physics

for scientists and engineers

Paul A. Tipler

physics for scientists and engineers

Fourth Edition

W.H. FREEMAN AND COMPANY/WORTH PUBLISHERS

For Claudia

Physics for Scientists and Engineers
Fourth Edition
Paul A. Tipler

Manufactured in the United States of America
Library of Congress Catalog Card Number: 98-60168
ISBN: 1-57259-673-2
Volume 1 (Chapters 1–21) paperback ISBN: 1-57259-491-8
Volume 1 (Chapters 1–21) hardcover ISBN: 1-57259-812-3
Volume 2 (Chapters 22–35) paperback ISBN: 1-57259-492-6
Volume 2 (Chapters 22–35) hardcover ISBN: 1-57259-813-1
Volume 3 (Chapters 36–41) paperback ISBN: 1-57259-490-X
Volume 3 (Chapters 36–41) hardcover ISBN: 1-57259-814-X
Volumes 1 and 2, ISBN: 1-57259-614-7
Volumes 1, 2, and 3, ISBN: 1-57259-615-5

Printing: 2 3 4 5 02 01 00

Executive Editors: Anne C. Duffy and Susan Finnemore Brennan
Development Editors: Steven Tenney and Morgan Ryan, with Richard Mickey
Marketing Managers: Kimberly Manzi and John Britch
Design: Malcolm Grear Designers
Art Director: George Touloumes
Production Editors: Margaret Comaskey and Elizabeth Geller
Production Managers: Patricia Lawson and Sarah Segal
Layout: Fernando Quiñones and Lee Mahler
Picture Editor: Elyse Rieder
Graphic Arts Manager: Demetrios Zangos
Three-dimensional art by DreamLight Incorporated
Illustrations: DreamLight Incorporated and Mel Erikson Art Services
Composition: Compset, Inc.
Separations: Creative Graphic Services
Printing and Binding: R. R. Donnelley and Sons
Cover Image: Sand atop a vertically driven shaker table spontaneously forms a roughly sinusoidal outline. Image by Max Aguilera-Hellweg.

Illustration credits begin on page IC-1 and constitute an extension of the copyright page.

W.H. Freeman and Company
41 Madison Avenue
New York, NY 10010 U.S.A.

preface

In this fourth edition I have worked toward four goals:

1. To help students increase their experience and ability in problem solving
2. To make the reading of the text easier and more fun for students
3. To bring the presentation of physics up to date to reflect the importance of the role of quantum theory
4. To make the text more flexible for the instructor in a wide variety of course formats

Enhanced Problem Solving

To help students learn how to solve problems, the number of worked *Examples* that correspond to intermediate-level problems has been greatly increased. Especially notable is a new two-column side-by-side example format that has been developed to better display the text and equations in worked examples. Care has been taken to show the students a logical method of solving problems. Examples begin with strategies, and often diagrams, in a *Picture the Problem* prologue. When possible, the first step gives an equation relating the quantity asked for to other quantities. This is usually followed by a statement of the general physical principle that applies. For example, this step may be "Apply Newton's second law" or "Use conservation of energy." Examples usually conclude with *Remarks* that discuss the problem and solution, and in many cases there are additional *Check the Result* sections that teach the student how to check the answer, as well as *Exercises* that present additional related problems, which students can solve on their own.

Also new are innovative, interactive types of examples, each labeled *Try it yourself*. In these, students are told in the left column how to proceed with each step of the problem-solving process, but in the right column are given only the answer. Thus, students are guided through the problem, but must independently work through the actual derivations and calculations.

A *Problem-Solving Guide* appears at the end of each chapter in the form of a summary of the worked examples in the chapter. The Problem-Solving Guide is designed to help students recognize types of problems and find the right conceptual strategy for solving them. Here again, general principles such as applying Newton's second law or the conservation of energy are emphasized.

Concluding each chapter is a selection of approximately one hundred *Problems*. The problems are grouped by type, which may or may not coincide with the section titles in the chapter. Each problem is designated easy, intermediate, or challenging. Qualitative questions and problems are integrated with quantitative problems within each group, in the hope that this organization will elevate the stature of qualitative problems in the minds of students (and instructors). At the back of the book, *Answers* are given to the odd-numbered problems. Preceding the answers for each chapter is a *Problem Map* that charts which odd-numbered intermediate-level problems correspond with worked examples in the text. Complete solutions to every other odd-numbered problem, worked out in the two-column example format, are available in the *Solutions Manual for Students.*

I do not believe that students can be given too much help in solving problems. Students learn best when they are successful at the tasks they are given. The hierarchy of worked examples, "Try it yourself" examples, Problem-Solving Guide, and Problem Map gives the student and the instructor maximum flexibility by leading the student through progressive levels of independence. "Try it yourself" problems take students step by step through a problem without doing the math for them. The Problem-Solving Guide gives an overview of the techniques that have been demonstrated in the chapter. The Problem Map shows students who are having difficulty where help may lie in the chapter but gives no other assistance.

Student Interest

Much effort has gone into making the written text more lively and informal. Students build their understanding of physics on the physics they've already learned, each concept serving as a building block that will provide the foundation for further inquiry. Over one hundred enthusiastic student reviews indicate that the changes in the fourth edition will successfully reach the widest range of students and will help them to enjoy learning and doing physics rather than focusing on the difficulty of the subject. To further stimulate the interest of students, supplemental, brief *"Exploring ..."* sections offer essays on various topics of interest to science and engineering undergraduates.

Modern Physics in the Introductory Course

Although quantum theory revolutionized the way we describe the physical world more than 70 years ago, we have been slow to integrate it into our introductory physics courses. To make physics more relevant to today's students, the mass–energy relationship and energy quantization sections are included in the conservation of energy chapter, and the quantization of angular momentum is discussed in the chapter on the conservation of angular momentum. These ideas are then used throughout the text, for example, in Chapter 19 to explain the failure of the equipartition theorem.

In addition, two optional chapters, "Wave–Particle Duality and Quantum Physics" (Chapter 17) and "The Microscopic Theory of Electrical Conduction" (Chapter 27), have been written so that instructors who choose to do so can integrate them into a two-semester course along with the usual topics in classical physics. These chapters offer something completely new—support for professors who choose to introduce quantum physics earlier in the course. Chapter 17 on the wave–particle duality of nature is the concluding chapter in Part II, immediately following the chapter on superposition and standing waves. This chapter introduces the idea of the wave–particle duality of light and matter and uses the frequency quantization of standing waves, just studied in the previous chapter, to introduce energy quantization of confined systems. Many students have heard of quantum theory and are curious about it. Having just studied frequency quantization that arises in standing waves, students can easily grasp energy quantization from standing electron waves, once they have seen from diffraction and interference patterns that electrons have wave properties. Because there is little time to cover even the usual material in the introductory course, some instructors are reluctant to consider adding even one more chapter such as Chapter 17. I would argue that quantum physics is at least as important as many of the other topics we teach.

Chapter 27 on the quantum explanation for electrical conduction is positioned so that it can be covered immediately after the discussion of electric current and dc circuits. The classical model of conduction is developed, concluding with the relation between resistivity and the average speed v_{av} and mean free path λ of electrons. The classical and quantum interpretations of v_{av} and λ are then discussed using the particle-in-a-box problem, discussed in the optional Chapter 17, to introduce the Fermi energy. Simple band theory is discussed to show why materials are conductors, insulators, or semiconductors. My hope in offering these optional chapters is that, given the choice, instructors will take advantage of the means to incorporate simple quantum theory into their elementary physics course.

Flexibility

To accommodate professors in a wide variety of course formats and to respond to the preferences of previous users of this text, there has been some revision in the order of material. With this new edition, instructors can give their students a brief exposure to modern physics integrated with the classical topics, or they can choose to skip the optional chapters on quantum physics entirely, perhaps returning to them in the final part of the course when this material is traditionally taught. To make room for these optional quantum chapters, some traditional material may be deleted from the course. To aid the instructor, material that can be skipped without jeopardizing coverage in other sections has been placed in optional sections. There are also two optional chapters in addition to Chapters 17 and 27. Chapter 12, "Static Equilibrium and Elasticity," and Chapter 21, "Thermal Properties and Processes," gather material that instructors sometimes choose to skip over or offer as added reading. The "optional" labeling of sections and chapters enables the instructor to pick and choose among topics with confidence that no material in nonoptional sections depends on previous coverage of an optional topic. Optional sections and chapters are clearly marked by gray borders down the side of the page. Some optional material, such as numerical methods and the use of complex numbers to solve the driven oscillator equation, is presented in "Exploring ..." essays.

Acknowledgments

Many people have contributed to this edition. I would like to thank everyone who used the earlier editions and offered comments and suggestions.

Gene Mosca, James Garland, Robert Lieberman, and Murray Scureman provided detailed reviews of nearly every chapter. Gene Mosca also wrote the student study guide along with Ron Gautreau. Robert Leiberman and Brooke Pridmore class-tested parts of the book, and assisted in obtaining student reviews and feedback. Howard McAllister was instrumental in the development of a standard approach to problem solving in the examples.

Many new problems were provided by Frank Blatt and Boris Korsunsky. Frank Blatt wrote the solutions manuals and offered many helpful suggestions. Jeff Culbert helped to enliven the problem sets with his story problems. Several of the graphs at the ends of the examples were provided by Robert Hollebeek.

I received invaluable help in manuscript checking from Murray Scureman, Thor Stromberg, and Howard Miles, and in checking problems and solutions from Thor Stromberg, Howard Miles, Robert Detenbeck, Daniel G. Tekleab, Jeannette Myers, Scott Sinawi, John Pratte, Yuriy Zhestkov, Huidong Guo, Fred Watts, Ilon Joseph, Monwhea Jeng, Harry Chu, and Roy Wood. Any errors remaining are of course my responsibility.

I would particularly like to thank the more than one hundred students who read and studied from various chapters and provided detailed and valuable comments. Many instructors have provided extensive and invaluable reviews of one or more chapters. They have all made fundamental contributions to the quality of this revision. I would therefore like to thank:

Michael Arnett, *Iowa State University*
William Bassichis, *Texas A&M*
Joel C. Berlinghieri, *The Citadel*
Frank Blatt, *Michigan State University*
John E. Byrne, *Gonzaga University*
Wayne Carr, *Stevens Institute of Technology*
George Cassidy, *University of Utah*
I. V. Chivets, *Trinity College, University of Dublin*
Harry T. Chu, *University of Akron*
Jeff Culbert, *London, Ontario*
Paul Debevec, *University of Illinois*
Robert W. Detenbeck, *University of Vermont*
Bruce Doak, *Arizona State University*
John Elliott, *University of Manchester, England*
James Garland, *Retired*
Ian Gatland, *Georgia Institute of Technology*
Ron Gautreau, *New Jersey Institute of Technology*
David Gavenda, *University of Texas at Austin*
Newton Greenburg, *SUNY Binghamton*
Huidong Guo, *Columbia University*
Richard Haracz, *Drexel University*
Michael Harris, *University of Washington*
Randy Harris, *University of California at Davis*
Dieter Hartmann, *Clemson University*
Robert Hollebeek, *University of Pennsylvania*
Madya Jalil, *University of Malaya*
Monwhea Jeng, *University of California, Santa Barbara*
Ilon Joseph, *Columbia University*
David Kaplan, *University of California, Santa Barbara*
John Kidder, *Dartmouth College*
Boris Korsunsky, *Northfield Mt. Hermon School*
Andrew Lang (graduate student), *University of Missouri*
David Lange, *University of California, Santa Barbara*
Isaac Leichter, *Jerusalem College of Technology*
William Lichten, *Yale University*
Robert Lieberman, *Cornell University*
Fred Lipschultz, *University of Connecticut*
Graeme Luke, *Columbia University*
Howard McAllister, *University of Hawaii*
M. Howard Miles, *Washington State University*
Matthew Moelter, *University of Puget Sound*
Eugene Mosca, *United States Naval Academy*
Aileen O'Donughue, *St. Lawrence University*
Jack Ord, *University of Waterloo*
Richard Packard, *University of California*
George W. Parker, *North Carolina State University*
Edward Pollack, *University of Connecticut*
John M. Pratte, *Clayton College & State University*
Brooke Pridmore, *Clayton State College*
David Roberts, *Brandeis University*
Lyle D. Roelofs, *Haverford College*
Larry Rowan, *University of North Carolina at Chapel Hill*
Lewis H. Ryder, *University of Kent, Canterbury*
Bernd Schuttler, *University of Georgia*

Cindy Schwarz, *Vassar College*
Murray Scureman, *Amdahl Corporation*
Scott Sinawi, *Columbia University*
Wesley H. Smith, *University of Wisconsin*
Kevork Spartalian, *University of Vermont*
Kaare Stegavik, *University of Trondheim, Norway*
Jay D. Strieb, *Villanova University*
Martin Tiersten, *City College of New York*
Oscar Vilches, *University of Washington*
Fred Watts, *College of Charleston*
John Weinstein, *University of Mississippi*
David Gordon Wilson, *MIT*
David Winter, *Columbia University*
Frank L. H. Wolfe, *University of Rochester*
Roy C. Wood, *New Mexico State University*
Yuriy Zhestkov, *Columbia University*

Focus Group Participants

Cherry Hill, New Jersey, July 15, 1997

John DiNardo, *Drexel University*
Eduardo Flores, *Rowan College*
Jeff Martoff, *Temple University*
Anthony Novaco, *Lafayette College*
Jay Strieb, *Villanova University*
Edward Whittaker, *Stevens Institute of Technology*

Denver, Colorado, August 15, 1997

Edward Adelson, *Ohio State University*
David Bartlett, *University of Colorado at Boulder*
David Elmore, *Purdue University*
Colonel Rolf Enger, *United States Air Force Academy*
Kendal Mallory, *University of Northern Colorado*
Samuel Milazzo, *University of Colorado at Colorado Springs*
Anders Schenstrom, *Milwaukee School of Engineering*
Daniel Schroeder, *Weber State University*
Ashley Schultz, *Fort Lewis College*

Student Reviewers

For this edition we invited the input of student reviewers at all stages of manuscript development. A number of the student reviews were blind submissions. The reviews of the following students were especially helpful:

Jesper Anderson, *Haverford College*
Anthony Bak, *Haverford College*
Luke Benes, *Cornell University*
Deborah Brown, *Northwestern University*
Andrew Burgess, *University of Kent, Canterbury*
Sarah Burnett, *Cornell University*
Sara Ellison, *University of Kent, Canterbury*
Ilana Greenstein, *Haverford College*
Sharon Hovey, *Northwestern University*
Samuel LaRoque, *Cornell University*
Valerie Larson, *Northwestern University*
Jonathan McCoy, *Haverford College*
Aaron Todd, *Cornell University*
Katalin Varju, *University of Kent, Canterbury*
Ryan Walker, *Haverford College*
Matthew Wolpert, *Haverford College*
Julie Zachiariadis, *Haverford College*

I would also like to thank the reviewers of previous editions, whose contributions are part of the foundation of this edition:

Walter Borst, *Texas Technological University*
Edward Brown, *Manhattan College*
James Brown, *The Colorado School of Mines*
Christopher Cameron, *University of Southern Mississippi*
Roger Clapp, *University of South Florida*
Bob Coakley, *University of Southern Maine*
Andrew Coates, *University College, London*
Miles Dresser, *Washington State University*
Manuel Gómez-Rodríguez, *University of Puerto Rice, Río Piedras*
Allin Gould, *John Abbott College C.E.G.E.P., Canada*

Dennis Hall, *University of Rochester*
Grant Hart, *Brigham Young University*
Jerold Izatt, *University of Alabama*
Alvin Jenkins, *North Carolina State University*
Lorella Jones, *University of Illinois, Urbana-Champaign*
Michael Kambour, *Miami-Dade Junior College*
Patrick Kenealy, *California State University at Long Beach*
Doug Kurtze, *Clarkson University*
Lui Lam, *San Jose State University*
Chelcie Liu, *City College of San Francisco*
Robert Luke, *Boise State University*
Stefan Machlup, *Case Western Reserve University*
Eric Matthews, *Wake Forest University*
Konrad Mauersberger, *University of Minnesota, Minneapolis*
Duncan Moore, *University of Rochester*
Elizabeth Nickles, *Albany College of Pharmacy*
Harry Otteson, *Utah State University*
Jack Overley, *University of Oregon*
Larry Panek, *Widener University*
Malcolm Perry, *Cambridge University, United Kingdom*
Arthur Quinton, *University of Massachusetts, Amherst*
John Risley, *North Carolina State University*
Robert Rundel, *Mississippi State University*
John Russell, *Southeastern Massachusetts University*
Michael Simon, *Housatonic Community College*
Jim Smith, *University of Illinois, Urbana-Champaign*
Richard Smith, *Montana State University*
Larry Sorenson, *University of Washington*
Thor Stromberg, *New Mexico State University*
Edward Thomas, *Georgia Institute of Technology*
Colin Thomson, *Queens University, Canada*
Gianfranco Vidali, *Syracuse University*
Brian Watson, *St. Lawrence University*
Robert Weidman, *Michigan Technological University*
Stan Williams, *Iowa State University*
Thad Zaleskiewicz, *University of Pittsburgh, Greensburg*
George Zimmerman, *Boston University*

Finally, I would like to thank everyone at Worth and W. H. Freeman Publishers for their help and encouragement. I was fortunate to work with two talented developmental editors. Steve Tenney worked on the beginning phases of the book and is responsible for many of the innovative ideas, such as the example format, summary format, problem-solving guide, and problem map. Morgan Ryan worked on the final stages, including the entire art program, and made significant improvements in the entire book. I am grateful also for the contributions of Kerry Baruth, Anne Duffy, Margaret Comaskey, Elizabeth Geller, Yuna Lee, Sarah Segal, Patricia Lawson, and George Touloumes.

Berkeley, California
December 1997

Paul Tipler

supplements

For Students

Study Guide

Volume 1 (Chapters 1–21) ISBN: 1-57259-511-6
Volumes 2 and 3 (Chapters 22–41) ISBN: 1-57259-512-4

Each chapter contains a description of key ideas, potential pitfalls, true-false questions that test essential definitions and relations, questions and answers that require qualitative reasoning, and problems and solutions.

Solutions Manual for Students

Volume 1 (Chapters 1–21) ISBN: 1-57259-513-2
Volumes 2 and 3 (Chapters 22–41) ISBN: 1-57259-524-8

The *Solutions Manual for Students* provides answers to every other odd end-of-chapter problem, presented in the same format and with the same level of detail as the *Instructor's Solutions Manual* (see below).

For Instructors

Instructor's Solutions Manual

Volume 1 (Chapters 1–21) ISBN: 1-57259-514-0
Volumes 2 and 3 (Chapters 22–41) ISBN: 1-57259-515-9

Complete solutions to all problems in the text are worked out in the same two-column format as the examples.

Test Bank

Approximately 3500 multiple-choice questions span all sections of the text. Each question is identified by topic and noted as factual, conceptual, or numerical. ISBN: 1-57259-517-5

Computerized Test-Generation System

A database comprises the questions in the *Test Bank.* Instructors can custom design their tests with the *Computerized Test Bank.* For Windows, ISBN: 1-57259-519-1; for Macintosh, ISBN: 1-57259-520-5

Instructor's Resource Manual

Demonstrations and a film and video cassette guide are included. ISBN: 1-57259-516-7

Transparencies

Approximately 150 full-color acetates of figures and tables from the text are included, with type enlarged for projection. Volume 1, ISBN: 1-57259-521-3; Volumes 2 and 3, ISBN: 1-57259-674-0

about the author

Paul Tipler was born in the small farming town of Antigo, Wisconsin, in 1933. He graduated from high school in Oshkosh, Wisconsin, where his father was superintendent of the Public Schools. He received his B.S. from Purdue University in 1955 and his Ph.D. at the University of Illinois in 1962, where he studied the structure of nuclei. He taught for one year at Wesleyan University in Connecticut while writing his thesis, then moved to Oakland University in Michigan, where he was one of the original members of the physics department, playing a major role in developing the physics curriculum. During the next 20 years, he taught nearly all the physics courses and wrote the first and second editions of his widely used textbooks *Modern Physics* (1969, 1978) and *Physics* (1976, 1982). In 1982, he moved to Berkeley, California, where he now resides, and where he wrote *College Physics* (1987) and the third edition of *Physics* (1991). In addition to physics, his interests include music, hiking, and camping, and he is an accomplished jazz pianist and poker player.

The author as a student, 1954

For over 20 years, the formula has been

Tipler = Quality

Tipler *Physics for Scientists and Engineers*, continues to be the best resource a student have for learning physics. Dynamic features like these guide the student to mastery . . .

EXAMPLES

- Text includes a greater number of intermediate-level worked examples.
- Each example has a **"Picture the Problem"** section that teaches students how to solve the problem conceptually before solving it mathematically. By learning how to find and organize the relevant information in a problem, students learn to think like a physicist.
- A major innovation is the potent **two-column side-by-side format** for the solutions to examples. Concepts are explained on the left, and the math is presented on the right. This format allows students to make the connections between the equation and what it means.
- Most examples conclude with a **"Remark"** that supplies additional information, discussion of common errors, and advice on solving problems as a physicist would.

Example 6-8

You ski downhill on waxed skis that are nearly frictionless. (*a*) What work is done on you as you ski a distance *s* down the hill? (*b*) What is your speed on reaching the bottom of the run? Assume the length of the ski run is *s*, its angle of incline is *θ*, and your mass is *m*. The height of the hill is then $h = s \sin\theta$.

Picture the Problem We assume that you are a particle. Two forces act on you: gravity, $m\vec{g}$, and the normal force exerted by the hill, $\vec{F}_n$ (Figure 6-15*a*). Only gravity does work on you, because the normal force is perpendicular to the hill, and hence has no component in the direction of your motion. The work–kinetic energy theorem with $v_i = 0$ gives the final speed v.

Figure 6-15*b* shows a free-body diagram for you on skis. The net force is $mg \sin\theta$, which is the component of the weight in the direction of the displacement Δs.

Figure 6-15*a*

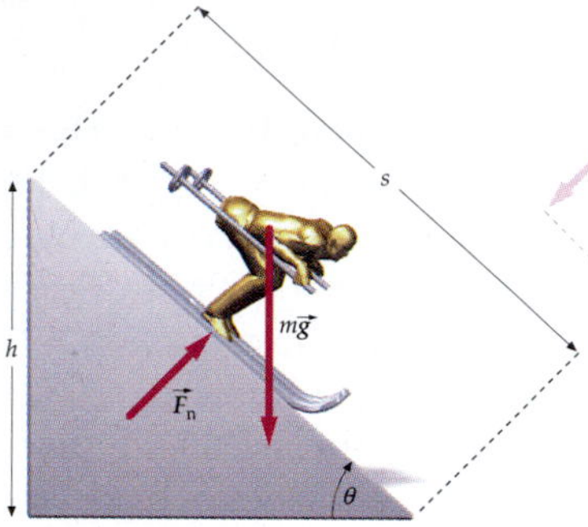

Figure 6-15*b*

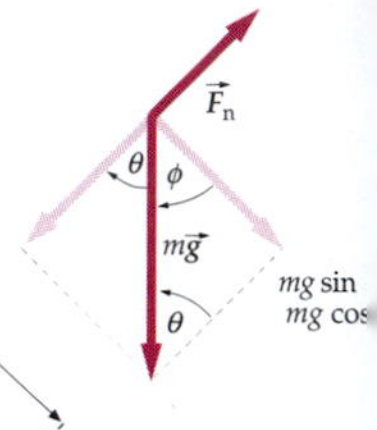

(*a*)1. The work done by gravity as you traverse the slope is $m\vec{g}\cdot\vec{s}$:	$W = m\vec{g}\cdot\vec{s} = mgs\cos\phi = mgs\sin\theta$
2. From Figure 6-15*a*, the angle *θ* is related to *h* and *s*:	$\sin\theta = \dfrac{h}{s}$
3. Substitute *h* for *s* sin *θ*:	$W = mgh$
(*b*) Apply the work–kinetic energy theorem to find the final speed *v*:	$W = mgh = \dfrac{1}{2}mv^2 - 0$ or $v = \sqrt{2g}$

Remarks $mg\sin\theta = mg\cos\phi$ is the component of the weight in the direction of the displacement. This is the component that does work on you. The final speed is independent of the angle *θ*, and the same as if the skier had dropped vertically a height *h* with acceleration *g*. If *θ* were smaller, the skier would travel a greater distance to drop the same vertical distance *h*, but the component of the force of gravity in the direction of motion would be less. The two effects cancel, and the work done by gravity is *mgh* independent of the angle of the slope. Figure 6-16 shows that for a hill of arbitrary shape, the work done by the earth on the skier is *mgh*.

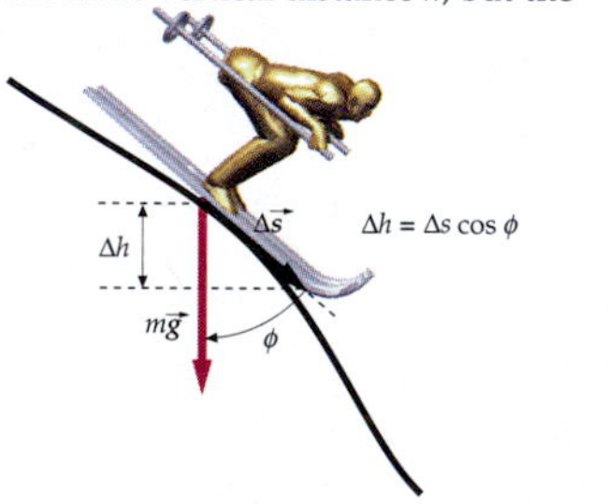

Figure 6-16 Skier skiing down a hill of arbitrary shape. The work done by the e during a displacement $\Delta\vec{s}$ is $m\vec{g}\cdot\Delta\vec{s} = mg\,\Delta s\cos\phi = mg\,\Delta h$, where Δh is the vertical distance dropped. The total work done by the earth when the skier skis down a vertical distance *h* is $W = \int_0^s m\vec{g}\cdot d\vec{s} = mg\int_0^s \cos\phi\, ds = mg\int_0^h dh = mgh$, independent of the shape of the hill.

- When appropriate, **"Check the Result"** sections teach students how to check their own work.
- Many examples are followed by one or more related **exercises.** Answers are given, but it is up to the student to relate the exercise to the worked-out example.

Check the Result The component of *A* along *B* is $A\cos\phi = (\sqrt{13}\text{ m})\cos 70.6° = 1.2$ m.

Exercise (*a*) Find $\vec{A}\cdot\vec{B}$ for $\vec{A} = (3\text{ m})\hat{i} + (4\text{ m})\hat{j}$ and $\vec{B} = (2\text{ m})\hat{i} + (8\text{ m})\hat{j}$. (*b*) Find *A*, *B*, and the angle between $\vec{A}$ and $\vec{B}$ for these vectors. (*Answers* (*a*) 38 m^2, (*b*) $A = 5$ m, $B = 8.25$ m, $\phi = 23°$)

Example 2-15

A car is speeding at 25 m/s (≈ 90 km/h ≈ 56 mi/h) in a school zone. A police car starts from rest just as the speeder passes and accelerates at a constant rate of 5 m/s². (*a*) When does the police car catch the speeding car? (*b*) How fast is the police car traveling when it catches up with the speeder?

Picture the Problem To determine when the two cars will be at the same position, we write the positions x_s of the speeder and x_p of the police car as functions of time and solve for the time t when $x_s = x_p$.

(*a*) 1. Write the position functions for the speeder and the police car: $x_s = v_s t$ and $x_p = \frac{1}{2}a_p t^2$

2. Set $x_s = x_p$ and solve for the time t: $v_s t = \frac{1}{2}a_p t^2$; $t = 0$ (initial condition)

$$t = \frac{2v_s}{a_p} = \frac{2(25\text{ m/s})}{5\text{ m/s}^2} = 10\text{ s}$$

(*b*) The velocity of the police car is given by $v = v_0 + at$ with $v_0 = 0$: $v_p = a_p t = (5\text{ m/s}^2)(10\text{ s}) = 50\text{ m/s}$

Remark The final speed of the police car in (*b*) is exactly twice that of the speeder. Since the two cars covered the same distance in the same time, they must have had the same average speed. The speeder's average speed, of course, is 25 m/s. For the police car to start from rest and have an average speed of 25 m/s, it must reach a final speed of 50 m/s.

Exercise How far have the cars traveled when the police car catches the speeder? (*Answer* 250 m)

Remark In Figure 2-13 the solid lines depict the speeder and the police car in this example. The dashed lines are variations on the example. The smaller acceleration depicted by the lower dashed line means the police car takes longer to reach the speeder. In the higher dashed line, the acceleration is the same as in the example, but the police car does not start accelerating until 4 s after the speeder passes by.

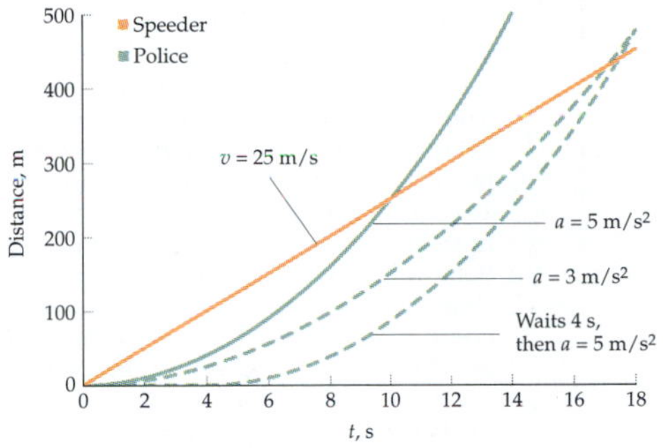

Figure 2-13

- Instructive illustrations accompany nearly all examples.
- Students are involved in every step of the complete solution for each example.

"TRY IT YOURSELF" EXAMPLES

- **"Try it yourself"** examples follow the same format as regular examples, except that the student is given only brief instructions on the left side, and only the relevant formulas and the answer on the right side, without being shown the mathematical steps in between.
- **"Try it yourself"** examples test understanding, identify weaknesses, and reinforce proper problem-solving procedures.

Example 6-7 ***try it yourself***

A particle is given a displacement $\Delta\vec{s} = 2\text{ m}\,\hat{i} - 5\text{ m}\,\hat{j}$ along a straight line. During the displacement, a constant force $\vec{F} = 3\text{ N}\,\hat{i} + 4\text{ N}\,\hat{j}$ acts on the particle (Figure 6-14). Find (*a*) the work done by the force, and (*b*) the component of the force in the direction of the displacement.

Picture the Problem The work W is found by computing $W = \vec{F}\cdot\Delta\vec{s} = F_x\Delta x + F_y\Delta y + F_z\Delta z$. Since $\vec{F}\cdot\Delta\vec{s} = F\cos\phi|\Delta\vec{s}|$, we can find the component of $\vec{F}$ in the direction of the displacement from

$$F\cos\phi = \frac{(\vec{F}\cdot\Delta\vec{s})}{|\Delta\vec{s}|} = \frac{W}{|\Delta\vec{s}|}$$

Figure 6-14

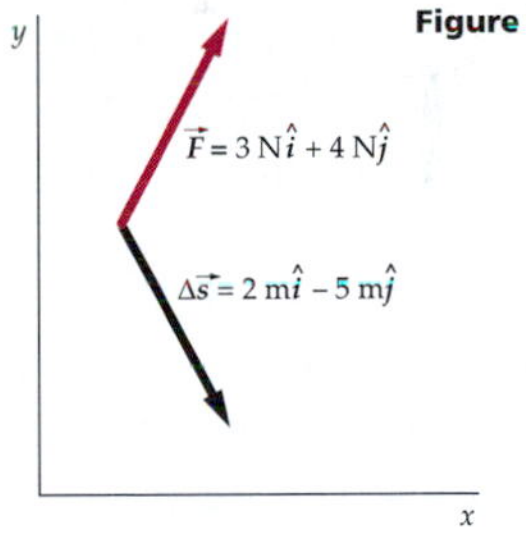

Cover the column to the right and try these on your own before looking at the answers.

Steps	*Answers*
(*a*) Compute the work done W.	$W = -14\text{ N}\cdot\text{m}$
(*b*) 1. Compute $\Delta\vec{s}\cdot\Delta\vec{s}$ and use your result to find the distance $\|\Delta\vec{s}\|$.	$\|\Delta\vec{s}\| = \sqrt{29}\text{ m}$
2. Compute $F\cos\phi = W/\|\Delta\vec{s}\|$.	$F\cos\phi = -2.60\text{ N}$

Remark The component of the force in the direction of the displacement is negative, so the work done is negative.

Exercise Find the magnitude of $\vec{F}$, and the angle ϕ between $\vec{F}$ and $\Delta\vec{s}$. (*Answer* $F = 5\text{ N}$, $\phi = 121°$)

"EXPLORING" ESSAYS

- "Exploring" essays enrich students by covering topics that are outside the core part of the course—but are too interesting to pass over completely.

Exploring

Numerical Methods: Euler's Method

If a particle moves under the influence of a *constant* force, its acceleration is constant and we can find its velocity and position from the constant-acceleration formulas in Chapter 2. But consider a particle moving through space where the force on it, and therefore its acceleration, depends on its position and velocity. The velocity and acceleration of the particle at one instant determine its position and velocity at the next instant, which then determines its acceleration at that instant. The actual position, velocity, and acceleration of an object all change continuously with time. We can approximate this by replacing the continuous time variations with small time steps of duration Δt. The simplest approximation is to assume constant acceleration during each step. This approximation is called **Euler's method**. If the time interval is sufficiently short, the change in acceleration during the interval will be small and can be neglected.

Let x_0, v_0, and a_0 be the known position, veloc-

$$x_2 = x_1 + v_1\Delta t$$

In general, the connection between the position and velocity at time t_n and time $t_{n+1} = t_n + \Delta t$ is given by

$$v_{n+1} = v_n + a_n\Delta t \qquad 3$$

and

$$x_{n+1} = x_n + v_n\Delta t \qquad 4$$

To find the velocity and position at some time t, we therefore divide the time interval $t - t_0$ into a large number of smaller intervals Δt and apply Equations 3 and 4, beginning at the initial time t_0. This involves a large number of simple, repetitive calculations that are easily done on a computer. The technique of breaking the time interval into small steps and computing the acceleration, velocity, and position at each step using the values from the previous step is called numerical integration.

Drag Forces

To illustrate the use of numerical methods, let us consider a problem in which a sky diver is dropped from rest at some height under the influences of gravity and a drag force that is proportional to the square of the speed. We will find the velocity v and the distance traveled x as functions of time.

The equation describing the motion of an object of mass m dropped from rest is Equation 5-7 with $n = 2$:

$$\sum F_y = mg - bv^n = ma_y$$

SUMMARY

- Detailed **Summary** sections reinforce the lessons of the chapter. The summary is not a substitute for attentive study of the text, but a skeletal outline to help students organize what they have learned.

Summary

1. Work, kinetic energy, potential energy, and power are important derived dynamic quantities.
2. The work–kinetic energy theorem is an important relation derived from Newton's laws applied to a particle.
3. The dot product of vectors is a mathematical definition that is useful throughout physics.

Topic	Remarks and Relevant Equations	
1. Work		
Constant force	The work done by a constant force is the product of the component of the force in the direction of motion and the displacement of the force: $W = F\cos\theta\,\Delta x = F_x\,\Delta x$	**6-1**
Variable force	$W = \int_{x_1}^{x_2} F_x\,dx$ = area under the F_x-versus-x curve	**6-9**
Force in three dimensions	$W = \int_1^2 \vec{F}\cdot d\vec{s}$	**6-14**
Units	The SI unit of work and energy is the joule (J): $1\text{ J} = 1\text{ N}\cdot\text{m}$	**6-2**
2. Kinetic Energy	$K = \frac{1}{2}mv^2$	**6-6**

PROBLEMS

- Types of problems are denoted by color swatches: **yellow** denotes conceptual problems and a **gray** band indicates optional or exploring sections.
- The difficulty level is denoted by bullets.
- Qualitative problems are included in context with related quantitative problems.

Problems

Conceptual Problems

Problems from Optional and Exploring sections

• Single-concept, single-step, relatively easy

•• Intermediate-level, may require synthesis of concepts

••• Challenging, for advanced students

In a few problems, you are given more data than you actually need; in a few other problems, you are required to supply data from your general knowledge, outside sources, or informed estimates.

Conditions for Equilibrium

1 • True or false:

(*a*) $\Sigma\vec{F} = 0$ is sufficient for static equilibrium to exist.
(*b*) $\Sigma\vec{F} = 0$ is necessary for static equilibrium to exist.
(*c*) In static equilibrium, the net torque about any point is zero.
(*d*) An object is in equilibrium only when there are no forces acting on it.

2 • A seesaw consists of a 4-m board pivoted at the center. A 28-kg child sits on one end of the board. Where should a 40-kg child sit to balance the seesaw?

3 • In Figure 12-23, Misako is about to do a push-up. Her center of gravity lies directly above point *P* on the floor, which is 0.9 m from her feet and 0.6 m from her hands. If her mass is 54 kg, what is the force exerted by the floor on her hands?

Figure 12-23 Problem 3

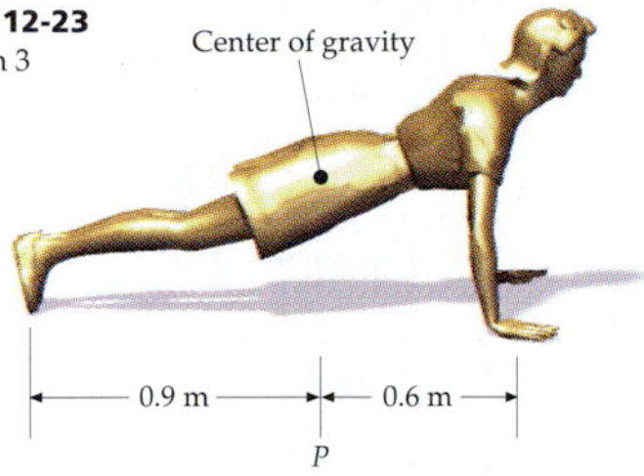

4 • Juan and Bettina are carrying a 60-kg block on a 4-m board as shown in Figure 12-24. The mass of the board is 10 kg. Since Juan spends most of his time reading cookbooks, whereas Bettina regularly does push-ups, they place the block 2.5 m from Juan and 1.5 m from Bettina. Find the force in newtons exerted by each to carry the block.

Figure 12-24 Problem 4

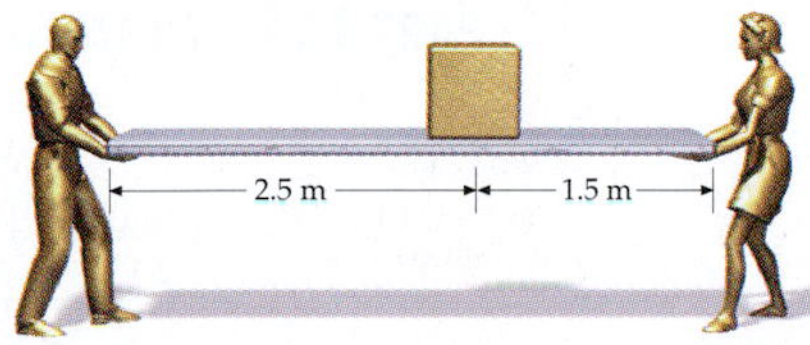

Figure 12-25 Problem 5

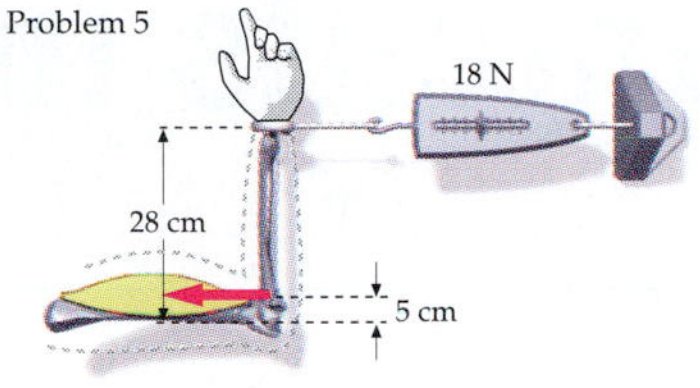

the pivot point. If the scale reads 18 N when she exerts her maximum force, what force is exerted by the biceps muscle?

6 • A crutch is pressed against the sidewalk with a force $\vec{F}_c$ along its own direction as in Figure 12-26. This force is balanced by a normal force $\vec{F}_n$ and a frictional force $\vec{f}_s$. (*a*) Show that when the force of friction is at its maximum value, the coefficient of friction is related to the angle θ by $\mu_s = \tan\theta$. (*b*) Explain how this result applies to the forces on your foot when you are not using a crutch. (*c*) Why is it advantageous to take short steps when walking on ice?

Figure 12-26 Problem 6

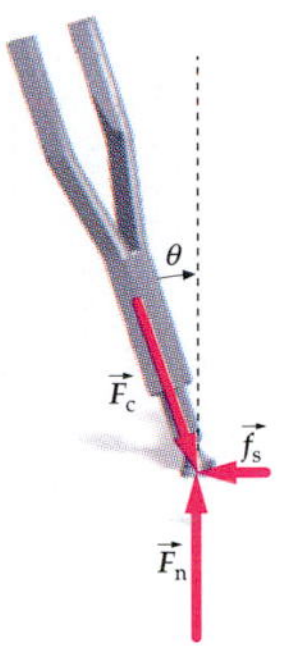

The Center of Gravity

7 • True or false: The center of gravity is always at the geometric center of a body.

8 • Must there be any material at the center of gravity of an object?

9 • If the acceleration of gravity is not constant over an object, is it the center of mass or the center of gravity that is the pivot point when the object is balanced?

10 • Two spheres of radius *R* rest on a horizontal table with their centers a distance 4*R* apart. One sphere has twice the weight of the other sphere. Where is the center of gravity of this system?

11 • An automobile has 58% of its weight on the front wheels. The front and back wheels are separated by 2 m.

General Problems

66 • If the net torque about some point is zero, must it be zero about any other point? Explain.

67 • The horizontal bar in Figure 12-52 will remain horizontal if

(*a*) $L_1 = L_2$ and $R_1 = R_2$. (*b*) $L_1 = L_2$ and $M_1 = M_2$.
(*c*) $R_1 = R_2$ and $M_1 = M_2$. (*d*) $L_1M_1 = L_2M_2$.
(*e*) $R_1L_1 = R_2L_2$.

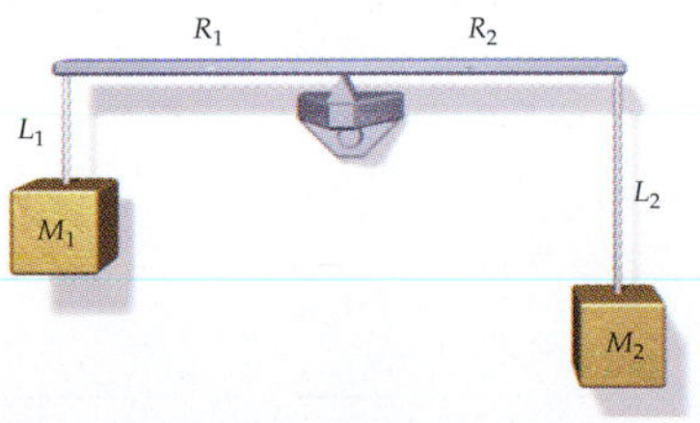

68 • Which of the following could not have units of N/m^2?

(*a*) Young's modulus (*b*) Shear modulus
(*c*) Stress (*d*) Strain

69 •• Sit in a chair with your back straight. Now try to stand up without leaning forward. Explain why you cannot do it.

70 • A 90-N board 12 m long rests on two supports, each 1 m from the end of the board. A 360-N block is placed on the board 3 m from one end as shown in Figure 12-53. Find the force exerted by each support on the board.

Figure 12-53 Problem 70

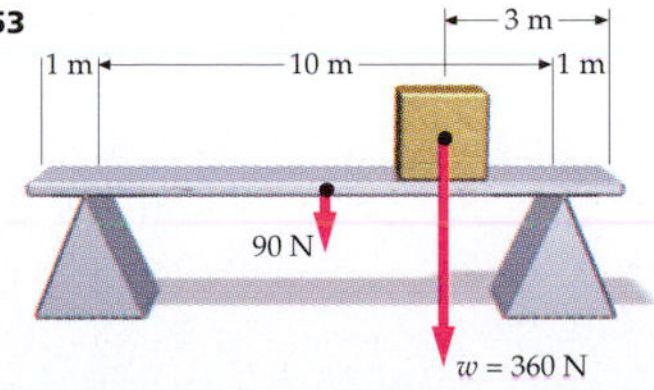

contents in brief

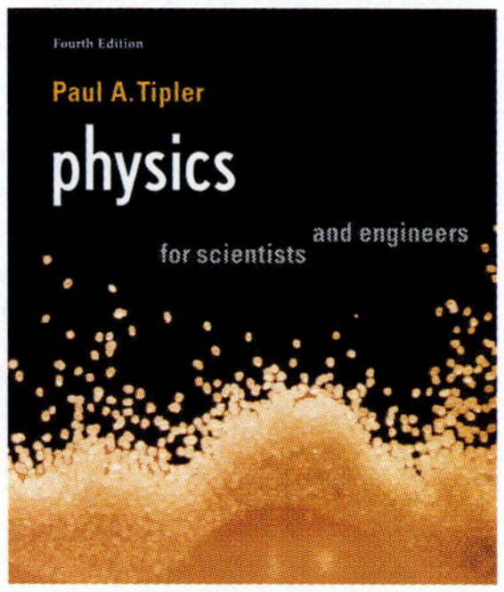

PART IV electricity and magnetism

PART V light

PART VI modern physics: quantum mechanics, relativity, and the structure of matter

Appendix

OPTIONAL SECTIONS: These sections contain material that can be omitted without jeopardizing coverage of other sections.

contents

PART I mechanics

PART III thermodynamics

PART IV electricity and magnetism

PART IV electricity and magnetism

physics

for scientists and engineers

CHAPTER 1

Systems of Measurement

Tycho Brahe (1546–1601) with his large brass quadrant for measuring the positions of planets and stars.

We have always been curious about the world around us. Since the beginnings of recorded thought, we have sought ways to impose order on the bewildering diversity of events that we observe. This search for order has taken a variety of forms, and has given birth to religion, art, and science. Although the word "science" has its origins in a Latin verb meaning "to know," science has come to mean not simply knowledge in general, but more specifically, knowledge of the natural world. Most importantly, science is a body of knowledge organized in a specific and rational way.

Today we think of science divided into separate fields, although this division occurred only in the last century or so. The separation of complex systems into smaller categories that can be more easily studied is one of the great successes of science. Biology, for example, is the study of living organisms. Chemistry deals with the interaction of elements and compounds. Geology is the study of the earth. Astronomy is the study of the solar system, the stars and galaxies, and the universe as a whole. Physics is the science of matter and energy, and includes the principles that govern the motion of particles and waves, the interactions of particles, and the properties of molecules, atoms, and atomic nuclei, as well as larger-scale systems such as gases, liquids, and solids. Some consider physics the most fundamental science because its principles supply the foundation of the other scientific fields.

Physics is the science of the exotic and the science of everyday life. At the exotic extreme, black holes boggle the imagination. In everyday life, engineers, musicians, architects, chemists, biologists, doctors, and many others routinely command such subjects as heat transfer, fluid flow, sound waves, radioactivity, and stresses in buildings or bones to perform their daily work. Countless questions about our world can be answered with a basic knowledge of physics. Why must a helicopter have two rotors? Why do astronauts float in space? Why does sound travel around corners while light appears to travel in straight lines? Why does an oboe sound different from a flute? How do CD players work? Why is there no hydrogen in the atmosphere? Why do metal objects feel colder than wood objects at the same temperature? Why is copper an electrical conductor while wood is an insulator? Why is lithium, with its three electrons, extremely reactive, whereas helium, with two electrons, is chemically inert?

Classical and Modern Physics

The earliest recorded efforts to systematically assemble knowledge concerning motion came from ancient Greece. In the system of natural philosophy set forth by Aristotle (384–322 B.C.), explanations of physical phenomena were deduced from assumptions about the world, rather than derived from experimentation. For example, it was a fundamental assumption that every substance had a "natural place" in the universe. Motion was ruled to be the result of a substance trying to reach its natural place. Because of the agreement between the deductions of Aristotelian physics and the motions observed throughout the physical universe, and because there was no tradition of experimentation that could overturn the ancient physics, the Greek view was accepted for nearly 2000 years. It was the Italian scientist Galileo Galilei (1564–1642) whose brilliant experiments on motion established for all time the absolute necessity of experimentation in physics and initiated the disintegration of Aristotelian physics. Within 100 years, Isaac Newton had generalized the results of Galileo's experiments into his three spectacularly successful laws of motion, and the natural philosophy of Aristotle was gone.

Experimentation during the next 200 years brought a flood of discoveries, inspiring the development of physical theories to explain them. By the end of the nineteenth century, Newton's laws for the motions of mechanical systems had been joined by equally impressive laws from Maxwell, Joule, Carnot, and others to describe electromagnetism and thermodynamics. The subjects that occupied physical scientists through the end of the nineteenth century—mechanics, light, heat, sound, electricity, and magnetism—are usually referred to as *classical physics*.

The remarkable success of classical physics led many scientists to believe that the description of the physical universe was complete. However, the discoveries of X rays by Roentgen in 1895 and of nuclear radioactivity by Becquerel in 1896 seemed to be outside the framework of classical physics. The theory of special relativity proposed by Albert Einstein in 1905 contradicted the ideas of space and time of Galileo and Newton. In the same year, Einstein suggested that light energy is quantized; that is, that light comes in discrete packets rather than being wavelike and continuous as had been assumed in classical physics. The generalization of this insight to the quantization of all types of energy is a central idea of quantum mechanics, one that has many amazing and important consequences. The application of special relativity and, particularly, quantum theory to such microscopic systems as atoms, molecules, and nuclei has led to a detailed understanding of solids, liquids, and gases and is often referred to as *modern physics*.

Except for the interiors of atoms and for motions at speeds near the speed of light, classical physics correctly and precisely describes the behavior of the physical world. It is classical physics we must master to understand the macroscopic world we live in, and classical physics is the main subject of this book. Modern physics is itself built on the concepts of classical physics. It is not possible to understand quantum theory without a knowledge of such classical concepts as energy, momentum, angular momentum, wave functions, and standing waves. We thus begin our study of physics with the study of classical topics. However, we will look ahead from time to time and note the relationship between classical and modern physics. When we discuss velocity in Chapter 2, for example, we will take a moment to consider velocities near the speed of light as we cross over to the relativistic universe first imagined by Einstein. After discussing the conservation of energy in Chapter 7, we will discuss the quantization of energy and Einstein's famous relation between mass and energy, $E = mc^2$. After introducing classical waves, we show that light, which is treated as a wave in classical physics, also has particle properties, and electrons, which are particles in the classical view, also have wave properties. Building on the classical concepts of wave functions, interference, diffraction, and standing waves, we show in Chapter 17 that the application of these concepts to electron waves leads to the quantization of energy. With a rudimentary knowledge of quantum theory, we are in position to understand the world around us. We are also prepared to understand when and why classical physics applies in our later studies of physics, and at what times it must be augmented or replaced by quantum physics.

1-1 Units

We all know of things that cannot be measured—the beauty of a flower or of a Bach fugue. As certain as our knowledge of these things may be, we readily admit that this knowledge is not science. The ability not only to define but also to measure is a requisite of science, and in physics, more than in any other field of knowledge, the precise definition of terms and the accurate measurement of quantities have led to great discoveries. We begin our study of physics by establishing a few basic definitions, introducing units, and showing how units are dealt with in equations. The fun comes later.

Measurement of any quantity involves comparison with some precisely defined unit value of the quantity. For example, to measure the distance between two points, we need a standard unit, such as the meter. The statement that a certain distance is 25 meters means that it is 25 times the length of the unit meter. That is, a standard meterstick fits into that distance 25 times. It is important to include the unit, in this case meters, along with the number 25 in expressing this distance, because there are other units of distance such as kilometers or miles in common use. To say that a distance is 25 is meaningless. The magnitude of any physical quantity must include both a number and a unit.

The International System of Units

A small number of fundamental units are sufficient to express all physical quantities. Many of the quantities that we shall be studying, such as velocity, force, momentum, work, energy, and power, can be expressed in terms of three fundamental measures: length, time, and mass. The choice of standard units for these fundamental quantities determines a system of units. The system used universally in the scientific community is called SI (for *Système International*). The standard SI unit for length is the meter, the standard unit of time is the second, and the standard mass is the kilogram. Complete definitions of the SI units are given in Appendix A.

The standard unit of length, the meter (abbreviated m), was originally indicated by two scratches on a bar made of a platinum–iridium alloy kept at the International Bureau of Weights and Measures in Sèvres, France. This length was chosen so that the distance between the equator and the North Pole along the meridian through Paris would be 10 million meters (Figure 1-1). The meter is now defined in terms of the speed of light—the meter is the distance light travels through empty space in 1/299,792,458 second. (This makes the speed of light exactly 299,792,458 m/s.)

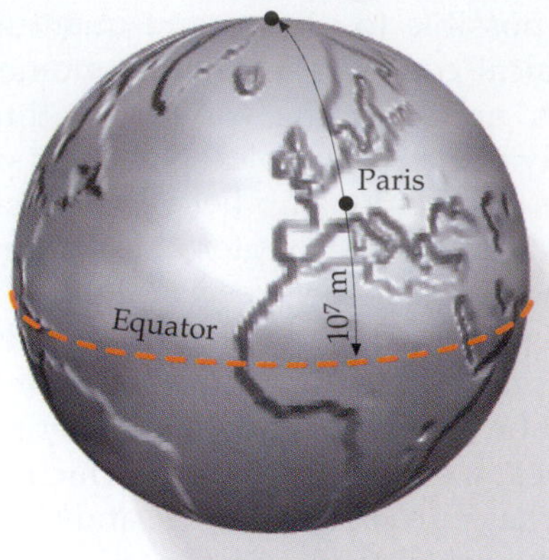

Figure 1-1 The meter was originally chosen so that the distance from the equator to the North Pole along the meridian through Paris would be 10^7 m.

The unit of time, the second (s), was originally defined in terms of the rotation of the earth and was equal to $(\frac{1}{60})(\frac{1}{60})(\frac{1}{24})$ of the mean solar day. The second is now defined in terms of a characteristic frequency associated with the cesium atom. All atoms, after absorbing energy, emit light with wavelengths and frequencies characteristic of the particular element. There is a set of wavelengths and frequencies for each element, with a particular frequency and wavelength associated with each energy transition within the atom. As far as we know, these frequencies remain constant. The second is defined so that the frequency of the light from a certain transition in cesium is 9,192,631,770 cycles per second. With these definitions, the fundamental units of length and time are accessible to laboratories throughout the world.

The unit of mass, the kilogram (kg), which equals 1000 grams (g), is defined to be the mass of a standard body, also kept at Sèvres. A duplicate of the standard 1-kg body is kept at the National Institute of Standards and Technology in Gaithersburg, Maryland. We shall discuss the concept of mass in detail in Chapter 4, where we will see that the weight of an object at a given point on earth is proportional to its mass. Thus, masses of ordinary size can be compared by weighing them.

In our study of thermodynamics and electricity, we shall need three more fundamental physical units: one for temperature, the kelvin (K) (formerly the degree Kelvin); one for the amount of a substance, the mole (mol); and one for electrical current, the ampere (A). There is another fundamental unit, the candela (cd) for luminous intensity, which we shall have no occasion to use in this book. These seven fundamental units, the meter (m), second (s), kilogram (kg), kelvin (K), ampere (A), mole (mol), and candela (cd), constitute the international system of units, or SI units.

The unit of every physical quantity can be expressed in terms of the fundamental SI units. Some frequently used combinations are given special names. For example, the SI unit of force, $\text{kg}\cdot\text{m/s}^2$, is called a newton (N). Similarly, the SI unit of power, $1\ \text{kg}\cdot\text{m}^2/\text{s}^3 = \text{N}\cdot\text{m/s}$, is called a watt (W).

Prefixes for common multiples and submultiples of SI units are listed in Table 1-1. These multiples are all powers of 10. Such a system is called a deci-

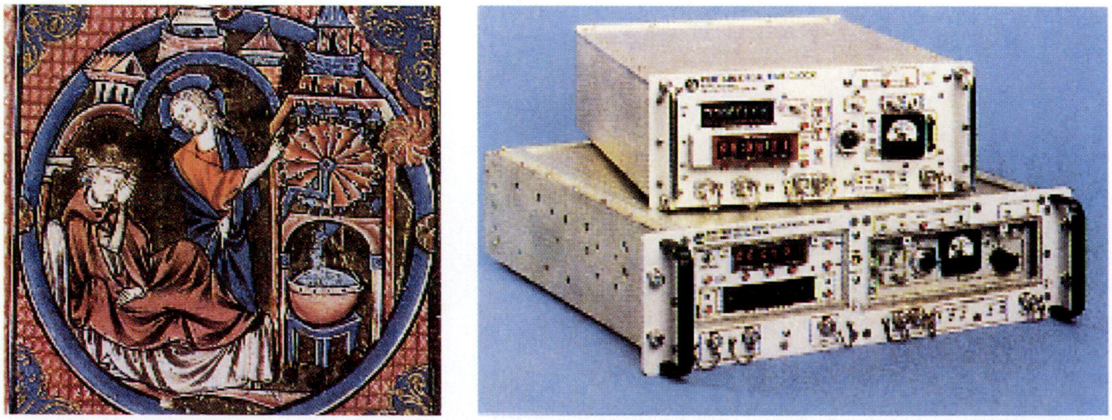

(*a*) (*b*)

(*a*) Water clock used to measure time intervals in the thirteenth century. (*b*) Modern portable cesium clock.

mal system. The decimal system based on the meter is called the metric system. The prefixes can be applied to any SI unit; for example, 0.001 second is 1 millisecond (ms); 1,000,000 watts is 1 megawatt (MW).

Table 1-1

Prefixes for Powers of 10

Multiple	Prefix	Abbreviation
10^{18}	exa	E
10^{15}	peta	P
10^{12}	tera	T
10^{9}	giga	G
10^{6}	mega	M
10^{3}	kilo	k
10^{2}	hecto	h
10^{1}	deka	da
10^{-1}	deci	d
10^{-2}	centi	c
10^{-3}	milli	m
10^{-6}	micro	μ
10^{-9}	nano	n
10^{-12}	pico	p
10^{-15}	femto	f
10^{-18}	atto	a

The prefixes hecto (h), deka (da), and deci (d) are not multiples of 10^3 or 10^{-3} and are rarely used. The other prefix that is not a multiple of 10^3 or 10^{-3} is centi (c), now used only with the meter, as in 1 cm = 10^{-2} m.

Other Systems of Units

Another decimal system still in use but gradually being replaced by SI units is the cgs system, based on the centimeter, gram, and second. The centimeter is defined as 0.01 m. The gram is now defined as 0.001 kg. Originally the gram was defined as the mass of one cubic centimeter of water. (The kilogram is then the mass of 1000 cubic centimeters or one liter of water.)

In another system of units, the U.S. customary system, a unit of force, the pound, is chosen to be a fundamental unit. In this system, the unit of mass is then defined in terms of the fundamental unit of force. The pound is defined in terms of the gravitational attraction of the earth at a particular place for a standard body. The fundamental unit of length in this system is the foot and the unit of time is the second, which has the same definition as the SI unit. The foot is defined as exactly one-third of a yard, which is now defined in terms of the meter:

$$1\text{ yd} = 0.9144\text{ m} \tag{1-1}$$

$$1\text{ ft} = \tfrac{1}{3}\text{ yd} = 0.3048\text{ m} \tag{1-2}$$

making the inch exactly 2.54 cm. This scheme is not a decimal system. It is less convenient than the SI or other decimal systems because common multiples of the unit are not powers of 10. For example, 1 yd = 3 ft and 1 ft = 12 in. We will see in Chapter 4 that mass is a better choice for a fundamental unit than force because mass is an intrinsic property of an object independent of its location. Relations between the U.S. customary system and SI units are given in Appendix A.

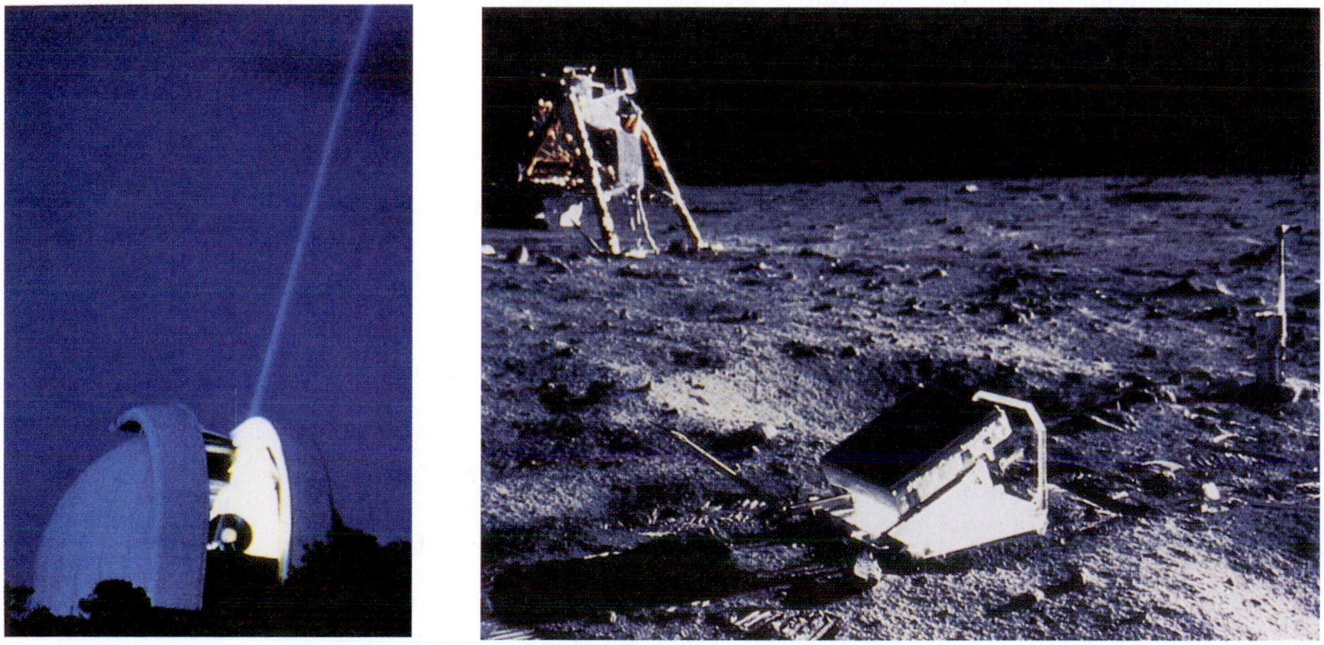

(*a*) (*b*)

(*a*) Laser beam from the Macdonald Observatory used to measure the distance to the moon. The distance can be measured to within a few centimeters by measuring the time required for the beam to go to the moon and back after reflecting off a mirror (*b*) placed on the moon by the Apollo 14 astronauts.

1-2 Conversion of Units

All physical quantities contain both a number and a unit. When such quantities are added, subtracted, multiplied, or divided in an algebraic equation, the unit can be treated like any other algebraic quantity. For example, suppose you wish to find the distance traveled in 3 hours (h) by a car moving at a constant rate of 80 kilometers per hour (km/h). The distance is the product of the speed v and the time t:

$$x = vt = \frac{80\text{ km}}{\cancel{\text{h}}} \times 3\cancel{\text{h}} = 240\text{ km}$$

We cancel the unit of time, the hours, just as we would any algebraic quantity to obtain the distance in the proper unit of length, the kilometer. This method of treating units makes it easy to convert from one unit of distance to another. Suppose we want to convert our answer of 240 km to miles (mi). Using the fact that 1 mi = 1.61 km, we divide each side of this equality by 1.61 km to obtain

$$\frac{1\text{ mi}}{1.61\text{ km}} = 1$$

Since any quantity can be multiplied by 1 without changing its value, we can now change 240 km to miles by multiplying by the factor (1 mi)/(1.61 km):

$$240\text{ km} = 240\,\cancel{\text{km}} \times \frac{1\text{ mi}}{1.61\,\cancel{\text{km}}} = 149\text{ mi}$$

The factor (1 mi)/(1.61 km) is called a **conversion factor**. All conversion factors have a value of 1 and are used to convert a quantity expressed in one unit of measure into its equivalent in another unit of measure. By writing out the units explicitly and canceling them, we need not think about whether we multiply by 1.61 or divide by 1.61 to change kilometers to miles because the units tell us whether we have chosen the correct—or incorrect—factor.

Example 1-1

If you drive 90 km/h, how fast are you going in meters per second and in miles per hour?

Picture the Problem We use the facts that 1000 m = 1 km, 60 s = 1 min, and 60 min = 1 h to convert to meters per second. The quantity 90 km/h is multiplied by a set of conversion factors each having the value 1, so the value of the speed is not changed. To convert to miles per hour, we use the conversion factor (1 mi)/(1.61 km) = 1.

1. Multiply 90 km/h by a set of conversion factors that convert kilometers to meters and hours to seconds:

$$\frac{90\,\cancel{\text{km}}}{\cancel{\text{h}}} \times \frac{1000\text{ m}}{1\,\cancel{\text{km}}} \times \frac{1\,\cancel{\text{h}}}{60\,\cancel{\text{min}}} \times \frac{1\,\cancel{\text{min}}}{60\text{ s}} = 25\text{ m/s}$$

2. Multiply 90 km/h by 1 mi/1.61 km:

$$\frac{90\,\cancel{\text{km}}}{\text{h}} \times \frac{1\text{ mi}}{1.61\,\cancel{\text{km}}} = 55.9\text{ mi/h}$$

Exercise What is the equivalent of 65 mi/h in meters per second? (*Answer* 29.1 m/s)

1-3 Dimensions of Physical Quantities

The area of a surface is found by multiplying one length by another. For example, the area of a rectangle of sides 2 m and 3 m is $A = (2\text{ m})(3\text{ m}) = 6\text{ m}^2$. The units of this area are square meters. Because area is the product of two lengths, it is said to have the dimensions of length times length, or length squared, often written L^2. The idea of dimensions is easily extended to other nongeometric quantities. For example, speed is said to have the dimension length divided by time, or L/T. The dimensions of other quantities such as force and energy are written in terms of the fundamental quantities of length, time, and mass. Adding two physical quantities makes sense only if the quantities have the same dimensions. For example, we cannot add an area to a speed to obtain a meaningful sum. If we have an equation like

$$A = B + C$$

the quantities A, B, and C must all have the same dimensions. Addition of B and C also requires that these quantities be in the same units. For example, if B is an area of 500 in^2 and C is 4 ft^2, we must either convert B into square feet or C into square inches in order to find the sum of the two areas.

We can often find mistakes in a calculation by checking the dimensions or units of the quantities in our result. Suppose, for example, that we mistakenly use the formula $A = 2\pi r$ for the area of a circle. We can see immediately that this cannot be correct because $2\pi r$ has dimensions of length, whereas area must have dimensions of length squared. Dimensional consistency is a necessary but not a sufficient condition for an equation to be correct. An equation can have the correct dimensions in each term without describing any physical situation. Table 1-2 gives the dimensions of some quantities we encounter in physics.

Table 1-2

Quantity	Symbol	Dimension
Area	A	L^2
Volume	V	L^3
Speed	v	L/T
Acceleration	a	L/T^2
Force	F	ML/T^2
Pressure (F/A)	p	M/LT^2
Density (M/V)	ρ	M/L^3
Energy	E	ML^2/T^2
Power (E/T)	P	ML^2/T^3

Example 1-2

The pressure in a fluid in motion depends on its density ρ and its speed v. Find a simple combination of density and speed that gives the correct dimensions of pressure.

Picture the Problem We note from Table 1-2 that both pressure and density have units of mass in the numerator, whereas speed does not contain M. We therefore divide the units of pressure by those of density and inspect the result.

1. Divide the units of pressure by those of density:

$$\frac{[p]}{[\rho]} = \frac{M/LT^2}{M/L^3} = \frac{L^2}{T^2}$$

2. By inspection, we note that the result has dimensions of v^2. The dimensions of pressure are thus the same as the dimensions of density times speed squared:

$$[p] = [\rho][v^2] = (M/L^3)(L/T)^2 = M/LT^2$$

Remarks When we study fluids in motion in Chapter 13, we will see from Bernoulli's law that for a fluid moving at a constant height, $p + \frac{1}{2}\rho v^2$ is constant where p is the pressure in the fluid. This is also known as the Venturi effect.

1-4 Scientific Notation

Handling very large or very small numbers is simplified by using scientific notation. In this notation, the number is written as a product of a number between 1 and 10 and a power of 10, such as 10^2 (= 100) or 10^3 (= 1000). For example, the number 12,000,000 is written 1.2×10^7; the distance from the earth to the sun, about 150,000,000,000 m, is written 1.5×10^{11} m. The number 11 in 10^{11} is called the **exponent**. For numbers smaller than 1, the exponent is negative. For example, $0.1 = 10^{-1}$, and $0.0001 = 10^{-4}$. The diameter of a virus, which is about 0.00000001 m, is written 1×10^{-8} m.

When numbers in scientific notation are multiplied, the exponents are added; when numbers are divided, the exponents are subtracted. These rules can be seen from some simple examples:

$$10^2 \times 10^3 = 100 \times 1000 = 100{,}000 = 10^5$$

Similarly,

$$\frac{10^2}{10^3} = \frac{100}{1000} = \frac{1}{10} = 10^{2-3} = 10^{-1}$$

In this notation, 10^0 is defined to be 1. To see why, suppose we divide 1000 by itself. We have

$$\frac{1000}{1000} = \frac{10^3}{10^3} = 10^{3-3} = 10^0 = 1$$

Example 1-3

In 12 g of carbon, there are $N_A = 6.02 \times 10^{23}$ carbon atoms (Avogadro's number). If you could count 1 atom per second, how long would it take to count the atoms in 1 g of carbon? Express your answer in years.

Picture the Problem We need to find the total number of atoms to be counted, N, and then use the fact that the number counted equals the counting rate R multiplied by the time t.

1. The time is equal to the total number of atoms N divided by the rate of counting R = 1 atom/s:

$$t = \frac{N}{R}$$

2. Find the number of carbon atoms in 1 g:

$$N = \frac{6.02 \times 10^{23}\ \text{atoms}}{12\ \text{g}} \times 1\ \text{g} = 5.02 \times 10^{22}\ \text{atoms}$$

3. Calculate the number of seconds it takes to count these at 1 per second:

$$t = \frac{N}{R} = \frac{5.02 \times 10^{22}\ \text{atoms}}{1\ \text{atom/s}} = 5.02 \times 10^{22}\ \text{s}$$

4. Calculate the number of seconds in a year:

$$n = \frac{365\ \text{d}}{1\ \text{y}} \times \frac{24\ \text{h}}{\text{d}} \times \frac{60\ \text{min}}{\text{h}} \times \frac{60\ \text{s}}{\text{min}} = 3.15 \times 10^7\ \text{s/y}$$

5. Use the conversion factor 3.15×10^7 s/y (a handy quantity to remember) to convert the answer in step 3 to years:

$$t = 5.02 \times 10^{22}\ \text{s} \times \frac{1\ \text{y}}{3.15 \times 10^7\ \text{s/y}}$$

$$= \frac{5.02}{3.15} \times 10^{22-7}\ \text{y} = 1.59 \times 10^{15}\ \text{y}$$

Remark The time required is about 100,000 times the age of the universe.

Exercise If you divided the task so that each person counted different atoms, how long would it take for 5 billion (5×10^9) people to count the atoms in 1 g of carbon? (*Answer* 3.19×10^5 y)

Example 1-4

A liter (L) is the volume of a cube that is 10 cm by 10 cm by 10 cm. If you drink 1 L of beer, how much volume in cubic centimeters and in cubic meters would it occupy in your stomach?

Picture the Problem The volume of a cube of side ℓ is $V = \ell^3$. The volume in cm^3 is found directly from $\ell = 10$ cm. To find the volume in m^3, convert cm^3 to m^3 using the conversion factor 1 cm $= 10^{-2}$ m.

1. Calculate the volume in cm^3: $V = \ell^3 = (10\text{ cm})^3 = 10^3\text{ cm}^3$

2. Convert to m^3: $$10^3\text{ cm}^3 = 10^3\text{ cm}^3 \times \left(\frac{10^{-2}\text{ m}}{1\text{ cm}}\right)^3$$ $$= 10^3\text{ cm}^3 \times \frac{10^{-6}\text{ m}^3}{1\text{ cm}^3} = 10^{-3}\text{ m}^3$$

Remark Note that the conversion factor (which equals 1) can be raised to the third power without changing its value, enabling us to cancel units.

Care is required when adding or subtracting numbers written in scientific notation when their exponents don't match. Consider, for example,

$$1.200 \times 10^2 + 8 \times 10^{-1} = 120.0 + 0.8 = 120.8$$

To find the sum without converting both numbers into ordinary decimal form, it is sufficient to rewrite either of the numbers so that its power of 10 is the same as that of the other. For example, we can find the sum by writing $1.200 \times 10^2 = 1200 \times 10^{-1}$ and then adding

$$1200 \times 10^{-1} + 8 \times 10^{-1} = 1208 \times 10^{-1} = 120.8$$

When the exponents are very different, one of the numbers is much smaller than the other. The smaller number can often be neglected in addition or subtraction. For example,

$$(2 \times 10^6) + (9 \times 10^{-3}) = 2{,}000{,}000 + 0.009$$
$$= 2{,}000{,}000.009 \approx 2 \times 10^6$$

where the symbol $\approx$ means "is approximately equal to."

When raising a power to another power, the exponents are multiplied. For example,

$$(10^2)^4 = 10^2 \times 10^2 \times 10^2 \times 10^2 = 10^8$$

1-5 Significant Figures and Order of Magnitude

Many of the numbers in science are the result of measurement and are therefore known only to within some degree of experimental uncertainty. The magnitude of the uncertainty depends on the skill of the experimenter and the apparatus used, and often can only be estimated. A rough indication of the uncertainty in a measurement is implied by the number of digits used. For example, if we say that a table is 2.50 m long, we are implying that its length is between 2.495 m and 2.505 m. That is, we know the length to about ± 0.005 m $= \pm 0.5$ cm. If we used a meterstick with millimeter markings and measured the table length carefully, we might estimate that we could measure the length to ± 0.5 mm rather than ± 0.5 cm. We would indicate this precision when giving the length by using four digits, such as 2.503 m. A reliably known digit (other than a zero used to locate the decimal point) is called a significant figure. The number 2.50 has three significant figures; 2.503 m has four. The number 0.00103 has three significant figures. (The first three zeroes are not significant figures but merely locate the decimal point.) In scientific notation, the number 0.00103 is written 1.03×10^{-3}. A common student error is to carry more digits than the certainty of measurement warrants. Suppose, for example, that you measure the area of a circular playing field by pacing off the radius and using the formula for the area of a circle, $A = \pi r^2$. If you estimate the radius to be 8 m and use a 10-digit calculator to compute the area, you obtain $\pi(8\text{ m})^2 = 201.0619298\text{ m}^2$. The digits after the decimal point give a false indication of the accuracy with which you know the area. If you found the radius by pacing, you might expect that your measurement was accurate to only about 0.5 m. That is, the radius could be as great as 8.5 m or as small as 7.5 m. If the radius is 8.5 m, the area is $\pi(8.5\text{ m})^2 = 226.9800692\text{ m}^2$, whereas if it is 7.5 m, the area is $\pi(7.5\text{ m})^2 = 176.714587\text{ m}^2$. There is a general rule to guide you when combining several numbers in multiplication or division:

> The number of significant figures in the result of multiplication or division is no greater than the least number of significant figures in any of the factors.

In the previous example, the radius is known to only one significant figure, so the area is also known only to one significant figure. It should be written as $2 \times 10^2\text{ m}^2$, which implies that the area is somewhere between 150 m^2 and 250 m^2.

The precision of the sum or difference of two measurements is only as good as the precision of the least precise of the two measurements. A general rule is

> The result of addition or subtraction of two numbers has no significant figures beyond the last decimal place where both of the original numbers had significant figures.

Example 1-5

Find the sum of 1.040 and 0.21342.

Picture the Problem The first number, 1.040, has only three significant figures beyond the decimal point, whereas the second, 0.21342 has five. According to the rule stated above, the sum can have only three significant figures beyond the decimal point.

Sum the numbers keeping only 3 digits beyond the decimal point: $1.040 + 0.21342 = 1.253$

Exercise Apply the appropriate rule for significant figures to calculate the following: (*a*) 1.58×0.03 (*b*) $1.4 + 2.53$ (*c*) $2.34 \times 10^2 + 4.93$. (*Answers* (*a*) 0.05, (*b*) 3.9, (*c*) 2.39×10^2)

Most examples and exercises in this book will be done with data to three (or sometimes four) significant figures, but occasionally we will say, for example, that a table top is 3 ft by 8 ft rather than taking the time and space to say it is 3.00 ft by 8.00 ft. Any data you see in an example or exercise can be assumed to be known to three significant figures unless otherwise indicated. In doing rough calculations or comparisons, we sometimes round off a number to the nearest power of 10. Such a number is called an **order of magnitude**. For example, the height of a small insect, say an ant, might be 8×10^{-4} m $\approx 10^{-3}$ m. We would say that the order of magnitude of the height of an ant is 10^{-3} m. Similarly, though the height of most people is about 2 m, we might round that off and say that the order of magnitude of the height of a person is 10^0 m. We do not mean to imply that a typical height is really 1 m but that it is closer to 1 m than to 10 m or to $10^{-1} = 0.1$ m. We might say that a typical human being is three orders of magnitude taller than a typical ant, meaning that the ratio of heights is about 1000 to 1. An order of magnitude does not provide any digits that are reliably known. It may be thought of as having no significant figures. Table 1-3 gives typical order-of-magnitude values for a variety of masses, sizes, and time intervals encountered in physics.

Table 1-3

The Universe by Orders of Magnitude

Size or Distance	(m)	Mass	(kg)	Time Interval	(s)
Proton	10^{-15}	Electron	10^{-30}	Time for light to cross nucleus	10^{-23}
Atom	10^{-10}	Proton	10^{-27}	Period of visible light radiation	10^{-15}
Virus	10^{-7}	Amino acid	10^{-25}	Period of microwaves	10^{-10}
Giant amoeba	10^{-4}	Hemoglobin	10^{-22}	Half-life of muon	10^{-6}
Walnut	10^{-2}	Flu virus	10^{-19}	Period of highest audible sound	10^{-4}
Human	10^{0}	Giant amoeba	10^{-8}	Period of human heartbeat	10^{0}
Highest mountain	10^{4}	Raindrop	10^{-6}	Half-life of free neutron	10^{3}
Earth	10^{7}	Ant	10^{-2}	Period of earth's rotation	10^{5}
Sun	10^{9}	Human	10^{2}	Period of earth's revolution around sun	10^{7}
Distance from earth to sun	10^{11}	Saturn V rocket	10^{6}	Lifetime of human	10^{9}
Solar system	10^{13}	Pyramid	10^{10}	Half-life of plutonium-239	10^{12}
Distance to nearest star	10^{16}	Earth	10^{24}	Lifetime of mountain range	10^{15}
Milky Way galaxy	10^{21}	Sun	10^{30}	Age of earth	10^{17}
Visible universe	10^{26}	Milky Way galaxy	10^{41}	Age of universe	10^{18}
		Universe	10^{52}		

Order of Magnitude

A Benzene molecules of the order of 10^{-10} m in diameter as seen in a scanning electron microscope.

B Chromosomes measuring of the order of 10^{-6} m across as seen in a scanning electron microscope.

C Distances familiar in our everyday world. The height of the girl is of the order of 10^{0} m and that of the mountain is of the order of 10^{4} m.

D Earth with a diameter of the order of 10^{7} m as seen from space.

E The diameter of the Andromeda galaxy is of the order of 10^{21}m.

In many cases the order of magnitude of a quantity can be estimated using reasonable assumptions and simple calculations. The physicist Enrico Fermi was a master at using cunning order-of-magnitude estimations to generate answers for questions that seemed impossible to calculate because of lack of information. Problems like these are often called **Fermi problems**. The following is an example of a Fermi problem.

Example 1-6

What thickness of rubber tread is worn off the tire of an automobile as it travels 1 km (0.6 mi)?

Picture the Problem We assume the tread thickness of a new tire is 1 cm. This may be off by a factor of 2 or so, but 1 mm is certainly too small and 10 cm is too large. Since tires have to be replaced after about 60,000 km (about 37,000 mi), we will assume that the tread is completely worn off after 60,000 km. In other words, the rate of wear is 1 cm of tire per 60,000 km of travel.

Use 1 cm wear per 60,000 km travel to compute the thickness worn after 1 km of travel:

$$\frac{1\text{ cm wear}}{60{,}000\text{ km travel}} = \frac{1.7\times10^{-5}\text{ cm wear}}{1\text{ km travel}}$$

$$\approx 0.2\ \mu\text{m wear/km of travel}$$

Summary

Topic	Remarks and Relevant Equations
1. Units	The magnitude of a physical quantity (for example, length, time, force, and energy) is expressed as a number times a unit.
Fundamental units	The fundamental SI units (short for *Système Internationale*) are the meter (m), the second (s), the kilogram (kg), the kelvin (K), the ampere (A), the mole (mol), and the candela (cd). The unit of every physical quantity can be expressed in terms of these fundamental units.
Units in equations	Units in equations are treated just like any other algebraic quantity.
Conversion	Conversion factors, which are always equal to 1, provide a convenient method for converting from one kind of unit to another.
2. Dimensions	The two sides of an equation must have the same dimensions.
3. Scientific Notation	For convenience, very small and very large numbers are generally written as a factor times a power of 10.
4. Exponents	
Multiplication	When multiplying two numbers, the exponents are added.
Division	When dividing two numbers, the exponents are subtracted.
Raising to a power	When a number containing an exponent is itself raised to an exponent, the exponents are multiplied.

5. Significant Figures		
Multiplication and division	The number of significant figures in the result of multiplication or division is no greater than the least number of significant figures in any of the numbers.	
Addition and subtraction	The result of addition or subtraction of two numbers has no significant figures beyond the last decimal place where both of the original numbers had significant figures.	
6. Order of Magnitude	A number rounded to the nearest power of 10 is called an order of magnitude. The order of magnitude of a quantity can often be estimated using reasonable assumptions and simple calculations.	

Problem-Solving Guide

Summary of Worked Examples

Type of Calculation	Procedure and Relevant Examples	
1. Units		
Convert a quantity from one set of units to another.	Multiply by a conversion factor that has the value 1.	**Examples 1-1, 1-3**
2. Dimensions		
Analyze the dimensions of quantities to determine how one quantity depends on two or more other quantities.	Inspect the dimensions and try various combinations.	**Example 1-2**
3. Scientific Notation		
Multiply two numbers.	Write the numbers in powers of 10 notation, then add the exponents.	**Example 1-3**
Divide two numbers.	Write the numbers in powers of 10 notation, then subtract the exponents.	**Example 1-3**
Add or subtract two numbers.	Write both numbers so they have the same power of 10.	
4. Significant Figures		
Multiply or divide two numbers.	Round off the answer to the lesser number of significant figures of either factor.	
Add or subtract two numbers.	Round off the answer to the lesser number of significant figures beyond the decimal point of either of the factors.	**Example 1-5**
5. Fermi Problems	Supply needed data by order-of-magnitude estimates.	**Example 1-6**

Problems

Conceptual Problems

Problems from Optional and Exploring sections

In a few problems, you are given more data than you actually need; in a few other problems, you are required to supply data from your general knowledge, outside sources, or informed estimates.

• Single-concept, single-step, relatively easy
•• Intermediate-level, may require synthesis of concepts
••• Challenging, for advanced students

Units

1 • Which of the following is *not* one of the fundamental physical quantities in the SI system?
(*a*) mass (*b*) length (*c*) force (*d*) time
(*e*) All of the above are fundamental physical quantities.

2 • In doing a calculation, you end up with m/s in the numerator and m/s^2 in the denominator. What are your final units?
(*a*) m^2/s^3 (*b*) 1/s (*c*) s^3/m^2 (*d*) s (*e*) m/s

3 • Write the following using the prefixes listed in Table 1-1 and the abbreviations listed on the inside cover; for example, 10,000 meters = 10 km. (*a*) 1,000,000 watts, (*b*) 0.002 gram, (*c*) 3×10^{-6} meter, (*d*) 30,000 seconds.

4 • Write each of the following without using prefixes: (*a*) 40 μW, (*b*) 4 ns, (*c*) 3 MW, (*d*) 25 km.

5 • Write out the following (which are not SI units) without using any abbreviations. For example, 10^3 meters = 1 kilometer. (*a*) 10^{-12} boo, (*b*) 10^9 low, (*c*) 10^{-6} phone, (*d*) 10^{-18} boy, (*e*) 10^6 phone, (*f*) 10^{-9} goat, (*g*) 10^{12} bull.

6 •• In the following equations, the distance x is in meters, the time t is in seconds, and the velocity v is in meters per second. What are the SI units of the constants C_1 and C_2?
(*a*) $x = C_1 + C_2 t$ (*b*) $x = \frac{1}{2} C_1 t^2$ (*c*) $v^2 = 2C_1 x$
(*d*) $x = C_1 \cos C_2 t$ (*e*) $v = C_1 e^{-C_2 t}$
(*Hint:* The arguments of trigonometric functions and exponentials must be dimensionless. The "argument" of $\cos\theta$ is θ and that of e^x is x.)

7 •• If x is in feet, t is in seconds, and v is in feet per second, what are the units of the constants C_1 and C_2 in each part of Problem 6?

Conversion of Units

8 • From the original definition of the meter in terms of the distance from the equator to the North Pole, find in meters (*a*) the circumference of the earth, and (*b*) the radius of the earth. (*c*) Convert your answers for (*a*) and (*b*) from meters into miles.

9 • The speed of sound in air is 340 m/s. What is the speed of a supersonic plane that travels at twice the speed of sound? Give your answer in kilometers per hour and miles per hour.

10 • A basketball player is 6 ft $10\frac{1}{2}$ in tall. What is his height in centimeters?

11 • Complete the following:
(*a*) 100 km/h = ______mi/h. (*b*) 60 cm = ______in.
(*c*) 100 yd = ______m.

12 • The main span of the Golden Gate Bridge is 4200 ft. Express this distance in kilometers.

13 • Find the conversion factor to convert from miles per hour into kilometers per hour.

14 • Complete the following:
(*a*) 1.296×10^5 km/h^2 = ______km/h·s.
(*b*) 1.296×10^5 km/h^2 = ______m/s^2
(*c*) 60 mi/h = ______ft/s. (*d*) 60 mi/h = ______m/s.

15 • There are 1.057 quarts in a liter and 4 quarts in a gallon. (*a*) How many liters are there in a gallon? (*b*) A barrel equals 42 gallons. How many cubic meters are there in a barrel?

16 • There are 640 acres in a square mile. How many square meters are there in one acre?

17 •• A right circular cylinder has a diameter of 6.8 in and a height of 2 ft. What is the volume of the cylinder in (*a*) cubic feet, (*b*) cubic meters, (*c*) liters?

18 •• In the following, x is in meters, t is in seconds, v is in meters per second, and the acceleration a is in meters per second squared. Find the SI units of each combination:
(*a*) v^2/x (*b*) $\sqrt{x/a}$ (*c*) $\frac{1}{2}at^2$

Dimensions of Physical Quantities

19 • What are the dimensions of the constants in each part of Problem 6?

20 •• The law of radioactive decay is $N(t) = N_0 e^{-\lambda t}$, where N_0 is the number of radioactive nuclei at $t = 0$, $N(t)$ is the number remaining at time t, and λ is a quantity known as the decay constant. What is the dimension of λ?

21 •• The SI unit of force, the kilogram-meter per second squared (kg·m/s^2) is called the newton (N). Find the dimensions and the SI units of the constant G in Newton's law of gravitation $F = Gm_1m_2/r^2$.

22 •• An object on the end of a string moves in a circle. The force exerted by the string has units of ML/T^2 and depends on the mass of the object, its speed, and the radius of the circle. What combination of these variables gives the correct dimensions?

23 •• Show that the product of mass, acceleration, and speed has the dimension of power.

24 •• The momentum of an object is the product of its velocity and mass. Show that momentum has the dimension of force multiplied by time.

25 •• What combination of force and one other physical quantity has the dimension of power?

26 •• When an object falls through air, there is a drag force that depends on the product of the surface area of the object and the square of its velocity, i.e., $F_{air} = CAv^2$, where C is a constant. Determine the dimension of C.

27 •• Kepler's third law relates the period of a planet to its radius r, the constant G in Newton's law of gravitation ($F = Gm_1m_2/r^2$), and the mass of the sun M_S. What combination of these factors gives the correct dimensions for the period of a planet?

Scientific Notation and Significant Figures

28 • The prefix giga means ______.

(*a*) 10^3 (*b*) 10^6 (*c*) 10^9 (*d*) 10^{12} (*e*) 10^{15}

29 • The prefix mega means ______.

(*a*) 10^{-9} (*b*) 10^{-6} (*c*) 10^{-3} (*d*) 10^6 (*e*) 10^9

30 • The prefix pico means ______.

(*a*) 10^{-12} (*b*) 10^{-6} (*c*) 10^{-3} (*d*) 10^6 (*e*) 10^9

31 • The number 0.0005130 has ______ significant figures.

(*a*) one (*b*) three (*c*) four (*d*) seven (*e*) eight

32 • The number 23.0040 has ______ significant figures.

(*a*) two (*b*) three (*c*) four (*d*) five (*e*) six

33 • Express as a decimal number without using powers of 10 notation:

(*a*) 3×10^4 (*b*) 6.2×10^{-3} (*c*) 4×10^{-6} (*d*) 2.17×10^5

34 • Write the following in scientific notation.

(*a*) 3.1 GW = ______W. (*b*) 10 pm = ______m.
(*c*) 2.3 fs = ______s. (*d*) 4 μs = ______s.

35 • Calculate the following, round off to the correct number of significant figures, and express your result in scientific notation.

(*a*) $(1.14)(9.99 \times 10^4)$ (*b*) $(2.78 \times 10^{-8}) - (5.31 \times 10^{-9})$
(*c*) $12\pi/(4.56 \times 10^{-3})$ (*d*) $27.6 + (5.99 \times 10^2)$

36 • Calculate the following, round off to the correct number of significant figures, and express your result in scientific notation.

(*a*) $(200.9)(569.3)$ (*b*) $(0.000000513)(62.3 \times 10^7)$
(*c*) $28{,}401 + (5.78 \times 10^4)$ (*d*) $63.25/(4.17 \times 10^{-3})$

37 • A cell membrane has a thickness of about 7 nm. How many cell membranes would it take to make a stack 1 in high?

38 • Calculate the following, round off to the correct number of significant figures, and express your result in scientific notation.

(*a*) $(2.00 \times 10^4)(6.10 \times 10^{-2})$ (*b*) $(3.141592)(4.00 \times 10^5)$
(*c*) $(2.32 \times 10^3)/(1.16 \times 10^8)$ (*d*) $(5.14 \times 10^3) + (2.78 \times 10^2)$
(*e*) $(1.99 \times 10^2) + (9.99 \times 10^{-5})$

39 • Perform the following calculations and round off the answers to the correct number of significant figures:

(*a*) $3.141592654 \times (23.2)^2$ (*b*) $2 \times 3.141592654 \times 0.76$
(*c*) $4/3\pi \times (1.1)^3$ (*d*) $(2.0)^5/(3.141592654)$

40 •• The sun has a mass of 1.99×10^{30} kg and is composed mostly of hydrogen, with only a small fraction being heavier elements. The hydrogen atom has a mass of 1.67×10^{-27} kg. Estimate the number of hydrogen atoms in the sun.

General Problems

41 • What are the advantages and disadvantages of using the length of your arm for a standard length?

42 • A certain clock is known to be consistently 10% fast compared with the standard cesium clock. A second clock varies in a random way by 1%. Which clock would make a more useful secondary standard for a laboratory? Why?

43 • True or false:

(*a*) Two quantities to be added must have the same dimensions.
(*b*) Two quantities to be multiplied must have the same dimensions.
(*c*) All conversion factors have the value 1.

44 • On many of the roads in Canada the speed limit is 100 km/h. What is the speed limit in miles per hour?

45 • If one could count \$1 per second, how many years would it take to count 1 billion dollars (1 billion = 10^9)?

46 • Sometimes a conversion factor can be derived from the knowledge of a constant in two different systems. (*a*) The speed of light in vacuum is 186,000 mi/s = 3×10^8 m/s. Use this fact to find the number of kilometers in a mile. (*b*) The weight of 1 ft^3 of water is 62.4 lb. Use this and the fact that 1 cm^3 of water has a mass of 1 g to find the weight in pounds of a 1-kg mass.

47 •• The mass of one uranium atom is 4.0×10^{-26} kg. How many uranium atoms are there in 8 g of pure uranium?

48 •• During a thunderstorm, a total of 1.4 in of rain falls. How much water falls on one acre of land? (640 acre = 1 mi^2.)

49 •• The angle subtended by the moon's diameter at a point on the earth is about 0.524° (Figure 1-2). Use this and the fact that the moon is about 384 Mm away to find the diameter of the moon. (The angle subtended by the moon θ is approximately D/r_m, where D is the diameter of the moon and r_m is the distance to the moon.)

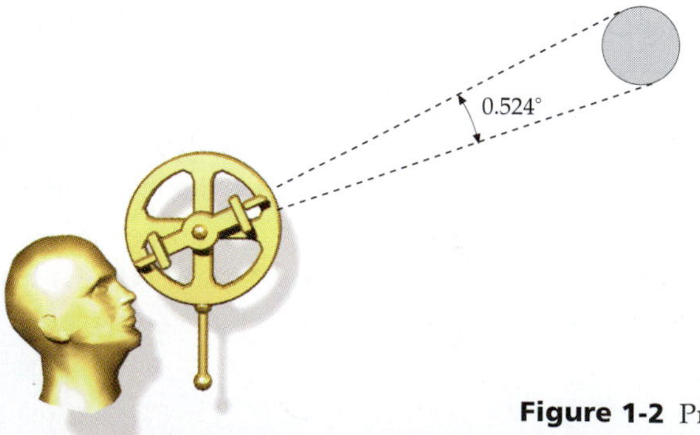

Figure 1-2 Problem 49

50 •• The United States imports 6 million barrels of oil per day. This imported oil provides about one-fourth of our total energy. A barrel fills a drum that stands about 1 m high. (*a*) If the barrels are laid end to end, what is the length in kilometers of barrels of oil imported each day? (*b*) The largest tankers hold about a quarter-million barrels. How many tanker loads per year would supply our imported oil? (*c*) If oil costs \$20 a barrel, how much do we spend for oil each year?

51 •• Every year the United States generates 160 million tons of municipal solid waste and a grand total of 10 billion tons of solid waste of all kinds. If one allows one cubic meter of volume per ton, how many square miles of area at an average height of 10 m is needed for landfill each year?

52 •• An iron nucleus has a radius of 5.4×10^{-15} m and a mass of 9.3×10^{-26} kg. (*a*) What is its mass per unit volume in $\mathrm{kg/m^3}$? (*b*) If the earth had the same mass per unit volume, what would its radius be? (The mass of the earth is 5.98×10^{24} kg.)

53 •• Evaluate the following expressions.

(*a*) $(5.6 \times 10^{-5})(0.0000075)/(2.4 \times 10^{-12})$
(*b*) $(14.2)(6.4 \times 10^{7})(8.2 \times 10^{-9}) - 4.06$
(*c*) $(6.1 \times 10^{-6})^2(3.6 \times 10^{4})^3/(3.6 \times 10^{-11})^{1/2}$
(*d*) $(0.000064)^{1/3}/[(12.8 \times 10^{-3})(490 \times 10^{-1})^{1/2}]$

54 •• The astronomical unit is defined in terms of the distance from the earth to the sun, namely 1.496×10^{11} m. The parsec is the radial length that one astronomical unit of arc length subtends at an angle of 1 s. The light-year is the distance that light travels in one year. (*a*) How many parsecs are there in one astronomical unit? (*b*) How many meters are in a parsec? (*c*) How many meters in a light-year? (*d*) How many astronomical units in a light-year? (*e*) How many light-years in a parsec?

55 •• If the average density of the universe is at least $6 \times 10^{-27}\ \mathrm{kg/m^3}$, then the universe will eventually stop expanding and begin contracting. (*a*) How many electrons are needed in a cubic meter to produce the critical density? (*b*) How many protons per cubic meter would produce the critical density? (The mass of an electron and of a proton can be found on the endpapers of this book.)

56 •• Observational estimates of the density of the universe yield an average of about $2 \times 10^{-28}\ \mathrm{kg/m^3}$. (*a*) If a 100-kg football player had his mass uniformly spread out in a sphere to match the estimate for the average mass density of the universe, what would be the radius of the sphere? (*b*) Compare this radius with the earth–moon distance of 3.82×10^{8} m.

57 •• Beer and soft drinks are sold in aluminum cans. The mass of a typical can is about 0.018 kg. (*a*) Estimate the number of aluminum cans used in the United States in one year. (*b*) Estimate the total mass of aluminum in a year's consumption from these cans. (*c*) If aluminum returns \$1/kg at a recycling center, how much is a year's accumulation of aluminum cans worth?

58 •• An aluminum rod is 8.00024 m long at 20.00°C. If the rod's temperature increases, it expands such that it lengthens by 0.0024% per degree temperature rise. Determine the rod's length at 28.00°C and at 31.45°C.

59 ••• The table below gives experimental results for a measurement of the period of motion T of an object of mass m suspended on a spring versus the mass of the object. These data are consistent with a simple equation expressing T as a function of m of the form $T = Cm^n$, where C and n are constants and n is not necessarily an integer. (*a*) Find n and C. (There are several ways to do this. One is to guess the value of n and check by plotting T versus m^n on graph paper. If your guess is right, the plot will be a straight line. Another is to plot log T versus log m. The slope of the straight line on this plot is n.) (*b*) Which data points deviate the most from a straight-line plot of T versus m^n?

Mass *m*, kg	0.10	0.20	0.40	0.50	0.75	1.00	1.50
Period *T*, s	0.56	0.83	1.05	1.28	1.55	1.75	2.22

60 ••• The table below gives the period T and orbit radius r for the motions of four satellites orbiting a dense, heavy asteroid. (*a*) These data can be fitted by the formula $T = Cr^n$. Find C and n. (*b*) A fifth satellite is discovered to have a period of 6.20 y. Find the radius for the orbit of this satellite, which fits the same formula.

Period *T*, y	0.44	1.61	3.88	7.89
Radius *r*, Gm	0.088	0.208	0.374	0.600

61 ••• The period T of a simple pendulum depends on the length L of the pendulum and the acceleration of gravity g (dimensions L/T^2). (*a*) Find a simple combination of L and g which has the dimensions of time. (*b*) Check the dependence of the period T on the length L by measuring the period (time for a complete swing back and forth) of a pendulum for two different values of L. (*c*) The correct formula relating T to L and g involves a constant that is a multiple of π, and cannot be obtained by the dimensional analysis of part (*a*). It can be found by experiment as in (*b*) if g is known. Using the value $g = 9.81\ \mathrm{m/s^2}$ and your experimental results from (*b*), find the formula relating T to L and g.

62 ••• The weight of the earth's atmosphere pushes down on the surface of the earth with a force of 14.7 lb for each square inch of earth's surface. What is the weight in pounds of the earth's atmosphere? (The radius of the earth is about 6370 km.)

63 ••• Each binary digit is termed a bit. A series of bits grouped together is called a word. An eight-bit word is called a byte. Suppose a computer hard disk has a capacity of 2 gigabytes. (*a*) How many bits can be stored on the disk? (*b*) Estimate the number of typical books that can be stored on the disk.

PART I

mechanics

Modern version of Galileo's legendary experiment in which he dropped a heavy ball and a light ball from the Leaning Tower of Pisa. In this demonstration, a feather and an apple in a large vacuum chamber fall with the same acceleration due to gravity.

CHAPTER 2

Motion in One Dimension

The most coveted records in top fuel dragstrip racing are the elapsed time to complete a quarter mile from a standing start, and the top speed, measured over the last 66 feet of a quarter mile sprint. The record for elapsed time as of early 1998 is 4.564 seconds, and the top speed is 321.78 miles per hour.

We begin our study of the physical universe by examining objects in motion. The study of motion, whose measurement more than 400 years ago gave birth to physics, is called **kinematics**. We start with the simplest case, the motion of a particle along a straight line, like that of a car moving along a flat, straight, narrow road. A particle is an object whose position can be described by a single point. Anything can be considered to be a particle—a molecule, a person, or a galaxy—as long as we can reasonably ignore its internal structure.

2-1 Displacement, Velocity, and Speed

Figure 2-1 shows a car at position x_1 at time t_1 and at position x_2 at a later time t_2. The change in the car's position, called the **displacement,** is given by $x_2 - x_1$. We use the Greek letter Δ (uppercase delta) to indicate the change in a quantity; thus, the change in x is written Δx:

$$\Delta x = x_2 - x_1 \qquad 2\text{-}1$$

Definition—Displacement

The notation Δx (read "delta x") stands for a single quantity, the change in x. It is not a product of Δ and x any more than $\cos\theta$ is a product of cos and θ. By convention, the change in a quantity is always its final value minus its initial value.

Velocity is the rate at which the position changes. The **average velocity** of the particle is defined as the ratio of the displacement Δx to the time interval $\Delta t = t_2 - t_1$:

$$v_{av} = \frac{\Delta x}{\Delta t} = \frac{x_2 - x_1}{t_2 - t_1} \qquad \text{2-2}$$

Definition—Average velocity

Displacement and average velocity may be positive or negative. A positive value indicates motion in the positive x direction. The SI unit of velocity is meters per second (m/s), and the U.S. customary unit is feet per second (ft/s).

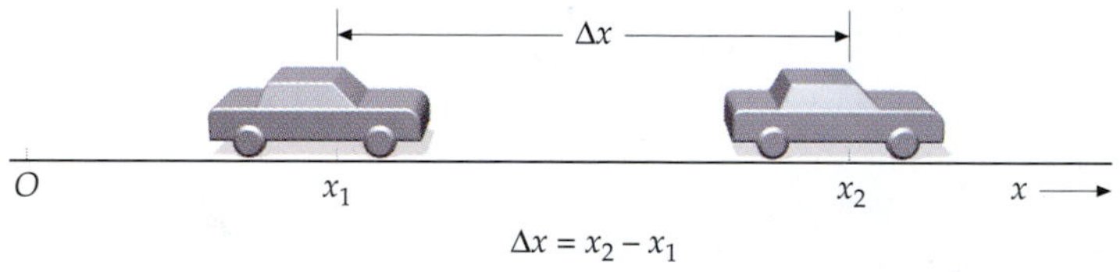

Figure 2-1 A car is moving in a coordinate system consisting of a line with a point chosen to be the origin O. Other points on the line are assigned a number x. The value of x depends on its distance from O. Points to the right of O are positive and points to the left are negative. When the car travels from point x_1 to point x_2, its displacement is $\Delta x = x_2 - x_1$.

Example 2-1

An errant space probe is traveling directly toward the sun. At time t_1 it is at $x_1 = 3.0 \times 10^{12}$ m relative to the sun (Figure 2-2). Exactly one year later, it is at $x_2 = 2.1 \times 10^{12}$ m. Find its displacement and average velocity.

Picture the Problem We are given x_1 and x_2. If we choose $t_1 = 0$, then $t_2 = 1\text{ y} = 3.16 \times 10^7$ s. The average velocity is $\Delta x/\Delta t$.

Figure 2-2

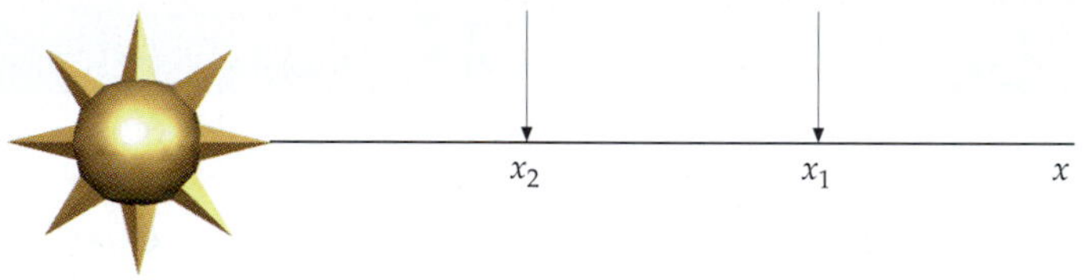

1. The displacement is found from its definition: $\Delta x = x_2 - x_1 = 2.1 \times 10^{12}\text{ m} - 3.0 \times 10^{12}\text{ m} = -9 \times 10^{11}\text{ m}$

2. The average velocity is the displacement divided by the time interval: $v_{av} = \frac{\Delta x}{\Delta t} = \frac{-9 \times 10^{11}\text{ m}}{3.16 \times 10^7\text{ s}}$
$= -2.85 \times 10^4\text{ m/s} = -28.5\text{ km/s}$

Remark Both displacement and average velocity are negative, because the probe moved toward smaller values of x. Note that the units, m for Δx, and m/s or km/s for v_{av}, are essential parts of the answers. It is meaningless to say "the displacement is -9×10^{11}" or "the average velocity of a particle is −28.5."

Exercise A jet plane leaves the gate in Detroit at 2:15 P.M. It arrives at the gate in Chicago, 438 km away, having completed the journey with an average velocity of 500 km/h. What is the arrival time of the flight in Chicago? (*Answer* 3:08 P.M. Detroit time, which is actually 2:08 P.M. Chicago time.)

Example 2-2 *try it yourself*

In a 100-m footrace, you cover the first 50 m with an average velocity of 10 m/s and the second 50 m with an average velocity of 8 m/s. What is your average velocity for the entire 100 m?

Picture the Problem The total displacement is $\Delta x = 100$ m. Find the total time by adding the times for each part of the race. Then compute v_{av} from its definition, $v_{av} = (\text{total displacement})/(\text{total time}) = \Delta x/\Delta t$.

Cover the column to the right and try these on your own before looking at the answers.

Steps	*Answers*
1. Write the average velocity in terms of the total displacement and total time interval.	$v_{av} = \dfrac{\Delta x}{\Delta t}$
2. Compute the time for the first 50 m.	$\Delta t_1 = 5$ s
3. Compute the time for the second 50 m.	$\Delta t_2 = 6.25$ s
4. Calculate the total time.	$\Delta t = \Delta t_1 + \Delta t_2 = 11.25$ s
5. Use your result for the total time to compute the average velocity.	$v_{av} = \dfrac{\Delta x}{\Delta t} = 8.89$ m/s

Check the Result Given the statement of the problem, the answer should be between 8 m/s and 10 m/s, which it is. Since more time is spent running at 8 m/s than at 10 m/s, the average velocity is closer to the lower value.

Exercise A car travels 80 km in a straight line. If the first 40 km is covered with an average velocity of 80 km/h, and the total trip takes 1.2 h, what was the average velocity during the second 40 km? (*Answer* 57.1 km/h)

The **average speed** of a particle is the ratio of the total distance traveled to the total time from start to finish:

$$\text{Average speed} = \frac{\text{total distance}}{\text{total time}} = \frac{\Delta s}{\Delta t} \qquad \text{2-3}$$

Since the total distance and total time are both always positive, the average speed is always positive.

Example 2-3

You run 100 m in 12 s, then turn around and jog 50 m back toward the starting point in 30 s (Figure 2-3). Calculate (*a*) your average speed, and (*b*) your average velocity for the total trip.

Figure 2-3

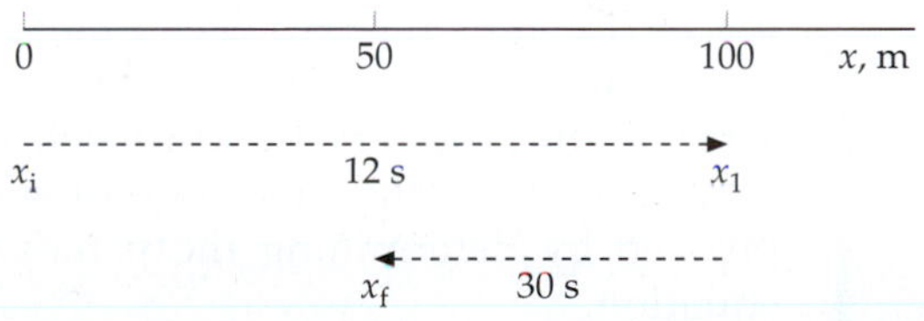

Picture the Problem We use the definitions of average speed and average velocity, noting that average *speed* is the total *distance* divided by Δt, whereas average *velocity* is the *net displacement* divided by Δt.

(*a*) 1. Your average speed equals the total distance divided by the total time: $\text{Average speed} = \dfrac{\Delta s}{\Delta t}$

2. Calculate the total distance traveled and the total time: $\Delta s = 100\text{ m} + 50\text{ m} = 150\text{ m}$

$\Delta t = 12\text{ s} + 30\text{ s} = 42\text{ s}$

3. Use s and t to find your average speed: $\text{Average speed} = \dfrac{150\text{ m}}{42\text{ s}} = 3.57\text{ m/s}$

(*b*) 1. Your average velocity is the ratio of the net displacement Δx to the time interval Δt: $v_{av} = \dfrac{\Delta x}{\Delta t}$

2. Your net displacement is $x_f - x_i$, where $x_i = 0$ is your initial position and $x_f = 50$ m is your final position: $\Delta x = x_f - x_i = 50\text{ m} - 0 = 50\text{ m}$

3. Use Δx and Δt to find your average velocity: $v_{av} = \dfrac{\Delta x}{\Delta t} = \dfrac{50\text{ m}}{42\text{ s}} = 1.19\text{ m/s}$

Check the Result The world record for a 100-m race is just under 10 s, so 10 m/s is about the maximum speed obtainable. The result of 3.57 m/s for the average speed in part (*a*) is reasonable, given that you merely jogged for one-third of the distance. If you had obtained 35.7 m/s for the average speed, that would have been a clue that something was wrong with the calculation.

Remark Note that your average speed is greater than your average velocity because the total distance traveled is greater than the total displacement.

Example 2-4

Two trains 75 km apart approach each other on parallel tracks, each moving at 15 km/h. A bird flies back and forth between the trains at 20 km/h until the trains pass each other. How far does the bird fly?

Picture the Problem This problem seems difficult at first, but viewed in the right way it is actually quite simple. We approach it by first writing an equation for the quantity to be found, the total distance Δs flown by the bird.

1. The total distance equals the average speed times the time: $\Delta s = \text{average speed} \times \Delta t = (20\text{ km/h})\Delta t$

2. The time that the bird is in the air is the time taken for the trains to meet. Since they are moving toward each other at 15 km/h, the distance between them decreases at 30 km/h. Use that to calculate how long it will take for the distance between them to go from 75 km to zero: $\Delta t = \dfrac{75\text{ km}}{30\text{ km/h}} = 2.5\text{ h}$

3. The total distance traveled by the bird is therefore: $\Delta s = (20\text{ km/h})(2.5\text{ h}) = 50\text{ km}$

Remark Some try to solve this problem by finding and summing the distances flown by the bird each time it moves from one train to the other. This makes a relatively easy problem quite difficult. It is important to develop a thoughtful, systematic approach to solving problems. Begin by writing an equation for the unknown quantity in terms of other quantities. Then proceed by determining the values for each of the other quantities in the equation.

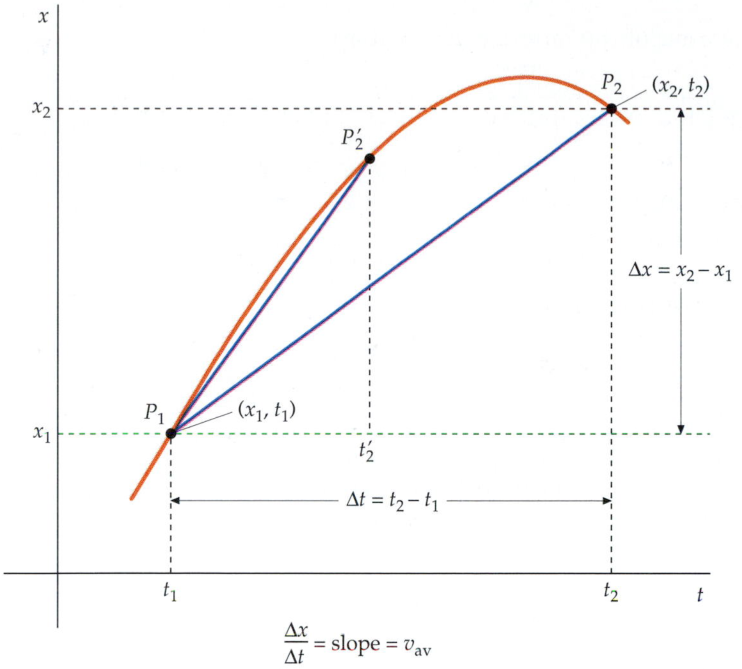

Figure 2-4 Graph of x versus t for a particle moving in one dimension. Each point on the curve represents the location x at a particular time t. We have drawn a straight line between positions P_1 and P_2. The displacement $\Delta x = x_2 - x_1$ and the time interval $\Delta t = t_2 - t_1$ between these points are indicated. The straight line between P_1 and P_2 is the hypotenuse of the triangle having sides Δx and Δt, and the ratio $\Delta x/\Delta t$ is its slope. In geometric terms, the slope is a measure of the line's steepness.

Figure 2-4 depicts average velocity graphically. A straight line connects points P_1 and P_2 and forms the hypotenuse of the triangle having sides Δx and Δt. The ratio $\Delta x/\Delta t$ is the line's **slope,** which gives us a geometric interpretation of average velocity:

> The average velocity is the slope of the straight line connecting the points (t_1, x_1) and (t_2, x_2).

Generally, average velocity depends on the time interval on which it is based. In Figure 2-4, for example, the smaller time interval indicated by t_2' and P_2' gives a larger average velocity, as shown by the greater steepness of the line connecting points P_1 and P_2'.

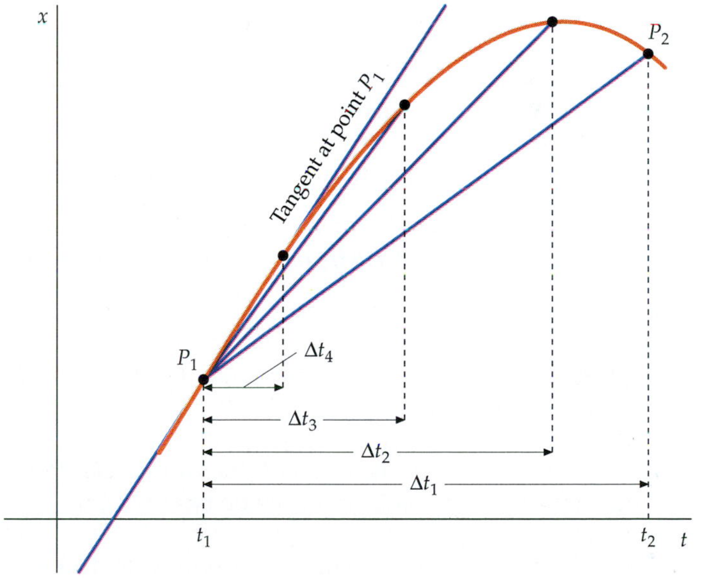

Figure 2-5 Graph of x versus t. Note the sequence of successively smaller time intervals, $\Delta t_1, \Delta t_2, \Delta t_3, \ldots$. The average velocity of each interval is the slope of the straight line for that interval. As the time intervals become smaller, these slopes approach the slope of the tangent to the curve at point t_1. The slope of this line is defined as the instantaneous velocity at time t_1.

Instantaneous Velocity

On first consideration, defining the velocity of a particle at a single instant seems impossible. At a given instant, a particle is at a single point. If it is at a single point, how can it be moving? If it is not moving, how can it have a velocity? This age-old paradox is resolved when we realize that observing and defining motion requires that we look at the position of the object at more than one time. Consider Figure 2-5. As we consider successively shorter time intervals beginning at t_1, the average velocity for the interval approaches the slope of the tangent at t_1. We define the slope of this tangent as the **instantaneous velocity** at t_1. This tangent is the limit of the ratio $\Delta x/\Delta t$ as Δt, and therefore Δx, approaches zero. So we can say,

The instantaneous velocity is the limit of the ratio $\Delta x/\Delta t$ as Δt approaches zero:

$$v(t) = \lim_{\Delta t \to 0} \frac{\Delta x}{\Delta t}$$
$$= \text{slope of the line tangent to the } x\text{-versus-}t \text{ curve*} \qquad 2\text{-}4$$

Definition—Instantaneous velocity

This limit is called the **derivative** of x with respect to t. In the usual calculus notation, the derivative is written dx/dt:

$$v(t) = \lim_{\Delta t \to 0} \frac{\Delta x}{\Delta t} = \frac{dx}{dt} \qquad 2\text{-}5$$

A line's slope may be positive, negative, or zero; consequently, instantaneous velocity (in one-dimensional motion) may be positive (x increasing), negative (x decreasing), or zero (no motion). The magnitude of the instantaneous velocity is the **instantaneous speed.**

* The slope of the line tangent to a curve is often referred to more simply as the "slope of the curve."

Example 2-5 *try it yourself*

The position of a particle as a function of time is given by the curve shown in Figure 2-6. Find the instantaneous velocity at time $t = 2$ s. When is the velocity greatest? When is it zero? Is it ever negative?

Picture the Problem In the figure, we have sketched the line tangent to the curve at $t = 2$ s. The tangent line's slope is the instantaneous velocity of the particle at the given time. You can use this figure to measure the slope of the tangent line.

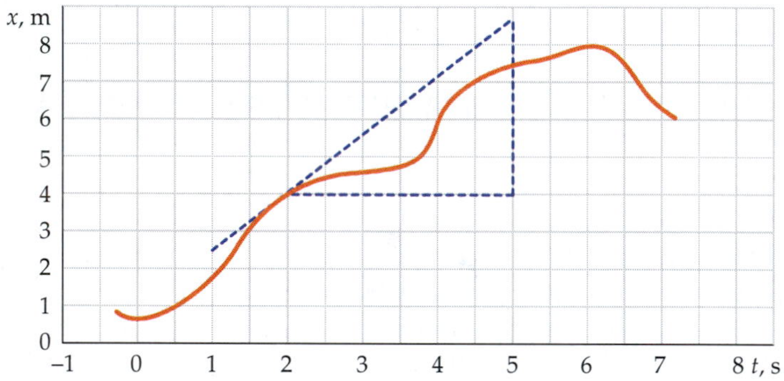

Figure 2-6

Cover the column to the right and try these on your own before looking at the answers.

Steps	*Answers*
1. Find the values x_1 and x_2 on the tangent line at times $t_1 = 2$ s and $t_2 = 5$ s.	$x_1 \approx 4$ m, $x_2 \approx 8.5$ m
2. Compute the slope of the tangent line from these values. This slope equals the instantaneous velocity at $t = 2$ s.	$v = \text{slope} = 1.5$ m/s
3. From the figure, the slope (and therefore velocity) is greatest at about $t = 4$ s. The slope and velocity are zero at $t = 0$ and $t = 6$ s and are negative before 0 and after 6 s.	

Exercise What is the average velocity of this particle between $t = 2$ s and $t = 5$ s? (*Answer* 1.17 m/s)

Example 2-6

The position of a stone dropped from a cliff is described approximately by $x = 5t^2$, where x is in meters measured downward from the original position at $t = 0$, and t is in seconds. Find the velocity at any time t. (We omit explicit indication of units to simplify the notation.)

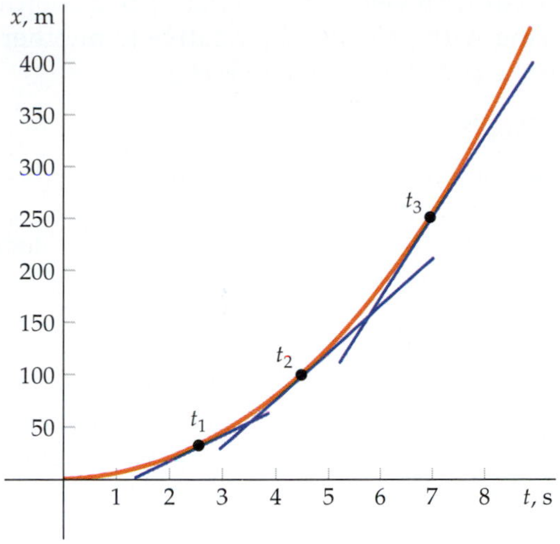

Figure 2-7

Picture the Problem We can compute the velocity at some time t by computing the derivative dx/dt directly from the definition in Equation 2-4. The corresponding curve giving x versus t is shown in Figure 2-7. Tangent lines are drawn at times t_1, t_2, and t_3. The slopes of these tangent lines increase steadily, indicating that the instantaneous velocity increases steadily with time.

1. By definition the instantaneous velocity is: $$v(t) = \lim_{\Delta t \to 0} \frac{\Delta x}{\Delta t} = \lim_{\Delta t \to 0} \frac{x(t + \Delta t) - x(t)}{\Delta t}$$

2. We compute the displacement Δx from the position function $x(t)$: $$x(t) = 5t^2$$

3. At a later time $t + \Delta t$ the position is $x(t + \Delta t)$, given by: $$x(t + \Delta t) = 5(t + \Delta t)^2 = 5[t^2 + 2t\Delta t + (\Delta t)^2]$$ $$= 5t^2 + 10t\Delta t + 5(\Delta t)^2$$

4. The displacement for this time interval is thus: $$\Delta x = x(t + \Delta t) - x(t)$$ $$= [5t^2 + 10t\Delta t + 5(\Delta t)^2] - 5t^2$$ $$= 10t\Delta t + 5(\Delta t)^2$$

5. Divide Δx by Δt to find the average velocity for this time interval: $$v_{\text{av}} = \frac{\Delta x}{\Delta t} = \frac{10t\,\Delta t + 5(\Delta t)^2}{\Delta t} = 10t + 5\Delta t$$

6. As we consider shorter and shorter time intervals, Δt approaches zero and the second term, $5\,\Delta t$, approaches zero, though the first term, $10t$, remains unchanged. The instantaneous velocity at time t is thus: $$v(t) = \lim_{\Delta t \to 0} \frac{\Delta x}{\Delta t} = 10t$$

Remark If we had set $\Delta t = 0$ in steps 4 and 5, the displacement would be $\Delta x = 0$, in which case the ratio $\Delta x/\Delta t$ would be undefined. Instead, we leave Δt as a variable until the final step, when the limit $\Delta t \to 0$ is well defined.

To find derivatives quickly, we use rules based on this limiting process (see Appendix Table D-4). A particularly useful rule is

$$\text{If } x = Ct^n, \quad \text{then } \frac{dx}{dt} = Cnt^{n-1} \tag{2-6}$$

where C and n are any constants. Using this rule in Example 2-6, we have $x = 5t^2$, and $v = dx/dt = 10t$, in agreement with our previous results.

Relative Velocity

If a particle moves with velocity v_{pA} relative to a coordinate system A, which is in turn moving with velocity v_{AB} relative to another coordinate system B, the velocity of the particle relative to B is

$$v_{pB} = v_{pA} + v_{AB} \qquad \text{2-7}a$$

For example, if you swim in a river parallel to the direction of the flow, your velocity relative to the shore, v_{ys}, equals your velocity relative to the water, v_{yw}, plus the velocity of the water relative to the shore, v_{ws}:

$$v_{ys} = v_{yw} + v_{ws}$$

The velocities add or subtract depending on whether you are swimming with the current or against it. For example, if you are swimming at 2 m/s against the current, and the water speed is 1.2 m/s relative to the shore, then your velocity relative to the shore is $v_{ys} = -2\text{ m/s} + 1.2\text{ m/s} = -0.8\text{ m/s}$, where we have chosen the direction of the water flow to be the positive direction. Other common instances when we might want to know the relative velocity are an airplane flying with or against the wind, or a traveler walking on a moving sidewalk at an airport.

Midair refueling. Each aircraft is nearly at rest relative to the other, though both are moving with very large velocities relative to the earth.

A great surprise of twentieth-century physics was the discovery that Equation 2-7*a* is only an approximation. A study of the theory of relativity shows that the exact expression for relative velocities is

$$v_{ys} = \frac{v_{yw} + v_{ws}}{1 + v_{yw}v_{ws}/c^2} \qquad \text{2-7}b$$

where $c = 3 \times 10^8$ m/s is the velocity of light in a vacuum. In all everyday cases with macroscopic objects, v_{yw} and v_{ws} are both much smaller than c, so Equations 2-7*a* and 2-7*b* are the same, but for very high speeds, such as the speed of an electron or the speed at which distant galaxies are receding from the earth, the difference between these two equations becomes significant. Equation 2-7*b* has the interesting property that if $v_{yw} = c$, then v_{ys} also equals c, which is a postulate of relativity, namely that the speed of light is the same in all reference frames moving with constant velocity relative to each other (Figure 2-8).

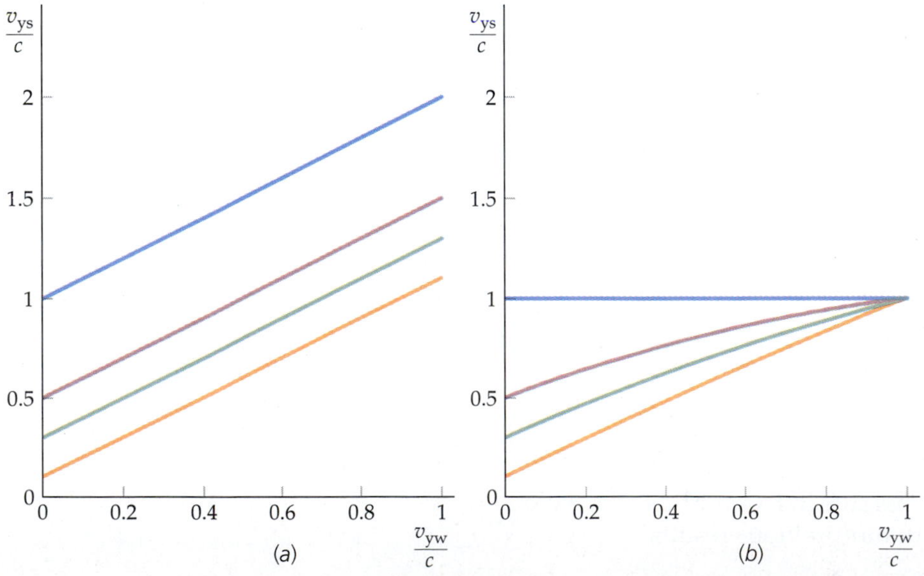

Figure 2-8 An boat moving down a river in the everyday and relativistic worlds. (*a*) Velocity addition in our everyday world. Each line shows the boat's velocities relative to shore (y coordinate) and water (x coordinate). Each of the four different lines corresponds to different relative velocities of water to shore. The uppermost line denotes water flowing relative to the shore at the speed of light, c. Note that all velocities are expressed as velocities divided by c—that is, as dimensionless units. (*b*) Relativistic velocity addition. Again, each line shows a boat's velocities relative to shore and water for a different relative velocity of water to shore. Unlike part (*a*), the lines are not straight, since the relative velocities combine according to Equation 2-7*b* rather than 2-7*a*. A boat moving at the speed of light c relative to the shore also moves at speed c relative to the river independent of the velocity of the river relative to the shore.

2-2 Acceleration

Acceleration is the rate of change of the instantaneous velocity. The **average acceleration** for a particular time interval $\Delta t = t_2 - t_1$ is defined as the ratio $\Delta v/\Delta t$, where $\Delta v = v_2 - v_1$:

$$a_{av} = \frac{\Delta v}{\Delta t} \tag{2-8}$$

Definition—Average acceleration

Acceleration has dimensions of length divided by time squared; the SI unit is meters per second squared (m/s^2). For example, if a particle at rest accelerates at 5.1 m/s^2, its velocity after 1 s is 5.1 m/s, its velocity after 2 s is 10.2 m/s, and so on.

Instantaneous acceleration is the limit of the ratio $\Delta v/\Delta t$ as Δt approaches zero. On a plot of velocity versus time, the instantaneous acceleration at time t is the slope of the line tangent to the curve at that time:

$$a = \lim_{\Delta t \to 0} \frac{\Delta v}{\Delta t} \tag{2-9}$$
$$= \text{slope of the line tangent to the } v \text{ versus } t \text{ curve}$$

Definition—Instantaneous acceleration

Thus, acceleration is the derivative of velocity with respect to time, dv/dt. Since velocity is the derivative of the position x with respect to t, acceleration is the second derivative of x with respect to t, d^2x/dt^2. We can see the reason for this notation when we write the acceleration as dv/dt and replace v with dx/dt:

$$a = \frac{dv}{dt} = \frac{d(dx/dt)}{dt} = \frac{d^2x}{dt^2} \tag{2-10}$$

If acceleration is zero, there is no change in velocity over time—velocity is constant. In this case, the curve of x versus t is a straight line. If acceleration is nonzero and constant, as in Example 2-4, then velocity varies linearly with time and the curve of x versus t is quadratic in t.

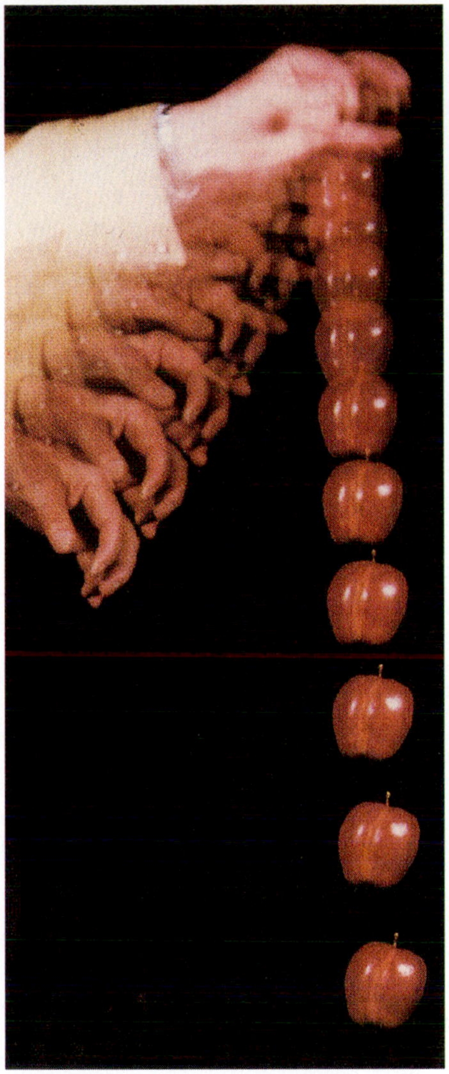

A falling apple captured by strobe photography at 60 flashes per second. The acceleration of the apple is indicated by the widening spaces between the images.

Example 2-7

A cheetah can accelerate from 0 to 96 km/h (60 mi/h) in 2 s, whereas a Corvette requires 4.5 s. Compute the average accelerations for the cheetah and Corvette and compare them with the free-fall acceleration due to gravity, $g = 9.81$ m/s^2.

1. Find the average acceleration from the information given:

$$\text{Cheetah } a_{av} = \frac{\Delta v}{\Delta t} = \frac{96\text{ km/h} - 0}{2\text{ s}} = 48\text{ km/h}\cdot\text{s}$$

$$\text{Corvette } a_{av} = \frac{\Delta v}{\Delta t} = \frac{96\text{ km/h} - 0}{4.5\text{ s}} = 21.3\text{ km/h}\cdot\text{s}$$

2. Convert to m/s^2 using 1 h = 3600 s = 3.6 ks:

$$\text{Cheetah } \frac{48\text{ km}}{\text{h}\cdot\text{s}} \times \frac{1\text{ h}}{3.6\text{ ks}} = 13.3\text{ m/s}^2$$

$$\text{Corvette } \frac{21.3\text{ km}}{\text{h}\cdot\text{s}} \times \frac{1\text{ h}}{3.6\text{ ks}} = 5.92\text{ m/s}^2$$

3. To compare the result with the acceleration due to gravity, multiply each by the conversion factor $g/(9.81\text{ m/s}^2)$:

$$\text{Cheetah } 13.3\text{ m/s}^2 \times \frac{g}{9.81\text{ m/s}^2} = 1.36g$$

$$\text{Corvette } 5.92\text{ m/s}^2 \times \frac{g}{9.81\text{ m/s}^2} = 0.60g$$

Remark Note that by expressing the time in kiloseconds, the k's in km and ks cancel.

Exercise A car is traveling at 45 km/h at time $t = 0$. It accelerates at a constant rate of 10 km/h·s. (*a*) How fast is it traveling at $t = 2$ s? (*b*) At what time is the car traveling at 70 km/h? (*Answers* (*a*) 65 km/h, (*b*) 2.5 s)

Exercise in Dimensional Analysis If a car starts from rest at $x = 0$ with constant acceleration a, its velocity v depends on a and the distance traveled x. Which of the following equations has the correct dimensions and therefore could be a possible equation relating x, a, and v?

(*a*) $v = 2ax$
(*b*) $v^2 = 2a/x$
(*c*) $v = 2ax^2$
(*d*) $v^2 = 2ax$

(*Answer* Only (*d*) has the same dimensions on both sides of the equation. Although we cannot obtain the exact equation from dimensional analysis, we can often obtain the functional dependence.)

Example 2-8

The position of a particle is given by $x = Ct^3$, where C is a constant having units of m/s^3. Find the velocity and acceleration as functions of time.

1. We find the velocity by applying $dx/dt = Cnt^{n-1}$ (Equation 2-6):

$$x = Ct^3$$

$$v = \frac{dx}{dt} = 3Ct^2$$

2. The time derivative of velocity gives the acceleration:

$$a = \frac{dv}{dt} = 6Ct$$

Check the Result We can check the units of our answers. For velocity, $[v] = [C][t^2] = (\text{m/s}^3)(\text{s}^2) = \text{m/s}$. For acceleration, $[a] = [C][t] = (\text{m/s}^3)(\text{s}) = \text{m/s}^2$.

2-3 Motion With Constant Acceleration

"It goes from zero to 60 in about 3 seconds."
© Sydney Harris

The motion of a particle that has constant acceleration is common in nature. For example, near the earth's surface all unsupported objects fall vertically with constant acceleration (provided air resistance is negligible).

If a particle has a constant acceleration a, it follows that the average acceleration for any time interval is also a. Thus,

$$a_{av} = \frac{\Delta v}{\Delta t} = a$$

If the velocity is v_0 at time $t = 0$, and v at some later time t, the corresponding acceleration is

$$a = \frac{\Delta v}{\Delta t} = \frac{v - v_0}{t - 0} = \frac{v - v_0}{t}$$

Rearranging yields v as a function of time.

$$v = v_0 + at \qquad \text{2-11}$$

Constant acceleration, v versus t

This is the equation for a straight line in a v-versus-t plot (Figure 2-9). The line's slope is the acceleration a, and its v intercept is the initial velocity v_0.

The displacement $\Delta x = x - x_0$ in the time interval $\Delta t = t - 0$ is

$$\Delta x = v_{av}\Delta t = v_{av}t \qquad \text{2-12a}$$

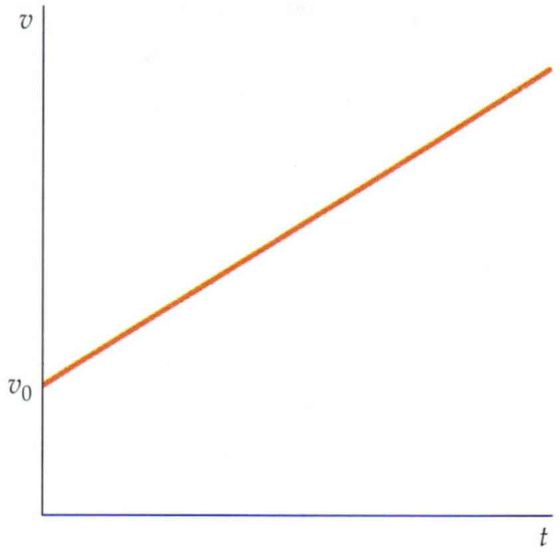

Figure 2-9 Graph of velocity versus time for constant acceleration.

For constant acceleration, the velocity varies linearly with time, and the average velocity is the mean value of the initial and final velocities. (This relation, which will be derived in Section 2-4, holds only if the acceleration is constant.) If v_0 is the initial velocity and v is the final velocity, the average velocity is

$$v_{av} = \tfrac{1}{2}(v_0 + v) \qquad \text{2-12b}$$

Constant acceleration, v_{av}

The displacement is then

$$\Delta x = x - x_0 = v_{av}t = \tfrac{1}{2}(v_0 + v)t \qquad \text{2-13}$$

We can eliminate v by substituting $v = v_0 + at$ from Equation 2-11:

$$\Delta x = \tfrac{1}{2}(v_0 + v)t = \tfrac{1}{2}(v_0 + v_0 + at)t = v_0t + \tfrac{1}{2}at^2$$

The displacement is thus

$$\Delta x = x - x_0 = v_0t + \tfrac{1}{2}at^2 \qquad \text{2-14}$$

Constant acceleration, Δx versus t

The first term on the right, v_0t, is the displacement that would occur if a were zero, and the second term, $\frac{1}{2}at^2$, is the additional displacement due to the constant acceleration.

Let's eliminate t from Equations 2-11 and 2-12a and find a relation between Δx, a, v, and v_0. From Equation 2-11, $t = (v - v_0)/a$. Substituting this into Equation 2-12a, we get

$$\Delta x = v_{av}t = \frac{1}{2}(v_0 + v)t = \frac{1}{2}(v_0 + v)\frac{v - v_0}{a} = \frac{v^2 - v_0^2}{2a}$$

or

$$v^2 = v_0^2 + 2a\,\Delta x \qquad \text{2-15}$$

Constant acceleration, v versus x

Equation 2-15 is useful, for example, if we want to find the final velocity of a ball dropped from rest at some height x and we are not interested in the time the fall takes.

Problems With One Object

Many practical problems deal with objects falling freely due to gravity. The magnitude of the acceleration caused by gravity is designated by g, which has the approximate value

$$g = 9.81\ \text{m/s}^2 = 32.2\ \text{ft/s}^2$$

By convention, g is always positive. If downward is the positive direction, the acceleration due to gravity is $a = g$; if upward is positive, then $a = -g$.

Example 2-9

Upon graduation, a joyful physics student throws his cap upward with an initial speed of 14.7 m/s (Figure 2-10). Given that its acceleration is 9.81 m/s² downward (we neglect air resistance), (*a*) how long does it take to reach its highest point? (*b*) What is the distance to the highest point? (*c*) What is the total time the cap is in the air?

Figure 2-10

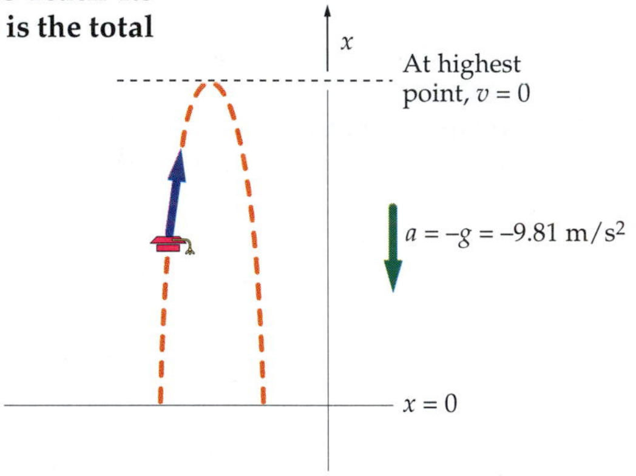

Picture the Problem When the cap is at its highest point, its instantaneous velocity is zero. Thus we translate the statement "highest point" into the mathematical condition $v = 0$. Similarly, "total time in the air" means the time t following the toss when $x = x_0$. We choose the origin to be at the initial position of the cap, and we designate upward as the positive direction. Then $x_0 = 0$, $v_0 = 14.7$ m/s, and the acceleration, which is downward, is $a = -g = -9.81\ \text{m/s}^2$.

(*a*) 1. The time is related to the velocity and acceleration:
$$v = v_0 + at$$

2. To find the time at which the cap reaches its greatest height, set $v = 0$, and solve for t:
$$t = \frac{v - v_0}{a} = \frac{0 - 14.7\ \text{m/s}}{-9.81\ \text{m/s}^2} = 1.50\ \text{s}$$

(*b*) We can find the distance traveled from the time t and the average velocity:
$$\Delta x = v_{av}t = \tfrac{1}{2}(v_0 + v)t = \tfrac{1}{2}(14.7\ \text{m/s})(1.50\ \text{s}) = 11.0\ \text{m}$$

(*c*) 1. To find total time, set $x = x_0$ in Equation 2-14 and solve for t:
$$x_0 - x_0 = v_0 t + \tfrac{1}{2}at^2$$
$$0 = t(v_0 + \tfrac{1}{2}at)$$

2. There are two solutions for t when $x = x_0$. The first corresponds to the time at which the cap is thrown, and the second corresponds to the time at which the cap lands:
$$t = 0 \qquad \text{(first solution)}$$
$$t = -\frac{2v_0}{a} = -\frac{2(14.7\ \text{m/s})}{-9.81\ \text{m/s}^2} = 3\ \text{s} \qquad \text{(second solution)}$$

Remark The $t = 3$ s solution also follows from a symmetry in the system: It takes the same time for the cap to fall from its greatest height as to rise to that height. In reality, however, the cap will not have a constant acceleration because air resistance has a significant effect on a light object like a cap.

Exercise Find Δx using (*a*) Equation 2-13 and (*b*) Equation 2-14. (*c*) Find the velocity of the cap when it returns to its starting point. (*Answers* (*a*) and (*b*) $\Delta x = 11.0$ m, (*c*) -14.7 m/s; notice that the final speed is the same as the initial speed.)

Exercise What is the velocity of the cap (*a*) 0.1 s before it reaches its highest point and (*b*) 0.1 s after it reaches its highest point? (*c*) Compute $\Delta v/\Delta t$ for this 0.2-s time interval. (*Answers* (*a*) 0.981 m/s, (*b*) -0.981 m/s, (*c*) $(-0.981 \text{ m/s} - 0.981 \text{ m/s})/(0.2 \text{ s}) = -9.81 \text{ m/s}^2$)

Exercise A car accelerates from rest at a constant rate of 8 m/s^2. (*a*) How fast is it going after 10 s? (*b*) How far has it gone after 10 s? (*c*) What is its average velocity for the interval $t = 0$ to $t = 10$ s? (*Answers* (*a*) 80 m/s, (*b*) 400 m, (*c*) 40 m/s)

Remark Figure 2-11 shows (*a*) x versus t and (*b*) v versus t for two tosses of a graduate's cap with different initial velocities. Notice that for both tosses the velocity is zero when the cap is at its maximum height. Notice also that the two velocity curves are parallel. This is because the slope of each velocity curve is equal to $-g = -9.81 \text{ m/s}^2$.

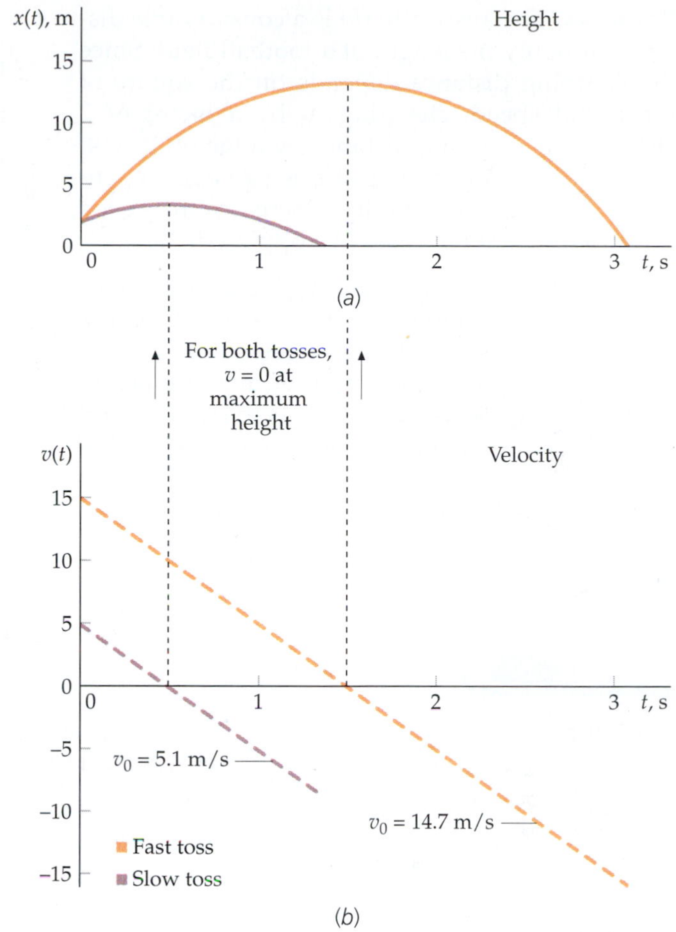

Figure 2-11

The next example concerns a car's **stopping distance**—how far it travels while coming to a halt.

Example 2-10

On a highway at night you see a stalled vehicle and brake your car to a stop with an acceleration of magnitude 5 m/s². (An acceleration that reduces the speed is often called a deceleration.) What is the car's stopping distance if its initial speed is (*a*) 15 m/s (about 34 mi/h) or (*b*) 30 m/s?

Picture the Problem If we choose the direction of motion to be positive, the stopping distance and the initial velocity are positive, but the acceleration is negative. Thus, the initial velocity is $v_0 = 15$ m/s, the final velocity is $v = 0$, and the acceleration is $a = -5 \text{ m/s}^2$. We seek the distance traveled, Δx. We do not need to know the time it takes for the car to stop, so Equation 2-15 is the most convenient formula to use.

(*a*) 1. Set $v = 0$ in Equation 2-15:

$$v^2 = v_0^2 + 2a\,\Delta x = 0$$

2. Solve for Δx:

$$\Delta x = \frac{v^2 - v_0^2}{2a} = \frac{-v_0^2}{2a} = \frac{-(15 \text{ m/s})^2}{2(-5 \text{ m/s}^2)} = 22.5 \text{ m}$$

(*b*) Use the above result for Δx, but replace v_0 with $v_0' = 2v_0$:

$$\Delta x' = \frac{-(2v_0)^2}{2a} = \frac{-4v_0^2}{2a} = 4\Delta x = 4(22.5 \text{ m}) = 90 \text{ m}$$

Remarks The answer to (b) is a considerable distance, roughly the length of a football field. Since the stopping distance depends on the square of the initial speed, changing v_0 by a factor of 2 changes the stopping distance by a factor $2^2 = 4$. The practical implication of this squared dependence is that even modest increases in speed cause significant increases in stopping distance.

Remark Figure 2-12 shows stopping distance as a function of the initial velocity. The middle curve shows the case where the acceleration is $a = -5.0\ \mathrm{m/s^2}$; the rightmost and leftmost points on the middle curve are the solutions to parts (a) and (b). Also shown are cases of larger deceleration ($a = -9.81\ \mathrm{m/s^2}$) with a shorter stopping distance (bottom curve), and of smaller deceleration with a larger stopping distance (top curve).

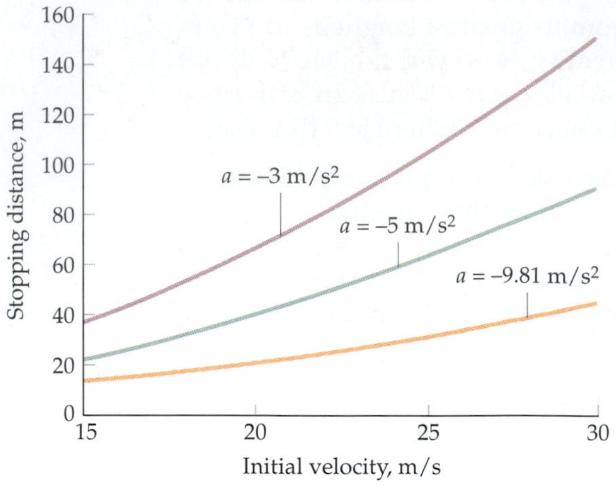

Figure 2-12

Example 2-11 *try it yourself*

In Example 2-10, (a) how long does it take for the car to stop if its initial velocity is 30 m/s? (b) How far does the car travel in the last second?

Picture the Problem (a) You can find the total time t from $v_f = v_0 + at = 0$, with $v_0 = 30\ \mathrm{m/s}$ and $a = -5\ \mathrm{m/s}$. (b) Since the velocity decreases by 5 m/s each second, the velocity 1 s before the car stops must be 5 m/s. Find the average velocity during the last second and use that to find the distance traveled.

Cover the column to the right and try these on your own before looking at the answers.

Steps	Answers
(a) Find the total stopping time t.	$t = 6\ \mathrm{s}$
(b) 1. Find the average velocity during the last second.	$v_{av} = 2.5\ \mathrm{m/s}$
2. Compute the distance traveled from $\Delta x = v_{av}\Delta t$.	$\Delta x_1 = v_{av}\Delta t = 2.5\ \mathrm{m}$

Sometimes valuable insight can be gained about the motion of an object by assuming that the constant-acceleration formulas still apply even when the acceleration is not constant.

Example 2-12

In a crash test, a car traveling 100 km/h (about 62 mi/h) hits an immovable concrete wall. (a) How soon does the car stop? (b) What is its acceleration?

Picture the Problem In this example, it is not accurate to treat the car as a particle because different parts of the vehicle will have different accelerations as the car crumples to a halt. Moreover, the accelerations are not constant. Nevertheless, we can approximate an answer by assuming constant acceleration for a point particle located in the center of the car. We need more information to solve this problem—either the stopping distance or the time to stop. We can estimate the stopping distance using common sense. Upon impact, the center of the car will certainly move forward less than half the length of the car. We'll choose 0.75 m as a reasonable estimate of the stopping distance.

(*a*) 1. The time needed for the car to stop is related to the stopping distance and the average velocity:

$$\Delta x = v_{av}\Delta t$$

2. The average velocity is found from the initial and final velocities (since we are only estimating, two significant figures are sufficient):

$$v_{av} = \frac{1}{2}(v_0 + v) = \frac{1}{2}(100\text{ km/h} + 0) = 50\text{ km/h}$$
$$= 14\text{ m/s}$$

3. The time taken to stop the car is thus:

$$\Delta t = \frac{\Delta x}{v_{av}} = \frac{0.75\text{ m}}{14\text{ m/s}} = 0.054\text{ s}$$

(*b*) 1. The average acceleration equals the ratio of the change in velocity and the time interval:

$$a = \frac{\Delta v}{\Delta t}$$

2. Since the car is brought from $v_0 = 100$ km/h $= 28$ m/s to rest in this time, the average acceleration is:

$$a = \frac{\Delta v}{\Delta t} = \frac{0 - 28\text{ m/s}}{0.054\text{ s}} = -520\text{ m/s}^2$$

Remark The magnitude of this acceleration is greater than $50g$.

Example 2-13 *try it yourself*

An electron in a cathode-ray tube accelerates from rest with a constant acceleration of 5.33×10^{12} m/s^2 for 0.15 μs (1 μs = 10^{-6} s). The electron then drifts with constant velocity for 0.2 μs. Finally, it comes to rest with an acceleration of -2.67×10^{13} m/s^2. How far does the electron travel?

Picture the Problem The equations for constant acceleration do not apply to this problem directly because the acceleration of the electron varies with time. Divide the electron's motion into three intervals, each with a different constant acceleration, and use the final position and velocity for one interval as the initial conditions for the next interval. Choose the origin to be at the electron's starting position, and the positive direction to be the direction of motion.

Cover the column to the right and try these on your own before looking at the answers.

Steps	*Answers*
1. Find the displacement and final velocity for the first 0.15-μs interval.	$\Delta x_1 = 6.00$ cm; $v_1 = 8.00 \times 10^5$ m/s
2. Use this final velocity as the constant velocity to find the displacement while it drifts at constant velocity.	$\Delta x_2 = 16$ cm
3. Use this same velocity as the initial velocity and Equation 2-15 with $v = 0$ to find the displacement for the third interval in which the electron slows down.	$\Delta x_3 = 1.20$ cm
4. Add the displacements found in steps 1, 2, and 3 to find the total displacement.	$\Delta x = 23.2$ cm

Remark In your television, electrons are accelerated from the cathode to the anode, after which they are focused and passed through deflecting plates. They then drift toward the screen and crash into it, coming abruptly to rest.

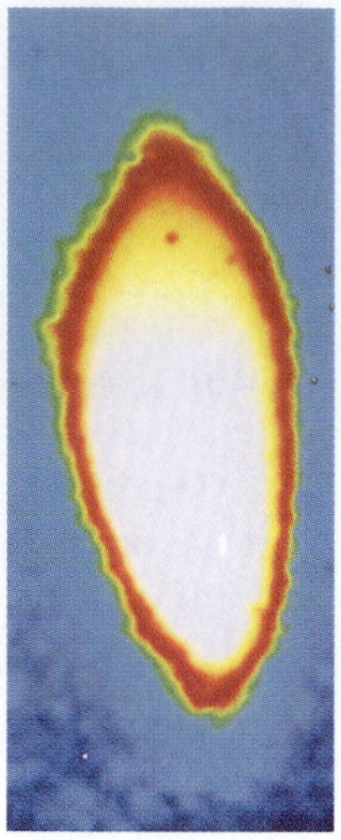

(*a*) The two-mile-long linear accelerator at Stanford University, used to accelerate electrons and positrons in a straight line to nearly the speed of light. (*b*) Cross section of the accelerator's electron beam as shown on a video monitor.

Example 2-14 ***try it yourself***

John climbs a tree to get a better view of the speaker at an outdoor graduation ceremony. Unfortunately, he leaves his binoculars behind. Marsha throws them up to John, but her strength is greater than her accuracy. The binoculars pass John's outstretched hand after 0.69 s and again 1.68 s later. How high is John?

Picture the Problem There are two unknowns in this problem, John's height h and the initial velocity of the binoculars v_0. We know that $x = h$ at $t_1 = 0.69$ s and $x = h$ at $t_2 = 0.69\text{ s} + 1.68\text{ s} = 2.37$ s. Expressing h as a function of time t gives us two equations from which the two unknowns can be determined.

Cover the column to the right and try these on your own before looking at the answers.

Steps	Answers
1. Write $x(t)$ for times t_1 and t_2, noting that $x_0 = 0$ and $x = h$ in each case.	$h = v_0t_1 - \frac{1}{2}gt_1^2$ and $h = v_0t_2 - \frac{1}{2}gt_2^2$
2. Eliminate h from these two equations and solve for v_0 in terms of the times t_1 and t_2.	$v_0 = \dfrac{\frac{1}{2}g(t_1^2 - t_2^2)}{t_1 - t_2} = \frac{1}{2}g(t_1 + t_2)$
3. Substitute your result for v_0 into either of the equations for h.	$h = \frac{1}{2}gt_1t_2 = 8.02$ m

Remark We have two unknowns, h and v_0, but are given two times, t_1 and t_2, so we can write two equations and solve them for the two unknowns.

Exercise Find the initial velocity of the binoculars and the velocity as they pass John on the way down. (*Answers* $v_0 = 15.0$ m/s and $v_2 = -8.24$ m/s)

Problems With Two Objects

We now give some examples of problems involving two objects moving with constant acceleration.

Example 2-15

A car is speeding at 25 m/s (≈ 90 km/h ≈ 56 mi/h) in a school zone. A police car starts from rest just as the speeder passes and accelerates at a constant rate of 5 m/s². (*a*) When does the police car catch the speeding car? (*b*) How fast is the police car traveling when it catches up with the speeder?

Picture the Problem To determine when the two cars will be at the same position, we write the positions x_s of the speeder and x_p of the police car as functions of time and solve for the time t when $x_s = x_p$.

(*a*) 1. Write the position functions for the speeder and the police car: $x_s = v_s t$ and $x_p = \frac{1}{2}a_p t^2$

2. Set $x_s = x_p$ and solve for the time t: $v_s t = \frac{1}{2}a_p t^2; \quad t = 0$ (initial condition)

$$t = \frac{2v_s}{a_p} = \frac{2(25\ \text{m/s})}{5\ \text{m/s}^2} = 10\ \text{s}$$

(*b*) The velocity of the police car is given by $v = v_0 + at$ with $v_0 = 0$: $v_p = a_p t = (5\ \text{m/s}^2)(10\ \text{s}) = 50\ \text{m/s}$

Remark The final speed of the police car in (*b*) is exactly twice that of the speeder. Since the two cars covered the same distance in the same time, they must have had the same average speed. The speeder's average speed, of course, is 25 m/s. For the police car to start from rest and have an average speed of 25 m/s, it must reach a final speed of 50 m/s.

Exercise How far have the cars traveled when the police car catches the speeder? (*Answer* 250 m)

Remark In Figure 2-13 the solid lines depict the speeder and the police car in this example. The dashed lines are variations on the example. The smaller acceleration depicted by the lower dashed line means the police car takes longer to reach the speeder. In the higher dashed line, the acceleration is the same as in the example, but the police car does not start accelerating until 4 s after the speeder passes by.

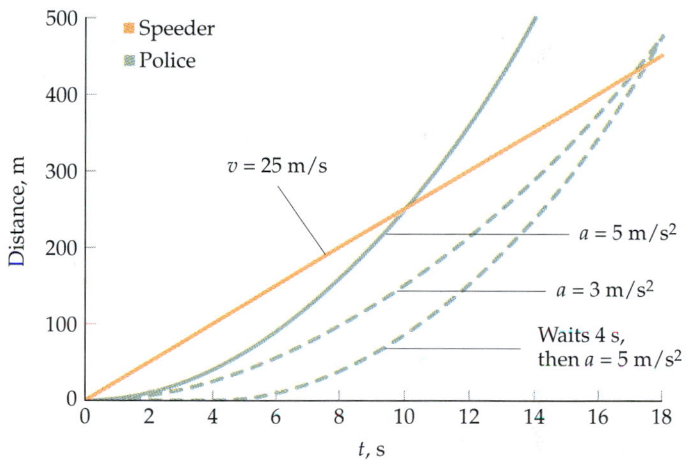

Figure 2-13

Example 2-16 *try it yourself*

How fast is the police car in Example 2-15 traveling when it is 25 m behind the speeding car?

Picture the Problem The speed is given by $v_p = at_1$, where t_1 is the time at which $x_s - x_p = 25$ m.

Cover the column to the right and try these on your own before looking at the answers.

Steps	*Answers*
1. Using the equations for x_p and x_s from Example 2-15, solve for t_1 when $x_s - x_p = 25$ m.	$t_1 = 5\ \text{s} \pm \sqrt{15}\ \text{s} = 1.13\ \text{s}$ or $8.87\ \text{s}$
2. Use $v_p = a_p t$ to compute the speed of the police car at $t = t_1$.	$v = 5.65$ m/s at 1.13 s and 44.4 m/s at $t = 8.87$ s

Remark We see from Figure 2-13 that the distance between the cars starts at zero, increases to a maximum value, and then decreases. The separation at any time is $D = x_s - x_p = (25\text{ m/s})t - \frac{1}{2}(5\text{ m/s}^2)t^2$. At maximum separation, $dD/dt = 0$, which occurs at $t = 5$ s. At equal intervals before and after $t = 5$ s, the separations are the same.

Example 2-17

Figure 2-14

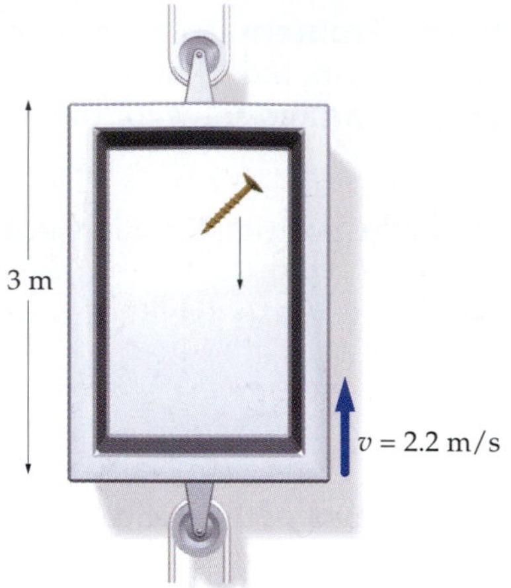

While standing in an elevator, you see a screw fall from the ceiling. The ceiling is 3 m above the floor (Figure 2-14). (*a*) If the elevator is moving upward with a constant speed of 2.2 m/s, how long does it take for the screw to hit the floor? (*b*) How long is the screw in the air if the elevator starts from rest when the screw falls, and moves upward with a constant acceleration of $a_e = 4.0\text{ m/s}^2$?

Picture the Problem Write the position as a function of time for both the screw, x_s, and the floor, x_f. The screw hits the floor when $x_s = x_f$. Choose the origin to be the initial position of the floor, and designate upward as the positive direction.

(*a*) 1. Write the position functions for the elevator floor and screw:

$$x_f - x_{0f} = v_{0f}t + \frac{1}{2}a_f t^2$$

$$x_s - x_{0s} = v_{0s}t + \frac{1}{2}a_s t^2$$

2. Identify the initial conditions and the accelerations:

$$x_{0f} = 0, \quad v_{0f} = 2.2\text{ m/s}, \quad a_f = 0$$

$$x_{0s} = h = 3\text{ m}, \quad v_{0s} = 2.2\text{ m/s}, \quad a_s = -g$$

3. Substitute these values into the position functions:

$$x_f = (2.2\text{ m/s})t$$

$$x_s = h + (2.2\text{ m/s})t - \frac{1}{2}gt^2$$

4. Set $x_s = x_f$ and solve for t:

$$h + (2.2\text{ m/s})t - \frac{1}{2}gt^2 = (2.2\text{ m/s})t$$

$$t = \sqrt{\frac{2h}{g}} = \sqrt{\frac{2(3\text{ m})}{9.81\text{ m/s}^2}} = 0.78\text{ s}$$

Note that this result does not depend on the velocity of the elevator.

(*b*) 1. Now the elevator floor moves upward from rest with constant acceleration. The initial conditions are then:

$$x_{0f} = 0, \quad v_{0f} = 0, \quad a_f = 4.0\text{ m/s}^2$$

$$x_{0s} = 3\text{ m}, \quad v_{0s} = 0, \quad a_s = -g$$

2. Use the initial conditions to write the position functions for this case:

$$x_f = \frac{1}{2}a_f t^2 = \frac{1}{2}(4.0\text{ m/s}^2)t^2$$

$$x_s = 3\text{ m} - \frac{1}{2}gt^2$$

3. Set $x_s = x_w$ and solve for t:

$$\frac{1}{2}a_f t^2 = 3\text{ m} - \frac{1}{2}gt^2$$

$$t = \sqrt{\frac{2(3\text{ m})}{g + a_f}} = \sqrt{\frac{2(3\text{ m})}{(9.81 + 4.0)\text{ m/s}^2}} = 0.66\text{ s}$$

Remark The time in the air is independent of the speed of the elevator, as long as the elevator does not accelerate. If the elevator has acceleration a_f you and the screw experience an "effective gravity" with acceleration $g' = g + a_f$. For the case in which the elevator accelerates downward with $a_f = -g$, the time of fall becomes infinite and the screw appears weightless.

Example 2-18 *try it yourself*

A probe launches vertically from the surface of Mars and reaches a height of 320 m and a velocity of 80 m/s at time $t = 0$ when its thrusters cut out. It then continues moving upward under the influence of Martian gravity, which is approximately constant and equal to $g_m = 3.72$ m/s^2. At the moment that the probe's thrusters cut out, $t = 0$, you are in a spacecraft 1500 m from the Martian surface, approaching the probe nearly head on, moving downward at 25 m/s and slowing down at a rate of 0.80 m/s^2. (*a*) When do you reach the probe? (*b*) How high above the planet's surface will the first rendezvous occur? (*c*) What is the velocity of each craft when they meet, assuming there are no course adjustments? (*d*) What is the velocity of the probe relative to the ship?

Figure 2-15

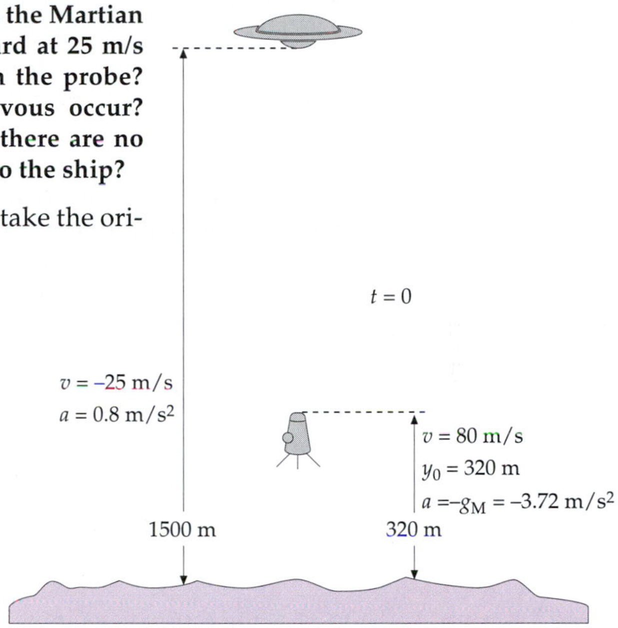

Picture the Problem Let upward be the positive direction, and take the origin to be at the surface of Mars (Figure 2-15). Then for the probe,

$y_0 = 320$ m

$v_0 = 80$ m/s

$a = -3.72$ m/s^2

and for your ship,

$y_0 = 1500$ m

$v_0 = -25$ m/s

$a = +0.8$ m/s^2

The ship reaches the probe when $y_p = y_s$.

Cover the column to the right and try these on your own before looking at the answers.

Steps	***Answers***
(*a*) 1. Write equations for y_p and y_s as functions of time.	$y_p = 320 + 80t - 1.86t^2$, $y_s = 1500 - 25t + 0.4t^2$
2. Set $y_p = y_s$ and solve for the time t. Note that you get two solutions.	$t_1 = 19.0$ s, $t_2 = 27.4$ s
(*b*) Substitute the result for t_1 into either equation in step 1 and solve for y.	$y_p(t_1) = y_s(t_1) = 1.17$ km
(*c*) 1. Write general equations for the velocities and substitute the time(s) found from part (*a*), step 2.	$v_p(t_1) = 9.32$ m/s $v_s(t_1) = -9.80$ m/s $v_p(t_2) = -21.9$ m/s $v_s(t_2) = -3.08$ m/s
(*d*) Find the relative velocity $v_p - v_s$.	$v_r(t_1) = 19.1$ m/s; $v_r(t_2) = -18.8$ m/s

Remark Your ship intercepts the probe twice, once when the probe is moving up and once when it is moving down, as shown in Figure 2-16.

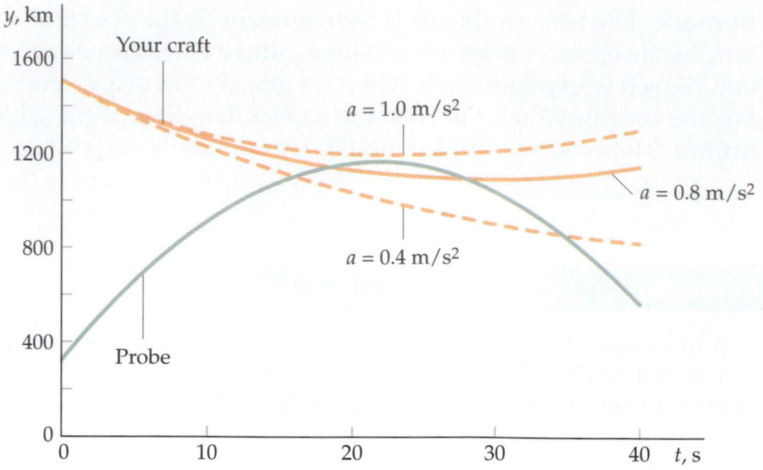

Figure 2-16 The solid curves depict the situation in Example 2-18. Also shown are two other cases: (lower dashed line) A spacecraft with $a = 0.4\ \text{m/s}^2$ crashes into the probe! (higher dashed line) A spacecraft with $a = 1.0\ \text{m/s}^2$ narrowly misses the probe. *Suggestion:* Find the conditions for the optimum encounter in which relative velocity is at a minimum.

2-4 Integration

To find the velocity from a given acceleration, we note that the velocity is the function $v(t)$ whose derivative is the acceleration $a(t)$:

$$\frac{dv(t)}{dt} = a(t)$$

If the acceleration is constant, the velocity is that function of time which, when differentiated, equals this constant. One such function is

$$v = at, \qquad a = \text{constant}$$

More generally, we can add any constant to at without changing the time derivative. Calling this constant v_0, we have

$$v = at + v_0$$

When $t = 0$, $v = v_0$. Thus, v_0 is the initial velocity.

Similarly, the position function $x(t)$ is that function whose derivative is the velocity:

$$\frac{dx}{dt} = v = v_0 + at$$

We can treat each term separately. The function whose derivative is a constant v_0 is $v_0 t$ plus any constant. The function whose derivative is at is $\frac{1}{2}at^2$ plus any constant. Writing x_0 for the combined arbitrary constants, we have

$$x = x_0 + v_0 t + \frac{1}{2}at^2$$

When $t = 0$, $x = x_0$. Thus, x_0 is the initial position.

Whenever we find a function from its derivative, we must include an arbitrary constant in the general function. Since we go through the integration process twice to find $x(t)$ from the acceleration, two constants arise. These constants are usually determined from the velocity and position at some given time, which is usually chosen to be $t = 0$. They are therefore called the **initial conditions.** A common problem, called the **initial-value problem,** takes the form "given $a(t)$ and the initial values of x and v, find $x(t)$." This problem is particularly important in physics because the acceleration of a particle is determined by the forces acting on it. Thus, if we know the forces acting on a particle and the position and velocity of the particle at some par-

A function $F(t)$ whose derivative (with respect to t) equals a function $f(t)$ is called the **antiderivative** of $f(t)$. Finding the antiderivative of a function is related to the problem of finding the area under a curve. Consider motion with a constant velocity v_0. The change in position Δx during an interval Δt is

$$\Delta x = v_0 \Delta t$$

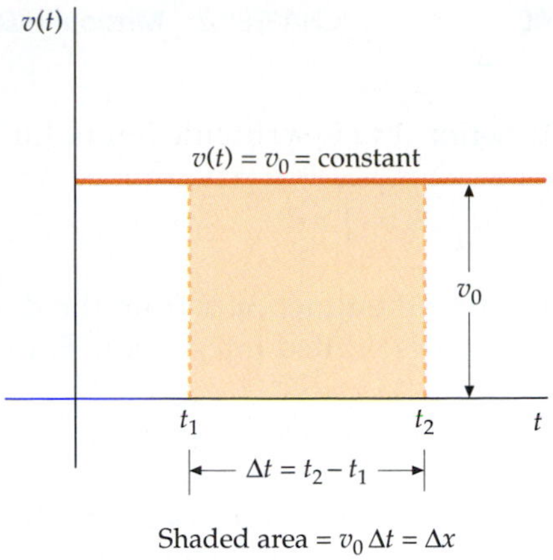

Figure 2-17 The displacement Δx during the time interval $\Delta t = t_2 - t_1$ is equal to the area of the shaded region. By the definition of average velocity, $\Delta x = v_{av}\,\Delta t$. This is just the area of a rectangle of height v_{av} and width Δt. Thus, the rectangular area $v_{av}\,\Delta t$ and the area under the v-versus-t curve must be equal.

This is the area under the v-versus-t curve (Figure 2-17). The geometric interpretation of the displacement as the area under the v-versus-t curve is true not only for constant velocity, but also in general, as illustrated in Figure 2-18. There, the area under the curve is approximated by first dividing the time interval into several smaller intervals, Δt_1, Δt_2, and so on, and drawing a set of rectangles. The area of the rectangle corresponding to the ith time interval Δt_i (shaded in the figure) is $v_i \Delta t_i$, which is approximately equal to the displacement Δx_i during the interval Δt_i. The sum of the rectangular areas is therefore approximately the sum of the displacements during the time intervals and is approximately equal to the total displacement from time t_1 to t_2. Mathematically, we write this as

$$\Delta x \approx \sum_i v_i \Delta t_i$$

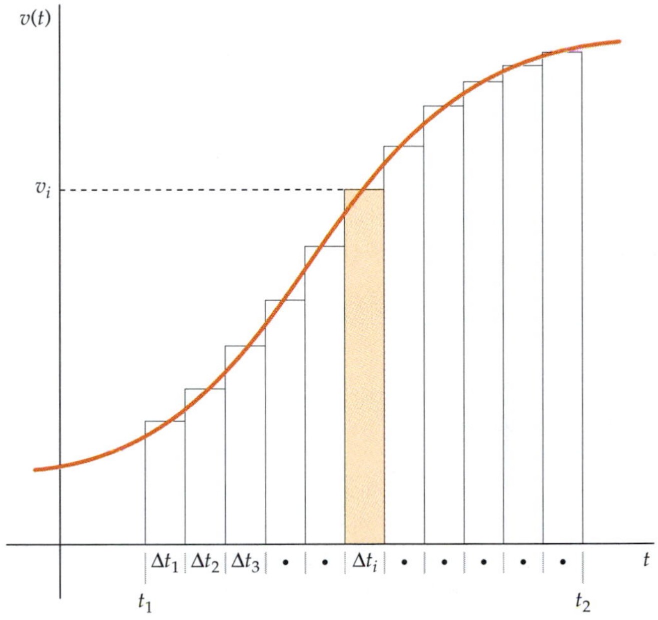

Figure 2-18 Graph of a general $v(t)$-versus-t curve. The total displacement from t_1 to t_2 is the area under the curve for this interval, which can be approximated by summing the areas of the rectangles.

where the Greek letter Σ (uppercase sigma) stands for "sum." We can make the approximation as accurate as we wish by choosing enough rectangles under the curve, each having a small value for Δt. For the limit of smaller and smaller time intervals, the resulting sum equals the area under the curve, which equals the displacement. This limit is called the **integral** and is written

$$\Delta x = x(t_2) - x(t_1) = \lim_{\Delta t \to 0} \sum_i v_i \Delta t_i = \int_{t_1}^{t_2} v\, dt \qquad \text{2-16}$$

It is helpful to think of the integral sign $\int$ as an elongated S indicating a sum. The limits t_1 and t_2 indicate the initial and final values of the variable t. The displacement is thus the area under the v-versus-t curve. Figure 2-19 demonstrates that the average velocity has a simple geometric interpretation in terms of the area under a curve.

The process of computing an integral is called **integration.** In Equation 2-16, v is the derivative of x, and x is the antiderivative of v. This is an example of the fundamental theorem of calculus, whose formulation in the seventeenth century greatly accelerated the mathematical development of physics:

$$\text{If } f(t) = \frac{dF(t)}{dt}, \quad \text{then} \quad F(t_2) - F(t_1) = \int_{t_1}^{t_2} f(t)\, dt \qquad \text{2-17}$$

Fundamental theorem of calculus

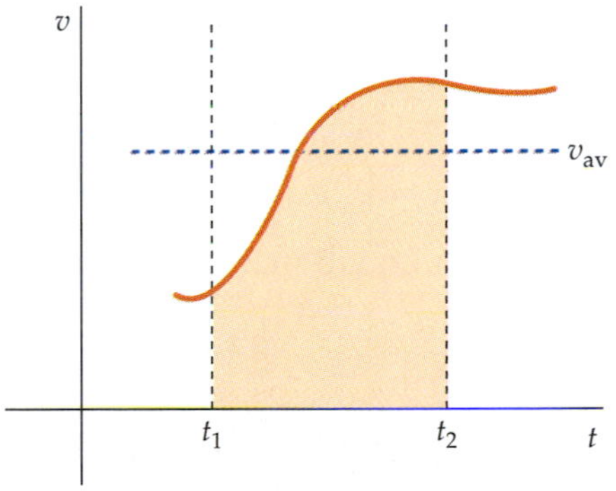

Figure 2-19 The displacement Δx during the time interval $\Delta t = t_2 - t_1$ is equal to the area of the shaded region. We know from the definition of average velocity that $\Delta x = v_{av}\,\Delta t$. This is just the area of a rectangle of height v_{av} and width Δt. Thus, the rectangular area $v_{av}\,\Delta t$ and the area under the v-versus-t curve must be equal.

The antiderivative of a function is also called the indefinite integral of the function and is written without limits on the integral sign:

$$x = \int v\, dt$$

Finding the function x from the derivative v (that is, finding the antiderivative) is also called integration. For example, if $v = v_0$, a constant, then

$$x = \int v_0\, dt = v_0 t + x_0$$

where x_0 is the arbitrary constant of integration. We can find a general rule for the integration of a power of t from Equation 2-6, which gives the general rule for the derivative of a power. The result is

$$\int t^n\, dt = \frac{t^{n+1}}{n+1} + C, \qquad n \neq -1 \tag{2-18}$$

where C is an arbitrary constant. This can easily be checked by differentiating the right side using the rule of Equation 2-6. (For the special case $n = -1$, $\int t^{-1}\, dt = \ln t + C$, where $\ln t$ is the natural logarithm of t.)

The change in velocity for some time interval can similarly be interpreted as the area under the a-versus-t curve for that interval. This is written

$$\Delta v = \lim_{\Delta t \to 0} \sum_i a_i \Delta t_i = \int_{t_1}^{t_2} a\, dt \tag{2-19}$$

We can now derive the constant-acceleration equations by computing the indefinite integrals of the acceleration and velocity. If a is constant, we have

$$v = \int a\, dt = v_0 + at \tag{2-20}$$

where we have written the constant of integration, v_0, first. Integrating again and writing x_0 for the constant of integration gives

$$x = \int (v_0 + at) dt = x_0 + v_0 t + \frac{1}{2} a t^2 \tag{2-21}$$

Having derived the constant-acceleration equations without any reference to average velocity, we can now show that the average velocity for the special case of constant acceleration is the mean value between the initial and final velocities as given by Equation 2-12. Let v_0 be the initial velocity at $t = 0$, and let v be the final velocity at time t. According to the definition of average velocity, the displacement is

$$\Delta x = v_{av} \Delta t = v_{av}(t - 0) = v_{av} t \tag{2-22}$$

Also, from Equation 2-21, we have

$$\Delta x = v_0 t + \frac{1}{2} a t^2$$

We can eliminate the acceleration using $a = (v - v_0)/t$ from Equation 2-20. Then

$$\Delta x = v_0 t + \frac{1}{2}\left(\frac{v - v_0}{t}\right) t^2 = v_0 t + \frac{1}{2} vt - \frac{1}{2} v_0 t = \frac{1}{2}(v + v_0) t \tag{2-23}$$

Comparing this with the definition of average velocity (Equation 2-22), we have

$$v_{av} = \frac{1}{2}\left(v_0 + v_f\right)$$

which is Equation 2-12*b*.

Example 2-19

A ferry boat moves with constant velocity $v_0 = 8$ m/s for 60 s. It then shuts off its engines and coasts. Its coasting velocity is given by $v = v_0 t_1^2/t^2$, where $t_1 = 60$ s. What is the displacement of the boat from $t = 0$ to $t \to \infty$?

Picture the Problem This velocity function is shown in Figure 2-20. The total displacement is calculated as the sum of the displacement Δx_1 from $t = 0$ to $t = 60$ s and the displacement Δx_2 from $t = 60$ s to $t \to \infty$.

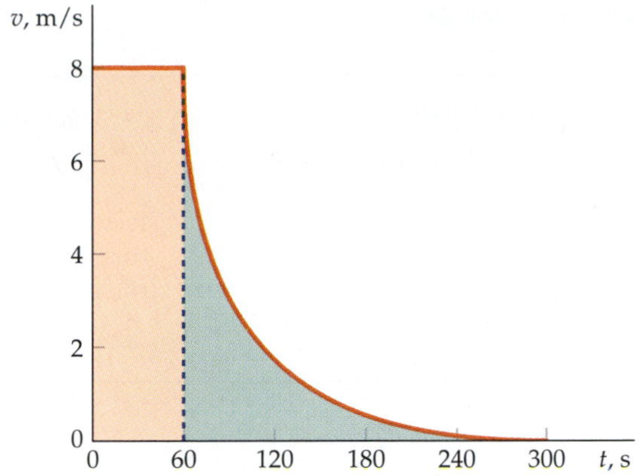

Figure 2-20

1. The velocity of the boat is constant during the first 60 seconds; thus, the displacement is simply the velocity times the elapsed time:

$$\Delta x_1 = v\Delta t = (8\text{ m/s})(60\text{ s}) = 480\text{ m}$$

2. The remaining displacement is given by the integral of the velocity from $t = 60$ s to $t = \infty$. We use Equation 2-18 to calculate the integral:

$$\Delta x_2 = \int_{60\text{ s}}^{\infty} v\,dt = \int_{60\text{ s}}^{\infty} \frac{v_0 t_1^2}{t^2}\,dt = v_0 t_1^2 \int_{60\text{ s}}^{\infty} t^{-2}\,dt = v_0 t_1^2 \left.\frac{t^{-1}}{-1}\right|_{60\text{ s}}^{\infty}$$

$$= \frac{v_0 t_1^2}{60\text{ s}} = \frac{(8\text{ m/s})(60\text{ s})^2}{60\text{ s}} = 480\text{ m}$$

3. The total displacement is the sum of the displacements found above:

$$\Delta x = \Delta x_1 + \Delta x_2 = 480\text{ m} + 480\text{ m} = 960\text{ m}$$

Remark Note that the area under the v-versus-t curve is finite. Thus, even though the boat never stops moving, it travels only a finite distance. A better representation of the velocity of a coasting boat might be an exponentially decreasing function. In that case, the boat would also coast a finite distance in the interval $60\text{ s} \le t \le \infty$.

Summary

Displacement, velocity, and acceleration are important *defined* kinematics quantities.

Topic	Remarks and Relevant Equations	
I. General Relations		
1. Displacement	$\Delta x = x_2 - x_1$	**2-1**
Graphical interpretation	Displacement is the area under the v-versus-t curve.	
2. Velocity		
Average velocity	$v_{av} = \dfrac{\Delta x}{\Delta t}$	**2-2**

Instantaneous velocity	$v(t) = \lim_{\Delta t \to 0} \frac{\Delta x}{\Delta t} = \frac{dx}{dt}$	**2-5**
Graphical interpretation	The instantaneous velocity is represented graphically as the slope of the x-versus-t curve.	
Relative velocity	If a particle moves with velocity v_{pA} relative to a coordinate system A, which is in turn moving with velocity v_{AB} relative to another coordinate system B, the velocity of the particle relative to B is $v_{pB} = v_{pA} + v_{AB}$	**2-7**

3. Speed

Average speed	$\text{Average speed} = \frac{\text{total distance}}{\text{total time}} = \frac{\Delta s}{\Delta t}$	**2-3**
Instantaneous speed	Instantaneous speed is the magnitude of the instantaneous velocity.	

4. Acceleration

Average acceleration	$a_{av} = \frac{\Delta v}{\Delta t}$	**2-8**
Instantaneous acceleration	$a = \frac{dv}{dt} = \frac{d^2x}{dt^2}$	**2-10**
Graphical interpretation	The instantaneous acceleration is represented graphically as the slope of the v-versus-t curve.	
Acceleration due to gravity	The acceleration of an object near the surface of the earth in free fall under the influence of gravity is directed downward and has the magnitude $g = 9.81\ \text{m/s}^2 = 32.2\ \text{ft/s}^2$	

5. Displacement and Velocity as Integrals

Displacement is represented graphically as the area under the v-versus-t curve. This area is the integral of v over time from some initial time t_1 to some final time t_2 and is written

$$\Delta x = \lim_{\Delta t \to 0} \sum_i v_i \Delta t_i = \int_{t_1}^{t_2} v\, dt \qquad \textbf{2-16}$$

Similarly, the change in velocity for some time is represented graphically as the area under the a-versus-t curve:

$$\Delta v = \lim_{\Delta t \to 0} \sum_i a_i \Delta t_i = \int_{t_1}^{t_2} a\, dt \qquad \textbf{2-17}$$

II. Constant-Acceleration Equations

Velocity	$v = v_0 + at$	**2-12*b***
Displacement in terms of v_{av}	$\Delta x = x - x_0 = v_{av} t = \frac{1}{2}(v_0 + v)t$	**2-13**
Displacement in terms of a	$\Delta x = x - x_0 = v_0 t + \frac{1}{2} a t^2$	**2-14**
v in terms of a and Δx	$v^2 = v_0^2 + 2a\,\Delta x$	**2-15**

Problem-Solving Guide

The following is applicable to all types of problems:

1. Begin by drawing a neat diagram that includes the important features of the problem.
2. Choose a convenient coordinate system and indicate it on your diagram. Show the origin and positive directions. When possible, choose the origin to be the location of the particle at $t = 0$ so that $x_0 = 0$.
3. Show known quantities on your diagram.
4. When possible, write an equation for the quantity to be found in terms of other quantities that are known or that can be found. Then proceed to find the other quantities in your equation.
5. When possible, solve the problem two different ways to check your solution.
6. Examine your answer to see if it is reasonable.

Summary of Worked Examples

Type of Calculation	Procedure and Relevant Examples
1. Displacement, Velocity, and Acceleration	
Find the average velocity for some time interval.	Find the displacement Δx for that time interval; then $v_{av} = \Delta x/\Delta t$. **Examples 2-1, 2-2, 2-3**
Find the total distance traveled.	Find the total time Δt and the average speed. The total distance is the average speed times the total time. **Example 2-3**
Find the instantaneous velocity from a graph of *x* versus *t*.	Draw the tangent line at the point in question. The slope of this line is v. **Example 2-5**
Find the instantaneous velocity and acceleration from a given function *x*(*t*).	Compute the derivatives $v = dx/dt$ and $a = dv/dt = d^2x/dt^2$. **Examples 2-6, 2-8**
Find the average acceleration for some time interval.	Find the total change in velocity Δv; then $a_{av} = \Delta v/\Delta t$. **Example 2-7**
2. Constant Acceleration—One Object	
Find the greatest height reached by a thrown object and the time needed to reach it.	Set $v = 0$ in the constant acceleration equation $v^2 = v_0^2 + 2a\Delta x$ and solve for Δx. To find the rise time, set $v = 0$ in $v = v_0 + at$, and solve for t. **Example 2-9**
Find the stopping distance for a braking automobile.	Use $v^2 = v_0^2 + 2a\Delta x$ with $v = 0$. **Example 2-10**
Find the speed after a given time.	Use $v = v_0 + at$. **Example 2-11**
Find the distance traveled during a given time interval.	Find the initial and final time and compute $x_2 - x_1$, or find the average velocity for the interval and compute $\Delta x = v_{av}\,\Delta t$. **Example 2-11**
Estimate the stopping time and average acceleration for a car crashing into a wall.	Estimate the stopping distance, then use $\Delta t = \Delta x/v_{av}$ and $a = \Delta v/\Delta t$ **Examples 2-10, 2-12**
Find the total distance when the acceleration is different for different intervals.	Find Δx_i for each interval using the final position and velocity for the previous interval as initial conditions for the next interval. **Example 2-13**

3. Constant Acceleration—Two Objects

Find the time of collision of two moving objects.	Write $x(t)$ for each object, then set $x_1(t) = x_2(t)$ and solve for t.	**Examples 2-15, 2-16, 2-17**
Find the speed of one object given the separation of two objects.	Write $x(t)$ for both objects and solve for t for the given separation. Then find v at that time.	**Example 2-18**

4. Integration

Find the displacement of a particle given $v(t)$.	Compute the integral of $v(t)\,dt$.	**Example 2-19**

Problems

Conceptual Problems

Problems from Optional and Exploring sections

In a few problems, you are given more data than you actually need; in a few other problems, you are required to supply data from your general knowledge, outside sources, or informed estimates.

• Single-concept, single-step, relatively easy
•• Intermediate-level, may require synthesis of concepts
••• Challenging, for advanced students

For all problems, use $g = 9.81$ m/s^2 for the acceleration due to gravity and neglect friction and air resistance unless instructed to do otherwise.

Speed, Displacement, and Velocity

1 • What is the approximate average velocity of the race cars during the Indianapolis 500?

2 • Does the following statement make sense? "The average velocity of the car at 9 A.M. was 60 km/h."

3 • Is it possible for the average velocity of an object to be zero during some interval even though its average velocity for the first half of the interval is not zero? Explain.

4 • The diagram in Figure 2-21 tracks the path of an object moving in a straight line. At which point is the object farthest from its starting point?

(*a*) A
(*b*) B
(*c*) C
(*d*) D
(*e*) E

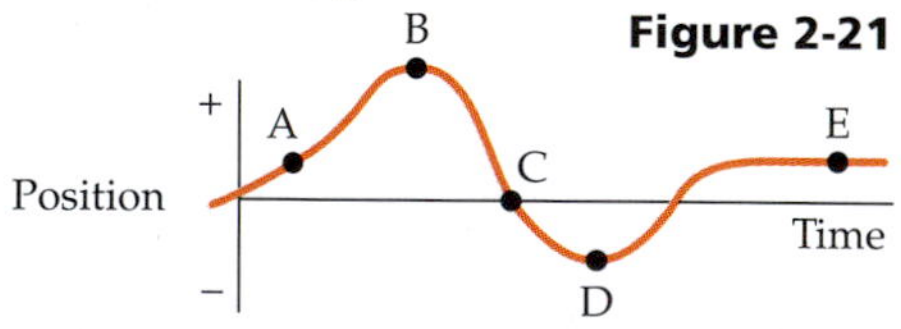

Figure 2-21 Problem 4

5 • (*a*) An electron in a television tube travels the 16-cm distance from the grid to the screen at an average speed of 4×10^7 m/s. How long does the trip take? (*b*) An electron in a current-carrying wire travels at an average speed of 4×10^{-5} m/s. How long does it take to travel 16 cm?

6 • A runner runs 2.5 km in 9 min and then takes 30 min to walk back to the starting point. (*a*) What is the runner's average velocity for the first 9 min? (*b*) What is the average velocity for the time spent walking? (*c*) What is the average velocity for the whole trip? (*d*) What is the average speed for the whole trip?

7 • A car travels in a straight line with an average velocity of 80 km/h for 2.5 h and then with an average velocity of 40 km/h for 1.5 h. (*a*) What is the total displacement for the 4-h trip? (*b*) What is the average velocity for the total trip?

8 • One busy air route across the Atlantic Ocean is about 5500 km. (*a*) How long does it take for a supersonic jet flying at 2.4 times the speed of sound to make the trip? Use 340 m/s for the speed of sound. (*b*) How long does it take a subsonic jet flying at 0.9 times the speed of sound to make the same trip? (*c*) Allowing 2 h at each end of the trip for ground travel, check-in, and baggage handling, what is your average speed door to door when traveling on the supersonic jet? (*d*) What is your average speed taking the subsonic jet?

9 • As you drive down a desert highway at night, an alien spacecraft passes overhead, causing malfunctions in your speedometer, wristwatch, and short-term memory. When you return to your senses, you can't tell where you are, where you are going, or even how fast you are traveling. The passenger sleeping next to you never woke up during this incident. Although your pulse is racing, hers is a steady 55 beats per minute. (*a*) If she has 45 beats between the mile markers posted along the road, determine your speed. (*b*) If you want to travel at 120 km/h, how many heartbeats should there be between mile markers?

10 • The speed of light, c, is 3×10^8 m/s. (*a*) How long does it take for light to travel from the sun to the earth, a distance of 1.5×10^{11} m? (*b*) How long does it take light to travel from the moon to the earth, a distance of 3.84×10^8 m? (*c*) A light-year is a unit of distance equal to that traveled by light in 1 year. Convert 1 light-year into kilometers and miles.

11 • The nearest star, Proxima Centauri, is 4.1×10^{15} km away. From the vicinity of this star, Gregor places an order at Tony's Pizza in Hoboken, New Jersey, communicating via light signals. Tony's fastest delivery craft travels at $10^{-4}c$ (see Problem 10). (*a*) How long does it take for Gregor's order to reach Tony's pizza? (*b*) How long does Gregor wait between sending the signal and receiving the pizza? If Tony's has a 1000-years-or-it's-yours-free delivery policy, does Gregor have to pay for the pizza?

12 • A car making a 100-km journey travels 40 km/h for the first 50 km. How fast must it go during the second 50 km to average 50 km/h?

13 •• John can run 6.0 m/s. Marcia can run 15% faster than John. (*a*) By what distance does Marcia beat John in a 100-m race? (*b*) By what time does Marcia beat John in a 100-m race?

14 •• Figure 2-22 shows the position of a particle versus time. Find the average velocities for the time intervals *a*, *b*, *c*, and *d* indicated in the figure.

Figure 2-22 Problem 14

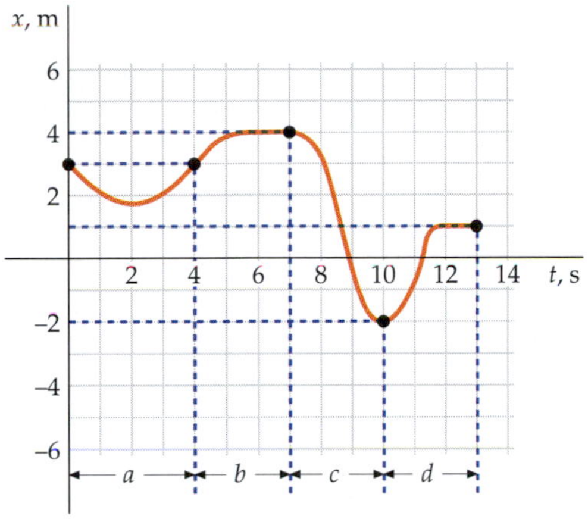

15 •• It has been found that galaxies are moving away from the earth at a speed that is proportional to their distance from the earth. This discovery is known as Hubble's law. The speed of a galaxy at distance r from the earth is given by $v = Hr$, where H is the Hubble constant, equal to $1.58 \times 10^{-18}\ \mathrm{s}^{-1}$. What is the speed of a galaxy (*a*) 5×10^{22} m from earth and (*b*) 2×10^{25} m from earth? (*c*) If each of these galaxies has traveled with constant speed, how long ago were they both located at the same place as the earth?

16 •• Cupid fires an arrow that strikes St. Valentine, producing the usual sounds of harp music and bird chirping as Valentine swoons into a fog of love. If Cupid hears these telltale sounds exactly one second after firing the arrow and the average speed of the arrow was 40 m/s, what was the distance separating them? Take 340 m/s for the speed of sound.

Instantaneous Velocity

17 • If the instantaneous velocity does not change, will the average velocities for different intervals differ?

18 • If $v_{av} = 0$ for some time interval Δt, must the instantaneous velocity v be zero at some point in the interval? Support your answer by sketching a possible x-versus-t curve that has $\Delta x = 0$ for some interval Δt.

19 •• An object moves along the x axis as shown in Figure 2-23. At which point or points is the magnitude of its velocity at a minimum?

(*a*) A and E
(*b*) B, D, and E
(*c*) C only
(*d*) E only
(*e*) None of these is correct.

Figure 2-23 Problem 19

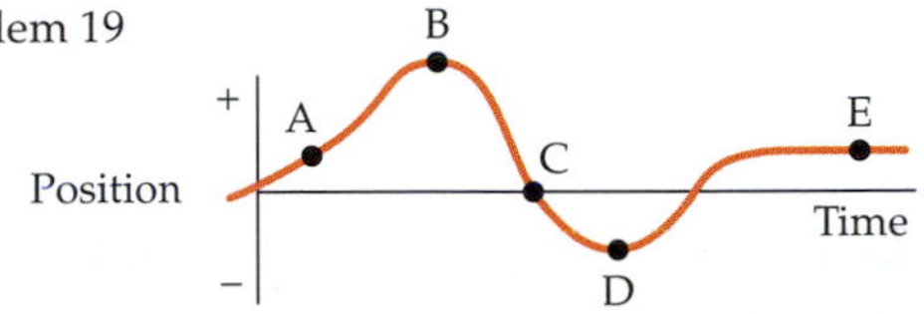

20 •• For each of the four graphs of x versus t in Figure 2-24, answer the following questions. (*a*) Is the velocity at time t_2 greater than, less than, or equal to the velocity at time t_1? (*b*) Is the speed at time t_2 greater than, less than, or equal to the speed at time t_1?

Figure 2-24 Problem 20

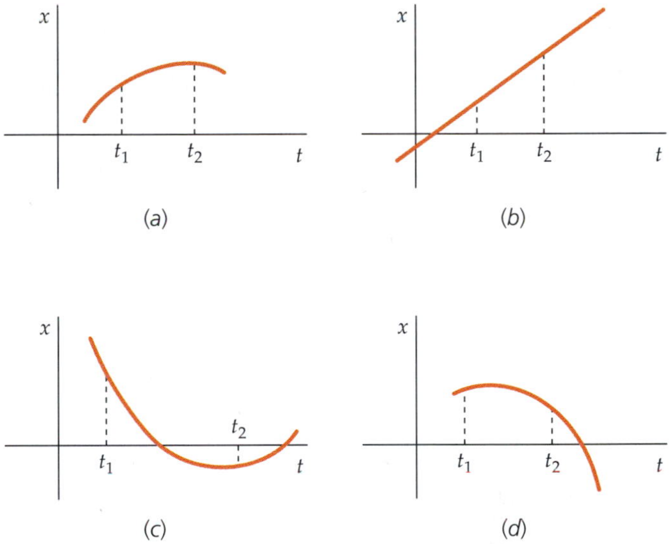

21 • Using the graph of x versus t in Figure 2-25, (*a*) find the average velocity between the times $t = 0$ and $t = 2$ s. (*b*) Find the instantaneous velocity at $t = 2$ s by measuring the slope of the tangent line indicated.

Figure 2-25 Problem 21

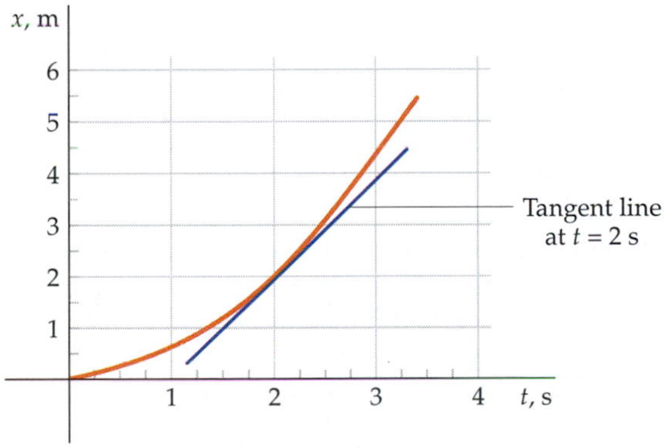

22 • Using the graph of x versus t in Figure 2-26, find (*a*) the average velocity for the time intervals $\Delta t = t_2 - 0.75$ s when t_2 is 1.75, 1.5, 1.25, and 1.0 s; (*b*) the instantaneous ve-

Figure 2-26 Problem 22

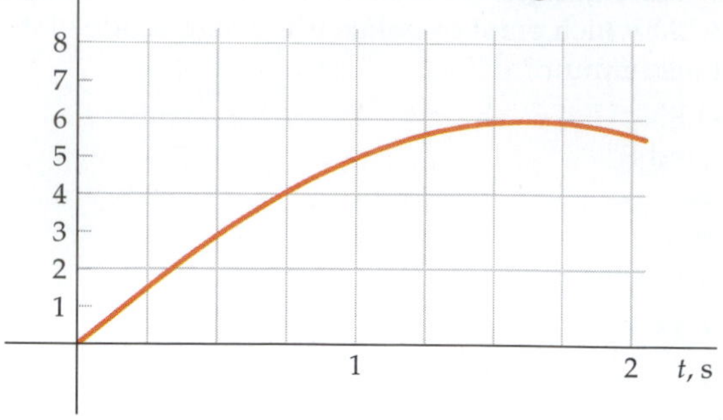

locity at $t = 0.75$ s; (c) the approximate time when the instantaneous velocity is zero.

23 •• The position of a certain particle depends on the time according to $x = (1\ \text{m/s}^2)t^2 - (5\ \text{m/s})t + 1$ m. (*a*) Find the displacement and average velocity for the interval $3\ \text{s} \leq t \leq 4$ s. (*b*) Find a general formula for the displacement for the time interval from t to $t + \Delta t$. (*c*) Use the limiting process to obtain the instantaneous velocity for any time t.

24 •• The height of a certain projectile is related to time by $y = -5(t - 5)^2 + 125$, where y is in meters and t is in seconds. (*a*) Sketch y versus t for $0 \leq t \leq 10$ s. (*b*) Find the average velocity for each of the 1-s time intervals between integral time values from $0 \leq t \leq 10$ s. Sketch v_{av} versus t. (*c*) Find the instantaneous velocity as a function of time.

25 ••• The position of a body oscillating on a spring is given by $x = A \sin \omega t$, where A and ω are constants with values $A = 5$ cm and $\omega = 0.175\ \text{s}^{-1}$. (*a*) Sketch x versus t for $0 \leq t \leq 36$ s. (*b*) Measure the slope of your graph at $t = 0$ to find the velocity at this time. (*c*) Calculate the average velocity for a series of intervals beginning at $t = 0$ and ending at $t = 6, 3, 2, 1, 0.5$, and 0.25 s. (*d*) Compute dx/dt and find the velocity at time $t = 0$.

Relative Velocity

26 • To avoid falling too fast during a landing, an airplane must maintain a minimum airspeed (the speed of the plane relative to the air). However, the slower the ground speed (speed relative to the ground) during a landing, the safer the landing. Is it safer for an airplane to land with the wind or against the wind?

27 •• Two cars are traveling along a straight road. Car A maintains a constant speed of 80 km/h; car B maintains a constant speed of 110 km/h. At $t = 0$, car B is 45 km behind car A. How far will car A travel from $t = 0$ before it is overtaken by car B?

28 •• A car traveling at a constant speed of 20 m/s passes an intersection at time $t = 0$, and 5 s later another car traveling 30 m/s passes the same intersection in the same direction. (*a*) Sketch the position functions $x_1(t)$ and $x_2(t)$ for the two cars. (*b*) Determine when the second car will overtake the first. (*c*) How far from the intersection will the two cars be when they pull even?

29 •• Margaret has just enough gas in her speedboat to get to the marina, an upstream journey that takes 4.0 h. Finding it closed for the season, she spends the next 8.0 h floating back downstream to her shack. The entire trip took 12.0 h; how long would it have taken if she had bought gas at the marina?

30 •• Joe and Sally tend to argue when they travel. Just as they reached the moving sidewalk at the airport, their struggle for itinerary-making powers peaked. Though they stepped on the moving belt at the same time, Joe chose to stand and ride, while Sally opted to keep walking. Sally reached the end in 1 min, while Joe took 2 min. How long would it have taken Sally if she had walked twice as fast?

Acceleration

31 • Walk across the room in such a way that, after getting started, your velocity is negative, but your acceleration is positive. (*a*) Describe how you did it. (*b*) Sketch a graph of v versus t for your motion.

32 • Give an example of a motion for which both the acceleration and the velocity are negative.

33 • Is it possible for a body to have zero velocity and nonzero acceleration?

34 • True or false:

(*a*) If the acceleration is zero, the body cannot be moving.
(*b*) If the acceleration is zero, the x-versus-t curve must be a straight line.

35 •• State whether the acceleration is positive, negative, or zero for each of the position functions $x(t)$ in Figure 2-27.

Figure 2-27 Problem 35

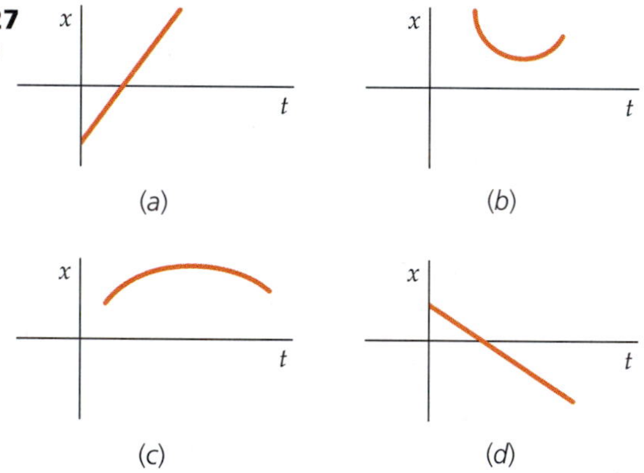

36 •• Answer the following questions for each of the graphs in Figure 2-28: (*a*) At what times are the accelerations of the objects positive, negative, and zero? (*b*) At what times are the accelerations constant? (*c*) At what times are the instantaneous velocities zero?

Figure 2-28 Problem 36

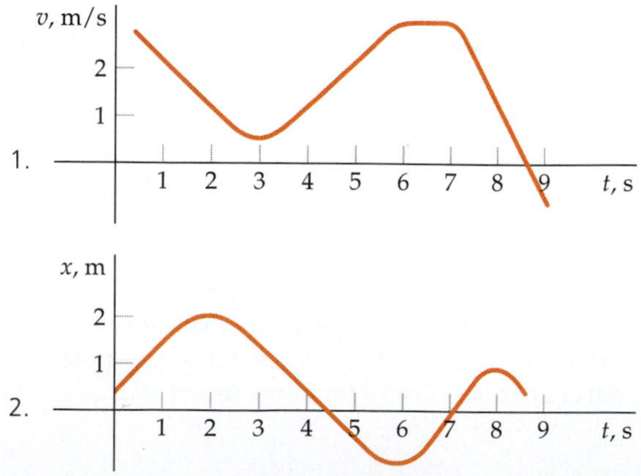

37 • A BMW M3 sports car can accelerate in third gear from 48.3 km/h (30 mi/h) to 80.5 km/h (50 mi/h) in 3.7 s. (*a*) What is the average acceleration of this car in m/s^2? (*b*) If the car continued at this acceleration for another second, how fast would it be moving?

38 • At $t = 5$ s, an object at $x = 3$ m is traveling at 5 m/s. At $t = 8$ s, it is at $x = 9$ m and its velocity is -1 m/s. Find the average acceleration for this interval.

39 •• A particle moves with velocity $v = 8t - 7$, where v is in meters per second and t is in seconds. (*a*) Find the average acceleration for the one-second intervals beginning at $t =$ 3 s and $t = 4$ s. (*b*) Sketch v versus t. What is the instantaneous acceleration at any time?

40 •• The position of an object is related to time by $x = At^2 - Bt + C$, where $A = 8\ m/s^2$, $B = 6$ m/s, and $C = 4$ m. Find the instantaneous velocity and acceleration as functions of time.

Motion With Constant Acceleration

41 • Identical twin brothers standing on a bridge each throw a rock straight down into the water below. They throw the rocks at exactly the same time, but one hits the water before the other. How can this occur if the rocks have the same acceleration?

42 • A ball is thrown straight up. What is the velocity of the ball at the top of its flight? What is its acceleration at that point?

43 • An object thrown straight up falls back to the ground T seconds later. Its maximum height is H meters. Its average velocity during those T seconds is (*a*) H/T, (*b*) 0, (*c*) $H/2T$, (*d*) $2H/T$.

44 • For an object thrown straight up, which of the following is true while it is in the air? (*a*) The acceleration is always opposite to the velocity. (*b*) The acceleration is always directed downward. (*c*) The acceleration is always in the direction of motion. (*d*) The acceleration is zero at the top of the trajectory.

45 • An object projected up with initial velocity v attains a height H. Another object projected up with initial velocity $2v$ will attain a height of (*a*) $4H$, (*b*) $3H$, (*c*) $2H$, (*d*) H.

46 • A ball is thrown upward. While it is in the air, its acceleration is (*a*) decreasing, (*b*) constant, (*c*) zero, (*d*) increasing.

47 • At $t = 0$, object A is dropped from the roof of a building. At the same instant, object B is dropped from a window 10 m below the roof. During their descent to the ground the distance between the two objects

(*a*) is proportional to t.
(*b*) is proportional to t^2.
(*c*) decreases.
(*d*) remains 10 m throughout.

48 •• A Porsche accelerates uniformly from 80.5 km/h (50 mi/h) at $t = 0$ to 113 km/h (70 mi/h) at $t = 9$ s. Which graph in Figure 2-29 best describes the motion of the car?

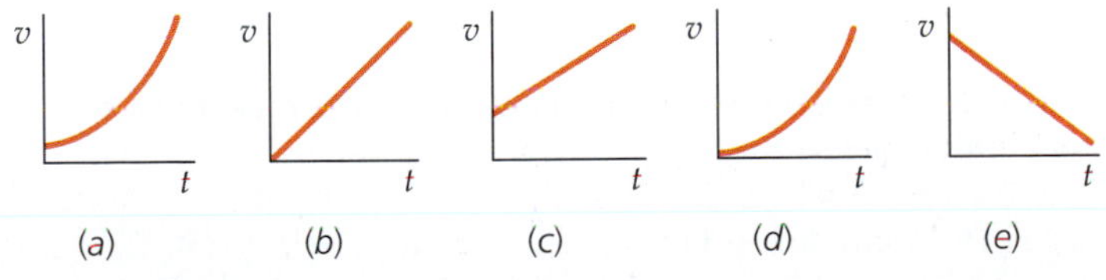

Figure 2-29 Problem 48

49 •• An object is dropped from rest. If the time during which it falls is doubled, the distance it falls will (*a*) double, (*b*) decrease by one-half, (*c*) increase by a factor of four, (*d*) decrease by a factor of four, (*e*) remain the same.

50 •• A ball is thrown upward with an initial velocity v_0. Its velocity halfway to its highest point is (*a*) $0.5v_0$, (*b*) $0.25v_0$, (*c*) v_0, (*d*) $0.707v_0$, (*e*) cannot be determined from the information given.

51 • A car starting at $x = 50$ m accelerates from rest at a constant rate of $8\ m/s^2$. (*a*) How fast is it going after 10 s? (*b*) How far has it gone after 10 s? (*c*) What is its average velocity for the interval $0 \le t \le 10$ s?

52 • An object with an initial velocity of 5 m/s has a constant acceleration of $2\ m/s^2$. When its speed is 15 m/s, how far has it traveled?

53 • An object with constant acceleration has velocity $v = 10$ m/s when it is at $x = 6$ m and $v = 15$ m/s when it is at $x = 10$ m. What is its acceleration?

54 • An object has constant acceleration $a = 4\ m/s^2$. At $t = 0$, its velocity is 1 m/s and it is at $x = 7$ m. How fast is it moving when it is at $x = 8$ m? What is t at that point?

55 • If a rifle fires a bullet straight up with a muzzle speed of 300 m/s, how high will the bullet rise? (Ignore air resistance.)

56 • A test of the prototype of a new automobile shows that the minimum distance for a controlled stop from 98 km/h to zero is 50 m. Find the acceleration, assuming it to be constant, and express your answer as a fraction of the free-fall acceleration due to gravity. How long does the car take to stop?

57 •• A ball is thrown upward with initial velocity of 20 m/s. (*a*) How long is the ball in the air? (*b*) What is the greatest height reached by the ball? (*c*) When is the ball 15 m above the ground?

58 •• A particle moves with a constant acceleration of $3\ m/s^2$. At $t = 4$ s, it is at $x = 100$ m; at $t = 6$ s, it has a velocity $v = 15$ m/s. Find its position at $t = 6$ s.

59 •• A bullet traveling at 350 m/s strikes a telephone pole and penetrates a distance of 12 cm before stopping. (*a*) Estimate the average acceleration by assuming it to be constant. (*b*) How long did it take for the bullet to stop?

60 •• A plane landing on an aircraft carrier has just 70 m to stop. If its initial speed is 60 m/s, (*a*) what is the acceleration of the plane during landing, assuming it to be constant? (*b*) How long does it take for the plane to stop?

61 •• An automobile accelerates from rest at $2\ m/s^2$ for 20 s. The speed is then held constant for 20 s, after which there is an acceleration of $-3\ m/s^2$ until the automobile stops. What is the total distance traveled?

62 •• In the Blackhawk landslide in California, a mass of rock and mud fell 460 m down a mountain and then traveled 8 km across a level plain on a cushion of compressed air. Assume that the mud dropped with the free-fall acceleration due to gravity and then slid horizontally with constant deceleration. (*a*) How long did the mud take to drop the 460 m? (*b*) How fast was it traveling when it reached the bottom? (*c*) How long did the mud take to slide the 8 km horizontally?

63 •• A load of bricks is being lifted by a crane at a steady velocity of 5 m/s when one brick falls off 6 m above the ground. (*a*) Sketch $x(t)$ to show the motion of the free brick. (*b*) What is the greatest height the brick reaches above the ground? (*c*) How long does it take to reach the ground? (*d*) What is its speed just before it hits the ground?

64 •• An egg with a mass of 50 g rolls off a table at a height of 1.2 m and splatters on the floor. Estimate the average acceleration of the egg while it is in contact with the floor.

65 •• To win publicity for her new CD release, Sharika, the punk queen, jumps out of an airplane without a parachute. She expects a stack of loose hay to break her fall. If she reaches a speed of 120 km/h prior to impact, and if a $35g$ deceleration is the greatest deceleration she can withstand, how high must the stack of hay be for her to survive? Assume uniform acceleration while she is in contact with the hay.

66 •• A bolt comes loose from underneath an elevator that is moving upward at a speed of 6 m/s. The bolt reaches the bottom of the elevator shaft in 3 s. (*a*) How high up was the elevator when the bolt came loose? (*b*) What is the speed of the bolt when it hits the bottom of the shaft?

67 •• An object is dropped from a height of 120 m. Find the distance it falls during its final second in the air.

68 •• An object is dropped from a height H. During the final second of its fall, it traverses a distance of 38 m. What was H?

69 •• A stone is thrown vertically from a cliff 200 m tall. During the last half-second of its flight the stone travels a distance of 45 m. Find the initial velocity of the stone.

70 •• An object in free fall from a height H traverses $0.4H$ during the first second of its descent. Determine the average speed of the object during free fall.

71 •• A bus accelerates at 1.5 m/s^2 from rest for 12 s. It then travels at constant speed for 25 s, after which it slows to a stop with an acceleration of -1.5 m/s^2. (*a*) How far did the bus travel? (*b*) What was its average velocity?

72 •• A basketball is dropped from a height of 3 m and rebounds from the floor to a height of 2 m. (*a*) What is the velocity of the ball just as it reaches the floor? (*b*) What is its velocity just as it leaves the floor? (*c*) Estimate the magnitude and direction of its average acceleration during this interval.

73 •• A rocket is fired vertically with an upward acceleration of 20 m/s^2. After 25 s, the engine shuts off and the rocket continues as a free particle until it reaches the ground. Calculate (*a*) the highest point the rocket reaches, (*b*) the total time the rocket is in the air, (*c*) the speed of the rocket just before it hits the ground.

74 •• A flowerpot falls from the ledge of an apartment building. A person in an apartment below, coincidentally holding a stopwatch, notices that it takes 0.2 s for the pot to fall past his window, which is 4 m high. How far above the top of the window is the ledge from which the pot fell?

75 •• Sharika arrives home late from a gig, only to find herself locked out. Her roommate and bass player Chico is practicing so loudly that he can't hear Sharika's pounding on the door downstairs. One of the band's props is a small trampoline, which Sharika places under Chico's window. She bounces progressively higher trying to get Chico's attention. Propelling herself furiously upward, she miscalculates on the last bounce and flies past the window and out of sight. Chico sees her face for 0.2 s as she moves a distance of 2.4 m from the bottom to the top of the window. (*a*) How long until she reappears? (*b*) What is her greatest height above the top of the window? (Treat Sharika as a point-particle punk.)

76 •• In a classroom demonstration, a glider moves along an inclined air track with constant acceleration a. It is projected from the start of the track ($x = 0$) with an initial velocity v_0. At time $t = 8$ s, it is at $x = 100$ cm and is moving along the track at velocity $v = -15$ cm/s. Find the initial speed v_0 and the acceleration a.

77 •• A rock dropped from a cliff falls one-third of its total distance to the ground in the last second of its fall. How high is the cliff?

78 ••• A typical automobile has a maximum deceleration of about 7 m/s^2; the typical reaction time to engage the brakes is 0.50 s. A school board sets the speed limit in a school zone to meet the condition that all cars should be able to stop in a distance of 4 m. (*a*) What maximum speed should be allowed for a typical automobile? (*b*) What fraction of the 4 m is due to the reaction time?

Constant Acceleration With Two Objects

79 •• Two trains face each other on adjacent tracks. They are initially at rest 40 m apart. The train on the left accelerates rightward at 1.4 m/s^2. The train on the right accelerates leftward at 2.2 m/s^2. How far does the train on the left travel before the two trains pass?

80 •• Two stones are dropped from the edge of a 60-m cliff, the second stone 1.6 s after the first. How far below the cliff is the second stone when the separation between the two stones is 36 m?

81 •• A motorcycle policeman hidden at an intersection observes a car that ignores a stop sign, crosses the intersection, and continues on at constant speed. The policeman starts off in pursuit 2.0 s after the car has passed the stop sign, accelerates at 6.2 m/s^2 until his speed is 110 km/h, and then continues at this speed until he catches the car. At that instant, the car is 1.4 km from the intersection. How fast was the car traveling?

82 •• At $t = 0$, a stone is dropped from a cliff above a lake; 1.6 s later another stone is thrown downward from the same point with an initial speed of 32 m/s. Both stones hit the water at the same instant. Find the height of the cliff.

83 •• A passenger train is traveling at 29 m/s when the engineer sees a freight train 360 m ahead traveling on the same track in the same direction. The freight train is moving at a speed of 6 m/s. If the reaction time of the engineer is 0.4 s, what must be the deceleration of the passenger train if a collision is to be avoided? If your answer is the maximum deceleration of the passenger train but the engineer's reaction time is 0.8 s, what is the relative speed of the two trains at the instant of collision and how far will the passenger train have traveled in the time between the sighting of the freight train and the collision?

84 •• After being forced out of farming, Lou has given up on trying to find work locally and is about to "ride the rails" to look for a job. Running at his maximum speed of 8 m/s, he is a distance d from the train when it begins to accelerate from rest at 1.0 m/s^2. (*a*) If d = 30 m and Lou keeps running, will he be able to jump into the train? (*b*) Sketch the position function $x(t)$ for the train, with $x = 0$ at $t = 0$. On the same graph, sketch $x(t)$ for various distances d, including d = 30 m and the critical separation distance d_c, the distance at which he just catches the train. (*c*) For the situation $d = d_c$, what is the speed of the train when Lou catches it? What is the train's average speed for the time interval between $t = 0$ and the moment Lou catches the train? What is the exact value of d_c?

85 •• A train pulls away from a station with a constant acceleration of 0.40 m/s^2. A passenger arrives at the track 6.0 s after the end of the train has passed the very same point. What is the slowest constant speed at which she can run and catch the train? Sketch curves for the motion of the passenger and the train as functions of time.

86 ••• Lou applies for a job as a perfume salesman. He tries to convince the boss to try his daring, aggressive promotional gimmick: dousing perspective customers as they wait at bus stops. A hard ball is to be thrown straight upward with an initial speed of 24 m/s. A thin-skinned ball filled with perfume is then thrown straight upward along the same path with a speed of 14 m/s. The balls are to collide when the perfume ball is at the high point of its trajectory, so that it breaks open and everyone gets a free sample. If $t = 0$ when the first ball is thrown, find the time when the perfume ball should be thrown.

87 ••• Ball A is dropped from the top of a building at the same instant that ball B is thrown vertically upward from the ground. When the balls collide, they are moving in opposite directions, and the speed of A is twice the speed of B. At what fraction of the height of the building does the collision occur?

88 ••• Solve Problem 87 if the collision occurs when the balls are moving in the same direction and the speed of A is 4 times that of B.

89 ••• The Sprint missile, designed to destroy incoming ballistic missiles, can accelerate at 100g. If an ICBM is detected at an altitude of 100 km moving straight down at a constant speed of 3×10^4 km/h and the Sprint missile is launched to intercept it, at what time and altitude will the interception take place? (*Note:* You can neglect the acceleration due to gravity in this problem. Why?)

90 ••• When a car traveling at speed v_1 rounds a corner, the driver sees another car traveling at a slower speed v_2 a distance d ahead. (*a*) If the maximum acceleration the driver's brakes can provide is a, show that the distance d must be greater than $(v_1 - v_2)^2/2a$ if a collision is to be avoided. (*b*) Evaluate this distance for v_1 = 90 km/h, v_2 = 45 km/h, and a = 6 m/s^2. (*c*) Estimate or measure your reaction time and calculate the effect it would have on the distance found in part (*b*).

Integration

91 • The velocity of a particle is given by $v = 6t + 3$, where t is in seconds and v is in meters per second. (*a*) Sketch $v(t)$ versus t, and find the area under the curve for the interval $t = 0$ to $t = 5$ s. (*b*) Find the position function $x(t)$. Use it to calculate the displacement during the interval $t = 0$ to $t = 5$ s.

92 • Figure 2-30 shows the velocity of a particle versus time. (*a*) What is the magnitude in meters of the area of the rectangle indicated? (*b*) Find the approximate displacement of the particle for the one-second intervals beginning at t = 1 s and t = 2 s. (*c*) What is the approximate average velocity for the interval from 1 s $\le t \le$ 3 s?

Figure 2-30 Problem 92

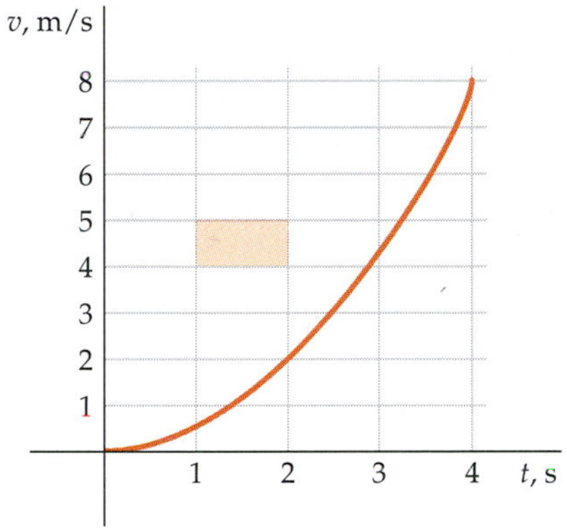

93 •• The velocity of a particle is given by $v = 7t^2 - 5$, where t is in seconds and v is in meters per second. Find the general position function $x(t)$.

94 •• The equation of the curve shown in Figure 2-30 is $v = 0.5t^2$ m/s. Find the displacement of the particle for the interval 1 s $\le t \le$ 3 s by integration, and compare this answer with your answer for Problem 92. Is the average velocity equal to the mean of the initial and final velocities for this case?

95 •• Figure 2-31 shows the acceleration of a particle versus time. (*a*) What is the magnitude of the area of the rectangle indicated? (*b*) The particle starts from rest at t = 0. Find the velocity at t = 1 s, 2 s, and 3 s by counting the rectangles under the curve. (*c*) Sketch the curve $v(t)$ versus t from your results for part (*b*), and estimate how far the particle travels in the interval t = 0 to t = 3 s.

Figure 2-31 Problem 95

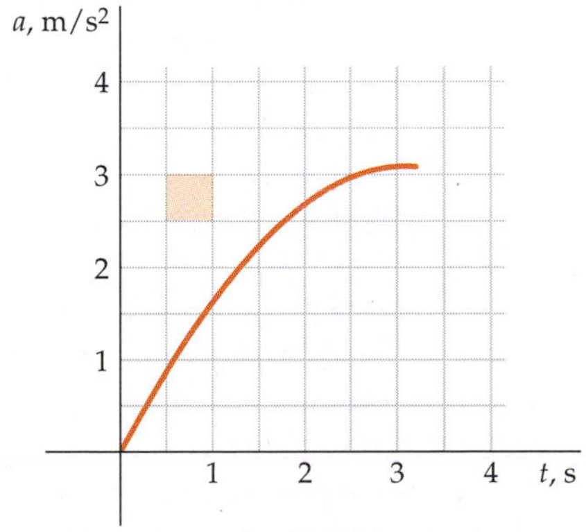

96 •• Figure 2-32 is a graph of v versus t for a particle moving along a straight line. The position of the particle at time $t = 0$ is $x_0 = 5$ m. (*a*) Find x for various times t by counting squares, and sketch x versus t. (*b*) Sketch the acceleration a versus t.

Figure 2-32 Problem 96

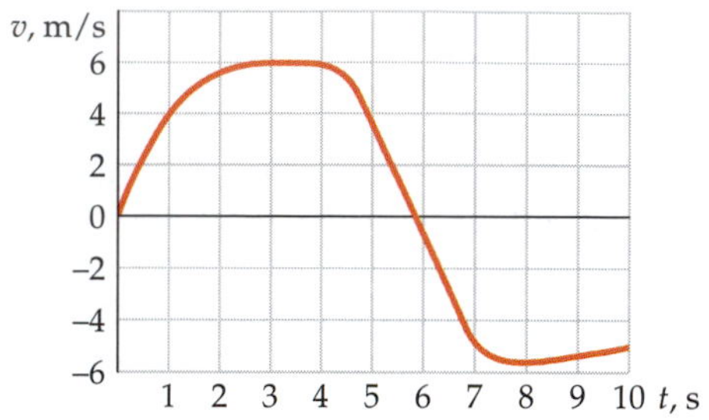

97 ••• Figure 2-33 shows a plot of x versus t for a body moving along a straight line. Sketch rough graphs of v versus t and a versus t for this motion.

Figure 2-33 Problem 97

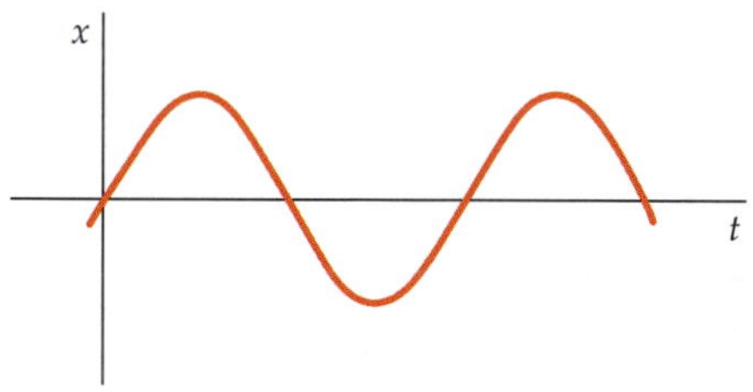

General Problems

98 • True or false:

(*a*) The equation $\Delta x = v_0 t + \frac{1}{2}at^2$ is valid for all particle motion in one dimension.
(*b*) If the velocity at a given instant is zero, the acceleration at that instant must also be zero.
(*c*) The equation $\Delta x = v_{av}\,\Delta t$ holds for all motion in one dimension.

99 • If an object is moving at constant acceleration in a straight line, its instantaneous velocity halfway through any time interval is

(*a*) greater than its average velocity.
(*b*) less than its average velocity.
(*c*) equal to its average velocity.
(*d*) half of its average velocity.
(*e*) twice its average velocity.

100 • On a graph showing position on the vertical axis and time on the horizontal axis, a straight line with a negative slope represents

(*a*) a constant positive acceleration.
(*b*) a constant negative acceleration.
(*c*) zero velocity.
(*d*) a constant positive velocity.
(*e*) a constant negative velocity.

101 •• On a graph showing position on the vertical axis and time on the horizontal axis, a parabola that opens upward represents

(*a*) a positive acceleration.
(*b*) a negative acceleration.
(*c*) no acceleration.
(*d*) a positive followed by a negative acceleration.
(*e*) a negative followed by a positive acceleration.

102 •• On a graph showing velocity on the vertical axis and time on the horizontal axis, zero acceleration is represented by

(*a*) a straight line with a positive slope.
(*b*) a straight line with a negative slope.
(*c*) a straight line with zero slope.
(*d*) either (*a*), (*b*), or (*c*).
(*e*) none of the above.

103 •• On a graph showing velocity on the vertical axis and time on the horizontal axis, constant acceleration is represented by

(*a*) a straight line with a positive slope.
(*b*) a straight line with a negative slope.
(*c*) a straight line with zero slope.
(*d*) either (*a*), (*b*), or (*c*).
(*e*) none of the above.

104 •• Which graph of v versus t in Figure 2-34 best describes the motion of a particle with positive velocity and negative acceleration?

Figure 2-34 Problems 104 and 105

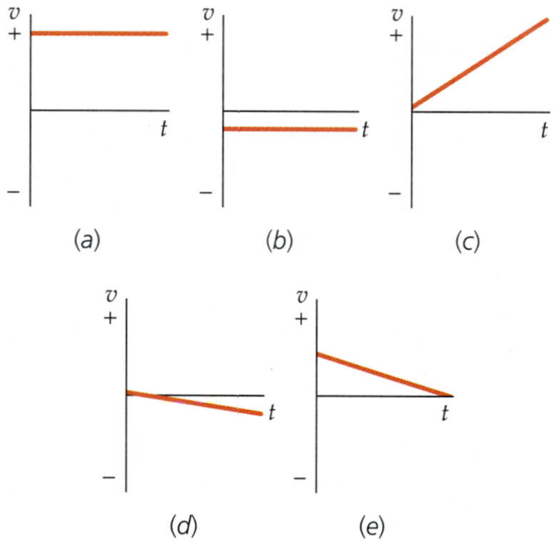

105 •• Which graph of v versus t in Figure 2-34 best describes the motion of a particle with negative velocity and negative acceleration?

106 •• A graph of the motion of an object is plotted with the velocity on the vertical axis and time on the horizontal axis. The graph is a straight line. Which of these quantities *cannot* be determined from this graph?

(*a*) The displacement from time $t = 0$
(*b*) The initial velocity at $t = 0$
(*c*) The acceleration of the object
(*d*) The average velocity of the object
(*e*) None of the above

(*e*) a constant negative velocity.

101 •• On a graph showing position on the vertical axis and time on the horizontal axis, a parabola that opens upward represents
(*a*) a positive acceleration.
(*b*) a negative acceleration.
(*c*) no acceleration.
(*d*) a positive followed by a negative acceleration.
(*e*) a negative followed by a positive acceleration.

Figure 2-35 Problem 107

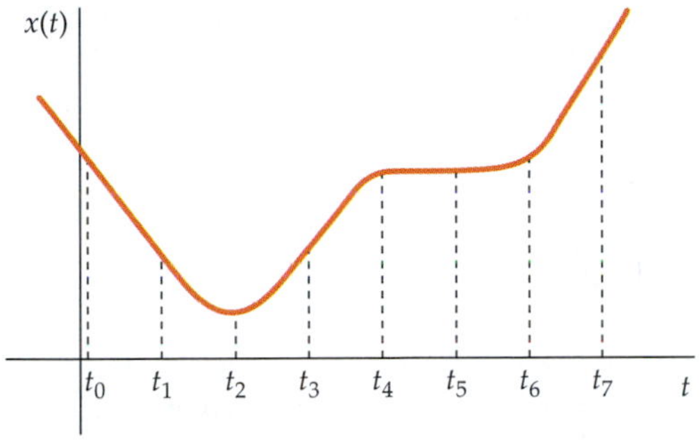

102 •• On a graph showing velocity on the vertical axis and time on the horizontal axis, zero acceleration is represented by
(*a*) a straight line with a positive slope.
(*b*) a straight line with a negative slope.
(*c*) a straight line with zero slope.
(*d*) either (*a*), (*b*), or (*c*).
(*e*) none of the above.

103 •• On a graph showing velocity on the vertical axis and time on the horizontal axis, constant acceleration is represented by
(*a*) a straight line with a positive slope.
(*b*) a straight line with a negative slope.
(*c*) a straight line with zero slope.
(*d*) either (*a*), (*b*), or (*c*).
(*e*) none of the above.

104 •• Which graph of v versus t in Figure 2-34 best describes the motion of a particle with positive velocity and negative acceleration?

105 •• Which graph of v versus t in Figure 2-34 best describes the motion of a particle with negative velocity and negative acceleration?

106 •• A graph of the motion of an object is plotted with the velocity on the vertical axis and time on the horizontal axis. The graph is a straight line. Which of these quantities *cannot* be determined from this graph?
(*a*) The displacement from time $t = 0$
(*b*) The initial velocity at $t = 0$
(*c*) The acceleration of the object
(*d*) The average velocity of the object
(*e*) None of the above

Figure 2-36 Problem 109

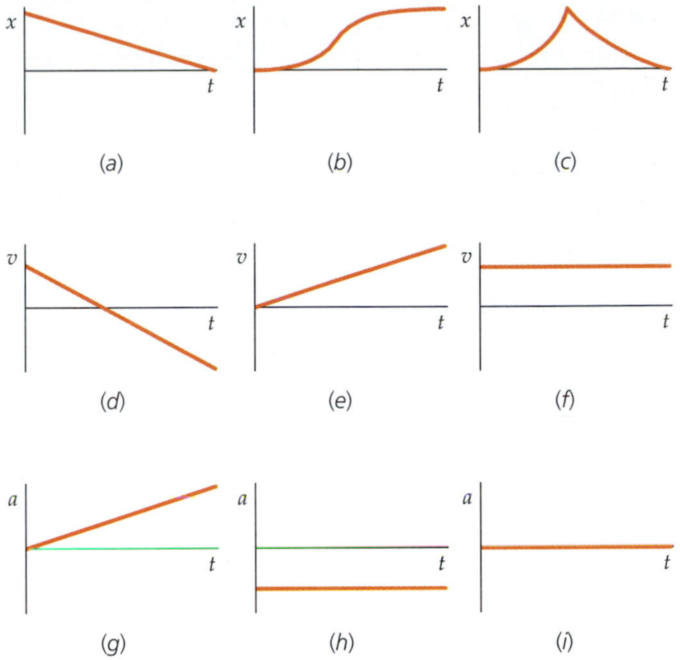

107 •• Figure 2-35 shows the position of a car plotted as a function of time. At which of the times t_0 to t_7 is the velocity
(*a*) negative?
(*b*) positive?
(*c*) zero?

At which times is the acceleration

(*d*) negative?
(*e*) positive?
(*f*) zero?

108 •• Sketch v-versus-t curves for each of the following conditions:
(*a*) Acceleration is zero and constant while velocity is not zero.
(*b*) Acceleration is constant but not zero.
(*c*) Velocity and acceleration are both positive.
(*d*) Velocity and acceleration are both negative.
(*e*) Velocity is positive and acceleration is negative.
(*f*) Velocity is negative and acceleration is positive.
(*g*) Velocity is zero but acceleration is not.

109 •• Figure 2-36 shows nine graphs of position, velocity, and acceleration for objects in linear motion. Indicate the graphs that meet the following conditions:
(*a*) Velocity is constant.
(*b*) Velocity has reversed its direction.
(*c*) Acceleration is constant.
(*d*) Acceleration is not constant.

Which graphs of velocity and acceleration are mutually consistent?

Figure 2-37 Problem 116

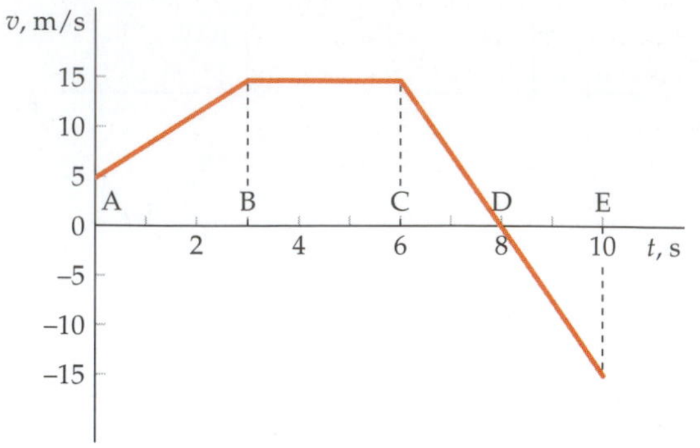

117 •• Consider the velocity graph in Figure 2-38. Assuming $x = 0$ at $t = 0$, write correct algebraic expressions for $x(t)$, $v(t)$, and $a(t)$ with appropriate numerical values inserted for all constants.

Figure 2-38 Problem 117

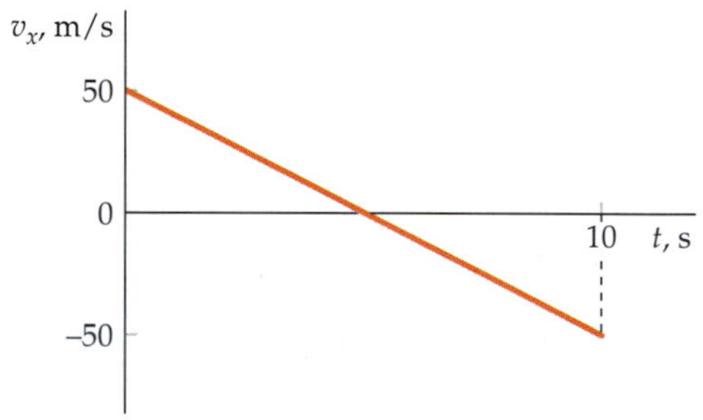

118 •• Starting at one station, a subway train accelerates from rest at a constant rate of $1.0\ \text{m/s}^2$ for half the distance to the next station, then slows down at the same rate for the second half of the journey. The total distance between stations is 900 m. (*a*) Sketch a graph of the velocity v as a function of time over the full journey. (*b*) Sketch a graph of the distance covered as a function of time over the full journey. Place appropriate numerical values along both axes.

119 •• The acceleration of a certain rocket is given by $a = Ct$, where C is a constant. (*a*) Find the general position function $x(t)$. (*b*) Find the position and velocity at $t = 5$ s if $x = 0$ and $v = 0$ at $t = 0$ and $C = 3\ \text{m/s}^3$.

120 •• A physics professor demonstrates his new "antigravity parachute" by exiting from a helicopter at an altitude of 1500 m with zero initial velocity. For 8 s he falls freely. Then he switches on the "parachute" and falls with a constant upward acceleration of $15\ \text{m/s}^2$ until his downward speed reaches 5 m/s, whereupon he adjusts his controls to maintain that speed until he reaches the ground. (*a*) On a single graph, sketch his acceleration and velocity as functions of time. (Take upward to be positive.) (*b*) What is his speed at the end of the first 8 s? (*c*) For how long does he maintain the constant upward acceleration of $15\ \text{m/s}^2$? (*d*) How far does he travel during the upward acceleration in part (c)? (*e*) How many seconds are required for the entire trip from the helicopter to the ground? (*f*) What is his average velocity for the entire trip?

121 •• Without telling Sally, Joe made travel arrangements that include a stopover in Toronto to visit Joe's old buddy. Sally doesn't like Joe's buddy and wants to change their tickets. She hops on a courtesy motor scooter and begins accelerating at $0.9\ \text{m/s}^2$ toward the ticket counter to make the arrangements. As she begins moving, Joe is 40 m behind her, running at a constant speed of 9 m/s. (*a*) How long does it take for Joe to catch up with her? (*b*) What is the time interval during which Sally remains ahead of Joe?

122 •• A speeder races past at 125 km/h. A patrol car pursues from rest with a constant acceleration of 8 km/h·s until it reaches its maximum speed of 190 km/h, which it maintains until it catches up with the speeder. (*a*) How long until the patrol car catches the speeder if it starts moving just as the speeder passes? (*b*) How far does each car travel? (*c*) Sketch $x(t)$ for each car.

123 •• When the patrol car in Problem 122 (traveling at 190 km/h) pulls to within 100 m behind the speeder (traveling at 125 km/h), the speeder sees the police car and slams on his brakes, locking the wheels. (*a*) Assuming that each car can brake at $6\ \text{m/s}^2$ and that the driver of the police car brakes instantly as she sees the brake lights of the speeder (reaction time = 0 s), show that the cars collide. (*b*) At what time after the speeder applies his brakes do the two cars collide? (*c*) Discuss how reaction time affects this problem.

124 •• The speed of a good base runner is 9.5 m/s. The distance between bases is 26 m, and the pitcher is about 18.5 m from home plate. If a runner on first base edges 2 m off the base and takes off for second the instant the ball leaves the pitcher's hand, what is the likelihood that the runner will steal second base safely?

125 •• Repeat Problem 124, but with the runner attempting to steal third base, starting from second base with a lead of 3 m.

126 •• Urgently needing the cash prize, Lou enters the Rest-to-Rest auto competition, in which each contestant's car begins and ends at rest, covering a distance L in as short a time as possible. The intention is to demonstrate mechanical and driving skills, and to consume the largest amount of fossil fuels in the shortest time possible. The course is designed so that maximum speeds of the cars are never reached. If Lou's car has a maximum acceleration of a, and a maximum deceleration of $2a$, then at what fraction of L should Lou move his foot from the gas pedal to the brake? What fraction of the time for the trip has elapsed at that point?

127 ••• The acceleration of a badminton birdie falling under the influence of gravity and a resistive force, such as air resistance, is given by $a = dv/dt = g - bv$, where g is the free-fall acceleration due to gravity and b is a constant that depends on the mass and shape of the birdie and on the properties of the medium. Suppose the birdie begins with zero velocity at time $t = 0$. (*a*) Discuss qualitatively how the speed v varies with time from your knowledge of the rate of change dv/dt given by this equation. What is the velocity when the acceleration is zero? This is called the *terminal velocity*. (*b*)

Sketch the solution $v(t)$ versus t without solving the equation. This can be done as follows: At $t = 0$, v is zero and the slope is g. Sketch a straight-line segment, neglecting any change in slope for a short time interval. At the end of the interval, the velocity is not zero, so the slope is less than g. Sketch another straight-line segment with a smaller slope. Continue until the slope is zero and the velocity equals the terminal velocity.

128 ••• Suppose acceleration is a function of x, where $a(x) = (2\ \mathrm{s}^{-2})x$. (*a*) If the velocity at $x = 1$ m is zero, what is the velocity at $x = 3$ m? (*b*) How long does it take to travel from $x = 1$ m to $x = 3$ m?

129 ••• Suppose that a particle moves in a straight line such that, at any time t, its position, velocity, and acceleration all have the same numerical value. Give the position x as a function of time.

130 ••• An object moving in a straight line doubles its velocity each second for the first 10 s. Let the initial speed be 2 m/s. (*a*) Sketch a smooth function $v(t)$ that gives the velocity. (*b*) What is the average velocity over the first 10 s?

131 ••• In a dream, you find that you can run at superhuman speeds, but there is also a resistant force that reduces your speed by one-half for each second that passes. Assume that the laws of physics still hold in your dreamworld, and that your initial speed is 1000 m/s. (*a*) Sketch a smooth function $v(t)$ that gives your velocity. (*b*) What is your average velocity over the first 10 s?

CHAPTER 3

Motion in Two and Three Dimensions

Illuminated fountains, St. Louis, Missouri. The jets follow parabolic paths like those followed by projectiles.

3-1 The Displacement Vector

When motion occurs in two or three dimensions, the displacement of a particle has a direction in space as well as a magnitude. The quantity that gives the direction and the straight-line distance between two points in space is a line segment called the **displacement vector.** It is represented graphically by an arrow whose direction is the same as the direction of the displacement and whose length is proportional to the magnitude of the displacement. We denote vectors by boldface italic letters with an overhead arrow, $\vec{A}$. The magnitude of $\vec{A}$ is written $|\vec{A}|$ or simply A.

Addition of Displacement Vectors

Figure 3-1 shows the path of a particle that moves from point P_1 to a second point P_2 and then to a third point P_3. The displacement from P_1 to P_2 is represented by the vector $\vec{A}$, and the displacement from P_2 to P_3 is represented by $\vec{B}$. Note that the displacement vectors depend only on the endpoints and not on the

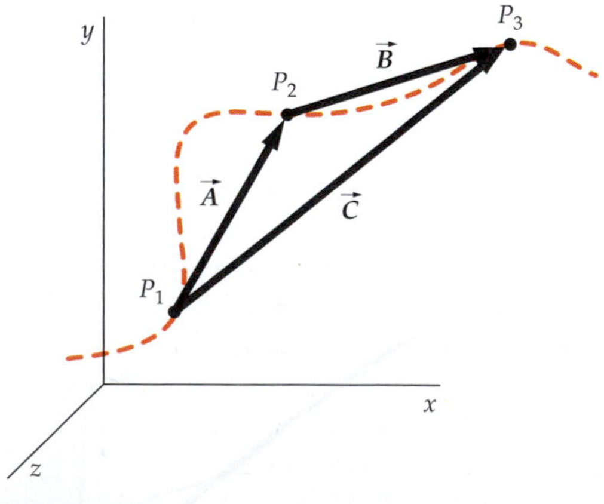

Figure 3-1

actual path of the particle. The *resultant* displacement from P_1 to P_3, labeled $\vec{C}$, is the sum of the two successive displacements $\vec{A}$ and $\vec{B}$:

$$\vec{C} = \vec{A} + \vec{B} \qquad \text{3-1}$$

Two displacement vectors are added graphically by placing the tail of one at the head of the other (Figure 3-2). The resultant vector extends from the tail of the first to the head of the second. Note that C does not equal $A + B$ unless $\vec{A}$ and $\vec{B}$ are in the same direction. That is, $\vec{C} = \vec{A} + \vec{B}$ does not imply that $C = A + B$.

An equivalent way of adding vectors, called the **parallelogram method,** is to move $\vec{B}$ so that it is tail to tail with $\vec{A}$. The diagonal of the parallelogram formed by $\vec{A}$ and $\vec{B}$ then equals $\vec{C}$. From Figure 3-3 we can see that it makes no difference in which order we add two vectors; that is, $\vec{A} + \vec{B} = \vec{B} + \vec{A}$.

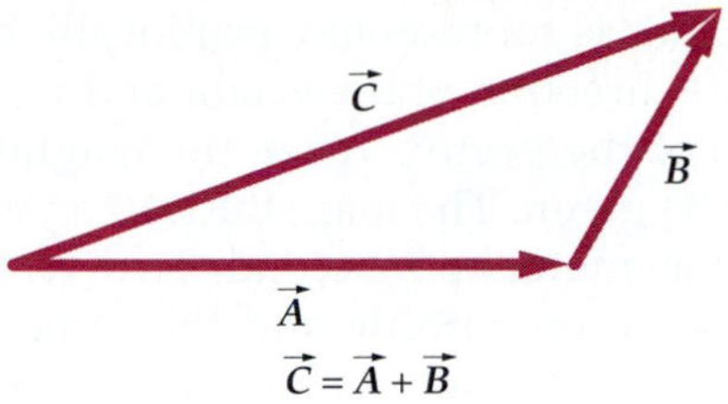

Figure 3-2 Vector addition.

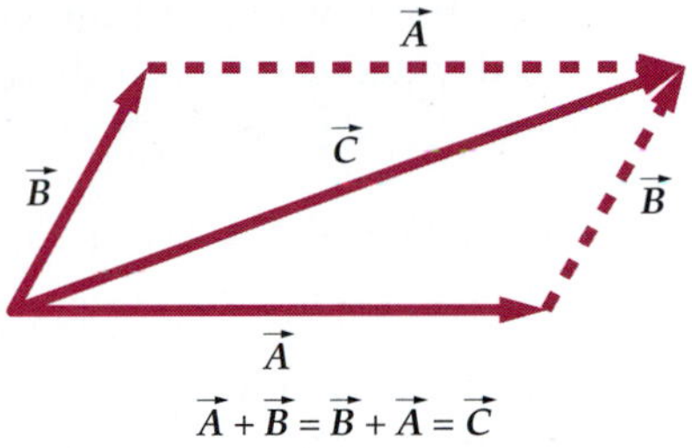

Figure 3-3 Parallelogram method for adding vectors.

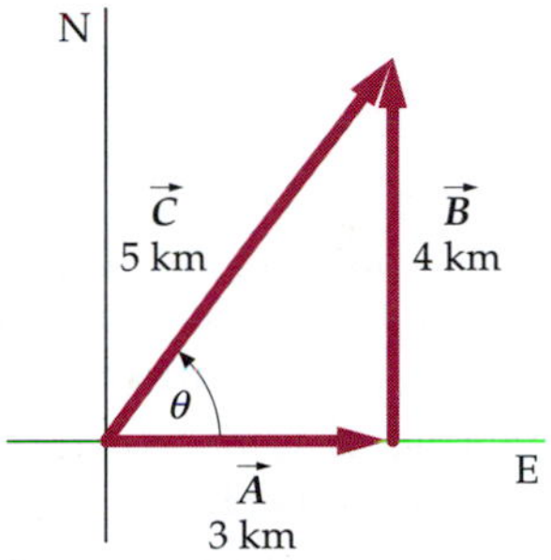

Figure 3-4

Example 3-1

You walk 3 km east and then 4 km north. What is your resultant displacement?

Picture the Problem The two displacements $\vec{A}$ and $\vec{B}$ and the resultant displacement $\vec{C}$ are shown in Figure 3-4. $\vec{A}$ and $\vec{B}$ are at right angles to each other, and $\vec{C} = \vec{A} + \vec{B}$ is the hypotenuse of the corresponding right triangle. The magnitude C can be found from the Pythagorean theorem. The direction of $\vec{C}$ is found using trigonometry.

1. The magnitude of the resultant displacement is related to the magnitudes of the two displacements by the Pythagorean theorem:

$$C^2 = A^2 + B^2$$
$$= (3\text{ km})^2 + (4\text{ km})^2$$
$$= 25\text{ km}^2$$
$$C = \sqrt{25\text{ km}^2} = 5\text{ km}$$

2. Let θ be the angle from the east axis to the resultant displacement $\vec{C}$. From the figure we find $\tan\theta$; using a calculator with trigonometric functions yields θ:

$$\tan\theta = \frac{4\text{ km}}{3\text{ km}} = 1.33$$
$$\theta = \tan^{-1}1.33 = 53.1°$$

Remarks A vector is described by its magnitude and its direction. Your resultant displacement is a vector of length 5 km in a direction 53.1° north of east.

3-2 General Properties of Vectors

Many quantities in physics have magnitude and direction, and add like displacements. Examples include velocity, acceleration, momentum, and force. Such quantities are called **vectors.** Quantities with magnitude but no associated direction—for example, distance and speed—are called **scalars**.

> Vectors are quantities with magnitude and direction that add and subtract like displacements.

Definition—Vectors

A vector is represented graphically by an arrow whose direction is the same as the direction of the vector and whose length is proportional to the magnitude of the vector. When the magnitude of a vector is given, its units must also be given. The magnitude of a velocity vector, for example, requires units such as meters per second. Two vectors are defined to be equal if they have the same magnitude and the same direction. Graphically, this means they have the same length and are parallel to one another. A consequence of this definition is that moving a vector so that it remains parallel to itself does not change it. Thus, all the vectors in Figure 3-5 are equal. If we translate or rotate the coordinate system, all the vectors in Figure 3-5 remain equal. A vector does not depend on the coordinate system used to represent it.

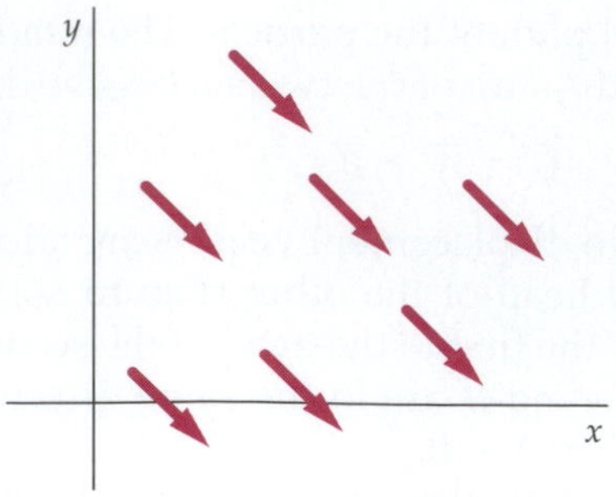

Figure 3-5 Vectors are equal if their magnitudes and directions are the same. All vectors in this figure are equal.

Multiplying a Vector by a Scalar

A vector $\vec{A}$ multiplied by a scalar s is the vector $\vec{B} = s\vec{A}$, which has magnitude $|s|A$ and is parallel to $\vec{A}$ if s is positive and antiparallel if s is negative. Thus the vector $-\vec{A}$ has the same magnitude as $\vec{A}$ but points in the opposite direction so that $\vec{A} + (-\vec{A}) = 0$. The dimensions of $s\vec{A}$ are those of s multiplied by those of A.

Subtracting Vectors

We subtract vector $\vec{B}$ from vector $\vec{A}$ by adding $-\vec{B}$, which has the same magnitude as $\vec{B}$. The result is $\vec{C} = \vec{A} + (-\vec{B}) = \vec{A} - \vec{B}$ (Figure 3-6*a*). An equivalent way of subtracting $\vec{B}$ from $\vec{A}$ is to draw them tail to tail and then draw a vector $\vec{C}$ from $\vec{B}$ to $\vec{A}$. That is, $\vec{C}$ is the vector that must be added to $\vec{B}$ to obtain the resultant vector $\vec{A}$ (Figure 3-6*b*). The rules for adding or subtracting any two vectors, such as two velocity vectors or two acceleration vectors, are the same as for adding displacement vectors.

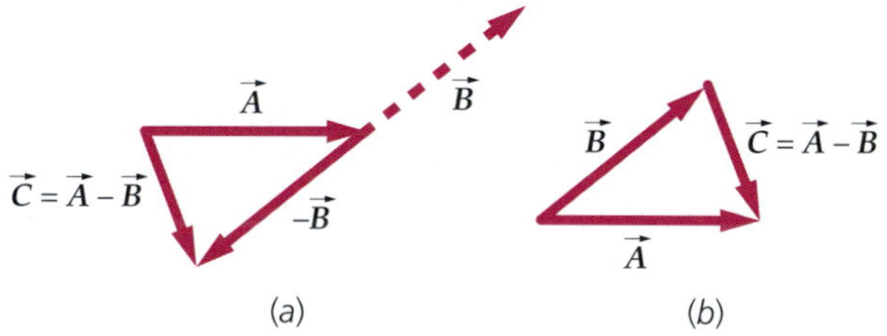

Figure 3-6

Components of Vectors

The component of a vector along a line in space is the length of the projection of the vector on that line. It is found by dropping a perpendicular from the head of the vector to the line, as shown in Figure 3-7. The components of a vector along the x, y, and z directions, illustrated in Figure 3-8 for a vector in the xy plane, are called rectangular components. Note that the components of a vector *do* depend on the coordinate system used to represent the vector, although the vector itself does not.

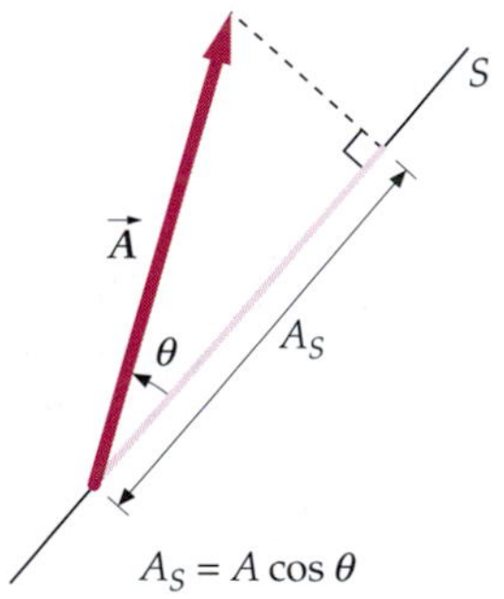

Figure 3-7 Definition of the component of a vector. The component of the vector $\vec{A}$ along the line S is A_S.

Rectangular components are useful for the addition or subtraction of vectors. If θ is the angle between $\vec{A}$ and the x axis, then

$$A_x = A\cos\theta \tag{3-2}$$

The x component of a vector

and

$$A_y = A\sin\theta \tag{3-3}$$

The y component of a vector

where A is the magnitude of $\vec{A}$.

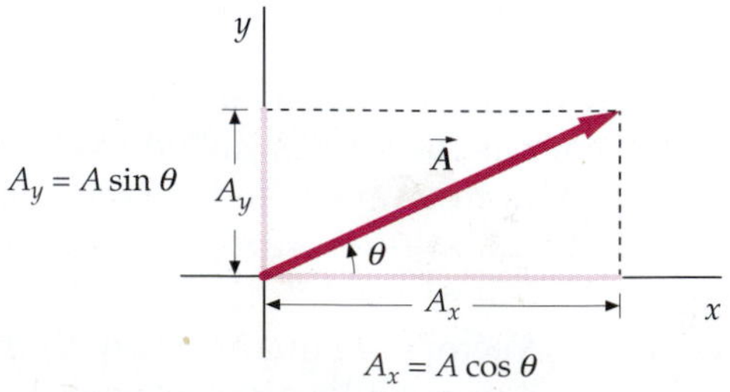

Figure 3-8 The rectangular components of a vector. $A_x = A\cos\theta$, $A_y = A\sin\theta$.

If we know A_x and A_y, we can find the angle θ from

$$\tan\theta = \frac{A_y}{A_x}, \qquad \theta = \tan^{-1}\frac{A_y}{A_x} \qquad \text{3-4}$$

and the magnitude A from the Pythagorean theorem:

$$A = \sqrt{A_x^2 + A_y^2} \qquad \text{3-5a}$$

In three dimensions,

$$A = \sqrt{A_x^2 + A_y^2 + A_z^2} \qquad \text{3-5b}$$

Components can be positive or negative. For example, if $\vec{A}$ points in the negative x direction, A_x is negative. Consider two vectors $\vec{A}$ and $\vec{B}$ that lie in the xy plane. The rectangular components of each vector and those of the sum $\vec{C} = \vec{A} + \vec{B}$ are shown in Figure 3-9. We see that $\vec{C} = \vec{A} + \vec{B}$ is equivalent to both

$$C_x = A_x + B_x \qquad \text{3-6a}$$

and

$$C_y = A_y + B_y \qquad \text{3-6b}$$

Figure 3-9

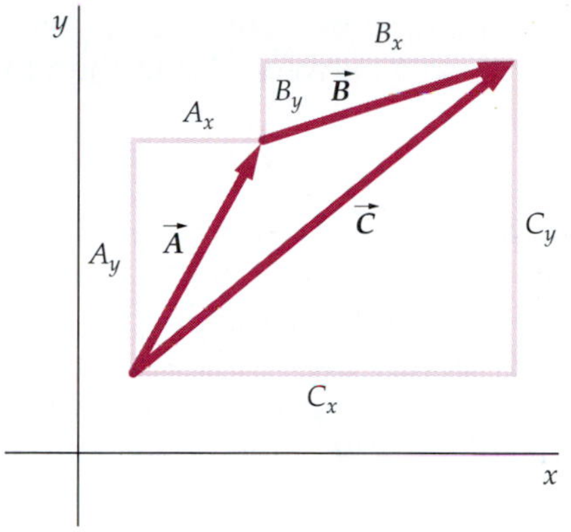

Exercise A car travels 20 km in a direction 30° north of west. Let the x axis point east and the y axis point north as in Figure 3-10. Find the x and y components of the displacement vector of the car. (*Answer* $A_x = -17.3$ km, $A_y = +10$ km)

Figure 3-10

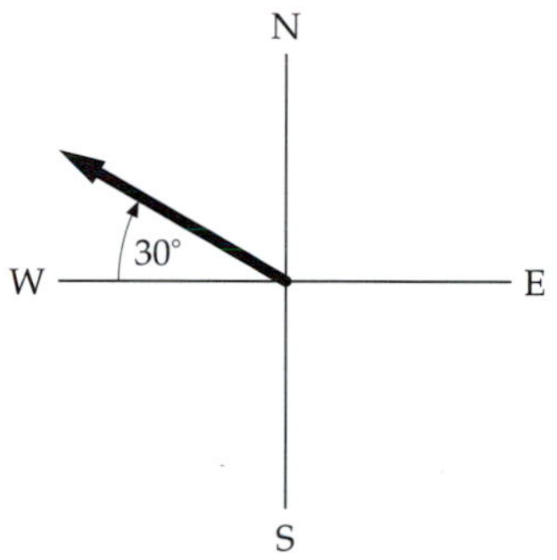

Example 3-2

You walk 3 km west and then 4 km headed 60° north of east (Figure 3-11). Find your resultant displacement (*a*) graphically and (*b*) using vector components.

Picture the Problem The triangle formed by the three vectors is not a right triangle, so the magnitudes of the vectors are not related by the Pythagorean theorem. We find the resultant graphically by drawing each of the displacements to scale and measuring the resultant displacement.

Figure 3-11

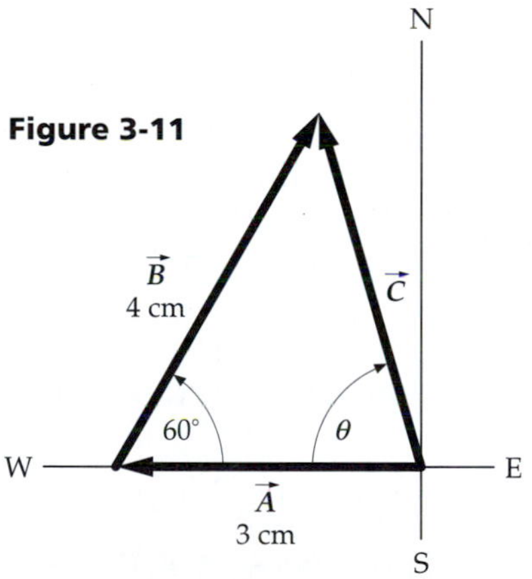

(*a*) If we draw the first displacement vector 3 cm long and the second one 4 cm long, we find the resultant vector to be about 3.5 cm long. Thus, the magnitude of the resultant displacement is 3.5 km. The angle θ made between the resultant displacement and the west direction can then be measured with a protractor. It is about 75°.

(*b*)1. Let $\vec{A}$ be the first displacement and choose the x axis to be in the easterly direction. Compute A_x and A_y from Equations 3-2 and 3-3:

$A_x = -3$ km and $A_y = 0$

2. Similarly, compute the components of the second displacement $\vec{B}$:

$$B_x = (4\text{ km})\cos 60^\circ = 2\text{ km}$$

$$B_y = (4\text{ km})\sin 60^\circ = 3.46\text{ km}$$

3. The components of the resultant displacement $\vec{C} = \vec{A} + \vec{B}$ are found by addition:

$$C_x = A_x + B_x = -3\text{ km} + 2\text{ km} = -1\text{ km}$$

$$C_y = A_y + B_y = 0 + 3.46\text{ km} = 3.46\text{ km}$$

4. The Pythagorean theorem gives the magnitude of $\vec{C}$:

$$C^2 = C_x^2 + C_y^2 = (-1\text{ km})^2 + (3.46\text{ km})^2 = 13.0\text{ km}^2$$

$$C = \sqrt{13.0\text{ km}^2} = 3.61\text{ km}$$

5. The ratio of C_y to C_x gives the tangent of the angle θ between $\vec{C}$ and the x axis:

$$\tan\theta = \frac{C_y}{C_x} = \frac{3.46\text{ km}}{-1\text{ km}} = -3.46$$

$$\theta = \tan^{-1} -3.46 = -74^\circ$$

Remarks Since the displacement (which is a vector) was asked for, the answer must include either the magnitude *and* direction, or both components. In (*b*) we could have stopped at step 3 because the x and y components completely define the displacement vector. We converted to the magnitude and direction to compare with the answer to part (*a*). Note that in step 5 of (*b*), a calculator gives the angle as -74°. But the calculator can't distinguish whether the x or y component is negative. We noted on the figure that the resultant displacement makes an angle of about 75° with the negative x axis and an angle of about 105° with the positive x axis. This agrees with the results in (*a*) within the accuracy of our measurement.

Unit Vectors

A **unit vector** is a *dimensionless* vector with unit magnitude. The vector $A^{-1}\vec{A}$ is an example of a unit vector that points in the direction of $\vec{A}$. Unit vectors are often written boldface italic with an overhead caret as in $\hat{A} = A^{-1}\vec{A}$. Unit vectors that point in the x, y, and z directions are convenient for expressing vectors in terms of their rectangular components. They are usually written $\hat{i}$, $\hat{j}$, and $\hat{k}$, respectively. Then the vector $A_x\hat{i}$ has a magnitude A_x and points in the positive x direction (or negative x direction if A_x is negative). A general vector $\vec{A}$ can be written as the sum of three vectors, each of which is parallel to a coordinate axis (Figure 3-12):

$$\vec{A} = A_x\hat{i} + A_y\hat{j} + A_z\hat{k} \qquad \text{3-7}$$

The addition of two vectors $\vec{A}$ and $\vec{B}$ can be written in terms of unit vectors as

$$\vec{A} + \vec{B} = (A_x\hat{i} + A_y\hat{j} + A_z\hat{k}) + (B_x\hat{i} + B_y\hat{j} + B_z\hat{k})$$
$$= (A_x + B_x)\hat{i} + (A_y + B_y)\hat{j} + (A_z + B_z)\hat{k} \qquad \text{3-8}$$

The general properties of vectors are summarized in Table 3-1.

Exercise Given two vectors

$$\vec{A} = (4\text{ m})\hat{i} + (3\text{ m})\hat{j} \qquad \text{and} \qquad \vec{B} = (2\text{ m})\hat{i} - (3\text{ m})\hat{j}$$

find (*a*) A, (*b*) B, (*c*) $\vec{A} + \vec{B}$, and (*d*) $\vec{A} - \vec{B}$. (*Answers* (*a*) $A = 5$ m, (*b*) $B = 3.61$ m, (*c*) $\vec{A} + \vec{B} = (6\text{ m})\hat{i}$, (*d*) $\vec{A} - \vec{B} = (2\text{ m})\hat{i} + (6\text{ m})\hat{j}$)

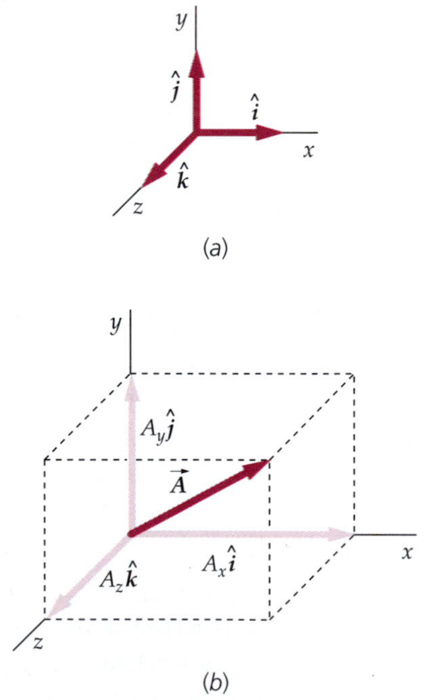

Figure 3-12 (*a*) The unit vectors $\hat{i}$, $\hat{j}$, and $\hat{k}$ in a rectangular coordinate system. (*b*) The vector $\vec{A}$ in terms of unit vectors: $\vec{A} = A_x\hat{i} + A_y\hat{j} + A_z\hat{k}$.

Table 3-1

Properties of Vectors

Property	Explanation	Figure	Component representation
Equality	$\vec{A} = \vec{B}$ if $\lvert\vec{A}\rvert = \lvert\vec{B}\rvert$ and their directions are the same		$A_x = B_x$ $A_y = B_y$ $A_z = B_z$
Addition	$\vec{C} = \vec{A} + \vec{B}$	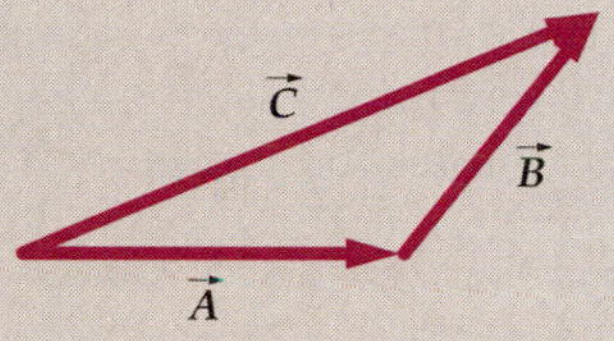	$C_x = A_x + B_x$ $C_y = A_y + B_y$ $C_z = A_z + B_z$
Negative of a vector	$\vec{A} = -\vec{B}$ if $\lvert\vec{B}\rvert = \lvert\vec{A}\rvert$ and their directions are opposite		$A_x = -B_x$ $A_y = -B_y$ $A_z = -B_z$
Subtraction	$\vec{C} = \vec{A} - \vec{B}$	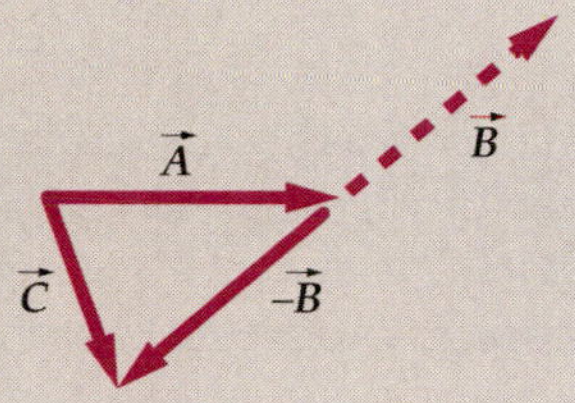	$C_x = A_x - B_x$ $C_y = A_y - B_y$ $C_z = A_z - B_z$
Multiplication by a scalar	$\vec{B} = s\vec{A}$ has magnitude $\lvert\vec{B}\rvert = s\lvert\vec{A}\rvert$ and has the same direction as $\vec{A}$ if s is positive or $-\vec{A}$ if s is negative	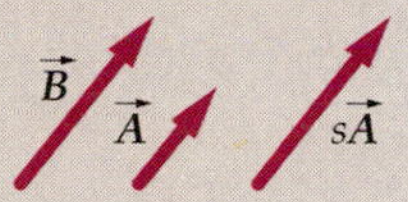	$B_x = sA_x$ $B_y = sA_y$ $B_z = sA_z$

3-3 Position, Velocity, and Acceleration

Position and Velocity Vectors

The **position vector** of a particle is a vector drawn from the origin of a reference frame to the xy position of the particle. For a particle at the point (x, y), its position vector $\vec{r}$ is

$$\vec{r} = x\hat{i} + y\hat{j} \qquad 3\text{-}9$$

Definition—Position vector

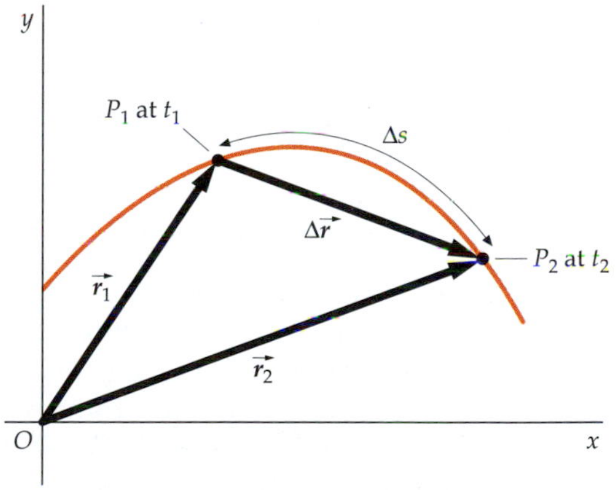

Figure 3-13 The displacement vector $\Delta\vec{r}$ is the difference in the position vectors, $\Delta\vec{r} = \vec{r}_2 - \vec{r}_1$. Equivalently, $\Delta\vec{r}$ is the vector that, when added to $\vec{r}_1$, yields the new position vector $\vec{r}_2$.

Figure 3-13 shows the actual path or trajectory of the particle. (Don't confuse the trajectory with the x-versus-t plots of the previous chapter.) At time t_1, the particle is at P_1, with position vector $\vec{r}_1$; by t_2, the particle has moved to P_2, with position vector $\vec{r}_2$. The particle's change in position is the displacement vector $\Delta\vec{r}$:

$$\Delta\vec{r} = \vec{r}_2 - \vec{r}_1 \qquad 3\text{-}10$$

Definition—Displacement vector

The ratio of the displacement vector to the time interval $\Delta t = t_2 - t_1$ is the **average-velocity vector:**

$$\vec{v}_{av} = \frac{\Delta \vec{r}}{\Delta t} \qquad \text{3-11}$$

Definition—Average-velocity vector

This vector points in the direction of the displacement.

The magnitude of the displacement vector is less than the distance traveled along the curve unless the particle moves in a straight line. However, if we consider smaller and smaller intervals, the magnitude of the displacement approaches the distance along the curve, and the direction of $\Delta \vec{r}$ approaches the tangent to the curve at the beginning of the interval (Figure 3-14). We define the **instantaneous-velocity vector** as the limit of the average-velocity vector as Δt approaches zero:

$$\vec{v} = \lim_{\Delta t \to 0} \frac{\Delta \vec{r}}{\Delta t} = \frac{d\vec{r}}{dt} \qquad \text{3-12}$$

Definition—Instantaneous-velocity vector

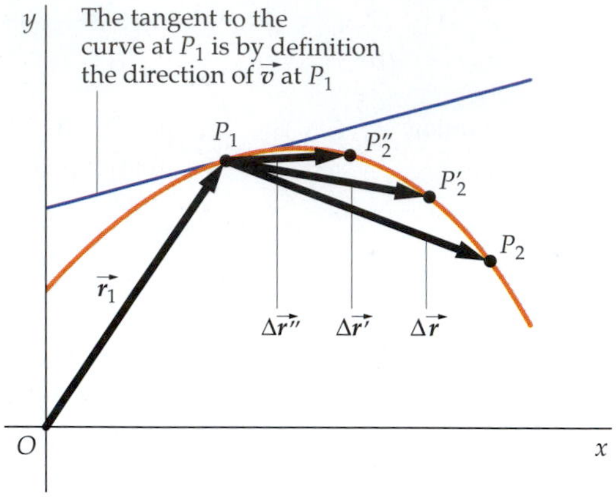

Figure 3-14 As the time interval is made smaller, the direction of the displacement vector approaches the tangent to the curve.

The instantaneous-velocity vector is the derivative of the position vector with respect to time. Its magnitude is the speed ds/dt, and its direction is the direction of motion of the particle along the line tangent to the curve.

To calculate the derivative in Equation 3-12, we write the position vector in terms of its components:

$$\Delta \vec{r} = \vec{r}_2 - \vec{r}_1 = (x_2 - x_1)\hat{i} + (y_2 - y_1)\hat{j} = \Delta x\hat{i} + \Delta y\hat{j}$$

Then

$$\vec{v} = \lim_{\Delta t \to 0} \frac{\Delta \vec{r}}{\Delta t} = \lim_{\Delta t \to 0} \frac{\Delta x\hat{i} + \Delta y\hat{j}}{\Delta t} = \lim_{\Delta t \to 0} \frac{\Delta x}{\Delta t}\hat{i} + \lim_{\Delta t \to 0} \frac{\Delta y}{\Delta t}\hat{j}$$

or

$$\vec{v} = \frac{dx}{dt}\hat{i} + \frac{dy}{dt}\hat{j} = v_x\hat{i} + v_y\hat{j} \qquad \text{3-13}$$

Example 3-3

A sailboat has coordinates $(x_1, y_1) = (110\text{ m}, 218\text{ m})$ at $t_1 = 60$ s. Two minutes later, at time t_2, it has the coordinates $(x_2, y_2) = (130\text{ m}, 205\text{ m})$. (*a*) Find the average velocity for this two-minute interval. Express v_{av} in terms of its rectangular components. (*b*) Find the magnitude and direction of this average velocity. (*c*) For $t \geq 20$ s, the position of the sailboat as a function of time is $x(t) = 100\text{ m} + (\frac{1}{6}\text{ m/s})t$ and $y(t) = 200\text{ m} + (1080\text{ m·s})t^{-1}$. Find the instantaneous velocity at a general time $t \geq 20$ s.

Picture the Problem The initial and final positions of the sailboat are shown in Figure 3-15. (*a*) The average velocity vector points from the initial to the final position. (*b*) The instantaneous velocity components are calculated from Equation 3-13: $v_x = dx/dt$ and $v_y = dy/dt$.

Figure 3-15

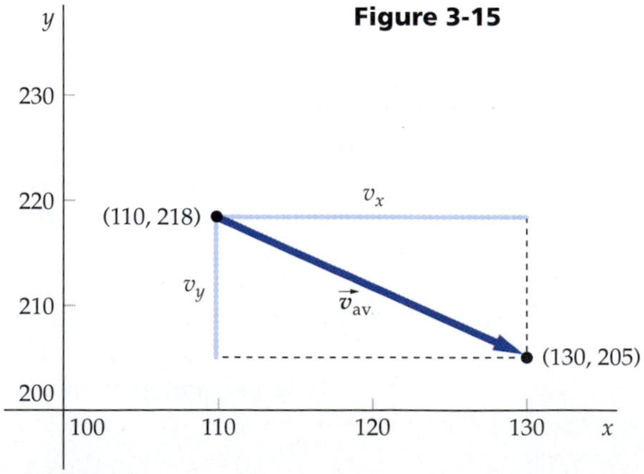

(*a*) The *x* and *y* components of the average velocity $\vec{v}_{av}$ are calculated directly from their definitions:

$$v_{x,av} = \frac{x_2 - x_1}{\Delta t} = \frac{130\text{ m} - 110\text{ m}}{120\text{ s}} = 0.167\text{ m/s}$$

$$v_{y,av} = \frac{y_2 - y_1}{\Delta t} = \frac{205\text{ m} - 218\text{ m}}{120\text{ s}} = -0.108\text{ m/s}$$

(*b*)1. The magnitude of $\vec{v}_{av}$ is found from the Pythagorean theorem:

$$v_{av} = \sqrt{(v_{x,av})^2 + (v_{y,av})^2} = 0.199\text{ m/s}$$

2. The ratio of $v_{y,av}$ to $v_{x,av}$ gives the tangent of the angle θ between $\vec{v}_{av}$ and the *x* axis:

$$\tan\theta = \frac{v_{y,av}}{v_{x,av}} = \frac{-0.108\text{ m/s}}{0.167\text{ m/s}} = -0.65$$

$$\theta = \tan^{-1}(-0.65) = -33.0°$$

(*c*) We find the instantaneous velocity $\vec{v}$ by calculating dx/dt and dy/dt:

$$\vec{v} = \frac{dx}{dt}\hat{i} + \frac{dy}{dt}\hat{j} = \left(\frac{1}{6}\text{ m/s}\right)\hat{i} - (1080\text{ m}\cdot\text{s})t^{-2}\hat{j}$$

Remark The magnitude of $\vec{v}$ can be found from $v = \sqrt{v_x^2 + v_y^2}$ and its direction can be found from $\tan\theta = v_y/v_x$.

Exercise Find the *x* and *y* components and the magnitude and direction of the instantaneous velocity of the sailboat at time $t_1 = 60$ s. (*Answers* $\vec{v}_1 = (\frac{1}{6}\text{ m/s})\hat{i} - (0.30\text{ m/s})\hat{j}$, $v_1 = 0.34$ m/s, $\theta_2 = -60.9°$)

Relative Velocity

Relative velocities in two and three dimensions combine just as they do in one dimension, except that the velocity vectors are not necessarily along the same line. If a particle moves with velocity $\vec{v}_{pA}$ relative to a coordinate system A, which is in turn moving with velocity $\vec{v}_{AB}$ relative to another coordinate system B, the velocity of the particle relative to B is

$$\vec{v}_{pB} = \vec{v}_{pA} + \vec{v}_{AB} \qquad \text{3-14}$$

Relative velocity

For example, if you are on a railroad car moving with velocity $\vec{v}_{cg}$ relative to the ground (Figure 3-16*a*), and you start walking with a velocity relative to the car of $\vec{v}_{pc}$ (Figure 3-16*b*), then your velocity relative to the ground is the sum of these two velocities: $\vec{v}_{pg} = \vec{v}_{pc} + \vec{v}_{cg}$ (Figure 3-16*c*).

Figure 3-16 Relative velocity in two dimensions.

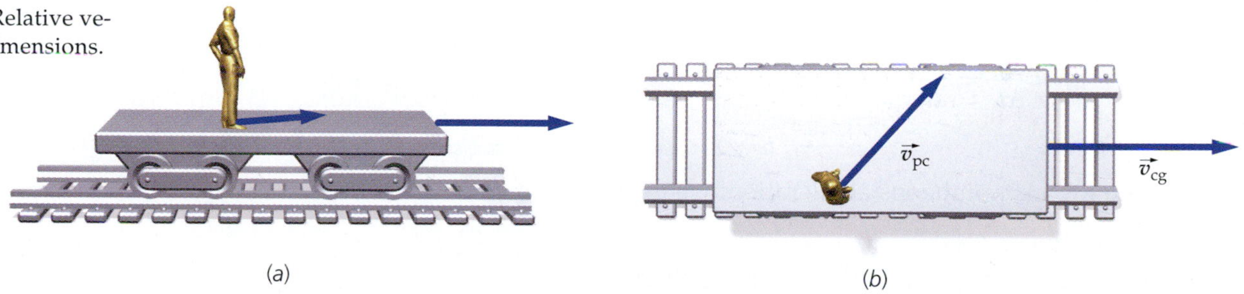

(*a*) (*b*)

The velocity of object A relative to object B is equal in magnitude and opposite in direction to the velocity of object B relative to object A. For example, $\vec{v}_{pc} = -\vec{v}_{cp}$, where $\vec{v}_{cp}$ is the velocity of the car relative to the person. The addition of relative velocities is done in the same way as the addition of displacements; either graphically, by placing the velocity vectors head to tail, or analytically, using vector components.

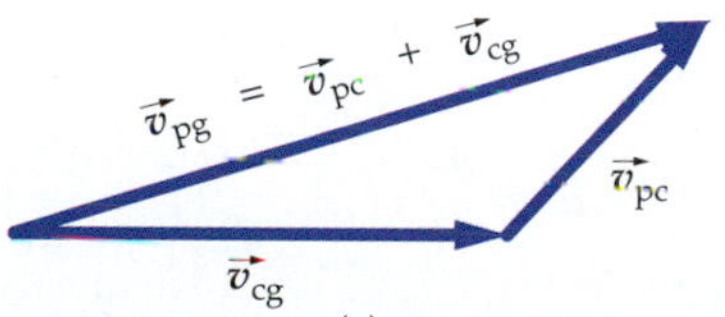

(*c*)

Example 3-4

A plane is to fly due north. The speed of the plane relative to the air is 200 km/h, and the wind is blowing from west to east at 90 km/h. (*a*) In which direction should the plane head? (*b*) How fast does the plane travel relative to the ground?

Picture the Problem Since the wind is blowing toward the east, the plane must head west of north as shown in Figure 3-17. The velocity of the plane relative to the ground $\vec{v}_{pg}$ will be the sum of the velocity of the plane relative to the air $\vec{v}_{pa}$ and the velocity of the air relative to the ground $\vec{v}_{ag}$.

Figure 3-17

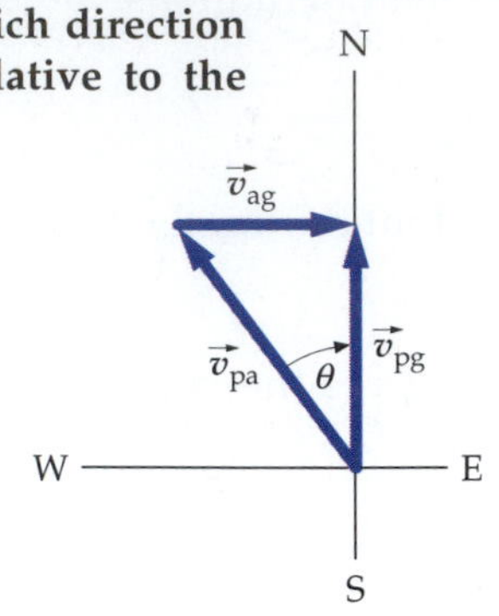

(*a*) 1. The velocity of the plane relative to the ground is given by Equation 3-14:

$$\vec{v}_{pg} = \vec{v}_{pa} + \vec{v}_{ag}$$

2. The sine of the angle θ between the velocity of the plane and north equals the ratio of v_{ag} and v_{pa}:

$$\sin\theta = \frac{v_{ag}}{v_{pa}} = \frac{90 \text{ km/h}}{200 \text{ km/h}} = 0.45$$

$$\theta = 26.7°$$

(*b*) Since v_{ag} and v_{pg} are perpendicular, we can use the Pythagorean theorem to find the magnitude of $\vec{v}_{pg}$:

$$v_{pa}^2 = v_{ag}^2 + v_{pg}^2$$

$$v_{pg} = \sqrt{v_{pa}^2 - v_{ag}^2}$$

$$= \sqrt{(200 \text{ km/h})^2 - (90 \text{ km/h})^2} = 179 \text{ km/h}$$

The Acceleration Vector

The **average-acceleration vector** is the ratio of the change in the instantaneous-velocity vector $\Delta\vec{v}$ to the time interval Δt:

$$\vec{a}_{av} = \frac{\Delta\vec{v}}{\Delta t} \qquad 3\text{-}15$$

Definition—Average-acceleration vector

The **instantaneous-acceleration vector** is the limit of this ratio as Δt approaches zero; in other words, it is the derivative of the velocity vector with respect to time:

$$\vec{a} = \lim_{\Delta t \to 0} \frac{\Delta\vec{v}}{\Delta t} = \frac{d\vec{v}}{dt} \qquad 3\text{-}16$$

Definition—Instantaneous-acceleration vector

To calculate the instantaneous acceleration, we express $\vec{v}$ in rectangular coordinates:

$$\vec{v} = v_x\hat{i} + v_y\hat{j} + v_z\hat{k} = \frac{dx}{dt}\hat{i} + \frac{dy}{dt}\hat{j} + \frac{dz}{dt}\hat{k}$$

Then

$$\vec{a} = \frac{dv_x}{dt}\hat{i} + \frac{dv_y}{dt}\hat{j} + \frac{dv_z}{dt}\hat{k} = \frac{d^2x}{dt^2}\hat{i} + \frac{d^2y}{dt^2}\hat{j} + \frac{d^2z}{dt^2}\hat{k}$$

$$= a_x\hat{i} + a_y\hat{j} + a_z\hat{k} \qquad 3\text{-}17$$

Example 3-5

The position of a thrown baseball is given by $\vec{r} = 1.5\text{ m}\,\hat{i} + (12\text{ m/s}\,\hat{i} + 16\text{ m/s}\,\hat{j})t - 4.9\text{ m/s}^2\,\hat{j}\,t^2$. Find its velocity and acceleration.

1. The x and y components of the velocity are found by differentiating x and y:

$$v_x = \frac{dx}{dt} = \frac{d}{dt}[1.5\text{ m} + (12\text{ m/s})t] = 12\text{ m/s}$$

$$v_y = \frac{dy}{dt} = \frac{d}{dt}[(16\text{ m/s})t - (4.9\text{ m/s}^2)t^2]$$

$$= 16\text{ m/s} - 2(4.9\text{ m/s}^2)t = 16\text{ m/s} - (9.8\text{ m/s}^2)t$$

2. We differentiate again to obtain the components of the acceleration:

$$a_x = \frac{dv_x}{dt} = 0$$

$$a_y = \frac{dv_y}{dt} = -9.8\text{ m/s}^2$$

3. In vector notation, the velocity and acceleration are:

$$\vec{v} = (12\text{ m/s})\hat{i} + [16\text{ m/s} - (9.8\text{ m/s}^2)t]\hat{j}$$

$$\vec{a} = -9.8\text{ m/s}^2\,\hat{j}$$

Remark This is an example of projectile motion, a topic we study in the next section.

For a vector to be constant, both its magnitude and direction must remain constant. If either changes, the vector changes. Thus, if a car rounds a curve in the road at constant speed, it is accelerating because the velocity is changing due to the change in direction of the velocity vector.

Example 3-6

A car is traveling east at 60 km/h. It rounds a curve, and 5 s later it is traveling north at 60 km/h. Find the average acceleration of the car.

Picture the Problem The initial and final velocity vectors are shown in Figure 3-18. We choose the unit vector $\hat{i}$ to be east and $\hat{j}$ to be north, and we calculate the average acceleration from its definition, $\vec{a} = \Delta\vec{v}/\Delta t$. Note that $\Delta\vec{v}$ is the vector that, when added to $\vec{v}_i$, results in $\vec{v}_f$.

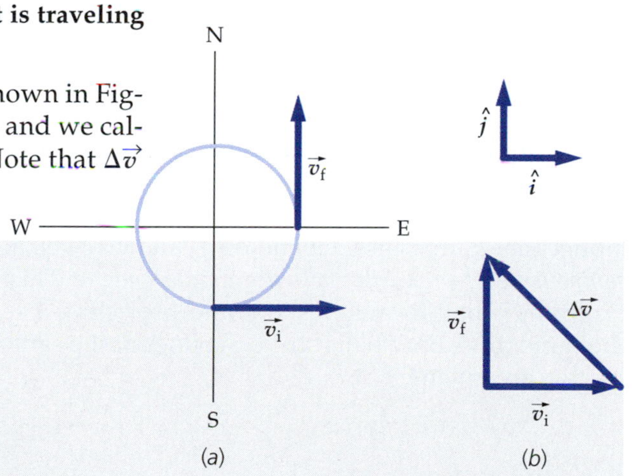

Figure 3-18

1. The average acceleration is the ratio of the velocity change to the time interval:

$$\vec{a}_{av} = \frac{\Delta\vec{v}}{\Delta t}$$

2. The change in velocity is related to the initial and final velocities:

$$\Delta\vec{v} = \vec{v}_f - \vec{v}_i$$

3. Express the initial and final velocities as vectors:

$$\vec{v}_i = (60\text{ km/h})\hat{i}$$

$$\vec{v}_f = (60\text{ km/h})\hat{j}$$

4. Substitute the above results to find the average acceleration:

$$\vec{a}_{av} = \frac{\vec{v}_f - \vec{v}_i}{\Delta t} = \frac{(60\text{ km/h})\hat{j} - (60\text{ km/h})\hat{i}}{5\text{ s}}$$

$$= -(12\text{ km/h}\cdot\text{s})\hat{i} + (12\text{ km/h}\cdot\text{s})\hat{j}$$

Remark Note that the car accelerates, even though its speed does not change.

Exercise Find the magnitude and direction of the average acceleration vector. (*Answers* $a = 17.0\ \text{km/h·s}$, $\theta = 135°$).

The motion of an object traveling in a circle is a common example of motion in which the velocity of an object changes even as its speed remains constant. We discuss circular motion in Chapter 5.

3-4 Projectile Motion

Figure 3-19 shows a particle launched with initial speed v_0 at angle θ with the horizontal axis. Let the launch point be at (x_0, y_0); y is positive upward, and x is positive to the right. The initial velocity then has components

$$v_{0x} = v_0 \cos \theta_0 \tag{3-18a}$$

$$v_{0y} = v_0 \sin \theta_0 \tag{3-18b}$$

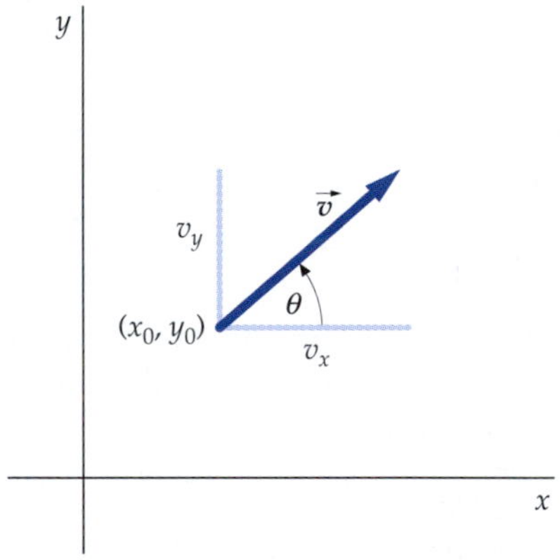

Figure 3-19

In the absence of air resistance, the acceleration is that of gravity, vertically downward:

$$a_x = 0 \tag{3-19a}$$

and

$$a_y = -g \tag{3-19b}$$

Since the acceleration is constant, we can use the kinematics equations discussed in Chapter 2. The x component of the velocity is constant because there is no horizontal acceleration:

$$v_x = v_{0x}$$

The y component varies with time according to Equation 2-11, with $a = -g$:

$$v_y = v_{0y} - gt$$

Notice that v_x does not depend on v_y and vice versa: *The horizontal and vertical components of projectile motion are independent.* This can be demonstrated by dropping a ball from a desktop and projecting a second ball horizontally at the same time. Both balls strike the floor at the same time. The displacements x and y are given by

$$x(t) = x_0 + v_{0x}t \tag{3-20a}$$

$$y(t) = y_0 + v_{0y}t - \frac{1}{2}gt^2 \tag{3-20b}$$

Equations of motion for a projectile

(See Equation 2-14.) The notation $x(t)$ and $y(t)$ simply emphasizes that x and y are functions of time. If the y component of the initial velocity is known, the time t for which the particle is at height y can be found from Equation 3-20*b*. The horizontal position at that time can then be found using Equation 3-20*a*. The total horizontal distance a projectile travels is called its **range.**

Example 3-7

Another joyful physics student throws his cap into the air with an initial velocity of 24.5 m/s at 36.9° from the horizontal. Find (*a*) the total time the cap is in the air, and (*b*) the total horizontal distance traveled.

Picture the Problem We choose the origin to be the initial position of the cap so that $x_0 = y_0 = 0$. The total time the cap is in the air is found by setting $y = 0$ in Equation 3-20*b*. We can then use this result in Equation 3-20*a* to find the total horizontal distance traveled.

(*a*) 1. Set $y = 0$ in Equation 3-20*b* and solve for t:

$$y = v_{0y}t - \tfrac{1}{2}gt^2 = t(v_{0y} - \tfrac{1}{2}gt) = 0$$

2. There are two solutions for t:

$$t = 0 \quad \text{(initial conditions)}$$

$$t = \frac{2v_{oy}}{g}$$

3. Compute the vertical component of the initial velocity vector:

$$v_{0y} - (24.5\ \text{m/s})\sin 36.9° = 14.7\ \text{m/s}$$

4. Use this result to find the total time t:

$$t = \frac{2v_{0y}}{g} = \frac{2(14.7\ \text{m/s})}{9.81\ \text{m/s}^2} = 3.0\ \text{s}$$

(*b*) Use this value for the time to calculate the total horizontal distance traveled:

$$x = v_{0x}t = (v_0 \cos\theta)t = (24.5\ \text{m/s})\cos 36.9°(3\ \text{s})$$

$$= (19.6\ \text{m/s})(3\ \text{s}) = 58.8\ \text{m}$$

Remarks The time the cap is in the air is the same as in Example 2-7, where the cap was thrown straight up with $v_0 = 14.7$ m/s. Figure 3-20 shows the height y versus t for the cap. This curve is identical to Figure 2-11 (Example 2-7) because the caps each have the same vertical acceleration and vertical velocity. The figure can be reinterpreted as a graph of y versus x if its time scale is converted to a distance scale. This can be done by multiplying the time values by 19.6 m/s, because the cap moves 19.6 m/s horizontally. The curve y versus x is a parabola.

Figure 3-21 shows graphs of the vertical heights versus the horizontal distances for projectiles with an initial speed of 24.5 m/s and several different initial angles. The angles drawn are 45°, which has the maximum range, and pairs of angles of equal amounts above and below 45°. Notice that the paired angles have the same range. The blue curve has an initial angle of 36.9° (0.64 rad), as in this example.

Figure 3-20

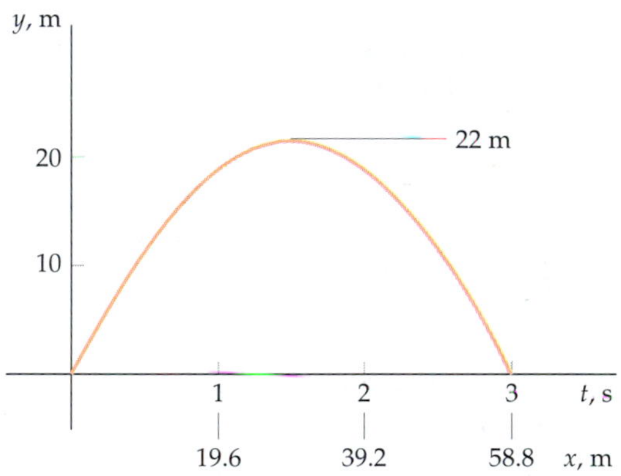

Figure 3-21

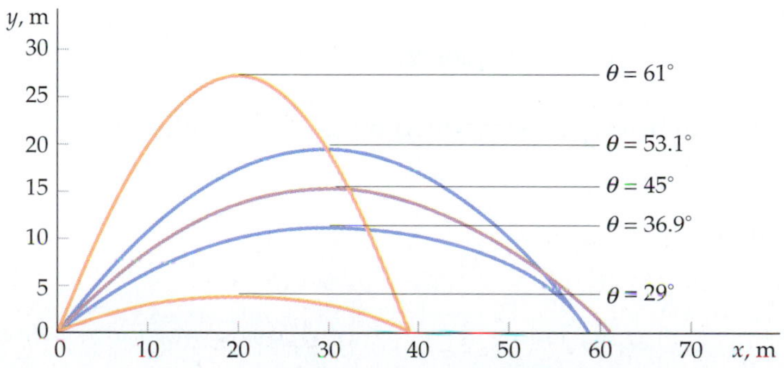

The general equation for the path $y(x)$ can be obtained from Equations 3-20*a* and 3-20*b* by eliminating the variable t. Choosing $x_0 = 0$ and $y_0 = 0$, we obtain $t = x/v_{0x}$ from Equation 3-20*a*. Substituting this into Equation 3-20*b* gives

$$y(x) = v_{0y}\left(\frac{x}{v_{0x}}\right) - \frac{1}{2}g\left(\frac{x}{v_{0x}}\right)^2 = \left(\frac{v_{0y}}{v_{0x}}\right)x - \left(\frac{g}{2v_{0x}^2}\right)x^2$$

Writing out the velocity components yields

$$y(x) = (\tan\theta_0)x - \left(\frac{g}{2v_0^2\cos^2\theta_0}\right)x^2 \qquad \text{3-21}$$

Path of a projectile

for the projectile's path. This is of the form $y = ax + bx^2$, the equation for a parabola passing through the origin. Figure 3-22 shows the path of a projectile with its velocity vector and components at several points.

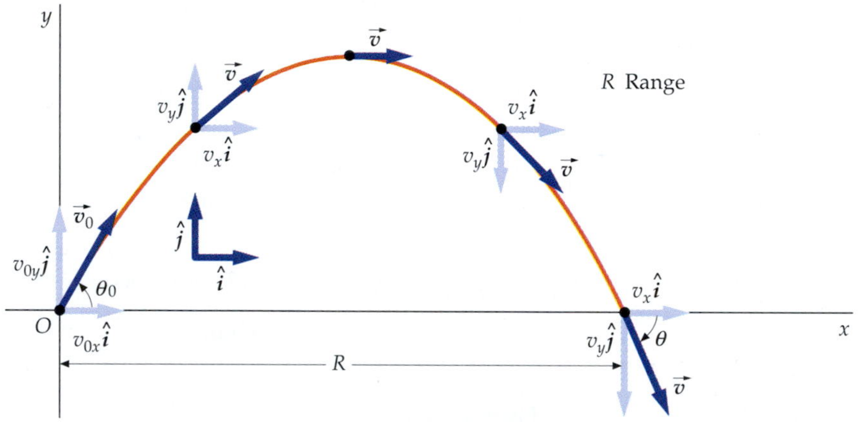

Figure 3-22 Path of a projectile showing velocity vectors.

If the initial and final elevations are equal, the range of a projectile can be written in terms of its initial speed and the angle of projection. As in the above examples, we find the range by multiplying the x component of the velocity by the total time that the projectile is in the air. The total flight time T is obtained by setting $y_0 = 0$ and $y = 0$ in Equation 3-20*b*:

$$y = v_{0y}t - \frac{1}{2}gt^2 = 0$$

$$t\left(v_{0y} - \frac{1}{2}gt\right) = 0$$

The flight time of the projectile is thus

$$T = \frac{2v_{0y}}{g} = \frac{2v_0}{g}\sin\theta_0$$

and the range is

$$R = v_{0x}T = (v_0\cos\theta_0)\left(\frac{2v_0}{g}\sin\theta_0\right) = \frac{2v_0^2}{g}\sin\theta_0\cos\theta_0$$

This can be further simplified by using the following trigonometric identity:

$$\sin 2\theta = 2\sin\theta\cos\theta$$

Thus,

$$R = \frac{v_0^2}{g}\sin 2\theta_0 \qquad \text{3-22}$$

Range of a projectile for equal initial and final elevations

Exercise Use Equation 3-21 for the path to derive Equation 3-22. (*Answer* Set $y(x) = 0$ and solve for x.)

Equation 3-22 is useful if you want to find the range for many projectiles with equal initial and final elevations. More importantly, this equation shows how the range depends on θ. Since the maximum value of $\sin 2\theta$ is 1 when $2\theta = 90°$ or $\theta = 45°$, the range is greatest when $\theta = 45°$. In many practical applications, the initial and final elevations may not be equal, and other considerations are important. For example, in the shot put, the ball ends its flight when it hits the ground, but it is projected from an initial height of about 2 m above the ground. This causes the range to be maximum at an angle somewhat lower than 45°, as shown in Figure 3-23. Studies of the best shot-putters show that maximum range occurs at an initial angle of about 42°. When calculating the range of artillery shells, air resistance must be taken into account to predict the range accurately. As expected, air resistance reduces the range for a given angle of projection. It also decreases the optimum angle of projection slightly.

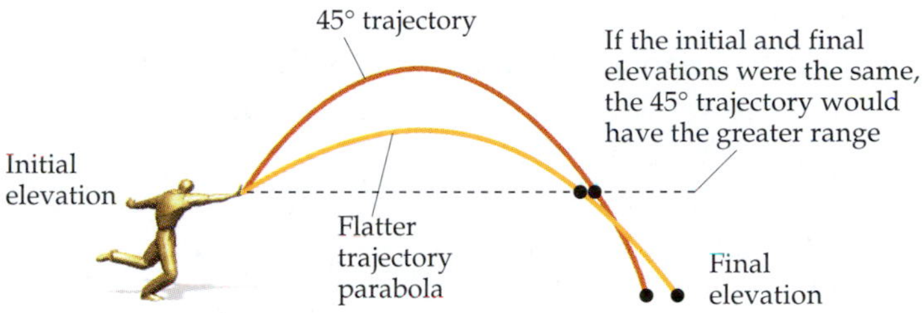

Figure 3-23 If a projectile lands at an elevation lower than the elevation of projection, maximum range is achieved when the projection angle is somewhat lower than 45°.

Example 3-8 *try it yourself*

A policeman chases a master jewel thief across city rooftops. They are both running at 5 m/s when they come to a gap between buildings that is 4 m wide and has a drop of 3 m (Figure 3-24). The thief, having studied a little physics, leaps at 5 m/s and at 45° and clears the gap easily. The policeman did not study physics and thinks he should maximize his horizontal velocity, so he leaps at 5 m/s horizontally. (*a*) Does he clear the gap? (*b*) By how much does the thief clear the gap?

Picture the Problem The time in the air depends only on the vertical motion. Choose the origin at the launch point, with upward positive so that Equations 3-20*a* and 3-20*b* apply. Use Equation 3-20*b* for $y(t)$ and solve for the time when $y = -3$ m. The horizontal distance traveled is the value of x at this time. (*a*) For the policeman, $\theta_0 = 0$, so the equations of motion are $x(t) = v_0 t$ and $y(t) = -\frac{1}{2}gt^2$. (*b*) For the thief, $\theta_0 = 45°$, so $x(t) = v_0 \cos 45° t$ and $y(t) = v_0 \sin 45°\, t - \frac{1}{2}gt^2$.

Figure 3-24

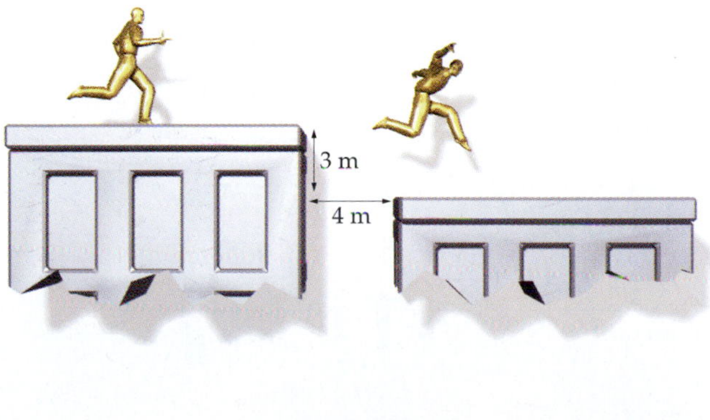

Cover the column to the right and try these on your own before looking at the answers.

Steps	***Answers***
(*a*) 1. Write $y(t)$ for the policeman and solve for t when $y = -3$ m.	$y(t) = -\frac{1}{2}gt^2 = -3\text{ m}$ $t = 0.782\text{ s}$
2. Find the horizontal distance traveled during this time.	$x = v_0 t = 3.91\text{ m}$ Since this is less than 4 m, the policeman fails to make it across the gap between buildings.
(*b*) 1. Write $y(t)$ for the thief and set $y = -3$ m.	$y(t) = v_{0y}t - \frac{1}{2}gt^2 = -3\text{ m}$ or $\frac{1}{2}gt^2 - v_{0y}t - 3\text{ m} = 0$
2. Find the two solutions for t.	$t = \frac{v_{0y}}{g} \pm \frac{1}{g}\sqrt{v_{0y}^2 - (-6\text{m})(g)}$ $t = -0.5\text{ s}$ or $t = 1.22\text{ s}$
3. Find the horizontal distance covered for the positive value of t.	$x = v_{0x}t = 4.31\text{ m}$
4. Subtract 4.0 m from this distance.	0.31 m

Remark The thief probably knew that he should jump at slightly less than 45°, but he didn't have time to solve the problem exactly.

Example 3-9

A helicopter drops a supply package to soldiers in a jungle clearing. When the package is dropped, the helicopter is 100 m above the clearing and flying at 25 m/s at an angle $\theta_0 = 36.9°$ above the horizontal (Figure 3-25). (*a*) Where does the package land? (*b*) If the helicopter flies at constant velocity, where is it when the package lands?

Picture the Problem The horizontal distance traveled by the package is given by Equation 3-20*a*, where t is the time the package is in the air. The value of t can be found from Equation 3-20*b*. Choose the origin to be directly below the helicopter when the package is dropped. The initial velocity of the package is the initial velocity of the helicopter.

Figure 3-25

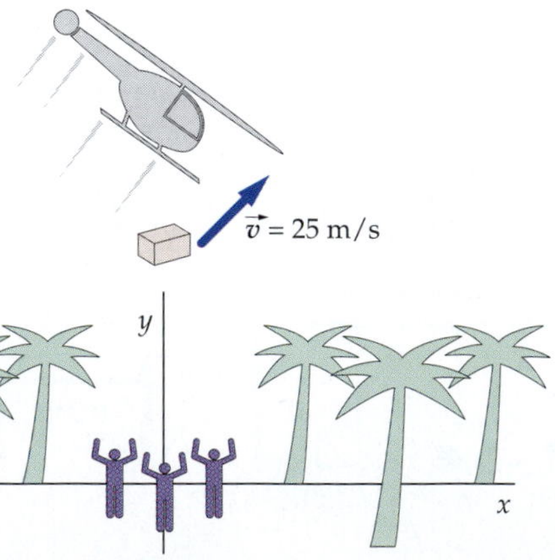

(*a*) 1. The point of impact for the package, x, is given by horizontal velocity times the time:

$$x = v_{0x}t$$

2. Find the horizontal velocity of the package:

$$v_{0x} = v_0 \cos \theta = (25 \text{ m/s})\cos 36.9° = 20 \text{ m/s}$$

3. Write $y(t)$ and solve for t when $y = 0$:

$$y(t) = y_0 + v_{0y}t - \tfrac{1}{2}gt^2$$
$$= 100 \text{ m} + (25 \text{ m/s})(\sin 36.9°)t - \tfrac{1}{2}(9.81 \text{ m/s}^2)t^2$$
$$= 100 \text{ m} + (15 \text{ m/s})t - 4.9t^2$$

$$y = 0 \text{ at } t = 6.30 \text{ s and } t = -3.24 \text{ s}$$

4. Use the positive time to find the range x:

$$x = v_{0x}t = (20 \text{ m/s})(6.30 \text{ s}) = 126 \text{ m}$$

(*b*) The coordinates of the helicopter at the time of impact are:

$$x_h = v_{0x}t = (20 \text{ m/s})(6.30 \text{ s}) = 126 \text{ m}$$
$$y_h = y_0 + v_{0y}t = 100 \text{ m} + (15 \text{ m/s})(6.30 \text{ s})$$
$$= 100 \text{ m} + 94.5 \text{ m} = 194.5 \text{ m}$$

Remark The positive time is appropriate because it corresponds to a time after the package is dropped (which occurs at $t = 0$). The negative time is when the package would have been at $y = 0$ if its motion had started earlier, as shown in Figure 3-26. Note that the helicopter is directly above the package when the package hits the ground (and at all other times before then).

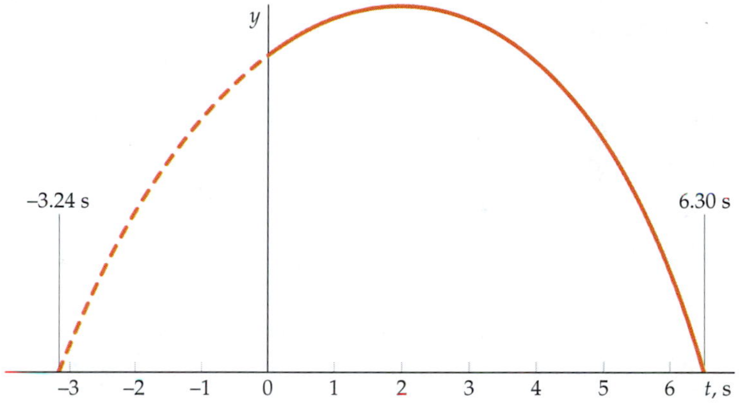

Figure 3-26

Remark Figure 3-27 shows a graph of y versus x for supply packages dropped at various initial angles and with an initial speed of 25 m/s. The green curve is the initial angle of 36.9° given in this example. Note that the maximum range no longer occurs at 45°.

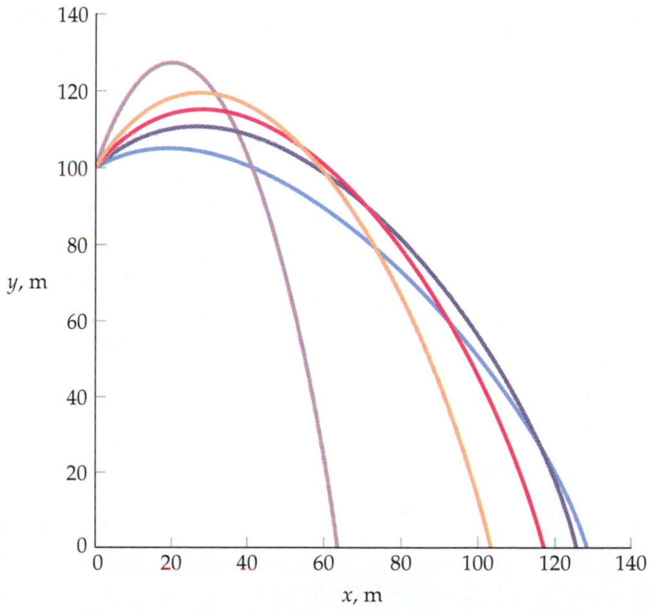

Figure 3-27

Example 3-10 *try it yourself*

In Example 3-9, (*a*) find the time t_1 for the package to reach its greatest height h, (*b*) find its greatest height h, and (*c*) find the time t_2 for the package to fall from its greatest height.

Cover the column to the right and try these on your own before looking at the answers.

Steps	**Answers**
(*a*) 1. Write $v_y(t)$ for the package.	$v_y(t) = v_{y0} - gt = 15\text{ m/s} - (9.81\text{ m/s}^2)t$
2. Set $v_y(t_1) = 0$ and solve for t_1.	$t_1 = 1.53\text{ s}$
(*b*) 1. Find $v_{y,\text{av}}$ during the time the package is moving up.	$v_{y,\text{av}} = 7.5\text{ m/s}$
2. Use $v_{y,\text{av}}$ to find the distance traveled up. Then find h.	$\Delta y = 11.5\text{ m}, \quad h = 111.5\text{ m}$
(*c*) Find the time for the package to fall a distance h.	$t_2 = 4.77\text{ s}$

Remark Note that $t_1 + t_2 = 6.3$ s, in agreement with Example 3-9.

Example 3-11

A park ranger with a tranquilizer dart gun intends to shoot a monkey hanging from a branch (Figure 3-28). The ranger aims directly at the monkey, not realizing that the dart will follow a parabolic path that will pass below the present position of the creature. The monkey, seeing the gun discharge, lets go of the branch and drops out of the tree, expecting to avoid the dart. Show that the monkey will be hit regardless of the initial speed of the dart so long as it is great enough for the dart to travel the horizontal distance to the tree before hitting the ground. Assume that the reaction time of the monkey is negligible.

Picture the Problem We choose the origin to be at the muzzle of the gun and let $\vec{r}_0$ be the initial position vector of the monkey. Since the gun is aimed at the original position of the monkey, the initial velocity of the dart $\vec{v}_0$ is parallel to $\vec{r}_0$. We find the position vectors for both the monkey and dart as functions of time, then solve for the time t when they are equal.

Figure 3-28

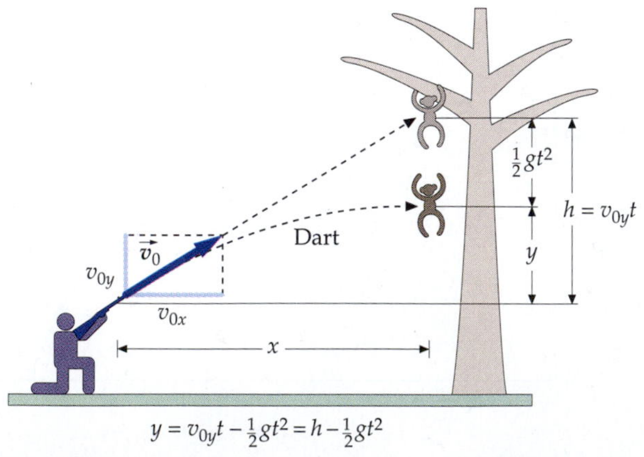

1. Write the position vector of the monkey at time t: $\vec{r}_m = \vec{r}_0 - \frac{1}{2}gt^2\hat{j}$

2. Write the position vector of the dart at some time t in terms of the initial velocity $\vec{v}_0$: $\vec{r}_d = \vec{v}_0 t - \frac{1}{2}gt^2\hat{j}$

3. When the dart hits the monkey, the position vectors are equal to each other:
$$\vec{r}_m = \vec{r}_d$$
$$\vec{r}_0 - \frac{1}{2}gt^2\hat{j} = \vec{v}_0 t - \frac{1}{2}gt^2\hat{j}$$
$$\vec{r}_0 = \vec{v}_0 t$$

4. We can solve for t in terms of the distance x and initial speed v_0 by taking the x component of the above equation:
$$r_{0x} = x = v_{0x}t$$
$$t = \frac{x}{v_{0x}}$$

Remarks According to these equations, the dart always hits the monkey (Figure 3-29). However, if the dart (and monkey) hits the ground at some time $t < t_1$, the equations for $\vec{r}_m$ and $\vec{r}_d$ are no longer valid. In a familiar lecture demonstration, a target is suspended by an electromagnet. When the dart leaves the gun, the circuit to the magnet is broken and the target falls. The initial velocity of the dart is varied so that for large v_0 the target is hit very near its original height and for small v_0 it is hit just before it reaches the floor.

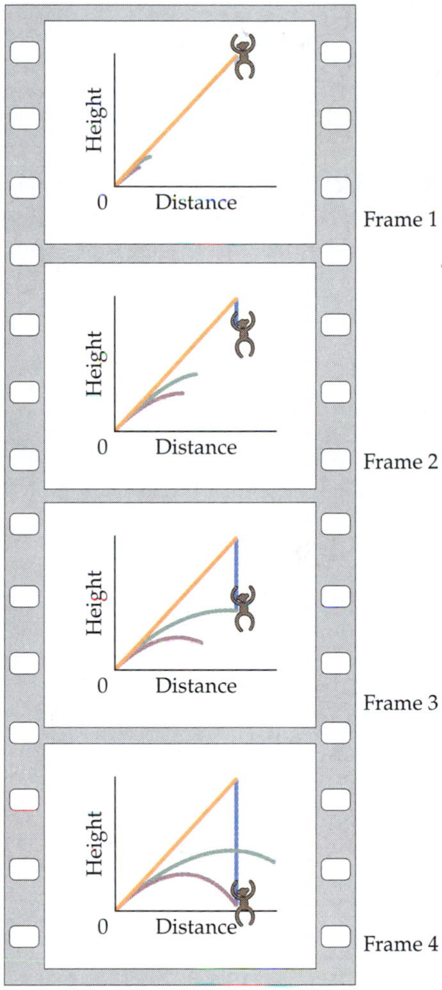

Figure 3-29 In the classic ranger and monkey problem, the dart and the monkey both fall with the same acceleration g. As a result they will meet, independent of the speed of the dart as long as the dart was aimed at the monkey. In *frame 1*, the dart is traveling so fast that it hits the monkey before the monkey has fallen very far. In *frame 2*, the monkey has fallen about a half meter from the tree branch and the dart has fallen about a half meter from a straight line of flight. In *frame 3*, the monkey has fallen about half the distance to the ground before the dart hits. The slower moving dart has also fallen away from a straight line of flight by exactly the same amount. In *frame 4*, the slowest dart and the monkey reach a point just above the ground at the same time.

Example 3-12 *try it yourself*

Your slapshot in hockey is wickedly fast but not very accurate. The puck, struck at ice level, misses the net and just clears the top of the Plexiglas wall of height h = 2.80 m. The flight time at the moment the puck clears the wall is t_1 = 0.650 s, and the horizontal distance is x_1 = 12.0 m. (*a*) Find the initial speed and direction of the puck. (*b*) When does the puck reach its maximum height? (*c*) What is the maximum height of the puck?

Cover the column to the right and try these on your own before looking at the answers.

Steps / ***Answers***

(*a*) 1. Find the horizontal component of the initial velocity. $v_{0x} = \frac{x_1}{t_1} = 18.5\text{ m/s}$

2. Write the equation for $y(t)$ and solve for v_{0y} using $y = h$ and $t = t_1$. $\quad y = v_{0y}t - \frac{1}{2}gt^2, \qquad v_{0y} = 7.49\text{ m/s}$

3. Find v from the components and θ_0 from $\tan\theta_0 = v_{0y}/v_{0x}$. $\quad v = \sqrt{v_{0x}^2 + v_{0y}^2} = 20.0\text{ m/s}, \qquad \theta_0 = 22.0°$

(*b*) Write the general equation for $v_y(t)$ and solve for t when $v_y = 0$. $\quad v_y = v_{0y} - gt, \qquad t = 0.764\text{ s}$

(*c*) Find the maximum height from $\Delta y = v_{y,\text{av}}\, t$. $\quad \Delta y = v_{y,\text{av}}t = 2.86\text{ m}$

Remark In this example, a hockey puck clears a glass wall 2.8 m high and 12 m distant. The puck reaches its maximum height after clearing the wall. Figure 3-30 shows several other cases of initial velocity and angle for which the puck would also just clear the wall.

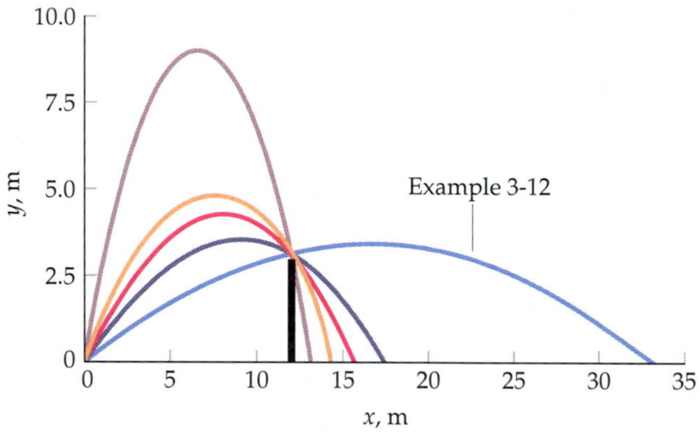

Figure 3-30

Summary

Topic	Remarks and Relevant Equations	
1. Vectors		
Definition	Vectors are quantities that have both magnitude and direction. Vectors add like displacements.	
Components	The component of a vector along a line in space is its projection on the line. If $\vec{A}$ makes an angle θ with the x axis, its x and y components are	
	$A_x = A\cos\theta$	**3-2**
	$A_y = A\sin\theta$	**3-3**
Magnitude	$A = \sqrt{A_x^2 + A_y^2}$	**3-5**
Adding vectors graphically	Any two vectors whose magnitudes have the same units may be added graphically by placing the tail of one arrow at the head of the other.	

Adding vectors using components	If $\vec{C} = \vec{A} + \vec{B}$, then $$C_x = A_x + B_x$$ and $$C_y = A_y + B_y$$	3-6a 3-6b
Unit vectors	A vector $\vec{A}$ can be written in terms of unit vectors $\hat{i}$, $\hat{j}$, and $\hat{k}$, which have unit magnitude and lie along the x, y, and z axes, respectively $$\vec{A} = A_x\hat{i} + A_y\hat{j} + A_z\hat{k}$$	3-7
Position vector	The position vector $\vec{r}$ points from the origin of the coordinate system to the particle's position.	
Instantaneous-velocity vector	The velocity vector $\vec{v}$ is the rate of change of the position vector. Its magnitude is the speed and it points in the direction of motion. $$\vec{v} = \lim_{\Delta t \to 0} \frac{\Delta \vec{r}}{\Delta t} = \frac{d\vec{r}}{dt}$$	3-12
Instantaneous-acceleration vector	$$\vec{a} = \lim_{\Delta t \to 0} \frac{\Delta \vec{v}}{\Delta t} = \frac{d\vec{v}}{dt}$$	3-16
2. Relative Velocity	If a particle moves with velocity $\vec{v}_{\text{pA}}$ relative to a coordinate system A, which is in turn moving with velocity $\vec{v}_{\text{AB}}$ relative to another coordinate system B, the velocity of the particle relative to B is $$\vec{v}_{\text{pB}} = \vec{v}_{\text{pA}} + \vec{v}_{\text{AB}}$$	3-14
3. Projectile Motion		
Independence of motion	In projectile motion, the horizontal and vertical motions are independent. The horizontal motion has constant velocity. The vertical motion is the same as motion in one dimension with constant acceleration due to gravity g downward.	
Equations	$$v_x(t) = v_{0x}$$ $$x(t) = x_0 + v_{0x}t = x_0 + (v_0 \cos\theta)t$$ $$v_y(t) = v_{0y} - gt = v_0 \sin\theta - gt$$ $$y(t) = y_0 + v_{0y}t - \frac{1}{2}gt^2 = y_0 + (v_0 \sin\theta)t - \frac{1}{2}gt^2$$	3-20a 3-20b
Path	$$y(x) = (\tan\theta_0)x - \left(\frac{g}{2v_0^2\cos^2\theta_0}\right)x^2$$	3-21
Range	The range is found by multiplying v_x by the total time the projectile is in the air.	
Range when initial and final elevations are equal	$$R = \frac{v_0^2}{g}\sin 2\theta_0$$	3-22

Problem-Solving Guide

The following are applicable to all types of problems:

1. Begin by drawing a neat diagram that includes the important features of the problem.
2. Choose a convenient coordinate system and indicate it on your sketch.
3. Indicate the given information on your sketch.

Summary of Worked Examples

Type of Calculation	Procedure and Relevant Examples
1. Vectors Add (or subtract) vectors.	Add (or subtract) the components of individual vectors to find the components of the resultant vector. **Example 3-2**
Find the direction of a resultant vector.	The angle made with the positive x direction is found from $\tan\theta = v_y/v_x$. **Examples 3-1, 3-2, 3-3**
Take derivatives of vectors.	Express the vector in component form using unit vectors, and take the derivative of each component separately. **Examples 3-3, 3-5**
2. Relative Velocity	
Express the velocity of a particle relative to a coordinate system that is itself moving relative to another coordinate system.	Use $\vec{v}_{\rm pB} = \vec{v}_{\rm pA} + \vec{v}_{\rm AB}$. **Example 3-4**
3. Projectile Motion	
Find a projectile's time of flight to various positions.	Calculate the time at which a vertical position is reached by using constant acceleration formulas. Calculate the time at which a horizontal position is reached by using constant velocity formulas. **Examples 3-7, 3-8, 3-9, 3-10, 3-11**
Find speeds and angles along a projectile's trajectory	Find the x and y components of the velocity from the constant acceleration formulas. Find the speed from $v = \sqrt{v_x^2 + v_y^2}$. The angle of a projectile's trajectory is the angle of its resultant velocity vector at that moment. **Examples 3-8, 3-12**
Find the position of a projectile	Use $x = v_{0x}\,t$ and $y = y_0 + v_{0y}t - \frac{1}{2}gt^2$ where t is the time of flight. **Examples 3-7, 3-8, 3-9, 3-12**

Problems

In a few problems, you are given more data than you actually need; in a few other problems, you are required to supply data from your general knowledge, outside sources, or informed estimates.

Conceptual Problems

Problems from Optional and Exploring sections

• Single-concept, single-step, relatively easy

•• Intermediate-level, may require synthesis of concepts

••• Challenging, for advanced students

For all problems, use $g = 9.81$ m/s^2 for the acceleration due to gravity and neglect friction and air resistance unless instructed to do otherwise.

Vectors and Vector Addition

1 • Can the magnitude of the displacement of a particle be less than the distance traveled by the particle along its path? Can its magnitude be more than the distance traveled? Explain.

2 • Give an example in which the distance traveled is a significant amount yet the corresponding displacement is zero.

3 • The magnitude of the displacement of a particle is __________ the distance the object has traveled.

(*a*) larger than
(*b*) smaller than
(*c*) either larger or smaller than
(*d*) the same as
(*e*) smaller than or equal to

4 • A bear walks northeast for 12 m and then east for 12 m. Show each displacement graphically, and find the resultant displacement vector.

5 • (*a*) A man walks along a circular arc from the position $x = 5$ m, $y = 0$ to a final position $x = 0$, $y = 5$ m. What is his displacement? (*b*) A second man walks from the same initial position along the x axis to the origin and then along the y axis to $y = 5$ m and $x = 0$. What is his displacement?

6 • A circle of radius 8 m has its center on the y axis at $y = 8$ m. You start at the origin and walk along the circle at a steady speed, returning to the origin exactly 1 min after you started. (*a*) Find the magnitude and direction of your displacement from the origin 15, 30, 45, and 60 s after you start. (*b*) Find the magnitude and direction of your displacement for each of the four successive 15-s intervals of your walk. (*c*) How is your displacement for the first 15 s related to that for the second 15 s? (*d*) How is your displacement for the second 15-s interval related to that for the last 15-s interval?

7 • For the two vectors $\vec{A}$ and $\vec{B}$ in Figure 3-31, find the following graphically: (*a*) $\vec{A} + \vec{B}$, (*b*) $\vec{A} - \vec{B}$, (*c*) $2\vec{A} + \vec{B}$, (*d*) $\vec{B} - \vec{A}$, (*e*) $2\vec{B} - \vec{A}$.

Figure 3-31 Problem 7

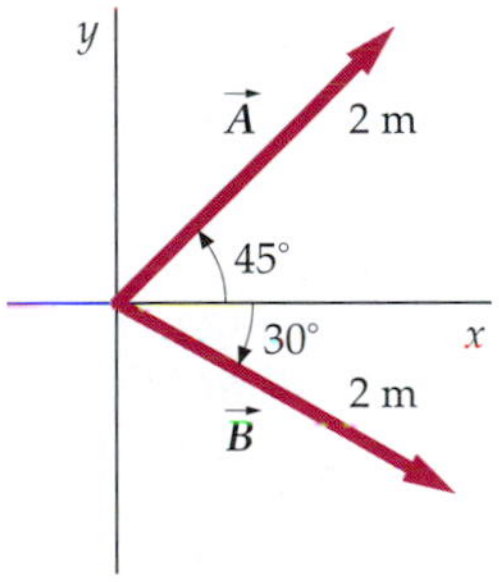

8 • A scout walks 2.4 km due east from camp, then turns left and walks 2.4 km along the arc of a circle centered at the campsite, and finally walks 1.5 km directly toward camp. (*a*) How far is the scout from camp at the end of his walk? (*b*) In what direction is the scout's position relative to the campsite? (*c*) What is the ratio of the final magnitude of the displacement to the total distance walked?

Adding Vectors by Components

9 • Can a component of a vector have a magnitude greater than the magnitude of the vector? Under what circumstances can a component of a vector have a magnitude equal to the magnitude of the vector?

10 • Can a vector be equal to zero and still have one or more components not equal to zero?

11 • Are the components of $\vec{C} = \vec{A} + \vec{B}$ necessarily larger than the corresponding components of either $\vec{A}$ or $\vec{B}$?

12 • The components of a vector are $A_x = -10$ m and $A_y = 6$ m. What angle does this vector make with the positive x axis?

(*a*) 31°
(*b*) −31°
(*c*) 180° − 31°
(*d*) 180° + 31°
(*e*) 90° − 31°

13 • A velocity vector has an x component of +5.5 m/s and a y component of −3.5 m/s. Which diagram in Figure 3-32 gives the direction of the vector?

Figure 3-32 Problem 13

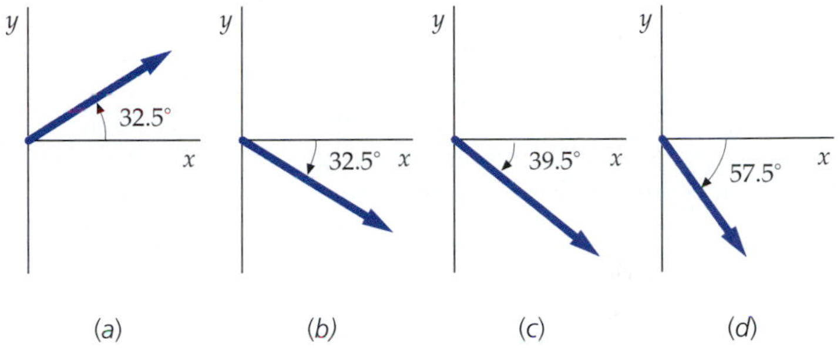

(*e*) None of the above.

14 • Three vectors $\vec{A}$, $\vec{B}$, and $\vec{C}$ have the following x and y components:

	$\vec{A}$	$\vec{B}$	$\vec{C}$
x component	+6	−3	+2
y component	−3	+4	+5

The magnitude of $\vec{A} + \vec{B} + \vec{C}$ is __________.

(*a*) 3.3
(*b*) 5.0
(*c*) 11
(*d*) 7.8
(*e*) 14

15 • Find the rectangular components of the following vectors $\vec{A}$, which lie in the xy plane, and make an angle θ with the x axis (Figure 3-33) if (*a*) $A = 10$ m, $\theta = 30°$; (*b*) $A = 5$ m, $\theta = 45°$; (*c*) $A = 7$ km, $\theta = 60°$; (*d*) $A = 5$ km, $\theta = 90°$; (*e*) $A = 15$ km/s, $\theta = 150°$; (*f*) $A = 10$ m/s, $\theta = 240°$; and (*g*) $A = 8$ m/s^2, $\theta = 270°$.

Figure 3-33 Problem 15

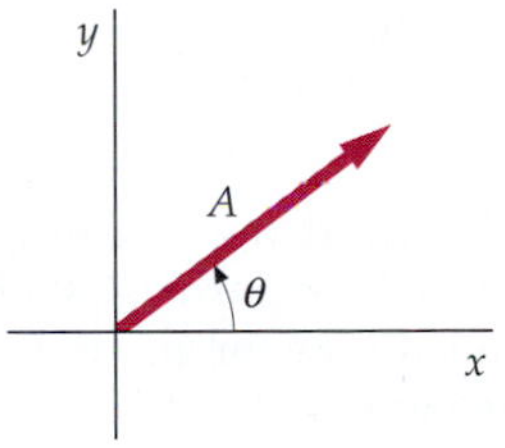

16 • Vector $\vec{A}$ has a magnitude of 8 m at an angle of 37° with the x axis; vector $\vec{B} = 3\text{ m}\,\hat{i} - 5\text{ m}\,\hat{j}$; vector $\vec{C} = -6\text{ m}\,\hat{i} + 3\text{ m}\,\hat{j}$. Find the following vectors: (*a*) $\vec{D} = \vec{A} + \vec{C}$; (*b*) $\vec{E} = \vec{B} - \vec{A}$; (*c*) $\vec{F} = \vec{A} - 2\vec{B} + 3\vec{C}$; (*d*) A vector $\vec{G}$ such that $\vec{G} - \vec{B} = \vec{A} + 2\vec{C} + 3\vec{G}$.

Unit Vectors

17 • Find the magnitude and direction of the following vectors: (*a*) $\vec{A} = 5\hat{i} + 3\hat{j}$; (*b*) $\vec{B} = 10\hat{i} - 7\hat{j}$; (*c*) $\vec{C} = -2\hat{i} - 3\hat{j} + 4\hat{k}$.

18 • Find the magnitude and direction of $\vec{A}$, $\vec{B}$, and $\vec{A} + \vec{B}$ for (*a*) $\vec{A} = -4\hat{i} - 7\hat{j}, \vec{B} = 3\hat{i} - 2\hat{j}$, and (*b*) $\vec{A} = 1\hat{i} - 4\hat{j}, \vec{B} = 2\hat{i} + 6\hat{j}$.

19 • Describe the following vectors using the unit vectors $\hat{i}$ and $\hat{j}$: (*a*) a velocity of 10 m/s at an angle of elevation of 60°; (*b*) a vector $\vec{A}$ of magnitude $A = 5$ m and $\theta = 225°$; (*c*) a displacement from the origin to the point $x = 14$ m, $y = -6$ m.

20 • For the vector $\vec{A} = 3\hat{i} + 4\hat{j}$, find any three other vectors $\vec{B}$ that also lie in the xy plane and have the property that $A = B$ but $\vec{A} \neq \vec{B}$. Write these vectors in terms of their components and show them graphically.

21 • If $\vec{A} = 5\hat{i} - 4\hat{j}$ and $\vec{B} = -7.5\hat{i} + 6\hat{j}$, write an equation relating $\vec{A}$ to $\vec{B}$.

22 •• The faces of a cube of side 3 m are parallel to the coordinate planes with one corner at the origin. A fly begins at the origin and walks along three edges until it is at the far corner. Write the displacement vector of the fly using the unit vectors $\hat{i}$, $\hat{j}$, and $\hat{k}$, and find the magnitude of this displacement.

Velocity and Acceleration Vectors

23 • For an arbitrary motion of a given particle, does the direction of the velocity vector have any particular relation to the direction of the position vector?

24 • Give examples in which the directions of the velocity and position vectors are (*a*) opposite, (*b*) the same, and (*c*) mutually perpendicular.

25 • How is it possible for a particle moving at constant speed to be accelerating? Can a particle with constant velocity be accelerating at the same time?

26 • If an object is moving toward the west, in what direction is its acceleration?

(*a*) North
(*b*) East
(*c*) West
(*d*) South
(*e*) May be any direction.

27 •• Consider the path of a particle as it moves in space. (*a*) How is the velocity vector related geometrically to the path of the particle? (*b*) Sketch a curved path and draw the velocity vector for the particle for several positions along the path.

28 •• A dart is thrown straight up. After it leaves the player's hand, it steadily loses speed as it gains altitude until it lodges in the ceiling of the game room. Draw the dart's velocity vector at times t_1 and t_2, where $\Delta t = t_2 - t_1$ is small. From your drawing find the direction of the change in velocity $\Delta\vec{v} = \vec{v}_2 - \vec{v}_1$, and thus the direction of the acceleration vector.

29 •• As a bungee jumper approaches the lowest point in her drop, she loses speed as she continues to move downward. Draw the velocity vectors of the jumper at times t_1 and t_2, where $\Delta t = t_2 - t_1$ is small. From your drawing find the direction of the change in velocity $\Delta\vec{v} = \vec{v}_2 - \vec{v}_1$, and thus the direction of the acceleration vector.

30 •• After reaching the lowest point in her jump at time t_{low}, the bungee jumper in the previous problem then moves upward, gaining speed for a short time until gravity again dominates her motion. Draw her velocity vectors at times t_1 and t_2, where $\Delta t = t_2 - t_1$ is small and $t_1 < t_{low} < t_2$. From your drawing find the direction of the change in velocity $\Delta\vec{v} = \vec{v}_2 - \vec{v}_1$, and thus the direction of the acceleration vector.

31 • A stationary radar operator determines that a ship is 10 km south of him. An hour later the same ship is 20 km southeast. If the ship moved at constant speed and always in the same direction, what was its velocity during this time?

32 • A particle's position coordinates (x, y) are (2 m, 3 m) at $t = 0$; (6 m, 7 m) at $t = 2$ s; and (13 m, 14 m) at $t = 5$ s. (*a*) Find v_{av} from $t = 0$ to $t = 2$ s. (*b*) Find v_{av} from $t = 0$ to $t = 5$ s.

33 • A particle moving at 4.0 m/s in the positive x direction is given an acceleration of 3.0 m/s^2 in the positive y direction for 2.0 s. The final speed of the particle is ________.

(*a*) −2.0 m/s
(*b*) 7.2 m/s
(*c*) 6.0 m/s
(*d*) 10 m/s
(*e*) None of the above

34 • A ball is thrown directly upward. Consider the 2-s time interval $\Delta t = t_2 - t_1$, where t_1 is 1 s before the ball reaches its highest point and t_2 is 1 s after it reaches its highest point. For the time interval Δt, find (*a*) the change in speed, (*b*) the change in velocity, and (*c*) the average acceleration.

35 • Initially, a particle is moving due west with a speed of 40 m/s; 5 s later it is moving north with a speed of 30 m/s. (*a*) What was the change in the magnitude of the particle's velocity during this time? (*b*) What was the change in the direction of the velocity? (*c*) What are the magnitude and direction of $\Delta\vec{v}$ for this interval? (*d*) What are the magnitude and direction of $\vec{a}_{av}$ for this interval?

36 • At $t = 0$, a particle located at the origin has a velocity of 40 m/s at $\theta = 45°$. At $t = 3$ s, the particle is at $x = 100$ m and $y = 80$ m with a velocity of 30 m/s at $\theta = 50°$. Calculate (*a*) the average velocity and (*b*) the average acceleration of the particle during this interval.

37 •• A particle moves in an xy plane with constant acceleration. At time zero, the particle is at $x = 4$ m, $y = 3$ m, and has velocity $\vec{v} = 2\text{ m/s}\,\hat{i} - 9\text{ m/s}\,\hat{j}$. The acceleration is given by the vector $\vec{a} = 4\text{ m/s}^2\,\hat{i} + 3\text{ m/s}^2\,\hat{j}$. (*a*) Find the velocity vector at $t = 2$ s. (*b*) Find the position vector at $t = 4$ s. Give the magnitude and direction of the position vector.

38 •• A particle has a position vector given by $\vec{r} = 30t\hat{i} + (40t - 5t^2)\hat{j}$, where r is in meters and t in seconds. Find the instantaneous-velocity and instantaneous-acceleration vectors as functions of time t.

39 •• A particle has a constant acceleration of $\vec{a} = (6\hat{i} + 4\hat{j})\text{ m/s}^2$. At time $t = 0$, the velocity is zero and the position vector is $\vec{r}_0 - (10\text{ m})\,\hat{i}$. (*a*) Find the velocity and position vectors at any time t. (*b*) Find the equation of the particle's path in the xy plane, and sketch the path.

40 ••• Mary and Robert decide to rendezvous on Lake Michigan. Mary departs in her boat from Petoskey at 9:00 A.M. and travels due north at 8 mi/h. Robert leaves from his home on the shore of Beaver Island, 26 mi 30° west of north of Petoskey, at 10:00 A.M. and travels at a constant speed of 6 mi/h. In what direction should Robert be heading to intercept Mary, and where and when will they meet?

Relative Velocity

41 • A river is 0.76 km wide. The banks are straight and parallel (Figure 3-34). The current is 5.0 km/h and is parallel to the banks. A boat has a maximum speed of 3 km/h in still water. The pilot of the boat wishes to go on a straight line from A to B, where AB is perpendicular to the banks. The pilot should

(*a*) head directly across the river.
(*b*) head 68° upstream from the line AB.
(*c*) head 22° upstream from the line AB.
(*d*) give up—the trip from A to B is not possible with this boat.
(*e*) do none of the above.

Figure 3-34 Problem 41

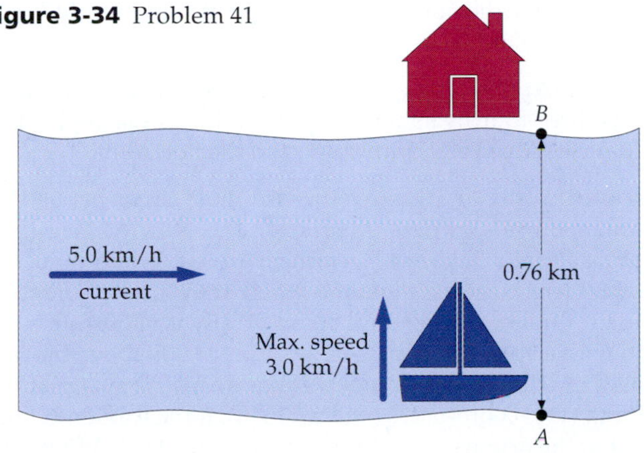

42 •• A plane flies at a speed of 250 km/h relative to still air. There is a wind blowing at 80 km/h in the northeast direction at exactly 45° to the east of north. (*a*) In what direction should the plane head so as to fly due north? (*b*) What is the speed of the plane relative to the ground?

43 •• A swimmer heads directly across a river, swimming at 1.6 m/s relative to still water. She arrives at a point 40 m downstream from the point directly across the river, which is 80 m wide. (*a*) What is the speed of the river current? (*b*) What is the swimmer's speed relative to the shore? (*c*) In what direction should the swimmer head so as to arrive at the point directly opposite her starting point?

44 •• A small plane departs from point A heading for an airport at point B 520 km due north. The airspeed of the plane is 240 km/h and there is a steady wind of 50 km/h blowing northwest to southeast. Determine the proper heading for the plane and the time of flight.

45 •• Two boat landings are 2.0 km apart on the same bank of a stream that flows at 1.4 km/h. A motorboat makes the round trip between the two landings in 50 min. What is the speed of the boat relative to the water?

46 •• A model airplane competition has the following rules: Each plane must fly to a point 1 km from the start and then back again. The winner is the plane with the shortest round-trip time. The contestants are free to launch their planes in any direction, so long as the plane travels exactly 1 km out and then returns. On the day of the race, a steady wind blows from the north at 5 m/s. Your plane can maintain an airspeed (speed relative to the air) of 15 m/s, and you know that starting, stopping, and turning times will be negligible. The question: Should you plan to fly into the wind and against the wind on your round-trip, or across the wind flying east and west? Make a reasoned choice by working out the following round-trip times: (1) The plane goes 1 km due north and then back; (2) the plane goes to point 1 km due east of the start, and then back.

47 •• Car A is traveling east at 20 m/s. As car A crosses the intersection shown in Figure 3-35, car B starts from rest 40 m north of the intersection and moves south with a constant acceleration of 2 m/s^2. (*a*) What is the position of B relative to A 6 s after A crosses the intersection? (*b*) What is the velocity of B relative to A for $t = 6$ s? (*c*) What is the acceleration of B relative to A for $t = 6$ s?

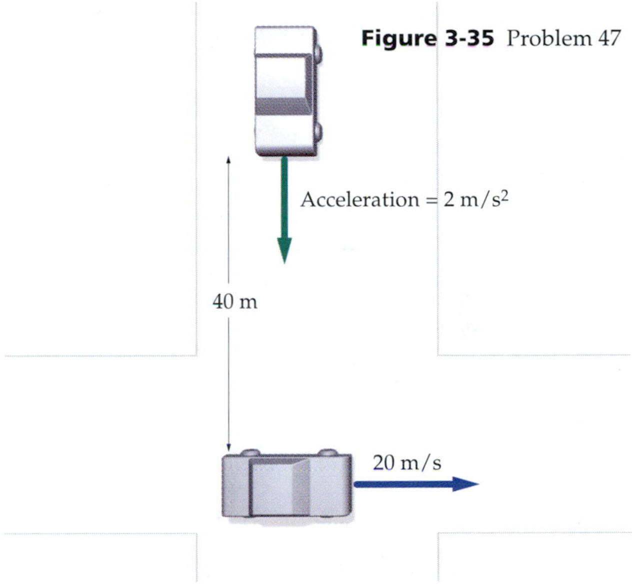

Figure 3-35 Problem 47

48 •• Bernie is showing Margaret his new boat and its autonavigation feature, of which he is particularly proud. "That island is 1 km east and 3 km north of this dock. So I just punch in the numbers like this, and we get ourselves a refreshment and enjoy the scenery." Forty-five minutes later, they find themselves due east of the island. "OK, something went wrong. I'll just reverse the instructions, and we'll go back to the dock and try again." But 45 min later, the boat is 6 km east of their original position at the dock. "Did you allow for the current?" asks Margaret. "For the what?" (*a*) What is the velocity of the current in the waterway where Bernie and Margaret are boating? (*b*) What is the velocity of the boat, relative to the water, for the first 45 min? (*c*) What is the velocity of the boat relative to the island for the first 45 min?

49 ••• Airports A and B are on the same meridian, with B 624 km south of A. Plane P departs airport A for B at the same time that an identical plane, Q, departs airport B for A. A steady 60 km/h wind is blowing from the south 30′ east of north. Plane Q arrives at airport A 1 h before plane P arrives at airport B. Determine the airspeeds of the two planes (assuming that they are the same) and the heading of each plane.

Projectiles

50 • What is the acceleration of a projectile at the top of its flight?

51 • True or false: When a bullet is fired horizontally, it takes the same amount of time to reach the ground as a bullet dropped from rest from the same height.

52 • A golfer drives her ball from the tee a distance of 240 yards down the fairway in a high arcing shot. When the ball is at the highest point of its flight,

(*a*) its velocity and acceleration are both zero.
(*b*) its velocity is zero but its acceleration is nonzero.
(*c*) its velocity is nonzero but its acceleration is zero.
(*d*) its velocity and acceleration are both nonzero.
(*e*) insufficient information is given to answer correctly.

53 • A projectile was fired at 35° above the horizontal. At the highest point in its trajectory, its speed was 200 m/s. The initial velocity had a horizontal component of

(*a*) 0.
(*b*) 200 cos(35°) m/s.
(*c*) 200 sin(35°) m/s.
(*d*) (200 m/s)/cos(35°).
(*e*) 200 m/s.

54 • Figure 3-36 represents the parabolic trajectory of a ball going from *A* to *E*. What is the direction of the acceleration at point *B*?

(*a*) Up and to the right
(*b*) Down and to the left
(*c*) Straight up
(*d*) Straight down
(*e*) The acceleration of the ball is zero.

Figure 3-36 Problems 54 and 55

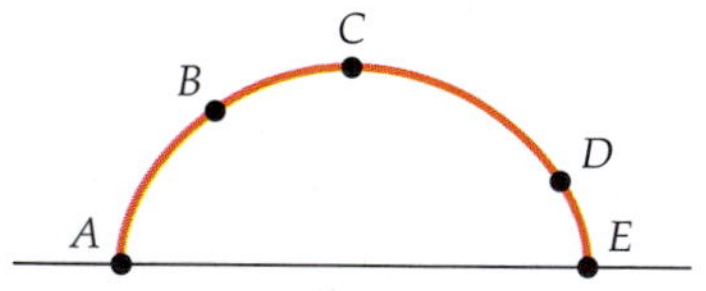

55 • Referring to Figure 3-36, (*a*) at which point(s) is the speed the greatest? (*b*) At which point(s) is the speed the lowest? (*c*) At which two points is the speed the same? Is the velocity the same at those points?

56 • A bullet is fired horizontally with an initial velocity of 245 m/s. The gun is 1.5 m above the ground. How long is the bullet in the air?

57 • A pitcher throws a fastball at 140 km/h toward home plate, which is 18.4 m away. Neglecting air resistance (not a good idea if you are the batter), find how far the ball drops because of gravity by the time it reaches home plate.

58 • A projectile is launched with speed v_0 at an angle of θ_0 with the horizontal. Find an expression for the maximum height it reaches above its starting point in terms of v_0, θ_0, and g.

59 • A projectile is fired with an initial velocity of 30 m/s at 60° above horizontal. At the projectile's highest point, what is its velocity? Its acceleration?

60 •• A projectile is fired with initial speed v at an angle 30° above the horizontal from a height of 40 m above the ground. The projectile strikes the ground with a speed of $1.2v$. Find v.

61 •• If the tree in Example 3-11 is 50 m away and the monkey hangs from a branch 10 m above the muzzle position, what is the minimum initial speed of the dart if it is to hit the monkey before hitting the ground?

62 •• A projectile is fired with an initial speed of 53 m/s. Find the angle of projection such that the maximum height of the projectile is equal to its horizontal range.

63 •• A ball thrown into the air lands 40 m away 2.44 s later. Find the direction and magnitude of the initial velocity.

64 •• Show that if an object is thrown with speed v_0 at an angle θ above the horizontal, its speed at some height h is independent of θ.

65 •• At half its maximum height, the speed of a projectile is three-fourths its initial speed. What is the angle of the initial velocity vector with respect to the horizontal?

66 ••• Wally and Luke advertise their circus act as "The Human Burrs—Trapeze Artists for the New Millennium." Their specialty involves wearing padded Velcro suits that cause them to stick together when they make contact in midair. While working on their act, Wally is shot from a cannon with a speed of 20 m/s at an angle of 30° above the horizontal. At the same moment, Luke drops from a platform having (x, y) coordinates of (8 m, 16 m), if the cannon is taken to sit at the origin. (*a*) Will they make contact? (*b*) What is the minimum distance separating Wally and Luke during their flight paths? (*c*) At what time does this minimum separation occur? (*d*) Give the coordinates of each daredevil at that time.

Projectile Range

67 • A cargo plane is flying horizontally at an altitude of 12 km with a speed of 900 km/h when a battle tank falls out of the rear loading ramp. (*a*) How long does it take the tank to hit the ground? (*b*) How far horizontally is the tank from where it fell off when it hits the ground? (*c*) How far is the tank from the aircraft when the tank hits the ground, assuming that the plane continues to fly with constant velocity?

68 • A cannon barrel is elevated at an angle of 45°. The cannon fires a ball with a speed of 300 m/s. (*a*) What height does the ball reach? (*b*) How long is the ball in the air? (*c*) What is the horizontal range of the cannon?

69 •• A stone thrown horizontally from the top of a 24-m tower hits the ground at a point 18 m from the base of the tower. (*a*) Find the speed at which the stone was thrown. (*b*) Find the speed of the stone just before it hits the ground.

70 •• A projectile is fired into the air from the top of a 200-m cliff above a valley (Figure 3-37). Its initial velocity is 60 m/s at 60° above the horizontal. Where does the projectile land?

Figure 3-37 Problem 70

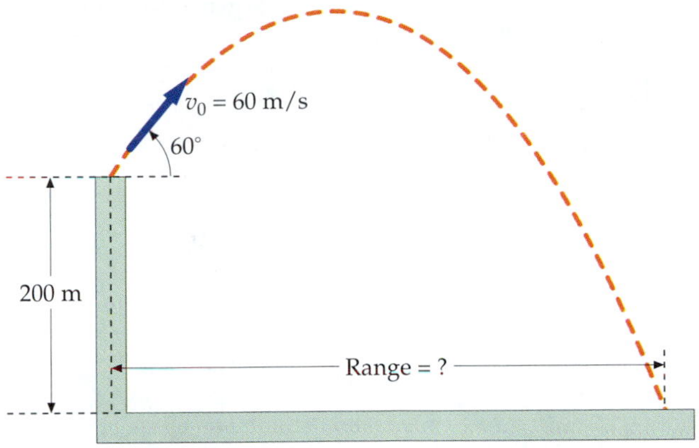

71 •• The range of a projectile fired horizontally from a cliff is equal to the height of the cliff. What is the direction of the velocity vector when the projectile strikes the ground?

72 •• Find the range of the projectile of Problem 60.

73 •• Compute $dR/d\theta$ from Equation 3-22 and show that setting $dR/d\theta = 0$ gives $\theta = 45°$ for the maximum range.

74 •• A rock is thrown from the top of a 20-m building at an angle of 53° above the horizontal. If the horizontal range of the throw is equal to the height of the building, with what speed was the rock thrown? What is the velocity of the rock just before it strikes the ground?

75 •• A stone is thrown horizontally from the top of an incline that makes an angle ϕ with the horizontal. If the stone's initial speed is v, how far down the incline will it land?

76 •• A flock of seagulls has decided to mount an organized response to the human overpopulation of their favorite beach. One tactic popular among the innovative radicals is bombing the sunbathers with clams. A gull dives with a speed of 16 m/s, at an angle of 40° below the horizontal. He releases a projectile when his vertical distance above his target, a sunbather's bronzed tummy, is 8.5 m, and scores a bull's-eye. (*a*) Where is the sunbather in relation to the gull at the instant of release? (*b*) How long is the projectile in the air? (*c*) What is the velocity of the projectile upon impact?

77 ••• A girl throws a ball at a vertical wall 4 m away (Figure 3-38). The ball is 2 m above the ground when it leaves the girl's hand with an initial velocity of $\vec{v}_0 = (10\hat{i} + 10\hat{j})$ m/s. When the ball hits the wall, the horizontal component of its velocity is reversed; the vertical component remains unchanged. Where does the ball hit the ground?

Jumping Gaps; Hitting Targets; Clearing Fences

78 • A boy uses a slingshot to project a pebble at a shoulder-height target 40 m away. He finds that to hit the target he must aim 4.85 m above the target. Determine the velocity of the pebble on leaving the slingshot and the time of flight.

79 •• The distance from the pitcher's mound to home plate is 18.4 m. The mound is 0.2 m above the level of the field. A pitcher throws a fast ball with an initial speed of 37.5 m/s. At the moment the ball leaves the pitcher's hand, it is 2.3 m above the mound. What should the angle between $\vec{v}$ and the horizontal be so that the ball crosses the plate 0.7 m above ground?

80 •• Suppose the puck in Example 3-12 is struck in such a way that it just clears the Plexiglas wall when it is at its highest point. Find v_{0y}, the time t to reach the wall, and v_{0x}, v_0, and θ_0 for this case.

81 •• The coach throws a baseball to a player with an initial speed of 20 m/s at an angle of 45° with the horizontal. At the moment the ball is thrown, the player is 50 m from the coach. At what speed and in what direction must the player run to catch the ball at the same height at which it was released?

82 •• Carlos is on his trail bike, approaching a creek bed that is 7 m wide. A ramp with an incline of 10° has been built for daring people who try to jump the creek. Carlos is traveling at his bike's maximum speed, 40 km/h. (*a*) Should Carlos attempt the jump or emphatically hit the brakes? (*b*) What is the minimum speed a bike must have to make this jump? (Assume equal elevations on either side of the creek.)

83 •• It's the bottom of the ninth with two outs and the winning runs on base. You hit a knee-high fastball that just clears the leaping third baseman's glove. He is standing 28 m from you and his glove reaches to 3.2 m above the ground. The flight time to that point is 0.64 s. Assume that the ball's initial height was 0.6 m. Find (*a*) the initial speed and direction of the ball; (*b*) the time at which the ball reaches its maximum height; (*c*) the maximum height of the ball.

Figure 3-38 Problem 77

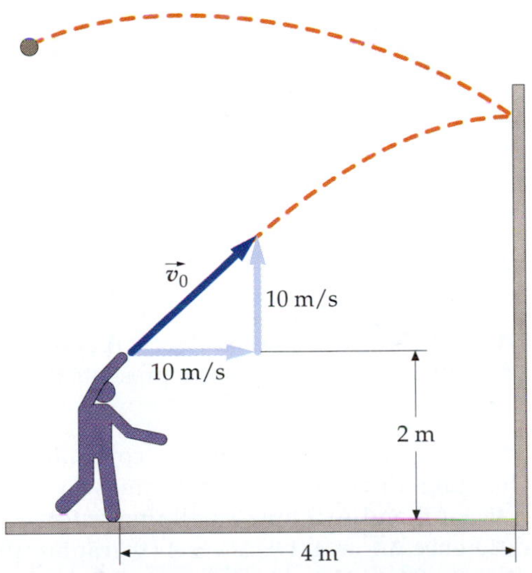

84 •• Noobus is a death-defying squirrel with miraculous jumping abilities. Running to the edge of a flat rooftop, she leaps horizontally with a speed of 6 m/s. If she just clears the 3-m gap between the houses and lands on the neighbor's roof, what is her speed upon landing?

85 ••• If a bullet that leaves the muzzle of a gun at 250 m/s is to hit a target 100 m away at the level of the muzzle, the gun must be aimed at a point above the target. How far above the target is this point?

86 ••• A baseball just clears a 3-m wall that is 120 m from home plate. If the ball leaves the bat at 45° and 1.2 m above the ground, what must its initial speed be?

87 ••• A baseball is struck by a bat, and 3 s later it is caught 30 m away. (*a*) If the baseball was 1 m above the ground when it was struck and caught, what was the greatest height it reached above the ground? (*b*) What were the horizontal and vertical components of its velocity when it was struck? (*c*) What was its speed when it was caught? (*d*) At what angle with the horizontal did it leave the bat?

88 ••• A baseball player hits a baseball that drops into the stands 22 m above the playing field. The ball lands with a velocity of 50 m/s at an angle of 35° below the horizontal. (*a*) If the batter contacted the ball 1.2 m above the playing field, what was the velocity of the ball upon leaving the bat? (*b*) What was the horizontal distance traveled by the ball? (*c*) How long was the ball in the air?

General Problems

89 • True or false:

(*a*) The magnitude of the sum of two vectors must be greater than the magnitude of either vector.
(*b*) If the speed is constant, the acceleration must be zero.
(*c*) If the acceleration is zero, the speed must be constant.

90 • The initial and final velocities of an object are as shown in Figure 3-39. Indicate the direction of the average acceleration.

Figure 3-39 Problem 90

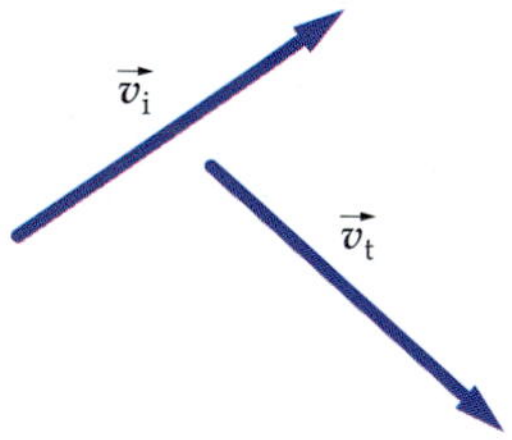

Figure 3-40 Problem 91

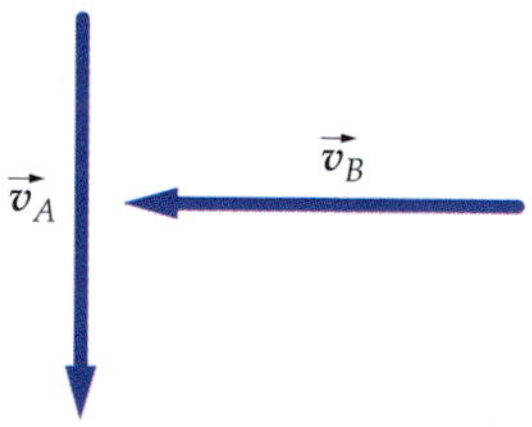

91 • The velocities of objects A and B are shown in Figure 3-40. Draw a vector that represents the velocity of B relative to A.

92 •• A vector $\vec{A}(t)$ has a constant magnitude but is changing direction in a uniform way. Draw the vectors $\vec{A}(t + \Delta t)$ and A(t) for a small time interval Δt, and find the difference $\Delta\vec{A} = \vec{A}(t + \Delta t) - \vec{A}(t)$ graphically. How is the direction of $\Delta\vec{A}$ related to $\vec{A}$ for small time intervals?

93 •• The automobile path shown in Figure 3-41 is made up of straight lines and arcs of circles. The automobile starts from rest at point A. After it reaches point B, it travels at constant speed until it reaches point E. It comes to rest at point F. (*a*) At the middle of each segment (AB, BC, CD, DE, and EF), what is the direction of the velocity vector? (*b*) At which of these points does the automobile have an acceleration? In those cases, what is the direction of the acceleration? (*c*) How do the magnitudes of the acceleration compare for segments BC and DE?

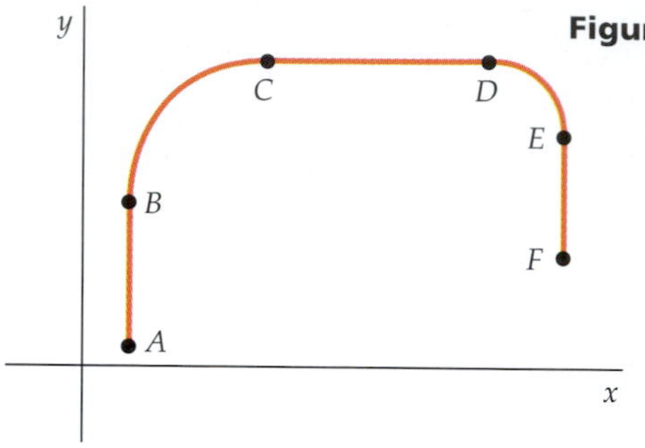

Figure 3-41 Problem 93

94 • The displacement vectors $\vec{A}$ and $\vec{B}$ in Figure 3-42 both have a magnitude of 2 m. (*a*) Find their x and y components. (*b*) Find the components, magnitude, and direction of the sum $\vec{A} + \vec{B}$. (*c*) Find the components, magnitude, and direction of the difference $\vec{A} - \vec{B}$.

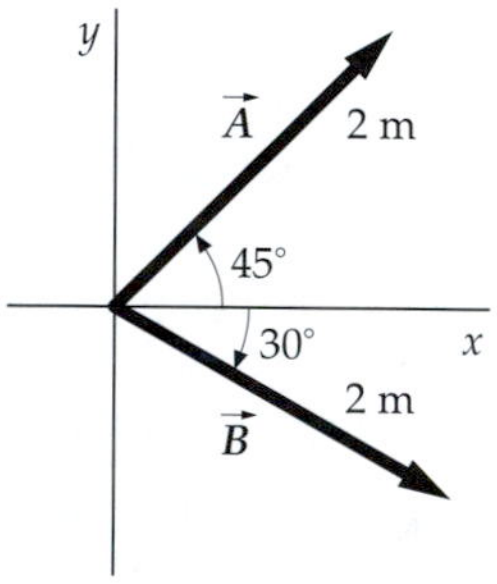

Figure 3-42 Problem 94

95 • A plane is inclined at an angle of 30° from the horizontal. Choose the x axis pointing down the slope of the plane and the y axis perpendicular to the plane. Find the x and y components of the acceleration of gravity, which has the magnitude 9.81 m/s^2 and points vertically down.

96 • Two vectors $\vec{A}$ and $\vec{B}$ lie in the xy plane. Under what conditions does the ratio A/B equal A_x/B_x?

97 • The position vector of a particle is given by $\vec{r} = 5t\hat{i} + 10t\hat{j}$, where t is in seconds and $\vec{r}$ is in meters. (*a*) Draw the path of the particle in the xy plane. (*b*) Find $\vec{v}$ in component form and then find its magnitude.

98 • Off the coast of Chile, a spotter plane sees a school of tuna swimming at a steady 5 km/h northwest (Figure 3-43). The pilot informs a fishing trawler located 100 km due south of the fish. The trawler sails at full steam along the best straight-line course and intercepts the tuna after 4 h. How fast did the trawler move?

Figure 3-43 Problem 98

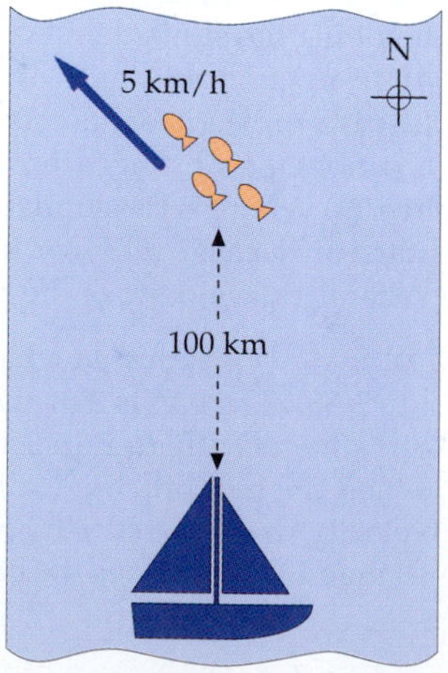

99 •• A worker on the roof of a house drops her hammer, which slides down the roof at a constant speed of 4 m/s. The roof makes an angle of 30° with the horizontal, and its lowest point is 10 m from the ground. What is the horizontal distance traveled by the hammer after it leaves the roof of the house and before it hits the ground?

100 •• A freight train is moving at a constant speed of 10 m/s. A man standing on a flatcar throws a ball into the air and catches it as it falls. Relative to the flatcar, the initial velocity of the ball is 15 m/s straight up. (*a*) What are the magnitude and direction of the initial velocity of the ball as seen by a second man standing next to the track? (*b*) How long is the ball in the air according to the man on the train? According to the man on the ground? (*c*) What horizontal distance has the ball traveled by the time it is caught according to the man on the train? According to the man on the ground? (*d*) What is the minimum speed of the ball during its flight according to the man on the train? According to the man on the ground? (*e*) What is the acceleration of the ball according to the man on the train? According to the man on the ground?

101 •• Estimate how far you can throw a ball if you throw it (*a*) horizontally while standing on level ground; (*b*) at $\theta = 45°$ while standing on level ground; (*c*) horizontally from the top of a building 12 m high; (*d*) at $\theta = 45°$ from the top of a building 12 m high.

102 •• A stunt motorcyclist wants to jump over 10 cars parked side by side below a horizontal launching ramp, as shown in Figure 3-44. With what minimum horizontal speed v_0 must the cyclist leave the ramp in order to clear the top of the last car?

Figure 3-44 Problem 102

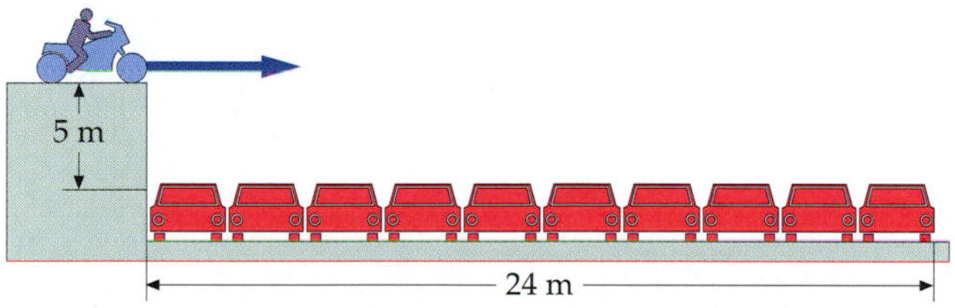

103 •• In 1978, Geoff Capes of Great Britain threw a heavy brick a horizontal distance of 44.5 m. Find the velocity of the brick at the highest point of its flight.

104 •• In 1940, Emanuel Zacchini flew about 53 m as a human cannonball, a record that remains unbroken. His initial velocity was 24.2 m/s at an angle θ. Find θ and the maximum height h Emanuel achieved during the record flight.

105 •• A particle moves in the xy plane with constant acceleration. At $t = 0$ the particle is at $\vec{r}_1 = 4\text{ m}\,\hat{i} + 3\text{ m}\,\hat{j}$, with velocity $\vec{v}_1$. At $t = 2$ s the particle has moved to $\vec{r}_2 = 10\text{ m}\,\hat{i} - 2\text{ m}\,\hat{j}$, and its velocity has changed to $\vec{v}_2 = 5\text{ m/s}\,\hat{i} - 6\text{ m/s}\,\hat{j}$. (*a*) Find $\vec{v}_1$. (*b*) What is the acceleration of the particle? (*c*) What is the velocity of the particle as a function of time? (*d*) What is the position vector of the particle as a function of time?

106 •• A small steel ball is projected horizontally off the top landing of a long rectangular staircase (Figure 3-45). The initial speed of the ball is 3 m/s. Each step is 0.18 m high and 0.3 m wide. Which step does the ball strike first?

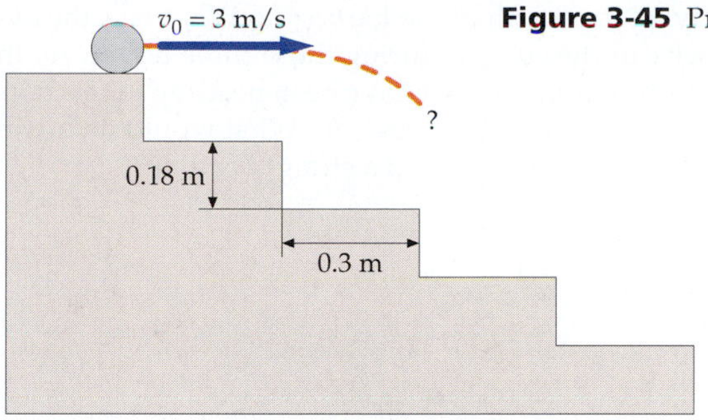

Figure 3-45 Problem 106

107 •• As a car travels down a highway at 25 m/s, a passenger flips out a can at a 45° angle of elevation in a plane perpendicular to the motion of the car. The initial speed of the can relative to the car is 10 m/s. The can is released at a height 1.2 m above the road. (*a*) Write the initial velocity of the can (relative to the road) in terms of the unit vectors $\hat{i}$, $\hat{j}$, and $\hat{k}$. (*b*) Where does the can land?

108 •• Suppose you can throw a ball a distance x_0 when standing on level ground. How far can you throw it from a building of height $h = x_0$ if you throw it at (*a*) 0°? (*b*) 30°? (*c*) 45°?

109 •• A baseball hit toward center field will land 72 m away unless it is caught first. At the moment the ball is hit, the center fielder is 98 m away. He uses 0.5 s to judge the flight of the ball, then races to catch it. The ball's speed as it leaves the bat is 35 m/s. Can the center fielder catch the ball before it hits the ground?

110 ••• Darlene is a stunt motorcyclist in a traveling circus. For the climax of her show, she takes off from the ramp at angle θ, clears a fiery ditch of width x, and lands on an elevated platform (height H) on the other side (Figure 3-46). Darlene notices, however, that night after night, the circus owner keeps raising the height of the platform and the flames to make the jump more spectacular. She is beginning to worry about how far this trend can be taken before she becomes a spectacular casualty, so she decides that it is time for some calculations. (*a*) For a given angle θ and distance x, what is the upper limit H_{max} such that the bike can make the jump? (*b*) For H less than H_{max}, what is the minimum takeoff speed necessary for a successful jump? (Neglect the size of the bike.)

Figure 3-46 Problem 110

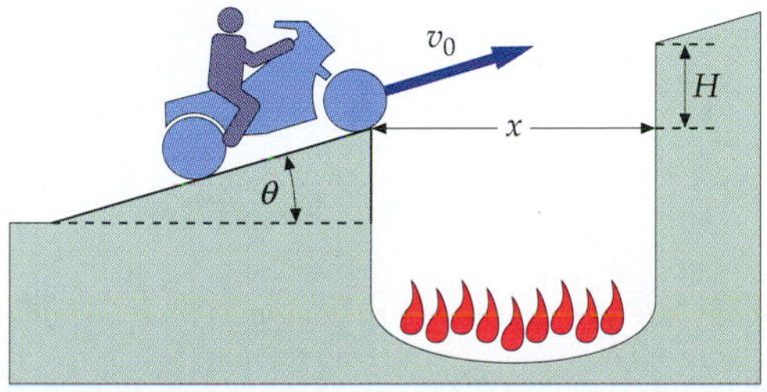

111 ••• A small boat is headed for a harbor 32 km northwest of its current position when it is suddenly engulfed in heavy fog. The captain maintains a compass bearing of northwest and a speed of 10 km/h relative to the water. Three hours later, the fog lifts and the captain notes that he is now exactly 4.0 km south of the harbor. (*a*) What was the average velocity of the current during those three hours? (*b*) In what direction should the boat have been heading to reach its destination along a straight course? (*c*) What would its travel time have been if it had followed a straight course?

112 ••• Galileo showed that, if air resistance is neglected, the ranges for projectiles whose angles of projection exceed or fall short of 45° by the same amount are equal. Prove Galileo's result.

113 ••• Two balls are thrown with equal speeds from the top of a cliff of height H. One ball is thrown upward at an angle α above the horizontal. The other ball is thrown downward at an angle β below the horizontal. Show that each ball strikes the ground with the same speed, and find that speed in terms of H and the initial speed v_0.

CHAPTER 4

Newton's Laws

Isaac Newton

Classical mechanics is a theory of motion based on mass and force. It describes phenomena using Newton's three laws, which relate an object's acceleration to its mass and the forces acting on it. A modern wording of Newton's laws follows.

First law. An object at rest stays at rest unless acted on by an external force. An object in motion continues to travel with constant velocity unless acted on by an external force.

Newton's first law

Second law. The acceleration of an object is in the direction of the net external force acting on it. It is proportional to the net external force, and is inversely proportional to the mass of the object:

$$\vec{a} = \frac{\vec{F}_{\text{net}}}{m}$$

or

$$\vec{F}_{\text{net}} = m\vec{a}$$

The net force acting on an object, also called the resultant force, is the vector sum of all the forces acting on it: $\vec{F}_{\text{net}} = \Sigma\vec{F}$. Thus,

$$\sum \vec{F} = \vec{F}_{\text{net}} = m\vec{a} \qquad 4\text{-}1$$

Newton's second law

Third law. Forces always occur in equal and opposite pairs. If object A exerts a force on object B, an equal but opposite force is exerted by object B on object A.

Newton's third law

4-1 Newton's First Law: The Law of Inertia

Friction is greatly reduced by a cushion of air that supports the hovercraft.

Push a piece of ice on a counter top: It slides, then stops. If the counter is wet, the ice will travel farther before stopping. A piece of dry ice riding on a cushion of carbon dioxide slides quite far with little change in velocity. Before Galileo, it was thought that a force, such as a push or pull, was always needed to keep an object moving with constant velocity. Galileo, and later Newton, recognized that the slowing of objects in everyday experience is due to friction. If friction is reduced, the change in velocity is reduced. A water slick or carbon dioxide cushion is especially effective at reducing friction, allowing the object to slide a great distance with little change in velocity. Remove all external forces on an object, Galileo reasoned, and its velocity will never change—a property of matter he described as its **inertia.** This conclusion, restated by Newton as his first law, is also called the **law of inertia.**

Inertial Reference Frames

Newton's first law makes no distinction between an object at rest and an object moving with constant velocity. Whether an object is at rest or is moving with constant velocity depends on the reference frame in which the object is observed. A **reference frame** is a set of coordinate systems at rest relative to each other. Consider a ball sitting in the aisle of an airplane cruising along a horizontal path. In a coordinate system attached to the plane (that is, in the reference frame of the plane) the ball is at rest. It will remain at rest relative to the plane as long as the plane flies with constant velocity. In a coordinate system attached to the earth, the ball is moving with the velocity of the plane. According to Newton's first law, the ball will continue to move with constant velocity in the reference frame of the earth, and will remain at rest in the reference frame of the plane unless it is acted on by a net force.

A reference frame in which the law of inertia holds exactly is called an **inertial reference frame.** Both the cruising plane and the ground are, to a good approximation, inertial reference frames. Any reference frame moving with constant velocity relative to an inertial reference frame is also an inertial reference frame.

Now suppose that the plane accelerates forward relative to the ground. The ball will roll backward, accelerating relative to the plane even though there is no net force acting on it. The ball accelerates *in the plane's frame of ref-*

erence despite there being no net external force acting on it. Also, the back of your seat will exert a horizontal forward force on you, but you do not accelerate *relative to the plane*. The law of inertia does not hold in the reference frame of the accelerating plane. A reference frame accelerating relative to an inertial reference frame is not an inertial reference frame. *Newton's first law thus gives us the criterion for determining if a reference frame is an inertial frame.*

A reference frame attached to the surface of the earth is not quite an inertial reference frame because of the small acceleration of the surface of the earth (relative to the center of the earth) due to the rotation of the earth, and the small acceleration of the earth itself due to its revolution around the sun. However, these accelerations are of the order of 0.01 m/s^2 or less, so to a good approximation, a reference frame attached to the surface of the earth is an inertial reference frame.

4-2 Force, Mass, and Newton's Second Law

Newton's first and second laws allow us to define force. A **force** is an external influence on an object that causes it to accelerate relative to an inertial reference frame. (We assume there are no other forces acting.) The direction of the force is the direction of the acceleration it causes if it is the only force acting on the object. The magnitude of the force is the product of the mass of the object and the magnitude of its acceleration. This definition of force is in accord with our intuitive idea of a force as a push or pull like that exerted by our muscles.

Mass is an intrinsic property of an object that measures its resistance to acceleration. That is, it is a measure of the object's inertia. The ratio of two masses is defined quantitatively by applying the same force to each and comparing their accelerations. If a force F produces acceleration a_1 when applied to an object of mass m_1, and the same force produces acceleration a_2 when applied to an object of mass m_2, then the ratio of the masses is defined by

$$\frac{m_2}{m_1} = \frac{a_1}{a_2} \qquad 4\text{-}2$$

Definition—Mass

This definition agrees with our intuitive idea of mass. If the same force is applied to two objects, the object with more mass will accelerate less. The ratio a_1/a_2 produced by an identical force acting on two objects is found experimentally to be independent of the magnitude, direction, or type of force used. Mass is an intrinsic property of an object that does not depend on its location—it remains the same whether the object is on the earth, on the moon, or in outer space.

If a direct comparison shows that $m_2/m_1 = 2$ and $m_3/m_1 = 4$, then m_3 will be twice m_2 when the two objects are compared with each other. We can therefore establish a mass scale by choosing a standard object and assigning it a mass of 1 unit. As we noted in Chapter 1, the object chosen as the international standard for mass is a platinum–iridium alloy cylinder carefully preserved at the International Bureau of Weights and Measures at Sèvres, France. The mass of the standard object is 1 **kilogram,** the SI unit of mass.* The force required to produce an acceleration of 1 m/s^2 on the standard object is defined to be 1 **newton** (N). The force that produces an acceleration of 2 m/s^2 on the standard object is 2 N, and so on.

* The standard kilogram was originally intended to be equal to the mass of 1000 cm^3 = 1 liter of water.

Example 4-1

A given force produces an acceleration of 5 m/s² on the standard object of mass m_1. When the same force is applied to a carton of ice cream of mass m_2, it produces an acceleration of 11 m/s². (*a*) What is the mass of the carton of ice cream? (*b*) What is the magnitude of the force?

(*a*)1. The ratio of the masses varies inversely as the ratio of the accelerations under the same applied force:
$$\frac{m_2}{m_1} = \frac{a_1}{a_2} = \frac{5\ \text{m/s}^2}{11\ \text{m/s}^2}$$

2. Solve for m_2 in terms of m_1, which is 1 kg:
$$m_2 = \frac{5}{11}m_1 = \frac{5}{11}(1\ \text{kg}) = 0.45\ \text{kg}$$

(*b*) The magnitude of the force F is found by using the mass and acceleration of either object:
$$F = m_1 a_1 = (1\ \text{kg})(5\ \text{m/s}^2) = 5\ \text{N}$$

Exercise A force of 3 N produces an acceleration of 2 m/s² on an object of unknown mass. (*a*) What is the mass of the object? (*b*) If the force is increased to 4 N, what is the acceleration? (*Answers* (*a*) 1.5 kg, (*b*) 2.67 m/s²)

It is found experimentally that two or more forces acting on an object accelerate it as if the object were acted on by a single force equal to the vector sum of the individual forces. That is, forces combine as vectors. Newton's second law is thus

$$\sum \vec{F} = \vec{F}_{\text{net}} = m\vec{a}$$

Example 4-2

You're stranded in space away from your spaceship. Fortunately, you have a propulsion unit that provides a constant force $\vec{F}$ for 3 s. After 3 s you have moved 2.25 m. If your mass is 68 kg, find $\vec{F}$.

Picture the Problem The force acting on you is constant, so your acceleration $\vec{a}$ is also constant. Hence, we use the kinematic equations of Chapter 2 to find $\vec{a}$, and then obtain the force from $\Sigma\vec{F} = m\vec{a}$. Choose $\vec{F}$ to be along the x axis, so that $\vec{F} = F_x\hat{i}$ (Figure 4-1). The component of Newton's second law along the x axis is then $F_x = ma_x$.

Figure 4-1

1. Apply $\Sigma\vec{F} = m\vec{a}$ to relate the net force to the mass and the acceleration:
$$F_x = ma_x$$

2. To find the acceleration, we use Equation 2-14 with $v_0 = 0$:
$$\Delta x = x - x_0 = v_0 t + \frac{1}{2}a_x t^2 = \frac{1}{2}a_x t^2$$
$$a_x = \frac{2\Delta x}{t^2} = \frac{2(2.25\ \text{m})}{(3\ \text{s})^2} = 0.500\ \text{m/s}^2$$

3. Substitute $a_x = 0.500\ \text{m/s}^2$ and $m = 68$ kg to find the force:
$$F_x = ma_x = (68\ \text{kg})(0.500\ \text{m/s}^2) = 34.0\ \text{N}$$

Example 4-3 *try it yourself*

A particle of mass 0.4 kg is subjected simultaneously to two forces $\vec{F}_1 = 2\,\text{N}\,\hat{i} - 4\,\text{N}\,\hat{j}$ and $\vec{F}_2 = -2.6\,\text{N}\,\hat{i} + 5\,\text{N}\,\hat{j}$. If the particle is at the origin and starts from rest at $t = 0$, find (*a*) its position vector $\vec{r}$ and (*b*) its velocity $\vec{v}$ at $t = 1.6$ s.

Picture the Problem Since $\vec{F}_1$ and $\vec{F}_2$ are constant, the acceleration of the particle is constant. Hence, you can use the kinematic equations of Chapter 2 to determine the particle's position and velocity as functions of time.

Cover the column to the right and try these on your own before looking at the answers.

Steps	*Answers*
(*a*)1. Write the general equation for the position vector $\vec{r}$ as a function of time t for constant acceleration $\vec{a}$ in terms of $\vec{r}_0$, $\vec{v}$, and $\vec{a}$, and substitute $\vec{r}_0 = 0, \vec{v}_0 = 0$.	$\vec{r} = \vec{r}_0 + \vec{v}_0 t + \frac{1}{2}\vec{a}t^2 = \frac{1}{2}\vec{a}t^2$
2. Use $\Sigma\vec{F} = m\vec{a}$ to write the acceleration $\vec{a}$ in terms of the resultant force $\Sigma\vec{F}$ and the mass m.	$\vec{a} = \dfrac{\Sigma\vec{F}}{m}$
3. Compute $\Sigma\vec{F}$ from the given forces.	$\Sigma\vec{F} = \vec{F}_1 + \vec{F}_2 = -0.6\,\text{N}\,\hat{i} + 1.0\,\text{N}\,\hat{j}$
4. Find the acceleration vector $\vec{a}$.	$\vec{a} = \dfrac{\Sigma\vec{F}}{m} = -1.5\,\text{m/s}^2\,\hat{i} + 2.5\,\text{m/s}^2\,\hat{j}$
5. Find the position vector $\vec{r}$ for a general time t.	$\vec{r} = \frac{1}{2}\vec{a}t^2 = \frac{1}{2}a_x t^2\hat{i} + \frac{1}{2}a_y t^2\hat{j}$ $= -0.75\,\text{m/s}^2\,t^2\hat{i} + 1.25\,\text{m/s}^2\,t^2\hat{j}$
6. Find $\vec{r}$ at $t = 1.6$ s.	$\vec{r} = -1.92\,\text{m}\,\hat{i} + 3.20\,\text{m}\,\hat{j}$
(*b*) Write the velocity vector $\vec{v}$ in terms of the acceleration and time and compute its components for the time $t = 1.6$ s.	$\vec{v} = \vec{a}t = (-1.5\,\text{m/s}^2\,\hat{i} + 2.5\,\text{m/s}^2\,\hat{j})t$ $= -2.4\,\text{m/s}\,\hat{i} + 4.00\,\text{m/s}\,\hat{j}$

4-3 The Force Due to Gravity: Weight

If we drop an object near the earth's surface, it accelerates toward the earth. If we neglect air resistance, all objects have the same acceleration, called the acceleration due to gravity $\vec{g}$ at any given point in space. The force causing this acceleration is the force of gravity on the object, called its weight $\vec{w}$. If $\vec{w}$ is the only force acting on an object, the object is said to be in **free fall.** If its mass is m, Newton's second law defines the weight $\vec{w}$:

$$\vec{w} = m\vec{g} \qquad \text{4-3}$$

Weight

Since $\vec{g}$ is the same for all objects at a given point, the weight of an object must be proportional to its mass. The vector $\vec{g}$ is the force per unit mass exerted by the earth on any object and is called the **gravitational field** of the earth. It is equal to the free-fall acceleration experienced by an object. Near the surface of the earth, g has the value

$$g = 9.81\ \text{N/kg} = 9.81\ \text{m/s}^2$$

Careful measurements show that $\vec{g}$ varies with location. In particular, at points above the surface of the earth, $\vec{g}$ points toward the center of the earth and varies inversely with the square of the distance to the center of the earth. Thus, an object weighs slightly less at very high altitudes than it does at sea level. The gravitational field also varies slightly with latitude because the earth is not exactly spherical but is slightly flattened at the poles. Thus, weight, unlike mass, is *not* an intrinsic property of an object. Although the weight of an object varies from place to place because of changes in g, this variation is too small to be noticed in most practical applications on or near the surface of the earth.

An example should help clarify the difference between mass and weight. Consider a bowling ball near the moon. Its weight is the force exerted on it by the moon, but that force is a mere one-sixth of the force exerted on the bowling ball when it is similarly positioned on earth. The ball weighs about one-sixth as much on the moon, and lifting the ball on the moon requires one-sixth the force. However, because the mass of the ball is the same on the moon as on the earth, throwing the ball with some horizontal acceleration requires the same force on the moon as on the earth or in free space.

Though an object's weight may vary from one place to another, at any particular location its weight is proportional to its mass. Thus, we can conveniently compare the masses of two objects at a given location by comparing their weights.

Our awareness of our own weight comes from other forces that balance it. When you sit on a chair, you feel a force exerted by the chair that balances your weight and prevents you from falling to the floor. When you stand on a spring scale, your feet feel the force exerted by the scale. The scale is calibrated to read the force it must exert (by the compression of its springs) to balance your weight. This force is called your **apparent weight.** If there is no force to balance your weight, as in free fall, your apparent weight is zero. This condition, called **weightlessness,** is experienced by astronauts in orbiting satellites. As will be discussed in Chapter 5, when an object travels in a circle, the direction of its velocity vector is constantly changing, and the object is therefore accelerating. A satellite in a circular orbit near the surface of the earth is accelerating toward the earth. The only force acting on the satellite is gravity (its weight), so it is in free fall with the acceleration due to gravity. Astronauts in the satellite are also in free fall. The only force on them is their weight, which produces the acceleration g. Since there is no force balancing the force of gravity, the astronauts have zero apparent weight.

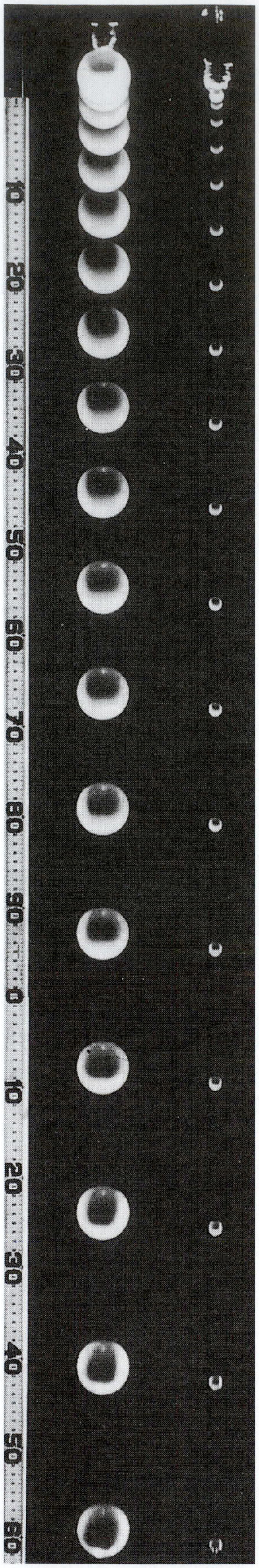

When air resistance can be neglected, objects of different mass fall with the same acceleration due to gravity.

Units of Force and Mass

Like the second and the meter, the SI unit of mass, the kilogram, is a fundamental unit. The unit of force, the newton, and the units for other quantities that we will study such as momentum and energy, are derived from the three fundamental units second, meter, and kilogram.

As noted above, the newton is defined as the force that produces an acceleration of 1 m/s^2 when it acts on 1 kg. Then Newton's second law gives

$$1\text{ N} = (1\text{ kg})(1\text{ m/s}^2) = 1\text{ kg}\cdot\text{m/s}^2 \qquad 4\text{-}4$$

A convenient standard unit for mass in atomic and nuclear physics is the unified mass unit (u), which is defined as one-twelfth the mass of the neutral carbon-12 (^{12}C) atom. The unified mass unit is related to the kilogram by

$$1\text{ u} = 1.660\,540 \times 10^{-27}\text{ kg} \qquad 4\text{-}5$$

The mass of a hydrogen atom is approximately 1 u.

Although we generally use SI units in this book, we need to know another system, the U.S. customary system, still used in the United States, which is based on the foot, the second, and the pound. The U.S. customary system differs from SI in that it uses a unit of force, the pound, as a fundamental unit rather than using a unit of mass. The **pound** was originally defined as the weight of a particular standard object at a particular location. It is now defined as 4.448222 N. Rounding to three places, we have 1 lb $\approx$ 4.45 N. Since 1 kg weighs 9.81 N, its weight in pounds is

$$9.81 \text{ N} \times \frac{1 \text{ lb}}{4.45 \text{ N}} = 2.20 \text{ lb} \tag{4-6}$$

Weight of 1 kg

The unit of mass in the U.S. customary system is the rarely encountered slug, defined as the mass of an object that weighs 32.2 lb. When working problems in the U.S. customary system, we substitute w/g for mass m, where w is the weight in pounds and g is the acceleration due to gravity in feet per second per second:

$$g = 32.2 \text{ ft/s}^2 \tag{4-7}$$

Example 4-4

The net force acting on a 130-lb student is 25 lb. What is her acceleration?

According to Newton's second law, her acceleration is the force divided by her mass:

$$a = \frac{F}{m} = \frac{F}{w/g} = \frac{25 \text{ lb}}{(130 \text{ lb})/(32.2 \text{ ft/s}^2)}$$
$$= 6.19 \text{ ft/s}^2$$

Exercise What force is needed to give an acceleration of 3 ft/s^2 to a 5-lb block? (*Answer* 0.466 lb)

4-4 Newton's Third Law

The word *force* is used to describe the interaction between two objects. When two objects interact, they exert forces on each other. Newton's third law states that these forces are equal in magnitude and opposite in direction. If object A exerts a force on object B, object B exerts a force on A that is equal in magnitude and opposite in direction. Thus, forces always occur in pairs. It is common to refer to one force in the pair as an action and the other as a reaction. This terminology is unfortunate because it sounds like one force "reacts" to the other, which is not true. Both forces occur simultaneously. Either can be called the action and the other the reaction. Action and reaction forces can never balance *each other* because they act on *different objects*. In Figure 4-2, a block rests on a table. The force acting downward on the block is the weight $\vec{w}$ due to the attraction of the earth. An equal and opposite force $\vec{w}' = -\vec{w}$ is exerted by the block on the earth. These forces are an action–reaction pair. If they were the only forces present, the block would accelerate downward,

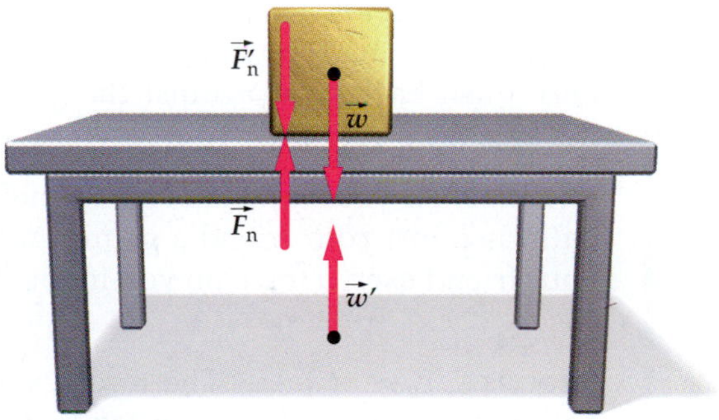

Figure 4-2

because it would have only a single force acting on it (and the earth would accelerate upward toward the block). However, the table exerts an upward force $\vec{F}_n$ on the block that balances the block's weight. The block also exerts a force $\vec{F}'_n = -\vec{F}_n$ downward on the table. The forces $\vec{F}_n$ and $\vec{F}'_n$ are also an action–reaction pair.

Example 4-5

A horse refuses to pull a cart (Figure 4-3). The horse reasons, "According to Newton's third law, whatever force I exert on the cart, the cart will exert an equal and opposite force on me, so the net force will be zero and I will have no chance of accelerating the cart." What is wrong with this reasoning?

Picture the Problem Since we are interested in the motion of the cart, we have enclosed it with a dashed line and have indicated the forces acting on it. The force exerted by the horse is labeled $\vec{T}$. It is exerted by the horse on the harness. (Since the harness is attached to the cart, we are considering it part of the cart.) Other forces acting on the cart are its weight $\vec{w}$, the vertical support force of the ground $\vec{F}_n$, and the horizontal force exerted by the ground, labeled $\vec{f}$ (for friction).

In the idealized diagram, the cart is drawn as a particle with the forces acting on it. The vertical forces $\vec{w}$ and $\vec{F}_n$ sum to zero. (We know this because we know the cart does not accelerate vertically.) The horizontal forces are $\vec{T}$ to the right and $\vec{f}$ to the left. The cart will accelerate if $\vec{T}$ is greater than $\vec{f}$.

Note that the reaction force to $\vec{T}$, which we call $\vec{T}'$, is exerted on the horse, not on the cart. It has no effect on the motion of the cart, but it does affect the motion of the horse. If the horse is to accelerate to the right, there must be a force $\vec{F}$ (to the right) exerted by the ground on the horse's feet that is greater than $\vec{T}'$.

Remark This example illustrates the importance of drawing a simple diagram when solving mechanics problems. Had the horse sketched a simple diagram, he would have seen that he need only push back hard against the ground so that the ground would push him forward.

Exercise As you stand facing a friend, place your palms against your friend's palms and push. Can your friend exert a force on you if you do not exert a force back? Try it.

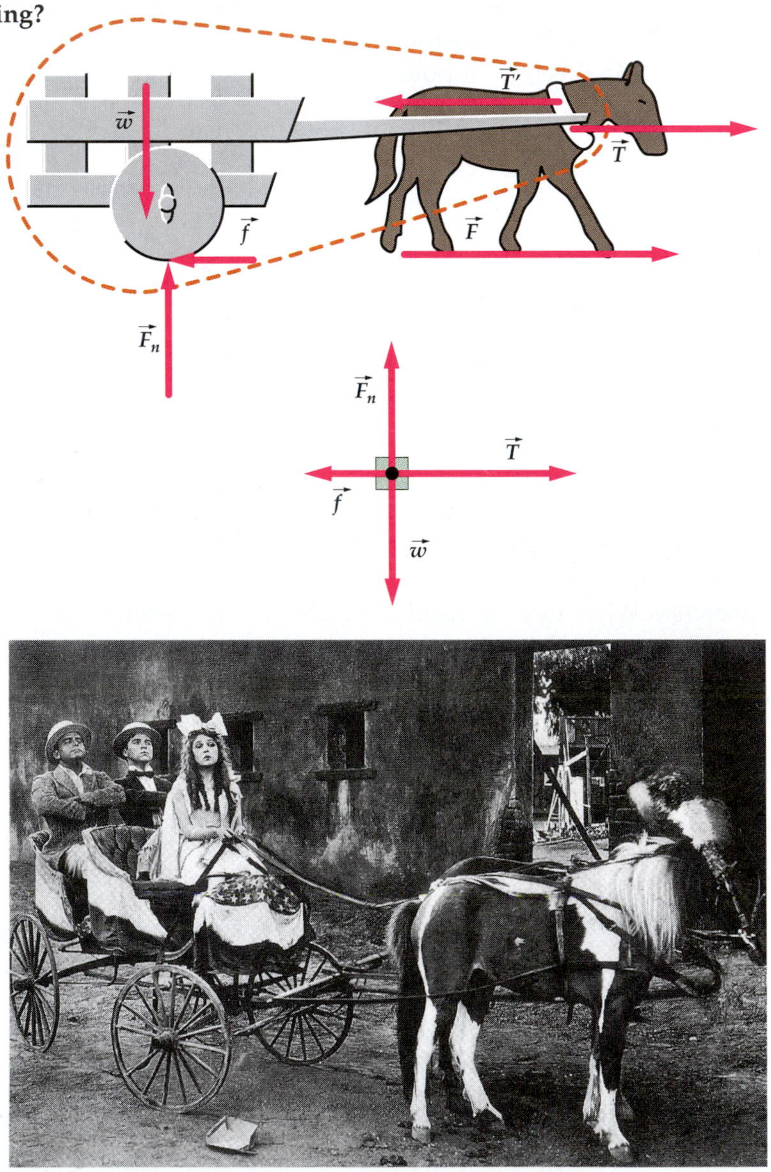

Figure 4-3

Exercise True or false: The force exerted by the cart on the horse is equal and opposite to the force exerted by the horse on the cart only when the cart is not accelerating. (*Answer* False! An action–reaction pair of forces describes the interaction of two objects. One force cannot exist without the other. They are always equal and opposite.)

4-5 Forces in Nature

The full power of Newton's second law emerges when it is combined with the force laws that describe the interactions of objects. For example, Newton's law for gravitation, which we study in Chapter 11, gives the gravitational force exerted by one object on another in terms of the distance between the objects and the masses of each. This, combined with Newton's second law, enables us to calculate the orbits of planets around the sun, the motion of the moon, and variations with altitude of g, the acceleration due to gravity.

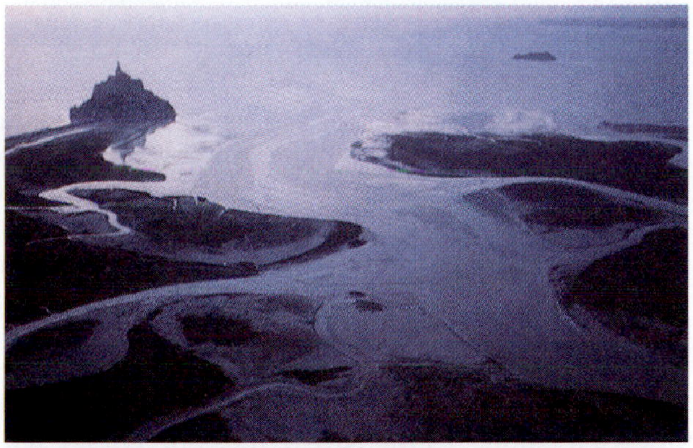

(*a*)

The Fundamental Forces

All the different forces observed in nature can be explained in terms of four basic interactions that occur between elementary particles:

1. The gravitational force
2. The electromagnetic force
3. The strong nuclear force (also called the hadronic force)
4. The weak nuclear force

The everyday forces that we observe between macroscopic objects are due to either the gravitational force or the electromagnetic force.

(*b*)

(*c*)

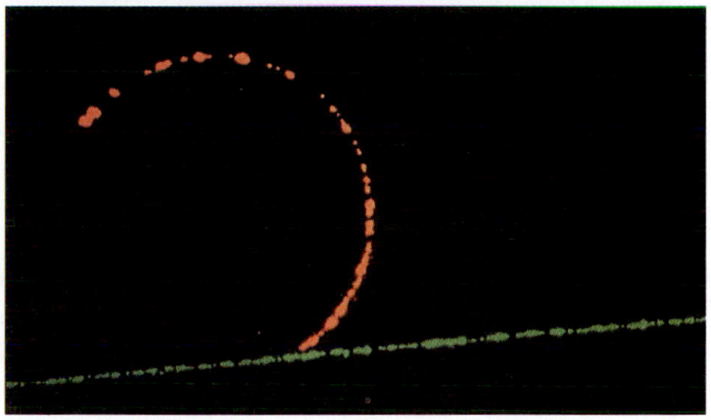

(*d*)

(*a*) The gravitational force between the earth and an object near the earth's surface is the weight of the object. The gravitational force exerted by the sun keeps the planets in their orbits. Similarly, the gravitational force exerted by the earth keeps the moon in its nearly circular orbit around the earth. The gravitational forces exerted by the moon and sun on the oceans of the earth cause the tides. Mont-Saint-Michel, France, shown in the photo, is an island when the tide is in. (*b*) The electromagnetic force includes both the electric and the magnetic forces. A familiar example of the electric force is the attraction between bits of paper and a comb that is electrified after being run through hair. The magnetic force between a magnet and iron arises when electric charges are in motion. The electromagnetic force between charged elementary particles is vastly greater than the gravitational force between them. For example, the electrostatic force of repulsion between two protons is of the order of 10^{36} times the gravitational attraction between them. The lightning bolts shown in the photo are the result of the electromagnetic force. (*c*) The strong nuclear force occurs between elementary particles called hadrons, which include protons and neutrons. The strong force results from the interaction of quarks, the building blocks of hadrons, and is responsible for holding nuclei together. The magnitude of the strong force decreases rapidly with distance and is negligible beyond a few nuclear diameters. The hydrogen bomb explosion shown in the photo illustrates the strong nuclear force. (*d*) The weak nuclear force, which also has a short range, occurs between leptons (which include electrons and muons) and between hadrons (which include protons and neutrons). This false-color cloud chamber photograph illustrates the weak interaction between a cosmic ray muon (green) and an electron (red) knocked out of an atom.

Action at a Distance

The fundamental forces of gravity and electromagnetism act between particles that are separated in space. This creates a philosophical problem referred to as **action at a distance.** Newton perceived action at a distance as a flaw in his theory of gravitation but avoided giving any other hypothesis. Today the problem is avoided by introducing the concept of a field, which acts as an intermediary agent. For example, we consider the attraction of the earth by the sun in two steps. The sun creates a condition in space that we call the gravitational field. This field then exerts a force on the earth. Similarly, the earth produces a gravitational field that exerts a force on the sun. Your weight is the force exerted by the gravitational field of the earth on you. When we study electricity and magnetism (Chapters 22–32) we will study electric fields, which are produced by electrical charges, and magnetic fields, which are produced by electrical charges in motion.

Contact Forces

Many forces we encounter are exerted by objects in direct contact. These forces are electromagnetic in origin and are exerted between the molecules of each object.

Solids Consider a book on a table. The weight of the book pulls it downward, pressing it against the molecules in the table's surface, which resist compression and exert a force upward on the book. Such a force, *perpendicular* to the surface, is called a **normal force** (one meaning of the word *normal* is "perpendicular"). A supporting surface bends slightly in response to a load, though this is rarely noticeable to the naked eye.

Normal forces can vary over a wide range of magnitude. A table, for instance, will exert an upward force on any object resting on it. As long as the table doesn't break, the normal force will balance the weight of the object. Furthermore, if you press down on the object, the table will exert a support force that counters the extra force, preventing the object from accelerating downward.

Objects in contact can also exert forces on each other that are *parallel* to the surfaces in contact. The parallel component of a contact force is called a **frictional force.** We will consider frictional forces in the next chapter.

Springs When a spring is compressed or extended by a small amount Δx, the force it exerts is found experimentally to be

$$F_x = -k\,\Delta x \qquad \text{4-8}$$

Hooke's law

where k is the force constant, a measure of the stiffness of the spring (Figure 4-4). The negative sign in Equation 4-8 signifies that when the spring is stretched or compressed, the force it exerts back is in the opposite direction. This relation, known as Hooke's law, turns out to be quite important. An object at rest under the influence of forces that balance is said to be in static equilibrium. If a small displacement results in a net restoring force toward the equilibrium position, the equilibrium is called stable equilibrium. For small displacements, nearly all restoring forces obey Hooke's law.

The molecular force of attraction between atoms in a molecule or solid varies approximately linearly with the change in separation (for small changes); the force varies much like that of a spring. We can therefore use

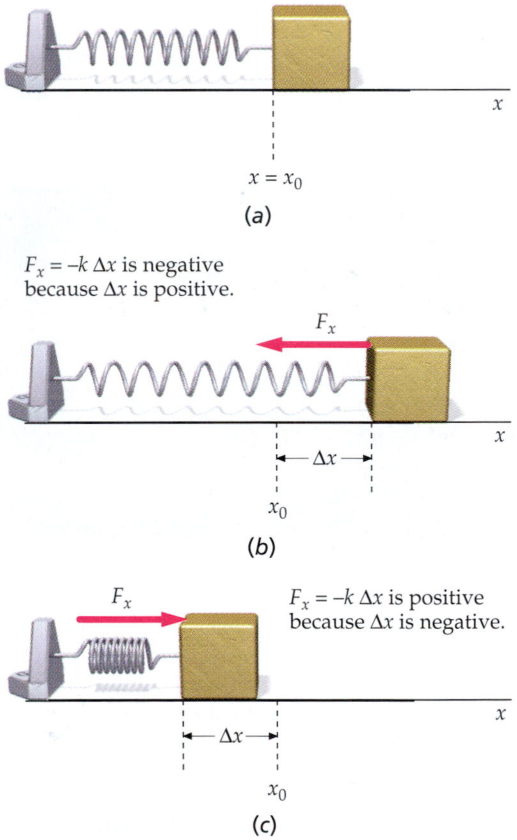

Figure 4-4 A horizontal spring. (*a*) When the spring is unstretched, it exerts no force on the block. (*b*) When the spring is stretched so that Δx is positive, it exerts a force of magnitude $k\,\Delta x$ in the negative x direction. (*c*) When the spring is compressed so that Δx is negative, the spring exerts a force of magnitude $k\,\Delta x$ in the positive direction.

two masses on a spring to model a diatomic molecule, or a set of masses connected by springs to model a solid, as shown in Figure 4-5.

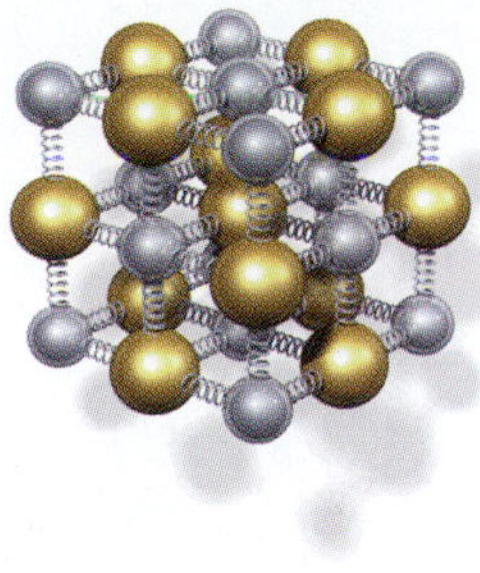

(*a*)

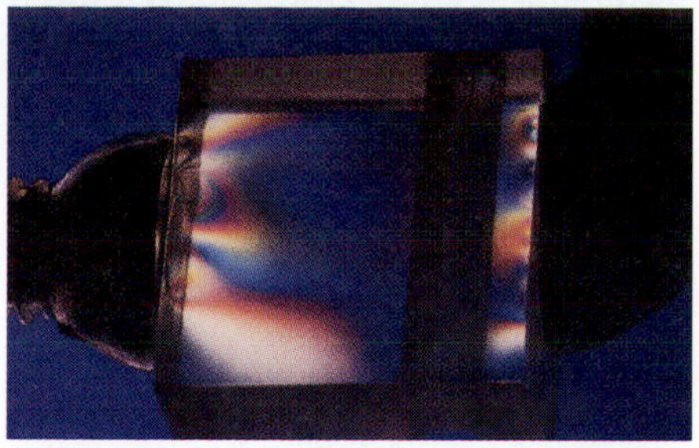

(*b*)

Figure 4-5 (*a*) Model of a solid consisting of atoms connected to each other by springs. The springs are very stiff (large force constant) so that when a weight is placed on the solid its deformation is not visible. However, compression such as that produced by the C clamp on the plastic block in (*b*) leads to stress patterns that are visible when viewed with polarized light.

Example 4-6

A 110-kg basketball player hangs on the rim following a slam dunk. When he comes to rest, the rim is bent down a distance of 15 cm. Assume that the rim can be approximated by a spring and calculate the force constant k.

Picture the Problem Since the acceleration of the player is zero, the net force exerted on him must be zero. The upward force exerted by the rim balances his weight (Figure 4-6). Let $y = 0$ be the original position of the rim and choose down to be positive. Then Δy is positive, the weight mg is positive, and the force exerted by the rim, $-k\,\Delta y$ is negative.

Figure 4-6

1. Apply $\Sigma \vec{F} = m\vec{a}$ to the player: $\sum F_y = mg + (-k\,\Delta y) = ma_y = 0$

2. Solve for k: $k = \dfrac{mg}{\Delta y} = \dfrac{(110\text{ kg})(9.81\text{ N/kg})}{0.15\text{ m}}$
$= 7.19 \times 10^3\text{ N/m}$

Remark Although a basketball rim doesn't look much like a spring, when the displacement is small, the force it exerts is proportional to the displacement and oppositely directed. Note that we used N/kg for the units of g so that kg cancels, giving N/m for the units of k. We can use either 9.81 N/kg or 9.81 m/s^2 for g, whichever is more convenient, because 1 N/kg = 1 m/s^2.

Exercise A 4-kg bunch of bananas is suspended motionless from a spring balance whose force constant is $k = 300$ N/m. By how much is the spring stretched? (*Answer* 13.1 cm)

Exercise A spring of force constant 400 N/m is attached to a 3-kg block that rests on a horizontal air track that renders friction negligible. What extension of the spring is needed to give the block an acceleration of 4 m/s^2 upon release? (*Answer* 3.0 cm)

Exercise in Dimensional Analysis An object of mass m oscillates at the end of an ideal spring of force constant k. The time for one complete oscillation is the period T. Assuming that T depends on m and k, use dimensional analysis to find the form of the relationship $T = f(m, k)$, ignoring numerical constants. This is most easily found by looking at the units. Note that the units of k are N/m = $(\text{kg}\cdot\text{m/s}^2)/\text{m} = \text{kg/s}^2$, and the units of m are kg. (*Answer* $T = C\sqrt{m/k}$, where C is some dimensionless constant. The correct expression for the period, as we will see in Chapter 14, is $T = 2\pi\sqrt{m/k}$.)

Strings If we pull on a string, the string stretches slightly and pulls back with an equal but opposite force (unless the string breaks). We can think of a string as a spring with such a large force constant that the extension of the string is negligible. The string is flexible, however, so we cannot exert a force of compression on it. When we push on a string, it merely flexes or bends.

4-6 Problem Solving: Free-Body Diagrams

Imagine a dogsled being pulled across icy ground. The dog in front pulls on a light rope attached to the sled with a force $\vec{F}$ (Figure 4-7*a*). The taut rope then pulls the sled forward. What forces act on the sled? Both the rope and ice touch the sled, so we know the rope and ice exert contact forces on it. We also know that the earth exerts a gravitational force on the sled (the sled's weight). Thus, three forces act on the sled (assuming that friction is negligible):

1. The weight of the sled, $\vec{w}$
2. The contact force $\vec{F}_n$ exerted by the ice (without friction, the contact force is perpendicular to the ice)
3. The contact force $\vec{T}$ exerted by the rope

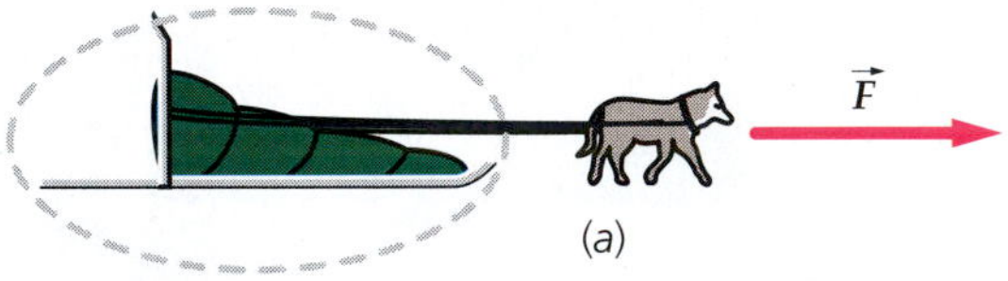

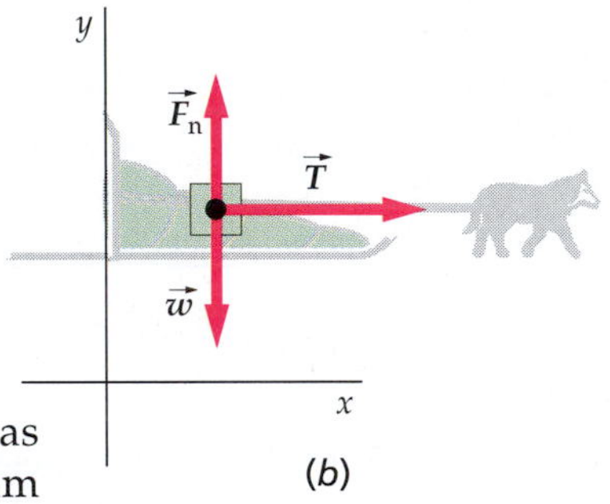

Figure 4-7 (*a*) A dog pulling a sled. The first step in problem solving is to isolate the system to be analyzed. In this case, an oval isolates the sled from its surroundings. (*b*) The forces acting on the sled of (*a*).

A diagram that shows schematically all the forces acting on a system, such as Figure 4-7*b*, is called a **free-body diagram.** Note that $\vec{F}_n$ and $\vec{w}$ in the diagram have equal magnitudes. The magnitudes must be equal because the sled doesn't accelerate vertically. Conditions on the motion of an object, such as the requirement that the sled remain on the ice, are called **constraints.**

The y component of Newton's second law gives

$$\sum F_y = F_{n,y} + w_y + T_y = ma_y$$

$$F_n - w + 0 = 0$$

or

$$F_n = w$$

The x component of Newton's second law gives

$$\sum F_x = F_{n,x} + w_x + T_x = ma_x$$

or

$$a_x = \frac{T}{m}$$

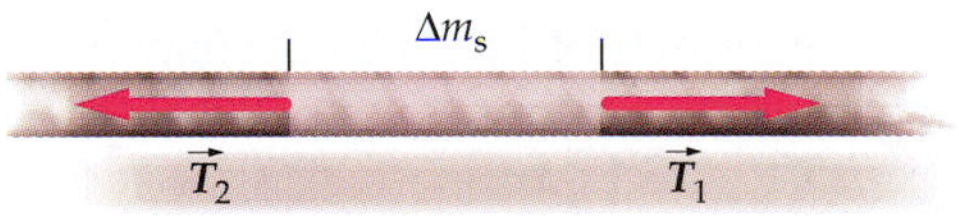

Figure 4-8 Free-body diagram for a segment Δm_s of the rope. Newton's second law applied to this segment gives $T_1 - T_2 = (\Delta m_s)a_x$. If the mass of the segment is negligible, $T_1 = T_2$. The tension T is the magnitude of the force each segment exerts on each adjacent segment. These forces act along the rope, so a light rope connecting two points has a tension that has a constant magnitude throughout.

The tension is constant along the rope, so the force $\vec{F}$ exerted by the dog on the rope equals the force $\vec{T}$ exerted by the rope on the sled (Figure 4-8). Constant tension in a string or rope also holds for a string that passes over a frictionless peg or pulley of negligible mass as long as there are no tangential forces acting on the string between the two points considered.

In this simple example, we found two things: the horizontal acceleration ($a_x = T/m = F/m$), and the vertical force $\vec{F}_n$ exerted by the ice ($F_n = w$). According to Newton's third law, forces always act in pairs. Figure 4-7 shows only those forces that act on the *sled*. Figure 4-9 shows the reaction forces to those in Figure 4-7. These are the gravitational force $\vec{w}'$ exerted by the sled *on the earth*, the force $\vec{F}'_n$ exerted by the sled *on the ice*, and the force $\vec{T}'$ exerted by the sled *on the rope*. Since these forces are not exerted on the sled, they have nothing to do with its motion. Therefore, they are not part of the application of Newton's second law to the motion of the sled.

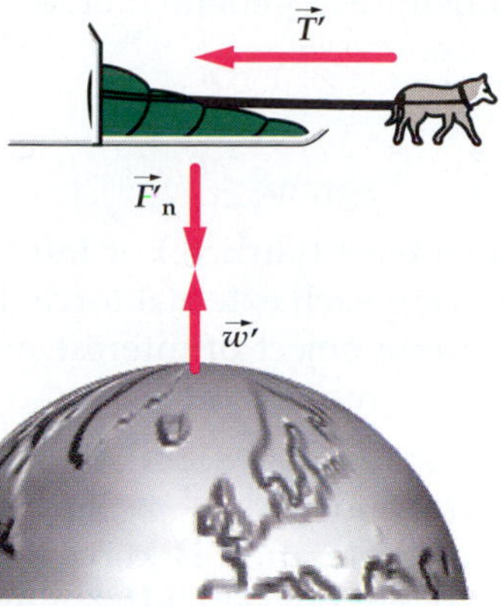

Figure 4-9 The reaction forces corresponding to the three forces shown in Figure 4-7. These forces do *not* act on the sled.

Example 4-7

During your winter break, you enter a dogsled race in which students replace the dogs. Wearing cleats for traction, you begin the race by pulling on a rope attached to the sled with a force of 150 N at 25° with the horizontal. The mass of the sled is 80 kg and there is negligible friction between the sled and ice (Figure 4-10). Find (*a*) the acceleration of the sled and (*b*) the normal force $\vec{F}_n$ exerted by the surface on the sled.

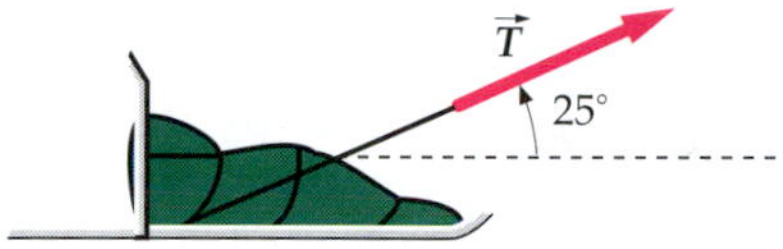

Figure 4-10

Picture the Problem Three forces act on the sled: its weight, $m\vec{g}$, which acts downward; the normal force, $\vec{F}_n$, which acts upward; and the tension in the rope, $\vec{T}$, directed 25° above the horizontal. Since the forces do not lie along a line, we study the system by applying Newton's second law to the x and y directions separately. We choose x to be in the direction of motion, and y to be perpendicular to the ice. Then we draw a free-body diagram for the sled.

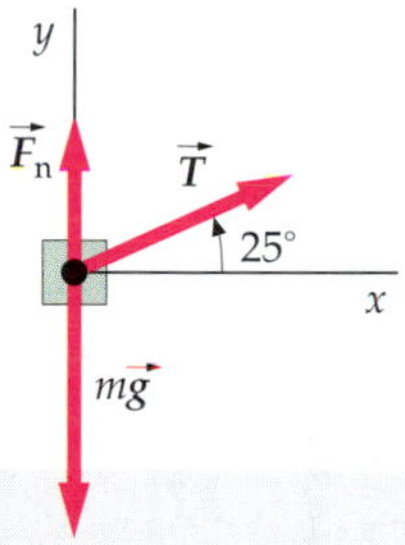

(*a*) Apply $\Sigma\vec{F} = m\vec{a}$ to motion along the x axis to determine the acceleration of the sled, a_x:

$$\sum F_x = T\cos\theta = ma_x$$

$$a_x = \frac{T\cos\theta}{m} = \frac{(150\text{ N})(\cos 25°)}{80\text{ kg}} = 1.70\text{ m/s}^2$$

(*b*) There is no acceleration in the y direction. Apply $\Sigma\vec{F} = m\vec{a}$ to motion along the y axis to determine F_n:

$$\sum F_y = T\sin\theta + F_n - mg = ma_y = 0$$

$$F_n = mg - T\sin\theta$$

$$= (80\text{ kg})(9.81\text{ N/kg}) - (150\text{ N})(\sin 25°) = 721\text{ N}$$

Remarks Note that only the x component of the tension, $T\cos\theta$, causes the sled to accelerate. Also note that the ice supports less than the full weight of the sled, since part of the weight, $T\sin\theta$, is supported by the rope.

Check the Result If $\theta = 0$, the sled is accelerated by a force T and the ice supports all the weight of the sled. Our results agree, giving $a_x = T/m$ and $F_n = mg$. For $\theta = 90°$, $a_x = 0$ and $F_n = mg - T$, as expected.

Exercise What is the greatest tension that can be applied to the rope without lifting the sled off the surface? (*Answer* $T = 1.86$ kN)

Example 4-7 illustrates a general method for solving problems using Newton's laws:

1. Draw a neat diagram.
2. Isolate the object (particle) of interest, and draw a free-body diagram showing each external force that acts on the object. If there is more than one object of interest in the problem, draw a separate free-body diagram for each.
3. Choose a convenient coordinate system for each object and apply Newton's second law, $\Sigma\vec{F} = m\vec{a}$, in component form. If the direction of the acceleration is known, choose a coordinate axis to be parallel to it. For objects sliding along a surface, choose one coordinate axis parallel to the surface and the other perpendicular to it.
4. Solve for the unknowns in the resulting equations.
5. Check to see whether your results have the correct units and seem reasonable. Substituting extreme values into your solution is a good way to check your work for errors.

Solving problems using Newton's laws

Example 4-8

You unload a moving van by sliding its cargo down a ramp that has rollers (i.e., the ramp is approximately frictionless). The ramp is inclined at an angle θ to the horizontal. For a box of mass m, find both the acceleration of the box as it slides down the ramp and the normal force exerted by the ramp on the box.

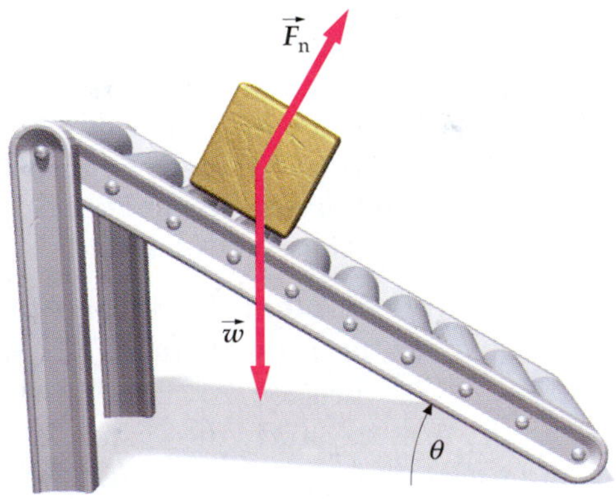

Figure 4-11

Picture the Problem Two forces act on the box, the weight $\vec{w}$ and the normal force $\vec{F}_n$ (Figure 4-11). Since these forces act along different lines, they cannot sum to zero, hence there is a net force on the box causing it to accelerate. The ramp constrains the box to move parallel to its surface, so we choose a coordinate system aligned with the ramp, as shown in Figure 4-12. Then the acceleration has only one nonzero component, a_x.

Note that $\vec{w}$ is perpendicular to the horizontal, and the negative y axis is perpendicular to the incline, so the angle between $\vec{w}$ and the negative y axis is the same as the angle θ of the incline.

1. Draw a free-body diagram for the box:

Figure 4-12

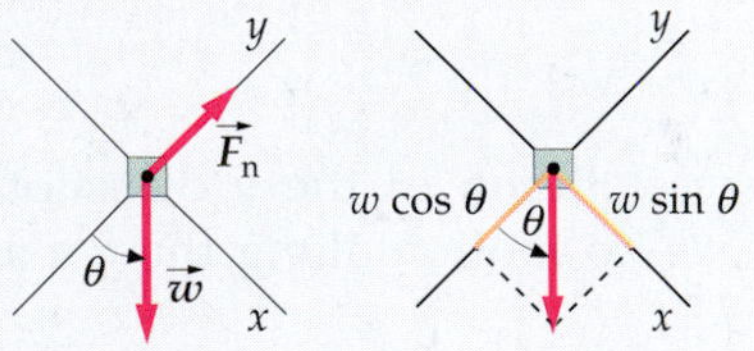

2. Apply $\Sigma\vec{F} = m\vec{a}$ to the box:

$$\sum \vec{F} = m\vec{a}$$
$$\vec{F}_n + \vec{w} = m\vec{a}$$

3. The normal force $\vec{F}_n$ is in the y direction:

$$F_{n,x} = 0, \quad F_{n,y} = F_n$$

4. The weight $\vec{w}$ has both x and y components:

$$w_x = w\sin\theta = mg\sin\theta$$
$$w_y = -w\cos\theta = -mg\cos\theta$$

5. Apply $\Sigma\vec{F} = m\vec{a}$ in component form:

$$\sum F_x = ma_x, \quad 0 + mg\sin\theta = ma_x$$
$$\sum F_y = ma_y, \quad F_n - mg\cos\theta = ma_y = 0$$

6. Solve for a_x and F_n:

$$a_x = g \sin\theta$$
$$F_n = mg\cos\theta$$

Remark The acceleration down the incline is constant and equal to $g \sin\theta$.

Check the Result It is useful to check our results at the extreme values of inclination, $\theta = 0$ and $\theta = 90°$. At $\theta = 0$, the surface is horizontal. The weight has only a y component, which is balanced by the normal force $\vec{F}_n$. The acceleration is zero: $a_x = g \sin 0° = 0$. At the opposite extreme, $\theta = 90°$, the incline is vertical. Then the weight has only an x component along the incline, and the normal force is zero: $F_n = mg\cos 90° = 0$. The acceleration is $a_x = g \sin 90° = g$. That is, the box is in free fall.

Example 4-9 *try it yourself*

A picture weighing 8 N is supported by two wires with tension $\vec{T}_1$ and $\vec{T}_2$, as shown in Figure 4-13. Find each tension.

Figure 4-13

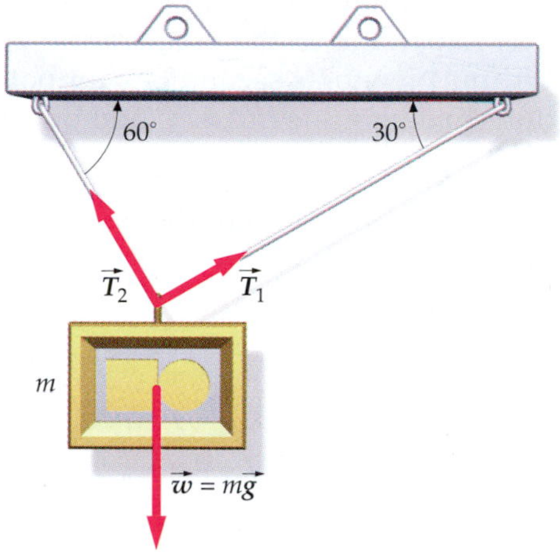

Picture the Problem Since the picture does not accelerate, the net force acting on it must be zero. The three forces acting on the picture, its weight $m\vec{g}$, the tension $\vec{T}_1$ in one wire, and the tension $\vec{T}_2$ in the other wire, must therefore sum to zero.

Cover the column to the right and try these on your own before looking at the answers.

Steps	***Answers***
1. Draw a free-body diagram for the picture. On your diagram show the x and y components of the tensions.	**Figure 4-14** 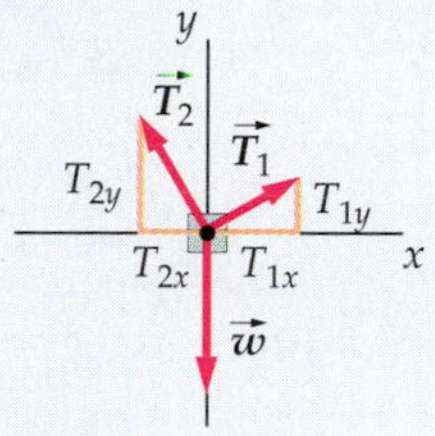
2. Apply $\Sigma\vec{F} = m\vec{a}$ in vector form to the picture.	$\sum\vec{F} = \vec{T}_1 + \vec{T}_2 + \vec{w} = m\vec{a} = 0$
3. Resolve each force into its x and y components. This gives you two equations for the two unknowns T_1 and T_2.	$\sum F_x = T_1\cos 30° - T_2\cos 60° = 0$ $\sum F_y = T_1\sin 30° + T_2\sin 60° - mg = 0$
4. Solve the x component equation for T_2 in terms of T_1.	$T_2 = T_1\dfrac{\cos 30°}{\cos 60°} = T_1\sqrt{3}$
5. Substitute your result for T_2 from step 3 into the y-component equation and solve for T_1.	$T_1\sin 30° + (T_1\sqrt{3})\sin 60° - mg = 0$ $T_1 = \dfrac{1}{2}mg = 4\text{ N}$
6. Use your result for T_1 to find T_2.	$T_2 = \sqrt{3}T_1 = \dfrac{\sqrt{3}}{2}mg = 6.93\text{ N}$

Remarks Note that the more vertical of the two wires supports the greater share of the load, as one might expect. Also, we see that $T_1 + T_2 > 8$ N. The "extra" force is due to the wires pulling to the right and left.

Example 4-10

As your jet plane speeds down the runway on takeoff, you decide to determine its acceleration, so you take out your yo-yo and note that when you suspend it, the string makes an angle of 22° with the vertical (Figure 4-15). (*a*) What is the acceleration of the plane? (*b*) If the mass of the yo-yo is 40 g, what is the tension in the string?

Picture the Problem Since the yo-yo accelerates in the horizontal direction, it must be acted on by a net horizontal force. This force is supplied by the horizontal component of the tension $\vec{T}$. The vertical component of $\vec{T}$ balances the weight of the yo-yo. We choose a coordinate system in which the x direction is parallel to the acceleration vector $\vec{a}$, and the y direction is vertical. Writing Newton's second law for both the x and y directions gives two equations to determine the two unknowns, a and T.

Figure 4-15

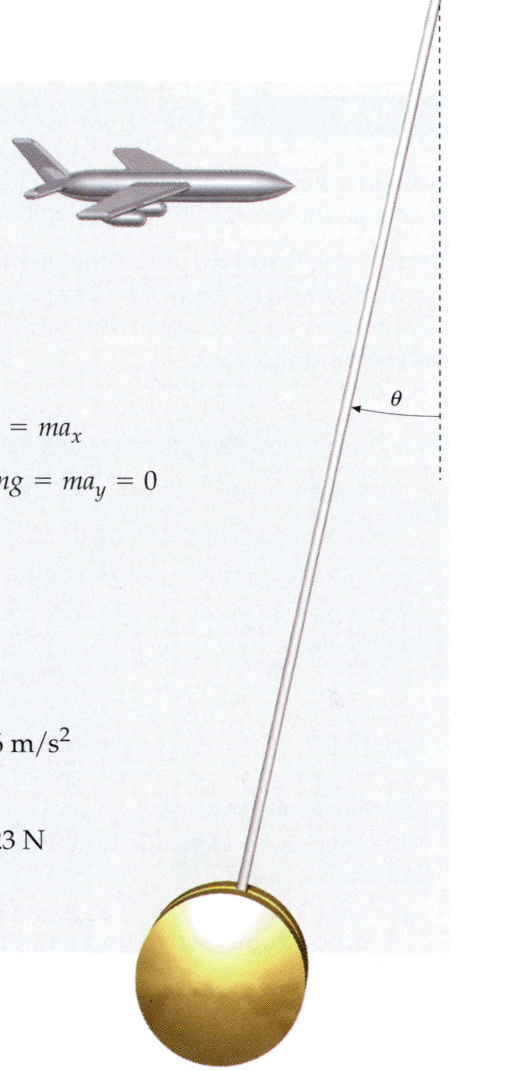

(*a*)1. Draw a free-body diagram for the yo-yo:

Figure 4-16

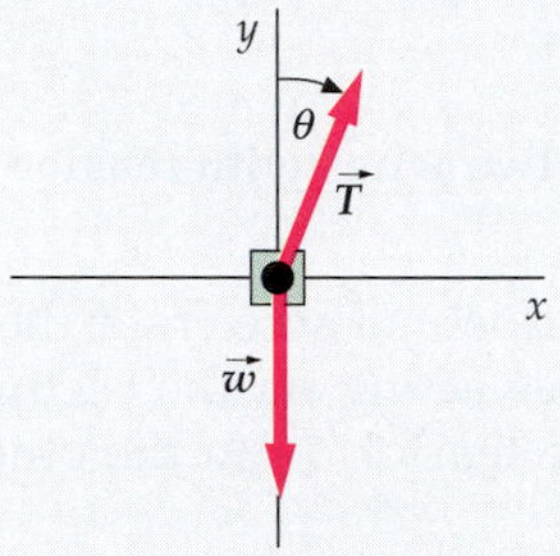

2. Apply $\Sigma\vec{F} = m\vec{a}$ in component form to the yo-yo:

$$\sum F_x = ma_x, \quad T_x + w_x = T\sin\theta + 0 = ma_x$$

$$\sum F_y = ma_y, \quad T_y + w_y = T\cos\theta - mg = ma_y = 0$$

3. Simplify:

$$T\sin\theta = ma_x, \quad T\cos\theta = mg$$

4. Dividing one of these equations by the other eliminates T and allows us to determine a:

$$\frac{T\sin\theta}{T\cos\theta} = \frac{ma_x}{mg}$$

or

$$a_x = g\tan\theta = (9.81\ \mathrm{m/s^2})\tan 22° = 3.96\ \mathrm{m/s^2}$$

(*b*) T can be found directly from the y component of Newton's second law:

$$T = \frac{mg}{\cos\theta} = \frac{(0.04\ \mathrm{kg})(9.81\ \mathrm{m/s^2})}{\cos 22°} = 0.423\ \mathrm{N}$$

Remark Notice that T is greater than the weight of the yo-yo ($mg = 0.392$ N), since the cord not only keeps the yo-yo from falling but also accelerates it in the horizontal direction. Here we use the units $\mathrm{m/s^2}$ for g because we are calculating acceleration.

Check the Result Note that the x equation gives the same value for the tension: $T = ma/\sin\theta = 0.423$ N. At $\theta = 0$, we find that $T = mg$ and $a = 0$.

Exercise For what acceleration a would the tension in the string be equal to $3mg$? What is θ in this case? (*Answers* $a = 27.8\ \mathrm{m/s^2}$, $\theta = 70.5°$)

Our next example applies Newton's laws to objects that are at rest relative to a reference frame that is itself accelerating.

Example 4-11

An 80-kg man stands on a scale fastened to the floor of an elevator. The scale is calibrated in newtons. What does the scale read when (*a*) the elevator is moving with upward acceleration *a*; (*b*) the elevator is moving with downward acceleration *a*′; (*c*) the elevator is moving upward at 20 m/s while its speed is decreasing at a rate of 8 m/s²?

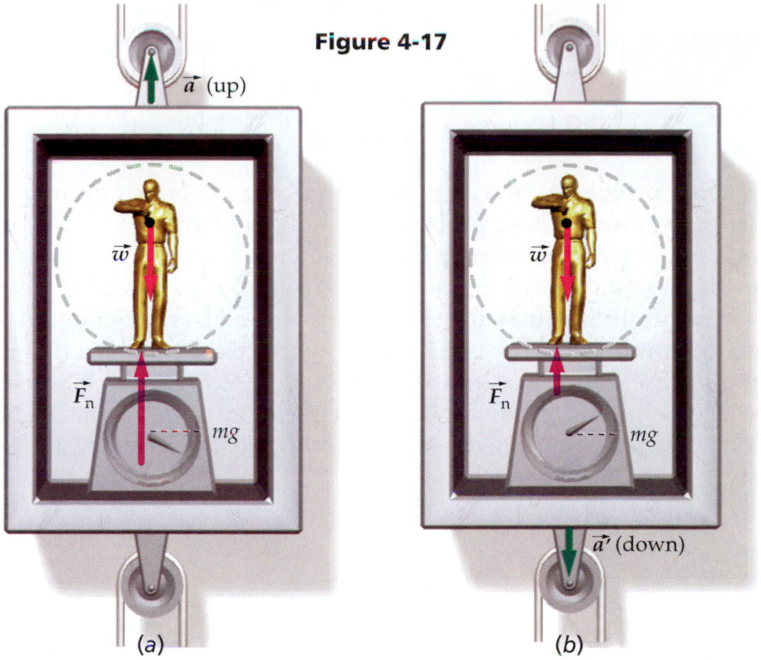

Figure 4-17

Picture the Problem The scale reading is the magnitude of the normal force $\vec{F}_n$ exerted by the scale on the man (Figure 4-17). Since the man is at rest relative to the elevator, he and the elevator have the same acceleration. Two forces act on the man; the downward force of gravity, $m\vec{g}$, and the upward normal force from the scale, $\vec{F}_n$. The sum of these forces gives the man the observed acceleration. In what follows, we choose upward to be the positive direction.

(*a*) 1. Draw a free-body diagram for the man:

Figure 4-18

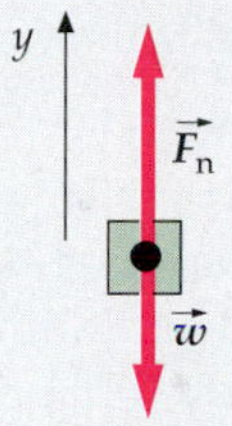

2. Apply $\Sigma\vec{F} = m\vec{a}$ in the y direction: $\sum F_y = ma_y$

3. Solve for F_n. This is the reading on the scale (the man's apparent weight): $F_n - mg = ma_y = ma$

$F_n = mg + ma$

(*b*) 1. Apply $\Sigma\vec{F} = m\vec{a}$ in the y direction for the case in which the elevator accelerates downward with magnitude a': $\sum F_y = ma_y$

$F_n - mg = ma_y = m(-a')$

2. Solve for F_n: $F_n = mg - ma'$

(*c*) 1. Apply $\Sigma\vec{F} = m\vec{a}$ in the y direction. Note that the acceleration of the elevator is downward: $\sum F_y = ma_y$

2. Solve for F_n: $F_n - mg = ma_y = (80\text{ kg})(-8\text{ m/s}^2)$

$F_n = (80\text{ kg})(-8\text{ m/s}^2) + (80\text{ kg})(9.81\text{ m/s}^2) = 145\text{ N}$

Remarks When the elevator accelerates upward, the man's apparent weight is greater than mg by the amount ma. For the man, it is as if gravity were increased from g to $g + a$. When the elevator accelerates downward, the man's apparent weight is less than mg by the amount ma'. He feels lighter, as if gravity were $g - a'$. If $a' = g$, the elevator is in free fall, and the man experiences weightlessness. Note that our conclusions are independent of the speed and direction of motion of the elevator.

Exercise An elevator descending to the ground floor comes to a stop with an acceleration of magnitude 4 m/s². If your mass is 70 kg and you are standing on a scale in the elevator, what does the scale read as the elevator is stopping? (*Answer* 967 N)

4-7 Problems With Two or More Objects

In some problems, two or more objects are in contact or are connected by a string or spring. These problems are solved by drawing a free-body diagram for each object and then applying Newton's second law to each object. The resultant equations, together with any equations describing constraints, are solved simultaneously for the unknown forces or accelerations. An example of a constraint would be two objects connected by a string that is always taut. In that case, the objects must have equal speeds and their accelerations must be equal in magnitude. If the objects are in direct contact, the forces they exert on each other must be equal and opposite, as stated in Newton's third law.

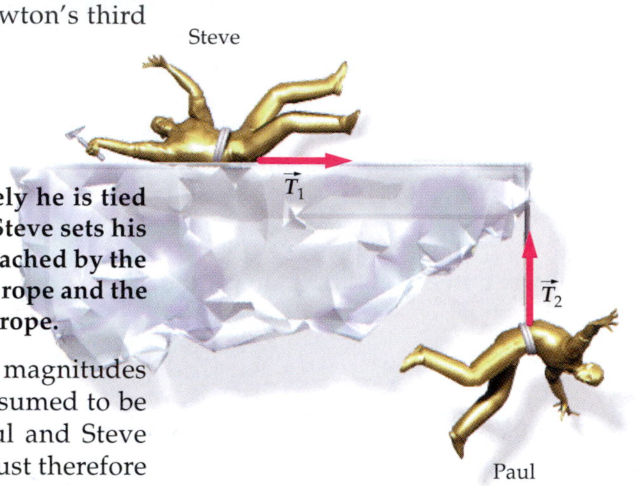

Figure 4-19

Example 4-12

Paul (mass m_P) accidentally falls off the edge of a cliff. Fortunately he is tied by a long rope to Steve (mass m_S), who has a climbing ax. Before Steve sets his ax to stop them, he slides without friction along the level snow, attached by the rope to Paul (Figure 4-19). Assume there is no friction between the rope and the cliff. Find the acceleration of each person and the tension T in the rope.

Picture the Problem The rope tensions $\vec{T}_1$ and $\vec{T}_2$ have equal magnitudes T because the rope is assumed to be massless, and the cliff is assumed to be frictionless. The rope does not stretch or become slack, so Paul and Steve have the same speed at all times. Their accelerations a_S and a_P must therefore be equal in magnitude (but not in direction). Because Steve has no vertical acceleration, the vertical forces F_n and $m_S g$ must balance. The acceleration of each person is related to the forces acting on him by Newton's second law.

1. Draw free-body diagrams for both Steve and Paul:

Figure 4-20

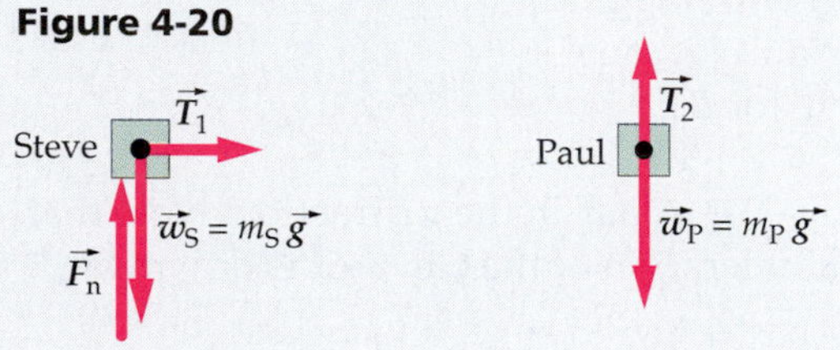

2. Apply $\Sigma \vec{F} = m\vec{a}$ in the horizontal direction to Steve, using $T_1 = T$: $\quad \sum F_x = ma_x, \qquad T = m_S a_S$

3. Apply $\Sigma \vec{F} = m\vec{a}$ to Paul. His acceleration is downward, and the forces acting on him are his weight $m_P g$ downward and $T_2 = T$ upward. Take the downward direction to be positive: $\quad \sum F_y = ma_y, \qquad m_P g - T = m_P a_P$

4. Because they are connected by the length of rope, the accelerations of Paul and Steve must be equal in magnitude. Let $a = a_S = a_P$: $\quad T = m_S a, \qquad m_P g - T = m_P a$

5. We solve these two equations for a and T by first eliminating one of the unknowns. T is eliminated by substituting $m_S a$ for T in the second equation: $\quad m_P g - m_S a = m_P a$

6. Solve for a: $a = \dfrac{m_P}{m_S + m_P} g$

7. The expression for a can be substituted into the first equation in step 4 to find T: $T = m_S a = \dfrac{m_S m_P}{m_S + m_P} g$

Remark In step 3, we chose downward to be positive to keep the solution as simple as possible. With this choice, when Steve moves in the positive direction (to the right), Paul also moves in the positive direction (downward). Note that the acceleration a is the same as that for a mass $m = m_S + m_P$ acted on by a force $m_P g$.

Check the Result If m_P is very much greater than m_S, we expect the acceleration to be approximately g and the tension to be approximately zero. Substituting $m_S = 0$ does indeed give $a = g$ and $T = 0$ in this case. If m_P is much less than m_S, we expect the acceleration to be approximately zero and the tension to be $m_P g$. If we neglect m_P in the denominator in steps 6 and 7, we indeed obtain $a \approx (m_P/m_S)\, g \approx 0$ and $T \approx m_P g$.

Exercise (*a*) Find the acceleration if the masses are $m_S = 78$ kg and $m_P = 92$ kg. (*b*) Find the acceleration if these two masses are interchanged. (*Answers* (*a*) $a = 0.541g$, (*b*) $a = 0.459g$)

Example 4-13 *try it yourself*

While constructing a space station, you push on a box of mass m_1 with a force $\vec{F}$. The box is in direct contact with a second box of mass m_2 (Figure 4-21). (*a*) What is the acceleration of the boxes? (*b*) What is the magnitude of the force exerted by one box on the other?

Figure 4-21

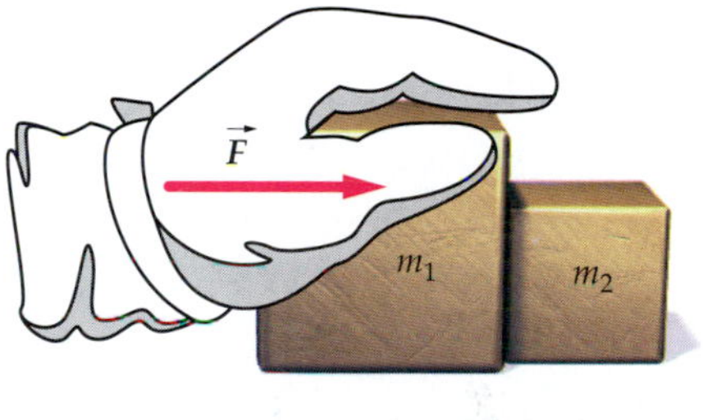

Picture the Problem Let $\vec{F}_{2,1}$ be the contact force exerted by m_2 on m_1 and $\vec{F}_{1,2}$ be the force exerted by m_1 on m_2. These forces are equal and opposite so $\vec{F}_{2,1} = -\vec{F}_{1,2}$ and $F_{2,1} = F_{1,2}$. Apply Newton's second law to each box separately and use the fact that the accelerations a_1 and a_2 are equal.

Cover the column to the right and try these on your own before looking at the answers.

Steps	***Answers***
(*a*) 1. Draw free-body diagrams for the two boxes.	(Figure 4-22)
2. Apply $\Sigma \vec{F} = m\vec{a}$ to the first box.	$F - F_{2,1} = m_1 a_1 = m_1 a$
3. Apply $\Sigma \vec{F} = m\vec{a}$ to the second box.	$F_{1,2} = m_2 a_2 = m_2 a$
4. Add these equations to eliminate $F_{2,1}$ and $F_{1,2}$ and solve for $a = a_1 = a_2$.	$a = \dfrac{F}{m_1 + m_2}$
(*b*) Substitute your value for a into the equation in step 3 and solve for $F_{1,2}$.	$F_{1,2} = \dfrac{m_2}{m_1 + m_2} F$

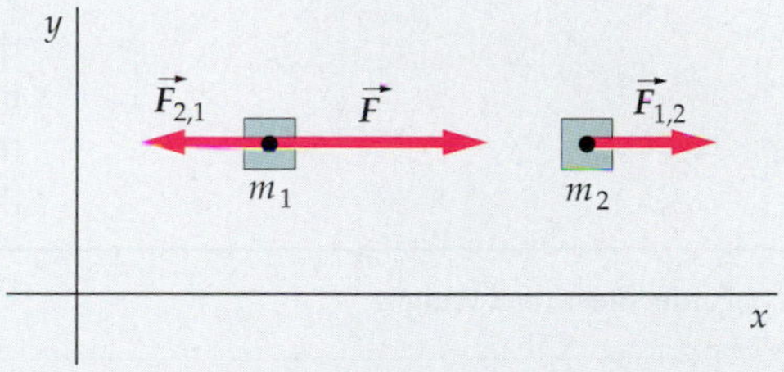

Figure 4-22

Remark Note that the result in step (*a*) 3 is the same as if the force F had acted on a single mass equal to the sum of the masses of the two boxes. In fact, since the two boxes have the same acceleration, we can consider them to be a single system with mass $m_1 + m_2$.

Exercise (*a*) Find the acceleration and the contact force if $m_1 = 2$ kg, $m_2 = 3$ kg, and $F = 12$ N. (*b*) Find the contact force for these values if the blocks are interchanged so that the first block has a mass of 3 kg and the second block has a mass of 2 kg (*Answer* (*a*) $a = 2.4\ \text{m/s}^2$, $F_{1,2} = 7.2$ N; (*b*) $F_{1,2} = 4.8$ N)

Summary

1. Newton's laws of motion are fundamental laws of nature that serve as the basis for our understanding of mechanics.
2. Mass is an *intrinsic* property of an object.
3. Force is an important *derived* dynamic quantity.

Topic	Remarks and Relevant Equations	
1. Newton's Laws		
First law	An object at rest stays at rest unless acted on by an external force. An object in motion continues to travel with constant velocity unless acted on by an external force.	
Second law	The acceleration of an object is inversely proportional to its mass and directly proportional to the net external force acting on it: $\vec{a} = \frac{\vec{F}_{\text{net}}}{m}$ or $\sum \vec{F} = \vec{F}_{\text{net}} = m\vec{a}$	4-1
Third law	Forces always occur in equal and opposite pairs. If object A exerts a force on object B, an equal but opposite force is exerted by object B on object A.	
2. Inertial Reference Frames	Newton's laws hold only in an inertial reference frame—a reference frame for which an object at rest remains at rest if no net force acts on the object. Any reference frame that is moving with constant velocity relative to an inertial reference frame is itself an inertial reference frame. A reference frame that is accelerating relative to an inertial frame is not an inertial reference frame. A reference frame attached to the earth is approximately an inertial reference frame.	
3. Force, Mass, and Weight		
Force	Force is defined in terms of the acceleration it produces on a given object. A force of 1 newton (N) is that force which produces an acceleration of 1 m/s^2 on a mass of 1 kilogram (kg).	
Mass	Mass is the intrinsic property of an object that measures its resistance to acceleration. Mass does not depend on the location of the object. The masses of two objects can be compared by applying the same force to each object and measuring their accelerations. The ratio of the masses of the objects is then equal to the inverse ratio of the accelerations of the objects produced by the same force: $\frac{m_2}{m_1} = \frac{a_1}{a_2}$	4-2

	Weight	The weight $\vec{w}$ of an object is the force of gravitational attraction exerted by the earth on the object. It is proportional to the mass m of the object and the gravitational field $\vec{g}$, which also equals the free-fall acceleration due to gravity: $$\vec{w} = m\vec{g} \qquad \textbf{4-3}$$ Weight is not an intrinsic property of an object; it depends on the location of the object.
4.	**Fundamental Forces**	All the forces observed in nature can be explained in terms of four basic interactions: 1. The gravitational force 2. The electromagnetic force 3. The strong nuclear force (also called the hadronic force) 4. The weak nuclear force
5.	**Contact Forces**	Contact forces of support and friction and those exerted by springs and strings are due to molecular forces that arise from the basic electromagnetic force.
	Hooke's law	When a spring is compressed or extended by a small amount Δx, the force it exerts is proportional to Δx. $$F_x = -k\,\Delta x \qquad \textbf{4-8}$$

Problem-Solving Guide

1. Begin by drawing a neat diagram that includes the important features of the problem.
2. Isolate the object (particle) of interest, and draw a free-body diagram that shows how each external force acts on the object. If there is more than one object of interest in the problem, draw a separate free-body diagram for each object.
3. Choose a convenient coordinate system for each object, and apply Newton's second law, $\Sigma \vec{F} = m\vec{a}$, in component form.
4. Solve the resulting equations for the unknowns.
5. Check to see whether your results are reasonable.

Summary of Worked Examples

Type of Calculation	Procedure and Relevant Examples
1. Compare an unknown mass with a known one.	Find the ratio of the accelerations of the objects under the influence of the same force. **Example 4-1**
2. Find the motions (a, v, or x) of objects using Newton's laws. **To find:**	
the motion of a single object under the influence of a single force	Draw a free-body diagram. Apply $\Sigma \vec{F} = m\vec{a}$ to the object. If the force is exerted by a spring, use Hooke's law, $F_x = -kx$. If the force is constant, the position and velocity of the object can be found from the constant-acceleration equations in Chapter 2. **Examples 4-2, 4-6, 4-10**
the motion of a single object under the influence of several forces	Draw a free-body diagram. Choose a coordinate system so that the acceleration is along one axis. Apply $\Sigma \vec{F} = m\vec{a}$ in component form to the object, and solve the equations for the acceleration. If the force is constant, the position and velocity of the object can be found from the constant-acceleration equations in Chapter 2. If the object is constrained to move along a surface, the normal force must balance the other forces perpendicular to that surface. **Examples 4-7, 4-8**

	the motions of two objects under the influence of several forces with constraints	Draw a free-body diagram for each object. Choose a coordinate system so that the acceleration is along one axis. Apply $\Sigma \vec{F} = m\vec{a}$ in component form to each object separately. Use the constraints to obtain information relating the magnitudes of the acceleration or the magnitudes of the forces (e.g., constant tension in a string). Solve the equations simultaneously to determine each force and acceleration. **Examples 4-12, 4-13**
3.	**Apply Newton's laws in the U.S. customary system.**	Draw a free-body diagram. Use $m = w/g$ for the mass. **Example 4-4**
4.	**Find the forces acting on static objects.**	Draw a free-body diagram. Choose a coordinate system in which one or more of the forces is along one of the axes. Apply $\Sigma \vec{F} = 0$ in component form, and solve for the desired quantities. **Example 4-9**
5.	**Apply Newton's laws to an object at rest in an accelerated system (such as a boxcar or elevator).**	Draw a free-body diagram. Choose a coordinate system in which the acceleration is along one axis. Apply $\Sigma \vec{F} = m\vec{a}$ in component form, noting that the acceleration is the acceleration of the object in an inertial reference frame. Since the object is at rest in a noninertial frame, $\vec{a}$ is the acceleration of the accelerated frame. Solve for the forces. **Examples 4-10, 4-11**

Problems

In a few problems, you are given more data than you actually need; in a few other problems, you are required to supply data from your general knowledge, outside sources, or informed estimates.

- Conceptual Problems
- Problems from Optional and Exploring sections

- • Single-concept, single-step, relatively easy
- •• Intermediate-level, may require synthesis of concepts
- ••• Challenging, for advanced students

For all problems, use $g = 9.81\ m/s^2$ for the acceleration due to gravity and neglect friction and air resistance unless instructed to do otherwise.

Newton's First Law: The Law of Inertia

1 •• How can you tell if a particular reference frame is an inertial reference frame?

2 •• Suppose you find that an object in a particular frame has an acceleration $\vec{a}$ when there are no forces acting on it. How can you use this information to find an inertial reference frame?

Force, Mass, and Newton's Second Law

3 • If an object has no acceleration in an inertial reference frame, can you conclude that no forces are acting on it?

4 • If only a single force acts on an object, must the object accelerate in an inertial reference frame? Can it ever have zero velocity?

5 • If an object is acted upon by a single known force, can you tell in which direction the object will move using no other information?

6 • An object is observed to be moving at constant velocity in an inertial reference frame. It follows that

(*a*) no forces act on the object.
(*b*) a constant force acts on the object in the direction of motion.
(*c*) the net force acting on the object is zero.
(*d*) the net force acting on the object is equal and opposite to its weight.

7 • A body moves with constant speed in a straight line in an inertial reference frame. Which of the following statements must be true?

(*a*) No force acts on the body.
(*b*) A single constant force acts on the body in the direction of motion.
(*c*) A single constant force acts on the body in the direction opposite to the motion.
(*d*) A net force of zero acts on the body.
(*e*) A constant net force acts on the body in the direction of motion.

8 •• Figure 4-23 shows the position x versus time t of a particle moving in one dimension. During what time intervals is there a net force acting on the particle? Give the direction (+ or −) of the net force during these time intervals.

Figure 4-23 Problem 8

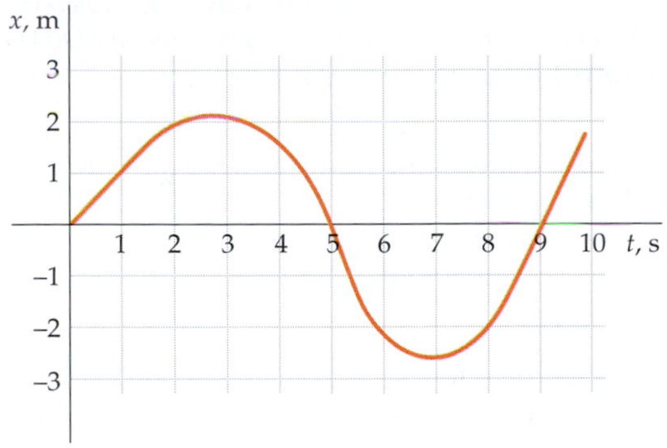

9 • A particle of mass m is traveling at an initial speed $v_0 = 25.0$ m/s. It is brought to rest in a distance of 62.5 m when a net force of 15.0 N acts on it. What is m?

(*a*) 37.5 kg (*b*) 3.00 kg (*c*) 1.50 kg
(*d*) 6.00 kg (*e*) 3.75 kg

10 • (*a*) An object experiences an acceleration of 3 m/s^2 when a certain force F_0 acts on it. What is its acceleration when the force is doubled? (*b*) A second object experiences an acceleration of 9 m/s^2 under the influence of the force F_0. What is the ratio of the masses of the two objects? (*c*) If the two objects are tied together, what acceleration will the force F_0 produce?

11 • A tugboat tows a ship with a constant force F_1. The increase in the ship's speed in a 10-s interval is 4 km/h. When a second tugboat applies a second constant force F_2 in the same direction, the speed increases by 16 km/h in a 10-s interval. How do the magnitudes of the two forces compare? (Neglect water resistance.)

12 • A force F_0 causes an acceleration of 3 m/s^2 when it acts on an object of mass m sliding on a frictionless surface. Find the acceleration of the same object in the circumstances shown in Figure 4-24*a* and *b*.

Figure 4-24 Problem 12

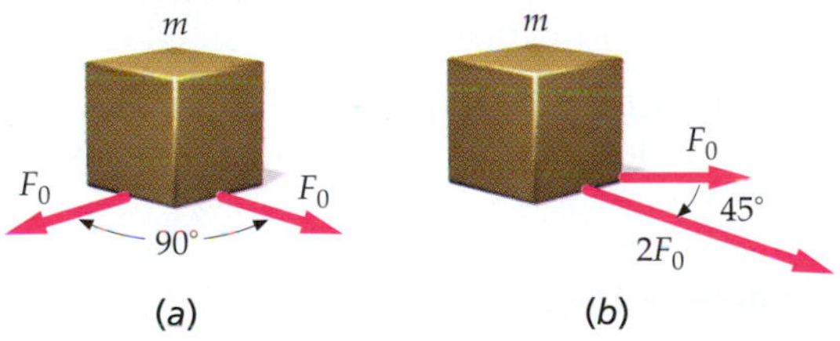

13 • A force $\vec{F} = 6\text{ N}\,\hat{i} - 3\text{ N}\,\hat{j}$ acts on an object of mass 1.5 kg. Find the acceleration $\vec{a}$. What is the magnitude a?

14 • A single force of 12 N acts on a particle of mass m. The particle starts from rest and travels in a straight line a distance of 18 m in 6 s. Find m.

15 • To drag a 75-kg log along the ground at constant velocity, you have to pull on it with a horizontal force of 250 N. (*a*) What is the resistive force exerted by the ground? (*b*) What force must you exert if you want to give the log an acceleration of 2 m/s^2?

Figure 4-25 Problem 16

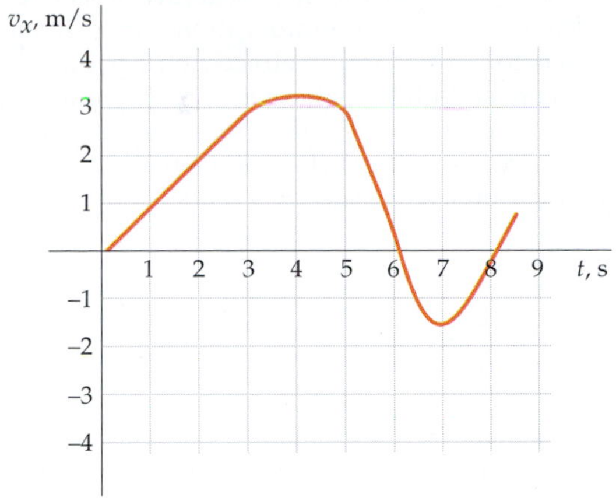

16 • Figure 4-25 shows a plot of v_x versus t for an object of mass 8 kg moving in a straight line. Make a plot of the net force acting on the object as a function of time.

17 •• A 4-kg object is subjected to two forces, $\vec{F}_1 = 2\text{ N}\,\hat{i} - 3\text{ N}\,\hat{j}$ and $\vec{F}_2 = 4\text{ N}\,\hat{i} - 11\text{ N}\,\hat{j}$. The object is at rest at the origin at time $t = 0$. (*a*) What is the object's acceleration? (*b*) What is its velocity at time $t = 3$ s? (*c*) Where is the object at time $t = 3$ s?

Weight and Mass

18 • Suppose an object were sent far out in space, away from galaxies, stars, or other bodies. How would its mass change? Its weight?

19 • How would an astronaut in apparent weightlessness be aware of her mass?

20 • Under what circumstances would your apparent weight be greater than your true weight?

21 • On the moon, the acceleration due to gravity is only about 1/6 of that on earth. An astronaut whose weight on earth is 600 N travels to the lunar surface. His mass as measured on the moon will be

(*a*) 600 kg. (*b*) 100 kg. (*c*) 61.2 kg.
(*d*) 9.81 kg. (*e*) 360 kg.

22 • Find the weight of a 54-kg girl in (*a*) newtons and (*b*) pounds.

23 • Find the mass of a 165-lb man in kilograms.

24 • After watching a space documentary, Lou speculates that there is money to be made by combining the phenomenon of weightlessness in space with the widespread longing for weight loss in the general population. Researching the matter, he learns that the gravitational force on a mass m at a height h above the earth's surface is given by $F = mgR_E^2/(R_E + h)^2$, where R_E is the radius of the earth (about 6370 km) and g is the acceleration due to gravity at the earth's surface. (*a*) Using this expression, find the weight in newtons and pounds of an 83-kg person at the earth's surface. (*b*) If this person were weight-conscious and rich, and Lou managed to sell the person a trip to a height of 400 km above the earth's surface, how much weight would the person lose? (*c*) What is the person's mass at this altitude?

25 •• Caught without a map again, Hayley lands her spacecraft on an unknown planet. Visibility is poor, but she finds someone on a local communications channel and asks for directions to Earth. "You're already on Earth," is the reply, "Wait there and I'll be right over." Hayley is suspicious, however, so she drops a lead ball of mass 76.5 g from the top of her ship, 18 m above the surface of the planet. It takes 2.5 s to reach the ground. (*a*) If Hayley's mass is 68.5 kg, what is her weight on this planet? (*b*) Is she on Earth?

Newton's Third Law

26 • True or false:

(*a*) Action–reaction forces never act on the same object.
(*b*) Action equals reaction only if the objects are not accelerating.

27 • An 80-kg man on ice skates pushes a 40-kg boy also on skates with a force of 100 N. The force exerted by the boy on the man is

(*a*) 200 N.
(*b*) 100 N.
(*c*) 50 N.
(*d*) 40 N.

28 • A boy holds a bird in his hand. The reaction force to the force exerted on the bird by the boy's hand is

(*a*) the force of the earth on the bird.
(*b*) the force of the bird on the earth.
(*c*) the force of the hand on the bird.
(*d*) the force of the bird on the hand.
(*e*) the force of the earth on the hand.

The reaction force to the weight of the bird is

(*a*) the force of the earth on the bird.
(*b*) the force of the bird on the earth.
(*c*) the force of the hand on the bird.
(*d*) the force of the bird on the hand.
(*e*) the force of the earth on the hand.

29 • A baseball player hits a ball with a bat. If the force with which the bat hits the ball is considered the action force, what is the reaction force?

(*a*) The force the bat exerts on the batter's hands.
(*b*) The force on the ball exerted by the glove of the person who catches it.
(*c*) The force the ball exerts on the bat.
(*d*) The force the pitcher exerts on the ball while throwing it.
(*e*) Friction, as the ball rolls to a stop.

30 •• Dean reads in his physics book that when two people pull on the end of a rope in a tug-of-war, the forces exerted by each on the other are equal and opposite, according to Newton's third law. Misunderstanding the law tragically, Dean runs out to challenge Hugo the Large, convinced that the laws of physics guarantee a tie. Hugo lumbers over, picks up the rope, pulls Dean off his feet, and then drags him through a puddle, across the road, and up the steps of the physics building. Use a force diagram to show Dean that, in spite of Newton's third law, it is possible for one side to win a tug-of-war.

31 • A 2.5-kg object hangs at rest from a string attached to the ceiling. (*a*) Draw a diagram showing all forces acting on the object and indicate each reaction force. (*b*) Do the same for each force acting on the string.

32 • A 9-kg box rests on a 12-kg box that rests on a horizontal table. (*a*) Draw a diagram showing all forces acting on the 9-kg box and indicate each reaction force. (*b*) Do the same for all forces acting on the 12-kg box.

Contact Forces

33 • A vertical spring of force constant 600 N/m has one end attached to the ceiling and the other to a 12-kg block resting on a horizontal surface so that the spring exerts an upward force on the block. The spring is stretched by 10 cm. (*a*) What force does the spring exert on the block? (*b*) What is the force that the surface exerts on the block?

34 • A 6-kg box on a frictionless horizontal surface is attached to a horizontal spring with a force constant of 800 N/m. If the spring is stretched 4 cm from its equilibrium length, what is the acceleration of the box?

35 •• The acceleration a versus spring length L observed when a 0.5-kg mass is pulled along a frictionless table by a single spring is shown in the following table:

L, cm	4	5	6	7	8	9	10	11	12	13	14
a, m/s^2	0	2.0	3.8	5.6	7.4	9.2	11.2	12.8	14.0	14.6	14.6

(*a*) Make a plot of the force exerted by the spring versus length L. (*b*) If the spring is extended to 12.5 cm, what force does it exert? (*c*) How much does the spring extend when the mass is suspended from it at rest near sea level, where $g = 9.81$ N/kg?

Problem Solving

36 • A picture is supported by two wires as in Example 4-9. Do you expect the tension in the wire that is more nearly vertical to be greater than or less than the tension in the other wire?

37 • A clothesline is stretched taut between two poles. Then a wet towel is hung at the center of the line. Can the line remain horizontal? Explain.

38 •• Which of the free-body diagrams in Figure 4-26 represents a block sliding down a frictionless inclined surface?

Figure 4-26 Problem 38

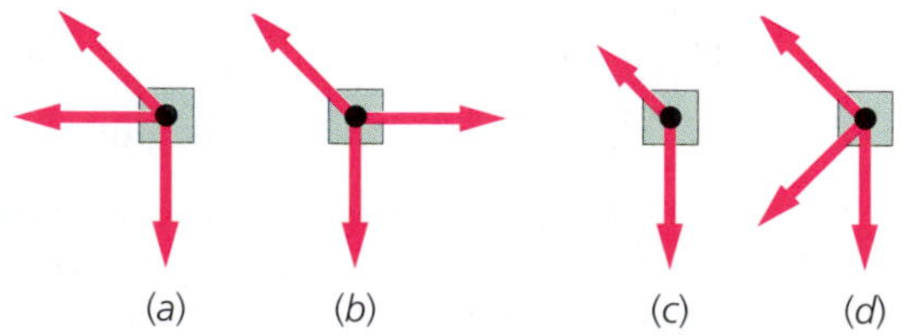

39 • A lamp with a mass $m = 42.6$ kg is hanging from wires as shown in Figure 4-27. The tension T_1 in the vertical handle is

(*a*) 209 N.
(*b*) 418 N.
(*c*) 570 N.
(*d*) 360 N.
(*e*) 730 N.

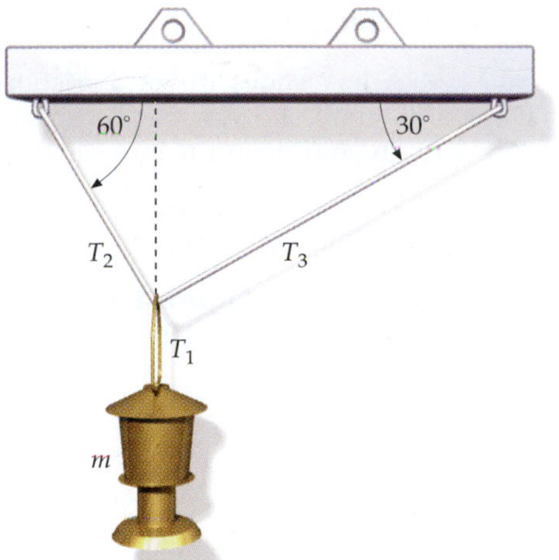

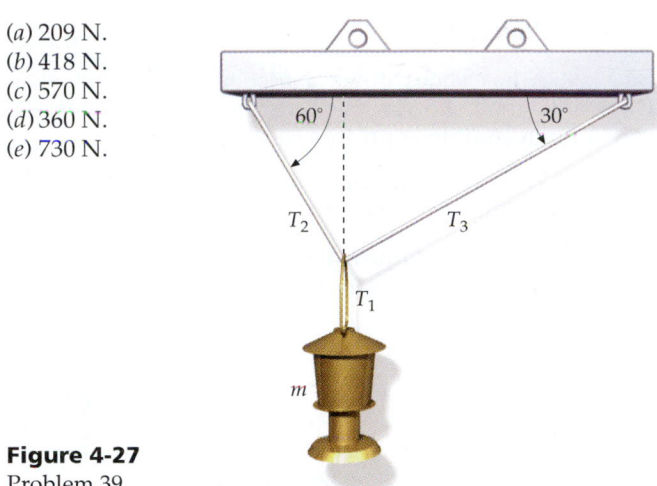

Figure 4-27
Problem 39

40 • A 40.0-kg object supported by a vertical rope is initially at rest. The object is then accelerated upward. The tension in the rope needed to give the object an upward speed of 3.50 m/s in 0.700 s is

(*a*) 590 N. (*b*) 390 N. (*c*) 200 N.
(*d*) 980 N. (*e*) 720 N.

41 • A hovering helicopter of mass m_h is lowering a truck of mass m_t. If the truck's downward speed is increasing at the rate $0.1g$, what is the tension in the supporting cable?

(*a*) $1.1m_t\,g$ (*b*) $m_t\,g$ (*c*) $0.9m_t\,g$
(*d*) $1.1(m_h + m_t)g$ (*e*) $0.9(m_h + m_t)g$

42 • A 10-kg object on a frictionless table is subjected to two horizontal forces, $\vec{F}_1$ and $\vec{F}_2$, with magnitudes $F_1 = 20$ N and $F_2 = 30$ N, as shown in Figure 4-28. (*a*) Find the acceleration $\vec{a}$ of the object. (*b*) A third force $\vec{F}_3$ is applied so that the object is in static equilibrium. Find $\vec{F}_3$.

Figure 4-28 Problem 42

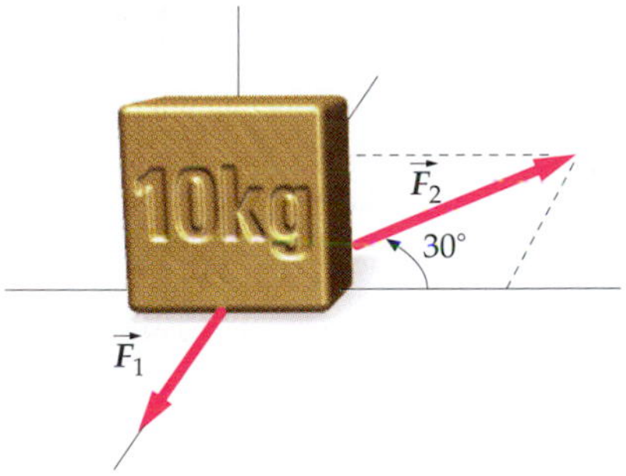

Figure 4-29 Problem 43

43 • A vertical force $\vec{T}$ is exerted on a 5-kg body near the surface of the earth, as shown in Figure 4-29. Find the acceleration of the body if (*a*) $T = 5$ N, (*b*) $T = 10$ N, and (*c*) $T = 100$ N.

44 •• To compensate for a distinct lack of personality, Herbert relies on the Grand Entrance technique when he attends parties. His latest plan for appearing at a pool party is to arrive by helicopter and then slide down a nylon rope as the helicopter hovers above poolside. However, as the helicopter approaches its destination, the pilot tells Herbert that the rope will break if the tension exceeds 300 N. Herbert, whose mass is 61.2 kg, realizes that the rope won't hold him unless he slides down with an appropriate acceleration. What must his acceleration be if the rope is not to break and ruin the whole effect?

45 •• A student has to escape from his girlfriend's dormitory through a window that is 15.0 m above the ground. He has a 24-m rope, but it will break when the tension exceeds 360 N, and the student weighs 600 N. The student will be injured if he hits the ground with a speed greater than 8 m/s. (*a*) Show that he cannot safely slide down the rope. (*b*) Find a strategy using the rope that will permit the student to reach the ground safely.

46 •• A rifle bullet of mass 9 g starts from rest and exits from the 0.6-m barrel at 1200 m/s. Find the force exerted on the bullet, assuming it to be constant, while the bullet is in the barrel.

47 •• A 2-kg picture is hung by two wires of equal length. Each makes an angle of θ with the horizontal, as shown in Figure 4-30. (*a*) Find the general equation for tension T, given θ and weight w for the picture. For what angle θ is T the least? The greatest? (*b*) If $\theta = 30°$, what is the tension in the wires?

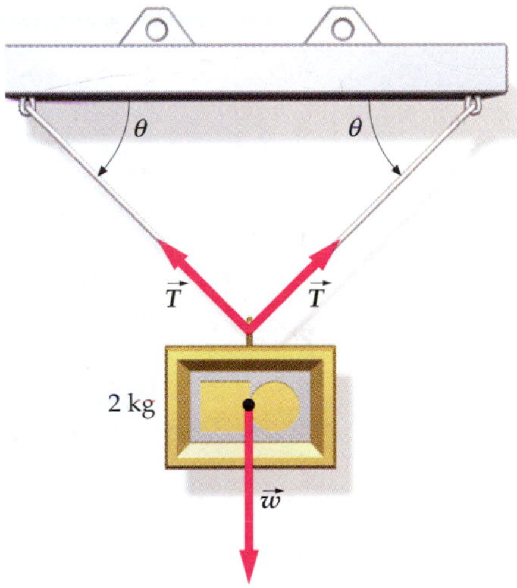

Figure 4-30 Problem 47

48 •• A bullet of mass 1.8×10^{-3} kg moving at 500 m/s impacts with a large fixed block of wood and travels 6 cm before coming to rest. Assuming that the deceleration of the bullet is constant, find the force exerted by the wood on the bullet.

49 •• A 1000-kg load is being moved by a crane. Find the tension in the cable that supports the load as (*a*) it is accelerated upward at 2 m/s^2, (*b*) it is lifted at constant speed, and (*c*) it moves upward with speed decreasing by 2 m/s each second.

50 •• A horse-drawn coach is decelerating at 3.0 m/s^2 while moving in a straight line. A lamp of mass 0.844 kg is hanging from the ceiling of the coach on a string 0.6 m long. The angle that the string makes with the vertical is

(*a*) 8.5° toward the front of the coach.
(*b*) 17° toward the front of the coach.
(*c*) 17° toward the back of the coach.
(*d*) 2.5° toward the front of the coach.
(*e*) 0° or straight down.

51 •• For the systems in equilibrium in Figure 4-31, find the unknown tensions and masses.

Figure 4-31 Problem 51

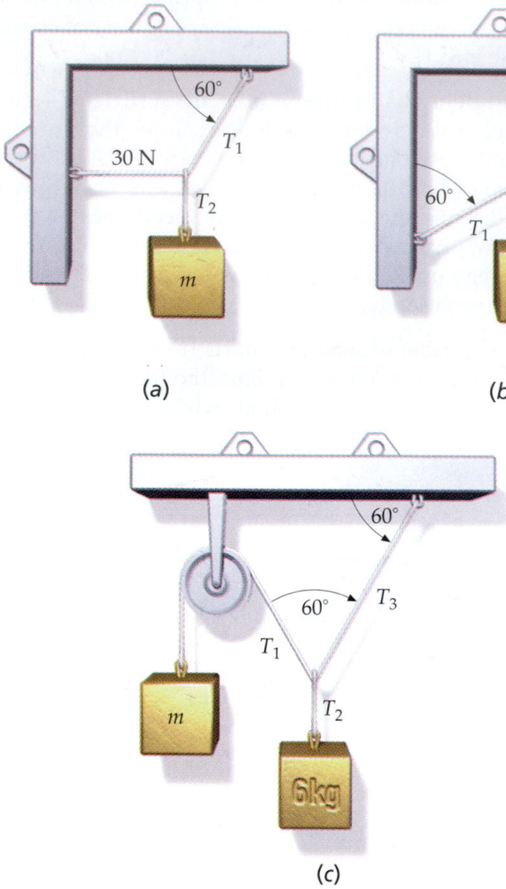

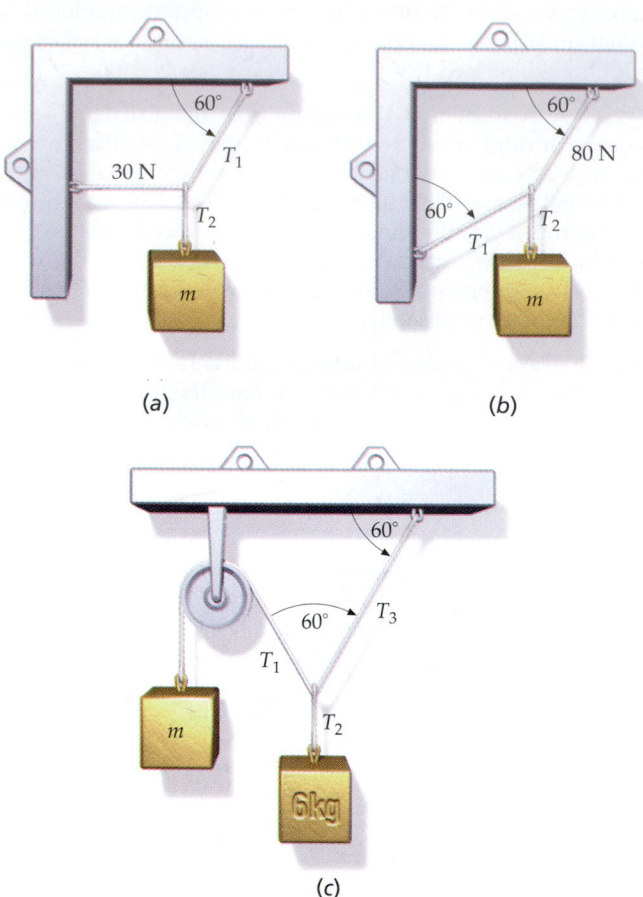

52 •• Your car is stuck in a mudhole. You are alone, but you have a long, strong rope. Having studied physics, you tie the rope tautly to a telephone pole and pull on it sideways, as shown in Figure 4-32. (*a*) Find the force exerted by the rope on the car when the angle θ is 3° and you are pulling with a force of 400 N but the car does not move. (*b*) How strong must the rope be if it takes a force of 600 N to move the car when $\theta = 4°$?

Figure 4-32 Problem 52

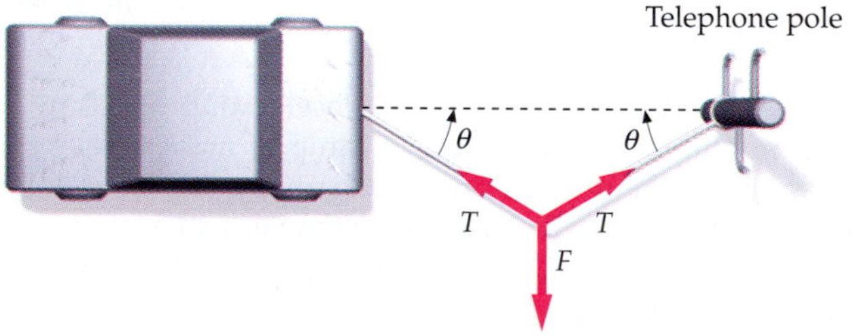

Inclined Planes

53 • A box slides down a frictionless inclined plane. Draw a diagram showing the forces acting on the box. For each force in your diagram, indicate the reaction force.

54 • The system shown in Figure 4-33 is in equilibrium. It follows that the mass m is

(*a*) 3.5 kg.
(*b*) 3.5 sin 40° kg.
(*c*) 3.5 tan 40° kg.
(*d*) none of the above.

Figure 4-33 Problem 54

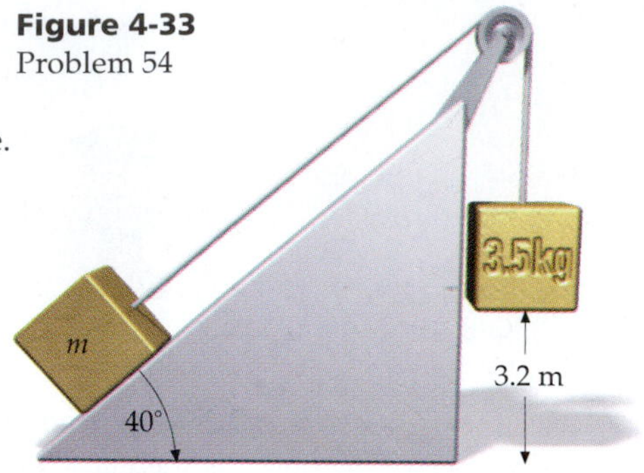

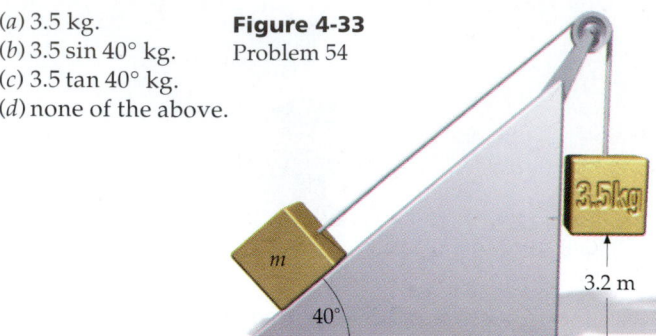

55 • In Figure 4-34, the objects are attached to spring balances calibrated in newtons. Give the readings of the balances in each case, assuming that the strings are massless and the incline is frictionless.

Figure 4-34 Problem 55

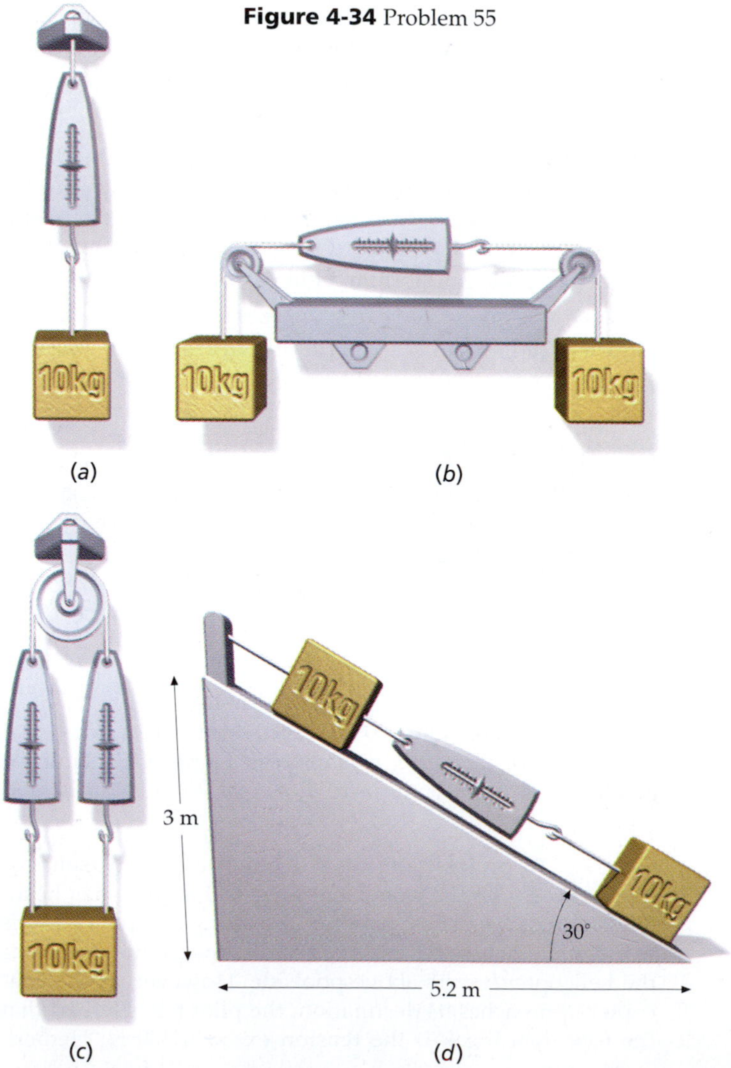

56 •• A box is held in position by a cable along a frictionless incline (Figure 4-35). (*a*) If $\theta = 60°$ and $m = 50$ kg, find the tension in the cable and the normal force exerted by the incline. (*b*) Find the tension as a function of θ and m, and check your result for $\theta = 0°$ and $\theta = 90°$.

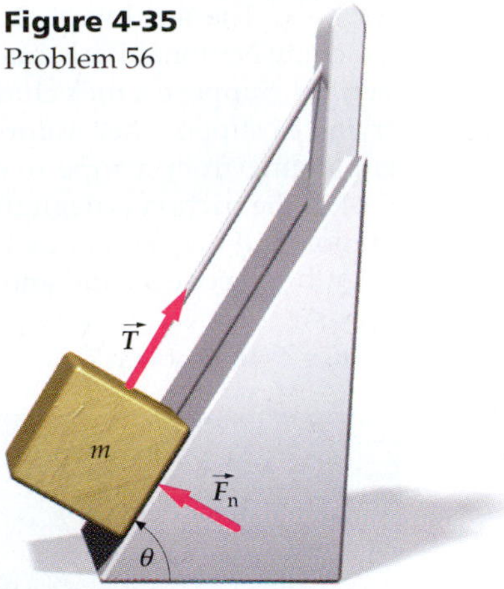

Figure 4-35 Problem 56

57 •• A horizontal force of 100 N pushes a 12-kg block up a frictionless incline that makes an angle of 25° with the horizontal. (*a*) What is the normal force that the incline exerts on the block? (*b*) What is the acceleration of the block?

58 •• A 65-kg boy weighs himself by standing on a scale mounted on a skateboard that is rolling down an incline, as shown in Figure 4-36. Assume there is no friction so that the force exerted by the incline on the skateboard is perpendicular to the incline. What is the reading on the scale if $\theta = 30°$?

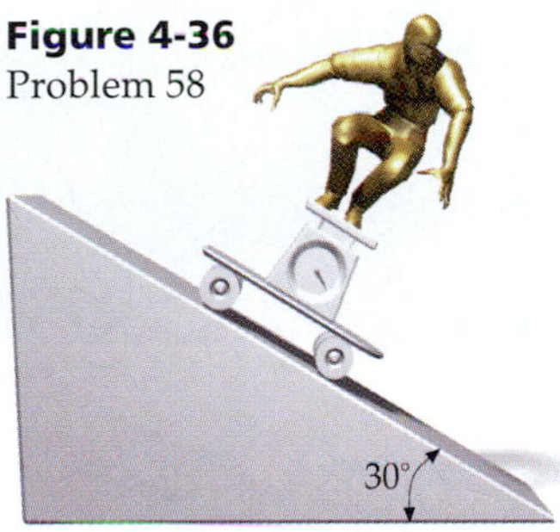

Figure 4-36 Problem 58

Elevators

59 • An object is suspended from the ceiling of an elevator that is descending at a constant speed of 9.81 m/s. The tension in the string holding the object is

(*a*) equal to the weight of the object.
(*b*) less than the weight of the object but not zero.
(*c*) greater than the weight of the object.
(*d*) zero.

60 • What effect does the velocity of an elevator have on the apparent weight of a person in the elevator?

61 • Suppose you are standing on a scale in a descending elevator as it comes to a stop on the ground floor. Will the scale's report of your weight be high, low, or correct?

62 • A person of weight w is in an elevator going up when the cable suddenly breaks. What is the person's apparent weight immediately after the cable breaks?

(*a*) w (*b*) Greater than w (*c*) Less than w
(*d*) $9.8w$ (*e*) Zero

63 • A person in an elevator is holding a 10-kg block by a cord rated to withstand a tension of 150 N. When the elevator starts up, the cord breaks. What was the minimum acceleration of the elevator?

64 • A 60-kg girl weighs herself by standing on a scale in an elevator. What does the scale read when (*a*) the elevator is descending at a constant rate of 10 m/s; (*b*) the elevator is descending at 10 m/s and gaining speed at a rate of 2 m/s^2; (*c*) the elevator is ascending at 10 m/s but its speed is decreasing by 2 m/s in each second?

65 •• A 2-kg block hangs from a spring balance calibrated in newtons that is attached to the ceiling of an elevator (Figure 4-37). What does the balance read when (*a*) the elevator is moving up with a constant velocity of 30 m/s; (*b*) the elevator is moving down with a constant velocity of 30 m/s; (*c*) the elevator is ascending at 20 m/s and gaining speed at a rate of 10 m/s^2?

From $t = 0$ to $t = 2$ s, the elevator moves upward at 10 m/s. Its velocity is then reduced uniformly to zero in the next 2 s, so that it is at rest at $t = 4$ s. Describe the reading on the balance during the interval $0 < t < 4$ s.

Figure 4-37 Problem 65

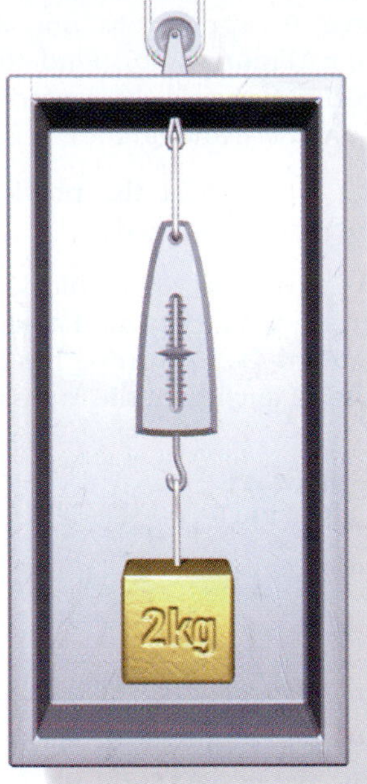

66 •• A man stands on a scale in an elevator that has an upward acceleration a. The scale reads 960 N. When he picks up a 20-kg box, the scale reads 1200 N. Find the mass of the man, his weight, and the acceleration a.

Two or More Objects

67 • Two boxes of mass m_1 and m_2 connected together by a massless string are accelerated uniformly on a frictionless surface, as shown in Figure 4-38. The ratio of the tensions T_1/T_2 is given by

(*a*) m_1/m_2. (*b*) m_2/m_1. (*c*) $(m_1 + m_2)/m_2$.
(*d*) $m_1/(m_1 + m_2)$. (*e*) $m_2/(m_1 + m_2)$.

Figure 4-38 Problem 67

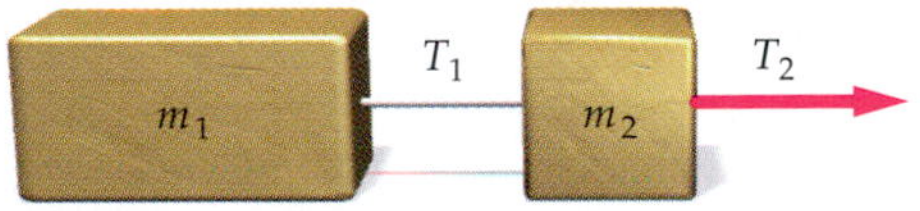

68 • A box of mass $m_2 = 3.5$ kg rests on a frictionless horizontal shelf and is attached by strings to boxes of masses $m_1 = 1.5$ kg and $m_3 = 2.5$ kg, which hang freely, as shown in Figure 4-39. Both pulleys are frictionless and massless. The system is initially held at rest. After it is released, find (*a*) the acceleration of each of the boxes, and (*b*) the tension in each string.

Figure 4-39 Problem 68

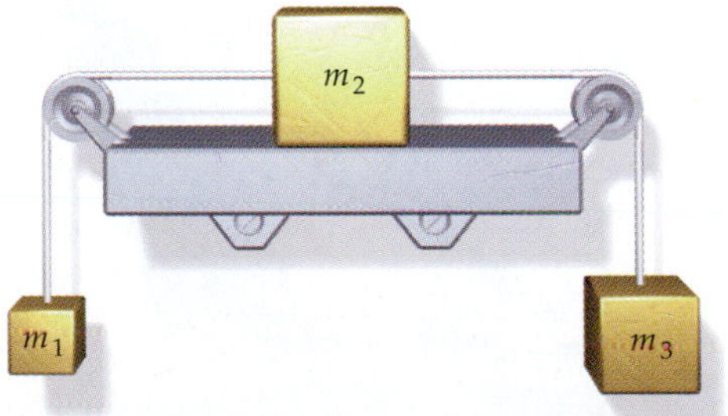

69 •• Two blocks are in contact on a frictionless, horizontal surface. The blocks are accelerated by a horizontal force $\vec{F}$ applied to one of them (Figure 4-40). Find the acceleration and the contact force for (*a*) general values of F, m_1, and m_2, and (*b*) for $F = 3.2$ N, $m_1 = 2$ kg, and $m_2 = 6$ kg.

Figure 4-40 Problem 69

70 •• Repeat the previous problem, but with the two blocks interchanged.

71 •• Two 100-kg blocks are dragged along a frictionless surface with a constant acceleration of 1.6 m/s^2, as shown in Figure 4-41. Each rope has a mass of 1 kg. Find the force F and the tension in the ropes at points A, B, and C.

Figure 4-41 Problem 71

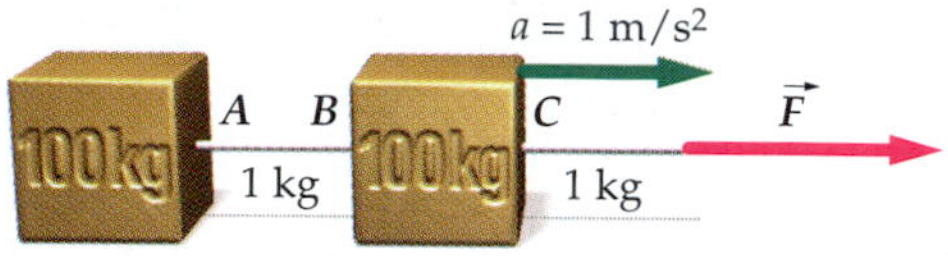

72 •• Two objects are connected by a massless string, as shown in Figure 4-42. The incline and pulley are frictionless. Find the acceleration of the objects and the tension in the string for (*a*) general values of θ, m_1, and m_2, and (*b*) $\theta = 30°$ and $m_1 = m_2 = 5$ kg.

Figure 4-42 Problem 72

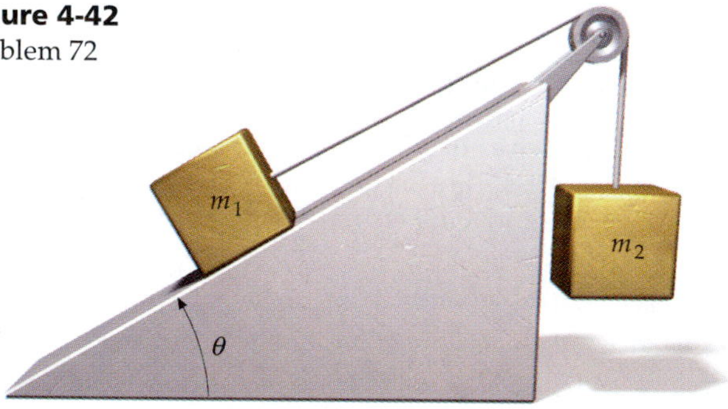

73 •• Two climbers on an icy (frictionless) slope, tied together by a 30-m rope, are in the predicament shown in Figure 4-43. At time $t = 0$, the speed of each is zero, but the top climber, Paul (mass 52 kg), has taken one step too many and his friend Jay (mass 74 kg) has dropped his pick. (*a*) Find the tension in the rope as Paul falls and his speed just before he hits the ground. (*b*) If Paul unhooks his rope after hitting the ground, find Jay's speed as he hits the ground.

Figure 4-43 Problem 73

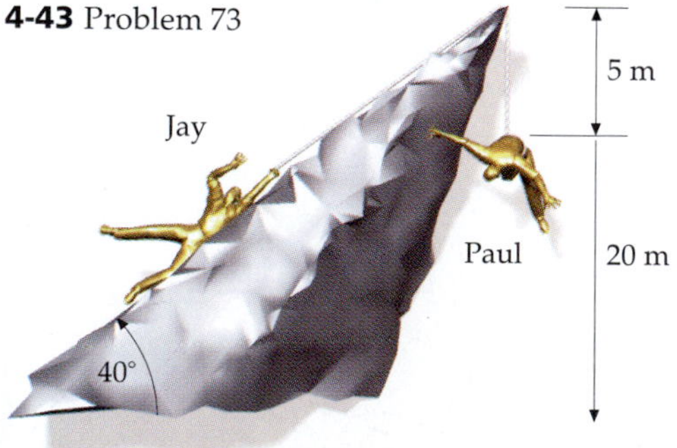

74 • The northwest face of Half Dome, a large rock in Yosemite National Park, makes an angle of $\theta = 7.0°$ with the vertical. Suppose a rock climber lying horizontal on the top is trying to support her unfortunate friend of equal mass who is hanging from a rope over the edge, as shown in Figure 4-44. If the friction is negligible (the top is icy!), at what acceleration will they slide down before the top partner manages to grab someone's hand and stop?

Figure 4-44 Problem 74

75 •• In a stage production of Peter Pan, the 50-kg actor playing Peter has to fly in vertically, and to be in time with the music, he must be lowered a distance of 3.2 m in 2.2 s. Backstage, a smooth surface sloped at 50° supports a counterweight of mass m, as shown in Figure 4-45. Show the calculations that the stage manager must perform to find (*a*) the mass of the counterweight that must be used, and (*b*) the tension in the wire.

Figure 4-45 Problem 75

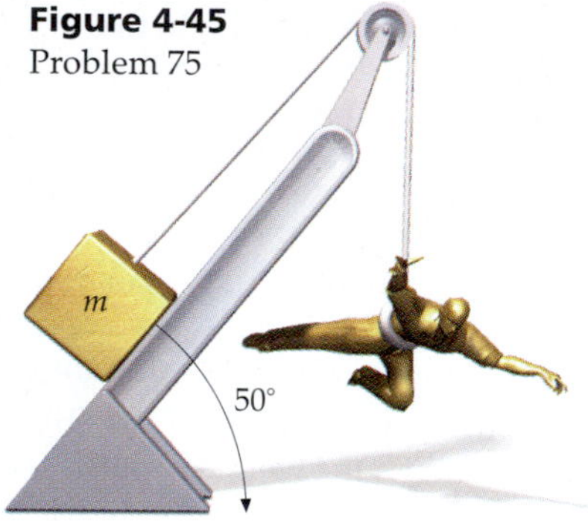

76 •• An 8-kg block and a 10-kg block connected by a rope that passes over a frictionless peg slide on frictionless inclines, as shown in Figure 4-46. (*a*) Find the acceleration of the blocks and the tension in the rope. (*b*) The two blocks are replaced by two others of mass m_1 and m_2 such that there is no acceleration. Find whatever information you can about the mass of these two new blocks.

Figure 4-46 Problem 76

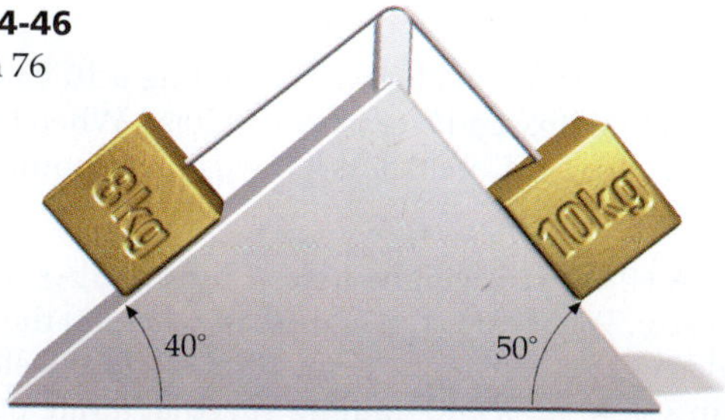

77 •• A heavy rope of length 5 m and mass 4 kg lies on a frictionless horizontal table. One end is attached to a 6-kg block. At the other end of the rope, a constant horizontal force of 100 N is applied. (*a*) What is the acceleration of the system? (*b*) Give the tension in the rope as a function of position along the rope.

78 •• A 60-kg housepainter stands on a 15-kg aluminum platform. The platform is attached to a rope that passes through an overhead pulley, which allows the painter to raise herself and the platform (Figure 4-47). (*a*) To accelerate herself and the platform at a rate of 0.8 m/s^2, with what force must she pull on the rope? (*b*) When her speed reaches 1 m/s, she pulls in such a way that she and the platform go up at a constant speed. What force is she exerting on the rope? (Ignore the mass of the rope.)

Figure 4-47 Problem 78

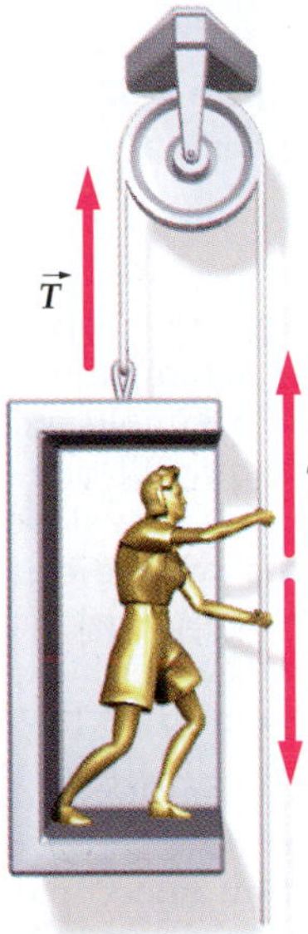

79 ••• Figure 4-48 shows a 20-kg block sliding on a 10-kg block. All surfaces are frictionless. Find the acceleration of each block and the tension in the string that connects the blocks.

Figure 4-48 Problem 79

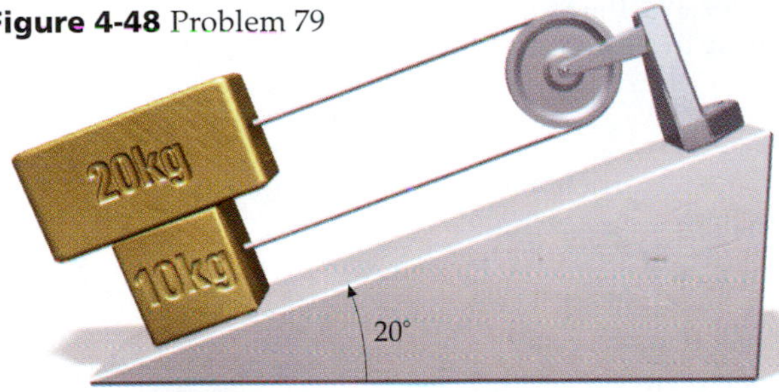

80 ••• A 20-kg block with a pulley attached slides along a frictionless ledge. It is connected by a massless string to a 5-kg block via the arrangement shown in Figure 4-49. Find the acceleration of each block and the tension in the connecting string.

Figure 4-49 Problem 80

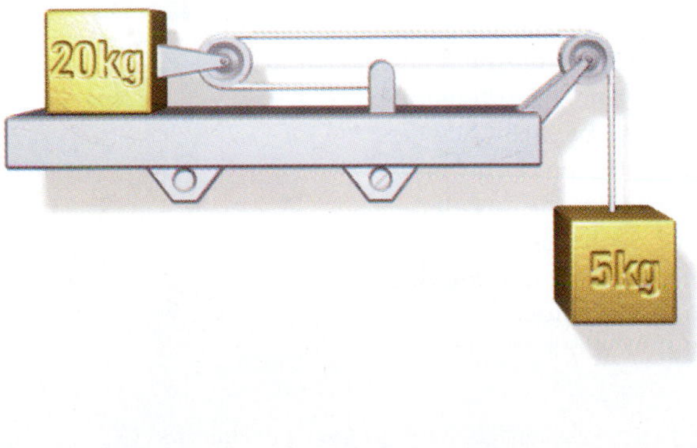

Atwood's Machine

81 •• The apparatus in Figure 4-50 is called an *Atwood's machine* and is used to measure the acceleration due to gravity g by measuring the acceleration of the two blocks. Assuming a massless, frictionless pulley and a massless string, show that the magnitude of the acceleration of either body and the tension in the string are

$$a = \frac{m_1 - m_2}{m_1 + m_2}g \quad \text{and} \quad T = \frac{2m_1 m_2 g}{m_1 + m_2}$$

Figure 4-50 Problems 81–84

82 •• If one of the masses of the Atwood's machine in Figure 4-50 is 1.2 kg, what should the other mass be so that the displacement of either mass during the first second following release is 0.3 m?

83 •• A small pebble of mass m rests on the block of mass m_2 of the Atwood's machine in Figure 4-50. Find the force exerted by the pebble on m_2.

84 •• Find the force exerted by the Atwood's machine on the hanger to which the pulley is attached, as shown in Figure 4-50, while the blocks accelerate. Neglect the mass of the pulley. Check your answer by considering appropriate variations for m_1 and/or m_2.

85 ••• The acceleration of gravity g can be determined by measuring the time t it takes for a mass m_2 in an Atwood's machine to fall a distance L, starting from rest. (*a*) Find an expression for g in terms of m_1, m_2, L, and t. (*b*) Show that if there is a small error in the time measurement dt, it will lead to an error in the determination of g by an amount dg given by $dg/g = -2\,dt/t$. If $L = 3$ m and m_1 is 1 kg, find the value of m_2 such that g can be measured with an accuracy of ±5% with a time measurement that is accurate to 0.1 s. Assume that the only significant uncertainty in the measurement is the time of fall.

General Problems

86 • True or false:

(*a*) If there are no forces acting on an object, it will not accelerate.
(*b*) If an object is not accelerating, there must be no forces acting on it.
(*c*) The motion of an object is always in the direction of the resultant force.
(*d*) The mass of an object depends on its location.

87 • A skydiver of weight w is descending near the surface of the earth. What is the magnitude of the force exerted by her body *on the earth*?

(*a*) w (*b*) Greater than w (*c*) Less than w (*d*) $9.8w$
(*e*) 0 (*f*) It depends on the air resistance.

88 • The net force on a moving object is suddenly reduced to zero. As a consequence, the object

(*a*) stops abruptly.
(*b*) stops during a short time interval.
(*c*) changes direction.
(*d*) continues at constant velocity.
(*e*) changes velocity in an unknown manner.

89 • A force of 12 N is applied to an object of mass m. The object moves in a straight line, with its speed increasing by 8 m/s every 2 s. Find m.

90 • A certain force F_1 gives an object an acceleration of $6 \times 10^6\ \mathrm{m/s^2}$. Another force F_2 gives the same object an acceleration of $15 \times 10^6\ \mathrm{m/s^2}$. What is the acceleration of the object if (*a*) the two forces act together on the object in the same direction; (*b*) the two forces act in opposite directions on the object; (*c*) the two forces act on the object at 90° to each other?

91 • A certain force applied to a particle of mass m_1 gives it an acceleration of $20\ \mathrm{m/s^2}$. The same force applied to a particle of mass m_2 gives it an acceleration of $50\ \mathrm{m/s^2}$. If the two particles are tied together and the same force is applied to the pair, find the acceleration.

92 • A 6-kg object is pulled along a frictionless horizontal surface by a horizontal force of 10 N. (*a*) If the object is at rest at $t = 0$, how fast is it moving after 3 s? (*b*) How far does it travel during these 3 s?

93 • If you weigh 125 lb on the earth, what would your weight be in pounds on the moon, where the free-fall acceleration due to gravity is $5.33\ \mathrm{ft/s^2}$?

94 • A redheaded woodpecker hits the bark of a tree extremely hard—the speed of its head reaches approximately $v = 3.5$ m/s before impact. If the mass of the bird's head is 0.060 kg, and the average force acting on the head during impact is $F = 6.0$ N, find (*a*) the acceleration of the head (assuming it is constant); (*b*) the depth of penetration into the bark; (*c*) the time t it takes the woodpecker's head to stop.

95 •• A simple accelerometer can be made by suspending a small object from a string attached to a fixed point on an accelerating object—to the ceiling of a passenger car, for example. When there is an acceleration, the object will deflect and the string will make some angle with the vertical. (*a*) How is the direction in which the suspended object deflects related to the direction of the acceleration? (*b*) Show that the acceleration a is related to the angle θ that the string makes by $a = g \tan\theta$; (*c*) Suppose the accelerometer is attached to the ceiling of an automobile that brakes to rest from 50 km/h in a distance of 60 m. What angle will the accelerometer make? Will the object swing forward or backward?

96 •• The mast of a sloop is supported at bow and stern by stainless steel wires, the forestay and backstay, anchored 10 m apart (Figure 4-51). The 12-m-long mast weighs 800 N and stands vertically on the deck of the sloop. The mast is positioned 3.6 m behind where the forestay is attached. The tension in the forestay is 500 N. Find the tension in the backstay and the force that the mast exerts on the deck of the sloop.

Figure 4-51
Problem 96

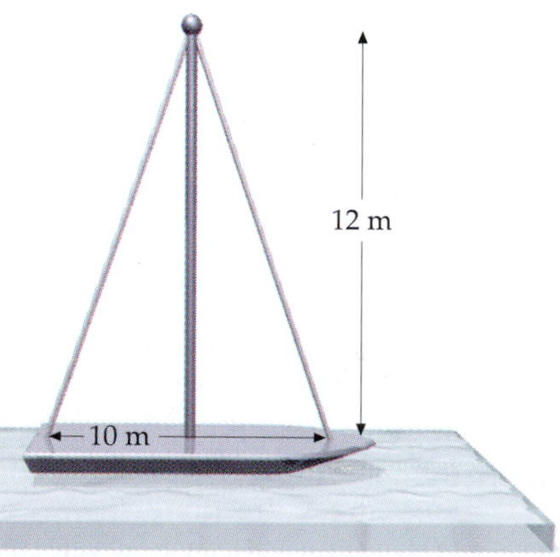

97 •• A block of mass m_1 is pulled along a smooth horizontal surface by a force $\vec{F}$ exerted at the end of a rope that has a much smaller mass m_2, as shown in Figure 4-52. (*a*) Find the acceleration of the rope and block, assuming them to be one object. (*b*) What is the net force acting on the rope? (*c*) Find the tension in the rope at the point where it is attached to the block. (*d*) The diagram, with the rope perfectly horizontal along its length, is not quite accurate. Correct the diagram, and state how this correction affects your solution.

Figure 4-52
Problem 97

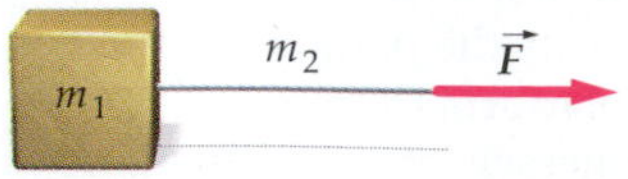

98 •• Joe and Sal are in a rollerbladers' club that is building a ramp to reach new levels of extremeness. The ramp is to be a simple incline, so that after coasting horizontally, a skater will ride up the slope at some angle θ. Sal suggests making the slope as steep as possible to maximize the height that will be reached. Joe whips out a pencil and paper to prove to Sal that, if the surfaces are smooth, the height reached is independent of the angle of the slope. Sal acknowledges that even though Joe is being smug and obnoxious, his argument is sound. Show Joe's proof.

99 •• A car traveling 90 km/h crashes into the rear end of an unoccupied stalled vehicle. Fortunately, the driver is wearing a seat belt. Using reasonable values for the mass of the driver and the stopping distance, estimate the force (assuming it to be constant) exerted on the driver by the seat belt.

100 •• A 2-kg body rests on a frictionless wedge that has an inclination of 60° and an acceleration a to the right such that the mass remains stationary relative to the wedge (Figure 4-53). (*a*) Find a. (*b*) What would happen if the wedge were given a greater acceleration?

Figure 4-53
Problem 100

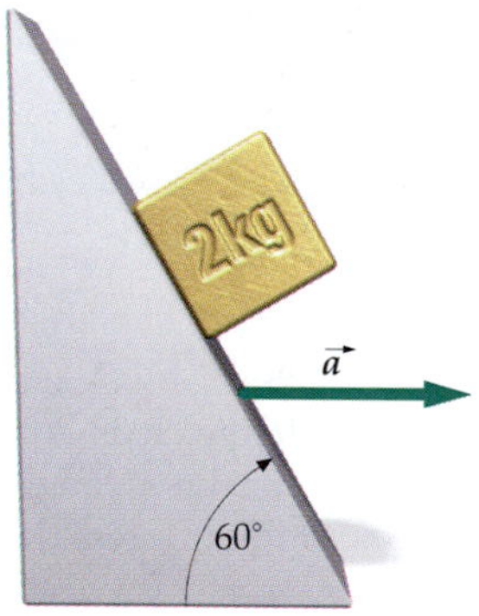

101 •• The masses attached to each side of an Atwood's machine consist of a stack of five washers each of mass m, as shown in Figure 4-54. The tension in the string is T_0. When one of the washers is removed from the left side, the remaining washers accelerate and the tension decreases by 0.3 N. (*a*) Find m. (*b*) Find the new tension and the acceleration of each mass when a second washer is removed from the left side.

Figure 4-54
Problem 101

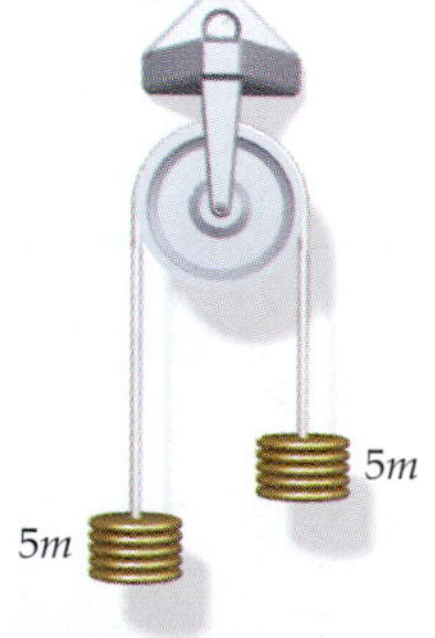

102 •• Consider the Atwood's machine in Figure 4-54. When N washers are transferred from the left side to the right side, the right side drops 47.1 cm in 0.40 s. Find N.

103 •• Blocks of mass m and $2m$ are connected by a string (Figure 4-55). (*a*) If the forces are constant, find the tension in the connecting string. (*b*) If the forces vary with time as $F_1 = Ct$ and $F_2 = 2Ct$, where C is a constant and t is time, find the time t_0 at which the tension in the string is T_0.

Figure 4-55 Problem 103

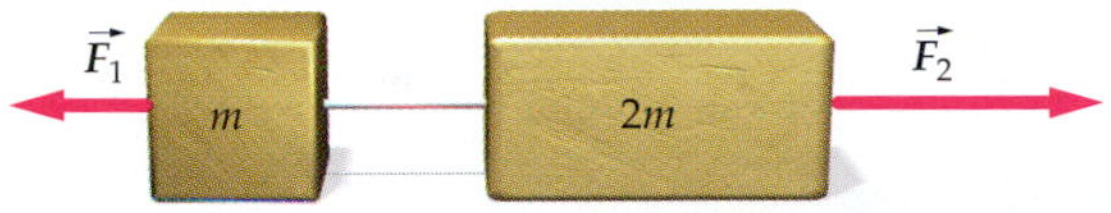

104 ••• Find the normal force and the tangential force exerted by the road on the wheels of your bicycle (*a*) as you climb an 8% grade at constant speed, (*b*) as you descend the 8% grade at constant speed. (An 8% grade means that the angle of inclination θ is given by $\tan \theta = 0.08$).

105 ••• The pulley in an Atwood's machine is given an upward acceleration $\vec{a}$, as shown in Figure 4-56. Find the acceleration of each mass and the tension in the string that connects them.

Figure 4-56 Problem 105

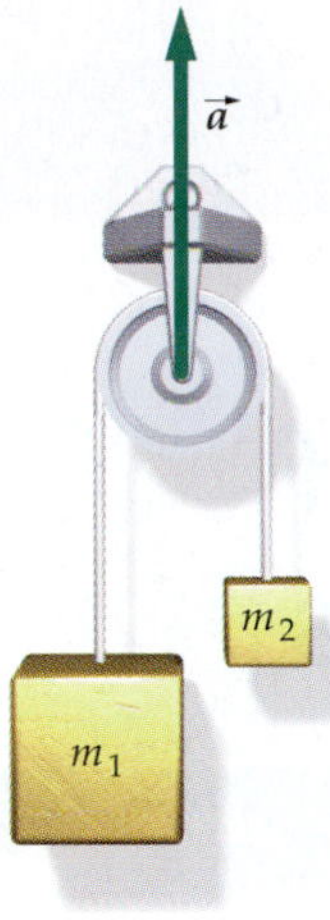

Figure 4-57 Problem 106

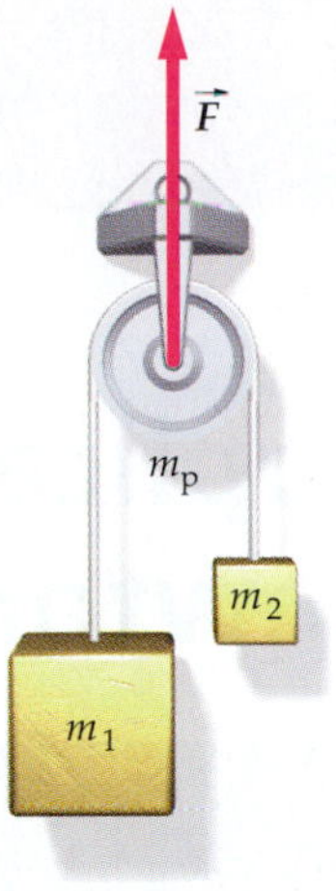

106 ••• The pulley in an Atwood's machine has a mass m_p. A force $\vec{F}$ is exerted on the pulley, as shown in Figure 4-57. Find the acceleration of each mass and the tension in the string that connects them.

CHAPTER 5

Applications of Newton's Laws

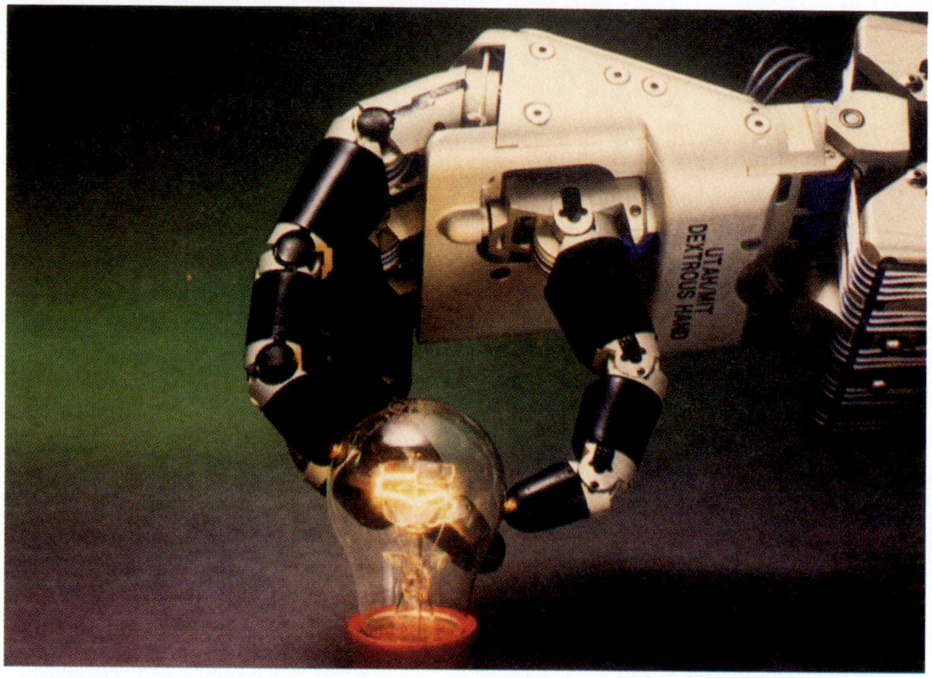

The Utah/MIT Dextrous Hand is a tendon-operated, multidegree-of-freedom dextrous hand that has multichannel touch sensing capability.

In this chapter we give examples of the application of Newton's laws to problems that involve frictional forces and to problems involving circular motion. We will also briefly discuss the motion of an object under the influence of drag forces, which are not constant but depend on the velocity of the object.

5-1 Friction

Static Friction

Friction is a complicated, incompletely understood phenomenon that arises due to the bonding of molecules between two surfaces that are in close contact. This bonding is the same as the molecular bonding that holds an object together. When you apply a small horizontal force to a large box resting on the floor, the box may not move because the force of **static friction**, $\vec{f}_s$, exerted by the floor on the box, balances the force you are applying (Figure 5-1). The force of static fric-

Figure 5-1

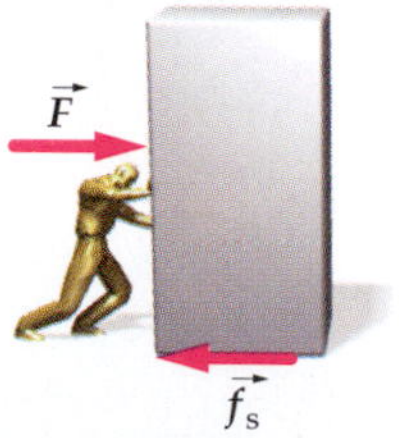

tion, which opposes the applied force, can adjust from zero to some maximum force $f_{s,max}$, depending on how hard you push. You might expect $f_{s,max}$ to be proportional to the area of contact between the two surfaces, but this is not the case. To a good approximation, $f_{s,max}$ is independent of the area of contact and is simply proportional to the normal force exerted by one surface on the other:

$$f_{s,max} = \mu_s F_n \qquad 5\text{-}1$$

Definition—Coefficient of static friction

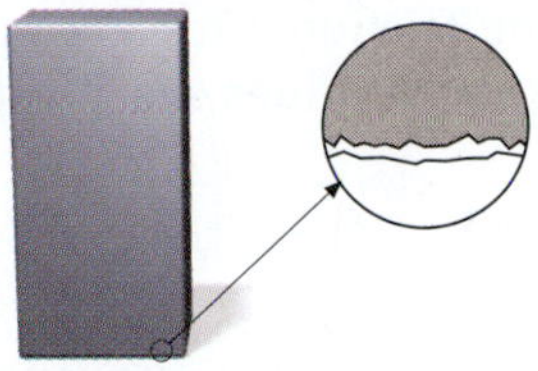

Figure 5-2 The microscopic area of contact between box and floor is only a small fraction of the macroscopic area of the box. This fraction is proportional to the normal force exerted between the surfaces. If the box rests on its side, the macroscopic area is increased, but the force per unit area is decreased by the same factor, so the microscopic area of contact is not changed.

where μ_s is called the **coefficient of static friction,** a dimensionless quantity that depends on the nature of the surfaces in contact (Figure 5-2). If you exert a horizontal force smaller than $f_{s,max}$ on the box, the frictional force will just balance this horizontal force. In general, we can write

$$f_s \leq \mu_s F_n \qquad 5\text{-}2$$

Kinetic Friction

If you push the box in Figure 5-1 hard enough, it will slide across the floor. When the box is sliding, molecular bonds are continually being formed and ruptured, and small pieces of the surfaces are being broken off. The result is a force of **kinetic friction,** $\vec{f}_k$ (also called sliding friction) that opposes the motion. To keep the box sliding with constant velocity, you must exert a force on the box that is equal in magnitude and opposite in direction to the force of kinetic friction exerted by the floor.

The **coefficient of kinetic friction,** μ_k, is defined as the ratio of the magnitudes of the kinetic frictional force f_k and the normal force F_n:

$$f_k = \mu_k F_n \qquad 5\text{-}3$$

Definition—Coefficient of kinetic friction

where μ_k depends on the nature of the surfaces in contact. Experimentally, it is found that μ_k is less than μ_s, and is approximately constant for speeds ranging from about 1 cm/s to several meters per second, the only situations we will consider. To a good approximation μ_k, like μ_s, is independent of the (macroscopic) area of contact.

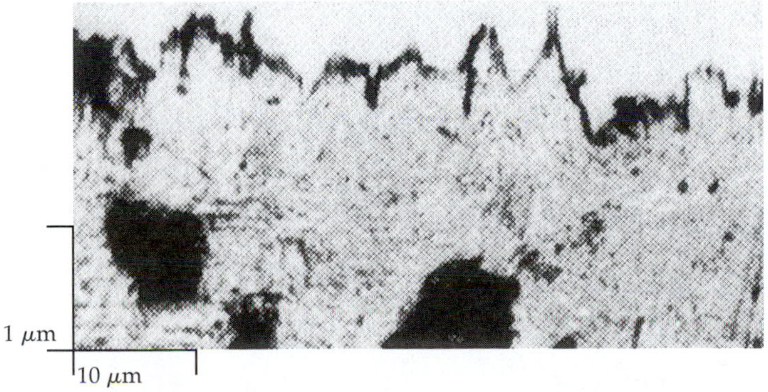

Magnified section of a polished steel surface showing surface irregularities. The irregularities are about 5×10^{-5} cm high, corresponding to several thousand atomic diameters.

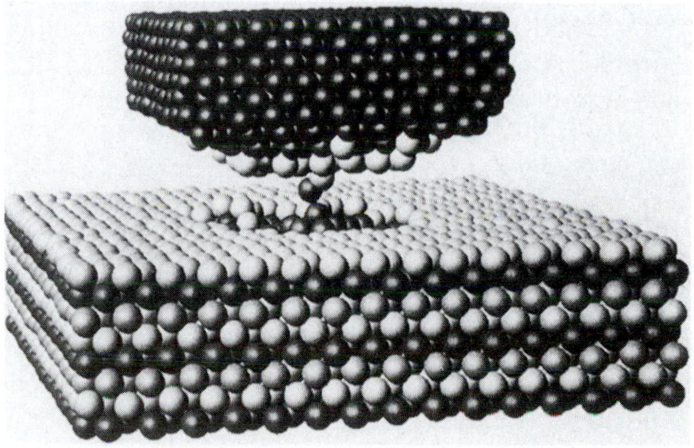

Computer graphic showing gold atoms (bottom) adhering to the fine point of a nickel probe (top) that has been in contact with the gold surface.

Figure 5-3 shows the frictional force exerted on the box by the floor as a function of the applied force. The force of friction balances the applied force until the applied force equals $\mu_s F_n$, at which point the box begins to slide. Then the frictional force is constant and equal to $\mu_k F_n$. Table 5-1 lists some approximate values of μ_s and μ_k for various surfaces.

Figure 5-3

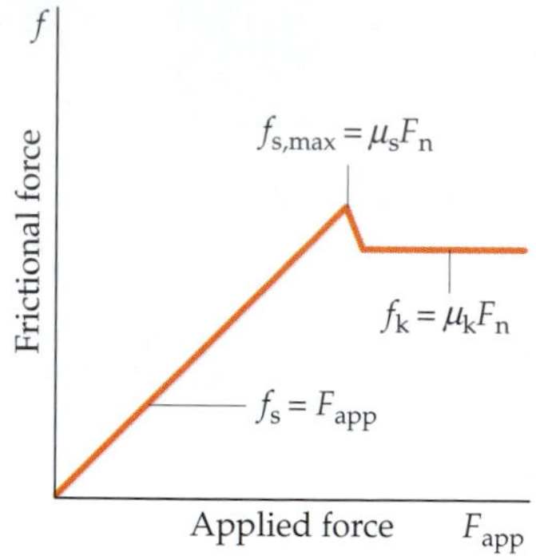

Table 5-1

Approximate Values of Frictional Coefficients

Materials	μ_s	μ_k
Steel on steel	0.7	0.6
Brass on steel	0.5	0.4
Copper on cast iron	1.1	0.3
Glass on glass	0.9	0.4
Teflon on Teflon	0.04	0.04
Teflon on steel	0.04	0.04
Rubber on concrete (dry)	1.0	0.80
Rubber on concrete (wet)	0.30	0.25
Waxed ski on snow (0°C)	0.10	0.05

Example 5-1

A bartender slides a beer stein of mass 0.45 kg horizontally along the bar with an initial speed of 3.5 m/s. The stein comes to rest near the customer after sliding 2.8 m. Find the coefficient of kinetic friction.

Picture the Problem The force of kinetic friction is the only horizontal force acting on the stein (Figure 5-4). Since the frictional force is constant, the acceleration is constant. We can find a_c from the constant-acceleration equations of Chapter 2 and relate it to μ_k using $\Sigma F_x = ma_x$. Choose the direction of motion of the stein to be positive.

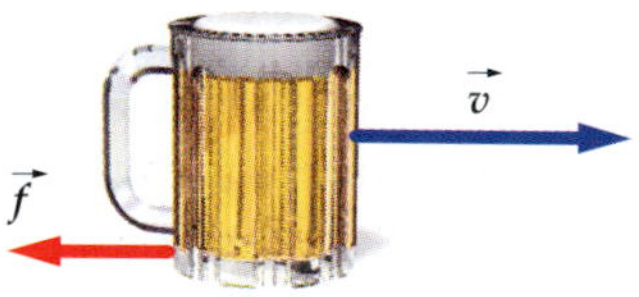

Figure 5-4

1. Draw a free-body diagram for the beer stein:

Figure 5-5

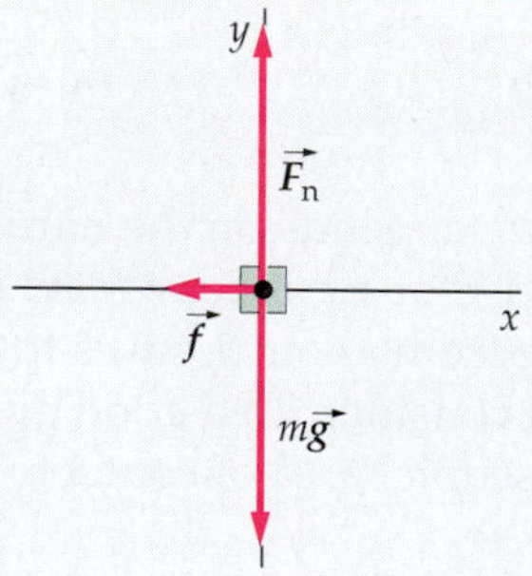

2. The coefficient of friction is related to the frictional force: $f_k = \mu_k F_n$

3. Apply $\Sigma\vec{F} = m\vec{a}$ in component form to the beer stein:

$$\sum F_y = ma_y, \qquad F_n - mg = 0$$

$$\sum F_y = ma_y, \qquad -f_k = ma_x$$

4. Substitute mg for F_n to write the frictional force in terms of mg:

$$f_k = \mu_k F_n = \mu_k mg = -ma_x$$

$$\mu_k = \frac{-a_x}{g}$$

5. Relate the constant acceleration to the total distance traveled and the initial velocity using Equation 2–14:

$$v^2 = v_0^2 + 2a_x\Delta x = 0$$

$$a_x = -\frac{v_0^2}{2\,\Delta x} = -\frac{(3.5\text{ m/s})^2}{2(2.8\text{ m})} = -2.19\text{ m/s}^2$$

6. Substitute this value of a_x to calculate μ_k:

$$\mu_k = -\frac{a_x}{g} = -\frac{-2.19\text{ m/s}^2}{9.81\text{ m/s}^2} = 0.223$$

Remark The mass m of the beer stein cancels. The greater the mass, the harder it is to stop the stein, but the greater mass is also accompanied by greater friction. The net result is that mass has no effect.

Example 5-2

A block rests on an inclined plane surface. The angle of inclination is increased until it reaches a critical angle θ_c, after which the block begins to slide. Find the coefficient of static friction μ_s.

Picture the Problem The forces acting on the block are its weight mg, the normal force F_n exerted by the plane, and the force of friction f (Figure 5-6). At angles less than the critical angle θ_c, the frictional force balances the component $mg \sin\theta$ down the incline. At the critical angle, $f_s = \mu_s F_n$. If we write $\Sigma\vec{F} = 0$, we can relate μ_s to the angle θ_c. Choose the x axis to be parallel to the plane and the y axis to be perpendicular to the plane.

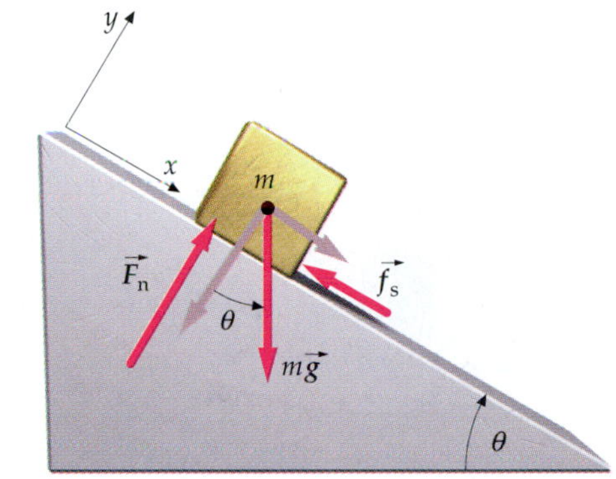

Figure 5-6

1. Draw a free-body diagram for the block:

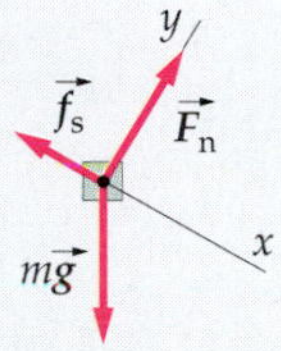

Figure 5-7

2. Apply $\Sigma\vec{F} = m\vec{a}$ in component form to the block:

$$\sum F_y = F_n - mg\cos\theta = ma_y = 0$$

$$\sum F_x = mg\sin\theta - f_s = ma_x = 0$$

3. Substitute $\mu_s F_n$ for f_s in the x equation:

$$mg\sin\theta - \mu_s F_n = 0$$

4. Solve for μ_s using $F_n = mg\cos\theta$ from the y equation:

$$\mu_s = \frac{mg\sin\theta}{F_n} = \frac{mg\sin\theta}{mg\cos\theta} = \tan\theta$$

Exercise The coefficient of static friction between a car's tires and the road on a particular day is 0.7. What is the steepest angle of inclination of the road for which the car can be parked with its wheels locked and not slide down the hill? (*Answer* 35°)

From Example 5-2 we see that the coefficient of static friction is related to the critical angle θ_c at which an object begins to slip by

$$\mu_s = \tan\theta_c \qquad 5\text{-}4$$

Example 5-3

Two children are pulled on a sled over snow-covered ground. The sled, which is initially at rest, is pulled by a rope that makes an angle of 40° with the horizontal. The children have a combined mass of 45 kg and the sled has a mass of 5 kg. The coefficients of static and kinetic friction are $\mu_s = 0.2$ and $\mu_k = 0.15$. Find the frictional force exerted by the ground on the sled and the acceleration of the children and sled, starting from rest, if the tension in the rope is (*a*) 100 N and (*b*) 140 N.

Picture the Problem First we need to find out whether the frictional force is static or kinetic. To do this we compare the maximum frictional force with the horizontal force exerted by the tension in the rope. We choose the coordinate system shown, where $\vec{f}$ is the frictional force, $\vec{F}_n$ is the vertical force exerted by the ground, and $\vec{T}$ is the tension in the rope (Figure 5-8).

Figure 5-8

(*a*) 1. Draw a free-body diagram for the sled:	**Figure 5-9** 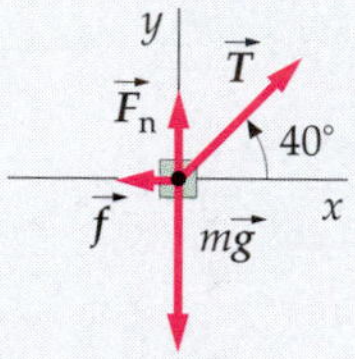
2. The maximum possible force of static friction is related to the magnitude of the normal force F_n:	$f_{s,max} = \mu_s F_n$
3. Apply $\Sigma\vec{F} = m\vec{a}$ in component form to the sled to obtain equations for f and F_n:	$\sum F_x = T_x - f = ma_x$ $\sum F_y = F_n + T_y - mg = ma_y$
4. Compute the horizontal and vertical components of the tension in the rope:	$T_x = T\cos 40° = (100\text{ N})(0.766) = 76.6\text{ N}$ $T_y = T\sin 40° = (100\text{ N})(0.643) = 64.3\text{ N}$
5. F_n can be determined by setting a_y equal to zero:	$F_n + T_y - mg = 0$ $F_n = mg - T_y = (50\text{ kg})(9.81\text{ m/s}^2) - 64.3\text{ N}$ $= 490\text{ N} - 64.3\text{ N} = 426\text{ N}$
6. Substitution of this result into step 1 above gives the maximum frictional force:	$f_{s,max} = \mu_s F_n = 0.2(426\text{ N}) = 85.2\text{ N}$
7. Since the applied horizontal force T_x does not exceed the maximum possible force of static friction, the sled remains at rest. The horizontal component of Newton's second law then gives:	$T_x - f = ma_x = 0$ $f = T_x = 76.6\text{ N}$

(*b*)1. Now we find T_x and T_y for $T = 140$ N:

$$T_x = (140\text{ N})(\cos 40°) = 107\text{ N}$$
$$T_y = (140\text{ N})(\sin 40°) = 90.0\text{ N}$$

2. Apply $\Sigma\vec{F} = m\vec{a}$ in the y direction:

$$\sum F_y = F_n + T_y - mg = ma_y$$

3. Set a_y equal to zero and solve for F_n:

$$F_n + T_y - mg = ma_y = 0$$
$$F_n = mg - T_y = 490\text{ N} - 90.0\text{ N} = 400\text{ N}$$

4. Use this new value of F_n to find the new maximum possible force of static friction:

$$f_{s,max} = \mu_s F_n = 0.2(400\text{ N}) = 80.0\text{ N}$$

5. Since this value for T_x is greater than the maximum force of static friction, the sled will slide. The frictional force on the sled will thus be due to kinetic friction:

$$f_k = \mu_k f_n = 0.15(400\text{ N}) = 60.0\text{ N}$$

6. Apply $\Sigma\vec{F} = m\vec{a}$ in the x direction to find the acceleration of the sled:

$$\sum F_x = T_x - f_k = ma_x$$
$$a_x = \frac{T_x - f_k}{m} = \frac{107\text{ N} - 60.0\text{ N}}{50\text{ kg}} = 0.940\text{ m/s}^2$$

Remark There are two important points in this example: (1) The normal force is not equal to the weight of the children and the sled because the vertical component of the tension helps lift the sled off the ground. (2) In part (*a*), the force of static friction is not equal to $\mu_s F_n$; it is less than this maximum possible limiting value.

Example 5-4 *try it yourself*

The mass m_2 in Figure 5-10 has been adjusted so that the block of mass m_1 is on the verge of sliding. (*a*) If $m_1 = 7$ kg and $m_2 = 5$ kg, what is the coefficient of static friction between the shelf and the block? (*b*) With a slight nudge, the blocks move with acceleration a. Find a if the coefficient of kinetic friction between the shelf and the block is $\mu_k = 0.54$.

Figure 5-10

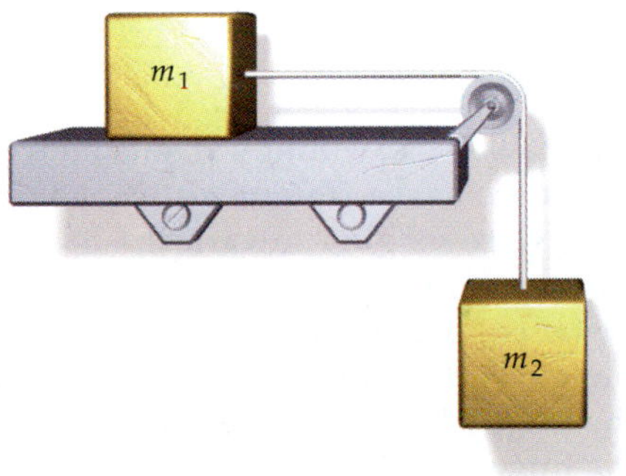

Picture the Problem Apply Newton's second law to each block, using the fact that T has the same magnitude throughout the rope, so $T_1 = T_2$, and that the accelerations have the same magnitude because the rope does not stretch. Choose the positive direction to be rightward for m_1 and downward for m_2.

To find the coefficient of static friction μ_s, as required in part (*a*), set the force of static friction on m_1 equal to its maximum value $f_{max} = \mu_s g$, and set the acceleration equal to zero.

Cover the column to the right and try these on your own before looking at the answers.

Steps | ***Answers***

(*a*)1. Draw a free-body diagram for each block:

Figure 5-11

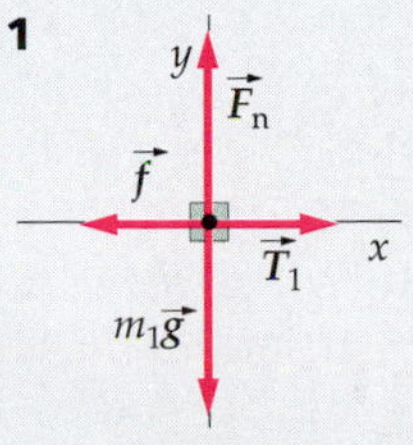

Block 1

Figure 5-12

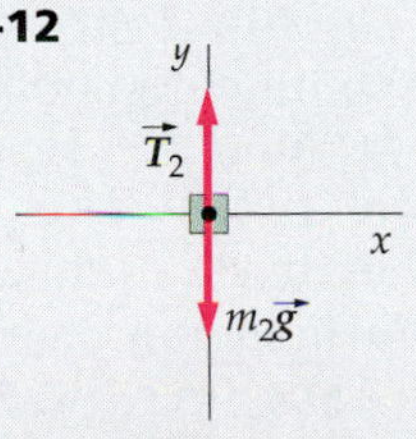

Block 2

2. Apply $\Sigma\vec{F} = m\vec{a}$ in component form to block 1 using $T_1 = T_2 = T$.

$\sum F_x = ma_{1x}, \quad T - f = 0$

$\sum F_y = ma_{1y}, \quad F_n - m_1 g = 0$

3. Apply $\Sigma\vec{F} = m\vec{a}$ to block 2.

$\sum F_y = ma_{2y}, \quad m_2 g - T = 0$

4. Add the two equations containing the tension T to eliminate T, and then solve for f.

$f = m_2 g$

5. Relate f to μ_s and the weight of block 1.

$f = \mu_s F_n = \mu_s m_1 g$

6. Combine your two equations for f and solve for μ_s.

$f = m_2 g = \mu_s m_1 g$

$$\mu_s = \frac{m_2}{m_1} = 0.714$$

(*b*)1. Apply $\Sigma\vec{F} = m\vec{a}$ to the horizontal motion of block 1 and the vertical motion of block 2. Use $f = \mu_k m_1 g$ for the frictional force on block 1 and note that T and a are the same for both blocks.

$\sum F_x = ma_{1x}, \quad T - \mu_k m_1 g = m_1 a$

$\sum F_x = ma_{1y}, \quad m_2 g - T = m_2 a$

2. Eliminate T from your equations in step 1 of part (*b*) and solve for a.

$$a = \frac{m_2 - \mu_k m_1}{m_1 + m_2} g = 0.997 \text{ m/s}^2$$

Check the Result Note that $\mu_k = 0$ gives the result derived in Example 4-12.

Exercise What is the tension in the rope when the blocks are sliding? (*Answer* $T = m_2(g - a) = m_1(a + \mu_k g) = 44.1$ N)

Example 5-5

A runaway baby buggy is sliding without friction across a frozen pond toward a hole in the ice. You race after the buggy on skates. As you grab it, you and the buggy are moving toward the hole at speed v_0. The coefficient of friction between your skates and the ice as you turn out the blades to brake is μ_k. D is the distance to the hole when you reach the buggy, M is the total mass of the buggy, and m is your mass. (*a*) What is the least value of D such that you stop the buggy before it reaches the hole in the ice? (*b*) What force do you exert on the buggy?

Picture the Problem Initially, you and the buggy are moving toward the hole with speed v_0, which we take to be in the positive x direction. If F is the magnitude of the force you exert on the buggy, the net force on the buggy is $-F$ and that on you is $F - f$, where $f = \mu_k mg$, the force of kinetic friction. The minimum value of D is that for which your speed is zero just as you reach the hole (Figure 5-13).

Figure 5-13

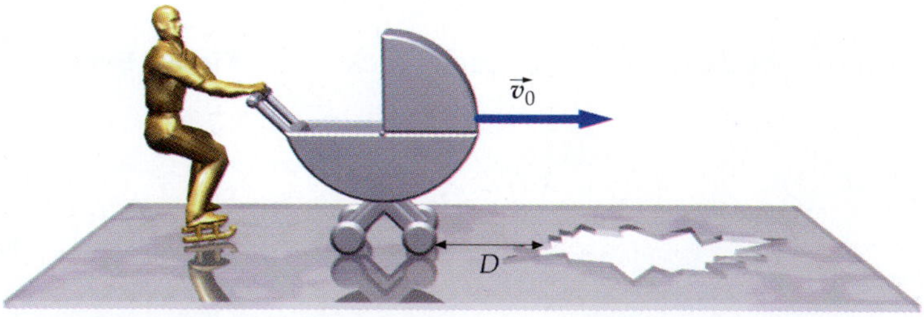

(*a*) 1. Draw free-body diagrams for you and the buggy:

Figure 5-14

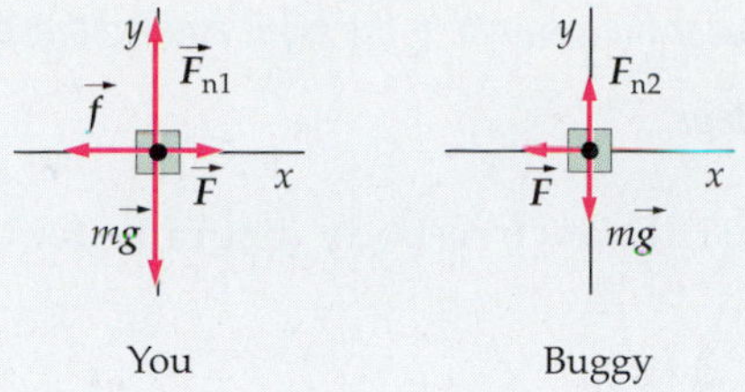

2. The minimum distance D is related to the initial speed v_0 and the final speed $v = 0$:

$$v^2 = v_0^2 + 2a_xD = 0$$

$$D = -\frac{v_0^2}{2a_x}$$

3. Apply $\Sigma\vec{F} = m\vec{a}$ to the motion of both you and the buggy in the x direction to relate the acceleration of each object to the forces acting upon them:

You $\sum F_x = ma_x,\quad F - f = F - \mu_k mg = ma_x$

Buggy $\sum F_x = Ma_x,\quad -F = Ma$

4. Add these two equations to eliminate F, then solve for a_x:

$$-\mu_k mg = (M + m)a_x$$

The acceleration is negative, as expected:

$$a_x = -\left(\frac{m}{M+m}\right)\mu_k g$$

5. Substitute this result for a_x into the equation for D in step 2:

$$D = -\frac{v_0^2}{2a_x} = \left(\frac{M+m}{m}\right)\frac{v_0^2}{2\mu_k g}$$

(*b*) F is found from Newton's second law applied to the buggy:

$$F = -Ma_x = \left(\frac{Mm}{M+m}\right)\mu_k g$$

Remark The minimum value of D is proportional to v_0^2 and inversely proportional to μ_k. Figure 5-15 shows the stopping distance D versus initial velocity squared for values of M/m equal to 0.1, 0.3, and 1.0, with $\mu_k = 0.5$. Note that when the mass of the buggy is larger, a greater stopping distance is required for a given initial velocity. This is somewhat like stopping a car while the car is pulling a trailer that does not have its own brakes. The mass of the trailer increases the stopping distance for a given speed.

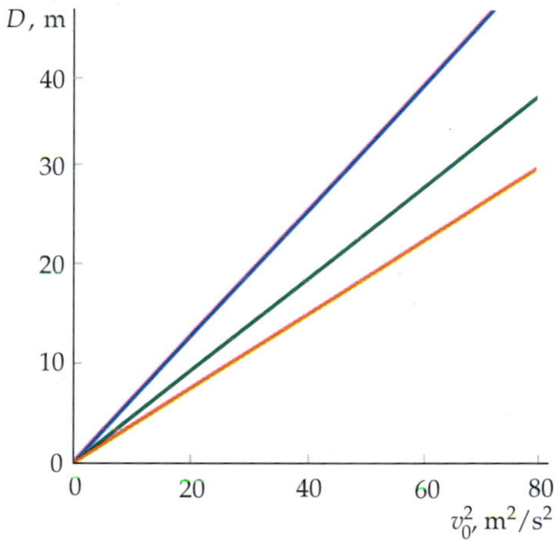

Figure 5-15

Example 5-6 *try it yourself*

A girl of mass m_g sits on a toboggan of mass m_t, which in turn sits on a frozen pond assumed to be frictionless. The toboggan is pulled with a horizontal force $\vec{F}$ (Figure 5-16). The coefficients of static and sliding friction between the girl and toboggan are μ_s and μ_k. (*a*) Find the maximum value of F for which the girl will not slide relative to the toboggan. (*b*) Find the acceleration of the toboggan and girl when F is greater than this value.

Picture the Problem The forces acting on the girl are her weight $\vec{w}_g$, friction $\vec{f}$, and the normal force $\vec{F}_{n1}$ exerted by the toboggan. The forces on the toboggan are its weight $\vec{w}_t$, friction $\vec{f}'$, the applied force $\vec{F}$, and the normal force $\vec{F}'_{n1}$ exerted by the girl. The forces $\vec{f}'$ and $\vec{f}$ are action–reaction forces, as are $\vec{F}_{n1}$ and $\vec{F}'_{n1}$. The vertical forces balance in both cases. The only horizontal force on the girl is that of friction. If static friction is great enough, the girl has the same acceleration as the toboggan. Otherwise, the girl starts to slide, and then kinetic friction causes her acceleration.

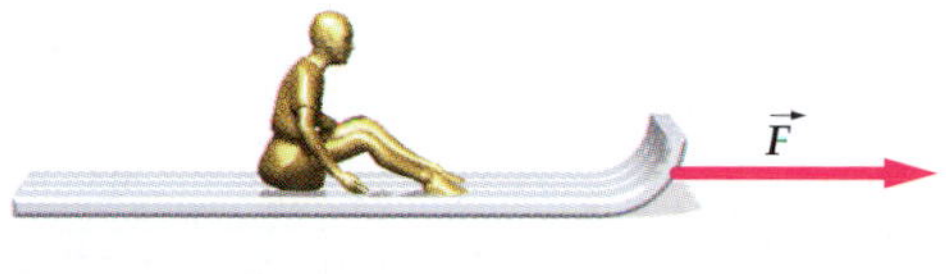

Figure 5-16

Cover the column to the right and try these on your own before looking at the answers.

Steps	***Answers***
(*a*) 1. Draw free-body diagrams for each object.	See Figure 5-17
2. Apply $\Sigma\vec{F} = m\vec{a}$ to the horizontal motions of the girl and the toboggan, using $f' = f$ for the frictional force.	$f = m_g a_1, \quad F - f = m_t a_2$
3. Eliminate the frictional force from your equations to obtain an expression for the force F in terms of the masses and the common acceleration a.	$F = (m_g + m_t)a$
4. The maximum acceleration of the girl occurs when the force of static friction on her is maximum. Find this maximum frictional force and use it to find her maximum acceleration a_{max}.	$f_{s,max} = \mu_s F_{n1} = \mu_s m_g g$ $a_{max} = \mu_s g$
5. Use your result for a_{max} to find the maximum force F_{max} for which the girl does not slide.	$F_{max} = (m_g + m_t)\mu_s g$
(*b*) 1. When the girl slides, the horizontal force on her is that of kinetic friction, f_k. Find f_k and the acceleration of the girl, a_g.	$f_k = \mu_k m_g g, \quad a_g = \mu_k g$
2. Write the horizontal component of $\Sigma\vec{F} = m\vec{a}$ for the toboggan.	$F - f_k = m_t a$
3. Solve for the acceleration of the toboggan a_t using the value of f_k found previously.	$a_t = \dfrac{F - \mu_k m_g g}{m_t}$

Figure 5-17

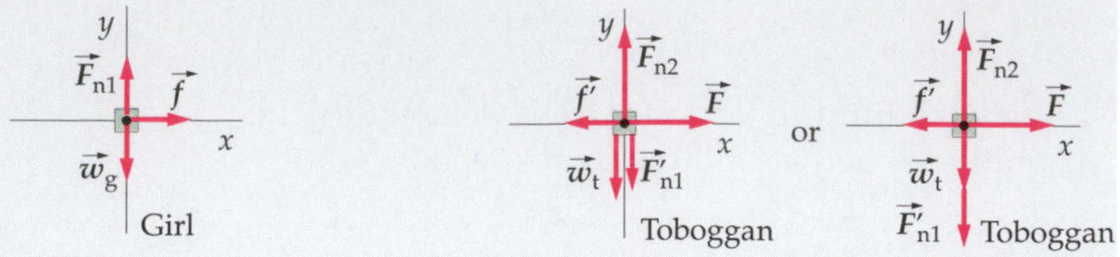

Remarks Two alternative styles for the free-body diagram of the toboggan are shown. In the first, the two downward forces are displaced slightly so they can be seen. In the second, they are drawn head to tail. In Figure 5-18, the accelerations of the toboggan and the girl are shown as functions of the applied force F.

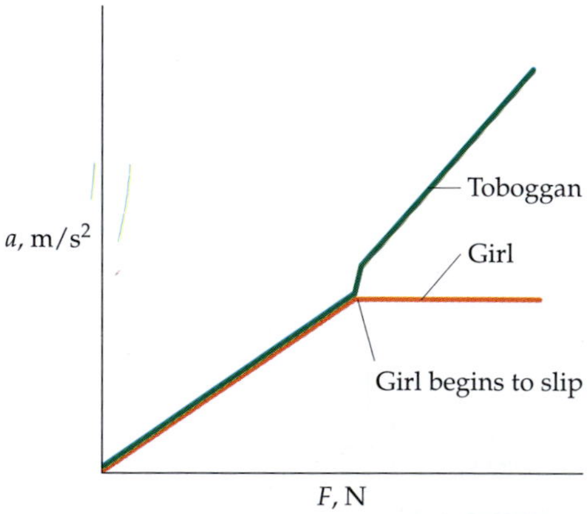

Figure 5-18 The accelerations of the toboggan and the girl as functions of the applied force F. When the friction is static, they move together with the same acceleration. When the girl begins to slip, the frictional force is smaller ($\mu_k m_g g$ rather than $\mu_s m_g g$), so the net force on the toboggan is greater and the slope of a_t versus F is greater. While the girl slips on the toboggan, the net force on the girl is constant ($\mu_k m_g g$), so the acceleration of the girl is constant.

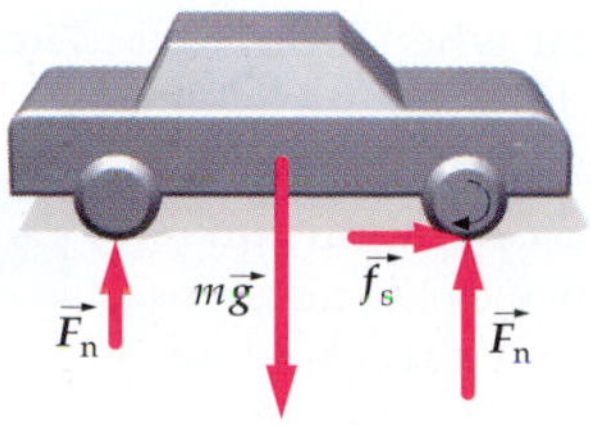

Figure 5-19 Forces acting on a car with front-wheel drive. The normal forces $\vec{F}_n$ are not generally equal on the front and rear tires.

Figure 5-19 shows the forces acting on a car that is just starting to move from rest. The weight of the car is balanced by the normal force F_n exerted on the tires. To start the car moving, the engine delivers power to the axle that makes the wheels rotate (we discuss power in Chapter 6). If the road were perfectly frictionless, the wheels would merely spin. When friction is present, the frictional force exerted by the road on the tires is in the forward direction and provides the force needed to accelerate the car. If the power delivered by the engine is small enough so that the force exerted by the tire surface on the road surface is not too great, the two surfaces do not slip. Then the wheels roll without slipping and the tire tread touching the road is at rest relative to it (Figure 5-20). The friction between the road and the tire is then static friction. The largest frictional force that the tires can exert on the road (and that the road can exert on the tires) is $\mu_s F_n$.

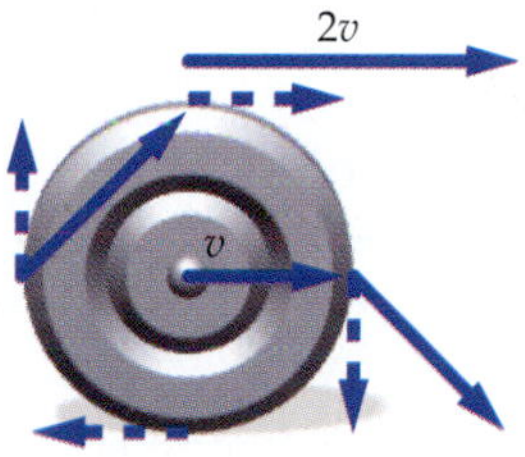

Figure 5-20 When a wheel rolls without slipping, each point on the rim has a velocity of magnitude v relative to the center of the wheel, where v is the speed of the center of the wheel relative to the ground. The velocity of the point on the tire in contact with the ground is zero relative to the ground. In this figure, dashed lines represent velocities relative to the center of the wheel, and solid lines represent velocities relative to the ground.

If the power delivered by the engine is too great, the surfaces in contact will slip and the wheels will spin. Then the force that accelerates the car is the force of kinetic friction, which is less than the force of static friction. If we are stuck on ice or snow, our chances of getting free are better if we use a light touch on the accelerator pedal. Similarly, when braking a car to a stop, the force exerted by the road on the tires may be either static friction or kinetic friction, depending on how the brakes are applied. If the brakes are applied so hard that the wheels lock, the tires will slide along the road and the stopping force will be that of kinetic friction. If the brakes are applied gently, so that no slipping occurs between the tires and the road, the stopping force will be that of static friction. Antilock braking systems in cars allow you to brake hard without locking the wheels to provide maximum friction for stopping.

When an ideal, rigid wheel rolls *at constant speed* along a horizontal road without slipping, no force accelerates it. But because a real tire continually deforms, and the tread and road are continually peeled apart, a small force is needed to maintain the constant velocity. The **coefficient of rolling friction,** μ_r, is the ratio of the force needed to keep a wheel rolling at constant velocity on a level surface to the normal force exerted by the surface on the wheel. Typical values of μ_r are 0.01 to 0.02 for rubber tires on concrete, and 0.001 to 0.002 for steel wheels on steel rails.

Example 5-7

A car is traveling at 30 m/s along a horizontal road. The coefficients of friction between the road and the tires are $\mu_s = 0.5$ and $\mu_k = 0.3$. How far does the car travel before stopping if (*a*) the car is braked with an antilock braking system so that the wheels do not slip, and (*b*) the car is braked hard with no antilock braking system so that the wheels lock?

Picture the Problem The force that stops a car when it brakes is the force of friction exerted by the road on the tires (Figure 5-21). Since frictional force exerted by the road is constant, the acceleration is constant, and we can use the constant acceleration equations of Chapter 2 to relate the stopping distance to the acceleration. We then find the acceleration from Newton's second law. We take the direction of motion to be the positive x direction.

Figure 5-21

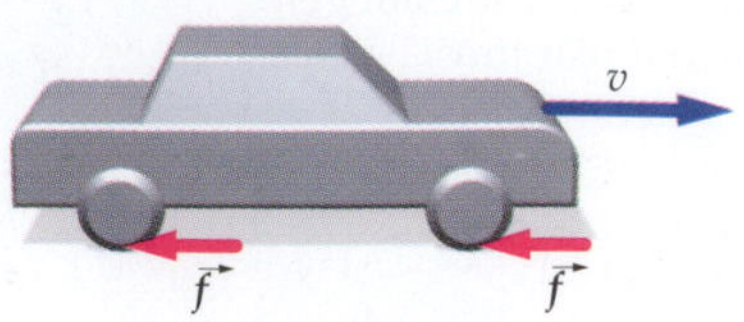

(*a*)1. Draw a free-body diagram for the car. Treat all four wheels as if they were a single point of contact with the ground. Assume further that the brakes are applied to all four wheels:

Figure 5-22

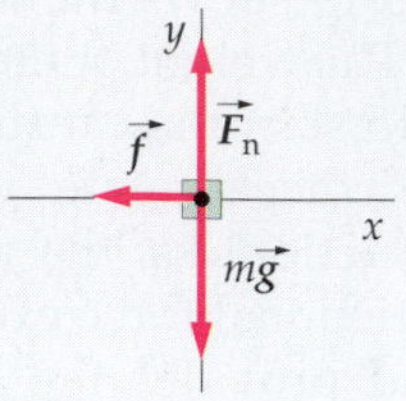

2. Equation 2-14 relates the stopping distance Δx to the initial speed v_0:

$$v^2 = v_0^2 + 2a_x\Delta x = 0$$

$$\Delta x = -\frac{v_0^2}{2a_x}$$

3. Apply $\Sigma\vec{F} = m\vec{a}$ to the car. Since the wheels do not slip, the horizontal force exerted by the road is that of static friction:

$$\sum F_x = ma_x, \qquad -\mu_s F_n = ma_x$$

$$\sum F_y = ma_y, \qquad F_n - mg = m(0) = 0$$

4. Since there is no vertical acceleration, the normal force F_n exerted by the road balances the weight mg of the car. Substitute mg for F_n and solve for a_x:

$$F_n = mg$$

$$a_x = \frac{-\mu_s F_n}{m} = \frac{-\mu_s mg}{m} = -\mu_s g$$

$$= -(0.5)(9.81\text{ m/s}^2) = -4.90\text{ m/s}^2$$

5. Substituting these results in the equation for Δx in step 2 gives the stopping distance:

$$\Delta x = -\frac{v_0^2}{2a}$$

$$= -\frac{(30\text{ m/s})^2}{2(-4.90\text{ m/s}^2)}$$

$$= 91.8\text{ m}$$

(*b*)1. When the wheels lock, the force exerted by the road on the car is that of kinetic friction. Using reasoning similar to that in part (*a*), we obtain for the acceleration:

$$a_x = -\mu_k g = -(0.3)(9.81\text{ m/s}^2)$$

$$= -2.94\text{ m/s}^2$$

2. The stopping distance is then:

$$\Delta x = -\frac{v_0^2}{2a}$$

$$= -\frac{(30\text{ m/s})^2}{2(-2.94\text{ m/s}^2)} = 153\text{ m}$$

Remark The stopping distance is more than 50% greater when the wheels are locked. This is why antilock braking systems were developed for automobiles. Also note that the stopping distance is independent of the car's mass—the stopping distance is the same for a subcompact or a large truck, provided the coefficients of friction are the same.

Exercise What must the coefficient of static friction be between the road and the tires of a four-wheel-drive car if the car is to accelerate from rest to 25 m/s in 8 s? (*Answer* 0.319)

5-2 Circular Motion

Centripetal Acceleration

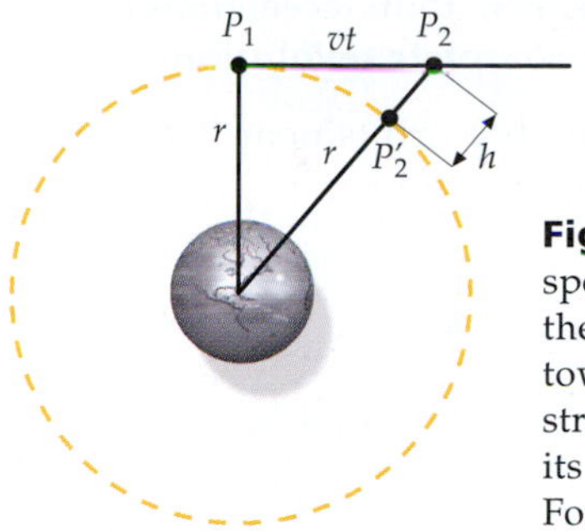

Figure 5-23 The satellite is moving with speed v in a circular orbit of radius r about the earth. If the satellite did not accelerate toward the earth, it would move in a straight line from point P_1 to P_2. Because of its acceleration, it instead falls a distance h. For small time t, $h = \frac{1}{2}(v^2/r)\,t^2 = \frac{1}{2}\,at^2$.

Newton showed that a particle moving with constant speed v in a circle of radius r has an acceleration of magnitude v^2/r directed radially inward toward the center of the circle. This acceleration, called **centripetal acceleration**, requires a net force directed toward the center of the circle. Figure 5-23 shows a satellite moving in a circular orbit around the earth. At an altitude of 200 km, the gravitational force on the satellite is just slightly less than at the earth's surface. Why doesn't the satellite fall toward the earth? Actually, the satellite does "fall." But because of its horizontal velocity, it continually misses the earth. If the satellite in Figure 5-23 were not accelerating, it would move from point P_1 to P_2 in some time t. Instead, it arrives at point P_2' on its circular orbit. In a sense, the satellite "falls" the distance h shown in Figure 5-23. If t is small, P_2 and P_2' are nearly on a radial line. In that case we can calculate h from the right triangle of sides vt, r, and $r + h$. Since $r + h$ is the hypotenuse of the right triangle, the Pythagorean theorem gives

$$(r + h)^2 = (vt)^2 + r^2$$

$$r^2 + 2hr + h^2 = v^2t^2 + r^2$$

or

$$h(2r + h) = v^2t^2$$

For very short times, h will be much less than r, so we can neglect h compared with $2r$ for the term in parentheses. Then

$$2rh \approx v^2t^2$$

or

$$h \approx \frac{1}{2}\left(\frac{v^2}{r}\right)t^2$$

Comparing this with the constant-acceleration expression $h = \frac{1}{2}at^2$, we see that the magnitude of the acceleration of the satellite is

$$a = \frac{v^2}{r} \qquad 5\text{-}5$$

Centripetal acceleration

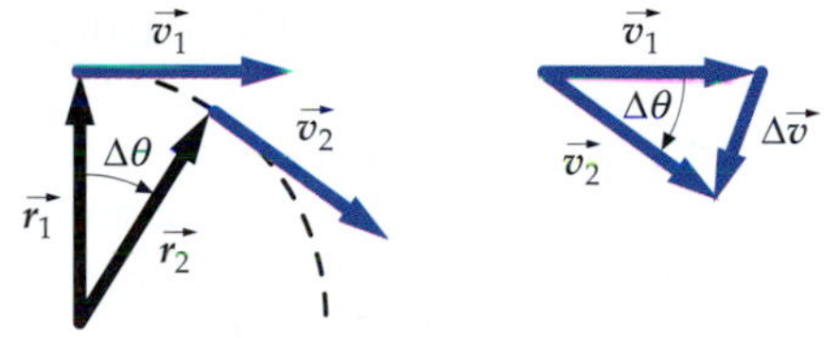

Figure 5-24 Position and velocity vectors for a particle moving in a circle at constant speed. The angle $\Delta\theta$ between $\vec{v}_1$ and $\vec{v}_2$ is the same as that between $\vec{r}_1$ and $\vec{r}_2$ because the position and velocity vectors must both move through equal angles to remain mutually perpendicular. For very small time intervals, the velocity change $\Delta\vec{v}$ is approximately perpendicular to $\vec{v}$ and points inward toward the center of the circle. The magnitude of the acceleration can be found from $\Delta\theta \approx v\,\Delta t/r$ and $\Delta\theta = \Delta v/v$. Then $\Delta v/\Delta t \approx v^2/r$.

From Figure 5-23 we see that the direction is inward toward the center of the circle. A geometric proof that this result holds in general for circular motion with constant speed is given in Figure 5-24. An algebraic proof is outlined in Problem 108.

The motion of a particle moving in a circle with constant speed is often described in terms of the time required for one complete revolution T, called the **period.** During one period, the particle travels a distance of $2\pi r$ (where r is the radius of the circle), so its speed is related to r and T by

$$v = \frac{2\pi r}{T} \qquad 5\text{-}6$$

Example 5-8

A satellite moves at constant speed in a circular orbit about the center of the earth and near the surface of the earth. If its acceleration is 9.81 m/s^2, find (*a*) its speed and (*b*) the time for one complete revolution.

Picture the Problem Since the satellite orbits near the surface of the earth, we take the radius of the orbit to be the radius of the earth, $r = 6370$ km.

(*a*) Setting the centripetal acceleration v^2/r equal to g yields the speed v:

$$a = \frac{v^2}{r} = g \quad \text{or}$$

$$v = \sqrt{rg} = \sqrt{(6370\text{ km})(9.81\text{ m/s}^2)}$$

$$= 7.91\text{ km/s} = 17{,}700\text{ mi/h}$$

(*b*) We use Equation 5-6 to get the period T:

$$T = \frac{2\pi r}{v} = \frac{2\pi(6370\text{ km})}{7.91\text{ km/s}} = 5060\text{ s} = 84.3\text{ min}$$

Remark For satellites in orbit a few hundred kilometers above the earth's surface, the orbital radius r is slightly greater than 6370 km. As a result, the centripetal acceleration is slightly less than 9.81 m/s^2 because of the decrease in the gravitational force with distance from the center of the earth. Many satellites are launched into such orbits, and their periods are roughly 90 min.

Exercise A car rounds a curve of radius 40 m at 48 km/h. What is its centripetal acceleration? (*Answer* 4.44 m/s^2)

A particle moving in a circle with *varying speed* has a component of acceleration tangent to the circle, dv/dt, as well as the radially inward centripetal acceleration, v^2/r. For general motion along a curve, we can treat a portion of the curve as an arc of a circle (Figure 5-25). The particle then has acceleration v^2/r toward the center of curvature, and if the speed is changing, it has tangential acceleration dv/dt.

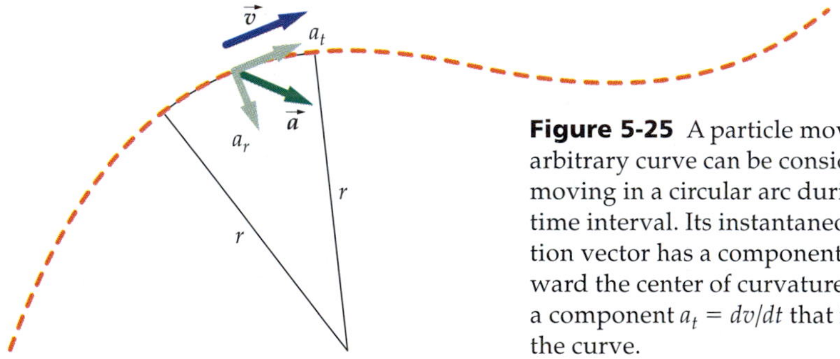

Figure 5-25 A particle moving along an arbitrary curve can be considered to be moving in a circular arc during a small time interval. Its instantaneous acceleration vector has a component $a_r = v^2/r$ toward the center of curvature of the arc and a component $a_t = dv/dt$ that is tangential to the curve.

Centripetal Force

As with any acceleration, there must be a net force in the direction of the acceleration to produce it. For centripetal accelerations, this force is called the **centripetal force.** It is *not* a new kind of force, but merely a name for the force needed for circular motion. The centripetal force may be due to a string, spring, or other contact force such as a normal force or friction; it may be an action-at-a-distance type of force such as a gravitational force, or it may be any combination of these. It is always directed inward, toward the center of the circle of motion.

Example 5-9

You swing a pail of water in a vertical circle of radius r. The speed of the pail is v_t at the top of the circle. (*a*) Find the force exerted on the water by the pail at the top of the circle. (*b*) Find the minimum value of v_t for the water to remain in the pail. (*c*) Find the force exerted by the pail on the water at the bottom of the circle, where the pail's speed is v_b.

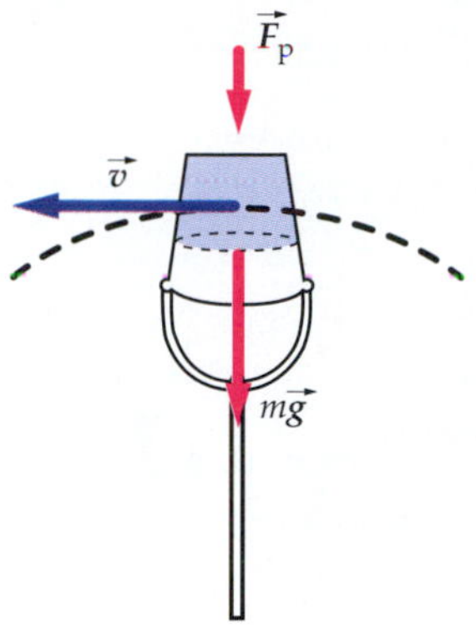

Figure 5-26

Picture the Problem We apply Newton's second law to find the force exerted by the pail. Two forces act on the water: gravity, $m\vec{g}$, and the force of the pail on the water, $\vec{F}_p$. Since the water moves in a circular path, it has a centripetal acceleration toward the center of the circle (Figure 5-26). We choose up for the positive y direction. Then at the top of the circle the acceleration is downward, $a_{y,\text{top}} = -v_t^2/r$; at the bottom it is upward, $a_{y,\text{bottom}} = +v_b^2/r$.

(*a*)1. Draw free-body diagrams for the water at the top and bottom of the circle.

Figure 5-27

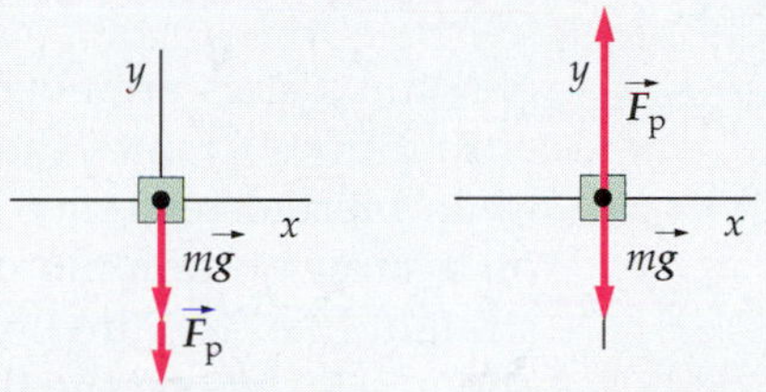

2. Apply $\Sigma\vec{F} = m\vec{a}$ to the water at the top of the circle. Both $\vec{F}_p$ and the weight are toward the center of the circle in the negative y direction:

$$\sum F_y = ma_y = m\left(-\frac{v_t^2}{r}\right)$$

$$-F_p - mg = m\left(-\frac{v_t^2}{r}\right) \quad \text{or} \quad F_p = m\frac{v_t^2}{r} - mg$$

(*b*) The pail cannot exert an upward force on the water at the top of the circle. The minimum force it can exert is zero. Set $F_p = 0$ and solve for the minimum speed, $v_{t,\min}$:

$$0 = m\frac{v_{t,\min}^2}{r} - mg \quad \text{or} \quad v_{t,\min} = \sqrt{rg}$$

(*c*) Apply $\Sigma\vec{F} = m\vec{a}$ to the water at the bottom of the circle where $\vec{F}_p$ and the acceleration are upward:

$$\sum F_y = ma_y = m\left(\frac{v_b^2}{r}\right)$$

$$F_p - mg = m\frac{v_b^2}{r} \quad \text{or} \quad F_p = m\frac{v_b^2}{r} + mg$$

Remark Note that there is no arrow for centripetal force in the free-body diagrams. Centripetal force is not a kind of force exerted by some agent; it is just the name for the resultant force that must point toward the center of the circle to provide the centripetal acceleration. When a whirling bucket is at the top of its circle, both gravity and the contact force of the pail contribute to the necessary centripetal force on the water. When the water is moving at the minimum speed at the top of the circle, its acceleration is $\vec{g}$, the free-fall acceleration due to gravity, and the only force acting on it at this point is its weight, $m\vec{g}$. At the bottom of the circle, F_p must be greater than the weight mg by enough to provide the necessary centripetal force.

Check the Result When $v = 0$ at the bottom, $F_p = mg$.

Exercise Estimate (*a*) the minimum speed at the top of the circle, and (*b*) the maximum period of revolution that will keep you from getting wet if you swing a pail of water in a vertical circle at constant speed. (*Answers* (*a*) Assuming $r \sim 1$ m, we find $v_{t,\min} \sim 3$ m/s, (*b*) $T = (2\pi r/v) \sim 2$ s)

Exploring

Noninertial Reference Frames, Pseudoforces, and Cyclones

Newton's laws are not valid in noninertial reference frames (frames that are accelerating relative to an inertial reference frame). Imagine standing on a subway train moving at constant speed in a straight line. You are holding onto a vertical pole, and a ball rests on the floor. When the train slows for the next station, the ball immediately accelerates forward until it hits the front of the car. You would also accelerate forward if you didn't hold onto the pole. Since there is no apparent force on the ball, its acceleration violates Newton's second law. Similarly, you feel the force of the pole pulling you backward even though, relative to the train, you are not accelerating. There seems to be a very real force acting in the forward direction trying to push you, the ball, and everything else toward the front of the car.

For an observer standing in the station, the ball continues to move forward with constant speed (since no net force acts on it) and the train slows (accelerates backward) so the ball hits the front of the car. The backward force exerted by the pole on you causes you to decelerate along with the train. For the observer in the station, who is in an inertial reference frame, Newton's laws are obeyed.

Your frame, which is accelerating relative to the station, is a noninertial reference frame. You can use $\Sigma\vec{F} = m\vec{a}$ in your frame if you introduce a so-called **pseudoforce** that acts on each object, $\vec{F}_p = -m\vec{a}_t$, where m is the mass of the object. No agent exerts this pseudoforce; it is a fictitious force that exists only in your noninertial frame.

Now imagine riding on the rim of a merry-go-round, again holding onto a vertical pole. You have to grip the pole to keep from flying off. It seems like there is an outward force acting on you and everything else in this frame. This pseudoforce is called a centrifugal ("center-fleeing")

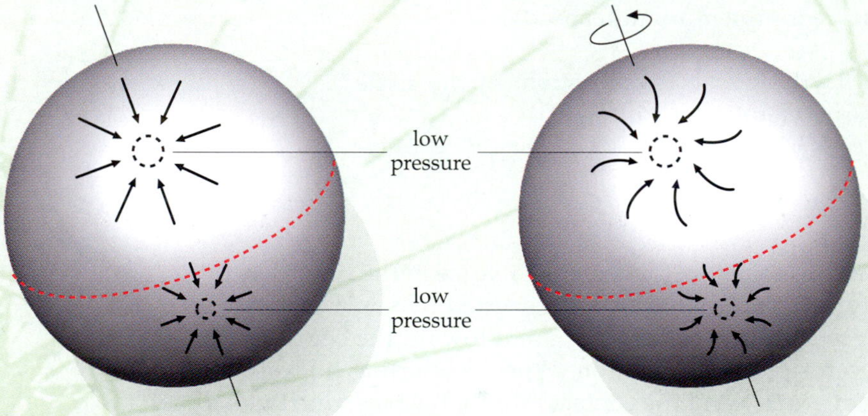

Figure 1 Suppose a low-pressure system develops at some middle latitude in the Northern Hemisphere. Since the pressure is low, air flows toward that region from all directions. Air moving down from the north is like the ball moving from the center of the platform toward the rim; it is deflected sideways to the west. Air moving up from the south is like a ball thrown from the rim toward the center, and the earth's rotation causes the air to move more rapidly eastward than the low-pressure region; thus, it outraces the low and deflects to the east. The net effect of the deflection of the winds sideways is a counterclockwise circulation pattern established around the low-pressure area, as indicated by the red arrows in the figure. This weather pattern is known as a cyclone; hurricanes are particularly fierce cyclones. Counterclockwise circulation is characteristic of cyclones in the Northern Hemisphere; in the Southern Hemisphere, cyclones rotate in a clockwise direction. On a smaller scale, the Coriolis effect causes water draining in a bathtub to rotate counterclockwise in the Northern Hemisphere and clockwise in the Southern Hemisphere.

Typhoon Pat photographed over the Western Pacific by astronauts in the Space Shuttle *Discovery*. Shown in the photo is the counterclockwise rotation of the typhoon due to the Coriolis effect. Similar phenomena occur elsewhere in the universe with respect to other rotating bodies. An example is the Great Red Spot on Jupiter, a cyclone so vast that the earth could fit comfortably inside. The Great Red Spot has endured for hundreds of years, unlike the relatively short-lived hurricanes on earth.

force. Again, no agent exerts this fictitious force; it exists only in the noninertial rotating frame. From the frame of reference of an observer on the ground, you are hanging onto the pole so that it will exert the centripetal force required to make you move in a circular path.

There is another pseudoforce that occurs in rotating frames that is interesting because of its connection with weather patterns. Suppose you toss a ball from the center of the merry-go-round to a friend who rides at the rim. You throw the ball straight to your friend, but because your friend is moving sideways, the ball misses its target. From your point of view, the ball was deflected sideways away from your friend, as if a sideways force had acted on it. This sideways pseudoforce, which depends on the velocity of an object in a rotating frame, is called the Coriolis force, and the sideways deflection is called the Coriolis effect. The same effect has dramatic consequences for weather on the earth (Figure 1).

Example 5-10 *try it yourself*

A tether ball of mass m is suspended from a rope of length L and travels at constant speed v in a horizontal circle of radius r. The rope makes an angle θ given by $\sin\theta = r/L$, as shown in Figure 5-28. Find (*a*) the tension in the rope, and (*b*) the speed of the ball.

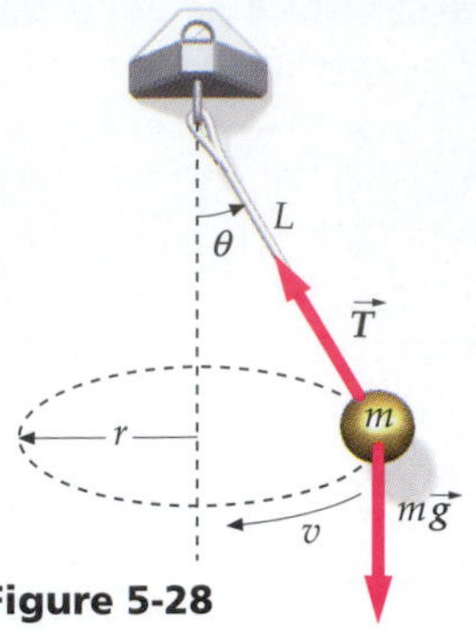

Figure 5-28

Picture the Problem Two forces act on the ball: its weight, $m\vec{g}$, and the tension in the rope, $\vec{T}$. Because of the ball's circular motion, it has a horizontal acceleration of magnitude v^2/r directed toward the center of the circle. Hence, the vertical component of the tension balances the ball's weight, and the horizontal component is the centripetal force. Choose y to be vertical and x to be directed toward the center of the circle.

Cover the column to the right and try these on your own before looking at the answers.

Steps	*Answers*
(*a*) 1. Draw a free-body diagram for the ball.	
2. Apply $\Sigma\vec{F} = m\vec{a}$ in component form to the ball.	$\sum F_x = T\sin\theta = ma_x = \dfrac{mv^2}{r}$ $\sum F_y = T\cos\theta - mg = ma_y = 0$
3. Solve the y equation for T.	$T = \dfrac{mg}{\cos\theta}$
(*b*) Substitute the result for T in the x equation and solve for v.	$v = \sqrt{rg\tan\theta}$

Figure 5-29

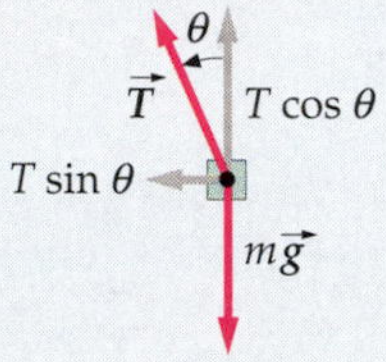

Remarks An object attached to a string and moving in a horizontal circle so that the string makes an angle θ with the vertical is called a *conical pendulum*.

When a car rounds a curve on a horizontal road, the centripetal force is provided by the force of friction exerted by the road on the tires of the car. If the car does not slide radially, the friction is static friction.

As the motorcycle rounds the curve, it is tilted so that the resultant of the normal and frictional forces exerted by the road acts along the plane of the cycle.

Example 5-11

In a skid test, a recent-model BMW 530i was able to travel in a circle of radius 45.7 m in 15.2 s without skidding. (*a*) What was its average speed? (*b*) Assuming v to be constant, what was the centripetal acceleration? (*c*) Again assuming v to be constant, what is the minimum value for the coefficient of static friction?

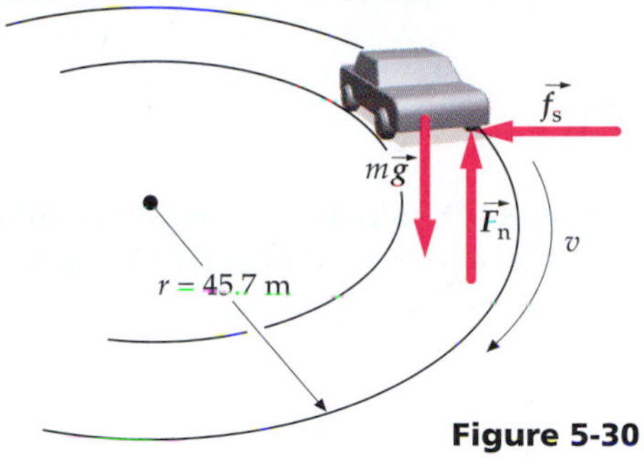

Figure 5-30

Picture the Problem Figure 5-30 shows the forces acting on the car. The normal force F_n balances the downward force due to gravity mg. The horizontal force is the force of static friction, which provides the centripetal force. The faster the car travels, the greater the required centripetal force. The average speed can be found from the circumference of the circle and the period T. This average speed puts a lower limit on the maximum value of the coefficient of static friction.

(*a*) 1. Draw a free-body diagram for the car:

Figure 5-31

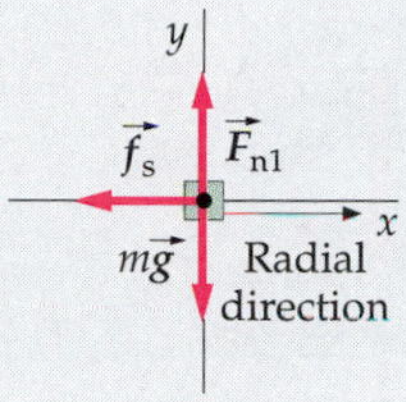

2. The average speed is the total circumference $2\pi r$ divided by the period T:

$$v = \frac{2\pi r}{T} = \frac{2\pi(45.7\text{ m})}{15.2\text{ s}} = 18.9\text{ m/s}$$

(*b*) Use v to calculate the centripetal acceleration:

$$a_c = \frac{v^2}{r} = \frac{(18.9\text{ m/s})^2}{45.7\text{ m}} = 7.82\text{ m/s}^2$$

(*c*) 1. Apply $\Sigma\vec{F} = m\vec{a}$ to the vertical and radial motions of the car. Choose the outward radial direction to be positive:

$$\sum F_y = ma_y, \qquad F_n - mg = 0$$

$$\sum F_r = ma_r, \qquad -f_s = -m\frac{v^2}{r}$$

2. The maximum value for static friction is proportional to the normal force:

$$f_{s,\max} = \mu_s F_n = \mu_s mg$$

3. Substituting $f_{s,\max}$ into $\Sigma\vec{F} = m\vec{a}$ relates μ_s to the speed:

$$\mu_s mg = m\frac{v^2}{r}$$

$$\mu_s = \frac{v^2}{rg} = \frac{(18.9\text{ m/s})^2}{(45.7\text{ m})(9.81\text{ m/s}^2)} = 0.797$$

Check the Result If μ_s were equal to 1, the inward force would be equal to mg and the centripetal acceleration would be g. Here μ_s is about 0.8 and the centripetal acceleration is about $0.8g$.

Banked Curves

If a curved road is not horizontal but banked, the normal force of the road will have a component directed inward toward the center of the circle that will contribute to the centripetal force. The banking angle can be chosen such that, for a given speed, no friction is needed for a car to handle the curve without sliding.

A large centrifuge used for research at Sandia National Laboratories.

optional

Example 5-12

A curve of radius 30 m is banked at an angle θ. Find θ for which a car can round the curve at 40 km/h even if the road is frictionless.

Figure 5-32

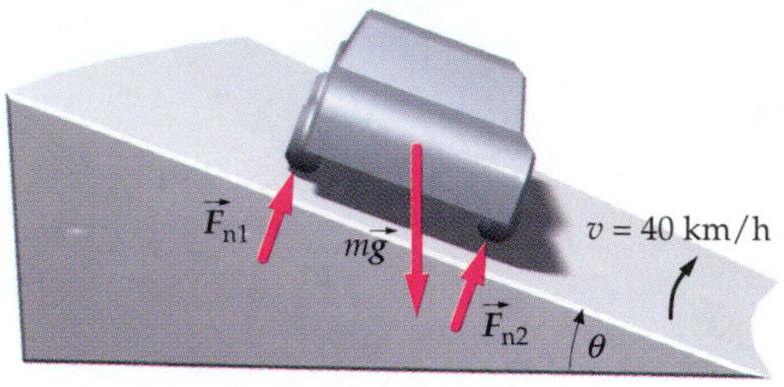

Figure 5-33

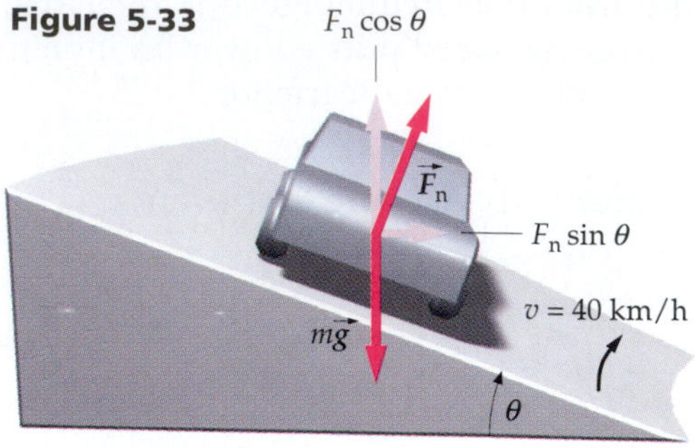

Picture the Problem In this case only two forces act on the car: gravity and the normal force. Since the road is banked, the normal force has a horizontal component that supplies the needed centripetal force. In Figure 5-32, the forces exerted by the road on the car are represented by $\vec{F}_{n1}$ and $\vec{F}_{n2}$. These forces are combined into $\vec{F}_n$ in Figure 5-33. We see that the angle between the normal force F_n and the vertical is θ, the same as the banking angle. We find this angle by applying Newton's second law. The total normal force has the component $F_n \sin\theta$ directed toward the center of the curve, which provides the centripetal acceleration of the car.

1. Draw a free-body diagram for the car. We choose y to be the vertical direction and x to be the horizontal direction toward the center of the circle. We call the total normal force exerted by the road F_n:

Figure 5-34

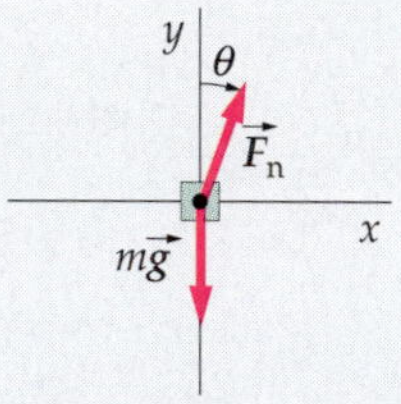

2. Apply $\Sigma\vec{F} = m\vec{a}$ in component form:

$$\sum F_x = ma_x, \quad F_n \sin\theta = ma_x = m\frac{v^2}{r}$$
$$\sum F_y = ma_y, \quad F_n \cos\theta - mg = ma_y = 0$$

3. Solve the y equation for the normal force F_n:

$$F_n = \frac{mg}{\cos\theta}$$

4. Substitute this result for F_n into the x equation:

$$\frac{mg}{\cos\theta}\sin\theta = m\frac{v^2}{r}$$

5. Solve for θ:

$$\tan\theta = \frac{v^2}{rg}$$

6. Substitute v = 40 km/h = 11.1 m/s, r = 30 m, and g = 9.81 m/s^2 and find θ:

$$\tan\theta = \frac{(11.1\text{ m/s})^2}{(30\text{ m})(9.81\text{ m/s}^2)} = 0.419$$
$$\theta = 22.7°$$

Remark The banking angle θ depends on v and r, but not the mass m; θ increases with increasing v, and decreases with increasing r. When the banking angle, speed, and radius satisfy $\tan\theta = v^2/rg$, the car rounds the curve smoothly, with no tendency to slide either inward or outward. If the car speed is greater than $\sqrt{rg \tan \theta}$, the road will exert a frictional force down the incline (Figure 5-35). This force has an inward horizontal component, which provides the additional centripetal force needed to keep the car from moving outward (sliding up the incline). If the car speed is less than this amount, the road must exert a frictional force up the incline.

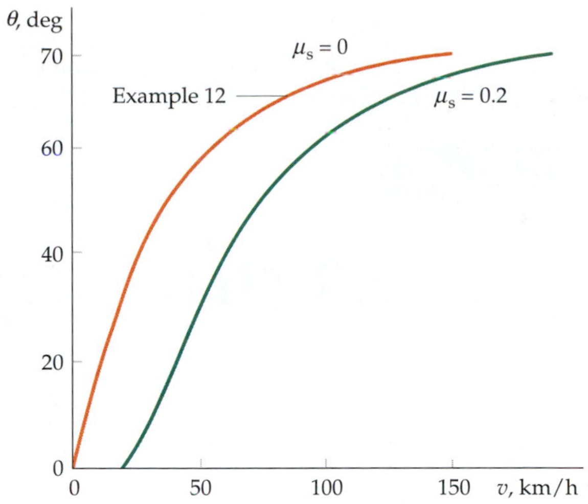

Figure 5-35 The addition of friction allows the car in this example to travel a bit faster around the curve without slipping. The two plots show banking angles θ versus the maximum velocities that can be attained without slipping, both without friction (red; $\mu_s = 0$) and with friction (green; $\mu_s = 0.2$). When the banking angle is zero, the maximum speed through the turn without slipping is zero with no friction, and about 30 km/h with $\mu_s = 0.2$.

5-3 Drag Forces

When an object moves through a fluid such as air or water, the fluid exerts a **drag force** or retarding force that tends to reduce the speed of the object. The drag force depends on the shape of the object, the properties of the fluid, and the speed of the object relative to the fluid. Unlike ordinary friction, the drag force increases as the speed of the object increases. At low speeds, the drag force is approximately proportional to the speed of the object; at higher speeds, it is more nearly proportional to the square of the speed.

Consider an object dropped from rest and falling under the influence of the force of gravity, which we assume to be constant. Now add a drag force of magnitude bv^n, where b is a constant that depends on the shape of the object and the properties of the air, and the exponent n is approximately 1 at low speeds and approximately 2 at higher speeds. We then have a constant downward force mg and an upward force bv^n (Figure 5-36). If we take the downward direction to be positive, we obtain from Newton's second law

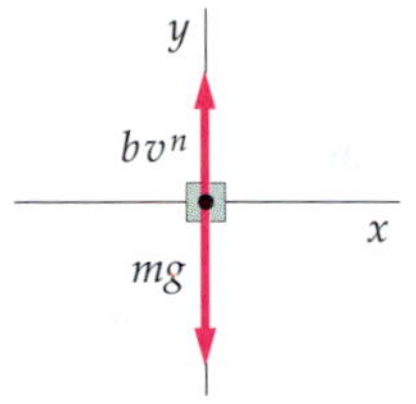

Figure 5-36 A free-body diagram showing forces on an object falling with air resistance.

$$\sum F_y = mg - bv^n = ma_y \qquad 5\text{-}7$$

At $t = 0$, the instant when the object is dropped, the speed is zero, so the retarding force is zero and the acceleration is g downward. As the speed of the object increases, the drag force increases and the acceleration becomes less than g. Eventually, the speed is great enough for the drag force bv^n to equal the force of gravity mg, at which point the acceleration is zero. The object then continues moving at a constant speed v_t, called its **terminal speed.** Setting the acceleration a in Equation 5-7 equal to zero, we obtain

$$bv_t^n = mg$$

Solving for the terminal speed, we get

$$v_t = \left(\frac{mg}{b}\right)^{1/n} \qquad 5\text{-}8$$

The larger the constant b, the lower the terminal speed. A parachute is designed to maximize b so that the terminal speed will be small. Cars, on the other hand, are designed to minimize b to reduce the effect of wind resistance.

Exploring

Numerical Methods: Euler's Method

If a particle moves under the influence of a *constant* force, its acceleration is constant and we can find its velocity and position from the constant-acceleration formulas in Chapter 2. But consider a particle moving through space where the force on it, and therefore its acceleration, depends on its position and velocity. The velocity and acceleration of the particle at one instant determine its position and velocity at the next instant, which then determines its acceleration at that instant. The actual position, velocity, and acceleration of an object all change continuously with time. We can approximate this by replacing the continuous time variations with small time steps of duration Δt. The simplest approximation is to assume constant acceleration during each step. This approximation is called **Euler's method**. If the time interval is sufficiently short, the change in acceleration during the interval will be small and can be neglected.

Let x_0, v_0, and a_0 be the known position, velocity, and acceleration of a particle at some initial time t_0. If we assume constant acceleration during Δt, velocity at time $t_1 = t_0 + \Delta t$ is given by

$$v_1 = v_0 + a\,\Delta t \qquad 1$$

Similarly, if we neglect the change in velocity during the time interval, the new position is given by

$$x_1 = x_0 + v_0 \Delta t \qquad 2$$

(Other methods of numerical integration are more accurate but less simple to use. For example, the accuracy is improved if a and v are computed at the midpoint of the interval rather than at the beginning.)

We can use the values v_1 and x_1 to compute the new acceleration a_1 from Newton's second law, and then use a_1 for the next time interval to compute v_2 and x_2.

$$v_2 = v_1 + a_1 \Delta t$$

$$x_2 = x_1 + v_1 \Delta t$$

In general, the connection between the position and velocity at time t_n and time $t_{n+1} = t_n + \Delta t$ is given by

$$v_{n+1} = v_n + a_n \Delta t \qquad 3$$

and

$$x_{n+1} = x_n + v_n \Delta t \qquad 4$$

To find the velocity and position at some time t, we therefore divide the time interval $t - t_0$ into a large number of smaller intervals Δt and apply Equations 3 and 4, beginning at the initial time t_0. This involves a large number of simple, repetitive calculations that are easily done on a computer. The technique of breaking the time interval into small steps and computing the acceleration, velocity, and position at each step using the values from the previous step is called numerical integration.

Drag Forces

To illustrate the use of numerical methods, let us consider a problem in which a sky diver is dropped from rest at some height under the influences of gravity and a drag force that is proportional to the square of the speed. We will find the velocity v and the distance traveled x as functions of time.

The equation describing the motion of an object of mass m dropped from rest is Equation 5-7 with $n = 2$:

$$\sum F_y = mg - bv^n = ma_y$$

The acceleration is thus

$$a = g - \left(\frac{b}{m}\right)v^2 \qquad 5$$

It is convenient to write the constant b/m in terms of the terminal speed v_t. Setting $a = 0$ in Equation 5 we obtain

$$0 = g - \left(\frac{b}{m}\right)v_t^2$$

$$\frac{b}{m} = \frac{g}{v_t^2}$$

Substituting g/v_t^2 for b/m in Equation 5 gives

$$a = g\left(1 - \frac{v^2}{v_t^2}\right) \qquad 6$$

To solve Equation 6 numerically, we need numerical values for g and v_t. A reasonable terminal speed for a sky diver is 60 m/s. Using this and $g = 9.81\ \mathrm{m/s^2}$, we obtain

$$a = 9.81\left(1 - \frac{v^2}{3600}\right) \qquad 7$$

We have omitted the units in this equation. Since we are using SI units, the unit for v is meters per second, and the unit for x is meters. If we choose $x_0 = 0$ for the initial position, the initial values are $x_0 = 0$, $v_0 = 0$, and $a_0 = g = 9.81$. To find the velocity v and position x after some time, say $t = 20$ s, we divide the time interval $0 < t < 20$ s into many small intervals Δt and apply Equations 3 and 4. We do this by writing a computer program or by using a computer spreadsheet. Figure 1 shows graphs of v versus t and x versus t based on data found using a spreadsheet with $\Delta t = 0.5$ s. At $t = 20$ s, the computed values are $v = 59.97$ m/s and $x = 957.5$ m.

But how accurate are our computations? We can estimate the accuracy by running the program again using a smaller time interval. If we use $\Delta t = 0.25$ s, one-half of the value we originally used, we obtain $v = 59.92$ m/s and $x = 952.0$ m at $t = 20$ s. The difference in v is about 0.1% and that in x is about 0.5%. These are our estimates of the accuracy of the original computations.

Since the difference between the value of a_{av} for some time interval Δt and the value of a_i at the beginning of the interval becomes smaller as the time interval becomes smaller, we might expect that it would be better to use very small time intervals, say $\Delta t = 0.000\,000\,001$ s. But there are two reasons for not using very small time intervals. First, the smaller the time interval, the larger the number of calculations required, and the longer the program takes to run. Second, the computer keeps only a fixed number of digits at each step of the calculation, so that at each step there is a round-off error. These round-off errors add up. The larger the number of calculations, the more significant the total round-off error becomes. When we first decrease the time interval, the accuracy improves because a_i more nearly approximates a_{av} for the interval. However, as the time interval is decreased further, the round-off errors build up and the accuracy of the computation decreases. A good rule of thumb to follow is to use no more than about 10^4 or 10^5 time intervals for a typical numerical integration.

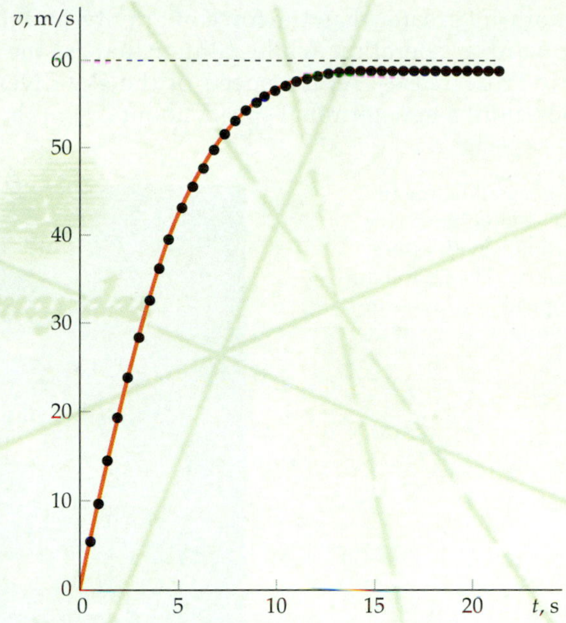

(a)

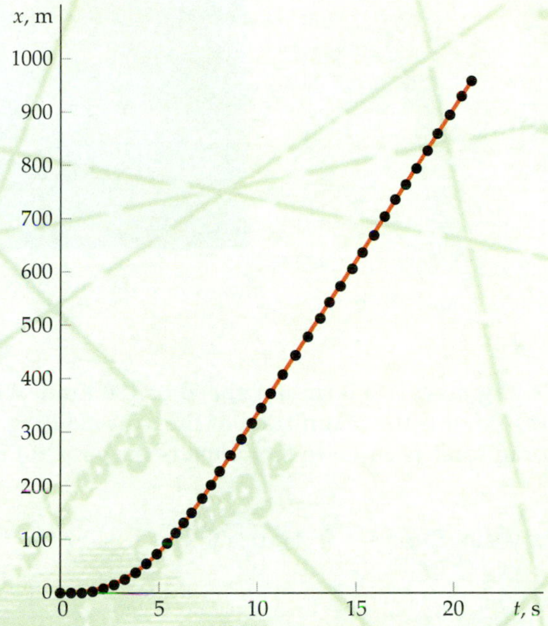

(b)

Figure 1 (a) Graph of v versus t for a sky diver found by numerical integration using $\Delta t = 0.5$ s. The horizontal dashed line is the terminal speed $v_t = 60$ m/s. (b) Graph of x versus t using $\Delta t = 0.5$ s.

optional

The terminal speed of a sky diver before release of the parachute is about 60 m/s, or ≈ 200 km/h. When the parachute is opened, the drag force shoots up to become temporarily greater than the force of gravity, and the sky diver experiences an upward acceleration while falling; that is, the downward speed of the sky diver decreases. As the speed of the sky diver drops, the drag force decreases, until a new terminal speed, about 20 km/h, is reached.

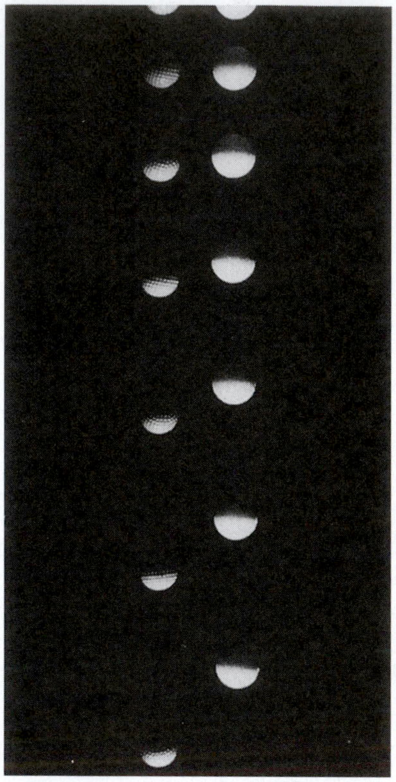

A golf ball and Styrofoam ball falling in air. The air resistance is negligible for the heavier golf ball, which falls with essentially constant acceleration. The Styrofoam ball reaches terminal speed quickly, as indicated by the nearly equal spacing at the bottom.

Example 5-13

A sky diver of mass 64 kg reaches a terminal speed of 180 km/h with her arms and legs outspread. (*a*) What is the magnitude of the upward drag force on the sky diver? (*b*) If the drag force is equal to bv^2, what is the value of b?

(*a*) 1. Draw a free-body diagram:

Figure 5-37

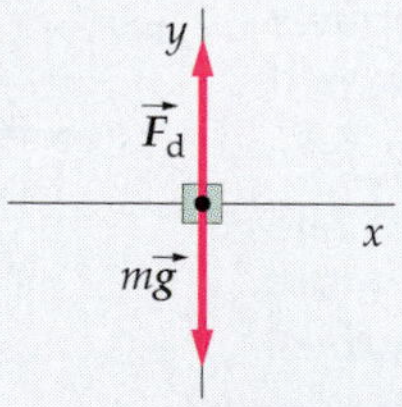

2. Apply $\Sigma\vec{F} = m\vec{a}$. Since the sky diver is moving with constant velocity, the acceleration is zero:

$$\sum F_y = ma_y$$

$$mg - F_d = 0$$

$$F_d = mg = (64\text{ kg})(9.81\text{ N/kg}) = 628\text{ N}$$

(*b*) To find b we set $F_d = bv^2$:

$$F_d = mg = bv^2$$

$$b = \frac{mg}{v^2} = 0.251\text{ N}\cdot\text{s}^2/\text{m}^2 = 0.251\text{ kg/m}$$

Summary

1. The acceleration due to circular motion ($a = v^2/r$) is an important *derived* result in kinematics.
2. Friction and drag forces are complex phenomena empirically approximated by simple equations.

Topic	Remarks and Relevant Equations	
1. Friction	Two objects in contact exert frictional forces on each other. These forces are parallel to the surfaces of the objects at the points of contact and directed opposite to the direction of sliding or tendency to slide.	
Static friction	$f_s \leq \mu_s F_n$	5-2
	where F_n is the normal force of contact and μ_s is the coefficient of static friction.	
Kinetic friction	$f_k = \mu_k F_n$	5-3
	where μ_k is the coefficient of kinetic friction. The coefficient of kinetic friction is slightly less than the coefficient of static friction.	
2. Circular Motion	When an object moves in a circle with constant speed, it is accelerating, because its velocity is changing in direction. The acceleration is called centripetal acceleration and is directed toward the center of the circle. There must be a net inward force to provide the centripetal acceleration.	
Centripetal acceleration	$a = \dfrac{v^2}{r}$	5-5
Speed and period	$v = \dfrac{2\pi r}{T}$	5-6
3. General Motion Along a Curve in Space	A particle moving along an arbitrary curve can be considered to be moving in a circular arc during a small time interval. Its instantaneous acceleration vector has a component $a_r = v^2/r$ toward the center of curvature of the arc and a component $a_t = dv/dt$ that is tangential to the curve.	
4. Drag Forces (optional)	When an object moves through a fluid, it experiences a drag force that opposes its motion. The drag force increases with increasing speed. If the body is dropped from rest, its speed increases until the drag force equals the force of gravity, after which it moves with a constant speed called its terminal speed. The terminal speed depends on the shape of the body and on the medium through which it falls.	

Problem-Solving Guide

1. Begin by drawing a neat diagram that includes the important features of the problem.
2. Isolate the object of interest and draw a free-body diagram that shows how each external force acts on the object. If there is more than one object of interest in the problem, draw a separate free-body diagram for each object.
3. Choose a convenient coordinate system for each object and apply Newton's second law, $\vec{F}_{net} = m\vec{a}$, in component form.
4. Solve the resulting equations for the unknowns.
5. Check to see whether your results are reasonable.

Summary of Worked Examples

Type of Calculation	Procedure and Relevant Examples
1. Problems With Friction	Remember that the force of static friction may not be equal to its limiting value of $\mu_s F_n$ and that the normal force between the surfaces of two objects is not necessarily equal to the weight of the upper object. **Example 5-3**
Find the stopping distance of an object.	Apply $\Sigma\vec{F} = m\vec{a}$ in component form to the object. If the object is sliding, the frictional force on it is $\mu_k F_n$, where F_n is the normal force. If the friction is static (as can be the case in braking a car), the maximum frictional force is $\mu_s F_n$. In both cases, the force is constant, so the constant acceleration equations of Chapter 2 can be used. **Examples 5-1, 5-4**
Determine how an object on an inclined plane will move.	Apply $\Sigma\vec{F} = m\vec{a}$ in component form to the object. The frictional force is $f_s \le \mu_s F_n$ if the object is static and $\mu_k F_n$ if it is sliding, where F_n is the normal force. **Example 5-2**
Determine if friction is static or kinetic and find the force of static friction.	Determine what the frictional force would have to be to prevent sliding. If this value is greater than $\mu_s F_n$, sliding will occur. Apply $\Sigma\vec{F} = m\vec{a}$ in component form to the object. **Examples 5-3, 5-4, 5-10**
Find the acceleration of one or more objects, taking into account various forces, including friction.	Determine if the frictional force is static or kinetic. Apply $\Sigma\vec{F} = m\vec{a}$ in component form to each object. **Examples 5-3, 5-4, 5-5, 5-6**
2. Circular Motion	Apply $\Sigma\vec{F} = m\vec{a}$ in component form to the object. When an object is moving in a circle, resolve the forces acting on the object into radial and tangential components. The inward radial component of the resultant force is the centripetal force, which equals mv^2/r. The tangential component of the resultant force equals $m\,dv/dt$.
Find the period of an object undergoing circular motion at constant speed.	Apply $\Sigma\vec{F} = m\vec{a}$ in component form, using v^2/r for the acceleration. Then use $T = 2\pi r/v$. **Example 5-8**
Find the forces on an object in circular motion.	Apply $\Sigma\vec{F} = m\vec{a}$. Resolve the forces into radial and tangential components. Set the net inward radial component equal to $ma = mv^2/r$. **Examples 5-9, 5-10, 5-11**
Find the forces on a car traveling on a banked curve. (optional)	Apply $\Sigma\vec{F} = m\vec{a}$. The normal force has an inward component $F_n \sin\theta$. If there is a frictional force, its radial component is $\pm f\cos\theta$, depending on whether the car is sliding or tending to slide up or down the bank. **Example 5-12**
3. Drag Force and Terminal Speed (optional)	Set $F_d = mg$ for the drag force on an object falling at terminal speed. **Example 5-13**

Problems

Conceptual Problems

Problems from Optional and Exploring sections

• Single-concept, single-step, relatively easy

•• Intermediate-level, may require synthesis of concepts

••• Challenging, for advanced students

In a few problems, you are given more data than you actually need; in a few other problems, you are required to supply data from your general knowledge, outside sources, or informed estimates.

Friction

1 • Various objects lie on the floor of a truck moving along a horizontal road. If the truck accelerates, what force acts on the objects to cause them to accelerate?

2 • Any object resting on the floor of a truck will slide if the truck's acceleration is too great. How does the critical acceleration at which a light object slips compare with that at which a much heavier object slips?

3 • True or false:

(*a*) The force of static friction always equals $\mu_s F_n$.
(*b*) The force of friction always opposes the motion of an object.
(*c*) The force of friction always opposes sliding.
(*d*) The force of kinetic friction always equals $\mu_k F_n$.

4 • A block of mass m rests on a plane inclined at an angle θ with the horizontal. It follows that the coefficient of static friction between the block and plane is

(*a*) $\mu_s \geq g$.
(*b*) $\mu_s = \tan\theta$.
(*c*) $\mu_s \leq \tan\theta$.
(*d*) $\mu_s \geq \tan\theta$.

5 • A block of mass m is at rest on a plane inclined at an angle of 30° with the horizontal, as in Figure 5-38. Which of the following statements about the force of static friction is true?

(*a*) $f_s > mg$
(*b*) $f_s > mg\cos 30°$
(*c*) $f_s = mg\cos 30°$
(*d*) $f_s = mg\sin 30°$
(*e*) None of these statements is true.

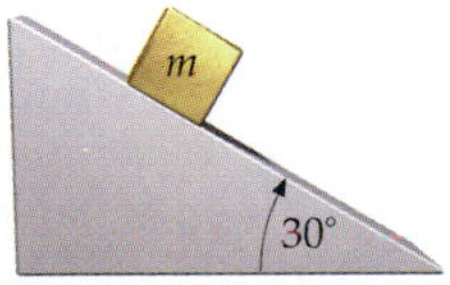

Figure 5-38 Problem 5

6 • A block of mass m slides at constant speed down a plane inclined at an angle θ with the horizontal. It follows that

(*a*) $\mu_k = mg\sin\theta$.
(*b*) $\mu_k = \tan\theta$.
(*c*) $\mu_k = 1 - \cos\theta$.
(*d*) $\mu_k = \cos\theta - \sin\theta$.

7 • A block of wood is pulled by a horizontal string across a horizontal surface at a constant velocity with a force of 20 N. The coefficient of kinetic friction between the surfaces is 0.3. The force of friction is

(*a*) impossible to determine without knowing the mass of the block.
(*b*) impossible to determine without knowing the speed of the block.
(*c*) 0.3 N.
(*d*) 6 N.
(*e*) 20 N.

8 • A 20-N block rests on a horizontal surface. The coefficients of static and kinetic friction between the surface and the block are $\mu_s = 0.8$ and $\mu_k = 0.6$. A horizontal string is attached to the block and a constant tension T is maintained in the string. What is the force of friction acting on the block if (*a*) $T = 15$ N or (*b*) $T = 20$ N.

9 • A block of mass m is pulled at a constant velocity across a horizontal surface by a string as in Figure 5-39. The magnitude of the frictional force is

(*a*) $\mu_k mg$.
(*b*) $T\cos\theta$.
(*c*) $\mu_k(T - mg)$.
(*d*) $\mu_k T\sin\theta$.
(*e*) $\mu_k(mg + T\sin\theta)$.

Figure 5-39 Problem 9

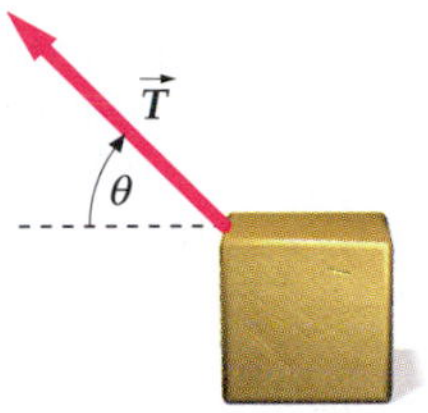

10 • A tired worker pushes with a horizontal force of 500 N on a 100-kg crate resting on a thick pile carpet. The coefficients of static and kinetic friction are 0.6 and 0.4, respectively. Find the frictional force exerted by the surface.

11 • A box weighing 600 N is pushed along a horizontal floor at constant velocity with a force of 250 N parallel to the floor. What is the coefficient of kinetic friction between the box and the floor?

12 • The coefficient of static friction between the tires of a car and a horizontal road is $\mu_s = 0.6$. If the net force on the car is the force of static friction exerted by the road, (*a*) what is the maximum acceleration of the car when it is braked? (*b*) What is the least distance in which the car can stop if it is initially traveling at 30 m/s?

13 • The force that accelerates a car along a flat road is the frictional force exerted by the road on the car's tires. (*a*) Explain why the acceleration can be greater when the wheels do not spin. (*b*) If a car is to accelerate from 0 to 90 km/h in 12 s at constant acceleration, what is the minimum coefficient of friction needed between the road and tires? Assume that half the weight of the car is supported by the drive wheels.

14 • On the current tour of the rock band Dead Wait, the show opens with a dark stage. Suddenly there is the sound of a large automobile accident. Lead singer Sharika comes sliding to the front of the stage on her knees. Her initial speed is 3 m/s. After sliding 2 m, she comes to rest in a dry ice fog as flash pots explode on either side. What is the coefficient of kinetic friction between Sharika and the stage?

15 • A 5-kg block is held at rest against a vertical wall by a horizontal force of 100 N. (*a*) What is the frictional force exerted by the wall on the block? (*b*) What is the minimum horizontal force needed to prevent the block from falling if the coefficient of friction between the wall and the block is $\mu_s = 0.40$?

16 • On a snowy day with the temperature near the freezing point, the coefficient of static friction between a car's tires and an icy road is 0.08. What is the maximum incline that this four-wheel-drive vehicle can climb with zero acceleration?

17 • A 50-kg box that is resting on a level floor must be moved. The coefficient of static friction between the box and the floor is 0.6. One way to move the box is to push down on it at an angle θ with the horizontal. Another method is to pull up on the box at an angle θ with the horizontal. (*a*) Explain why one method is better than the other. (*b*) Calculate the force necessary to move the box by each method if $\theta = 30°$ and compare the answer with the results when $\theta = 0°$.

18 • A 3-kg box resting on a horizontal shelf is attached to a 2-kg box by a light string as in Figure 5-40. (*a*) What is the minimum coefficient of static friction such that the objects remain at rest? (*b*) If the coefficient of static friction is less than that found in part (*a*), and the coefficient of kinetic friction between the box and the shelf is 0.3, find the time for the 2-kg mass to fall 2 m to the floor if the system starts from rest.

Figure 5-40
Problem 18

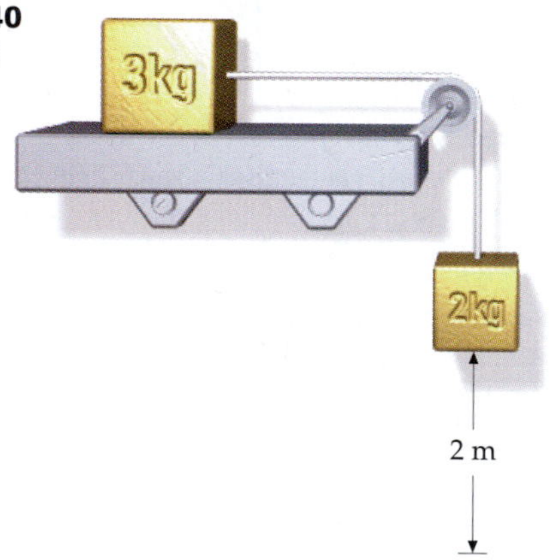

19 •• A block on a horizontal plane is given an initial velocity v. It comes to rest after a displacement d. The coefficient of kinetic friction between the block and the plane is given by

(*a*) $\mu_k = v^2 d/2g$. (*b*) $\mu_k = v^2/2dg$.
(*c*) $\mu_k = v^2 g/d^2$. (*d*) none of the above.

20 •• A block of mass $m_1 = 250$ g is at rest on a plane that makes an angle $\theta = 30°$ above the horizontal (Figure 5-41). The coefficient of kinetic friction between the block and the plane is $\mu_k = 0.100$. The block is attached to a second block of mass $m_2 = 200$ g that hangs freely by a string that passes over a frictionless and massless pulley. When the second block has fallen 30.0 cm, its speed is

(*a*) 83 cm/s.
(*b*) 48 cm/s.
(*c*) 160 cm/s.
(*d*) 59 cm/s.
(*e*) 72 cm/s.

Figure 5-41 Problems 20–22

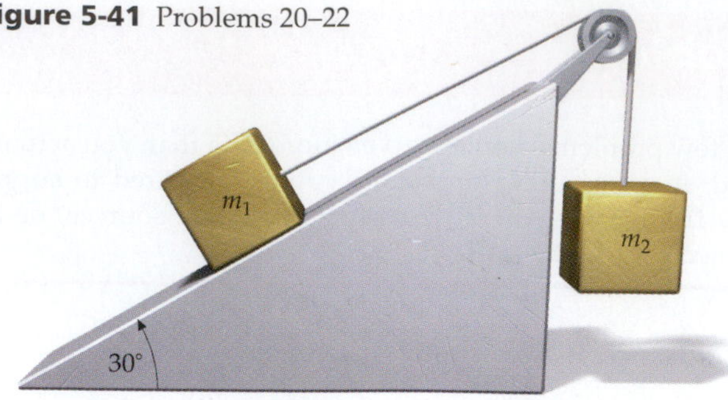

21 •• Returning to Figure 5-41, this time $m_1 = 4$ kg. The coefficient of static friction between the block and the incline is 0.4. (*a*) Find the range of possible values for m_2 for which the system will be in static equilibrium. (*b*) What is the frictional force on the 4-kg block if $m_2 = 1$ kg?

22 •• Returning once again to Figure 5-41, this time $m_1 = 4$ kg, $m_2 = 5$ kg, and the coefficient of kinetic friction between the inclined plane and the 4-kg block is $\mu_k = 0.24$. Find the acceleration of the masses and the tension in the cord.

23 •• The coefficient of static friction between the bed of a truck and a box resting on it is 0.30. The truck is traveling at 80 km/h along a horizontal road. What is the least distance in which the truck can stop if the box is not to slide?

24 •• A 4.5-kg mass is given an initial velocity of 14 m/s up an incline that makes an angle of 37° with the horizontal. When its displacement is 8.0 m, its upward velocity has diminished to 5.2 m/s. Find (*a*) the coefficient of kinetic friction between the mass and plane, (*b*) the displacement of the mass from its starting point at the time when it momentarily comes to rest, and (*c*) the speed of the block when it again reaches its initial position.

25 •• An automobile is going up a grade of 15° at a speed of 30 m/s. The coefficient of static friction between the tires and the road is 0.7. (*a*) What minimum distance does it take to stop the car? (*b*) What minimum distance would it take if the car were going down the grade?

26 •• A block of mass m slides with initial speed v_0 on a horizontal surface. If the coefficient of kinetic friction between the block and the surface is μ_k, find the distance d that the block moves before coming to rest.

27 •• A rear-wheel-drive car supports 40% of its weight on its two drive wheels and has a coefficient of static friction of 0.7. (*a*) What is the vehicle's maximum acceleration? (*b*) What is the shortest possible time in which this car can achieve a speed of 100 km/h? (Assume that the engine has unlimited power.)

28 •• Lou bets an innocent stranger that he can place a 2-kg block against the side of a cart, as in Figure 5-42, and that the block will not fall to the ground, even though Lou will use no hooks, ropes, fasteners, magnets, glue, or adhesives of any kind. When the stranger accepts the bet, Lou begins to push the cart in the direction shown. The coefficient of static friction between the block and the cart is 0.6. (*a*) Find the minimum acceleration for which Lou will win the bet. (*b*) What is the magnitude of the frictional force in this case?

(*c*) Find the force of friction on the block if a is twice the minimum needed for the block not to fall. (*d*) Show that, for a block of any mass, the block will not fall if the acceleration is $a \geq g/\mu_s$, where μ_s is the coefficient of static friction.

Figure 5-42 Problem 28

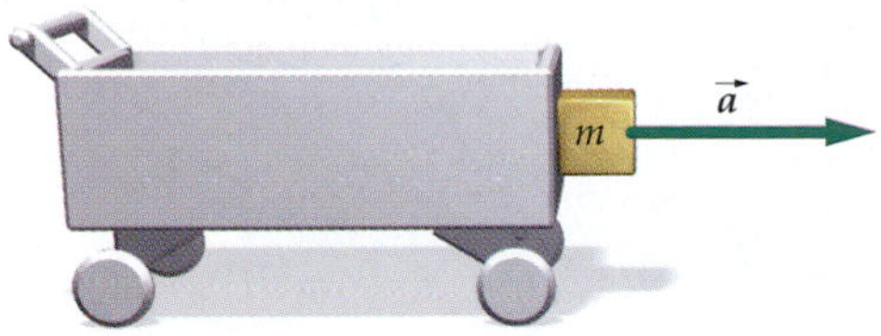

29 •• Two blocks attached by a string slide down a 20° incline. The lower block has a mass of $m_1 = 0.25$ kg and a coefficient of kinetic friction $\mu_k = 0.2$. For the upper block, $m_2 = 0.8$ kg and $\mu_k = 0.3$. Find (*a*) the acceleration of the blocks and (*b*) the tension in the string.

30 •• Two blocks attached by a string are at rest on an inclined surface. The lower block has a mass of $m_1 = 0.2$ kg and a coefficient of static friction $\mu_s = 0.4$. The upper block has a mass $m_2 = 0.1$ kg and $\mu_s = 0.6$. (*a*) At what angle θ_c do the blocks begin to slide? (*b*) What is the tension in the string just before sliding begins?

31 •• Two blocks connected by a massless, rigid rod slide on a surface inclined at an angle of 20°. The lower block has a mass $m_1 = 1.2$ kg, and the upper block's mass is $m_2 = 0.75$ kg. (*a*) If the coefficients of kinetic friction are $\mu_k = 0.3$ for the lower block and $\mu_k = 0.2$ for the upper block, what is the acceleration of the blocks? (*b*) Determine the force transmitted by the rod.

32 •• A block of mass m rests on a horizontal surface (Figure 5-43). The block is pulled by a massless rope with a force $\vec{F}$ at an angle θ. The coefficient of static friction is 0.6. The minimum value of the force needed to move the block depends on the angle θ. (*a*) Discuss qualitatively how you would expect this force to depend on θ. (*b*) Compute the force for the angles $\theta = 0°, 10°, 20°, 30°, 40°, 50°$, and $60°$, and make a plot of F versus θ for $mg = 400$ N. From your plot, at what angle is it most efficient to apply the force to move the block?

Figure 5-43 Problem 32

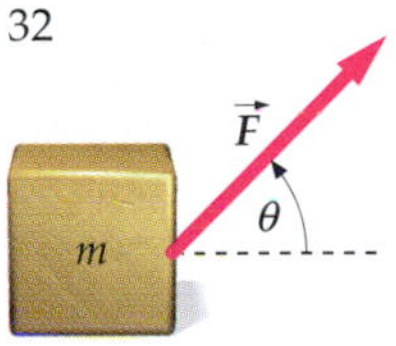

33 •• Answer the same questions as in Problem 32, only this time with a force $\vec{F}$ that pushes down on the block in Figure 5-44 at an angle θ with the horizontal.

Figure 5-44 Problem 33

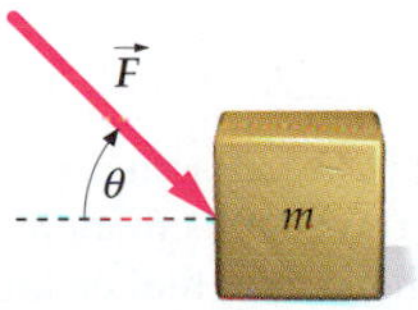

34 •• A 100-kg mass is pulled along a frictionless surface by a horizontal force $\vec{F}$ such that its acceleration is 6 m/s^2 (Figure 5-45). A 20-kg mass slides along the top of the 100-kg mass and has an acceleration of 4 m/s^2. (It thus slides backward relative to the 100-kg mass.) (*a*) What is the frictional force exerted by the 100-kg mass on the 20-kg mass? (*b*) What is the net force acting on the 100-kg mass? What is the force F? (*c*) After the 20-kg mass falls off the 100-kg mass, what is the acceleration of the 100-kg mass? (Assume that the force F does not change.)

Figure 5-45 Problem 34

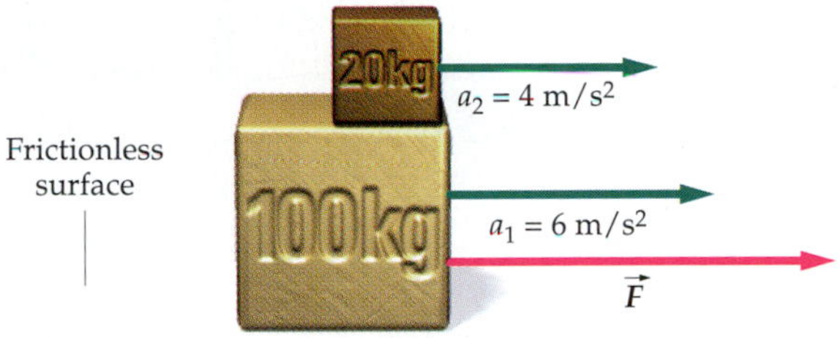

35 •• A 60-kg block slides along the top of a 100-kg block with an acceleration of 3 m/s^2 when a horizontal force $\vec{F}$ of 320 N is applied, as in Figure 5-46. The 100-kg block sits on a horizontal frictionless surface, but there is friction between the two blocks. (*a*) Find the coefficient of kinetic friction between the blocks. (*b*) Find the acceleration of the 100-kg block during the time that the 60-kg block remains in contact.

Figure 5-46 Problem 35

36 •• The coefficient of static friction between a rubber tire and the road surface is 0.85. What is the maximum acceleration of a 1000-kg four-wheel-drive truck if the road makes an angle of 12° with the horizontal and the truck is (*a*) climbing, and (*b*) descending?

37 •• A 2-kg block sits on a 4-kg block that is on a frictionless table (Figure 5-47). The coefficients of friction between the blocks are $\mu_s = 0.3$ and $\mu_k = 0.2$. (*a*) What is the maximum force F that can be applied to the 4-kg block if the 2-kg block is not to slide? (*b*) If F is half this value, find the acceleration of each block and the force of friction acting on each block. (*c*) If F is twice the value found in (*a*), find the acceleration of each block.

Figure 5-47 Problem 37

38 •• In Figure 5-48, the mass $m_2 = 10$ kg slides on a frictionless shelf. The coefficients of static and kinetic friction between m_2 and $m_1 = 5$ kg are $\mu_s = 0.6$ and $\mu_k = 0.4$. (*a*) What is the maximum acceleration of m_1? (*b*) What is the maximum value of m_3 if m_1 moves with m_2 without slipping? (*c*) If $m_3 = 30$ kg, find the acceleration of each body and the tension in the string.

Figure 5-48 Problem 38

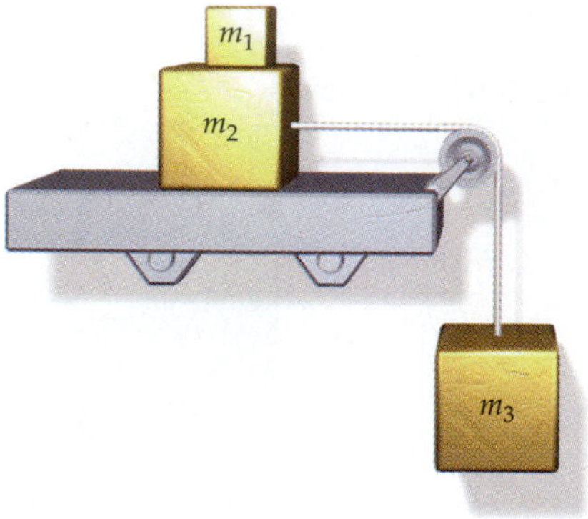

39 ••• A box of mass m rests on a horizontal table. The coefficient of static friction is μ_s. A force $\vec{F}$ is applied at an angle θ as shown in Problem 5-32. (*a*) Find the force F needed to move the box as a function of angle θ. (*b*) At the angle θ for which this force is minimum, the slope $dF/d\theta$ of the curve F versus θ is zero. Compute $dF/d\theta$ and show that this derivative is zero at the angle θ that obeys $\tan\theta = \mu_s$. Compare this general result with that obtained in Problem 5-32.

40 ••• A 10-kg block rests on a 5-kg bracket like the one in Figure 5-49. The 5-kg bracket sits on a frictionless surface. The coefficients of friction between the 10-kg block and the bracket on which it rests are $\mu_s = 0.40$ and $\mu_k = 0.30$. (*a*) What is the maximum force F that can be applied if the 10-kg block is not to slide on the bracket? (*b*) What is the corresponding acceleration of the 5-kg bracket?

Figure 5-49 Problem 40

41 ••• Lou has set up a kiddie ride at the Winter Ice Fair. He builds a right-angle triangular wedge, which he intends to push along the ice with a child sitting on the hypotenuse. If he pushes too hard, the kid will slide up and over the top, and Lou could be looking at a lawsuit. If he doesn't push hard enough, the kid will slide down the wedge, and the parents will want their money back. If the angle of inclination of the wedge is 40°, what are the minimum and maximum values for the acceleration that Lou must achieve? Use m for the child's mass, and μ_s for the coefficient of static friction between the child and the wedge.

Figure 5-50 Problem 42

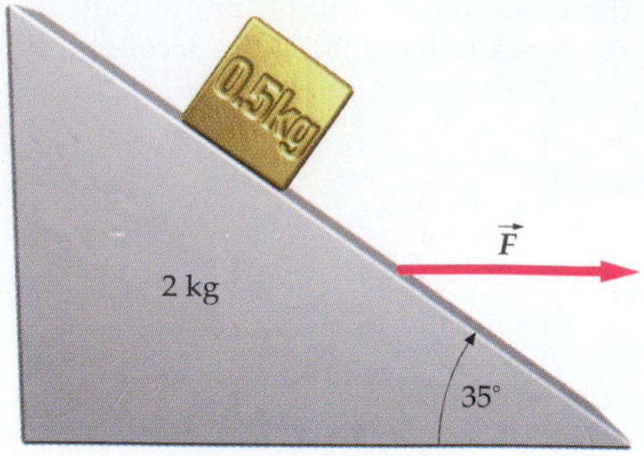

42 ••• A block of mass 0.5 kg rests on the inclined surface of a wedge of mass 2 kg, as in Figure 5-50. The wedge is acted on by a horizontal force $\vec{F}$ and slides on a frictionless surface. (*a*) If the coefficient of static friction between the wedge and the block is $\mu_s = 0.8$, and the angle of the incline is 35°, find the maximum and minimum values of F for which the block does not slip. (*b*) Repeat part (*a*) with $\mu_s = 0.4$.

Circular Motion

43 True or false: An object cannot move in a circle unless there is a net force acting on it.

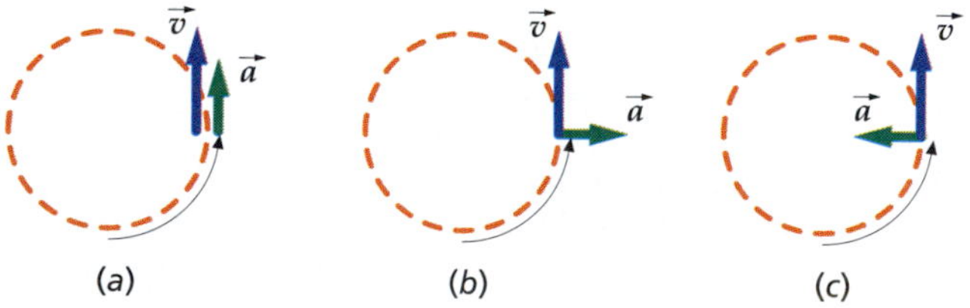

Figure 5-51 Problem 44

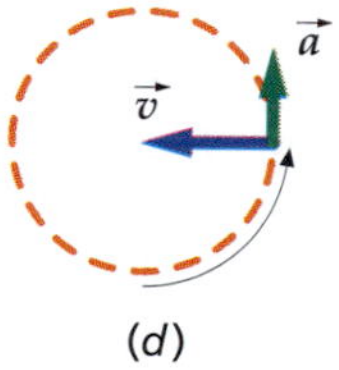

44 • An object moves in a circle counterclockwise with constant speed (Figure 5-51). Which figure shows the correct velocity and acceleration vectors?

45 • A particle is traveling in a vertical circle at constant speed. One can conclude that the ________ is constant.

(*a*) velocity (*b*) acceleration (*c*) net force
(*d*) apparent weight (*e*) None of the above.

46 • An object travels with a constant speed v in a circular path of radius r. (*a*) If v is doubled, how is the acceleration a affected? (*b*) If r is doubled, how is a affected? (*c*) Why is it impossible for an object to travel around a perfectly sharp angular turn?

47 • A boy whirls a ball on a string in a horizontal circle of radius 0.8 m. How many revolutions per minute must the ball make if the magnitude of its centripetal acceleration is to be the same as the free-fall acceleration due to gravity g?

48 • A 0.20-kg stone attached to a 0.8-m long string is rotated in the horizontal plane. The string makes an angle of 20° with the horizontal. Determine the speed of the stone.

49 • A 0.75-kg stone attached to a string is whirled in a horizontal circle of radius 35 cm as in the conical pendulum of Example 5-10. The string makes an angle of 30° with the vertical. (*a*) Find the speed of the stone. (*b*) Find the tension in the string.

50 •• A stone with a mass $m = 95$ g is being whirled in a horizontal circle on the end of a string that is 85 cm long. The length of time required for the stone to make one complete revolution is 1.22 s. The angle that the string makes with the horizontal is ________.
(*a*) 52° (*b*) 46° (*c*) 26° (*d*) 23° (*e*) 3°

51 •• A pilot of mass 50 kg comes out of a vertical dive in a circular arc such that her upward acceleration is $8.5g$. (*a*) What is the magnitude of the force exerted by the airplane seat on the pilot at the bottom of the arc? (*b*) If the speed of the plane is 345 km/h, what is the radius of the circular arc?

52 •• A 65-kg airplane pilot pulls out of a dive by following the arc of a circle whose radius is 300 m. At the bottom of the circle, where her speed is 180 km/h, (*a*) what are the direction and magnitude of her acceleration? (*b*) What is the net force acting on her at the bottom of the circle? (*c*) What is the force exerted on the pilot by the airplane seat?

53 •• Mass m_1 moves with speed v in a circular path of radius R on a frictionless horizontal table (Figure 5-52). It is attached to a string that passes through a frictionless hole in the center of the table. A second mass m_2 is attached to the other end of the string. Derive an expression for R in terms of m_1, m_2, and v.

Figure 5-52 Problem 53

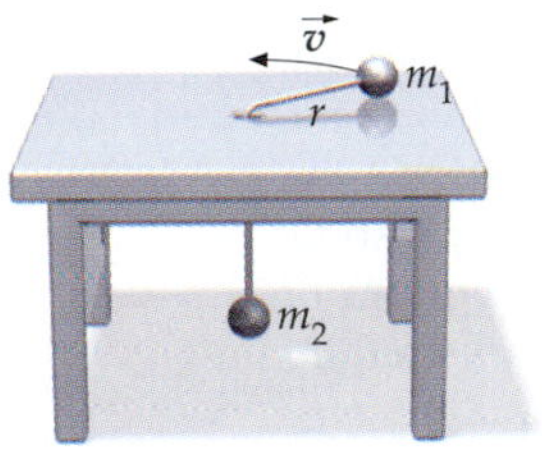

54 •• In Figure 5-53 particles are shown traveling counterclockwise in circles of radius 5 m. The acceleration vectors are indicated at three specific times. Find the values of v and dv/dt for each of these times.

Figure 5-53 Problem 54

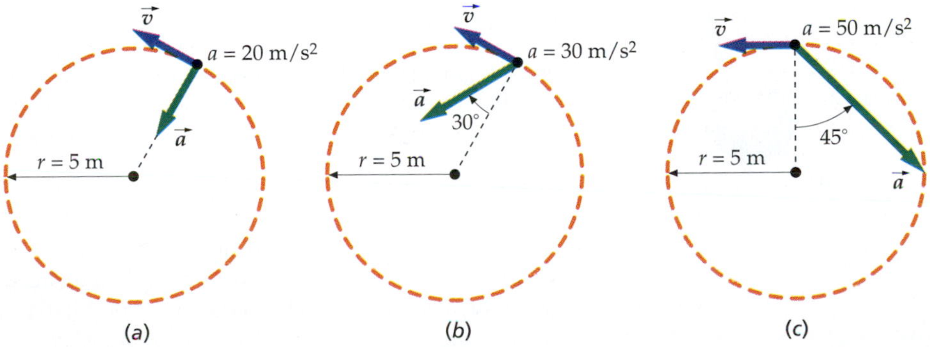

55 •• A block of mass m_1 is attached to a cord of length L_1, which is fixed at one end. The block moves in a horizontal circle on a frictionless table. A second block of mass m_2 is attached to the first by a cord of length L_2 and also moves in a circle, as shown in Figure 5-54. If the period of the motion is T, find the tension in each cord.

Figure 5-54 Problem 55

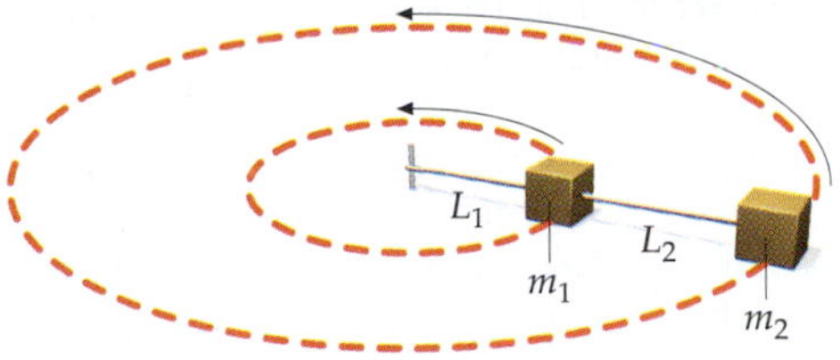

56 •• A particle moves with constant speed in a circle of radius 4 cm. It takes 8 s to make a complete trip. Draw the path of the particle to scale, and indicate the particle's position at 1-s intervals. Draw displacement vectors for each interval. These vectors also indicate the directions for the average-velocity vectors for each interval. Find graphically the change in the average velocity $\Delta\vec{v}$ for two consecutive 1-s intervals. Compare $\Delta\vec{v}/\Delta t$, measured in this way, with the instantaneous acceleration computed from $a = v^2/r$.

57 •• A man swings his child in a circle of radius 0.75 m, as shown in the photo. If the mass of the child is 25 kg and the child makes one revolution in 1.5 s, what are the magnitude and direction of the force that must be exerted by the man on the child? (Assume the child to be a point particle.)

58 •• The string of a conical pendulum is 50 cm long and the mass of the bob is 0.25 kg. Find the angle between the string and the horizontal when the tension in the string is six times the weight of the bob. Under those conditions, what is the period of the pendulum?

59 •• Frustrated with his inability to make a living through honest channels, Lou sets up a deceptive weight-loss scam. The trick is to make insecure customers believe that they can "think those extra pounds away" if they will only take a ride in a van that Lou claims to be "specially equipped to enhance mental-mass fluidity." The customer sits on a platform scale in the back of the van, and Lou drives off at a constant speed of 14 m/s. Lou then asks the customer to "think heavy" as he drives through the bottom of a dip in the road having a radius of curvature of 80 m. Sure enough, the scale's reading increases, until Lou says, "Now think light," and drives over the crest of a hill having a radius of curvature of 100 m. If the scale reads 800 N when the van is on level ground, what is the range of readings for the trip described here?

60 •• A 100-g disk sits on a horizontally rotating turntable. The turntable makes one revolution each second. The disk is located 10 cm from the axis of rotation of the turntable. (*a*) What is the frictional force acting on the disk? (*b*) The disk will slide off the turntable if it is located at a radius larger than 16 cm from the axis of rotation. What is the coefficient of static friction?

61 •• A tether ball of mass 0.25 kg is attached to a vertical pole by a cord 1.2 m long. Assume the cord attaches to the center of the ball. If the cord makes an angle of 20° with the vertical, then (*a*) what is the tension in the cord? (b) What is the speed of the ball?

62 •• An object on the equator has an acceleration toward the center of the earth because of the earth's rotation and an acceleration toward the sun because of the earth's motion along its orbit. Calculate the magnitudes of both accelerations, and express them as fractions of the free-fall acceleration due to gravity g.

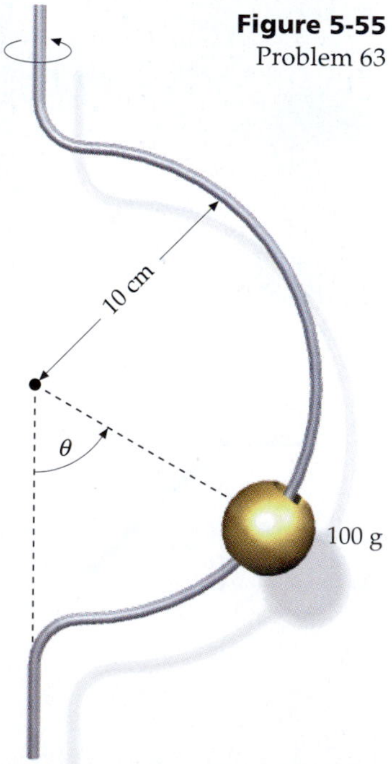

Figure 5-55 Problem 63

63 •• A small bead with a mass of 100 g slides along a semicircular wire with a radius of 10 cm that rotates about a vertical axis at a rate of 2 revolutions per second, as in Figure 5-55. Find the values of θ for which the bead will remain stationary relative to the rotating wire.

64 ••• Consider a bead of mass m that is free to move on a thin, circular wire of radius r. The bead is given an initial speed v_0, and there is a coefficient of kinetic friction μ_k. The experiment is performed in a spacecraft drifting in space. Find the speed of the bead at any subsequent time t.

65 ••• Revisiting the previous problem, (*a*) find the centripetal acceleration of the bead. (*b*) Find the tangential acceleration of the bead. (*c*) What is the magnitude of the resultant acceleration?

Loop-the-Loop

66 • A block is sliding on a frictionless surface along a loop-the-loop, as in Figure 5-56. The block is moving fast enough that it never loses contact with the track. Match the points along the track to the appropriate free-body diagrams (Figure 5-57).

Figure 5-56
Problem 66

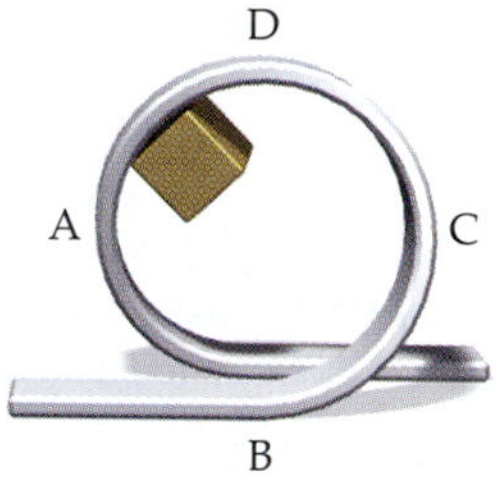

Figure 5-57
Problem 66

Point A
Point B
Point C
Point D

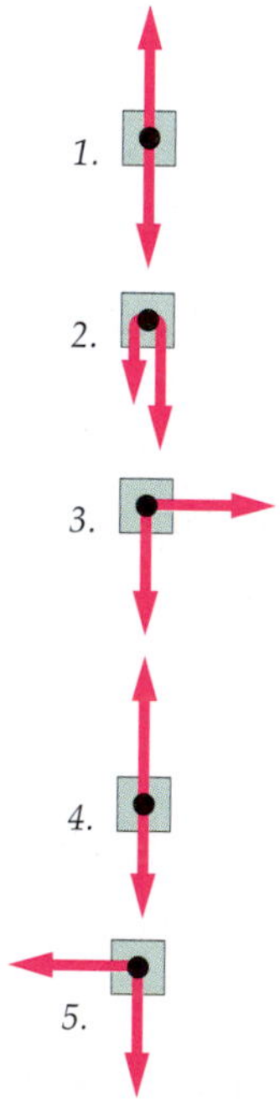

67 • A person rides a loop-the-loop at an amusement park. The cart circles the track at a constant speed. At the top of the loop, the normal force exerted by the seat equals the person's weight, mg. At the bottom of the loop, the force exerted by the seat will be ________.

(*a*) 0
(*b*) mg
(*c*) $2mg$
(*d*) $3mg$
(*e*) greater than mg, but the exact value cannot be calculated from the information given

68 • The radius of curvature of a loop-the-loop roller coaster is 12.0 m. At the top of the loop, the force that the seat exerts on a passenger of mass m is $0.4mg$. Find the speed of the roller coaster at the top of the loop.

Unbanked Curves

69 • Realizing that he has left the gas stove on, Aaron races for his car to drive home. He lives at the other end of a long, unbanked curve in the highway, and he knows that when he is traveling alone in his car at 40 km/h, he can just make it around the curve without skidding. He yells at his friends, "Get in the car! With greater mass, I can take the curve at a higher speed!" Carl says, "No, that will make you skid at an even *lower* speed." Bonita says, "The mass doesn't matter. Just get going!" Who is right?

70 • A car speeds along the curved exit ramp of a freeway. The radius of the curve is 80 m. A 70-kg passenger holds the arm rest of the car door with a 220-N force to keep from sliding across the front seat of the car. (Assume that the exit ramp is not banked and ignore friction with the car seat.) What is the car's speed?

(*a*) 16 m/s
(*b*) 57 m/s
(*c*) 18 m/s
(*d*) 50 m/s
(*e*) 28 m/s

71 ••• Suppose you ride a bicycle on a horizontal surface in a circle with a radius of 20 m. The resultant force exerted by the road on the bicycle (normal force plus frictional force) makes an angle of 15° with the vertical. (*a*) What is your speed? (*b*) If the frictional force is half its maximum value, what is the coefficient of static friction?

Banked Curves (optional)

optional

72 • A 750-kg car travels at 90 km/h around a curve with a radius of 160-m. What should the banking angle of the curve be so that the only force between the pavement and tires of the car is the normal reaction force?

73 •• A curve of radius 150 m is banked at an angle of 10°. An 800-kg car negotiates the curve at 85 km/h without skidding. Find (*a*) the normal force on the tires exerted by the pavement, (*b*) the frictional force exerted by the pavement on the tires of the car, and (*c*) the minimum coefficient of static friction between the pavement and tires.

74 •• On another occasion, the car in the previous problem negotiates the curve at 38 km/h. Find (*a*) the normal force exerted on the tires by the pavement, and (*b*) the frictional force exerted on the tires by the pavement.

75 ••• A civil engineer is asked to design a curved section of roadway that meets the following conditions: With ice on the road, when the coefficient of static friction between the road and rubber is 0.08, a car at rest must not slide into the ditch and a car traveling less than 60 km/h must not skid to the outside of the curve. What is the minimum radius of curvature of the curve and at what angle should the road be banked?

76 ••• A curve of radius 30 m is banked so that a 950-kg car traveling 40 km/h can round it even if the road is so icy that the coefficient of static friction is approximately zero. Find the range of speeds at which a car can travel around this curve without skidding if the coefficient of static friction between the road and the tires is 0.3.

Drag Forces (optional)

77 • How would you expect the value of b for air resistance to depend on the density of air?

78 • True or false: The terminal speed of an object depends on its shape.

79 • As a skydiver falls through the air, her terminal speed

(*a*) depends on her mass.
(*b*) depends on her orientation as she falls.
(*c*) depends on the density of the air.
(*d*) depends on all of the above.

80 • What are the dimensions and SI units of the constant b in the retarding force bv^n if (*a*) $n = 1$, and (*b*) $n = 2$?

81 • A small pollution particle settles toward the earth in still air with a terminal speed of 0.3 mm/s. The particle has a mass of 10^{-10} g and a retarding force of the form bv. What is the value of b?

82 • A Ping-Pong ball has a mass of 2.3 g and a terminal speed of 9 m/s. The retarding force is of the form bv^2. What is the value of b?

83 • A sky diver of mass 60 kg can slow herself to a constant speed of 90 km/h by adjusting her form. (*a*) What is the magnitude of the upward drag force on the sky diver? (*b*) If the drag force is equal to bv^2, what is the value of b?

84 • Newton showed that the air resistance on a falling object with a circular cross section should be approximately $\frac{1}{2}\rho\pi r^2 v^2$, where $\rho = 1.2\ \text{kg/m}^3$, the density of air. Find the terminal speed for a 56-kg sky diver, assuming that his cross-sectional area is equivalent to that of a disk of radius 0.30 m.

85 •• An 800-kg car rolls down a very long 6° grade. The drag force for motion of the car has the form $F_d = 100\ \text{N} + (1.2\ \text{N}\cdot\text{s}^2/\text{m}^2)v^2$. What is the terminal velocity for the car rolling down this grade?

86 •• While claims of hailstones the size of golf balls may be a slight exaggeration, hailstones are often substantially larger than raindrops. Estimate the terminal velocity of a raindrop and a large hailstone. (See Problem 84.)

87 •• (*a*) A parachute creates enough air resistance to keep the downward speed of an 80-kg sky diver to a constant 6.0 m/s. Assuming that the force of air resistance is given by $f = bv^2$, calculate b for this case. (*b*) A sky diver free-falls until his speed is 60 m/s before opening his parachute. If the parachute opens instantaneously, calculate the initial upward force exerted by the chute on the sky diver moving at 60 m/s. Explain why it is important that the parachute takes a few seconds to open.

88 ••• An object falls under the influence of gravity and a drag force $F_d = -bv$. (*a*) By applying Newton's second law, show that the acceleration of the object can be written

$$a = \frac{dv}{dt} = g - \frac{b}{m}v$$

(*b*) Rearrange this equation to obtain

$$\frac{dv}{v - v_t} = -\frac{g}{v_t}\,dt$$

where $v_t = mg/b$.

(*c*) Integrate this equation to obtain the exact solution

$$v = \frac{mg}{b}\left(1 - e^{-bt/m}\right) = v_t\left(1 - e^{-gt/v_t}\right)$$

(*d*) Plot v versus t for $v_t = 60$ m/s.

89 ••• Small spherical particles experience a viscous drag force given by Stokes' law: $F_d = 6\pi\eta r v$, where r is the radius of the particle, v is its speed, and η is the viscosity of the fluid medium. (*a*) Estimate the terminal speed of a spherical pollution particle of radius 10^{-5} m and density of 2000 kg/m^3. (*b*) Assuming that the air is still and that η is 1.8×10^{-5} N·s/m^2, estimate the time it takes for such a particle to fall from a height of 100 m.

90 ••• An air sample containing pollution particles of the size and density given in Problem 89 is captured in a test tube 8.0 cm long. The test tube is then placed in a centrifuge with the midpoint of the test tube 12 cm from the center of the centrifuge. The centrifuge spins at 800 revolutions per minute. Estimate the time required for nearly all of the pollution particles to sediment at the end of the test tube and compare this to the time required for a pollution particle to fall 8.0 cm under the action of gravity and subject to the viscous drag of air.

General Problems

91 • The mass of the moon is about 1% that of the earth. The centripetal force that keeps the moon in its orbit around the earth

(*a*) is much smaller than the gravitational force exerted on the moon by the earth.
(*b*) depends on the phase of the moon.
(*c*) is much greater than the gravitational force exerted on the moon by the earth.
(*d*) is the same as the gravitational force exerted on the moon by the earth.
(*e*) I cannot answer; we haven't studied Newton's law of gravity yet.

92 • True or false: Centripetal force is one of the four fundamental forces.

93 • On an icy winter day, the coefficient of friction between the tires of a car and a roadway might be reduced to one-half of its value on a dry day. As a result, the maximum speed at which a curve of radius R can be safely negotiated is

(*a*) the same as on a dry day.
(*b*) reduced to 71% of its value on a dry day.
(*c*) reduced to 50% of its value on a dry day.
(*d*) reduced to 37% of its value on a dry day.
(*e*) reduced by an unknown amount depending on the car's mass.

94 • A 4.5-kg block slides down an inclined plane that makes an angle of 28° with the horizontal. Starting from rest, the block slides a distance of 2.4 m in 5.2 s. Find the coefficient of kinetic friction between the block and plane.

95 • A model airplane of mass 0.4 kg is attached to a horizontal string and flies in a horizontal circle of radius 5.7 m. (The weight of the plane is balanced by the upward "lift" force of the air on the wings of the plane.) The plane makes 1.2 revolutions every 4 s. (*a*) Find the speed v of the plane. (*b*) Find the tension in the string.

96 •• Show with a force diagram how a motorcycle can travel in a circle on the inside vertical wall of a hollow cylinder. Assume reasonable parameters (coefficient of friction, radius of the circle, mass of the motorcycle, or whatever is required), and calculate the minimum speed needed.

97 •• An 800-N box rests on a plane inclined at 30° to the horizontal. A physics student finds that she can prevent the box from sliding if she pushes on it with a force of at least 200 N parallel to the surface. (*a*) What is the coefficient of static friction between the box and the surface? (*b*) What is the greatest force that can be applied to the box parallel to the incline before the box slides up the incline?

98 •• The position of a particle is given by the vector $\vec{r} = -10 \text{ m} \cos \omega t \hat{i} + 10 \text{ m} \sin \omega t \hat{j}$, where $\omega = 2 \text{ s}^{-1}$. (*a*) Show that the path of the particle is a circle. (*b*) What is the radius of the circle? (*c*) Does the particle move clockwise or counterclockwise around the circle? (*d*) What is the speed of the particle? (*e*) What is the time for one complete revolution?

99 •• A crate of books is to be put on a truck with the help of some planks sloping up at 30°. The mass of the crate is 100 kg, and the coefficient of sliding friction between it and the planks is 0.5. You and your friends push *horizontally* with a force $\vec{F}$. Once the crate has started to move, how large must F be to keep the crate moving at constant speed?

100 •• Brother Bernard is a very large dog with a taste for tobogganing. Ernie gives him a ride down Idiots' Hill—so named because it is a steep slope that levels out at the bottom for 10 m, and then drops into a river. When they reach the level ground at the bottom, their speed is 40 km/h, and Ernie, sitting in front, starts to dig in his heels to make the toboggan stop. He knows, however, that if he brakes too hard, he will be mashed by Brother Bernard. If the coefficient of static friction between the dog and the toboggan is 0.8, what is the minimum stopping distance that will keep Brother Bernard off Ernie's back?

101 •• An object with a mass of 5.5 kg is allowed to slide from rest down an inclined plane. The plane makes an angle of 30° with the horizontal and is 72 m long. The coefficient of kinetic friction between the plane and the object is 0.35. The speed of the object at the bottom of the plane is

(*a*) 5.3 m/s.
(*b*) 15 m/s.
(*c*) 24 m/s.
(*d*) 17 m/s.
(*e*) 11 m/s.

102 •• A brick slides down an inclined plank at constant speed when the plank is inclined at an angle θ_0. If the angle is increased to θ_1, the block accelerates down the plank with acceleration a. The coefficient of kinetic friction is the same in both cases. Given θ_0 and θ_1, calculate a.

103 •• One morning, Lou was in a particularly deep and peaceful slumber. Unfortunately, he had spent the night in the back of a dump truck, and Barry, the driver, was keen to go off to work and start dumping things. Rather than risk a ruckus with Lou, Barry simply raised the back of the truck, and when it reached an angle of 30°, Lou slid down the 4-m incline in 2 s, plopped onto a pile of sand, rolled over, and continued to sleep. Calculate the coefficients of static and kinetic friction between Lou and the truck.

104 •• In a carnival ride, the passenger sits on a seat in a compartment that rotates with constant speed in a vertical circle of radius $r = 5$ m. The heads of the seated passengers always point toward the axis of rotation. (*a*) If the carnival ride completes one full circle in 2 s, find the acceleration of the passenger. (*b*) Find the slowest rate of rotation (in other words, the longest time T to complete one full circle) if the seat belt is to exert no force on the passenger at the top of the ride.

105 •• A flat-topped toy cart moves on frictionless wheels, pulled by a rope under tension T. The mass of the cart is m_1. A load of mass m_2 rests on top of the cart with a coefficient of static friction μ_s. The cart is pulled up a ramp that is inclined at angle θ above the horizontal. The rope is parallel to the ramp. What is the maximum tension T that can be applied without making the load slip?

106 •• A sled weighing 200 N rests on a 15° incline, held in place by static friction (Figure 5-58). The coefficient of static friction is 0.5. (*a*) What is the magnitude of the normal force on the sled? (*b*) What is the magnitude of the static friction on the sled? (*c*) The sled is now pulled up the incline at constant speed by a child. The child weighs 500 N and pulls on the rope with a constant force of 100 N. The rope makes an angle of 30° with the incline and has negligible weight. What is the magnitude of the kinetic friction force on the sled? (*d*) What is the coefficient of kinetic friction between the sled and the incline? (*e*) What is the magnitude of the force exerted on the child by the incline?

Figure 5-58 Problem 106

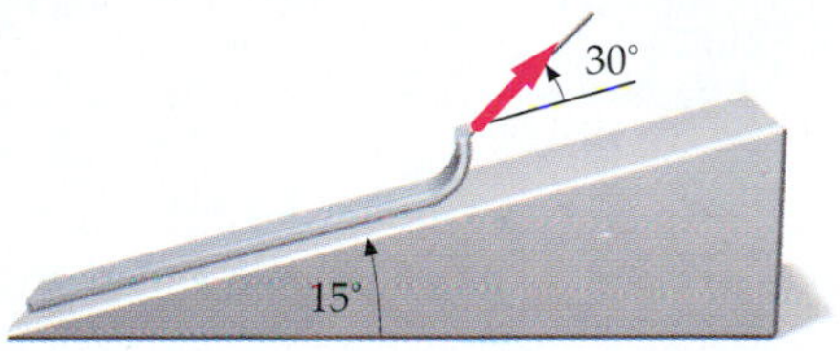

107 •• A child slides down a slide inclined at 30° in time t_1. The coefficient of kinetic friction between her and the slide is μ_k. She finds that if she sits on a small cart with frictionless wheels, she slides down the same slide in time $t_1/2$. Find μ_k.

108 •• The position of a particle of mass $m = 0.8$ kg as a function of time is

$$\vec{r} = x\hat{i} + y\hat{j} = R \sin \omega t\hat{i} + R \cos \omega t\hat{j}$$

where $R = 4.0$ m, and $\omega = 2\pi\ \mathrm{s}^{-1}$. (*a*) Show that the path of this particle is a circle of radius R with its center at the origin. (*b*) Compute the velocity vector. Show that $v_x/v_y = -y/x$. (*c*) Compute the acceleration vector and show that it is in the radial direction and has the magnitude v^2/r. (*d*) Find the magnitude and direction of the net force acting on the particle.

109 •• In an amusement-park ride, riders stand with their backs against the wall of a spinning vertical cylinder. The floor falls away and the riders are held up by friction. If the radius of the cylinder is 4 m, find the minimum number of revolutions per minute necessary to prevent the riders from dropping when the coefficient of static friction between a rider and the wall is 0.4.

110 •• Some bootleggers race from the police down a road that has a sharp, level curve with a radius of 30 m. As they go around the curve, the bootleggers squirt oil on the road behind them, reducing the coefficient of static friction from 0.7 to 0.2. When taking this curve, what is the maximum safe speed of (*a*) the bootleggers' car, and (*b*) the police car?

111 •• A mass m_1 on a horizontal shelf is attached by a thin string that passes over a frictionless peg to a 2.5-kg mass m_2 that hangs over the side of the shelf 1.5 m above the ground (Figure 5-59). The system is released from rest at $t = 0$ and the 2.5-kg mass strikes the ground at $t = 0.82$ s. The system is now placed in its initial position and a 1.2-kg mass is placed on top of the block of mass m_1. Released from rest, the 2.5-kg mass now strikes the ground 1.3 s later. Determine the mass m_1 and the coefficient of kinetic friction between m_1 and the shelf.

Figure 5-59 Problem 111

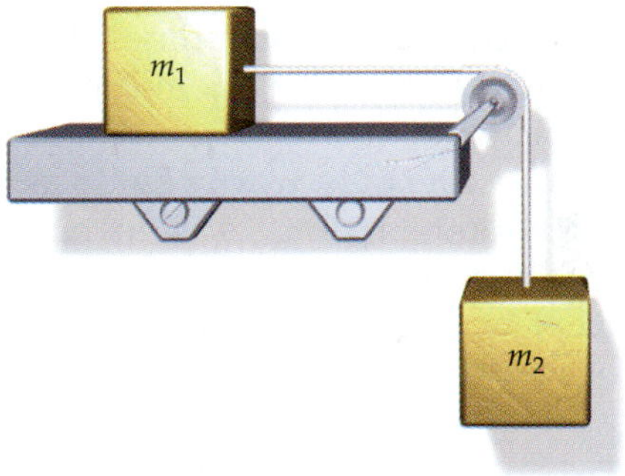

112 ••• (*a*) Show that a point on the surface of the earth at latitude θ has an acceleration relative to a reference frame not rotating with the earth with a magnitude of $3.37 \cos \theta\ \mathrm{cm/s^2}$. What is the direction of this acceleration? (*b*) Discuss the effect of this acceleration on the apparent weight of an object near the surface of the earth. (*c*) The free-fall acceleration of an object at sea level measured *relative to the earth's surface* is $9.78\ \mathrm{m/s^2}$ at the equator and $9.81\ \mathrm{m/s^2}$ at latitude $\theta = 45°$. What are the values of the gravitational field g at these points?

CHAPTER 6

Work and Energy

Work and energy are important concepts in physics as well as in our everyday life. In physics, a force does **work** when it acts on an object that moves through a distance, and there is a component of the force along the line of motion. For a constant force in one dimension, the work done equals the force times the distance. (This differs somewhat from the everyday use of the word work. When you study hard for an exam, the only work you do as the term is understood in physics is in moving your pencil or turning the pages of your book.)

The concept of **energy** is closely associated with that of work. When work is done by one system on another, energy is transferred between the two systems. For example, when you do work pushing a swing, chemical energy in your body is transferred to the swing and appears as kinetic energy of motion or gravitational potential energy of the earth–swing system. There are many forms of energy. Kinetic energy is associated with the motion of an object. Potential energy is associated with the configuration of a system, such as the separation distance between some object and the earth. Thermal energy is associated with the random motion of the molecules within a system and is closely connected with the temperature of the system.

The action of the pole vaulter shown here demonstrates several kinds of energy. First the vaulter transforms the internal chemical energy of his body into kinetic energy as he runs. Some of this kinetic energy is then converted into elastic potential energy, represented by the deformation of the pole. The rest of the vaulter's kinetic energy is eventually converted into gravitational potential energy, which in turn is converted into kinetic energy as the vaulter drops. Mechanical energy is finally converted into thermal energy when the athlete drops onto the mat.

6-1 Work and Kinetic Energy

Motion in One Dimension With Constant Forces

The work W done by a constant force $\vec{F}$ whose point of application moves through a distance Δx is defined to be

$$W = F\cos\theta\,\Delta x = F_x\Delta x \qquad \text{6-1}$$

Definition—Work by a constant force

where θ is the angle between $\vec{F}$ and the x axis, and Δx is the displacement of the force as shown in Figure 6-1.

Work is a scalar quantity that is positive if Δx and F_x have the same signs and negative if they have opposite signs. The dimensions of work are those of force times distance. The SI unit of work and energy is the **joule** (J), which equals the product of a newton and a meter*:

$$1\text{ J} = 1\text{ N}\cdot\text{m} \tag{6-2}$$

A convenient unit of work and energy in atomic and nuclear physics is the electron volt (eV):

$$1\text{ eV} = 1.6 \times 10^{-19}\text{ J} \tag{6-3}$$

Commonly used multiples are keV (1000 eV) and MeV (10^6 eV). The work required to remove an electron from an atom is of the order of several eV, whereas the work needed to remove a proton or neutron from an atomic nucleus is of the order of several MeV.

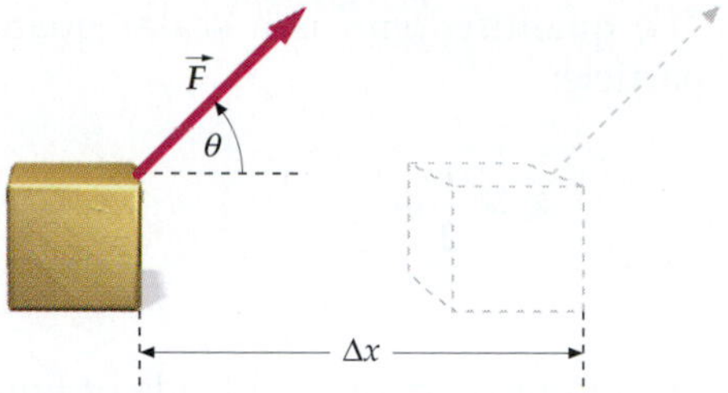

Figure 6-1 When a constant force $\vec{F}$ moves through a distance Δx, the work done is $F\cos\theta\,\Delta x = F_x\,\Delta x$.

Exercise A force of 12 N is exerted on a box at an angle of $\theta = 20°$, as in Figure 6-1. How much work is done by the force as the box moves along the table a distance of 3 m? (*Answer* 33.8 J)

When there are several forces that do work, the total work is found by computing the work done by each force and summing:

$$W_{\text{total}} = F_{1x}\Delta x_1 + F_{2x}\Delta x_2 + F_{3x}\Delta x_3 + \cdots$$

When the forces do work on a *particle*, the displacement of the force Δx_i is the same for each force and is equal to the displacement of the particle Δx:

$$\begin{aligned} W_{\text{total}} &= F_{1x}\Delta x + F_{2x}\Delta x + F_{3x}\Delta x + \cdots \\ &= (F_{1x} + F_{2x} + F_{3x})\Delta x \\ &= F_{\text{net }x}\Delta x \end{aligned} \tag{6-4}$$

Thus for a particle, the total work can be found by summing all the forces to find the net force and then computing the work done by the net force.

The Work–Kinetic Energy Theorem

There is an important relation between the total work done on a particle and the initial and final speeds of the particle. If F_x is the net force acting on a particle, Newton's second law gives

$$F_x = ma_x$$

Since the work done by the net force equals the total work done on the particle,

$$W_{\text{total}} = F_x\Delta x = ma_x\,\Delta x$$

For a constant force, the acceleration is constant, and we can relate the distance the particle moves to its initial speed v_{i} and final speed v_{f} by using the constant-acceleration formula (Equation 2-15):

$$v_{\text{f}}^2 = v_{\text{i}}^2 + 2a_x\Delta x$$

Substituting $\frac{1}{2}(v_{\text{f}}^2 - v_{\text{i}}^2)$ for $a_x\,\Delta x$ yields

$$W_{\text{total}} = \frac{1}{2}mv_{\text{f}}^2 - \frac{1}{2}mv_{\text{i}}^2 \tag{6-5}$$

* In the U.S. customary system, the unit of work is the foot-pound: 1 ft-lb = 1.356 J.

The quantity $\frac{1}{2}mv^2$ is a scalar quantity called the **kinetic energy** K of the particle:

$$K = \frac{1}{2}mv^2 \qquad 6\text{-}6$$

Definition—Kinetic energy

The quantity on the right side of Equation 6-5 is the change in the kinetic energy of the particle. Thus,

The *total* work done on a particle is equal to the *change* in its kinetic energy:

$$W_{\text{total}} = \Delta K = \frac{1}{2}mv_{\text{f}}^2 - \frac{1}{2}mv_{\text{i}}^2 \qquad 6\text{-}7$$

Work–kinetic energy theorem

This result is known as the **work–kinetic energy theorem.** It holds whether the net force is constant or variable, as we will see in the next section.

Exercise A girl of mass 50 kg is running at 3.5 m/s. What is her kinetic energy? (*Answer* 306 J)

Example 6-1

A truck of mass 3000 kg is to be loaded onto a ship by a crane that exerts an upward force of 31 kN on the truck. This force, which is just strong enough to get the truck started upward, is applied over a distance of 2 m. Find (*a*) the work done by the crane, (*b*) the work done by gravity, and (*c*) the upward speed of the truck after 2 m.

Picture the Problem The applied force is in the direction of motion, so the work it does is positive. On the other hand, gravity is opposite in direction to the motion, so the work done by gravity is negative (Figure 6-2). The final speed of the truck can be obtained from its final kinetic energy, which equals the total work done on the truck because it starts from rest. The total work is the sum of the results for (*a*) and (*b*).

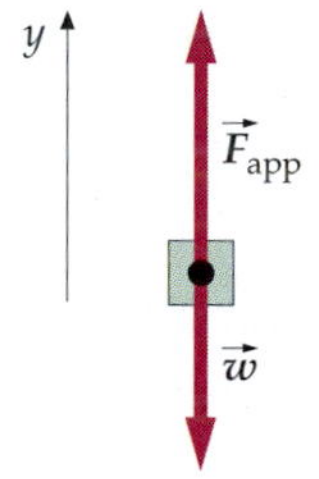

Figure 6-2

(*a*) Calculate the work done by the applied force: $W_{\text{app}} = F_{\text{app}} \cos 0° \, \Delta y = (31 \text{ kN})(1)(2 \text{ m}) = 62 \text{ kJ}$

(*b*) Calculate the work done by gravity:

$$W_{\text{g}} = mg \cos 180° \, \Delta y$$
$$= (3000 \text{ kg})(9.81 \text{ N/kg})(-1)(2 \text{ m}) = -59 \text{ kJ}$$

(*c*) 1. The final speed is related to the final kinetic energy:

$$K_{\text{f}} = \frac{1}{2}mv_{\text{f}}^2$$
$$v_{\text{f}} = \sqrt{\frac{2K_{\text{f}}}{m}}$$

2. Apply the work–kinetic energy theorem, with $v_{\text{i}} = 0$:

$$W_{\text{total}} = \Delta K = K_{\text{f}} - K_{\text{i}} = K_{\text{f}}$$

3. The total work is the sum of the applied work and the work done by gravity:

$$W_{\text{total}} = W_{\text{app}} + W_{\text{g}} = 62 \text{ kJ} - 59 \text{ kJ} = 3.0 \text{ kJ}$$

4. Substitute $K_{\text{f}} = 3.0$ J to obtain the final speed of the truck:

$$v_{\text{f}} = \sqrt{\frac{2K_{\text{f}}}{m}} = \sqrt{\frac{2(3.0 \text{ kJ})}{3000 \text{ kg}}} = 1.4 \text{ m/s}$$

Remark We treat each force separately when calculating the work done. We could also find the total work by first adding the forces to obtain the net force, then applying $W_{\text{total}} = F_{\text{net }x}\,\Delta x$. In either case, the work–kinetic energy theorem applies only to the total work. We could have also found the speed using Newton's second law.

Exercise Find the final speed of the truck if the same upward force were applied for 2 m after it was already moving upward at 1 m/s. (*Answer* 1.73 m/s. Note that the answer is *not* 1.4 m/s + 1 m/s. Why not?)

Example 6-2

In a television tube, an electron is accelerated from rest to a kinetic energy of 2.5 keV over a distance of 80 cm. (The force that accelerates the electron is an electric force due to the electric field in the tube.) Find the force on the electron, assuming it to be constant and in the direction of motion.

Picture the Problem Since the electron starts from rest, the work done equals the final kinetic energy. To find the force in newtons, we must convert the energy from keV to joules.

1. Set the work done to be equal to the change in kinetic energy:

$$W = F\,\Delta x = \Delta K = K_f - K_i = K_f = 2.5\text{ keV}$$

2. Solve for F and convert the energy to joules:

$$F = \frac{W}{\Delta x} = \frac{2.5\text{ keV}}{0.8\text{ m}} \times \frac{1.6 \times 10^{-19}\text{ J}}{1\text{ eV}} = 5.0 \times 10^{-16}\text{ N}$$

Remark When we discuss electricity we will see that the work done per charge is called the potential difference and is measured in volts. Thus, 1 eV is the energy acquired or lost by a particle of charge e (an electron or proton, for example) when its potential difference changes by 1 V.

Example 6-3

Your professor enters the dogsled race during winter break. To get started, he pulls his sled (total mass 80 kg) with a force of 180 N at 20° to the horizontal. Find (*a*) the work he does, and (*b*) the final speed of the sled after it moves $\Delta x = 5$ m, assuming that it starts from rest and there is no friction.

Figure 6-3

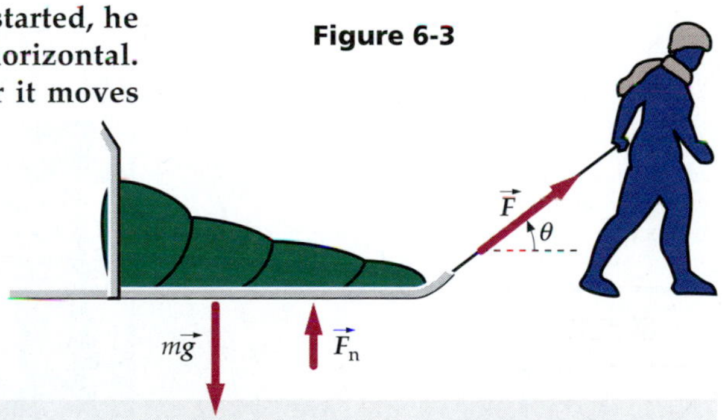

Picture the Problem The work done by the professor is $F_x\,\Delta x$ since there is no motion in the y direction. This is also the total work done on the sled because the other forces, mg and F_n, have no x components (Figure 6-3). The final speed of the sled is found by applying the work–kinetic energy theorem with $v_i = 0$.

(*a*) The work done by the professor is $F_x\,\Delta x$. This is also the total work done on the sled:

$$W = F_x\Delta x = (F\cos 20°)\Delta x$$
$$= (180\text{ N})(\cos 20°)(5\text{ m}) = 846\text{ J} = W_{\text{total}}$$

(*b*) Apply the work–kinetic energy theorem and solve for the final speed:

$$W_{\text{total}} = \frac{1}{2}mv_f^2 - \frac{1}{2}mv_i^2 = \frac{1}{2}mv_f^2$$

$$v_f = \sqrt{\frac{2W_{\text{total}}}{m}} = \sqrt{\frac{2(846\text{ J})}{80\text{ kg}}} = 4.60\text{ m/s}$$

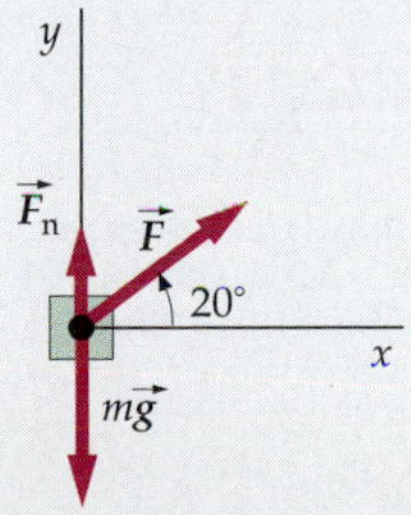

Remark We do not need to work out the units. If we have a correct equation, and all quantities are in SI units, the result will be in the correct SI units. However, as a check on the equation, we can show that $1\ \mathrm{J/kg} = 1\ \mathrm{m^2/s^2}$. We have $1\ \mathrm{J/kg} = 1\ \mathrm{N{\cdot}m/kg} = (1\ \mathrm{kg{\cdot}m/s^2})\ \mathrm{m{\cdot}kg} = 1\ \mathrm{m^2/s^2}$.

Exercise What force did your professor exert if the sled starts with a speed of 2 m/s, and its final speed is 4.5 m/s after he pulls it through a distance of 5 m? (*Answer* 138 N)

What if you hold a weight in a fixed position? You are expending energy, but are you doing work? According to the definition of work, you are not doing work *on the weight* because the weight does not move (Figure 6-4). But your muscles are continually contracting and relaxing as you hold the weight. Molecular assemblies in your muscle *do* move, and work is done. In the process, internal chemical energy in your body is converted to thermal energy (Figure 6-5).

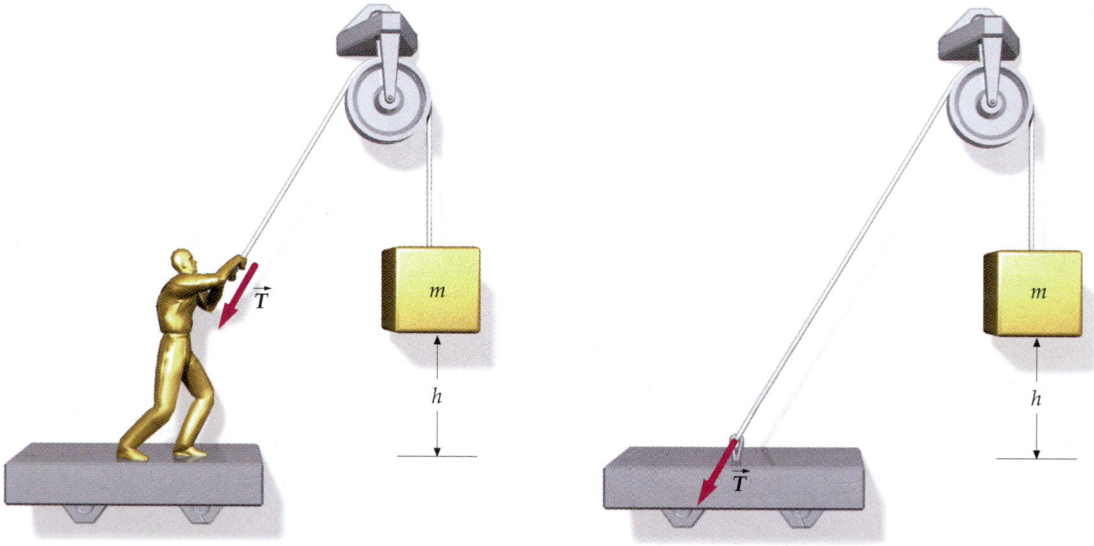

Figure 6-4 The man standing on a ledge does not do work on the weight when holding it at a fixed position. The same task could be accomplished by tying the rope to a fixed point.

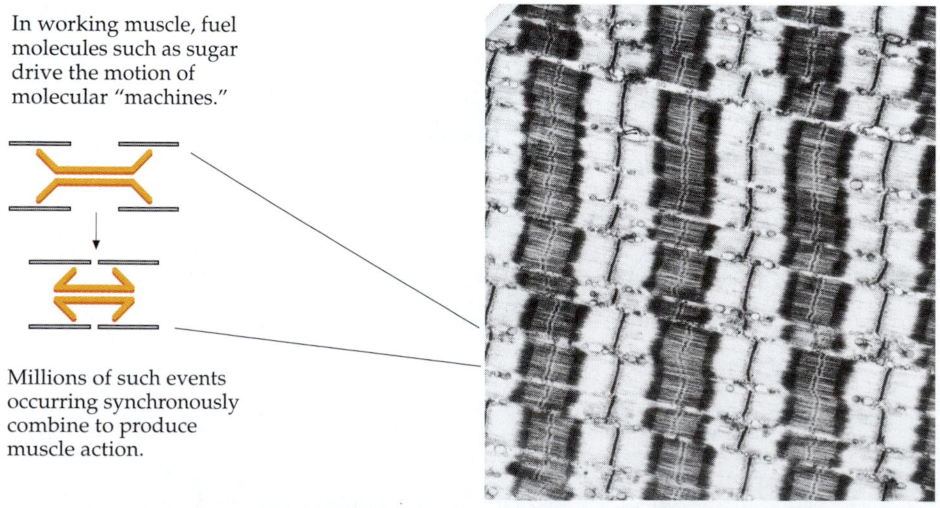
In working muscle, fuel molecules such as sugar drive the motion of molecular "machines."
Millions of such events occurring synchronously combine to produce muscle action.

Figure 6-5 Muscle work. While the man holding the weight in Figure 6-4 may be doing no work on the weight, his body is putting out work on the molecular level, as structures within the muscle slide over each other during muscular extension and contraction.

Work Done by a Variable Force

In Figure 6-6, we plot a constant force F_x as a function of position x. The work done on a particle whose displacement is Δx is represented by the area under the force-versus-position curve, indicated by the shading in Figure 6-6.

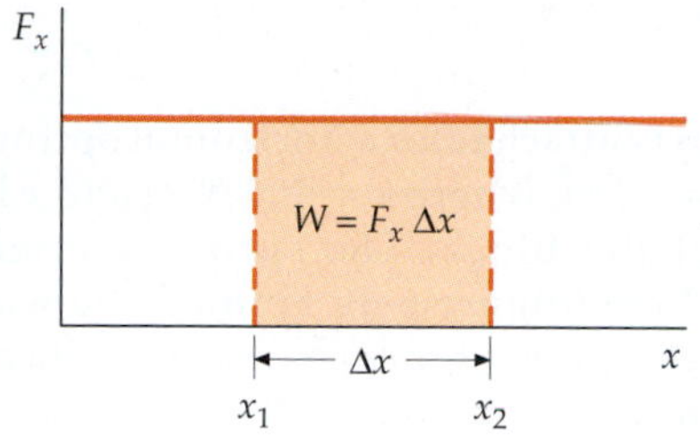

Figure 6-6 The work done by a constant force is represented graphically as the area under the F_x-versus-x curve.

Many forces vary with distance. For example, a spring exerts a force proportional to the distance it is stretched or compressed. And the gravitational force the earth exerts on a spaceship varies inversely with the square of the distance between the two bodies. We can approximate a variable force by a series of constant forces (Figure 6-7). The work done by a variable force is then

$$W = \lim_{\Delta x_i \to 0} \sum_i F_x \Delta x_i = \text{area under the } F_x\text{-versus-}x \text{ curve} \qquad 6\text{-}8$$

This limit is the integral of F_x over x. So the work done by a variable force F_x acting on a particle as it moves from x_1 to x_2 is

$$W = \int_{x_1}^{x_2} F_x\, dx = \text{area under the } F_x\text{-versus-}x \text{ curve} \qquad 6\text{-}9$$

Definition—Work by a variable force

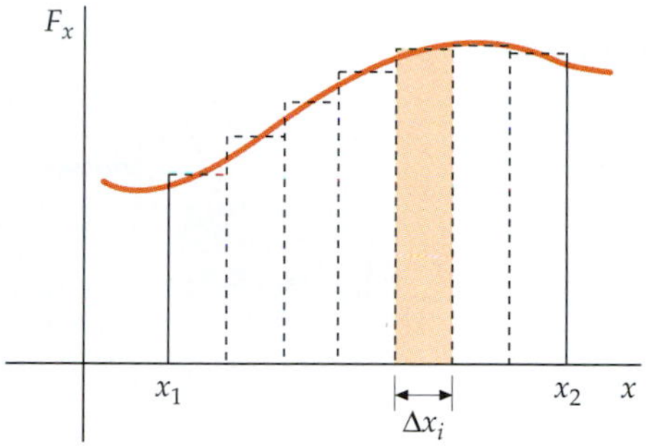

Figure 6-7 A variable force can be approximated by a series of constant forces over small intervals. The work done by the constant force in each interval is the area of the rectangle beneath the force curve. The sum of these rectangular areas is the sum of the work done by the set of constant forces that approximates the varying force. In the limit of infinitesimally small Δx_i, the sum of the areas of the rectangles equals the area under the complete curve.

For each rectangular area, the force is constant, so the work done equals the change in the kinetic energy over that interval. The total work done is the sum of the areas over all intervals, which equals the change in kinetic energy over the complete interval. Thus, $W_{\text{total}} = \Delta K$ holds for variable forces as well as for constant forces.

Exercise in Dimensional Analysis A spring is characterized by its force constant k, which has dimensions N/m. How does the work required to stretch a spring by an amount x_0 depend on k and x_0? (*Answer* Since work has dimensions of N·m, the work must depend on k and x_0 in the combination kx_0^2. We will see in Example 6-5 that the actual expression is $W = \frac{1}{2}kx_0^2$. The factor $\frac{1}{2}$ arises because the force varies from 0 to a maximum value of kx_0, and has the average value $\frac{1}{2}kx_0$.)

Example 6-4

A force F_x varies with x as shown in Figure 6-8. Find the work done by the force on a particle as the particle moves from $x = 0$ to $x = 6$ m.

1. We find the work done by calculating the area under the F_x-versus-x curve: $W = A$

2. This area is the sum of the two areas shown:

$$W = A = A_1 + A_2$$
$$= (5\text{ N})(4\text{ m}) + \frac{1}{2}(5\text{ N})(2\text{ m})$$
$$= 20\text{ J} + 5\text{ J} = 25\text{ J}$$

Figure 6-8

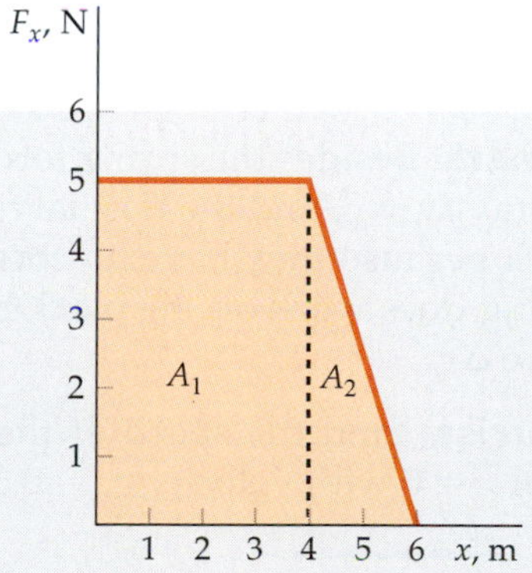

Exercise The force shown is the only force that acts on a particle of mass 3 kg. If the particle starts from rest at $x = 0$, how fast is it moving when it reaches $x = 6$ m? (*Answer* 4.08 m/s)

Example 6-5

A 4-kg block on a frictionless table is attached to a horizontal spring that obeys Hooke's law and exerts a force $\vec{F} = -kx\hat{i}$, where $k = 400$ N/m and x is measured from the equilibrium position of the block. The spring is originally compressed with the block at $x_1 = -5$ cm (Figure 6-9). Find (*a*) the work done by the spring on the block as the block moves from $x_1 = -5$ cm to its equilibrium position $x_2 = 0$, and (*b*) the speed of the block at $x_2 = 0$.

Figure 6-9

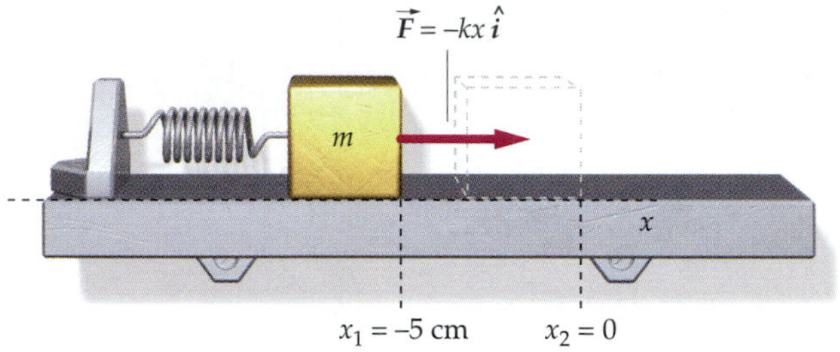

Figure 6-10

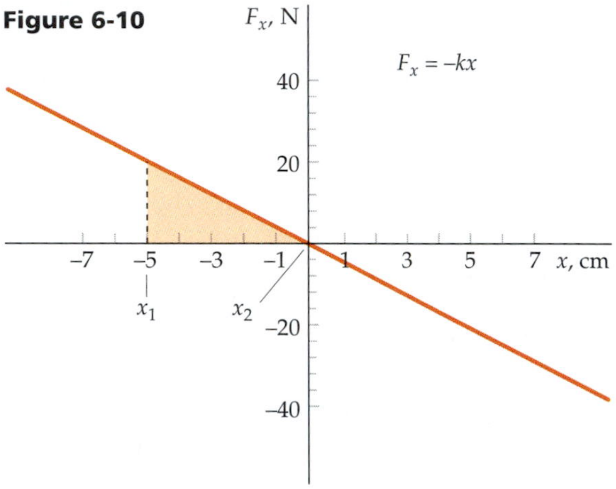

Picture the Problem The work done on the block as it moves from x_1 to $x_2 = 0$ equals the area under the F_x-versus-x curve between these limits (the shaded area in Figure 6-10), which can be calculated by integrating the force over the distance. The work done equals the change in kinetic energy, which is just the final kinetic energy since the initial kinetic energy is zero. The speed of the block at $x = 0$ is found from the kinetic energy of the block.

(*a*) The work W done by the spring on the block is the integral of $F_x\,dx$ from $x_1 = -5$ cm to $x_2 = 0$:

$$W = \int_{x_1}^{x_2} F_x\,dx = \int_{x_1}^{0} -kx\,dx = -k\int_{x_1}^{0} x\,dx = -\frac{1}{2}kx^2\Big|_{x_1}^{0}$$

$$= \frac{1}{2}kx_1^2 = \frac{1}{2}(400\text{ N/m})(0.05\text{ m})^2 = 0.500\text{ J}$$

(*b*) Apply the work–kinetic energy theorem with $v_1 = 0$ and solve for v_2:

$$W = \frac{1}{2}mv_2^2 - \frac{1}{2}mv_1^2 = \frac{1}{2}mv_2^2$$

$$v_2 = \sqrt{\frac{2W}{m}} = \sqrt{\frac{2(0.500\text{ J})}{4\text{ kg}}} = 0.50\text{ m/s}$$

Remark Besides the spring force, two other forces act on the block; the force of gravity, $m\vec{g}$, and the normal force of the table, $\vec{F}_n$. These latter forces do no work because they have no component in the direction of motion. Only the spring does work on the block because the force it exerts acts through a distance Δx.

Exercise Find the speed of the block when it reaches $x = 3$ cm if it starts from $x = 0$ with velocity $v_x = 0.5$ m/s. (*Answer* 0.4 m/s)

Note that we could not have solved Example 6-5 by finding the acceleration and then using the constant-acceleration equations. The force exerted by the spring on the block, $F_x = -kx$, varies with position, so the acceleration also varies.

6-2 Work and Energy in Three Dimensions

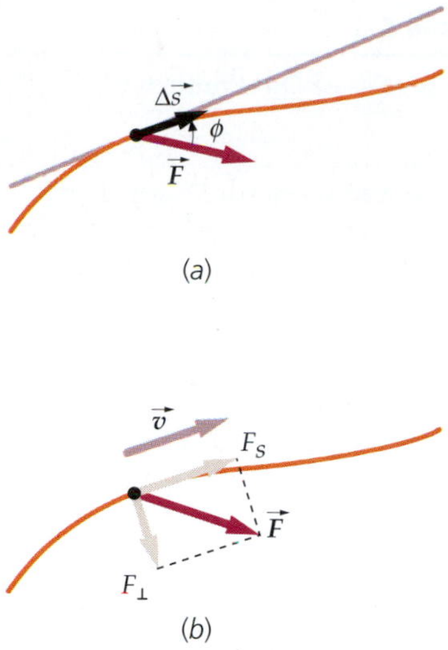

Figure 6-11 (*a*) A particle moving along an arbitrary curve in space. (*b*) The perpendicular component of the force $F_\perp$ changes the direction of the particle's motion but not its speed. The tangential component F_s changes the particle's speed but not its direction. F_s equals the mass m times the tangential acceleration dv/dt. Only this component does work.

Figure 6-11 shows a particle of mass m acted on by a force $\vec{F}$ as it moves along a curve in space. Consider a small displacement Δs, where s is the distance measured along the curve. $\vec{F}$ has components F_s parallel to and $F_\perp$ perpendicular to the displacement. The component $F_\perp$ provides the centripetal force needed for the particle to round the curve, but since it is perpendicular to the motion, it does not contribute to the work done on the particle by $\vec{F}$, which is

$$\Delta W = F_s \Delta s$$

To find the work done as the particle moves along the curve from point 1 to point 2, we compute $F_s\,\Delta s$ for each element of the path and sum. In the limit of smaller and smaller displacement elements, this sum becomes an integral:

$$W = \int_{s_1}^{s_2} F_s\,ds$$

From Newton's second law,

$$F_s = m\frac{dv}{dt}$$

If we think of the speed as a function of the distance s, we can apply the chain rule for derivatives:

$$\frac{dv}{dt} = \frac{dv}{ds}\frac{ds}{dt} = v\frac{dv}{ds}$$

where we have used $ds/dt = v$, the speed. The work done by the net force is then

$$W_{\text{total}} = \int_{s_1}^{s_2} F_s\,ds = \int_{s_1}^{s_2} m\frac{dv}{dt}\,ds = \int_{s_1}^{s_2} mv\frac{dv}{ds}\,ds = \int_{v_1}^{v_2} mv\,dv$$

or

$$W_{\text{total}} = \int_{s_1}^{s_2} F_s\,ds = \frac{1}{2}mv_2^2 - \frac{1}{2}mv_1^2 \qquad \text{6-10}$$

Work–kinetic energy theorem in three dimensions

Equation 6-10, along with its one-dimensional counterpart, Equation 6-7, follows directly from the definition of work and from Newton's second law of motion.

The Dot Product

The component F_s in Figure 6-11 is related to the angle ϕ between $\vec{F}$ and $\Delta\vec{s}$ by $F_s = F\cos\phi$, so the work done by $\vec{F}$ for a displacement $\Delta\vec{s}$ is

$$\Delta W = F_s \Delta s = (F\cos\phi)\,\Delta s$$

This combination of two vectors and the cosine of the angle between them is called the **dot product** (or **scalar product**) of the vectors. The dot product of two general vectors $\vec{A}$ and $\vec{B}$ is written $\vec{A}\cdot\vec{B}$ and is defined by

$$\vec{A}\cdot\vec{B} = AB\cos\phi \qquad \text{6-11}$$

Definition—Dot product

Table 6-1

Properties of Dot Products

If	*then*
$\vec{A}$ and $\vec{B}$ are perpendicular,	$\vec{A}\cdot\vec{B} = 0$ (since $\phi = 90°$, $\cos 90° = 0$)
$\vec{A}$ and $\vec{B}$ are parallel,	$\vec{A}\cdot\vec{B} = AB$ (since $\phi = 0°$, $\cos 0° = 1$)
$\vec{A}\cdot\vec{B} = 0$,	Either $\vec{A} = 0$ or $\vec{B} = 0$ or $\vec{A}$ and $\vec{B}$ are perpendicular
Furthermore,	
$\vec{A}\cdot\vec{A} = A^2$	Since $\vec{A}$ is parallel to itself
$\vec{A}\cdot\vec{B} = \vec{B}\cdot\vec{A}$	Commutative rule of multiplication
$(\vec{A} + \vec{B})\cdot\vec{C} = \vec{A}\cdot\vec{C} + \vec{B}\cdot\vec{C}$	Distributive rule of multiplication

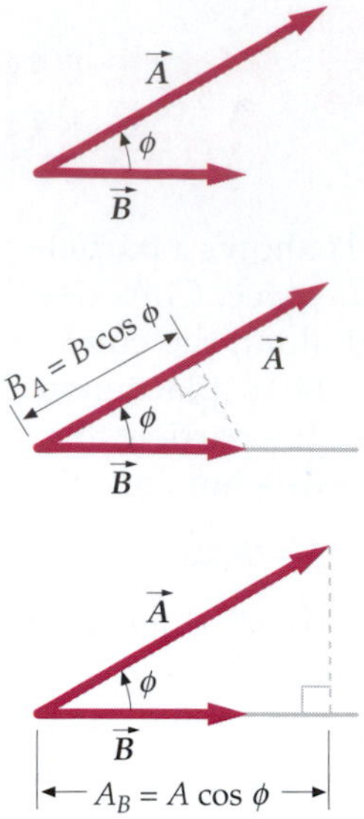

Figure 6-12 The dot product $\vec{A}\cdot\vec{B}$ is the product of A and the projection of $\vec{B}$ on $\vec{A}$ or the product of B and the projection of $\vec{A}$ on $\vec{B}$.

where ϕ is the angle between $\vec{A}$ and $\vec{B}$. The dot product $\vec{A}\cdot\vec{B}$ can be thought of as A times the component of $\vec{B}$ in the direction of $\vec{A}$ (that is, A times $B\cos\phi$), or as B times the component of $\vec{A}$ in the direction of $\vec{B}$ (that is, B times $A\cos\phi$). Figure 6-12 shows a geometric representation of the dot product $\vec{A}\cdot\vec{B}$. Properties of the dot product are summarized in Table 6-1. We can use unit vectors to write the dot product in terms of the rectangular components of the two vectors:

$$\vec{A}\cdot\vec{B} = (A_x\hat{i} + A_y\hat{j} + A_z\hat{k})\cdot(B_x\hat{i} + B_y\hat{j} + B_z\hat{k})$$

Since the unit vectors $\hat{i}$, $\hat{j}$, and $\hat{k}$ are mutually perpendicular, $\hat{i}\cdot\hat{j} = \hat{i}\cdot\hat{k} = \hat{j}\cdot\hat{k} = 0$. So the cross terms like $A_x\hat{i}\cdot B_y\hat{j}$ are zero. In addition, the dot product of a unit vector with itself is 1; $\hat{i}\cdot\hat{i} = \hat{j}\cdot\hat{j} = \hat{k}\cdot\hat{k} = 1$, so a term like $A_x\hat{i}\cdot B_x\hat{i}$ equals A_xB_x. The result is

$$\vec{A}\cdot\vec{B} = A_xB_x + A_yB_y + A_zB_z \qquad 6\text{-}12$$

The component of a vector along some axis can be written as the dot product of the vector and the unit vector along that axis. For example, the component A_x is found from

$$\vec{A}\cdot\hat{i} = (A_x\hat{i} + A_y\hat{j} + A_z\hat{k})\cdot\hat{i} = A_x \qquad 6\text{-}13$$

Example 6-6

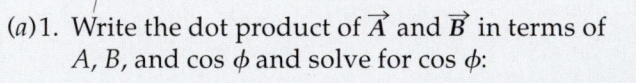

(*a*) Find the angle between the vectors $\vec{A} = (3\text{ m})\hat{i} + (2\text{ m})\hat{j}$ and $\vec{B} = (4\text{ m})\hat{i} - (3\text{ m})\hat{j}$ (Figure 6-13). (*b*) Find the component of $\vec{A}$ in the direction of $\vec{B}$.

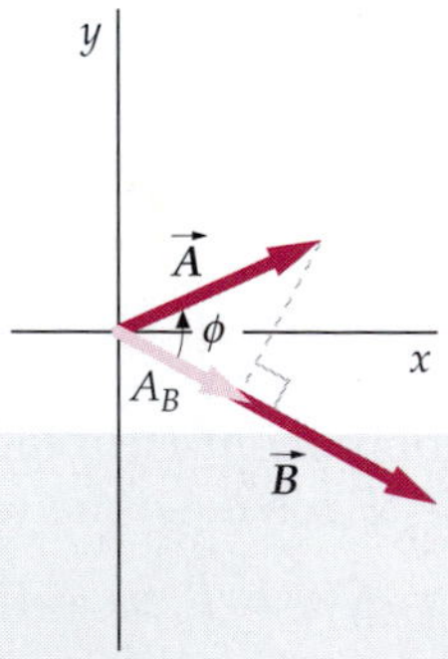

Figure 6-13

Picture the Problem We find the angle ϕ from the definition of the dot product. The component of $\vec{A}$ in the direction of $\vec{B}$ is found from the dot product of $\vec{A}$ with the unit vector $\vec{B}/B$.

(*a*)1. Write the dot product of $\vec{A}$ and $\vec{B}$ in terms of A, B, and $\cos\phi$ and solve for $\cos\phi$:

$$\vec{A}\cdot\vec{B} = AB\cos\phi$$

$$\cos\phi = \frac{\vec{A}\cdot\vec{B}}{AB}$$

2. Find $\vec{A}\cdot\vec{B}$ from their components:

$$\vec{A}\cdot\vec{B} = A_xB_x + A_yB_y$$
$$= (3\text{ m})(4\text{ m}) + (2\text{ m})(-3\text{ m})$$
$$= 12\text{ m}^2 - 6\text{ m}^2 = 6\text{ m}^2$$

3. The magnitudes of the vectors are obtained from the dot product of the vector with itself:

$$\vec{A}\cdot\vec{A} = A^2 = A_x^2 + A_y^2$$
$$= (2\text{ m})^2 + (3\text{ m})^2 = 13\text{ m}^2$$
$$A = \sqrt{13}\text{ m}$$

and

$$\vec{B}\cdot\vec{B} = B^2 = B_x^2 + B_y^2$$
$$= (4\text{ m})^2 + (-3\text{ m})^2 = 25\text{ m}^2$$
$$B = 5\text{ m}$$

4. Substitute these values into the equation in step 1 for $\cos\phi$ to find ϕ:

$$\cos\phi = \frac{\vec{A}\cdot\vec{B}}{AB} = \frac{6\text{ m}^2}{(\sqrt{13}\text{ m})(5\text{ m})} = 0.333$$
$$\phi = 70.6°$$

(*b*) The component of $\vec{A}$ along $\vec{B}$ is the dot product of $\vec{A}$ with the unit vector $\vec{B}/B$:

$$A_B = \vec{A}\cdot\frac{\vec{B}}{B} = \frac{\vec{A}\cdot\vec{B}}{B} = \frac{6\text{ m}^2}{5\text{ m}} = 1.2\text{ m}$$

Check the Result The component of A along B is $A\cos\phi = (\sqrt{13}\text{ m})\cos 70.6° = 1.2\text{ m}$.

Exercise (*a*) Find $\vec{A}\cdot\vec{B}$ for $\vec{A} = (3\text{ m})\hat{i} + (4\text{ m})\hat{j}$ and $\vec{B} = (2\text{ m})\hat{i} + (8\text{ m})\hat{j}$. (*b*) Find A, B, and the angle between $\vec{A}$ and $\vec{B}$ for these vectors. (*Answers* (*a*) 38 m^2, (*b*) $A = 5\text{ m}$, $B = 8.25\text{ m}$, $\phi = 23°$)

In dot-product notation, the work dW done by a force $\vec{F}$ on a particle undergoing a displacement $d\vec{s}$ is

$$dW = F\cos\phi\, ds = \vec{F}\cdot d\vec{s} \qquad \text{6-14}$$

and the work done on the particle as it moves from point 1 to point 2 is

$$W = \int_{s_1}^{s_2} \vec{F}\cdot d\vec{s} \qquad \text{6-15}$$

The general definition of work

When several forces $\vec{F}_i$ act on a particle whose displacement is $d\vec{s}$, the total work is

$$dW_{\text{total}} = \vec{F}_1\cdot d\vec{s} + \vec{F}_2\cdot d\vec{s} + \cdots = \left(\sum_i \vec{F}_i\right)\cdot d\vec{s} \qquad \text{6-16}$$

Example 6-7 *try it yourself*

A particle is given a displacement $\Delta\vec{s} = 2\text{ m}\,\hat{i} - 5\text{ m}\,\hat{j}$ along a straight line. During the displacement, a constant force $\vec{F} = 3\text{ N}\,\hat{i} + 4\text{ N}\,\hat{j}$ acts on the particle (Figure 6-14). Find (*a*) the work done by the force, and (*b*) the component of the force in the direction of the displacement.

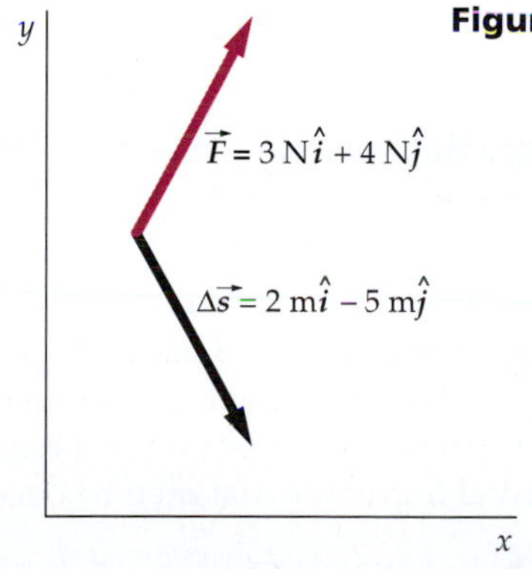

Figure 6-14

Picture the Problem The work W is found by computing $W = \vec{F}\cdot\Delta\vec{s} = F_x\Delta x + F_y\Delta y + F_z\Delta z$. Since $\vec{F}\cdot\Delta\vec{s} = F\cos\phi|\Delta\vec{s}|$, we can find the component of $\vec{F}$ in the direction of the displacement from

$$F\cos\phi = \frac{(\vec{F}\cdot\Delta\vec{s})}{|\Delta\vec{s}|} = \frac{W}{|\Delta\vec{s}|}$$

Cover the column to the right and try these on your own before looking at the answers.

Steps	**Answers**
(*a*) Compute the work done W.	$W = -14\ \text{N}\cdot\text{m}$
(*b*) 1. Compute $\Delta\vec{s}\cdot\Delta\vec{s}$ and use your result to find the distance $\lvert\Delta\vec{s}\rvert$.	$\lvert\Delta\vec{s}\rvert = \sqrt{29}\ \text{m}$
2. Compute $F\cos\phi = W/\lvert\Delta\vec{s}\rvert$.	$F\cos\phi = -2.60\ \text{N}$

Remark The component of the force in the direction of the displacement is negative, so the work done is negative.

Exercise Find the magnitude of $\vec{F}$, and the angle ϕ between $\vec{F}$ and $\Delta\vec{s}$. (*Answer* $F = 5\ \text{N}$, $\phi = 121°$)

Example 6-8

You ski downhill on waxed skis that are nearly frictionless. (*a*) What work is done on you as you ski a distance *s* down the hill? (*b*) What is your speed on reaching the bottom of the run? Assume the length of the ski run is *s*, its angle of incline is θ, and your mass is *m*. The height of the hill is then $h = s\sin\theta$.

Picture the Problem We assume that you are a particle. Two forces act on you: gravity, $m\vec{g}$, and the normal force exerted by the hill, $\vec{F}_n$ (Figure 6-15*a*). Only gravity does work on you, because the normal force is perpendicular to the hill, and hence has no component in the direction of your motion. The work–kinetic energy theorem with $v_i = 0$ gives the final speed v.

Figure 6-15*b* shows a free-body diagram for you on skies. The net force is $mg\sin\theta$, which is the component of the weight in the direction of the displacement Δs.

Figure 6-15*a*

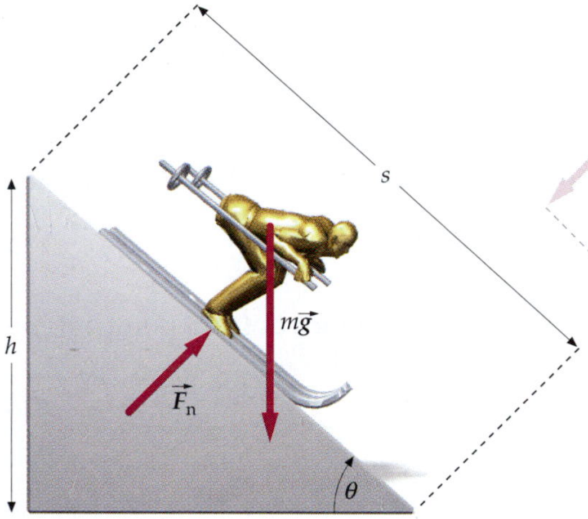

Figure 6-15*b*

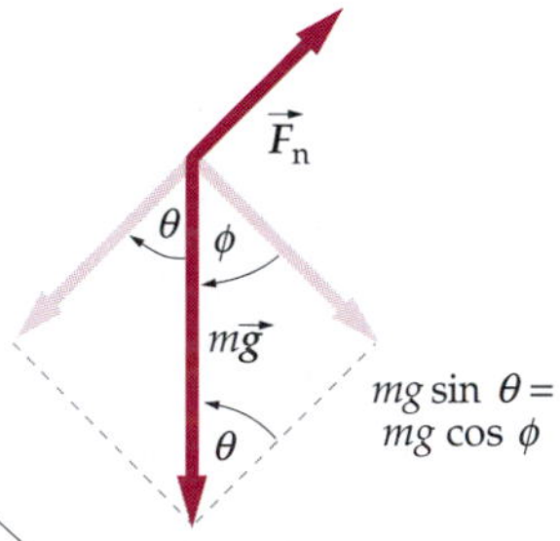

(*a*) 1. The work done by gravity as you traverse the slope is $m\vec{g}\cdot\vec{s}$:	$W = m\vec{g}\cdot\vec{s} = mgs\cos\phi = mgs\sin\theta$
2. From Figure 6-15*a*, the angle θ is related to h and s:	$\sin\theta = \dfrac{h}{s}$
3. Substitute h for $s\sin\theta$:	$W = mgh$
(*b*) Apply the work–kinetic energy theorem to find the final speed v:	$W = mgh = \frac{1}{2}mv^2 - 0$ or $v = \sqrt{2gh}$

Remarks $mg\sin\theta = mg\cos\phi$ is the component of the weight in the direction of the displacement. This is the component that does work on you. The final speed is independent of the angle θ, and the same as if the skier had dropped vertically a height h with acceleration g. If θ were smaller, the skier would travel a greater distance to drop the same vertical distance h, but the component of the force of gravity in the direction of motion would be less. The two effects cancel, and the work done by gravity is mgh independent of the angle of the slope. Figure 6-16 shows that for a hill of arbitrary shape, the work done by the earth on the skier is mgh.

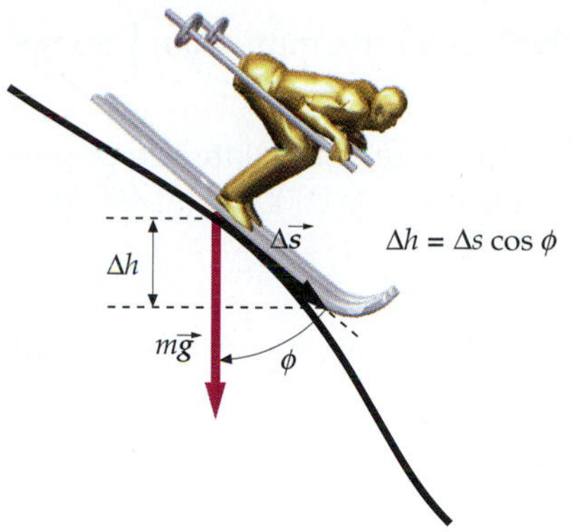

Figure 6-16 Skier skiing down a hill of arbitrary shape. The work done by the earth during a displacement $\Delta\vec{s}$ is $m\vec{g}\cdot\Delta\vec{s} = mg\,\Delta s\cos\phi = mg\,\Delta h$, where Δh is the vertical distance dropped. The total work done by the earth when the skier skis down a vertical distance h is $W = \int_0^s m\vec{g}\cdot d\vec{s} = mg\int_0^s \cos\phi\, ds = mg\int_0^h dh = mgh$, independent of the shape of the hill.

6-3 Power

The **power** P supplied by a force is the rate at which the force does work. Consider a particle moving with instantaneous velocity $\vec{v}$. In a short time interval dt, the particle has displacement $d\vec{s} = \vec{v}\,dt$. The work done by a force $\vec{F}$ acting on the particle during this time interval is

$$dW = \vec{F}\cdot d\vec{s} = \vec{F}\cdot\vec{v}\,dt$$

The power delivered to the particle is then

$$P = \frac{dW}{dt} = \vec{F}\cdot\vec{v} \qquad \text{6-17}$$

Definition—Power

The SI unit of power, one joule per second, is called a watt (W):

$$1\text{ W} = 1\text{ J/s}$$

Note the difference between power and work. Two motors that lift a given load a given distance do the same amount of work, but the one that does it in the least time supplies more power. Gas and electric companies charge for energy, not power, usually by the kilowatt-hour (kW·h). A kilowatt-hour of energy is

$$1\text{ kW}\cdot\text{h} = (10^3\text{ W})(3600\text{ s}) = 3.6\times10^6\text{ W}\cdot\text{s} = 3.6\text{ MJ}$$

In the U.S. customary system, the unit of energy is the foot-pound and the unit of power is the foot-pound per second. A commonly used multiple of this unit, called a horsepower (hp), is defined as

$$1\text{ hp} = 550\text{ ft}\cdot\text{lb/s} = 746\text{ W}$$

Example 6-9

A small motor is used to operate a lift that raises a load of bricks weighing 800 N to a height of 10 m in 20 s. What is the minimum power the motor must produce?

Picture the Problem Assuming that the bricks are lifted without acceleration, the upward force exerted by the motor is equal to the weight of the bricks, $F = 800$ N. The speed of the bricks is $v = 10\text{ m}/(20\text{ s}) = 0.5\text{ m/s}$.

Figure 6-17

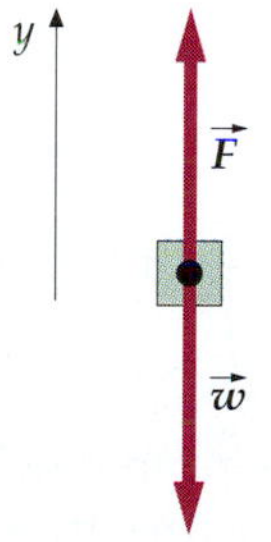

The power is the product of the speed v and the component of force in the direction of motion, which is simply F in this case (Figure 6-17):

$$P = Fv = (800\text{ N})(0.5\text{ m/s})$$
$$= 400\text{ N}\cdot\text{m/s} = 400\text{ J/s} = 400\text{ W}$$

Remark This minimum power output of 400 W is a little more than $\frac{1}{2}$ horsepower.

Exercise (*a*) Find the total work done by the force. (*b*) Calculate the power by dividing the total work by the total time. (*Answers* (*a*) 8000 J, (*b*) 400 W)

Consider a net force F_x acting on a particle in one dimension. The rate at which this force does work is

$$P = F_x v_x$$

Substituting $F_x = ma_x$ we have

$$P = F_x v_x = ma_x v_x \qquad \text{6-18}$$

or

$$a_x = \frac{P}{mv_x} \qquad \text{6-19}$$

Thus, for a constant power P, the acceleration varies inversely as the speed. A familiar example is the difficulty in passing an automobile at high speeds. For a given power, the acceleration at high speeds is smaller than at lower speeds. Alternatively, it takes more power to give the same acceleration to an automobile moving at 80 km/h than one moving at 60 km/h.

If we write $a_x = dv_x/dt$ in Equation 6-18, then

$$P = ma_x v_x = mv_x \frac{dv_x}{dt} = \frac{d}{dt}\left(\frac{1}{2}mv_x^2\right) = \frac{dK}{dt}$$

Assuming that P is constant, and integrating over some time interval, we get

$$P\,\Delta t = \Delta K \qquad \text{(constant power)} \qquad \text{6-20}$$

So the time it takes an automobile or airplane at constant power to accelerate from one speed to another speed is proportional to the change in the kinetic energy.

Example 6-10

A new Cadillac can accelerate from 0 to 96 km/h in 6.5 s. How quickly would you expect it to be able to accelerate from 80 km/h to 112 km/h?

Picture the Problem According to Equation 6-20, the time is related to the power and change in kinetic energy $\Delta t = \Delta K/P$. If we assume constant power, the time required is proportional to the change in kinetic energy. Since we only need to calculate ratios, we do not need to convert the units. Let m be the mass of the Cadillac.

1. The time Δt_1 for a change in kinetic energy ΔK_1 is: $\quad \Delta t_1 = \dfrac{\Delta K_1}{P}$

2. If Δt_2 is the time needed for a change in kinetic energy ΔK_2, the times Δt_1 and Δt_2 are related by: $\quad \dfrac{\Delta t_2}{\Delta t_1} = \dfrac{\Delta K_2}{\Delta K_1} = \dfrac{\frac{1}{2}mv_{2f}^2 - \frac{1}{2}mv_{2i}^2}{\frac{1}{2}mv_{1f}^2 - \frac{1}{2}mv_{1i}^2} = \dfrac{v_{2f}^2 - v_{2i}^2}{v_{1f}^2 - v_{1i}^2}$

3. Substitute the given values for the speeds: $\quad \dfrac{\Delta t_2}{\Delta t_1} = \dfrac{v_{2f}^2 - v_{2i}^2}{v_{1f}^2 - v_{1i}^2} = \dfrac{(112\text{ km/h})^2 - (80\text{ km/h})^2}{(96\text{ km/h})^2 - 0} = 0.667$

4. Solve for Δt_2: $\quad \Delta t_2 = 0.667\,\Delta t_1 = (0.667)(6.5\text{ s}) = 4.33\text{ s}$

Remark The actual time to accelerate from 80 km/h to 112 km/h as measured in tests was 4.0 s. Air resistance, which we have neglected, is greater at higher speeds, but automobile engines actually have somewhat greater power at higher speeds.

Exercise A car accelerates from 0 to 40 km/h in T seconds. If the power output of the car is constant, how long does it take for the car to accelerate from 40 km/h to 80 km/h? (*Answer* $3T$ seconds; see Figure 6-18.)

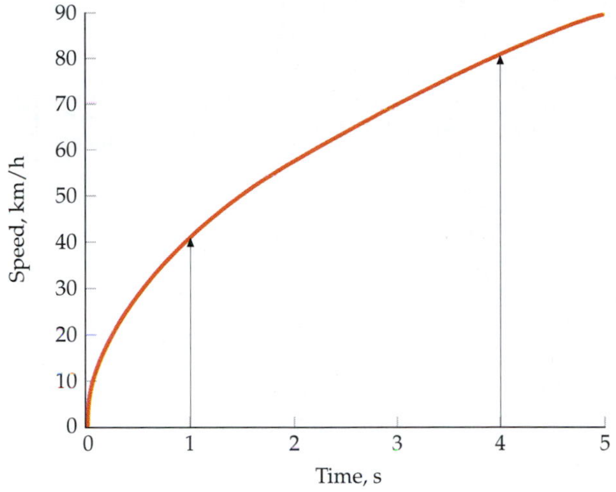

Figure 6-18 The arrows indicate the times at which the speed is 40 km/h and 80 km/h.

Example 6-11 *try it yourself*

A truck of mass m is accelerated from rest at $t = 0$ with constant power P along a level road. (a) Find the speed of the truck as a function of time. (b) Show that if $x = 0$ at time $t = 0$, the position function $x(t)$ is given by

$$x = \sqrt{\frac{8P}{9m}}\, t^{3/2}$$

Picture the Problem You can calculate the velocity function by integrating the acceleration $a = dv/dt = P/mv$, as given by Equation 6-19. The position function $x(t)$ can then be obtained by integrating the velocity.

Cover the column to the right and try these on your own before looking at the answers.

Steps	*Answers*
(a)1. Show that Equation 6-19 can be written as $v\,dv = (\text{constant})\,dt$.	$v\,dv = \frac{P}{m}\,dt$
2. Integrate to obtain v^2 using the fact that $v_0 = 0$.	$\frac{v^2}{2} = \frac{P}{m}t$
3. Solve for v.	$v = \left(\frac{2P}{m}\right)^{1/2} t^{1/2}$
(b)1. Set $v = dx/dt$ and solve for dx.	$dx = \left(\frac{2P}{m}\right)^{1/2} t^{1/2}\,dt$
2. Integrate to obtain $x(t)$.	$x = \int dx = \int \left(\frac{2P}{m}\right)^{1/2} t^{1/2}\,dt = \left(\frac{8P}{9m}\right)^{1/2} t^{3/2}$

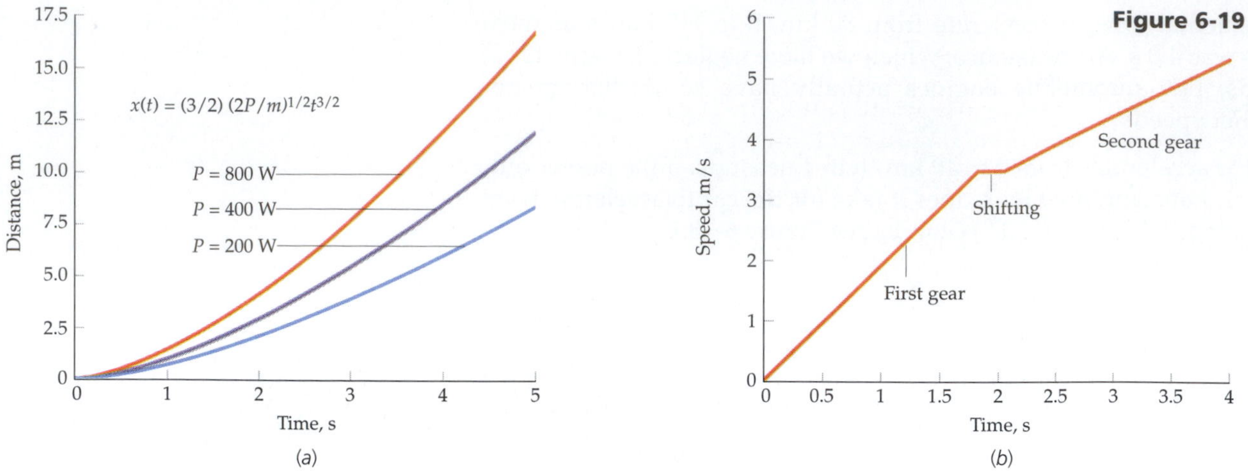

Figure 6-19

Remark Figure 6-19 shows two graphical views of power. Figure 6-19*a* shows distance versus time for various constant powers for a truck of mass $m = 1600$ kg. Figure 6-19*b* shows a vehicle's speed increasing approximately linearly with time up to a maximum speed, and then climbing with a lesser slope after gears are shifted at approximately 2 s.

6-4 Potential Energy

The total work done on a particle equals the change in its kinetic energy. But we are frequently interested in the work done on a *system* of two or more particles.* Often, the work done by external forces on a *system* does not increase the kinetic energy *of the system*, but instead is stored as **potential energy.**

Consider lifting a barbell of mass m to a height h. The work you do on the barbell is mgh. The kinetic energy of the barbell does not increase because the earth does negative work $-mgh$, so the total work on the barbell is zero. Now consider the barbell and the planet Earth (but not you) to be a *system* of particles. The external forces on the earth–barbell system are the gravitational attraction you exert on the earth, w, the force your feet exert on the earth, $w + mg$, and the force mg exerted by your hands on the barbell (Figure 6-20). (We can neglect the gravitational force you exert on the barbell.) The barbell moves, but the earth doesn't, so the only external force exerted on the system that does work is the force you exert on the barbell. The total work done on the earth–barbell system by forces *external* to the system is mgh. This work is stored as potential energy, which is associated with the configuration of the earth–barbell system.

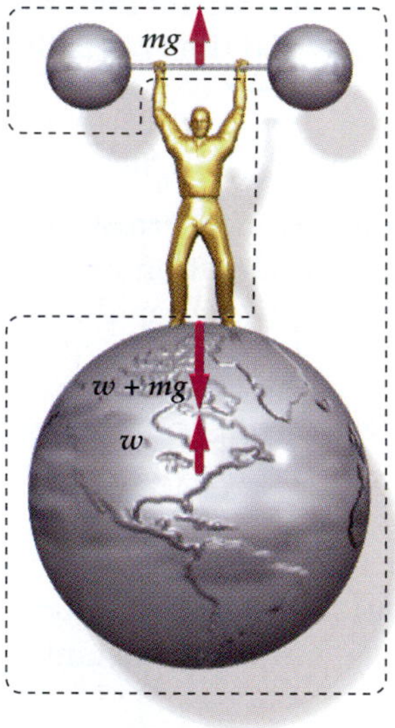

Figure 6-20 A system consisting of a barbell and the earth. When you lift the barbell, you do work on this system.

Consider another system consisting of a dart and spring in a toy dart gun. You compress the spring by pushing the dart into the gun. The work you do on this system is stored as potential energy in the dart–spring system. Its configuration has been changed because the spring has been compressed. Figure 6-21 shows a schematic description of such a system. The spring is compressed by the two forces $\vec{F}_1$ and $\vec{F}_2$, which are equal and opposite. Note that even though each force does (positive) work on the two-mass system, the net external force on the system is zero.

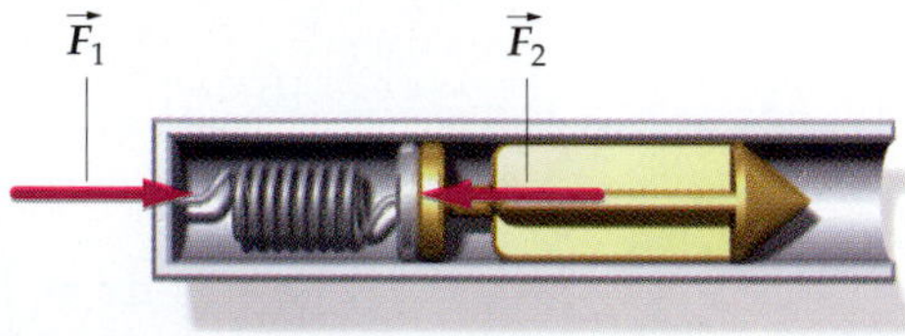

Figure 6-21 Potential energy of a dart gun.

* Systems of particles are discussed more thoroughly in Chapter 8.

Conservative Forces

When you ride a ski lift to the top of a hill of height h, the work done by the lift on you is mgh and that done by gravity is $-mgh$. When you ski down the hill to the bottom, the work done by gravity is $+mgh$ independent of the shape of the hill. The total work done by gravity on you during the round trip is zero independent of the path you take. The force of gravity exerted by the earth on you is called a **conservative force.**

> A force is conservative if the total work it does on a particle is zero when the particle moves around any closed path returning to its initial position.

Definition—Conservative force

From Figure 6-22 we see that this definition implies the following:

> The work done by a conservative force on a particle is independent of the path taken as the particle moves from one point to another.

Alternative definition—Conservative force

Now consider you and the earth to be a *two-particle system*. When a ski lift raises you to the top of the hill, it does work mgh on the system. This work is stored as potential energy of the system. When you ski down the hill, this potential energy is converted to kinetic energy of motion.

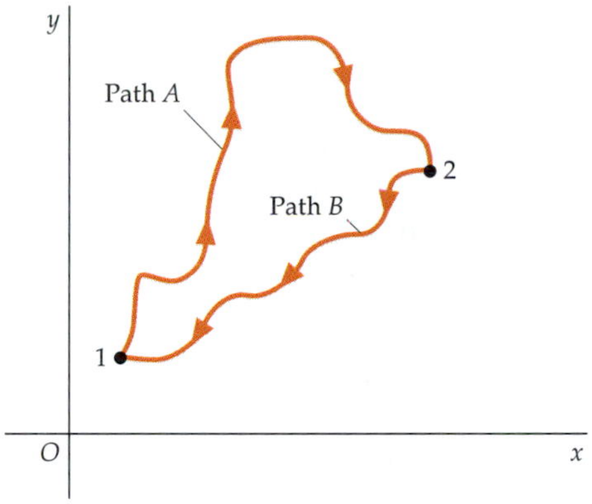

Figure 6-22 Two paths in space connecting the points 1 and 2. If the work done by a conservative force along path A from 1 to 2 is W, then the work done on the return trip along path B must be $-W$ because the round-trip work is zero. When traversing path B from 1 to 2, the force is the same at each point, but the displacement is opposite that when going from 2 to 1. Then the work done along path B from 1 to 2 must also be W. It follows that the work done as a particle going from point 1 to 2 is the same along any path connecting the two points.

Potential-Energy Functions

Since the work done by a conservative force on a particle does not depend on the path, it can depend only on the endpoints 1 and 2. We can use this property to define the **potential-energy function** U that is associated with a conservative force. Note that when the skier skis down the hill, the work done by gravity *decreases* the potential energy of the system. In general, we define the potential energy function such that the work done by a conservative force equals the decrease in the potential-energy function:

$$W = \int \vec{F}\cdot d\vec{s} = -\Delta U$$

or

$$\Delta U = U_2 - U_1 = -\int_{s_1}^{s_2} \vec{F}\cdot d\vec{s} \qquad \text{6-21}a$$

Definition—Potential-energy function

For infinitesimal displacement, we have

$$dU = -\vec{F}\cdot d\vec{s} \qquad \text{6-21}b$$

Gravitational Potential Energy Near the Earth's Surface We can calculate the potential-energy function associated with the gravitational force near the surface of the earth from Equation 6-21*b*. For the force $\vec{F} = -mg\hat{j}$, we have

$$dU = -\vec{F}\cdot d\vec{s} = -(-mg\hat{j})\cdot(dx\hat{i} + dy\hat{j} + dz\hat{k}) = +mg\,dy$$

Integrating, we obtain

$$U = \int mg\,dy = mgy + U_0$$

$$U = U_0 + mgy \qquad \text{6-22}$$

Gravitational potential energy near the earth's surface

where U_0, the arbitrary constant of integration, is the value of the potential energy at $y = 0$. Since only a change in the potential energy is defined, the actual value of U is not important. We are free to choose U to be zero at any convenient reference point. For example, if the gravitational potential energy of the earth–skier system is chosen to be zero when the skier is at the bottom of the hill, its value when the skier is at a height h above that level is mgh. Or we could choose the potential energy to be zero when the skier is at sea level, in which case its value at any other point would be mgy, where y is measured from sea level.

Exercise A 55-kg girl stands on a ledge that is 8 m above the ground. What is the potential energy U of the girl–earth system if (*a*) U is chosen to be zero on the ground and (*b*) U is chosen to be zero 4 m above the ground and (*c*) U is chosen to be zero 10 m above the ground? (*Answers* (*a*) 4.32 kJ, (*b*) 2.16 kJ, (*c*) −1.08 kJ)

Example 6-12

A bottle of mass 0.350 kg falls from rest from a shelf that is 1.75 m above the floor. Find the original potential energy of the bottle–earth system relative to the floor, and the kinetic energy of the bottle just before it hits the floor.

Picture the Problem We choose the potential energy of the bottle–earth system to be zero when the bottle is on the floor. The work done by the earth on the bottle as it falls equals the change in its kinetic energy.

1. The original potential energy U at $y = 1.75$ m is:
   $U = mgy = (0.350\text{ kg})(9.81\text{ N/kg})(1.75\text{ m})$
   $= 6.01\text{ J}$

2. Set the work done to be equal to the change in kinetic energy. The total work on the bottle is the work done by the earth:
   $\Delta K = W_{\text{total}} = mgy = 6.01\text{ J}$

3. Since the original kinetic energy is zero, the final kinetic energy equals the change in kinetic energy:
   $K = \Delta K = 6.01\text{ J}$

Remark In this example, the potential energy lost by the bottle–earth system is converted entirely to kinetic energy of the bottle as it falls. Note that in step 1 we used the definition 1 J = 1 N·m.

Potential Energy of a Spring Another example of a conservative force is that of a stretched spring. Suppose we pull a block attached to a spring from a position $x = 0$ (equilibrium) to x_1 (Figure 6-23). The spring does negative work because its force is opposite the direction of motion. If we then release the block, the spring does positive work as it accelerates the block toward its initial position. The total work done by the spring when the block reaches its initial position is zero independent of how far we stretched the spring (assuming we did not stretch the spring so far that it was damaged). The force exerted by the spring is therefore a conservative force. We can calculate the potential-energy function associated with this force from Equation 6-21*b*:

$$dU = -\vec{F}\cdot d\vec{s} = -F_x\,dx = -(-kx)dx = +kx\,dx$$

Then

$$U = \int kx\,dx = \frac{1}{2}kx^2 + U_0$$

where U_0 is the potential energy when $x = 0$, that is, when the spring is unstretched. Choosing U_0 to be zero gives

$$U = \frac{1}{2}kx^2 \qquad \text{6-23}$$

Potential energy of a spring

When we pull the block from $x = 0$ to x_1, we must exert an applied force $F_{app} = +kx$ to balance the spring force. The work we do is

$$W_{app} = \int_0^{x_1} kx\,dx = \frac{1}{2}kx_1^2$$

This work is stored as potential energy in the spring–block system.

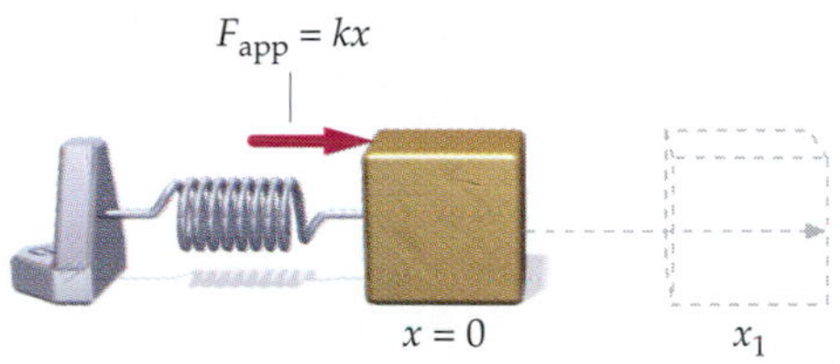

Figure 6-23 To stretch the spring, a force $F_{app} = +kx$ must be applied to the block.

Example 6-13

Find the total potential energy of the basketball player hanging on the rim in Example 4-6 (Figure 6-24). Assume the player can be described as a point mass of 110 kg at 2 m above the floor and the force constant of the rim is 7.2 kN/m. The rim is displaced a distance $s = 15$ cm.

Picture the Problem The potential energy consists of gravitational potential energy, $U_g = mgy$, and energy stored in the displaced rim, whose potential energy is assumed to be the same as if it were a spring: $U_s = \frac{1}{2}ks^2$. Choose $y = 0$ at the floor for the gravitational potential energy.

Figure 6-24

The total potential energy is the sum of gravitational potential energy and potential energy of the rim (see Figure 6-25):

$$U = U_g + U_s = mgy + \frac{1}{2}ks^2$$

$$= (110\ \text{kg})(9.81\ \text{N/kg})(2\ \text{m}) + \frac{1}{2}(7.2\ \text{kN/m})(0.15\ \text{m})^2$$

$$= 2158\ \text{J} + 81.0\ \text{J} = 2239\ \text{J}$$

Remark Nearly all of the potential energy is gravitational in this case, because even though the force constant of the "spring" is very large, the displacement is very small.

Exercise A 3-kg block is hung vertically from a spring with a force constant of 600 N/m. (*a*) By how much is the spring stretched when the block is in equilibrium? (*b*) How much potential energy is stored in the spring–block system? (*Answers* (*a*) 4.9 cm, (*b*) 0.72 J)

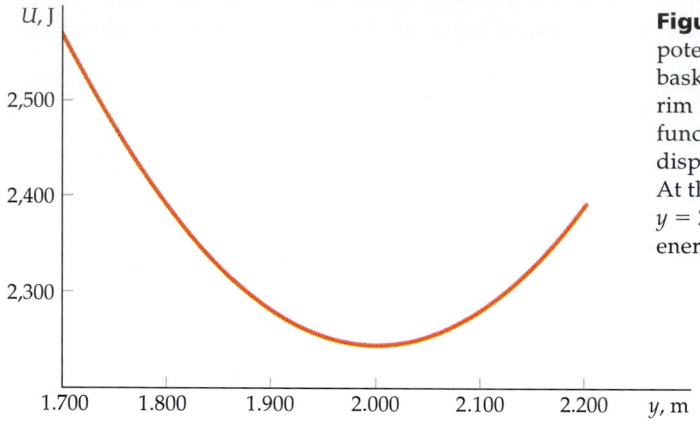

Figure 6-25 Total potential energy of the basketball player plus rim $U = U_s + U_g$ as a function of the vertical displacement of the player. At the equilibrium position $y = 2.00$ m, the potential energy is at a minimum.

Potential energy is associated with the configuration of a *system of particles*, but we sometimes have systems such as the earth–skier system, in which only one particle moves (the earth's motion is negligible). For brevity, then, we sometimes refer to the potential energy of the earth–skier system as simply the potential energy of the skier.

Nonconservative Forces

Not all forces are conservative. An example of a nonconservative force is kinetic friction. Suppose you push a box around some closed path on a rough table so that the box ends up at its original position. The force of kinetic friction is always opposite the direction of motion, so the work it does is always negative, and the total round-trip work it does cannot be zero. Another example of a nonconservative force is a force applied by a human agent. The work that you do in pushing a box around a closed path on a rough table is not generally zero. It depends on how great a force you decide to exert on the box. Thus, neither the force you exert nor the force of kinetic friction is conservative, and no potential-energy function can be defined for either.

Potential Energy and Equilibrium in One Dimension

For a general conservative force in one dimension, $\vec{F} = F_x\hat{i}$, Equation 6-21*b* is

$$dU = -\vec{F}\cdot d\vec{s} = -F_x\,dx$$

The force is therefore the negative derivative of the potential-energy function:

$$F_x = -\frac{dU}{dx} \qquad \text{6-24}$$

We can illustrate this general relation for a block–spring system by differentiating the function $U = \frac{1}{2}kx^2$. We obtain

$$F_x = -\frac{dU}{dx} = -\frac{d}{dx}\left(\frac{1}{2}kx^2\right) = -kx$$

Figure 6-26 shows a plot of $U = \frac{1}{2}kx^2$ versus x for a block and spring. The derivative of this function is represented graphically as the slope of the line tangent to the curve. The force is thus equal to the negative of the slope of the curve. At $x = 0$, the force $F_x = -dU/dx$ is zero and the block is in equilibrium.

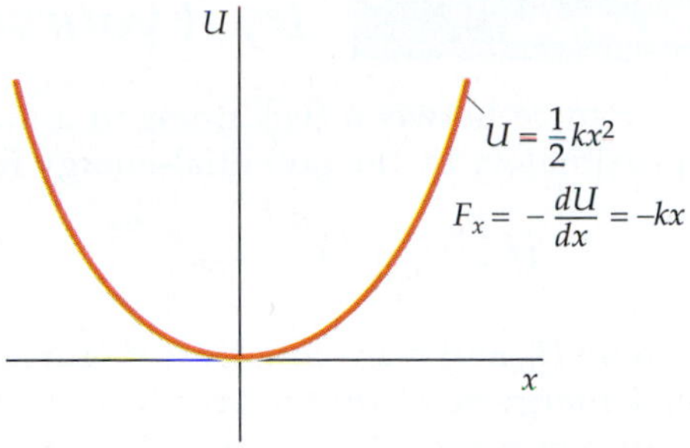

Figure 6-26 Plot of the potential-energy function U versus x for an object on a spring. A minimum in a potential energy curve is a point of stable equilibrium. Displacement in either direction results in a force directed toward the equilibrium position.

A particle is in equilibrium if the net force acting on it is zero.

Condition for equilibrium

When x is positive in Figure 6-26, the slope is positive and the force F_x is negative. When x is negative the slope is negative and the force F_x is positive. In either case, the force is in the direction that will accelerate the block toward lower potential energy. If the block is displaced slightly from $x = 0$, the force is directed back toward $x = 0$. The equilibrium at $x = 0$ is thus **stable equilibrium**.

In stable equilibrium, a small displacement results in a restoring force that accelerates the particle back toward its equilibrium position.

Figure 6-27 shows a potential-energy curve with a maximum rather than a minimum at the equilibrium point $x = 0$. Such a curve could represent the potential energy of a skier at the top of a hill. For this curve, when x is positive, the slope is negative and the force F_x is positive, and when x is negative, the slope is positive and the force F_x is negative. Again, the force is in the direction that will accelerate the particle toward lower potential energy, but this time the force is away from the equilibrium position. The maximum at $x = 0$ in Figure 6-27 is a point of **unstable equilibrium**.

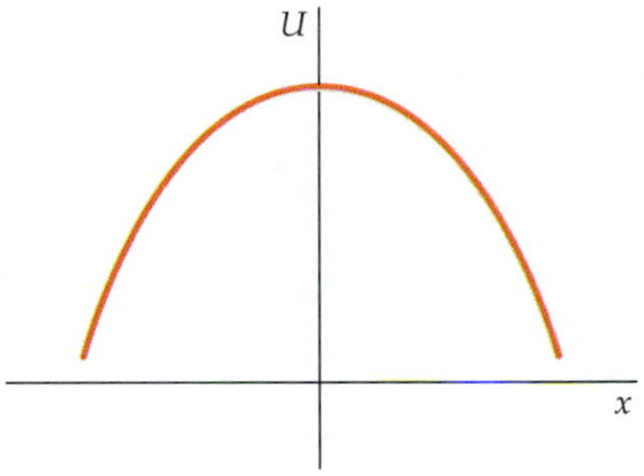

Figure 6-27 A particle at $x = 0$ on this potential-energy curve will be in unstable equilibrium because a displacement in either direction results in a force directed away from the equilibrium position.

In unstable equilibrium, a small displacement results in a force that accelerates the particle away from its equilibrium position.

Figure 6-28 shows a potential-energy curve that is flat in the region near $x = 0$. No force acts on a particle at $x = 0$, and hence the particle is at equilibrium; furthermore, there will be no resulting force if the particle is displaced slightly in either direction. This is an example of **neutral equilibrium.**

In neutral equilibrium, a small displacement results in zero force and the particle remains in equilibrium.

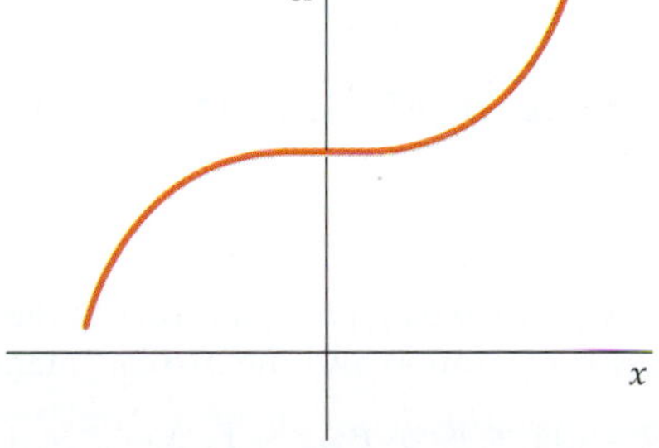

Figure 6-28 Neutral equilibrium. The force $Fx = -dU/dx$ is zero at $x = 0$ and at neighboring points, so displacement away from $x = 0$ results in no force, and the system remains in equilibrium.

Example 6-14 ***try it yourself***

The force between two atoms in a diatomic molecule can be represented approximately by the potential-energy function

$$U = U_0\left[\left(\frac{a}{x}\right)^{12} - 2\left(\frac{a}{x}\right)^{6}\right]$$

where U_0 and a are constants (Figure 6-29). (*a*) At what value of x is the potential energy zero? (*b*) Find the force F_x. (*c*) At what value of x is the potential energy a minimum? Show that $U_{min} = -U_0$.

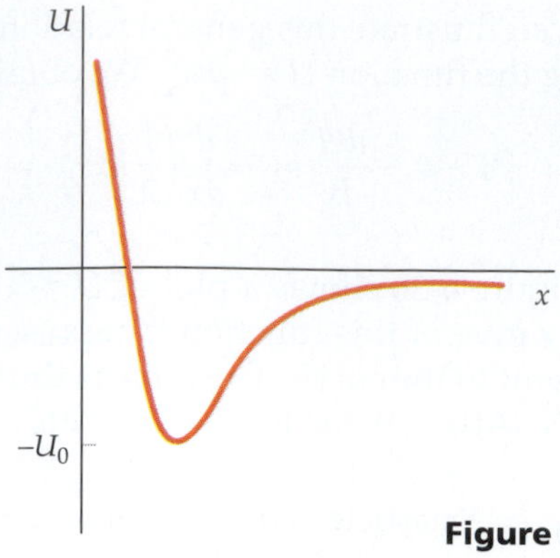

Figure 6-29

Picture the Problem The force is the negative derivative of the potential-energy function. The potential energy has its minimum value when its slope is zero; that is, when the force is zero.

Cover the column to the right and try these on your own before looking at the answers.

Steps	*Answers*
(*a*) Set $U = 0$ and solve for x.	$x = \frac{a}{\sqrt[6]{2}}$
(*b*) Compute $F_x = -dU/dx$	$F_x = \frac{12U_0}{a}\left[\left(\frac{a}{x}\right)^{13} - \left(\frac{a}{x}\right)^{7}\right]$
(*c*) 1. Set F_x equal to zero and solve for x.	$x = a$
2. Use your result to find U_{min}.	$U_{min} = -U_0$

Remark This potential-energy function is generally known as the "Lennard-Jones" or "6–12" potential. The minimum occurs at $x = a$, which is the average spacing between atoms in such a molecule. The lowest energy of the molecule is slightly greater than the minimum $-U_0$, so the energy needed to separate the atoms is slightly less than U_0.

Summary

1. Work, kinetic energy, potential energy, and power are important derived dynamic quantities.
2. The work–kinetic energy theorem is an important relation derived from Newton's laws applied to a particle.
3. The dot product of vectors is a mathematical definition that is useful throughout physics.

Topic	Remarks and Relevant Equations
1. Work	
Constant force	The work done by a constant force is the product of the component of the force in the direction of motion and the displacement of the force: $W = F \cos\theta\,\Delta x = F_x\,\Delta x$ **6-1**

	Variable force	$W = \int_{x_1}^{x_2} F_x\,dx$ = area under the F_x-versus-x curve	6-9
	Force in three dimensions	$W = \int_1^2 \vec{F} \cdot d\vec{s}$	6-14
	Units	The SI unit of work and energy is the joule (J): $1\text{ J} = 1\text{ N}\cdot\text{m}$	6-2
2.	**Kinetic Energy**	$K = \frac{1}{2} mv^2$	6-6
3.	**Work–Kinetic Energy Theorem**	$W_{\text{total}} = \Delta K = \frac{1}{2} mv_f^2 - \frac{1}{2} mv_i^2$	6-7
4.	**Dot Product**	$\vec{A} \cdot \vec{B} = AB \cos \phi$ where ϕ is the angle between the vectors.	6-11
	In terms of components	$\vec{A} \cdot \vec{B} = A_x B_x + A_y B_y + A_z B_z$	6-12
	Vector component	$\vec{A} \cdot \hat{i} = A_x$	6-13
5.	**Power**	$P = \frac{dW}{dt} = \vec{F} \cdot \vec{v}$	6-17
6.	**Conservative Force**	A force is conservative if the total work it does on a particle is zero when the particle moves along any path that returns it to its initial position. The work done by a conservative force on a particle is independent of the path taken by the particle as it moves from one point to another.	
7.	**Potential Energy**	The potential energy of a system is the energy associated with the configuration of the system. The change in the potential energy of a system is defined as the negative of the work done by conservative forces acting on the system.	
	Definition	$\Delta U = U_2 - U_1 = -W = -\int_1^2 \vec{F} \cdot d\vec{s}$	6-21*a*
		$dU = -\vec{F} \cdot d\vec{s}$	6-21*b*
	Gravitational	$U = U_0 + mgy$	6-22
	Spring	$U = \frac{1}{2} kx^2$	6-23
	Conservative force	In one dimension, a conservative force equals the negative derivative of the potential-energy function associated with it: $F_x = -\frac{dU}{dx}$	6-24
	Potential-energy curve	At a minimum on the curve of the potential-energy function versus the displacement, the force is zero and the system is in stable equilibrium. At a maximum, the force is zero and the system is in unstable equilibrium. A conservative force always tends to accelerate a particle toward a position of lower potential energy.	

Problem-Solving Guide

1. Begin by drawing a neat diagram that includes the important features of the problem.
2. The work–kinetic energy theorem relates the initial and final speeds of a particle to the total work done on the particle.

Summary of Worked Examples

Type of Calculation	Procedure and Relevant Examples	
1. Work		
Find the work done by a constant force.	$W = F_x \Delta x$	**Examples 6-1, 6-2, 6-3**
Find the work done by a force that varies with position.	The work is given by $W = \int_{x_1}^{x_2} F\,dx$. This integral equals the area under the F-versus-x curve. For a spring, $W = \int_{x_1}^{x_2} -kx\,dx = -\frac{1}{2}k\left(x_2^2 - x_1^2\right)$.	**Examples 6-4,6-5**
Find the work done by a constant force when $\vec{F}$ and $\Delta\vec{s}$ are given in terms of unit vectors.	Compute $W = \vec{F} \cdot \Delta\vec{s}$.	**Example 6-7**
Find the work done by gravity.	The work is $+mgh$ if the object moves downward and $-mgh$ if it moves upward, independent of the path.	**Example 6-8**
2. Work–Kinetic Energy Theorem		
Find the final speed of an object.	The final speed is found from the kinetic energy, which is obtained from the work–kinetic energy theorem. $W_{total} = \Delta K$.	**Examples 6-1, 6-2, 6-3, 6-5, 6-8, 6-12**
3. Find the angle between two vectors or the component of one vector along another.	Find the dot product $\vec{A}\cdot\vec{B} = AB\cos\phi$. Then divide by the magnitudes obtained from $A = \sqrt{\vec{A}\cdot\vec{A}}$ and $B = \sqrt{\vec{B}\cdot\vec{B}}$. The component of $\vec{B}$ along $\vec{A}$ is $B_A = \vec{B}\cdot(\vec{A}/A) = B\cos\phi$.	**Examples 6-6, 6-7**
4. Power		
Find the power supplied by a force.	The instantaneous power is $P = \vec{F}\cdot\vec{v}$.	**Example 6-9**
Find the time needed to accelerate from one speed to another at constant power.	Find the change in kinetic energy and use $P\,\Delta t = \Delta K$.	**Example 6-10**
Find v and x at a given time t.	If $x_0 = 0$ and $v_0 = 0$, use the results of Example 6-2. Otherwise, derive expressions for v from $a = dv/dt = P/mv$ and $dx/dt = v$.	**Example 6-11**
5. Potential Energy		
Find the gravitational potential energy of an object.	Use $U_g = mgy$.	**Example 6-12**
Find the potential energy of a stretched spring.	Use $U_s = \frac{1}{2}kx^2$.	**Example 6-13**
Find the force from the potential energy function.	Compute $F_x = -dU/dx$.	**Example 6-14**

Problems

Conceptual Problems

Problems from Optional and Exploring sections

In a few problems, you are given more data than you actually need; in a few other problems, you are required to supply data from your general knowledge, outside sources, or informed estimates.

• Single-concept, single-step, relatively easy

•• Intermediate-level, may require synthesis of concepts

••• Challenging, for advanced students

Take $g = 9.81$ N/kg $= 9.81$ m/s^2 and neglect friction in all problems unless otherwise stated.

Work and Kinetic Energy

1 • True or false:

(*a*) Only the net force acting on an object can do work.
(*b*) No work is done on a particle that remains at rest.
(*c*) A force that is always perpendicular to the velocity of a particle never does work on the particle.

2 • A heavy box is to be moved from the top of one table to the top of another table of the same height on the other side of the room. Is work required to do this?

3 • To get out of bed in the morning, do you have to do work?

4 • By what factor does the kinetic energy of a car change when its speed is doubled?

5 • An object moves in a circle at constant speed. Does the force that accounts for its acceleration do work on it? Explain.

6 • An object initially has kinetic energy K. The object then moves in the opposite direction with three times its initial speed. What is the kinetic energy now?

(*a*) K
(*b*) $3K$
(*c*) $-3K$
(*d*) $9K$
(*e*) $-9K$

7 • A 15-g bullet has a speed of 1.2 km/s. (*a*) What is its kinetic energy in joules? (*b*) What is its kinetic energy if its speed is halved? (*c*) What is its kinetic energy if its speed is doubled?

8 • Find the kinetic energy in joules of (*a*) a 0.145-kg baseball moving with a speed of 45 m/s and (*b*) a 60-kg jogger running at a steady pace of 9 min/mi.

9 • A 6-kg box is raised from rest a distance of 3 m by a vertical force of 80 N. Find (*a*) the work done by the force, (*b*) the work done by gravity, and (*c*) the final kinetic energy of the box.

10 • A constant force of 80 N acts on a box of mass of 5.0 kg that is moving in the direction of the applied force with a speed of 20 m/s. Three seconds later the box is moving with a speed of 68 m/s. Determine the work done by this force.

11 •• You run a race with your girlfriend. At first you each have the same kinetic energy, but you find that she is beating you. When you increase your speed by 25%, you are running at the same speed she is. If your mass is 85 kg, what is her mass?

Work Done by a Variable Force

12 • How does the work required to stretch a spring 2 cm from its natural length compare with that required to stretch it 1 cm from its natural length?

13 •• A 3-kg particle is moving with a speed of 2 m/s when it is at $x = 0$. It is subjected to a single force F_x that varies with position as shown in Figure 6-30. (*a*) What is the kinetic energy of the particle when it is at $x = 0$? (*b*) How much work is done by the force as the particle moves from $x = 0$ to $x = 4$ m? (*c*) What is the speed of the particle when it is at $x = 4$ m?

Figure 6-30
Problem 13

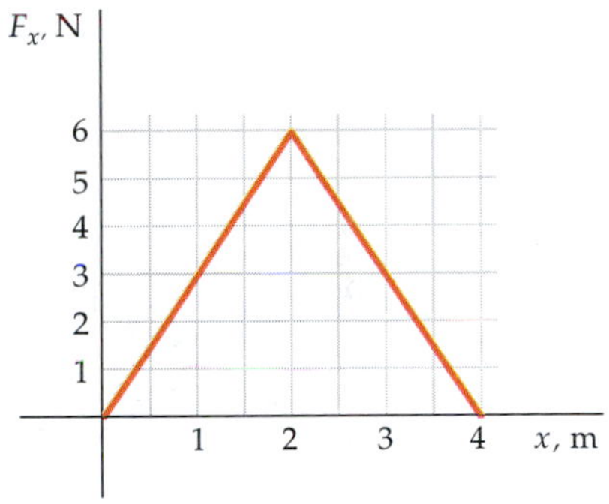

14 •• A 4-kg particle is initially at rest at $x = 0$. It is subjected to a single force F_x that varies with position as shown in Figure 6-31. Find the work done by the force as the particle moves (*a*) from $x = 0$ to $x = 3$ m, and (*b*) from $x = 3$ m to $x = 6$ m. Find the kinetic energy of the particle when it is at (*c*) $x = 3$ m and (*d*) $x = 6$ m.

Figure 6-31
Problem 14

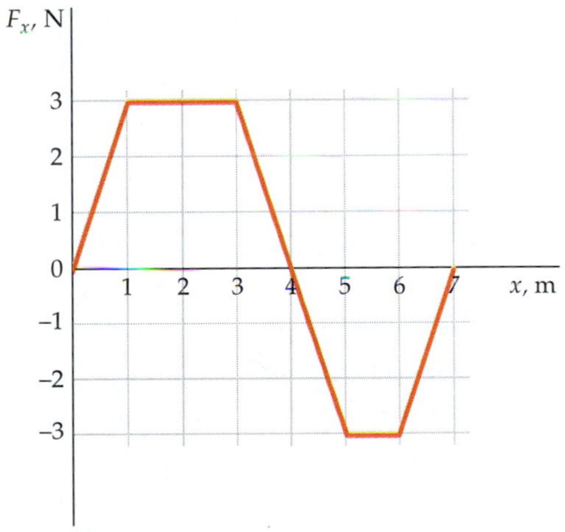

15 •• A force F_x acts on a particle. The force is related to the position of the particle by the formula $F_x = Cx^3$, where C is a constant. Find the work done by this force on the particle when the particle moves from $x = 1.5$ m to $x = 3$ m.

16 •• Lou's latest invention, aimed at urban dog owners, is the X-R-Leash. It is made of a rubber-like material that exerts a force $F_x = -kx - ax^2$ when it is stretched a distance x, where k and a are constants. The ad claims, "You'll never go back to your old dog leash after you've had the thrill of an X-R-Leash experience. And you'll see a new look of respect in the eyes of your proud pooch." Find the work done on a dog by the leash if the person remains stationary and the dog bounds off, stretching the X-R-Leash from $x = 0$ to $x = x_0$.

17 •• A 3-kg object is moving with a speed of 2.40 m/s in the x direction when it passes the origin. It is acted on by a single force F_x that varies with x as shown in Figure 6-32. (*a*) What is the work done by the force from $x = 0$ to $x = 2$ m? (*b*) What is the kinetic energy of the object at $x = 2$ m? (*c*) What is the speed of the object at $x = 2$ m? (*d*) What is the work done on the object from $x = 0$ to $x = 4$ m? (*e*) What is the speed of the object at $x = 4$ m?

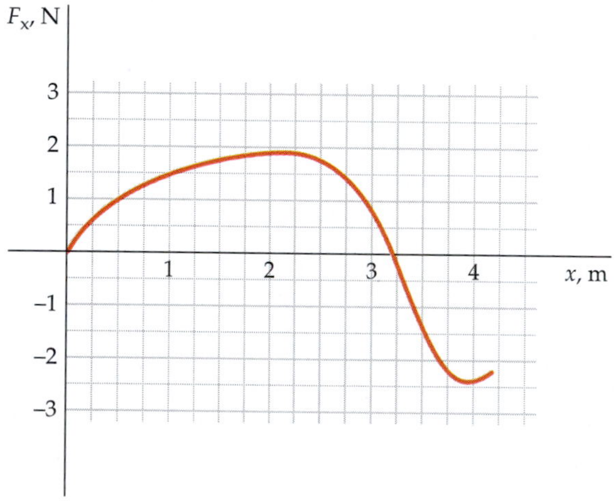

Figure 6-32 Problem 17

18 •• Near Margaret's cabin is a 20-m water tower that attracts many birds during the summer months. During a hot spell last year, the tower went dry, and Margaret had to have her water hauled in. She got lonesome without the birds visiting, so she decided to carry some water up the tower to attract them back. Her bucket has a mass of 10 kg and holds 30 kg of water when it is full. However, the bucket has a hole, and as Margaret climbed at a constant speed, water leaked out at a constant rate. Several birds took advantage of the shower below, but when she got to the top, only 10 kg of water remained for the birdbath. (*a*) Write an expression for the mass of the bucket plus water as a function of the height y climbed. (*b*) Find the work done by Margaret on the bucket.

Work and Energy in Three Dimensions

19 • Suppose there is a net force acting on a particle but it does no work. Can the particle be moving in a straight line?

20 • A 6-kg block slides down a frictionless incline making an angle of 60° with the horizontal. (*a*) List all the forces acting on the block, and find the work done by each force when the block slides 2 m (measured along the incline). (*b*) What is the total work done on the block? (*c*) What is the speed of the block after it has slid 1.5 m if it starts from rest? (*d*) What is its speed after 1.5 m if it starts with an initial speed of 2 m/s?

21 • An 85-kg cart is deposited on a 1.5-m platform after being rolled up an incline formed by a plank of length L that has been laid from the lower level to the top of the platform. (Assume that the rolling is equivalent to sliding without friction.) (*a*) Find the force parallel to the incline needed to push the cart up without acceleration for $L = 3$, 4, and 5 m. (*b*) Calculate directly from Equation 6-15 the work needed to push the cart up the incline for each value of L. (*c*) Since the work found in (*b*) is the same for each value of L, what advantage, if any, is there in choosing one length over another?

22 • A 2-kg object attached to a horizontal string moves with a speed of 2.5 m/s in a circle of radius 3 m on a frictionless horizontal surface. (*a*) Find the tension in the string. (*b*) List the forces acting on the object, and find the work done by each force during one revolution.

Dot Products

23 • What is the angle between the vectors $\vec{A}$ and $\vec{B}$ if $\vec{A}\cdot\vec{B} = -AB$?

24 • Two vectors $\vec{A}$ and $\vec{B}$ have magnitudes of 6 m and make an angle of 60° with each other. Find $\vec{A}\cdot\vec{B}$.

25 • Find $\vec{A}\cdot\vec{B}$ for the following vectors: (*a*) $\vec{A} = 3\hat{i} - 6\hat{j}$, $\vec{B} = -4\hat{i} + 2\hat{j}$; (*b*) $\vec{A} = 5\hat{i} + 5\hat{j}$, $\vec{B} = 2\hat{i} - 4\hat{j}$; and (*c*) $\vec{A} = 6\hat{i} + 4\hat{j}$, $\vec{B} = 4\hat{i} - 6\hat{j}$.

26 • Find the angles between the vectors $\vec{A}$ and $\vec{B}$ in Problem 25.

27 • A 2-kg object is given a displacement $\Delta\vec{s} = 3\text{ m}\,\hat{i} + 3\text{ m}\,\hat{j} - 2\text{ m}\,\hat{k}$ along a straight line. During the displacement, a constant force $\vec{F} = 2\text{ N}\,\hat{i} - 1\text{ N}\,\hat{j} + 1\text{ N}\,\hat{k}$ acts on the object. (*a*) Find the work done by $\vec{F}$ for this displacement. (*b*) Find the component of $\vec{F}$ in the direction of the displacement.

28 •• (*a*) Find the unit vector that is parallel to the vector $\vec{A} = A_x\hat{i} + A_y\hat{j} + A_z\hat{k}$. (*b*) Find the component of the vector $\vec{A} = 2\hat{i} - \hat{j} - \hat{k}$ in the direction of the vector $\vec{B} = 3\hat{i} + 4\hat{j}$.

29 •• When a particle moves in a circle with constant speed, the magnitudes of its position vector and velocity vectors are constant. (*a*) Differentiate $\vec{r}\cdot\vec{r} = r^2 =$ constant with respect to time to show that $\vec{v}\cdot\vec{r} = 0$ and therefore $\vec{v} \perp \vec{r}$. (*b*) Differentiate $\vec{v}\cdot\vec{v} = v^2 =$ constant with respect to time to show that $\vec{a}\cdot\vec{v} = 0$ and therefore $\vec{a} \perp \vec{v}$. What do the results of (*a*) and (*b*) imply about the direction of $\vec{a}$? (*c*) Differentiate $\vec{v}\cdot\vec{r} = 0$ with respect to time and show that $\vec{a}\cdot\vec{r} + v^2 = 0$ and therefore $a_r = -v^2/r$.

30 •• Vectors $\vec{A}$, $\vec{B}$, and $\vec{C}$ form a triangle as shown in Figure 6-33. The angle between $\vec{A}$ and $\vec{B}$ is θ, and the vectors are related by $\vec{C} = \vec{A} - \vec{B}$. Compute $\vec{C}\cdot\vec{C}$ in terms of A, B, and θ, and derive the law of cosines, $C^2 = A^2 + B^2 - 2AB\cos\theta$.

Figure 6-33 Problem 30

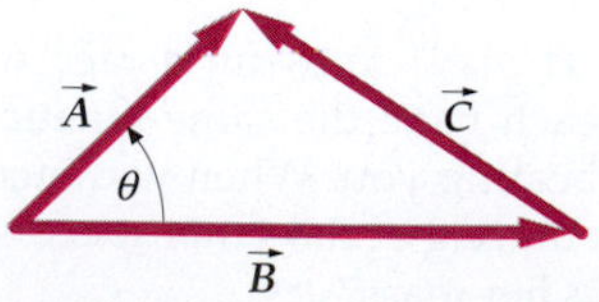

Power

31 • The dimension of power is ______.

(a) $[M][L]^2[T]^2$
(b) $[M][L]^2/[T]$
(c) $[M][L]^2/[T]^2$
(d) $[M][L]^2/[T]^3$

32 • True or false: A kilowatt-hour is a unit of power.

33 • The engine of a car operates at constant power. The ratio of acceleration of the car at a speed of 60 km/h to that at 30 km/h (neglecting air resistance) is ______.

(a) $\frac{1}{2}$ (b) $1/\sqrt{2}$ (c) $\sqrt{2}$ (d) 2

34 •• A car starts from rest and travels at constant acceleration. Which of the following statements are true?

(a) The power delivered by the engine is constant.
(b) The power delivered by the engine increases as the car gains speed.
(c) The power delivered by the engine decreases as the car gains speed.
(d) Both (b) and (c) are correct.

35 •• Force A does 5 J of work in 10 s. Force B does 3 J of work in 5 s. Which force delivers greater power?

36 • A 5-kg box is lifted by a force equal to the weight of the box. The box moves upward at a constant velocity of 2 m/s. (a) What is the power input of the force? (b) How much work is done by the force in 4 s?

37 • Fluffy has just caught a mouse, and decides that the only decent thing to do is to bring it to the bedroom so that his human roommate can admire it when she wakes up. A constant horizontal force of 3 N is enough to drag the mouse across the rug at a constant speed v. If Fluffy's force does work at the rate of 6 W, (a) what is her speed, v? (b) How much work does Fluffy do in 4 s?

38 • A single force of 5 N acts in the x direction on an 8-kg object. (a) If the object starts from rest at $x = 0$ at time $t = 0$, find its velocity v as a function of time. (b) Write an expression for the power input as a function of time. (c) What is the power input of the force at time $t = 3$ s?

39 • Find the power input of a force $\vec{F}$ acting on a particle that moves with a velocity $\vec{v}$ for (a) $\vec{F} = 4\text{ N}\,\hat{i} + 3\text{ N}\,\hat{k}$, $\vec{v} = 6\text{ m/s}\,\hat{i}$; (b) $\vec{F} = 6\text{ N}\,\hat{i} - 5\text{ N}\,\hat{j}$, $\vec{v} = -5\text{ m/s}\,\hat{i} + 4\text{ m/s}\,\hat{j}$; and (c) $\vec{F} = 3\text{ N}\,\hat{i} + 6\text{ N}\,\hat{j}$, $\vec{v} = 2\text{ m/s}\,\hat{i} + 3\text{ m/s}\,\hat{j}$.

40 •• A particle of mass m moves from rest at $t = 0$ under the influence of a single force of magnitude F. Show that the power delivered by the force at time t is $P = F^2t/m$.

41 •• At a speed of 20 km/h, a 1200-kg car accelerates at 3 m/s^2 using 20 kW of power. How much power must be expended to accelerate the car at 2 m/s^2 at a speed of 40 km/h?

42 •• A car manufacturer claims that his car can accelerate from rest to 100 km/h in 8 s. The car's mass is 800 kg. (a) Assuming that this performance is achieved at constant power, determine the power developed by the car's engine. (b) What is the car's speed after 4 s? (Neglect friction and air resistance.)

43 •• Show that the position of the truck in Example 6-11 is related to its speed by $x = (m/3P)v^3$.

44 •• A 700-kg car accelerates from rest under constant power. At the end of 8.0 s, its speed is 90 km/h and it is located 133 m from its starting point. If the car continues to accelerate using the same power, what will its speed be at the end of 10 s, and how far will the car be from the starting point?

45 •• A 4.0-kg object initially at rest at $x = 0$ is accelerated at constant power of 8.0 W. At $t = 9.0$ s, it is at $x = 36.0$ m. Find its speed at $t = 6.0$ s and its position at that instant.

46 •• A 700-kg car accelerates from rest under constant power at $t = 0$. At $t = 9$ s it is 117.7 m from its starting point and its acceleration is then 1.09 m/s^2. Find the power expended by the car's engine, neglecting frictional losses.

Potential Energy

47 • Two knowledge seekers decide to ascend a mountain. Sal chooses a short, steep trail, while Joe, who weighs the same as Sal, goes up via a long, gently sloped trail. At the top, they get into an argument about who gained more potential energy. Which of the following is true?

(a) Sal gains more gravitational potential energy than Joe.
(b) Sal gains less gravitational potential energy than Joe.
(c) Sal gains the same gravitational potential energy as Joe.
(d) To compare energies, we must know the height of the mountain.
(e) To compare energies, we must know the length of the two trails.

48 • The gravitational potential energy of an object changes by −6 J. It follows that the work done by the gravitational force on this object is

(a) −6 J and the elevation of the object is increased.
(b) −6 J and the elevation of the object is decreased.
(c) +6 J and the elevation of the object is increased.
(d) +6 J and the elevation of the object is decreased.

49 • A woman runs up a flight of stairs. The gain in her gravitational potential energy is U. If she runs up the same stairs with twice the speed, what will be her gain in potential energy?

(a) U
(b) $2U$
(c) $U/2$
(d) $4U$
(e) $U/4$

50 • Which of the following statements is true?

(a) The kinetic and potential energies of an object must always be positive quantities.
(b) The kinetic and potential energies of an object must always be negative quantities.
(c) Kinetic energy can be negative, but potential energy cannot.
(d) Potential energy can be negative, but kinetic energy cannot.
(e) None of the preceding statements is true.

51 • A block slides a certain distance down an incline. The work done by gravity is W. What is the work done by gravity if this block slides the same distance up the incline?
(*a*) W
(*b*) Zero
(*c*) $-W$
(*d*) Gravity can't do work; some other force does work.
(*e*) Cannot be determined unless given the distance traveled.

52 • True or false:
(*a*) Only conservative forces can do work.
(*b*) If only conservative forces act, the kinetic energy of a particle does not change.
(*c*) The work done by a conservative force equals the decrease in the potential energy associated with that force.

53 • When you climb a mountain, is the work done on you by gravity different if you take a short, steep trail instead of a long, gentle trail? If not, why do you find one trail easier?

54 • Which of the following forces are conservative and which are nonconservative?
(*a*) the frictional force exerted on a sliding box
(*b*) the force exerted by a linear spring that obeys Hooke's law
(*c*) the force of gravity
(*d*) the wind resistance on a moving car

55 • An 80-kg man climbs up a 6-m high flight of stairs. What is the increase in gravitational potential energy?

56 • One of the highlights of Sharika's concert is her daredevil swan dive into the audience from a height of 2 m above the crowd's outstretched hands. If her mass is 60 kg, and the time of her dive is defined as $t = 0$, (*a*) what is her initial potential energy relative to $U = 0$ at the position of the crowd's hands? (*b*) From Newton's laws, find the distance she has fallen and her speed at $t = 0.20$ s. (*c*) Find her potential and kinetic energy at $t = 0.40$ s. (*d*) Find her kinetic energy and speed just as she reaches the hands of the crowd in the mosh pit. Assume her horizontal speed is negligible.

57 • Water flows over Victoria Falls, which is 128 m high, at an average rate of 1.4×10^6 kg/s. If half the potential energy of this water were converted into electric energy, how much power would be produced by these falls?

58 • A 2-kg box slides down a long, frictionless incline of angle 30°. It starts from rest at time $t = 0$ at the top of the incline at a height of 20 m above the ground. (*a*) What is the original potential energy of the box relative to the ground? (*b*) From Newton's laws, find the distance the box travels in 1 s and its speed at $t = 1$ s. (*c*) Find the potential energy and the kinetic energy of the box at $t = 1$ s. (*d*) Find the kinetic energy and the speed of the box just as it reaches the bottom of the incline.

59 • A force $F_x = 6$ N is constant. (*a*) Find the potential-energy function U associated with this force for an arbitrary reference position x_0 at which $U = 0$. (*b*) Find U such that $U = 0$ at $x = 4$ m. (*c*) Find U such that $U = 14$ J at $x = 6$ m.

60 • A spring has a force constant of $k = 10^4$ N/m. How far must it be stretched for its potential energy to be (*a*) 50 J and (*b*) 100 J?

61 •• A simple Atwood's machine uses two masses, m_1 and m_2 (Figure 6-34). Starting from rest, the speed of the two masses is 4.0 m/s at the end of 3.0 s. At that instant, the kinetic energy of the system is 80 J and each mass has moved a distance of 6.0 m. Determine the values of m_1 and m_2.

Figure 6-34 Problem 61

62 •• A straight rod of negligible mass is mounted on a frictionless pivot as in Figure 6-35. Masses m_1 and m_2 are suspended at distances l_1 and l_2. (*a*) Write an expression for the gravitational potential energy of the masses as a function of the angle θ made by the rod and the horizontal. (*b*) For what angle θ is the potential energy a minimum? Is the statement "systems tend to move toward a configuration of minimum potential energy" consistent with your result? (*c*) Show that if $m_1 l_1 = m_2 l_2$, the potential energy is the same for all values of θ. (When this holds, the system will balance at any angle θ. This result is known as *Archimedes' law of the lever*.)

Figure 6-35 Problem 62

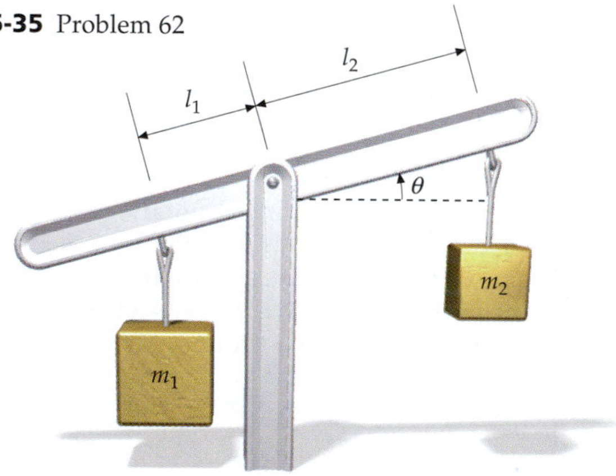

Force, Potential Energy, and Equilibrium

63 •• Figure 6-36 shows the plot of a potential-energy function U versus x. (*a*) At each point indicated, state whether the force F_x is positive, negative, or zero. (*b*) At which point does the force have the greatest magnitude? (*c*) Identify any equilibrium points, and state whether the equilibrium is stable, unstable, or neutral.

Figure 6-36 Problem 63

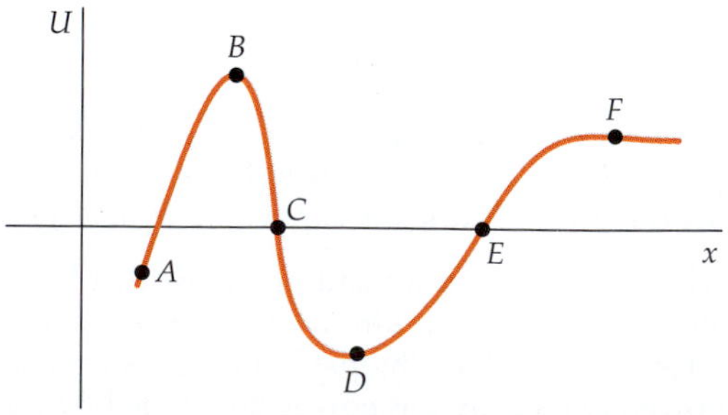

64 • (*a*) Find the force F_x associated with the potential-energy function $U = Ax^4$, where A is a constant. (*b*) At what point(s) is the force zero?

65 •• A potential-energy function is given by $U = C/x$, where C is a positive constant. (*a*) Find the force F_x as a function of x. (*b*) Is this force directed toward the origin or away from it? (*c*) Does the potential energy increase or decrease as x increases? (*d*) Answer parts (*b*) and (*c*) where C is a negative constant.

66 •• On the potential-energy curve for U versus y shown in Figure 6-37, the segments AB and CD are straight lines. Sketch a plot of the force F_y versus y.

Figure 6-37 Problem 66

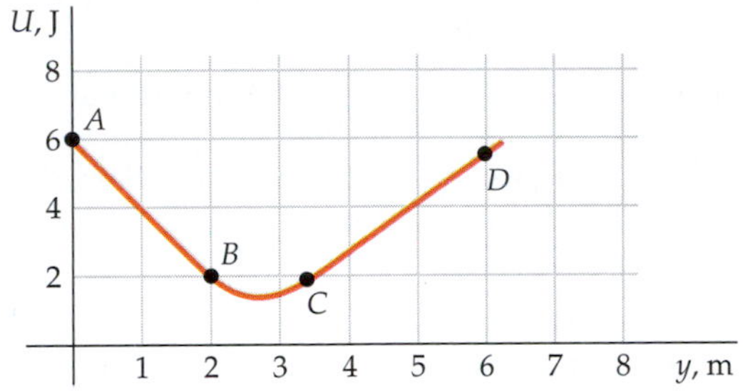

67 •• The force acting on an object is given by $F_x = a/x^2$. Determine the potential energy of the object as a function of x.

68 •• The potential energy of an object is given by $U(x) = 3x^2 - 2x^3$, where U is in joules and x is in meters. (*a*) Determine the force acting on this object. (*b*) At what positions is this object in equilibrium? (*c*) Which of these equilibrium positions are stable and which are unstable?

69 •• During a Dead Wait concert, Sharika and Chico, each of mass M, are attached to the ends of a light rope that is hung over two frictionless pulleys, as shown in Figure 6-38. A large gong of mass m is attached to the middle of the rope, between the pulleys, and Sharika and Chico beat it madly in lieu of the usual guitar solo. (*a*) Find the potential energy of the system as a function of the distance y to the center of the gong. (*b*) Find the value of y for which the potential energy function of the system is a minimum. (*c*) Check your answer by applying Newton's laws to the gong.

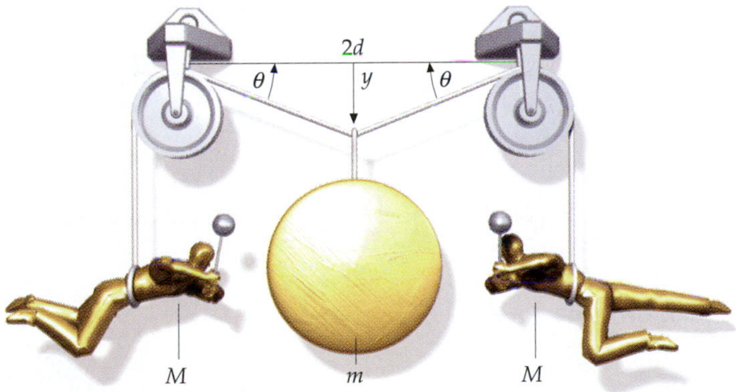

Figure 6-38 Problem 69

70 ••• The potential energy of an object is given by $U(x) = 8x^2 - x^4$, where U is in joules and x is in meters. (*a*) Determine the force acting on this object. (*b*) At what positions is this object in equilibrium? (*c*) Which of these equilibrium positions are stable and which are unstable?

71 ••• The force acting on an object is given by $F(x) = x^3 - 4x$. Locate the positions of unstable and stable equilibrium and show that at these points $U(x)$ is a local maximum or minimum, respectively.

72 ••• The potential energy of a 4-kg object is given by $U = 3x^2 - x^3$ for $x \leq 3$ m, and $U = 0$ for $x \geq 3$ m, where U is in joules and x is in meters. (*a*) At what positions is this object in equilibrium? (*b*) Sketch a plot of U versus x. (*c*) Discuss the stability of the equilibrium for the values of x found in (*a*). (*d*) If the total energy of the particle is 12 J, what is its speed at $x = 2$ m?

73 ••• A force is given by $F_x = Ax^{-3}$, where $A = 8\ \text{N}\cdot\text{m}^3$. (*a*) For positive values of x, does the potential energy associated with this force increase or decrease with increasing x? (You can determine the answer to this question by imagining what happens to a particle that is placed at rest at some point x and is then released.) (*b*) Find the potential-energy function U associated with this force such that U approaches zero as x approaches infinity. (*c*) Sketch U versus x.

General Problems

74 • True or false:

(*a*) The gravitational force cannot do work because it acts at a distance.
(*b*) Work is the area under the force-versus-time curve.

75 • Negative work by an applied force implies that

(*a*) the kinetic energy of the object increases.
(*b*) the applied force is variable.
(*c*) the applied force is perpendicular to the displacement.
(*d*) the applied force has a component that is opposite to the displacement.
(*e*) nothing; there is no such thing as negative work.

76 •• A movie crew is in the Badlands when their car overheats. After they stop to let it cool down, an argument breaks out. They agree that they must go easy on the engine, but they disagree about when the engine works the hardest, and therefore about how they should drive for the rest of the trip. Carolyn claims that the work done by the car in accelerating from 0 to 20 km/h is less than that required to accelerate from 20 to 30 km/h, meaning they should drive more slowly. Ted says no, the work done between 0 and 20 km/h is more than the work done between 20 and 30 km/h. Ernie says it all depends on the mass of the car, and Bloop says it all depends on how long you take to change from one speed to another. Who is right?

77 • Figure 6–39 shows two pulleys arranged to help lift a heavy load. A rope runs around two massless, frictionless pulleys and the weight $\vec{w}$ hangs from one pulley. You exert a force of magnitude F on the free end of the cord. (*a*) If the weight is to move up a distance h, through what distance must the force move? (*b*) How much work is

Figure 6-39 Problem 77

done by the ropes on the weight? (*c*) How much work do you do? (This is an example of a simple machine in which a small force F_1 moves through a large distance x_1 to exert a large force F_2 (= w) through a smaller distance $x_2 = h$.)

78 • In February 1995, a total of 54.3 billion kW-h of electrical energy was generated by nuclear power plants in the United States. At the same time, the population of the United States was about 255 million people. If the average American has a mass of 60 kg, and if the entire energy output of all nuclear power plants was diverted to supplying energy for a single giant elevator, estimate the height h at which the entire population of the country could be lifted by the elevator. In your calculations, assume that 25% of the energy goes into lifting the people; assume also that g is constant over the entire height h.

79 • One of the most powerful cranes in the world, operating in Switzerland, can slowly raise a load of M = 6000 tonne to a height of h = 12.0 m (1 tonne = 1000 kg). (*a*) How much work is done by the crane? (*b*) If it takes 1.00 min to lift the load at constant velocity to this height, find the power developed by the crane.

80 • In Australia, there used to be a ski lift of length 5.6 km. It took about 60 min for a gondola to travel all the way up. If there were 12 gondolas going up at once, each of mass 550 kg, and the angle of ascent was 30°, estimate the power P of the engine needed to operate the ski lift.

81 • A 2.4-kg object attached to a horizontal string moves with constant speed in a circle of radius R on a frictionless horizontal surface. The kinetic energy of the object is 90 J and the tension in the string is 360 N. Find R.

82 • How high must an 800-kg Ford Escort be lifted to gain an amount of potential energy equal to the kinetic energy it has when it is moving at 100 km/h?

83 • The movie crew arrives in the Badlands ready to shoot a scene. The script calls for a car to crash into a vertical rock face at 100 km/h. Unfortunately, the car won't start, and there is no mechanic in sight. They are about to skulk back to the studio to face the producer's wrath when the cameraman gets an idea. They use a crane to lift the car by its rear end and then drop it, filming at an angle that makes the car appear to be traveling horizontally. How high should the 800-kg car be lifted so that it reaches a speed of 100 km/h in the fall?

84 •• The force acting on a particle that is moving along the x axis is given by $F_x = -ax^2$, where a is a constant. Calculate the potential-energy function U relative to $U = 0$ at $x = 0$, and sketch a graph of U versus x.

85 •• Water from behind a dam flows through a large turbine at a rate of 1.5×10^6 kg/min. The turbine is located 50 m below the surface of the reservoir, and the water leaves the turbine with a speed of 5 m/s. (*a*) Neglecting any energy dissipation, what is the power output of the turbine? (*b*) How many U.S. citizens would be supplied with energy by this dam if each citizen uses 3×10^{11} J of energy per year?

86 •• A force acts on a cart of mass m in such a way that the speed v of the cart increases with distance x as $v = Cx$, where C is a constant. (*a*) Find the force acting on the cart as a function of position. (*b*) What is the work done by the force in moving the cart from $x = 0$ to $x = x_1$?

87 •• A force $\vec{F} = (2\ \text{N/m}^2)x^2\,\hat{i}$ is applied to a particle. Find the work done on the particle as it moves a total distance of 5 m (*a*) parallel to the y axis from point (2 m, 2 m) to point (2 m, 7 m) and (*b*) in a straight line from (2 m, 2 m) to (5 m, 6 m).

88 •• A particle of mass m moves along the x axis. Its position varies with time according to $x = 2t^3 - 4t^2$ where x is in meters and t is in seconds. Find (*a*) the velocity and acceleration of the particle at any time t; (*b*) the power delivered to the particle at any time t; and (*c*) the work done by the force from $t = 0$ to $t = t_1$.

89 •• A 3-kg particle starts from rest at $x = 0$ and moves under the influence of a single force $F_x = 6 + 4x - 3x^2$ where F_x is in newtons and x is in meters. (*a*) Find the work done by the force as the particle moves from $x = 0$ to $x = 3$ m. (*b*) Find the power delivered to the particle when it is at $x = 3$ m.

90 •• The initial kinetic energy imparted to a 20-g bullet is 1200 J. Neglecting air resistance, find the range of this projectile when it is fired at an angle such that the range equals the maximum height attained.

91 •• A force F_x acting on a particle is shown as a function of x in Figure 6-40. (*a*) From the graph, calculate the work done by the force when the particle moves from $x = 0$ to the following values of x: −4, −3, −2, −1, 0, 1, 2, 3, and 4 m. (*b*) Plot the potential energy U versus x for the range of values of x from −4 m to +4 m, assuming that $U = 0$ at $x = 0$.

Figure 6-40 Problem 91

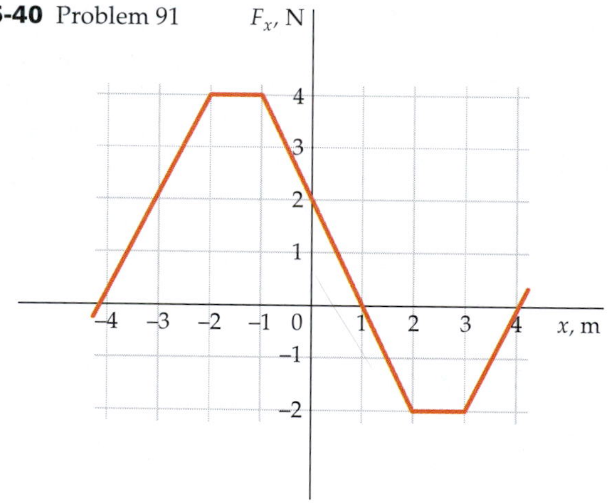

92 •• Repeat Problem 91 for the force F_x shown in Figure 6-41.

Figure 6-41 Problem 92

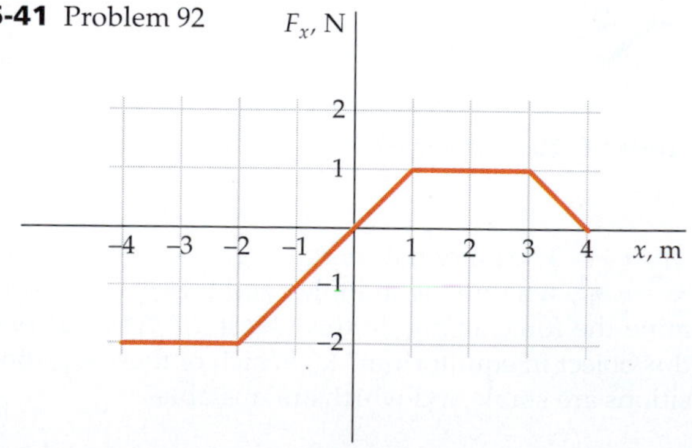

93 •• A rope of length L and mass per unit length of μ lies coiled on the floor. (*a*) What force F is required to hold one end of the rope a distance $y < L$ above the floor as shown in Figure 6-42? (*b*) Find the work required to lift one end of the rope from the floor to a height $l < L$ by integrating $F\,dy$ from $y = 0$ to $y =$ ell.

Figure 6-42 Problem 93

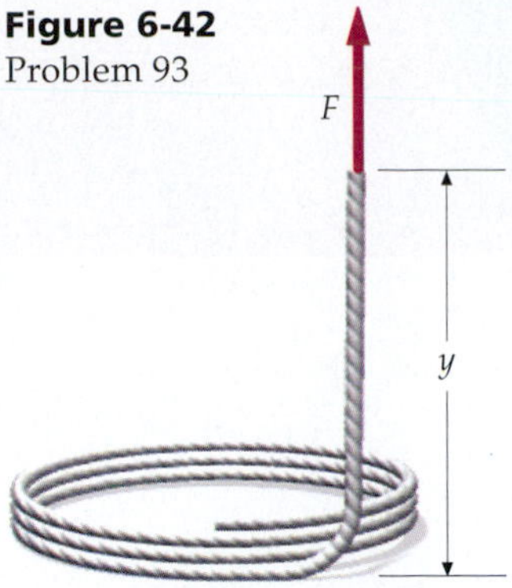

94 ••• A box of mass M is at the bottom of a frictionless inclined plane (Figure 6-43). The box is attached to a string that pulls with a constant tension T. (*a*) Find the work done by the tension T when the box has moved a distance x along the plane. (*b*) Find the speed of the box as a function of x and θ. (*c*) Determine the power produced by the tension in the string as a function of x and θ.

Figure 6-43 Problem 94

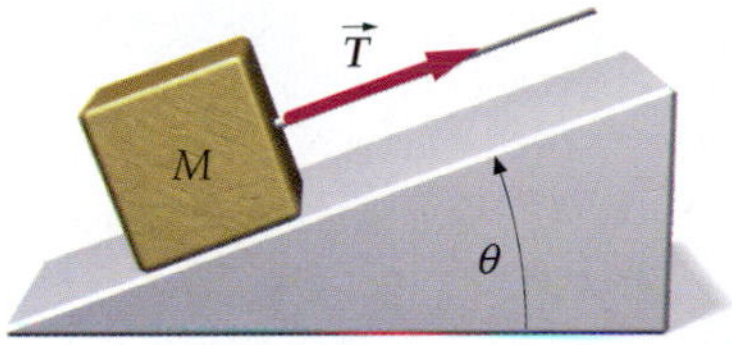

95 ••• A force in the xy plane is given by $\vec{F} = (F_0/r)(y\hat{i} - x\hat{j})$, where F_0 is a constant and $r = \sqrt{x^2 + y^2}$. (*a*) Show that the magnitude of this force is F_0 and that its direction is perpendicular to $\vec{r} = x\hat{i} + y\hat{j}$. (*b*) Find the work done by this force on a particle that moves in a circle of radius 5 m centered at the origin. Is this force conservative?

96 ••• A theoretical formula for the potential energy associated with the nuclear force between two protons, two neutrons, or a neutron and a proton is the *Yukawa potential:*

$$U = -U_0\left(\frac{a}{x}\right)e^{-x/a}$$

where U_0 and a are constants. (*a*) Sketch U versus x using $U_0 = 4$ pJ (a picojoule, pJ, is 1×10^{-12} J) and $a = 2.5$ fm (a femtometer, fm, is 1×10^{-15} m). (*b*) Find the force F_x. (*c*) Compare the magnitude of the force at the separation $x = 2a$ to that at $x = a$. (*d*) Compare the magnitude of the force at the separation $x = 5a$ to that at $x = a$.

CHAPTER 7

Conservation of Energy

The waterfall in this 1961 lithograph by the Dutch artist M. C. Escher violates the law of conservation of energy. As the water falls, part of its potential energy is converted into the kinetic energy of the waterwheel. How then does the water get back to the top of the waterfall?

The potential energy of a system is defined in such a way that the work done by an internal conservative force on the system equals the *decrease* in potential energy. If the conservative force is the only force that does work, the work it does also equals the *increase* in kinetic energy. Since the decrease in potential energy equals the increase in kinetic energy, the sum of potential and kinetic energy, or the total mechanical energy, does not change. This is known as the law of conservation of mechanical energy. It follows from Newton's laws, and presents a useful alternative to Newton's laws for solving many problems in mechanics. The use of conservation of energy is limited, however, because there are usually nonconservative forces present, such as friction. When friction is present, the mechanical energy of the system decreases.

Since mechanical energy is often not conserved, the importance of energy was not realized until the nineteenth century, when it was discovered that the disappearance of macroscopic mechanical energy is always accompanied by the appearance of some other kind of energy, often thermal energy, which is usually indicated by an increase in temperature. We now know that, on the microscopic scale, this thermal energy consists of the kinetic and potential energies of the molecules in the system.

There are other forms of energy, such as the internal chemical energy in your body, the energy of sound, and electromagnetic energy. Whenever the energy of a system changes, we can account for the change by the appearance or disappearance of energy somewhere else. This experimental observation is the law of conservation of energy, one of the most fundamental and important laws in all of science. Although energy changes from one form to another, it is never created or destroyed.

We begin by considering systems in which mechanical energy is conserved. We then extend the discussion to include thermal and chemical energy, and we develop specific methods to deal with dissipative systems, wherein kinetic friction converts mechanical energy to thermal energy. After discussing Einstein's famous relation between mass and energy, we conclude by considering the quantization of energy, the surprising result, discovered in the first quarter of the twentieth century, that energy changes in a system are not continuous, but occur in lumps or quanta. Although the quantum of energy is so small that the incremental quality of energy goes unnoticed in the macroscopic world, the quantization of energy has profound consequences for microscopic systems such as atoms and molecules.

7-1 The Conservation of Mechanical Energy

Consider a system for which the only forces acting are internal, conservative forces. The skier–earth system discussed in Chapter 6 is such a system. The total work done on each particle in the system equals the increase in the kinetic energy of that particle, so the total work done by all the forces equals the increase in the total kinetic energy of the system:

$$W_{\text{total}} = \sum \Delta K_{\text{i}} = \Delta K \qquad \text{7-1}$$

Since each internal force is conservative, the work it does decreases the potential energy associated with that force. So the total work done by all the internal forces equals the total decrease in potential energy of the system:

$$W_{\text{total}} = \sum - \Delta U_{\text{i}} = -\Delta U \qquad \text{7-2}$$

Thus, $\Delta K = -\Delta U$ or

$$\Delta K + \Delta U = \Delta(K + U) = 0 \qquad \text{7-3}$$

The sum of the kinetic energy K and the potential energy U of a system is called the **total mechanical energy** E_{mech}:

$$E_{\text{mech}} = K + U \qquad \text{7-4}$$

Definition—Total mechanical energy

We have just shown that when only internal, conservative forces do work on a system of two or more particles, the total mechanical energy of the system does not change:

$$\Delta(K + U) = \Delta E_{\text{mech}} = 0$$

$$E_{\text{mech}} = K + U = \text{constant} \qquad \text{7-5}$$

Conservation of mechanical energy

This is the **law of conservation of mechanical energy** and is the origin of the expression "conservative force."

If $E_i = K_i + U_i$ is the initial mechanical energy of the system and $E_f = K_f + U_f$ is the final mechanical energy, conservation of mechanical energy implies that

$$E_f = E_i$$
$$K_f + U_f = K_i + U_i \tag{7-6}$$

Conservation of mechanical energy

Many mechanics problems can be solved by setting the final mechanical energy of a system equal to its initial mechanical energy.

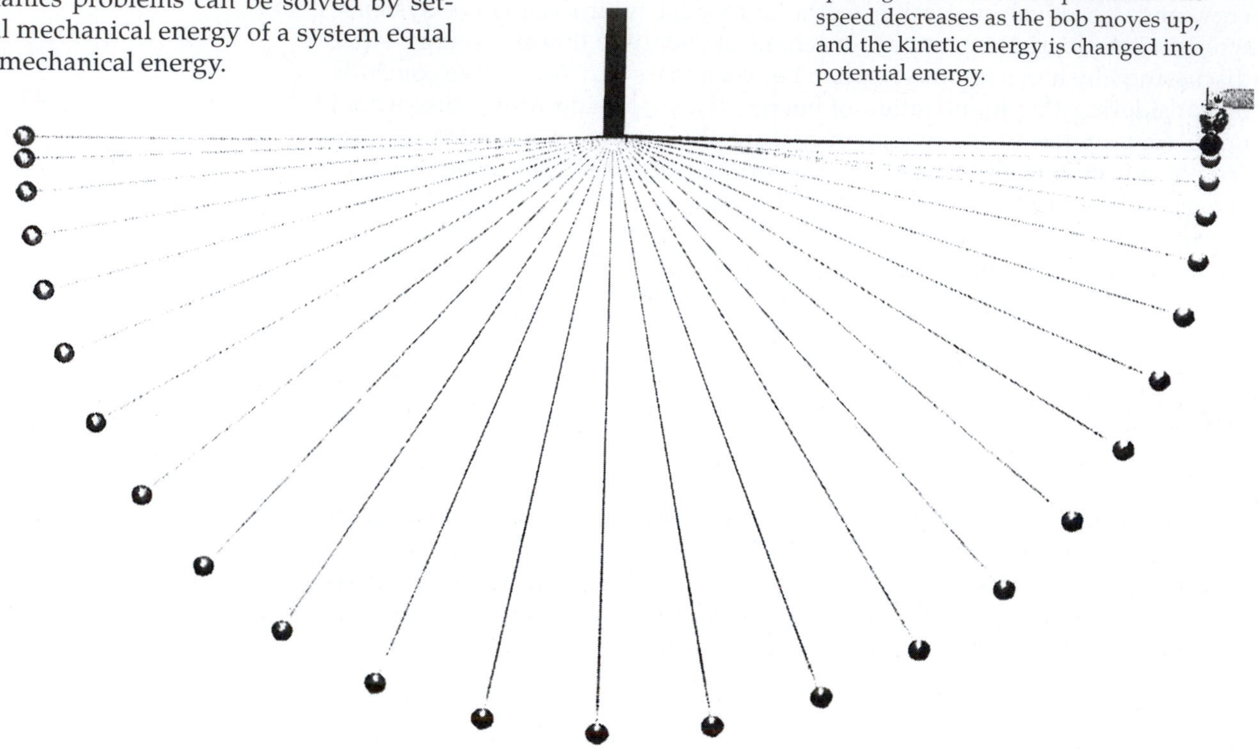

Multiflash photograph of a simple pendulum. As the bob descends, gravitational potential energy is converted into kinetic energy, and the speed increases as indicated by the increased spacing of the recorded positions. The speed decreases as the bob moves up, and the kinetic energy is changed into potential energy.

Applications

Consider a downhill skier who starts at rest from a height h above the bottom of a hill, which we assume to be frictionless. What is the skier's speed at a height y above the bottom of the hill? The mechanical energy of the earth–skier system is conserved because the only force doing work is the internal, conservative force of gravity. If we choose $U = 0$ at the bottom of the hill, the original potential energy is mgh. This is also the total mechanical energy because the initial kinetic energy is zero. Thus,

$$E_i = K_i + U_i = 0 + mgh = mgh$$

At the height y, the potential energy is mgy and the speed is v. Hence,

$$E_f = K_f + U_f = \frac{1}{2}mv^2 + mgy$$

Setting $E_f = E_i$ we find

$$\frac{1}{2}mv^2 + mgy = mgh$$

or

$$v = \sqrt{2g(h - y)}$$

The speed of the skier is the same as if she had undergone free fall through a distance $h - y$.

Example 7-1

Standing near the edge of the roof of a 12-m high building, you kick a ball with an initial speed of $v_i = 16$ m/s at an angle of 60° above the horizontal. Neglecting air resistance, find (*a*) how high above the building the ball rises, and (*b*) its speed just before it hits the ground.

Picture the Problem Since gravity is the only force that does work on the ball–earth system, mechanical energy is conserved. At the top of its flight, the ball is moving horizontally with its initial horizontal velocity $v_{top} = v_i \cos 60°$. We choose $U = 0$ at the top of the building (Figure 7-1).

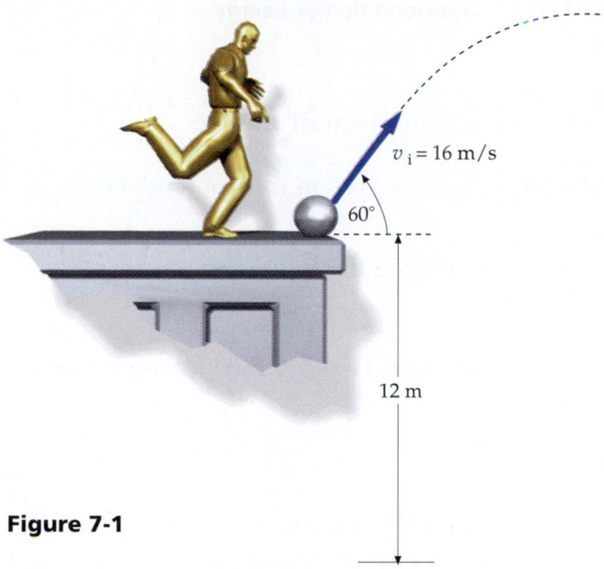

Figure 7-1

(*a*)1. Conservation of mechanical energy relates the height h to the initial velocity v_i and the velocity at the top of its flight v_{top}:

$$E_{top} = E_i$$

$$\frac{1}{2}mv_{top}^2 + mgh = \frac{1}{2}mv_i^2$$

2. Solve for h:

$$h = \frac{v_i^2 - v_{top}^2}{2g}$$

3. The velocity at the top of its flight equals its initial horizontal velocity:

$$v_{top} = v_i \cos\theta = (16\text{ m/s})\cos 60° = 8\text{ m/s}$$

4. Substitute this value for v_t and $v_i = 16$ m/s and solve for h:

$$h = \frac{v_i^2 - v_{top}^2}{2g} = \frac{(16\text{ m/s})^2 - (8\text{ m/s})^2}{2(9.81\text{ m/s}^2)} = 9.79\text{ m}$$

(*b*)1. If v_f is the speed of the ball just before it hits the ground, its energy is:

$$E_f = \frac{1}{2}mv_f^2 + mgy$$

2. Apply conservation of mechanical energy:

$$\frac{1}{2}mv_f^2 + mgy = \frac{1}{2}mv_i^2$$

3. Solve for v_f, and set $y = -12$ m to find the final velocity:

$$v_f = \sqrt{v_i^2 - 2gy}$$
$$= \sqrt{16\text{ (m/s)}^2 - 2(9.81\text{ m/s}^2)(-12\text{ m})} = 22.2\text{ m/s}$$

Example 7-2

A pendulum consists of a bob of mass m attached to a string of length L. The bob is pulled aside so that the string makes an angle θ_0 with the vertical and is released from rest. Find expressions for (*a*) the speed v at the bottom of the swing, and (*b*) the tension in the string at that time.

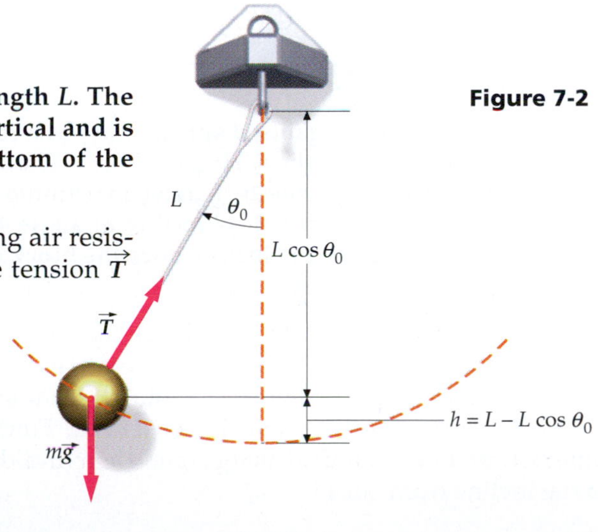

Figure 7-2

Picture the Problem The two forces acting on the bob (neglecting air resistance) are the force of gravity $m\vec{g}$, which is conservative, and the tension $\vec{T}$ (Figure 7-2). Since $\vec{T}$ is perpendicular to the motion, it does no work. Hence, the mechanical energy of the bob–earth system is conserved. To find the speed of the bob, equate the initial and final energies. The tension in the string is found from Newton's second law. We choose $U = 0$ at the bottom of the swing. The initial height h above the bottom is related to the initial angle θ_0 by $h = L - L\cos\theta_0$.

(*a*)1. Apply conservation of mechanical energy:	$E_f = E_i$
2. The initial energy E_i of the system is its potential energy:	$E_i = K_i + U_i = 0 + mgh$
3. At the bottom, the energy E_f is all kinetic:	$E_f = K_f + U_f = \frac{1}{2}mv^2 + 0 = \frac{1}{2}mv^2$
4. Conservation of mechanical energy thus relates the speed v to the height h:	$\frac{1}{2}mv^2 = mgh$
5. Solve for the speed v:	$v = \sqrt{2gh}$
6. To express speed in terms of the initial angle θ_0, we need to relate h to θ_0. This is done in Figure 7-2:	$h = L - L\cos\theta_0 = L(1 - \cos\theta_0)$
7. Substitute this value for h to express the speed at the bottom in terms of θ_0:	$v = \sqrt{2gh} = \sqrt{2gL(1 - \cos\theta_0)}$
(*b*)1. The forces on the bob are its weight $m\vec{g}$ down and $\vec{T}$, which is up when the bob is at the bottom of the circle. Choose up to be the positive y direction, and apply $\Sigma F_y = ma_y$:	$T - mg = ma$
2. At the bottom, the bob has a centripetal acceleration v^2/L toward the center of the circle, which is upward at this point:	$a = \frac{v^2}{L} = \frac{2gL(1 - \cos\theta_0)}{L} = 2g(1 - \cos\theta_0)$
3. Substitute this value of a into the equation in (*b*)1 to find T:	$T = mg + ma = mg + 2mg(1 - \cos\theta_0)$ $= mg(3 - 2\cos\theta_0)$

Remarks The tension at the bottom is greater than the weight of the bob because the bob is accelerating upward. If the bob is released at $\theta_0 = 90°$, the tension at the bottom is $3mg$. The speed of the bob can also be found using Newton's laws (see Problem 97), but the solution is difficult because the acceleration tangential to the curve varies with the angle θ and therefore with time, so the constant-acceleration formulas do not apply. Finally, step 4 in part (*a*) shows that the speed at the bottom is the same as if the bob had dropped in free fall from a height h.

Example 7-3 *try it yourself*

A 2-kg block is pushed against a spring that has a force constant of 500 N/m, compressing the spring by 20 cm. The block is then released, and the spring projects it along a frictionless horizontal surface and then up a frictionless incline of angle 45° as shown in Figure 7-3. How far up the incline does the block travel before momentarily coming to rest?

Picture the Problem After the block is released, the only forces that do work are the conservative forces exerted by the spring and the force of gravity. The total mechanical energy of the block–spring–earth system is conserved. Find h from the conservation of mechanical energy, and then find the distance s up the incline from $\sin 45° = h/s$.

Figure 7-3

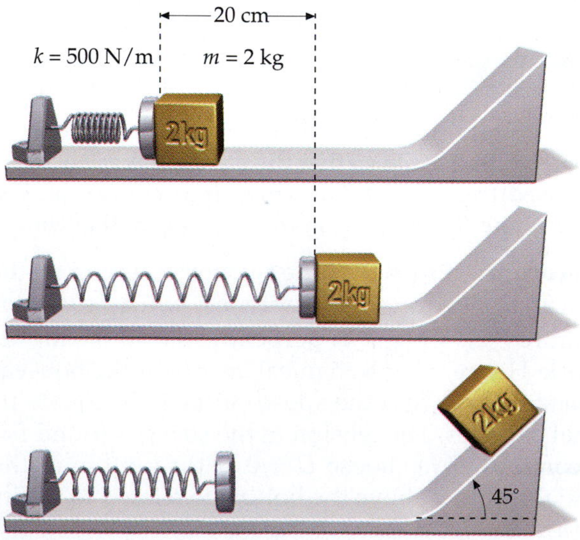

Cover the column to the right and try these on your own before looking at the answers.

Steps	**Answers**
1. Write the initial mechanical energy in terms of the compression distance x.	$E_i = \frac{1}{2}kx^2$
2. Write the final mechanical energy in terms of the height h.	$E_f = mgh$
3. Apply conservation of mechanical energy, and solve for h.	$mgh = \frac{1}{2}kx^2$ $h = \frac{kx^2}{2mg} = 0.51\text{ m}$
4. Find the distance s from $h = s\sin\theta$.	$s = 0.721\text{ m}$

Remark In this problem, the initial potential energy in the spring is converted first into kinetic energy and then into gravitational potential energy.

Exercise Find the speed of the block just after it leaves the spring. (*Answer* 3.16 m/s)

Example 7-4

A spring with a force constant of k hangs vertically. A block of mass m is attached to the unstretched spring and allowed to fall from rest. Find an expression for the maximum distance the block falls before it begins moving upward.

Picture the Problem As the block drops, its speed first increases, then reaches some maximum value, and then decreases until it is again zero when the block is at its lowest point (Figure 7-4). Only conservative forces are present, so we apply the conservation of mechanical energy to the earth–spring–block system. The initial and final positions of the block are shown. Choose the gravitational potential energy of the block to be zero at the original position $y = 0$. The initial potential energy of the spring is zero because the spring is unstretched at this position. Since the block is at rest at this point, the total mechanical energy is zero. Let d be the distance the block falls.

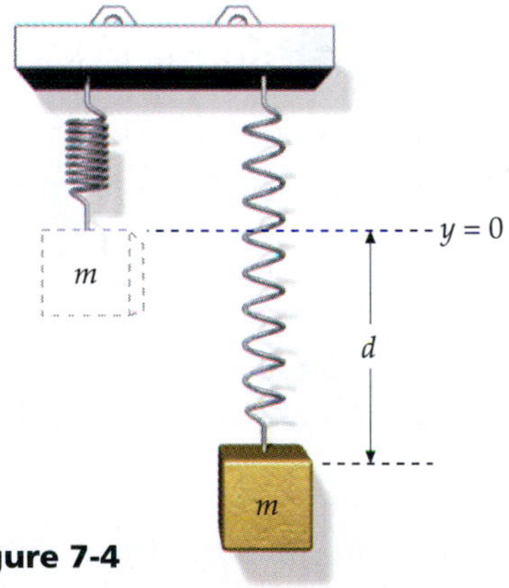

Figure 7-4

1. At a general point y, the total mechanical energy is the sum of the kinetic energy K, the gravitational potential energy U_g, which is equal to $-mgy$, and the spring potential energy U_s, which is equal to $\frac{1}{2}ky^2$:

$$E = K + U_g + U_s = \frac{1}{2}mv^2 - mgy + \frac{1}{2}ky^2$$

2. Apply conservation of mechanical energy:

$$E = E_i = 0$$

$$\frac{1}{2}mv^2 - mgy + \frac{1}{2}ky^2 = 0$$

3. Set $v = 0$ and solve for y. There are two solutions. One gives the initial position and the other is the one we want:

$$0 - mgy + \frac{1}{2}ky^2 = 0$$

$$y = 0 \quad \text{(initial position)}$$

$$y = d = \frac{2mg}{k}$$

Remarks Gravitational potential energy is converted into the kinetic energy of the block plus the potential energy of the spring. At the lowest point, where the block is momentarily at rest, the gain in potential energy of the spring equals the loss in gravitational potential energy of the block. In bungee jumping from a bridge, you are attached to an elastic cord of length L less than the height of the bridge. When you have dropped a distance L, the cord is unstretched and you have kinetic energy equal to mgL. You therefore fall farther than $2mg/k$ (assuming the cord acts like a spring of constant k) before you come back up. If mechanical energy were perfectly conserved, you would come up and strike the bridge from which you had leapt. Instead, some mechanical energy is dissipated because of nonelastic properties of the cord.

Example 7-5 *try it yourself*

Two blocks are attached to a light string that passes over a massless, frictionless pulley. The two blocks have masses m_1 and m_2 and are initially at rest. Find the speed of either block when the heavier one falls a distance h.

Figure 7-5

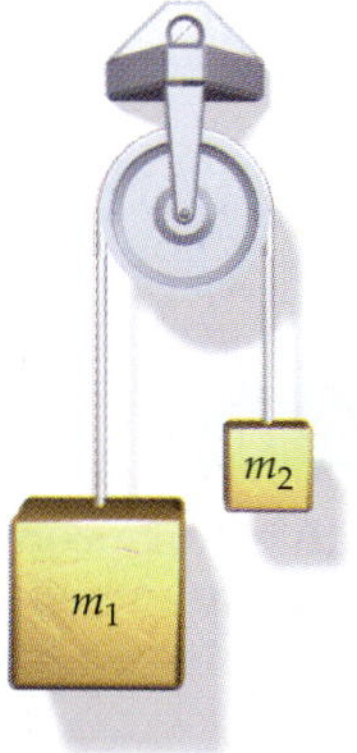

Picture the Problem Mechanical energy is conserved. The net work done by the string tension is zero; it does positive work lifting the lighter object and an equal amount of negative work as the heavier object moves downward the same distance in the opposite direction (Figure 7-5). If we choose $U_i = 0$ when the blocks are at rest, the total energy is zero. Because the string does not stretch, both blocks move with the same speed v.

Cover the column to the right and try these on your own before looking at the answers.

Steps	***Answers***
1. Write the total kinetic energy of the system when the blocks are moving with speed v.	$K = \frac{1}{2}m_1v^2 + \frac{1}{2}m_2v^2$
2. Write the total potential energy of the system when m_1 has moved down a distance h and m_2 has moved up the same distance.	$U = -m_1gh + m_2gh$
3. Add U and K to obtain the total energy E.	$E = K + U = \frac{1}{2}m_1v^2 + \frac{1}{2}m_2v^2 - m_1gh + m_2gh$
4. Apply conservation of mechanical energy.	$E = E_i = 0$ $\frac{1}{2}m_1v^2 + \frac{1}{2}m_2v^2 - m_1gh + m_2gh = 0$
5. Solve for v.	$v = \sqrt{\frac{2(m_1 - m_2)}{m_1 + m_2}gh}$

Remarks This device, called an *Atwood's machine*, is analyzed in terms of forces in Problems 81–85 in Chapter 4. If a pulley has mass, it has kinetic energy when it rotates. Our pulley is massless, so we can neglect its energy of rotation. We consider the more complicated problem of a pulley with mass in Chapter 9.

Since all the forces are constant, the acceleration of the blocks is constant. From the constant-acceleration equation $v^2 = a\,\Delta x$, we see that the acceleration is given by $a = [(m_1 - m_2)/(m_1 + m_2)]g$, so $g = [(m_1 + m_2)/(m_1 - m_2)]a$. If m_1 and m_2 are not too different, the acceleration of either object is a small fraction of g. It was easily measured with the rather crude timing devices available in the eighteenth century, whereas a direct measurement of g was difficult if not impossible.

Exercise What is the magnitude of the acceleration of either block if the masses are $m_1 = 3$ kg and $m_2 = 5$ kg? (*Answer* $a = 0.25g = 2.45\text{ m/s}^2$)

We've seen that the law of conservation of mechanical energy can be used as an alternative to Newton's laws for solving certain problems in mechanics. When we are not interested in the time t, the conservation of mechanical energy is often much easier to use than Newton's second law (Figure 7-6). Since the conservation of mechanical energy was derived from Newton's laws, any problem that can be solved using it can also be solved directly from Newton's laws, though often with much more difficulty.

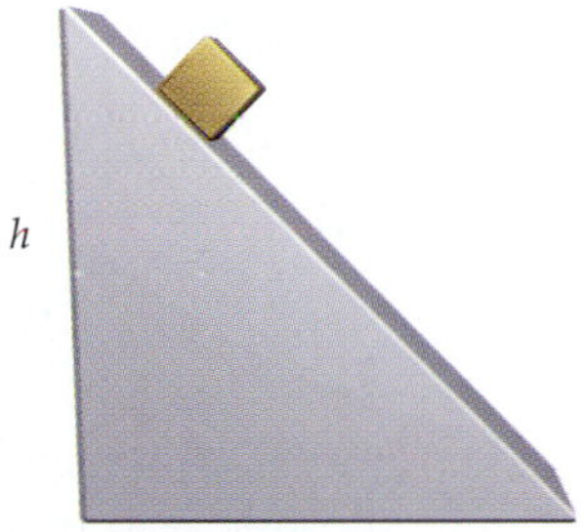

(*a*)

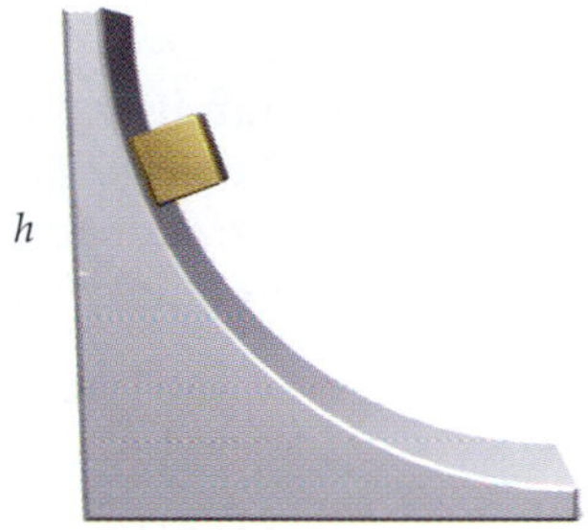

(*b*)

Figure 7-6 (*a*) One can easily find the speed of a block sliding down a frictionless incline of constant slope by applying Newton's second law or by using conservation of mechanical energy. However, if the incline is frictionless but not of constant slope, as in (*b*), the problem can still be solved easily using conservation of mechanical energy, whereas it can be solved using Newton's second law only if the slope of the incline is known at each point, and then the calculation is quite tedious.

7-2 The Conservation of Energy

In the macroscopic world, nonconservative forces are always present to some extent, the most common being frictional forces, which decrease the mechanical energy of a system. However, the decrease in mechanical energy is found to be equal to the increase in thermal energy produced by the frictional forces. Another type of nonconservative force is that involved in the deformations of objects. When you bend a coat hanger back and forth, you do work on the coat hanger, but the work you do does not appear as mechanical energy. Instead, the coat hanger becomes warm. The work done in deforming the hanger is dissipated as thermal energy. Similarly, when a ball of putty is dropped to the floor, it warms as it deforms on impact, and the original potential energy appears as thermal energy. If thermal energy is added to mechanical energy, the total energy is conserved even when there are frictional forces or forces of deformation.

A third type of nonconservative force is associated with chemical reactions. When we include systems in which chemical reactions take place, the sum of mechanical energy plus thermal energy is not conserved. For example, suppose that you begin running from rest. Originally you have no kinetic energy. When you begin to run, internal chemical energy in your muscle is converted to kinetic energy of your body, and thermal energy is produced. It is possible to identify and measure the chemical energy that is used. In this case, the sum of mechanical, thermal, and chemical energy is conserved.

Even when thermal energy and chemical energy are included, the total energy of the system does not always remain constant. The energy of a system can change because of some form of radiation, such as sound waves or electromagnetic waves. However, *the increase or decrease in the total energy of a system can always be accounted for by the appearance or disappearance of energy somewhere else.* This experimental result is known as the **law of conservation of energy**. Let E_{sys} be the total energy of a given system, E_{in} be the energy that enters the system, and E_{out} be the energy that leaves the system. The law of conservation of energy then states:

$$E_{in} - E_{out} = \Delta E_{sys} \qquad 7\text{-}7$$

Law of conservation of energy

Alternatively,

> The total energy of the universe is constant. Energy can be converted from one form to another, or transmitted from one region to another, but energy can never be created or destroyed.

Law of conservation of energy

The total energy E of many systems familiar from everyday life can be accounted for completely by mechanical energy E_{mech}, thermal energy E_{therm}, and chemical energy E_{chem}. To be comprehensive and include other possible forms of energy, such as electromagnetic or nuclear energy, we include E_{other}, and write generally

$$E_{sys} = E_{mech} + E_{therm} + E_{chem} + E_{other} \qquad 7\text{-}8$$

The Work–Energy Theorem

A common way to transfer energy into or out of a system is to do work on the system from the outside. If this is the only source of energy transferred,* the law of conservation of energy becomes

$$W_{ext} = \Delta E_{sys} \qquad 7\text{-}9$$

Work–energy theorem

where W_{ext} is the work done on the system by external forces, and ΔE_{sys} is the change in the system's total energy. This work–energy theorem for systems, which we will call simply the work–energy theorem, is a powerful tool for studying a wide variety of systems. Note that if the system is just a single particle, its energy can only be kinetic, so Equation 7-9 is equivalent to the work–kinetic energy theorem studied in Chapter 6.

*Energy can also be transferred when heat is exchanged between a system and its surroundings. Exchanges of heat energy, which occur when there is a temperature difference between a system and its surroundings, are discussed in Chapter 19.

Example 7-6

A ball of putty of mass m is released from rest from a height h and falls to the hard floor (Plop!). Discuss the application of the law of conservation of energy to (*a*) the system consisting of the ball alone, and (*b*) the system consisting of the earth and ball.

Picture the Problem Two forces act on the ball: gravity and the force of the floor. Since the floor does not move, the force it exerts does no work. There are no chemical or other energy changes, so we can neglect E_{chem} and E_{other}. If we neglect the sound energy radiated when the ball hits the floor, the only energy transferred to or from the ball is the work done by gravity, so we can use the work–energy theorem.

(*a*)1. Write the work–energy theorem: $W_{\text{ext}} = \Delta E_{\text{sys}} = \Delta E_{\text{mech}} + \Delta E_{\text{therm}}$

2. The two external forces on the system are gravity and the force exerted by the floor. The floor does not move and therefore does no work. The only work done on the ball is by gravity: $W_{\text{ext}} = mgh$

3. Since the ball alone is our system, its mechanical energy is entirely kinetic, which is zero both initially and finally. Thus, the change in mechanical energy is zero: $\Delta E_{\text{mech}} = 0$

4. Substitute mgh for W_{ext} and 0 for ΔE_{mech} in step 1: $W_{\text{ext}} = \Delta E_{\text{therm}} = mgh$

(*b*)1. There are no external forces acting on this system (the force of gravity and the force of the floor are now internal to the system), so there is no external work done: $W_{\text{ext}} = 0$

2. Write the work–energy theorem with $W_{\text{ext}} = 0$:

$$W_{\text{ext}} = \Delta E_{\text{sys}} = \Delta E_{\text{mech}} + \Delta E_{\text{therm}} = 0$$

$$\Delta E_{\text{therm}} = -\Delta E_{\text{mech}}$$

3. The original mechanical energy of the ball–earth system is the original gravitational potential energy, and the final mechanical energy is zero:

$$E_{\text{i}} = mgh$$

$$E_{\text{f}} = 0$$

4. The change in mechanical energy of the ball–earth system is thus: $\Delta E_{\text{mech}} = E_{\text{f}} - E_{\text{i}} = 0 - mgh = -mgh$

5. The work–energy theorem thus gives the same result found in (*a*): $\Delta E_{\text{therm}} = -\Delta E_{\text{mech}} = mgh$

Remarks In (*a*), energy is transferred to the ball by the work done on it by gravity. This energy appears as the kinetic energy of the ball before it hits the floor and as thermal energy after. The ball warms slightly and the energy is eventually transferred to the surroundings as heat. In (*b*), the original potential energy of the ball–earth system is converted to kinetic energy of the ball just before it hits and then into thermal energy.

(a)

(b)

(c)

(a) In this power plant in Kansas, energy stored in the fossil fuel coal (the black mound at lower right) is released by burning the coal to produce steam; the steam is then used to drive turbines to produce electricity. The excess heat is dissipated by cooling towers. (b) The potential energy of the water at the top of Niagara Falls is used to produce electrical energy. (c) This wind farm at Altamont Pass in California uses hundreds of windmills to convert wind energy into electrical energy.

Problems Involving Kinetic Friction

Kinetic frictional forces exerted by one surface on another when the surfaces slide across each other decrease the total mechanical energy of a system and increase the thermal energy. Consider a block that begins with initial velocity v_i and slides on a rough table until it stops (Figure 7-7). We choose the block and table to be our system. Then $\Delta E_{chem} = \Delta E_{other} = 0$ and no external work is done on this system. The work–energy theorem gives

$$0 = \Delta E_{mech} + \Delta E_{therm}$$

The mechanical energy lost is the initial kinetic energy of the block

$$\Delta E_{mech} = -\frac{1}{2}mv_i^2 \qquad 7\text{-}10$$

We can relate the loss in mechanical energy to frictional force. If f is the magnitude of the frictional force, Newton's second law gives

$$-f = ma$$

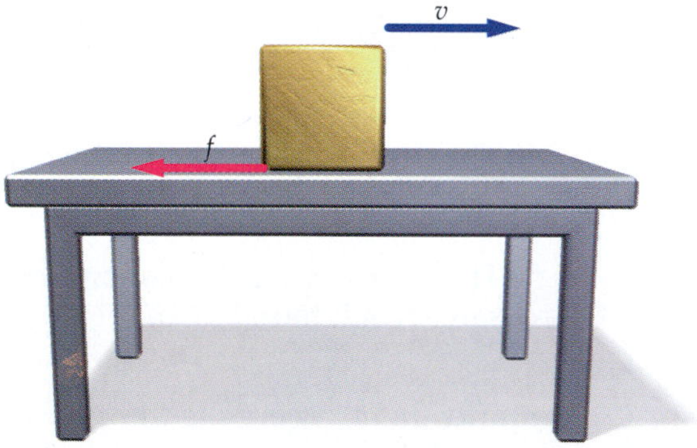

Figure 7-7 A block sliding on a rough table. The force of friction reduces the mechanical energy of the block–table system.

Multiplying both sides of this equation by Δs, we find

$$-f\,\Delta s = ma\,\Delta s = m\left(\frac{1}{2}v_f^2 - \frac{1}{2}v_i^2\right) = -\frac{1}{2}mv_i^2 \qquad \text{7-11}$$

where we have used the constant-acceleration formula $2a\,\Delta s = v_f^2 - v_i^2$, and $v_f = 0$. Comparing Equations 7-10 and 7-11 we find

$$f\,\Delta s = -\Delta E_{mech} \qquad \text{7-12}$$

Note that the quantity $-f\,\Delta s$ is *not* the work done by friction on the sliding block, because the displacement of the frictional force is not, in general, equal to the displacement of the block. However, it can be shown that $f\,\Delta s$ does equal the increase in thermal energy due to the dissipation of mechanical energy on the surfaces as they slide across one another.* Thus,

$$f\,\Delta s = \Delta E_{therm} \qquad \text{7-13}$$

Energy dissipated by friction

Substituting this result into the work–energy theorem (with $E_{chem} = E_{other} = 0$), we obtain

$$W_{ext} = \Delta E_{mech} + \Delta E_{therm} = \Delta E_{mech} + f\,\Delta s \qquad \text{7-14}$$

Work–energy theorem for problems with friction

When there is no external work done on the system, the energy dissipated by friction equals the decrease in mechanical energy:

$$\Delta E_{therm} = f\,\Delta s = -\Delta E_{mech} \qquad (W_{ext} = 0) \qquad \text{7-15}$$

*The work done by kinetic friction is examined in detail in "Work and Heat Transfer in the Presence of Sliding Friction" by B. A. Shewood and W. H. Bernard, *American Journal of Physics*, **52**, 1001 (1984).

Example 7-7

A horizontal force of 25 N is applied to a 4-kg block, which is initially at rest on a horizontal table. The coefficient of kinetic friction μ_k between the block and table is 0.35. Find (*a*) the external work done on the block–table system, (*b*) the energy dissipated by friction, (*c*) the kinetic energy of the block after it has been pushed 3 m, and (*d*) the speed of the block after it has been pushed 3 m.

Figure 7-8

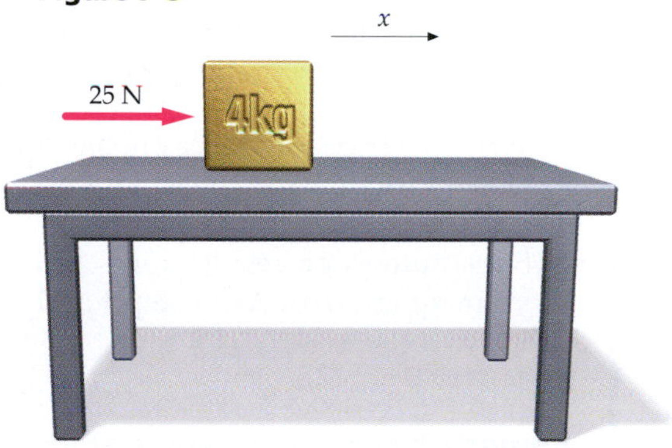

Picture the Problem We choose the block plus table as our system (Figure 7-8). The speed of the block is found from its final kinetic energy, which we find using the work–energy theorem with $\Delta E_{chem} = 0$ and $\Delta E_{therm} = f\,\Delta s$. The mechanical energy is increased by the external work and decreased by the energy dissipated by friction.

(*a*) The external work done is the product of the external force and the distance traveled:

$$W_{ext} = F_{ext}\,\Delta x = (25\text{ N})(3\text{ m}) = 75\text{ J}$$

(*b*) The energy dissipated by friction is $f\,\Delta x$:

$$\Delta E_{therm} = f\,\Delta x = \mu_k mg\,\Delta x = (0.35)(4\text{ kg})(9.81\text{ N/kg})(3\text{ m}) = 41.2\text{ J}$$

(*c*)1. Apply the work–energy theorem to find the final kinetic energy:

$$W_{ext} = \Delta E_{mech} + f\,\Delta x$$

2. Since the initial kinetic energy is zero and there is no change in the potential energy, the final kinetic energy equals the change in mechanical energy:

$$\Delta E_{mech} = \Delta K = K_f - K_i = K_f$$

3. Substitute this result into the work–energy theorem:

$$W_{ext} = K_f + f\,\Delta x$$
$$K_f = W_{ext} - f\,\Delta x = 75\text{ J} - 41.2\text{ J} = 33.8\text{ J}$$

(*d*)1. The final speed of the block is related to its kinetic energy:

$$K_f = \frac{1}{2}mv^2$$

2. Solve for the final speed of the block:

$$v = \sqrt{\frac{2K_f}{m}} = \sqrt{\frac{2(33.8\text{ J})}{4\text{ kg}}} = 4.11\text{ m/s}$$

Example 7-8 *try it yourself*

A 5-kg toboggan is sliding with an initial speed of 4 m/s. If the coefficient of friction between the toboggan and the snow is 0.14, how far will the toboggan go before coming to rest?

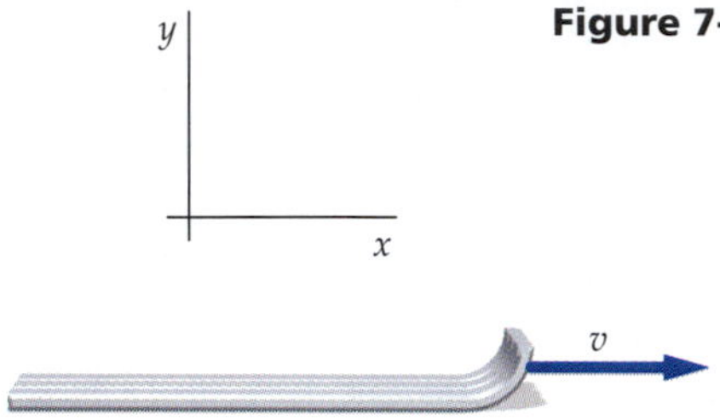

Figure 7-9

Picture the Problem We choose the toboggan and snow as our system (Figure 7-9). Then $W_{ext} = 0$, and the work–energy theorem implies that the energy dissipated by friction equals the change in mechanical energy.

Cover the column to the right and try these on your own before looking at the answers.

Steps	***Answers***
1. Write the work–energy theorem with $W_{ext} = 0$:	$W_{ext} = \Delta E_{mech} + f\,\Delta x = 0$
2. Solve for Δx.	$\Delta x = -\dfrac{\Delta E_{mech}}{f}$
3. Write the change in mechanical energy in terms of the initial speed of the toboggan.	$\Delta E_{mech} = -\dfrac{1}{2}mv^2$
4. Write the frictional force f in terms of the coefficient of friction and the weight of the toboggan.	$f = \mu_k mg$
5. Substitute your results for f and ΔE_{mech} into your equation for Δx in step 1 and calculate Δx.	$\Delta x = -\dfrac{-\frac{1}{2}mv^2}{\mu_k\,mg} = 5.82\text{ m}$

Remark Figure 7-10 shows stopping distance versus initial speed of the toboggan for three different values of the coefficient of kinetic friction.

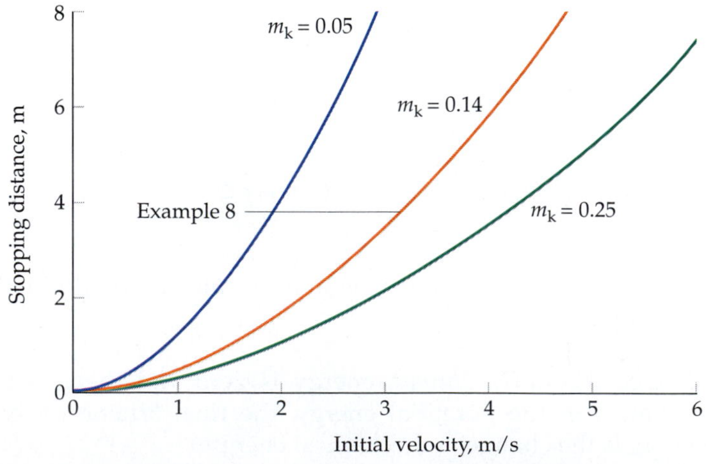

Figure 7-10

Example 7-9

A child of mass 40 kg goes down a rough slide inclined at 30°. The coefficient of kinetic friction between the child and the slide is $\mu_k = 0.2$. If the child starts from rest at the top of the slide, a height 4 m above the bottom, how fast is she traveling when she reaches the bottom?

Picture the Problem As the child slides down, some of her potential energy is converted into kinetic energy and some into thermal energy because of friction. We choose the child–slide–earth as our system (Figure 7-11). Then $W_{ext} = 0$, and the work–energy theorem implies that the energy dissipated by friction equals the change in mechanical energy. We choose $y = 0$ at the bottom of the slide so the final potential energy of the child is zero.

Figure 7-11

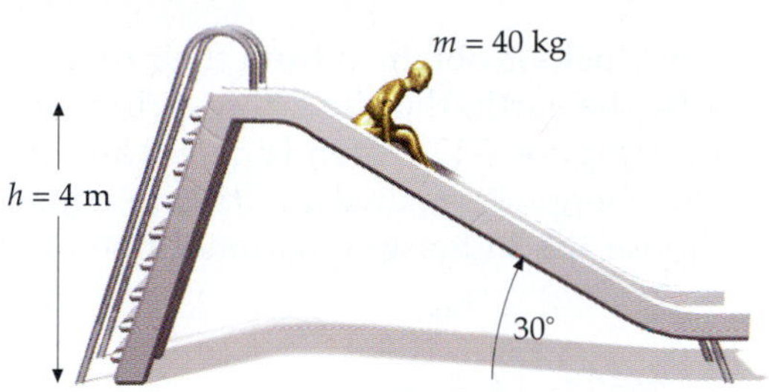

1. The speed at the bottom is related to the final kinetic energy: $v = \sqrt{\frac{2K_f}{m}}$

2. The final kinetic energy equals the initial potential energy minus the energy dissipated by friction: $K_f = mgh - f\,\Delta s$

3. The frictional force is related to the normal force F_n: $f = \mu_k F_n = \mu_k mg\cos 30°$

4. The total distance traveled is related to the height h: $\Delta s = \frac{h}{\sin 30°}$

5. Substitute these results to calculate the final kinetic energy:
$$K_f = mgh - f\,\Delta s = mgh - \mu_k mg\cos 30°\frac{h}{\sin 30°} = mgh(1 - \mu_k\cot 30°)$$

6. Use this final kinetic energy to find v:
$$v = \sqrt{\frac{2K_f}{m}} = \sqrt{2gh(1 - \mu_k\cot 30°)}$$
$$= \sqrt{2(9.81\text{ m/s}^2)(4\text{ m})[1 - 0.2\,(1.73)]} = 7.16\text{ m/s}$$

Remark The energy dissipated by friction decreases the total mechanical energy of the child–slide–earth system from mgh to its final value $mgh(1 - \cot 30°)$, which equals the kinetic energy at the bottom. Note that the result is independent of the mass of the child.

Exercise For the earth–child–slide system, calculate (*a*) the initial mechanical energy, (*b*) the final mechanical energy, and (*c*) the energy dissipated by friction. (*Answers* (*a*) 1570 J, (*b*) 1026 J, (*c*) 544 J)

Example 7-10 *try it yourself*

A 4-kg block hangs by a light string that passes over a massless, frictionless pulley and is connected to a 6-kg block that rests on a rough shelf. The coefficient of kinetic friction is $\mu_k = 0.2$. The 6-kg block is pushed against a spring to which it is not attached. The spring has a force constant of 180 N/m, and it is compressed 30 cm. Find the speed of the blocks after the spring is released and the 4-kg block has fallen a distance of 40 cm.

Picture the Problem The speed of the blocks is obtained from their final kinetic energy. Consider the system to be the earth, the shelf, the spring, and the two blocks $m_1 = 6$ kg and $m_2 = 4$ kg (Figure 7-12). Then $W_{ext} = 0$ and the work–energy theorem implies that the energy dissipated by friction equals the change in mechanical energy. Choose the initial gravitational potential energy to be zero.

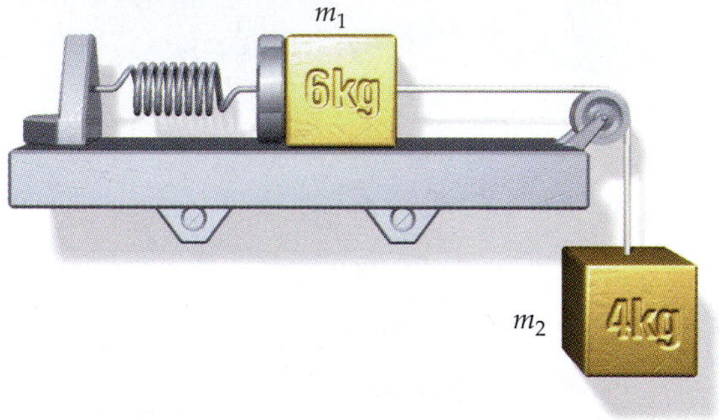

Figure 7-12

Cover the column to the right and try these on your own before looking at the answers.

Steps	*Answers*
1. Write expressions for the initial mechanical energy and the final mechanical energy when each block has moved a distance Δs.	$E_i = \frac{1}{2}kx^2$ $E_f = K_f + U_f = \frac{1}{2}(m_1 + m_2)v^2 - m_2 g\,\Delta s$
2. Write an expression for the energy dissipated by friction in terms of the coefficient of friction and Δs.	$f\,\Delta s = \mu_k m_1 g\,\Delta s$
3. Set the energy dissipated by friction equal to the loss in mechanical energy of the system.	$\mu_k m_1 g\,\Delta s = \frac{1}{2}kx^2 - \left[\frac{1}{2}(m_1 + m_2)v^2 - m_2 g\,\Delta s\right]$
4. Solve your equation for v^2 and substitute the numerical values.	$v^2 = \frac{kx^2 + 2m_2 g\,\Delta s - 2\mu_k m_1 g\,\Delta s}{m_1 + m_2} = 3.82\ \text{m}^2/\text{s}^2$
5. Solve for v.	$v = 1.95$ m/s

Remarks This solution assumes that the string remains taut at all times. This will be true if the acceleration of m_1 is less than g, that is, if the net force on m_1 is less than $m_1 g$. Initially, the force exerted by the spring on m_1 has the magnitude $k\,\Delta x_1 = (180\ \text{N/m})(0.3\ \text{m}) = 54$ N, which is less than $m_1 g = (6\ \text{kg})(9.81\ \text{N/kg}) = 58.9$ N. Since the spring force decreases as block m_1 moves forward, and the frictional force decreases the net force, the acceleration of the 6-kg block will always be less than g, and the string will remain taut.

Systems With Chemical Energy

Sometimes a system's internal chemical energy is converted into mechanical energy and thermal energy with no work being done by an outside agent. For example, to walk forward, you push back on the floor and the floor pushes forward on you with the force of static friction. This force accelerates you, but it does *not* do work. The displacement of the point of application of the force is zero (assuming your shoes do not slip on the floor), therefore no work is done and no energy is transferred from the floor to your body. The kinetic energy of your body comes from the conversion of chemical energy in your body derived from the food you eat. We consider a similar case in the next example.

This pizza contains about 16 megajoules of energy, approximately the same as the energy in a gallon (3.78 L) of gasoline.

Example 7-11

A man of mass m walks with a small constant speed up a flight of stairs to a height h. Discuss the application of energy conservation to the system consisting of the man alone.

Picture the Problem Two forces act on the man: gravity and the force of the stairs. Since the stairs do not move, they do no work. In this case we *cannot* neglect changes in chemical energy.

1. Write the work–energy theorem: $W_{ext} = \Delta E_{sys} = \Delta E_{mech} + \Delta E_{therm} + \Delta E_{chem}$

2. The only work done on the man is done by gravity. This work is negative because the force is in the opposite direction of the displacement: $W_{ext} = -mgh$

3. Since the man alone is our system, his mechanical energy is entirely kinetic, which is zero both initially and finally: $\Delta E_{mech} = 0$

4. Substitute these results into the work–energy theorem: $-mgh = \Delta E_{therm} + \Delta E_{chem}$

Remark If there were no change in thermal energy, the chemical energy of the man would decrease by mgh. Because the body is relatively inefficient, the amount of chemical energy converted in the man's body will be considerably greater than mgh. The amount of energy above mgh appears as thermal energy, which is eventually transferred from the man to his surroundings as heat.

Exercise Discuss the energy conservation for the system of man plus earth. (*Answer* For this system no external work is done, so the total energy, which now includes potential energy, is conserved. The change in mechanical energy is mgh, so the work–energy theorem gives $0 = mgh + \Delta E_{therm} + \Delta E_{chem}$.)

Example 7-12

A 1000-kg car travels at a constant speed of 100 km/h = 28 m/s = 62 mi/h up a 10% grade. (A 10% grade means that the road rises 1 m for each 10 m of horizontal distance—that is, the angle of inclination θ is given by $\tan\theta = 0.1$ [Figure 7-13].) What is the minimum power that must be delivered by the car's engine? (Neglect rolling friction and air drag.)

Picture the Problem The power delivered by the car's engine is the rate of decrease of its chemical energy. Some of it goes into increasing the potential energy of the car as it climbs the hill, and some goes into an increase in thermal energy, which is expelled as exhaust. For a 10% grade, $\tan\theta = 0.10$ is given, and $\sin\theta \approx \tan\theta$ because the angle is small (Figure 7-13). For the car–earth system, $W_{ext} = 0$, so the total energy is conserved.

Figure 7-13

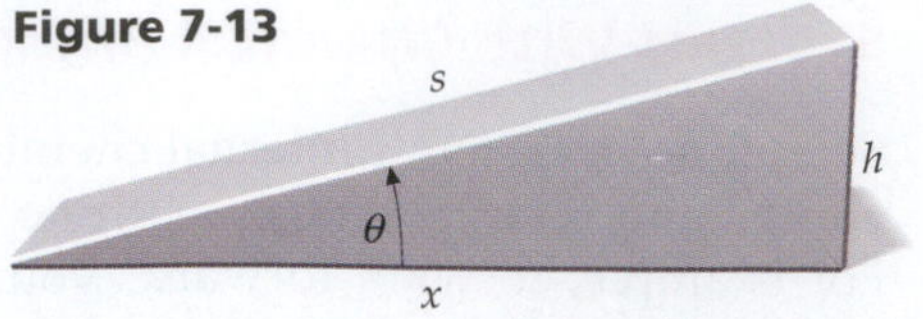

$\tan\theta = h/x \sim \sin\theta = h/s$

1. The power input by the engine is the rate of decrease of its chemical energy: $P = -\dfrac{dE_{chem}}{dt}$

2. The chemical energy change is found from the work–energy theorem:
$$W_{ext} = \Delta E_{mech} + \Delta E_{therm} + \Delta E_{chem} = 0$$
$$\Delta E_{chem} = -\Delta E_{mech} - \Delta E_{therm}$$

3. Convert the changes to time derivatives:
$$P = -\frac{dE_{chem}}{dt} = \frac{dE_{mech}}{dt} + \frac{dE_{therm}}{dt}$$

4. Since the speed $v = ds/dt$ is constant, the rate of change of the mechanical energy is just the rate of change of potential energy:
$$\frac{dE_{mech}}{dt} = \frac{dU}{dt} = \frac{d(mgh)}{dt} = mg\frac{dh}{dt}$$

5. From Figure 7-13 we can see that when the car travels a distance s along the road, it climbs a height h, which is related to s by:
$$h = s\sin\theta$$

6. We can use the approximation $\tan\theta \approx \sin\theta$ because the angle is small:
$$h = s\sin\theta \approx s\tan\theta = 0.1s$$

7. We can now relate the rate of change of mechanical energy to the speed:
$$\frac{dE_{mech}}{dt} = mg\frac{dh}{dt} = 0.1\,mg\frac{ds}{dt} = 0.1\,mgv$$

8. Substitute these results into the equation for power in step 3:
$$P = \frac{dE_{mech}}{dt} + \frac{dE_{therm}}{dt}$$
$$= 0.1mgv + \frac{dE_{therm}}{dt}$$
$$= (0.1)(1000\text{ kg})(9.81\text{ N/kg})(28\text{ m/s}) + \frac{dE_{therm}}{dt}$$
$$= 27.5\text{ kW} + \frac{dE_{therm}}{dt}$$

9. The minimum power occurs when $dE_{therm}/dt = 0$: $P_{min} = 27.5\text{ kW}$

Remarks The actual power needed by a car is considerably greater than our result because cars are typically only about 15% efficient. About 85% of the power generated by a car's engine goes to internal thermal energy that is expelled as heat exhaust plus thermal energy created by rolling friction and wind resistance.

Remark Figure 7-14 shows the speed of a car versus power for various incline angles. The speed is larger at a fixed power for smaller inclines. The blue curve shows the approximate effect of a term in the required power proportional to v (from rolling friction) and v^3 (from air drag).

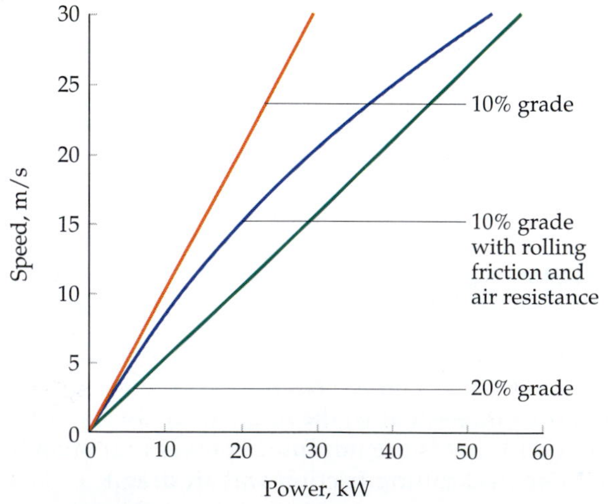

Figure 7-14

exploring

Transducers

Devices that convert one form of energy to another are called transducers. Those shown here convert nonelectrical energy to electrical energy.

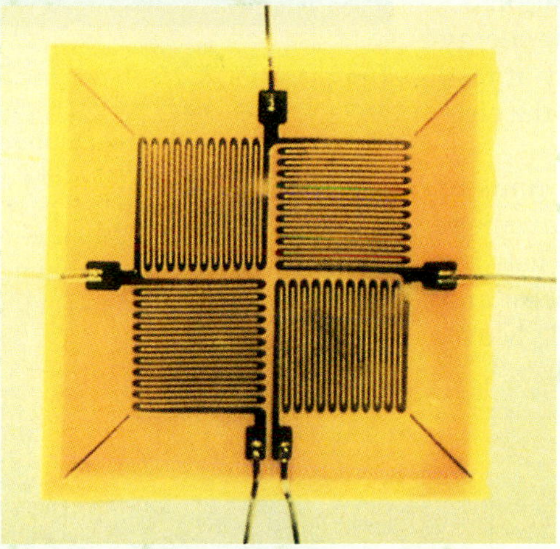

The strain gauge shown above consists of a grid of very fine wires or foils of a substance such as carbon that changes its electrical resistance when mechanically stressed. The wire is bonded to a thin insulating backing, which is attached by adhesive to an object. Stresses that distort the object deform the attached strain gauge as well. The degree of deformation is measured by the change in resistance of the gauge. If a fixed voltage is applied across the ends of the gauge wire, a varying resistance will produce variations in the current.

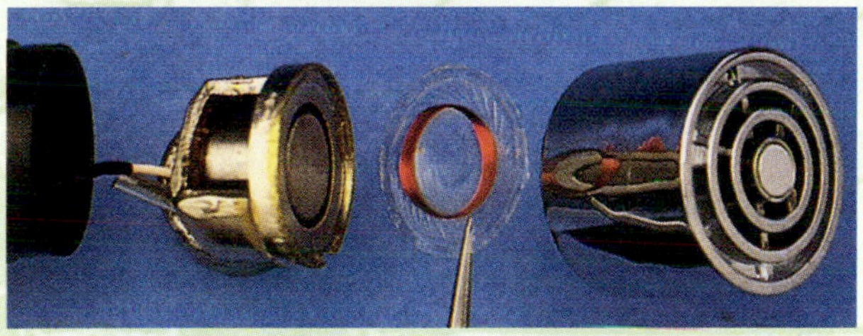

A microphone converts sound energy to electrical energy. In the kind shown here, a copper ring is attached to a thin plastic membrane. Sound waves hitting the membrane cause it and the ring to vibrate. The ring is mounted in the field of a permanent magnet. Motion back and forth across the magnetic field, caused by the vibration, induces an alternating current in the ring. This current causes a secondary alternating magnetic field to arise, which in turn creates a secondary alternating current, this time in a wire coil (connected to the output leads) positioned behind the ring. These processes are discussed more fully in Chapters 28 (Section 28-2), 29 (Section 29-2), and 30 (Section 30-7).

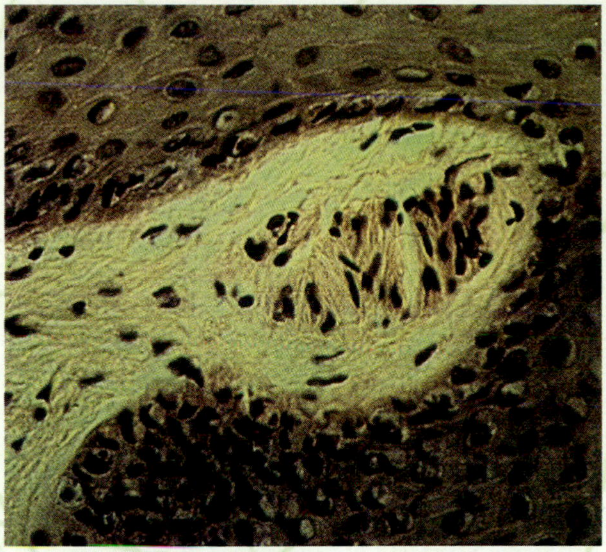

Our sense of touch arises from encapsulated nerve fibers called Meissner's corpuscles, shown above, that are located directly under the outer layer of skin. When skin overlying them is touched, the corpuscles are deformed, triggering electrical impulses in the nerve fibers. A stronger touch produces greater deformation and increases the frequency of impulses. The system is in some respects a biological counterpart to a strain gauge; in both transducers, the energy of mechanical stress is used to modulate changes in electrical conduction.

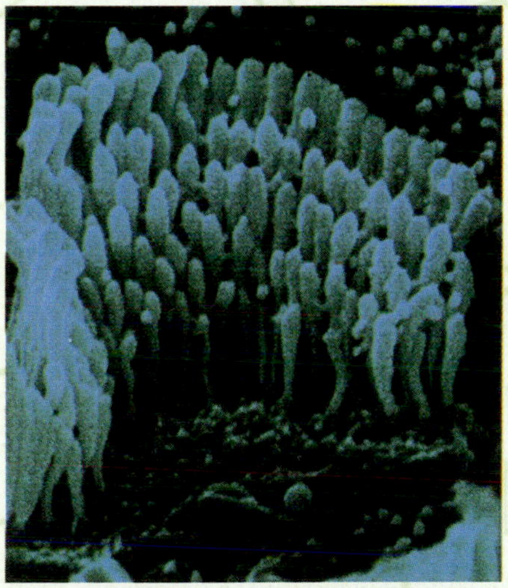

Sound waves transmitted to the spiral chamber of the inner ear cause the fluid there to vibrate. Sensory receptor cells (above, magnified 3500 times) are attached to the walls of the chamber. The receptor cells, stimulated by the vibrating fluid, cause neurons connected to their base to transmit electrical impulses. The impulses, traveling along a chain of neurons, eventually register in the brain as the sensation "sound." This system for converting sound energy to electical energy is a biological counterpart to a microphone.

7-3 Mass and Energy

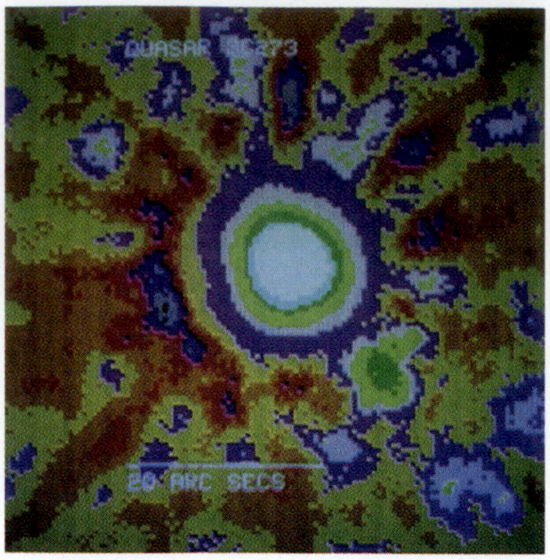

The quasar 3C 273 is shown imaged via X-ray energy. The X-ray energy emitted by this quasar is more than a million times that emitted by the entire Milky Way galaxy. The mechanism that generates this enormous energy is not known. It is conjectured to be caused by the annihilation of vast amounts of matter and antimatter.

In 1905, Albert Einstein published his special theory of relativity, a result of which is the famous equation

$$E_0 = mc^2 \qquad \text{7-16}$$

where $c = 3 \times 10^8$ m/s is the speed of light in vacuum. We will study this theory in some detail in Chapter 39.

According to Equation 7-16, a particle or system of mass m has "rest" energy mc^2. This energy is intrinsic to the particle. Consider the positron, a particle emitted in a nuclear process called beta decay. Positrons and electrons have identical masses, but equal and opposite electrical charge. When a positron encounters an electron in matter, electron–positron annihilation occurs, a process in which the electron and positron disappear and their energy appears as electromagnetic radiation. If the two particles are initially at rest, the energy of the electromagnetic radiation equals the rest energy of the electron plus that of the positron.

Energies in atomic and nuclear physics are usually expressed in units of electron volts (eV) or mega-electron-volts (1 MeV = 10^6 eV). A convenient unit for the masses of atomic particles is eV/c^2 or MeV/c^2. Table 7-1 lists the rest energies (and therefore the masses) of some elementary particles and light nuclei. The total rest energy of a positron plus electron is 2(0.511 MeV), which is the radiation energy emitted upon annihilation.

Table 7-1

Rest Energies of Some Elementary Particles and Light Nuclei

Particle		Rest Energy (MeV)
Electron	e^-	0.5110
Positron	e^+	0.5110
Proton	p	938.280
Neutron	n	939.573
Deuteron	d	1875.628
Triton	t	2808.944
Helium-3	^{3}He	2808.41
Alpha particle	α	3727.409

The rest energy of a *system* can consist of the potential energy of the system, or other internal energies of the system in addition to the intrinsic rest energies of the particles in the system. If the system at rest absorbs energy, ΔE, its rest energy increases, and its mass increases by

$$\Delta M = \frac{\Delta E}{c^2} \qquad \text{7-17}$$

Consider two 1-kg blocks connected by a spring of force constant k. If we stretch the spring a distance A, the potential energy of the system increases by $\Delta U = \frac{1}{2}kA^2$. According to Equation 7-17, the mass of the system has also increased by $\Delta M = \Delta U/c^2$. Because c is such a large number, this increase in mass cannot be observed in macroscopic systems. For example, suppose $k = 800$ N/m, and $A = 10$ cm $= 0.1$ m. The potential energy of the spring system is then $\frac{1}{2}kA^2 = \frac{1}{2}(800\text{ N/m})(0.1\text{ m})^2 = 4$ J. The increase in mass of the system is

$$\Delta M = \frac{\Delta U}{c^2} = \frac{4\text{ J}}{(3 \times 10^8\text{ m/s})^2} = 4.44 \times 10^{-17}\text{ kg}$$

The relative mass increase $\Delta M/M \approx 2 \times 10^{-17}$ is much too small to be observed.

Nuclear Energy

In nuclear reactions, the energy changes are often an appreciable fraction of the rest energy of the system. Consider the deuteron, which is the nucleus of deuterium, an isotope of hydrogen called heavy hydrogen. The deuteron consists of a proton and neutron bound together. From Table 7-1, we see that the mass of the proton is 938.28 MeV/c^2 and the mass of the neutron is 939.57 MeV/c^2. The sum of these two masses is 1877.85 MeV/c^2. But the mass of the deuteron is 1875.63 MeV/c^2, which is less than the sum of the masses of the proton and neutron by 2.22 MeV/c^2. Note that this mass difference is about 0.12%, much greater than the uncertainties inherent in the measurement of these masses, and very much greater than the unobservable 10^{-17} relative mass increase discussed above for a macroscopic system. So where do we find the missing mass of 2.22 MeV/c^2?

Deuterons can be produced by letting neutrons from a reactor collide with protons. When a neutron is captured to form a deuteron, 2.22 MeV of energy is released, usually in the form of electromagnetic radiation. Thus, the mass of the proton–neutron system decreases by 2.22 MeV/c^2 when the particles combine to form a deuteron. Similarly, to break up a deuteron into its constituent parts, a proton plus a neutron, 2.22 MeV of energy must be put into the system. The energy needed to break up a nucleus into its constituent parts is called the **binding energy** of the nucleus. Deuterons can be broken up by bombarding them with energetic particles or electromagnetic radiation possessing energy of at least 2.22 MeV. If the energy is greater, 2.22 MeV is converted to the excess mass of the neutron and proton over that of the deuteron, and the rest appears as kinetic energy of the outgoing particles.

The deuteron is an example of a bound system. Its rest energy is less than the rest energy of its parts, so energy must be put into the system to break it apart. If the rest energy of a system is greater than the rest energy of its parts, the system is unbound. An example is uranium-236, which breaks apart or **fissions** into two smaller nuclei.* The sum of the masses of the resultant parts is less than the mass of the original nucleus. Thus the mass of the system decreases, and energy is released.

In nuclear fusion, two very light nuclei, such as a deuteron and a triton (nucleus of the hydrogen isotope tritium), fuse together. The mass of the resultant nucleus is less than that of the original parts and again energy is released. In a chemical reaction that produces energy, such as coal burning, the mass decrease is of the order of 1 eV/c^2 per atom. This is more than a million times smaller than the mass changes in nuclear reactions, and is not readily observable.

* Uranium-236, written ^{236}U, is made in a nuclear reactor when the stable isotope uranium-235 absorbs a neutron.

Example 7-13

A hydrogen atom consisting of a proton and an electron has a binding energy of 13.6 eV. By what percentage is the mass of the proton plus the electron greater than that of the hydrogen atom?

1. The percentage difference between the mass of the hydrogen atom and the masses of its parts is the binding energy E_b divided by $m_e + m_p$:

$$\% \text{ difference} = \frac{E_b}{m_e + m_p} = \frac{13.6 \text{ eV}/c^2}{m_e + m_p}$$

2. Obtain the rest masses of the proton and electron from Table 7-1:

$$m_p = 938.28 \text{ MeV}/c^2, \qquad m_e = 0.511 \text{ MeV}/c^2$$

3. Add to find the sum of these masses:

$$m_p + m_e = 938.79 \text{ MeV}/c^2$$

4. The rest mass of the hydrogen atom is less than this by 13.6 eV$/c^2$. The percentage difference is:

$$\% \text{ difference} = \frac{13.6 \text{ eV}/c^2}{938.79 \times 10^6 \text{ eV}/c^2} = 1.45 \times 10^{-8}$$

$$= 1.45 \times 10^{-6}\ \%$$

Remark This mass difference is too small to be measured directly. However, binding energies can be accurately measured, so the mass difference can be found from $E_b = \Delta m/c^2$.

Example 7-14 *try it yourself*

In a typical nuclear fusion reaction, a tritium nucleus (^{3}H) and a deuterium nucleus (^{2}H) fuse together to form a helium nucleus (^{4}He) plus a neutron (Figure 7-15). The reaction is written $^2\text{H} + {^3\text{H}} \rightarrow {^4\text{He}} + \text{n}$. If the initial kinetic energy of the particles is negligible, how much energy is released in this fusion reaction?

Figure 7-15

Picture the Problem Since energy is released, the total rest energy of the initial particles must be greater than that of the final particles. This difference equals the energy released.

Cover the column to the right and try these on your own before looking at the answers.

Steps	***Answers***
1. Write down the rest energies of ^{2}H and ^{3}H from Table 7-1 and add to find the total initial rest energy.	$E_0\,(\text{initial}) = 1875.628 \text{ MeV} + 2808.944 \text{ MeV} = 4684.572 \text{ MeV}$
2. Do the same for ^{4}He and n to find the final rest energy.	$E_0\,(\text{final}) = 3727.409 \text{ MeV} + 939.573 \text{ MeV} = 4666.982 \text{ MeV}$
3. Find the energy released from $E_{\text{released}} = E_0\,(\text{initial}) - E_0\,(\text{final})$.	$E_{\text{released}} = 4684.572 \text{ MeV} - 4666.982 \text{ MeV} = 17.59 \text{ MeV} \approx 17.6 \text{ MeV}$

Remarks This and other fusion reactions occur in the sun. The energy that is released bathes the earth and is ultimately responsible for all life on the planet. The energy constantly pouring out from the sun is matched by a continuous decrease in the sun's rest mass.

Newtonian Mechanics and Relativity

When the speed of a particle approaches the speed of light, Newton's second law breaks down, and we must modify Newtonian mechanics according to Einstein's theory of relativity.* The criterion for the validity of Newtonian mechanics can also be stated in terms of the energy of a particle. In Newtonian mechanics, the kinetic energy of a particle moving with speed v is

$$K = \frac{1}{2}mv^2 = \frac{1}{2}mc^2\frac{v^2}{c^2} = \frac{1}{2}E_0\frac{v^2}{c^2}$$

where $E_0 = mc^2$ is the rest energy of the particle. Then

$$\frac{v}{c} = \sqrt{\frac{2K}{E_0}}$$

Newtonian mechanics is valid if the speed of the particle is much less than the speed of light, or, alternatively, if the kinetic energy of the particle is much less than its rest energy.

7-4 Quantization of Energy

When energy is put into a system that remains at rest, the internal energy of the system increases. It would seem that we could choose to put any amount of energy into a system. However, this is not true for microscopic systems such as atoms or molecules. The internal energy of a microscopic system can increase only by discrete increments.

If we have two blocks attached to a spring and we pull the blocks apart, we do work on the block–spring system, and its potential energy increases. If we then release the blocks, they oscillate back and forth. The energy of oscillation E, which is the kinetic energy of motion of the blocks plus the potential energy due to the stretching of the spring, equals the original potential energy. In time, the energy of the system decreases because of various damping effects such as friction and air resistance. As close as we can measure, the energy decreases continuously. Eventually all the energy is dissipated and the energy of oscillation is zero.

Now consider a diatomic molecule such as molecular oxygen, O_2. The force of attraction between the two oxygen atoms varies approximately linearly with the change in separation (for small changes) much like that of a spring. If a diatomic molecule is set oscillating with some energy E, the energy decreases with time as the molecule radiates, or interacts with its surroundings, but the decrease is *not continuous*. The energy decreases in finite steps, and the lowest energy state, called the ground state, is not zero. The vibrational energy of a diatomic molecule is said to be **quantized;** that is, the molecule can possess energies only in certain amounts, known as quanta.

When blocks on a spring or diatomic molecules oscillate, the time for one oscillation is called the period T. The reciprocal of the period is the frequency of oscillation $f = 1/T$. We will see in Chapter 12 that the period and frequency of an oscillator do not depend on the energy of oscillation. As the en-

* Einstein published two very different theories of relativity. His special theory of relativity, which applies to our discussion here, was published in 1905 and applies to particles moving at speeds near the speed of light. Einstein's general theory of relativity, published in 1916, deals with gravity.

ergy decreases, the frequency remains the same. Figure 7-16 shows an **energy-level diagram** for an oscillator. The allowed energies are approximately equally spaced, and are given by*

$$E_n = \left(n + \frac{1}{2}\right)hf, \qquad n = 0, 1, 2, 3, \ldots \qquad \text{7-18}$$

where f is the frequency of oscillation and h is a fundamental constant of nature called Planck's constant†:

$$h = 6.626 \times 10^{-34}\ \text{J·s} \qquad \text{7-19}$$

The integer n in Equation 7-18 is called a **quantum number.** The lowest possible energy is the **ground-state energy** $E_0 = \frac{1}{2}hf$.

Microscopic systems often gain or lose energy by absorbing or emitting electromagnetic radiation. By conservation of energy, if E_i and E_f are the initial and final energies of a system, the energy of the radiation emitted or absorbed is

$$E_{rad} = E_i - E_f$$

Since the system energies E_i and E_f are quantized, the radiated energy is also quantized. Historically, the quantization of electromagnetic radiation, as proposed by Max Planck and Albert Einstein, was the first "discovery" of energy quantization. The quantum of radiation energy is called a **photon**. The energy of a photon is given by

$$E_{photon} = hf \qquad \text{7-20}$$

where f is the frequency of the electromagnetic radiation. Electromagnetic radiation includes light, microwaves, radio waves, television waves, X rays, and gamma rays. These differ from one another only in their range of frequencies and thereby in the range in energy of their photons.

As far as we know, all systems exhibit energy quantization. For macroscopic systems, the steps between energy levels are so small as to be unobservable. For example, typical oscillation frequencies for two blocks on a spring are 1 to 10 times per second. If $f = 10$ oscillations per second, the spacing between allowed levels is $hf = (6.626 \times 10^{-34}\ \text{J·s}) \times (10/\text{s}) \approx 6 \times 10^{-33}$ J. Since the energy of a macroscopic system is of the order of joules, a quantum step of 10^{-33} J is too small to be noticed. To put it another way, if the energy of a system is 1 J, the value of n is of the order of 10^{32} and changes of one or two quantum units will not be observable.

For a diatomic molecule, a typical frequency of vibration is 10^{14} vibrations per second, and a typical energy is 10^{-19} J. The spacing between allowed levels is then $E_{n+1} - E_n = hf \approx (6.63 \times 10^{-34}\ \text{J·s})(10^{14}\ \text{s}) \approx 6 \times 10^{-20}$ J. Thus, changes in the energy of oscillation are on the same order as the energy of the molecule, and quantization is definitely noticeable.

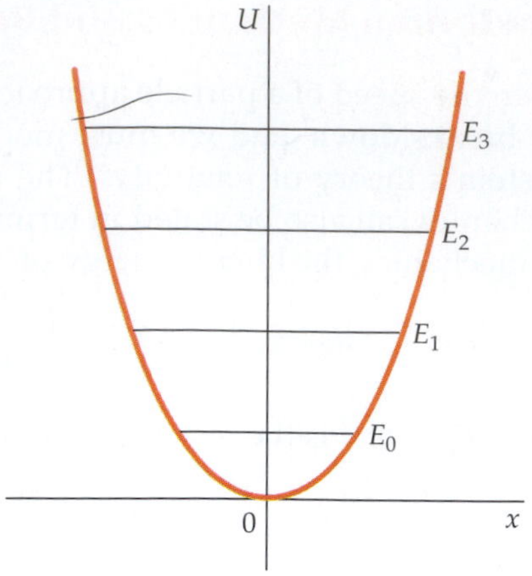

Figure 7-16 Energy-level diagram for an oscillator.

* A diatomic molecule can also have rotational energy. The rotational energy is quantized, as expected, but the energy levels are not equally spaced, and the lowest possible energy is zero. We study rotational energy in Chapters 9 and 10.

† In 1900 the German physicist Max Planck had introduced this constant as a calculational device to explain discrepancies between theory and experiment on the spectrum of blackbody radiation. The significance of Planck's constant was not appreciated by Planck or anyone else until Einstein postulated in 1905 that the energy of electromagnetic radiation is not continuous, but occurs in packets of size hf, where f is the frequency of the radiation and h is the constant discovered by Planck.

Summary

1. The conservation of mechanical energy is an important relation derived from Newton's laws for conservative forces. It is useful in solving many problems.
2. The work–energy theorem and the conservation of energy are fundamental laws of nature that have applications in all areas of physics.
3. Einstein's equation $E_0 = mc^2$ is a fundamental relation between mass and energy.
4. Quantization is a fundamental property of the energy in bound systems.

Topic	Remarks and Relevant Equations	
1. Mechanical Energy	The sum of the kinetic and potential energy of a system is called the total mechanical energy	
	$E_{mech} = K + U$	7-4
Conservation of mechanical energy	If no external forces do work on a system, and the internal forces are all conservative, the total mechanical energy of the system remains constant	
	$E_{mech} = K + U = \text{constant}$	7-5
	$K_f + U_f = K_i + U_i$	7-6
2. Total Energy of a System	The energy of a system consists of mechanical energy E_{mech}, thermal energy E_{therm}, chemical energy E_{chem}, and other types of energy E_{other}, such as sound radiation and electromagnetic radiation.	
	$E_{sys} = E_{mech} + E_{therm} + E_{chem} + E_{other}$	7-8
3. Conservation of Energy		
Universe	The total energy of the universe is constant. Energy can be converted from one form to another, or transmitted from one region to another, but energy can never be created or destroyed.	
System	The energy of a system can be changed by various means such as work done on the system, heat transfer, and emission or absorption of radiation. The increase or decrease in the energy of the system can always be accounted for by the appearance or disappearance of some kind of energy somewhere else.	
	$E_{in} - E_{out} = \Delta E_{sys}$	7-7
Work–energy theorem	$W_{ext} = \Delta E_{sys} = \Delta E_{mech} + \Delta E_{therm} + \Delta E_{chem} + \Delta E_{other}$	7-9
4. Energy Dissipated by Friction	For a system that involves a pair of sliding surfaces, the total energy dissipated by friction on both surfaces equals the increase in thermal energy of the system and is given by	
	$f\,\Delta s = \Delta E_{therm}$	7-13
	where Δs is the displacement of one surface relative to the other. If there is no external work done, the increase in thermal energy equals the decrease in mechanical energy of the system:	
	$\Delta E_{therm} = f\,\Delta s = -\Delta E_{mech}$	7-14
5. Problem Solving	The conservation of mechanical energy and the work–energy theorem can be used as an alternative to Newton's laws to solve mechanics problems that require the determination of the speed of a particle as a function of its position.	

6.	**Mass and Energy**	A particle with mass m has an intrinsic rest energy E_0 given by $E_0 = mc^2$ **7-15** where $c = 3 \times 10^8$ m/s is the speed of light in vacuum. A system with mass M also has a rest energy $E_0 = Mc^2$. If a system gains or loses internal energy ΔE, it simultaneously gains or loses mass $\Delta M = \Delta E/c^2$.
7.	**Binding Energy**	The energy required to separate a system into its constituent parts is called its binding energy. The binding energy is ΔMc^2, where ΔM is the increase in mass of the parts over the mass of the system.
8.	**Newtonian Mechanics and the Theory of Relativity**	When the speed of a particle approaches the speed of light c, or the kinetic energy of the particle approaches its rest energy, Newtonian mechanics breaks down and must be replaced by Einstein's special theory of relativity.
9.	**Energy Quantization**	The internal energy of a microscopic system is found to have only a discrete set of possible values. For a system oscillating with frequency f, the allowed energy values are separated by an amount hf, where h is Planck's constant: $h = 6.626 \times 10^{-34}\ \text{J}\cdot\text{s}$ **7-19**
	Photons	Microscopic systems often exchange energy with their surroundings by emitting or absorbing electromagnetic radiation, which is also quantized. The quantum of radiation energy is called the photon: $E_{\text{photon}} = hf$ **7-20** where f is the frequency of the electromagnetic radiation.

Problem-Solving Guide

1. Begin by drawing a neat diagram with a suitable coordinate system.
2. Choose your system and indicate any external forces that act on it. Include in your system objects for which we have a potential-energy function, such as springs, and the earth. If there is sliding friction, make sure your system includes both surfaces.
3. Determine whether mechanical energy is conserved. If it is, write expressions for the initial and final mechanical energy and set them equal to each other. Choose a convenient point for the zero of potential energy.
4. If there is sliding friction, equate $f\Delta s$ to ΔE_{therm}.
5. Apply the work–energy theorem.

Summary of Worked Examples

Type of Calculation	Procedure and Relevant Examples
1. Mechanical Energy Conserved	
Find the speed of an object that is falling in the earth's gravitational field.	Use conservation of mechanical energy to find the final kinetic energy. The gravitational potential energy is mgy, where y is measured from an arbitrary point. **Examples 7-1, 7-2**
Find the distance traveled or the final velocity of an object attached to a spring moving in the earth's gravitational field.	Use conservation of mechanical energy. The potential energy of a spring that is stretched or compressed by an amount x is $\frac{1}{2}kx^2$. **Examples 7-3, 7-4**

	Find the final energy of two objects connected by a string.	Use conservation of mechanical energy for the two-object system. The work done by the string tension is internal to the system. **Example 7-5**
2.	**Analyzing Energy Changes When Thermal or Chemical Energies Are Involved**	Choose your system and apply conservation of energy as expressed in the work–energy theorem. **Examples 7-6, 7-7, 7-8**
	Find the power needed to climb a hill.	Use conservation of energy as expressed in the work–energy theorem. Convert each term to a time rate of change. **Example 7-12**
3.	**Problems With Sliding Friction**	If W_{ext} is zero, the energy dissipated by friction equals the decrease in mechanical energy. If there is work done by external forces, use conservation of energy as expressed in the work–energy theorem with $\Delta E_{therm} = f\Delta s$. **Examples 7-8, 7-9, 7-11**
4.	**Find the Energy Released in a Nuclear Reaction**	Compute the initial and final rest energies. Then $E_{released} = E_0(\text{initial}) - E_0(\text{final})$. **Example 7-14**

Problems

In a few problems, you are given more data than you actually need; in a few other problems, you are required to supply data from your general knowledge, outside sources, or informed estimates.

Conceptual Problems

Problems from Optional and Exploring sections

• Single-concept, single-step, relatively easy

•• Intermediate-level, may require synthesis of concepts

••• Challenging, for advanced students

Take $g = 9.81$ N/kg $= 9.81$ m/s^2 and neglect friction in all problems unless otherwise stated.

The Conservation of Mechanical Energy

1 •• What are the advantages and disadvantages of using the conservation of mechanical energy rather than Newton's laws to solve problems?

2 •• Two objects of unequal mass are connected by a massless cord passing over a frictionless peg. After the objects are released from rest, which of the following statements are true? (U = gravitational potential energy, K = kinetic energy of the system.)

(*a*) $\Delta U < 0$ and $\Delta K > 0$
(*b*) $\Delta U = 0$ and $\Delta K > 0$
(*c*) $\Delta U < 0$ and $\Delta K = 0$
(*d*) $\Delta U = 0$ and $\Delta K = 0$
(*e*) $\Delta U > 0$ and $\Delta K < 0$

3 •• Two stones are thrown with the same initial speed at the same instant from the roof of a building. One stone is thrown at an angle of 30° above the horizontal; the other is thrown horizontally. (Neglect air resistance.) Which statement below is true?

(*a*) The stones strike the ground at the same time and with equal speeds.
(*b*) The stones strike the ground at the same time with different speeds.
(*c*) The stones strike the ground at different times with equal speeds.
(*d*) The stones strike the ground at different times with different speeds.

4 • A block of mass m is pushed up against a spring, compressing it a distance x, and the block is then released. The spring projects the block along a frictionless horizontal surface, giving the block a speed v. The same spring projects a second block of mass $4m$, giving it a speed $3v$. What distance was the spring compressed in the second case?

5 • A woman on a bicycle traveling at 10 m/s on a horizontal road stops pedaling as she starts up a hill inclined at 3.0° to the horizontal. Ignoring friction forces, how far up the hill will she travel before stopping?

(*a*) 5.1 m (*b*) 30 m
(*c*) 97 m (*d*) 10.2 m
(*e*) The answer depends on the mass of the woman.

6 • A pendulum of length L with a bob of mass m is pulled aside until the bob is a distance $L/4$ above its equilibrium position. The bob is then released. Find the speed of the bob as it passes the equilibrium position.

7 • When she hosts a garden party, Julie likes to launch bagels to her guests with a spring device that she has devised. She places one of her 200-g bagels against a horizontal spring mounted on her gazebo. The force constant of the spring is 300 N/m, and she compresses it 9 cm. (*a*) Find the work done by Julie and the spring when Julie launches a bagel. (*b*) If the released bagel leaves the spring at the spring's equilibrium position, find the speed of the bagel at that point. (*c*) If the bagel launcher is 2.2 m above the grass, what is Julie's horizontal range firing 200-g bagels?

8 • A 3-kg block slides along a frictionless horizontal surface with a speed of 7 m/s (Figure 7-17). After sliding a distance of 2 m, the block makes a smooth transition to a frictionless ramp inclined at an angle of 40° to the horizontal.

Figure 7-17 Problem 8

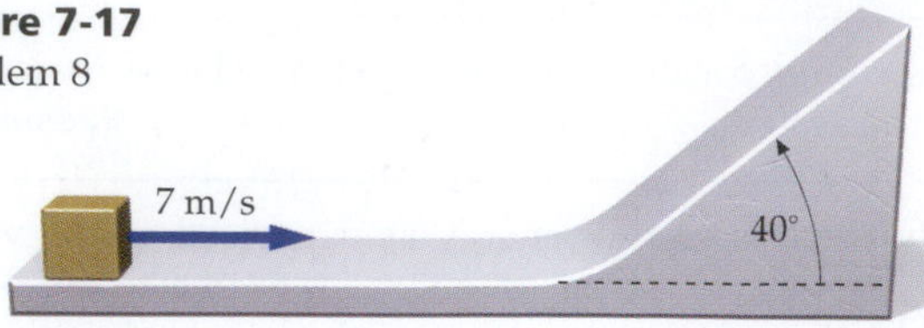

How far up the ramp does the block slide before coming momentarily to rest?

9 • The 3-kg object in Figure 7-18 is released from rest at a height of 5 m on a curved frictionless ramp. At the foot of the ramp is a spring of force constant $k = 400$ N/m. The object slides down the ramp and into the spring, compressing it a distance x before coming momentarily to rest. (*a*) Find x. (*b*) What happens to the object after it comes to rest?

Figure 7-18 Problem 9

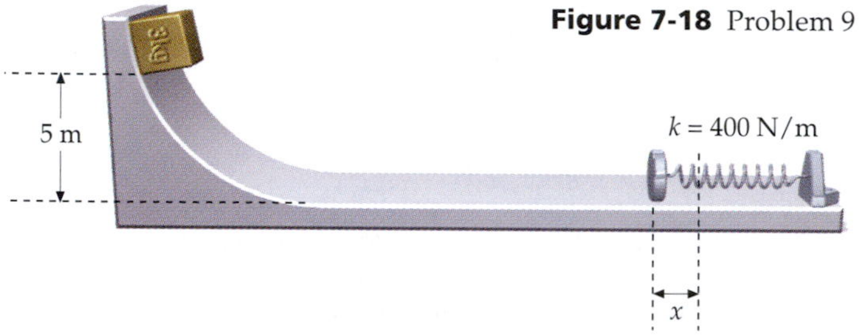

10 • A vertical spring compressed a distance x sits on a concrete floor. When a block of mass m_1 is placed on the spring and the spring is released, the block is projected upward to a height h. If a block of mass $m_2 = 2m_1$ is placed on the spring and the spring is again compressed a distance x and released, to what height will the block rise?

(*a*) $h/4$ (*b*) $h/2$ (*c*) $h/\sqrt{2}$ (*d*) h

11 • If the spring in Problem 10 is compressed an amount $2x$ when the block of mass m_2 is placed on it, to what height will the block rise when the spring is released?

(*a*) $2h$ (*b*) $\sqrt{2}\,h$ (*c*) h (*d*) $h/\sqrt{2}$

12 • A 15-g ball is shot from a spring gun whose spring has a force constant of 600 N/m. The spring can be compressed 5 cm. How high will the ball go if the gun is aimed vertically?

13 • A stone is projected horizontally with a speed of 20 m/s from a bridge 16 m above the surface of the water. What is the speed of the stone as it strikes the water?

14 • At a dock, a crane lifts a 4000-kg container 30 m, swings it out over the deck of a freighter, and lowers the container into the hold of the freighter, which is 8 m below the level of the dock. How much work is done by the crane? (Neglect friction losses.)

15 • A 16-kg child on a playground swing moves with a speed of 3.4 m/s when the 6-m-long swing is at its lowest point. What is the angle that the swing makes with the vertical when the child is at the highest point?

16 •• In 1983, Jacqueline De Creed, driving a 1967 Ford Mustang, made a jump of 71 m, taking off from a ramp inclined at 30° with the horizontal. If the mass of the car and driver was about 900 kg, find the kinetic energy K and potential energy U of De Creed's vehicle at the top point of her flight.

17 •• The system in Figure 7-19 is initially at rest when the lower string is cut. Find the speed of the objects when they are at the same height.

Figure 7-19 Problem 17

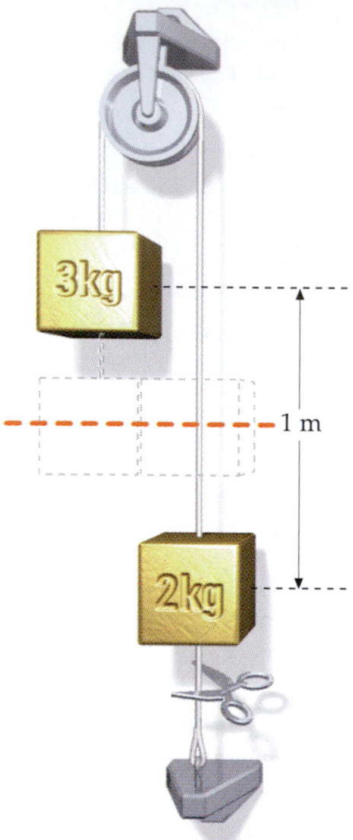

18 •• While traveling in the far north, one of your companions gets snow blindness, and you have to lead him along by the elbow. Looking back, you see your other companion, Sandy, fall and slide along the frictionless surface of the frozen river valley shown in Figure 7-20. If point Q is 4.5 m higher than point P, and your hapless companion fell at point P with a velocity v_0 down the slope, describe his motion to your snow-blind friend if (*a*) $v_0 = 2$ m/s and (*b*) $v_0 = 5$ m/s. (*c*) What is the minimum initial speed required for the fall to carry your partner past point Q?

Figure 7-20 Problem 18

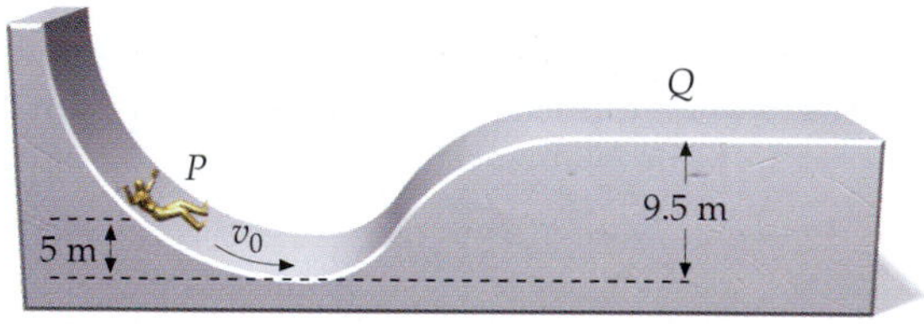

19 •• A block rests on an inclined plane as in Figure 7-21. A spring to which it is attached via a pulley is being pulled downward with gradually increasing force. The value of μ_s is known. Find the potential energy U of the spring at the moment when the block begins to move.

Figure 7-21 Problem 19

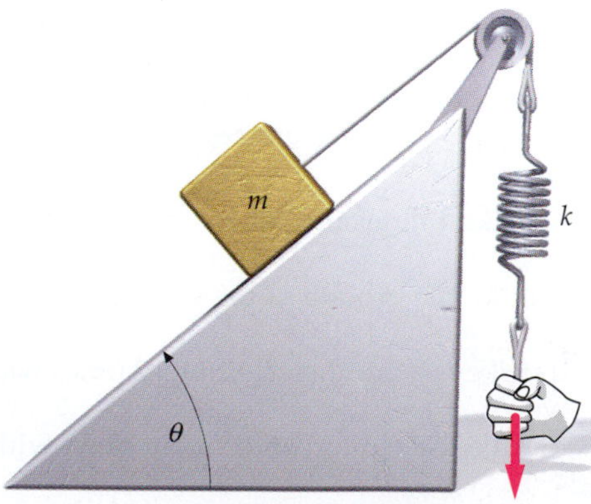

20 •• Sandy is sliding helplessly across the frictionless ice with her climbing rope trailing behind (Figure 7-22). Racing after her, you get hold of her rope just as she goes over the edge of a cliff. You manage to grab a tree branch in time to keep from going over yourself. Let $U = 0$ for the position of Sandy dangling in midair at the other end of the rope. Snap! The branch to which you are clinging breaks. (*a*) Write an expression for the total mechanical energy of this two-body system after Sandy has fallen a distance y. (*b*) There is another tree branch 2 m closer to the cliff edge than the first. What is your speed as you reach it?

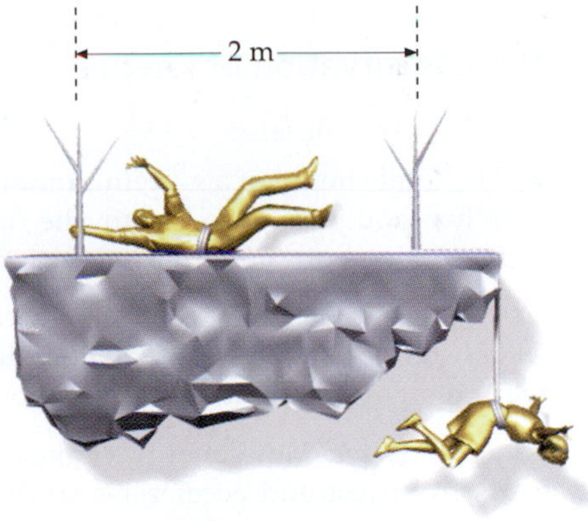

Figure 7-22 Problem 20

21 •• A 2.4-kg block is dropped from a height of 5.0 m onto a spring of spring constant 3955 N/m. When the block is momentarily at rest, the spring has compressed by 25 cm. Find the speed of the block when the compression of the spring is 15.0 cm.

22 •• Red is a girl of mass m who is taking a picnic lunch to her grandmother. She ties a rope of length R to a tree branch over a creek and starts to swing from rest at point A, which is a distance $R/2$ lower than the branch (Figure 7-23). What is the minimum breaking tension for the rope if it is not to break and drop Red into the creek?

Figure 7-23 Problem 22
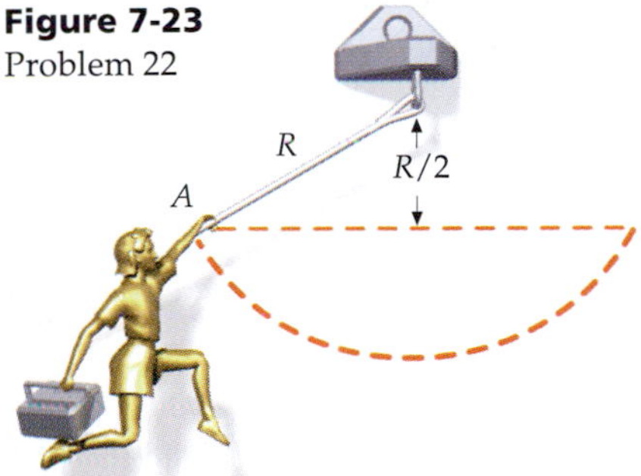

23 •• A ball at the end of a string moves in a vertical circle with constant energy E. What is the difference between the tension at the bottom of the circle and the tension at the top?

24 •• A roller coaster car of mass 1500 kg starts a distance $H = 23$ m above the bottom of a loop 15 m in diameter (Figure 7-24). If friction is negligible, the downward force of the rails on the car when it is upside down at the top of the loop is __________.

(*a*) 4.6×10^4 N (*b*) 3.1×10^4 N (*c*) 1.7×10^4 N
(*d*) 980 N (*e*) 1.6×10^3 N

Figure 7-24 Problem 24
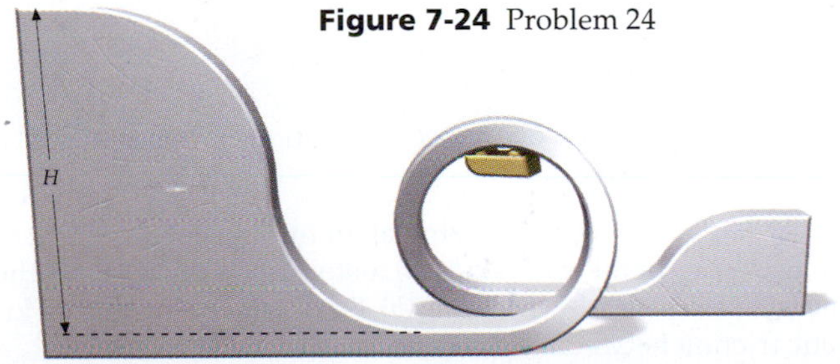

25 •• A stone is thrown upward at an angle of 53° above the horizontal. Its maximum height during the trajectory is 24 m. What was the stone's initial speed?

26 •• A baseball of mass 0.17 kg is thrown from the roof of a building 12 m above the ground. Its initial velocity is 30 m/s at an angle of 40° above the horizontal. (*a*) What is the maximum height of the ball? (*b*) What is the work done by gravity as the ball moves from the roof to its maximum height? (*c*) What is the speed of the ball as it strikes the ground?

27 •• An 80-cm-long pendulum with a 0.6-kg bob is released from rest at initial angle θ_0 with the vertical. At the bottom of the swing, the speed of the bob is 2.8 m/s. (*a*) What was the initial angle of the pendulum? (*b*) What angle does the pendulum make with the vertical when the speed of the bob is 1.4 m/s?

28 •• The Royal Gorge bridge over the Arkansas River is about $L = 310$ m high. A bungee jumper of mass 60 kg has an elastic cord of length $d = 50$ m attached to her feet. Assume that the cord acts like a spring of force constant k. The jumper leaps, barely touches the water, and after numerous ups and downs comes to rest at a height h above the water. (*a*) Find h. (*b*) Find the maximum speed of the jumper.

29 •• A pendulum consists of a 2-kg bob attached to a light string of length 3 m. The bob is struck horizontally so that it has an initial horizontal velocity of 4.5 m/s. For the point at which the string makes an angle of 30° with the vertical, what is (*a*) the speed? (*b*) the potential energy? (*c*) the tension in the string? (*d*) What is the angle of the string with the vertical when the bob reaches its greatest height?

30 •• Lou is trying to kill mice by swinging a clock of mass m attached to one end of a light (massless) stick 1.4 m in length hanging on a nail in the wall (Figure 7-25). The clock end of the stick is free to rotate around its other end in a vertical circle. Lou raises the clock until the stick is horizontal, and when mice peek their heads out from the hole to their den, he gives it an initial downward velocity v. The clock misses a mouse and continues on its circular path with just enough energy to complete the circle and bonk Lou on the

Figure 7-25 Problem 30
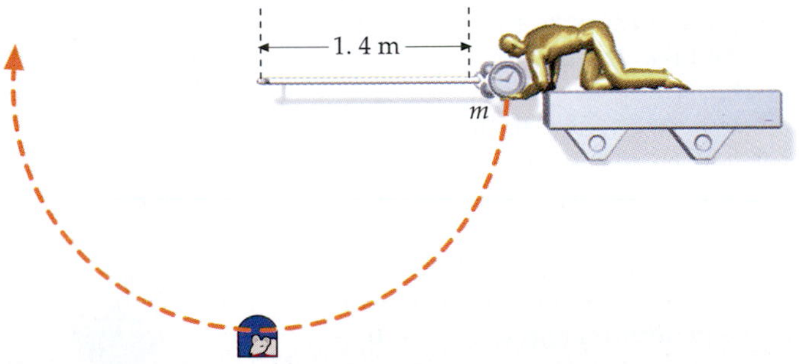

back of his head, to the sound of cheering mice. (*a*) What was the value of v? (*b*) What was the clock's speed at the bottom of its swing?

31 •• A pendulum consists of a string of length L and a bob of mass m. The string is brought to a horizontal position and the bob is given the minimum initial speed enabling the pendulum to make a full turn in the vertical plane. (*a*) What is the maximum kinetic energy K of the bob? (*b*) What is the tension in the string when the kinetic energy is maximum?

32 •• A child whose weight is 360 N swings out over a pool of water using a rope attached to the branch of a tree at the edge of the pool. The branch is 12 m above ground level and the surface of the pool is 1.8 m below ground level. The child holds onto the rope at a point 10.6 m from the branch and moves back until the angle between the rope and the vertical is 23°. When the rope is in the vertical position, the child lets go and drops into the pool. Find the speed of the child at the surface of the pool.

33 •• Walking by a pond, you find a rope attached to a tree limb 5.2 m off the ground. You decide to use the rope to swing out over the pond. The rope is a bit frayed but supports your weight. You estimate that the rope might break if the tension is 80 N greater than your weight. You grab the rope at a point 4.6 m from the limb and move back to swing out over the pond. (*a*) What is the maximum safe initial angle between the rope and the vertical so that it will not break during the swing? (*b*) If you begin at this maximum angle, and the surface of the pond is 1.2 m below the level of the ground, with what speed will you enter the water if you let go of the rope when the rope is vertical?

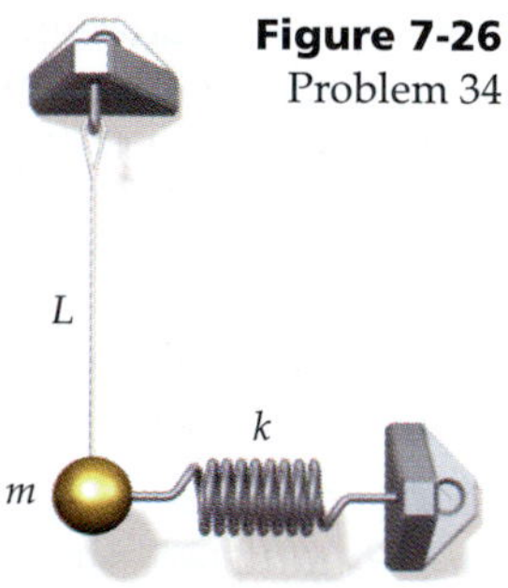

Figure 7-26 Problem 34

34 •• A pendulum of length L has a bob of mass m attached to a light string, which is attached to a spring of force constant k. With the pendulum in the position shown in Figure 7-26, the spring is at its unstretched length. If the bob is now pulled aside so that the string makes a *small* angle θ with the vertical, what is the speed of the bob after release as it passes through the equilibrium position?

Figure 7-27 Problem 35

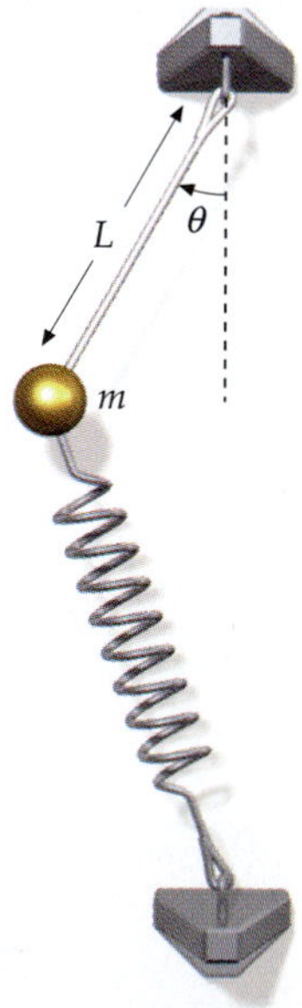

35 ••• A pendulum is suspended from the ceiling and attached to a spring fixed at the opposite end directly below the pendulum support (Figure 7-27). The mass of the pendulum bob is m, the length of the pendulum is L, and the spring constant is k. The unstretched length of the spring is $L/2$ and the distance between the bottom of the spring and the ceiling is $1.5L$. The pendulum is pulled aside so that it makes a small angle θ with the vertical and is then released from rest. Obtain an expression for the speed of the pendulum bob when $\theta = 0$.

The Conservation of Energy

36 • True or false:

(*a*) The total energy of a system cannot change.
(*b*) When you jump into the air, the floor does work on you, increasing your potential energy.

37 • A man stands on roller skates next to a rigid wall. To get started, he pushes off against the wall. Discuss the energy changes pertinent to this situation.

38 • Discuss the energy changes involved when a car starts from rest and accelerates so that the car's wheels do not slip. What external force accelerates the car? Does this force do work?

39 • A body falling through the atmosphere (air resistance is present) gains 20 J of kinetic energy. The amount of gravitational potential energy that is lost is

(*a*) 20 J. (*b*) more than 20 J. (*c*) less than 20 J.
(*d*) impossible to tell without knowing the mass of the body.
(*e*) impossible to tell without knowing how far the body falls.

40 • Assume that you can expend energy at a constant rate of 250 W. Estimate how fast you can run up four flights of stairs, with each flight 3.5 m high.

41 • A 70-kg skater pushes off the wall of a skating rink, acquiring a speed of 4 m/s. (*a*) How much work is done on the skater? (*b*) What is the change in the mechanical energy of the skater? (*c*) Discuss the conservation of energy as applied to the skater.

42 • In a volcanic eruption, 4 km^3 of mountain with a density of 1600 kg/m^3 was lifted an average height of 500 m. (*a*) How much energy in joules was released in this eruption? (*b*) The energy released by thermonuclear bombs is measured in megatons of TNT, where 1 megaton of TNT $= 4.2 \times 10^{15}$ J. Convert your answer for (*a*) to megatons of TNT.

43 •• An 80-kg physics student climbs a 120-m hill. (*a*) What is the increase in the gravitational potential energy of the student? (*b*) Where does this energy come from? (*c*) The student's body is 20% efficient; that is, for every 20 J that are converted to mechanical energy, 100 J of internal energy are expended, with 80 J going into thermal energy. How much chemical energy is expended by the student during the climb?

44 •• In 1993, Carl Fentham of Great Britain raised a full keg of beer (mass 62 kg) to a height of about 2 m 676 times in 6 h. Assuming that work was done only as the keg was going up, estimate how many such kegs of beer he would have to drink to reimburse his energy expenditure. (1 liter of beer is approximately 1 kg and provides about 1.5 MJ of energy; in your calculations, neglect the mass of the empty keg.)

Kinetic Friction

45 • Discuss the energy considerations when you pull a box along a rough road.

46 • A 2000-kg car moving at an initial speed of 25 m/s along a horizontal road skids to a stop in 60 m. (*a*) Find the energy dissipated by friction. (*b*) Find the coefficient of kinetic friction between the tires and the road.

47 • An 8-kg sled is initially at rest on a horizontal road. The coefficient of kinetic friction between the sled and the road is 0.4. The sled is pulled a distance of 3 m by a force of 40 N applied to the sled at an angle of 30° to the horizontal. (*a*) Find the work done by the applied force. (*b*) Find the energy dissipated by friction. (*c*) Find the change in the kinetic energy of the sled. (*d*) Find the speed of the sled after it has traveled 3 m.

48 • Returning to Problem 8, suppose the surfaces described are not frictionless and that the coefficient of kinetic friction between the block and the surfaces is 0.30. Find (*a*) the speed of the block when it reaches the ramp, and (*b*) the distance that the block slides up the ramp before coming momentarily to rest. (Neglect the energy dissipated along the transition curve.)

49 • The 2-kg block in Figure 7-28 slides down a frictionless curved ramp, starting from rest at a height of 3 m. The block then slides 9 m on a rough horizontal surface before coming to rest. (*a*) What is the speed of the block at the bottom of the ramp? (*b*) What is the energy dissipated by friction? (*c*) What is the coefficient of friction between the block and the horizontal surface?

Figure 7-28 Problem 49

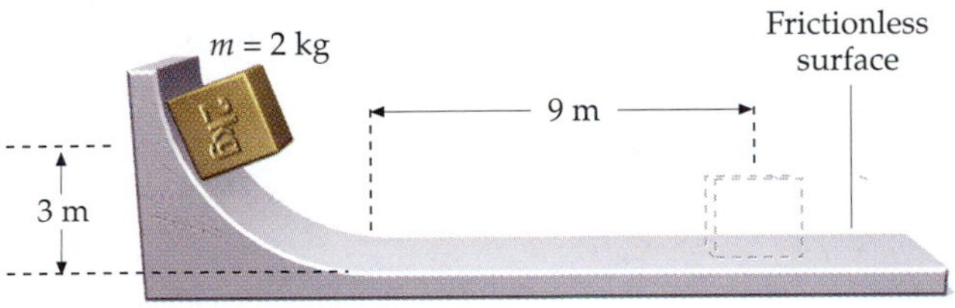

50 •• A 20-kg girl slides down a playground slide that is 3.2 m high. When she reaches the bottom of the slide, her speed is 1.3 m/s. (*a*) How much energy was dissipated by friction? (*b*) If the slide is inclined at 20°, what is the coefficient of friction between the girl and the slide?

Figure 7-29 Problem 51

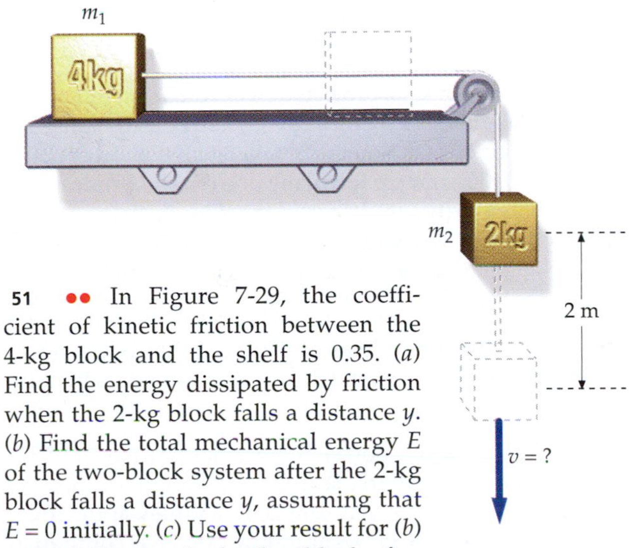

51 •• In Figure 7-29, the coefficient of kinetic friction between the 4-kg block and the shelf is 0.35. (*a*) Find the energy dissipated by friction when the 2-kg block falls a distance y. (*b*) Find the total mechanical energy E of the two-block system after the 2-kg block falls a distance y, assuming that $E = 0$ initially. (*c*) Use your result for (*b*) to find the speed of either block after the 2-kg block falls 2 m.

52 •• Nils Lied, an Australian meteorologist, once played golf on the ice in Antarctica and drove a ball a horizontal distance of 2400 m. For a rough estimate, let us assume that the ball took off at $\theta = 45°$, flew a horizontal distance of 200 m without air resistance, and then slid on the ice without bouncing, its velocity being equal to the horizontal component of the initial velocity. Estimate the coefficient of kinetic friction μ_k between the ice and ball.

53 •• A particle of mass m moves in a horizontal circle of radius r on a rough table. It is attached to a horizontal string fixed at the center of the circle. The speed of the particle is initially v_0. After completing one full trip around the circle, the speed of the particle is $\frac{1}{2}v_0$. (*a*) Find the energy dissipated by friction during that one revolution in terms of m, v_0, and r. (*b*) What is the coefficient of kinetic friction? (*c*) How many more revolutions will the particle make before coming to rest?

54 •• In 1987, British skier Graham Wilkie achieved a speed of v = 211 km/h going downhill. Assuming that he reached the maximum speed at the end of the hill and then continued on the horizontal surface, find the maximum distance d he *could have* covered on the horizontal surface. Take the coefficient of kinetic friction μ_k to be constant throughout the run; neglect air resistance. Assume the hill is 225 m high with a constant slope of 30° with the horizontal.

55 •• During a move, Kate and Lou have to push Kate's 80-kg stove up a rough loading ramp, pitched at an angle of 10°, to get it into a truck. They push it along the horizontal floor to pick up speed and give it one last push at the bottom of the ramp, hoping for the best. Unfortunately, the stove stops short and then slides down the ramp, sending them leaping to the side. (*a*) If the stove has a speed of 3.0 m/s at the bottom of the ramp, and a speed of 0.8 m/s when it is 2 m up the ramp, what is the maximum height reached by the stove? (*b*) What is the stove's speed when it passes the 2-m spot again? (*c*) What is the energy dissipated by friction during the complete round trip back to the bottom of the ramp?

56 •• A 2.4-kg box has an initial velocity of 3.8 m/s upward along a rough plane inclined at 37° to the horizontal. The coefficient of kinetic friction between the box and plane is 0.30. How far up the incline does the box travel? What is its speed when it passes its starting point on its way down the incline?

57 ••• A block of mass m rests on a rough plane inclined at θ with the horizontal (Figure 7-30). The block is attached to a spring of constant k near the top of the plane. The coefficients of static and kinetic friction between the block and plane are μ_s and μ_k, respectively. The spring is slowly pulled upward along the plane until the block starts to move. (*a*) Obtain an expression for the extension d of the spring the

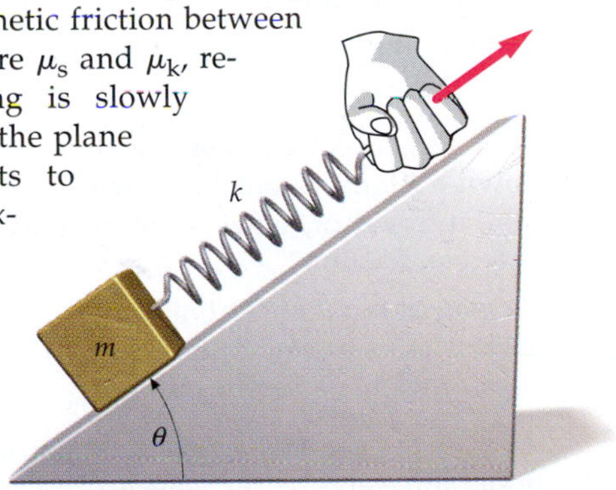

Figure 7-30 Problem 57

instant the block moves. (*b*) Determine the value of μ_k such that the block comes to rest just as the spring is in its unstressed condition, i.e., neither extended nor compressed.

Mass and Energy

58 • How much rest mass is consumed in the core of a nuclear-fueled electric generating plant in producing (*a*) one joule of thermal energy? (*b*) enough energy to keep a 100-W light bulb burning for 10 years?

59 • (*a*) Calculate the rest energy in 1 g of dirt. (*b*) If you could convert this energy into electrical energy and sell it for 10 cents per kilowatt-hour, how much money would you get? (*c*) If you could power a 100-W light bulb with this energy, for how long could you keep the bulb lit?

60 • A muon has a rest energy of 105.7 MeV. Calculate its rest mass in kilograms.

61 • For the fusion reaction in Example 7-14, calculate the number of reactions per second that are necessary to generate 1 kW of power.

62 • How much energy is needed to remove one neutron from ^{4}He, leaving ^{3}He plus a neutron? (The rest energy of ^{3}He is 2808.41 MeV.)

63 • A free neutron at rest decays into a proton plus an electron:

$$\mathrm{n} \rightarrow \mathrm{p} + \mathrm{e}$$

Use Table 7-1 to calculate the energy released in this reaction.

64 •• In one nuclear fusion reaction, two ^{2}H nuclei combine to produce ^{4}He. (*a*) How much energy is released in this reaction? (*b*) How many such reactions must take place per second to produce 1 kW of power?

65 •• A large nuclear power plant produces 3000 MW of power by nuclear fission, which converts matter into energy. (*a*) How many kilograms of matter does the plant consume in one year? (*b*) In a coal-burning power plant, each kilogram of coal releases 31 MJ of energy when burned. How many kilograms of coal are needed each year for a 3000-MW plant?

General Problems

66 •• A block of mass *m*, starting from rest, is pulled by a string up a frictionless inclined plane that makes an angle θ with the horizontal. The tension in the string is *T* and the string is parallel to the plane. After traveling a distance *L*, the speed of the block is *v*. The work done by the tension *T* is

(*a*) $mgL \sin \theta$ (*b*) $mgL \cos \theta + \frac{1}{2} mv^2$ (*c*) $mgL \sin \theta + \frac{1}{2}mv^2$
(*d*) $mgL \cos \theta$ (*e*) $TL \cos \theta$

67 •• A block of mass *m* slides with constant velocity *v* down a plane inclined at θ with the horizontal. During the time interval Δt, what is the magnitude of the energy dissipated by friction?

(*a*) $mgv\ \Delta t \tan \theta$ (*b*) $mgv\ \Delta t \sin \theta$ (*c*) $\frac{1}{2}mv^3\Delta t$
(*d*) The answer cannot be determined without knowing the coefficient of kinetic friction.

68 •• Assume that on applying the brakes a constant frictional force acts on the wheels of a car. If that is so, it follows that

(*a*) the distance the car travels before coming to rest is proportional to the speed of the car before the brakes are applied.
(*b*) the car's kinetic energy diminishes at a constant rate.
(*c*) the kinetic energy of the car is inversely proportional to the time that has elapsed since the application of the brakes.
(*d*) none of the above apply.

69 • Our bodies convert internal chemical energy into work and heat at the rate of about 100 W, which is called our metabolic rate. (*a*) How much internal chemical energy do we use in 24 h? (*b*) The energy comes from the food that we eat and is usually measured in kilocalories, where 1 kcal = 4.184 kJ. How many kilocalories of food energy must we ingest per day if our metabolic rate is 100 W?

70 • A 3.5-kg box rests on a horizontal frictionless surface in contact with a spring of spring constant 6800 N/m. The spring is fixed at its other end and is initially at its uncompressed length. A constant horizontal force of 70 N is applied to the box so that the spring compresses. Determine the distance the spring is compressed when the box is momentarily at rest.

71 • The average energy per unit time per unit area that reaches the upper atmosphere of the earth from the sun, called the solar constant, is 1.35 kW/m^2. Because of absorption and reflection by the atmosphere, about 1 kW/m^2 reaches the surface of the earth on a clear day. How much energy is collected in 8 h of daylight by a solar panel 1 m by 2 m on a rotating mount that is always perpendicular to the sun's rays.

72 • When the jet-powered car *Spirit of America* went out of control during a test drive at Bonneville Salt Flats, Utah, it left skid marks about 9.5 km long. (*a*) If the car was moving initially at a speed of $v = 708$ km/h, estimate the coefficient of kinetic friction μ_k. (*b*) What was the kinetic energy *K* of the car at time $t = 60$ s after the brakes were applied? Take the mass of the car to be 1250 kg.

73 •• A T-bar tow is required to pull 80 skiers up a 600-m slope inclined at 15° above horizontal at a speed of 2.5 m/s. The coefficient of kinetic friction is 0.06. Find the motor power required if the mass of the average skier is 75 kg.

74 •• A 2-kg box is projected with an initial speed of 3 m/s up a rough plane inclined at 60° above horizontal. The coefficient of kinetic friction is 0.3. (*a*) List all the forces acting on the box. (*b*) How far up the plane does the box slide before it stops momentarily? (*c*) What is the energy dissipated by friction as the box slides up the plane? (*d*) What is the speed of the box when it again reaches its initial position?

75 •• A 1200-kg elevator driven by an electric motor can safely carry a maximum load of 800 kg. What is the power provided by the motor when the elevator ascends with a full load at a speed of 2.3 m/s?

76 •• To reduce the power requirement of elevator motors, elevators are counterbalanced with weights connected to the elevator by a cable that runs over a pulley at the top of the elevator shaft. If the elevator in Problem 75 is counterbalanced with a mass of 1500 kg, what is the power provided by the motor when the elevator ascends fully loaded at a speed of 2.3 m/s? How much power is provided by the motor when the elevator ascends without a load at 2.3 m/s?

77 •• The spring constant of a toy dart gun is 5000 N/m. To cock the gun the spring is compressed 3 cm. The 7-g dart, fired straight upward, reaches a maximum height of 24 m. Determine the energy dissipated by air friction during the dart's ascent. Estimate the speed of the projectile when it returns to its starting point.

78 •• A 0.050-kg dart is fired vertically from a spring gun that has a spring constant of 4000 N/m. Prior to release, the spring was compressed by 10.7 cm. When the dart is 6.8 m above the gun, its upward speed is 28 m/s. Determine the maximum height reached by the dart.

79 •• In a volcanic eruption, a 2-kg piece of porous volcanic rock is thrown vertically upward with an initial speed of 40 m/s. It travels upward a distance of 50 m before it begins to fall back to the earth. (*a*) What is the initial kinetic energy of the rock? (*b*) What is the increase in thermal energy due to air friction during ascent ? (*c*) If the increase in thermal energy due to air friction on the way down is 70% of that on the way up, what is the speed of the rock when it returns to its initial position?

80 •• A block of mass m starts from rest at a height h and slides down a frictionless plane inclined at θ with the horizontal as shown in Figure 7-31. After sliding a distance L, the block strikes a spring of force constant k. Find the compression of the spring when the block is momentarily at rest.

Figure 7-31 Problem 80

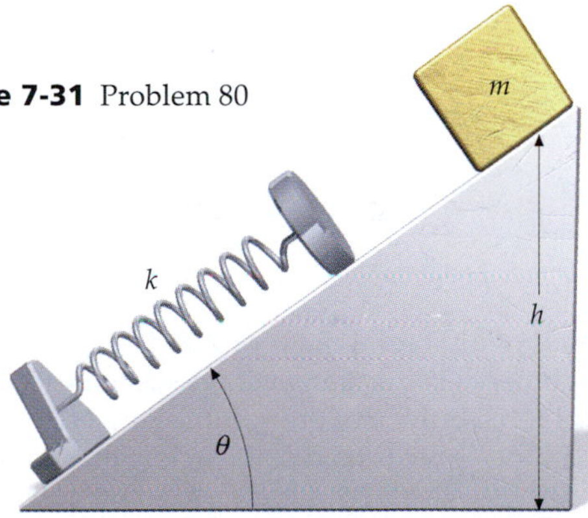

81 •• A car of mass 1500 kg traveling at 24 m/s is at the foot of a hill that rises 120 m in 2.0 km. At the top of the hill, the speed of the car is 10 m/s. Find the average power delivered by the car's engine, neglecting any frictional losses.

82 •• In a new ski jump event, a loop is installed at the end of the ramp as shown in Figure 7-32. This problem addresses the physical requirement such an event would place on the skiers. Neglect friction.

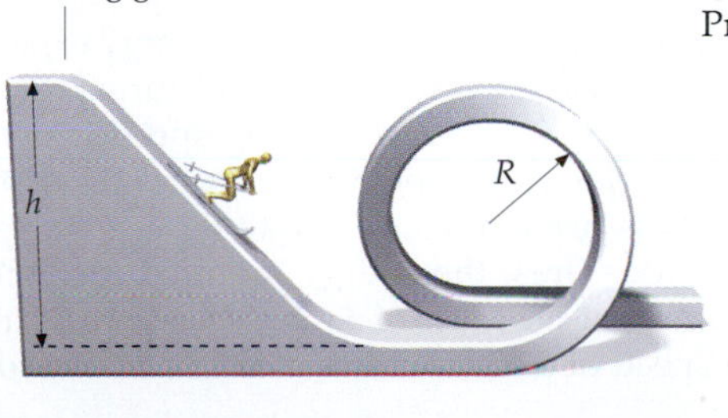

Figure 7-32 Problem 82

(*a*) Along the track of the ramp and the loop, where do the legs of the skier have to support the maximum weight?
(*b*) If the loop has a radius R, where should the starting gate (indicated by h) be placed so that the maximum force on the skier's legs is 4 times the skier's body weight?
(*c*) With the starting gate at the height found in part (*b*), will the skier be able to make it completely around the loop? Why or why not?
(*d*) What is the minimum height h such that the skier can make it around the loop? What is the maximum force on the skier's legs for this height?

83 •• A mass m is suspended from the ceiling by a spring and is free to move vertically in the y direction as indicated in Figure 7-33. We are given that the potential energy as a function of position is $U = \frac{1}{2}ky^2 - mgy$.

(*a*) Sketch U as a function of y. What value of y corresponds to the *unstretched* condition of the spring, y_0?
(*b*) From the given expression for U, find the net downward force acting on m at any position y.
(*c*) The mass is released from rest at $y = 0$; if there is no friction, what is the maximum value y_{max} that will be reached by the mass? Indicate y_{max} on your sketch.
(*d*) Now consider the effect of friction. The mass ultimately settles down into an equilibrium position y_{eq}. Find this point on your sketch.
(*e*) Find the amount of thermal energy produced by friction from the start of the operation to the final equilibrium.

Figure 7-33 Problem 83

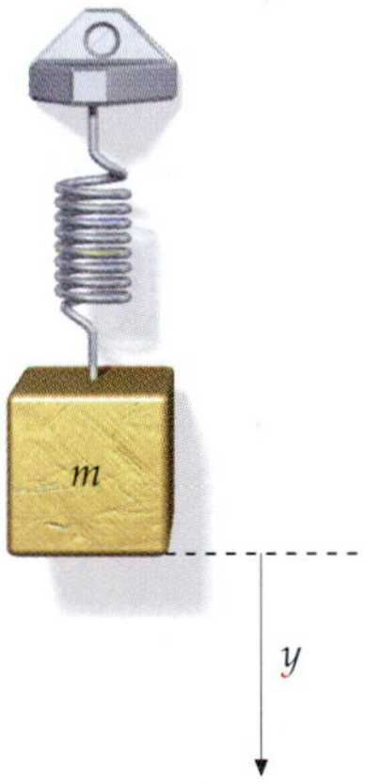

84 •• A spring-loaded gun is cocked by compressing a short, strong spring by a distance d. The gun fires a signal flare of mass m directly upward. The flare has speed v_0 as it leaves the spring and is observed to rise to a maximum height h above the point where it leaves the spring. After it leaves the spring, effects of drag force by the air on the packet are significant. (Express answers in terms of m, v_0, d, h, and g, the acceleration due to gravity.) (*a*) How much work is done on the spring in the course of the compression? (*b*) What is the value of the spring constant k? (*c*) How much mechanical energy is converted to thermal energy because of the drag force of the air on the flare between the time of firing and the time at which maximum elevation is reached?

85 •• A roller-coaster car having a total mass (including passengers) of 500 kg travels freely along the winding frictionless track in Figure 7-34. Points A, E, and G are horizontal straight sections, all at the same height of 10 m above ground. Point C is at a height of 10 m above the ground on a section sloped at an angle of 30°. Point B is at the top of a hill, while point D is at ground level at the bottom of a valley. The radius of curvature at each of these points is 20 m. Point F is at the middle of a banked horizontal curve of radius of curvature of 30 m, and at the same height of 10 m above the ground as points A, E, and G. At point A the speed of the car is 12 m/s.

(*a*) If the car is just barely able to make it over the hill at point B, what is the height of that point above the ground?
(*b*) If the car is just barely able to make it over the hill at point B, what is the magnitude of the total force exerted on the car by the track at that point?
(*c*) What is the acceleration of the car at point C?
(*d*) What are the magnitude and direction of the total force exerted on the car by the track at point D?
(*e*) What are the magnitude and direction of the total force exerted on the car by the track at point F?
(*f*) At point G, a constant braking force is applied to the car, bringing the car to a halt in a distance of 25 m. What is the braking force?

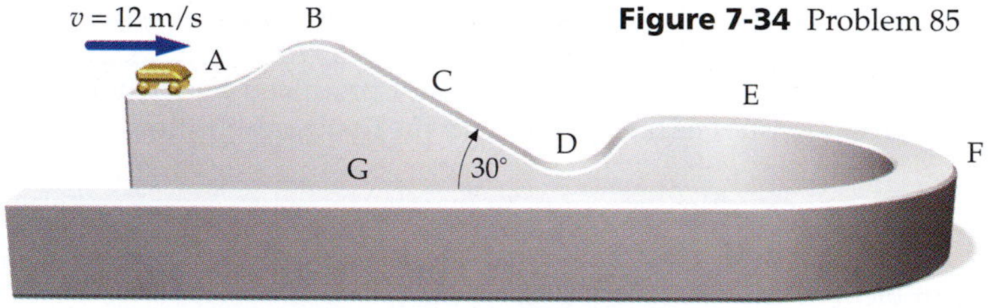

Figure 7-34 Problem 85

86 • An elevator (mass $M = 2000$ kg) is moving downward at $v_0 = 1.5$ m/s. A braking system prevents the downward speed from increasing. (*a*) At what rate (in J/s) is the braking system converting mechanical energy to thermal energy? (*b*) While the elevator is moving downward at $v_0 = 1.5$ m/s, the braking system fails and the elevator is in free fall for a distance $d = 5$ m before hitting the top of a large safety spring with force constant $k = 1.5 \times 10^4$ N/m. After the elevator cage hits the top of the spring, we want to know the distance Δy that the spring is compressed before the cage is brought to rest. Write an algebraic expression for the value of Δy in terms of the known quantities M, v_0, g, k, and d, and substitute the given values to find Δy.

87 • To measure the force of friction on a moving car, engineers turn off the engine and allow the car to coast down hills of known steepness. The engineers collect the following data:

1. On a 2.87° hill, the car can coast at a steady 20 m/s.
2. On a 5.74° hill, the steady coasting speed is 30 m/s.

The total mass of the car is 1000 kg.

(*a*) What is the force of friction at 20 m/s (F_{20}) and at 30 m/s (F_{30})?
(*b*) How much useful power must the engine deliver to drive the car on a level road at steady speeds of 20 m/s (P_{20}) and 30 m/s (P_{30})?
(*c*) At full throttle, the engine delivers 40 kW. What is the angle of the steepest incline up which the car can maintain a steady 20 m/s?
(*d*) Assume that the engine delivers the same total useful work from each liter of gas, no matter what speed. At 20 m/s on a level road, the car goes 12.7 km/L. How many kilometers per liter does it get if it goes 30 m/s instead?

88 •• A 50,000-kg barge is pulled along a canal at a constant speed of 3 km/h by a heavy tractor. The towrope makes an angle of 18° with the velocity vector of the barge. The tension in the towrope is 1200 N. If the towrope breaks, how far will the barge move before coming to rest? Assume that the drag force between the barge and water is independent of velocity.

89 •• A 2-kg block is released 4 m from a massless spring with a force constant $k = 100$ N/m that is fixed along a frictionless plane inclined at 30°, as shown in Figure 7-35. (*a*) Find the maximum compression of the spring. (*b*) If the plane is rough rather than frictionless, and the coefficient of kinetic friction between the plane and the block is 0.2, find the maximum compression. (*c*) For the rough plane, how far up the incline will the block travel after leaving the spring?

Figure 7-35 Problem 89

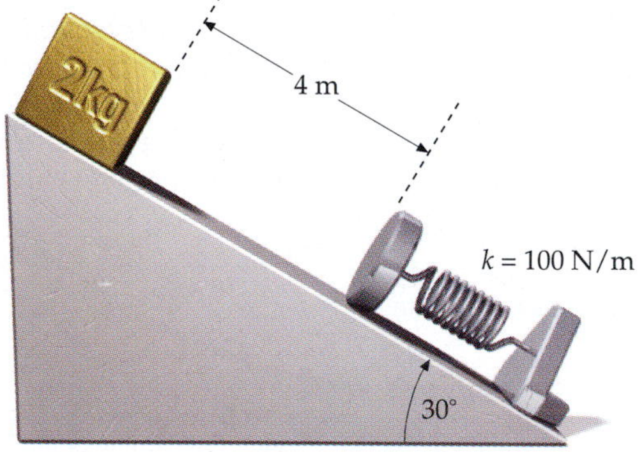

90 •• A train with a total mass of 2×10^6 kg rises 707 m in a travel distance of 62 km at an average speed of 15.0 km/h. If the frictional force is 0.8% of the weight, find (*a*) the kinetic energy of the train, (*b*) the total change in its potential energy, (*c*) the energy dissipated by kinetic friction, and (*d*) the power output of the train's engines.

91 •• While driving, one expects to spend more energy accelerating than driving at a constant speed. (*a*) Neglecting friction, calculate the energy required to give a 1200-kg car a speed of 50 km/h. (*b*) If friction results in a retarding force of 300 N at a speed of 50 km/h, what is the energy needed to move the car a distance of 300 m at a constant speed of 50 km/h? (*c*) Assuming that the energy losses due to friction in part (*a*) are 75% of those found in part (*b*), estimate the ratio of the energy consumption for the two cases considered.

92 •• In one model of jogging, the energy expended is assumed to go into accelerating and decelerating the legs. If the mass of the leg is m and the running speed is v, the energy needed to accelerate the leg from rest to v is $\frac{1}{2}mv^2$, and the same energy is needed to decelerate the leg back to rest for the next stride. Thus, the energy required for each stride is mv^2. Assume that the mass of a man's leg is 10 kg and that he runs at a speed of 3 m/s with 1 m between one footfall and the next. Therefore, the energy he must provide to his legs in each second is $3 \times mv^2$. Calculate the rate of the man's energy expenditure using this model and assuming that his muscles have an efficiency of 25%.

93 •• On July 31, 1994, Sergei Bubka pole-vaulted over a height of 6.14 m. If his body was momentarily at rest at the top of the leap, and all of the energy required to raise his body derived from his kinetic energy just prior to planting his pole, how fast was he moving just before takeoff? Neglect the mass of the pole. If he could maintain that speed for a 100-m sprint, how fast would he cover that distance? Since the world record for the 100-m dash is just over 9.8 s, what do you conclude about world-class pole-vaulters?

94 •• A 5-kg block is held against a spring of force constant 20 N/cm, compressing it 3 cm. The block is released and the spring extends, pushing the block along a rough horizontal surface. The coefficient of friction between the surface and the block is 0.2. (*a*) Find the work done on the block by the spring as it extends from its compressed position to its equilibrium position. (*b*) Find the energy dissipated by friction while the block moves the 3 cm to the equilibrium position of the spring. (*c*) What is the speed of the block when the spring is at its equilibrium position? (*d*) If the block is not attached to the spring, how far will it slide along the rough surface before coming to rest?

95 •• A pendulum of length L has a bob of mass m. It is released from some angle θ_1. The string hits a peg at a distance x directly below the pivot as in Figure 7-36, effectively shortening the length of the pendulum. Find the maximum angle θ_2 between the string and the vertical when the bob is to the right of the peg.

Figure 7-36 Problem 95

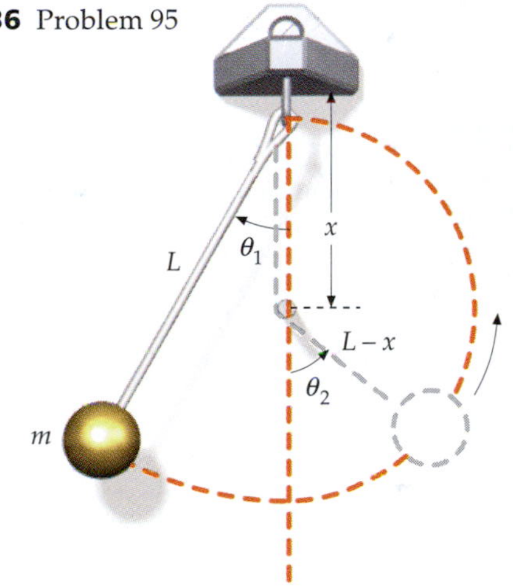

96 ••• A block of mass m is dropped onto the top of a vertical spring whose force constant is k. If the block is released from a height h above the top of the spring, (*a*) what is the maximum kinetic energy of the block? (*b*) What is the maximum compression of the spring? (*c*) At what compression is the block's kinetic energy half its maximum value?

97 ••• The bob of a pendulum of length L is pulled aside so the string makes an angle θ_0 with the vertical, and the bob is then released. In Example 7-2, the conservation of energy was used to obtain the speed of the bob at the bottom of its swing. In this problem, you are to obtain the same result using Newton's second law. (*a*) Show that the tangential component of Newton's second law gives $dv/dt = -g \sin \theta$, where v is the speed and θ is the angle made by the string and the vertical. (*b*) Show that v can be written $v = L\, d\theta/dt$. (*c*) Use this result and the chain rule for derivatives to obtain

$$\frac{dv}{dt} = \frac{dv}{d\theta}\frac{d\theta}{dt} = \frac{dv}{d\theta}\frac{v}{L}$$

(*d*) Combine the results of (*a*) and (*c*) to obtain

$$v\, dv = -gL \sin \theta\, d\theta$$

(*e*) Integrate the left side of the equation in part (*d*) from $v = 0$ to the final speed v and the right side from $\theta = \theta_0$ to $\theta = 0$, and show that the result is equivalent to $v = \sqrt{2gh}$, where h is the original height of the bob above the bottom.

CHAPTER 8

Systems of Particles and Conservation of Momentum

This break shot illustrates the transfer of momentum in two dimensions from the white cue ball to the other balls on a pool table.

We've discussed Newton's laws in terms of the motion of point particles, but many applications concern extended objects—cars, rockets, people. We will justify these applications by showing that there is one point of a system, the **center of mass,** that moves as if all the mass of the system were concentrated at that point, and all the external forces acting on the system were acting exclusively on that point. The motion of any object or system of particles can be described in terms of the motion of the center of mass (which may be thought of as the bulk motion of the system) plus the motion of individual particles in the system relative to the center of mass.

The mass of a particle times its velocity is called the **momentum** of the particle. The momentum of a system is the sum of the momenta of the individual particles in the system. When the net external force acting on a system is zero, the system's total momentum remains constant. Momentum in an isolated system is a conserved quantity, just like energy. We will use conservation of momentum to analyze collisions between billiard balls, cars, and subatomic particles, and we will apply it to the decay of radioactive nuclei.

8-1 The Center of Mass

We first consider a simple system of two particles in one dimension. If two point masses, m_1 and m_2, have coordinates x_1 and x_2 on the x axis, then the center-of-mass coordinate x_{cm} is defined by

$$Mx_{cm} = m_1x_1 + m_2x_2 \tag{8-1}$$

where $M = m_1 + m_2$ is the total mass of the system. In the case of just two particles, the center of mass lies at some point on the line between the particles; if the particles have equal masses, the center of mass is midway between them (Figure 8-1).

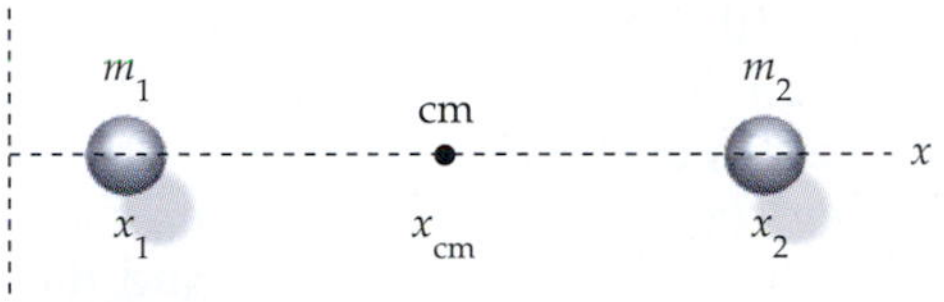

Figure 8-1

If the particles are of unequal mass, the center of mass is closer to the more massive particle (Figure 8-2).

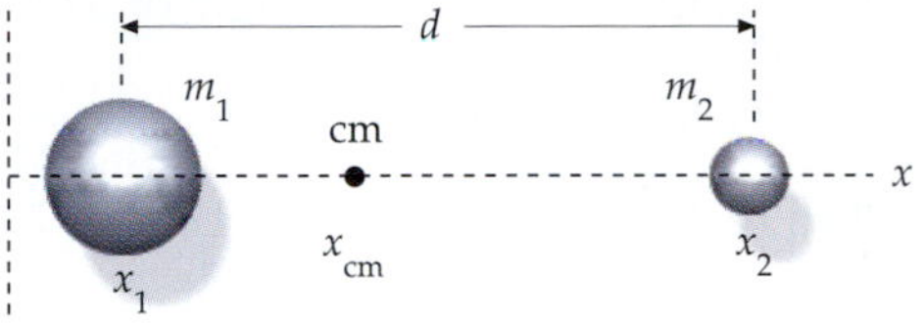

Figure 8-2

If we choose the position of m_1 to be the origin, x_2 is the distance d between the particles (Figure 8-3) and the center of mass is given by

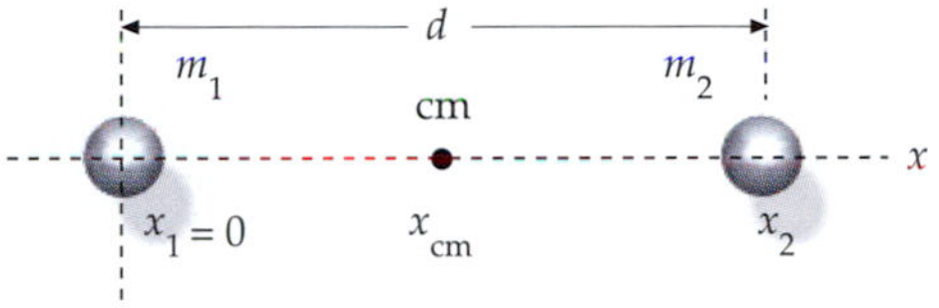

Figure 8-3

$$Mx_{cm} = m_1x_1 + m_2x_2 = m_1(0) + m_2d$$

$$x_{cm} = \frac{m_2}{M}d = \frac{m_2}{m_1 + m_2}d \tag{8-2}$$

Exercise A 4-kg mass is at the origin and a 2-kg mass is at $x = 6$ cm. Find x_{cm}. (*Answer* $x_{cm} = 2$ cm)

We can generalize from two particles in one dimension to a system of many particles in three dimensions. For N particles,

$$Mx_{cm} = m_1x_1 + m_2x_2 + m_3x_3 + \cdots + m_Nx_N = \sum_i m_ix_i \tag{8-3a}$$

where again $M = \Sigma m_i$ is the total mass of the system. Similarly,

$$My_{cm} = \sum_i m_iy_i \quad \text{and} \quad Mz_{cm} = \sum_i m_iz_i \tag{8-3b}$$

In vector notation, $\vec{r}_i = x_i\hat{i} + y_i\hat{j} + z_i\hat{k}$ is the position vector of the ith particle. The position vector of the **center of mass,** $\vec{r}_{cm}$, is defined by

$$M\vec{r}_{cm} = \sum_i m_i\vec{r}_i \tag{8-4}$$

Definition—Center of mass, system of particles

where $\vec{r}_{cm} = x_{cm}\hat{i} + y_{cm}\hat{j} + z_{cm}\hat{k}$.

To find the center of mass of a continuous object, we replace the sum in Equation 8-4 with an integral:

$$M\vec{r}_{cm} = \int \vec{r}\, dm \tag{8-5}$$

Definition—Center of mass, continuous object

where dm is an element of mass located at position $\vec{r}$, as shown in Figure 8-4. Examples showing how to calculate the center of mass using integration are given in Section 8-2. For regularly shaped objects, we can use symmetry to find the center of mass. For example, the center of mass of a uniform cylinder or sphere is at the geometric center of the object.

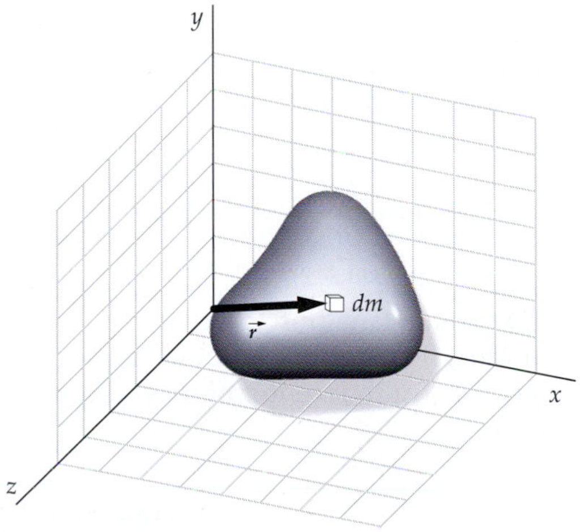

Figure 8-4 Mass element dm located at $\vec{r}$ for finding the center of mass by integration.

Example 8-1

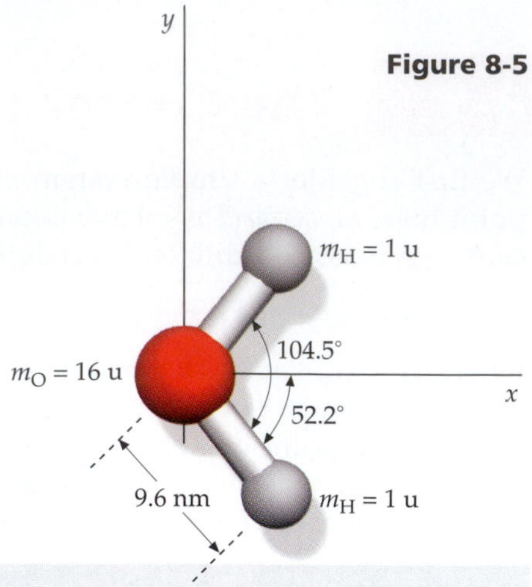

Figure 8-5

Find the center of mass of a water molecule.

Picture the Problem A water molecule consists of an oxygen atom and two hydrogen atoms (Figure 8-5). Oxygen has a mass of 16 unified mass units (u) and each hydrogen has a mass of 1 u. The hydrogen atoms are each at an average distance of 9.6 nm (9.6×10^{-9} m) from the oxygen atom, and are separated from one another by an angle of 104.5°. The calculation is simplified if we place the origin at the location of the oxygen atom, with the x axis bisecting the angle between the hydrogen atoms. Then, given the symmetries of the molecule, the center of mass will be on the x axis, and the line from the oxygen atom to each hydrogen atom will make an angle of 52.2°.

1. The location of the center of mass is given by its coordinates, x_{cm} and y_{cm}:

$$x_{cm} = \frac{\Sigma m_i x_i}{M}, \qquad y_{cm} = \frac{\Sigma m_i y_i}{M}$$

2. By symmetry, the center of mass is on the x axis:

$$y_{cm} = 0$$

3. Write out the expressions for x_{cm} explicitly:

$$x_{cm} = \frac{m_H x_{H1} + m_H x_{H2} + m_O x_O}{m_H + m_H + m_O}$$

4. We have chosen the origin to be the location of oxygen, so the x coordinate of oxygen is zero. The x coordinates of the hydrogen atoms are calculated from the 52.2° angle each hydrogen atom makes with the x axis:

$$x_O = 0$$

$$x_{H1} = x_{H2} = 9.6 \text{ nm} \cos 52.2° = 5.9 \text{ nm}$$

5. Substituting the x coordinates and the mass values into step 3 gives x_{cm}:

$$x_{cm} = \frac{(1 \text{ u})5.9 \text{ nm} + (1 \text{ u})5.9 \text{ nm} + (16 \text{ u})0}{1 \text{ u} + 1 \text{ u} + 16 \text{ u}}$$

$$= 0.66 \text{ nm}$$

Example 8-1 can also be solved by first finding the center of mass of just the two hydrogen atoms. For a system of three particles, Equation 8-4 is

$$M\vec{r}_{cm} = m_1\vec{r}_1 + m_2\vec{r}_2 + m_3\vec{r}_3$$

The first two terms on the right side of this equation are related to the center of mass of the first two particles $\vec{r}\,'_{cm}$:

$$m_1\vec{r}_1 + m_2\vec{r}_2 = (m_1 + m_2)\vec{r}\,'_{cm}$$

The center of mass of the three-particle system can then be written

$$M\vec{r}_{cm} = (m_1 + m_2)\vec{r}\,'_{cm} + m_3\vec{r}_3$$

So we can first find the center of mass for two of the particles, the hydrogen atoms, for example, and then replace them with a single particle of total mass $m_1 + m_2$ at that center of mass (Figure 8-6). The same technique enables us to

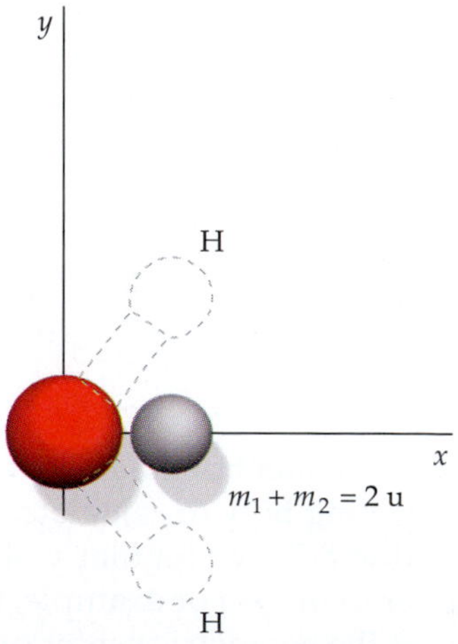

Figure 8-6 Example 8-1 with the two H atoms replaced by a single particle of mass $m_1 + m_2 = 2$ u on the x axis at the center of mass of the original atoms. The center of mass then falls between the oxygen atom at the origin and the calculated center of mass of the two hydrogen atoms.

calculate centers of mass for more complex systems, such as two uniform sticks (Figure 8-7). The center of mass of each stick separately is at the center of the stick. The center of mass of the system is found by treating each stick as a point particle at its individual center of mass.

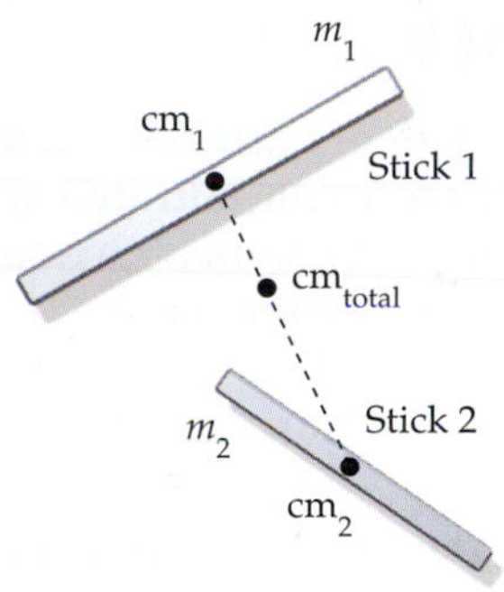

Figure 8-7

Example 8-2 *try it yourself*

Find the center of mass of the uniform sheet of plywood in Figure 8-8.

Picture the Problem The sheet can be divided into two symmetrical parts. The center of mass of each part is at its geometric center. Let m_1 be the mass of part 1 and m_2 be the mass of part 2. The total mass is $M = m_1 + m_2$. The masses are proportional to the areas.

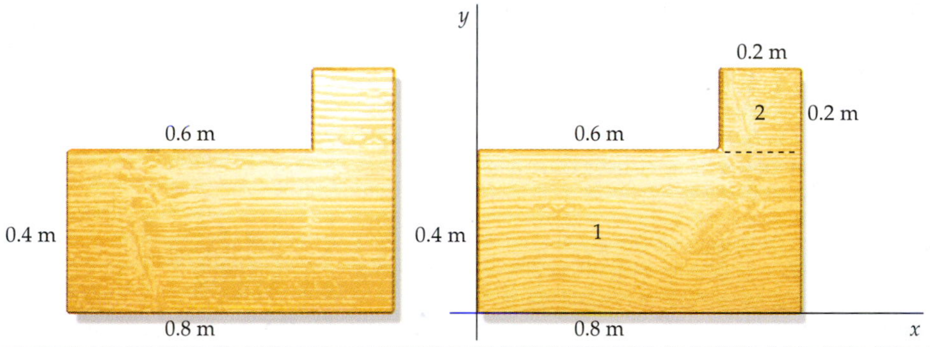

Figure 8-8

Cover the column to the right and try these on your own before looking at the answers.

Steps	***Answers***
1. Write the x and y coordinates of the center of mass in terms of m_1 and m_2.	$mx_{cm} = m_1x_{cm,1} + m_2x_{cm,2}$ $my_{cm} = m_1y_{cm,1} + m_2y_{cm,2}$
2. The mass of each part is proportional to its area. Calculate the areas A_1 and A_2 and the ratio A_1/A_2.	$A_1 = 0.32\ \text{m}^2$, $A_2 = 0.04\ \text{m}^2$, $\frac{A_1}{A_2} = 8$
3. Express the masses m_1 and $M = m_1 + m_2$ in terms of the smallest mass m_2.	$m_1 = 8m_2$, $M = 9m_2$
4. Write the x and y coordinates of the center of mass coordinates for each part by inspection of the figure.	$x_1 = 0.4\ \text{m}$, $y_1 = 0.2\ \text{m}$ $x_2 = 0.7\ \text{m}$, $y_2 = 0.5\ \text{m}$
5. Substitute these results to calculate x_{cm} and y_{cm}.	$x_{cm} = 0.433\ \text{m}$, $y_{cm} = 0.233\ \text{m}$

Remark The center of mass is very near the center of mass of part 1 because $m_1 = 8m_2$.

Gravitational Potential Energy of a System

The gravitational potential energy of a system of particles in a uniform gravitational field is the same as if all the mass were concentrated at the center of mass. Let h_i be the height of the ith particle in a system above some reference level. The gravitational potential energy of the system is

$$U = \sum_i m_i g h_i = g \sum_i m_i h_i$$

But, by definition of the center of mass, the height of the center of mass is given by

$$Mh_{\text{cm}} = \sum_i m_i h_i$$

so

$$U = Mgh_{\text{cm}} \tag{8-6}$$

We can use this result to locate the center of mass of an object experimentally. For example, two objects connected by a light rod will balance if the pivot is at the center of mass (Figure 8-9). If we pivot the system at any other point, the system will rotate until the potential energy is at a minimum, which occurs when the center of mass is at its lowest possible point directly below the pivot (Figure 8-10).

If we suspend any irregular object from a pivot, the object will hang so that its center of mass lies somewhere on the vertical line drawn directly downward from the pivot. Now suspend the object from another point and note where the vertical line now passes across the object. The center of mass will lie at the intersection of the two lines (Figure 8-11).

Figure 8-9

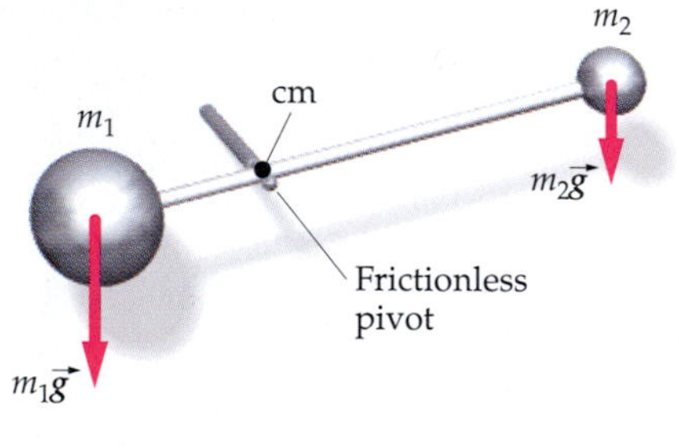

Figure 8-10

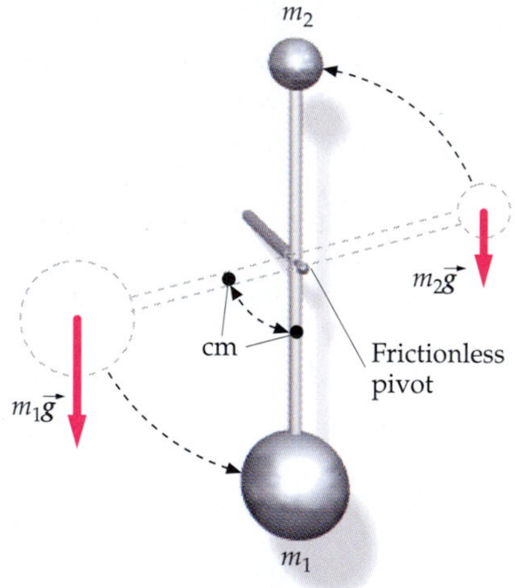

Figure 8-11 The center of mass of an irregular object can be found by suspending it from two points.

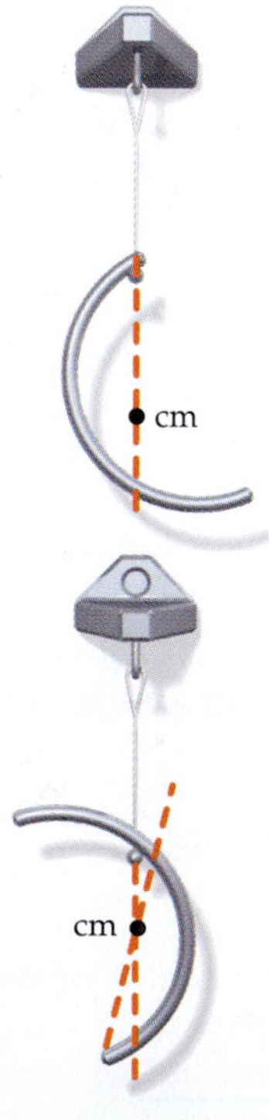

8-2 Finding the Center of Mass by Integration

optional

In this section we illustrate finding the center of mass by integration (Equation 8-5):

$$M\vec{r}_{\text{cm}} = \int \vec{r}\, dm$$

Uniform Stick

This problem, whose answer we can guess from symmetry considerations, illustrates the technique for setting up the integration. We first choose a coordinate system with the x axis along the stick and one end of the stick at the origin (Figure 8-12). Let the mass per unit length of the stick be λ. Since the stick is uniform, $\lambda = M/L$. In Figure 8-12, we have indicated a mass element dm of length dx at a distance x from the origin. The mass of an element of length dx is

$$dm = M\frac{dx}{L} = \frac{M}{L}dx = \lambda\, dx$$

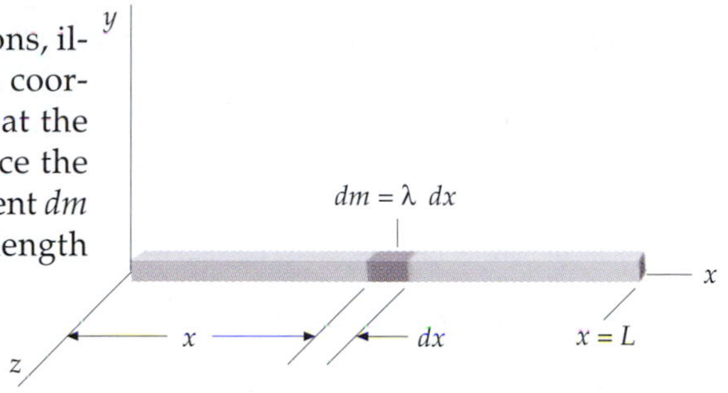

Figure 8-12

Equation 8-5 thus gives

$$Mx_{\text{cm}} = \int x\, dm = \int_0^L x\lambda\, dx = \left.\frac{\lambda x^2}{2}\right|_0^L$$

Using $\lambda = M/L$, we obtain the expected result:

$$x_{\text{cm}} = \frac{\lambda L^2}{2M} = \frac{M}{L}\left(\frac{L^2}{2M}\right) = \frac{1}{2}L$$

Semicircular Hoop

The calculation for determining the center of mass of a semicircular hoop is simplest with the origin at the center of curvature and the y axis on the hoop's line of symmetry (Figure 8-13). Then $x_{\text{cm}} = 0$ because of symmetry. However, $y_{\text{cm}} > 0$, since all of the mass is at positive values of y. In Figure 8-13, we indicate a mass element of length $ds = R\, d\theta$. Since the total length of the hoop is πR, the mass per unit length is $\lambda = M/\pi R$, where M is the total mass. The mass of the element is thus

$$dm = \lambda\, ds = \lambda R\, d\theta$$

The y coordinate of the mass element is related to the angle θ by $y = R \sin\theta$. The angle θ varies from 0 to π. We thus have

$$My_{\text{cm}} = \int y\, dm = \int y\lambda\, ds = \int y\lambda R\, d\theta$$

$$= \int_0^\pi (R\sin\theta)\lambda R\, d\theta = R^2\lambda \int_0^\pi \sin\theta\, d\theta$$

$$= R^2\lambda(-\cos\theta)\Big|_0^\pi = 2R^2\lambda$$

Using $\lambda = M/\pi R$, we have

$$My_{\text{cm}} = 2R^2\frac{M}{\pi R}, \qquad y_{\text{cm}} = \frac{2R}{\pi}$$

In this case, the center of mass is not within the body of the object.

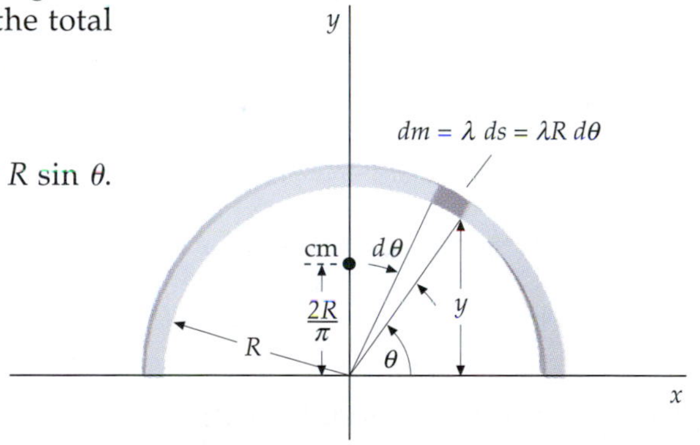

Figure 8-13 Geometry for calculating the center of mass of a semicircular hoop by integration. The center of mass lies on the y axis.

8-3 Motion of the Center of Mass

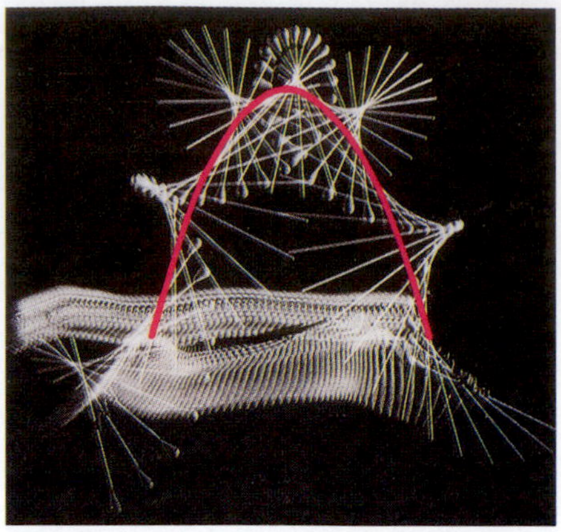

Figure 8-14 Multiflash photo of a baton thrown into the air. The center of mass follows the same simple parabolic path it would if it were a single point particle.

Figure 8-14 is a multiflash photograph of a baton thrown into the air. Although the motion of the baton is complicated, the motion of the center of mass is simple. While the baton is in the air, the center of mass follows a parabolic path, the same path that would be followed by a point particle. We will show in general that the acceleration of the center of mass of a system of particles equals the net external force acting on the system divided by the total mass of the system. For the baton thrown into the air, the acceleration of the center of mass is $\vec{g}$ downward.

To find the acceleration of the center of mass, we first find its velocity by differentiating Equation 8-4 with respect to time:

$$M\frac{d\vec{r}_{\text{cm}}}{dt} = m_1\frac{d\vec{r}_1}{dt} + m_2\frac{d\vec{r}_2}{dt} + \cdots = \sum_i m_i\frac{d\vec{r}_i}{dt}$$

or

$$M\vec{v}_{\text{cm}} = m_1\vec{v}_1 + m_2\vec{v}_2 + \cdots = \sum_i m_i\vec{v}_i \qquad \text{8-7}$$

We differentiate again to obtain the acceleration of the center of mass:

$$M\vec{a}_{\text{cm}} = m_1\vec{a}_1 + m_2\vec{a}_2 + \cdots = \sum_i m_i\vec{a}_i \qquad \text{8-8}$$

According to Newton's second law, we can replace the quantity $m_i\vec{a}_i$ with $\vec{F}_i$, the net force acting on the ith particle. Forces acting on a particle fall into two categories: *internal* forces due to interactions with other particles within the system, and *external* forces due to agents outside the system:

$$\vec{F}_i = m_i\vec{a}_i = \vec{F}_{i,\text{int}} + \vec{F}_{i,\text{ext}}$$

Substituting this into Equation 8-8 gives

$$M\vec{a}_{\text{cm}} = \sum_i \vec{F}_{i,\text{int}} + \sum_i \vec{F}_{i,\text{ext}} \qquad \text{8-9}$$

According to Newton's third law, for each internal force acting on one particle, there is an equal but opposite force acting on another particle. The internal forces thus occur in pairs of equal and opposite forces. When we sum over all the particles in the system, the internal forces cancel, $\Sigma\vec{F}_{i,\text{int}} = 0$, leaving only the external forces. Equation 8-9 then becomes

$$\vec{F}_{\text{net,ext}} = \sum_i \vec{F}_{i,\text{ext}} = M\vec{a}_{\text{cm}} \qquad \text{8-10}$$

Newton's second law for a system

That is, the net external force acting on the system equals the total mass M of the system times the acceleration of the center of mass $\vec{a}_{\text{cm}}$. Thus,

> The center of mass of a system moves like a particle of mass $M = \Sigma\, m_i$ under the influence of the net external force acting on the system.

This theorem is important because it describes the motion of the center of mass for any system of particles: The center of mass behaves just like a single point particle acted on by the external forces. The motions of the individual particles of the system are usually much more complex and are not described by Equation 8-10. The baton thrown into the air in Figure 8-14 is an example. The only external force acting is gravity, so the center of mass of the baton moves in a simple parabolic path, as would a point particle. (The rotation of the baton about its center of mass is not described by Equation 8-10.)

Exercise A cylinder rests on a sheet of paper on a table. You pull the paper to the right, causing the cylinder to roll backward relative to the paper (Figure 8-15). How does the cylinder's center of mass move? (*Answer* It accelerates to the right, because the net external force acting on the cylinder is that of friction to the right. Try it. The cylinder may *appear* to accelerate to the left, because it rolls backward on the paper. But mark its original position on the *table*: While the cylinder is on the paper, the motion of the center of mass is to the right.)

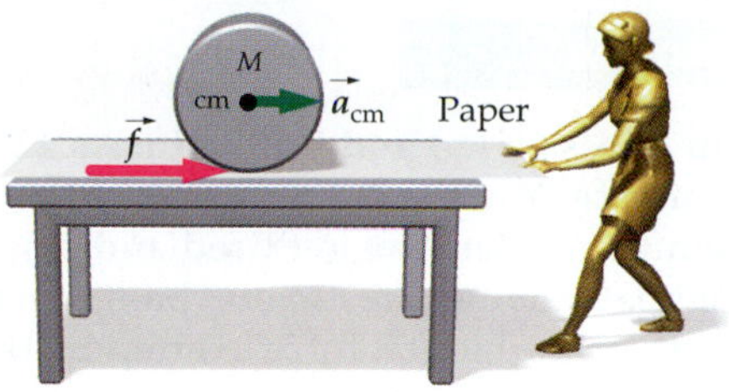

Figure 8-15

Example 8-3

A projectile is fired into the air over level ground with an initial velocity of 24.5 m/s at 36.9° to the horizontal. At its highest point, it explodes into two fragments of equal mass. One fragment falls straight down to the ground. Where does the other fragment land?

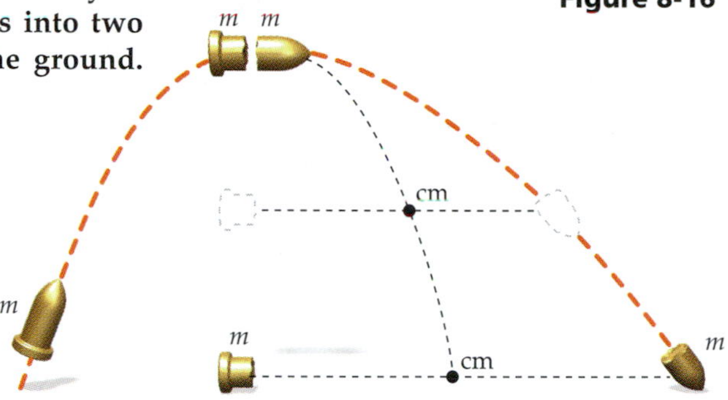

Figure 8-16

Picture the Problem Since the only *external* force acting on the system is gravity, the center of mass, which is midway between the fragments, continues on its parabolic path as if there had been no explosion (Figure 8-16). It lands at $x = R$, where R is the range. The fragment that falls straight down lands at a point $x_1 = 0.5R$. The other fragment must then land at $x_2 = 1.5R$. Let m be the mass of each fragment.

1. The landing positions x_1 and x_2 of the fragments are related to the final position of the center of mass by: $mx_1 + mx_2 = (2m)x_{cm}$

2. Set $x_{cm} = R$, $x_1 = 0.5R$, and solve for x_2: $x_2 = 2x_{cm} - x_1 = 2R - 0.5R = 1.5R$

3. Find the range for the given initial velocity: $R = \frac{v_0^2}{g}\sin 2\theta = \frac{(24.5\ \text{m/s})^2}{9.81\ \text{m/s}^2}\sin(73.8°) = 58.8\ \text{m}$

4. Substitute this value of R to find x_2: $x_2 = 1.5R = 88.2\ \text{m}$

Remarks Figure 8-17 plots the height versus distance for exploding projectiles when the first fragment has a horizontal velocity of half the initial horizontal velocity. As in the original example, in which the first fragment falls straight down, the center of mass follows a normal parabolic trajectory.

Exercise If one of the fragments lands back at the initial position of the projectile, where does the other one land? (*Answer* $2R$)

Remarks If both fragments have no vertical component of velocity after the explosion, they land at the same time. If one fragment is moving downward after the explosion, the other fragment will have an upward component of velocity. The downward-moving fragment will then hit the ground first, and since the ground exerts a force on it before the other fragment lands, our analysis breaks down because there is an unbalanced force on the system.

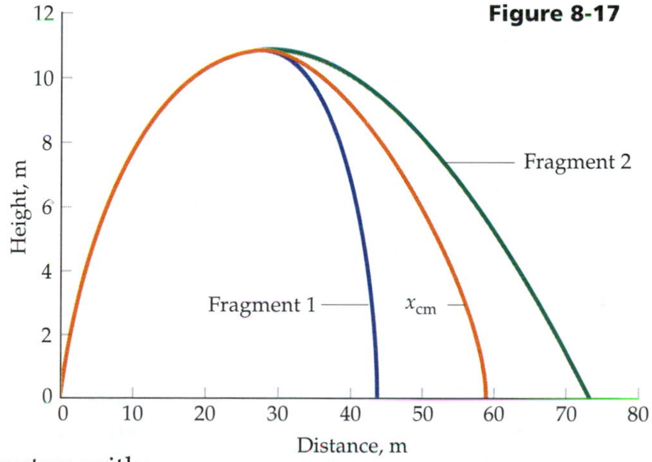

Figure 8-17

A special case of a system's center of mass in motion is the system with zero net external force acting on it. Here, $\vec{a}_{cm} = 0$, so the center of mass is at rest or moves with constant velocity. The internal forces and motion may be complex, but the behavior of the center of mass is simple.

Example 8-4

You (mass 80 kg) and Bubba (mass 120 kg) are in a rowboat (mass 60 kg) on a calm lake. You are at the center of the boat, rowing, and he is at the back, 2 m from the center. You get tired and stop rowing. Bubba offers to row, and after the boat comes to rest, you change places. How far does the boat move? (Neglect any horizontal force exerted by the water.)

Picture the Problem Since there are no external forces in the horizontal direction, the center of mass does not move. This determines the distance the boat must move (Figure 8-18). Choose the origin at the center of the boat. Then $x_{\text{Bubba}} = 2$ m, $x_{\text{you}} = 0$, and $x_{\text{boat}} = 0$. After you and Bubba switch places, the center of mass will have a different x value because the origin moves. The difference in the x values for the center of mass is the distance the origin moves.

Figure 8-18

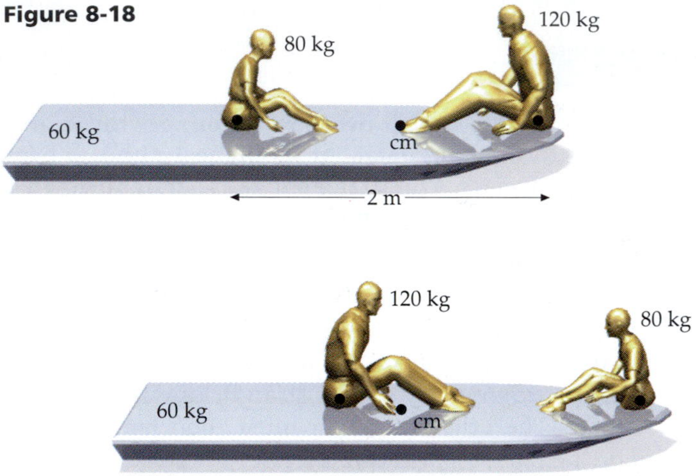

1. Find the initial x value of the center of mass:

$$x_{\text{cm}} = \frac{\Sigma m_i x_i}{M}$$

$$= \frac{(80\text{ kg})(0) + (60\text{ kg})(0) + (120\text{ kg})(2\text{ m})}{80\text{ kg} + 60\text{ kg} + 120\text{ kg}} = 0.923\text{ m}$$

2. Now compute the new coordinate x'_{cm}:

$$x'_{\text{cm}} = \frac{(120\text{ kg})(0) + (60\text{ kg})(0) + (80\text{ kg})(2\text{ m})}{80\text{ kg} + 60\text{ kg} + 120\text{ kg}} = 0.615\text{ m}$$

3. The difference $x_{\text{cm}} - x'_{\text{cm}}$ is the distance that the boat has moved:

$$\Delta x = x_{\text{cm}} - x'_{\text{cm}} = 0.923\text{ m} - 0.615\text{ m} = 0.308\text{ m}$$

Remark Choosing the origin at the center of the boat rather than at one end simplifies the calculation of the center of mass position because two of the masses are then located at $x = 0$.

Example 8-5

A wedge of mass m_2 sits at rest on a scale as shown in Figure 8-19. A small block of mass m_1 slides down the frictionless incline of the wedge. Find the scale reading while the block slides.

Picture the Problem We choose the wedge plus block to be the system. Since the block accelerates down the wedge, the center of mass has acceleration components to the right and downward. The forces on the system are the weights of the block and wedge, the force F_x exerted by the scale on the wedge to the right, and the normal force F_n exerted upward by the scale. The scale reading equals the magnitude of F_n.

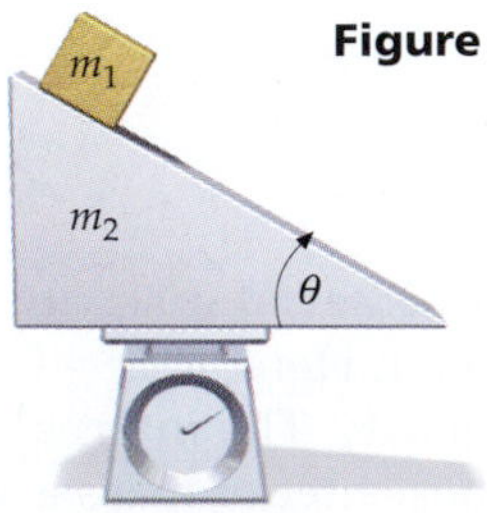

Figure 8-19

1. Draw a free-body diagram for the wedge–block system:

Figure 8-20

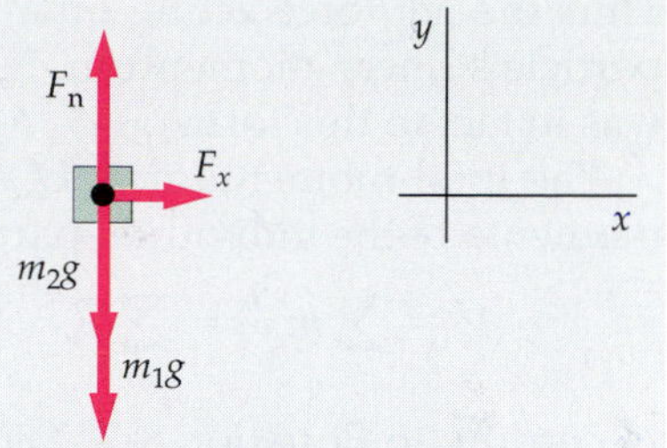

2. Write the vertical component of Newton's second law for the system and solve for F_n:

$$F_n - m_1g - m_2g = Ma_{cm,y} = (m_1 + m_2)a_{cm,y}$$

$$F_n = (m_1 + m_2)g + (m_1 + m_2)a_{cm,y}$$

3. Write the vertical component of the acceleration of the center of mass in terms of the acceleration of the block:

$$Ma_{cm,y} = m_1a_{1y} + m_2a_{2y} = m_1a_{1y}$$

$$a_{cm,y} = \frac{m_1}{m_1 + m_2}a_{1y}$$

4. From Example 4-8, a block sliding down a stationary incline has acceleration $g \sin\theta$ down the incline. Find the y component of this acceleration and use it to find $a_{cm,y}$:

$$a_{1y} = -a_1 \cos\theta = -g \sin^2\theta$$

$$a_{cm,y} = \frac{m_1}{m_1 + m_2}a_{1y} = -\frac{m_1}{m_1 + m_2}g \sin^2\theta$$

5. Substitute this value for $a_{cm,y}$ and calculate F_n:

$$F_n = (m_1 + m_2)g + (m_1 + m_2)a_{cm,y}$$

$$= (m_1 + m_2)g - m_1g \sin^2\theta$$

Exercise Find the force F_x exerted on the wedge by the scale (*Answer* $F_x = m_1g \sin\theta \cos\theta$)

8-4 Conservation of Momentum

A particle's **momentum** $\vec{p}$ is defined as the product of its mass and velocity:

$$\vec{p} = m\vec{v} \qquad \text{8-11}$$

Definition—Momentum of a particle

Momentum is a vector quantity that may be thought of as a measurement of the effort needed to bring a particle to rest.* For example, a heavy truck has more momentum than a light car traveling at the same speed. It takes a greater force to stop the truck in a given time than it does to stop the car.

Newton's second law can be written in terms of the momentum of a particle. Differentiating Equation 8-11 with respect to time, we obtain

$$\frac{d\vec{p}}{dt} = \frac{d(m\vec{v})}{dt} = m\frac{d\vec{v}}{dt} = m\vec{a}$$

Then substituting the force $\vec{F}_{net}$ for $m\vec{a}$, we get

$$\vec{F}_{net} = \frac{d\vec{p}}{dt} \qquad \text{8-12}$$

* The quantity $m\vec{v}$ is sometimes referred to as the *linear momentum* of a particle to distinguish it from the *angular momentum*, which is discussed in Chapter 10.

Thus the net force acting on a particle equals the time rate of change of the particle's linear momentum. Newton's original statement of his second law was in fact in this form.

The total momentum $\vec{P}$ of a system of many particles is the sum of the momenta of the individual particles:

$$\vec{P} = \sum_i m_i\vec{v}_i = \sum_i \vec{p}_i$$

According to Equation 8-7, $\Sigma m_i\vec{v}_i$ equals the total mass M times the velocity of the center of mass:

$$\vec{P} = \sum_i m_i\vec{v}_i = M\vec{v}_{cm} \qquad 8\text{-}13$$

Total momentum of a system

Differentiating this equation with respect to time, we obtain

$$\frac{d\vec{P}}{dt} = M\frac{d\vec{v}_{cm}}{dt} = M\vec{a}_{cm}$$

But according to Newton's second law (Equation 8-10), $M\vec{a}_{cm}$ equals the net external force acting on the system. Thus,

$$\sum_i \vec{F}_{ext} = \vec{F}_{net,ext} = \frac{d\vec{P}}{dt} \qquad 8\text{-}14$$

When the net external force acting on a system of particles is zero, the rate of change of the total momentum is zero, and the total momentum of the system remains constant:

$$\vec{P} = \sum_i m_i\vec{v}_i = M\vec{v}_{cm} = \text{constant} \qquad (\vec{F}_{net,ext} = 0) \qquad 8\text{-}15$$

Conservation of momentum

This result is known as the **law of conservation of momentum:**

If the net external force on a system is zero, the total momentum of the system remains constant.

This law is one of the most important in physics. It is more widely applicable than the law of conservation of mechanical energy, because internal forces exerted by one particle in a system on another are often not conservative. Thus, these internal forces can change the total mechanical energy of the system, though they have no effect on the system's total momentum. We see from Equation 8-15 that if the total momentum is constant, the velocity of the center of mass of the system is constant.

Example 8-6

During repair of the Hubble Space Telescope, an astronaut replaces two solar panels whose frames are bent. Pushing the detached panels away into space, she is propelled in the opposite direction. The astronaut's mass is 60 kg and the panel's mass is 80 kg. The astronaut is at rest relative to her spaceship when she shoves away the panel, and she shoves it at 0.3 m/s relative to the spaceship. What is her subsequent velocity relative to the space ship? (During this operation the astronaut is tethered to the ship; for our calculation, assume that the tether remains slack.)

Picture the Problem The velocity of the astronaut can be found from the velocity of the panel using conservation of momentum. Choose the direction of motion of the panel to be positive.

1. Apply conservation of momentum to find the velocity of the astronaut. Since the total momentum is initially zero, it remains zero:

$$p_p + p_a = m_p v_p + m_a v_a = 0$$

2. Solve for the astronaut's velocity:

$$v_a = -\frac{m_p}{m_a} v_p = -\frac{80 \text{ kg}}{60 \text{ kg}}(0.3 \text{ m/s}) = -0.4 \text{ m/s}$$

Remark Although momentum is conserved, the mechanical energy of this system increased because chemical energy of the astronaut was converted to kinetic energy.

Exercise Find the final kinetic energy of the astronaut–panel system. (*Answer* 8.4 J)

Example 8-7

A runaway 14,000-kg railroad car is rolling at 4 m/s toward a switchyard. A sudden downpour fills the open-topped car with 2000 kg of rainwater. After the rainstorm, how long does it take the car to cover the 500-m distance to the switchyard? Assume that the rain comes straight down and that slowing due to friction is negligible.

Figure 8-21

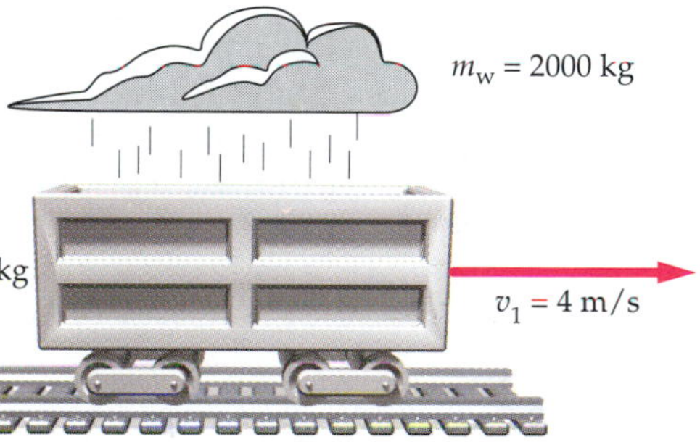

Picture the Problem We find the travel time that we seek from the distance traveled and the speed of the car. Consider the car and the water falling into the car as our system (Figure 8-21). No horizontal external forces act on this system, so the horizontal component of the momentum of the system is conserved. The final speed of the rain-filled car is found from its final momentum, which equals the car's initial momentum. The water initially has no horizontal momentum. Let m_c and m_w be the masses of the car and water, respectively.

1. The time from the end of the storm until the car reaches the yard is the distance d to the yard divided by the car's final speed v_f:

$$\Delta t = \frac{d}{v_f} = \frac{500 \text{ m}}{v_f}$$

2. Apply conservation of momentum to relate the final speed v_f to the initial speed v_i:

$$m_c v_i + m_w(0) = (m_c + m_w)v_f$$

3. Solve for v_f:

$$v_f = \frac{m_c v_i}{m_c + m_w} = \frac{(14{,}000 \text{ kg})(4 \text{ m/s})}{14{,}000 \text{ kg} + 2000 \text{ kg}} = 3.5 \text{ m/s}$$

4. Substitute the result for v_f into step 1:

$$\Delta t = \frac{500 \text{ m}}{v_f} = \frac{500 \text{ m}}{3.5 \text{ m/s}} = 143 \text{ s}$$

Remark Mechanical energy of the system is converted to thermal energy. Let K_w be the kinetic energy of the rainwater just as it hits the car. The initial mechanical energy is $K_w + \frac{1}{2}m_c v_i^2 = K_w + \frac{1}{2}(14{,}000 \text{ kg})(4 \text{ m/s})^2 = K_w + 112 \text{ kJ}$, whereas the final energy is $\frac{1}{2}(m_c + m_w)v_f^2 = \frac{1}{2}(16{,}000 \text{ kg})(3.5 \text{ m/s})^2 = 98 \text{ kJ}$.

Exercise Suppose that there is a small hole in the bottom of the car so that the water leaks out at 10 kg/s. Assume that the car is full when the rain stops. How long does it take the car to cover the 500 m? (*Answer* 143 s. The water leaking out does not impart any momentum to the rest of the system. If the ground were frictionless and nonporous, all of the water initially in the car would arrive at the switchyard along with the car.)

Example 8-8

A 40-kg skateboarder on a 3-kg board is training with two 5-kg weights. Beginning from rest, she throws the weights horizontally one at a time from her board. The velocity of each weight is 7 m/s relative to her and the board after it is thrown. How fast is she propelled in the opposite direction after throwing the second weight? Assume that the board rolls without friction.

Picture the Problem No external horizontal forces act on the system, so the horizontal component of momentum is conserved. We need to find the velocity of the skateboarder after throwing each weight (Figure 8-22). Choose the direction of her motion to be the positive direction. If v_1 is her velocity relative to the ground after throwing the first weight, the weight travels at $v_1 - 7$ m/s relative to the ground. After finding v_1 from conservation of momentum, we use it to find v_2.

Figure 8-22

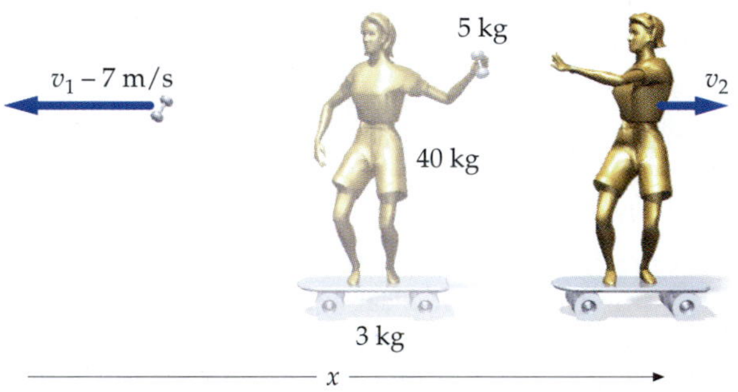

1. Her final velocity v_f is related to her final momentum: $p_f = mv_f$

2. Apply conservation of momentum to relate the final momentum p_f to the initial momentum p_i: $p_i = p_f$

3. The initial momentum p_i is zero. Let v_1 and p_1 be her velocity and momentum after throwing the first weight. The momentum p_1 is that of the skateboard plus girl (43 kg), and one weight (5 kg) with velocity v_1, plus the other weight (5 kg) with velocity $v_1 - 7$ m/s:

$$0 = (48 \text{ kg})v_1 + (5 \text{ kg})(v_1 - 7 \text{ m/s})$$

$$v_1 = \frac{35 \text{ kg·m/s}}{53 \text{ kg}} = 0.660 \text{ m/s}$$

4. When the second weight is thrown, the initial momentum of the girl, skateboard, and weight is $(48 \text{ kg})v_1$. Apply conservation of momentum and solve for v_2:

$$(48 \text{ kg})v_1 = (43 \text{ kg})v_2 + (5 \text{ kg})(v_2 - 7 \text{ m/s})$$

$$v_2 = \frac{(48 \text{ kg})v_1 + 35 \text{ kg·m/s}}{48 \text{ kg}} = 1.73 \text{ m/s}$$

Remark This example illustrates the principle of the rocket; a rocket moves forward by throwing its fuel out the back in the form of exhaust gases.

Exercise How fast is the skateboarder moving if, starting from rest, she throws both weights together, and the weights have velocity 7 m/s relative to her and the board *after they are thrown*? (*Answer* 1.32 m/s)

Example 8-9 *try it yourself*

A thorium-227 nucleus at rest decays into a radium-223 nucleus (mass 223 u) by emitting an α particle (mass 4 u) (Figure 8-23). The kinetic energy of the α particle is found to be 6.00 MeV. What is the kinetic energy of the recoiling radium nucleus?

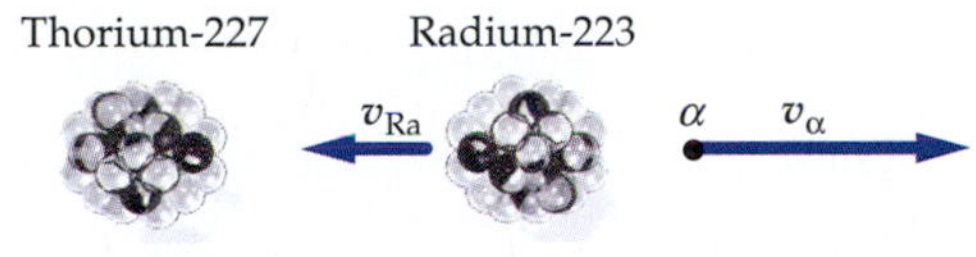

Figure 8-23

Picture the Problem Since the thorium nucleus before decay is at rest, its total momentum is zero. You can therefore relate the velocity of the radium nucleus to that of the α particle using conservation of momentum.

Cover the column to the right and try these on your own before looking at the answers.

Steps	*Answers*
1. Write the kinetic energy of the radium nucleus K_{Ra} in terms of its mass m_{Ra} and speed v_{Ra}.	$K_{Ra} = \frac{1}{2} m_{Ra} v_{Ra}^2$
2. Use conservation of momentum to relate v_{Ra} to the speed of the α particle v_α.	$m_\alpha v_\alpha = m_{Ra} v_{Ra}$
3. Substitute your expression for v_α into the kinetic energy expression in step 1.	$K_{Ra} = \frac{1}{2} m_{Ra} \times \left(\frac{m_\alpha v_\alpha}{m_{Ra}}\right)^2$
4. Factor $K_{Ra} = \frac{1}{2} m_\alpha v_\alpha^2$ from your expression in step 3.	$K_{Ra} = \frac{m_\alpha}{m_{Ra}}\left(\frac{1}{2} m_\alpha v_\alpha^2\right)$
5. Substitute the given values to calculate K_{Ra}.	$K_{Ra} = 0.107$ MeV

Remark In this process, rest energy of the thorium nucleus is converted into kinetic energy of the α particle plus radium nucleus. The mass of the thorium nucleus is greater than that of the α particle plus radium nucleus by about 6.1 MeV$/c^2$.

8-5 Kinetic Energy of a System

Although the total momentum of a system of particles must be constant if the net external force on the system is zero, the total mechanical energy of the system can change. As we saw in the examples of the previous section, internal forces that cannot change the total momentum may be nonconservative and thus change the total mechanical energy of the system. There is an important theorem concerning the kinetic energy of a system of particles that allows us to treat the energy of complex systems more easily and gives us insight into energy changes within a system:

> The kinetic energy of a system of particles can be written as the sum of two terms: (1) the kinetic energy associated with the motion of the center of mass, $\frac{1}{2}Mv_{cm}^2$, where M is the total mass of the system; and (2) the kinetic energy associated with the motion of the particles of the system relative to the center of mass, $\Sigma\frac{1}{2}m_i u_i^2$, where $\vec{u}_i$ is the velocity of the ith particle relative to the center of mass.

The kinetic energy of a system of particles is the sum of the kinetic energies of the individual particles:

$$K = \sum_i K_i = \sum_i \frac{1}{2} m_i v_i^2 = \sum_i \frac{1}{2} m_i (\vec{v}_i \cdot \vec{v}_i)$$

The velocity of each particle can be written as the sum of the velocity of the center of mass, $\vec{v}_{cm}$, and the velocity of the particle relative to the center of mass, $\vec{u}_i$:

$$\vec{v}_i = \vec{v}_{cm} + \vec{u}_i \qquad 8\text{-}16$$

Then

$$K = \sum_i \frac{1}{2} m_i(\vec{v}_i \cdot \vec{v}_i) = \sum_i \frac{1}{2} m_i(\vec{v}_{cm} + \vec{u}_i)\cdot(\vec{v}_{cm} + \vec{u}_i)$$

$$= \sum_i \frac{1}{2} m_i v_{cm}^2 + \sum_i \frac{1}{2} m_i u_i^2 + \vec{v}_{cm} \cdot \sum_i m_i \vec{u}_i$$

where in the last term we have removed $\vec{v}_{cm}$ from the sum because it is the same for each particle; that is, it refers to the system and not to any particular particle. The quantity $\Sigma m_i \vec{u}_i$ is the total momentum of the system *relative to the center of mass*. This quantity is necessarily zero. Relative to the center of mass, the velocity of the center of mass, $\vec{u}_{cm}$, is zero, and the total momentum $M\vec{u}_{cm}$ is also zero. Then

$$K = \sum_i \frac{1}{2} m_i v_{cm}^2 + \sum_i \frac{1}{2} m_i u_i^2 = \frac{1}{2} M v_{cm}^2 + K_{rel} \qquad 8\text{-}17$$

Kinetic energy of a system of particles

where M is the total mass and K_{rel} is the kinetic energy of the particles relative to the center of mass. When there are no external forces, v_{cm} is constant and the kinetic energy associated with bulk motion ($\frac{1}{2}Mv_{cm}^2$) does not change. Only the relative kinetic energy can change in an isolated system.

8-6 Collisions

In a collision, two objects approach and interact strongly for a very short time. During the brief time of collision, any external forces are much smaller than the forces of interaction between the objects. Thus, the only important forces acting on the two-object system are the interaction forces, which are equal and opposite, so the total momentum of the system remains unchanged. The collision time is usually so small that the displacement of the objects during the collision can be neglected. Before and after the collision, the interaction of the two objects is small compared with the interaction during the collision. Examples are a cue ball hitting an object ball, a baseball being hit by a bat, a dart colliding with a dart board, and a comet swinging around the sun.

When the total kinetic energy of the two objects is the same after the collision as before, the collision is called an **elastic collision;** otherwise, it is called an **inelastic collision.** An extreme case is the **perfectly inelastic collision**, in which all of the kinetic energy relative to the center of mass is converted to thermal or internal energy of the system, and the two objects stick together after the collision.

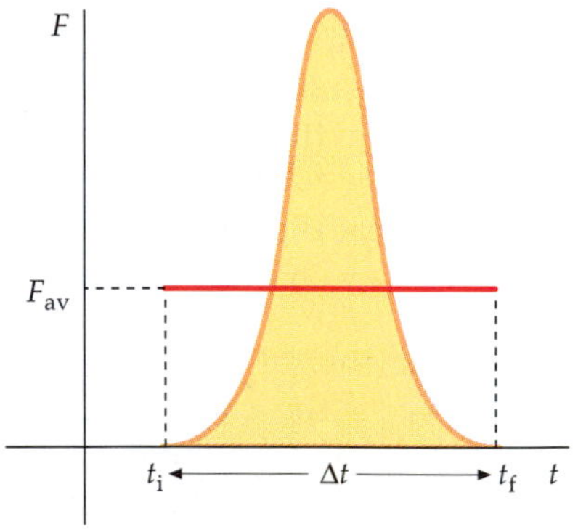

Figure 8-24 Typical time variation of force during a collision. The area under the F-versus-t curve is the magnitude of the impulse, I. F_{av} is the average force for time interval Δt. The rectangular area $F_{av}\,\Delta t$ is the same as the area under the F-versus-t curve.

Impulse and Average Force

Figure 8-24 shows the time variation of the magnitude of a typical force exerted by one object on another during a collision. During the collision time $\Delta t = t_f - t_i$, the force is large. For other times, the force is negligibly small.

The **impulse** $\vec{I}$ of the force is a vector defined as

$$\vec{I} = \int_{t_i}^{t_f} \vec{F}\,dt \qquad 8\text{-}18$$

Definition—Impulse

The magnitude of the impulse of the force is the area under its F-versus-t curve. The unit of impulse is N·s. If $\vec{F}$ is the net force acting on a particle, it is related to the rate of change of momentum of the particle by Newton's second law, $\vec{F} = d\vec{p}/dt$. Then the impulse of the net force equals the total change in momentum during the time interval:

$$\vec{I}_{net} = \int_{t_i}^{t_f} \vec{F}_{net}\,dt = \int_{t_i}^{t_f} \frac{d\vec{p}}{dt}\,dt = \vec{p}_f - \vec{p}_i = \Delta\vec{p} \qquad 8\text{-}19$$

The **average force** for the interval $\Delta t = t_f - t_i$ is defined as

$$\vec{F}_{av} = \frac{1}{\Delta t}\int_{t_i}^{t_f} \vec{F}\,dt = \frac{\vec{I}}{\Delta t} \qquad 8\text{-}20$$

Definition—Average force

The average force is the constant force that gives the same impulse as the actual force in the time interval Δt, as shown by the rectangle in Figure 8-24. The average force can be calculated from the change in momentum if a collision time is known. This time is often estimated using the distance traveled by one of the objects during the collision.

Example 8-10

With an expert karate blow, you shatter a concrete block. Consider your fist to have a mass 0.70 kg, to be moving 5.0 m/s as it strikes the block, and to stop within 6 mm of the point of contact. (*a*) What impulse does the block exert on your fist? (*b*) What is the approximate collision time and the average force the block exerts on your fist?

Picture the Problem The impulse equals the change in momentum $\Delta\vec{p}$. We find $\Delta\vec{p}$ from the mass and velocity of the fist. The time of collision for part (*b*) comes from the given distance $\Delta y = 6$ mm and the average velocity during the collision, which we can estimate by assuming constant acceleration. We will choose upward as the positive direction.

(*a*) 1. Set the impulse equal to the change in momentum: $\vec{I} = \Delta\vec{p} = \vec{p}_f - \vec{p}_i$

2. The initial momentum is that of the fist just before it hits the block with speed v, and the final momentum is zero: $\vec{p}_i = -(0.7\text{ kg})(5.0\text{ m/s})\,\hat{j} = -3.5\text{ kg·m/s}\,\hat{j}$

$\vec{p}_f = 0$

3. Find the impulse exerted by the block on the fist: $\vec{I} = \vec{p}_f - \vec{p}_i = 0 - (-3.5\text{ kg·m/s}\,\hat{j}) = 3.5\text{ kg·m/s}\,\hat{j}$

(*b*) 1. The collision time is the distance moved divided by the average speed: $\Delta t = \dfrac{\Delta y}{v_{av}}$

2. Assuming constant acceleration, $v_{av} = \frac{1}{2}v$. Since we have chosen upward to be positive, both Δy and v_{av} are negative. Calculate Δt: $\Delta t = \dfrac{\Delta y}{\frac{1}{2}v} = \dfrac{-0.006\text{ m}}{-2.5\text{ m/s}} = 0.0024\text{ s} = 2.4\text{ ms}$

3. The average force is the impulse divided by the collision time. It is upward, as expected: $\vec{F}_{av} = \dfrac{\vec{I}}{\Delta t} = \dfrac{3.5\text{ N·s}\,\hat{j}}{0.0024\text{ s}} = 1.46\text{ kN}\,\hat{j}$

Remark The average force is large—about 116 times the weight of the fist.

Example 8-11 *try it yourself*

A car equipped with an 80-kg crash-test dummy drives into a wall at 25 m/s (about 56 mi/h). Estimate the force that the seat belt exerts on the dummy upon impact.

Picture the Problem Assume that the car and dummy travel about 1 m as the car comes to rest, and that the acceleration is constant during the crash (Figure 8-25). This means that the average speed of the car during the collision is one-half the initial speed, or $v_{av} = 12.5$ m/s. To find the force, calculate the impulse I, then divide by the collision time Δt.

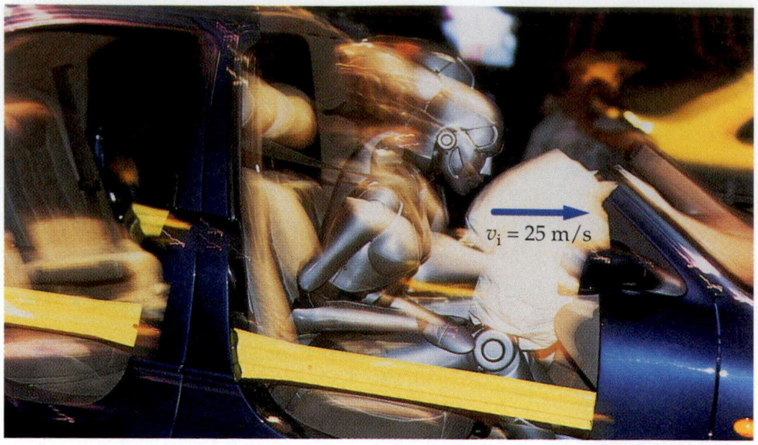

Figure 8-25

Cover the column to the right and try these on your own before looking at the answers.

Steps	*Answers*
1. Find the dummy's initial momentum.	$mv = 2000$ kg·m/s
2. Set the impulse equal to the change in momentum to find the magnitude of the impulse exerted by the seat belt on the dummy.	$I = 2000$ N·s
3. Estimate the collision time using $\Delta x = 1$ m and $v_{av} = 12.5$ m/s.	$\Delta t = 0.08$ s
4. Compute the average force.	$F_{av} = 25{,}000$ N

Remark The average acceleration is $a_{av} = \Delta v/\Delta t = 313\ \text{m/s}^2$, or roughly 32 times the acceleration due to gravity. A large acceleration means a large force, as step 4 of the example reveals. 25,000 N (about 5600 lb) is clearly enough to cause serious injuries. An air bag increases the stopping distance somewhat, which helps to prevent injury. The air bag also allows the force to be distributed over a much larger area.

Remark (*a*) Figure 8-26 shows the average force exerted by the seat belt on the dummy as a function of the stopping distance x. With no seat belt or air bag, you either fly through the windshield or are stopped in a fraction of a meter by the dashboard or steering wheel. (*b*) The plot shows the force as a function of the initial velocity for three stopping distances: 2 m, 1.5 m, and 1 m.

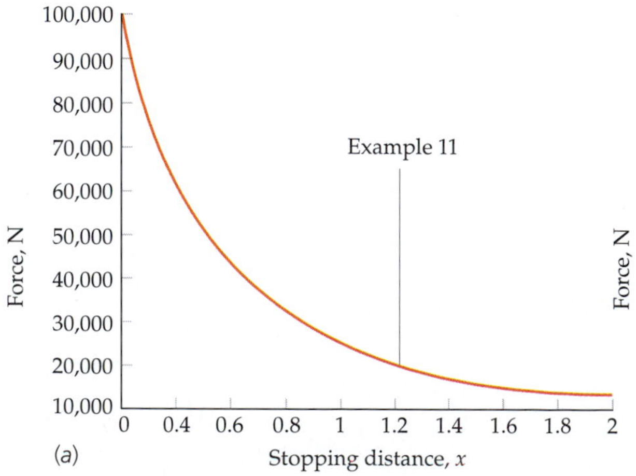

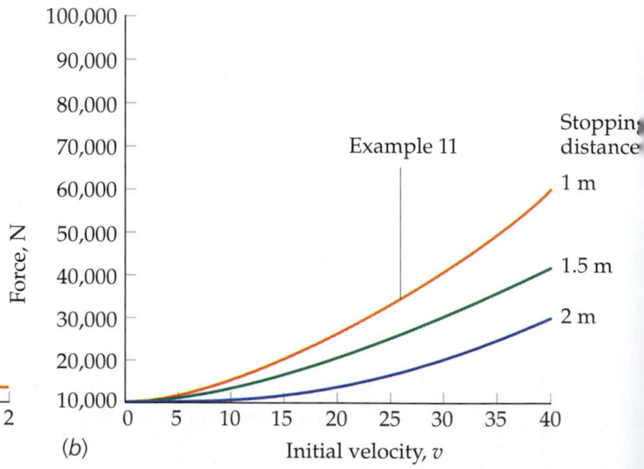

Figure 8-26

Example 8-12

You strike a golf ball with a club. What are reasonable estimates for (*a*) the impulse I, (*b*) the collision time Δt, and (*c*) the average force F_{av}? A typical golf ball has a mass $m = 45$ g and a radius $r = 2$ cm (Figure 8-27). For a typical drive, the range is roughly $R = 160$ m (about 175 yards).

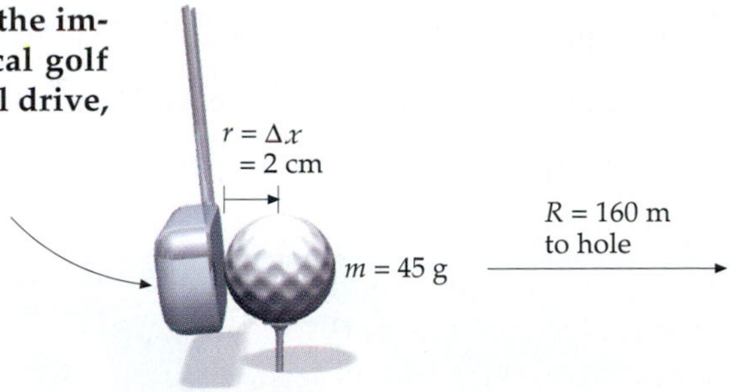

Figure 8-27

Picture the Problem The impulse equals the change in momentum of the ball, which is mv_0. We estimate the initial speed v_0 from the range. We estimate the collision time from the distance traveled Δx and the average speed $\frac{1}{2}v_0$, assuming constant acceleration. We use $\Delta x = 2$ cm, the radius of the ball. The average force is then obtained from the impulse I and collision time Δt.

(*a*) 1. Set the impulse equal to the change in momentum of the ball:

$$I = mv_0$$

2. The initial speed is related to the range R, which is given by Equation 2-22:

$$R = \frac{v_0^2}{g}\sin 2\theta_0$$

3. Take $\theta_0 = 45°$ corresponding to maximum range ($\sin 2\theta_0 = 1$), and calculate the initial speed:

$$v_0 = \sqrt{Rg} = \sqrt{(160\text{ m})(9.81\text{ m/s}^2)} = 40\text{ m/s}$$

4. Use this value of v_0 to calculate the magnitude of the impulse:

$$I = mv_0 = (0.045\text{ kg})(40\text{ m/s})$$
$$= 1.8\text{ kg}\cdot\text{m/s} = 1.8\text{ N}\cdot\text{s}$$

(*b*) Calculate the collision time Δt using $x = 2$ cm, and $v_{av} = \frac{1}{2}v_0$:

$$\Delta t = \frac{\Delta x}{v_{av}} = \frac{\Delta x}{\frac{1}{2}v_0} = \frac{0.02\text{ m}}{20\text{ m/s}} = 0.001\text{ s}$$

(*c*) Use the calculated values of I and Δt to find the magnitude of the average force:

$$f_{av} = \frac{I}{\Delta t} = \frac{1.8\text{ N}\cdot\text{s}}{0.001\text{ s}} = 1800\text{ N}$$

Remark Again we see very large forces exerted during a collision. Here the force exerted on the golf ball by the club is roughly 4000 times the weight of the ball, giving it a brief acceleration of $4000g$. In comparison, the force of friction exerted on the ball during the collision is negligible.

Collisions in One Dimension

Consider an object of mass m_1 with initial velocity v_{1i} approaching a second object of mass m_2 that is moving in the same direction with initial velocity v_{2i}. If $v_{2i} < v_{1i}$, the objects collide. Let v_{1f} and v_{2f} be their final velocities after the collision. (The velocities can be positive or negative, depending on whether the objects are moving to the right or left.) Conservation of momentum gives one relation between the two unknown velocities v_{1f} and v_{2f}:

$$m_1v_{1f} + m_2v_{2f} = m_1v_{1i} + m_2v_{2i} \qquad \text{8-21}$$

To determine v_{1f} and v_{2f}, we must have a second relation. That second relation, which we shall now develop, depends on the type of collision.

Perfectly Inelastic Collisions in One Dimension In perfectly inelastic collisions, the particles stick together after the collision. The second relation between the final velocities is that they are equal to each other and to the velocity of the center of mass:

$$v_{1f} = v_{2f} = v_{cm}$$

This result combined with conservation of momentum gives

$$(m_1 + m_2)v_{cm} = m_1v_{1i} + m_2v_{2i} \qquad 8\text{-}22$$

Perfectly inelastic collision of two cars.

Example 8-13

An astronaut of mass 60 kg is on a space walk to repair a communications satellite. Suddenly she needs to consult her physics book. You happen to have it with you, so you throw it to her with speed 4 m/s relative to your spacecraft. She is at rest relative to the spacecraft just before catching the 3.0-kg book (Figure 8-28). Find (*a*) her velocity just after she catches the book, (*b*) the initial and final mechanical energy of the book–astronaut system, and (*c*) the impulse exerted by the book on the astronaut.

Picture the Problem (*a*) The final velocity of the book and astronaut is the velocity of the center of mass. We find this using conservation of momentum, as expressed in Equation 8-21. The initial and final kinetic energies are calculated from the initial and final velocities. Since the book and astronaut move with the same final velocity, the collision is perfectly inelastic. (*b*) The mechanical energies of the book and astronaut are calculated directly from their masses and speeds. (*c*) The impulse exerted by the book on the astronaut equals the change in momentum of the astronaut.

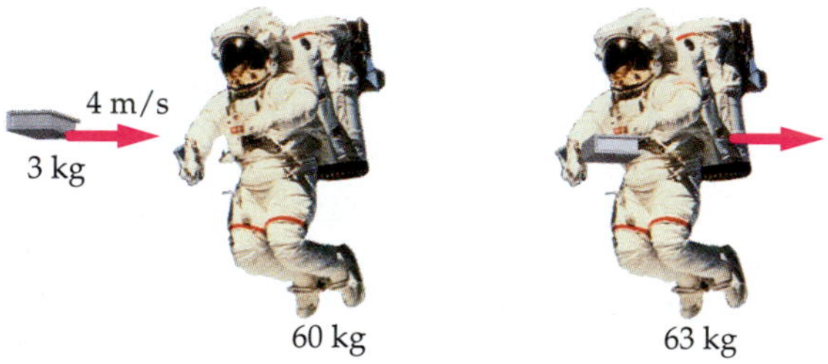

Figure 8-28

(*a*)1. Use conservation of momentum to relate the final velocity of the system, v_{cm}, to the initial velocities:

$$m_bv_b + m_av_a = (m_b + m_a)v_{cm}$$

$$(3.0\text{ kg})(4\text{ m/s}) + (60\text{ kg})(0) = (63\text{ kg})v_{cm}$$

2. Solve for v_{cm}:

$$v_{cm} = \frac{(3.0\text{ kg})(4\text{ m/s})}{63\text{ kg}} = 0.19\text{ m/s}$$

(*b*)1. The initial mechanical energy of the book–astronaut system is the kinetic energy of the book:

$$E_1 = K_b = \frac{1}{2}m_bv_b = \frac{1}{2}(3.0\text{ kg})(4\text{ m/s})^2 = 24\text{ J}$$

2. The final mechanical energy is the kinetic energy of the book and astronaut moving together:

$$E_f = K_f = \frac{1}{2}(m_b + m_a)v_{cm}^2$$

$$= \frac{1}{2}(63\text{ kg})(0.19\text{ m/s})^2 = 1.14\text{ J}$$

(*c*) Set the impulse exerted on the astronaut equal to the change in momentum of the astronaut:

$I = \Delta p_{ast} = m_{ast}\Delta v_{ast} = (60\text{ kg})(0.19\text{ m/s})$
$= 11.4\text{ kg}\cdot\text{m/s} = 11.4\text{ N}\cdot\text{s}$

Remark Most of the initial mechanical energy in this collision is lost by conversion to thermal energy. The impulse exerted by the book on the astronaut is equal and opposite to that exerted by the astronaut on the book, so the total change in momentum is zero.

It is useful to express the kinetic energy K of a particle in terms of its momentum p. For a mass m moving with speed v, we have

$$K = \frac{1}{2}mv^2 = \frac{(mv)^2}{2m}$$

Since $\vec{p} = m\vec{v}$,

$$K = \frac{p^2}{2m} \qquad \text{8-23}$$

We can apply this to a perfectly inelastic collision where one object is initially at rest. The momentum of the system is that of the incoming object:

$$P = m_1 v_{1i}$$

The initial kinetic energy is

$$K_i = \frac{P^2}{2m_1} \qquad \text{8-24}$$

After colliding, the objects move together as a single mass $m_1 + m_2$ with v_{cm}. Momentum is conserved, so the final momentum equals P. The final kinetic energy is then

$$K_f = \frac{P^2}{2(m_1 + m_2)} \qquad \text{8-25}$$

Comparing Equations 8-24 and 8-25, we see that the final energy is less than the initial energy.

Example 8-14

In a feat of public marksmanship, you fire a bullet into a hanging target (Figure 8-29). The target, with bullet embedded, swings upward. Noting the height reached at the top of the swing, you immediately inform the crowd of the bullet's speed. For arbitrary masses m_1 and m_2, and height h, how would you calculate the speed?

Picture the Problem The initial speed of the bullet, v_{1i}, is related to the speed of the bullet–block system, v_f, just after the inelastic collision by conservation of momentum. The speed v_f is related to the height h by conservation of mechanical energy. Let m_1 be the mass of the bullet and m_2 be the mass of the target.

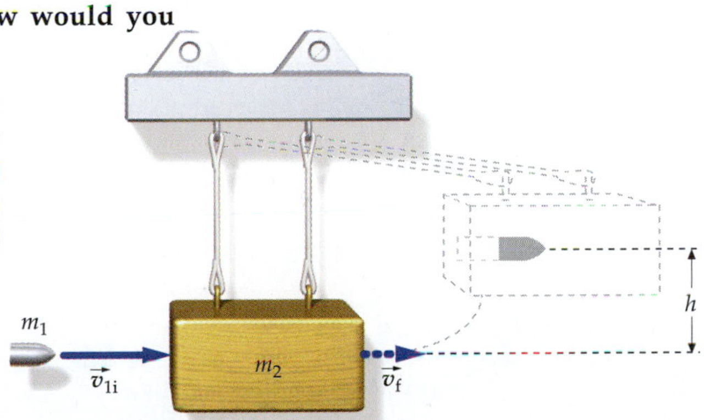

Figure 8-29

1. Use conservation of momentum during the collision to find v_{1i} in terms of v_f:

$$m_1 v_{1i} + m_2(0) = (m_1 + m_2)v_f$$

$$v_{1i} = \frac{m_1 + m_2}{m_1} v_f$$

2. Use conservation of mechanical energy after the collision to find v_f in terms of the height h:

$$\frac{1}{2}(m_1 + m_2)v_f^2 = (m_1 + m_2)gh$$

$$v_f = \sqrt{2gh}$$

3. Substituting v_f into the equation in step 1, we can solve for v_{1i}:

$$v_{1i} = \frac{m_1 + m_2}{m_1} v_f = \frac{m_1 + m_2}{m_1}\sqrt{2gh}$$

Remark In this problem, as in all collision problems, we assume that the time of the collision is so short that the displacement of the block during the collision is negligible. Devices such as the one pictured are called *ballistic pendulums*.

Exercise If the mass of the bullet is 12 g, the mass of the block on the ballistic pendulum is 2 kg, and the final height is 10.4 cm, what speed did you announce to the crowd? (*Answer* 240 m/s)

Exercise A 2000-kg car moving 25 m/s runs head-on into a 1500-kg car initially at rest. If the collision is perfectly inelastic, find (*a*) each car's speed after the collision, and (*b*) the ratio of the system's final kinetic energy to its initial kinetic energy. (*Answers* (*a*) 14.3 m/s, (*b*) 0.57)

Example 8-15 *try it yourself*

You repeat your feat of Example 8-14, this time with an empty box as a target. The bullet strikes the target and passes through it completely. A laser ranging device indicates that the bullet emerged with half its initial velocity. Hearing this, you correctly report how high the target must have swung. How high did it swing?

Picture the Problem The height h is related to the box's speed after colliding, v_2, by conservation of mechanical energy (Figure 8-30). This speed can be determined using conservation of momentum.

Figure 8-30

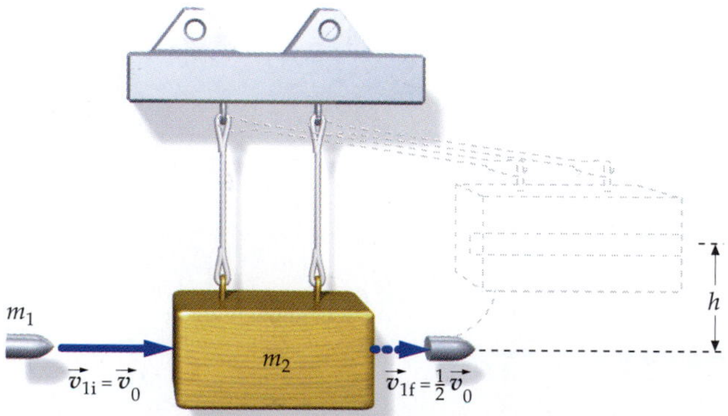

Cover the column to the right and try these on your own before looking at the answers.

Steps	Answers
1. Use conservation of mechanical energy to relate the final height h to the speed v_2 of the box after the collision.	$m_2gh = \frac{1}{2}m_2v_2^2$
2. Use conservation of momentum to write an equation relating the speed v_2 of the box to v_0.	$m_2v_2 + m_1\left(\frac{1}{2}v_0\right) = m_1v_0$
3. Solve for v_2.	$v_2 = \frac{m_1}{2m_2}v_0$
4. Substitute this value of v_2 into your equation in step 1 for h.	$h = \frac{v_2^2}{2g} = \frac{m_1^2v_0^2}{8m_2^2g}$

Remark Inelastic collisions also occur in microscopic systems. For example, when an electron is scattered by an atom, the atom is sometimes excited to a higher internal energy state. As a result, the total kinetic energy of the atom and the electron is lower after the collision than it was before.

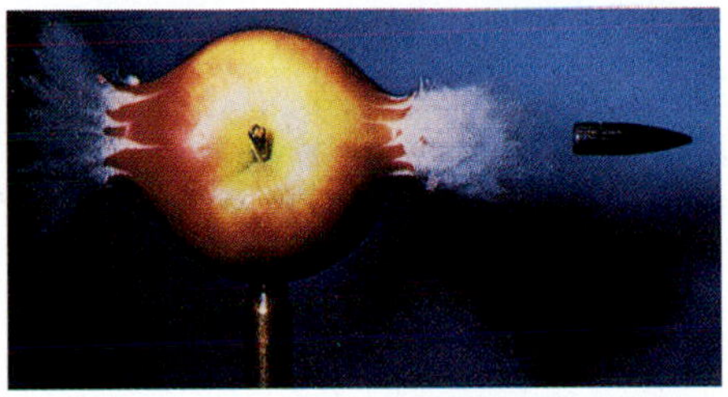

A bullet traveling 850 m/s collides inelastically with an apple, which moments later disintegrates completely. Exposure time is less than a millionth of a second.

Elastic Collisions in One Dimension For elastic collisions, the initial and final kinetic energies are equal:

$$\frac{1}{2}m_1v_{1f}^2 + \frac{1}{2}m_2v_{2f}^2 = \frac{1}{2}m_1v_{1i}^2 + \frac{1}{2}m_2v_{2i}^2 \qquad 8\text{-}26$$

This, together with conservation of momentum (Equation 8-21), is sufficient to determine the final velocities of the two objects. However, the quadratic nature of Equation 8-26 often complicates the solution of an elastic collision problem. Such problems can be treated more easily if we express the relative velocity of the two particles after the collision in terms of the relative velocity before the collision. Rearranging Equation 8-26 gives

$$m_2(v_{2f}^2 - v_{2i}^2) = m_1(v_{1i}^2 - v_{1f}^2)$$

or

$$m_2(v_{2f} - v_{2i})(v_{2f} + v_{2i}) = m_1(v_{1i} - v_{1f})(v_{1i} + v_{1f}) \qquad 8\text{-}27$$

From conservation of momentum, we know that

$$m_1v_{1f} + m_2v_{2f} = m_1v_{1i} + m_2v_{2i}$$

so that

$$m_2(v_{2f} - v_{2i}) = m_1(v_{1i} - v_{1f}) \qquad 8\text{-}28$$

Then dividing Equation 8-27 by Equation 8-28, we get

$$v_{2f} + v_{2i} = v_{1i} + v_{1f}$$

or

$$v_{2f} - v_{1f} = -(v_{2i} - v_{1i}) \qquad 8\text{-}29$$

Relative velocities in an elastic collision

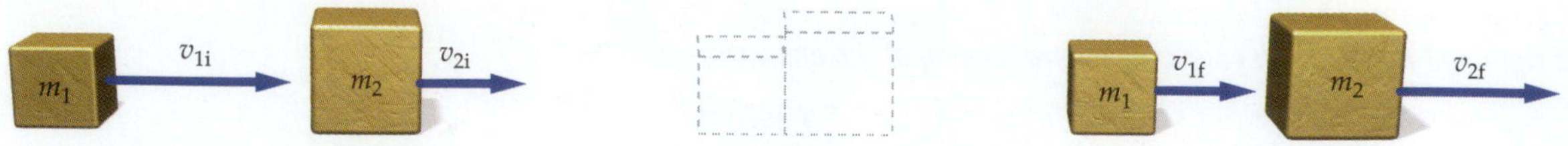

Figure 8-31 Approach and recession in an elastic collision.

If two objects are to collide, $v_{2i} - v_{1i}$ must be negative (Figure 8-31), making their **speed of approach** $-(v_{2i} - v_{1i})$. After colliding, the objects' **speed of recession** is $v_{2f} - v_{1f}$, which is positive. Equation 8-29 states

> In elastic collisions, the speed of recession equals the speed of approach.

Solving elastic-collision problems is usually easier using Equation 8-29 than Equation 8-26. But remember, Equation 8-29 depends on conservation of mechanical energy, so it applies only to *elastic* collisions.

Example 8-16

A 4-kg block moving right at 6 m/s collides elastically with a 2-kg block moving right at 3 m/s (Figure 8-32). Find their final velocities.

Picture the Problem Conservation of momentum (Equation 8-18) and conservation of energy (expressed as a reversal of relative velocities, Equation 8-24) give two equations for the two unknown final velocities. Let subscript 1 denote the 4-kg block, and subscript 2 denote the 2-kg block.

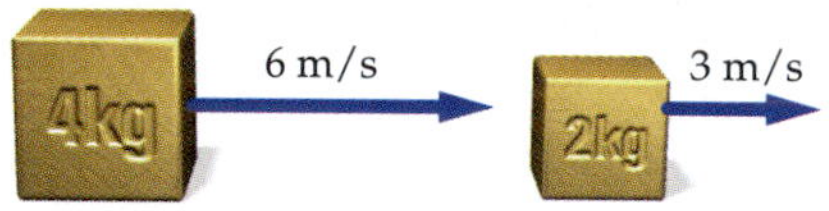

Figure 8-32

1. Apply conservation of momentum: $p_i = m_1v_{1i} + m_2v_{2i} = 30\ \text{kg}\cdot\text{m/s}$

   $p_f = m_1v_{1f} + m_2v_{2f} = p_i$

2. Substituting numerical values in step 1 relates v_{1f} and v_{2f}: $4v_{1f} + 2v_{2f} = 30\ \text{m/s}$

3. Calculate the velocity of approach: $v_{2i} - v_{1i} = -3\ \text{m/s}$

4. Use conservation of mechanical energy to set the velocity of recession equal to the negative of the velocity of approach: $v_{2f} - v_{1f} = -(v_{2i} - v_{1i}) = 3\ \text{m/s}$

5. With the two relations for two unknowns in steps 2 and 4, we solve for the final velocities: $v_{1f} = 4\ \text{m/s}$ and $v_{2f} = 7\ \text{m/s}$

Check the Result As a check, we calculate the initial and final kinetic energies: $K_i = \frac{1}{2}(4\ \text{kg})(6\ \text{m/s})^2 + \frac{1}{2}(2\ \text{kg})(3\ \text{m/s})^2 = 72\ \text{J} + 9\ \text{J} = 81\ \text{J}$; $K_f = \frac{1}{2}(4\ \text{kg})(4\ \text{m/s})^2 + \frac{1}{2}(2\ \text{kg})(7\ \text{m/s})^2 = 32\ \text{J} + 49\ \text{J} = 81\ \text{J} = K_i$.

Example 8-17

A neutron of mass m_1 and speed v_{1i} collides elastically with a carbon nucleus of mass m_2 at rest (Figure 8-33). (*a*) What are the final velocities of both particles? (*b*) What fraction of its initial energy does the neutron lose?

Figure 8-33

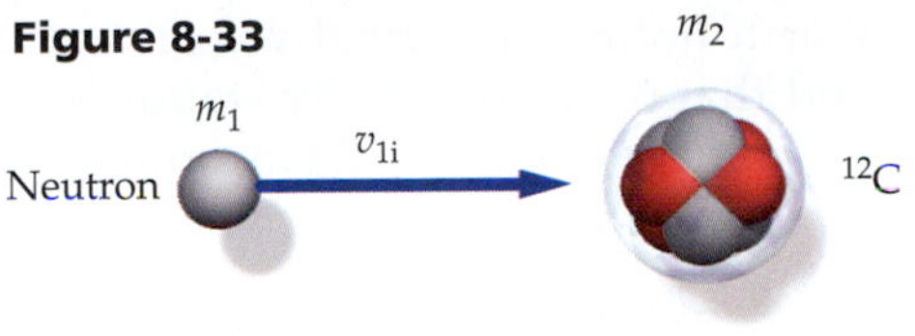

Picture the Problem Conservation of momentum and conservation of energy allow us to find the final velocities. Since the initial energy of the carbon nucleus is zero, its final energy equals the energy lost by the neutron.

(*a*)1. Use conservation of momentum to obtain one relation for the final velocities:

$$m_1v_{1i} = m_1v_{1f} + m_2v_{2f}$$

2. Use conservation of mechanical energy to set the velocity of recession equal to the negative of the velocity of approach:

$$v_{2f} - v_{1f} = -(v_{2i} - v_{1i}) = v_{1i}$$

3. To eliminate v_{2f}, solve step 2 for v_{2f}, and substitute the answer in step 1:

$$v_{2f} = v_{1i} + v_{1f}$$

$$m_1v_{1i} = m_1v_{1f} + m_2(v_{1i} + v_{1f})$$

4. Solve for v_{1f} (Note that v_{1f} is negative because the mass of the carbon nucleus m_2 is greater than the mass of the neutron.):

$$v_{1f} = \frac{m_1 - m_2}{m_1 + m_2}v_{1i}$$

5. Use v_{1f} to find v_{2f}:

$$v_{2f} = v_{1i} + v_{1f} = \frac{2m_1}{m_1 + m_2}v_{1i}$$

(*b*)1. The energy lost by the neutron is the final energy of the carbon nucleus:

$$-\Delta K_n = K_{2f} = \frac{1}{2}m_2v_{2f}^2 = \frac{2m_2m_1^2v_{1i}^2}{(m_1 + m_2)^2}$$

$$= \frac{4m_1m_2}{(m_1 + m_2)^2}\left(\frac{1}{2}\right)m_1v_{1i}^2$$

2. Divide $-\Delta K_n$ by K_n to find the fraction of the initial energy of the neutron lost:

$$f = \frac{-\Delta K_n}{K_n} = \frac{4m_1m_2}{(m_1 + m_2)^2}$$

Remark An important application of energy transfer in elastic collisions is the slowing down of neutrons in a nuclear reactor. High-energy neutrons are emitted in the fission of a uranium nucleus. If these neutrons are to cause another uranium nucleus to fission, their energy must be reduced, that is, they must be slowed down or "moderated." One mechanism for slowing down neutrons is the elastic scattering of the neutrons with the nuclei in the reactor. The fractional energy loss $f = -\Delta K_n/K_n$ depends on the ratio of the mass of the moderator nucleus to that of the neutron, as shown in Figure 8-34. For uranium, $m_2 \approx 235m_1$ and $f \approx 0.017 = 1.7\%$. For carbon, $m_2 \approx 12m_1$ and $f = 0.28 = 28\%$; for hydrogen, $m_1 \approx m_1$ and $f = 1 = 100\%$. A moderator such as graphite or water is added to a reactor to slow down the neutrons so that they can be captured by uranium nuclei.

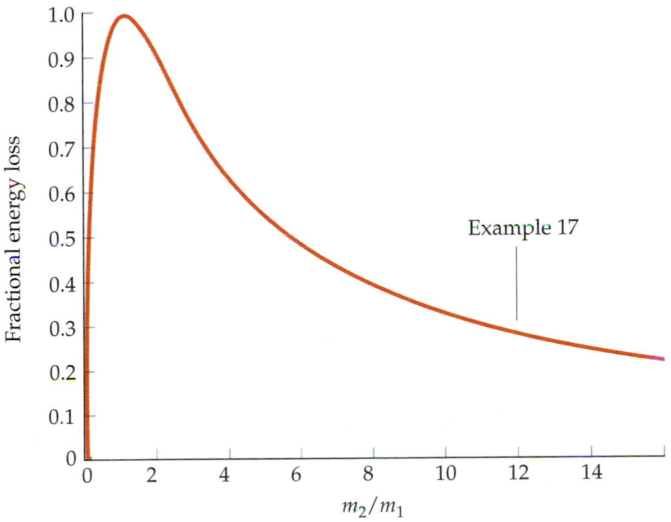

Figure 8-34 Fractional energy loss as a function of the ratio of the two masses. The maximum energy loss occurs when $m_1 = m_2$.

Exercise A 2-kg box moving at 3 m/s makes an elastic collision with a stationary 4-kg box. (*a*) What is the original mechanical energy? (*b*) How much energy is transferred to the 4-kg box? (*Answers* (*a*) 9 J, (*b*) 8 J)

The results of Example 8-17 for the final velocities of an incoming particle colliding with a second particle initially at rest are worth noting. The final velocity of the incoming particle, v_{1f}, and that of the originally stationary particle, v_{2f}, are related to the initial velocity of the incoming particle by

$$v_{1f} = \frac{m_1 - m_2}{m_1 + m_2} v_{1i} \tag{8-30a}$$

and

$$v_{2f} = \frac{2m_1}{m_1 + m_2} v_{1i} \tag{8-30b}$$

When a very massive object (say a bowling ball) collides with a light stationary object (say a Ping-Pong ball), the massive object is essentially unaffected. Before the collision, the relative velocity of approach is v_{1i}. If the massive object continues with a velocity that is essentially v_{1i} after the collision, the velocity of the smaller object must be $2v_{1i}$ so that the speed of recession is equal to the speed of approach. This result also follows from Equations 8-30*a* and 8-30*b* if we take m_2 to be much smaller than m_1, in which case $v_{1f} \approx v_{1i}$ and $v_{2f} \approx 2v_{1i}$, as expected. An example of such a collision, shown in Figure 8-35, is that between a golf ball and a golf club (whose mass is augmented by the mass of the golfer swinging the club).

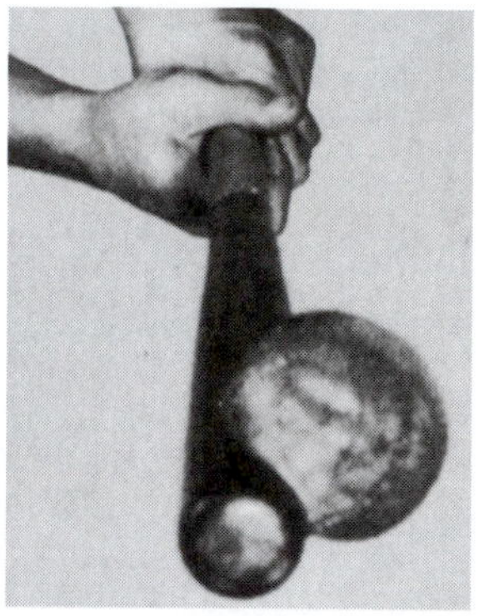

(*left*) A baseball and bat collide. The ball briefly deforms due to the large force the bat exerts during contact. Rebounding, the ball springs back to its original shape, converting elastic potential energy of deformation into kinetic energy.

Figure 8-35 Multiflash photograph of a golfer hitting a ball. The ball travels at approximately twice the speed of the club, as can be seen by comparing the distances Δs_c traveled by the club and Δs_b traveled by the ball between flashes.

optional

The Coefficient of Restitution Most collisions lie somewhere between the extreme cases of elastic, in which the relative velocities are reversed, and perfectly inelastic, in which there is no relative velocity after the collision. The **coefficient of restitution,** *e*, is a measure of the elasticity of a collision. It is defined as the ratio of the relative speed of recession to the relative speed of approach:

$$e = \frac{|v_{2f} - v_{1f}|}{|v_{2i} - v_{1i}|} = \frac{v_{rec}}{v_{app}} \tag{8-31}$$

Definition—Coefficient of restitution

For an elastic collision, $e = 1$; for a perfectly inelastic collision, $e = 0$.

Collisions in Three Dimensions

Perfectly Inelastic Collisions in Three Dimensions For collisions in three dimensions, the total initial momentum is the sum of the initial momentum vectors of each object involved in the collision. For perfectly inelastic collisions, the objects stick together, and since their final momentum equals the initial momentum, they move off in the direction of the resultant total momentum with velocity $\vec{v}_{cm}$ given by

$$\vec{v}_{cm} = \frac{\vec{P}}{m_1 + m_2} \tag{8-32}$$

where $\vec{P} = \vec{p}_1 + \vec{p}_2$ is the total momentum of the system. Since $\vec{P}$ is in the plane formed by $\vec{p}_1$ and $\vec{p}_2$, the collision takes place in this plane.

Example 8-18 ***try it yourself***

A small car of mass 1.2 Mg (1.2 × 10³ kg) traveling east at 60 km/h collides at an intersection with a truck of mass 3 Mg traveling north at 40 km/h, as shown. The car and truck stick together. Find the velocity of the wreckage just after the collision.

Picture the Problem Choose our coordinate system so that initially the car is traveling in the x direction and the truck is traveling in the y direction (Figure 8-36). Then write the momentum of each object in vector form, and use conservation of momentum.

Figure 8-36

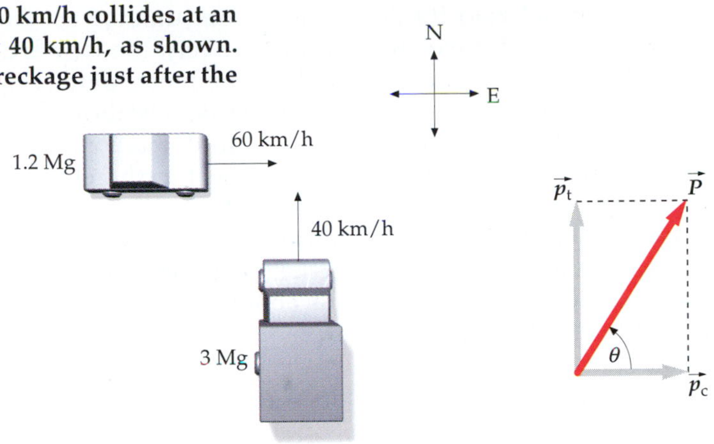

Cover the column to the right and try these on your own before looking at the answers.

Steps	***Answers***
1. Write the initial momentum vectors for the car and truck, $\vec{p}_c$ and $\vec{p}_t$, using the unit vectors $\hat{i}$ and $\hat{j}$.	$\vec{p}_c = (72\ \text{Mg·km/h})\,\hat{i}$, $\vec{p}_t = (120\ \text{Mg·km/h})\,\hat{j}$
2. Add the vectors in step 1 to obtain the total initial momentum $\vec{P}$, which is also the final momentum.	$\vec{P} = (72\ \text{Mg·km/h})\,\hat{i} + (120\ \text{Mg·km/h})\,\hat{j}$
3. Divide by the total mass M to find the velocity of the center of mass, which is the final velocity of the wreckage.	$\vec{v}_{cm} = \vec{P}/M = (17.1\ \text{km/h})\,\hat{i} + (28.6\ \text{km/h})\,\hat{j}$
4. Find the magnitude of the final velocity.	$v_{cm} = 33.3\ \text{km/h}$
5. Find the direction of the final velocity from $\tan\theta = P_y/P_x$.	$\theta = 59°$

Elastic Collisions in Three Dimensions Elastic collisions in three dimensions are more complicated than those we have covered previously. Figure 8-37 shows an off-center collision between an object of mass m_1 moving with velocity $\vec{v}_{1i}$ parallel to the x axis toward an object of mass m_2 that is initially at rest at the origin. The distance b between the centers measured perpendicular to the direction of $\vec{v}_{1i}$ is called the **impact parameter**. After the collision, object 1 moves off with velocity $\vec{v}_{1f}$, making an angle θ_1 with its initial velocity, and object 2 moves with velocity $\vec{v}_{2f}$, making an angle θ_2 with $\vec{v}_{1f}$. Conservation of momentum gives

$$\vec{P} = m_1\vec{v}_{1i} = m_1\vec{v}_{1f} + m_2\vec{v}_{2f}$$

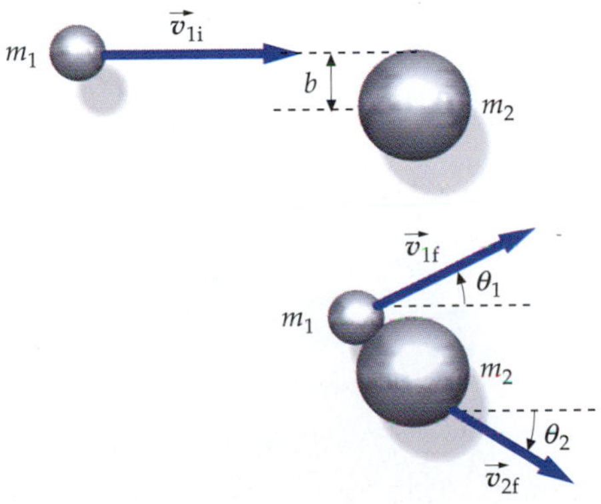

Figure 8-37 Off-center collision. The final velocities depend on the impact parameter b and on the type of force exerted by one object on the other.

We can see from this equation that the vector $\vec{v}_{2f}$ must lie in the plane formed by $\vec{v}_{1i}$ and $\vec{v}_{1f}$, which we will take to be the xy plane. Assuming that we know the initial velocity $\vec{v}_{1i}$, we have four unknowns: the x and y components of both final velocities; or alternatively, the two final speeds and the two angles of deflection. The x and y components of the conservation-of-momentum equation give us two of the needed relations among these quantities. Conservation of energy gives a third relation. To find the four unknowns, we need another relation. The fourth relation depends on the impact parameter b and on the type of interacting force exerted by the bodies on each other. In practice, the fourth relation is often found experimentally, by measuring the angle of deflection or the angle of recoil. Such a measurement can then give us information about the type of interacting force between the bodies.

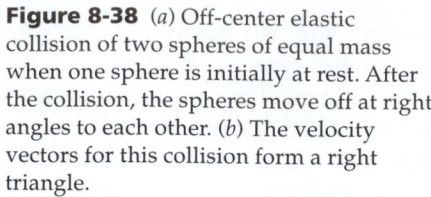

Figure 8-38 (*a*) Off-center elastic collision of two spheres of equal mass when one sphere is initially at rest. After the collision, the spheres move off at right angles to each other. (*b*) The velocity vectors for this collision form a right triangle.

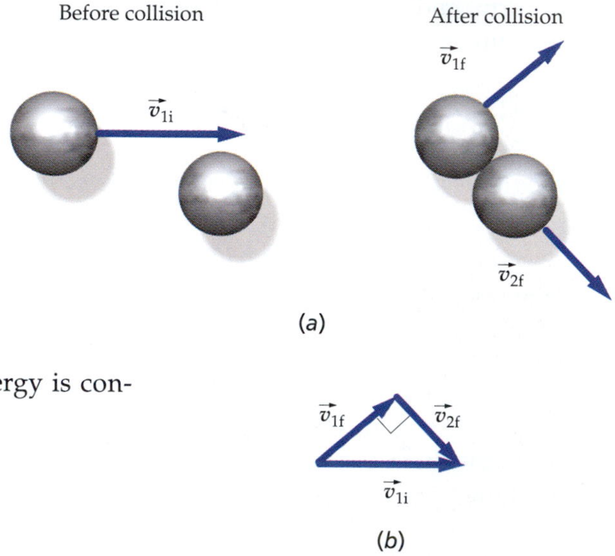

We omit further discussion of elastic collisions in three dimensions except for the interesting special case of the off-center elastic collision of two objects *of equal mass* when one is initially at rest (Figure 8-38*a*). If $\vec{v}_{1i}$ and $\vec{v}_{1f}$ are the initial and final velocities of object 1 and $\vec{v}_{2f}$ is the final velocity of object 2, conservation of momentum gives

$$m\vec{v}_{1i} = m\vec{v}_{1f} + m\vec{v}_{2f}$$

or

$$\vec{v}_{1i} = \vec{v}_{1f} + \vec{v}_{2f}$$

These vectors form the triangle shown in Figure 8-38*b*. Since energy is conserved in the collision,

$$\frac{1}{2}mv_{1i}^2 = \frac{1}{2}mv_{1f}^2 + \frac{1}{2}mv_{2f}^2$$

or

$$v_{1i}^2 = v_{1f}^2 + v_{2f}^2 \qquad \text{8-33}$$

Equation 8-33 is the Pythagorean theorem for a right triangle formed by the vectors $\vec{v}_{1f}$, $\vec{v}_{2f}$, and $\vec{v}_{1i}$ with the hypotenuse of the triangle being $\vec{v}_{1i}$. So for this special case, the final velocity vectors $\vec{v}_{1f}$ and $\vec{v}_{2f}$ are perpendicular to each other, as shown in Figure 8-38*b*.

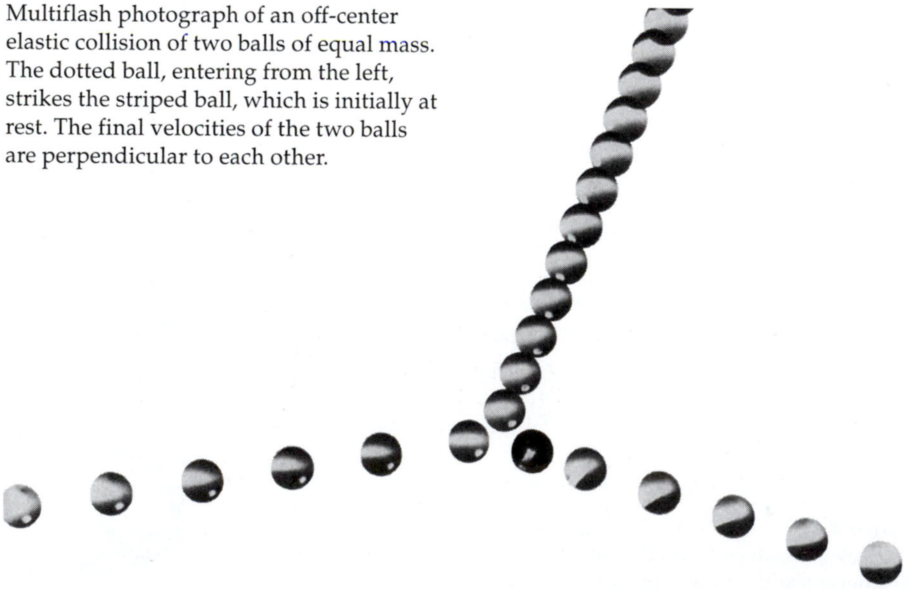

Multiflash photograph of an off-center elastic collision of two balls of equal mass. The dotted ball, entering from the left, strikes the striped ball, which is initially at rest. The final velocities of the two balls are perpendicular to each other.

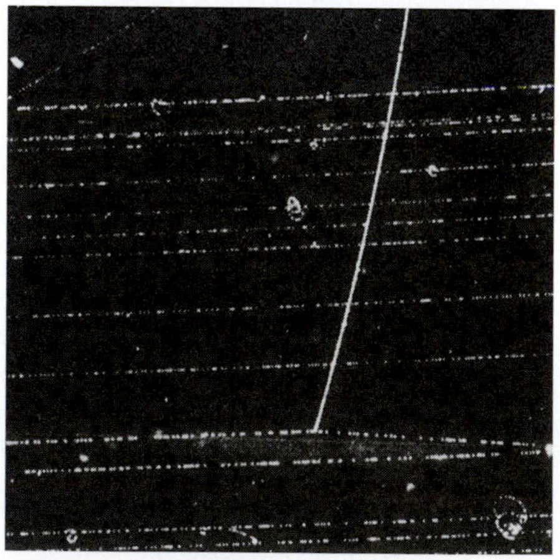

Proton–proton collision in a liquid-hydrogen bubble chamber. A proton entering from the left interacts with a stationary proton. The two then move off at right angles. The slight curvature of the tracks is due to a magnetic field.

exploring

Collisions on Extreme Scales

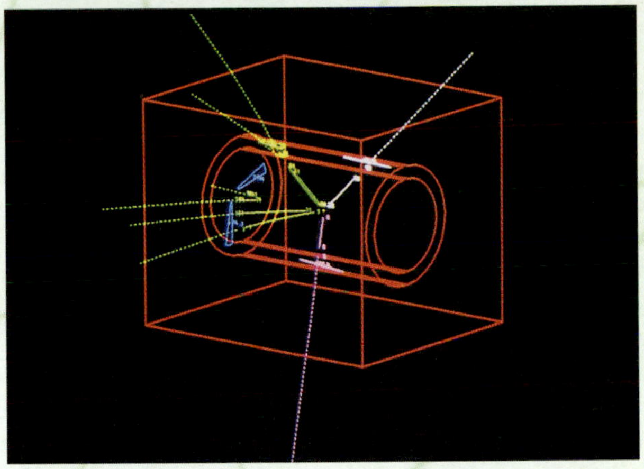

A proton and an antiproton collide, producing a shower of other particles, including the rare Z particle. The electrically neutral Z leaves no track, but quickly decays into other particles that are charged and do leave tracks. Here the Z decays into an electron (pink track) and a positron (white track). The energy of the electron and positron is measured by the curvature of their tracks in a magnetic field. The total energy of these decay particles equals $m_Z c^2$, where m_Z is the predicted mass of the Z particle.

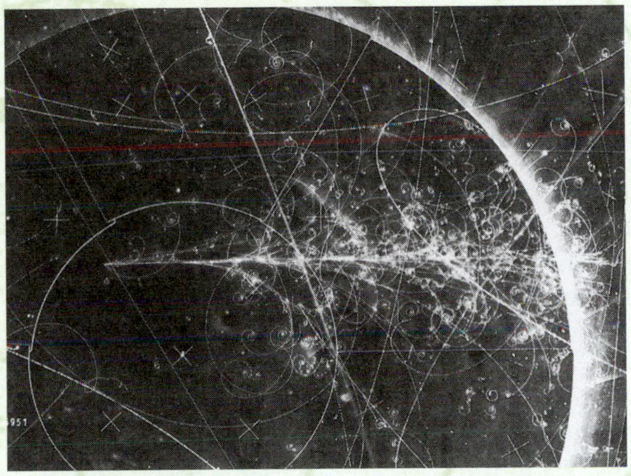

Tracks of a complicated spray of particles emitted when a neutrino (entering from the left) collides with a proton in the Big European Bubble Chamber at CERN. Neutrinos were first discovered when it was observed that the decay of a neutron into a proton and electron appeared to violate conservation of energy and momentum.

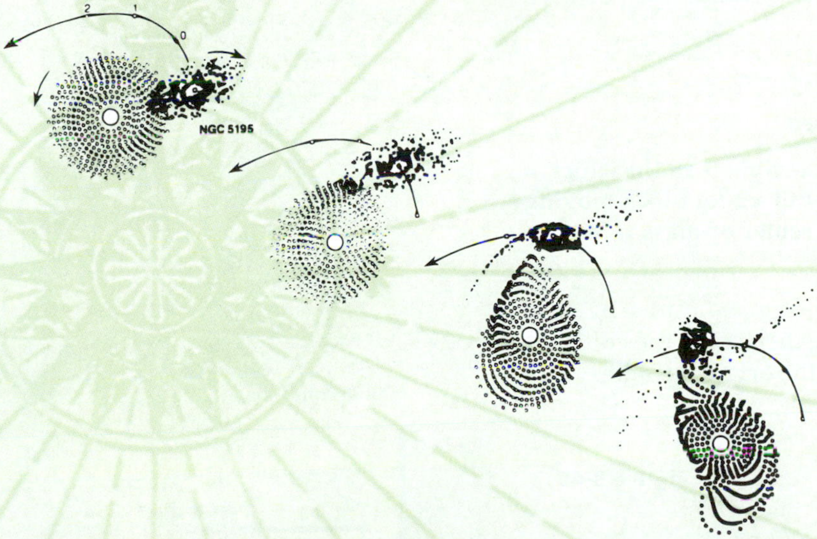

Computer simulation of the collision of two galaxies. Each step represents 100 million years.

The collision of the large spiral galaxy NGC 5194 (bottom), which has a mass of about 10^{11} solar masses, and a much younger galaxy, NGC 5195, which is about one-third as massive.

optional

8-7 The Center-of-Mass Reference Frame

When the net external force on a system is zero, the velocity of the center of mass is constant. It is often convenient to choose a coordinate system with the origin at the center of mass. Then, relative to the original coordinate system, this coordinate system moves with a constant velocity $\vec{v}_{cm}$. The frame of reference attached to the center of mass is called the **center-of-mass reference frame.** If a particle has velocity $\vec{v}$ in the original reference frame, its velocity relative to the center of mass is $\vec{u}_1 = \vec{v}_1 - \vec{v}_{cm}$. In the center-of-mass frame, the velocity of the center of mass is zero. Since the total momentum of a system equals the total mass times the velocity of the center of mass, the total momentum is also zero in the center-of-mass frame. Thus, the center-of-mass reference frame is also called the **zero-momentum reference frame.**

The mathematics of collisions are greatly simplified when considered within the center-of-mass reference frame. The momenta of the two incoming objects are equal and opposite. After a perfectly inelastic collision, the objects remain at rest. All of the original energy is lost to thermal energy. A perfectly elastic collision in one dimension reverses the velocity of each object but does not change the magnitude of v (you will derive this in Problem 93).

Consider a simple two-particle system in a reference frame in which one particle of mass m_1 is moving with a velocity $\vec{v}_1$ and a second particle of mass m_2 is moving with a velocity $\vec{v}_2$ (Figure 8-39). In this frame, the velocity of the center of mass is

$$\vec{v}_{cm} = \frac{m_1\vec{v}_1 + m_2\vec{v}_2}{m_1 + m_2}$$

We can transform the velocities of the two particles to their velocities in the center-of-mass reference frame by subtracting $\vec{v}_{cm}$. The velocities of the particles in the center-of-mass frame are $\vec{u}_1$ and $\vec{u}_2$, given by

$$\vec{u}_1 = \vec{v}_1 - \vec{v}_{cm} \qquad \text{8-34}a$$

and

$$\vec{u}_2 = \vec{v}_2 - \vec{v}_{cm} \qquad \text{8-34}b$$

Since the total momentum is zero in the center-of-mass frame, the particles have equal and opposite momenta in this frame.

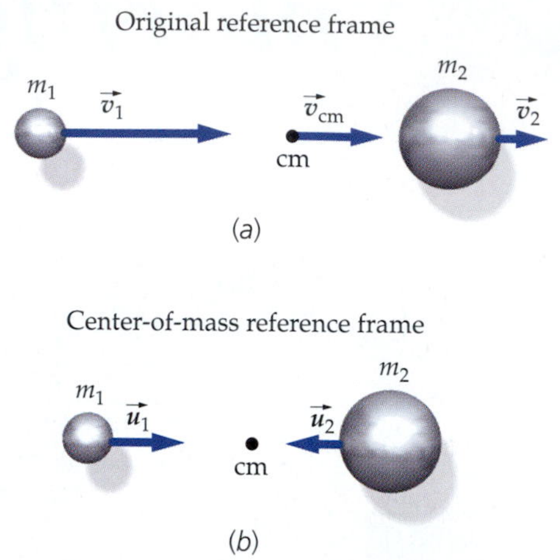

Figure 8-39 (*a*) Two particles moving in a general reference frame in which the center of mass has a velocity $\vec{v}_{cm}$. (*b*) In the center-of-mass reference frame, the center of mass is at rest and the particles have equal and opposite momenta. The velocities in the two frames are related by $\vec{u}_1 = \vec{v}_1 - \vec{v}_{cm}$ and $\vec{u}_2 = \vec{v}_2 - \vec{v}_{cm}$.

Example 8-19

Find the final velocities for the elastic collision in Example 8-16 (in which a 4-kg block moving right at 6 m/s collides elastically with a 2-kg block moving right at 3 m/s) by transforming their velocities to the center-of-mass reference frame.

Picture the Problem We transform to the center-of-mass frame by first finding v_{cm} and subtracting it from each velocity. We then solve the collision by reversing the velocities and transforming back to the original frame.

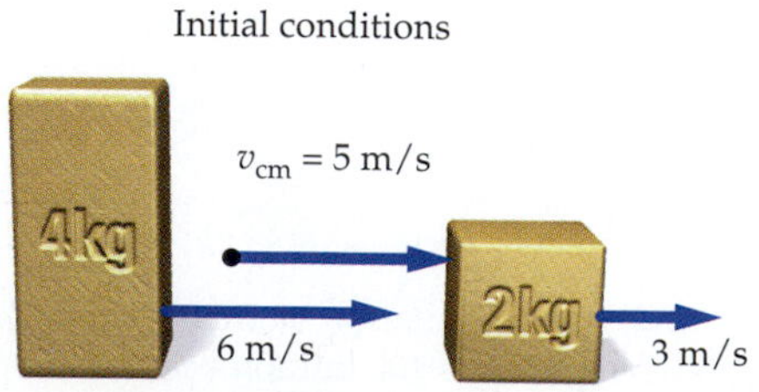

Figure 8-40

1. Calculate the velocity of the center of mass, v_{cm}:

$$v_{cm} = \frac{m_1 v_{1i} + m_2 v_{2i}}{m_1 + m_2} = \frac{(4\text{ kg})(6\text{ m/s}) + (2\text{ kg})(3\text{ m/s})}{4\text{ kg} + 2\text{ kg}} = 5\text{ m/s}$$

2. Transform the initial velocities to the center-of-mass reference frame by subtracting v_{cm} from the initial velocities:

$$u_{1i} = v_{1i} - v_{cm} = 6\ \text{m/s} - 5\ \text{m/s} = 1\ \text{m/s}$$

$$u_{2i} = v_{2i} - v_{cm} = 3\ \text{m/s} - 5\ \text{m/s} = -2\ \text{m/s}$$

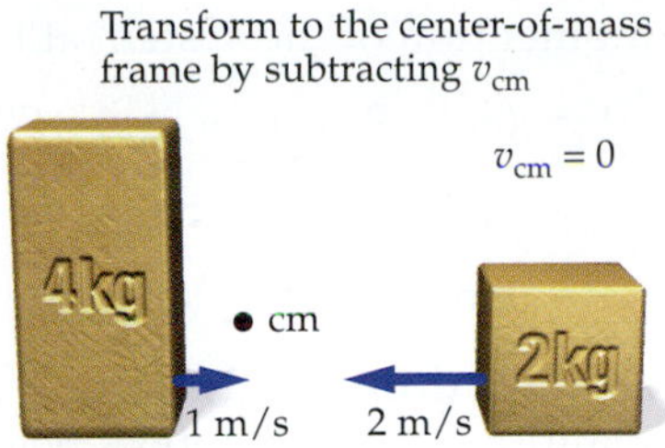

Figure 8-41

3. Solve the collision in the center-of-mass reference frame by reversing the velocity of each object:

$$u_{1f} = -u_{1i} = -1\ \text{m/s}$$

$$u_{2f} = -u_{2i} = +2\ \text{m/s}$$

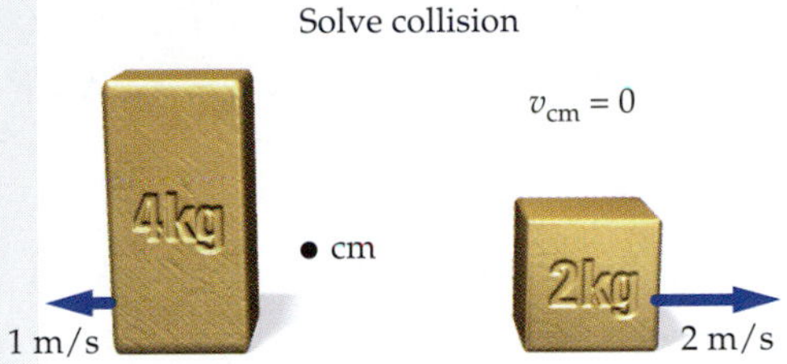

Figure 8-42

4. To find the final velocities in the original frame, add v_{cm} to each final velocity:

$$v_{1f} = u_{1f} + v_{cm} = -1\ \text{m/s} + 5\ \text{m/s} = 4\ \text{m/s}$$

$$v_{2f} = u_{2f} + v_{cm} = 2\ \text{m/s} + 5\ \text{m/s} = 7\ \text{m/s}$$

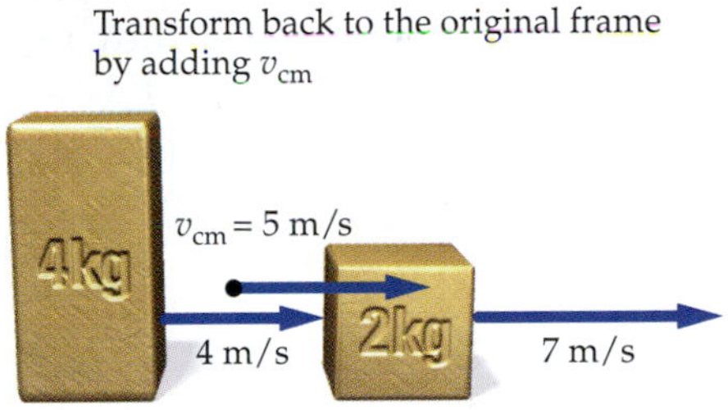

Figure 8-43

Remark This is the same result found in Example 8-16.

Exercise Show that the total momentum both before and after the collision is zero in the center-of-mass reference frame. (*Answer* Before the collision: $P_i = (4\ \text{kg})(1\ \text{m/s}) + (2\ \text{kg})(-2\ \text{m/s}) = 0$; after the collision: $P_f = (4\ \text{kg}) \times (-1\ \text{m/s}) + (2\ \text{kg})(2\ \text{m/s}) = 0$)

8-8 Rocket Propulsion

Rocket propulsion is a striking example of the conservation of momentum in action. The mathematical description of rocket propulsion can become quite complex because the mass of the rocket changes continuously as it burns fuel and expels exhaust gas. The easiest approach is to compute the change in the momentum of the total system (including the exhaust gas) for some time interval and use Newton's law in the form $F_{ext} = dP/dt$, where F_{ext} is the net force acting on the rocket.

Consider a rocket moving with speed v relative to the earth (Figure 8-44). If the fuel is burned at a constant rate, $R = |dm/dt|$, the rocket's mass at time t is

$$m = m_0 - Rt \qquad \text{8-35}$$

where m_0 is the initial mass of the rocket. The momentum of the system at time t is

$$P_i = mv$$

At a later time $t + \Delta t$, the rocket has expelled gas of mass $R\,\Delta t$. If the gas is exhausted at a speed u_{ex} *relative to the rocket*, the velocity of the gas relative to the earth is $v - u_{ex}$. The rocket then has a mass $m - R\,\Delta t$ and is moving at a speed $v + \Delta v$ (Figure 8-45).

Figure 8-44

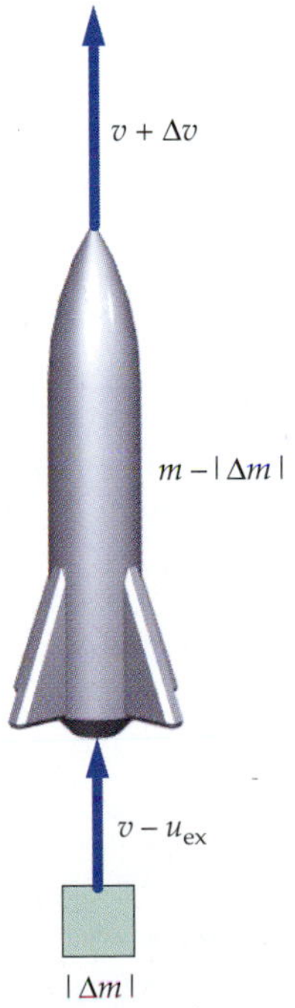

Figure 8-45

The momentum of the system at $t + \Delta t$ is

$$P_f = (m - R\,\Delta t)(v + \Delta v) + R\,\Delta t(v - u_{ex})$$
$$= mv + m\,\Delta v - v\,R\,\Delta t - R\,\Delta t\,\Delta v + v\,R\,\Delta t - u_{ex}\,R\,\Delta t$$
$$\approx mv + m\,\Delta v - u_{ex}\,R\,\Delta t$$

where we have dropped the term $R\,\Delta t\,\Delta v$, which is the product of two very small quantities, and therefore negligible compared with the others. The change in momentum is

$$\Delta P = P_f - P_i = m\,\Delta v - u_{ex}\,R\,\Delta t$$

and

$$\frac{\Delta P}{\Delta t} = m\frac{\Delta v}{\Delta t} - u_{ex}R \qquad \text{8-36}$$

As Δt approaches zero, $\Delta v/\Delta t$ approaches the derivative dv/dt, which is the acceleration. For a rocket moving upward near the surface of the earth, $F_{ext} = -mg$. Setting $dP/dt = F_{ext} = -mg$ gives us the **rocket equation**:

$$m\frac{dv}{dt} = Ru_{ex} + F_{ext} = Ru_{ex} - mg \qquad \text{8-37}$$

Rocket equation

or

$$\frac{dv}{dt} = \frac{Ru_{ex}}{m} - g = \frac{Ru_{ex}}{m_0 - Rt} - g \qquad \text{8-38}$$

The quantity Ru_{ex} is the force exerted on the rocket by the exhausting fuel. This is called the **thrust**:

$$F_{th} = Ru_{ex} = \left|\frac{dm}{dt}\right|u_{ex} \qquad \text{8-39}$$

Definition—Rocket thrust

Equation 8-38 is solved by integrating both sides with respect to time. For a rocket starting at rest at $t = 0$, the result is

$$v = -u_{ex}\ln\left(\frac{m_0 - Rt}{m_0}\right) - gt \qquad \text{8-40}$$

as can be verified by taking the time derivative of v. The **payload** of a rocket is the final mass, m_f, after all the fuel has been burned. The **burn time** t_b is given by $m_f = m_0 - Rt_b$, or

$$t_b = \frac{m_0 - m_f}{R} \qquad \text{8-41}$$

Thus, a rocket starting at rest with mass m_0, and payload of m_f, attains a final speed

$$v_f = -u_{ex}\ln\frac{m_f}{m_0} - gt_b \qquad \text{8-42}$$

Final speed of rocket

assuming the acceleration of gravity to be constant.

Example 8-20 *try it yourself*

The Saturn V rocket used in the Apollo moon-landing program had an initial mass m_0 of 2.85×10^6 kg, a payload of 27%, a burn rate R of 13.84×10^3 kg/s, and a thrust F_{th} of 34×10^6 N. Find (*a*) the exhaust speed, (*b*) the burn time t_b, (*c*) the acceleration at liftoff, (*d*) the acceleration at burnout t_b, and (*e*) the final speed of the rocket.

Picture the Problem (*a*) The exhaust speed can be found from the thrust and burn rate. (*b*) To find the burn time, you need to find the total mass of fuel burned, which is the initial mass minus the payload. (*c*) The acceleration is found from Equation 8-38. (*d*) The final speed is given by Equation 8-42.

Cover the column to the right and try these on your own before looking at the answers.

Steps	Answers
(*a*) Calculate u_{ex} from the given thrust and burn rate.	$u_{ex} = 2.46$ km/s
(*b*)1. Calculate the final mass m_f of the rocket.	$m_f = (0.27)m_0 = 7.70 \times 10^5$ kg
2. Use your result to calculate the burn time t_b.	$t_b = \frac{m_0 - m_f}{R} = 150$ s
(*c*) Calculate dv/dt for $m = m_0$.	Initially, $dv/dt = 2.14$ m/s^2
(*d*) Calculate dv/dt for $m = m_f$.	Finally, $dv/dt = 34.4$ m/s^2
(*e*) Calculate the final speed from Equation 8-42.	$v_f = 1.75$ km/s

Remarks The initial acceleration is small—only $0.21g$. At burnout, the rocket's acceleration has increased to $3.5\ g$. The speed of the rocket at burnout, after two and a half minutes of burning, is roughly 6300 km/h (3900 mi/h).

Summary

1. The conservation of momentum for an isolated system is a fundamental law of nature that has applications in all areas of physics.

Topic	Remarks and Relevant Equations	
1. Center of Mass		
Position for multiple objects	$M\vec{r}_{cm} = m_1\vec{r}_1 + m_2\vec{r}_2 + \cdots = \sum_i m_i\vec{r}_i$	8-4
Position for continuous objects	$M\vec{r}_{cm} = \int \vec{r}\, dm$	8-5
Motion of center of mass	$\vec{F}_{net,ext} = \sum_i \vec{F}_{i,ext} = M\vec{a}_{cm}$	8-10

2. Momentum		
Definition for a particle	$\vec{p} = m\vec{v}$	**8-11**
Kinetic energy of a particle in terms of momentum	$K = \dfrac{p^2}{2m}$	**8-23**
Definition for a system of particles	$\vec{P} = \sum_i m_i\vec{v}_i = M\vec{v}_{cm}$	**8-13**
Newton's second law for systems	$\sum_i \vec{F}_{ext} = \vec{F}_{net,ext} = \dfrac{d\vec{P}}{dt}$	**8-14**
Law of conservation of momentum	If the net external force acting on a system is zero, the total momentum of the system is conserved.	
3. Energy of a System		
Kinetic energy	The kinetic energy of a system of particles can be written as the sum of two terms: (1) the kinetic energy associated with the motion of the center of mass, $\frac{1}{2}Mv_{cm}^2$, and (2) the kinetic energy associated with the motion of the particles of the system relative to the center of mass, $\Sigma\frac{1}{2}m_iu_i^2$, where $\vec{u}_i$ is the velocity of the ith particle relative to the center of mass. $K = \sum_i \frac{1}{2} m_i v_{cm}^2 + \sum_i \frac{1}{2} m_i u_i^2 = \frac{1}{2} M v_{cm}^2 + K_{rel}$	**8-17**
Gravitational potential energy	$U = Mgh_{cm}$	**8-6**
4. Collisions		
Impulse	The impulse of a force is defined as the integral of the force over the time interval during which the force acts. It equals the total change in momentum of the particle. $\vec{I} = \int_{t_i}^{t_f} \vec{F}\,dt = \Delta\vec{p}$	**8-19**
Average force	$\vec{F}_{av} = \dfrac{1}{\Delta t}\int_{t_i}^{t_f} \vec{F}\,dt = \dfrac{\vec{I}}{\Delta t}$	**8-20**
Elastic collisions	An elastic collision is one in which the total kinetic energy of the two objects is the same before and after the collision.	
Relative speeds of approach and recession	The relative speed of recession of the objects after an elastic collision in one dimension equals the relative speed of approach before the collision. $v_{2f} - v_{1f} = -(v_{2i} - v_{1i})$	**8-29**
Perfectly inelastic collisions	In a perfectly inelastic collision, the objects stick together and move with the velocity of the center of mass.	
Coefficient of restitution (optional)	The coefficient of restitution e is a measure of the elasticity of a collision and is defined as the ratio of the relative speed of recession to the relative speed of approach: $e = \dfrac{\lvert v_{2f} - v_{1f}\rvert}{\lvert v_{2i} - v_{1i}\rvert} = \dfrac{v_{rec}}{v_{app}}$ For an elastic collision, $e = 1$; for a perfectly inelastic collision, $e = 0$.	**8-31**

5. Center-of-Mass Reference Frame (optional)	The center-of-mass reference frame is one that moves with the velocity of the center of mass. In this frame, the total momentum of a system is zero. In perfectly inelastic collisions, the particles remain at rest within the center-of-mass reference frame after the collision. In elastic collisions in one dimension, the velocity of each particle is reversed.	
Transforming into the center-of-mass reference frame	The velocity of a particle in the center-of-mass frame, $\vec{u}_i$, is related to the velocity in the original frame, $\vec{v}_i$, by $\vec{u}_i = \vec{v}_i - \vec{v}_{cm}$	8-34
6. Rocket Propulsion (optional)	Rockets achieve thrust by burning fuel and exhausting the resulting gases. The force exerted by the exhaust gases on the rocket propels the rocket forward.	
Rocket equation	$m\frac{dv}{dt} = Ru_{ex} + F_{ext} = Ru_{ex} - mg$ where u_{ex} is the exhaust speed and R is the thrust.	8-37
Thrust	$F_{th} = Ru_{ex} = \left\|\frac{dm}{dt}\right\| u_{ex}$	8-39
Final speed	$v_f = -u_{ex} \ln \frac{m_f}{m_0} - gt_b$	8-42

Problem-Solving Guide

1. Begin by drawing a neat diagram with a suitable coordinate system.
2. If possible, choose a system for which the net external force is zero, and then apply conservation of momentum.
3. In collision problems, choose a system that includes both objects. Any external forces will generally be negligible compared to the collision forces during the collision. Apply conservation of momentum. If the collision is elastic, apply conservation of mechanical energy by setting the velocity of recession equal to the velocity of approach.

Summary of Worked Examples

Type of Calculation	Procedure and Relevant Examples	
1. Find the Center of Mass	Draw a coordinate system and show the locations of all the masses in the system. A thoughtful choice for the coordinate system can greatly simplify the calculation. For symmetric systems, the center of mass is at the geometric center.	
System of particles	Determine x_{cm}, y_{cm}, and z_{cm} separately using $Mx_{cm} = \sum_i m_i x_i$, etc.	Example 8-1
Continuous system	The center of mass of a uniform symmetric object is at its geometric center. Divide nonsymmetric objects into sets of symmetric objects if possible. In general, x_{cm} can be found from $x_{cm} = \int x\, dm$.	Example 8-2
2. Find the Forces Acting on a System	Use $\Sigma\vec{F}_{i,ext} = M\vec{a}_{cm}$. Calculate a_{cm} from $M\vec{a}_{cm} = \Sigma m_i\vec{a}_i$	Example 8-5
3. Problems With Variable Mass	Choose a system that includes all of the masses and apply conservation of momentum to the directions where there is no external force.	Examples 8-3, 8-7, 8-8

4. Collisions	Sketch the system before and after the collision. Indicate masses and velocities in your sketch. Apply conservation of momentum.
Find the impulse	Set the impulse equal to the change in momentum. **Examples 8-10, 8-11, 8-12, 8-13**
Perfectly inelastic	Use conservation of momentum, $m_1\vec{v}_1 + m_2\vec{v}_2 = (m_1 + m_2)\vec{v}_{cm}$. The final velocity equals the velocity of the center of mass. The loss in mechanical energy can be found by writing the kinetic energy as $K = P^2/2m$, and using the fact that the momentum P is the same before and after the collision. **Examples 8-7, 8-13, 8-14, 8-15, 8-18**
Elastic, one dimension	Use conservation of momentum and conservation of energy as expressed by the fact that the relative speed of recession equals the relative speed of approach. **Examples 8-16, 8-17**
In center-of-mass reference frame (optional)	Find the velocity of each particle in the center-of-mass reference frame by subtracting the velocity of the center of mass in the original frame. In the center-of-mass frame, the velocity of each particle is reversed by the collision. The final velocities in the original frame can then be found by adding $\vec{v}_{cm}$ to each velocity. **Example 8-19**
5. Estimate Time of Collision, Impulse, Acceleration, and Average Force (optional)	Use a reasonable guess for the distance traveled during the collision, Δs, and assume constant acceleration to find v_{av}. Then the time of collision is $\Delta t = \Delta s/v_{av}$, the acceleration is $a = \Delta v/\Delta t$, the impulse is $\vec{I} = \Delta\vec{p}$, and the average force is $\vec{F}_{av} = \vec{I}/\Delta t = \Delta\vec{p}/\Delta t$. **Examples 8-10, 8-11, 8-12**
6. Rocket Propulsion (optional)	
Find the final speed of the rocket	Use $$v_f = -u_{ex} \ln \frac{m_f}{m_0} - gt_b$$ where u_{ex} is the exhaust speed, m_f is the mass of the rocket without fuel, m_0 is the initial mass with fuel, and t_b is the burn time. **Example 8-20**

Problems

Conceptual Problems

Problems from Optional and Exploring sections

In a few problems, you are given more data than you actually need; in a few other problems, you are required to supply data from your general knowledge, outside sources, or informed estimates.

• Single-concept, single-step, relatively easy

•• Intermediate-level, may require synthesis of concepts

••• Challenging, for advanced students

Take $g = 9.81$ N/kg $= 9.81$ m/s^2 and neglect friction in all problems unless otherwise stated.

The Center of Mass

1 • Give an example of a three-dimensional object that has no mass at its center of mass.

2 • Three point masses of 2 kg each are located on the x axis at the origin, $x = 0.20$ m, and $x = 0.50$ m. Find the center of mass of the system.

3 • A 24-kg child is 20 m from an 86-kg adult. Where is the center of mass of this system?

4 • Three objects of 2 kg each are located in the xy plane at points (10 cm, 0), (0, 10 cm), and (10 cm, 10 cm). Find the location of the center of mass.

5 • Find the center of mass x_{cm} of the three masses in Figure 8-46.

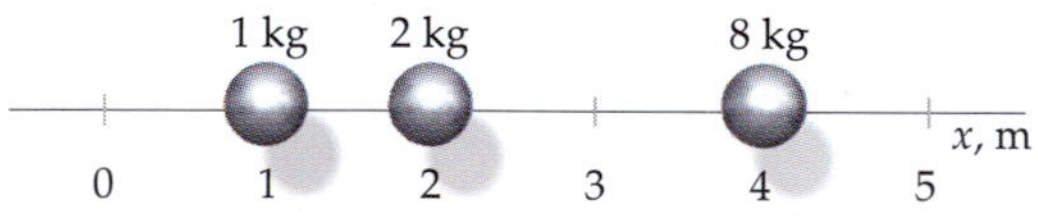

Figure 8-46 Problem 5

6 • Alley Oop's club-ax consists of a symmetrical 8-kg stone attached to the end of a uniform 2.5-kg stick that is 98 cm long. The dimensions of the club-ax are shown in Figure 8-47. How far is the center of mass from the handle end of the club-ax?

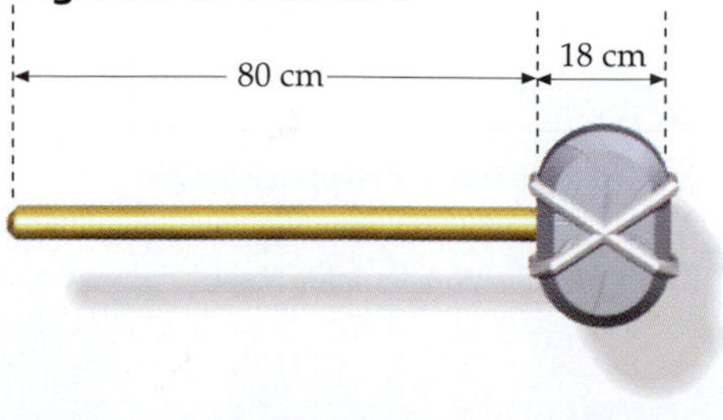

Figure 8-47 Problem 6

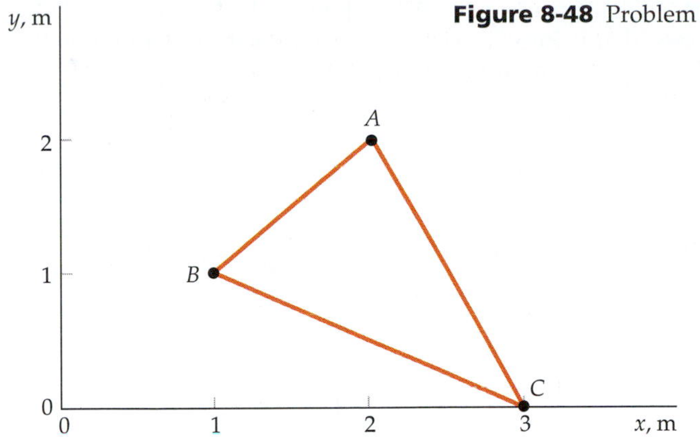

Figure 8-48 Problem 7

7 • Three balls A, B, and C, with masses of 3 kg, 1 kg, and 1 kg, respectively, are connected by massless rods. The balls are located as in Figure 8-48. What are the coordinates of the center of mass?

8 • By symmetry, locate the center of mass of an equilateral triangle of side length a located with one vertex on the y axis and the others at $(-a/2, 0)$ and $(+a/2, 0)$.

9 •• The uniform sheet of plywood in Figure 8-49 has a mass of 20 kg. Find its center of mass.

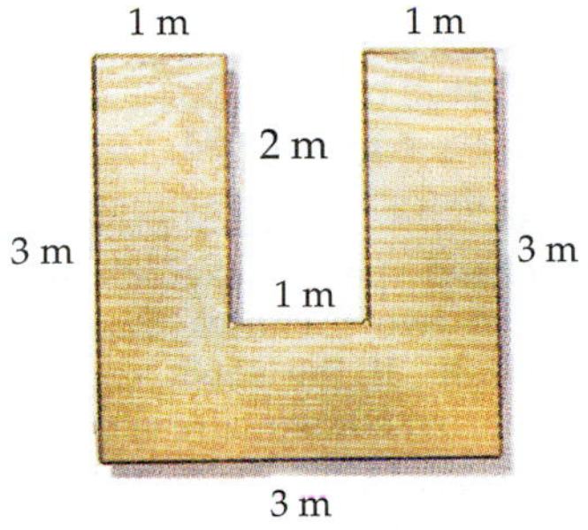

Figure 8-49 Problem 9

Finding the Center of Mass by Integration (optional)

10 •• Show that the center of mass of a uniform semicircular disk of radius R is at a point $(4/3\pi)\,R$ from the center of the circle.

11 •• A baseball bat of length L has a peculiar linear density (mass per unit length) given by $\lambda = \lambda_0(1 + x^2/L^2)$. Find the x coordinate of the center of mass in terms of L.

12 ••• Find the center of mass of a homogeneous solid hemisphere of radius R and mass M.

13 ••• Find the center of mass of a thin hemispherical shell.

14 ••• A sheet of metal is cut in the shape of a parabola. The edge of the sheet is given by the expression $y = ax^2$, and y ranges from $y = 0$ to $y = b$. Find the center of mass in terms of a and b.

Motion of the Center of Mass of a System

15 • On the night before your physics exam, you hear a banging on your door, and in walks Kelly. She says, "There's a big problem here. According to Newtonian physics, only external forces can cause the center of mass of a system to accelerate. But a car accelerates because of its own engine, so obviously Newton was wrong." She crosses her arms in a way that suggests that she is not going anywhere until she gets a satisfactory explanation. How can you explain Kelly's error to her in order to rescue Newton and get back to your studying?

16 • Two pucks of mass m_1 and m_2 lie unconnected on a frictionless table. A horizontal force F_1 is exerted on m_1 only. What is the magnitude of the acceleration of the center of mass of the pucks?

(*a*) F_1/m_1
(*b*) $F_1/(m_1 + m_2)$
(*c*) F_1/m_2
(*d*) $(m_1 + m_2)\,F_1/m_1m_2$

17 • The two pucks in Problem 16 are lying on a frictionless table and connected by a spring of force constant k. A horizontal force F_1 is again exerted only on m_1 along the spring away from m_2. What is the magnitude of the acceleration of the center of mass?

(*a*) F_1/m_1
(*b*) $F_1/(m_1 + m_2)$
(*c*) $(F_1 + k\,\Delta x)/m_1m_2$, where Δx is the amount the spring is stretched
(*d*) $(m_1 + m_2)F_1/m_1m_2$

18 • Two 3-kg masses have velocities $\vec{v}_1 = 2\text{ m/s}\,\hat{i} + 3\text{ m/s}\,\hat{j}$ and $\vec{v}_2 = 4\text{ m/s}\,\hat{i} - 6\text{ m/s}\,\hat{j}$. Find the velocity of the center of mass for the system.

19 • A 1500-kg car is moving westward with a speed of 20 m/s, and a 3000-kg truck is traveling east with a speed of 16 m/s. Find the velocity of the center of mass of the system.

20 • A force $\vec{F} = 12\,N\,\hat{i}$ is applied to the 3-kg ball in Problem 7. What is the acceleration of the center of mass?

21 •• A block of mass m is attached to a string and suspended inside a hollow box of mass M. The box rests on a scale that measures the system's weight. (*a*) If the string breaks, does the reading on the scale change? Explain your reasoning. (*b*) Assume that the string breaks and mass m falls with constant acceleration g. Find the acceleration of the center of mass, giving both direction and magnitude. (*c*) Using the result from (*b*), determine the reading on the scale while m is in free fall.

22 •• A vertical spring of force constant k is attached at the bottom to a platform of mass m_p, and at the top to a massless cup, as in Figure 8-50. The platform rests on a scale. A ball of mass m_b is placed in the cup. What is the reading on the scale when (*a*) the spring is compressed an amount $d = mg/k$? (*b*) the ball comes to rest momentarily with the spring compressed? (*c*) the ball again comes to rest in its original position?

Figure 8-50 Problem 22

23 •• In the Atwood's machine in Figure 8-51, the string passes over a fixed, frictionless cylinder of mass m_c. *(a)* Find the acceleration of the center of mass of the two-block-and-cylinder system. *(b)* Use Newton's second law for systems to find the force F exerted by the support. *(c)* Find the tension in the string connecting the blocks and show that $F = m_c g + 2T$.

Figure 8-51 Problem 23

24 •• Repeat Problems 22*a* and 22*b* with the ball dropped into the cup from a height h above the cup.

The Conservation of Momentum

25 • True or false:

(a) The momentum of a heavy object is greater than that of a light object moving at the same speed.
(b) The momentum of a system may be conserved even when mechanical energy is not.
(c) The velocity of the center of mass of a system equals the total momentum of the system divided by its total mass.

26 • How is the recoil of a rifle related to momentum conservation?

27 • A man is stranded in the middle of an ice rink that is perfectly frictionless. How can he get to the edge?

28 • A girl jumps from a boat to a dock. Why does she have to jump with more energy than she would need if she were jumping the same distance from one dock to another?

29 •• Much early research in rocket motion was done by Robert Goddard, physics professor at Clark College in Worcester, Massachusetts. A quotation from a 1921 editorial in the *New York Times* illustrates the public acceptance of his work: "That Professor Goddard with his 'chair' at Clark College and the countenance of the Smithsonian Institution does not know the relation between action and reaction, and the need to have something better than a vacuum against which to react—to say that would be absurd. Of course, he only seems to lack the knowledge ladled out daily in high schools." The belief that a rocket needs something to push against was a prevalent misconception before rockets in space were commonplace. Explain why that belief is wrong.

30 •• Liz, Jay, and Tara discover that sinister chemicals are leaking at a steady rate from a hole in the bottom of a railway car. To collect evidence of a potential environmental mishap, they videotape the car as it rolls without friction at an initial speed v_0. Tara claims that careful analysis of the videotape will show that the car's speed is increasing, because it is losing mass as it drains. The increase in speed will help to prove that the leak is occurring. Liz says no, that with a loss of mass, the car's speed will be decreasing. Jay says the speed will remain the same. *(a)* Who is right? *(b)* What forces are exerted on the system of the car plus chemical cargo?

31 • A girl of mass 55 kg jumps off the bow of a 75-kg canoe that is initially at rest. If her velocity is 2.5 m/s to the right, what is the velocity of the canoe after she jumps?

32 • Two masses of 5 kg and 10 kg are connected by a compressed spring and rest on a frictionless table. After the spring is released, the smaller mass has a velocity of 8 m/s to the left. What is the velocity of the larger mass?

33 • Figure 8-52 shows the behavior of a projectile just after it has broken up into three pieces. What was the speed of the projectile the instant before it broke up?

(a) v_3
(b) $v_3/3$
(c) $v_3/4$
(d) $4v_3$
(e) $(v_1 + v_2 + v_3)/4$

Figure 8-52 Problem 33

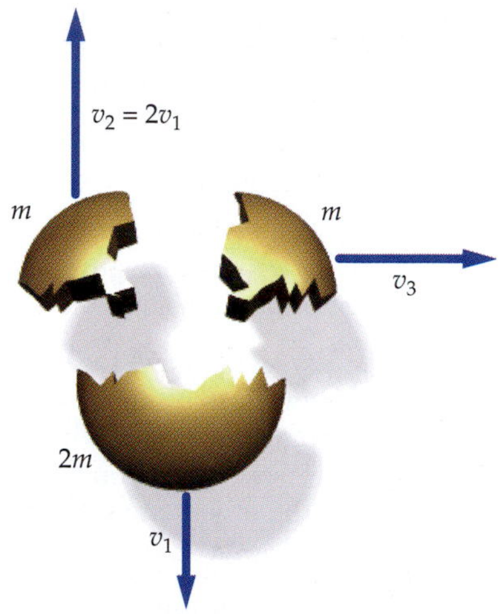

34 • A shell of mass m and speed v explodes into two identical fragments. If the shell was moving horizontally with respect to the earth, and one of the fragments is subsequently moving vertically with the speed v, find the velocity $\vec{v}\,'$ of the other fragment.

35 •• In a circus act, Marcello (mass 70.0 kg) is shot from a cannon with a muzzle velocity of 24.0 m/s at an angle of 30° above horizontal. His partner, Tina (mass 50.0 kg), stands on an elevated platform located at the top of his trajectory. He grabs her as he flies by and the two fly off together. They land in a net at the same elevation as the cannon a horizontal distance x away. Find x.

36 •• A block and a handgun loaded with one bullet are firmly affixed to opposite ends of a massless cart that rests on a level frictionless air table (Figure 8-53). The mass of the handgun is m_g, the mass of the block is m_{bk}, and the mass of the bullet is m_{bt}. The gun is aimed so that when fired, the bullet will go into the block. When the bullet leaves the barrel of the handgun, it has a velocity v_b as measured by an observer at rest with the table. Take the fall of the bullet to be negligible and its penetration into the block to be small. *(a)* What is the velocity of the cart immediately after the bullet leaves the gun barrel? *(b)* What is the velocity of the cart immediately after the bullet comes to rest in the block? *(c)* How far has the block moved from its initial position at the moment when the bullet comes to rest in the block?

Figure 8-53 Problem 36

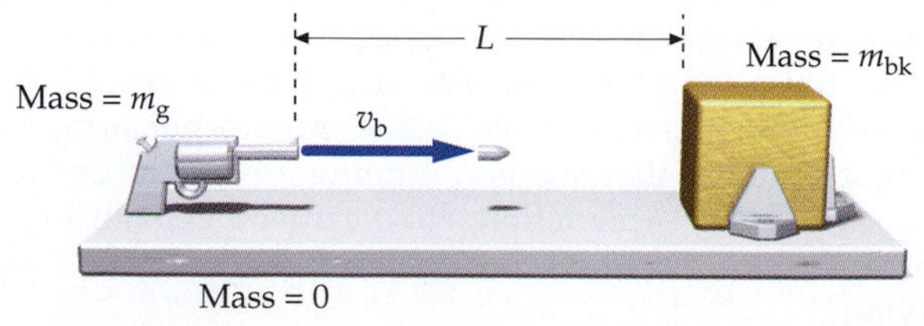

37 •• A small object of mass m slides down a wedge of mass $2m$ and exits smoothly onto a frictionless table. The wedge is initially at rest on the table. If the object is initially at rest at a height h above the table, find the velocity of the wedge when the object leaves it.

Kinetic Energy of a System

38 • Describe how a basketball is moving when (*a*) its total kinetic energy is just the energy of motion of its center of mass, and (*b*) its total kinetic energy is the energy of its motion relative to its center of mass.

39 • Two bowling balls are moving with the same velocity, but one just slides down the alley, whereas the other rolls down the alley. Which ball has more energy?

40 • A 3-kg block is traveling to the right at 5 m/s, and a second 3-kg block is traveling to the left at 2 m/s. (*a*) Find the total kinetic energy of the two blocks in this reference frame. (*b*) Find the velocity of the center of mass of the two-body system. (*c*) Find the velocities of the two blocks relative to the center of mass. (*d*) Find the kinetic energy of the motion of the blocks relative to the center of mass. (*e*) Show that your answer for part (*a*) is greater than your answer for part (*d*) by an amount equal to the kinetic energy of the center of mass.

41 • Repeat Problem 40 with the second, 3-kg block replaced by a block having a mass of 5 kg and moving to the right at 3 m/s.

Impulse and Average Force

42 • Explain why a safety net can save the life of a circus performer.

43 • How might you estimate the collision time of a baseball and bat?

44 • Why does a wine glass survive a fall onto a carpet but not onto a concrete floor?

45 • A soccer ball of mass 0.43 kg leaves the foot of the kicker with an initial speed of 25 m/s. (*a*) What is the impulse imparted to the ball by the kicker? (*b*) If the foot of the kicker is in contact with the ball for 0.008 s, what is the average force exerted by the foot on the ball?

46 • A 0.3-kg brick is dropped from a height of 8 m. It hits the ground and comes to rest. (*a*) What is the impulse exerted by the ground on the brick? (*b*) If it takes 0.0013 s from the time the brick first touches the ground until it comes to rest, what is the average force exerted by the ground on the brick?

47 • A meteorite of mass 30.8 tonne (1 tonne = 1000 kg) is exhibited in the Hayden Planetarium in New York. Suppose the kinetic energy of the meteorite as it hit the ground was 617 MJ. Find the impulse I experienced by the meteorite up to the time its kinetic energy was halved (which took about 3.0 s). Find also the average force F exerted on the meteorite during this time interval.

48 •• When a 0.15-kg baseball is hit, its velocity changes from +20 m/s to −20 m/s. (*a*) What is the magnitude of the impulse delivered by the bat to the ball? (*b*) If the baseball is in contact with the bat for 1.3 ms, what is the average force exerted by the bat on the ball?

49 •• A 300-g handball moving with a speed of 5.0 m/s strikes the wall at an angle of 40° and then bounces off with the same speed at the same angle. It is in contact with the wall for 2 ms. What is the average force exerted by the ball on the wall?

50 •• A 2000-kg car traveling at 90 km/h crashes into a concrete wall that does not give at all. (*a*) Estimate the time of the collision, assuming that the center of the car travels halfway to the wall with constant deceleration. (Use any reasonable length for the car.) (*b*) Estimate the average force exerted by the wall on the car.

51 •• You throw a 150-g ball to a height of 40 m. (*a*) Use a reasonable value for the distance the ball moves while it is in your hand to calculate the average force exerted by your hand and the time the ball is in your hand while you throw it. (*b*) Is it reasonable to neglect the weight of the ball while it is being thrown?

52 •• A handball of mass 300 g is thrown straight against a wall with a speed of 8 m/s. It rebounds with the same speed. (*a*) What impulse is delivered to the wall? (*b*) If the ball is in contact with the wall for 0.003 s, what average force is exerted on the wall by the ball? (*c*) The ball is caught by a player who brings it to rest. In the process, her hand moves back 0.5 m. What is the impulse received by the player? (*d*) What average force was exerted on the player by the ball?

53 ••• The great limestone caverns were formed by dripping water. (*a*) If water droplets of 0.03 mL fall from a height of 5 m at a rate of 10 per minute, what is the average force exerted on the limestone floor by the droplets of water? (*b*) Compare this force to the weight of a water droplet.

54 ••• A favorite game at picnics is the egg toss. Two people toss a raw egg back and forth as they move farther apart. If the force required to break the egg's shell is about 5 N and the mass of the egg is 50 g, estimate the maximum separation distance for the egg throwers. Make whatever assumptions seem reasonable.

Collisions in One Dimension

55 • True or false:

(*a*) In any perfectly inelastic collision, all the kinetic energy of the bodies is lost.
(*b*) In a head-on elastic collision, the relative speed of recession after the collision equals the relative speed of approach before the collision.
(*c*) Kinetic energy is conserved in an elastic collision.

56 •• Under what conditions can all the initial kinetic energy of colliding bodies be lost in a collision?

57 •• Consider a perfectly inelastic collision of two objects of equal mass. (*a*) Is the loss of kinetic energy greater if the two objects have oppositely directed velocities of equal magnitude $v/2$, or if one of the two objects is initially at rest and the other has an initial velocity of v? (b) In which situation is the percentage loss in kinetic energy the greatest?

58 •• A mass m_1 traveling with a speed v makes a head-on elastic collision with a stationary mass m_2. In which scenario will the energy imparted to m_2 be greatest?
(*a*) $m_2 \ll m_1$
(*b*) $m_2 = m_1$
(*c*) $m_2 \gg m_1$
(*d*) None of the above.

59 • Joe and Sal decide that little Ronny is well-behaved enough to sit at the table with the family for Thanksgiving dinner. They are wrong. Ronny throws a 150-g handful of mashed potatoes horizontally with a speed of 5 m/s. It strikes a 1.2-kg gravy boat that is initially at rest on the frictionless table. If the potatoes stick to the gravy boat, what is the speed of the combined system as it slides down the table toward Grandpa?

60 • A 2000-kg car traveling to the right at 30 m/s is chasing a second car of the same mass that is traveling to the right at 10 m/s. (*a*) If the two cars collide and stick together, what is their speed just after the collision? (*b*) What fraction of the initial kinetic energy of the cars is lost during this collision? Where does it go?

61 • An 85-kg running back moving at 7 m/s makes a perfectly inelastic collision with a 105-kg linebacker who is initially at rest. What is the speed of the players just after their collision?

62 • A 5.0-kg object with a speed of 4.0 m/s collides head-on with a 10-kg object moving toward it with a speed of 3.0 m/s. The 10-kg object stops dead after the collision. (*a*) What is the final speed of the 5-kg object? (*b*) Is the collision elastic?

63 • A ball of mass m moves with speed v to the right toward a much heavier bat that is moving to the left with speed v. Find the speed of the ball after it makes an elastic collision with the bat.

64 •• During the Great Muffin Wars of ’98, students from rival residences became familiar with the characteristics of various muffins. Mushy Pumpkin Surprise, for example, was good for temporarily blinding an attacker, while Mrs. O’Brien’s Bran Muffins, having the density of lacrosse balls, were used more sparingly, and mainly as a deterrent. According to the rules, all muffins must have a mass of 0.3 kg. During one of the more memorable battles, a muffin moving to the right at 5 m/s collides with a muffin moving to the left at 2 m/s. Find the final velocities if (*a*) it is a perfectly inelastic collision of two pumpkin muffins, and (*b*) it is an elastic collision of two bran muffins.

65 •• Repeat Problem 64 with a second (illegal) muffin having a mass of 0.5 kg and moving to the right at 3 m/s.

66 •• A proton of mass m undergoes a head-on elastic collision with a stationary carbon nucleus of mass $12m$. The speed of the proton is 300 m/s. (*a*) Find the velocity of the center of mass of the system. (*b*) Find the velocity of the proton after the collision.

67 •• A 3-kg block moving at 4 m/s makes an elastic collision with a stationary block of mass 2 kg. Use conservation of momentum and the fact that the relative velocity of recession equals the relative velocity of approach to find the velocity of each block after the collision. Check your answer by calculating the initial and final kinetic energies of each block.

68 •• Night after night, Lucy is tormented by nocturnal wailing from the house next door. One day she seizes a revolver, stalks to the neighbors’ window with a crazed look in her eye, takes aim, and fires a 10-g bullet into her target: a 1.2-kg saxophone that rests on a frictionless surface. The bullet passes right through and emerges on the other side with a speed of 100 m/s, and the saxophone is given a speed of 4 m/s. Find the initial speed of the bullet, and the amount of energy dissipated in its trip through the saxophone.

69 •• A block of mass $m_1 = 2$ kg slides along a frictionless table with a speed of 10 m/s. Directly in front of it, and moving in the same direction with a speed of 3 m/s, is a block of mass $m_2 = 5$ kg. A massless spring with spring constant $k = 1120$ N/m is attached to the second block as in Figure 8-54. (*a*) Before m_1 runs into the spring, what is the velocity of the center of mass of the system? (*b*) After the collision, the spring is compressed by a maximum amount Δx. What is the value of Δx? (*c*) The blocks will eventually separate again. What are the final velocities of the two blocks measured in the reference frame of the table?

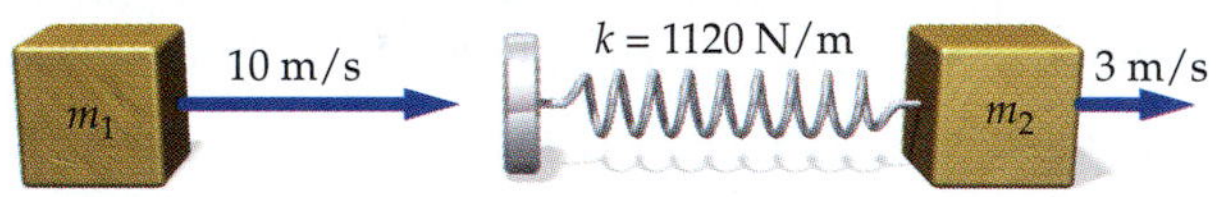

Figure 8-54 Problem 69

70 •• A bullet of mass m is fired vertically from below into a block of wood of mass M that is initially at rest, supported by a thin sheet of paper. The bullet blasts through the block, which rises to a height of H above its initial position before falling back down. The bullet continues rising to height h. (*a*) Express the upward velocity of the bullet and the block immediately after the bullet exits the block in terms of h and H. (*b*) Use conservation of momentum to express the speed of the bullet before it enters the block of wood in terms of given parameters. (*c*) Obtain expressions for the mechanical energies of the system before and after the inelastic collision. (*d*) Express the energy dissipated in the block of wood in terms of m, h, M, and H.

71 •• A proton of mass m is moving with initial speed v_0 toward an α particle of mass $4m$, which is initially at rest. Because both particles carry positive electrical charge, they repel each other. Find the speed v' of the α particle (*a*) when the distance between the two particles is least, and (*b*) when the two particles are far apart.

Ballistic Pendulums

72 •• A 16-g bullet is fired into the bob of a ballistic pendulum of mass 1.5 kg. When the bob is at its maximum height, the strings make an angle of 60° with the vertical. The length of the pendulum is 2.3 m. Find the speed of the bullet.

73 •• A bullet of mass m_1 is fired with a speed v into the bob of a ballistic pendulum of mass m_2. The bob is attached to a very light rod of length L that is pivoted at the other end. The bullet is stopped in the bob. Find the minimum v such that the bob will swing through a complete circle.

74 •• A bullet of mass m_1 is fired with a speed v into the bob of a ballistic pendulum of mass m_2. Find the maximum height h attained by the bob if the bullet passes through the bob and emerges with a speed $v/2$.

Exploding Objects and Radioactive Decay

75 •• A 3-kg bomb slides along a frictionless horizontal plane in the x direction at 6 m/s. It explodes into two pieces, one of mass 2 kg and the other of mass 1 kg. The 1-kg piece moves along the horizontal plane in the y direction at 4 m/s. (*a*) Find the velocity of the 2-kg piece. (*b*) What is the velocity of the center of mass after the explosion?

76 •• The beryllium isotope ^{4}Be is unstable and decays into two α particles (helium nuclei of mass $m = 6.68 \times 10^{-27}$ kg) with the release of 1.5×10^{-14} J of energy. Determine the velocities of the two α particles that arise from the decay of a ^{4}Be nucleus at rest.

77 •• The light isotope of lithium, ^{5}Li, is unstable and breaks up spontaneously into a proton (hydrogen nucleus) and an α particle (helium nucleus). In this process, a total energy of 3.15×10^{-13} J is released, appearing as the kinetic energy of the two reaction products. Determine the velocities of the proton and α particle that arise from the decay of a ^{5}Li nucleus at rest. (*Note:* The masses of the proton and alpha particle are $m_p = 1.67 \times 10^{-27}$ kg and $m_\alpha = 4m_p = 6.68 \times 10^{-27}$ kg.)

78 •• Jay and Dave decide that the best way to protest the opening of a new incinerator is to launch a stink bomb into the middle of the ceremony. They calculate that a 6-kg projectile launched with an initial speed of 40 m/s at an angle of 30° will do the trick. The bomb will explode on impact, no one will get hurt, but everyone will stink. Perfect. However, at the top of its flight, the bomb explodes into two fragments, each having a horizontal trajectory. To top it off—this really isn't their day—the 2-kg fragment lands right at the feet of Dave and Jay. (*a*) Where does the 4-kg fragment land? (*b*) Find the energy of the explosion by comparing the kinetic energy of the projectiles just before and just after the explosion.

79 •• A projectile of mass $m = 3$ kg is fired with initial speed of 120 m/s at an angle of 30° with the horizontal. At the top of its trajectory, the projectile explodes into two fragments of masses 1 kg and 2 kg. The 2-kg fragment, lands on the ground directly below the point of explosion 3.6 s after the explosion. (*a*) Determine the velocity of the 1-kg fragment immediately after the explosion. (*b*) Find the distance between the point of firing and the point at which the 1-kg fragment strikes the ground. (*c*) Determine the energy released in the explosion.

80 ••• The boron isotope ^{9}B is unstable and disintegrates into a proton and two α particles. The total energy released as kinetic energy of the decay products is 4.4×10^{-14} J. In one such event, with the ^{9}B nucleus at rest prior to decay, the velocity of the proton is measured to be 6.0×10^6 m/s. If the two α particles have equal energies, find the magnitude and the direction of their velocities with respect to that of the proton.

The Coefficient of Restitution (optional)

81 • The coefficient of restitution for steel on steel is measured by dropping a steel ball onto a steel plate that is rigidly attached to the earth. If the ball is dropped from a height of 3 m and rebounds to a height of 2.5 m, what is the coefficient of restitution?

82 • According to the official rules of racquetball, a ball acceptable for tournament play must bounce to a height of between 173 and 183 cm when dropped from a height of 254 cm at room temperature. What is the acceptable range of values for the coefficient of restitution for the racquetball–floor system?

83 • A ball bounces to 80% of its original height. (*a*) What fraction of its mechanical energy is lost each time it bounces? (*b*) What is the coefficient of restitution of the ball–floor system?

84 •• A 2-kg object moving at 6 m/s collides with a 4-kg object that is initially at rest. After the collision, the 2-kg object moves backward at 1 m/s. (*a*) Find the velocity of the 4-kg object after the collision. (*b*) Find the energy lost in the collision. (*c*) What is the coefficient of restitution for this collision?

85 •• A 2-kg block moving to the right with speed 5 m/s collides with a 3-kg block that is moving in the same direction at 2 m/s, as in Figure 8-55. After the collision, the 3-kg block moves at 4.2 m/s. Find (*a*) the velocity of the 2-kg block after the collision, and (*b*) the coefficient of restitution for the collision.

Figure 8-55 Problem 85

Collisions in Three Dimensions (optional)

86 • In a pool game, the cue ball, which has an initial speed of 5 m/s, makes an elastic collision with the eight ball, which is initially at rest. After the collision, the eight ball moves at an angle of 30° with the original direction of the cue ball. (*a*) Find the direction of motion of the cue ball after the collision. (*b*) Find the speed of each ball. Assume that the balls have equal mass.

87 •• An object of mass $m_1 = m$ collides with velocity $v_0\hat{i}$ into an object of mass $m_2 = 2m$ with velocity $\frac{1}{2}v_0\hat{j}$. Following the collision, the mass m_2 has a velocity $\frac{1}{4}v_0\hat{i}$. (*a*) Determine the velocity of the mass m_1 after the collision. (*b*) Was this an elastic collision? If not, express the energy change in terms of m and v_0.

88 •• A puck of mass 0.5 kg approaches a second, similar puck that is stationary on frictionless ice. The initial speed of the moving puck is 2 m/s. After the collision, one puck leaves with a speed v_1 at 30° to the original line of motion; the

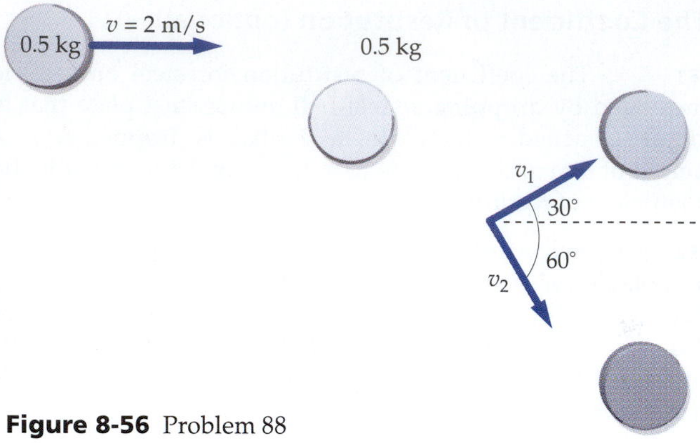

Figure 8-56 Problem 88

second puck leaves with speed v_2 at 60°, as in Figure 8-56. (*a*) Calculate v_1 and v_2. (*b*) Was the collision elastic?

89 •• Figure 8-57 shows the result of a collision between two objects of unequal mass. (*a*) Find the speed v_2 of the larger mass after the collision and the angle θ_2. (*b*) Show that the collision is elastic.

Figure 8-57 Problem 89

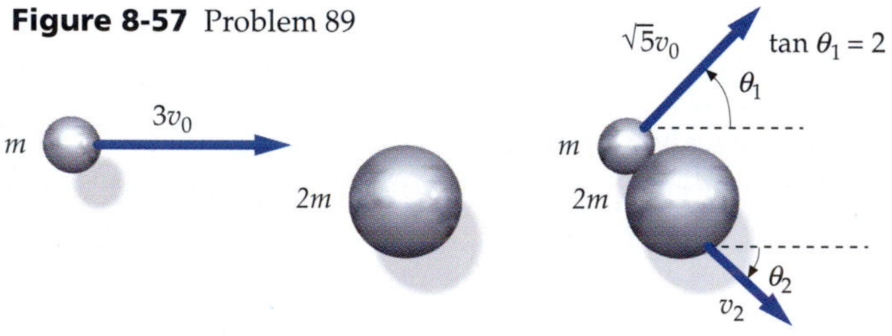

90 •• A ball moving at 10 m/s makes an off-center elastic collision with another ball of equal mass that is initially at rest. The incoming ball is deflected at an angle of 30° from its original direction of motion. Find the velocity of each ball after the collision.

91 ••• A particle has an initial speed v_0. It collides with a second particle that is at rest and is deflected through an angle ϕ. Its speed after the collision is v. The second particle recoils. Its velocity makes an angle θ with the initial direction of the first particle. (*a*) Show that

$$\tan\theta = \frac{v\sin\phi}{v_0 - v\cos\phi}$$

(*b*) Do you have to assume that the collision is either elastic or inelastic to get the result in part (*a*)?

The Center-of-Mass Reference Frame (optional)

92 • Describe a perfectly inelastic collision as viewed in the center-of-mass reference frame.

93 •• A particle with momentum p_1 in one dimension makes an elastic collision with a second particle of momentum $p_2 = -p_1$ in the center-of-mass reference frame. After the collision its momentum is p_1'. Write the total initial and final energies in terms of p_1 and p_1' and show that $p_1' = \pm p_1$. If $p_1' = -p_1$, the particle is merely turned around by the collision and leaves with the speed it had initially. What is the significance of the plus sign in your solution?

94 •• A 3-kg block is traveling to the right at 5 m/s, and a 1-kg block is traveling to the left at 3 m/s. (*a*) Find the velocity v_{cm} of the center of mass. (*b*) Subtract v_{cm} from the velocity of each block to find the velocity of each block in the center-of-mass reference frame. (*c*) After they make an elastic collision, the velocity of each block is reversed in this frame. Find the velocity of each block after an elastic collision. (*d*) Transform back into the original frame by adding v_{cm} to the velocity of each block. (*e*) Check your result by finding the initial and final kinetic energies of the blocks in the original frame.

95 •• Repeat Problem 94 with a second block having a mass of 5 kg and moving to the right at 3 m/s.

Rocket Propulsion (optional)

96 •• A rocket burns fuel at a rate of 200 kg/s and exhausts the gas at a relative speed of 6 km/s. Find the thrust of the rocket.

97 •• The payload of a rocket is 5% of its total mass, the rest being fuel. If the rocket starts from rest and moves with no external forces acting on it, what is its final velocity if the exhaust velocity of its gas is 5 km/s?

98 •• A rocket moves in free space with no external forces acting on it. It starts from rest and has an exhaust speed of 3 km/s. Find the final velocity if the payload is (*a*) 20%, (*b*) 10%, (*c*) 1%.

99 •• A rocket has an initial mass of 30,000 kg, of which 20% is the payload. It burns fuel at a rate of 200 kg/s and exhausts its gas at a relative speed of 1.8 km/s. Find (*a*) the thrust of the rocket, (*b*) the time until burnout, and (*c*) its final speed assuming it moves upward near the surface of the earth where the gravitational field g is constant.

General Problems

100 • Why can friction and the force of gravity usually be neglected in collision problems?

101 • The condition necessary for the conservation of momentum of a given system is that

(*a*) energy is conserved.
(*b*) one object is at rest.
(*c*) no external force acts.
(*d*) internal forces equal external forces.
(*e*) the net external force is zero.

102 • As a pendulum bob swings back and forth, is the momentum of the bob conserved? Explain why or why not.

103 • A model-train car of mass 250 g traveling with a speed of 0.50 m/s links up with another car of mass 400 g that is initially at rest. What is the speed of the cars immediately after they have linked together? Find the initial and final kinetic energies.

104 • (*a*) Find the total kinetic energy of the two model-train cars of Problem 103 before they couple. (*b*) Find the initial velocities of the two cars relative to the center of mass of the system, and use them to calculate the initial kinetic energy of the system relative to the center of mass. (*c*) Find the kinetic energy of the center of mass. (*d*) Compare your answers for (*b*) and (*c*) with that for (*a*).

105 • A 4-kg fish is swimming at 1.5 m/s to the right. He swallows a 1.2-kg fish swimming toward him at 3 m/s. Neglecting water resistance, what is the velocity of the larger fish immediately after his lunch?

106 • A 3-kg block moves at 6 m/s to the right while a 6-kg block moves at 3 m/s to the right. Find (*a*) the total kinetic energy of the two-block system, (*b*) the velocity of the center of mass, (*c*) the center-of-mass kinetic energy, and (d) the kinetic energy relative to the center of mass.

107 • A 1500-kg car traveling north at 70 km/h collides at an intersection with a 2000-kg car traveling west at 55 km/h. The two cars stick together. (*a*) What is the total momentum of the system before the collision? (*b*) Find the magnitude and direction of the velocity of the wreckage just after the collision.

108 • The great white shark can have a mass as great as 3000 kg. Suppose such a shark is cruising the ocean when it spots a meal below it: a 200.0-kg fish swimming horizontally at 8.00 m/s. The shark rushes vertically downward at 3.00 m/s and swallows the prey at once. At what angle to the vertical θ will the shark be moving immediately after the snack? What is the final speed of the shark? (Neglect any drag effects of the water.)

109 • Repeat Problem 106 for a 3-kg block moving at 6 m/s to the right and a 6-kg block moving at 3 m/s to the left.

110 • Repeat Problem 106 for a 3-kg block moving at 10 m/s to the right and a 6-kg block moving at 1 m/s to the right.

111 •• A 60-kg woman stands on the back of a 6-m-long, 120-kg raft that is floating at rest in still water with no friction. The raft is 0.5 m from a fixed pier, as in Figure 8-58.

(*a*) The woman walks to the front of the raft and stops. How far is the raft from the pier now?

(*b*) While the woman walks, she maintains a constant speed of 3 m/s relative to the raft. Find the total kinetic energy of the system (woman plus raft), and compare with the kinetic energy if the woman walked at 3 m/s on a raft *tied to the pier*.

(*c*) Where does this energy come from, and where does it go when the woman stops at the front of the raft?

(*d*) On land, the woman can put a lead shot 6 m. She stands at the back of the raft, aims forward, and puts the shot so that just after it leaves her hand, it has the same velocity *relative to her* as it does when she throws it from the ground. Where does the shot land?

Figure 8-58 Problem 111

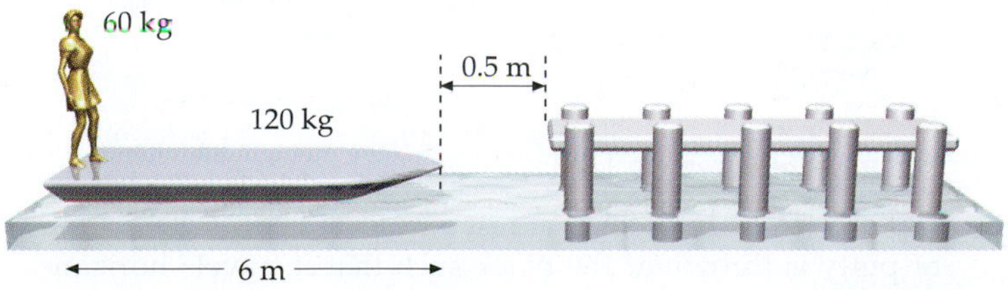

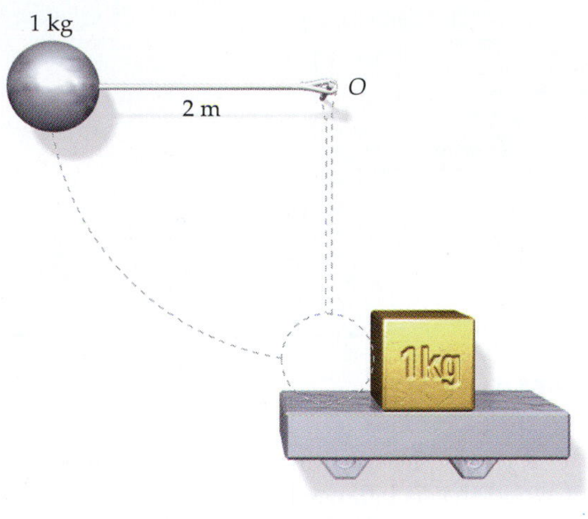

Figure 8-59 Problem 112

112 •• A 1-kg steel ball and a 2-m cord of negligible mass make up a simple pendulum that can pivot without friction about the point *O*, as in Figure 8-59. This pendulum is released from rest in a horizontal position and when the ball is at its lowest point it strikes a 1-kg block sitting at rest on a rough shelf. Assume that the collision is perfectly elastic and take the coefficient of friction between the block and shelf to be 0.1. (*a*) What is the velocity of the block just after impact? (*b*) How far does the block move before coming to rest?

113 •• In World War I, the most awesome weapons of war were huge cannons mounted on railcars. Figure 8-60 shows such a cannon, mounted so that it will project a shell at an angle of 30°. With the car initially at rest, the cannon fires a 200-kg projectile at 125 m/s. Now consider a system composed of a cannon, shell, and railcar, all rolling on the track without frictional losses. (*a*) Will the total vector momentum of that system be the same (i.e., "conserved") before and after the shell is fired? Explain your answer in a few words. (*b*) If the mass of the railcar plus cannon is 5000 kg, what will be the recoil velocity of the car along the track after the firing? (*c*) The shell is observed to rise to a maximum height of 180 m as it moves through its trajectory. At this point, its speed is 80 m/s. On the basis of this information, calculate the amount of thermal energy produced by air friction on the shell on its way from firing to this maximum height.

Figure 8-60 Problem 113

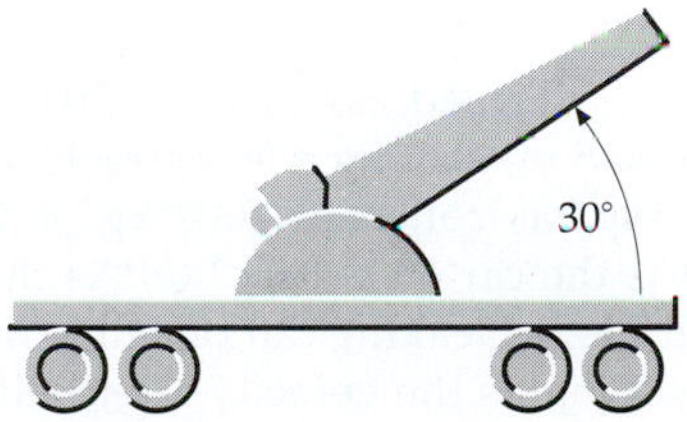

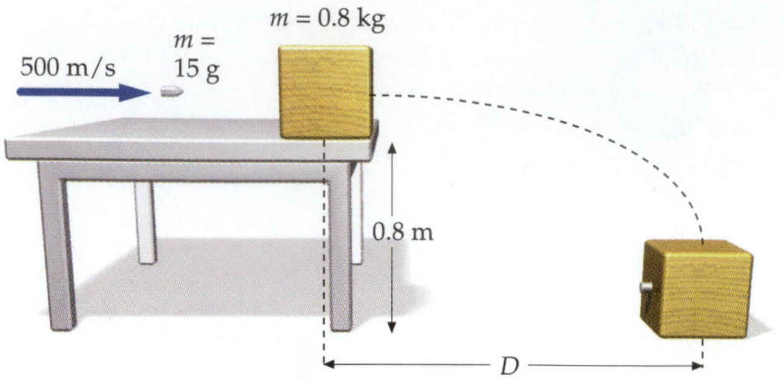

Figure 8-61 Problem 114

114 •• A 15-g bullet traveling at 500 m/s strikes an 0.8-kg block of wood that is balanced on a table edge 0.8 m above the ground (Figure 8-61). If the bullet buries itself in the block, find the distance D at which the block hits the floor.

115 •• In hand-pumped railcar races, a speed of 32 km/h has been achieved by teams of four. A car of mass 350 kg is moving at that speed toward a river when Carlos, the chief pumper, notices that the bridge ahead is out. All four people (of mass 75 kg each) jump simultaneously backward off the car with a velocity that has a horizontal component of 4 m/s *relative to the car after jumping*. The car proceeds off the bank and falls in the water a distance 25.0 m off the bank. (*a*) Estimate the time of the fall of the railcar. (*b*) What happens to the team of pumpers?

116 •• A constant force $\vec{F} = 12\,\text{N}\,\vec{i}$ is applied to the 8-kg mass of Problem 5 at $t = 0$. (*a*) What is the velocity of the center of mass of the three-particle system at $t = 5$ s? (*b*) What is the location of the center of mass at $t = 5$ s?

117 •• Two particles of mass m and $4m$ are moving in a vacuum at right angles as in Figure 8-62. A force $\vec{F}$ acts on both particles for a time T. As a result, the velocity of the particle m is $4v$ in its original direction. Find the new velocity $\vec{v}\,'$ of the particle of mass $4m$.

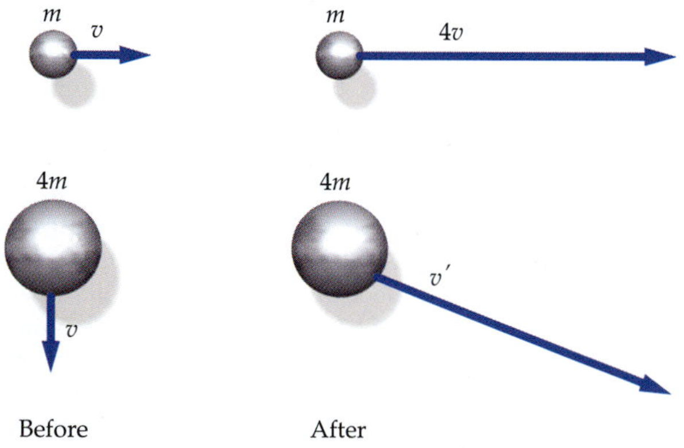

Figure 8-62 Problem 117

118 •• An open railroad car of mass 20,000 kg is rolling without friction at 5 m/s along a level track when it starts to rain. After the car has collected 2000 kg of water, the rain stops. (*a*) What is the car's velocity? (*b*) As the car is rolling along, the water begins leaking out of a hole in the bottom at a rate of 5 kg/s. What is the velocity after half the water has leaked out? (*c*) What is the velocity after all the water has leaked out?

119 •• In the "slingshot effect," the transfer of energy in an elastic collision is used to boost the energy of a space probe so that it can escape from the solar system. Figure 8-63 shows a space probe moving at 10.4 km/s (relative to the sun) toward Saturn, which is moving at 9.6 km/s (relative to the sun) toward the probe. Because of the gravitational attraction between Saturn and the probe, the probe swings around Saturn and heads back in the opposite direction with speed v_f. (*a*) Assuming this collision to be a one-dimensional elastic collision with the mass of Saturn much greater than that of the probe, find v_f. (*b*) By what factor is the kinetic energy of the probe increased? Where does this energy come from?

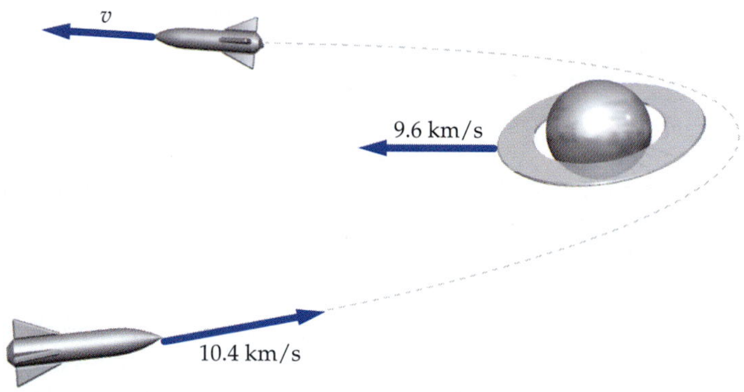

Figure 8-63 Problem 119

120 •• You (mass 80 kg) and your friend (mass unknown) are in a rowboat (mass 60 kg) on a calm lake. You are at the center of the boat rowing and she is at the back, 2 m from the center. You get tired and stop rowing. She offers to row and after the boat comes to rest, you change places. You notice that after changing places the boat has moved 20 cm relative to a fixed log. What is your friend's mass?

121 •• A small car of mass 800 kg is parked behind a small truck of mass 1600 kg on a level road (Figure 8-64). The brakes of both the car and the truck are off so that they are free to roll with negligible friction. A man sitting on the tailgate of the truck shoves the car away by exerting a constant force on the car with his feet. The car accelerates at 1.2 m/s^2. (*a*) What is the acceleration of the truck? (*b*) What is the magnitude of the force exerted on either the truck or the car?

Figure 8-64 Problem 121

122 •• A 13-kg block is at rest on a level floor. A 400-g glob of putty is thrown at the block such that it travels horizontally, hits the block, and sticks to it. The block and putty slide 15 cm along the floor. If the coefficient of sliding friction is 0.4, what is the initial speed of the putty?

123 •• A careless driver rear-ends a car that is halted at a stop sign. Just before impact, the driver slams on his brakes, locking the wheels. The driver of the struck car also has his foot solidly on the brake pedal, locking his brakes. The mass of the struck car is 900 kg, and that of the initially moving vehicle is 1200 kg. On collision, the bumpers of the two cars mesh. Police determine from the skid marks that after the collision the two cars moved 0.76 m together. Tests revealed that the coefficient of sliding friction between the tires and pavement was 0.92. The driver of the moving car claims that he was traveling at less than 15 km/h as he approached the intersection. Is he telling the truth?

124 •• A pendulum consists of a 0.4-kg bob attached to a string of length 1.6 m. A block of mass m rests on a horizontal frictionless surface (Figure 8-65). The pendulum is released from rest at an angle of 53° with the vertical and the bob collides elastically with the block. Following the collision, the maximum angle of the pendulum with the vertical is 5.73°. Determine the mass m.

Figure 8-65 Problem 124

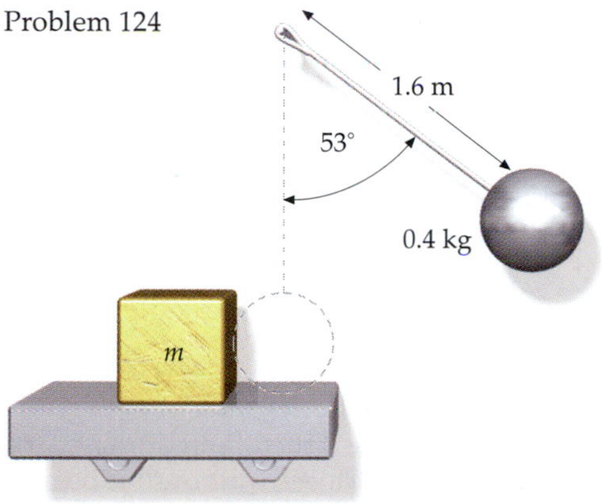

125 •• Initially, mass m = 1.0 kg and mass M are both at rest on a frictionless inclined plane (Figure 8-66). Mass M rests against a spring that has a spring constant of 11,000 N/m. The distance along the plane between m and M is 4.0 m. Mass m is released, makes an elastic collision with mass M, and rebounds a distance of 2.56 m back up the inclined plane. Mass M comes to rest momentarily 4.0 cm from its initial position. Find the mass M.

Figure 8-66 Problem 125

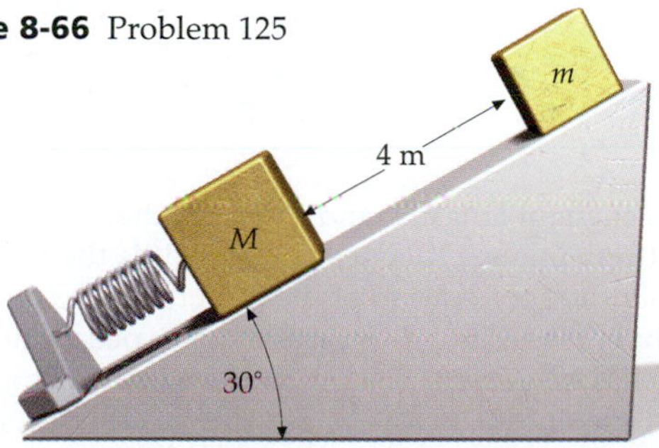

126 •• A circular plate of radius r has a circular hole cut out of it having radius $r/2$ (Figure 8-67). Find the center of mass of the plate. *Hint:* The hole can be represented by two disks superimposed, one of mass m and the other of mass $-m$.

Figure 8-67 Problem 126

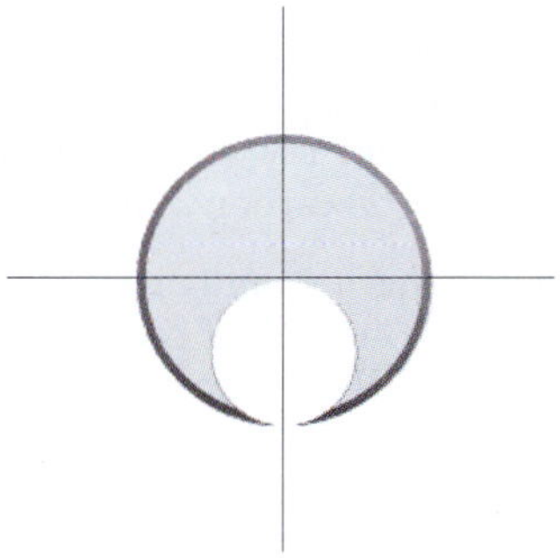

127 •• Using the hint from Problem 126, find the center of mass of a solid sphere of radius r that has a spherical cavity of radius $r/2$, as in Figure 8-68.

Figure 8-68 Problem 127

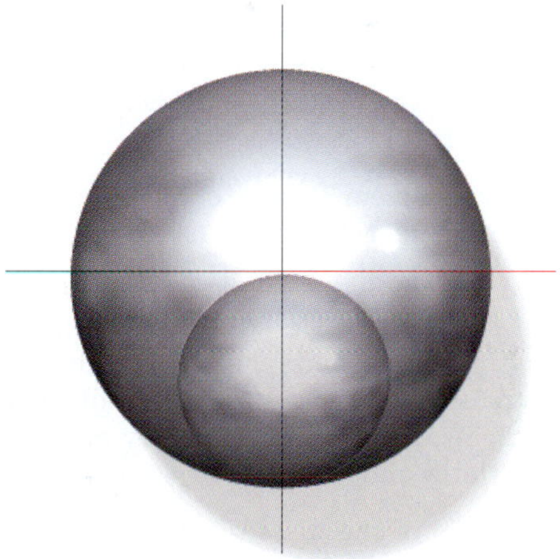

128 •• A neutron of mass m makes an elastic head-on collision with a stationary nucleus of mass M. (*a*) Show that the energy of the nucleus after the collision is $K_{\text{nucleus}} = [4mM/(m+M)^2]K_n$, where K_n is the initial energy of the neutron. (*b*) Show that the fraction of energy lost by the neutron in this collision is

$$\frac{-\Delta K_n}{K_n} = \frac{4mM}{(m+M)^2} = \frac{4(m/M)}{(1+m/M)^2}$$

129 •• The mass of a carbon nucleus is approximately 12 times the mass of a neutron. (*a*) Use the results of Problem 128 to show that after N head-on collisions of a neutron with carbon nuclei at rest, the energy of the neutron is approximately $0.716^N E_0$, where E_0 is its original energy. Neutrons emitted in the fission of a uranium nucleus have an energy of about 2 MeV. For such a neutron to cause the fission of another uranium nucleus in a reactor, its energy must be reduced to about 0.02 eV. (*b*) How many head-on collisions are needed to reduce the energy of a neutron from 2 MeV to 0.02 eV, assuming elastic head-on collisions with stationary carbon nuclei?

130 •• On average, a neutron loses 63% of its energy in an elastic collision with a hydrogen atom and 11% of its energy in an elastic collision with a carbon atom. The numbers are lower than the ones we have been using in earlier problems because most collisions are not head-on. Calculate the number of collisions, on average, needed to reduce the energy of a neutron from 2 MeV to 0.02 eV (a desirable outcome for reasons explained in Problem 129) if the neutron collides with (*a*) hydrogen atoms and (*b*) carbon atoms.

131 •• A rope of length L and mass M lies coiled on a table. Starting at $t = 0$, one end of the rope is lifted from the table with a force F such that it moves with a constant velocity v. (*a*) Find the height of the center of mass of the rope as a function of time. (*b*) Differentiate your result in (*a*) twice to find the acceleration of the center of mass. (*c*) Assuming that the force exerted by the table equals the weight of the rope still there, find the force F you exert on the top of the rope.

132 •• A tennis ball of mass m_t is held a small distance above a basketball of mass m_b. Both are dropped from a height h above the floor. (Take h to be the distance to the center of the basketball.) The basketball collides elastically with the floor. Find the speed v_t of the tennis ball after it then collides elastically with the basketball. Calculate the height reached by the tennis ball if $m_b = 0.480$ kg, $m_t = 0.060$ kg, and $h = 2$ m. (*Caution:* If you try this experimentally, get out of the way of the tennis ball!)

133 •• Repeat Problem 24 if the cup has a mass m_c and the ball collides with it inelastically.

134 •• Two astronauts at rest face each other in space. One, with mass m_1, throws a ball of mass m_b to the other, whose mass is m_2. She catches the ball and throws it back to the first astronaut. If they each throw the ball with a speed of v relative to themselves, how fast are they moving after each has made one throw and one catch?

135 •• The ratio of the mass of the earth to the mass of the moon is $M_e/m_m = 81.3$. The radius of the earth is about 6370 km, and the distance from the earth to the moon is about 384,000 km. (*a*) Locate the center of mass of the earth–moon system relative to the surface of the earth. (*b*) What external forces act on the earth–moon system? (*c*) In what direction is the acceleration of the center of mass of this system? (*d*) Assume that the center of mass of this system moves in a circular orbit around the sun. How far must the center of the earth move in the radial direction (toward or away from the sun) during the 14 days between the time the moon is farthest from the sun (full moon) and the time it is closest to the sun (new moon)?

136 •• You wish to enlarge a skating surface so you stand on the ice at one end and aim a hose horizontally to spray water on the schoolyard pavement. Water leaves the hose at 2.4 kg/s with speed 30 m/s. If your mass is 75 kg, what is your recoil acceleration? (Neglect friction and the mass of the hose.)

137 •• A neutron at rest decays into a proton plus an electron. The conservation of momentum implies that the electron and proton should have equal and opposite momentum. However, experimentally they do not. This apparent nonconservation of momentum led Wolfgang Pauli to suggest in 1931 that there was a third, unseen particle emitted in the decay. This particle is called a neutrino, and it was finally observed directly in 1957. Suppose that the electron has momentum $p = 4.65 \times 10^{-22}$ kg·m/s along the negative x direction and the proton ($m = 1.67 \times 10^{-27}$ kg) moves with speed 2.93×10^5 m/s at an angle 17.9° above the x axis. Find the momentum of the neutrino. (The kinetic energy of the electron is comparable to its rest energy, so its energy and momentum are related relativistically rather than classically. However, the rest energy of the proton is large compared with its kinetic energy so the classical relation $E = \frac{1}{2}mv^2 = p^2/2m$ is valid.)

138 ••• A stream of glass beads, each with a mass of 0.5 g, comes out of a horizontal tube at a rate of 100 per second (Figure 8-69). The beads fall a distance of 0.5 m to a balance pan and bounce back to their original height. How much mass must be placed in the other pan of the balance to keep the pointer at zero?

Figure 8-69 Problem 138

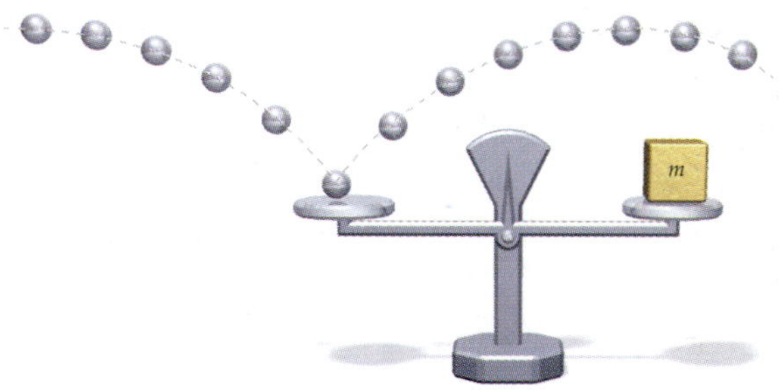

139 ••• A dumbbell consisting of two balls of mass m connected by a massless rod of length L rests on a frictionless floor against a frictionless wall until it begins to slide down the wall as in Figure 8-70. Find the speed v of the bottom ball at the moment when it equals the speed of the top one.

Figure 8-70 Problem 139

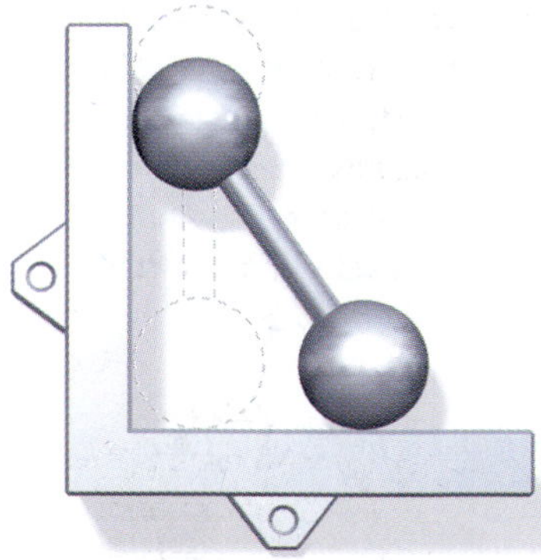

140 ••• A chain of length L and mass m is held vertically so that the bottom link just touches the floor. It is then dropped. (*a*) Show that the acceleration of the top end of the chain is g. (*b*) If the chain is moving downward with speed v at time t, and speed $v + \Delta v$ at time $t + \Delta v$, find an expression for the change in momentum of the chain during the interval Δt. (*c*) Find the force exerted on the chain by the floor.

CHAPTER 9

Rotation

Star tracks in a time exposure of the night sky.

Rotational motion is all around us from molecules to galaxies. The earth rotates about its axis. Wheels, gears, propellers, motors, the drive shaft in a car, a CD in its player, a pirouetting ice skater, all rotate. Our study of rotation is simplified by analogies between linear motion and rotational motion. In this chapter, we consider rotation about an axis that is fixed in space, or one that is moving parallel to itself as in a rolling ball. More general examples of rotational motion are discussed in Chapter 10.

9-1 Angular Velocity and Angular Acceleration

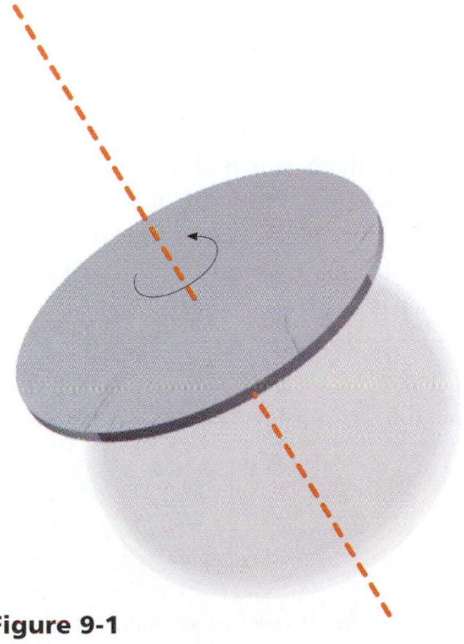

Figure 9-1

Imagine a disk rotating about a fixed axis perpendicular to the disk and through its center (Figure 9-1). Points near the rim move faster than points near the axis. But when a point near the rim moves through a complete circle, so does any other point on the disk. As the disk rotates through a given angle, *all* points on the disk rotate through the same angle. The angle through which a disk rotates is a characteristic of the disk as a whole, as is the rate at which the angle changes. As the disk turns, the distance between any two particles that make up the disk remains fixed. Such a system is called a **rigid body.**

Consider a typical particle in the disk (Figure 9-2). Let r_i be the distance from the center of the disk to the ith particle, and θ_i be the angle measured counterclockwise from a fixed reference line in space to a line from the center to the particle. As the disk rotates through an angle $d\theta$, the particle moves through a circular arc of length

$$ds_i = r_i|d\theta| \tag{9-1}$$

where $d\theta$ is measured in radians. The distance ds_i varies from particle to particle, but the angle $d\theta$, called the **angular displacement,** is the same for all particles of the disk. For one complete revolution, the arc length Δs_i is $2\pi r$ and the angular displacement $\Delta\theta$ is

$$\Delta\theta = \frac{2\pi r_i}{r_i} = 2\pi \text{ rad} = 360° = 1 \text{ rev}$$

Figure 9-2

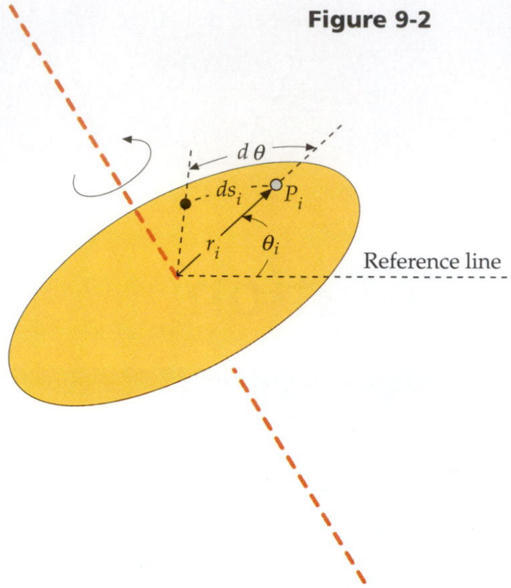

The time rate of change of the angle, $d\theta/dt$, is the same for all particles of the disk, and is called the **angular velocity** ω of the disk:

$$\omega = \frac{d\theta}{dt} \tag{9-2}$$

Definition—Angular velocity

For counterclockwise rotation, θ increases, so ω is positive. For clockwise rotation, θ decreases and ω is negative. The units of ω are radians per second. Since radians are dimensionless, the dimensions of angular velocity are those of reciprocal time (T^{-1}). The magnitude of the angular velocity is called the **angular speed.** We often use revolutions per minute (rev/min) to describe rotation. To convert between revolutions, radians, and degrees, we use

$$1 \text{ rev} = 2\pi \text{ rad} = 360°$$

Exercise A CD-ROM disc is rotating at 3000 revolutions per minute. What is its angular speed in radians per second? (*Answer* 314 rad/s)

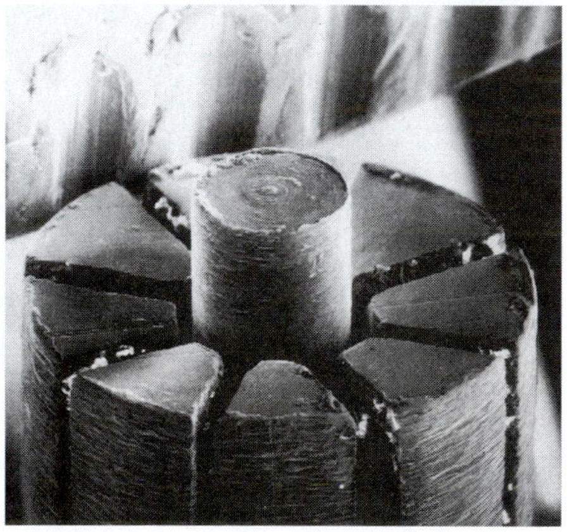

A tiny device called a wobble motor. This motor, which has a diameter of the order of a millimeter, has achieved angular speeds in excess of 120,000 rev/min. The edge of a dime is visible in the background.

The time rate of change of angular velocity is called the **angular acceleration** α:

$$\alpha = \frac{d\omega}{dt} = \frac{d^2\theta}{dt^2} \tag{9-3}$$

Definition—Angular acceleration

The units of α are radians per second per second (rad/s^2). If ω is increasing, α is positive; if ω is decreasing, α is negative.

The linear velocity of a particle on the disk is tangent to the circular path of the particle and has magnitude $v_{it} = ds_i/dt$. We can relate this *tangential velocity* to the angular velocity of the disk using Equations 9-1 and 9-2:

$$v_{it} = \frac{ds_i}{dt} = \frac{r_i d\theta}{dt} = r_i\omega \tag{9-4}$$

Similarly, the tangential acceleration of a particle on the disk is

$$a_{it} = \frac{dv_{it}}{dt} = r_i\frac{d\omega}{dt}$$

so

$$a_{it} = r_i\alpha \tag{9-5}$$

Each particle of the disk also has a radial acceleration, the centripetal acceleration, which points inward along the radial line, and has the magnitude

$$a_{ic} = \frac{v_{it}^2}{r_i} = \frac{(r_i\omega)^2}{r_i} = r_i\omega^2 \qquad \text{9-6}$$

Exercise A point on the rim of a compact disc is 6.0 cm from the axis of rotation. Find the tangential speed v_t, tangential acceleration a_t, and centripetal acceleration a_c of the point when the disc is rotating at a constant angular speed of 300 rev/min. (*Answers* $v_t = 188$ cm/s, $a_t = 0$, $a_c = 5.92 \times 10^3$ cm/s^2)

The angular displacement θ, angular velocity ω, and angular acceleration α are analogous to the linear displacement x, linear velocity v, and linear acceleration a in one-dimensional motion. If the angular acceleration α is constant, we can integrate Equation 9-3 to find ω:

$$\omega = \omega_0 + \alpha t \qquad \text{9-7}$$

where the constant of integration ω_0 is the initial angular velocity. This is the rotational analog of $v = v_0 + at$. Integrating again, we obtain

$$\theta = \theta_0 + \omega_0 t + \frac{1}{2}\alpha t^2 \qquad \text{9-8}$$

which is the rotational analog of $x = x_0 + v_0 t + \frac{1}{2}at^2$ with θ replacing x, ω replacing v, and α replacing a. Similarly, by eliminating t from Equations 9-7 and 9-8, we get

$$\omega^2 = \omega_0^2 + 2\alpha(\theta - \theta_0) \qquad \text{9-9}$$

which is the rotational analog of $v^2 = v_0^2 + 2a(x - x_0)$. The equations for constant angular acceleration have the same form as those for constant linear acceleration.

Example 9-1

A compact disc rotates from rest to 500 rev/min in 5.5 s. (*a*) What is its angular acceleration, assuming that it is constant? (*b*) How many revolutions does it make in 5.5 s? (*c*) How far does a point on the rim 6 cm from the center travel during the 5.5 s it takes to get to 500 rev/min?

Picture the Problem Part (*a*) is analogous to the linear problem of finding the acceleration given the final velocity. To find α in rad/s^2 we need to convert ω to rad/s. Part (*b*) is analogous to finding the distance traveled in a given time.

(*a*)1. The angular acceleration is related to the initial and final angular velocities:

$$\omega = \omega_0 + \alpha t = 0 + \alpha t$$

2. Solve for α:

$$\alpha = \frac{\omega}{t} = \frac{500\ \text{rev/min}}{5.5\ \text{s}} \times \frac{2\pi\ \text{rad}}{1\ \text{rev}} \times \frac{1\ \text{min}}{60\ \text{s}} = 9.52\ \text{rad/s}^2$$

(*b*)1. The angular displacement is related to the time by Equation 9-8:

$$\theta - \theta_0 = \omega_0 t + \frac{1}{2}\alpha t^2$$

$$= 0 + \frac{1}{2}(9.52\ \text{rad/s}^2)(5.5\ \text{s})^2 = 144\ \text{rad}$$

2. Convert radians to revolutions:

$$144\ \text{rad} \times \frac{1\ \text{rev}}{2\pi\ \text{rad}} = 22.9\ \text{rev}$$

(c) The distance traveled Δs is r times the angular displacement: $\Delta s = r\,\Delta\theta = (0.06\text{ m})(144\text{ rad}) = 8.64\text{ m}$

Check the Result The average angular velocity in revolutions per minute is 250 rev/min. In 5.5 s, the compact disc rotates (250 rev/60 s)(5.5 s) = 22.9 rev.

Remarks A compact disc is scanned by a laser that begins at the inner radius of about 2.4 cm and moves out to the edge at 6.0 cm. As the laser moves outward, the angular velocity of the disc decreases from 500 rev/min to 200 rev/min so that the linear (tangential) velocity of the disc at the point where the laser beam strikes remains constant.

Exercise Convert 500 rev/min to rad/s. (*Answer* 500 rev/min = 52.4 rad/s)

Exercise Check the result of part (*b*) in the example using $\omega^2 = \omega_0^2 + 2\alpha(\theta - \theta_0)$.

Exercise Find the linear speed of a point on the disc at (*a*) $r = 2.4$ cm when the disc rotates at 500 rev/min, and (*b*) $r = 6.0$ cm when the disc rotates at 200 rev/min. (*Answers* (*a*) 126 cm/s, (*b*) 126 cm/s)

9-2 Torque, Moment of Inertia, and Newton's Second Law for Rotation

To set a top spinning, you twist it. In Figure 9-3, a disk is set spinning by the forces $\vec{F}_1$ and $\vec{F}_2$ exerted at the edges of the disk. The points at which these forces are applied is important. The same forces applied so that their lines of action pass through the center of the disk, as in Figure 9-4, will not spin the disk. Figure 9-5 shows a single force $\vec{F}_i$ acting on the ith particle of a disk. The perpendicular distance between the line of action of a force and the axis of rotation is called the **lever arm** ℓ of the force. A force times its lever arm is the magnitude of the **torque** τ_i:

$$\tau_i = F_i\ell \qquad \text{9-10}$$

Torque can be thought of as a twist, just as a force is a push or a pull. It is the torque that affects the angular velocity of the object.

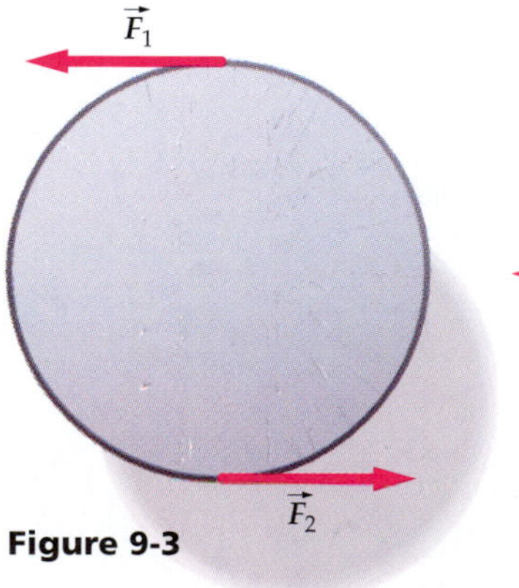

Figure 9-3

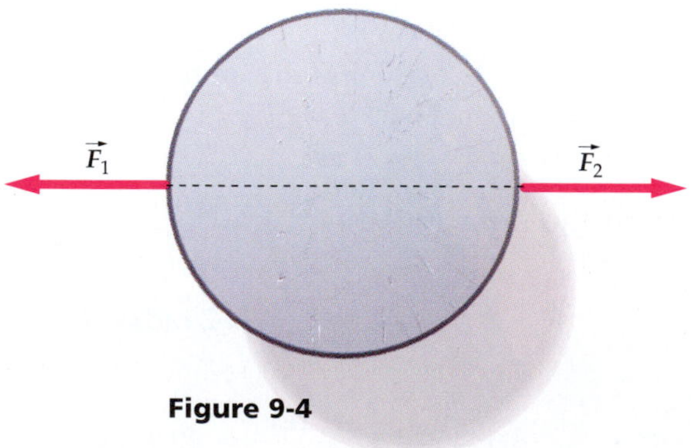

Figure 9-4

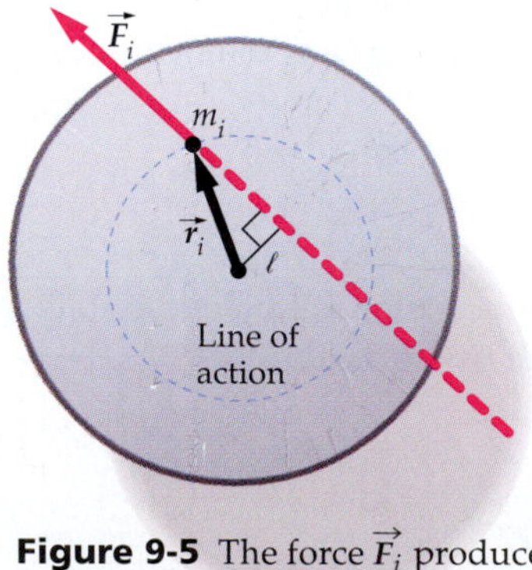

Figure 9-5 The force $\vec{F}_i$ produces a torque $F_i\ell$ about the center.

The lever arm of the force in Figure 9-5 is $\ell = |r_i \sin\phi|$, where ϕ is the angle between $\vec{F}_i$ and the position vector $\vec{r}_i$ from the center of the disk to the point of application of the force. The torque exerted by this force is

$$\tau_i = F_i r_i \sin\phi \qquad 9\text{-}11$$

The torque is taken to be positive if it tends to rotate the disk counterclockwise, and negative if it tends to rotate the disk clockwise. In Figure 9-6, $\vec{F}_i$ is resolved into two components, $F_{ir} = F_i \cos\phi$ along the radial line $\vec{r}_i$ and $F_{it} = F_i \sin\phi$ perpendicular to the radial line. The radial component has no effect on the rotation of the disk. The torque exerted by $\vec{F}_i$ can be written in terms of F_{it}. From Equation 9-10,

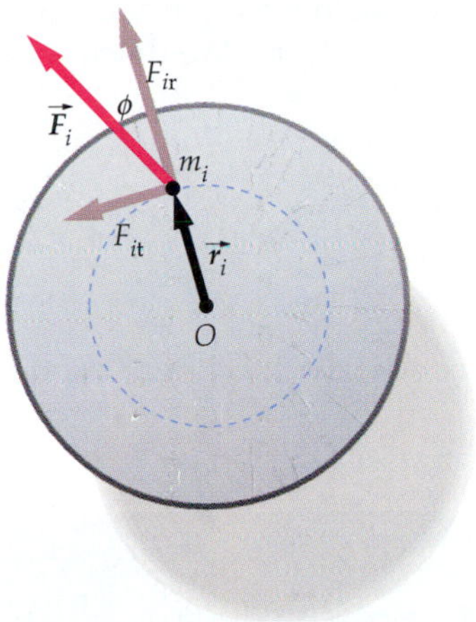

Figure 9-6

$$\tau_i = F_i\ell = F_i r_i \sin\phi = F_{it} r_i \qquad 9\text{-}12$$

We now show that a rigid body's angular acceleration is proportional to the net torque acting on it. Let $\vec{F}_i$ be the net external force acting on the ith particle. The tangential acceleration of the ith particle is, by Newton's second law,

$$F_{it} = m_i a_{it} = m_i r_i \alpha \qquad 9\text{-}13$$

where we have used $a_{it} = r_i\alpha$ (Equation 9-5). Multiply each side by r_i:

$$r_i F_{it} = m_i r_i^2 \alpha \qquad 9\text{-}14$$

The left side of Equation 9-14 is the torque τ_i exerted by the force $\vec{F}_i$ about the pivot O. So

$$\tau_i = m_i r_i^2 \alpha \qquad 9\text{-}15$$

Summing over all the particles in the object gives

$$\sum_i \tau_i = \sum_i m_i r_i^2 \alpha \qquad 9\text{-}16$$

$\Sigma\tau_i$ is the net torque acting on the object. For a rigid body, the angular acceleration is the same for all the particles of the object and can therefore be taken out of the sum. The quantity $\Sigma\, m_i r_i^2$ is called the **moment of inertia** I. For a continuous object, the sum is replaced by an integral.

$$I = \sum_i m_i r_i^2 \quad \text{(system of particles)} \qquad 9\text{-}17$$

$$I = \int r^2 dm \quad \text{(continuous object)} \qquad 9\text{-}18$$

Definition—Moment of inertia

In Equation 9-17, r_i is the distance of the ith particle from the axis of rotation, and in Equation 9-18, r is the distance of the mass element dm from the axis of rotation. For a disk with the origin at the center on the axis, this is the same as the distance to the origin.

Writing I for the moment of inertia, Equation 9-16 becomes

$$\sum_i \tau_i = I\alpha \qquad 9\text{-}19$$

In Chapter 8, we saw that the net force acting on a system of particles is equal to the net *external* force acting on the system because the internal forces (those exerted by the particles within the system on one another) cancel in pairs. The treatment of internal torques exerted by the particles within a system on one another leads to a similar result, that is, the net torque acting on a system equals the net *external* torque acting on the system. (We discuss this further in Chapter 10.) We can thus write Equation 9-19 as

$$\tau_{\text{net,ext}} = \sum_i \tau_{i,\text{ext}} = I\alpha \qquad \text{9-20}$$

Newton's second law for rotation

This is the rotational analog of Newton's second law for linear motion, $\Sigma \vec{F} = m\vec{a}$.

The torque exerted by a wrench on a nut is proportional to the force and to the lever arm. Charlie could exert a greater torque with the same force if he held the wrenches nearer their ends.

9-3 Calculating the Moment of Inertia

The moment of inertia is a measure of the resistance of an object to changes in its rotational motion. It is the rotational analog of mass. The moment of inertia depends on the distribution of mass within the object relative to the axis of rotation. The farther the mass from the axis, the greater the moment of inertia. Thus, unlike the mass of an object, which is a property of the object itself, the moment of inertia of an object also depends on the location of the axis of rotation.

Systems of Particles

For systems consisting of discrete particles, we can compute the moment of inertia about a given axis directly from Equation 9-17.

Example 9-2

Four particles of mass m are connected by massless rods to form a rectangle of sides $2a$ and $2b$ as shown. The system rotates about an axis in the plane of the figure through the center (Figure 9-7). Find the moment of inertia about this axis.

Picture the Problem Since we are given that the objects are particles, we use Equation 9-17. In that equation, r_i is the perpendicular distance from the particle of mass m_i to the axis (line) of rotation.

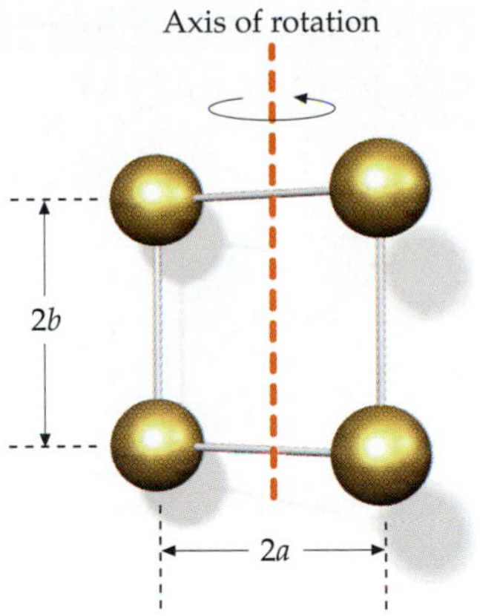

Figure 9-7

1. Apply the definition of moment of inertia for discrete particles (Equation 9-17): $I = \sum_i m_i r_i^2 = m_1 r_1^2 + m_2 r_2^2 + m_3 r_3^2 + m_4 r_4^2$

2. The masses m_i and the distances r_i are given: $m_1 = m_2 = m_3 = m_4 = m$

   $r_1 = r_2 = r_3 = r_4 = a$

3. Substitution gives the moment of inertia: $I = ma^2 + ma^2 + ma^2 + ma^2 = 4ma^2$

Remark Notice that I is independent of the length b, which has no effect on how far the masses are from the axis of rotation. This example and exercise illustrate the fact that the moment of inertia depends on the location of the axis of rotation. In Figure 9-8, the moment of inertia, represented by the dimensionless quantity I/ma^2, is plotted versus the distance from the two left particles to the axis of rotation. Note that the moment of inertia is a minimum when the axis is directly in the middle.

Exercise Find the moment of inertia of this system for rotation about an axis parallel to the first axis but passing through two of the particles as shown in Figure 9-9. (*Answer* $I = 8ma^2$)

Figure 9-9

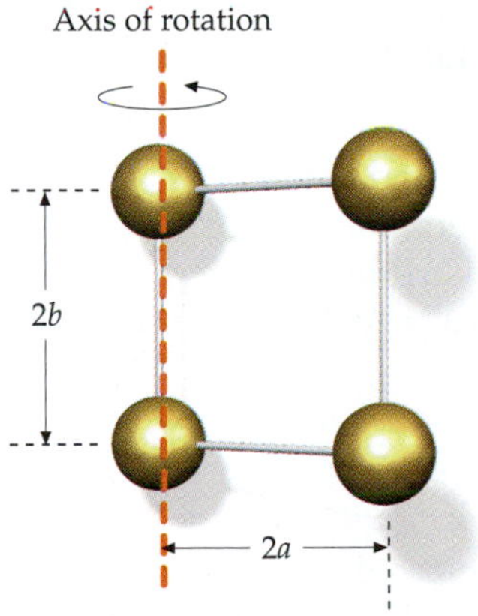

Figure 9-8

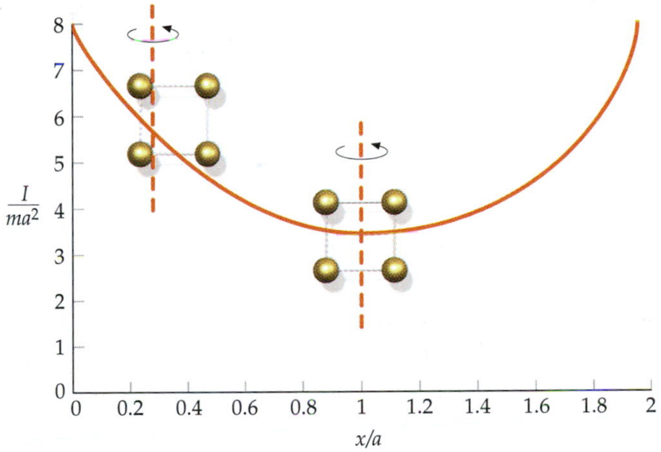

Continuous Objects

To calculate the moment of inertia for continuous objects, we use Equation 9-18 where r is the distance of the mass element dm from the axis of rotation. Table 9-1 lists the moments of inertia of various uniform objects. We now give several more examples of calculating I for continuous objects.

Table 9-1

Moments of Inertia of Uniform Bodies of Various Shapes

Cylindrical shell about axis

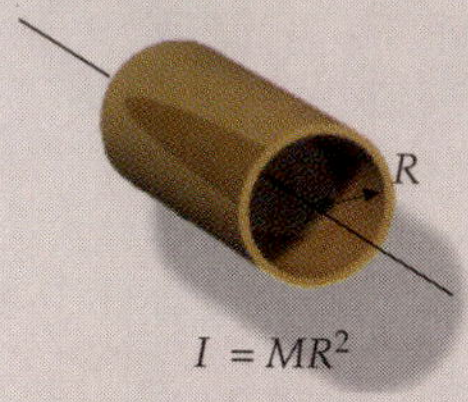

$I = MR^2$

Solid cylinder about axis

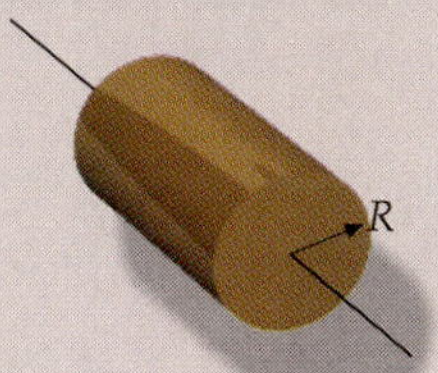

$I = \frac{1}{2}MR^2$

Hollow cylinder about axis

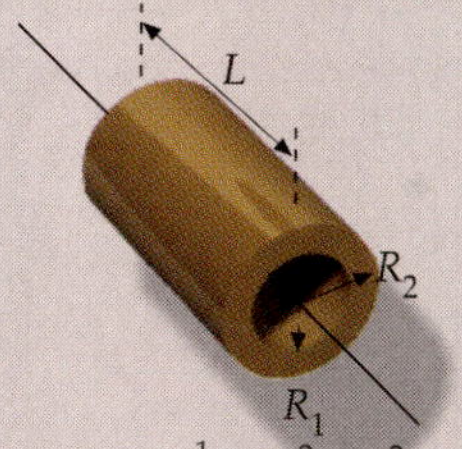

$I = \frac{1}{2}M(R_1^2 + R_2^2)$

Cylindrical shell about diameter through center

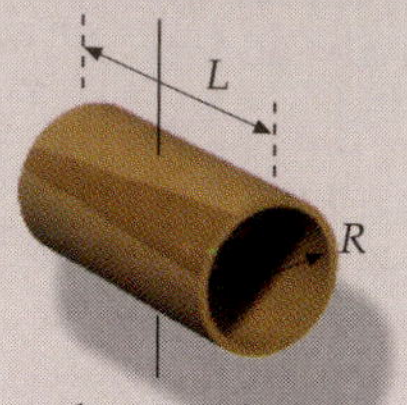

$I = \frac{1}{2}MR^2 + \frac{1}{12}ML^2$

Solid cylinder about diameter through center

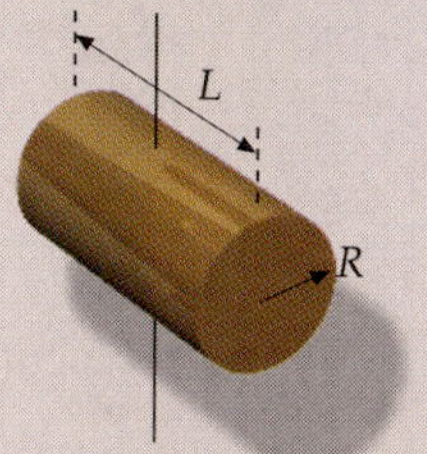

$I = \frac{1}{4}MR^2 + \frac{1}{12}ML^2$

Thin rod about perpendicular line through center

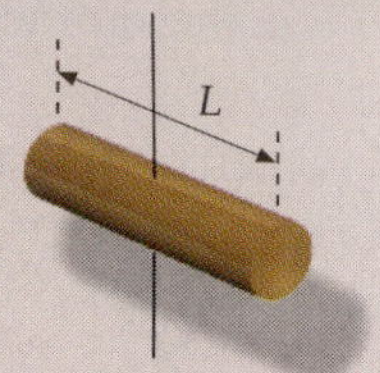

$I = \frac{1}{12}ML^2$

Thin rod about perpendicular line through one end

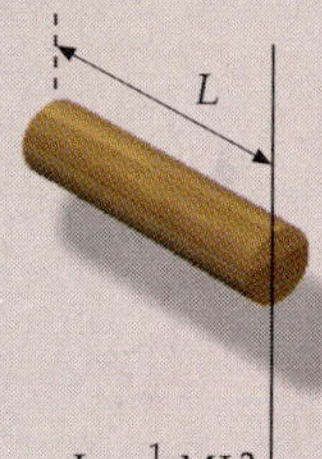

$I = \frac{1}{3}ML^2$

Thin spherical shell about diameter

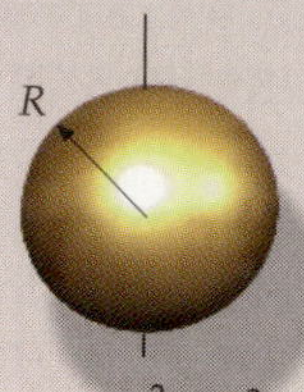

$I = \frac{2}{3}MR^2$

Solid sphere about diameter

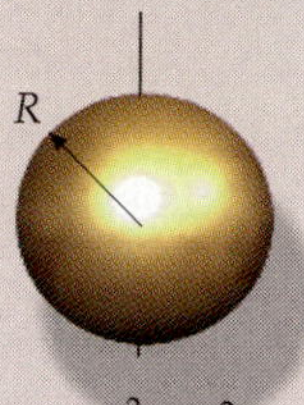

$I = \frac{2}{5}MR^2$

Solid rectangular parallelepiped about axis through center perpendicular to face

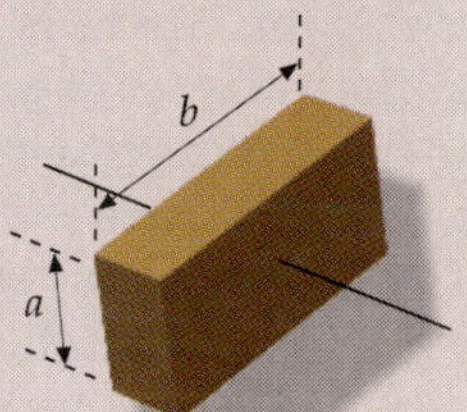

$I = \frac{1}{12}M(a^2 + b^2)$

Example 9-3

Find the moment of inertia of a uniform stick of length L and mass M about an axis perpendicular to the stick and through one end (Figure 9-10). Assume that the stick has negligible thickness.

Picture the Problem Let the stick lie along the x axis with its end at the origin. To calculate I_y about the y axis, we choose a mass element dm at a distance x from the axis. Since the total mass M is uniformly distributed along the length L, the mass per unit length (linear mass density) is $\lambda = M/L$.

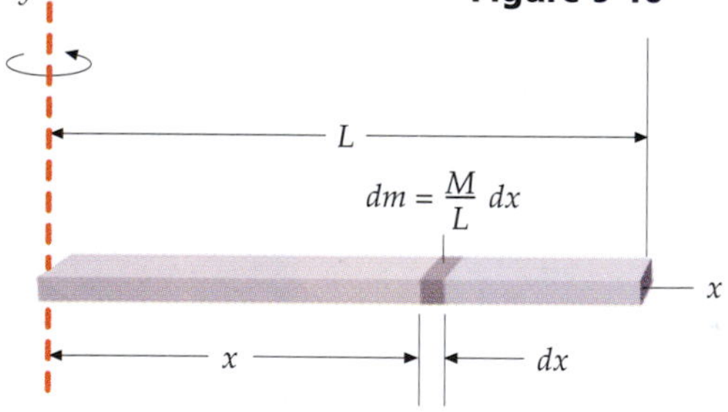

Figure 9-10

1. The moment of inertia is given by the integral: $I = \int_0^L x^2\,dm$

2. Write dm in terms of the mass density λ and dx: $dm = \lambda\,dx = \frac{M}{L}\,dx$

3. Substitute and perform the integration:

$$I_y = \int_0^L x^2\,dm = \int_0^L x^2 \frac{M}{L}\,dx = \frac{M}{L}\int_0^L x^2\,dx$$

$$= \frac{M}{L}\frac{1}{3}x^3\Big|_0^L = \frac{M}{L}\frac{L^3}{3} = \frac{1}{3}ML^2$$

Remark The moment of inertia about the z axis is also $\frac{1}{3}ML^2$ and that about the x axis is zero, assuming that all of the mass is right on the x axis.

Hoop About a Perpendicular Axis Through Its Center Assume that a hoop has mass M and radius R (Figure 9-11). The axis of rotation is the axis of the hoop, which is perpendicular to the plane of the hoop. All the mass is at a distance $r = R$, and the moment of inertia is

$$I = \int r^2\, dm = R^2 \int dm = MR^2$$

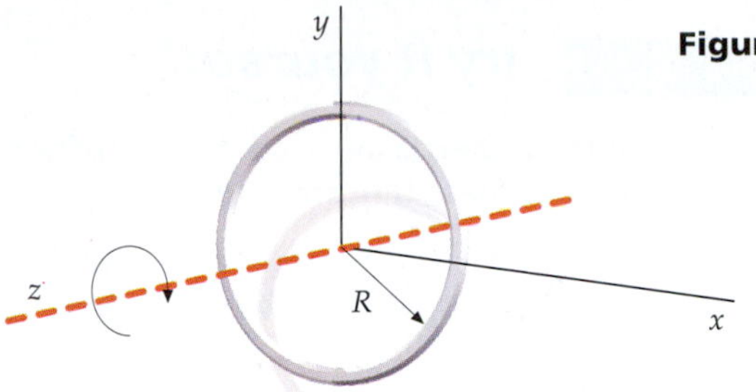

Figure 9-11

Uniform Disk About a Perpendicular Axis Through Its Center For the case of a uniform disk, we expect that I will be smaller than MR^2 since the mass is uniformly distributed from $r = 0$ to $r = R$ rather than being concentrated at $r = R$ as it is in a hoop. In Figure 9-12, each mass element is a hoop of radius r and thickness dr. The moment of inertia of any given mass element is $r^2\, dm$. Since the area of each element is $dA = 2\pi r\, dr$, the mass of each element is

$$dm = \frac{M}{A}\, dA = \frac{M}{A}\, 2\pi r\, dr$$

where $A = \pi R^2$ is the area of the disk. We thus have

$$I = \int r^2\, dm = \int_0^R r^2 \frac{M}{A}\, 2\pi r\, dr$$

$$= \frac{2\pi M}{\pi R^2} \int_0^R r^3\, dr = \frac{2M}{R^2}\frac{R^4}{4} = \frac{1}{2} MR^2$$

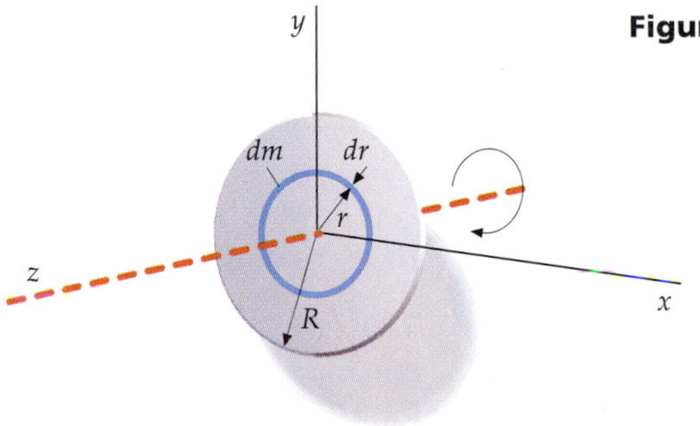

Figure 9-12

Uniform Cylinder About Its Axis We consider a cylinder to be a set of disks, each with mass dm and moment of inertia $\frac{1}{2} dm\, R^2$ (Figure 9-13). The moment of inertia of the complete cylinder is then

$$I = \int \frac{1}{2}\, dm\, R^2 = \frac{1}{2} R^2 \int dm = \frac{1}{2} MR^2$$

where M is the total mass of the cylinder.

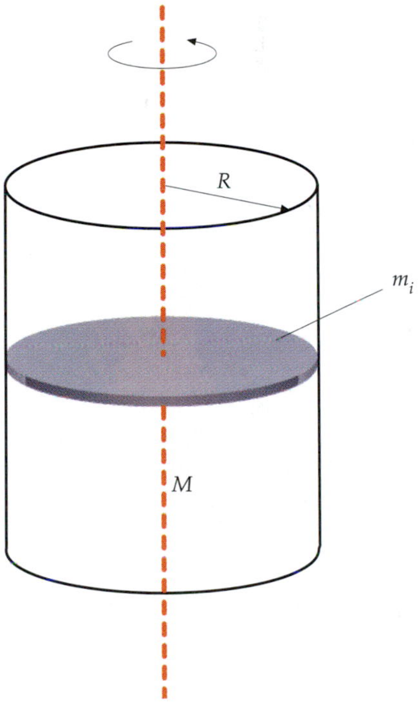

Figure 9-13

The Parallel-Axis Theorem

We can often simplify the calculation of moments of inertia for various bodies by using the **parallel-axis theorem,** which relates the moment of inertia about an axis through the center of mass of an object to the moment of inertia about a second, parallel axis (Figure 9-14). Let I_{cm} be the moment of inertia about an axis through the center of mass of an object of total mass M, and let I be that about a parallel axis a distance h away. The parallel-axis theorem states that

$$I = I_{cm} + Mh^2 \qquad \text{9-21}$$

Parallel-axis theorem

Example 9-2 and the exercise following it illustrated a special case of this theorem with $h = a$, $M = 4m$, and $I_{cm} = 4ma^2$. A proof of the parallel-axis theorem is given at the end of this section.

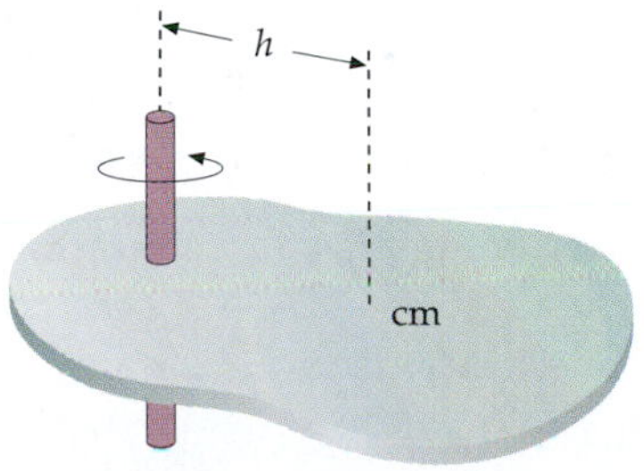

Figure 9-14 An object rotating about an axis parallel to an axis through the center of mass and a distance h from it.

optional

Example 9-4 *try it yourself*

Find the moment of inertia of a stick of uniform density about the y' axis through the center of mass (Figure 9-15).

Picture the Problem Here you know $I = \frac{1}{3}ML^2$ about one end and wish to find I_{cm}. Use the parallel-axis theorem with $h = \frac{1}{2}L$.

Figure 9-15

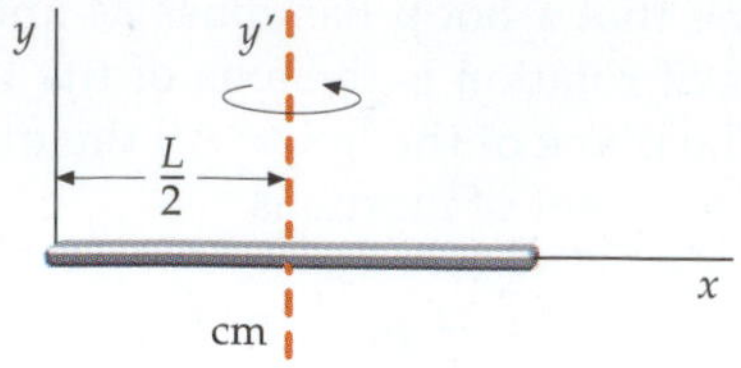

Cover the column to the right and try these on your own before looking at the answers.

Steps	*Answers*
1. Apply the parallel-axis theorem to write I about the end in terms of I_{cm}.	$I = Mh^2 + I_{cm}$
2. Substitute $I = \frac{1}{3}ML^2$ about the end and solve for I_{cm}.	$I_{cm} = \dfrac{1}{12}ML^2$

Remark The moment of inertia is least when an object is rotated about its center of mass.

optional

Proof of the Parallel-Axis Theorem

We can prove the parallel-axis theorem using the result developed in Chapter 8 that the kinetic energy of a system of particles is the sum of the kinetic energy of the motion of the center of mass plus the kinetic energy of the motion relative to the center of mass:

$$K = \frac{1}{2}MV_{cm}^2 + K_{rel} \qquad 9\text{-}22$$

Consider a rigid object rotating with an angular velocity ω about an axis a distance h from a parallel axis through the center of mass as shown in Figure 9-16. When the body rotates through an angle $d\theta$ measured about the axis of rotation, it rotates through the same angle $d\theta$ measured about any other parallel axis. The motion of the object relative to the center of mass is thus a rotation about the center-of-mass axis with the same angular velocity ω. The kinetic energy of this relative motion is

Figure 9-16

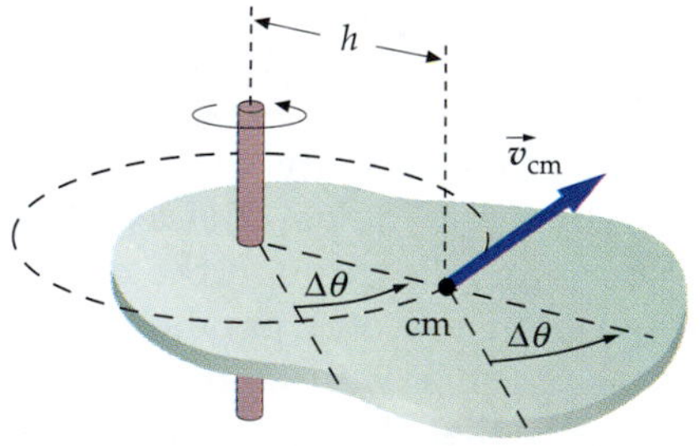

$$K_{rel} = \frac{1}{2}I_{cm}\omega^2$$

The velocity of the center of mass relative to any point on the axis of rotation is $v_{cm} = h\omega$. The kinetic energy of the motion of the center of mass is thus

$$\frac{1}{2}Mv_{cm}^2 = \frac{1}{2}M(h\omega)^2 = \frac{1}{2}M\omega^2h^2$$

The relative energy is $\frac{1}{2}I\omega^2$. Equation 9-21 then becomes

$$K = \frac{1}{2}M\omega^2h^2 + \frac{1}{2}I_{cm}\omega^2 = \frac{1}{2}(Mh^2 + I_{cm})\omega^2 = \frac{1}{2}I\omega^2$$

with

$$I = Mh^2 + I_{cm}$$

which is the parallel-axis theorem.

9-4 Applications of Newton's Second Law for Rotation

In this section, we give several applications of Newton's second law for rotation as expressed in Equation 9-20.

Example 9-5

To get some exercise without going anywhere, you set your bike on a stand so that the rear wheel is free to turn (Figure 9-17). As you pedal, the chain applies a force of 18 N to the sprocket at a distance of $r_s = 7$ cm from the axle of the wheel. Consider the wheel to be a hoop ($I = MR^2$) of radius $R = 35$ cm and mass 2.4 kg. What is the angular velocity of the wheel after 5 s?

Figure 9-17

Picture the Problem The angular velocity is found from the angular acceleration, which is found from Newton's second law for rotation. Since the forces are constant, the torques are constant and the constant angular acceleration equations apply. Note that $\vec{F}$ acts in the direction of the chain, so the line of force is tangent to the wheel and the lever arm is the radius r_s of the sprocket.

1. The angular velocity is related to the angular acceleration and the time: $\omega = \omega_0 + \alpha t = 0 + \alpha t$

2. Apply Newton's second law for rotational motion, $\Sigma\tau_{i,\text{ext}} = I\alpha$, to relate α to the net torque and the moment of inertia: $\alpha = \dfrac{\tau_{\text{net}}}{I}$

3. The only torque acting on the system is the applied force F with lever arm r_s: $\tau_{\text{net}} = Fr_s$

4. Substitute this value for the torque and $I = MR^2$ for the moment of inertia: $\alpha = \dfrac{\tau_{\text{net}}}{I} = \dfrac{Fr_s}{MR^2}$

5. Substitute the given values to obtain α: $\alpha = \dfrac{Fr_s}{MR^2} = \dfrac{(18\text{ N})(0.07\text{ m})}{(2.4\text{ kg})(0.35\text{ m})^2} = 4.29\text{ rad/s}^2$

6. Use this value of α to find the angular velocity after 5 s: $\omega = \alpha t = (4.29\text{ rad/s}^2)5\text{ s} = 21.4\text{ rad/s}$

Rotation Under Nonslip Conditions

There are many physical situations in which a string is wrapped around a rotating cylinder. If the string doesn't slip, its linear velocity must equal the tangential velocity of the rim of the cylinder:

$$v_t = R\omega \qquad \text{9-23}$$

Nonslip condition for v and ω

ation expressed in Equation 9-23 is called a nonslip condition. If we iate it with respect to time, we can relate the tangential acceleration eel to the linear acceleration of the chain:

$$R\alpha \qquad \text{9-24}$$

Nonslip condition for a and α

Example 9-6

An object of mass m is tied to a light string wound around a wheel that has a moment of inertia I and radius R. The wheel bearing is frictionless, and the string does not slip on the rim. Find the tension in the string and the acceleration of the object.

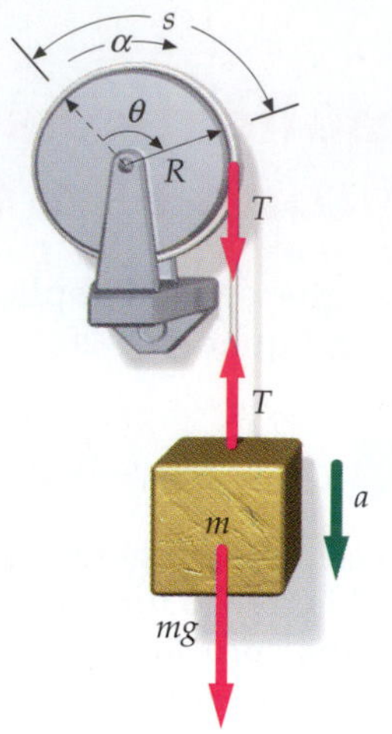

Figure 9-18

Picture the Problem In this system, the object descends with a constant downward acceleration a, while the wheel turns with a constant angular acceleration α (Figure 9-18). Because the string unwinds from the wheel without slipping, $a = R\alpha$. We apply Newton's law for rotation to the wheel to determine α, and apply Newton's second law to the object to obtain a. Since the object moves downward and the wheel rotates clockwise, we take these directions to be positive.

1. The only force that exerts a torque on the wheel is the tension T, which has lever arm R. Apply Newton's second law for rotational motion, $\Sigma\tau_{i,\text{ext}} = I\alpha$ to relate T and the angular acceleration α:

$$TR = I\alpha$$

2. Draw a free-body diagram for the suspended object (Figure 9-19), and apply $\Sigma\vec{F} = m\vec{a}$ to relate T to the linear acceleration a:

$$mg - T = ma$$

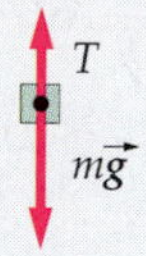

Figure 9-19

3. We have two equations for three unknowns, T, a, and α. A third equation is the nonslip condition relating a and α:

$$a = R\alpha$$

4. We now have three equations to determine T, a, and α. Use $a = R\alpha$ in the equation in step 1 to eliminate α, and solve for a:

$$TR = I\alpha = I\frac{a}{R} \quad \text{or} \quad a = \frac{TR^2}{I}$$

5. Substitute this result for a into the linear Newton's second-law equation, and solve for T:

$$mg - T = m\frac{TR^2}{I}$$

$$T = \frac{mg}{1 + mR^2/I} = \frac{I}{I + mR^2}mg$$

6. Substituting this value for T in step 4 yields a:

$$a = \frac{mR^2}{I + mR^2}g$$

Check the Result Let's check a couple extreme limits. If $I = 0$, the object should fall freely, and the string should be slack; our results give $T = 0$, $a = g$. What happens if $I \to \infty$? For $I \gg mR^2$, our equations give $T \approx mg$, and $a \approx 0$.

Remark We see that the tension and acceleration depend on the dimensionless quantity I/mR^2. A plot of the acceleration in units of g (a/g), versus I/mR^2 is shown in Figure 9-20.

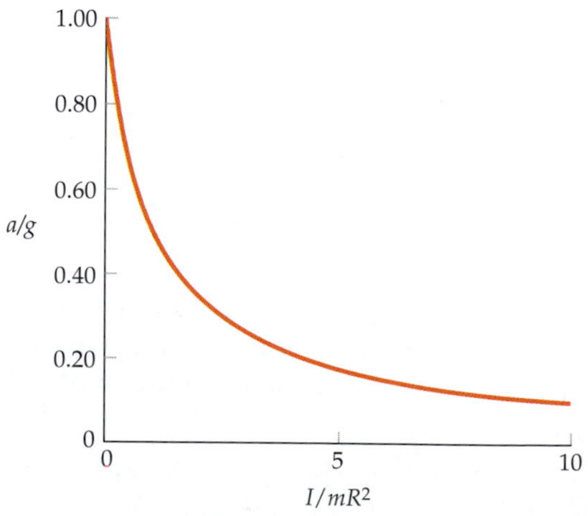

Figure 9-20

Example 9-7 *try it yourself*

Two blocks are connected by a string that passes over a pulley of radius R and moment of inertia I. The block of mass m_1 slides on a frictionless, horizontal surface; the block of mass m_2 is suspended from the string (Figure 9-21). Find the acceleration a of the blocks and the tensions T_1 and T_2 assuming that the string does not slip on the pulley.

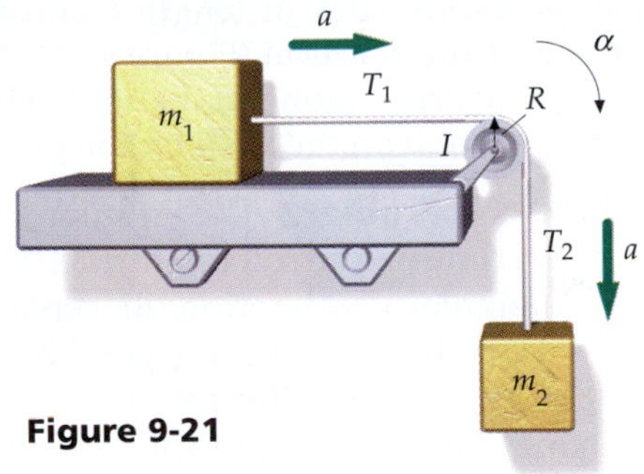

Figure 9-21

Picture the Problem In this problem, the tensions T_1 and T_2 are not equal because there is friction between the string and the pulley. (Otherwise, the pulley would not turn.) Note that T_2 exerts a clockwise torque and T_1 exerts a counterclockwise torque on the pulley. Use Newton's second law for each block and $\tau = I\alpha$ for the pulley, then relate α and a by $a = R\alpha$.

Cover the column to the right and try these on your own before looking at the answers.

Steps	***Answers***
1. Draw a free-body diagram for each block and for the pulley (Figure 9-22). Note that the pulley does not accelerate, so the support must exert a force on the axle F_s that balances the forces exerted by the string.	**Figure 9-22**
2. Apply $\Sigma\vec{F} = m\vec{a}$ to each block.	$T_1 = m_1 a, \quad m_2 g - T_2 = m_2 a$
3. Add the two equations in step 2 and rearrange to get an equation for $T_2 - T_1$.	$T_2 - T_1 = m_2 g - (m_1 + m_2)a$
4. Apply Newton's second law for rotational motion, $\Sigma\tau_{i,\text{ext}} = I\alpha$, to the pulley and obtain another equation for $T_2 - T_1$. Use the nonslip condition to eliminate α.	$(T_2 - T_1)R = I\alpha, \quad T_2 - T_1 = \left(\frac{I}{R^2}\right)a$
5. Set the equations for $T_2 - T_1$ in steps 3 and 4 equal and solve for a in terms of the masses, I, and R^2.	$a = \frac{m_2}{m_1 + m_2 + I/R^2} g$
6. Substitute your result for a into each of the equations in step 2 and solve for T_1 and T_2.	$T_1 = \frac{m_1}{m_1 + m_2 + I/R^2} m_2 g$ $T_2 = \frac{(m_1 + I/R^2)}{m_1 + m_2 + I/R^2} m_2 g$

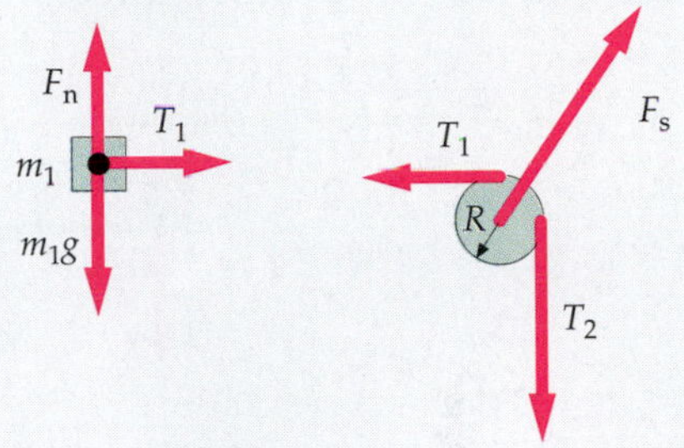

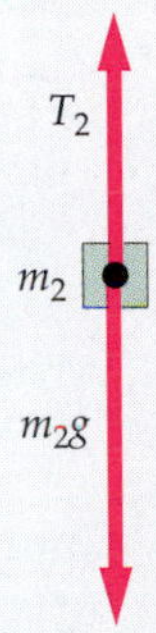

Check the Result If $I = 0$, $T_1 = T_2$, and the acceleration is $a = m_2 g/(m_1 + m_2)$, as expected. If I is very large, $I/R^2 >> (m_1 + m_2)$, then $T_1 \approx 0$, $T_2 \approx m_2 g$, and $a \approx 0$.

Remarks In general, the acceleration can be written in dimensionless form

$$\frac{a}{g} = \frac{m_2/m_1}{1 + m_2/m_1 + I/m_1R^2}$$

We see from this expression that I/m_1R^2 is a convenient dimensionless parameter to characterize the moment of inertia for this problem. A plot of a/g versus m_2/m_1 for various values of I/m_1R^2 is shown in Figure 9-23. As m_2/m_1 becomes large, a/g approaches 1. The approach is more rapid for small I/m_1R^2.

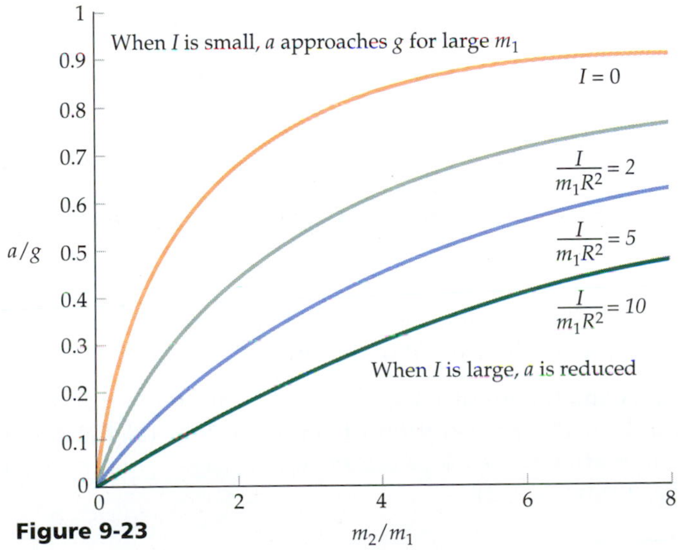

Figure 9-23

Example 9-8

A uniform thin stick of length L and mass M is pivoted at one end. It is held horizontal and released (Figure 9-24). Assume the pivot is frictionless. Find (*a*) the angular acceleration of the stick immediately after it is released, and (*b*) the force F_0 exerted on the stick by the pivot at this time.

Picture the Problem The angular acceleration is found from $\tau = I\alpha$, where τ is the torque on the rod relative to the pivot exerted by gravity. Since the rod has an angular acceleration, its center of mass has a tangential acceleration $a_{cm} = \frac{1}{2}L\alpha$. The initial centripetal acceleration of the rod is zero because its velocity is zero just after release. The force exerted by the pivot is found by applying Newton's second law to the rod. Since the acceleration is downward and the weight is downward, F_0 must be vertical. Assume that it is upward and take the positive direction to be downward.

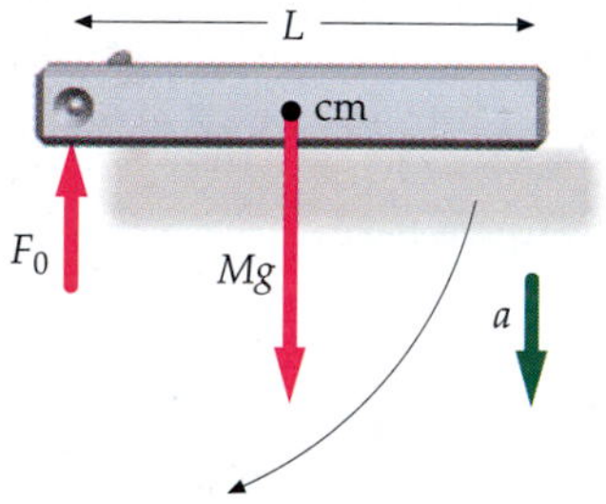

Figure 9-24

(*a*)1. Write Newton's second law for rotation: $\sum \tau_{i,\text{ext}} = I\alpha$

2. Compute the torque about the end of the stick: $\tau = Mg\dfrac{L}{2}$

3. Find the moment of inertia about the end of the stick from Table 9-1: $I = \dfrac{1}{3}ML^2$

4. Substitute these values to compute α: $\alpha = \dfrac{\tau}{I} = \dfrac{MgL/2}{ML^2/3} = \dfrac{3}{2}\dfrac{g}{L}$

(*b*)1. Write $\Sigma\vec{F} = m\vec{a}$ for the stick: $Mg - F_0 = Ma_{cm}$

2. Relate a_{cm} to α: $a_{cm} = r\alpha = \dfrac{L}{2}\alpha = \dfrac{L}{2}\dfrac{3}{2}\dfrac{g}{L} = \dfrac{3}{4}g$

3. Substitute a_{cm} into Newton's second law and solve for F_0: $F_0 = Mg - Ma_{cm} = Mg - M\left(\dfrac{3}{4}g\right) = \dfrac{1}{4}Mg$

Remark Just after the stick is released, the pivot exerts an upward force equal to one-fourth the weight of the stick.

Exercise: A small coin of mass $m \ll M$ is placed on top of the stick at its center. Find *(a)* the acceleration of the coin, and *(b)* the force it exerts on the stick just after the stick is released. (*Answers* *(a)* $a = 3g/4$ downward, *(b)* $f = mg/4$ downward)

9-5 Rotational Kinetic Energy

The kinetic energy of a rotating object is the sum of the kinetic energies of the individual particles in the object. The kinetic energy of a mass element m_i is

$$K = \frac{1}{2} m_i v_i^2$$

Summing over all the elements and using $v_i = r_i \omega$ gives

$$K_{\text{rot}} = \sum_i \frac{1}{2} m_i v_i^2 = \sum_i \frac{1}{2} m_i (r_i \omega)^2 = \frac{1}{2}\left(\sum_i m_i r_i^2\right)\omega^2$$

The term in the second set of parentheses is the moment of inertia I relative to the axis of rotation. The kinetic energy is thus

$$K_{\text{rot}} = \frac{1}{2} I\omega^2 \qquad \text{9-25}$$

Kinetic energy of rotation

Equation 9-25 is the rotational analog of $K = \frac{1}{2}mv^2$ for linear motion.

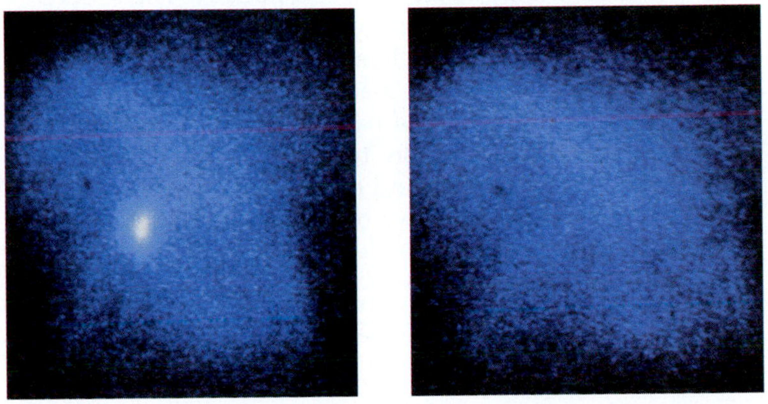

The Crab Pulsar is one of the fastest-rotating neutron stars known, but it is slowing down. It appears to blink on (left) and off (right) like the rotating lamp in a lighthouse, at the fast rate of about 30 times per second, but the period is increasing by about 10^{-5} s per year. The loss in rotational energy, which is equivalent to the power output of 100,000 suns, appears as light emitted by electrons accelerated in the magnetic field of the pulsar.

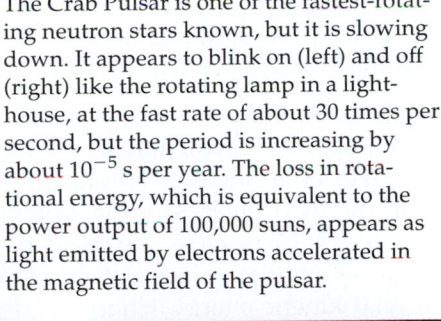

Example 9-9

A flywheel used for storing energy consists of a uniform disk of mass 1.5×10^5 kg and radius 2.2 m that rotates at 3000 rev/min about its center of mass. Find its kinetic energy.

Picture the Problem The kinetic energy is calculated directly from $K = \frac{1}{2}I\omega^2$. To obtain K in joules, we must express the angular velocity in radians per second.

1. The kinetic energy of rotation is: $K_{\text{rot}} = \frac{1}{2}I\omega^2$

2. Calculate the moment of inertia of the disk: $I = \frac{1}{2}mR^2 = \frac{1}{2}(1.5 \times 10^5\ \text{kg})(2.2\ \text{m})^2$
$= 3.63 \times 10^5\ \text{kg}\cdot\text{m}^2$

3. Convert ω to rad/s: $\omega = \frac{3000\ \text{rev}}{60\ \text{s}} \times \frac{2\pi\ \text{rad}}{1\ \text{rev}} = 314\ \text{rad/s}$

4. Substitute these values to find the kinetic energy: $K_{\text{rot}} = \frac{1}{2}I\omega^2 = \frac{1}{2}(3.63 \times 10^5\ \text{kg}\cdot\text{m}^2)(314\ \text{rad/s})^2$
$= 1.79 \times 10^{10}\ \text{J}$

Remark To use $K = \frac{1}{2}I\omega^2$, we must express ω in radians per second. Since a radian is dimensionless, the units of step 4 are $\text{kg}\cdot\text{m}^2/\text{s}^2 = \text{J}$. This energy is about 5000 kW·h.

Example 9-10

The stick of Example 9-8 is again released from rest when it is horizontal. Assuming the pivot to be frictionless, find (*a*) the angular velocity of the stick when it reaches its vertical position, and (*b*) the force exerted by the pivot at this time. (*c*) What initial angular velocity is needed for the stick to reach a vertical position at the top of its swing?

Picture the Problem (*a*) As the stick swings down, its potential energy decreases, and its kinetic energy of rotation about the pivot point increases (Figure 9-25). As it swings up, its kinetic energy decreases and its potential energy increases. Since the pivot is frictionless, we use conservation of mechanical energy. The angular velocity of the stick is then found from its rotational kinetic energy. Choose $U = 0$ initially. Then use $U = -MgL/2$ when the stick hangs vertically. (*b*) When the stick is vertical, there is no torque on it, so the stick has no angular acceleration, and the center of mass has no tangential acceleration. But the center of mass has a tangential velocity, so it has a centripetal acceleration toward the pivot. We apply $\Sigma\vec{F}_{\text{ext}} = m\vec{a}_{\text{cm}}$ to the stick to find the force exerted by the pivot. (*c*) We find the initial angular velocity from conservation of mechanical energy.

Figure 9-25

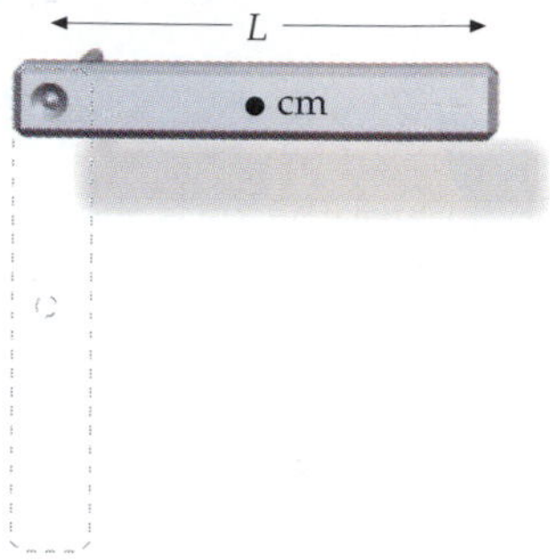

(*a*)1. The angular velocity of the stick is related to its kinetic energy of rotation *K*:

$$K_f = \frac{1}{2}I\omega_f^2$$

2. Apply conservation of mechanical energy with $E_f = K_f - MgL/2$, and $E_i = 0$:

$$E_f = E_i$$

$$\frac{1}{2}I\omega_f^2 - Mg\frac{L}{2} = 0$$

3. Solve for ω_f:

$$\omega_f = \sqrt{\frac{MgL}{I}} = \sqrt{\frac{MgL}{\frac{1}{3}ML^2}} = \sqrt{\frac{3g}{L}}$$

(*b*)1. Let $\vec{F}_p$ be the force exerted by the pivot when the stick is vertical. Apply $\Sigma\vec{F}_{\text{ext}} = m\vec{a}_{\text{cm}}$, taking the upward direction to be positive:

$$\sum\vec{F}_{\text{ext}} = m\vec{a}_{\text{cm}}$$

$$F_p - Mg = Ma_{\text{cm}}$$

2. a_{cm} is the centripetal acceleration:

$$a_{\text{cm}} = \frac{v_{\text{cm}}^2}{r} = \frac{(\frac{1}{2}L\omega_f)^2}{\frac{1}{2}L} = \frac{L}{2}\omega_f^2 = \frac{L}{2}\frac{3g}{L} = \frac{3}{2}g$$

3. Substitute and calculate F_p:

$$F_p = Mg + Ma_{\text{cm}} = M\left(g + \frac{3}{2}g\right) = \frac{5}{2}Mg$$

(*c*)1. The initial angular velocity ω_i is related to the initial kinetic energy:

$$K_i = \frac{1}{2}I\omega_i^2 = \frac{1}{2}\left(\frac{1}{3}ML^2\right)\omega_i^2 = \frac{1}{6}ML^2\omega_i^2$$

2. Apply conservation of mechanical energy with $K_f = 0$ and $U_i = 0$ to relate the initial kinetic energy to the final position:

$$K_f + U_f = K_i + U_i$$

$$0 + U_f = K_i + 0$$

$$Mg\frac{L}{2} = K_i = \frac{1}{6}ML^2\omega_i^2$$

3. Solve for the initial angular velocity:

$$\omega_i = \sqrt{\frac{3g}{L}}$$

Remarks We could not easily have used Newton's laws to solve this problem because the acceleration is not constant. The angular velocity ω versus angle θ is shown in Figure 9-26 for a stick of length 1 m and various values of the initial angular velocity ω_i. For no initial angular velocity (bottom curve), θ varies from 0 to π (180°). For any initial velocity, ω is maximum at $\theta = \pi/2$, corresponding to the stick being vertical and below the pivot. If the initial angular velocity is great enough, the stick will swing completely around. Then ω oscillates, as shown in the top curve.

Figure 9-26

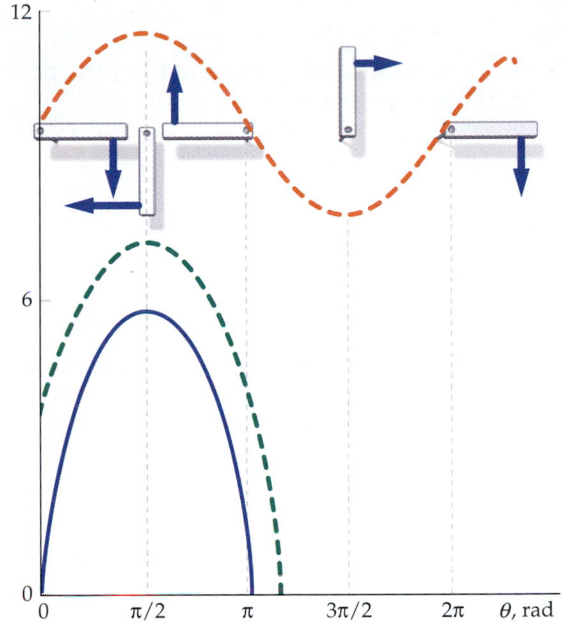

Example 9-11 *try it yourself*

The drum of a winch has mass M and radius R. A cable wound around the drum suspends a load of mass m. The entire cable has a length L and density (mass per unit length) λ, with a total mass $m_c = L\lambda$. The load begins to fall toward the ground, unwinding cable as it goes. How fast is the load moving after it has fallen a distance d?

Figure 9-27

Picture the Problem As the load falls, mechanical energy is conserved. Choose the initial potential energy to be zero. Then the total mechanical energy is zero. When the load has fallen a distance d, its potential energy is $-mgd$ (Figure 9-27). In addition, the center of mass of the hanging cable (mass λd) has dropped a distance $d/2$, so the potential energy of the cable is $-(\lambda d)g(d/2)$. When the load is moving at speed v, the drum is rotating at angular speed $\omega = v/R$. Since the hanging part of the cable moves with speed v and the cable does not stretch or become slack, the entire cable must move at speed v. We find v from the conservation of mechanical energy. Assume that the drum is a uniform cylinder of moment of inertia $\frac{1}{2}MR^2$.

Cover the column to the right and try these on your own before looking at the answers.

Steps	***Answers***
1. Apply conservation of mechanical energy.	$E_f = E_i = 0$ $K_f + U_f = 0$
2. Write an expression for the total potential energy of the load and cable when the load has fallen a distance d.	$U = -mgd - (\lambda d)g\frac{d}{2} = -mgd - \frac{m_c g d^2}{2L}$
3. Express the kinetic energy of the winch in terms of its moment of inertia I and angular speed ω.	$K_w = \frac{1}{2}I\omega^2$
4. Use the nonslip condition and $I = \frac{1}{2}MR^2$ to express the kinetic energy of the winch in terms of M and v.	$K_w = \frac{1}{2}I\omega^2 = \frac{1}{4}Mv^2$
5. Write an expression for the kinetic energy of the cable and of the load.	$K_c + K_{load} = \frac{1}{2}m_c v^2 + \frac{1}{2}mv^2$

6. Find the total final kinetic energy plus potential energy and set it equal to zero. $\frac{1}{4}Mv^2 + \frac{1}{2}m_c v^2 + \frac{1}{2}mv^2 - mgd - \frac{m_c g d^2}{2L} = 0$

7. Solve for v. $v = \sqrt{\frac{4mgd + 2m_c g d^2/L}{M + 2m + 2m_c}}$

Power

When you spin an object you do work on it, increasing its kinetic energy. Consider a force F_i acting on the ith particle of a rotating object. As the object turns through an angle $d\theta$, the ith particle moves a distance $ds_i = r_i\, d\theta$, and the force does work:

$$dW_i = F_{it}\, ds_i = F_{it} r_i\, d\theta = \tau_i\, d\theta$$

where τ_i is the torque exerted by the force F_i. In general, the work done by a torque τ when an object turns through a small angle $d\theta$ is

$$dW = \tau\, d\theta \qquad 9\text{-}26$$

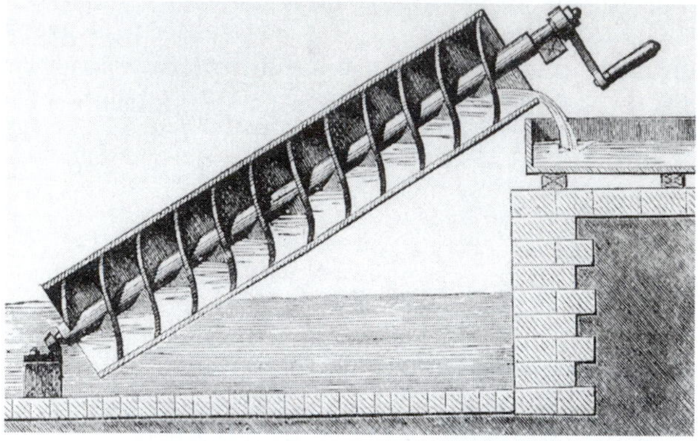

The Archimedes screw is a device for lifting water. The rotational work done by the torque exerted at the handle is converted into increased potential energy of the water.

The rate at which the torque does work is the power input of the torque:

$$P = \frac{dW}{dt} = \tau\frac{d\theta}{dt}$$

or

$$P = \tau\omega \qquad 9\text{-}27$$

Power

Equations 9-26 and 9-27 are the rotational analogs of $dW = F_s\, ds$ and $P = F_s v_s$.

Table 9-2 compares rotational and linear motion. (Angular momentum, which appears in the last two table entries, is discussed in Chapter 10.)

Table 9-2

Analogs in Rotational and Linear Motion

Rotational Motion		Linear Motion	
Angular displacement	$\Delta\theta$	Displacement	Δx
Angular velocity	$\omega = \frac{d\theta}{dt}$	Velocity	$v = \frac{dx}{dt}$
Angular acceleration	$\alpha = \frac{d\omega}{dt} = \frac{d^2\theta}{dt^2}$	Acceleration	$a = \frac{dv}{dt} = \frac{d^2x}{dt^2}$
Constant angular acceleration equations	$\omega = \omega_0 + \alpha t$ $\Delta\theta = \omega_{av}\,\Delta t$ $\omega_{av} = \frac{1}{2}(\omega_0 + \omega)$ $\theta = \theta_0 + \omega_0 t + \frac{1}{2}\alpha t^2$ $\omega^2 = \omega_0^2 + 2\alpha\,\Delta\theta$	Constant acceleration equations	$v = v_0 + at$ $\Delta x = v_{av}\,\Delta t$ $v_{av} = \frac{1}{2}(v_0 + v)$ $x = x_0 + v_0 t + \frac{1}{2}at^2$ $v^2 = v_0^2 + 2a\,\Delta x$
Torque	τ	Force	F
Moment of inertia	I	Mass	m
Work	$dW = \tau\, d\theta$	Work	$dW = F_s\, ds$
Kinetic energy	$K = \frac{1}{2}I\omega^2$	Kinetic energy	$K = \frac{1}{2}mv^2$
Power	$P = \tau\omega$	Power	$P = Fv$
Angular momentum	$L = I\omega$	Momentum	$p = mv$
Newton's second law	$\tau_{net} = I\alpha = \frac{dL}{dt}$	Newton's second law	$F_{net} = ma = \frac{dp}{dt}$

Example 9-12 ***try it yourself***

A Dodge Neon delivers 175 N·m of torque at 5000 rev/min. Find the power output of the car at that engine speed.

Picture the Problem The power equals the product of the torque and angular velocity, which are given. You must express ω in rad/s to obtain the power in watts.

Cover the column to the right and try these on your own before looking at the answers.

Steps	*Answers*
1. Write the power in terms of τ and ω.	$P = \tau\omega$
2. Convert ω to rad/s.	$\omega = 523$ rad/s
3. Calculate the power.	$P = 91.5$ kW

Remark This power output is about 123 horsepower.

9-6 Rolling Objects

Rolling Without Slipping

Consider a ball of radius R rolling without slipping along a plane surface. As the ball turns through the angle ϕ (Figure 9-28), the point of contact between the ball and the plane moves a distance s that is related to ϕ by

$$s = R\phi \qquad \text{9-28}$$

Nonslip condition for displacement

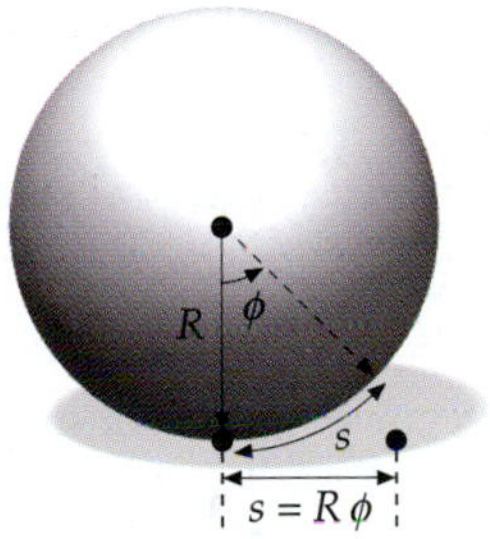

Figure 9-28

Since the ball's center of mass lies directly over the point of contact, it also moves through s. The velocity of the center of mass is therefore

$$v_{\text{cm}} = \frac{ds}{dt} = R\frac{d\phi}{dt}$$

or

$$v_{\text{cm}} = R\omega \qquad \text{9-29}$$

Nonslip condition for velocity

Differentiating each side again gives

$$a_{\text{cm}} = R\alpha \qquad \text{9-30}$$

Nonslip condition for acceleration

These conditions are the same as the nonslip conditions for a string wrapped around a pulley or peg.

When a ball rotates with angular velocity ω, the top and bottom of the ball move with speed $v = R\omega$ relative to the center of the ball (Figure 9-29*a*). When the ball rolls with speed v without slipping, the top of the ball moves with speed $2v$ and the bottom of the ball in contact with the surface is instantaneously at rest (Figure 9-29*b*). If a frictional force is exerted by the surface on the ball, it is static friction and no energy is dissipated.

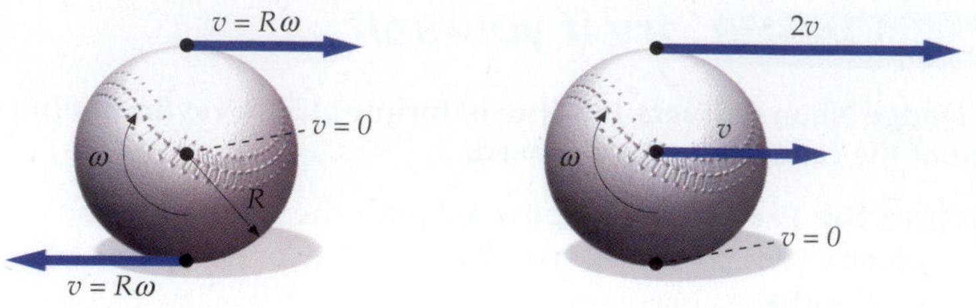

Figure 9-29 (*a*) Rotation without translation. The top of the ball moves to the right with a speed $v = R\omega$ relative to the center, which is at rest. The bottom moves to the left with the same speed relative to the center. (*b*) Rolling without slipping. If the center moves with speed v, the top moves with speed $2v$ and the bottom of the ball is momentarily at rest.

We saw in Chapter 8 that the kinetic energy of a system can be written as the sum of the kinetic energy of motion of the center of mass plus the kinetic energy relative to the center of mass. For a rolling object, the relative kinetic energy is $\frac{1}{2}I_{\text{cm}}\,\omega^2$. Thus, the kinetic energy of a rolling object is

$$K = \frac{1}{2}I_{\text{cm}}\omega^2 + \frac{1}{2}Mv_{\text{cm}}^2 \qquad \text{9-31}$$

Kinetic energy of a rolling object

Example 9-13 *try it yourself*

A bowling ball of radius 11 cm and mass $m = 7.2$ kg is rolling without slipping on a horizontal ball return at 2 m/s. It then rolls without slipping up a hill to a height h before momentarily coming to rest. Find h.

Picture the Problem Mechanical energy is conserved. The initial kinetic energy, which is the translational kinetic energy of the center of mass, $\frac{1}{2}mv_{\text{cm}}^2$, plus the kinetic energy of rotation about the center of mass, $\frac{1}{2}I_{\text{cm}}\omega^2$, is converted to potential energy mgh. Since the sphere rolls without slipping, the linear and angular speeds are related by $v_{\text{cm}} = R\omega$.

Cover the column to the right and try these on your own before looking at the answers.

Steps	*Answers*
1. Apply conservation of mechanical energy with $U_{\text{i}} = 0$ and $K_{\text{f}} = 0$.	$E_{\text{f}} = E_{\text{i}}$ $U_{\text{f}} = K_{\text{i}}$
2. Write the total initial kinetic energy K_{i} in terms of the speed v_{cm} and angular speed ω.	$K_{\text{i}} = \frac{1}{2}mv_{\text{cm}}^2 + \frac{1}{2}I_{\text{cm}}\omega^2$
3. Substitute $\omega = R/v_{\text{cm}}$ and $I_{\text{cm}} = \frac{2}{5}mR^2$ and solve for K_{i} in terms of the mass m and v_{cm}.	$K_{\text{i}} = \frac{1}{2}mv_{\text{cm}}^2 + \frac{1}{2}\left(\frac{2}{5}mR^2\right)\left(\frac{v_{\text{cm}}}{R}\right)^2 = \frac{7}{10}mv_{\text{cm}}^2$
4. Set this initial kinetic energy equal to the final potential energy mgh.	$\frac{7}{10}mv_{\text{cm}}^2 = mgh$
5. Solve for h.	$h = \frac{7v_{\text{cm}}^2}{10g} = 0.285\text{ m} = 28.5\text{ cm}$

Remark The height is independent of the mass or radius of the ball.

Exercise Find the total mechanical energy of the ball. (*Answer* 20.2 J)

Example 9-14

A cue stick hits a cue ball horizontally a distance x above the center of the ball (Figure 9-30). Find the value of x for which the cue ball will roll without slipping from the beginning. Express your answer in terms of the radius R of the ball.

Picture the Problem If the stick hits at the level of the ball's center, the ball initially translates with no rotation. If the stick hits below the center, the ball initially has back spin. At a certain value of x, the ball has just the right forward spin and forward acceleration to satisfy the nonslip condition. The value of x determines the torque exerted on the ball, and hence its angular acceleration α. The linear acceleration a is F/m independent of x. For the ball to roll without slipping from the start, we find α and a, then set $a = R\alpha$ (nonslip condition) to find x. The weight and normal force act through the center of mass and thus exert no torque about it. The frictional force is much smaller than the collision force of the stick and can be neglected.

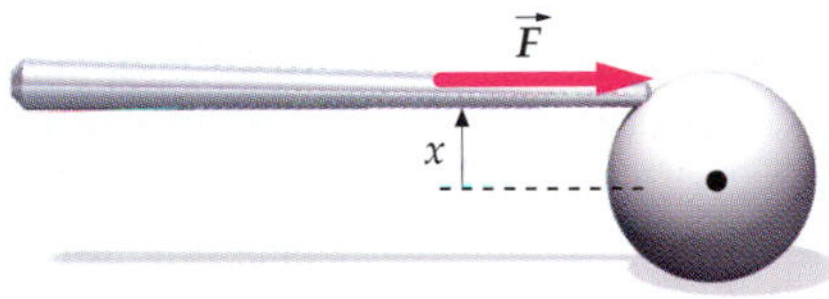

Figure 9-30

1. The torque about the center of the ball equals F times x: $\tau = Fx$
2. Apply Newton's second law $\Sigma\vec{F} = m\vec{a}$ and Newton's second law for rotational motion about the center of the ball, $\Sigma\tau = I\alpha$: $F = ma$ and $\tau = Fx = I\alpha$
3. The nonslip condition relates a and α: $a = R\alpha$
4. Express a and α in terms of the force F from step 2: $\frac{F}{m} = R\frac{Fx}{I}$
5. Solve for x: $x = \frac{I}{mR}$
6. For a sphere, $I = \frac{2}{5}mR^2$: $x = \frac{\frac{2}{5}mR^2}{mR} = \frac{2}{5}R$

Remark If the ball is struck at a point higher than $2R/5$ or lower than $2R/5$ from the center, it will have topspin or backspin and slip. Rolling with slipping is discussed in the next subsection.

When an object rolls down an incline, its center of mass is accelerated. The analysis of such a problem is simplified by an important theorem concerning the center of mass:

> If the torques are computed relative to the center of mass, Newton's second law for rotation holds for rotation about an axis through the center of mass, no matter how the center of mass is moving.

$$\tau_{\text{net,cm}} = I_{\text{cm}}\alpha \qquad \text{9-32}$$

This is the same as Equation 9-19 except that here the torque is computed relative to the center of mass rather than relative to some fixed point, and the moment of inertia is computed about an axis through the center of mass. It is often convenient to compute torques about the center of mass. When the center of mass is accelerating (a ball rolling down an incline, for example), its reference frame is a noninertial one, where we would not necessarily expect Newton's second law for rotation to hold. Nevertheless, it does.*

*A proof is given in most intermediate-level mechanics books—for example, G. R. Fowles, *Analytical Mechanics*, Holt, Rinehart & Winston, New York, 1993.

Example 9-15

A uniform solid ball of mass m and radius R rolls without slipping down a plane inclined at an angle θ. Find the acceleration of the center of mass.

Picture the Problem From Newton's second law, the acceleration of the center of mass equals the net force divided by the mass. The forces acting are the weight $m\vec{g}$ downward, the normal force $\vec{F}_n$ that balances the normal component of the weight, and the force of friction $\vec{f}$ acting up the incline (Figure 9-31). As the object accelerates down the incline, the angular velocity of rotation must increase to maintain the nonslip condition. We apply Newton's second law for rotation about a horizontal axis through the center of mass to find α, which is related to the acceleration by the nonslip condition. The only torque about the cm is due to $\vec{f}$. (Both $m\vec{g}$ and $\vec{F}_n$ act through the center of mass.) Choose the positive direction to be down the incline:

Figure 9-31

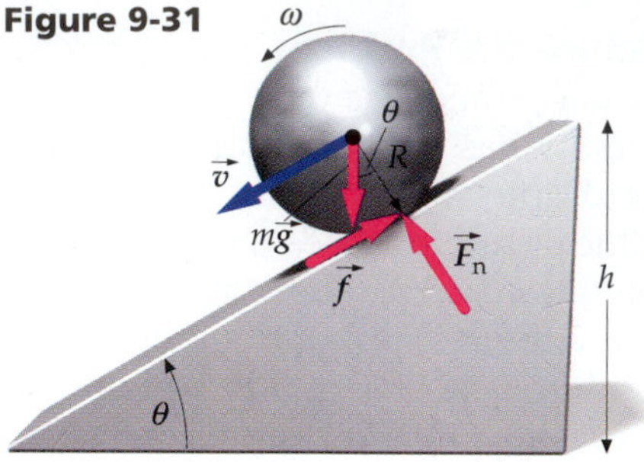

1. Apply $\Sigma\vec{F} = m\vec{a}$ along the incline: $\quad mg \sin\theta - f = ma_{cm}$

2. Apply $\Sigma\tau = I_{cm}\alpha$: $\quad fR = I_{cm}\alpha$

3. Use the nonslip condition to eliminate α and solve for f: $\quad fR = I_{cm}\dfrac{a_{cm}}{R}$

$$f = \frac{I_{cm}}{R^2}a_{cm}$$

4. Substitute this result for f and solve for a_{cm}: $\quad mg\sin\theta - \dfrac{I_{cm}}{R^2}a_{cm} = ma_{cm}$

$$a_{cm} = \frac{1}{1 + I_{cm}/mR^2}g\sin\theta$$

5. Substitute $I_{cm} = \frac{2}{5}mR^2$ for a sphere: $\quad a_{cm} = \dfrac{1}{1 + 2/5}g\sin\theta = \dfrac{5}{7}g\sin\theta$

Remarks Since the ball rolls without slipping, the friction is static friction. Note that the result is independent of the coefficient of static friction as long as it is great enough so the ball does not slip.

The results of steps 3 and 4 in the above example apply equally to any rolling object:

$$f = \frac{I_{cm}}{R^2}a_{cm} \qquad \text{9-33}$$

$$a_{cm} = \frac{1}{1 + I_{cm}/mR^2}g\sin\theta \qquad \text{9-34}$$

For a cylinder, $I_{cm} = \frac{1}{2}mR^2$, and the acceleration is $\frac{2}{3}g\sin\theta$. For a hoop, $I_{cm} = mR^2$, and the acceleration is $\frac{1}{2}g\sin\theta$. The linear acceleration of any object rolling down an incline is less than $g\sin\theta$ because of the frictional force directed up the incline. Note that these accelerations are independent of both the mass and radius of the objects. If we release a sphere, a cylinder, and a hoop at the top of an incline, and if they all roll without slipping, the sphere will reach the bottom first because it has the greatest acceleration. The cylinder will be second and the hoop third (Figure 9-32). If any object could slide down the incline without friction, it would reach the bottom before any of the rolling objects.

Since the friction is static, it does no work, and there is no dissipation of mechanical energy. We can therefore use the conservation of mechanical energy to find the speed of an object rolling without slipping down an incline. At the top of the incline, the total energy is the potential energy mgh. At the bottom, the total energy is kinetic energy. Conservation of mechanical energy therefore gives

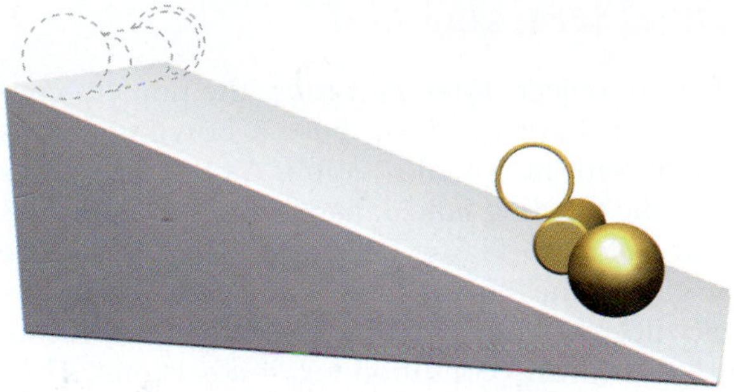

Figure 9-32 A sphere, a cylinder, and a hoop are started together from rest at the top of an incline. The sphere reaches the bottom first, followed by the cylinder and then the hoop.

$$\frac{1}{2}mv_{\text{cm}}^2 + \frac{1}{2}I_{\text{cm}}\omega^2 = mgh$$

We can use the nonslip condition to eliminate either v_{cm} or ω. Substituting $\omega = v_{\text{cm}}/R$, we obtain

$$\frac{1}{2}mv_{\text{cm}}^2 + \frac{1}{2}I_{\text{cm}}\left(\frac{v_{\text{cm}}}{R}\right)^2 = mgh$$

$$v_{\text{cm}}^2 = \frac{1}{1 + I_{\text{cm}}/mR^2}2gh \qquad \text{9-35}$$

For a cylinder, with $I_{\text{cm}} = \frac{1}{2}mR^2$, we obtain $v_{\text{cm}} = \sqrt{\frac{4}{3}gh}$. Note that the speed is independent of the mass and radius of the cylinder, and is less than $\sqrt{2gh}$, the speed of an object sliding with no friction down the incline.

We can find the force of static friction f for an object rolling down an incline from Equations 9-33 and 9-34:

$$f = \frac{I_{\text{cm}}}{R^2}a_{\text{cm}} = \frac{I_{\text{cm}}}{R^2}\frac{1}{1 + I_{\text{cm}}/R^2}g\sin\theta$$

or

$$f = \frac{1}{1 + mR^2/I_{\text{cm}}}mg\sin\theta \qquad \text{9-36}$$

For a cylinder, for example, $I_{\text{cm}} = \frac{1}{2}mR^2$, so $f = \frac{1}{3}mg\sin\theta$. Note that we have found f without considering the coefficient of static friction, μ_s. For an object rolling down an incline without slipping, f is less than its maximum value; that is,

$$f \le \mu_s F_n = \mu_s mg\cos\theta$$

Then for a cylinder,

$$f = \frac{1}{3}mg\sin\theta \le \mu_s mg\cos\theta$$

or

$$\tan\theta \le 3\mu_s \qquad \text{(cylinder)} \qquad \text{9-37}$$

If the tangent of the angle of incline is greater than $3\mu_s$, the cylinder will slip as it moves down the incline.

Exercise A cylinder rolls down a plane inclined at $\theta = 50°$. What is the minimum value of the coefficient of static friction for which the cylinder will roll without slipping? (*Answer* 0.40)

Exercise For a hoop rolling down an incline, (*a*) what is the force of friction, and (*b*) what is the maximum value of $\tan\theta$ for which the hoop will roll without slipping? (*Answers* (*a*) $f = \frac{1}{2}mg\sin\theta$; (*b*) $\tan\theta \le 2\mu_s$)

Rolling With Slipping

When an object slides as it rolls, the nonslip condition does not hold. Suppose a bowling ball is thrown with no initial rotation. As the ball slides along the bowling lane, kinetic friction reduces its linear velocity (Figure 9-33). The frictional force also causes the ball to start rotating. The linear velocity decreases and the angular velocity increases until the nonslip condition $v_{cm} = R\omega$ is met. Then the ball rolls without slipping.

Another example of rolling with slipping is a ball with topspin, such as a cue ball struck at a point greater than $2R/5$ above the center (see Example 9-14). Then the frictional force increases v and reduces ω until the nonslip condition is met (Figure 9-34).

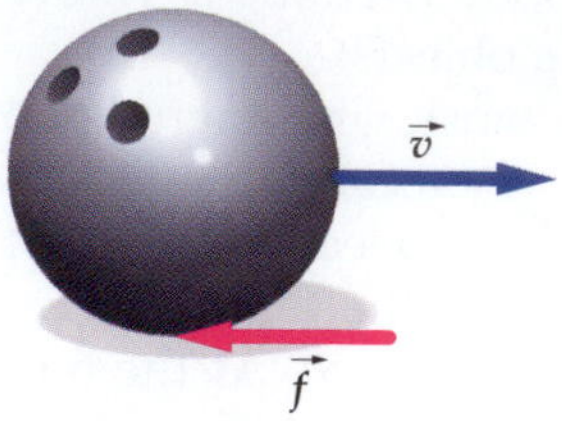

Figure 9-33 A bowling ball moving with no initial rotation. The frictional force $\vec{f}$ exerted by the floor reduces the speed v and increases the angular speed ω until $v = R\omega$.

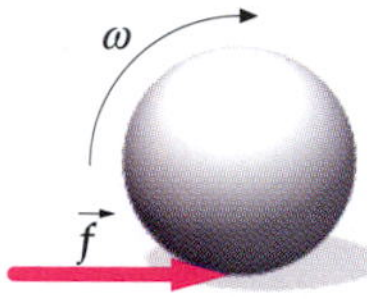

Figure 9-34 Ball with topspin. The frictional force accelerates the ball in the direction of motion.

Example 9-16

A bowling ball of mass M and radius R is thrown such that the instant it touches the floor it is moving horizontally with speed $v_0 = 5$ m/s and is not rotating. The coefficient of kinetic friction between the ball and the floor is $\mu_k = 0.08$. Find (*a*) the time the ball slides before the nonslip condition is met, and (*b*) the distance the ball slides before it rolls without slipping.

Picture the Problem We calculate v_{cm} and ω as functions of time, set $v_{cm} = R\omega$, and solve for t. The linear and angular accelerations are found from $\Sigma\vec{F} = m\vec{a}$, and $\tau = I\alpha$. Let the direction of motion be positive.

(*a*)1. The net force on the ball is the force of kinetic friction, f_k, which acts in the negative direction. Apply $\Sigma\vec{F} = m\vec{a}$:

$$f_k = -\mu_k Mg = Ma_{cm} \quad \text{or}$$
$$a_{cm} = -\mu_k g$$

2. The linear velocity is related to the acceleration and the time:

$$v_{cm} = v_0 + a_{cm}t = v_0 - \mu_k g t$$

3. The torque about the center of the ball is the frictional force times the lever arm R. Apply $\tau = I\alpha$ with $I = \frac{2}{5}MR^2$:

$$\tau = \mu_k MgR = I_{cm}\alpha \quad \text{or}$$
$$\alpha = \frac{\mu_k MgR}{I_{cm}} = \frac{\mu_k MgR}{\frac{2}{5}MR^2} = \frac{5}{2}\left(\frac{\mu_k g}{R}\right)$$

4. The angular velocity is related to the angular acceleration and the time:

$$\omega = \omega_0 + \alpha t = 0 + \alpha t = \frac{5}{2}\left(\frac{\mu_k g}{R}\right)t$$

5. Solve for time t_1 at which $v_{cm} = R\omega$:

$$v_{cm} = v_0 - \mu_k g t_1 = R\omega = \frac{5}{2}\mu_k g t_1$$
$$t_1 = \frac{2v_0}{7\mu_k g} = \frac{2(5\text{ m/s})}{7(0.08)(9.81\text{ m/s}^2)} = 1.82\text{ s}$$

(*b*)1. The distance traveled in time t_1 is:

$$\Delta x = v_0 t_1 + \frac{1}{2}a_{cm}t_1^2$$
$$= v_0\frac{2v_0}{7\mu_k g} + \frac{1}{2}(-\mu_k g)\left(\frac{2v_0}{7\mu_k g}\right)^2 = \frac{12}{49}\frac{v_0^2}{\mu_k g}$$

2. Substitute the given values:

$$\Delta x = \frac{12}{49}\frac{v_0^2}{\mu_k g} = \frac{12}{49}\frac{(5\text{ m/s})^2}{(0.08)(9.81\text{ m/s}^2)} = 7.80\text{ m}$$

Exercise Find the speed of the bowling ball when it begins to roll without slipping. (*Answer* $v_{cm} = \frac{5}{7}v_0$. This result is independent of the coefficient of kinetic friction. The rolling speed is $\frac{5}{7}v_0$ whether friction is large or small. The total mechanical energy lost is thus independent of μ_k. The time and distance are sensitive to the value of μ_k, however.)

Exercise Find the total kinetic energy of the ball after it begins to roll without slipping. (*Answer* $K = \frac{5}{14}Mv_0^2$)

Remarks In a well-maintained bowling alley, the lanes are lightly oiled and very slick, so that the ball slides over a great distance, which gives the bowler added control. Figure 9-35 shows the sliding distance Δx versus initial velocity for the bowling ball for $\mu_k = 0.08$, as in this example (slippery floor), and for $\mu_k = 0.24$ (sticky floor).

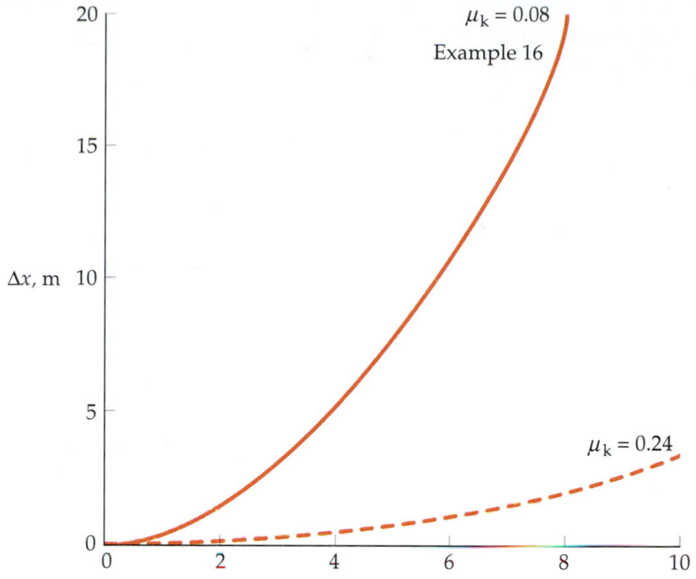

Figure 9-35

Summary

1. Angular displacement, angular velocity, and angular acceleration are fundamental defined quantities in rotational kinematics.
2. Torque and moment of inertia are important derived dynamic concepts. Torque is a measure of the ability of a force to cause an object to start or stop rotating. Moment of inertia is the measure of an object's inertial resistance to angular accelerations. The moment of inertia depends on the distribution of the mass relative to the rotation axis.
3. The parallel-axis theorem, which follows from the definition of the moment of inertia, often simplifies the calculation of I.
4. Newton's second law for rotation, $\Sigma\tau_{i,\text{ext}} = I\alpha$, is derived from Newton's second law and the definitions of τ, I, and α. It is an important relation for problems involving the rotation of a rigid object about an axis of fixed direction.

Topic	Remarks and Relevant Equations	
1. Angular Velocity and Angular Acceleration		
Angular velocity	$\omega = \dfrac{d\theta}{dt}$ (Definition)	9-2
Angular acceleration	$\alpha = \dfrac{d\omega}{dt} = \dfrac{d^2\theta}{dt^2}$ (Definition)	9-3
Tangential speed	$v_{it} = r_i\omega$	9-4
Tangential acceleration	$a_{it} = r_i\alpha$	9-5
Centripetal acceleration	$a_{ic} = \dfrac{v_i^2}{r_i} = r_i\omega^2$	9-6

2.	**Equations for Rotation With Constant Angular Acceleration**	$\omega = \omega_0 + \alpha t$	**9-7**
		$\theta = \theta_0 + \omega_0 t + \frac{1}{2}\alpha t^2$	**9-8**
		$\omega^2 = \omega_0^2 + 2\alpha(\theta - \theta_0)$	**9-9**
3.	**Torque**	The magnitude of the torque exerted by a force on an object is defined as the product of the force and the lever arm.	
		$\tau = F\ell$ (Definition)	**9-10**
4.	**Moment of Inertia**		
	System of particles	$I = \sum m_i r_i^2$ (Definition)	**9-17**
	Continuous object	$I = \int r^2\, dm$ (Definition)	**9-18**
	Parallel-axis theorem	The moment of inertia about an axis a distance h from a parallel axis through the center of mass is	
		$I = I_{cm} + Mh^2$	**9-21**
		where I_{cm} is the moment of inertia about the axis through the center of mass and M is the total mass of the object.	
5.	**Newton's Second Law for Rotation**	$\tau_{net,ext} = \sum_i \tau_{i,ext} = I\alpha$	**9-20**
		If the torques are computed relative to the center of mass, Newton's second law for rotation holds for rotation about an axis through the center of mass, no matter how the center of mass is moving.	
6.	**Nonslip Conditions**	When a string that is wrapped around a pulley or disk does not slip, the linear and angular quantities are related by	
		$v = R\omega$	**9-19**
		$a = R\alpha$	**9-20**
7.	**Energy**		
	Kinetic energy of rotation	$K = \frac{1}{2}I\omega^2$	**9-25**
	K for rotation plus translation	$K = \frac{1}{2}I_{cm}\omega^2 + \frac{1}{2}Mv_{cm}^2$	**9-31**
	Power	$P = \tau\omega$	**9-27**
8.	**Rolling Objects**		
	Rolling without slipping	$v_{cm} = R\omega$	**9-29**
	Rolling with slipping (optional)	When an object rolls and slips, $v_{cm} \neq R\omega$. Kinetic friction exerts a force that tends to change v_{cm}, and also exerts a torque that changes ω until $v_{cm} = R\omega$ and rolling without slipping sets in.	

Problem-Solving Guide

1. Begin by drawing a neat diagram including the important features of the problem. Draw a free-body diagram for each object showing the forces acting, the chosen coordinate system, and appropriate axes of rotation.
2. Write $\Sigma \vec{F} = m\vec{a}$ for each translating object and $\Sigma \tau_i = I\alpha$ for each rotating object, and relate the translational and rotational velocities and accelerations by nonslip conditions whenever applicable.
3. If there is no energy dissipation, conservation of mechanical energy provides a useful method to find final velocities or angular velocities.

Summary of Worked Examples

Type of Calculation	Procedure and Relevant Examples	
1. Kinematics		
Find $\Delta\theta$, ω, or α for rotation due to a constant torque	Use $\tau = I\alpha$ and constant angular acceleration equations.	**Examples 9-1, 9-5**
2. Moment of Inertia		
Calculate I for a system of particles	Use $I = \Sigma m_i r_i^2$	**Example 9-2**
Calculate I for a continuous object	Use $I = \int r^2\,dm$	**Example 9-3**
Calculate I about an axis parallel to one through the center of mass	Use the parallel axis theorem $I = I_{cm} + Mh^2$ where h is the distance between the axes	**Example 9-4**
3. Finding Forces and Accelerations		
Find the acceleration of an object hanging from a string wrapped around a disk or pulley	Use $\Sigma \vec{F} = m\vec{a}$ for the object and $\Sigma \tau_i = I\alpha$ for the disk or pulley, and the nonslip condition $a = R\alpha$.	**Examples 9-6, 9-7**
Find the acceleration of an object rotating about a pivot and the forces exerted by the pivot	Apply $\Sigma\tau = I\alpha$ to find α, then use it to find a_{cm}. Apply $\Sigma\vec{F} = m\vec{a}$ to find the forces.	**Example 9-8**
4. Energy and Power		
Find the kinetic energy of a rotating object	Use $K = \frac{1}{2}I\omega^2$. Remember that ω must be in rad/s.	**Example 9-9**
Find the power of a motor	Use $P = \tau\omega$. Remember that ω must be in rad/s.	**Example 9-12**
Find the final speed or angular speed of a system for which there is no energy dissipation	Use conservation of mechanical energy with $K = \frac{1}{2}I\omega^2$ for rotating the object. Remember that ω must be in rad/s.	**Examples 9-15, 9-11**
5. Rolling Objects		
Hit a cue ball, or throw a ball such that it rolls without slipping	The torque and force must be such that the acceleration and angular acceleration are related by $a = R\alpha$.	**Example 9-14**
Find the acceleration of an object rolling down an incline	Apply $\Sigma\tau_i = I\alpha$ about the center of mass, and $\Sigma\vec{F} = m\vec{a}_{cm}$ to eliminate the frictional force, and use the nonslip condition $a_{cm} = R\alpha$ to write α in terms of a_{cm}.	**Example 9-15**

Find the kinetic energy of a rolling object	Add $\frac{1}{2}Mv_{cm}^2$ and $\frac{1}{2}I\omega^2$ and use the nonslip condition $v_{cm} = R\omega$.	**Example** 9-13
Rolling and sliding objects (optional)	While sliding, $v_{cm} \neq R\omega$. Kinetic friction exerts a force that accelerates the center of mass and also a torque that produces angular acceleration. Then v_{cm} and ω change until $v_{cm} = R\omega$, at which time rolling without slipping begins.	**Example 9-16**

Problems

Conceptual Problems

Problems from Optional and Exploring sections

• Single-concept, single-step, relatively easy
•• Intermediate-level, may require synthesis of concepts
••• Challenging, for advanced students

In a few problems, you are given more data than you actually need; in a few other problems, you are required to supply data from your general knowledge, outside sources, or informed estimates.

Take $g = 9.81$ N/kg $= 9.81$ m/s^2 and neglect friction in all problems unless otherwise stated. Assume that all objects are points unless otherwise indicated.

Angular Velocity and Angular Acceleration

1 • Two points are on a disk turning at constant angular velocity, one point on the rim and the other halfway between the rim and the axis. Which point moves the greater distance in a given time? Which turns through the greater angle? Which has the greater speed? The greater angular velocity? The greater tangential acceleration? The greater angular acceleration? The greater centripetal acceleration?

2 • True or false:

(*a*) Angular velocity and linear velocity have the same dimensions.
(*b*) All parts of a rotating wheel must have the same angular velocity.
(*c*) All parts of a rotating wheel must have the same angular acceleration.

3 •• Starting from rest, a disk takes 10 revolutions to reach an angular velocity ω. At constant angular acceleration, how many additional revolutions are required to reach an angular velocity 2ω?

(*a*) 10 rev (*b*) 20 rev (*c*) 30 rev (*d*) 40 rev (*e*) 50 rev

4 • A particle moves in a circle of radius 90 m with a constant speed of 25 m/s. (*a*) What is its angular velocity in radians per second about the center of the circle? (*b*) How many revolutions does it make in 30 s?

5 • A wheel starts from rest with a constant angular acceleration of 2.6 rad/s^2 and rolls for 6 s. At the end of that time, (*a*) what is its angular velocity? (*b*) Through what angle has the wheel turned? (*c*) How many revolutions has it made? (*d*) What is the speed and acceleration of a point 0.3 m from the axis of rotation?

6 • When a turntable rotating at $33\frac{1}{3}$ rev/min is shut off, it comes to rest in 26 s. Assuming constant angular acceleration, find (*a*) the angular acceleration, (*b*) the average angular velocity of the turntable, and (*c*) the number of revolutions it makes before stopping.

7 • A disk of radius 12 cm, initially at rest, begins rotating about its axis with a constant angular acceleration of 8 rad/s^2. At $t = 5$ s, what are (*a*) the angular velocity of the disk, and (*b*) the tangential acceleration a_t and the centripetal acceleration a_c of a point on the edge of the disk?

8 • Radio announcers who still play vinyl records have to be careful when cueing up live recordings. While studio albums have blank spaces between the songs, live albums have audiences cheering. If the volume levels are left up when the turntable is turned on, it sounds as though the audience has suddenly burst through the wall. If a turntable begins at rest and rotates through 10° in 0.5 s, how long must an announcer wait before the record reaches the required angular speed of $33\frac{1}{3}$ rev/min? Assume constant angular acceleration.

9 • A Ferris wheel of radius 12 m rotates once in 27 s. (*a*) What is its angular velocity in radians per second? (*b*) What is the linear speed of a passenger? (*c*) What is the centripetal acceleration of a passenger?

10 • A cyclist accelerates from rest. After 8 s, the wheels have made 3 rev. (*a*) What is the angular acceleration of the wheels? (*b*) What is the angular velocity of the wheels after 8 s?

11 • What is the angular velocity of the earth in rad/s as it rotates about its axis?

12 • A wheel rotates through 5.0 rad in 2.8 s as it is brought to rest with constant angular acceleration. The initial angular velocity of the wheel before the braking began was

(*a*) 0.6 rad/s. (*b*) 0.9 rad/s. (*c*) 1.8 rad/s.
(*d*) 3.6 rad/s. (*e*) 7.2 rad/s.

13 • A circular space station of radius 5.10 km is a long way from any star. Its rotational speed is controllable to some degree, and so the apparent gravity changes according to the tastes of those who are making the decisions. Dave the Earthling puts in a request for artificial gravity of 9.8 m/s^2 at the circumference. His secret agenda is to give the Earthlings a home-gravity advantage in the upcoming interstellar basketball tournament. Dave's request would require an angular speed of

(*a*) 4.4×10^{-2} rad/s. (*b*) 7.0×10^{-3} rad/s.
(*c*) 0.28 rad/s. (*d*) −0.22 rad/s.
(*e*) 1300 rad/s.

14 • A bicycle has wheels of 1.2 m diameter. The bicyclist accelerates from rest with constant acceleration to 24 km/h in 14.0 s. What is the angular acceleration of the wheels?

15 •• The tape in a standard VHS videotape cassette has a length $L = 246$ m; the tape plays for $t = 2.0$ h (Figure 9-36). As the tape starts, the full reel has an outer radius of about $R = 45$ mm, and an inner radius of about $r = 12$ mm. At some point during the play, both reels have the same angular speed. Calculate this angular speed in rad/s and in rev/min.

Figure 9-36 Problem 15

Torque, Moment of Inertia, and Newton's Second Law for Rotation

16 • The dimension of torque is the same as that of

(*a*) impulse. (*b*) energy.
(*c*) momentum. (*d*) none of the above.

17 • The moment of inertia of an object of mass M

(*a*) is an intrinsic property of the object.
(*b*) depends on the choice of axis of rotation.
(*c*) is proportional to M regardless of the choice of axis.
(*d*) Both (*b*) and (*c*) are correct.

18 • Can an object continue to rotate in the absence of torque?

19 • Does an applied net torque always increase the angular speed of an object?

20 • True or false:

(*a*) If the angular velocity of an object is zero at some instant, the net torque on the object must be zero at that instant.
(*b*) The moment of inertia of an object depends on the location of the axis of rotation.
(*c*) The moment of inertia of an object depends on the angular velocity of the object.

21 • A disk is free to rotate about an axis. A force applied a distance d from the axis causes an angular acceleration α. What angular acceleration is produced if the same force is applied a distance $2d$ from the axis?

(*a*) α (*b*) 2α (*c*) $\alpha/2$ (*d*) 4α (*e*) $\alpha/4$

22 • A disk-shaped grindstone of mass 1.7 kg and radius 8 cm is spinning at 730 rev/min. After the power is shut off, a woman continues to sharpen her ax by holding it against the grindstone for 9 s until the grindstone stops rotating. (*a*) What is the angular acceleration of the grindstone? (*b*) What is the torque exerted by the ax on the grindstone? (Assume constant angular acceleration and a lack of other frictional torques.)

23 • A 2.5-kg cylinder of radius 11 cm is initially at rest. A rope of negligible mass is wrapped around it and pulled with a force of 17 N. Find (*a*) the torque exerted by the rope, (*b*) the angular acceleration of the cylinder, and (*c*) the angular velocity of the cylinder at $t = 5$ s.

24 •• A wheel mounted on an axis that is not frictionless is initially at rest. A constant external torque of 50 N·m is applied to the wheel for 20 s, giving the wheel an angular velocity of 600 rev/min. The external torque is then removed, and the wheel comes to rest 120 s later. Find (*a*) the moment of inertia of the wheel, and (*b*) the frictional torque, which is assumed to be constant.

25 •• A pendulum consisting of a string of length L attached to a bob of mass m swings in a vertical plane. When the string is at an angle θ to the vertical, (*a*) what is the tangential component of acceleration of the bob? (*b*) What is the torque exerted about the pivot point? (*c*) Show that $\tau = I\alpha$ with $a_t = L\alpha$ gives the same tangential acceleration found in part (*a*).

26 ••• A uniform rod of mass M and length L is pivoted at one end and hangs as in Figure 9-37 so that it is free to rotate without friction about its pivot. It is struck by a horizontal force F_0 for a short time Δt at a distance x below the pivot as shown. (*a*) Show that the speed of the center of mass of the rod just after being struck is given by $v_0 = 3F_0 x\ \Delta t/2ML$. (*b*) Find the force delivered by the pivot, and show that this force is zero if $x = 2L/3$. (*Note:* The point $x = 2L/3$ is called the *center of percussion* of the rod.)

Figure 9-37 Problem 26

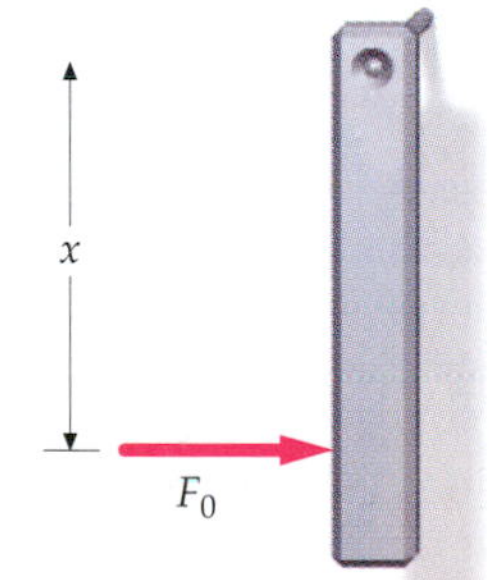

27 ••• A uniform horizontal disk of mass M and radius R is rotating about its vertical axis with an angular velocity ω. When it is placed on a horizontal surface, the coefficient of kinetic friction between the disk and surface is μ_k. (*a*) Find the torque $d\tau$ exerted by the force of friction on a circular element of radius r and width dr. (*b*) Find the total torque exerted by friction on the disk. (*c*) Find the time required to bring the disk to a halt.

Calculating the Moment of Inertia

28 • The moment of inertia of an object about an axis that does not pass through its center of mass is ________ the moment of inertia about a parallel axis through its center of mass.

(*a*) always less than (*b*) sometimes less than
(*c*) sometimes equal to (*d*) always greater than

29 • A tennis ball has a mass of 57 g and a diameter of 7 cm. Find the moment of inertia about its diameter. Assume that the ball is a thin spherical shell.

30 • Four particles at the corners of a square with a side length $L = 2$ m are connected by massless rods (Figure 9-38). The particle masses are $m_1 = m_4 = 3$ kg and $m_2 = m_3 = 4$ kg. Find the moment of inertia of the system about the z axis.

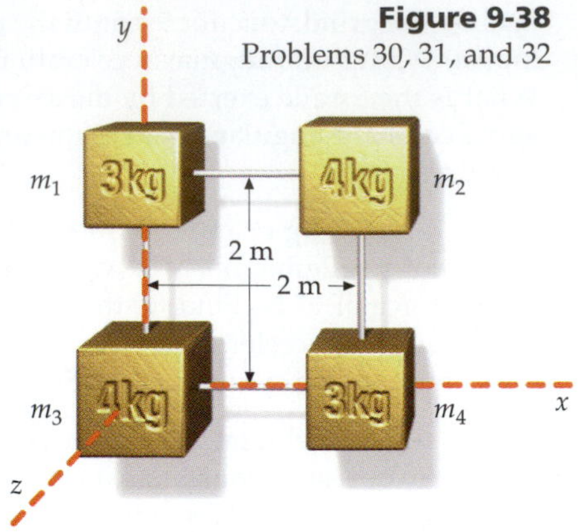

Figure 9-38 Problems 30, 31, and 32

31 • Use the parallel-axis theorem and your results for Problem 30 to find the moment of inertia of the four-particle system in Figure 9-38 about an axis that is perpendicular to the plane of the masses and passes through the center of mass of the system. Check your result by direct computation.

32 • For the four-particle system of Figure 9-38, (*a*) find the moment of inertia I_x about the x axis, which passes through m_3 and m_4, and (*b*) find I_y about the y axis, which passes through m_1 and m_3.

33 • Use the parallel-axis theorem to find the moment of inertia of a solid sphere of mass M and radius R about an axis that is tangent to the sphere (Figure 9-39).

Figure 9-39 Problem 33

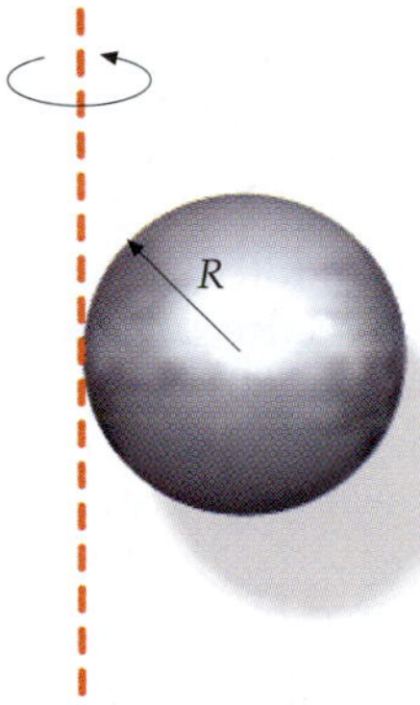

34 •• A wagon wheel 1.0 m in diameter consists of a thin rim having a mass of 8 kg and six spokes each having a mass of 1.2 kg. Determine the moment of inertia of the wagon wheel for rotation about its axis.

35 •• Two point masses m_1 and m_2 are separated by a massless rod of length L. (*a*) Write an expression for the moment of inertia about an axis perpendicular to the rod and passing through it at a distance x from mass m_1. (*b*) Calculate dI/dx and show that I is at a minimum when the axis passes through the center of mass of the system.

36 •• A uniform rectangular plate has mass m and sides a and b. (*a*) Show by integration that the moment of inertia of the plate about an axis that is perpendicular to the plate and passes through one corner is $\frac{1}{3}m(a^2 + b^2)$. (*b*) What is the moment of inertia about an axis that is perpendicular to the plate and passes through its center of mass?

37 •• Tracey and Corey are doing intensive research on theoretical baton twirling. Each is using "The Beast" as a model baton: two uniform spheres, each of mass 500 g and radius 5 cm, mounted at the ends of a 30-cm uniform rod of mass 60 g (Figure 9-40). Tracey and Corey want to calculate the moment of inertia of The Beast about an axis perpendicular to the rod and passing through its center. Corey uses the approximation that the two spheres can be treated as point particles that are 20 cm from the axis of rotation, and that the mass of the rod is negligible. Tracey, however, makes her calculations without approximations. (*a*) Compare the two results. (*b*) If the spheres retained the same mass but were hollow, would the rotational inertia increase or decrease? Justify your choice with a sentence or two. It is not necessary to calculate the new value of I.

Figure 9-40 Problem 37

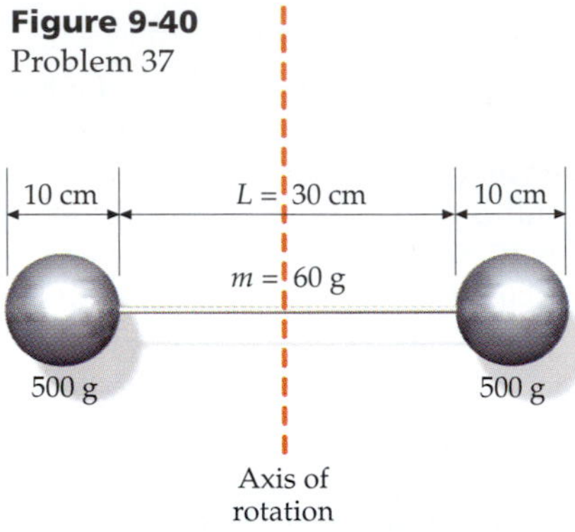

Figure 9-41 Methane (Problem 38)

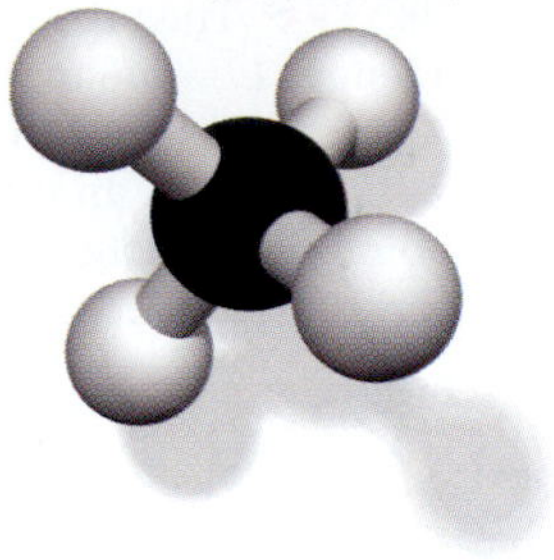

38 •• The methane molecule (CH_4) has four hydrogen atoms located at the vertices of a regular tetrahedron of side length 1.4 nm, with the carbon atom at the center of the tetrahedron (Figure 9-41). Find the moment of inertia of this molecule for rotation about an axis that passes through the carbon atom and one of the hydrogen atoms.

39 ••• A hollow cylinder has mass m, an outside radius R_2, and an inside radius R_1. Show that its moment of inertia about its symmetry axis is given by $I = \frac{1}{2}m(R_2^2 + R_1^2)$.

40 ••• Show that the moment of inertia of a spherical shell of radius R and mass m is $\frac{2}{3}mR^2$. This can be done by direct integration or, more easily, by finding the increase in the moment of inertia of a solid sphere when its radius changes. To do this, first show that the moment of inertia of a solid sphere of density ρ is $I = \frac{8}{15}\pi\rho R^5$. Then compute the change dI in I for a change dR, and use the fact that the mass of this shell is $m = 4\pi R^2\rho\, dR$.

41 ••• The density of the earth is not quite uniform. It varies with the distance r from the center of the earth as $\rho = C\,(1.22 - r/R)$, where R is the radius of the earth and C is a constant. (*a*) Find C in terms of the total mass M and the radius R. (*b*) Find the moment of inertia of the earth. (See Problem 40.)

42 ••• Use integration to determine the moment of inertia of a right circular homogeneous cone of height H, base radius R, and mass density ρ about its symmetry axis.

43 ••• Use integration to determine the moment of inertia of a hollow, thin-walled, right circular cone of mass M, height H, and base radius R about its symmetry axis.

44 ••• Use integration to determine the moment of inertia of a thin uniform disk of mass M and radius R for rotation about a diameter. Check your answer by referring to Table 9-1.

45 ••• Use integration to determine the moment of inertia of a thin circular hoop of radius R and mass M for rotation about a diameter. Check your answer by referring to Table 9-1.

46 ••• A roadside ice-cream stand uses rotating cones to catch the eyes of travelers. Each cone rotates about an axis perpendicular to its axis of symmetry and passing through its apex. The sizes of the cones vary, and the owner wonders if it would be more energy-efficient to use several smaller cones or a few big ones. To answer this, he must calculate the moment of inertia of a homogeneous right circular cone of height H, base radius R, and mass density ρ. What is the result?

Rotational Kinetic Energy

47 • A constant torque acts on a merry-go-round. The power input of the torque is

(*a*) constant.
(*b*) proportional to the angular speed of the merry-go-round.
(*c*) zero.
(*d*) none of the above.

48 • The particles in Figure 9-42 are connected by a very light rod whose moment of inertia can be neglected. They rotate about the y axis with angular velocity $\omega = 2$ rad/s. (*a*) Find the speed of each particle, and use it to calculate the kinetic energy of this system directly from $\Sigma\, \frac{1}{2}m_i v_i^2$. (*b*) Find the moment of inertia about the y axis, and calculate the kinetic energy from $K = \frac{1}{2}I\omega^2$.

Figure 9-42 Problem 48

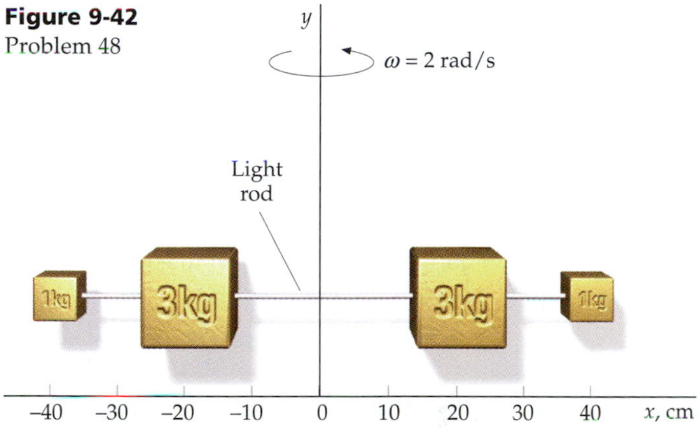

49 • Four 2-kg particles are located at the corners of a rectangle of sides 3 m and 2 m as shown in Figure 9-43. (*a*) Find the moment of inertia of this system about the z axis. (*b*) The system is set rotating about this axis with a kinetic energy of 124 J. Find the number of revolutions the system makes per minute.

Figure 9-43 Problem 49

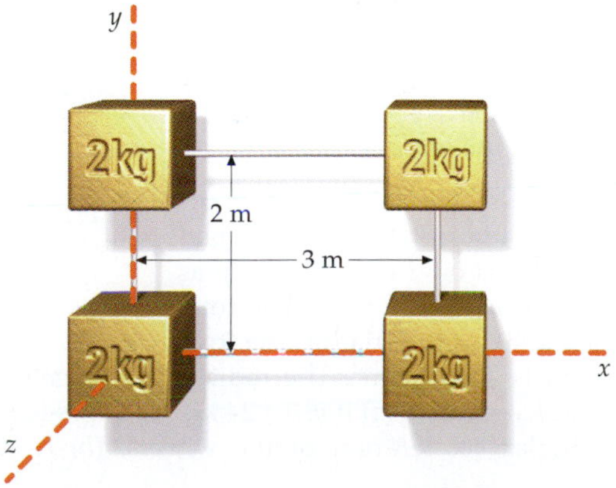

50 • A solid ball of mass 1.4 kg and diameter 15 cm is rotating about its diameter at 70 rev/min. (*a*) What is its kinetic energy ? (*b*) If an additional 2 J of energy are supplied to the rotational energy, what is the new angular speed of the ball?

51 • An engine develops 400 N·m of torque at 3700 rev/min. Find the power developed by the engine.

52 •• Two point masses m_1 and m_2 are connected by a massless rod of length L to form a dumbbell that rotates about its center of mass with angular velocity ω. Show that the ratio of kinetic energies of the masses is $K_1/K_2 = m_2/m_1$.

53 •• Calculate the kinetic energy of rotation of the earth, and compare it with the kinetic energy of motion of the earth's center of mass about the sun. Assume the earth to be a homogeneous sphere of mass 6.0×10^{24} kg and radius 6.4×10^6 m. The radius of the earth's orbit is 1.5×10^{11} m.

54 •• A 2000-kg block is lifted at a constant speed of 8 cm/s by a steel cable that passes over a massless pulley to a motor-driven winch (Figure 9-44). The radius of the winch drum is 30 cm. (*a*) What force must be exerted by the cable? (*b*) What torque does the cable exert on the winch drum? (*c*) What is the angular velocity of the winch drum? (*d*) What power must be developed by the motor to drive the winch drum?

Figure 9-44 Problem 54

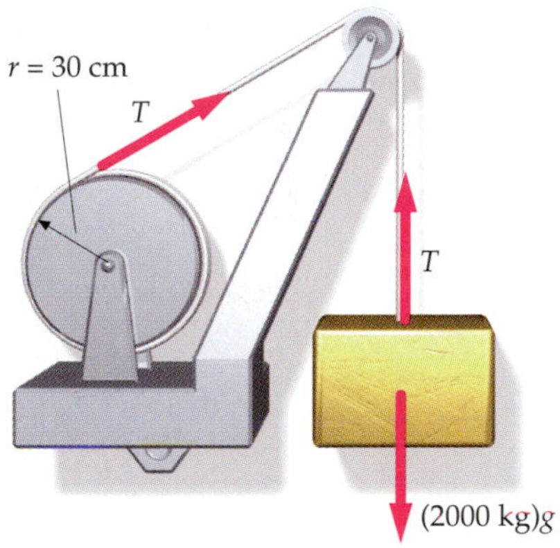

55 •• A uniform disk of mass M and radius R is pivoted such that it can rotate freely about a horizontal axis through its center and perpendicular to the plane of the disk. A small particle of mass m is attached to the rim of the disk at the top, directly above the pivot. The system is given a gentle start, and the disk begins to rotate. (*a*) What is the angular velocity of the disk when the particle is at its lowest point? (*b*) At this point, what force must be exerted on the particle by the disk to keep it on the disk?

56 •• A ring 1.5 m in diameter is pivoted at one point on its circumference so that it is free to rotate about a horizontal axis. Initially, the line joining the support and center is horizontal. (*a*) If released from rest, what is its maximum angular velocity? (*b*) What must its initial angular velocity be if it is to just make a complete revolution?

57 •• You set out to design a car that uses the energy stored in a flywheel consisting of a uniform 100-kg cylinder of radius R. The flywheel must deliver an average of 2 MJ of mechanical energy per kilometer, with a maximum angular velocity of 400 rev/s. Find the least value of R such that the car can travel 300 km without the flywheel having to be recharged.

58 •• A ladder that is 8.6 m long and has mass 60 kg is placed in a nearly vertical position against the wall of a building. You stand on a rung with your center of mass at the top of the ladder. Assume that your mass is 80 kg. As you lean back slightly, the ladder begins to rotate about its base away from the wall. Is it better to quickly step off the ladder and drop to the ground or to hold onto the ladder and step off just before the top end hits the ground?

59 ••• Consider the situation in Problem 58 with a ladder of length L and mass M. Find the ratio of your speed as you hit the ground if you hang on to the ladder to your speed if you immediately step off as a function of the mass ratio M/m, where m is your mass.

Pulleys, Yo-Yos, and Hanging Things

60 •• A 4-kg block resting on a frictionless horizontal ledge is attached to a string that passes over a pulley and is attached to a hanging 2-kg block (Figure 9-45). The pulley is a uniform disk of radius 8 cm and mass 0.6 kg. (*a*) Find the speed of the 2-kg block after it falls from rest a distance of 2.5 m. (*b*) What is the angular velocity of the pulley at this time?

Figure 9-45 Problems 60–63

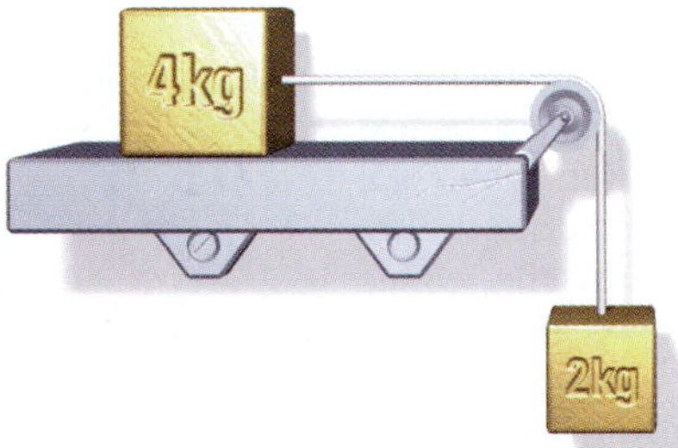

61 •• For the system in Problem 60, find the linear acceleration of each block and the tension in the string.

62 •• Work Problem 60 for the case in which the coefficient of friction between the ledge and the 4-kg block is 0.25.

63 •• Work Problem 61 for the case in which the coefficient of friction between the ledge and the 4-kg block is 0.25.

64 •• In 1993, a giant yo-yo of mass 400 kg and measuring about 1.5 m in radius was dropped from a crane 57 m high. Assuming that the axle of the yo-yo had a radius of $r = 0.1$ m, find the velocity of the descent v at the end of the fall.

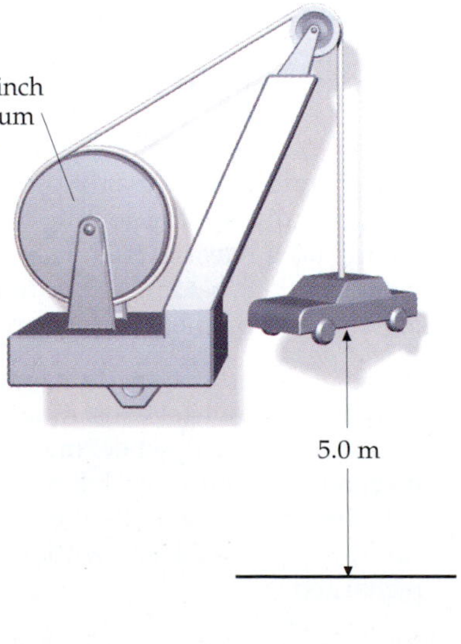

Figure 9-46 Problem 65

65 •• A 1200-kg car is being unloaded by a winch. At the moment shown in Figure 9-46, the gearbox shaft of the winch breaks, and the car falls from rest. During the car's fall, there is no slipping between the (massless) rope, the pulley, and the winch drum. The moment of inertia of the winch drum is 320 kg·m^2 and that of the pulley is 4 kg·m^2. The radius of the winch drum is 0.80 m and that of the pulley is 0.30 m. Find the speed of the car as it hits the water.

66 •• The system in Figure 9-47 is released from rest. The 30-kg block is 2 m above the ledge. The pulley is a uniform disk with a radius of 10 cm and mass of 5 kg. Find (*a*) the speed of the 30-kg block just before it hits the ledge, (*b*) the angular speed of the pulley at that time, (*c*) the tensions in the strings, and (*d*) the time it takes for the 30-kg block to reach the ledge. Assume that the string does not slip on the pulley.

Figure 9-47 Problem 66

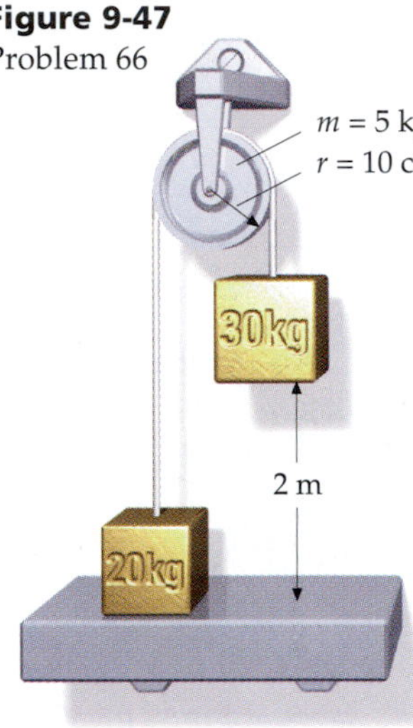

Figure 9-48 Problem 67

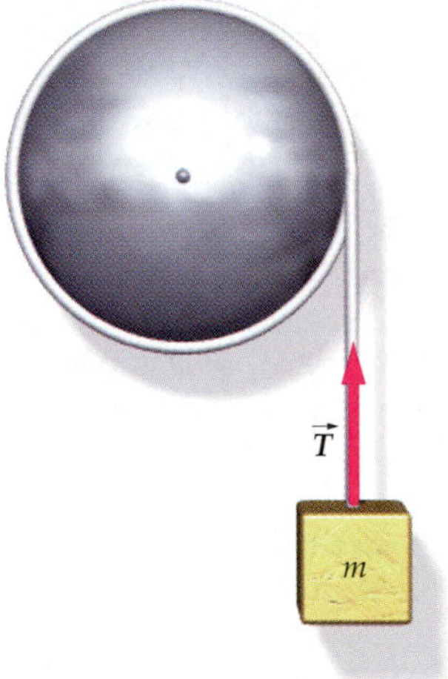

67 •• A uniform sphere of mass M and radius R is free to rotate about a horizontal axis through its center. A string is wrapped around the sphere and is attached to an object of mass m as shown in Figure 9-48. Find (*a*) the acceleration of the object, and (*b*) the tension in the string.

68 •• An Atwood's machine has two objects of mass m_1 = 500 g and m_2 = 510 g, connected by a string of negligible mass that passes over a frictionless pulley (Figure 9-49). The pulley is a uniform disk with a mass of 50 g and a radius of 4 cm. The string does not slip on the pulley. (*a*) Find the acceleration of the objects. (*b*) What is the tension in the string supporting m_1? In the string supporting m_2? By how much do they differ? (*c*) What would your answers have been if you had neglected the mass of the pulley?

Figure 9-49 Problem 68

69 •• Two objects are attached to ropes that are attached to wheels on a common axle as shown in Figure 9-50. The total moment of inertia of the two wheels is 40 kg·m^2. The radii of the wheels are R_1 = 1.2 m and R_2 = 0.4 m. (*a*) If m_1 = 24 kg, find m_2 such that there is no angular acceleration of the wheels. (*b*) If 12 kg is gently

added to the top of m_1, find the angular acceleration of the wheels and the tensions in the ropes.

Figure 9-50 Problem 69

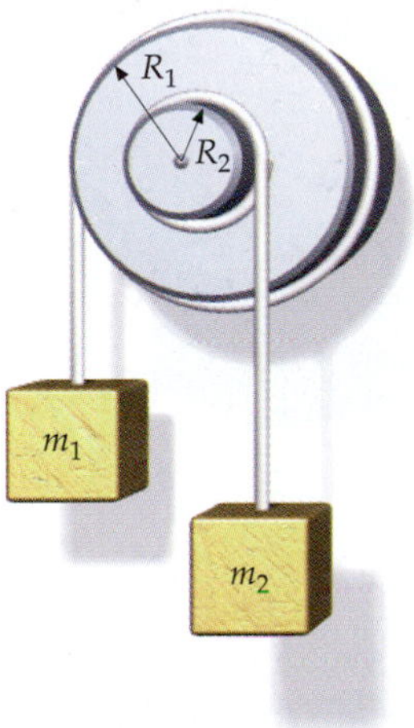

Figure 9-51 Problems 70 and 71

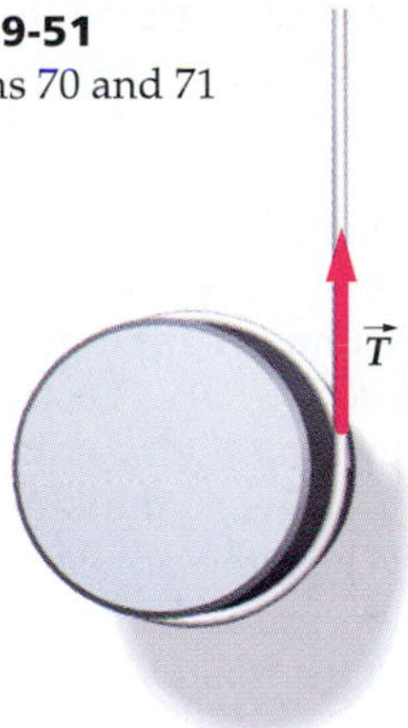

70 •• A uniform cylinder of mass M and radius R has a string wrapped around it. The string is held fixed, and the cylinder falls vertically as shown in Figure 9-51. (*a*) Show that the acceleration of the cylinder is downward with a magnitude $a = 2g/3$. (*b*) Find the tension in the string.

71 •• The cylinder in Figure 9-51 is held by a hand that is accelerated upward so that the center of mass of the cylinder does not move. Find (*a*) the tension in the string, (*b*) the angular acceleration of the cylinder, and (*c*) the acceleration of the hand.

72 •• A 0.1-kg yo-yo consists of two solid disks of radius 10 cm joined together by a massless rod of radius 1 cm and a string wrapped around the rod. One end of the string is held fixed and is under constant tension T as the yo-yo is released. Find the acceleration of the yo-yo and the tension T.

73 •• A uniform cylinder of mass m_1 and radius R is pivoted on frictionless bearings. A massless string wrapped around the cylinder connects to a mass m_2, which is on a frictionless incline of angle θ as shown in Figure 9-52. The system is released from rest with m_2 a height h above the bottom of the incline. (*a*) What is the acceleration of m_2? (*b*) What is the tension in the string? (*c*) What is the total energy of the system when m_2 is at height h? (*d*) What is the total energy when m_2 is at the bottom of the incline and has a speed v? (*e*) What is the speed v? (*f*) Evaluate your answers for the extreme cases of $\theta = 0°$, $\theta = 90°$, and $m_1 = 0$.

Figure 9-52 Problem 73

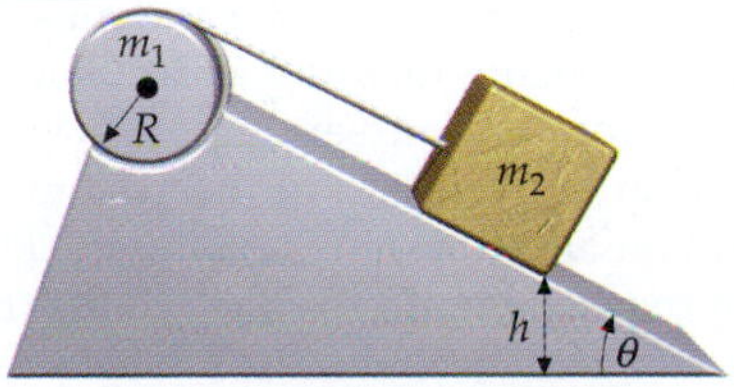

74 •• A device for measuring the moment of inertia of an object is shown in Figure 9-53. A circular platform has a concentric drum of radius 10 cm about which a string is wound. The string passes over a frictionless pulley to a weight of mass M. The weight is released from rest, and the time for it to drop a distance D is measured. The system is then rewound, the object placed on the platform, and the system again released from rest. The time required for the weight to drop the same distance D then provides the data needed to calculate I. With $M = 2.5$ kg, and $D = 1.8$ m, the time is 4.2 s. (*a*) Find the combined moment of inertia of the platform, drum, shaft, and pulley. (*b*) With an object placed on the platform, the time is 6.8 s for $D = 1.8$ m. Find I of that object about the axis of the platform.

Figure 9-53 Problem 74

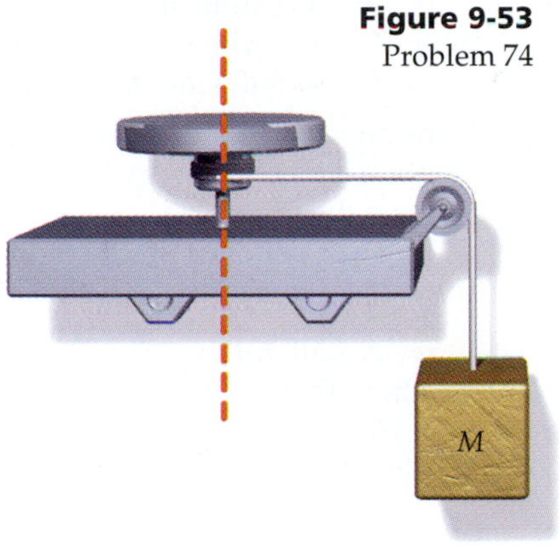

Objects Rolling Without Slipping

75 • True or false: When an object rolls without slipping, friction does no work on the object.

76 • A wheel of radius R is rolling without slipping. The velocity of the point on the rim that is in contact with the surface, relative to the surface, is

(*a*) equal to $R\omega$ in the direction of motion of the center of mass.
(*b*) equal to $R\omega$ opposite the direction of motion of the center of mass.
(*c*) zero.
(*d*) equal to the velocity of the center of mass and in the same direction.
(*e*) equal to the velocity of the center of mass but in the opposite direction.

77 •• A solid cylinder and a solid sphere have equal masses. Both roll without slipping on a horizontal surface. If their kinetic energies are the same, then

(*a*) the translational speed of the cylinder is greater than that of the sphere.
(*b*) the translational speed of the cylinder is less than that of the sphere.
(*c*) the translational speeds of the two objects are the same.
(*d*) (*a*), (*b*), or (*c*) could be correct depending on the radii of the objects.

78 •• Starting from rest at the same time, a coin and a ring roll down an incline without slipping. Which of the following is true?

(*a*) The ring reaches the bottom first.
(*b*) The coin reaches the bottom first.
(*c*) The coin and ring arrive at the bottom simultaneously.
(*d*) The race to the bottom depends on their relative masses.
(*e*) The race to the bottom depends on their relative diameters.

79 •• For a hoop of mass M and radius R that is rolling without slipping, which is larger, its translational kinetic energy or its rotational kinetic energy?

(*a*) Translational kinetic energy is larger.
(*b*) Rotational kinetic energy is larger.
(*c*) Both are the same size.
(*d*) The answer depends on the radius.
(*e*) The answer depends on the mass.

80 •• For a disk of mass M and radius R that is rolling without slipping, which is larger, its translational kinetic energy or its rotational kinetic energy?

(*a*) Translational kinetic energy is larger.
(*b*) Rotational kinetic energy is larger.
(*c*) Both are the same size.
(*d*) The answer depends on the radius.
(*e*) The answer depends on the mass.

81 •• A ball rolls without slipping along a horizontal plane. Show that the frictional force acting on the ball must be zero. *Hint:* Consider a possible direction for the action of the frictional force and what effects such a force would have on the velocity of the center of mass and on the angular velocity.

82 • A homogeneous solid cylinder rolls without slipping on a horizontal surface. The total kinetic energy is K. The kinetic energy due to rotation about its center of mass is

(*a*) $\frac{1}{2}K$. (*b*) $\frac{1}{3}K$. (*c*) $\frac{4}{7}K$.
(*d*) none of the above.

83 • A homogeneous cylinder of radius 18 cm and mass 60 kg is rolling without slipping along a horizontal floor at 5 m/s. How much work is needed to stop the cylinder?

84 • Find the percentages of the total kinetic energy associated with rotation and translation, respectively, for an object that is rolling without slipping if the object is (*a*) a uniform sphere, (*b*) a uniform cylinder, or (*c*) a hoop.

85 • A hoop of radius 0.40 m and mass 0.6 kg is rolling without slipping at a speed of 15 m/s toward an incline of slope 30°. How far up the incline will the hoop roll, assuming that it rolls without slipping?

86 • A ball rolls without slipping down an incline of angle θ. The coefficient of static friction is μ_s. Find (*a*) the acceleration of the ball, (*b*) the force of friction, and (*c*) the maximum angle of the incline for which the ball will roll without slipping.

87 •• An empty can of total mass $3M$ is rolling without slipping. If its mass is distributed as in Figure 9-54, what is the value of the ratio of kinetic energy of translation to the kinetic energy of rotation about its center of mass?

Figure 9-54 Problem 87

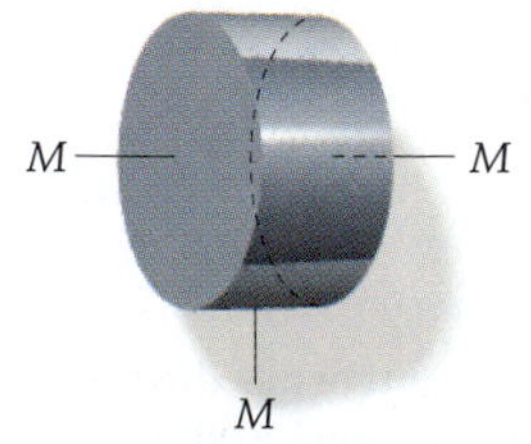

88 •• A bicycle of mass 14 kg has 1.2-m-diameter wheels, each of mass 3 kg. The mass of the rider is 38 kg. Estimate the fraction of the total kinetic energy of bicycle and rider associated with rotation of the wheels.

89 •• A hollow sphere and uniform sphere of the same mass m and radius R roll down an inclined plane from the same height H without slipping (Figure 9-55). Each is moving horizontally as it leaves the ramp. When the spheres hit the ground, the range of the hollow sphere is L. Find the range L' of the uniform sphere.

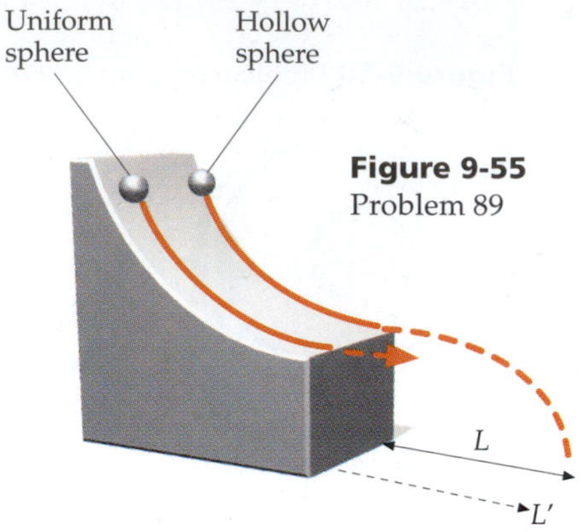

Figure 9-55 Problem 89

90 •• A hollow cylinder and a uniform cylinder are rolling horizontally without slipping. The speed of the hollow cylinder is v. The cylinders encounter an inclined plane that they climb without slipping. If the maximum height they reach is the same, find the initial speed v' of the uniform cylinder.

91 •• A hollow, thin-walled cylinder and a solid sphere start from rest and roll without slipping down an inclined plane of length 3 m. The cylinder arrives at the bottom of the plane 2.4 s after the sphere. Determine the angle between the inclined plane and the horizontal.

92 •• A uniform solid sphere of radius r starts from rest at a height h and rolls without slipping along the loop-the-loop track of radius R as shown in Figure 9-56. (*a*) What is the smallest value of h for which the sphere will not leave the track at the top of the loop? (*b*) What would h have to be if, instead of rolling, the ball slides without friction?

Figure 9-56 Problem 92

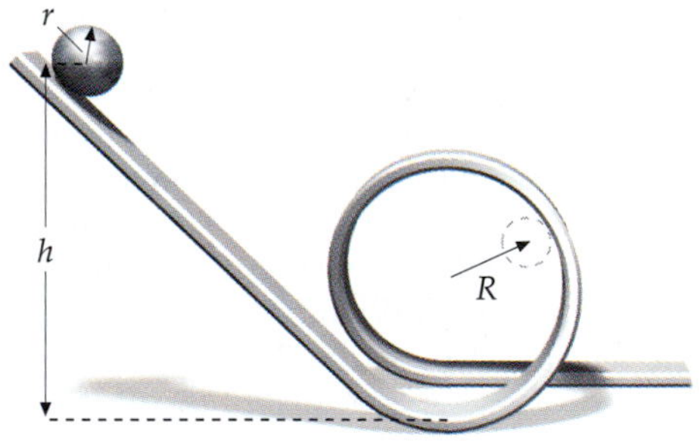

93 ••• A wheel has a thin 3.0-kg rim and four spokes each of mass 1.2 kg. Find the kinetic energy of the wheel when it rolls at 6.0 m/s on a horizontal surface.

94 ••• Two uniform 20-kg disks of radius 30 cm are connected by a short rod of radius 2 cm and mass 1 kg. When the rod is placed on a plane inclined at 30°, such that the disks hang over the sides, the assembly rolls without slipping. Find (*a*) the linear acceleration of the system, and (*b*) the angular acceleration of the system. (*c*) Find the kinetic energy of translation of the system after it has rolled 2 m down the incline starting from rest. (*d*) Find the kinetic energy of rotation of the system at the same point.

95 ••• A wheel of radius R rolls without slipping at a speed V. The coordinates of the center of the wheel are X, Y. (*a*) Show that the x and y coordinates of point P in Figure 9-57 are $X + r_0 \cos\theta$ and $R + r_0 \sin\theta$, respectively. (*b*) Show that the total velocity v of point P has the components $v_x = V + (r_0 V \sin\theta)/R$ and $v_y = -(r_0 V \cos\theta)/R$. (*c*) Show that at the instant that $X = 0$, v and r are perpendicular to each other by calculating $\vec{v} \cdot \vec{r}$. (*d*) Show that $v = r\omega$, where $\omega = V/R$ is the angular velocity of the wheel. These results demonstrate that, in the case of rolling without slipping, the motion is the same as if the rolling object were instantaneously rotating about the point of contact with an angular speed $w = V/R$.

Figure 9-57 Problem 95

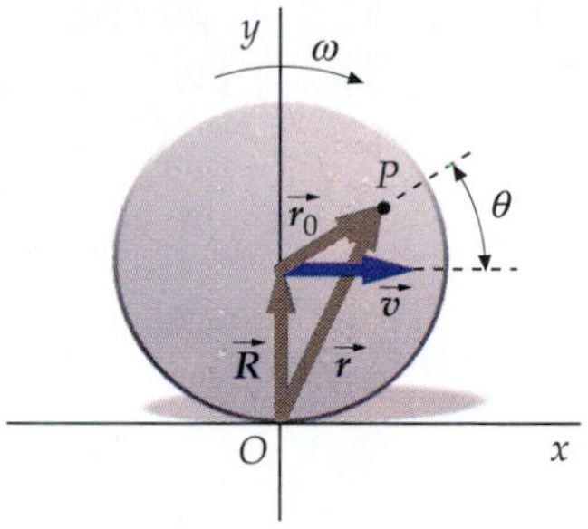

96 ••• A uniform cylinder of mass M and radius R rests on a block of mass m, which in turn is at rest on a horizontal, frictionless table (Figure 9-58). If a horizontal force $\vec{F}$ is applied to the block, it accelerates and the cylinder rolls without slipping. Find the acceleration of the block.

Figure 9-58 Problems 96, 97, and 98

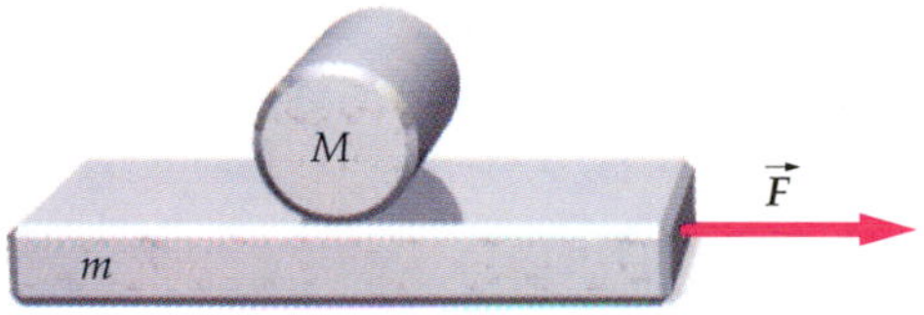

97 ••• (*a*) Find the angular acceleration of the cylinder in Problem 96. Is the cylinder rotating clockwise or counterclockwise? (*b*) What is the cylinder's linear acceleration relative to the table? Let the direction of $\vec{F}$ be the positive direction. (*c*) What is the linear acceleration of the cylinder relative to the block?

98 ••• If the force in Problem 96 acts over a distance d, find (*a*) the kinetic energy of the block, and (*b*) the kinetic energy of the cylinder. (*c*) Show that the total kinetic energy is equal to the work done on the system.

99 ••• A marble of radius 1 cm rolls from rest without slipping from the top of a large sphere of radius 80 cm, which is held fixed (Figure 9-59). Find the angle from the top of the sphere to the point where the marble breaks contact with the sphere.

Figure 9-59 Problem 99

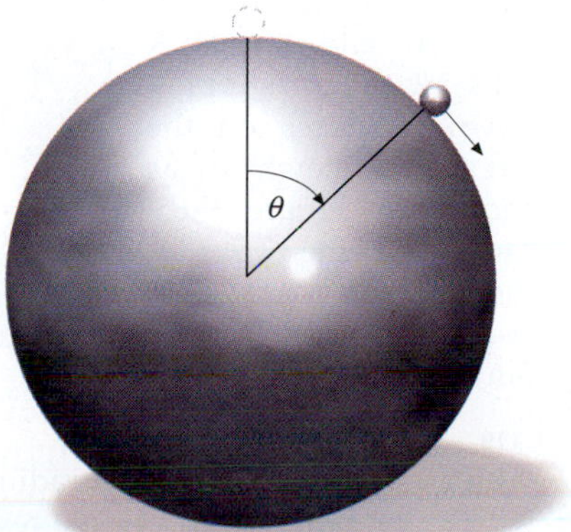

Rolling With Slipping (optional)

100 • True or false: When a sphere rolls and slips on a rough surface, mechanical energy is dissipated.

101 • A cue ball is hit very near the top so that it starts to move with topspin. As it slides, the force of friction

(*a*) increases v_{cm}.
(*b*) decreases v_{cm}.
(*c*) has no effect on v_{cm}.

102 •• A bowling ball of mass M and radius R is thrown such that at the instant it touches the floor it is moving horizontally with a speed v_0 and is not rotating. It slides for a time t_1 a distance s_1 before it begins to roll without slipping. (*a*) If μ_k is the coefficient of sliding friction between the ball and the floor, find s_1, t_1, and the final speed v_1 of the ball. (*b*) Evaluate these quantities for $v_0 = 8$ m/s and $\mu_k = 0.06$. (*c*) Find the ratio of the final mechanical energy to the initial mechanical energy of the ball.

103 •• A cue ball of radius r is initially at rest on a horizontal pool table (Figure 9-60). It is struck by a horizontal cue stick that delivers a force of magnitude P_0 for a very short time Δt. The stick strikes the ball at a point h above the ball's point of contact with the table. (*a*) Show that the ball's initial angular velocity ω_0 is related to the initial linear velocity of its center of mass v_0 by $\omega_0 = 5v_0(h - r)/2r^2$.

Figure 9-60 Problem 103

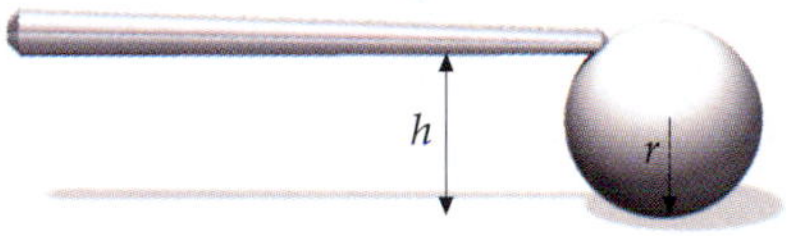

104 ••• A uniform spherical ball is set rotating about a horizontal axis with an angular speed ω_0 and is placed on the floor. If the coefficient of sliding friction between the ball and the floor is μ_k, find the speed of the center of mass of the ball when it begins to roll without slipping.

105 ••• A uniform solid ball resting on a horizontal surface has a mass of 20 g and a radius of 5 cm. A sharp force is applied to the ball in the horizontal direction 9 cm above the horizontal surface. The force increases linearly from 0 to a peak value of 40,000 N in 10^{-4} s and then decreases linearly to 0 in 10^{-4} s. (*a*) What is the velocity of the ball after impact? (*b*) What is the angular velocity of the ball after impact? (*c*) What is the velocity of the ball when it begins to roll without sliding? (*d*) For how long does the ball slide on the surface? Assume that $\mu_k = 0.5$.

106 ••• A 0.3-kg billiard ball of radius 3 cm is given a sharp blow by a cue stick. The applied force is horizontal and passes through the center of the ball. The initial velocity of the ball is 4 m/s. The coefficient of kinetic friction is 0.6. (*a*) For how many seconds does the ball slide before it begins to roll without slipping? (*b*) How far does it slide? (*c*) What is its velocity once it begins rolling without slipping?

107 ••• A billiard ball initially at rest is given a sharp blow by a cue stick. The force is horizontal and is applied at a distance $2R/3$ below the centerline, as shown in Figure 9-61. The initial speed of the ball is v_0, and the coefficient of kinetic friction is μ_k. (*a*) What is the initial angular speed ω_0? (*b*) What is the speed of the ball once it begins to roll without slipping? (*c*) What is the initial kinetic energy of the ball? (*d*) What is the frictional work done as it slides on the table?

Figure 9-61 Problem 107

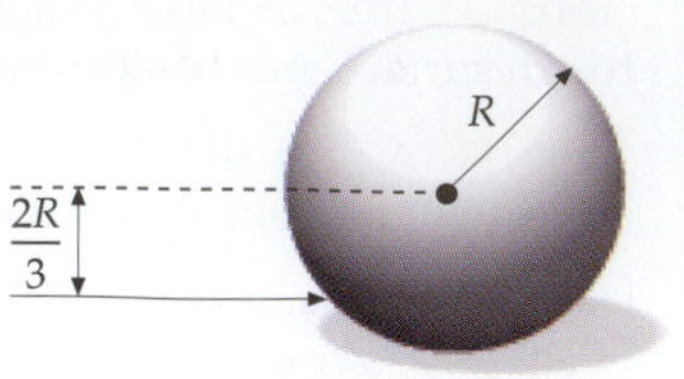

108 ••• A bowling ball of radius R is given an initial velocity v_0 down the lane and a forward spin $\omega_0 = 3v_0/R$. The coefficient of kinetic friction is μ_k. (*a*) What is the speed of the ball when it begins to roll without slipping? (*b*) For how long does the ball slide before it begins to roll without slipping? (*c*) What distance does the ball slide down the lane before it begins rolling without slipping?

109 ••• A solid cylinder of mass M resting on its side on a horizontal surface is given a sharp blow by a cue stick. The applied force is horizontal and passes through the center of the cylinder so that the cylinder begins translating with initial velocity v_0. The coefficient of sliding friction between the cylinder and surface is μ_k. (*a*) What is the translational velocity of the cylinder when it is rolling without slipping? (*b*) How far does the cylinder travel before it rolls without slipping? (*c*) What fraction of its initial mechanical energy is dissipated in friction?

General Problems

110 • The torque exerted on an orbiting communications satellite by the gravitational pull of the earth is

(*a*) directed toward the earth.
(*b*) directed parallel to the earth's axis and toward the north pole.
(*c*) directed parallel to the earth's axis and toward the south pole.
(*d*) directed toward the satellite.
(*e*) zero.

111 • The moon rotates as it revolves around the earth so that we always see the same side. Use this fact to find the angular velocity (in rad/s) of the moon about its axis. (The period of revolution of the moon about the earth is 27.3 days.)

112 • Find the moment of inertia of a hoop about an axis perpendicular to the plane of the hoop and through its edge.

113 •• The radius of a park merry-go-round is 2.2 m. To start it rotating, you wrap a rope around it and pull with a force of 260 N for 12 s. During this time, the merry-go-round makes one complete rotation. (*a*) Find the angular acceleration of the merry-go-round. (*b*) What torque is exerted by the rope on the merry-go-round? (*c*) What is the moment of inertia of the merry-go-round?

114 •• A uniform disk of radius 0.12 m and mass 5 kg is pivoted such that it rotates freely about its central axis (Figure 9-62). A string wrapped around the disk is pulled with a force of 20 N. (*a*) What is the torque exerted on the disk? (*b*) What is the angular acceleration of the disk? (*c*) If the disk starts from rest, what is its angular velocity after 5 s? (*d*) What is its kinetic energy after 5 s? (*e*) What is the total angle θ that the disk turns through in 5 s? (*f*) Show that the work done by the torque $\tau\,\Delta\theta$ equals the kinetic energy.

Figure 9-62 Problem 114

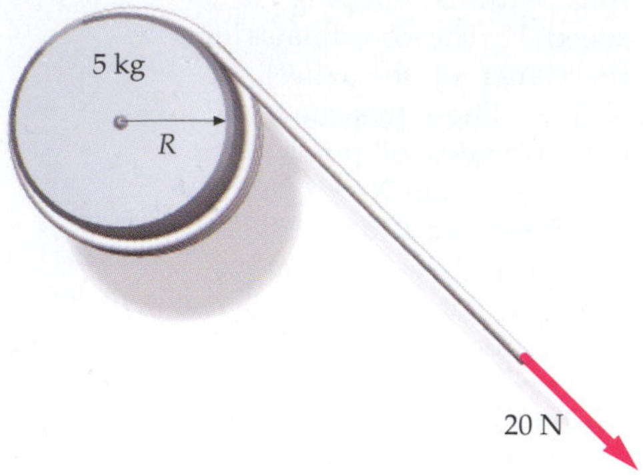

115 •• A 0.25-kg rod of length 80 cm is suspended by a frictionless pivot at one end. It is held horizontal and released. Immediately after it is released, what is (*a*) the acceleration of the center of the rod, and (*b*) the initial acceleration of a point on the end of the rod? (*c*) Find the linear velocity of the center of mass of the rod when it is vertical.

116 •• A uniform rod of length $3L$ is pivoted as shown in Figure 9-63 and held in a horizontal position. What is the initial angular acceleration α of the rod upon release?

117 •• A uniform rod of length L and mass m is pivoted at the middle as shown in Figure 9-64. It has a load of mass $2m$ attached to one of the ends. If the system is released from a horizontal position, what is the maximum velocity of the load?

Figure 9-63 Problem 116

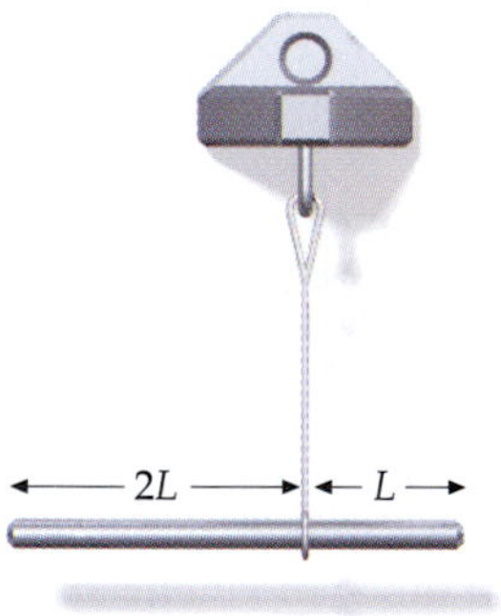

Figure 9-64 Problem 117

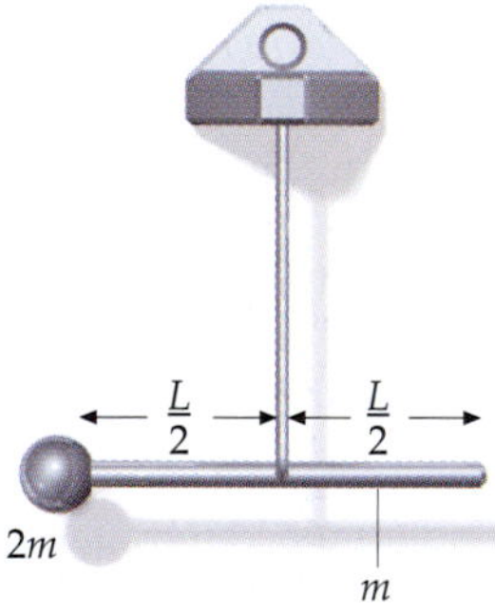

118 •• A marble of mass M and radius R rolls without slipping down the track on the left from a height h_1 as shown in Figure 9-65. The marble then goes up the *frictionless* track on the right to a height h_2. Find h_2.

Figure 9-65 Problem 118

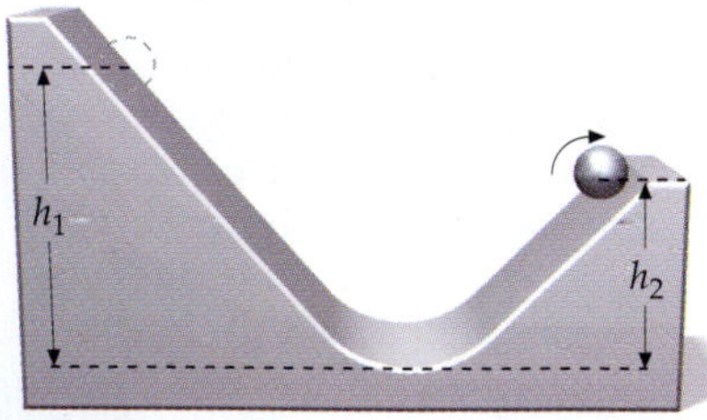

119 •• A uniform disk with a mass of 120 kg and a radius of 1.4 m rotates initially with an angular speed of 1100 rev/min. (*a*) A constant tangential force is applied at a radial distance of 0.6 m. What work must this force do to stop the wheel? (*b*) If the wheel is brought to rest in 2.5 min, what torque does the force produce? What is the magnitude of the force? (*c*) How many revolutions does the wheel make in these 2.5 min?

120 •• A park merry-go-round consists of a 240-kg circular wooden platform 4.00 m in diameter. Four children running alongside push tangentially along the platform's circumference until, starting from rest, the merry-go-round reaches a steady speed of one complete revolution every 2.8 s. (*a*) If each child exerts a force of 26 N, how far does each child run? (*b*) What is the angular acceleration of the merry-go-round? (*c*) How much work does each child do? (*d*) What is the kinetic energy of the merry-go-round?

121 •• A hoop of mass 1.5 kg and radius 65 cm has a string wrapped around its circumference and lies flat on a horizontal frictionless table. The string is pulled with a force of 5 N. (*a*) How far does the center of the hoop travel in 3 s? (*b*) What is the angular velocity of the hoop about its center of mass after 3 s?

122 •• A vertical grinding wheel is a uniform disk of mass 60 kg and radius 45 cm. It has a handle of radius 65 cm of negligible mass. A 25-kg load is attached to the handle when it is in the horizontal position. Neglecting friction, find (*a*) the initial angular acceleration of the wheel, and (*b*) the maximum angular velocity of the wheel.

123 •• In this problem, you are to derive the perpendicular-axis theorem for planar objects, which relates the moments of inertia about two perpendicular axes in the plane of Figure 9-66 to the moment of inertia about a third axis that is perpendicular to the plane of figure. Consider the mass element dm for the figure shown in the xy plane. (*a*) Write an expression for the moment of inertia of the figure about the z axis in terms of dm and r. (*b*) Relate the distance r of dm to the distances x and y, and thus show that $I_z = I_y + I_x$. (*c*) Apply your result to find the moment of inertia of a uniform disk of radius R about a diameter of the disk.

Figure 9-66 Problem 123

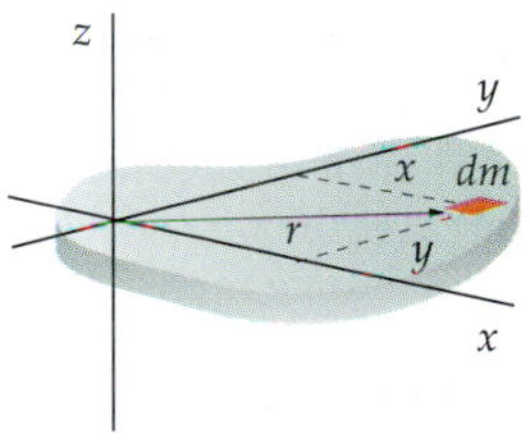

124 •• A uniform disk of radius r and mass M is pivoted about a horizontal axis parallel to its symmetry axis and passing through its edge such that it can swing freely in a vertical plane (Figure 9-67). It is released from rest with its center of mass at the same height as the pivot. (*a*) What is the angular velocity of the disk when its center of mass is directly below the pivot? (*b*) What force is exerted by the pivot at this time?

Figure 9-67 Problem 124

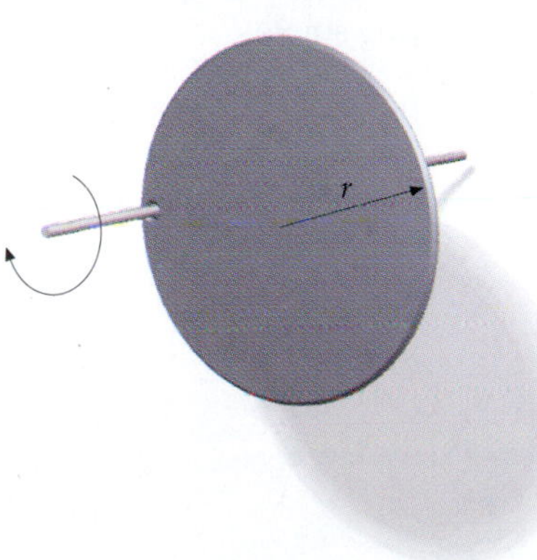

125 •• A spool of mass M rests on an inclined plane at a distance D from the bottom. The ends of the spool have radius R, the center has radius r, and the moment of inertia of the spool about its axis is I. A long string of negligible mass is wound many times around the center of the spool. The other end of the string is fastened to a hook at the top of the inclined plane such that the string always pulls parallel to the slope as shown in Figure 9-68. (*a*) Suppose that initially the slope is so icy that there is *no* friction. How does the spool move as it slips down the slope? Use energy considerations to determine the speed of the center of mass of the spool when it reaches the bottom of the slope. Give your answer in terms of M, I, r, R, g, D, and θ. (*b*) Now suppose that the ice is gone and that when the spool is set up in the same way, there is enough friction to keep it from slipping on the slope. What is the direction and magnitude of the friction force in this case?

Figure 9-68 Problem 125

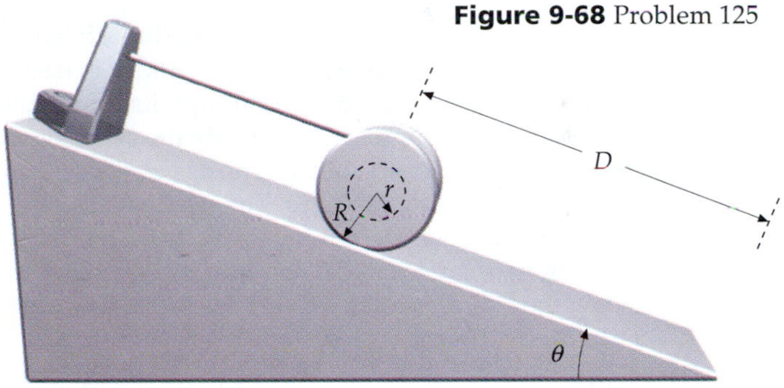

126 •• Ian has suggested another improvement for the game of hockey. Instead of the usual two-minute penalty, he would like to see an offender placed in a barrel at mid-ice and then spun in a circle by the other team. When the offender is silly with dizziness, he is put back into the game. Assume that a penalized player in a barrel approximates a uniform, 100-kg cylinder of radius 0.60 m, and that the ice is smooth (Figure 9-69). Ropes are wound around the barrel, so that pulling them causes rotation. If two players simultaneously pull the ropes with forces of 40 N and 60 N for 6 s, describe the motion of the barrel. Give its acceleration, velocity, and the position of its center of mass as functions of time.

Figure 9-69 Problem 126

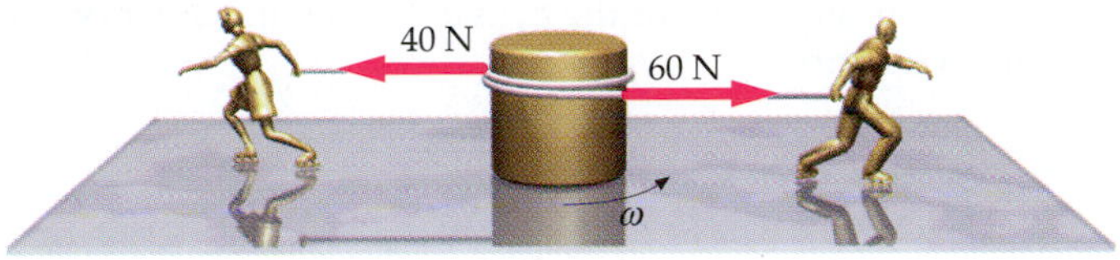

127 •• A solid metal rod 1.5 m long is free to rotate without friction about a fixed, horizontal axis perpendicular to the rod and passing through one end. The other end is held in a horizontal position. Small coins of mass m are placed on the rod 25 cm, 50 cm, 75 cm, 1 m, 1.25 m, and 1.5 m from the bearing. If the free end is now released, calculate the initial force exerted on each coin by the rod. Assume that the mass of the coins may be neglected in comparison to the mass of the rod.

128 •• A thin rod of length L and mass M is supported in a horizontal position by two strings, one attached to each end as shown in Figure 9-70. If one string is cut, the rod begins to rotate about the point where it connects to the other string (point A in the figure). (*a*) Find the initial acceleration of the center of mass of the rod. (*b*) Show that the initial tension in the string is $mg/4$ and that the initial angular acceleration of the rod about an axis through the point A is $3g/2L$. (*c*) At what distance from point A is the initial linear acceleration equal to g?

Figure 9-70 Problem 128

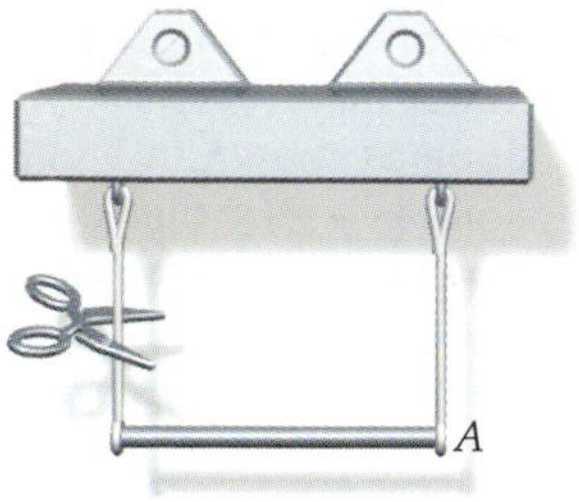

129 •• Figure 9-71 shows a hollow cylinder of length 1.8 m, mass 0.8 kg, and radius 0.2 m. The cylinder is free to rotate about a vertical axis that passes through its center and is perpendicular to the cylinder's axis. Inside the cylinder are two masses of 0.2 kg each, attached to springs of spring constant k and unstretched lengths 0.4 m. The inside walls of the cylinder are frictionless. (*a*) Determine the value of the spring constant if the masses are located 0.8 m from the center of the cylinder when the cylinder rotates at 24 rad/s. (*b*) How much work was needed to bring the system from $\omega = 0$ to $\omega = 24$ rad/s?

Figure 9-71 Problems 129 and 130

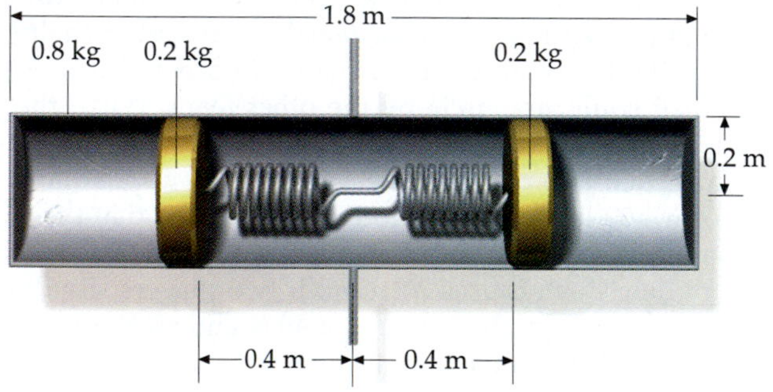

130 •• Suppose that for the system described in Problem 129, the spring constants are each $k = 60$ N/m. The system starts from rest and slowly accelerates until the masses are 0.8 m from the center of the cylinder. How much work was done in the process?

131 •• A string is wrapped around a uniform cylinder of radius R and mass M that rests on a horizontal frictionless surface. The string is pulled horizontally from the top with force F. (*a*) Show that the angular acceleration of the cylinder is twice that needed for rolling without slipping, so that the bottom point on the cylinder slides backward against the table. (*b*) Find the magnitude and direction of the frictional force between the table and cylinder needed for the cylinder to roll without slipping. What is the acceleration of the cylinder in this case?

132 •• Figure 9-72 shows a solid cylinder of mass M and radius R to which a hollow cylinder of radius r is attached. A string is wound about the hollow cylinder. The solid cylinder rests on a horizontal surface. The coefficient of static friction between the cylinder and surface is μ_s. If a light tension is applied to the string in the vertical direction, the cylinder will roll to the left; if the tension is applied with the string horizontally, the cylinder rolls to the right. Find the angle of the string with the horizontal that will allow the cylinder to remain stationary when a small tension is applied to the string.

Figure 9-72 Problem 132

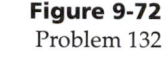

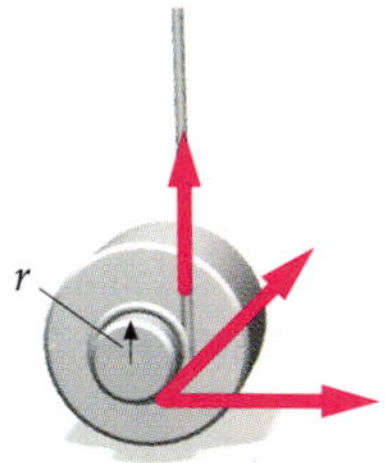

133 ••• A heavy, uniform cylinder has a mass m and a radius R (Figure 9-73). It is accelerated by a force $\vec{T}$, which is applied through a rope wound around a light drum of radius r that is attached to the cylinder. The coefficient of static friction is sufficient for the cylinder to roll without slipping. (*a*) Find the frictional force. (*b*) Find the acceleration a of the center of the cylinder. (*c*) Is it possible to choose r so that a is greater than T/m? How? (*d*) What is the direction of the frictional force in the circumstances of part (*c*)?

Figure 9-73 Problem 133

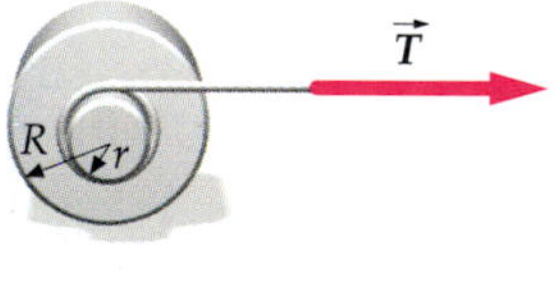

134 ••• A uniform stick of length L and mass M is hinged at one end. It is released from rest at an angle θ_0 with the vertical. Show that when the angle with the vertical is θ, the hinge exerts a force F_r along the stick and a force F_t perpendicular to the stick given by $F_r = \frac{1}{2}Mg(5\cos\theta - 3\cos\theta_0)$ and $F_t = \frac{1}{4}Mg\sin\theta$.

CHAPTER 10

Conservation of Angular Momentum

Waterspouts off the Grand Bahama Islands offer a stunning visualization of rotational motion.

In this chapter, we extend our study of rotational motion to situations in which the direction of the axis of rotation may change. We begin with an examination of the vector properties of angular velocity and torque and then we introduce the concept of angular momentum, which is the rotational analog of linear momentum. We then show that the net torque acting on a system equals the rate of change of its angular momentum, a result that is equivalent to Newton's second law for rotational motion. Angular momentum is therefore conserved in systems with zero net torque. Like conservation of linear momentum, conservation of angular momentum is a fundamental law of nature, applying even in the atomic domain where Newtonian mechanics fails.

10-1 The Vector Nature of Rotation

To indicate the direction of rotation about a *fixed* axis, we assigned plus and minus signs to the angular velocity vector, just as we used them to indicate the direction of the velocity vector in one-dimensional motion. When the direction of the axis of rotation is not fixed in space, the vector nature of angular velocity becomes important. Consider, for example, the rotating disk in Figure 10-1. We describe the direction of rotation by giving the direction of

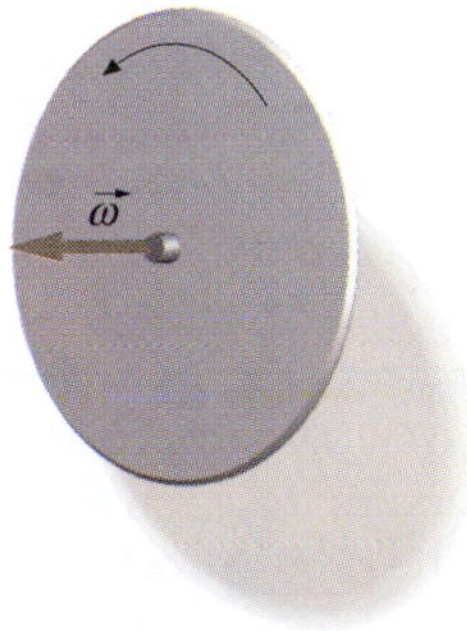

Figure 10-1 A disk rotating about an axis through its center and perpendicular to its plane.

the axis of rotation. (By symmetry, all directions in the plane of the disk are equivalent.) We therefore choose the angular velocity vector $\vec{\omega}$ to be along the axis of rotation, and we determine the direction of $\vec{\omega}$ by a convention known as the **right-hand rule**, which is illustrated in Figure 10-2. Thus, if the rotation is counterclockwise, as in Figure 10-1, $\vec{\omega}$ is outward; if it is clockwise, $\vec{\omega}$ is inward.

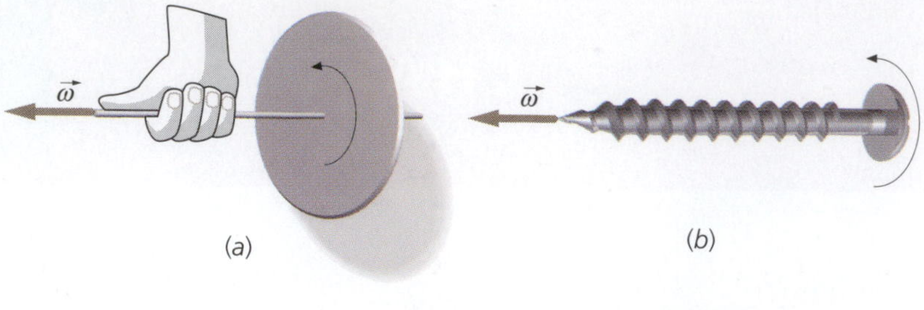

Figure 10-2 The right-hand rule for determining the direction of the angular velocity $\vec{\omega}$. (*a*) When the fingers of the right hand curl in the direction of rotation, the thumb points in the direction of $\vec{\omega}$. (*b*) Looked at another way, the direction of $\vec{\omega}$ is that of the advance of a rotating right-hand screw.

We apply similar considerations to the torque. Figure 10-3 shows a force $\vec{F}$ acting on a particle at some position $\vec{r}$ relative to the origin O. The torque $\vec{\tau}$ exerted by this force relative to the origin O is defined as a vector that is perpendicular to the plane formed by $\vec{F}$ and $\vec{r}$, and has magnitude $Fr \sin \phi$, where ϕ is the angle between $\vec{F}$ and $\vec{r}$. If $\vec{F}$ and $\vec{r}$ are in the xy plane, as in Figure 10-3, the torque is along the z axis. If $\vec{F}$ is applied to the rim of a disk of radius r, as shown in Figure 10-4, the torque has the magnitude Fr, and is along the axis of rotation as shown.

Figure 10-3 Force $\vec{F}$ acting on a particle at position $\vec{r}$.

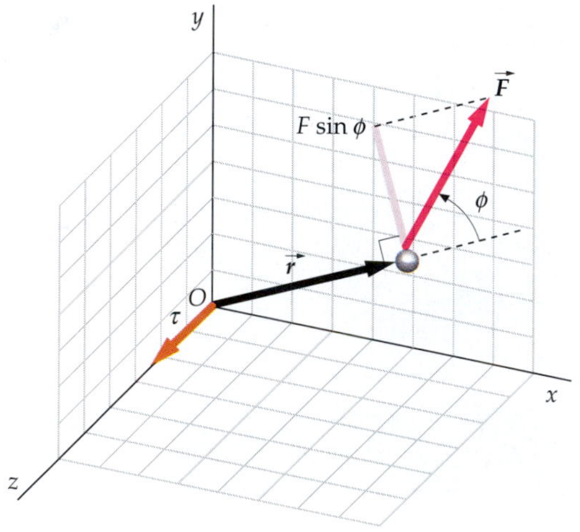

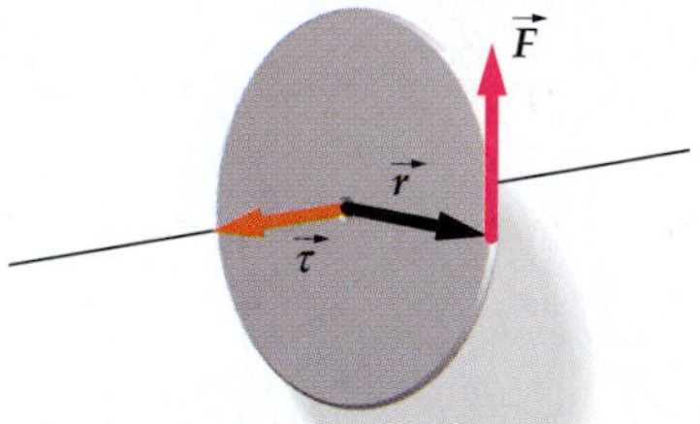

Figure 10-4 The force $\vec{F}$, which is tangent to the disk, exerts a torque along the axis of rotation.

The Cross Product

Torque is expressed mathematically as the **cross product** (or **vector product**) of $\vec{r}$ and $\vec{F}$:

$$\vec{\tau} = \vec{r} \times \vec{F} \qquad \text{10-1}$$

The cross product of two vectors $\vec{A}$ and $\vec{B}$ is defined to be a vector $\vec{C} = \vec{A} \times \vec{B}$ whose magnitude equals the area of the parallelogram formed by the two vectors (Figure 10-5). The vector $\vec{C}$ is perpendicular to the plane containing $\vec{A}$ and $\vec{B}$ in the direction given by the right-hand rule as $\vec{A}$ is rotated into $\vec{B}$ through the smaller angle between these vectors (Figure 10-6). If ϕ is the angle between the two vectors and $\hat{n}$ is a unit vector that is perpendicular to each vector in the direction described, the cross product of $\vec{A}$ and $\vec{B}$ is

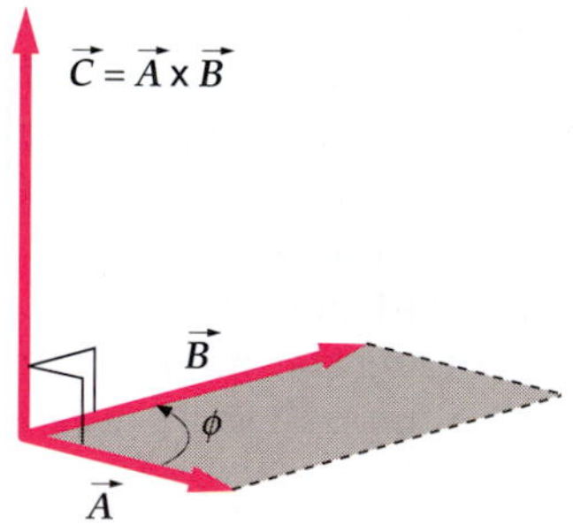

Figure 10-5 The cross product $\vec{A} \times \vec{B}$ is a vector $\vec{C}$ that is perpendicular to both $\vec{A}$ and $\vec{B}$ and has a magnitude $AB \sin \phi$, which equals the area of the parallelogram shown.

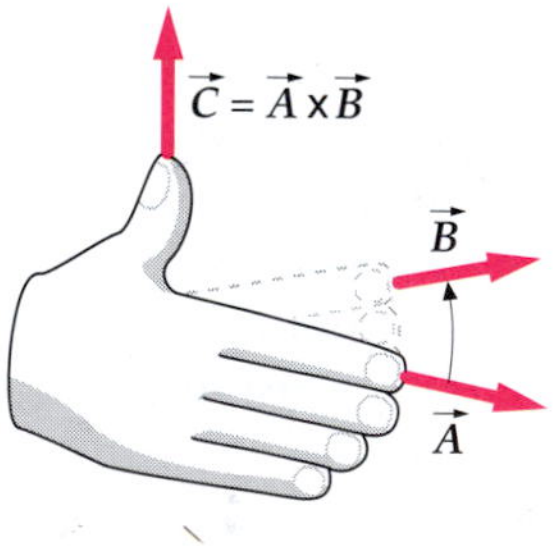

Figure 10-6 The direction of $\vec{A} \times \vec{B}$ is given by the right-hand rule when $\vec{A}$ is rotated into $\vec{B}$ through the angle ϕ.

$$\vec{A} \times \vec{B} = AB \sin \phi \, \hat{n} \qquad \text{10-2}$$

Definition—Cross product

If $\vec{A}$ and $\vec{B}$ are parallel, $\vec{A} \times \vec{B}$ is zero. It follows from the definition of the cross product that

$$\vec{A} \times \vec{A} = 0 \qquad \text{10-3}$$

and

$$\vec{A} \times \vec{B} = -\vec{B} \times \vec{A} \qquad \text{10-4}$$

Note that the order in which two vectors are multiplied in a cross product is significant. Below are some properties of the cross product of two vectors:

1. The cross product obeys a distributive law under addition:

$$\vec{A} \times (\vec{B} + \vec{C}) = \vec{A} \times \vec{B} + \vec{A} \times \vec{C} \qquad \text{10-5}$$

2. If $\vec{A}$ and $\vec{B}$ are functions of some variable such as t, the derivative of $\vec{A} \times \vec{B}$ follows the usual product rule for derivatives:

$$\frac{d}{dt}(\vec{A} \times \vec{B}) = \vec{A} \times \frac{d\vec{B}}{dt} + \frac{d\vec{A}}{dt} \times \vec{B} \qquad \text{10-6}$$

3. The unit vectors $\hat{i}$, $\hat{j}$, and $\hat{k}$ (Figure 10-7), which are mutually perpendicular, have cross products given by

$$\hat{i} \times \hat{j} = \hat{k}, \qquad \hat{j} \times \hat{k} = \hat{i}, \qquad \hat{k} \times \hat{i} = \hat{j} \qquad \text{10-7a}$$

Furthermore,

$$\hat{i} \times \hat{i} = \hat{j} \times \hat{j} = \hat{k} \times \hat{k} = 0 \qquad \text{10-7b}$$

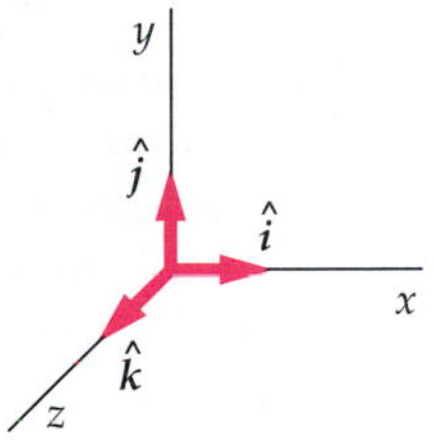

Figure 10-7 The unit vectors $\hat{i}$, $\hat{j}$, and $\hat{k}$ are mutually perpendicular and have magnitude 1.

10-2 Angular Momentum

Figure 10-8 shows a particle of mass m moving with a velocity $\vec{v}$ at a position $\vec{r}$ relative to the origin O. The linear momentum of the particle is $\vec{p} = m\vec{v}$. The **angular momentum** $\vec{L}$ of the particle relative to the origin O is defined to be the cross product of $\vec{r}$ and $\vec{p}$:

$$\vec{L} = \vec{r} \times \vec{p} \qquad \text{10-8}$$

Definition—Angular momentum of a particle

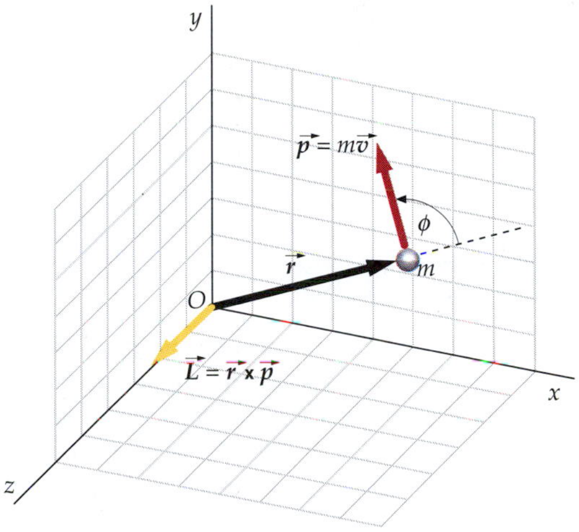

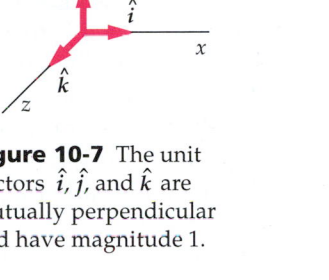

Figure 10-8 A particle with a momentum $\vec{p}$ at position $\vec{r}$ relative to the origin O.

If $\vec{r}$ and $\vec{p}$ are in the xy plane, as in Figure 10-8, then $\vec{L}$ is along the z axis and is given by $\vec{L} = \vec{r} \times \vec{p} = mvr \sin \phi \, \hat{k}$. Like torque, angular momentum is defined *relative to a point in space.*

Figure 10-9 shows a particle moving in a circle in the xy plane with the center of the circle at the origin. The speed v of the particle and the magnitude of its angular velocity ω are related by $v = r\omega$. The angular momentum of the particle relative to the center of the circle is

$$\vec{L} = \vec{r} \times \vec{p} = \vec{r} \times m\vec{v} = rmv \sin 90^\circ \, \hat{k} = rmv\hat{k} = mr^2\omega\hat{k} = mr^2\vec{\omega}$$

The angular momentum is in the same direction as the angular velocity.

Since mr^2 is the moment of inertia for a single particle about the z axis, we have

$$\vec{L} = mr^2\vec{\omega} = I\vec{\omega}$$

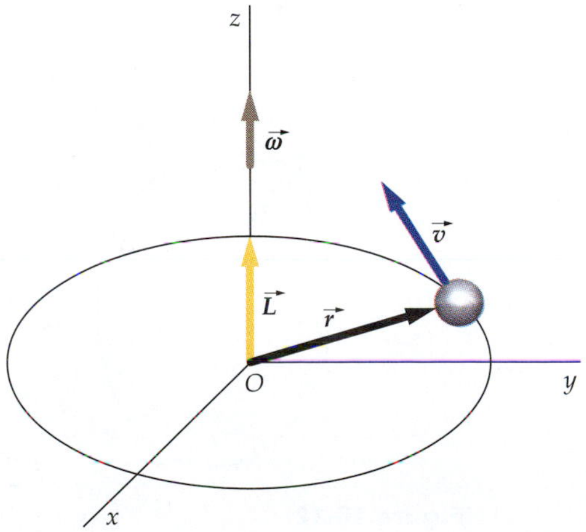

Figure 10-9 A particle moving in a circle has angular momentum relative to the center of the circle $\vec{L} = \vec{r} \times \vec{p} = I\vec{\omega}$.

This result does not hold for a general point. Figure 10-10 shows the angular-momentum vector $\vec{L}'$ for the same particle moving in the same circle but with $\vec{L}'$ computed relative to a point on the z axis that is not at the center of the circle. In this case, the angular momentum is not parallel to the angular velocity $\vec{\omega}$, which is along the z axis.

In Figure 10-11, we add a second particle of equal mass moving in the same circle. The angular-momentum vectors $\vec{L}_1'$ and $\vec{L}_2'$ are shown relative to the same point as in Figure 10-10. The total angular momentum $\vec{L}_1' + \vec{L}_2'$ of the two-particle system is again parallel to the angular velocity $\vec{\omega}$. In this case, the axis of rotation, the z axis, passes through the center of mass of the two-particle system, and the mass distribution is symmetric about this axis. Such an axis is called a **symmetry axis**. For any system of particles that rotates about a symmetry axis, the total angular momentum (which is the sum of the angular momenta of the individual particles) is parallel to the angular velocity and is given by

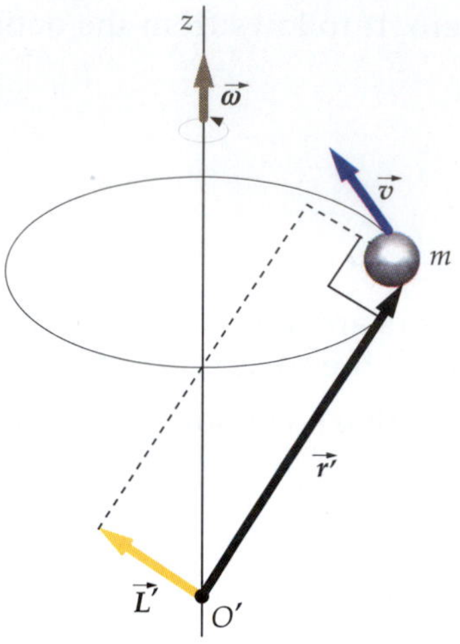

Figure 10-10

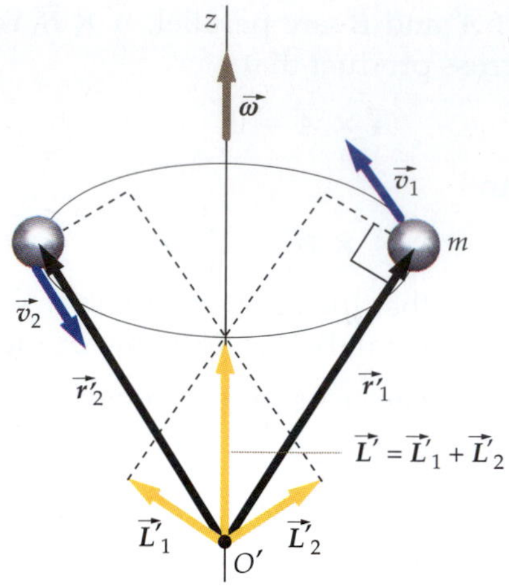

Figure 10-11

$$\vec{L} = I\vec{\omega} \qquad 10\text{-}9$$

Angular momentum of a system rotating about a symmetry axis

Example 10-1

Find the angular momentum about the origin for the following situations. (*a*) A car of mass 1200 kg moves counterclockwise in a circle of radius 20 m with speed 15 m/s. (*b*) The same car moves with velocity $\vec{v} = -(15\text{ m/s})\hat{i}$ along the line $y = y_0 = 20$ m parallel to the x axis. (*c*) A disk in the xy plane of radius 20 m and mass 1200 kg rotates at 0.75 rad/s about its axis, which is the z axis.

Picture the Problem

(*a*) $\vec{r}$ and $\vec{p}$ are perpendicular, and $\vec{r} \times \vec{p}$ is in the z direction:

$$\vec{L} = \vec{r} \times \vec{p} = rmv\hat{k} = (20\text{ m})(1200\text{ kg})(15\text{ m/s})\hat{k}$$
$$= 3.6 \times 10^5\text{ kg}\cdot\text{m}^2/\text{s }\hat{k}$$

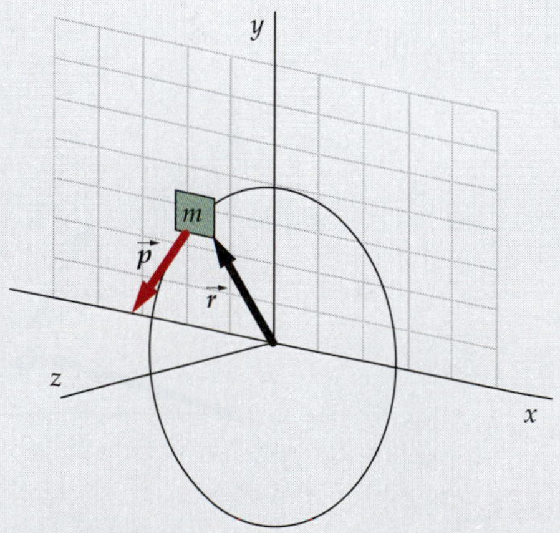

Figure 10-12

(*b*) 1. For the same car moving to the left along the line $y = y_0 =$ 20 m, we express $\vec{r}$ and $\vec{p}$ in terms of unit vectors:

$$\vec{r} = x\hat{i} + y\hat{j} = x\hat{i} + y_0\hat{j}$$
$$\vec{p} = m\vec{v} = -p\hat{i}$$

Figure 10-13

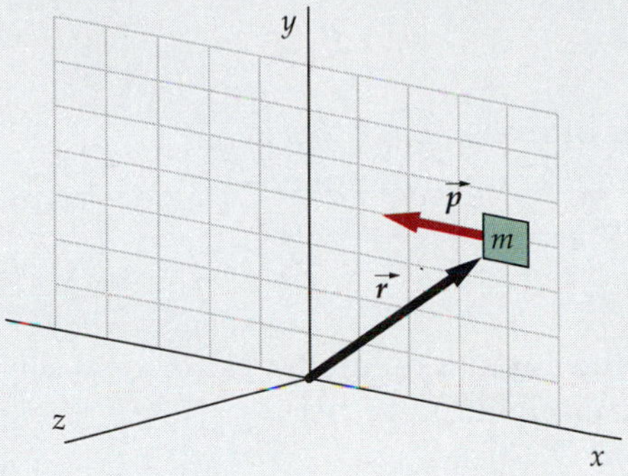

2. Now compute $\vec{r} \times \vec{p}$:

$$\begin{aligned}\vec{L} = \vec{r} \times \vec{p} &= (x\hat{i} + y_0\hat{j}) \times (-p\hat{i})\\ &= -xp(\hat{i} \times \hat{i}) - y_0p(\hat{j} \times \hat{i})\\ &= -xp(0) - y_0p(-\hat{k}) = y_0p\hat{k}\\ &= (20\text{ m})(1200\text{ kg})(15\text{ m/s})\hat{k}\\ &= 3.6 \times 10^5\text{ kg·m}^2\text{/s }\hat{k}\end{aligned}$$

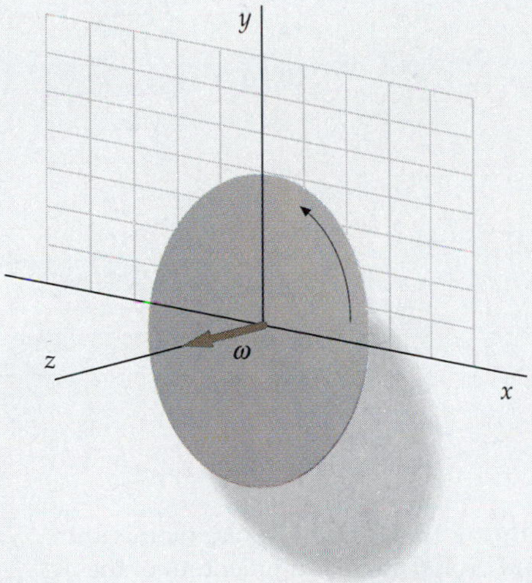

(*c*) Use $\vec{L} = I\vec{\omega}$:

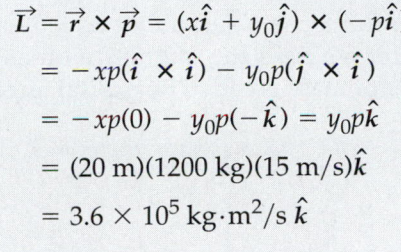

$$\vec{L} = I\vec{\omega} = I\omega\hat{k} = \left(\frac{1}{2}mR^2\right)\omega\hat{k}$$
$$= \frac{1}{2}(1200\text{ kg})(20\text{ m})^2(0.75\text{ rad/s})\hat{k}$$
$$= 1.8 \times 10^5\text{ kg·m}^2\text{/s}\hat{k}$$

Figure 10-14

Remarks The angular momentum of the car moving in a circle in (*a*) is the same as that of the car moving along a straight line in (*b*). In (*c*), the velocity of a point on the rim is $v = R\omega = (20\text{ m})(0.75\text{ rad/s}) = 15\text{ m/s}$, the same as the velocity of the car in parts (*a*) and (*b*). The moment of inertia of a 1200-kg disk of radius 20 m is less than that of a 1200-kg car at 20 m from the axis because much of the mass of the disk is closer to the axis of rotation.

Figure 10-15 shows a disk rotating about an axis parallel to its symmetry axis a distance h away. The angular momentum $\vec{L}_i$ of the ith particle on the disk relative to point O' is in the z direction, parallel to $\vec{\omega}$, and given by

$$\vec{L}_i = \vec{r}_i \times m_i\vec{v}_i = m_iv_ir_i\hat{k} = m_ir_i^2\omega\hat{k}$$

Summing over all the particles yields

$$\vec{L} = \sum \vec{L}_i = \sum m_ir_i^2\omega\hat{k} = I'\omega\hat{k} = I'\vec{\omega}$$

where I' is the moment of inertia of the disk about the axis through O'. Equation 10-9 thus also holds for rotation about an axis parallel to a symmetry axis. I' and I are related by the parallel-axis theorem (Equation 9-20):

$$I' = I_{cm} + Mh^2$$

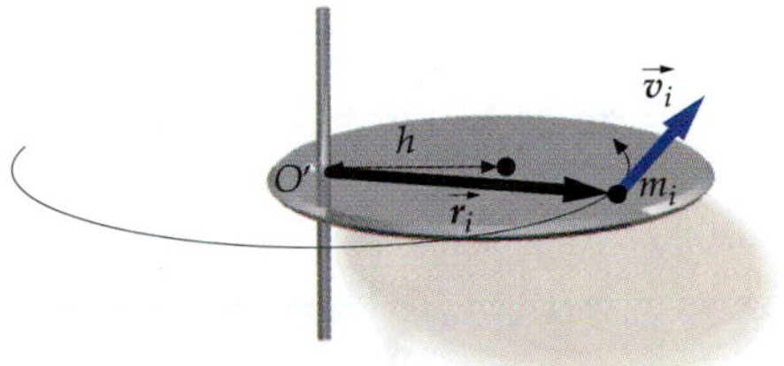

Figure 10-15 A disk rotating about an axis through point O' a distance h from the parallel symmetry axis.

Then

$$\vec{L} = I\omega\hat{k} = (I_{cm} + Mh^2)\omega\hat{k} = I_{cm}\omega\hat{k} + Mh^2\omega\hat{k}$$

or

$$\vec{L} = \vec{L}_{cm} + Mv_{cm}h\hat{k}$$

where $\vec{L}_{cm}$ is the angular momentum about the center of mass, $v_{cm} = h\omega$ is the velocity of the center of mass, and $Mv_{cm}h\hat{k}$ is the angular momentum relative to O' of a particle of mass M moving with speed v_{cm}. This result, obtained for rotation about a fixed axis, holds in general:

> The angular momentum about any point O' is the angular momentum about the center of mass plus the angular momentum associated with the motion of the center of mass about O'.

The angular momentum of an object about its center of mass is called its **spin angular momentum,** whereas the angular momentum associated with the motion of the center of mass is called its **orbital angular momentum.**

$$\vec{L} = \vec{L}_{orbit} + \vec{L}_{spin} = \vec{r}_{cm} \times M\vec{v}_{cm} + \sum_i \vec{r}_i' \times m_i\vec{u}_i \qquad \text{10-10}$$

where $\vec{u}_i$ is the velocity of the ith particle relative to the center of mass. The earth has spin angular momentum due to its rotation about its axis and orbital angular momentum relative to the sun due to its revolving around the sun (Figure 10-16). The total angular momentum of the earth relative to the sun is the vector sum of its spin and orbital angular momenta.

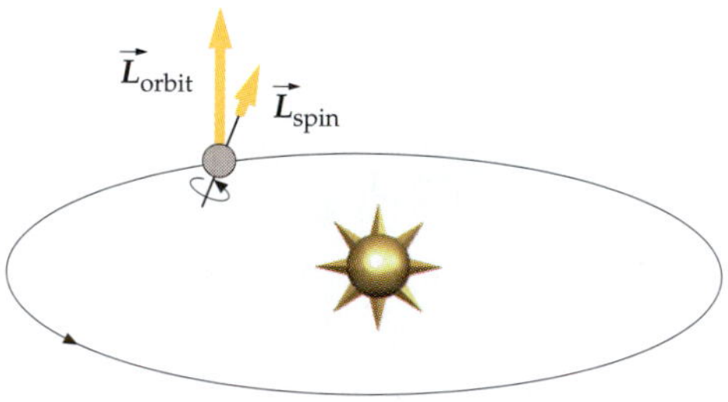

Figure 10-16 Spin angular momentum and orbital angular momentum of the earth.

10-3 Torque and Angular Momentum

We will now show that Newton's second law implies that the rate of change of the angular momentum of a particle equals the net torque acting on the particle. For a number of forces acting on a particle, the net torque relative to the origin O is the sum of the torques due to each force:

$$\vec{\tau}_{net} = \vec{r} \times \vec{F}_1 + \vec{r} \times \vec{F}_2 + \cdots = \vec{r} \times \sum_i \vec{F}_i = \vec{r} \times \vec{F}_{net}$$

According to Newton's second law, the net force equals the rate of change of the linear momentum $d\vec{p}/dt$. Thus,

$$\vec{\tau}_{net} = \vec{r} \times \vec{F}_{net} = \vec{r} \times \frac{d\vec{p}}{dt} \qquad \text{10-11}$$

We now compare this with the rate of change of the angular momentum. We can compute $d\vec{L}/dt$ using the product rule for derivatives:

$$\frac{d\vec{L}}{dt} = \frac{d}{dt}(\vec{r} \times \vec{p}) = \frac{d\vec{r}}{dt} \times \vec{p} + \vec{r} \times \frac{d\vec{p}}{dt}$$

The first term on the right of this equation is zero because

$$\frac{d\vec{r}}{dt} \times \vec{p} = \vec{v} \times m\vec{v} = 0$$

Thus

$$\frac{d\vec{L}}{dt} = \vec{r} \times \frac{d\vec{p}}{dt} \qquad 10\text{-}12$$

Comparing this result with Equation 10-11 gives

$$\vec{\tau}_{\text{net}} = \frac{d\vec{L}}{dt} \qquad 10\text{-}13$$

The net torque acting on a system of particles is the sum of the individual torques. The generalization of Equation 10-13 to a system of particles is then

$$\sum_i \vec{\tau} = \sum_i \frac{d\vec{L}_i}{dt} = \frac{d}{dt}\sum_i \vec{L}_i = \frac{d\vec{L}}{dt}$$

In this equation, the sum of the torques may include internal as well as external torques. We show in Section 10-4 that the sum of the internal torques must be zero. Therefore,

$$\sum_i \vec{\tau}_{i,\text{ext}} = \frac{d\vec{L}}{dt} \qquad 10\text{-}14$$

The net external torque acting on a system equals the rate of change of the angular momentum of the system.

Newton's second law for rotation

Equation 10-14 is the rotational analog of $\vec{F}_{\text{net,ext}} = d\vec{p}/dt$ for linear motion. It holds for a general system of particles, rotating about any axis, whether or not the moment of inertia is constant. For a rigid body rotating about a fixed axis, the moment of inertia is constant and Equation 10-14 becomes

$$\sum_i \vec{\tau}_{i,\text{ext}} = \frac{d\vec{L}}{dt} = \frac{d(I\vec{\omega})}{dt} = I\frac{d\vec{\omega}}{dt} = I\vec{\alpha} \qquad 10\text{-}15$$

where $\vec{\alpha} = d\vec{\omega}/dt$ is the angular acceleration vector. Equation 10-15 is the same as Equation 9-20.

The direction of rotation is changed by this bevel gear in a diesel engine.

Example 10-2

An Atwood's machine has two blocks of mass m_1 and m_2 ($m_1 > m_2$), connected by a string of negligible mass that passes over a pulley with frictionless bearings (Figure 10-17). The pulley is a uniform disk of mass M and radius R. The string does not slip on the pulley. Apply Equation 10-14 to the system consisting of both blocks plus pulley to find the angular acceleration of the pulley and the acceleration of the blocks.

Picture the Problem Let the pulley and blocks be in the xy plane with the z axis out of the page. We compute the torques and angular momenta about the center of the pulley. Since m_1 is greater than m_2, the disk will rotate counterclockwise corresponding to $\vec{\omega}$ out of the page in the positive z direction.

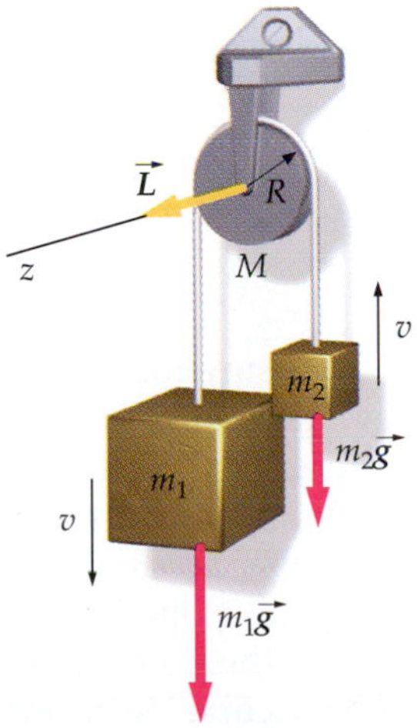

Figure 10-17

The weight of m_1 exerts a torque m_1gR out of the page, and the weight of m_2 exerts a torque m_2gR into the page. Since m_1 moves down and m_2 moves up, they both have angular momentum about the center of the pulley directed outward. Since the torque, angular-velocity, and angular-momentum vectors are all along the z axis, we can forget about their vector nature and treat the problem like a one-dimensional problem with positive assigned to counterclockwise motion and negative to clockwise motion. The speed v of the blocks is related to the angular speed of the pulley ω by the nonslip condition $v = R\omega$.

1. Use $\Sigma\,\vec{\tau}_{i,\text{ext}} = d\vec{L}/dt$:

$$\sum \vec{\tau}_{i,\text{ext}} = \frac{d\vec{L}}{dt}$$

2. The total angular momentum about the center of the pulley equals the angular momentum of the pulley plus the angular momentum of the blocks, each of which is in the positive z direction:

$$L_z = L_p + L_1 + L_2 = I\omega + m_1vR + m_2vR$$

3. The weight $m_1\vec{g}$ exerts a torque in the positive z direction, whereas $m_2\vec{g}$ exerts a torque in the negative z direction. The lever arm for each force is R. The net torque is:

$$\tau_{z,\text{net}} = m_1gR - m_2gR$$

4. Substitute these results into Newton's second law for rotation in step 1:

$$m_1gR - m_2gR = \frac{dL_z}{dt} = \frac{d}{dt}(I\omega + m_1vR + m_2vR)$$

$$= I\alpha + (m_1 + m_2)Ra$$

5. Relate I to M and R, and use the nonslip condition to relate α to a:

$$m_1gR - m_2gR = \frac{1}{2}MR^2\frac{a}{R} + (m_1 + m_2)Ra$$

6. Solve for a:

$$a = \frac{m_1 - m_2}{\frac{1}{2}M + m_1 + m_2}g$$

Remarks This problem could be solved by writing the tensions T_1 on the left and T_2 on the right and using $\tau = I\alpha$ for the pulley and $\Sigma\vec{F} = m\vec{a}$ for each block. If you do not need to know the tensions, using angular momentum is easier. Note that the linear momenta of the blocks are in opposite directions, but their angular momenta are in the same direction.

There are many problems in which the forces, position vectors, and velocities all lie in a plane, so that the torques, angular velocities, and angular-momentum vectors are all along the axis of rotation that remains fixed in space. In such cases, we can assign positive and negative values to counterclockwise or clockwise rotations, as we did in Example 10-2, and treat the case like a one-dimensional problem. However, there are other situations, such as the motion of a gyroscope (discussed in the "Exploring . . . Motion of a Gyroscope" section) where the vector natures of torque, angular velocity, and angular momentum are important.

Exploring

Motion of a Gyroscope

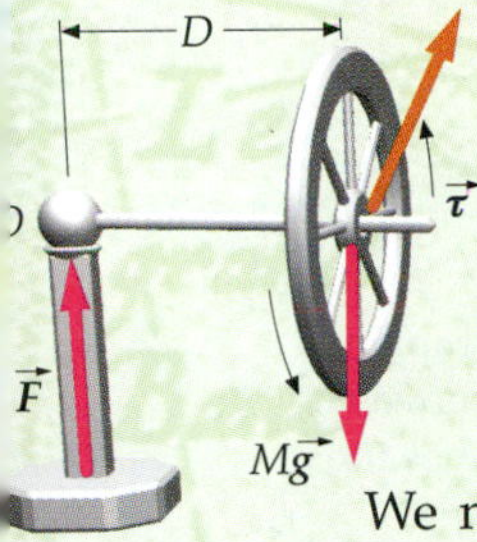

Figure 1

A gyroscope is a common example of motion in which the axis of rotation changes direction. Figure 1 shows a gyroscope that is free to turn on its axle. The axle is pivoted at a point a distance D from the center of the wheel, and is free to turn in any direction. We now give a qualitative understanding of the complex motion of such a system by using Newton's second law for rotation,

$$\vec{\tau}_{\text{net}} = \frac{d\vec{L}}{dt} \quad \text{or} \quad d\vec{L} = \vec{\tau}_{\text{net}}dt \qquad 1$$

along with the relations

$$\vec{\tau}_{\text{net}} = \vec{r} \times M\vec{g} \qquad 2$$

and

$$\vec{L} = I_s\vec{\omega}_s \qquad 3$$

where I_s and $\vec{\omega}_s$ are the moment of inertia and angular velocity of the wheel about its spin axis. All we really need to remember in order to describe the motion of a gyroscope is that the *change* in angular momentum of the wheel must be in the direction of the net torque acting on it.

Suppose the axle is held horizontally and then released. If the wheel isn't spinning, it simply falls, rotating about a horizontal axis through O and perpendicular to $\vec{r}$. The torque is horizontal, into the page. For this case, the *change* in angular momentum equals the angular momentum itself, which, in our example, is just $\vec{r} \times M\vec{v}_{\text{cm}}$. However, if the wheel *is* spinning and has a large angular momentum along its axle, it does not fall when the axle is released. If it were to fall, the axle would point downward, resulting in a large component of angular momentum in the downward direction. But there is no torque in the downward direction; the torque is horizontal. What actually happens is that the axle moves horizontally (into the paper in Figure 1). The wheel must move this way so that the *change* in angular momentum is in the direction of the net torque. This is illustrated in Figure 2*a*, where we see a large angular momentum along the axis of the wheel and a change in angular momentum $d\vec{L}$ in the direction of the torque. This motion, which is always surprising when first encountered, is called **precession**. We can calculate the angular velocity of precession. In a small time interval dt, the change in the angular momentum has a magnitude dL:

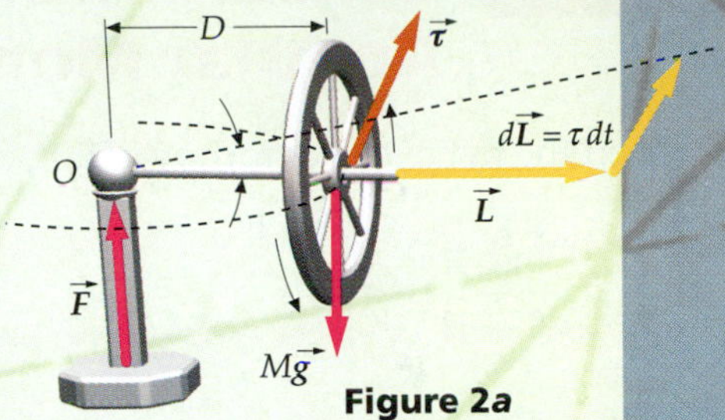

Figure 2*a*

$$dL = \tau\, dt = MgD\, dt$$

where MgD is the magnitude of the torque about the pivot point. From Figure 2*b*, the angle $d\phi$ through which the axle moves is

$$d\phi = \frac{dL}{L} = \frac{MgD\, dt}{L}$$

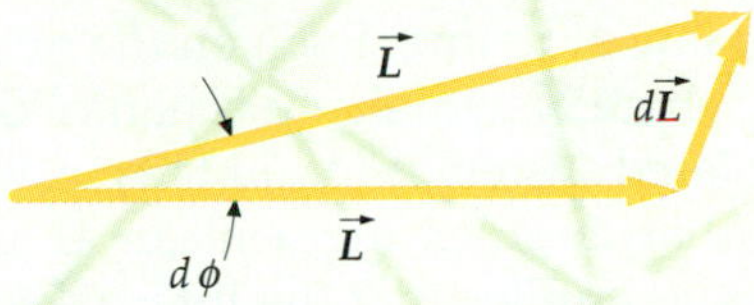

Figure 2*b*

The angular velocity of the precession is thus

$$\omega_p = \frac{d\phi}{dt} = \frac{MgD}{L} = \frac{MgD}{I_s\omega_s} \qquad 4$$

If the angular momentum due to the spin of the wheel is large, the precession can be very slow.

If you perform this experiment, you will notice a small up-and-down oscillation of the axle. This motion is called **nutation**. Suppose the wheel is simply released at the beginning. The bulk motion of the wheel as it precesses results in a component of angular momentum in the upward direction. Since there is no torque in that direction, the axle must initially dip down slightly to make the vertical component of angular momentum zero. Close analysis shows that it first dips down, then overshoots and moves upward, and then continues oscillating (nutating). The nutation can be avoided by giving the wheel a slight push in the direction of precession when released. This provides the upward torque needed for the upward angular momentum. If it is pushed too hard, the axle will move upward initially and then nutate.

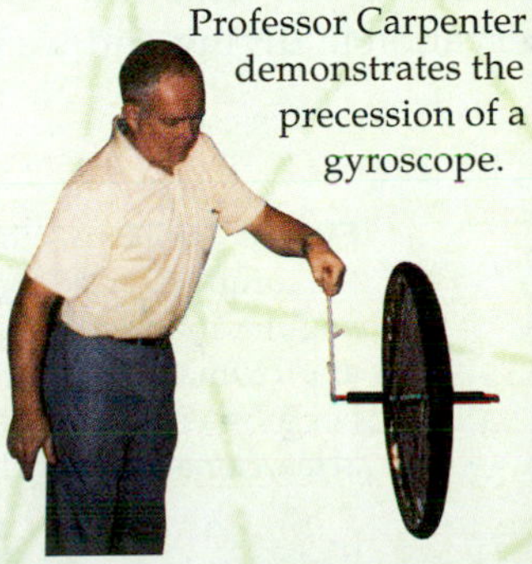

Professor Carpenter demonstrates the precession of a gyroscope.

10-4 Conservation of Angular Momentum

When the net external torque acting on a system is zero, we have

$$\frac{d\vec{L}}{dt} = 0$$

or

$$\vec{L} = \text{constant} \qquad 10\text{-}16$$

Equation 10-16 is a statement of the **law of conservation of angular momentum**:

> If the net external torque acting on a system is zero, the total angular momentum of the system is constant.

Conservation of angular momentum

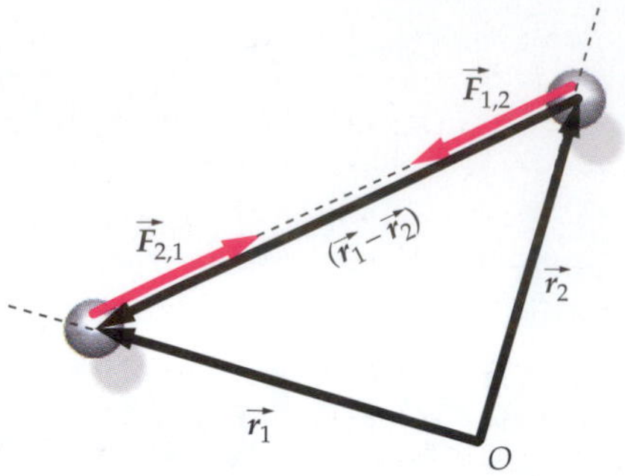

Figure 10-18 If the internal forces exerted by one particle on another are along the line joining the particles, the net torque exerted by the forces is zero about any point O.

This is the rotational analog of the law of conservation of linear momentum. If a system is isolated from its surroundings, so that there are no external forces or torques acting on it, three quantities are conserved: energy, linear momentum, and angular momentum. The law of conservation of angular momentum is a fundamental law of nature. Even on the microscopic scale of atomic and nuclear physics, where Newtonian mechanics does not hold, the angular momentum of an isolated system is found to be constant over time.

The experimental result that angular momentum is conserved in the absence of a net external torque implies that the internal torques must sum to zero. This fact is suggested by Newton's third law. Consider the two particles shown in Figure 10-18. Let $\vec{F}_{1,2}$ be the force exerted by particle 1 on particle 2, and $\vec{F}_{2,1}$ be that exerted by particle 2 on particle 1. By Newton's third law, $\vec{F}_{2,1} = -\vec{F}_{1,2}$. The sum of the torques exerted by these forces about the origin O is

$$\vec{\tau}_1 + \vec{\tau}_2 = \vec{r}_1 \times \vec{F}_{2,1} + \vec{r}_2 \times \vec{F}_{1,2}$$
$$= \vec{r}_1 \times \vec{F}_{2,1} + \vec{r}_2 \times (-\vec{F}_{1,2}) = (\vec{r}_1 - \vec{r}_2) \times \vec{F}_{2,1}$$

The vector $\vec{r}_1 - \vec{r}_2$ is along the line joining the two particles. If $\vec{F}_{2,1}$ acts parallel to the line joining m_1 and m_2, $\vec{F}_{2,1}$ and $\vec{r}_1 - \vec{r}_2$ are either parallel or antiparallel and

$$(\vec{r}_1 - \vec{r}_2) \times \vec{F}_{2,1} = 0$$

If this is true for all the internal forces, the internal torques cancel in pairs.

There are many examples of the conservation of angular momentum in everyday life. Figures 10-19 and 10-20 illustrate angular momentum conservation in diving and ice skating.

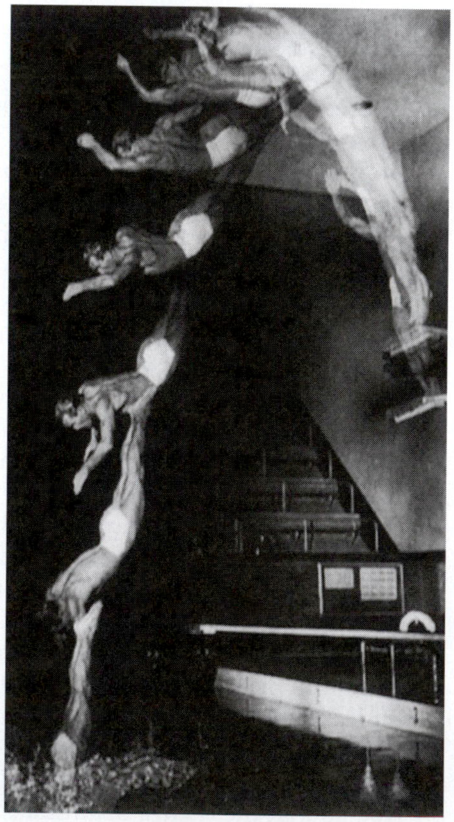

Figure 10-19 Multiflash photograph of a diver. The diver's center of mass moves along a parabolic path after he leaves the board. The angular momentum is provided by the initial external torque due to the force of the board, which does not pass through the diver's center of mass if he leans forward as he jumps. If the diver wanted to undergo one or more somersaults in the air, he would draw in his arms and legs, decreasing his moment of inertia to increase his angular velocity.

Figure 10-20 A spinning skater. Because the torque exerted by the ice is small, the angular momentum of the skater is approximately constant. When she reduces her moment of inertia by drawing in her arms, her angular velocity increases.

Example 10-3

A disk is rotating with an initial angular speed ω_i about a frictionless shaft through its symmetry axis as shown in Figure 10-21. Its moment of inertia about this axis is I_1. It drops onto another disk of moment of inertia I_2 that is initially at rest on the same shaft. Because of surface friction, the two disks eventually attain a common angular speed ω_f. Find ω_f.

Picture the Problem We find the final angular speed from the final angular momentum, which is equal to the initial angular momentum because there are no external torques acting on the system. Note that we do *not* use conservation of mechanical energy. The angular speed of the upper disk is reduced while that of the lower disk is increased by the force of kinetic friction that acts between the surfaces. We therefore expect that the total mechanical energy is decreased.

Figure 10-21

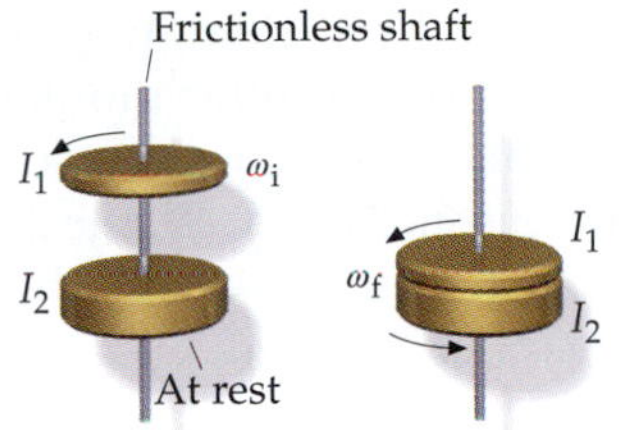

1. The final angular speed is related to the initial angular speed by conservation of angular momentum:	$L_f = L_i$ $(I_1 + I_2)\omega_f = I_1\omega_i$
2. Solve for the final angular speed:	$\omega_f = \dfrac{I_1}{I_1 + I_2}\omega_i$

Check the Result If $I_2 \ll I_1$, the collision should have little effect on disk 1. Our results agree, and give $\omega_f \to \omega_i$. If $I_2 \gg I_1$, then disk 1 should slow to a stop without causing disk 2 to move. Our results give $\omega_f \to 0$, as expected.

In the collision of the two disks in Example 10-3, mechanical energy is not conserved. We can see this by writing the energy in terms of the angular momentum. An object rotating with an angular velocity ω has kinetic energy

$$K = \frac{1}{2}I\omega^2 = \frac{(I\omega)^2}{2I}$$

Using $L = I\omega$, we get

$$K = \frac{L^2}{2I}$$

Compare this result with that for linear motion, $K = p^2/2m$ (Equation 8-23). The initial kinetic energy in Example 10-3 is

$$K_i = \frac{L_i^2}{2I_1}$$

and the final kinetic energy is

$$K_f = \frac{L_f^2}{2(I_1 + I_2)}$$

Since $L_f = L_i$, the final kinetic energy is less than the initial kinetic energy by the factor $I_1/(I_1 + I_2)$. This interaction of the disks is analogous to a one-dimensional perfectly inelastic collision of two objects.

10-17 The rotating plates in the transmission of a truck make inelastic collisions when engaged.

Example 10-4

A merry-go-round of radius 2 m and moment of inertia 500 kg·m^2 is rotating about a frictionless pivot, making one revolution every 5 s. A child of mass 25 kg originally standing at the center walks out to the rim (Figure 10-22). Find the new angular speed of the merry-go-round.

Picture the Problem The new angular speed ω_f is related to the final angular momentum of the system, which is the sum of the angular momentum of the child, L_c, and the angular momentum of the merry-go-round, L_m. No external torques act on the system, so the final angular momentum equals the initial angular momentum. When the child (considered as a particle of mass m) is a distance r from the center, her moment of inertia is $I_c = mr^2$. Initially, $r = 0$ and $I_{ci} = 0$. At the rim, $I_{cf} = mR^2$.

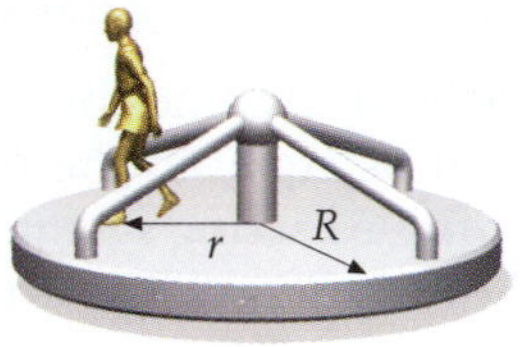

Figure 10-22

1. The final angular velocity is related to the initial angular velocity by conservation of angular momentum:

$$\vec{L}_f = \vec{L}_i$$

$$I_{sys,f}\,\omega_f = I_{sys,i}\,\omega_i$$

2. The moment of inertia of the system is the moment of inertia of the merry-go-round plus that of the child:

$$I_{sys} = I_m + I_c = I_m + mr^2$$

3. Initially $r = 0$, finally $r = R$. Substitute initial and final expressions for I_{sys}:

$$(I_m + mR^2)\omega_f = I_m\omega_i$$

4. Solve for ω_f:

$$\omega_f = \frac{I_m}{mR^2 + I_m}\,\omega_i = \frac{500\text{ kg}\cdot\text{m}^2}{(25\text{ kg})(2\text{ m})^2 + 500\text{ kg}\cdot\text{m}^2}\,\omega_i$$

$$= \frac{5}{6}\,\omega_i = \frac{5}{6}\left(\frac{1\text{ rev}}{5\text{ s}}\right) = \frac{1\text{ rev}}{6\text{ s}}$$

When the child reaches the rim, the merry-go-round rotates once every 6 s.

Remarks To calculate the value of L_i or L_f, or to find the kinetic energy of the system, it is necessary to convert the angular speed to radians per second.

When the child is at the center of the merry-go-round, she is at rest. As she walks outward, she begins to move in a circle. The force that accelerates the child is the friction between her shoes and the merry-go-round. This force exerts a torque on the child, increasing her angular momentum. The child exerts an equal and opposite frictional force on the merry-go-round. The torque associated with this force decreases the angular momentum of the merry-go-round.

From Equation 10-17 we can see that the kinetic energy of the child and merry-go-round system decreases since the angular momentum is constant, but the moment of inertia increases as the child walks toward the rim. At each step, the child makes an inelastic collision with a part of the merry-go-round that is farther out and therefore moving faster than she is. Mechanical energy is lost in each of these inelastic collisions. If the child walks inward, the moment of inertia of the child–merry-go-round system decreases, hence the total kinetic energy of the system must increase. This energy comes from the child's internal chemical energy.

Example 10-5 ***try it yourself***

The same child as in Example 10-4 runs with an initial speed 2.5 m/s along a path tangential to the rim of the merry-go-round, which is initially at rest, and then jumps on (Figure 10-23). Find the final angular velocity of the child and the merry-go-round together.

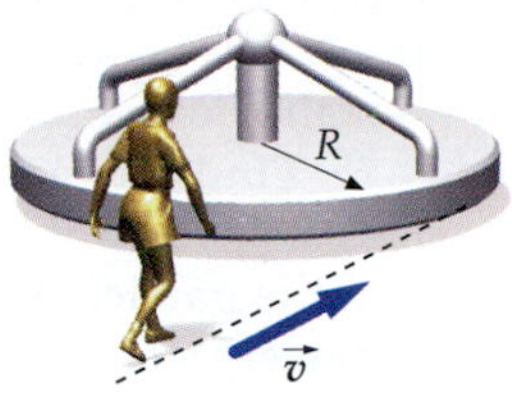

Figure 10-23

Picture the Problem Once the child's feet leave the ground, no external torques act on the child–merry-go-round system, hence the total angular momentum of the system is conserved. The mass of the child is $m = 25$ kg, her initial speed is $v = 2.5$ m/s, and the radius of the merry-go-round is $R = 2.0$ m. The initial angular speed of the merry-go-round is $w_i = 0$.

Cover the column to the right and try these on your own before looking at the answers.

Steps	***Answers***
1. Write an expression for the initial angular momentum of the running child relative to the center of the merry-go-round.	$L_i = mvR$
2. Write an expression for the total final angular momentum of the child–merry-go-round system in terms of the final angular velocity ω_f.	$L_f = (mr^2 + I_m)\omega_f$
3. Set your expressions in 1 and 2 equal and solve for ω_f.	$\omega_f = \dfrac{mvR}{mR^2 + I_m} = 0.208 \text{ rad/s}$

Exercise Calculate the initial and final kinetic energies of the child–merry-go-round system. (*Answer* $K_i = 78.1$ J, $K_f = 13.0$ J)

The Hubble Space Telescope is aimed by regulating the spin rates of 45-kg flywheels arranged off-axis from each other and spinning at up to 3000 rpm. Software-controlled changes in the spin rates create angular momentum that causes the satellite to slew into new positions. This aiming mechanism can achieve and hold a target to within 0.005 arcsec—equivalent to holding a flashlight beam in Los Angeles on a dime in San Francisco.

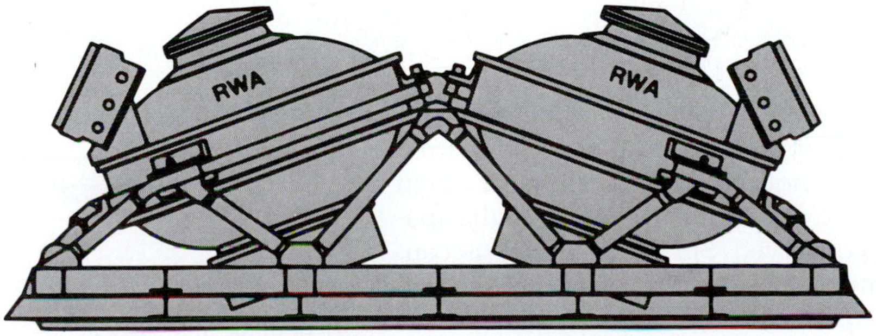

Example 10-6

A particle of mass m moves with speed v_0 in a circle of radius r_0 on a frictionless tabletop. The particle is attached to a string that passes through a hole in the table, as in Figure 10-24. The string is slowly pulled downward so that the particle moves in a smaller circle of radius r_f. (*a*) Find the final velocity in terms of r_0, v_0, and r_f. (*b*) Find the tension when the particle is moving in a circle of radius r in terms of m, r, and the angular momentum $L_0 = mv_0r_0$. (*c*) Calculate the work done on the particle by the tension T by integrating $\vec{T}\cdot d\vec{r}$ from r_0 to r_f. Express your answer in terms of r and L_0.

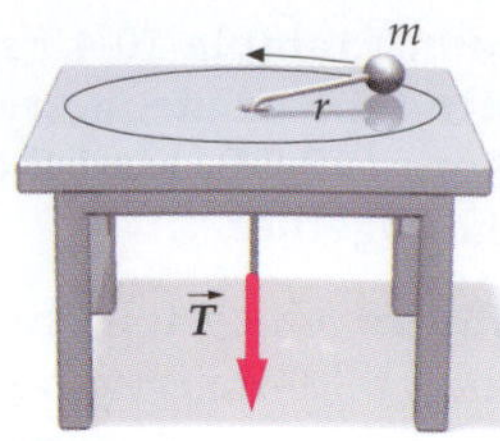

Figure 10-24

Picture the Problem The speed of the particle is related to its angular momentum. Since the net force acting on the particle is the tension in the string T, which is always directed toward the hole, the torque about the hole is zero. Thus the angular momentum remains constant, $L = mvr = L_0 = mv_0r_0$.

(*a*) 1. Conservation of angular momentum relates the final speed to the initial speed and the initial and final radii:

$$mv_0r_0 = mv_fr_f$$

2. Solve for v_f:

$$v_f = \frac{r_0}{r_f}v_0$$

(*b*) 1. Apply $\Sigma\vec{F} = m\vec{a}$ to relate T to v and r:

$$T = m\frac{v^2}{r}$$

2. Apply conservation of angular momentum to obtain a second relation between v and r:

$$mvr = mv_0r_0 = L_0$$

3. Eliminate v and solve for T:

$$T = m\frac{v^2}{r} = m\frac{(L_0/mr)^2}{r} = \frac{L_0^2}{mr^3}$$

(*c*) 1. Write $dW = \vec{T}\cdot d\vec{r} = T_r\,dr$ with $T_r = -L_0^2/mr^3$ from part (*b*):

$$dW = \vec{T}\cdot d\vec{r} = T_r\,dr = -\frac{L_0^2}{mr^3}\,dr = -\frac{L_0^2}{m}r^{-3}\,dr$$

2. Integrate from r_0 to r_f:

$$W = \int_{r_0}^{r_f} T_r\,dr = -\frac{L_0^2}{m}\int_{r_0}^{r_f} r^{-3}\,dr = -\frac{L_0^2}{m}\frac{r^{-2}}{-2}\bigg|_{r_0}^{r_f}$$

$$= \frac{L_0^2}{2m}(r_f^{-2} - r_0^{-2})$$

Check the Result Note that work must be done to pull the string downward. Since r_f is less than r_0, the work is positive. This work is converted into an increased kinetic energy. We can calculate the change in kinetic energy of the particle directly. Using $K = L^2/2I$, with $L_i = L_f$, and $I = mr^2$, the change in kinetic energy is $K_f - K_i = (L_0^2/2mr_f^2) - (L_0^2/2mr_0^2) = (L_0^2/2m)(r_f^{-2} - r_0^{-2})$, which is the same as found by direct integration.

In Figure 10-25, a puck on a frictionless plane is given an initial speed v_0. The puck is attached to a string that wraps around a vertical post. This situation looks similar to Example 10-6, but it is not the same. There is no agent that can do work on the puck, nor is there any mechanism for energy dissipation. Thus, mechanical energy must be conserved. Since $K = L^2/2I$ is constant and I decreases as r_0 decreases, L must also decrease. Note that the tension does not act toward the center of the post. The tension produces a torque about the center of the post in the downward direction, which reduces the angular momentum of the puck, which is in the upward direction.

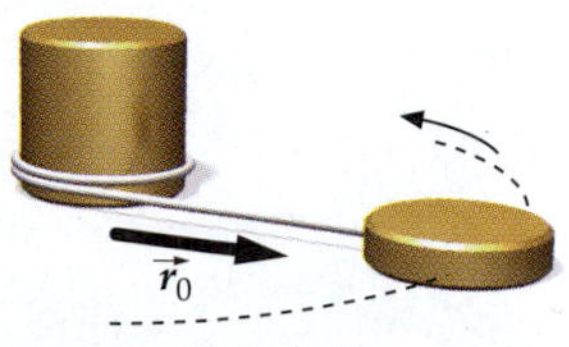

Figure 10-25 A puck sliding around a post on a frictionless table. As the puck slides, the string wraps around the post.

Example 10-7 ***try it yourself***

A thin stick of mass M and length d is attached to a pivot at the top. A piece of clay of mass m and speed v hits the stick a distance x from the pivot and sticks to it (Figure 10-26). Find the ratio of the final energy to the initial energy.

Picture the Problem The collision is inelastic, so we do not expect mechanical energy to be conserved. During the collision, the pivot exerts a large force on the stick, so linear momentum is also not conserved. However, there are no external torques about the pivot point on the clay–stick system, so angular momentum is conserved. The kinetic energy after the inelastic collision can be written in terms of the angular momentum L_f and the moment of inertia I' of the combined clay–stick system. Conservation of angular momentum allows you to relate L_f to the mass m and velocity v of the clay.

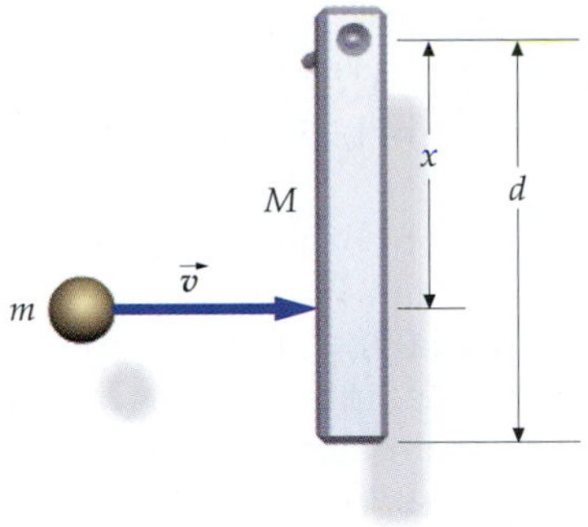

Figure 10-26

Cover the column to the right and try these on your own before looking at the answers.

Steps	***Answers***
1. Write the kinetic energy after the collision in terms of the angular momentum L_f and the moment of inertia I' of the combined stick–clay system.	$E_f = \dfrac{L_f^2}{2I'}$
2. Apply conservation of angular momentum to relate L_f to m, v, and x.	$L_f = L_i = mvx$
3. Write I' in terms of m, x, M, and d.	$I' = mx^2 + \dfrac{1}{3}Md^2$
4. Substitute these expressions for L_f and I' into your equation for E_f.	$E_f = \dfrac{L_f^2}{2I'} = \dfrac{(mvx)^2}{2(mx^2 + \frac{1}{3}Md^2)} = \dfrac{3}{2}\dfrac{m^2x^2v^2}{3mx^2 + Md^2}$
5. Divide the energy after the collision by the initial energy of the clay.	$\dfrac{E_f}{E_i} = \dfrac{\frac{3}{2}m^2x^2v^2/(3mx^2 + Md^2)}{\frac{1}{2}mv^2} = \dfrac{3mx^2}{3mx^2 + Md^2}$

Remark This example is the rotational analog of the ballistic pendulum discussed in Example 8-13. In that example, we used conservation of linear momentum to find the energy of the pendulum after the collision.

10-5 Quantization of Angular Momentum

Angular momentum plays an important role in the description of atoms, molecules, nuclei, and elementary particles. Like energy, angular momentum is **quantized,** that is, changes in angular momentum occur only in discrete amounts.

The angular momentum of a particle due to its motion is its orbital angular momentum. The magnitude of the orbital angular momentum of a particle can have only the values

$$L = \sqrt{\ell(\ell + 1)}\,\hbar, \qquad \ell = 0, 1, 2, \ldots$$

where $\hbar$ (read "h-bar") is the **fundamental unit of angular momentum**, which is related to another fundamental constant of nature, Planck's constant h:

$$\hbar = \frac{h}{2\pi} = 1.05 \times 10^{-34}\,\text{J·s} \qquad \text{10-18}$$

The component of orbital angular momentum along any line in space is also quantized and can have only the values $\pm m\hbar$ where m is an integer that is less than or equal to ℓ. For example, if $\ell = 2$, m can be 2, 1, or 0.

Because the quantum of angular momentum, $\hbar$, is so small, the quantization of angular momentum is not noticed in the macroscopic world. Consider a particle of mass 1 g ($= 10^{-3}$ kg) moving in a circle of radius 1 cm with a period of 1 s. Its orbital angular momentum is

$$L = mvr = mr^2\omega = mr^2\frac{2\pi}{T} = (10^{-3}\,\text{kg})(10^{-2}\,\text{m})^2\frac{2\pi}{1\,\text{s}} = 6.28 \times 10^{-7}\,\text{J·s}$$

If we divide by $\hbar$, we obtain

$$\frac{L}{\hbar} = \frac{6.28 \times 10^{-7}\,\text{J·s}}{1.05 \times 10^{-34}\,\text{J·s}} = 6 \times 10^{27}$$

The angular momentum of this macroscopic system is equal to 6×10^{27} units of the fundamental unit of angular momentum. Even if we could measure L to one part in a billion, we would never notice the quantization of this macroscopic angular momentum.

The quantization of orbital angular momentum leads to the quantization of rotational energy. Consider a molecule rotating about its center of mass with angular momentum L (Figure 10-27). Let I be its moment of inertia. Its kinetic energy is

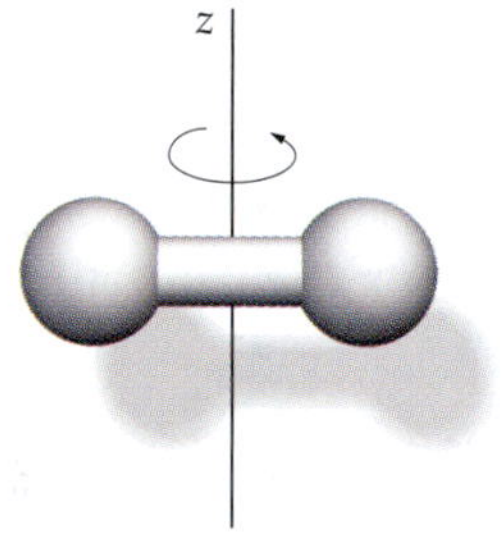

Figure 10-27 Model of a rigid diatomic molecule rotating about the z axis.

$$K = \frac{L^2}{2I} \qquad \text{10-19}$$

But L^2 is quantized to the values $L^2 = \ell(\ell + 1)\hbar^2$ with $\ell = 0, 1, 2, \ldots$. Thus the kinetic energy is quantized to the values K_ℓ given by

$$K_\ell = \frac{L^2}{2I} = \frac{\ell(\ell + 1)\hbar^2}{2I} = \ell(\ell + 1)E_{0r} \qquad \text{10-20}a$$

where

$$E_{0r} = \frac{\hbar^2}{2I} \qquad \text{10-20}b$$

Figure 10-28 shows an energy-level diagram for a rotating molecule with constant moment of inertia I. Note that, unlike the energy levels for a vibrating system (Section 7-4), the rotational energy levels are not equally spaced, and the lowest level is zero.

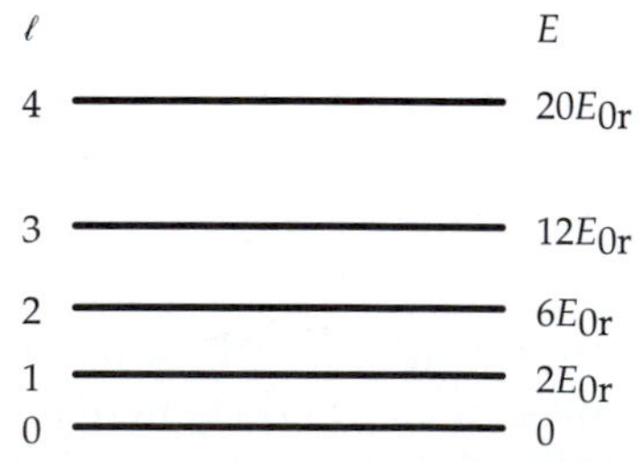

Figure 10-28 Energy level diagram for a rotating molecule.

Stable matter contains just three kinds of particles: electrons, protons, and neutrons. In addition to its orbital angular momentum, each of these particles also has an intrinsic angular momentum called its **spin**. The spin angular momentum of a particle, like its mass and electric charge, is a fundamental property of the particle that cannot be changed. It does not have anything to do with the particle's motion. The magnitude of the spin angular momentum vector for these particles is $s = \sqrt{\frac{1}{2}(\frac{1}{2} + 1)}\,\hbar$ and the component along any line in space can have just two values, $+\frac{1}{2}\hbar$ and $-\frac{1}{2}\hbar$. Such particles are called "spin-one-half" particles.*

* Electrons, protons, and neutrons and other spin-$\frac{1}{2}$ particles are also called fermions. There are other particles called bosons, such as a photons and α particles, that have zero spin or integral spin.

The picture of an electron as a spinning ball that orbits the nucleus in an atom (like the spinning earth orbiting the sun) is a useful visualization but is not entirely accurate. Spin is a quantum mechanical property, analogous but not identical to "spin" in the macroscopic sense. Whereas the angular momentum of a spinning ball can be increased or decreased, the spin of the electron is a fixed property that does not change. Furthermore, as far as we know, an electron is a point particle that has no size, and hence can be compared only figuratively with a spinning ball.

Summary

1. Angular momentum is an important derived dynamic quantity in macroscopic physics. In microscopic physics, an elementary particle has orbital angular momentum due to its motion, and spin angular momentum, which is an intrinsic, fundamental property of the particle.
2. Conservation of angular momentum is a fundamental law of nature.
3. Quantization of angular momentum is a fundamental law of nature.

Topic	Remarks and Relevant Equations	
1. Vector Nature of Rotation	When the axis of rotation changes direction in space, the vector nature of rotational quantities is important.	
Angular velocity $\vec{\omega}$	The direction of the angular velocity $\vec{\omega}$ is along the axis of rotation in the sense given by the right-hand rule.	
Torque $\vec{\tau}$	$\vec{\tau} = \vec{r} \times \vec{F}$	10-1
2. Vector Product	$\vec{A} \times \vec{B} = AB \sin\phi\, \hat{n}$	10-2
	where ϕ is the angle between the vectors and $\hat{n}$ is a unit vector perpendicular to the plane of $\vec{A}$ and $\vec{B}$ in the sense given by the right-hand rule as $\vec{A}$ is rotated into $\vec{B}$.	
Properties	$\vec{A} \times \vec{B} = -\vec{B} \times \vec{A}$	10-4
	$\frac{d}{dt}(\vec{A} \times \vec{B}) = \vec{A} \times \frac{d\vec{B}}{dt} + \frac{d\vec{A}}{dt} \times \vec{B}$	10-6
	$\hat{i} \times \hat{j} = \hat{k}, \quad \hat{j} \times \hat{k} = \hat{i}, \quad \hat{k} \times \hat{i} = \hat{j}$	10-7*a*
	$\hat{i} \times \hat{i} = \hat{j} \times \hat{j} = \hat{k} \times \hat{k} = 0$	10-7*b*
3. Angular Momentum		
For a particle	$\vec{L} = \vec{r} \times \vec{p}$	10-8
For a system rotating about a symmetry axis	$\vec{L} = I\vec{\omega}$	10-9
For a system rotating and translating	The angular momentum about any point O' is the angular momentum about the center of mass (spin angular momentum) plus the angular momentum associated with center of mass motion about O' (orbital angular momentum).	
	$\vec{L} = \vec{L}_{\text{orbit}} + \vec{L}_{\text{spin}} = \vec{r}_{\text{cm}} \times M\vec{v}_{\text{cm}} + \sum_i \vec{r}_i' \times m_i\vec{u}_i$	10-10

Newton's second law for rotation	$\vec{\tau}_{\text{net,ext}} = \frac{d\vec{L}}{dt}$	**10-13**
Kinetic energy of a rotating object	$K = \frac{L^2}{2I}$	**10-16**
Conservation of angular momentum	If the net external torque is zero, the angular momentum of the system is conserved.	
Quantization of angular momentum	The magnitude of the orbital angular momentum of a particle can have only the values $L = \sqrt{\ell(\ell+1)}\,\hbar, \quad \ell = 0, 1, 2, \ldots$ where	
Fundamental unit of angular momentum	$\hbar = \frac{h}{2\pi} = 1.05 \times 10^{-34}\,\text{J·s}$ is the fundamental unit of angular momentum, and h is Planck's constant.	**10-17**
Quantization of any component of orbital angular momentum	The component of orbital angular momentum along any line in space is also quantized and can have only the values $\pm m\hbar$, where m is an integer that is less than or equal to ℓ.	
Spin	Electrons, protons, and neutrons have an intrinsic angular momentum called spin. The magnitude of the spin angular momentum vector for these particles is $s = \sqrt{\frac{1}{2}(\frac{1}{2}+1)}\,\hbar$ and the component along any line in space can have just two values, $+\frac{1}{2}\hbar$ and $-\frac{1}{2}\hbar$.	

Problem-Solving Guide

1. Begin by drawing a neat diagram that includes the important features of the problem.
2. If there is a net torque acting on the system, write an expression for the total angular momentum of the system and apply Newton's second law in the form $\vec{\tau}_{\text{net,ext}} = d\vec{L}/dt$.
3. If the net external torque acting on the system is zero, use conservation of angular momentum to relate the final angular velocity to the initial angular velocity.

Summary of Worked Examples

Type of Calculation	Procedure and Relevant Examples
1. Calculate the Angular Momentum	For a particle, use the definition $\vec{L} = \vec{r} \times \vec{p}$. For a system rotating about a symmetry axis or an axis parallel to a symmetry axis, use $\vec{L} = I\vec{\omega}$. **Examples 10-1, 10-2**
2. Applying Newton's Second Law	
Find the angular acceleration of a pulley with masses hanging from it.	Write an expression for the angular momentum of the pulley and each object in the system. Use the nonslip condition to relate v and ω, or a and α. Apply $\Sigma\vec{\tau}_{i,\text{ext}} = d\vec{L}/dt$. **Example 10-2**
3. Conservation of Angular Momentum	
Find the final angular speed for a system in which I changes (inelastic collision of rotating disks, walking or jumping on merry-go-round).	Apply conservation of angular momentum to the system. The initial and final energies can be most easily compared if you write $K = L^2/2I$, since L is constant. **Examples 10-3, 10-4, 10-5**

Find the work done by a force that exerts no torque.	Apply conservation of angular momentum to find the final velocity or angular velocity, then use the work–energy theorem. **Example 10-6**
Find the energy lost in an inelastic collision when there are no external torques acting.	Apply conservation of angular momentum to find the final velocity or angular velocity. **Example 10-7**

Problems

Conceptual Problems

Problems from Optional and Exploring sections

In a few problems, you are given more data than you actually need; in a few other problems, you are required to supply data from your general knowledge, outside sources, or informed estimates.

• Single-concept, single-step, relatively easy
•• Intermediate-level, may require synthesis of concepts
••• Challenging, for advanced students

The Vector Nature of Rotation

1 • True or false:

(*a*) If two vectors are parallel, their cross product must be zero.
(*b*) When a disk rotates about its symmetry axis, $\vec{\omega}$ is along the axis.
(*c*) The torque exerted by a force is always perpendicular to the force.

2 • Two vectors $\vec{A}$ and $\vec{B}$ have equal magnitude. Their cross product has the greatest magnitude if $\vec{A}$ and $\vec{B}$ are

(*a*) parallel.
(*b*) equal.
(*c*) perpendicular.
(*d*) antiparallel.
(*e*) at an angle of 45° to each other.

3 • A force of magnitude F is applied horizontally in the negative x direction to the rim of a disk of radius R as shown in Figure 10-29. Write $\vec{F}$ and $\vec{r}$ in terms of the unit vectors $\hat{i}$, $\hat{j}$, and $\hat{k}$, and compute the torque produced by the force about the origin at the center of the disk.

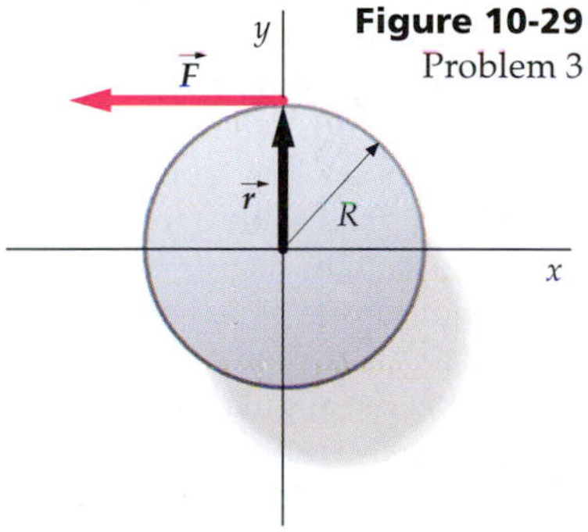

Figure 10-29 Problem 3

4 • Compute the torque about the origin for the force $\vec{F} = -mg\hat{j}$ acting on a particle at $\vec{r} = x\hat{i} + y\hat{j}$, and show that this torque is independent of the y coordinate.

5 • Find $\vec{A} \times \vec{B}$ for (*a*) $\vec{A} = 4\hat{i}$ and $\vec{B} = 6\hat{i} + 6\hat{j}$, (*b*) $\vec{A} = 4\hat{i}$ and $\vec{B} = 6\hat{i} + 6\hat{k}$, and (*c*) $\vec{A} = 2\hat{i} + 3\hat{j}$ and $\vec{B} = -3\hat{i} + 2\hat{j}$.

6 • Under what conditions is the magnitude of $\vec{A} \times \vec{B}$ equal to $\vec{A} \cdot \vec{B}$?

7 •• A particle moves in a circle of radius $\vec{r}$ with an angular velocity $\vec{\omega}$. (*a*) Show that its velocity is $\vec{v} = \vec{\omega} \times \vec{r}$. (*b*) Show that its centripetal acceleration is $\vec{a}_c = \vec{\omega} \times \vec{v} = \vec{\omega} \times (\vec{\omega} \times \vec{r})$.

8 •• If $\vec{A} = 4\hat{i}$, $B_z = 0$, $|\vec{B}| = 5$, and $\vec{A} \times \vec{B} = 12\hat{k}$, determine $\vec{B}$.

9 •• If $\vec{A} = 3\hat{j}$, $\vec{A} \times \vec{B} = 9\hat{i}$, and $\vec{A} \cdot \vec{B} = 12$, find $\vec{B}$.

Angular Momentum

10 • What is the angle between a particle's linear momentum $\vec{p}$ and its angular momentum $\vec{L}$?

11 • A particle of mass m is moving with speed v along a line that passes through point P. What is the angular momentum of the particle about point P?

(*a*) mv
(*b*) zero
(*c*) It changes sign as the particle passes through point P.
(*d*) It depends on the distance of point P from the origin of coordinates.

12 •• A particle travels in a circular path. (*a*) If its linear momentum p is doubled, how is its angular momentum affected? (*b*) If the radius of the circle is doubled but the speed is unchanged, how is the angular momentum of the particle affected?

13 •• A particle moves along a straight line at constant speed. How does its angular momentum about any point vary over time?

14 • A particle moving at constant velocity has zero angular momentum about a particular point. Show that the particle either has passed through that point or will pass through it.

15 • A 2-kg particle moves at a constant speed of 3.5 m/s around a circle of radius 4 m. (*a*) What is its angular momentum about the center of the circle? (*b*) What is its moment of inertia about an axis through the center of the circle and perpendicular to the plane of the motion? (*c*) What is the angular speed of the particle?

16 • A 2-kg particle moves at constant speed of 4.5 m/s along a straight line. (*a*) What is the magnitude of its angular momentum about a point 6 m from the line? (*b*) Describe qualitatively how its angular speed about that point varies with time.

17 •• A particle is traveling with a constant velocity $\vec{v}$ along a line that is a distance b from the origin O (Figure 10-30). Let dA be the area swept out by the position vector from O to the particle in time dt. Show that dA/dt is constant in time and equal to $\frac{1}{2}L/m$, where L is the angular momentum of the particle about the origin.

Figure 10-30 Problem 17

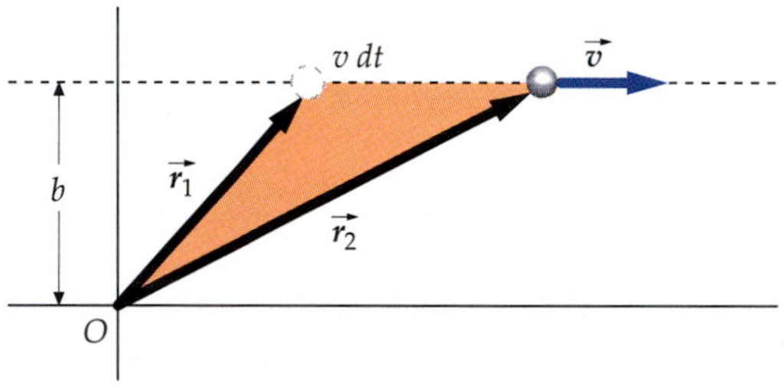

18 •• A 15-g coin of diameter 1.5 cm is spinning at 10 rev/s about a vertical diameter at a fixed point on a tabletop. (*a*) What is the angular momentum of the coin about its center of mass? (*b*) What is its angular momentum about a point on the table 10 cm from the coin? If the coin spins about a vertical diameter at 10 rev/s while its center of mass travels in a straight line across the tabletop at 5 cm/s, (*c*) what is the angular momentum of the coin about a point on the line of motion? (*d*) What is the angular momentum of the coin about a point 10 cm from the line of motion? (There are two answers to this question. Explain why and give both.)

19 •• Two particles of masses m_1 and m_2 are located at $\vec{r}_1$ and $\vec{r}_2$ relative to some origin O as in Figure 10-31. They exert equal and opposite forces on each other. Calculate the resultant torque exerted by these internal forces about the origin O and show that it is zero if the forces $\vec{F}_1$ and $\vec{F}_2$ lie along the line joining the particles.

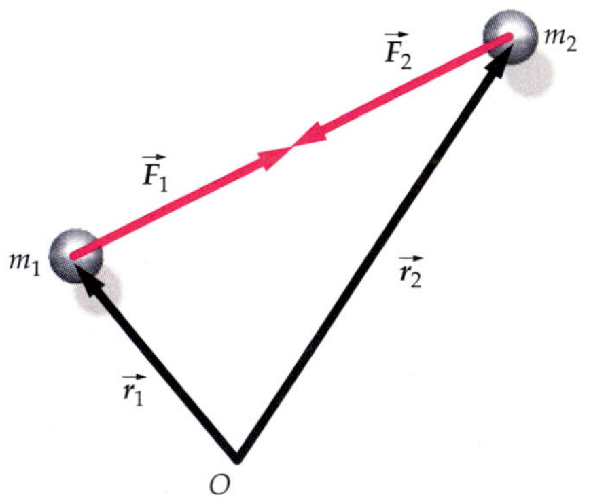

Figure 10-31 Problem 19

Torque and Angular Momentum

20 • True or false:

(*a*) The rate of change of a system's angular momentum is always parallel to the net external torque.

(*b*) If the net torque on a body is zero, the angular momentum must be zero.

21 • A 1.8-kg particle moves in a circle of radius 3.4 m. The magnitude of its angular momentum relative to the center of the circle depends on time according to $L = (4\ \text{N·m})t$. (*a*) Find the magnitude of the torque acting on the particle. (*b*) Find the angular speed of the particle as a function of time.

22 •• A uniform cylinder of mass 90 kg and radius 0.4 m is mounted so that it turns without friction on its fixed symmetry axis. It is rotated by a drive belt that wraps around its perimeter and exerts a constant torque. At time $t = 0$, its angular velocity is zero. At time $t = 25$ s, its angular velocity is 500 rev/min. (*a*) What is its angular momentum at $t = 25$ s? (*b*) At what rate is the angular momentum increasing? (*c*) What is the torque acting on the cylinder? (*d*) What is the magnitude of the force acting on the rim of the cylinder?

23 •• In Figure 10-32, the incline is frictionless and the string passes through the center of mass of each block. The pulley has a moment of inertia I and a radius r. (*a*) Find the net torque acting on the system (the two masses, string, and pulley) about the center of the pulley. (*b*) Write an expression for the total angular momentum of the system about the center of the pulley when the masses are moving with a speed v. (*c*) Find the acceleration of the masses from your results for parts (*a*) and (*b*) by setting the net torque equal to the rate of change of the angular momentum of the system.

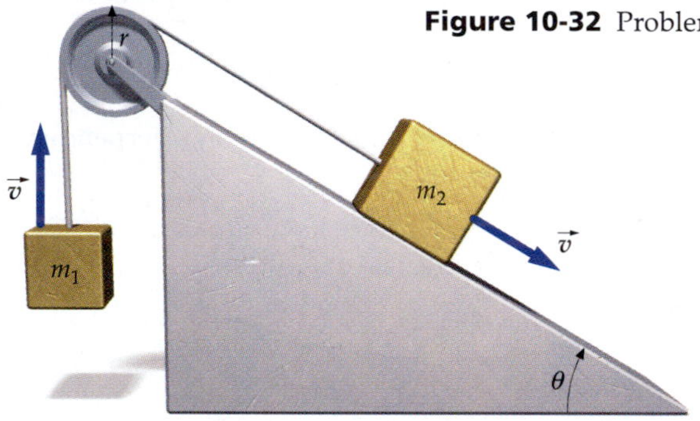

Figure 10-32 Problem 23

24 •• From her elevated DJ booth at a dance club, Caroline is lowering a 2-kg speaker using a 0.6-kg disk of radius 8 cm as a pulley (Figure 10-33). The speaker wire runs straight up from the speaker, over the pulley, and then horizontally across the table. She attaches the wire to the 4-kg amplifier on her tabletop, and then turns to get the other speaker. The table, however, is nearly frictionless, and the whole system begins to move when she lets go. (*a*) What is the net torque about the center of the pulley? (*b*) What is the total angular momentum of the system 3.5 s after release? (*c*) What is the angular momentum of the pulley at this time? (*d*) What is the ratio of the angular momentum of each piece of equipment to the angular momentum of the pulley?

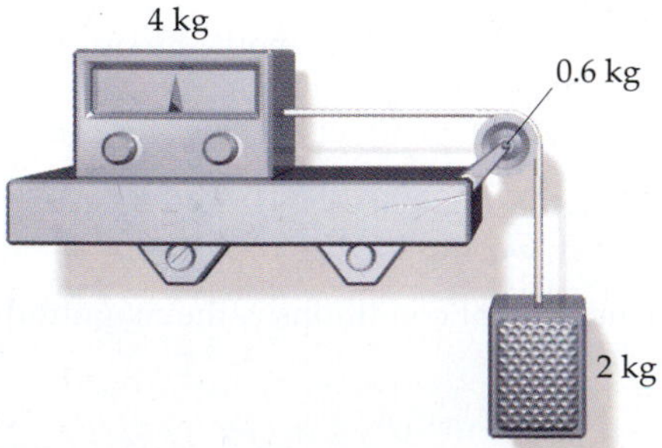

Figure 10-33 Problems 24 and 25

25 •• Work Problem 24 for the case in which the coefficient of friction between the table and the 4-kg amplifier is 0.25.

26 •• Figure 10-34 shows the rear view of a spaceship that is rotating about its longitudinal axis at 6 rev/min. The occupants wish to stop this rotation. They have small jets mounted tangentially, at a distance $R = 3$ m from the axis, as indicated, and can eject 10 g/s of gas from each jet with a nozzle velocity of 800 m/s. For how long must they turn on these jets to stop the rotation? The rotational inertia of the ship around its axis (assumed to be constant) is 4000 kg·m^2.

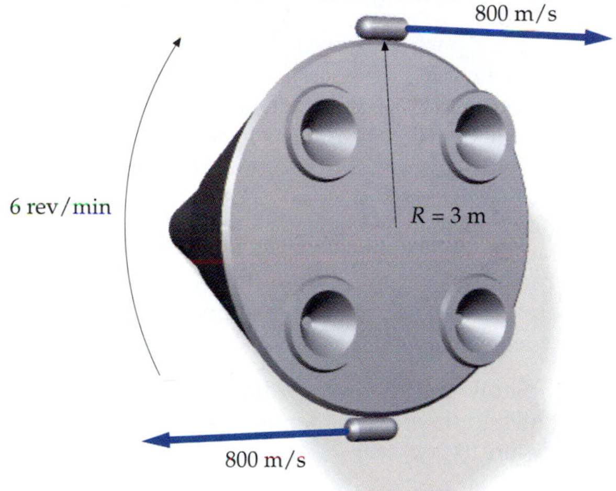

Figure 10-34 Problem 26

Conservation of Angular Momentum

27 • True or false: If the net torque on a rotating system is zero, the angular velocity of the system cannot change.

28 • Folk wisdom says that a cat always lands on its feet. If a cat starts falling with its feet up, how can it land on its feet without violating the law of conservation of angular momentum?

29 • If the angular momentum of a system is constant, which of the following statements must be true?

(*a*) No torque acts on any part of the system.
(*b*) A constant torque acts on each part of the system.
(*c*) Zero net torque acts on each part of the system.
(*d*) A constant external torque acts on the system.
(*e*) Zero net torque acts on the system.

30 • Two identical cylindrical disks have a common axis. Initially, one of the disks is spinning. When the two disks are brought into contact they stick together. Which of the following statements is true?

(*a*) The total kinetic energy and the total angular momentum are unchanged from their initial values.
(*b*) Both the total kinetic energy and the total angular momentum are reduced to half of their original values.
(*c*) The total angular momentum is unchanged, but the total kinetic energy is reduced to half its original value.
(*d*) The total angular momentum is reduced to half of its original value, but the total kinetic energy is unchanged.
(*e*) The total angular momentum is unchanged, and the total kinetic energy is reduced to one-quarter of its original value.

31 •• In Example 10-4, does force exerted by the merry-go-round on the child do work ?

32 •• Is it easier to crawl radially outward or radially inward on a rotating merry-go-round? Why?

33 •• A block sliding on a frictionless table is attached to a string that passes through a hole in the table. Initially, the block is sliding with speed v_0 in a circle of radius r_0. A student under the table pulls slowly on the string. What happens as the block spirals inward? Give supporting arguments for your choice.

(*a*) Its energy and angular momentum are conserved.
(*b*) Its angular momentum is conserved, and its energy increases.
(*c*) Its angular momentum is conserved, and its energy decreases.
(*d*) Its energy is conserved, and its angular momentum increases.
(*e*) Its energy is conserved, and its angular momentum decreases.

34 • A planet moves in an elliptical orbit about the sun with the sun at one focus of the ellipse as in Figure 10-35. (*a*) What is the torque produced by the gravitational force of attraction of the sun for the planet? (*b*) At position *A*, the planet is a distance r_1 from the sun and is moving with a speed v_1 perpendicular to the line from the sun to the planet. At position *B*, it is at distance r_2 and is moving with speed v_2, again perpendicular to the line from the sun to the planet. What is the ratio of v_1 to v_2 in terms of r_1 and r_2?

Figure 10-35 Problem 34

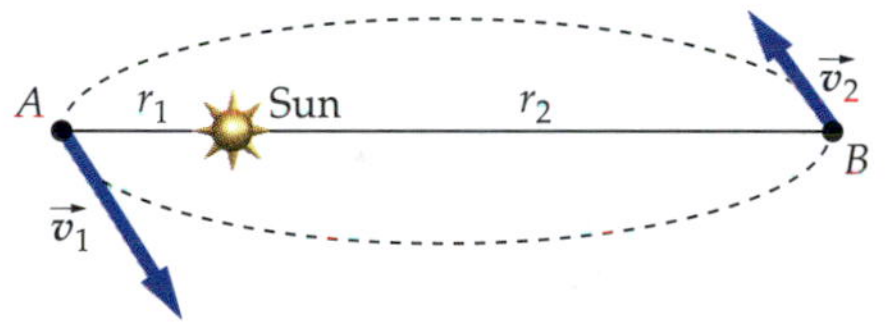

35 • Under gravitational collapse (all forces on various pieces are inward toward the center), the radius of a spinning spherical star of uniform density shrinks by a factor of 2, with the resulting increased density remaining uniform throughout as the star shrinks. What will be the ratio of the final angular speed ω_2 to the initial angular speed ω_1?

(*a*) 2 (*b*) 0.5 (*c*) 4
(*d*) 0.25 (*e*) 1.0

36 •• A man stands at the center of a platform that rotates without friction with an angular speed of 1.5 rev/s. His arms are outstretched, and he holds a heavy weight in each hand. The moment of inertia of the man, the extended weights, and the platform is 6 kg·m^2. When the man pulls the weights inward toward his body, the moment of inertia decreases to 1.8 kg·m^2. (*a*) What is the resulting angular speed of the platform? (*b*) What is the change in kinetic energy of the system? (*c*) Where did this increase in energy come from?

37 •• A small blob of putty of mass *m* falls from the ceiling and lands on the outer rim of a turntable of radius *R* and moment of inertia I_0 that is rotating freely with angular speed ω_i about its vertical fixed symmetry axis. (*a*) What is the postcollision angular speed of the turntable plus putty?

(*b*) After several turns, the blob flies off the edge of the turntable. What is the angular speed of the turntable after the blob flies off?

38 •• Two disks of identical mass but different radii (r and $2r$) are spinning on frictionless bearings at the same angular speed ω_0 but in opposite directions (Figure 10-36). The two disks are brought slowly together. The resulting frictional force between the surfaces eventually brings them to a common angular velocity. What is the magnitude of that final angular velocity in terms of ω_0?

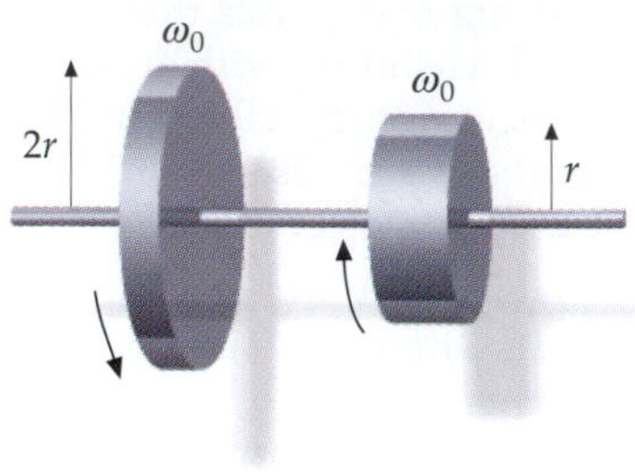

Figure 10-36 Problem 38

39 •• A block of mass m sliding on a frictionless table is attached to a string that passes through a hole in the table. Initially, the block is sliding with speed v_0 in a circle of radius r_0. Find (*a*) the angular momentum of the block, (*b*) the kinetic energy of the block, and (*c*) the tension in the string. A student under the table now pulls slowly on the string. How much work is required to reduce the radius of the circle from r_0 to $r_0/2$?

40 •• At the beginning of each term, a physics professor named Dr. Zeus shows the class his expectations of them through a demonstration that he calls "Lesson #1." He stands at the center of a turntable that can rotate without friction. He then takes a 2-kg globe of the earth and swings it around his head at the end of a 0.8-m chain. The world revolves around him every 3 s, and the professor and the platform have a moment of inertia of 0.5 kg·m^2. (*a*) What is the angular speed of the professor? (*b*) What is the total kinetic energy of the globe, professor, and platform?

41 •• The sun's radius is 6.96×10^8 m, and it rotates with a period of 25.3 d. Estimate the new period of rotation of the sun if it collapses with no loss of mass to become a neutron star of radius 5 km.

42 •• Arriving at the baggage claim area in a small airport, Alan (mass m) discovers a large turntable (radius R and moment of inertia I) that is spinning out of control. Not wanting to pass up an opportunity for magnificence, Alan leaps onto the edge of the turntable, which continues to spin freely with an angular speed of 7.5 rad/s. He struggles on his hands and knees to the center, and then rises up into a pose that resembles a hood ornament and spins like a figure skater in finale. Security is notified, but passengers applaud. Assume that $mR^2 = 2.8I$, and that Alan has a moment of inertia of $I/10$ in his final pose. What is his final angular speed if friction is neglected?

43 •• A 0.2-kg point mass moving on a frictionless horizontal surface is attached to a rubber band whose other end is fixed at point P. The rubber band exerts a force $F = bx$ toward P, where x is the length of the rubber band and b is an unknown coefficient. The mass moves along the dotted line in Figure 10-37. When it passes point A, its velocity is 4 m/s directed as shown. The distance AP is 0.6 m and BP is 1.0 m. (*a*) Find the velocity of the mass at points B and C. (*b*) Find b.

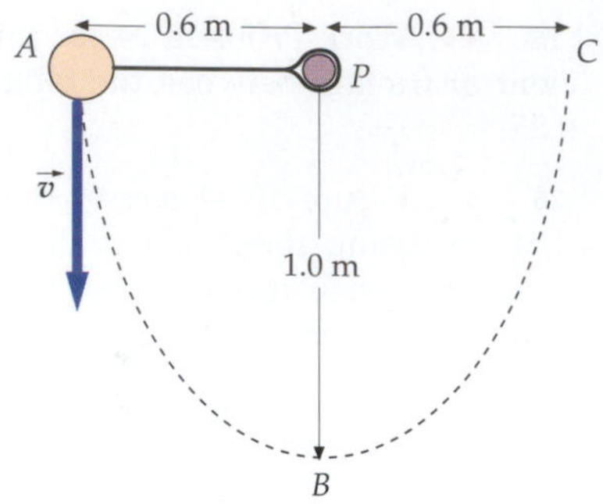

Figure 10-37 Problem 43

Quantization of Angular Momentum

44 • A 2-g particle moves at a constant speed of 3 mm/s around a circle of radius 4 mm. (*a*) Find the magnitude of the angular momentum of the particle. (*b*) If $L = \sqrt{\ell(\ell + 1)}\,\hbar$, find the value of $\ell(\ell + 1)$ and the approximate value of ℓ. (*c*) Explain why the quantization of angular momentum is not noticed in macroscopic physics.

45 • The z component of the spin of an electron is $\frac{1}{2}\hbar$, but the magnitude of the spin vector is $\sqrt{0.75}\,\hbar$. What is the angle between the electron's spin angular momentum vector and the z axis?

46 •• Show that the energy difference between one rotational state and the next higher state is proportional to $\ell + 1$ (see Equation 10-20*a*).

47 •• In the HBr molecule, the mass of the bromine nucleus is 80 times that of the hydrogen nucleus (a single proton); consequently, in calculating the rotational motion of the molecule, one may, to a good approximation, assume that the Br nucleus remains stationary as the H atom (mass 1.67×10^{-27} kg) revolves around it. The separation between the H atom and bromine nucleus is 0.144 nm. Calculate (*a*) the moment of inertia of the HBr molecule about the bromine nucleus, and (*b*) the rotational energies for $\ell = 1$, $\ell = 2$, and $\ell = 3$.

48 •• The equilibrium separation between the nuclei of the nitrogen molecule is 0.11 nm. The mass of each nitrogen nucleus is 14 u, where u $= 1.66 \times 10^{-27}$ kg. We wish to calculate the energies of the three lowest angular momentum states of the nitrogen molecule. (*a*) Approximate the nitrogen molecule as a rigid dumbbell of two equal point masses, and calculate the moment of inertia about its center of mass. (*b*) Find the rotational energy levels using the relation $E_\ell = \ell(\ell + 1)\hbar^2/2I$.

Collision Problems

49 •• A 16.0-kg, 2.4-m-long rod is supported on a knife edge at its midpoint. A 3.2-kg ball of clay is dropped from rest from a height of 1.2 m and makes a perfectly inelastic collision with the rod 0.9 m from the point of support (Figure 10-38). Find the angular momentum of the rod-and-clay system immediately after the inelastic collision.

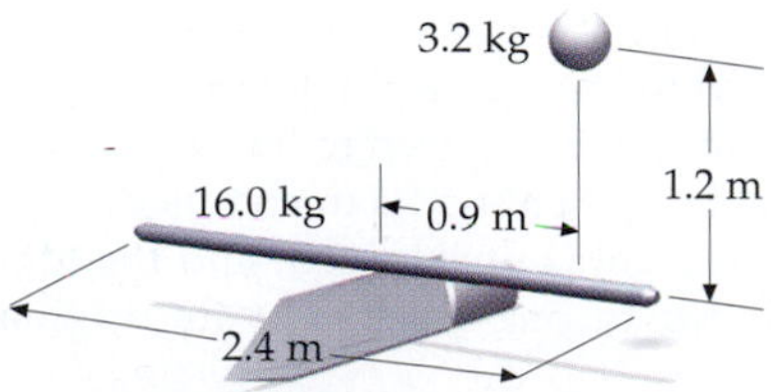

Figure 10-38 Problem 49

50 •• Figure 10-39 shows a thin bar of length L and mass M, and a small blob of putty of mass m. The system is supported on a frictionless horizontal surface. The putty moves to the right with velocity v, strikes the bar at a distance d from the center of the bar, and sticks to the bar at the point of contact. Obtain expressions for the velocity of the system's center of mass and for the angular velocity of the system about its center of mass.

Figure 10-39 Problems 50 and 51

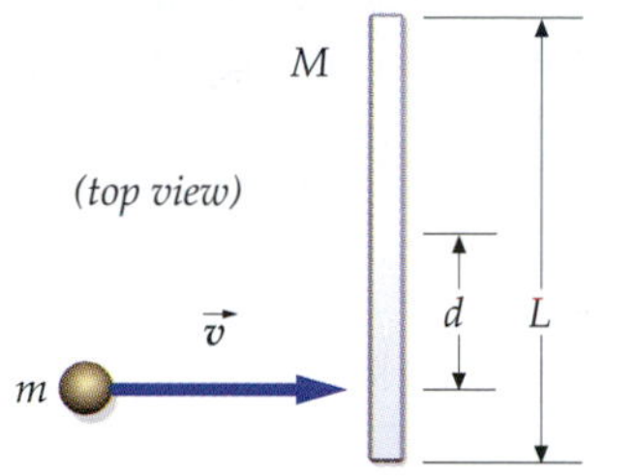

51 •• In Problem 50, replace the blob of putty with a small hard sphere of negligible size that collides elastically with the bar. Find d such that the sphere is at rest after the collision.

52 •• Figure 10-40 shows a uniform rod of length L and mass M pivoted at the top. The rod, which is initially at rest, is struck by a particle of mass m at a point $d = 0.8L$ below the pivot. Assume that the collision is perfectly inelastic. What must be the speed v of the particle so that the maximum angle between the rod and the vertical is 90°?

Figure 10-40 Problems 52 and 53

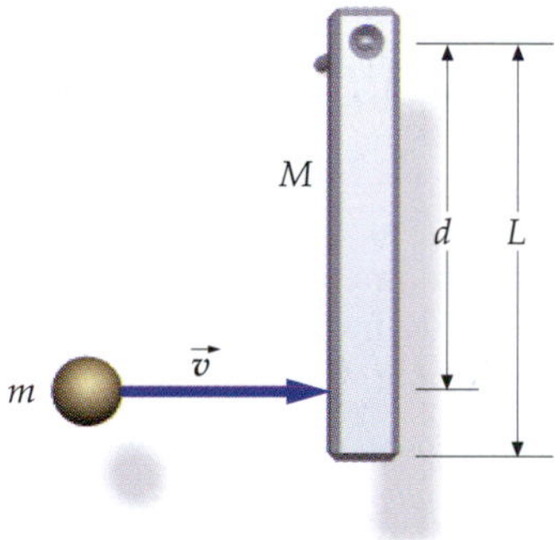

53 •• If, for the system in Problem 52, $L = 1.2$ m, $M = 0.8$ kg, and $m = 0.3$ kg, and the maximum angle between the rod and the vertical is 60°, find the speed of the particle before impact.

54 •• A projectile of mass m_p is traveling at a constant velocity $\vec{v}_0$ toward a stationary disk of mass M and radius R that is free to rotate about a pivot through its axis O (Figure 10-41). Before impact, the projectile is traveling along a line displaced a distance b below the axis. The projectile strikes the disk and sticks to point B. Treat the projectile as a point mass. (*a*) Before impact, what is the total angular momentum L_0 of the projectile and disk about the O axis? (*b*) What is the angular speed ω of the disk and projectile system just after the impact? (*c*) What is the kinetic energy of the disk and projectile system after impact? (*d*) How much mechanical energy is lost in this collision?

Figure 10-41 Problem 54

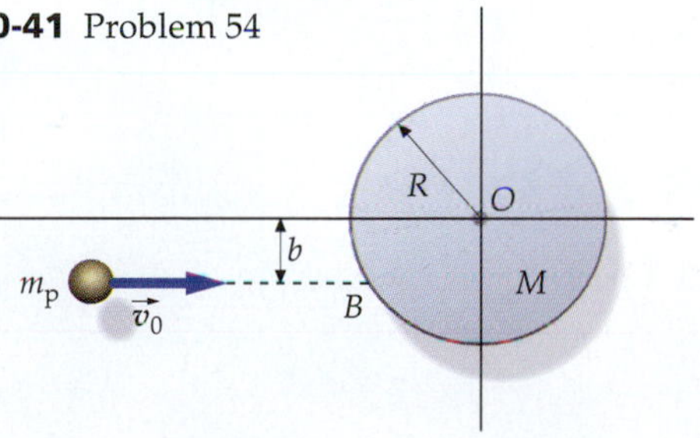

55 •• A uniform rod of length L_1 and mass $M = 0.75$ kg is supported by a hinge at one end and is free to rotate in the vertical plane (Figure 10-42). The rod is released from rest in the position shown. A particle of mass $m = 0.5$ kg is supported by a thin string of length L_2 from the hinge. The particle sticks to the rod on contact. What should be the ratio L_2/L_1 so that $\theta_{max} = 60°$ after the collision?

Figure 10-42 Problems 55–58

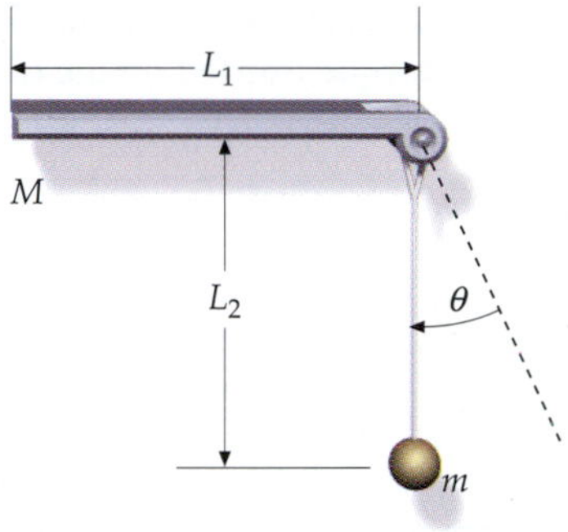

56 •• Returning to Figure 10-42, this time set $L_1 = 1.2$ m, $M = 2.0$ kg, and $L_2 = 0.8$ m. After the inelastic collision, $\theta_{max} = 37°$. Find m. How much energy is dissipated in this inelastic collision?

57 ••• Suppose that in Figure 10-42, $m = 0.4$ kg, $M = 0.75$ kg, $L_1 = 1.2$ m, and $L_2 = 0.8$ m. What minimum initial angular velocity must be imparted to the rod so that the system will revolve completely about the hinge following the inelastic collision? How much energy is then dissipated in the inelastic collision?

58 ••• Repeat Problem 56 if the collision between the rod and particle is elastic.

Exploring . . . Motion of a Gyroscope

59 • True or false:

(*a*) Nutation and precession are the same phenomenon.
(*b*) The direction of precession is the direction of the net torque.
(*c*) When the gyroscope is not spinning, $\vec{\tau} = d\vec{L}/dt$ does not hold.

60 •• The angular momentum vector for a spinning wheel lies along its axle and is pointed east. To make this vector point south, it is necessary to exert a force on the east end of the axle in which direction?

(*a*) Up
(*b*) Down
(*c*) North
(*d*) South
(*e*) East

61 •• A man is walking north carrying a suitcase that contains a spinning gyroscope mounted on an axle attached to the front and back of the case. The angular velocity of the gyroscope points north. The man now begins to turn to walk east. As a result, the front end of the suitcase will

(*a*) resist his attempt to turn and will try to remain pointed north.
(*b*) fight his attempt to turn and will pull to the west.
(*c*) rise upward.
(*d*) dip downward.
(*e*) cause no effect whatsoever.

62 •• The angular momentum of the propeller of a small airplane points forward. (*a*) As the plane takes off, the nose lifts up and the airplane tends to veer to one side. To which side does it veer and why? (*b*) If the plane is flying horizontally and suddenly turns to the right, does the nose of the plane tend to move up or down? Why?

63 •• A car is powered by the energy stored in a single flywheel with an angular momentum $\vec{L}$. Discuss the problems that would arise for various orientations of $\vec{L}$ and various maneuvers of the car. For example, what would happen if $\vec{L}$ points vertically upward and the car travels over a hilltop or through a valley? What would happen if $\vec{L}$ points forward or to one side and the car attempts to turn to the left or right? In each case that you examine, consider the direction of the torque exerted on the car by the road.

64 •• A bicycle wheel of radius 28 cm is mounted at the middle of an axle 50 cm long. The tire and rim weigh 30 N. The wheel is spun at 12 rev/s, and the axle is then placed in a horizontal position with one end resting on a pivot. (*a*) What is the angular momentum due to the spinning of the wheel? (Treat the wheel as a hoop.) (*b*) What is the angular velocity of precession? (*c*) How long does it take for the axle to swing through 360° around the pivot? (*d*) What is the angular momentum associated with the motion of the center of mass, that is, due to the precession? In what direction is this angular momentum?

65 •• A uniform disk of mass 2.5 kg and radius 6.4 cm is mounted in the center of a 10-cm axle and spun at 700 rev/min. The axle is then placed in a horizontal position with one end resting on a pivot. The other end is given an initial horizontal velocity such that the precession is smooth with no nutation. (*a*) What is the angular velocity of precession? (*b*) What is the speed of the center of mass during the precession? (*c*) What are the magnitude and direction of the acceleration of the center of mass? (*d*) What are the vertical and horizontal components of the force exerted by the pivot?

General Problems

66 • An object of mass M is rotating about a fixed axis with angular momentum L. Its moment of inertia about this axis is I. What is its kinetic energy?

(*a*) $IL^2/2$
(*b*) $L^2/2I$
(*c*) $ML^2/2$
(*d*) $IL^2/2M$

67 • Explain why a helicopter with just one main rotor has a second smaller rotor mounted on a horizontal axis at the rear as in Figure 10-43. Describe the resultant motion of the helicopter if this rear rotor fails during flight.

Figure 10-43 Problem 67

68 •• A woman sits on a spinning piano stool with her arms folded. When she extends her arms out to the side, her kinetic energy

(*a*) increases.
(*b*) decreases.
(*c*) remains the same.

69 •• In tetherball, a ball is attached to a string that is attached to a pole. When the ball is hit, the string wraps around the pole and the ball spirals inward. Neglecting air resistance, what happens as the ball swings around the pole? Give supporting arguments for your choice.

(*a*) The mechanical energy and angular momentum of the ball are conserved.
(*b*) The angular momentum of the ball is conserved, but the mechanical energy of the ball increases.
(*c*) The angular momentum of the ball is conserved, and the mechanical energy of the ball decreases.
(*d*) The mechanical energy of the ball is conserved and the angular momentum of the ball increases.
(*e*) The mechanical energy of the ball is conserved and the angular momentum of the ball decreases.

70 •• A uniform rod of mass M and length L lies on a horizontal frictionless table. A piece of putty of mass $m = M/4$ moves along a line perpendicular to the rod, strikes the rod near its end, and sticks to the rod. Describe qualitatively the subsequent motion of the rod and putty.

71 • A particle of mass 3 kg moves with velocity $\vec{v} = 3\text{ m/s}\,\hat{i}$ along the line $z = 0$, $y = 5.3$ m. (*a*) Find the angular momentum $\vec{L}$ relative to the origin when the particle is at $x = 12$ m, $y = 5.3$ m. (*b*) A force $\vec{F} = -3\text{ N}\,\hat{i}$ is applied to the particle. Find the torque relative to the origin due to this force.

72 • The position vector of a particle of mass 3 kg is given by $\vec{r} = 4\hat{i} + 3t^2\hat{j}$, where $\vec{r}$ is in meters and t is in seconds. Determine the angular momentum and the torque acting on the particle about the origin.

73 •• An ice skater starts her pirouette with arms outstretched, rotating at 1.5 rev/s. Estimate her rotational speed (in revolutions per second) when she brings her arms flat against her body.

74 •• Two ice skaters hold hands and rotate, making one revolution in 2.5 s. Their masses are 55 kg and 85 kg, and they are separated by 1.7 m. Find (*a*) the angular momentum of the system about their center of mass, and (*b*) the total kinetic energy of the system.

75 •• A 2-kg ball attached to a string of length 1.5 m moves in a horizontal circle as a conical pendulum (Figure 10-44). The string makes an angle $\theta = 30°$ with the vertical. (*a*) Show that the angular momentum of the ball about the point of support *P* has a horizontal component toward the center of the circle as well as a vertical component, and find these components. (*b*) Find the magnitude of $d\vec{L}/dt$, and show that it equals the magnitude of the torque exerted by gravity about the point of support.

Figure 10-44 Problem 75

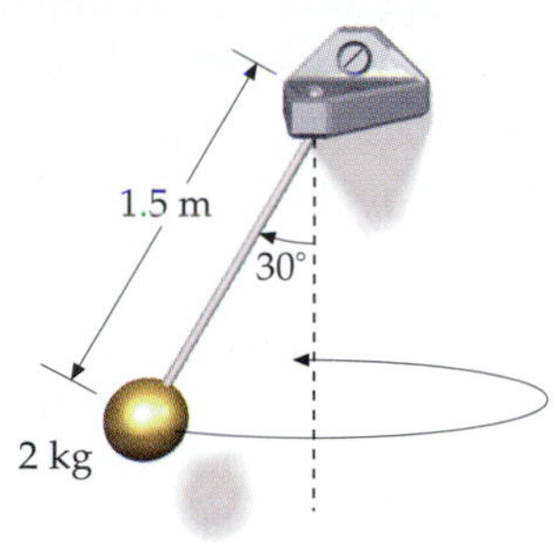

76 •• A mass *m* on a horizontal, frictionless surface is attached to a string that wraps around a vertical cylindrical post so that when it is set into motion it follows a path that spirals inward. (*a*) Is the angular momentum of the mass conserved? (*b*) Is the energy of the mass conserved? (*c*) If the speed of the mass is v_0 when the length of the string is *r*, what is its speed when the unwrapped length has shortened to $r/2$?

77 •• Figure 10-45 shows a hollow cylindrical tube of mass *M*, length *L*, and moment of inertia $ML^2/10$. Inside the cylinder are two masses *m*, separated a distance ℓ and tied to a central post by a thin string. The system can rotate about a vertical axis through the center of the cylinder. With the system rotating at ω, the strings holding the masses suddenly break. When the masses reach the end of the cylinder, they stick. Obtain expressions for the final angular velocity and the initial and final energies of the system. Assume that the inside walls of the cylinder are frictionless.

Figure 10-45 Problems 77–81

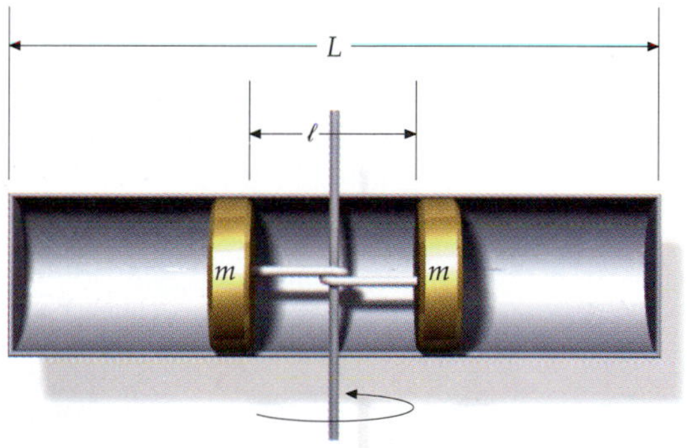

78 •• Repeat Problem 77, this time adding friction between the masses and walls of the cylinder. However, the coefficient of friction is not enough to prevent the masses from reaching the ends of the cylinder. Can the final energy of the system be determined without knowing the coefficient of kinetic friction?

79 •• Suppose that in Figure 10-45, $\ell = 0.6$ m, $L = 2.0$ m, $M = 0.8$ kg, and $m = 0.4$ kg. The system rotates at ω such that the tension in the string is 108 N just before it breaks. Determine the initial and final angular velocities and initial and final energies of the system. Assume that the inside walls of the cylinder are frictionless.

80 •• For Problem 77, determine the radial velocity of each mass just before it reaches the end of the cylinder.

81 •• Given the numerical values of Problem 79, suppose the coefficient of friction between the masses and the walls of the cylinder is such that the masses cease sliding 0.2 m from the ends of the cylinder. Determine the initial and final angular velocities of the system and the energy dissipated in friction.

82 •• Kepler's second law states: *The radius vector from the sun to a planet sweeps out equal areas in equal times.* Show that this law follows directly from the law of conservation of angular momentum and the fact that the force of gravitational attraction between a planet and the sun acts along the line joining the two celestial objects.

83 •• Figure 10-46 shows a hollow cylinder of length 1.8 m, mass 0.8 kg, and radius 0.2 m that is free to rotate about a vertical axis through its center and perpendicular to the cylinder's axis. Inside the cylinder are two thin disks of 0.2 kg each, attached to springs of spring constant *k* and unstretched lengths 0.4 m. The system is brought to a rotational speed of 8 rad/s with the springs clamped so they do not stretch. The springs are then suddenly unclamped. When the disks have stopped their radial motion due to friction between the disks and the wall, they come to rest 0.6 m from the central axis. What is the angular velocity of the cylinder when the disks have stopped their radial motion? How much energy was dissipated in friction between the disks and cylinder wall?

Figure 10-46 Problem 83

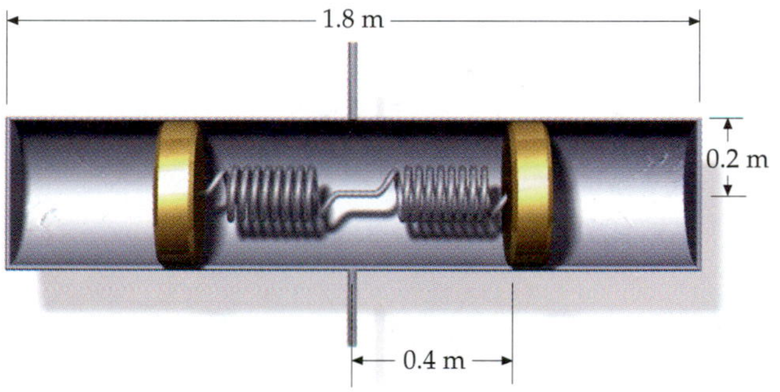

84 •• (*a*) Assuming the earth to be a homogeneous sphere of radius *r* and mass *m*, show that the period *T* of the earth's rotation about its axis is related to its radius by $T = (4\pi m/5L)r^2$, where *L* is the angular momentum of the earth due to its rotation. (*b*) Suppose that the radius *r* changes by a very small amount Δr due to some internal effect such as thermal expansion. Show that the fractional change in the period ΔT is given approximately by $\Delta T/T = 2\,\Delta r/r$. *Hint*: Use the differentials *dr* and *dT* to approximate the changes in these quantities. (*c*) By how many kilometers would the earth need to expand for the period to change by $\frac{1}{4}d/y$ so that leap years would no longer be necessary?

85 •• The polar ice caps contain about 2.3×10^{19} kg of ice. This mass contributes negligibly to the moment of inertia of the earth because it is located at the poles, close to the axis of rotation. Estimate the change in the length of the day that would be expected if the polar ice caps were to melt and the water were distributed uniformly over the surface of the earth. (The moment of inertia of a spherical shell of mass m and radius r is $\frac{2}{3}mr^2$.)

86 ••• Figure 10-47 shows a hollow cylinder of mass $M = 1.2$ kg and length $L = 1.6$ m that is free to rotate about a vertical axis through its center. Inside the cylinder are two disks, each of mass 0.4 kg that are tied to a central post by a thin string and separated by a distance $\ell = 0.8$ m. The string breaks if the tension exceeds 100 N. Starting from rest, a torque is applied to the system until the string breaks. Assuming the disks are point masses and the radius of the cylinder is negligible, find the amount of work done up to that instant. Suppose that at that instant, the applied torque is removed, and that the walls of the cylinder are frictionless. Obtain an expression for the angular velocity of the system as a function of x for $x < L/2$, where x is the distance between each mass and the central post.

87 ••• For the system of Problem 86, find the angular velocity of the system just before and just after the point masses pass the ends of the cylinder.

Figure 10-47 Problems 86–88

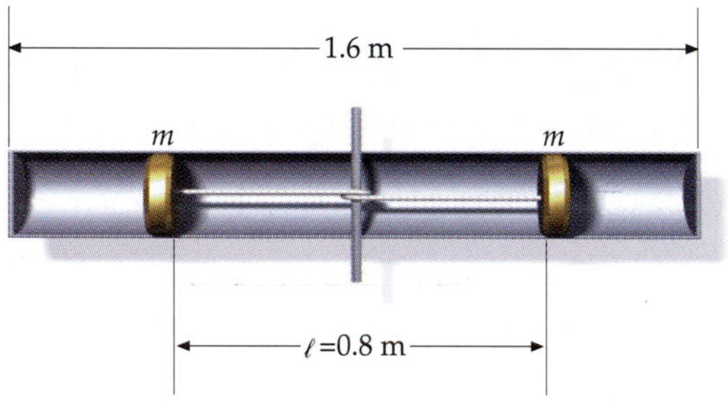

88 ••• Repeat Problem 86 with the radius of the hollow cylinder as 0.4 m and the masses treated as thin disks rather than point masses.

89 ••• Figure 10-48 shows a pulley in the shape of a uniform disk with a heavy rope hanging over it. The circumference of the pulley is 1.2 m and its mass is 2.2 kg. The rope is 8.0 m long and its mass is 4.8 kg. At the instant shown in the figure, the system is at rest and the difference in height of the two ends of the rope is 0.6 m. (*a*) What is the angular velocity of the pulley when the difference in height between the two ends of the rope is 7.2 m? (*b*) Obtain an expression for the angular momentum of the system as a function of time while neither end of the rope is above the center of the pulley. There is no slippage between rope and pulley.

Figure 10-48 Problem 89

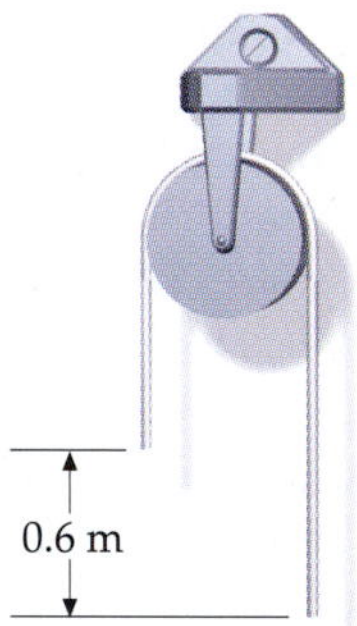

CHAPTER 11

Gravity

A mechanical model of the solar system, called an orrery, in the collection of Historical Scientific Instruments at Harvard University.

Gravity is the weakest of the four basic forces. It is negligible in the interactions of elementary particles, and thus plays no role in molecules, atoms, and nuclei. The gravitational attraction between objects of ordinary size, such as the gravitational force exerted by a building on a car, is too small to be noticed. Yet when we consider very large objects, such as moons, planets, and stars, gravity is of primary importance. The gravitational force exerted by the earth on us and on the objects around us is a fundamental part of our experience. It is gravity that binds us to the earth and keeps the earth and the other planets on course within the solar system. The gravitational force plays an important role in the life history of stars and in the behavior of galaxies. On the largest of all scales, it is gravity that controls the evolution of the universe.

Johannes Kepler (1571–1630)

11-1 Kepler's Laws

The nighttime sky with its myriad stars and shining planets has always fascinated humankind. Toward the end of the sixteenth century, the astronomer Tycho Brahe studied the motions of the planets and made observations that were considerably more accurate than those previously available. Using

Figure 11-1 Orbits of the planets around the sun.

Brahe's data, Johannes Kepler discovered that the paths of the planets about the sun are ellipses (Figure 11-1). He also showed that the planets move faster when their orbit brings them closer to the sun and slower when their orbit takes them farther away. Finally, Kepler developed a precise mathematical relation between the period of a planet and its average distance from the sun. Kepler stated his results in three empirical laws of planetary motion. Ultimately, these laws provided the basis for Newton's discovery of the law of gravity. Kepler's three laws are

> Law 1. All planets move in elliptical orbits with the sun at one focus.
> Law 2. A line joining any planet to the sun sweeps out equal areas in equal times.
> Law 3. The square of the period of any planet is proportional to the cube of the semimajor axis of its orbit.

Figure 11-2 An ellipse is the locus of points for which $r_1 + r_2 =$ constant. The distance a is called the *semimajor* axis, and b is the *semiminor* axis. You can draw an ellipse with a piece of string by fixing each end at a focus F and using it to guide the pencil. Circles are special cases in which the two foci coincide.

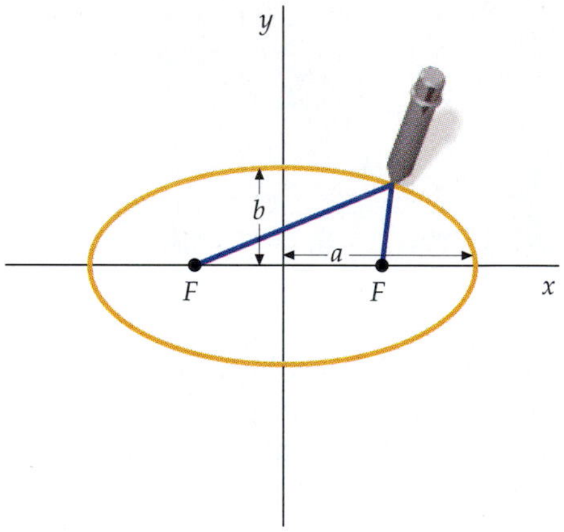

An ellipse is the locus of points for which the sum of the distances from two foci F is constant, as shown in Figure 11-2. Figure 11-3 shows a planet following an elliptical path with the sun at one focus. The earth's orbit is nearly circular, with the distance to the sun at perihelion (closest point) being 1.48×10^{11} m, and at aphelion (farthest point) being 1.52×10^{11} m. The semimajor axis equals the average of these distances, which is 1.50×10^{11} m (93 million miles) for the earth's orbit. The mean earth–sun distance defines the astronomical unit (AU):

$$1 \text{ AU} = 1.50 \times 10^{11} \text{ m} = 93.0 \times 10^{6} \text{ mi} \qquad \text{11-1}$$

The AU is used frequently in problems dealing with the solar system.

Figure 11-4 illustrates Kepler's second law, the law of equal areas. A planet moves faster when it is closer to the sun than when it is farther away, so that the area swept out by the radius vector in a given time interval is the same throughout the orbit. The law of equal areas is related to the conservation of angular momentum, as we will see in the next section.

Kepler's third law relates the period of any planet to its mean distance from the sun, which equals the semimajor axis of its elliptical path. In algebraic form, if r is the mean distance

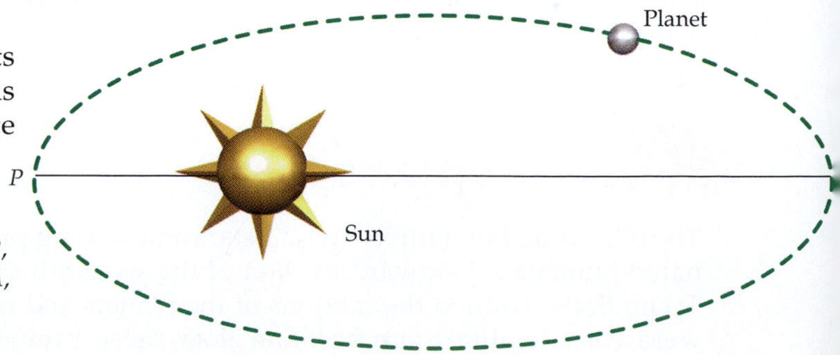

Figure 11-3 Elliptical path of a planet with the sun at one focus. Point P, where the planet is closest to the sun, is called the perihelion, and point A, where it is farthest, is called the aphelion. The average distance between the planet and the sun is equal to the semimajor axis.

between a planet and the sun and T is the planet's period of revolution, Kepler's third law states that

$$T^2 = Cr^3 \tag{11-2}$$

where the constant C has the same value for all the planets. This law is a consequence of the fact that the force exerted by the sun on a planet varies inversely with the square of the distance from the sun to the planet. We will demonstrate this in the next section for the special case of a circular orbit.

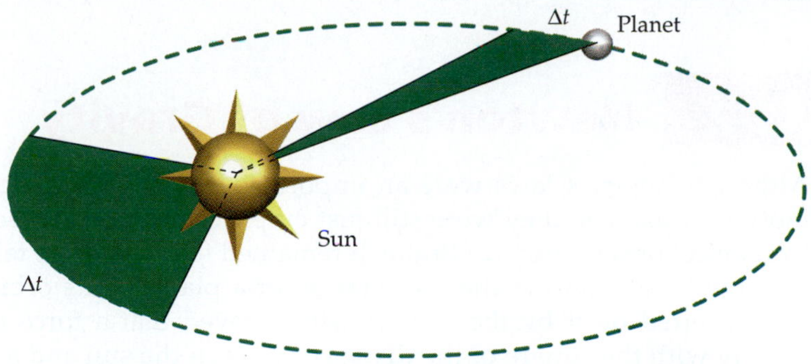

Figure 11-4 When a planet is close to the sun, it moves faster than when it is farther away. The areas swept out by the radius vector in a given time interval are equal.

Example 11-1

The mean distance from the sun to Jupiter is 5.20 AU. What is the period of Jupiter's orbit around the sun?

Picture the Problem We use Kepler's third law to relate the period of Jupiter to its mean distance from the sun. The constant C can be obtained from the known mean distance and period of the earth. Let $T_E = 1$ y be the period of the earth and let $r_E = 1$ AU be the mean distance from the earth to the sun. Let T_J and $r_J = 5.20$ AU be the period and mean distance for Jupiter.

1. Kepler's third law relates Jupiter's period T_J and mean distance r_J: $\quad T_J^2 = Cr_J^3$

2. Apply Kepler's third law to the earth to find C in terms of T_E and r_E: $\quad T_E^2 = Cr_E^3 \quad \text{or} \quad C = \dfrac{T_E^2}{r_E^3}$

3. Substitute this value of C and solve for T_J: $\quad T_J^2 = Cr_J^3 = \dfrac{T_E^2}{r_E^3}r_J^3$

$$T_J = \left(\frac{r_J}{r_E}\right)^{3/2} T_E = \left(\frac{5.20\ \text{AU}}{1\ \text{AU}}\right)^{3/2} (1\ \text{y}) = 11.9\ \text{y}$$

Exercise The period of Neptune is 164.8 y. Calculate its mean distance from the sun. (*Answer* 30.1 AU)

Remark Figure 11-5 shows the periods of the planets Earth, Jupiter, and Neptune as functions of their mean distances from the sun. In (*a*), periods are plotted on an arithmetic scale. The same data plotted on a log–log scale (*b*) fall on the straight line $\log T = \frac{1}{2}\log C + \frac{3}{2}\log R$.

Figure 11-5

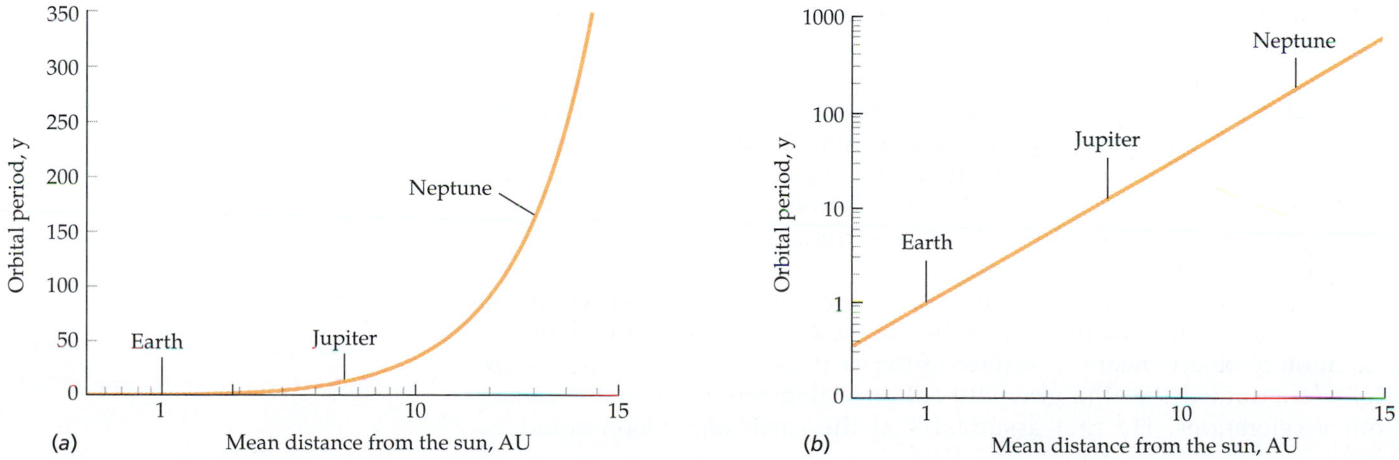

11-2 Newton's Law of Gravity

Although Kepler's laws were an important first step in understanding the motion of planets, they were still just empirical rules obtained from the astronomical observations of Brahe. It remained for Newton to take the next giant step by attributing the acceleration of a planet in its orbit to a specific force exerted on it by the sun. Newton proved that a force that varies inversely with the square of the distance between the sun and a planet results in elliptical orbits, as observed by Kepler. He then made the bold assumption that this force acts between any two objects in the universe. Before Newton, it was not even generally accepted that the laws of physics observed on earth were applicable to the heavenly bodies. **Newton's law of gravity** postulates that there is a force of attraction between each pair of objects that is proportional to the product of the masses of the objects and inversely proportional to the square of the distance separating them. The magnitude of the gravitational force exerted by a particle of mass m_1 on another particle of mass m_2 a distance r away is thus given by

$$F = \frac{Gm_1m_2}{r^2} \tag{11-3}$$

where G is the **universal gravitational constant**, which has the value

$$G = 6.67 \times 10^{-11}\ \mathrm{N \cdot m^2/kg^2} \tag{11-4}$$

Newton published his theory of gravitation in 1686, but it was not until a century later that an accurate experimental determination of G was made by Cavendish, whose findings will be discussed in the next section. If m_1 is at position $\vec{r}_1$ and m_2 is at $\vec{r}_2$ (Figure 11-6*a*), the force $\vec{F}_{1,2}$ exerted by mass m_1 on m_2 is

$$\vec{F}_{1,2} = -\frac{Gm_1m_2}{r_{1,2}^2}\hat{r}_{1,2} \tag{11-5}$$

Newton's law of gravity

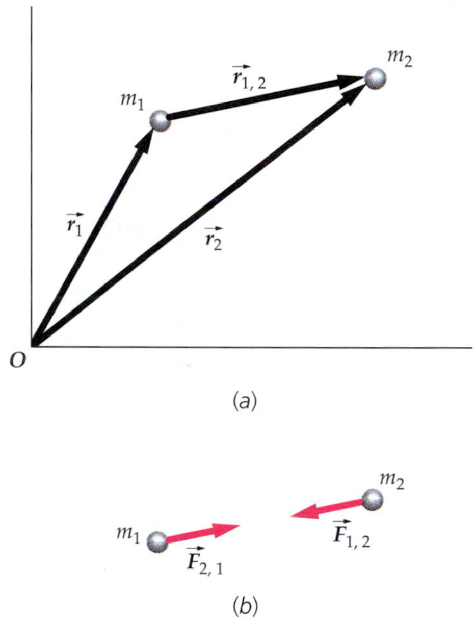

Figure 11-6 (*a*) Particles at $\vec{r}_1$ and $\vec{r}_2$. (*b*) The particles exert equal and opposite forces on each other.

where $\vec{r}_{1,2}$ is the vector pointing from mass m_1 to m_2 and $\hat{r}_{1,2} = \vec{r}_{1,2}/r_{1,2}$ is a unit vector point from m_1 to m_2. The force $\vec{F}_{2,1}$ exerted by m_2 on m_1 is the negative of $\vec{F}_{1,2}$, according to Newton's third law (Figure 11-6*b*).

We can use the known value of G to compute the gravitational attraction between two ordinary objects.

Exercise Find the gravitational force that attracts a 65-kg boy to a 50-kg girl when they are 0.5 m apart. Assume that they are point masses. (*Answer* 8.67×10^{-7} N)

This exercise demonstrates that the gravitational force exerted by an object of ordinary size on another such object is extremely small. For example, the weight of a 50-kg person is 491 N, about half a billion times the force of attraction calculated in the exercise. The gravitational attraction can be easily noticed only if at least one of the objects is extremely massive, as with an ordinary object and the earth.

To check the validity of the inverse-square nature of the gravitational force, Newton compared the acceleration of the moon in its orbit with the acceleration of objects near the surface of the earth (such as the legendary apple). He assumed that the gravitational attraction due to the earth causes both accelerations. He first assumed that the earth and moon could be treated as point particles with their total masses concentrated at their centers. The force on a particle of mass m a distance r from the center of the earth is

$$F = \frac{GM_E m}{r^2} \qquad 11\text{-}6$$

From Newton's second law, the acceleration is

$$a = \frac{F}{m} = \frac{GM_E}{r^2} \qquad 11\text{-}7$$

For objects near the surface of the earth, $r = R_E$ and the acceleration is g:

$$g = \frac{GM_E}{R_E^2} \qquad 11\text{-}8$$

Since the distance to the moon is about 60 times the radius of the earth, the acceleration of objects near the surface of the earth ($g = 9.81\ \text{m/s}^2$) should be $60^2 = 3600$ times the acceleration of the moon. The moon's centripetal acceleration can be calculated from its known distance from the center of the earth $r = 3.84 \times 10^8$ m, and its period $T = 27.3$ days $= 2.36 \times 10^6$ s:

$$a_m = \frac{v^2}{r} = \frac{(2\pi r/T)^2}{r} = \frac{4\pi^2 r}{T^2} = \frac{4\pi^2(3.84 \times 10^8\ \text{m})}{(2.36 \times 10^6\ \text{s})^2} = 2.72 \times 10^{-3}\ \text{m/s}^2$$

Then

$$\frac{g}{a_m} = \frac{9.81\ \text{m/s}^2}{2.72 \times 10^{-3}\ \text{m/s}^2} = 3607 \approx 60^2$$

In Newton's words, "I thereby compared the force requisite to keep the Moon in her orb with the force of gravity at the surface of the Earth, and found them answer pretty nearly."

The assumption that the earth and moon can be treated as point particles in the calculation of the force on the moon is reasonable because the moon is far from the earth compared with the radius of either the earth or the moon, but such an assumption is certainly questionable when applied to an object near the earth's surface. After considerable effort, Newton was able to prove that the force exerted by any spherically symmetric object on a point mass either on or outside its surface is the same as if all the mass of the object were concentrated at its center. The proof involves integral calculus, which Newton developed to solve this problem.

Since $g = 9.81\ \text{m/s}^2$ is easily measured and the radius of the earth is known, Equation 11-8 can be used to determine either the constant G or the mass of the earth M_E if one of these quantities is known. Newton estimated the value of G from an approximation of the mass of the earth. When Cavendish determined G some 100 years later by measuring the force between small spheres of known mass and separation, he called his experiment "weighing the earth."

Cavendish used the apparatus shown in Figure 11-7. His measurement of G has been repeated by other experimenters with various improvements and refinements. All measurements of G are difficult because of the extreme weakness of the gravitational attraction. Consequently, the value of G is known today only to about 1 part in 10,000. Although G was one of the first physical constants ever measured, it remains one of the least accurately known.

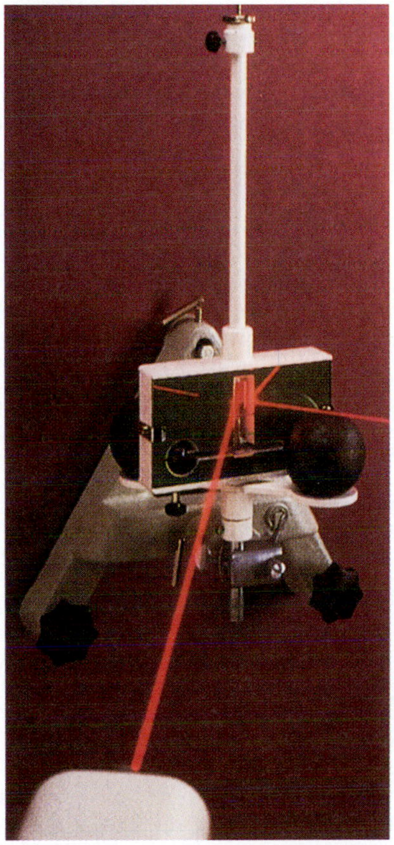

Gravitational torsion balance used in student labs for the measurement of G. A tiny angular deflection of the balance results in a large angular deflection of the laser beam that reflects from a mirror on the balance.

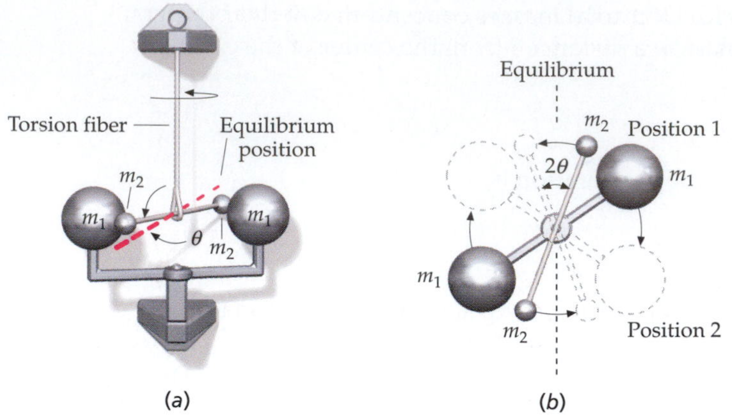

Figure 11-7 (*a*)Two small spheres, each of mass m_2, are at the ends of a light rod that is suspended by a fine fiber. Careful measurements determine the torque required to turn the fiber through a given angle. Two large spheres, each of mass m_1, are then placed near the small spheres. Because of the gravitational attraction of the large spheres of mass m_1 for the small spheres, the fiber is turned through a very small angle θ from its equilibrium position. (*b*) The apparatus as seen from above. After the apparatus comes to rest, the positions of the large spheres are reversed, as shown by the dashed lines, so that they are at the same distance from the equilibrium position of the balance but on the other side. If the apparatus is again allowed to come to rest, the fiber will turn through angle 2θ in response to the reversal of the torque. Once the torsion constant has been determined, the forces between the masses m_1 and m_2 can be determined from the measurement of this angle. Since the masses and their separations are known, G can be calculated. Cavendish obtained a value for G within about 1% of the presently accepted value given by Equation 11-4.

Example 11-2

What is the free-fall acceleration of an object at the altitude of the space shuttle's orbit, about 400 km above the earth's surface?

Picture the Problem The force is given by Equation 11-6 with $r = R_E + 400$ km.

1. The acceleration is given by $a = F/m$, where F is given by Newton's law of gravity:

$$a = \frac{F}{m} = \frac{GmM_E/r^2}{m} = \frac{GM_E}{r^2}$$

2. The distance r is related to the radius of the earth R_E and the altitude h:

$$r = R_E + h = 6370\text{ km} + 400\text{ km} = 6770\text{ km}$$

3. The acceleration is then:

$$a = \frac{GM_E}{r^2} = \frac{(6.67 \times 10^{-11}\text{ N}\cdot\text{m}^2/\text{kg}^2)(5.98 \times 10^{24}\text{ kg})}{(6770\text{ km})^2}$$

$$= 8.70\text{ m/s}^2$$

Remark This is also the acceleration of the "weightless" shuttle astronauts as they accelerate in their circular orbit.

The calculation in Example 11-2 could have been simplified by using Equation 11-8 to write

$$GM_E = gR_E^2 \qquad 11\text{-}9$$

Then the acceleration at a distance r is

$$a = \frac{F}{m} = \frac{GM_E}{r^2} = g\frac{R_E^2}{r^2} \qquad 11\text{-}10$$

Exercise At what distance h above the surface of the earth is the acceleration of gravity half its value at sea level? (*Answer* $r = \sqrt{2}\,R_E = R_E + h$, $h = (\sqrt{2} - 1)R_E = 2639$ km)

Derivation of Kepler's Laws

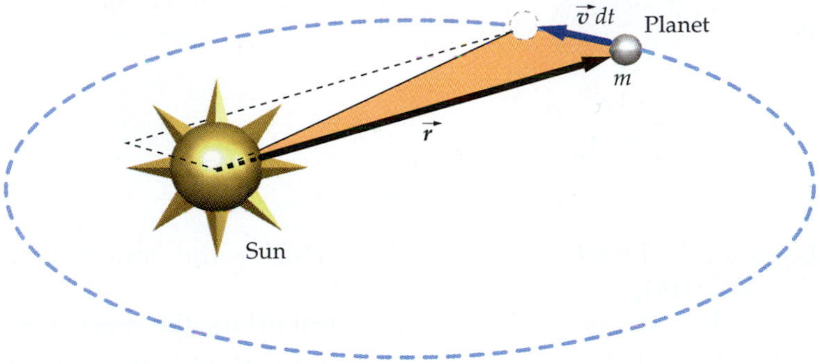

Figure 11-8 Area swept out by the radius vector of a planet orbiting the sun is proportional to the angular momentum of the planet around the sun.

Newton showed that when an object such as a planet or comet moves around a $1/r^2$ force center such as the sun, the object's path is an ellipse, a parabola, or a hyperbola.* The parabolic and hyperbolic paths apply to objects that make one pass by the sun and never return. Such orbits are not closed. The only closed orbits in an inverse-square force field are ellipses. Thus, Kepler's first law is a direct consequence of Newton's law of gravity. Kepler's second law, the law of equal areas, follows from the fact that the force exerted by the sun on a planet is directed toward the sun. Such a force is called a **central force.** Figure 11-8 shows a planet moving in an elliptical orbit around the sun. In time dt, the planet moves a distance $v\,dt$ and sweeps out the area indicated in the figure. This is half the area of the parallelogram formed by the vectors $\vec{r}$ and $\vec{v}\,dt$, which is $|\vec{r} \times \vec{v}\,dt|$. Thus, the area dA swept out by the radius vector $\vec{r}$ in time dt is

$$dA = \frac{1}{2}|\vec{r} \times \vec{v}\,dt| = \frac{1}{2m}|\vec{r} \times m\vec{v}|\,dt$$

or

$$dA = \frac{1}{2m}L\,dt \qquad \text{11-11}$$

where $\vec{L} = \vec{r} \times m\vec{v}$ is the angular momentum of the planet relative to the sun. The area swept out in a given time interval dt is therefore proportional to the angular momentum L. Since the force on a planet is along the line from the planet to the sun, it has no torque about the sun. Thus, the angular momentum of the planet is conserved, that is, L is constant. Therefore, the area swept out in a given time interval dt is the same for all parts of the orbit, which is Kepler's second law.

We will now show that Newton's law of gravity implies Kepler's third law for the special case of a circular orbit. Consider a planet moving with speed v in a circular orbit of radius r about the sun. The gravitational force of attraction between the sun and the planet provides the centripetal acceleration v^2/r. Newton's second law gives

$$F = m_p a$$

$$\frac{GM_s m_p}{r^2} = m_p\frac{v^2}{r} \qquad \text{11-12}$$

where M_s is the mass of the sun and m_p is that of the planet. Solving for v^2, we find

$$v^2 = \frac{GM_s}{r} \qquad \text{11-13}$$

Since the planet moves a distance $2\pi r$ in time T, its speed is related to the period by

$$v = \frac{2\pi r}{T} \qquad \text{11-14}$$

*These are paths produced by slicing a cone and are therefore called "conic sections." A circle is a special case of an ellipse.

Substituting this expression for v in Equation 11-13, we obtain

$$v^2 = \frac{4\pi^2 r^2}{T^2} = \frac{GM_s}{r}$$

or

$$T^2 = \frac{4\pi^2}{GM_s} r^3 \qquad \text{11-15}$$

Kepler's third law

Equation 11-15 is Kepler's third law, which is the same as Equation 11-2 with $C = 4\pi^2/GM_s$.

For the more general case of elliptical orbits, the proof is more difficult. In such cases, the distance r is the mean distance from the sun, which also equals the semimajor axis a. Equation 11-15 also applies to the orbits of the satellites of any planet if we replace the mass of the sun M_s with the mass of the planet.* Finally, since G is known, we can determine the mass of a planet by measuring the period T and the mean orbital radius r of a moon orbiting it.

Example 11-3

A satellite travels in a circular orbit around the earth. Find its period if (*a*) the satellite is just above the surface of the earth, and (*b*) the satellite is at the space shuttle's altitude of 400 km. (Assume that air resistance can be neglected.)

Picture the Problem We use Kepler's third law with M_s in Equation 11-15 replaced by the mass of the earth M_E. The numerical calculation is simplified by using $GM_E = R_E^2 g$ from Equation 11-9.

(*a*)1. Apply Kepler's third law to the satellite:

$$T^2 = \frac{4\pi^2}{GM_E} r^3$$

2. Substitute $r = R_E$ for a satellite just above the earth's surface:

$$T^2 = \frac{4\pi^2}{GM_E} R_E^3$$

3. Use $GM_E = R_E^2 g$ to write T in terms of g:

$$T^2 = \frac{4\pi^2}{GM_E} R_E^3 = \frac{4\pi^2}{R_E^2 g} R_E^3 = \frac{4\pi^2 R_E}{g}$$

$$T = 2\pi\sqrt{\frac{R_E}{g}} = 2\pi\sqrt{\frac{6.37 \times 10^6 \text{ m}}{9.81 \text{ m/s}^2}} = 84.4 \text{ min}$$

(*b*) At an altitude $h = 400$ km, $r = R_E + h = 6770$ km. We find the period at this altitude by noting that T is proportional to $r^{3/2}$:

$$T = (84.4 \text{ min})\left(\frac{r}{R_E}\right)^{3/2}$$

$$= (84.4 \text{ min})\left(\frac{6.77 \times 10^6 \text{ m}}{6.37 \times 10^6 \text{ m}}\right)^{3/2} = 92.5 \text{ min}$$

Exercise Find the radius of the circular orbit of a satellite that orbits the earth with a period of one day. (*Answer* $r = 6.63R_E = 4.22 \times 10^7$ m $= 26{,}200$ mi. If such a satellite is in orbit over the equator and moves in the same direction as the rotation of the earth, it appears stationary relative to the earth. Most satellites are parked in such an orbit, which is called a geosynchronous orbit.)

* For example, it applies to the earth's moon and to all the artificial satellites orbiting the earth if the sun's mass M_s is replaced by the earth's mass M_E.

Exploring

Gravitational and Inertial Mass

The property of an object responsible for the gravitational force it exerts on another object is its *gravitational* mass, whereas the property of an object that measures its resistance to acceleration is its *inertial* mass. We have used the same symbol m for these two properties because, experimentally, they are equal. The fact that the gravitational force exerted on an object is proportional to its inertial mass is a characteristic unique to the force of gravity. One consequence is that all objects near the surface of the earth fall with the same acceleration if air resistance is neglected. This fact has seemed surprising to all since it was discovered. The famous story of Galileo demonstrating it by dropping objects from the Tower of Pisa is just one example of the excitement this discovery aroused in the sixteenth century.

We could easily imagine that the gravitational and inertial masses of an object were not the same. Suppose we write m_G for the gravitational mass and m for the inertial mass. The force exerted by the earth on an object near its surface would then be

$$F = \frac{GM_E m_G}{R_E^2} \qquad 1$$

where M_E is the gravitational mass of the earth. The free-fall acceleration of the object near the earth's surface would then be

$$a = \frac{F}{m} = \left(\frac{GM_E}{R_E^2}\right)\frac{m_G}{m} \qquad 2$$

If gravity were just another property of matter, like color or hardness, it might be reasonable to expect that the ratio m_G/m would depend on such things as the chemical composition of the object or its temperature. The free-fall acceleration would then be different for different objects. The experimental fact, however, is that a is the same for all objects. Thus, we need not maintain the distinction between m_G and m and can set $m_G = m$. We must keep in mind, however, that the equivalence of gravitational and inertial mass is an experimental law, one that is limited by the accuracy of experiment. Experiments testing this equivalence were carried out by Simon Stevin in the 1580s. Galileo publicized the law widely, and his contemporaries made considerable improvements in the experimental accuracy with which the law was established.

The most precise early comparisons of gravitational and inertial mass were made by Newton. From experiments using simple pendulums rather than falling bodies, Newton was able to establish the equivalence between gravitational and inertial mass to an accuracy of about 1 part in 1000. Experiments comparing gravitational and inertial mass have improved steadily over the years. Their equivalence is now established to about 1 part in 10^{12}. The equivalence of gravitational and inertial mass is therefore one of the most well established of all physical laws. It is the basis for the principle of equivalence, which is the foundation of Einstein's general theory of relativity.

11-3 Gravitational Potential Energy

Near the surface of the earth, the gravitational force exerted by the earth on an object is constant because the distance to the center of the earth $r = R_E + h$ is always approximately R_E for $h \ll R_E$. The potential energy of an object near the earth's surface is $mg(r - R_E) = mgh$, where we have chosen $U = 0$ at the earth's surface, $r = R_E$. When we are far from the surface of the earth, we must take into account the fact that the gravitational force exerted by the earth is not uniform but decreases as $1/r^2$. The general definition of potential energy (Equation 5–20*b*) gives

$$dU = -\vec{F}\cdot d\vec{s}$$

where $\vec{F}$ is the force on a particle and $d\vec{s}$ is a general displacement of the particle. For the radial gravitational force $\vec{F}$ given by Equation 11-6 we have

$$dU = -\vec{F}\cdot d\vec{s} = -F_r\,dr = -\left(-\frac{GM_E m}{r^2}\right)dr = +\frac{GM_E m}{r^2}\,dr \qquad \text{11-16}$$

Integrating both sides of this equation we obtain

$$U = -\frac{GM_E m}{r} + U_0 \qquad \text{11-17}$$

where U_0 is a constant of integration. Since only changes in potential energy are important, we can choose the potential energy to be zero at any position. The earth's surface is a good choice for many everyday problems, but it is not always a convenient choice. For example, when considering the potential energy associated with a planet and the sun, there is no reason to want the potential energy to be zero at the surface of the sun. In fact, it is nearly always more convenient to choose the gravitational potential energy of a two-object system to be zero when the separation of the objects is infinite. Thus, $U_0 = 0$ is often a convenient choice. Then

$$U(r) = -\frac{GMm}{r}, \qquad U = 0 \text{ at } r = \infty \qquad \text{11-18}$$

Gravitational potential energy with $U = 0$ at infinite separation

Figure 11-9 is a plot of $U(r)$ versus r for this choice of $U = 0$ at $r = \infty$ for an object of mass m and the earth of mass M_E. This function begins at the negative value $U = -GM_E\,m/R_E = -mgR_E$ at the earth's surface and increases as r increases, approaching zero at infinite r.

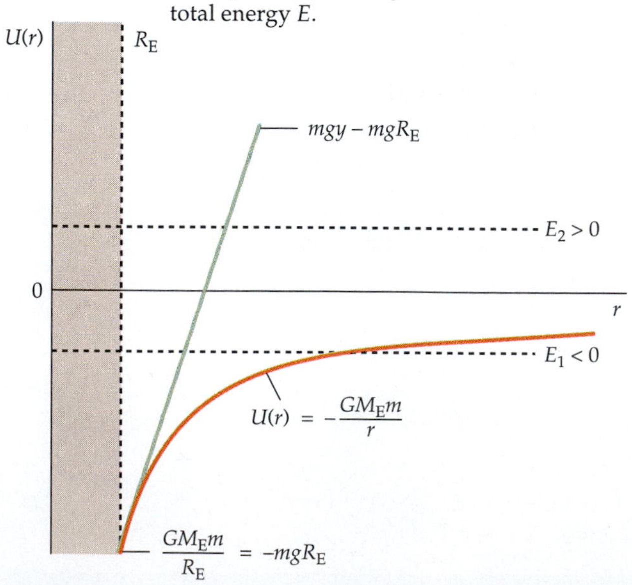

Figure 11-9 Gravitational potential energy $U(r)$ versus distance r from the center of mass. For r just slightly greater than R_E, $U(r)$ is approximately $mgy - mgR_E$, where y is the distance above the earth's surface. Horizontal dashed lines show positive and negative values for the total energy E.

Escape Speed

In the past few decades, the idea of escaping from the earth's gravity has changed from fantasy to reality. Space probes have been sent out to the far reaches of the solar system. Some of these probes are expected to orbit the sun, others will leave the solar system and drift on into outer space. We will see that there is a minimum initial speed, called the **escape speed**, that is required for an object to escape from the earth.

If we project an object upward from the earth with some initial kinetic energy, the kinetic energy decreases and the potential energy increases as the object rises. The maximum increase in potential energy is GM_Em/R_E, as can be seen from Figure 11-10. Therefore, this is the most that the kinetic energy can decrease. If the initial kinetic energy is greater than GM_Em/R_E, the total energy E will be greater than zero (E_2 in Figure 11-10), and the object will still have some kinetic energy when r is very great (or even when r is infinite). Thus, the object will escape from the earth if the initial kinetic energy is greater than GM_Em/R_E. Since the potential energy at the earth's surface is $-GM_Em/R_E$, the total energy $E = K + U$ must be greater than or equal to zero for the object to escape. The speed near the earth's surface corresponding to zero total energy is called the escape speed v_e. It is found by setting the total energy at the surface of the earth equal to zero.

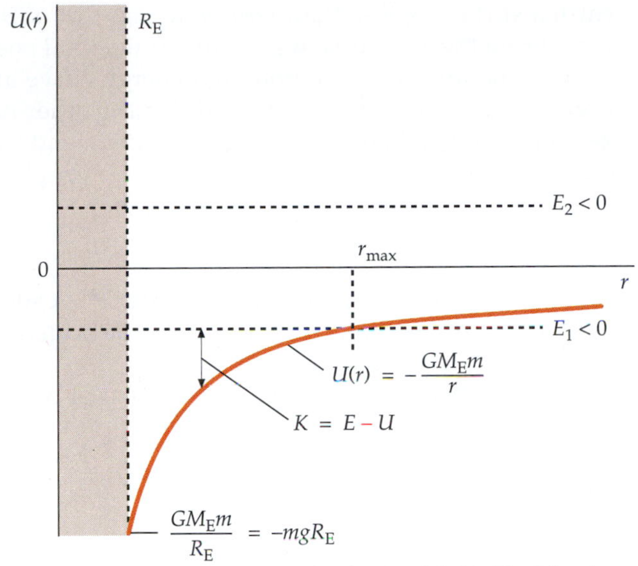

Figure 11-10 The kinetic energy of an object at a distance r from the center of the earth is $E - U(r)$. When the total energy is less than zero (E_1 in the figure), the kinetic energy is zero at $r = r_{max}$ and the object is bound to the earth. When the total energy is greater than zero (E_2 in the figure), the object can escape the earth.

$$E = K + U = 0$$

$$\frac{1}{2}mv_e^2 - \frac{GM_Em}{R_E} = 0$$

$$v_e = \sqrt{\frac{2GM_E}{R_E}} = \sqrt{2gR_E} \qquad 11\text{-}19$$

Escape speed

Using $g = 9.81\ \text{m/s}^2$ and $R_E = 6.37 \times 10^6$ m, we obtain

$$v_e = \sqrt{2(9.81\ \text{m/s}^2)(6.37 \times 10^6\ \text{m})} = 11.2\ \text{km/s}$$

This is about 6.95 mi/s or 25,000 mi/h. An object with this speed will just escape the earth. (However, it will not escape the solar system, because we have neglected the gravitational attraction of the sun and other planets; see Problem 57.)

The escape speed for a planet or moon relative to the thermal speeds of gas molecules determines the kind of atmosphere a planet or moon can have. The average kinetic energy of gas molecules, $(\frac{1}{2}mv^2)_{av}$, is proportional to the absolute temperature T (Chapter 18). At the surface of the earth, the speeds of nearly all of the oxygen and nitrogen molecules are much lower than the escape speed, so these gases are retained in our atmosphere. For the lighter molecules hydrogen and helium, however, a considerable fraction of the molecules have speeds greater than the escape speed. Hydrogen and helium gases are therefore not found in our atmosphere. The escape speed at the surface of the moon is 2.3 km/s, which can be calculated from Equation 11-19 with the mass and radius of the moon replacing M_E and R_E. This is considerably smaller than the escape speed for earth, and in fact is too small for any atmosphere to exist.

Earth as seen from Apollo 11 orbiting the moon on July 31, 1969.

Exercise Find the escape speed at the surface of Mercury, which has a mass $M = 3.31 \times 10^{23}$ kg and a radius $R = 2440$ km. (*Answer* $v_e = \sqrt{2GM/R} = 4.25$ km/s)

Classification of Orbits by Energy

In Figures 11-9 and 11-10, two possible values for the total energy E are indicated on a graph of U versus r: E_1, which is negative, and E_2, which is positive. A negative total energy merely means that the kinetic energy at the earth's surface is less than GM_Em/R_E, so that $K + U$ is never greater than zero. From these figures, we see that if the total energy is negative, the total-energy line intersects the potential-energy curve at some maximum separation r_{max} and the system is bound. On the other hand, if the total energy is zero or positive, there is no such intersection and the system is unbound. The criteria for a bound or unbound system are simply stated.

If $E < 0$, the system is bound.
If $E \geq 0$, the system is unbound.

When E is negative, its absolute value $|E|$ is called the binding energy. The binding energy is the energy that must be added to the system to bring the total energy up to zero.

The potential energy of an object such as a planet or comet of mass m at a distance r from the sun is

$$U(r) = -\frac{GM_sm}{r} \qquad \text{11-20}$$

where M_s is the mass of the sun. The kinetic energy of the object is $\frac{1}{2}mv^2$. If the total energy, kinetic plus potential, is less than zero, the orbit will be an ellipse (or a circle), and the object will be bound to the sun. That is, it cannot escape from the sun. On the other hand, if the total energy is positive, the orbit will be a hyperbola, and the object will make one swing around the sun and leave, never to return again. If the total energy is exactly zero, the orbit will be a parabola, and again the object will escape. That is, when the total energy is zero or positive the object is not bound to the sun.

Example 11-4

A projectile is fired straight up from the surface of the earth with an initial speed v_i = 8 km/s. Find the maximum height the projectile reaches, neglecting air resistance.

Picture the Problem The maximum height is found using energy conservation. We take the surface of the earth as the initial point, with $U_i = -GM_Em/R_E$ and $K_i = \frac{1}{2}mv_i^2$. At the greatest height, $K_f = 0$.

1. Apply conservation of mechanical energy:

$$U_i + K_i = U_f + K_f$$

$$-\frac{GM_Em}{R_E} + \frac{1}{2}mv_i^2 = -\frac{GM_Em}{r} + 0$$

2. Cancel the common term m, use $g = GM_E/R_E^2$, and solve for r:

$$\frac{1}{2}v_i^2 = \frac{GM_E}{R_E}\left(1 - \frac{R_E}{r}\right) = gR_E\left(1 - \frac{R_E}{r}\right)$$

$$1 - \frac{R_E}{r} = \frac{v_i^2}{2gR_E}$$

$$r = \frac{R_E}{1 - v_i^2/2gR_E}$$

3. Substitute numerical values to find r and $h = r - R_E$:

$$\frac{v_i^2}{2gR_E} = \frac{(8000\text{ m/s})^2}{2(9.81\text{ m/s}^2)(6.37 \times 10^6\text{ m})} = 0.512$$

$$r = \frac{R_E}{1 - 0.512} = 2.05R_E$$

$$h = r - R_E = 1.05R_E$$

Example 11-5 *try it yourself*

A projectile is fired straight up from the surface of the earth with an initial speed $v_i = 15$ km/s. Find the speed of the projectile when it is very far from the earth, neglecting air resistance.

Picture the Problem The initial speed is greater than the escape speed, so the total energy of the projectile is positive, and the projectile will escape the earth with some final kinetic energy. Use conservation of mechanical energy to find this kinetic energy and then solve for the final speed.

Cover the column to the right and try these on your own before looking at the answers.

Steps	*Answers*
1. Apply conservation of mechanical energy, noting that $r_f = \infty$, so $U_f = 0$.	$U_i + K_i = U_f + K_f$ $-\frac{GM_Em}{R_E} + \frac{1}{2}mv_i^2 = 0 + \frac{1}{2}mv_f^2$
2. Solve for v_f^2 using $GM_E/R_E^2 = g$ to simplify.	$v_f^2 = v_i^2 - \frac{2GM_E}{R_E} = v_i^2 - 2gR_E$
3. Substitute known values for g and R_E to calculate v_f.	$v_f^2 = 10^8\ \text{m}^2/\text{s}^2$ $v_f = 10^4\ \text{m/s} = 10\ \text{km/s}$

Remark In Figure 11-11, the speed of the projectile in kilometers per second is plotted versus h/R_E, where h is the height above the earth's surface. At very large values of h/R_E, the speed approaches the dashed line $v_f = 10$ km/s.

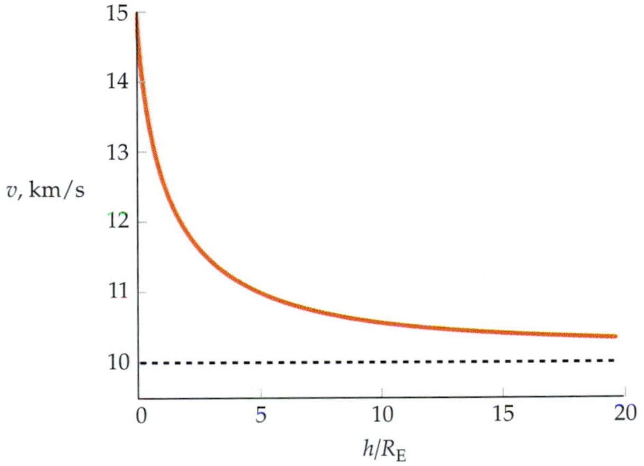

Figure 11-11

Example 11-6 *try it yourself*

Show that the total energy of a satellite in a circular orbit is half its potential energy.

Picture the Problem The total energy of a satellite is the sum of its potential and kinetic energy, $E = U + K$. The kinetic energy depends on the satellite's speed, which can be determined by equating the gravitational force with the centripetal force needed for the circular orbit. Assume that the mass of the earth is much greater than that of the satellite so the center of mass is essentially at the center of the earth.

Cover the column to the right and try these on your own before looking at the answers.

Steps	*Answers*
1. Write the potential energy U of the satellite in terms of the separation distance r.	$U = -\frac{GM_Em}{r}$

2. Write the kinetic energy K in terms of the mass of the satellite and its velocity v. $K = \frac{1}{2}mv^2$

3. Set the gravitational force on the satellite equal to its mass times its centripetal acceleration. $\frac{mv^2}{r} = \frac{GM_E m}{r^2}$

4. Substitute mv^2 from step 3 into the expression for K in step 2. This will give you an expression for K in terms of the distance r. $K = \frac{GM_E m}{2r}$

5. Write the total energy $E = K + U$ as a function of r. Compare it with U in step 1. $E = U + K = -\frac{1}{2}\frac{GM_E m}{r} = \frac{1}{2}U$

Exercise A satellite of mass 450 kg orbits the earth in a circular orbit at 6830 km above the earth's surface. Find (*a*) the potential energy, (*b*) the kinetic energy, and (*c*) the total energy of the satellite. (*Answers* Note that $r = R_E + h = 13{,}200$ km. (*a*) $U = -13.6$ GJ, (*b*) $K = 6.80$ GJ, (*c*) $E = -6.80$ GJ)

11-4 The Gravitational Field $\vec{g}$

The gravitational force exerted by a point mass m_1 on a second mass m_2 a distance $r_{1,2}$ away is given by

$$\vec{F}_{1,2} = -\frac{Gm_1m_2}{r_{1,2}^2}\hat{r}_{1,2}$$

where $\hat{r}_{1,2} = \vec{r}_{1,2}/r_{1,2}$ is a unit vector pointing from m_1 to m_2. The gravitational force on a small test mass m divided by m is called the **gravitational field** $\vec{g}$.

$$\vec{g} = \frac{\vec{F}}{m} \qquad \text{11-21}$$

Definition—Gravitational field

The gravitational field at a point due to a set of point masses is the vector sum of the fields due to the individual masses at that point.

$$\vec{g} = \sum_i \vec{g}_i \qquad \text{11-22}a$$

To find the gravitational field at a point due to a continuous object, we find the field $d\vec{g}$ due to a small mass element dm, assuming it to be a point mass, and integrate over the entire object.

$$\vec{g} = \int d\vec{g} \qquad \text{11-22}b$$

The gravitational field of the earth at a distance $r \geq R_E$ points toward the earth and is given by

$$\vec{g} = -\frac{GM_E}{r^2}\hat{r} \qquad \text{11-23}$$

Gravitational field of the earth

Example 11-7

Two particles each of mass M are fixed on the y axis at $y = +a$ and $y = -a$ (Figure 11-12). Find the gravitational field at a point P on the x axis.

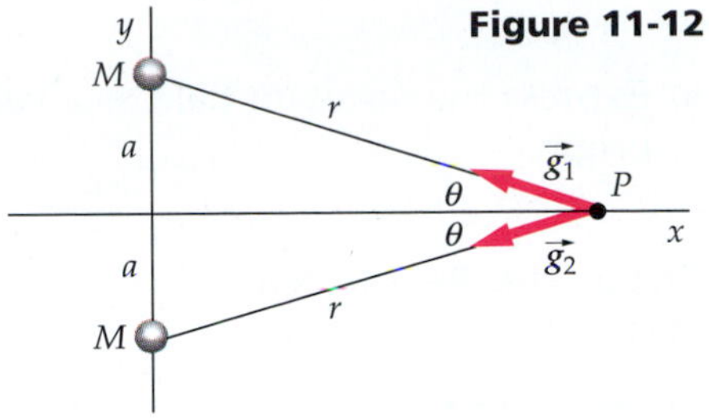

Figure 11-12

Picture the Problem Two particles of mass M produce a gravitational field at point P. The distance between P and either particle is $r = \sqrt{x^2 + a^2}$. The field is the vector sum of the fields $\vec{g}_1$ and $\vec{g}_2$ due to each mass.

1. Calculate the magnitude of either $\vec{g}_1$ or $\vec{g}_2$:
$$g_1 = g_2 = \frac{GM}{r^2} = \frac{GM}{x^2 + a^2}$$

2. By symmetry, the y component of the resultant field is zero. The x component is the sum of g_{1x} and g_{2x}:
$$g_x = g_{1x} + g_{2x} = 2g_{1x} = -2g_1 \cos\theta$$

3. Express $\cos\theta$ in terms of x and a from the figure:
$$\cos\theta = \frac{x}{r} = \frac{x}{\sqrt{x^2 + a^2}}$$

4. Combining the last two results yields $\vec{g}$:
$$\vec{g} = g_x\hat{i} = -\frac{2GMx}{(x^2 + a^2)^{3/2}}\,\hat{i}$$

Check the Result If $x = 0$, we find that $\vec{g} = 0$; the fields due to m_1 and m_2 are equal and opposite at $x = 0$, and hence they cancel. For $x >> a$, $\vec{g} \approx (2GM/x^2)\hat{i}$. The field is the same as if a single mass of $2M$ were at the origin.

Example 11-8

A uniform stick of mass M and length L is centered on the origin and lies along the x axis. Find the gravitational field due to the stick at a point x_0 on the x axis, where $x_0 > L/2$.

Picture the Problem We choose a mass element dm at a distance dx (Figure 11-13). All such elements produce a gravitational field at P that points toward the origin. Thus, we can calculate the total field by integrating the magnitude of the field produced by dm from $x = -L/2$ to $x = +L/2$.

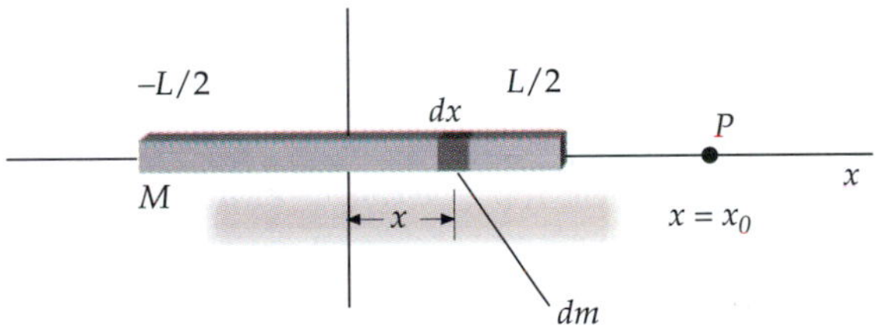

Figure 11-13

1. Find the magnitude of the field at P due to the element dm:
$$dg = \frac{G\,dm}{r^2}$$

2. The mass dm is proportional to the size of the element dx:
$$dm = \frac{M}{L}\,dx$$

3. Write the distance r between dm and point P in terms of x and x_0:
$$r = x_0 - x$$

4. Substitute these results to express dg in terms of x:
$$dg = \frac{G\,dm}{r^2} = \frac{G(M/L)\,dx}{(x_0 - x)^2}$$

5. Integrate to find the total field:
$$g = \int dg = \frac{GM}{L}\int_{-L/2}^{L/2}\frac{dx}{(x_0 - x)^2}$$
$$= \frac{GM}{L}\left[\frac{1}{x_0 - x}\right]_{-L/2}^{L/2}$$
$$= \frac{GM}{L}\left(\frac{1}{x_0 - L/2} - \frac{1}{x_0 + L/2}\right) = \frac{GM}{x_0^2 - L^2/4}$$

6. Express the resultant field as a vector that points toward the origin:

$$\vec{g} = -\frac{GM}{x_0^2 - L^2/4}\hat{i}$$

Check the Result For $x_0 >> L/2$, the field approaches the field of a point mass $\vec{g} = -(GM/x_0^2)\hat{i}$.

$\vec{g}$ of a Spherical Shell and of a Solid Sphere

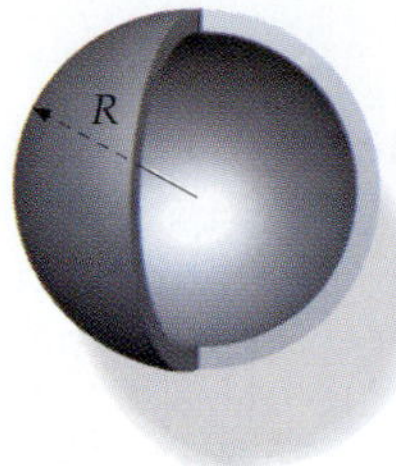

Figure 11-14 A uniform spherical shell of mass M and radius R.

One of Newton's motivations for developing calculus was to prove that the gravitational field outside a solid sphere is the same as if all the mass of the sphere were concentrated at its center. In the next section, we will show that the gravitational field at a distance r from the center of a uniform spherical shell of mass M and radius R (Figure 11-14) is given by

$$\vec{g} = -\frac{GM}{r^2}\hat{r} \qquad \text{for } r > R \qquad \text{11-24}a$$

$$\vec{g} = 0 \qquad \text{for } r < R \qquad \text{11-24}b$$

Gravitational field of a spherical shell

We can understand the result that $\vec{g} = 0$ inside the shell from Figure 11-15, which shows a point mass m_0 inside a spherical shell. In this figure, the masses of the shell segments m_1 and m_2 are related by

$$m_2 = m_1\left(\frac{r_2^2}{r_1^2}\right)$$

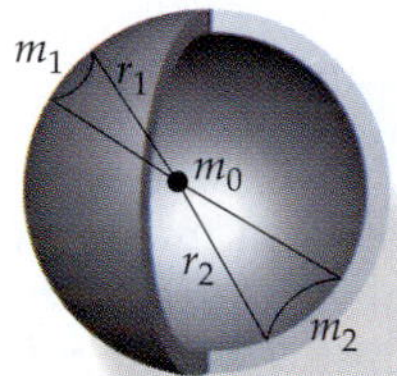

Figure 11-15 A point mass m_0 inside a uniform spherical shell feels no net force.

Since the force due to each mass is proportional to $1/r^2$, the force due to the smaller mass on the left is exactly balanced by that due to the more distant larger mass on the right.

The gravitational field outside a solid sphere is a simple extension of Equation 11-24*a*. We merely consider the solid sphere to consist of a continuous set of spherical shells. Since the field due to each shell is the same as if its mass were concentrated at the center of the shell, the field due to the entire sphere is the same as if the entire mass of the sphere were concentrated at its center:

$$\vec{g} = -\frac{GM}{r^2}\hat{r} \qquad \text{for } r > R \qquad \text{11-25}$$

This result holds whether or not the sphere has a constant density, as long as the density depends only on r so that spherical symmetry is maintained.

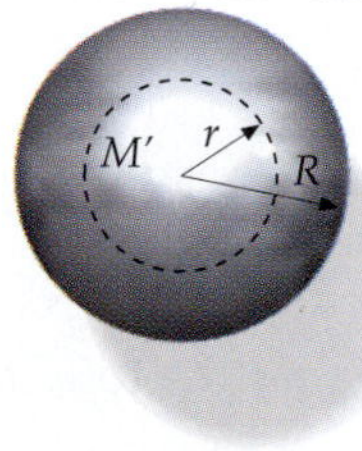

Figure 11-16 A uniform solid sphere of radius R and mass M. Only the mass M', which is inside the sphere of radius r, contributes to the gravitational field at the distance r.

$\vec{g}$ Inside a Solid Sphere We now use Equations 11-24*a* and 11-24*b* to find the gravitational field inside of a solid sphere of constant density at a point a distance r from the center, where r is less than the radius R of the sphere. This would apply, for example, to finding the weight of an object at the bottom of a deep mine shaft. As we have seen, the field inside a spherical shell is zero. Thus, in Figure 11-16, the mass of the sphere outside r exerts no force at or inside r. Therefore, only the mass M' within the radius r contributes to the gravitational field at r. This mass produces a field equal to that of a point mass M' at the center of the sphere. The fraction of the total mass of the sphere within r is equal to the ratio of the volume of a sphere of radius r to that of a sphere of radius R. Thus, for a uniform mass distribution, if M is the total mass of the sphere, M' is given by

$$M' = \frac{\frac{4}{3}\pi r^3}{\frac{4}{3}\pi R^3} M = \frac{r^3}{R^3} M \qquad \text{11-26}$$

The gravitational field at the distance r is thus

$$g_r = -\frac{GM'}{r^2} = -\frac{GMr^3/R^3}{r^2}$$

or

$$\vec{g} = -\frac{GM}{R^3}\vec{r} \qquad \text{for } r < R \qquad \text{11-27}$$

The magnitude of the field increases with distance r inside the sphere. Figure 11-17 shows a plot of the field g_r as a function of r for a solid sphere of constant mass density.

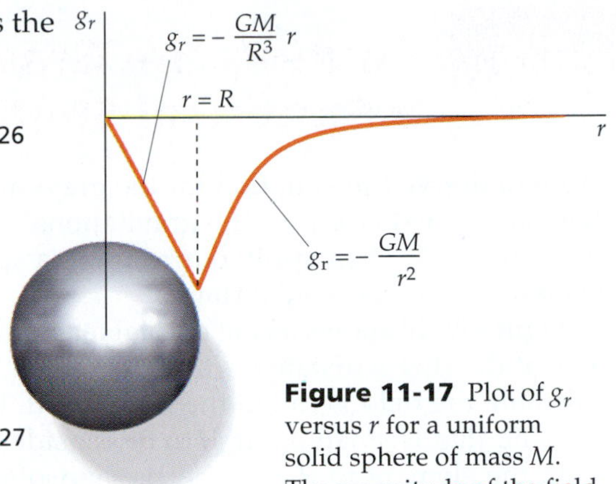

Figure 11-17 Plot of g_r versus r for a uniform solid sphere of mass M. The magnitude of the field increases with r inside the sphere and decreases as $1/r^2$ outside the sphere.

Example 11-9 *try it yourself*

A solid sphere of radius R and mass M is spherically symmetric but not uniform. Its density ρ, defined as its mass per unit volume, is proportional to the distance from the center r for $r \le R$. That is, $\rho = Cr$ for $r \le R$, and $\rho = 0$ for $r > R$, where C is a constant. (*a*) Find C. (*b*) Find g_r for $r \ge R$. (*c*) Find g_r at $r = R/2$.

Picture the Problem (*a*) You can find C by integrating the density over the volume of the sphere and setting the result equal to M. For a volume element, take a spherical shell of radius r and thickness dr. Its volume is $4\pi r^2\,dr$, and its mass is $dm = \rho\,dV = Cr\,(4\pi r^2\,dr)$. (*b*) The field at $r \ge R$ is the same as if the total mass M were at the center of the sphere. (*c*) The field at $r = R/2$ is the same as if mass M' were at the center of the sphere, where M' is the amount of mass within the sphere of radius $R/2$. The mass between $r = R/2$ and $r = R$ produces zero field at $r = R/2$.

Cover the column to the right and try these on your own before looking at the answers.

Steps	*Answers*
(*a*) 1. Integrate dm from $r = 0$ to $r = R$.	$\int_0^R \rho\,dV = C\pi R^4$
2. Set your result equal to M and solve for C in terms of the given quantities M and R.	$C = \frac{M}{\pi R^4}$
(*b*) Write an expression for the field outside the sphere in terms of the mass M and the distance $r \ge R$.	$g_r = \frac{GM}{r^2}$
(*c*) 1. Compute the mass M' that is within the radius $R/2$ by integrating dm from $r = 0$ to $r = R/2$ and use the value of C found in step 2.	$M' = \int_0^{R/2} \rho\,dV = \frac{C\pi R^4}{16} = \frac{M}{16}$
2. Write an expression for the field at $r = R/2$ in terms of M and R.	$g_r = \frac{GM'}{(R/2)^2} = \frac{GM}{4R^2}$

Check the Result The units for C are kg/m^4, so the units for ρ are kg/m^3, which is mass per volume.

optional

11-5 Finding the Gravitational Field of a Spherical Shell by Integration

We will derive the equation for the gravitational field of a spherical shell in two steps. First, we find the gravitational field on the axis of a ring of uniform mass. We then apply our result to a spherical shell, which we can consider to be a set of coaxial rings.

Figure 11-18 shows a ring of total mass m and radius a and a point P on the axis of the ring a distance x from its center. The point P at which we wish to calculate the field is called the **field point.** We choose a mass element dm on the ring that is small enough to be considered a point particle. The distance from the element to P is s, and the line joining the element and P makes an angle α with the axis of the ring.

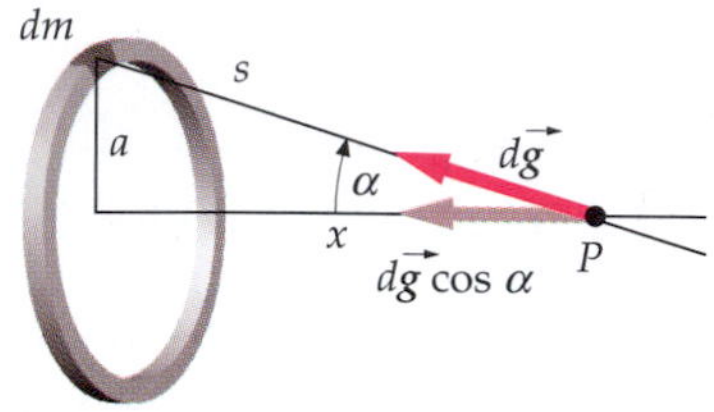

Figure 11-18 The gravitational field at a point P a distance x from a uniform ring. The field due to the element dm points toward the element. The total field due to the ring is along the axis of the ring.

The field at P due to the element dm is toward the element and has the magnitude dg given by

$$dg = \frac{G(dm)}{s^2}$$

From the symmetry of the figure, we can see that when we sum over all the elements of the ring, the net field will be along the axis of the ring; that is, the perpendicular components will sum to zero. For example, the perpendicular component of the field shown in the figure will be canceled by the perpendicular component due to another element of the ring directly opposite the one shown. The net field will therefore be in the negative x direction. The x component of the field due to the element dm is

$$dg_x = -dg \cos \alpha = -\frac{G(dm)}{s^2} \cos \alpha$$

We obtain the total field by summing over all the elements of the ring:

$$g_x = -\int \frac{G\,dm}{s^2} \cos \alpha$$

Since s and α are the same for all points on the ring, they are constants as far as the integration is concerned. Thus,

$$g_x = -\frac{G}{s^2} \cos \alpha \int dm = -\frac{Gm}{s^2} \cos \alpha \qquad \text{11-28}$$

where $m = \int dm$ is the total mass of the ring.

We now use this result to calculate the gravitational field of a spherical shell of mass M and radius R at a point a distance r from the center of the shell. We first consider the case in which the field point P is outside the shell, as in Figure 11-19. By symmetry, the field must be radial. We choose for our element of mass the strip shown, which can be considered to be a ring of mass dM. The field due to this strip is given by Equation 11-28 with m replaced by dM:

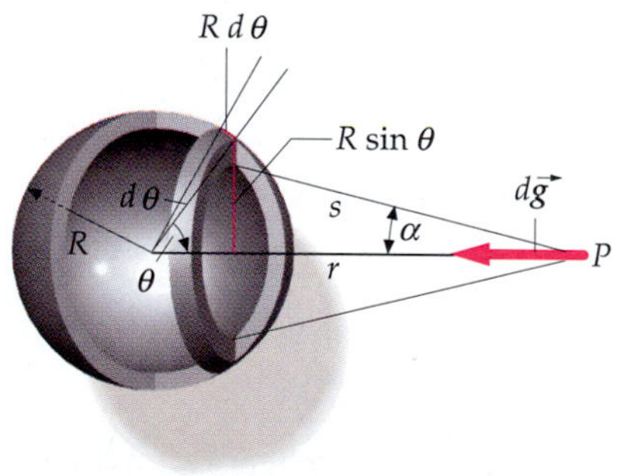

Figure 11-19 A uniform spherical shell of radius R and total mass M. The strip shown can be considered to be a ring of width $R\,d\theta$ and circumference $2\pi R \sin \theta$.

$$dg_r = -\frac{G\,dM}{s^2} \cos \alpha \qquad \text{11-29}$$

The mass dM is proportional to the area of the strip dA, which equals the circumference times the width. The radius of the strip is $R \sin \theta$ so the circumference is $2\pi R \sin \theta$. The width is $R\,d\theta$. If M is the total mass of the shell, and $A = 4\pi R^2$ is its total area, the mass of the strip of area dA is

$$dM = M\frac{dA}{A} = M\frac{(2\pi R \sin \theta)(R\,d\theta)}{4\pi R^2} = \frac{M}{2} \sin \theta\, d\theta \qquad \text{11-30}$$

Substituting this result into Equation 11-29 gives

$$dg_r = -\frac{G\,dM}{s^2}\cos\alpha = -\frac{GM\sin\theta\,d\theta}{2s^2}\cos\alpha \qquad \text{11-31}$$

Before integrating over the entire shell, we must eliminate two of the three related variables s, θ, and α. It turns out to be easiest to write everything in terms of s, which varies from $s = r - R$ at $\theta = 0$ to $s = r + R$ at $\theta = 180°$. By the law of cosines, we have

$$s^2 = r^2 + R^2 - 2rR\cos\theta$$

Differentiating gives

$$2s\,ds = +2rR\sin\theta\,d\theta$$

or

$$\sin\theta\,d\theta = s\frac{ds}{rR}$$

An expression for $\cos\alpha$ can be obtained by applying the law of cosines to the same triangle. We have

$$R^2 = s^2 + r^2 - 2sr\cos\alpha$$

or

$$\cos\alpha = \frac{s^2 + r^2 - R^2}{2sr}$$

Substituting these results into Equation 11-31 gives

$$dg_r = -\frac{GM}{2s^2}\frac{s\,ds}{rR}\frac{s^2 + r^2 - R^2}{2sr}$$

$$= -\frac{GM}{4r^2R}\left(1 + \frac{r^2 - R^2}{s^2}\right)ds \qquad \text{11-32}$$

The field due to the entire shell is found by integrating from $s = r - R$ $(\theta = 0)$ to $s = r + R$ $(\theta = 180°)$.

$$g_r = -\frac{GM}{4r^2R}\int_{r-R}^{r+R}\left(1 + \frac{r^2 - R^2}{s^2}\right)ds$$

$$= -\frac{GM}{4r^2R}\left[s - \frac{r^2 - R^2}{s}\right]_{r-R}^{r+R}$$

Substitution of the upper and lower limits yields $4R$ for the quantity in brackets. Thus,

$$g_r = -\frac{GM}{r^2}$$

which is the same result as in Equation 11-24*a*.

If the field point P is inside the shell, the calculation is identical except that s now varies from $R - r$ to $R + r$. Thus,

$$g_r = -\frac{GM}{4r^2R}\left[s - \frac{r^2 - R^2}{s}\right]_{R-r}^{R+r}$$

Substitution of these upper and lower limits yields 0. Therefore,

$$g_r = 0$$

which is the same as Equation 11-24*b*.

Exploring

Tidal Forces and the Roche Limit

Because the gravitational field of a spherical object is not uniform, but varies as $1/r^2$, the force exerted on an extended object varies across the object. For example, the force exerted by the moon is stronger on the parts of the earth nearest the moon than on the parts farthest away. Figure 1 shows the earth a distance r from the moon.

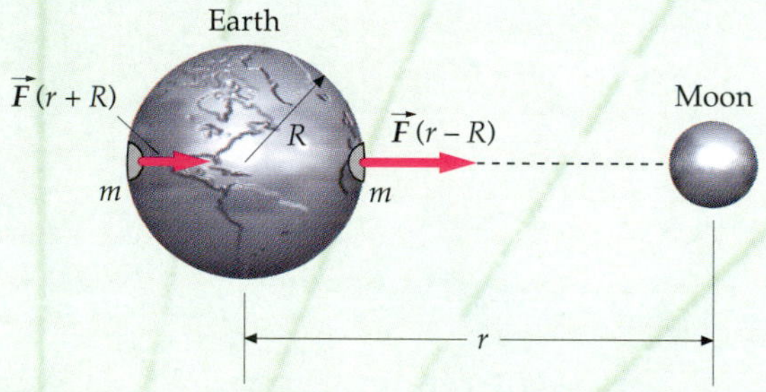

Figure 1

The difference between the force exerted by the moon on a mass m on the near side of the earth at $r - R$ and that exerted on the far side at $r + R$ is

$$\Delta F = F(r-R) - F(r+R)$$
$$= \frac{GMm}{(r-R)^2} - \frac{GMm}{(r+R)^2}$$
$$= \frac{GMm[(r+R)^2 - (r-R)^2]}{(r-R)^2(r+R)^2}$$
$$= \frac{4GMmrR}{(r^2-R^2)^2} \approx \frac{4GMmR}{r^3} \qquad 1$$

where we have neglected R^2 compared with r^2 in the denominator. Although the sun exerts a much greater force on the earth's oceans than does the moon (see Problem 111), the *difference* in the force exerted by the moon when the ocean is closest compared to when it is farthest is much greater than the corresponding differential force exerted by the sun. Because this differential force is responsible for the observed tides, it is called a tidal force.

Most large astronomical objects are held together by gravity. If the tidal force on such an object is greater than the gravitational forces holding the object together, the object will fly apart. Consider a planet of mass M. Because the tidal forces exerted by the planet vary as M/r^3, there is a minimum distance r_m at which a satellite can exist. This minimum distance is called the Roche limit after the French scientist Edouard Roche, who investigated this problem in 1848. We can estimate the Roche limit from a simple calculation. We consider an object of mass $2m$ consisting of two uniform spheres each of mass m and radius a (Figure 2). These objects exert an attractive force on each other as if each were a point mass a distance $2a$ from the other. We consider the force of attraction of these spheres $Gmm/(2a)^2$ to be the force that keeps the object together. When this object is at a distance r from a large object of mass M, as shown in Figure 1, the tidal force is given by Equation 1 with $R = a$. At the Roche limit $r = r_m$, the tidal force equals the force of attraction:

$$\frac{4GMma}{r_m^3} = \frac{Gm^2}{(2a)^2}$$

$$r_m^3 = \frac{16Ma^3}{m}$$

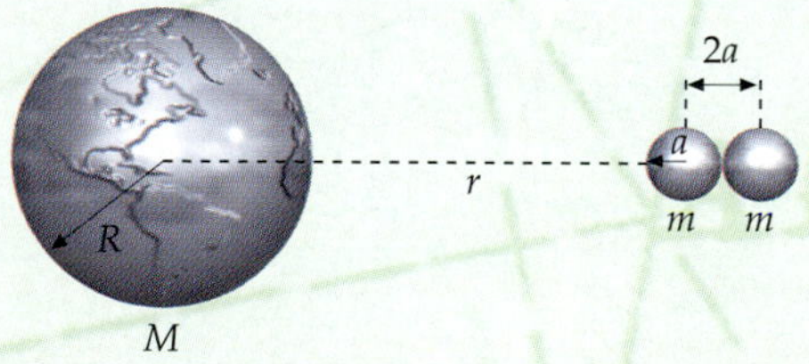

Figure 2

Let ρ_o be the density of the large object, whose radius is R, and ρ_s be the density of the satellite. Then $M = \frac{4}{3}\rho_o R^3$, $m = \frac{4}{3}\rho_s a^3$, and the Roche limit is

$$r_m = \left(\frac{16\rho_o}{\rho_s}\right)^{1/3} R \qquad 2$$

If the densities are equal, the Roche limit is about 2.5 times the radius of the planet. Natural satellites can exist only outside the Roche limit of a planet. Around Saturn, we find that inside the

Roche limit are rings of small particles that cannot form a satellite held together by gravity. Artificial satellites can, of course, exist within the Roche limit of a planet because they are held together by nuts and bolts rather than by gravitational attraction.

If the central object in our system is a black hole, its density ρ_0 will be much greater than that of ordinary objects. The Roche limit for a black hole is many times greater than the radius R of the black hole.

The rings and three moons of Saturn. Close to the planet the tidal forces are too great for a satellite to exist. Instead, we find rings of small particles orbiting close to Saturn. The three moons shown orbit at distances greater than the Roche limit, where the tidal forces are much smaller. (Planetary rings are not uncommon. Faint rings have also been observed around Jupiter, Uranus, and Neptune.)

Summary

1. Kepler's laws are *empirical* observations that were later derived from Newton's laws.
2. Newton's law of gravity is a fundamental law of physics.
3. The gravitational potential energy of a two-mass system relative to $U = 0$ at infinite separation is given by $U = -Gm_1m_2/r$. If the system is bound, its total energy is negative.
4. The gravitational field is a physical concept that describes the condition in space set up by a mass distribution.

Topic	Remarks and Relevant Equations
1. Kepler's Three Laws	Law 1. All planets move in elliptical orbits with the sun at one focus. Law 2. A line joining any planet to the sun sweeps out equal areas in equal times. Law 3. The square of the period of any planet is proportional to the cube of the planet's mean distance from the sun: $T^2 = Cr^3$ **11-2** where C has the same value for all planets; from Newton's law of gravity, C can be shown to be $4\pi^2/GM_S$. Then $T^2 = \frac{4\pi^2}{GM_s}r^3$ **11-15** Kepler's laws can be derived from Newton's law of gravity. The first and third laws follow from the fact that the force exerted by the sun on the planets varies inversely as the square of the separation distance. The second law follows from the fact that the force exerted by the sun on a planet is along the line joining them, so the angular momentum of the planet is conserved. Kepler's laws also hold for any object orbiting another in an inverse-square field, such as a satellite orbiting a planet.
2. Newton's Law of Gravity	Every object exerts on every other object an attractive force that is proportional to the masses of the two objects and inversely proportional to the square of the distance separating them: $\vec{F}_{1,2} = -\frac{Gm_1m_2}{r_{1,2}^2}\hat{r}_{1,2}$ **11-5**
Universal gravitational constant	$G = 6.67 \times 10^{-11}\ \mathrm{N{\cdot}m^2/kg^2}$ **11-4**
3. Gravitational Potential Energy	The gravitational potential energy for any two objects of mass M and m is $U(r) = -\frac{GMm}{r}, \quad U = 0 \text{ at } r = \infty$ **11-18**
4. Escape Speed	If an object of mass m is given an initial kinetic energy at the earth's surface greater than or equal to GM_Em/R_E so that the total energy $E \geq 0$, the object will escape from the earth. In the absence of air resistance, the speed needed for escape is $v_e = \sqrt{\frac{2GM_E}{R_E}} = \sqrt{2gR_E} \approx 11.2\ \mathrm{km/s}$ **11-19**
5. Classification of Orbits	If $E < 0$, the system is bound and the orbit is an ellipse (or circle). If $E \geq 0$, the system is unbound and the orbit is a hyperbola (or parabola for $E = 0$).
6. Gravitational Field	
Definition	$\vec{g} = \frac{\vec{F}}{m}$ **11-21**

Field due to the earth	$\vec{g} = -\frac{GM_E}{r^2}\hat{r}$	**11-23**
Field of a spherical shell	The gravitational field of a uniform spherical shell outside the shell is the same as if all the mass of the shell were concentrated at the center. The field inside the shell is zero.	
	$\vec{g} = -\frac{GM}{r^2}\hat{r}$ for $r > R$	**11-24*a***
	$\vec{g} = 0$ for $r < R$	**11-24*b***
Field of a solid sphere	$\vec{g} = -\frac{GM}{r^2}\hat{r}$ for $r > R$	**11-25**
	$\vec{g} = -\frac{GM}{R^3}\vec{r}$ for $r < R$	**11-27**

Problem-Solving Guide

1. Begin by drawing a neat diagram including the important features of the problem.
2. Kepler's third law, $T^2 \propto R^3$, is useful in relating periods and orbital radii of objects orbiting the same force center.
3. Numerical calculations can be simplified using $GM_E = gR_E^2$.

Summary of Worked Examples

Type of Calculation	Procedure and Relevant Examples
1. Kepler's Laws	
Find the period of a planet or comet around the sun.	Use Kepler's third law and compare with the period of the earth, which is 1 y. **Example 11-1**
Find the period of a satellite orbiting the earth or another planet at a given height.	Use Kepler's third law with M = mass of earth or planet to relate the period to the radius of the orbit and the mass of the planet. For a satellite orbiting the earth, use $GM_E = gR_E^2$. **Example 11-3**
2. Newton's Law of Gravity	
Find the free-fall acceleration of a particle a given distance above the earth's surface.	Use Newton's law of gravity and write $GM_E = gR_E^2$. **Example 11-2**
3. Gravitational Potential Energy	
Find the maximum height of a projectile or find the final speed of a projectile fired with speed greater than the escape speed.	Use conservation of mechanical energy with $U(r) = -GMm/r$. If projected from the earth's surface, use $U(r) = -GMm/R_E = -mgR_E$ **Examples 11-4, 11-5**
Find the escape speed at the surface of a planet.	Use $v_e = \sqrt{2GM_p/R_p}$, which is derived from the conservation of mechanical energy. If the free-fall acceleration at the surface of the planet is known, you can use $GM_p = gR_p^2$.
Compare the kinetic energy, potential energy, and total energy of a particle in a circular orbit.	Use $U(r) = -GMm/r$. Set the gravitational force equal to the mass times the centripetal acceleration to obtain an expression for K in terms of r. **Example 11-6**

4. Gravitational Field

1. Due to a set of objects	Find $\vec{g}_i$ due to each object separately from Newton's law of gravitation and sum the vectors. **Example 11-7**
2. Due to a continuous object	Find $d\vec{g}$ due to a mass element dm and integrate. **Example 11-8**
3. At some point r' inside a spherically symmetric mass distribution	Use $g_r = -GM'/r^2$ where M' is the total mass inside $r \le r'$. The mass M' is found by integrating $\rho\, dV$ where ρ is the mass per unit volume and $dV = 4\pi r^2\, dr$ is the volume of a shell of radius r and thickness dr. **Example 11-9**

Problems

Conceptual Problems

Problems from Optional and Exploring sections

In a few problems, you are given more data than you actually need; in a few other problems, you are required to supply data from your general knowledge, outside sources, or informed estimates.

• Single-concept, single-step, relatively easy
•• Intermediate-level, may require synthesis of concepts
••• Challenging, for advanced students

Take $g = 9.81$ N/kg $= 9.81$ m/s^2 and neglect friction and air resistance in all problems unless otherwise stated.

Kepler's Laws

1 • True or false:

(*a*) Kepler's law of equal areas implies that gravity varies inversely with the square of the distance.
(*b*) The planet closest to the sun, on the average, has the shortest period of revolution about the sun.

2 • If the mass of a satellite is doubled, the radius of its orbit can remain constant if the speed of the satellite

(*a*) increases by a factor of 8.
(*b*) increases by a factor of 2.
(*c*) does not change.
(*d*) is reduced by a factor of 8.
(*e*) is reduced by a factor of 2.

3 • One night, Lucy picked up a strange message on her ham radio. "Help! We ran away from earth to live in peace and serenity, and we got disoriented. All we know is that we are orbiting the sun with a period of 5 years. Where are we?" Lucy did some calculations and told the travelers their mean distance from the sun. What is it?

4 • Halley's comet has a period of about 76 y. What is its mean distance from the sun?

5 • A comet has a period estimated to be about 4210 y. What is its mean distance from the sun? (4210 y was the estimated period of the comet Hale–Bopp, which was seen in the Northern Hemisphere in early 1997. Gravitational interactions with the major planets that occurred during this apparition of the comet greatly changed its period, which is now expected to be about 2380 y.)

6 • The radius of the earth's orbit is 1.496×10^{11} m and that of Uranus is 2.87×10^{12} m. What is the period of Uranus?

7 • The asteroid Hektor, discovered in 1907, is in a nearly circular orbit of radius 5.16 AU about the sun. Determine the period of this asteroid.

8 •• The asteroid Icarus, discovered in 1949, was so named because its highly eccentric elliptical orbit brings it close to the sun at perihelion. The eccentricity e of an ellipse is defined by the relation $d_p = a(1 - e)$, where d_p is the perihelion distance and a is the semimajor axis. Icarus has an eccentricity of 0.83. The period of Icarus is 1.1 years. (*a*) Determine the semimajor axis of the orbit of Icarus. (*b*) Find the perihelion and aphelion distances of the orbit of Icarus.

Newton's Law of Gravity

9 • Why don't you feel the gravitational attraction of a large building when you walk near it?

10 • Astronauts orbiting in a satellite 300 km above the surface of the earth feel weightless. Why? Is the force of gravity exerted by the earth on them negligible at this height?

11 •• The distance from the center of the earth to a point where the acceleration due to gravity is $g/4$ is

(*a*) R_E.
(*b*) $4R_E$.
(*c*) $\frac{1}{2}R_E$.
(*d*) $2R_E$.
(*e*) none of the above.

12 •• At the surface of the moon, the acceleration due to the gravity of the moon is a. At a distance from the center of the moon equal to four times the radius of the moon, the acceleration due to the gravity of the moon is

(*a*) $16a$.
(*b*) $a/4$.
(*c*) $a/3$.
(*d*) $a/16$.
(*e*) None of the above.

13 • One of Jupiter's moons, Io, has a mean orbital radius of 4.22×10^8 m and a period of 1.53×10^5 s. (*a*) Find the mean orbital radius of another of Jupiter's moons, Callisto, whose period is 1.44×10^6 s. (*b*) Use the known value of G to compute the mass of Jupiter.

14 • The mass of Saturn is 5.69×10^{26} kg. (*a*) Find the period of its moon Mimas, whose mean orbital radius is 1.86×10^8 m. (*b*) Find the mean orbital radius of its moon Titan, whose period is 1.38×10^6 s.

15 • Calculate the mass of the earth from the period of the moon $T = 27.3$ d, its mean orbital radius $r_m = 3.84 \times 10^8$ m, and the known value of G.

16 • Use the period of the earth (1 y), its mean orbital radius (1.496×10^{11} m), and the value of G to calculate the mass of the sun.

17 • An object is dropped from a height of 6.37×10^6 m above the surface of the earth. What is its initial acceleration?

18 • Suppose you leave the solar system and arrive at a planet that has the same mass per unit volume as the earth but has 10 times the earth's radius. What would you weigh on this planet compared with what you weigh on earth?

19 • Suppose that the earth retained its present mass but was somehow compressed to half its present radius. What would be the value of g, the acceleration due to gravity, at the surface of this new, compact planet?

20 • A planet moves around a massive sun with constant angular momentum. When the planet is at perihelion, it has a speed of 5×10^4 m/s and is 1.0×10^{15} m from the sun. The orbital radius increases to 2.2×10^{15} m at aphelion. What is the planet's speed at aphelion?

21 • A comet orbits the sun with constant angular momentum. It has a maximum radius of 150 AU, and at aphelion its speed is 7×10^3 m/s. The comet's closest approach to the sun is 0.4 AU. What is its speed at perihelion?

22 •• The speed of an asteroid is 20 km/s at perihelion and 14 km/s at aphelion. Determine the ratio of the aphelion to perihelion distance.

23 •• A satellite with a mass of 300 kg moves in a circular orbit 5×10^7 m above the earth's surface. (*a*) What is the gravitational force on the satellite? (*b*) What is the speed of the satellite? (*c*) What is the period of the satellite?

24 •• At the airport, a physics student weighs 800 N. The student boards a jet plane that rises to an altitude of 9500 m. What is the student's loss in weight?

25 •• Suppose that Kepler had found that the period of a planet's circular orbit is proportional to the square of the orbit radius. What conclusion would Newton have drawn concerning the dependence of the gravitational attraction on distance between two masses?

26 •• A superconducting gravity meter can measure changes in gravity of the order $\Delta g/g = 10^{-11}$. (*a*) Estimate the maximum range at which an 80-kg person can be detected by this gravity meter. Assume that the gravity meter is stationary, and that the person's mass can be considered to be concentrated at his or her center of gravity. (*b*) What vertical change in the position of the gravity meter in the earth's gravitational field is detectable?

27 •• During a solar eclipse, when the moon is between the earth and the sun, the gravitational pull of the moon and the sun on a student are in the same direction. (*a*) If the pull of the earth on the student is 800 N, what is the force of the moon on the student? (*b*) What is the force of the sun on the student? (*c*) What percentage correction due to the sun and moon when they are directly overhead should be applied to the reading of a very accurate scale to obtain the student's weight?

28 •• Suppose that the attractive interaction between a star of mass M and a planet of mass $m \ll M$ were of the form $F = KMm/r$, where K is the gravitational constant. What would be the relation between the radius of the planet's circular orbit and its period?

29 •• The mass of the earth is 5.98×10^{24} kg and its radius is 6370 km. The radius of the moon is 1738 km. The acceleration of gravity at the surface of the moon is 1.62 m/s^2. What is the ratio of the average density of the moon to that of the earth?

30 ••• A plumb bob near a large mountain is slightly deflected from the vertical by the gravitational attraction of the mountain. Estimate the order of magnitude of the angle of deflection using any assumptions you like.

Measurement of *G*

31 • Why is G so difficult to measure?

32 • The masses in a Cavendish apparatus are $m_1 = 10$ kg and $m_2 = 10$ g, the separation of their centers is 6 cm, and the rod separating the two small masses is 20 cm long. (*a*) What is the force of attraction between the large and small masses? (*b*) What torque must be exerted by the suspension to balance these forces?

33 • The masses in a Cavendish apparatus are $m_1 = 12$ kg and $m_2 = 15$ g, and the separation of their centers is 7 cm. (*a*) What is the force of attraction between these two masses? (*b*) If the rod separating the two small masses is 18 cm long, what torque must be exerted by the suspension to balance the torque exerted by gravity?

Exploring . . . Gravitational and Inertial Mass

34 •• How would everyday life change if gravitational and inertial mass were not identical?

35 •• If gravitational and inertial mass were not identical, what would change for

(*a*) an offensive lineman on a football team?
(*b*) a car?
(*c*) a paperweight?

36 • A standard object defined as having a mass of exactly 1 kg is given an acceleration of 2.6587 m/s^2 when a certain force is applied to it. A second object of unknown mass acquires an acceleration of 1.1705 m/s^2 when the same force is applied to it. (*a*) What is the mass of the second object? (*b*)

Is the mass that you determined in part (*a*) gravitational or inertial mass?

37 • The weight of a standard object defined as having a mass of exactly 1 kg is measured to be 9.81 N. In the same laboratory, a second object weighs 56.6 N. (*a*) What is the mass of the second object? (*b*) Is the mass you determined in part (*a*) gravitational or inertial mass?

Gravitational Potential Energy

38 • (*a*) Taking the potential energy to be zero at infinite separation, find the potential energy of a 100-kg object at the surface of the earth. (Use 6.37×10^6 m for the earth's radius.) (*b*) Find the potential energy of the same object at a height above the earth's surface equal to the earth's radius. (*c*) Find the escape speed for a body projected from this height.

39 • A point mass m_0 is initially at the surface of a large sphere of mass M and radius R. How much work is needed to remove it to a very large distance away from the large sphere?

40 • Suppose that in space there is a duplicate earth, except that it has no atmosphere, is not rotating, and is not in motion around any sun. What initial velocity must a spacecraft on its surface have to travel vertically upward a distance above the surface of the planet equal to one earth radius?

41 •• An object is dropped from rest from a height of 4×10^6 m above the surface of the earth. If there is no air resistance, what is its speed when it strikes the earth?

42 •• An object is projected upward from the surface of the earth with an initial speed of 4 km/s. Find the maximum height it reaches.

43 •• A spherical shell has a radius R and a mass M. (*a*) Write expressions for the force exerted by the shell on a point mass m_0 when m_0 is outside the shell and when it is inside the shell. (*b*) What is the potential-energy function $U(r)$ for this system when the mass m_0 is at a distance r ($r \geq R$) if $U = 0$ at $r = \infty$? Evaluate this function at $r = R$. (*c*) Using the general relation for $dU = -\vec{F}\cdot d\vec{r} = -F_r\, dr$, show that U is constant everywhere inside the shell. (*d*) Using the fact that U is continuous everywhere, including at $r = R$, find the value of the constant U inside the shell. (*e*) Sketch $U(r)$ versus r for all possible values of r.

44 ••• Our galaxy can be considered to be a large disk of radius R and mass M of approximately uniform mass density. (*a*) Consider a ring element of radius r and thickness dr of such a disk. Find the gravitational potential energy of a 1-kg mass on the axis of this element a distance x from its center. (*b*) Integrate your result for part (*a*) to find the total gravitational potential energy of a 1-kg mass at a distance x due to the disk. (*c*) From $F_x = -dU/dx$ and your result for part (*b*), find the gravitational field g_x on the axis of the disk.

45 ••• The assumption of uniform mass density in Problem 44 is rather unrealistic. For most galaxies, the mass density increases greatly toward the center of the galaxy. Repeat Problem 44 using a surface mass density of the form $\sigma(r) = C/r$, where $\sigma(r)$ is the mass per unit area of the disk at a distance r from the center. First determine the constant C in terms of R and M; then proceed as in Problem 44.

Escape Speed

46 • What is the effect of air resistance on the escape speed near the earth's surface?

47 • Would it be possible in principle for the earth to escape from the solar system?

48 • If the mass of a planet is doubled with no increase in its size, the escape speed for that planet will be

(*a*) increased by a factor of 1.4.
(*b*) increased by a factor of 2.
(*c*) unchanged.
(*d*) reduced by a factor of 1.4.
(*e*) reduced by a factor of 2.

49 • The planet Saturn has a mass 95.2 times that of the earth and a radius 9.47 times that of the earth. Find the escape speed for objects near the surface of Saturn.

50 • Find the escape speed for a rocket leaving the moon. The acceleration of gravity on the moon is 0.166 times that on earth, and the moon's radius is $0.273R_E$.

51 •• A particle is projected from the surface of the earth with a speed equal to twice the escape speed. When it is very far from the earth, what is its speed?

52 •• What initial speed should a particle be given if it is to have a final speed when it is very far from the earth equal to its escape speed?

53 •• A space probe launched from the earth with an initial speed v_i is to have a speed of 60 km/s when it is very far from the earth. What is v_i?

54 •• (*a*) Calculate the energy in joules necessary to launch a 1-kg mass from the earth at escape speed. (*b*) Convert this energy to kilowatt-hours. (*c*) If energy can be obtained at 10 cents per kilowatt-hour, what is the minimum cost of giving an 80-kg astronaut enough energy to escape the earth's gravitational field?

55 •• Show that the escape speed from a planet is related to the speed of a circular orbit just above the surface of the planet by $v_e = \sqrt{2}\, v_c$, where v_c is the speed of an object in the circular orbit.

56 •• Find the speed of the earth v_c as it orbits the sun, assuming a circular orbit. Use this and the result of Problem 55 to calculate the speed v_{eS} needed by the earth to escape from the sun.

57 •• If an object has just enough energy to escape from the earth, it will not escape from the solar system because of the attraction of the sun. Use Equation 11-19 with M_S replacing M_E and the distance to the sun r_S replacing R_E to calculate the speed v_{eS} needed to escape from the sun's gravitational field for an object at the surface of the earth. Neglect the attraction of the earth. Compare your answer with that in Problem 56. Show that if v_e is the speed needed to escape from the earth, neglecting the sun, then the speed of an object at the earth's surface needed to escape from the solar system is given by $v_{e,\text{solar}}^2 = v_e^2 + v_{eS}^2$, and calculate $v_{e,\text{solar}}$.

58 •• Why is it reasonable to neglect the other planets in calculating the speed needed to escape from the solar system? Would you expect the actual value of this speed to be greater or less than that calculated in Problem 57?

59 •• An object is projected vertically from the surface of the earth. Show that the maximum height reached by the object is $H = R_E H'/(R_E - H')$, where H' is the height that it would reach if the gravitational field were constant.

Orbits

60 •• An object (say, a newly discovered comet) enters the solar system and makes a pass around the sun. How can we tell if the object will return many years later, or if it will never return?

61 •• A spacecraft of 100 kg mass is in a circular orbit about the earth at a height $h = 2R_E$. (*a*) What is the period of the spacecraft's orbit about the earth? (*b*) What is the spacecraft's kinetic energy? (*c*) Express the angular momentum L of the spacecraft about the earth in terms of its kinetic energy K and find its numerical value.

62 •• Many satellites orbit the earth about 1000 km above the earth's surface. Geosynchronous satellites orbit at a distance of 4.22×10^7 m from the center of the earth. How much more energy is required to launch a 500-kg satellite into a geosynchronous orbit than into an orbit 1000 km above the surface of the earth?

63 •• It is theoretically possible to place a satellite at a position between the earth and the sun on the line joining them, where the gravitational forces of the sun and the earth on the satellite combine in such a way that the satellite will execute a circular orbit around the sun that is synchronous with the earth's orbit around the sun. (In other words, the satellite and the earth have the same orbital period about the sun, even though they are at different distances from the sun. The satellite always remains on the line joining the earth and the sun.) Write an expression that relates the appropriate circular orbital speed v of a satellite in such a situation to its distance r from the sun. Your expression may also contain quantities shown in Figure 11-20 plus the gravitational constant G.

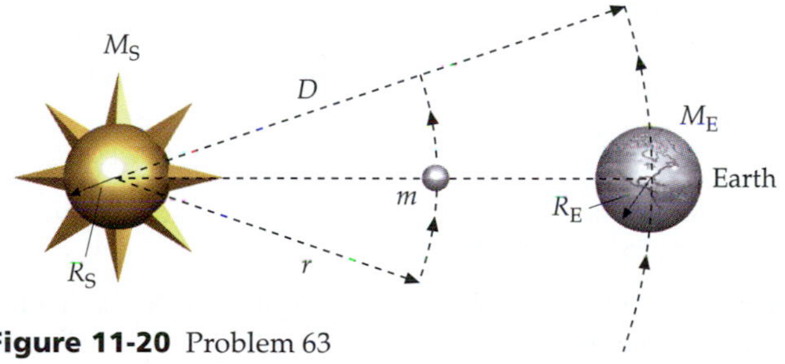

Figure 11-20 Problem 63

The Gravitational Field $\vec{g}$

64 • A 3-kg mass experiences a gravitational force of $12\text{ N}\,\hat{i}$ at some point P. What is the gravitational field at that point?

65 • The gravitational field at some point is given by $\vec{g} = 2.5 \times 10^{-6}\text{ N/kg}\,\hat{j}$. What is the gravitational force on a mass of 4 g at that point?

66 •• A point mass m is on the x axis at $x = L$ and a second equal point mass m is on the y axis at $y = L$. (*a*) Find the gravitational field at the origin. (*b*) What is the magnitude of this field?

67 •• Five equal masses M are equally spaced on the arc of a semicircle of radius R as in Figure 11-21. A mass m is located at the center of curvature of the arc. (*a*) If M is 3 kg, m is 2 kg, and R is 10 cm, what is the force on m due to the five masses? (*b*) If m is removed, what is the gravitational field at the center of curvature of the arc?

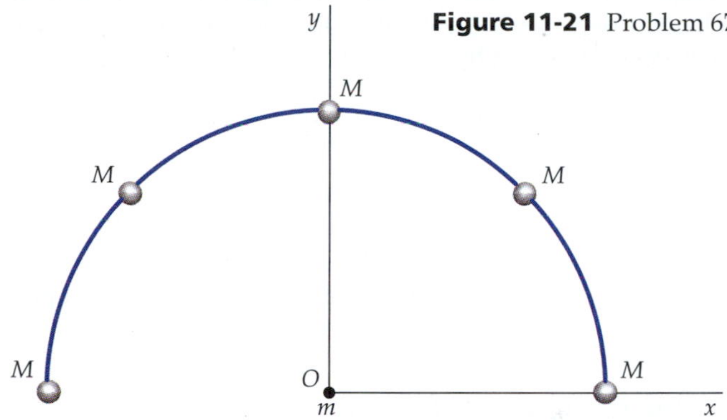

Figure 11-21 Problem 67

68 •• A point mass $m_1 = 2$ kg is at the origin and a second point mass $m_2 = 4$ kg is on the x axis at $x = 6$ m. Find the gravitational field at (*a*) $x = 2$ m, and (*b*) $x = 12$ m. (*c*) Find the point on the x axis for which $g = 0$.

69 •• (*a*) Show that the gravitational field of a ring of uniform mass is zero at the center of the ring.
(*b*) Figure 11-22 shows a point P in the plane of the ring but not at its center. Consider two elements of the ring of length s_1 and s_2 at distances of r_1 and r_2, respectively.
1. What is the ratio of the masses of these elements?
2. Which produces the greater gravitational field at point P?
3. What is the direction of the field at point P due to these elements?

(*c*) What is the direction of the gravitational field at point P due to the entire ring?
(*d*) Suppose that the gravitational field varied as $1/r$ rather than $1/r^2$. What would be the net gravitational field at point P due to the two elements?
(*e*) How would your answers to parts (*b*) and (*c*) differ if point P were inside a spherical shell of uniform mass rather than inside a plane circular ring?

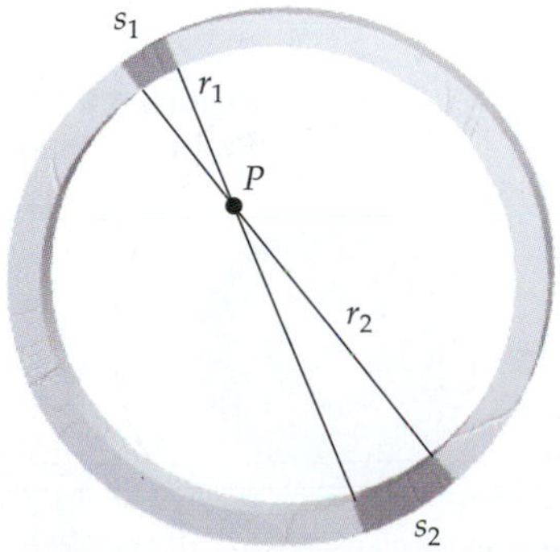

Figure 11-22 Problem 69

70 •• Show that the maximum value of $|g_x|$ for the field of Example 11-7 occurs at the points $x = \pm a/\sqrt{2}$.

71 •• A nonuniform stick of length L lies on the x axis with one end at the origin. Its mass density λ (mass per unit length) varies as $\lambda = Cx$, where C is a constant. (Thus, an element of the stick has mass $dm = \lambda\,dx$.) (*a*) What is the total mass of the stick? (*b*) Find the gravitational field due to the stick at a point $x_0 > L$.

72 ••• A uniform rod of mass M and length L lies along the x axis with its center at the origin. Consider an element of length dx at a distance x from the origin. (*a*) Show that this element produces a gravitational field at a point x_0 on the x axis ($x_0 > \frac{1}{2}L$) given by

$$dg_x = -\frac{GM}{L(x_0 - x)^2}\,dx$$

(*b*) Integrate this result over the length of the rod to find the total gravitational field at the point x_0 due to the rod. (*c*) What is the force on an object of mass m_0 at x_0? (*d*) Show that for $x_0 >> L$, the field is approximately equal to that of a point mass M.

$\vec{g}$ due to Spherical Objects

73 •• Explain why the gravitational field increases with r rather than decreasing as $1/r^2$ as one moves out from the center inside a solid sphere of uniform mass.

74 • A spherical shell has a radius of 2 m and a mass of 300 kg. What is the gravitational field at the following distances from the center of the shell: (*a*) 0.5 m; (*b*) 1.9 m; (*c*) 2.5 m?

75 • A spherical shell has a radius of 2 m and a mass of 300 kg, and its center is located at the origin of a coordinate system. Another spherical shell with a radius of 1 m and mass 150 kg is inside the larger shell with its center at 0.6 m on the x axis. What is the gravitational force of attraction between the two shells?

76 • Two spheres, S_1 and S_2, have equal radii R and equal masses M. The density of sphere S_1 is constant, whereas that of sphere S_2 depends on the radial distance according to $\rho(r) = C/r$. If the acceleration of gravity at the surface of sphere S_1 is g_1, what is the acceleration of gravity at the surface of sphere S_2?

77 •• Two homogeneous spheres, S_1 and S_2, have equal masses but different radii, R_1 and R_2. If the acceleration of gravity on the surface of sphere S is g_1, what is the acceleration of gravity on the surface of sphere S_2?

78 •• Two concentric uniform spherical shells have masses M_1 and M_2 and radii a and $2a$ as in Figure 11-23. What is the magnitude of the gravitational force on a point mass m located (*a*) a distance $3a$ from the center of the shells? (*b*) a distance $1.9a$ from the center of the shells? (*c*) a distance $0.9a$ from the center of the shells?

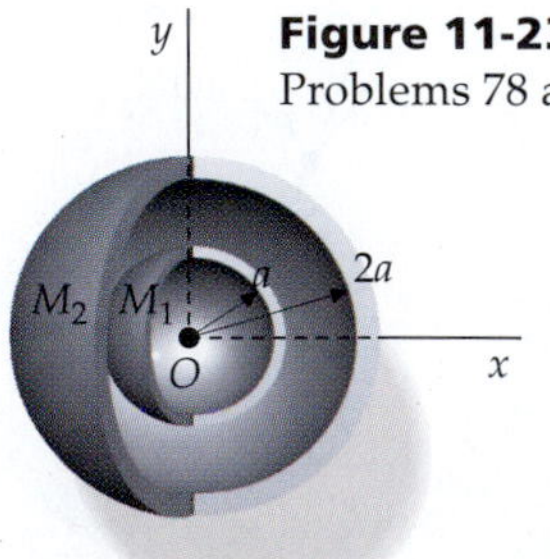

Figure 11-23 Problems 78 and 79

79 •• The inner spherical shell in Problem 78 is shifted such that its center is now at $x = 0.8a$. The points $3a$, $1.9a$, and $0.9a$ lie along the same radial line from the center of the larger spherical shell. (*a*) What is the force on m at $x = 3a$? (*b*) What is the force on m at $x = 1.9a$? (*c*) What is the force on m at $x = 0.9a$?

$\vec{g}$ Inside Solid Spheres

80 •• Suppose the earth were a sphere of uniform mass. If there were a deep elevator shaft going 15,000 m into the earth, what would be the loss in weight at the bottom of this deep shaft for a student who weighs 800 N at the surface of the earth?

81 •• A sphere of radius R has its center at the origin. It has a uniform mass density ρ_0, except that there is a spherical cavity in it of radius $r = \frac{1}{2}R$ centered at $x = \frac{1}{2}R$ as in Figure 11-24. Find the gravitational field at points on the x axis for $|x| > R$. (*Hint*: The cavity may be thought of as a sphere of mass $m = \frac{4}{3}\pi r^3 \rho_0$ plus a sphere of mass $-m$.)

82 ••• For the sphere with the cavity in Problem 81, show that the gravitational field inside the cavity is uniform, and find its magnitude and direction.

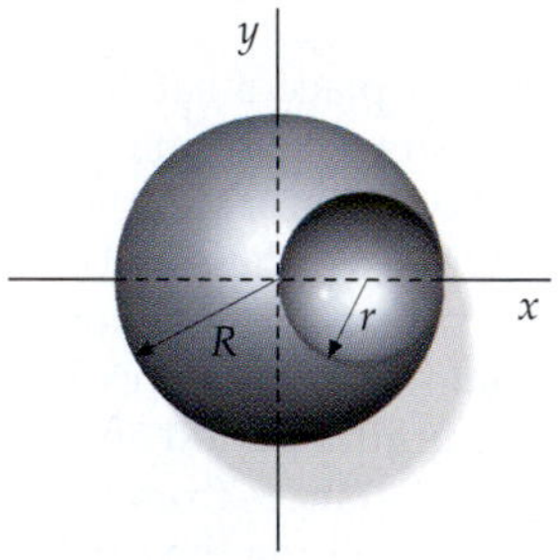

Figure 11-24 Problems 81 and 82

83 ••• A straight, smooth tunnel is dug through a spherical planet whose mass density ρ_0 is constant. The tunnel passes through the center of the planet and is perpendicular to the planet's axis of rotation, which is fixed in space. The planet rotates with an angular velocity ω such that objects in the tunnel have no acceleration relative to the tunnel. Find ω.

84 ••• The density of a sphere is given by $\rho(r) = C/r$. The sphere has a radius of 5 m and a mass of 1011 kg. (*a*) Determine the constant C. (*b*) Obtain expressions for the gravitational field for (1) $r > 5$ m, and (2) $r < 5$ m.

85 ••• A hole is drilled into the sphere of Problem 84 toward the center of the sphere to a depth of 2 km below the sphere's surface. A small mass is dropped from the surface into the hole. Determine the speed of the small mass as it strikes the bottom of the hole.

86 ••• The solid surface of the earth has a density of about 3000 kg/m^3. A spherical deposit of heavy metals with a density of 8000 kg/m^3 and a radius of 1000 m is centered 2000 m below the surface. Find $\Delta g/g$ at the surface directly above this deposit, where Δg is the increase in the gravitational field due to the deposit.

87 ••• Two identical spherical hollows are made in a lead sphere of radius R. The hollows have a radius $R/2$. They touch the outside surface of the sphere and its center as in Figure 11-25. The mass of the lead sphere before hollowing was M. (*a*) Find the force of attraction of a small sphere of mass m to the lead sphere at the position shown in the figure below. (*b*) What is the attractive force if m is located right at the surface of the lead sphere?

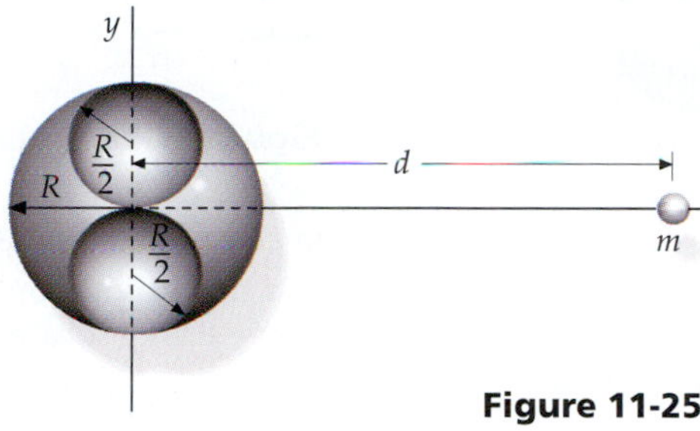

Figure 11-25 Problem 87

General Problems

88 • If K is the kinetic energy of the moon in its orbit around the earth, and U is the potential energy of the earth–moon system, what is the relationship between K and U?

89 •• A woman whose weight on earth is 500 N is lifted to a height two earth radii above the surface of the earth. Her weight will

(*a*) decrease to one-half of the original amount.
(*b*) decrease to one-quarter of the original amount.
(*c*) decrease to one-third of the original amount.
(*d*) decrease to one-ninth of the original amount.

90 • The mean distance of Pluto from the sun is 39.5 AU. Find the period of Pluto.

91 • The semimajor axis of Ganymede, a moon of Jupiter discovered by Galileo, is 1.07×10^6 km, and its period is 7.155 days. Determine the mass of Jupiter.

92 • Calculate the mass of the earth using the known values of G, g, and R_E.

93 • Uranus has a moon, Umbriel, whose mean orbital radius is 2.67×10^8 m and whose period is 3.58×10^5 s. (*a*) Find the period of another of Uranus's moons, Oberon, whose mean orbital radius is 5.86×10^8 m. (*b*) Use the known value of G to find the mass of Uranus.

94 •• Joe and Sally learn that there is a point between the earth and the moon where the gravitational effects of the two bodies balance each other. Being of a New Age bent, they decide to try to conceive a child free from the bondage of gravity, so they book an earth-to-moon trip. How far from the center of the earth should they try to conceive Zerog, the first zero-gravity baby?

95 •• The force exerted by the earth on a particle of mass m a distance r from the center of the earth has the magnitude $GM_Em/r^2 = mgR_E^2/r^2$. (*a*) Calculate the work you must do against gravity to move the particle from a distance r_1 to r_2. (*b*) Show that when $r_1 = R_E$ and $r_2 = R_E + h$, the result can be written

$$W = mgR_E^2\left(\frac{1}{R_E} - \frac{1}{R_E + h}\right)$$

(*c*) Show that when $h \ll R_E$, the work is given approximately by $W = mgh$.

96 •• Suppose that the gravitational force of attraction depended not on $1/r^2$ but was proportional to the distance between the two masses, like the force of a spring. In a planetary system like the solar system, what would then be the relation between the period of a planet and its orbit radius, assuming all orbits were circular?

97 •• A uniform sphere of radius 100 m and density 2000 kg/m^3 is in free space far from other massive objects. (*a*) Find the gravitational field outside of the sphere as a function of r. (*b*) Find the gravitational field inside the sphere as a function of r.

98 •• Two spherical planets have identical mass densities. Planet P_1 has a radius R_1, and planet P_2 has a radius R_2. If the acceleration of gravity at the surface of planet P_1 is g_1, what is the acceleration of gravity at the surface of planet P_2?

99 •• Jupiter has a mass 320 times that of Earth and a volume 1320 times that of Earth. A "day" on Jupiter is 9 h 50 min long. Find the height h above Jupiter at which a satellite must be revolving to have a period equal to 9 h 50 min.

100 •• The average density of the moon is $\rho = 3340$ kg/m^3. Find the minimum possible period T of a spacecraft orbiting the moon.

101 •• A satellite is circling around the moon (radius 1700 km) close to the surface at a speed v. A projectile is launched from the moon vertically up at the same initial speed v. How high will it rise?

102 •• Two space colonies of equal mass orbit a star (Figure 11-26). The Yangs in m_1 move in a circular orbit of radius 10^{11} m with a period of 2 y. The Yins in m_2 move in an elliptical orbit with a closest distance $r_1 = 10^{11}$ m and a farthest distance $r_2 = 1.8 \times 10^{11}$ m. (*a*) Using the fact that the mean radius of an elliptical orbit is the length of the semimajor axis, find the length of the Yin year. (*b*) What is the mass of the star? (*c*) Which colony moves faster at point P in Figure 11-26? (*d*) Which colony has the greater total energy? (*e*) How does the speed of the Yins at point P compare with their speed at point A?

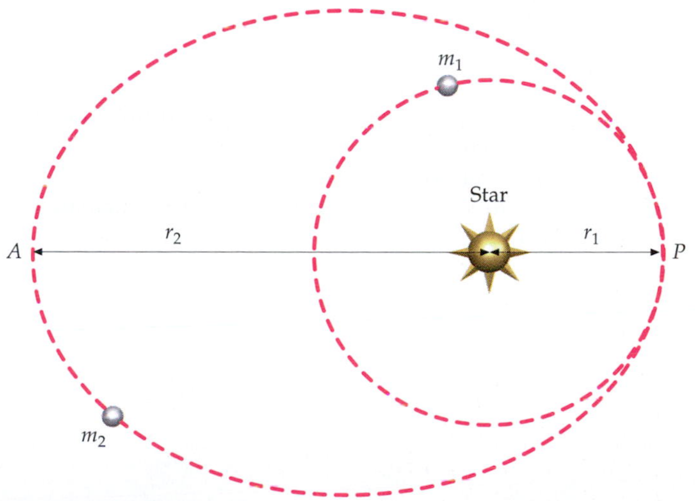

Figure 11-26 Problem 102

103 •• In a binary star system, two stars orbit about their common center of mass. If the stars have masses m_1 and m_2 and are separated by a distance r, show that the period of rotation is related to r by

$$T^2 = \frac{4\pi^2}{G(m_1 + m_2)} r^3$$

104 •• Two particles of mass m_1 and m_2 are released from rest with infinite separation. Find their speeds v_1 and v_2 when their separation distance is r.

105 •• A hole is drilled from the surface of the earth to its center as in Figure 11-27. Ignore the earth's rotation and air resistance. (*a*) How much work is required to lift a particle of mass m from the center of the earth to the earth's surface? (*b*) If the particle is dropped from rest at the surface of the earth, what is its speed when it reaches the center of the earth? (*c*) What is the escape speed for a particle projected from the center of the earth? Express your answers in terms of m, g, and R_E.

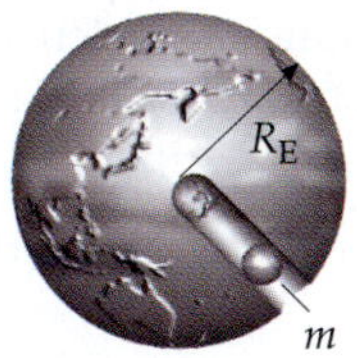

Figure 11-27 Problem 105

106 •• A thick spherical shell of mass M and uniform density has an inner radius R_1 and an outer radius R_2. Find the gravitational field g_r as a function of r for all possible values of r. Sketch a graph of g_r versus r.

107 •• (*a*) Sketch a plot of the gravitational field g_x versus x due to a uniform ring of mass M and radius R whose axis is the x axis. (*b*) At what points is the magnitude of g_x maximum?

108 ••• In this problem, you are to find the gravitational potential energy of the stick in Example 11-8 and a point mass m_0 that is on the x axis at x_0. (*a*) Show that the potential energy of an element of the stick dm and m_0 is given by

$$dU = -\frac{Gm_0\,dm}{x_0 - x} = \frac{GMm_0}{L(x_0 - x)}\,dx$$

where $U = 0$ at $x_0 = \infty$. (*b*) Integrate your result for part (*a*) over the length of the rod to find the total potential energy for the system. Write your result as a general function $U(x)$ by setting x_0 equal to a general point x. (*c*) Compute the force on m_0 at a general point x from $F_x = -dU/dx$ and compare your result with $m_0 g$, where g is the field at x_0 calculated in Example 11-8.

109 ••• A uniform sphere of mass M is located near a thin, uniform rod of mass m and length L as in Figure 11-28. Find the gravitational force of attraction exerted by the sphere on the rod. (See Problem 72.)

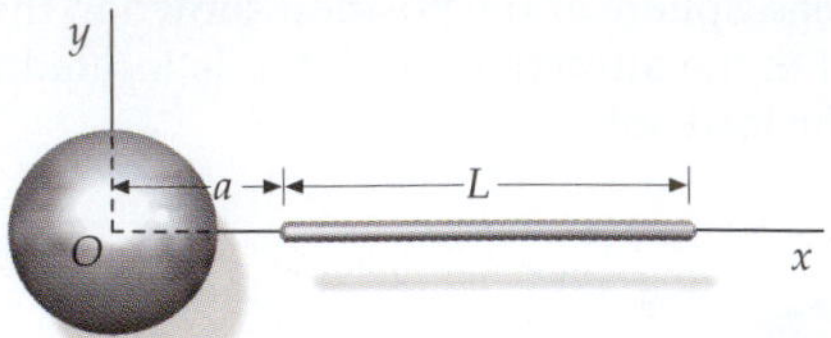

Figure 11-28 Problem 109

110 ••• A uniform rod of mass $M = 20$ kg and length $L = 5$ m is bent into a semicircle. What is the gravitational force exerted by the rod on a point mass $m = 0.1$ kg located at the center of the circular arc?

111 ••• Both the sun and the moon exert gravitational forces on the oceans of the earth, causing tides. (*a*) Show that the ratio of the force exerted by the sun to that exerted by the moon is $M_s r_m^2 / M_m r_s^2$, where M_s and M_m are the masses of the sun and moon and r_s and r_m are the distances from the earth to the sun and to the moon. Evaluate this ratio. (*b*) Even though the sun exerts a much greater force on the oceans than the moon does, the moon has a greater effect on the tides because it is the difference in the force from one side of the earth to the other that is important. Differentiate the expression $F = Gm_1m_2/r^2$ to calculate the change in F due to a small change in r. Show that $dF/F = (-2\,dr)/r$. (*c*) During one full day, the rotation of the earth can cause the distance from the sun or moon to an ocean to change by at most the diameter of the earth. Show that for a small change in distance, the change in the force exerted by the sun is related to the change in the force exerted by the moon by

$$\frac{\Delta F_s}{\Delta F_m} \approx \frac{M_s r_m^3}{M_m r_s^3}$$

and calculate this ratio.

CHAPTER 12

Static Equilibrium and Elasticity

This gymnast, standing on one hand on a balance beam, is in static equilibrium.

If an object is stationary and remains stationary, it is said to be in static equilibrium. Being able to determine the forces acting on an object in static equilibrium has many important applications. For example, the forces exerted by the cables of a suspension bridge must be known so that the cables can be designed to be strong enough to support the bridge. Similarly, cranes must be designed so that they do not topple over when lifting a weight.

The forces exerted by the cables and beams in a structure are called elastic forces. They are the result of slight deformations—the stretching or compression of solid objects under stress from bearing loads. We will first study the equilibrium of a rigid body, an ideal object whose deformation can be neglected. We then briefly consider the deformations and elastic forces that arise when real solids are under stress.

12-1 Conditions for Equilibrium

A necessary condition for a particle at rest to remain at rest is that the net force acting on it must be zero. Similarly, the center of mass of a rigid object remains at rest if the net force acting on the object is zero. However, even if its center of mass is at rest, an object may still rotate. If there is rotation about any point, the object is not in static equilibrium. Therefore, for static equilibrium to exist, the net torque acting on an object in equilibrium must be zero about *any* point. This condition gives us the freedom to choose any point when calculating torques, which is often useful in solving problems.

The two necessary conditions for a rigid body to be in static equilibrium are therefore

1. The net external force acting on the body must be zero:

$$\sum_i \vec{F}_i = 0 \qquad \text{12-1}$$

2. The net external torque about any point must be zero:

$$\sum_i \vec{\tau}_i = 0 \qquad \text{12-2}$$

Conditions for equilibrium

As we have seen, we can describe the vector nature of rotation about a fixed axis as being positive or negative. We will choose counterclockwise torques* to be positive and clockwise torques to be negative.

* A counterclockwise torque is one that tends to produce rotation in the counterclockwise sense.

Example 12-1

A board of length L = 3 m and mass M = 2 kg is supported by scales on either end, as in Figure 12-1. A 6-kg mass m rests on the board a distance x_1 = 2.5 m from the left end and x_2 = 0.5 m from the right end. Find the readings on the scales.

Picture the Problem Let $\vec{F}_1$ and $\vec{F}_2$ be the forces exerted by the scales on the left and right ends of the board, respectively (Figure 12-2). Since the board exerts an equal but opposite force on each scale, the magnitudes of $\vec{F}_1$ and $\vec{F}_2$ are the readings on the scales. To find F_1 and F_2, we apply the two conditions for equilibrium. We take upward to be positive.

Figure 12-1

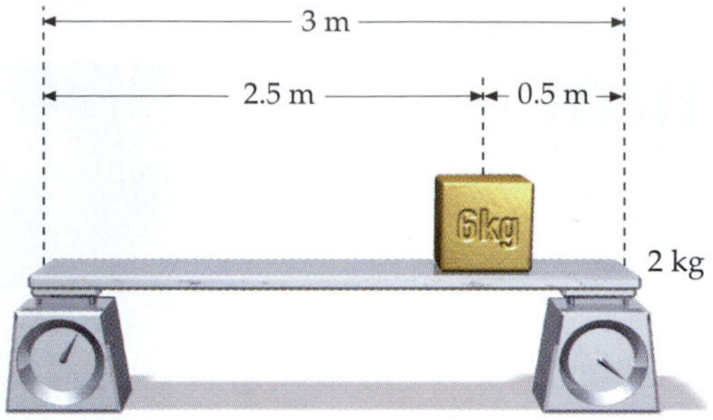

Figure 12-2

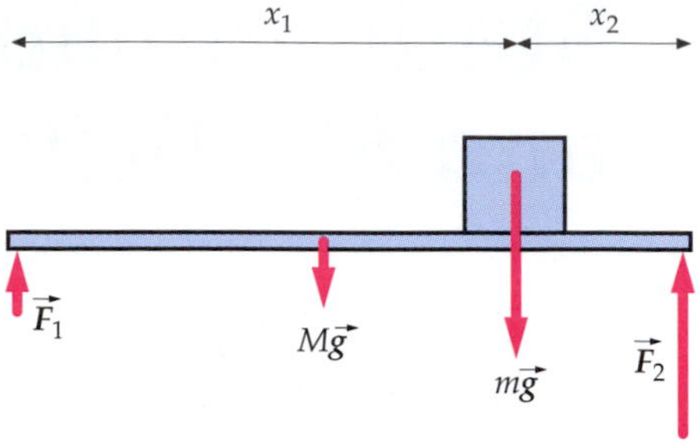

1. Set the net force equal to zero: $F_1 + F_2 - Mg - mg = 0$

2. Set the net torque about the right scale equal to zero: $-F_1L + Mg\frac{L}{2} + mgx_2 + F_2 \times (0) = 0$

3. The torque equation yields F_1: $F_1 = \frac{1}{2}Mg + \frac{x_2}{L}mg$

4. Substitute this result for F_1 into step 1: $F_2 = Mg + mg - F_1 = \frac{1}{2}Mg + mg - \frac{x_2}{L}mg$

5. Substitute numerical values to obtain F_1 and F_2: $F_1 = 19.6\text{ N}$

   $F_2 = 58.9\text{ N}$

Remarks Note that the right scale supports the greater weight, as expected.

Exercise Repeat this example choosing the left scale to be the point about which you set the net torque equal to zero. (*Answer* In this case, the torque equation is $F_2L - MgL/2 - mgx_1 = 0$. Again, we find that $F_1 = 19.6$ N and $F_2 =$ 58.9 N.)

Example 12-1 can be solved using a pivot point located at the mass m, but in this case both F_1 and F_2 occur in the torque equation, hence the algebra is a bit more complex. In general, a statics problem can be simplified by computing the torques about a point on the line of action of one of the unknown forces, as when we chose the left or right scale above.

12-2 The Center of Gravity

Figure 12-3 shows an object divided into many smaller objects, which we can consider to be particles. The weight of each particle is $\vec{w}_i$, and the total weight of the object is $\vec{W} = \Sigma\vec{w}_i$. We can imagine this total weight concentrated at a single point in the object such that if the object were supported at that point, it would be in static equilibrium. This point is the **center of gravity**, defined so that the torque produced by $\vec{W}$ about any point is the same as that produced by the weights of the particles. If X_{cg} is the x coordinate of the center of gravity relative to some origin O, the magnitude of the torque about O is

$$X_{cg}W = \sum_i w_i x_i \qquad 12\text{-}3$$

Center of gravity defined

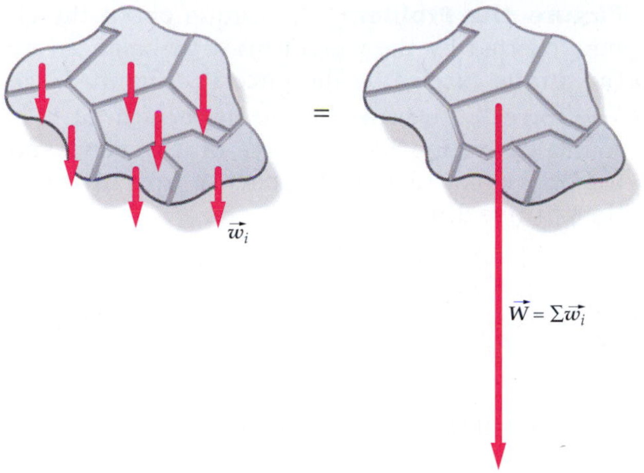

Figure 12-3 The weights of all the particles of an object can be replaced by the total weight $\vec{W}$ of the object acting at the center of gravity.

If the acceleration of gravity is constant over the object (as is nearly always the case), we can write $w_i = m_i g$ and $W = Mg$ and cancel the common factor g. Then

$$X_{cg}Mg = \sum_i m_i g x_i$$

or

$$MX_{cg} = \sum_i m_i x_i \qquad 12\text{-}4$$

This is the same as Equation 8-3, which gives the x coordinate for the center of mass. Thus, the center of gravity and the center of mass coincide when the gravitational field is uniform.

If we choose our origin to be at the center of gravity, $X_{cg} = 0$, then

$$X_{cg}W = \sum_i w_i x_i = 0$$

The center of gravity is that point about which the forces of gravity acting on all the particles of an object produce zero torque. We can use the methods discussed in Chapter 8 for locating the center of mass to locate the center of gravity (see Equation 8-6 and the discussion following). For example, the center of gravity of a stick is the point at which it balances on a pivot.

12-3 Some Examples of Static Equilibrium

Example 12-2

A 60-N weight is held in the hand with the forearm making a 90° angle with the upper arm, as in Figure 12-4. The biceps muscle exerts a force $\vec{F}_m$ that acts 3.4 cm from the pivot point O at the elbow joint. Neglecting the weight of the arm and hand, (*a*) find the magnitude of $\vec{F}_m$ if the distance from the weight to the pivot point is 30 cm, and (*b*) find the force exerted on the elbow joint by the upper arm.

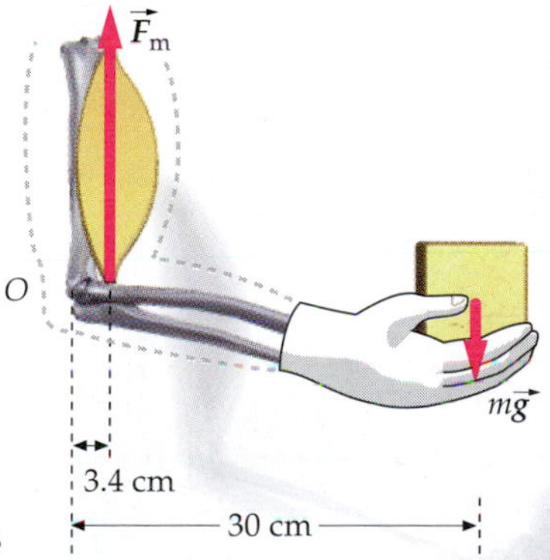

Figure 12-4

Picture the Problem The torque about the elbow exerted by the weight must be balanced by the torque exerted by the force $\vec{F}_m$ (Figure 12-5). The force $\vec{F}_{ua}$ exerted by the upper arm at O is found by setting the net force on the hand and forearm equal to zero. We choose upward to be the positive direction.

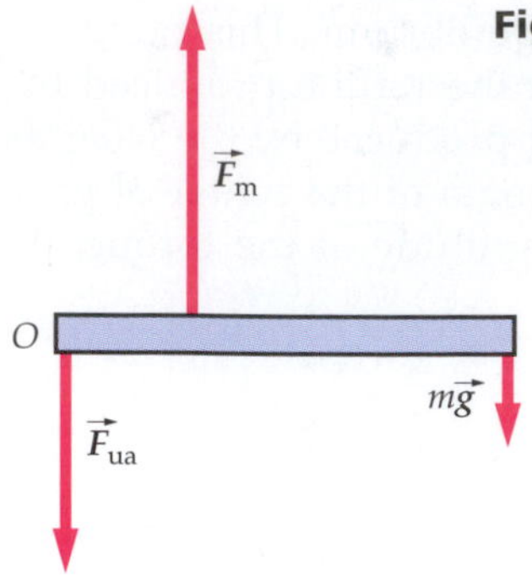

Figure 12-5

(*a*) $\Sigma \vec{\tau} = 0$ about point O gives F_m:

$$F_{ua} \times (0) + F_m \times (3.4 \text{ cm}) - (60 \text{ N})(30 \text{ cm}) = 0$$

$$F_m = \frac{(30 \text{ cm})(60 \text{ N})}{3.4 \text{ cm}} = 529 \text{ N}$$

(*b*) $\Sigma \vec{F} = 0$ gives F_{ua}:

$$-F_{ua} + F_m - 60 \text{ N} = 0$$

$$F_{ua} = F_m - 60 \text{ N} = 469 \text{ N}$$

Remarks The force that must be exerted by the muscle is 8.8 times the weight of the object! In addition, as the muscle pulls upward, the upper arm must push downward to keep the forearm in equilibrium. The force exerted by the upper arm is also several times greater than the object's weight.

Exercise Show that F_{ua} can be found in one step by choosing the pivot point to be where the biceps attaches to the forearm. (*Answer* Setting net torque equal to zero gives $(3.4 \text{ cm})F_{ua} + (0)F_m - (30 \text{ cm} - 3.4 \text{ cm})(60 \text{ N}) = 0$. This yields $F_{ua} = (60 \text{ N})(26.6 \text{ cm}/3.4 \text{ cm}) = 469 \text{ N}$.)

Remarks This example and this exercise show that we can choose the pivot point wherever it is convenient for our calculation.

Example 12-3 *try it yourself*

A sign of mass 20 kg hangs from the end of a rod of length 2 m and mass 4 kg. A wire is attached to the end of the rod and to a point 1 m above point *O* (Figure 12-6). Find the tension $\vec{T}$ in the wire and the force $\vec{F}$ exerted by the wall on the rod at point *O*.

Picture the Problem We have three unknowns: T, and the components F_x and F_y of the force exerted by the wall on the rod. We can find T_y by setting the net torque about O equal to zero. Then T_x is found from $\tan\theta = T_y/T_x = \frac{1}{2}$. F_x and F_y can then be found by applying the zero-force condition to both the x and y directions.

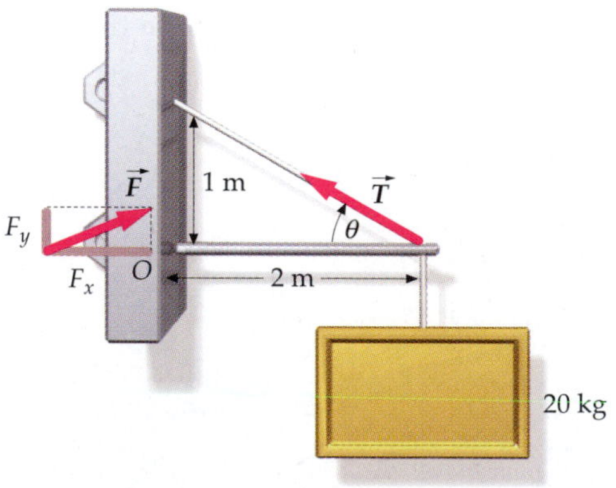

Figure 12-6

Cover the column to the right and try these on your own before looking at the answers.

Steps	Answers
1. Draw a free-body diagram for the rod.	**Figure 12-7** 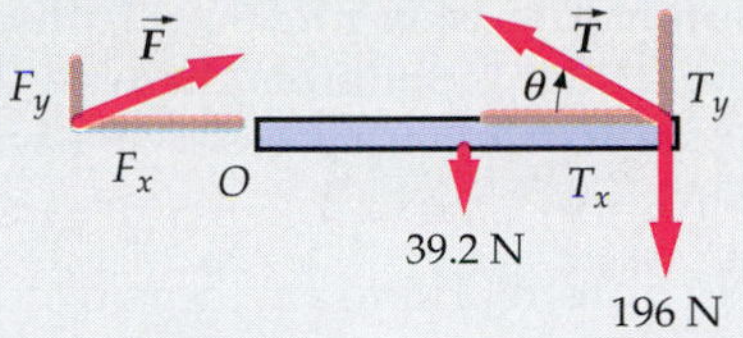
2. Set $\Sigma\vec{\tau} = 0$ about point O.	$T_y\,(2\text{ m}) - 39.2\text{ N}(1\text{ m}) - 196\text{ N}(2\text{ m}) = 0$
3. Solve the torque equation for T_y.	$T_y = 216\text{ N}$
4. Using your result for T_y and the value of $\tan\theta$ from the figure, find T_x.	$T_x = \dfrac{-T_y}{\tan\theta} = -2T_y = -432\text{ N}$
5. Set $\Sigma F_x = 0$ and $\Sigma F_y = 0$.	$F_x + T_x = 0$ $F_y + T_y - 39.2\text{ N} - 196\text{ N} = 0$
6. Use your results for T_x and T_y to find the force components F_x and F_y in step 1.	$F_x = 432\text{ N}, \quad F_y = 19.2\text{ N}$

Example 12-4 *try it yourself*

A wheel of mass M and radius R rests on a horizontal surface against a step of height h ($h < R$). The wheel is to be raised over the step by a horizontal force $\vec{F}$ applied to the axle of the wheel as shown in Figure 12-8. Find the force $\vec{F}$ necessary to raise the wheel over the step.

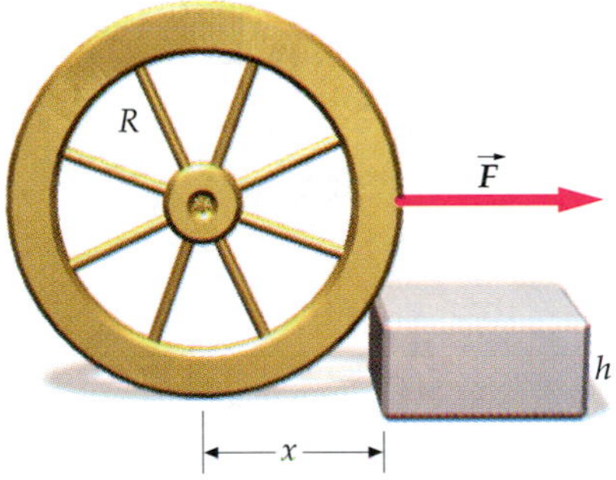

Figure 12-8

Picture the Problem When F is very small, the forces exerted on the wheel are $\vec{F}$, the upward normal force $\vec{F}_n$ exerted by the surface at the bottom of the wheel, its weight $M\vec{g}$, and the force $\vec{F}'$ exerted at the corner in contact with the wheel. As F is increased, F_n decreases. Set F_n equal to zero and take torques about the corner to eliminate $\vec{F}'$.

Cover the column to the right and try these on your own before looking at the answers.

Steps	Answers
1. Express the clockwise torque exerted by $\vec{F}$ about the corner in terms of F, h, and R.	$\tau_1 = F(R - h)$
2. Express the counterclockwise torque exerted by the weight about the corner in terms of x, the horizontal distance from the center of the wheel to the corner.	$\tau_2 = Mgx$
3. Use trigonometry to express x in terms of h and R.	$x^2 + (R - h)^2 = R^2$ $x = \sqrt{h(2R - h)}$
4. Set the magnitudes of the torques equal to each other and solve for F.	$F = \dfrac{Mgx}{R - h} = \dfrac{Mg\sqrt{h(2R - h)}}{R - h}$

Example 12-5

A uniform, 5-m ladder weighing 60 N leans against a frictionless vertical wall. The foot of the ladder is 3 m from the wall (Figure 12-9). What is the minimum coefficient of static friction necessary between the ladder and the floor if the ladder is not to slip?

Figure 12-9

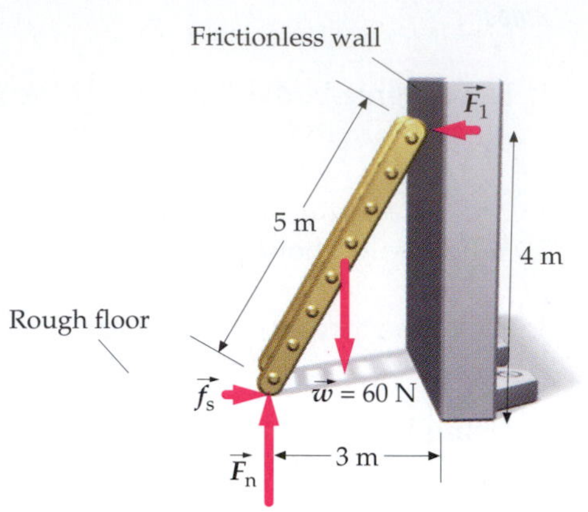

Picture the Problem There are three conditions for the ladder to be in equilibrium; $F_{net,x} = 0$, $F_{net,y} = 0$, and $\vec{\tau}_{net} = 0$. The forces acting on the ladder are the force $\vec{w}$ due to gravity acting downward at the ladder's center of gravity, the force $\vec{F}_1$ exerted horizontally by the wall (since the wall is frictionless, it exerts only a normal force), and the force exerted by the floor, which consists of a normal force $\vec{F}_n$ and a horizontal force of static friction $\vec{f}_s$. Thus, our three conditions determine F_1, f_s, and F_n. We choose the foot of the ladder for our pivot point so that the torque equation contains only one unknown, F_1.

1. The coefficient of static friction is related to the frictional force f_s and normal force F_n: $\quad f_s \le \mu_s F_n \quad \text{or} \quad \mu_s \ge \dfrac{f_s}{F_n}$

2. Set $\Sigma F_x = 0$ and $\Sigma F_y = 0$: $\quad F_{net,x} = 0 = f_s - F_1 \quad \text{and} \quad F_{net,y} = 0 = F_n - w$

3. Solve for f_s and F_n: $\quad f_s = F_1 \quad \text{and} \quad F_n = w = 60\text{ N}$

4. Set $\Sigma \vec{\tau} = 0$ about the foot of the ladder: $\quad \tau_{net} = F_1(4\text{ m}) - w(1.5\text{ m}) = 0$

5. Solve for the force F_1: $\quad F_1 = \dfrac{w(15\text{ m})}{4\text{ m}} = \dfrac{(60\text{ N})(1.5\text{ m})}{4\text{ m}} = 22.5\text{ N}$

6. Use this result for F_1 and $f_s = F_1$ from step 3 to find f_s: $\quad f_s = F_1 = 22.5\text{ N}$

7. Using these results for f_s and F_n, we obtain the minimum value of μ_s from step 1: $\quad \mu_s \ge \dfrac{f_s}{F_n} = \dfrac{22.5\text{ N}}{60\text{ N}} = 0.375$

Remarks There is another way to look at this problem. Whenever an object is in static equilibrium under the influence of three nonparallel forces, the lines of action of the forces must intersect at one point. In the free-body diagram for the ladder shown in Figure 12-10, the lines of action of the weight $\vec{w}$ and the force $\vec{F}_1$ exerted by the wall intersect at point P. The line of action of the resultant force exerted by the ground $\vec{f} + \vec{F}_n$ must also go through point P or there would be an unbalanced torque about this point. The cotangent of the angle made by this resultant force equals $1.5\text{ m}/4\text{ m} = 0.375 = f/F_n$.

Figure 12-10

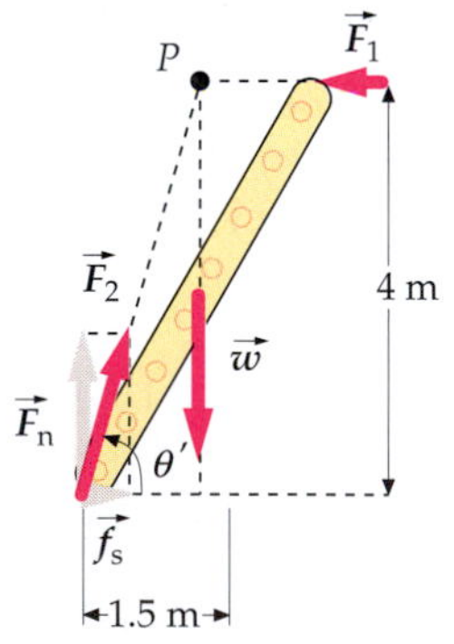

12-4 Couples

The forces $\vec{F}_n$ and $\vec{w}$ in Figures 12-9 and 12-10 of Example 12-5 are equal and opposite. Such a pair of forces, called a couple, tends to produce rotation, but its net force is zero. The forces $\vec{f}_s$ and $\vec{F}_1$ in those figures also constitute a couple. Figure 12-11 shows a couple consisting of forces $\vec{F}_1$ and

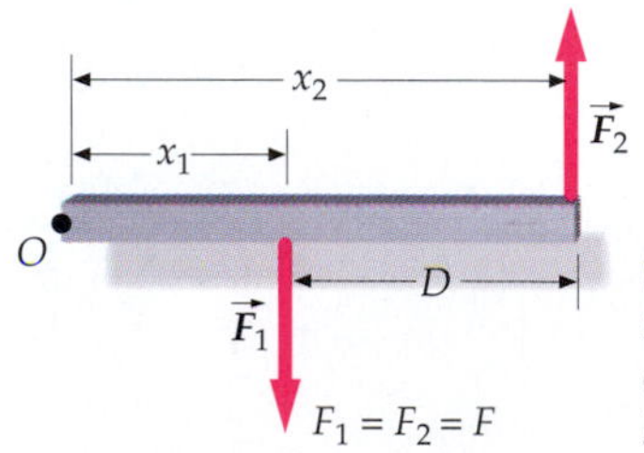

Figure 12-11 Two equal and opposite forces constitute a couple. The torque exerted by a couple has the same value *FD* about any point in space.

$\vec{F}_2$ a distance D apart. The torque produced by this couple about an arbitrary point O is

$$\tau = Fx_2 - Fx_1 = F(x_2 - x_1) = FD$$

where F is the magnitude of either force and D is the distance between them. This result does not depend on the choice of the point O.

> The torque produced by a couple is the same about all points in space.

12-5 Static Equilibrium in an Accelerated Frame

If we consider an object at rest in an accelerated frame of reference, the net force is not zero. If the object is to be at rest relative to the accelerated frame, the object must have the same acceleration as the frame. The two conditions for an object to be in static equilibrium in an accelerated reference frame are

1. $\Sigma\vec{F} = m\vec{a}_{\text{cm}}$

   where $\vec{a}_{\text{cm}}$ is the acceleration of the center of mass, which is the acceleration of the reference frame.

2. $\Sigma\vec{\tau}_{\text{cm}} = 0$

   The sum of the torques about the center of mass must be zero.

The second condition follows from the fact that Newton's second law for rotation, $\Sigma\vec{\tau}_{\text{cm}} = I_{\text{cm}}\vec{\alpha}$, holds for torques about the center of mass whether or not the center of mass is accelerating.* (Newton's second law for rotation does not hold for any other point that is accelerating.)

* See the discussion surrounding Equation 9-32.

Example 12-6

A truck carries a uniform box of mass m, height h, and square cross section of side L (Figure 12-12). What is the greatest acceleration the truck can have without tipping over the box? Assume that the box tips before it slides.

Picture the Problem The acceleration of the box is caused by the frictional force, as shown in Figure 12-13. This force exerts a counterclockwise torque about the center of mass of the box. The only other force that exerts a torque about the center of mass of the box is the normal force. When the box is not accelerating, $f = 0$, and the normal force is exerted through the center of the box. As the acceleration increases, the normal force moves to the left to provide a balance torque about the center of mass. The greatest balancing torque this force can exert is when it is at the edge of the box.

Figure 12-13

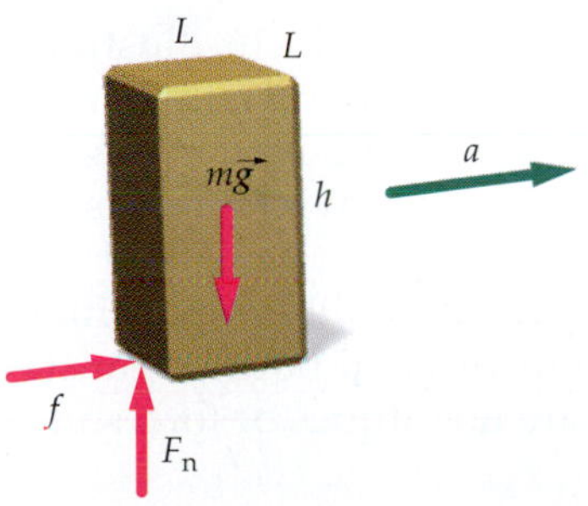

Figure 12-12

1. Apply $\Sigma\vec{F} = m\vec{a}$ to the box:

$$F_{\text{net},x} = f = ma$$

$$F_{\text{net},y} = F_{\text{n}} - mg = 0 \quad \text{or} \quad F_{\text{n}} = mg$$

2. Apply $\Sigma\vec{\tau}_{\text{cm}} = I_{\text{cm}}\vec{\alpha} = 0$:

$$f\frac{h}{2} - F_{\text{n}}\frac{L}{2} = 0$$

3. Substitute $f = ma$ and $F_{\text{N}} = mg$ and solve for a:

$$mah - mgL = 0$$

$$a = \frac{L}{h}g$$

Remarks The maximum acceleration is proportional to L/h. This maximum acceleration is small for a tall, narrow box (L/h small) and large for a short, wide box (L/h large). Thus, a short, wide box is more stable.

12-6 Stability of Rotational Equilibrium

There are three categories of rotational equilibrium for an object: stable, unstable, or neutral. **Stable rotational equilibrium** occurs when the torques that arise from a small angular displacement of the object urge the object back toward its equilibrium position. Stable equilibrium is illustrated in Figure 12-14. When the box is rotated slightly about one end, the resulting torque about the pivot point tends to restore the box to its original position. Note that this slight rotation lifts the center of gravity, increasing the potential energy of the box.

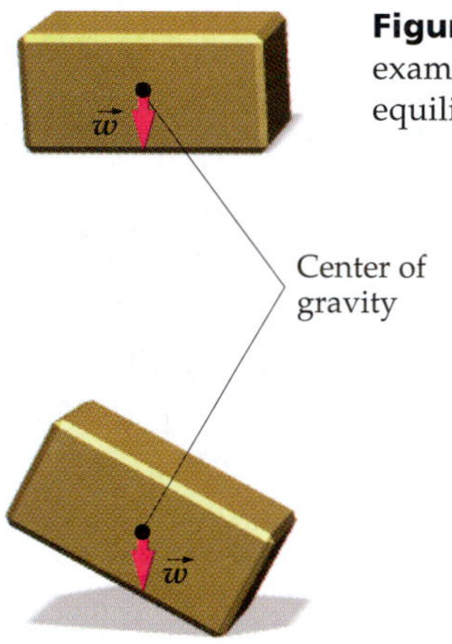

Figure 12-14 An example of stable equilibrium.

Unstable rotational equilibrium, illustrated in Figure 12-15, occurs when the torques that arise from a small angular displacement of the object urge the object away from its equilibrium position. A slight rotation of the narrow stick causes it to fall over because the torque due to its weight urges it away from its original position. Here the rotation lowers the center of gravity and decreases the potential energy of the stick.

Figure 12-15 An example of unstable equilibrium.

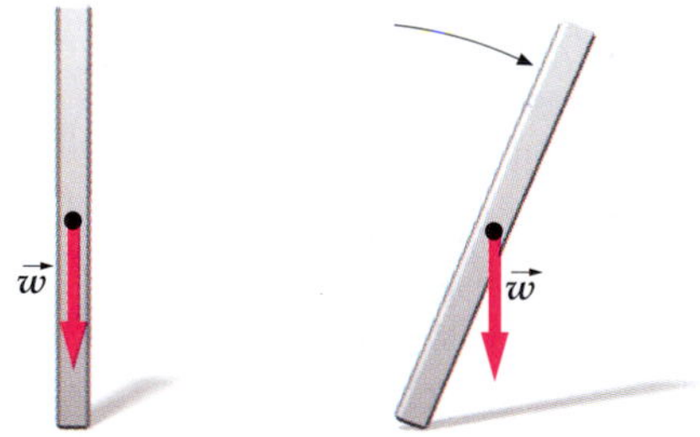

The cylinder resting on a horizontal surface in Figure 12-16 illustrates **neutral rotational equilibrium.** If the cylinder is rotated slightly, there is no torque or force that urges it either back toward its original position or away from it. As the cylinder rotates, the height of the center of gravity remains unchanged, so the potential energy does not change.

In summary, if a system is disturbed slightly from its equilibrium position, the equilibrium is stable if the system returns to its original position, unstable if it moves farther away, and neutral if there are no torques or forces urging it in either direction.

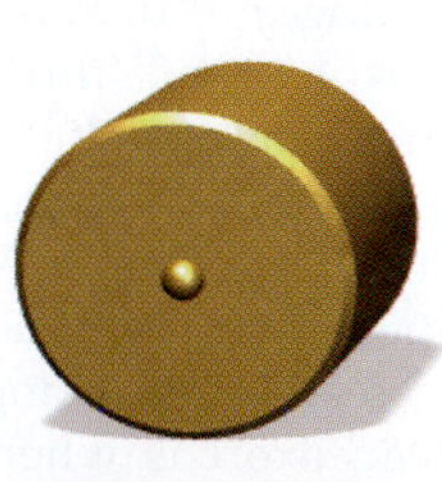

Figure 12-16 An example of neutral equilibrium.

Since "disturbed slightly" is a relative term, stability is also relative. One example of equilibrium may be more or less stable than another. Figure 12-17*a* shows a stick balanced on end that is not as narrow as that in Figure 12-15. Here, if the disturbance is very small (Figure 12-17*b*), the stick will move back toward its original position, but if the disturbance is great enough so that the center of gravity no longer lies over the base of support (Figure 12-17*c*), the stick will fall.

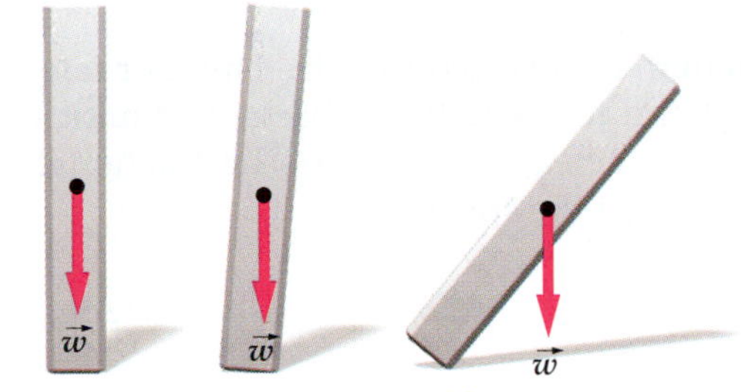

Figure 12-17 Stability of equilibrium is relative. If the stick in (*a*) is rotated slightly, as in (*b*), it returns to its original equilibrium position as long as the center of gravity lies over the base of support. (*c*) If the rotation is too great, the center of gravity is no longer over the base of support, and the stick falls over.

We can improve the stability of a system by either lowering the center of gravity or widening the base of support. Figure 12-18 shows a nonuniform stick that is loaded so that its center of gravity is near one end. If it stands on its heavy end so that the center of gravity is low (Figure 12-18*a*), it is much more stable than if it stands on the other end so that the center of gravity is high (Figure 12-18*b*).

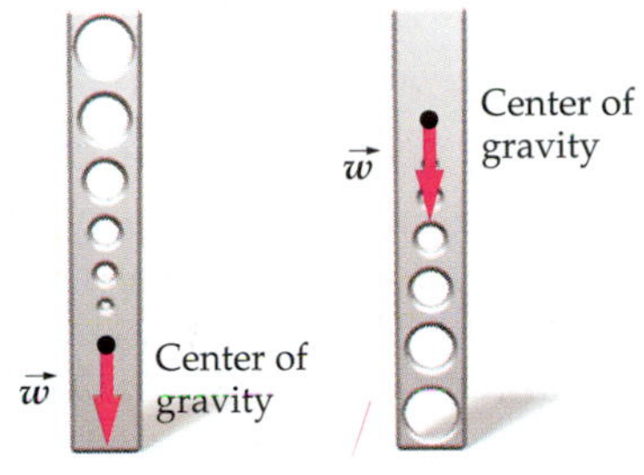

Figure 12-18 When a nonuniform stick rests on its heavy end with its center of gravity low (*a*), the equilibrium is more stable than when its center of gravity is high (*b*).

In Figure 12-19 the center of gravity lies below the point of support of the system. This system is stable for any displacement because the resulting torque always rotates the system back toward its equilibrium position.

Standing or walking upright presents a challenge for humans because the center of gravity is high and must be kept in balance over a relatively small base of support, the feet. Human infants take about a year to learn to walk. A four-footed creature has a much easier time because its base of support is larger and its center of gravity is lower. Newborn kittens can walk almost immediately.

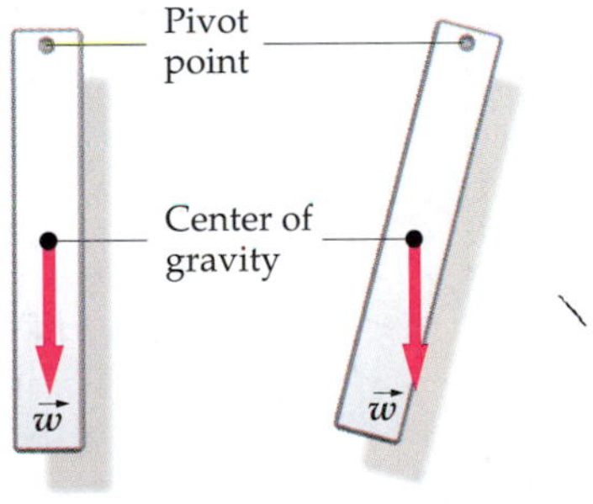

Figure 12-19 When a stick is pivoted so that its center of gravity is below the pivot point, the equilibrium is stable no matter how far the stick is displaced from equilibrium.

12-7 Indeterminate Problems

When objects are not rigid, but deformable, we need more information to determine the forces required for equilibrium. Consider a car resting on a horizontal surface. Suppose there is a very heavy object on one side of the trunk. We wish to find the vertical support force exerted by the road on each tire. Let the road be in the xy plane. If we choose one of the tires as our origin, the torque exerted by all the forces about that point has x and y components, but no z component because there are no horizontal forces. We thus obtain two equations by setting the net torque equal to zero, and a third equation by setting the net vertical force equal to zero. We need another equation to find the force exerted by the road on each of the four tires. If we let air out of one of the tires and pump up another tire to a greater pressure, the car remains in equilibrium, but the force exerted on each tire changes. Clearly, the forces on the tires in this problem are not determined by the information given. The tires are not rigid bodies. To some extent every object is deformable, a concept that is the foundation of the next section.

12-8 Stress and Strain

If a solid object is subjected to forces that tend to stretch, shear, or compress the object, its shape changes. If the object returns to its original shape when the forces are removed, it is said to be **elastic.** Most objects are elastic for forces up to a certain maximum, called the **elastic limit.** If the forces exceed the elastic limit, the object does not return to its original shape but is permanently deformed.

Figure 12-20 shows a solid bar of length L subjected to a stretching or **tensile force** F acting equally to the right and to the left. The bar is in equilibrium, but the forces acting on it tend to increase its length. The fractional change in the length of the bar $\Delta L/L$ is called the **strain**:

$$\text{Strain} = \frac{\Delta L}{L} \qquad \text{12-5}$$

The ratio of the force F to the cross-sectional area A is called the **tensile stress**:

$$\text{Stress} = \frac{F}{A} \qquad \text{12-6}$$

Figure 12-21 shows a graph of strain versus stress for a typical solid bar. The graph is linear until point A. Up to this point, known as the proportional limit, the strain is proportional to the stress. The result that strain varies linearly with stress is known as Hooke's law.* Point B in Figure 12-21 is the elastic limit of the material. If the bar is stretched beyond this point, it is permanently deformed. If an even greater stress is applied, the material eventually breaks, shown happening at point C. The ratio of stress to strain in the linear region of the graph is a constant called **Young's modulus** Y:

$$Y = \frac{\text{stress}}{\text{strain}} = \frac{F/A}{\Delta L/L} \qquad \text{12-7}$$

Definition—Young's modulus

The units of Young's modulus are newtons per square meter (or pounds per square inch). Approximate values of Young's modulus for various materials are listed in Table 12-1.

Exercise A certain person's biceps muscle has a maximum cross-sectional area of 12 $cm^2 = 1.2 \times 10^{-3}\ m^2$. What is the stress in the muscle if it exerts a force of 300 N? (*Answer* Stress $= F/A = 2.5 \times 10^5\ N/m^2$. The maximum stress that can be exerted is approximately the same for all human muscles. Greater forces can be exerted by muscles with greater cross-sectional areas.)

If a bar is subjected to forces that tend to compress it rather than stretch it, the stress is called **compressive stress.** For many materials, Young's modulus for compressive stress is the same as that for tensile stress. Note that ΔL in Equation 12-7 is then taken to be the *decrease* in the length of the bar. If the tensile or compressive stress is too great, the bar breaks. The stress at which breakage occurs is called the **tensile strength,** or in the case of compression, the **compressive strength.** Approximate values of the tensile and compressive strengths of various materials are listed in Table 12-1. Note that Young's modulus for bone is quite different for compressive and tensile stress, unlike

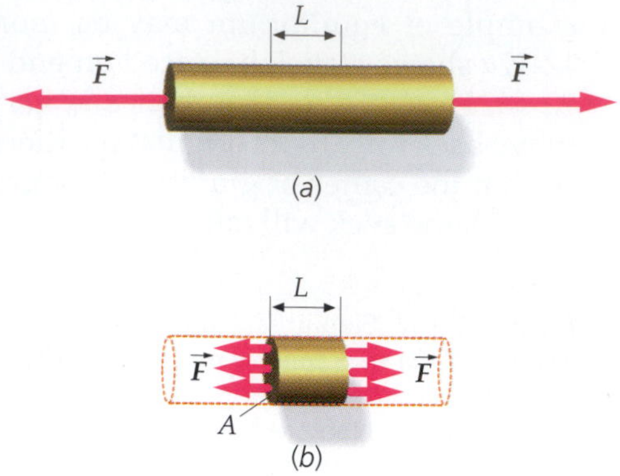

Figure 12-20 (*a*) A solid bar subjected to stretching forces $\vec{F}$ acting on each end. (*b*) A small section of the bar of length L. The elements of the bar to the left and right of this section exert forces on this section. If the section is not too close to the end, these forces are distributed equally over the cross-sectional area. The force per unit area is the stress.

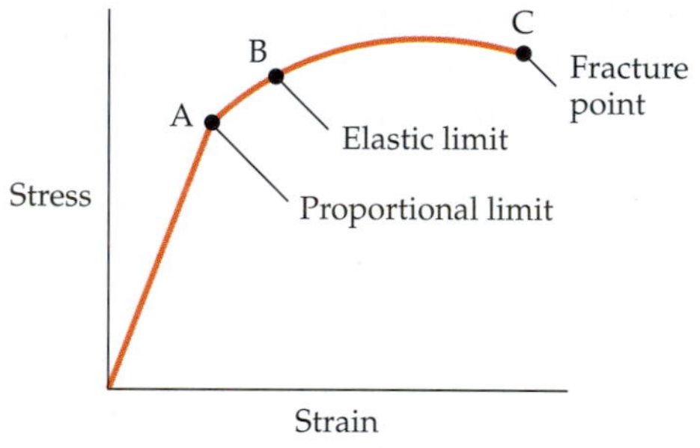

Figure 12-21 A graph of stress versus strain. Up to point A, the strain is proportional to the stress. Beyond the elastic limit at point B, the bar will not return to its original length when the stress is removed. At point C, the bar fractures.

* This is the same behavior as that of a coiled spring for small stretching.

many other materials. This fact has biological significance, because the major job of bone is to resist the compressive load exerted by contracting muscles.

Table 12-1

Young's Modulus *Y* and Strengths of Various Materials*

Material	*Y*, GN/m^2†	Tensile strength, MN/m^2	Compressive strength, MN/m^2
Aluminum	70	90	
Bone			
Tensile	16	200	
Compressive	9		270
Brass	90	370	
Concrete	23	2	17
Copper	110	230	
Iron (wrought)	190	390	
Lead	16	12	
Steel	200	520	520

*These values are representative. Actual values for particular samples may differ.
†1 GN = 10^3 MN = 10^9 N.

Example 12-7

A 500-kg mass is hung from a 3-m steel wire with a cross-sectional area of 0.15 cm². How much does the wire stretch?

Picturing the Problem L is the unstretched length of the wire, F is the force acting on it, and A is its cross-sectional area. The stretch in the wire ΔL is related to Young's modulus by $Y = (F/A)/(\Delta L/L)$. From Table 12-1 we find the numerical value of Young's modulus for steel, $Y = 2.0 \times 10^{11}\ \text{N/m}^2$.

1. The amount the wire is stretched, ΔL, is found from Young's modulus:

$$Y = \frac{F/A}{\Delta L/L}$$

$$\Delta L = L\frac{F/A}{Y}$$

2. The force acting on the wire is the weight of the 500-kg mass:

$$F = mg = (500\ \text{kg})(9.81\ \text{m/s}^2) = 4.90 \times 10^3\ \text{N}$$

3. Convert the area to m²:

$$A = 0.15\ \text{cm}^2 \times \frac{10^{-4}\ \text{m}^2}{1\ \text{cm}^2} = 1.5 \times 10^{-5}\ \text{m}^2$$

4. Substituting numerical values yields ΔL:

$$\Delta L = L\frac{F/A}{Y}$$

$$= (3\ \text{m})\frac{(4.9 \times 10^3\ \text{N})/(1.5 \times 10^{-5}\ \text{m}^2)}{2.0 \times 10^{11}\ \text{N/m}^2}$$

$$= 0.49\ \text{cm}$$

Exercise A wire 1.5 m long has a cross-sectional area of 2.4 mm². It is hung vertically and stretches 0.32 mm when a 10-kg block is attached to it. Find (*a*) the stress, (*b*) the strain, and (*c*) Young's modulus for the wire. (*Answers* (*a*) $4.09 \times 10^7\ \text{N/m}^2$, (*b*) 2.13×10^{-4}, (*c*) 192 GN/m^2)

In Figure 12-22, a force $\vec{F}_s$ is applied tangentially to the top of a book. Such a force is called a **shear force.** The ratio of the shear force F_s to the area A is called the **shear stress:**

$$\text{Shear stress} = \frac{F_s}{A} \qquad \text{12-8}$$

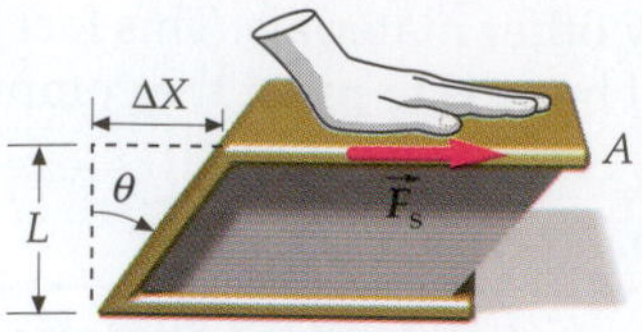

Figure 12-22 The application of the horizontal force F_s to the book causes a shear stress defined as the force per unit area. The ratio $\Delta X/L = \tan\theta$ is the shear strain.

A shear stress tends to deform an object, as shown in Figure 12-22. The ratio $\Delta X/L$ is called the **shear strain:**

$$\text{Shear strain} = \frac{\Delta X}{L} = \tan\theta \qquad \text{12-9}$$

where θ is the shear angle shown in the figure. The ratio of the shear stress to the shear strain is called the **shear modulus** M_s:

$$M_s = \frac{\text{shear stress}}{\text{shear strain}} = \frac{F_s/A}{\Delta X/L} = \frac{F_s/A}{\tan\theta} \qquad \text{12-10}$$

Definition—Shear modulus

The shear modulus is also known as the **torsion modulus.** The torsion modulus is approximately constant for small stresses, which implies that the shear strain varies linearly with the shear stress. This observation is known as Hooke's law for torsional stress. In a torsion balance, such as that used in Cavendish's apparatus for measuring the universal gravitational constant G, the torque (which is related to the stress) is proportional to the angle of twist (which equals the strain for small angles). Approximate values of the shear modulus for various materials are listed in Table 12-2.

Table 12-2

Approximate Values of the Shear Modulus M_s of Various Materials

Material	M_s, GN/m^2
Aluminum	30
Brass	36
Copper	42
Iron	70
Lead	5.6
Steel	84
Tungsten	150

Summary

Topic	Remarks and Relevant Equations
1. Equilibrium of a Rigid Object	
Conditions for	1. The net external force acting on the body must be zero: $$\sum_i \vec{F}_i = 0 \qquad \text{12-1}$$ 2. The net external torque about any point must be zero: $$\sum_i \vec{\tau}_i = 0 \qquad \text{12-2}$$ An alternative statement of the second condition is that the sum of the torques that tend to produce clockwise rotation about any point must equal the sum of the torques that tend to produce counterclockwise rotation about that point.
Stability	The equilibrium of an object can be classified as stable, unstable, or neutral. An object resting on some surface will be in equilibrium if its center of gravity lies over its base of support. Stability can be improved by lowering the center of gravity or by increasing the size of the base.
2. Center of Gravity	The force of gravity exerted on the various parts of an object can be replaced by a single force, the total weight of the object W, acting at the center of gravity. The x coordinate of the center of gravity X_{cg} relative to some origin is given by $$X_{cg}W = \sum_i w_i x_i \qquad \text{12-3}$$

When the acceleration of gravity is the same at all points of an object, the center of gravity coincides with the center of mass.

3. Couples	A pair of equal and opposite forces constitutes a couple. The torque produced by a couple is the same about any point in space.	
4. Stress and Strain		
Strain	$\text{Strain} = \frac{\Delta L}{L}$	**12-5**
Stress	$\text{Stress} = \frac{F}{A}$	**12-6**
Young's modulus	$Y = \frac{\text{stress}}{\text{strain}} = \frac{F/A}{\Delta L/L}$	**12-7**
Shear stress	$\text{Shear stress} = \frac{F_s}{A}$	**12-8**
Shear strain	$\text{Shear strain} = \frac{\Delta X}{L} = \tan\theta$	**12-9**
Shear modulus	$M_s = \frac{\text{shear stress}}{\text{shear strain}} = \frac{F_s/A}{\Delta X/L} = \frac{F_s/A}{\tan\theta}$	**12-10**

Problem-Solving Guide

1. Begin by drawing a neat diagram that includes the important features of the problem.
2. To solve problems involving the equilibrium of rigid bodies, set the net force and net torque equal to zero. A judicious choice for the point about which to calculate the torques can simplify the calculation.

Summary of Worked Examples

Type of Calculation	Procedure and Relevant Examples
1. Equilibrium of Rigid Object	
Find the forces exerted on an object in equilibrium.	Use $\Sigma\vec{F} = 0$ and $\Sigma\vec{\tau} = 0$. Choose a point for the torque equation that is on the line of action of one or more of the unknown forces so those forces do not appear in the equation. **Examples 12-1, 12-2, 12-3, 12-4, 12-5**
Equilibrium in an accelerated frame	Use $\Sigma\vec{F} = m\vec{a}_{cm}$ and $\Sigma\vec{\tau}_{cm} = 0$ where the torques must be computed about the center of mass. A box in an accelerated truck will not tip if the normal force can exert a great enough torque to balance that exerted by static friction. **Example 12-6**
2. Stress and Strain	
Find the amount that a wire stretches under a given load.	Use Young's modulus. **Example 12-7**

optional

Problems

Conceptual Problems

Problems from Optional and Exploring sections

In a few problems, you are given more data than you actually need; in a few other problems, you are required to supply data from your general knowledge, outside sources, or informed estimates.

• Single-concept, single-step, relatively easy

•• Intermediate-level, may require synthesis of concepts

••• Challenging, for advanced students

Conditions for Equilibrium

1 • True or false:

(*a*) $\Sigma\vec{F} = 0$ is sufficient for static equilibrium to exist.

(*b*) $\Sigma\vec{F} = 0$ is necessary for static equilibrium to exist.

(*c*) In static equilibrium, the net torque about any point is zero.

(*d*) An object is in equilibrium only when there are no forces acting on it.

2 • A seesaw consists of a 4-m board pivoted at the center. A 28-kg child sits on one end of the board. Where should a 40-kg child sit to balance the seesaw?

3 • In Figure 12-23, Misako is about to do a push-up. Her center of gravity lies directly above point *P* on the floor, which is 0.9 m from her feet and 0.6 m from her hands. If her mass is 54 kg, what is the force exerted by the floor on her hands?

Figure 12-23 Problem 3

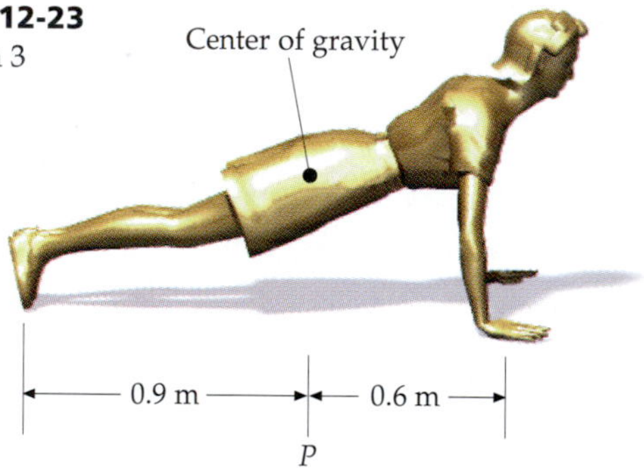

4 • Juan and Bettina are carrying a 60-kg block on a 4-m board as shown in Figure 12-24. The mass of the board is 10 kg. Since Juan spends most of his time reading cookbooks, whereas Bettina regularly does push-ups, they place the block 2.5 m from Juan and 1.5 m from Bettina. Find the force in newtons exerted by each to carry the block.

Figure 12-24 Problem 4

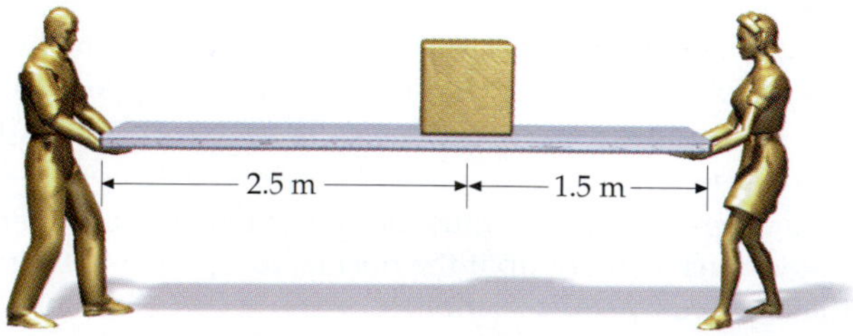

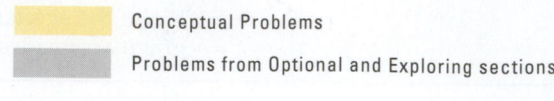

5 • Misako wishes to measure the strength of her biceps muscle by exerting a force on a test strap as shown in Figure 12-25. The strap is 28 cm from the pivot point at the elbow, and her biceps muscle is attached at a point 5 cm from the pivot point. If the scale reads 18 N when she exerts her maximum force, what force is exerted by the biceps muscle?

Figure 12-25 Problem 5

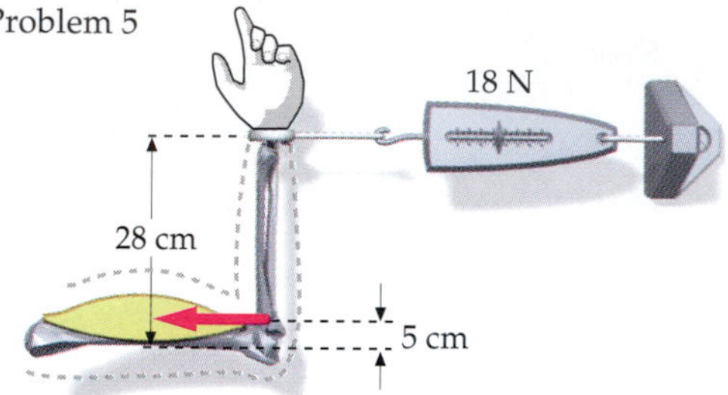

6 • A crutch is pressed against the sidewalk with a force $\vec{F}_c$ along its own direction as in Figure 12-26. This force is balanced by a normal force $\vec{F}_n$ and a frictional force $\vec{f}_s$. (*a*) Show that when the force of friction is at its maximum value, the coefficient of friction is related to the angle θ by $\mu_s = \tan\theta$. (*b*) Explain how this result applies to the forces on your foot when you are not using a crutch. (*c*) Why is it advantageous to take short steps when walking on ice?

Figure 12-26 Problem 6

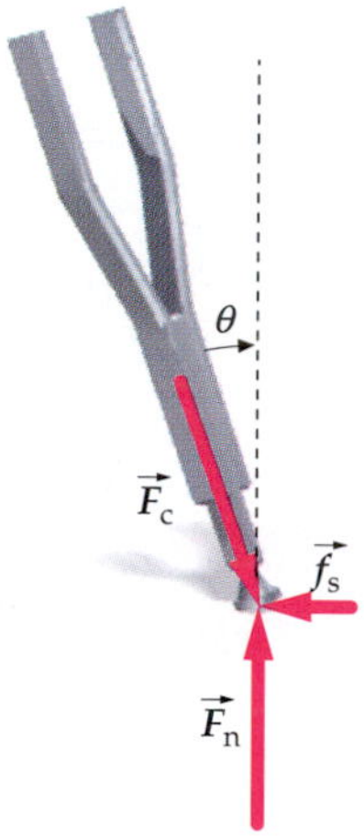

The Center of Gravity

7 • True or false: The center of gravity is always at the geometric center of a body.

8 • Must there be any material at the center of gravity of an object?

9 • If the acceleration of gravity is not constant over an object, is it the center of mass or the center of gravity that is the pivot point when the object is balanced?

10 • Two spheres of radius *R* rest on a horizontal table with their centers a distance 4*R* apart. One sphere has twice the weight of the other sphere. Where is the center of gravity of this system?

11 • An automobile has 58% of its weight on the front wheels. The front and back wheels are separated by 2 m. Where is the center of gravity located with respect to the front wheels?

12 • Each of the objects shown in Figure 12-27 is suspended from the ceiling by a thread attached to the point

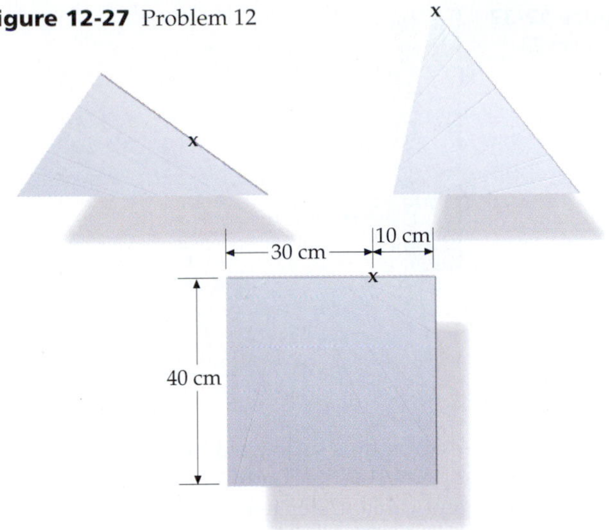

Figure 12-27 Problem 12

marked × on the object. Describe the orientation of each suspended object with a diagram.

13 •• A square plate is produced by welding together four smaller square plates, each of side a as shown in Figure 12-28. Plate 1 weighs 40 N; plate 2, 60 N; plate 3, 30 N; and plate 4, 50 N. Find the center of gravity (x_{cg}, y_{cg}).

Figure 12-28 Problem 13

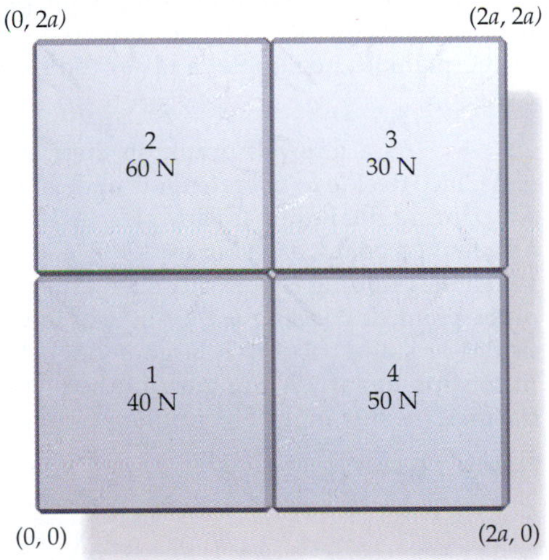

14 •• A uniform rectangular plate has a circular section of radius R cut out as shown in Figure 12-29. Find the center of gravity of the system. *Hint:* Do not integrate. Use superposition of a rectangular plate minus a circular plate.

Figure 12-29 Problem 14

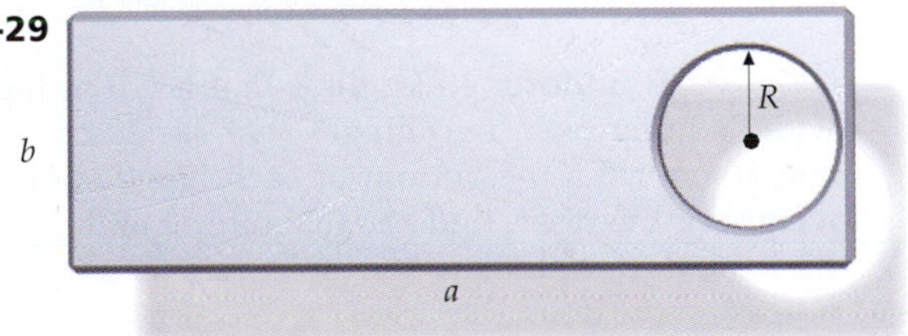

Some Examples of Static Equilibrium

15 • When the tree in front of his house was cut down to widen the road, Jay did not want it to go without ceremony, so he hauled out his electric guitar and amplifier. All that remained was a uniform 10-m log of mass 100 kg resting on two supports, waiting to be cut up and taken away the next day. One support was 2 m from the left end, and the other was 4 m from the right end. Find the forces exerted on the log by the supports as Jay played his ear-splitting "Requiem for a Fallen Tree."

16 • Bubba uses a crowbar that is 1 m long to lift a heavy crate off the ground. The crowbar rests on a rigid fulcrum 10 cm from one end as shown in Figure 12-30. (*a*) If Bubba exerts a downward force of 600 N on one end of the crowbar, what is the upward force exerted on the crate by the other end? (*b*) The ratio of the forces at the ends of the crowbar is called the mechanical advantage of the crowbar. What is the mechanical advantage here?

Figure 12-30 Problem 16

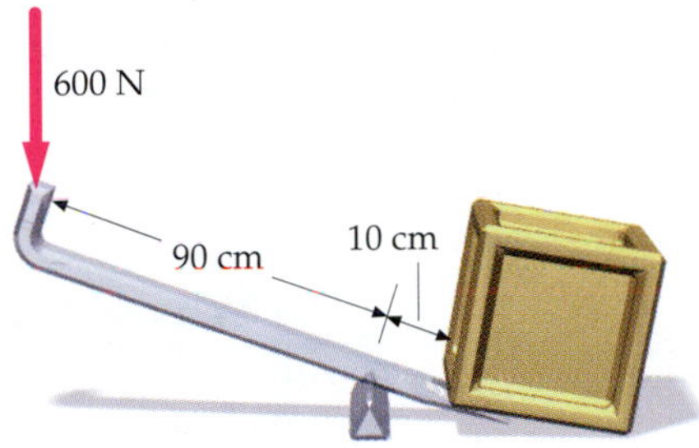

17 • Figure 12-31 shows a 25-foot sloop. The mast is a uniform pole of 120 kg and is supported on the deck and held fore and aft by wires as shown. The tension in the forestay (wire leading to the bow) is 1000 N. Determine the tension in the backstay and the force that the deck exerts on the mast. Is there a tendency for the mast to slide forward or aft? If so, where should a block be placed to prevent the mast from moving?

Figure 12-31 Problem 17

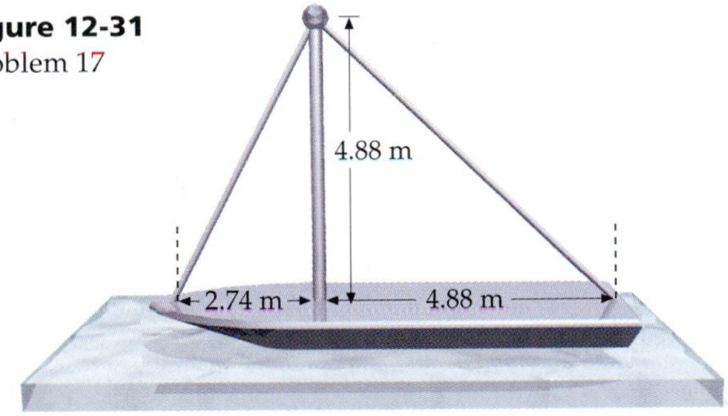

18 • The sloop in Figure 12-32 is rigged slightly differently from the one in Problem 17. The mass of the mast is 150 kg and the tension in the forestay is again 1000 N. Find the tension in the backstay and the force that the deck exerts on the mast. Is there a tendency for the mast to slide forward or aft? If so, where should a block be placed on the deck to prevent the mast from moving?

Figure 12-32 Problem 18

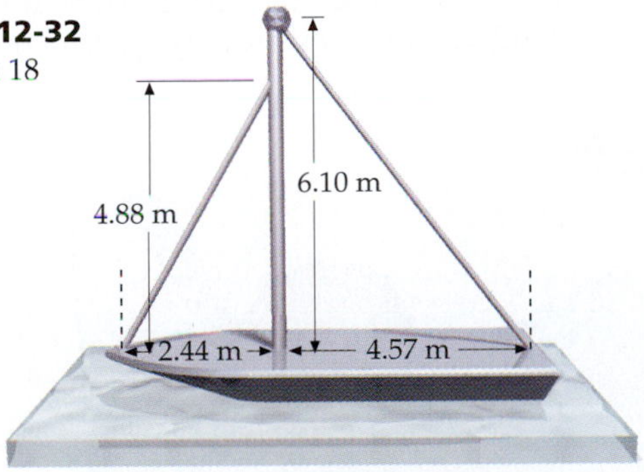

19 •• A 10-m beam of mass 300 kg extends over a ledge as in Figure 12-33. The beam is not attached, but simply rests on the surface. A 60-kg student intends to position the beam so that he can walk to the end of it. How far from the edge of the ledge can the beam extend?

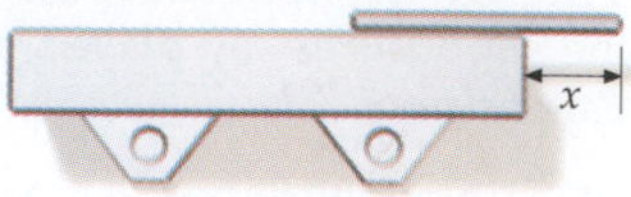

Figure 12-33 Problem 19

20 •• A gravity board for locating the center of gravity of a person consists of a horizontal board supported by a fulcrum at one end and by a scale at the other end. A physics student lies horizontally on the board with the top of his head above the fulcrum point as shown in Figure 12-34. The scale is 2 m from the fulcrum. The student has a mass of 70 kg, and when he is on the gravity board, the scale advances 250 N. Where is the center of gravity of the student?

Figure 12-34 Problem 20

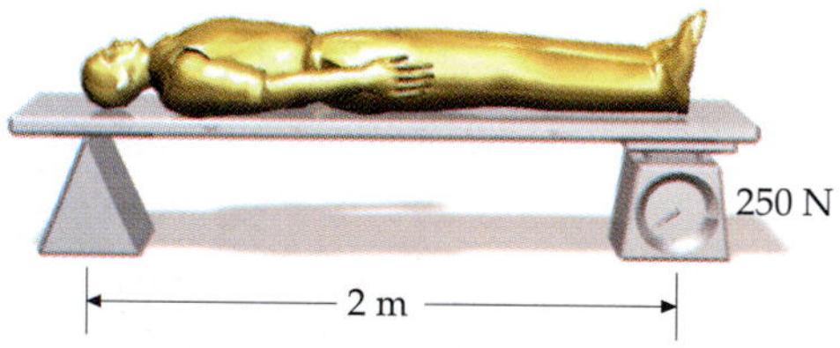

21 •• A 3-m board of mass 5 kg is hinged at one end. A force $\vec{F}$ is applied vertically at the other end to lift a 60-kg block, which rests on the board 80 cm from the hinge, as shown in Figure 12-35. (*a*) Find the magnitude of the force needed to hold the board stationary at $\theta = 30°$. (*b*) Find the force exerted by the hinge at this angle. (*c*) Find the magnitude of the force $\vec{F}$ and the force exerted by the hinge if $\vec{F}$ is exerted perpendicular to the board when $\theta = 30°$.

Figure 12-35 Problem 21

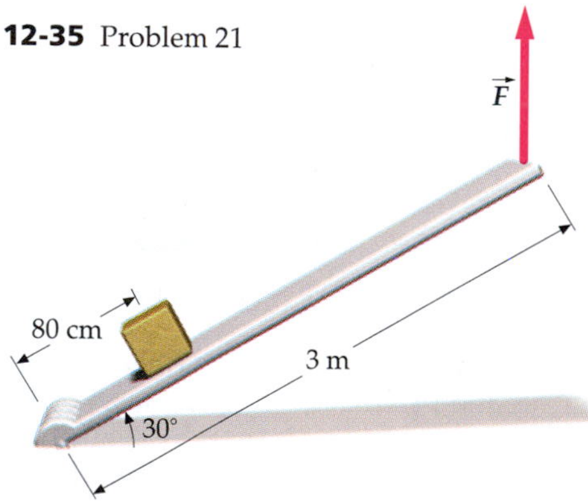

22 •• A cylinder of weight W is supported by a frictionless trough formed by a plane inclined at 30° to the horizontal on the left and one inclined at 60° on the right as shown in Figure 12-36. Find the force exerted by each plane on the cylinder.

Figure 12-36 Problem 22

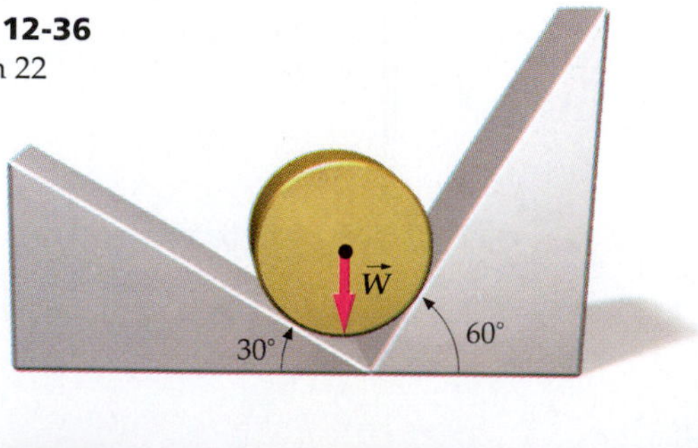

Figure 12-37 Problem 23

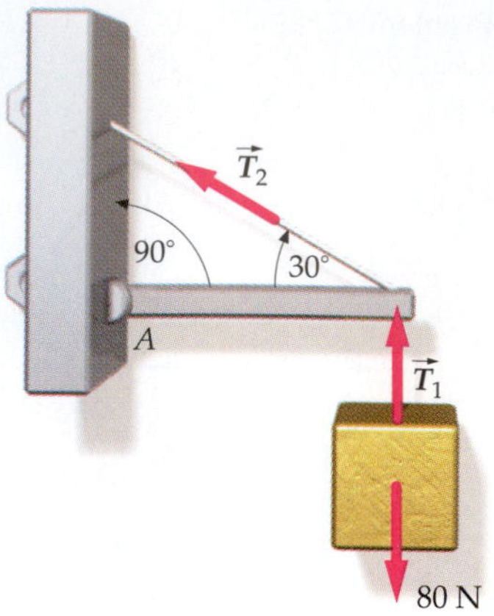

23 •• An 80-N weight is supported by a cable attached to a strut hinged at point A as in Figure 12-37. The strut is supported by a second cable under tension $\vec{T}_2$. The mass of the strut is negligible. (*a*) What are the three forces acting on the strut? (*b*) Show that the vertical component of the tension $\vec{T}_2$ must equal 80 N. (*c*) Find the force exerted on the strut by the hinge.

24 •• A horizontal board 8.0 m long is used by pirates to make their victims walk the plank. A pirate of mass 105 kg stands on the shipboard end of the plank to prevent it from tipping. Find the maximum distance the plank can overhang for a 63-kg victim to be able to walk to the end if (*a*) the mass of the plank is negligible, and (*b*) the mass of the plank is 25 kg.

25 •• As a farewell prank on their alma mater, Sharika and Chico decide to liberate thousands of marbles in the hallway during final exams. They place a 2-m × 1-m × 1-m box on a hinged board, as in Figure 12-38, and fill it with marbles. When the building is perfectly silent, they slowly lift one end of the plank, increasing θ, the angle of the incline. If the coefficient of static friction is large enough to prevent the box from slipping, at what angle will the box tip? (Assume that the marbles stay in the box until it tips over.)

Figure 12-38 Problem 25

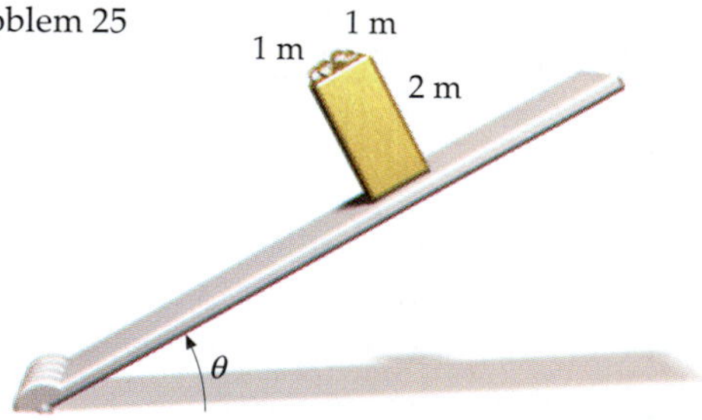

26 •• A uniform 18-kg door that is 2.0 m high by 0.8 m wide is hung from two hinges that are 20 cm from the top and 20 cm from the bottom. If each hinge supports half the weight of the door, find the magnitude and direction of the horizontal components of the forces exerted by the two hinges on the door.

27 •• Find the force exerted by the corner on the wheel in Example 12-4, just as the wheel lifts off the surface.

28 •• Lou is promoting the grand opening of Roswell's, a new nightclub with an alien theme. One end of a uniform 100-kg beam, 10 m long, is hinged to a wall, and the other end sticks out horizontally over the dance floor. A cable connects to the beam 6 m from the wall, as in Figure 12-39. Lou

Figure 12-39 Problems 28 and 32

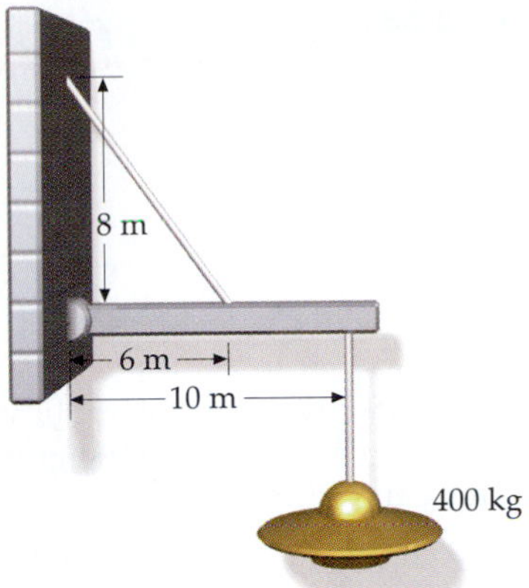

sits at the controls of a mock UFO, which hangs from the free end of the beam. From there, he sends down abduction beams, hypnotic light effects, and spaceship noises to the patrons below. If the combined weight of Lou and his UFO is 400 kg, (*a*) what is the tension in the cable? (*b*) What is the horizontal force on the hinge? (*c*) What is the vertical force of the beam on the hinge?

29 •• The diving board shown in Figure 12-40 has a mass of 30 kg. Find the force on the supports when a 70-kg diver stands at the end of the diving board. Give the direction of each support force as a tension or a compression.

Figure 12-40 Problem 29

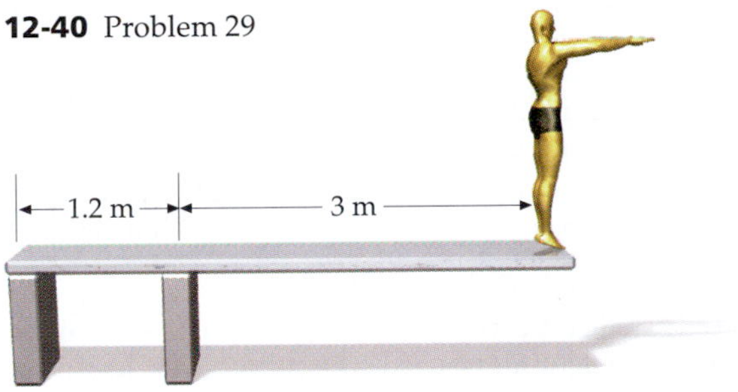

30 •• Find the force exerted on the strut by the hinge at *A* for the arrangement in Figure 12-41 if (*a*) the strut is weightless, and (*b*) the strut weighs 20 N.

Figure 12-41 Problem 30

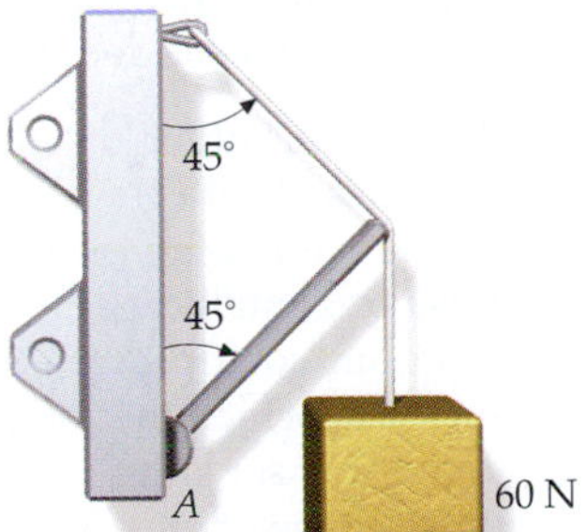

31 •• Julie has been hired to help paint the trim of a building, but she is not convinced of the safety of the apparatus. A 5.0-m plank is suspended horizontally from the top of the building by ropes attached at each end. She knows from previous experience, however, that the ropes being used will break if the tension exceeds 1 kN. Her 80-kg boss dismisses Julie's worries and begins painting while standing 1 m from the end of the plank. If Julie's mass is 60 kg and the plank has a mass of 20 kg, then over what range of positions can Julie stand if a colorful plummet is to be avoided?

32 •• The cable in Figure 12-39 must remain attached to the wall 8 m above the hinge, but its length can vary so that it can be connected to the beam at various distances *x* from the wall. How far from the wall should it be attached so that the force on the hinge has no vertical component?

33 •• A cylinder of mass *M* and radius *R* rolls against a step of height *h* as shown in Figure 12-42. When a horizontal force $\vec{F}$ is applied to the top of the cylinder, the cylinder remains at rest. (*a*) What is the normal force exerted by the floor on the cylinder? (*b*) What is the horizontal force exerted by the corner of the step on the cylinder? (*c*) What is the vertical component of the force exerted by the corner on the cylinder?

Figure 12-42 Problems 33 and 34

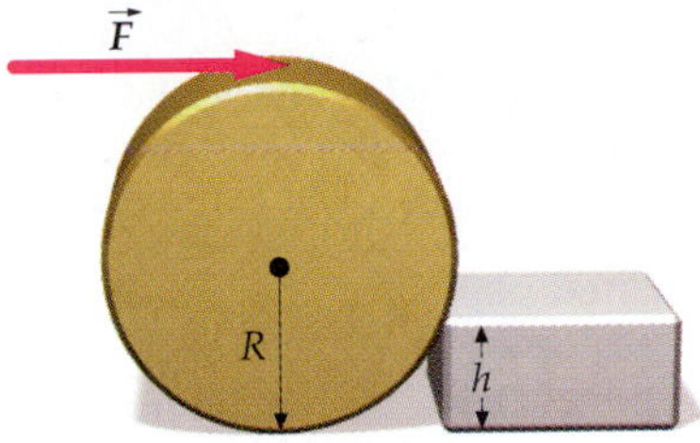

34 •• For the cylinder in Problem 33, find the minimum horizontal force $\vec{F}$ that will roll the cylinder over the step if the cylinder does not slide on the corner.

35 •• A strong man holds one end of a 3-m rod of mass 5 kg at rest in a horizontal position. (*a*) What total force does the man exert on the rod? (*b*) What total torque does the man exert on the rod? (*c*) If you approximate the effort of the man with two forces that act in opposite directions and are separated by the width of the man's hand, which is taken to be 10 cm, what are the magnitudes and directions of the two forces?

Figure 12-43 Problem 35

36 •• A large gate weighing 200 N is supported by hinges at the top and bottom and is further supported by a wire as shown in Figure 12-44. (*a*) What must the tension in the wire be for the force on the upper hinge to have no horizontal component? (*b*) What is the horizontal force on the lower hinge? (*c*) What are the vertical forces on the hinges?

Figure 12-44 Problem 36

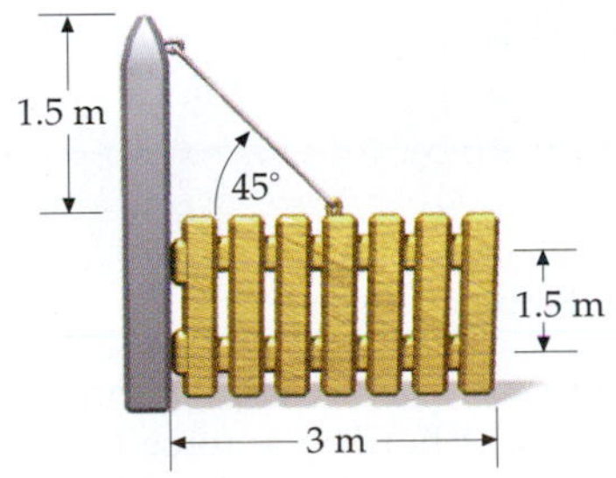

37 ••• A uniform log with a mass of 100 kg, a length of 4 m, and a radius of 12 cm is held in an inclined position, as shown in Figure 12-45. The coefficient of static friction between the log and the horizontal surface is 0.6. The log is on

Figure 12-45 Problem 37

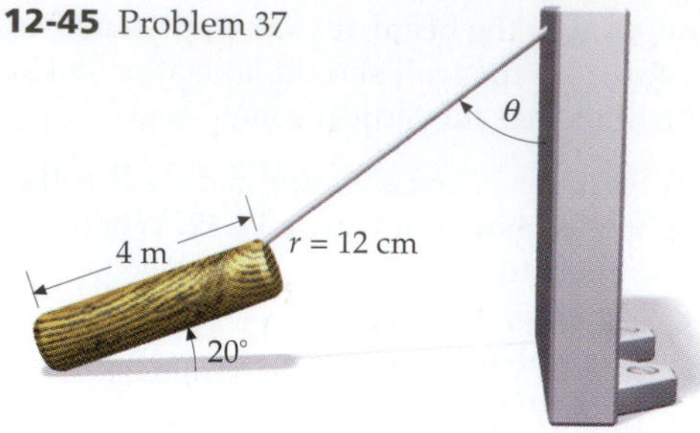

the verge of slipping to the right. Find the tension in the support wire and the angle the wire makes with the vertical wall.

38 ••• A tall, uniform, rectangular block sits on an inclined plane as shown in Figure 12-46. A cord is attached to the top of the block to prevent it from falling down the incline. What is the maximum angle θ for which the block will not slide on the incline? Let b/a be 4 and $\mu_s = 0.8$.

Figure 12-46 Problem 38

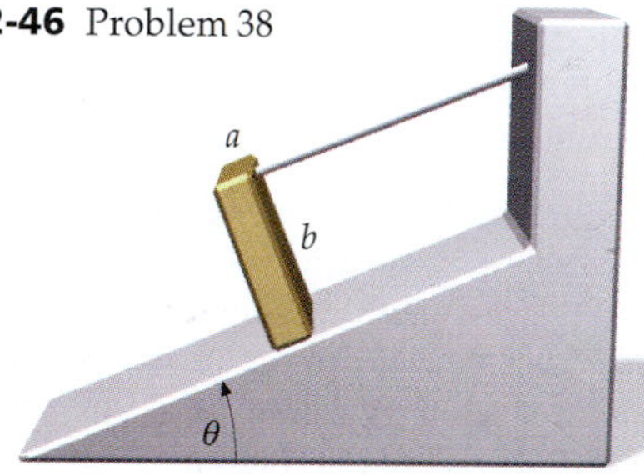

39 ••• A thin rail of length 10 m and mass 20 kg is supported at a 30° incline. One support is 2 m and the other is 6 m from the lower end of the rail. Friction prevents the rail from sliding off the supports. Find the force (magnitude and direction) exerted on the rail by each support.

Couples

40 • Two 80-N forces are applied to opposite corners of a rectangular plate as shown in Figure 12-47. Find the torque produced by this couple.

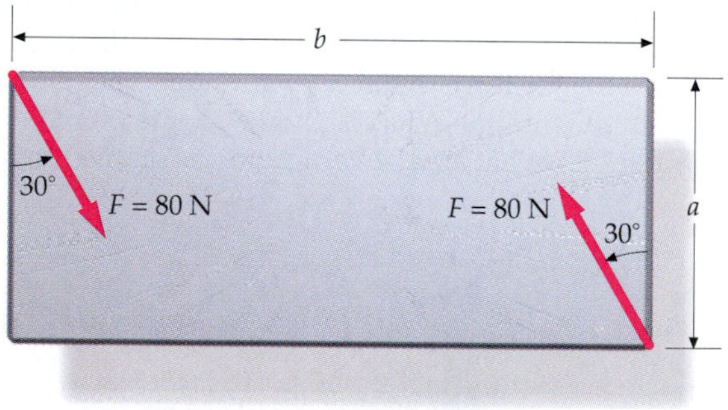

Figure 12-47 Problems 40 and 42

41 • A uniform cube of side a and mass M rests on a horizontal surface. A horizontal force $\vec{F}$ is applied to the top of the cube as in Figure 12-48. This force is not sufficient to move or tip the cube. (*a*) Show that the force of static friction exerted by the surface and the applied force constitute a couple, and find the torque exerted by the couple. (*b*) This couple is balanced by the couple consisting of the normal force exerted by the surface and the weight of the cube. Use this fact to find the effective point of application of the normal force when $F = Mg/3$. (*c*) What is the greatest magnitude of $\vec{F}$ for which the cube will not tip?

Figure 12-48 Problem 41

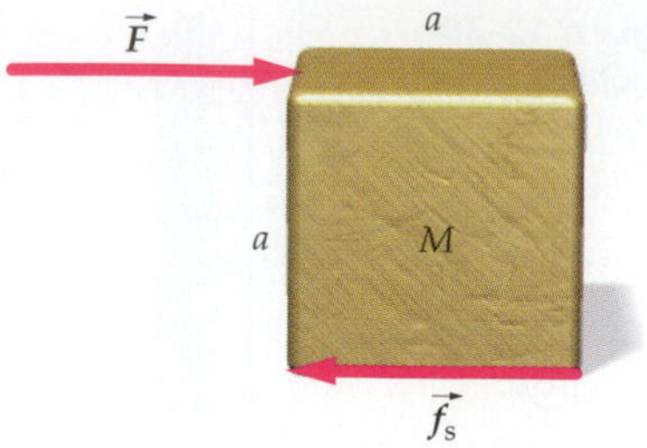

42 •• Resolve each force in Problem 40 into its horizontal and vertical components, producing two couples. The algebraic sum of the two component couples equals the resultant couple. Use this result to find the perpendicular distance between the lines of action of the two forces.

Ladder Problems

43 • Is it possible to climb a ladder placed against a wall where the ground is frictionless but the wall is not? Explain.

44 •• Romeo takes a uniform 10-m ladder and leans it against the smooth wall of the Capulet residence. The ladder's mass is 22.0 kg, and the bottom rests on the ground 2.8 m from the wall. When Romeo, whose mass is 70 kg, gets 90% of the way to the top, the ladder begins to slip. What is the coefficient of static friction between the ground and the ladder?

45 •• A massless ladder of length L leans against a smooth wall making an angle θ with the horizontal floor. The coefficient of friction between the ladder and the floor is μ_s. A man of mass M climbs the ladder. What height h can he reach before the ladder slips?

46 •• A uniform ladder of length L and mass m leans against a frictionless vertical wall with its lower end on the ground. It makes an angle of 60° with the horizontal ground. The coefficient of static friction between the ladder and ground is 0.45. If your mass is four times that of the ladder, how far up the ladder can you climb before it begins to slip?

47 •• A ladder of mass m and length L leans against a frictionless, vertical wall making an angle θ with the horizontal. The center of mass is at a height h from the floor. A force F pulls horizontally against the ladder at the midpoint. Find the minimum coefficient of static friction μ_s for which the top end of the ladder will separate from the wall while the lower end does not slip.

48 •• A 900-N boy sits on top of a ladder of negligible weight that rests on a frictionless floor as in Figure 12-49. There is a cross brace halfway up the ladder. The angle at the apex is $\theta = 30°$. (*a*) What is the force exerted by the floor on each leg of the ladder? (*b*) Find the tension in the cross brace. (*c*) If the cross brace is moved down toward the bottom of the

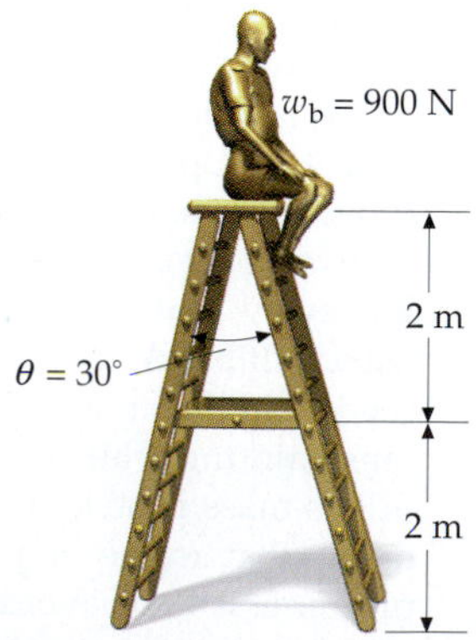

Figure 12-49 Problem 48

ladder (maintaining the same angle θ), will its tension be greater or less?

49 •• A ladder rests against a frictionless vertical wall. The coefficient of static friction between the ladder and the floor is 0.3. What is the smallest angle at which the ladder will remain stationary?

50 ••• Having failed in his first attempt, Romeo acquires a new ladder to try once again to get to Juliet's window. This one has a length L and a weight of 200 N. He tries placing it on the other side of the window, where the coefficients of static friction are 0.4 between the ladder and the wall, and 0.7 between the ladder and the ground. Because of bruises suffered in his last fall, Romeo wears heavy padding, which gives him a total mass of 80 kg. Sure enough, when he is $\frac{4}{5}$ of the way up the ladder, it begins to slip. What was the angle between the ladder and the ground when Romeo was making his ascent?

51 ••• A ladder leans against a large smooth sphere of radius R that is fixed in place on a horizontal surface. The ladder makes an angle of 60° with the horizontal surface and has a length $5R/2$. (*a*) What is the force that the sphere exerts on the ladder? (*b*) What is the frictional force that prevents the ladder from slipping? (*c*) What is the normal force that the horizontal surface exerts on the ladder?

Stress and Strain

52 • An aluminum wire and a steel wire of the same length L and diameter d are joined to form a wire of length $2L$. The wire is fastened to the roof and a weight W is attached to the other end. Neglecting the mass of the wires, which of the following statements is true?

(*a*) The aluminum portion will stretch by the same amount as the steel portion.
(*b*) The tensions in the aluminum portion and the steel portion are the same.
(*c*) The tension in the aluminum portion is greater than that in the steel portion.
(*d*) None of the above statements is true.

53 • A 50-kg ball is suspended from a steel wire of length 5 m and radius 2 mm. By how much does the wire stretch?

54 • Copper has a breaking stress of about 3×10^8 N/m^2. (*a*) What is the maximum load that can be hung from a copper wire of diameter 0.42 mm? (*b*) If half this maximum load is hung from the copper wire, by what percentage of its length will it stretch?

55 • A 4-kg mass is supported by a steel wire of diameter 0.6 mm and length 1.2 m. How much will this wire stretch under this load?

56 • As a runner's foot touches the ground, the shearing force acting on an 8-mm-thick sole is as shown in Figure 12-50. If the force of 25 N is distributed over an area of 15 cm^2, find the angle of shear θ shown, given that the shear modulus of the sole is 1.9×10^5 N/m^2.

Figure 12-50
Problem 56

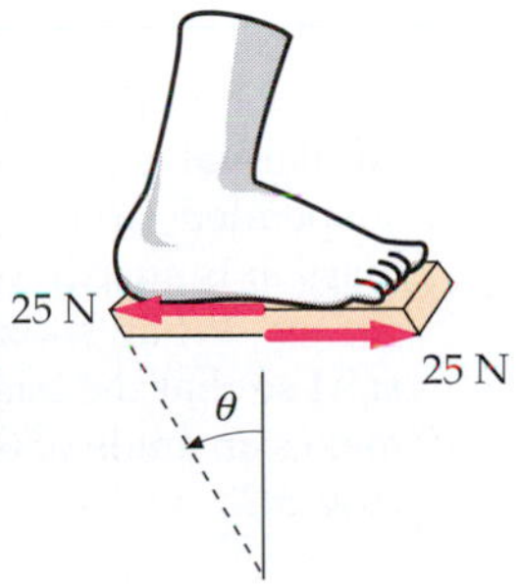

57 •• A steel wire of length 1.5 m and diameter 1 mm is joined to an aluminum wire of identical dimensions to make a composite wire of length 3.0 m. What is the length of the composite wire if it is used to support a mass of 5 kg?

58 •• A force F is applied to a long wire of length L and cross-sectional area A. Show that if the wire is considered to be a spring, the force constant k is given by $k = AY/L$ and the energy stored in the wire is $U = \frac{1}{2}F\,\Delta L$, where Y is Young's modulus and ΔL is the amount the wire has stretched.

59 •• The steel E string of a violin is under a tension of 53 N. The diameter of the string is 0.20 mm, and its length under tension is 35.0 cm. Find (*a*) the unstretched length of this string, and (*b*) the work needed to stretch the string. (See Problem 58.)

60 •• When a rubber strip with a cross section of 3 mm × 1.5 mm is suspended vertically and various masses are attached to it, a student obtains the following data for length versus load:

Load, g	0	100	200	300	300	500
Length, cm	5.0	5.6	6.2	6.9	7.8	10.0

(*a*) Find Young's modulus for the rubber strip for small loads. (*b*) Find the energy stored in the strip when the load is 150 g. (See Problem 58.)

61 •• A building is to be demolished by a 400-kg steel ball swinging on the end of a 30-m steel wire of diameter 5 cm hanging from a tall crane. As the ball is swung through an arc from side to side, the wire makes an angle of 50° with the vertical at the top of the swing. Find the amount by which the wire is stretched at the bottom of the swing.

62 •• A large mirror is hung from a nail as shown in Figure 12-51. The supporting steel wire has a diameter of 0.2 mm and an unstretched length of 1.7 m. The distance between the points of support at the top of the mirror's frame is 1.5 m. The mass of the mirror is 2.4 kg. What is the distance between the nail and the top of the frame when the mirror is hung?

Figure 12-51
Problems 62 and 91

63 •• Two masses, M_1 and M_2, are supported by wires of equal length when unstretched. The wire supporting M_1 is an aluminum wire 0.7 mm in diameter, and the one supporting M_2 is a steel wire 0.5 mm in diameter. What is the ratio M_1/M_2 if the two wires stretch by the same amount?

64 •• A mass of 0.5 kg is attached to an aluminum wire having a diameter of 1.6 mm and an unstretched length of 0.7 m. The other end of the wire is fixed to a post. The mass

rotates about the post in a horizontal plane at a rotational speed such that the angle between the wire and the horizontal is 5.0°. Find the tension in the wire and its length.

65 ••• It is apparent from Table 12-2 that the tensile strength of most materials is two to three orders of magnitude smaller than Young's modulus. Consequently, these materials, e.g., aluminum, will break before the strain exceeds even 1%. For nylon, however, the tensile strength and Young's modulus are approximately equal. If a nylon line of unstretched length L_0 and cross section A_0 is subjected to a tension T, the cross section may be substantially less than A_0 before the line breaks. Under these conditions, the tensile stress T/A may be significantly greater than T/A_0. Derive an expression that relates the area A to the tension T, A_0, and Young's modulus Y.

General Problems

66 • If the net torque about some point is zero, must it be zero about any other point? Explain.

67 • The horizontal bar in Figure 12-52 will remain horizontal if

(*a*) $L_1 = L_2$ and $R_1 = R_2$. (*b*) $L_1 = L_2$ and $M_1 = M_2$.
(*c*) $R_1 = R_2$ and $M_1 = M_2$. (*d*) $L_1M_1 = L_2M_2$.
(*e*) $R_1L_1 = R_2L_2$.

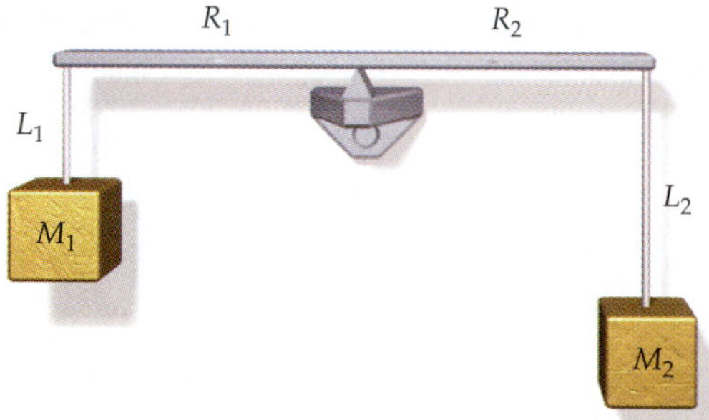

Figure 12-52 Problem 67

68 • Which of the following could not have units of N/m^2?

(*a*) Young's modulus (*b*) Shear modulus
(*c*) Stress (*d*) Strain

69 •• Sit in a chair with your back straight. Now try to stand up without leaning forward. Explain why you cannot do it.

70 • A 90-N board 12 m long rests on two supports, each 1 m from the end of the board. A 360-N block is placed on the board 3 m from one end as shown in Figure 12-53. Find the force exerted by each support on the board.

Figure 12-53 Problem 70

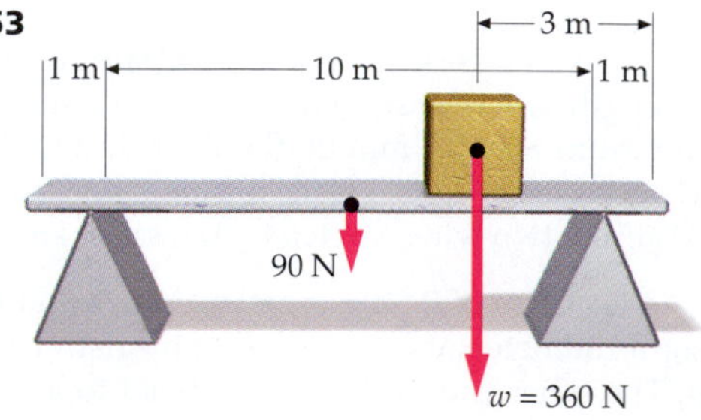

71 • The height of the center of gravity of a man standing erect is determined by weighing the man as he lies on a board of negligible weight supported by two scales as shown in Figure 12-54. If the man's height is 188 cm and the left scale reads 445 N while the right scale reads 400 N, where is his center of gravity relative to his feet?

Figure 12-54 Problem 71

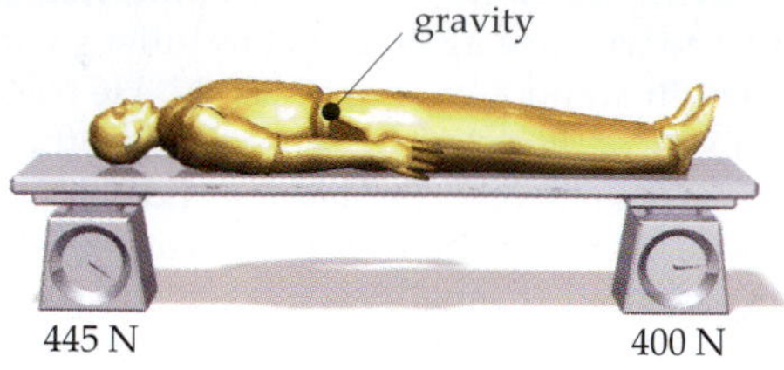

72 • Figure 12-55 shows a mobile consisting of four weights hanging on three rods of negligible mass. Find the value of each of the unknown weights if the mobile is to balance. *Hint:* Find weight w_1 first.

Figure 12-55 Problem 72

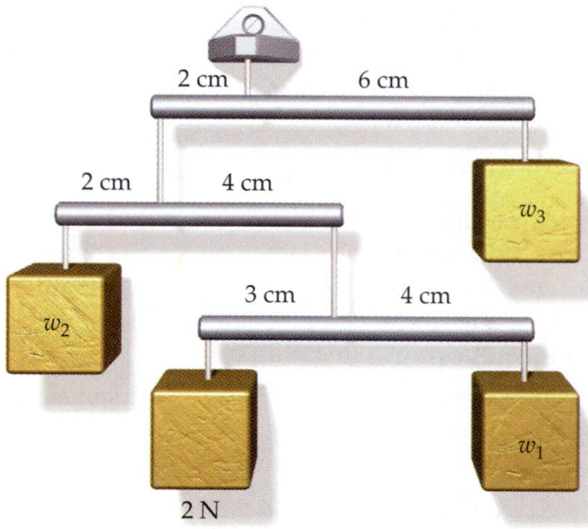

73 • A block and tackle is used to support a mass of 120 kg as shown in Figure 12-56. (*a*) What is the tension in the rope? (*b*) What is the mechanical advantage of this device?

Figure 12-56 Problem 73

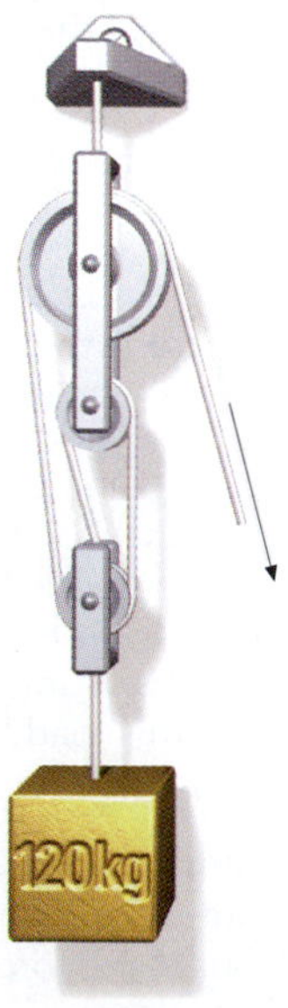

74 •• A plate in the shape of an equilateral triangle of mass M is suspended from one corner and a mass m is suspended from another corner. What should be the ratio m/M so that the base of the triangle makes an angle of 6.0° with the horizontal?

75 •• A standard six-sided pencil is placed on a paper pad (Figure 12-57). Find the minimum coefficient of static friction μ_s such that it rolls down rather than slides if the pad is inclined.

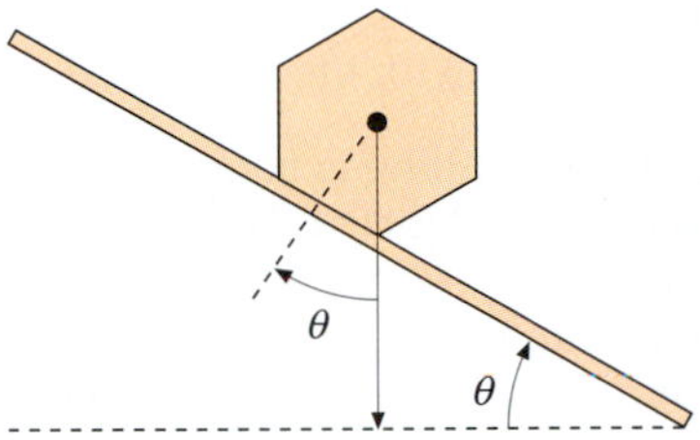

Figure 12-57 Problem 75

76 •• Having lost his job at the post office, Barry decides to explore the possibility that he might be a brilliant sculptor. Not one to start at the bottom, he borrows the money for a marble slab 3 m × 1 m × 1 m. After loading the marble onto the back of his truck, he drives off with the slab resting on its square end. But on the way home, a confused squirrel runs into his path, and Barry slams on the brakes. What deceleration will cause the uniform slab to tip over?

77 •• A uniform box of mass 8 kg that is twice as tall as it is wide rests on the floor of a truck. What is the maximum coefficient of static friction between the box and floor so that the box will slide toward the rear of the truck rather than tip when the truck accelerates on a level road?

78 •• Barry's art exhibit contains many tiny marble sculptures placed around a central piece called "Politics." The central piece consists of three identical bars, each of length L and mass m, joined as in Figure 12-58. Two bars form a fixed T, and the third bar is suspended on a hinge. Asked to explain the name, Barry said, "It swings from left to right with hinges flapping, but no matter where you start it from, you end up in the same place." When the system is in equilibrium, what is the value of θ?

Figure 12-58 Problem 78

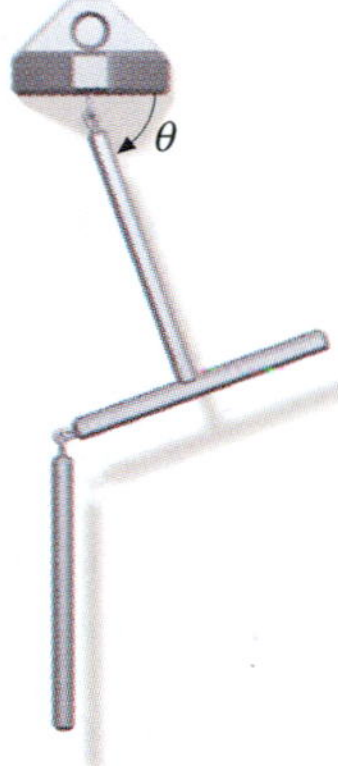

79 •• In the 1996 Olympics, the Russian super-heavyweight weightlifter Andrei Chemerkin broke the world record with a lift of mass 260 kg. Suppose his grip was slightly asymmetrical as shown in Figure 12-59. Find the maximum mass of the barbell Chemerkin could have handled with a symmetrical grip, assuming that his arms are equally strong.

Figure 12-59 Problem 79

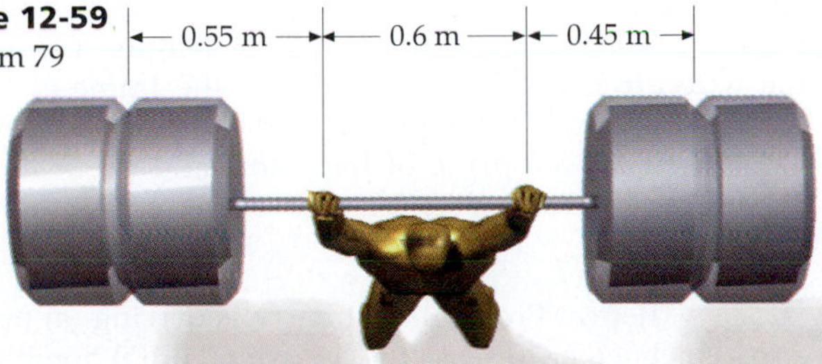

Figure 12-60 Problem 80

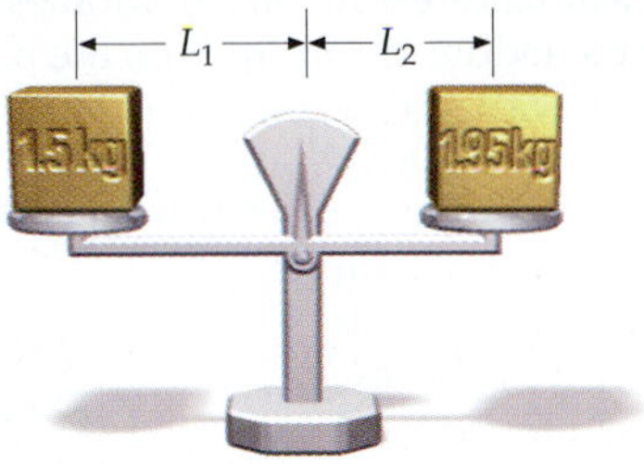

80 •• A balance scale has unequal arms. A 1.5-kg block appears to have a mass of 1.95 kg on the left pan of the scale (Figure 12-60). Find its apparent mass if the block is placed on the right pan.

81 •• A cube of mass M leans against a frictionless wall making an angle θ with the floor as shown in Figure 12-61. Find the minimum coefficient of static friction μ_s between the cube and the floor that allows the cube to stay at rest.

Figure 12-61 Problem 81

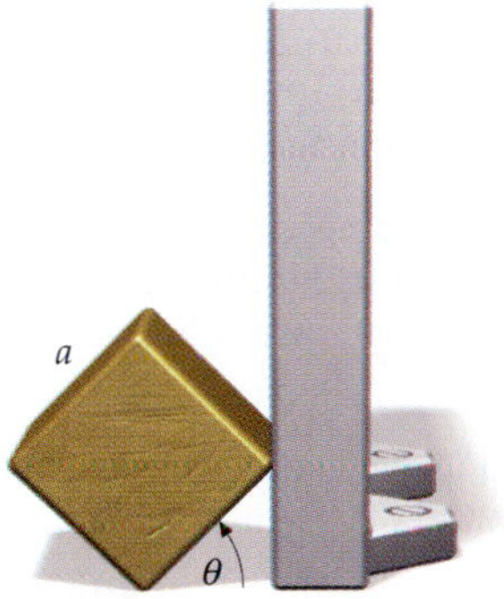

82 •• Figure 12-62 shows a steel meter stick hinged to a vertical wall and supported by a thin wire. The wire and meter stick make angles of 45° with the vertical. The mass of the meter stick is 5.0 kg. When a mass M = 10.0 kg is suspended from the midpoint of the meter stick, the tension T in the supporting wire is 52 N. If the wire will break should the tension exceed 75 N, what is the maximum distance along the meter stick at which the 10.0-kg mass can be suspended?

Figure 12-62 Problem 82

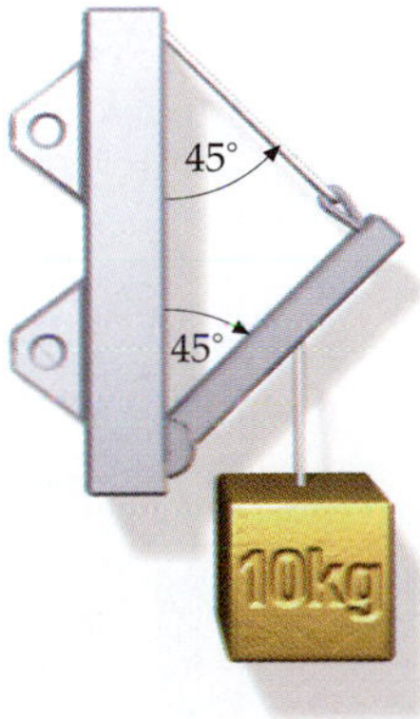

83 •• Figure 12-63 shows a 20-kg ladder leaning against a frictionless wall and resting on a frictionless horizontal surface. To keep the ladder from slipping, the bottom of the ladder is tied to the wall with a thin wire; the tension in the wire is 29.4 N. The wire will break if the tension exceeds 200 N. (*a*) If an 80-kg person climbs halfway up the ladder, what force will be exerted by the ladder against the wall? (*b*) How far up can an 80-kg person climb this ladder?

Figure 12-63 Problem 83

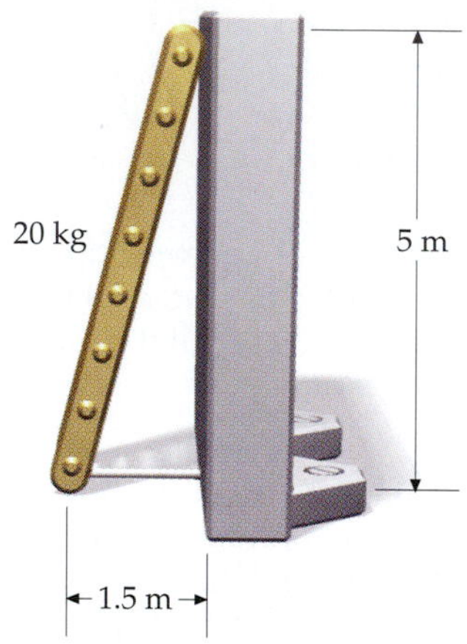

84 •• Suppose that the bar hanging from the end of the T in Problem 78 is of a different length $\ell \neq L$, although its mass per unit length is the same as that of the bars of the T. Find the ratio L/ℓ such that $\theta = 75°$.

85 •• A uniform cube can be moved along a horizontal plane either by pushing the cube so that it slips or by turning it over ("rolling"). What coefficient of static friction μ_k between the cube and the floor makes both ways equal in terms of the work needed?

86 •• A tall, uniform, rectangular block sits on an inclined plane as shown in Figure 12-64. If $\mu_s = 0.4$, does the block slide or fall over as the angle θ is slowly increased?

Figure 12-64 Problem 86

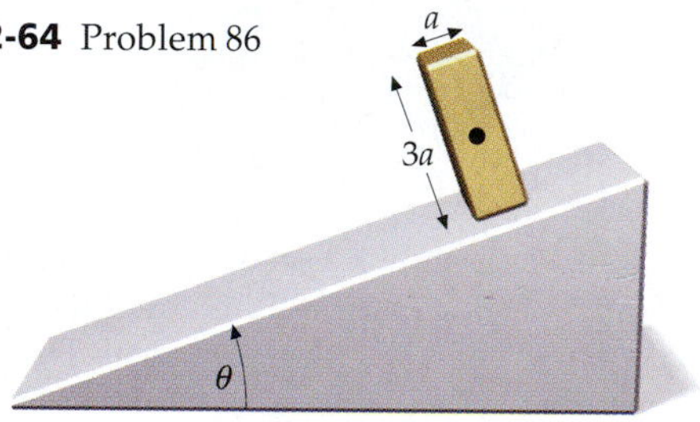

87 •• A 360-kg mass is supported on a wire attached to a 15-m-long steel bar that is pivoted at a vertical wall and supported by a cable as shown in Figure 12-65. The mass of the bar is 85 kg. (*a*) With the cable attached to the bar 5.0 m from the lower end as shown, find the tension in the cable and the force exerted by the wall on the steel bar. (*b*) Repeat if a somewhat longer cable is attached to the steel bar 5.0 m from its upper end, maintaining the same angle between the bar and the wall.

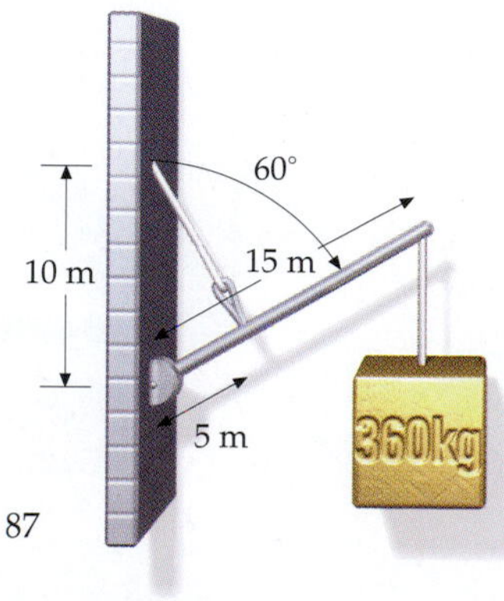

Figure 12-65 Problem 87

88 •• Repeat Problem 77 if the truck accelerates up a hill that makes an angle of 9.0° with the horizontal.

89 •• A thin rod 60 cm long is balanced 20 cm from one end when a mass of $2m + 2$ g is at the end nearest the pivot and a mass of m at the opposite end (Figure 12-66*a*). Balance is again achieved if the mass $2m + 2$ g is replaced by the mass m and no mass is placed at the other end (Figure 12-66*b*). Determine the mass M of the rod and the mass m.

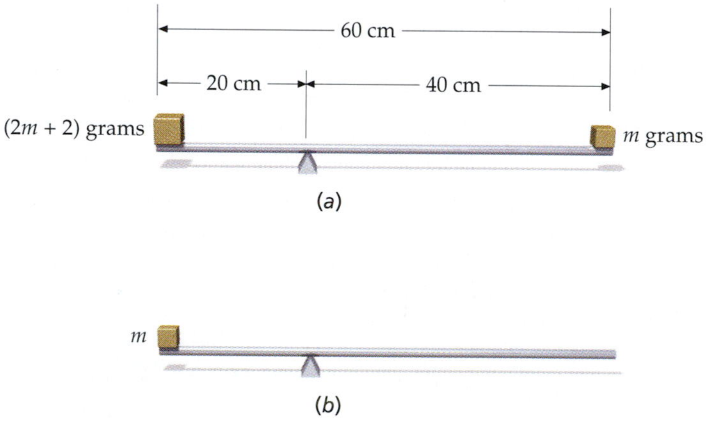

Figure 12-66 Problem 89

90 •• The planet Mars has two satellites, Phobos and Deimos, in nearly circular orbits. The orbit radii of Phobos and Deimos are 9.38×10^3 km and 23.46×10^3 km, respectively. The mass of Mars is 6.42×10^{23} kg, that of Phobos is 9.63×10^{15} kg, and that of Deimos is 1.93×10^{15} kg. Find the center of gravity and the center of mass of the two-satellite system using the center of Mars as the origin when (*a*) the satellites are in opposition (i.e., on exactly opposite sides of Mars), and (*b*) the satellites are in conjunction (i.e., in line on the same side of Mars).

91 •• When a picture is hung on a smooth vertical wall using a wire and a nail, as in Figure 12-51, the picture almost always tips slightly forward, i.e., the plane of the picture makes a small angle with the vertical. (*a*) Explain why pictures supported in this manner generally do not hang flush against the wall. (*b*) A framed picture 1.5 m wide and 1.2 m high and having a mass of 8.0 kg is hung as in Figure 12-51 using a wire of 1.7 m length. The ends of the wire are fastened to the sides of the frame at the rear and 0.4 m below the top. When the picture is hung, the angle between the plane of the frame and the wall is 5.0°. Determine the force that the wall exerts on the bottom of the frame.

92 •• Repeat Problem 76 if Barry is driving (*a*) uphill on a road inclined at 10° with the horizontal or (*b*) downhill on a road inclined at 10° with the horizontal.

93 •• If a train travels around a bend in the railbed too fast, the freight cars will tip over. Assume that the cargo portions of the freight cars are regular parallelepipeds of uniform density and 1.5×10^4 kg mass, 10 m long, 3.0 m high, and 2.20 m wide, and that their base is 0.65 m above the rails. The axles are 7.6 m apart, each 1.2 m from the ends of the boxcar. The separation between the rails is 1.55 m. Find the maximum safe speed of the train if the radius of curvature of the bend is (*a*) 150 m, and (*b*) 240 m.

94 •• For balance, a tightrope walker uses a thin rod 8 m long and bowed in a circular arc shape. At each end of the rod is a lead mass of 20 kg. The tightrope walker, whose mass is 58 kg and whose center of gravity is 0.90 m above the rope, holds the rod tightly at its center 0.65 m above the rope. What should the radius of curvature of the arc of the rod be so that he will be in stable equilibrium as he slowly makes his way across the rope? Neglect the mass of the rod.

95 •• A large crate weighing 4500 N rests on four 12-cm-high blocks on a horizontal surface (Figure 12-67). The crate is 2 m long, 1.2 m high, and 1.2 m deep. You are asked to lift one end of the crate using a long steel bar. The coefficient of static friction between the blocks and the supporting surface is 0.4. Estimate the length of the steel bar you will need to lift the end of the crate.

Figure 12-67 Problem 95

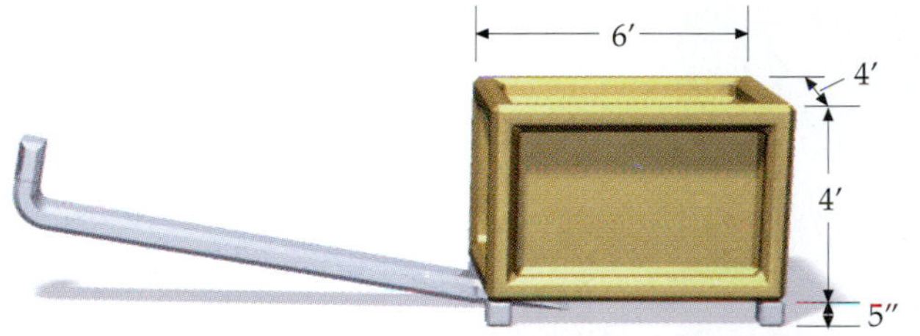

96 ••• Six identical bricks are stacked one on top of the other lengthwise and slightly offset to produce a stepped tower with the maximum offset that will still allow the tower to stand. (*a*) Starting from the top, give the maximum possible offset for each successive brick. (*b*) What is the total protrusion or offset of the six bricks?

97 ••• A uniform sphere of radius R and mass M is held at rest on an inclined plane of angle θ by a horizontal string, as shown in Figure 12-68. Let $R = 20$ cm, $M = 3$ kg, and $\theta = 30°$. (*a*) Find the tension in the string. (*b*) What is the normal force exerted on the sphere by the inclined plane? (*c*) What is the frictional force acting on the sphere?

Figure 12-68 Problem 97

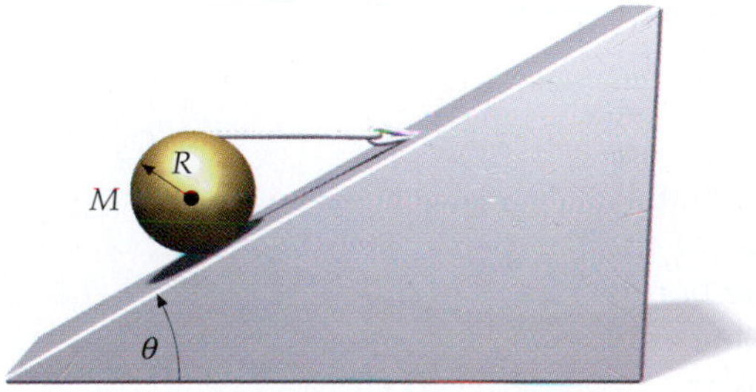

98 ••• The legs of a tripod make equal angles of 90° with each other at the apex, where they join together. A 100-kg block hangs from the apex. What are the compressional forces in the three legs?

99 ••• Figure 12-69 shows a 20-cm-long uniform beam resting on a cylinder of 4 cm radius. The mass of the beam is 5.0 kg, and that of the cylinder is 8.0 kg. The coefficient of friction between beam and cylinder is zero. (*a*) Find the forces that act on the beam and on the cylinder. (*b*) What must the minimum coefficients of static friction be between beam and floor and between the cylinder and floor to prevent slipping?

Figure 12-69
Problem 99

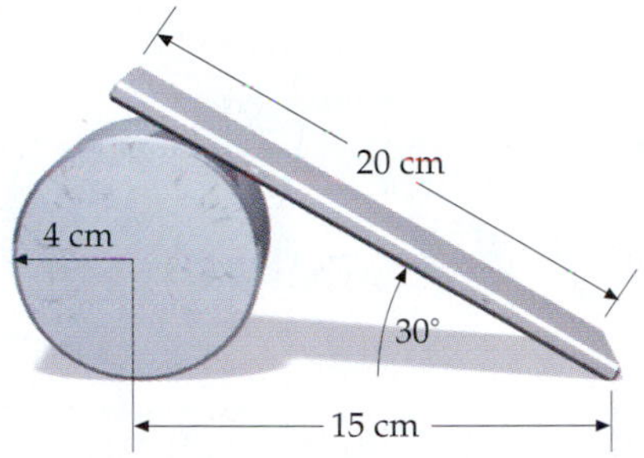

100 ••• Two solid smooth spheres of radius r are placed inside a cylinder of radius R as in Figure 12-70. The mass of each sphere is m. Find the force exerted by the bottom of the cylinder, the force exerted by the wall of the cylinder, and the force exerted by one sphere on the other.

Figure 12-70 Problem 100

101 ••• A solid cube of side length a balanced atop a cylinder of diameter d is in unstable equilibrium if $d \ll a$ and is in stable equilibrium if $d \gg a$ (Figure 12-71). Determine the minimum value of d/a for which the cube is in stable equilibrium.

Figure 12-71
Problem 101

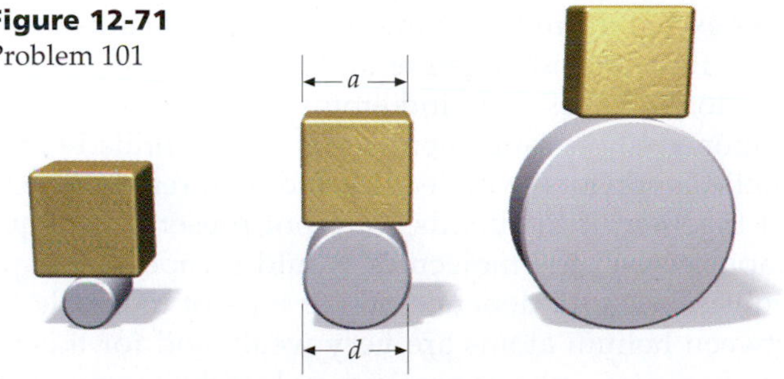

CHAPTER 13

Fluids

Airflow above and below the wing of this Indy race car creates greater pressure above the wing, increasing the effective weight of the car for better control at high speeds. An airplane wing is designed so that the flow creates greater pressure below the wing to lift the plane.

Fluids include both liquids and gases. Liquids flow under gravity until they occupy the lowest possible regions of their containers. Gases expand to fill their containers regardless of the containers' shapes.

In a gas, the average distance between two molecules is large compared with the size of a molecule. The molecules have little influence on one another except during their frequent but brief collisions. In a liquid or solid, the molecules are close together and exert forces on one another that are comparable to the forces that bind atoms into molecules. Molecules in a liquid form temporary short-range bonds that are continually broken due to the thermal kinetic energy of the molecules and then reformed. These bonds hold the liquid together; if the bonds were not present, the liquid would immediately evaporate and the molecules would escape as a vapor. The strength of the bonds in a liquid depends on the type of molecule. For example, the bonds between helium atoms are very weak, and for this reason, helium does not liquefy at atmospheric pressure unless the temperature is 4.2 K or lower.

13-1 Density

An important property of a substance is the ratio of its mass to its volume, which is called its **density**:

$$\text{Density} = \frac{\text{mass}}{\text{volume}}$$

The Greek letter ρ (rho) is usually used to denote density:

$$\rho = \frac{m}{V} \tag{13-1}$$

Definition—Density

The gram was originally defined as the mass of one cubic centimeter of water, so the density of water has served as a benchmark. In cgs units, the density of water is 1 g/cm^3. Converting to SI units, we obtain for the density of water

$$\rho_w = \frac{1\ \text{g}}{\text{cm}^3} \times \frac{\text{kg}}{10^3\ \text{g}} \times \left(\frac{100\ \text{cm}}{\text{m}}\right)^3 = 10^3\ \text{kg/m}^3 \tag{13-2}$$

The densities of most materials, including water, vary with temperature. Equation 13-2 gives the maximum value for the density of water, which occurs at 4°C.

A convenient unit of volume for fluids is the **liter** (L):

$$1\ \text{L} = 10^3\ \text{cm}^3 = 10^{-3}\ \text{m}^3$$

In terms of this unit, the density of water at 4°C is 1.00 kg/L. When an object's density is greater than that of water, it sinks in water. When its density is less, it floats. The ratio of the density of a substance to that of water is called the **specific gravity** of the substance. For example, the specific gravity of aluminum is 2.7, meaning that a volume of aluminum has 2.7 times the mass of an equal volume of water. The specific gravities of objects that sink in water range from 1 to about 22.5 (for the densest element, osmium).

Most solids and liquids expand only slightly when heated, and contract slightly when subjected to an increase in external pressure. Since these changes in volume are relatively small, we often treat the densities of solids and liquids as approximately independent of temperature and pressure. The density of a gas, on the other hand, depends strongly on the pressure and temperature, so these variables must be specified when reporting the densities of gases. By convention, **standard conditions** are atmospheric pressure at sea level and a temperature of 0°C. Figure 13-1 gives the densities for a variety of substances under these conditions. Note that the densities of liquids and solids are considerably greater than those of gases. For example, the density of water is about 800 times that of air under standard conditions.

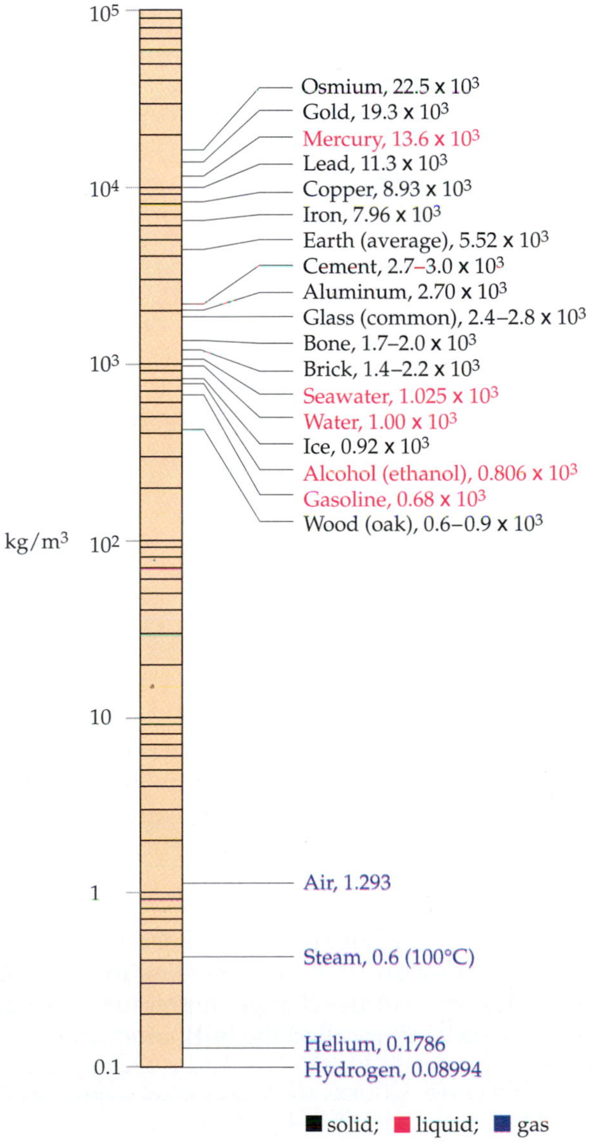

Figure 13-1 Densities of selected substances. Unless otherwise indicated, t = 0°C and P = 1 atm.

Example 13-1

A 200-mL flask is filled with water at 4°C. When the flask is heated to 80°C, 6 g of water spill out. What is the density of water at 80°C? (Assume that the expansion of the flask is negligible.)

Picture the Problem The density of water at 80°C is $\rho' = m'/V$, where $V = 200\text{ mL} = 200\text{ cm}^3$ is the volume of the flask, and m' is the mass remaining in the flask after 6 g spills out. We find m' by first finding the mass of water originally in the flask.

1. Calculate the original mass of water in the flask at 4°C using $\rho = 1\text{ g/cm}^3$: $m = \rho V = (1\text{ g/cm}^3)(200\text{ cm}^3) = 200\text{ g}$

2. Calculate the mass of water remaining, m', after 6 g spill out: $m' = m - 6\text{ g} = 200\text{ g} - 6\text{ g} = 194\text{ g}$

3. Use this value of m' to find the density of water at 80°: $\rho' = \dfrac{m'}{V} = \dfrac{194\text{ g}}{200\text{ cm}^3} = 0.97\text{ g/cm}^3$

Exercise A solid metal cube 8 cm on a side has a mass of 4.08 kg. (*a*) What is the density of the cube? (*b*) If the cube is made from a single element listed in Figure 13-1, what is the element? (*Answers* (*a*) 7.97 kg/L, (*b*) iron. Note that the slight difference between the answer in (*a*) and Figure 13-1 is due to a difference in the number of significant digits used to derive the two values.)

Exercise A gold brick is 5 cm × 10 cm × 20 cm. How much does it weigh? (*Answer* 189 N ≈ 42.6 lb)

13-2 Pressure in a Fluid

When a body is submerged in a fluid such as water, the fluid exerts a force perpendicular to the surface of the body at each point on the surface. This force per unit area is called the **pressure** P of the fluid:

$$P = \frac{F}{A} \qquad \text{13-3}$$

Definition—Pressure

The SI unit of pressure is the newton per square meter (N/m^2), which is called the **pascal** (Pa):

$$1\text{ Pa} = 1\text{ N/m}^2 \qquad \text{13-4}$$

In the U.S. customary system, pressure is usually given in pounds per square inch (lb/in^2). Another common unit of pressure is the atmosphere (atm), which is approximately the air pressure at sea level. One atmosphere is now defined to be 101.325 kilopascals, which is approximately 14.70 lb/in^2:

$$1\text{ atm} = 101.325\text{ kPa} = 14.70\text{ lb/in}^2 \qquad \text{13-5}$$

Other units of pressure in common use are discussed later.

The pressure due to a fluid pressing in on an object tends to compress the object. The ratio of the change in pressure (ΔP) to the fractional decrease in volume ($-\Delta V/V$) is called the **bulk modulus***:

* The minus sign in Equation 13-6 is introduced to make B positive since all materials decrease in volume when subjected to external pressure.

$$B = -\frac{\Delta P}{\Delta V/V} \qquad \text{13-6}$$

Definition—Bulk modulus

The more difficult a material is to compress, the smaller is $\Delta V/V$ for a given pressure, and hence the greater the bulk modulus. Liquids, gases, and solids all have a bulk modulus. Since liquids and solids are relatively incompressible, they have large values of B, and these values are relatively independent of temperature and pressure. Gases, on the other hand, are easily compressed, and their values for B depend strongly on the pressure and temperature. Figure 13-2 charts values for the bulk modulus of various materials.

Exercise Water is contained in a cylindrical iron container sealed with an iron piston. The pressure on the piston is increased to 100 atm. (*a*) What is the percentage change in the volume of the water? (*b*) What is the percentage change in the volume of the iron? (*Answer* $-\Delta V/V \approx 0.5\%$ for water and 0.01% for iron.)

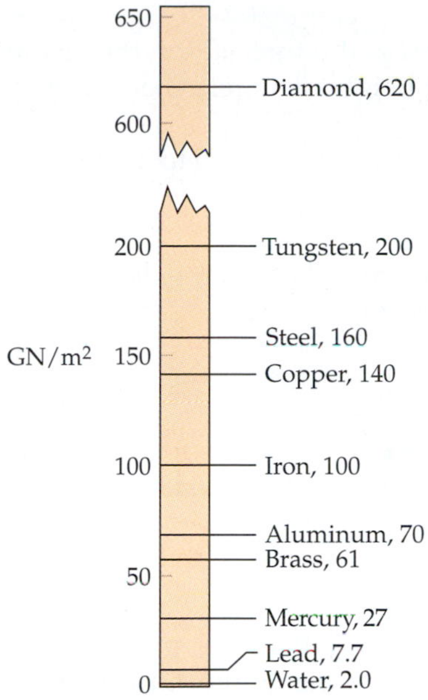

Figure 13-2 Approximate values for the bulk modulus B of various materials.

As any scuba diver knows, the pressure in a lake or ocean increases with depth. Similarly, the pressure of the atmosphere decreases with altitude. For a liquid such as water, whose density is approximately constant throughout, the pressure increases linearly with depth. We can see this by considering a column of liquid of height h and cross-sectional area A, as shown in Figure 13-3. To support the weight of the column, the pressure at the bottom must be greater than the pressure at the top. The weight of this liquid column is

$$w = mg = \rho V g = \rho A h g$$

If P_0 is the pressure at the top and P is the pressure at the bottom, the net upward force exerted by this pressure difference is $PA - P_0A$. Setting this net upward force equal to the weight of the column, we obtain

$$PA - P_0 A = \rho A h g$$

or

$$P = P_0 + \rho g h \qquad (\rho \text{ constant}) \qquad \text{13-7}$$

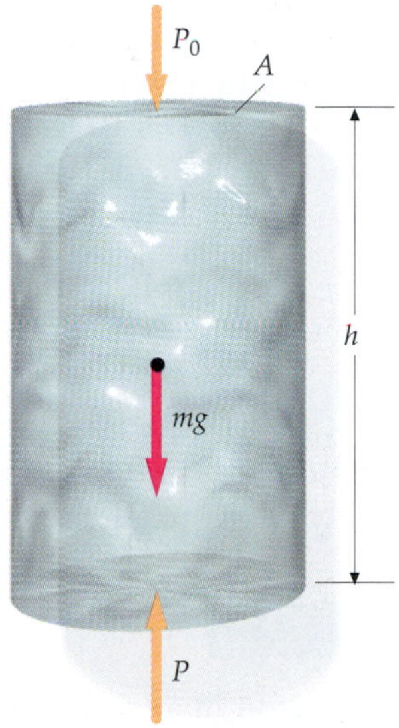

Figure 13-3 Column of water of height h and cross-sectional area A. The pressure P at the bottom must be greater than the pressure P_0 at the top to balance the weight of the water.

Exercise Find the pressure at a depth of 10 m below the surface of a lake if the pressure at the surface is 1 atm. (*Answer* With $P_0 = 1$ atm $= 101$ kPa, $\rho = 10^3\ \text{kg/m}^3$, and $g = 9.81$ N/kg, we have $P = P_0 + \rho g h = 1.97$ atm. The pressure at a depth of 10 m is nearly twice that at the surface.)

Example 13-2

A rectangular dam 30 m wide supports a body of water to a height of 25 m (Figure 13-4). Find the total horizontal force on the dam.

Figure 13-4

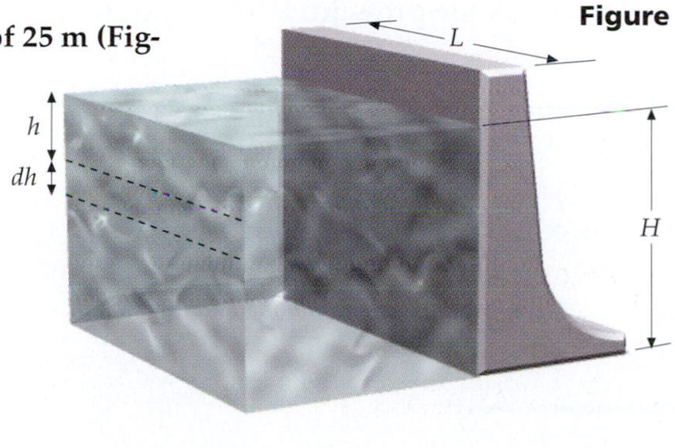

Picture the Problem Because the pressure varies with depth, we cannot merely multiply the pressure times the area of the dam to find the force exerted by the water. We therefore consider the force exerted on a strip of width $L = 30$ m, height dh, and area $dA = L\,dh$ at a depth h, and integrate from $h = 0$ to $h = H = 25$ m. The water pressure at depth h is $P_{atm} + \rho g h$. We can omit the atmospheric pressure because it is exerted on each side of the wall.

1. Express the force dF on the element of length L and height dh in terms of the net pressure ρhg:

$$dF = P\,dA = \rho ghL\,dh$$

2. Integrate from $h = 0$ to $h = H$:

$$F = \int_{h=0}^{h=H} dF = \int_0^H \rho ghL\,dh = \rho gL\frac{h^2}{2}\bigg|_0^H = \frac{1}{2}\rho gLH^2$$

3. Substitute the given values to find the numerical result:

$$F = \frac{1}{2}\rho gLH^2 = \frac{1}{2}(10^3\ \text{kg/m}^3)(9.81\ \text{N/kg})(30\ \text{m})(25\ \text{m})^2$$

$$= 9.20 \times 10^7\ \text{N}$$

Remarks Because the force on an element of the dam is proportional to the depth of the water, the dam must be designed to be thicker at the bottom.

The result that the pressure at a depth h is greater than that at the surface of the liquid by the amount ρgh holds for a liquid in any container, independent of the shape of the container. Furthermore, the pressure is the same at all points at the same depth. If we increase the pressure by pressing down on the top surface with a piston, the increase in pressure is the same throughout the liquid. This is known as **Pascal's principle,** named after Blaise Pascal (1623–1662):

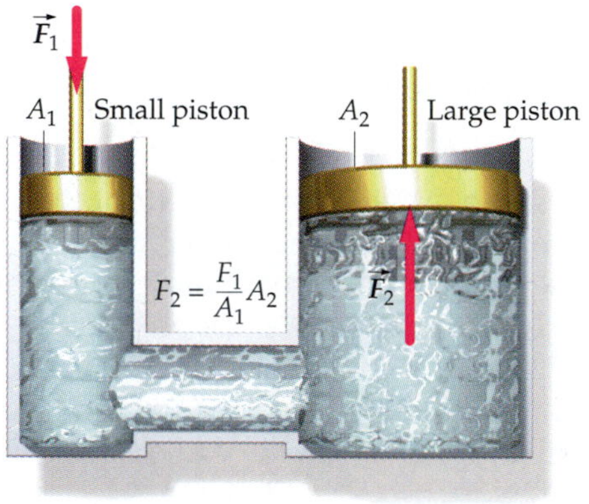

Figure 13-5 Hydraulic lift. A small force $\vec{F}_1$ on the small piston produces a change in pressure F_1/A_1 that is transmitted by the liquid to the large piston. Since the pressures at the small and large pistons are the same, the forces are related by $F_2/A_2 = F_1/A_1$. Since the area of the large piston is much greater than that of the small piston, the force on the large piston $F_2 = (A_2/A_1)F_1$ is much greater than F_1.

> Pressure applied to an enclosed liquid is transmitted undiminished to every point in the fluid and to the walls of the container.

Pascal's principle

A common application of Pascal's principle is the hydraulic lift shown in Figure 13-5.

Example 13-3

The large piston in a hydraulic lift has a radius of 20 cm. What force must be applied to the small piston of radius 2 cm to raise a car of mass 1500 kg?

Picture the Problem The pressure P times the area A_2 of the large piston must equal the weight mg of the car. The force that must be exerted on the small piston F_1 is this same pressure times the area A_1 (see Figure 13-5).

1. The force F_1 is the pressure P times the area A_1:

$$F_1 = PA_1$$

2. The pressure times the area A_2 equals the weight of the car:

$$PA_2 = F_2 = mg, \qquad \text{so} \quad P = \frac{mg}{A_2}$$

3. Use this result for P to calculate F_1:

$$F_1 = PA_1 = \frac{mg}{A_2}A_1 = mg\frac{\pi r_1^2}{\pi r_2^2}$$

$$= (1500\ \text{kg})(9.81\ \text{N/kg})\frac{(2\ \text{cm})^2}{(20\ \text{cm})^2} = 147\ \text{N}$$

Figure 13-6 shows water in a container with sections of different shapes. At first glance, it might seem that the pressure in the largest section of the container would be greatest and that water would therefore be forced to a greater height in the smaller sections of the container. The fact that this does not happen is known as the **hydrostatic paradox.** The pressure depends only on the depth of the water, not on the shape of the container, so at the same depth the pressure is the same in all parts of the container, a finding that can be shown experimentally. Although the water in the largest section of the container weighs more than that in the smaller sections, some of this weight is supported by the normal force exerted by the sides of the container. In fact, the shaded portion of the water is completely supported by the sides of the container.

Figure 13-6 The hydrostatic paradox. The water level is the same regardless of the shape of the vessel. The shaded portion of the water is supported by the sides of the container.

We can use the fact that the pressure difference is proportional to the depth of a fluid to measure unknown pressures. Figure 13-7 shows a simple pressure gauge, the open-tube manometer, consisting of a U-tube containing a liquid, which is designed to measure small pressure differences between its two arms. The top of the tube is open to the atmosphere at pressure P_{at}. The other end of the tube is at pressure P, which is to be measured. The difference $P - P_{at}$, called the **gauge pressure** P_{gauge}, is equal to ρgh, where ρ is the density of the liquid in the tube. The pressure you measure in your automobile tire is gauge pressure. When the tire is entirely flat, the gauge pressure is zero, and the absolute pressure in the tire is atmospheric pressure. The absolute pressure P is obtained from gauge pressure by adding atmospheric pressure to it:

$$P = P_{gauge} + P_{at} \qquad \text{13-8}$$

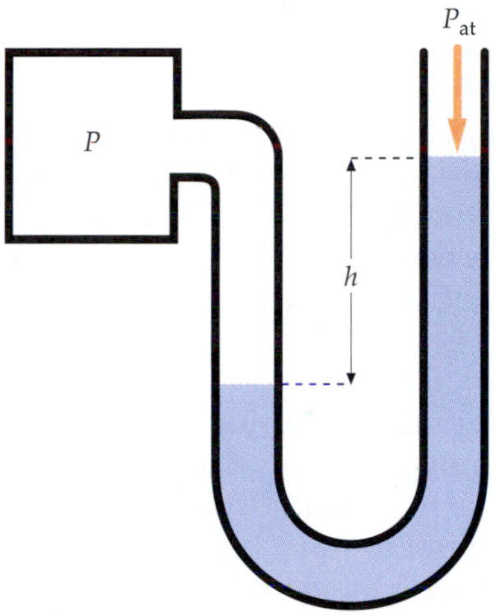

Figure 13-7 Open-tube manometer for measuring an unknown pressure P. The difference $P - P_{at}$ equals ρgh.

Figure 13-8 shows a mercury U-tube barometer, which is used to measure atmospheric pressure. The top end of the tube has been closed off and evacuated so that the pressure there is zero. The other end is open to the atmosphere at pressure P_{at}. The pressure P_{at} is given by $P_{at} = \rho gh$, where ρ is the density of mercury.

Exercise At 0°C, the density of mercury is 13.595×10^3 kg/m^3. What is the height of the mercury column in a U-tube barometer if the pressure is 1 atm = 101.325 kPa? (*Answer* $h = P/\rho g = 0.760$ m = 760 mm)

In practice, pressure is often measured in millimeters of mercury (mmHg), a unit called the **torr,** after the Italian physicist Torricelli, or in inches of mercury (written inHg). The various units of pressure are related as follows:

$$1 \text{ atm} = 760 \text{ mmHg} = 760 \text{ torr} = 29.9 \text{ inHg}$$
$$= 101.325 \text{ kPa} = 14.7 \text{ lb/in}^2 \qquad \text{13-9}$$

Other units commonly used on weather maps are the **bar** and the **millibar,** which are defined as

$$1 \text{ bar} = 10^3 \text{ millibars} = 100 \text{ kPa} \qquad \text{13-10}$$

A pressure of 1 bar is slightly less than 1 atm.

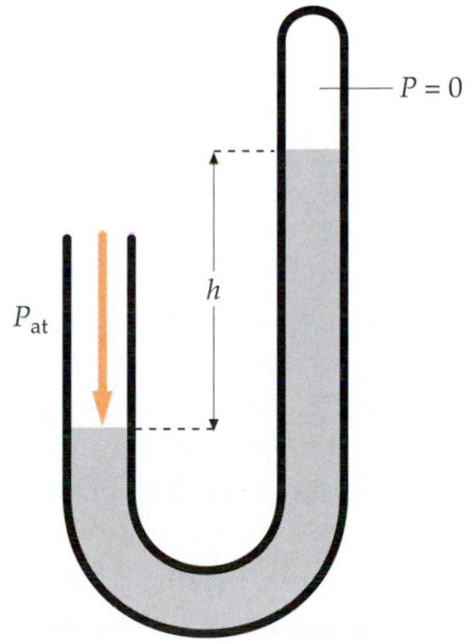

Figure 13-8 U-tube barometer for measuring atmospheric pressure P_{at}.

Example 13-4

The average (gauge) pressure in the aorta is about 100 mmHg. Convert this average blood pressure to pascals.

We use the conversion factors implied in Equation 13-9:

$$P = 100\ \text{mmHg}\left(\frac{101.325\ \text{kPa}}{760\ \text{mmHg}}\right) = 13.3\ \text{kPa}$$

Exercise Convert a pressure of 45 kPa to (*a*) mmHg, and (*b*) atmospheres. (*Answers* (*a*) 338 mmHg, (*b*) 0.444 atm)

The relation between pressure and altitude (or depth) is more complicated for a gas than for a liquid because the density of a gas is not constant like that of a liquid, but is approximately proportional to the pressure. As you go up from the surface of the earth, pressure in a column of air decreases, just as the pressure would decrease as you go up from the bottom in a water column. But the decrease in air pressure is not linear with distance. Instead, the air pressure decreases by a constant fraction for a given increase in height, as shown in Figure 13-9. At a height of about 5.5 km (18,000 ft), the air pressure is half its value at sea level. If we go up another 5.5 km to an altitude of 11 km (a typical altitude for airliners), the pressure is again halved so that it is one-fourth its value at sea level, and so on. This example of an *exponential decrease* is called the law of atmospheres. At the high altitudes at which commercial jets fly, the cabins must be pressurized. The density of air is proportional to the pressure, so the density of air decreases with altitude. There is less oxygen available on a mountain than at normal elevations, which makes exercising in the Rockies difficult and climbing in the Himalayas dangerous.

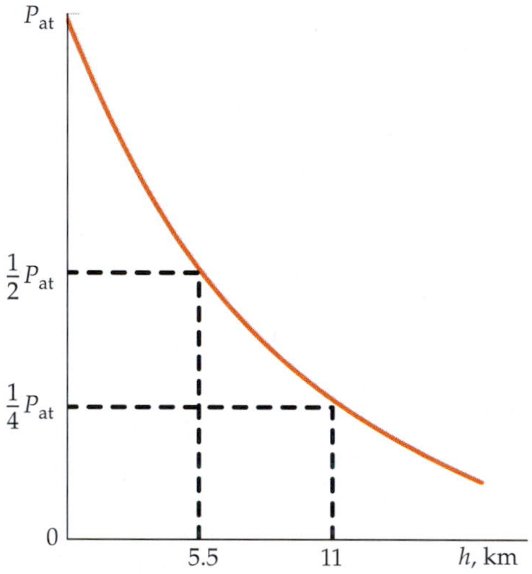

Figure 13-9 Variation in pressure with height above the earth's surface. For each 5.5-km increase in height, the pressure decreases by half.

13-3 Buoyancy and Archimedes' Principle

If a heavy object submerged in water is "weighed" by suspending it from a spring scale, the reading on the scale is less than when the object is weighed in air (Figure 13-10*a*). This is because the water exerts an upward force that partially balances the force of gravity. The force is even more evident when we submerge a piece of cork. When completely submerged, the cork experiences an upward force from the water pressure that is greater than the force of gravity, so it accelerates up toward the surface. The force exerted by a fluid on a body submerged in it is called the **buoyant force.** It is equal to the weight of the fluid displaced by the body.

> A body wholly or partially submerged in a fluid is buoyed up by a force equal to the weight of the displaced fluid.

Archimedes' principle

This result is known as **Archimedes' principle.**

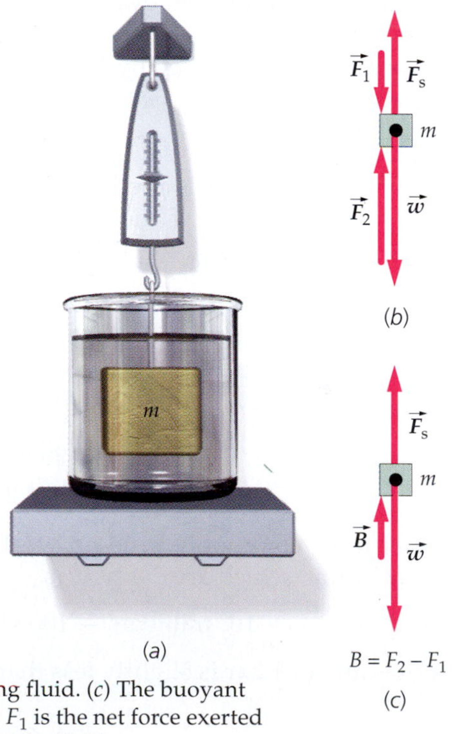

Figure 13-10 (*a*) Weighing an object submerged in a fluid. (*b*) Free-body diagram showing the weight $\vec{w}$, the force $\vec{F}_s$ of the spring, and the forces $\vec{F}_1$ and $\vec{F}_2$ exerted by the surrounding fluid. (*c*) The buoyant force $B = F_2 - F_1$ is the net force exerted on the object by the fluid.

We can derive Archimedes' principle from Newton's laws by considering the forces acting on a portion of a fluid and noting that in static equilibrium the net force must be zero. Figure 13-10*b* shows the vertical forces acting on an object being weighed while submerged. These are the force of gravity $\vec{w}$ down, the force $\vec{F}_s$ of the spring scale acting up, a force $\vec{F}_1$ acting down because of the fluid pressing on the top surface of the object, and a force $\vec{F}_2$ acting up because of the fluid pressing on the bottom surface of the object. Since the spring scale reads a force less than the weight, the force $\vec{F}_2$ must be greater in magnitude than the force $\vec{F}_1$. The difference in magnitude of these two forces is the buoyant force $B = F_2 - F_1$. The buoyant force occurs because the pressure of the fluid at the bottom of the object is greater than that at the top.

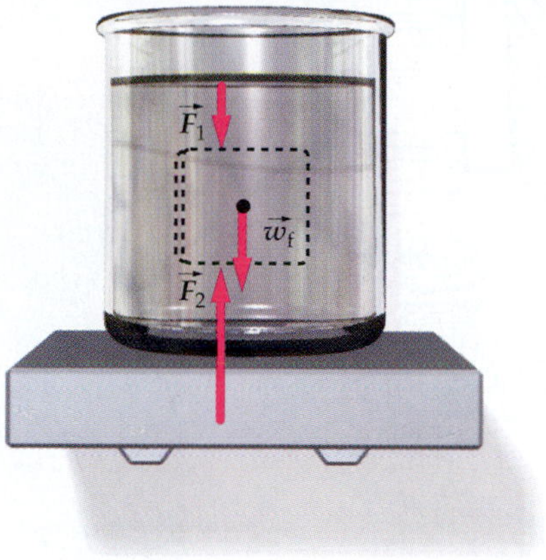

Figure 13-11 Figure 13-10 with the submerged body replaced by an equal volume of fluid. The forces $\vec{F}_1$ and $\vec{F}_2$ due to the pressure of the fluid are the same as in Figure 13-10. The buoyant force is thus equal to the weight w_f of the displaced fluid.

In Figure 13-11, the spring scale has been eliminated and the submerged object has been replaced by an equal volume of fluid (indicated by the dotted lines). The buoyant force $B = F_2 - F_1$ acting on this volume of fluid is the same as the buoyant force that acted on our original object since the fluid surrounding the space is the same. Because this volume of fluid is in equilibrium, the net force acting on it must be zero. The upward buoyant force thus equals the downward weight of this volume of fluid:

$$B = w_f \tag{13-11}$$

Note that this result does not depend on the shape of the submerged object. If we consider any irregularly shaped portion of fluid, there must be a buoyant force acting on it due to the surrounding fluid that is equal to the weight of that portion. Thus, we have derived Archimedes' principle.

Hot-air balloons rising in the night sky over Albuquerque during a balloon festival.

Archimedes (287–212 B.C.) had been given the task of determining whether a crown made for King Hieron II was of pure gold or had been adulterated with some cheaper metal, such as silver. The problem was to determine the density of the irregularly shaped crown without destroying it. As the story goes, Archimedes came upon the solution while sinking himself into a bath and immediately rushed naked through the streets of Syracuse shouting "Eureka!" ("I have found it!"). This flash of insight preceded Newton's laws, which we used to derive Archimedes' principle, by some 1900 years. What Archimedes found was a simple and accurate way to determine the specific gravity of the crown, which he could then compare with the specific gravity of gold.

The specific gravity of an object is the weight of the object in air divided by the weight of an equal volume of water:

$$\text{Specific gravity} = \frac{\text{weight of object in air}}{\text{weight of equal volume of water}} = \frac{w_o}{w_w}$$

But according to Archimedes' principle, the weight of an equal volume of water equals the buoyant force on the object when it is submerged. It is therefore equal to the loss in weight of the object when it is weighed while submerged in water. Thus,

$$\text{Specific gravity} = \frac{\text{weight of object in air}}{\text{weight loss when submerged in water}} = \frac{w_o}{w_{loss}} \tag{13-12}$$

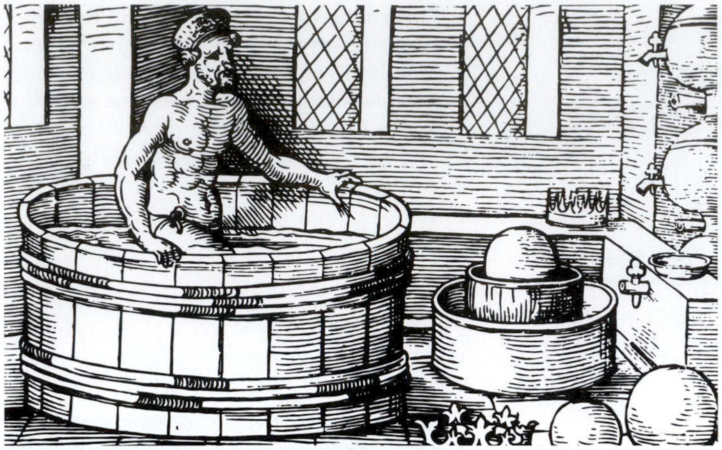

(*a*)

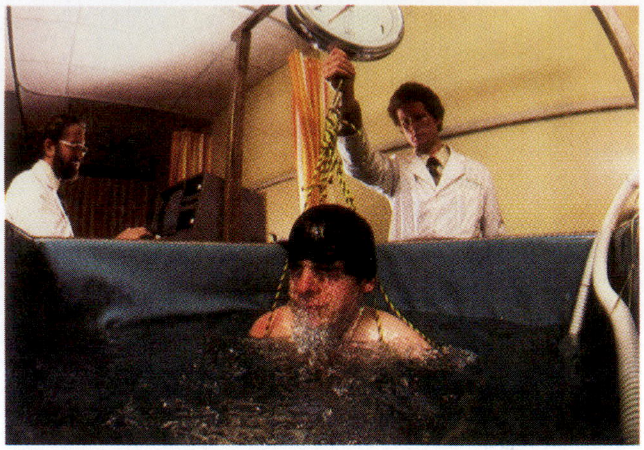

(*b*)

(*a*) Archimedes in his bath at the legendary moment when he discovered how to measure the density of an object. (*b*) To determine the percentage of fat in this man's body, his density is measured by weighing him while he is submerged under water.

The purity of King Hieron's crown could thus be determined by weighing the crown in air and then again when it was submerged in water, and comparing the results to those for a quantity of gold known to be pure.

Example 13-5

The specific gravity of gold is 19.3. If a crown made of pure gold weighs 8 N in air, what will its weight be when it is submerged in water?

1. The submerged weight equals the weight in air minus the weight loss in water: $w_{\text{sub}} = w - w_{\text{loss}}$
2. Equation 13-12 relates the weight loss to the weight and specific gravity: $w_{\text{loss}} = \dfrac{\text{weight in air}}{\text{specific gravity}} = \dfrac{8\text{ N}}{19.3} = 0.415\text{ N}$
3. The weight is thus: $w = 8\text{ N} - 0.415\text{ N} = 7.59\text{ N}$

Exercise A block of an unknown material weighs 3 N in air and 1.89 N when submerged in water. What is the specific gravity of the material? (*Answer* 2.70; the material is aluminum.)

The measured weight F_s of an object submerged in a fluid is the difference between the true weight w and the buoyant force B:

$$F_s = w - B$$

If the density of the object is ρ, its volume is V, and the density of the fluid is ρ_f, the weight is $w = \rho g V$, and the buoyant force is $B = \rho_f g V$. The measured weight is then

$$F_s = w - B = \rho g V - \rho_f g V = \rho g V\left(1 - \frac{\rho_f}{\rho}\right) \quad \text{13-13}$$

Exercise An aluminum block weighs 3 N in air. What is the true weight of the block? (*Answer* $w_o = F_s/(1 - \rho_f/\rho)$, where $F_s = 3$ N, $\rho_f = 1.293\text{ kg/m}^3$, and $\rho = \rho_{\text{alum}} = 2.7 \times 10^3\text{ kg/m}^3$. Thus, $w_o = 3.0014$ N, which is only 0.048% greater than the weight in air. Clearly, buoyancy due to air can usually be ignored.)

Exercise A piece of lead (specific gravity = 11.3) weighs 80 N in air. What does it weigh when submerged in water? (*Answer* 72.9 N)

Example 13-6

A raft of area A, thickness h, and mass 600 kg floats in still water with 7 cm submerged. When Bubba stands on the raft, 8.4 cm are submerged (Figure 13-12). What is Bubba's mass?

Figure 13-12

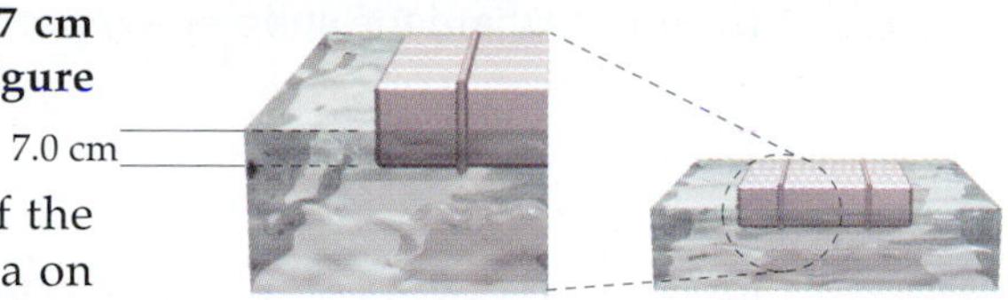

Picture the Problem Taking A to be the area of the raft, the weight of the displaced fluid is then $\rho_w A d_1 g$ with just the raft and $\rho_w A d_2 g$ with Bubba on the raft, where ρ_w is the density of the water, $d_1 = 7$ cm, and $d_2 = 8.4$ cm. If we set the weight of displaced fluid equal to the weight of the floating objects in each case, we can eliminate A and ρ_w and solve for Bubba's mass m in terms of the mass of the raft, $M = 600$ kg.

1. Set the buoyant force with $d_1 = 7$ cm submerged equal to the weight of the raft and with $d_2 = 8.4$ cm submerged equal to the weight of the raft plus Bubba:

$$\rho_w A d_1 g = Mg$$
$$\rho_w A d_2 g = (M + m)g$$

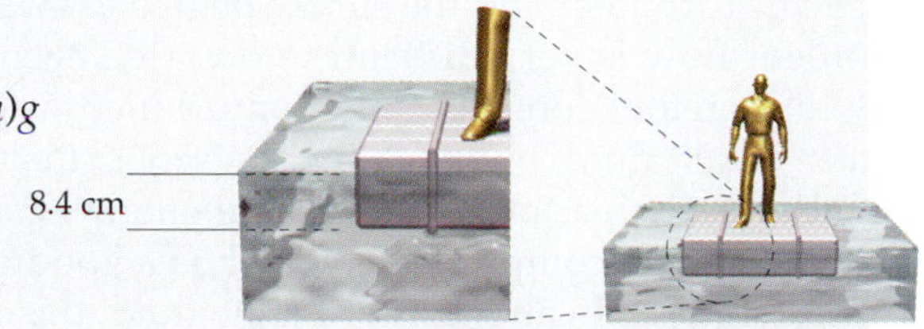

2. Divide these two equations to eliminate the unknowns, A and ρ_w:

$$\frac{d_2}{d_1} = \frac{M + m}{M}$$

3. Solve for m:

$$m = \left(\frac{d_2}{d_1} - 1\right)M = \left(\frac{8.4\text{ cm}}{7.0\text{ cm}} - 1\right)(600\text{ kg}) = 120\text{ kg}$$

Example 13-7

A cork has a density of 200 kg/m^3. Find the fraction of the volume of the cork that is submerged when the cork floats in water.

Picture the Problem Let V be the volume of the cork and V' be the volume that is submerged when it floats. The weight of the cork is $\rho_c V g$, and the buoyant force due to the water is $\rho_w V' g$.

1. Since the cork is in equilibrium, the buoyant force equals the weight:

$$\rho_c V g = \rho_w V' g$$

2. Solve for V'/V:

$$\frac{V'}{V} = \frac{\rho_c}{\rho_w} = \frac{200\text{ kg/m}^3}{1000\text{ kg/m}^3} = \frac{1}{5}$$

Remarks We see that only one-fifth of the cork is submerged. This result is independent of the shape of the cork.

If we replace ρ_w in the calculation above with ρ_f, the density of the fluid, we can determine the submerged fraction of an object floating in any fluid. From the example above, the fraction of a floating object that is submerged equals the ratio of the density of the object to the density of the fluid.

$$\frac{V'}{V} = \frac{\rho}{\rho_f} \qquad \text{13-14}$$

Since the density of ice is 920 kg/m^3 and that of seawater is 1025 kg/m^3, the fraction of an iceberg that is submerged in seawater is

$$\frac{V'}{V} = \frac{\rho}{\rho_f} = \frac{920\text{ kg/m}^3}{1025\text{ kg/m}^3} = 0.898$$

The great danger of icebergs to ships arises directly from the fact that only about 10% of an iceberg is visible above the water, and the visible portion gives little hint of where the submerged portion may extend.

13-4 Fluids in Motion

The behavior of a fluid in motion can be complex, as illustrated in the photographs at the right. Consider, for example, the rise of smoke from a burning cigarette. At first the smoke rises in a regular stream, but the simple streamlined flow quickly becomes turbulent and the smoke begins to swirl irregularly. Turbulent flow is very difficult to describe even qualitatively. We will therefore consider only nonturbulent, steady-state flow of an "ideal" fluid, one that is nonviscous; that is, one that flows with no dissipation of mechanical energy. We also assume that the fluid is incompressible, which is a good approximation for most liquids. In an incompressible fluid, the density is constant throughout the fluid.

Figure 13-13 shows a fluid flowing in a tube of varying cross-sectional area. The shading on the left indicates the volume of fluid flowing into the tube at point 1 in some time Δt. If the speed of the fluid at this point is v_1 and the cross-sectional area of the tube is A_1, the volume flowing into the tube in time Δt is

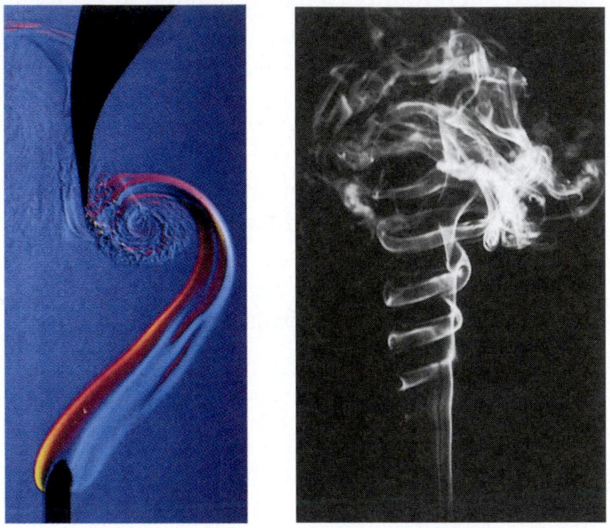

(a) (b)

(*a*) Smoke from a burning cigarette. At first the smoke rises in a regular stream, but the simple streamlined flow quickly becomes turbulent and the smoke begins to swirl irregularly. (*b*) Vortex of turbulent flow formed at the tip of a fan blade by heated air above an alcohol lamp.

$$\Delta V = A_1 v_1 \,\Delta t$$

Since we assume the fluid to be incompressible, an equal volume of fluid must flow out of the tube at point 2, as indicated by the shading on the right. If the speed of the fluid at this point is v_2 and the cross-sectional area is A_2, the volume is $\Delta V = A_2 v_2 \,\Delta t$. Since these volumes must be equal, we have

$$A_1 v_1 \,\Delta t = A_2 v_2 \,\Delta t \; A_1 v_1 = A_2 v_2 \qquad \text{13-15}$$

The quantity Av is called the **volume flow rate** I_V. The dimensions of I_V are those of volume per time. In the steady flow of an incompressible fluid, the volume flow rate is the same at any point in the fluid:

$$I_V = Av = \text{constant} \qquad \text{13-16}$$

Continuity equation

Equation 13-16 is called the **continuity equation.**

Exercise Blood flows in an aorta of radius 1.0 cm at 30 cm/s. What is the volume flow rate? (*Answer* $I_V = Av = 9.42 \times 10^{-5}\ \text{m}^3/\text{s}$. It is customary to give the pumping rate of the heart in liters per minute. Using 1 L = $10^{-3}\ \text{m}^3$, we have $I_V = 5.65$ L/min.)

Exercise Blood flows from an artery of radius 0.3 cm, where its speed is 10 cm/s, into a region where the radius has been reduced to 0.2 cm because of thickening of the arterial walls (arteriosclerosis). What is the speed of the blood in the narrower region? [*Answer* If v_1 and v_2 are the initial and final speeds and A_1 and A_2 are the initial and final areas, Equation 13-16 gives

$$v_2 = \frac{A_1}{A_2} v_1 = \frac{\pi (0.3\ \text{cm})^2}{\pi (0.2\ \text{cm})^2} (10\ \text{cm/s}) = 22.5\ \text{cm/s}$$]

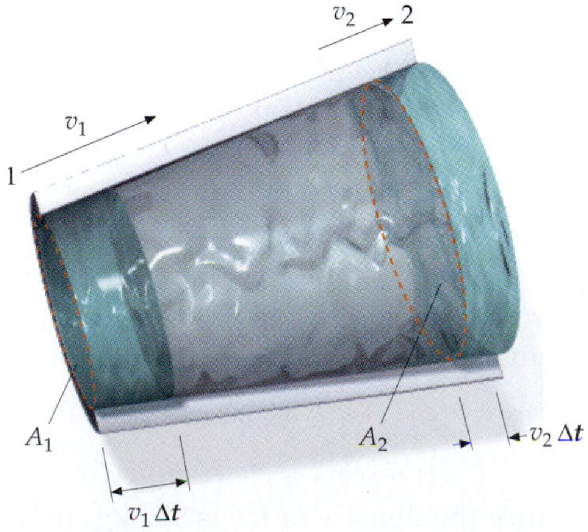

Figure 13-13 An incompressible fluid flowing in a tube of varying cross section. The shaded volumes are equal.

Bernoulli's Equation

Bernoulli's equation relates the pressure, elevation, and speed of an incompressible fluid in steady flow. It follows from Newton's laws and is most easily derived by applying the work–energy theorem to a segment of the fluid.

Consider a fluid flowing in a tube that varies in elevation as well as in cross-sectional area, as shown in Figure 13-14. We apply the work–energy theorem to the fluid that is contained initially between points 1 and 2 in Figure 13-14*a*. After some time Δt this fluid moves along the tube to the region between points 1′ and 2′ in Figure 13-14*b*. Let $\Delta m = \rho\,\Delta V$ be the mass of this fluid. The net effect on the fluid in time Δt is that its mass Δm is lifted from height y_1 to height y_2 and its speed is changed from v_1 to v_2. The change in the potential energy of the fluid is thus

$$\Delta U = \Delta m\, gy_2 - \Delta m\, gy_1 = \rho\,\Delta V\, g(y_2 - y_1)$$

and the change in its kinetic energy is

$$\Delta K = \frac{1}{2}(\Delta m)v_2^2 - \frac{1}{2}(\Delta m)v_1^2 = \frac{1}{2}\rho\,\Delta V(v_2^2 - v_1^2)$$

The fluid behind the mass of fluid in the pipe (to the left of the shaded portion in Figure 13-14*a*) exerts a force on the fluid ahead of magnitude $F_1 = P_1A_1$, where P_1 is the pressure at point 1. This force does work:

$$W_1 = F_1\,\Delta x_1 = P_1A_1\,\Delta x_1 = P_1\,\Delta V$$

At the same time, the fluid in front of it (to the right) exerts a force $F_2 = P_2A_2$ on it to the left in the figure. This force does negative work because it opposes the motion:

$$W_2 = -F_2\Delta x_2 = -P_2A_2\Delta x_2 = -P_2\Delta V$$

The total work done by these forces is

$$W_{\text{total}} = P_1\Delta V - P_2\Delta V = (P_1 - P_2)\Delta V$$

The work–energy theorem gives

$$W_{\text{total}} = \Delta U + \Delta K$$

so

$$(P_1 - P_2)\Delta V = \rho\,\Delta V g(y_2 - y_1) + \frac{1}{2}\rho\,\Delta V(v_2^2 - v_1^2)$$

If we divide by ΔV, we obtain

$$P_1 - P_2 = \rho g y_2 - \rho g y_1 + \frac{1}{2}\rho v_2^2 - \frac{1}{2}\rho v_1^2$$

When we collect all the quantities having a subscript 1 on one side and those having a subscript 2 on the other, this equation becomes

$$P_1 + \rho g y_1 + \frac{1}{2}\rho v_1^2 = P_2 + \rho g y_2 + \frac{1}{2}\rho v_2^2 \qquad \text{13-17}a$$

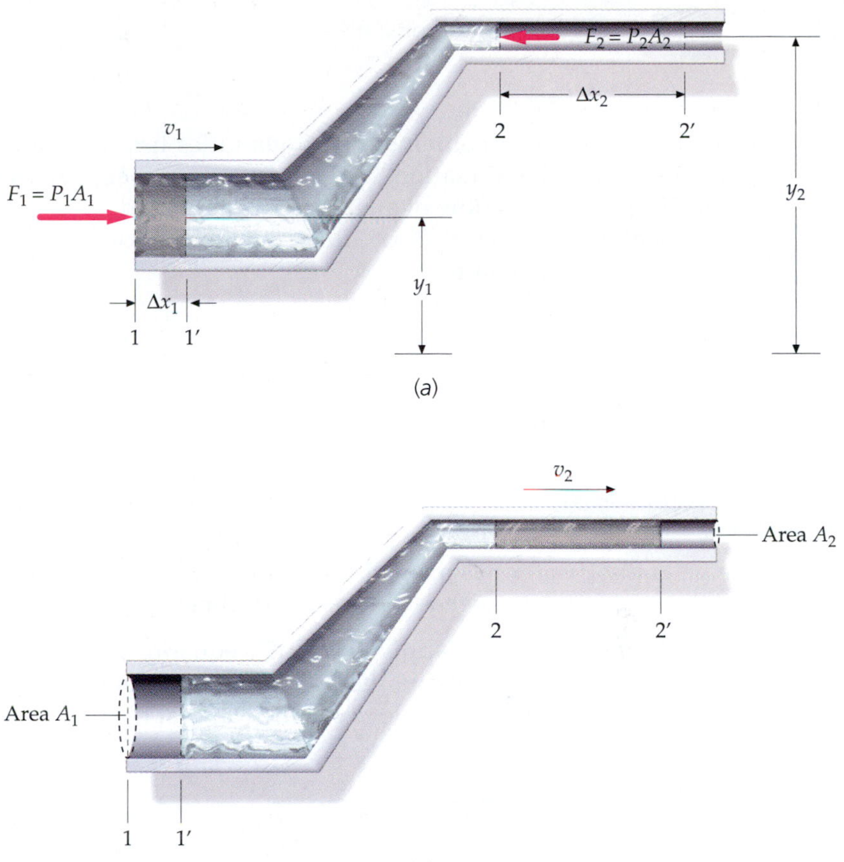

Figure 13-14 Fluid moving in a pipe that varies in both height and cross-sectional area. The total work done by the forces $F_1 = P_1A_1$ and $F_2 = P_2A_2$ has the effect of raising the shaded portion of the fluid from height y_1 to y_2 and changing its speed from v_1 to v_2.

This result can be restated as

$$P + \rho g y + \frac{1}{2}\rho v^2 = \text{constant} \qquad \text{13-17}b$$

Bernoulli's equation

which means that this combination of quantities has the same value at any point along the tube. Equation 13-17*b* is known as **Bernoulli's equation** for the steady, nonviscous flow of an incompressible fluid.

A special application of Bernoulli's equation is for a fluid at rest. Then $v_1 = v_2 = 0$, and we obtain

$$P_1 - P_2 = \rho g(y_2 - y_1) = \rho g h$$

where $h = y_2 - y_1$ is the difference in height between points 2 and 1. This is the same as Equation 13-7.

Example 13-8

A large tank of water has a small hole a distance *h* below the water surface. Find the speed of the water as it flows out the hole.

Picture the Problem We apply Bernoulli's equation to points *a* and *b* in Figure 13-15. Since the diameter of the hole is much smaller than the diameter of the tank, we can neglect the velocity of the water at the top (point *a*).

Figure 13-15

1. Bernoulli's equation with $v_a = 0$ gives: $P_a + \rho g y_a + 0 = P_b + \rho g y_b + \frac{1}{2}\rho v_b^2$

2. The pressure at point *a* and point *b* is the same, P_{at}, since both points are open to the air: $P_a = P_b = P_{at}$

   or

   $P_{at} + \rho g y_a + 0 = P_{at} + \rho g y_b + \frac{1}{2}\rho v_b^2$

3. Solve for the speed v_b of the water flowing from the hole: $v_b^2 = 2g(y_a - y_b) = 2gh$ or $v_b = \sqrt{2gh}$

Exercise If the water flowing out of the hole is directed vertically upward, how high does it rise? (*Answer* The water shoots upward a distance *h*; that is, to the same level as the water in the tank.)

In Example 13-8, the water emerges from the hole with a speed equal to the speed it would have if it dropped in free fall a distance h. This finding is known as **Torricelli's law.**

In Figure 13-16, water is flowing through a horizontal pipe that has a constricted section. Since both sections of the pipe are at the same elevation, $y_1 = y_2$ in Equation 13-17*a*. Then Bernoulli's equation becomes

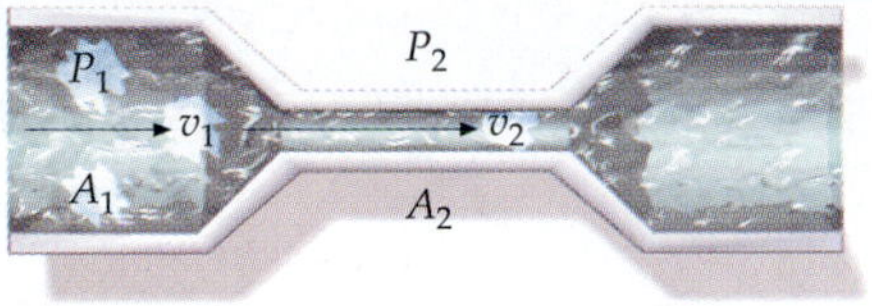

Figure 13-16 Constriction in a pipe carrying a moving fluid. The pressure is lower in the narrow section of the pipe where the fluid is moving faster.

$$P + \frac{1}{2}\rho v^2 = \text{constant} \qquad \text{13-18}$$

Bernoulli's equation for constant elevation

When the fluid moves into the constriction, the area A gets smaller, so the speed v must get larger since Av remains constant. But since $P + \frac{1}{2}\rho v^2$ is constant, if the speed gets larger the pressure must get smaller. Thus, the pressure in the constriction is reduced.

When the speed of a fluid increases, the pressure drops.

Venturi effect

This result is often referred to as the **Venturi effect.** Equation 13-18 is an important result that applies to many situations in which we can ignore changes in height.

Example 13-9 *try it yourself*

A *Venturi meter* is a device used to measure the flow rate of a fluid. The fluid of density ρ_F passes through a pipe of cross-sectional area A_1 that has a constriction of cross-sectional area A_2 (Figure 13-17). The two parts of the pipe are connected with a U-tube manometer partially filled with a liquid of density ρ_L. Since the velocity of flow is greater in the constricted region, the pressure in that section is less than in the other portion of the pipe. The pressure difference is measured by the difference h in the level of the liquid in the U-tube. Express the velocity v_1 in terms of the measured height h and the known quantities ρ_F, ρ_L, and $r = A_1/A_2$.

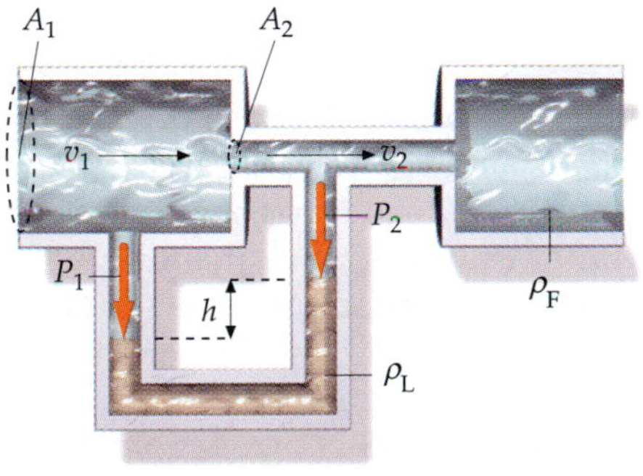

Figure 13-17 A Venturi meter.

Picture the Problem The pressures P_1 and P_2 in the wider and narrower regions of the pipe are related to the speeds v_1 and v_2 by Bernoulli's equation. The pressure difference is related to the height h by $P_1 - P_2 = \rho_L gh$. You can express v_2 in terms of v_1 and the areas A_1 and A_2 by the continuity equation.

Cover the column to the right and try these on your own before looking at the answers.

Steps	*Answers*
1. Write Bernoulli's equation for constant elevation for the wider and narrower regions of the pipe.	$P_1 + \frac{1}{2}\rho_F v_1^2 = P_2 + \frac{1}{2}\rho_F v_2^2$
2. Write the continuity equation for the two regions, and solve for v_2 in terms of v_1 and the areas A_1 and A_2.	$v_2 = \frac{A_1}{A_2} v_1 = r v_1$
3. Substitute your result for v_2 into the equation in step 1 and obtain an equation for $P_1 - P_2$.	$P_1 - P_2 = \frac{1}{2}\rho_F(r^2 - 1)v_1^2$
4. Write $P_1 - P_2$ in terms of the difference in height h of the water in the arms of the U-tube.	$P_1 - P_2 = \rho_L gh$

5. Equate your two expressions for $P_1 - P_2$, and solve for v_1 in terms of h.

$$v_1 = \sqrt{\frac{2\rho_L g h}{\rho_F(r^2 - 1)}}$$

Exercise Find v_1 if $h = 3$ cm, $r = 4$, the fluid is air ($\rho_F = 1.29\ \text{kg/m}^3$), and the liquid in the U-tube portion of the Venturi meter is water ($\rho_w = 10^3\ \text{kg/m}^3$). (*Answer* $v_1 = 5.52$ m/s)

Remark This calculation is not very accurate for air because Bernoulli's equation and the continuity equation hold only for incompressible fluids.

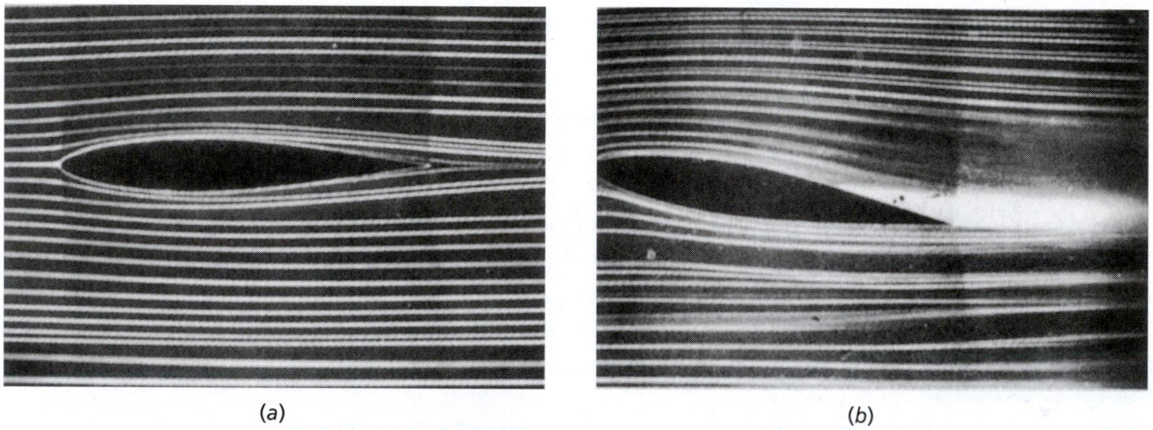

Air flow around a wing foil. (*a*) When the wing is horizontal, the flow is uniform and the pressure above the wing is the same as below it. (*b*) When the wing is tilted, the pressure is greater at the bottom of the wing than at the top of the wing, creating lift.

The Venturi effect can be used to give a qualitative understanding of the lift of an airplane wing and the path of a pitcher's curveball. An airplane wing is designed so that air moves faster over the top of the wing than it does under the wing, thus making the air pressure less on top than underneath. This difference in pressure results in a net force upward on the wing. Figure 13-18*a* shows a top view of the motion of a curveball. As the ball spins, it tends to drag air around with it. Figure 13-18*b* is drawn from the point of view of the ball being stationary (but spinning) with the air rushing past it.

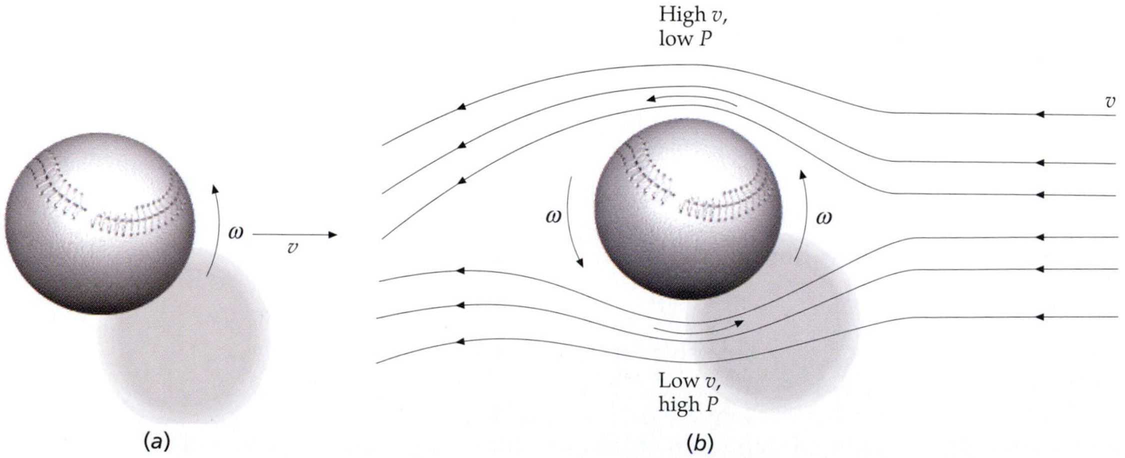

Figure 13-18 (*a*) Top view of a baseball thrown with a counterclockwise spin ω, like a curveball thrown by a right-handed pitcher. (*b*) In the frame of the ball, the ball is stationary (but spinning) and the air rushes past it. Because of its rough cover, the spinning ball drags the air around with it, making the air speed higher on the left side and lower on the right. The pressure is therefore lower on the left side, so the ball curves to the left.

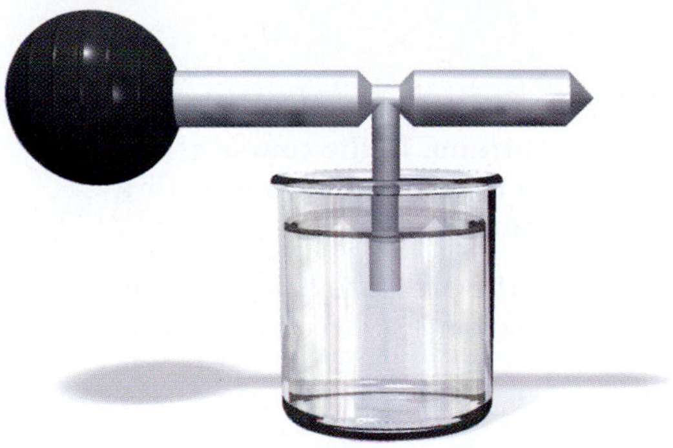

Figure 13-19 When the bulb of an atomizer is squeezed, air is forced through the constriction in the horizontal tube, which reduces the pressure there below atmospheric pressure. Because of the resulting pressure difference, the liquid in the jar is pumped up through the vertical tube, enters the airstream, and emerges from the nozzle. A similar effect occurs in the carburetor of a gasoline engine.

The air movement caused by the spinning ball adds to the velocity of the air rushing by on the left side of the ball and subtracts from it on the right. Thus, the air speed is greater on the left side of the ball than on the right, causing the pressure on the left to be less than that on the right. The ball therefore curves to the left. The atomizer in Figure 13-19 also works on the principle of the Venturi effect.

Although Bernoulli's equation is very useful for qualitative descriptions of many features of fluid flow, such descriptions are often grossly inaccurate when compared with the quantitative results of experiments. Prominent reasons for the discrepancies are that gases like air are hardly incompressible, and liquids like water have viscosity, which invalidates the assumption of the conservation of mechanical energy. In addition, it is often difficult to maintain steady-state, streamlined flow without turbulence, and the introduction of turbulence can greatly affect the results.

Viscous Flow

According to Bernoulli's equation, when a fluid flows steadily through a long, narrow, horizontal pipe of constant cross section, the pressure along the pipe will be constant. In practice, however, we observe a pressure drop as we move along the direction of the flow. Put another way, a pressure difference is required to push a fluid through a horizontal pipe. This pressure difference is needed because of the drag force that is exerted by the pipe on the layer of fluid in contact with the pipe, and because of the drag force exerted by each layer of the fluid on an adjacent layer that is moving with a different velocity. These drag forces are called **viscous forces.** As a result of these viscous forces, the velocity of the fluid is not constant across the diameter of the pipe. Instead, it is greatest near the center of the pipe and least near the edge; at the point of contact between the fluid and the walls of the pipe, the fluid is actually stationary (Figure 13-20). Let P_1 be the pressure at point 1 and P_2 be that at point 2, a distance L downstream from point 1. The pressure drop $\Delta P = P_1 - P_2$ is proportional to the volume flow rate:

$$\Delta P = P_1 - P_2 = I_V R \qquad \text{13-19}$$

where $I_V = vA$ is the volume flow rate, and the proportionality constant R is the resistance to flow, which depends on the length L of the pipe, the radius r, and the viscosity of the fluid.

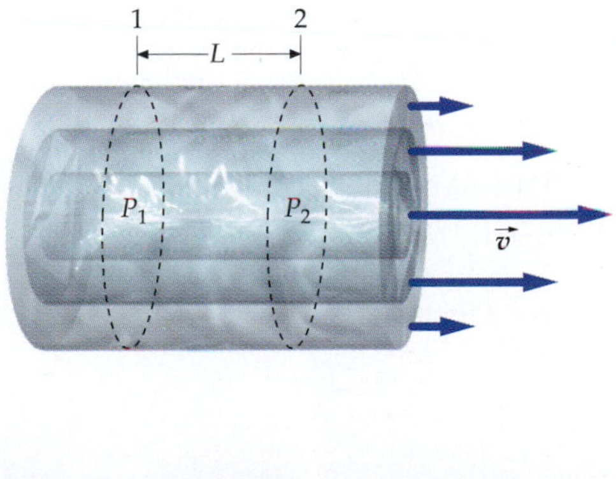

Figure 13-20 When a viscous fluid flows through a pipe, the speed is greatest at the center of the pipe. At the walls of the pipe, the fluid tends to remain stationary.

optional

Example 13-10

Blood flows from the aorta through the major arteries, the small arteries, the capillaries, and the veins until it reaches the right atrium. In the course of that flow, the (gauge) pressure drops from about 100 torr to zero. If the volume flow rate is 0.8 L/s, find the total resistance of the circulatory system.

Picture the Problem The resistance is related to the pressure drop and volume flow rate by Equation 13-19. We can use Equation 13-9 to convert from torr to N/m^2.

Write the resistance in terms of the pressure drop and volume flow rate, and convert all terms to SI units:

$$R = \frac{\Delta P}{I_V} = \frac{100\ \text{torr}}{0.8\ \text{L/s}} \times \frac{133.3\ \text{Pa}}{1\ \text{torr}} \times \frac{1\ \text{L}}{10^3\ \text{cm}^3} \times \frac{1\ \text{cm}^3}{10^{-6}\ \text{m}^3}$$

$$= 1.66 \times 10^7\ \text{Pa}\cdot\text{s/m}^3$$

Remarks We could have used $1\ \text{Pa} = 1\ \text{N/m}^2$ to write the result as $1.66 \times 10^7\ \text{N}\cdot\text{s/m}^5$.

To define the coefficient of viscosity of a fluid, we consider a fluid that is confined between two parallel plates, each of area A, separated by a distance z as shown in Figure 13-21. The upper plate is pulled at a constant speed v by a force $\vec{F}$ while the bottom plate is held at rest. A force is needed to pull the upper plate because the fluid next to the plate exerts a viscous drag force opposing its motion. The speed of the fluid between the plates is essentially v near the upper plate and zero near the lower plate, and it varies linearly with separation between the plates. The force $\vec{F}$ is found to be directly proportional to v and A and inversely proportional to the plate separation z. The proportionality constant is the **coefficient of viscosity** η:

$$F = \eta \frac{vA}{z} \qquad \text{13-20}$$

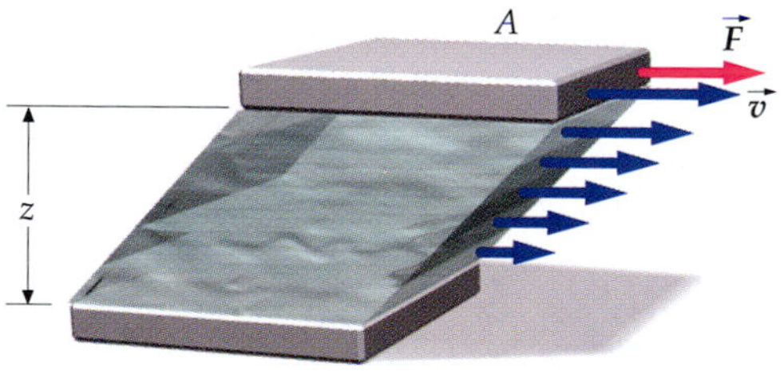

Figure 13-21 Two plates of equal area with a viscous fluid between them. When the upper plate is moved relative to the lower one, each layer of fluid exerts a drag force on adjacent layers. The force needed to pull the upper plate is directly proportional to v and the area A and inversely proportional to z, the separation between the plates.

The SI unit of viscosity is the $\text{N}\cdot\text{s/m}^2 = \text{Pa}\cdot\text{s}$. An older cgs unit still in common use is the **poise**, named after the French physicist Poiseuille. These units are related by

$$1\ \text{Pa}\cdot\text{s} = 10\ \text{poise} \qquad \text{13-21}$$

Table 13-1 lists the coefficients of viscosity for several fluids at various temperatures. Generally, the viscosity of a liquid increases as the temperature decreases. Thus, in cold climates, a thinner grade of oil is used to lubricate automobile engines in the winter than in summer.

Table 13-1

Coefficients of Viscosity for Various Fluids

Fluid	t, °C	η, mPa·s
Water	0	1.8
	20	1.00
	60	0.65
Blood (whole)	37	4.0
Engine oil (SAE 10)	30	200
Glycerin	0	10,000
	20	1,410
	60	81
Air	20	0.018

Poiseuille's Law The resistance to flow, R, in Equation 13-19 for steady flow through a circular tube of radius r can be shown to be

$$R = \frac{8\eta L}{\pi r^4} \qquad \text{13-22}$$

Equations 13-19 and 13-22 can be combined to give the pressure drop over a length L of a circular tube of radius r:

$$\Delta P = \frac{8\eta L}{\pi r^4} I_V \qquad \text{13-23}$$

Poiseuille's law

Equation 13-23 is known as **Poiseuille's law.** Note the inverse r^4 dependence of the pressure drop. If the radius of the tube is halved, the pressure drop for a given volume flow rate is increased by a factor of 16; or a pressure 16 times as great is needed to pump the fluid through the tube at the original volume flow rate. Thus, if the diameter of the blood vessels or arteries is reduced for some reason, either the volume flow rate of the blood is greatly reduced, or the heart must work much harder to maintain the volume flow rate. For water flowing through a long garden hose, the pressure drop is fixed. It equals the difference in pressure between that at the water source and atmospheric pressure at the open end. The volume flow rate is then proportional to the fourth power of the radius. Thus, if the radius is halved, the volume flow rate drops by a factor of 16.

Poiseuille's law applies only to the laminar (nonturbulent) flow of a fluid of constant viscosity. In some fluids, viscosity changes with velocity, violating Poiseuille's law. Blood, for example, is a complex fluid consisting of solid particles of various shapes suspended in a liquid. Red blood cells are disk-shaped objects that are randomly oriented at low velocities but tend to become oriented at high velocities to facilitate the flow. Thus, the viscosity of blood decreases as the flow velocity increases, so Poiseuille's law cannot be strictly applied. Nevertheless, Poiseuille's law is a good approximation that is very useful for obtaining a qualitative understanding of blood flow.

Turbulence: Reynolds Number When the flow velocity of a fluid becomes sufficiently great, laminar flow breaks down and turbulence sets in. The critical velocity above which the flow through a tube is turbulent depends on the density and viscosity of the fluid and on the radius of the tube. The flow of a fluid can be characterized by a dimensionless number called the **Reynolds number**, N_R, which is defined by

$$N_R = \frac{2r\rho v}{\eta} \qquad \text{13-24}$$

where v is the average velocity of the fluid. Experiments have shown that the flow will be laminar if the Reynolds number is less than about 2000 and turbulent if it is greater than 3000. Between these values, the flow is unstable and may change from one type to the other.

optional

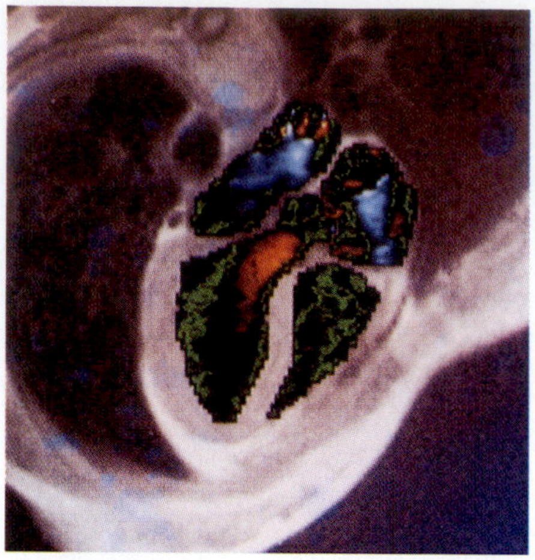

False-color view of turbulence of blood flowing into and out of the heart as seen by magnetic resonance imaging (MRI). Systolic ejection from the left ventricle into the aorta is seen in red, and diastolic filling of the ventricles is in blue.

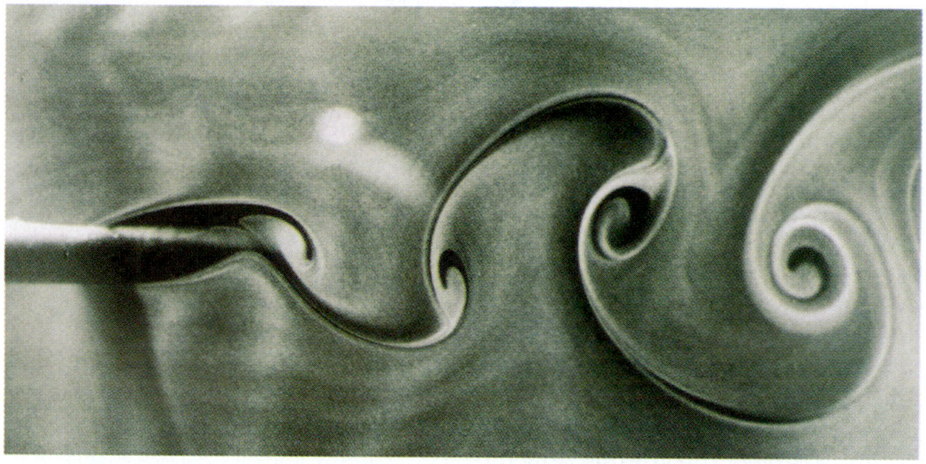

This turbulent motion of air around a vibrating cylinder in a wind tunnel is called *vortex shedding*. The cylinder was oscillating normal to the oncoming flow of air and the vortices were shed periodically in unison with the oscillations. The Reynolds number was 200. A sheet of minute aerosol particles was introduced into the wind tunnel upstream of the cylinder to make the flow pattern visible.

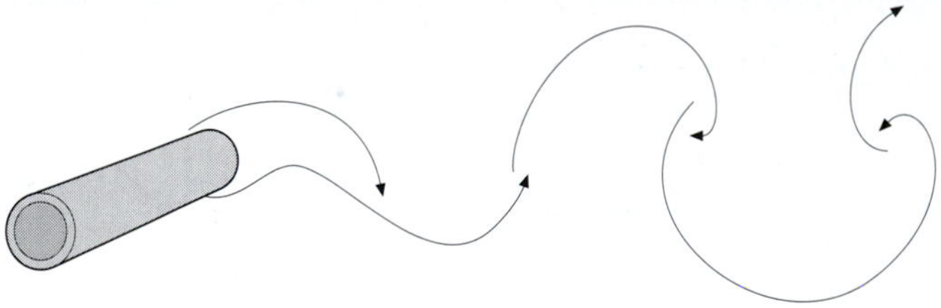

Example 13-11

Calculate the Reynolds number for blood flowing at 30 cm/s through an aorta of radius 1.0 cm. Assume that blood has a viscosity of 4 mPa·s and a density of 1060 kg/m³.

Picture the Problem Since N_R is dimensionless, we can use any set of units as long as we are consistent.

Write Equation 13-24 for the Reynolds number expressing each quantity in SI units:

$$N_R = \frac{2r\rho v}{\eta} = \frac{2(0.01\text{ m})(1060\text{ kg/m}^3)(0.3\text{ m/s})}{4 \times 10^{-3}\text{ Pa}\cdot\text{s}} = 1590$$

Remarks Since the Reynolds number is less than 2000, this flow will be laminar rather than turbulent.

Summary

1. Density, specific gravity, and pressure are defined quantities that are important in fluid statics and dynamics.
2. Pascal's principle and Archimedes' principle are derived from Newton's laws.
3. Bernoulli's equation is derived from the conservation of mechanical energy.
4. The Venturi effect is a special case of Bernoulli's equation.

Topic	Remarks and Relevant Equations	
1. Density	The density of a substance is the ratio of its mass to its volume:	
	$\rho = \frac{m}{V}$	**13-1**
	The densities of most solids and liquids are approximately independent of temperature and pressure, whereas those of gases depend strongly on these quantities.	
2. Specific Gravity	The specific gravity of a substance is the ratio of its density to that of water. An object sinks or floats in a given fluid depending on whether its density is greater than or less than that of the fluid.	
3. Pressure	$P = \frac{F}{A}$	**13-3**
Units	$1\text{ Pa} = 1\text{ N/m}^2$	**13-4**
	$1\text{ atm} = 760\text{ mmHg} = 760\text{ torr} = 29.9\text{ inHg}$ $= 101.325\text{ kPa} = 14.7\text{ lb/in}^2$	**13-9**
	$1\text{ bar} = 10^3\text{ millibars} = 100\text{ kPa}$	**13-10**
Gauge pressure	Gauge pressure is the difference between the absolute pressure and atmospheric pressure:	
	$P = P_{\text{gauge}} + P_{\text{at}}$	**13-8**
In a liquid	$P = P_0 + \rho g h$ (ρ constant)	**13-7**
In a gas	In a gas such as air, pressure decreases exponentially with altitude.	
Bulk modulus	$B = -\frac{\Delta P}{\Delta V/V}$	**13-6**
4. Pascal's Principle	Pressure applied to an enclosed liquid is transmitted undiminished to every point in the fluid and to the walls of the container.	
5. Archimedes' Principle	A body wholly or partially submerged in a fluid is buoyed up by a force equal to the weight of the displaced fluid.	
6. Fluid Flow		
Volume flow rate	$I_V = Av$	
Continuity equation	$I_V = Av = \text{constant}$	**13-16**

Bernoulli's equation	$P + \rho g y + \frac{1}{2}\rho v^2 = \text{constant}$	**13-17*b***
Bernoulli's equation for constant elevation	$P + \frac{1}{2}\rho v^2 = \text{constant}$	**13-18**
Venturi effect	When the speed of a fluid increases, the pressure drops.	
Resistance to fluid flow (optional)	$\Delta P_2 = I_V R$	**13-19**
Coefficient of viscosity (optional)	$F = \eta \frac{vA}{z}$	**13-20**
Poiseuille's law for viscous flow (optional)	$\Delta P = \frac{8\eta L}{\pi r^4} I_V$	**13-23**
Laminar flow, turbulent flow, and the Reynolds number (optional)	The flow will be laminar if the Reynolds number N_R is less than about 2000 and turbulent if it is greater than 3000, where N_R is given by $N_R = \frac{2r\rho v}{\eta}$	**13-24**

Problem-Solving Guide

1. Begin by drawing a neat diagram that includes the important features of the problem.
2. Fluid statics problems are usually solved using Archimedes' principle.
3. To solve fluid dynamics problems, use Bernoulli's equation plus the continuity equation.

Summary of Worked Examples

Type of Calculation	Procedure and Relevant Examples	
1. Density		
Find the density of a substance.	Use $\rho = m/V$.	**Example 13-1**
2. Pressure in a Fluid		
Find the force exerted by water on a vertical wall.	Find the force dF on an area element dA of height dh and integrate.	**Example 13-2**
Find the force in a hydraulic lift.	Use $F_1 A_1 = F_2 A_2$.	**Example 13-3**
Change the units of pressure.	Use the conversion factors implied by Equations 13-9 and 13-10.	**Example 13-4**
3. Archimedes' Principle		
Find the weight of a submerged object.	Use the definition of specific gravity (weight divided by the weight of an equal volume of water) plus Archimedes' principle.	**Example 13-5**

Find the mass of a floating object given the additional amount submerged when a weight is added.	Use Archimedes' principle.	**Example 13-6**
Find the fraction of the volume of an object that is submerged.	Set the buoyant force equal to the weight $\rho_c V g = \rho_w V' g$ and solve for V'/V.	**Example 13-7**

4. Fluid Flow

Find the speed of a fluid at some point.	Use Bernoulli's equation and the continuity equation.	**Examples 13-8, 13-9**
Find the resistance in viscous flow (optional).	Use the definition of resistance $R = \Delta P / I_V$.	**Example 13-10**
Calculate the Reynolds number (optional).	Use $N_R = 2r\rho v/\eta$ with all quantities in SI units.	**Example 13-11**

Problems

Conceptual Problems

Problems from Optional and Exploring sections

• Single-concept, single-step, relatively easy

•• Intermediate-level, may require synthesis of concepts

••• Challenging, for advanced students

In a few problems, you are given more data than you actually need; in a few other problems, you are required to supply data from your general knowledge, outside sources, or informed estimates.

Density

1 • A copper cylinder is 6 cm long and has a radius of 2 cm. Find its mass.

2 • Find the mass of a lead sphere of radius 2 cm.

3 • Find the mass of air in a room 4 m × 5 m × 4 m.

4 • A solid oak door is 200 cm high, 75 cm wide, and 4 cm thick. How much does it weigh?

5 •• A 60-mL flask is filled with mercury at 0°C (Figure 13-22). When the temperature rises to 80°C, 1.47 g of mercury spills out of the flask. Assuming that the volume of the flask is constant, find the density of mercury at 80°C if its density at 0°C is 13,645 kg/m^3.

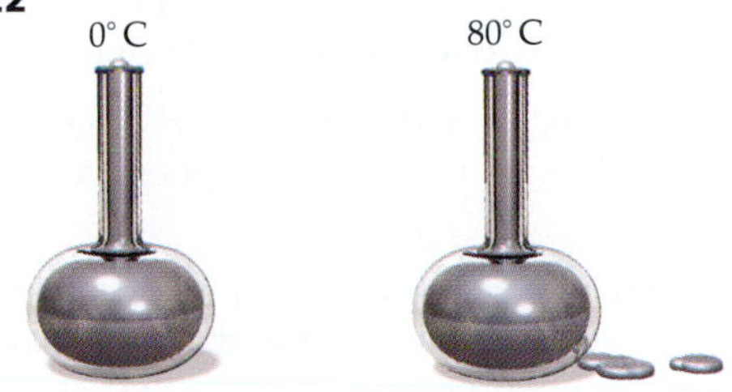

Figure 13-22 Problem 5

Pressure

6 • If the gauge pressure is doubled, the absolute pressure will be

(*a*) halved.
(*b*) doubled.
(*c*) unchanged.
(*d*) squared.
(*e*) Not enough information is given to determine the effect.

7 • Barometer readings are commonly given in inches of mercury. Find the pressure in inches of mercury equal to 101 kPa.

8 • The pressure on the surface of a lake is atmospheric pressure P_{at} = 101 kPa. (*a*) At what depth is the pressure twice atmospheric pressure? (*b*) If the pressure at the top of a deep pool of mercury is P_{at}, at what depth is the pressure $2P_{at}$?

9 • Find (*a*) the absolute pressure and (*b*) the gauge pressure at the bottom of a swimming pool of depth 5.0 m.

10 • When a woman in high heels takes a step, she momentarily places her entire weight on one heel of her shoe, which has a radius of 0.4 cm. If her mass is 56 kg, what is the pressure exerted on the floor by her heel?

11 • A hydraulic lift is used to raise an automobile of mass 1500 kg. The radius of the shaft of the lift is 8 cm and that of the piston is 1 cm. How much force must be applied to the piston to raise the automobile?

12 • Blood flows into the aorta through a circular opening of radius 0.9 cm. If the blood pressure is 120 torr, how much force must be exerted by the heart?

13 • What pressure is required to reduce the volume of 1 kg of water from 1.00 L to 0.99 L?

14 • A 1500-kg car rests on four tires, each of which is inflated to a gauge pressure of 200 kPa. What is the area of contact of each tire with the road, if the four tires support the weight equally?

15 • In the seventeenth century, Pascal performed the experiment shown in Figure 13-23. A wine barrel filled with water was coupled to a long tube. Water was added to the

Figure 13-23 Problem 15

tube until the barrel burst. (*a*) If the radius of the lid was 20 cm and the height of the water in the tube was 12 m, calculate the force exerted on the lid. (*b*) If the tube had an inner radius of 3 mm, what mass of water in the tube caused the pressure that burst the barrel?

16 •• Blood plasma flows from a bag through a tube into a patient's vein, where the blood pressure is 12 mmHg. The specific gravity of blood plasma at 37°C is 1.03. What is the minimum elevation the bag must have so that the pressure of the plasma as it flows into the vein is at least 12 mmHg?

17 •• Many people have imagined that if they were to float the top of a flexible snorkel tube out of the water, they would be able to breathe through it while walking underwater (Figure 13-24). However, they generally do not reckon with just how much water pressure opposes the expansion of the chest and the inflation of the lungs. Suppose you can just breathe while lying on the floor with a 400-N weight on your chest. How far below the surface of the water could your chest be for you still to be able to breathe, assuming your chest has a frontal area of 0.09 m^2?

Figure 13-24 Problem 17

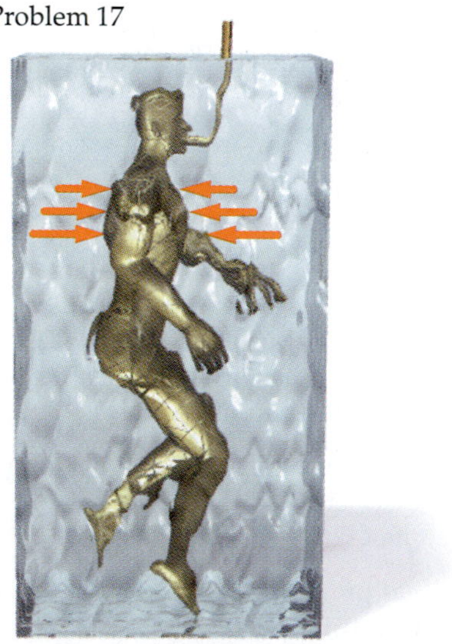

18 •• When the ground becomes saturated with water during floods, pressure develops similar to the pressure of water contained in a vessel of the same volume. This pressure forces water through the joints in concrete-block cellar walls. If this happens quickly enough to fill up the cellar with water, there may be no further damage. Otherwise, the upward pressure on the cellar floor may float the house like a ship. What upward force would be exerted on a 10-m × 10-m basement floor if the floor were 2 m below the surface of the water?

19 •• In Example 13-3, a force of 147 N is applied to a small piston to lift a car that weighs 14,700 N. Demonstrate that this does not violate the law of conservation of mechanical energy by showing that, when the car is lifted some distance h, the work done by the force acting on the small piston equals the work done by the large piston on the car.

20 •• A hollow cube with edge a is half-filled with water of density ρ. Find the force exerted on a side of the cube by the water.

21 ••• The volume of a cone of height h and base radius r is $V = \frac{1}{3}\pi r^2 h$. A conical vessel of height 25 cm resting on its base of radius 15 cm is filled with water. (*a*) Find the volume and weight of the water in the vessel. (*b*) Find the force exerted by the water on the base of the vessel. Explain how this force can be greater than the weight of the water.

Buoyancy

22 • Does Archimedes' principle hold in a satellite orbiting the earth in a circular orbit? Explain.

23 •• A rock of mass M with a density twice that of water is sitting on the bottom of an aquarium tank filled with water. The normal force exerted on the rock by the bottom of the tank is

(*a*) $2Mg$.
(*b*) Mg.
(*c*) $Mg/2$.
(*d*) zero.
(*e*) impossible to determine from the given information.

24 •• A rock is thrown into a swimming pool filled with water of uniform temperature. Which of the following statements is true?

(*a*) The buoyant force on the rock is zero as it sinks.
(*b*) The buoyant force on the rock increases as it sinks.
(*c*) The buoyant force on the rock decreases as it sinks.
(*d*) The buoyant force on the rock is constant as it sinks.
(*e*) The buoyant force on the rock as it sinks is nonzero at first but becomes zero once the terminal velocity is reached.

25 •• A fishbowl rests on a scale. The fish suddenly swims upward to get food. What happens to the scale reading?

26 •• Two objects are balanced as in Figure 13-25. The objects have identical volumes but different masses. Will the equilibrium be disturbed if the entire system is completely immersed in water? Explain.

Figure 13-25 Problem 26

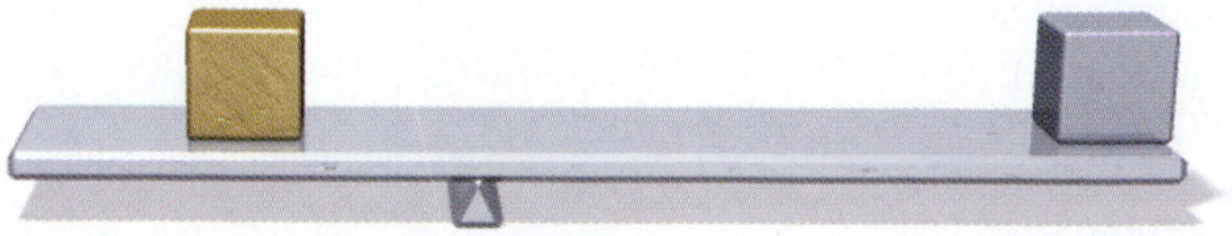

27 •• A 200-g block of lead and a 200-g block of copper rest at the bottom of an aquarium filled with water. Which of the following is true?

(*a*) The buoyant force is greater on the lead than on the copper.
(*b*) The buoyant force is greater on the copper than on the lead.
(*c*) The buoyant force is the same on both blocks.
(*d*) More information is needed to choose from the above.

28 •• A 20-cm^3 block of lead and a 20-cm^3 block of copper rest at the bottom of an aquarium filled with water. Which of the following is true?

(*a*) The buoyant force is greater on the lead than on the copper.
(*b*) The buoyant force is greater on the copper than on the lead.
(*c*) The buoyant force is the same on both blocks.
(*d*) More information is needed to choose from the above.

29 • A 500-g piece of copper (specific gravity 9.0) is suspended from a spring scale and is submerged in water (Figure 13-26). What force does the spring scale read?

Figure 13-26 Problem 29

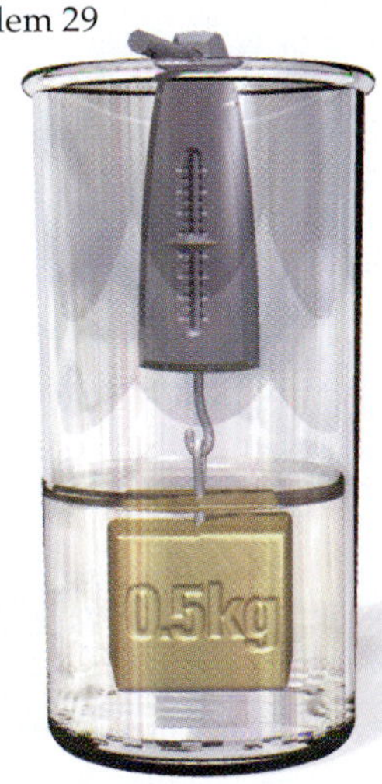

30 • When a 60-N stone is attached to a spring scale and is submerged in water, the scale reads 40 N. What is the specific gravity of the stone?

31 • A block of an unknown material weighs 5 N in air and 4.55 N when submerged in water. (*a*) What is the density of the material? (*b*) Of what material is the block made?

32 • A solid piece of metal weighs 90 N in air and 56.6 N when submerged in water. Determine the specific gravity of this metal.

33 •• An object floats on water with 80% of its volume below the surface. The same object when placed in another liquid floats on that liquid with 72% of its volume below the surface. Determine the density of the object and the specific gravity of the liquid.

34 •• A 5-kg iron block is attached to a spring scale and is submerged in a fluid of unknown density. The spring scale reads 6.16 N. What is the density of the fluid?

35 •• A large piece of cork weighs 0.285 N in air. When held submerged underwater by a spring scale as shown in Figure 13-27, the spring scale reads 0.855 N. Find the density of the cork.

Figure 13-27 Problem 35

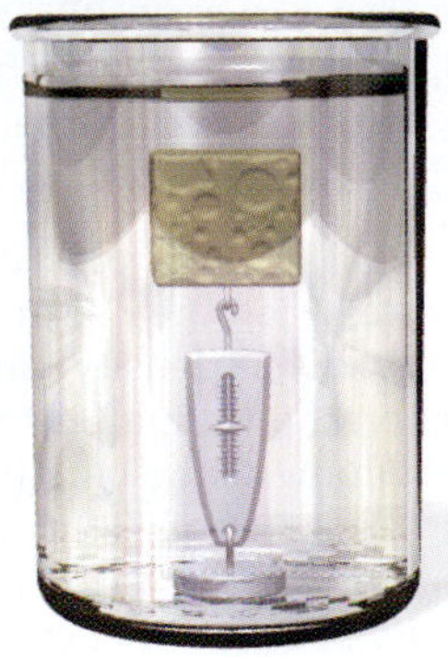

36 •• As you step onto the *Icarus* spacecraft, you are supposed to give a very accurate report of your weight. As you approach the front of the line, you realize that you've forgotten to subtract the buoyant force exerted on you by the earth's atmosphere. Estimate the correction that you'll have to make to the spring-scale reading of your weight.

37 •• A helium balloon lifts a basket and cargo of total weight 2000 N under standard conditions, in which the density of air is 1.29 kg/m^3 and the density of helium is 0.178 kg/m^3. What is the minimum volume of the balloon?

38 •• Zoe is packing her belongings and moving in with Margaret. Her books are in boxes, which she plans to float down the river to Margaret's shack on a square raft that is 3 m on each side and 11 cm thick. It is made of wood having a specific gravity of 0.6. If each box has a mass of 20 kg, how many boxes can be placed on the raft if the books are to remain dry? Assume that the water remains calm.

39 •• An object has neutral buoyancy when its density equals that of the liquid in which it is submerged, which means that it neither floats nor sinks. If the average density of an 85-kg diver is 0.96 kg/L, what mass of lead should be added to give him neutral buoyancy?

40 •• A beaker of mass 1 kg containing 2 kg of water rests on a scale. A 2-kg block of aluminum (specific gravity 2.70) suspended from a spring scale is submerged in the water as in Figure 13-28. Find the readings of both scales.

Figure 13-28 Problem 40

41 ••• A ship sails from seawater (specific gravity 1.025) into fresh water and therefore sinks slightly. When its load of 600,000 kg is removed, it returns to its original level. Assuming that the sides of the ship are vertical at the water line, find the mass of the ship before it was unloaded.

42 ••• The hydrometer shown in Figure 13-29 is a device for measuring the density of liquids. The bulb contains lead shot, and the density can be read directly from the liquid level on the stem after the hydrometer has been calibrated. The volume of the bulb is 20 mL, the stem is 15 cm long and has a diameter of 5.00 mm, and the mass of the glass is 6.0 g. (*a*) What mass of lead shot must be added so that the least density of liquid that can be measured is 0.9 kg/L? (*b*) What is the maximum density of liquid that can be measured?

Figure 13-29 Problem 42

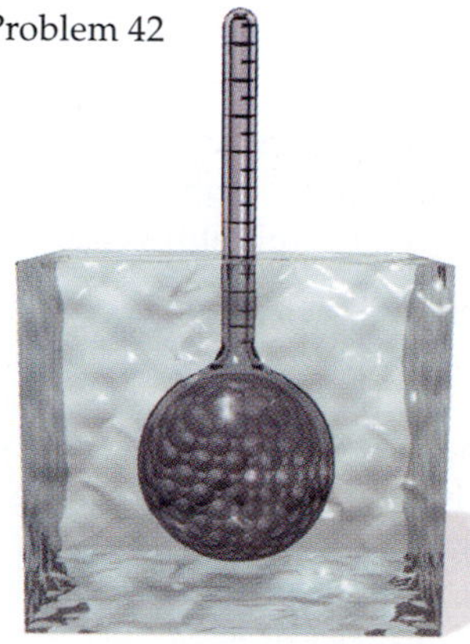

Continuity and Bernoulli's Equation

For the problems in this set, neglect the viscosity of the fluids.

43 • In a department store, a beach ball is supported by the airstream from a hose connected to the exhaust of a vacuum cleaner. Does the air blow under or over the ball to support it? Why?

44 • A horizontal pipe narrows from a diameter of 10 cm to 5 cm. For a fluid flowing from the larger diameter to the smaller,

(*a*) the velocity and pressure both increase.
(*b*) the velocity increases and the pressure decreases.
(*c*) the velocity decreases and the pressure increases.
(*d*) the velocity and pressure both decrease.
(*e*) either the velocity or pressure changes but not both.

45 •• When water emerges from a faucet, the stream narrows as the water falls. Explain why.

46 • Water flows at 0.65 m/s through a hose with a diameter of 3 cm. The diameter of the nozzle is 0.30 cm. (*a*) At what speed does the water pass through the nozzle? (*b*) If the pump at one end of the hose and the nozzle at the other end are at the same height, and if the pressure at the nozzle is atmospheric pressure, what is the pressure at the pump?

47 • Water is flowing at 3 m/s in a horizontal pipe under a pressure of 200 kPa. The pipe narrows to half its original diameter. (*a*) What is the speed of flow in the narrow section? (*b*) What is the pressure in the narrow section? (*c*) How do the volume flow rates in the two sections compare?

48 • The pressure in a section of horizontal pipe with a diameter of 2 cm is 142 kPa. Water flows through the pipe at 2.80 L/s. If the pressure at a certain point is to be reduced to 101 kPa by constricting a section of the pipe, what should the diameter of the constricted section be?

49 •• Blood flows in an aorta of radius 9 mm at 30 cm/s. (*a*) Calculate the volume flow rate in liters per minute. (*b*) Although the cross-sectional area of a capillary is much smaller than that of the aorta, there are many capillaries, so their total cross-sectional area is much larger. If all the blood from the aorta flows into the capillaries and the speed of flow through the capillaries is 1.0 mm/s, calculate the total cross-sectional area of the capillaries.

50 •• Dorothy is up on her 15-m × 15-m roof enjoying the view of Kansas. Suddenly, a strong wind blows down her ladder, leaving her stranded. She knows that a high wind reduces the air pressure on the roof, and that there is a danger that the atmospheric pressure inside the house will blow the roof off. As the wind reaches a speed of 30 m/s, she calls to her Auntie Em for help. Calculate the force on the roof.

51 •• A large tank of water is tapped a distance h below the water surface by a small pipe as in Figure 13-30. Find the distance x reached by the water flowing out the pipe.

Figure 13-30 Problems 51 and 56

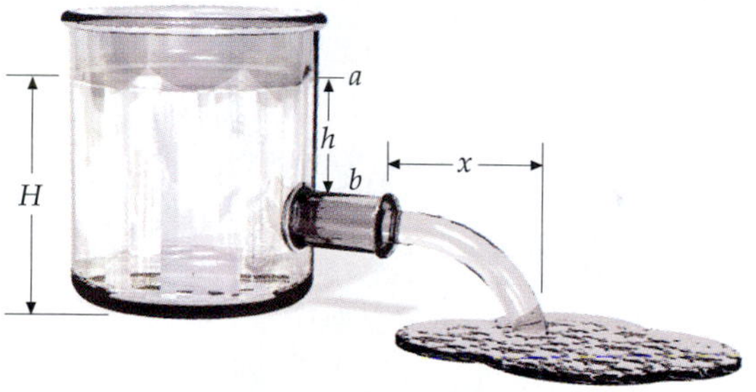

52 •• The \$8-billion, 800-mile long Alaskan Pipeline has a capacity of 240,000 m^3 of oil per day. It has a standard radius of 60 cm. Find the pressure P' at a point where the pipe has half the standard radius. Take the standard pressure to be $P = 180$ kPa and the density of oil to be 800 kg/m^3.

53 •• Water flows through a Venturi meter like that in Example 13-9 with a pipe diameter of 9.5 cm and a constriction diameter of 5.6 cm. The U-tube manometer is partially filled with mercury. Find the flow rate of the water in the pipe of 9.5-cm diameter if the difference in the mercury level in the U-tube is 2.40 cm.

54 •• A firefighter holds a hose with a bend in it as in Figure 13-31. Water is expelled from the hose in a stream of radius 1.5 cm at a speed of 30 m/s. (*a*) What mass of water emerges from the hose in 1 s? (*b*) What is the horizontal momentum of this water? (*c*) Before reaching the bend, the water has momentum upward, whereas afterward, its momen-

Figure 13-31 Problem 54

tum is horizontal. Draw a vector diagram of the initial and final momentum vectors, and find the change in the momentum of the water at the bend in 1 s. From this, find the force exerted on the water by the hose.

55 •• A fountain designed to spray a column of water 12 m into the air has a 1-cm-diameter nozzle at ground level. The water pump is 3 m below the ground. The pipe to the nozzle has a diameter of 2 cm. Find the necessary pump pressure.

56 ••• In Figure 13-30, (*a*) find the distance x at which the water strikes the ground as a function of h and H. (*b*) Show that there are two values of h that are equidistant from the point $h = \frac{1}{2}H$ that give the same distance x. (*c*) Show that x is a maximum when $h = \frac{1}{2}H$. What is the value of this maximum distance x?

Viscous Flow (optional)

57 • A horizontal tube with an inside diameter of 1.2 mm and a length of 25 cm has water flowing through it at 0.30 mL/s. Find the pressure difference required to drive this flow if the viscosity of water is 1.00 mPa·s.

58 • Find the diameter of a tube that would give double the flow rate for the pressure difference in Problem 57.

59 • Blood takes about 1.0 s to pass through a 1-mm-long capillary of the human circulatory system. If the diameter of the capillary is 7 μm and the pressure drop is 2.60 kPa, find the viscosity of blood.

General Problems

60 • True or false: The buoyant force on a submerged object depends on the shape of the object.

61 • A glass of water has ice cubes floating in it. What happens to the water level when the ice melts?

62 • Why is it easier to float in salt water than in fresh water?

63 • Smoke usually rises from a smokestack, but it may sink on a very humid day. What can be concluded about the relative densities of humid air and dry air?

64 •• A certain object has a density just slightly less than that of water so that it floats almost completely submerged. However, the object is more compressible than water. What happens if the floating object is given a slight push to submerge it?

65 •• In Example 13-9, the fluid is accelerated to a greater speed as it enters the narrow part of the pipe. Identify the forces that act on the fluid to produce this acceleration.

66 •• A glass of water is accelerating to the right along a horizontal surface. What is the origin of the force that produces the acceleration on a small element of water in the middle of the glass? Explain with a picture.

67 •• A 0.5-kg mass of lead is submerged in a container filled to the brim with water and a block of wood floats on top. The lead mass is slowly lifted from the container by a thin wire, and as it emerges into air the level of the water in the container drops a bit. The lead mass is now placed on the block of wood, which remains afloat. As the lead is placed on the block of wood,

(*a*) some water spills over the edge of the container.
(*b*) the water level rises exactly to the brim as before.
(*c*) the water level rises but does not reach the brim of the container.
(*d*) there is not enough information provided to decide between the three options.

68 •• You are sitting in a boat floating on a very small pond. You take the anchor out of the boat and drop it into the water. What happens to the water level in the pond?

69 • The top of a card table is 80 cm × 80 cm. What is the force exerted on it by the atmosphere? Why doesn't the table collapse?

70 • A 4.0-g Ping-Pong ball is attached by a thread to the bottom of a container. When the container is filled with water so that the ball is totally submerged, the tension in the thread is 2.8×10^{-2} N. Determine the diameter of the ball.

71 • Seawater has a bulk modulus of 2.3×10^9 N/m^2. Find the density of seawater at a depth where the pressure is 800 atm if the density at the surface is 1025 kg/m^3.

72 • Your car misses a turn and sinks into a small lake to a depth of 8 m. Your quick mind tells you that the chances of driving out are slim, so you'd better swim for it. The car door, however, is not budging, even though it seems undamaged. (*a*) If the outside area of the car door is 0.9 m^2, what is the force exerted on the door by the water? (*b*) What is the force exerted on the inside of the door by the air, assuming atmospheric pressure inside? (*c*) What will you have to do to get the door open and save your skin?

73 • A solid cubical block of side length 0.6 m is suspended from a spring balance. When the block is in water, the spring balance reads 80% of the reading when the block is in air. Determine the density of the block.

74 • When submerged in water, a block of copper has an apparent weight of 56 N. What fraction of this copper block will be submerged if floated on a pool of mercury?

75 • A 4.5-kg block of material floats on ethanol with 10% of its volume above the liquid surface. What fraction of this block will be submerged if floated on water?

76 • What is the buoyant force on your body when floating (*a*) in a freshwater lake (specific gravity = 1.0) and (*b*) in the ocean (specific gravity = 1.03)?

77 • Suppose that when you are floating in fresh water, 96% of your body is submerged. What is the volume of water your body displaces when it is fully submerged?

78 •• A block of wood of 1.5-kg mass floats on water with 68% of its volume submerged. A lead block is placed on the wood and the wood is then fully submerged. Find the mass of the lead block.

79 •• A Styrofoam cube, 25 cm on a side, is weighed on a simple beam balance. The balance is in equilibrium when a 20-g mass of brass is placed on the opposite pan of the balance. Find the true mass of the Styrofoam cube.

80 •• A spherical shell of copper with an outer diameter of 12 cm floats on water with half its volume above the water surface. Determine the inner diameter of the shell.

81 •• A beaker filled with water is balanced on the left cup of a scale. A cube 4 cm on a side is attached to a string and lowered into the water so that it is completely submerged. The cube is not touching the bottom of the beaker. A weight m is added to the system to retain equilibrium. What is the weight m and on which cup of the balance is it added?

82 •• Crude oil has a viscosity of about 0.8 Pa·s at normal temperature. A 50-km pipeline is to be constructed from an oil field to a tanker terminal. The pipeline is to deliver oil at the terminal at a rate of 500 L/s and the flow through the pipeline is to be laminar to minimize the pressure needed to push the fluid through the pipeline. Estimate the diameter of the pipeline that should be used.

83 •• Water flows through the pipe in Figure 13-32 and exits to the atmosphere at C. The diameter of the pipe is 2.0 cm at A, 1.0 cm at B, and 0.8 cm at C. The gauge pressure in the pipe at A is 1.22 atm and the flow rate is 0.8 L/s. The vertical pipes are open to the air. Find the level of the liquid–air interfaces in the two vertical pipes.

Figure 13-32
Problems 83 and 84

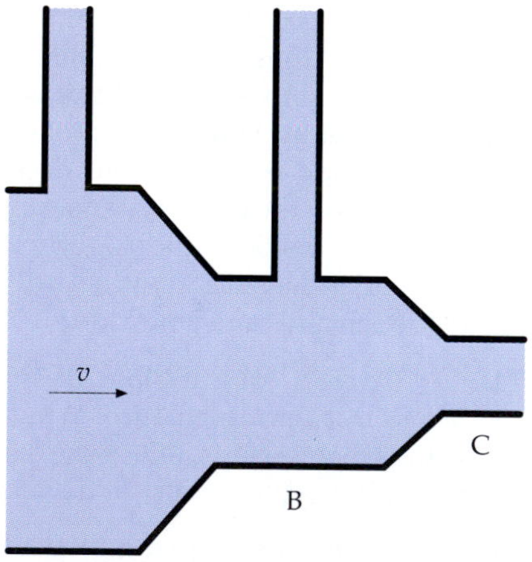

84 •• Repeat Problem 83 with the flow rate reduced to 0.6 L/s and the size of the opening at C reduced so that the pressure in the pipe at A remains unchanged.

85 •• Figure 13-33 is a sketch of an *aspirator*, a simple device that can be used to achieve a partial vacuum in a reservoir connected to the vertical tube at B. An aspirator attached to the end of a garden hose may be used to deliver soap or fertilizer from the reservoir. Suppose that the diameter at A is 2.0 cm and at C, where the water exits to the atmosphere, it is 1.0 cm. If the flow rate is 0.5 L/s and the gauge pressure at A is 0.187 atm, what diameter of the constriction at B will achieve a pressure of 0.1 atm in the container?

Figure 13-33
Problem 85

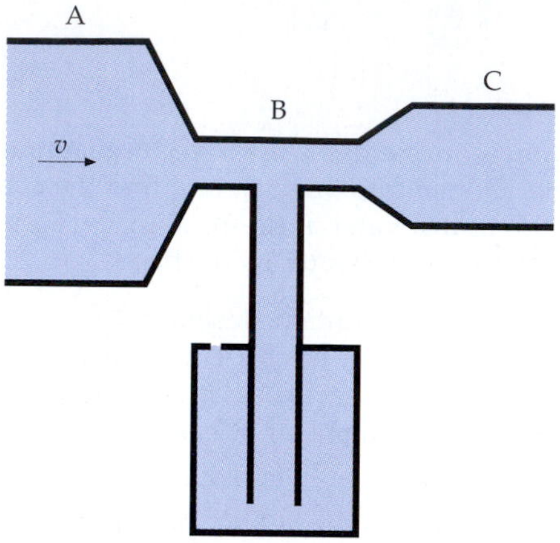

86 •• A cylindrical buoy at the entrance of a harbor has a diameter of 0.9 m and a height of 2.6 m. The mass of the buoy is 600 kg. It is attached to the bottom of the sea with a nylon cable of negligible mass. The specific gravity of the seawater is 1.025. (*a*) How much of the buoy is visible when the cable is slack? (*b*) If a tidal wave completely submerges the buoy, what is the tension in the taut cable? (*c*) If the cable breaks, what is the initial upward acceleration of the buoy?

87 •• Two communicating vessels contain a liquid of density ρ_0 (Figure 13-34). The cross-sectional areas of the vessels are A and $3A$. Find the change in elevation of the liquid level if an object of mass m and density $\rho' = 0.8\rho_0$ is put into one of the vessels.

Figure 13-34 Problem 87

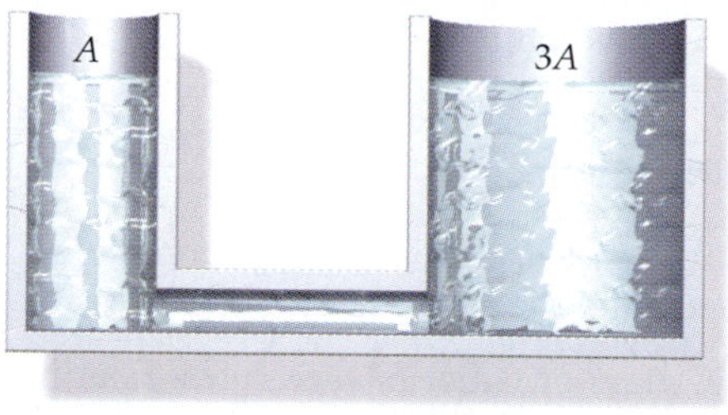

88 •• If an oil-filled manometer ($\rho = 900\ \text{kg/m}^3$) can be read to ±0.05 mm, what is the smallest pressure change that can be detected?

89 •• A rectangular dam 30 m wide supports a body of water to a height of 25 m. (*a*) Neglecting atmospheric pressure, find the total force due to water pressure acting on a thin strip of the dam of height dy located at a depth y. (*b*) Integrate your result in part (*a*) to find the total horizontal force exerted by the water on the dam. (*c*) Why is it reasonable to neglect atmospheric pressure?

90 •• A U-tube is filled with water until the liquid level is 28 cm above the bottom of the tube. An oil of specific gravity 0.78 is now poured into one arm of the U-tube until the level of the water in the other arm of the tube is 34 cm above the bottom of the tube. Find the level of the oil–water and oil–air interfaces in the other arm of the tube.

91 •• A U-tube contains liquid of unknown specific gravity. An oil of density 800 kg/m^3 is poured into one arm of the tube until the oil column is 12 cm high. The oil–air interface is then 5.0 cm above the liquid level in the other arm of the U-tube. Find the specific gravity of the liquid.

92 •• A lead block is suspended from the underside of a 0.5-kg block of wood of specific gravity of 0.7. If the upper surface of the wood is just level with the water, what is the mass of the lead block?

93 •• A helium balloon can just lift a load of 750 N. The skin of the balloon has a mass of 1.5 kg. *(a)* What is the volume of the balloon? *(b)* If the volume of the balloon is twice that found in part *(a)*, what is the initial acceleration of the balloon when it carries a load of 900 N?

94 •• A hollow sphere with an inner radius R and an outer radius $2R$ is made of material of density ρ_0 and is floating in a liquid of density $2\rho_0$. The interior is now filled with material of density ρ' so that the sphere just floats completely submerged. Find ρ'.

95 •• A balloon is filled with helium at atmospheric pressure. The skin of the balloon has a mass of 2.8 kg and the volume of the balloon is 16 m^3. What is the greatest weight that this balloon can lift?

96 •• As mentioned in the discussion of *the law of atmospheres*, the fractional decrease in atmospheric pressure is proportional to the change in altitude. Expressed in mathematical terms we have

$$\frac{dP}{P} = -C\,dh$$

where C is a constant. *(a)* Show that $P(h) = P_0e^{-Ch}$ is a solution of the differential equation. *(b)* Show that if $\Delta h \ll h_0$, then $P(h + \Delta h) \approx P(h)(1 - \Delta h/h_0)$, where $h_0 = 1/C$. *(c)* Given that the pressure at $h = 5.5$ km is half that at sea level, find the constant C.

97 •• A submarine has a total mass of 2.4×10^6 kg, including crew and equipment. The vessel consists of two parts, the pressure hull, which has a volume of 2×10^3 m^3, and the diving tanks, which have a volume of 4×10^2 m^3. When the sub cruises on the surface, the diving tanks are filled with air; when cruising below the surface, seawater is admitted into the tanks. *(a)* What fraction of the submarine's volume is above the water surface when the tanks are filled with air? *(b)* How much water must be admitted into the tanks to give the submarine neutral buoyancy? Neglect the mass of air in the tanks and use 1.025 as the specific gravity of seawater.

98 •• A marine salvage crew raises a crate that measures 1.4 m $\times$ 0.75 m $\times$ 0.5 m. The average density of the empty crate is the same as seawater, 1.025×10^3 kg/m^3, and its mass when empty is 32 kg. The crate contains gold bullion that fills 36% of its volume; the remaining volume is filled with seawater. *(a)* What is the tension in the cable that raises the crate and bullion while the crate is below the surface of the sea? *(b)* What is the tension in the cable while the crate is lifted to the deck of the ship if (1) none of the seawater leaks out of the crate, and (2) the crate is lifted so slowly that all of the seawater leaks out of the crate?

99 ••• When the hydrometer in Problem 42 is placed in a liquid whose specific gravity is greater than some minimum value, the device floats with part of the glass tube above the liquid level. Consider a hydrometer that has a spherical bulb 2.4 cm in diameter. The glass tube attached to the bulb is 20 cm long and has a diameter of 7.5 mm. The mass of the hydrometer before lead pellets are dropped into the bulb and the tube is sealed is 7.28 g. *(a)* What mass of lead should be placed in the bulb so that the hydrometer just floats in a liquid of specific gravity 0.78? *(b)* If the hydrometer is now placed in water, what is the length of the tube that shows above the surface of the water? *(c)* The hydrometer is placed in a liquid of unknown specific gravity; the length of the tube above the surface of the liquid is 12.2 cm. Determine the specific gravity of the liquid.

100 ••• A large beer keg of height H and cross-sectional area A_1 is filled with beer. The top is open to atmospheric pressure. At the bottom is a spigot opening of area A_2, which is much smaller than A_1. *(a)* Show that when the height of the beer is h, the speed of the beer leaving the spigot is approximately $\sqrt{2gh}$. *(b)* Show that for the approximation $A_2 \ll A_1$, the rate of change of the height h of the beer is given by

$$\frac{dh}{dt} = -\frac{A_2}{A_1}(2gh)^{1/2}$$

(c) Find h as a function of time if $h = H$ at $t = 0$. *(d)* Find the total time needed to drain the keg if $H = 2$ m, $A_1 = 0.8$ m^2, and $A_2 = (10^{-4})A_1$.

PART II

Vibration patterns in a violin. The vibrations of the violin strings are transmitted via the bridge to the body of the violin. The vibrations of the body, made visible here by interference effects, resonate with the vibrations of the strings, giving the violin its tone.

oscillations and waves

CHAPTER 14

Oscillations

The swaying of the Citicorp Building in New York during high winds is reduced by this tuned-mass damper mounted on an upper floor. It consists of a 400-ton sliding block connected to the building by a spring. The spring constant is chosen so that the natural frequency of the spring–block system is the same as the natural sway frequency of the building. Set into motion by winds, the building and damper oscillate 180° out of phase with each other, thereby significantly reducing the swaying.

Oscillation occurs when a system is disturbed from a position of stable equilibrium. There are many familiar examples: boats bob up and down, clock pendulums swing back and forth, and the strings and reeds of musical instruments vibrate. Other, less familiar examples are the oscillations of air molecules in a sound wave and the oscillations of electric currents in radios and television sets.

14-1 Simple Harmonic Motion

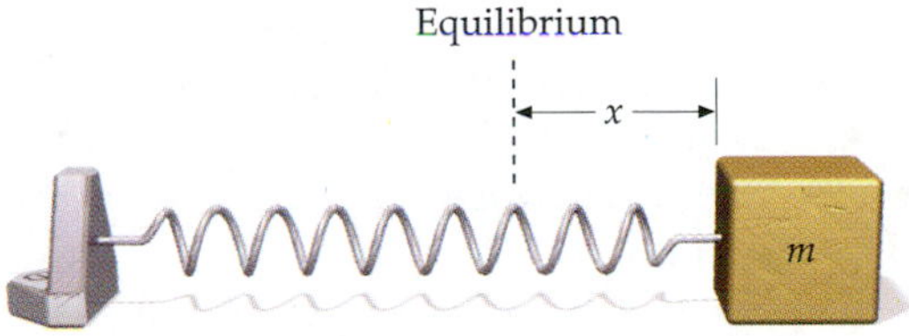

Figure 14-1 An object and spring on a frictionless surface. The displacement x, measured from the equilibrium position, is positive if the spring is stretched and negative if the spring is compressed.

A common and very important kind of oscillatory motion is **simple harmonic motion**, such as the motion of an object attached to a spring (Figure 14-1). In equilibrium, the spring exerts no force on the object. When the object is displaced an amount x from its equilibrium position, the spring exerts a force $-kx$, as given by Hooke's law:

$$F_x = -kx \qquad 14\text{-}1$$

where k is the force constant of the spring, a measure of the spring's stiffness.

The minus sign indicates that the force is a restoring force; that is, it is opposite to the direction of the displacement. Combining Equation 14-1 with Newton's second law, we have

$$F_x = ma_x$$

$$-kx = m\frac{d^2x}{dt^2}$$

or

$$a = \frac{d^2x}{dt^2} = -\frac{k}{m}x \qquad \text{14-2}$$

The acceleration is proportional to the displacement and is oppositely directed. This is a general characteristic of simple harmonic motion and can be used to identify systems that will exhibit it:

> Whenever the acceleration of an object is proportional to its displacement and is oppositely directed, the object will move with simple harmonic motion.

Conditions for simple harmonic motion in terms of acceleration

Since the acceleration is proportional to the net force, whenever the net force on an object is proportional to its displacement and is oppositely directed, the object will move with simple harmonic motion.

The time it takes for a displaced object to make a complete oscillation back and forth about its equilibrium position is called the **period** T. The reciprocal of the period is the **frequency** f, which is the number of oscillations per second:

$$f = \frac{1}{T} \qquad \text{14-3}$$

The unit of frequency is the reciprocal second (s^{-1}), which is called a **hertz** (Hz). For example, if the time for one complete oscillation is 0.25 s, the frequency is 4 Hz.

Figure 14-2 shows how we can experimentally obtain x versus t for a mass on a spring. The general equation for such a curve is

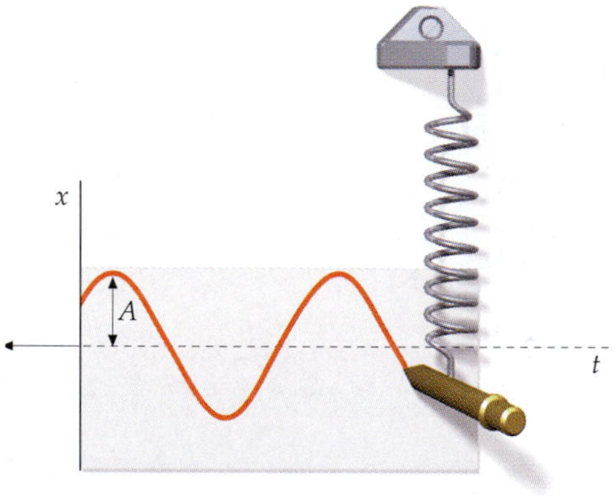

Figure 14-2 A marking pen is attached to a mass on a spring, and the paper is pulled to the left. As the paper moves with constant speed, the pen traces out the displacement x as a function of time t. (Here we have chosen x to be positive when the spring is compressed.)

$$x = A\cos(\omega t + \delta) \qquad \text{14-4}$$

Position in simple harmonic motion

where A, ω, and δ are constants.* The maximum displacement from equilibrium is called the **amplitude** A. The argument of the cosine function, $\omega t + \delta$, is called the **phase** of the motion, and the constant δ is called the **phase constant**. The phase constant depends on the choice of $t = 0$. If we have just one oscillating system, we can always choose $t = 0$ such that $\delta = 0$. If we have two systems oscillating with the same amplitude and frequency but different phase, we can choose $\delta = 0$ for one of them. The equations for the two systems are then

$$x_1 = A\cos(\omega t)$$

and

$$x_2 = A\cos(\omega t + \delta)$$

If the phase difference δ is 0 or an integer times 2π, then $x_2 = x_1$ and the systems are said to be in phase. If the phase difference δ is π or an odd integer times π, then $x_2 = -x_1$ and the systems are said to be out of phase.

---

* Note that $\cos(\omega t + \delta) = \sin(\omega t + \delta + \pi/2)$; thus, whether the equation is expressed as a cosine function or a sine function simply depends on the phase of the oscillation at the moment we designate to be $t = 0$.

We can show that Equation 14-4 is a solution of Equation 14-2 by differentiating x twice with respect to time. The first derivative of x gives the velocity v:

$$v = \frac{dx}{dt} = -A\omega \sin(\omega t + \delta) \qquad 14\text{-}5$$

Velocity in simple harmonic motion

Differentiating the velocity with respect to time gives the acceleration:

$$a = \frac{dv}{dt} = \frac{d^2x}{dt^2} = -\omega^2 A \cos(\omega t + \delta) \qquad 14\text{-}6$$

or

$$a = -\omega^2 x \qquad 14\text{-}7$$

Acceleration in simple harmonic motion

Comparing $a = -\omega^2 x$ with $a = -(k/m)x$ (Equation 14-2), we see that $x = A\cos(\omega t + \delta)$ is a solution of $d^2x/dt^2 = a = -(k/m)x$ if

$$\omega = \sqrt{\frac{k}{m}} \qquad 14\text{-}8$$

The amplitude A and the phase constant δ can be determined from the initial position x_0 and the initial velocity v_0 of the system. Setting $t = 0$ in $x = A\cos(\omega t + \delta)$ gives

$$x_0 = A\cos\delta \qquad 14\text{-}9$$

Similarly, setting $t = 0$ in $v = dx/dt = -A\omega\sin(\omega t + \delta)$ gives

$$v_0 = -A\omega\sin\delta \qquad 14\text{-}10$$

These equations can be solved for A and δ in terms of x_0 and v_0.

The period T is the time after which x repeats. Then

$$x(t) = x(t + T)$$

$$A\cos(\omega t + \delta) = A\cos[\omega(t + T) + \delta] = A\cos(\omega t + \delta + \omega T)$$

The cosine (and sine) function repeats in value when the phase increases by 2π, so

$$\omega T = 2\pi$$

or

$$\omega = \frac{2\pi}{T}$$

The constant ω is called the **angular frequency.** It has units of radians per second and dimensions of inverse time, the same as angular velocity, which is also designated by ω.

The frequency is the reciprocal of the period:

$$f = \frac{1}{T} = \frac{\omega}{2\pi} \qquad 14\text{-}11$$

Definition—Frequency, period, and angular frequency

Since $\omega = \sqrt{k/m}$, the frequency and period of an object on a spring are related to the force constant k and the mass m by

$$f = \frac{1}{T} = \frac{1}{2\pi}\sqrt{\frac{k}{m}} \qquad 14\text{-}12$$

Frequency and period for an object on a spring

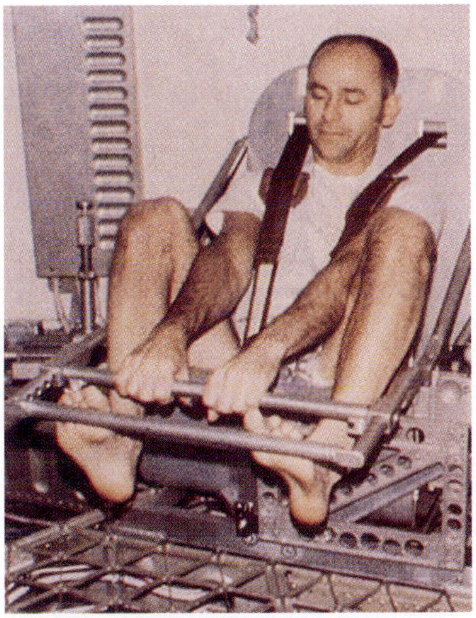

Astronaut Alan L. Bean measures his body mass during the second Skylab mission by sitting in a seat attached to a spring and oscillating back and forth. The total mass of the astronaut plus the seat is related to his frequency of vibration by Equation 14-12.

The frequency increases with increasing k (spring stiffness) and decreases with increasing mass.

Example 14-1

You are on a boat that is bobbing up and down. The boat's vertical displacement y is given by

$$y = (1.2\ \text{m})\cos\left(\frac{t}{2\ \text{s}} + \frac{\pi}{6}\right)$$

(*a*) Find the amplitude, angular frequency, phase constant, frequency, and period of the motion. (*b*) Where is the boat at $t = 1$ s? (*c*) Find the velocity and acceleration at any time t. (*d*) Find the initial position, velocity, and acceleration of the boat.

Picture the Problem We find the quantities asked for in (*a*) by comparing the equation of motion with the standard equation for simple harmonic motion, Equation 14-4. The velocity and acceleration are found by differentiating $y(t)$.

(*a*)1. Compare the equation for the boat's vertical displacement with Equation 14-4, $y = A\cos(\omega t + \delta)$, to get A, ω, and δ:

$$y = (1.2\ \text{m})\cos\left(\frac{t}{2\ \text{s}} + \frac{\pi}{6}\right)$$

$$A = 1.2\ \text{m}, \qquad \omega = 1/2\ \text{rad/s}, \qquad \delta = \pi/6\ \text{rad}$$

2. The frequency and period are found from ω:

$$f = \frac{\omega}{2\pi} = 0.0796\ \text{Hz}, \qquad T = \frac{1}{f} = 12.6\ \text{s}$$

(*b*) Set $t = 1$ s to find the boat's position:

$$y = (1.2\ \text{m})\cos\left[\frac{1}{2\ \text{s}}(1\ \text{s}) + \frac{\pi}{6}\right] = 0.624\ \text{m}$$

(*c*) The velocity and acceleration are obtained from the position by differentiation with respect to time:

$$v_y = \frac{dy}{dt} = -(1.2\ \text{m})\sin\left(\frac{1}{2\ \text{s}}t + \frac{\pi}{6}\right)\frac{d(t/2\ \text{s})}{dt}$$

$$= -(0.6\ \text{m/s})\sin\left(\frac{1}{2\ \text{s}}t + \frac{\pi}{6}\right)$$

$$a_y = \frac{dv_y}{dt} = -(0.6\ \text{m/s})\cos\left(\frac{1}{2\ \text{s}}t + \frac{\pi}{6}\right)\frac{d(t/2\ \text{s})}{dt}$$

$$= -(0.3\ \text{m/s}^2)\cos\left(\frac{1}{2\ \text{s}}t + \frac{\pi}{6}\right)$$

(*d*) Set $t = 0$ to find y_0, v_{y0}, and a_{y0}:

$$y_0 = (1.2\ \text{m})\cos\frac{\pi}{6} = 1.04\ \text{m}$$

$$v_{y0} = -(0.6\ \text{m/s})\sin\frac{\pi}{6} = -0.300\ \text{m/s}$$

$$a_{y0} = -(0.3\ \text{m/s}^2)\cos\frac{\pi}{6} = -0.260\ \text{m/s}^2$$

Exercise A 0.8-kg object is attached to a spring of force constant $k = 400$ N/m. Find the frequency and period of motion of the object when it is displaced from equilibrium. (*Answer* $f = 3.56$ Hz, $T = 0.281$ s)

Figure 14-3 shows two identical masses attached to identical springs and resting on a frictionless surface. One spring is stretched 10 cm and the other 5 cm. If they are released at the same time, which object reaches the equilibrium position first? According to Equation 14-12, the period depends only on k and m and not on the amplitude. Since k and m are the same for both systems, the periods are the same. Thus, the objects reach the equilibrium position at the same time. The second object has twice as far to go to reach equilibrium, but it will also have twice the average speed. Figure 14-4 shows a sketch of the position functions for the two objects. This illustrates an important general property of simple harmonic motion:

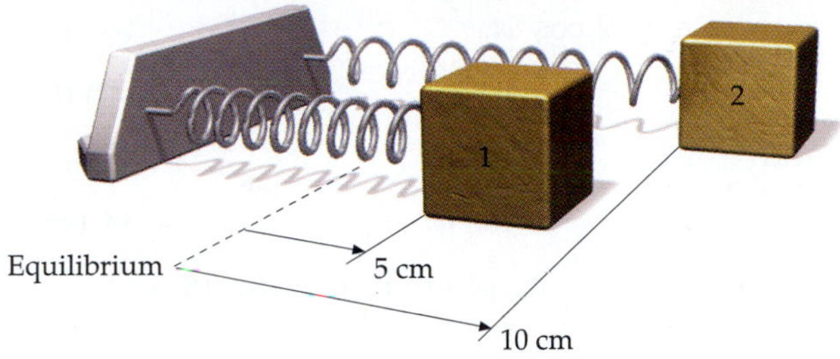

Figure 14-3 Two identical mass–spring systems.

> The frequency and period of simple harmonic motion are independent of the amplitude.

The fact that the frequency in simple harmonic motion is independent of the amplitude has important consequences in many fields. In music, for example, it means that when a note is struck on the piano, the pitch (which corresponds to the frequency) does not depend on how loudly the note is played (which corresponds to the amplitude).* If changes in amplitude had a large effect on the frequency, musical instruments would be unplayable.

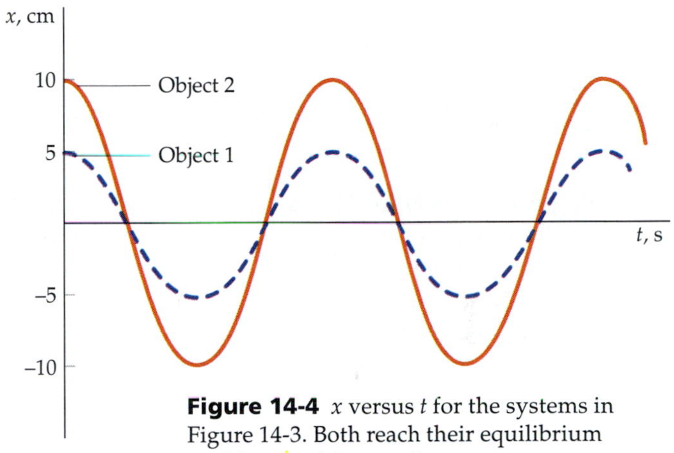

Figure 14-4 x versus t for the systems in Figure 14-3. Both reach their equilibrium positions at the same time.

Example 14-2

An object oscillates with angular frequency $\omega = 8.0$ rad/s. At $t = 0$, the object is at $x_0 = 4$ cm with an initial velocity $v_0 = -25$ cm/s. (*a*) Find the amplitude and phase constant for the motion. (*b*) Write x as a function of time.

Picture the Problem The initial position and velocity give us two equations from which to determine the amplitude A and the phase constant δ.

(*a*) 1. The initial position and velocity are related to the amplitude and phase constant:

$$x_0 = A\cos\delta \quad \text{and} \quad v_0 = -\omega A\sin\delta$$

2. Divide these equations to eliminate A:

$$\frac{v_0}{x_0} = \frac{-\omega A\sin\delta}{A\cos\delta} = -\omega\tan\delta$$

3. Substituting numerical values yields δ:

$$\tan\delta = -\frac{v_0}{\omega x_0} = -\frac{-25\text{ cm/s}}{(8.0\text{ rad/s})(4\text{ cm})} = 0.78$$

$$\delta = \tan^{-1}(0.78) = 0.66\text{ rad}$$

4. The amplitude can be found using either the x_0 or v_0 equation. Here we use x_0:

$$A = \frac{x_0}{\cos\delta} = \frac{4\text{ cm}}{\cos 0.66} = 5.06\text{ cm}$$

(*b*) Comparing with Equation 14-4 yields x:

$$x = (5.06\text{ cm})\cos(8.0t + 0.66)$$

---

* For many musical instruments, there is a slight dependence of frequency on amplitude. The vibration of an oboe reed, for example, is not exactly simple harmonic, thus its pitch depends slightly on how hard it is blown. This effect is corrected for by skilled musicians.

When the phase constant is $\delta = 0$, Equations 14-4, 14-5, and 14-6 then become

$$x = A \cos \omega t \qquad \text{14-13}a$$

$$v = -\omega A \sin \omega t \qquad \text{14-13}b$$

and

$$a = -\omega^2 A \cos \omega t \qquad \text{14-13}c$$

These functions are plotted in Figure 14-5.

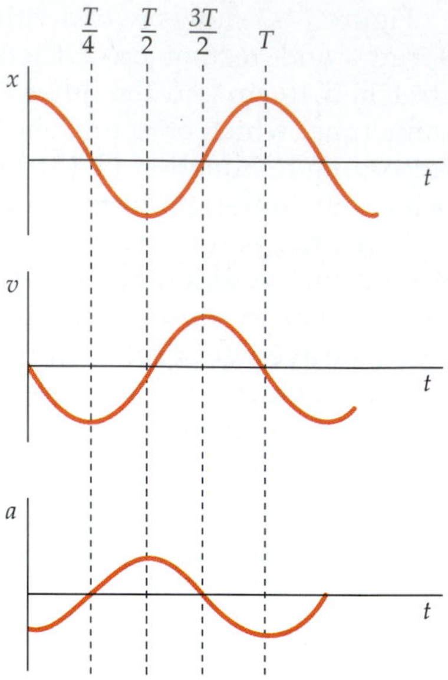

Figure 14-5 Plots of x, v, and a as functions of time t for $\delta = 0$. At $t = 0$, the displacement is maximum, the velocity is zero, and the acceleration is negative and equal to $-\omega^2 A$. The velocity becomes negative as the object moves back toward its equilibrium position. After one quarter-period ($t = T/4$), the object is at equilibrium, $x = 0$, $a = 0$, and the speed has its maximum value of ωA. At $t = T/2$, the displacement is $-A$, the velocity is again zero, and the acceleration is $+\omega^2 A$. At $t = 3T/4$, $x = 0$, $a = 0$, and $v = +\omega A$.

Example 14-3 *try it yourself*

A 2-kg object is attached to a spring as in Figure 14-1. The force constant of the spring is $k = 196$ N/m. The object is held a distance 5 cm from the equilibrium position and is released at $t = 0$. (*a*) Find the angular frequency ω, the frequency f, and the period T. (*b*) Write x as a function of time.

Cover the column to the right and try these on your own before looking at the answers.

Steps	***Answers***
(*a*)1. Calculate ω from $\omega = \sqrt{k/m}$.	$\omega = 9.90$ rad/s
2. Use your result to find f and T.	$f = 1.58$ Hz, $T = 0.633$ s
3. Find A and δ from the initial conditions.	$A = 5$ cm, $\delta = 0$
(*b*) Write $x(t)$ using your results for A, ω, and δ.	$x = (5\text{ cm}) \cos(9.90\text{ s}^{-1}\,t)$

Example 14-4

Consider an object on a spring whose position is given by the equation $x = (5\text{ cm})\cos(9.90\text{ s}^{-1}\,t)$. (*a*) What is the maximum speed of the object? (*b*) When does this maximum speed first occur? (*c*) What is the maximum acceleration of the object? (*d*) When does maximum acceleration first occur?

Picture the Problem Since the object is released from rest, $\delta = 0$, and the velocity and acceleration are given by Equations 14-13*b* and *c*.

(*a*)1. Equation 14-13*b* gives the velocity for $\delta = 0$: $\quad v = -\omega A \sin \omega t$

2. Maximum speed occurs when $\sin \omega t = \pm 1$: $\quad |v_{max}| = \omega A = (9.90\text{ rad/s})(5\text{ cm}) = 49.5\text{ cm/s}$

(*b*)1. Sin $\omega t = \pm 1$ first occurs when $\omega t = \pi/2$: $\quad \sin \omega t = \pm 1$

$$\omega t = \frac{\pi}{2}, \frac{3\pi}{2}, \frac{5\pi}{2}, \ldots$$

2. Solve for t when $\omega t = \pi/2$: $\quad t = \dfrac{\pi}{2\omega} = \dfrac{\pi}{2(9.90\text{ s}^{-1})} = 0.159\text{ s}$

(*c*) 1. The acceleration is given by Equation 14-13*c*: $a = -\omega^2 A \cos \omega t$

2. Maximum acceleration corresponds to $\cos \omega t = \pm 1$: $|a_{max}| = \omega^2 A = 490\ \mathrm{cm/s^2} \approx \frac{1}{2} g$

(*d*) The maximum acceleration occurs first at: $t = 0$

Remark The maximum speed first occurs after one quarter-period,

$$t = \frac{\pi}{2\omega} = \frac{\pi}{2(2\pi/T)} = \frac{1}{4}T$$

Simple Harmonic Motion and Circular Motion

There is a relation between simple harmonic motion and circular motion with constant speed. Imagine a particle moving with constant speed v in a circle of radius A (Figure 14-6). Its angular displacement relative to the positive x axis is

$$\theta = \omega t + \delta$$

where δ is the angular displacement at time $t = 0$ and $\omega = v/A$ is the angular velocity of the particle. The x component of the particle's position is

$$x = A \cos \theta = A \cos(\omega t + \delta)$$

which is the same as Equation 14-4 for simple harmonic motion.

> When a particle moves with constant speed in a circle, its projection on a diameter of the circle moves with simple harmonic motion.

The y component of the particle's position is

$$y = A \sin \theta = A \sin(\omega t + \delta) = A \cos(\omega t + \delta - \pi/2)$$

The y component also describes simple harmonic motion. If we compare the phase of the x and y components, we see that they differ by $\pi/2$ or 90°. Circular motion is therefore the combination of perpendicular simple harmonic motions having the same amplitude and frequency but with a relative phase difference of $\pi/2$. Figure 14-7 gives two examples demonstrating the relation of circular motion and simple harmonic motion.

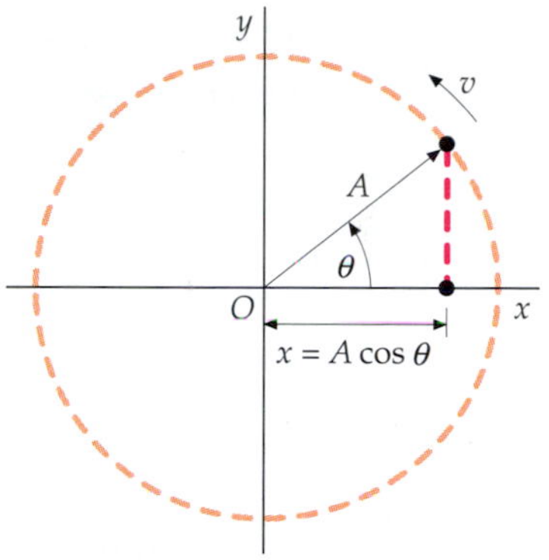

Figure 14-6 When a particle moves in a circular path with constant speed, its x component of position describes simple harmonic motion.

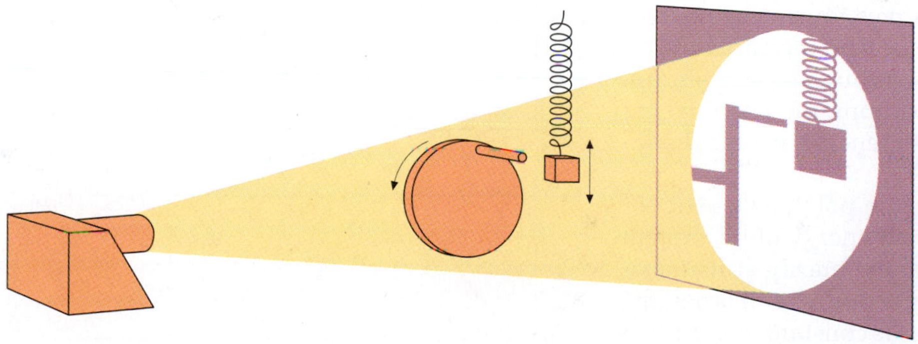

(*a*)

(*b*)

Figure 14-7 The relation between circular motion and simple harmonic motion. (*a*) The projected shadows of a rotating peg and an object on a spring move in unison when the period of the rotating turntable equals that of the oscillating object and the radius of the turntable equals the amplitude of the spring system. (*b*) Bubbles foaming off the edge of a rotating propeller that is moving through water produce a sinusoidal pattern.

14-2 Energy in Simple Harmonic Motion

When an object undergoes simple harmonic motion, the system's potential and kinetic energies vary with time. Their sum, the total energy $E = K + U$, is constant. Consider an object a distance x from equilibrium, acted on by a restoring force $-kx$. The system's potential energy is

$$U = \tfrac{1}{2}kx^2$$

This is Equation 6-23. For simple harmonic motion, $x = A\cos(\omega t + \delta)$. Substituting gives

$$U = \tfrac{1}{2}kA^2\cos^2(\omega t + \delta) \qquad \text{14-14}$$

Potential energy in simple harmonic motion

The kinetic energy of the system is

$$K = \tfrac{1}{2}mv^2$$

where m is the object's mass and v is its speed. For simple harmonic motion, $v = -A\omega\sin(\omega t + \delta)$. Substituting gives

$$K = \tfrac{1}{2}mA^2\omega^2\sin^2(\omega t + \delta)$$

Then using $\omega^2 = k/m$,

$$K = \tfrac{1}{2}kA^2\sin^2(\omega t + \delta) \qquad \text{14-15}$$

Kinetic energy in simple harmonic motion

The total energy is the sum of the potential and kinetic energies:

$$E_{\text{total}} = U + K = \tfrac{1}{2}kA^2\cos^2(\omega t + \delta) + \tfrac{1}{2}kA\sin^2(\omega t + \delta)$$
$$= \tfrac{1}{2}kA^2[\cos^2(\omega t + \delta) + \sin^2(\omega t + \delta)]$$

Since $\sin^2(\omega t + \delta) + \cos^2(\omega t + \delta) = 1$,

$$E_{\text{total}} = \tfrac{1}{2}kA^2 \qquad \text{14-16}$$

Total energy in simple harmonic motion

This equation reveals an important general property of simple harmonic motion:

> The total energy in simple harmonic motion is proportional to the square of the amplitude.

Figure 14-8 Plots of U and K versus t.

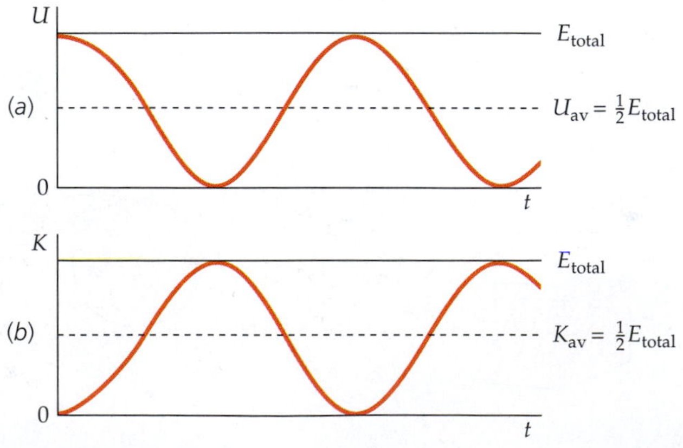

For an object at its maximum displacement, the total energy is all potential energy. As the object moves toward its equilibrium position, the kinetic energy of the system increases and its potential energy decreases. As it moves through its equilibrium position, the speed of the object is maximum, the potential energy of the system is zero, and the total energy equals the kinetic energy.

As the object moves past the equilibrium point, its kinetic energy begins to decrease, and the potential energy of the system increases until the object again stops momentarily at its maximum displacement (now in the other direction). At all times, the sum of the potential and kinetic energies is constant. Figure 14-8 shows plots of U and K versus time. These curves have the same shape except that one is zero when the other is maximum. Their average values over one or more cycles are equal, and since $U + K = E$, their average values are given by

$$U_{av} = K_{av} = \tfrac{1}{2}E_{total} \qquad \text{14-17}$$

In Figure 14-9, the potential energy U is graphed as a function of x. The total energy E_{total} is constant and is therefore plotted as a horizontal line. This line intersects the potential-energy curve at $x = A$ and $x = -A$, called the **turning points,** since these are the points at which oscillating objects reverse direction and head back toward the equilibrium position.

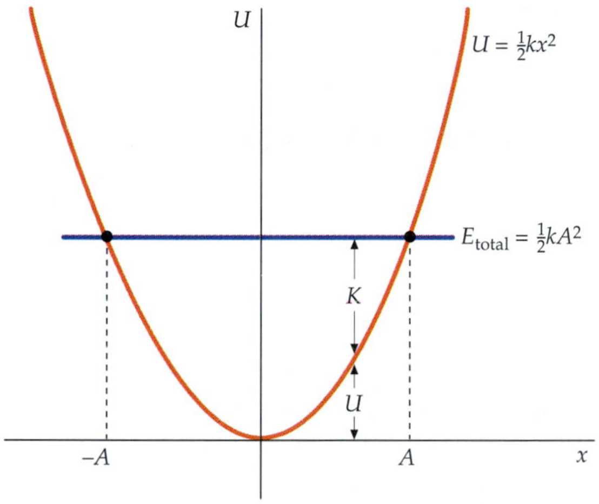

Figure 14-9 The potential-energy function $U = \frac{1}{2}kx^2$ for an object of mass m on a (massless) spring of force constant k. The horizontal blue line represents the total energy E_{total} for an amplitude of A. The kinetic energy K is represented by the vertical distance $K = E_{total} - U$. Since $E_{total} \geq U$, the motion is restricted to $-A \leq x \leq +A$.

Example 14-5

A 3-kg object attached to a spring oscillates with an amplitude of 4 cm and a period of 2 s. (*a*) What is the total energy? (*b*) What is the maximum speed of the object? (*c*) At what position x_1 is the speed equal to half its maximum value?

Picture the Problem (*a*) The total energy can be found from the amplitude and force constant, which can be found from the mass and period. (*b*) The maximum speed occurs when the kinetic energy equals the total energy. (*c*) We can relate the position to the speed by using conservation of energy.

(*a*) 1. Write the total energy E in terms of the force constant k and amplitude A:

$$E = \tfrac{1}{2}kA^2$$

2. The force constant is related to the period and mass:

$$k = m\omega^2 = \frac{4\pi^2 m}{T^2} = \frac{4\pi^2(3\text{ kg})}{(2\text{ s})^2} = 29.6\text{ N/m}$$

3. Substitute $k = 29.6$ N/m, and $A = 0.04$ m to find E:

$$E = \tfrac{1}{2}kA^2 = \tfrac{1}{2}(29.6\text{ N/m})(0.04\text{ m})^2 = 2.37 \times 10^{-2}\text{ J}$$

(*b*) To find v_{max}, set the kinetic energy equal to the total energy and solve for v:

$$\tfrac{1}{2}mv_{max}^2 = E$$

$$v_{max} = \sqrt{\frac{2E}{m}} = \sqrt{\frac{2(2.37 \times 10^{-2}\text{ J})}{3\text{ kg}}} = 0.126\text{ m/s}$$

(*c*) 1. Conservation of energy relates the position x to the speed v:

$$E = \tfrac{1}{2}mv^2 + \tfrac{1}{2}kx^2$$

2. Substitute $v = \frac{1}{2}v_{max}$ and solve for x_1. It is convenient to find x in terms of E and then write $E = \frac{1}{2}kA^2$ to obtain an expression for x in terms of A:

$$E = \tfrac{1}{2}m(\tfrac{1}{2}v_{max})^2 + \tfrac{1}{2}kx_1^2 = \tfrac{1}{4}(\tfrac{1}{2}mv_{max}^2) + \tfrac{1}{2}kx_1^2$$

$$E = \tfrac{1}{4}E + \tfrac{1}{2}kx_1^2$$

$$\tfrac{1}{2}kx_1^2 = E - \tfrac{1}{4}E = \tfrac{3}{4}E = \tfrac{3}{4}(\tfrac{1}{2}kA^2)$$

$$x_1 = \frac{\sqrt{3}}{2}A = \frac{\sqrt{3}}{2}(4\text{ cm}) = 3.46\text{ cm}$$

Exercise Calculate ω for this example and find v_{max} from $v_{max} = \omega A$. (*Answer* $\omega = 3.14$ rad/s, $v_{max} = 0.126$ m/s)

Exercise An object of mass 2 kg is attached to a spring of force constant 40 N/m. The object is moving at 25 cm/s when it is at its equilibrium position. (*a*) What is the total energy of the object? (*b*) What is the amplitude of the motion? (*Answers* (*a*) $E_{total} = \frac{1}{2}mv_{max}^2 = 0.0625$ J, (*b*) $A = \sqrt{2E_{total}/k} = 5.59$ cm).

optional

General Motion Near Equilibrium

Simple harmonic motion is important because it occurs whenever a particle is displaced slightly from a position of stable equilibrium. Figure 14-10 is a graph of the potential energy U versus x for a force that has a position of stable equilibrium and a position of unstable equilibrium. As discussed in Chapter 6, the maximum at x_2 on Figure 14-10 corresponds to unstable equilibrium, whereas the minimum at x_1 corresponds to stable equilibrium. Any smooth curve that has a minimum like the one in Figure 14-10 can be approximated near the minimum by a parabola. The dashed curve in this figure is a parabolic curve that approximately fits U near the stable equilibrium point. The general equation for a parabola that has a minimum at point x_1 can be written

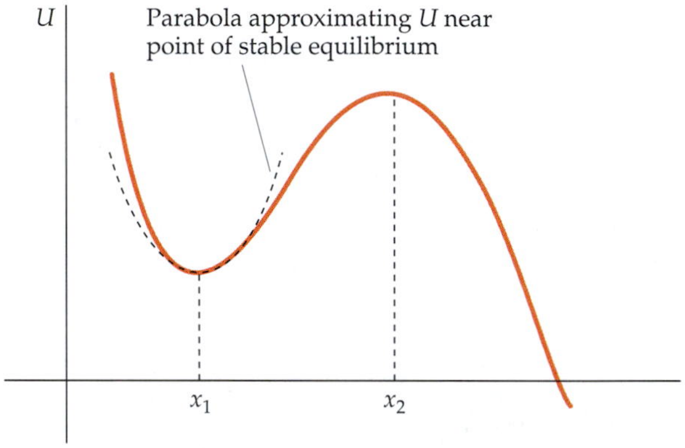

Figure 14-10 Plot of U versus x for a force that has a position of stable equilibrium (x_1) and a position of unstable equilibrium (x_2).

$$U = A + B(x - x_1)^2 \qquad 14\text{-}18$$

where A and B are constants. The constant A is the value of U at the equilibrium position $x = x_1$. The force is related to the potential energy curve by $F_x = -dU/dx$. Then

$$F_x = -\frac{dU}{dx} = -2B(x - x_1)$$

If we set $2B = k$, this equation reduces to

$$F_x = -\frac{dU}{dx} = -k(x - x_1) \qquad 14\text{-}19$$

According to Equation 14-19, the force is proportional to the displacement and oppositely directed, so the motion will be simple harmonic. Figure 14-11 shows a real potential energy function that has a position r_0 of stable equilibrium—$U(r)$ versus separation r for two hydrogen atoms.

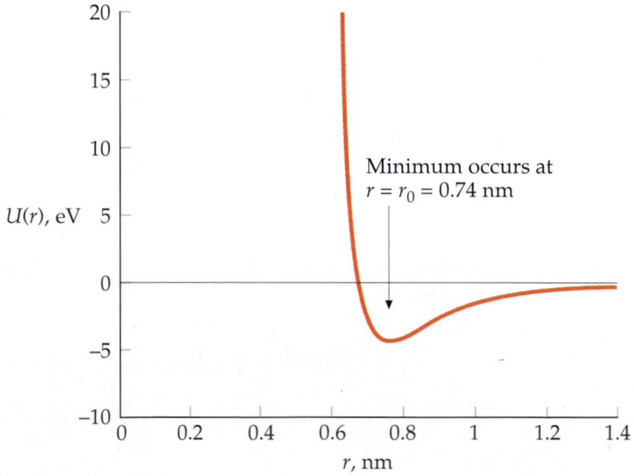

Figure 14-11 Potential energy function $U(r)$ versus separation r for two hydrogen atoms. The minimum value of the potential energy $U = -4.48$ eV occurs at the equilibrium separation $r = r_0 = 0.74$ nm. The atoms in the H_2 molecule oscillate about this point. If the energy of oscillation is not too great, so that $|r - r_0|$ is small, the oscillation is simple harmonic.

14-3 Some Oscillating Systems

Object on a Vertical Spring

When an object hangs from a vertical spring, there is a force mg downward in addition to the force of the spring (Figure 14-12). If we choose the downward direction to be positive, the spring's force on the object is $F_s = -ky$, where y is the difference in position between the end of the spring when it is unstretched and when it is stretched by the weight of the object. Then Newton's second law gives

$$m\frac{d^2y}{dt^2} = -ky + mg \qquad 14\text{-}20$$

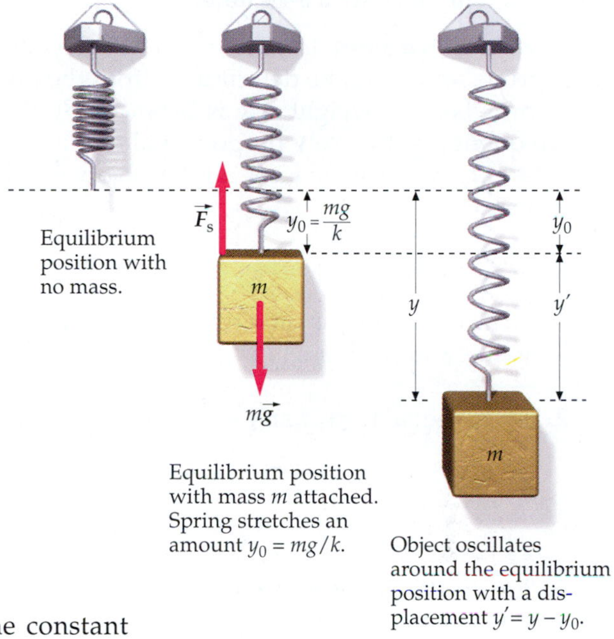

Equilibrium position with no mass.
$\vec{F}_s$
$y_0 = \frac{mg}{k}$
m
$m\vec{g}$
y
y_0
y'
m
Equilibrium position with mass m attached. Spring stretches an amount $y_0 = mg/k$.
Object oscillates around the equilibrium position with a displacement $y' = y - y_0$.

Figure 14-12 The problem of a mass on a vertical spring is simplified if the displacement (y') is measured from the equilibrium position of the spring with the mass attached.

Equation 14-20 differs from Equation 14-2 by the addition of the constant term mg. We handle this extra term by changing to a new variable $y' = y - y_0$, where $y_0 = mg/k$ is the amount the spring is stretched when the object is in equilibrium. Substituting $y = y' + y_0$ into Equation 14-20 gives

$$m\frac{d^2(y' + y_0)}{dt^2} = -k(y' + y_0) + mg = -ky' - ky_0 + mg$$

But $ky_0 = mg$. Using this and the fact that the derivative of a constant is zero, we have

$$m\frac{d^2y'}{dt^2} = -ky'$$

which has the familiar solution

$$y' = A\cos(\omega t + \delta)$$

Thus, the effect of the gravitational force mg is merely to shift the equilibrium position from $y = 0$ to $y' = 0$. When the object is displaced from this equilibrium position by the amount y', the unbalanced force is $-ky'$. The object oscillates about this equilibrium position with an angular frequency $\omega = \sqrt{k/m}$, the same as that for an object on a horizontal spring.

When a mass hangs on a vertical spring, there is both gravitational potential energy U_g and spring potential energy U_s. At the equilibrium point, the spring is stretched and has potential energy $\frac{1}{2}ky_0^2$ relative to its unstretched position ($y = 0$), and the gravitational potential energy is $-mgy_0$ relative to $y = 0$. We will show below that if we choose the total potential energy (including gravitational potential energy) to be zero at the equilibrium position $y' = 0$, the total potential energy can be written

$$U = U_s + U_g = \frac{1}{2}ky'^2 \qquad (U = 0 \text{ at } y' = 0) \qquad 14\text{-}21$$

So if we measure the displacement from the equilibrium position, we can forget about the effect of gravity.

Example 14-6

A 3-kg object stretches a spring 16 cm when it hangs vertically in equilibrium. The spring is then stretched from its equilibrium position and the object is released. (*a*) Find the frequency of the motion. (*b*) Find the frequency if the 3-kg object is replaced with a 6-kg object.

Picture the Problem (*a*) The frequency depends on the force constant for the spring, which can be determined from the position of the spring at equilibrium, when the weight m_1g is balanced by the spring force ky_0. (*b*) Since the frequency is inversely proportional to the square root of the mass (Equation 14-12), the frequency for a 6-kg object will be $1/\sqrt{2}$ times that for the 3-kg object.

(*a*) 1. Write the frequency in terms of the force constant k and the mass m_1:

$$f = \frac{\omega}{2\pi} = \frac{1}{2\pi}\sqrt{\frac{k}{m_1}}$$

2. Set ky_0 equal to m_1g and solve for k:

$$ky_0 = m_1g$$

$$k = \frac{m_1g}{y_0} = \frac{(3\text{ kg})(9.81\text{ N/kg})}{0.16\text{ m}} = 184\text{ N/m}$$

3. Substitute this value for k and the known values for m_1, g, and y_0:

$$f = \frac{1}{2\pi}\sqrt{\frac{k}{m_1}} = \frac{1}{2\pi}\sqrt{\frac{m_1g/y_0}{m_1}} = \frac{1}{2\pi}\sqrt{\frac{g}{y_0}}$$

$$= \frac{1}{2\pi}\sqrt{\frac{9.81\text{ m/s}^2}{0.16\text{ m}}} = 1.25\text{ Hz}$$

(*b*) Replace m_1 by $m_2 = 2m_1$:

$$f = \frac{1}{2\pi}\sqrt{\frac{k}{m_2}} = \frac{1}{2\pi}\sqrt{\frac{k}{2m_1}} = \frac{1}{\sqrt{2}}\left(\frac{1}{2\pi}\sqrt{\frac{k}{m_1}}\right)$$

$$= \frac{1}{\sqrt{2}}f_1 = \frac{1}{\sqrt{2}}1.25\text{ Hz} = 0.884\text{ Hz}$$

Remark Note that in (*a*) we did not need to use the value of m or k because the force constant is mg/y_0, and k/m is therefore equal to g/y_0.

Example 14-7

A block rests on a spring and oscillates vertically with a frequency of 4 Hz and an amplitude of 7 cm. A tiny bead is placed on top of the oscillating block just as it reaches its lowest point (Figure 14-13). Assume that the bead has no effect on the oscillation. (*a*) At what distance from the block's equilibrium position does the bead lose contact with the block? (*b*) What is the speed of the bead when it leaves the block?

Picture the Problem (*a*) Let y' be positive upward with $y' = 0$ at the equilibrium position of the block. The equation of motion for the block is $y' = -A\cos\omega t$ with $A = 0.07$ m and $\omega = 2\pi f = 8\pi$. The forces on the bead are its weight mg downward and the upward normal force exerted by the block. As the block moves upward *from equilibrium*, its acceleration and that of the bead are *downward* and increasing in magnitude. When the acceleration reaches $-g$, the normal force on the bead is zero and the bead leaves the block. (*b*) The velocity of the bead when it leaves the block is the same as that of the block at this time.

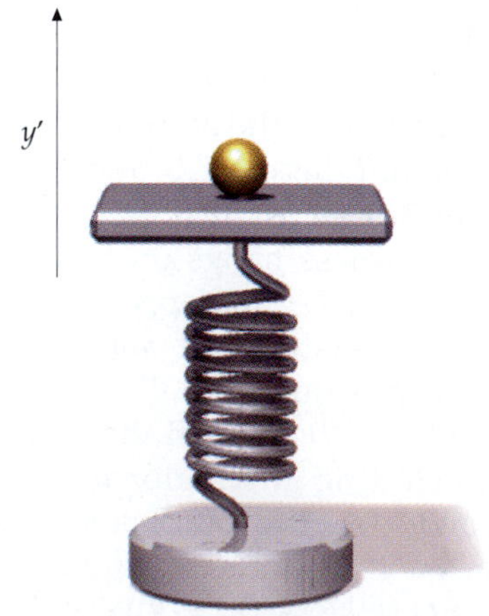

Figure 14-13

(a)1. Write the equation of motion for the block:	$y' = -A\cos\omega t$
2. Compute the acceleration of the block and set it equal to $-g$:	$a = -\omega^2 y' = +A\omega^2\cos\omega t = -g$ $\cos\omega t = \dfrac{-g}{\omega^2 A}$
3. Find the displacement y at this time:	$y' = -A\cos\omega t = \dfrac{g}{\omega^2} = \dfrac{9.81\text{ m/s}^2}{(8\pi\text{ rad/s})^2} = 0.0155\text{ m} = 1.55\text{ cm}$
(b)1. Find the velocity of the block (and bead) at any time:	$v = \dfrac{dy'}{dt} = +\omega A\sin\omega t = \omega A\sqrt{1-\cos^2\omega t}$
2. Compute v when $y' = 1.55$ cm:	$\cos\omega t = -\dfrac{y'}{A} = -\dfrac{1.55\text{ cm}}{7\text{ cm}} = -0.221$ $v = \omega A\sqrt{1-\cos^2\omega t}$ $= (8\pi\text{ rad/s})(0.07\text{ m})\sqrt{1-(0.221)^2}$ $= 1.72\text{ m/s}$

Remark The bead leaves when y' is positive, as expected.

Example 14-8

The 3-kg object in Example 14-6 stretches a spring 16 cm when it hangs vertically in equilibrium. The spring is then stretched 5 cm from its equilibrium position and the object is released. Find the total energy and the potential energy of the spring when the mass is at its maximum displacement.

Picture the Problem The total energy including gravitational energy is $\frac{1}{2}kA^2$. The potential energy of the spring at maximum displacment is the total energy minus the gravitational potential energy, which is $-mgA$. The value of k was calculated to be 184 N/m in Example 14-6.

1. The total energy at maximum displacement is the potential energy given by Equation 14-21 with $y' = A = 0.05$ m:	$E = U_{max} = \frac{1}{2}kA^2$ $= \frac{1}{2}(184\text{ N/m})(0.05\text{ m})^2 = 0.23\text{ N}\cdot\text{m} = 0.23\text{ J}$
2. To find the potential energy of the spring, subtract the gravitational potential energy:	$U_s = U_{total} - U_g = 0.23\text{ J} - (-m_1 gA)$ $= 0.23\text{ J} + (3\text{ kg})(9.81\text{ N/kg})(0.05\text{ m})$ $= 0.23\text{ J} + 1.47\text{ J} = 1.70\text{ J}$

Check the Result When the spring is stretched by 16 cm, its potential energy is $\frac{1}{2}ky_0^2 = \frac{1}{2}(184\text{ N/m})(0.16\text{ m})^2 = 2.36\text{ J}$ *relative to its unstretched position*. When the mass is at the maximum displacement of 5 cm from equilibrium, the spring is stretched by 5 cm + 16 cm = 21 cm and the potential energy of the spring *relative to its unstretched position* is $\frac{1}{2}ky^2 = \frac{1}{2}(184\text{ N/m})(0.21\text{ m})^2 = 4.06\text{ J}$. The potential energy of the spring *relative to its equilibrium position* is then 4.06 J − 2.36 J = 1.70 J.

optional

Derivation of Potential Energy for a Vertical Spring At the equilibrium point $y = y_0$, the gravitational potential energy is $-mgy_0$ relative to the unstretched length of the spring ($y = 0$), and the potential energy of the mass–spring system is $\frac{1}{2}ky_0^2$. Let us now choose $U_s = U_g = 0$ at the *equilibrium* position $y' = y - y_0 = 0$. Then at a general point y, the potential energy of the spring is $\frac{1}{2}ky^2 - \frac{1}{2}ky_0^2$ and the gravitational potential energy is $-mgy + mgy_0 = -mgy'$. The total potential energy is then

$$\begin{aligned} U = U_s + U_g &= (\tfrac{1}{2}ky^2 - \tfrac{1}{2}ky_0^2) - mgy' \\ &= \tfrac{1}{2}k(y' + y_0)^2 - \tfrac{1}{2}ky_0^2 - mgy' \\ &= (\tfrac{1}{2}ky'^2 + ky'y_0 + \tfrac{1}{2}ky_0^2) - \tfrac{1}{2}ky_0^2 - mgy' \\ &= \tfrac{1}{2}ky'^2 + ky'y_0 - mgy' \end{aligned}$$

The terms $+ky'y_0$ and $-mgy'$ cancel because $ky_0 = mg$. The total potential energy is therefore

$$U = U_s + U_g = \tfrac{1}{2}ky'^2$$

which is Equation 14-21.

The Simple Pendulum

A simple pendulum consists of a string of length L and a bob of mass m. When the bob is released from an initial angle ϕ_0 with the vertical, it swings back and forth with some period T. We wish to find the period T.

Exercise in Dimensional Analysis We might expect the period of a simple pendulum to depend on the mass m of a pendulum bob, the length L of the pendulum, the acceleration g due to gravity, and the initial angle ϕ_0. Find a simple combination of these quantities that gives the correct dimensions for the period. (*Answer* $\sqrt{L/g}$. The units of length, mass, and g are m, kg, and m/s^2, respectively. The angle ϕ_0 is dimensionless. If we divide L by g, the meters cancel and we are left with seconds squared, suggesting the form $\sqrt{L/g}$. If the formula for period contained the mass, the unit kg must be canceled by some other quantity. But there is no combination of L and g that can cancel mass units. So the period cannot depend on the mass of the bob. Since the initial angle ϕ_0 is dimensionless, we cannot tell whether or not it is a factor in the period. We will see below that for small ϕ_0, the period is given by $T = 2\pi\sqrt{L/g}$.)

The forces on the bob are its weight $m\vec{g}$ and the string tension $\vec{T}$ (Figure 14-14). At an angle ϕ with the vertical, the weight has components $mg \cos \phi$ along the string and $mg \sin \phi$ tangential to the circular arc in the direction of decreasing ϕ. Let s be the arc length measured from the bottom of the circle. Then

$$s = L\phi \qquad \text{14-22}$$

where ϕ is in radians. The tangential component of Newton's second law gives

$$\sum F_t = -mg \sin \phi = m\frac{d^2s}{dt^2} = mL\frac{d^2\phi}{dt^2}$$

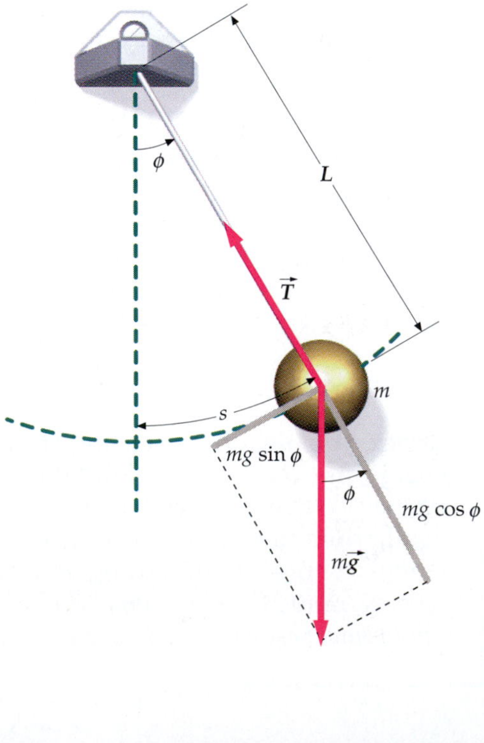

Figure 14-14
Forces on a pendulum bob.

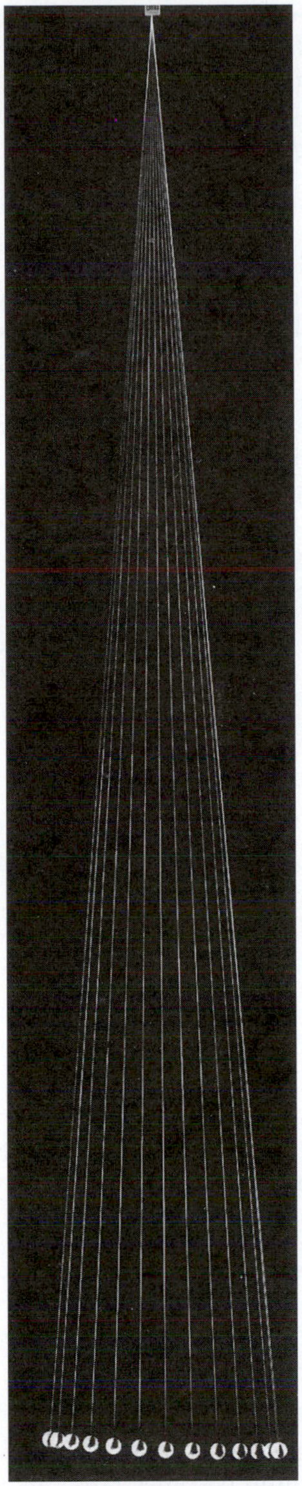

Motion of a simple pendulum. The bob is shown at equal time intervals. It moves faster at the bottom, as shown by the greater spacing of the images.

or

$$\frac{d^2\phi}{dt^2} = -\frac{g}{L}\sin\phi \qquad \text{14-23}$$

Note that the mass m does not appear in Equation 14-23—the motion of a pendulum does not depend on its mass. For small ϕ, $\sin\phi \approx \phi$, and

$$\frac{d^2\phi}{dt^2} = -\frac{g}{L}\phi \qquad \text{14-24}$$

Equation 14-24 is of the same form as Equation 14-2 for an object on a spring. The motion of a pendulum is thus approximately simple harmonic motion for small angular displacements.

Equation 14-24 can be written

$$\frac{d^2\phi}{dt^2} = -\omega^2\phi \qquad \text{14-25}$$

where

$$\omega^2 = \frac{g}{L} \qquad \text{14-26}$$

The period of the motion is thus

$$T = \frac{2\pi}{\omega} = 2\pi\sqrt{\frac{L}{g}} \qquad \text{14-27}$$

Period of a simple pendulum

The solution of Equation 14-25 is

$$\phi = \phi_0\cos(\omega t + \delta)$$

where ϕ_0 is the maximum angular displacement.

According to Equation 14-27, the greater the length of a pendulum, the greater the period, which is consistent with experimental observation. The period, and therefore the frequency, are independent of the amplitude of oscillation (as long as the amplitude is small), a general feature of simple harmonic motion. Galileo observed this by timing the period of a swinging lamp while in church. (He went back the next Sunday and found that the period also doesn't depend on the mass!)

Exercise Find the period of a pendulum of length 1 m. (*Answer* $T = 2\pi\sqrt{L/g} = 2\pi\sqrt{(1\text{ m})/(9.81\text{ m/s}^2)} = 2.01$ s. The validity of this result can be easily demonstrated by swinging a weight on a 1-m string and timing it.)

The acceleration due to gravity can be measured using a simple pendulum. You need only measure its length L and period T, and using Equation 14-27, solve for g.*

* When finding T, one usually measures the time for n oscillations and then divides by n, which helps minimize measurement error.

Pendulum in an Accelerated Reference Frame Figure 14-15*a* shows a simple pendulum suspended from the ceiling of a boxcar that has acceleration $\vec{a}_0$ to the right. Relative to a nonaccelerated inertial frame, the bob has a horizontal acceleration $\vec{a}_0$. The vertical and horizontal components of Newton's law for the bob are

$$\sum F_x = T\sin\theta = ma_0$$

$$\sum F_y = T\cos\theta - mg = 0$$

The equilibrium angle is thus given by $\tan\theta = a_0/g$.

Relative to the boxcar, the bob appears to be acted on by a horizontal pseudoforce $-ma_0$ to the left in addition to the downward force of gravity mg. This pseudoforce, like the real gravitational force, is proportional to the mass of the bob. Relative to the boxcar, all objects will fall at an angle θ to the vertical with acceleration $\vec{g}\,' = \vec{g} - \vec{a}_0$. We can use Newton's laws relative to the accelerating boxcar if we add a pseudoforce $-m\vec{a}_0$ that acts on each object of mass m. This is equivalent to replacing the acceleration due to gravity $\vec{g}$ by $\vec{g}\,' = \vec{g} - \vec{a}_0$ (Figure 14-15*b*).

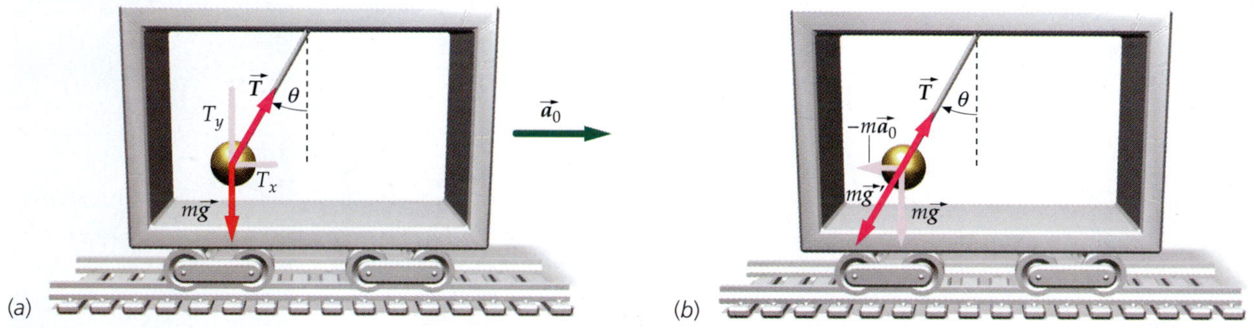

Figure 14-15 (*a*) Simple pendulum in apparent equilibrium in an accelerating boxcar. Forces are those as seen from a separate, stationary frame. (*b*) Forces on the bob as seen in the accelerated frame. Adding the pseudoforce $-ma_0$ is equivalent to replacing $\vec{g}$ by $\vec{g}\,'$.

If the bob is displaced slightly from equilibrium, it will oscillate with a period T given by Equation 14-27 with g replaced by g'.

Exercise A simple pendulum of length 1 m is in a boxcar that is accelerating horizontally with an acceleration a_0 = 3 m/s². Find g' and the period T. (*Answer* $g' = 10.3\ \text{m/s}^2$, $T = 1.96$ s)

Large-Amplitude Oscillations When the amplitude of a pendulum becomes large, its motion continues to be periodic, but it is no longer simple harmonic. A slight dependence on the amplitude must be accounted for when determining the period. For a general angular amplitude ϕ_0, the period can be shown to be

$$T = T_0\left[1 + \frac{1}{2^2}\sin^2\frac{1}{2}\phi_0 + \frac{1}{2^2}\left(\frac{3}{4}\right)^2\sin^4\frac{1}{2}\phi_0 + \cdots\right] \qquad 14\text{-}28$$

Period for large-amplitude oscillations

where $T_0 = 2\pi\sqrt{L/g}$ is the period for very small amplitudes. Figure 14-16 shows T/T_0 as a function of amplitude ϕ_0.

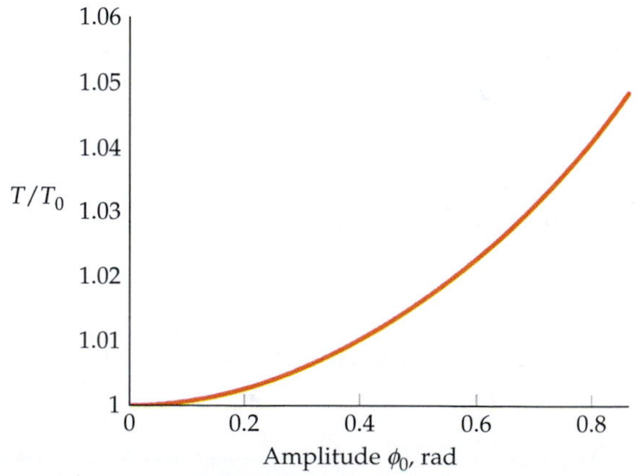

Figure 14-16 T/T_0 as a function of amplitude ϕ_0.

Example 14-9 ***try it yourself***

A simple pendulum clock is calibrated to keep accurate time at an angular amplitude of $\phi_0 = 10°$. When the amplitude has decreased to the point that it is very small, does the clock gain or lose time? How much time does the clock gain or lose in one day?

Cover the column to the right and try these on your own before looking at the answers.

Steps	***Answers***
1. Answer the first question by finding if the period increases or decreases.	T decreases as ϕ decreases so the clock gains time.
2. Use Equation 14-28 to find the percentage change $[(T - T_0)/T_0] \times 100\%$ for $\phi = 10°$. Use only the first correction term.	0.190%.
3. Find the number of minutes in a day.	There are 1440 minutes in a day.
4. Combine steps 2 and 3 to find the change in the number of minutes in a day.	The gain is 2.74 minutes per day.

The Physical Pendulum

A rigid object pivoted about a point other than its center of mass will oscillate when displaced from equilibrium. Such a system is called a **physical pendulum.** Consider a plane figure pivoted about a point a distance D from its center of mass and displaced from equilibrium by the angle ϕ (Figure 14-17). The torque about the pivot has a magnitude $MgD \sin \phi$ and tends to decrease ϕ. Newton's second law applied to rotation is

$$\tau = I\alpha = I\frac{d^2\phi}{dt^2}$$

where α is the angular acceleration, and I is the moment of inertia about the pivot point. Substituting $-MgD \sin \phi$ for the net torque, we have

$$-MgD \sin \phi = I\frac{d^2\phi}{dt^2}$$

or

$$\frac{d^2\phi}{dt^2} = -\frac{MgD}{I}\sin \phi \qquad \text{14-29}$$

Again, the motion is approximately simple harmonic if the angular displacements are small, so the approximation $\sin \phi \approx \phi$ holds. In this case, we have

$$\frac{d^2\phi}{dt^2} = -\frac{MgD}{I}\phi = -\omega^2\phi \qquad \text{14-30}$$

where $\omega^2 = MgD/I$. The period is therefore

$$T = \frac{2\pi}{\omega} = 2\pi\sqrt{\frac{I}{MgD}} \qquad \text{14-31}$$

Period of a physical pendulum

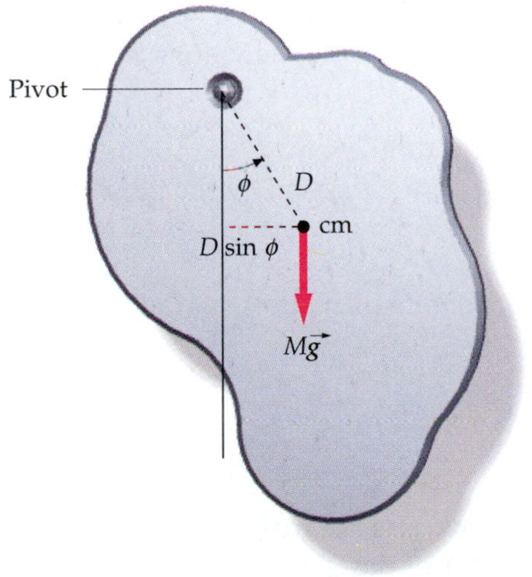

Figure 14-17 A physical pendulum.

optional

optional

For large amplitudes, the period is given by Equation 14-28, with T_0 given by Equation 14-31. For a simple pendulum of length L, the moment of inertia is $I = ML^2$ and $D = L$. Then Equation 14-31 gives $T = 2\pi\sqrt{ML^2/MgL} = 2\pi\sqrt{L/g}$, the same as Equation 14-27.

Example 14-10

A uniform stick of mass M and length L is pivoted at one end. (*a*) Find the period of oscillation for small angular displacements. (*b*) Find the period of oscillation if the stick is pivoted about point P a distance x from the center of mass.

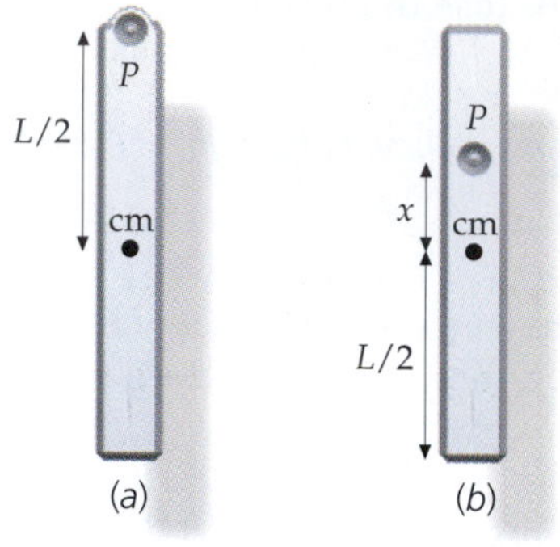

Figure 14-18

Picture the Problem (*a*) The period is given by $T = 2\pi\sqrt{I/MgD}$. The center of mass is at the center of the stick, so the distance from the center of mass to the pivot is $D = L/2$ (Figure 14-18*a*). The moment of inertia of a uniform stick about one end is $I = \frac{1}{3}ML^2$ (Table 9-1). (*b*) For a pivoting motion around point P, the distance D is given as x (Figure 14-18*b*), and the moment of inertia can be found from the parallel-axis theorem $I = I_{cm} + MD^2$, where $I_{cm} = \frac{1}{12}ML^2$.

(*a*)1. The period is given by Equation 14-31:
$$T = 2\pi\sqrt{\frac{I}{MgD}}$$

2. I about the end is known, and D is known:
$$I = \tfrac{1}{3}ML^2$$
$$D = \frac{L}{2}$$

3. Substitute the values for I and D to find T:
$$T = 2\pi\sqrt{\frac{\frac{1}{3}ML^2}{Mg\frac{1}{2}L}} = 2\pi\sqrt{\frac{2L}{3g}}$$

(*b*)1. About point P, $D = x$, and the moment of inertia is given by the parallel axis theorem:
$$D = x$$
$$I = I_{cm} + MD^2 = \tfrac{1}{12}ML^2 + Mx^2$$

2. Substitute these values to find T:
$$T = 2\pi\sqrt{\frac{I}{MgD}} = 2\pi\sqrt{\frac{\frac{1}{12}ML^2 + Mx^2}{Mgx}}$$
$$= 2\pi\sqrt{\frac{\frac{1}{12}L^2 + x^2}{gx}}$$

Check the Result When $x = 0$, the stick is pivoted about its center of mass and the period is infinite. When $x = L/2$, we get the same result as found in (*a*).

Exercise What is the period of oscillation for small angular displacements of a meter stick pivoted about one end? (*Answer* $T = 1.64$ s. Note that this is a smaller period than for a simple pendulum of length $L = 1$ m. The period of the simple pendulum is greater because its moment of inertia is mL^2 rather than $\frac{1}{3}mL^2$.)

Exercise Show that when $x = L/6$, the period is the same as when $x = L/2$.

Remark The period versus distance x from the center of mass for a stick of length 1 m is shown in Figure 14-19.

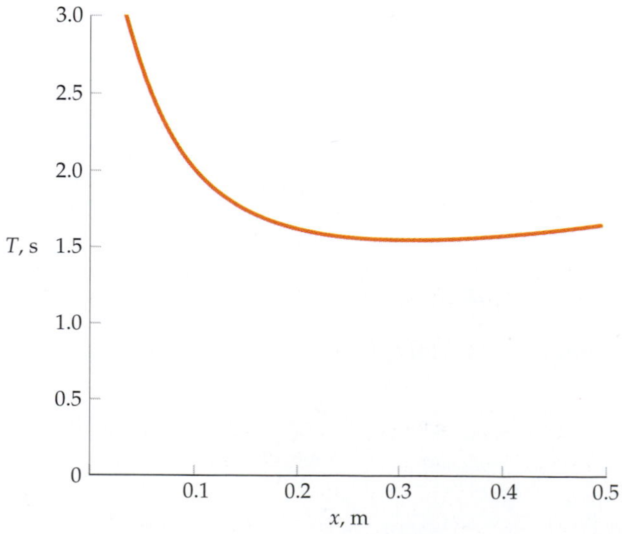

Figure 14-19

optional

Example 14-11 ***try it yourself***

Find the value of x in Example 14-10 such that the period is a minimum.

Picture the Problem At the value of x for which T is a minimum, $dT/dx = 0$. To simplify the calculation, let $Z = (L^2 + 12x^2)/x = x^{-1}(L^2 + 12x^2)$; then $T = (2\pi/\sqrt{12g})Z^{1/2}$ and $dT/dx = 0$ when $dZ/dx = 0$.

Cover the column to the right and try these on your own before looking at the answers.

Steps	***Answers***
1. Show that if $dZ/dx = 0$, then $dT/dx = 0$.	$\frac{dT}{dx} = \frac{2\pi}{\sqrt{12g}}\frac{1}{2}Z^{-1/2}\frac{dZ}{dx} = 0,$ only if $\frac{dZ}{dx} = 0$
2. Compute dZ/dx.	$\frac{dZ}{dx} = -x^{-2}(L^2 + 12x^2) + x^{-1}(24x)$ $= -\frac{L^2}{x^2} + 12$
3. Set $dZ/dx = 0$ and solve for x.	$x = \frac{L}{\sqrt{12}}$

14-4 Damped Oscillations

Left to itself, a spring or a pendulum eventually stops oscillating because the mechanical energy is dissipated by frictional forces. Such motion is said to be **damped.** If damping is small, the system oscillates with an amplitude that decreases slowly with time (Figure 14-20).

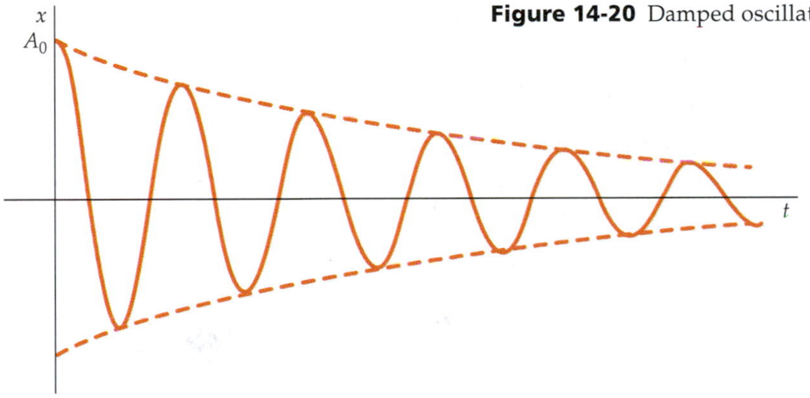

Figure 14-20 Damped oscillation.

Both the amplitude and the energy, which is proportional to the square of the amplitude, decrease by a constant percentage in a given time interval. This type of decrease is called exponential decrease. The force exerted on a damped oscillator such as the one shown in Figure 14-21 can be represented by the empirical expression

$$\vec{F}_{\rm d} = -b\vec{v}$$

where b is a constant. Since the damping force is opposite the direction of motion, it does negative work and causes the mechanical energy of the system to decrease. The motion of a damped system can be obtained from Newton's second law. For an object of mass m on a spring of force constant k the

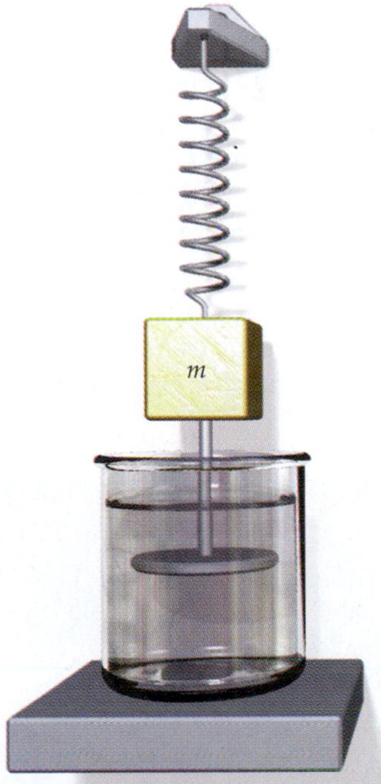

Figure 14-21 A damped oscillator. The motion is damped by the plunger immersed in the liquid.

net force is $-kx - b\,dx/dt$. Setting the net force equal to the mass times the acceleration d^2x/dt^2, we obtain

$$F_x = ma_x$$

$$-kx - b\frac{dx}{dt} = m\frac{d^2x}{dt^2} \qquad 14\text{-}32$$

Differential equation for a damped oscillator

The exact solution of this equation can be found using standard methods for solving differential equations.* The solution for the case of small damping is

$$x = A_0e^{-(b/2m)t}\cos(\omega' t + \delta) = A_0e^{-t/2\tau}\cos(\omega' t + \delta) \qquad 14\text{-}33$$

where A_0 is the maximum amplitude, and

$$\tau = \frac{m}{b} \qquad 14\text{-}34$$

is called the **decay time** or **time constant.** The frequency ω' is given by

$$\omega' = \sqrt{\omega_0^2 - (b/2m)^2} = \omega_0\sqrt{1 - (b/2m\omega_0)^2} \qquad 14\text{-}35$$

where ω_0 is the frequency with no damping ($\omega_0 = \sqrt{k/m}$ for a mass on a spring). For small damping, $b/2m\omega_0 \ll 1$ and ω' is nearly equal to ω_0. The dashed curves in Figure 14-20 correspond to $x = A$ and $x = -A$ where A is given by

$$A = A_0e^{-(b/2m)t} = A_0e^{-t/2\tau} \qquad 14\text{-}36$$

If the damping constant b is gradually increased, the angular frequency ω' decreases until it becomes zero at the critical value

$$b_c = 2m\omega_0 \qquad 14\text{-}37$$

When b is greater than or equal to b_c, the system does not oscillate. If $b = b_c$, the system is said to be **critically damped;** it returns to equilibrium with no oscillation in the shortest time possible. When b is greater than b_c, the system is **overdamped**. Figure 14-22 shows plots of the displacement versus time for a critically damped and an overdamped oscillator. We often use critical damping when we wish to have a system avoid oscillations and yet return to equilibrium quickly. One example is the use of shock absorbers to damp the oscillations of an automobile on its springs. You can test the damping of your shock absorbers by pushing down on the front or back of your car and releasing it. If the car returns to equilibrium with no oscillation, the system is critically damped or overdamped. Usually, you will note one or two oscillations, indicating that the damping is just under the critical value.

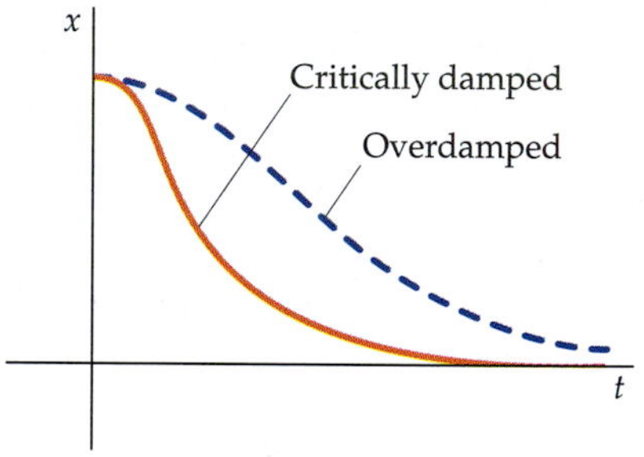

Figure 14-22 Plots of displacement versus time for a critically damped or overdamped oscillator.

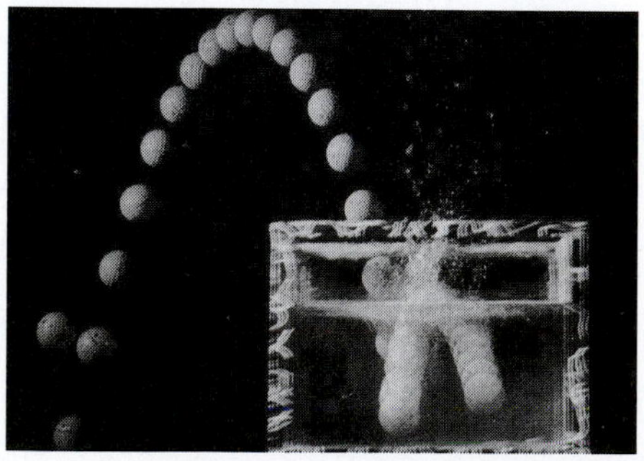

The ball's motion is damped—the energy of each bounce is less than that of the previous bounce because of the resistance of the water.

* A solution to this equation using complex numbers is given in the "Exploring . . ." section on page 428.

Because the energy of an oscillator is proportional to the square of the amplitude, the energy of an underdamped oscillator (averaged over a cycle) also decreases exponentially with time:

$$E = \tfrac{1}{2}m\omega^2A^2 = \tfrac{1}{2}m\omega^2(A_0\varepsilon^{-t/2\tau})^2 = E_0e^{-t/\tau} \qquad \text{14-38}$$

where $E_0 = \tfrac{1}{2}m\omega^2A_0^2$. We see that the decay time is the time for the energy to decrease to e^{-1} times its original value.

A damped oscillator is often described by its Q factor (for quality factor),

$$Q = \omega_0\tau = \frac{\omega_0 m}{b} \qquad \text{14-39}$$

Definition—Q factor

Shock absorbers (yellow cylinders) are used to damp the oscillations of this truck.

We can relate Q to the fractional energy loss per cycle. Differentiating Equation 14-38 gives

$$dE = -\frac{1}{\tau}E_0e^{-t/\tau}\,dt = -\frac{1}{\tau}E\,dt$$

If the energy loss per period is small, we can replace dE by ΔE, and dt by the period T. Then $|\Delta E|/E$ in one period is given by

$$\frac{|\Delta E|}{E} = \frac{T}{\tau} = \frac{2\pi}{\omega_0\tau} = \frac{2\pi}{Q} \qquad \text{14-40}$$

or

$$Q = \frac{2\pi}{(|\Delta E|/E)_{\text{cycle}}} \qquad \text{14-41}$$

Physical interpretation of Q for small damping

Q is thus inversely proportional to the fractional energy loss per cycle.

Example 14-12

When middle C on the piano (frequency 262 Hz) is struck, it loses half its energy after 4 s. (*a*) What is the decay time τ? (*b*) What is the Q factor for this piano wire? (c) What is the fractional energy loss per cycle?

Picture the Problem (*a*) We use $E = E_0e^{-t/\tau}$ and set E equal to $\tfrac{1}{2}E_0$. (*b*) The Q value can then be found from the decay time and the frequency.

(*a*)1. Set the energy at time $t = 4$ s equal to half the original energy:

$$E = E_0e^{-t/\tau} = E_0e^{-4\text{ s}/\tau} = \tfrac{1}{2}E_0$$
$$e^{4\text{ s}/\tau} = 2$$

2. Solve for the time t by taking the natural logarithm:

$$\frac{4\text{ s}}{\tau} = \ln 2$$
$$\tau = \frac{4\text{ s}}{\ln 2} = 5.77\text{ s}$$

(*b*) Calculate Q from τ and ω_0:

$$Q = \omega_0\tau = (2\pi f)\tau$$
$$= 2\pi(262\text{ Hz})(5.77\text{ s}) = 9.50 \times 10^3$$

(*c*) The fractional energy loss in a period is given by Equation 14-40:

$$\frac{|\Delta E|}{E} = \frac{T}{\tau} = \frac{1}{f\tau} = \frac{1}{(262\text{ Hz})(5.77\text{ s})} = 6.61 \times 10^{-4}$$

Check the Result Q can also be calculated from $Q = 2\pi/(\Delta E/E)_{\text{cycle}} = 2\pi/(6.61 \times 10^{-4}) = 9.50 \times 10^3$. Note that the fractional energy loss after 4 s is not just the number of cycles (4×262) times the fractional energy loss per cycle because the energy decrease is exponential, not constant.

Remarks Figure 14-23 shows amplitude versus time for the oscillation of a piano string after middle C is struck. After 4 s, the amplitude has decreased to about 0.7 times its initial value, and the energy, which is proportional to the amplitude squared, drops to about half its initial value.

Note that Q is quite large. You can estimate τ and Q of various oscillating systems. Tap a crystal wine glass and see how long it rings. The longer it rings, the greater the value of τ and Q, and the lower the damping. Glass beakers from your laboratory may also have a high Q. Now tap a plastic cup.

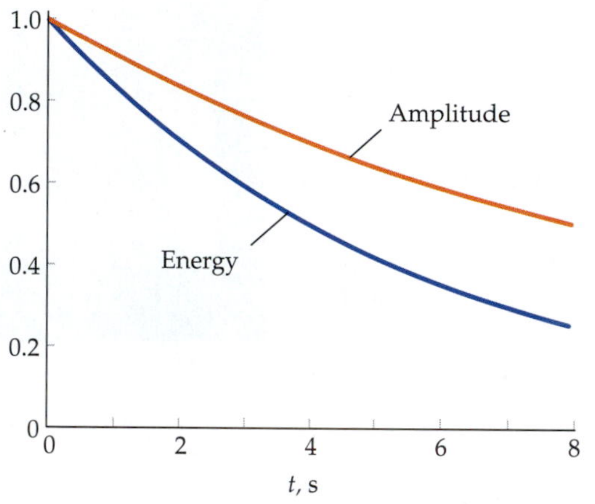

Figure 14-23

In terms of Q, the exact frequency of an underdamped oscillator is

$$\omega' = \omega_0\sqrt{1 - \left(\frac{b}{2m\omega_0}\right)^2} = \omega_0\sqrt{1 - \frac{1}{4Q^2}} \qquad \text{14-42}$$

Since Q is quite large for a slightly damped oscillator (Example 14-12), we see that ω' is nearly equal to ω_0.

We can understand the behavior of a damped oscillator qualitatively by considering its energy. The power dissipated by the damping force equals the instantaneous rate of change of the total mechanical energy

$$P = \frac{dE}{dt} = \vec{F}_d \cdot \vec{v} = -bv^2 \qquad \text{14-43}$$

For a slightly damped oscillator, the total mechanical energy decreases slowly with time. The average kinetic energy equals half the total energy

$$\left(\tfrac{1}{2}mv^2\right)_{\text{av}} = \tfrac{1}{2}E$$

If we substitute $(v^2)_{\text{av}} = E/m$ for v^2 in Equation 14-43, we have

$$\frac{dE}{dt} = -bv^2 \approx -b(v^2)_{\text{av}} = -\frac{b}{m}E \qquad \text{14-44}$$

Equation 14-44 may be solved by direct integration. Its solution is

$$E = E_0 e^{-(b/m)t} = E_0 e^{-t/\tau}$$

which is Equation 14-38.

14-5 Driven Oscillations and Resonance

To keep a damped system going, energy must be put into the system. When this is done, the oscillator is said to be driven or forced. When you keep a swing going by "pumping," that is, by moving your body and legs, you are driving an oscillator. If you put energy into the system faster than it is dissipated, the energy increases with time, and the amplitude increases. If you put energy in at the same rate it is being dissipated, the amplitude remains constant over time.

Figure 14-24 shows a system consisting of an object on a spring that is being driven by moving the point of support up and down with simple harmonic motion of frequency ω. At first the motion is complicated, but eventually a steady state is reached in which the system oscillates with the same frequency as that of the driver and with a constant amplitude and, therefore, a constant energy. In the steady state, the energy put into the system per cycle by the driving force equals the energy dissipated per cycle because of the damping.

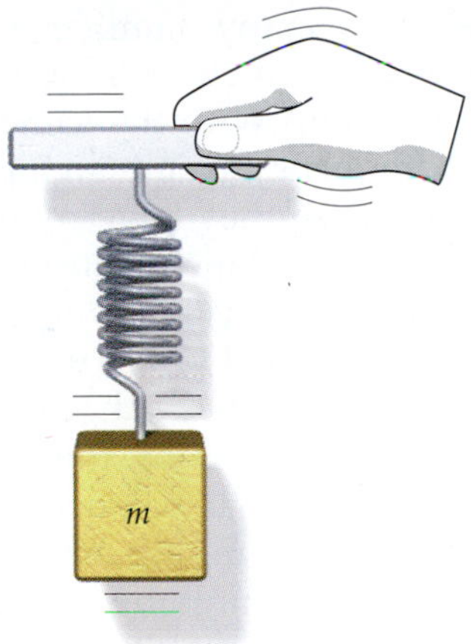

Figure 14-24 An object on a vertical spring can be driven by moving the support up and down.

The amplitude, and therefore the energy, of a system in the steady state depends not only on the amplitude of the driver, but also on its frequency. The **natural frequency** of an oscillator, ω_0, is the frequency when no driving or damping forces are present. (In the case of a spring, for example, $\omega_0 = \sqrt{k/m}$.) If the driving frequency is approximately equal to the natural frequency of the system, the system will oscillate with a very large amplitude. For example, if the support in Figure 14-24 oscillates with the natural frequency of the mass–spring system, the mass will oscillate with a much greater amplitude than that of the support. This phenomenon is called **resonance.** When the driving frequency equals the natural frequency of the oscillator, the energy absorbed by the oscillator is maximum. The natural frequency of the system is thus called the **resonance frequency** of the system.* The average rate at which energy is absorbed equals the average power delivered by the driving force. Figure 14-25 shows plots of the average power delivered to an oscillator as a function of the driving frequency for two different values of damping. These curves are called **resonance curves.** When the damping is small (large Q), the oscillator absorbs much more energy from the driving force at or near the resonance frequency than it does at other frequencies. The width of the peak of the resonance curve is correspondingly narrow, and we speak of the resonance as being sharp. When the damping is large, the resonance curve is broad. The width $\Delta\omega$ of each resonance curve is indicated in the figure. For relatively small damping, the ratio of the width of the resonance to the frequency can be shown to be equal to the reciprocal of the Q factor (see Problem 130):

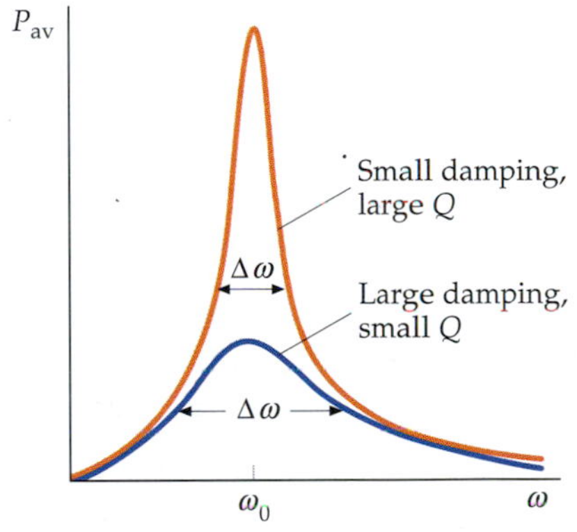

Figure 14-25 Resonance curves. Shown are plots of the average power delivered to an oscillator by a sinusoidal driving force versus the driving angular frequency ω for different values of damping. Resonance occurs when the (angular) frequency of the driving force equals the natural (angular) frequency of the system, ω_0. The resonance is sharp when the damping is small.

$$\frac{\Delta\omega}{\omega_0} = \frac{\Delta f}{f_0} = \frac{1}{Q} \qquad \text{14-45}$$

Resonance width for small damping

Thus, the Q factor is a direct measure of the sharpness of resonance.

You can do a simple experiment to demonstrate resonance. Hold a meter stick at one end so that it acts like a pendulum and then move your hand back and forth horizontally to drive it. Intuitively, you will move your hand back and forth with the natural frequency of the stick, and the amplitude of the oscillations of the stick will be much greater than the amplitude of oscillations of your hand. Now move your hand back and forth at much greater frequency and note the decrease in amplitude of the oscillating stick.

*Mathematically, the angular frequency ω is more convenient than the frequency $f = \omega/2\pi$. Since ω and f are proportional, most statements concerning angular frequency also hold for frequency. In verbal descriptions, we usually omit the word angular when the omission will not cause confusion.

There are many familiar examples of resonance. When you sit on a swing, you learn intuitively to pump with the same frequency as the natural frequency of the swing. Many machines vibrate because they have rotating parts that are not in perfect balance. (Observe a washing machine in the spin cycle for an example.) If such a machine is attached to a structure that can vibrate, the structure becomes a driven oscillatory system that is set in motion by the machine. Engineers pay great attention to balancing the rotary parts of such machines, damping their vibrations, and isolating them from building supports.

A glass with low damping can be broken by an intense sound wave at a frequency equal to or very nearly equal to the natural frequency of vibration of the glass. This is often done in physics demonstrations using an audio oscillator and an amplifier.

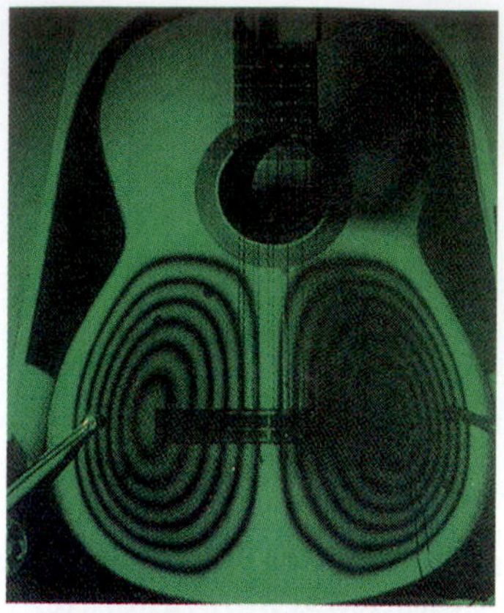

268 Hz ($Q = 52$)

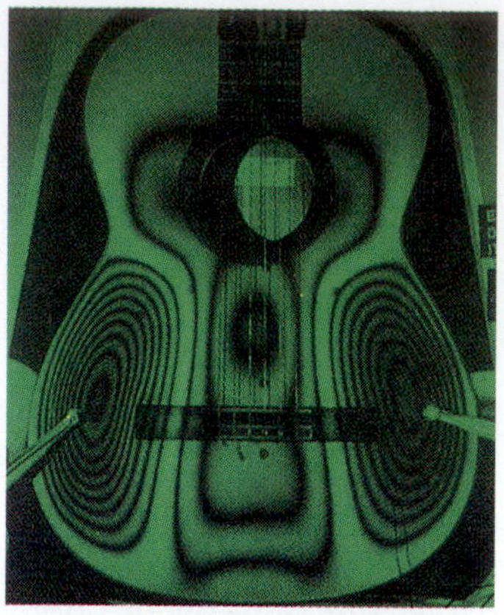

553 Hz ($Q = 66$)

672 Hz ($Q = 61$)

1010 Hz ($Q = 80$)

Extended objects have more than one resonance frequency. When plucked, a guitar string transmits its energy to the body of the guitar. The body's oscillations, coupled to those of the air mass it encloses, produce the resonance patterns shown.

Mathematical Treatment of Resonance

We can treat a driven oscillator mathematically by assuming that, in addition to the restoring force and a damping force, the oscillator is subject to an external driving force that varies harmonically with time:

$$F_{\text{ext}} = F_0 \cos \omega t \qquad \textbf{14-46}$$

where ω is the angular frequency of the driving force. This frequency is generally not related to the natural angular frequency of the system ω_0.

Newton's second law applied to an object of mass m attached to a spring of force constant k and subject to a damping force $-bv$ and an external force $F_0 \cos \omega t$ gives

$$\sum F = ma = m\frac{dv}{dt}$$

$$-kx - bv + F_0 \cos \omega t = m\frac{dv}{dt}$$

or

$$m\frac{d^2x}{dt^2} + b\frac{dx}{dt} + m\omega_0^2 x = F_0 \cos \omega t \qquad \textbf{14-47}$$

Differential equation for a driven oscillator

where we have used $k = m\omega_0^2$ and $dv/dt = d^2x/dt^2$.

We will discuss the general solution of Equation 14-47 qualitatively. It consists of two parts, the **transient solution** and the **steady-state solution.** The transient part of the solution is identical to that for a damped oscillator given in Equation 14-33. The constants in this part of the solution depend on the initial conditions. Over time, this part of the solution becomes negligible because of the exponential decrease of the amplitude. We are then left with the steady-state solution, which can be written as*

$$x = A\cos(\omega t - \delta) \qquad \text{14-48}$$

Position for a driven oscillator

where the angular frequency ω is the same as that of the driving force, and the amplitude A and phase constant δ are given by

$$A = \frac{F_0}{\sqrt{m^2(\omega_0^2 - \omega^2)^2 + b^2\omega^2}} \qquad \text{14-49}$$

Amplitude for a driven oscillator

and

$$\tan\delta = \frac{b\omega}{m(\omega_0^2 - \omega^2)} \qquad \text{14-50}$$

Phase constant for a driven oscillator

The steady-state solution does not depend on the initial conditions (Figure 14-26). Comparing Equations 14-46 and 14-48, we can see that the displacement and the driving force oscillate with the same frequency, but they differ in phase by δ. When the driving frequency ω is much less than the natural frequency ω_0, $\delta \approx 0$, as can be seen from Equation 14-50. At resonance, $\delta = \pi/2$. When ω is much greater than ω_0, $\delta \approx \pi$. In your simple experiment of driving a meter stick by moving your hand back and forth, you should be able to note that, at resonance, the oscillation of your hand is not in phase with the oscillation of the stick.

The velocity of the object in the steady state is obtained by differentiating x with respect to t:

$$v = \frac{dx}{dt} = -A\omega\sin(\omega t - \delta)$$

At resonance, $\delta = \pi/2$, and the velocity is in phase with the driving force:

$$v = -A\omega\sin\left(\omega t - \frac{\pi}{2}\right) = +A\omega\cos\omega t$$

Thus, at resonance, the object is always moving in the direction of the driving force, as would be expected for maximum power input. The speed is maximum at $\omega = \omega_0$.

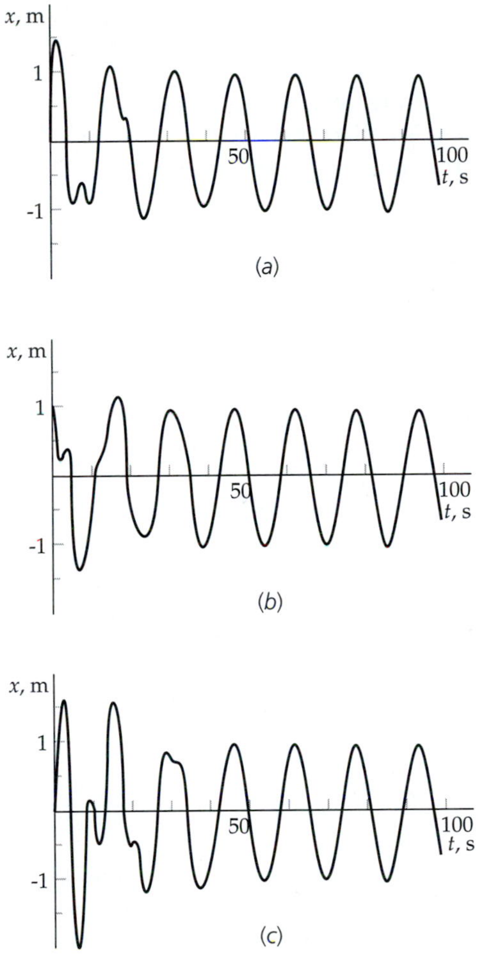

Figure 14-26 Three solutions of Equation 14-47 for the same driven oscillator under different initial conditions. For each solution, $\omega_0 = 1$ rad/s, $\omega = 0.4$ rad/s, and $Q = 8$. Note that the transient solutions are very different, but the steady-state solutions are all the same.

*The negative sign in the phase of Equation 14-48 is introduced so that the phase constant δ is positive.

Exploring

Using Complex Numbers to Solve the Oscillator Equations

Equations 14-32 and 14-47 for the damped and driven oscillator can be solved using complex numbers. A general complex number z can be written

$$z = \alpha + \beta i \qquad 1$$

where α and β are real numbers and $i = \sqrt{-1}$. The number α is called the real part of z, whereas β is called the imaginary part. An important relation for complex numbers is*

$$e^{i\phi} = \cos\phi + i\sin\phi \qquad 2$$

Using Equation 2 and referring to Figure 1, a general complex number can be written in polar form:

$$z = \alpha + \beta i = r\cos\phi + ir\sin\phi$$
$$= r(\cos\phi + i\sin\phi) = re^{i\phi} \qquad 3$$

where $r = |z| = \sqrt{\alpha^2 + \beta^2}$, and $\tan\phi = \beta/\alpha$.

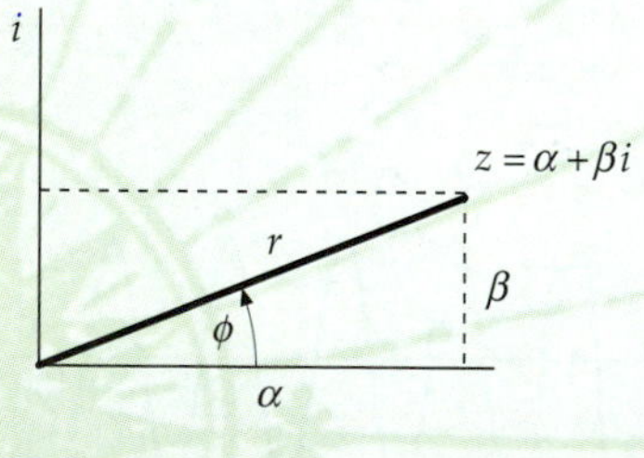

Figure 1 A general complex number $z = \alpha + \beta i$ is represented as a point in the complex plane in which the real part α is plotted along the horizontal axis and the imaginary part β is plotted along the vertical axis. The magnitude $|z| = r$ and the angle ϕ are then the polar coordinates of the point (α, β).

Damped Oscillator

Using $\omega_0^2 = k/m$, and replacing x with z, Equation 14-32 for the damped oscillator can be written

$$\frac{d^2z}{dt^2} + \frac{b}{m}\frac{dz}{dt} + \omega_0^2 z = 0 \qquad 4$$

Consider the complex function $z = x + yi$. Equation 4 is linear, so if the complex function z satisfies the equation, then the real part x (and imaginary part y) must also satisfy the equation. We try the complex function

$$z = Be^{i\omega t} \qquad 5$$

Differentiating with respect to t we obtain

$$\frac{dz}{dt} = i\omega Be^{i\omega t} = i\omega z \qquad 6$$

$$\frac{d^2z}{dt^2} = \frac{d}{dt}\frac{dz}{dt} = (i\omega)^2 Be^{i\omega t} = -\omega^2 z$$

Substituting these results into Equation 4 gives

$$-\omega^2 z + \frac{ib\omega}{m} z + \omega_0^2 z = 0$$

Canceling the common factor z and rearranging, we obtain

$$\omega^2 - \frac{ib}{m}\omega - \omega_0^2 = 0 \qquad 7$$

Equation 7 can be solved using the quadratic formula:

$$\omega = \frac{b}{2m}i \pm \frac{1}{2}\sqrt{\left(\frac{ib}{m}\right)^2 + 4\omega_0^2}$$
$$= \frac{b}{2m}i \pm \omega_0\sqrt{1 - \left(\frac{b}{2m\omega_0}\right)^2} = \frac{b}{2m}i \pm \omega' \qquad 8$$

where

$$\omega' = \omega_0\sqrt{1 - \left(\frac{b}{2m\omega_0}\right)^2}$$

Now substitute this value of ω into Equation 5:

$$z = Be^{i\omega t} = Be^{i(ib/2m \pm \omega')t} = Be^{-(b/2m)t}e^{\pm i\omega' t} \qquad 9$$

The real part of z is

$$x = \text{Re}(z) = Be^{-(b/2m)t}\cos\omega' t$$

which is the same as Equation 14-33 except for the phase constant δ, which is arbitrary.†

---

* See Appendix D for a brief discussion of complex numbers.

† We could have easily included the arbitrary phase constant by writing B as $B = Ae^{i\delta}$ where A is a real number. Then $z = Be^{i\omega t} = Ae^{i\delta}e^{i\omega t} = Ae^{i\omega t + \delta}$.

Driven Oscillator

To find the steady-state solution for the driven oscillator, we note that the right side of Equation 14-47 is the real part of $F_0e^{i\omega t}$. We then find the complex function z that satisfies

$$m\frac{d^2z}{dt^2} + b\frac{dz}{dt} + m\omega_0^2 z = F_0 e^{i\omega t} \qquad 10$$

Then the real part of z must satisfy Equation 14-47. From physical considerations we expect that in the steady state, x will oscillate with frequency ω, so we again try

$$z = Be^{i\omega t} \qquad 11$$

Here, the frequency ω is given, and we wish to determine B. Computing the derivatives as before and substituting them into Equation 10 gives

$$-m\omega^2 z + ib\omega z + m\omega_0^2 z = F_0 e^{i\omega t} = \frac{F_0}{B}z \qquad 12$$

Dividing out the common term z and solving for B, we obtain

$$B = \frac{F_0}{m(\omega_0^2 - \omega^2) + ib\omega} \qquad 13$$

Equation 13 is most easily solved by putting the denominator in polar form.

$$\alpha + \beta i = \sqrt{\alpha^2 + \beta^2}\, e^{i\delta}$$

where $\tan\delta = \beta/\alpha$. Then

$$m(\omega_0^2 - \omega^2) + ib\omega = \sqrt{m^2(\omega_0^2 - \omega^2)^2 + b^2\omega^2}\, e^{i\delta}$$

where

$$\tan\delta = \frac{b\omega}{m(\omega_0^2 - \omega^2)} \qquad 14$$

We can then write B as

$$B = \frac{F_0}{\sqrt{m^2(\omega_0^2 - \omega^2)^2 + b^2\omega^2}\, e^{i\delta}} = \frac{F_0 e^{-i\delta}}{\sqrt{m^2(\omega_0^2 - \omega^2)^2 + b^2\omega^2}}$$

The complex solution z is then

$$z = Be^{i\omega t} = \frac{F_0 e^{-i\delta}e^{i\omega t}}{\sqrt{m^2(\omega_0^2 - \omega^2)^2 + b^2\omega^2}} = \frac{F_0 e^{i(\omega t - \delta)}}{\sqrt{m^2(\omega_0^2 - \omega^2)^2 + b^2\omega^2}}$$

or

$$z = Ae^{i(\omega t - \delta)} \qquad 15$$

with

$$A = \frac{F_0}{\sqrt{m^2(\omega_0^2 - \omega^2)^2 + b^2\omega^2}} \qquad 16$$

The real part of Equation 15 is then

$$x = A\cos(\omega t - \delta)$$

with A given by Equation 16 and δ given by Equation 14.

Example 14-13 *try it yourself*

An object of mass 1.5 kg on a spring of force constant 600 N/m loses 3% of its energy in each cycle. The system is driven by a sinusoidal force with a maximum value of $F_0 = 0.5$ N. (*a*) What is Q for this system? (*b*) What is the resonance (angular) frequency? (*c*) If the driving frequency is varied, what is the width $\Delta\omega$ of the resonance? (*d*) What is the amplitude at resonance? (*e*) What is the amplitude if the driving frequency is $\omega = 19$ rad/s?

Picture the Problem We can find Q from $Q = 2\pi/(\Delta E/E)_{\text{cycle}}$ and then use this result to find the width of the resonance $\Delta\omega = \omega_0/Q$. The resonance frequency is the natural frequency. The amplitude can be found from Equation 14-49 both at resonance and off resonance, with the damping constant calculated from Q using Equation 14-41 in the form $b = \omega_0 m/Q$.

Cover the column to the right and try these on your own before looking at the answers.

Steps	***Answers***
(*a*)1. Relate Q to the fractional energy loss.	$Q = 2\pi\dfrac{E}{\lvert\Delta E\rvert}$
2. Substitute $\Delta E/E = 3\%$ and calculate Q.	$Q = 2\pi\dfrac{100}{3} = 209$
(*b*) Relate the resonance frequency to the natural frequency of the system.	$w_0 = \sqrt{k/m} = 20\ \text{rad/s}$
(*c*) Relate the width of the resonance $\Delta\omega$ to Q.	$\Delta w = w_0/Q = 0.0957\ \text{rad/s}$
(*d*)1. Write an expression for the amplitude A for any driving frequency ω.	$A = \dfrac{F_0}{\sqrt{m^2(\omega_0^2 - \omega^2)^2 + b^2\omega^2}}$
2. Substitute $\omega = \omega_0$ to calculate A at resonance.	$A = \dfrac{F_0}{b\omega_0}$
3. Use Equation 14-39 to relate the damping constant b to Q.	$b = \dfrac{m\omega_0}{Q} = \dfrac{(1.5\ \text{kg})(20\ \text{Hz})}{209} = 0.144\ \text{kg/s}$
4. Use the results of the previous two steps to calculate the amplitude at resonance.	$A = \dfrac{F_0}{b\omega_0} = \dfrac{0.5\ \text{N}}{(0.144\ \text{kg/s})(20\ \text{Hz})} = 0.174\ \text{m} = 17.4\ \text{cm}$
(*e*) Calculate the amplitude for $\omega = 19$ Hz. (We can omit the units to simplify the equation. Since all quantities are in SI units, A will be in meters.)	$A = \dfrac{F_0}{\sqrt{m^2(\omega_0^2 - \omega^2)^2 + b^2\omega^2}}$ $= 8.54 \times 10^{-3}\ \text{m} = 0.854\ \text{cm}$

Remarks At just 1 rad/s off resonance, the amplitude drops by a factor of 20. This is not surprising, because the width of the resonance is only 0.0957 rad/s. Note that off resonance the term $b^2\omega^2$ is negligible compared with the other term in the denominator of the expression for A. When $\omega - \omega_0$ is more than several times the half width $\Delta\omega$, we can neglect that term and calculate A from $A \approx F_0/m(\omega_0^2 - \omega^2)$. Figure 14-27 shows the amplitude versus driving frequency ω.

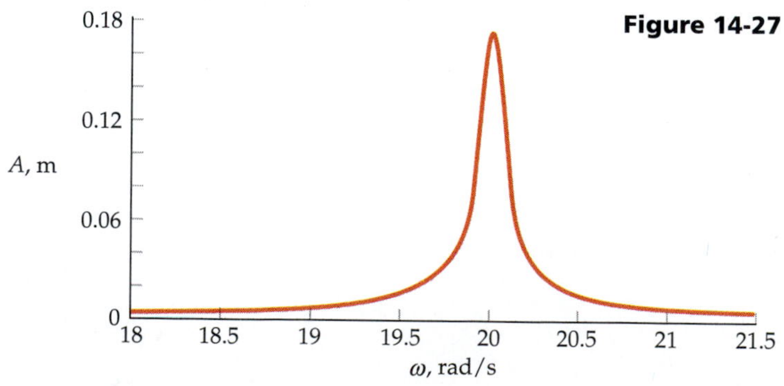

Figure 14-27

Summary

1. Simple harmonic motion occurs whenever the restoring force is proportional to the displacement from equilibrium. It has wide application in the study of oscillations, waves, and electrical circuits.
2. Resonance is an important phenomenon in many areas of physics. It occurs when the frequency of the driving force equals the natural frequency of the oscillating system.

Topic	Remarks and Relevant Equations	
1. Simple Harmonic Motion	In simple harmonic motion, the net force and acceleration are proportional to the displacement and oppositely directed.	
Position function	$x = A\cos(\omega t + \delta)$	**14-4**
Velocity	$v = -\omega A\sin(\omega t + \delta)$	**14-5**
Acceleration	$a_x = -\omega^2 x = -\omega^2 A\cos(\omega t + \delta)$	**14-6, 14-7**
Angular frequency	$\omega = 2\pi f = \frac{2\pi}{T}$	**14-11**
Total energy	$E_{\text{total}} = K + U = \frac{1}{2}kA^2$	**14-16**
Average kinetic or potential energy	$K_{\text{av}} = U_{\text{av}} = \frac{1}{2}E_{\text{total}}$	**14-17**
Circular motion	When a particle moves in a circle with constant speed, the x and y components of its position vary with simple harmonic motion.	
General motion near equilibrium	When an object is given a small displacement from any position of stable equilibrium, it oscillates about this position with simple harmonic motion	
2. Period for Various Systems		
Mass on a spring	$T = 2\pi\sqrt{\frac{m}{k}}$	**14-12**
Simple pendulum	$T = 2\pi\sqrt{\frac{L}{g}}$	**14-27**
Physical pendulum (optional)	$T = 2\pi\sqrt{\frac{I}{MgD}}$ where D is the distance of the center of mass to the pivot point and I is the moment of inertia about the pivot point.	**14-31**
3. Damped Oscillations	In the oscillations of real systems, the motion is damped because of dissipative forces. If the damping is greater than some critical value, the system does not oscillate when disturbed but merely returns to its equilibrium position. The motion of a slightly damped system is nearly simple harmonic with an amplitude that decreases exponentially with time.	
Frequency	$\omega' = \omega_0\sqrt{1 - \left(\frac{b}{2m\omega_0}\right)^2} = \omega_0\sqrt{1 - \frac{1}{4Q^2}}$	**14-35, 14-42**
Amplitude	$A = A_0 e^{-(b/2m)t} = A_0 e^{-t/2\tau}$	**14-36**

	Energy	$E = E_0 e^{-(b/m)t} = E_0 e^{-t/\tau}$	14-38
	Decay time	$\tau = \dfrac{m}{b}$	14-34
	Q Factor	$Q = \omega_0 \tau = \dfrac{\omega_0 m}{b} = \dfrac{2\pi}{(\lvert\Delta E\rvert/E)_{\text{cycle}}}$	14-39, 14-41
4.	**Driven Oscillations**	When a slightly damped system is driven by an external sinusoidal force $F_{\text{ext}} = F_0 \cos \omega t$, the system oscillates with a frequency ω equal to the driving frequency and an amplitude A that depends on the driving frequency.	
	Resonance frequency	$\omega = \omega_0$	
	Resonance width	$\dfrac{\Delta\omega}{\omega_0} = \dfrac{\Delta f}{f_0} = \dfrac{1}{Q}$	14-45
	Position function (optional)	$x = A \cos(\omega t - \delta)$	14-48
	Amplitude (optional)	$A = \dfrac{F_0}{\sqrt{m^2(\omega_0^2 - \omega^2)^2 + b^2\omega^2}}$	14-49
	Phase constant (optional)	$\tan \delta = \dfrac{b\omega}{m(\omega_0^2 - \omega^2)}$	14-50

Problem-Solving Guide

Begin by drawing a neat diagram that includes the important features of the problem.

Summary of Worked Examples

Type of Calculation	Procedure and Relevant Examples	
1. Simple Harmonic Motion		
Find amplitude, phase constant, angular frequency, frequency, and period of simple harmonic motion given the equation for *x*.	Compare the equation with the standard form $x = A \cos(\omega t + \delta)$.	**Example 14-1**
Find the velocity and acceleration given the position function *x*.	Compute the derivatives $v = dx/dt$, and $a = dv/dt = d^2x/dt^2$.	**Example 14-1**
Find the phase constant and amplitude given the initial position and velocity.	Use $x_0 = A \cos \delta$ and $v_0 = -\omega A \sin \delta$.	**Example 14-2**
Find the maximum speed and the time that it occurs given *x*(*t*).	Use $v = -\omega A \sin \omega t$ and solve for t such that $\sin \omega t = \pm 1$.	**Example 14-4**
Find *x* for some given *v*.	Use conservation of energy.	**Example 14-5**
2. Object on a Vertical Spring		
Find the frequency of oscillation given the equilibrium stretching of the spring when an object hangs from it.	Calculate $\omega = \sqrt{k/m}$ from $k\,\Delta y = mg$.	**Example 14-6**

3. Pendulums		
Find the time lost or gained by a pendulum clock.	If the change is due to an amplitude change, find the percentage change in the period using Equation 14-28.	Example 14-9
Find the period of oscillation of a rigid body about some point. (optional)	Use $T = \frac{2\pi}{\omega} = 2\pi\sqrt{\frac{I}{MgD}}$.	Example 14-10
4. Damped and Driven Oscillators		
Find the Q value for an oscillator that loses some fraction of its energy per cycle.	Use $Q = 2\pi E/\lvert\Delta E\rvert$.	Examples 14-12, 14-13
Find the amplitude of a driven oscillator. (optional)	Use Equation 14-49.	Example 14-13

Problems

Conceptual Problems

Problems from Optional and Exploring sections

In a few problems, you are given more data than you actually need; in a few other problems, you are required to supply data from your general knowledge, outside sources, or informed estimates.

• Single-concept, single-step, relatively easy
•• Intermediate-level, may require synthesis of concepts
••• Challenging, for advanced students

Simple Harmonic Motion

1 • Deezo the Clown slept in again. As he roller-skates out the door at breakneck speed on his way to a lunchtime birthday party, his superelastic suspenders catch on a fence post, and he flies back and forth, oscillating with an amplitude A. What distance does he move in one period? What is his displacement over one period?

2 • A neighbor takes a picture of the oscillating Deezo (from Problem 1) at a moment when his speed is zero. What is his displacement from the fence post at that time?

3 • What is the magnitude of the acceleration of an oscillator of amplitude A and frequency f when its speed is maximum? When its displacement is maximum?

4 • Can the acceleration and the displacement of a simple harmonic oscillator ever be in the same direction? The acceleration and the velocity? The velocity and the displacement? Explain.

5 • True or false:

(*a*) In simple harmonic motion, the period is proportional to the square of the amplitude.
(*b*) In simple harmonic motion, the frequency does not depend on the amplitude.
(*c*) If the acceleration of a particle is proportional to the displacement and oppositely directed, the motion is simple harmonic.

6 • The position of a particle is given by $x = (7\text{ cm}) \times \cos 6\pi t$, where t is in seconds. What is (*a*) the frequency, (*b*) the period, and (*c*) the amplitude of the particle's motion? (*d*) What is the first time after $t = 0$ that the particle is at its equilibrium position? In what direction is it moving at that time?

7 • (*a*) What is the maximum speed of the particle in Problem 6? (*b*) What is its maximum acceleration?

8 • What is the phase constant δ in Equation 14-4 if the position of the oscillating particle at time $t = 0$ is (*a*) 0, (*b*) $-A$, (*c*) A, (*d*) $A/2$?

9 • A particle of mass m begins at rest from $x = +25$ cm and oscillates about its equilibrium position at $x = 0$ with a period of 1.5 s. Write equations for (*a*) the position x versus the time t, (*b*) the velocity v versus t, and (*c*) the acceleration a versus t.

10 • Find (*a*) the maximum speed, and (*b*) the maximum acceleration of the particle in Problem 6. (*c*) What is the first time that the particle is at $x = 0$ and moving to the right?

11 •• Work Problem 9 with the particle initially at $x =$ 25 cm and moving with velocity $v_0 = +50$ cm/s.

12 •• The period of an oscillating particle is 8 s, and its amplitude is 12 cm. At $t = 0$, it is at its equilibrium position. Find the distance traveled during the interval (*a*) $t = 0$ to $t = 2$ s, (*b*) $t = 2$ s to $t = 4$ s, (*c*) $t = 0$ to $t = 1$ s, and (*d*) $t = 1$ s to $t = 2$ s.

13 •• The period of an oscillating particle is 8 s. At $t = 0$, the particle is at rest at $x = A = 10$ cm. (*a*) Sketch x as a function of t. (*b*) Find the distance traveled in the first second, the next second, the third second, and the fourth second after $t = 0$.

14 •• Military specifications often call for electronic devices to be able to withstand accelerations of $10g = 98.1\text{ m/s}^2$. To make sure that their products meet this specification, manufacturers test them using a shaking table that can vibrate a device at various specified frequencies and amplitudes. If a device is given a vibration of amplitude 1.5 cm, what should its frequency be in order to test for compliance with the 10g military specification?

15 •• The position of a particle is given by $x = 2.5 \cos \pi t$, where x is in meters and t is in seconds. (*a*) Find the maximum speed and maximum acceleration of the particle. (*b*) Find the speed and acceleration of the particle when $x = 1.5$ m.

16 •• The bow of a destroyer undergoes a simple harmonic vertical pitching motion with a period of 8.0 s and an amplitude of 2.0 m. (*a*) What is the maximum vertical velocity of the destroyer's bow? (*b*) What is its maximum acceleration? (*c*) An 80-kg sailor is standing on a scale in the bunkroom in the bow. What are the maximum and minimum readings on the scale in newtons?

Simple Harmonic Motion and Circular Motion

17 • A particle moves in a circle of radius 40 cm with a constant speed of 80 cm/s. Find (*a*) the frequency of the motion, and (*b*) the period of the motion. (*c*) Write an equation for the x component of the position of the particle as a function of time t, assuming that the particle is on the positive x axis at time $t = 0$.

18 • A particle moves in a circle of radius 15 cm, making 1 revolution every 3 s. (*a*) What is the speed of the particle? (*b*) What is its angular velocity ω? (*c*) Write an equation for the x component of the position of the particle as a function of time t, assuming that the particle is on the positive x axis at time $t = 0$.

Energy in Simple Harmonic Motion

19 • If the amplitude of a simple harmonic oscillator is tripled, by what factor is the energy changed?

20 •• An object attached to a spring has simple harmonic motion with an amplitude of 4.0 cm. When the object is 2.0 cm from the equilibrium position, what fraction of its total energy is potential energy?

(*a*) one-quarter (*b*) one-third (*c*) one-half
(*d*) two-thirds (*e*) three-quarters

21 • A 2.4-kg object is attached to a horizontal spring of force constant $k = 4.5$ kN/m. The spring is stretched 10 cm from equilibrium and released. Find its total energy.

22 • Find the total energy of a 3-kg object oscillating on a horizontal spring with an amplitude of 10 cm and a frequency of 2.4 Hz.

23 • A 1.5-kg object oscillates with simple harmonic motion on a spring of force constant $k = 500$ N/m. Its maximum speed is 70 cm/s. (*a*) What is the total energy? (*b*) What is the amplitude of the oscillation?

24 • A 3-kg object oscillating on a spring of force constant 2 kN/m has a total energy of 0.9 J. (*a*) What is the amplitude of the motion? (*b*) What is the maximum speed?

25 • An object oscillates on a spring with an amplitude of 4.5 cm. Its total energy is 1.4 J. What is the force constant of the spring?

26 •• A 3-kg object oscillates on a spring with an amplitude of 8 cm. Its maximum acceleration is 3.50 m/s^2. Find the total energy.

Springs

27 • True or false:

(*a*) For a given object on a given spring, the period is the same if the spring is vertical or horizontal.
(*b*) For a given object oscillating with amplitude A on a given spring, the maximum speed is the same if the spring is vertical or horizontal.

28 • Herb plans to ring in the new year by playing trombone while oscillating up and down on a spring that hangs down from a building at Times Square in New York City. He intends to oscillate with a period of one second to synchronize with the crowd as it counts down to midnight. If he uses a spring with a spring constant of 3000 N/m, Herb must be sure that the total of his vibrating mass adds up to

(*a*) 3000 kg. (*b*) $\sqrt{3000}$ kg. (*c*) $4\pi^2(3000)$ kg.
(*d*) $3000/4\pi^2$ kg. (*e*) none of the above.

29 • A 2.4-kg object is attached to a horizontal spring of force constant $k = 4.5$ kN/m. The spring is stretched 10 cm from equilibrium and released. Find (*a*) the frequency of the motion, (*b*) the period, (*c*) the amplitude, (*d*) the maximum speed, and (*e*) the maximum acceleration. (*f*) When does the object first reach its equilibrium position? What is its acceleration at this time?

30 • Answer the questions in Problem 29 for a 5-kg object attached to a spring of force constant $k = 700$ N/m when the spring is initially stretched 8 cm from equilibrium.

31 • A 3-kg object attached to a horizontal spring oscillates with an amplitude $A = 10$ cm and a frequency $f = 2.4$ Hz. (*a*) What is the force constant of the spring? (*b*) What is the period of the motion? (*c*) What is the maximum speed of the object? (*d*) What is the maximum acceleration of the object?

32 • An 85-kg person steps into a car of mass 2400 kg, causing it to sink 2.35 cm on its springs. Assuming no damping, with what frequency will the car and passenger vibrate on the springs?

33 • A 4.5-kg object oscillates on a horizontal spring with an amplitude of 3.8 cm. Its maximum acceleration is 26 m/s^2. Find (*a*) the force constant k, (*b*) the frequency, and (*c*) the period of the motion.

34 • An object oscillates with an amplitude of 5.8 cm on a horizontal spring of force constant 1.8 kN/m. Its maximum speed is 2.20 m/s. Find (*a*) the mass of the object, (*b*) the frequency of the motion, and (*c*) the period of the motion.

35 •• A 0.4-kg block attached to a spring of force constant 12 N/m oscillates with an amplitude of 8 cm. Find (*a*) the maximum speed of the block, (*b*) the speed and acceleration of the block when it is at $x = 4$ cm from the equilibrium position, and (*c*) the time it takes the block to move from $x = 0$ to $x = 4$ cm.

36 •• An object of mass m is supported by a vertical spring of force constant 1800 N/m. When pulled down 2.5 cm from equilibrium and released from rest, the object oscillates at 5.5 Hz. (*a*) Find m. (*b*) Find the amount the spring is stretched from its natural length when the object is in equilibrium. (*c*) Write expressions for the displacement x, the velocity v, and the acceleration a as functions of time t.

37 •• An object of unknown mass is hung on the end of an unstretched spring and is released from rest. If the object falls 3.42 cm before first coming to rest, find the period of the motion.

38 •• A spring of force constant $k = 250\ \mathrm{N/m}$ is suspended from a rigid support. An object of mass 1 kg is attached to the unstretched spring and the object is released from rest. (*a*) How far below the starting point is the equilibrium position for the object? (*b*) How far down does the object move before it starts up again? (*c*) What is the period of oscillation? (*d*) What is the speed of the object when it first reaches its equilibrium position? (*e*) When does it first reach its equilibrium position?

39 •• The St. Louis Arch has a height of 192 m. Suppose a stunt woman of mass 60 kg jumps off the top of the arch with an elastic band attached to her feet. She reaches the ground at zero speed. Find her kinetic energy K after 2.00 s of the flight. (Assume that the elastic band obeys Hooke's law, and neglect its length when relaxed.)

40 •• A 0.12-kg block is suspended from a spring. When a small stone of mass 30 g is placed on the block, the spring stretches an additional 5 cm. With the stone on the block, the spring oscillates with an amplitude of 12 cm. (*a*) What is the frequency of the motion? (*b*) How long does the block take to travel from its lowest point to its highest point? (*c*) What is the net force of the stone when it is at a point of maximum upward displacement?

41 •• In Problem 40, find the maximum amplitude of oscillation such that the stone will remain on the block.

42 •• An object of mass 2.0 kg is attached to the top of a vertical spring that is anchored to the floor. The uncompressed length of the spring is 8.0 cm, and the equilibrium position of the object on the spring is 5.0 cm from the floor. When the object is resting at its equilibrium position, it is given a downward impulse with a hammer such that its initial speed is 0.3 m/s. (*a*) To what maximum height above the floor does the object eventually rise? (*b*) How long does it take for the object to reach its maximum height the first time? (*c*) Does the spring ever become uncompressed? What minimum initial velocity must be given to the object for the spring to be uncompressed at some time?

43 •• Lou has devised a new kiddie ride and is testing it for safety. A child is placed on a large block that is attached to a horizontal spring. When pulled back and released, the child and block oscillate with a period of 2 s. (*a*) If the coefficient of static friction between the child and the block is 0.25, will an amplitude of 1 m cause the child to slip? (*b*) What is the maximum amplitude that will avoid slipping?

Energy of an Object on a Vertical Spring

44 •• A 2.5-kg object hanging from a vertical spring of force constant 600 N/m oscillates with an amplitude of 3 cm. When the object is at its maximum downward displacement, find (*a*) the total energy of the system, (*b*) the gravitational potential energy, and (*c*) the potential energy in the spring. (*d*) What is the maximum kinetic energy of the object? Choose $U = 0$ when the object is in equilibrium.

45 •• A 1.5-kg object that stretches a spring 2.8 cm from its natural length when hanging at rest oscillates with an amplitude of 2.2 cm. (*a*) Find the total energy of the system. (*b*) Find the gravitational potential energy at maximum downward displacement. (*c*) Find the potential energy in the spring at maximum downward displacement. (*d*) What is the maximum kinetic energy of the object? (Choose $U = 0$ when the object is in equilibrium.)

46 •• A 1.2-kg object hanging from a spring of force constant 300 N/m oscillates with a maximum speed of 30 cm/s. (*a*) What is its maximum displacement? When the object is at its maximum displacement, find (*b*) the total energy of the system, (*c*) the gravitational potential energy, and (*d*) the potential energy in the spring. (Choose $U = 0$ when the object is in equilibrium.)

Simple Pendulums

47 • True or false: The motion of a simple pendulum is simple harmonic for any initial angular displacement.

48 • True or false: The motion of a simple pendulum is periodic for any initial angular displacement.

49 •• The length of the string or wire supporting a pendulum increases slightly when its temperature is raised. How would this affect a clock operated by a simple pendulum?

50 • Find the length of a simple pendulum if the period is 5 s at a point where $g = 9.81\ \mathrm{m/s^2}$.

51 • What would be the period of the pendulum in Problem 50 if the pendulum were on the moon, where the acceleration due to gravity is one-sixth that on earth?

52 • If the period of a pendulum 70 cm long is 1.68 s, what is the value of g at the location of the pendulum?

53 • A pendulum set up in the stairwell of a 10-story building consists of a heavy weight suspended on a 34.0-m wire. If $g = 9.81\ \mathrm{m/s^2}$, what is the period of oscillation?

54 •• Show that the total energy of a simple pendulum undergoing oscillations of small amplitude ϕ_0 is approximately $E \approx 1/2mgL\phi_0^2$ (*Hint:* Use the approximation $\cos\phi \approx 1 - \phi^2/2$ for small ϕ.)

55 •• A simple pendulum of length L is attached to a cart that slides without friction down a plane inclined at angle θ with the horizontal as shown (Figure 14-28). Find the period of oscillation of the pendulum on the sliding cart.

Figure 14-28 Problem 55

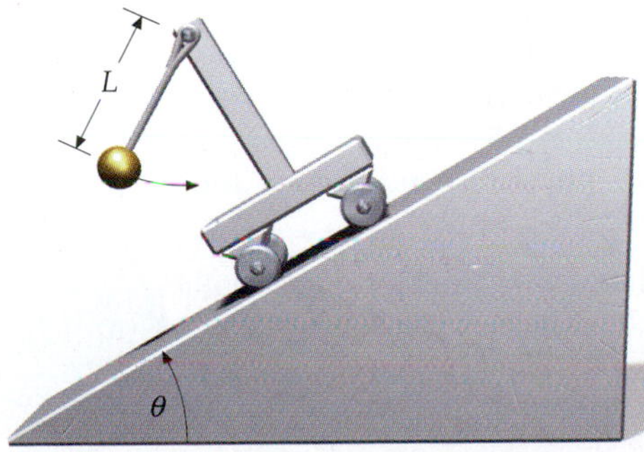

56 •• A simple pendulum of length L is released from rest from an angle ϕ_0. (*a*) Assuming that the pendulum undergoes simple harmonic motion, find its speed as it passes through $\phi = 0$. (*b*) Using the conservation of energy, find this speed exactly. (*c*) Show that your results for (*a*) and (*b*) are the same when ϕ_0 is small. (*d*) Find the difference in your results for $\phi_0 = 0.20$ rad and $L = 1$ m.

Physical Pendulums (optional)

57 • A thin disk of mass 5 kg and radius 20 cm is suspended by a horizontal axis perpendicular to the disk through its rim. The disk is displaced slightly from equilibrium and released. Find the period of the subsequent simple harmonic motion.

58 • A circular hoop of radius 50 cm is hung on a narrow horizontal rod and allowed to swing in the plane of the hoop. What is the period of its oscillation, assuming that the amplitude is small?

59 • A 3-kg plane figure is suspended at a point 10 cm from its center of mass. When it is oscillating with small amplitude, the period of oscillation is 2.6 s. Find the moment of inertia I about an axis perpendicular to the plane of the figure through the pivot point.

Figure 14-29 Problem 60

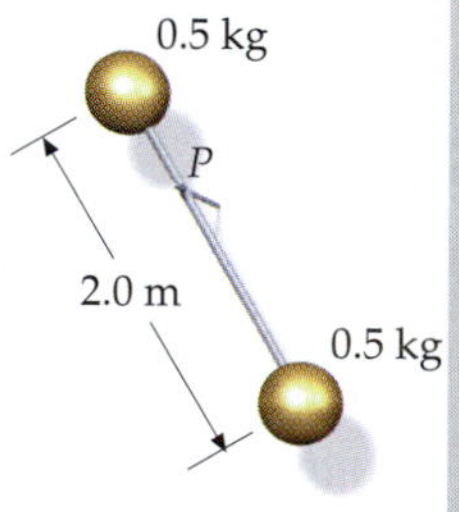

60 •• Figure 14-29 shows a dumbbell with two equal masses (to be considered as point masses) attached to a very thin (massless) rod of length L. (*a*) Show that the period of this pendulum is a minimum when the pivot point P is at one of the masses. (*b*) Find the period of this physical pendulum if the distance between P and the upper mass is $L/4$.

Figure 14-30 Problem 61

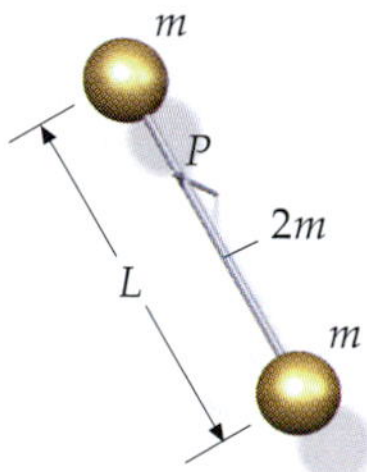

61 •• Suppose the rod in Problem 60 has a mass of $2m$ (Figure 14-30). Determine the distance between the upper mass and the pivot point P such that the period of this physical pendulum is a minimum.

62 •• You are given a meter stick and asked to drill a hole in it so that when pivoted about the hole the period of the pendulum will be a minimum. Where should you drill the hole?

Figure 14-31 Problem 63

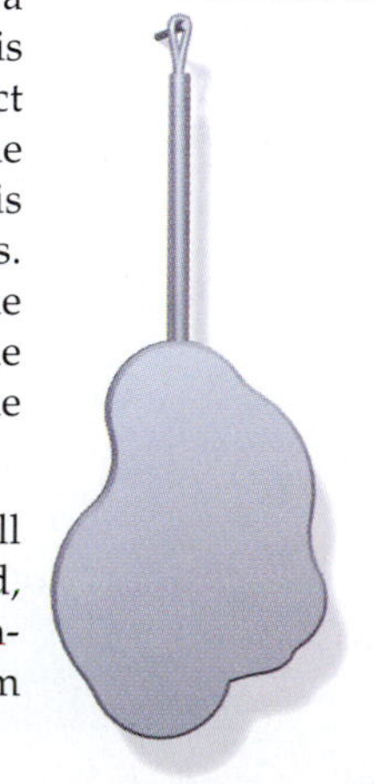

63 •• An irregularly shaped plane object of mass 3.2 kg is suspended by a massless rod of adjustable length and is free to swing in the plane of the object (Figure 14-31). When the length of the supporting rod is 1.0 m, the period of this pendulum for small oscillations is 2.6 s. When the rod is shortened to 0.8 m, the period decreases to 2.5 s. What will be the period of this physical pendulum if the length of the rod is 0.5 m?

64 •• When a short person and a tall person walk together at the same speed, the short person will take more steps. Consider the leg to be a physical pendulum that swings about the hip joint. Estimate the natural frequency of this pendulum for a person of your height, and compare the result with the rate at which you take steps when walking in a leisurely manner.

Figure 14-32 Problem 65

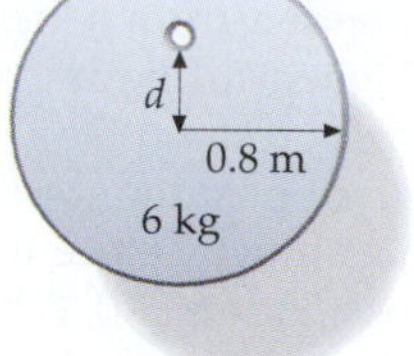

65 •• Figure 14-32 shows a uniform disk of radius $R = 0.8$ m and a 6-kg mass with a small hole a distance d from the disk's center that can serve as a pivot point. (*a*) What should be the distance d so that the period of this physical pendulum is 2.5 s? (*b*) What should be the distance d so that this physical pendulum will have the shortest possible period? What is this period?

Figure 14-33 Problem 66

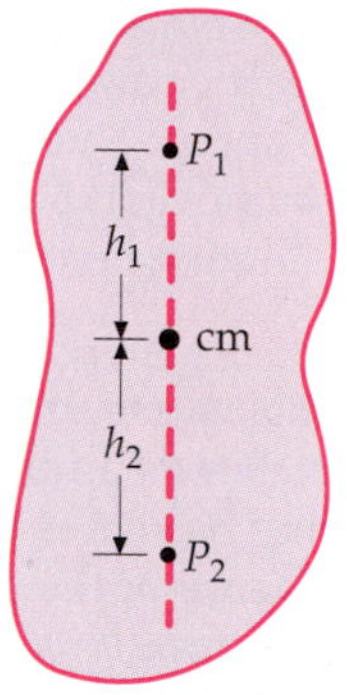

66 ••• A plane object has moment of inertia I about its center of mass. When pivoted at point P_1, as shown in Figure 14-33, it oscillates about the pivot with a period T. There is a second point P_2 on the opposite side of the center of mass about which the object can be pivoted so that the period of oscillation is also T. Show that $h_1 + h_2 = gT^2/4\pi^2$.

Figure 14-34 Problem 67

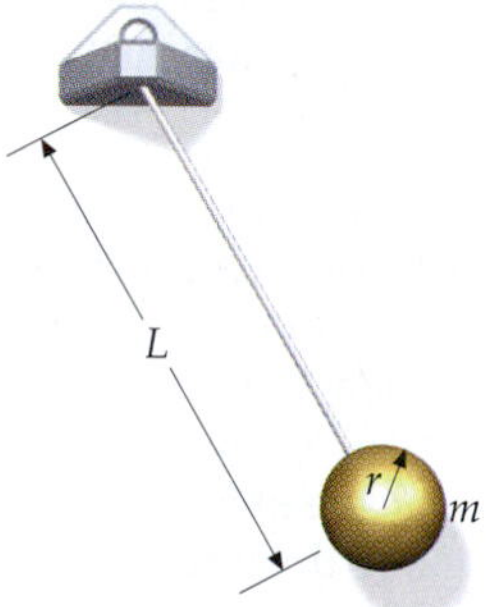

67 ••• A physical pendulum consists of a spherical bob of radius r and mass m suspended from a string (Figure 14-34). The distance from the center of the sphere to the point of support is L. When r is much less than L, such a pendulum is often treated as a simple pendulum of length L. (*a*) Show that the period for small oscillations is given by

$$T = T_0\sqrt{1 + \frac{2r^2}{5L^2}}$$

where $T_0 = 2\pi\sqrt{L/g}$ is the period of a simple pendulum of length L. (*b*) Show that when r is much smaller than L, the period is approximately $T \approx T_0(1 + r^2/5L^2)$. (*c*) If $L = 1$ m and $r = 2$ cm, find the error when the approximation $T = T_0$ is used for this pendulum. How large must the radius of the bob be for the error to be 1%?

Figure 14-35 Problem 68

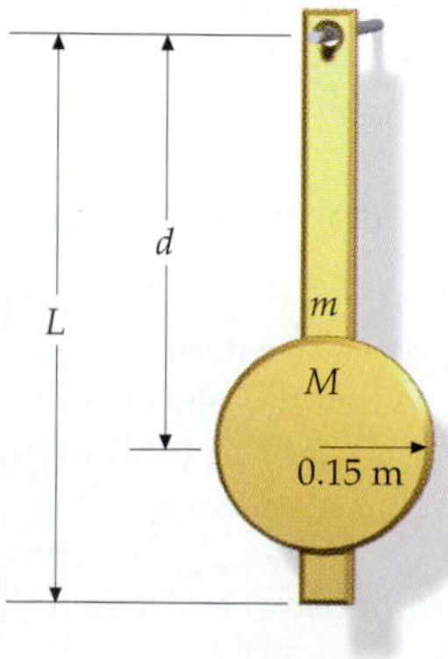

68 ••• Figure 14-35 shows the pendulum of a clock. The uniform rod of length $L = 2.0$ m has a mass $m = 0.8$ kg. Attached to the rod is a disk of mass $M = 1.2$ kg and radius 0.15 m. The clock is constructed to keep perfect time if the period of the pendulum is exactly 3.50 s. (*a*) What should be the distance d so that the period of this pendulum is 2.50 s? (*b*) Suppose that the pendulum clock loses 5.0 min per day. How far and in what direction should the disk be moved to ensure that the clock will keep perfect time?

Clocks

69 •• Two clocks have simple pendulums of identical lengths L. The pendulum of clock A swings through an arc of 10°; that of clock B swings through an arc of 5°. When the two clocks are compared one will find that

(*a*) A runs slow compared to B.
(*b*) A runs fast compared to B.
(*c*) both clocks keep the same time.
(*d*) the answer depends on the length L.

70 •• A simple-pendulum clock keeps accurate time when its length is L. If the length is increased a small amount, how will the accuracy of the clock be affected?

(*a*) The clock will lose time.
(*b*) The clock will gain time.
(*c*) The clock will continue to keep accurate time.
(*d*) The answer cannot be determined without knowing the original length of the pendulum.
(*e*) The answer cannot be determined without knowing the percent increase in the length of the pendulum.

71 •• A pendulum clock loses 48 s per day when the amplitude of the pendulum is 8.4°. What should be the amplitude of the pendulum so that the clock keeps perfect time?

72 •• A pendulum clock that has run down to a very small amplitude gains 5 min each day. What angular amplitude should the pendulum have to keep the correct time?

Damped Oscillations

73 • True or false: The energy of a damped, undriven oscillator decreases exponentially with time.

74 • Show that the dampening constant, b, has units of kg/s.

75 • An oscillator has a Q factor of 200. By what percentage does its energy decrease during one period?

76 • A 2-kg object oscillates with an initial amplitude of 3 cm on a spring of force constant $k = 400$ N/m. Find (*a*) the period, and (*b*) the total initial energy. (*c*) If the energy decreases by 1% per period, find the damping constant b and the Q factor.

77 •• Show that the ratio of the amplitudes for two successive oscillations is constant for a damped oscillator.

78 •• An oscillator has a period of 3 s. Its amplitude decreases by 5% during each cycle. (*a*) By how much does its energy decrease during each cycle? (*b*) What is the time constant τ? (*c*) What is the Q factor?

79 •• An oscillator has a Q factor of 20. (*a*) By what fraction does the energy decrease during each cycle? (*b*) Use Equation 14-35 to find the percentage difference between ω' and ω_0. (*Hint:* Use the approximation $(1 + x)^{1/2} \approx 1 + \frac{1}{2}x$ for small x.)

80 •• For a child on a swing, the amplitude drops by a factor of $1/e$ in about eight periods if no energy is fed in. Estimate the Q factor for this system.

81 •• A damped mass–spring system oscillates at 200 Hz. The time constant of the system is 2.0 s. At $t = 0$, the amplitude of oscillation is 6.0 cm and the energy of the oscillating system is then 60 J. (*a*) What are the amplitudes of oscillation at $t = 2.0$ s and at $t = 4.0$ s? (*b*) How much energy is dissipated in the first 2-s interval and in the second 2-s interval?

82 •• It has been stated that the vibrating earth has a resonance period of 54 min and a Q factor of about 400 and that after a large earthquake, the earth "rings" (continues to vibrate) for about 2 months. (*a*) Find the percentage of the energy of vibration lost to damping forces during each cycle. (*b*) Show that after n periods, the energy is $E_n = (0.984)^n E_0$, where E_0 is the original energy. (*c*) If the original energy of vibration of an earthquake is E_0, what is the energy after 2 days?

83 ••• A 3-kg sphere dropped through air has a terminal speed of 25 m/s. (Assume that the drag force is $-bv$.) Now suppose the sphere is attached to a spring of force constant $k = 400$ N/m, and that it oscillates with an initial amplitude of 20 cm. (*a*) What is the time constant τ? (*b*) When will the amplitude be 10 cm? (*c*) How much energy will have been lost when the amplitude is 10 cm?

Driven Oscillations and Resonance

84 • True or false:

(*a*) Resonance occurs when the driving frequency equals the natural frequency.
(*b*) If the Q value is high, the resonance is sharp.

85 • Give some examples of common systems that can be considered to be driven oscillators.

86 • A crystal wineglass shattered by an intense sound is an example of

(*a*) resonance.
(*b*) critical damping.
(*c*) an exponential decrease in energy.
(*d*) overdamping.

87 • Find the resonance frequency for each of the three systems shown in Figure 14-36.

Figure 14-36 Problem 87

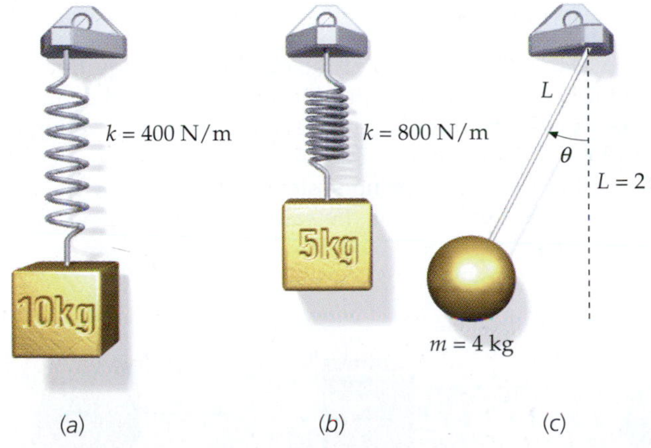

88 • A damped oscillator loses 2% of its energy during each cycle. (*a*) What is its Q factor? (*b*) If its resonance frequency is 300 Hz, what is the width of the resonance curve $\Delta\omega$ when the oscillator is driven?

89 •• A 2-kg object oscillates on a spring of force constant $k = 400$ N/m. The damping constant has a value of $b = 2.00$ kg/s. The system is driven by a sinusoidal force of maximum value 10 N and angular frequency $\omega = 10$ rad/s. (*a*) What is the amplitude of the oscillations? (*b*) If the driving frequency is varied, at what frequency will resonance occur? (*c*) What is the amplitude of oscillation at resonance? (*d*) What is the width of the resonance curve $\Delta\omega$?

90 •• A damped oscillator loses 3.5% of its energy during each cycle. (*a*) How many cycles elapse before half of its original energy is dissipated? (*b*) What is its Q factor? (*c*) If the natural frequency is 100 Hz, what is the width of the resonance curve when the oscillator is driven?

91 •• Tarzan is depressed again. He ties a vine to his ankle and swings upside-down with a period of 3 s as he contemplates his troubles. Cheetah the chimpanzee pushes him so that the amplitude remains constant. Tarzan's mass is 90 kg and his speed at the bottom of the swing is 2.0 m/s. (*a*) What is Tarzan's total energy? (*b*) If $Q = 20$, how much energy is dissipated during each oscillation? (*c*) What is Cheetah's power input? (*Note:* Pushing a swing is usually not done sinusoidally. However, to maintain a steady amplitude, the energy lost per cycle due to damping must be replaced by an external energy source.)

Collisions

92 •• Peter lays his jack-in-the-box on its side with the lid open, so that Jack, a painted 0.4-kg clown, sticks out horizontally at the end of a spring. Peter then takes a 0.6-kg wad of putty, places it in his favorite slingshot, and fires it at the top of Jack's head. The putty sticks to the clown's head, and the clown and putty oscillate with an amplitude of 16 cm and a frequency of 0.38 Hz. Assuming that the box remains immobile, determine (*a*) the putty's speed before the collision, and (*b*) the spring constant.

93 ••• Figure 14-37 shows a vibrating mass–spring system supported on a frictionless surface and a second equal mass that is moving toward the vibrating mass with velocity v. The motion of the vibrating mass is given by

$$x(t) = (0.1\text{ m})\cos(40\text{ s}^{-1}t)$$

where x is the displacement of the mass from its equilibrium position. The two masses collide elastically just as the vibrating mass passes through its equilibrium position traveling to the right. (*a*) What should be the velocity v of the second mass so that the mass–spring system is at rest following the elastic collision? (*b*) What is the velocity of the second mass after the elastic collision?

Figure 14-37 Problem 93

94 ••• Following the elastic collision in Problem 93, the energy of the recoiling mass is 8.0 J. Find the masses m and the spring constant k.

95 ••• An object of mass 2 kg resting on a frictionless horizontal surface is attached to a spring of force constant 600 N/m. A second object of mass 1 kg slides along the surface toward the first object at 6 m/s. (*a*) Find the amplitude of oscillation if the objects make a perfectly inelastic collision and remain together on the spring. What is the period of oscillation? (*b*) Find the amplitude and period of oscillation if the collision is elastic. (*c*) For each type of collision, write an expression for the position x as a function of time t for the object attached to the spring, assuming that the collision occurs at time $t = 0$.

General Problems

96 • The effect of the mass of a spring on the motion of an object attached to it is usually neglected. Describe qualitatively its effect when it is not neglected.

97 •• A lamp hanging from the ceiling of the club car in a train oscillates with period T_0 when the train is at rest. The period will be (match left and right columns)

1. greater than T_0 when
2. less than T_0 when
3. equal to T_0 when

A. the train moves horizontally with constant velocity.
B. the train rounds a curve of radius R with speed v.
C. the train climbs a hill of inclination θ at constant speed.
D. the train goes over a hill of radius of curvature R with constant speed.

98 •• Two mass–spring systems oscillate at frequencies f_A and f_B. If $f_A = 2f_B$ and the spring constants of the two springs are equal, it follows that the masses are related by

(*a*) $M_A = 4M_B$. (*b*) $M_A = M_B/\sqrt{2}$.
(*c*) $M_A = M_B/2$. (*d*) $M_A = M_B/4$.

99 •• Two mass–spring systems A and B oscillate so that their energies are equal. If $M_A = 2M_B$, then which formula below relates the amplitudes of oscillation?

(*a*) $A_A = A_B/4$ (*b*) $A_A = A_B/\sqrt{2}$ (*c*) $A_A = A_B$
(*d*) Not enough information is given to determine the ratio of the amplitudes.

100 •• Two mass–spring systems A and B oscillate so that their energies are equal. If $k_A = 2k_B$, then which formula below relates the amplitudes of oscillation?

(*a*) $A_A = A_B/4$ (*b*) $A_A = A_B/\sqrt{2}$ (*c*) $A_A = A_B$
(*d*) Not enough information is given to determine the ratio of the amplitudes.

101 •• Pendulum A has a bob of mass M_A and a length L_A; pendulum B has a bob of mass M_B and a length L_B. If the period of A is twice that of B, then

(*a*) $L_A = 2L_B$ and $M_A = 2M_B$.
(*b*) $L_A = 4L_B$ and $M_A = M_B$.
(*c*) $L_A = 4L_B$ whatever the ratio M_A/M_B.
(*d*) $L_A = \sqrt{2}L_B$ whatever the ratio M_A/M_B.

102 • A particle has a displacement $x = 0.4\cos(3t + \pi/4)$, where x is in meters and t is in seconds. (*a*) Find the frequency f and period T of the motion. (*b*) Where is the particle at $t = 0$? (*c*) Where is the particle at $t = 0.5$ s?

103 • (*a*) Find an expression for the velocity of the particle whose position is given in Problem 102. (*b*) What is the velocity at time $t = 0$? (*c*) What is the maximum velocity? (*d*) At what time after $t = 0$ does this maximum velocity first occur?

104 • An object on a horizontal spring oscillates with a period of 4.5 s. If the object is suspended from the spring vertically, by how much is the spring stretched from its natural length when the object is in equilibrium?

105 •• A small particle of mass m slides without friction in a spherical bowl of radius r. (*a*) Show that the motion of the particle is the same as if it were attached to a string of length r. (*b*) Figure 14-38 shows a particle of mass m_1 that is displaced a small distance s_1 from the bottom of the bowl, where s_1 is much smaller than r. A second particle of mass m_2 is displaced in the opposite direction a distance $s_2 = 3s_1$, where s_2 is also much smaller than r. If the particles are released at the same time, where do they meet? Explain.

Figure 14-38 Problem 105

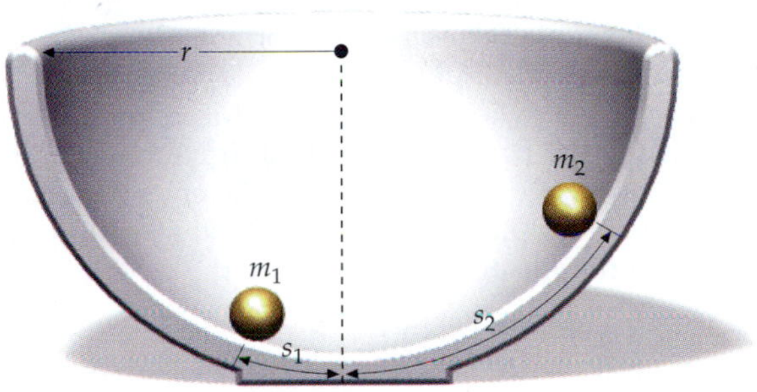

106 •• As your jet plane speeds down the runway on takeoff, you measure its acceleration by suspending your yo-yo as a simple pendulum and noting that when the bob (mass 40 g) is at rest relative to you, the string (length 70 cm) makes an angle of 22° with the vertical. Find the period T for small oscillations of this pendulum.

107 •• Two identical blocks placed one on top of the other rest on a frictionless horizontal air track. The lower block is attached to a spring of spring constant $k = 600$ N/m. When displaced slightly from its equilibrium position, the system oscillates with a frequency of 1.8 Hz. When the amplitude of oscillation exceeds 5 cm, the upper block starts to slide relative to the lower one. (*a*) What are the masses of the two blocks? (*b*) What is the coefficient of static friction between the two blocks?

108 •• Two atoms are bound together in a molecule. The potential energy U resulting from their interaction is shown in Figure 14-39. The variable r is the distance between the atom centers, and E_0 is the lowest (ground-state) energy.

(*a*) As a result of a collision, the molecule acquires a kinetic energy of vibration whose maximum value is 1.0 eV. With this kinetic energy, over what range of separation distance will the molecule vibrate?

(*b*) Give an approximate value for the force $F(r)$ between the two atoms at $r = 0.4$ nm. Express your answer in units appropriate to those used on the graph.

(*c*) Calculate the force in (*b*) in newtons. Is this force attractive or repulsive?

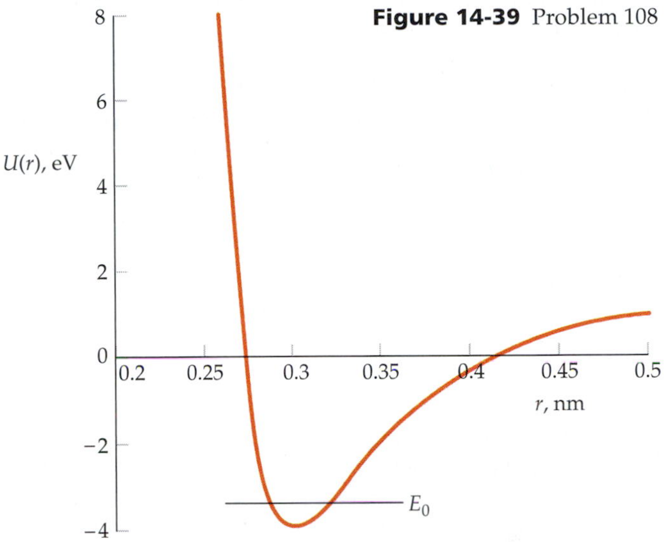

Figure 14-39 Problem 108

109 •• A wooden cube with edge a and mass m floats in water with one of its faces parallel to the water surface. The density of the water is ρ. Find the period of oscillation in the vertical direction if it is pushed down slightly.

110 •• A spider of mass 0.36 g sits in the middle of its horizontal web, which sags 3.00 mm under its weight. Estimate the frequency of vertical vibration for this system.

111 •• A clock with a pendulum keeps perfect time on the earth's surface. In which case will the error be greater: if the clock is placed in a mine of depth h or if the clock is elevated to a height h? Assume that $h \ll R_E$.

112 •• Figure 14-40 shows a pendulum of length L with a bob of mass M. The bob is attached to a spring of spring constant k as shown. When the bob is directly below the pendulum support, the spring is at its equilibrium length. (*a*) Derive an expression for the period of this oscillating system for small amplitude vibrations. (*b*) Suppose that $M = 1$ kg and L is such that in the absence of the spring the period is 2.0 s. What is the spring constant k if the period of the oscillating system is 1.0 s?

Figure 14-40 Problem 112

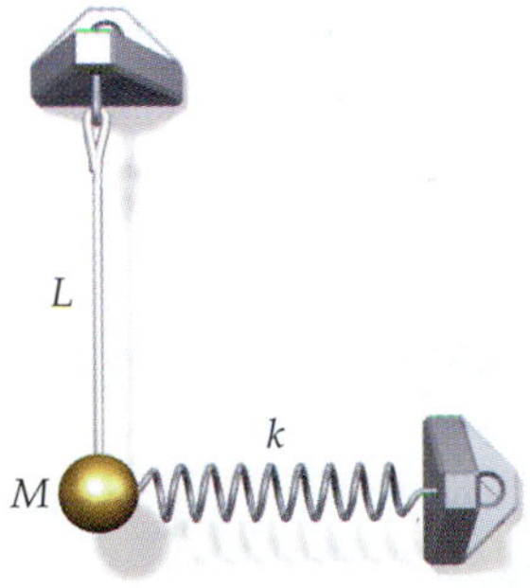

113 •• An object of mass m_1 sliding on a frictionless horizontal surface is attached to a spring of force constant k and oscillates with an amplitude A. When the spring is at its greatest extension and the mass is instantaneously at rest, a second object of mass m_2 is placed on top of it. (*a*) What is the smallest value for the coefficient of static friction μ_s such that the second object does not slip on the first? (*b*) Explain how

the total energy E, the amplitude A, the angular frequency ω, and the period T of the system are changed by placing m_2 on m_1, assuming that the friction is great enough so that there is no slippage.

114 •• The acceleration due to gravity g varies with geographical location because of the earth's rotation and because the earth is not exactly spherical. This was first discovered in the seventeenth century, when it was noted that a pendulum clock carefully adjusted to keep correct time in Paris lost about 90 s per day near the equator. (*a*) Show that a small change in the acceleration of gravity Δg produces a small change in the period ΔT of a pendulum given by

$$\frac{\Delta T}{T} \approx -\frac{1}{2}\frac{\Delta g}{g}$$

(Use differentials to approximate ΔT and Δg.) (*b*) How great a change in g is needed to account for a change in the period of 90 s per day?

115 •• Figure 14-41 shows two equal masses of 0.6 kg glued to each other and connected to a spring of spring constant $k = 240$ N/m. The masses, which rest on a frictionless horizontal surface, are displaced 0.6 m from their equilibrium position and released. Before being released, a few drops of solvent are deposited on the glue. (*a*) Find the frequency of vibration and total energy of the vibrating system before the glue has dissolved. (*b*) Find the frequency, amplitude of vibration, and energy of the vibrating system if the glue dissolves when the spring is (1) at maximum compression and (2) at maximum extension.

Figure 14-41
Problem 115

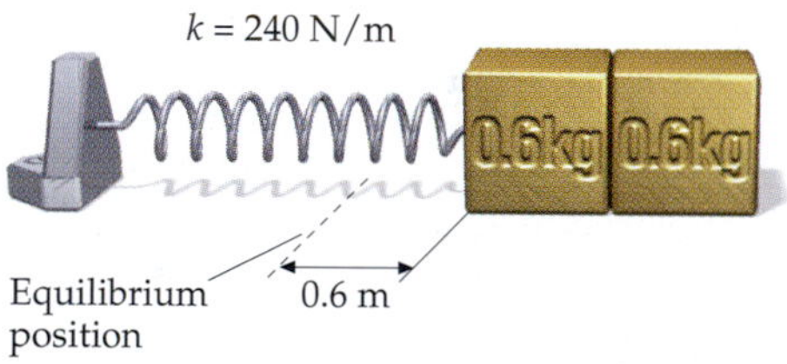

116 •• Show that for the situations in Figures 14-42*a* and *b*, the object oscillates with a frequency $f = [1/(2\pi)]\sqrt{k_{\text{eff}}/m}$, where k_{eff} is given by (*a*) $k_{\text{eff}} = k_1 + k_2$ and (*b*) $1/k_{\text{eff}} = 1/k_1 + 1/k_2$. (*Hint:* Find the net force F on the object for a small displacement x and write $F = -k_{\text{eff}}x$. Note that in (*b*) the springs stretch by different amounts, the sum of which is x.)

Figure 14-42 Problem 116

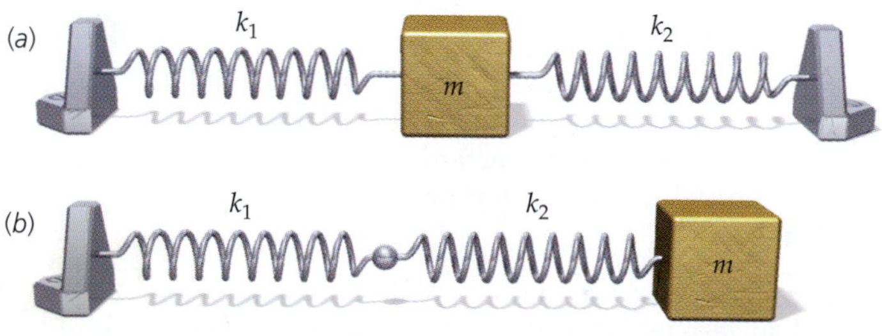

117 •• A small block of mass m_1 rests on a piston that is vibrating vertically with simple harmonic motion given by $y = A \sin \omega t$. (*a*) Show that the block will leave the piston if $\omega^2 A > g$. (*b*) If $\omega^2 A = 3g$ and $A = 15$ cm, at what time will the block leave the piston?

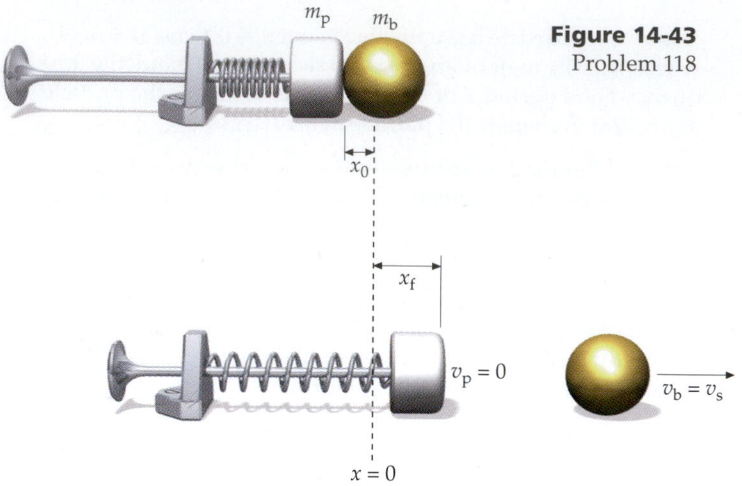

Figure 14-43
Problem 118

118 •• The plunger of a pinball machine has mass m_{p} and is attached to a spring of force constant k (Figure 14-43). The spring is compressed a distance x_0 from its equilibrium position $x = 0$ and released. A ball of mass m_{b} is next to the plunger. (*a*) Where does the ball leave the plunger? (*b*) What is the speed v_{s} of the ball when it separates? (*c*) At what distance x_{f} does the plunger come to rest momentarily? (Assume that the surface is horizontal and frictionless so that the ball slides rather than rolls.)

119 •• A level platform vibrates horizontally with simple harmonic motion with a period of 0.8 s. (*a*) A box on the platform starts to slide when the amplitude of vibration reaches 40 cm; what is the coefficient of static friction between the body and the platform? (*b*) If the coefficient of friction between the box and platform were 0.40, what would be the maximum amplitude of vibration before the box would slip?

120 ••• The potential energy of a mass m as a function of position is given by $U(x) = U_0(\alpha + 1/\alpha)$, where $\alpha = x/a$ and a is a constant. (*a*) Plot $U(x)$ versus x for $0.1a < x < 3a$. (*b*) Find the value of $x = x_0$ at stable equilibrium. (*c*) Write the potential energy $U(x)$ for $x = x_0 + \varepsilon$, where ε is a small displacement from the equilibrium position x_0. (*d*) Approximate the $1/x$ term using the binomial expansion

$$(1 + r)^n = 1 + nr + \frac{n(n-1)}{(2)(1)}r^2 + \frac{n(n-1)(n-2)}{(3)(2)(1)}r^3 + \cdots$$

with $r = \varepsilon/x_0 \ll 1$ and discarding all terms of power greater than r^2. (*e*) Compare your result with the potential for a simple harmonic oscillator. Show that the mass will undergo simple harmonic motion for small displacements from equilibrium and determine the frequency of this motion.

121 ••• Do Problem 120 with $U(x) = U_0(\alpha^2 + 1/\alpha^2)$.

122 ••• A solid cylindrical drum of mass 6.0 kg and diameter 0.06 m rolls without slipping on a horizontal surface (Figure 14-44). The axle of the drum is attached to a spring of spring constant $k = 4000$ N/m as shown. (*a*) Determine the frequency of oscillation of this system for small displacements from equilibrium. (*b*) What is the minimum value of the coefficient of static friction such that the drum will not slip when the vibrational energy is 5.0 J?

Figure 14-44 Problem 122

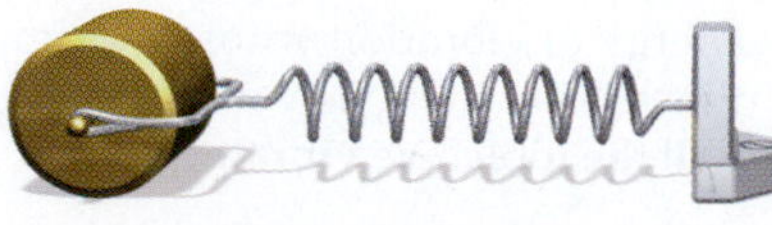

123 ••• Figure 14-45 shows a solid half-cylinder of mass M and radius R resting on a horizontal surface. If one side of this cylinder is pushed down slightly and then released, the object will oscillate about its equilibrium position. Determine the period of this oscillation.

124 ••• Repeat Problem 123 replacing the half-cylinder with a half-sphere.

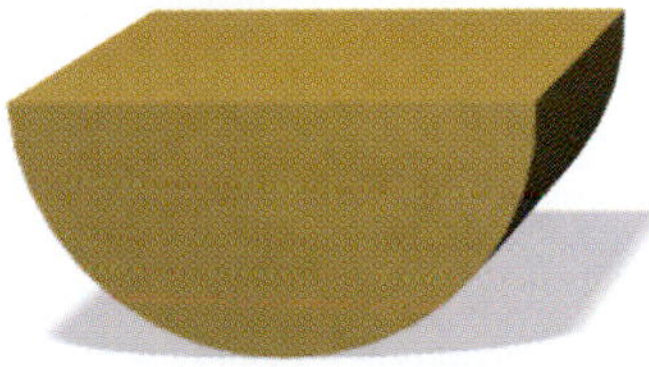

Figure 14-45 Problems 123 and 124

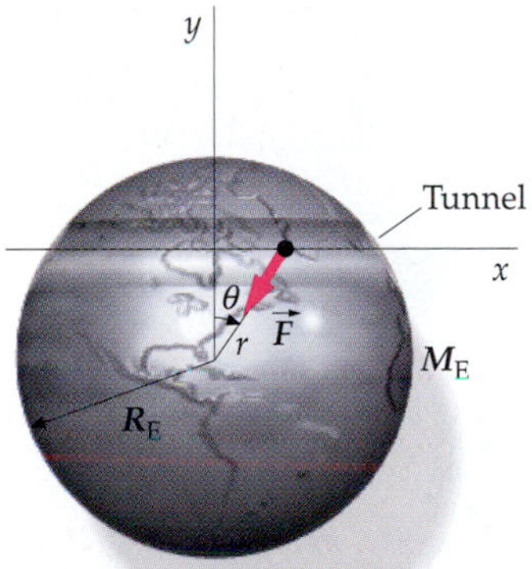

Figure 14-46 Problem 125

125 ••• A straight tunnel is dug through the earth as shown in Figure 14-46. Assume that the walls of the tunnel are frictionless. (*a*) The gravitational force exerted by the earth on a particle of mass m at a distance r from the center of the earth when $r < R_E$ is $F_r = -(GmM_E/R_E^3)r$, where M_E is the mass of the earth and R_E is its radius. Show that the net force on a particle of mass m at a distance x from the middle of the tunnel is given by $F_x = -(GmM_E/R_E^3)x$, and that the motion of the particle is therefore simple harmonic motion. (*b*) Show that the period of the motion is given by $T = 2\pi\sqrt{R_E/g}$ and find its value in minutes. (This is the same period as that of a satellite orbiting near the surface of the earth and is independent of the length of the tunnel.)

126 ••• A damped oscillator has a frequency ω' that is 10% less than its undamped frequency. (*a*) By what factor is the amplitude of the oscillator decreased during each oscillation? (*b*) By what factor is its energy reduced during each oscillation?

127 ••• Show by direct substitution that Equation 14-48 is a solution of Equation 14-47.

128 ••• A block of mass m on a horizontal table is attached to a spring of force constant k as shown in Figure 14-47. The coefficient of kinetic friction between the block and the table is μ_k. The spring is stretched a distance A and released. (*a*) Apply Newton's second law to the block to obtain an equation for its acceleration d^2x/dt^2 for the first half-cycle, during which the block is moving to the left. Show that the resulting equation can be written $d^2x'/dt^2 = -\omega^2 x'$, where $x = 0$ at the equilibrium position of the spring, and $x' = x - x_0$, with $x_0 = \mu_k mg/k = \mu_k g/\omega^2$. (*b*) Repeat part (*a*) for the second half-cycle as the block moves to the right, and show that $d^2x''/dt^2 = -\omega^2 x''$, where $x'' = x + x_0$ and x_0 has the same value. (*c*) Sketch $x(t)$ for the first few cycles for $A = 10x_0$.

Figure 14-47 Problem 128

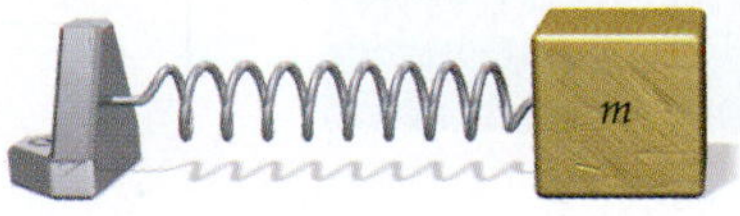

129 ••• In this problem, you will derive the expression for the average power delivered by a driving force to a driven oscillator (Figure 14-25).

(*a*) Show that the instantaneous power input of the driving force is given by

$$P = Fv = -A\omega F_0 \cos \omega t \sin(\omega t - \delta)$$

(*b*) Use the trigonometric identity $\sin(\theta_1 - \theta_2) = \sin\theta_1 \cos\theta_2 - \cos\theta_1 \sin\theta_2$ to show that the equation in (*a*) can be written

$$P = A\omega F_0 \sin\delta \cos^2 \omega t - A\omega F_0 \cos\delta \cos\omega t \sin\omega t$$

(*c*) Show that the average value of the second term in your result for (*b*) over one or more periods is zero and that therefore

$$P_{av} = \frac{1}{2} A\omega F_0 \sin\delta$$

(*d*) From Equation 14-50 for $\tan\delta$, construct a right triangle in which the side opposite the angle δ is $b\omega$ and the side adjacent is $m(\omega_0^2 - \omega^2)$, and use this triangle to show that

$$\sin\delta = \frac{b\omega}{\sqrt{m^2(\omega_0^2 - \omega^2)^2 + b^2\omega^2}} = \frac{b\omega A}{F_0}$$

(*e*) Use your result for (*d*) to eliminate ωA so that the average power input can be written

$$P_{av} = \frac{1}{2}\frac{F_0^2}{b}\sin^2\delta = \frac{1}{2}\left[\frac{b\omega^2 F_0^2}{m^2(\omega_0^2 - \omega^2)^2 + b^2\omega^2}\right] \qquad \text{14-51}$$

130 ••• In this problem, you are to use the result of Problem 129 to derive Equation 14-45, which relates the width of the resonance curve to the Q value when the resonance is sharp. At resonance, the denominator of the fraction in brackets in Equation 14-51 is $b^2\omega_0^2$ and P_{av} has its maximum value. For a sharp resonance, the variation in ω in the numerator in Equation 14-51 can be neglected. Then the power input will be half its maximum value at the values of ω, for which the denominator is $2b^2\omega_0^2$.

(*a*) Show that ω then satisfies

$$m^2(\omega - \omega_0)^2(\omega + \omega_0)^2 = b^2\omega_0^2$$

(*b*) Using the approximation $\omega + \omega_0 \approx 2\omega_0$, show that

$$\omega - \omega_0 \approx \pm\frac{b}{2m}$$

(*c*) Express b in terms of Q.

(*d*) Combine the results of (*b*) and (*c*) to show that there are two values of ω for which the power input is half that at resonance, and that they are given by

$$\omega_1 = \omega_0 - \frac{\omega_0}{2Q} \quad \text{and} \quad \omega_2 = \omega_0 + \frac{\omega_0}{2Q}$$

Therefore, $\omega_2 - \omega_1 = \Delta\omega = \omega_0/Q$, which is equivalent to Equation 14-45.

CHAPTER 15

Wave Motion

15-1 Simple Wave Motion

Transverse and Longitudinal Waves

Waves transport energy and momentum through space without transporting matter. In mechanical waves this happens via a disturbance in a medium. When a string under tension is given a flip, the bump that is produced travels down the string as a wave pulse. The disturbance in this case is the change in shape of the string from its equilibrium shape. Its propagation

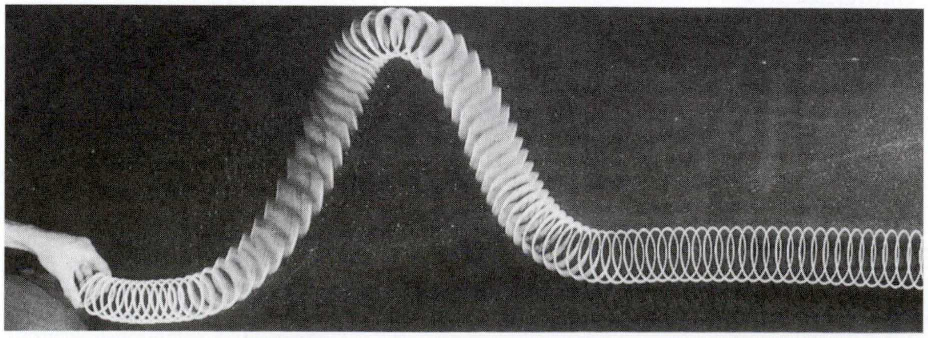

(a)

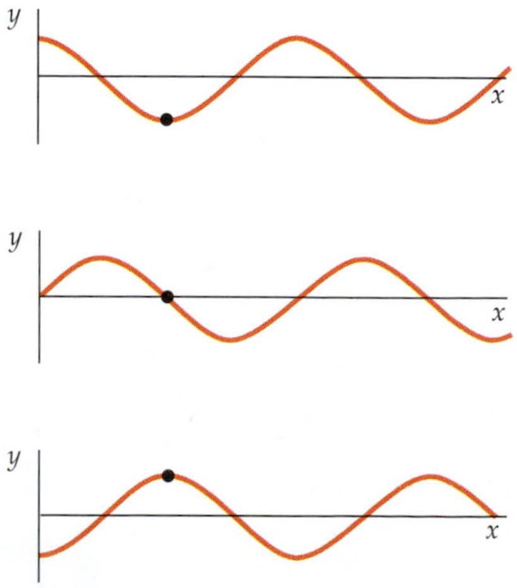

(b)

Figure 15-1 *(a)* Transverse wave pulse on a spring. The disturbance is perpendicular to the direction of the motion of the wave. *(b)* Three successive drawings of a transverse wave on a string traveling to the right. An element of the string moves up and down.

arises from the interaction of each string segment with the adjacent segments. The segments of the string (the medium) move in the direction perpendicular to the string as the pulse propagates down the string. Waves such as those on a string, in which the disturbance is perpendicular to the direction of propagation, are called **transverse** (Figure 15-1). Waves in which the disturbance is parallel to the propagation are called **longitudinal** (Figure 15-2). Sound waves are examples of longitudinal waves—the molecules of a gas, liquid, or solid through which sound travels move back and forth along the line of propagation, alternately compressing and rarefying the medium.

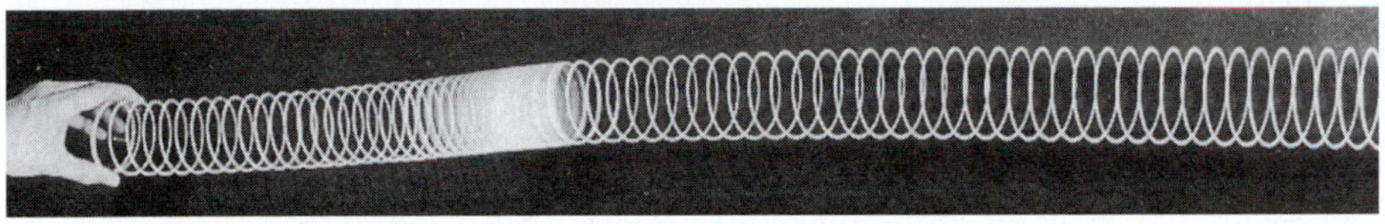

Figure 15-2 Longitudinal wave pulse on a spring. The disturbance is in the direction of the motion of the wave.

Wave Pulses

Figure 15-3 shows a pulse on a string at time $t = 0$. The shape of the string at this time can be represented by some function $y = f(x)$. At some later time, the pulse is farther down the string. In a new coordinate system with origin O' that moves with the speed of the pulse, the pulse is stationary. The string is described in this frame by $f(x')$ for all times. The coordinates of the two reference frames are related by

$$x = x' + vt$$

Thus, the shape of the string in the original frame is

$$y = f(x - vt), \qquad \text{wave moving right} \qquad 15\text{-}1$$

The same line of reasoning for a pulse moving to the left leads to

$$y = f(x + vt), \qquad \text{wave moving left} \qquad 15\text{-}2$$

In both expressions, v is the speed of propagation of the wave. The function $y = f(x - vt)$ is called the **wave function**. For waves on a string, the wave function represents the transverse displacement of the string. For sound waves in air, the wave function can be the longitudinal displacement of the air molecules or the pressure of the air. These wave functions are solutions of a differential equation called the wave equation that can be derived from Newton's laws.

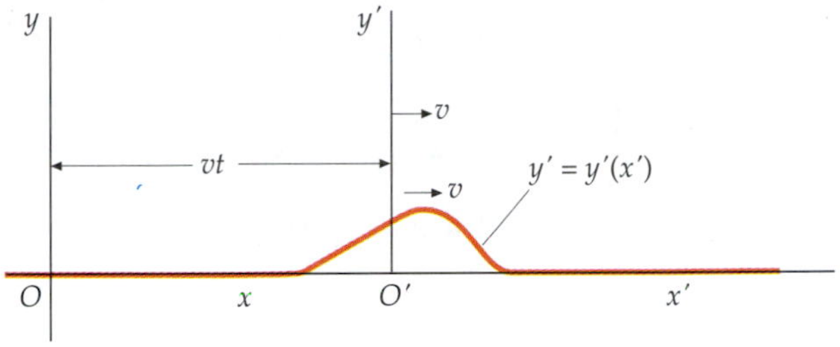

Figure 15-3 A wave pulse moving without change in shape in the positive x direction with a speed v relative to the origin O. In the primed coordinate system moving with the pulse, the wave function is $y' = f(x')$ at all times. In the original, unprimed system, the wave function is $y = f(x - vt)$.

Speed of Waves

A general property of waves is that their speed depends on the properties of the medium but is independent of the motion of the source of the waves. For example, the speed of a sound from a car horn depends only on the properties of air and not on the motion of the car. For wave pulses on a rope, we can easily demonstrate that the greater the tension, the faster the waves propagate. Furthermore, waves propagate faster in a light rope than a heavy rope under the same tension. We show below that if F is the tension* and μ is the linear mass density (mass per unit length), then the wave speed is

$$v = \sqrt{\frac{F}{\mu}} \qquad 15\text{-}3$$

Speed of waves on a string

*We use F for tension rather than T because we use T for the period.

Example 15-1

The tension in a string is provided by hanging an object of mass $m = 3$ kg at one end as shown in Figure 15-4. The length of the string is $L = 2.5$ m and its mass is $m_s = 50$ g. What is the speed of waves on the string?

Figure 15-4

1. The speed is related to the tension F and mass density μ: $v = \sqrt{\frac{F}{\mu}}$

2. Calculate the mass density and tension from the information given: $\mu = \frac{m_s}{L} = \frac{0.05\text{ kg}}{2.5\text{ m}} = 0.02\text{ kg/m}$

   $F = mg = (3\text{ kg})(9.81\text{ N/kg}) = 29.4\text{ N}$

3. Substitute these values to calculate the speed: $v = \sqrt{\frac{F}{\mu}} = \sqrt{\frac{29.4\text{ N}}{0.02\text{ kg/m}}} = 38.3\text{ m/s}$

Exercise If the 3-kg mass is replaced with a 6-kg mass, what is the speed of waves on the string? (*Answer* 54.2 m/s)

Exercise Show that the units of $\sqrt{F/\mu}$ are m/s when F is in newtons and μ is in kg/m.

For sound waves in a fluid such as air or water, the speed v can be shown to be given by

$$v = \sqrt{\frac{B}{\rho}} \tag{15-4}$$

where ρ is the equilibrium density of the medium and B is the bulk modulus (Equation 13-6). Comparing Equations 15-3 and 15-4, we can see that, in general, the speed of waves depends on an elastic property of the medium (the tension for string waves and the bulk modulus for sound waves) and on an inertial property of the medium (the linear mass density or the volume mass density).

For sound waves in a gas such as air, the bulk modulus is proportional to the pressure, which in turn is proportional to the density ρ and to the absolute temperature T of the gas. The ratio B/ρ is thus independent of density and is merely proportional to the absolute temperature T. In Chapter 19, we show that, in this case, Equation 15-4 is equivalent to

$$v = \sqrt{\frac{\gamma RT}{M}} \tag{15-5}$$

Speed of sound in a gas

In this equation T is the absolute temperature measured in kelvins (K), which is related to the Celsius temperature t_C by

$$T = t_C + 273 \tag{15-6}$$

The constant γ depends on the kind of gas. For diatomic molecules, such as O_2 and N_2, γ has the value 1.4, and, since O_2 and N_2 comprise 98% of the atmosphere, that is the value for air. (For monatomic molecules such as He, γ has the value 1.67.) The constant R is the universal gas constant,

$$R = 8.314 \text{ J/mol}\cdot\text{K} \qquad \text{15-7}$$

and M is the molar mass of the gas (that is, the mass of 1 mol of the gas), which for air is

$$M = 29 \times 10^{-3} \text{ kg/mol}$$

Example 15-2 ***try it yourself***

Calculate the speed of sound in air at (*a*) 0°C and (*b*) 20°C.

Cover the column to the right and try these on your own before looking at the answers.

Steps	***Answers***
(*a*)1. Convert 0°C to kelvins.	$T = 273$ K
2. Substitute your result for T into Equation 15-5 to find v.	$v = 331$ m/s
(*b*)1. Convert 20°C to kelvins.	$T = 293$ K
2. Use the fact that v is proportional to $\sqrt{T}$ to write an expression for the ratio of the speed at 293 K to the speed at 273 K.	$\dfrac{v_{293}}{v_{273}} = \dfrac{\sqrt{293}}{\sqrt{273}} = 1.036$
3. Calculate v at 293 K.	$v = 343$ m/s

Remarks We see from this example that the speed of sound in air is about 340 m/s at normal temperatures.

Exercise For helium, $M = 4 \times 10^{-3}$ kg/mol and $\gamma = 1.67$. What is the speed of sound waves in helium at 20°C? (*Answer* 1.01 km/s)

Derivation of *v* for Waves on a String Equation 15-3 can be obtained from Newton's laws. Consider a pulse traveling along a string with a speed v to the right (Figure 15-5*a*). If the amplitude of the pulse is small compared to the length of the string, the tension F will be approximately constant along the string. In a reference frame moving with speed v to the right, the pulse is stationary and the string moves with a speed v to the left. Figure 15-5*b* shows a small segment of the string of length Δs. The segment forms part of a circular arc of radius R. Instantaneously the segment is moving with speed v in a circular path, so it has a centripetal acceleration v^2/R. The forces acting on the segment are the tension F at each end. The horizontal components of these forces are equal and opposite and thus cancel. The vertical components of these forces point radially inward toward the center of the circular arc. These radial forces provide the centripetal acceleration.

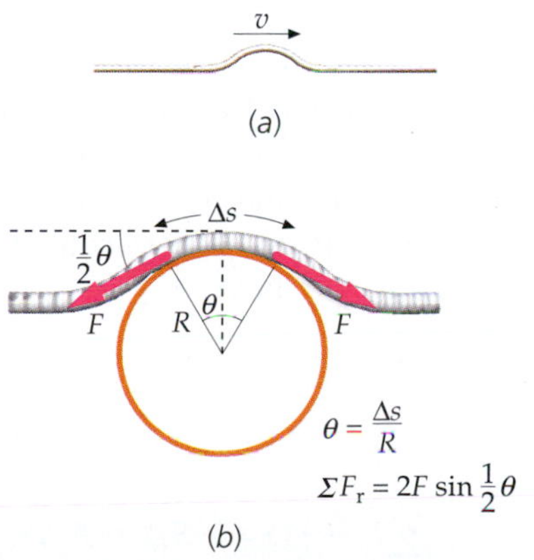

Figure 15-5 (*a*) Wave pulse moving with a speed v along a string. (*b*) In a frame in which the wave pulse of (*a*) is at rest, the string is moving with a speed v to the left. A small segment of the string of length Δs is moving in a circular arc of radius R. The centripetal acceleration of the segment is provided by the radial components of the tension.

Let the angle subtended by the string be θ: The net radial force acting on the segment is

$$\sum F_r = 2F \sin \tfrac{1}{2}\theta \approx 2F(\tfrac{1}{2}\theta) = F\theta$$

where we have used the approximation $\sin \frac{1}{2}\theta \approx \frac{1}{2}\theta$ for small θ. If μ is the mass per unit length of the string, the mass of a segment of length Δs is $m = \mu\, \Delta s$. The angle θ is related to Δs by

$$\theta = \frac{\Delta s}{R}$$

The mass of the element is thus

$$m = \mu\, \Delta s = \mu R\theta$$

Setting the net radial force equal to the mass times the centripetal acceleration gives

$$F\theta = \mu R\theta \frac{v^2}{R}$$

Solving for v, we obtain $v = \sqrt{F/\mu}$. Since v is independent of R and θ, this result holds for all segments of the string. However, the derivation depends on θ being small, which is true if the height of the pulse is small compared to its length.

In the original frame, the string is fixed, and the pulse moves with speed $v = \sqrt{F/\mu}$, which is Equation 15-3.

optional

The Wave Equation

We can apply Newton's laws to a segment of the string to derive a differential equation known as the wave equation, which relates the spatial derivatives of $y(x, t)$ to its time derivatives. Figure 15-6 shows one segment of a string. We consider only small vertical displacements. Then the length of the segment is approximately Δx and its mass is $m = \mu\, \Delta x$, where μ is the string's mass per unit length. The segment moves vertically, and the net force in this direction is

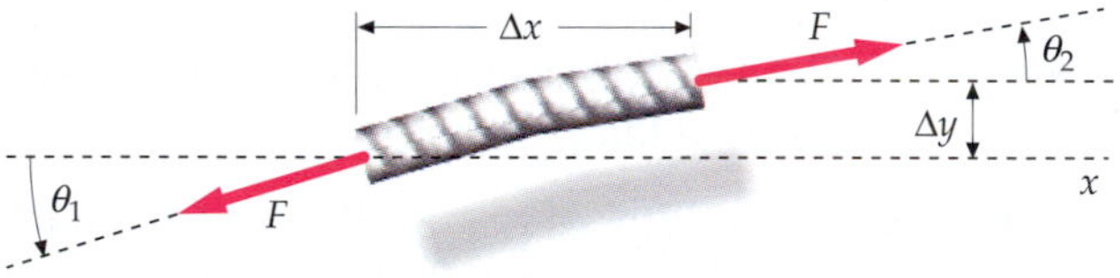

Figure 15-6 Segment of a stretched string used for the derivation of the wave equation. The net vertical force on the segment is $F \sin \theta_2 - F \sin \theta_1$, where F is the tension in the string. The wave equation is derived by applying Newton's second law to the segment.

$$\sum F = F \sin \theta_2 - F \sin \theta_1$$

where θ_2 and θ_1 are the angles shown, and F is the tension in the string. Since the angles are assumed to be small, we may approximate $\sin \theta$ by $\tan \theta$. Then the net vertical force on the string segment can be written

$$\sum F = F(\sin \theta_2 - \sin \theta_1) \approx F(\tan \theta_2 - \tan \theta_1)$$

The tangent of the angle made by the string with the horizontal is the slope of the curve formed by the string. The slope S is the first derivative of $y(x, t)$ with respect to x for constant t. A derivative of a function of two variables with respect to one of the variables with the other held constant is called a **partial derivative.** The partial derivative of y with respect to x is written $\partial y/\partial x$. Thus, we have

$$S = \tan \theta = \frac{\partial y}{\partial x}$$

Then

$$\sum F = F(S_2 - S_1) = F\, \Delta S$$

where S_1 and S_2 are the slopes of either end of the string segment, and ΔS is the change in the slope. Setting this net force equal to the mass $\mu\,\Delta x$ times the acceleration $\partial^2 y/\partial t^2$ gives

$$F\,\Delta S = \mu\,\Delta x\,\frac{\partial^2 y}{\partial t^2}$$

or

$$F\frac{\Delta S}{\Delta x} = \mu\frac{\partial^2 y}{\partial t^2} \qquad \text{15-8}$$

In the limit $\Delta x \to 0$, we have

$$\lim_{\Delta x \to 0}\frac{\Delta S}{\Delta x} = \frac{\partial S}{\partial x} = \frac{\partial}{\partial x}\frac{\partial y}{\partial x} = \frac{\partial^2 y}{\partial x^2}$$

Thus, Equation 15-8 becomes

$$\frac{\partial^2 y}{\partial x^2} = \frac{\mu}{F}\frac{\partial^2 y}{\partial t^2} \qquad \text{15-9}a$$

Wave equation for a stretched string

Equation 15-9*a* is the **wave equation** for a stretched string.

We now show that the wave equation is satisfied by any function $y(x - vt)$. Let $\alpha = x - vt$ and consider any wave function

$$y = y(x - vt) = y(\alpha)$$

We will use y' for the derivative of y with respect to α. Then, by the chain rule for derivatives,

$$\frac{\partial y}{\partial x} = \frac{\partial y}{\partial \alpha}\frac{\partial \alpha}{\partial x} = y'\frac{\partial \alpha}{\partial x}$$

and

$$\frac{\partial y}{\partial t} = \frac{\partial y}{\partial \alpha}\frac{\partial \alpha}{\partial t} = y'\frac{\partial \alpha}{\partial t}$$

Using $\partial\alpha/\partial x = 1$ and $\partial\alpha/\partial t = -v$ gives

$$\frac{\partial y}{\partial x} = y' \quad \text{and} \quad \frac{\partial y}{\partial t} = -vy'$$

Taking the second derivatives, we obtain

$$\frac{\partial^2 y}{\partial x^2} = y''$$

$$\frac{\partial^2 y}{\partial t^2} = -v\frac{\partial y'}{\partial t} = -v\frac{\partial y'}{\partial \alpha}\frac{\partial \alpha}{\partial t} = +v^2 y''$$

Thus,

$$\frac{\partial^2 y}{\partial x^2} = \frac{1}{v^2}\frac{\partial^2 y}{\partial t^2} \qquad \text{15-9}b$$

General wave equation

The same result can be obtained for any function of $x + vt$. Comparing Equations 15-9*a* and 15-9*b*, we see that the speed of propagation of the wave is $v = \sqrt{F/\mu}$, which is Equation 15-3.

optional

Example 15-3

Show by explicitly calculating the derivatives that the function $y(x,t) = A\sin(kx - \omega t)$ satisfies Equation 15-9*b*.

1. Take two partial derivatives of y with respect to x:

$$\frac{\partial y}{\partial x} = \frac{\partial}{\partial x}[A\sin(kx - \omega t)] = A\cos(kx - \omega t)\frac{\partial(kx)}{\partial x} = kA\cos(kx - \omega t)$$

$$\frac{\partial^2 y}{\partial x^2} = -k^2 A\sin(kx - \omega t)$$

2. Similarly, the two partial derivatives with respect to t are:

$$\frac{\partial y}{\partial t} = \frac{\partial}{\partial t}[A\sin(kx - \omega t)] = A\cos(kx - \omega t)\frac{\partial(-\omega t)}{\partial t} = -\omega A\cos(kx - \omega t)$$

$$\frac{\partial^2 y}{\partial t^2} = -\omega^2 A\sin(kx - \omega t)$$

3. Substitute these results in Equation 15-9*b*. The equation is satisfied provided that $v = \omega/k$:

$$-k^2 A\sin(kx - \omega t) = \frac{1}{v^2}[-\omega^2 A\sin(kx - \omega t)]$$

Remarks We have shown that the function $y = A\sin(kx - \omega t)$ is a solution to the wave equation. It describes a sinusoidal wave with speed $v = \omega/k$.

Exercise Show that any function $y(x + vt)$ satisfies Equation 15-9*b*.

A wave equation for sound waves can also be derived using Newton's laws. In one dimension, this equation is

$$\frac{\partial^2 s}{\partial x^2} = \frac{1}{v_s^2}\frac{\partial^2 s}{\partial t^2}$$

where s is the displacement of the medium in the x direction and v_s is the speed of sound.

15-2 Harmonic Waves

Harmonic Waves on a String

If one end of a string is attached to a vibrating tuning fork that is moving up and down with simple harmonic motion, a sinusoidal wave propagates along the string. The shape of the string at some instant in time is that of a sine function,* as shown in Figure 15-7. A sinusoidal wave such as that shown is called a **harmonic wave**. The distance after which the wave repeats (the distance between crests, for example) is the **wavelength** λ.

As the wave propagates, each point on the string moves up and down—perpendicular to the direction of propagation—in simple harmonic motion

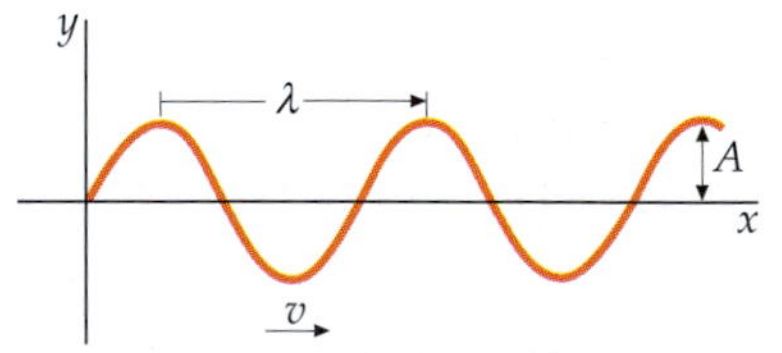

Figure 15-7 Harmonic wave at some instant in time. A is the amplitude and λ is the wavelength. For waves on a string, this figure can be obtained by taking a snapshot of the string.

* Whether this is a sine function or a cosine function depends on where the origin is chosen on the x axis.

with the frequency f of the tuning fork. During one period $T = 1/f$, the wave moves a distance of one wavelength, so its speed is given by

$$v = \frac{\lambda}{T} = f\lambda \qquad \text{15-10}$$

Since this relation arises from the definitions of wavelength and frequency, it applies to all harmonic waves.

A harmonic wave has a single frequency and wavelength. Other waves, such as wave pulses, can be considered to be made up of many harmonic waves of different frequencies. The superposition of harmonic waves is discussed in Chapter 16.

The sine function that describes the displacement in Figure 15-7 is

$$y(x) = A\sin(kx + \delta) \qquad \text{15-11}$$

where A is the **amplitude,** k is a constant called the **wave number,** and δ is a phase constant that depends on the choice of the origin $x = 0$. When dealing with a single harmonic wave, we are free to choose the origin anywhere, so we usually choose it to make $\delta = 0$.

Consider a point x_1 and another, x_2, one wavelength away, such that $x_2 = x_1 + \lambda$. The displacement at each point is the same: $y(x_1) = y(x_2)$. So,

$$\sin kx_1 = \sin kx_2 = \sin k(x_1 + \lambda) = \sin(kx_1 + k\lambda)$$

Therefore,

$$k\lambda = 2\pi$$

or

$$k = \frac{2\pi}{\lambda} \qquad \text{15-12}$$

Note that k has dimensions of m^{-1}. (Because the angle must be in radians, we sometimes write the units of k as rad/m.) Since $1/\lambda$ is the number of waves in a length of 1 meter, $k = 2\pi/\lambda$ is the number of waves in a distance of 2π meters.

For a wave traveling to the right with speed v, replace x in Equation 15-11 with $x - vt$ (see "Wave Pulses" in Section 15-1). With δ chosen to be zero, this gives

$$y(x, t) = A\sin k(x - vt) = A\sin(kx - kvt)$$

or

$$y(x, t) = A\sin(kx - \omega t) \qquad \text{15-13}$$

Harmonic wave function

where

$$\omega = kv \qquad \text{15-14}$$

is the angular frequency, which is related to the frequency f and period T by

$$\omega = 2\pi f = \frac{2\pi}{T} \qquad \text{15-15}$$

Substituting $\omega = 2\pi f$ into Equation 15-14 and using $k = 2\pi/\lambda$, we obtain

$$2\pi f = kv = \frac{2\pi}{\lambda}v$$

or $v = f\lambda$, which is Equation 15-10.

Example 15-4

The wave function for a harmonic wave on a string is $y(x,t) = (0.03\ \text{m}) \times \sin(2.2\ \text{m}^{-1}x - 3.5\ \text{s}^{-1}t)$. (*a*) In what direction does this wave travel, and what is its speed? (*b*) Find the wavelength, frequency, and period of this wave. (*c*) What is the maximum displacement of any string segment? (*d*) What is the maximum speed of any string segment?

Picture the Problem (*a*) To find the wave speed and direction, express $y(x,t)$ as a function of $x \pm vt$. (*b*) The wavelength, frequency, and period can be found from the wave number k and the angular frequency ω. (*c*) The maximum displacement of a string segment is the amplitude A. (*d*) The velocity of any string segment is dy/dt.*

(*a*)1. Express $y(x,t)$ as a function of $x \pm vt$:

$$y(x,t) = (0.03\ \text{m})\sin(2.2\ \text{m}^{-1}x - 3.5\ \text{s}^{-1}t)$$
$$= (0.03\ \text{m})\sin\left[(2.2\ \text{m}^{-1})\left(x - \frac{3.5\ \text{m}}{2.2\ \text{s}}t\right)\right]$$
$$= (0.03\ \text{m})\sin[(2.2\ \text{m}^{-1})(x - (1.59\ \text{m/s})t)]$$

2. Since the form is $x - vt$, the wave is moving to the right:

$$v = 1.59\ \text{m/s to the right}$$

(*b*)1. The wavelength λ is related to the wave number k, and the frequency and period are related to ω:

$$\lambda = \frac{2\pi}{k}, \quad f = \frac{\omega}{2\pi}, \quad T = \frac{1}{f} = \frac{2\pi}{\omega}$$

2. We find k, ω, and A by comparing the given equation with Equation 15-13:

$$y(x,t) = (0.03\ \text{m})\sin(2.2\ \text{m}^{-1}x - 3.5\ \text{s}^{-1}t)$$
$$y(x,t) = A\sin(kx - \omega t)$$
$$k = 2.2\ \text{m}^{-1}, \quad \omega = 3.5\ \text{s}^{-1}, \quad A = 0.03\ \text{m}$$

3. Substitute these values to calculate the wavelength λ, frequency f, and period T:

$$\lambda = \frac{2\pi}{k} = \frac{2\pi}{2.2\ \text{m}^{-1}} = 2.86\ \text{m}$$
$$f = \frac{\omega}{2\pi} = \frac{3.5\ \text{s}^{-1}}{2\pi} = 0.557\ \text{Hz}$$
$$T = \frac{1}{f} = \frac{1}{0.557\ \text{Hz}} = 1.80\ \text{s}$$

(*c*) The maximum displacement of a string segment is the amplitude A:

$$A = 0.03\ \text{m}$$

(*d*)1. Compute dy/dt to find the velocity of a string segment:

$$v_y = \frac{dy}{dt} = (0.03\ \text{m})\frac{d[\sin(2.2\ \text{m}^{-1}x - 3.5\ \text{s}^{-1}t)]}{dt}$$
$$= (0.03\ \text{m})(-3.5\ \text{s}^{-1})\cos(2.2\ \text{m}^{-1}x - 3.5\ \text{s}^{-1}t)$$
$$= -(0.105\ \text{m/s})\cos(2.2\ \text{m}^{-1}x - 3.5\ \text{s}^{-1}t)$$

2. The maximum speed occurs when the cosine function has the value of ± 1:

$$v_{y,\text{max}} = 0.105\ \text{m/s}$$

Remarks We have included the units to show how they work out. Often we will omit the units for simplicity.

*The derivative of a function of two variables, such as x and t, with one held constant is called a partial derivative and is usually written $\partial y/\partial t$.

Energy of Waves on a String Consider a string attached to a tuning fork. As the fork vibrates, it imparts energy to the segment of the string attached to it. For example, as the fork moves through its equilibrium position, it stretches the segment, increasing its potential energy, and the fork imparts a transverse speed to the segment, increasing its kinetic energy. As a wave moves along the string, energy is imparted to the other segments of the string.

We can calculate the kinetic energy of a segment from the wave function. Consider a segment of length Δx and mass $\mu\,\Delta x$. Its displacement from equilibrium is the wave function $y = A\sin(kx - \omega t)$. Its speed is dy/dt where x is considered to be fixed. The kinetic energy ΔK of the segment is then

$$\Delta K = \frac{1}{2}(\Delta m)v_y^2 = \frac{1}{2}(\mu\,\Delta x)\left(\frac{dy}{dt}\right)^2$$

Using $y = A\sin(kx - \omega t)$, we obtain $dy/dt = -\omega A\cos(kx - \omega t)$, so the kinetic energy of the segment is

$$\Delta K = \frac{1}{2}\mu\omega^2 A^2 \Delta x \cos^2(kx - \omega t) \qquad \text{15-16}a$$

The potential energy of a segment is the work done in stretching the string, and depends on the slope dy/dx. For small slopes, it can be shown to be related to the slope and tension F by*

$$\Delta U \approx \frac{1}{2}F\left(\frac{dy}{dx}\right)^2 \Delta x$$

Using $dy/dx = kA\cos(kx - \omega t)$ and $F = \mu v^2 = \mu\omega^2/k^2$ (Equation 15-3), we obtain for the potential energy

$$\Delta U = \frac{1}{2}\left(\frac{\mu\omega^2}{k^2}\right)k^2A^2\,\Delta x\cos^2(kx - \omega t)$$

or

$$\Delta U = \frac{1}{2}\mu\omega^2 A^2\,\Delta x \cos^2(kx - \omega t) \qquad \text{15-16}b$$

which is the same as the kinetic energy. The total energy of a string segment carrying a harmonic wave is

$$\Delta E = \Delta K + \Delta U = \mu\omega^2 A^2\,\Delta x\cos^2(kx - \omega t) \qquad \text{15-17}$$

We see that the energy of a segment varies with time. Since the average value of $\cos^2(kx - wt)$ at any point is $\frac{1}{2}$, the average energy is

$$\Delta E_{\text{av}} = \frac{1}{2}\mu\omega^2 A^2\,\Delta x \qquad \text{15-18}$$

This is the same result as for a mass $\mu\,\Delta x$ attached to a spring and oscillating with simple harmonic motion. However, for a mass attached to a spring, the potential energy is maximum when the displacement is maximum; for a string segment, the potential energy depends on the slope of the string and is maximum when the slope is maximum, which is at the equilibrium position of the segment, the same position for which the kinetic energy is maximum.

Consider a wave that has just reached some point P_1 at time t_1. The string to the left of P_1 has energy due to the simple harmonic motion of its elements,

* You will derive this result in Problem 123.

Figure 15-8

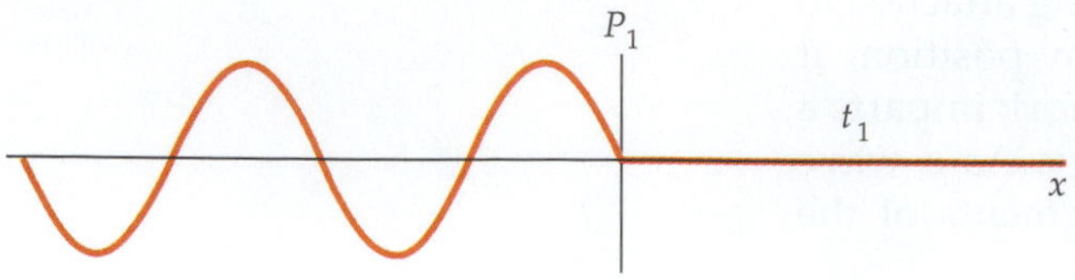

Figure 15-9

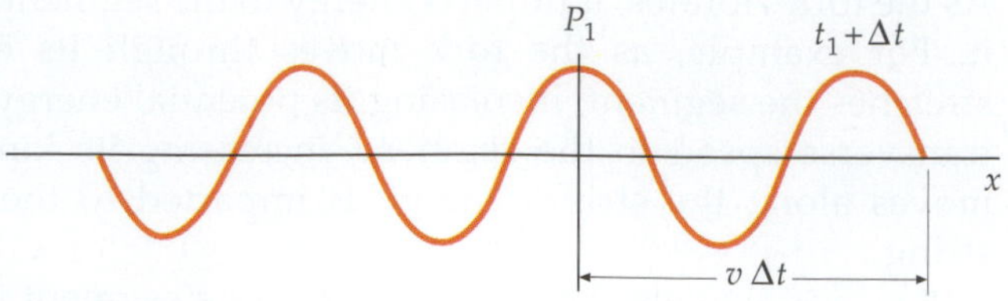

whereas the string to the right of P_1 has no energy (Figure 15-8). In the time interval Δt, the wave travels an additional distance $v\,\Delta t$ (Figure 15-9). The average energy transmitted past point P_1 during this time is the average energy in $\Delta x = v\,\Delta t$, which is

$$\Delta E_{\text{av}} = \frac{1}{2}\mu\omega^2 A^2 v\,\Delta t$$

The average rate at which energy is transmitted is the average power transmitted:

$$P_{\text{av}} = \frac{dE_{\text{av}}}{dt} = \frac{1}{2}\mu\omega^2 A^2 v \qquad \text{15-19}$$

The average energy and the power transmitted are proportional to the square of the amplitude of the wave.

Example 15-5

Waves of wavelength 35 cm and amplitude 1.2 cm move along a 15-m string that has a mass of 80 g and a tension of 12 N. (*a*) What are the speed and angular frequency of the waves? (*b*) What is the average total energy of the waves in the string?

Picture the Problem The speed of the waves is $v = \sqrt{F/\mu}$, where F is given and $\mu = m/L$. We find ω from $\omega = 2\pi f$, where $f = v/\lambda$. The energy is found using Equation 15-18.

(*a*)1. The speed is related to the tension and mass density:
$$v = \sqrt{F/\mu}$$

2. Calculate the linear mass density:
$$\mu = m/L = (80\text{ g}/15\text{ m}) = 5.33 \times 10^{-3}\text{ kg/m}$$

3. Substitute to find the speed:
$$v = \sqrt{\frac{F}{\mu}} = \sqrt{\frac{12\text{ N}}{5.33 \times 10^{-3}\text{ kg/m}}} = 47.4\text{ m/s}$$

4. The angular frequency is found from the frequency, which is found from the speed and wavelength:
$$\omega = 2\pi f = 2\pi\frac{v}{\lambda} = 2\pi\frac{47.4\text{ m/s}}{0.35\text{ m}} = 851\text{ rad/s}$$

(*b*) The average total energy of waves on the string is given by Equation 15-18 with $\mu\,\Delta x = \Delta m = 80$ g
$$\Delta E_{\text{av}} = \frac{1}{2}\mu\omega^2 A^2\,\Delta x = \frac{1}{2}(\Delta m)\omega^2 A^2$$
$$= \frac{1}{2}(8 \times 10^{-2}\text{ kg})(851\text{ rad/s})^2(0.012\text{ m})^2 = 4.17\text{ J}$$

Exercise: Calculate the rate at which energy is transmitted along the string. (*Answer* 13.2 W)

Harmonic Sound Waves

Harmonic sound waves can be generated by a tuning fork or loudspeaker that is vibrating with simple harmonic motion. The vibrating source causes the air molecules next to it to oscillate with simple harmonic motion about their equilibrium positions. These molecules collide with neighboring molecules, causing them to oscillate, thereby propagating the sound wave. Equation 15-13 describes a harmonic sound wave if the wave function $y(x, t)$ is replaced by $s(x, t)$, the displacement of the molecules from equilibrium.

$$s(x, t) = s_0 \sin(kx - \omega t) \qquad \text{15-20}$$

These displacements are along the direction of the motion of the wave, and lead to variations in the density and pressure of the air. Figure 15-10 shows the displacement of air molecules and the density changes caused by a sound wave at some fixed time. Since the pressure in a gas is proportional to its density, the change in pressure is maximum where the change in density is maximum. We can see from these figures that the pressure or density wave is 90° out of phase with the displacement wave.* When the displacement is zero, the pressure and density changes are either maximum or minimum. When the displacement is a maximum or minimum, the pressure and density changes are zero. A displacement wave given by Equation 15-20 thus implies a pressure wave given by

$$p = p_0 \sin(kx - \omega t - \pi/2) \qquad \text{15-21}$$

where p stands for the *change* in pressure from the equilibrium pressure and p_0 is the maximum value of this change. It can be shown that the maximum pressure amplitude p_0 is related to the maximum displacement amplitude s_0 by

$$p_0 = \rho \omega v s_0 \qquad \text{15-22}$$

where v is the speed of propagation and ρ is the equilibrium density of the gas. Thus, as a sound wave moves in time, the displacement of air molecules, the pressure, and the density all vary sinusoidally with the frequency of the vibrating source.

Exercise We can hear the sound of frequencies from about 20 Hz to about 20,000 Hz (though many people have rather limited hearing above 15,000 Hz). If the speed of sound in air is 340 m/s, what are the wavelengths that correspond to these extreme frequencies? (*Answer* λ = 17 m at 20 Hz, 1.7 cm at 20,000 Hz)

Energy of Sound Waves The average energy of a harmonic sound wave in a volume element ΔV is given by Equation 15-18 with A replaced by s_0, and the mass element $\Delta m = \mu\, \Delta x$ replaced by $\rho\, \Delta V$, where ρ is the average density of the medium.

$$\Delta E_{av} = \frac{1}{2}\rho\omega^2 s_0^2\, \Delta V \qquad \text{15-23}$$

The energy per unit volume is the average **energy density** η_{av}.

$$\eta_{av} = \frac{\Delta E_{av}}{\Delta V} = \frac{1}{2}\rho\omega^2 s_0^2 \qquad \text{15-24}$$

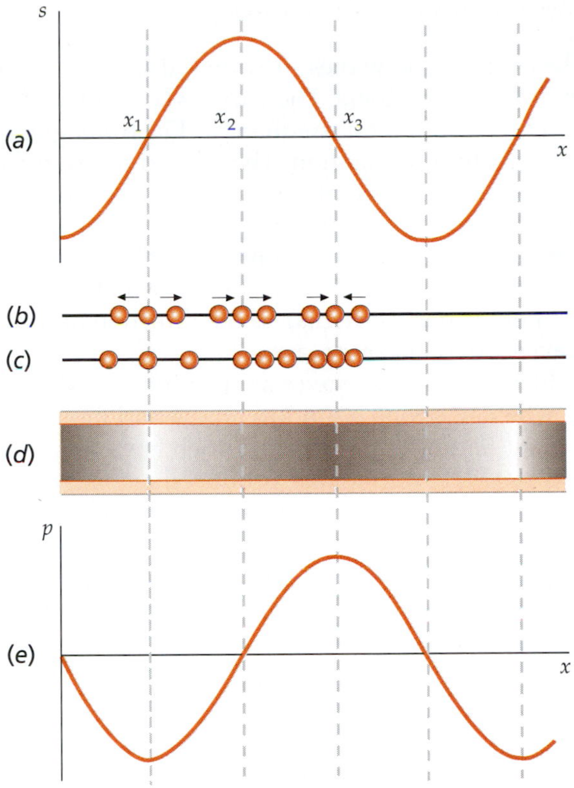

Figure 15-10 (*a*) Displacement from equilibrium of air molecules in a harmonic sound wave versus position at some instant. Points x_1 and x_3 are points of zero displacement. (*b*) Some representative molecules equally spaced at their equilibrium positions before the sound wave arrives. The arrows indicate the direction of the displacement that will be caused by the sound wave. (*c*) Molecules near points x_1, x_2, and x_3 after the sound wave arrives. Just to the left of x_1, the displacement is negative, indicating that the gas molecules are displaced to the left, away from point x_1, at this time. Just to the right of x_1, the displacement is positive, indicating that the molecules are displaced to the right, which is again away from point x_1. So at point x_1, the density is a minimum because the gas molecules on both sides are displaced away from that point. At point x_3, the density is a maximum because the molecules on both sides of that point are displaced toward point x_3. At point x_2 the density does not change because the gas molecules on both sides of that point have equal displacements in the same direction. (*d*) Density of the air at this time. The density is maximum at x_3 and minimum at x_1, which are both points of zero displacement. It is zero at point x_2, which is a maximum in displacement. (*e*) Pressure change, which is proportional to the density change versus position. The pressure change and displacement are 90° out of phase.

* In the arguments of sine or cosine functions we will always express phase angles in radians. However, in verbal descriptions we usually say that "two waves are 90° out of phase" rather than "two waves are out of phase by $\pi/2$ rad."

Electromagnetic Waves

Electromagnetic waves include light, radio waves, x rays, gamma rays, microwaves, and others. The various types of electromagnetic waves differ only in wavelength and frequency. Electromagnetic waves do not require a medium for propagation. They travel through a vacuum with speed c, which is a universal constant $c \approx 3 \times 10^8$ m/s. The wave function for electromagnetic waves is an electric field associated with the wave, $\vec{E}(x, t)$.* It is a solution of a wave equation similar to that for string waves and sound waves, which can be derived from the laws of electricity and magnetism.† The electric field is perpendicular to the direction of propagation, so electromagnetic waves are transverse waves.

Electromagnetic waves are produced when free electric charges accelerate, or when electrons bound to atoms and molecules make transitions to lower energy states. Radio waves, which have frequencies of about 1 MHz for AM and 100 MHz for FM, are produced by macroscopic electric currents oscillating in radio antennas. The frequency of the emitted waves equals the frequency of oscillation of the charges. Light waves, which have frequencies of the order of 10^{14} Hz, are generally produced by transitions of bound atomic charges. The spectrum of electromagnetic waves is discussed in Chapter 33.

15-3 Waves in Three Dimensions

Figure 15-11 shows two-dimensional circular waves on the surface of water in a ripple tank. These waves are generated by a point source moving up and down with simple harmonic motion. The wavelength is the distance between successive wave crests, which in this case are concentric circles. These circles are called **wavefronts.** For a point source of sound, the waves move out in three dimensions, and the wavefronts are concentric spherical surfaces.

The motion of any set of wavefronts can be indicated by **rays**, which are directed lines perpendicular to the wavefronts (Figure 15-12). For circular or spherical waves, the rays are radial lines.

Figure 15-11 Circular wavefronts diverging from a point source in a ripple tank.

Sound waves from a telephone handset spreading out in air. The waves have been made visible by sweeping out the space in front of the handset with a light source whose brightness is controlled by a microphone.

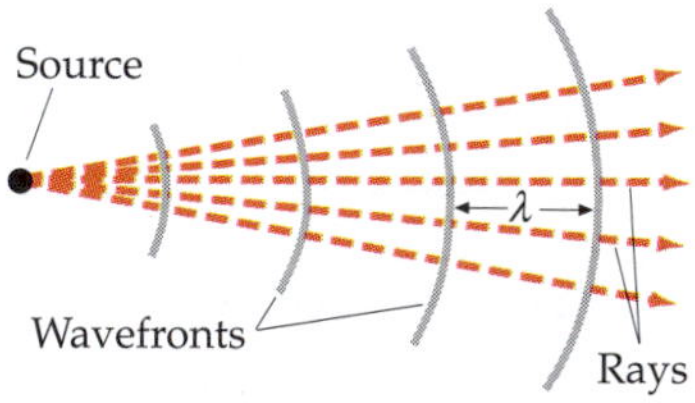

Figure 15-12 The motion of wavefronts can be represented by rays drawn perpendicular to the wavefronts. For a point source, the rays are radial lines diverging from the source.

* Electric fields are discussed in Chapter 22. There is also a magnetic field vector $\vec{B}$ that is perpendicular to $\vec{E}$ and to the direction of propagation. Either $\vec{E}$ or $\vec{B}$ can be considered to be the wave function for an electromagnetic wave. Magnetic fields are discussed in Chapter 29.

† The wave equation for electromagnetic waves is derived in Chapter 32.

In a homogeneous medium, such as air at constant density, a wave travels in straight lines in the direction of the rays, much like a beam of particles. At a great distance from a point source, a small part of the wavefront can be approximated by a plane, and the rays are approximately parallel lines; such a wave is called a **plane wave** (Figure 15-13). The two-dimensional analog of a plane wave is a line wave, which is a small part of a circular wavefront at a great distance from the source. Such waves can also be produced in a ripple tank by a line source, as in Figure 15-14.

Figure 15-13 Plane waves. At great distances from a point source, the wavefronts are approximately parallel planes, and the rays are approximately parallel lines perpendicular to the wavefronts.

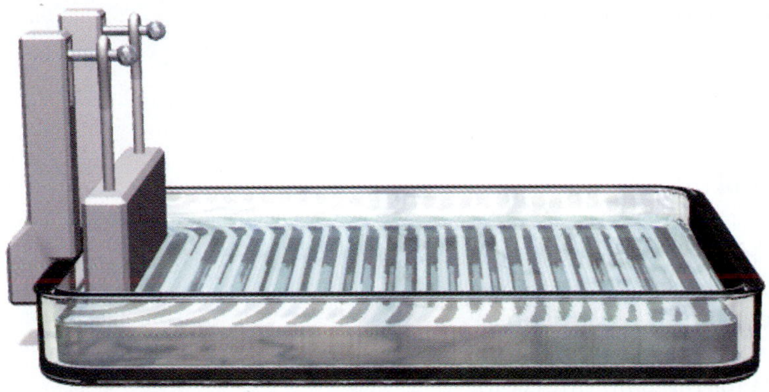

Figure 15-14 A two-dimensional analog of a plane wave can be generated in a ripple tank by a flat board that oscillates up and down in the water to produce wavefronts that are straight lines.

Wave Intensity

If a point source emits waves uniformly in all directions, the energy at a distance r from the source is distributed uniformly on a spherical surface of radius r and area $A = 4\pi r^2$. If P is the power emitted by the source, the power per unit area at a distance r from the source is $P/4\pi r^2$. The average power per unit area that is incident perpendicular to the direction of propagation is called the **intensity:**

$$I = \frac{P_{av}}{A} \qquad 15\text{-}25$$

The units of intensity are watts per square meter. At a distance r from a point source, the intensity is

$$I = \frac{P_{av}}{4\pi r^2} \qquad 15\text{-}26$$

Intensity due to a point source

The intensity of a three-dimensional wave varies inversely with the square of the distance from a point source.

There is a simple relation between the intensity of a wave and the energy density in the medium through which it propagates. Figure 15-15 shows a spherical wave that has just reached the radius r_1. The volume inside the radius r_1 contains energy because the particles in that region are oscillating with simple harmonic motion. The region outside r_1 contains no energy because the wave has not yet reached that region. After a short time Δt, the wave moves out past r_1 a short distance $\Delta r = v\,\Delta t$. The average energy in the spherical shell of surface area A, thickness $v\,\Delta t$, and volume $\Delta V = Av\,\Delta t$ is

Figure 15-15

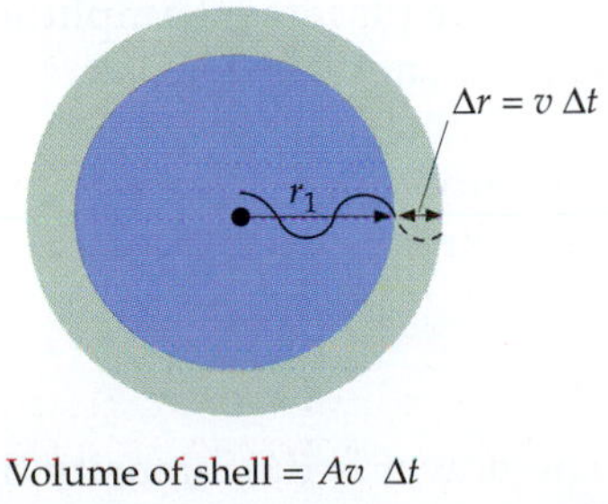

$$\Delta E_{av} = \eta_{av}\Delta V = \eta_{av}Av\,\Delta t$$

The rate of increase of energy is the power passing into the shell. The average incident power is

$$P_{av} = \frac{(\Delta E)_{av}}{\Delta t} = \eta_{av} A v$$

and the intensity of the wave is

$$I = \frac{P_{av}}{A} = \eta_{av} v \tag{15-27}$$

Thus, the intensity equals the product of the wave speed v and the average energy density η_{av}. This result applies to all waves. Substituting $\eta_{av} = \frac{1}{2}\rho\omega^2 s_0^2$ from Equation 15-24 for the energy density in a sound wave, we obtain

$$I = \eta_{av} v = \frac{1}{2}\rho\omega^2 s_0^2 v = \frac{1}{2}\frac{p_0^2}{\rho v} \tag{15-28}$$

where we have used $s_0 = p_0/\rho\omega v$ from Equation 15-22. This result—that the intensity of a sound wave is proportional to the square of the amplitude—is a general property of harmonic waves.

The human ear can accommodate a large range of sound-wave intensities, from about 10^{-12} W/m^2 (which is usually taken to be the threshold of hearing) to about 1 W/m^2 (a volume so loud it produces pain in most people). The pressure variations that correspond to these extreme intensities are about 3×10^{-5} Pa for the hearing threshold and 30 Pa for the pain threshold. (Recall that a pascal is a newton per square meter.) These very small pressure variations add or subtract to the normal atmospheric pressure of about 101,000 Pa.

Example 15-6

A loudspeaker diaphragm 30 cm in diameter is vibrating at 1 kHz with an amplitude of 0.020 mm. Assuming that the air molecules in the vicinity have this same amplitude of vibration, find (*a*) the pressure amplitude immediately in front of the diaphragm, (*b*) the sound intensity in front of the diaphragm, and (*c*) the acoustic power being radiated. (*d*) If the sound is radiated uniformly into the forward hemisphere, find the intensity at 5 m from the loudspeaker.

Picture the Problem (*a* and *b*) The pressure amplitude is calculated directly from $p_0 = \rho\omega v s_0$ (Equation 15-22), and the intensity from $I = \frac{1}{2}\rho\omega^2 s_0^2 v$ (Equation 15-28). (*c*) The power radiated is the intensity times the area of the diaphragm. (*d*) The area of a hemisphere of radius r is $2\pi r^2$. We can use Equation 15-26 if we replace the area $4\pi r^2$ by $2\pi r^2$.

(*a*) Equation 15-22 relates the pressure amplitude to the displacement amplitude, frequency, wave velocity, and air density:

$$p_0 = \rho\omega v s_0 = (1.29\ \text{kg/m}^3)2\pi(10^3\ \text{Hz})(340\ \text{m/s})(2 \times 10^{-5}\ \text{m})$$
$$= 55.1\ \text{N/m}^2$$

(*b*) Equation 15-28 relates the intensity to these same known quantities:

$$I = \tfrac{1}{2}\rho\omega^2 s_0^2 v$$
$$= (0.5)(1.29\ \text{kg/m}^3)[2\pi(10^3\ \text{Hz})]^2(2 \times 10^{-5}\ \text{m})^2(340\ \text{m/s})$$
$$= 3.46\ \text{W/m}^2$$

(*c*) The power is the intensity times the area of the diaphragm:

$$P = IA = (3.46\ \text{W/m}^2)\pi(0.15\ \text{m})^2 = 0.245\ \text{W}$$

(*d*) Calculate the intensity at $r = 5$ m assuming uniform radiation into the forward hemisphere:

$$I = \frac{P_{av}}{2\pi r^2} = \frac{0.245\ \text{W}}{2\pi\,(5\ \text{m})^2} = 1.56 \times 10^{-3}\ \text{W/m}^2$$

Remarks The assumption of uniform radiation in the forward hemisphere is not very good because the wavelength in this case ($\lambda = v/f = (340\ \text{m/s})/(1000\ \text{s}^{-1}) = 34$ cm) is not large compared with the speaker diameter. There is also some radiation in the backward direction as can be observed if you stand behind a loudspeaker.

Loudspeakers at a rock concert may put out more than 100 times as much power as the speaker in this example.

optional

Intensity Level and Loudness The psychological sensation of loudness varies approximately logarithmically rather than directly with intensity. We therefore use a logarithmic scale to describe the **intensity level** of a sound wave β, which is measured in **decibels** (dB) and defined by

$$\beta = 10 \log \frac{I}{I_0} \qquad \text{15-29}$$

Definition—Intensity level in dB

Here I is the intensity of the sound and I_0 is a reference level, which we take to be the threshold of hearing:

$$I_0 = 10^{-12}\ \text{W/m}^2 \qquad \text{15-30}$$

On this scale, the threshold of hearing is $\beta = 10 \log(I_0/I_0) = 0$ dB and the pain threshold ($I = 1\ \text{W/m}^2$) is $\beta = 10 \log(1/10^{-12}) = 10 \log 10^{12} = 120$ dB. Thus, the range of sound intensities from 10^{-12} W/m^2 to 1 W/m^2 corresponds to a range of intensity levels from 0 dB to 120 dB. Table 15-1 lists the intensity levels of some common sounds.

Table 15-1

Intensity and Intensity Level of Some Common Sounds ($I_0 = 10^{-12}$ W/m^2)

Source	I/I_0	dB	Description
	10^0	0	Hearing threshold
Normal breathing	10^1	10	Barely audible
Rustling leaves	10^2	20	
Soft whisper (at 5 m)	10^3	30	Very quiet
Library	10^4	40	
Quiet office	10^5	50	Quiet
Normal conversation (at 1 m)	10^6	60	
Busy traffic	10^7	70	
Noisy office with machines; average factory	10^8	80	
Heavy truck (at 15 m); Niagara Falls	10^9	90	Constant exposure endangers hearing
Old subway train	10^{10}	100	
Construction noise (at 3 m)	10^{11}	110	
Rock concert with amplifiers (at 2 m); jet takeoff (at 60 m)	10^{12}	120	Pain threshold
Pneumatic riveter; machine gun	10^{13}	130	
Jet takeoff (nearby)	10^{15}	150	
Large rocket engine (nearby)	10^{18}	180	

optional

Example 15-7

A sound absorber attenuates the sound level by 30 dB. By what factor is the intensity decreased?

From Table 15-1, we can see that for every 10-dB drop in the intensity level, the intensity decreases by a factor of 10. Thus, if the sound level drops 30 dB then the intensity drops by a factor of $10^3 = 1000$.

The sensation of loudness depends on the frequency as well as the intensity of a sound. Figure 15-16 is a plot of intensity level versus frequency for sounds of equal loudness to the human ear. (In this figure, the frequency is plotted on a logarithmic scale to display the wide range of frequencies from 20 Hz to 10 kHz.) We note from this figure that the human ear is most sensitive at about 4 kHz for all intensity levels.

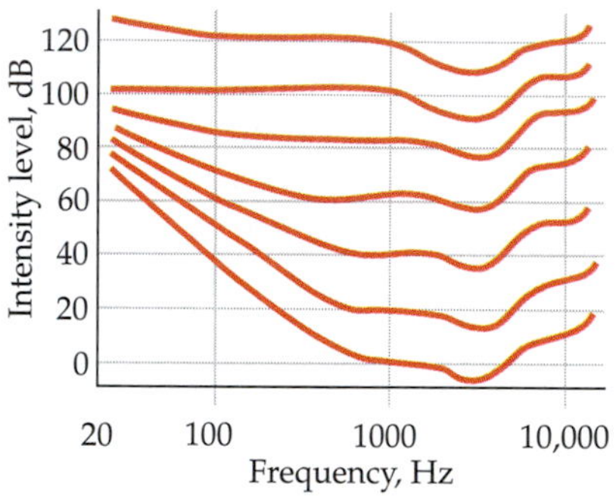

Figure 15-16 Intensity level versus frequency for sounds perceived to be of equal loudness. The lowest curve is below the threshold for hearing of all but about 1% of the population. The second lowest curve is approximately the hearing threshold for about 50% of the population.

Example 15-8

A barking dog delivers about 1 mW of power. (*a*) If this power is uniformly distributed in all directions, what is the sound intensity level at a distance of 5 m? (*b*) What would be the intensity level of two dogs barking at the same time if each delivered 1 mW of power?

Picture the Problem The intensity level is found from the intensity, which is found from $I = P/(4\pi r^2)$. For two dogs, the intensities add.

(*a*) 1. The intensity level is related to the intensity:

$$\beta = 10 \log \frac{I}{I_0}$$

2. Calculate the intensity at $r = 5$ m:

$$I = \frac{P}{4\pi r^2} = \frac{10^{-3}\text{ W}}{4\pi (5\text{ m})^2} = 3.18 \times 10^{-6}\text{ W/m}^2$$

3. Use your result to find the intensity level at 5 m:

$$\beta = 10 \log \frac{I}{I_0} = 10 \log \frac{3.18 \times 10^{-6}}{10^{-12}} = 10 \log(3.18 \times 10^6)$$

$$= 10(\log 3.18 + \log 10^6) = 10(0.50 + 6) = 65.0\text{ dB}$$

(*b*) If I_1 is the intensity for one dog, the intensity for two dogs is $I_2 = 2I_1$. The intensity level for two dogs is then:

$$\beta_2 = 10 \log \frac{I_2}{I_0} = 10 \log \frac{2I_1}{I_0} = 10 \log 2 + 10 \log \frac{I_1}{I_0}$$

$$= 10 \log 2 + \beta_1 = 3.01 + 65.0 = 68.0\text{ dB}$$

Remark We can see from this example that whenever the intensity is doubled, the intensity level increases by 3 dB.

15-4 Waves Encountering Barriers

Reflection and Refraction

When a wave is incident on a boundary that separates two regions of differing wave speed, part of the wave is reflected and part is transmitted. Figure 15-17*a* shows a pulse on a light string that is attached to a heavier string. In this case, the pulse reflected at the boundary is inverted. If the second string is lighter than the first (Figure 15-14*b*), the reflected pulse is not inverted. In either case, the transmitted pulse is not inverted. If the string is tied to a fixed point the pulse is reflected and inverted.

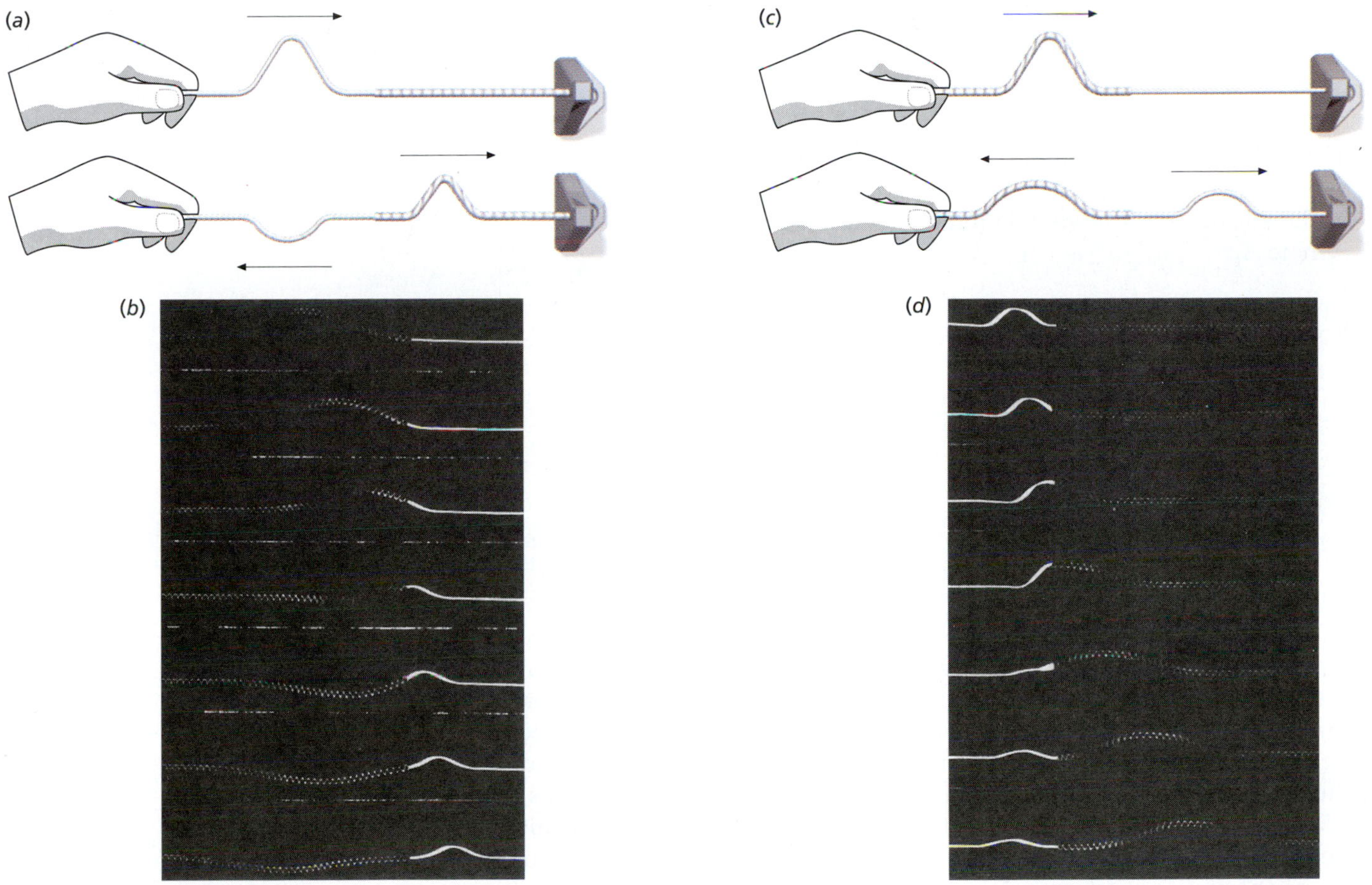

Figure 15-17 (*a*) A wave pulse traveling on a light string attached to a heavier string in which the wave speed is smaller. The reflected pulse is inverted, whereas the transmitted pulse is not. (*b*) Photograph of a similar pulse on a light spring attached to a heavier spring. (*c*) A wave pulse traveling on a heavy string attached to a light string in which the wave speed is greater. In this case, the reflected pulse is not inverted. (*d*) Photograph of a similar pulse on a heavy spring attached to a lighter spring.

Example 15-9

Two wires of different linear mass densities are soldered together end to end and then stretched under a tension F (the tension is the same in both wires). The wave speed in the first wire is twice that in the second wire. When a harmonic wave traveling in the first wire is reflected at the junction of the wires, the reflected wave has half the amplitude of the transmitted wave. (*a*) If the amplitude of the incident wave is A, what are the amplitudes of the reflected and transmitted waves? (*b*) Assuming no loss in the wire, what fraction of the incident power is reflected at the junction and what fraction is transmitted?

Picture the Problem By conservation of energy, the power incident on the junction equals the power reflected plus the power transmitted. Each power is expressed in Equation 15-19 as a function of the density μ, amplitude A, frequency ω, and wave speed v (Figure 15-18). The angular frequencies of all the waves are equal. Since the reflected wave and incident wave are in the same medium, they have the same wave speed v_1. We are given that the speed in the second wire is $v_2 = \frac{1}{2}v_1$. Then $\mu_2 = F/v_2^2 = 4\mu_1$.

Figure 15-18

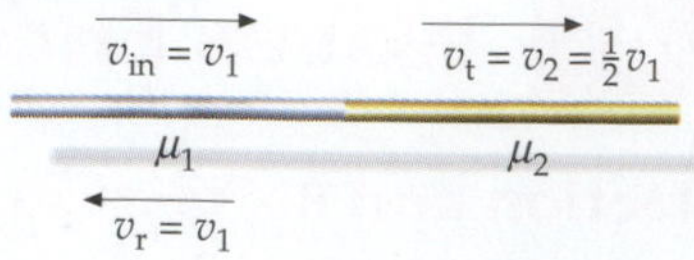

(*a*)1. By conservation of energy, the incident power equals the transmitted power plus the reflected power:

$$P_{in} = P_t + P_r$$

2. Express the incident, reflected, and transmitted power in terms of the mass density, amplitude, angular frequency, and wave speed:

$$\tfrac{1}{2}\mu_1\omega^2A_{in}^2v_1 = \tfrac{1}{2}\mu_2\omega^2A_t^2v_2 + \tfrac{1}{2}\mu_1\omega^2A_r^2v_1$$

3. Express the mass densities in terms of the wave speeds, and cancel the common terms:

$$\tfrac{1}{2}(F/v_1^2)\omega^2A_{in}^2v_1 = \tfrac{1}{2}(F/v_2^2)\omega^2A_t^2v_2 + \tfrac{1}{2}(F/v_1^2)\omega^2A_r^2v_1$$

$$A_{in}^2/v_1 = A_t^2/v_2 + A_r^2/v_1$$

4. The following relations are given:

$$v_2 = \tfrac{1}{2}v_1, \qquad A_r = \tfrac{1}{2}A_t$$

5. Substitute the known relations, and eliminate A_t:

$$A_{in}^2/v_1 = (2A_r)^2/(v_1/2) + A_r^2/v_1$$

$$A_{in}^2 = 8\,A_r^2 + A_r^2 = 9A_r^2$$

6. Solve for A_r:

$$A_r = \tfrac{1}{3}A_{in}$$

7. Use this result to find A_t:

$$A_t = 2A_r = \tfrac{2}{3}A_{in}$$

(*b*)1. Write the reflected power in terms of the incident power:

$$P_r = \tfrac{1}{2}\mu_1\omega^2A_r^2v_1 = \tfrac{1}{2}\mu_1\omega^2(\tfrac{1}{3}A_{in})^2v_1 = \tfrac{1}{9}P_{in}$$

2. Write the transmitted power in terms of the incident power:

$$P_t = \tfrac{1}{2}\mu_2\omega^2A_t^2v_2 = \tfrac{1}{2}(4\mu_1)\omega^2(\tfrac{2}{3}A_{in})^2\tfrac{1}{2}v_1 = \tfrac{8}{9}P_{in}$$

Remarks The reflected wave is inverted relative to the incident wave, so it is 180° out of phase with it. When the displacement of the wire to the left of the junction is y_1 due to the incident wave, it is $-(y_1/3)$ due to the reflected wave. These add (according to the principle of superposition, to be studied in the next chapter) giving a total displacement of $2y_1/3$, which equals the displacement that occurs to the right of the junction due to the transmitted wave. It can be shown that, given the ratio of the wave speeds, the amplitudes of the transmitted and reflected waves can be determined from the conditions that the displacement and slope of the wire must be continuous at the junction.

In three dimensions, a boundary between two regions of differing wave speed is a surface. Figure 15-19 shows a ray incident on such a boundary surface. This example could be a sound wave in air striking a solid or liquid surface. The reflected ray makes an angle with the normal to the surface equal to that of the incident ray, as shown.

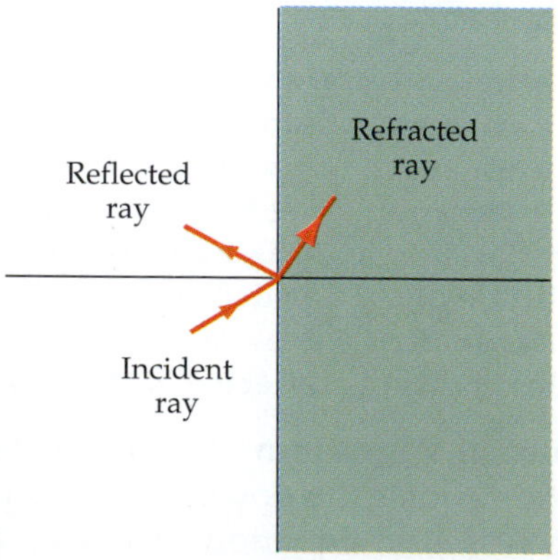

Figure 15-19 A wave striking a boundary surface between two media in which the wave speed differs. Part of the wave is reflected and part is transmitted. The change in direction of the transmitted ray is called refraction.

The transmitted ray is bent toward or away from the normal—depending on whether the wave speed in the second medium is less or greater than that in the incident medium. The bending of the transmitted ray is called **refraction**. When the wave speed in the second medium is greater than that in the incident medium (as occurs when a light wave in glass or water is refracted into the air), the ray describing the direction of propagation is bent away from the normal, as shown in Figure 15-20. As the angle of incidence is increased, the angle of refraction increases, until a critical angle of incidence is reached for which the angle of refraction is 90°. For incident angles greater than the critical angle, there is no refracted ray, a phenomenon known as **total internal reflection**.

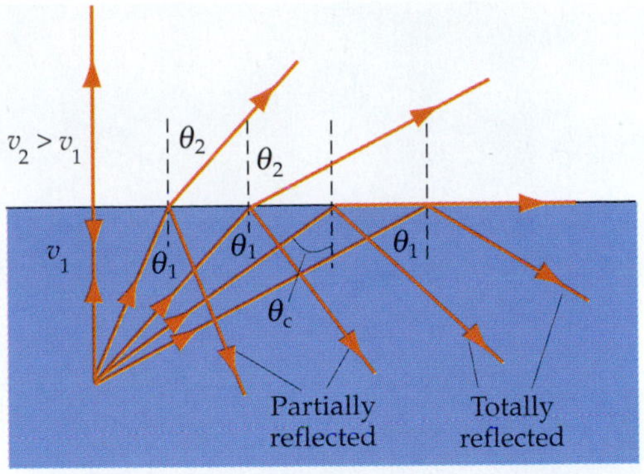

Figure 15-20 Light from a source in the water is bent away from the normal when it enters air. For angles of incidence above a critical angle, there is no transmitted ray, a condition known as total internal reflection.

The amount of energy reflected from a surface depends on the surface. Flat walls, floors, and ceilings make good reflectors for sound waves, whereas less rigid and porous materials, such as cloth in draperies and furniture coverings, absorb much of the incident sound. The reflection of sound waves plays an important role in the design of a lecture hall, a library, or a music auditorium. If a lecture hall has many flat reflecting surfaces, speech is difficult to understand because of the many echoes that arrive at different times at the listener's ear. Absorbent material is often placed on the walls and ceiling to reduce such reflections. In a concert hall, a reflecting shell is placed behind the orchestra, and reflecting panels are hung from the ceiling to reflect and direct the sound back toward the listeners.

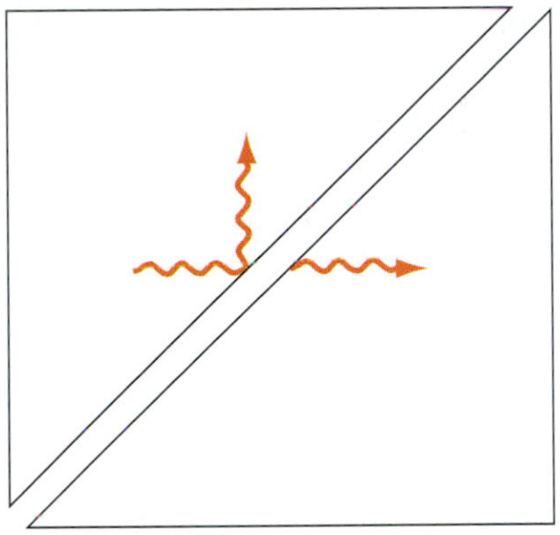

Figure 15-21 Penetration of optical barrier. If the pieces of glass are close enough, some light gets through the barrier.

Tunneling In total internal reflection, the wave function does not immediately drop to zero at the surface but instead decreases exponentially and becomes negligible within a few wavelengths of the surface. In Figure 15-21, light is totally reflected at the right surface of the glass. When another piece of glass is brought near the surface, some of the light is transmitted across the barrier. This is called **barrier penetration** or **tunneling**. Figure 15-22 shows a barrier penetration by water waves in a ripple tank.

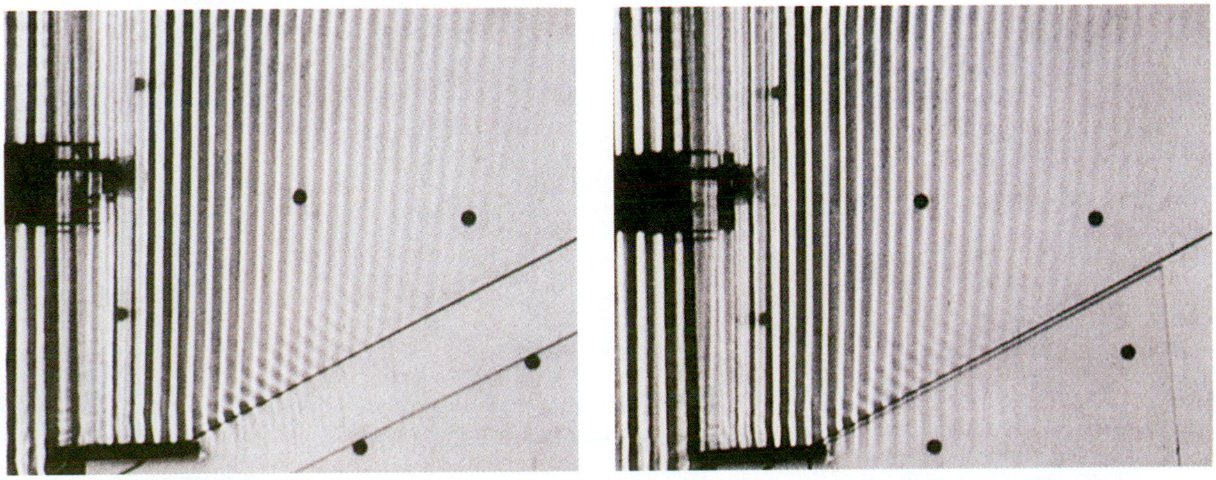

(*a*) (*b*)

Figure 15-22 Penetration of a barrier by water waves in a ripple tank. In (*a*), the waves are totally reflected from a large gap in deeper water. When the gap is very narrow, as in (*b*), a transmitted wave appears.

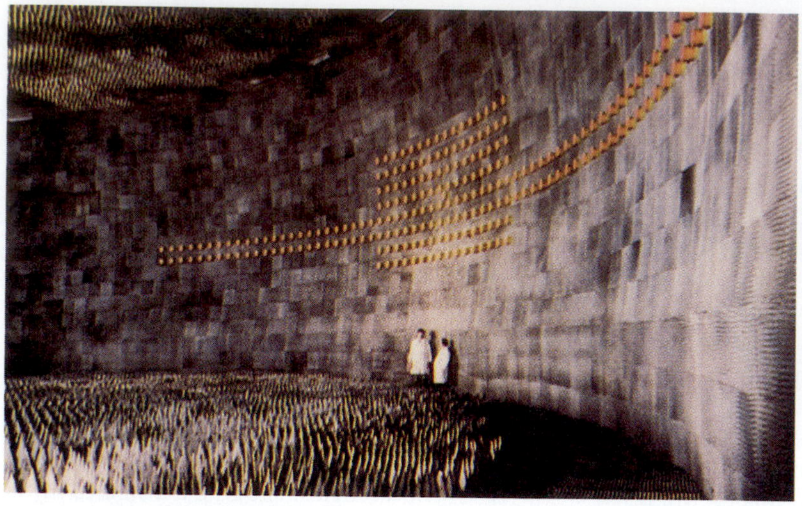

(Left) This anechoic chamber at the Naval Research Laboratory is used in the testing of electronic equipment. *(Right)* Davies Symphony Hall in San Francisco. The plastic reflectors above the orchestra reflect the sound waves out toward the audience.

Diffraction

A wave encountering an obstacle tends to bend around the obstacle. When the wave encounters a barrier with a small aperture, the wave bends and spreads out as a spherical or circular wave (Figure 15-23). This bending of the wavefront is called **diffraction**. In contrast, a beam of particles falling upon a barrier with an aperture either is halted by the barrier or passes through cleanly with no change in the direction of the particles (Figure 15-24). Diffraction is one of the key characteristics that distinguishes waves from particles.*

Though waves encountering an obstacle or aperture always bend, or diffract, to some extent, the amount of diffraction depends on whether the wavelength is small or large relative to the size of the obstacle or aperture. If the wavelength is large relative to the aperture, as in Figure 15-23, the diffraction effects are large. There the waves spread out as they pass through the aperture—as if the waves were originating from a point source. On the other hand, if the wavelength is small relative to the aperture, the effect of diffraction is small, as shown in Figure 15-25. Near the edges of the aperture the

Figure 15-23 Plane waves in a ripple tank meeting a barrier with an opening that is small compared to the wavelength λ. To the right of the barrier are circular waves that are concentric about the opening, just as if there were a point source at the opening.

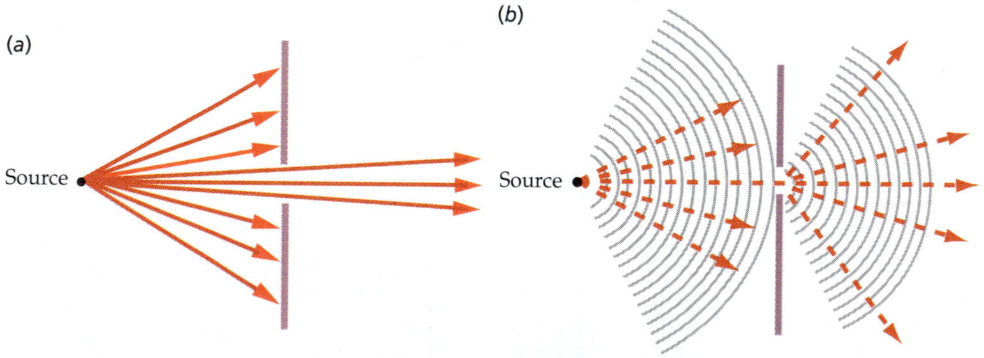

Figure 15-24 Comparison of particles and waves passing through a narrow opening in a barrier. (*a*) Transmitted particles are confined to a narrow angle. (*b*) Transmitted waves radiate widely from the aperture, which acts like a point source of circular waves.

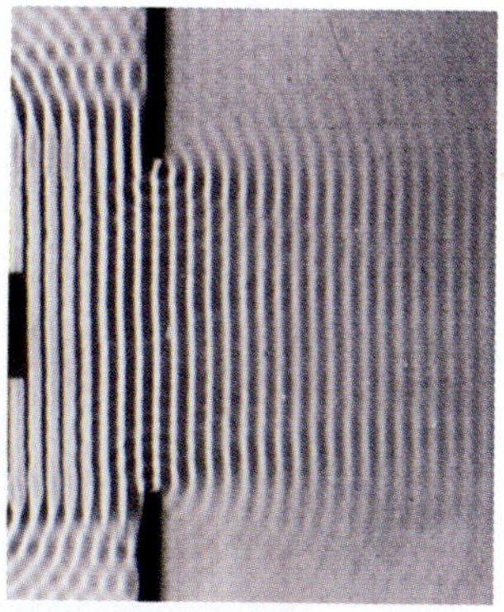

Figure 15-25 Plane waves in a ripple tank meeting a barrier with an opening that is large compared to λ. The barrier has a noticeable effect only near the edges of the opening.

* Diffraction is closely related to interference, which we discuss in Chapter 16. We show how diffraction arises when we study the interference and diffraction of light in Chapter 35.

wavefronts are distorted and the waves appear to bend slightly. For the most part, however, the wavefronts are not affected and the waves propagate in straight lines in the direction of the rays, much like a beam of particles. The approximation that these waves propagate in straight lines in the direction of the rays with no diffraction is known as the **ray approximation.**

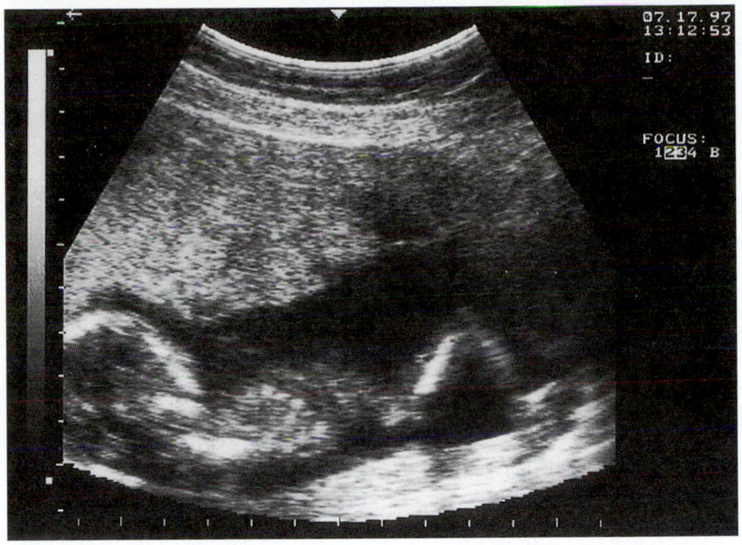

Sonogram of a pregnant woman showing the fetus in the womb.

Because the wavelengths of audible sound (which range from a few centimeters to several meters) are generally large compared with apertures and obstacles (doors or windows, for example), diffraction of sound waves is a common phenomenon. On the other hand, the wavelengths of visible light (4×10^{-7} to 7×10^{-7} m) are so small compared with the size of ordinary objects and apertures that the diffraction of light is not easily noticed; light appears to travel in straight lines. Nevertheless, the diffraction of light is an important phenomenon, which we study in detail in Chapter 35.

Diffraction places a limitation on how accurately small objects can be located by reflecting waves off of them, and on how well details of the objects can be resolved. Waves are not reflected appreciably from objects smaller than the wavelength, so detail cannot be observed on a scale smaller than the wavelength used. If waves of wavelength λ are used to locate an object, its position can be known only to within $\pm\lambda$. Since the smallest wavelength of audible sound is about 2 cm, the location of an object cannot be fixed to better than ± 2 cm using audible sound.

Sound waves with frequencies above 20,000 Hz are called **ultrasonic waves.** Because of their very small wavelengths, narrow beams of ultrasonic waves can be sent out and reflected from small objects. Bats can emit and detect frequencies up to about 120,000 Hz, corresponding to a wavelength of 2.8 mm, which they use to locate small prey such as moths. Ships use a device called sonar (from *so*und *n*avigation *a*nd *r*anging) to detect the outlines of submarines and other submerged objects with ultrasonic waves. In medicine, ultrasonic waves are used for diagnostic purposes. To create a sonogram, ultrasonic waves are passed through the body and information about the frequency and intensity of the transmitted and reflected waves is processed to construct a three-dimensional picture of the body's interior.

15-5 The Doppler Effect

When a wave source and a receiver are moving relative to each other, the frequency observed is not the same as that emitted. When they are moving toward each other, the observed frequency is greater than the source frequency; when they are moving away from each other, the observed frequency is less than the source frequency. This is called the **Doppler effect**. A familiar example is the change in pitch of a car horn as the car approaches or recedes.

The change in frequency of a sound wave is slightly different depending on whether the source or the receiver moves relative to the medium. When the source moves, the wavelength changes, and the new frequency f' is found by first finding the new wavelength λ' and then computing $f' = v/\lambda'$. When the source is stationary and the receiver moves, the wavelength is unchanged, and the frequency is different simply because the receiver moves past more or fewer waves in a given time.

Consider a source of frequency f_0 moving with speed u_s relative to the medium. The waves in front of the source are compressed, whereas behind the source, they are farther apart, as shown in Figure 15-26. Let v be the speed of the waves relative to the medium. This speed depends only on the properties of the medium and not on the motion of the source. In some time Δt, the source emits a number of waves $N = f_0\,\Delta t$. The first wavefront moves a distance $v\,\Delta t$, while the source moves a distance $u_s\,\Delta t$. The wavelength λ' is the distance $(v \pm u_s)\Delta t$ divided by the number of waves:

$$\lambda' = \frac{(v \pm u_s)\Delta t}{N} = \frac{(v \pm u_s)\Delta t}{f_0 \Delta t} = \frac{v \pm u_s}{f_0} \qquad \text{15-31}$$

In front of the source, the wavelength decreases, so the minus sign in Equation 15-31 applies. Behind the source, the plus sign applies.

The number of waves that pass a receiver in time Δt is the number of waves in the distance $v_r \Delta t$, where v_r is the speed of the waves relative to the receiver (Figure 15-27):

$$N = \frac{v_r \Delta t}{\lambda'} = \frac{(v \pm u_r)\Delta t}{\lambda'}$$

The frequency observed is this number of waves divided by the time interval:

$$f' = \frac{N}{\Delta t} = \frac{v \pm u_r}{\lambda'} \qquad \text{15-32}$$

If the receiver is stationary, $u_r = 0$, and the frequency is

$$f' = \frac{v}{\lambda'} = \frac{v}{v \pm u_s} f_0 = \frac{1}{1 \pm u_s/v} f_0 \qquad \text{(moving source)} \qquad \text{15-33}$$

When the source is moving toward the receiver, the frequency increases so the minus sign applies.

If the source is stationary, $\lambda' = \lambda_0 = v/f_0$, and the observed frequency is

$$f' = \frac{v \pm u_r}{v/f_0} = \left(1 \pm \frac{u_r}{v}\right) f_0 \qquad \text{(moving receiver)} \qquad \text{15-34}$$

We can combine Equations 15-33 and 15-34 to cover the general case of either source, receiver, or both moving:

$$f' = \frac{v \pm u_r}{\lambda'} = \frac{v \pm u_r}{v \pm u_s} f_0 = \frac{1 \pm u_r/v}{1 \pm u_s/v} f_0 \qquad \text{15-35}$$

The correct choices for the plus or minus signs are most easily determined by remembering that the frequency increases when the source and receiver are moving toward each other, whereas it decreases when they are moving away from each other. Thus, for example, if the source is moving toward the receiver and the receiver is moving toward the source, the plus sign is used in the numerator and the minus sign is used in the denominator.

It can be shown (see Problem 80) that if both u_s and u_r are much smaller than the wave speed v, then the shift in frequency is given approximately by

$$\frac{\Delta f}{f_0} \approx \pm \frac{u}{v} \qquad (u << v) \qquad \text{15-36}$$

where $u = u_s \pm u_r$ is the relative speed of the source and receiver.

If the medium is moving (for example, if air is the medium and there is a wind blowing), the wave speed v is replaced by $v' = v \pm u_w$, where u_w is the speed of the wind.

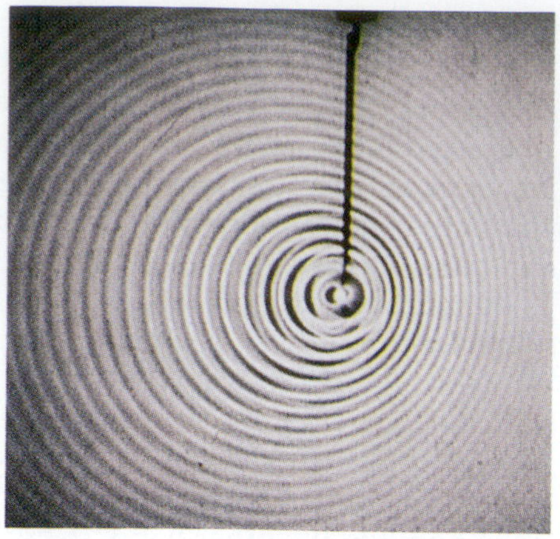

(a)

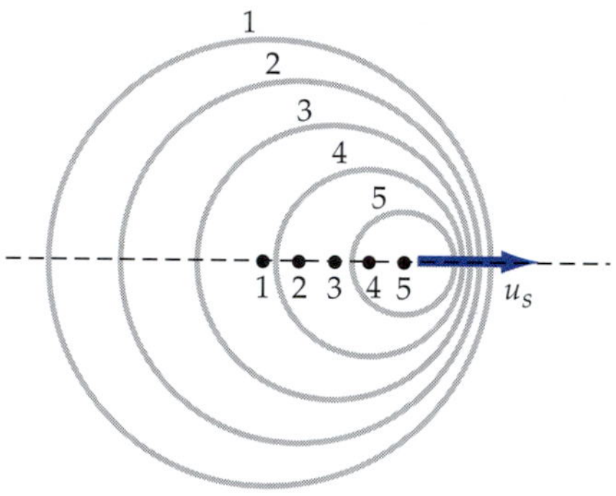

(b)

Figure 15-26 (*a*) Waves in a ripple tank produced by a point source moving to the right. The wavefronts are closer together in front of the source and farther apart behind the source. (*b*) Successive wavefronts emitted by a point source moving with speed u_s to the right. The numbers of the wavefronts correspond to the positions of the source when the wave was emitted.

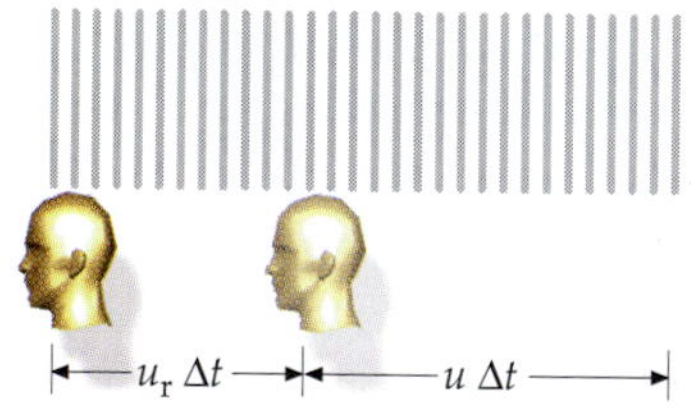

Figure 15-27 The number of waves passing a stationary receiver in time Δt is the number in the distance $v\,\Delta t$, where v is the wave speed. If the receiver moves toward the source with speed u_r, he passes the additional number of waves in the distance $u_r\,\Delta t$.

A familiar example of the Doppler effect is the radar used by police to measure the speed of a car. Electromagnetic waves emitted by the radar transmitter strike the moving car. The car acts as both a moving receiver and a moving source as the waves reflect off it back to the radar receiver. Since electromagnetic waves travel at the speed of light, $c = 3 \times 10^8$ m/s, the approximation $u \ll v$ is always valid and Equation 15-36 can be used to calculate the Doppler shift.

Example 15-10

The frequency of a car horn is 400 Hz. Find (*a*) the wavelength of the sound, and (*b*) the frequency observed if the car moves with a speed $u_s = 34$ m/s (about 122 km/h) through still air toward a stationary receiver. Take the speed of sound in air to be 340 m/s. (*c*) Find the frequency observed if the car is stationary and a receiver moves with a speed $u_s = 34$ m/s toward the car.

Picture the Problem (*a*) The waves in front of the source are compressed, so we use the minus sign in Equation 15-30. (*b*) We calculate the frequency from $f' = v/\lambda'$. (*c*) For a moving receiver, the wavelength does not change; the receiver merely passes more waves. We use the plus sign in Equation 15-34.

(*a*) Calculate the wavelength in front of the car, λ':

$$\lambda' = \frac{v - u_s}{f_0} = \frac{340\text{ m/s} - 34\text{ m/s}}{400\text{ s}^{-1}} = 0.765\text{ m}$$

(*b*) Use your result to find the observed frequency:

$$f = \frac{v}{\lambda'} = \frac{340\text{ m/s}}{0.765\text{ m}} = 444\text{ Hz}$$

(*c*) For a moving receiver, the observed frequency is given by Equation 15-34:

$$f' = f_0\left(1 + \frac{u_r}{v}\right) = f_0\left(1 + \frac{34}{340}\right) = f_0(1.10) = 440\text{ Hz}$$

Exercise As a train moving at 90 km/h is approaching a stationary listener, it blows its horn, which has a frequency of 630 Hz. (*a*) What is the wavelength of the sound waves in front of the train? (*b*) What frequency is heard by the listener? (Use 340 m/s for the speed of sound.) (*Answers* (*a*) $\lambda_f = 0.5$ m, (*b*) $f' = 680$ Hz)

Example 15-11 *try it yourself*

The ratio of the frequency of a note to the frequency of the semitone above it on the diatonic scale is about 15:16. How fast is a car going if its horn drops a semitone as it passes you? (Use $v = 340$ m/s for the speed of sound.)

Picture the Problem Let u be the speed of the car and f_0 be the original frequency. The frequency observed as the car approaches f' is greater than f_0, and the frequency observed as the car recedes f'' is less than f_0. Set the ratio $f''/f' = 15/16$ and solve for u.

Cover the column to the right and try these on your own before looking at the answers.

Steps	***Answers***
1. Write the frequency observed as the car approaches in terms of f_0.	$f' = f_0(1 + u/v)$

2. Write the frequency observed as the car recedes in terms of f_0. — $f'' = f_0(1 - u/v)$

3. Set the ratio f''/f' equal to 15/16. — $\frac{f''}{f'} = \frac{f_0(1 - u/v)}{f_0(1 + u/v)} = \frac{15}{16}$

4. Solve for u. — $u = \frac{v}{31} = 11.0\text{ m/s} = 39.6\text{ km/h}$

Example 15-12 *try it yourself*

The radar unit in a stationary police car sends out electromagnetic waves of frequency f_0 that travel at the speed of light c. The waves reflect from a speeding car moving at speed u away from the police car. A frequency difference Δf between the emitted radar and the waves reflected from the speeding car is detected at the police car.* Find u in terms of f_0 and Δf.

Picture the Problem The frequency of the radar that strikes the speeding car f' is less than f_0 because of the Doppler shift given by Equation 15-36. The car then acts as a moving source emitting waves of frequency f'. The police unit detects waves of frequency $f'' < f'$ because of the Doppler shift due to the moving source. The frequency difference is $f_0 - f''$.

Cover the column to the right and try these on your own before looking at the answers.

Steps	Answers
1. Write the frequency f' received by the moving car in terms of f_0, u, and c.	$f' = (1 - u/c)f_0$
2. Write the frequency f'' received by the police car in terms of f', u, and c.	$f'' = (1 - u/c)f'$
3. Use your result in step 1 to eliminate f'. Simplify using the fact that u is much less than c.	$f'' = (1 - u/c)^2 f_0 = (1 - 2u/c + u^2/c^2)f_0 \approx (1 - 2u/c)f_0$
4. Calculate Δf and solve for u.	$\Delta f = f_0 - f'' = \frac{2u}{c}f_0, \quad u = \frac{\Delta f}{2f_0}c$

Exercise Calculate Δf if $f_0 = 1.5 \times 10^9$ Hz, $c = 3 \times 10^8$ m/s, and $u = 50$ m/s. (*Answer* $\Delta f = 500$ Hz)

*The difference in frequency between two waves of nearly equal frequency is easy to detect because the two waves interfere to produce a wave whose amplitude oscillates with frequency Δf, which is called the beat frequency. Interference and beats are discussed in Chapter 16.

The Doppler Shift and Relativity We see from Example 15-10 (and Equations 15-33, 15-34, and 15-35) that the magnitude of the Doppler shift in frequency depends on whether it is the source or the receiver that is moving relative to the medium. For sound, these two situations are physically different. For example, if you move relative to still air, you feel air rushing past you. In your reference frame, there is a wind. For sound waves in air, therefore, we can tell whether the source or receiver is moving by noting if there is a wind in the reference frame of the source or the receiver. However, light and other electromagnetic waves propagate through empty space in which there is no medium. There is no "wind" to tell us whether the source or receiver is moving. According to Einstein's theory of relativity, absolute motion cannot be detected, and all observers measure the same speed c for light independent of their motion relative to the source. Thus, Equation 15-35 cannot be correct for the Doppler shift for light. Two modifications must be made in calculating the relativistic Doppler effect for light. First, the speed of waves passing a receiver is c independent of the motion of the receiver. Second, the time interval between the emission of the first wave and the Nth wave, which is $\Delta t = N/f_0$ in the reference frame of the source, is different in the reference frame of the receiver when they are in relative motion because of relativistic time dilation. (We discuss time dilation and the relativistic Doppler effect in Chapter 39.) The result is that the frequency received depends only on the relative speed of approach or recession u, and is related to the frequency emitted by

$$f = \sqrt{\frac{1 + u/c}{1 - u/c}}\, f_0 \qquad \text{(approaching)} \qquad 15\text{-}37a$$

$$f = \sqrt{\frac{1 - u/c}{1 + u/c}}\, f_0 \qquad \text{(receding)} \qquad 15\text{-}37b$$

where c is the speed of light. Again, when $u \ll c$, $f/f_0 \approx 1 \pm u/c$, as given by Equation 15-36.

Shock waves from a supersonic airplane.

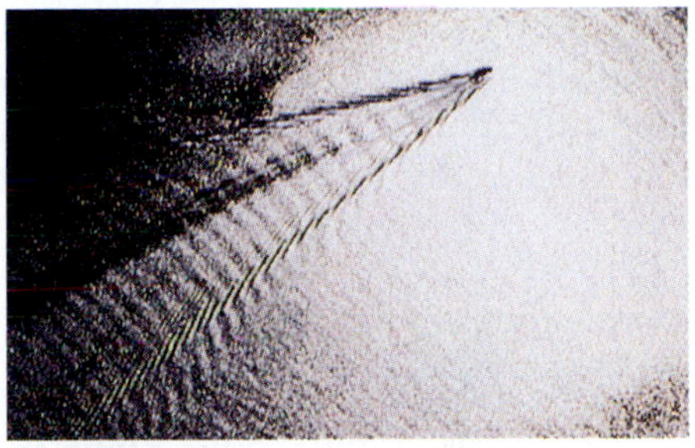

Bow waves from a boat.

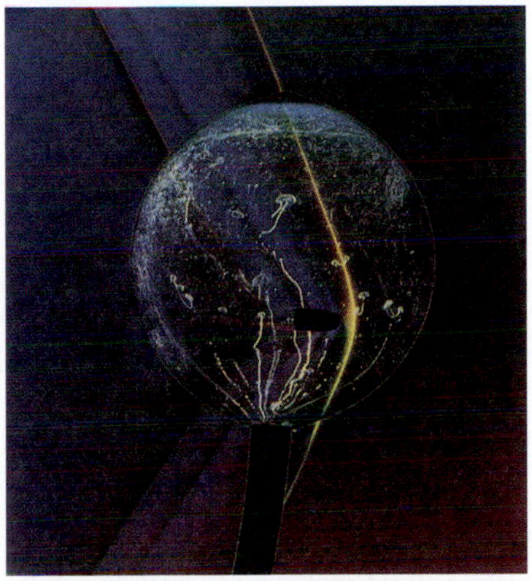

Shock waves produced by a bullet traversing a helium balloon.

Shock Waves

In our derivations of the Doppler-shift expressions, we assumed that the speed u of the source or receiver was less than the wave speed v. If the receiver moves faster than the wave speed, but toward the source, our derivation is still valid—Equation 15-34 continues to hold for the observed frequency. If the receiver moves faster than the wave speed, but away from the source, then the waves never reach the receiver. If a source moves with speed greater than the wave speed, there will be no waves in front of the source. Instead, the waves pile up behind the source to form a shock wave. In the case of sound waves, this shock wave is heard as a sonic boom when it arrives at the receiver.

Figure 15-28a shows a source originally at point P_1 moving to the right with speed u. After some time t, the wave emitted from point P_1 has traveled

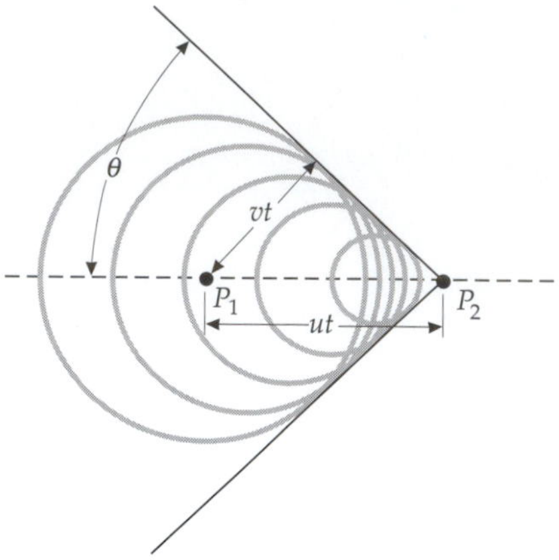

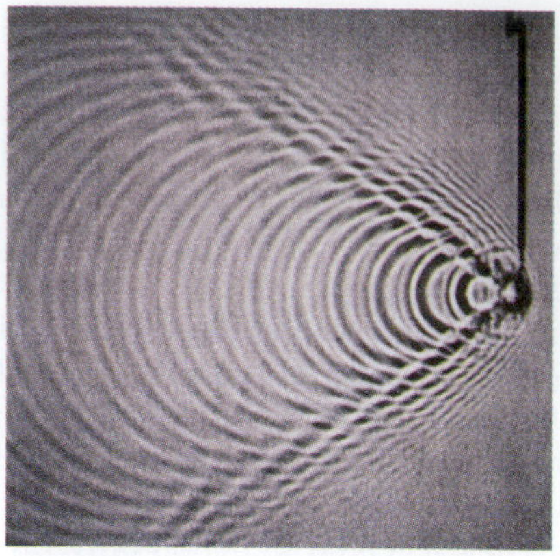

Figure 15-28 (*Left*) Source moving with a speed u that is greater than the wave speed v. The envelope of the wavefronts forms a cone with the source at the apex. (*Right*) Waves in a ripple tank produced by a source moving with a speed $u > v$.

a distance vt. The source has traveled a distance ut and will be at point P_2. The line from this new position of the source to the wavefront emitted when the source was at P_1 makes an angle θ with the path of the source, given by

$$\sin\theta = \frac{vt}{ut} = \frac{v}{u} \qquad \text{15-38}$$

Thus, the shock wave is confined to a cone that narrows as u increases. The ratio of the source speed u to the wave speed v is called the Mach number:

$$\text{Mach number} = \frac{u}{v} \qquad \text{15-39}$$

Equation 15-38 also applies to the electromagnetic radiation called Cerenkov radiation, which is given off when a charged particle moves in a medium with speed u that is greater than the speed of light v in that medium.* The blue glow surrounding the fuel elements used in nuclear reactors is an example of Cerenkov radiation.

Glow produced by Cerenkov radiation in the Waste Encapsulation and Storage Facility at the Hanford weapons complex.

Example 15-13 ***try it yourself***

At time $t = 0$, a supersonic plane is directly over point P flying due east at an altitude of 15 km. The sonic boom is heard at point P when the plane is 22 km east of point P (Figure 15-29). What is the speed of the supersonic plane?

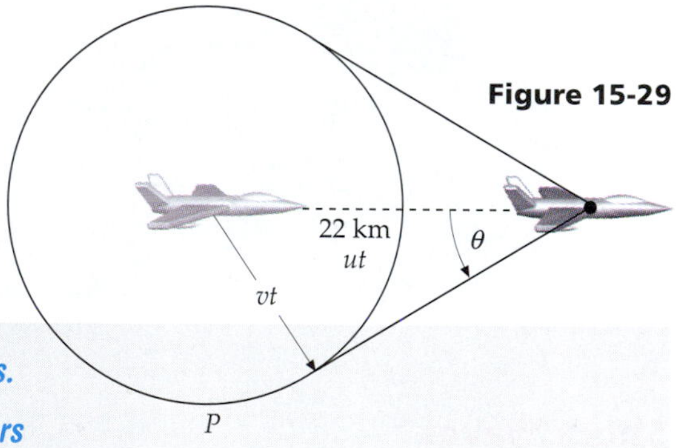

Figure 15-29

Cover the column to the right and try these on your own before looking at the answers.

Steps	***Answers***
1. Sketch the position of the plane when the sonic boom is heard at point P, and calculate $\tan\theta$ from the altitude and distance of the plane; then solve for θ.	$\tan\theta = \frac{v}{u} = \frac{vt}{ut} = \frac{15\text{ km}}{22\text{ km}}$ $\theta = 34.3°$
2. From your result and Equation 15-38, calculate u.	$u = \frac{v}{\sin\theta} = 604\text{ m/s}$

* According to the special theory of relativity, it is impossible for a particle to move faster than c, the speed of light in vacuum. In a medium such as glass, however, electrons and other particles can move faster than the speed of light in that medium.

Summary

1. In wave motion, energy and momentum are transported from one point in space to another without the transport of matter.
2. The relation $v = f\lambda$ holds for all harmonic waves.

Topic	Remarks and Relevant Equations	
1. Transverse and Longitudinal Waves	In transverse waves, such as waves on a string, the disturbance is perpendicular to the direction of propagation. In longitudinal waves, such as sound waves, the disturbance is along the direction of propagation.	
2. Speed of Waves	The speed of a wave depends on the density and elastic properties of the medium. It is independent of the motion of the wave source.	
Waves on a string	$v = \sqrt{F/\mu}$	**15-3**
Sound waves	$v = \sqrt{B/\rho}$	**15-4**
Sound waves in a gas	$v = \sqrt{\gamma RT/M}$	**15-5**
	where T is the absolute temperature, $T = t_C + 273$	**15-6**
	and R is the universal gas constant, $R = 8.314\ \text{J/mol}\cdot\text{K}$	**15-7**
	M is the molar mass of the gas, which for air is 29×10^{-3} kg/mol, and γ is a constant that depends on the kind of gas. For a diatomic gas such as air, $\gamma = 1.4$.	
Electromagnetic waves	The speed of electromagnetic waves such as light in vacuum is a universal constant. $c \approx 3 \times 10^8$ m/s.	
3. Wave Equation (optional)	$\frac{\partial^2 y}{\partial x^2} = \frac{1}{v^2}\frac{\partial^2 y}{\partial t^2}$	**15-9*b***
4. Harmonic Waves		
Wave function	$y(x, t) = A \sin(kx - \omega t)$, wave traveling to the right; $y(x, t) = A \sin(kx + \omega t)$, wave traveling to the left	**15-13**
	where A is the amplitude, k is the wave number, and ω is the angular frequency.	
Wave number	$k = \frac{2\pi}{\lambda}$	**15-12**
Angular frequency	$\omega = 2\pi f = \frac{2\pi}{T}$	**15-15**
Speed	$v = f\lambda = \frac{\omega}{\kappa}$	**15-10, 15-14**
Energy	The energy in a harmonic wave is proportional to the square of the amplitude.	
Energy of waves on a string	$\Delta E_{\text{av}} = \frac{1}{2}\mu\omega^2 y_0^2\,\Delta x$	**15-18**

5. Harmonic Sound Waves	Sound waves can be considered to be either displacement waves or pressure waves. The human ear is sensitive to sound waves of frequencies from about 20 Hz to 20 kHz. In a harmonic sound wave, the pressure and displacement are 90° out of phase.	
Amplitudes	The pressure and displacement amplitudes are related by $p_0 = \rho\omega v s_0$ where ρ is the density of the medium.	**15-22**
Energy density	$\eta_{\text{av}} = \dfrac{\Delta E_{\text{av}}}{\Delta V} = \dfrac{1}{2}\rho\omega^2 s_0^2$	**15-24**
6. Intensity	The intensity of a wave is the average energy on a unit area per unit time.	
Due to a point source	$I = \dfrac{P_{\text{av}}}{4\pi r^2}$	**15-26**
Average energy density η_{av} of a sound wave	$I = \eta_{\text{av}}v = \dfrac{1}{2}\rho\omega^2 s_0^2 v = \dfrac{1}{2}\dfrac{p_0^2}{\rho v}$	**15-28**
Intensity level β in dB (optional)	Sound intensity levels are measured on a logarithmic scale. $\beta = 10 \log \dfrac{I}{I_0}$ where $I_0 = 10^{-12}\ \text{W/m}^2$ is approximately the threshold of hearing.	**15-29**
7. Reflection and Refraction	When a wave is incident on a boundary surface that separates two regions of differing wave speed, part of the wave is reflected and part is transmitted.	
8. Diffraction	Diffraction is the bending of a wave around an obstacle or aperture that occurs when the wavefront is limited.	
Ray approximation	If a wavelength is small compared with obstacles or apertures, diffraction is negligible and the wave propagates in straight lines in the direction of the rays.	
9. Doppler Effect	When a sound source and receiver are in relative motion, the observed frequency f' is higher than the emitted frequency f_0 if they move toward each other, and lower if they move away from each other.	
Moving source	$\lambda' = \dfrac{v \pm u_s}{f_0}$	**15-31**
	$f' = \dfrac{1}{1 \pm u_s/v} f_0$	**15-33**
Moving receiver	$f' = \left(1 \pm \dfrac{u_r}{v}\right) f_0$	**15-34**
Either source or receiver moving	$f' = \dfrac{1 \pm u_r/v}{1 \pm u_s/v} f_0$	**15-35**
Small speeds of source or receiver	$\dfrac{\Delta f}{f_0} \approx \pm \dfrac{u}{v} \quad (u \ll v)$	**15-36**

Relativistic Doppler shift	$f = \sqrt{\frac{1 + u/c}{1 - u/c}}\, f_0$ (approaching)	**15-37*a***
	$f = \sqrt{\frac{1 - u/c}{1 + u/c}}\, f_0$ (receding)	**15-37*b***
10. Shock Waves	When the source speed is greater than the wave speed, the waves behind the source are confined to a cone of angle θ given by $\sin\theta = \frac{v}{u}$	**15-38**
Mach number	$\text{Mach number} = \frac{u}{v}$	**15-39**

Problem-Solving Guide

Begin by drawing a neat diagram including the important features of the problem.

Summary of Worked Examples

Type of Calculation	Procedure and Relevant Examples	
1. Find the Speed of a Wave		
When the wave is on a string whose tension is provided by a hanging weight.	The speed is $v = \sqrt{F/\mu}$, where the tension $F = mg$ and the linear mass density is $\mu = m/L$.	**Example 15-1**
For sound waves in a gas at a given temperature.	The speed is related to the absolute temperature T by $v = \sqrt{\gamma RT/M}$, where $\gamma = 1.4$ for air, $R = 8.314\ \text{J/mol·K}$, M is the molecular mass of the gas ($M = 29 \times 10^{-3}$ kg/mol for air), and T is the absolute temperature, which is related to the Celsius temperature by $T = t_C + 273$	**Example 15-2**
2. Harmonic Waves		
Find the wavelength, wave number, or frequency given the wave function.	Compare with the standard form $y(x,t) = A\sin(kx - \omega t)$, where $k = 2\pi/\lambda$, and $\omega = 2\pi f$.	**Example 15-4**
Find the speed of a string segment.	Compute dy/dt from the wave function.	**Example 15-4**
Find the total energy of a wave in a string.	Use $\Delta E_{av} = \frac{1}{2}(\Delta m)\omega^2 A^2$.	**Example 15-5**
3. Intensity		
Find the intensity given the frequency and displacement of the molecules.	Use $I = \frac{1}{2}\rho\omega^2 s_0^2 v$.	**Example 15-6**
Find the intensity given the average power of a point source.	If the energy is radiated uniformly in all directions, use $I = P_{av}/4\pi r^2$. If it is radiated uniformly into a hemisphere, use $I = P_{av}/2\pi r^2$.	**Example 15-6**
Find the intensity level from the intensity. (optional)	The intensity level is $\beta = 10\log(I/I_0)$, where $I_0 = 10^{-12}\ \text{W/m}^2$.	**Example 15-8**

Find the change in intensity from the change in intensity level. (optional)	Use Table 15-1. A change of ($n \times 10$) dB is equivalent to an intensity change by a factor of 10^n.	**Example 15-7**
4. Reflection and Refraction		
Find the fraction of power transmitted and reflected given information about the amplitudes, speeds, and densities.	Use $P = \frac{1}{2}\mu\omega^2 A^2 v$ and conservation of energy.	**Example 15-9**
5. The Doppler Effect		
Find the frequency of waves from a moving source.	Find the change in wavelength from $\lambda' = (v \pm u_s)/f_0$ and then find the frequency from $f' = v/\lambda'$.	**Examples 15-10, 15-11**
Find the frequency observed by a moving receiver.	Use $f' = f_0(1 \pm u_r/v)$.	**Example 15-10**
Find the frequency when $u_s, u_r \ll v$.	Use $\Delta f \approx \pm(u/v)f_0$, where $u = u_s \pm u_r$ is the relative speed of the source and receiver.	**Example 15-12**

Problems

In a few problems, you are given more data than you actually need; in a few other problems, you are required to supply data from your general knowledge, outside sources, or informed estimates.

Conceptual Problems

Problems from Optional and Exploring sections

• Single-concept, single-step, relatively easy

•• Intermediate-level, may require synthesis of concepts

••• Challenging, for advanced students

Use $v = 340$ m/s for the speed of sound in air unless otherwise indicated.

Speed of Waves

1 • A rope hangs vertically from the ceiling. Do waves on the rope move faster, slower, or at the same speed as they move from bottom to top? Explain.

2 • (*a*) The bulk modulus for water is 2.0×10^9 N/m^2. Use it to find the speed of sound in water. (*b*) The speed of sound in mercury is 1410 m/s. What is the bulk modulus for mercury ($\rho = 13.6 \times 10^3$ kg/m^3)?

3 • Calculate the speed of sound waves in hydrogen gas at $T = 300$ K. (Take $M = 2$ g/mol and $\gamma = 1.4$.)

4 • A steel wire 7 m long has a mass of 100 g. It is under a tension of 900 N. What is the speed of a transverse wave pulse on this wire?

5 • Transverse waves travel at 150 m/s on a wire of length 80 cm that is under a tension of 550 N. What is the mass of the wire?

6 • A wave pulse propagates along a wire in the positive x direction at 20 m/s. What will the pulse velocity be if we (*a*) double the length of the wire but keep the tension and mass per unit length constant? (*b*) double the tension while holding the length and mass per unit length constant? (*c*) double the mass per unit length while holding the other variables constant?

7 • A steel piano wire is 0.7 m long and has a mass of 5 g. It is stretched with a tension of 500 N. (*a*) What is the speed of transverse waves on the wire? (*b*) To reduce the wave speed by a factor of 2 without changing the tension, what mass of copper wire would have to be wrapped around the steel wire?

8 • The cable of a ski lift runs 400 m up a mountain and has a mass of 80 kg. When the cable is struck with a transverse blow at one end, the return pulse is detected 12 s later. (*a*) What is the speed of the wave? (*b*) What is the tension in the cable?

9 •• A common method for estimating the distance to a lightning flash is to begin counting when the flash is observed and continue until the thunder clap is heard. The number of seconds counted is then divided by 3 to get the distance in kilometers. (*a*) What is the velocity of sound in kilometers per second? (*b*) How accurate is this procedure? (*c*) Is a correction for the time it takes for the light to reach you important? (The speed of light is 3×10^8 m/s.)

10 •• A method for measuring the speed of sound using an ordinary watch with a second hand is to stand some distance from a large flat wall and clap your hands rhythmically in such a way that the echo from the wall is heard halfway between every two claps. (*a*) Show that the speed of sound is given by $v = 4LN$, where L is the distance to the wall and N is the number of claps per second. (*b*) What is a reasonable value for L for this experiment to be feasible? (If you have access to a flat wall outdoors somewhere, try this method and compare your result with the standard value for the speed of sound.)

11 •• A man drops a stone from a high bridge and hears it strike the water below exactly 4 s later. (*a*) Estimate the distance to the water based on the assumption that the travel time for the sound to reach the man is negligible. (*b*) Improve your estimate by using your result from part (*a*) for the distance to the water to estimate the time it takes for sound to travel this distance and then calculate the distance the rock falls in 4 s minus this time. (*c*) Calculate the exact distance and compare your result with your previous estimates.

12 •• (*a*) Compute the derivative of the speed of a wave on a string with respect to the tension dv/dF, and show that the differentials dv and dF obey $dv/v = \frac{1}{2}dF/F$. (*b*) A wave moves with a speed of 300 m/s on a wire that is under a tension of 500 N. Using dF to approximate a change in tension, determine how much the tension must be changed to increase the speed to 312 m/s.

13 •• (*a*) Compute the derivative of the velocity of sound with respect to the absolute temperature, and show that the differentials dv and dT obey $dv/v = \frac{1}{2}dT/T$. (*b*) Use this result to compute the percentage change in the velocity of sound when the temperature changes from 0 to 27°C. (*c*) If the speed of sound is 331 m/s at 0°C, what is it (approximately) at 27°C? How does this approximation compare with the result of an exact calculation?

14 ••• In this problem, you will derive a convenient formula for the speed of sound in air at temperature t in Celsius degrees. Begin by writing the temperature as $T = T_0 + \Delta T$, where $T_0 = 273$ K corresponds to 0°C and $\Delta T = t$, the Celsius temperature. The speed of sound is a function of T, $v(T)$. To a first-order approximation, you can write

$$v(T) \approx v(T_0) + (dv/dT)_{T_0}\,\Delta T$$

where $(dv/dt)_{T_0}$ is the derivative evaluated at $T = T_0$. Compute this derivative, and show that the result leads to

$$v = (331\ \text{m/s})\left(1 + \frac{t}{2T_0}\right) = (331 + 0.606t)\text{m/s}$$

15 ••• While studying physics in her dorm room, a student is listening to a live radio broadcast of a baseball game. She is 1.6 km due south of the baseball field. Over her radio, the student hears a noise generated by the electromagnetic pulse of a lightning bolt. Two seconds later, she hears over the radio the thunder picked up by the microphone at the baseball field. Four seconds after she hears the noise of the electromagnetic pulse over the radio, thunder rattles her windows. Where, relative to the ballpark, did the lightning bolt occur?

16 ••• A coiled spring, such as a Slinky, is stretched to a length L. It has a force constant k and a mass m. (*a*) Show that the velocity of longitudinal compression waves along the spring is given by $v = L\sqrt{k/m}$. (*b*) Show that this is also the velocity of transverse waves along the spring if the natural length of the spring is much less than L.

The Wave Equation (optional)

17 • Show explicitly that the following functions satisfy the wave equation: (*a*) $y(x, t) = k(x + vt)^3$; (*b*) $y(x, t) = Ae^{ik(x-vt)}$, where A and k are constants and $i = \sqrt{-1}$; and (*c*) $y(x, t) = \ln k(x - vt)$.

18 • Show that the function $y = A \sin kx \cos \omega t$ satisfies the wave equation.

19 ••• Consider the following equation:

$$\frac{\partial^2 y}{\partial x^2} + i\alpha\frac{\partial y}{\partial t} = 0, \qquad i = \sqrt{-1}$$

where α is a constant. Show that $y(x, t) = A \sin(kx - \omega t)$ is not a solution of this equation but that the functions $y(x, t) = Ae^{i(kx-\omega t)}$ and $y(x, t) = Ae^{i(kx+\omega t)}$ do satisfy that equation.

Harmonic Waves on a String

20 • A traveling wave passes a point of observation. At this point, the time between successive crests is 0.2 s. Which of the following is true?

(*a*) The wavelength is 5 m.
(*b*) The frequency is 5 Hz.
(*c*) The velocity of propagation is 5 m/s.
(*d*) The wavelength is 0.2 m.
(*e*) There is not enough information to justify any of these statements.

21 • True or false: The energy in a wave is proportional to the square of the amplitude of the wave.

22 • A rope hangs vertically. You shake the bottom back and forth, creating a sinusoidal wave train. Is the wavelength at the top the same as, less than, or greater than the wavelength at the bottom?

23 • One end of a string 6 m long is moved up and down with simple harmonic motion at a frequency of 60 Hz. The waves reach the other end of the string in 0.5 s. Find the wavelength of the waves on the string.

24 • Equation 15-13 expresses the displacement of a harmonic wave as a function of x and t in terms of the wave parameters k and ω. Write the equivalent expressions that contain the following pairs of parameters instead of k and ω: (*a*) k and v, (*b*) λ and f, (*c*) λ and T, (*d*) λ and v, and (*e*) f and v.

25 • Equation 15-10 applies to all types of periodic waves, including electromagnetic waves such as light waves and microwaves, which travel at 3×10^8 m/s in a vacuum.

(*a*) The range of wavelengths of light to which the eye is sensitive is about 4×10^{-7} to 7×10^{-7} m. What are the frequencies that correspond to these wavelengths? (*b*) Find the frequency of a microwave that has a wavelength of 3 cm.

26 • A harmonic wave on a string with a mass per unit length of 0.05 kg/m and a tension of 80 N has an amplitude of 5 cm. Each section of the string moves with simple harmonic motion at a frequency of 10 Hz. Find the power propagated along the string.

27 • A rope 2 m long has a mass of 0.1 kg. The tension is 60 N. A power source at one end sends a harmonic wave with an amplitude of 1 cm down the rope. The wave is extracted at the other end without any reflection. What is the frequency of the power source if the power transmitted is 100 W?

28 •• The wave function for a harmonic wave on a string is $y(x, t) = (0.001 \text{ m}) \sin(62.8 \text{ m}^{-1}x + 314 \text{ s}^{-1} t)$. (*a*) In what direction does this wave travel, and what is its speed? (*b*) Find the wavelength, frequency, and period of this wave. (*c*) What is the maximum speed of any string segment?

29 •• A harmonic wave with a frequency of 80 Hz and an amplitude of 0.025 m travels along a string to the right with a speed of 12 m/s. (*a*) Write a suitable wave function for this wave. (*b*) Find the maximum speed of a point on the string. (*c*) Find the maximum acceleration of a point on the string.

30 •• Waves of frequency 200 Hz and amplitude 1.2 cm move along a 20-m string that has a mass of 0.06 kg and a tension of 50 N. (*a*) What is the average total energy of the waves on the string? (*b*) Find the power transmitted past a given point on the string.

31 •• In a real string, a wave loses some energy as it travels down the string. Such a situation can be described by a wave function whose amplitude $A(x)$ depends on x:

$$y = A(x) \sin(kx - \omega t) = (A_0 e^{-bx}) \sin(kx - \omega t)$$

(*a*) What is the original power carried by the wave at the origin? (*b*) What is the power transported by the wave at point x?

32 •• Power is to be transmitted along a stretched wire by means of transverse harmonic waves. The wave speed is 10 m/s, and the linear mass density of the wire is 0.01 kg/m. The power source oscillates with an amplitude of 0.50 mm. (*a*) What average power is transmitted along the wire if the frequency is 400 Hz? (*b*) The power transmitted can be increased by increasing the tension in the wire, the frequency of the source, or the amplitude of the waves. How would each of these quantities have to be changed to effect an increase in power by a factor of 100 if it is the only quantity changed? (*c*) Which of the quantities would probably be the easiest to change?

Harmonic Sound Waves

33 • A sound wave in air produces a pressure variation given by

$$p(x, t) = 0.75 \cos \frac{\pi}{2}(x - 340t)$$

where p is in pascals, x is in meters, and t is in seconds. Find (*a*) the pressure amplitude of the sound wave, (*b*) the wavelength, (*c*) the frequency, and (*d*) the speed.

34 • (*a*) Middle C on the musical scale has a frequency of 262 Hz. What is the wavelength of this note in air? (*b*) The frequency of the C an octave above middle C is twice that of middle C. What is the wavelength of this note in air?

35 • (*a*) What is the displacement amplitude for a sound wave having a frequency of 100 Hz and a pressure amplitude of 10^{-4} atm? (*b*) The displacement amplitude of a sound wave of frequency 300 Hz is 10^{-7} m. What is the pressure amplitude of this wave?

36 • (*a*) Find the displacement amplitude of a sound wave of frequency 500 Hz at the pain-threshold pressure amplitude of 29 Pa. (*b*) Find the displacement amplitude of a sound wave with the same pressure amplitude but a frequency of 1 kHz.

37 • A typical loud sound wave with a frequency of 1 kHz has a pressure amplitude of about 10^{-4} atm. (*a*) At $t = 0$, the pressure is a maximum at some point x_1. What is the displacement at that point at $t = 0$? (*b*) What is the maximum value of the displacement at any time and place? (Take the density of air to be 1.29 kg/m^3.)

38 • (*a*) Find the displacement amplitude of a sound wave of frequency 500 Hz at the threshold-of-hearing pressure amplitude of 2.9×10^{-5} Pa. (*b*) Find the displacement amplitude of a wave of the same pressure amplitude but a frequency of 1 kHz.

Waves in Three Dimensions: Intensity

39 • A piston at one end of a long tube filled with air at room temperature and normal pressure oscillates with a frequency of 500 Hz and an amplitude of 0.1 mm. The area of the piston is 100 cm^2. (*a*) What is the pressure amplitude of the sound waves generated in the tube? (*b*) What is the intensity of the waves? (*c*) What average power is required to keep the piston oscillating (neglecting friction)?

40 • A spherical source radiates sound uniformly in all directions. At a distance of 10 m, the sound intensity level is 10^{-4} W/m^2. (*a*) At what distance from the source is the intensity 10^{-6} W/m^2? (*b*) What power is radiated by this source?

41 • A loudspeaker at a rock concert generates 10^{-2} W/m^2 at 20 m at a frequency of 1 kHz. Assume that the speaker spreads its energy uniformly in three dimensions. (*a*) What is the total acoustic power output of the speaker? (*b*) At what distance will the intensity be at the pain threshold of 1 W/m^2? (*c*) What is the intensity at 30 m?

42 •• When a pin of mass 0.1 g is dropped from a height of 1 m, 0.05% of its energy is converted into a sound pulse with a duration of 0.1 s. (*a*) Estimate the range at which the dropped pin can be heard if the minimum audible intensity is 10^{-11} W/m^2. (*b*) Your result in (*a*) is much too large in practice because of background noise. If you assume that the intensity must be at least 10^{-8} W/m^2 for the sound to be heard, estimate the range at which the dropped pin can be heard. (In both parts, assume that the intensity is $P/4\pi r^2$.)

Intensity Level (optional)

43 • True or false: A 60-dB sound has twice the intensity of a 30-dB sound.

44 • What is the intensity level in decibels of a sound wave of intensity (*a*) 10^{-10} W/m^2, and (*b*) 10^{-2} W/m^2?

45 • Find the intensity of a sound wave if (*a*) $\beta = 10$ dB, and (*b*) $\beta = 3$ dB. (*c*) Find the pressure amplitudes of sound waves in air for each of these intensities.

46 • The sound level of a dog's bark is 50 dB. The intensity of a rock concert is 10,000 times that of the dog's bark. What is the sound level of the rock concert?

47 • Two sounds differ by 30 dB. The intensity of the louder sound is I_L and that of the softer sound is I_S. The value of the ratio I_L / I_S is

(*a*) 1000.
(*b*) 30.
(*c*) 9.
(*d*) 100.
(*e*) 300.

48 • Show that if the intensity is doubled, the intensity level increases by 3.0 dB.

49 • What fraction of the acoustic power of a noise would have to be eliminated to lower its sound intensity level from 90 to 70 dB?

50 •• Normal human speech has a sound intensity level of about 65 dB at 1 m. Estimate the power of human speech.

51 •• A spherical source radiates sound uniformly in all directions. At a distance of 10 m, the sound intensity level is 80 dB. (*a*) At what distance from the source is the intensity level 60 dB? (*b*) What power is radiated by this source?

52 •• A spherical source of intensity I_0 radiates sound uniformly in all directions. Its intensity level is β_1 at a distance r_1, and β_2 at a distance r_2. Find β_2 / β_1.

53 •• A loudspeaker at a rock concert generates 10^{-2} W/m^2 at 20 m at a frequency of 1 kHz. Assume that the speaker spreads its energy uniformly in all directions. (*a*) What is the intensity level at 20 m? (*b*) What is the total acoustic power output of the speaker? (*c*) At what distance will the intensity level be at the pain threshold of 120 dB? (*d*) What is the intensity level at 30 m?

54 •• An article on noise pollution claims that sound intensity levels in large cities have been increasing by about 1 dB annually. (*a*) To what percentage increase in intensity does this correspond? Does this increase seem reasonable? (*b*) In about how many years will the intensity of sound double if it increases at 1 dB annually?

55 •• Three noise sources produce intensity levels of 70, 73, and 80 dB when acting separately. When the sources act together, their intensities add. (*a*) Find the sound intensity level in decibels when the three sources act at the same time. (*b*) Discuss the effectiveness of eliminating the two least intense sources in reducing the intensity level of the noise.

56 •• The equation $I = P_{av}/4\pi r^2$ is predicated on the assumption that the transmitting medium does not absorb any energy. It is known that absorption of sound by dry air results in a decrease of intensity of approximately 8 dB/km. The intensity of sound at a distance of 120 m from a jet engine is 130 dB. Find the intensity at 2.4 km from the jet engine (*a*) assuming no absorption of sound by air, and (*b*) assuming a diminution of 8 dB/km. (Assume that the sound radiates uniformly in all directions.)

57 ••• Everyone at a party is talking equally loudly. If only one person were talking, the sound level would be 72 dB. Find the sound level when all 38 people are talking.

58 ••• When a violinist pulls the bow across a string, the force with which the bow is pulled is fairly small, about 0.6 N. Suppose the bow travels across the A string, which vibrates at 440 Hz, at 0.5 m/s. A listener 35 m from the performer hears a sound of 60 dB intensity. With what efficiency is the mechanical energy of bowing converted to sound energy? (Assume that the sound radiates uniformly in all directions.)

59 ••• The noise level in an empty examination hall is 40 dB. When 100 students are writing an exam, the sounds of heavy breathing and pens traveling rapidly over paper cause the noise level to rise to 60 dB (not counting the occasional groans). Assuming that each student contributes an equal amount of noise power, find the noise level to the nearest decibel when 50 students have left.

The Doppler Effect

60 • If the source and receiver are at rest relative to each other but the wave medium is moving relative to them, will there be any Doppler shift in frequency?

61 • The frequency of a car horn is f_0. What frequency is observed if both the car and the observer are at rest, but a wind blows toward the observer?

(*a*) f_0
(*b*) Greater than f_0
(*c*) Less than f_0
(*d*) It could be either greater or less than f_0.
(*e*) It could be f_0 or greater than f_0, depending on how wind speed compares to speed of sound.

62 •• Stars often occur in pairs revolving around their common center of mass. If one of the stars is a black hole, it is invisible. Explain how the existence of such a black hole might be inferred from the light observed from the other, visible star.

63 • A conveyor belt moves to the right with a speed $v = 300$ m/min. A very fast piemaker puts pies on the belt at a rate of 20 per minute, and they are received at the other end by a pie eater. (*a*) If the piemaker is stationary, find the spacing λ between the pies and the frequency f with which they are received by the stationary pie eater. (*b*) The piemaker now walks with a speed of 30 m/min toward the receiver while continuing to put pies on the belt at 20 per minute. Find the spacing of the pies and the frequency with which they are received by the stationary pie eater. (*c*) Repeat your

calculations for a stationary piemaker and a pie eater who moves toward the piemaker at 30 m/min.

64 • For the situation described in Problem 63, derive general expressions for the spacing of the pies λ and the frequency f with which they are received by the pie eater in terms of the speed of the belt v, the speed of the sender u_s, the speed of the receiver u_r, and the frequency f_0 with which the piemaker places pies on the belt.

In Problems 65 through 70, a source emits sounds of frequency 200 Hz that travel through still air at 340 m/s.

65 • The sound source described above moves with a speed of 80 m/s relative to still air toward a stationary listener. (*a*) Find the wavelength of the sound between the source and the listener. (*b*) Find the frequency heard by the listener.

66 • Consider the situation in Problem 65 from the reference frame in which the source is at rest. In this frame, the listener moves toward the source with a speed of 80 m/s, and there is a wind blowing at 80 m/s from the listener to the source. (*a*) What is the speed of the sound from the source to the listener in this frame? (*b*) Find the wavelength of the sound between the source and the listener. (*c*) Find the frequency heard by the listener.

67 • The source moves away from the stationary listener at 80 m/s. (*a*) Find the wavelength of the sound waves between the source and the listener. (*b*) Find the frequency heard by the listener.

68 • The listener moves at 80 m/s relative to still air toward the stationary source. (*a*) What is the wavelength of the sound between the source and the listener? (*b*) What is the frequency heard by the listener?

69 • Consider the situation in Problem 68 in a reference frame in which the listener is at rest. (*a*) What is the wind velocity in this frame? (*b*) What is the speed of the sound from the source to the listener in this frame, that is, relative to the listener? (*c*) Find the wavelength of the sound between the source and the listener in this frame. (*d*) Find the frequency heard by the listener.

70 • The listener moves at 80 m/s relative to the still air away from the stationary source. Find the frequency heard by the listener.

71 • A jet is traveling at Mach 2.5 at an altitude of 5000 m. (*a*) What is the angle that the shock wave makes with the track of the jet? (Assume that the speed of sound at this altitude is still 340 m/s.) (*b*) Where is the jet when a person on the ground hears the shock wave?

72 • If you are running at top speed toward a source of sound at 1000 Hz, estimate the frequency of the sound that you hear. Suppose that you can recognize a change in frequency of 3%. Can you use your sense of pitch to estimate your running speed?

73 •• A radar device emits microwaves with a frequency of 2.00 GHz. When the waves are reflected from a car moving directly away from the emitter, a frequency difference of 293 Hz is detected. Find the speed of the car.

74 •• A stationary destroyer is equipped with sonar that sends out pulses of sound at 40 MHz. Reflected pulses are received from a submarine directly below with a time delay of 80 ms at a frequency of 39.958 MHz. If the speed of sound in seawater is 1.54 km/s, find (*a*) the depth of the submarine, and (*b*) its vertical speed.

75 •• Two airplanes, one flying due east and the other due west, are on a near collision course separated by 15 km when the pilot of one plane, traveling at 900 km/h, observes the other on his Doppler radar. The radar unit emits electromagnetic waves of frequency 3×10^{10} Hz. The radar readout indicates that the other plane's speed is 750 km/h. Determine the frequency of the signal received by the pilot's radar.

76 •• A police radar unit transmits microwaves of frequency 3×10^{10} Hz. The speed of these waves in air is 3.0×10^8 m/s. Suppose a car is receding from the stationary police car at a speed of 140 km/h. What is the frequency difference between the transmitted signal and the signal received from the receding car?

77 •• Suppose the police car of Problem 76 is moving in the same direction as the other vehicle at a speed of 60 km/h. What then is the difference in frequency between the emitted and the reflected signals?

78 •• At time $t = 0$, a supersonic plane is directly over point P flying due west at an altitude of 12 km and a speed of Mach 1.6. Where is the plane when the sonic boom is heard?

79 •• A small radio of 0.10 kg mass is attached to one end of an air track by a spring. The radio emits a sound of 800 Hz. A listener at the other end of the air track hears a sound whose frequency varies between 797 and 803 Hz. (*a*) Determine the energy of the vibrating mass–spring system. (*b*) If the spring constant is 200 N/m, what is the amplitude of vibration of the mass and what is the period of the oscillating system?

80 •• A sound source of frequency f_0 moves with speed u_s relative to still air toward a receiver who is moving with speed u_r relative to still air away from the source. (*a*) Write an expression for the received frequency f'. (*b*) Use the result that $(1 - x)^{-1} \approx 1 + x$ to show that if both u_s and u_r are small compared to v, then the received frequency is approximately

$$f' \approx \left(1 + \frac{u_s - u_r}{v}\right) f_0 = \left(1 + \frac{u_{rel}}{v}\right) f_0$$

where u_{rel} is the relative velocity of the source and receiver.

81 •• Two students with vibrating 440-Hz tuning forks walk away from each other with equal speeds. How fast must they walk so that they each hear a frequency of 438 Hz from the other fork?

82 •• A physics student walks down a long hall carrying a vibrating 512-Hz tuning fork. The end of the hall is closed so that sound reflects from it. The student hears a sound of 516 Hz from the wall. How fast is the student walking?

83 •• A small speaker radiating sound at 1000 Hz is tied to one end of an 0.8-m-long rod that is free to rotate about its other end. The rod rotates in the horizontal plane at 4.0 rad/s. Derive an expression for the frequency heard by a stationary observer far from the rotating speaker.

84 •• You have won a free trip on the *Queen Elizabeth II* and are in mid-Atlantic steaming due east at 45 km/h as the Concorde passes directly overhead flying due west at Mach 1.6 at an altitude of 12,500 m. Where is the Concorde relative to the *QEII* when you hear the sonic boom?

85 •• A balloon driven by a 36-km/h wind emits a sound of 800 Hz as it approaches a tall building. (*a*) What is the frequency of the sound heard by an observer at the window of this building? (*b*) What is the frequency of the reflected sound heard by a person riding in the balloon?

86 •• A car is approaching a reflecting wall. A stationary observer behind the car hears a sound of frequency 745 Hz from the car horn and a sound of frequency 863 Hz from the wall. (*a*) How fast is the car traveling? (*b*) What is the frequency of the car horn? (*c*) What frequency does the car driver hear reflected from the wall?

87 •• The driver of a car traveling at 100 km/h toward a vertical cliff briefly sounds the horn. Exactly one second later she hears the echo and notes that its frequency is 840 Hz. How far from the cliff was the car when the driver sounded the horn and what is the frequency of the horn?

88 •• You are on a transatlantic flight traveling due west at 800 km/h. A Concorde flying at Mach 1.6 and 3 km to the north of your plane is also on an east-to-west course. What is the distance between the two planes when you hear the sonic boom from the Concorde?

89 ••• Astronomers can deduce the existence of a binary star system even if the two stars cannot be visually resolved by noting an alternating Doppler shift of a spectral line. Suppose that an astronomical observation shows that the source of light is eclipsed once every 18 h. The wavelength of the spectral line observed changes from a maximum of 563 nm to a minimum of 539 nm. Assume that the double star system consists of a very massive, dark object and a relatively light star that radiates the observed spectral line. Use the data to determine the separation between the two objects (assume that the light object is in a circular orbit about the massive one) and the mass of the massive object. (Use the approximation $\Delta f/f_0 \approx v/c$.)

90 ••• A physics student drops a vibrating 440-Hz tuning fork down the elevator shaft of a tall building. When the student hears a frequency of 400 Hz, how far has the tuning fork fallen?

General Problems

91 • When a guitar string is plucked, is the wavelength of the wave it produces in air the same as the wavelength of the wave on the string?

92 • A wave pulse travels along a light string that is attached to a heavier string in which the wave speed is smaller. The reflected pulse is _____, and the transmitted pulse is _____.

(*a*) inverted/inverted
(*b*) inverted/not inverted
(*c*) not inverted/not inverted
(*d*) not inverted/inverted
(*e*) nonexistent/not inverted

93 • True or false:

(*a*) Wave pulses on strings are transverse waves.
(*b*) Sound waves in air are transverse waves of compression and rarefaction.
(*c*) The speed of sound at 20°C is twice that at 5°C.

94 • Sound travels at 340 m/s in air and 1500 m/s in water. A sound of 256 Hz is made under water. In the air, the frequency will be

(*a*) the same, but the wavelength will be shorter.
(*b*) higher, but the wavelength will stay the same.
(*c*) lower, but the wavelength will be longer.
(*d*) lower, and the wavelength will be shorter.
(*e*) the same, and the wavelength too will stay the same.

95 •• Figure 15-30 shows a wave pulse at time $t = 0$ moving to the right. At this particular time, which segments of the string are moving up? Which are moving down? Is there any segment of the string at the pulse that is instantaneously at rest? Answer these questions by sketching the pulse at a slightly later time and a slightly earlier time to see how the segments of the string are moving.

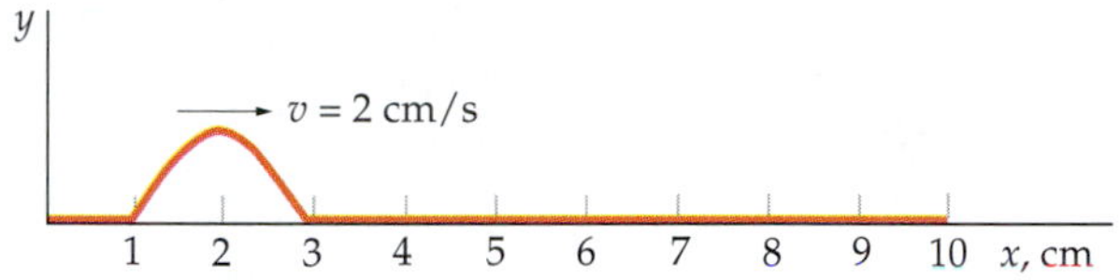

Figure 15-30 Problems 95 and 96

96 •• Make a sketch of the velocity of each string segment versus position for the pulse shown in Figure 15-30.

97 •• Consider a long line of cars equally spaced by one car length and moving slowly with the same speed. One car suddenly slows to avoid a dog and then speeds up until it is again one car length behind the car ahead. Discuss how the space between cars propagates back along the line. How is this like a wave pulse? Is there any transport of energy? What does the speed of propagation depend on?

98 • At time $t = 0$, the shape of a wave pulse on a string is given by the function

$$y(x, 0) = \frac{0.12\ \text{m}^3}{(2.00\ \text{m})^2 + x^2}$$

where x is in meters. (*a*) Sketch $y(x, 0)$ versus x. Give the wave function $y(x, t)$ at a general time t if (*b*) the pulse is moving in the positive x direction with a speed of 10 m/s, and (*c*) the pulse is moving in the negative x direction with a speed of 10 m/s.

99 • A wave with frequency of 1200 Hz propagates along a wire that is under a tension of 800 N. The wavelength of the wave is 24 cm. What will be the wavelength if the tension is decreased to 600 N and the frequency is kept constant?

100 • In a common lecture demonstration of wave pulses, a piece of rubber tubing is tied at one end to a fixed post and is passed over a pulley to a weight hanging at the other end. Suppose that the distance from the fixed support to the pulley is 10 m, the mass of this length of tubing is 0.7 kg, and the suspended weight is 110 N. If the tubing is given a transverse blow at one end, how long will it take the resulting pulse to reach the other end?

101 • The following wave functions represent traveling waves:

(*a*) $y_1(x, t) = A \cos k[x + (34\text{ m/s})t]$,
(*b*) $y_2(x, t) = Ae^{k[x-(20\text{m/s})t]}$,
(*c*) $y_3(x, t) = BC + \{k[x - (10\text{ m/s})t]\}^2$,

where x is in meters, t is in seconds, and A, k, B, and C are constants that have the proper units for y to be in meters. Give the direction of propagation and the speed of the wave for each wave function.

102 • A boat traveling at 10 m/s on a still lake makes a bow wave at an angle of 20° with its direction of motion. What is the speed of the bow wave?

103 • If a wavelength is much larger than the diameter of a loudspeaker, the speaker radiates in all directions, much like a point source. On the other hand, if the wavelength is much smaller than the diameter, the sound travels in an approximately straight line in front of the speaker. Find the frequency of a sound wave that has a wavelength (*a*) 10 times the diameter of a 30-cm speaker, and (*b*) one-tenth the diameter of a 30-cm speaker. (*c*) Repeat this problem for a 6-cm speaker.

104 • A whistle of frequency 500 Hz moves in a circle of radius 1 m at 3 rev/s. What are the maximum and minimum frequencies heard by a stationary listener in the plane of the circle and 5 m away from its center?

105 • Ocean waves move toward the beach with a speed of 8.9 m/s and a crest-to-crest separation of 15.0 m. You are in a small boat anchored off shore. (*a*) What is the frequency of the ocean waves? (*b*) You now lift anchor and head out to sea at a speed of 15 m/s. What frequency of the waves do you observe?

106 •• Two connected wires with linear mass densities that are related by $\mu_1 = 3\mu_2$ are under the same tension. When the wires oscillate at a frequency of 120 Hz, waves of wavelength 10 cm travel down the first wire with the linear density of μ_1. (*a*) What is the wave speed in the first wire? (*b*) What is the wave speed in the second wire? (*c*) What is the wavelength in the second wire?

107 •• A 12.0-m wire of mass 85 g is stretched under a tension of 180 N. A pulse is generated at the left end of the wire, and 25 ms later a second pulse is generated at the right end of the wire. Where do the pulses first meet?

108 •• A harmonic wave moves down a string with speed 12.4 m/s. A particle on the string has a maximum displacement of 4.5 cm and a maximum speed of 9.4 m/s. Find (*a*) the wavelength of the wave, and (*b*) the frequency. (*c*) Write an equation for the wave function.

109 •• Find the speed of a car for which the tone of its horn will drop by 10% as it passes you.

110 •• A loudspeaker diaphragm 20 cm in diameter is vibrating at 800 Hz with an amplitude of 0.025 mm. Assuming that the air molecules in the vicinity have this same amplitude of vibration, find (*a*) the pressure amplitude immediately in front of the diaphragm, (*b*) the sound intensity, and (*c*) the acoustic power being radiated.

111 •• A plane, harmonic, acoustical wave that oscillates in air with an amplitude of 10^{-6} m has an intensity of 10^{-2} W/m^2. What is the frequency of the sound wave?

112 •• Water flows at 7 m/s in a pipe of radius 5 cm. A plate having an area equal to the cross-sectional area of the pipe is suddenly inserted to stop the flow. Find the force exerted on the plate. Take the speed of sound in water to be 1.4 km/s. (*Hint:* When the plate is inserted, a pressure wave propagates through the water at the speed of sound v_s. The mass of water brought to a stop in time Δt is the water in a length of tube equal to $v_s\,\Delta t$.)

113 •• Two wires of different linear mass densities are soldered together end to end and are then stretched under a tension F (the tension is the same in both wires). The wave speed in the second wire is three times that in the first wire. When a harmonic wave traveling in the first wire is reflected at the junction of the wires, the reflected wave has half the amplitude of the incident wave. (*a*) If the amplitude of the incident wave is A, what are the amplitudes of the reflected and transmitted waves? (*b*) Assuming no loss in the wire, what fraction of the incident power is reflected at the junction and what fraction is transmitted? (*c*) Show that the displacement just to the left of the junction equals that just to the right of the junction.

114 •• A column of precision marchers keeps in step by listening to the band positioned at the head of the column. The beat of the music is for 100 paces per minute. A television camera shows that only the marchers at the front and the rear of the column are actually in step. The marchers in the middle section are striding forward with the left foot when those at the front and rear are striding forward with the right foot. The marchers are so well trained, however, that they are all certain that they are in proper step with the music. Explain the source of the problem, and calculate the length of the column.

115 •• Hovering over the pit of hell, the devil observes that as a student falls past (with terminal velocity), the frequency of his scream decreases from 842 to 820 Hz. (*a*) Find the speed of descent of the student. (*b*) The student's scream reflects from the bottom of the pit. Find the frequency of the echo as heard by the student. (*c*) Find the frequency of the echo as heard by the devil.

116 •• A bat flying toward an obstacle at 12 m/s emits brief, high-frequency sound pulses at a repetition frequency of 80 Hz. What is the time interval between the echo pulses heard by the bat?

117 •• A tuning fork attached to a stretched wire generates transverse waves. The vibration of the fork is perpendicular to the wire. Its frequency is 400 Hz, and the amplitude of its oscillation is 0.50 mm. The wire has linear mass density of 0.01 kg/m and is under a tension of 1 kN. Assume that there are no reflected waves. (*a*) Find the period and frequency of waves on the wire. (*b*) What is the speed of the waves? (*c*) What are the wavelength and wave number? (*d*) Write a suitable wave function for the waves on the wire. (*e*) Calculate the maximum speed and acceleration of a point on the wire. (*f*) At what average rate must energy be supplied to the fork to keep it oscillating at a steady amplitude?

118 •• A very long wire can be vibrated up and down with a mechanical motor to produce waves traveling down the wire. At the far end of the wire, the traveling waves are absorbed by a clever device that allows no reflected waves to be returned to the motor. The wave speed is observed to be 240 m/s, the maximum transverse displacement of the wire is 1 cm, and the distance between maxima is 3.0 m. (*a*) Write a wave function to represent the wave propagating down this wire. (*b*) What is the frequency of vibration of the motor? (*c*) What is the period of the transverse oscillations of the wire? (*d*) What is the maximum transverse velocity of a small insect clinging to the wire?

119 ••• If a loop of chain is spun at high speed, it will roll like a hoop without collapsing. Consider a chain of linear mass density μ that is rolling without slipping at a high speed v_0. (*a*) Show that the tension in the chain is $F = \mu v_0^2$. (*b*) If the chain rolls over a small bump, a transverse wave pulse will be generated in the chain. At what speed will it travel along the chain? (*c*) How far around the loop (in degrees) will a transverse wave pulse travel in the time the hoop rolls through one complete revolution?

120 ••• A long rope with a mass per unit length of 0.1 kg/m is under a constant tension of 10 N. A motor at the point $x = 0$ drives one end of the rope with harmonic motion at 5 oscillations per second and an amplitude of 4 cm. (*a*) What is the wave speed? (*b*) What is the wavelength? (*c*) What is the maximum transverse linear momentum of a 1-mm segment of the rope? (*d*) What is the maximum net force on a 1-mm segment of the rope?

121 ••• A heavy rope 3 m long is attached to the ceiling and is allowed to hang freely. (*a*) Show that the speed of transverse waves on the rope is independent of its mass and length but does depend on the distance y from the bottom according to the formula $v = \sqrt{gy}$. (*b*) If the bottom end of the rope is given a sudden sideways displacement, how long does it take the resulting wave pulse to go to the ceiling, reflect, and return to the bottom of the rope?

122 ••• The linear mass density of a nonuniform wire under constant tension decreases gradually along the wire so that an incident wave is transmitted without reflection. The wire is uniform for $-\infty \le x \le 0$. In this region, a transverse wave has the form $y(x, t) = 0.003\cos(25x - 50t)$, where y and x are in meters and t is in seconds. From $x = 0$ to $x = 20$ m, the linear mass density decreases gradually from μ_1 to $\mu_1/4$. For $20 \le x \le \infty$, the linear mass density is $\mu = \mu_1/4$. (*a*) Find the wave velocity for large values of x. (*b*) Find the amplitude of the wave for large values of x. (*c*) Give $y(x, t)$ for $20 \le x \le \infty$.

123 ••• In this problem you will derive an expression for the potential energy of a segment of a string carrying a traveling wave (Figure 15-31). The potential energy of a segment equals the work done by the tension in stretching the string, which is $\Delta U = F(\Delta\ell - \Delta x)$, where F is the tension, $\Delta\ell$ is the length of the stretched segment, and Δx is its original length. From the figure we see that

$$\Delta\ell \approx \sqrt{(\Delta x)^2 + (\Delta y)^2} = \Delta x\{1 + (\Delta y/\Delta x)^2\}^{1/2}$$

(*a*) Use the binomial expansion to show that $\Delta\ell - \Delta x \approx \frac{1}{2}(\Delta y/\Delta x)^2\,\Delta x$, and therefore $\Delta U \approx \frac{1}{2}F\,(\Delta y/\Delta x)^2\,\Delta x$. (*b*) Compute dy/dx from the wave function in Equation 15-13 and show that $\Delta U \approx \frac{1}{2}Fk^2A^2\cos^2(kx - \omega t)\,\Delta x$. (*c*) Use $F = \mu v^2$ and $v = \omega/k$ to show that your result for (*b*) is the same as Equation 15-16*b*.

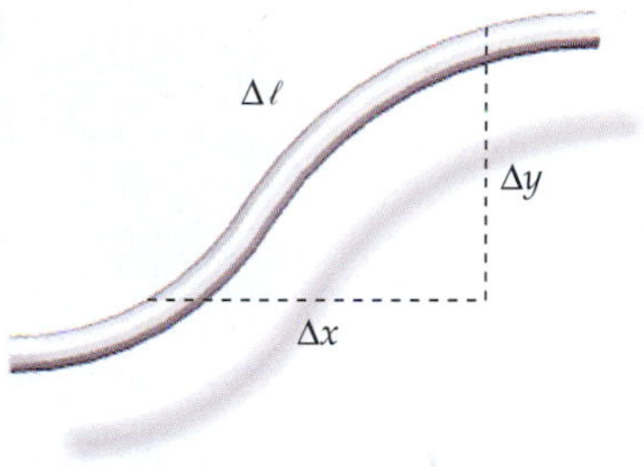

Figure 15–31
Problem 123

CHAPTER 16

Superposition and Standing Waves

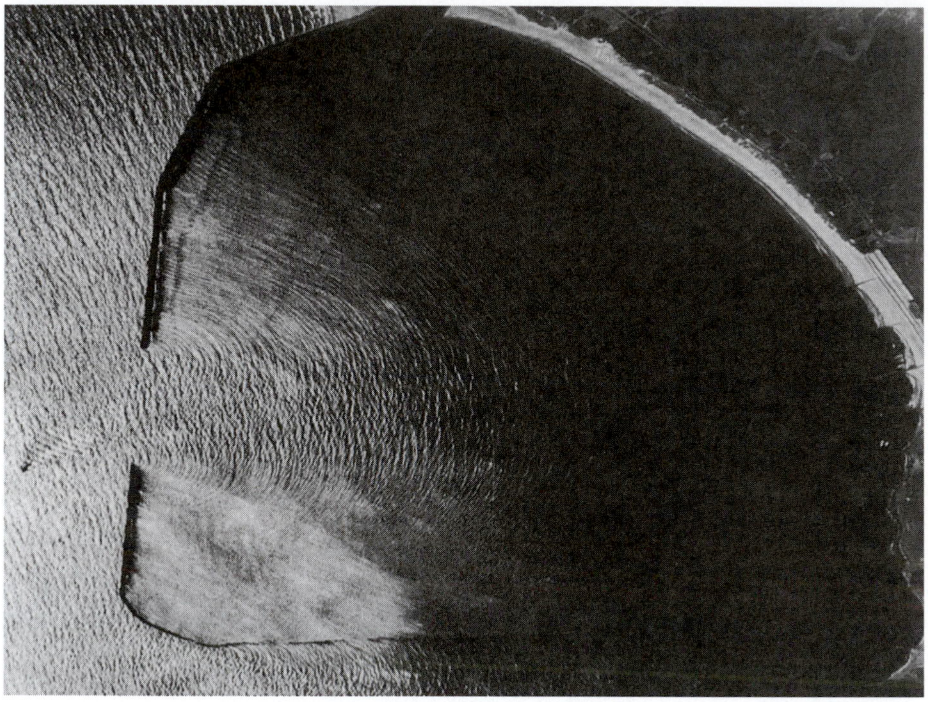

Water waves on the surface of the ocean diffract as they encounter an aperture formed by the breakwater.

When two waves meet in space, their individual disturbances (represented mathematically by their wave functions) superimpose and add algebraically, creating a new wave. The superposition of harmonic waves is called interference. Interference, like diffraction, is an important wave phenomenon. It was the observation of interference of light by Young in 1801 that led to the understanding that light propagates as a wave motion, not a particle motion as had been proposed by Newton. (It was the inability to observe interference or diffraction of light in Newton's time that led Newton to his particle model of light.) The observation of interference of electron waves by Davisson and Germer in 1927 led to our understanding of the wave nature of electrons and eventually to quantum physics, which we will study in Chapter 17.

In this chapter we begin by studying the superposition of wave pulses on a string and then consider the superposition and interference of harmonic waves. We will examine the phenomenon of beats, which result from the interference of two waves of slightly different frequencies, and we will then study standing waves, which result from the interference of two harmonic waves of the same frequency traveling in opposite directions in a confined space. Finally, we will consider the analysis of complex musical tones in terms of their component harmonic waves, and the inverse problem of the synthesis of harmonic waves to produce complex tones. We will conclude with a qualitative discussion of the extension of harmonic analysis to nonperiodic waves such as wave pulses.

16-1 Superposition of Waves

Figure 16-1 shows small wave pulses moving in opposite directions on a string. The shape of the string when they meet can be found by adding the displacements produced by each pulse separately. The **principle of superposition** is a property of wave motion that states

> When two or more waves combine, the resultant wave is the algebraic sum of the individual waves.

Principle of superposition

Mathematically, when there are two pulses on the string, the total wave function is the algebraic sum of the individual wave functions.

In the special case of two pulses that are identical except that one is inverted relative to the other, as in Figure 16-1*b*, there will be a moment in time when the pulses exactly overlap and add to zero. At this time the string is horizontal, but it is not at rest. Just to the right of the overlap region the string is moving up whereas just to the left it is moving down. A short time later the pulses emerge, each continuing in its original direction.

Superposition is a characteristic and unique property of wave motion. There is no analogous situation in particle motion; that is, two particles never overlap or add together in this way.

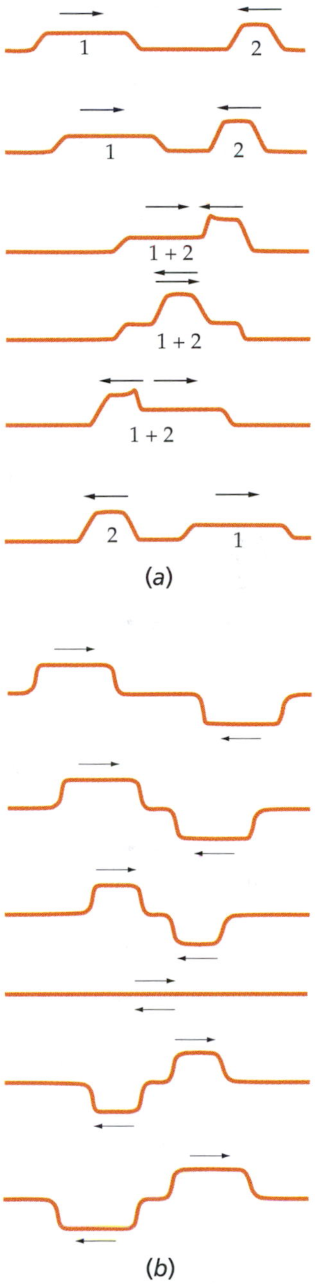

Figure 16-1 Wave pulses moving in opposite directions on a string. The shape of the string when the pulses meet is found by adding the displacements of each separate pulse. (*a*) Superposition of pulses having displacements in the same direction. (*b*) Superposition of pulses having opposite displacements. Here the algebraic addition of the displacements amounts to the subtraction of the magnitudes.

Superposition and the Wave Equation

The principle of superposition follows from the fact that the wave equation (Equation 15-9) is linear for small transverse displacements. That is, the function $y(x, t)$ and its derivatives occur only to the first power. An important property of linear equations is that if y_1 and y_2 are two solutions of the wave equation, the linear combination

$$y_3 = C_1y_1 + C_2y_2 \qquad \text{16-1}$$

is also a solution, where C_1 and C_2 are any constants. This can be shown by the direct substitution of y_3 into the wave equation. This result is the mathematical statement of the principle of superposition. If any two waves satisfy a wave equation, their sum also satisfies the same wave equation.

Exercise Show that the function y_3 given by Equation 16-1 satisfies Equation 15-9*b* if y_1 and y_2 both satisfy Equation 15-9*b*.

Interference of Harmonic Waves

The result of the superposition of harmonic waves depends on the phase difference between the waves. Let y_1 be the wave function for a harmonic wave traveling to the right with amplitude y_0, angular frequency ω, and wave number k:

$$y_1 = y_0 \sin(kx - \omega t) \qquad \text{16-2}$$

optional

For this wave function, we have chosen $t = 0$ when the displacement is zero at $x = 0$.* If we have another harmonic wave also traveling to the right with the same amplitude, frequency, and wave number, the general equation for its wave function can be written

$$y_2 = y_0 \sin(kx - \omega t + \delta) \qquad \text{16-3}$$

where δ is the phase constant. The two waves described by Equations 16-2 and 16-3 differ in phase by δ. Figure 16-2 shows a plot of the two wave functions versus position for a fixed time. The resultant wave is the sum

$$y_1 + y_2 = y_0 \sin(kx - \omega t) + y_0 \sin(kx - \omega t + \delta) \qquad \text{16-4}$$

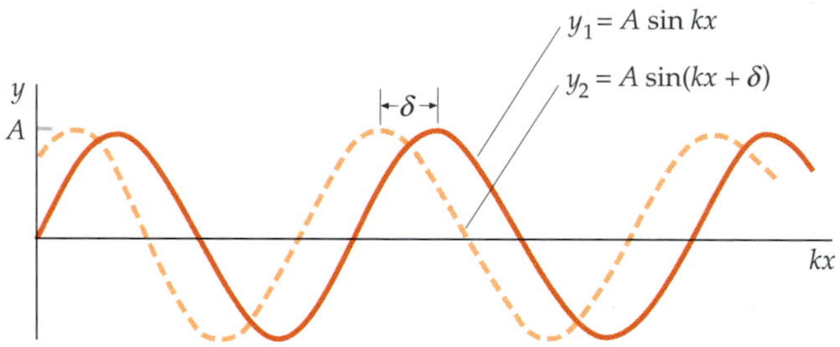

Figure 16-2 Displacement versus position for two harmonic waves having the same amplitude, frequency, and wavelength, but differing in phase by δ.

We can simplify Equation 16-4 by using the trigonometric identity

$$\sin \theta_1 + \sin \theta_2 = 2 \cos \tfrac{1}{2}(\theta_1 - \theta_2) \sin \tfrac{1}{2}(\theta_1 + \theta_2) \qquad \text{16-5}$$

For this case, $\theta_1 = kx - \omega t$ and $\theta_2 = kx - \omega t + \delta$, so that

$$\tfrac{1}{2}(\theta_1 - \theta_2) = -\tfrac{1}{2}\delta$$

and

$$\tfrac{1}{2}(\theta_1 + \theta_2) = kx - \omega t + \tfrac{1}{2}\delta$$

Thus, Equation 16-4 becomes

$$y_1 + y_2 = (2y_0 \cos \tfrac{1}{2}\delta) \sin(kx - \omega t + \tfrac{1}{2}\delta) \qquad \text{16-6}$$

Superposition of two waves of the same amplitude and frequency

where we have used $\cos(-\tfrac{1}{2}\delta) = \cos \tfrac{1}{2}\delta$. We see that the result of the superposition of two harmonic waves of equal wave number and frequency is a harmonic wave having the same wave number and frequency. The resultant wave differs in phase from both of the original waves, and its amplitude is $|2y_0 \cos \tfrac{1}{2}\delta|$. If the two waves are in phase, $\delta = 0$, $\cos 0 = 1$, and the amplitude of the resultant wave is $2A$. The interference of two waves in phase is called **constructive interference** (Figure 16-3). If the two waves are 180° out of phase, $\delta = \pi$ rad, $\cos(\pi/2) = 0$, and the amplitude of the resultant wave is zero. The interference of two waves 180° out of phase is called **destructive interference** (Figure 16-4).

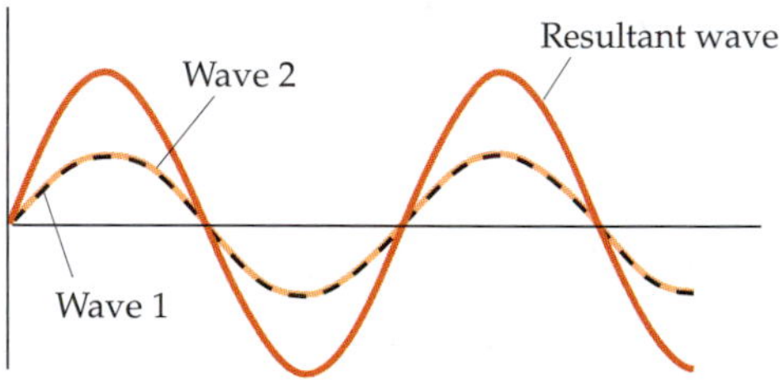

Figure 16-3 Constructive interference. When two waves are in phase, the amplitude of the resultant wave is the sum of the amplitudes of the individual waves.

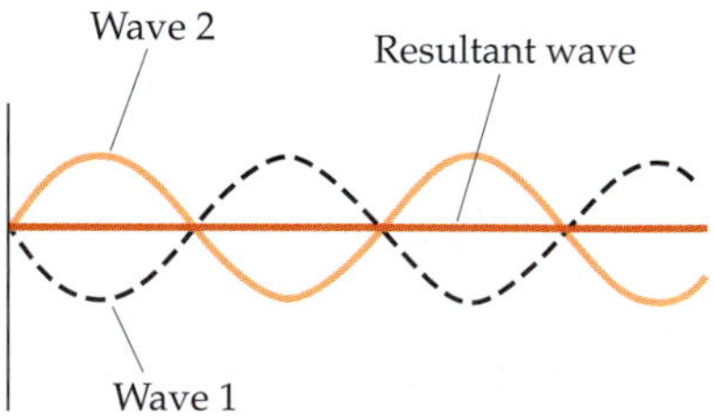

Figure 16-4 Destructive interference. When two waves have a phase difference of π, the amplitude of the resultant wave is the difference between the amplitudes of the individual waves. If the original waves have equal amplitudes, they cancel completely.

Exercise Two waves with the same frequency, wavelength, and amplitude are traveling in the same direction. (*a*) If they differ in phase by $\pi/2$ and each has an amplitude of 4.0 cm, what is the amplitude of the resultant wave? (*b*) For what phase difference δ will the resultant amplitude be equal to 4.0 cm? (*Answers* (*a*) 5.66 cm, (*b*) 120° or 240°)

---

* This choice is convenient but not mandatory. If, for example, we chose $t = 0$ when the displacement was maximum at $x = 0$, we would write $y_1 = A \cos(kx - \omega t) = A \sin(kx - \omega t - \pi/2)$.

Beats The interference of two sound waves with slightly different frequencies produces the interesting phenomenon known as **beats.** Consider two sound waves that have angular frequencies of ω_1 and ω_2 and the same pressure amplitude p_0. What do we hear? At a fixed point, the spatial dependence of the wave merely contributes a phase constant so we can neglect it. The pressure at the ear due to either wave acting alone will be a simple harmonic function of the type

$$p_1 = p_0 \sin \omega_1 t$$

and

$$p_2 = p_0 \sin \omega_2 t$$

where we have chosen sine functions for convenience and have assumed that the waves are in phase at time $t = 0$. Using Equation 16-5 for the sum of two sine functions, we obtain for the resultant wave

$$p = p_0 \sin \omega_1 t + p_0 \sin \omega_2 t = 2p_0 \cos \tfrac{1}{2}(\omega_1 - \omega_2)t \sin \tfrac{1}{2}(\omega_1 + \omega_2)t$$

If we write $\omega_{av} = \frac{1}{2}(\omega_1 + \omega_2)$ for the average angular frequency and $\Delta\omega = \omega_1 - \omega_2$ for the difference in angular frequencies, the resultant wave function is

$$p = 2p_0 \cos(\tfrac{1}{2}\Delta\omega\, t) \sin \omega_{av} t = 2p_0 \cos(2\pi\tfrac{1}{2}\Delta f\, t) \sin 2\pi f_{av} t \qquad \text{16-7}$$

where $\Delta f = \Delta\omega/2\pi$ and $f_{av} = \omega_{av}/2\pi$.

Figure 16-5 shows a plot of pressure variations as a function of time. The waves are originally in phase and add constructively at time $t = 0$. Because their frequencies differ, the waves gradually become out of phase, and at time t_1 they are 180° out of phase and interfere destructively.* An equal time interval later (time t_2 in the figure), the two waves are again in phase and interfere constructively. The greater the difference in frequencies of the two waves, the more rapidly they oscillate in and out of phase.

The tone we hear has a frequency of $f_{av} = \frac{1}{2}(f_1 + f_2)$ and amplitude $2p_0 \cos(2\pi\frac{1}{2}\Delta f\, t)$. The amplitude oscillates with the frequency $\frac{1}{2}\Delta f$. Since the sound intensity is proportional to the square of the amplitude, the sound is loud whenever the amplitude is either maximum or minimum. The frequency of this variation in intensity, which is twice $\frac{1}{2}\Delta f$, is called the **beat frequency.**

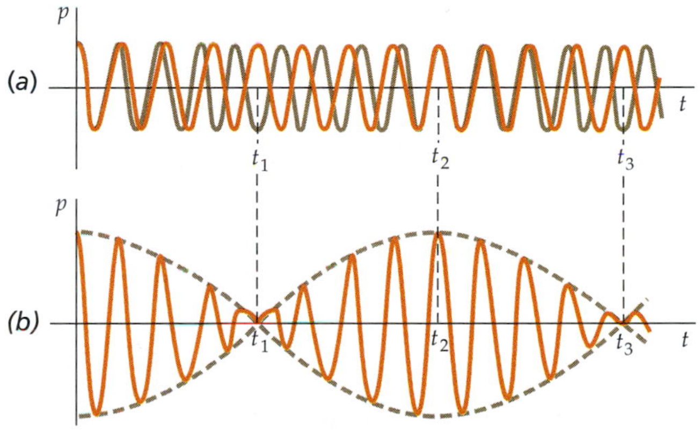

Figure 16-5 Beats. (*a*) Two waves of different but nearly equal frequencies that are in phase at $t_0 = 0$ are 180° out of phase at some later time t_1. At a still later time t_2, they are back in phase. (*b*) The resultant of the two waves shown in (*a*). The frequency of the resultant wave is about the same as those of the original waves, but the amplitude is modulated as indicated by the dashed envelope. The amplitude is maximum at times t_0 and t_2 and zero at times t_1 and t_3.

$$f_{beat} = \Delta f \qquad \text{16-8}$$

Beat frequency

The beat frequency equals the difference in the individual frequencies of the two waves: If we simultaneously strike two tuning forks having the frequencies 241 and 243 Hz, we will hear a pulsating tone at the average frequency of 242 Hz that has a maximum intensity 2 times per second; that is, the beat frequency is 2 Hz. The ear can detect up to about 15 to 20 beats per second. Above this frequency, the fluctuations in loudness are too rapid to be distinguished.

* Complete cancellation occurs only when the pressure amplitudes of the two waves are equal.

The phenomenon of beats is often used to compare an unknown frequency with a known frequency, as when a tuning fork is used to tune a piano string. Pianos are tuned by ringing the tuning fork and the piano note at the same time and tuning the piano string until the beats are far apart, indicating that the difference in frequency of the two sound generators is very small.

Example 16-1

When a 440-Hz (concert A) tuning fork is struck simultaneously with the playing of the A string of a slightly out-of-tune guitar, 3 beats per second are heard. After the guitar string is tightened a bit to increase its frequency, the beat frequency increases to 6 beats per second. What is the frequency of the guitar string after it is tightened?

Since 3 beats per second were heard initially, the original frequency of the guitar string was either 443 or 437 Hz. Had it been 437 Hz, increasing the string's frequency by tightening it would decrease the beat frequency. Since the beat frequency increases to 6 beats per second, the original frequency must have been 443 Hz, and hence the new frequency is 446 Hz.

Phase Difference due to Path Difference A common cause of a phase difference between two waves is different path lengths between the sources of the waves and the point of interference. Suppose that two sources oscillate in phase (for example, positive crests leave the sources at the same time) and emit harmonic waves of the same frequency and wavelength. Now consider a point in space for which the path lengths to the two sources differ. If the path difference is one wavelength, as is the case in Figure 16-6*a*, or is an integral number of wavelengths, the interference is constructive. If the path difference is one-half of a wavelength or an odd number of half-wavelengths, as in Figure 16-6*b*, the maximum of one wave falls at the minimum of the other and the interference is destructive.

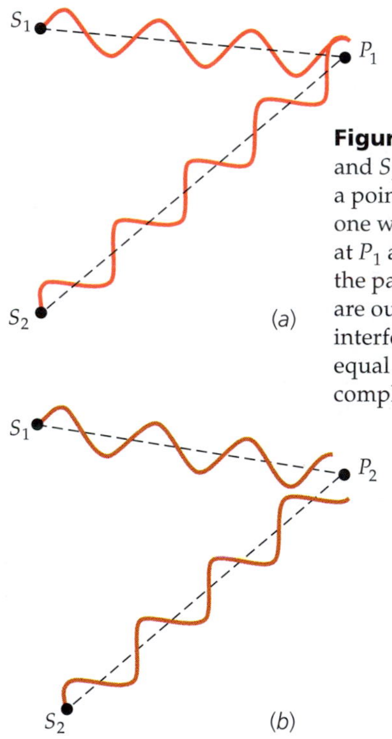

Figure 16-6 Waves from two sources S_1 and S_2 that are in phase when they meet at a point P_1. (*a*) When the path difference is one wavelength λ, the waves are in phase at P_1 and interfere constructively. (*b*) When the path difference is $\frac{1}{2}\lambda$, the waves at P_2 are out of phase by 180° and therefore interfere destructively. If the waves are of equal amplitude at P_2, they will cancel completely at this point.

The wave functions for waves from two sources oscillating in phase can be written

$$p_1 = p_0 \sin(kx_1 - \omega t)$$

and

$$p_2 = p_0 \sin(kx_2 - \omega t)$$

The phase difference for these two wave functions is

$$\delta = (kx_2 - \omega t) - (kx_1 - \omega t) = k(x_2 - x_1) = k\,\Delta x$$

Using $k = 2\pi/\lambda$, we have

$$\delta = 2\pi\frac{\Delta x}{\lambda} = (360°)\frac{\Delta x}{\lambda} \qquad \text{16-9}$$

Phase difference due to path difference

Example 16-2

Two sound sources oscillate in phase. At a point 5.00 m from one source and 5.17 m from the other, the amplitude of the sound from each source separately is p_0. Find the amplitude of the resultant wave if the frequency of the sound waves is (*a*) 1000 Hz, (*b*) 2000 Hz, and (*c*) 500 Hz. (Use 340 m/s for the speed of sound.)

Picture the Problem The amplitude of the resultant wave due to superposition of two waves differing in phase by δ is given by $A = |2p_0 \cos \frac{1}{2}\delta|$ (Equation 16-6), where p_0 is the amplitude of either wave, and $\delta = 2\pi\Delta x/\lambda$ is the phase difference. We are given the path difference, $\Delta x = 5.17\text{ m} - 5\text{ m} = 0.17\text{ m}$, so all that is needed is the wavelength λ.

(*a*)1. The wavelength equals the speed divided by the frequency. Calculate λ for $f = 1000$ Hz:

$$\lambda = \frac{v}{f} = \frac{340\text{ m/s}}{1000\text{ Hz}} = 0.34\text{ m}$$

2. For $\lambda = 0.34$ m, the given path difference ($\Delta x = 0.17$ m) is $\frac{1}{2}\lambda$, so we expect destructive interference. Use this value of λ to calculate the phase difference δ, and use δ to calculate the amplitude A:

$$\delta = 2\pi\frac{\Delta x}{\lambda} = 2\pi\frac{0.17\text{ m}}{0.34\text{ m}} = \pi$$

$$A = \left|2p_0 \cos\frac{1}{2}\delta\right| = \left|2p_0 \cos\frac{\pi}{2}\right| = 0$$

(*b*)1. Calculate λ for $f = 2000$ Hz:

$$\lambda = \frac{v}{f} = \frac{340\text{ m/s}}{2000\text{ Hz}} = 0.17\text{ m}$$

2. For $\lambda = 0.17$ m, the path difference equals λ, so we expect constructive interference. Calculate the phase difference and amplitude:

$$\delta = 2\pi\frac{\Delta x}{\lambda} = 2\pi\frac{0.17\text{ m}}{0.17\text{ m}} = 2\pi$$

$$A = \left|2p_0 \cos\frac{1}{2}\delta\right| = |2p_0 \cos \pi| = 2p_0$$

(*c*)1. Calculate λ for $f = 500$ Hz:

$$\lambda = \frac{v}{f} = \frac{340\text{ m/s}}{500\text{ Hz}} = 0.68\text{ m}$$

2. Calculate the phase difference and amplitude:

$$\delta = 2\pi\frac{\Delta x}{\lambda} = 2\pi\frac{0.17\text{ m}}{0.68\text{ m}} = \frac{\pi}{2}$$

$$A = \left|2p_0 \cos\frac{1}{2}\delta\right| = \left|2p_0 \cos\frac{\pi}{4}\right| = 1.41\,p_0$$

Example 16-3

Two speakers face each other at a distance of 90 cm and are driven by a common audio oscillator at 680 Hz (Figure 16-7). Locate the points between the speakers along a line joining them for which the sound intensity is (*a*) maximum and (*b*) minimum. (Neglect the variation in intensity from either speaker with distance, and use 340 m/s for the speed of sound.)

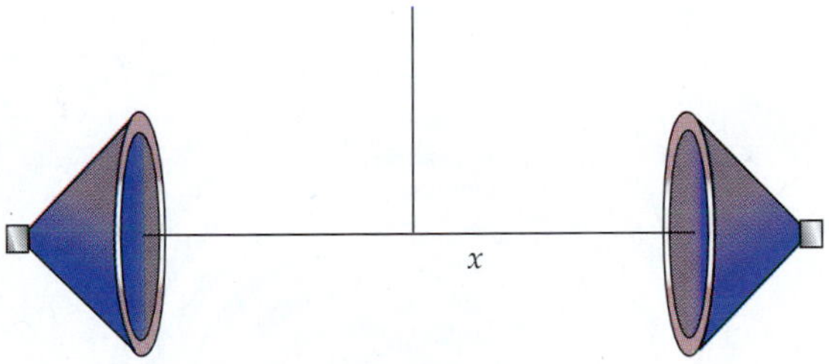

Figure 16-7

Picture the Problem We choose the origin to be at the midpoint between the speakers. Since this point is equidistant from the speakers, it is a point of maximum intensity. When we move a distance x toward one of the speakers, the path difference is $2x$. The intensity will be maximum when $2x = \lambda, 2\lambda, 3\lambda, \ldots$ and minimum when $2x = \frac{1}{2}\lambda, \frac{3}{2}\lambda, \frac{5}{2}\lambda, \ldots$.

(*a*)1. The intensity will be maximum when $2x$ equals an integral number of wavelengths: $2x = \pm\lambda,\ \pm 2\lambda,\ \pm 3\lambda,\ \ldots$

2. Calculate the wavelength: $\lambda = \dfrac{v}{f} = \dfrac{340\text{ m/s}}{680\text{ Hz}} = 0.5\text{ m} = 50\text{ cm}$

3. Solve for x using the calculated wavelength: $x = \pm\frac{1}{2}\lambda,\ \pm\lambda,\ \pm\frac{3}{2}\lambda,\ \ldots$

$= \pm 25\text{ cm},\ \pm 50\text{ cm},\ \pm 75\text{ cm}$

(*b*)1. The intensity will be minimum when $2x$ equals an odd number of half wavelengths: $2x = \pm\frac{1}{2}\lambda,\ \pm\frac{3}{2}\lambda,\ \pm\frac{5}{2}\lambda,\ \ldots$

2. Solve for x using the calculated wavelength: $x = \pm\frac{1}{4}\lambda,\ \pm\frac{3}{4}\lambda,\ \pm\frac{5}{4}\lambda,\ \ldots$

$= \pm 12.5\text{ cm},\ \pm 37.5\text{ cm},\ \pm 62.5\text{ cm}$

Remarks The maxima and minima will be relative maxima and relative minima because the amplitude from the near speaker will be slightly greater than that from the far speaker. Only three values for x were used in each case because the fourth value for x would be at a distance beyond one speaker.

Figure 16-8*a* shows the wave pattern produced by two point sources in a ripple tank that are oscillating in phase and are separated by a small distance. Each source produces circular waves of wavelength λ. We can construct a similar pattern with a compass by drawing circular arcs representing the wave crests from each source at some particular time (Figure 16-8*b*). Where the crests from each source overlap, the waves interfere constructively. At these points, the paths for the waves from the two sources are either equal in length or they differ by an integral number of wavelengths. The dashed lines indicate the points that are equidistant from the sources or whose path differences are one wavelength, two wavelengths, or three wavelengths. Between each pair of interference maxima is an interference minimum for which the path difference is an odd number of half-wavelengths. The lines along which the waves completely cancel are called **nodal lines**. At any point between the maxima and minima, the amplitude of the resultant wave is given by $A = 2p_0 \cos\frac{1}{2}\lambda$, where p_0 is the amplitude of each wave separately, and δ is related to the path difference by Equation 16-9.

Figure 16-8 (*a*) Water waves in a ripple tank produced by two nearby sources oscillating in phase. (*b*) Drawing of wave crests for the sources in (*a*). The dashed lines indicate points for which the path difference is an integral number of wavelengths.

(*a*)

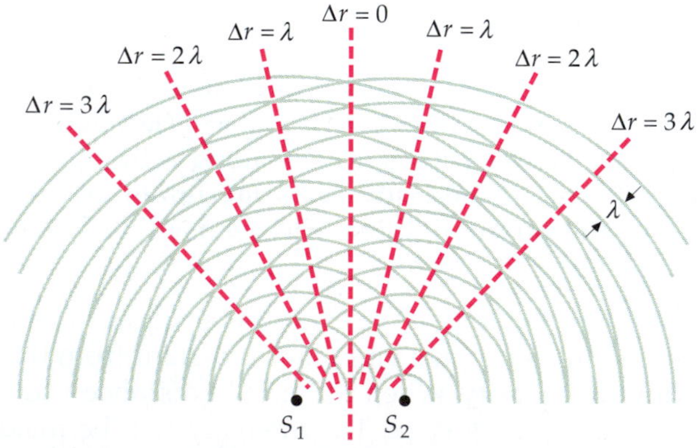

(*b*)

Figure 16-9 shows the intensity of the resultant wave from two sources as a function of path difference. At points where the interference is constructive, the amplitude of the resultant wave is twice that of either wave alone, and since the intensity is proportional to the square of the amplitude, the intensity is 4 times that due to either source alone. At points of destructive interference, the intensity is zero. The average intensity, shown by the dashed line in the figure, is twice the intensity due to either source alone, a result required by the conservation of energy. The interference of the waves from the two sources thus redistributes the energy in space. The interference of two sound sources can be demonstrated by driving two separated speakers with the same amplifier (so that they are always in phase) fed by an audio-signal generator. Moving about the room, one can detect by ear the positions of constructive and destructive interference.*

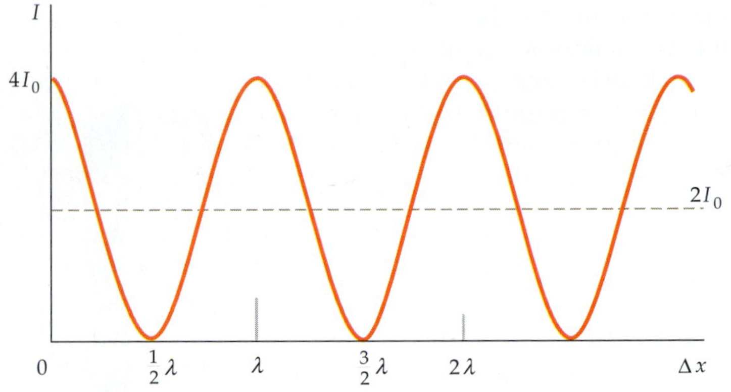

Figure 16-9 Intensity versus path difference for two sources that are in phase. I_0 is the intensity due to each source individually.

Coherence Two sources need not be in phase to produce an interference pattern. Consider two sources that are 180° out of phase.† The intensity pattern is the same as that in Figure 16-9 except that the maxima and minima are interchanged. At points for which the distance differs by an integral number of wavelengths, the interference is destructive because the waves are 180° out of phase. At points where the path difference is an odd number of half-wavelengths, the waves are now in phase because the 180° phase difference of the sources is offset by the 180° phase difference due to the path difference.

Similar interference patterns will be produced by any two sources whose phase difference is constant over time. Two sources that are in phase or have a constant phase difference are said to be **coherent**. Coherent sources of water waves in a ripple tank are easy to produce by driving both sources with the same motor. Coherent sound sources are obtained by driving two speakers with the same signal source and amplifier.

Wave sources whose phase difference is not constant over time but varies randomly are said to be **incoherent sources**. There are many examples of incoherent sources, such as two speakers driven by different amplifiers or two violins bowed by different players. For incoherent sources, the interference at a particular point varies rapidly back and forth from constructive to destructive, and no interference pattern is observed. At every point the resultant intensity of waves from two or more incoherent sources is simply the sum of the intensities due to the individual sources.

The Double-Slit Experiment Because a light beam is usually the result of millions of atoms radiating independently, the phase difference between the waves from such sources fluctuates randomly many times per second, so two light sources are usually not coherent. Coherence in optics is commonly achieved by splitting the light beam from a single source into two or more beams that can then be combined to produce an interference pattern. One

* In this demonstration, the sound intensity will be not quite zero at the points of destructive interference because of sound reflections from the walls or objects in the room.

† Two speakers that are in phase can be made to be out of phase by 180° merely by switching the leads to one of the speakers.

method of achieving this splitting is by the diffraction of a light beam by two small openings or slits in an opaque barrier (Figure 16-10). The intensity pattern on a screen far from the slits is shown in Figure 16-10*b*. This method was used by Thomas Young in 1801 to demonstrate that light exhibits interference and is therefore a wave phenomenon, rather than a particle motion as proposed by Newton. The demonstration is now known as Young's experiment. The intensity is maximum when the difference in path from the point on the screen to the two slits is an integer times the wavelength of the light. When this path difference is $\frac{1}{2}\lambda$, $\frac{3}{2}\lambda$, . . . , the intensity is zero. We discuss Young's experiment further in Chapter 17.

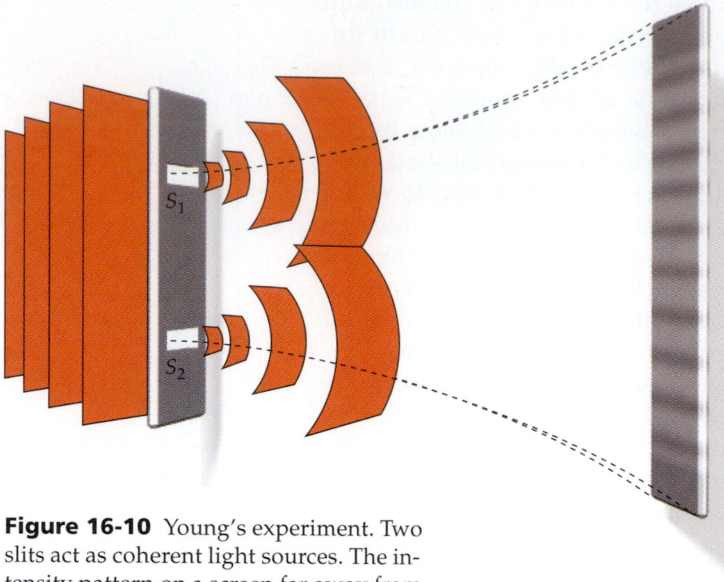

Figure 16-10 Young's experiment. Two slits act as coherent light sources. The intensity pattern on a screen far away from the slits exhibits interference.

16-2 Standing Waves

When waves are confined in space, like the waves on a piano string or sound waves in an organ pipe, reflections at both ends cause the waves to travel in both directions. These waves combine according to the principle of superposition. For a given string or pipe, there are certain frequencies for which superposition results in a stationary vibration pattern called a **standing wave.** Standing waves have important applications in musical instruments and in quantum theory.

String Fixed at Both Ends

If we fix both ends of a string and move a portion of the string up and down with simple harmonic motion of small amplitude, we find that at certain frequencies, standing-wave patterns such as those shown in Figure 16-11 are produced. The frequencies that produce these patterns are called the **resonance frequencies** of the string system. Each such frequency with its accompanying wave function is called a **mode of vibration.** The lowest resonance frequency is called the fundamental frequency f_1. It produces the standing-wave pattern shown in Figure 16-11*a*, which is called the **fundamental mode** of vibration or the **first harmonic.** The second lowest frequency f_2 produces the pattern shown in Figure 16-11*b*. This mode of vibration has a frequency twice that of the fundamental frequency and is called the second harmonic.* The third lowest frequency f_3

Figure 16-11 Standing waves on a string that is fixed at both ends. Points labeled A are antinodes and those labeled N are nodes. In general, the *n*th harmonic has *n* antinodes.

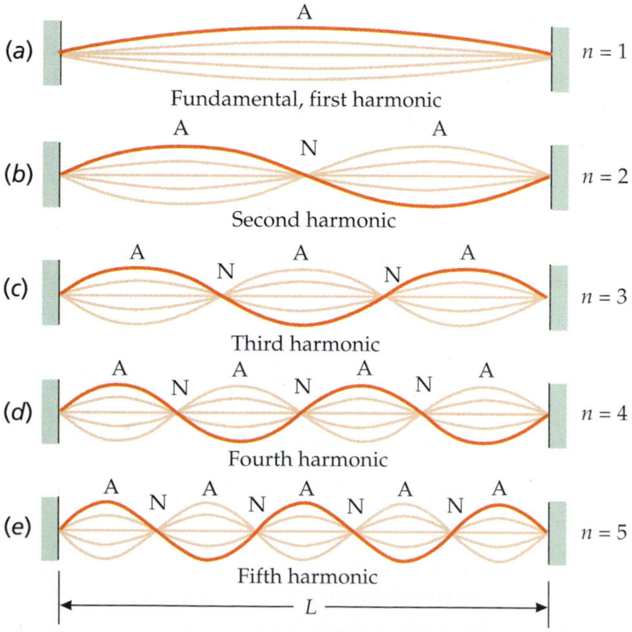

* In the terminology often used in music, the second harmonic is called the first **overtone,** the third harmonic is called the second overtone, and so on.

is three times the fundamental frequency, and it produces the third harmonic pattern shown in Figure 16-11*c*.

We note from Figure 16-11 that for each harmonic there are certain points on the string (the midpoint in Figure 16-11*b*, for example) that do not move. Such points are called **nodes.** Midway between each pair of nodes is a point of maximum amplitude of vibration called an **antinode.** Both fixed ends of the string are, of course, nodes.* We note that the first harmonic has one antinode, the second harmonic has two antinodes, and so on.

We can relate the resonance frequencies to the wave speed in the string and the length of the string. The length L of the string equals one-half the wavelength in the fundamental mode of vibration (Figure 16-12), and as Figure 16-11 reveals, L equals two half-wavelengths for the second harmonic, $\frac{3}{2}\lambda$ for the third harmonic, and so forth. In general, if λ_n is the wavelength for the nth harmonic, we have

Figure 16-12

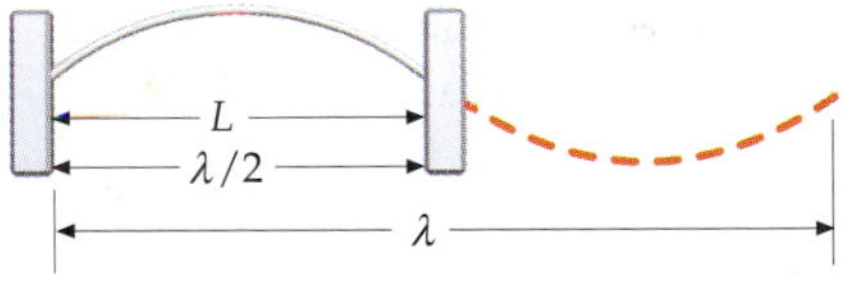

$$L = n\frac{\lambda_n}{2}, \qquad n = 1, 2, 3, \ldots \tag{16-10}$$

Standing-wave condition, both ends fixed

This result is known as the **standing-wave condition.** We can find the frequency of the nth harmonic from the fact that the wave speed v equals the frequency f times the wavelength. Thus,

$$f_n = \frac{v}{\lambda_n} = \frac{v}{2L/n}$$

or

$$f_n = n\frac{v}{2L} = nf_1, \qquad n = 1, 2, 3, \ldots \tag{16-11}$$

Resonance frequencies, both ends fixed

where $f_1 = v/2L$ is the fundamental frequency. You don't need to memorize Equation 16-11. Just sketch Figure 16-11 to remind yourself of the standing-wave condition, $L = n\lambda_n/2$, and then use $f = v/\lambda$.

We can understand standing waves in terms of resonance. Consider a string of length L that is attached at one end to a vibrating tuning fork and is fixed at the other end. The first wave sent out by the tuning fork travels down the string a distance L to the fixed end, where it is reflected and inverted. It then travels back a distance L and is again reflected and inverted at the tuning fork. The total time for the round trip is $2L/v$. If this time equals the period of the vibrating fork, the twice-reflected wave exactly overlaps the second wave produced by the fork, and the two waves interfere constructively, producing a wave with twice the original amplitude. The combined wave travels down the string and back and adds to the third wave produced by the fork, increasing the amplitude threefold, and so on. Thus, the tuning fork is in resonance with the string. Resonance also occurs if the time it takes for the first wave to travel the distance $2L$ is twice the period of the vibrating

A classic Steinway grand piano. The strings vibrate when struck by the hammers, which are controlled by the keys. The longer strings (left) vibrate at lower frequencies than the shorter strings (right).

* If one end is attached to a tuning fork rather than being fixed, it will still be approximately a node because the amplitude of the vibration at that end is so much smaller than the amplitude at the antinodes.

tuning fork or any integer n times the period. We can thus write the resonance condition as

$$\frac{2L}{v} = nT = \frac{n}{f}$$

or

$$f = n\frac{v}{2L}$$

where $T = 1/f$ is the period. This is the same result we found by fitting an integral number of half-wavelengths into the distance L. Various damping effects, such as the loss of energy during reflection and the imperfect flexibility of the string, put a limit on the maximum amplitude that can be reached.

The resonance frequencies given by Equation 16-11 are also called the **natural frequencies** of the string. When the frequency of the tuning fork is not one of the natural frequencies of the vibrating string, standing waves are not produced. After the first wave travels the distance $2L$ and is reflected from the fork, it differs in phase from the wave being generated at the fork (Figure 16-13). When this resultant wave has traveled the distance $2L$ and is again reflected at the fork, it will differ in phase from the next wave generated. In some cases, the new resultant wave will have a greater amplitude than the previous wave; in other cases, the new amplitude will be less. On the average, the amplitude will not increase but will remain on the order of the amplitude of the first wave generated, which is the amplitude of the tuning fork. This amplitude is very small compared with the amplitudes attained at resonance frequencies.

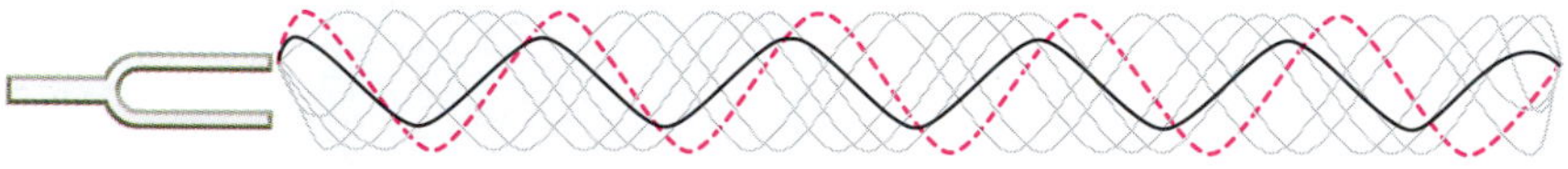

Figure 16-13

Figure 16-13 Waves on a string produced by a tuning fork whose frequency is not in resonance with the natural frequencies of the string. The wave leaving the tuning fork for the first time (dashed line) is not in phase with the waves that have been reflected two or more times (gray lines), and these waves are not in phase with each other, so there is no buildup in amplitude. The resultant wave (black line) has about the same amplitude as the individual waves, which is about the amplitude of the fork.

This resonance of standing waves is analogous to the resonance of a simple harmonic oscillator with a harmonic driving force. However, an oscillator has just one natural frequency, whereas a vibrating string has a sequence of natural frequencies that are integral multiples of the fundamental frequency. This sequence is called a **harmonic series**.

Turbulent winds set up standing waves in the Tacoma Narrows suspension bridge, leading to its collapse on November 7, 1940, just four months after it had been opened for traffic.

Example 16-4

A string is stretched between two fixed supports 0.7 m apart and the tension is adjusted until the fundamental frequency of the string is concert A, 440 Hz. What is the speed of transverse waves on the string?

1. The wave speed is related to the frequency and wavelength: $v = f\lambda$

2. The wavelength for the fundamental is twice the length of the string: $\lambda_1 = 2L$

3. Use this wavelength and the given frequency to find the speed: $v = f\lambda = (440\text{ Hz})(2L)$ $= (440\text{ Hz})(2)(0.7\text{ m}) = 616\text{ m/s}$

Exercise The speed of transverse waves on a stretched string is 200 m/s. If the string is 5 m long, find the frequencies of the fundamental and the second and third harmonics. (*Answer* $f_1 = 20\text{ Hz}, f_2 = 40\text{ Hz}, f_3 = 60\text{ Hz}$)

Example 16-5

A string of length 3 m and linear mass density 0.0025 kg/m is fixed at both ends. One of its resonance frequencies is 252 Hz. The next higher resonance frequency is 336 Hz. (*a*) Which harmonic is 252 Hz? (*b*) What is the fundamental frequency? (*c*) What is the tension in the string?

Picture the Problem Let 252 Hz be the *n*th harmonic. Then $f_n = nf_1$ and $f_{n+1} = (n+1)f_1$ are given. We can solve these two equations for n and f_1. The tension F is found from $v = \sqrt{F/\mu}$, where the speed v can be found from $v = f\lambda$ using any harmonic.

(*a*) 1. The consecutive harmonics f_n and f_{n+1} are related to the fundamental frequency f_1:

$$f_n = nf_1 = 252\text{ Hz}$$

$$f_{n+1} = (n+1)f_1 = 336\text{ Hz}$$

2. Dividing these equations eliminates f_1 and allows us to determine n:

$$\frac{f_n}{f_{n+1}} = \frac{n}{n+1} = \frac{252\text{ Hz}}{336\text{ Hz}} = 0.75 = \frac{3}{4}$$

$$4n = 3n + 3$$

$$n = 3$$

(*b*) Solve for f_1:

$$f_1 = \frac{f_n}{n} = \frac{f_3}{3} = \frac{252\text{ Hz}}{3} = 84\text{ Hz}$$

(*c*) 1. The tension is related to the wave speed:

$$v = \sqrt{\frac{F}{\mu}} \quad \text{or} \quad F = \mu v^2$$

2. Use the fundamental, with $\lambda_1 = 2L$, to obtain the speed:

$$v = f_1\lambda_1 = f_1(2L) = (84\text{ Hz})(6\text{ m}) = 504\text{ m/s}$$

3. Substitute this value of v to find the tension:

$$F = \mu v^2 = (0.0025\text{ kg/m})(504\text{ m/s})^2 = 635\text{ N}$$

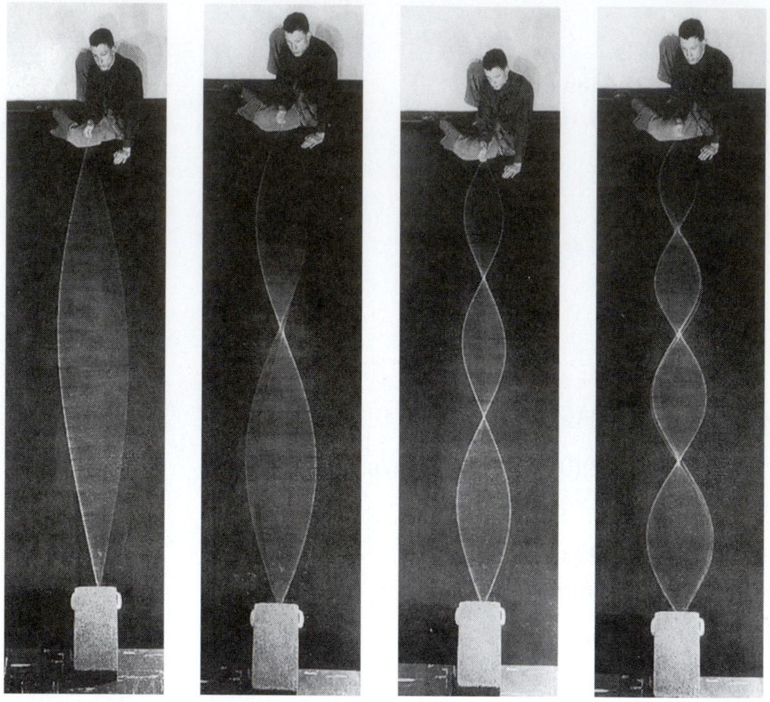

The first four harmonics of standing waves on a string fixed at both ends.

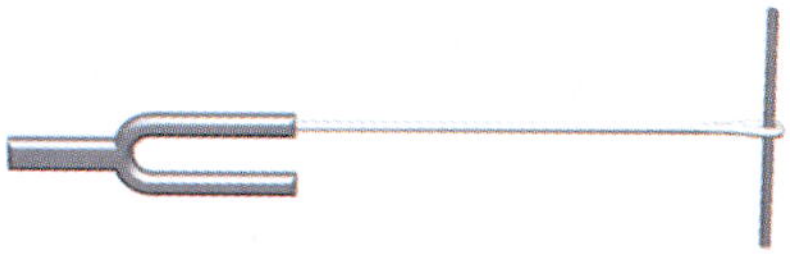

Figure 16-14 An approximation of a string fixed at one end and free at the other end can be produced by connecting the "free" end of the string to a ring that is free to move on a post. The end attached to the fork is approximately fixed because the amplitude of the tuning fork is very small.

String Fixed at One End

Figure 16-14 shows a string that has one end fixed and one end free. The standing-wave patterns for such a string are shown in Figure 16-15. Note that the free end is an antinode. In the fundamental mode of vibration for a string fixed at one end only, the length L of the string equals $\lambda_1/4$ (Figure 16-16).

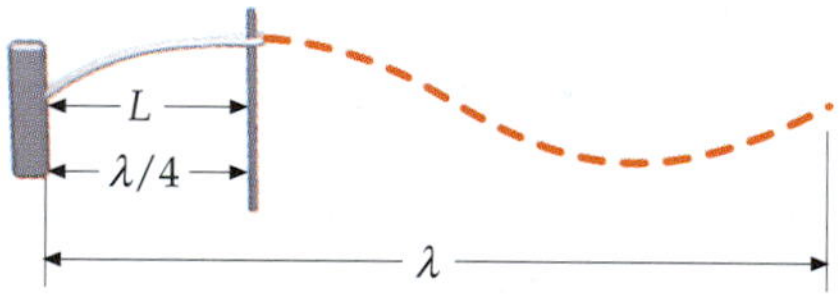

Figure 16-16

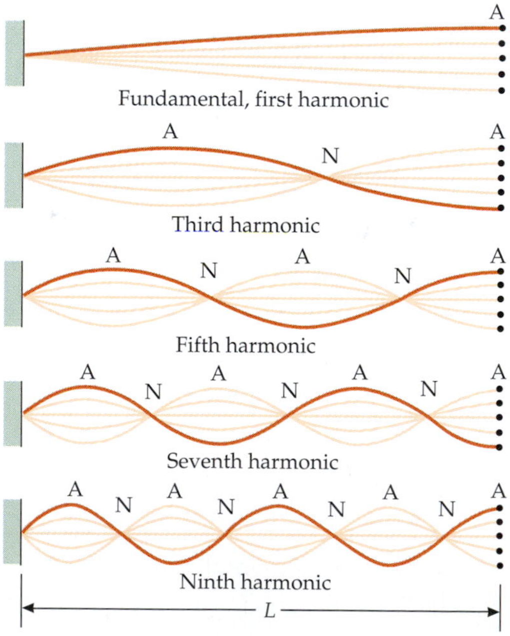

Figure 16-15 Standing waves on a string fixed at only one end. The free end is an antinode.

In the next highest mode, $L = 3\lambda_3/4$. The standing-wave condition can thus be written

$$L = n\frac{\lambda_n}{4}, \qquad n = 1, 3, 5, \ldots \qquad \text{16-12}$$

Standing-wave condition, one end free

The resonance frequencies are therefore given by

$$f_n = n\frac{v}{4L} = nf_1, \qquad n = 1, 3, 5, \ldots \qquad \text{16-13}$$

Resonance frequencies, one end free

where

$$f_1 = \frac{v}{4L} \qquad \text{16-14}$$

is the fundamental frequency. The natural frequencies of this system occur in the ratios 1:3:5:7: $\cdots$, which means that the even harmonics are missing. Again, an easy way to remember the resonance frequencies given by Equation 16-13 is to sketch Figure 16-15 to remind yourself of the standing-wave condition and use $f = v/\lambda$.

Wave Functions for Standing Waves

When a string vibrates in its nth mode, a segment of string moves with simple harmonic motion. Its displacement $y_n(x, t)$ is given by

$$y_n(x, t) = A_n(x)\cos(\omega_n t + \delta_n)$$

where ω_n is the angular frequency, δ_n is the phase constant, which depends on the initial conditions, and $A_n(x)$ is the amplitude, which depends on the location of the segment. The function $A_n(x)$ is the shape of the string at any instant of time. At the instant that the vibration has its maximum amplitude, the shape of a string vibrating in its nth harmonic is

$$A_n(x) = A_n \sin k_n x \qquad \text{16-15}$$

where $k_n = 2\pi/\lambda_n$ is the wave number. The wave function for a standing wave in the nth harmonic can thus be written

$$y_n(x, t) = A_n \sin k_n x \cos(\omega_n t + \delta_n) \qquad \text{16-16}$$

Example 16-6 *try it yourself*

(*a*) The wave functions for two waves that have equal amplitude, frequency, and wavelength, but that travel in opposite directions, are given by $y_1 = y_0 \sin(kx - \omega t)$ and $y_2 = y_0 \sin(kx + \omega t)$. Show that the sum of these two waves is a standing wave. (*b*) A standing wave on a string that is fixed at both ends is given by $y(x, t) = (0.024\ \text{m}) \sin(52.3\ \text{m}^{-1} x) \cos 480\ \text{s}^{-1} t$. Find the speed of waves on the string, and find the distance between nodes for the standing waves.

Cover the column to the right and try these on your own before looking at the answers.

Steps	***Answers***
(*a*) Add the two wave functions using Equation 16-5: $\sin\theta_1 + \sin\theta_2 = 2\sin\frac{1}{2}(\theta_1 + \theta_2)\cos\frac{1}{2}(\theta_1 - \theta_2)$.	$y = y_1 + y_2 = 2y_0 \sin kx \cos \omega t$
(*b*)1. Calculate the speed from $v = \omega/k$.	$v = 9.18$ m/s
2. Find the wavelength $\lambda = 2\pi/k$, and use it to find the distance between nodes.	$\Delta x = 0.5\lambda = 0.06$ m $= 6.0$ cm

Remarks Note that it doesn't matter which you call θ_1 and which you call θ_2 because $\cos(-\omega t) = \cos\omega t$. The resulting wavefunction is of the form given by Equation 16-16 with $A = 2y_0$.

Standing Sound Waves

A familiar example of standing waves are the waves in the air column of an organ pipe. In the flue-type organ pipe, a stream of air is directed against the sharp edge of an opening (point A in Figure 16-17). The complicated swirling motion of the air near the edge sets up vibrations in the air column. The resonance frequencies of the pipe depend on the length of the pipe and on whether the top is closed or open.

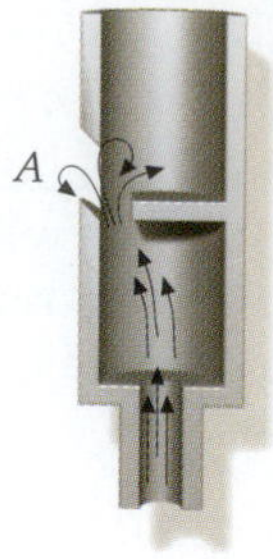

Figure 16-17 Flue-type organ pipe. Air is blown against the edge, causing a swirling motion of the air near point A that excites standing waves in the pipe. There is a pressure node near point A, which is open to the atmosphere.

In an open organ pipe, the pressure at both ends of the pipe is the same as atmospheric pressure and does not vary. Thus, there is a pressure node at both ends of the pipe. (This result is based on the assumption that the sound wave in the tube is a one-dimensional wave, which is approximately true if the tube diameter is much smaller than the wavelength. In practice, the pressure nodes lie slightly beyond the ends of the tube. The effective length of the pipe is $L_{eff} = L + \Delta L$, where ΔL is the end correction, a distance somewhat smaller than the tube diameter.) The standing-wave condition for this system is the same as that for a string fixed at both ends; after L is replaced by L_{eff}, the effective length of the tube, all the same equations apply.

In a closed organ pipe (open at one end, closed at the other), there is a pressure node near the opening (point A in Figure 16-17) and a pressure antinode at the closed end. The standing-wave condition for this system is the same as that for a string with one end fixed and one end free. The effective length of the tube must equal an odd integer times $\lambda/4$. That is, the wavelength of the fundamental mode is 4 times the effective length of the tube, and only the odd harmonics are present.

As we have seen, a sound wave can be thought of as either a pressure wave or a displacement wave. The pressure and displacement variations in a sound wave are 90° out of phase. Thus, in a standing sound wave, the pressure nodes are displacement antinodes and vice versa. The open end of an organ pipe is a pressure node and displacement antinode, whereas the closed end is a pressure antinode and a displacement node.

Example 16-7 *try it yourself*

If the speed of sound is 340 m/s, what are the allowed frequencies and wavelengths for standing sound waves in an open tube whose effective length is 1 m?

Cover the column to the right and try these on your own before looking at the answers.

Steps	*Answers*
1. Calculate the fundamental wavelength from $\lambda_1 = 2L_{eff}$.	$\lambda_1 = 2\text{ m}$
2. Use your value of λ_1 to calculate the fundamental frequency f_1.	$f_1 = 170\text{ Hz}$
3. Write expressions for the frequencies f_n and wavelengths λ_n of the other harmonics in terms of n:	$f_n = n(170\text{ Hz}), \quad n = 1, 2, 3, \ldots$ $\lambda_n = \dfrac{2\text{ m}}{n}, \quad n = 1, 2, 3, \ldots$

Example 16-8

When a tuning fork of frequency 500 Hz is held above the tube in Figure 16-18 as shown, resonances are found when the water level is at distances 16.0, 50.5, 85.0, and 119.5 cm from the top of the tube. (*a*) What is the speed of sound in air? (*b*) How far from the end of the tube is the pressure node?

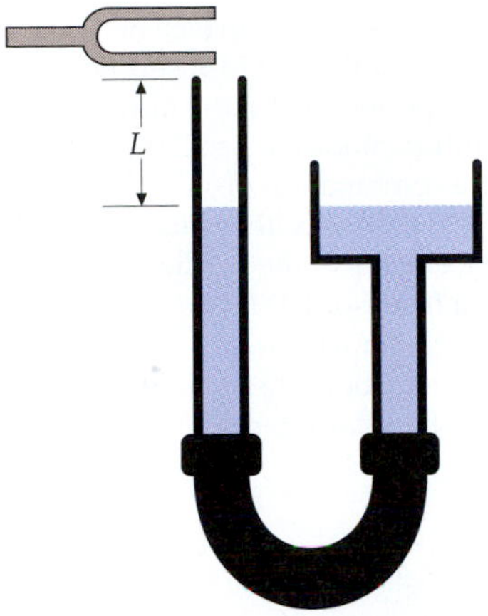

Figure 16-18

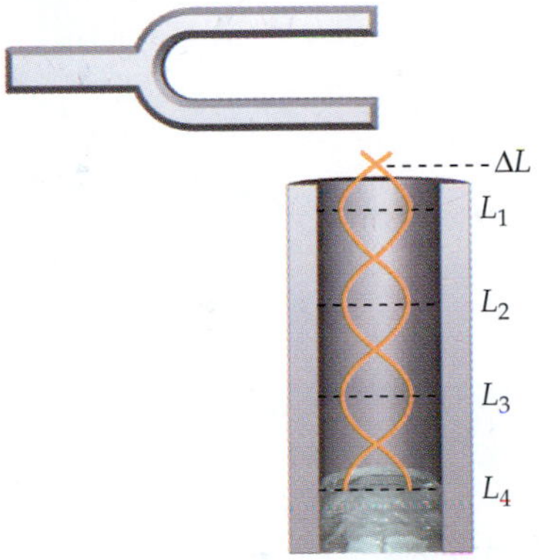

Figure 16-19

Picture the Problem Sound waves of the frequency 500 Hz are excited in the tube, whose length L can be adjusted. Resonance occurs when the effective length of the tube $L_{eff} = L + \Delta L$ equals $\frac{1}{4}\lambda$, $\frac{3}{4}\lambda$, $\frac{5}{4}\lambda$, and so on, where λ is the wavelength of the sound. Even though the pressure node is not exactly at the end of the tube because of the end correction, the wavelength can be found from the fact that the distance between water levels for successive resonances ($L_2 - L_1$ in Figure 16-19) is half the wavelength. The speed is then found from $v = f\lambda$, where f is 500 Hz. The end correction ΔL can be found from the fact that, for the fundamental, $L_{eff} = \frac{1}{4}\lambda = L_1 + \Delta L$.

(*a*) 1. The speed of sound in air is related to the frequency and wavelength:

$$v = f\lambda$$

2. The wavelength is twice the distance between successive water levels Δx:

$$\lambda = 2\,\Delta x$$

3. Δx is found from the data given in the problem:

$$\Delta x = 50.5\text{ cm} - 16.0\text{ cm} = 85\text{ cm} - 50.5\text{ cm}$$
$$= 119.5\text{ cm} - 85\text{ cm} = 34.5\text{ cm}$$

4. Substitute the values of f and λ to determine v:

$$\lambda = 2\,\Delta x = 2(34.5\text{ cm}) = 69\text{ cm} = 0.69\text{ m}$$
$$v = f\lambda = (500\text{ Hz})(0.69\text{ m}) = 345\text{ m/s}$$

(*b*) For the fundamental, the water level is at $L_1 = 16$ cm and the pressure node is $\frac{1}{4}\lambda = (69\text{ cm})/4 = 17.25$ cm. The difference is the end correction:

$$\Delta L = \tfrac{1}{4}\lambda - L_1$$
$$= \tfrac{1}{4}(69.0\text{ cm}) - (16.0\text{ cm}) = 1.25\text{ cm}$$

Most musical instruments are much more complicated than simple cylindrical tubes. The conical tube, which is the basis for the oboe, bassoon, English horn, and saxophone, has a complete harmonic series with its fundamental wavelength equal to twice the length of the cone. Brass instruments are combinations of cones and cylinders. The analysis of these instruments is extremely complex. The fact that they have nearly harmonic series is a triumph of educated trial and error rather than mathematical calculation.

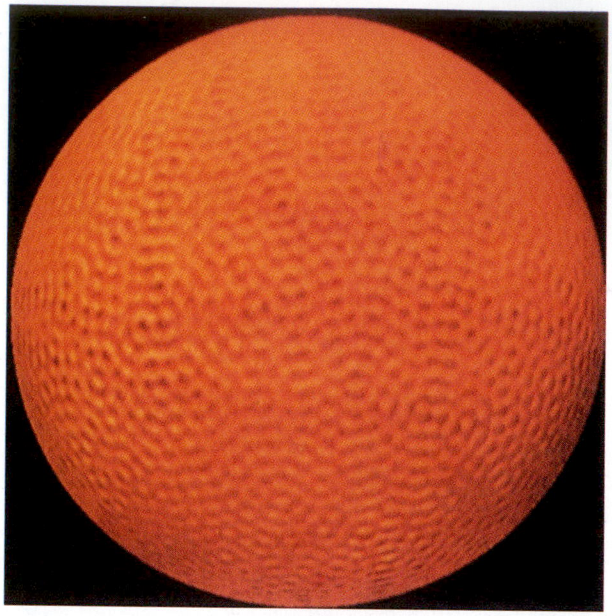

Standing sound waves on the surface of the sun. The surface of the sun is observed to oscillate with a period of about 5 min. Out of the 10 million modes of oscillation known to exist, a combination of approximately 100 modes is illustrated here. The displacements of the surface are exaggerated by a factor of 1000. The period of oscillation for each mode contains information about the structure and dynamics of the solar interior.

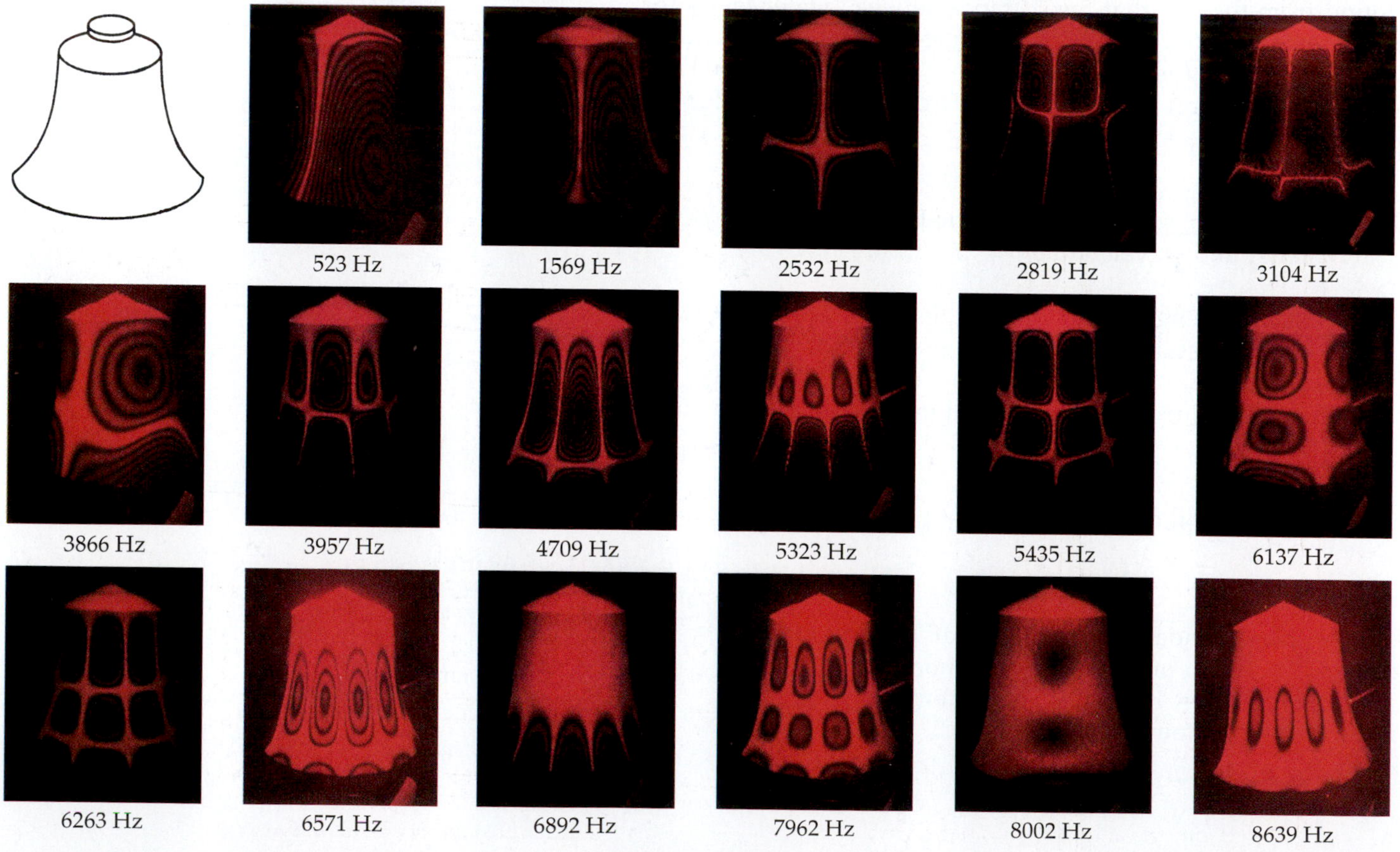

Holographic interferograms showing standing waves in a handbell. The "bull's eyes" locate the antinodes.

16-3 The Superposition of Standing Waves

optional

We have just seen that there is a set of natural resonance frequencies that produce standing waves for sound waves in air columns or vibrating strings that are fixed at one or both ends. For example, for a string fixed at both ends, the frequency of the fundamental mode of vibration is $f_1 = v/2L$, where L is the length of the string and v is the wave speed, and the wave function is Equation 16-16:

$$y_1(x,t) = A_1 \sin k_1 x \cos(\omega_1 t + \delta_1)$$

In general, a vibrating system does not vibrate in a single harmonic mode. Instead, the motion consists of a mixture of the allowed harmonics. The wave function is a linear combination of the harmonic wave functions:

$$y(x,t) = \sum_n A_n \sin k_n x \cos(\omega_n t + \delta_n) \qquad \text{16-17}$$

where $k_n = 2\pi/\lambda_n$, $\omega_n = 2\pi f_n$, and A_n and δ_n are constants. Since the energy in a wave is proportional to the square of the amplitude, the quantity A_n^2 describes the fraction of the energy associated with the nth harmonic. The constants A_n and δ_n depend on the initial position and velocity of the string. If a harp string, for example, is plucked at the center and released, as in Figure 16-20, the initial shape of the string is symmetric about the point $x = \frac{1}{2}L$. The motion of the string after it has been released will remain symmetric about this point. Only the odd harmonics, which are also symmetric about $x = \frac{1}{2}L$, will be excited. The even harmonics, which are antisymmetric about $x = \frac{1}{2}L$, are not excited; that is, the constant A_n is zero for all even n. The shapes of the first four harmonics are shown in Figure 16-21. Most of the energy of the plucked string is associated with the fundamental, but small amounts of energy are associated with the third and higher odd harmonic modes. Figure 16-22 shows an approximation to the initial shape of the string using the superposition of only the first three odd harmonics.

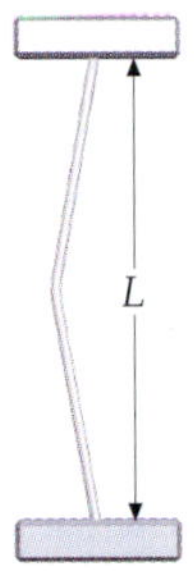

Figure 16-20 A string plucked at the center. When it is released, its vibration is a linear combination of standing waves.

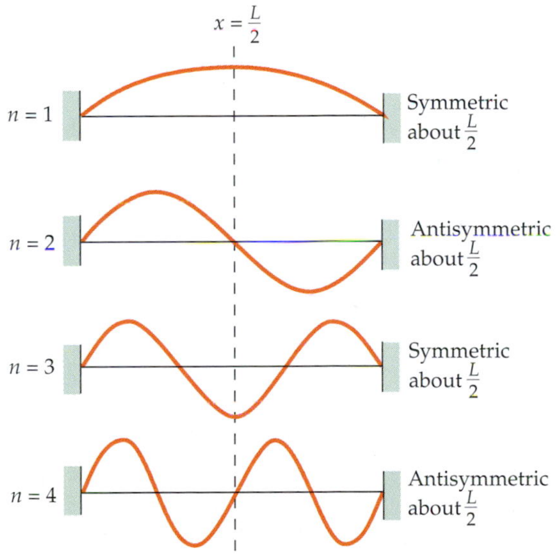

Figure 16-21 The first four harmonics for a string fixed at both ends. The odd harmonics are symmetrical about the center of the string, whereas the even harmonics are not. When a string is plucked at the center, it vibrates only in its odd harmonics.

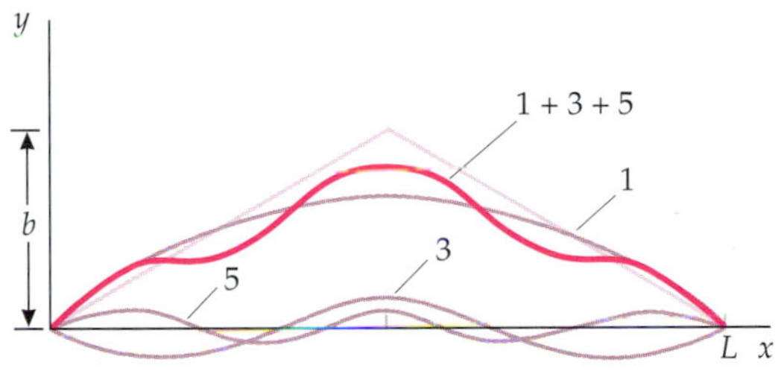

Figure 16-22 Approximating the shape of a string plucked at the center, as in Figure 16-20, using harmonics. The red line is an approximation of the original shape of the string based on the first three odd harmonics. The height of the string is exaggerated in this drawing to show the relative amplitudes of the harmonics. Most of the energy is associated with the fundamental, but there is some energy in the third, fifth, and other odd harmonics.

16-4 Harmonic Analysis and Synthesis

When a clarinet and an oboe play the same note, say, concert A, they sound quite different. Both notes have the same **pitch,** a physiological sensation of the highness or lowness of the note that is strongly correlated with frequency. However, the notes differ in what is called **tone quality.** The principal reason for the difference in tone quality is that, although both the clarinet and oboe are producing vibrations at the same fundamental frequency, each instrument is also producing harmonics whose relative intensities depend on the instrument and how it is played. If all instruments produced only the fundamental frequency, they would sound the same.

Figure 16-23 shows plots of the pressure variations versus time for a tuning fork, a clarinet, and an oboe, each playing the same note. These patterns are called **waveforms.** The waveform for the tuning fork is nearly a pure sine wave, but those for the clarinet and the oboe are clearly more complex.

Waveforms can be analyzed in terms of the harmonics that comprise them by means of **harmonic analysis.*** Figure 16-24 shows a plot of the relative intensities of the harmonics of the waveforms in Figure 16-23. The waveform of the tuning fork contains only the fundamental frequency. That of the clarinet contains the fundamental, large amounts of the third, fifth, and seventh harmonics, and lesser amounts of the second, fourth, and sixth harmonics. For the oboe, there is more energy in the second and fourth harmonics than in the fundamental.

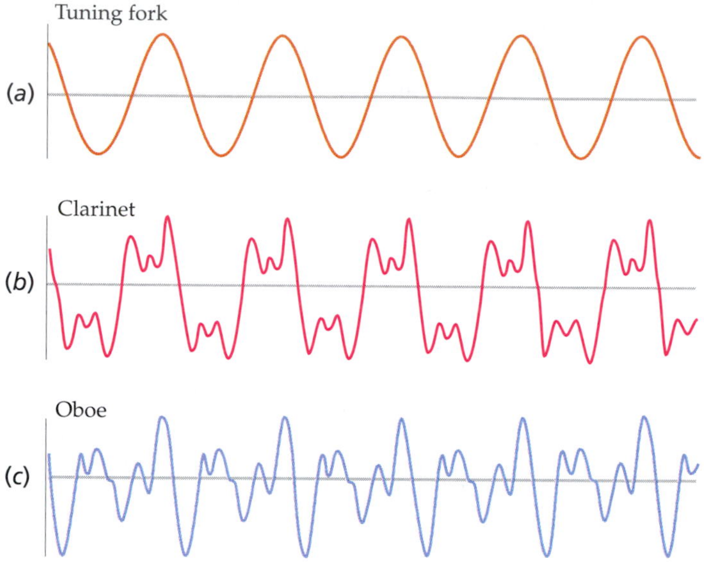

Figure 16-23 Waveforms of (*a*) a tuning fork, (*b*) a clarinet, and (*c*) an oboe, each at a fundamental frequency of 440 Hz and at approximately the same intensity.

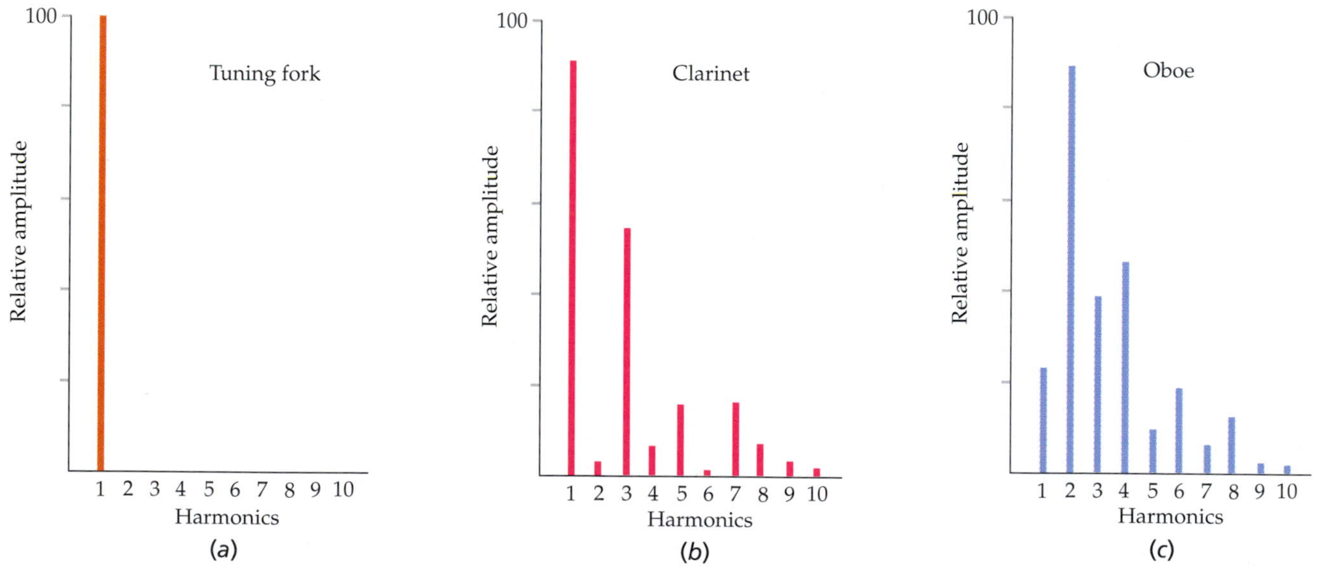

Figure 16-24 Relative intensities of the harmonics in the waveforms shown in Figure 16-23 for (*a*) the tuning fork, (*b*) the clarinet, and (*c*) the oboe.

* Harmonic analysis is also sometimes called **Fourier analysis** after the French mathematician Fourier, who developed the techniques for analyzing periodic functions.

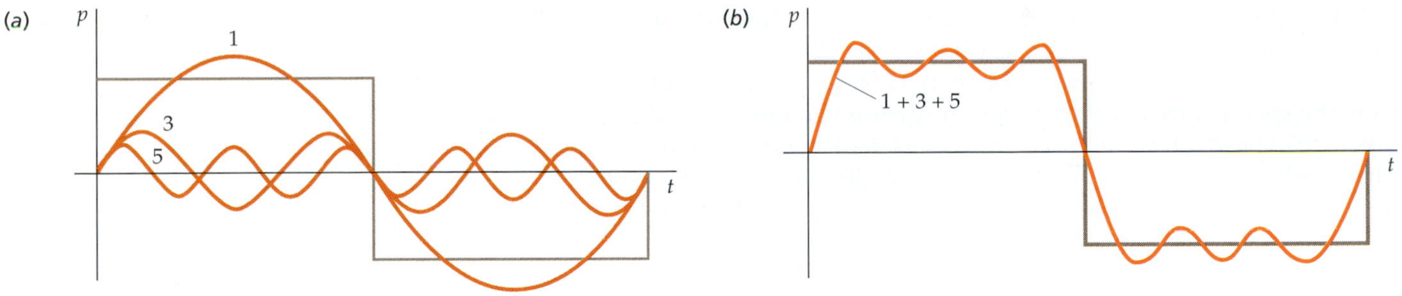

Figure 16-25 (*a*) The first three odd harmonics of a simple sine wave, used to synthesize a square wave. (*b*) The approximation of a square wave that results from summing the first three odd harmonics in (*a*).

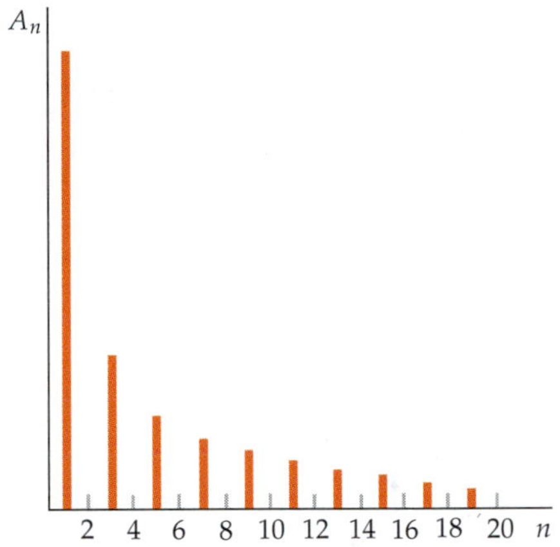

Figure 16-26 Relative amplitudes A_n of the first ten harmonics needed to synthesize a square wave. The more harmonics that are used, the closer the approximation is to the square wave.

The inverse of harmonic analysis is **harmonic synthesis,** which is the construction of a periodic wave from harmonic components. Figure 16-25*a* shows the first three odd harmonics used to synthesize a square wave (the dark gray line in the figure) and 16-25*b* shows the square wave that results from the sum of the three harmonics. The more harmonics used in a synthesis, the closer the approximation will be to the intended waveform. The relative amplitudes of the harmonics needed to synthesize a square wave are shown in Figure 16-26.

16-5 Wave Packets and Dispersion

The waveforms discussed in the previous section are periodic in time. Pulses, which are not periodic, can also be represented by a group of sinusoidal waves of different frequencies. However, the synthesis of a pulse requires a continuous distribution of frequencies rather than a discrete set of harmonics as in Figure 16-26. Such a group is called a **wave packet**. The characteristic feature of a wave pulse is that it has a beginning and an end, whereas a harmonic wave repeats over and over. If the duration Δt of the pulse is very short, the range of frequencies $\Delta\omega$ needed to describe the pulse is very large. The general relation between Δt and $\Delta\omega$ is

$$\Delta\omega\,\Delta t \sim 1 \qquad \text{16-18}$$

The exact value of this product depends on just how the quantities $\Delta\omega$ and Δt are defined. For any reasonable definitions, $\Delta\omega$ is of the order of $1/\Delta t$. A wave pulse produced by a source of short duration Δt has a narrow width in space $\Delta x = v\,\Delta t$, where v is the wave speed. Each harmonic wave of frequency ω has a wave number $k = \omega/v$. A range of frequencies $\Delta\omega$ implies a range of wave numbers $\Delta k = \Delta\omega/v$. Substituting $v\,\Delta k$ for $\Delta\omega$ in Equation 16-18 gives $v\,\Delta k\,\Delta t \sim 1$, or

$$\Delta k\,\Delta x \sim 1 \qquad \text{16-19}$$

If a wave packet is to maintain its shape as it travels, all of the component harmonic waves that make up the packet must travel with the same speed. This occurs if the speed of the component waves in a given medium is independent of wavelength and frequency. Such a medium is called a **nondispersive medium**. Air is a nondispersive medium for sound waves, but solids and liquids generally are not.*

* Probably the most familiar example of dispersion is the rainbow, which arises because the velocity of light waves in water depends slightly on the frequency and wavelength of the light, so the different colors corresponding to different wavelengths have slightly different angles of refraction.

optional

optional

When the wave speed in a dispersive medium depends only slightly on the frequency and wavelength, a wave packet changes shape very slowly as it travels, and it covers a considerable distance as a recognizable entity. However, the speed of this packet, called the **group velocity,** is not the same as the (average) speed of the individual component harmonic waves, called the **phase velocity.** For example, the group velocity of surface waves in deep water is half the average phase velocity of the component harmonic waves.

Summary

1. The principle of superposition, which holds for all electromagnetic waves and for string waves and sound waves of small amplitude, follows from the linearity of the corresponding wave equations.
2. Interference is an important wave phenomenon that applies to all waves. It follows from the principle of superposition.
3. The standing-wave conditions can be recalled by sketching a string or tube and drawing waves that have nodes at a fixed or closed end and antinodes at a free or open end.

	Remarks and Relevant Equations	
1. Superposition and Interference	The superposition of two harmonic waves of equal wave number and frequency but phase difference δ results in a harmonic wave of the same wave number and frequency, but differing in phase from each of the two waves: $y_1 + y_2 = (2y_0 \cos \frac{1}{2}\delta) \sin(kx - \omega t + \frac{1}{2}\delta)$	**16-6**
Constructive interference	If waves are in phase or differ in phase by an integer times 2π, the amplitudes of the waves add and the interference is constructive.	
Destructive interference	If waves differ in phase by π or by an odd integer times π, the amplitudes subtract and the interference is destructive.	
Beats	Beats are the result of the interference of two waves of slightly different frequencies. The beat frequency equals the difference in the frequencies of the two waves: $f_{\text{beat}} = \Delta f$	
Phase difference δ due to path difference Δx	$\delta = 2\pi \frac{\Delta x}{\lambda} = (360°)\frac{\Delta x}{\lambda}$	**16-9**
2. Coherent and Incoherent Sources	Two sources that are in phase or have a constant phase difference are coherent. Interference is observed only for waves from coherent sources. Waves from incoherent sources have a phase difference that varies randomly in time, so the interference at a particular point varies back and forth from constructive to destructive and no interference pattern is observed.	
3. Standing Waves	When waves are confined in space, there are certain frequencies for which superposition results in stationary vibration patterns called standing waves.	

String fixed at both ends	For a string fixed at both ends, there is a node at each end. Since an integral number of half-wavelengths must fit into the length of the string, the standing-wave condition in this case is $$L = n\frac{\lambda_n}{2}, \quad n = 1, 2, 3, \ldots \qquad \textbf{16-10}$$ The allowed waves form a harmonic series, with the frequencies given by $$f_n = n\frac{v}{2L} = nf_1, \quad n = 1, 2, 3, \ldots$$ where $f_1 = v/2L$ is the lowest frequency, the fundamental.
Organ pipe open at both ends	Standing sound waves in a pipe that is open at both ends have a pressure node at each end so that the standing-wave condition is the same as for a string fixed at both ends.
String fixed at one end and free at the other	If a string has one end fixed and one end free, there is a node at one end and an antinode at the other. The standing-wave condition in this case is $$L = n\frac{\lambda_n}{4}, \quad n = 1, 3, 5, \ldots \qquad \textbf{16-12}$$ Only the odd harmonics are present. Their frequencies are given by $$f_n = n\frac{v}{4L} = nf_1, \quad n = 1, 3, 5, \ldots$$ where $f_1 = v/4L$.
Organ pipe open at one end and closed at the other	Standing sound waves in a pipe that is open at one end and closed at the other end have a pressure node at the open end and an antinode at the closed end. The standing-wave condition is the same as for a string fixed at one end.
4. Standing Wave Functions	$$y_n(x, t) = A_n \sin k_n x \cos(\omega_n t + \delta_n) \qquad \textbf{16-16}$$ where $k_n = 2\pi/\lambda_n$ and $\omega_n = 2\pi f_n$.
5. Superposition of Standing Waves (optional)	In general, a vibrating system does not vibrate in a single harmonic mode but in a mixture of the allowed harmonics.
6. Harmonic Analysis and Synthesis (optional)	Sounds of different tone quality contain different mixtures of harmonics. The analysis of a particular tone in terms of its harmonic content is called harmonic analysis. Harmonic synthesis is the construction of a tone by the addition of harmonics.
7. Wave Packets (optional)	A wave pulse can be represented by a continuous distribution of harmonic waves called a wave packet. The range of frequencies $\Delta\omega$ is related to the width in time Δt, and the range of wave numbers Δk is related to the width in space Δx.
Frequency and time ranges	$\Delta\omega\,\Delta t \sim 1$ **16-18**
Wave number and space ranges	$\Delta k\,\Delta x \sim 1$ **16-19**
8. Dispersion (optional)	In a nondispersive medium, the wave speed v is independent of f and λ, and a pulse travels without change in shape. In a dispersive medium, v depends on f and λ, and the pulse changes shape as it moves. Then the velocity of the pulse, called the group velocity, is not the same as the phase velocity, which is the average velocity of the harmonic components of the pulse.

Problem-Solving Guide

1. Begin by drawing a neat diagram that includes the important features of the problem.
2. The standing-wave conditions can be recalled by sketching the string or tube and drawing waves that have nodes at a fixed or closed end and antinodes at a free or open end.

Summary of Worked Examples

Type of Calculation	Procedure and Relevant Examples
1. Superposition and Interference	
Relate the frequencies of two sources to their beat frequency.	The beat frequency is the difference in the two frequencies. **Example 16-1**
Find the amplitude or intensity due to two coherent sources in phase.	The amplitude is given by $A = \lvert 2p_0 \cos(\frac{1}{2}\delta) \rvert$ where the phase difference is $\delta = 2\pi(\Delta x/\lambda)$. The intensity is proportional to the square of the amplitude. **Example 16-2**
Find the positions of maximum and minimum interference of waves from two coherent sources in phase.	Maxima occur when the path difference is an integral number of wavelengths. Minima occur when the path difference is an odd number of half-wavelengths. **Example 16-3**
2. Standing Waves	
Find the speed of waves on a string given a particular mode of vibration and frequency.	Find the wavelength λ from the standing-wave condition, then find the speed from $v = f\lambda$. **Example 16-4**
Given the frequency of two successive harmonics, find which ones they are.	Use $f_n = nf_1$ and $f_{n+1} = (n+1)f_1$. Divide the equations and solve for n. **Example 16-5**
Find the allowed frequencies of standing waves in a tube.	If the tube is open at both ends or closed at both ends, the standing wave condition is $L = n(\lambda_n/2)$, $n = 1, 2, 3, 4, \ldots$. If it is closed at one end and open at the other end, the condition is $L = n(\lambda_n/4)$, $n = 1, 3, 5, 7, \ldots$. **Examples 16-6, 16-7, 16-8**

Problems

Conceptual Problems

Problems from Optional and Exploring sections

In a few problems, you are given more data than you actually need; in a few other problems, you are required to supply data from your general knowledge, outside sources, or informed estimates.

• Single-concept, single-step, relatively easy
•• Intermediate-level, may require synthesis of concepts
••• Challenging, for advanced students

Use $v = 340$ m/s for the speed of sound in air unless otherwise indicated.

Superposition and Interference

1 • True or false:

(*a*) The waves from two coherent sources that are radiating in phase interfere constructively everywhere in space.
(*b*) Two wave sources that are out of phase by 180° are incoherent.
(*c*) Interference patterns are observed only for coherent sources.

2 • Two violinists standing a few feet apart are playing the same notes. Are there places in the room where certain notes are not heard because of destructive interference? Explain.

3 •• Two rectangular wave pulses are traveling in opposite directions along a string. At $t = 0$, the two pulses are as

Figure 16-27 Problems 3 and 4

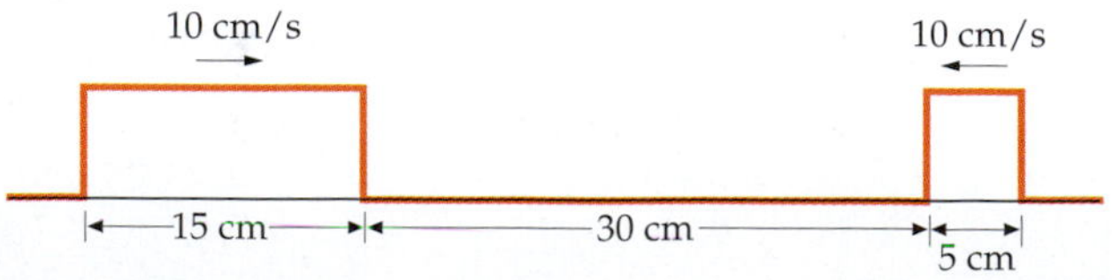

shown in Figure 16-27. Sketch the wave functions for $t = 1, 2,$ and 3 s.

4 •• Repeat Problem 3 for the case in which the pulse on the right is inverted.

5 • Two waves traveling on a string in the same direction both have a frequency of 100 Hz, a wavelength of 2 cm, and an amplitude of 0.02 m. What is the amplitude of the resultant wave if the original waves differ in phase by (*a*) $\pi/6$, and (*b*) $\pi/3$?

6 • What is the phase difference between the two waves of Problem 5 if the amplitude of the resultant wave is 0.02 m, the same as the amplitude of each original wave?

7 • Two waves having the same frequency, wavelength, and amplitude are traveling in the same direction. If they differ in phase by $\pi/2$ and each has an amplitude of 0.05 m, what is the amplitude of the resultant wave?

8 • Two sound sources oscillate in phase with the same amplitude A. They are separated in space by $\frac{1}{3}\lambda$. What is the amplitude of the resultant wave from the two sources at a point that is on the line that passes through the sources but is not between the sources?

9 • Two sound sources oscillate in phase with a frequency of 100 Hz. At a point 5.00 m from one source and 5.85 m from the other, the amplitude of the sound from each source separately is A. (*a*) What is the phase difference in the sound waves from the two sources at that point? (*b*) What is the amplitude of the resultant wave at that point?

10 • With a compass, draw circular arcs representing wave crests originating from each of two point sources a distance $d = 6$ cm apart for $\lambda = 1$ cm. Connect the intersections corresponding to points of constant path difference, and label the path difference for each line. (See Figure 16-8.)

11 • Two loudspeakers are separated by a distance of 6 m. A listener sits directly in front of one speaker at a distance of 8 m so that the two speakers and the listener form a right triangle. (*a*) Find the two lowest frequencies for which the path difference from the speakers to the listener is an odd number of half-wavelengths. (*b*) Why can these frequencies be heard even if the speakers are driven in phase by the same amplifier?

12 • Two speakers separated by some distance emit sound waves of the same frequency. At some point P, the intensity due to each speaker separately is I_0. The path distance from P to one of the speakers is $\frac{1}{2}\lambda$ greater than that from P to the other speaker. What is the intensity at P if (*a*) the speakers are coherent and in phase, (*b*) the speakers are incoherent, and (*c*) the speakers are coherent but have a phase difference of π rad?

13 • Answer the questions of Problem 12 for a point P' for which the distance to the far speaker is 1λ greater than the distance to the near speaker. Assume that the intensity at point P' due to each speaker separately is again I_0.

14 • Two speakers separated by some distance emit sound waves of the same frequency, but the speakers are out of phase by 90°. Let r_1 be the distance from some point to speaker 1 and r_2 be the distance from that point to speaker 2. Find the smallest value of $r_2 - r_1$ such that the sound at that point will be (*a*) maximum, and (*b*) minimum. (Express your answers in terms of the wavelength.)

15 •• Show that if the separation between two sound sources radiating coherently in phase is less than half a wavelength, complete destructive interference will not be observed in any direction.

16 •• A transverse wave of frequency 40 Hz propagates down a string. Two points 5 cm apart are out of phase by $\pi/6$. (*a*) What is the wavelength of the wave? (*b*) At a given point, what is the phase difference between two displacements for times 5 ms apart? (*c*) What is the wave velocity?

17 •• It is thought that the brain determines the direction toward the source of a sound by sensing the phase difference between the sound waves striking the eardrums. A distant source emits sound of frequency 680 Hz. When you are facing directly toward a sound source there should be no phase difference. Estimate the change in phase difference between the sounds received by the ears as you turn from facing directly toward the source through 90°.

18 •• Sound source A is located at $x = 0$, $y = 0$, and sound source B is placed at $x = 0$, $y = 2.4$ m. The two sources radiate coherently in phase. An observer at $x = 40$ m, $y = 0$ notes that as she walks along in either the positive or negative y direction away from $y = 0$, the sound intensity diminishes. What is the lowest and the next higher frequency of the sources that can account for that observation?

19 •• Suppose that the observer in Problem 18 finds herself at a point of minimum intensity at $x = 40$ m, $y = 0$. What is then the lowest and next higher frequency of the sources consistent with this observation?

20 •• Two point sources that are in phase are separated by a distance d. An interference pattern is detected along a line parallel to the line through the sources and a large distance D from the sources, as shown in Figure 16-28.

(*a*) Show that the path difference from the two sources to some point on the line at a small angle θ is given approximately by $\Delta s = d \sin \theta$. (*Hint:* Assume that the lines from the sources to P are approximately parallel.)

(*b*) Show that the distance y_m from the central maximum point to the mth interference maximum is given approximately by

$$y_m = m\frac{D\lambda}{d}$$

Figure 16-28 Problems 20–23

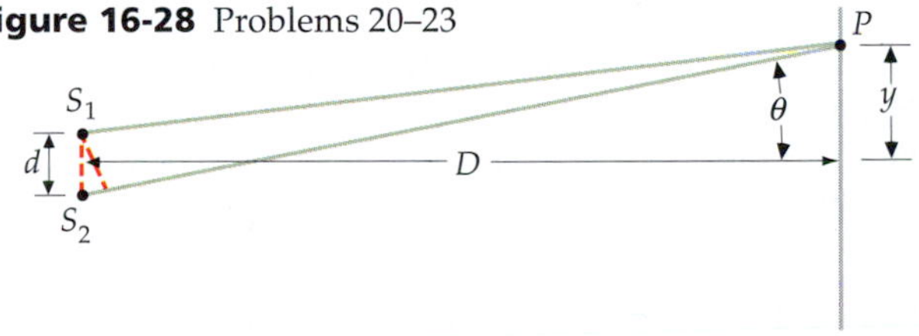

21 •• Two sound sources radiating in phase at a frequency of 480 Hz interfere such that maxima are heard at angles of 0 and 23° from a line perpendicular to that joining the two sources. Find the separation between the two sources and any other angles at which a maximum intensity will be heard. (Use the result of Problem 20.)

22 ••• Two loudspeakers are driven in phase by an audio amplifier at a frequency of 600 Hz. The speakers are on the y axis, one at $y = +1.00$ m and the other at $y = -1.00$ m. A listener begins at $y = 0$ a very large distance D away and walks along a line parallel to the y axis. (See Problem 20.) (*a*) At what angle θ will she first hear a minimum in the sound intensity? (*b*) At what angle will she first hear a maximum (after $\theta = 0$)? (*c*) How many maxima can she possibly hear if she keeps walking in the same direction?

23 ••• Two sound sources driven in phase by the same amplifier are 2 m apart on the y axis. At a point a very large distance from the y axis, constructive interference is first heard at an angle $\theta_1 = 0.140$ rad with the x axis and is next heard at $\theta_2 = 0.283$ rad (see Figure 16-28). (*a*) What is the wavelength of the sound waves from the sources? (*b*) What is the frequency of the sources? (*c*) At what other angles is constructive interference heard? (*d*) What is the smallest angle for which the sound waves cancel?

24 ••• Two identical sound sources have a frequency of 500 Hz. The coordinates of the sources are (0, 1 m) and (0, −1 m). A detector 80 m from the origin is free to revolve in the xy plane with a radius of 80 m. The first maximum in intensity is detected when the detector is at (80, 0) m. (*a*) Find the coordinates of the detector for the first five maxima in order of decreasing positive values of x. (*b*) Find the coordinates for the first four minima in order of decreasing positive values of x subject to the condition $\sqrt{x^2 + y^2} = 80$ m.

25 ••• A radio telescope consists of two antennas separated by a distance of 200 m. Both antennas are tuned to a particular frequency, such as 20 MHz. The signals from each antenna are fed into a common amplifier, but one signal first passes through a phase adjuster that delays its phase by a chosen amount so that the telescope can "look" in different directions. When the phase delay is zero, plane radio waves that are incident vertically on the antennas produce signals that add constructively at the amplifier. What should the phase delay be so that signals coming from an angle $\theta = 10°$ with the vertical (in the plane formed by the vertical and the line joining the antennas) will add constructively at the amplifier?

Beats

26 • Beats are produced by the superposition of two harmonic waves only if

(*a*) their amplitudes and frequencies are equal.
(*b*) their amplitudes are the same but their frequencies differ slightly.
(*c*) their frequencies differ slightly even if their amplitudes are not equal.
(*d*) their frequencies are equal but their amplitudes differ slightly.

27 • True or false: The beat frequency between two sound waves of nearly equal frequencies equals the difference in the frequencies of the individual sound waves.

28 •• About how accurately do you think you can tune a piano string to a tuning fork?

29 • Two tuning forks have frequencies of 256 and 260 Hz. If the forks are vibrating at the same time, what is the beat frequency?

30 • When a violin string is played (without fingering) simultaneously with a tuning fork of frequency 440 Hz, beats are heard at the rate of 3 per second. When the tension in the string is increased slightly, the beat frequency decreases. What was the initial frequency of the violin string?

31 • When two tuning forks are struck simultaneously, 4 beats per second are heard. The frequency of one fork is 500 Hz. (*a*) What are the possible values for the frequency of the other fork? (*b*) A piece of wax is placed on the 500-Hz fork to lower its frequency slightly. Explain how the measurement of the new beat frequency can be used to determine which of your answers to part (*a*) is the correct frequency of the second fork.

Standing Waves

32 • True or false:

(*a*) The frequency of the third harmonic is three times that of the first harmonic.
(*b*) The frequency of the fifth harmonic is five times that of the fundamental.
(*c*) In a pipe that is open at one end and closed at the other, the even harmonics are not excited.

33 • Standing waves result from the superposition of two waves of

(*a*) the same amplitude, frequency, and direction of propagation.
(*b*) the same amplitude and frequency and opposite directions of propagation.
(*c*) the same amplitude, slightly different frequency, and the same direction of propagation.
(*d*) the same amplitude, slightly different frequency, and opposite directions of propagation.

34 • An organ pipe open at both ends has a fundamental frequency of 400 Hz. If one end of this pipe is now closed, the fundamental frequency will be

(*a*) 200 Hz.
(*b*) 400 Hz.
(*c*) 546 Hz.
(*d*) 800 Hz.

35 •• A string fixed at both ends resonates at a fundamental frequency of 180 Hz. Which of the following will reduce the fundamental frequency to 90 Hz?

(*a*) Double the tension and double the length.
(*b*) Halve the tension and keep the length fixed.
(*c*) Keep the tension fixed and double the length.
(*d*) Keep the tension fixed and halve the length.

36 •• How do the resonance frequencies of an organ pipe change when the air temperature increases?

37 • A string fixed at both ends is 3 m long. It resonates in its second harmonic at a frequency of 60 Hz. What is the speed of transverse waves on the string?

38 • A string 3 m long and fixed at both ends is vibrating in its third harmonic. The maximum displacement of any point on the string is 4 mm. The speed of transverse waves on this string is 50 m/s. (*a*) What are the wavelength and frequency of this wave? (*b*) Write the wave function for this wave.

39 • Calculate the fundamental frequency for a 10-m organ pipe that is (*a*) open at both ends, and (*b*) closed at one end.

40 • A steel wire having a mass of 5 g and a length of 1.4 m is fixed at both ends and has a tension of 968 N. (*a*) Find the speed of transverse waves on the wire. (*b*) Find the wavelength and frequency of the fundamental. (*c*) Find the frequencies of the second and third harmonics.

41 • A rope 4 m long is fixed at one end; the other end is attached to a light string so that it is free to move. The speed of waves on the rope is 20 m/s. Find the frequency of (*a*) the fundamental, (*b*) the second harmonic, and (*c*) the third harmonic.

42 • A piano wire without windings has a fundamental frequency of 200 Hz. When it is wound with wire, its linear mass density is doubled. What is its new fundamental frequency, assuming that the tension is unchanged?

43 • The normal range of hearing is about 20 to 20,000 Hz. What is the greatest length of an organ pipe that would have its fundamental note in this range if (*a*) it is closed at one end, and (*b*) it is open at both ends?

44 •• The length of the B string on a certain guitar is 60 cm. Its fundamental is at 247 Hz. (*a*) What is the speed of transverse waves on the string? (*b*) If the linear mass density of the guitar string is 0.01 g/cm, what should its tension be when it is in tune?

45 •• The wave function $y(x, t)$ for a certain standing wave on a string fixed at both ends is given by $y(x, t) = 4.2 \sin 0.20x \cos 300t$, where y and x are in centimeters and t is in seconds. (*a*) What are the wavelength and frequency of this wave? (*b*) What is the speed of transverse waves on this string? (*c*) If the string is vibrating in its fourth harmonic, how long is it?

46 •• The wave function $y(x, t)$ for a certain standing wave on a string fixed at both ends is given by $y(x, t) = (0.05 \text{ m}) \sin 2.5 \text{ m}^{-1}x \cos 500 \text{ s}^{-1}t$. (*a*) What are the speed and amplitude of the two traveling waves that result in this standing wave? (*b*) What is the distance between successive nodes on the string? (*c*) What is the shortest possible length of the string?

47 •• A 2.51-m-long string has the wave function given in Problem 46. (*a*) Sketch the position of the string at the times $t = 0$, $t = \frac{1}{4}T$, $t = \frac{1}{2}T$, and $t = \frac{3}{4}T$, where $T = 1/f$ is the period of the vibration. (*b*) Find T in seconds. (*c*) At a time t when the string is horizontal, that is, $y(x) = 0$ for all x, what has become of the energy in the wave?

48 •• A string fixed at one end only is vibrating in its fundamental mode. The wave function is $y(x, t) = 0.02 \times \sin 2.36x \cos 377t$, where y and x are in meters and t is in seconds. (*a*) What is the wavelength of the wave? (*b*) What is the length of the string? (*c*) What is the speed of transverse waves on the string?

49 •• Three successive resonance frequencies for a certain string are 75, 125, and 175 Hz. (*a*) Find the ratios of each pair of successive resonance frequencies. (*b*) How can you tell that these frequencies are for a string fixed at one end only rather than for a string fixed at both ends? (*c*) What is the fundamental frequency? (*d*) Which harmonics are these resonance frequencies? (*e*) If the speed of transverse waves on this string is 400 m/s, find the length of the string.

50 •• The space above the water in a tube like that shown in Example 16-8 is 120 cm long. Near the open end, there is a loudspeaker that is driven by an audio oscillator whose frequency can be varied from 10 to 5000 Hz. (*a*) What is the lowest frequency of the oscillator that will produce resonance within the tube? (*b*) What is the highest frequency that will produce resonance? (*c*) How many different frequencies of the oscillator will produce resonance? (Neglect the end correction.)

51 •• A 460-Hz tuning fork causes resonance in the tube in Example 16-8 when the top of the tube is 18.3 and 55.8 cm above the water surface. (*a*) Find the speed of sound in air. (*b*) What is the end correction to adjust for the fact that the antinode does not occur exactly at the end of the open tube?

52 •• At 16°C, the fundamental frequency of an organ pipe is 440.0 Hz. What will be the fundamental frequency of the pipe if the temperature increases to 32°C? Would it be better to construct the pipe with a material that expands substantially as the temperature increases or should the pipe be made of a material that maintains the same length at all normal temperatures?

53 •• A violin string of length 40 cm and mass 1.2 g has a frequency of 500 Hz when it is vibrating in its fundamental mode. (*a*) What is the wavelength of the standing wave on the string? (*b*) What is the tension in the string? (*c*) Where should you place your finger to increase the frequency to 650 Hz?

54 •• The G string on a violin is 30 cm long. When played without fingering, it vibrates at a frequency of 196 Hz. The next higher notes on the C-major scale are A (220 Hz), B (247 Hz), C (262 Hz), and D (294 Hz). How far from the end of the string must a finger be placed to play each of these notes?

55 •• A string with a mass density of 4×10^{-3} kg/m is under a tension of 360 N and is fixed at both ends. One of its resonance frequencies is 375 Hz. The next higher resonance frequency is 450 Hz. (*a*) What is the fundamental frequency of this string? (*b*) Which harmonics are the ones given? (*c*) What is the length of the string?

56 •• A string fastened at both ends has successive resonances with wavelengths of 0.54 m for the nth harmonic and 0.48 m for the $(n + 1)$th harmonic. (*a*) Which harmonics are these? (*b*) What is the length of the string?

57 •• A rubber band with an unstretched length of 0.80 m and a mass of 6×10^{-3} kg stretches to 1.20 m when under a tension of 7.60 N. What is the fundamental frequency of

oscillation of this band when stretched between two fixed posts 1.20 m apart?

58 •• The strings of a violin are tuned to the tones G, D, A, and E, which are separated by a fifth from one another. That is, $f(\text{D}) = 1.5f(\text{G})$, $f(\text{A}) = 1.5f(\text{D}) = 440$ Hz, and $f(\text{E}) = 1.5f(\text{A})$. The distance between the two fixed points, the bridges at the scroll and over the body of the instrument, is 30 cm. The tension on the E string is 90 N. (*a*) What is the mass per meter of the E string? (*b*) To prevent distortion of the instrument over time, it is important that the tension on all strings be the same. Find the masses per meter of the other strings.

59 •• To tune a violin, the violinist first tunes the A string to the correct pitch of 440 Hz and then bows two adjoining strings simultaneously and listens for a beat pattern. While bowing the A and E strings, the violinist hears a beat frequency of 3 Hz and notes that the beat frequency increases as the tension on the E string is increased. (The E string is to be tuned to 660 Hz.) (*a*) Why is a beat produced by these two strings bowed simultaneously? (*b*) What is the frequency of the E string vibration when the beat frequency is 3 Hz? (*c*) If the tension on the E string is 80.0 N when the beat frequency is 3 Hz, what tension corresponds to perfect tuning of that string?

60 •• (*a*) For the wave function given in Problem 48, find the velocity of a string segment at some point x as a function of time. (*b*) Which point has the greatest speed at any time? What is the maximum speed of this point? (*c*) Find the acceleration of a string segment at some point x as a function of time. (*d*) Which point has the greatest acceleration? What is the maximum acceleration of this point?

61 •• A student carries a small oscillator and speaker as she walks very slowly down a long hall. The speaker emits a sound of frequency 680 Hz which is reflected from the walls at each end of the hall. The student notes that as she walks along, the sound intensity she hears passes through successive maxima and minima. What distance must she walk to pass from one maximum to the next?

62 •• Assume that the rubber band of Problem 57 behaves like an ideal spring. The band is attached to two posts whose separation D can be varied. (*a*) Derive an expression for the frequency of the fundamental vibration of this system. (*b*) What should be the separation between the fixed ends of the rubber band so that it will vibrate with a fundamental frequency of 21 Hz?

63 •• A 2-m string is fixed at one end and is vibrating in its third harmonic with amplitude 3 cm and frequency 100 Hz. (*a*) Write the wave function for this vibration. (*b*) Write an expression for the kinetic energy of a segment of the string of length dx at a point x at some time t. At what time is this kinetic energy maximum? What is the shape of the string at this time? (*c*) Find the maximum kinetic energy of the string by integrating your expression for part (*b*) over the total length of the string.

Wave Packets (optional)

64 • Information for use by computers is transmitted along a cable in the form of short electric pulses at the rate of 10^7 pulses per second. (*a*) What is the maximum duration of each pulse if no two pulses overlap? (*b*) What is the range of frequencies to which the receiving equipment must respond?

65 • A tuning fork of frequency f_0 begins vibrating at time $t = 0$ and is stopped after a time interval Δt. The waveform of the sound at some later time is shown as a function of x. Let N be the (approximate) number of cycles in this waveform. (*a*) How are N, f_0, and Δt related? (*b*) If Δx is the length in space of this wave packet, what is the wavelength in terms of Δx and N? (*c*) What is the wave number k in terms of N and Δx? (*d*) The number N is uncertain by about ± 1 cycle. Use Figure 16-29 to explain why. (*e*) Show that the uncertainty in the wave number due to the uncertainty in N is $2\pi/\Delta x$.

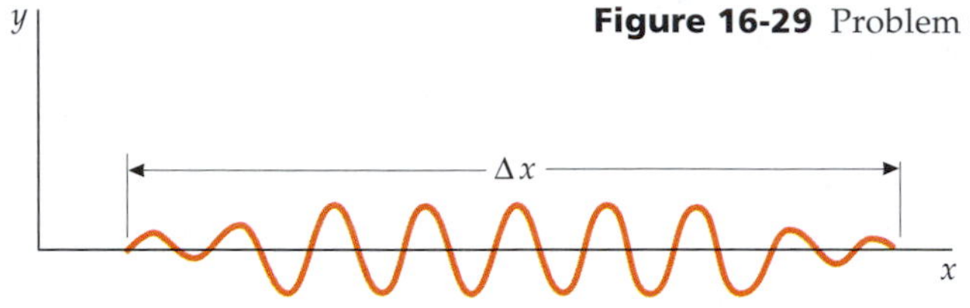

Figure 16-29 Problem 65

General Problems

66 • When two waves moving in opposite directions superimpose as in Figure 16-1, does either impede the progress of the other?

67 • When a guitar string is plucked, is the wavelength of the wave it produces in air the same as the wavelength of the wave on the string?

68 • When two waves interfere constructively or destructively, is there any gain or loss in energy? Explain.

69 • A musical instrument consists of drinking glasses partially filled with water that are struck with a small mallet. Explain how this works.

70 •• During an organ recital, the air compressor that drives the organ pipes suddenly fails. An enterprising physics student in the audience comes to the rescue by connecting a tank of pure nitrogen gas under high pressure to the output of the compressor. What effect, if any, will this change have on the operation of the organ? What if the tank contained helium?

71 •• When the tension on a piano wire is increased, which of the following occurs?

(*a*) Its wavelength decreases.
(*b*) Its wavelength remains the same while its frequency increases.
(*c*) Its wavelength and frequency increase.
(*d*) None of the above.

72 •• The following instructions are given for connecting stereo speakers to an amplifier so that they are in phase: "After both speakers are connected, play a monophonic record or program with the bass control turned up and the treble control turned down. While listening to the speakers, turn the balance control so that first one speaker is heard separately, then the two together, and then the other separately. If the bass is stronger when both speakers play together, they are connected properly. If the bass is weaker when both play together than when each plays separately, reverse the connections on one speaker." Explain why this method works. In particular, explain why a stereo source is not used and why only the bass is compared.

73 •• The constant γ for helium (and all monatomic gases) is 1.67. If a man inhales helium and then commences to speak, he sounds like Alvin of the Chipmunks. Why?

74 • Middle C on the equal-temperament scale used by modern instrument makers has a frequency of 261.63 Hz. If a 7-g piano wire that is 80 cm long is to be tuned so that 261.63 Hz is its fundamental frequency, what should the tension in the wire be?

75 • The ear canal, which is about 2.5 cm long, roughly approximates a pipe that is open at one end and closed at the other. (*a*) What are the resonance frequencies of the ear canal? (*b*) Describe the possible effect of the resonance modes of the ear canal on the threshold of hearing.

76 • A 4-m-long, 160-g rope is fixed at one end and is tied to a light string at the other end. Its tension is 400 N. (*a*) What are the wavelengths of the fundamental and the next two harmonics? (*b*) What are the frequencies of these standing waves?

77 • The shortest pipes used in organs are about 7.5 cm long. (*a*) What is the fundamental frequency of a pipe this long that is open at both ends? (*b*) For such a pipe, what is the highest harmonic that is within the audible range (see Problem 43)?

78 •• Two waves from two coherent sources have the same wavelength λ, frequency ω, and amplitude A. What is the path difference if the resultant wave at some point has amplitude A?

79 •• A 35-m string has a linear mass density of 0.0085 kg/m and is under a tension of 18 N. Find the frequencies of the lowest four harmonics if (*a*) the string is fixed at both ends, and (*b*) the string is fixed at one end and attached to a long, thin, massless thread at the other end.

80 •• You find an abandoned mine shaft and wish to measure its depth. Using an audio oscillator of variable frequency, you note that you can produce successive resonances at frequencies of 63.58 and 89.25 Hz. What is the depth of the shaft?

81 •• A string 5 m long that is fixed at one end only is vibrating in its fifth harmonic with a frequency of 400 Hz. The maximum displacement of any segment of the string is 3 cm. (*a*) What is the wavelength of this wave? (*b*) What is the wave number k? (*c*) What is the angular frequency? (*d*) Write the wave function for this standing wave.

82 •• The wave function for a standing wave on a string is described by

$$y(x, t) = 0.02 \sin 4\pi x \cos 60\pi t$$

where y and x are in meters and t is in seconds. Determine the maximum displacement and maximum speed of a point on the string at (*a*) $x = 0.10$ m, (*b*) $x = 0.25$ m, (*c*) $x = 0.30$ m, and (*d*) $x = 0.50$ m.

83 •• A 2.5-m-long wire having a mass of 0.10 kg is fixed at both ends and is under tension of 30 N. When the nth harmonic is excited, there is a node 0.50 m from one end. (*a*) What is n? (*b*) What are the frequencies of the first three allowed modes of vibration?

84 •• In an early method of determining the speed of sound in gases, powder was spread along the bottom of a horizontal, cylindrical glass tube. One end of the tube was closed by a piston that oscillated at a known frequency f. The other end was closed by a movable piston whose position was adjusted until resonance occurred. At resonance, the powder collected in equally spaced piles along the bottom of the tube. (*a*) Explain why the powder collects in this way. (*b*) Derive a formula that gives the speed of sound in the gas in terms of f and the distance between the piles of powder. (*c*) Give suitable values for the frequency f and the distance between the piles of powder. (*d*) Give suitable values for the frequency f and the length L of the tube for which the speed of sound could be measured in either air or helium.

85 •• In a lecture demonstration of standing waves, a string is attached to a tuning fork that vibrates at 60 Hz and sets up transverse waves of that frequency on the string. The other end of the string passes over a pulley, and the tension is varied by attaching weights to that end. The string has approximate nodes at the tuning fork and at the pulley. (*a*) If the string has a linear mass density of 8 g/m and is 2.5 m long (from the tuning fork to the pulley), what must the tension be for the string to vibrate in its fundamental mode? (*b*) Find the tensions necessary for the string to vibrate in its second, third, and fourth harmonic.

86 •• Three successive resonance frequencies in an organ pipe are 1310, 1834, and 2358 Hz. (*a*) Is the pipe closed at one end or open at both ends? (*b*) What is the fundamental frequency? (*c*) What is the length of the pipe?

87 •• A wire of mass 1 g and length 50 cm is stretched with a tension of 440 N. It is then placed near the open end of the tube in Example 16-8 and stroked with a violin bow so that it oscillates at its fundamental frequency. The water level in the tube is then lowered until a resonance is obtained, which occurs at 18 cm below the top of the tube. Use the data given to determine the speed of sound in air. Why is this method not very accurate?

88 •• On a windy day, a drain pipe will sometimes resonate. Estimate the resonance frequency of a drain pipe on a single-story house. How much might this frequency change from winter to summer in your region?

89 •• A 50-cm-long wire fixed at both ends vibrates with a fundamental frequency f_0 when the tension is 50 N. If the tension is increased to 60 N, the fundamental frequency increases by 5 Hz, and a further increase in tension to 70 N results in a fundamental frequency of $(f_0 + 9.6)$ Hz. Determine the mass of the wire.

90 •• A standing wave on a rope is represented by the following wave function:

$$y(x, t) = 0.02 \sin \frac{\pi x}{2} \cos 40\pi t$$

where x and y are in meters and t is in seconds. (*a*) Write wave functions for two traveling waves that when superimposed will produce the resultant standing-wave pattern. (*b*) What is the distance between the nodes of the standing wave? (*c*) What is the velocity of a segment of the rope at $x = 1$ m? (*d*) What is the acceleration of a segment of the rope at $x = 1$ m?

91 •• Two identical speakers emit sound waves of frequency 680 Hz uniformly in all directions. The total audio output of each speaker is 1 mW. A point P is 2.00 m from one speaker and 3.00 m from the other. (*a*) Find the intensities I_1 and I_2 from each speaker separately at point P. (*b*) If the speakers are driven coherently and are in phase, what is the intensity at point P? (*c*) If they are driven coherently but are 180° out of phase, what is the intensity at point P? (*d*) If the speakers are incoherent, what is the intensity at point P?

92 •• Three waves with the same frequency, wavelength, and amplitude are traveling in the same direction. The three waves are given by

$$y_1(x, t) = 0.05 \sin\left(kx - \omega t - \frac{\pi}{3}\right)$$

$$y_2(x, t) = 0.05 \sin(kx - \omega t)$$

$$y_3(x, t) = 0.05 \sin\left(kx - \omega t + \frac{\pi}{3}\right)$$

Find the resultant wave.

93 •• (*a*) Show that if the temperature changes by a small amount ΔT, the fundamental frequency of an organ pipe changes by approximately Δf, where $\Delta f/f = \frac{1}{2}\Delta T/T$. (*b*) Suppose an organ pipe that is closed at one end has a fundamental frequency of 200 Hz when the temperature is 20°C. What will its fundamental frequency be when the temperature is 30°C? (Ignore any change in the length of the pipe due to thermal expansion.)

94 •• Two traveling wave pulses on a string are represented by the wave functions

$$y_1(x, t) = \frac{0.02 \text{ m}^3}{2 \text{ m}^2 + (x - 2t)^2}$$

and

$$y_2(x, t) = \frac{-0.02 \text{ m}^3}{2 \text{ m}^2 + (x + 2t)^2}$$

where x is in meters and t is in seconds. (*a*) Sketch each wave function separately as a function of x at $t = 0$, and describe the behavior of each as time increases. (*b*) Find the resultant wave function at $t = 0$. (*c*) Find the resultant wave function at $t = 1$ s. (*d*) Sketch the resultant wave function at $t = 1$ s.

95 •• The kinetic energy of a segment Δm of a vibrating string is given by

$$\Delta K = \frac{1}{2}\Delta m\left(\frac{\partial y}{\partial t}\right)^2 = \frac{1}{2}\mu\left(\frac{\partial y}{\partial t}\right)^2 \Delta x$$

(*a*) Find the total kinetic energy of the nth mode of vibration of a string of length L fixed at both ends. (*b*) Give the maximum kinetic energy of the string. (*c*) What is the wave function when the kinetic energy has its maximum value? (*d*) Show that the maximum kinetic energy in the nth mode is proportional to $n^2A_n^2$.

96 •• (*a*) Show that when the tension in a string fixed at both ends is changed by a small amount dF, the frequency of the fundamental is changed by approximately df, where $df/f = \frac{1}{2}\, dF/F$. Does this result apply to all harmonics? (*b*) Use this result to find the percentage change in the tension needed to increase the frequency of the fundamental of a piano wire from 260 to 262 Hz.

97 ••• Two sources have a phase difference δ_0 that is proportional to time: $\delta_0 = Ct$, where C is a constant. The amplitude of the wave from each source at some point P is A_0. (*a*) Write the wave functions for each of the two waves at point P, assuming this point to be a distance x_1 from one source and $x_1 + \Delta x$ from the other. (*b*) Find the resultant wave function, and show that its amplitude is $2A_0 \cos\frac{1}{2}(\delta + \delta_0)$, where δ is the phase difference at P due to the path difference. (*c*) Sketch the intensity at point P versus time for a zero path difference. (Let I_0 be the intensity due to each wave separately.) What is the time average of the intensity? (*d*) Make the same sketch for the intensity at a point for which the path difference is $\frac{1}{2}\lambda$.

98 ••• The wave functions of two standing waves on a string of length L are

$$y_1(x, t) = A_1 \cos \omega_1 t \sin k_1 x$$

and

$$y_2(x, t) = A_2 \cos\omega_2 t \sin k_2 x$$

where $k_n = n\pi/L$, and $\omega_n = n\omega_1$. The wave function of the resultant wave is

$$y_r(x, t) = y_1(x, t) + y_2(x, t)$$

(*a*) Find the velocity of a segment dx of the string. (*b*) Find the kinetic energy of this segment. (*c*) By integration, find the total kinetic energy of the resultant wave. Notice the disappearance of the cross terms so that the total kinetic energy is proportional to $(n_1A_1)^2 + (n_2A_2)^2$.

99 ••• A 2-m wire fixed at both ends is vibrating in its fundamental mode. The tension in the wire is 40 N and the mass of the wire is 0.1 kg. At the midpoint of the wire, the amplitude is 2 cm. (*a*) Find the maximum kinetic energy of the wire. (*b*) At the instant the transverse displacement is given by (0.02 m) $\sin(\pi x/2)$, what is the kinetic energy of the wire? (*c*) At what position on the wire does the kinetic energy per unit length have its largest value? (*d*) Where does the potential energy per unit length have its maximum value?

100 ••• A string 3.2 m long and with a linear mass density of 0.008 kg/m is kept under tension so that traveling waves propagate at 48 m/s along the string. The ends of the string are clamped and the string vibrates in its third harmonic with an amplitude of 5.0 cm. How much energy is stored in this vibrating system at that time? If the amplitude of the standing wave diminishes to 3.0 cm in 1.0 s, what is the Q of this vibrating system?

CHAPTER 17

Wave–Particle Duality and Quantum Physics

We have seen that the propagation of waves through space is quite different from the propagation of particles. Waves bend around corners (diffraction) and interfere with one another, producing interference patterns. When a wave encounters a small aperture, the wave spreads out on the other side as if the aperture were a point source (Figure 15-23). When two waves of equal intensity I_0 and originating from coherent sources meet in space, the result can be a wave of intensity $4I_0$ (constructive interference), an intensity of zero (destructive interference), or a wave of intensity between zero and $4I_0$, depending on the phase difference between the waves at their meeting point.

The propagation of particles is quite different. Particles travel along well-defined paths. When two particles meet in space, they never produce an interference pattern.

Particles and waves also exchange energy differently. Particles exchange energy in collisions that occur at specific points in space and time. The energy of waves, on the other hand, is spread out in space and deposited continuously as the wave front interacts with matter.

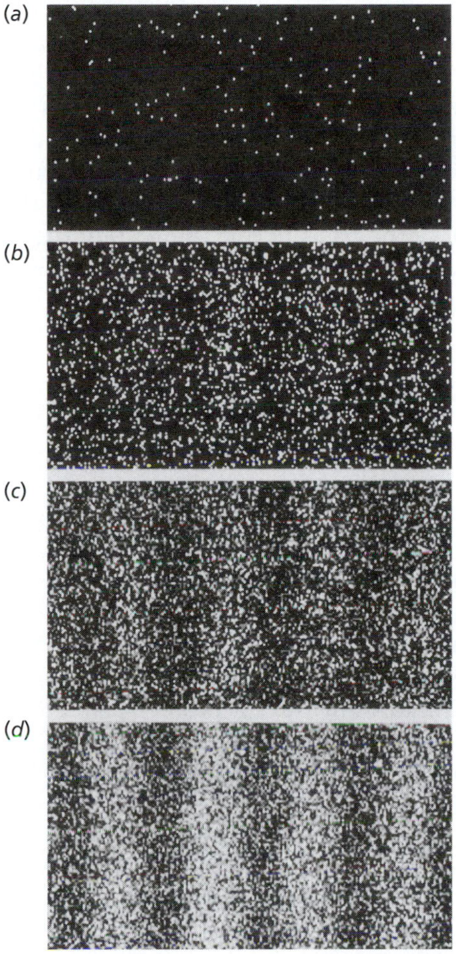

Electron interference pattern produced by electrons incident on a barrier containing two slits: (*a*) 10 electrons; (*b*) 100 electrons; (*c*) 3,000 electrons; (*d*) 70,000 electrons. The maxima and minima demonstrate the wave nature of the electron as it traverses the slits. Individual dots on the screen indicate the particle nature of the electron as it exchanges energy with the detector. The pattern is the same whether electrons or photons (particles of light) are used.

Sometimes the propagation of a wave cannot be distinguished from that of a beam of particles. When the wavelength λ is very small compared with the sizes of apertures and obstacles, diffraction effects are negligible and the wave appears to travel along a well-defined path. Also, interference maxima and minima are so close together in space as to be unobservable. Similarly, when there are very many small particles each exchanging a small amount of energy, the exchange cannot be distinguished from that of a wave. For example, you do not observe the individual air molecules bouncing off your face when the wind blows on it. Instead, the interaction with billions of particles is perceived to be continuous, as if the particles were a wave.

At the beginning of the twentieth century, it was thought that sound, light, and other electromagnetic radiation such as radio were waves, whereas electrons, protons, atoms, and similar constituents of nature were understood to be particles. The first 30 years of the new century revealed such startling developments in theoretical and experimental physics as the finding that light, thought to be a wave, actually exchanges energy in discrete lumps, or quanta, just like particles, and that an electron, thought to be a particle, exhibits diffraction and interference as it propagates through space, just like a wave.

The fact that light exchanges energy like a particle implies that light energy is not continuous but is *quantized*. Similarly, the wave nature of the electron, along with the fact that the standing-wave condition requires a discrete set of frequencies, implies that the energy of an electron in a confined region of space is not continuous, but is quantized to a discrete set of values.

In this chapter we begin by discussing some basic properties of light and electrons, examining both their wave and particle characteristics. We then consider some of the detailed properties of matter waves, showing, in particular, how standing waves imply the quantization of energy. We will then discuss some of the important features of the theory of quantum physics, which was developed in the 1920s and which has been extremely successful in describing nature. Quantum physics is now the basis of our understanding of the microscopic world.

17-1 Light

The question of whether light consists of a beam of particles or waves in motion is one of the most interesting in the history of science. Newton used a particle theory of light to explain the laws of reflection and refraction, but for refraction he needed to assume that light travels faster in water or glass than in air, an assumption later shown to be false. The chief early proponents of the wave theory were Robert Hooke and Christian Huygens, who explained refraction by assuming that light travels more slowly in glass or water than in air. Newton rejected the wave theory because in his time light was believed to travel only in straight lines—diffraction had not yet been observed.

Because of Newton's great reputation and authority, his particle theory of light was accepted for more than a century. Then in 1801 Thomas Young demonstrated the wave nature of light in a famous experiment in which two coherent light sources are produced by illuminating a pair of narrow, parallel slits with a single source (Figure 17-1). We saw in Chapter 16 that when light encounters a small opening, the opening acts as a point source of waves (Figure 16-10). In Young's experiment, each slit acts as a line source, which is equivalent to a point source in two dimensions. The interference pattern is observed on a screen placed behind the slits. Interference maxima occur at angles such that the path difference is an integral number of wavelengths. Similarly, interference minima occur when the path difference is one half-wavelength or any odd number of

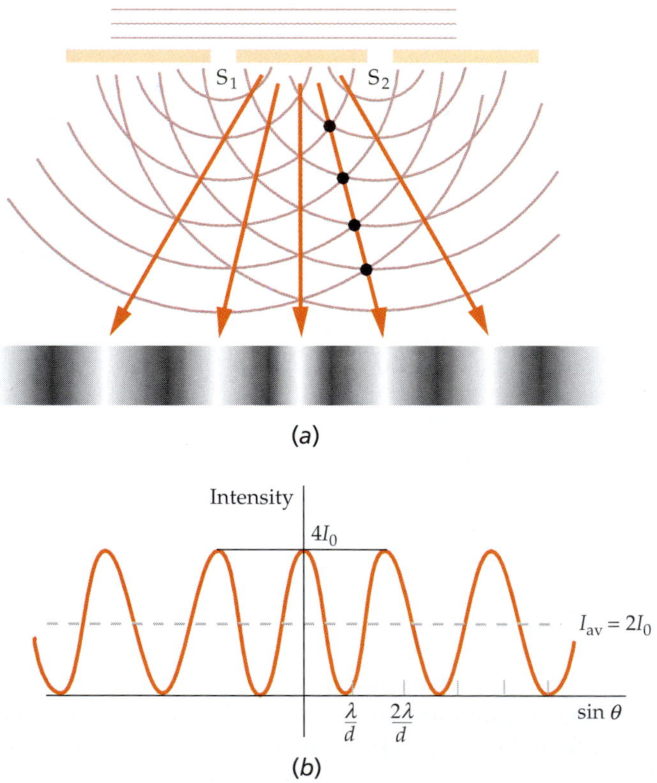

Figure 17-1 (*a*) Two slits act as coherent sources of light for the observation of interference in Young's experiment. Waves from the slits overlap and produce an interference pattern on a screen far away. (*b*) Graph of the intensity pattern produced in (*a*). The intensity is maximum at points where the path difference is an integral number of wavelengths and zero where the path difference is an odd number of half-wavelengths.

half-wavelengths. Figure 17-1*b* shows a graph of the intensity pattern seen on the screen. This and many other experiments demonstrate that light propagates like a wave.

In the early nineteenth century, the French physicist Augustin Fresnel (1788–1827) performed extensive experiments on interference and diffraction and put the wave theory on a mathematical basis. Among his results, he showed that the observed straight-line propagation of light is a result of the very short wavelengths of visible light.

The classical wave theory of light culminated in 1860 when James Clerk Maxwell published his mathematical theory of electromagnetism. This theory yielded a wave equation that predicted the existence of electromagnetic waves that propagate with a speed that can be calculated from the laws of electricity and magnetism. The fact that the result of this calculation was $c \approx 3 \times 10^8$ m/s, the same as the measured value for the speed of light, suggested to Maxwell that light is an electromagnetic wave. The eye is sensitive to electromagnetic waves with wavelengths in the range from about 400 nm (1 nm $= 10^{-9}$ m) to about 700 nm. This range is called visible light. Other electromagnetic waves such as microwaves, radio, television, and X rays differ from light only in wavelength and frequency.

17-2 The Particle Nature of Light: Photons

The diffraction of light and the existence of an interference pattern in the two-slit experiment give clear evidence that light has wave properties. However, early in the twentieth century it was found that light energy is exchanged in discrete amounts—an unwavelike property.

The Photoelectric Effect

The quantum nature of light and the quantization of energy were suggested by Einstein in 1905 in his explanation of the photoelectric effect. Einstein's work marked the beginning of quantum theory, and for it he received the Nobel prize for physics. Figure 17-2 shows a schematic diagram of the basic apparatus for studying the photoelectric effect. When light is incident on a clean metal surface C, electrons are emitted. Some of these electrons strike the second metal plate A, constituting an electric current between the plates. The maximum energy of the emitted electrons can be measured. Experiments give the surprising result that the maximum kinetic energy of the ejected electrons is *independent of the intensity* of the incident light. From the wave theory of light, we would expect that increasing the rate at which light energy falls on the metal surface would increase the amount of energy absorbed by individual electrons and therefore would increase the maximum kinetic energy of the electrons emitted. This is not what happens. The maximum kinetic energy of the ejected electrons is the same for a given wavelength of incident light, no matter how intense the light. Einstein suggested that this experimental result can be explained if light energy is quantized in small bundles called **photons**. The energy of each photon is given by

$$E = hf = \frac{hc}{\lambda} \qquad \text{17-1}$$

Einstein equation for photon energy

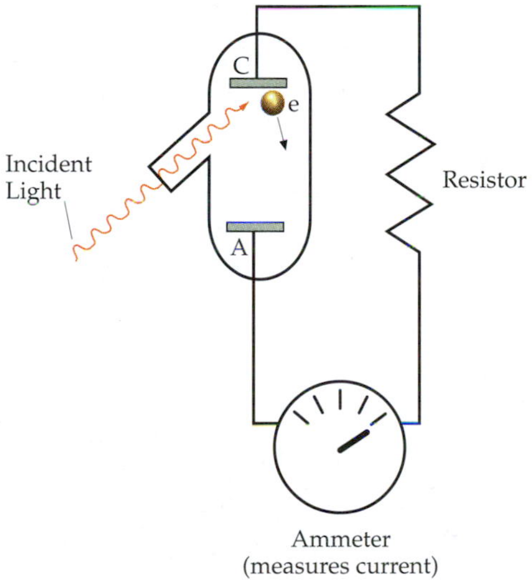

Figure 17-2 Schematic drawing of the apparatus for studying the photoelectric effect. Light strikes the cathode C and ejects electrons. The number of electrons that reach the anode A is measured by the current in an ammeter placed in a circuit between A and C.

optional

where f is the frequency, and h is a constant now known as **Planck's constant**.* The accepted value of this constant is now

$$h = 6.626 \times 10^{-34}\ \text{J}\cdot\text{s} = 4.136 \times 10^{-15}\ \text{eV}\cdot\text{s} \tag{17-2}$$

Equation 17-1 is sometimes called the **Einstein equation**.

In this picture, a light beam consists of a beam of particles—photons, each having energy hf. The intensity of a light beam (energy per unit area per unit time) is the number of photons per unit area per unit of time times the energy of each photon. The interaction of the light beam with the metal surface consists of collisions between photons and electrons. In these collisions, the photon disappears, giving all of its energy to the electron. An electron emitted from a metal surface exposed to light thus receives its energy from a single photon. When the intensity of light is increased, more photons fall on the surface per unit time, and more electrons are ejected. However, each photon still has the same energy hf, so the energy absorbed by each electron is the same.

If ϕ is the minimum energy necessary to remove an electron from a metal surface, the maximum kinetic energy of the electrons emitted is given by

$$K_{max} = (\tfrac{1}{2}mv^2)_{max} = hf - \phi \tag{17-3}$$

Einstein's photoelectric equation

The quantity ϕ, called the **work function,** is a characteristic of the particular metal. (Some electrons will have kinetic energies less than $hf - \phi$ because of the loss of energy from traveling through the metal.)

According to Einstein's photoelectric equation, a plot of K_{max} versus frequency f should be a straight line with the slope h. This was a bold prediction, for at the time there was no evidence that Planck's constant had any application outside of the then-mysterious phenomenon of blackbody radiation. In addition, there were no experimental data on K_{max} versus frequency because no one before that time had even suspected that the frequency of the light was related to K_{max}. This prediction was difficult to verify experimentally, but eventually careful experiments by R. A. Millikan about ten years later showed that Einstein's equation is correct. Figure 17-3 shows a plot of Millikan's data.

Photons with frequencies less than a **threshold frequency** f_t, and therefore with wavelengths greater than a **threshold wavelength** $\lambda_t = c/f_t$, do not have enough energy to eject an electron from a particular metal. The threshold frequency and the corresponding threshold wavelength can be related to the work function ϕ by setting the maximum kinetic energy of the electrons equal to zero in Equation 17-3. Then

$$\phi = hf_t = \frac{hc}{\lambda_t} \tag{17-4}$$

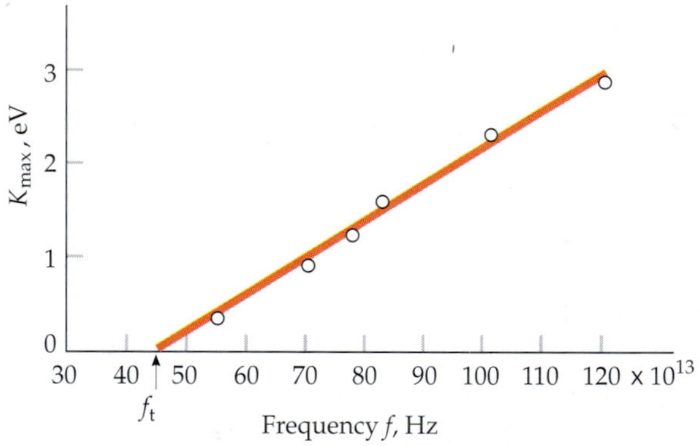

Figure 17-3 Millikan's data for the maximum kinetic energy K_{max} versus frequency f for the photoelectric effect. The data fall on a straight line that has a slope h, as predicted by Einstein a decade before the experiment was performed.

Work functions for metals are typically a few electron volts. Since light wavelengths are usually given in nanometers and energies in electron volts, it is useful to have the value of hc in electron volt-nanometers:

$$hc = (4.136 \times 10^{-15}\ \text{eV}\cdot\text{s})(2.997 \times 10^8\ \text{m/s}) = 1.240 \times 10^{-6}\ \text{eV}\cdot\text{m}$$

or

$$hc = 1240\ \text{eV}\cdot\text{nm} \tag{17-5}$$

* In 1900, the German physicist Max Planck had introduced this constant to explain discrepancies between theoretical curves and experimental data related to the spectrum of blackbody radiation. Planck also assumed that the radiation was emitted and absorbed by a blackbody in quanta of energy hf, but he considered his assumption to be just a calculational device rather than a fundamental property of electromagnetic radiation. We discuss blackbody radiation in Chapter 21.

Example 17-1

Calculate the photon energies for light of wavelengths 400 nm (violet) and 700 nm (red). (These are the approximate wavelengths at the two extremes of the visible spectrum.)

1. The energy is related to the wavelength by Equation 17-1: $E = hf = \frac{hc}{\lambda}$
2. For $\lambda = 400$ nm, the energy is: $E = \frac{hc}{\lambda} = \frac{1240\ \text{eV}\cdot\text{nm}}{400\ \text{nm}} = 3.10\ \text{eV}$
3. For $\lambda = 700$ nm, the energy is: $E = \frac{hc}{\lambda} = \frac{1240\ \text{eV}\cdot\text{nm}}{700\ \text{nm}} = 1.77\ \text{eV}$

Remark We can see from these calculations that visible light contains photons with energies that range from about 1.8 to 3.1 eV. X rays, which have much shorter wavelengths, contain photons with energies of the order of keV. Gamma rays emitted by nuclei have even shorter wavelengths and photons with energies of the order of MeV.

Example 17-2 ***try it yourself***

The intensity of sunlight at the earth's surface is approximately 1400 W/m^2. Assuming that the average photon energy is 2 eV (corresponding to a wavelength of about 600 nm), calculate the number of photons that strike an area of 1 cm^2 in 1 s.

Cover the column to the right and try these on your own before looking at the answers.

Steps	***Answers***
1. The number N of photons is related to the total energy.	$E = Nhf = N(2\ \text{eV})$
2. Use 1 W = 1 J/s to find the energy in joules striking an area of 1 cm^2 in 1 s.	$E = 0.14$ J
3. Use the conversion factor 1 eV = 1.6×10^{-19} J to find the energy in eV striking an area of 1 cm^2 in 1 s.	$E = 8.75 \times 10^{17}$ eV
4. Use this value of E to solve for N.	$N = 4.38 \times 10^{17}$ photons

Remark This is an enormous number of photons. In most everyday situations, the number of photons is so great that the quantization of light is not noticeable.

Exercise Find the energy of a photon corresponding to electromagnetic radiation in the FM radio band of wavelength 3 m. (*Answer* 4.13×10^{-7} eV)

Exercise Find the wavelength of a photon whose energy is (*a*) 0.1 eV, (*b*) 1 keV, and (*c*) 1 MeV. (*Answers* (*a*) 12.4 μm, (*b*) 1.24 nm, (*c*) 1.24 pm)

optional

Compton Scattering

Further evidence of the correctness of the photon concept was furnished by Arthur H. Compton, who measured the scattering of X rays by electrons in 1923. According to classical theory, when an electromagnetic wave of frequency f_1 is incident on material containing charges, the charges will oscillate with this frequency and reradiate electromagnetic waves of the same frequency. Compton pointed out that if the scattering process were considered to be a collision between a photon and an electron, the electron would recoil and thus absorb energy. The scattered photon would then have less energy and therefore a lower frequency and larger wavelength than the incident photon.

According to classical wave theory, the energy and momentum of an electromagnetic wave are related by

$$E = pc \tag{17-6}$$

If a photon has energy $E = hf = hc/\lambda$, its momentum should then be $p = E/c = hf/c = h/\lambda$:

$$p = \frac{h}{\lambda} \tag{17-7}$$

Momentum of a photon

Compton applied the laws of conservation of momentum and energy to the collision of a photon and an electron to calculate the momentum p_2 and thus the wavelength $\lambda_2 = h/p_2$ of a scattered photon (Figure 17-4). Because the calculation requires Einstein's theory of special relativity, we present only the result here. The wavelengths λ_1, associated with the incoming photon, and λ_2, associated with the scattered photon, are related to each other and to the scattering angle θ by

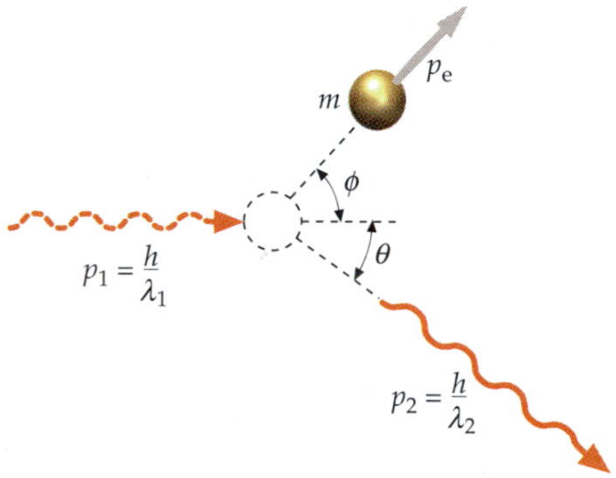

Figure 17-4 The scattering of light by an electron is considered as a collision of a photon of momentum h/λ_1 and a stationary electron. The scattered photon has less energy and therefore a greater wavelength.

$$\lambda_2 - \lambda_1 = \frac{h}{m_e c}(1 - \cos\theta) \tag{17-8}$$

where m_e is the mass of the electron. The change in wavelengths is independent of the original wavelength. The quantity $h/m_e c$ depends only on the mass of the electron. It has dimensions of length and is called the Compton wavelength. Its value is

$$\lambda_C = \frac{h}{m_e c} = \frac{hc}{m_e c^2} = \frac{1240 \text{ eV}\cdot\text{nm}}{5.11 \times 10^5 \text{ eV}} = 2.43 \times 10^{-12} \text{ m} \tag{17-9}$$
$$= 2.43 \text{ pm}$$

where 1 pm $= 10^{-12}$ m $= 10^{-3}$ nm. Because $\lambda_2 - \lambda_1$ is small, it is difficult to observe unless λ_1 is so small that the fractional change $(\lambda_2 - \lambda_1)/\lambda_1$ is appreciable. Compton used X rays of wavelength 71.1 pm. The energy of a photon of this wavelength is $E = hc/\lambda = (1240 \text{ eV}\cdot\text{nm})/(0.0711 \text{ nm}) = 17.4$ keV. Compton's experimental results for $\lambda_2 - \lambda_1$ as a function of scattering angle θ agreed with Equation 17-8, thereby confirming the correctness of the photon concept, that is, of the particle nature of light.

Example 17-3

An X-ray photon of wavelength 6 pm makes a head-on collision with an electron so that it is scattered by an angle of 180° (Figure 17-5). (*a*) What is the change in wavelength of the photon? (*b*) What is the kinetic energy of the recoiling electron?

Picture the Problem We can calculate the change in wavelength and the new wavelength from Equation 17-8. We then use the new wavelength to find the energy of the scattered photon, and we use conservation of energy to find the energy of the recoiling electron.

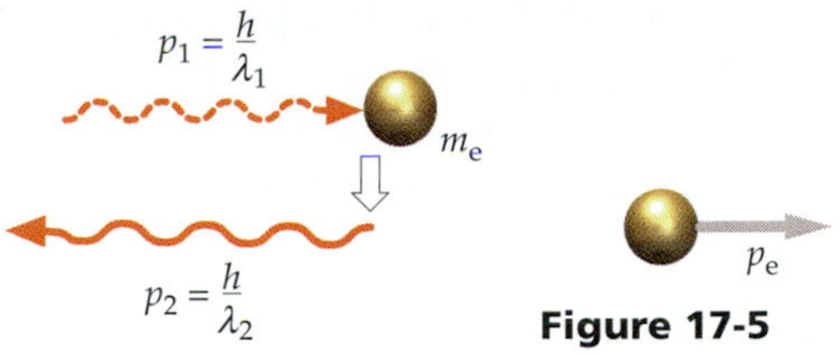

Figure 17-5

(*a*) Use Equation 17-8 to calculate the change in wavelength:

$$\Delta\lambda = \lambda_2 - \lambda_1$$
$$= \frac{h}{m_e c}(1 - \cos\theta) = 2.43\text{ pm}\,(1 - \cos 180°)$$
$$= 2.43\text{ pm}\,(1 - (-1)) = 4.86\text{ pm}$$

(*b*)1. The energy of the recoiling electron equals the energy of the incident photon E_1 minus the energy of the scattered photon E_2:

$$K_e = E_1 - E_2 = \frac{hc}{\lambda_1} - \frac{hc}{\lambda_2}$$

2. Calculate the energy of the incident photon:

$$E_1 = \frac{hc}{\lambda_1} = \frac{1240\text{ eV}\cdot\text{nm}}{6.0\text{ pm}} = \frac{1.24\text{ keV}\cdot\text{nm}}{6.0 \times 10^{-3}\text{ nm}} = 207\text{ keV}$$

3. Calculate λ_2 from the given wavelength of the incident photon and the change found in (*a*):

$$\lambda_2 = \lambda_1 + \Delta\lambda = 6\text{ pm} + 4.86\text{ pm} = 10.86\text{ pm}$$

4. Use this result to find E_2:

$$E_2 = \frac{hc}{\lambda_2} = \frac{1240\text{ eV}\cdot\text{nm}}{10.86\text{ pm}} = \frac{1.24\text{ keV}\cdot\text{nm}}{10.86 \times 10^{-3}\text{ nm}} = 114\text{ keV}$$

5. Substitute the calculated values of E_1 and E_2 to find the energy of the recoiling electron:

$$K_e = E_1 - E_2 = 207\text{ keV} - 114\text{ keV} = 93\text{ keV}$$

17-3 Energy Quantization in Atoms

Ordinary white light has a continuous spectrum, that is, it contains all the wavelengths in the visible spectrum. But when atoms in a gas at low pressure are excited by an electric discharge, they emit light of specific wavelengths that are characteristic of the type of atom. Since the energy of a photon is related to its wavelength by $E = hf = hc/\lambda$, a discrete set of wavelengths implies a discrete set of energies. Conservation of energy then implies that when an atom radiates, its internal energy changes by a discrete amount. This led Niels Bohr in 1913 to postulate that the internal energy of an atom can have only a discrete set of values. That is, the internal energy of an atom is **quantized.** When an atom radiates light of frequency f, the atom makes a transition from one allowed level to another level that is lower in energy by $\Delta E = hf$. Bohr was able to construct a model of the hydrogen atom that had a discrete set of energy levels consistent with the observed spectrum of emitted light.* However, the *reason* for the quantization of energy levels in atoms and other systems remained a mystery until the wave nature of electrons was discovered a decade later.

*We study the Bohr model in Chapter 37.

17-4 Electrons and Matter Waves

In 1897, J. J. Thomson showed that the rays of a cathode-ray tube (Figure 17-6) consist of electrically charged particles, and he showed that all the particles have the same charge-to-mass ratio q/m. He also showed that particles with this charge-to-mass ratio can be obtained using any material for the cathode, which means that these particles, now called **electrons,** are a fundamental constituent of all matter.

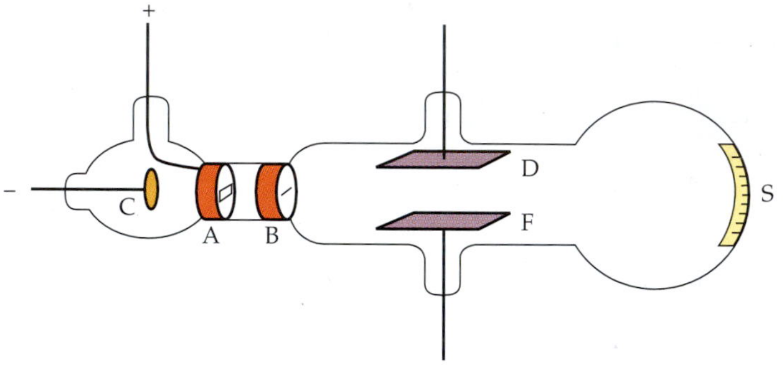

Figure 17-6 Schematic diagram of the cathode-ray tube Thomson used to measure q/m for the particles that comprise cathode rays (electrons). Electrons from the cathode C pass through the slits at A and B and strike a phosphorescent screen S. The beam can be deflected by an electric field between plates D and F or by a magnetic field (not shown).

The de Broglie Hypothesis

Since light seems to have both wave and particle properties, it is natural to ask whether matter—electrons, protons, etc.—might also have both wave and particle characteristics. In 1924, a French physics student, Louis de Broglie, suggested this idea in his doctoral dissertation. de Broglie's work was highly speculative since there was no evidence at that time of any wave aspects of matter.

For the wavelength of electron waves, de Broglie chose

$$\lambda = \frac{h}{p} \qquad \text{17-10}$$

de Broglie relation for the wavelength of electron waves

where p is the momentum of the electron. Note that this is the same as Equation 17-7 for a photon. For the frequency of electron waves de Broglie chose the Einstein equation relating the frequency and energy of a photon:

$$f = \frac{E}{h} \qquad \text{17-11}$$

de Broglie relation for the frequency of electron waves

These equations are thought to apply to all matter. However, for macroscopic objects, the wavelengths calculated from Equation 17-10 are so small that it is impossible to observe the usual wave properties of interference or diffraction. Even a dust particle as small as 1 μg is much too massive for any wave characteristics to be noticed, as we see in the following example.

Example 17-4 ***try it yourself***

Find the de Broglie wavelength of a particle of mass 10^{-6} g moving with a speed of 10^{-6} m/s.

Cover the column to the right and try this on your own before looking at the answer.

Step	*Answer*
Write down the definition of the de Broglie wavelength and substitute the given data.	$\lambda = \frac{h}{p} = \frac{h}{mv} = \frac{6.63 \times 10^{-34}\ \text{J}\cdot\text{s}}{(10^{-9}\ \text{kg})(10^{-6}\ \text{m/s})}$ $= 6.63 \times 10^{-19}\ \text{m}$

Remark This wavelength is much smaller than the diameter of the atomic nucleus, which is about 10^{-15} m.

Since the wavelength found in Example 17-4 is much smaller than any possible apertures or obstacles, diffraction or interference of such waves cannot be observed. In fact, the propagation of waves of very small wavelengths is indistinguishable from the propagation of particles. The momentum of the particle in Example 17-4 was only 10^{-15} kg·m/s. A macroscopic particle with a greater momentum would have an even smaller de Broglie wavelength. We therefore do not observe the wave properties of such macroscopic objects as baseballs and billiard balls.

Exercise Find the de Broglie wavelength of a baseball of mass 0.17 kg moving at 100 km/h. (*Answer* 1.4×10^{-34} m)

The situation is different for low-energy electrons and other microscopic particles. Consider a particle with kinetic energy K. Its momentum is found from

$$K = \frac{p^2}{2m}$$

or

$$p = \sqrt{2mK}$$

Its wavelength is then

$$\lambda = \frac{h}{p} = \frac{h}{\sqrt{2mK}}$$

If we multiply both numerator and denominator by c we obtain*

$$\lambda = \frac{hc}{\sqrt{2mc^2K}} = \frac{1240\ \text{eV}\cdot\text{nm}}{\sqrt{2mc^2K}} \qquad \text{17-12}$$

Wavelength associated with a particle of mass m

where we have used $hc = 1240$ eV·nm. For electrons, $mc^2 = 0.511$ MeV. Then

$$\lambda = \frac{1240\ \text{eV}\cdot\text{nm}}{\sqrt{2mc^2K}} = \frac{1240\ \text{eV}\cdot\text{nm}}{\sqrt{2(0.511 \times 10^6\ \text{eV})K}}$$

* Equations 17-12 and 17-13 do not hold for relativistic particles whose kinetic energies are comparable to their rest energies mc^2.

or

$$\lambda = \frac{1.23}{\sqrt{K}}\,\text{nm} \qquad (K \text{ in electron volts}) \tag{17-13}$$

Electron wavelength

Exercise Find the wavelength of an electron whose kinetic energy is 10 eV. (*Answer* 0.388 nm. This is on the same order of magnitude as the size of the atom and the spacing of atoms in a crystal.)

Electron Interference and Diffraction

The observation of diffraction and interference of electron waves would provide the crucial evidence that electrons have wave properties. This evidence was obtained accidentally in 1927 by C. J. Davisson and L. H. Germer as they were studying electron scattering from a nickel target at the Bell Telephone Laboratories. After an accidental break in the vacuum system they were using, they were obliged to heat the target to remove an oxide coating that had accumulated. Afterward, they found that the scattered-electron intensity as a function of the scattering angle showed maxima and minima. By chance they had observed electron diffraction. After realizing that the scattering pattern had changed because the heating procedure had caused the target to crystallize, they prepared a target consisting of a single crystal of nickel and investigated this phenomenon extensively. Figure 17-7*a* illustrates their experiment. Electrons from an electron gun are directed at a crystal and detected at some angle ϕ that can be varied. Figure 17-7*b* shows a typical pattern observed. There is a strong scattering maximum at an angle of 50°. The angle for maximum scattering of waves from a crystal depends on the wavelength of the waves and the spacing of the atoms in the crystal. Using the known spacing of the atoms in their crystal, Davisson and Germer calculated the wavelength that could produce such a maximum and found that it agreed with the de Broglie equation (Equation 17-10) for the electron energy they were using. By varying the energy of the incident electrons, they could vary the electron wavelengths and produce maxima and minima at different locations in the diffraction patterns. In all cases, the measured wavelengths agreed with de Broglie's hypothesis.

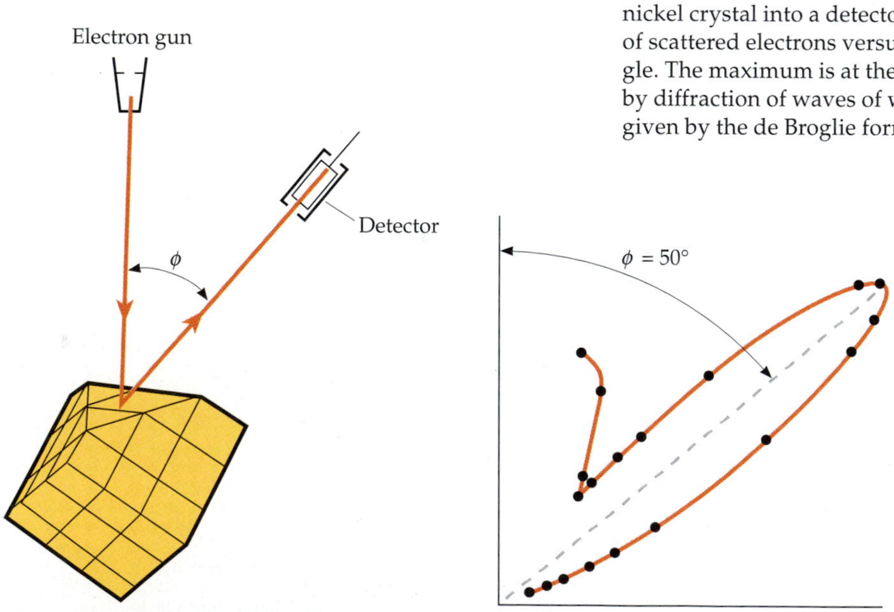

(*a*) (*b*)

Figure 17-7 The Davisson–Germer experiment. (*a*) Electrons are scattered from a nickel crystal into a detector. (*b*) Intensity of scattered electrons versus scattering angle. The maximum is at the angle predicted by diffraction of waves of wavelength λ given by the de Broglie formula.

Another demonstration of the wave nature of electrons was provided in the same year by G. P. Thomson (son of J. J. Thomson), who observed electron diffraction in the transmission of electrons through thin metal foils. A metal foil consists of tiny, randomly oriented crystals. The diffraction pattern resulting from such a foil is a set of concentric circles. Figure 17-8*a* and *b* shows the diffraction pattern observed using X rays and electrons on an aluminum-foil target. Figure 17-8*c* shows the diffraction patterns of neutrons on a copper-foil target. Note the similarity of the patterns. The diffraction of hydrogen and helium atoms was observed in 1930. In all cases, the measured wavelengths agree with the de Broglie predictions. Figure 17-8*d* shows a diffraction pattern produced by electrons incident on two narrow slits. This experiment is equivalent to Young's famous double-slit experiment with light; the pattern is identical to that observed with photons of the same wavelength (compare with Figure 17-1).

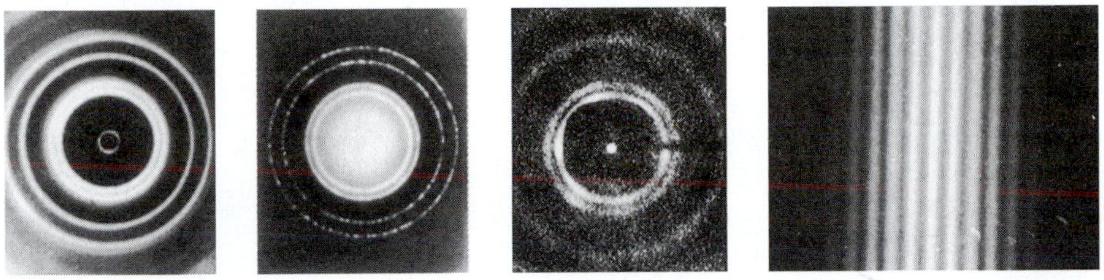

(*a*) (*b*) (*c*) (*d*)

Figure 17-8 (*a*) Diffraction pattern produced by X rays of wavelength 0.071 nm on an aluminum-foil target; (*b*) diffraction pattern produced by 600-eV electrons (λ = 0.050 nm) on an aluminum-foil target; and (*c*) diffraction of 0.0568 eV neutrons (λ = 0.12 nm) incident on a copper foil. (*d*) A two-slit electron diffraction–interference pattern.

Shortly after the wave properties of the electron were demonstrated, it was suggested that electrons rather than light might be used to "see" small objects. As was mentioned in Chapter 15, reflected waves can resolve details of objects only when the details are larger than the wavelength of the reflected wave. Beams of electrons, which can be focused electrically, can have very small wavelengths—much shorter than visible light. Today, the electron microscope is an important research tool used to visualize specimens at scales far smaller than those possible with a light microscope (Figure 17-9).

Standing Waves and Energy Quantization

Given that electrons have wavelike properties, it should be possible to produce standing electron waves. We saw in Chapter 16 that standing waves on a string or standing sound waves occur only for a discrete set of wavelengths and frequencies. If energy is associated with the frequency of a standing wave, as in $E = hf$ (Equation 17-11), then standing waves imply a discrete set of energies. In other words, standing waves imply that energy is quantized.

The idea that the discrete energy states in atoms could be explained by standing waves led to the development by Erwin Schrödinger and

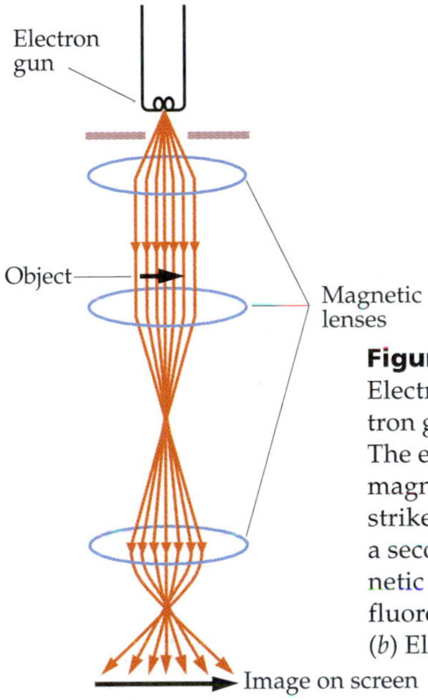

(*a*)

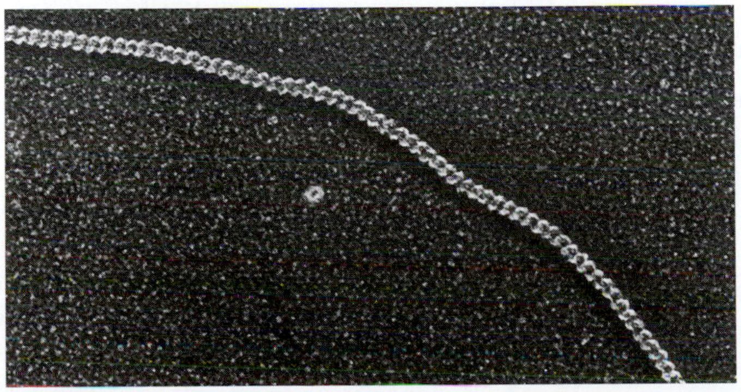

(*b*)

Figure 17-9 (*a*) Electron microscope. Electrons from a heated filament (the electron gun) are accelerated to a high energy. The electron beam is made parallel by a magnetic focusing lens. The electrons strike a thin target and are then focused by a second magnetic lens. The third magnetic lens projects the electron beam onto a fluorescent screen to produce the image. (*b*) Electron micrograph of DNA.

optional

others in 1928 of a detailed mathematical theory known as quantum theory, quantum mechanics, or wave mechanics. In this theory, the electron is described by a wave function that obeys a wave equation called the Schrödinger equation. The form of the Schrödinger equation for a particular situation depends on the forces acting on the particle, which are described by the potential energy functions associated with those forces. In Chapter 36 we discuss this equation, which is somewhat similar to the classical wave equations for sound or light. Schrödinger solved the standing-wave problem for the hydrogen atom, the simple harmonic oscillator, and other systems of interest. He found that the allowed frequencies, combined with $E = hf$, resulted in the set of energy levels found experimentally for the hydrogen atom, thereby demonstrating that quantum theory provides a general method of finding the quantized energy levels for a given system. Quantum theory is the basis for our understanding of the modern world, from the inner workings of the atomic nucleus to the radiation spectra of distant galaxies.

17-5 The Interpretation of the Wave Function

The wave function for waves on a string is the string displacement y. The wave function for sound waves can be either the displacement s of the air molecules or the density ρ. The wave function for light and other electromagnetic waves is the electric field $\vec{E}$. What is the wave function for electron waves? The symbol we use for this wave function is ψ (the Greek letter psi). When Schrödinger published his wave equation, neither he nor anyone else knew just how to interpret the wave function ψ. We can get a hint about how to interpret ψ by considering the quantization of light waves. For sound or light waves, the energy per unit volume in the wave is proportional to the square of the wave function. Since the energy of a light wave is quantized, the energy per unit volume is proportional to the number of photons per unit volume. We might therefore expect the square of the photon's wave function to be proportional to the number of photons per unit volume in a light wave. But suppose we have a very low-energy source of light that emits just one photon at a time. In any unit volume, there is either one photon or none. The square of the wave function must then describe the *probability* of finding a photon in some unit volume.

The Schrödinger equation describes a single particle. The square of the wave function for a particle must then describe the *probability* of finding the particle in some unit volume. The probability of finding the particle in some volume element must also be proportional to the size of the volume element dV. Thus, in one dimension, the probability of finding a particle in a region dx at the position x is $\psi^2(x)\,dx$. If we call this probability $P(x)\,dx$, where $P(x)$ is the **probability density,** we have

$$P(x) = \psi^2(x) \qquad \text{17-14}$$

Probability density

Generally, the wave function depends on time as well as position and is written $\Psi(x,t)$, with an uppercase psi. However, for standing waves, the probability density is independent of time. Since we will be concerned mostly with standing waves in this chapter, we omit the time dependence of the wave function, and write it $\psi(x)$ or just ψ.

The probability of finding the particle in dx at point x_1 *or* at point x_2 is the sum of the separate probabilities $P(x_1)\,dx + P(x_2)\,dx$. If we have a particle at all, the probability of finding the particle somewhere must be 1. Then the sum of the probabilities over all the possible values of x must equal 1. That is,

$$\int_{-\infty}^{\infty} \psi^2\,dx = 1 \qquad \text{17-15}$$

Normalization condition

Equation 17-15 is called the **normalization condition.** If ψ is to satisfy the normalization condition, it must approach zero as x approaches infinity. This places a restriction on the possible solutions of the Schrödinger equation.

Example 17-5

A classical point particle moves back and forth with constant speed between two walls at $x = 0$ and $x = 8$ cm (Figure 17-10). (*a*) What is the probability density $P(x)$? (*b*) What is the probability of finding the particle at $x = 2$ cm? (*c*) What is the probability of finding the particle between $x = 3.0$ cm and $x = 3.4$ cm?

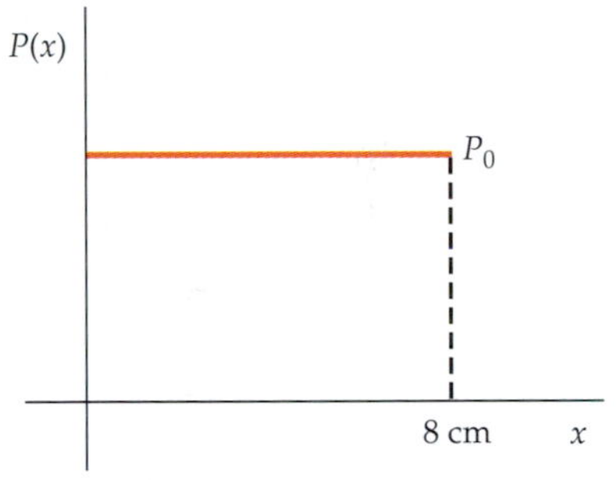

Figure 17-10 Probability function $P(x)$.

Picture the Problem The probability of finding a classical particle in some region dx is proportional to the time spent in that region, dx/v, where v is the speed. Since the speed is constant, the probability density $P(x)$ is constant, independent of x, for $0 < x < 8$ cm. Outside of this range, $P(x)$ is zero. We can find the constant by normalization, that is, by requiring that the probability of finding the particle somewhere between $x = 0$ and $x = 8$ cm is 1.

(*a*)1. The probability density $P(x)$ is constant between the walls:

$$P(x) = P_0, \quad 0 < x < 8\text{ cm}$$

$$P(x) = 0, \quad x < 0 \text{ or } x > 8\text{ cm}$$

2. Apply the normalization condition:

$$\int_{-\infty}^{+\infty} P(x)\,dx = \int_0^{8\text{ cm}} P_0\,dx = P_0(8\text{ cm}) = 1$$

3. Solve for P_0:

$$P(x) = P_0 = \frac{1}{8\text{ cm}}$$

(*b*) The probability of finding the particle in some range dx is proportional to dx. Since $dx = 0$, the probability of finding the particle at the point $x = 2$ cm is 0.

(*c*) Since the probability density is constant, the probability of a particle being in some range Δx in the region $0 < x < 8$ cm is $P_0\,\Delta x$. The probability of the particle being in the region $3.0\text{ cm} < x < 3.4$ cm is thus:

$$P_0\,\Delta x = \frac{1}{8\text{ cm}}\,0.4\text{ cm} = 0.05$$

Remark Note in step 2 of part (*a*) that we need only integrate from 0 to 8 cm because $P(x)$ is zero outside this range.

optional

17-6 Wave–Particle Duality

We have seen that light, which we ordinarily think of as a wave, exhibits particle properties when it interacts with matter, as in the photoelectric effect or in Compton scattering. Electrons, which we usually think of as particles, exhibit the wave properties of interference and diffraction. All carriers of momentum and energy, such as electrons, atoms, light, or sound, have both particle and wave characteristics. It might be tempting to say that an electron, for example, is both a wave and a particle, but what does this mean? In classical physics, the concepts of waves and particles are mutually exclusive. A **classical particle** behaves like a piece of shot; it can be localized and scattered, it exchanges energy suddenly at a point in space, and it obeys the laws of conservation of energy and momentum in collisions. It does *not* exhibit interference or diffraction. A **classical wave,** on the other hand, behaves like a water wave; it exhibits diffraction and interference, and its energy is spread out continuously in space and time. Nothing can be both a classical particle and a classical wave at the same time.

After Thomas Young observed the two-slit interference pattern with light in 1801, light was thought to be a classical wave. On the other hand, the electrons discovered by J. J. Thomson were thought to be classical particles. We now know that these classical concepts of waves and particles do not adequately describe the complete behavior of any phenomenon.

> All carriers of energy and momentum, such as light and electrons, propagate like a wave and exchange energy like a particle.

Often the concepts of the classical particle and the classical wave give the same results. When the wavelength is very small, diffraction effects are negligible, so the waves travel in straight lines like classical particles. Also, interference is not seen for waves of very small wavelength because the interference maxima and minima are too closely spaced to be observed. It then makes no difference which concept we use. When diffraction is negligible, we can think of light as a wave propagating along rays or as a beam of photon particles. Similarly, we can think of an electron as a wave propagating in straight lines along rays or, more commonly, as a particle.

We can also use either the wave or particle concept to describe exchanges of energy if we have a large number of particles and we are interested only in the average values of energy and momentum exchanges.

The Two-Slit Experiment Revisited The wave–particle duality of nature is illustrated by the analysis of the experiment in which an electron is incident on a barrier with two slits. The analysis is the same whether we use electrons or photons (light). To describe the propagation of the electron, we must use wave theory. Consider an electron wave that traverses both slits of the two-slit barrier. The two slits act as point sources of spherical electron waves. The wave function at a point on a screen or film far from the slits depends on the path difference from the two slits. At points for which the path difference is 0 or an integral number of wavelengths, the wave function ψ is maximum. Since the probability of detecting the electron is proportional to ψ^2, the electron is most likely to arrive at these points. At points for which the path difference is a half-wavelength or an odd number of half-wavelengths, the wave function ψ is zero, implying that there is zero probability of the electron arriving at such a point. The chapter opening photo on page 509 shows the interference pattern produced by 10 electrons, 100 electrons, 3,000 electrons, and 70,000 electrons. Note that, although the electron propagates through the slits like a wave, it interacts with the screen at a single point like a particle.

The Uncertainty Principle An important consequence of the wave–particle duality of nature is the uncertainty principle, which states that it is impossible in principle to simultaneously measure both the position and momentum of a particle with unlimited precision. A common way to measure the position of an object is to look at it with light. When we do this, we scatter light from the object and determine the position by the direction of the scattered light. If we use light of wavelength λ, we can measure the position only to an uncertainty of the order of λ because of diffraction effects:

$$\Delta x \sim \lambda$$

To reduce the uncertainty in position we therefore use light of very short wavelength, perhaps even X rays. In principle, there is no limit to the accuracy of such a position measurement because there is no limit on how small the wavelength λ can be.

If we know the mass of a particle, we can determine its momentum by measuring its position at two nearby times and computing its velocity. If we use light of wavelength λ, the photons carry momentum h/λ. When these photons are scattered by the particle under scrutiny, the momentum of the particle is changed by the scattering in an uncontrollable way. Each photon carries momentum h/λ, so the uncertainty in the momentum of the particle, introduced by looking at it, is of the order of h/λ:

$$\Delta p \sim \frac{h}{\lambda}$$

When the wavelength of the radiation is small, the momentum of each photon will be large and the momentum measurement will have a large uncertainty. This uncertainty cannot be eliminated by reducing the intensity of light; such a reduction merely reduces the number of photons in the beam. To "see" the particle we must scatter at least one photon. Therefore, the uncertainty in the momentum measurement of the particle will be large if λ is small, and the uncertainty in the position measurement of the particle will be large if λ is large.

Of course we could always "look at" the particles by scattering electrons instead of photons, but the same difficulty remains. If we use low-momentum electrons to reduce the uncertainty in the momentum measurement, we have a large uncertainty in the position measurement because of diffraction of the electrons. The relation between the wavelength and momentum $\lambda = h/p$ is the same for electrons as for photons.

The product of the intrinsic uncertainties in position and momentum is

$$\Delta x\, \Delta p \sim \lambda \times \frac{h}{\lambda} = h$$

If we define precisely what we mean by uncertainties in measurement, we can give a precise statement of the uncertainty principle. If Δx and Δp are defined to be the standard deviations in the measurements of position and momentum, it can be shown that their product must be greater than or equal to $\hbar/2$:

$$\Delta x\, \Delta p \geq \tfrac{1}{2}\hbar \qquad \text{17-16}$$

where $\hbar$ (read h bar) $= h/2\pi$.*

Equation 17-16 provides a statement of the uncertainty principle first enunciated by Werner Heisenberg in 1927. In practice, the experimental uncertainties are usually much greater than the intrinsic lower limit that results from wave–particle duality.

*The combination $h/2\pi$ occurs so often that it is given a special symbol, somewhat analogous to giving the special symbol ω for $2\pi f$, which occurs often in oscillations.

optional

17-7 A Particle in a Box

We can illustrate many of the important features of quantum physics by considering the simple problem of a particle of mass m confined to a one-dimensional box of length L, like the particle in Example 17-5. This situation is analogous to an electron confined within an atom, or a proton confined within a nucleus. When a classical particle bounces back and forth between the walls of the box, its energy and momentum can have any values. However, according to quantum theory, the particle is described by a wave function ψ, whose square describes the probability of finding the particle in some region. Since we are assuming that the particle is indeed inside the box, the wave function must be zero everywhere outside the box. If the box is between $x = 0$ and $x = L$, we have

$$\psi = 0 \qquad \text{for } x \leq 0 \quad \text{and} \quad \text{for } x \geq L$$

In particular, since the wave function is continuous, it must be zero at the end points of the box $x = 0$ and $x = L$. This is the same condition as that for standing waves on a string fixed at $x = 0$ and $x = L$, and the results are the same. The allowed wavelengths for a particle in the box are those such that the length L equals an integral number of half-wavelengths (Figure 17-11).

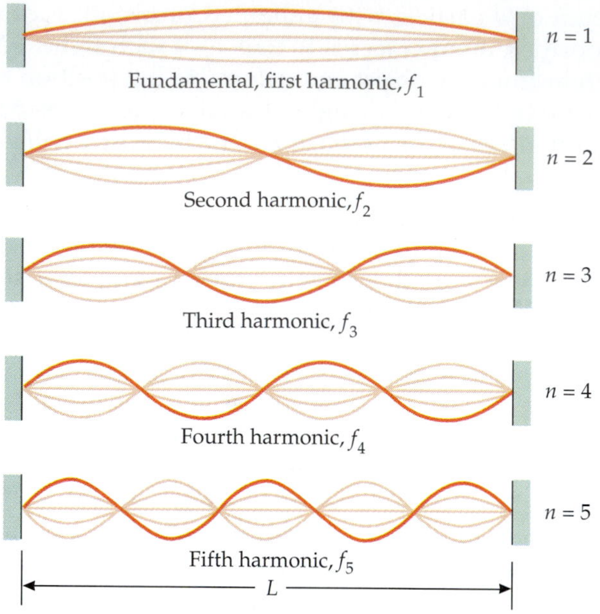

Figure 17-11 Standing waves on a string fixed at both ends. The standing-wave condition is the same as for standing electron waves in a box.

$$L = n\frac{\lambda_n}{2}, \qquad n = 1, 2, 3, \ldots \qquad \text{17-17}$$

Standing-wave condition, particle in a box of length L

The total energy of a particle in a box is its kinetic energy:

$$E = \frac{1}{2}mv^2 = \frac{p^2}{2m}$$

Substituting the de Broglie relation $p_n = h/\lambda_n$, we get

$$E_n = \frac{p_n^2}{2m} = \frac{(h/\lambda_n)^2}{2m}$$

Then the standing-wave condition $\lambda_n = 2L/n$ gives the allowed energies:

$$E_n = \frac{h^2}{2m\lambda_n^2} = n^2\frac{h^2}{8mL^2} = n^2E_1 \qquad \text{17-18}$$

Allowed energies for a particle in a box

where

$$E_1 = \frac{h^2}{8mL^2} \qquad \text{17-19}$$

Ground-state energy for a particle in a box

is the energy of the lowest state, the ground state.

The condition $\psi = 0$ at $x = 0$ and $x = L$ is called a **boundary condition.** Boundary conditions in quantum theory lead to energy quantization. Figure 17-12 shows the energy-level diagram for a particle in a box. Note that the lowest energy is not zero. This result is a general feature of quantum theory.

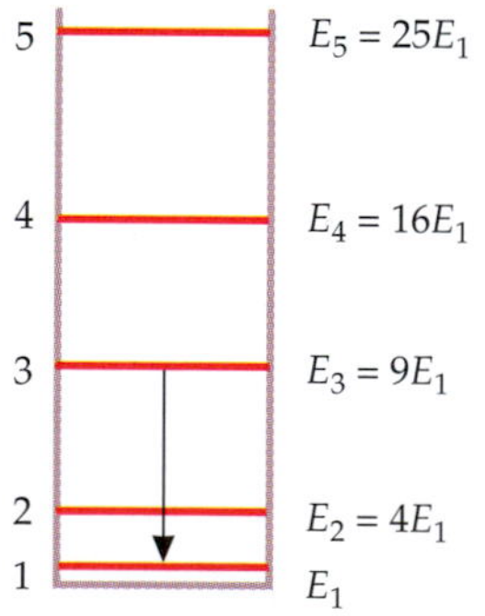

Figure 17-12 Energy-level diagram for a particle in a box. Classically, a particle can have any energy value. Quantum mechanically, only those energy values given by Equation 17-18 are allowed. A transition between the state $n = 3$ and the ground state $n = 1$ is indicated by the vertical arrow.

When a particle is confined to some region of space, it has a minimum energy, which is called the **zero-point energy.** The smaller the region of space, the greater the zero-point energy, as indicated by the fact that E_1 varies as $1/L^2$ in Equation 17-19.

If an electron is in some energy state E_i, it can make a transition to another energy state E_f with the emission of a photon (if $E_f < E_i$) or the absorption of a photon (if $E_f > E_i$). The transition from state 3 to the ground state is indicated in Figure 17-12 by the vertical arrow. The frequency of the emitted photon is found from conservation of energy*

$$hf = E_i - E_f \qquad 17\text{-}20$$

The wavelength of the photon is then

$$\lambda = \frac{c}{f} = \frac{hc}{E_i - E_f} \qquad 17\text{-}21$$

Wave Functions for Standing Waves

The instantaneous shape of a vibrating string fixed at $x = 0$ and $x = L$ is given by Equation 16-15:

$$y_n = A_n \sin k_n x$$

where A_n is a constant, and $k_n = 2\pi/\lambda_n$ is the wave number. The wave functions for a particle in a box (which can be obtained by solving the Schrödinger equation, as we will see in Chapter 36) are the same:

$$\psi_n(x) = A_n \sin k_n x$$

where $k_n = 2\pi/\lambda_n$. Using $\lambda_n = 2L/n$, we have

$$k_n = \frac{2\pi}{\lambda_n} = \frac{2\pi}{2L/n} = \frac{n\pi}{L}$$

The wave functions can thus be written

$$\psi_n(x) = A_n \sin \frac{n\pi x}{L}$$

The constant A_n is determined by the normalization condition (Equation 17-15):

$$\int_{-\infty}^{\infty} \psi^2\, dx = \int_0^L A_n^2 \sin^2 \frac{n\pi x}{L}\, dx = 1$$

Note that we need integrate only from $x = 0$ to $x = L$ because $\psi(x)$ is zero everywhere else. The integration can be done using tables. The result is

$$A_n = \sqrt{\frac{2}{L}}$$

independent of n. The normalized wave functions for a particle in a box are thus

$$\psi_n(x) = \sqrt{\frac{2}{L}} \sin \frac{n\pi x}{L} \qquad 17\text{-}22$$

Wave functions for a particle in a box

---

* This equation was first proposed by Niels Bohr in his model of the hydrogen atom in 1913, about 10 years before de Broglie's suggestion that electrons have wave properties. We study the Bohr model in Chapter 37.

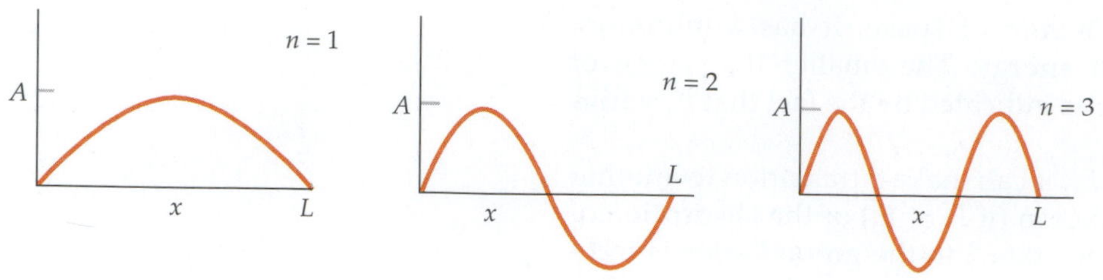

Figure 17-13 Standing-wave functions for $n = 1, 2$, and 3.

These functions for $n = 1, 2$, and 3 are shown in Figure 17-13.

The number n is called a **quantum number.** It characterizes the wave function for a particular state and the energy of that state. In our one-dimensional problem, it arises from the boundary condition on the wave function that it must be zero at $x = 0$ and $x = L$. In three-dimensional problems, three quantum numbers arise, one associated with a boundary condition in each dimension.

Figure 17-14 shows plots of ψ^2 for the ground state $n = 1$, the first excited state $n = 2$, the second excited state $n = 3$, and the state $n = 10$. In the ground state, the particle is most likely to be found near the center of the box, as indicated by the maximum value of ψ^2 at $x = L/2$. In the first excited state, the particle is never found exactly in the center of the box because ψ^2 is zero at $x = L/2$. For very large values of n, the maxima and minima of ψ^2 are very close together, as illustrated for $n = 10$. The average value of ψ^2 is indicated in this figure by the dashed line. For very large values of n, the maxima are so closely spaced that ψ^2 cannot be distinguished from its average value. The fact that $(\psi^2)_{av}$ is constant across the whole box means that the particle is equally likely to be found anywhere in the box—the same as the classical result. This is an example of **Bohr's correspondence principle**:

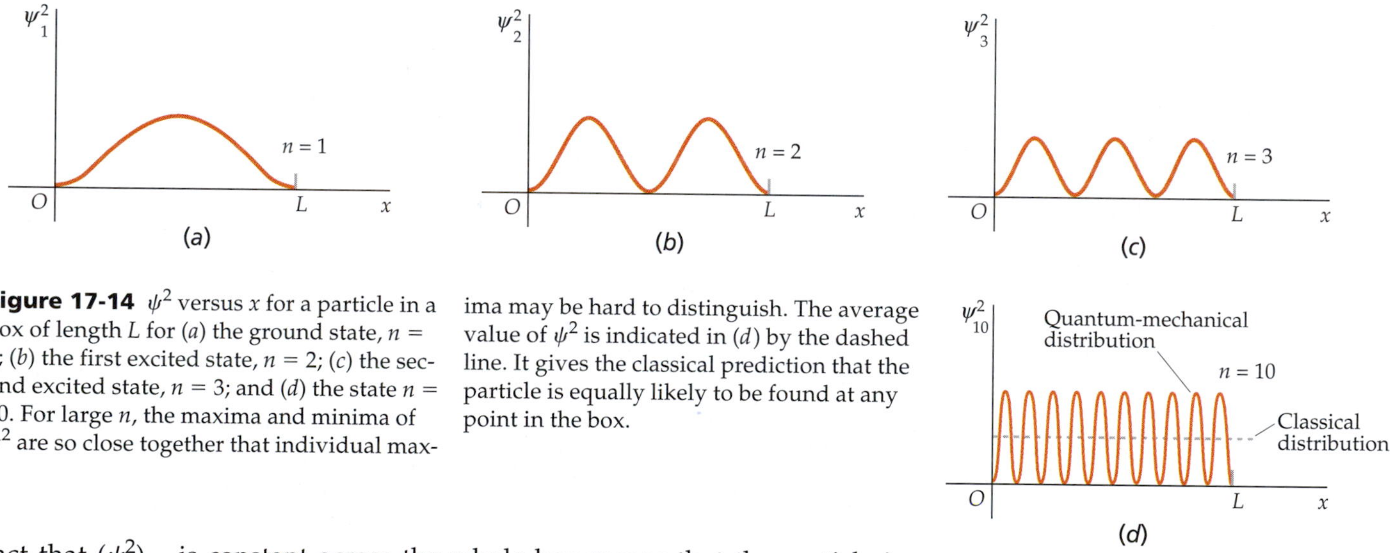

Figure 17-14 ψ^2 versus x for a particle in a box of length L for (*a*) the ground state, $n = 1$; (*b*) the first excited state, $n = 2$; (*c*) the second excited state, $n = 3$; and (*d*) the state $n = 10$. For large n, the maxima and minima of ψ^2 are so close together that individual maxima may be hard to distinguish. The average value of ψ^2 is indicated in (*d*) by the dashed line. It gives the classical prediction that the particle is equally likely to be found at any point in the box.

> In the limit of very large quantum numbers, the classical calculation and the quantum calculation must yield the same results.

Bohr's correspondence principle

When the quantum numbers are very large, the energy is very large. For large energies, the percentage change in energy between adjacent quantum states is very small, so energy quantization is not important (see Problem 83).

We are so accustomed to thinking of the electron as a classical particle that we tend to think of an electron in a box as a particle bouncing back and forth between the walls. But the probability distributions shown in Figure 17-14 are stationary; that is, they do not depend on time. A better picture for an electron in a bound state is a cloud of charge with the charge density proportional to ψ^2. The graphs in Figure 17-14 can then be thought of as plots of the charge density versus x for the various states. In the ground state, $n = 1$, the electron cloud is centered in the middle of the box and is spread out over most of the box, as indicated in Figure 17-14*a*. In the first excited state, $n = 2$, the charge density of the electron cloud has two maxima, as indicated in Figure 17-14*b*. For very large values of n, there are many closely spaced maxima and minima in the charge density resulting in an average charge density that is approximately uniform throughout the box. This probability-cloud picture of an electron is very useful in understanding the structure of atoms and molecules. However, it should be noted that whenever an electron is observed to interact with matter or radiation, it is always observed as a whole unit charge.

Example 17-6

An electron is in a one-dimensional box of length 0.1 nm. (*a*) Find the ground-state energy. (*b*) Find the energy in electron volts of the five lowest states and make an energy-level diagram. (*c*) Find the wavelength of the photon emitted for each transition from the state $n = 3$ to a lower-energy state.

Picture the Problem For (*a*) and (*b*), the energies are given by $E_n = n^2 E_1$, where $E_1 = h^2/8mL^2 = (hc)^2/8(mc^2)L^2$. For (*c*), the photon wavelengths are given by $\lambda = hc/(E_i - E_f)$.

(*a*) Use $hc = 1240$ eV·nm, and $mc^2 = 5.11 \times 10^5$ eV to calculate E_1:

$$E_1 = \frac{(hc)^2}{8(mc^2)L^2} = \frac{(1240\ \text{eV}\cdot\text{nm})^2}{8(5.11 \times 10^5\ \text{eV})(0.1\ \text{nm})^2} = 37.6\ \text{eV}$$

(*b*) Calculate $E_n = n^2 E_1$ for $n = 2, 3, 4$, and 5:

$$E_2 = (2)^2(37.6\ \text{eV}) = 150\ \text{eV}$$

$$E_3 = (3)^2(37.6\ \text{eV}) = 338\ \text{eV}$$

$$E_4 = (4)^2(37.6\ \text{eV}) = 602\ \text{eV}$$

$$E_5 = (5)^2(37.6\ \text{eV}) = 940\ \text{eV}$$

(*c*)1. Use the energies in (*b*) to calculate the wavelength for a transition from state 3 to state 2:

$$\lambda = \frac{hc}{E_3 - E_2} = \frac{1240\ \text{eV}\cdot\text{nm}}{338\ \text{eV} - 150\ \text{eV}} = 6.60\ \text{nm}$$

2. Then use the energies in (*a*) and (*b*) to calculate the wavelength for a transition from state 3 to state 1:

$$\lambda = \frac{hc}{E_3 - E_1} = \frac{1240\ \text{eV}\cdot\text{nm}}{338\ \text{eV} - 37.6\ \text{eV}} = 4.13\ \text{nm}$$

Remarks The energy-level diagram is shown in Figure 17-15. The transitions from $n = 3$ to $n = 2$ and from $n = 3$ to $n = 1$ are indicated by the vertical arrows. The ground-state energy of 37.6 eV is of the same order of magnitude as the kinetic energy of the electron in the ground state of the hydrogen atom, which is 13.6 eV. In the hydrogen atom, the electron also has potential energy of −27.2 eV in the ground state, giving a total ground-state energy of −13.6 eV.

Exercise Calculate the wavelength of the photon emitted if the electron makes a transition from $n = 4$ to $n = 3$. (*Answer* 4.70 nm)

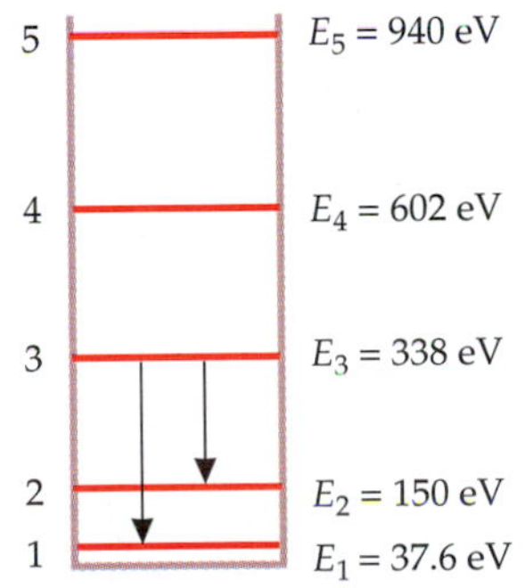

Figure 17-15

optional

17-8 Expectation Values

The solution of a classical mechanics problem is typically specified by giving the position of a particle as a function of time. But the wave nature of matter prevents us from doing this for microscopic systems. The most that we can know is the probability of measuring a certain value of position x. If we measure the position for a large number of identical systems, we get a range of values corresponding to the probability distribution. The average value of x obtained from such measurements is called the **expectation value** and is written $\langle x \rangle$. The expectation value of x is the same as the average value of x that we would expect to obtain from a measurement of the positions of a large number of particles with the same wave function $\psi(x)$.

Since $\psi^2(x)\,dx$ is the probability of finding a particle in the region dx, the expectation value of x is

$$\langle x \rangle = \int x\psi^2(x)\,dx \qquad \text{17-23}$$

Expectation value of x defined

The expectation value of any function $f(x)$ is given by

$$\langle f(x) \rangle = \int f(x)\psi^2(x)\,dx \qquad \text{17-24}$$

Expectation value of f(x) defined

Calculating Probabilities and Expectation Values*

The problem of a particle in a box allows us to illustrate the calculation of the probability of finding the particle in various regions of the box, and the expectation values for various energy states. We give two examples, using the wave functions given by Equation 17-22.

*These calculations are somewhat complicated and may be skipped over on a first reading. Students required to perform similar calculations in problems will find these examples helpful.

Example 17-7

A particle in a one-dimensional box of length L is in the ground state. Find the probability of finding the particle (*a*) in the region $\Delta x = 0.01L$ at $x = \frac{1}{2}L$, and (*b*) in the region $0 < x < \frac{1}{4}L$.

Picture the Problem The probability of finding the particle in some range dx is $\psi^2\,dx$. For (*a*) (Figure 17-16*a*), the region $\Delta x = 0.01L$ is so small that we can neglect the variation in $\psi(x)$ and just compute $\psi^2\,\Delta x$. For (*b*) (Figure 17-16*b*), we must take into account the variation of $\psi(x)$ and integrate from 0 to $L/4$. These probabilities are indicated by the shaded regions in the figures.

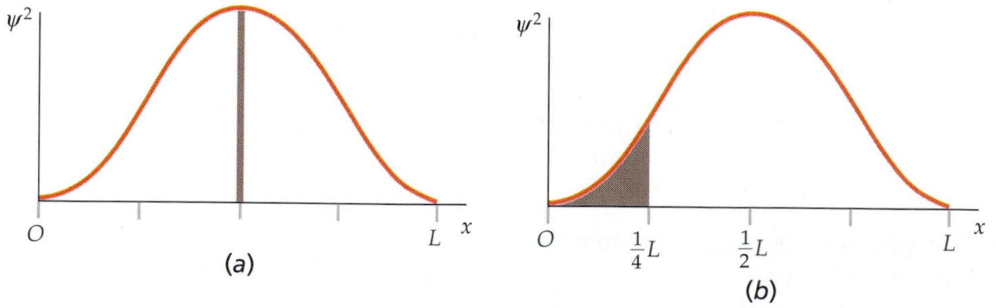

Figure 17-16

(*a*)1. The probability of finding the particle in some range dx is $\psi^2\,dx$: $\quad P(x)\,dx = \psi^2(x)\,dx = \frac{2}{L}\sin^2\frac{\pi x}{L}\,dx$

2. Since ψ^2 does not vary rapidly near $x = L/2$, and the region $\Delta x = 0.01L$ is very small compared with L, we do not need to integrate. The approximate probability is $\psi^2(x)\,\Delta x$. Substitute $x = \frac{1}{2}L$ and $\Delta x = 0.01L$: $\quad P = \frac{2}{L}\left(\sin^2\frac{\pi}{2}\right)(0.01L) = \frac{2}{L}(1.0)(0.01L) = 0.02$

(*b*)1. For the region $0 < x < L/4$, integrate from $x = 0$ to $x = L/4$: $\quad P = \int_0^{L/4}\frac{2}{L}\sin^2\frac{\pi x}{L}\,dx$

2. Change the integration variables to $\theta = \pi x/L$: $\quad P = \frac{2}{L}\frac{L}{\pi}\int_0^{\pi/4}\sin^2\theta\,d\theta$

3. The integral can be found in tables: $\quad \int_0^{\pi/4}\sin^2\theta\,d\theta = \left(\frac{\theta}{2} - \frac{\sin 2\theta}{4}\right)\Big|_0^{\pi/4} = \frac{\pi}{8} - \frac{1}{4}$

4. Use this result to calculate the probability: $\quad P = \frac{2}{\pi}\int_0^{\pi/4}\sin^2\theta\,d\theta = \frac{2}{\pi}\left(\frac{\pi}{8} - \frac{1}{4}\right) = 0.091$

Remarks: The chance of finding the particle in the region $\Delta x = 0.01L$ at $x = \frac{1}{2}L$ is approximately 2%. The chance of finding the particle in the region $0 < x < L/4$ is about 9.1%.

Example 17-8

(*a*) Find $\langle x\rangle$ for a particle in its ground state in a box of length L, and (*b*) find $\langle x^2\rangle$.

Picture the Problem We use $\langle f(x)\rangle = \int f(x)\psi^2(x)\,dx$ with

$$\psi_n(x) = \sqrt{\frac{2}{L}}\sin\frac{n\pi x}{L}$$

(*a*)1. Write $\langle x\rangle$ using the ground-state wave function given by Equation 17-22 with $n = 1$: $\quad \langle x\rangle = \int_{-\infty}^{+\infty} x\psi^2(x)\,dx = \int_0^L x\frac{2}{L}\sin^2\frac{\pi x}{L}\,dx$

2. Substitute $\theta = \pi x/L$: $\quad \langle x\rangle = \frac{2}{L}\left(\frac{L}{\pi}\right)^2\int_0^{\pi}\theta\sin^2\theta\,d\theta = \frac{2L}{\pi^2}\int_0^{\pi}\theta\sin^2\theta\,d\theta$

3. Evaluate the integral by looking it up in tables: $\quad \int_0^{\pi}\theta\sin^2\theta\,d\theta = \left[\frac{\theta^2}{4} - \frac{\theta\sin 2\theta}{4} - \frac{\cos 2\theta}{8}\right]_0^{\pi} = \frac{\pi^2}{4}$

4. Substitute this value into the expression in step 2: $\quad \langle x\rangle = \frac{2L}{\pi^2}\int_0^{\pi}\theta\sin^2\theta\,d\theta = \frac{2L}{\pi^2}\frac{\pi^2}{4} = \frac{L}{2}$

(*b*)1. Repeat step 1 for $\langle x^2\rangle$: $\quad \langle x^2\rangle = \int_{-\infty}^{+\infty} x^2\psi^2(x)\,dx = \int_0^L x^2\frac{2}{L}\sin^2\left(\frac{\pi x}{L}\right)dx$

2. Again, substitute $\theta = \pi x/L$: $\quad \langle x^2\rangle = \frac{2}{L}\left(\frac{L}{\pi}\right)^3\int_0^{\pi}\theta^2\sin^2\theta\,d\theta = \frac{2L^2}{\pi^3}\int_0^{\pi}\theta^2\sin^2\theta\,d\theta$

optional

3. Evaluate the integral by looking it up in tables:

$$\int_0^{\pi} \theta^2 \sin^2 \theta \, d\theta = \left[\frac{\theta^3}{6} - \left(\frac{\theta^2}{4} - \frac{1}{8}\right)\sin 2\theta - \frac{\theta \cos 2\theta}{4}\right]_0^{\pi} = \frac{\pi^3}{6} - \frac{\pi}{4}$$

4. Substitute this value into the expression in step 2 of part (*b*):

$$\langle x^2 \rangle = \frac{2L^2}{\pi^3}\int_0^{\pi} \theta^2 \sin^2 \theta \, d\theta = \frac{2L^2}{\pi^3}\left(\frac{\pi^3}{6} - \frac{\pi}{4}\right) = L^2\left(\frac{1}{3} - \frac{1}{2\pi^2}\right) = 0.283L^2$$

Remarks The expectation value of x is $L/2$, as we would expect, because the probability distribution is symmetric about the midpoint of the box. Note that $\langle x^2 \rangle$ is not equal to $\langle x \rangle^2$.

17-9 Energy Quantization in Other Systems

The quantized energies of a system are generally determined by solving the Schrödinger equation for that system. The form of the Schrödinger equation depends on the potential energy of the particle. The potential energy for a one-dimensional box from $x = 0$ to $x = L$ is shown in Figure 17-17. This potential-energy function is called an **infinite square-well potential** and is described mathematically by

$$U(x) = 0, \qquad 0 < x < L$$
$$U(x) = \infty, \qquad x < 0 \quad \text{or} \quad x > L \qquad \text{17-25}$$

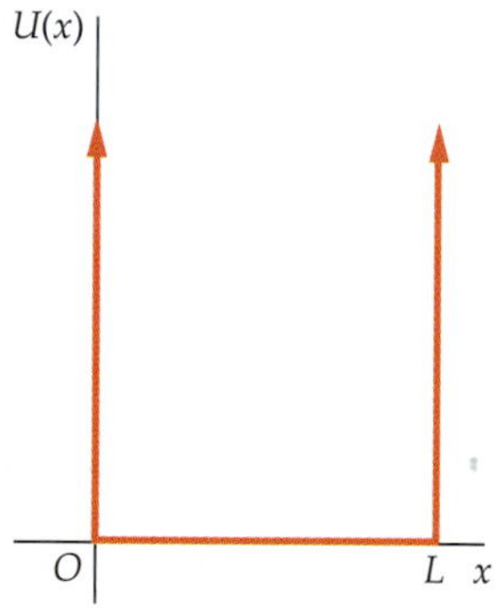

Figure 17-17 The infinite square-well potential energy. For $x < 0$ and $x > L$, the potential energy $U(x)$ is infinite. The particle is confined to the region in the well $0 < x < L$.

Inside the box, the particle moves freely so the potential energy is zero. Outside the box, the potential energy is infinite, so the particle cannot exist outside the box no matter what its energy. We did not need to solve the Schrödinger equation for this potential because the wave functions and quantized frequencies are the same as for a string fixed at both ends, which we studied in Chapter 16. Although this problem seems artificial, it is actually useful in dealing with some physical problems, such as a neutron inside a nucleus.

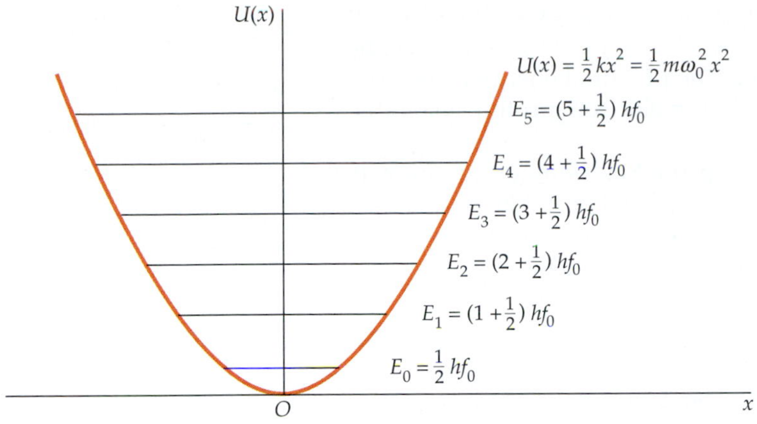

Figure 17-18 Harmonic oscillator potential-energy function. The allowed energy levels are indicated by the equally spaced horizontal lines.

optional

The Harmonic Oscillator

More realistic than the particle in a box is the harmonic oscillator, which applies to an object of mass m on a spring of force constant k or to any system undergoing small oscillations about a stable equilibrium. Figure 17-18 shows the potential-energy function

$$U(x) = \tfrac{1}{2}kx^2 = \tfrac{1}{2}m\omega_0^2 x^2$$

where $\omega_0 = \sqrt{k/m}$ is the natural frequency of the oscillator. Classically, the object oscillates between $x = +A$ and $x = -A$. Its total energy is $E = \frac{1}{2}m\omega_0^2 A^2$, which can have any nonnegative value, including zero.

In quantum theory, the particle is represented by the wave function $\psi(x)$, which is determined by solving the Schrödinger equation for this potential. Normalizable wave functions $\psi_n(x)$ occur only for discrete values of the energy E_n given by

$$E_n = (n + \tfrac{1}{2})hf_0, \qquad n = 0, 1, 2, 3, \ldots \qquad \text{17-26}$$

where $f_0 = \omega_0/2\pi$ is the frequency of the oscillator. Note that the energy levels of a harmonic oscillator are evenly spaced with separation hf as indicated in Figure 17-18. Compare this with the uneven spacing of the energy levels for the particle in a box, shown in Figure 17-12. When a harmonic oscillator makes a transition from energy level n to the next lowest energy level $n - 1$, the energy of the photon emitted is

$$E_n - E_{n-1} = (n + \tfrac{1}{2})hf_0 - (n - 1 + \tfrac{1}{2})hf_0 = hf_0$$

The frequency of the emitted photon is therefore equal to the classical frequency of the oscillator.

The Hydrogen Atom

In the hydrogen atom, an electron is bound to a proton by the electrostatic force of attraction, which we shall study in Chapter 22. This force varies inversely as the square of the separation distance (exactly like the gravitational attraction of the earth and sun). The potential energy of the electron–proton system therefore varies inversely with separation distance like the gravitational potential energy described by Equation 11-18. As in the case of gravitational potential energy, the potential energy of the electron–proton system is chosen to be zero when the electron is an infinite distance from the proton. Then for all finite distances the potential energy is negative. Like the case of an object orbiting the earth, the electron–proton system is a bound system when its total energy is negative.

The allowed energies obtained by solving the Schrödinger equation for the hydrogen atom are described by a quantum number n, like the energies of a particle in a box and of a harmonic oscillator. As we shall see in Chapter 37, the allowed energies of the hydrogen atom are given by

$$E_n = -\frac{13.6\text{ eV}}{n^2}, \qquad n = 1, 2, 3, \ldots \tag{17-27}$$

The lowest energy corresponds to $n = 1$. The ground-state energy is thus -13.6 eV. The energy of the first excited state is $-(13.6\text{ eV}/2^2) = -3.40$ eV. Figure 17-19 shows the energy-level diagram for the hydrogen atom. Transitions from a higher state to a lower state with the emission of electromagnetic radiation are indicated by the vertical arrows. Only those transitions ending at the first excited state ($n = 2$) involve energy differences in the range of visible light of 1.77 to 3.1 eV, as calculated in Example 17-1.

Other atoms are more complicated than the hydrogen atom, but their energy levels are similar to those of hydrogen. The ground-state energies are of the order of -1 to -10 eV, and many transitions involve energies corresponding to photons in the visible range.

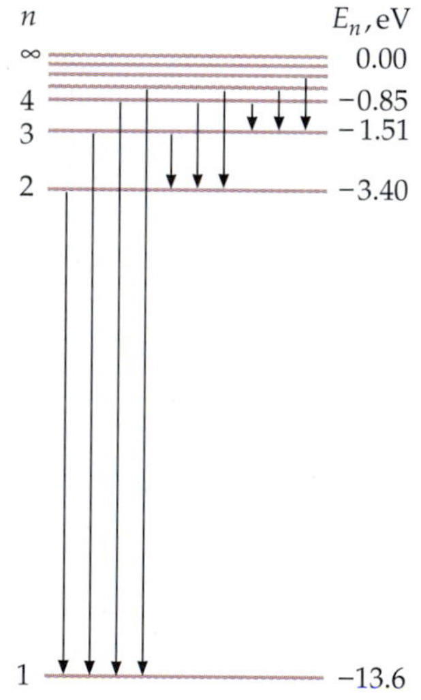

Figure 17-19 Energy-level diagram for the hydrogen atom. The energy of the ground state is -13.6 eV. As n approaches ∞, the energy approaches 0, the highest energy state.

Summary

1. All carriers of energy and momentum propagate like waves and exchange energy like particles.
2. The quantum of light is called a photon. It has energy $E = hf$, where h is Planck's constant.
3. The wavelength of electrons and other "particles" is given by the de Broglie relation $\lambda = h/p$.
4. Energy quantization in bound systems arises from standing wave conditions, which are equivalent to boundary conditions on the wave function.
5. The uncertainty principle is a fundamental law of nature that places theoretical restrictions on the precision of a simultaneous measurement of the position and momentum of a particle. The uncertainty principle follows from the general properties of waves.

Topic	Remarks and Relevant Equations	
1. Wave Nature of Light	The wave nature of light can be demonstrated by the interference of light from two narrow slits illuminated by a single source.	
2. Quantization of Radiation	Light and other electromagnetic energy is not continuous, but instead is quantized. The quantum of light energy is called a photon.	
Einstein equation for photon energy	$E = hf = \dfrac{hc}{\lambda}$	17-1
Planck's constant	$h = 6.626 \times 10^{-34}\ \text{J}\cdot\text{s} = 4.136 \times 10^{-15}\ \text{eV}\cdot\text{s}$	17-2
Einstein's photoelectric equation	$K_{max} = (\frac{1}{2}mv^2)_{max} = hf - \phi$ where ϕ is the work function of the cathode	17-3
hc	$hc = 1240\ \text{eV}\cdot\text{nm}$	17-5
Momentum of a photon	$p = \dfrac{h}{\lambda}$	17-7
Compton scattering equation	$\lambda_2 - \lambda_1 = \dfrac{h}{m_e c}(1 - \cos\theta)$	17-8
3. de Broglie Hypothesis	Electrons and other "particles" have wave properties.	
de Broglie wavelength	$\lambda = \dfrac{h}{p}$	17-10
de Broglie frequency	$f = \dfrac{E}{h}$	17-11
λ for nonrelativistic particles	$\lambda = \dfrac{hc}{\sqrt{2mc^2K}}$	17-12
λ for nonrelativistic electrons	$\lambda = \dfrac{1.23}{\sqrt{K}}\ \text{nm}$ (K in electron volts)	17-13

optional

4. Quantum Mechanics	The state of a particle such as an electron is described by its wave function ψ, which is the solution of the Schrödinger wave equation.
Probability density	The probability of finding the particle in some region of space dx is given by $$P(x)\,dx = \psi^2(x)\,dx \qquad \textbf{17-14}$$
Normalization condition	$$\int_{-\infty}^{\infty} \psi^2\,dx = 1 \qquad \textbf{17-15}$$
Quantum number	The wave function for a particular energy state is characterized by a quantum number n. In three dimensions there are three quantum numbers, one associated with each dimension.
Bohr correspondence principle	In the limit of very large quantum numbers, the classical calculation and the quantum calculation must yield the same results.
Expectation value	The expectation value of x is the average value of x that we would expect to obtain from a measurement of the positions of a large number of particles with the same wave function $\psi(x)$. $$\langle x \rangle = \int x\psi^2(x)\,dx \qquad \textbf{17-23}$$ $$\langle f(x) \rangle = \int f(x)\psi^2(x)\,dx \qquad \textbf{17-24}$$
5. Wave–Particle Duality	Light, electrons, neutrons, and all carriers of energy and momentum exhibit both wave and particle properties. Each propagates like a classical wave, exhibiting diffraction and interference, yet exchanges energy in discrete lumps like a classical particle. Because the wavelength of macroscopic objects is so small, diffraction and interference are not observed. Also, when a macroscopic amount of energy is exchanged, so many quanta are involved that the particle nature of the energy is not evident.
Uncertainty principle	The wave–particle duality of nature leads to the uncertainty principle, which states that the product of the uncertainty in a measurement of position and the uncertainty in a measurement of momentum must be greater than $\frac{1}{2}\hbar$. $$\Delta x\,\Delta p \geq \tfrac{1}{2}\hbar \quad \text{where } \hbar = h/2\pi \qquad \textbf{17-16}$$
6. Particle in a Box	
Allowed energy	$$E_n = \frac{p_n^2}{2m} = n^2\frac{h^2}{8mL^2} = n^2E_1 \qquad \textbf{17-18}$$
Ground-state energy	$$E_1 = \frac{h^2}{8mL^2} \qquad \textbf{17-19}$$
Wave function	$$\psi_n(x) = \sqrt{\frac{2}{L}}\sin\frac{n\pi x}{L} \qquad \textbf{17-22}$$
Transitions between energy states	A system in state of energy E_i can make a transition to a state of energy E_f by emitting or absorbing a photon of frequency f given by $$hf = \lvert E_i - E_f \rvert \qquad \textbf{17-20}$$ $$\lambda = \frac{c}{f} = \frac{hc}{\lvert E_i - E_f \rvert} \qquad \textbf{17-21}$$

7. Energy Quantization in Other Systems		
Harmonic oscillator	$E_n = (n + \frac{1}{2})hf_0, \quad n = 0, 1, 2, 3, \ldots$	**17-26**
Hydrogen atom	$E_n = -\dfrac{13.6\text{eV}}{n^2}, \quad n = 1, 2, 3, \ldots$	**17-27**

optional

Problem-Solving Guide

1. Begin by drawing a neat diagram that includes the important features of the problem.
2. Numerical calculations of the energies of photons or the wavelengths of electrons can often be simplified by using the combination $hc = 1240$ eV·nm.

Summary of Worked Examples

Type of Calculation	Procedure and Relevant Examples	
1. Photons		
Find the energy of a photon from its wavelength or the wavelength from its energy.	Use $E = hf = hc/\lambda$ with $hc = 1240$ eV·nm.	**Example 17-1**
Find the number of photons in a light beam.	The intensity gives the energy per second per unit area. The energy is Nhf.	**Example 17-2**
2. Matter Waves		
Find the de Broglie wavelength of an electron.	Use $\lambda = h/p = h/mv$.	**Example 17-4**
3. Probability		
Calculate the classical probability density.	Classically, $P(x)\,dx$ is proportional to the time spent in dx which is dx/v. If the particle is confined, the total probability of finding it in the confined region must be 1.	**Example 17-5**
Calculate the probability of finding a particle in some region of space Δx.	The probability of a particle being in dx is $\psi^2\,dx$. If Δx is very small, just replace dx with Δx. Otherwise integrate $\psi^2\,dx$ over the region of interest.	**Example 17-7**
Calculate the expectation value of x or $f(x)$ for a particular state.	Use $\langle f(x)\rangle = \int f(x)\psi^2(x)\,dx$, where $\psi(x)$ is the wave function for that state.	**Example 17-8**
4. Energy Quantization		
Find the energy levels for a particle in a box.	Use $E_n = n^2h^2/8mL^2 = n^2(hc)^2/8mc^2L^2$.	**Example 17-6**
Find the energy of a photon emitted by a system making a transition between two energy levels.	Use $\lambda = \dfrac{hc}{E_i - E_f}$	**Example 17-6**

optional

Problems

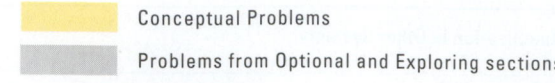

In a few problems, you are given more data than you actually need; in a few other problems, you are required to supply data from your general knowledge, outside sources, or informed estimates.

Conceptual Problems

Problems from Optional and Exploring sections

• Single-concept, single-step, relatively easy
•• Intermediate-level, may require synthesis of concepts
••• Challenging, for advanced students

Photons

1 • The quantized character of electromagnetic radiation is revealed by

(*a*) the Young double-slit experiment.
(*b*) diffraction of light by a small aperture.
(*c*) the photoelectric effect.
(*d*) the J. J. Thomson cathode-ray experiment.

2 •• Two monochromatic light sources, A and B, emit the same number of photons per second. The wavelength of A is $\lambda_A = 400$ nm, and that of B is $\lambda_B = 600$ nm. The power radiated by source B is

(*a*) equal to that of source A.
(*b*) less than that of source A.
(*c*) greater than that of source A.
(*d*) cannot be compared to that from source A using the available data.

3 • Find the photon energy in joules and in electron volts for an electromagnetic wave of frequency (*a*) 100 MHz in the FM radio band, and (*b*) 900 kHz in the AM radio band.

4 • An 80-kW FM transmitter operates at a frequency of 101.1 MHz. How many photons per second are emitted by the transmitter?

5 • What are the frequencies of photons having the following energies? (*a*) 1 eV, (*b*) 1 keV, and (*c*) 1 MeV.

6 • Find the photon energy for light of wavelength (*a*) 450 nm, (*b*) 550 nm, and (*c*) 650 nm.

7 • Find the photon energy if the wavelength is (*a*) 0.1 nm (about 1 atomic diameter), and (*b*) 1 fm (1 fm = 10^{-15} m, about 1 nuclear diameter).

8 •• The wavelength of light emitted by a 3-mW He-Ne laser is 632 nm. If the diameter of the laser beam is 1.0 mm, what is the density of photons in the beam?

The Photoelectric Effect

9 • True or false: In the photoelectric effect,

(*a*) the current is proportional to the intensity of the incident light.
(*b*) the work function of a metal depends on the frequency of the incident light.
(*c*) the maximum kinetic energy of electrons emitted varies linearly with the frequency of the incident light.
(*d*) the energy of a photon is proportional to its frequency.

10 • In the photoelectric effect, the number of electrons emitted per second is

(*a*) independent of the light intensity.
(*b*) proportional to the light intensity.
(*c*) proportional to the work function of the emitting surface.
(*d*) proportional to the frequency of the light.

11 • The work function of a surface is ϕ. The threshold wavelength for emission of photoelectrons from the surface is

(*a*) hc/ϕ.
(*b*) ϕ/hf.
(*c*) hf/ϕ.
(*d*) none of the above.

12 •• When light of wavelength λ_1 is incident on a certain photoelectric cathode, no electrons are emitted no matter how intense the incident light is. Yet when light of wavelength $\lambda_2 < \lambda_1$ is incident, electrons are emitted even when the incident light has low intensity. Explain.

13 • The work function for tungsten is 4.58 eV. (*a*) Find the threshold frequency and wavelength for the photoelectric effect. (*b*) Find the maximum kinetic energy of the electrons if the wavelength of the incident light is 200 nm, and (*c*) 250 nm.

14 • When light of wavelength 300 nm is incident on potassium, the emitted electrons have maximum kinetic energy of 2.03 eV. (*a*) What is the energy of an incident photon? (*b*) What is the work function for potassium? (*c*) What would be the maximum kinetic energy of the electrons if the incident light had a wavelength of 430 nm? (*d*) What is the threshold wavelength for the photoelectric effect with potassium?

15 • The threshold wavelength for the photoelectric effect for silver is 262 nm. (*a*) Find the work function for silver. (*b*) Find the maximum kinetic energy of the electrons if the incident radiation has a wavelength of 175 nm.

16 • The work function for cesium is 1.9 eV. (*a*) Find the threshold frequency and wavelength for the photoelectric effect. Find the maximum kinetic energy of the electrons if the wavelength of the incident light is (*b*) 250 nm, and (*c*) 350 nm.

17 •• When a surface is illuminated with light of wavelength 512 nm, the maximum kinetic energy of the emitted electrons is 0.54 eV. What is the maximum kinetic energy if the surface is illuminated with light of wavelength 365 nm?

Compton Scattering

18 • Find the shift in wavelength of photons scattered at $\theta = 60°$.

19 • When photons are scattered by electrons in carbon, the shift in wavelength is 0.33 pm. Find the scattering angle.

20 • The wavelength of Compton-scattered photons is measured at $\theta = 90°$. If $\Delta\lambda/\lambda$ is to be 1.5%, what should the wavelength of the incident photons be?

21 • Compton used photons of wavelength 0.0711 nm. (*a*) What is the energy of these photons? (*b*) What is the wavelength of the photon scattered at $\theta = 180°$? (*c*) What is the energy of the photon scattered at this angle?

22 • For the photons used by Compton, find the momentum of the incident photon and that of the photon scattered at 180°, and use the conservation of momentum to find the momentum of the recoil electron for this case (see Problem 21).

23 •• An X-ray photon of wavelength 6 pm that collides with an electron is scattered by an angle of 90°. (*a*) What is the change in wavelength of the photon? (*b*) What is the kinetic energy of the scattered electron?

24 •• How many head-on Compton scattering events are necessary to double the wavelength of a photon having initial wavelength 200 pm?

Matter Waves

25 • True or false:

(*a*) The de Broglie wavelength of an electron varies inversely with its momentum.
(*b*) Electrons can be diffracted.
(*c*) Neutrons can be diffracted.
(*d*) An electron microscope is used to look at electrons.

26 • If the de Broglie wavelength of an electron and a proton are equal, then

(*a*) the velocity of the proton is greater than that of the electron.
(*b*) the velocity of the proton and electron are equal.
(*c*) the velocity of the proton is less than that of the electron.
(*d*) the energy of the proton is greater than that of the electron.
(*e*) both (*a*) and (*d*) are correct.

27 • A proton and an electron have equal kinetic energies. It follows that the de Broglie wavelength of the proton is

(*a*) greater than that of the electron.
(*b*) equal to that of the electron.
(*c*) less than that of the electron.

28 • Use Equation 17-13 to calculate the de Broglie wavelength for an electron of kinetic energy (*a*) 2.5 eV, (*b*) 250 eV, (*c*) 2.5 keV, and (*d*) 25 keV.

29 • An electron is moving at $v = 2.5 \times 10^5$ m/s. Find its de Broglie wavelength.

30 • An electron has a wavelength of 200 nm. Find (*a*) its momentum, and (*b*) its kinetic energy.

31 • Find the energy of an electron in electron volts if its de Broglie wavelength is (*a*) 5 nm, and (*b*) 0.01 nm.

32 • A neutron in a reactor has kinetic energy of about 0.02 eV. Calculate the de Broglie wavelength of this neutron from Equation 17-12, where $mc^2 = 940$ MeV is the rest energy of the neutron.

33 • Use Equation 17-12 to find the de Broglie wavelength of a proton (rest energy $mc^2 = 938$ MeV) that has a kinetic energy of 2 MeV.

34 • A proton is moving at $v = 0.003c$, where c is the speed of light. Find its de Broglie wavelength.

35 • What is the kinetic energy of a proton whose de Broglie wavelength is (*a*) 1 nm, and (*b*) 1 fm?

36 • Find the de Broglie wavelength of a baseball of mass 0.145 kg moving at 30 m/s.

37 • The energy of the electron beam in Davisson and Germer's experiment was 54 eV. Calculate the wavelength for these electrons.

38 • The distance between Li^+ and Cl^- ions in a LiCl crystal is 0.257 nm. Find the energy of electrons that have a wavelength equal to this spacing.

39 • An electron microscope uses electrons of energy 70 keV. Find the wavelength of these electrons.

40 • What is the de Broglie wavelength of a neutron with speed 10^6 m/s?

Wave–Particle Duality

41 • Suppose you have a spherical object of mass 4 g moving at 100 m/s. What size aperture is necessary for the object to show diffraction? Show that no common objects would be small enough to squeeze through such an aperture.

42 • A neutron has a kinetic energy of 10 MeV. What size object is necessary to observe neutron diffraction effects? Is there anything in nature of this size that could serve as a target to demonstrate the wave nature of 10-MeV neutrons?

43 • What is the de Broglie wavelength of an electron of kinetic energy 200 eV? What are some common targets that could demonstrate the wave nature of such an electron?

Particle in a Box

44 •• Sketch the wave function $\psi(x)$ and the probability distribution $\psi^2(x)$ for the state $n = 4$ of a particle in a box.

45 •• (*a*) Find the energy of the ground state ($n = 1$) and the first two excited states of a proton in a one-dimensional box of length $L = 10^{-15}$ m = 1 fm. (These are of the order of magnitude of nuclear energies.) Make an energy-level diagram for this system and calculate the wavelength of electromagnetic radiation emitted when the proton makes a transition from (*b*) $n = 2$ to $n = 1$, (*c*) $n = 3$ to $n = 2$, and (*d*) $n = 3$ to $n = 1$.

46 •• (*a*) Find the energy of the ground state ($n = 1$) and the first two excited states of a proton in a one-dimensional box of length 0.2 nm (about the diameter of a H_2 molecule). Calculate the wavelength of electromagnetic radiation emitted when the proton makes a transition from (*b*) $n = 2$ to $n = 1$, (*c*) $n = 3$ to $n = 2$, and (*d*) $n = 3$ to $n = 1$.

optional

47 •• (*a*) Find the energy of the ground state and the first two excited states of a small particle of mass 1 μg confined to a one-dimensional box of length 1 cm. (*b*) If the particle moves with a speed of 1 mm/s, calculate its kinetic energy and find the approximate value of the quantum number n.

Calculating Probabilities and Expectation Values

48 •• A particle is in the ground state of a box of length L. Find the probability of finding the particle in the interval $\Delta x = 0.002L$ at (*a*) $x = L/2$, (*b*) $x = 2L/3$, and (*c*) $x = L$. (Since Δx is very small, you need not do any integration because the wave function is slowly varying.)

49 •• Do Problem 48 for a particle in the first excited state ($n = 2$).

50 •• Do Problem 48 for a particle in the second excited state ($n = 3$).

51 •• The classical probability distribution function for a particle in a box of length L is given by $P(x) = 1/L$. Use this to find $\langle x \rangle$ and $\langle x^2 \rangle$ for a classical particle in such a box.

52 •• (*a*) Find $\langle x \rangle$ for the first excited state ($n = 2$) for a particle in a box of length L, and (*b*) find $\langle x^2 \rangle$.

53 •• (*a*) Find $\langle x \rangle$ for the second excited state ($n = 3$) for a particle in a box of length L, and (*b*) find $\langle x^2 \rangle$.

54 •• A particle in a one-dimensional box is in the first excited state ($n = 2$). (*a*) Sketch $\psi^2(x)$ versus x for this state. (*b*) What is the expectation value $\langle x \rangle$ for this state? (*c*) What is the probability of finding the particle in some small region dx centered at $x = \frac{1}{2}L$? (*d*) Are your answers for (*b*) and (*c*) contradictory? If not, explain.

55 •• A particle of mass m has a wave function given by $\psi(x) = Ae^{-|x|/a}$, where A and a are constants. (*a*) Find the normalization constant A. (*b*) Calculate the probability of finding the particle in the region $-a \leq x \leq a$.

56 •• A particle in a one-dimensional box of length L is in its ground state. Calculate the probability that the particle will be found in the region (*a*) $0 < x < \frac{1}{2}L$, (*b*) $0 < x < \frac{1}{3}L$, and (*c*) $0 < x < \frac{3}{4}L$.

57 •• Repeat Problem 56 for a particle in the first excited state of the box.

58 •• (*a*) For the wave functions

$$\psi_n(x) = \sqrt{\frac{2}{L}} \sin \frac{n\pi x}{L}, \qquad n = 1, 2, 3, \ldots$$

corresponding to a particle in the nth state of a one-dimensional box of length L, show that

$$\langle x^2 \rangle = \frac{L^2}{3} - \frac{L^2}{2n^2\pi^2}$$

(*b*) Compare this result for $n >> 1$ with your answer for the classical distribution of Problem 51.

59 •• The wave functions for a particle of mass m in a one-dimensional box of length L *centered at the origin* (so that the ends are at $x = \pm L/2$) are given by

$$\psi_n(x) = \sqrt{\frac{2}{L}} \cos \frac{n\pi x}{L}, \qquad n = 1, 3, 5, 7, \ldots$$

and

$$\psi_n(x) = \sqrt{\frac{2}{L}} \sin \frac{n\pi x}{L}, \qquad n = 2, 4, 6, 8, \ldots$$

Calculate $\langle x \rangle$ and $\langle x^2 \rangle$ for the ground state.

60 •• Calculate $\langle x \rangle$ and $\langle x^2 \rangle$ for the first excited state of the box described in Problem 59.

General Problems

61 • Can the expectation value of x ever equal a value that has zero probability of being measured?

62 • Explain why the maximum kinetic energy of electrons emitted in the photoelectric effect does not depend on the intensity of the incident light, but the total number of electrons emitted does.

63 •• A six-sided die has the number 1 painted on three sides and the number 2 painted on the other three sides. (*a*) What is the probability of a 1 coming up when the die is thrown? (*b*) What is the expectation value of the number that comes up when the die is thrown?

64 •• True or false:

(*a*) It is impossible in principle to know precisely the position of an electron.
(*b*) A particle that is confined to some region of space cannot have zero energy.
(*c*) All phenomena in nature are adequately described by classical wave theory.
(*d*) The expectation value of a quantity is the value that you expect to measure.

65 •• It was once believed that if two identical experiments are done on identical systems under the same conditions, the results must be identical. Explain why this is not true, and how it can be modified so that it is consistent with quantum physics.

66 • A light beam of wavelength 400 nm has an intensity of 100 W/m^2. (*a*) What is the energy of each photon in the beam? (*b*) How much energy strikes an area of 1 cm^2 perpendicular to the beam in 1 s? (*c*) How many photons strike this area in 1 s?

67 • A mass of 10^{-6} g is moving with a speed of about 10^{-1} cm/s in a box of length 1 cm. Treating this as a one-dimensional particle in a box, calculate the approximate value of the quantum number n.

68 • (*a*) For the classical particle of Problem 67, find Δx and Δp, assuming that these uncertainties are given by $\Delta x/L = 0.01\%$ and $\Delta p/p = 0.01\%$. (*b*) What is $(\Delta x\, \Delta p)/\hbar$?

69 • In 1987, a laser at Los Alamos National Laboratory produced a flash that lasted 1.0×10^{-12} s and had a power of 5.0×10^{15} W. Estimate the number of emitted photons if their wavelength was 400 nm.

70 • You can't see anything smaller than the wavelength λ used. What is the minimum energy of an electron needed in an electron microscope to see an atom, which has a diameter of about 0.1 nm?

71 • A common flea that has a mass of 0.008 g can jump vertically as high as 20 cm. Estimate the de Broglie wavelength for the flea immediately after takeoff.

72 • The work function for sodium is $\phi = 2.3$ eV. Find the minimum de Broglie wavelength for the electrons emitted by a sodium cathode illuminated by violet light with a wavelength of 420 nm.

73 •• Suppose that a 100-W source radiates light of wavelength 600 nm uniformly in all directions and that the eye can detect this light if only 20 photons per second enter a dark-adapted eye having a pupil 7 mm in diameter. How far from the source can the light be detected under these rather extreme conditions?

74 •• Data for maximum kinetic energy of the electrons versus wavelength for the photoelectric effect using sodium are

λ, nm	200	300	400	500	600
K_{max}, eV	4.20	2.06	1.05	0.41	0.03

Plot these data so as to obtain a straight line and from your plot find (*a*) the work function, (*b*) the threshold frequency, and (*c*) the ratio h/e.

75 •• The diameter of the pupil of the eye under room-light conditions is about 5 mm. (It can vary from about 1 to 8 mm.) Find the intensity of light of wavelength 600 nm such that 1 photon per second passes through the pupil.

76 •• A light bulb radiates 90 W uniformly in all directions. (*a*) Find the intensity at a distance of 1.5 m. (*b*) If the wavelength is 650 nm, find the number of photons per second that strike a surface of area 1 cm^2 oriented so that the line to the bulb is perpendicular to the surface.

77 •• When light of wavelength λ_1 is incident on the cathode of a photoelectric tube, the maximum kinetic energy of the emitted electrons is 1.8 eV. If the wavelength is reduced to $\lambda_1/2$, the maximum kinetic energy of the emitted electrons is 5.5 eV. Find the work function ϕ of the cathode material.

78 •• A photon of energy E is scattered at an angle of θ. Show that the energy E' of the scattered photon is given by

$$E' = \frac{E}{(E/m_ec^2)(1 - \cos\theta) + 1}$$

79 •• A particle is confined to a one-dimensional box. In making a transition from the state n to the state $n - 1$, radiation of 114.8 nm is emitted; in the transition from the state $n - 1$ to the state $n - 2$, radiation of wavelength 147.6 nm is emitted. The ground-state energy of the particle is 1.2 eV. Determine n.

80 •• A particle confined to a one-dimensional box has a ground-state energy of 0.4 eV. When irradiated with light of 206.7 nm it makes a transition to an excited state. When decaying from this excited state to the next lower state it emits radiation of 442.9 nm. What is the quantum number of the state to which the particle has decayed?

81 •• When a surface is illuminated with light of wavelength λ the maximum kinetic energy of the emitted electrons is 1.2 eV. If wavelength $\lambda' = 0.8\,\lambda$ is used the maximum kinetic energy increases to 1.76 eV, and for wavelength $\lambda'' = 0.6\lambda$ the maximum kinetic energy of the emitted electrons is 2.7 eV. Determine the work function of the surface and the wavelength λ.

82 •• A simple pendulum of length 1 m has a bob of mass 0.3 kg. The energy of this oscillator is quantized to the values $E_n = (n + \frac{1}{2})hf_0$, where n is an integer and f_0 is the frequency of the pendulum. (*a*) Find n if the angular amplitude is 10°. (*b*) Find Δn if the energy changes by 0.01%.

83 •• (*a*) Show that for large n, the fractional difference in energy between state n and state $n + 1$ for a particle in a box is given approximately by

$$\frac{E_{n+1} - E_n}{E_n} \approx \frac{2}{n}$$

(*b*) What is the approximate percentage energy difference between the states $n_1 = 1000$ and $n_2 = 1001$? (*c*) Comment on how this result is related to Bohr's correspondence principle.

84 •• In 1985, a light pulse of 1.8×10^{12} photons was produced in an AT&T laboratory during a time interval of 8×10^{-15} s. The wavelength of the produced light was $\lambda = 2400$ nm. Suppose all of the light was absorbed by the black surface of a screen. Estimate the force exerted by the photons on the screen.

85 •• This problem is one of estimating the time lag (expected classically but not observed) in the photoelectric effect. Let the intensity of the incident radiation be 0.01 W/m^2. (*a*) If the area of the atom is 0.01 nm^2, find the energy per second falling on an atom. (*b*) If the work function is 2 eV, how long would it take classically for this much energy to fall on one atom?

PART III

Two red intersecting laser beams are used here to study the combustion of coal–water slurries in a conventional power conversion device. The test material is injected into the combustion reactor (blue flame), giving off a yellow-orange emission as it ignites and burns. The laser light is used to measure the particle size of combustible material.

thermodynamics

CHAPTER 18

Temperature and the Kinetic Theory of Gases

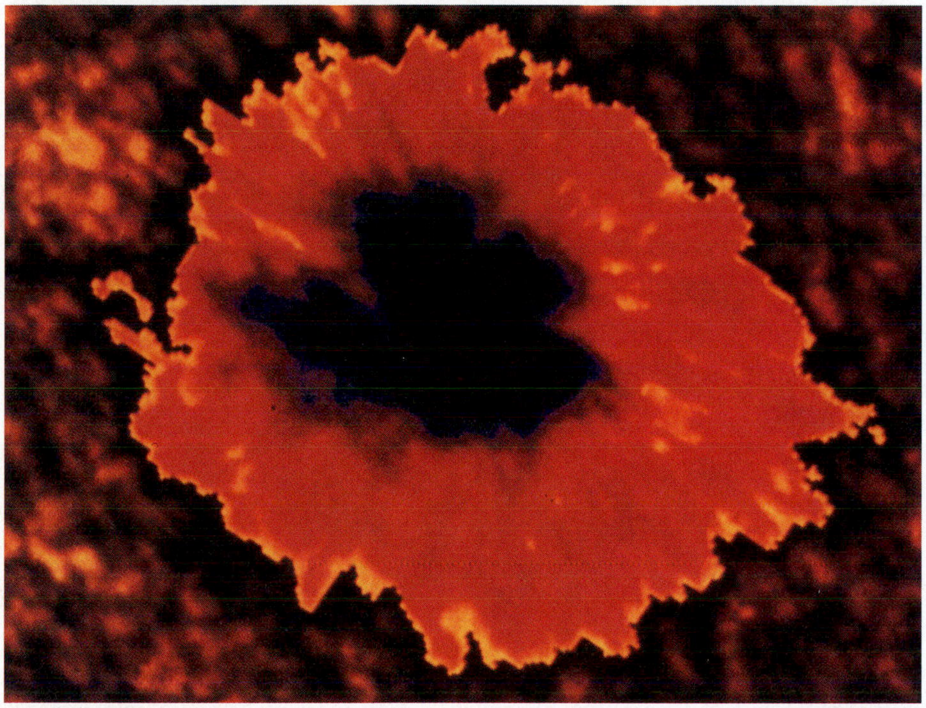

Sunspots appear on the surface of the sun when streams of gases slowly erupt from deep within the star. The solar "flower" is 10,000 miles in diameter. The temperature variation, indicated by computer-enhanced color changes, is not fully understood. The central portion of the sunspot is cooler than the outer regions as indicated by the dark area. The temperature at the sun's core is of the order of 10^7 K, whereas at the surface the temperature is only about 6000 K.

Temperature is familiar to us all as the measure of the hotness or coldness of objects. In this chapter we show that a consistent temperature scale can be defined in terms of the properties of gases at low densities, and that temperature is a measure of the average internal molecular kinetic energy of an object.

18-1 Thermal Equilibrium and Temperature

Our sense of touch can usually tell us if an object is hot or cold. Early in childhood we learn that to make a cold object warmer, we place it in contact with a hot object. To make a warm object cool, we place it in contact with a cold object.

When an object is heated or cooled, some of its physical properties change. Most solids and liquids expand when they are heated. A gas, if permitted, will also expand when it is heated, or, if its volume is kept constant, its pressure will rise. If an electrical conductor is heated, its electrical resistance changes. A physical property that changes with temperature is called a **thermometric property.** A change in a thermometric property indicates a change in the temperature of the object.

Suppose we place a warm copper bar in close contact with a cold iron bar so that the copper bar cools and the iron bar warms. We say the two bars are in **thermal contact**. The copper bar contracts slightly as it cools, and the iron bar expands slightly as it warms. Eventually this process stops and the lengths of the bars remain constant. The two bars are then in **thermal equilibrium** with each other.

Suppose instead we place the warm copper bar in a cool lake. The bar cools until it and the water are in thermal equilibrium. (We assume the lake is large enough so that the warming of its water will be negligible.) Next we place a cold iron bar in the lake far away from the copper bar. The iron bar will warm until it and the lake water are also in thermal equilibrium. If we remove the bars and place them in thermal contact with each other, we find their lengths do not change. They are in thermal equilibrium with each other. Though it is common sense, there is no logical way to deduce this fact, which is called the **zeroth law of thermodynamics** (Figure 18-1):

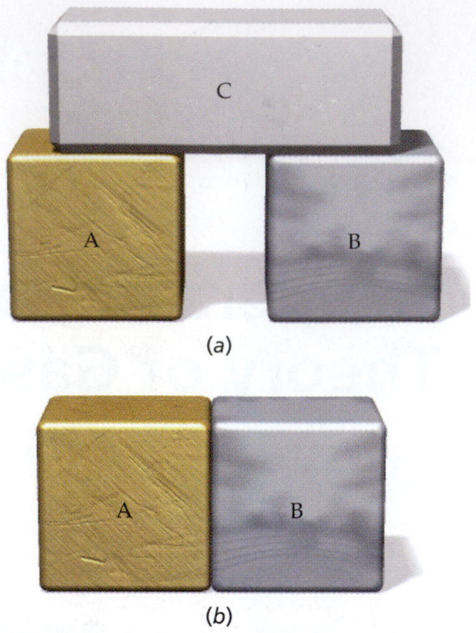

Figure 18-1 The zeroth law of thermodynamics. (*a*) Systems A and B are in thermal contact with system C but not with each other. When A and B are each in thermal equilibrium with C, they are in thermal equilibrium with each other, which can be checked by placing them in contact with each other as in (*b*).

> If two objects are in thermal equilibrium with a third, then they are in thermal equilibrium with each other.

Zeroth law of thermodynamics

Two objects are defined to have the **same temperature** if they are in thermal equilibrium with each other. The zeroth law, as we will see, enables us to define a temperature scale.

18-2 The Celsius and Fahrenheit Temperature Scales

Any thermometric property can be used to establish a temperature scale. The common mercury thermometer consists of a glass bulb and tube containing a fixed amount of mercury. When this thermometer is put in contact with a warmer body, the mercury expands, increasing the length of the mercury column (the glass expands too, but by a negligible amount). We can create a scale along the glass tube as follows. First the thermometer is placed in ice and water in equilibrium* at a pressure of 1 atm. When the thermometer is in thermal equilibrium with the ice water, the position of the mercury column is marked on the glass tube. This is the **ice-point temperature** (also called the **normal freezing point** of water). Next, the thermometer is placed in boiling water at a pressure of 1 atm. When the thermometer is in thermal equilibrium with the boiling water, the new position of the column is marked. This is the **steam-point temperature** (also called the **normal boiling point** of water).

The **Celsius temperature scale** defines the ice-point temperature as zero degrees Celsius (0°C) and the steam-point temperature as 100°C. The space between the 0 and 100° marks is divided into 100 equal intervals (degrees). Degree markings are also extended below and above these points. If L_t is the length of the mercury column, the Celsius temperature t_C is given by

$$t_C = \frac{L_t - L_0}{L_{100} - L_0} \times 100° \qquad \text{18-1}$$

where L_0 is the length of the mercury column when the thermometer is in an ice bath and L_{100} is its length when the thermometer is in a steam bath. The

*Water and ice in equilibrium provide a constant-temperature bath. When ice is placed in warm water, the water cools as some of the ice melts. Eventually, thermal equilibrium is reached and no more ice melts. If the system is heated slightly, some more of the ice melts, but the temperature does not change as long as some ice remains.

normal temperature of the human body measured on the Celsius scale is about 37°C.

The **Fahrenheit temperature scale** (which is used in the United States) defines the ice-point temperature as 32°F and the steam-point temperature as 212°F.* To convert temperatures between Fahrenheit and Celsius, we note there are 100 Celsius degrees and 180 Fahrenheit degrees between the ice and steam points. A temperature change of one Celsius degree therefore equals a change of $1.8 = \frac{9}{5}$ Fahrenheit degrees. To convert a temperature from one scale to the other, we must also take into account the fact that the zero temperatures of the two scales are not the same. The general relation between a Fahrenheit temperature t_F and Celsius temperature t_C is

$$t_C = \tfrac{5}{9}(t_F - 32°) \qquad \text{18-2}$$

Fahrenheit–Celsius conversion

Example 18-1

(*a*) Find the temperature on the Celsius scale equivalent to 41°F. (*b*) Find the temperature on the Fahrenheit scale equivalent to −10°C.

(*a*) Apply Equation 18-2 with $t_F = 41°F$: $t_C = \frac{5}{9}(t_F - 32°) = \frac{5}{9}(41° - 32°) = \frac{5}{9}(9°) = 5°C$

(*b*)1. Solve Equation 18-2 for t_F in terms of t_C: $t_F = \frac{9}{5}t_C + 32°$

2. Substitute $t_C = -10°C$: $t_F = \frac{9}{5}(-10°) + 32° = -18° + 32° = 14°F$

Exercise (*a*) Find the Celsius temperature equivalent to 68°F. (*b*) Find the Fahrenheit temperature equivalent to −40°C. (*Answers* (*a*) 20°C, (*b*) −40°F)

Other thermometric properties can be used to set up thermometers and construct temperature scales. Figure 18-2 shows a bimetallic strip consisting of two different metals bonded together. When the strip is heated or cooled, it bends to accommodate the difference in the thermal expansion of the two metals. Figure 18-3 shows a thermometer consisting of a bimetallic coil with

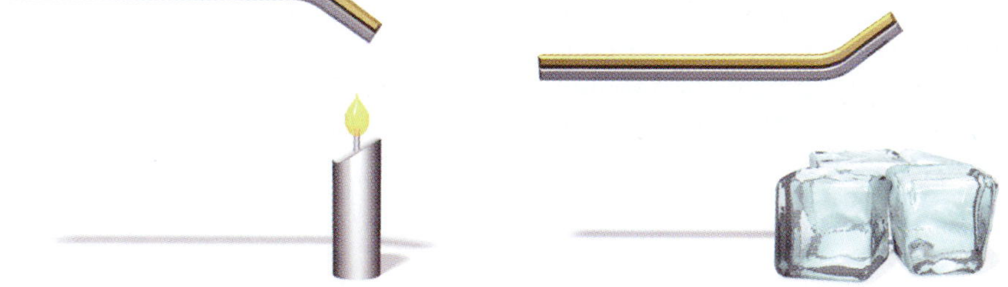

Figure 18-2 A bimetallic strip. When heated or cooled the two metals expand or contract by different amounts, causing the strip to bend.

* When the German physicist Daniel Fahrenheit devised his temperature scale, he wanted all measurable temperatures to be positive. Originally, he chose 0°F for the coldest temperature he could obtain with a mixture of ice and saltwater, and 96°F (a convenient number with many factors for subdivision) for the temperature of the human body. He then modified his scale slightly to make the ice-point and steam-point temperatures whole numbers. This resulted in the average temperature of the human body being between 98 and 99°F.

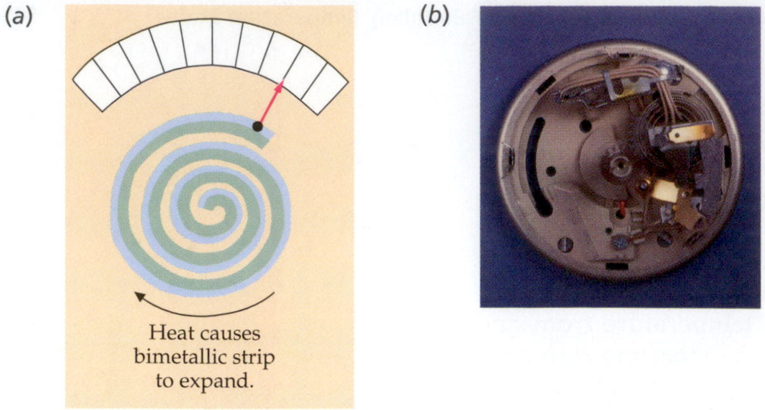

Figure 18-3 (*a*) A thermometer using a bimetallic strip in the form of a coil. When the coil is heated or cooled, the two different metals expand or shrink at different rates. (*b*) A home thermostat. The coil on the right controls the air conditioner. When the room gets warmer, the coil expands, the tube mounted on it tilts, and mercury in the tube slides to close an electrical switch, turning on the air conditioning. A slide lever (at the lower right) mounted on the coil is used to set the desired temperature. The circuit will be broken when the cooler air causes the bimetallic coil to contract.

a pointer attached to indicate the temperature. When the thermometer is heated, the coil bends and the pointer moves to the right. Like mercury thermometers, it is calibrated by dividing the interval between the ice point and steam point into 100 Celsius degrees (or 180 Fahrenheit degrees).

18-3 Gas Thermometers and the Absolute Temperature Scale

When different types of thermometers are calibrated in ice water and steam, they agree (by definition) at 0°C and at 100°C, but they give slightly different readings at points in between. Discrepancies increase markedly above the steam point and below the ice point. However, in one group of thermometers, gas thermometers, the measured temperatures agree closely even far from the calibration points. In a **constant-volume gas thermometer**, the gas volume is kept constant, and change in gas pressure is used to measure a change in temperature (Figure 18-4). An ice-point pressure P_0 and steam-point pressure P_{100} are determined by placing the thermometer in ice–water and water–steam baths, and the interval between is divided into 100 equal degrees (for the Celsius scale). If the pressure is P_t in a bath whose temperature is to be determined, that temperature in degrees Celsius is defined to be

$$t_C = \frac{P_t - P_0}{P_{100} - P_0} \times 100° \qquad 18\text{-}3$$

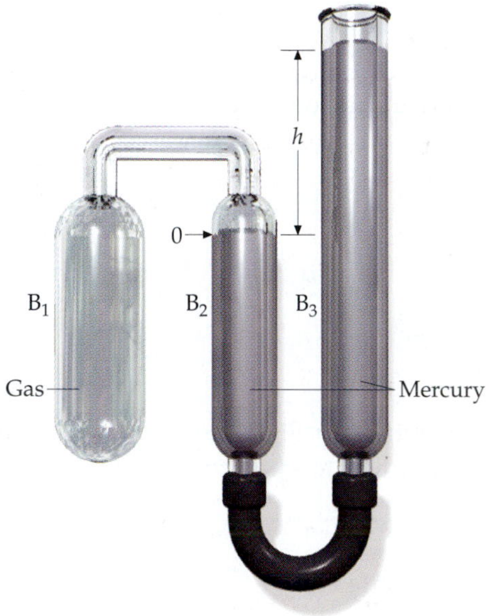

Figure 18-4 A constant-volume gas thermometer. The volume is kept constant by raising or lowering tube B_3 so that the mercury in tube B_2 remains at the zero mark. The temperature is chosen to be proportional to the pressure of the gas in tube B_1, which is indicated by the height *h* of the mercury column in tube B_3.

Figure 18-5 shows the results of measurements of the boiling point of sulfur using constant-volume gas thermometers filled with various gases. The measured temperature is plotted as a function of the steam-point pressure P_{100}, which is varied by changing the amount of gas in the thermometer. As the amount of gas is reduced, its density and the steam-point pressure both decrease. We see that agreement among the thermometers is very close at low gas densities. In the limit as gas density goes to zero, all gas thermometers give the same value for any temperature. Because this temperature measurement is independent of the properties of any particular gas, low-density gas thermometers can be used to define temperature.

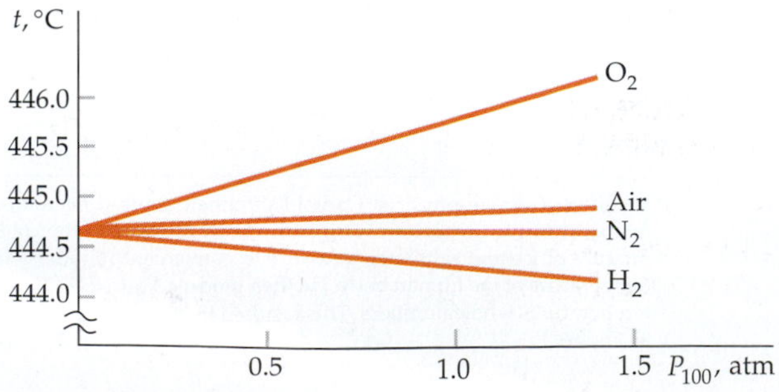

Figure 18-5 Temperature of the boiling point of sulfur measured with constant-volume gas thermometers filled with various gases. The pressure of the steam point of water, P_{100}, is varied by varying the amount of gas in the thermometers. As the amount of gas is reduced, the temperatures measured by all the thermometers approach the value 444.60°C.

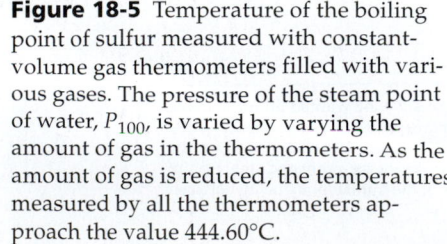

Let us now consider a series of temperature measurements with a constant-volume gas thermometer that has a very small but fixed amount of gas. According to Equation 18-3, the pressure P_t in the thermometer varies linearly with the measured temperature t_C. Figure 18-6 shows a plot of measured temperature versus pressure in a constant-volume gas thermometer. When we extrapolate this straight line to zero pressure, the temperature approaches −273.15°C. This limit is the same no matter what kind of gas is used.

A reference state that is much more precisely reproducible than either the ice or steam points is the **triple point of water**—the unique temperature and pressure at which water, water vapor, and ice coexist in equilibrium. This equilibrium state occurs at 4.58 mmHg and 0.01°C. The **ideal-gas temperature scale** is defined so that the temperature of the triple point is 273.16 kelvins (K).* The temperature T of any other state is defined to be proportional to the pressure in a constant-volume gas thermometer:

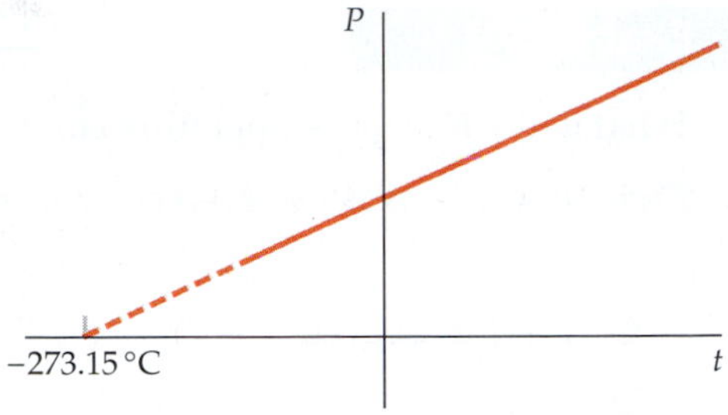

Figure 18-6 Plot of pressure versus temperature as measured by a constant-volume gas thermometer. When extrapolated to zero pressure, the plot intersects the temperature axis at the value −273.15°C.

$$T = \frac{273.16\text{ K}}{P_3} P \qquad \text{18-4}$$

Ideal-gas temperature scale

where P is the observed pressure of the gas in the thermometer, and P_3 is the pressure when the thermometer is immersed in a water–ice–vapor bath at its triple point. The value of P_3 depends on the amount of gas in the thermometer. The ideal-gas temperature scale, defined by Equation 18-4, has the advantage that any measured temperature does not depend on the properties of the particular gas that is used, but depends only on the general properties of gases.

The lowest temperature that can be measured with a gas thermometer is about 1 K, and requires helium for the gas. Below this temperature helium liquefies; all other gases liquefy at higher temperatures. In Chapter 20, we will see that the second law of thermodynamics can be used to define the **absolute temperature scale** independent of the properties of any substance, and with no limitations on the range of temperatures that can be measured. Temperatures as low as a millionth of a kelvin have been measured. The absolute scale so defined is identical to that defined by Equation 18-4 for the range of temperatures for which gas thermometers can be used. The symbol T is used when referring to absolute temperature.

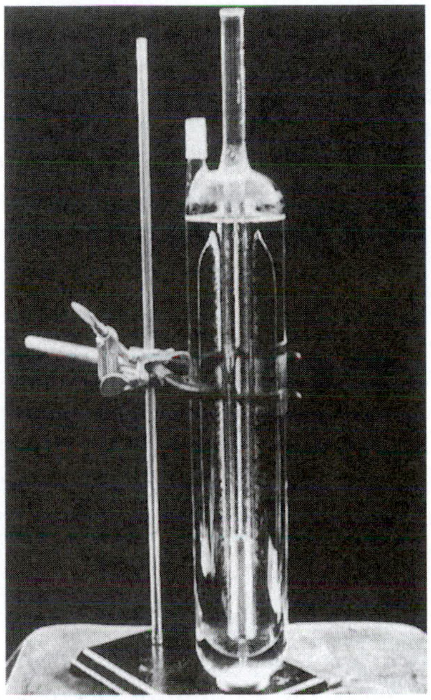

H_2O at its triple point. The container is a hollow cylindrical shell that is sealed and evacuated. It contains water, ice, and water vapor in equilibrium. The cylindrical well in the center is filled with water and contains an aluminum bushing for inserting a thermometer. The container, stored at the National Institute of Standards and Technology, is used in an ice-water bath that is just slightly below the triple-point temperature so that the water in the cell freezes very slowly.

Because the Celsius degree and the kelvin are the same size, temperature *differences* are the same on both the Celsius and the absolute temperature scale (also called the **Kelvin scale**). That is, a temperature *change* of 1 K is identical to a temperature *change* of 1 C°.† The two scales differ only in the choice of zero temperature. To convert from degrees Celsius to kelvins, we merely add 273.15:‡

$$T = t_C + 273.15\text{ K} \qquad \text{18-5}$$

Celsius–Kelvin conversion

Although the Celsius and Fahrenheit scales are convenient for everyday use, the absolute scale is much more convenient for scientific purposes, partly because many formulas are more simply expressed in it, and partly because the absolute temperature can be given a more fundamental interpretation.

---

* The kelvin is a degree unit that is the same size as the Celsius degree.

† We write 1 C° to indicate a *temperature change* of one Celsius degree, in contrast to 1°C, which means a temperature of one degree Celsius.

‡ For most purposes, we can round off the temperature of absolute zero to −273°C.

Example 18-2

What is the Kelvin temperature corresponding to 70°F?

Picture the Problem First convert to degrees Celsius, then to kelvins.

1. Convert to degrees Celsius: $t_C = \frac{5}{9}(70° - 32°) = 21.1°C$

2. To find the Kelvin temperature we add 273: $T = t_C + 273 = 21.1 + 273 = 294\ K$

Exercise The "high-temperature" superconductor $YBa_2Cu_3O_7$ becomes superconducting when the temperature is lowered to 92 K. Find this superconducting temperature in degrees Fahrenheit. (*Answer* −294°F)

18-4 The Ideal-Gas Law

The properties of gases at low densities allow the definition of the ideal-gas temperature scale. If we compress such a gas while keeping its temperature constant, the pressure increases. Similarly, if a gas expands at constant temperature, its pressure decreases. To a good approximation, the product of the pressure and volume of a low-density gas is constant at a constant temperature. This result was discovered experimentally by Robert Boyle (1627–1691), and is known as **Boyle's law:**

$$PV = \text{constant} \qquad \text{(constant temperature)}$$

A more general law exists that reproduces Boyle's law as a special case. According to Equation 18-4, the absolute temperature of a low-density gas is proportional to its pressure at constant volume. In addition—a result discovered experimentally by Jacques Charles (1746–1823) and Gay–Lussac (1778–1850)—the absolute temperature of a low-density gas is proportional to its volume at constant pressure. We can combine these two results by stating

$$PV = CT \tag{18-6}$$

where C is a constant of proportionality. We can see that this constant is proportional to the amount of gas by considering combining two identical containers, each holding the same volume of the same kind of gas at the same temperature. Then twice the amount of gas occupies twice the volume at the same pressure P and temperature T. We therefore write C as a constant k times the number of molecules in the gas N:

$$C = kN$$

Equation 18-6 then becomes

$$PV = NkT \tag{18-7}$$

The constant k is called **Boltzmann's constant.** It is found experimentally to have the same value for any kind of gas:

$$k = 1.381 \times 10^{-23}\ \text{J/K} = 8.617 \times 10^{-5}\ \text{eV/K} \tag{18-8}$$

An amount of gas is often expressed in moles. A **mole** (mol) of any substance is the amount of that substance that contains Avogadro's number N_A

of atoms or molecules, defined as the number of carbon atoms in 12 g of ^{12}C:

$$N_A = 6.022 \times 10^{23} \qquad 18\text{-}9$$

Avogadro's number

If we have n moles of a substance, the number of molecules is then

$$N = nN_A \qquad 18\text{-}10$$

Equation 18-7 is then

$$PV = nN_AkT = nRT \qquad 18\text{-}11$$

where $R = N_Ak$ is called the **universal gas constant**. Its value, which is the same for all gases, is

$$R = N_Ak = 8.314\ \text{J/mol}\cdot\text{K} = 0.08206\ \text{L}\cdot\text{atm/mol}\cdot\text{K} \qquad 18\text{-}12$$

Figure 18-7 shows plots of PV/nT versus the pressure P for several gases. For all gases, PV/nT is nearly constant over a large range of pressures. Even oxygen, which varies the most in this graph, changes by only about 1% between 0 and 5 atm. An **ideal gas** is defined as one for which PV/nT is constant for all pressures. The pressure, volume, and temperature of an ideal gas are related by

$$PV = nRT \qquad 18\text{-}13$$

Ideal-gas law

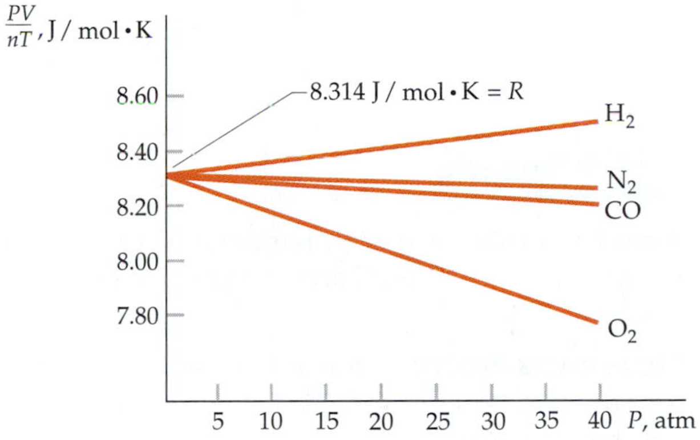

Figure 18-7 Plot of PV/nT versus P for real gases. In these plots, the pressure is varied by varying the amount of gas. As we reduce the density of the gas and thereby the pressure, the ratio PV/nT approaches the same value, 8.314 J/mol·K, for all gases. This value is the universal gas constant R.

The variables P, V, and T describe the **macroscopic state** of a gas at a given instant. (The **microscopic state** of the gas would be described by giving the position and velocity of each of the molecules in the gas.) An equation such as Equation 18-13, which relates these **macroscopic state variables**, is called an **equation of state**. For any gas at any density, there is an equation of state relating P, V, and T for a given amount of gas. Thus, the macroscopic state of a given amount of gas is determined by any two of the three state variables P, V, and T. Equation 18-13 describes the properties of real gases with low densities (and therefore low pressures). At higher densities, corrections must be made to Equation 18-13. In Chapter 21 we discuss another equation of state, the van der Waals equation, that includes such corrections.

Example 18-3

What volume is occupied by 1 mol of gas at a temperature of 0°C and a pressure of 1 atm?

We can find the volume using the ideal-gas law, with $T = 273$ K:

$$V = \frac{nRT}{P} = \frac{(1\ \text{mol})(0.0821\ \text{L}\cdot\text{atm/mol}\cdot\text{K})(273\ \text{K})}{1\ \text{atm}} = 22.4\ \text{L}$$

Remarks Note that by writing R in L·atm/mol, we could write P in atmospheres to get V in liters.

Exercise Find (*a*) the number of moles, n, and (*b*) the number of molecules, N, in 1 cm^3 of a gas at 0°C and 1 atm. (*Answers* (*a*) $n = 4.46 \times 10^{-5}$ mol, (*b*) $N = 2.68 \times 10^{19}$ molecules)

The temperature of 0°C (273 K) and the pressure of 1 atm are often referred to as **standard conditions.** We see from Example 18-3 that under standard conditions, 1 mol of an ideal gas occupies a volume of 22.4 L.

Figure 18-8 shows plots of P versus V at several constant temperatures T. These curves are called **isotherms**. The isotherms for an ideal gas are hyperbolas. For a fixed amount of gas, we can see from Equation 18-13 that the quantity PV/T is constant. Using the subscripts 1 for the initial values and 2 for the final values, we have

$$\frac{P_2V_2}{T_2} = \frac{P_1V_1}{T_1} \qquad \text{18-14}$$

Ideal-gas law for fixed amount of gas

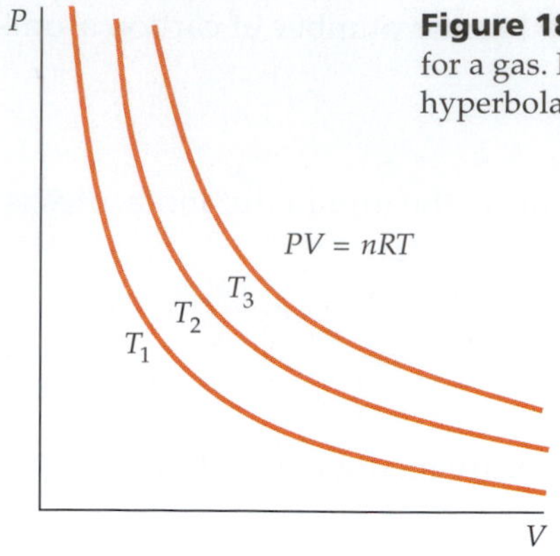

Figure 18-8 Isotherms on the PV diagram for a gas. For an ideal gas, these curves are hyperbolas given by $PV = nRT$.

Example 18-4

A gas has a volume of 2 L, a temperature of 30°C, and a pressure of 1 atm. When the gas is heated to 60°C and compressed to a volume of 1.5 L, what is its new pressure?

Picture the Problem Since the amount of gas is fixed, the pressure can be found using Equation 18-14. Let subscripts 1 and 2 refer to the initial and final state, respectively.

1. Express the pressure P_2 in terms of P_1 and the initial and final volumes and temperatures:

$$\frac{P_1V_1}{T_1} = \frac{P_2V_2}{T_2}$$

$$P_2 = \frac{T_2V_1}{T_1V_2}P_1$$

2. Calculate the initial and final absolute temperatures:

$$T_1 = 273 + 30 = 303\text{ K}$$

$$T_2 = 273 + 60 = 333\text{ K}$$

3. Substitute numerical values in step 1 to find P_2:

$$P_2 = \frac{(333\text{ K})(2\text{ L})}{(303\text{ K})(1.5\text{ L})}(1\text{ atm}) = 1.47\text{ atm}$$

Exercise How many moles of gas are in the system described in this example? (*Answer* $n = 0.0804$ mol)

The mass per mole of a substance is called its **molar mass** M.* The molar mass of ^{12}C is, by definition, 12 g/mol or 12×10^{-3} kg/mol. Molar masses of the elements are given in the periodic table in Appendix C. The molar mass of a molecule, such as CO_2, is the sum of the molar masses of the elements in the molecule. Since the molar mass of oxygen is 16 g/mol (actually 15.999 g/mol), the molar mass of O_2 is 32 g/mol and that of CO_2 is 12 + 32 = 44 g/mol.

* The terms *molecular weight* and *molecular mass* are also sometimes used.

The mass of n moles of gas is given by

$$m = nM$$

and the density ρ of an ideal gas is

$$\rho = \frac{m}{V} = \frac{nM}{V}$$

Using $n/V = P/RT$ from Equation 18-13, we have

$$\rho = \frac{M}{RT}P \qquad \text{18-15}$$

Density of an ideal gas

At a given temperature, the density of an ideal gas is proportional to its pressure.

Example 18-5

The molar mass of hydrogen is 1.008 g/mol. What is the mass of one hydrogen atom?

Picture the Problem Let m be the mass of a hydrogen atom. Since there are N_A atoms in a mole, the molar mass M is given by $M = mN_A$. We can use this to solve for m.

The mass of a hydrogen atom is the molar mass divided by Avogadro's number:

$$m = \frac{M}{N_A} = \frac{1.008\ \text{g/mol}}{6.022 \times 10^{23}\ \text{atoms/mol}}$$

$$= 1.67 \times 10^{-24}\ \text{g/atom}$$

Remark Note that Avogadro's number is approximately the reciprocal of the mass of the hydrogen atom measured in grams.

Example 18-6 *try it yourself*

One hundred grams of CO_2 occupies a volume of 55 L at a pressure of 1 atm. (*a*) What is the temperature? (*b*) If the volume is increased to 80 L and the temperature is kept constant, what is the new pressure?

Picture the Problem Both questions can be answered using the ideal-gas law (Equation 18-13) if we first find the number of moles, n.

Cover the column to the right and try these on your own before looking at the answers.

Steps	*Answers*
(*a*) 1. Use $M = 44$ g/mol to find the number of moles of CO_2.	$n = m/M = 2.27$ mol
2. Find the temperature T from the ideal-gas law.	$T = PV/nR = 295$ K
(*b*) Use PV = constant to find the new pressure for $V = 80$ L.	$P_2 = 0.688$ atm

18-5 The Kinetic Theory of Gases

The description of the behavior of a gas in terms of the macroscopic state variables P, V, and T can be related to simple averages of microscopic quantities such as the mass and speed of the molecules in the gas. The resulting theory is called **the kinetic theory of gases.**

From the point of view of kinetic theory, a gas consists of a large number of molecules making elastic collisions with each other and with the walls of a container. In the absence of external forces (we may neglect gravity), there is no preferred position for a molecule in the container,* and no preferred direction for its velocity vector. The molecules are separated, on the average, by distances that are large compared with their diameters, and they exert no forces on each other except when they collide. (This final assumption is equivalent to assuming a very low gas density, which as we saw in the last section is the same as assuming that the gas is an ideal gas. Since momentum is conserved, the collisions the molecules make with each other have no effect on the total momentum in any direction—thus such collisions can be neglected.)

Calculating the Pressure Exerted by a Gas The pressure that a gas exerts on its container is due to collisions between gas molecules and the container walls. This pressure is a force per unit area and, by Newton's second law, this force is the rate of change of momentum of the gas molecules colliding with the wall.

Consider a rectangular container of volume V containing N molecules, each of mass m moving with a speed v. Let us calculate the force exerted by these molecules on the right wall, which is perpendicular to the x axis and has area A. The number of molecules hitting this wall in a time interval Δt is the number that are within distance $v_x\,\Delta t$ of the wall (Figure 18-9) and are moving to the right. This is the number of molecules per unit volume N/V times the volume $v_x\,\Delta t\,A$ times $\frac{1}{2}$ because, on the average, only half the molecules are moving to the right. Thus, the number that hit the wall in Δt is

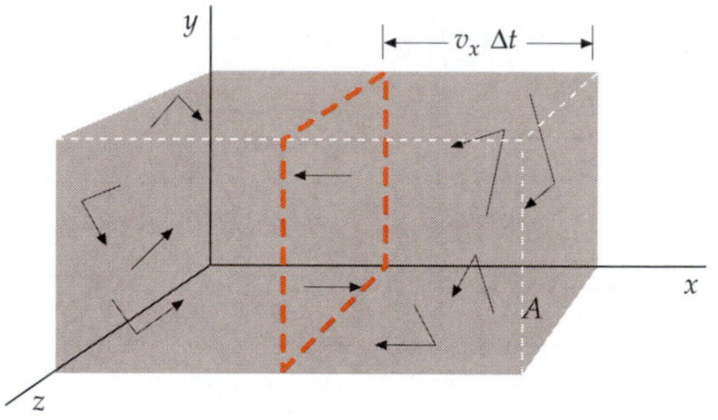

Figure 18-9 Gas molecules in a rectangular container. In a time interval Δt, the molecules at a distance $v_x\,\Delta t$ from the right wall will hit the right wall if they are moving to the right.

$$\text{Molecules that hit the wall} = \frac{1}{2}\frac{N}{V}v_x\Delta t\,A$$

The x component of momentum of a molecule is $+mv_x$ before it hits the wall, and $-mv_x$ after an elastic collision with the wall. The change in momentum has the magnitude $2mv_x$. The magnitude of the total change in momentum Δp of all molecules during a time interval Δt is $2mv_x$ times the number of molecules that hit the wall during this interval:

$$\Delta p = (2mv_x) \times \left(\frac{1}{2}\frac{N}{V}v_x\Delta t\,A\right) = \frac{N}{V}mv_x^2A\,\Delta t \qquad \text{18-16}$$

The force exerted by the wall on the molecules, and by the molecules on the wall, is this change in momentum divided by the time interval Δt. The pressure is this force divided by the area:

$$P = \frac{F}{A} = \frac{1}{A}\frac{\Delta p}{\Delta t} = \frac{N}{V}mv_x^2$$

* Because of gravity, the density of molecules at the bottom of the container is slightly greater than at the top. As discussed in Chapter 13, the density of air decreases by half at a height of about 5.5 km, so the variation over a normal-sized container is negligible.

or

$$PV = Nmv_x^2 \qquad \text{18-17}$$

To allow for the fact that all the molecules in a container do not have the same speed, we merely replace v_x^2 with the average value $(v_x^2)_{av}$. Then, writing Equation 18-17 in terms of the kinetic energy $\frac{1}{2}mv_x^2$ associated with motion along the x axis, we have

$$PV = 2N(\tfrac{1}{2}mv_x^2)_{av} \qquad \text{18-18}$$

The Molecular Interpretation of Temperature Comparing Equation 18-18 with Equation 18-7, which was obtained experimentally for any gas at very low densities, we can see that

$$PV = NkT = 2N(\tfrac{1}{2}mv_x^2)_{av}$$

or

$$(\tfrac{1}{2}mv_x^2)_{av} = \tfrac{1}{2}kT \qquad \text{18-19}$$

The average energy associated with motion in the x direction

Thus, the average kinetic energy associated with motion along the x axis is $\frac{1}{2}kT$. But there is nothing special about the x direction. On the average,

$$(v_x^2)_{av} = (v_y^2)_{av} = (v_z^2)_{av} \qquad \text{18-20}$$

and

$$(v^2)_{av} = (v_x^2)_{av} + (v_y^2)_{av} + (v_z^2)_{av} = 3(v_x^2)_{av}$$

Writing $(v_x^2)_{av} = \frac{1}{3}(v^2)_{av}$ and K_{av} for the average kinetic energy of the molecules, Equation 18-19 becomes

$$K_{av} = (\tfrac{1}{2}mv^2)_{av} = \tfrac{3}{2}kT \qquad \text{18-21}$$

Average kinetic energy of a molecule

The absolute temperature is thus a measure of the average translational kinetic energy of the molecules.* The total translational kinetic energy of n moles of a gas containing N molecules is

$$K = N(\tfrac{1}{2}mv^2)_{av} = \tfrac{3}{2}NkT = \tfrac{3}{2}nRT \qquad \text{18-22}$$

Kinetic energy of translation for n moles of a gas

where we've used $Nk = nN_Ak = nR$. Thus, the translational kinetic energy is $\frac{3}{2}kT$ per molecule and $\frac{3}{2}RT$ per mole.

We can use these results to estimate the order of magnitude of the speeds of the molecules in a gas. The average value of v^2 is, by Equation 18-21,

$$(v^2)_{av} = \frac{3kT}{m} = \frac{3N_AkT}{N_Am} = \frac{3RT}{M}$$

where $M = N_Am$ is the molar mass. The square root of $(v^2)_{av}$ is referred to as the **root mean square** (rms) speed:

$$v_{rms} = \sqrt{(v^2)_{av}} = \sqrt{\frac{3kT}{m}} = \sqrt{\frac{3RT}{M}} \qquad \text{18-23}$$

* We include the word "translational" because the molecules may also have rotational or vibrational kinetic energy. Only the translational kinetic energy is relevant to the calculation of the pressure exerted by a gas on the walls of its container.

Note that Equation 18-23 is similar to Equation 15-5 for the speed of sound in a gas:

$$v_{\text{sound}} = \sqrt{\frac{\gamma RT}{M}} \qquad \text{18-24}$$

where $\gamma = 1.4$ for air. This is not surprising since a sound wave in air is a pressure disturbance propagated by collisions between air molecules.

Example 18-7

Oxygen gas (O_2) has a molar mass of about 32 g/mol and hydrogen gas (H_2) has a molar mass of about 2 g/mol. Calculate (*a*) the rms speed of an oxygen molecule when the temperature is 300 K, and (*b*) the rms speed of a hydrogen molecule at the same temperature.

Picture the Problem (a) We find v_{rms} using Equation 18-23. For the units to work out right we use $R = 8.31$ J/mol·K and we express the molecular mass of O_2 in kg/mol. (*b*) Since v_{rms} is proportional to $1\sqrt{M}$, and the molar mass of hydrogen is one-sixteenth that of oxygen, the rms speed of hydrogen is 4 times that of oxygen.

(*a*) Substitute the given values into Equation 18-23:

$$v_{\text{rms}}(O_2) = \sqrt{\frac{3RT}{M}} = \sqrt{\frac{3(8.31\ \text{J/mol}\cdot\text{K})(300\ \text{K})}{32 \times 10^{-3}\ \text{kg/mol}}}$$
$$= 483\ \text{m/s}$$

(*b*) Use $v_{\text{rms}} \propto 1/\sqrt{M}$ to calculate v_{rms} for hydrogen:

$$v_{\text{rms}}(H_2) = \frac{\sqrt{M_{O_2}}}{\sqrt{M_{H_2}}}\, v_{\text{rms}}(O_2) = \sqrt{\frac{32\ \text{g/mol}}{2\ \text{g/mol}}}\,(483\ \text{m/s})$$
$$= 1.93\ \text{km/s}$$

Remarks The rms speed of oxygen molecules is of the same order of magnitude as the speed of sound in air, which at 300 K is about 347 m/s.

Exercise Find the rms speed of a nitrogen molecule ($M = 28$ g/mol) at 300 K. (*Answer* 516 m/s)

The Equipartition Theorem

We have seen that the average kinetic energy associated with translational motion in any direction is $\frac{1}{2}kT$ per molecule (Equation 18-19) (or, equivalently, $\frac{1}{2}RT$ per mole), where k is Boltzmann's constant. If the energy of a molecule associated with its motion in one direction is momentarily increased, say, by a collision between the molecule and a moving piston during a compression, collisions between that molecule and other molecules will quickly redistribute the additional energy. When the gas is again in equilibrium, the energy will be equally partitioned among the translational kinetic energies associated with motion in the x, y, and z directions. This sharing of the energy equally between the three terms in the translational kinetic energy is a special case of the **equipartition theorem**, a result that follows from classical statistical mechanics. Each component of position and momentum (including angular position and angular momentum) that appears as a squared term in the expression for the energy of the system is called a **degree of freedom**. Typical degrees of freedom are associated with the kinetic energy of translation, rotation, and vibration, and with the potential energy of vibration. The equipartition theorem states that

> When a substance is in equilibrium, there is an average energy of $\frac{1}{2}kT$ per molecule or $\frac{1}{2}RT$ per mole associated with each degree of freedom.

Equipartition theorem

In Chapter 19, we use the equipartition theorem to relate the measured heat capacities of gases to their molecular structure.

Mean Free Path

The average speed of molecules in a gas at normal pressures is several hundred meters per second, yet if somebody across a room from you opens a perfume bottle, you don't detect the odor instantly. In fact, if it were not for the bulk flow of air in the room, you wouldn't detect the odor for weeks. The transmission of the odor is slow because perfume molecules, speedy as they are, do not travel directly toward you, but instead travel a zigzag path, as often back as forward, due to collisions with air molecules. The average distance traveled by a molecule between collisions λ is called its **mean free path**.

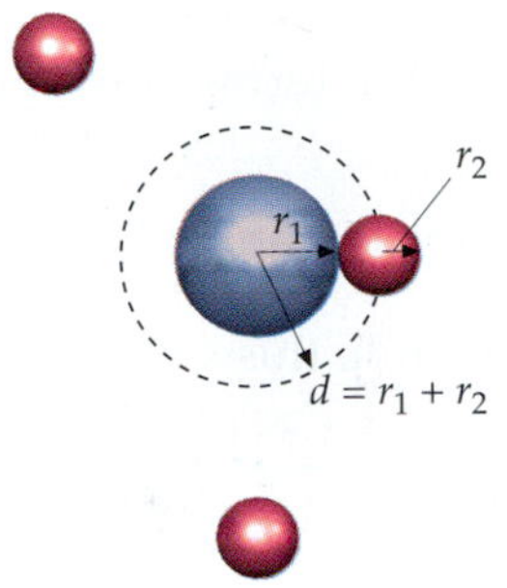

Figure 18-10 Model of a molecule (center sphere) moving in a gas. The molecule of radius r_1 will collide with any molecule of radius r_2 if their centers are a distance $d = r_1 + r_2$ apart, which is any molecule whose center is in a circle of radius $d = r_1 + r_2$ centered about the molecule.

The mean free path of a molecule is related to its size, the size of the surrounding gas molecules, and the density of the gas. Consider one molecule of radius r_1 moving with speed v through a region of stationary molecules (Figure 18-10). The molecule will collide with another molecule of radius r_2 if the centers of the two molecules come within a distance $d = r_1 + r_2$ from each other. (If all the molecules are the same type, d is the molecular diameter.) As the molecule moves, it will collide with any molecule whose center is in a circle of radius d (Figure 18-11). In some time t, the molecule moves a distance vt and collides with every molecule in the cylindrical volume $\pi d^2 vt$. The number of molecules in this volume is $n_v \pi d^2 vt$, where $n_v = N/V$ is the number of molecules per unit volume. (After each collision, the direction of the molecule changes, so the path actually zigs and zags.) The total path length divided by the number of collisions is the mean free path:

$$\lambda = \frac{vt}{n_v \pi d^2 vt} = \frac{1}{n_v \pi d^2}$$

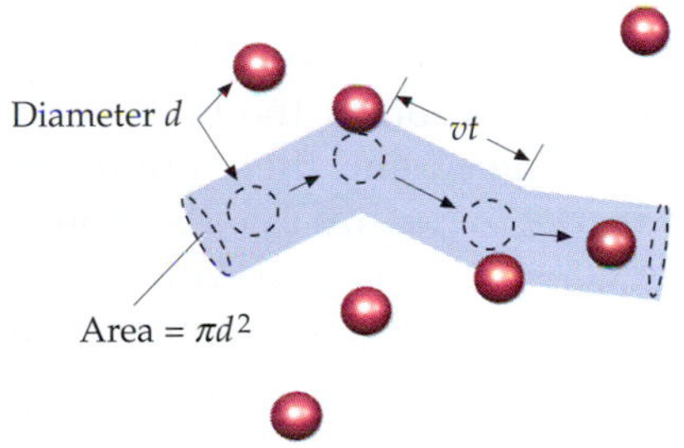

Figure 18-11 Model of a molecule moving with speed v in a gas of similar molecules. In time t the molecule of diameter d will collide with any similar molecule whose center is in a cylinder of volume $\pi d^2 vt$. In this picture, all the molecules but one are assumed to be at rest.

This calculation of the mean free path assumes that all but one of the gas molecules are stationary. When the motion of all the molecules is taken into account, the correct expression for the mean free path is given by

$$\lambda = \frac{1}{\sqrt{2}\, n_v \pi d^2} \qquad \text{18-25}$$

The average time between collisions is called the collision time τ. The reciprocal of the collision time, $1/\tau$, is approximately equal to the average number of collisions per second, or the collision frequency. If v_{av} is the average speed, the average distance traveled between collisions is

$$\lambda = v_{av}\tau \qquad \text{18-26}$$

Example 18-8 *try it yourself*

The diameter of a nitrogen molecule is about 3.75×10^{-10} m, and the molar mass of nitrogen is 28 g/mol. (*a*) Calculate the mean free path of a nitrogen molecule at 300 K and 1 atm. (*b*) Estimate the time between collisions, assuming that the average speed equals v_{rms}.

Picture the Problem (*a*) Since d is given, you need only n_v to calculate λ from $\lambda = 1/(\sqrt{2}n_v\pi d^2)$ (Equation 18-25). You can use the ideal-gas law to find $n_v = N/V$. (Be sure to convert 1 atm to pascals so the units come out right.) (*b*) Use Equation 18-26 to relate τ to λ and v_{av}, and estimate v_{av} by $v_{av} \approx v_{rms} = \sqrt{3RT/M}$ (Equation 18-23).

Cover the column to the right and try these on your own before looking at the answers.

Steps	*Answers*
(*a*)1. Write λ in terms of the number density n_v and the molecular diameter d.	$\lambda = \dfrac{1}{\sqrt{2}n_v\pi d^2}$
2. Use the equation $PV = NkT$ to calculate $n_v = N/V$.	$n_v = 2.45 \times 10^{25}$ molecules/m^3
3. Substitute this value of n_v and the given value of d to calculate λ.	$\lambda = 6.53 \times 10^{-8}$ m
(*b*)1. Write τ in terms of the mean free path λ.	$\tau = \dfrac{\lambda}{v_{av}}$
2. Estimate v_{av} by calculating v_{rms}.	$v_{rms} = 517$ m/s
3. Use $v_{av} \approx v_{rms}$ to estimate τ.	$\tau \approx 1.3 \times 10^{-10}$ s

Remark Note that the mean free path is about 2000 times the diameter of the molecule, and that the collision frequency is about $1/\tau \approx 8 \times 10^9$ collisions per second.

optional

The Distribution of Molecular Speeds

We would not expect all of the molecules in a gas to have the same speed. The calculation of the pressure of a gas allows us to calculate the average square speed and therefore the average energy of molecules in a gas, but it does not yield any details about the *distribution* of molecular speeds. Before we consider this problem, we discuss the idea of distribution functions in general with some elementary examples from common experience.

Distribution Functions Suppose a teacher gave a 25-point quiz to a large number N of students. To describe the results, the teacher might give the average score, but this would not be a complete description. If all the students received a score of 12.5, for example, that would be quite different from half the students receiving 25 and the other half zero, but the average score would be the same in both cases. A complete description of the results would be to give the number n_i of students that received a score s_i for all the scores received. Alternatively, one could give the fraction $f_i = n_i/N$ of the students that received the score s_i. Both n_i and f_i, which are functions of the variable s, are called **distribution functions.** The fractional distribution is somewhat more convenient to use. The probability that one of the N students selected at random received the score s_i equals the total number of students that received that score n_i divided by N, that is, the probability equals f_i. Note that

$$\sum_i f_i = \sum_i \frac{n_i}{N} = \frac{1}{N}\sum_i n_i$$

and since $\Sigma n_i = N$,

$$\sum_i f_i = 1 \qquad \text{18-27}$$

Equation 18-27 is called the **normalization condition** for fractional distributions.

To find the average score, we add all the scores and divide by N. Since each score s_i was obtained by $n_i = Nf_i$ students, this is equivalent to

$$s_{av} = \frac{1}{N}\sum_i n_i s_i = \sum_i s_i f_i \qquad \text{18-28}$$

Similarly, the average of any function $g(s)$ is defined by

$$g(s)_{av} = \frac{1}{N}\sum_i g(s_i) n_i = \sum_i g(s_i) f_i \qquad \text{18-29}$$

In particular, the average square score is

$$(s^2)_{av} = \frac{1}{N}\sum_i s_i^2 n_i = \sum_i s_i^2 f_i \qquad \text{18-30}$$

The square root of $(s^2)_{av}$ is called the **root mean square score,** or rms score. A possible distribution function is shown in Figure 18-12. For this distribution, the most probable score (that obtained by the most students) is 16, the average score is 14.2, and the rms score is 14.9.

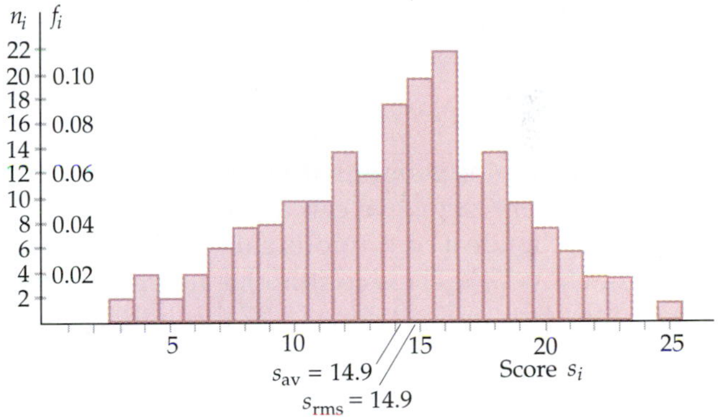

Figure 18-12 Grade distribution for a 25-point quiz given to 200 students. n_i is the number of students receiving grade s_i, and $f_i = n_i/N$ is the fractional distribution function.

Example 18-9

Fifteen students took a 25-point quiz. Their scores were 25, 22, 22, 20, 20, 20, 18, 18, 18, 18, 18, 15, 15, 15, and 10. Find the average score and the rms score.

Picture the Problem The distribution function for this problem is $n_{25} = 1$, $n_{22} = 2$, $n_{20} = 3$, $n_{18} = 5$, $n_{15} = 3$, and $n_{10} = 1$. To find the average score, we use Equation 18-28. To find the rms score, we use Equation 18-30 and then take the square root.

(*a*) By definition, s_{av} is:

$$s_{av} = \frac{1}{N}\sum_i n_i s_i$$
$$= \tfrac{1}{15}(1(25) + 2(22) + 3(20) + 5(18) + 3(15) + 1(10))$$
$$= \tfrac{1}{15}(274) = 18.3$$

(*b*)1. To calculate v_{rms}, first find the average of s^2:

$$(s^2)_{av} = \frac{1}{N}\sum_i n_i s_i^2$$
$$= \tfrac{1}{15}(1(25)^2 + 2(22)^2 + 3(20)^2 + 5(18)^2 + 3(15)^2 + 1(10)^2)$$
$$= \tfrac{1}{15}(5188) = 346$$

2. Take the square root of $(s^2)_{av}$:

$$s_{rms} = \sqrt{(s^2)_{av}} = 18.6 \text{ m/s}$$

optional

Now consider the case of a continuous distribution, for example, the distribution of heights in a population. For any finite number N, the number of people that are exactly 2 m tall is zero. If we assume that height can be determined to any desired accuracy, there are an infinite number of possible heights, so the chance that anybody has a particular exact height is zero. We therefore divide the heights into intervals Δh (for example, Δh might be 1 cm or 0.5 cm) and ask what fraction of people have heights that fall in any particular interval. For very large N, this number is proportional to the size of the interval. We define the distribution function $f(h)$ as the fraction of the number of people with heights in the interval between h and $h + \Delta h$. Then for N people, $Nf(h)\,\Delta h$ is the number of people whose height is between h and $h + \Delta h$. Figure 18-13 shows a possible height distribution.

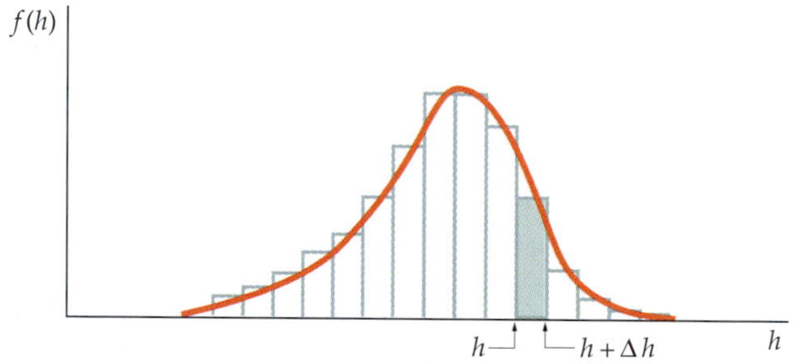

Figure 18-13 A possible height distribution function. The fraction of the number of heights between h and $h + \Delta h$ equals the shaded area $f(h)\,\Delta h$. The histogram can be approximated by a continuous curve as shown.

The fraction of people with heights in a given interval Δh is the area $f(h)\,\Delta h$. If N is very large, we can choose Δh to be very small, and the histogram will approximate a continuous curve. We can therefore consider the distribution function $f(h)$ to be a continuous function, write the interval as dh, and replace the sums in Equations 18-27 through 18-30 with integrals:

$$\int f(h)\,dh = 1 \qquad \text{18-31}$$

$$h_{av} = \int hf(h)\,dh \qquad \text{18-32}$$

$$g(h)_{av} = \int g(h)f(h)\,dh \qquad \text{18-33}$$

$$(h^2)_{av} = \int h^2 f(h)\,dh \qquad \text{18-34}$$

The probability of a person selected at random having a height between h and $h + dh$ is $f(h)\,dh$. We have already encountered a continuous distribution function in Chapter 17, where the probability of measuring the position of an electron described by the wave function $\psi(x)$ is $\psi^2(x)\,dx$.

A useful quantity characterizing a distribution is the **standard deviation** σ defined by

$$\sigma^2 = [(x - x_{av})^2]_{av} \qquad \text{18-35a}$$

Expanding the square on the right, we obtain

$$\sigma^2 = (x^2 - 2xx_{av} + x_{av}^2)_{av} = (x^2)_{av} - 2x_{av}x_{av} + x_{av}^2$$

or

$$\sigma^2 = (x^2)_{av} - x_{av}^2 \qquad \text{18-35b}$$

The standard deviation measures the spread of the values about the average value. For most distributions there will be few values that differ from x_{av} by more than a few multiples of σ. For the familiar bell-shaped distribution (called a normal distribution), two-thirds of the values are expected to fall within $x_{av} \pm \sigma$.

In Example 18-7, we found that the rms value was greater than the average value. This is a general feature for any distribution (unless all the values are identical, in which case $x_{rms} = x_{av}$). According to Equation 18-35*b*, the square of the rms value $(x^2)_{av}$ minus the square of the average value (x^2_{av}) is σ^2, which is by definition positive.

The Maxwell–Boltzmann Distribution The distribution of the molecular speeds of a gas can be measured directly using the apparatus illustrated in Figure 18-14*a*. In Figure 18-14*b*, these speeds are shown for two different temperatures. The quantity $f(v)$ in Figure 18-14*b* is called the **Maxwell–Boltzmann speed distribution function.** In a gas of N molecules, the number that have speeds in the range between v and $v + dv$ is dN, given by

$$dN = N f(v)\, dv \qquad \text{18-36}$$

(*a*)

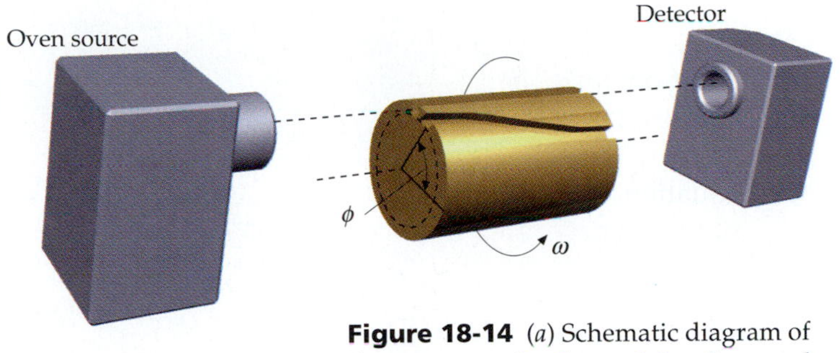

(*b*)

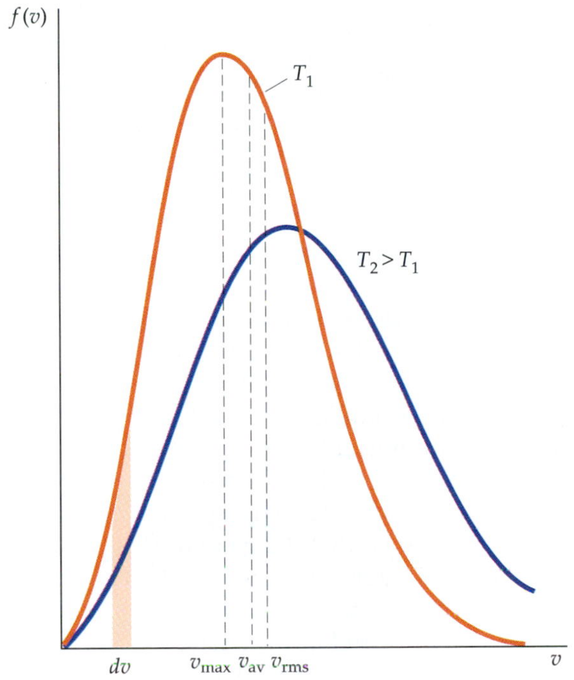

Figure 18-14 (*a*) Schematic diagram of the apparatus for determining the speed distribution of the molecules of a gas. A substance is vaporized in an oven, and the vapor molecules are allowed to escape through a hole in the oven wall into a vacuum chamber. The molecules are collimated into a narrow beam by a series of slits (not shown). The beam is aimed at a detector that counts the number of molecules that are incident on it in a given period of time. Most of the beam is stopped by a rotating cylinder. Small helical slits in the cylinder (only one of which is depicted here) allow the passage of molecules that have a narrow range of speeds that is determined by the angular velocity of rotation of the cylinder. The number of molecules in each range of speeds is measured by varying the angular velocity of the cylinder and counting the number of molecules that reach the detector for each angular velocity. (*b*) Distributions of molecular speeds in a gas at two temperatures, T_1 and $T_2 > T_1$. The shaded area $f(v)\, dv$ equals the fraction of the number of molecules having a particular speed in a narrow range of speeds dv. The mean speed v_{av} and the rms speed v_{rms} are both slightly greater than the most probable speed v_{max}.

The fraction $dN/N = f(v)\, dv$ in a particular range dv is illustrated by the shaded region in the figure. The Maxwell–Boltzmann speed distribution function can be derived using statistical mechanics. The result is

$$f(v) = \frac{4}{\sqrt{\pi}}\left(\frac{m}{2kT}\right)^{3/2} v^2 e^{-mv^2/2kT} \qquad \text{18-37}$$

Maxwell–Boltzmann speed distribution function

The most probable speed v_{max} is that speed for which $f(v)$ is maximum. It is left as a problem to show that

$$v_{max} = \sqrt{\frac{2kT}{m}} = \sqrt{\frac{2RT}{M}} \qquad \text{18-38}$$

Comparing Equation 18-38 with Equation 18-23, we see that the most probable speed is slightly less than the rms speed.

optional

Example 18-10

Use the Maxwell–Boltzmann distribution function to calculate the average value of v^2 for the molecules in a gas.

Picture the Problem The average value of v^2 is calculated from Equation 18-34 with v replacing h and $f(v)$ given by Equation 18-37.

1. By definition, $(v^2)_{av}$ is:

$$(v^2)_{av} = \int_0^\infty v^2 f(v)\, dv$$

2. Use Equation 18-37 for $f(v)$:

$$(v^2)_{av} = \int_0^\infty v^2 \frac{4}{\sqrt{\pi}}\left(\frac{m}{2kT}\right)^{3/2} v^2 e^{-mv^2/2kT}\, dv$$

$$= \frac{4}{\sqrt{\pi}}\left(\frac{m}{2kT}\right)^{3/2} \int_0^\infty v^4 e^{-mv^2/2kT}\, dv$$

3. The integral in step 2 can be found in standard integral tables:

$$\int_0^\infty v^4 e^{-mv^2/2kT}\, dv = \frac{3}{8}\sqrt{\pi}\left(\frac{2kT}{m}\right)^{5/2}$$

4. Use this result to calculate $(v^2)_{av}$:

$$(v^2)_{av} = \frac{4}{\sqrt{\pi}}\left(\frac{m}{2kT}\right)^{3/2} \frac{3}{8}\sqrt{\pi}\left(\frac{2kT}{m}\right)^{5/2} = \frac{3kT}{m}$$

Remarks Note that our result agrees with $v_{rms} = \sqrt{3kT/m}$ from Equation 18-23.

In Example 18-7, we found that the rms speed of hydrogen molecules is about 1.93 km/s. This is about one-sixth of the escape speed at the surface of the earth, which we found to be 11.2 km/s in Section 11-3. So why is there no free hydrogen in the earth's atmosphere? As we can see from Figure 18-14*b*, a considerable fraction of the molecules of a gas in equilibrium have speeds greater than the rms speed. When the rms speed of the molecules of a particular gas is as great as 15 to 20% of the escape speed for a planet, enough of the molecules have speeds greater than the escape speed so that the gas cannot exist in the atmosphere of that planet. Thus, no hydrogen. The rms speed of oxygen molecules, on the other hand, is about one-fourth that of hydrogen molecules, which makes it only about 4% of the escape speed at the surface of the earth. Therefore, few oxygen molecules have speeds greater than the escape speed, and oxygen is found in the earth's atmosphere.

The Energy Distribution The Maxwell–Boltzmann speed distribution as given by Equation 18-37 can also be written as an energy distribution. We write the number of molecules with energy E in the range between E and $E + dE$ as

$$dN = NF(E)\, dE$$

where $F(E)$ is the energy distribution function. This will be the same number as given by Equation 18-37 if the energy E is related to the speed v by $E = \frac{1}{2}mv^2$. Then

$$dE = mv\, dv$$

and

$$Nf(v)\, dv = NF(E)\, dE$$

We can write

$$f(v)\,dv = Cv^2 e^{-mv^2/2kT}\,dv = Cve^{-E/kT}\,v\,dv = C\left(\frac{2E}{m}\right)^{1/2} e^{-E/kT}\frac{dE}{m}$$

where $C = (4/\sqrt{\pi})(m/2kT)^{3/2}$ from Equation 18-37. The energy distribution function $F(E)$ is thus given by

$$F(E) = \frac{4}{\sqrt{\pi}}\left(\frac{m}{2kT}\right)^{3/2}\left(\frac{2}{m}\right)^{1/2}\frac{1}{m}E^{1/2}e^{-E/kT}$$

Simplifying, we obtain the **Maxwell–Boltzmann energy distribution function:**

$$F(E) = \frac{2}{\sqrt{\pi}}\left(\frac{1}{kT}\right)^{3/2} E^{1/2}e^{-E/kT} \qquad \text{18-39}$$

Maxwell–Boltzmann energy distribution function

In the language of statistical mechanics, the energy distribution is considered to be the product of two factors: one, called the **density of states,** is proportional to $E^{1/2}$; the other is the probability of a state being occupied, which is $e^{-E/kT}$ and is called the **Boltzmann factor.**

Jupiter as seen from about 12 million miles. Because the escape speed at the surface of Jupiter is about 600 km/s, Jupiter easily retains hydrogen in its atmosphere.

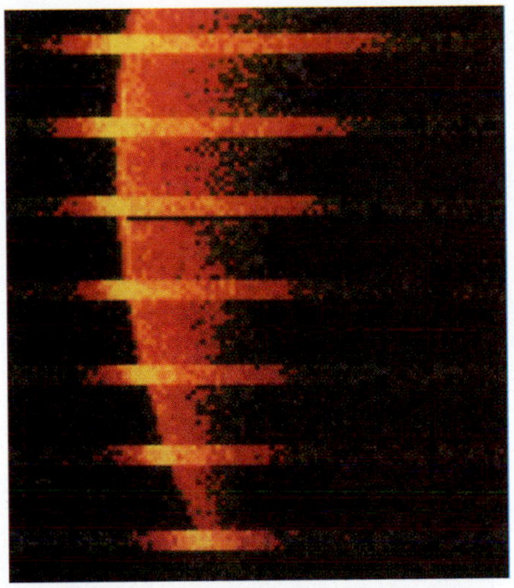

The atmosphere of Venus is almost entirely CO_2. However, measurements by the Pioneer Venus Orbiter show an atomic hydrogen cloud surrounding Venus. The crescent-shaped image shows atomic oxygen, and the bars show hydrogen extending far above the atmosphere. Since the escape speed on Venus is 10.3 km/s, slightly smaller than the escape speed on earth, and since the Venusian atmosphere is considerably warmer than earth's atmosphere, all of the hydrogen in the atmosphere at the time of the formation of Venus should have escaped by now.

Summary

Topic	Remarks and Relevant Equations
1. Celsius and Fahrenheit Scales	On the Celsius scale, the ice point is defined to be 0°C and the steam point is 100°C. On the Fahrenheit scale, the ice point is 32°F and the steam point is 212°F. Temperatures on the Fahrenheit and Celsius scales are related by $t_C = \frac{5}{9}(t_F - 32°)$ **18-2**
2. Gas Thermometers	Gas thermometers have the property that they all agree with each other in the measurement of any temperature as long as the density of the gas is very low. The ideal-gas temperature T is defined by $T = \frac{273.16\text{ K}}{P_3}P$ **18-4** where P is the observed pressure of the gas in the thermometer, and P_3 is the pressure when the thermometer is immersed in a water–ice–vapor bath at its triple point.

3. Kelvin Temperature Scale	The absolute temperature or temperature in kelvins is related to the Celsius temperature by $T = t_C + 273.15\text{ K}$	**18-5**
4. Ideal Gas		
Equation of state	At low densities, all gases obey the ideal-gas law: $PV = nRT$	**18-13**
Universal gas constant	$R = kN_A = 8.314\text{ J/mol}\cdot\text{K} = 0.08206\text{ L}\cdot\text{atm/mol}\cdot\text{K}$	**18-12**
Boltzmann's constant	$k = 1.381 \times 10^{-23}\text{ J/K} = 8.617 \times 10^{-5}\text{ eV/K}$	**18-8**
Avogadro's number	$N_A = 6.022 \times 10^{23}$	**18-9**
Equation for a fixed amount of gas	A form of the ideal-gas law that is useful for solving problems involving a fixed amount of gas is $\frac{P_2V_2}{T_2} = \frac{P_1V_1}{T_1}$	**18-14**
5. Kinetic Theory of Gases		
Molecular interpretation of temperature	The absolute temperature T is a measure of the average molecular energy.	
Equipartition theorem	When a system is in equilibrium, there is an average energy of $\frac{1}{2}kT$ per molecule or $\frac{1}{2}RT$ per mole associated with each degree of freedom.	
Average kinetic energy	For an ideal gas, the average translational kinetic energy of the molecules is $K_{av} = (\frac{1}{2}mv^2)_{av} = \frac{3}{2}kT$	**18-21**
Total kinetic energy	The total translational kinetic energy of n moles of a gas containing N molecules is given by $K = N(\frac{1}{2}mv^2)_{av} = \frac{3}{2}NkT = \frac{3}{2}nRT$	**18-22**
rms speed of molecules	The rms speed of a molecule of a gas is related to the absolute temperature by $v_{rms} = \sqrt{(v^2)_{av}} = \sqrt{\frac{3kT}{m}} = \sqrt{\frac{3RT}{M}}$ where m is the mass of the molecule and M is the molar mass.	**18-23**
Mean free path	The mean free path of a molecule is related to its diameter d and the number of molecules per unit volume n_v by $\lambda = \frac{1}{\sqrt{2}n_v\pi d^2}$	**18-25**
6. Maxwell–Boltzmann Distribution Functions (optional)		
Speed distribution function	$f(v) = \frac{4}{\sqrt{\pi}}\left(\frac{m}{2kT}\right)^{3/2} v^2 e^{-mv^2/2kT}$	**18-37**
Energy distribution function	$F(E) = \frac{2}{\sqrt{\pi}}\left(\frac{1}{kT}\right)^{3/2} E^{1/2}e^{-E/kT}$	**18-39**

Problem-Solving Guide

Summary of Worked Examples

Type of Calculation	Procedure and Relevant Examples	
1. Temperature		
Convert temperatures from one scale to another.	To convert between °F and °C, use $t_C = \frac{5}{9}(t_F - 32°)$. To convert to kelvins, first convert to Celsius, then use $T = t_C + 273$	**Examples 18-1, 18-2**
2. Ideal gas		
Find the volume, pressure, or temperature of an ideal gas.	Use $PV = nRT$ or $\frac{P_1V_1}{T_1} = \frac{P_2V_2}{T_2}$	**Examples 18-3, 18-4**
3. Molecules		
Find the mass of a molecule given the molar mass.	Use $M = mN_A$	**Example 18-5**
4. Kinetic Theory		
Find the rms speed of molecules.	For a gas at temperature T use $v_{rms} = \sqrt{\frac{3kT}{m}} = \sqrt{\frac{3RT}{M}}$	**Example 18-7**
Find the mean free path of molecules.	Use $\lambda = \frac{1}{\sqrt{2}n_v\pi d^2}$ and calculate $n_v = N/V$ from $PV = NkT$	**Example 18-8**
Find v_{av} and v_{rms} (optional).	For a small number of objects with given speeds, use $Nv_{av} = \Sigma n_i v_i$ and $v_{rms} = \sqrt{(v^2)_{av}}$ where $N(v^2)_{av} = \Sigma n_i v_i^2$	**Example 18-9**
	For a gas at temperature T use $v_{av} = \int vf(v)\,dv$ and $(v^2)_{av} = \int v^2 f(v)\,dv$ where $f(v)$ is the Maxwell–Boltzmann distribution function.	**Example 18-10**

Problems

In a few problems, you are given more data than you actually need; in a few other problems, you are required to supply data from your general knowledge, outside sources, or informed estimates.

Conceptual Problems

Problems from Optional and Exploring sections

• Single-concept, single-step, relatively easy

•• Intermediate-level, may require synthesis of concepts

••• Challenging, for advanced students

Temperature Scales

1 • True or false:

(*a*) Two objects in thermal equilibrium with each other must be in thermal equilibrium with a third object.

(*b*) The Fahrenheit and Celsius temperature scales differ only in the choice of the zero temperature.

(*c*) The kelvin is the same size as the Celsius degree.

(*d*) All thermometers give the same result when measuring the temperature of a particular system.

2 • How can you determine if two bodies are in thermal equilibrium with each other if it is impossible to put them into thermal contact with each other?

3 • Which is greater, an increase in temperature of 1 C° or of 1 F°?

4 • "One day I woke up and it was 20°F in my bedroom," said Mert to his old friend Mort. "That's nothing" replied Mort. "My room was once −5°C." Which room was colder?

5 • A certain ski wax is rated for use between −12 and −7°C. What is this temperature range on the Fahrenheit scale?

6 • The melting point of gold (Au) is 1945.4°F. Express this temperature in degrees Celsius.

7 • The highest and lowest temperatures ever recorded in the United States are 134°F (in California in 1913) and −80°F (in Alaska in 1971). Express these temperatures using the Celsius scale.

8 • What is the Celsius temperature corresponding to the normal temperature of the human body, 98.6°F?

9 • The length of the column of mercury in a thermometer is 4.0 cm when the thermometer is immersed in ice water and 24.0 cm when the thermometer is immersed in boiling water. (*a*) What should the length be at room temperature, 22.0°C? (*b*) If the mercury column is 25.4 cm long when the thermometer is immersed in a chemical solution, what is the temperature of the solution?

10 • The temperature of the interior of the sun is about 10^7 K. What is this temperature on (*a*) the Celsius scale, and (*b*) the Fahrenheit scale?

11 • The boiling point of nitrogen N_2 is 77.35 K. Express this temperature in degrees Fahrenheit.

12 • The pressure of a constant-volume gas thermometer is 0.400 atm at the ice point and 0.546 atm at the steam point. (*a*) When the pressure is 0.100 atm, what is the temperature? (*b*) What is the pressure at 444.6°C, the boiling point of sulfur?

13 • A constant-volume gas thermometer reads 50 torr at the triple point of water. (*a*) What will the pressure be when the thermometer measures a temperature of 300 K? (*b*) What ideal-gas temperature corresponds to a pressure of 678 torr?

14 • A constant-volume gas thermometer has a pressure of 30 torr when it reads a temperature of 373 K. (*a*) What is its triple-point pressure P_3? (*b*) What temperature corresponds to a pressure of 0.175 torr?

15 • At what temperature do the Fahrenheit and Celsius temperature scales give the same reading?

16 • Sodium melts at 371 K. What is the melting point of sodium on the Celsius and Fahrenheit temperature scales?

17 • The boiling point of oxygen at one atmosphere is 90.2 K. What is the boiling point of oxygen on the Celsius and Fahrenheit scales?

18 •• On the Réaumur temperature scale, the melting point of ice is 0° R and the boiling point of water is 80° R. Derive expressions for converting temperatures on the Réaumur scale to the Celsius and Fahrenheit scales.

19 ••• A thermistor is a solid-state device whose resistance varies greatly with temperature. Its temperature dependence is given approximately by $R = R_0 e^{B/T}$, where R is in ohms (Ω), T is in kelvins, and R_0 and B are constants that can be determined by measuring R at calibration points such as the ice point and the steam point. (*a*) If $R = 7360\ \Omega$ at the ice point and 153 Ω at the steam point, find R_0 and B. (*b*) What is the resistance of the thermistor at $t = 98.6$°F? (*c*) What is the rate of change of the resistance with temperature (dR/dT) at the ice point and the steam point? (*d*) At which temperature is the thermistor most sensitive?

The Ideal-Gas Law

20 •• Two identical vessels contain different ideal gases at the same pressure and temperature. It follows that

(*a*) the number of gas molecules is the same in both vessels.
(*b*) the total mass of gas is the same in both vessels.
(*c*) the average speed of the gas molecules is the same in both vessels.
(*d*) none of the above is correct.

21 •• Figure 18-15 shows a plot of volume versus temperature for a process that takes an ideal gas from point A to point B. What happens to the pressure of the gas?

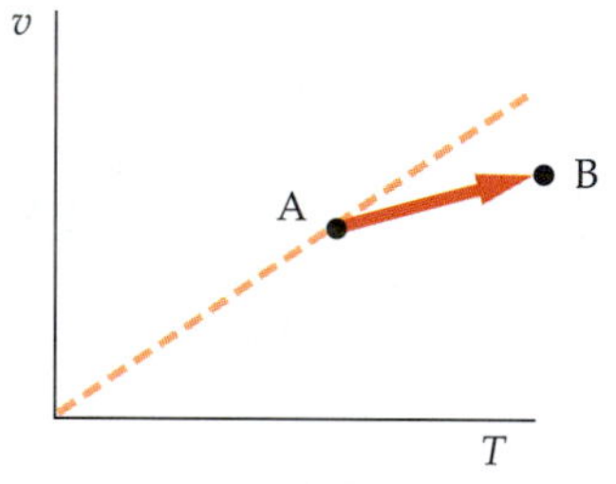

Figure 18-15 Problem 21

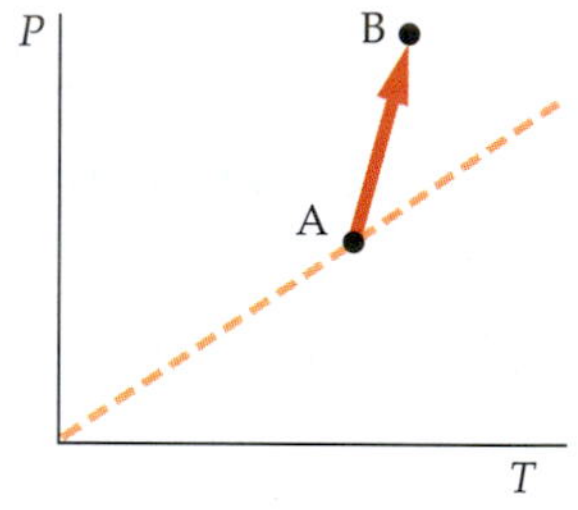

Figure 18-16 Problem 22

22 •• Figure 18-16 shows a plot of pressure versus temperature for a process that takes an ideal gas from point A to point B. What happens to the volume of the gas?

23 • A gas is kept at constant pressure. If its temperature is changed from 50 to 100°C, by what factor does the volume change?

24 • A 10-L vessel contains gas at a temperature of 0°C and a pressure of 4 atm. How many moles of gas are in the vessel? How many molecules?

25 •• A pressure as low as 1×10^{-8} torr can be achieved using an oil diffusion pump. How many molecules are there in 1 cm^3 of a gas at this pressure if its temperature is 300 K?

26 •• A motorist inflates the tires of her car to a pressure of 180 kPa on a day when the temperature is −8.0°C. When she arrives at her destination, the tire pressure has increased to 245 kPa. What is the temperature of the tires if we assume

that (*a*) the tires do not expand, or (*b*) that the tires expand by 7%?

27 •• A room is 6 m by 5 m by 3 m. (*a*) If the air pressure in the room is 1 atm and the temperature is 300 K, find the number of moles of air in the room. (*b*) If the temperature rises by 5 K and the pressure remains constant, how many moles of air leave the room?

28 •• A seafood restaurant hires Lou to run its advertising campaign. Lou figures that snorklers are a great pool of potential customers for seafood, so he prints ads on Mylar balloons that he ties to the coral of an underwater reef. Each balloon has a volume of 4 L and is filled with air at 20°C. At 15 m below the ocean surface, the volume has diminished to 1.60 L. What is the temperature of the water at this depth?

29 •• The boiling point of helium at one atmosphere is 4.2 K. What is the volume occupied by helium gas due to evaporation of 10 g of liquid helium at 1 atm pressure and a temperature of (*a*) 4.2 K, and (*b*) 293 K?

30 •• A container with a volume of 6.0 L holds 10 g of liquid helium. As the container warms to room temperature, what is the pressure exerted by the gas on its walls?

31 •• An automobile tire is filled to a gauge pressure of 200 kPa when its temperature is 20°C. (Gauge pressure is the difference between the actual pressure and atmospheric pressure.) After the car has been driven at high speeds, the tire temperature increases to 50°C. (*a*) Assuming that the volume of the tire does not change, and that air behaves as an ideal gas, find the gauge pressure of the air in the tire. (*b*) Calculate the gauge pressure if the volume of the tire expands by 10%.

32 •• A scuba diver is 40 m below the surface of a lake, where the temperature is 5°C. He releases an air bubble with a volume of 15 cm^3. The bubble rises to the surface, where the temperature is 25°C. What is the volume of the bubble right before it breaks the surface? *Hint:* Remember that the pressure also changes.

33 ••• A helium balloon is used to lift a load of 110 N. The weight of the balloon's skin is 50 N, and the volume of the balloon when fully inflated is 32 m^3. The temperature of the air is 0°C and the atmospheric pressure is 1 atm. The balloon is inflated with sufficient helium gas so that the net upward force on the balloon and its load is 30 N. Neglect changes of temperature with altitude.

(*a*) How many moles of helium gas are contained in the balloon?
(*b*) At what altitude will the balloon be fully inflated?
(*c*) Does the balloon ever reach the altitude at which it is fully inflated?
(*d*) If the answer to (*c*) is affirmative, what is the maximum altitude attained by the balloon?

Kinetic Theory of Gases

34 • True or false: The absolute temperature of a gas is a measure of the average translational kinetic energy of the gas molecules.

35 • By what factor must the absolute temperature of a gas be increased to double the rms speed of its molecules?

36 • How does the average translational kinetic energy of a molecule of a gas change if the pressure is doubled while the volume is kept constant? If the volume is doubled while the pressure is kept constant?

37 • A mole of He molecules is in one container and a mole of CH_4 molecules is in a second container, both at standard conditions. Which molecules have the greater mean free path?

38 •• A vessel holds an equal number of moles of helium and methane, CH_4. The ratio of the rms speeds of the helium atoms to the CH_4 molecules is ________.

(*a*) 1
(*b*) 2
(*c*) 4
(*d*) 16

39 • (*a*) Find v_{rms} for an argon atom if 1 mol of the gas is confined to a 1-L container at a pressure of 10 atm. (For argon, $M = 40 \times 10^{-3}$ kg/mol.) (*b*) Compare this with v_{rms} for a helium atom under the same conditions. (For helium, $M = 4 \times 10^{-3}$ kg/mol.)

40 • Find the total translational kinetic energy of 1 L of oxygen gas held at a temperature of 0°C and a pressure of 1 atm.

41 • Find the rms speed and the average kinetic energy of a hydrogen atom at a temperature of 10^7 K. (At this temperature, which is of the order of the temperature in the interior of a star, the hydrogen is ionized and consists of a single proton.)

42 • In one model of a solid, the material is assumed to consist of a regular array of atoms in which each atom has a fixed equilibrium position and is connected by springs to its neighbors. Each atom can vibrate in the *x*, *y*, and *z* directions. The total energy of an atom in this model is

$$E = \tfrac{1}{2}mv_x^2 + \tfrac{1}{2}mv_y^2 + \tfrac{1}{2}mv_z^2 + \tfrac{1}{2}Kx^2 + \tfrac{1}{2}Ky^2 + Kz^2$$

What is the average energy of an atom in the solid when the temperature is *T*? What is the total energy of one mole of such a solid?

43 • Show that the mean free path for a molecule in an ideal gas at temperature *T* and pressure *P* is given by

$$\lambda = \frac{kT}{\sqrt{2}P\pi d^2}$$

44 •• The escape velocity on Mars is 5.0 km/s, and the surface temperature is typically 0°C. Calculate the rms speeds for (*a*) H_2, (*b*) O_2, and (*c*) CO_2 at this temperature. (*d*) If the rms speed of a gas is greater than about 15 to 20% of the escape velocity of a planet, virtually all of the molecules of that gas will escape the atmosphere of the planet. Based on this criterion, are H_2, O_2, and CO_2 likely to be found in Mars's atmosphere?

45 •• Repeat Problem 44 for Jupiter, whose escape velocity is 60 km/s and whose temperature is typically −150°C.

46 •• A pressure as low as $P = 7 \times 10^{-11}$ Pa has been obtained. Suppose a chamber contains helium at this pressure and at room temperature (300 K). Estimate the mean free

path λ and the collision time τ for helium in the chamber. Take the diameter of a helium molecule to be 10^{-10} m.

47 •• Oxygen (O_2) is confined to a cubic container 15 cm on a side at a temperature of 300 K. Compare the average kinetic energy of a molecule of the gas to the change in its gravitational potential energy if it falls from the top of the container to the bottom.

The Distribution of Molecular Speeds

48 •• The class in Room 101 prepares their traditional greeting for a substitute teacher. Ten toy cars are wound up and released as the teacher arrives. The cars have the following speeds:

Speed, m/s	2	5	6	8
Number of cars	3	3	3	1

Calculate (*a*) the average speed, and (*b*) the rms speed of the cars.

49 •• Show that $f(v)$ given by Equation 18–37 is maximum when $v = \sqrt{2kT/m}$. *Hint:* Set $df/dv = 0$ and solve for v.

50 •• Since $f(v)\,dv$ gives the fraction of molecules that have speeds in the range dv, the integral of $f(v)\,dv$ over all the possible ranges of speeds must equal 1. Given the integral

$$\int_0^\infty v^2 e^{-av^2}\,dv = \frac{\sqrt{\pi}}{4}a^{-3/2}$$

show that $\int_0^\infty f(v)\,dv = 1$, where $f(v)$ is given by Equation 18–37.

51 •• Given the integral

$$\int_0^\infty v^3 e^{-av^2}\,dv = \frac{a^{-2}}{2}$$

calculate the average speed v_{av} of molecules in a gas using the Maxwell–Boltzmann distribution function.

52 ••• In Chapter 11, we found that the escape speed at the surface of a planet of radius R is $v_e = \sqrt{2gR}$, where g is the acceleration due to gravity. (*a*) At what temperature is v_{rms} for O_2 equal to the escape speed for the earth? (*b*) At what temperature is v_{rms} for H_2 equal to the escape speed for the earth? (*c*) Temperatures in the upper atmosphere reach 1000 K. How does this account for the low abundance of hydrogen in the earth's atmosphere? (*d*) Compute the temperatures for which the rms speeds of O_2 and H_2 are equal to the escape velocity at the surface of the moon, where g is about one-sixth of its value on earth and R = 1738 km. How does this account for the absence of an atmosphere on the moon?

General Problems

53 • True or false: If the pressure of a gas increases, the temperature must increase.

54 • What is the difference between 1°C and 1 C°?

55 • Why might the Celsius and Fahrenheit scales be more convenient than the absolute scale for ordinary, nonscientific purposes?

56 • The temperature of the interior of the sun is said to be about 10^7 degrees. Do you think that this is degrees Celsius or kelvins, or does it matter?

57 • If the temperature of an ideal gas is doubled while maintaining constant pressure, the average speed of the molecules

(*a*) remains constant.
(*b*) increases by a factor of 4.
(*c*) increases by a factor of 2.
(*d*) increases by a factor of $\sqrt{2}$.

58 • If both temperature and volume of an ideal gas are halved, the pressure

(*a*) diminishes by a factor of 2.
(*b*) remains constant.
(*c*) increases by a factor of 2.
(*d*) diminishes by a factor of $\sqrt{2}$.

59 • The average translational kinetic energy of the molecules of an ideal gas depends on

(*a*) the number of moles of the gas and its temperature.
(*b*) the pressure of the gas and its temperature.
(*c*) the pressure of the gas only.
(*d*) the temperature of the gas only.

60 • If a vessel contains equal amounts, by weight, of helium and argon, which of the following are true?

(*a*) The pressure exerted by the two gases on the walls of the container is the same.
(*b*) The average speed of a helium atom is the same as that of an argon atom.
(*c*) The number of helium atoms and argon atoms in the vessel are equal.
(*d*) None of the above statements is correct.

61 • Two different gases are at the same temperature. What can you say about the rms speeds of the gas molecules? What can you say about the average kinetic energies of the molecules?

62 •• Explain in terms of molecular motion why the pressure on the walls of a container increases when a gas is heated at constant volume.

63 •• Explain in terms of molecular motion why the pressure on the walls of a container increases when the volume of a gas is reduced at constant temperature.

64 •• Oxygen has a molar mass of 32 g/mol, and nitrogen has a molar mass of 28 g/mol. The oxygen and nitrogen molecules in a room have

(*a*) equal average kinetic energies, but the oxygen molecules are faster.
(*b*) equal average kinetic energies, but the oxygen molecules are slower.
(*c*) equal average kinetic energies and speeds.
(*d*) equal average speeds, but the oxygen molecules have a higher average kinetic energy.
(*e*) equal average speeds, but the oxygen molecules have a lower average kinetic energy.
(*f*) None of the above is correct.

65 • At what temperature will the rms speed of an H_2 molecule equal 331 m/s?

66 • A solid-state temperature transducer is essentially a linear amplifier whose amplification is linearly temperature dependent. If the amplification is 25 times at 20°C and 60 times at 70°C, what is the temperature when the amplification is 45 times?

67 •• (*a*) If 1 mol of a gas in a container occupies a volume of 10 L at a pressure of 1 atm, what is the temperature of the gas in kelvins? (*b*) The container is fitted with a piston so that the volume can change. When the gas is heated at constant pressure, it expands to a volume of 20 L. What is the temperature of the gas in kelvins? (*c*) The volume is fixed at 20 L, and the gas is heated at constant volume until its temperature is 350 K. What is the pressure of the gas?

68 •• A cubic metal box with sides of 20 cm contains air at a pressure of 1 atm and a temperature of 300 K. The box is sealed so that the volume is constant, and it is heated to a temperature of 400 K. Find the net force on each wall of the box.

69 •• Water, H_2O, can be converted into H_2 and O_2 gas by electrolysis. How many moles of these gases result from the electrolysis of 2 L of water?

70 •• A massless cylinder 40 cm long rests on a horizontal frictionless table. The cylinder is divided into two equal sections by a membrane. One section contains nitrogen and the other contains oxygen. The pressure of the nitrogen is twice that of the oxygen. How far will the cylinder move if the membrane is removed?

71 •• A cylinder contains a mixture of nitrogen gas (N_2) and hydrogen gas (H_2). At a temperature T_1 the nitrogen is completely dissociated but the hydrogen does not dissociate at all, and the pressure is P_1. If the temperature is doubled to $T_2 = 2T_1$, the pressure is tripled due to complete dissociation of hydrogen. If the mass of hydrogen is m_H, find the mass of nitrogen m_N.

72 •• A vertical closed cylinder of cross-sectional area A is divided into two equal parts by a heavy insulating movable piston of mass m_p. The top part contains nitrogen at a temperature T_1 and pressure P_1, and the bottom part is filled with oxygen at a temperature $2T_1$. The cylinder is turned upside-down. To keep the piston in the middle, the oxygen must be cooled to $T_2 = T_1/3$, with the temperature of the nitrogen remaining at T_1. Find the initial pressure of oxygen P_i.

73 •• Three insulated vessels of equal volumes V are connected by thin tubes that can transfer gas but do not transfer heat. Initially all vessels are filled with the same type of gas at a temperature T_0 and pressure P_0. Then the temperature in the first vessel is doubled and the temperature in the second vessel is tripled. The temperature in the third vessel remains unchanged. Find the final pressure P' in the system in terms of the initial pressure P_0.

74 •• At the surface of the sun, the temperature is about 6000 K, and all the substances present are gaseous. From data given by the light spectrum of the sun, it is known that most elements are present. (*a*) What is the average kinetic energy of translation of an atom at the surface of the sun? (*b*) What is the range of rms speeds at the surface of the sun if the atoms present range from hydrogen (M = 1 g/mol) to uranium (M = 238 g/mol)?

75 •• A constant-volume gas thermometer with a triple-point pressure $P_3 = 500$ torr is used to measure the boiling point of some substance. When the thermometer is placed in thermal contact with the boiling substance, its pressure is 734 torr. Some of the gas in the thermometer is then allowed to escape so that its triple-point pressure is 200 torr. When it is again placed in thermal contact with the boiling substance, its pressure is 293.4 torr. Again, some of the gas is removed from the thermometer so that its triple-point pressure is 100 torr. When the thermometer is placed in thermal contact with the boiling substance once again, its pressure is 146.65 torr. Find the ideal-gas temperature of the boiling substance.

76 ••• A cylinder 2.4 m tall is filled with 0.1 mol of an ideal gas at standard temperature and pressure (Figure 18-17). The top of the cylinder is then closed with a tight-fitting piston whose mass is 1.4 kg and the piston is allowed to drop until it is in equilibrium. (*a*) Find the height of the piston, assuming that the temperature of the gas does not change as it is compressed. (*b*) Suppose that the piston is pushed down below its equilibrium position by a small amount and then released. Assuming that the temperature of the gas remains constant, find the frequency of vibration of the piston.

Figure 18-17 Problem 76

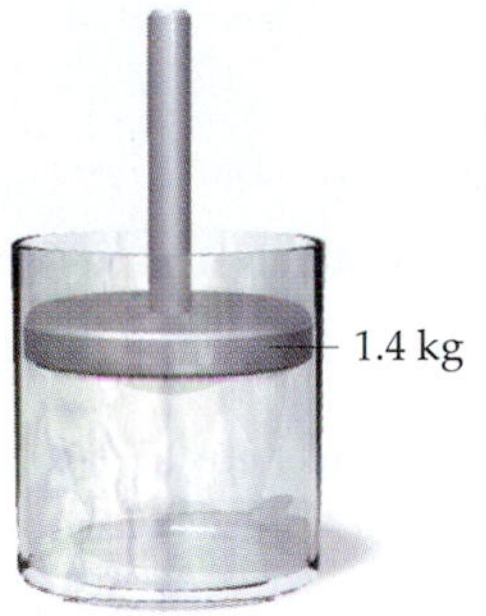

CHAPTER 19

Heat and the First Law of Thermodynamics

Steel ingots in a twin-tube tunnel furnace. The three 53-cm diameter carbon steel ingots seen here have been heated for about 7 hours to approximately 1340°C. Each 3200-kg ingot sits on a furnace car that transports it through the 81-m furnace, which is divided into twelve separate heating zones so that the temperature of the ingot is increased gradually to prevent cracking. The ingots, glowing a yellow-whitish color, exit the furnace to be milled into large, heavy-walled pipes.

Heat is energy that is transferred from one system to another because of a difference in temperature. In the seventeenth century, Galileo, Newton, and other scientists generally supported the theory of the ancient Greek atomists who considered heat to be a manifestation of molecular motion. In the next century, methods were developed for making quantitative measurements of the amount of heat that leaves or enters an object, and it was found that when objects are in thermal contact, the amount of heat that leaves one object equals the amount that enters the other. This discovery led to the caloric theory of heat as a conserved material substance. In this theory, an invisible fluid called "caloric" flowed out of one object and into another and could be neither created nor destroyed.

The caloric theory reigned until the nineteenth century, when it was found that friction between objects could generate an unlimited amount of heat, deposing the idea that caloric was a substance present in a fixed amount. The modern theory of heat did not emerge until the 1840s, when James Joule (1818–1889) showed that the gain or loss of a given amount of heat was accompanied by the disappearance or appearance of an equivalent quantity of mechanical energy. Heat, therefore, is not itself conserved. Instead, heat is a form of energy, and it is energy that is conserved.

19-1 Heat Capacity and Specific Heat

When heat energy flows into a substance, the temperature of the substance usually rises.* The amount of heat energy Q needed to raise the temperature of a substance is proportional to the temperature change and to the mass of the substance:

$$Q = C\,\Delta T = mc\,\Delta T \qquad 19\text{-}1$$

where C is the **heat capacity,** which is defined as the heat energy needed to raise the temperature of a substance by one degree, and c is the **specific heat**, the heat capacity per unit mass:

$$c = \frac{C}{m} \qquad 19\text{-}2$$

The historical unit of heat energy, the **calorie,** was originally defined to be the amount of heat energy needed to raise the temperature of one gram of water one Celsius degree.† Since we now recognize that heat is just another form of energy, we do not need any special units for it. The calorie is now defined in terms of the SI unit of energy, the joule:

$$1\text{ cal} = 4.184\text{ J} \qquad 19\text{-}3$$

The U.S. customary unit of heat is the **Btu** (for British thermal unit), which was originally defined to be the amount of energy needed to raise the temperature of one pound of water by one Fahrenheit degree. The Btu is related to the calorie and to the joule by

$$1\text{ Btu} = 252\text{ cal} = 1.054\text{ kJ} \qquad 19\text{-}4$$

The original definition of the calorie implies that the specific heat of water is‡

$$c_{\text{water}} = 1\text{ cal/g}\cdot\text{C}^\circ = 1\text{ kcal/kg}\cdot\text{C}^\circ$$
$$= 1\text{ kcal/kg}\cdot\text{K} = 4.184\text{ kJ/kg}\cdot\text{K} \qquad 19\text{-}5a$$

Similarly, from the definition of the Btu, the specific heat of water in U.S. customary units is

$$c_{\text{water}} = 1\text{ Btu/lb}\cdot\text{F}^\circ \qquad 19\text{-}5b$$

The heat capacity per mole is called the **molar specific heat** c',

$$c' = \frac{C}{n}$$

where n is the number of moles. Since $C = mc$, the molar specific heat c' and specific heat c are related by

$$c' = \frac{C}{n} = \frac{mc}{n} = Mc \qquad 19\text{-}6$$

where $M = m/n$ is the molar mass. Table 19-1 lists the specific heats and molar specific heats of some solids and liquids. Note

Table 19-1

Specific Heats and Molar Specific Heats of Some Solids and Liquids

Substance	c, kJ/kg·K	c, kcal/kg·K or Btu/lb·F°	c', J/mol·K
Aluminum	0.900	0.215	24.3
Bismuth	0.123	0.0294	25.7
Copper	0.386	0.0923	24.5
Glass	0.840	0.20	
Gold	0.126	0.0301	25.6
Ice (−10°C)	2.05	0.49	36.9
Lead	0.128	0.0305	26.4
Silver	0.233	0.0558	24.9
Tungsten	0.134	0.0321	24.8
Zinc	0.387	0.0925	25.2
Alcohol (ethyl)	2.4	0.58	111
Mercury	0.140	0.033	28.3
Water	4.18	1.00	75.2

* An exception occurs during a change in phase, as when water freezes or evaporates. Changes of phase are discussed in Section 19-2.

† The kilocalorie is then the amount of heat energy needed to raise the temperature of one kilogram of water by one Celsius degree. The "calorie" used in measuring the energy equivalent of foods is actually the kilocalorie.

‡ Careful measurement shows that the specific heat of water varies by about 1% over the temperature range from 0 to 100°C. We will usually neglect this small variation.

that the molar heats of all the metals are about the same. We discuss the significance of this in Section 19-7.

Example 19-1

How much heat is needed to raise the temperature of 3 kg of copper by 20 C°?

The required heat is given by Equation 19-1 with $c = 0.386$ kJ/kg·K from Table 19-1:

$$Q = mc\,\Delta T = (3\text{ kg})(0.386\text{ kJ/kg·K})(20\text{ K})$$
$$= 23.2\text{ kJ}$$

Remarks Note that we use $\Delta T = 20$ C° $= 20$ K. Alternatively, we could express the specific heat as 0.386 kJ/kg·C° and write the temperature change as 20 C°.

Exercise A 2-kg aluminum block is originally at 10°C. If 36 kJ of heat energy are added to the block, what is its final temperature? (*Answer* 30°C)

We see from Table 19-1 that the specific heat of water is considerably larger than that of the other substances. Thus, water is an excellent material for storing thermal energy, as in a solar heating system. It is also an excellent coolant, as in a car engine. Large bodies of water, such as lakes or oceans, tend to moderate variations of temperature nearby because they can absorb or release large quantities of thermal energy while undergoing only very small changes in temperature.

Calorimetry

The specific heat of an object can be conveniently measured by heating the object to some temperature, placing it in a water bath of known mass and temperature, and measuring the final equilibrium temperature. If the system is isolated from its surroundings, the heat leaving the object equals the heat entering the water and its container. This procedure is called **calorimetry,** and the insulated water container is called a **calorimeter.**

Let m be the mass of the object, let c be its specific heat, and let T_{io} be the object's initial temperature. If T_{f} is the final temperature of the object in its water bath, the heat leaving the object is

$$Q_{\text{out}} = mc(T_{\text{io}} - T_{\text{f}})$$

Similarly, if T_{iw} is the initial temperature of the water and container, and T_{f} is their final equilibrium temperature, the heat absorbed by the water and container is

$$Q_{\text{in}} = m_{\text{w}}c_{\text{w}}(T_{\text{f}} - T_{\text{iw}}) + m_{\text{c}}c_{\text{c}}(T_{\text{f}} - T_{\text{iw}})$$

where m_{w} and $c_{\text{w}} = 4.18$ kJ/kg·K are the mass and specific heat of the water, and m_{c} and c_{c} are the mass and specific heat of the container. (Note that we have chosen the temperature differences so that the heat in and heat out are both positive quantities.) Setting these amounts of heat equal yields the specific heat c of the object:

$$Q_{\text{out}} = Q_{\text{in}}$$

$$mc(T_{\text{io}} - T_{\text{f}}) = m_{\text{w}}c_{\text{w}}(T_{\text{f}} - T_{\text{iw}}) + m_{\text{c}}c_{\text{c}}(T_{\text{f}} - T_{\text{iw}}) \qquad \text{19-7}$$

Since only temperature differences occur in Equation 19-7, and since the kelvin and Celsius degree are the same size, it doesn't matter whether kelvins or Celsius degrees are used.

Example 19-2

To measure the specific heat of lead, you heat 600 g of lead shot to 100°C and place it in an aluminum calorimeter of mass 200 g that contains 500 g of water initially at 17.3°C. If the final temperature of the mixture is 20.0°C, what is the specific heat of lead? (The specific heat of the aluminum container is 0.900 kJ/kg·K.)

Picture the Problem We set the heat out of the lead equal to the heat into the water and container and solve for the specific heat of lead c_{Pb}.

1. Write the heat given off by the lead in terms of its specific heat:

$$Q_{\text{Pb}} = mc_{\text{Pb}}(T_{\text{io}} - T_{\text{f}}) = (0.6\ \text{kg})c_{\text{Pb}}(100°\text{C} - 20°\text{C}) = (0.6\ \text{kg})c_{\text{Pb}}(80\ \text{K})$$

2. Find the heat absorbed by the water:

$$Q_{\text{w}} = m_{\text{w}}c_{\text{w}}\,\Delta T_{\text{w}} = m_{\text{w}}c_{\text{w}}(20.0°\text{C} - 17.3°\text{C}) = (0.5\ \text{kg})(4.18\ \text{kJ/kg·K})(2.7\ \text{K}) = 5.64\ \text{kJ}$$

3. Find the heat absorbed by the container:

$$Q_{\text{c}} = m_{\text{c}}c_{\text{c}}\,\Delta T_{\text{c}} = (0.2\ \text{kg})(0.900\ \text{kJ/kg·K})(2.7\ \text{K}) = 0.486\ \text{kJ}$$

4. Set the heat out equal to the heat in:

$$Q_{\text{Pb}} = Q_{\text{w}} + Q_{\text{c}}$$

$$(0.6\ \text{kg})c_{\text{Pb}}(80\ \text{K}) = 5.64\ \text{kJ} + 0.486\ \text{kJ} = 6.13\ \text{kJ}$$

5. Solve for c_{Pb}:

$$c_{\text{Pb}} = \frac{6.13\ \text{kJ}}{(0.6\ \text{kg})(80.0\ \text{K})} = 0.128\ \text{kJ/kg·K}$$

Remarks Note that the specific heat of lead is considerably less than that of water.

19-2 Change of Phase and Latent Heat

When heat is added to ice at 0°C, the temperature of the ice does not change. Instead, the ice melts.This is an example of a **phase change**. Common types of phase changes include fusion (liquid to solid), melting (solid to liquid), vaporization (liquid to vapor or gas), condensation (gas or vapor to liquid), and sublimation (solid directly to vapor, as when solid carbon dioxide (dry ice) changes to vapor). There are other types of phase changes as well, such as the change of a solid from one crystalline form to another. Carbon under intense pressure, for example, becomes diamond.

The fact that the temperature remains constant during a phase change can be understood in terms of molecular theory. The molecules in a liquid are close together and exert attractive forces on each other. The molecules in a gas are far apart. Changing the substance from a liquid to a vapor requires energy to overcome the attraction between the molecules of the liquid. The energy put into the liquid to vaporize it thus increases the potential energy of the molecules, not their kinetic energy. Since temperature measures the kinetic energy of the molecules, the temperature remains the same during a phase change.

For a pure substance, a change in phase at a given pressure occurs only at a particular temperature. For example, pure water at a pressure of 1 atm changes from solid to liquid at 0°C (the normal melting point of water) and from liquid to gas at 100°C (the normal boiling point of water).

The heat energy required to melt a substance of mass m with no change in its temperature is proportional to the mass of the substance:

$$Q_f = mL_f \tag{19-8}$$

where L_f is called the **latent heat of fusion** of the substance. At a pressure of 1 atm, the latent heat of fusion for water is 333.5 kJ/kg = 79.7 kcal/kg. When the phase change is from liquid to gas, the heat required is

$$Q_v = mL_v \tag{19-9}$$

where L_v is the **latent heat of vaporization**. For water at a pressure of 1 atm, the latent heat of vaporization is 2.26 MJ/kg = 540 kcal/kg. Table 19-2 gives the normal melting and boiling points, and the latent heats of fusion and vaporization at 1 atm, for various substances.

Table 19-2

Normal Melting Point (MP), Latent Heat of Fusion L_f, Normal Boiling Point (BP), and Latent Heat of Vaporization L_v for Various Substances at 1 atm

Substance	MP, K	L_f, kJ/kg	BP, K	L_v, kJ/kg
Alcohol, ethyl	159	109	351	879
Bromine	266	67.4	332	369
Carbon dioxide	—	—	194.6[a]	573[a]
Copper	1356	205	2839	4726
Gold	1336	62.8	3081	1701
Helium	—	—	4.2	21
Lead	600	24.7	2023	858
Mercury	234	11.3	630	296
Nitrogen	63	25.7	77.35	199
Oxygen	54.4	13.8	90.2	213
Silver	1234	105	2436	2323
Sulfur	388	38.5	717.75	287
Water	273.15	333.5	373.15	2257
Zinc	692	102	1184	1768

[a] These values are for sublimation. Carbon dioxide does not have a liquid state at 1 atm.

Example 19-3 try it yourself

How much heat do you need to heat 1.5 kg of ice at a pressure of 1 atm from −20°C until all the ice has been changed into steam?

Picture the Problem The heat required consists of four parts: Q_1, the heat needed to warm the ice from −20 to 0°C; Q_2, the heat needed to melt the ice; Q_3, the heat needed to warm the water from 0 to 100°C; and Q_4, the heat needed to vaporize the water. In calculating Q_1 and Q_3, we will assume that the specific heats are constant, with the values 2.05 kJ/kg·K for ice and 4.18 kJ/kg·K for water.

Cover the column to the right and try these on your own before looking at the answers.

Steps	***Answers***
1. Use $Q_1 = mc\,\Delta T$ to find the heat needed to warm the ice to 0°C.	$Q_1 = 61.5\text{ kJ} = 0.0615\text{ MJ}$
2. Use L_f from Table 19-2 to find the heat Q_2 needed to melt the ice.	$Q_2 = 500\text{ kJ} = 0.500\text{ MJ}$
3. Find the heat Q_3 needed to warm the water from 0 to 100°C.	$Q_3 = 627\text{ kJ} = 0.627\text{ MJ}$
4. Use L_v from Table 19-2 to find the heat Q_4 needed to vaporize the water.	$Q_4 = 3.39\text{ MJ}$
5. Sum your results to find the total heat Q.	$Q = 4.58\text{ MJ}$

Remarks Notice that most of the heat was needed to vaporize the water, and that the amount needed to melt the ice was almost as much as that needed to raise the temperature of the water by 100 C°. Figure 19-1 shows a graph of temperature versus time for the case in which the heat is added at a constant rate of 1.5 kJ/s. Note that it takes considerably longer to vaporize the water than it does to melt the ice or to raise the temperature of the water. When all of the water has vaporized, the temperature again rises as heat is added.

Exercise An 830-g piece of lead is heated to its melting point of 600 K. How much additional heat energy must be added to melt the lead? (*Answer* 20.5 kJ)

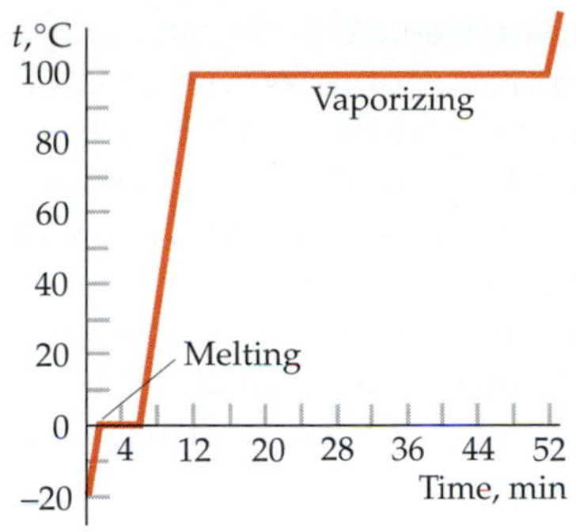

Figure 19-1

Example 19-4

Lemonade has been sitting on the picnic table all day at 33°C. You pour 0.24 kg into a Styrofoam cup and add 2 ice cubes (each 0.025 kg at 0°C). (*a*) Assuming no heat lost to the surroundings, what is the final temperature of the lemonade? (*b*) What is the final temperature if you add 6 ice cubes? Assume that the lemonade has the same heat capacity as water.

Picture the Problem We set the heat lost by the lemonade equal to that gained by the ice cubes. Let t_f be the final temperature of the lemonade and water.

(*a*)1. Write the heat lost by the lemonade in terms of the final temperature t_f:

$$Q_{out} = m_L c\, \Delta t = (0.24\text{ kg})(4.18\text{ kJ/kg}\cdot\text{C}°)(33°\text{C} - t_f)$$
$$= 33\text{ kJ} - (1.00\text{ kJ/C}°)t_f$$

2. Write the heat gained by the ice cubes and resulting water in terms of the final temperature:

$$Q_{in} = L_f m_{ice} + m_{ice} c\, \Delta t$$
$$= (0.05\text{ kg})(333.5\text{ kJ/kg})$$
$$+ (0.05\text{ kg})(4.18\text{ kJ/kg}\cdot\text{C}°)(t_f - 0)$$
$$= 16.7\text{ kJ} + (0.209\text{ kJ/C}°)t_f$$

3. Set the heat lost equal to the heat gained and solve for t_f:

$$33\text{ kJ} - (1.00\text{ kJ/C}°)t_f = 16.7\text{ kJ} + (0.209\text{ kJ/C}°)t_f$$
$$t_f = 13.5°\text{C}$$

(*b*)1. For 6 ice cubes, $m_{ice} = 0.15$ kg. Step 1 is the same. Find the heat gained by the ice as in step 2 of part (*a*):

$$Q_{in} = L_f m_{ice} + m_{ice} c\, \Delta t$$
$$= (0.15\text{ kg})(333.5\text{ kJ/kg})$$
$$+ (0.15\text{ kg})(4.18\text{ kJ/kg}\cdot\text{C}°)(t_f - 0)$$
$$= 50.0\text{ kJ} + (0.627\text{ kJ/C}°)t_f$$

2. Set the heat lost equal to the heat gained and solve for t_f:

$$33\text{ kJ} - (1.00\text{ kJ/C}°)t_f = 50.0\text{ kJ} + (0.627\text{ kJ/C}°)t_f$$
$$t_f = -10°\text{C}$$

But this cannot be correct! We know that if we add ice at 0°C to lemonade at 33°C that the final temperature of the mixture cannot be −10°C. What's wrong? The heat given off by the lemonade as it cools from 33 to 0°C is not enough to melt all of the ice, contrary to our assumption that all of the ice melts. The final temperature is thus

$$t_f = 0°\text{C}$$

Check the Result Let's calculate how much ice is melted in part (*b*). For the lemonade to cool from 33 to 0°C, it must give off heat in the amount $Q_{\text{out}} = (0.24\text{ kg})(4.18\text{ kJ/kg}\cdot\text{C°})(33\text{°C}) = 33.1\text{ kJ}$. The mass of ice that this amount of heat will melt is $m_{\text{ice}} = Q_{\text{in}}/L_{\text{f}} = 33.1\text{ kJ}/(333.5\text{ kJ/kg}) = 0.10\text{ kg}$. This is about 4 ice cubes. Adding more than 4 ice cubes does not lower the temperature below 0°C. It merely increases the amount of ice in the ice–lemonade mixture. In problems like this one, we should first find out how much ice must be melted to reduce the temperature of the liquid to 0°C. If less than that amount is added, we can proceed as in part (*a*). If more ice is added, the final temperature is 0°C.

19-3 Joule's Experiment and the First Law of Thermodynamics

We can raise the temperature of a system by adding heat, but we can also raise its temperature by just doing work on it. Figure 19-2*a* is a diagram of the apparatus Joule used in his famous experiment for determining the amount of work needed to raise the temperature of one gram of water by one Celsius degree. Joule's machine converts the potential energy of falling weights into work done on the water. Joule found that he could raise the temperature of his water sample by 1 Fahrenheit degree when he drove his machine by the fall of 772 pounds of weight through one foot. In modern units, Joule found that it takes about 4.184 J (the energy units adopted by the scientific community in 1948) to raise the temperature of 1 g of water by 1 C°. The result that 4.184 J of mechanical energy is equivalent to 1 cal of heat energy is known as the mechanical equivalence of heat.

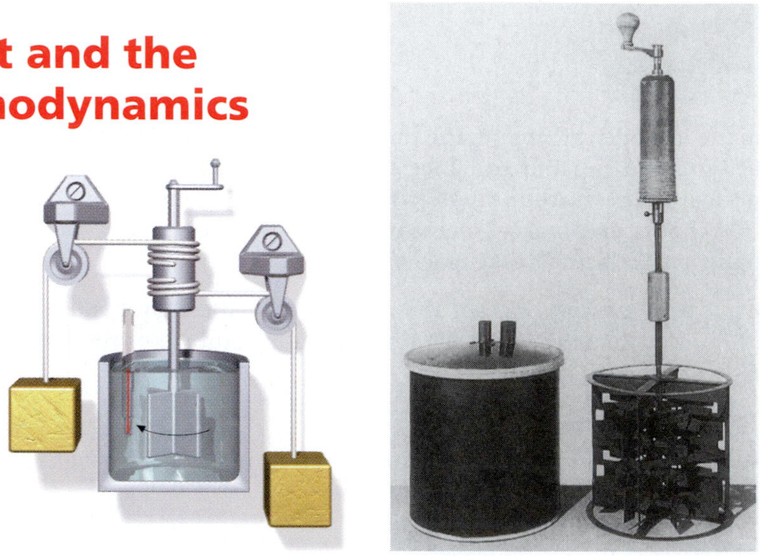

(*a*) (*b*)

Figure 19-2 (*a*) Schematic diagram for Joule's experiment. Water is enclosed by insulating walls to prevent heat transfer. As the weights fall at constant speed, they turn a paddle wheel, which does work on the water. If friction is negligible, the work done by the paddle wheel against the water equals the loss of mechanical energy of the weights, which is determined by calculating the loss in the potential energy of the weights. (*b*) Photograph of the apparatus for Joule's experiment.

There are other ways of doing work on this system. For example, we could drop the insulated container of water from some height h, letting the system make an inelastic collision with the ground, or we could do mechanical work to generate electricity and then use the electricity to heat the water (Figure 19-3). In all such experiments, the same amount of work is required to produce a given temperature change. By the conservation of energy, the work done must go into an increase in internal energy of the system.

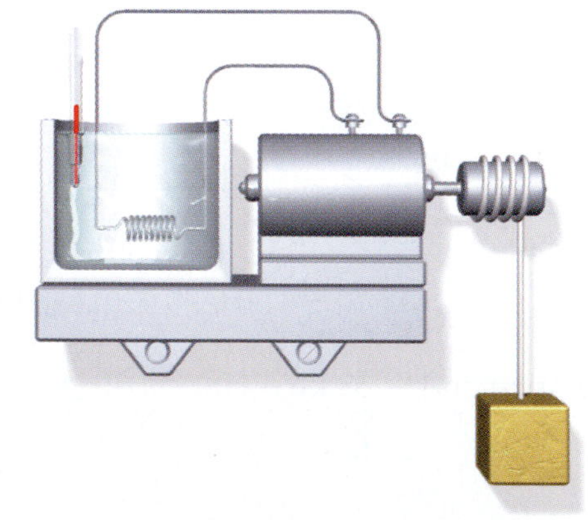

Figure 19-3 Another method of doing work on a thermally insulated container of water. Electrical work is done on the system by the generator, which is driven by the falling weight.

Example 19-5

To demonstrate the equivalence of heat and energy, you drop a thermally insulated container of water from a height *h* to the ground. If the collision is perfectly inelastic and all of the energy lost goes into internal energy of the water, what must *h* be for the temperature of the water to increase by 1 C°?

Picture the Problem The kinetic energy of the water just before it hits the ground equals its original potential energy mgh. During the collision, this energy is converted into thermal energy Q, which in turn causes a rise in temperature given by $Q = mc\,\Delta T$.

1. Set the potential energy equal to the thermal energy: $mgh = mc\,\Delta T$

2. Solve for the height h: $h = c\,\Delta T/g$

3. Substitute $c = 4.18\text{ kJ/kg·K}$ and $\Delta T = 1\text{ C}° = 1\text{ K}$: $h = \dfrac{(4.18\text{ kJ/kg·K})(1\text{ K})}{9.81\text{ N/kg}} = 0.426\text{ km} = 426\text{ m}$

Remarks Note that h is independent of the mass of the water. It is also rather large, which illustrates one of the difficulties with Joule's experiment—a large amount of work must be done to produce a measurable change in the temperature of the water.

Now suppose we perform Joule's experiment but replace the insulating walls of the container with conducting walls. We find that the work needed to produce a given change in the temperature of the system depends on how much heat is added to or subtracted from the system by conduction through the walls. However, if we sum the work done on the system and the net heat added to or subtracted from the system, the result is always the same for a given temperature change. That is, the sum of the heat added and the work done on the system equals the change in the internal energy of the system. This is the **first law of thermodynamics**, which is simply a statement of the conservation of energy.

It is customary to write W for the work done *by* the system on its surroundings.* Then $-W$ is the work done *on* the system. For example, if a gas expands against a piston, doing work on the surroundings, W is positive. The heat Q is taken to be positive if it is put *into* the system, and negative if it is taken *out of* the system (Figure 19-4). Using these conventions, and denoting the internal energy by U, the first law of thermodynamics is written†

$$Q = \Delta U + W \qquad \text{19-10}$$

The heat added to a system equals the change in the internal energy of the system plus the work done by the system.

First law of thermodynamics

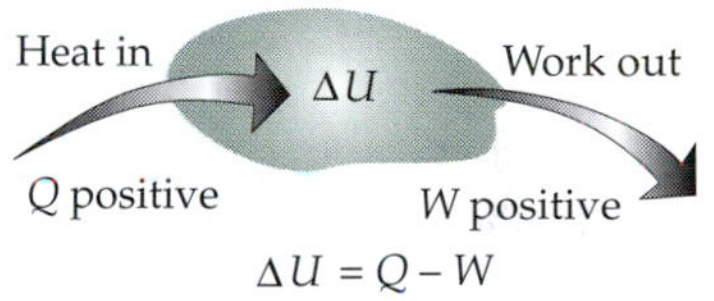

Figure 19-4 Sign convention for the first law of thermodynamics.

Equation 19-10 is the same as the work–energy theorem, $W_{ext} = \Delta E_{sys}$ of Chapter 7 (Equation 7-9), except that we have added the heat term Q, changed the sign convention for W, and called the energy of the system U.

*We choose this so that the work done by an expanding gas is positive, and the work done by a heat engine, which we study in the next chapter, is positive.

†The symbol U, which we used in previous chapters to denote potential energy, is now used for the internal energy of a system, which may include both kinetic and potential energy of the molecules in the system.

Example 19-6

You do 25 kJ of work on a system consisting of 3 kg of water by stirring it with a paddle wheel. During this time, 15 kcal of heat is removed. What is the change in the internal energy of the system?

Picture the Problem We express all energies in joules and apply the first law of thermodynamics.

1. ΔU is found from the first law of thermodynamics: $Q = \Delta U + W$

2. Heat is *removed* from the system, thus the heat *added* is negative: $Q = -15\text{ kcal} = -(15\text{ kcal})\left(\dfrac{4.18\text{ kJ}}{1\text{ kcal}}\right) = -62.7\text{ kJ}$

3. Work is done *on* the system, thus the work done *by* the system is negative: $W = -25\text{ kJ}$

4. Substitute these quantities and solve for ΔU: $\Delta U = Q - W = (-62.7\text{ kJ}) - (-25\text{ kJ}) = -37.7\text{ kJ}$

Remark The internal energy decreases because the system loses more energy as heat than it gains from the work done on it.

It is important to understand that the internal energy U is a function of the state of the system, just as P, V, and T are functions of the state of the system. Consider a gas in some initial state (P_1, V_1). The temperature T_1 is determined by the equation of state. For example, if the gas is ideal, $T_1 = P_1V_1/nR$. The internal energy U_1 also depends only on the state of the gas, which is determined by any two state variables, such as P and V. If we compress the gas or let it expand, add or remove heat from it, do work on it or let it do work, the gas will move through a sequence of states; that is, it will have different values of the state functions P, V, T, and U. If the gas is then returned to its original state (P_1, V_1), the temperature T and the internal energy U must equal their original values.

On the other hand, the net heat input Q and the work W done by the gas are not functions of the state of the system. There are no functions Q or W associated with any particular state of the gas. We could take the gas through a sequence of states beginning and ending at state (P_1, V_1) during which the gas did positive work and absorbed an equal amount of heat. Or we could take it through a different sequence of states such that work was done on the gas and heat was removed from the gas. It is correct, then, to say that a system has a large amount of internal energy, but it is not correct to say that a system has a large amount of heat or a large amount of work. Heat is not something that is contained in a system. Rather, it is a measure of the energy that flows from one system to another because of a difference in temperature.

For very small amounts of heat added, work done, or changes in internal energy, it is customary to write Equation 19-10 as

$$dQ = dU + dW \tag{19-11}$$

In this equation, dU is the differential of the internal-energy function. However, neither dQ nor dW is a differential of any function. Instead, dQ merely represents a small amount of heat added to the system, and dW represents a small amount of work done by the system.

19-4 The Internal Energy of an Ideal Gas

The translational kinetic energy K of the molecules in an *ideal* gas is related to the absolute temperature T by Equation 18-22:

$$K = \tfrac{3}{2}nRT$$

where n is the number of moles of gas and R is the universal gas constant. If the internal energy of a gas is just this translational kinetic energy, then $U = K$, and

$$U = \tfrac{3}{2}nRT \qquad 19\text{-}12$$

Then the internal energy will depend only on the temperature of the gas and not on its volume or pressure. If the molecules have other types of energy in addition to translational energy, such as energy of rotation, the internal energy will be greater than that given by Equation 19-12. But according to the equipartition theorem, the average energy associated with any degree of freedom will be $\frac{1}{2}kT$ per molecule or $\frac{1}{2}RT$ per mole, so again, the internal energy will depend only on the temperature and not on the volume or pressure.

We can imagine that the internal energy of a *real* gas might include other kinds of energy that depend on the pressure and volume of the gas. Suppose, for example, that nearby gas molecules exert attractive forces on each other. Work is then required to increase the separation of the molecules. Then, when the average distance between the molecules is increased, the potential energy associated with the molecular attraction will increase. The internal energy of the gas will then depend on the volume of the gas as well as on its temperature.

Joule, using an apparatus like that shown in Figure 19-5, performed an interesting experiment to determine whether the internal energy of a gas depends on its volume. Initially, the compartment on the left in Figure 19-5 contains a gas and the compartment on the right is evacuated. The compartments are connected by a stopcock that is closed. The whole system is thermally insulated from its surroundings—no heat can go into or out of the system and no work can be done. When the stopcock is opened, the gas rushes into the evacuated chamber. This process is called a **free expansion.** Eventually, the gas reaches thermal equilibrium with itself. Since no work has been done and no heat has been transferred, the final internal energy of the gas must equal its initial internal energy. If the gas molecules exert attractive forces on one another, the potential energy associated with these forces will increase when the volume increases. Since energy is conserved, the kinetic energy of translation must therefore decrease, which will result in a decrease in the temperature of the gas.

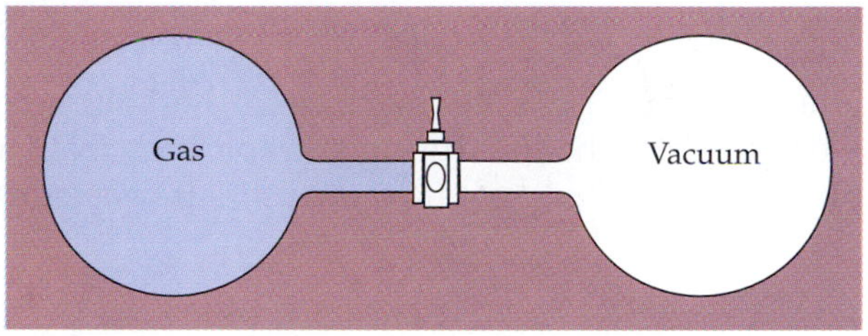

Figure 19-5 Free expansion of a gas. When the stopcock is opened, the gas expands rapidly into the evacuated chamber. Since no work is done and the whole system is thermally insulated, the initial and final internal energies of the gas are equal.

When Joule did this experiment, he found the final temperature to be equal to the initial temperature. Subsequent experiments verified this result for low gas densities. This implies that for a gas at low density—that is, for an ideal gas—the temperature depends only on the internal energy, or as we usually think of it, the internal energy depends only on the temperature. However, when the experiment is done with a large amount of gas initially in the left compartment so that the density is high, the temperature after expansion is slightly lower than that before the expansion. This indicates that there is a small attraction between the gas molecules of a real gas.

19-5 Work and the *PV* Diagram for a Gas

In many types of engines, work is done by a gas expanding against a movable piston. For example, in a steam engine, water is heated in a boiler to produce steam. The steam then does work as it expands and drives a piston. In an automobile engine, a mixture of gasoline vapor and air is ignited, causing it to explode. The resulting high temperatures and pressures cause the gas to expand rapidly, driving a piston and doing work.

Quasi-static Processes

Figure 19-6 shows an ideal gas confined in a container with a tightly fitting piston that we assume to be frictionless. When the piston moves, the volume of the gas changes. The temperature or pressure or both must also change since these three variables are related by the equation of state $PV = nRT$. If we suddenly push the piston in to compress the gas, the pressure will initially be greater near the piston than far from it. Eventually the gas will settle down to a new equilibrium pressure and temperature. Until equilibrium is restored in the gas, we cannot determine such macroscopic variables as T, P, or U for the entire gas system. However, if we move the piston slowly in small steps and allow equilibrium to be reestablished after each step, we can compress or expand the gas in such a way that the gas is never far from an equilibrium state. In this kind of process, called a **quasi-static process,** the gas moves through a series of equilibrium states. In practice, it is possible to approximate quasi-static processes fairly well.

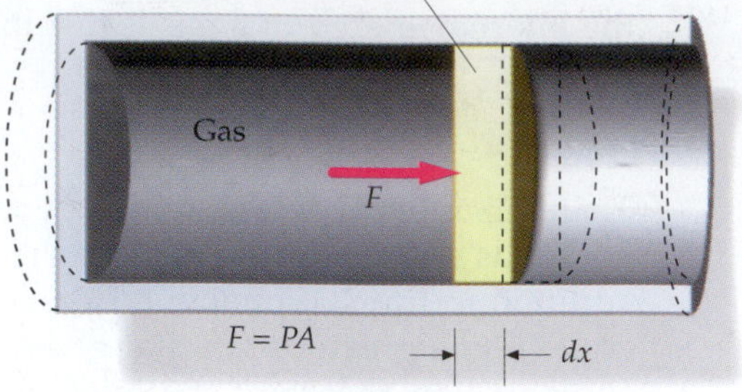

Figure 19-6 Gas confined in a thermally insulated cylinder with a movable piston. When the piston moves a distance dx, the volume of the gas changes by $dV = A\,dx$. The work done by the gas is $PA\,dx = P\,dV$, where P is the pressure.

Let us begin with a gas at a high pressure, and let it expand quasi-statically. The force exerted by the gas on the piston is PA, where A is the area of the piston and P is the gas pressure. If the piston moves a small distance dx, the work done by the gas on the piston is

$$dW = F\,dx = PA\,dx = P\,dV \qquad \text{19-13}$$

where $dV = A\,dx$ is the increase in the volume of the gas. To calculate the work done by the gas during an expansion from a volume of V_1 to a volume of V_2, we need to know how the pressure varies during the expansion.

PV Diagrams

We can represent the states of a gas on a diagram of P versus V. Each point on the PV diagram indicates a particular state of the gas. Figure 19-7 shows a PV diagram with a horizontal line representing a series of states that all have the same value of P. This line represents an expansion at constant pressure. Such a process is called an **isobaric expansion.** For a volume change of ΔV, the work done is $P\,\Delta V$, which is equal to the shaded area under the curve in the figure. In general, the work done by the gas is equal to the area under the P-versus-V curve:

$$W = \int P\,dV = \text{area under the } P\text{-versus-}V \text{ curve} \qquad \text{19-14}$$

Work done by a gas

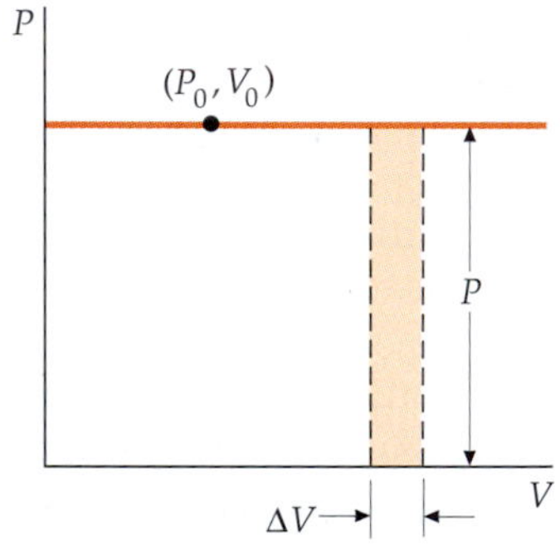

Figure 19-7 Each point on a PV diagram, such as (P_0,V_0), represents a particular state of the gas. The horizontal line represents states with a constant pressure P_0. The work done by a gas as it expands an amount ΔV is represented by the shaded area, $P_0\,\Delta V$.

Since pressures are often given in atmospheres and volumes are often given in liters, it is convenient to have a conversion factor between liter-atmospheres and joules:

$$1\text{ L}\cdot\text{atm} = (10^{-3}\text{ m}^3)(101.3 \times 10^3\text{ N/m}^2) = 101.3\text{ J} \qquad \text{19-15}$$

Exercise If 3 L of an ideal gas at a pressure of 2 atm is heated so that it expands at constant pressure until its volume is 5 L, what is the work done by the gas? (*Answer* 405.2 J)

Figure 19-8 shows three different possible paths on a PV diagram for a gas that is initially in state (P_1,V_1) and is finally in state (P_2, V_2). We assume that the gas is ideal and have chosen the original and final states to have the same temperature so that $P_1V_1 = P_2V_2 = nRT$. Since the internal energy depends only on the temperature, the initial and final internal energies are also the same.

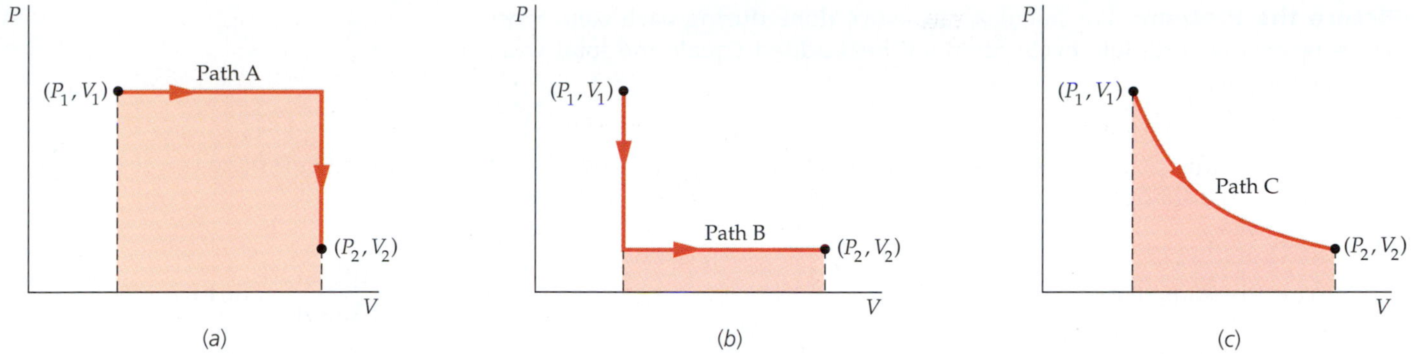

Figure 19-8 Three paths on *PV* diagrams connecting an initial state (P_1, V_1) and a final state (P_2, V_2). The work done along each path is indicated by the shaded area.

In Figure 19-8*a*, the gas is heated at constant pressure until its volume is V_2, after which it is cooled at constant volume until its pressure is P_2. The work done along path A is $P_1(V_2 - V_1)$ for the horizontal part of the path and zero for the constant-volume part.

In Figure 19-8*b*, the gas is first cooled at constant volume until its pressure is P_2, after which it is heated at constant pressure until its volume is V_2. The work done along path B is $P_2(V_2 - V_1)$, which is much less than that done along path A as can be seen by comparing the shaded regions in Figure 19-8*a* and *b*.

In Figure 19-8*c*, path C represents an **isothermal expansion**, meaning that the temperature remains constant. We can calculate the work done along path C by using $P = nRT/V$.

$$dW = P\,dV = \frac{nRT}{V}\,dV$$

Hence, the work done by the gas as it expands from V_1 to V_2 is

$$W = \int_{V_1}^{V_2} P\,dV = \int_{V_1}^{V_2} \frac{nRT}{V}\,dV$$

Since T is constant for an isothermal process, we can remove it from the integral. We then have

$$W_{\text{isothermal}} = nRT\int_{V_1}^{V_2} \frac{dV}{V} = nRT \ln \frac{V_2}{V_1} \qquad \text{19-16}$$

We see that the amount of work done by the gas is different for each process illustrated. Since $U_2 = U_1$ for these states, the net amount of heat added must also be different for each of the processes. This discussion illustrates the fact that both the work done and the heat added depend on just how a system moves from one state to another, but the change in the internal energy of the system does not.

Example 19-7

An ideal gas undergoes a cyclic process from point A to point B to point C to point D and back to point A as shown in Figure 19-9. The gas begins at a volume of 1 L and a pressure of 2 atm and expands at constant pressure until the volume is 2.5 L, after which it is cooled at constant volume until its pressure is 1 atm. It is then compressed at constant pressure until its volume is again 1 L, after which it is heated at constant volume until it is back in its original state. Find the total work done by the gas and the total heat added during the cycle.

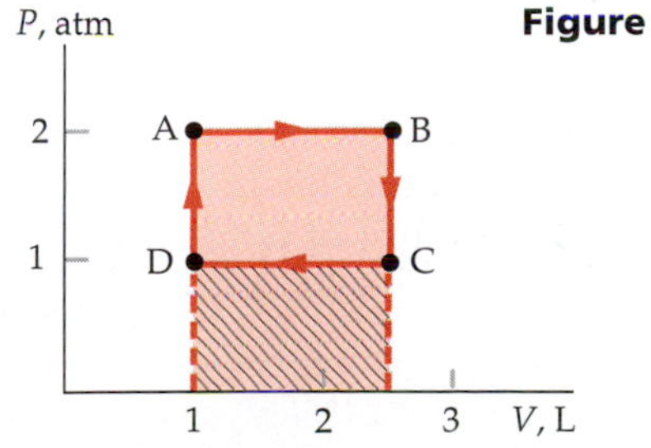

Figure 19-9

Picture the Problem We calculate the work done during each step. Since $\Delta U = 0$ for any complete cycle, the total heat added equals the total work done by the gas.

1. From A to B the pressure is constant. The work done by the gas equals the shaded area under the curve AB in the figure:

$$W_{AB} = P(V_B - V_A) = (2\text{ atm})(2.5\text{ L} - 1\text{ L}) = 3\text{ L}\cdot\text{atm}$$

2. Convert the units to joules:

$$W_{AB} = 3\text{ L}\cdot\text{atm} \times \frac{101.3\text{ J}}{1\text{ L}\cdot\text{atm}} = 304\text{ J}$$

3. From B to C the gas cools at constant volume so the work done is zero:

$$W_{BC} = 0$$

4. As the gas is compressed at constant pressure from point C to point D, it does negative work. The magnitude of the work is the area under the CD curve indicated by cross hatching:

$$W_{CD} = P(V_D - V_C) = (1\text{ atm})(1\text{ L} - 2.5\text{ L}) = -1.5\text{ L}\cdot\text{atm} = -152\text{ J}$$

5. As the gas is heated back to its original state A, the volume is again constant, so no work is done:

$$W_{DA} = 0$$

6. The total work done by the gas is the sum of the work done along each step:

$$W_{\text{total}} = W_{AB} + W_{BC} + W_{CD} + W_{DA} = 304\text{ J} + 0 + (-152\text{ J}) + 0 = 152\text{ J}$$

7. Since the gas is back in its original state, the total change in internal energy is zero:

$$\Delta U = 0$$

8. The heat added is found from the first law:

$$Q = \Delta U + W = W = 152\text{ J}$$

Remarks The net work done by the gas is represented by the shaded area enclosed by the cycle in Figure 19-9 (without cross hatching). Such cyclic processes have important applications for heat engines, as we will see in the next chapter.

19-6 Heat Capacities of Gases

The determination of the heat capacity of a substance provides information about its internal energy, which is related to its molecular structure. For all substances that expand when heated, the heat capacity at constant pressure C_p is greater than the heat capacity at constant volume C_v. When heat is added at constant pressure, the substance expands and does work so it takes more heat for a given temperature change than if heated at constant volume. The expansion is usually negligible for solids and liquids, so for them $C_p \approx C_v$. But a gas heated at constant pressure readily expands and does a significant amount of work, so $C_p - C_v$ is not negligible.

When heat is added to a gas at constant volume, no work is done, so the heat added equals the increase in the internal energy of the gas. Writing Q_v for the heat added at constant volume, we have

$$Q_v = C_v \Delta T$$

Since $W = 0$, we have from the first law of thermodynamics

$$Q_v = \Delta U + W = \Delta U$$

Thus,

$$\Delta U = C_v \Delta T$$

Taking the limit as ΔT approaches zero, we obtain

$$dU = C_v\, dT \qquad \text{19-17}$$

and

$$C_v = \frac{dU}{dT} \qquad \text{19-18}$$

The heat capacity at constant volume is the rate of change of the internal energy with temperature. Since U and T are state functions, Equations 19-17 and 19-18 hold for any process.

Now let's calculate the difference $C_p - C_v$ for an ideal gas. From the definition of C_p, the heat added at constant pressure is

$$Q_p = C_p \Delta T$$

From the first law of thermodynamics,

$$Q_p = \Delta U + W = \Delta U + P\,\Delta V$$

Then

$$C_p \Delta T = \Delta U + P\,\Delta V$$

For infinitesimal changes, this becomes

$$C_p\, dT = dU + P\, dV$$

Using Equation 19-17 for dU, we obtain

$$C_p\, dT = C_v dT + P\, dV \qquad \text{19-19}$$

The pressure, volume, and temperature for an ideal gas are related by

$$PV = nRT$$

Take the differentials of both sides of the ideal-gas law, with $dP = 0$ for constant pressure.

$$P\, dV + V\, dP = P\, dV = nR\, dT$$

Substituting this into Equation 19-19 gives

$$C_p\, dT = C_v\, dT + nR\, dT$$

Therefore,

$$C_p = C_v + nR \qquad \text{19-20}$$

For an ideal gas, the heat capacity at constant pressure is greater than that at constant volume by the amount nR.

Table 19-3

Molar Heat Capacities (J/mol·K) of Various Gases at 25°C

Gas	c'_p	c'_v	c'_v/R	$c'_p - c'_v$	$(c'_p - c'_v)/R$
Monatomic					
He	20.79	12.52	1.51	8.27	0.99
Ne	20.79	12.68	1.52	8.11	0.98
Ar	20.79	12.45	1.50	8.34	1.00
Kr	20.79	12.45	1.50	8.34	1.00
Xe	20.79	12.52	1.51	8.27	0.99
Diatomic					
N_2	29.12	20.80	2.50	8.32	1.00
H_2	28.82	20.44	2.46	8.38	1.01
O_2	29.37	20.98	2.52	8.39	1.01
CO	29.04	20.74	2.49	8.30	1.00
Polyatomic					
CO_2	36.62	28.17	3.39	8.45	1.02
N_2O	36.90	28.39	3.41	8.51	1.02
H_2S	36.12	27.36	3.29	8.76	1.05

Table 19-3 lists measured molar heat capacities c'_p and c'_v for several gases. We note from this table that the ideal gas prediction, $c'_p - c'_v = R$, holds quite well for all gases. The table also shows that c'_v is approximately $1.5R$ for all monatomic gases, $2.5R$ for all diatomic gases, and greater than $2.5R$ for gases consisting of more complex molecules. We can understand these results by considering the molecular model of a gas discussed in Chapter 18. The total translational kinetic energy of n moles of a gas is $K = \frac{3}{2}nRT$ (Equation 18-22). Thus, if the internal energy of a gas consists of translational kinetic energy only, we have

$$U = \frac{3}{2}nRT \qquad 19\text{-}21$$

The heat capacities are then

$$C_v = \frac{dU}{dT} = \frac{3}{2}nR \qquad 19\text{-}22$$

C_v for an ideal monatomic gas

and

$$C_p = C_v + nR = \frac{5}{2}nR \qquad 19\text{-}23$$

C_p for an ideal monatomic gas

The results in Table 19-3 agree well with these predictions for monatomic gases, but for other gases, the heat capacities are greater than those predicted by Equations 19-22 and 19-23. The internal energy for a gas consisting of diatomic or more complicated molecules is evidently greater than $\frac{3}{2}nRT$. The reason is that such molecules can have other types of energy, such as rotational or vibrational energy, in addition to translational kinetic energy.

Example 19-8 ***try it yourself***

A system consisting of 0.32 mol of a monatomic ideal gas with $c'_{\mathrm{v}} = \frac{3}{2}R$ occupies a volume of 2.2 L at a pressure of 2.4 atm, as represented by point A in Figure 19-10. The system is carried through a cycle consisting of three processes:

1. The gas is heated at constant pressure until its volume is 4.4 L at point B.
2. The gas is cooled at constant volume until the pressure decreases to 1.2 atm (point C).
3. The gas undergoes an isothermal compression back to point A.

(*a*) What is the temperature at points A, B, and C? (*b*) Find W, Q, and ΔU for each process and for the entire cycle.

Figure 19-10

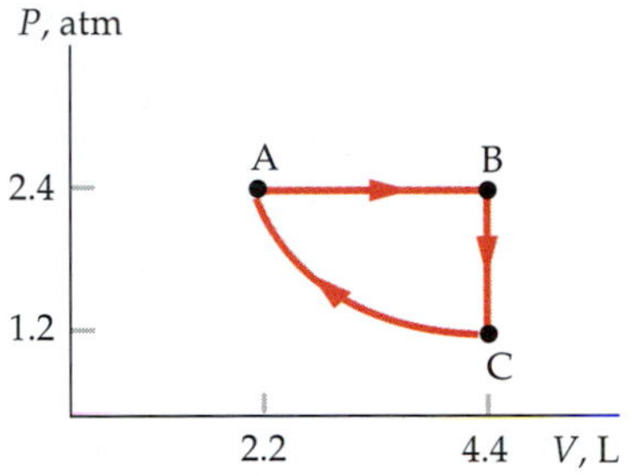

Picture the Problem You can find the temperatures at all points from the ideal-gas law. You can find the work for each process by finding the area under the curve, and the heat exchanged from the given heat capacity and the initial and final temperatures for each process. In process 3, T is constant, so $\Delta U = 0$ and the heat input equals the work done.

Cover the column to the right and try these on your own before looking at the answers.

Steps	***Answers***
(*a*) Find the temperatures at points A, B, and C using the ideal-gas law.	$T_A = T_C = 201$ K, $T_B = 402$ K
(*b*)1. For process 1, use $W_1 = P_C\,\Delta V$ to calculate the work, and $C_p = \frac{5}{2}nR$ to calculate the heat Q_1. Then use W_1 and Q_1 to calculate ΔU_2.	$W_1 = 5.28\ \mathrm{L{\cdot}atm} = 535$ J, $Q_1 = 1337$ J $\Delta U_1 = Q_1 - W_1 = 802$ J
2. For process 2, use $C_v = \frac{3}{2}nR$ and $T_C - T_B$ from step 1 to find Q_2. Then, since $W_2 = 0$, $\Delta U = Q_2$.	$W_2 = 0$, $Q_2 = -802$ J, $\Delta U_2 = -802$ J
3. Calculate W_1 from $W = nRT\ln(V_A/V_C)$ in the isothermal compression. Then, since $\Delta U_1 = 0$, $Q_3 = W_3$.	$W_3 = -371$ J, $Q_3 = -371$ J, $\Delta U_3 = 0$
5. Find the total work W, the total heat Q, and the total change ΔU by summing the quantities found in steps 2, 3, and 4.	$W_{\mathrm{total}} = W_1 + W_2 + W_3 = 535\ \mathrm{J} + 0 + (-371\ \mathrm{J}) = 164$ J $Q_{\mathrm{total}} = Q_1 + Q_2 + Q_3 = 1337\ \mathrm{J} + (-802\ \mathrm{J}) + (-371\ \mathrm{J}) = 164$ J $\Delta U_{\mathrm{total}} = \Delta U_1 + \Delta U_2 + \Delta U_3 = 802\ \mathrm{J} + (-802\ \mathrm{J}) + 0 = 0$

Remarks The total change in internal energy is zero, as it must be for a cyclic process. The work done by the system equals the heat absorbed. This work equals the area under the AB curve minus the area under the CA curve, which equals the area enclosed by the complete figure.

Heat Capacities and the Equipartition Theorem

According to the equipartition theorem stated in Chapter 18, the internal energy of n moles of a gas should equal $\frac{1}{2}nRT$ for each degree of freedom of the gas molecule. The heat capacity at constant volume of a gas should then be $\frac{1}{2}nR$ times the number of degrees of freedom of the molecule. From Table 19-2, nitrogen, oxygen, hydrogen, and carbon monoxide all have molar heat capacities at constant volume of about $\frac{5}{2}R$. Thus, the molecules in each of these gases have five degrees of freedom (Figure 19-11). About 1880, Clausius speculated that these gases must consist of diatomic molecules that can rotate about two axes, giving them two additional degrees of freedom. The two degrees of freedom besides the three for translation are now known to be associated with their rotation about each of the two axes, x' and y', perpendicular to the line joining the atoms. The kinetic energy of a diatomic molecule is therefore

$$K = \tfrac{1}{2}mv_x^2 + \tfrac{1}{2}mv_y^2 + \tfrac{1}{2}mv_z^2 + \tfrac{1}{2}I_{x'}\omega_{x'}^2 + \tfrac{1}{2}I_{y'}\omega_{y'}^2$$

The total internal energy of n moles of such a gas is then

$$U = 5 \times (\tfrac{1}{2}nRT) = \tfrac{5}{2}nRT \qquad 19\text{-}24$$

and the heat capacity at constant volume is

$$C_V = \tfrac{5}{2}nR \qquad 19\text{-}25$$

Apparently, diatomic gases do not rotate about the line joining the two atoms—if they did, there would be six degrees of freedom and C_V would be $\frac{6}{2}nR = 3nR$, contrary to experiment. Furthermore, monatomic gases apparently do not rotate at all. We will see in Section 19-8 that these puzzling facts are easily explained when we take into account the quantization of energy.

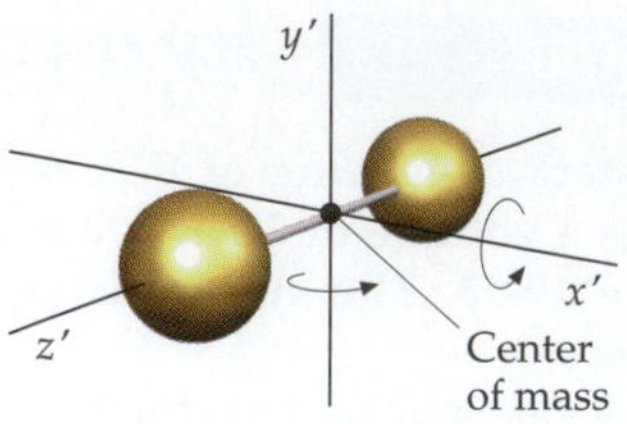

Figure 19-11 Rigid-dumbbell model of a diatomic molecule.

Example 19-9

One mole of oxygen gas is heated from a temperature of 20°C and a pressure of 1 atm to a temperature of 100°C. Assume that oxygen is an ideal gas. (*a*) How much heat must be supplied if the volume is kept constant during the heating? (*b*) How much heat must be supplied if the pressure is kept constant? (*c*) How much work does the gas do in part (*b*)?

Picture the Problem The heat needed for constant-volume heating is $Q_V = C_V\,\Delta T$, where $C_V = \frac{5}{2}nR = \frac{5}{2}R$ since oxygen is a diatomic gas, and $n = 1$ mole. For constant-pressure heating, $Q_P = C_P\,\Delta T$, where $C_P = C_V + R$. Finally, the amount of work done can be found from $W = Q - \Delta U$, or from $W = P\,\Delta V$.

(*a*) 1. Write the heat needed for constant volume in terms of C_V and ΔT: $\quad Q_V = C_V\,\Delta T$

2. Calculate the heat capacity at constant volume: $\quad C_V = \frac{5}{2}nR = \frac{5}{2}(1\text{ mol})(8.31\text{ J/mol}\cdot\text{K}) = 20.8\text{ J/K}$

3. Calculate the heat for $\Delta T = 80\text{ C}° = 80\text{ K}$: $\quad Q_V = C_V\,\Delta T = (20.8\text{ J/K})(80\text{ K}) = 1.66\text{ kJ}$

(*b*) 1. Write the heat needed for constant pressure in terms of C_P and ΔT: $\quad Q_P = C_P\,\Delta T$

2. Calculate the heat capacity at constant pressure: $\quad C_P = C_V + nR = 20.8\text{ J/K} + (1\text{ mol})(8.31\text{ J/mol}\cdot\text{K}) = 29.1\text{ J/K}$

3. Calculate the heat added at constant pressure for $\Delta T = 80\text{ K}$: $\quad Q_P = C_P\,\Delta T = (29.1\text{ J/K})(80\text{ K}) = 2.33\text{ kJ}$

(*c*) 1. The work W can be found from the first law of thermodynamics: $\quad W = Q - \Delta U$

2. The internal energy change equals the heat added at constant volume in (a), since there was no work done: $\Delta U = Q_v = C_V \Delta T = 1.66 \text{ kJ}$

3. The work done at constant pressure is then: $W = Q_p - \Delta U = 2.33 \text{ kJ} - 1.66 \text{ kJ} = 0.67 \text{ kJ}$

Remarks Note that the change in internal energy is the same independent of the process. It depends only on the initial and final states.

Exercise Find the initial and final volumes of this gas from the ideal-gas law, and use them to calculate the work done when the heat is added at constant pressure from $W = P\,\Delta V$. (*Answers* $V_i = 24.0$ L, $V_f = 30.6$ L, $W = 6.6$ L·atm $= 0.67$ kJ)

19-7 Heat Capacities of Solids

In Section 19-1, we noted that all of the metals listed in Table 19-1 have approximately equal molar specific heats. Experimentally, most solids have molar heat capacities approximately equal to $3R$:

$$c' = 3R = 24.9 \text{ J/mol·K} \qquad \text{19-26}$$

This result is known as the **Dulong–Petit law.** We can understand it by applying the equipartition theorem to the simple model for a solid shown in Figure 19-12. According to this model, a solid consists of a regular array of atoms in which each of the atoms has a fixed equilibrium position and is connected by springs to its neighbors. Each atom can vibrate in the x, y, and z directions. The total energy of an atom in a solid is thus

$$E = \tfrac{1}{2}mv_x^2 + \tfrac{1}{2}mv_y^2 + \tfrac{1}{2}mv_z^2 + \tfrac{1}{2}k_{\text{eff}}x^2 + \tfrac{1}{2}k_{\text{eff}}y^2 + \tfrac{1}{2}k_{\text{eff}}z^2$$

where k_{eff} is the effective force constant of the hypothetical springs. Each atom thus has six degrees of freedom. The equipartition theorem states that a substance in equilibrium has an average energy of $\frac{1}{2}RT$ per mole for each degree of freedom. Thus, the internal energy of a mole of a solid is

$$U_m = 6 \times \tfrac{1}{2}RT = 3RT \qquad \text{19-27}$$

which means that c' is equal to $3R$.

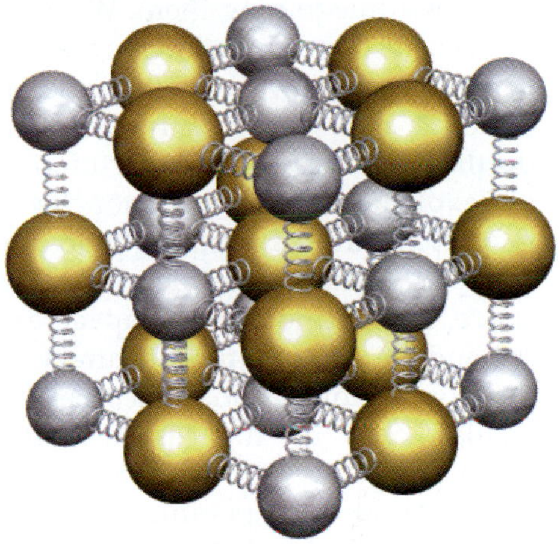

Figure 19-12 Model of a solid in which the atoms are connected to each other by springs. The internal energy of the molecule consists of the kinetic and potential energies of vibration.

Example 19-10

The molar mass of copper is 63.5 g/mol. Use the Dulong–Petit law to calculate the specific heat of copper.

Picture the Problem The Dulong–Petit law gives the molar specific heat of a solid, c'. The specific heat is then $c = c'/M$ (Equation 19-6), where M is the molar mass.

1. The Dulong–Petit law gives c': $c' = 3R = 3(8.31 \text{ J/mol·K}) = 24.9 \text{ J/mol·K}$

2. Using $M = 63.5$ g/mol for copper, the specific heat is: $c = \dfrac{c'}{M} = \dfrac{24.9 \text{ J/mol·K}}{63.5 \text{ g/mol}} = 0.392 \text{ J/g·K} = 0.392 \text{ kJ/kg·K}$

Remarks This solution agrees fairly closely with the measured value of 0.386 kJ/kg·K given in Table 19-1.

Exercise The specific heat of a certain metal is measured to be 1.02 kJ/kg·K. (*a*) Calculate the molar mass of this metal, assuming that the metal obeys the Dulong–Petit law. (*b*) What is the metal? (*Answers* (*a*) $M = 24.4$ g/mol. (*b*) The metal must be magnesium, which has a molar mass of 24.31 g/mol.)

19-8 Failure of the Equipartition Theorem

Although the equipartition theorem had spectacular successes in explaining the heat capacities of gases and solids, it had equally spectacular failures. For example, if a diatomic gas molecule like the one in Figure 19-11 rotates about the line joining the atoms, there should be an additional degree of freedom. Similarly, if a diatomic molecule is not rigid, the two atoms should vibrate along the line joining them. We would then have two more degrees of freedom corresponding to kinetic and potential energies of vibration. According to the measured values of the molar heat capacities in Table 19-2, however, diatomic gases apparently do not rotate about the line joining them, nor do they vibrate. The equipartition theorem gives no explanation for this, nor for the fact that monatomic atoms apparently do not rotate about any of the three possible perpendicular axes in space. Furthermore, heat capacities are found to depend on temperature, contrary to the predictions of the equipartition theorem. The most spectacular case of the temperature dependence of heat capacity is that of H_2, shown in Figure 19-13. At low temperatures, H_2 behaves like a monatomic molecule that does not rotate. At very high temperatures, H_2 begins to vibrate, but the molecule dissociates before c'_v reaches $\frac{7}{2}R$. Finally, the equipartition theorem predicts a constant value of $3R$ for the heat capacity of solids. This result holds for many solids at high temperatures, although not all; however, it does not hold at very low temperatures.

Figure 19-13 Temperature dependence of the molar heat capacity of H_2.

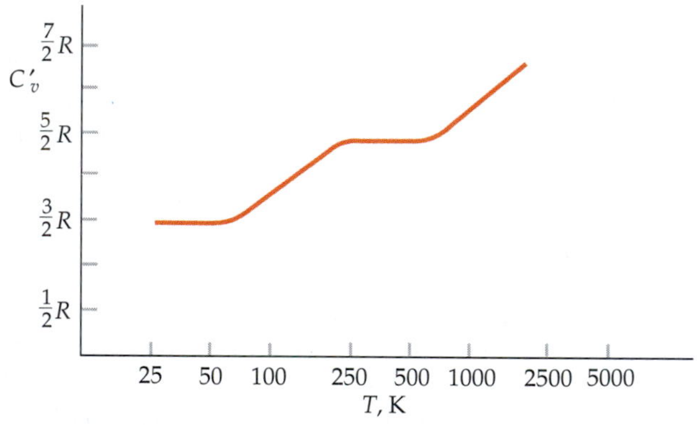

The equipartition theorem fails because energy is quantized. A molecule can have only certain values of energy, as illustrated schematically by the energy-level diagram in Figure 19-14. The molecule can gain or lose energy only if the gain or loss takes it to another allowed level. For example, in a gas the energy that can be transferred between the molecules in collisions is of the order of kT, the typical thermal energy of a molecule. The validity of the equipartition theorem depends on the relative size of kT and the spacing of the energy levels.

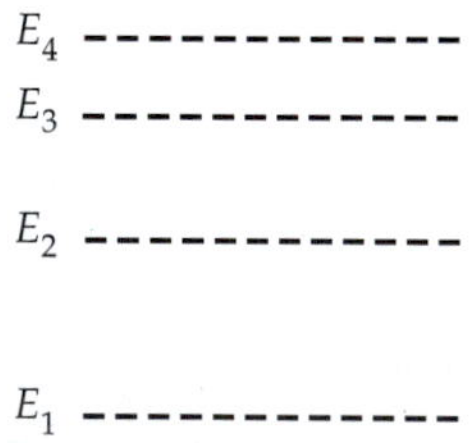

Figure 19-14 Energy-level diagram. A system can have only certain discrete energies.

> If the spacing of the levels is large compared with kT, energy cannot be transferred by collisions and the classical equipartition theorem is not valid. If the spacing of the levels is much smaller than kT, energy quantization will not be noticed and the equipartition theorem will hold.

Conditions for the validity of the equipartition theorem

Consider the rotation of a molecule. The energy of rotation is

$$E = \frac{1}{2}I\omega^2 = \frac{(I\omega)^2}{2I} = \frac{L^2}{2I} \qquad \text{19-28}$$

where I is the moment of inertia of the molecule, ω is its angular velocity, and $L = Iw$ is its angular momentum. In Section 10-5, we mentioned that angular momentum is quantized, and its magnitude is restricted to

$$L = \sqrt{\ell(\ell + 1)}\,\hbar, \qquad \ell = 0, 1, 2, \ldots \qquad \text{19-29}$$

where $\hbar = h/2\pi$, and h is Planck's constant. The energy of a rotating molecule is therefore quantized to the values

$$E = \frac{L^2}{2I} = \frac{\ell(\ell + 1)\hbar^2}{2I} = \ell(\ell + 1)E_{0r} \qquad \text{19-30}$$

where

$$E_{0r} = \frac{\hbar^2}{2I} \qquad \text{19-31}$$

is characteristic of the energy gap between levels. If this energy is much less than kT, we expect classical physics and the equipartition theorem to hold. Let us define a critical temperature T_c by

$$kT_c = E_{0r} = \frac{\hbar^2}{2I} \qquad \text{19-32}$$

When T is much greater than this critical temperature, kT will be much greater than the spacing of the energy levels, which is of the order of kT_c, and we expect classical physics and the equipartition theorem to be valid. When T is less than or of the order of T_c, kT will not be much greater than the energy-level spacing, and we expect classical physics and the equipartition theorem to break down. We will estimate T_c for some cases of interest.

1. *Rotation of H_2 about an axis perpendicular to the line joining the H atoms and through the center of mass (see Figure 19-11):* The moment of inertia of H_2 is

$$I_H = 2M_H\left(\frac{r_s}{2}\right)^2 = \frac{1}{2}M_H r_s^2$$

where M_H is the mass of a H atom, and r_s is the separation distance. For hydrogen, $M_H = 1.67 \times 10^{-27}$ kg, and $r_s \approx 8 \times 10^{-11}$ m. The critical temperature is then

$$T_c = \frac{\hbar^2}{2kI_H} = \frac{\hbar^2}{kM_H r_s^2}$$

$$\approx \frac{(1.05 \times 10^{-34}\,\text{J}\cdot\text{s})^2}{(1.38 \times 10^{-23}\,\text{J/K})(1.67 \times 10^{-27}\,\text{kg})(8 \times 10^{-11}\text{m})^2} \approx 75\ \text{K}$$

As can be seen from Figure 19-13, this is approximately the temperature below which the rotational energy does not contribute to the heat capacity.

2. *O_2*: Since the mass of O_2 is about 16 times that of H_2, and the separation is about the same, the critical temperature for O_2 should be about $75/16 \approx 4.7$ K. For all temperatures for which O_2 exists as a gas, $T >> T_c$, so kT is much greater than the energy level spacing, and we expect the equipartition theorem of classical physics to apply.

3. *Rotation of a monatomic gas:* We consider the He atom. The mass of the electron is about 2000 times smaller than that of the nucleus. But the radius of the nucleus is about 100,000 times smaller than the distance to the electron. So, the moment of inertia of the atom is almost entirely due to its electrons. The distance from the nucleus to the two electrons in He is about half the separation distance of the H atoms in H_2. Thus, using $m_e = M_H/2000$ and $r = r_s/2$, the moment of inertia of the two electrons in He is roughly

$$I_{He} = 2m_e r^2 \approx 2\left(\frac{M_H}{2000}\right)\left(\frac{r_s}{2}\right)^2 = \frac{I_H}{2000}$$

The critical temperature for He is thus about 2000 times that of H_2 or about 150,000 K. This is much higher than the dissociation temperature (the temperature at which electrons are stripped from their nuclei) for helium. So the gap between allowed energy levels is always much greater than kT and the He molecules cannot be induced to rotate by collisions occurring in the gas. Other monatomic gases have slightly greater moments of inertia because they have more electrons, but their critical temperatures are still tens of thousands of kelvins, so their molecules also cannot be induced to rotate by collisions occurring in the gas.

4. *Rotation of a diatomic gas about the axis joining the atoms:* From our discussion of monatomic gases, we see that the moment of inertia for the diatomic case will also be due mainly to the electrons and will be of the same order of magnitude as for a monatomic gas. Again, the critical temperature T_c calculated in order for this rotation to occur due to collisions between molecules in the gas exceeds the gas's dissociation temperature, making rotation under those circumstances impossible.

It is interesting to note that the successes of the equipartition theorem in explaining the measured heat capacities of gases and solids led to the first real understanding of molecular structure in the nineteenth century, whereas its failures played an important role in the development of quantum mechanics in the twentieth century.

Example 19-11

(*a*) Estimate the lowest (nonzero) rotational energy for the hydrogen atom and compare it to kT at room temperature, $T = 300$ K. (*b*) Calculate the critical temperature T_c.

Picture the Problem From Equation 19-29, the lowest rotational energy is for $\ell = 1$. Then $E = 2E_{0r} = \hbar^2/I$. Since we can neglect the moment of inertia of the nucleus (because its radius is 100,000 times smaller than the radius of the atom), the moment of inertia for the atom is essentially the moment of inertia of the electron. Then $I = m_e r^2$, where $r \approx 5 \times 10^{-11}$ m is the distance to the electron.

(*a*)1. The lowest energy greater than zero occurs for $\ell = 1$:

$$E = \frac{\ell(\ell + 1)\hbar^2}{2I} = \frac{1(2)\hbar^2}{2I} = \frac{\hbar^2}{m_e r^2}$$

2. The numerical values are:

$$\hbar = 1.05 \times 10^{-34}\ \text{J}\cdot\text{s}$$
$$m_e = 9.11 \times 10^{-31}\ \text{kg}$$
$$r = 5 \times 10^{-11}\ \text{m}$$

3. Substitute the numerical values:

$$E_1 = \frac{\hbar^2}{m_e r^2} = \frac{(1.05 \times 10^{-34}\ \text{J}\cdot\text{s})^2}{(9.11 \times 10^{-31}\ \text{kg})(5 \times 10^{-11}\ \text{m})^2} = 4.8 \times 10^{-18}\ \text{J}$$

4. The value of kT at $T = 300$ K is:

$$kT = (1.38 \times 10^{-23}\ \text{J/K})(300\ \text{K}) = 4.1 \times 10^{-21}\ \text{J}$$

(*b*) Set $kT_c = E_1$ and solve for T_c:

$$kT_c = E_1 = 4.8 \times 10^{-18}\ \text{J}$$
$$T_c = \frac{E_1}{k} = \frac{4.8 \times 10^{-18}\ \text{J}}{1.38 \times 10^{-23}\ \text{J/K}} = 3.48 \times 10^5\ \text{K}$$

Remarks The lowest rotational energy is about 1000 times kT at room temperature. The critical temperature of a hydrogen atom is so high that the atom would be ionized well before the critical temperature could be reached.

19-9 The Quasi-static Adiabatic Expansion of a Gas

A process in which no heat flows into or out of a system is called an **adiabatic process.** Consider the quasi-static adiabatic expansion of a gas in which the gas in a thermally insulated container expands slowly against a piston, doing work on it. Since no heat enters or leaves the gas, the work done by the gas equals the decrease in the internal energy of the gas, and the temperature of the gas decreases. The curve representing this process on a PV diagram is shown in Figure 19-15.

We can find the equation for the adiabatic curve for an ideal gas using the equation of state and the first law of thermodynamics. We have

$$dQ = dU + dW = C_v\,dT + P\,dV = 0 \qquad \text{19-33}$$

where we have used $dU = C_v\,dT$ from Equation 19-11. Then, using $P = nRT/V$,

$$C_v dT + nRT\frac{dV}{V} = 0$$

Rearranging gives

$$\frac{dT}{T} + \frac{nR}{C_v}\frac{dV}{V} = 0 \qquad \text{19-34}$$

Equation 19-34 can be simplified by noting that $C_p - C_v = nR$, so

$$\frac{nR}{C_v} = \frac{C_p - C_v}{C_v} = \frac{C_p}{C_v} - 1 = \gamma - 1$$

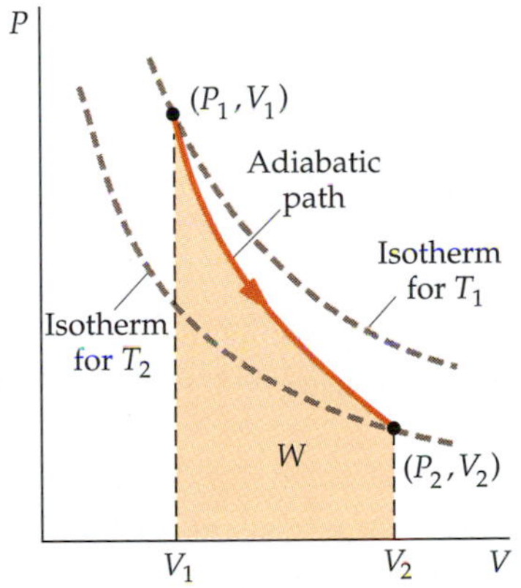

Figure 19-15 Quasi-static adiabatic expansion of an ideal gas. The dashed lines are the isotherms for the initial and final temperatures. The curve connecting the initial and final states of the adiabatic expansion is steeper than the isotherms because the temperature drops during the expansion.

where γ is the ratio of the heat capacities:

$$\gamma = \frac{C_p}{C_v} \tag{19-35}$$

We then have

$$\frac{dT}{T} + (\gamma - 1)\frac{dV}{V} = 0$$

Integration gives

$$\ln T + (\gamma - 1) \ln V = \text{constant}$$

Using the properties of logarithms (Appendix D), we obtain

$$\ln(TV^{\gamma-1}) = \text{constant}$$

or

$$TV^{\gamma-1} = \text{constant} \tag{19-36}$$

Quasi-static adiabatic process

Clouds form when rising moist air cools due to adiabatic expansion of the air and then condenses into liquid droplets.

where the constants in the two preceding equations are not the same. We can eliminate T from Equation 19-36 using $PV = nRT$. We then have

$$\frac{PV}{nR} V^{\gamma-1} = \text{constant}$$

or

$$PV^{\gamma} = \text{constant} \tag{19-37}$$

Quasi-static adiabatic process

Equation 19-37 relates P and V for adiabatic expansions and compressions.

The work done by the gas in an adiabatic expansion can be calculated from the first law of thermodynamics:

$$dQ = dU + dW = dU + p\,dV$$

Since $dQ = 0$, we have

$$dW = -dU = -C_v\,dT$$

Then

$$W_{\text{adiabatic}} = \int dW = \int -C_v\,dT = -C_v\,\Delta T \tag{19-38}$$

Adiabatic work

where we have assumed that C_v is constant.* We note that the work done by the gas depends only on the change in the absolute temperature of the gas. In an adiabatic expansion, the gas does work, and its internal energy and temperature decrease. In an adiabatic *compression*, work is done *on* the gas, and the internal energy and temperature increase.

* For an ideal gas, U is proportional to the absolute temperature, and therefore $C_v = dU/dT$ is a constant.

We can use the ideal-gas law to write Equation 19-38 in terms of the initial and final values of the pressure and volume. If T_1 is the initial temperature and T_2 is the final temperature, we have for the work done

$$W_{\text{adiabatic}} = -C_v\,\Delta T = -C_v(T_2 - T_1) = C_v(T_1 - T_2)$$

Using $PV = nRT$, we obtain

$$W_{\text{adiabatic}} = C_v\left(\frac{P_1V_1}{nR} - \frac{P_2V_2}{nR}\right) = \frac{C_v}{C_p - C_v}(P_1V_1 - P_2V_2)$$

where we have used $nR = C_p - C_v$. Dividing the numerator and denominator by C_v and writing γ for C_p/C_v, we obtain

$$W_{\text{adiabatic}} = \frac{P_1V_1 - P_2V_2}{\gamma - 1} \qquad \text{19-39}$$

Adiabatic work

Example 19-12

A quantity of air ($\gamma = 1.4$) expands adiabatically and quasi-statically from an initial pressure of 2 atm and volume of 2 L at temperature 20°C to twice its original volume. Find (*a*) the final pressure, (*b*) the final temperature, and (*c*) the work done by the gas.

Picture the Problem Since the process is adiabatic, we know that $PV^{\gamma} =$ constant, and $TV^{\gamma-1} =$ constant. These relations yield the final pressure and final temperature, respectively. The work done is found from Equation 19-39. Let subscript 1 refer to initial values, and subscript 2 to final values. Then $P_1 = 2$ atm, $V_1 = 2$ L, $V_2 = 4$ L, and $T_1 = 20°\text{C} = 293$ K.

(*a*)1. Write $PV^{\gamma} =$ constant in terms of initial and final values:

$$P_1V_1^{\gamma} = P_2V_2^{\gamma}$$

2. Solve for P_2:

$$P_2 = P_1\left(\frac{V_1}{V_2}\right)^{\gamma} = (2\text{ atm})\left(\frac{2\text{ L}}{4\text{ L}}\right)^{1.4} = 0.758\text{ atm}$$

(*b*)1. Write $TV^{\gamma-1} =$ constant in terms of initial and final values:

$$T_1V_1^{\gamma-1} = T_2V_2^{\gamma-1}$$

2. Solve for T_2:

$$T_2 = T_1\left(\frac{V_1}{V_2}\right)^{\gamma-1} = (293\text{ K})\left(\frac{2\text{ L}}{4\text{ L}}\right)^{0.4} = 222\text{ K} = -51°\text{C}$$

(*c*) Equation 19-39 gives the work done:

$$W_{\text{adiabatic}} = \frac{P_1V_1 - P_2V_2}{\gamma - 1} = \frac{(2\text{ atm})(2\text{ L}) - (0.758\text{ atm})(4\text{ L})}{1.4 - 1}$$

$$= 2.42\text{ L}\cdot\text{atm} = 245\text{ J}$$

Speed of Sound Waves Let's use Equation 19-37 to calculate the adiabatic bulk modulus of an ideal gas, which is related to the speed of sound waves in air. We first compute the differential of both sides of Equation 19-37:

$$P\,d(V^{\gamma}) + V^{\gamma}\,dP = 0$$

or

$$P\gamma V^{\gamma-1}\,dV + V^{\gamma}\,dP = 0$$

Then

$$dP = -\frac{\gamma P\,dV}{V}$$

Referring to Equation 13-6, the adiabatic bulk modulus* is then:

$$B_{\text{adiab}} = -\frac{dP}{dV/V} = \gamma P \qquad \textbf{19-40}$$

The speed of sound (Equation 15-4) is given by

$$v = \sqrt{\frac{B_{\text{adiab}}}{\rho}}$$

where the mass density ρ is related to the number of moles n and the molecular mass M by $\rho = m/V = nM/V$. Using the ideal gas law, $PV = nRT$, we can eliminate V from the density:

$$\rho = \frac{nM}{V} = \frac{nM}{nRT/P} = \frac{MP}{RT}$$

Using this result and γP for B_{adiab}, we obtain

$$v = \sqrt{\frac{B_{\text{adiab}}}{\rho}} = \sqrt{\frac{\gamma P}{MP/RT}} = \sqrt{\frac{\gamma RT}{M}}$$

which is Equation 15-5 for the speed of sound in a gas.

*The bulk modulus is the negative ratio of the pressure change to the fractional change in volume, $B = -\Delta P/(\Delta V/V)$ (Chapter 13). The isothermal bulk modulus, which describes changes that occur at constant temperature, differs from the adiabatic bulk modulus, which describes changes with no heat transfer. For sound waves at audible frequencies the changes occur too rapidly for appreciable heat flow, so the appropriate bulk modulus is the adiabatic bulk modulus.

Summary

1. The first law of thermodynamics is a statement of the conservation of energy.
2. The equipartition theorem is a fundamental law of classical physics. It breaks down when the typical thermal energy kT is small compared to the spacing of quantized energy levels.

Topic	Remarks and Relevant Equations	
1. Heat	Heat is energy that is transferred from one object to another because of a temperature difference.	
Calorie	The calorie, originally defined as the heat necessary to raise the temperature of one gram of water by one Celsius degree, is now defined to be 4.184 joules.	
2. Heat Capacity	Heat capacity is the amount of heat needed to raise the temperature of a substance by one degree.	
	$c = \dfrac{Q}{\Delta T}$	**19-1**

At constant volume	$C_v = \frac{Q_v}{\Delta T}$	
At constant pressure	$C_p = \frac{Q_p}{\Delta T}$	
Specific heat (heat capacity per unit mass)	$c' = \frac{C}{m}$	19-2
Molar specific heat (heat capacity per mole)	$c' = \frac{C}{n}$	19-6
Heat capacity related to internal energy	$C_v = \frac{dU}{dT}$	19-18
Of ideal gas	$C_p - C_v = +nR$	19-20
Of monatomic ideal gas	$C_v = \frac{3}{2}nR$	19-22
Of diatomic ideal gas	$C_v = \frac{5}{2}nR$	19-25
3. Fusion and Vaporization	During melting and vaporization, the temperature does not change.	
Latent heat of fusion	The heat needed to melt a substance is the product of the mass of the substance and its latent heat of fusion L_f: $Q_f = mL_f$	19-8
L_f of water	333.5 kJ/kg	
Latent heat of vaporization	The heat needed to vaporize a liquid is the product of the mass of the liquid and its latent heat of vaporization L_v: $Q_v = mL_v$	19-9
L_v of water	2257 kJ/kg	
4. First Law of Thermodynamics	The net heat added to a system equals the change in the internal energy of the system plus the work done by the system: $Q = \Delta U + W$	19-10
5. Internal Energy *U*	The internal energy of a system is a property of the state of the system, as are the pressure, volume, and temperature. Heat and work are not properties of state.	
Ideal gas	U depends only on the temperature T.	
Monatomic ideal gas	$U = \frac{3}{2}nRT$	19-12
Internal energy related to heat capacity	$dU = C_v dT$	19-18
6. Quasi-static Process	A quasi-static process is one that occurs slowly so that the system moves through a series of equilibrium states	
Isobaric	$P =$ constant	
Isothermal	$T =$ constant	

Adiabatic	$Q = 0$	
Adiabatic, ideal gas	$TV^{\gamma-1} = \text{constant}$	19-36
	or $PV^{\gamma} = \text{constant}$	19-37
	where $\gamma = \frac{C_P}{C_V}$	19-35
7. Work Done by a Gas in a Quasi-static Process	$W = \int P\,dV$	19-14
Isothermal	$W_{\text{isothermal}} = nRT \ln \frac{V_2}{V_1}$	19-16
Adiabatic	$W_{\text{adiabatic}} = -C_V\,\Delta T = \frac{P_1V_1 - P_2V_2}{\gamma - 1}$	19-38, 19-39
8. Equipartition Theorem	The equipartition theorem states that when a system is in equilibrium, there is an average energy of $\frac{1}{2}kT$ per molecule or $\frac{1}{2}RT$ per mole associated with each degree of freedom.	
Failure of the equipartition theorem	The equipartition theorem fails when the thermal energy ($\sim kT$) that can be transferred in collisions is smaller than the energy gap ΔE between quantized energy levels. For example, monatomic gases cannot rotate because the first nonzero energy permitted is much greater than kT.	
9. Dulong–Petit Law	The molar specific heat of most solids is $3R$. This is predicted by the equipartition theorem, assuming that a solid atom has six degrees of freedom.	

Problem-Solving Guide

Summary of Worked Examples

Type of Calculation	Procedure and Relevant Examples	
1. Heat Capacity and Specific Heat		
Find the heat needed to raise the temperature of an object.	Use $Q = C\,\Delta T = mc\,\Delta T$.	Example 19-1
Calorimeter problems	Use $Q_{\text{in}} = Q_{\text{out}}$.	Examples 19-2, 19-4
Find the heat needed to change the temperature of an ideal gas.	Use $C_V = \frac{3}{2}nR$ for a monatomic gas, $C_V = \frac{5}{2}nR$ for a diatomic gas, and $C_P = C_V + nR$.	Examples 19-3, 19-7
Find the specific heat of a metal, or find the molar mass given the specific heat.	Use the Dulong–Petit law, $C_V = 6nR$.	Example 19-6

2.	**Latent Heat**		
	Find the heat needed to melt a solid or vaporize a liquid.	Use $Q_f = mL_f$ or $Q_v = mL_v$.	**Example** 19-3
3.	**First Law of Thermodynamics**		
	Find the work needed for a given temperature change.	Set the work equal to the thermal energy needed, $mc\,\Delta T$.	**Example 19-4**
	Find the internal energy change of a system.	Use the first law, $\Delta U = Q - W$.	**Example 19-5**
	Find the work done and the heat added during a complete cycle.	Calculate the work from $dW = p\,dV$. Since $\Delta U = 0$, $Q = W$.	**Example 19-6**
	Find W, Q, and ΔU for a quasi-static process.	Use $dW = p\,dV$, $Q = C\,\Delta T$, and $dQ = \Delta U + dW$.	**Example 19-7**
4.	**Calculate Work**		
	Find the work done by a gas during expansion or compression.	Use $W = \int p\,dV$. For isothermal expansion of a gas, $W = nRT\ln(V_2/V_1)$.	**Example** 19-8
		For adiabatic expansion, $$W = \frac{P_1V_1 - P_2V_2}{\gamma - 1} = -C_v\,\Delta T$$	**Example 19-9**
5.	**Equipartition Theorem**		
	Determine if the equipartition theorem holds for a system at a certain temperature.	Compare the spacing of the quantized energy levels with kT.	**Example 19-11**
6.	**Quasi-static Adiabatic Expansion**		
	Find P, V, or T after an adiabatic expansion.	Use $P_1V_1^\gamma = P_2V_2^\gamma$ to relate P and V. Use $T_1V_1^{\gamma-1} = T_2V_2^{\gamma-1}$ to relate T and V.	**Example 19-10**

Problems

In a few problems, you are given more data than you actually need; in a few other problems, you are required to supply data from your general knowledge, outside sources, or informed estimates.

Conceptual Problems

Problems from Optional and Exploring sections

• Single-concept, single-step, relatively easy

•• Intermediate-level, may require synthesis of concepts

••• Challenging, for advanced students

Heat Capacity; Specific Heat; Latent Heat

1 • Body A has twice the mass and twice the specific heat of body B. If they are supplied with equal amounts of heat, how do the subsequent changes in their temperatures compare?

2 • The temperature change of two blocks of masses M_A and M_B is the same when they absorb equal amounts of heat. It follows that the specific heats are related by

(*a*) $c_A = (M_A/M_B)c_B$. (*b*) $c_A = (M_B/M_A)c_B$.
(*c*) $c_A = c_B$. (*d*) none of the above.

3 • The specific heat of aluminum is more than twice that of copper. Identical masses of copper and aluminum, both at 20°C, are dropped into a calorimeter containing water at 40°C. When thermal equilibrium is reached,

(*a*) the aluminum is at a higher temperature than the copper.
(*b*) the aluminum has absorbed less energy than the copper.
(*c*) the aluminum has absorbed more energy than the copper.
(*d*) both (*a*) and (*c*) are correct statements.

4 • Sam the shepherd's partner, Bernard, who is a working dog, consumes 2500 kcal of food each day. (*a*) How many joules of energy does Bernard consume each day? (*b*) Sam and Bernard often find themselves sleeping out in the cold night. If the energy consumed by Bernard is dissipated as heat at a steady rate over 24 h, what is his power output in watts as a heater for Sam?

5 • A solar home contains 10^5 kg of concrete (specific heat = 1.00 kJ/kg·K). How much heat is given off by the concrete when it cools from 25 to 20°C?

6 • How many calories must be supplied to 60 g of ice at −10°C to melt it and raise the temperature of the water to 40°C?

7 •• How much heat must be removed when 100 g of steam at 150°C is cooled and frozen into 100 g of ice at 0°C? (Take the specific heat of steam to be 2.01 kJ/kg·K.)

8 •• A 50-g piece of aluminum at 20°C is cooled to −196°C by placing it in a large container of liquid nitrogen at that temperature. How much nitrogen is vaporized? (Assume that the specific heat of aluminum is constant and is equal to 0.90 kJ/kg·K.)

9 •• If 500 g of molten lead at 327°C is poured into a cavity in a large block of ice at 0°C, how much of the ice melts?

10 •• A 30-g lead bullet initially at 20°C comes to rest in the block of a ballistic pendulum. Assume that half the initial kinetic energy of the bullet is converted into thermal energy within the bullet. If the speed of the bullet was 420 m/s, what is the temperature of the bullet immediately after coming to rest in the block?

11 •• A 1400-kg car traveling at 80 km/h is brought to rest by applying the brakes. If the specific heat of steel is 0.11 cal/g·K, what total mass of steel must be contained in the steel brake drums if the temperature of the brake drums is not to rise by more than 120°C?

Calorimetry

12 • A 200-g piece of lead is heated to 90°C and is then dropped into a calorimeter containing 500 g of water that is initially at 20°C. Neglecting the heat capacity of the container, find the final temperature of the lead and water.

13 • The specific heat of a certain metal can be determined by measuring the temperature change that occurs when a piece of the metal is heated and then placed in an insulated container made of the same material and containing water. Suppose a piece of metal has a mass of 100 g and is initially at 100°C. The container has a mass of 200 g and contains 500 g of water at an initial temperature of 20.0°C. The final temperature is 21.4°C. What is the specific heat of the metal?

14 •• A 25-g glass tumbler contains 200 mL of water at 24°C. If two 15-g ice cubes each at a temperature of −3°C are dropped into the tumbler, what is the final temperature of the drink? Neglect thermal conduction between the tumbler and the room.

15 •• A 200-g piece of ice at 0°C is placed in 500 g of water at 20°C. The system is in a container of negligible heat capacity and is insulated from its surroundings. (*a*) What is the final equilibrium temperature of the system? (*b*) How much of the ice melts?

16 •• A 3.5-kg block of copper at a temperature of 80°C is dropped into a bucket containing a mixture of ice and water whose total mass is 1.2 kg. When thermal equilibrium is reached the temperature of the water is 8°C. How much ice was in the bucket before the copper block was placed in it? (Neglect the heat capacity of the bucket.)

17 •• A well-insulated bucket contains 150 g of ice at 0°C. (*a*) If 20 g of steam at 100°C is injected into the bucket, what is the final equilibrium temperature of the system? (*b*) Is any ice left afterward?

18 •• A calorimeter of negligible mass contains 1 kg of water at 303 K and 50 g of ice at 273 K. Find the final temperature *T*. Solve the same problem if the mass of ice is 500 g.

19 •• A 200-g aluminum calorimeter contains 500 g of water at 20°C. A 100-g piece of ice cooled to −20°C is placed in the calorimeter. (*a*) Find the final temperature of the system, assuming no heat loss. (Assume that the specific heat of ice is 2.0 kJ/kg·K.) (*b*) A second 200-g piece of ice at −20°C is added. How much ice remains in the system after it reaches equilibrium? (*c*) Would you give a different answer for (*b*) if both pieces of ice were added at the same time?

20 •• The specific heat of a 100-g block of material is to be determined. The block is placed in a 25-g copper calorimeter that also holds 60 g of water. The system is initially at 20°C. Then 120 mL of water at 80°C are added to the calorimeter vessel. When thermal equilibrium is attained, the temperature of the water is 54°C. Determine the specific heat of the block.

21 •• Between innings at his weekly softball game, Stan likes to have a sip or two of beer. He usually consumes about 6 cans, which he prefers at exactly 40°F. His wife Bernice puts a six-pack of 12-ounce aluminum cans of beer (1 ounce has a mass of 28.4 g) originally at 80°F in a well-insulated Styrofoam container and begins adding ice. How many 30-g ice cubes must she add to the container so that the final temperature is 40°F? (Neglect heat losses through the container and the heat removed from the aluminum and assume that the beer is mostly water.)

22 •• A 100-g piece of copper is heated in a furnace to a temperature *t*. The copper is then inserted into a 150-g copper calorimeter containing 200 g of water. The initial temperature of the water and calorimeter is 16°C, and the final temperature after equilibrium is established is 38°C. When the calorimeter and its contents are weighed, 1.2 g of water are found to have evaporated. What was the temperature *t*?

23 •• A 200-g aluminum calorimeter contains 500 g of water at 20°C. Aluminum shot of mass 300 g is heated to 100°C and is then placed in the calorimeter. (*a*) Using the value of the specific heat of aluminum given in Table 19-1, find the final temperature of the system, assuming that no heat is lost to the surroundings. (*b*) The error due to heat transfer between the system and its surroundings can be minimized if the initial temperature of the water and calorimeter is chosen to be $\frac{1}{2}\Delta t_w$ below room temperature, where Δt_w is the temperature change of the calorimeter and water during the measurement. Then the final temperature is $\frac{1}{2}\Delta t_w$ above room temperature. What should the initial temperature of the water and container be if the room temperature is 20°C?

First Law of Thermodynamics

24 • Joule's experiment establishing the mechanical equivalence of heat involved the conversion of mechanical energy into internal energy. Give some examples of the internal energy of a system being converted into mechanical energy.

25 • Can a system absorb heat with no change in its internal energy?

26 • In the equation $Q = \Delta U + W$ (the formal statement of the first law of thermodynamics), the quantities Q and W represent

(*a*) the heat supplied to the system and the work done by the system.
(*b*) the heat supplied to the system and the work done on the system.
(*c*) the heat released by the system and the work done by the system.
(*d*) the heat released by the system and the work done on the system.

27 • A diatomic gas does 300 J of work and also absorbs 600 cal of heat. What is the change in internal energy of the gas?

28 • If 400 kcal is added to a gas that expands and does 800 kJ of work, what is the change in the internal energy of the gas?

29 • A lead bullet moving at 200 m/s is stopped in a block of wood. Assuming that all of the energy change goes into heating the bullet, find the final temperature of the bullet if its initial temperature is 20°C.

30 • (*a*) At Niagara Falls, the water drops 50 m. If the change in potential energy goes into the internal energy of the water, compute the increase in its temperature. (*b*) Do the same for Yosemite Falls, where the water drops 740 m. (These temperature rises are not observed because the water cools by evaporation as it falls.)

31 • When 20 cal of heat are absorbed by a gas, the system performs 30 J of work. What is the change in the internal energy of the gas?

32 •• A lead bullet initially at 30°C just melts upon striking a target. Assuming that all of the initial kinetic energy of the bullet goes into the internal energy of the bullet to raise its temperature and melt it, calculate the speed of the bullet upon impact.

33 •• A piece of ice is dropped from a height H. (*a*) Find the minimum value of H such that the ice melts when it makes an inelastic collision with the ground. Assume that all the mechanical energy lost goes into melting the ice. (*b*) Is it reasonable to neglect the variation in the acceleration of gravity in doing this problem? (*c*) Comment on the reasonableness of neglecting air resistance. What effect would air resistance have on your answer?

34 •• On a cold day you can warm your hands by rubbing them together. (*a*) Assume that the coefficient of friction between your hands is 0.5, that the normal force between your hands is 35 N, and that you rub them together at an average speed of 35 cm/s. What is the rate at which heat is generated? (*b*) Assume further that the mass of each of your hands is approximately 350 g, that the specific heat of your hands is about 4 kJ/kg·K, and that all the heat generated goes into raising the temperature of your hands. How long must you rub your hands together to produce a 5-C° increase in their temperature?

The Internal Energy of an Ideal Gas

35 • A real gas cools during a free expansion, though an ideal gas does not. Explain.

36 • An ideal gas at one atmosphere pressure and 300 K is confined to half of an insulated container by a thin partition. The partition is then removed and equilibrium is established. At that point, which of the following is correct?

(*a*) The pressure is half an atmosphere and the temperature is 150 K.
(*b*) The pressure is one atmosphere and the temperature is 150 K.
(*c*) The pressure is half an atmosphere and the temperature is 300 K.
(*d*) None of the above.

37 • A certain gas consists of ions that repel each other. The gas undergoes a free expansion with no heat exchange and no work done. How does the temperature of the gas change? Why?

Work and the *PV* Diagram for a Gas

38 • A gas changes its state reversibly from A to C (Figure 19-16). The work done by the gas is

(*a*) greatest for path A → B → C.
(*b*) least for path A → C.
(*c*) greatest for path A → D → C.
(*d*) the same for all three paths.

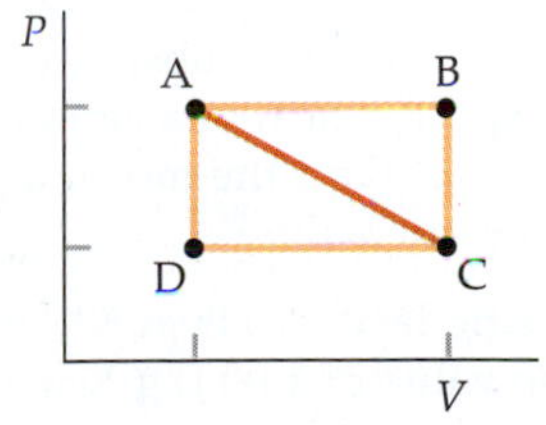

Figure 19-16 Problem 38

In Problems 39 through 42, the initial state of 1 mol of an ideal gas is $P_1 = 3$ atm, $V_1 = 1$ L, and $U_1 = 456$ J, and its final state is $P_2 = 2$ atm, $V_2 = 3$ L, and $U_2 = 912$ J.

39 • The gas is allowed to expand at constant pressure to a volume of 3 L. It is then cooled at constant volume until its pressure is 2 atm. (*a*) Show this process on a *PV* diagram, and calculate the work done by the gas. (*b*) Find the heat added during this process.

40 • The gas is first cooled at constant volume until its pressure is 2 atm. It is then allowed to expand at constant pressure until its volume is 3 L. (*a*) Show this process on a *PV* diagram, and calculate the work done by the gas. (*b*) Find the heat added during this process.

41 •• The gas is allowed to expand isothermally until its volume is 3 L and its pressure is 1 atm. It is then heated at constant volume until its pressure is 2 atm. (*a*) Show this process on a *PV* diagram, and calculate the work done by the gas. (*b*) Find the heat added during this process.

42 •• The gas is heated and is allowed to expand such that it follows a straight-line path on a *PV* diagram from its initial state to its final state. (*a*) Show this process on a *PV* diagram, and calculate the work done by the gas. (*b*) Find the heat added during this process.

43 •• One mole of the ideal gas is initially in the state $P_0 = 1$ atm, $V_0 = 25$ L. As the gas is slowly heated, the plot of its state on a *PV* diagram moves in a straight line to the state $P = 3$ atm, $V = 75$ L. Find the work done by the gas.

44 •• One mole of the ideal gas is heated so that $T = AP^2$, where A is a constant. The temperature changes from T_0 to $4T_0$. Find the work done by the gas.

45 •• One mole of an ideal gas initially at a pressure of 1 atm and a temperature of 0°C is compressed isothermally and quasi-statically until its pressure is 2 atm. Find (*a*) the work needed to compress the gas, and (*b*) the heat removed from the gas during the compression.

46 •• An ideal gas initially at 20°C and 200 kPa has a volume of 4 L. It undergoes a quasi-static, isothermal expansion until its pressure is reduced to 100 kPa. Find (*a*) the work done by the gas, and (*b*) the heat added to the gas during the expansion.

Heat Capacities of Gases and the Equipartition Theorem

47 • The heat capacity at constant volume of a certain amount of a monatomic gas is 49.8 J/K. (*a*) Find the number of moles of the gas. (*b*) What is the internal energy of the gas at $T = 300$ K? (*c*) What is the heat capacity of the gas at constant pressure?

48 • The Dulong–Petit law was originally used to determine the molecular mass of a substance from its measured heat capacity. The specific heat of a certain solid is measured to be 0.447 kJ/kg·K. (*a*) Find the molecular mass of the substance. (*b*) What element is this?

49 •• The specific heat of air at 0°C is listed in a handbook as having the value of 1.00 J/g·K measured at constant pressure. (*a*) Assuming that air is an ideal gas with a molar mass $M = 29.0$ g/mol, what is its specific heat at 0°C and constant volume? (*b*) How much internal energy is there in 1 L of air at 0°C and at 1 atm?

50 •• One mole of an ideal diatomic gas is heated at constant volume from 300 to 600 K. (*a*) Find the increase in internal energy, the work done, and the heat added. (*b*) Find the same quantities if this gas is heated from 300 to 600 K at constant pressure. Use the first law of thermodynamics and your results for (*a*) to calculate the work done. (*c*) Calculate the work done in (*b*) directly from $dW = P\,dV$.

51 •• A diatomic gas (molar mass *M*) is confined to a closed container of volume *V* at a pressure P_0. What amount of heat *Q* should be transferred to the gas in order to triple the pressure? (Express your answer in terms of P_0 and *V*.)

52 •• One mole of air ($c_v = 5R/2$) is confined at atmospheric pressure in a cylinder with a piston at 0°C. The initial volume, occupied by gas, is *V*. Find the volume of gas V' after the equivalent of 13,200 J of heat is transferred to it.

53 •• The heat capacity of a certain amount of a particular gas at constant pressure is greater than that at constant volume by 29.1 J/K. (*a*) How many moles of the gas are there? (*b*) If the gas is monatomic, what are C_v and C_p? (*c*) If the gas consists of diatomic molecules that rotate but do not vibrate, what are C_v and C_p?

54 •• One mole of a monatomic ideal gas is initially at 273 K and 1 atm. (*a*) What is its initial internal energy? (*b*) Find its final internal energy and the work done by the gas when 500 J of heat are added at constant pressure. (c) Find the same quantities when 500 J of heat are added at constant volume.

55 •• A certain molecule has vibrational energy levels that are equally spaced by 0.15 eV. Find the critical temperature T_c such that for $T \gg T_c$ you would expect the equipartition theorem to hold and for $T \ll T_c$ you would expect the equipartition theorem to fail.

Quasi-static Adiabatic Expansion of a Gas

56 • When an ideal gas is subjected to an adiabatic process,

(*a*) no work is done by the system.
(*b*) no heat is supplied to the system.
(*c*) the internal energy remains constant.
(*d*) the heat supplied to the system equals the work done by the system.

57 • One mole of an ideal gas ($\gamma = \frac{5}{3}$) expands adiabatically and quasi-statically from a pressure of 10 atm and a temperature of 0°C to a pressure of 2 atm. Find (*a*) the initial and final volumes, (*b*) the final temperature, and (*c*) the work done by the gas.

58 • An ideal gas at a temperature of 20°C is compressed quasi-statically and adiabatically to half its original volume. Find its final temperature if (*a*) $C_v = \frac{3}{2}nR$, and (*b*) $C_v = \frac{5}{2}nR$.

59 • Two moles of neon gas initially at 20°C and a pressure of 1 atm are compressed adiabatically to one-fourth of their initial volume. Determine the temperature and pressure following compression.

60 •• Half a mole of an ideal monatomic gas at a pressure of 400 kPa and a temperature of 300 K expands until the pressure has diminished to 160 kPa. Find the final temperature and volume, the work done, and the heat absorbed by the gas if the expansion is *(a)* isothermal, and *(b)* adiabatic.

61 •• Repeat Problem 60 for a diatomic gas.

62 •• One-half mole of helium is expanded adiabatically and quasi-statically from an initial pressure of 5 atm and temperature of 500 K to a final pressure of 1 atm. Find *(a)* the final temperature, *(b)* the final volume, *(c)* the work done by the gas, and *(d)* the change in the internal energy of the gas.

63 ••• A hand pump is used to inflate a bicycle tire to a gauge pressure of 482 kPa (about 70 lb/in^2). How much work must be done if each stroke of the pump is an adiabatic process? Atmospheric pressure is 1 atm, the air temperature is initially 20°C, and the volume of the air in the tire remains constant at 1 L.

64 ••• An ideal gas at initial volume V_1 and pressure P_1 expands quasi-statically and adiabatically to volume V_2 and pressure P_2. Calculate the work done by the gas directly by integrating $P\,dV$, and show that your result is the same as that given by Equation 19-39.

Cyclic Processes

65 •• One mole of N_2 ($C_v = \frac{5}{2}R$) gas is originally at room temperature (20°C) and a pressure of 5 atm. It is allowed to expand adiabatically and quasi-statically until its pressure equals the room pressure of 1 atm. It is then heated at constant pressure until its temperature is again 20°C. During this heating, the gas expands. After it reaches room temperature, it is heated at constant volume until its pressure is 5 atm. It is then compressed at constant pressure until it is back to its original state. *(a)* Construct an accurate *PV* diagram showing each process in the cycle. *(b)* From your graph, determine the work done by the gas during the complete cycle. *(c)* How much heat is added or subtracted from the gas during the complete cycle? *(d)* Check your graphical determination of the work done by the gas in *(b)* by calculating the work done during each part of the cycle.

66 •• Two moles of an ideal monatomic gas have an initial pressure $P_1 = 2$ atm and an initial volume $V_1 = 2$ L. The gas is taken through the following quasi-static cycle: It is expanded isothermally until it has a volume $V_2 = 4$ L. It is then heated at constant volume until it has a pressure $P_3 = 2$ atm. It is then cooled at constant pressure until it is back to its initial state. *(a)* Show this cycle on a *PV* diagram. *(b)* Calculate the heat added and the work done by the gas during each part of the cycle. *(c)* Find the temperatures T_1, T_2, and T_3.

67 ••• At point D in Figure 19-17 the pressure and temperature of 2 mol of an ideal monatomic gas are 2 atm and 360 K. The volume of the gas at point B on the *PV* diagram is three times that at point D and its pressure is twice that at point C. Paths AB and CD represent isothermal processes. The gas is carried through a complete cycle along the path DABCD. Determine the total amount of work done by the gas and the heat supplied to the gas along each portion of the cycle.

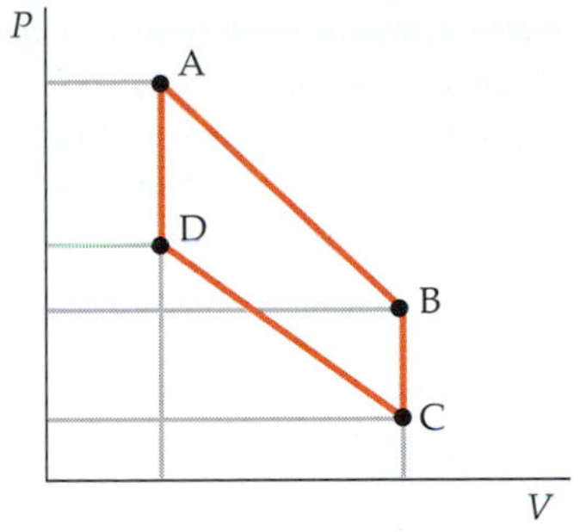

Figure 19-17
Problems 67–70

68 ••• Repeat Problem 67 with the paths AB and CD representing adiabatic processes.

69 ••• Repeat Problem 67 with a diatomic gas.

70 ••• Repeat Problem 68 with a diatomic gas.

71 ••• An ideal gas of *n* mol is initially at pressure P_1, volume V_1, and temperature T_h. It expands isothermally until its pressure and volume are P_2 and V_2. It then expands adiabatically until its temperature is T_c and its pressure and volume are P_3 and V_3. It is then compressed isothermally until it is at a pressure P_4 and a volume V_4, which is related to its initial volume V_1 by $T_c V_4^{\gamma-1} = T_h V_1^{\gamma-1}$. The gas is then compressed adiabatically until it is back in its original state. *(a)* Assuming that each process is quasi-static, plot this cycle on a *PV* diagram. (This cycle is known as the Carnot cycle for an ideal gas.) *(b)* Show that the heat Q_h absorbed during the isothermal expansion at T_h is $Q_h = nRT_h \ln(V_2/V_1)$. *(c)* Show that the heat Q_c given off by the gas during the isothermal compression at T_c is $Q_c = nRT_c \ln(V_3/V_4)$. *(d)* Using the result that $TV^{\gamma-1}$ is constant for an adiabatic expansion, show that $V_2/V_1 = V_3/V_4$. *(e)* The efficiency of a Carnot cycle is defined to be the net work done divided by the heat absorbed Q_h. Using the first law of thermodynamics, show that the efficiency is $1 - Q_c/Q_h$. *(f)* Using your results from the previous parts of this problem, show that $Q_c/Q_h = T_c/T_h$.

General Problems

72 • After a potato wrapped in aluminum foil has been baked in an oven, it is taken out and its foil removed. The foil cools much faster than the potato. Why?

73 • True or false:

(a) The heat capacity of a body is the amount of heat it can store at a given temperature.
(b) When a system goes from state 1 to state 2, the amount of heat added to the system is the same for all processes.
(c) When a system goes from state 1 to state 2, the work done on the system is the same for all processes.
(d) When a system goes from state 1 to state 2, the change in the internal energy of the system is the same for all processes.
(e) The internal energy of a given amount of an ideal gas depends only on its absolute temperature.
(f) A quasi-static process is one in which there is no motion.
(g) For any material that expands when heated, C_p is greater than C_v.

74 • If a system's volume remains constant while undergoing changes in temperature and pressure, then

(*a*) the internal energy of the system is unchanged.
(*b*) the system does no work.
(*c*) the system absorbs no heat.
(*d*) the change in internal energy equals the heat absorbed by the system.

75 • When an ideal gas is subjected to an isothermal process,
(*a*) no work is done by the system.
(*b*) no heat is supplied to the system.
(*c*) the heat supplied to the system equals the change in internal energy.
(*d*) the heat supplied to the system equals the work done by the system.

76 •• The 1-L fuel tank of a gas grill contains 600 g of propane (C_3H_8) at a pressure of 2 MPa. What can you say about the phase state of the propane?

77 •• An ideal gas undergoes a process during which $P\sqrt{V}$ = constant and the volume of the gas decreases. What happens to the temperature?

78 • The volume of three moles of a monatomic gas is increased from 50 to 200 L at constant pressure. The initial temperature of the gas is 300 K. How much heat must be supplied to the gas?

79 • In the process of compressing n moles of an ideal diatomic gas to one-fifth of its initial volume, 180 kJ of work is done on the gas. If this is accomplished isothermally at room temperature (293 K), how many calories of heat are removed from the gas?

80 • What is the number of moles n of the gas in Problem 79?

Figure 19-18 Problems 81–85

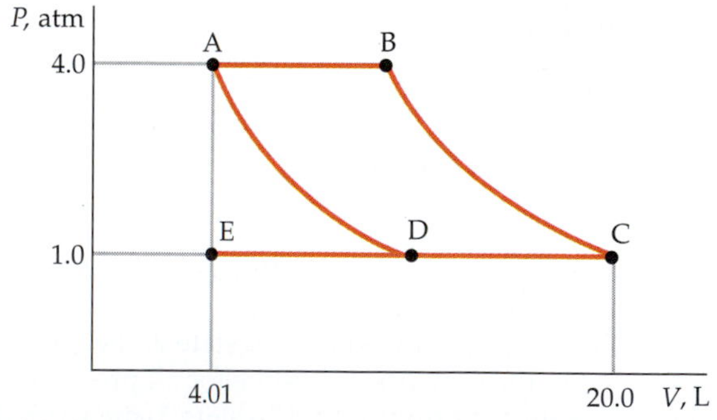

81 • The PV diagram in Figure 19-18 represents 3 mol of an ideal monatomic gas. The gas is initially at point A. The paths AD and BC represent isothermal changes. If the system is brought to point C along the path AEC, find (*a*) the initial and final temperatures, (*b*) the work done by the gas, and (*c*) the heat absorbed by the gas.

82 •• Repeat Problem 81 with the gas following path ABC.

83 •• Repeat Problem 82 with the gas following path ADC.

84 •• Suppose that the paths AD and BC represent adiabatic processes. What then are the work done by the gas and the heat absorbed by the gas in following the path ABC?

85 •• Repeat Problem 84 for the path ADC.

86 •• At very low temperatures, the specific heat of a metal is given by $c = aT + bT^3$. For the metal copper, $a = 0.0108\ \mathrm{J/kg{\cdot}K^2}$ and $b = 7.62 \times 10^{-4}\ \mathrm{J/kg{\cdot}K^4}$. (*a*) What is the specific heat of copper at 4 K? (*b*) How much heat is required to heat copper from 1 to 3 K?

87 •• Two moles of a diatomic ideal gas are compressed isothermally from 18 to 8 L. In the process, 170 calories escape from the system. Determine the amount of work done by the gas, the change in internal energy, and the initial and final temperatures of the gas.

88 •• Suppose the two moles of a diatomic ideal gas in Problem 87 are compressed from 18 to 8 L adiabatically. The work done on the gas is 820 J. Find the initial temperature and the initial and final pressures.

89 •• Repeat Problem 87 with the diatomic ideal gas replaced by a monatomic ideal gas.

90 •• Repeat Problem 88 with the diatomic ideal gas replaced by a monatomic ideal gas.

91 •• How much work must be done to 30 g of CO at standard temperature and pressure to compress it to a fifth of its initial volume if the process is (*a*) isothermal; (*b*) adiabatic?

92 •• Repeat Problem 91 if the gas is CO_2.

93 •• Repeat Problem 91 if the gas is argon.

94 •• A thermally insulated system consists of 1 mol of a diatomic ideal gas at 100 K and 2 mol of a solid at 200 K that are separated by a rigid insulating wall. Find the equilibrium temperature of the system after the insulating wall is removed, assuming that the solid obeys the Dulong–Petit law.

95 •• When an ideal gas undergoes a temperature change at constant volume, its energy changes by $\Delta U = C_\mathrm{v}\,\Delta T$. (*a*) Explain why this result holds for an ideal gas for any temperature change independent of the process. (*b*) Show explicitly that this result holds for the expansion of an ideal gas at constant pressure by first calculating the work done and showing that it can be written as $W = nR\,\Delta T$, and then by using $\Delta U = Q - W$, where $Q = C_\mathrm{P}\,\Delta T$.

96 •• One mole of an ideal monatomic gas is heated at constant volume from 300 to 600 K. (*a*) Find the heat added, the work done by the gas, and the change in its internal energy. (*b*) Find these same quantities if the gas is heated from 300 to 600 K at constant pressure.

97 •• Heat in the amount of 500 J is supplied to 2 mol of an ideal diatomic gas. (*a*) Find the change in temperature if the pressure is kept constant. (*b*) Find the work done by the gas. (*c*) Find the ratio of the final volume of the gas to the initial volume if the initial temperature is 20°C.

98 •• An insulated cylinder is fitted with a movable piston to maintain constant pressure. The cylinder initially contains 100 g of ice at −10°C. Heat is supplied to the contents at a constant rate by a 100-W heater. Make a graph showing the temperature of the cylinder contents as a function of time starting at $t = 0$, when the temperature is -10°C, and ending when the temperature is 110°C. (Use $c = 2.0$ kJ/kg·K for the average specific heat of ice from -10 to 0°C and of steam from 100 to 110°C.)

99 •• For the winter festival, a sculptor creates a 20-kg statue of a skier made of ice at 0°C. To show off the statue's stability, the sculptor hires someone to repeatedly slide the statue down a plane 8 m long and inclined at 30°. Unfortunately, the sculptor forgot about the thermal energy produced by friction. If the coefficient of sliding friction between the ice and the plane is 0.05, how much ice melts due to the friction on the first run? (Assume that all the mechanical energy lost goes into melting the ice.)

100 •• Two moles of a diatomic ideal gas expand adiabatically. The initial temperature of the gas is 300 K. The work done by the gas during the expansion is 3.5 kJ. What is the final temperature of the gas?

101 •• One mole of monatomic gas, initially at temperature T, undergoes a process in which its temperature is quadrupled and its volume is halved. Find the amount of heat Q transferred to the gas. It is known that in this process the pressure was never less than the initial pressure, and the work done on the gas was the minimum possible.

102 •• A vertical heat-insulated cylinder is divided into two parts by a movable piston of mass m. Initially the piston is held at rest. The top part is evacuated and the bottom part is filled with 1 mol of diatomic ideal gas at temperature 300 K. After the piston is released and the system comes to equilibrium, the volume, occupied by gas, is halved. Find the final temperature of the gas.

103 •• According to the Einstein model of a crystalline solid, the internal energy per mole is given by

$$U = \frac{3N_A k T_E}{e^{T_E/T} - 1}$$

where T_E is a characteristic temperature called the Einstein temperature, and T is the temperature of the solid in kelvins. Evaluate the molar internal energy of diamond ($T_E = 1060$ K) at 300 and 600 K, and thereby the increase in internal energy as diamond is heated from 300 to 600 K.

104 ••• In an isothermal expansion, an ideal gas at an initial pressure P_0 expands until its volume is twice its initial volume. (*a*) Find its pressure after the expansion. (*b*) The gas is then compressed adiabatically and quasi-statically back to its original volume, at which point its pressure is $1.32P_0$. Is the gas monatomic, diatomic, or polyatomic? (*c*) How does the translational kinetic energy of the gas change in these processes?

105 ••• Prove that the slope of the adiabatic curve passing through a point on the PV diagram for an ideal gas is γ times the slope of the isothermal curve passing through the same point.

Note: Although Problems 106–109 involve non-quasi-static processes, the state functions of the end products can be calculated assuming that the gases are ideal and using the first law of thermodynamics. For $T > 2000$ K, vibration of the atoms contribute to C_p of CO_2 and H_2O so that the C_p of these gases is 7.5 R at high temperatures. Assume the gases do not dissociate.

106 ••• The combustion of benzene is represented by the chemical reaction $2(C_6H_6) + 15(O_2) \rightarrow 12(CO_2) + 6(H_2O)$. The amount of energy released in the combustion of two mol of benzene is 1516 kcal. One mole of benzene and 7.5 mol of oxygen at 300 K are confined in an insulated enclosure at a pressure of 1 atm. (*a*) Find the temperature and volume following combustion if the pressure is maintained at 1 atm. (*b*) If, following combustion, the thermal insulation about the container is removed and the system is cooled to 300 K, what is the final pressure?

107 ••• Repeat Problem 106, parts (*a*) and (*b*), using as the combustible substance 1 mol of acetylene for which the combustion reaction is $2(C_2H_2) + 5(O_2) \rightarrow 4(CO_2) + 2(H_2O)$. The combustion of 1 mol of acetylene releases 300 kcal.

108 ••• Carbon monoxide and oxygen combine to form carbon dioxide with an energy release of 280 kJ/mol of CO according to the reaction $2(CO) + O_2 \rightarrow 2(CO_2)$. Two moles of CO and 1 mol of O_2 at 300 K are confined in an 80-L container; the combustion reaction is initiated with a spark. (*a*) What is the pressure in the container prior to the reaction? (*b*) If the reaction proceeds adiabatically, what are the final temperature and pressure? (*c*) If the resulting CO_2 gas is cooled to 0°C, what is the pressure in the container?

109 ••• Suppose that instead of pure oxygen, just enough air is mixed with the 2 mol of CO in the container of Problem 108 to permit complete combustion. Air is 80% N_2 and 20% O_2 by weight, and the nitrogen does not participate in the reaction. What then are the answers to parts (*a*), (*b*), and (*c*) of Problem 108?

110 ••• Use the expression given in Problem 103 for the internal energy per mole of a solid according to the Einstein model to show that the molar heat capacity at constant volume is given by

$$c'_v = 3R\left(\frac{T_E}{T}\right)^2 \frac{e^{T_E/T}}{(e^{T_E/T} - 1)^2}$$

111 ••• (*a*) Use the results of Problem 110 to show that the Dulong–Petit law, $c'_v \approx 3R$, holds for the Einstein model when $T > T_E$. (*b*) For diamond, T_E is approximately 1060 K. Numerically integrate $\Delta U = \int c'_v \, dT$ to find the increase in the internal energy if 1 mol of diamond is heated from 300 to 600 K. Compare your result to that obtained in Problem 103.

112 ••• A refinement of the Einstein model by Debye resulted in the following expression for the specific heat:

$$c'_v = 9R\left(\frac{T}{T_D}\right)^3 \int_0^y \frac{x^4 e^x}{(e^x - 1)^2}\, dx$$

where T_D is called the Debye temperature and $y = T_D/T$. (*a*) Show that when $T >> T_D$, the above expression reduces to the Dulong–Petit result $c_v = 3R$. (*Hint:* When $T >> T_D$, $y << 1$ and therefore x is always much less than 1. Then $e^x \approx 1 + x$.) (*b*) When $T << T_D$, the integral's upper limit may be approximated by infinity; the definite integral then has the value $4\pi^4/15$. Show that at very low temperatures the specific heat is given by $c_v = (12\pi^4/5)R(T/T_D)^3$.

CHAPTER 20

The Second Law of Thermodynamics

Solar energy is directed toward the solar oven at the center by this circular array of reflectors at Barstow, California.

We are often asked to conserve energy. But according to the first law of thermodynamics, energy is always conserved. What then does it mean to conserve energy if the total amount of energy in the universe does not change no matter what we do? The first law of thermodynamics does not tell the whole story. Energy is always conserved, but some forms of energy are more useful than others. The possibility or impossibility of putting energy to use is the subject of the second law of thermodynamics. For example, it is easy to convert mechanical work completely into thermal energy, but it is impossible to remove thermal energy from a system and convert it completely into mechanical work with no other changes. This experimental fact is one statement of the **second law of thermodynamics.**

> It is impossible to remove thermal energy from a system at a single temperature and convert it to mechanical work without changing the system or surroundings in some other way.

Second law of thermodynamics: Kelvin statement

We will encounter several other formulations of the same law in this chapter.

A common example of the conversion of mechanical energy into thermal energy is movement with friction. For example, when a block slides along a rough table, the initial mechanical (kinetic) energy of the block is converted into thermal energy as the block and the table are heated. The reverse process never occurs—a block and table that are warm will never spontaneously cool by converting their thermal energy into kinetic energy that sends the block sliding across the table. Yet such an amazing occurrence would not violate the first law of thermodynamics or any other physical laws we have encountered so far. It does, however, violate the second law of thermodynamics. There is thus a lack of symmetry in the roles played by heat and work that is not evident from the first law. This lack of symmetry is related to the fact that some processes are irreversible.

Irreversible processes take many forms, but all are related by the second law. For example, heat conduction is an irreversible process. If we place a hot body in contact with a cold body, heat will flow from the hot body to the cold body until they are at the same temperature. The reverse does not occur. Two bodies in contact at the same temperature remain at the same temperature; heat does not flow from one to the other making one colder and the other warmer. This experimental fact gives us another statement of the second law of thermodynamics.

> There can be no process whose only final result is to transfer thermal energy from a cooler object to a hotter one.

Second law of thermodynamics: Clausius statement

We will show in this chapter that the Kelvin and Clausius statements of the second law are equivalent.

20-1 Heat Engines and the Second Law of Thermodynamics

The study of the efficiency of heat engines gave rise to the first clear statements of the second law. A **heat engine** is a cyclic device whose purpose is to convert as much heat input into work as possible. Heat engines contain a **working substance** (water in a steam engine, air and gasoline vapor in an internal-combustion engine) that absorbs a quantity of heat $Q_{\text{in, h}}$, does work W, and gives off heat $|Q_{\text{out,c}}|$* as it returns to its initial state.

The earliest heat engines were steam engines, invented in the eighteenth century for pumping water from coal mines. Today steam engines are used to generate electricity. In a typical steam engine, water is heated under several hundred atmospheres pressure until it vaporizes at about 500°C (Figure 20-1). This steam expands against a piston, doing work, then exits at a much lower temperature and is further cooled until it condenses. The water is then pumped back into the boiler and heated again.

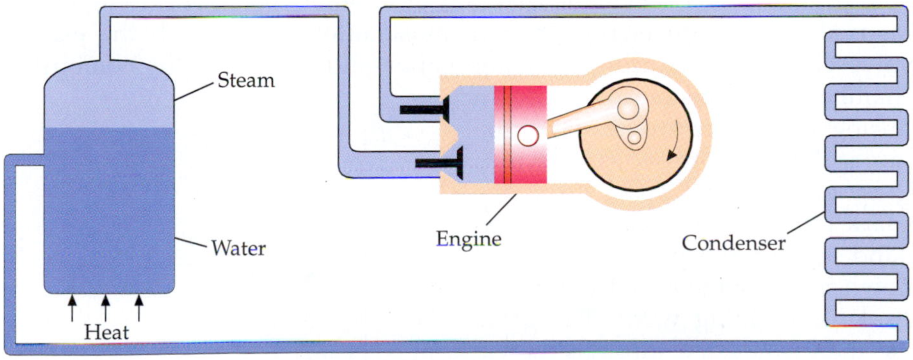

Figure 20-1 Schematic drawing of a steam engine. High-pressure steam does work against the piston.

* According to our sign convention for the first law, the heat given off by a system is negative. Since we are interested here only in the magnitudes of the heat absorbed or given off, we will use absolute-value signs where needed.

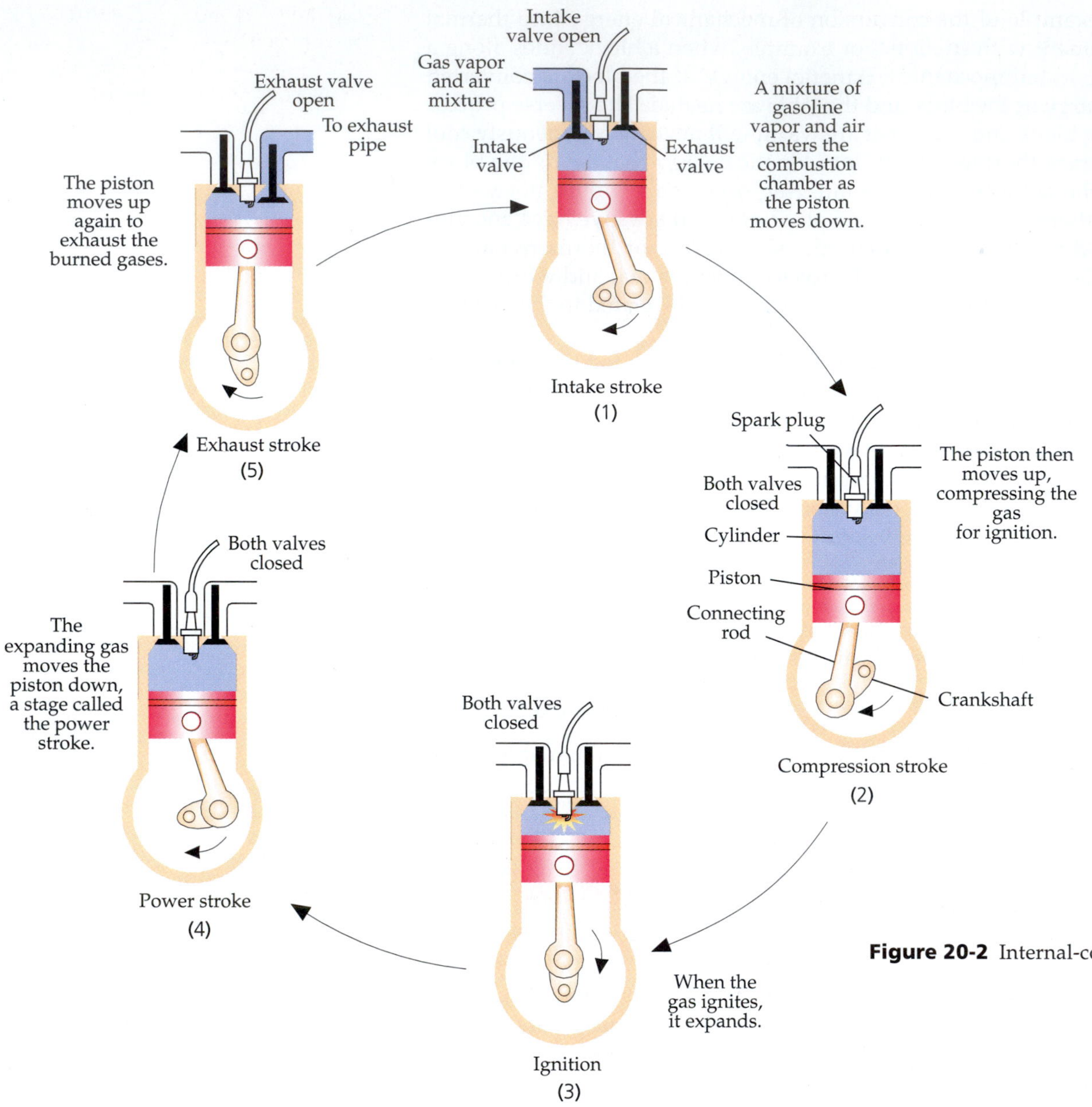

Figure 20-2 Internal-combustion engine.

Figure 20-2 is a schematic diagram of another type of heat engine, the internal-combustion engine used in most automobiles. With the exhaust valve closed, a mixture of gasoline vapor and air enters the combustion chamber as the piston moves down during the intake stroke. The mixture is then compressed, after which it is ignited by a spark from the spark plug. The hot gases then expand against the piston, driving it down in the stage called the power stroke and doing work. The gases are then exhausted through the exhaust valve, and the cycle repeats. An idealized model of the processes in the internal combustion engine is the *Otto cycle* shown in Figure 20-3.

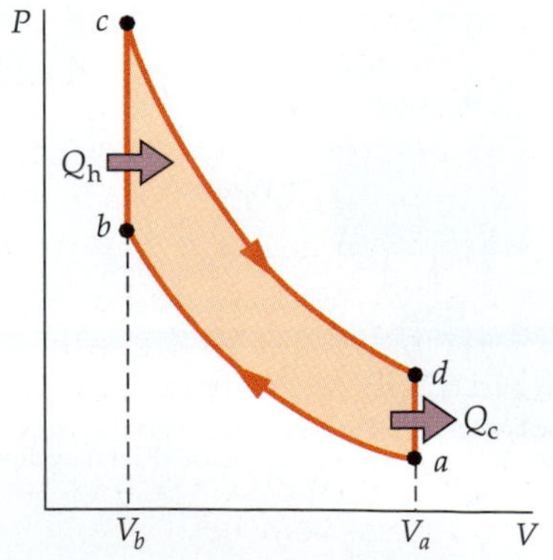

Figure 20-3 Otto cycle, representing the internal-combustion engine. The gasoline–air mixture enters at a and is adiabatically compressed to b. It is then heated (by ignition from the spark plug) at constant volume to c. The power stroke is represented by the adiabatic expansion from c to d. The cooling at constant volume from c to a represents the exhausting of the burned gases and the intake of a fresh gasoline–air mixture.

Figure 20-4 shows a schematic representation of a basic heat engine. The heat input is represented as coming from a **hot reservoir** at temperature T_h, and the exhaust goes into a **cold reservoir** at a lower temperature T_c. A hot or cold reservoir is an idealized body or system that has a very large heat capacity that allows it to absorb or give off thermal energy with no appreciable change in temperature. In practice the surrounding atmosphere or a lake often acts as a thermal reservoir. Since the initial and final states of the engine and working substance are the same, the initial and final internal energies of the engine are equal so $\Delta U = 0$. Then, according to the first law of thermodynamics, the work done equals the net heat absorbed:

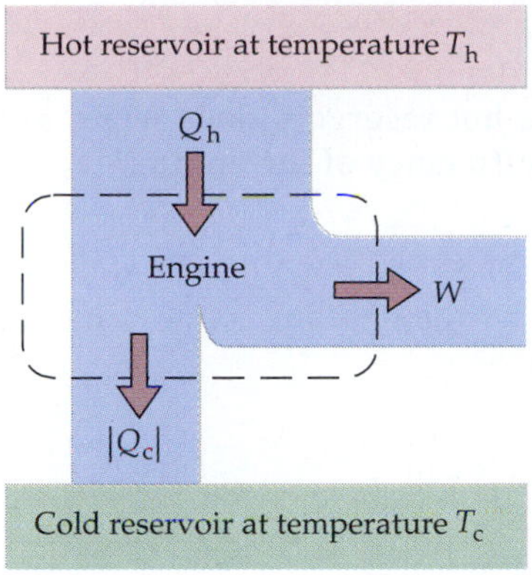

Figure 20-4 Schematic representation of a heat engine. The engine removes heat energy Q_h from a hot reservoir at a temperature T_h, does work W, and gives off heat $|Q_c|$ to a cold reservoir at a temperature T_c.

$$Q = \Delta U + W = W$$

or

$$W = Q_{in,h} - |Q_{out,c}| \qquad 20\text{-}1$$

The **efficiency** ε of a heat engine is defined as the ratio of the work done to the heat absorbed from the hot reservoir:

$$\varepsilon = \frac{W}{Q_{in,\,h}} = \frac{Q_{in,\,h} - |Q_{out,\,c}|}{Q_{in,\,h}} = 1 - \frac{|Q_{out,\,c}|}{Q_{in,\,h}} \qquad 20\text{-}2$$

Definition—Efficiency of a heat engine

Since the heat Q_{in} is usually produced by burning some fuel like coal or oil that must be paid for, heat engines are designed to have the greatest possible efficiency. The best steam engines operate near 40% efficiency; the best internal-combustion engines operate near 50%. At 100% efficiency ($\varepsilon = 1$), all the thermal energy absorbed from the hot reservoir would be converted into work and no thermal energy would be given off to the cold reservoir. However, *it is impossible to make a heat engine whose efficiency is 100%.* This experimental result is the **heat-engine statement of the second law of thermodynamics,** which is equivalent to the Kelvin statement above:

> It is impossible for a heat engine working in a cycle to produce no other effect than that of extracting thermal energy from a reservoir and performing an equivalent amount of work.

Second law of thermodynamics: Heat-engine statement

Exhaust manifolds in the 8-cylinder Dusenberg carry heat away from the engine to reduce its temperature.

The word "cycle" in this statement is important because it *is* possible to convert heat completely into work in a noncyclic process. An ideal gas undergoing an isothermal expansion does just this. But after the expansion, the gas is not in its original state. To bring the gas back to its original state, work must be done on the gas, and some heat will be exhausted.

The second law tells us that to do work with energy extracted from a heat reservoir, we must have a colder reservoir available to receive part of the energy as exhaust. If this were not true, we could design a ship with a heat engine that was powered by simply extracting thermal energy from the ocean. Unfortunately, the lack of a colder reservoir for exhaust makes this enormous reservoir of energy unavailable for such use.* The point is that in order to convert completely disordered thermal energy at a single temperature into the completely ordered energy of an object in motion (with no other changes in the source or object), a separate cold reservoir must be used.

*It is theoretically possible to run a heat engine between the warmer surface water of the ocean and the colder water at greater depths, but no practical scheme for using this temperature difference has yet emerged.

Example 20-1

A heat engine absorbs 200 J of heat from a hot reservoir, does work, and exhausts 160 J to a cold reservoir. What is the efficiency of the engine?

1. The efficiency is the work done divided by the heat in: $\varepsilon = \frac{W}{Q_{in}}$
2. The heat in is given: $Q_{in} = 200 \text{ J}$
3. The work is found from the first law: $W = Q_{in} - |Q_{out}| = 200 \text{ J} - 160 \text{ J} = 40 \text{ J}$
4. Substitute the values of Q_{in} and W to calculate the efficiency: $\varepsilon = \frac{W}{Q_{in}} = \frac{40 \text{ J}}{200 \text{ J}} = 0.20 = 20\%$

Exercise A heat engine has an efficiency of 35%. (*a*) How much work does it perform in a cycle if it extracts 150 J of thermal energy from a hot reservoir per cycle? (*b*) How much thermal energy is exhausted per cycle? (*Answers* (*a*) 52.5 J, (*b*) 97.5 J)

Example 20-2 *try it yourself*

(*a*) Find the efficiency of the Otto cycle shown in Figure 20-3. (*b*) Express your answer in terms of the ratio of the volumes $r = V_a/V_b = V_d/V_c$.

Picture the Problem (*a*) To find ε, you need to find Q_{in} and Q_{out}. Heat transfer occurs only during the two constant-volume processes, b to c and d to a. You can thus find Q_{in} and Q_{out} and therefore ε in terms of the temperatures T_a, T_b, T_c, and T_d. (*b*) The temperatures can be related to the volumes using $TV^{\gamma-1}$ = constant for adiabatic processes.

Cover the column to the right and try these on your own before looking at the answers.

Steps	***Answers***
(*a*)1. Write the efficiency in terms of Q_{in} and $\lvert Q_{out}\rvert$.	$\varepsilon = 1 - \frac{\lvert Q_{out}\rvert}{Q_{in}}$
2. The heat out occurs at constant volume from d to a. Write $\lvert Q_{out}\rvert$ in terms of C_v and the temperatures T_a and T_d.	$\lvert Q_{out}\rvert = \lvert Q_{da}\rvert = C_v\lvert T_a - T_d\rvert = C_v(T_d - T_a)$
3. The heat in occurs at constant volume from b to c. Write Q_{in} in terms of C_v and the temperatures T_c and T_b.	$Q_{in} = Q_{bc} = C_v(T_c - T_b)$
4. Substitute these values of $\lvert Q_{out}\rvert$ and Q_{in} to find the efficiency in terms of the temperatures T_a, T_b, T_c, and T_d.	$\varepsilon = 1 - \frac{T_d - T_a}{T_c - T_b}$

(*b*)1. Relate T_c to T_d using $TV^{\gamma-1}$ = constant, and $V_d/V_c = r$.
$$T_cV_c^{\gamma-1} = T_dV_d^{\gamma-1}; \qquad T_c = T_d\frac{V_d^{\gamma-1}}{V_c^{\gamma-1}} = T_dr^{\gamma-1}$$

2. Relate T_b to T_a as in step 1.
$$T_b = T_ar^{\gamma-1}$$

3. Use these relations to eliminate T_c and T_b from ε in (*a*) so that ε is expressed in terms of *r*.
$$\varepsilon = 1 - \frac{T_d - T_a}{T_dr^{\gamma-1} - T_ar^{\gamma-1}} = 1 - \frac{1}{r^{\gamma-1}}$$

Remark The ratio *r* (volume before compression/volume after compression) is called the compression ratio.

20-2 Refrigerators and the Second Law of Thermodynamics

Figure 20-5 is a schematic representation of a **refrigerator**, which is essentially a heat engine run backward. Work is put into the engine to extract thermal energy from the refrigerator (cold reservoir) and transfer it to the surroundings (hot reservoir). Experience shows that such a transfer always requires some work—a result known as the **refrigerator statement of the second law of thermodynamics,** which is equivalent to the Clausius statement:

> It is impossible for a refrigerator working in a cycle to produce no other effect than the transfer of thermal energy from a cold object to a hot object.

Second law of thermodynamics: Refrigerator statement

Were the above statement not true, we could cool our homes in the summer with refrigerators that pumped thermal energy to the outside without using any electricity or other energy.

A measure of a refrigerator's performance is the ratio $Q_{\text{in, c}}/W$, where $Q_{\text{in, c}}$ is the heat removed from a cold reservoir by the refrigerator doing work *W*. This ratio is called the **coefficient of performance,** COP:

$$\text{COP} = \frac{Q_c}{W} \qquad \text{20-3}$$

Definition—Coefficient of performance

The greater the coefficient of performance, the better the refrigerator. Typical refrigerators have coefficients of performance of about 5 or 6. In terms of this ratio, the refrigerator statement of the second law says that the coefficient of performance of a refrigerator cannot be infinite.

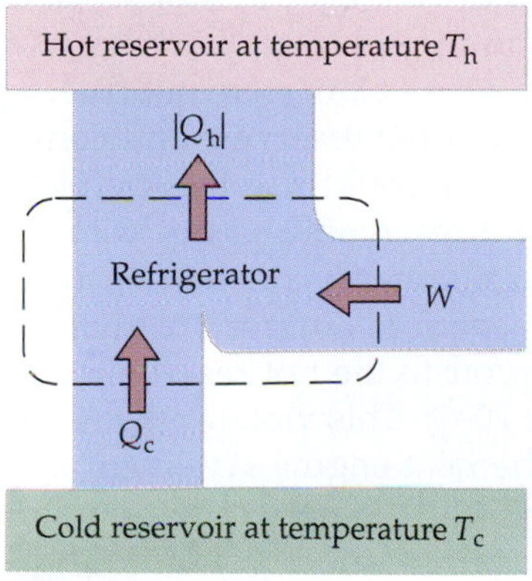

Figure 20-5 Schematic representation of a refrigerator. The refrigerator removes heat energy Q_c from a cold reservoir and gives off heat $|Q_h|$ to a hot reservoir using work *W*.

Example 20-3

A refrigerator has a coefficient of performance of 5.5. How much work is needed for this refrigerator to make ice cubes from 1 L of water at 10°C?

1. The work is related to the coefficient of performance and the heat extracted:

$$\text{COP} = \frac{Q_c}{W}$$

$$W = \frac{Q_c}{\text{COP}}$$

2. The total heat Q_c is the heat to be removed to cool the water and to freeze it:

$$Q_c = Q_{cool} + Q_{freeze}$$

3. The heat needed to cool 1 L of water (mass 1 kg) by 10 C° = 10 K is:

$$Q_{cool} = mc\,\Delta T = (1\text{ kg})(4.18\text{ kJ/kg·K})(10\text{ K}) = 41.8\text{ kJ}$$

4. The heat needed to freeze 1 L of water into ice cubes is:

$$Q_{freeze} = mL_f = (333.5\text{ kJ/kg})(1\text{ kg}) = 333.5\text{ kJ}$$

5. Add these heats to obtain Q_c:

$$Q_c = 41.8\text{ kJ} + 333.5\text{ kJ} = 375.3\text{ kJ} \approx 375\text{ kJ}$$

6. Substitute Q_c into step 1 to find the work W:

$$W = \frac{Q_c}{\text{COP}} = \frac{375\text{ kJ}}{5.5} = 68.2\text{ kJ}$$

Exercise A refrigerator has a coefficient of performance of 4.0. How much heat is exhausted to the hot reservoir when 200 kJ of heat are removed from the cold reservoir? (*Answer* 250 kJ)

20-3 Equivalence of the Heat-Engine and Refrigerator Statements

The heat-engine and refrigerator statements (or the Kelvin and Clausius statements) of the second law of thermodynamics seem quite different, but they are actually equivalent. We can prove this by showing that if either statement is assumed to be false, the other must also be false. We'll use a numerical example to show that if the heat-engine statement is false, then the refrigerator statement is false.

Figure 20-6*a* shows an ordinary refrigerator that uses 50 J of work to remove 100 J of energy from a cold reservoir and reject 150 J of energy to a hot reservoir. If the heat-engine statement of the second law were not true, a perfect heat engine could remove energy from the hot reservoir and convert it completely into work with 100% efficiency. We could use this perfect heat engine to remove 50 J of energy from the hot reservoir and do 50 J of work (Figure 20-6*b*). Then, using the perfect heat engine in conjunction with the ordinary refrigerator, we could construct a perfect refrigerator that would transfer 100 J of energy from the cold reservoir to the hot reservoir without requiring any work, as illustrated in Figure 20-6*c*. This violates the refrigerator statement of the second law. Thus, if the heat-engine statement is false, the refrigerator statement is also false. Similarly, if a perfect refrigerator existed, it could be used in conjunction with an ordinary heat engine to construct a perfect heat engine (see Problem 17). Thus, if the refrigerator statement is false, the heat-engine statement is also false. It then follows that if one statement is true, the other is also true. Therefore, the two statements are equivalent.

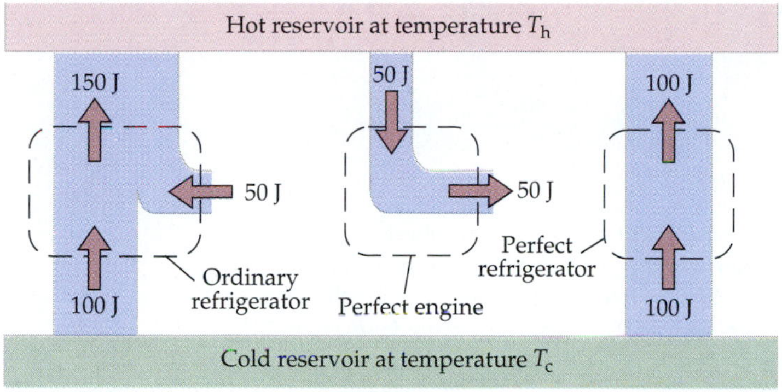

(*a*) An ordinary refrigerator removes 100 J from a cold reservoir, requiring the input of 50 J of work.

(*b*) A perfect heat engine violates the heat engine statement of the second law by removing 50 J from the hot reservoir and converting it completely into work.

(*c*) Putting the two together makes a perfect refrigerator that violates the refrigerator statement of the second law by transferring 100 J from the cold reservoir to the hot reservoir with no other effect.

Figure 20-6 Demonstration of the equivalence of the heat-engine and refrigerator statements of the second law of thermodynamics.

20-4 The Carnot Engine

According to the second law of thermodynamics, it is impossible for a heat engine working between two heat reservoirs to be 100% efficient. What, then, is the maximum possible efficiency for such an engine? This question was answered in 1824 by a young French engineer, Sadi Carnot, before either the first or second law of thermodynamics had been established. Carnot found that a *reversible engine* is the most efficient engine that can operate between any two given reservoirs. This result is known as the Carnot theorem:

> No engine working between two given heat reservoirs can be more efficient than a reversible engine working between those two reservoirs.

Carnot theorem

A reversible engine is thus an ideal engine, in that it operates with the greatest possible efficiency. A reversible engine working in a cycle between two heat reservoirs is called a **Carnot engine,** and its cycle is called a **Carnot cycle**. The Carnot engine is an ideal engine in that its efficiency cannot be improved upon. Figure 20-7 illustrates the Carnot theorem with a numerical example.

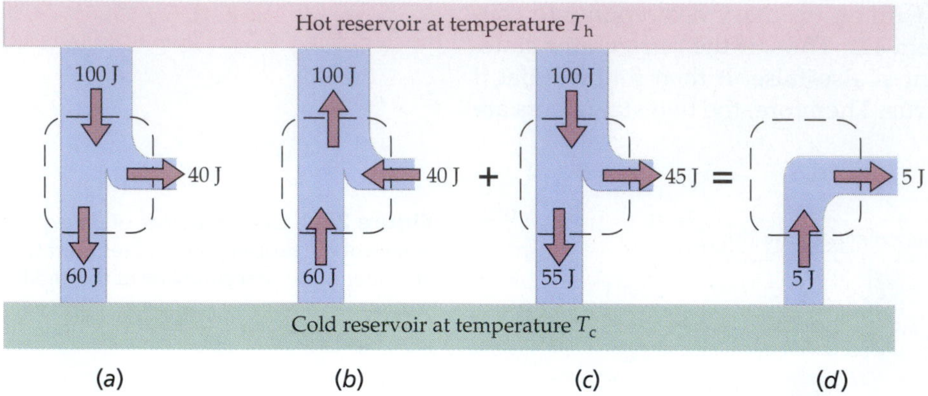

(a) A reversible heat engine with 40% efficiency removes 100 J from a hot reservoir, does 40 J of work, and exhausts 60 J to the cold reservoir.

(b) When the same engine runs backward as a refrigerator, 40 J of work are done to remove 60 J from the cold reservoir and exhaust 100 J to the hot reservoir.

(c) This assumed heat engine works between the same two reservoirs with a 45% efficiency, which is greater than that of the reversible engine in (*a*).

(d) The net effect of running the engine in (*c*) in conjunction with the refrigerator in (*b*) is the same as that of a perfect heat engine that removes 5 J from the cold reservoir and converts it completely into work with no other effect, violating the second law of thermodynamics. Thus, the reversible engine in (*a*) is the most efficient engine that can operate between these two reservoirs.

Figure 20-7 Demonstration of the Carnot theorem.

If no engine can have a greater efficiency than a Carnot engine, it follows that all Carnot engines working between the same two reservoirs have the same efficiency. This efficiency, called the **Carnot efficiency,** must be independent of the working substance of the engine and thus can depend only on the temperatures of the reservoirs.

Let us look at what makes a process reversible or irreversible. According to the second law, the conversion of mechanical energy into heat by friction is *not* reversible; nor is the conduction of heat from a hot object to a cold one. A third type of irreversibility occurs when a system passes through nonequilibrium states, such as when there is turbulence in a gas, or when a gas explodes. For a process to be reversible, we must be able to move the system back through the same equilibrium states in the reverse order.

From these considerations and our statements of the second law of thermodynamics, we can list some conditions that are necessary for a process to be reversible:

1. No work must be done by friction, viscous forces, or other dissipative forces that produce heat.
2. Heat conduction can only occur isothermally.
3. The process must be quasi-static so that the system is always in an equilibrium state (or infinitesimally near an equilibrium state).

Conditions for reversibility

Any process that violates any of the above conditions is irreversible. Most processes in nature are irreversible. To have a reversible process, great care must be taken to eliminate frictional and other dissipative forces and to make the process quasi-static. Since this can never be done completely, a reversible process is an idealization, similar to the idealization of motion without friction in mechanics problems. Nevertheless, reversibility can in practice be approximated quite closely.

We can now understand the features of a Carnot cycle, which is a reversible cycle between just two reservoirs. Since all heat transfer must be done isothermally for the process to be reversible, the heat absorbed from the hot reservoir must be absorbed isothermally. The next step must be a quasi-static adiabatic expansion to the lower temperature of the cold reservoir. Then heat is given off isothermally to the cold reservoir. Finally, there is a quasi-static, adiabatic compression to the higher temperature of the hot reservoir. The Carnot cycle thus consists of four reversible steps:

1. A quasi-static isothermal absorption of heat from a hot reservoir
2. A quasi-static adiabatic expansion to a lower temperature
3. A quasi-static isothermal exhaustion of heat to a cold reservoir
4. A quasi-static adiabatic compression back to the original state

Steps in a Carnot cycle

To calculate the efficiency of a Carnot engine, we choose as the working substance a material of which we have some knowledge—an ideal gas—and explicitly calculate the work done by it over a Carnot cycle (Figure 20-8). Since all Carnot cycles have the same efficiency independent of the working substance, our result will be valid in general.

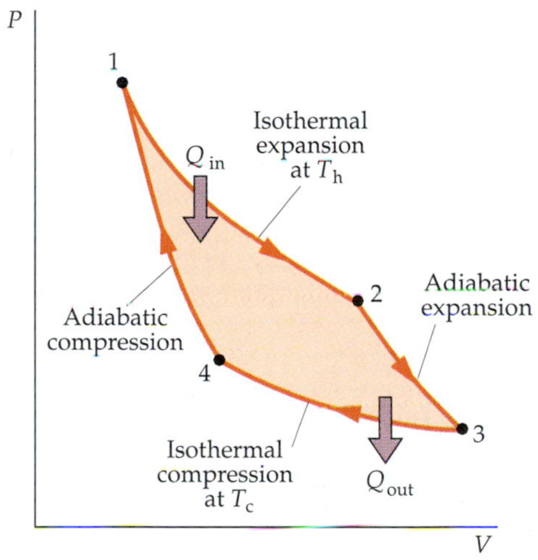

Figure 20-8 Carnot cycle for an ideal gas:
Step 1: Heat is absorbed from a hot reservoir at temperature T_h during an isothermal expansion from state 1 to state 2.
Step 2: The gas expands adiabatically from state 2 to state 3 and its temperature is reduced to T_c.
Step 3: The gas gives off heat to the cold reservoir as it is compressed isothermally at T_c from state 3 to state 4.
Step 4: The gas is compressed adiabatically until its temperature is again T_h.
Work is done on the gas or by the gas during each step. The net work done during the cycle is represented by the shaded area. All processes are reversible. All steps are quasi-static.

The efficiency of this cycle is

$$\varepsilon = 1 - \frac{|Q_{\text{out,c}}|}{Q_{\text{in,h}}}$$

The heat $Q_{\text{in,h}}$ is absorbed during the isothermal expansion from state 1 to state 2. Since $\Delta U = 0$ for an isothermal expansion of an ideal gas, $Q_{\text{in,h}}$ equals the work done by the gas.

$$Q_{\text{in,h}} = W = \int_1^2 P\,dV = \int_1^2 \frac{nRT_h}{V}\,dV = nRT_h \ln \frac{V_2}{V_1}$$

Similarly, the heat given off to the cold reservoir equals the work done on the gas during the isothermal compression at temperature T_c from state 3 to state 4. This work has the same magnitude as that done by the gas if it expands from state 4 to state 3. The heat rejected is thus

$$|Q_{out,c}| = nRT_c \ln \frac{V_3}{V_4}$$

The ratio of these heats is

$$\frac{|Q_{out,c}|}{Q_{in,h}} = \frac{T_c \ln(V_3/V_4)}{T_h \ln(V_2/V_1)} \qquad \text{20-4}$$

We can relate the volumes V_1, V_2, V_3, and V_4 using Equation 19-22 for a quasi-static adiabatic expansion:

$$TV^{\gamma-1} = \text{constant}$$

For the expansion from state 2 to state 3, we have

$$T_h V_2^{\gamma-1} = T_c V_3^{\gamma-1}$$

Similarly, for the adiabatic compression from state 4 to state 1, we have

$$T_h V_1^{\gamma-1} = T_c V_4^{\gamma-1}$$

Dividing these two equations, we obtain

$$\left(\frac{V_2}{V_1}\right)^{\gamma-1} = \left(\frac{V_3}{V_4}\right)^{\gamma-1}$$

and so $V_2/V_1 = V_3/V_4$. Therefore, $\ln(V_2/V_1) = \ln(V_3/V_4)$ and we can cancel the logarithmic terms in Equation 20-4 to obtain

$$\frac{|Q_{out,c}|}{Q_{in,h}} = \frac{T_c}{T_h} \qquad \text{20-5}$$

(*a*)

(*a*) Coal-fueled electric generating plant at Four Corners, New Mexico. (*b*) Power plant at Wairakei, New Zealand, that converts geothermal energy into electricity. (*c*) Solar energy is focused and collected individually to produce electricity by these heliostats being tested at Sandia National Laboratory. (*d*) Control rods are inserted into this nuclear reactor at Tihange, Belgium. (*e*) An experimental wind-powered electric generator at Sandia National Laboratory. The propeller is designed for optimum transfer of wind energy to mechanical energy.

The Carnot efficiency ε_C is thus

$$\varepsilon_C = 1 - \frac{T_c}{T_h} \qquad \text{20-6}$$

Carnot efficiency

Equation 20-6 demonstrates what we noted earlier, that the Carnot efficiency depends only on the temperatures of the two reservoirs.

(*b*)

(*c*)

(*d*)

(*e*)

Example 20-4

A steam engine works between a hot reservoir at 100°C = 373 K and a cold reservoir at 0°C = 273 K. (*a*) What is the maximum possible efficiency of this engine? (*b*) If the engine is run backward as a refrigerator, what is its maximum coefficient of performance?

Picture the Problem The maximum efficiency is the Carnot efficiency given by Equation 20-6. To find the coefficient of performance, it is easiest to choose some values for the heat absorbed and given off. If the engine absorbs 100 J from the hot reservoir, it does work $W = \varepsilon(100\text{ J})$ and gives off heat $Q_c = 100 - \varepsilon(100\text{ J})$. Then $\text{COP} = Q_c/W$.

(*a*) The maximum efficiency is the Carnot efficiency:
$$\varepsilon_C = 1 - \frac{T_c}{T_h} = 1 - \frac{273\text{ K}}{373\text{ K}} = 0.268 = 26.8\%$$

(*b*)1. Find the work done by the engine when it absorbs 100 J:
$$W = \varepsilon_C\, Q_{ih,h} = \; = 0.268(100\text{ J}) = 26.8\text{ J}$$

2. Find the heat given off:
$$Q_{out,c} = Q_{ih,h} - W = 100\text{ J} - 26.8\text{ J} = 73.2\text{ J}$$

3. Use these results to calculate COP when the engine is run backward:
$$\text{COP} = \frac{Q_{out,c}}{W} = \frac{73.2\text{ J}}{26.8\text{ J}} = 2.73$$

Remarks Even though this efficiency seems to be quite low, it is the greatest efficiency possible for any engine working between these temperatures. Real engines will have lower efficiencies because of friction, heat conduction, and other irreversible processes. Similarly, real refrigerators will have a lower coefficient of performance. In Section 20-5 it is shown that the coefficient of performance of a Carnot refrigerator is $T_c/\Delta T$ (Equation 20-9). For these reservoirs, $\Delta T = 100$ K and COP = 273 K/100 K = 2.73.

The Carnot efficiency is useful to know because it gives us an upper limit on possible efficiencies. It tells us, for example, that an engine working between reservoirs at 373 and 273 K with an efficiency of 25% is a very good engine. However much friction and other irreversible losses were reduced, the best efficiency obtainable between those temperatures is 26.8%, as calculated in Example 20-4.

Example 20-5

An engine removes 200 J from a hot reservoir at 373 K, does 48 J of work, and exhausts 152 J to a cold reservoir at 273 K. How much work is "lost" per cycle due to irreversible processes in this engine?

Picture the Problem The difference between maximum amount of work that could be done using a Carnot engine and 48 J is the work lost.

1. The work lost is the maximum amount of work that could be done minus the work actually done:
$$W_{lost} = W_{max} - W = W_{max} - 48\text{ J}$$

2. The maximum amount of work that could be done is: $W_{max} = \varepsilon_C Q_{in,h}$

3. The Carnot efficiency was found in Example 20-4: $\varepsilon_C = 1 - \frac{T_c}{T_h} = 1 - \frac{273\text{ K}}{373\text{ K}} = 0.268 = 26.8\%$

4. The maximum amount of work that could be done is then: $W_{max} = \varepsilon_C Q_{in,h} = (0.268)(200) = 53.6\text{ J}$

5. The work lost is then: $W_{lost} = W_{max} - W = 53.6\text{ J} - 48\text{ J} = 5.6\text{ J}$

Remarks The 5.6 J of energy in the answer is not "lost" to the universe—total energy is conserved. That 5.6 J of energy exhausted into the cold reservoir by the nonideal engine of the problem is only lost in that it would have been converted into useful work if an ideal (reversible) engine had been used.

Example 20-6

If 200 J of heat is conducted from a heat reservoir at 373 K to one at 273 K, with no engine between the reservoirs as in Example 20-5, how much work capability is "lost" in this process?

We saw in Example 20-5 that a Carnot engine working between these two reservoirs could do 53.6 J of work if it extracted 200 J from the 373-K reservoir and exhausted to a 273-K reservoir. Thus, if 200 J is conducted directly from the hot reservoir to the cold reservoir without any work being done, 53.6 J of this energy has been "lost" in the sense that it could have been converted into useful work.

Exercise A Carnot engine works between heat reservoirs at 500 and 300 K. (*a*) What is its efficiency? (*b*) If it removes 200 kJ of heat from the hot reservoir, how much work does it do? (*Answers* (*a*) 40%, (*b*) 80 kJ)

Exercise A real engine works between heat reservoirs at 500 and 300 K. It removes 500 kJ of heat from the hot reservoir and does 150 kJ of work during each cycle. What is its efficiency? (*Answer* 30%)

The Absolute Temperature Scale

In Chapter 18, the ideal-gas temperature scale was defined in terms of the properties of gases at low densities. Since the Carnot efficiency depends only on the temperatures of the two heat reservoirs, it can be used to define the ratio of the temperatures of the reservoirs independent of the properties of any substance. We define the ratio of the absolute temperatures of the hot and cold reservoirs to be

$$\frac{T_c}{T_h} = \frac{|Q_c|}{Q_h} \qquad \text{20-7}$$

where Q_h is the energy removed from the hot reservoir and $|Q_c|$ is the energy exhausted to the cold reservoir by a Carnot engine working between the two reservoirs. Thus, to find the ratio of two reservoir temperatures, we set up a reversible engine operating between them and measure the heat absorbed from or given off by each reservoir during one cycle. The absolute temperature is completely determined by Equation 20-7 and the choice of one fixed point. If the fixed point is defined to be 273.16 K for the triple point of water, then the absolute temperature scale matches the ideal-gas temperature scale for the range of temperatures over which a gas thermometer can be used.

optional

20-5 Heat Pumps

A **heat pump** is essentially a refrigerator that is used to pump thermal energy from a cold reservoir (for example, the cold air outside a house) to a hot reservoir (for example, the hot air inside the house). If work W is done to remove heat Q_c from the cold reservoir and reject heat $|Q_h| = W + Q_c$ to the hot reservoir, the coefficient of performance (Equation 20-3) is

$$\text{COP} = \frac{Q_c}{W}$$

Using $W = |Q_h| - Q_c$, this can be written

$$\text{COP} = \frac{Q_c}{|Q_h| - Q_c} = \frac{Q_c/|Q_h|}{1 - Q_c/|Q_h|} \qquad \text{20-8}$$

The maximum coefficient of performance is obtained using a Carnot heat pump. Then Q_c and Q_h are related by Equation 20-5. Substituting $Q_c/|Q_h| = T_c/T_h$ into Equation 20-8, we obtain for the maximum coefficient of performance

$$\text{COP}_{max} = \frac{T_c/T_h}{1 - T_c/T_h} = \frac{T_c}{T_h - T_c} = \frac{T_c}{\Delta T} \qquad \text{20-9}$$

where ΔT is the difference in temperature between the hot and cold reservoirs. Real heat pumps and refrigerators have COPs less than the COP_{max} because of losses due to friction, heat conduction, and other irreversible processes.

We are usually interested in the work that must be done to exhaust a given amount of heat $|Q_h|$ into the hot reservoir, which in the case of a home heat pump would be the hot-air supply for the heating fan of a house. Using $|Q_h| = Q_c + W$, we can write Equation 20-3 as

$$\text{COP} = \frac{Q_c}{W} = \frac{|Q_h| - W}{W}$$

or

$$W = \frac{|Q_h|}{1 + \text{COP}} \qquad \text{20-10}$$

Example 20-7 ***try it yourself***

An ideal heat pump is used to pump heat from the outside air at −5°C to the hot-air supply for the heating fan in a house, which is at 40°C. How much work is required to pump 1 kJ of heat into the house?

Picture the Problem Use Equation 20-10 with COP calculated from Equation 20-9 for $T_c = -5°\text{C} = 268$ K and $\Delta T = 45$ K.

Cover the column to the right and try these on your own before looking at the answers.

Steps	**Answers**
1. Calculate the COP from Equation 20-9.	$\text{COP}_{\text{max}} = \frac{T_c}{\Delta T} = 5.96$
2. Calculate the work needed from Equation 20-10.	$W = \frac{\lvert Q_c \rvert}{1 + \text{COP}} = 0.144 \text{ kJ}$

Remarks Only 0.144 kJ of work is needed to pump 1 kJ of heat into the hot-air supply in the house. We see that the heat pump essentially multiplies the energy needed to run the pump by 1 + COP. If we use 1 kJ to run a heat pump with a COP = 5.96, we can exhaust 6.96 kJ of heat into the house.

20-6 Irreversibility and Disorder

There are many irreversible processes that cannot be described by the heat-engine or refrigerator statements of the second law, such as a glass falling to the floor and breaking. However, all irreversible processes have one thing in common—the system plus its surroundings moves toward a less ordered state.

Suppose a box containing a gas of mass M at a temperature T is moving along a frictionless table with a velocity v_{cm} (Figure 20-9a). The total kinetic energy of the gas has two components: that associated with the movement of the center of mass $\frac{1}{2}Mv_{\text{cm}}^2$, and the energy of the motion of its molecules relative to its center of mass. The center-of-mass energy $\frac{1}{2}Mv_{\text{cm}}^2$ is ordered mechanical energy that could be converted completely into work.* The relative energy is the gas's internal thermal energy, which is related to its temperature T. It is random, nonordered energy that cannot be converted directly into work.

Now, suppose the box hits a fixed wall and stops (Figure 20-9b):

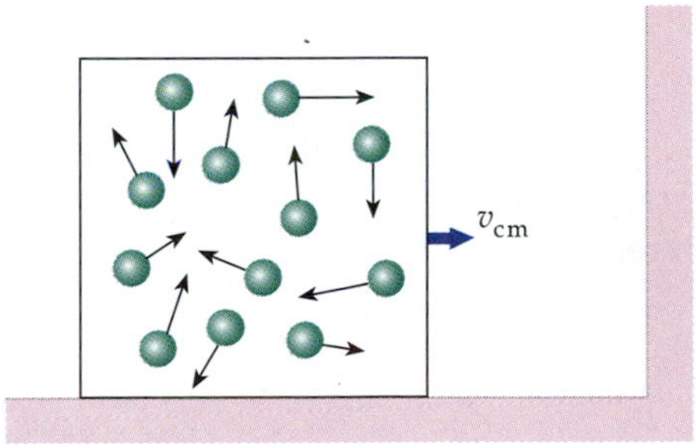

Figure 20-9*a*

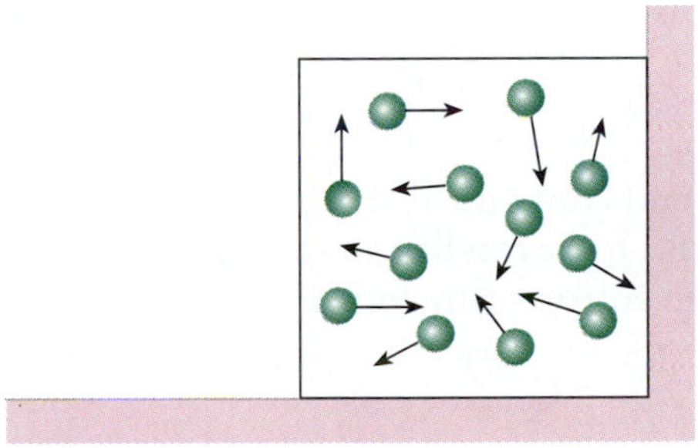

Figure 20-9*b*

This inelastic collision is clearly an irreversible process. The ordered mechanical energy of the gas is converted into random internal energy and the temperature of the gas rises. The gas still has the same total energy, but now all of it is associated with the random motion of its molecules about its center of mass, which is now at rest. Thus, the gas has become less ordered (or more disordered), and it has lost the ability to do work.

*For example, if a weight were attached with a string to the moving box, this energy could be used to lift the weight.

20-7 Entropy

There is a thermodynamic function called **entropy** S that is a measure of the disorder of a system. Like the pressure P, volume V, temperature T, and internal energy U, entropy S is a function of the state of a system. As with potential energy, it is the *change* in entropy that is important. The change in entropy dS of a system when it goes from one state to another is defined as

$$dS = \frac{dQ_{\text{rev}}}{T} \qquad \text{20-11}$$

Definition—Entropy change

where dQ_{rev} is the heat that must be added to the system in a *reversible process* that brings the system from the initial state to the final state. If dQ_{rev} is negative, the entropy change of the system is negative.

The term dQ_{rev} does not mean that a reversible heat transfer must take place for the entropy of a system to change. Indeed, there are many situations in which the entropy of a system changes when there is no transfer of heat whatsoever. Equation 20-11 simply gives us a method for *calculating* the entropy difference between two states of a system. Since entropy is a state function, its change when the system moves from one state to another depends only on the system's initial and final states, not on the process by which the change occurs.

Entropy of an Ideal Gas

We now illustrate that dQ/T is in fact the differential of a state function for an ideal gas even though dQ is not. Consider an arbitrary reversible quasi-static process in which a system consisting of an ideal gas absorbs an amount of heat dQ. According to the first law, dQ is related to the change in the internal energy dU of the gas and the work done $dW = P\,dV$ by

$$dQ = dU + dW = dU + P\,dV$$

For an ideal gas, we can write dU in terms of the heat capacity, $dU = C_v\,dT$, and we can substitute nRT/V for P from the equation of state. Then

$$dQ = C_v\,dT + nRT\frac{dV}{V} \qquad \text{20-12}$$

Equation 20-12 cannot be integrated unless we know how T depends on V. This is just another way of saying that dQ is not a differential of a state function Q. But if we divide each term by T, we obtain

$$\frac{dQ}{T} = C_v\frac{dT}{T} + nR\frac{dV}{V} \qquad \text{20-13}$$

Since C_v depends only on T, the first term on the right can be integrated and now the second term can be integrated also.* Thus dQ/T is the differential of a function, the entropy function S:

$$dS = \frac{dQ}{T} = C_v\frac{dT}{T} + nR\frac{dV}{V} \qquad \text{20-14}$$

For simplicity, we will assume that C_v is constant. Integrating Equation 20-14, we obtain

$$\Delta S = \int\frac{dQ}{T} = C_v\ln\frac{T_2}{T_1} + nR\ln\frac{V_2}{V_1} \qquad \text{20-15}$$

Equation 20-15 gives the entropy change of an ideal gas that undergoes a reversible expansion from an initial state of volume V_1 and temperature T_1 to a final state of volume V_2 and temperature T_2.

*Mathematically, the factor $1/T$ is called an integrating factor for Equation 20-12.

Entropy Changes for Various Processes

ΔS for an Isothermal Expansion of an Ideal Gas When an ideal gas undergoes an isothermal expansion, $T_2 = T_1$ and its entropy change is

$$\Delta S = \int \frac{dQ}{T} = nR \ln \frac{V_2}{V_1} \qquad \text{20-16}$$

The entropy change of the gas is positive because V_2 is greater than V_1. In this process, an amount of heat $|Q|$ leaves the reservoir and enters the gas. This amount of heat equals the work done by the gas:

$$Q = W = \int_1^2 P\,dV = nRT \int_{V_1}^{V_2} \frac{dV}{V} = nRT \ln \frac{V_2}{V_1} \qquad \text{20-17}$$

The entropy change of the gas is $+|Q|/T$. Since the same amount of heat leaves the reservoir at temperature T, the entropy change of the reservoir is $-|Q|/T$. The net entropy change of the gas plus the reservoir is zero. We will refer to the system under consideration plus its surroundings as the "universe." This example illustrates a general result:

> In a reversible process, the entropy change of the universe is zero.

ΔS for a Free Expansion of an Ideal Gas In the free expansion of a gas discussed in Section 19-4, a gas is initially confined in one compartment of a container, which is connected by a stopcock to another compartment that is evacuated. The whole system has rigid walls and is thermally insulated from its surroundings so that no heat can flow in or out, and no work can be done (Figure 19-5). When the stopcock is opened, the gas rushes into the evacuated chamber. Eventually, the gas reaches thermal equilibrium with itself. Since no work is done and no heat is transferred, the final internal energy of the gas must equal its initial internal energy. Let us assume that the gas is ideal. Then the final temperature T equals the initial temperature.

We might think that there is no entropy change of the gas because there is no heat transfer. But this process is not reversible, so we cannot use $\int dQ/T$ to find the change in entropy of the gas. However, the initial and final states of the gas in the free expansion are the same as those of the gas in the isothermal expansion just discussed. *Since the change in the entropy of a system for any process depends only on the initial and final states of the system, the entropy change of the gas for the free expansion is the same as that for the isothermal expansion.* If V_1 is the initial volume of the gas and V_2 is its final volume, the entropy change of the gas is given by Equation 20-16:

$$\Delta S_{\text{gas}} = nR \ln \frac{V_2}{V_1}$$

In this case, there is no change in the surroundings, so the entropy change of the gas is also the entropy change of the universe:

$$\Delta S_{\text{u}} = nR \ln \frac{V_2}{V_1} \qquad \text{20-18}$$

Note that because V_2 is greater than V_1, the change in entropy of the universe for this irreversible process is positive; that is, the entropy of the universe increases. This is also a general result:

> In an irreversible process, the entropy of the universe increases.

If the final volume in the free expansion were less than the initial volume, the entropy of the universe would decrease, but this does not happen. A gas does not freely contract by itself into a smaller volume. This leads us to yet another statement of the second law of thermodynamics:

> For any process, the entropy of the universe never decreases.

Example 20-8

Find the entropy change for the free expansion of 0.75 mol of an ideal gas from $V_1 = 1.5$ L to $V_2 = 3$ L.

Picture the Problem The entropy change for this irreversible process is equivalent to the entropy change for an isothermal process from V_1 to V_2. For the isothermal process, we first calculate $Q = W$, then set $\Delta S = Q/T$.

1. The entropy change is the same as for an isothermal expansion from V_1 to V_2: $\Delta S = \Delta S_{\text{isothermal}} = \dfrac{Q}{T}$

2. The heat Q that would enter the gas during an isothermal expansion at temperature T equals the work done by the gas during the expansion: $Q = W = nRT \ln \dfrac{V_2}{V_1}$

3. Substitute this value of Q to calculate ΔS: $\Delta S = \dfrac{Q}{T} = nR \ln \dfrac{V_2}{V_1} = (0.75 \text{ mol})(8.31 \text{ J/mol}\cdot\text{K}) \ln 2 = 4.32 \text{ J/K}$

ΔS for Constant-Pressure Processes When a substance is heated from temperature T_1 to temperature T_2 at constant pressure, the heat absorbed dQ is related to its temperature change dT by

$$dQ = C_p \, dT$$

We can approximate reversible heat conduction if we have a large number of heat reservoirs with temperatures ranging from T_1 to T_2 in very small steps. We then place our substance, whose initial temperature is T_1, in contact with the first reservoir at a temperature just slightly greater than T_1 and let the substance absorb a small amount of heat. Since the heat transfer is approximately isothermal, the process will be approximately reversible. We then place the substance in contact with the next reservoir at a slightly higher temperature, and so on, until the final temperature T_2 is reached. When heat dQ is absorbed reversibly, the entropy change of the substance is

$$dS = \frac{dQ}{T} = C_p \frac{dT}{T}$$

Integrating from T_1 to T_2, we obtain the total entropy change of the substance:

$$\Delta S = C_p \int_{T_1}^{T_2} \frac{dT}{T} = C_p \ln \frac{T_2}{T_1} \qquad \text{20-19}$$

This result gives the entropy change of a substance that is heated from T_1 to T_2 by any process, reversible or irreversible, as long as the final pressure equals the initial pressure. It also gives the entropy change of a substance that is cooled. In this case, T_2 is less than T_1, and $\ln(T_2/T_1)$ is negative, giving a negative entropy change.

Exercise Find the change in entropy of 1 kg of water that is heated from 0 to 100°C. (*Answer* $\Delta S = 1.31$ kJ/K)

Example 20-9 *try it yourself*

1 kg of water at temperature $T_1 = 30°C$ is mixed with 2 kg of water at $T_2 = 90°C$ in a calorimeter of negligible heat capacity at a constant pressure of 1 atm. Find the change in entropy of the system.

Picture the Problem When the two amounts of water are mixed, they eventually come to a final equilibrium temperature T_f that can be found by setting the heat lost equal to the heat gained. To calculate the entropy change of each mass of water, we consider a reversible isobaric heating of the 1-kg mass of water from 30°C to T_f and an isobaric cooling of the 2-kg mass from 90°C to T_f using Equation 20-19. The entropy change of the system is the sum of the entropy changes of each part.

Cover the column to the right and try these on your own before looking at the answers.

Steps	*Answers*
1. Write an expression for the entropy change ΔS_1 of the mass $m_1 = 1$ kg of water going from the state $T = T_1$ to the state $T = T_f$.	$\Delta S_1 = m_1 c_p \ln\frac{T_f}{T_1}$
2. Write an expression for the entropy change ΔS_2 of the mass $m_2 = 2$ kg of water going from the state $T = T_2$ to the state $T = T_f$.	$\Delta S_2 = m_2 c_p \ln\frac{T_f}{T_2}$
3. Calculate T_f by setting the heat lost equal to the heat gained.	$m_1 c_p(T_f - 30°C) = m_2 c_p(90°C - T_f)$ $T_f = 70°C = 343\text{ K}$
4. Use your result for T_f and the data given to calculate ΔS_1 and ΔS_2.	$\Delta S_1 = m_1 c_p \ln\frac{T_f}{T_1} = 0.518\text{ kJ/K}$ $\Delta S_2 = m_2 c_p \ln\frac{T_f}{T_2} = -0.474\text{ kJ/K}$
5. Add ΔS_1 and ΔS_2 to find the total entropy change of the system.	$\Delta S = \Delta S_1 + \Delta S_2 = +0.044\text{ kJ/K}$

Remarks Note that we had to convert the temperatures to the absolute scale to calculate the entropy changes. The entropy change of the universe is positive as expected.

ΔS for an Inelastic Collision Since mechanical energy is converted into thermal energy in an inelastic collision, such a process is clearly irreversible. The entropy of the universe must therefore increase. Consider a block of mass m falling from a height h and making an inelastic collision with the ground. Let the block, ground, and atmosphere all be at a temperature T, which is not significantly changed by the process. If we consider the block, ground, and atmosphere as our isolated system, there is no heat conducted into or out of the system. The state of the system has been changed because its internal energy has been increased by an amount mgh. This change is the same as if we added heat $Q = mgh$ to the system at constant temperature T.

To calculate the change in entropy of the system, we thus consider a reversible process in which heat $Q_{\text{rev}} = mgh$ is added at a constant temperature T. According to Equation 20-11, the change in entropy is then

$$\Delta S = \frac{Q}{T} = \frac{mgh}{T}$$

This positive entropy change is also the entropy change of the universe.

ΔS for Heat Conduction From One Reservoir to Another Heat conduction is also an irreversible process, so we expect the entropy of the universe to increase. Consider the simple case of heat $|Q|$ conducted from a hot reservoir at a temperature T_{h} to a cold reservoir at a temperature T_{c}. The state of a heat reservoir is determined only by its temperature and its internal energy. The change in entropy of a heat reservoir due to a heat exchange is the same whether the heat exchange is reversible or not. If heat $|Q|$ is put into a reservoir at temperature T, the entropy of the reservoir increases by $|Q|/T$. If the heat is removed, the entropy of the reservoir decreases by $-|Q|/T$. In the case of heat conduction, the hot reservoir loses heat, so its entropy change is

$$\Delta S_{\text{h}} = -\frac{|Q|}{T_{\text{h}}}$$

The cold reservoir absorbs heat, so its entropy change is

$$\Delta S_{\text{c}} = +\frac{|Q|}{T_{\text{c}}}$$

The net entropy change of the universe is

$$\Delta S_{\text{u}} = \Delta S_{\text{c}} + \Delta S_{\text{h}} = +\frac{|Q|}{T_{\text{c}}} - \frac{|Q|}{T_{\text{h}}} \qquad \text{20-20}$$

Note that since heat always flows from a hot reservoir to a cold reservoir, the change in entropy of the universe is positive.

ΔS for a Carnot Cycle Since a Carnot cycle is by definition reversible, the entropy change of the universe after a cycle must be zero. We demonstrate this by showing that the entropy change of the reservoirs is zero. (Since a Carnot engine works in a cycle, its entropy change is zero; the entropy change of the universe is just the sum of the entropy changes of the reservoirs.) The entropy change of the hot reservoir is $\Delta S_{\text{h}} = -|Q_{\text{h}}|/T_{\text{h}}$, where $|Q_{\text{h}}|$ is the heat removed. The entropy change of the cold reservoir is $\Delta S_{\text{c}} = +|Q_{\text{c}}|/T_{\text{c}}$, where $|Q_{\text{c}}|$ is the energy exhausted. These energies are related by Equation 20-7,

$$|Q_{\text{c}}| = |Q_{\text{h}}|\frac{T_{\text{c}}}{T_{\text{h}}}$$

The entropy change of the universe is thus*

$$\Delta S_{\text{u}} = \Delta S_{\text{h}} + \Delta S_{\text{c}} = -\frac{|Q_{\text{h}}|}{T_{\text{h}}} + \frac{|Q_{\text{c}}|}{T_{\text{c}}} = -\frac{|Q_{\text{h}}|}{T_{\text{h}}} + \frac{|Q_{\text{h}}|(T_{\text{c}}/T_{\text{h}})}{T_{\text{c}}} = 0$$

The entropy change of the universe is zero, as expected.

* Because the heat *out* of the hot reservoir is the heat *into* the Carnot engine, the sign conventions and the subscripts in and out are confusing here, so we are using absolute values signs for both Q_{h} and Q_{c}, and displaying the signs explicitly.

Example 20-10

During each cycle, a Carnot engine removes 100 J of energy from a reservoir at 400 K, does work, and exhausts heat to a reservoir at 300 K. Compute the entropy change of each reservoir for each cycle, and show explicitly that the entropy change of the universe is zero for this reversible process.

Picture the Problem Since the engine works in a cycle, its entropy change is zero. We therefore compute the entropy change of each reservoir and add them to obtain the entropy change of the universe.

1. The entropy change of the universe equals the sum of the entropy changes of the reservoirs:
$$\Delta S_u = \Delta S_{400} + \Delta S_{300}$$

2. Calculate the entropy change of the hot reservoir:
$$\Delta S_{400} = -\frac{|Q_h|}{T_h} = -\frac{100\ \text{J}}{400\ \text{K}} = -0.250\ \text{J/K}$$

3. The entropy change of the cold reservoir is $|Q_c|$ divided by T_c:
$$\Delta S_{300} = \frac{|Q_c|}{T_c} = \frac{|Q_c|}{300\ \text{K}}$$

4. The heat given off $|Q_c|$ is related to the efficiency:
$$\varepsilon_C = 1 - \frac{|Q_c|}{|Q_h|} = 1 - \frac{T_c}{T_h} = 1 - \frac{300\ \text{K}}{400\ \text{K}} = 0.25$$
$$|Q_c| = |Q_h|(1 - 0.25) = (100\ \text{J})(0.75) = 75\ \text{J}$$

5. Calculate the entropy change of the cold reservoir:
$$\Delta S_{300} = \frac{|Q_c|}{T_c} = \frac{75\ \text{J}}{300\ \text{K}} = +0.250\ \text{J/K}$$

6. Substitute these results into step 1 to find the entropy change of the universe:
$$\Delta S_u = \Delta S_{400} + \Delta S_{300}$$
$$= -0.250\ \text{J/K} + 0.250\ \text{J/K} = 0$$

Remarks Suppose an ordinary, nonreversible engine removed 100 J from the hot reservoir. Since its efficiency must be less than that of a Carnot engine, it would do less work and exhaust more heat to the cold reservoir. Then the entropy increase of the cold reservoir would be greater than the entropy decrease of the hot reservoir, and the entropy change of the universe would be positive.

Example 20-11

Since entropy is a state function, thermodynamic processes can be represented as *ST*, *SV*, or *SP* diagrams instead of the *PV* diagrams we have used so far. Make a sketch of the Carnot cycle on an *ST* plot.

Picture the Problem The Carnot cycle consists of a reversible isothermal expansion followed by a reversible adiabatic expansion, then a reversible isothermal compression followed by a reversible adiabatic compression. During the isothermal processes, heat is absorbed or expelled at constant temperature, so S increases or decreases at constant T. During the adiabatic processes, the temperature changes, but since $\Delta Q_r = 0$, S is constant.

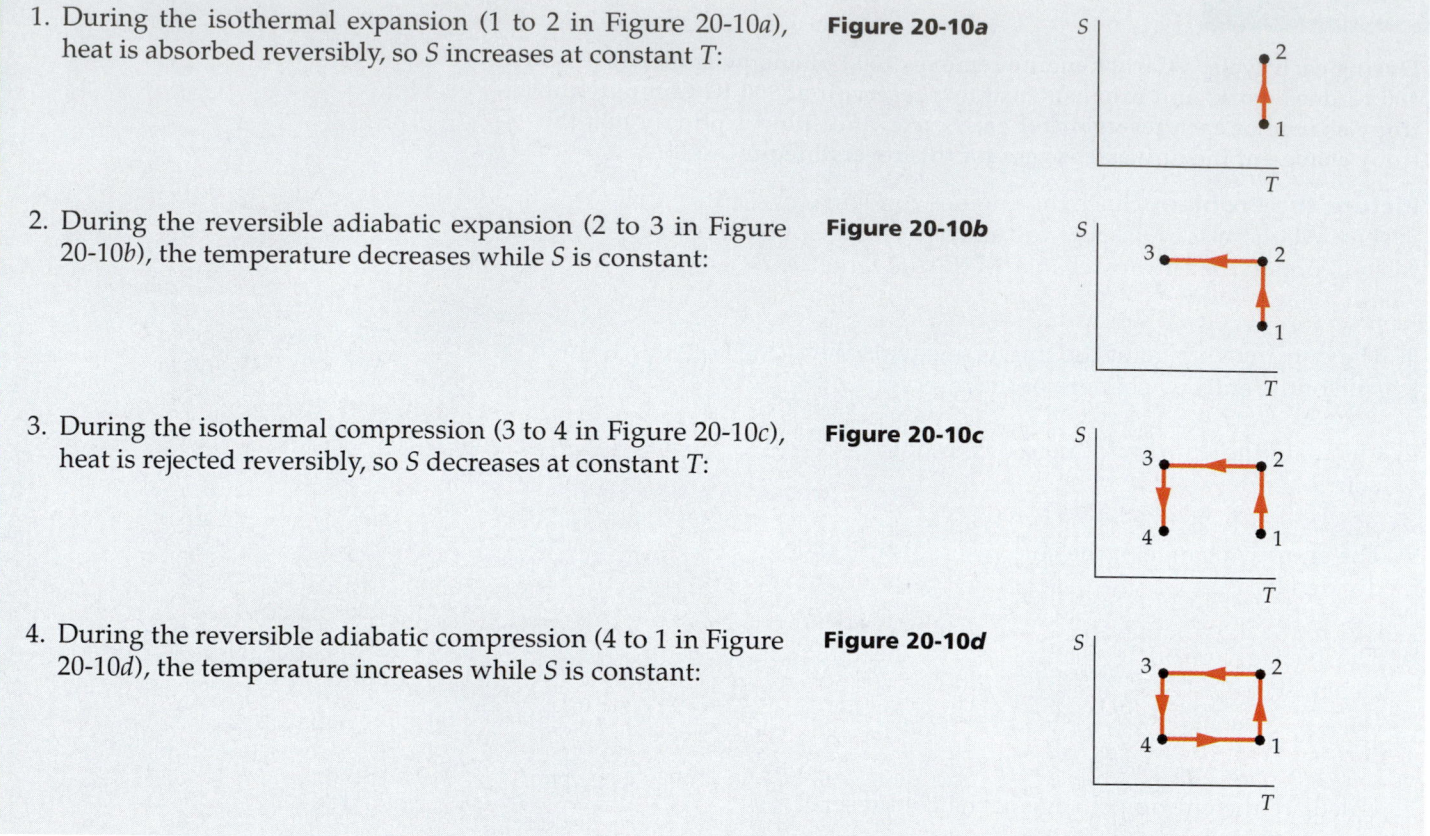

1. During the isothermal expansion (1 to 2 in Figure 20-10*a*), heat is absorbed reversibly, so S increases at constant T:

Figure 20-10*a*

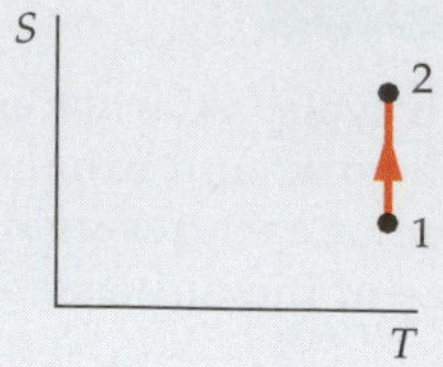

2. During the reversible adiabatic expansion (2 to 3 in Figure 20-10*b*), the temperature decreases while S is constant:

Figure 20-10*b*

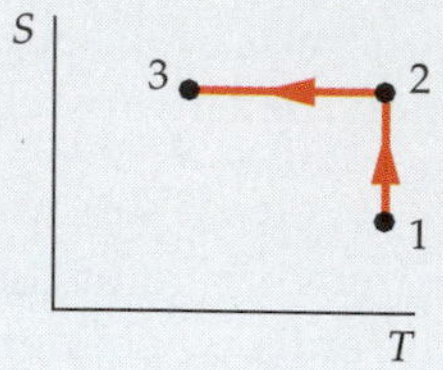

3. During the isothermal compression (3 to 4 in Figure 20-10*c*), heat is rejected reversibly, so S decreases at constant T:

Figure 20-10*c*

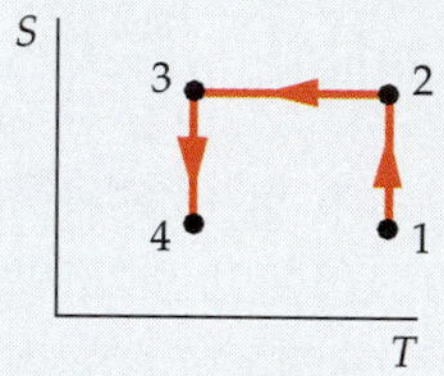

4. During the reversible adiabatic compression (4 to 1 in Figure 20-10*d*), the temperature increases while S is constant:

Figure 20-10*d*

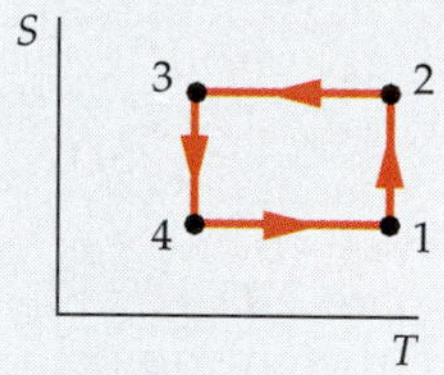

Remarks The Carnot cycle is a rectangle when plotted on an ST diagram.

20-8 Entropy and the Availability of Energy

When an irreversible process occurs, energy is conserved, but some of the energy is "wasted," meaning it becomes unavailable to do work. Consider a block falling to the ground. When the block was at a height h, its potential energy mgh could have been used to do useful work. After the inelastic collision of the block with the ground, this energy is no longer available because it has become the disordered internal energy of the block and its surroundings. The energy that has become unavailable is equal to $mgh = T\,\Delta S_u$. This is a general result:

> In an irreversible process, energy equal to $T\,\Delta S_u$ becomes unavailable for doing work, where T is the temperature of the coldest available reservoir.

For simplicity, we will call the energy that becomes unavailable for doing work the "work lost":

$$W_{lost} = T\,\Delta S_u$$ 20-21

Example 20-12

Suppose that the box discussed in Section 20-6 and shown in Figure 20-9*a* and *b* has a mass of 2.4 kg and slides with a speed of $v = 3$ m/s before crashing into a fixed wall and stopping. The temperature T of the block, table, and surroundings is 293 K and does not change appreciably as the block comes to rest. Find the entropy change of the universe.

Picture the Problem The initial mechanical energy of the block, $\frac{1}{2}mv^2$, is converted to thermal energy. The entropy change is equivalent to what would occur if the heat $Q = \frac{1}{2}mv^2$ were added to the system reversibly.

The entropy change of the universe is Q/T:

$$\Delta S_u = \frac{Q}{T} = \frac{\frac{1}{2}mv^2}{T} = \frac{\frac{1}{2}(2.4\text{ kg})(3\text{ m/s})^2}{293\text{ K}} = 0.0369\text{ J/K}$$

Remark Energy is conserved, but the energy $T\,\Delta S_u = \frac{1}{2}mv^2$ is no longer available to do work.

In the free expansion discussed earlier, the ability to do work was also lost. In that case, the entropy change of the universe was $nR\ln(V_2/V_1)$, so the work lost was $nRT\ln(V_2/V_1)$. This is the amount of work that could have been done if the gas had expanded quasi-statically and isothermally from V_1 to V_2, as given by Equation 19-16.

When heat is conducted from a hot reservoir to a cold reservoir, the change in entropy of the universe is given by Equation 20-20, and the work lost is

$$W_{\text{lost}} = T_c\,\Delta S_u = |Q|\left(1 - \frac{T_c}{T_h}\right)$$

We can see that this is just the work that could have been done by a Carnot engine running between these reservoirs, removing heat $|Q|$ from the hot reservoir and doing work $W = \varepsilon_C|Q|$, where $\varepsilon_C = 1 - T_c/T_h$.

20-9 Entropy and Probability

Entropy, which is a measure of the disorder of a system, is related to probability. Essentially, a state of high order has a low probability, whereas a state of low order has a high probability. Thus, in an irreversible process, the universe moves from a state of low probability to one of high probability.

Let us consider a free expansion in which a gas expands from an initial volume V_1 to a final volume $V_2 = 2V_1$. The entropy change of the universe for this process is given by Equation 20-18:

$$\Delta S = nR\ln\frac{V_2}{V_1} = nR\ln 2 \qquad \text{20-22}$$

Why is this process irreversible? Why can't the gas compress by itself back into its original volume? Since there is no energy change involved, a compression would not violate the first law of thermodynamics. The reason is merely that such a compression is extremely improbable. To see this, let us begin by assuming that the gas consists of only 10 molecules, and that, initially, these molecules occupy the entire volume of their container. Then the chance that any one particular molecule will be in the left half of the container at any given time is $\frac{1}{2}$. The chance that any two particular molecules will both be in the left half is $\frac{1}{2} \times \frac{1}{2} = \frac{1}{4}$.* The chance that three particular molecules will be in the left half is $\frac{1}{2} \times \frac{1}{2} \times \frac{1}{2} = (\frac{1}{2})^3 = \frac{1}{8}$. The chance that all 10 molecules will be in the left half is $(\frac{1}{2})^{10} = \frac{1}{1024}$. That is, there is 1 chance in 1024 that all 10 molecules will be in the left half of the container at any given time.

Though the probability of all 10 molecules being on one side of the container is small, we would not be completely surprised to see it occur. If we look at the gas once each second, we could expect to see it happen once in every 1024 seconds, or about once every 17 minutes. If we started with the 10 molecules randomly distributed and then found them all in the left half of the original volume, the entropy of the universe would have *decreased* by $nR \ln 2$. However, this decrease is extremely small, since the number of moles n corresponding to 10 molecules is only about 10^{-23}. Still, it would violate the entropy statement of the second law of thermodynamics, which says that for any process, the entropy of the universe never decreases. If we wish to apply the second law to microscopic systems, such as a small number of molecules, we should consider the second law to be a statement of probability.

We can relate the probability of a gas spontaneously compressing itself into a smaller volume to the change in its entropy. If the original volume is V_1, the probability p of finding N molecules in a smaller volume V_2 is

$$p = \left(\frac{V_2}{V_1}\right)^N$$

Taking the natural logarithm of both sides of this equation, we obtain

$$\ln p = N \ln\left(\frac{V_2}{V_1}\right) = nN_A \ln\left(\frac{V_2}{V_1}\right) \qquad \text{20-23}$$

where n is the number of moles and N_A is Avogadro's number. The entropy change of the gas is

$$\Delta S = nR \ln\left(\frac{V_2}{V_1}\right) \qquad \text{20-24}$$

(The entropy change is negative if V_2 is less than V_1.) Comparing Equations 20-23 and 20-24, we see that

$$\Delta S = \frac{R}{N_A} \ln p = k \ln p \qquad \text{20-25}$$

where k is Boltzmann's constant.

It may be disturbing to learn that events such as the spontaneous compression of a gas or the spontaneous conduction of heat from a cold body to a hot body (processes for which $\Delta S_u < 0$) are only improbable, not impossible. But, as we have just seen, there is a reasonable chance that such an event will occur only if the system consists of a very small number of molecules. However, *thermodynamics itself is applicable only to macroscopic systems*, that is, to systems that have a very large number of molecules. Consider trying to measure the pressure of a gas consisting of only 10 molecules. The pressure would vary wildly depending on whether no molecule, 2 molecules, or 10 molecules were colliding with the wall of the container at the time of measurement. The macroscopic variables of pressure and temperature are simply not applicable to a microscopic system with only 10 molecules.

As we increase the number of molecules in a system, the chance of an event occurring for which $\Delta S_u < 0$ decreases dramatically. For example, if we have

---

* This is the same as the chance that a coin flipped twice will come up heads both times.

50 molecules in a container, the chance that they will all be in the left half of the volume is $(\frac{1}{2})^{50} \approx 10^{-15}$. Thus, if we look at the gas once each second, we could expect to see all 50 molecules in the left half of the volume about once in every 10^{15} s or once in every 36 million years! For 1 mol $= 6 \times 10^{23}$ molecules, the chance that all will wind up in half of the volume is vanishingly small, essentially zero. For macroscopic systems, the probability of a process resulting in a decrease in the entropy of the universe is so extremely small that the distinction between improbable and impossible becomes blurred.

Summary

The second law of thermodynamics is a fundamental law of nature.

Topic	Remarks and Relevant Equations	
1. Efficiency of a Heat Engine	If the engine removes Q_h from a hot reservoir, does work W, and exhausts heat $\lvert Q_c \rvert$ to a cold reservoir, its efficiency is $\varepsilon = \frac{W}{Q_h} = 1 - \frac{\lvert Q_c \rvert}{Q_h}$	20-2
2. Coefficient of Performance of a Refrigerator	$\text{COP} = \frac{Q_c}{W}$	20-3
3. Equivalent Statements of the Second Law of Thermodynamics		
The Kelvin statement	It is impossible to remove thermal energy from a system at a single temperature and convert it to mechanical work without changing the system or surroundings in some other way.	
The heat-engine statement	It is impossible for a heat engine working in a cycle to remove heat from a reservoir and convert it completely into work with no other effects.	
The Clausius statement	A process whose only final result is to transfer thermal energy from a cooler object to a hotter one is impossible.	
The refrigerator statement	It is impossible for a refrigerator working in a cycle to produce no other effect than the transfer of thermal energy from a cold object to a hot object.	
The entropy statement	The entropy of the universe (system plus surroundings) can never decrease.	
4. Conditions for a Reversible Process	1. No work must be done by friction, viscous forces, or other dissipative forces that produce heat. 2. Heat conduction can occur only isothermally. 3. The process must be quasi-static so that the system is always in an equilibrium state (or infinitesimally near an equilibrium state).	
5. Carnot Engine	A Carnot engine is a reversible engine that works between two reservoirs. It uses a Carnot cycle, which consists of	
Carnot cycle	1. A quasi-static isothermal expansion absorbing heat at temperature T_h 2. A quasi-static adiabatic expansion 3. A quasi-static isothermal compression rejecting heat at temperature T_c 4. A quasi-static adiabatic compression back to the original state	

	Carnot efficiency	$\varepsilon_C = 1 - \frac{\lvert Q_c \rvert}{Q_h} = 1 - \frac{T_c}{T_h}$	20-6
6.	**Absolute Temperature Scale**	The ratio of the absolute temperatures of two reservoirs is defined to be the ratio of the heat exhausted to the heat intake of a Carnot engine running between the reservoirs. $\frac{T_c}{T_h} = \frac{\lvert Q_c \rvert}{Q_h}$	20-7
7.	**Entropy**	Entropy is a measure of the disorder of a system. The difference in entropy between two nearby states is given by $dS = \frac{dQ_{rev}}{T}$ where dQ_{rev} is the heat added in a reversible process connecting the states. The entropy change of a system can be positive or negative.	20-11
	Entropy and loss of work capability	During an irreversible process, the entropy of the universe S_u increases and a certain amount of energy $W_{lost} = T\,\Delta S_u$ becomes unavailable for doing work.	20-21
	Entropy and probability	Entropy is related to probability. A highly ordered system is one of low probability and low entropy. An isolated system moves toward a state of low order, high probability, and high entropy.	

Problem-Solving Guide

Summary of Worked Examples

Type of Calculation	Procedure and Relevant Examples	
1. Heat Engines		
Find the efficiency of a heat engine.	Use $\varepsilon = W/Q_{in} = 1 - Q_{out}/Q_{in}$.	**Examples 20-1, 20-2**
Find the maximum efficiency of an engine.	Use $\varepsilon_C = 1 - T_c/T_h$.	**Example 20-4**
Find the work lost due to irreversibility of an engine.	Compare the work done with the maximum work that could be done using a Carnot engine.	**Examples 20-4, 20-5, 20-6**
2. Refrigerators		
Find the work needed to run a refrigerator.	Use $W = Q_c/\text{COP}$.	**Example 20-3**
3. Entropy		
Find the change in entropy when two objects at different temperatures are put in contact.	First find the final temperature. Then use $\Delta S = mC_p \ln(T_f/T_i)$ for each object.	**Example 20-9**

Find the entropy change due to a free expansion of an ideal gas.	Use $\Delta S = Q/T$ where $Q = W = nRT\ln(V_2/V_1)$ is the heat input for an isothermal expansion to the same final volume.	**Example 20-8**
Find the entropy change when mechanical energy ΔE is converted into thermal energy.	Use $\Delta S = \Delta E/T$ where T is the temperature of the surroundings.	**Example 20-12**
Find the entropy change of a heat reservoir of temperature T.	Use $\Delta S = \pm Q/T$, where $+Q$ is for heat in and $-Q$ is for heat out.	**Example 20-10**
Plot a reversible cyclic process on an S versus T diagram.	During adiabatic expansions or compressions, T changes at constant S. During isothermal expansions or compressions, S changes at constant T.	**Example 20-11**

Problems

Conceptual Problems

Problems from Optional and Exploring sections

In a few problems, you are given more data than you actually need; in a few other problems, you are required to supply data from your general knowledge, outside sources, or informed estimates.

• Single-concept, single-step, relatively easy

•• Intermediate-level, may require synthesis of concepts

••• Challenging, for advanced students

Heat Engines and Refrigerators

1 • Where does the energy come from in an internal-combustion engine? In a steam engine?

2 • How does friction in an engine affect its efficiency?

3 • John is house-sitting for a friend who keeps delicate plants in her kitchen. She warns John not to let the room get too warm or the plants will wilt, but John forgets and leaves the oven on all day after his brownies are baked. As the plants begin to droop, John turns off the oven and opens the refrigerator door, intending to use the refrigerator to cool the kitchen. Explain why this doesn't work.

4 • Why do power-plant designers try to increase the temperature of the steam fed to engines as much as possible?

5 • An engine with 20% efficiency does 100 J of work in each cycle. (*a*) How much heat is absorbed in each cycle? (*b*) How much heat is rejected in each cycle?

6 • An engine absorbs 400 J of heat and does 120 J of work in each cycle. (*a*) What is its efficiency? (*b*) How much heat is rejected in each cycle?

7 • An engine absorbs 100 J and rejects 60 J in each cycle. (*a*) What is its efficiency? (*b*) If each cycle takes 0.5 s, find the power output of this engine in watts.

8 • A refrigerator absorbs 5 kJ of energy from a cold reservoir and rejects 8 kJ to a hot reservoir. (*a*) Find the coefficient of performance of the refrigerator. (*b*) The refrigerator is reversible and is run backward as a heat engine between the same two reservoirs. What is its efficiency?

9 •• An engine operates with 1 mol of an ideal gas for which $C_v = \frac{3}{2}R$ and $C_p = \frac{5}{2}R$ as its working substance. The cycle begins at $P_1 = 1$ atm and $V_1 = 24.6$ L. The gas is heated at constant volume to $P_2 = 2$ atm. It then expands at constant pressure until $V_2 = 49.2$ L. During these two steps, heat is absorbed by the gas. The gas is then cooled at constant volume until its pressure is again 1 atm. It is then compressed at constant pressure to its original state. During the last two steps, heat is rejected by the gas. All the steps are quasi-static and reversible. (*a*) Show this cycle on a *PV* diagram. Find the work done, the heat added, and the change in the internal energy of the gas for each step of the cycle. (*b*) Find the efficiency of the cycle.

10 •• An engine using 1 mol of a diatomic ideal gas performs a cycle consisting of three steps: (1) an adiabatic expansion from an initial pressure of 2.64 atm and an initial volume of 10 L to a pressure of 1 atm and a volume of 20 L, (2) a compression at constant pressure to its original volume of 10 L, and (3) heating at constant volume to its original pressure of 2.64 atm. Find the efficiency of this cycle.

11 •• An engine using 1 mol of an ideal gas initially at $V_1 = 24.6$ L and $T = 400$ K performs a cycle consisting of four steps: (1) an isothermal expansion at $T = 400$ K to twice its initial volume, (2) cooling at constant volume to $T = 300$ K, (3) an isothermal compression to its original volume, and (4) heating at constant volume to its original temperature of 400 K. Assume that $C_v = 21$ J/K. Sketch the cycle on a *PV* diagram and find its efficiency.

12 •• One mole of an ideal monatomic gas at an initial volume $V_1 = 25$ L follows the cycle shown in Figure 20-11. All the processes are quasi-static. Find (*a*) the temperature of each state of the cycle, (*b*) the heat flow for each part of the cycle, and (*c*) the efficiency of the cycle.

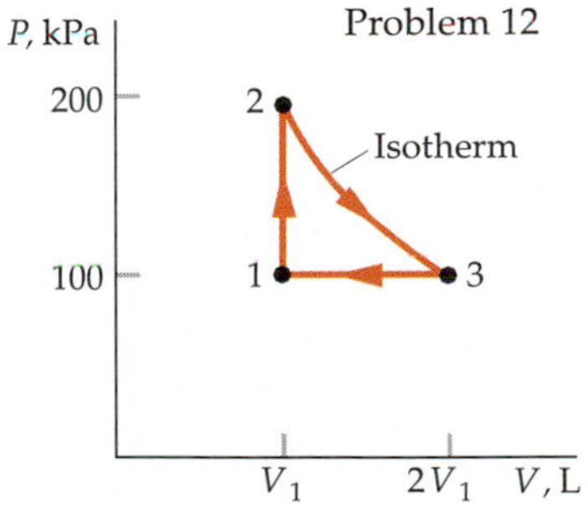

Figure 20-11 Problem 12

13 •• An ideal gas ($\gamma = 1.4$) follows the cycle shown in Figure 20-12. The temperature of state 1 is 200 K. Find (*a*) the temperatures of the other three states of the cycle and (*b*) the efficiency of the cycle.

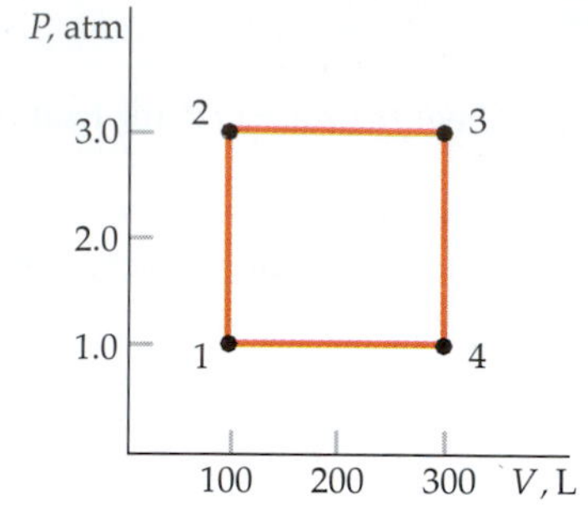

Figure 20-12
Problem 13

14 ••• The *diesel cycle* shown in Figure 20-13 approximates the behavior of a diesel engine. Process *ab* is an adiabatic compression, process *bc* is an expansion at constant pressure, process *cd* is an adiabatic expansion, and process *da* is cooling at constant volume. Find the efficiency of this cycle in terms of the volumes V_a, V_b, V_c, and V_d.

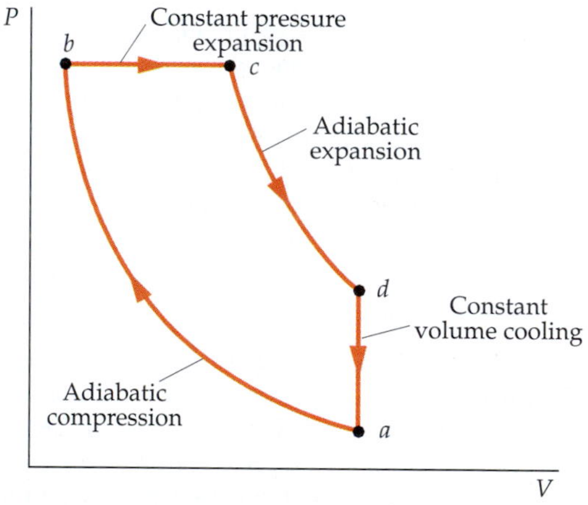

Figure 20-13
Diesel cycle for Problem 14

15 ••• In the *Stirling cycle* shown in Figure 20-14, process *ab* is an isothermal compression, process *bc* is heating at constant volume, process *cd* is an isothermal expansion, and process *da* is cooling at constant volume. Find the efficiency of the Stirling cycle in terms of the temperatures T_h and T_c and the volumes V_a and V_b.

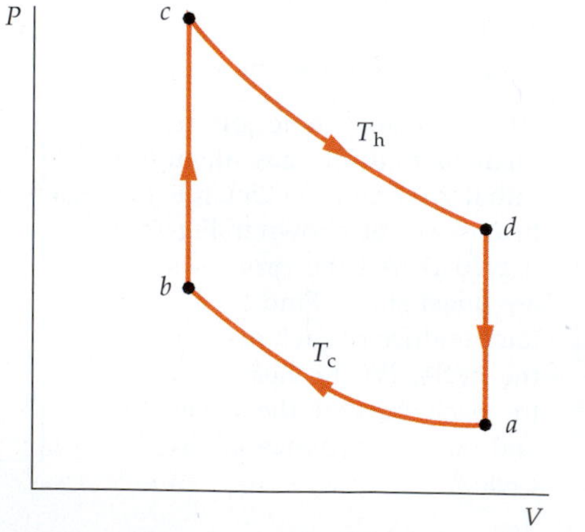

Figure 20-14
Stirling cycle for Problems 15 and 77

16 ••• The Clausius equation of state is $P(V - bn) = nRT$, where b is a constant. Show that the efficiency of a Carnot cycle is the same for a gas that obeys this equation of state as it is for one that obeys the ideal-gas equation of state, $PV = nRT$.

Second Law of Thermodynamics

17 •• A certain engine running at 30% efficiency draws 200 J of heat from a hot reservoir. Assume that the refrigerator statement of the second law of thermodynamics is false, and show how this engine combined with a perfect refrigerator can violate the heat-engine statement of the second law.

18 •• A certain refrigerator takes in 500 J of heat from a cold reservoir and gives off 800 J to a hot reservoir. Assume that the heat-engine statement of the second law of thermodynamics is false, and show how a perfect engine working with this refrigerator can violate the refrigerator statement of the second law.

19 •• If two adiabatic curves intersect on a *PV* diagram, a cycle could be completed by an isothermal path between the two adiabatic curves shown in Figure 20-15. Show that such a cycle could violate the second law of thermodynamics.

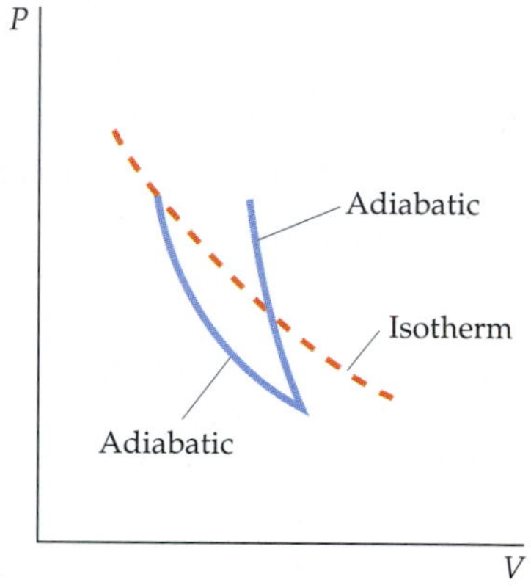

Figure 20-15
Problem 19

Carnot Engines

20 • A Carnot engine works between two heat reservoirs at temperatures $T_h = 300$ K and $T_c = 200$ K. (*a*) What is its efficiency? (*b*) If it absorbs 100 J from the hot reservoir during each cycle, how much work does it do? (*c*) How much heat does it give off during each cycle? (*d*) What is the COP of this engine when it works as a refrigerator between the same two reservoirs?

21 • A refrigerator works between an inside temperature of 0°C and a room temperature of 20°C. (*a*) What is the largest possible coefficient of performance it can have? (*b*) If the inside of the refrigerator is to be cooled to −10°C, what is the largest possible coefficient of performance it can have, assuming the same room temperature of 20°C?

22 • An engine removes 250 J from a reservoir at 300 K and exhausts 200 J to a reservoir at 200 K. (*a*) What is its efficiency? (*b*) How much more work could be done if the engine were reversible?

23 •• A reversible engine working between two reservoirs at temperatures T_h and T_c has an efficiency of 30%.

Working as a heat engine, it gives off 140 J of heat to the cold reservoir. A second engine working between the same two reservoirs also gives off 140 J to the cold reservoir. Show that if the second engine has an efficiency greater than 30%, the two engines working together would violate the heat-engine statement of the second law.

24 •• A reversible engine working between two reservoirs at temperatures T_h and T_c has an efficiency of 20%. Working as a heat engine, it does 100 J of work in each cycle. A second engine working between the same two reservoirs also does 100 J of work in each cycle. Show that if the efficiency of the second engine is greater than 20%, the two engines working together would violate the refrigerator statement of the second law.

25 •• A Carnot engine works between two heat reservoirs as a refrigerator. It requires 50 J of work to remove 100 J from the cold reservoir and gives off 150 J to the hot reservoir during each cycle. Its coefficient of performance COP = $Q_c/W = (100\text{ J})/(50\text{ J}) = 2$. (*a*) What is the efficiency of the Carnot engine when it works as a heat engine between the same two reservoirs? (*b*) Show that no other engine working as a refrigerator between the same two reservoirs can have a COP greater than 2.

26 •• A Carnot engine works between two heat reservoirs at temperatures $T_h = 300$ K and $T_c = 77$ K. (*a*) What is its efficiency? (*b*) If it absorbs 100 J from the hot reservoir during each cycle, how much work does it do? (*c*) How much heat does it give off in each cycle? (*d*) What is the coefficient of performance of this engine when it works as a refrigerator between these two reservoirs?

27 •• In the cycle shown in Figure 20-16, 1 mol of an ideal gas ($\gamma = 1.4$) is initially at a pressure of 1 atm and a temperature of 0°C. The gas is heated at constant volume to $t_2 = 150$°C and is then expanded adiabatically until its pressure is again 1 atm. It is then compressed at constant pressure back to its original state. Find (*a*) the temperature t_3 after the adiabatic expansion, (*b*) the heat entering or leaving the system during each process, (*c*) the efficiency of this cycle, and (*d*) the efficiency of a Carnot cycle operating between the temperature extremes of this cycle.

Figure 20-16
Problem 27

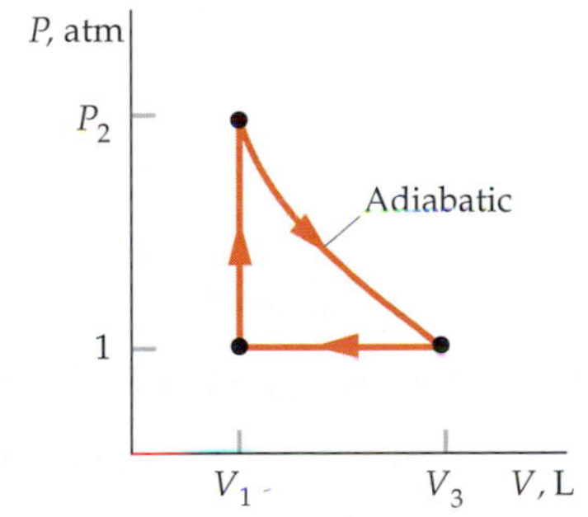

28 •• A steam engine takes in superheated steam at 270°C and discharges condensed steam from its cylinder at 50°C. Its efficiency is 30%. (*a*) How does this efficiency compare with the maximum possible efficiency for these temperatures? (*b*) If the useful power output of the engine is 200 kW, how much heat does the engine discharge to its surroundings in 1 h?

Heat Pumps (optional)

29 • A heat pump delivers 20 kW to heat a house. The outside temperature is −10°C and the inside temperature of the hot-air supply for the heating fan is 40°C. (*a*) What is the coefficient of performance of a Carnot heat pump operating between these temperatures? (*b*) What must be the minimum power of the engine needed to run the heat pump? (*c*) If the COP of the heat pump is 60% of the COP of an ideal pump, what must the minimum power of the engine be?

30 • Rework Problem 29 for an outside temperature of −20°C.

31 • A refrigerator is rated at 370 W. (*a*) What is the maximum amount of heat it can remove in 1 min if the inside temperature of the refrigerator is 0°C and it exhausts into a room at 20°C? (*b*) If the COP of the refrigerator is 70% of that of an ideal pump, how much heat can it remove in 1 min?

32 • Rework Problem 31 for a room temperature of 35°C.

Entropy Changes

33 •• On a humid day, water vapor condenses on a cold surface. During condensation, the entropy of the water

(*a*) increases.
(*b*) remains constant.
(*c*) decreases.
(*d*) may decrease or remain unchanged.

34 • What is the change in entropy of 1 mol of water at 0°C that freezes?

35 • Two moles of an ideal gas at $T = 400$ K expand quasi-statically and isothermally from an initial volume of 40 L to a final volume of 80 L. (*a*) What is the entropy change of the gas? (*b*) What is the entropy change of the universe for this process?

36 • The gas in Problem 35 is taken from the same initial state ($T = 400$ K, $V_1 = 40$ L) to the same final state ($T = 400$ K, $V_2 = 80$ L) by a process that is not quasi-static. (*a*) What is the entropy change of the gas? (*b*) What can be said about the entropy change of the universe?

37 • What is the change in entropy of 1.0 kg of water when it changes to steam at 100°C and a pressure of 1 atm?

38 • Jay approached his guru in a depressed mood. "I want to change the world, but I feel helpless," he said. The guru turned and pushed a 5-kg rock over a ledge. It hit the ground 6 m below and came to rest. "There," said the guru. "I have changed the world." If the rock, the ground, and the atmosphere are all initially at 300 K, calculate the entropy change of the universe.

39 • What is the change in entropy of 1.0 kg of ice when it changes to water at 0°C and a pressure of 1 atm?

40 •• A system absorbs 200 J of heat reversibly from a reservoir at 300 K and gives off 100 J reversibly to a reservoir at 200 K as it moves from state A to state B. During this process, the system does 50 J of work. (*a*) What is the change in the internal energy of the system? (*b*) What is the change in entropy of the system? (*c*) What is the change in entropy of the universe? (*d*) If the system goes from state A to state B by a nonreversible process, how would your answers for parts (*a*), (*b*), and (*c*) differ?

41 •• A system absorbs 300 J from a reservoir at 300 K and 200 J from a reservoir at 400 K. It then returns to its original state, doing 100 J of work and rejecting 400 J of heat to a reservoir at a temperature T. (*a*) What is the entropy change of the system for the complete cycle? (*b*) If the cycle is reversible, what is the temperature T?

42 •• Two moles of an ideal gas originally at $T = 400$ K and $V = 40$ L undergo a free expansion to twice their initial volume. What is (*a*) the entropy change of the gas, and (*b*) the entropy change of the universe?

43 •• A 200-kg block of ice at 0°C is placed in a large lake. The temperature of the lake is just slightly higher than 0°C, and the ice melts. (*a*) What is the entropy change of the ice? (*b*) What is the entropy change of the lake? (*c*) What is the entropy change of the universe (the ice plus the lake)?

44 •• A 100-g piece of ice at 0°C is placed in an insulated container with 100 g of water at 100°C. (*a*) When thermal equilibrium is established, what is the final temperature of the water? Ignore the heat capacity of the container. (*b*) Find the entropy change of the universe for this process.

45 •• A 1-kg block of copper at 100°C is placed in a calorimeter of negligible heat capacity containing 4 L of water at 0°C. Find the entropy change of (*a*) the copper block, (*b*) the water, and (*c*) the universe.

46 •• If a 2-kg piece of lead at 100°C is dropped into a lake at 10°C, find the entropy change of the universe.

47 •• A 1500-kg car traveling at 100 km/h crashes into a concrete wall. If the temperature of the air is 20°C, calculate the entropy change of the universe.

48 •• Find the net change in entropy of the universe when 10 g of steam at 100°C and a pressure of 1 atm are introduced into a calorimeter of negligible heat capacity containing 150 g of water and 150 g of ice at 0°C.

Entropy and Work Lost

49 •• If 500 J of heat is conducted from a reservoir at 400 K to one at 300 K, (*a*) what is the change in entropy of the universe, and (*b*) how much of the 500 J of heat conducted could have been converted into work using a cold reservoir at 300 K?

50 •• One mole of an ideal gas first undergoes a free expansion from $V_1 = 12.3$ L and $T_1 = 300$ K to $V_2 = 24.6$ L and $T_2 = 300$ K. It is then compressed isothermally and quasi-statically back to its original state. (*a*) What is the entropy change of the universe for the complete cycle? (*b*) How much work is wasted in this cycle? (*c*) Show that the work wasted is $T\,\Delta S_u$.

General Problems

51 • In a reversible adiabatic process,

(*a*) the internal energy of the system remains constant.
(*b*) no work is done by the system.
(*c*) the entropy of the system remains constant.
(*d*) the temperature of the system remains constant.

52 •• True or false:

(*a*) Work can never be converted completely into heat.
(*b*) Heat can never be converted completely into work.
(*c*) All heat engines have the same efficiency.
(*d*) It is impossible to transfer a given quantity of heat from a cold reservoir to a hot reservoir.
(*e*) The coefficient of performance of a refrigerator cannot be greater than 1.
(*f*) All Carnot engines are reversible.
(*g*) The entropy of a system can never decrease.
(*h*) The entropy of the universe can never decrease.

53 •• An ideal gas is taken reversibly from an initial state P_i, V_i, T_i to the final state P_f, V_f, T_f. Two possible paths are (A) an isothermal expansion followed by an adiabatic compression, and (B) an adiabatic compression followed by an isothermal expansion. For these two paths,

(*a*) $\Delta U_A > \Delta U_B$.
(*b*) $\Delta S_A > \Delta S_B$.
(*c*) $\Delta S_A < \Delta S_B$.
(*d*) none of the above is correct.

54 •• Figure 20-17 shows a thermodynamic cycle on an ST diagram. Identify this cycle and sketch it on a PV diagram.

Figure 20-17 Problems 54 and 65

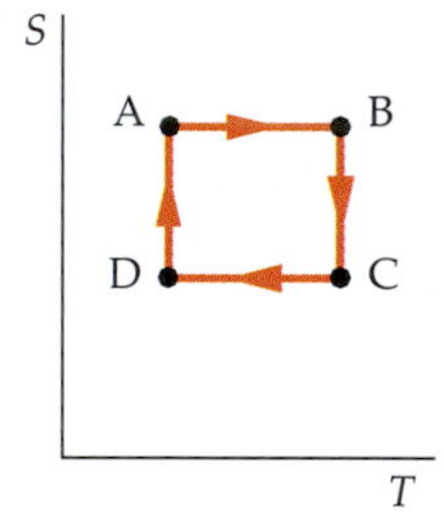

Figure 20-18 Problem 55

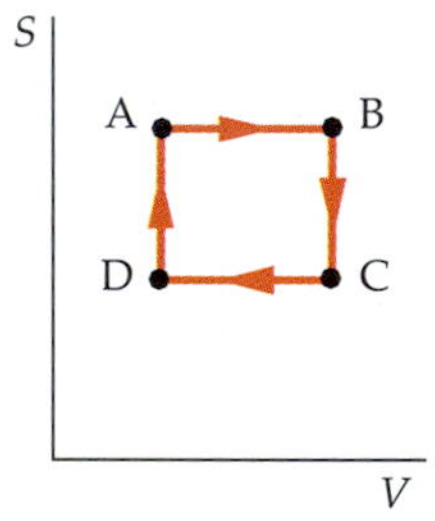

55 •• Figure 20-18 shows a thermodynamic cycle on an SV diagram. Identify the type of engine represented by this diagram.

56 •• Sketch an ST diagram of the Otto cycle.

57 •• Sketch an SV diagram of the Carnot cycle.

58 •• Sketch an SV diagram of the Otto cycle.

59 •• Figure 20-19 shows a thermodynamic cycle on an SP diagram. Make a sketch of this cycle on a PV diagram.

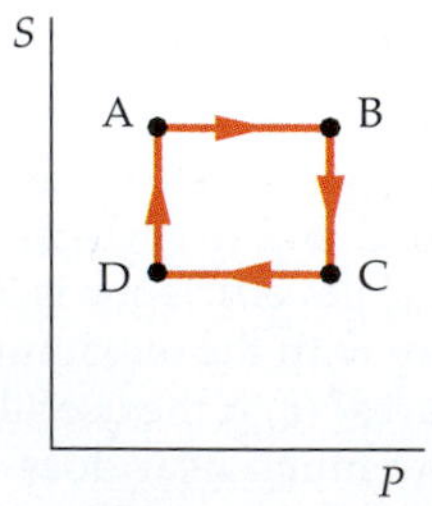

Figure 20-19 Problem 59

60 • An engine with an output of 200 W has an efficiency of 30%. It works at 10 cycles/s. (*a*) How much work is done in each cycle? (*b*) How much heat is absorbed and how much is given off in each cycle?

61 • Which has a greater effect on increasing the efficiency of a Carnot engine, a 5-K increase in the temperature of the hot reservoir or a 5-K decrease in the temperature of the cold reservoir?

62 • In each cycle, an engine removes 150 J from a reservoir at 100°C and gives off 125 J to a reservoir at 20°C. (*a*) What is the efficiency of this engine? (*b*) What is the ratio of its efficiency to that of a Carnot engine working between the same reservoirs? (This ratio is called the *second law efficiency*.)

63 • An engine removes 200 kJ of heat from a hot reservoir at 500 K in each cycle and exhausts heat to a cold reservoir at 200 K. Its efficiency is 85% of a Carnot engine working between the same reservoirs. (*a*) What is the efficiency of this engine? (*b*) How much work is done in each cycle? (*c*) How much heat is exhausted in each cycle?

64 • To maintain the temperature inside a house at 20°C, the power consumption of the electric baseboard heaters is 30 kW on a day when the outside temperature is −7°C. At what rate does this house contribute to the increase in the entropy of the universe?

65 •• The system represented in Figure 20-17 (Problem 54) is 1 mol of an ideal monatomic gas. The temperatures at points A and B are 300 and 750 K, respectively. What is the thermodynamic efficiency of the cyclic process ABCDA?

66 •• A sailor is in a tropical ocean on a boat. She has a 2-kg piece of ice at 0°C, and the temperature of the ocean is T_h = 27°C. Find the maximum work W that can be done using the fusion of ice.

67 •• (*a*) Which process is more wasteful: (1) a block moving with 500 J of kinetic energy being slowed to rest by friction when the temperature of the atmosphere is 300 K or (2) 1 kJ of heat being conducted from a reservoir at 400 K to one at 300 K? *Hint:* How much of the 1 kJ of heat could be converted into work in an ideal situation? (*b*) What is the change in entropy of the universe for each process?

68 •• Helium gas (γ = 1.67) is initially at a pressure of 16 atm, a volume of 1 L, and a temperature of 600 K. It is expanded isothermally until its volume is 4 L and is then compressed at constant pressure until its volume and temperature are such that an adiabatic compression will return the gas to its original state. (*a*) Sketch this cycle on a *PV* diagram. (*b*) Find the volume and temperature after the isobaric compression. (*c*) Find the work done during each cycle. (*d*) Find the efficiency of the cycle.

69 •• A heat engine that does the work of blowing up a balloon at a pressure of 1 atm extracts 4 kJ from a hot reservoir at 120°C. The volume of the balloon increases by 4 L, and heat is exhausted to a cold reservoir at a temperature T_c. If the efficiency of the heat engine is 50% of the efficiency of a Carnot engine working between the same reservoirs, find the temperature T_c.

70 •• Show that the COP of a Carnot refrigerator is related to the efficiency of a Carnot engine by COP = $T_c/(\varepsilon_C T_h)$.

71 •• A freezer has a temperature T_c = −23°C. The air in the kitchen has a temperature T_h = +27°C. Since the heat insulation is not perfect, some heat flows into the freezer at a rate of 50 W. Find the power of the motor that is needed to maintain the temperature in the freezer.

72 •• Two moles of a diatomic gas are taken through the cycle ABCA as shown on the *PV* diagram in Figure 20-20. At A the pressure and temperature are 5 atm and 600 K. The volume at B is twice that at A. The segment BC is an adiabatic expansion and the segment CA is an isothermal compression. (*a*) What is the volume of the gas at A? (*b*) What are the volume and temperature of the gas at B? (*c*) What is the temperature of the gas at C? (*d*) What is the volume of the gas at C? (*e*) How much work is done by the gas in each of the three segments of the cycle? (*f*) How much heat is absorbed by the gas in each segment of the cycle? (*g*) What is the thermodynamic efficiency of this cycle?

Figure 20-20
Problems 72 and 74

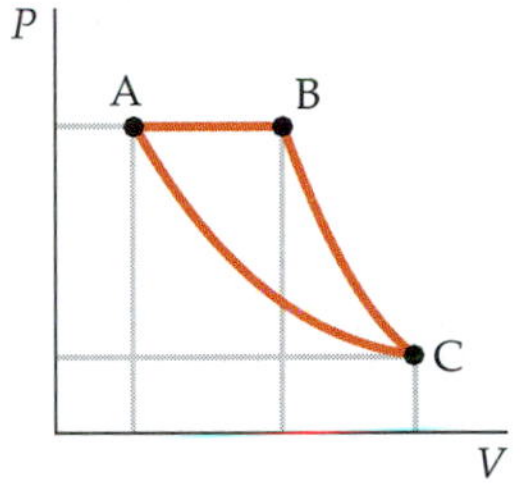

Figure 20-21
Problems 73 and 75

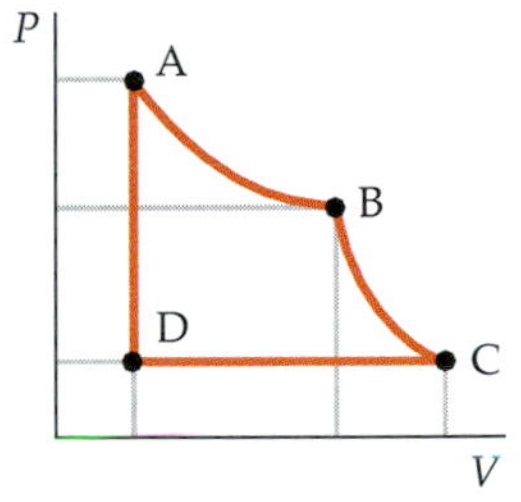

73 •• Two moles of a diatomic gas are carried through the cycle ABCDA shown in the *PV* diagram in Figure 20-21. The segment AB represents an isothermal expansion, the segment BC an adiabatic expansion. The pressure and temperature at A are 5 atm and 600 K. The volume at B is twice that at A. The pressure at D is 1 atm. (*a*) What is the pressure at B? (*b*) What is the temperature at C? (*c*) Find the work done by the gas in one cycle and the thermodynamic efficiency of this cycle.

74 •• Repeat Problem 72 for a monatomic gas.

75 •• Repeat Problem 73 for a monatomic gas.

76 •• Compare the efficiency of the Otto engine and the Carnot engine operating between the same maximum and minimum temperatures.

77 •• Compare the efficiency of the Stirling cycle (see Figure 20-14) and the Carnot engine operating between the same maximum and minimum temperatures.

78 ••• Using the equation for the entropy change of an ideal gas when the volume and temperature change and $TV^{\gamma-1}$ is a constant, show explicitly that the entropy change is zero for a quasi-static adiabatic expansion from state (V_1, T_1) to state (V_2, T_2).

79 ••• *(a)* Show that if the refrigerator statement of the second law of thermodynamics were not true, the entropy of the universe could decrease. *(b)* Show that if the heat-engine statement of the second law were not true, the entropy of the universe could decrease. *(c)* An alternative statement of the second law is that the entropy of the universe cannot decrease. Have you just proved that this statement is equivalent to the refrigerator and heat-engine statements?

80 ••• Suppose that two heat engines are connected in series, such that the heat exhaust of the first engine is used as the heat input of the second engine as shown in Figure 20-22. The efficiencies of the engines are ε_1 and ε_2, respectively. Show that the net efficiency of the combination is given by

$$\varepsilon_{\rm net} = \varepsilon_1 + (1 - \varepsilon_1)\varepsilon_2$$

Figure 20-22
Problems 80 and 81

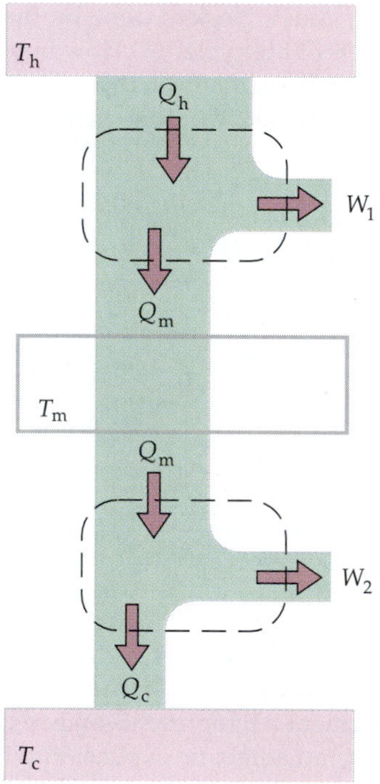

81 ••• Suppose that each engine in Figure 20-22 is an ideal reversible heat engine. Engine 1 operates between temperatures T_h and T_m and engine 2 operates between T_m and T_c, where $T_h > T_m > T_c$. Show that

$$\varepsilon_{\rm net} = 1 - \frac{T_c}{T_h}$$

This means that two reversible heat engines in series are equivalent to one reversible heat engine operating between the hottest and coldest reservoirs.

82 ••• The cooling compartment of a refrigerator and its contents are at 5°C and have an average heat capacity of 84 kJ/K. The refrigerator exhausts heat to the room, which is at 25°C. What minimum power will be required by the motor that runs the refrigerator if the temperature of the cooling compartment and its contents is to be reduced by 1 C° in 1 min?

83 ••• An insulated container is separated into two chambers of equal volume by a thin partition. On one side of the container there are twelve ^{131}Xe atoms, and on the other side there are twelve ^{132}Xe atoms. The partition is then removed. Calculate the change in entropy of the system after equilibrium has been established (that is, when the ^{131}Xe and ^{132}Xe atoms are evenly distributed throughout the total volume).

CHAPTER 21

Thermal Properties and Processes

The temperature in the interior of a furnace is measured by a radiation thermometer.

When an object absorbs thermal energy, various changes in the physical properties of the object may occur. The temperature of the object may rise, accompanied by an expansion or contraction of the object, or the object may liquefy or vaporize, during which the temperature remains constant. In this chapter, we examine some of the thermal properties of matter and some important processes involving thermal energy.

21-1 Thermal Expansion

When the temperature of an object increases, the object usually expands. Consider a long rod of length L at a temperature T. When the temperature changes by ΔT, the change ΔL in length is proportional to ΔT and to the original length L:

$$\Delta L = \alpha L \, \Delta T \qquad 21\text{-}1$$

where α, called the **coefficient of linear expansion,** is the ratio of the fractional change in length to the change in temperature:

$$\alpha = \frac{\Delta L/L}{\Delta T} \tag{21-2}$$

The units for the coefficient of linear expansion are reciprocal Celsius degrees (1/C°), which are the same as reciprocal kelvins (1/K). The value of α for a solid or liquid doesn't vary much with pressure, but it may vary significantly with temperature. Equation 21-2 gives the average value over the temperature interval ΔT. The coefficient of linear expansion at a particular temperature T is found by taking the limit as ΔT approaches zero:

$$\alpha = \lim_{\Delta T \to 0} \frac{\Delta L/L}{\Delta T} = \frac{1}{L}\frac{dL}{dT} \tag{21-3}$$

The accuracy obtained by using the average value of α over a wide temperature range is sufficient for most purposes.

The **coefficient of volume expansion** β is similarly defined as the ratio of the fractional change in volume to the change in temperature (at constant pressure):

$$\beta = \lim_{\Delta T \to 0} \frac{\Delta V/V}{\Delta T} = \frac{1}{V}\frac{dV}{dT} \tag{21-4}$$

Like α, β does not usually vary with pressure for solids and liquids, but may vary with temperature. Average values for α and β for various substances are given in Figure 21-1.

For a given material, $\beta = 3\alpha$. We can show this by considering a box of dimensions L_1, L_2, and L_3. Its volume at a temperature T is

$$V = L_1L_2L_3$$

The rate of change of the volume with respect to temperature is

$$\frac{dV}{dT} = L_1L_2\frac{dL_3}{dT} + L_1L_3\frac{dL_2}{dT} + L_2L_3\frac{dL_1}{dT}$$

Dividing each side of the equation by the volume, we obtain

$$\beta = \frac{1}{V}\frac{dV}{dT} = \frac{1}{L_3}\frac{dL_3}{dT} + \frac{1}{L_2}\frac{dL_2}{dT} + \frac{1}{L_1}\frac{dL_1}{dT}$$

Since each term on the right side of the above equation equals α, we have

$$\beta = 3\alpha \tag{21-5}$$

Similarly, the coefficient of *area* expansion is twice that of *linear* expansion.

The increase in size of any part of a body for a given temperature change is proportional to the original size of that part of the body. Thus, if we increase the temperature of a steel ruler, for example, the effect will be similar to that of a (very slight) photographic enlargement. The dimensions of the ruler itself will be larger, as will the distance between the equally spaced lines. If the

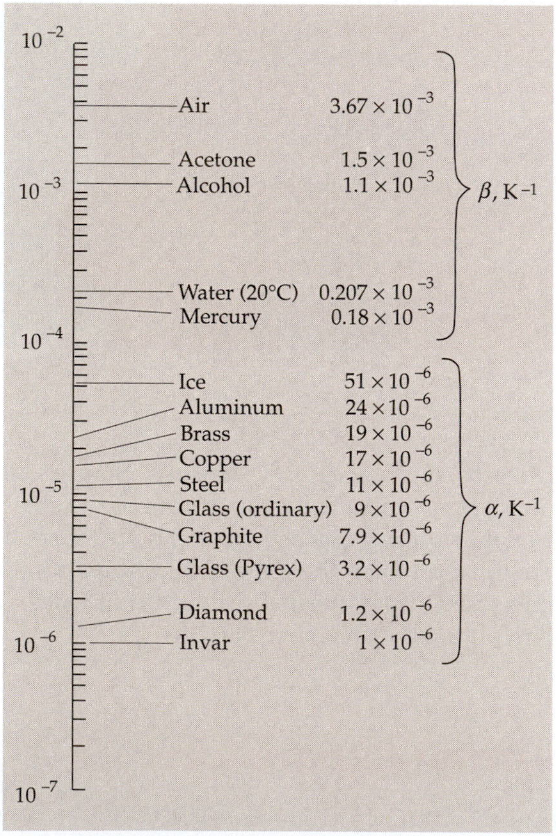

Figure 21-1 Approximate values of the coefficients of thermal expansion for various substances.

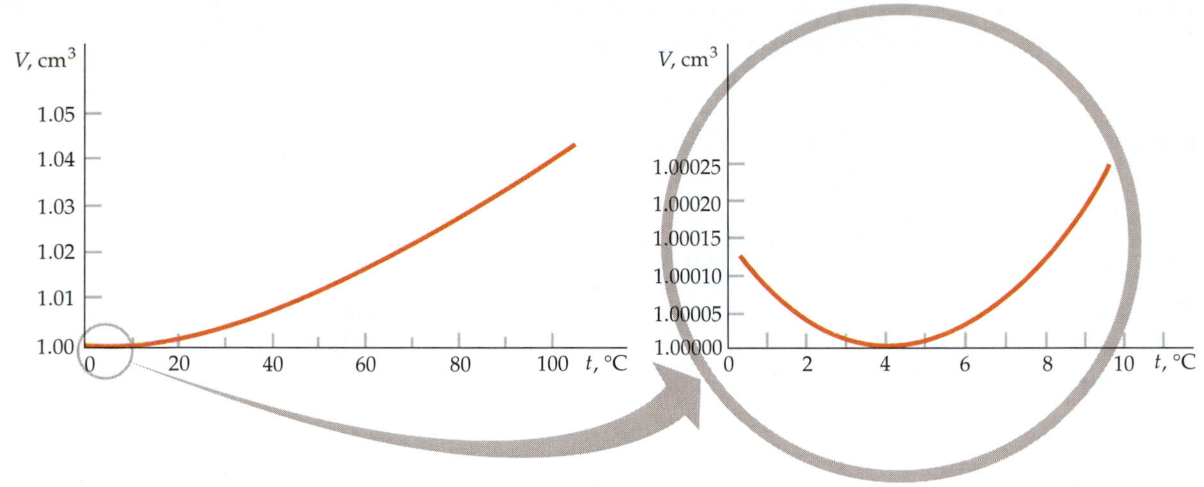

Figure 21-2 Volume of 1 g of water at atmospheric pressure versus temperature. The minimum volume, which corresponds to the maximum density, occurs at 4°C.

ruler has a hole in it, the hole will get larger, just as it would in a photographic enlargement.

Most materials expand when heated. Water, however, presents an important exception. Figure 21-2 shows the volume occupied by 1 g of water as a function of temperature. The volume is minimum, and therefore the density is maximum, at 4°C. Thus, when water is heated from temperatures below 4°C, it contracts rather than expands. This property has important consequences for the ecology of lakes. At temperatures above 4°C, the water in a lake becomes denser as it cools, and thus sinks to the bottom. But as the water cools below 4°C it becomes less dense, and rises to the surface. Ice therefore forms first on the surface of a lake and, being less dense than water, it remains there and acts as a thermal insulator for the water below. If water contracted when it froze, as most substances do, ice would sink and new water would be exposed at the surface to freeze. Lakes would fill with ice from the bottom up and would be much more likely to freeze completely in the winter, killing fish and other aquatic life.

Example 21-1

A steel bridge is 1000 m long. By how much does it expand when the temperature rises from 0 to 30°C?

Picture the Problem Use $\alpha = 11 \times 10^{-6}\,\text{K}^{-1}$ from Figure 21-1 and calculate ΔL from Equation 21-2.

The change in length for a 30-C° = 30-K change in temperature is the product of α, L, and ΔT:

$$\Delta L = \alpha L\,\Delta T = (11 \times 10^{-6} K^{-1})(1000\text{ m})(30\text{ K})$$

$$= 0.33\text{ m} = 33\text{ cm}$$

Expansion joints are included in bridges to relieve the enormous stresses that would occur without them. We can calculate the stress that would result in a steel bridge without expansion joints by using Young's modulus (Equation 12-7):

$$Y = \frac{\text{stress}}{\text{strain}} = \frac{F/A}{\Delta L/L}$$

optional

Then

$$\frac{F}{A} = Y\frac{\Delta L}{L} = Y\alpha\,\Delta T$$

For $\Delta T = 30$ K, $\Delta L/L = 0.33$ m/1000 m as found in Example 21-1. Then using $Y = 2 \times 10^{11}$ N/m^2 (from Table 12-1),

$$\frac{F}{A} = Y\frac{\Delta L}{L} = (2 \times 10^{11}\ \mathrm{N/m^2})\frac{0.33\ \mathrm{m}}{1000\ \mathrm{m}} = 6.6 \times 10^{7}\ \mathrm{N/m^2}$$

This stress is about one-third of the breaking stress for steel under compression. A compression stress of this magnitude would cause a steel bridge to buckle and become permanently deformed.

Example 21-2

Your 1-L glass flask is filled to the brim with alcohol at 10°C. If the temperature is raised to 30°C, how much alcohol spills out of the flask?

Picture the Problem The glass flask and the alcohol both expand when heated, but the alcohol expands more, so some spills out. We calculate the amount spilled by finding the changes in volume for $\Delta T = 20$ K using $\Delta V_a = \beta V\,\Delta T$ with $\beta = 1.1 \times 10^{-3}$ K^{-1} for alcohol (from Figure 21-1), and $\Delta V_g = \beta V\,\Delta T = 3\alpha V\,\Delta T$ with $\alpha = 9 \times 10^{-6}$ K^{-1} for glass. The difference in these volume changes equals the volume spilled.

1. The volume of alcohol spilled, ΔV_s, is the difference in the changes in volume of the alcohol and glass:
$$\Delta V_s = \Delta V_a - \Delta V_g$$

2. Find the increase in the volume of the alcohol:
$$\Delta V_a = \beta V\,\Delta T = (1.1 \times 10^{-3}\,\mathrm{K^{-1}})(1\ \mathrm{L})(20\ \mathrm{K})$$
$$= 2.2 \times 10^{-2}\ \mathrm{L} = 22.0\ \mathrm{mL}$$

3. Find the increase in the volume of the glass flask:
$$\Delta V_g = \beta V\,\Delta T = 3\alpha V\,\Delta T$$
$$= 3(9 \times 10^{-6}\,\mathrm{K^{-1}})(1\ \mathrm{L})(20\ \mathrm{K})$$
$$= 5.4 \times 10^{-4}\ \mathrm{L} = 0.54\ \mathrm{mL}$$

4. Subtract to find the amount of alcohol spilled:
$$\Delta V_s = \Delta V_a - \Delta V_g = 22.0\ \mathrm{mL} - 0.54\ \mathrm{mL} = 21.5\ \mathrm{mL}$$

Example 21-3

A copper bar is heated to 300°C and is then clamped rigidly between two fixed points so that it can neither expand nor contract. If the breaking stress of copper is 230 MN/m^2, at what temperature will the bar break as it cools?

Picture the Problem As the bar cools, the change ΔL in length that *would* occur if the bar contracted is offset by an equal stretching which is due to tensile stress in the bar. The stress F/A is related to the stretching ΔL by $Y = (F/A)/(\Delta L/L)$, where Young's modulus for copper is $Y = 110$ GN/m^2 (from Table 12-1). The maximum allowable stretching occurs when F/A equals 230 MN/m^2. We thus find the temperature change that would produce this maximum contraction.

1. Calculate the change ΔL in length that would occur if the bar were unclamped and cooled by ΔT:

$$\Delta L = \alpha L\,\Delta T$$

2. A tensile stress F/A stretches the bar by the same amount ΔL found in step 1:

$$\Delta L = L\frac{F/A}{Y} = \alpha L\,\Delta T$$

3. Solve for ΔT and set the stress equal to the breaking value:

$$\Delta T = \frac{F/A}{\alpha Y} = \frac{230 \times 10^6\ \text{N/m}^2}{(17 \times 10^{-6}\ \text{K}^{-1})(110 \times 10^9\ \text{N/m}^2)}$$

$$= 123\ \text{K} = 123\ \text{C}^\circ$$

4. Subtract this result from the original temperature to find the final temperature at which the bar breaks:

$$300^\circ\text{C} - 123^\circ\text{C} = 177^\circ\text{C}$$

21-2 The van der Waals Equation and Liquid–Vapor Isotherms

Although most gases behave like an ideal gas at ordinary pressures, this ideal behavior breaks down when the pressure is high enough or the temperature is low enough that the density of the gas is high and the molecules are, on average, close together. An equation of state called the **van der Waals equation** describes the behavior of many real gases over a wide range of pressures more accurately than does the ideal-gas equation of state ($PV = nRT$). The van der Waals equation for n moles of gas is

$$\left(P + \frac{an^2}{V^2}\right)(V - bn) = nRT \qquad 21\text{-}6$$

The van der Waals equation of state

The constant b in this equation arises because the gas molecules are not point particles but objects that have a finite size; therefore, the volume available to each molecule is reduced. The magnitude of b is the volume of one mole of gas molecules. The term an^2/V^2 arises from the attraction of the gas molecules for each other. As a molecule approaches the wall of the container, it is pulled back by the molecules surrounding it with a force that is proportional to the density of those molecules, n/V. Since the number of molecules that hit the wall in a given time is also proportional to the density of the molecules, the decrease in pressure due to the attraction of the molecules for each other is proportional to the square of the density and therefore to n^2/V^2. The constant a depends on the gas and is small for inert gases, which have very weak chemical interactions. The terms bn and an^2/V^2 are both negligible when the volume V is large, so at low densities the van der Waals equation approaches the ideal-gas law. At high densities the van der Waals equation provides a much better description than the ideal-gas law of the behavior of real gases.

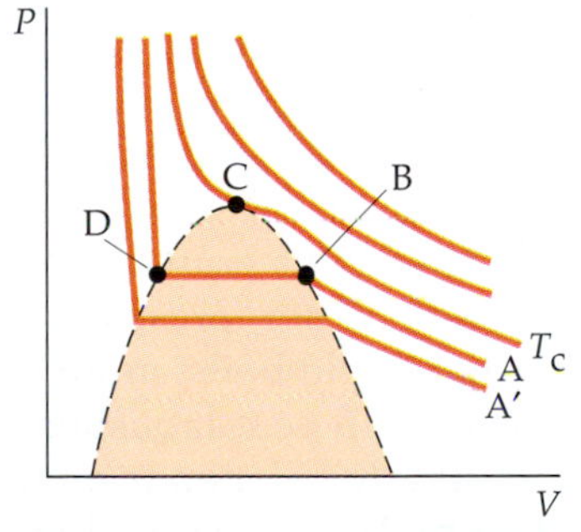

Figure 21-3 Isotherms on the PV diagram for a real substance. For temperatures above the critical temperature T_c, the substance remains a gas at all pressures and is described by the van der Waals equation. The pressure for the horizontal portions of the curves in the shaded region is the vapor pressure, which is the pressure at which the vapor and liquid are in equilibrium. To the left of the shaded region for temperatures below the critical temperature, the substance is a liquid and is nearly incompressible.

Figure 21-3 shows PV isothermal curves for a real substance at various temperatures. For temperatures above some critical temperature T_c, these curves are described quite accurately by the van der Waals equation and can be used to determine the constants a and b. For example, the values of these constants that give the best fit to the experimental curves for nitrogen are $a = 0.14\ \text{Pa}\cdot\text{m}^6/\text{mol}^2$ and $b = 39.1\ \text{cm}^3/\text{mol}$. This volume of 39.1 cm^3 per mole is about 0.2% of the volume of 22,400 cm^3 occupied by 1 mol of nitrogen under standard conditions. Since the molar mass of nitrogen is 28 g/mol, if 1 mol of nitrogen molecules were packed into a volume of 39.1 cm^3, the density would be

$$\rho = \frac{M}{V} = \frac{28\text{ g}}{39.1\text{ cm}^3} = 0.72\text{ g/cm}^3$$

which compares favorably with the density of liquid nitrogen, 0.80 g/cm^3.

The value of the constant b can be used to estimate the size of a molecule. Since 1 mol = N_A molecules of nitrogen has a volume of 39.1 cm^3, the volume of one nitrogen molecule is

$$V = \frac{b}{N_A} = \frac{39.1\text{ cm}^3/\text{mol}}{6.02 \times 10^{23}\text{ molecules/mol}}$$
$$= 6.50 \times 10^{-23}\text{ cm}^3/\text{molecule}$$

If we assume that each molecule is a sphere of diameter d occupying a cubic volume of side d, we obtain

$$d^3 = 6.50 \times 10^{-23}\text{ cm}^3$$

or

$$d = 4.0 \times 10^{-8}\text{ cm} = 4.0 \times 10^{-10}\text{ m}$$

which is a reasonable estimate for the diameter of a molecule.

At temperatures below T_c, the van der Waals equation describes those portions of the isotherms outside the shaded region in Figure 21-3 but not those portions inside the shaded region. Suppose we have a gas at a temperature below T_c that initially has a low pressure and a large volume. We begin to compress the gas while holding the temperature constant (isotherm A in the figure). At first the pressure rises, but when we reach point B on the dashed curve, the pressure ceases to rise and the gas begins to liquefy at constant pressure. Along the horizontal line BD in the figure, the gas and liquid are in equilibrium. As we continue to compress the gas, more and more gas liquefies until at point D on the dashed curve we have only liquid. Then, if we try to compress the substance further, the pressure rises sharply because a liquid is nearly incompressible.

Now consider putting a liquid such as water in a sealed evacuated container. As some of the water evaporates, water-vapor molecules fill the previously empty space in the container. Some of these molecules will hit the liquid surface and rejoin the liquid water in the process called condensation. Initially, the rate of evaporation will be greater than the rate of condensation, but eventually equilibrium will be reached. The pressure at which a liquid is in equilibrium with its own vapor is called the **vapor pressure.** If we now heat the container slightly, the liquid boils, more liquid will evaporate, and a new equilibrium will be established at a higher vapor pressure. Vapor pressure thus depends on the temperature. We can see this from Figure 21-3. If we had started compressing the gas at a lower temperature, as with isotherm A′ in Figure 21-3, the vapor pressure would be lower, as is indicated by the horizontal constant-pressure line for A′ at a lower value of pressure. The temperature for which the vapor pressure for a substance equals 1 atm is the **normal boiling point** of that substance. For example, the temperature at which the vapor pressure of water is 1 atm is 373 K = 100°C, so this temperature is the normal boiling point of water. At high altitudes, such as on the top of a mountain, the pressure is less than 1 atm, and water then boils at a temperature lower than 373 K. Figure 21-4 gives the vapor pressures of water at various temperatures.

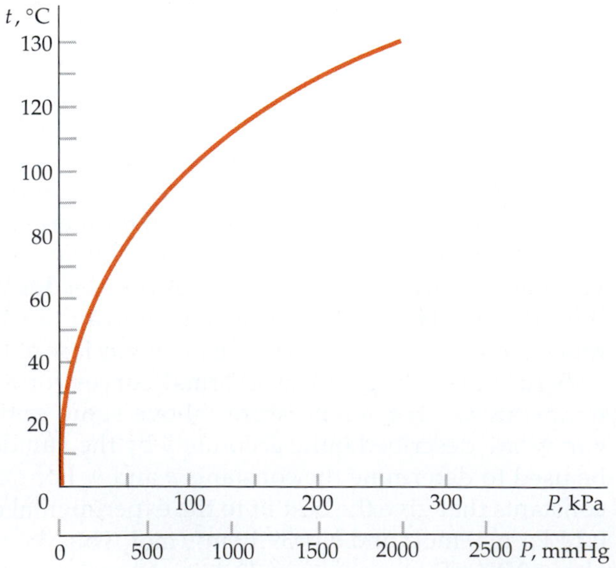

Figure 21-4 Vapor pressure of water versus temperature.

At temperatures greater than the critical temperature T_c, a gas will not liquefy at any pressure. The critical temperature for water vapor is 647 K = 374°C. The point at which the critical isotherm intersects the dashed curve in Figure 21-3 (point C) is called the critical point.

21-3 Phase Diagrams

Figure 21-5 is a plot of pressure versus temperature at a constant volume for water. Such a plot is called a **phase diagram.** The portion of the diagram between points O and C shows the vapor pressure versus the temperature. As we continue to heat the container, the density of the liquid decreases and the density of the vapor increases. At point C on the diagram, these densities are equal. Point C is called the **critical point.** At this point and above it, there is no distinction between the liquid and the gas.* Critical-point temperatures T_c for various substances are listed in Figure 21-6. At temperatures greater than the critical temperature a gas will not liquefy at any pressure.

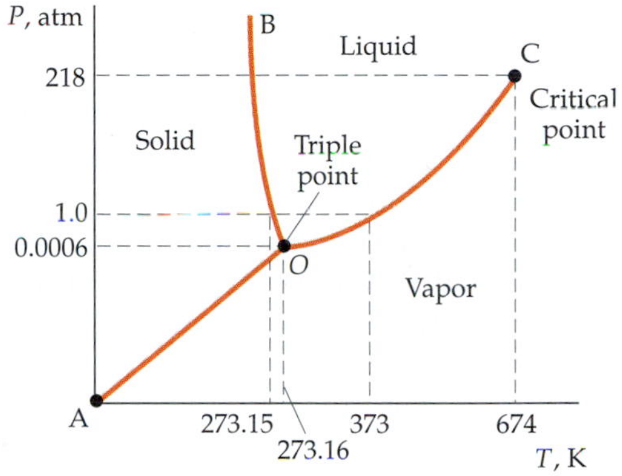

Figure 21-5 Phase diagram for water. The pressure and temperature scales are not linear but are compressed to show the interesting points. Curve OC is the curve of vapor pressure versus temperature. Curve OB is the melting curve, and curve OA is the sublimation curve.

If we now cool our container, some of the vapor condenses into a liquid as we move back down the curve OC until the substance reaches point O in Figure 21-5. At this point, the liquid begins to solidify. Point O is the triple point, that one point at which the vapor, liquid, and solid phases of a substance can coexist in equilibrium. Every substance has a unique triple point at a specific temperature and pressure. The triple-point temperature for water is 273.16 K = 0.01°C and the triple-point pressure is 4.58 mmHg.

At temperatures and pressures below the triple point, the liquid cannot exist. The curve OA in the phase diagram of Figure 21-5 is the locus of pressures and temperatures for which the solid and vapor coexist in equilibrium. The direct change from a solid to a vapor is called **sublimation.** You can observe sublimation if you put ice cubes in the freezer compartment of a refrigerator (especially a self-defrosting refrigerator). The ice cubes will eventually disappear due to sublimation. Because atmospheric pressure is well above the triple-point pressure of water, equilibrium is never established between the ice and water vapor. The triple-point temperature and pressure of carbon dioxide (CO_2) are 216.55 K and 3880 mmHg, which means that liquid CO_2 can exist only at pressures above 3880 mmHg = 5.1 atm. Thus, at ordinary atmospheric pressures, liquid carbon dioxide cannot exist at any temperature. When solid carbon dioxide "melts," it sublimates directly into gaseous CO_2 without going through the liquid phase, hence the name "dry ice."

The curve OB in Figure 21-5 is the melting curve separating the liquid and solid phases. For a substance like water for which the melting temperature decreases as the pressure increases, curve OB slopes upward to the left from the triple point, as in this figure. For most other substances, the melting temperature increases as the pressure increases. For such a substance, curve OB slopes upward to the right from the triple point.

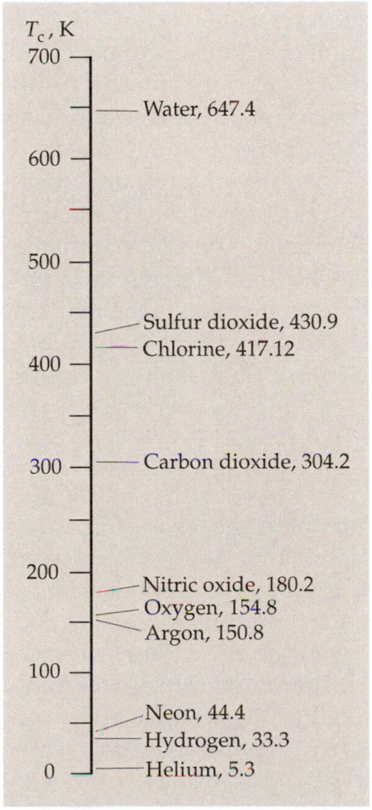

Figure 21-6 Critical temperatures T_c for various substances.

* Often the word "vapor" is used if the temperature is below the critical temperature, and the word "gas" is used if the temperature is above the critical temperature, though there is no need for such a distinction.

For a molecule to escape from a liquid, energy is required to break the molecular bonds at the liquid's surface. Vaporization therefore cools the liquid left behind. If water is brought to a boil over heat, the cooling effect keeps the temperature of the liquid constant at the boiling point. This is the reason that the boiling point of a substance can be used to calibrate thermometers. However, water can also be caused to boil without adding heat by evacuating the air above it, thereby lowering the applied pressure. The energy needed for vaporization is then taken from the liquid left behind. As a result, the liquid will cool down, even to the point that ice forms on top of the boiling water.

21-4 The Transfer of Thermal Energy

Thermal energy is transferred from one place to another by three processes: conduction, convection, and radiation. In **conduction,** thermal energy is transferred by interactions among atoms or molecules, though there is no transport of the atoms or molecules themselves. For example, if one end of a solid bar is heated, the atoms in the heated end vibrate with greater energy than those at the cooler end. Because of the interaction of the more energetic atoms with their neighbors, this energy is transported along the bar.*

In **convection,** heat is transported by direct mass transport. For example, warm air in a room expands and rises because of its lower density. Thermal energy is thus transported upward along with the mass of warm air.

In **radiation**, thermal energy is transported through space in the form of electromagnetic waves that move at the speed of light. Thermal radiation, light waves, radio waves, television waves, and X rays are all forms of electromagnetic radiation that differ from one another only in their wavelengths and frequencies.

In all mechanisms of heat transfer, the rate of cooling of a body is approximately proportional to the temperature difference between the body and its surroundings. This result is known as **Newton's law of cooling.**

In many real situations, all three mechanisms for heat transfer occur simultaneously, though one may be more effective than the others. For example, an ordinary space heater uses both radiation and convection. If the heating element is quartz, the main mechanism of heat transference is radiation. If the heating element is metal, which does not radiate as efficiently as quartz, convection is the main mechanism by which heat is transmitted, with the heated air rising to be replaced by cooler air. Often a fan is included in heaters with hot elements to speed the convection process.

Conduction

Figure 21-7 shows a uniform solid bar of cross-sectional area A. If we keep one end of the bar at a high temperature and the other end at a low temperature, thermal energy is conducted down the bar from the hot end to the cold end. In the steady state, the temperature varies uniformly from the hot end to the cold end. The rate of change of the temperature along the bar, $\Delta T/\Delta x$, is called a **temperature gradient**.

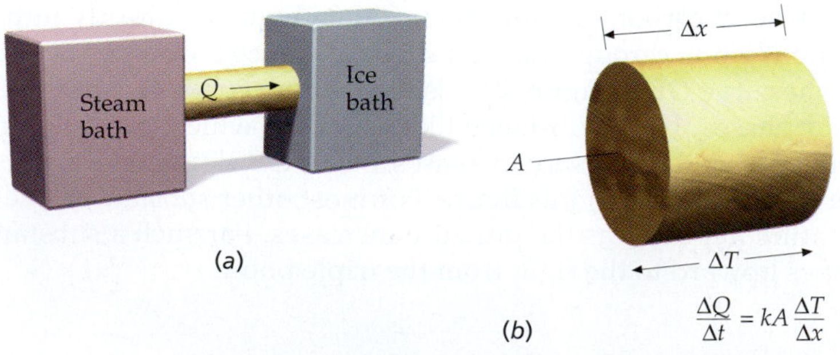

Figure 21-7 (*a*) A conducting bar with its ends at two different temperatures. (*b*) A segment of the bar of length Δx. The rate at which thermal energy is conducted across the segment is proportional to the cross-sectional area and the temperature difference and is inversely proportional to the thickness of the segment.

* If the solid is a metal, the transport of thermal energy is helped by free electrons, which move throughout the metal.

Let ΔT be the temperature difference across a small segment of thickness Δx (Figure 21-7*b*). If ΔQ is the amount of thermal energy conducted through the segment in some time Δt, the rate of conduction of thermal energy, $\Delta Q/\Delta t$, is called the thermal current I. Experimentally, it is found that the thermal current is proportional to the temperature gradient* and to the cross-sectional area A:

$$I = \frac{\Delta Q}{\Delta t} = kA\frac{\Delta T}{\Delta x} \qquad \text{21-7}$$

Thermal current

The proportionality constant k, called the **thermal conductivity**, depends on the composition of the bar.† In SI units, thermal current is expressed in watts, and the thermal conductivity has units of watts per meter-kelvin.‡ In practical calculations in the United States, the thermal current is usually expressed in Btu per hour, the area in square feet, the thickness in inches, and the temperature in Fahrenheit degrees. The thermal conductivity is then given in Btu·in/h·ft^2·F°. Figure 21-8 lists thermal conductivities for various materials in both SI and U.S. customary units.

If we solve Equation 21-7 for the temperature difference, we obtain

$$\Delta T = I\frac{\Delta x}{kA} \qquad \text{21-8}$$

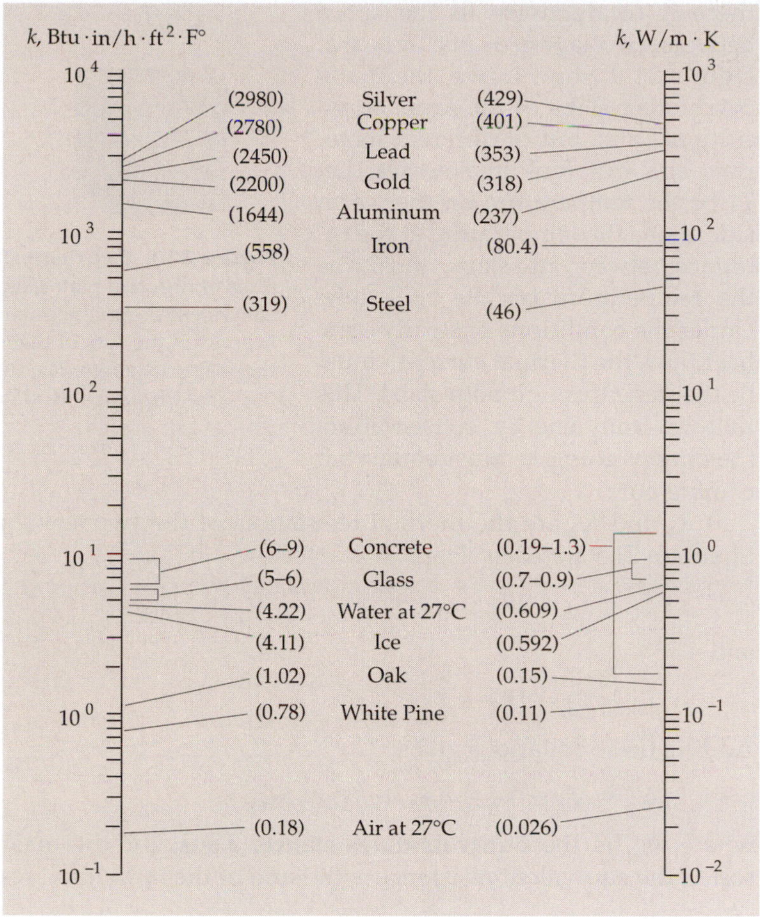

Figure 21-8 Thermal conductivities k for various materials.

or

$$\Delta T = IR \qquad \text{21-9}$$

Temperature change versus current

where $\Delta x/kA$ is the **thermal resistance** R:

$$R = \frac{\Delta x}{kA} \qquad \text{21-10}$$

Definition—Thermal resistance

Exercise Calculate the thermal resistance of an aluminum slab of cross-sectional area 15 cm^2 and thickness 2 cm. (*Answer* 0.0563 K/W)

Exercise What thickness of silver would be required to give the same thermal resistance as a 1-cm thickness of air of the same area? (*Answer* $\Delta x = (1\text{ cm})(429)/(0.026) = 16{,}500\text{ cm} = 165\text{ m}$)

In many practical problems, we are interested in the flow of heat through two or more conductors (or insulators) in series. For example, we may wish to know the effect of adding insulating material of a certain thickness and

* The thermal current is in the direction of the decrease in temperature.

† Don't confuse the thermal conductivity with Boltzmann's constant, which is also designated by k.

‡ In some tables, the energy may be given in calories or kilocalories and the thickness in centimeters.

thermal conductivity to the space between two layers of plasterboard. Figure 21-9 shows two thermally conducting slabs of the same cross-sectional area but of different materials and different thicknesses. Let T_1 be the temperature on the warm side, T_2 be the temperature at the interface between the slabs, and T_3 be the temperature on the cool side. Under the conditions of steady-state heat flow, the thermal current I must be the same through both slabs. This follows from energy conservation; the energy going in must equal that coming out.

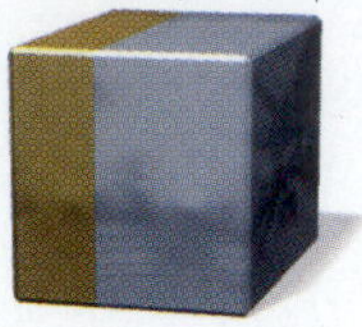

Figure 21-9 Two thermally conducting slabs of different materials in series. The equivalent thermal resistance of the slabs in series is the sum of their individual thermal resistances. The thermal current is the same through both slabs.

If R_1 and R_2 are the thermal resistances of the two slabs, we have from Equation 21-9 for each slab,

$$T_1 - T_2 = IR_1$$

and

$$T_2 - T_3 = IR_2$$

Adding these equations gives

$$\Delta T = T_1 - T_3 = I(R_1 + R_2) = IR_{eq} \qquad 21\text{-}11$$

where R_{eq} is the **equivalent resistance.** Thus, for thermal resistances in series, the equivalent resistance is the sum of the individual resistances:

$$R_{eq} = R_1 + R_2 + \cdots \qquad 21\text{-}12$$

Thermal resistances in series

This result can be applied to any number of resistances in series. In Chapter 26 we will find that the same formula applies to electrical resistances in series.

To calculate the amount of heat leaving a room by conduction in a given time, we need to know how much leaves through the walls, the windows, the floor, and the ceiling. For this type of problem, in which there are several paths for heat flow, the resistances are said to be in parallel. The temperature difference is the same for each path, but the thermal current is different. The total thermal current is the sum of the thermal currents through each of the independent or parallel paths:

$$I_{total} = I_1 + I_2 + \cdots = \frac{\Delta T}{R_1} + \frac{\Delta T}{R_2} + \cdots = \Delta T\left(\frac{1}{R_1} + \frac{1}{R_2} + \cdots\right)$$

or

$$I_{total} = \frac{\Delta T}{R_{eq}} \qquad 21\text{-}13$$

where the equivalent thermal resistance is given by

$$\frac{1}{R_{eq}} = \frac{1}{R_1} + \frac{1}{R_2} + \cdots \qquad 21\text{-}14$$

Thermal resistances in parallel

We will encounter this equation again when we study electrical conduction through parallel resistances. Note that I is proportional to ΔT, in agreement with Newton's law of cooling.

Example 21-4 try it yourself

Two metal bars, each of length 5 cm and rectangular cross section with sides 2 cm and 3 cm, are wedged between two walls, one held at 100°C and the other at 0°C (Figure 21-10). The bars are lead and silver. Find (*a*) the total thermal current through the bars, and (*b*) the temperature at the interface.

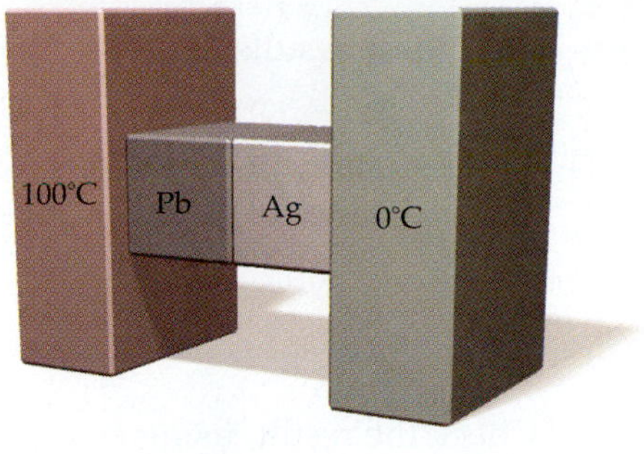

Figure 21-10

Picture the Problem (*a*) You can find the total thermal current I from $\Delta T = IR_{eq}$. The equivalent resistance R_{eq} is the sum of the individual resistances, which you can find from the thermal conductivities given in Figure 21-8. (*b*) You can find the temperature at the interface by applying $\Delta T = IR_1$ to the first cube only, and solving for ΔT in terms of the value for I found in (*a*).

Cover the column to the right and try these on your own before looking at the answers.

Steps	***Answers***
(*a*)1. Write the equivalent thermal resistance in terms of the thermal resistances of the two cubes.	$R_{eq} = R_{Pb} + R_{Ag}$
2. Calculate the thermal resistance from its definition for each cube.	$R_{Pb} = 0.236\ \text{K/W};\quad R_{Ag} = 0.194\ \text{K/W}$
3. Find the equivalent thermal resistance from your results in step 2.	$R_{eq} = 0.430\ \text{K/W}$
4. Substitute R_{eq} and $\Delta T = 100$ K into Equation 21-13 to find the thermal current.	$I = 232.6\ \text{W}$
(*b*)1. Calculate the temperature difference across the lead cube using the current and thermal resistance found in (*a*).	$\Delta T_{Pb} = IR_{Pb} = 54.9\ \text{K} = 54.9\ \text{C}°$
2. Use your result in the previous step to find the temperature at the interface.	$T_{if} = 100°\ \text{C} - \Delta T_{Pb} = 45.1°\text{C}$
3. Check your answer in (*b*)1 by finding the temperature difference across the silver cube.	$\Delta T_{Ag} = IR_{Ag} = 45.1\ \text{C}°$

Example 21-5

The metal bars in Example 21-4 are rearranged as shown in Figure 21-11. Find (*a*) the thermal current in each bar, (*b*) the total thermal current, and (*c*) the equivalent thermal resistance of the two-bar system.

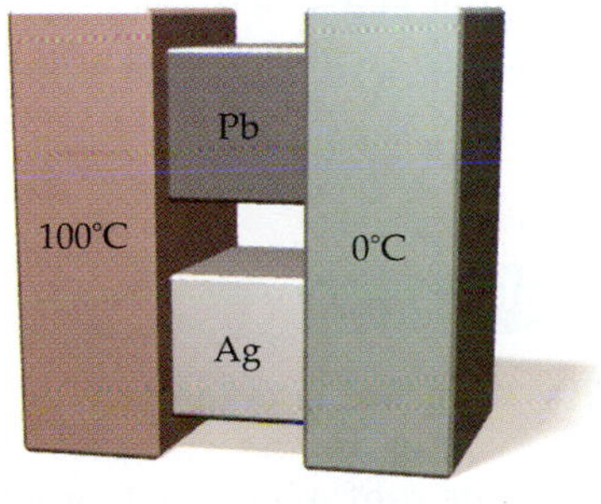

Figure 21-11

Picture the Problem The current in each bar is found from $I = \Delta T/R$, where R is the thermal resistance of the bar found in Example 21-4. The total current is the sum of the currents. The equivalent resistance can be found from Equation 21-14 or from $I_{total} = \Delta T/R_{eq}$.

(*a*) Calculate the thermal current for each cube:

$$I_{Pb} = \frac{\Delta T}{R_{Pb}} = \frac{100\ \text{K}}{0.236\ \text{K/W}} = 424\ \text{W}$$

$$I_{Ag} = \frac{\Delta T}{R_{Ag}} = \frac{100\ \text{K}}{0.194\ \text{K/W}} = 515\ \text{W}$$

optional

(*b*) Add these results to find the total current: $I_{total} = I_{Pb} + I_{Ag} = 424\text{ W} + 515\text{ W} = 939\text{ W}$

(*c*)1. Use Equation 21-14 to calculate the equivalent resistance of the two cubes in parallel:

$$\frac{1}{R_{eq}} = \frac{1}{R_{Pb}} + \frac{1}{R_{Ag}}$$

$$R_{eq} = \frac{R_{Pb}R_{Ag}}{R_{Pb} + R_{Ag}} = \frac{(0.236)(0.194)}{0.236 + 0.194} = 0.106\text{ K/W}$$

2. Check the result, using $I_{total} = \Delta T/R_{eq}$:

$$I_{total} = \frac{\Delta T}{R_{eq}}; \quad R_{eq} = \frac{\Delta T}{I_{total}} = \frac{100\text{ K}}{939\text{ W}} = 0.106\text{ K/W}$$

Remark Note that the equivalent resistance is less than either of the individual resistances. This is always the case for parallel resistors.

In the building industry, the thermal resistance in U.S. customary units for a square foot of material is called the ***R* factor,** R_f. The *R* factor is simply the thickness of the material divided by its thermal conductivity:

$$R_f = \frac{\Delta x}{k} = RA \qquad 21\text{-}15$$

Definition—R factor

Table 21-1 lists *R* factors for several materials. In terms of the *R* factor, Equation 21-9 for the thermal current is

$$\Delta T = IR = \frac{I}{A}R_f \qquad 21\text{-}16$$

For slabs of insulating material of the same area in series, R_f is replaced by the equivalent *R* factor $R_{f,eq}$

$$R_{f,eq} = R_{f1} + R_{f2} + \cdots$$

For parallel slabs, we calculate the thermal current through each slab and add to obtain the total current.

Table 21-1

R Factors $\Delta x/k$ for Various Building Materials

Material	Thickness, in	R_f, h·ft²·F°/Btu
Building board		
Gypsum or plasterboard	0.375	0.32
Plywood (Douglas fir)	0.5	0.62
Plywood or wood panels	0.75	0.93
Particle board, medium density	1.0	1.06
Finish flooring materials		
Carpet and fibrous pad	1.0	2.08
Tile		0.5
Wood, hardwood finish	0.75	0.68
Roof insulation	1.0	2.8
Roofing		
Asphalt roll roofing		0.15
Asphalt shingles		0.44
Windows		
Single-pane		0.9
Double-pane		1.8

Example 21-6

A 60 × 20-ft roof is made of 1-in pine board covered with asphalt shingles. (*a*) If the overlap in the shingles is neglected, at what rate is heat conducted through the roof when the temperature inside is 70°F and the temperature outside is 40°F? (*b*) Find the rate at which heat is conducted if 2 in of roof insulation is added.

Picture the Problem (*a*) The rate at which heat is lost, *I*, is given by $I = \Delta T/R = A\,\Delta T/R_{eq,f}$. For materials in series, the *R* factors add. R_f for pine board is found from its thermal conductivity, given in Figure 21-8. The R_f for asphalt shingles is given in Table 21-1. (*b*) The heat loss will be reduced by the ratio of the new *R* factor to the old *R* factor.

(*a*)1. The rate of heat loss is the thermal current: $I = \dfrac{A\,\Delta T}{R_{\text{f,eq}}}$

2. The equivalent R factor is the sum of the individual R factors: $R_{\text{f,eq}} = R_{\text{f,p}} + R_{\text{f,a}}$

3. We find the R factor for 1-in pine board using k from Figure 21-8:
$$R_{\text{f,p}} = \frac{\Delta x}{k} = \frac{1\text{ in}}{0.78\text{ Btu}\cdot\text{in/h}\cdot\text{ft}^2\cdot\text{F}^\circ}$$
$$= 1.28\,\frac{\text{h}\cdot\text{ft}^2\cdot\text{F}^\circ}{\text{Btu}}$$

4. We find the R factor for asphalt shingles from Table 21-1: $R_{\text{f,a}} = 0.44\text{ h}\cdot\text{ft}^2\cdot\text{F}^\circ/\text{Btu}$

5. The equivalent R factor is the sum of $R_{\text{f,p}}$ and $R_{\text{f,a}}$:
$$R_{\text{f,eq}} = R_{\text{f,p}} + R_{\text{f,a}} = (1.28 + 0.44)\text{h}\cdot\text{ft}^2\cdot\text{F}^\circ/\text{Btu}$$
$$= 1.72\text{ h}\cdot\text{ft}^2\cdot\text{F}^\circ/\text{Btu}$$

6. Calculate the area and temperature: $A = 60\text{ ft} \times 20\text{ ft} = 1200\text{ ft}^2$

and

$\Delta T = 70^\circ\text{F} - 40^\circ\text{F} = 30\text{ F}^\circ$

7. Substitute to find the thermal current:
$$I = \frac{A\Delta T}{R_{\text{f,eq}}} = \frac{(1200\text{ ft}^2)(30\text{ F}^\circ)}{1.72\text{ h}\cdot\text{ft}^2\cdot\text{F}^\circ/\text{Btu}} = 20{,}900\text{ Btu/h}$$

(*b*)1. The R factor for roof insulation, given in Table 21-1, is $2.8\text{ h}\cdot\text{ft}^2\cdot\text{F}^\circ/\text{Btu}$ for 1 in, twice that for 2 in:
$$R_{\text{f,in}} = 2 \times 2.8\text{ h}\cdot\text{ft}^2\cdot\text{F}^\circ/\text{Btu}$$
$$= 5.6\text{ h}\cdot\text{ft}^2\cdot\text{F}^\circ/\text{Btu}$$

2. Add the R factor for the insulation to that for the roof to find the new equivalent R factor: $R_{\text{f,eq}} = 1.72 + 5.6 = 7.32\text{ h}\cdot\text{ft}^2\cdot\text{F}^\circ/\text{Btu}$

3. Since the ratio of R factors in (*b*) and (*a*) is $7.32/1.72 = 4.26$, the rate of heat loss will be reduced by a factor of 4.26:
$$I_{\text{b}} = I_{\text{a}}\frac{R_{\text{f,eq a}}}{R_{\text{f,eq a}}} = (20{,}900\text{ Btu/h})\frac{1.72\text{ h}\cdot\text{ft}^2\cdot\text{F}^\circ/\text{Btu}}{7.32\text{ h}\cdot\text{ft}^2\cdot\text{F}^\circ/\text{Btu}}$$
$$= \frac{20{,}900\text{ Btu/h}}{4.26} = 4910\text{ Btu/h}$$

The thermal conductivity of air is very small compared with that of solid materials, so air is a very good insulator. However, when there is a large air gap—say, between a storm window and the inside window—the insulating efficiency of air is greatly reduced because of convection. Whenever there is a temperature difference between different parts of the air space, convection currents act quickly to equalize the temperature, so the effective conductivity is greatly increased. For storm windows, air gaps of about 1 to 2 cm are optimal. Wider air gaps actually reduce the thermal resistance of a double-pane window because of convection. The insulating properties of air are used most effectively when the air is trapped in small pockets that prevent convection from taking place. This is the principle underlying the excellent insulating properties of both goose down and Styrofoam.

If you touch the inside surface of a glass window when it is cold outside, you will observe that the surface is considerably colder than the inside air. The thermal resistance of windows is due mainly to thin films of insulating air that adhere to either side of the glass surface. The thickness of the glass has little effect on the overall thermal resistance. An air film typically adds an R factor of about 0.45. Thus, the R factor of a window with N separated glass layers is approximately $0.9N$ because of the two sides of each layer. Under windy conditions, the outside air film may be greatly decreased, leading to a smaller R factor for the window.

Convection

Convection is the transport of thermal energy by the transport of the medium itself. It is responsible for the great ocean currents as well as the global circulation of the atmosphere. In the simplest case, convection arises when a fluid (gas or liquid) is heated from below. The warm fluid then expands and rises as the cooler fluid sinks. The mathematical description of convection is very complex because the flow depends on the temperature difference in different parts of the fluid, and this temperature difference is affected by the flow itself.

The heat transferred from an object to its surroundings by convection is approximately proportional to the area of the object and to the difference in temperature between the object and the surrounding fluid. It is possible to write an equation for the thermal energy transported by convection and to define a coefficient of convection, but the analysis of practical problems involving convection is quite difficult and will not be discussed here.

Radiation

All objects emit and absorb electromagnetic radiation. When an object is in thermal equilibrium with its surroundings, it emits and absorbs energy at the same rate. The rate at which an object radiates energy is proportional to the area of the object and to the fourth power of its absolute temperature. This result, found empirically by Josef Stefan in 1879 and derived theoretically by Ludwig Boltzmann about five years later, is called the Stefan–Boltzmann law:

$$P_{\mathrm{r}} = e\sigma A T^4 \qquad \text{21-17}$$

Stefan–Boltzmann law

where P_{r} is the power radiated in watts, A is the area, σ is a universal constant called Stefan's constant, which has the value

$$\sigma = 5.6703 \times 10^{-8}\ \mathrm{W/m^2{\cdot}K^4} \qquad \text{21-18}$$

and e is the **emissivity** of the object, a number between 0 and 1 that depends on the composition of the surface of the object.

When radiation falls on an opaque object, part of the radiation is reflected and part is absorbed. Light-colored objects reflect most visible radiation, whereas dark objects absorb most of it. The rate at which an object absorbs radiation is given by

$$P_{\mathrm{a}} = e\sigma A T_0^4 \qquad \text{21-19}$$

where T_0 is the temperature of the surroundings.

If an object emits more radiation than it absorbs, it cools while the surroundings absorb radiation from the object and warm. If the object absorbs more than it emits, the object warms and the surroundings cool. The net power radiated by an object at temperature T in an environment at temperature T_0 is

$$P_{\mathrm{net}} = e\sigma A(T^4 - T_0^4) \qquad \text{21-20}$$

When an object is in thermal equilibrium with its surroundings, $T = T_0$, and the object emits and absorbs radiation at the same rate.

An object that absorbs all the radiation incident upon it has an emissivity equal to 1 and is called a blackbody. A blackbody is also an ideal radiator. The concept of a blackbody is important because the characteristics of the radiation emitted by such an ideal object can be calculated theoretically. Materials such as black velvet come close to being ideal blackbodies. The best practical approximation of an ideal blackbody is a small hole leading into a cavity, such as a keyhole in a closet door (Figure 21-12). Radiation incident on the hole has little chance of being reflected back out the hole before it is absorbed by the walls of the cavity. The radiation emitted through the hole is thus characteristic of the temperature of the object.

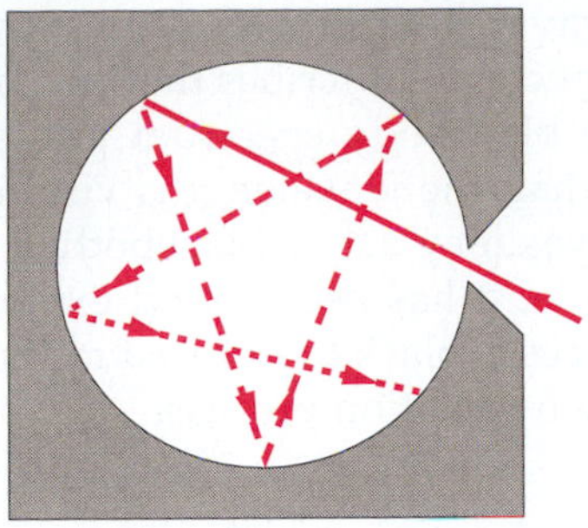

Figure 21-12 A hole in a cavity approximates an ideal blackbody. Radiation entering the cavity has little chance of leaving the cavity before it is completely absorbed. The radiation emitted through the hole is therefore characteristic of the temperature of the object.

Most of the radiation emitted by an object at temperatures below about 600°C is concentrated at wavelengths much longer than those of visible light.* As an object is heated, the rate of energy emission increases, and the energy radiated extends to shorter and shorter wavelengths. Between about 600 and 700°C, enough of the radiated energy is in the visible spectrum for the body to glow a dull red. At higher temperatures, it may become bright red or even "white hot." Figure 21-13 shows the power radiated by a blackbody as a function of wavelength for several different temperatures. The wavelength at which the power is a maximum varies inversely with the temperature, a result known as Wien's displacement law:

$$\lambda_{max} = \frac{2.898 \text{ mm}\cdot\text{K}}{T} \qquad \text{21-21}$$

Wien's displacement law

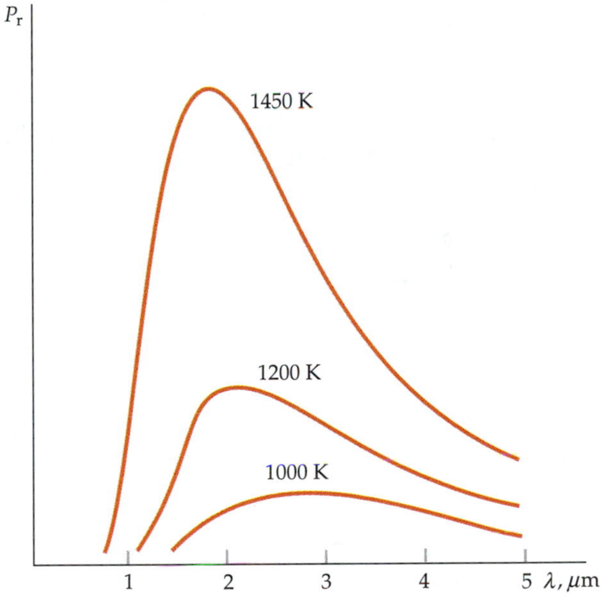

Figure 21-13 Radiated power versus wavelength for radiation emitted by a blackbody. The wavelength of the maximum power varies inversely with the absolute temperature of the blackbody.

This law is used to determine the temperatures of stars from analyses of their radiation. It can also be used to map the variation in temperature over different regions of the surfaces of an object. Such a map is called a thermograph. Thermographs can be used to detect cancer because cancerous material is at a slightly higher temperature than the surrounding tissue.

The spectral-distribution curves shown in Figure 21-13 played an important role in the history of physics. It was the discrepancy between theoretical calculations of what the blackbody spectral distribution should be using classical thermodynamics and the experimental measurements that led to Max Planck's first ideas about the quantization of energy in 1897.

* When we study light, we will see that visible light is electromagnetic radiation with wavelengths between about 400 and 700 nm.

Example 21-7

(*a*) The surface temperature of the sun is about 6000 K. If the sun is assumed to be a blackbody radiator, at what wavelength λ_{max} would its spectrum peak? (*b*) Calculate λ_{max} for a blackbody at room temperature, T = 300 K.

(*a*) We can find λ_{max} from Wien's displacement law:

$$\lambda_{max} = \frac{2.898 \text{ mm}\cdot\text{K}}{6000 \text{ K}} = 483 \times 10^{-9} \text{ m} = 483 \text{ nm}$$

(*b*) Apply Wien's displacement law for T = 300 K:

$$\lambda_{max} = \frac{2.898 \text{ mm}\cdot\text{K}}{300 \text{ K}} = 9.66 \times 10^{-6} \text{ m} = 9660 \text{ nm}$$

optional

Remarks The peak wavelength from the sun is in the visible spectrum. The blackbody radiation spectrum describes the sun's radiation fairly well.

For $T = 300$ K, the blackbody spectrum peaks in the infrared at wavelengths much longer than the wavelengths visible to the eye. Surfaces that are not black to our eyes may act as blackbodies for infrared radiation and absorption. For example, it has been found experimentally that the skin of human beings of all races is black to infrared radiation; hence, the emissivity of skin is 1.00 for its own radiation process.

Example 21-8 *try it yourself*

Calculate the net loss in radiated energy for a naked person in a room at 20°C, assuming the person to be a blackbody with a surface area of 1.4 m^2 and a surface temperature of 33°C = 306 K. (The surface temperature of the human body is slightly less than the internal temperature of 37°C because of the thermal resistance of the skin.)

Cover the column to the right and try this on your own before looking at the answer.

Step	*Answer*
Use $P_{net} = e\sigma A(T^4 - T_0^4)$ with $e = 1$, $T = 306$ K, and $T_0 = 293$ K.	$P_{net} = 111$ W

Remarks This large energy loss is approximately equal to the basal metabolic rate of about 120 W. We protect ourselves from this great loss of energy by wearing clothing, which, because of its low thermal conductivity, has a much lower outside temperature than the skin and therefore a much lower rate of thermal radiation.

When the temperature T of an object is not too different from the surrounding temperature T_0, a radiating object obeys Newton's law of cooling. We can see this by writing Equation 21-20 as

$$P_{net} = e\sigma A(T^4 - T_0^4) = e\sigma A(T^2 + T_0^2)(T^2 - T_0^2)$$
$$= e\sigma A(T^2 + T_0^2)(T + T_0)(T - T_0)$$

When $T - T_0$ is small, we can replace T_0 by T in the sums with little change in the result. Then

$$P_{net} = e\sigma A(T^4 - T_0^4) \approx e\sigma A(T^2 + T^2)(T + T)(T - T_0) = 4e\sigma AT^3\Delta T$$

The net power radiated is approximately proportional to the temperature difference, in agreement with Newton's law of cooling. This result can also be obtained by computing the differential dP for a small change in temperature dT. We have

$$P_r = e\sigma AT^4$$
$$dP_r = e\sigma A(4T^3\,dT) = 4e\sigma AT^3\,dT$$

Summary

Topic	Remarks and Relevant Equations	
1. Thermal Expansion		
Coefficient of linear expansion	$\alpha = \dfrac{\Delta L/L}{\Delta T}$	**21-2**
Coefficient of volume expansion	$\beta = \dfrac{\Delta V/V}{\Delta T} = 3\alpha$	**21-4, 21-5**
2. The van der Waals Equation of State	The van der Waals equation of state describes the behavior of real gases over a wide range of temperatures and pressures, taking into account the space occupied by the gas molecules themselves and the attraction of the molecules for one another. $\left(P + \dfrac{an^2}{V^2}\right)(V - bn) = nRT$	**21-6**
3. Vapor Pressure	Vapor pressure is the pressure at which the liquid and gas phases of a substance are in equilibrium at a given temperature. The liquid boils at that temperature for which the external pressure equals the vapor pressure.	
4. The Triple Point	The triple point is the unique temperature and pressure at which the gas, liquid, and solid phases of a substance can coexist in equilibrium. At temperatures and pressures below the triple point, the liquid phase of a substance cannot exist.	
5. Heat Transfer	The three mechanisms by which thermal energy is transferred are radiation, conduction, and convection.	
Newton's law of cooling	For all mechanisms of heat transfer, if the temperature difference between the body and its surroundings is small, the rate of cooling of a body is approximately proportional to the temperature difference.	
6. Heat Conduction		
Current	The rate of conduction of thermal energy is given by $I = \dfrac{\Delta Q}{\Delta t} = kA\dfrac{\Delta T}{\Delta x}$ where I is the thermal current and k is the coefficient of thermal conductivity.	**21-7**
Thermal resistance	$\Delta T = IR$	**21-9**
	where R is the thermal resistance: $R = \dfrac{\Delta x}{kA}$	**21-10**
Equivalent resistance:		
Series	$R_{eq} = R_1 + R_2 + \cdots$	**21-12**
Parallel	$\dfrac{1}{R_{eq}} = \dfrac{1}{R_1} + \dfrac{1}{R_2} + \cdots$	**21-14**

optional

R factor	The *R* factor is the thermal resistance for a unit area of a slab of material: $$R_f = \frac{\Delta x}{k} = RA \qquad \textbf{21-15}$$

7. Thermal Radiation

Rate of power radiated	$$P_r = e\sigma AT^4 \qquad \textbf{21-17}$$ where $\sigma = 5.6703 \times 10^{-8}\ \mathrm{W/m^2{\cdot}K^4}$ is Stefan's constant, and e is the emissivity, a number between 0 and 1 that depends on the composition of the surface of the object. Materials that are good heat absorbers are good heat radiators.
Net power radiated by an object at T to its environment at T_0	$$P_{net} = e\sigma A(T^4 - T_0^4) \qquad \textbf{21-20}$$
Blackbody	A blackbody has an emissivity of 1. It is a perfect radiator, and it absorbs all the radiation incident upon it.
Wien's law	The power spectrum of electromagnetic energy radiated by a blackbody has a maximum at a wavelength λ_{max}, which varies inversely with the absolute temperature of the body: $$\lambda_{max} = \frac{2.898\ \mathrm{mm{\cdot}K}}{T} \qquad \textbf{21-21}$$

Problem-Solving Guide

Summary of Worked Examples

Type of Calculation	Procedure and Relevant Examples	
1. Thermal Expansion		
Find the expansion due to temperature change.	Use definitions of coefficients α or β, and values from tables.	**Examples 21-1, 21-2**
Find the breaking temperature of an object that is constrained	Find ΔL for which stress equals breaking stress. Then find ΔT that would give ΔL if expansion were allowed.	**Example 21-3**
2. Thermal Conduction		
Find the thermal current through series conductors.	Use $\Delta T = IR$ with $R = R_1 + R_2 + \cdots$.	**Examples 21-4, 21-6**
Find the thermal current through parallel resistors.	Use $I_{total} = \frac{\Delta T}{R_{eq}}$ with $\frac{1}{R_{eq}} = \frac{1}{R_1} + \frac{1}{R_2} + \cdots$.	**Example 21-5**
Find the temperature at the interface of two series conductors.	First find the total current, then use $\Delta T_1 = IR_1$.	**Example 21-4**
3. Blackbody Radiation		
Find the peak wavelength of a blackbody spectrum.	Use $\lambda_{max} = \frac{2.898\ \mathrm{mm{\cdot}K}}{T}$.	**Example 21-7**
Find the net loss in radiated energy of a body.	Use $P_{net} = e\sigma A(T^4 - T_0^4)$.	**Example 21-8**

Problems

Conceptual Problems

Problems from Optional and Exploring sections

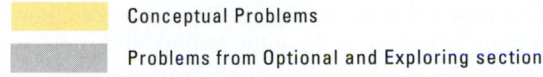

In a few problems, you are given more data than you actually need; in a few other problems, you are required to supply data from your general knowledge, outside sources, or informed estimates.

• Single-concept, single-step, relatively easy
•• Intermediate-level, may require synthesis of concepts
••• Challenging, for advanced students

Thermal Expansion

1 • Why does the mercury level first decrease slightly when a thermometer is placed in warm water?

2 • A large sheet of metal has a hole cut in the middle of it. When the sheet is heated, the area of the hole will

(*a*) not change.
(*b*) always increase.
(*c*) always decrease.
(*d*) increase if the hole is not in the exact center of the sheet.
(*e*) decrease only if the hole is in the exact center of the sheet.

3 • A steel ruler has a length of 30 cm at 20°C. What is its length at 100°C?

4 • A bridge 100 m long is built of steel. If it is built as a single, continuous structure, how much will its length change from the coldest winter days (−30°C) to the hottest summer days (40°C)?

5 •• (*a*) Define a coefficient of area expansion. (*b*) Calculate it for a square and a circle, and show that it is 2 times the coefficient of linear expansion.

6 •• The density of aluminum is $2.70 \times 10^3\ \text{kg/m}^3$ at 0°C. What is the density of aluminum at 200°C?

7 •• A copper collar is to fit tightly about a steel shaft whose diameter is 6.0000 cm at 20°C. The inside diameter of the copper collar at that temperature is 5.9800 cm. To what temperature must the copper collar be raised so that it will just slip on the steel shaft, assuming that the steel shaft remains at 20°C?

8 •• Repeat Problem 7 when the temperature of both the steel shaft and copper collar are raised simultaneously.

9 •• A container is filled to the brim with 1.4 L of mercury at 20°C. When the temperature of container and mercury is raised to 60°C, 7.5 mL of mercury spill over the brim of the container. Determine the linear expansion coefficient of the container.

10 •• A hole is drilled in an aluminum plate with a steel drill bit whose diameter at 20°C is 6.245 cm. In the process of drilling, the temperature of the drill bit and of the aluminum plate rise to 168°C. What is the diameter of the hole in the aluminum plate when it has cooled to room temperature?

11 •• Len sells trees that double in price when they are over 2.00 m high. To make a standard, he cuts an aluminum rod 2.00 m in length, as measured by a steel measuring tape. That day, the temperature of both the rod and the tape is 25°C. What will the tape indicate the length of the rod to be when both the tape and the rod are at (*a*) 0°C and (*b*) 50°C?

12 •• A rookie crew was left to put in the final 1 km of rail for a stretch of railroad track. When they finished, the temperature was 20°C, and they headed to town for some refreshments with their coworkers. After an hour or two, one of the old-timers noticed that the temperature had gone up to 25°C, so he said, "I hope you left some gaps to allow for expansion." By the look on their faces, he knew that they had not, and they all rushed back to the work site. The rail had buckled into an isosceles triangle. How high was the buckle?

13 •• A car has a 60-L steel gas tank filled to the top with gasoline when the temperature is 10°C. The coefficient of volume expansion of gasoline is $\beta = 0.900 \times 10^{-3}\ \text{K}^{-1}$. Taking the expansion of the steel tank into account, how much gasoline spills out of the tank when the car is parked in the sun and its temperature rises to 25°C?

14 •• A thermometer has an ordinary glass bulb and thin glass tube filled with 1 mL of mercury. A temperature change of 1 C° changes the level of mercury in the thin tube by 3.0 mm. Find the inside diameter of the thin glass tube.

15 •• A mercury thermometer consists of a 0.4-mm capillary tube connected to a glass bulb. The mercury level rises 7.5 cm as the temperature of the thermometer increases from 35 to 43°C. Find the volume of the thermometer bulb.

16 ••• A grandfather's clock is calibrated at a temperature of 20°C. (*a*) On a hot day, when the temperature is 30°C, does the clock run fast or slow? (*b*) How much does it gain or lose in a 24-h period? Assume that the pendulum is a thin brass rod of negligible mass with a heavy bob attached to the end.

17 ••• A steel tube has an outside diameter of 3.000 cm at room temperature (20°C). A brass tube has an inside diameter of 2.997 cm at the same temperature. To what temperature must the ends of the tubes be heated if the steel tube is to be inserted into the brass tube?

18 ••• What is the tensile stress in the copper collar of Problem 7 when its temperature returns to 20°C?

The van der Waals Equation, Liquid–Vapor Isotherms, and Phase Diagrams

19 • Mountaineers say that you cannot hard boil an egg on the top of Mount Rainier. This is true because

(*a*) the air is too cold to boil water.
(*b*) the air pressure is too low for stoves to burn.
(*c*) boiling water is not hot enough to hard boil the egg.
(*d*) the oxygen content of the air is too low.
(*e*) the eggs always break in their backpacks.

20 • Which gases in Figure 21-6 cannot be liquefied by applying pressure at 20°C?

optional

21 •• The phase diagram in Figure 21-14 can be interpreted to yield information on how the boiling and melting points of water change with altitude. (*a*) Explain how this information can be obtained. (*b*) How might this information affect cooking procedures in the mountains?

Figure 21-14 Problems 21 and 22

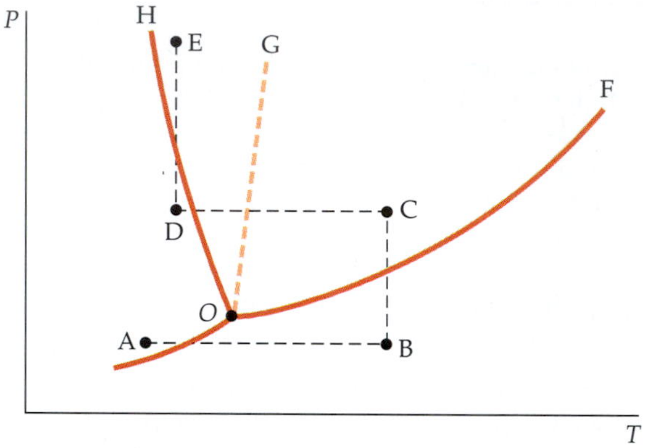

22 •• For the phase diagram given in Figure 21-14, state what changes (if any) occur for each line segment—AB, BC, CD, and DE—in (*a*) volume and (*b*) phase. (*c*) For what type of substance would *OH* be replaced by *OG*? (*d*) What is the significance of point F?

23 • (*a*) Calculate the volume of 1 mol of steam at 100°C and a pressure of 1 atm, assuming that it is an ideal gas. (*b*) Find the temperature at which the steam will occupy the volume found in part (*a*) if it obeys the van der Waals equation with $a = 0.55\ \text{Pa}\cdot\text{m}^6/\text{mol}^2$ and $b = 30\ \text{cm}^3/\text{mol}$.

24 •• From Figure 21-4, find (*a*) the temperature at which water boils on a mountain where the atmospheric pressure is 70 kPa, (*b*) the temperature at which water will boil in a container in which the pressure has been reduced to 0.5 atm, and (*c*) the pressure at which water will boil at 115°C.

25 •• The van der Waals constants for helium are $a = 0.03412\ \text{L}^2\cdot\text{atm}/\text{mol}^2$ and $b = 0.0237\ \text{L/mol}$. Use these data to find the volume in cubic centimeters occupied by one helium atom and to estimate the radius of the atom.

26 ••• (*a*) For a van der Waals gas, show that the critical temperature is $8a/27Rb$ and the critical pressure is $a/27b^2$. (*b*) Rewrite the van der Waals equation of state in terms of the reduced variable $V_r = V/V_c$, $P_r = P/P_c$, and $T_r = T/T_c$.

Heat Conduction

27 • A copper bar 2 m long has a circular cross section of radius 1 cm. One end is kept at 100°C and the other end is kept at 0°C. The surface of the bar is insulated so that there is negligible heat loss through it. Find (*a*) the thermal resistance of the bar, (*b*) the thermal current *I*, (*c*) the temperature gradient $\Delta T/\Delta x$, and (*d*) the temperature of the bar 25 cm from the hot end.

28 • A 20 × 30-ft slab of insulation has an *R* factor of 11. How much heat (in Btu per hour) is conducted through the slab if the temperature on one side is 68°F and that on the other side is 30°F?

29 •• Two metal cubes with 3-cm edges, one copper (Cu) and one aluminum (Al), are arranged as shown in Figure 21-15. Find (*a*) the thermal resistance of each cube, (*b*) the thermal resistance of the two-cube system, (*c*) the thermal current *I*, and (*d*) the temperature at the interface of the two cubes.

Figure 21-15 Problem 29

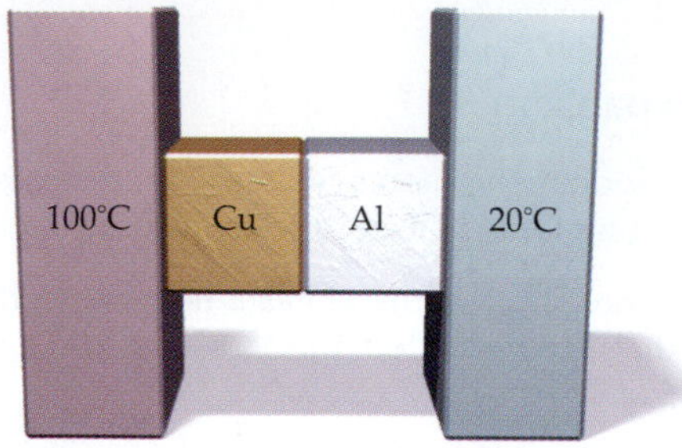

30 •• The cubes in Problem 29 are rearranged in parallel as shown in Figure 21-16. Find (*a*) the thermal current carried by each cube from one side to the other, (*b*) the total thermal current, and (*c*) the equivalent thermal resistance of the two-cube system.

Figure 21-16 Problem 30

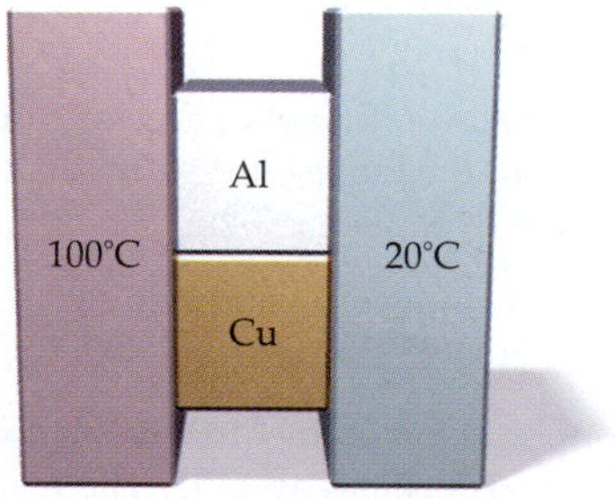

31 •• A spherical shell of thermal conductivity *k* has inside radius r_1 and outside radius r_2 (Figure 21-17). The inside of the shell is held at a temperature T_1, and the outside at temperature T_2. In this problem, you are to show that the thermal current through the shell is given by

$$I = \frac{4\pi k r_1 r_2}{r_2 - r_1}(T_2 - T_1) \qquad \text{21-22}$$

Consider a spherical element of the shell of radius *r* and thickness *dr*. (*a*) Why must the thermal current through each such element be the same? (*b*) Write the thermal current *I* through such a shell element in terms of the area $A = 4\pi r^2$, the thickness *dr*, and the temperature difference *dT* across the element. (*c*) Solve for *dT* in terms of *dr* and integrate from $r = r_1$ to $r = r_2$. (*d*) Show that when r_1 and r_2 are much larger than $r_2 - r_1$, Equation 21-22 is the same as Equation 21-7.

Figure 21-17
Problems 31 and 50

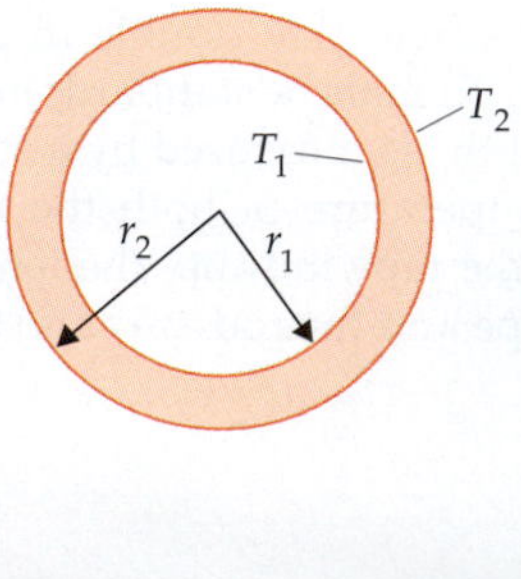

32 •• A group of anthropologists is staying in the high Arctic for a month, and they need accommodation. They are directed to a small company, Inuit Igloos."How thick do you want the walls?" asks Inuk, the head igloo maker. After some conferring, they reply that it should be 20°C inside when the temperature is −20°C outside. After looking the anthropologists over and poking them a bit, Inuk estimates that they would give off 38 MJ of heat per day. If the inside radius of the hemispherical igloo is to be 2 m, and the thermal conductivity of the compacted snow is 0.209 W/m·K, how thick should the walls be? (As an approximation, assume that the inner surface area of the igloo is equal to the outer surface area.)

33 •• For a boiler at a power station, heat must be transferred to boiling water at the rate of 3 GW. The boiling water passes through copper pipes having a wall thickness of 4.0 mm and a surface area of 0.12 m^2 per meter length of pipe. Find the total length of pipe (actually there are many pipes in parallel) that must pass through the furnace if the steam temperature is 225°C and the external temperature of the pipes is 600°C.

34 ••• A steam pipe of length L is insulated with a layer of material of thermal conductivity k. Find the rate of heat transfer if the temperature outside the insulation is t_1, the temperature inside is t_2, the outside radius of the insulation is r_1, and the inside radius is r_2.

35 ••• Brine at −16°C circulating through copper pipes with walls 1.5 mm thick is used to keep a cold room at 0°C. The diameter of each pipe is very large compared to the thickness of its walls. By what fraction is the transfer of heat reduced when the pipes are coated with a 5-mm layer of ice?

Radiation

36 • If the absolute temperature of an object is tripled, the rate at which it radiates thermal energy

(*a*) triples.
(*b*) increases by a factor of 9.
(*c*) increases by a factor of 27.
(*d*) increases by a factor of 81.
(*e*) depends on whether the absolute temperature is above or below zero.

37 • Calculate λ_{max} for a human blackbody radiator, assuming the surface temperature of the skin to be 33°C.

38 •• The heating wires of a 1-kW electric heater are red hot at a temperature of 900°C. Assuming that 100% of the heat output is due to radiation and that the wires act as blackbody radiators, what is the effective area of the radiating surface? (Assume a room temperature of 20°C.)

39 •• A blackened, solid copper sphere of radius 4.0 cm hangs in a vacuum in an enclosure whose walls have a temperature of 20°C. If the sphere is initially at 0°C, find the rate at which its temperature changes, assuming that heat is transferred by radiation only.

40 •• The surface temperature of the filament of an incandescent lamp is 1300°C. If the electric power input is doubled, what will the temperature become? *Hint:* Show that you can neglect the temperature of the surroundings.

41 •• Liquid helium is stored at its boiling point (4.2 K) in a spherical can that is separated by a vacuum space from a surrounding shield that is maintained at the temperature of liquid nitrogen (77 K). If the can is 30 cm in diameter and is blackened on the outside so that it acts as a blackbody, how much helium boils away per hour?

General Problems

42 • In a cool room, a metal or marble table top feels much colder to the touch than a wood surface does even though they are at the same temperature. Why?

43 • True or false:

(*a*) During a phase change, the temperature of a substance remains constant.
(*b*) The rate of conduction of thermal energy is proportional to the temperature gradient.
(*c*) The rate at which an object radiates energy is proportional to the square of its absolute temperature.
(*d*) All materials expand when they are heated.
(*e*) The vapor pressure of a liquid depends on the temperature.

44 • Conduction is a method of heat transfer that

(*a*) can proceed in vacuum.
(*b*) involves the transfer of mass.
(*c*) is dominant in solids.
(*d*) depends on the fourth power of the absolute temperature.

45 • The earth loses heat by

(*a*) conduction.
(*b*) convection.
(*c*) radiation.
(*d*) all of the above.

46 • Which heat-transfer mechanisms are most important in the warming effect of a fire in a fireplace?

47 • Which heat-transfer mechanism is important in the transfer of energy from the sun to the earth?

48 •• Two cylinders made of materials A and B have the same lengths; their diameters are related by $d_A = 2d_B$. When the same temperature difference is maintained between the ends of the cylinders they conduct heat at the same rate. Their thermal conductivities are related by

(*a*) $k_A = k_B/4$.
(*b*) $k_A = k_B/2$.
(*c*) $k_A = k_B$.
(*d*) $k_A = 2k_B$.
(*e*) $k_A = 4k_B$.

49 • A steel tape is placed around the earth at the equator when the temperature is 0°C. What will the clearance between the tape and the ground (assumed to be uniform) be if the temperature of the tape rises to 30°C? Neglect the expansion of the earth.

50 •• Use the result of Problem 31 (Equation 21-22) to calculate the wall thickness of the hemispherical igloo of Problem 32 without assuming that the inner surface area equals the outer surface area.

optional

51 •• Show that change in the density of an isotropic material due to an increase in temperature ΔT is given by $\Delta\rho = -\beta\rho\,\Delta T$.

52 •• The solar constant is the power received from the sun per unit area perpendicular to the sun's rays at the mean distance of the earth from the sun. Its value at the upper atmosphere of the earth is about 1.35 kW/m^2. Calculate the effective temperature of the sun if it radiates like a blackbody. (The radius of the sun is 6.96×10^8 m.)

53 •• Lou has patented a cooking timer, which he is marketing as "Nature's Way: Taking You Back To Simpler Times." The timer consists of a 28-cm copper rod having a 5.0-cm diameter. Just as the lower end is placed in boiling water, an ice cube is placed on the top of the rod. When the ice melts completely, the cooking time is up. A special ice cube tray makes cubes of various sizes to correspond to the boiling time required. What is the cooking time when a 30-g ice cube at −5.0°C is used?

54 •• To determine the R value of insulating material that comes in sheets of $\frac{1}{2}$-in thickness, you construct a cubical box of 12 in per side and place a thermometer and a 100-W heater inside the box. After thermal equilibrium has been attained, the temperature inside the box is 90°C when the external temperature is 20°C. Determine the R value of this material.

55 •• A 2-cm-thick copper sheet is pressed against a sheet of aluminum. What should be the thickness of the aluminum sheet so that the temperature of the copper–aluminum interface is $(T_1 + T_2)/2$, where T_1 and T_2 are the temperatures at the copper–air and aluminum–air interfaces?

56 •• At a temperature of 20°C, a steel bar of radius 2.2 cm and length 60 cm is jammed horizontally perpendicular between two vertical concrete walls. With a blowtorch, the temperature of the bar is raised to 60°C. Find the force exerted by the bar on each wall.

57 •• (*a*) From the definition of β, the coefficient of volume expansion (at constant pressure), show that $\beta = 1/T$ for an ideal gas. (*b*) The experimentally determined value of β for N_2 gas at 0°C is 0.003673 K^{-1}. Compare this value with the theoretical value $\beta = 1/T$, assuming that N_2 is an ideal gas.

58 •• One way to construct a device with two points whose separation remains the same in spite of temperature changes is to bolt together one end of two rods having different coefficients of linear expansion as in the arrangement shown in Figure 21-18. (*a*) Show that the distance L will not change with temperature if the lengths L_A and L_B are chosen such that $L_A/L_B = \alpha_B/\alpha_A$. (*b*) If material B is steel, material A is brass, and $L_A = 250$ cm at 0°C, what is the value of L?

Figure 21-18 Problem 58

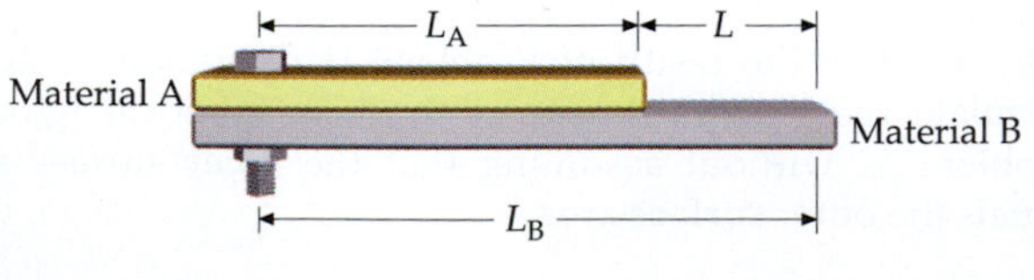

59 •• On the average, the temperature of the earth's crust increases 1.0 C° for every 30 m of depth. The average thermal conductivity of the earth's crust is 0.74 J/m·s·K. What is the heat loss of the earth per second due to conduction from the core? How does this heat loss compare with the average power received from the sun? (The solar constant is about 1.35 kW/m^2.)

60 •• A copper-bottomed saucepan containing 0.8 L of boiling water boils dry in 10 min. Assuming that all the heat flows through the flat copper bottom, which has a diameter of 15 cm and a thickness of 3.0 mm, calculate the temperature of the outside of the copper bottom while some water is still in the pan.

61 •• A hot-water tank of cylindrical shape has an inside diameter of 0.55 m and inside height of 1.2 m. The tank is enclosed with a 5-cm-thick insulating layer of glass wool whose thermal conductivity is 0.035 W/m·K. The metallic interior and exterior walls of the container have thermal conductivities that are much greater than that of the glass wool. How much power must be supplied to this tank to maintain the water temperature at 75°C when the external temperature is 1°C?

62 ••• The diameter of a rod is given by $d = d_0(1 + ax)$, where a is a constant and x is the distance from one end. If the thermal conductivity of the material is k what is the thermal resistance of the rod if its length is L?

63 ••• A solid disk of radius R and mass M is spinning in a frictionless environment with angular velocity ω_1 at temperature T_1. The temperature of the disk is then changed to T_2. Express the angular velocity ω_2, rotational kinetic energy E_2, and angular momentum L_2 in terms of their values at the temperature T_1 and the linear expansion coefficient α of the disk.

64 ••• A small pond has a layer of ice 1 cm thick floating on its surface. (*a*) If the air temperature is −10°C, find the rate in centimeters per hour at which ice is added to the bottom of the layer. The density of ice is 0.917 g/cm^3. (*b*) How long does it take for a 20-cm layer to be built up?

65 ••• A body initially at a temperature T_i cools by convection and radiation in a room where the temperature is T_0. The body obeys Newton's law of cooling, which can be written $dQ/dt = hA(T - T_0)$, where A is the area of the body and h is a constant called the surface coefficient of heat transfer. Show that the temperature T at any time t is given by $T = T_0 + (T_i - T_0)e^{-hAt/mc}$, where m is the mass of the body and c is its specific heat.

66 ••• Two 200-g copper containers, each holding 0.7 L of water, are connected by a 10-cm copper rod of cross-sectional area 1.5 cm^2. Initially, one container is at 60°C; the second is maintained at 0°C. (*a*) Show that the temperature t_c of the first container changes over time t according to

$$t_c = t_{c0}e^{-t/RC}$$

where t_{c0} is the initial temperature of the first container, R is the thermal resistance of the rod, and C is the total heat capacity of the container plus the water. (*b*) Evaluate R, C, and the "time constant" RC. (*c*) Show that the total amount of the "time constant" RC. (*c*) Show that the total amount of heat Q conducted in time t is

$$Q = Ct_{c0}(1 - e^{-t/RC})$$

(*d*) Find the time it takes for the temperature of the first container to be reduced to 30°C.

67 ••• Liquid helium is stored in containers fitted with 7-cm-thick "superinsulation" consisting of a large number of layers of very thin aluminized Mylar sheets. The rate of evaporation of liquid in a 200-L container is about 0.7 L per day. Assume that the container is spherical and that the external temperature is 20°C. The specific gravity of liquid helium is 0.125 and the latent heat of vaporization is 21 kJ/kg. Estimate the thermal conductivity of superinsulation.

PART IV

A small, permanent magnet levitates above a disk of the superconductor yttrium–barium copper oxide cooled to 77 K. The magnetic field of the cube sets up circulating electric currents in the superconducting disk, such that the resultant magnetic field in the superconductor is zero. These currents produce a magnetic field that repels the cube.

electricity and magnetism

CHAPTER 22

The Electric Field I: Discrete Charge Distributions

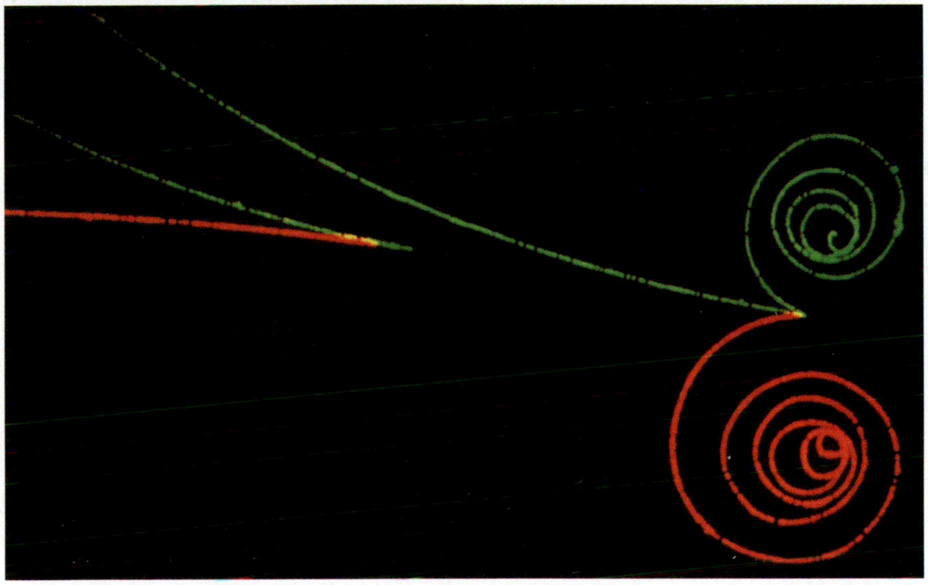

Pair production. An electron of charge $-e$ and a positron of charge $+e$ are created by the interaction of electromagnetic radiation with matter. The paths of the oppositely charged particles, made visible in a bubble chamber, are bent in opposite directions by a magnetic field.

We are extremely dependent on electricity in our daily lives, while just a century ago we had nothing more than a few electric lights. Yet the study of electricity has a history reaching long before the first electric lamp glowed. Observations of electrical attraction can be traced back to the ancient Greeks, who noticed that after amber has been rubbed, it attracts small objects such as straw or feathers. Indeed, the word "electric" comes from the Greek word for amber, *elektron.*

We begin our examination of electricity with **electrostatics**, the study of electrical charges at rest. After introducing the concept of electric charge, we briefly look at conductors and insulators and how conductors can be given a net charge. We then study Coulomb's law, which gives the force exerted by one electric charge on another. Next, we introduce the electric field and show how it can be described by electric field lines that indicate the magnitude and direction of the field. Finally, we discuss the behavior of point charges and electric dipoles in electric fields.

22-1 Electric Charge

Suppose we rub a plastic rod with fur and suspend it from a string so that it is free to rotate. Now we bring a second similarly rubbed plastic rod near it. The rods repel each other (Figure 22-1). We get the same results if we use two glass rods that have been rubbed with silk. But a plastic rod rubbed with fur and a glass rod rubbed with silk attract each other.

Rubbing a rod causes it to become electrically charged. If we repeat the experiment with various materials, we find that all charged objects fall into one of just two groups—those like the plastic rod rubbed with fur and those like the glass rod rubbed with silk. Benjamin Franklin explained this by a model in which every object has a "normal" amount of electricity that can be transferred from one object to the other when two objects are in close contact, as when they are rubbed together. This leaves one with an excess charge and the other with a deficiency of charge in the same amount as the excess. Franklin described the resulting charges with plus and minus signs, choosing positive to be the charge acquired by a glass rod when it is rubbed with a piece of silk. The piece of silk acquires a negative charge of equal magnitude during the procedure. Based on Franklin's convention, plastic rubbed with fur acquires a negative charge and the fur acquires a positive charge. Two objects that carry the same type of charge repel each other, and two objects that carry opposite charges attract each other (Figure 22-2).

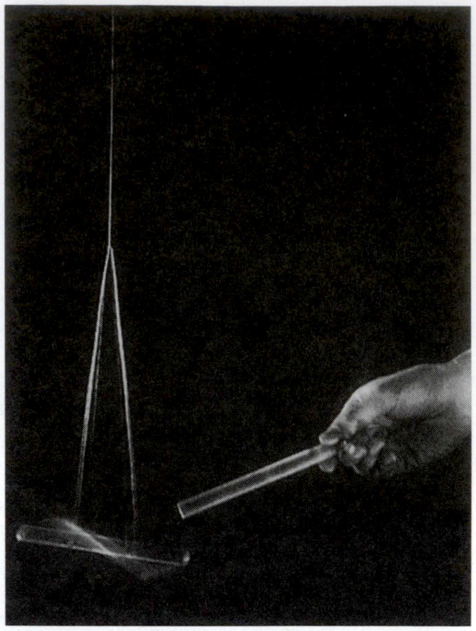

Figure 22-1 Two plastic rods that have been rubbed with fur repel each other.

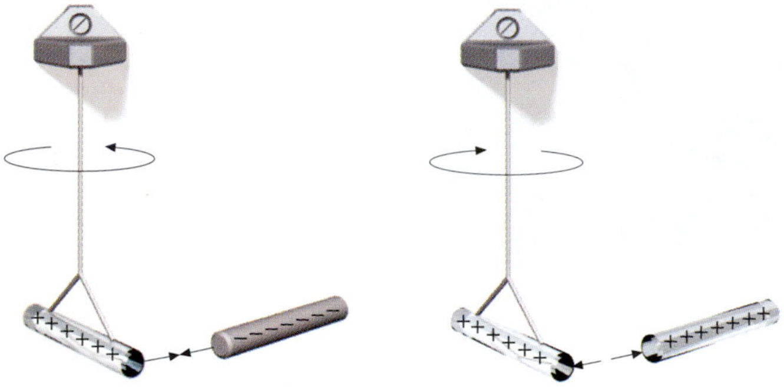

Figure 22-2 (*a*) Objects carrying charges of opposite sign attract each other. (*b*) Objects carrying charges of the same sign repel each other.

Today, we know that when glass is rubbed with silk, electrons are transferred from the glass to the silk. Since the silk is negatively charged (according to Franklin's classification, which we still use), electrons are said to carry a negative charge.

Charge Quantization

Matter consists of atoms that are electrically neutral. Each atom has a tiny but massive nucleus that contains protons and neutrons. Protons are positively charged, whereas neutrons are uncharged. The number of protons in the nucleus is the atomic number Z of the element. Surrounding the nucleus is an equal number of negatively charged electrons, leaving the atom with zero net charge. The electron is about 2000 times less massive than the proton, yet the charges of these two particles are exactly equal in magnitude. The charge of the proton is e and that of the electron is $-e$, where e is called the **funda-**

mental unit of charge. The charge of an electron or proton is an intrinsic property of the particle, just as mass and spin are intrinsic properties of these particles.

All observable charges occur in integral amounts of the fundamental unit of charge e. That is, *charge is quantized.* Any charge Q occurring in nature can be written $Q = \pm Ne$, where N is an integer.* For large systems, however, N is usually very large and charge appears to be continuous, just as air appears to be continuous even though air consists of many discrete molecules. To give an everyday example of N, charging a plastic rod by rubbing it with a piece of fur typically transfers 10^{10} or more electrons to the rod.

Charge Conservation

When objects are rubbed together, one object is left with an excess number of electrons and is therefore negatively charged; the other object is left lacking electrons and is therefore positively charged. The net charge of the two objects remains constant; that is, *charge is conserved.* The **law of conservation of charge** is a fundamental law of nature. In certain interactions among elementary particles, charged particles such as electrons are created or annihilated. However, in all these processes, equal amounts of positive and negative charge are produced or destroyed, so the net charge of the universe is unchanged.

The SI unit of charge is the coulomb, which is defined in terms of the unit of electric current, the ampere.† The **coulomb** (C) is the amount of charge flowing through a wire in one second when the current in the wire is one ampere. The fundamental unit of electric charge e is related to the coulomb by

$$e = 1.602\,177 \times 10^{-19}\,\text{C} \tag{22-1}$$

Fundamental unit of charge

Exercise A charge of magnitude 50 nC ($1\text{ nC} = 10^{-9}$ C) can be produced in the laboratory by simply rubbing two objects together. How many electrons must be transferred to produce this charge? (*Answer* $N = Q/e = (50 \times 10^{-9}\text{ C})/(1.6 \times 10^{-19}\text{ C}) = 3.12 \times 10^{11}$. Charge quantization cannot be detected in a charge of this size; even adding or subtracting a million electrons produces a negligibly small effect.)

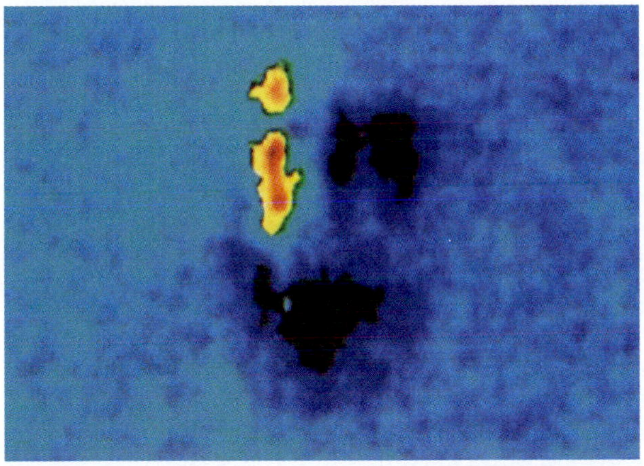

Charging by contact. A sample of plastic about 0.02 mm wide that was charged by contact with a piece of nickel. Although the plastic carries a net positive charge, regions of negative charge (dark) as well as regions of positive charge (yellow) are indicated. The photograph was taken by sweeping a charged needle of width 10^{-7} m over the sample and recording the electrostatic force on the needle.

* In the Standard Model of elementary particles, protons, neutrons, and some other elementary particles are made up of more fundamental particles called quarks that carry charges of $\pm\frac{1}{3}e$ or $\pm\frac{2}{3}e$. Only combinations that result in a net charge of $\pm Ne$ or 0 are known.

† The ampere (A) is the unit of current used in everyday electrical work. It will be defined later.

Example 22-1

A copper penny ($Z = 29$) has a mass of 3 g. What is the total charge of all the electrons in the penny?

Picture the Problem The electrons have a total charge given by the number of electrons in the penny, N_e, times the charge of an electron, $-e$. The number of electrons is 29 times the number of copper atoms N. To find N, we use the fact that one mole of any substance has Avogadro's number ($N_A = 6.02 \times 10^{23}$) of molecules, and the number of grams in a mole is the molecular mass M, which is 63.5 for copper. Since each molecule of copper is just one copper atom, we find the number of atoms per gram by dividing N_A atoms/mole by M grams/mole.

1. The total charge is the number of electrons times the electronic charge:

$$Q = N_e(-e)$$

2. The number of electrons is Z times the number of copper atoms N_a:

$$N_e = ZN_a$$

3. Compute the number of copper atoms in 3 g of copper:

$$N_a = (3\text{ g})\frac{6.02 \times 10^{23}\text{ atoms/mol}}{63.5\text{ g/mol}} = 2.84 \times 10^{22}\text{ atoms}$$

4. Compute the number of electrons N_e:

$$N_e = ZN_a = (29\text{ electrons/atom})(2.84 \times 10^{22}\text{ atoms})$$
$$= 8.24 \times 10^{23}\text{ electrons}$$

5. Use this value of N_e to find the total charge:

$$Q = N_e(-e)$$
$$= (8.24 \times 10^{23}\text{ electrons})(-1.6 \times 10^{-19}\text{ C/electron})$$
$$= -1.32 \times 10^{5}\text{ C}$$

Exercise If one million electrons are given to each man, woman, and child in the United States (about 250,000,000 people), what percentage of the number of electrons in a penny would this represent? (*Answer* About thirty billionths of a percent.)

22-2 Conductors and Insulators

In many materials, such as copper and other metals, some of the electrons are free to move about the entire material. Such materials are called **conductors.** In other materials, such as wood or glass, all the electrons are bound to nearby atoms and none can move freely. These materials are called **insulators.**

In a single atom of copper, 29 electrons are bound to the nucleus by the electrostatic attraction between the negatively charged electrons and the positively charged nucleus. The outer electrons are more weakly bound than the inner electrons because of their greater distance from the nucleus and because of the repulsive force exerted by the inner electrons. When a large number of copper atoms are combined in a piece of metallic copper, the binding of the electrons of each individual atom is reduced by interactions with neighboring atoms. One or more of the outer electrons in each atom is no longer bound but is free to move throughout the whole piece of metal, much as a gas molecule is free to move about in a box. The number of free electrons depends on the particular metal, but it is typically about one per atom. An atom with an electron removed or added, resulting in a net charge on the atom, is called an **ion**. In metallic copper, the copper ions are arranged in a regular array called a *lattice*. Normally, a conductor is electrically neutral because there is a lattice ion carrying a positive charge $+e$ for each free electron carrying a negative charge $-e$. A conductor can be given a net charge by adding or removing free electrons.

The Electroscope

Figure 22-3 shows a device for detecting electric charge called an **electroscope**. Two gold leaves are attached to a conducting post that has a conducting ball on top. The leaves are otherwise insulated from the container. When uncharged, the leaves hang together vertically. When the ball is touched by a negatively charged plastic rod, some of the negative charge from the rod is transferred to the ball and moves to the gold leaves, which then spread apart because of electrical repulsion between their negative charges. Touching the ball with a positively charged glass rod also causes the leaves to spread apart. In this case, the positively charged glass rod attracts electrons from the metal ball, leaving a net positive charge on the leaves.

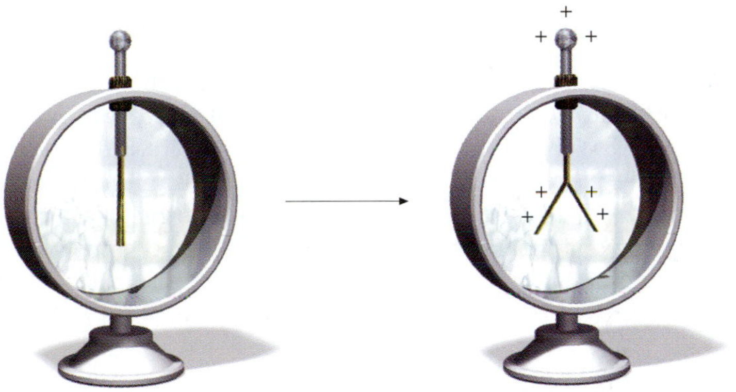

Figure 22-3 An electroscope. The two gold leaves are attached to a metal rod that has a metal ball on top. When a charge is placed on the metal ball, the charge is conducted to the leaves and they repel each other.

Charging by Induction

The conservation of charge is illustrated by a simple method of charging a conductor called **charging by induction,** as shown in Figure 22-4. Two uncharged metal spheres are in contact. When a charged rod is brought near one of the spheres, free electrons flow from one sphere to the other, toward a positively charged rod or away from a negatively charged rod. The positively charged rod in Figure 22-4*a* attracts the negatively charged electrons, and the sphere nearest the rod acquires electrons from the sphere farther away. This leaves the near sphere with a net negative charge and the far sphere with an equal net positive charge. A conductor that has separated equal and opposite charges is said to be **polarized**. If the spheres are separated before the rod is removed, they will be left with equal amounts of opposite charges (Figure 22-4*b*). A similar result would be obtained with a negatively charged rod, which would drive electrons from the near sphere to the far sphere.

Figure 22-4 Charging by induction. (*a*) Conductors in contact become oppositely charged when a charged rod attracts electrons to the left sphere. (*b*) If the spheres are separated before the rod is removed, they will retain their equal and opposite charges. (*c*) When the rod is removed and the spheres are far apart, the spheres are uniformly charged with equal and opposite charges.

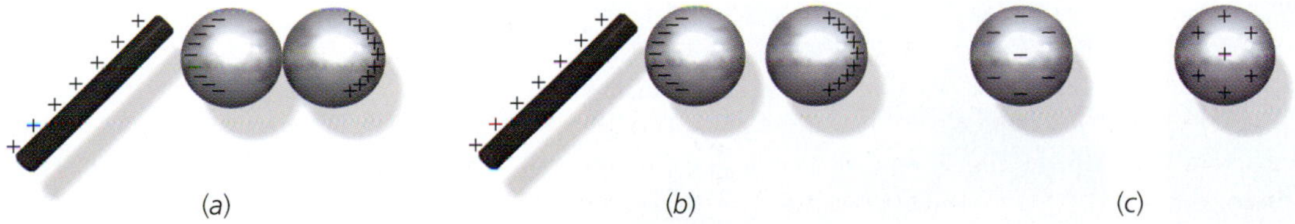

Exercise Two identical conducting spheres, one with an initial charge Q, the other initially uncharged, are brought into contact. (*a*) What is the new charge on each sphere? (*b*) While the spheres are in contact, a negatively charged rod is moved close to one sphere, causing it to have a charge of $2Q$. What is the charge on the other sphere? (*Answers* (*a*) $\frac{1}{2}Q$. Since the spheres are identical, they must share the total charge equally. (*b*) $-Q$, which is necessary to satisfy the conservation of charge)

Exercise Two identical spheres are charged by induction and then separated; sphere 1 has charge Q and sphere 2 has charge $-Q$. A third identical sphere is initially uncharged. If sphere 3 is touched to sphere 1 and separated, then touched to sphere 2 and separated, what is the final charge on each of the three spheres? (*Answer* $Q_1 = Q/2, Q_2 = -Q/4, Q_3 = -Q/4$)

For many purposes, the earth itself can be considered to be an infinitely large conductor with an abundant supply of free charge. When a conductor is connected to the earth, it is said to be **grounded** (indicated schematically in Figure 22-5*b* by a connecting wire ending in parallel horizontal lines). Figure 22-5 demonstrates how we can induce a charge in a single conductor by transferring charge from the earth through the ground wire and then breaking the connection to the ground.

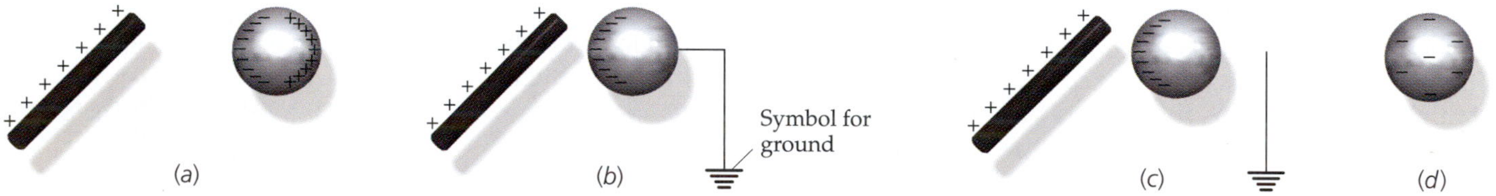

Figure 22-5 Induction via grounding. (*a*) The free charge on the single conducting sphere is polarized by the positively charged rod, which attracts negative charges on the sphere. (*b*) When the conductor is grounded by connecting it with a wire to a very large conductor, such as the earth, electrons from the ground neutralize the positive charge on the far face. The conductor is then negatively charged. (*c*) The negative charge remains if the connection to the ground is broken before the rod is removed. (*d*) When the rod is removed, the sphere has a uniform negative charge.

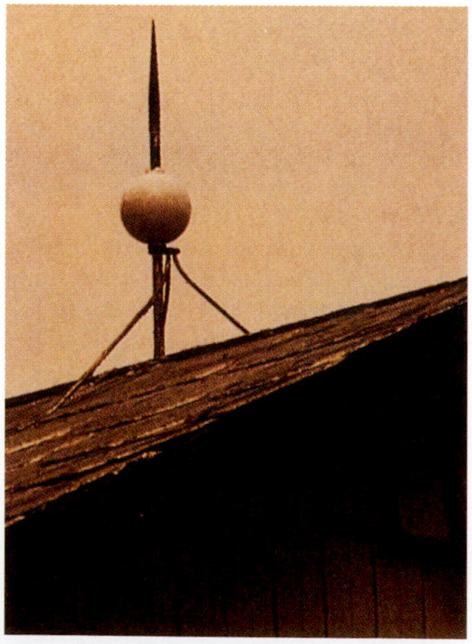

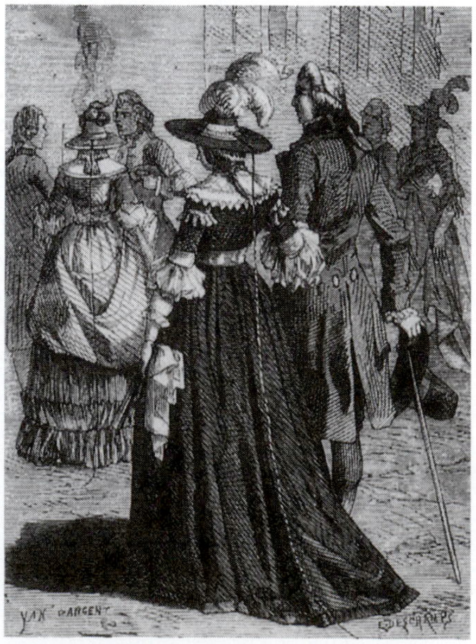

(*Left*) The lightning rod on this building is grounded so that it can conduct electrons from the ground to the positively charged clouds, thus neutralizing them. (*Right*) These fashionable ladies are wearing hats with metal chains that drag along the ground, which were supposed to protect them from lightning.

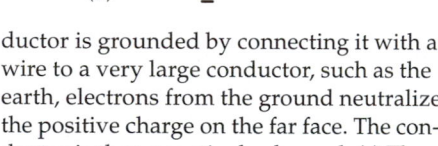

22-3 Coulomb's Law

The force exerted by one charge on another was studied by Charles Coulomb (1736–1806) using a torsion balance of his own invention.* In Coulomb's experiment, the charged spheres were much smaller than the distance between them so that the charges could be treated as point charges. Coulomb used the method of charging by induction to produce equally charged spheres and to vary the amount of charge on the spheres. For example, beginning with charge q_0 on each sphere, he could reduce the charge to $\frac{1}{2}q_0$ by temporarily grounding one sphere to discharge it and then placing the two spheres in contact. The results of the experiments of Coulomb and others are summarized in **Coulomb's law**:

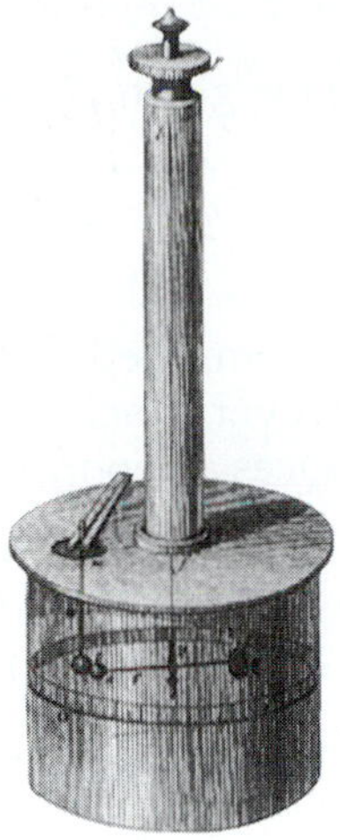

Coulomb's torsion balance.

> The force exerted by one point charge on another acts along the line between the charges. It varies inversely as the square of the distance separating the charges and is proportional to the product of the charges. The force is repulsive if the charges have the same sign and attractive if the charges have opposite signs.

Coulomb's law

The *magnitude* of the electric force exerted by a charge q_1 on another charge q_2 a distance r away is thus given by

$$F = \frac{k|q_1q_2|}{r^2} \qquad 22\text{-}2$$

where k is an experimentally determined constant called the **Coulomb constant,** which has the value

$$k = 8.99 \times 10^9\ \mathrm{N\cdot m^2/C^2} \qquad 22\text{-}3$$

If q_1 is at position $\vec{r}_1$ and q_2 is at $\vec{r}_2$ (Figure 22-6), the force $\vec{F}_{1,2}$ exerted by q_1 on q_2 is

$$\vec{F}_{1,2} = \frac{kq_1q_2}{r_{1,2}^2}\hat{r}_{1,2} \qquad 22\text{-}4$$

Coulomb's law for the force exerted by q_1 on q_2

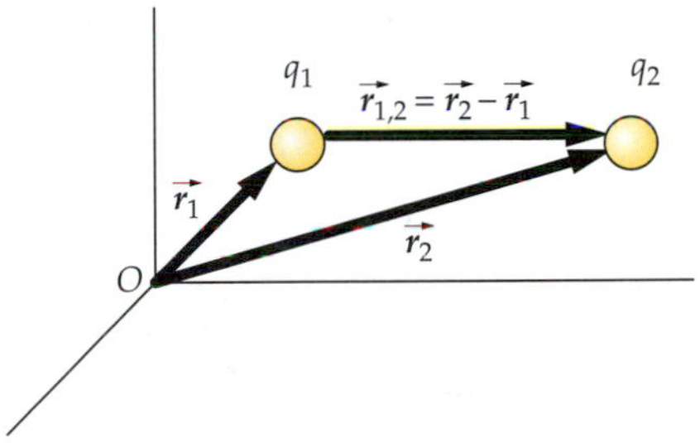

Figure 22-6 Charge q_1 at position $\vec{r}_1$ and charge q_2 at $\vec{r}_2$ relative to the origin O. The force exerted by q_1 on q_2 is in the direction of the vector $\vec{r}_{1,2} = \vec{r}_2 - \vec{r}_1$ if both charges have the same sign, and in the opposite direction if they have opposite signs.

where $\vec{r}_{1,2} = \vec{r}_2 - \vec{r}_1$ is the vector pointing from q_1 to q_2, and $\hat{r}_{1,2} = \vec{r}_{1,2}/r_{1,2}$ is a unit vector pointing from q_1 to q_2.

By Newton's third law, the force $\vec{F}_{2,1}$ exerted by q_2 on q_1 is the negative of $\vec{F}_{1,2}$. Note the similarity between Coulomb's law and Newton's law of gravity (Equation 11-3). Both are inverse-square laws. But the gravitational force between two particles is proportional to the masses of the particles and is always attractive, whereas the electric force is proportional to the charges of the particles and is repulsive if both charges have the same sign and attractive if they have opposite signs.

* Coulomb's experimental apparatus was essentially the same as that described for the Cavendish experiment in Chapter 11, with the masses replaced by small charged spheres. For the magnitudes of charges easily transferred by rubbing, the gravitational attraction of the spheres is completely negligible compared with their electric attraction or repulsion.

Example 22-2

In a hydrogen atom, the electron is separated from the proton by an average distance of about 5.3×10^{-11} m. Calculate the magnitude of the electrostatic force of attraction exerted by the proton on the electron.

Substitute the given values into Coulomb's law:

$$F = \frac{k|q_1q_2|}{r^2} = \frac{ke^2}{r^2} = \frac{(8.99 \times 10^9 \text{ N}\cdot\text{m}^2/\text{C}^2)(1.6 \times 10^{-19} \text{ C})^2}{(5.3 \times 10^{-11} \text{ m})^2}$$

$$= 8.19 \times 10^{-8} \text{ N}$$

Remarks Compared with macroscopic interactions, this is a very small force. However, since the mass of the electron is only about 10^{-30} kg, this force produces an enormous acceleration of $F/m = 8 \times 10^{22}$ m/s^2.

Exercise Two point charges of 0.05 μC each are separated by 10 cm. Find the magnitude of the force exerted by one point charge on the other. (*Answer* 2.25×10^{-3} N)

Since the electrical force and the gravitational force between any two particles both vary inversely with the square of the separation between the particles, the ratio of these forces is independent of separation. We can therefore compare the relative strengths of the electrical and gravitational forces for elementary particles such as the electron and proton.

Example 22-3

Compute the ratio of the electric force to the gravitational force exerted by a proton on an electron in a hydrogen atom.

Picture the Problem We use Coulomb's law with $q_1 = e$ and $q_2 = -e$ to find the electric force, and Newton's law of gravity with the mass of the proton, $m_p = 1.67 \times 10^{-27}$ kg, and the mass of the electron, $m_e = 9.11 \times 10^{-31}$ kg.

1. Express the magnitudes of the electric force F_e and the gravitational force F_g in terms of the charges, masses, separation distance r, and electrical and gravitational constants:

$$F_e = \frac{ke^2}{r^2}, \quad F_g = \frac{Gm_pm_e}{r^2}$$

2. Take the ratio. Note that the separation distance r cancels:

$$\frac{F_e}{F_g} = \frac{ke^2}{Gm_pm_e}$$

3. Substitute numerical values:

$$\frac{F_e}{F_g} = \frac{(8.99 \times 10^9 \text{ N}\cdot\text{m}^2/\text{C}^2)(1.6 \times 10^{-19} \text{ C})^2}{(6.67 \times 10^{-11} \text{ N}\cdot\text{m}^2/\text{kg}^2)(1.67 \times 10^{-27} \text{ kg})(9.11 \times 10^{-31} \text{ kg})}$$

$$= 2.27 \times 10^{39}$$

Remark This result shows why the effects of gravity are not considered when discussing atomic or molecular interactions.

Though the gravitational force is incredibly weak compared with the electric force and plays essentially no role on the atomic level, gravity is the dominant force between large objects such as planets and stars, because large objects contain almost equal numbers of positive and negative charges, and hence the attractive and repulsive electrical forces cancel. The net force between astronomical objects is therefore essentially the force of gravitational attraction alone.

Force Exerted by a System of Charges

In a system of charges, each charge exerts a force given by Equation 22-4 on every other charge. The net force on any charge is the vector sum of the individual forces exerted on that charge by all the other charges in the system. This follows from the principle of superposition of forces.

Example 22-4 *try it yourself*

Three point charges lie on the x axis; $q_1 = 25$ nC is at the origin, $q_2 = -10$ nC is at $x = 2$ m, and $q_0 = 20$ nC is at $x = 3.5$ m (Figure 22-7). Find the net force on q_0 due to q_1 and q_2.

Picture the Problem The net force on q_0 is the vector sum of the force $\vec{F}_{1,0}$ exerted by q_1, and the force $\vec{F}_{2,0}$ exerted by q_2. The individual forces are found using Coulomb's law. Note that $\hat{r}_{1,0} = \hat{r}_{2,0} = \hat{i}$ because q_0 is to the right of both q_1 and q_2.

Figure 22-7

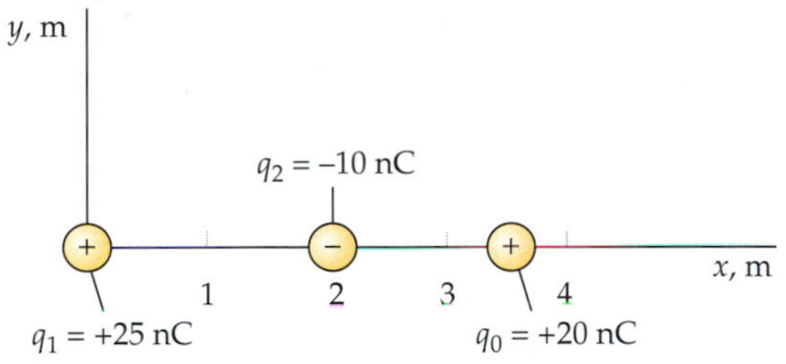

Cover the column to the right and try these on your own before looking at the answers.

Steps	***Answers***
1. Find the force $\vec{F}_{1,0}$ due to q_1.	$\vec{F}_{1,0} = (0.367\ \mu\text{N})\hat{i}$
2. Find the force $\vec{F}_{2,0}$ due to q_2.	$\vec{F}_{2,0} = (-0.799\ \mu\text{N})\hat{i}$
3. Combine your results to obtain the net force.	$\vec{F}_{\text{net}} = \vec{F}_{1,0} + \vec{F}_{2,0} = -(0.432\ \mu\text{N})\hat{i}$

Remark Figure 22-8 shows the force F_x on q_0 as a function of its position x. Near either of the other charges, the force on q_0 is essentially due to the nearest charge alone. Note that the force is undefined at the position of the charges q_1 at $x = 0$, and q_2 at $x = 2$.

Exercise If q_0 is at $x = 1$ m, find (*a*) $\hat{r}_{1,0}$, (*b*) $\hat{r}_{2,0}$, and (*c*) the net force acting on q_0. (*Answers* (*a*) $\hat{i}$, (*b*) $-\hat{i}$, (*c*) $(6.29\ \mu\text{N})\hat{i}$)

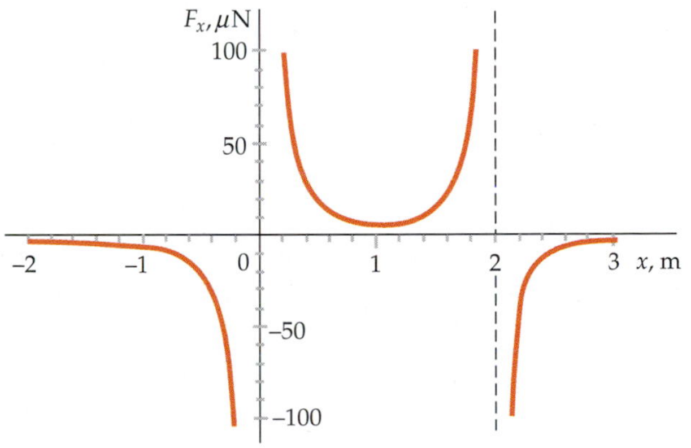

Figure 22-8

If a system of charges is to remain stationary, there must be other forces acting on the charges so that the net force from all sources acting on each charge is zero. In the preceding example and those that follow, we assume that there are such forces so that all the charges remain stationary.

Example 22-5

Charge $q_1 = +25$ nC is at the origin, charge $q_2 = -15$ nC is on the x axis at $x = 2$ m, and charge $q_0 = +20$ nC is at the point $x = 2$ m, $y = 2$ m as shown in Figure 22-9. Find the resultant force ΣF on q_0.

Picture the Problem The resultant force is the vector sum of the individual forces exerted by each charge on q_0. We compute each force from Coulomb's law and write it in terms of its rectangular components. Figure 22-9*a* shows the resultant force on charge q_0 as the vector sum of the forces $\vec{F}_{1,0}$ due to q_1 and $\vec{F}_{2,0}$ due to q_2. Figure 22-9*b* shows the net force in Figure 22-9*a* and its x and y components.

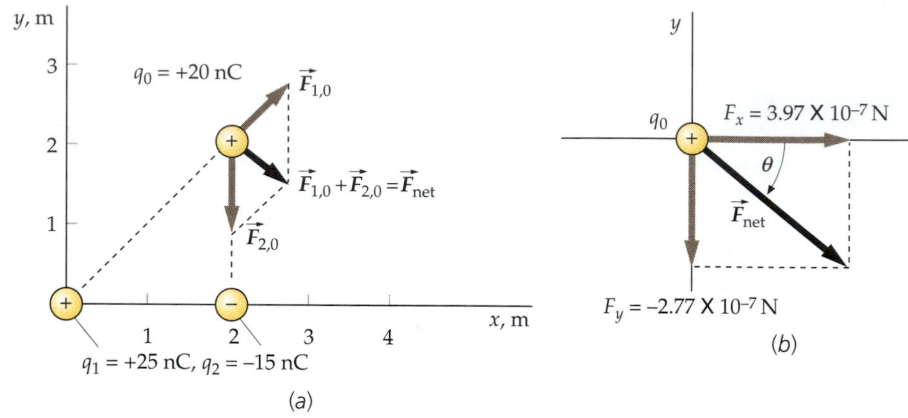

Figure 22-9

1. The resultant force $\Sigma\vec{F}$ on q_0 is the sum of the individual forces:

$$\sum \vec{F} = \vec{F}_{1,0} + \vec{F}_{2,0}$$
$$\sum F_x = F_{1,0x} + F_{2,0x}$$
$$\sum F_y = F_{1,0y} + F_{2,0y}$$

2. The force $\vec{F}_{1,0}$ is directed along the line from q_1 to q_0. Use $r_{1,0} = 2\sqrt{2}$ m for the distance between q_1 and q_0 to calculate its magnitude:

$$F_{1,0} = \frac{(8.99 \times 10^9\ \text{N}\cdot\text{m}^2/\text{C}^2)(25 \times 10^{-9}\ \text{C})(20 \times 10^{-9}\ \text{C})}{(2\sqrt{2}\ \text{m})^2}$$
$$= 5.62 \times 10^{-7}\ \text{N}$$

3. Since $\vec{F}_{1,0}$ makes an angle of 45° with the x and y axes, its x and y components are equal to each other:

$$F_{1,0x} = F_{1,0y} = \frac{F_{1,0}}{\sqrt{2}} = \frac{5.62 \times 10^{-7}\ \text{N}}{\sqrt{2}} = 3.97 \times 10^{-7}$$

4. The force $\vec{F}_{2,0}$ exerted by q_2 on q_0 is attractive and in the negative y direction as shown in Figure 22-9*a*:

$$\vec{F}_{2,0} = \frac{kq_2q_0}{r_{2,0}^2}\hat{r}_{2,0}$$
$$= \frac{(8.99 \times 10^9\ \text{N}\cdot\text{m}^2/\text{C}^2)(-15 \times 10^{-9}\ \text{C})(20 \times 10^{-9}\ \text{C})}{(2\ \text{m})^2}\hat{j}$$
$$= (-6.74 \times 10^{-7}\ \text{N})\hat{j}$$

5. Calculate the components of the resultant force:

$$\sum F_x = F_{1,0x} + F_{2,0x} = (3.97 \times 10^{-7}\ \text{N}) + 0 = 3.97 \times 10^{-7}\ \text{N}$$
$$\sum F_y = F_{1,0y} + F_{2,0y} = (3.97 \times 10^{-7}\ \text{N}) + (-6.74 \times 10^{-7}\ \text{N})$$
$$= -2.77 \times 10^{-7}\ \text{N}$$

6. The magnitude of the resultant force is found from its components:

$$F = \sqrt{F_x^2 + F_y^2} = \sqrt{(3.97 \times 10^{-7}\ \text{N})^2 + (-2.77 \times 10^{-7}\ \text{N})^2}$$
$$= 4.84 \times 10^{-7}\ \text{N}$$

7. The resultant force points to the right and downward as shown in Figure 22-9*b*, making an angle θ with the x axis given by:

$$\tan\theta = \frac{F_y}{F_x} = \frac{-2.77}{3.97} = -0.698$$
$$\theta = -34.9°$$

22-4 The Electric Field

The electric force exerted by one charge on another is an example of an action-at-a-distance force, similar to the gravitational force exerted by one mass on another. The idea of action at a distance presents a difficult conceptual problem. What is the mechanism by which one particle can exert a force on another across the empty space between the particles? Suppose that a charged particle at some point is suddenly moved. Does the force exerted on the second particle some distance r away change instantaneously? To avoid the problem of action at a distance, the concept of the **electric field** is introduced. One charge produces an electric field $\vec{E}$ everywhere in space, and this field exerts the force on the other charge. The force is thus exerted *by the field* at the position of the second charge, rather than by the first charge itself, which is some distance away. The field propagates through space with the speed of light, c. Thus, if a charge is suddenly moved, the force it exerts on another charge a distance r away does not change until a time r/c later.

Figure 22-10 shows a set of point charges, q_1, q_2, and q_3, arbitrarily arranged in space. These charges produce an electric field $\vec{E}$ everywhere in space. If we place a small **test charge** q_0 at some point nearby, there will be a force exerted on q_0 due to the other charges.* The net force on q_0 is the vector sum of the individual forces exerted on q_0 by each of the other charges in the system. Since each of these forces is proportional to q_0, the net force will be proportional to q_0. The electric field $\vec{E}$ at a point is this force divided by q_0†:

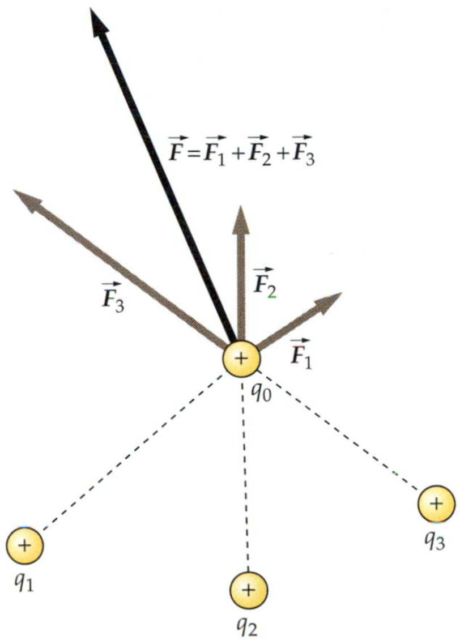

Figure 22-10 A small test charge q_0 in the vicinity of a system of charges q_1, q_2, q_3, . . . experiences a force $\vec{F}$ that is proportional to q_0. The ratio $\vec{F}/q_0$ is the electric field at that point.

$$\vec{E} = \frac{\vec{F}}{q_0} \quad (q_0 \text{ small}) \qquad \text{22-5}$$

Definition—Electric field

The SI unit of the electric field is the newton per coulomb (N/C). Table 22-1 lists the magnitudes of some of the electric fields found in nature.

The electric field describes the condition in space set up by the system of point charges. By moving a test charge q_0 from point to point, we can find $\vec{E}$ at all points in space (except at any point occupied by a charge q). The electric field $\vec{E}$ is thus a vector function of position. The force exerted on a test charge q_0 at any point is related to the electric field at that point by

$$\vec{F} = q_0\vec{E} \qquad \text{22-6}$$

Exercise When a 5-nC test charge is placed at a certain point, it experiences a force of 2×10^{-4} N in the x direction. What is the electric field $\vec{E}$ at that point? (*Answer* $\vec{E} = \vec{F}/q_0 = [(2 \times 10^{-4}\,\text{N})\hat{i}]/(5 \times 10^{-9}\,\text{C}) = (4 \times 10^4\,\text{N/C})\hat{i}$)

Exercise What is the force on an electron placed at a point where the electric field is $\vec{E} = (4 \times 10^4\,\text{N/C})\hat{i}$? (*Answer* $(-6.4 \times 10^{-15}\,\text{N})\hat{i}$)

Table 22-1

Some Electric Fields in Nature

	E, N/C
In household wires	10^{-2}
In radio waves	10^{-1}
In the atmosphere	10^2
In sunlight	10^3
Under a thundercloud	10^4
In a lightning bolt	10^4
In an X-ray tube	10^6
At the electron in a hydrogen atom	6×10^{11}
At the surface of a uranium nucleus	2×10^{21}

* The presence of the charge q_0 will generally change the original distribution of the other charges, particularly if the charges are on conductors. However, we may choose q_0 to be small enough so that its effect on the original charge distribution is negligible.

† This definition is similar to that for the gravitational field of the earth, which was defined in Section 4-3 as the force per unit mass exerted by the earth on an object.

The electric field due to a single point charge can be calculated from Coulomb's law. Consider a small, positive test charge q_0 at some point P a distance r_{i0} away from a charge q_i. The force on it is

$$\vec{F}_{i,0} = \frac{kq_i q_0}{r_{i,0}^2}\hat{r}_{i,0}$$

The electric field at point P due to charge q_i is thus

$$\vec{E}_i = \frac{kq_i}{r_{i,0}^2}\hat{r}_{i,0} \qquad \text{22-7}$$

Coulomb's law for $\vec{E}$ due to a point charge

where $\hat{r}_{i0}$ is a unit vector pointing from the charge to the **field point** P. The net electric field due to a distribution of point charges is found by summing the fields due to each charge separately:

$$\vec{E} = \sum_i \vec{E}_i = \sum_i \frac{kq_i}{r_{i,0}^2}\hat{r}_{i,0} \qquad \text{22-8}$$

Electric field due to a system of point charges

Example 22-6

A positive charge $q_1 = +8$ nC is at the origin, and a second positive charge $q_2 = +12$ nC is on the x axis at $a = 4$ m (Figure 22-11). Find the net electric field (a) at point P_1 on the x axis at $x = 7$ m, and (b) at point P_2 on the x axis at $x = 3$ m.

Picture the Problem Because point P_1 is to the right of both charges, each charge produces a field to the right at that point. At point P_2, which is between the charges, the 5-nC charge gives a field to the right and the 12-nC charge gives a field to the left. We calculate each field using

$$\vec{E} = \sum_i \frac{kq_i}{r_{i,0}^2}\hat{r}_{i,0}$$

At point P_1, both unit vectors point along the x axis in the positive direction, so $\hat{r}_{1,0} = \hat{r}_{2,0} = \hat{i}$. At point P_2, $\hat{r}_{1,0} = \hat{i}$, but the unit vector from the 12-nC charge points along the negative x direction, so $\hat{r}_{2,0} = -\hat{i}$.

Figure 22-11

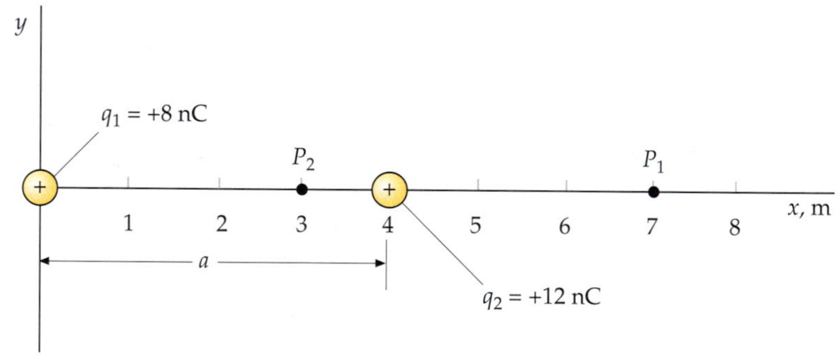

(a) Calculate $\vec{E}$ at point P_1, using $r_{1,0} = x = 7$ m and $r_{2,0} = (x - a) = 7\text{ m} - 4\text{ m} = 3$ m:

$$\vec{E} = \frac{kq_1}{r_{1,0}^2}\hat{r}_{1,0} + \frac{kq_2}{r_{2,0}^2}\hat{r}_{2,0} = \frac{kq_1}{x^2}\hat{i} + \frac{kq_2}{(x-a)^2}\hat{i}$$

$$= \frac{(8.99 \times 10^9\text{ N}\cdot\text{m}^2/\text{C}^2)(8 \times 10^{-9}\text{ C})}{(7\text{ m})^2}\hat{i}$$

$$+ \frac{(8.99 \times 10^9\text{ N}\cdot\text{m}^2/\text{C}^2)(12 \times 10^{-9}\text{ C})}{(3\text{ m})^2}\hat{i}$$

$$= (1.47\text{ N/C})\hat{i} + (12.0\text{ N/C})\hat{i} = (13.5\text{ N/C})\hat{i}$$

(*b*) Calculate $\vec{E}$ at point P_2, where $r_{1,0} = x = 3$ m and $r_{2,0} = a - x = 4\text{ m} - 3\text{ m} = 1$ m:

$$\vec{E} = \frac{kq_1}{r_{1,0}^2}\hat{r}_{1,0} + \frac{kq_2}{r_{2,0}^2}\hat{r}_{2,0} = \frac{kq_1}{x^2}\hat{i} + \frac{kq_2}{(a-x)^2}(-\hat{i})$$

$$= \frac{(8.99 \times 10^9\ \text{N}\cdot\text{m}^2/\text{C}^2)(8 \times 10^{-9}\ \text{C})}{(3\ \text{m})^2}\hat{i}$$

$$+ \frac{(8.99 \times 10^9\ \text{N}\cdot\text{m}^2/\text{C}^2)(12 \times 10^{-9}\ \text{C})}{(1\ \text{m})^2}(-\hat{i})$$

$$= (7.99\ \text{N/C})\hat{i} - (108\ \text{N/C})\hat{i} = (-100\ \text{N/C})\hat{i}$$

Remarks The electric field at point P_2 is in the negative x direction because the field due to the +12-nC charge, which is 1 m away, is larger than that due to the +8-nC charge, which is 3 m away. As we move toward the +8-nC charge, the magnitude of its field increases and that due to the +12-nC charge decreases. There is one point between the charges where the net electric field is zero. At this point, a test charge experiences no net force. A sketch of E_x versus x for this system is shown in Figure 22-12.

Exercise Find the point on the x axis where the electric field is zero. (*Answer* $x = 1.80$ m)

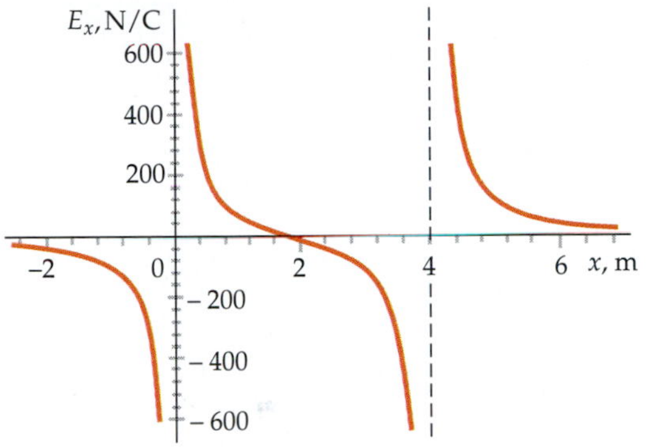

Figure 22-12

Example 22-7 *try it yourself*

Find the electric field on the y axis at $y = 3$ m for the charges in Example 22-6.

Picture the Problem On the y axis, the electric field $\vec{E}_1$ due to charge q_1 is directed along the y axis, and the field $\vec{E}_2$ due to charge q_2 makes an angle θ with the y axis (Figure 22-13*a*). To find the resultant field, we first find the x and y components of these fields, as shown in Figure 22-13*b*.

Figure 22-13

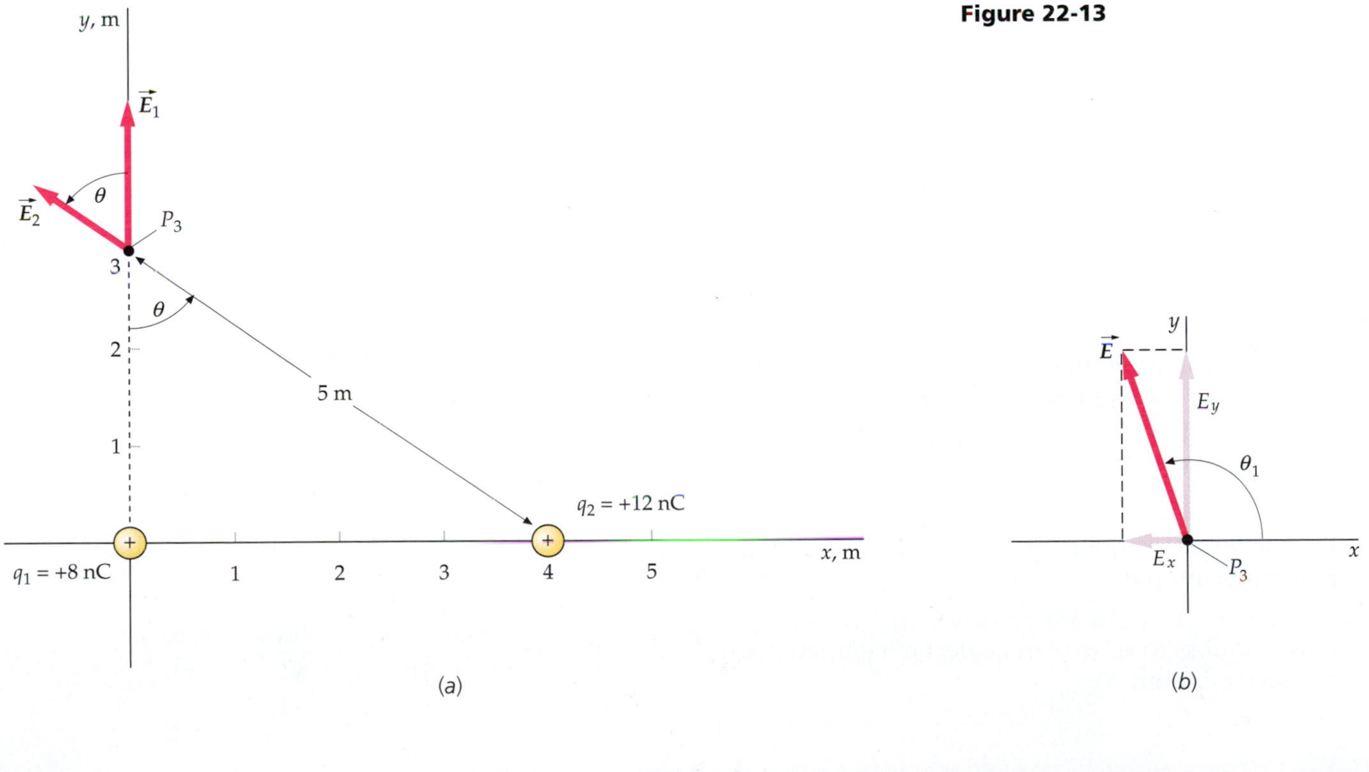

Cover the column to the right and try these on your own before looking at the answers.

Steps	Answers
1. Calculate the magnitude of the field $\vec{E}_1$ due to q_1. Find the x and y components of $\vec{E}_1$.	$E_1 = kq_1/y^2 = 7.99\ \text{N/C}$, $E_{1x} = 0$, $E_{1y} = 7.99\ \text{N/C}$
2. Calculate the magnitude of the field $\vec{E}_2$ due to q_2.	$E_2 = 4.32\ \text{N/C}$
3. Write the x and y components of $\vec{E}_2$ in terms of the angle θ.	$E_x = -E_2 \sin\theta$, $E_y = E_2 \cos\theta$
4. Compute $\sin\theta$ and $\cos\theta$.	$\sin\theta = 0.8$, $\cos\theta = 0.6$
5. Calculate E_{2x} and E_{2y}.	$E_{2x} = -3.46\ \text{N/C}$, $E_{2y} = 2.59\ \text{N/C}$
6. Find the x and y components of the resultant field $\vec{E}$.	$E_x = -3.46\ \text{N/C}$, $E_y = 10.6\ \text{N/C}$
7. Calculate the magnitude of $\vec{E}$ from its components.	$E = \sqrt{E_x^2 + E_y^2} = 11.2\ \text{N/C}$
8. Find the angle θ_1 made by $\vec{E}$ with the x axis.	$\theta_1 = \tan^{-1}\left(\frac{E_y}{E_x}\right) = 108°$

Example 22-8

A charge $+q$ is at $x = a$ and a second charge $-q$ is at $x = -a$ (Figure 22-14). (*a*) Find the electric field on the x axis at an arbitrary point $x > a$. (*b*) Find the limiting form of the electric field for $x >> a$.

Picture the Problem We calculate the electric field using

$$\vec{E} = \sum_i \frac{kq_i}{r_{i,0}^2}\hat{r}_{i,0}$$

(Equation 22-8). For $x > a$, the unit vector for each charge is $\hat{i}$. The distances are $x - a$ to the plus charge and $x - (-a) = x + a$ to the minus charge.

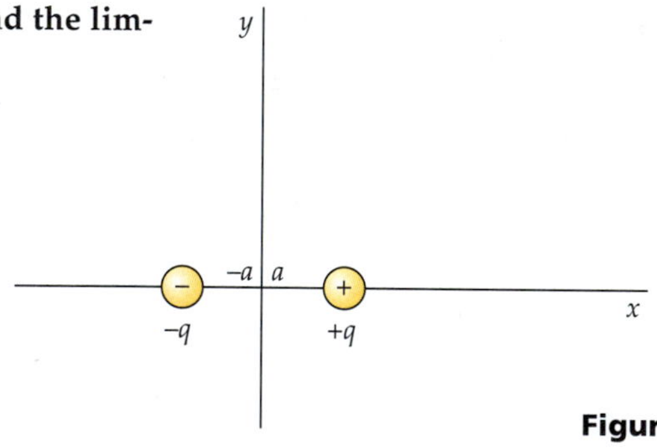

Figure 22-14

(*a*)1. Calculate $\vec{E}$ due to the two charges for $x > a$: [*Note:* The equation on the right holds only for $x > a$. For $x < a$, the signs of the two terms are reversed. For $-a < x < a$, both terms have negative signs.]

$$\vec{E} = \frac{kq}{(x-a)^2}\hat{i} + \frac{k(-q)}{(x+a)^2}\hat{i} = kq\left[\frac{1}{(x-a)^2} - \frac{1}{(x+a)^2}\right]\hat{i}$$

2. Put the terms in square brackets under a common denominator and simplify:

$$\vec{E} = kq\left[\frac{(x+a)^2 - (x-a)^2}{(x+a)^2(x-a)^2}\right]\hat{i} = kq\frac{4ax}{(x^2-a^2)^2}\hat{i}$$

(*b*) In the limit $x >> a$, we can neglect a^2 compared with x^2 in the denominator:

$$\vec{E} = kq\frac{4ax}{(x^2-a^2)^2}\hat{i} \approx kq\frac{4ax}{x^4}\hat{i} = \frac{4kqa}{x^3}\hat{i}$$

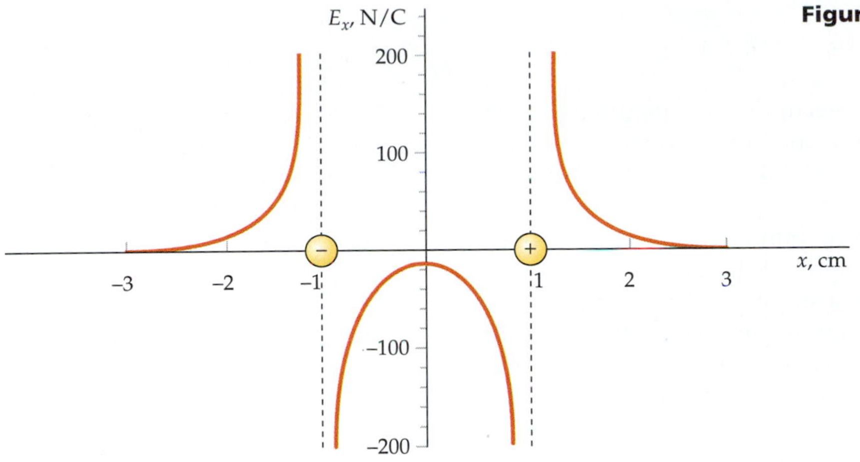

Figure 22-15

Remarks Figure 22-15 shows E_x versus x for all x, for $q = 1$ nC and $a = 1$ cm. Far from the charges, the field is given by

$$\vec{E} = \frac{4kqa}{|x^3|}\hat{i}$$

Between the charges, the contribution from each charge is in the negative direction. An expression that holds for all x is

$$\vec{E} = \frac{kq}{(x-a)^2}\left[\frac{(x-a)\hat{i}}{|x-a|}\right] + \frac{k(-q)}{(x+a)^2}\left[\frac{(x+a)\hat{i}}{|x+a|}\right]$$

Note that the unit vectors (quantities in square brackets) in this expression point in the proper direction for all x.

Electric Dipoles

A system of two equal and opposite charges q separated by a small distance L is called an **electric dipole.** Its strength and orientation are described by the **electric dipole moment** $\vec{p}$, which is a vector that points from the negative charge to the positive charge and has the magnitude qL (Figure 22-16).

$$\vec{p} = q\vec{L} \qquad \text{22-9}$$

Definition—Electric dipole moment

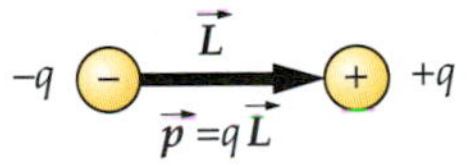

Figure 22-16 An electric dipole consists of two equal and opposite charges separated by some distance L. The dipole moment points from the negative charge to the positive charge and has the magnitude $p = qL$.

For the configuration in Figure 22-14, $\vec{L} = 2a\hat{i}$ and the electric dipole moment is

$$\vec{p} = 2aq\hat{i}$$

In terms of the dipole moment, the electric field on the axis of the dipole at a point a great distance x away is in the direction of the dipole moment and has the magnitude

$$E = \frac{2kp}{|x^3|} \qquad \text{22-10}$$

(See Example 22-8.) At any point far from a dipole, the electric field is proportional to the dipole moment and decreases with the cube of the distance. When a system has a net charge, the electric field decreases as $1/r^2$ at large distances. In a system with zero net charge, the electric field falls off more rapidly with distance. In the case of an electric dipole, the field falls off as $1/r^3$.

22-5 Electric Field Lines

We can picture the electric field by drawing lines to indicate its direction. At any given point, the field vector $\vec{E}$ is tangent to the lines. Electric field lines are also called **lines of force** because they show the direction of the force exerted on a positive test charge. At any point near a positive charge, the electric field points radially away from the charge. Similarly, the electric field lines converge toward a point occupied by a negative charge.

Figure 22-17 shows the electric field lines of a single positive point charge. The spacing of the lines is related to the strength of the electric field. As we move away from the charge, the field becomes weaker and the lines become farther apart. Consider a spherical surface of radius r with its center at the charge. Its area is $4\pi r^2$. If N lines diverge from the point charge, the number of lines per unit area on a spherical surface a distance r away is $N/4\pi r^2$. Thus, as the distance increases, the density of the field lines (the number of lines per unit area) decreases as $1/r^2$, the same rate of decrease as E. So, if we adopt the convention of drawing a fixed number of lines from a point charge, the number being proportional to the charge q, and if we draw the lines symmetrically about the point charge, the field strength is indicated by the density of the lines. The more closely spaced the lines, the stronger the electric field.

Figure 22-18 shows the electric field lines for two equal positive point charges q separated by a small distance. Near each charge, the field is approximately due to that charge alone because the other charge is far away. Consequently, the field lines near either charge are radial and equally spaced. Since the charges are equal, we draw an equal number of lines originating from each charge. At very large distances, the details of the system are not important and the system looks like a point charge of magnitude $2q$. (For example, if the two charges were 1 mm apart and we were looking at them from a point 100 km away, they would look like a single charge.) So far from the charges, the field is approximately the same as that due to a point charge $2q$ and the lines are approximately equally spaced. Looking at the figure, we see that the electric field in the space between the charges is weak because there are few lines in this region compared with the region just to the right or left of the charges, where the lines are more closely spaced. This information can, of course, also be obtained by direct calculation of the field at points in these regions.

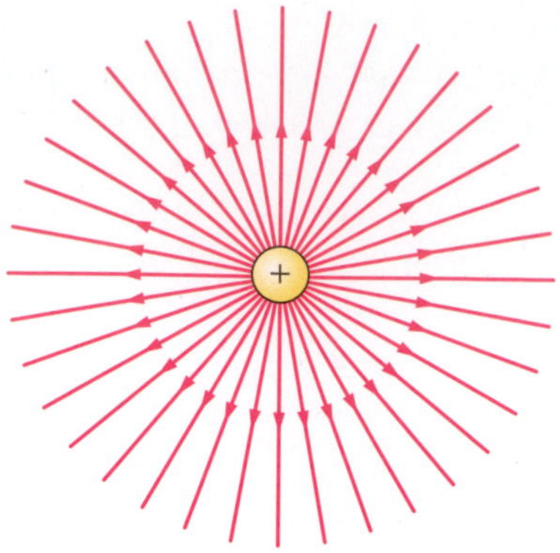

(a)

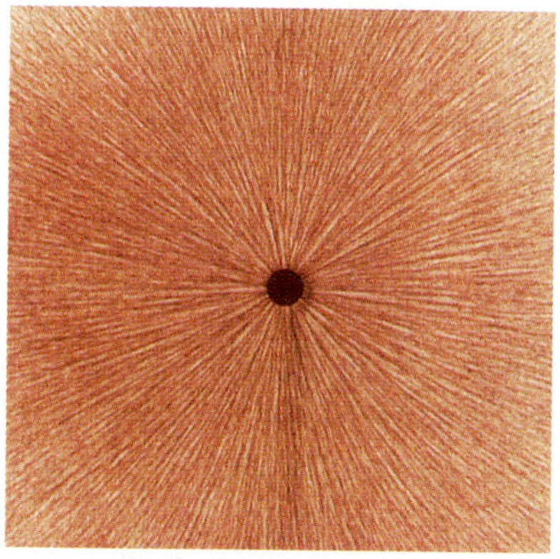

(b)

Figure 22-17 (a) Electric field lines of a single positive point charge. If the charge were negative, the arrows would be reversed. (b) The same electric field lines shown by bits of thread suspended in oil. The electric field of the charged object in the center induces opposite charges on the ends of each bit of thread, causing the threads to align themselves parallel to the field.

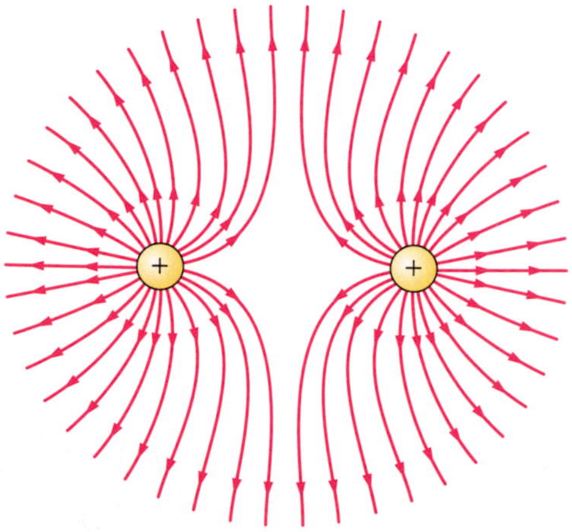

(a)

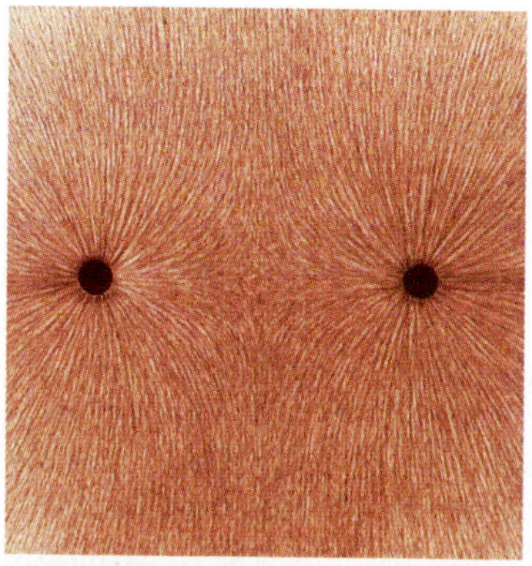

(b)

Figure 22-18 (a) Electric field lines due to two positive point charges. The arrows would be reversed if both charges were negative. (b) The same electric field lines shown by bits of thread in oil.

We can apply this reasoning to draw the electric field lines for any system of point charges. Very near each charge, the field lines are equally spaced and leave or enter the charge radially, depending on the sign of the charge. Very far from all the charges, the detailed structure of the system is not important so the field lines are just like those of a single point charge carrying the net charge of the system. For future reference, the rules for drawing electric field lines can be summarized as follows:

1. Electric field lines begin on positive charges (or at infinity) and end on negative charges (or at infinity).
2. The lines are drawn symmetrically entering or leaving an isolated charge.
3. The number of lines leaving a positive charge or entering a negative charge is proportional to the magnitude of the charge.
4. The density of the lines (the number of lines per unit area perpendicular to the lines) at any point is proportional to the magnitude of the field at that point.
5. At large distances from a system of charges, the field lines are equally spaced and radial, as if they came from a single point charge equal to the net charge of the system.
6. Field lines do not cross. (If two field lines crossed, that would indicate two directions for $\vec{E}$ at the point of intersection.)

Rules for drawing electric field lines

Figure 22-19 shows the electric field lines due to an electric dipole. Very near the positive charge, the lines are directed radially outward. Very near the negative charge, the lines are directed radially inward. Since the charges have equal magnitudes, the number of lines that begin at the positive charge equals the number that end at the negative charge. In this case, the field is strong in the region between the charges, as indicated by the high density of field lines in this region in the figure.

Figure 22-20 shows the electric field lines for a negative charge $-q$ at a small distance from a positive charge $+2q$. Twice as many lines leave the positive charge as enter the negative charge. Thus, half the lines beginning on the positive charge $+2q$ enter the negative charge $-q$; the rest leave the system. Very far from the charges, the lines leaving the system are approximately symmetrically spaced and point radially outward, just as they would for a single positive point charge $+q$.

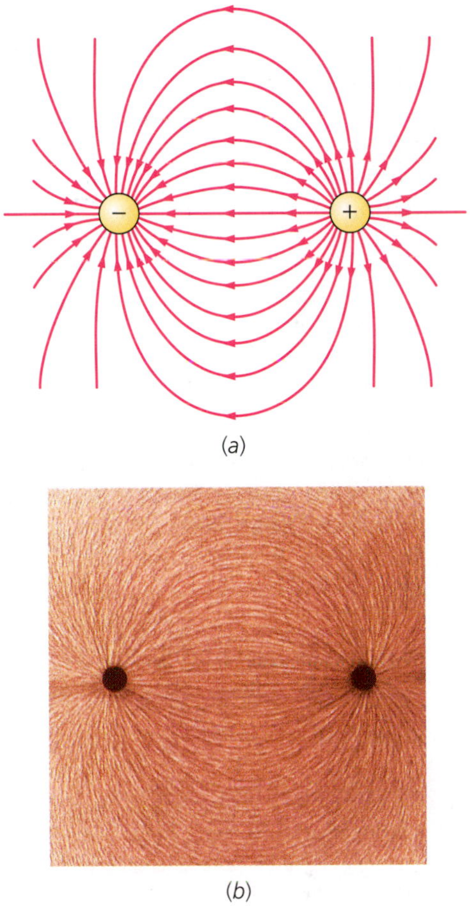

Figure 22-19 (*a*) Electric field lines for an electric dipole. (*b*) The same field lines shown by bits of thread in oil.

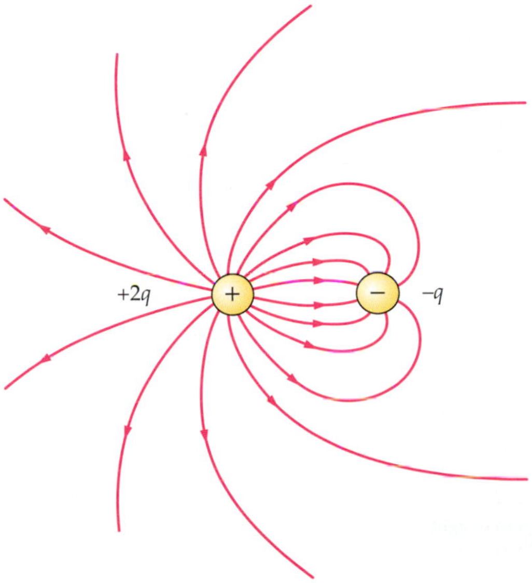

Figure 22-20 Electric field lines for a point charge $+2q$ and a second point charge $-q$. At great distances from the charges, the lines are the same as those for a single charge $+q$.

Example 22-9

The electric field lines for two conducting spheres are shown in Figure 22-21. What is the relative sign and magnitude of the charges on the two spheres?

Picture the Problem The charge on a sphere is positive if more lines leave than enter and negative if more enter than leave. The ratio of the magnitudes of the charges equals the ratio of the net number of lines entering or leaving.

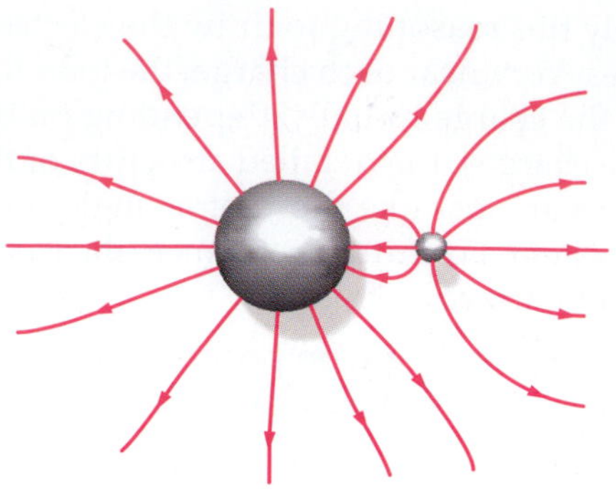

Figure 22-21

Since 11 electric field lines leave the large sphere on the left and 3 enter, the net number leaving is 8, so the charge on the large sphere is positive. For the small sphere on the right, 8 lines leave and none enter, so its charge is also positive. Since the net number of lines leaving each sphere is 8, the spheres carry equal positive charges. The charge on the small sphere creates an intense field at the nearby surface of the large sphere that causes a local accumulation of negative charge on the large sphere—indicated by the three entering field lines. Most of the large sphere's surface has positive charge, however, so its total charge is positive.

The convention relating the electric field strength to the electric field lines works because the electric field varies inversely as the square of the distance from a point charge. Since the gravitational field of a point mass also varies inversely as the square of the distance, field-line drawings are also useful for picturing the gravitational field. Near a point mass, the gravitational field lines converge toward the mass just as electric field lines converge toward a negative charge. However, there are no points in space where gravitational field lines diverge like electric field lines near a positive charge, because the gravitational force is always attractive, never repulsive.

22-6 Motion of Point Charges in Electric Fields

When a particle with a charge q is placed in an electric field $\vec{E}$, it experiences a force $q\vec{E}$. If the electric force is the only significant force acting on the particle, the particle has an acceleration

$$\vec{a} = \frac{\Sigma\vec{F}}{m} = \frac{q}{m}\vec{E}$$

where m is the mass of the particle.* If the electric field is known, the charge-to-mass ratio of the particle can be determined from the measured acceleration. The deflection of electrons in a uniform electric field was used by J. J. Thomson in 1897 to demonstrate the existence of electrons and to measure their charge-to-mass ratio. Familiar examples of devices that rely on the motion of electrons in electric fields are oscilloscopes, computer monitors, and television picture tubes.

* If the particle is an electron, its speed in an electric field is often a significant fraction of the speed of light. In such cases, Newton's laws of motion must be modified by Einstein's special theory of relativity.

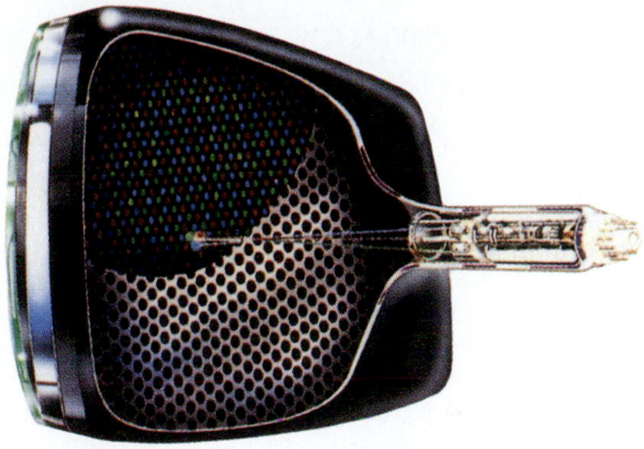

Schematic drawing of a cathode-ray tube used for color television. The beams of electrons from the electron gun on the right activate phosphors on the screen at the left, giving rise to bright spots whose colors depend on the relative intensity of each beam. Electric fields between deflection plates in the gun (or magnetic fields from coils within the gun) deflect the beams. The beams sweep across the screen in a horizontal line, are deflected downward, then sweep across again. The entire screen is covered in this way 30 times per second.

Example 22-10

An electron is projected into a uniform electric field $\vec{E} = (1000\ \text{N/C})\hat{i}$ with an initial velocity $\vec{v}_0 = (2 \times 10^6\ \text{m/s})\hat{i}$ in the direction of the field (Figure 22-22). How far does the electron travel before it is brought momentarily to rest?

Picture the Problem Since the charge of the electron is negative, the force $-e\vec{E}$ acting on the electron is in the direction opposite that of the field. Since $\vec{E}$ is constant, the force is constant and we can use constant acceleration formulas from Chapter 2. We choose the field to be in the positive x direction.

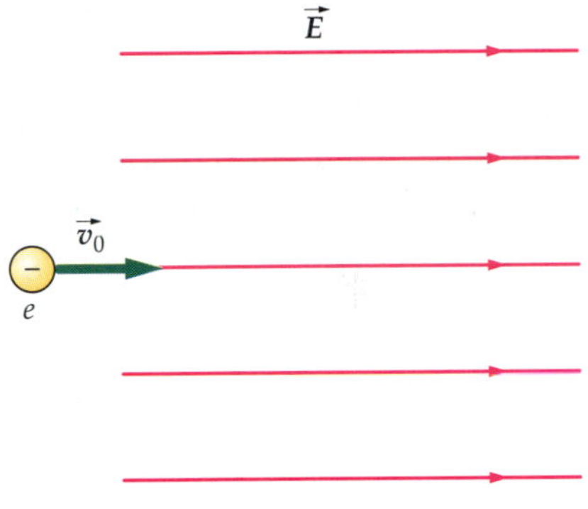

Figure 22-22

1. The displacement Δx is related to the initial and final velocities: $v^2 = v_0^2 + 2a\,\Delta x$

2. The acceleration is obtained from Newton's second law: $a = \dfrac{F}{m} = \dfrac{-eE}{m}$

3. When $v = 0$, the displacement is:

$$\Delta x = \frac{v^2 - v_0^2}{2a} = \frac{0 - v_0^2}{2(-eE/m)} = \frac{mv_0^2}{2eE}$$

$$= \frac{(9.11 \times 10^{-31}\ \text{kg})(2 \times 10^6\ \text{m/s})^2}{2(1.6 \times 10^{-19}\ \text{C})(1000\ \text{N/C})} = 1.14 \times 10^{-2}\ \text{m}$$

Example 22-11

An electron enters a uniform electric field $\vec{E} = (-2000\ \text{N/C})\hat{j}$ with an initial velocity $\vec{v}_0 = (10^6\ \text{m/s})\hat{i}$ perpendicular to the field (Figure 22-23). (*a*) Compare the gravitational force acting on the electron to the electric force acting on it. (*b*) By how much has the electron been deflected after it has traveled 1 cm in the x direction?

Picture the Problem (*a*) Calculate the ratio of the electric force $qE = -eE$ to the gravitational force mg. (*b*) Since mg is negligible, the force on the electron is $-eE$ vertically upward. The electron thus moves with constant horizontal velocity v_x and is deflected upward by an amount $y = \frac{1}{2}at^2$, where t is the time to travel 1 cm in the x direction.

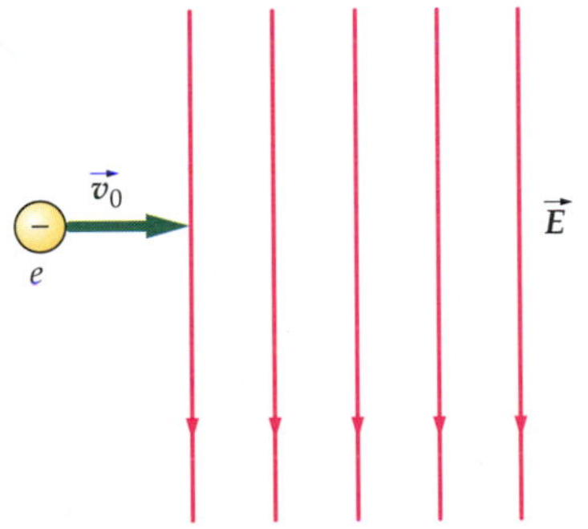

Figure 22-23

(*a*) Calculate the ratio of the magnitude of the electric force, F_e, to the magnitude of the gravitational force, F_g:

$$\frac{F_e}{F_g} = \frac{eE}{mg} = \frac{(1.6 \times 10^{-19}\ \text{C})(2000\ \text{N/C})}{(9.11 \times 10^{-31}\ \text{kg})(9.81\ \text{N/kg})} = 3.6 \times 10^{13}$$

(*b*)1. Express the vertical deflection in terms of the acceleration a and time t:

$$y = \frac{1}{2}at^2$$

2. The time required for the electron to travel a distance x with constant horizontal velocity v_0 is:

$$t = \frac{x}{v_0} = \frac{10^{-2}\ \text{m}}{10^6\ \text{m/s}} = 10^{-8}\ \text{s}$$

3. Use this result for t and eE/m for a to calculate y:

$$y = \frac{1}{2}\frac{eE}{m}t^2 = \frac{1}{2}\frac{(1.6 \times 10^{-19}\ \text{C})(2000\ \text{N/C})}{9.11 \times 10^{-31}\ \text{kg}}(10^{-8}\ \text{s})^2$$

$$= 1.76\ \text{cm}$$

Remarks (*a*) As is usually the case, the electric force is huge compared with the gravitational force. Thus, it is not necessary to consider gravity when designing a cathode-ray tube, for example, or when calculating the deflection in the problem above. In fact, a television picture tube works equally well upside down and right side up, as if gravity were not even present. (*b*) The path of an electron moving in a uniform electric field is a parabola, the same as the path of a mass moving in a uniform gravitational field.

22-7 Electric Dipoles in Electric Fields

In Example 22-6 we found the electric field produced by a dipole, a system of two equal and opposite point charges that are close together. Here we consider the behavior of an electric dipole in an external electric field. Some molecules have permanent electric dipole moments due to a nonuniform distribution of charge within the molecule. Such molecules are called **polar molecules**. An example is HCl, which is essentially a positive hydrogen ion of charge $+e$ combined with a negative chlorine ion of charge $-e$. The center of charge of the positive ion does not coincide with the center of charge for the negative ion, so the molecule has a permanent dipole moment. Another example is water (Figure 22-24). A uniform external electric field exerts no net force on a dipole, but it does exert a torque that tends to rotate the dipole into the direction of the field. We see in Figure 22-25 that the torque about the negative charge* has the magnitude $F_1L \sin\theta = qEL \sin\theta = pE \sin\theta$. The direction of the torque is into the paper such that it rotates the dipole moment $\vec{p}$ into the direction of $\vec{E}$. The torque can be conveniently written as the cross product of the dipole moment $\vec{p}$ and the electric field $\vec{E}$:

$$\vec{\tau} = \vec{p} \times \vec{E} \qquad 22\text{-}11$$

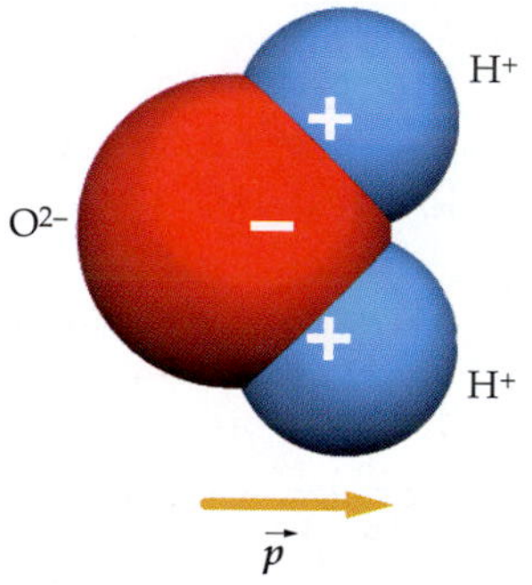

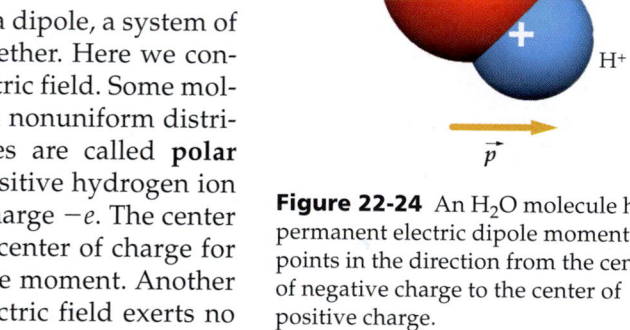

Figure 22-24 An H_2O molecule has a permanent electric dipole moment that points in the direction from the center of negative charge to the center of positive charge.

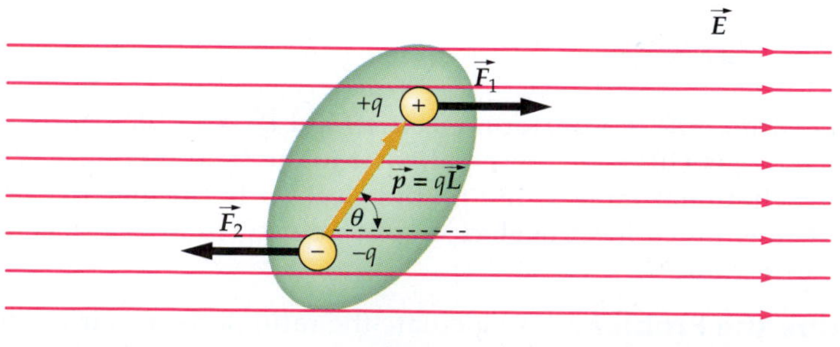

Figure 22-25 A dipole in a uniform electric field experiences equal and opposite forces that tend to rotate the dipole so that its dipole moment is aligned with the electric field.

* The torque produced by two equal and opposite forces (an arrangement called a couple) is the same about any point in space.

When the dipole rotates through $d\theta$, the electric field does work:

$$dW = -\tau\, d\theta = -pE \sin\theta\, d\theta$$

(The minus sign arises because the torque tends to decrease θ.) Setting this work equal to the decrease in potential energy, we have

$$dU = -dW = +pE \sin\theta\, d\theta$$

Integrating, we obtain

$$U = -pE \cos\theta + U_0$$

If we choose the potential energy U_0 to be zero when $\theta = 90°$, then the potential energy of the dipole is

$$U = -pE \cos\theta = -\vec{p}\cdot\vec{E} \qquad 22\text{-}12$$

Potential energy of a dipole in an electric field

Microwave ovens take advantage of the electric dipole moment of water molecules to cook food. Like all electromagnetic waves, microwaves have an oscillating electric field that can cause electric dipoles to vibrate. The microwaves in home ovens are tuned to the natural frequency of vibration of water molecules. The water molecules in food resonate with the oscillating electric field and absorb large amounts of energy, accounting for the rapid cooking times that make microwave ovens so convenient.

Nonpolar molecules have no permanent electric dipole moment. However, all neutral molecules contain equal amounts of positive and negative charge. In the presence of an external electric field $\vec{E}$, the charges become separated in space. The positive charges are pushed in the direction of $\vec{E}$ and the negative charges are pushed in the opposite direction. The molecule thus acquires an induced dipole moment parallel to the external electric field and is said to be **polarized**.

In a nonuniform electric field, an electric dipole experiences a net force because the electric field has different magnitudes at the positive and negative poles. Figure 22-26 shows how a positive point charge polarizes a nonpolar molecule and then attracts it. A familiar example is the attraction that holds an electrostatically charged balloon against a wall. The nonuniform field produced by the charge on the balloon polarizes molecules in the wall and attracts them. An equal and opposite force is exerted by the wall molecules on the balloon.

The diameter of an atom or molecule is of the order of 10^{-10} m = 0.1 nm. A convenient unit for the electric dipole moment of atoms and molecules is the fundamental electronic charge e times the distance 1 nm. For example, the dipole moment of H_2O in these units has a magnitude of about 0.04 e·nm.

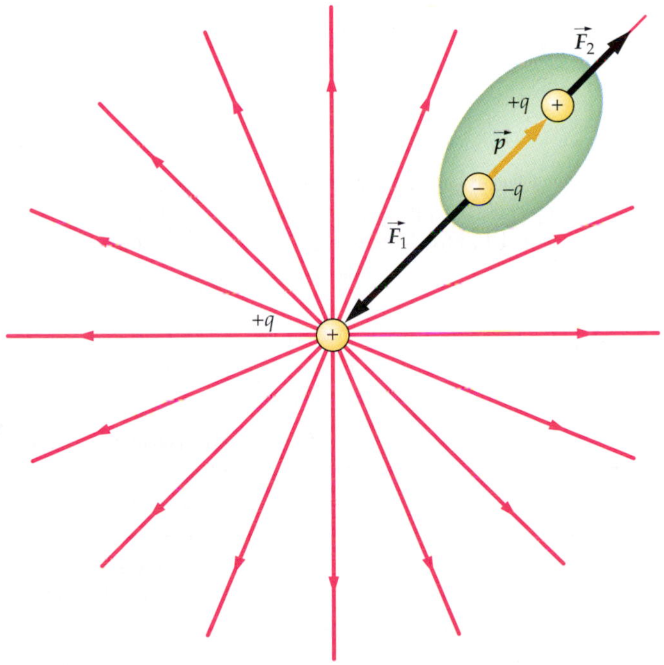

Figure 22-26 A nonpolar molecule in the nonuniform electric field of a positive point charge. The induced electric dipole moment $\vec{p}$ is parallel to the field of the point charge. Since the point charge is closer to the center of negative charge than to the center of positive charge, there is a net force of attraction between the dipole and the point charge. If the point charge were negative, the induced dipole moment would be reversed, and the molecule would again be attracted to the point charge.

Example 22-12

A dipole with a moment of magnitude 0.02 e·nm makes an angle of 20° with a uniform electric field of magnitude 3×10^3 N/C (Figure 22-27). Find (*a*) the magnitude of the torque on the dipole and (*b*) the potential energy of the system.

Figure 22-27

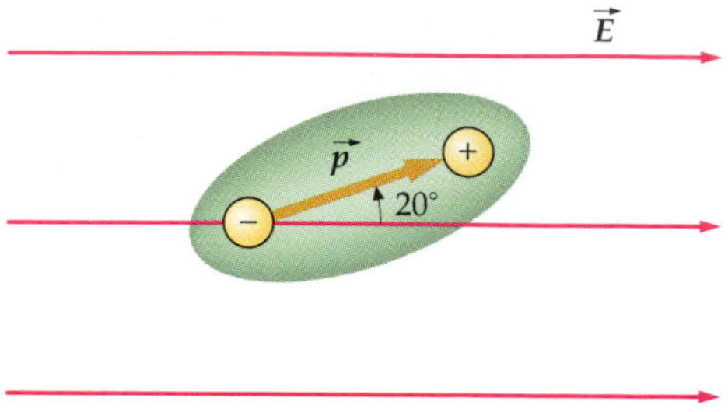

Picture the Problem The torque is found from $\vec{\tau} = \vec{p} \times \vec{E}$ and the potential energy is found from $U = -\vec{p}\cdot\vec{E}$.

(*a*) Calculate the magnitude of the torque:

$$\tau = |\vec{p} \times \vec{E}| = pE \sin\theta = (0.02\ \text{e}\cdot\text{nm})(3 \times 10^3\ \text{N/C})(\sin 20°)$$
$$= (0.02)(1.6 \times 10^{-19}\ \text{C})(10^{-9}\ \text{m})(3 \times 10^3\ \text{N/C})(\sin 20°)$$
$$= 3.28 \times 10^{-27}\ \text{N}\cdot\text{m}$$

(*b*) Calculate the potential energy:

$$U = -\vec{p}\cdot\vec{E} = -pE\cos\theta$$
$$= -(0.02)(1.6 \times 10^{-19}\ \text{C})(10^{-9}\ \text{m})(3 \times 10^3\ \text{N/C})\cos 20°$$
$$= -9.02 \times 10^{-27}\ \text{J}$$

Summary

1. Quantization and conservation are fundamental properties of electric charge.
2. Coulomb's law is the fundamental law of interaction between charges at rest.
3. The electric field describes the condition in space set up by a charge distribution.

Topic	Remarks and Relevant Equations	
1. Electric Charge	There are two kinds of electric charge, positive and negative.	
Quantization	Electric charge is quantized—it always occurs in integral multiples of the fundamental unit of charge e. The charge of the electron is $-e$ and that of the proton is $+e$.	
Magnitude	$e = 1.60 \times 10^{-19}$ C	22-1
Conservation	Charge is conserved. It is neither created nor destroyed in any process, but is merely transferred.	

2.	**Conductors and Insulators**	In conductors, about one electron per atom is free to move about the entire material. In insulators, all the electrons are bound to nearby atoms.	
	Ground	A very large conductor that can supply an unlimited amount of charge (such as the earth) is called a ground.	
3.	**Charging by Induction**	A conductor can be charged by holding a charge near it to attract or repel the free electrons and then grounding the conductor to drain off the far away charges.	
4.	**Coulomb's Law**	The force exerted by a charge q_1 on q_2 is given by $\vec{F}_{1,2} = \frac{kq_1q_2}{r_{1,2}^2}\hat{r}_{1,2}$ where $\hat{r}_{1,2}$ is a unit vector that points from q_1 to q_2	22-2
	Coulomb constant	$k = 8.99 \times 10^9\ \mathrm{N{\cdot}m^2/C^2}$	
5.	**Electric Field**	The electric field due to a system of charges at a point is defined as the net force exerted by those charges on a very small positive test charge q_0 divided by q_0: $\vec{E} = \frac{\vec{F}}{q_0}$	22-5
	Due to a point charge	$\vec{E}_i = \frac{kq_i}{r_{i,0}^2}\hat{r}_{i,0}$	22-7
	Due to a system of point charges	The electric field due to several charges is the vector sum of the fields due to the individual charges: $\vec{E} = \sum_i \vec{E}_i = \sum_i \frac{kq_i}{r_{i,0}^2}\hat{r}_{i,0}$	22-8
6.	**Electric Field Lines**	The electric field can be represented by electric field lines that originate on positive charges and end on negative charges. The strength of the electric field is indicated by the density of the electric field lines.	
7.	**Electric Dipole**	An electric dipole is a system of two equal but opposite charges separated by a small distance.	
	Dipole moment	$\vec{p} = q\vec{L}$ where $\vec{L}$ points from the negative charge to the positive charge.	
	Field due to dipole	The electric field far from a dipole is proportional to the dipole moment and decreases with the cube of the distance.	
	Torque on a dipole	In a uniform electric field, the net force on a dipole is zero, but there is a torque $\vec{\tau}$ that tends to align the dipole in the direction of the field. $\vec{\tau} = \vec{p} \times \vec{E}$	
	Potential energy of a dipole	$U = -\vec{p}\cdot\vec{E}$	
8.	**Polar and Nonpolar Molecules**	Polar molecules, such as H_2O, have permanent dipole moments because their centers of positive and negative charge do not coincide. They behave like simple dipoles in an electric field. Nonpolar molecules do not have permanent dipole moments, but they acquire induced dipole moments in the presence of an electric field.	

Problem-Solving Guide

Begin by drawing a neat diagram that includes the important features of the problem. Include the location, sign and magnitude of the charges, as well as the individual force vectors or electric field vectors in your sketch. Show the vector sum of these force or field vectors when appropriate.

Summary of Worked Examples

Type of Calculation	Procedure and Relevant Examples
1. Charges and Coulomb's Law	
Find the number of electrons in a given charge.	The number of electrons in a charge Q is $N = Q/e$. **Example 22-1**
Find the electric force between two point charges.	Use Coulomb's law. **Examples 22-2, 22-3**
Find the electric force on one point charge due to two or more other charges.	Use Coulomb's law to find the force due to each charge, then add the force vectors. **Examples 22-4, 22-5**
2. The Electric Field	
Find the electric field due to two or more point charges.	Find the electric field due to each charge and add the electric field vectors. **Examples 22-6, 22-7, 22-8**
3. Electric Field Lines	
Using electric field lines to determine the relative signs and magnitudes of charges.	The lines begin on positive charges and end on negative charges with the number of lines being proportional to the magnitude of the charge. **Example 22-9**
4. Point Charges and Dipoles in Electric Fields	
Motion of a charged particle in a constant electric field.	Find the acceleration of the particle using $\vec{a} = q\vec{E}/m$, then use constant-acceleration kinematics. **Examples 22-10 and 22-11**
Find the torque acting on an electric dipole.	Use $\vec{\tau} = \vec{p} \times \vec{E}$. **Example 22-12**
Find the potential energy of a dipole in an electric field.	Use $U = -pE \cos \theta = -\vec{p} \cdot \vec{E}$. **Example 22-12**

Problems

Conceptual Problems

Problems from Optional and Exploring sections

In a few problems, you are given more data than you actually need; in a few other problems, you are required to supply data from your general knowledge, outside sources, or informed estimates.

- • Single-concept, single-step, relatively easy
- •• Intermediate-level, may require synthesis of concepts
- ••• Challenging, for advanced students

Electric Charge

1 • If the sign convention for charge were changed so that the charge on the electron were positive and the charge on the proton were negative, would Coulomb's law still be written the same?

2 •• Discuss the similarities and differences in the properties of electric charge and gravitational mass.

3 • A plastic rod is rubbed against a wool shirt, thereby acquiring a charge of $-0.8\ \mu C$. How many electrons are transferred from the wool shirt to the plastic rod?

4 • A charge equal to the charge of Avogadro's number of protons ($N_A = 6.02 \times 10^{23}$) is called a *faraday*. Calculate the number of coulombs in a faraday.

5 • How many coulombs of positive charge are there in 1 kg of carbon? Twelve grams of carbon contain Avogadro's number of atoms, with each atom having six protons and six electrons.

Conductors, Insulators, and Charging by Induction

6 • Can insulators be charged by induction?

7 •• A metal rectangle B is connected to ground through a switch S that is initially closed (Figure 22-28). While the charge $+Q$ is near B, switch S is opened. The charge $+Q$ is then removed. Afterward, what is the charge state of the metal rectangle B?

(*a*) It is positively charged.
(*b*) It is uncharged.
(*c*) It is negatively charged.
(*d*) It may be any of the above depending on the charge on B before the charge $+Q$ was placed nearby.

Figure 22-28 Problem 7

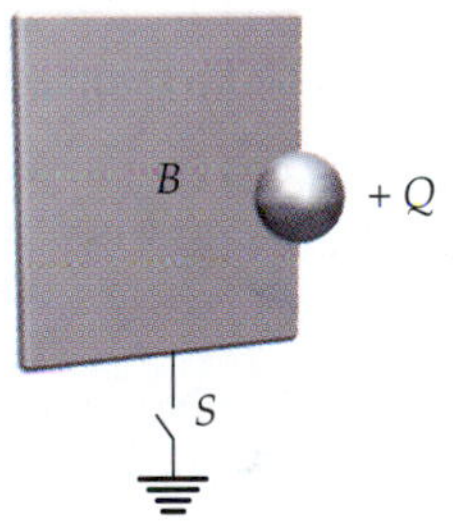

8 •• Explain, giving each step, how a positively charged insulating rod can be used to give a metal sphere (*a*) a negative charge, and (*b*) a positive charge. (*c*) Can the same rod be used to simultaneously give one sphere a positive charge and another sphere a negative charge without the rod having to be recharged?

9 •• Two uncharged conducting spheres with their conducting surfaces in contact are supported on a large wooden table by insulated stands. A positively charged rod is brought up close to the surface of one of the spheres on the side opposite its point of contact with the other sphere. (*a*) Describe the induced charges on the two conducting spheres, and sketch the charge distributions on them. (*b*) The two spheres are separated far apart and the charged rod is removed. Sketch the charge distributions on the separated spheres.

Coulomb's Law

10 • Three charges, $+q$, $+Q$, and $-Q$, are placed at the corners of an equilateral triangle as shown in Figure 22-29. The net force on charge $+q$ due to the other two charges is

(*a*) vertically up.
(*b*) vertically down.
(*c*) zero.
(*d*) horizontal to the left.
(*e*) horizontal to the right.

Figure 22-29 Problem 10

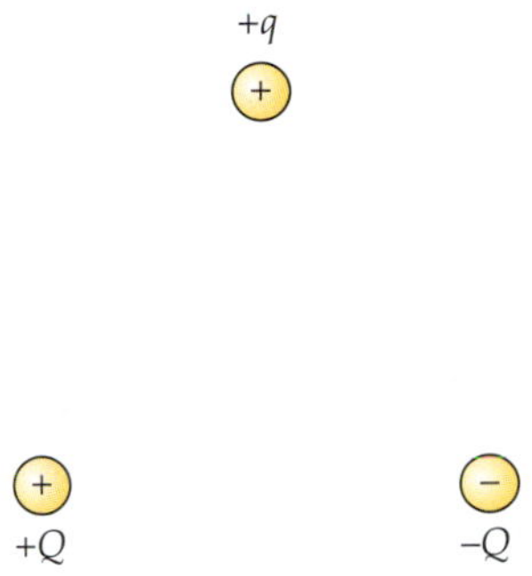

11 • A charge $q_1 = 4.0\ \mu C$ is at the origin, and a charge $q_2 = 6.0\ \mu C$ is on the x axis at $x = 3.0$ m. (*a*) Find the force on charge q_2. (*b*) Find the force on q_1. (*c*) How would your answers for parts (*a*) and (*b*) differ if q_2 were $-6.0\ \mu C$?

12 • Three point charges are on the x axis: $q_1 = -6.0\ \mu C$ is at $x = -3.0$ m, $q_2 = 4.0\ \mu C$ is at the origin, and $q_3 = -6.0\ \mu C$ is at $x = 3.0$ m. Find the force on q_1.

13 •• Two equal charges of 3.0 μC are on the y axis, one at the origin and the other at $y = 6$ m. A third charge $q_3 = 2\ \mu C$ is on the x axis at $x = 8$ m. Find the force on q_3.

14 •• Three charges, each of magnitude 3 nC, are at separate corners of a square of side 5 cm. The two charges at opposite corners are positive, and the other charge is negative. Find the force exerted by these charges on a fourth charge $q = +3$ nC at the remaining corner.

15 •• A charge of 5 μC is on the y axis at $y = 3$ cm, and a second charge of $-5\ \mu C$ is on the y axis at $y = -3$ cm. Find the force on a charge of 2 μC on the x axis at $x = 8$ cm.

16 •• A point charge of $-2.5\ \mu C$ is located at the origin. A second point charge of 6 μC is at $x = 1$ m, $y = 0.5$ m. Find the x and y coordinates of the position at which an electron would be in equilibrium.

17 •• A charge of $-1.0\ \mu C$ is located at the origin, a second charge of 2.0 μC is located at $x = 0$, $y = 0.1$ m, and a third charge of 4.0 μC is located at $x = 0.2$ m, $y = 0$. Find the forces that act on each of the three charges.

18 •• A charge of 5.0 μC is located at $x = 0$, $y = 0$ and a charge Q_2 is located at $x = 4.0$ cm, $y = 0$. The force on a 2-μC charge at $x = 8.0$ cm, $y = 0$ is 19.7 N, pointing in the negative x direction. When this 2-μC charge is positioned at $x = 17.75$ cm, $y = 0$, the force on it is zero. Determine the charge Q_2.

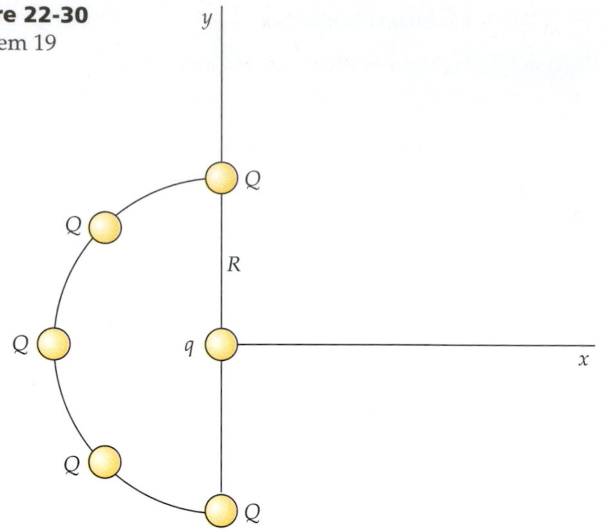

Figure 22-30 Problem 19

19 •• Five equal charges Q are equally spaced on a semicircle of radius R as shown in Figure 22-30. Find the force on a charge q located at the center of the semicircle.

20 ••• The configuration of the NH_3 molecule is approximately that of a regular tetrahedron, with three H^+ ions forming the base and an N^{3-} ion at the apex of the tetrahedron. The length of each side is 1.64×10^{-10} m. Calculate the force that acts on each ion.

The Electric Field

21 • A positive charge that is free to move but is at rest in an electric field $\vec{E}$ will

(*a*) accelerate in the direction perpendicular to $\vec{E}$.
(*b*) remain at rest.
(*c*) accelerate in the direction opposite to $\vec{E}$.
(*d*) accelerate in the same direction as $\vec{E}$.
(*e*) do none of the above.

22 • If four charges are placed at the corners of a square as shown in Figure 22-31, the field $\vec{E}$ is zero at

(*a*) all points along the sides of the square midway between two charges.
(*b*) the midpoint of the square.
(*c*) midway between the top two charges and midway between the bottom two charges.
(*d*) none of the above.

Figure 22-31 Problem 22

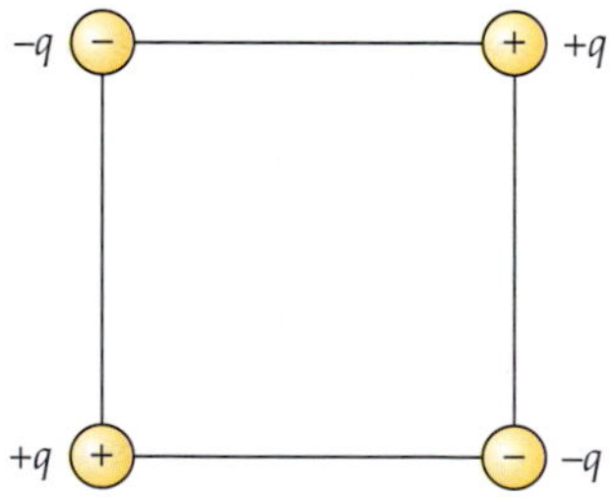

23 •• At a particular point in space, a charge Q experiences no net force. It follows that

(*a*) there are no charges nearby.
(*b*) if charges are nearby, they have the opposite sign of Q.
(*c*) if charges are nearby, the total positive charge must equal the total negative charge.
(*d*) none of the above need be true.

24 • A charge of 4.0 μC is at the origin. What is the magnitude and direction of the electric field on the x axis at (*a*) $x = 6$ m and (*b*) $x = -10$ m? (*c*) Sketch the function E_x versus x for both positive and negative values of x. (Remember that E_x is negative when $\vec{E}$ points in the negative x direction.)

25 • Two charges, each $+4\ \mu$C, are on the x axis, one at the origin and the other at $x = 8$ m. Find the electric field on the x axis at (*a*) $x = -2$ m, (*b*) $x = 2$ m, (*c*) $x = 6$ m, and (*d*) $x = 10$ m. (*e*) At what point on the x axis is the electric field zero? (*f*) Sketch E_x versus x.

26 • When a test charge $q_0 = 2$ nC is placed at the origin, it experiences a force of 8.0×10^{-4} N in the positive y direction. (*a*) What is the electric field at the origin? (*b*) What would be the force on a charge of -4 nC placed at the origin? (*c*) If this force is due to a charge on the y axis at $y = 3$ cm, what is the value of that charge?

27 • An oil drop has a mass of 4×10^{-14} kg and a net charge of 4.8×10^{-19} C. An upward electric force just balances the downward force of gravity so that the oil drop is stationary. What is the direction and magnitude of the electric field?

28 • The electric field near the surface of the earth points downward and has a magnitude of 150 N/C. (*a*) Compare the upward electric force on an electron with the downward gravitational force. (*b*) What charge should be placed on a penny of mass 3 g so that the electric force balances the weight of the penny near the earth's surface?

29 •• Two equal positive charges of magnitude $q_1 = q_2 = 6.0$ nC are on the y axis at $y_1 = +3$ cm and $y_2 = -3$ cm. (*a*) What is the magnitude and direction of the electric field on the x axis at $x = 4$ cm? (*b*) What is the force exerted on a third charge $q_0 = 2$ nC when it is placed on the x axis at $x = 4$ cm?

30 •• A point charge of $+5.0\ \mu$C is located at $x = -3.0$ cm, and a second point charge of $-8.0\ \mu$C is located at $x = +4.0$ cm. Where should a third charge of $+6.0\ \mu$C be placed so that the electric field at $x = 0$ is zero?

31 •• A point charge of $-5\ \mu$C is located at $x = 4$ m, $y = -2$ m. A second point charge of 12 μC is located at $x = 1$ m, $y = 2$ m. (*a*) Find the magnitude and direction of the electric field at $x = -1$ m, $y = 0$. (*b*) Calculate the magnitude and direction of the force on an electron at $x = -1$ m, $y = 0$.

32 •• Two equal positive charges q are on the y axis, one at $y = a$ and the other at $y = -a$. (*a*) Show that the electric field on the x axis is along the x axis with $E_x = 2kqx(x^2 + a^2)^{-3/2}$. (*b*) Show that near the origin, when x is much smaller than a, E_x is approximately $2kqx/a^3$. (*c*) Show that for values of x much larger than a, E_x is approximately $2kq/x^2$. Explain why you would expect this result even before calculating it.

33 •• A 5-μC point charge is located at $x = 1$ m, $y = 3$ m, and a -4-μC point charge is located at $x = 2$ m, $y = -2$ m. (*a*) Find the magnitude and direction of the electric field at $x = -3$ m, $y = 1$ m. (*b*) Find the magnitude and direction of the force on a proton at $x = -3$ m, $y = 1$ m.

34 •• (*a*) Show that the electric field for the charge distribution in Problem 32 has its greatest magnitude at the points $x = a/\sqrt{2}$ and $x = -a/\sqrt{2}$ by computing dE_x/dx and setting the derivative equal to zero. (*b*) Sketch the function E_x versus x using your results for part (*a*) of this problem and parts (*b*) and (*c*) of Problem 32.

35 ••• For the charge distribution in Problem 32, the electric field at the origin is zero. A test charge q_0 placed at the

origin will therefore be in equilibrium. (*a*) Discuss the stability of the equilibrium for a positive test charge by considering small displacements from equilibrium along the x axis and small displacements along the y axis. (*b*) Repeat part (*a*) for a negative test charge. (*c*) Find the magnitude and sign of a charge q_0 that when placed at the origin results in a net force of zero on each of the three charges. (*d*) What will happen if any of the charges is displaced slightly from equilibrium?

36 ••• Two positive point charges $+q$ are on the y axis at $y = +a$ and $y = -a$ as in Problem 32. A bead of mass m carrying a negative charge $-q$ slides without friction along a thread that runs along the x axis. (*a*) Show that for small displacements of $x \ll a$, the bead experiences a restoring force that is proportional to x and therefore undergoes simple harmonic motion. (*b*) Find the period of the motion.

Electric Field Lines

37 • Which of the following statements about electric field lines is (are) *not* true?

(*a*) The number of lines leaving a positive charge or entering a negative charge is proportional to the charge.
(*b*) The lines begin on positive charges and end on negative charges.
(*c*) The density of lines (the number per unit area perpendicular to the lines) is proportional to the magnitude of the field.
(*d*) Electric field lines cross midway between charges that have equal magnitude and sign.

38 • Figure 22-32 shows the electric field lines for a system of two point charges. (*a*) What are the relative magnitudes of the charges? (*b*) What are the signs of the charges? (*c*) In what regions of space is the electric field strong? In what regions is it weak?

Figure 22-32 Problem 38

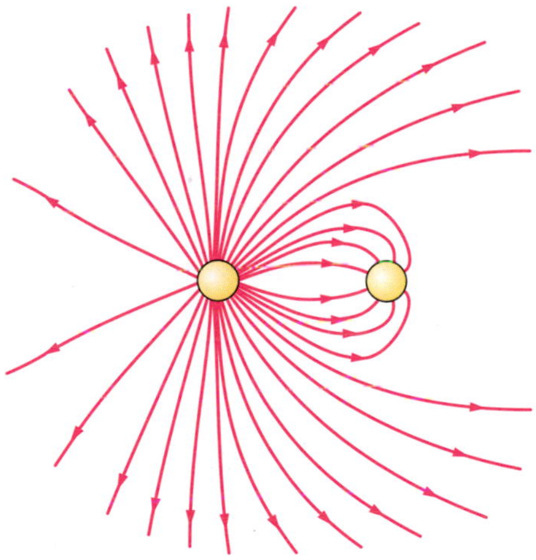

39 • Two charges $+4q$ and $-3q$ are separated by a small distance. Draw the electric field lines for this system.

40 • Two charges $+q$ and $-3q$ are separated by a small distance. Draw the electric field lines for this system.

41 • Three equal positive point charges are situated at the corners of an equilateral triangle. Sketch the electric field lines in the plane of the triangle.

Motion of Point Charges in Electric Fields

42 • The acceleration of a particle in an electric field depends on the ratio of the charge to the mass of the particle. (*a*) Compute e/m for an electron. (*b*) What is the magnitude and direction of the acceleration of an electron in a uniform electric field with a magnitude of 100 N/C? (*c*) When the speed of an electron approaches the speed of light c, relativistic mechanics must be used to calculate its motion, but at speeds significantly less than c, Newtonian mechanics applies. Using Newtonian mechanics, compute the time it takes for an electron placed at rest in an electric field with a magnitude of 100 N/C to reach a speed of $0.01c$. (*d*) How far does the electron travel in that time?

43 • (*a*) Compute e/m for a proton, and find its acceleration in a uniform electric field with a magnitude of 100 N/C. (*b*) Find the time it takes for a proton initially at rest in such a field to reach a speed of $0.01c$ (where c is the speed of light).

44 • An electron has an initial velocity of 2×10^6 m/s in the x direction. It enters a uniform electric field $\vec{E} = (400\text{ N/C})\hat{j}$, which is in the y direction. (*a*) Find the acceleration of the electron. (*b*) How long does it take for the electron to travel 10 cm in the x direction in the field? (*c*) By how much and in what direction is the electron deflected after traveling 10 cm in the x direction in the field?

45 •• An electron, starting from rest, is accelerated by a uniform electric field of 8×10^4 N/C that extends over a distance of 5.0 cm. Find the speed of the electron after it leaves the region of uniform electric field.

46 •• An electron moves in a circular orbit about a stationary proton. The centripetal force is provided by the electrostatic force of attraction between the proton and the electron. The electron has a kinetic energy of 2.18×10^{-18} J. (*a*) What is the speed of the electron? (*b*) What is the radius of the orbit of the electron?

47 •• A mass of 2 g located in a region of uniform electric field $\vec{E} = (300\text{ N/C})\hat{i}$ carries a charge Q. The mass, released from rest at $x = 0$, has a kinetic energy of 0.12 J at $x = 0.50$ m. Determine the charge Q.

48 •• A particle leaves the origin with a speed of 3×10^6 m/s at 35° to the x axis. It moves in a constant electric field $\vec{E} = E_y\hat{j}$. Find E_y such that the particle will cross the x axis at $x = 1.5$ cm if the particle is (*a*) an electron and (*b*) a proton.

49 •• An electron starts at the position shown in Figure 22-33 with an initial speed $v_0 = 5 \times 10^6$ m/s at 45° to the x axis. The electric field is in the positive y direction and has a magnitude of 3.5×10^3 N/C. On which plate and at what location will the electron strike?

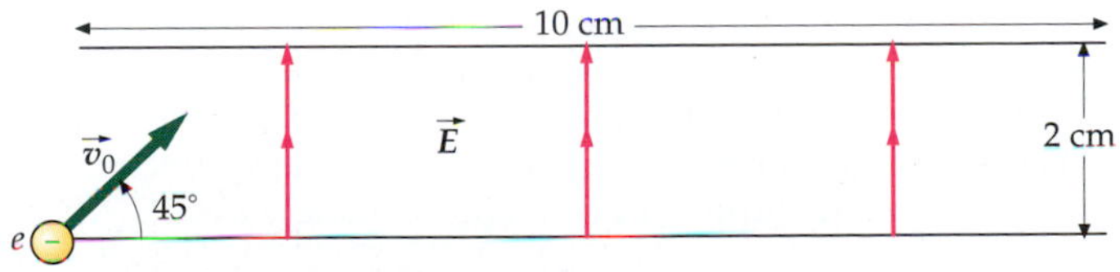

Figure 22-33 Problem 49

Figure 22-34 Problem 50

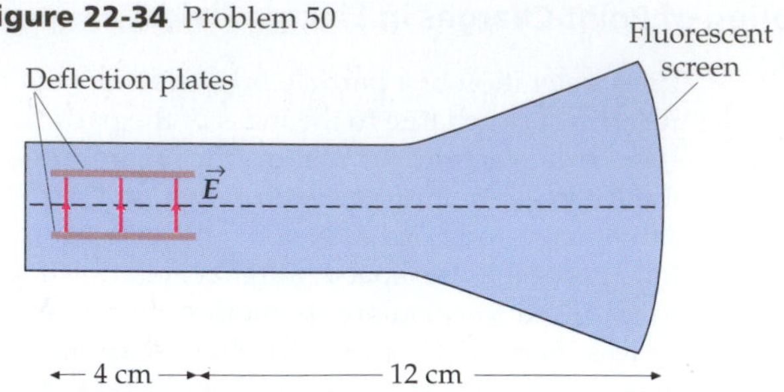

50 •• An electron with kinetic energy of 2×10^{-16} J is moving to the right along the axis of a cathode-ray tube as shown in Figure 22-34. There is an electric field $\vec{E} = (2 \times 10^4 \text{ N/C})\hat{j}$ in the region between the deflection plates. Everywhere else, $\vec{E} = 0$. (*a*) How far is the electron from the axis of the tube when it reaches the end of the plates? (*b*) At what angle is the electron moving with respect to the axis? (*c*) At what distance from the axis will the electron strike the fluorescent screen?

Electric Dipoles

51 • Two point charges, $q_1 = 2.0$ pC and $q_2 = -2.0$ pC, are separated by 4 μm. (*a*) What is the dipole moment of this pair of charges? (*b*) Sketch the pair, and show the direction of the dipole moment.

52 • A dipole of moment 0.5 e·nm is placed in a uniform electric field with a magnitude of 4.0×10^4 N/C. What is the magnitude of the torque on the dipole when (*a*) the dipole is parallel to the electric field, (*b*) the dipole is perpendicular to the electric field, and (*c*) the dipole makes an angle of 30° with the electric field? (*d*) Find the potential energy of the dipole in the electric field for each case.

53 •• For a dipole oriented along the x axis, the electric field falls off as $1/x^3$ in the x direction and $1/y^3$ in the y direction. Use dimensional analysis to prove that, in any direction, the field far from the dipole falls off as $1/r^3$.

54 •• A water molecule has its oxygen atom at the origin, one hydrogen nucleus at $x = 0.077$ nm, $y = 0.058$ nm, and the other hydrogen nucleus at $x = -0.077$ nm, $y = 0.058$ nm. If the hydrogen electrons are transferred completely to the oxygen atom so that it has a charge of $-2e$, what is the dipole moment of the water molecule? (Note that this characterization of the chemical bonds of water as totally ionic is simply an approximation that overestimates the dipole moment of a water molecule.)

55 •• An electric dipole consists of two charges $+q$ and $-q$ separated by a very small distance $2a$. Its center is on the x axis at $x = x_1$, and it points along the x axis in the positive x direction. The dipole is in a nonuniform electric field, which is also in the x direction, given by $\vec{E} = Cx\hat{i}$, where C is a constant. (*a*) Find the force on the positive charge and that on the negative charge, and show that the net force on the dipole is $Cp\hat{i}$. (*b*) Show that, in general, if a dipole of moment $\vec{p}$ lies along the x axis in an electric field in the x direction, the net force on the dipole is given approximately by $(dE_x/dx)p\hat{i}$.

56 ••• A positive point charge $+Q$ is at the origin, and a dipole of moment $\vec{p}$ is a distance r away and in the radial direction as in Figure 22-26. (*a*) Show that the force exerted by the electric field of the point charge on the dipole is attractive and has a magnitude of approximately $2kQp/r^3$ (see Problem 55). (*b*) Now assume that the dipole is centered at the origin and that a point charge Q is a distance r away along the line of the dipole. From your result for part (*a*) and Newton's third law, show that the magnitude of the electric field of the dipole along the line of the dipole a distance r away is approximately $2kp/r^3$.

57 ••• A quadrupole consists of two dipoles that are close together, as shown in Figure 22-35. The effective charge at the origin is $-2q$ and the other charges on the y axis at $y = a$ and $y = -a$ are each $+q$. (*a*) Find the electric field at a point on the x axis far away so that $x >> a$. (*b*) Find the electric field on the y axis far away so that $y >> a$.

Figure 22-35 Problem 57

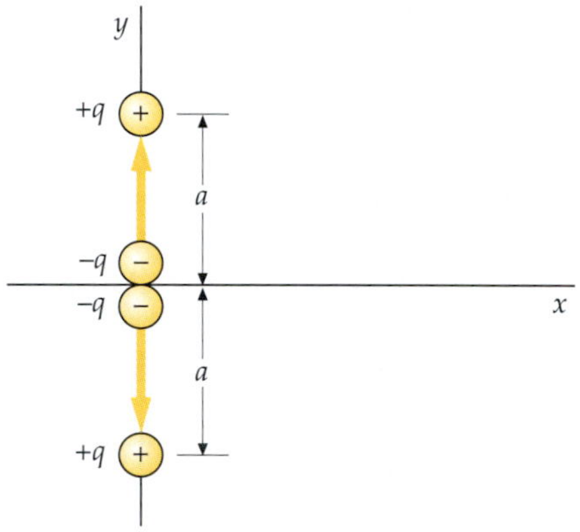

General Problems

58 • A charged insulator and an uncharged metal
(*a*) always repel one another.
(*b*) exert no electrostatic force on one another.
(*c*) always attract one another.
(*d*) may attract or repel, depending on the sign of the charge on the insulator.

59 • Which of the following statements are true?
(*a*) A positive charge experiences an attractive electrostatic force toward a nearby neutral conductor.
(*b*) A positive charge experiences no electrostatic force near a neutral conductor.
(*c*) A positive charge experiences a repulsive force, away from a nearby conductor.
(*d*) Whatever the force on a positive charge near a neutral conductor, the force on a negative charge is then oppositely directed.
(*e*) None of the above is correct.

60 • The electric field lines around an electrical dipole are best represented by which, if any, of the diagrams in Figure 22-36?

61 •• A molecule with electric dipole moment $\vec{p}$ is oriented so that $\vec{p}$ makes an angle θ with a uniform electric field $\vec{E}$. The dipole is free to move in response to the force from the field. Describe the motion of the dipole. Suppose the electric field is nonuniform and is larger in the x direction. How will the motion be changed?

Figure 22-36 Problem 60

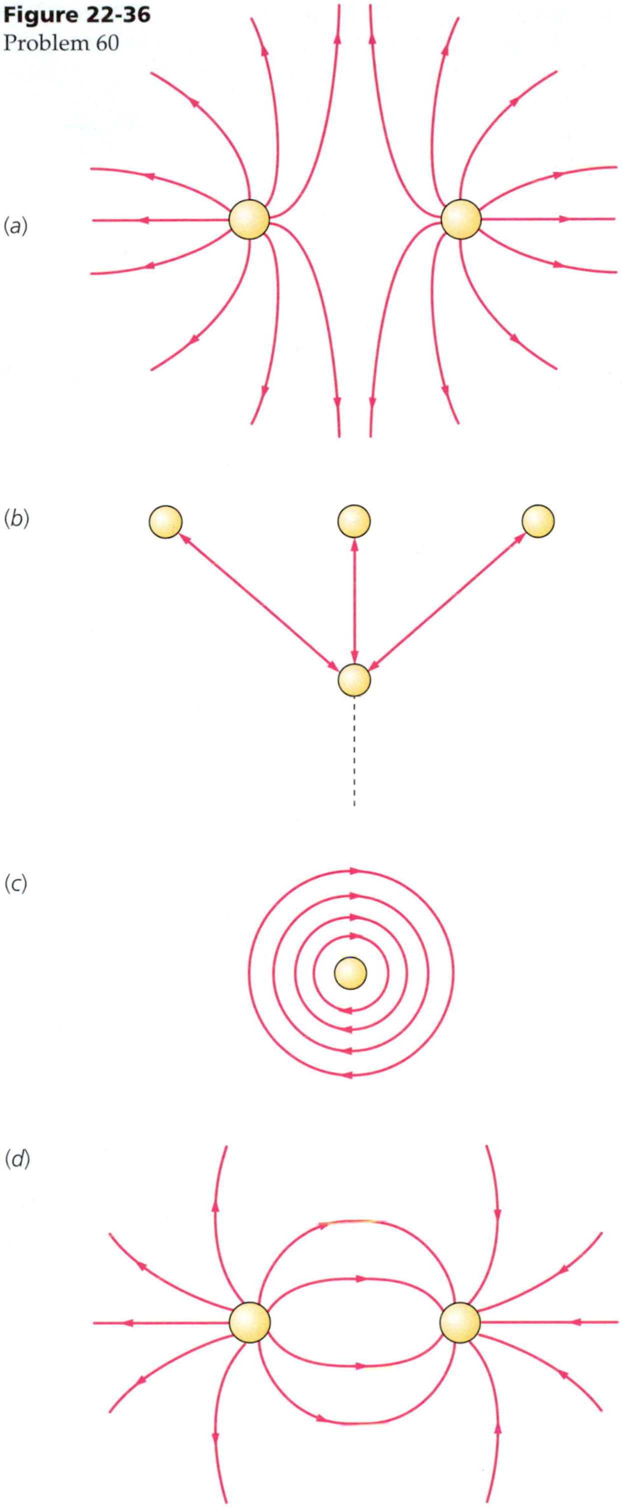

62 •• True or false:

(*a*) The electric field of a point charge always points away from the charge.
(*b*) All macroscopic charges Q can be written as $Q = \pm Ne$, where N is an integer and e is the charge of the electron.
(*c*) Electric field lines never diverge from a point in space.
(*d*) Electric field lines never cross at a point in space.
(*e*) All molecules have electric dipole moments in the presence of an external electric field.

63 •• A small, nonconducting ball with no net charge is suspended from a thread. When a positive charge is brought near the ball, the ball is attracted toward the charge. How does this come about? How would the situation be different if the charge brought near the ball were negative instead of positive?

64 •• Two metal balls have charges $+q$ and $-q$. How will the force on one of them change if (*a*) the balls are placed in water, the distance between them being unchanged, and (*b*) a third uncharged metal ball is placed between the first two? Explain.

65 •• A metal ball is positively charged. Is it possible for it to attract another positively charged ball? Explain.

66 • In interstellar space, two charged point-like objects, each of mass m and charge q, are separated by a distance d and released. They remain motionless at that separation. Find an expression for q in terms of m, G, and k.

67 •• Point charges of $-5.0\ \mu C$, $+3.0\ \mu C$, and $+5.0\ \mu C$ are located along the x axis at $x = -1.0$ cm, $x = 0$, and $x = +1.0$ cm, respectively. Calculate the electric field at $x = 3.0$ cm and at $x = 15.0$ cm. Is there some point on the x axis where the magnitude of the electric field is zero? Locate that point.

68 •• For the charge distribution of Problem 67, find the electric field at $x = 15.0$ cm as the vector sum of the electric field due to a dipole formed by the two 5.0-μC charges and a point charge of 3.0 μC, both located at the origin. Compare your result with the result obtained in Problem 67 and explain any difference between these two.

69 •• In copper, about one electron per atom is free to move about. A copper penny has a mass of 3 g. (*a*) What percentage of the free charge would have to be removed to give the penny a charge of 15 μC? (*b*) What would be the force of repulsion between two pennies carrying this charge if they were 25 cm apart? Assume that the pennies are point charges.

70 •• Two charges q_1 and q_2 have a total charge of 6 μC. When they are separated by 3 m, the force exerted by one charge on the other has a magnitude of 8 mN. Find q_1 and q_2 if (*a*) both are positive so that they repel each other, and (*b*) one is positive and the other is negative so that they attract each other.

71 •• Three charges, $+q$, $+2q$, and $+4q$, are connected by strings as shown in Figure 22-37. Find the tensions T_1 and T_2.

Figure 22-37 Problem 71

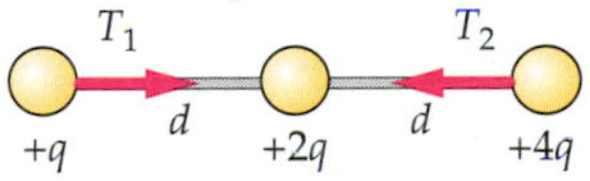

72 •• A positive charge Q is to be divided into two positive charges q_1 and q_2. Show that, for a given separation D, the force exerted by one charge on the other is greatest if $q_1 = q_2 = \frac{1}{2}Q$.

73 •• A charge Q is located at $x = 0$ and a charge $4Q$ is at $x = 12.0$ cm. The force on a charge of $-2\ \mu C$ is zero if that charge is placed at $x = 4.0$ cm and is 126.4 N in the positive x direction if placed at $x = 8.0$ cm. Determine the charge Q.

74 •• Two small spheres (point charges) separated by 0.60 m carry a total charge of 200 μC. (*a*) If the two spheres repel each other with a force of 80 N, what are the charges on each of the two spheres? (*b*) If the two spheres attract each other with a force of 80 N, what are the charges on the two spheres?

75 •• A ball of known charge q and unknown mass m, initially at rest, falls freely from a height h in a uniform electric field $\vec{E}$ that is directed vertically downward. The ball hits the ground at a speed $v = 2\sqrt{gh}$. Find m in terms of E, q, and g.

76 •• Charges of 3.0 μC are located at $x = 0$, $y = 2.0$ m and at $x = 0$, $y = -2.0$ m. Charges Q are located at $x = 4.0$ m, $y = 2.0$ m and at $x = 4.0$ m, $y = -2.0$ m (Figure 22-38). The electric field at $x = 0$, $y = 0$ is $(4.0 \times 10^3\ \text{N/C})\hat{i}$. Determine Q.

Figure 22-38 Problem 76

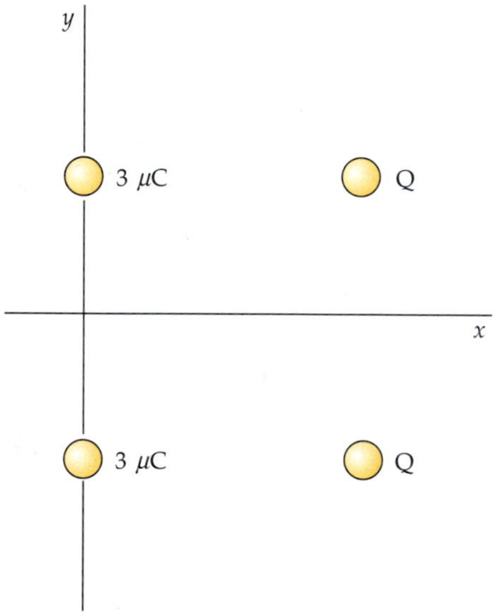

77 •• Two identical small spherical conductors (point charges), separated by 0.60 m, carry a total charge of 200 μC. They repel one another with a force of 120 N. (*a*) Find the charge on each sphere. (*b*) The two spheres are placed in electrical contact and then separated so that each carries 100 μC. Determine the force exerted by one sphere on the other when they are 0.60 m apart.

78 •• Repeat Problem 77 if the two spheres initially attract one another with a force of 120 N.

79 •• A charge of -3.0 μC is located at the origin; a charge of 4.0 μC is located at $x = 0.2$ m, $y = 0$; a third charge Q is located at $x = 0.32$ m, $y = 0$. The force on the 4.0-μC charge is 240 N, directed in the positive x direction. (*a*) Determine the charge Q. (*b*) With this configuration of three charges, where, along the x direction, is the electric field zero?

80 •• Two small spheres of mass m are suspended from a common point by threads of length L. When each sphere carries a charge q, each thread makes an angle θ with the vertical, as shown in Figure 22-39. (*a*) Show that the charge q is given by

$$q = 2L\sin\theta\sqrt{\frac{mg\tan\theta}{k}}$$

where k is the Coulomb constant. (*b*) Find q if $m = 10$ g, $L = 50$ cm, and $\theta = 10°$.

Figure 22-39 Problems 80 and 90

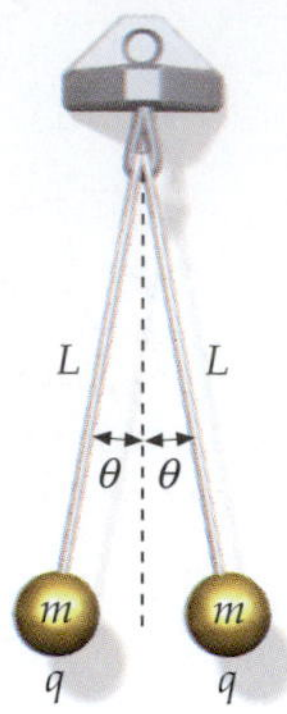

81 •• (*a*) Suppose that in Problem 80, $L = 1.5$ m, $m = 0.01$ kg, and $q = 0.75$ μC. What is the angle that each string makes with the vertical? (*b*) Find the angle that each string makes with the vertical if one mass carries a charge of 0.50 μC, the other a charge of 1.0 μC.

82 •• Four charges of equal magnitude are arranged at the corners of a square of side L as shown in Figure 22-40. (*a*) Find the magnitude and direction of the force exerted on the charge in the lower left corner by the other charges. (*b*) Show that the electric field at the midpoint of one of the sides of the square is directed along that side toward the negative charge and has a magnitude E given by

$$E = k\frac{8q}{L^2}\left(1 - \frac{\sqrt{5}}{25}\right)$$

Figure 22-40 Problem 82

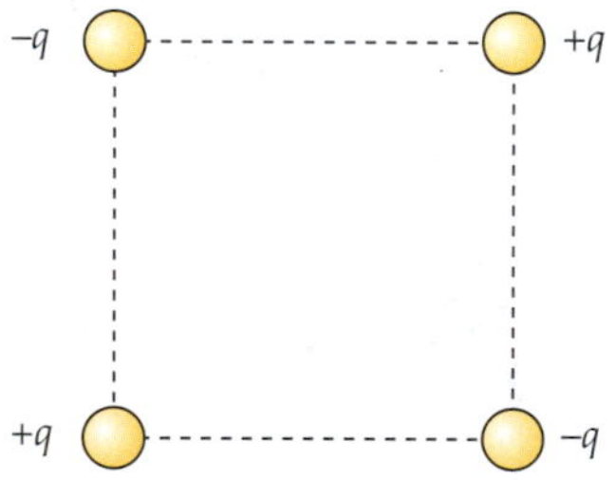

83 •• Figure 22-41 shows a dumbbell consisting of two identical masses m attached to the ends of a thin (massless) rod of length a that is pivoted at its center. The masses carry charges of $+q$ and $-q$ and the system is located in a uniform electric field $\vec{E}$. Show that for small values of the angle θ between the direction of the dipole and the electric field, the system displays simple harmonic motion, and obtain an expression for the period of that motion.

Figure 22-41 Problems 83 and 84

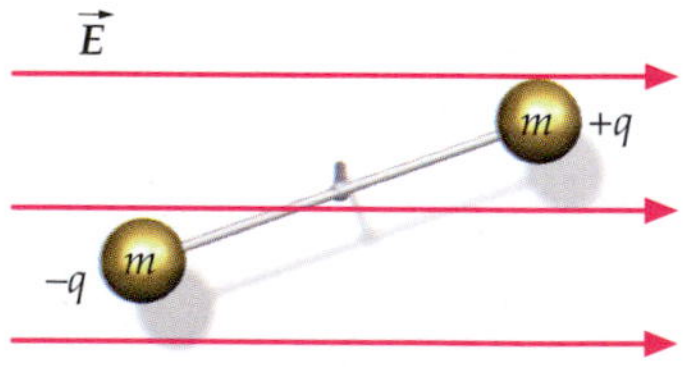

84 •• For the dumbbell in Figure 22-41, let $m = 0.02$ kg, $a = 0.3$ m, and $\vec{E} = (600\ \text{N/C})\hat{i}$. Initially the dumbbell is at rest and makes an angle of 60° with the x axis. The dumbbell is then released, and when it is momentarily aligned with the electric field, its kinetic energy is 5×10^{-3} J. Determine the magnitude of q.

85 •• An electron (charge $-e$, mass m) and a positron (charge $+e$, mass m) revolve around their common center of mass under the influence of their attractive coulomb force. Find the speed of each particle v in terms of e, m, k, and their separation r.

86 •• The equilibrium separation between the nuclei of the ionic molecule KBr is 0.282 nm. The masses of the two ions, K^+ and Br^-, are very nearly the same, 1.4×10^{-25} kg, and each of the two ions carries a charge of magnitude e. Use the result of Problem 83 to determine the frequency of oscillation of a KBr molecule in a uniform electric field of 1000 N/C.

87 ••• A small (point) mass m, which carries a charge q, is constrained to move vertically inside a narrow, frictionless cylinder (Figure 22-42). At the bottom of the cylinder is a point mass of charge Q having the same sign as q. (*a*) Show that the mass m will be in equilibrium at a height $y_0 = (kqQ/mg)^{1/2}$. (*b*) Show that if the mass m is displaced by a small amount from its equilibrium position and released, it will exhibit simple harmonic motion with angular frequency $\omega = (2g/y_0)^{1/2}$.

Figure 22-42 Problem 87

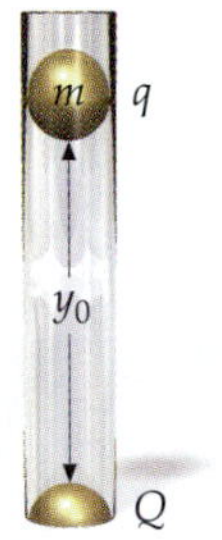

88 ••• A small bead of mass m and carrying a negative charge $-q$ is constrained to move along a thin frictionless rod (Figure 22-43). A distance L from this rod is a positive charge Q. Show that if the bead is displaced a distance x, where $x \ll L$, and released, it will exhibit simple harmonic motion. Obtain an expression for the period of this motion in terms of the parameters L, Q, q, and m.

Figure 22-43 Problem 88

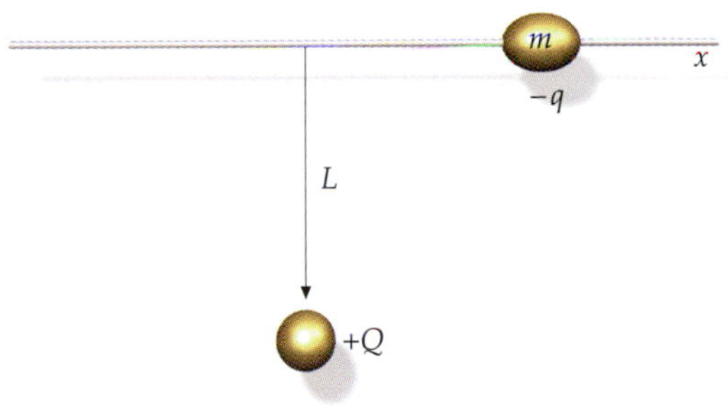

89 ••• Repeat Problem 81 with the system located in a uniform electric field of 1.0×10^5 N/C that points vertically downward.

90 ••• Suppose that the two masses in Problem 80 are not equal. One mass is 0.01 kg, the other is 0.02 kg. The charges on the two masses are 2.0 and 1.0 μC, respectively. Determine the angle that each of the strings supporting the masses makes with the vertical.

91 ••• A simple pendulum of length $L = 1.0$ m and mass $M = 5.0 \times 10^{-3}$ kg is placed in a uniform, vertically directed electric field $\vec{E}$. The bob carries a charge of -8.0 μC. The period of the pendulum is 1.2 s. What is the magnitude and direction of $\vec{E}$?

92 ••• Two neutral polar molecules attract each other. Suppose that each molecule has a dipole moment $\vec{p}$ and that these dipoles are aligned along the x axis and separated by a distance d. Derive an expression for the force of attraction in terms of p and d.

93 ••• A small bead of mass m, carrying a charge q, is constrained to slide along a thin rod of length L. Charges Q are fixed at each end of the rod (Figure 22-44). (*a*) Obtain an expression for the electric field due to the two charges Q as a function of x, where x is the distance from the midpoint of the rod. (*b*) Show that for $x \ll L$, the magnitude of the field is proportional to x. (*c*) Show that if q is of the same sign as Q, the force that acts on the object of mass m is always directed toward the center of the rod and is proportional to x. (*d*) Find the period of oscillation of the mass m if it is displaced by a small distance from the center of the rod and then released.

Figure 22-44 Problem 93

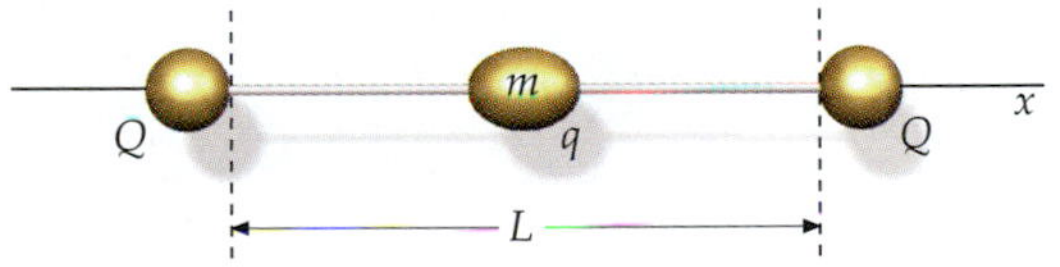

94 ••• Two equal positive charges Q are on the x axis at $x = \frac{1}{2}L$ and $x = -\frac{1}{2}L$. (*a*) Obtain an expression for the electric field as a function of y on the y axis. (*b*) A ring of mass m, which carries a charge q, moves on a thin frictionless rod along the y axis. Find the force that acts on the charge q as a function of y; determine the sign of q such that this force always points toward $y = 0$. (*c*) Show that for small values of y the ring exhibits simple harmonic motion. (*d*) If $Q = 5$ μC, $|q| = 2$ μC, $L = 24$ cm, and $m = 0.03$ kg, what is the frequency of the oscillation for small amplitudes?

CHAPTER 23

The Electric Field II: Continuous Charge Distributions

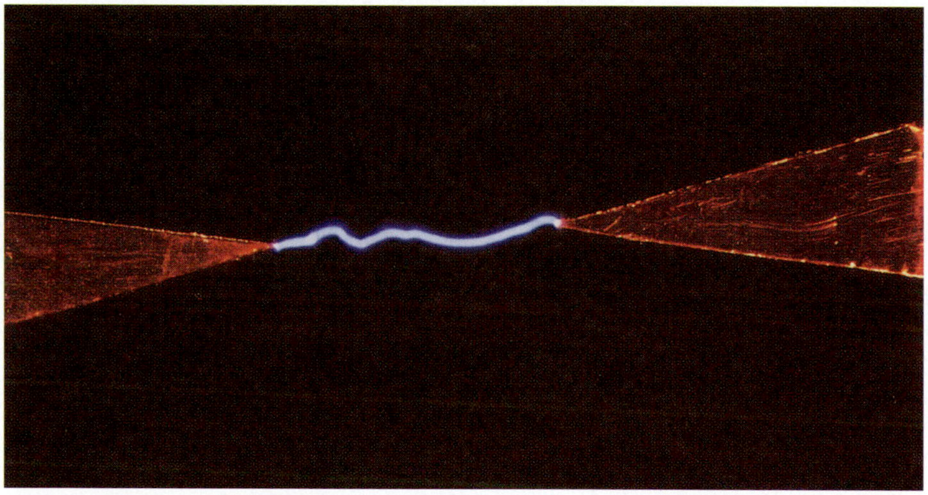

Electrical discharge between two charged conductors. The electric field near the sharp points is strong enough to strip the electrons from nearby air molecules, thus ionizing them and causing the air to conduct.

On a microscopic scale, electric charge is quantized. However, there are often situations in which many charges are so close together that they can be considered to be continuously distributed. The use of a continuous charge density to describe a large number of discrete charges is similar to the use of a continuous mass density to describe air, which actually consists of a large number of discrete molecules. In either case, it is usually easy to find a volume element ΔV that is large enough to contain a multitude of individual charges or molecules and yet is small enough that replacing ΔV by a differential dV and using calculus introduces negligible error.

We describe the charge per unit volume by the **volume charge density** ρ:

$$\rho = \frac{\Delta Q}{\Delta V} \qquad \text{23-1}$$

Often charge is distributed in a thin layer on the surface of an object. We define the **surface charge density** σ as the charge per unit area:

$$\sigma = \frac{\Delta Q}{\Delta A} \qquad \text{23-2}$$

Similarly, we sometimes encounter charge distributed along a line in space.

We define the **linear charge density** λ as the charge per unit length:

$$\lambda = \frac{\Delta Q}{\Delta L} \qquad 23\text{-}3$$

In this chapter, we show how Coulomb's law is used to calculate the electric field due to various types of continuous charge distributions. We then introduce Gauss's law, which relates the electric field on a closed surface to the net charge within the surface, and we use this relation to calculate the electric field for symmetric charge distributions.

23-1 Calculating $\vec{E}$ From Coulomb's Law

Figure 23-1 shows an element of charge $dq = \rho\, dV$ that is small enough to be considered a point charge. The electric field $d\vec{E}$ at a field point P due to this charge element is given by Coulomb's law:

$$d\vec{E} = \frac{k\, dq}{r^2}\hat{r}$$

where $\hat{r}$ is a unit vector that points from the charge element to the field point P. The total field at P is found by integrating this expression over the entire charge distribution. Thus,

$$\vec{E} = \int_V \frac{k\, dq}{r^2}\hat{r} \qquad 23\text{-}4$$

Electric field due to a continuous charge distribution

where $dq = \rho\, dV$. If the charge is distributed on a surface or line, we use $dq = \sigma\, dA$ or $dq = \lambda\, dL$ and integrate over the surface or line.

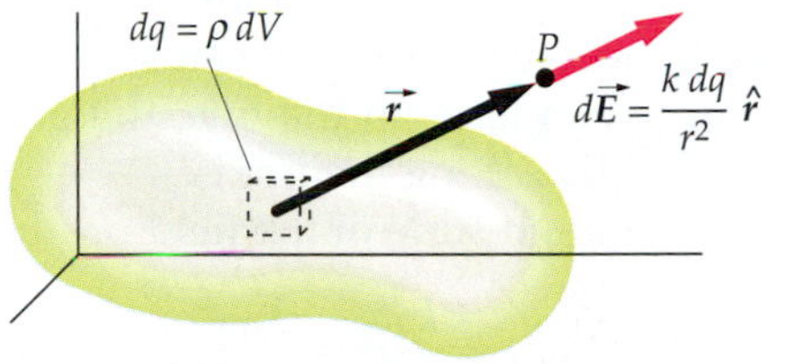

Figure 23-1 An element of charge dq produces a field $d\vec{E} = (k\,dq/r^2)\hat{r}$ at point P. The field at P due to the total charge is found by integrating over the entire charge distribution.

$\vec{E}$ on the Axis of a Finite Line Charge

A uniform charge Q lies along the x axis from $x = 0$ to $x = L$ as shown in Figure 23-2. The linear charge density for this charge is $\lambda = Q/L$. We wish to find the electric field produced by this line charge at some point P on the x axis at $x = x_0$, for $x_0 > L$. In the figure, we have chosen a small differential element dx at a distance x from the origin. The field point P is at a distance $r = x_0 - x$ from this charge element. The electric field due to this element of charge is given by Coulomb's law for a point charge. It is directed along the x axis and is given by

$$dE_x = \frac{k\, dq}{(x_0 - x)^2} = \frac{k\lambda\, dx}{(x_0 - x)^2}$$

We find the total field by integrating over the entire line charge from $x = 0$ to $x = L$:

$$E_x = k\lambda \int_0^L \frac{dx}{(x_0 - x)^2} = k\lambda \left[\frac{1}{x_0 - x}\right]_0^L$$

$$= k\lambda \left\{\frac{1}{x_0 - L} - \frac{1}{x_0}\right\} = k\lambda \left\{\frac{L}{x_0(x_0 - L)}\right\}$$

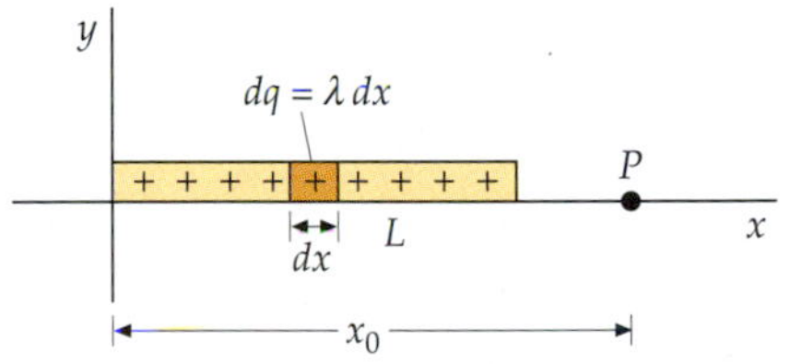

Figure 23-2 Geometry for the calculation of the electric field on the axis of a uniform line charge of length L, charge Q, and linear charge density $\lambda = Q/L$. An element $dq = \lambda\, dx$ is treated as a point charge.

Using $\lambda = Q/L$, we obtain

$$E_x = \frac{kQ}{x_0(x_0 - L)} \qquad \text{23-5}$$

We can see that if x_0 is much larger than L, the electric field at x_0 is approximately kQ/x_0^2. That is, if we are sufficiently far away from the line charge, it looks like a point charge.

$\vec{E}$ off the Axis of a Finite Line Charge

We first consider the simple case of the field at a point P on the perpendicular bisector of the line charge. The geometry is shown in Figure 23-3. We chose a coordinate system such that the charge is on the x axis with the origin at its center, and the field point P is on the y axis. A typical charge element $dq = \lambda\, dx$ that produces a field $d\vec{E}$ is shown in the figure.

The field has a component parallel to the line charge and a component perpendicular to it. We can see from symmetry that when we sum over all charge elements in the line, the parallel components sum to zero. Thus, the field $\vec{E}$ lies along the y axis.

The magnitude of the field produced by an element of charge $dq = \lambda\, dx$ is

$$|d\vec{E}| = \frac{k\,dq}{r^2} = \frac{k\lambda\, dx}{r^2}$$

The y component is

$$dE_y = \frac{k\lambda\, dx}{r^2}\cos\theta = \frac{k\lambda y\, dx}{r^3} \qquad \text{23-6}$$

where $\cos\theta = y/r$ and $r = \sqrt{x^2 + y^2}$. The total field E_y is computed by integrating from $x = -\frac{1}{2}L$ to $x = +\frac{1}{2}L$. Because of symmetry, the contribution of each half of the line charge to the total field is the same, so we can integrate from $x = 0$ to $x = \frac{1}{2}L$ and multiply by 2:

$$E_y = \int_{x=-\frac{1}{2}L}^{x=+\frac{1}{2}L} dE_y = 2\int_{x=0}^{x=\frac{1}{2}L} dE_y = 2k\lambda y \int_{x=0}^{x=\frac{1}{2}L} \frac{dx}{r^3} \qquad \text{23-7}$$

This integral is of a standard form found in tables:

$$\int \frac{dx}{r^3} = \frac{1}{y^2}\frac{x}{r} = \frac{1}{y^2}\sin\theta$$

We see from the figure that at $x = 0$, $\theta = 0$, so $\sin\theta = 0$ at the lower limit. At the upper limit at $x = L/2$, $\theta = \theta_0$, as shown in Figure 23-3. The field is thus

$$E_y = \frac{2k\lambda y}{y^2}\sin\theta_0$$

or

$$E_y = \frac{2k\lambda}{y}\sin\theta_0 = \frac{2k\lambda}{y}\frac{\frac{1}{2}L}{\sqrt{(\frac{1}{2}L)^2 + y^2}} \qquad \text{23-8}$$

$\vec{E}$ on the perpendicular bisector of a finite line charge

where (from Figure 23-3) $\sin\theta_0$ is related to L and y by

$$\sin\theta_0 = \frac{\frac{1}{2}L}{\sqrt{(\frac{1}{2}L)^2 + y^2}}$$

Exercise Show that when y is much greater than L, a finite line charge looks like a point charge, that is, Equation 23-8 reduces to $E_y \approx kQ/y^2$.

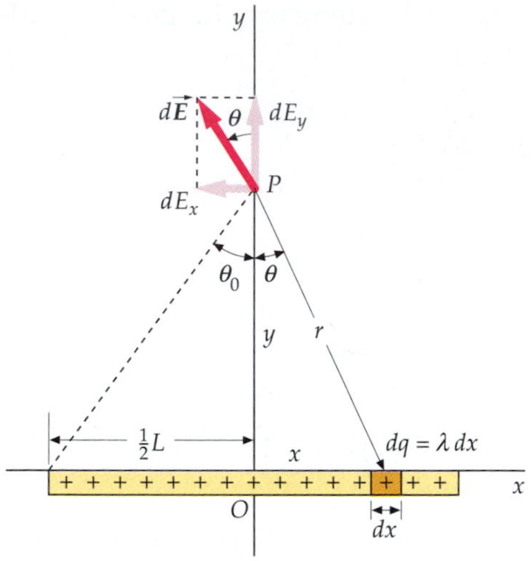

Figure 23-3 Geometry for the calculation of the electric field at a point on the perpendicular bisector of a uniform finite line charge. By symmetry, the net electric field is perpendicular to the line charge.

At a point a distance y from the line charge, but not on the perpendicular bisector, the field has both x and y components. Calculation of these components is left to the Problems section at the end of the chapter (see Problems 23-12 and 23-13).

$\vec{E}$ due to an Infinite Line Charge

When we are very close to a line charge (when $y \ll L$) or when the line charge is very long, the angle θ_0 in Figure 23-3 is approximately 90°, and $\sin\theta_0 \approx 1$. Then

$$E_y = \frac{2k\lambda}{y} \qquad 23\text{-}9$$

$\vec{E}$ at a distance y from an infinite line charge

Thus, $\vec{E}$ is directed away from the line (for a positive line charge) and its magnitude decreases as $1/y$.

Exercise Show that Equation 23-9 has the correct units for the electric field.

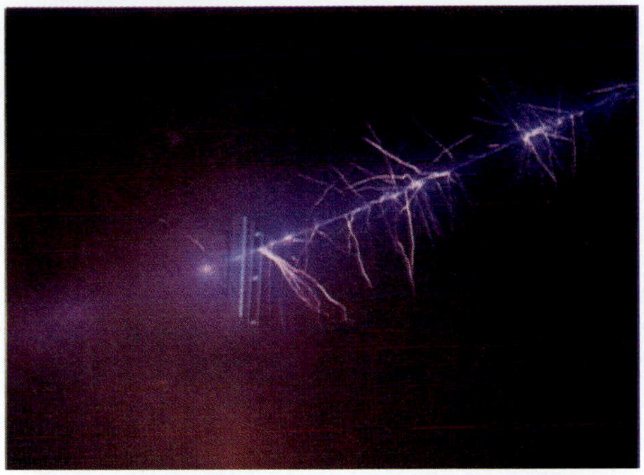

Electric field lines near a long wire. The electric field near a high-voltage power line can be large enough to strip the electrons from air molecules, thus ionizing them and making the air a conductor. The glow resulting from the recombination of free electrons with the ions is called corona discharge.

Example 23-1

A line charge of linear density $\lambda = 4.5$ nC/m lies on the x axis and extends from $x = -5$ cm to $x = 5$ cm. Calculate the electric field on the y axis using the exact expression

$$E_y = \frac{2k\lambda}{y}\sin\theta_0 = \frac{2k\lambda}{y}\frac{\frac{1}{2}L}{\sqrt{(\frac{1}{2}L)^2 + y^2}}$$

(Equation 23-8) at (*a*) $y = 1$ cm, (*b*) $y = 4$ cm, and (*c*) $y = 40$ cm. (*d*) Calculate the electric field on the y axis at $y = 1.0$ cm, assuming the line charge to be infinite. (*e*) Find the total charge and calculate the field at $y = 40$ cm, assuming the line charge to be a point charge.

Picture the Problem In the expression for $\sin\theta_0$, we can express L and y in centimeters because the units cancel. (*d*) To find the field very near the line charge, we use $E_y = 2k\lambda/y$. (*e*) To find the field very far from the charge, we use $E_y = kQ/y^2$ with $Q = \lambda L$.

(*a*) Calculate E_y at $y = 1$ cm for $\lambda = 4.5$ nC/m and $L = 0.5$ cm:

$$E_y = \frac{2k\lambda}{y}\frac{\frac{1}{2}L}{\sqrt{(\frac{1}{2}L)^2 + y^2}}$$

$$= \frac{2(8.99 \times 10^9\ \text{N}\cdot\text{m}^2/\text{C}^2)(4.5 \times 10^{-9}\ \text{C/m})}{0.01\ \text{m}}\frac{5\ \text{cm}}{\sqrt{(5\ \text{cm})^2 + (1\ \text{cm})^2}}$$

$$= \frac{80.9\ \text{N}\cdot\text{m/C}}{0.01\ \text{m}}\frac{5\ \text{cm}}{\sqrt{26}\ \text{cm}}$$

$$= 7.93 \times 10^3\ \text{N/C} = 7.93\ \text{kN/C}$$

(*b*) Repeat the calculation for $y = 4$ cm $= 0.04$ m using the result $2k\lambda = 80.9$ N·m/C to simplify the notation:

$$E_y = \frac{2k\lambda}{y}\frac{\frac{1}{2}L}{\sqrt{(\frac{1}{2}L)^2 + y^2}} = \frac{80.9\ \text{N}\cdot\text{m/C}}{0.04\ \text{m}}\frac{5\ \text{cm}}{\sqrt{(5\ \text{cm})^2 + (4\ \text{cm})^2}}$$

$$= 1.58 \times 10^3\ \text{N/C} = 1.58\ \text{kN/C}$$

(*c*) Repeat the calculation for y = 40 cm:

$$E_y = \frac{2k\lambda}{y}\frac{\frac{1}{2}L}{\sqrt{(\frac{1}{2}L)^2 + y^2}} = \frac{80.9\ \text{N}\cdot\text{m/C}}{0.4\ \text{m}}\frac{5\ \text{cm}}{\sqrt{(5\ \text{cm})^2 + (40\ \text{cm})^2}} = 25.1\ \text{N/C}$$

(*d*) Calculate the field at $y = 1$ cm $= 0.01$ m due to an infinite line charge:

$$E_y \approx \frac{2k\lambda}{y} = \frac{80.9\ \text{N}\cdot\text{m/C}}{0.01\ \text{m}} = 8.09\ \text{kN/m}$$

(*e*) Calculate the total charge λL for $L = 0.1$ m and use it to find the field of a point charge at $y = 4$ m:

$$Q = \lambda L = (4.5\ \text{nC/m})(0.1\ \text{m}) = 0.45\ \text{nC}$$

$$E_y \approx \frac{k\lambda L}{y^2} = \frac{kQ}{y^2} = \frac{(8.99 \times 10^9\ \text{N}\cdot\text{m}^2/\text{C}^2)(0.45 \times 10^{-9}\ \text{C})}{(0.40\ \text{m})^2} = 25.3\ \text{N/C}$$

Remarks At 1 cm from the 10-cm-long line charge, the appoximate value of 8.09 kN/C obtained by assuming an infinite line charge differs from the exact value of 7.93 calculated in (*a*) by about 2%. At 40 cm from the line charge, the approximate value of 25.3 N/C obtained by assuming the line charge to be a point charge differs from the exact value of 25.1 N/C obtained in (*c*) by about 1%. Figure 23-4 shows the exact result for this line segment of length 10 cm and charge density 4.5 nC/m, and for the limiting cases of an infinite line charge of the same charge density, and a point charge $Q = \lambda L$.

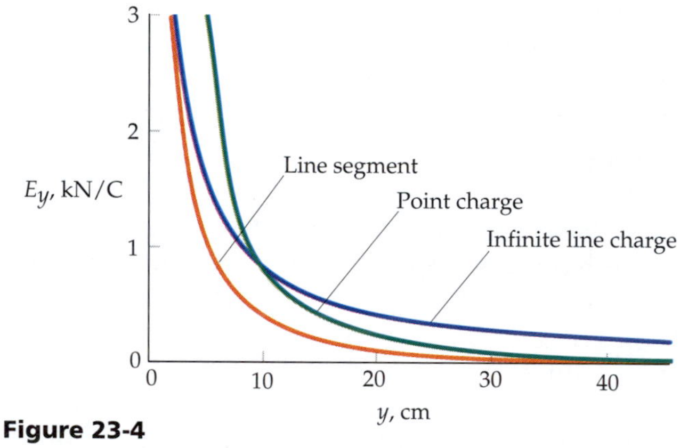

Figure 23-4

Example 23-2 *try it yourself*

An infinite line charge of linear charge density λ = 0.6 μC/m lies along the z axis, and a point charge q = 8 μC lies on the y axis at y = 3 m. Find the electric field at the point P on the x axis at x = 4 m.

Picture the Problem The electric field for this system is the superposition of the fields due to the infinite line charge and the point charge. The field of the line charge, $\vec{E}_L$, points radially away from the z axis (Figure 23-5). Thus, at point P on the x axis, $\vec{E}_L$ is in the positive x direction. The point charge produces a field $\vec{E}_P$ along the line connecting q and the point P. The distance from q to P is $r = \sqrt{(3\ \text{m})^2 + (4\ \text{m})^2} = 5$ m.

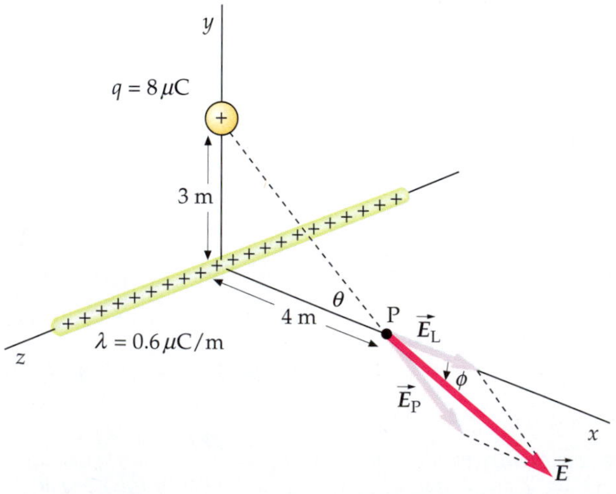

Figure 23-5

Cover the column to the right and try these on your own before looking at the answers.

Steps	***Answers***
1. Calculate the field $\vec{E}_L$ at point P due to the infinite line charge.	$\vec{E}_L = 2.70\ \text{kN/C}\ \hat{i}$
2. Find the field $\vec{E}_p$ at point P due to the point charge. Express $\vec{E}_p$ in terms of the unit vector $\hat{r}$ that points from q to P.	$\vec{E}_p = 2.88\ \text{kN/C}\ \hat{r}$
3. Find the x and y components of $\vec{E}_p$.	$E_{px} = E_p\,(0.8) = 2.30\ \text{kN/C}$ $E_{py} = E_p(-0.6) = -1.73\ \text{kN/C}$
4. Find the x and y components of the total field at point P.	$E_x = 5.00\ \text{kN/C}, \quad E_y = -1.73\ \text{kN/C}$
5. Use your result in step 4 to calculate the magnitude of the total field.	$E = \sqrt{E_x^2 + E_y^2} = 5.29\ \text{kN/C}$
6. Use your results in step 4 to find the angle ϕ between the field and the x axis.	$\phi = -19.1°$

$\vec{E}$ on the Axis of a Ring Charge

Figure 23-6 shows a uniform ring charge of radius a and total charge Q. The field $d\vec{E}$ at point P on the axis due to the charge element dq is shown in the figure. This field has a component dE_x directed along the axis of the ring and a component $dE_\perp$ directed perpendicular to the axis. From the symmetry of the charge distribution, we can see that the net field due to the entire ring must lie along the axis of the ring; that is, the perpendicular components sum to zero. The axial component of the field due to the charge element shown is

$$dE_x = \frac{k\,dq}{r^2}\cos\theta = \frac{k\,dq}{r^2}\frac{x}{r} = \frac{k\,dq\,x}{(x^2 + a^2)^{3/2}}$$

where

$$r^2 = x^2 + a^2$$

and

$$\cos\theta = \frac{x}{r} = \frac{x}{\sqrt{x^2 + a^2}}$$

The field due to the entire ring of charge is

$$E_x = \int \frac{kx\,dq}{(x^2 + a^2)^{3/2}}$$

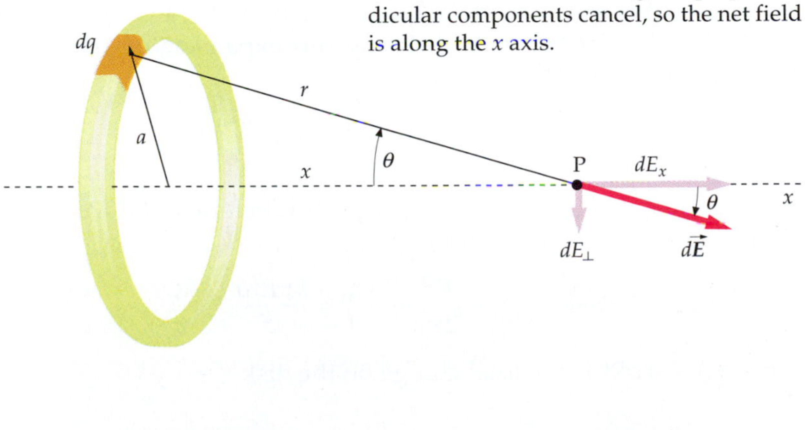

Figure 23-6 A ring charge of radius a. The electric field at point P on the x axis due to the charge element dq shown has one component along the x axis and one perpendicular to the x axis. When summed over the total ring, the perpendicular components cancel, so the net field is along the x axis.

Since x does not vary as we integrate over the elements of charge, we can remove it from the integral. Then

$$E_x = \frac{kx}{(x^2 + a^2)^{3/2}} \int dq$$

or

$$E_x = \frac{kQx}{(x^2 + a^2)^{3/2}} \qquad \text{23-10}$$

$\vec{E}$ on the axis of a ring charge

Exercise Find the field on the axis of a ring charge when (*a*) $x = 0$ and (*b*) x is much greater than a. (*Answers* (*a*) $E_x = 0$ at $x = 0$, (*b*) $E_x \approx kQ/x^2$ for $x >> a$)

$\vec{E}$ on the Axis of a Uniformly Charged Disk

Figure 23-7 shows a uniformly charged disk of radius R and total charge Q. We can calculate the field on the axis of the disk by treating the disk as a set of concentric ring charges. By symmetry, $\vec{E}$ on the axis of the disk is along the axis. A ring of radius a and width da is shown in the figure. The area of this ring is $dA = 2\pi a\, da$, and its charge is $dq = \sigma\, dA = 2\pi\sigma a\, da$, where $\sigma = Q/\pi R^2$ is the charge per unit area. The field produced by this ring is given by Equation 23-10 if we replace Q with $dq = 2\pi\sigma a\, da$.

$$dE_x = \frac{kx 2\pi\sigma a\, da}{(x^2 + a^2)^{3/2}}$$

The total field is found by integrating from $a = 0$ to $a = R$:

$$E_x = \int_0^R \frac{kx 2\pi\sigma a\, da}{(x^2 + a^2)^{3/2}} = kx\pi\sigma \int_0^R (x^2 + a^2)^{-3/2}\, 2a\, da$$

This integral is of the form $\int u^n\, du$, with $u = x^2 + a^2$ and $n = -\frac{3}{2}$. The integration thus gives

$$E_x = kx\pi\sigma \left[\frac{(x^2 + a^2)^{-1/2}}{-\frac{1}{2}}\right]_0^R = -2kx\pi\sigma\left(\frac{1}{\sqrt{x^2 + R^2}} - \frac{1}{x}\right)$$

or

$$E_x = 2\pi k\sigma\left(1 - \frac{x}{\sqrt{x^2 + R^2}}\right) \qquad \text{23-11}$$

$\vec{E}$ on the axis of a disk charge

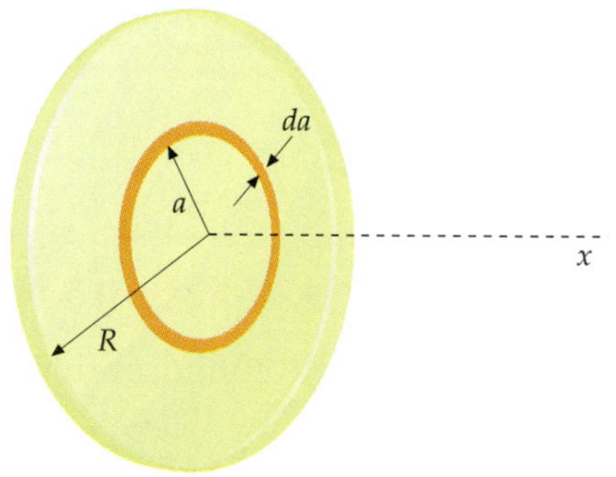

Figure 23-7 A uniform disk of charge can be treated as a set of ring charges, each of radius a and thickness da and carrying a charge $dq = \sigma\, dA = (Q/\pi R^2) 2\pi a\, da$.

When we are very far from the disk, we expect it to look like a point charge. If we merely replace $x^2 + R^2$ with x^2 for $x >> R$, we get $E_x \to 0$. Although this is correct, it does not tell us anything about how E_x depends on x for large x. We can find this dependence by using the binomial expansion, $(1 + \epsilon)^n \approx 1 + n\epsilon$, for $\epsilon \ll 1$. Using this expansion on the second term in Equation 23-11, we obtain

$$\frac{x}{\sqrt{x^2 + R^2}} = \frac{x}{x(1 + R^2/x^2)^{1/2}} = \left(1 + \frac{R^2}{x^2}\right)^{-1/2} \approx 1 - \frac{R^2}{2x^2} + \cdots$$

If we use just the first term, we get $E_x \approx 0$ as before. But if we use the first two terms, Equation 23-11 becomes

$$E_x \approx 2\pi k\sigma\left(1 - 1 + \frac{R^2}{2x^2} + \cdots\right) = \frac{k\pi R^2\sigma}{x^2} = \frac{kQ}{x^2}$$

where $Q = \sigma\pi R^2$ is the total charge on the disk.

$\vec{E}$ due to an Infinite Plane of Charge

The field of an infinite plane of charge can be obtained from Equation 23-11 by either letting R go to infinity or letting x go to zero. Then

$$E_x = 2\pi k\sigma, \qquad x > 0 \tag{23-12a}$$

$\vec{E}$ near an infinite plane of charge

Thus, the field due to an infinite-plane charge distribution is uniform; that is, the field does not depend on x. On the other side of the infinite plane, for negative values of x, the field points in the negative x direction, so

$$E_x = -2\pi k\sigma, \qquad x < 0 \tag{23-12b}$$

As we move along the x axis, the electric field jumps from $-2\pi k\sigma\hat{i}$ to $+2\pi k\sigma\hat{i}$ when we pass through an infinite plane of charge (Figure 23-8). There is thus a discontinuity in E_x in the amount $4\pi k\sigma$.

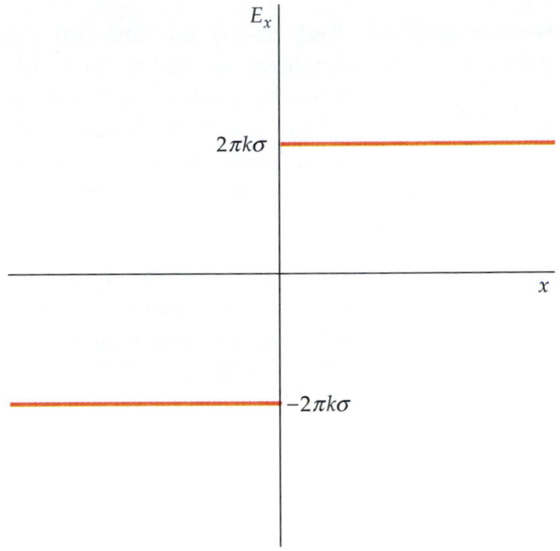

Figure 23-8 Graph showing the discontinuity of $\vec{E}$ at a plane charge.

Example 23-3

A disk of radius 5 cm carries a uniform surface charge density of 4 μC/m^2. Using reasonable approximations, find the electric field on the axis of the disk at distances of (*a*) 0.01 cm, (*b*) 0.03 cm, (*c*) 6 m, and (*d*) 6 cm.

Picture the Problem For (*a*) and (*b*), the field point is very near the disk compared with its radius, so we can approximate the disk as an infinite plane. For (*c*), the field point is far enough from the disk that we can approximate the disk as a point charge. (*d*) Since 6 cm is neither much less than nor much greater than the radius of 5 cm, we use the exact expression.

(*a*) The electric field near the disk is that due to an infinite plane charge:

$$E_x = 2\pi k\sigma = 2\pi(8.99 \times 10^9\ \mathrm{N{\cdot}m^2/C^2})(4 \times 10^{-6}\ \mathrm{C/m^2}) = 226\ \mathrm{kN/C}$$

(*b*) Since 0.03 cm is still very near the disk, the disk still looks like an infinite plane charge:

$$E_x = 2\pi k\sigma = 226\ \mathrm{kN/C}$$

(*c*) 1. Far from the disk, the field is approximately that due to a point charge:

$$E_x = \frac{kQ}{x^2}$$

2. The total charge Q is:

$$Q = \sigma\pi r^2 = (4\ \mu\mathrm{C/m^2})\pi(0.05\ \mathrm{m})^2 = 31.4\ \mathrm{nC}$$

3. Substitute this value into step 1:

$$E_x = \frac{(8.99 \times 10^9\ \mathrm{N{\cdot}m^2/C^2})(31.4 \times 10^{-9}\ \mathrm{C})}{(6\ \mathrm{m})^2} = 7.84\ \mathrm{N/C}$$

(*d*) For $x = 6$ cm, we use the exact expression for E_x:

$$E_x = 2\pi k\sigma\left[1 - \frac{x}{\sqrt{(x^2 + R^2)}}\right]$$

$$= (226\ \mathrm{kN/C})\left[1 - \frac{6\ \mathrm{cm}}{\sqrt{(6\ \mathrm{cm})^2 + (5\ \mathrm{cm})^2}}\right]$$

$$= (226\ \mathrm{kN/C})(1 - 0.768) = 52.4\ \mathrm{kN/C}$$

Remarks Note that in (*d*) we did not need to convert from centimeters to meters to find $x/\sqrt{x^2 + R^2}$ because the units cancel. Figure 23-9 shows E_x versus x for the disk charge in this example, for an infinite plane with the same charge density, and for a point charge.

Exercise Calculate E_x for parts (*a*) and (*b*) to five significant figures using the infinite plane approximation and then using the exact expression of Equation 23-11 and compare your results. (*Answers* (*a*) $E_x \approx 225.94$ kN/C and $E_x = 225.49$ kN/C, a difference of about 0.2%; (*b*) $E_x \approx 225.94$ kN/C and $E_x = 224.58$ kN/C, a difference of about 0.6%.)

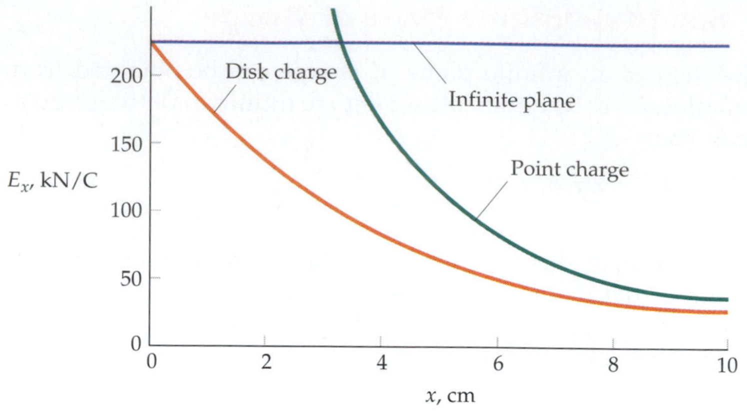

Figure 23-9

23-2 Gauss's Law

The qualitative description of the electric field using electric field lines, discussed in Chapter 22, is related to a mathematical equation known as Gauss's law, which relates the electric field on a closed surface to the net charge within the surface. Gauss's law is the basis for one of Maxwell's equations, the fundamental equations of electromagnetism, which is discussed in Chapter 32. For static charges, Gauss's law and Coulomb's law are equivalent. Electric fields arising from some symmetrical charge distributions, such as a spherical shell of charge or an infinite line of charge, can be easily calculated using Gauss's law. In this section, we give an argument for the validity of Gauss's law based on the properties of electric field lines. A rigorous derivation of Gauss's law is given in Section 23-6.

Figure 23-10 shows a surface of arbitrary shape enclosing a dipole. The number of electric field lines emanating from the positive charge and crossing the surface to the outside depends on where the surface is drawn, but it is exactly equal to the number of lines entering the surface and ending on the negative charge. If we count the number of lines leaving as positive and the number entering as negative, the net number leaving and entering is zero. For surfaces enclosing other types of charge distributions, such as that shown in Figure 23-11, *the net number of lines leaving any surface enclosing the charges is proportional to the net charge enclosed by the surface.* This rule is a qualitative statement of Gauss's law.

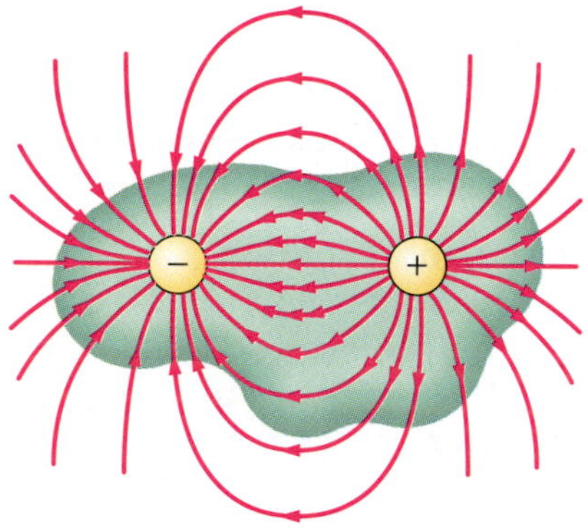

Figure 23-10 A surface of arbitrary shape enclosing an electric dipole. As long as the surface encloses both charges, the number of lines leaving the surface is exactly equal to the number of lines entering the surface no matter where the surface is drawn.

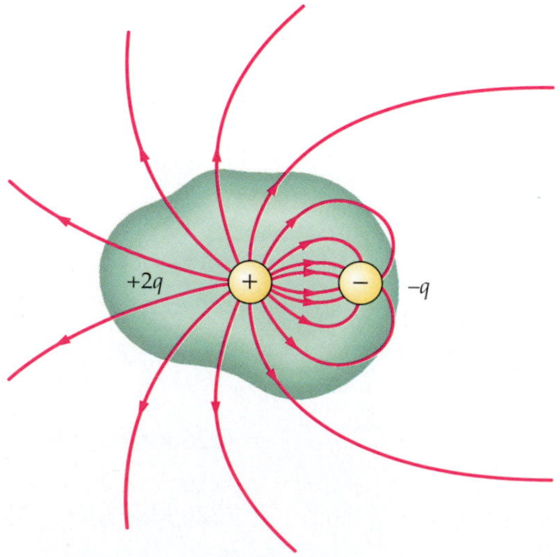

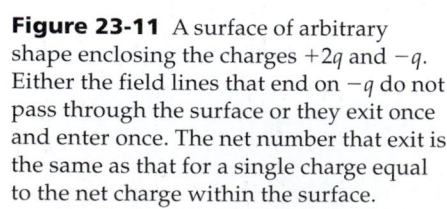

Figure 23-11 A surface of arbitrary shape enclosing the charges $+2q$ and $-q$. Either the field lines that end on $-q$ do not pass through the surface or they exit once and enter once. The net number that exit is the same as that for a single charge equal to the net charge within the surface.

Electric Flux

The mathematical quantity that corresponds to the number of field lines crossing a surface is called the **electric flux** ϕ, which for a surface perpendicular to $\vec{E}$ (Figure 23-12) is defined as the product of the magnitude of the field E and the area A:

$$\phi = EA$$

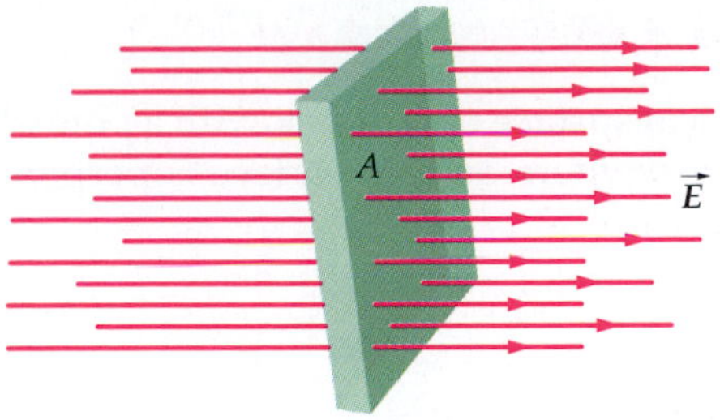

Figure 23-12 Electric field lines of a uniform field crossing an area A that is perpendicular to the field. The product EA is the electric flux through the area.

The units of flux are $\mathrm{N{\cdot}m^2/C}$. Since E is proportional to the number of lines per unit area, the flux is proportional to the number of field lines through the area.

In Figure 23-13, the surface of area A_2 is not perpendicular to the electric field $\vec{E}$. However, the number of lines that cross area A_2 is the same as the number that cross area A_1, which is perpendicular to $\vec{E}$. These areas are related by

$$A_2 \cos\theta = A_1 \qquad \text{23-13}$$

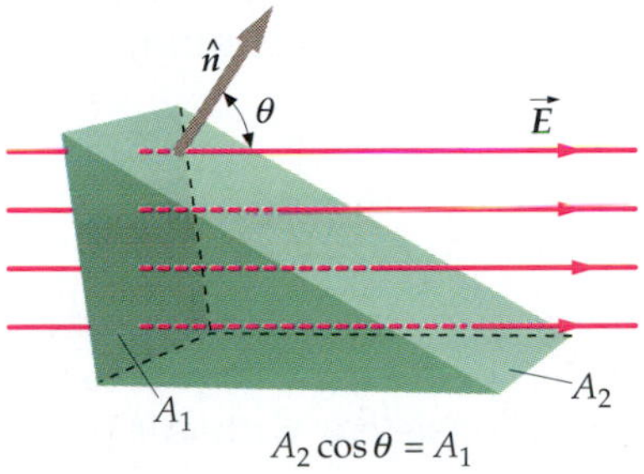

Figure 23-13 Electric field lines of a uniform electric field that is perpendicular to the area A_1 but makes an angle θ with the unit vector $\hat{n}$ that is normal to the area A_2. When $\vec{E}$ is not perpendicular to the area, the flux is E_nA, where $E_n = E\cos\theta$ is the component of $\vec{E}$ that is perpendicular to the area. The flux through A_2 is the same as the flux through A_1.

where θ is the angle between $\vec{E}$ and the unit vector $\hat{n}$ that is perpendicular to the surface A_2, as shown in the figure. The flux through a surface that is not perpendicular to $\vec{E}$ is defined to be

$$\phi = \vec{E}\cdot\hat{n}A = EA\cos\theta = E_nA$$

where $E_n = \vec{E}\cdot\hat{n}$ is the component of $\vec{E}$ that is perpendicular, or normal, to the surface.

Figure 23-14 shows an arbitrary surface over which $\vec{E}$ may vary. If the area element ΔA_i that we choose is small enough, it can be considered to be a plane, and the variation of the electric field across the element can be neglected. The flux of the electric field through this element is

$$\Delta\phi_i = \vec{E}\cdot\hat{n}_i\,\Delta A_i$$

where $\hat{n}_i$ is the unit vector perpendicular to the element. If the surface is curved, the unit vectors for different elements will have different directions. The total flux through the surface is the sum of $\Delta\phi_i$ over all the elements. In the limit as the number of elements approaches infinity and the area of each element approaches zero, this sum becomes an integral. The general definition of electric flux is thus

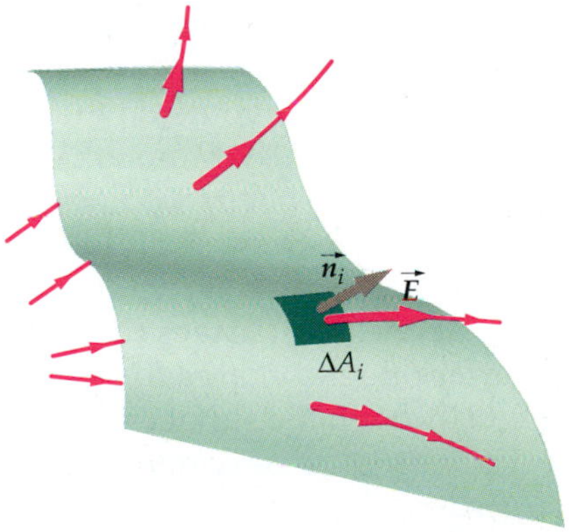

Figure 23-14 When $\vec{E}$ varies in either magnitude or direction, the area of the surface is divided into small elements ΔA_i. The flux through the area is computed by summing $\vec{E}\cdot\hat{n}\,\Delta A_i$ over all the area elements.

$$\phi = \lim_{\Delta A_i \to 0} \sum_i \hat{n}_i\,\Delta A_i = \int_S \vec{E}\cdot\hat{n}\,dA \qquad \text{23-14}$$

Definition—Electric flux

where the S reminds us that we are integrating over a surface.

On a *closed* surface, the normal unit vector $\hat{n}$ is defined to be outward at each point. The integral over a closed surface is indicated by the symbol $\oint$. The total or net flux through a closed surface is therefore written

$$\phi_{\text{net}} = \oint_S \vec{E}\cdot\hat{n}\,dA = \oint_S E_n\,dA \qquad \text{23-15}$$

The net flux ϕ_{net} through the closed surface is positive or negative, depending on whether $\vec{E}$ is predominantly outward or inward at the surface.

Quantitative Statement of Gauss's Law

Figure 23-15 shows a spherical surface of radius R with a point charge Q at its center. The electric field everywhere on this surface is perpendicular to the surface and has the magnitude

$$E_n = \frac{kQ}{R^2}$$

The net flux through this spherical surface is

$$\phi_{net} = \oint_S E_n\,dA = E_n \oint_S dA$$

where we have taken E_n out of the integral because it is constant everywhere on the surface. The integral of dA over the surface is just the total area of the surface, which equals $4\pi R^2$. Using this and substituting kQ/R^2 for E_n, we obtain

$$\phi_{net} = \frac{kQ}{R^2}4\pi R^2 = 4\pi kQ \qquad \text{23-16}$$

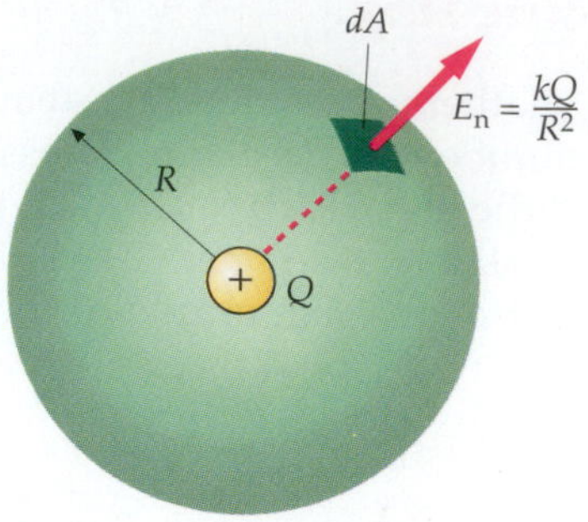

Figure 23-15 A spherical surface enclosing a point charge Q. The same number of electric field lines that pass through this surface will pass through any surface that encloses Q. The flux is easily calculated for a spherical surface. It equals E_n times the surface area, or $E_n \cdot 4\pi R^2$.

Thus, the net flux through a spherical surface with a point charge at its center is independent of the radius of the sphere and is equal to $4\pi k$ times the magnitude of the point charge. This is consistent with our previous observation that the net number of lines going out of a surface is proportional to the net charge inside the surface. *This number of lines is the same for all surfaces surrounding the charge, independent of the shape of the surface.* Thus, the net flux through *any surface* surrounding a point charge Q equals $4\pi kQ$.

We can extend this result to systems containing multiple charges. In Figure 23-16, the surface encloses two point charges q_1 and q_2, and there is a third point charge q_3 outside the surface. Since the electric field at any point on the surface is the vector sum of the electric fields produced by each of the three charges, the net flux $\phi_{net} = \oint \vec{E}\cdot\hat{n}\,dA$ through the surface is just the sum of the fluxes due to the individual charges. The flux due to charge q_3, which is outside the surface, is zero because every field line from q_3 that enters the surface at one point leaves the surface at some other point. The flux through the surface due to charge q_1 is $4\pi kq_1$ and that due to charge q_2 is $4\pi kq_2$. The net flux through the surface therefore equals $4\pi k(q_1 + q_2)$, which may be positive, negative, or zero, depending on the signs and magnitudes of the two charges.

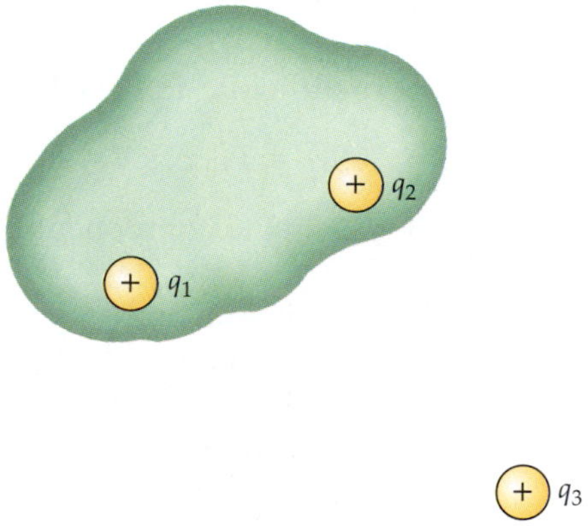

Figure 23-16 A surface enclosing point charges q_1 and q_2 but not q_3. The net flux through this surface is $4\pi k(q_1 + q_2)$.

The net flux through any surface equals $4\pi k$ times the net charge inside the surface:

$$\phi_{net} = \oint_S E_n\,dA = 4\pi kQ_{inside} \qquad \text{23-17}$$

Gauss's law

This is **Gauss's law.** Its validity depends on the fact that the electric field due to a single point charge varies inversely with the square of the distance from the charge. It was this property of the electric field that made it possible to draw a fixed number of electric field lines from a charge and have the density of lines be proportional to the field strength.

It is customary to write the Coulomb constant k in terms of another constant ϵ_0, which is called the **permittivity of free space:**

$$k = \frac{1}{4\pi\epsilon_0} \qquad \text{23-18}$$

Using this notation, Coulomb's law for $\vec{E}$ is written

$$\vec{E} = \frac{1}{4\pi\epsilon_0}\frac{q}{r^2}\hat{r} \qquad \text{23-19}$$

Coulomb's law in terms of ϵ_0

and Gauss's law is written

$$\phi_{\text{net}} = \oint_S E_n\,dA = \frac{1}{\epsilon_0}Q_{\text{inside}} \qquad \text{23-20}$$

Gauss's law in terms of ϵ_0

The value of ϵ_0 in SI units is

$$\epsilon_0 = \frac{1}{4\pi k} = \frac{1}{4\pi(8.99 \times 10^9\ \text{N}\cdot\text{m}^2/\text{C}^2)} = 8.85 \times 10^{-12}\ \text{C}^2/\text{N}\cdot\text{m}^2$$

Gauss's law is valid for all surfaces and all charge distributions. For charge distributions that have high degrees of symmetry, it can be used to calculate the electric field, as we illustrate in the next section. For static charge distributions, Gauss's law and Coulomb's law are equivalent. However, Gauss's law is more general in that it is always valid whether or not the charges are static.

Example 23-4

An electric field is $\vec{E}$ = (200 N/C) $\hat{i}$ for $x > 0$ and $\vec{E}$ = (−200 N/C) $\hat{i}$ for $x < 0$. A cylinder of length 20 cm and radius R = 5 cm has its center at the origin and its axis along the x axis, so that one end is at x = +10 cm and the other is at x = −10 cm (Figure 23-17). (*a*) What is the flux through each end? (*b*) What is the flux through the curved surface of the cylinder? (*c*) What is the net outward flux through the entire closed surface? (*d*) What is the net charge inside the cylinder?

Figure 23-17

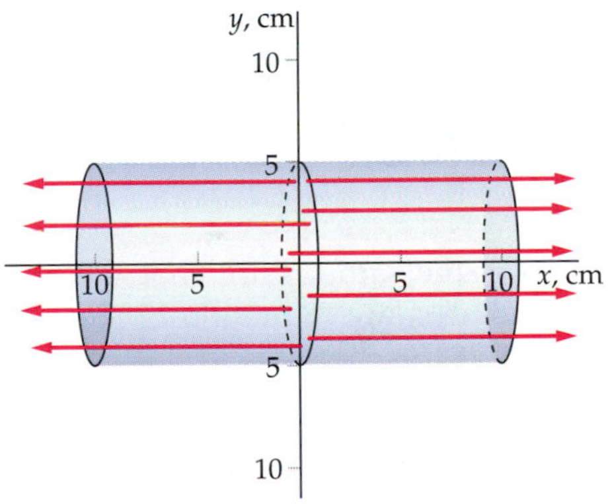

Picture the Problem The field at both circular faces of the cylinder is parallel to the outward vector normal to the surface, so the flux is just EA. There is no flux through the curved surface because the normal to that surface is perpendicular to $\vec{E}$. The net flux through the closed surface is related to the net charge inside by Gauss's law.

(*a*) 1. Calculate the flux through the right circular surface:

$$\phi_{\text{right}} = \vec{E}_{\text{right}}\cdot\hat{n}_{\text{right}}A = (200\ \text{N/C})\hat{i}\cdot\hat{i}\ (\pi)(0.05\ \text{m})^2 = 1.57\ \text{N}\cdot\text{m}^2/\text{C}$$

2. Calculate the flux through the left circular surface:

$$\phi_{\text{left}} = \vec{E}_{\text{left}}\cdot\hat{n}_{\text{left}}A = (-200\ \text{N/C})\hat{i}\cdot(-\hat{i})(\pi)(0.05\ \text{m})^2 = +1.57\ \text{N}\cdot\text{m}^2/\text{C}$$

(*b*) The flux through the curved surface is zero because $\vec{E}$ is perpendicular to $\hat{n}$:

$$\phi_{\text{curved}} = \vec{E}_{\text{curved}}\cdot\hat{n}_{\text{curved}}A = 0$$

(*c*) The total flux is the sum through all surfaces:

$$\phi_{\text{net}} = \phi_{\text{right}} + \phi_{\text{left}} + \phi_{\text{curved}} = 1.57\ \text{N}\cdot\text{m}^2/\text{C} + 1.57\ \text{N}\cdot\text{m}^2/\text{C} + 0 = 3.14\ \text{N}\cdot\text{m}^2/\text{C}$$

(*d*) Gauss's law relates the charge inside to the net flux:

$$Q_{\text{inside}} = \epsilon_0\phi_{\text{net}} = (8.85 \times 10^{-12}\ \text{C}^2/\text{N}\cdot\text{m}^2)(3.14\ \text{N}\cdot\text{m}^2/\text{C}) = 2.78 \times 10^{-11}\ \text{C}$$

23-3 Calculating $\vec{E}$ From Gauss's Law

The electric field due to a highly symmetrical charge distribution can often be easily calculated using Gauss's law. We first find a surface, called a **Gaussian surface**, on which the magnitude of the field E is constant. The flux through this surface will be proportional to the field E on the surface. Gauss's law then relates this field to the charge inside the surface.

Plane Geometry

$\vec{E}$ due to an Infinite Plane of Charge Figure 23-18 shows an infinite plane of charge of surface charge density σ. By symmetry, $\vec{E}$ must be perpendicular to the plane and can depend only on the distance from it. Also, $\vec{E}$ must have the same magnitude but the opposite direction at points the same distance above and below the plane. For our Gaussian surface, we choose a pillbox-shaped cylinder as shown. Let each face of the cylinder have an area A. Since $\vec{E}$ is parallel to the curved cylindrical surface, there is no flux through this surface. The flux through each face is $E_n A$, so the total flux is $2E_n A$. The net charge inside the surface is σA. Gauss's law then gives

$$\phi_{\text{net}} = \oint E_n\, dA = \frac{1}{\epsilon_0} Q_{\text{inside}}$$

$$2E_n A = \frac{1}{\epsilon_0}\sigma A$$

or

$$E_n = \frac{\sigma}{2\epsilon_0} = 2\pi k\sigma \qquad 23\text{-}21$$

$\vec{E}$ near an infinite plane of charge

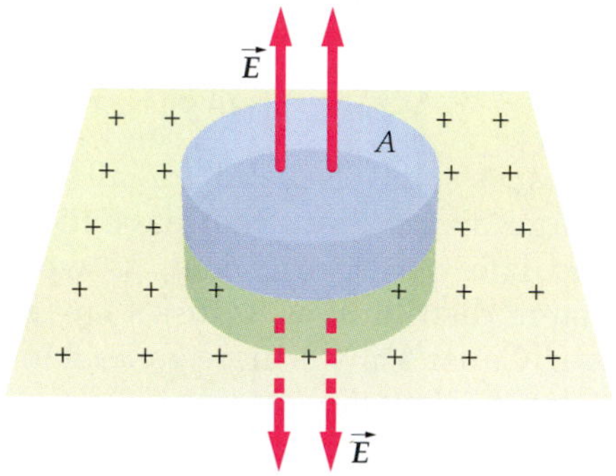

Figure 23-18 Gaussian surface for the calculation of $\vec{E}$ due to an infinite plane of charge. On the upper and lower faces of this pillbox, $\vec{E}$ is perpendicular to the surface and constant in magnitude.

This is the same result that we obtained with much more difficulty using Coulomb's law (Equation 23-12*a*). Note that the field is discontinuous at the plane. If the charge is in the xy plane, the field is $E_z = \sigma/2\epsilon_0$ just above the plane and $E_z = -\sigma/2\epsilon_0$ just below the plane. Thus, the field is discontinuous by $\Delta E_z = \sigma/2\epsilon_0 - (-\sigma/2\epsilon_0) = \sigma/\epsilon_0$.

Example 23-5

In Figure 23-19, an infinite plane of surface charge density $\sigma = +4.5\ \text{nC/m}^2$ lies in the yz plane, and a second infinite plane of surface charge density $\sigma = -4.5\ \text{nC/m}^2$ lies in a plane parallel to the yz plane at $x = 2$ m. Find the electric field at (*a*) $x = 1.8$ m and (*b*) $x = 5$ m.

Picture the Problem Each plane produces a uniform electric field of magnitude $E = \sigma/2\epsilon_0$. We use superposition to find the resultant field. Between the planes the fields add, producing a net field of magnitude σ/ϵ_0 in the positive x direction. For $x > 2$ m or $x < 0$, the fields point in opposite directions and cancel.

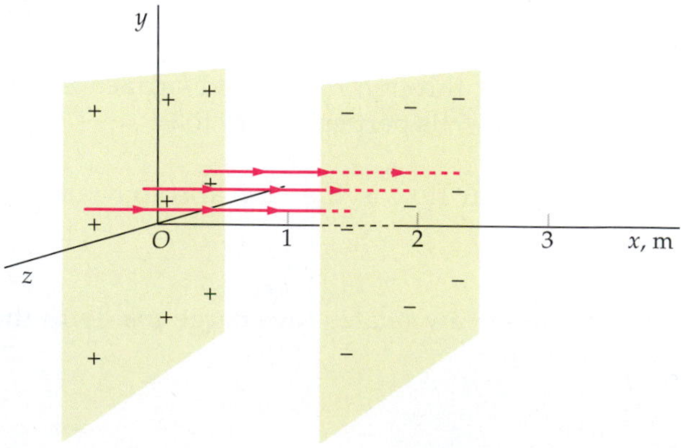

Figure 23-19

(*a*) 1. Calculate the magnitude of the field E produced by each plane:	$E = \dfrac{\sigma}{2\epsilon_0} = \dfrac{4.5 \times 10^{-9}\ \mathrm{C/m^2}}{2(8.85 \times 10^{-12}\ \mathrm{C^2/N \cdot m^2})} = 254\ \mathrm{N/C}$
2. At $x = 1.8$ m, between the planes, the field due to each plane points in the positive x direction:	$E_{x,\mathrm{net}} = E_1 + E_2 = 254\ \mathrm{N/C} + 254\ \mathrm{N/C} = 508\ \mathrm{N/C}$
(*b*) At $x = 5$ m, the fields due to the two planes are in opposite directions:	$E_{x,\mathrm{net}} = E_1 - E_2 = 0$

Remarks Because the two planes carry equal and opposite charge densities, the electric field lines begin on the positive plane and end on the negative plane. $\vec{E}$ is zero except between the planes. Note that $E_{x,\mathrm{net}} = 508$ N/C not just at $x = 1.8$ m but at any point between the planes.

Spherical Geometry

To calculate the electric field due to spherically symmetric charge distributions, we use a spherical surface for our Gaussian surface. We illustrate this by first finding the electric field at a distance r from a point charge q. We choose a spherical surface of radius r centered at the charge for our Gaussian surface. By symmetry, $\vec{E}$ must be radial, and its magnitude can depend only on the distance from the charge. The normal component of $\vec{E}$, $E_n = \vec{E} \cdot \hat{n} = E_r$, has the same value everywhere on the spherical surface. The net flux through this surface is thus

$$\phi_{\mathrm{net}} = \oint \vec{E} \cdot \hat{n}\, dA = \oint E_r\, dA = E_r \oint dA$$

But $\oint dA$ is equal to $4\pi r^2$, the total area of the spherical surface. Since the total charge inside the surface is just the point charge q, Gauss's law gives

$$E_r 4\pi r^2 = \frac{q}{\epsilon_0}$$

or

$$E_r = \frac{1}{4\pi\epsilon_0} \frac{q}{r^2}$$

which is Coulomb's law. We have thus derived Coulomb's law from Gauss's law. Since Gauss's law can also be obtained from Coulomb's law (see Section 23-6), we have shown that the two laws are equivalent for static charges.

$\vec{E}$ due to a Spherical Shell of Charge

Consider a uniformly charged spherical shell of radius R and total charge Q. By symmetry, $\vec{E}$ must be radial, and its magnitude can depend only on the distance r from the center of the sphere. In Figure 23-20, we have chosen a spherical Gaussian surface of radius $r > R$. Since $\vec{E}$ is perpendicular to this surface and constant in magnitude everywhere on it, the flux through the surface is

$$\phi_{\mathrm{net}} = \oint E_r\, dA = E_r 4\pi r^2$$

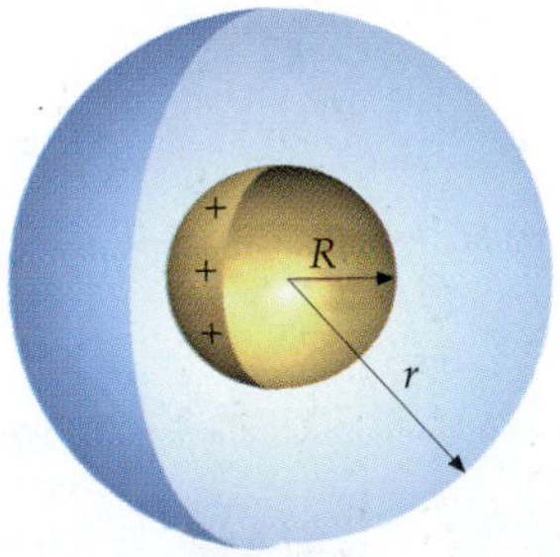

Figure 23-20 Spherical Gaussian surface of radius $r > R$ for the calculation of the electric field outside a uniformly charged spherical shell of radius R.

Since the total charge inside the Gaussian surface is the total charge on the shell Q, Gauss's law gives

$$E_r 4\pi r^2 = \frac{Q}{\epsilon_0}$$

or

$$E_r = \frac{1}{4\pi\epsilon_0}\frac{Q}{r^2}, \qquad r > R \qquad \text{23-22a}$$

$\vec{E}$ outside a spherical shell of charge

Thus, the electric field outside a uniformly charged spherical shell is the same as if all the charge were at the center of the shell.

If we choose a spherical Gaussian surface inside the shell, where $r < R$, the net flux is again $E_r 4\pi r^2$, but the total charge inside the surface is zero. Therefore, for $r < R$, Gauss's law gives

$$\phi_{\text{net}} = E_r 4\pi r^2 = 0$$

and

$$E_r = 0, \qquad r < R \qquad \text{23-22b}$$

$\vec{E}$ inside a spherical shell of charge

These results can also be obtained by direct integration of Coulomb's law, but that calculation is much more difficult.

Figure 23-21 shows E_r versus r for a spherical-shell charge distribution. Again, note that the electric field is discontinuous at $r = R$, where the surface charge density is σ. Just outside the shell at $r \approx R$, the electric field is $E_r = Q/4\pi\epsilon_0 R^2 = \sigma/\epsilon_0$ since $\sigma = Q/4\pi R^2$. Since the field just inside the shell is zero, the electric field is discontinuous by the amount σ/ϵ_0 as we pass through the shell.

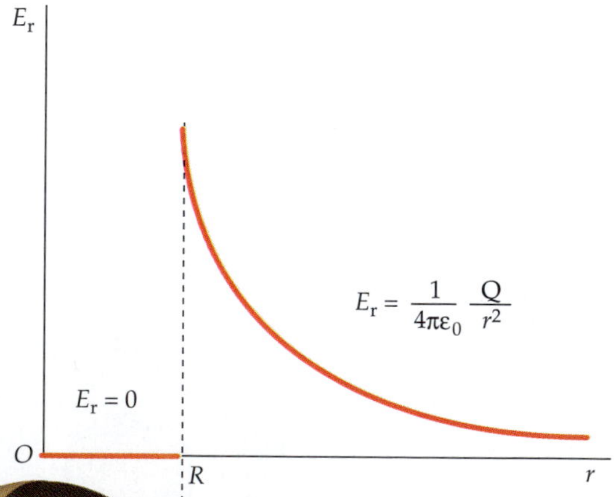

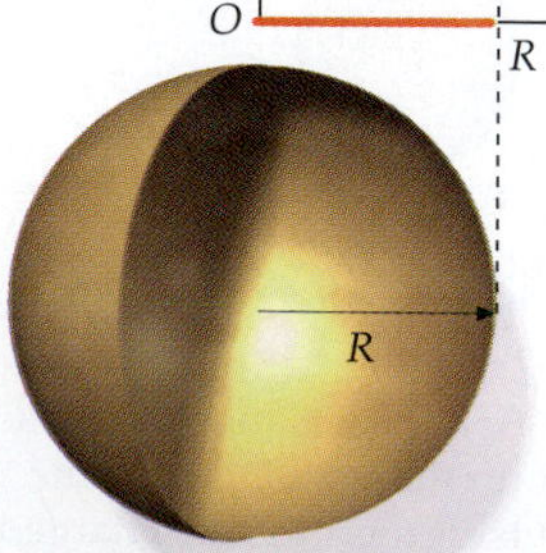

(a)

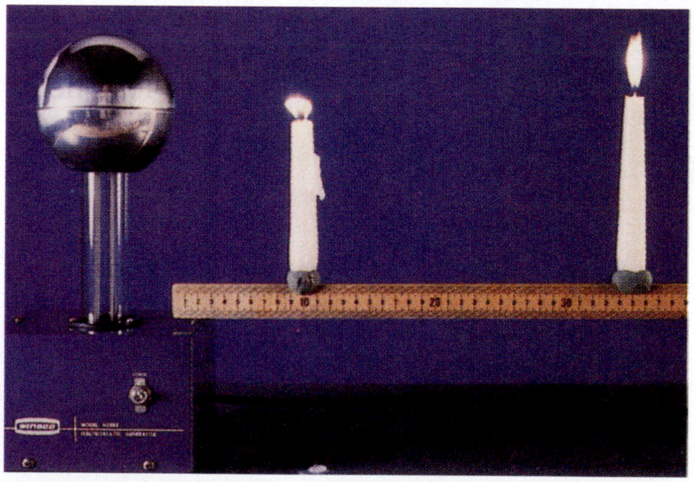

(b)

Figure 23-21 (*a*) A plot of E_r versus r for a spherical-shell charge distribution. The electric field is discontinuous at $r = R$, where there is a surface charge of density σ. (*b*) The decrease in E_r over distance due to a charged spherical shell is evident by the effect of the field on the flames of these two candles. The spherical shell at the left (part of a Van de Graaff generator, a device that is discussed in Chapter 24) carries a large negative charge that attracts the positive ions in the nearby candle flame. The flame at right, which is much farther away, is not affected.

Example 23-6

A spherical shell of radius $R = 3$ m has its center at the origin and carries a surface charge density of $\sigma = 3\ \text{nC/m}^2$. A point charge $q = 250$ nC is on the y axis at $y = 2$ m. Find the electric field on the x axis at (*a*) $x = 2$ m and (*b*) $x = 4$ m.

Picture the Problem We find the field due to the point charge and that due to the spherical shell and sum the field vectors. For (*a*), the field point is inside the shell, so the field is due only to the point charge (Figure 23-22*a*). For (*b*), the field point is outside the shell, so the shell can be considered to be a point charge at the origin. We then find the field due to two point charges (Figure 23-22*b*).

Figure 23-22

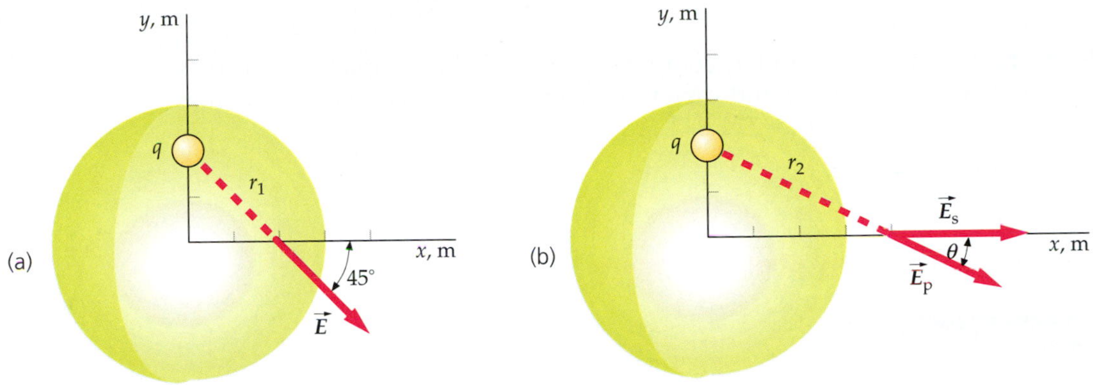

(*a*) 1. Inside the shell, $\vec{E}_1$ is due only to the point charge: $\vec{E}_1 = \dfrac{kq}{r_1^2}\hat{r}$

2. Calculate the distance r_1: $r_1 = \sqrt{(2\ \text{m})^2 + (2\ \text{m})^2} = \sqrt{8}\ \text{m}$

3. Use r_1 to calculate the magnitude of the field: $E_1 = \dfrac{kq}{r_1^2} = \dfrac{(8.99 \times 10^9\ \text{N}\cdot\text{m}^2/\text{C}^2)(250 \times 10^{-9}\ \text{C})}{(\sqrt{8}\ \text{m})^2} = 281\ \text{N/C}$

4. From Figure 23-22*a*, we can see that the field makes an angle of $-45°$ with the x axis: $\theta = -45°$

(*b*) 1. Outside of its perimeter, the shell can be treated as a point charge at the origin, and the field due to the shell $\vec{E}_s$ is therefore along the x axis: $\vec{E}_s = \dfrac{kQ}{x_2^2}\hat{i}$

2. Calculate the total charge Q on the shell: $Q = \sigma 4\pi R^2 = (3\ \text{nC/m}^2)4\pi(3\ \text{m})^2 = 339\ \text{nC}$

3. Use Q to calculate the field due to the shell: $E_s = \dfrac{(8.99 \times 10^9\ \text{N}\cdot\text{m}^2/\text{C}^2)(339 \times 10^{-9}\ \text{C})}{(4\ \text{m})^2} = 190\ \text{N/C}$

4. The field due to the point charge is: $\vec{E}_p = \dfrac{kq}{r_2^2}\hat{r}_2$

5. Calculate the distance from the point charge q on the y axis to the field point at $x = 4$ m: $r_2 = \sqrt{(2\ \text{m})^2 + (4\ \text{m})^2} = \sqrt{20}\ \text{m}$

6. Calculate the magnitude of the field due to the point charge: $E_p = \dfrac{kq}{r_2^2} = \dfrac{(8.99 \times 10^9\ \text{N}\cdot\text{m}^2/\text{C}^2)(250 \times 10^{-9}\ \text{C})}{(\sqrt{20}\ \text{m})^2} = 112\ \text{N/C}$

7. This field makes an angle θ with the x axis, where: $\cos\theta = \dfrac{4}{\sqrt{20}}$

$$\sin\theta = \frac{-2}{\sqrt{20}}$$

8. The x and y components of this field are thus:

$$E_{px} = E_p \cos\theta = (112\text{ N/C})\left(\frac{4}{\sqrt{20}}\right) = 100\text{ N/C}$$

$$E_{py} = E_p \sin\theta = (112\text{ N/C})\left(-\frac{2}{\sqrt{20}}\right) = -50\text{ N/C}$$

9. Find the x and y components of the net electric field:

$$E_x = E_{sx} + E_{px} = 190\text{ N/C} + 100\text{ N/C} = 290\text{ N/C}$$

$$E_y = E_{sy} + E_{py} = 0 - 50\text{ N/C} = -50\text{ N/C}$$

Remark Giving the x and y components of a field completely specifies the field. The magnitude and direction of the net field can be found from $E = \sqrt{E_x^2 + E_y^2}$ and $\tan\theta' = E_y/E_x$.

$\vec{E}$ due to a Uniformly Charged Sphere

Example 23-7

Find the electric field (*a*) outside and (*b*) inside a uniformly charged sphere of radius R carrying a total charge Q that is uniformly distributed throughout the volume of the sphere with charge density $\rho = Q/V$, where $V = \frac{4}{3}\pi R^3$ is the volume of the sphere.

Picture the Problem By symmetry, the electric field must be radial. (*a*) To find E_r outside the charged sphere, we choose a spherical Gaussian surface of radius $r > R$ (Figure 23-23a). (*b*) To find E_r inside the charge we choose a spherical Gaussian surface of radius $r < R$ (Figure 23-23b). On each of these surfaces, E_r is constant. Gauss's law then relates E_r to the total charge inside the surface.

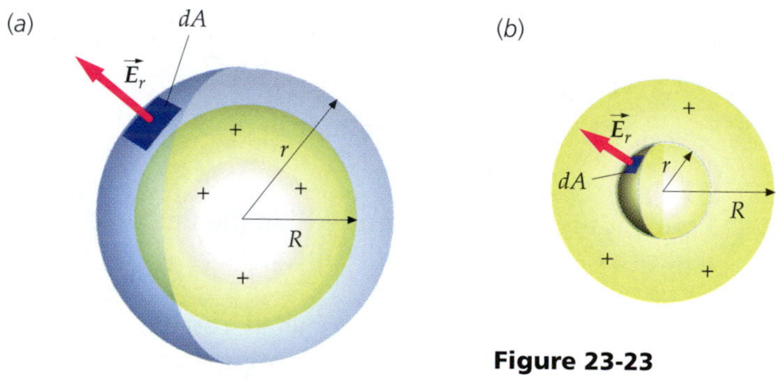

Figure 23-23

(*a*) 1. (Outside) Relate the flux through the Gaussian surface to the electric field E_r on the Gaussian surface at $r > R$:

$$\phi_{\text{net}} = E_r 4\pi r^2$$

2. Apply Gauss's law to relate the field to the total charge inside the surface, which is Q:

$$E_r 4\pi r^2 = \frac{Q}{\epsilon_0}$$

3. Solve for E_r:

$$E_r = \frac{1}{4\pi\epsilon_0}\frac{Q}{r^2}, \qquad r > R$$

(*b*) 1. (Inside) Relate the flux through the Gaussian surface to the electric field E_r on the Gaussian surface at $r < R$:

$$\phi_{\text{net}} = E_r 4\pi r^2$$

2. Apply Gauss's law to relate the field to the total charge inside the surface Q_{inside}:

$$E_r 4\pi r^2 = \frac{Q_{\text{inside}}}{\epsilon_0}$$

3. The total charge inside the surface is $\rho V'$, where $V' = \frac{4}{3}\pi r^3$ is the volume inside the Gaussian surface:

$$Q_{\text{inside}} = \rho V' = \left(\frac{Q}{V}\right)V' = \left(\frac{Q}{\frac{4}{3}\pi R^3}\right)\left(\frac{4}{3}\pi r^3\right) = Q\frac{r^3}{R^3}$$

4. Substitute this value for Q_{inside} and solve for E_r:

$$E_r 4\pi r^2 = \frac{Q_{\text{inside}}}{\epsilon_0} = \frac{1}{\epsilon_0} Q \frac{r^3}{R^3}$$

$$E_r = \frac{1}{4\pi\epsilon_0}\frac{Q}{R^3} r, \qquad r \leq R$$

Remarks Figure 23-24 shows E_r versus r for the charge distribution in this example. Inside a sphere of charge, E_r increases with r. Note that E_r is continuous at $r = R$. A uniformly charged sphere is sometimes used as a model to describe the electric field of an atomic nucleus.

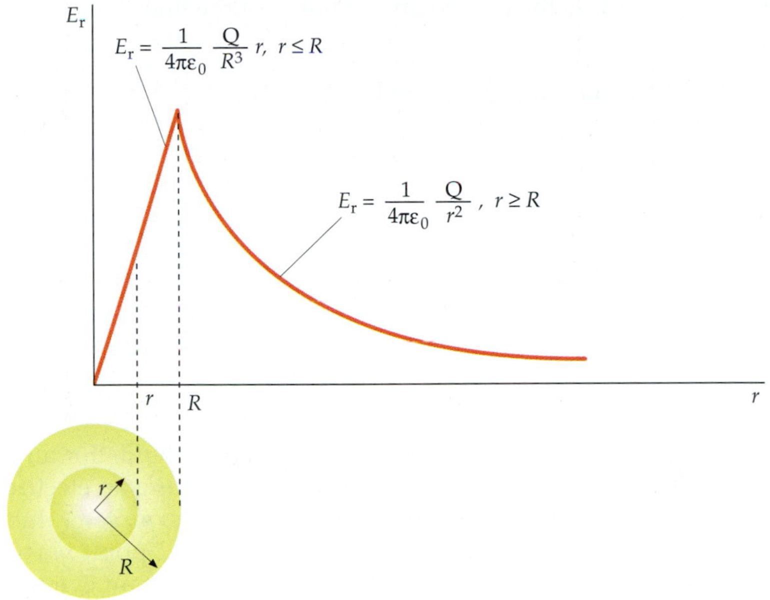

Figure 23-24

We see from Example 23-7 that the electric field at a distance r from the center of a uniformly charged sphere of radius R is given by

$$E_r = \frac{1}{4\pi\epsilon_0}\frac{Q}{r^2}, \qquad r \geq R \tag{23-23a}$$

$$E_r = \frac{1}{4\pi\epsilon_0}\frac{Q}{R^3} r, \qquad r \leq R \tag{23-23b}$$

where Q is the total charge of the sphere.

Cylindrical Geometry

To calculate the electric field due to cylindrically symmetric charge distributions, we use a cylindrical Gaussian surface. We illustrate this by calculating the electric field due to an infinitely long line charge of uniform linear charge density, a problem we have already solved using Coulomb's law.

Example 23-8

Use Gauss's law to find the electric field at a distance r from an infinitely long line charge of uniform charge density λ.

Picture the Problem By symmetry, the electric field lines radiate uniformly from the line, outward if λ is positive and inward if λ is negative. We therefore choose a cylindrical Gaussian surface of length L and radius r (Figure 23-25). The electric field is perpendicular to the cylindrical surface and has the same value E_r everywhere on the surface. The electric flux is then just the product of the electric field and the area of the cylindrical surface, which is $2\pi rL$. There is no flux through the flat surfaces at the ends of the cylinder because $\vec{E} \cdot \hat{n} = 0$ on these surfaces.

Figure 23-25

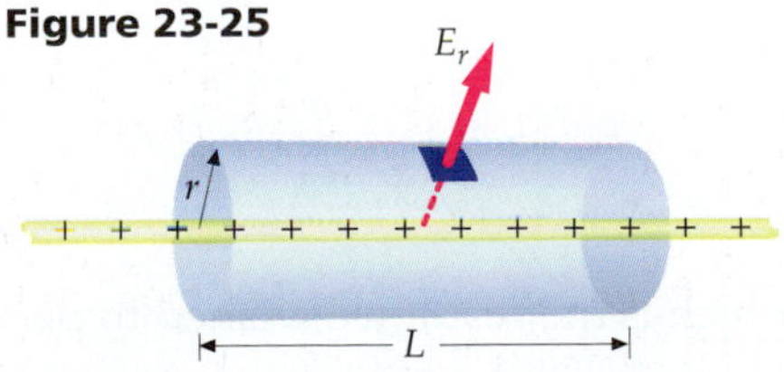

1. Relate the flux through the Gaussian surface to the electric field E_r on the Gaussian surface at $r > R$:
$$\phi_{\text{net}} = E_r \oint dA = E_r 2\pi rL$$
2. Apply Gauss's law to relate the field to the total charge inside the surface Q_{inside}:
$$E_r 2\pi rL = \frac{Q_{\text{inside}}}{\epsilon_0}$$
3. The charge inside is the charge on a length L of the line:
$$Q_{\text{inside}} = \lambda L$$
4. Substitute this value for Q_{inside} and solve for E_r:
$$E_r 2\pi rL = \frac{\lambda L}{\epsilon_0}$$
$$E_r = \frac{1}{2\pi\epsilon_0}\frac{\lambda}{r}$$

Remark Since $1/(2\pi\epsilon_0) = 2k$, the field is $2k\lambda/r$, the same as Equation 23-9 with $r = y$.

It is important to realize that although Gauss's law holds for any surface surrounding any charge distribution, its use in calculating an electric field is limited to charge distributions with a high degree of symmetry. In the preceding calculation, we needed to assume that the field point was very far from the ends of the line charge so that E_n would be constant everywhere on the cylindrical Gaussian surface. This is equivalent to assuming that, at the distance r from the line, the line charge appears to be infinitely long. If we are near the end of a finite line charge, we cannot assume that $\vec{E}$ is perpendicular to the cylindrical surface, or that E_n is constant everywhere on it, so we cannot use Gauss's law to calculate the electric field.

23-4 Discontinuity of E_n

We have seen that the electric field for an infinite plane of charge and a spherical shell of charge is discontinuous by the amount σ/ϵ_0 at a point where there is a surface charge density σ. We now show that this is a general result for the component of the electric field that is perpendicular to a surface carrying a charge density of σ.

Figure 23-26 shows a pill box-shaped Gaussian surface with faces of area A on each side of an arbitrary surface carrying a surface charge density σ. Let the normal component of the electric field be E_{n2} on one side of the surface and E_{n1} on the other side, as shown in the figure. If we make the height of the cylinder very small compared with the radius of the faces, we can neglect the flux through the sides of the cylinder compared with the flux through the faces of area πR^2. The net flux through the Gaussian surface is then $E_{n2}A - E_{n1}A$, and the charge inside the surface is σA. Gauss's law gives

$$E_{n2}A - E_{n1}A = \frac{\sigma A}{\epsilon_0}$$

or

$$E_{n2} - E_{n1} = \frac{\sigma}{\epsilon_0} \qquad 23\text{-}24$$

Discontinuity of E_n at a surface charge

which is the result we wished to prove.

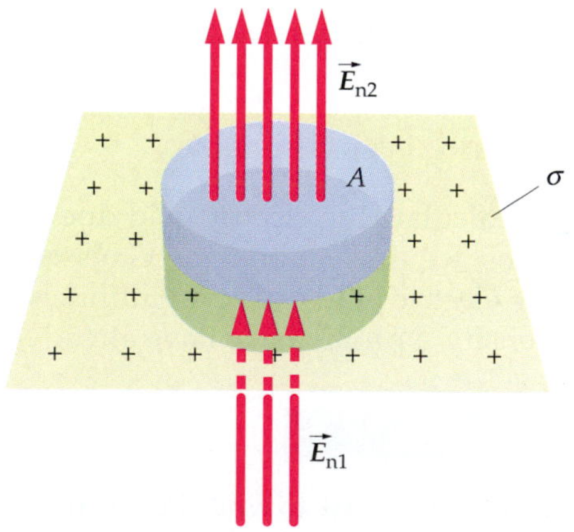

Figure 23-26 A Gaussian, pillbox-shaped surface with faces of area A on each side of a surface charge of charge density of σ. The net flux through the pillbox is $(E_{n2} - E_{n1})A$. The electric field E_{n2} on one side is greater than the electric field E_{n1} on the other side by the amount σ/ϵ_0.

Note that there is no discontinuity in the electric field at the surface of a uniform ball of charge. The field just inside the surface of a charged sphere is the same as the field just outside such a charge distribution, as can be seen from Figure 23-24.

23-5 Charge and Field at Conductor Surfaces

A conductor contains charge that is free to move about the volume of the conductor. If there is an electric field inside the conductor, there will be a net force on this charge causing a momentary electric current (electric currents are discussed in Chapter 26). However, unless there is a source of energy to maintain this current, the free charge in a conductor will merely redistribute itself to create an electric field that cancels the external field within the conductor. The conductor is then said to be in **electrostatic equilibrium.** Thus, in equilibrium, the electric field inside a conductor must be zero. The time to reach equilibrium depends on the conductor. For copper and other good conductors, the time is so small that, for all practical purposes, electrostatic equilibrium is reached instantaneously.

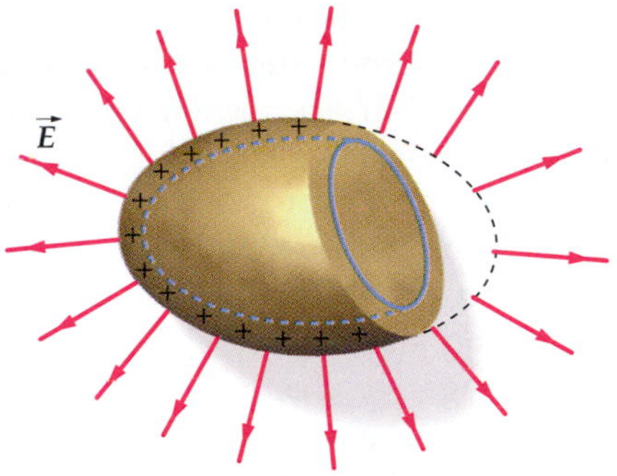

Figure 23-27 A Gaussian surface (the blue line) just inside the surface of a conductor. Since the electric field is zero inside a conductor in electrostatic equilibrium, the net flux through this surface must also be zero. Therefore, the net charge inside this surface must be zero.

We can use Gauss's law to show that any net electric charge on a conductor resides on the surface of the conductor. Consider a Gaussian surface just inside the actual conductor surface in electrostatic equilibrium (Figure 23-27). The electric field is zero everywhere on the Gaussian surface because the surface is completely within the conductor. The net flux through the surface must therefore be zero, and by Gauss's law, the net charge inside the surface must be zero. Thus there can be no net charge inside any surface lying completely within the conductor. If there is any net charge on the conductor, it must be on the conductor's surface. At the surface of a conductor in equilibrium, $\vec{E}$ must be perpendicular to the surface. If the electric field had a tangential component, the free charge would be accelerated tangential to the surface until equilibrium was established.

Since E_n is discontinuous at any surface by the amount σ/ϵ_0, and since $\vec{E}$ is zero inside a conductor, the field just outside a conductor is given by

$$E_n = \frac{\sigma}{\epsilon_0} \qquad 23\text{-}25$$

E_n just outside the surface of a conductor

This result is exactly twice the field produced by an infinite plane of charge. We can understand this result from Figure 23-28. The charge on the conductor consists of two parts: (1) the charge near point P, and (2) all the rest of the charge. The charge near point P looks like an infinite plane and produces a

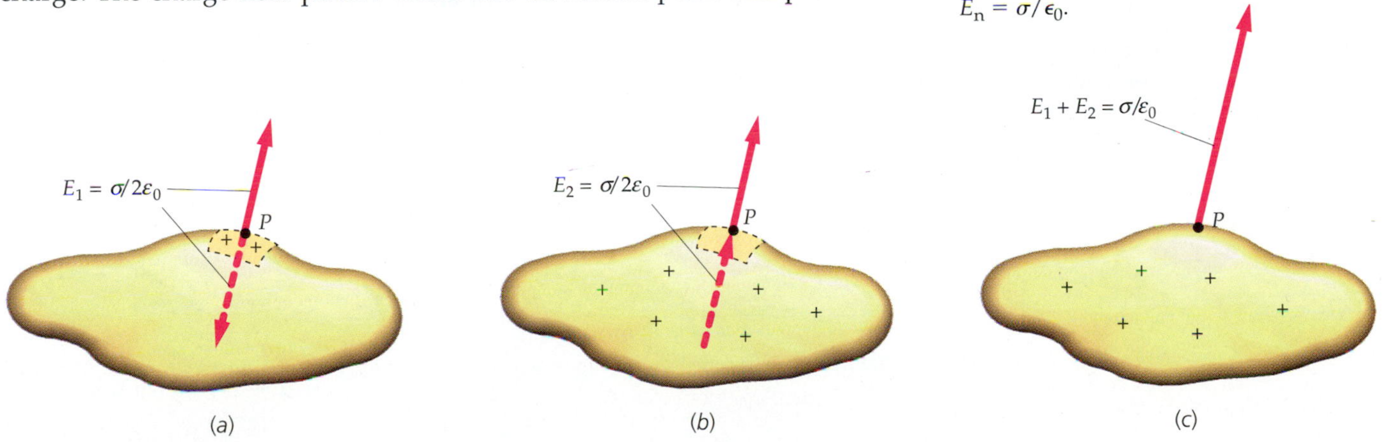

Figure 23-28 An arbitrarily shaped conductor carrying a charge on its surface. (*a*) The charge in the vicinity of point P near the surface looks like an infinite plane of charge, giving an electric field of magnitude $\sigma/2\epsilon_0$ pointing away from the surface both inside and outside the surface. Inside the conductor this field points down from point P. (*b*) Since the net field inside the conductor is zero, the rest of the charge must produce a field of magnitude $\sigma/2\epsilon_0$ in the upward direction. The field due to this charge is the same inside the surface as outside the surface. (*c*) Inside the surface, the fields shown in (*a*) and (*b*) cancel, but outside at point P they add to give $E_n = \sigma/\epsilon_0$.

field of magnitude $\sigma/2\epsilon_0$ just inside and just outside the conductor. The rest of the charge must produce a field of magnitude $\sigma/2\epsilon_0$ that cancels the field inside the conductor. This field due to the rest of the charge adds to the field due to the local charge just outside the conductor to give a total field of σ/ϵ_0.

Figure 23-29 shows a positive point charge q at the center of a spherical cavity inside a spherical conductor. Since the net charge must be zero within any surface drawn within the conductor, there must be a negative charge $-q$ induced in the inside surface. In Figure 23-30, the point charge has been moved so that it is no longer at the center of the cavity. The field lines in the cavity are altered, and the surface charge density of the induced negative charge on the inner surface is no longer uniform. However, the positive surface charge density on the outside surface is not disturbed—it is still uniform—because it is shielded from the cavity by the conductor.

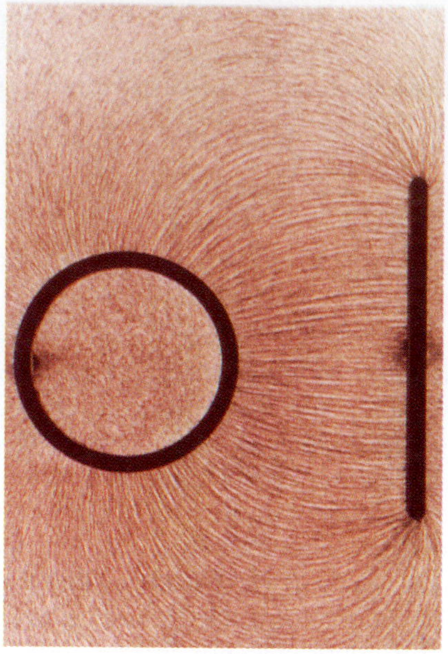

Electric field lines for an oppositely charged cylinder and plate, shown by bits of fine thread suspended in oil. Note that the field lines are perpendicular to the conductors and that there are no lines inside the cylinder.

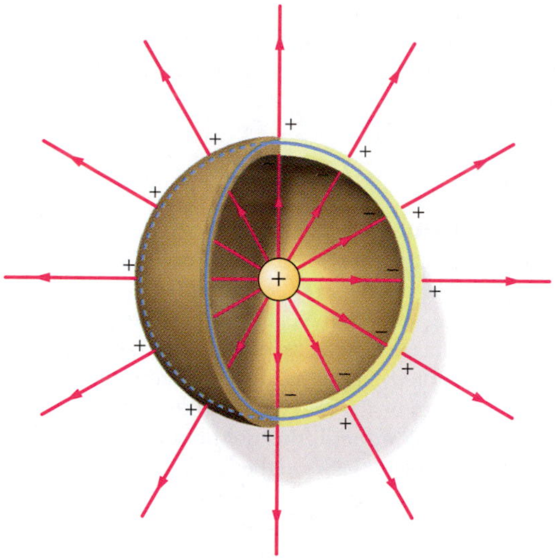

Figure 23-29 A point charge q in the cavity at the center of a thick spherical conducting shell. Since the net charge within the Gaussian surface (indicated in blue) must be zero, a surface charge $-q$ is induced on the inner surface of the shell, and since the conductor is neutral, an equal but opposite charge $+q$ is induced on the outer surface. Electric field lines begin on the point charge and end on the inner surface. The field lines begin again on the outer surface.

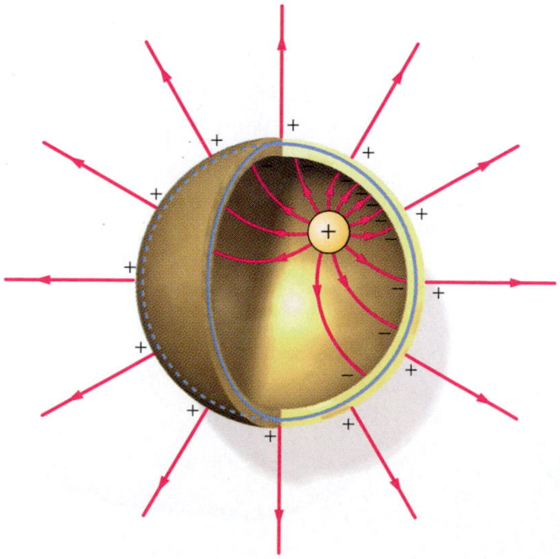

Figure 23-30 The same conductor as in Figure 23-29 with the point charge moved away from the center of the sphere. The charge on the outer surface and the electric field lines outside the sphere are not affected.

Example 23-9

An uncharged square conducting slab of negligible thickness and with 4-m sides is placed in an external uniform field $\vec{E} = (450\ \text{kN/C})\,\hat{i}$ that is perpendicular to the faces of the slab (Figure 23-31). (*a*) Find the charge density on each face of the slab. (*b*) A net charge of 96 μC is placed on the slab. Find the new charge density on each face and the electric field near each face but far from the edges of the slab.

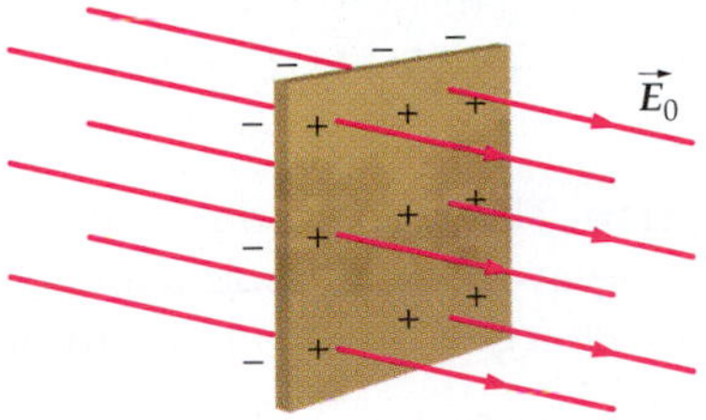

Figure 23-31

Picture the Problem (*a*) We find the charge density by using the value of E_n near each surface. (*b*) The additional charge of 96 μC must be distributed uniformly on each face of the slab so that the electric field inside the slab remains zero. The net charge density is the sum of the original density plus the added density.

(*a*)1. The charge density is related to the field just outside the face:

$$\sigma = \epsilon_0 E_n$$

2. At the right face, $\vec{E}$ points away from the slab so $E_n = 450$ kN/C. Use the value of E_n to calculate σ_R:

$$\sigma_R = \epsilon_0 E_n = (8.85 \times 10^{-12}\ \text{C}^2/\text{N}\cdot\text{m}^2)(450\ \text{kN/C}) = 3.98 \times 10^{-6}\ \text{C/m}^2 = 3.98\ \mu\text{C/m}^2$$

3. At the left face, the electric field points toward the slab, so $E_n = -450$ kN/C. Use this to calculate σ_L:

$$\sigma_L = \epsilon_0 E_n = (8.85 \times 10^{-12}\ \text{C}^2/\text{N}\cdot\text{m}^2)(-450\ \text{kN/C}) = -3.98\ \mu\text{C/m}^2$$

(*b*)1. The new charge density on a face equals the old charge density plus the additional charge density:

$$\sigma' = \sigma + \sigma_a$$

2. The additional charge density on each face equals the total charge on the face divided by its area:

$$\sigma_a = \frac{Q}{A}$$

3. The additional charge on each face is half the total additional charge (or 48 μC). Use this fact and the given dimensions of the slab to calculate σ_a:

$$\sigma_a = \frac{Q}{A} = \frac{48\ \mu\text{C}}{(4\ \text{m})^2} = 3.0\ \mu\text{C/m}^2$$

4. Add this additional charge density to the original charge density to find the new charge density:

$$\sigma_R = 3.98\ \mu\text{C/m}^2 + 3.0\ \mu\text{C/m}^2 = 6.98\ \mu\text{C/m}^2$$

$$\sigma_L = -3.98\ \mu\text{C/m}^2 + 3.0\ \mu\text{C/m}^2 = -0.98\ \mu\text{C/m}^2$$

5. Use these charge densities to calculate the normal component of the electric field just outside each face:

$$E_{nR} = \frac{\sigma_R}{\epsilon_0} = \frac{6.98\ \mu\text{C/m}^2}{8.85 \times 10^{-12}\ \text{C}^2/\text{N}\cdot\text{m}^2} = 789\ \text{kN/C}$$

$$E_{nL} = \frac{\sigma_R}{\epsilon_0} = \frac{-0.98\ \mu\text{C/m}^2}{8.85 \times 10^{-12}\ \text{C}^2/\text{N}\cdot\text{m}^2} = -111\ \text{kN/m}^2$$

Exercise Adding the positive charge density of 3.0 μC/m^2 to each face in this example is equivalent to adding two planes of positive charge. (*a*) Find the electric field due to these planes outside the slabs. (*b*) Combine this field with the original field to find the resultant electric fields on the right and left of the slab. (*Answers* (*a*) $E = \frac{1}{2}\sigma/\epsilon_0 + \frac{1}{2}\sigma/\epsilon_0 = \sigma/\epsilon_0 = 339$ kN/C (*b*) $E_R = 450$ kN/C + 339 kN/C = 789 kN/C; $E_L = 450$ kN/C − 339 kN/C = 111 kN/C)

Exercise The electric field just outside the surface of a certain conductor points away from the conductor and has a magnitude of 2000 N/C. What is the surface charge density on the surface of the conductor? (*Answer* 17.7 nC/m^2)

optional

23-6 Derivation of Gauss's Law from Coulomb's Law

Gauss's law can be derived mathematically using the concept of the **solid angle**. Consider an area element ΔA on a spherical surface. The solid angle $\Delta\Omega$ subtended by ΔA at the center of the sphere is defined to be

$$\Delta\Omega = \frac{\Delta A}{r^2}$$

where r is the radius of the sphere. Since ΔA and r^2 both have dimensions of length squared, the solid angle is dimensionless. The unit of the solid angle is the **steradian** (sr). Since the total area of a sphere is $4\pi r^2$, the total solid angle subtended by a sphere is

$$\frac{4\pi r^2}{r^2} = 4\pi \text{ steradians}$$

There is a close analogy between the solid angle and the ordinary plane angle $\Delta\theta$, which is defined to be the ratio of an element of arc length of a circle Δs to the radius of the circle:

$$\Delta\theta = \frac{\Delta s}{r} \text{ radians}$$

The total plane angle subtended by a circle is 2π radians.

In Figure 23-32, the area element ΔA is not perpendicular to the radial lines from point O. The unit vector $\hat{n}$ normal to the area element makes an angle θ with the radial unit vector $\hat{r}$. In this case, the solid angle subtended by ΔA at point O is defined to be

$$\Delta\Omega = \frac{\Delta A \hat{n} \cdot \hat{r}}{r^2} = \frac{\Delta A \cos\theta}{r^2} \qquad \text{23-26}$$

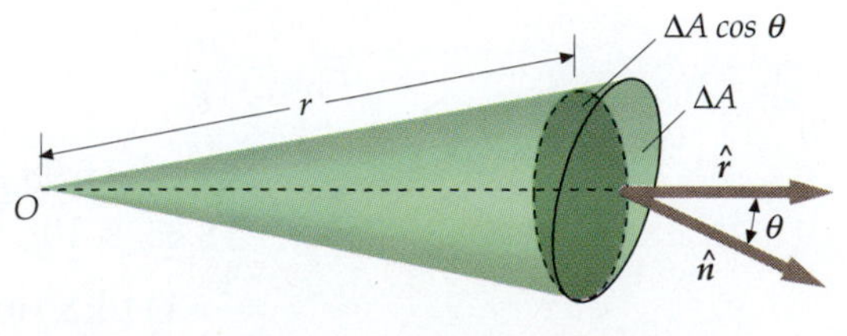

Figure 23-32 An area element ΔA whose normal is not parallel to the radial line from O to the center of the element. The solid angle subtended by this element at O is defined to be $(\Delta A \cos\theta)/r^2$.

Figure 23-33 shows a point charge q surrounded by a surface S of arbitrary shape. To calcuate the flux through this surface, we want to find $\vec{E}\cdot\hat{n}\,\Delta A$ for each element of area on the surface and sum over the entire surface. The flux through the area element shown is

$$\Delta\phi = \vec{E}\cdot\hat{n}\ \Delta A = \frac{kq}{r^2}\,\hat{r}\cdot\hat{n}\ \Delta A = kq\ \Delta\Omega$$

The solid angle $\Delta\Omega$ is the same as that subtended by the corresponding area element of a spherical surface of any radius. The sum of the flux through the entire surface is kq times the total solid angle subtended by the closed surface, which is 4π steradians:

$$\phi_{\rm net} = \oint_S \vec{E}\cdot\hat{n}\ dA = kq\oint d\Omega = 4\pi kq = \frac{q}{\epsilon_0} \qquad \text{23-27}$$

which is Gauss's law.

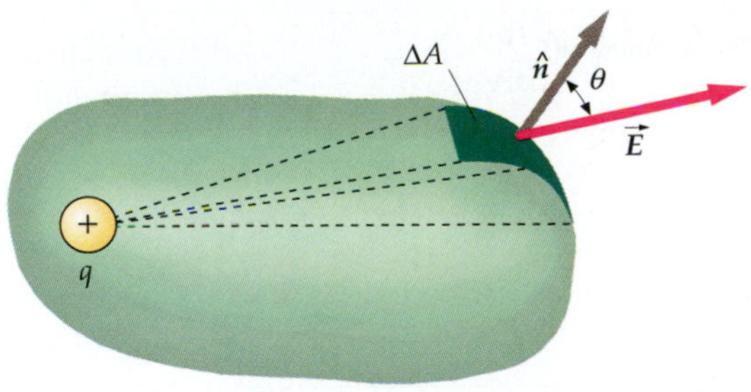

Figure 23-33 A point charge enclosed by an arbitrary surface S. The flux through an area element ΔA is proportional to the solid angle subtended by the area element at the charge. The net flux through the surface, found by summing over all the area elements, is proportional to the total solid angle 4π at the charge, which is independent of the shape of the surface.

Summary

1. Gauss's law is a fundamental law of physics that is equivalent to Coulomb's law for static charges.
2. For highly symmetric charge distributions, Gauss's law can be used to calculate the electric field.

Topic	Remarks and Relevant Equations	
Definitions and General Relations		
1. Electric field for a continuous charge distribution	$\vec{E} = \int_V \frac{k\,dq}{r^2}\,\hat{r}$ (Coulomb's law) where $dq = \rho\,dV$ for a charge distributed throughout a volume, $dq = \sigma\,dA$ for a charge distributed on a surface, and $dq = \lambda\,dL$ for a charge distributed along a line.	23-4
2. Electric flux	$\phi = \lim_{\Delta A_i \to 0} \sum_i \hat{n}_i\,\Delta A_i = \int_S \vec{E}\cdot\hat{n}\ dA$	23-14
3. Gauss's law	$\phi_{\rm net} = \oint_S E_{\rm n}\,dA = 4\pi k Q_{\rm inside}$ The net flux through a closed surface equals $4\pi k$ times the net charge within the surface.	23-17
4. Coulomb constant k and permittivity of free space ϵ_0	$k = \frac{1}{4\pi\epsilon_0} = 8.99 \times 10^9\ \mathrm{N\cdot m^2/C^2}$ $\epsilon_0 = 8.85 \times 10^{-12}\ \mathrm{C^2/N\cdot m^2}$	23-18
5. Coulomb's law and Gauss's law in terms of ϵ_0	$\vec{E} = \frac{1}{4\pi\epsilon_0}\frac{q}{r^2}\,\hat{r}$	23-19
	$\phi_{\rm net} = \oint_S E_{\rm n}\,dA = \frac{1}{\epsilon_0}Q_{\rm inside}$	23-20

6. Discontinuity of E_n	At a surface carrying a surface charge density σ, the component of the electric field perpendicular to the surface is discontinuous by σ/ϵ_0: $$E_{n2} - E_{n1} = \frac{\sigma}{\epsilon_0}$$	23-24
7. Charge on a conductor	In electrostatic equilibrium, the net electric charge on a conductor resides on the surface of the conductor.	
8. $\vec{E}$ just outside a conductor	The electric field just outside the surface of a conductor is perpendicular to the surface and has the magnitude σ/ϵ_0, where σ is the local surface charge density at that point on the conductor: $$E_n = \frac{\sigma}{\epsilon_0}$$ The force per unit area exerted on the charge on the surface of a conductor by all the other charges is called the electrostatic stress.	23-25

Electric Fields for Various Charge Distributions

At the bisector of a finite line charge	$E_r = \frac{2k\lambda}{r}\sin\theta_0$		23-8
At the bisector of an infinite line charge	$E_r = \frac{1}{2\pi\epsilon_0}\frac{\lambda}{r} = 2k\frac{\lambda}{r}$		23-9
On the axis of a ring charge	$E_x = \frac{kQx}{(x^2 + a^2)^{3/2}}$		23-10
On the axis of a disk charge	$E_x = 2\pi k\sigma\left(1 - \frac{x}{\sqrt{x^2 + R^2}}\right)$		23-11
Of an infinite plane of charge	$E_n = \frac{\sigma}{2\epsilon_0} = 2\pi k\sigma$		23-12, 23-21
Of a spherical shell of charge	$E_r = \frac{1}{4\pi\epsilon_0}\frac{Q}{r^2},$	$r > R$	23-22*a*
	$E_r = 0,$	$r < R$	23-22*b*
Of a solid sphere of charge	$E_r = \frac{1}{4\pi\epsilon_0}\frac{Q}{r^2},$	$r \geq R$	23-23*a*
	$E_r = \frac{1}{4\pi\epsilon_0}\frac{Q}{R^3}r,$	$r \leq R$	23-23*b*

Problem-Solving Guide

1. Begin by drawing a neat diagram that includes the important features of the problem. When using Coulomb's law, your sketch should show dq, the unit vector $\hat{r}$ from dq to the field point P, and the field element $d\vec{E}$. Resolve $d\vec{E}$ into components and use symmetry when possible. When finding $\vec{E}$ using superposition, show the individual $\vec{E}$ vectors in your sketch, along with an appropriate coordinate system.
2. For symmetrical charge distributions, $\vec{E}$ is most easily found using Gauss's law. Show the Gaussian surface in your sketch.

Summary of Worked Examples

Type of Calculation	Procedure and Relevant Examples
1. Coulomb's Law	
Calculate $\vec{E}$ using Coulomb's law.	Find $d\vec{E}$ for a given charge element dq, then integrate over the entire charge distribution. **(A sample calculation appears in the text of Section 23-1.)**
2. Gauss's Law	
Find the charge enclosed by a surface, given $\vec{E}$ on the surface.	Calculate the flux through the surface, then use Gauss's law. **Example 23-4**
Calculate $\vec{E}$ using Gauss's law.	Choose an appropriate Gaussian surface, calculate ϕ_{net} (in terms of E), calculate Q_{inside}, then use Gauss's law to solve for E. **Examples 23-7, 23-8**
3. Find $\vec{E}$ due to two or more charge distributions.	Determine $\vec{E}$ for each distribution separately, then do a vector sum to find the total field. **Examples 23-2, 23-3, 23-5, 23-6**

Problems

Conceptual Problems

Problems from Optional and Exploring sections

In a few problems, you are given more data than you actually need; in a few other problems, you are required to supply data from your general knowledge, outside sources, or informed estimates.

• Single-concept, single-step, relatively easy

•• Intermediate-level, may require synthesis of concepts

••• Challenging, for advanced students

Calculate $\vec{E}$ from Coulomb's Law

1 • A uniform line charge of linear charge density $\lambda = 3.5$ nC/m extends from $x = 0$ to $x = 5$ m. (*a*) What is the total charge? Find the electric field on the x axis at (*b*) $x = 6$ m, (*c*) $x = 9$ m, and (*d*) $x = 250$ m. (*e*) Find the field at $x = 250$ m, using the approximation that the charge is a point charge at the origin, and compare your result with that for the exact calculation in part (*d*).

2 • Two infinite vertical planes of charge are parallel to each other and are separated by a distance $d = 4$ m. Find the electric field to the left of the planes, to the right of the planes, and between the planes (*a*) when each plane has a uniform surface charge density $\sigma = +3\ \mu\text{C/m}^2$ and (*b*) when the left plane has a uniform surface charge density $\sigma = +3\ \mu\text{C/m}^2$ and that of the right plane is $\sigma = -3\ \mu\text{C/m}^2$. Draw the electric field lines for each case.

3 • A 2.75-μC charge is uniformly distributed on a ring of radius 8.5 cm. Find the electric field on the axis at (*a*) 1.2 cm, (*b*) 3.6 cm, and (*c*) 4.0 m from the center of the ring. (*d*) Find the field at 4.0 m using the approximation that the ring is a point charge at the origin, and compare your results with that for part (*c*).

4 • A disk of radius 2.5 cm carries a uniform surface charge density of 3.6 $\mu\text{C/m}^2$. Using reasonable approximations, find the electric field on the axis at distances of (*a*) 0.01 cm, (*b*) 0.04 cm, (*c*) 5 m, and (*d*) 5 cm.

5 • For the disk charge of Problem 4, calculate exactly the electric field on the axis at distances of (*a*) 0.04 cm and (*b*) 5 m, and compare your results with those for parts (*b*) and (*c*) of Problem 4.

6 • A uniform line charge extends from $x = -2.5$ cm to $x = +2.5$ cm and has a linear charge density of $\lambda = 6.0$ nC/m. (*a*) Find the total charge. Find the electric field on the y axis at (*b*) $y = 4$ cm, (*c*) $y = 12$ cm, and (*d*) $y = 4.5$ m. (*e*) Find the field at $y = 4.5$ m, assuming the charge to be a point charge, and compare your result with that for part (*d*).

7 • A disk of radius a lies in the yz plane with its axis along the x axis and carries a uniform surface charge density σ. Find the value of x for which $E_x = \frac{1}{2}\sigma/2\epsilon_0$.

8 • A ring of radius a with its center at the origin and its axis along the x axis carries a total charge Q. Find E_x at (*a*) $x = 0.2a$, (*b*) $x = 0.5a$, (*c*) $x = 0.7a$, (*d*) $x = a$, and (*e*) $x = 2a$. (*f*) Use your results to plot E_x versus x for both positive and negative values of x.

9 • Repeat Problem 8 for a disk of uniform surface charge density σ.

10 •• A disk of radius 30 cm carries a uniform charge density σ. (*a*) Compare the approximation $E = 2\pi k\sigma$ with the exact expression (Equation 23-11) for the electric field on the axis of the disk by computing the fractional difference $\Delta E/E \approx x/\sqrt{x^2 + R^2}$ for the distances $x = 0.1$, $x = 0.2$, and $x = 3$ cm. (*b*) At what distance is the neglected term 1% of $2\pi k\sigma$?

11 •• Show that E_x on the axis of a ring charge of radius a has its maximum and minimum values at $x = +a/\sqrt{2}$ and $x = -a/\sqrt{2}$. Sketch E_x versus x for both positive and negative values of x.

12 •• A line charge of uniform linear charge density λ lies along the x axis from $x = 0$ to $x = a$. (*a*) Show that the x component of the electric field at a point on the y axis is given by

$$E_x = -\frac{k\lambda}{y} + \frac{k\lambda}{\sqrt{y^2 + a^2}}$$

(*b*) Show that if the line charge extends from $x = -b$ to $x = a$, the x component of the electric field at a point on the y axis is given by

$$E_x = \frac{k\lambda}{\sqrt{y^2 + a^2}} - \frac{k\lambda}{\sqrt{y^2 + b^2}}$$

13 •• (*a*) A finite line charge of uniform linear charge density λ lies on the x axis from $x = 0$ to $x = a$. Show that the y component of the electric field at a point on the y axis is given by

$$E = \frac{k\lambda}{y} \sin\theta_1 = \frac{k\lambda}{y} \frac{a}{\sqrt{y^2 + a^2}}$$

where θ_1 is the angle subtended by the line charge at the field point. (*b*) Show that if the line charge extends from $x = -b$ to $x = a$, the y component of the electric field at a point on the y axis is given by

$$E_y = \frac{k\lambda}{y}(\sin\theta_1 + \sin\theta_2)$$

where $\sin\theta_2 = b/\sqrt{y^2 + b^2}$.

14 •• A semicircular ring of radius R carries a uniform line charge of λ. Find the electric field at the center of the semicircle.

15 ••• A hemispherical thin shell of radius R carries a uniform surface charge σ. Find the electric field at the center of the hemispherical shell ($r = 0$).

16 ••• A line charge of linear charge density λ with the shape of a square of side L lies in the yz plane with its center at the origin. Find the electric field on the x axis at an arbitrary distance x, and compare your result to that for the field on the axis of a charged ring of radius $r = \frac{1}{2}L$ with its center at the origin and carrying the same total charge. (*Hint:* Use Equation 23-8 for the field due to each segment of the square).

Gauss's Law

17 •• True or false:

(*a*) Gauss's law holds only for symmetric charge distributions.

(*b*) The result that $E = 0$ inside a conductor can be derived from Gauss's law.

18 •• What information in addition to the total charge inside a surface is needed to use Gauss's law to find the electric field?

19 ••• Is the electric field E in Gauss's law only that part of the electric field due to the charge inside a surface, or is it the total electric field due to all charges both inside and outside the surface?

20 • Consider a uniform electric field $\vec{E} = 2\text{ kN/C }\hat{i}$. (*a*) What is the flux of this field through a square of side 10 cm in a plane parallel to the yz plane? (*b*) What is the flux through the same square if the normal to its plane makes a 30° angle with the x axis?

21 • A single point charge $q = +2\ \mu$C is at the origin. A spherical surface of radius 3.0 m has its center on the x axis at $x = 5$ m. (*a*) Sketch electric field lines for the point charge. Do any lines enter the spherical surface? (*b*) What is the net number of lines that cross the spherical surface, counting those that enter as negative? (*c*) What is the net flux of the electric field due to the point charge through the spherical surface?

22 • An electric field is $\vec{E} = 300\text{ N/C }\hat{i}$ for $x > 0$ and $\vec{E} = -300\text{ N/C }\hat{i}$ for $x < 0$. A cylinder of length 20 cm and radius 4 cm has its center at the origin and its axis along the x axis such that one end is at $x = +10$ cm and the other is at $x = -10$ cm. (*a*) What is the flux through each end? (*b*) What is the flux through the curved surface of the cylinder? (*c*) What is the net outward flux through the entire cylindrical surface? (*d*) What is the net charge inside the cylinder?

23 • A positive point charge q is at the center of a cube of side L. A large number N of electric field lines are drawn from the point charge. (*a*) How many of the field lines pass through the surface of the cube? (*b*) How many lines pass through each face, assuming that none pass through the edges or corners? (*c*) What is the net outward flux of the electric field through the cubic surface? (*d*) Use symmetry arguments to find the flux of the electric field through one face of the cube. (*e*) Which, if any, of your answers would change if the charge were inside the cube but not at its center?

24 • Careful measurement of the electric field at the surface of a black box indicates that the net outward flux through the surface of the box is 6.0 kN·m²/C. (*a*) What is the net charge inside the box? (*b*) If the net outward flux through the surface of the box were zero, could you conclude that there were no charges inside the box? Why or why not?

25 • A point charge $q = +2\ \mu$C is at the center of a sphere of radius 0.5 m. (*a*) Find the surface area of the sphere. (*b*) Find the magnitude of the electric field at points on the surface of the sphere. (*c*) What is the flux of the electric field due to the point charge through the surface of the sphere? (*d*) Would your answer to part (*c*) change if the point charge were moved so that it was inside the sphere but not at its center? (*e*) What is the net flux through a cube of side 1 m that encloses the sphere?

26 • Since Newton's law of gravity and Coulomb's law have the same inverse-square dependence on distance, an expression analogous in form to Gauss's law can be found for gravity. The gravitational field $\vec{g}$ is the force per unit mass on a test mass m_0. Then for a point mass m at the origin, the gravitational field g at some position r is

$$\vec{g} = -\frac{Gm}{r^2}\hat{r}$$

Compute the flux of the gravitational field through a spherical surface of radius r centered at the origin, and show that the gravitational analog of Gauss's law is $\phi_{\text{net}} = -4\pi G m_{\text{inside}}$.

27 •• A charge of 2 μC is 20 cm above the center of a square of side length 40 cm. Find the flux through the square. (*Hint:* Don't integrate.)

28 •• In a particular region of the earth's atmosphere, the electric field above the earth's surface has been measured to be 150 N/C downward at an altitude of 250 m and 170 N/C downward at an altitude of 400 m. Calculate the volume charge density of the atmosphere, assuming it to be uniform between 250 and 400 m. (You may neglect the curvature of the earth. Why?)

Spherical Charge Distributions

29 •• Explain why the electric field increases with r rather than decreasing as $1/r^2$ as one moves out from the center inside a spherical charge distribution of constant volume charge density.

30 • A spherical shell of radius R_1 carries a total charge q_1 that is uniformly distributed on its surface. A second, larger spherical shell of radius R_2 that is concentric with the first carries a charge q_2 that is uniformly distributed on its surface. (*a*) Use Gauss's law to find the electric field in the regions $r < R_1$, $R_1 < r < R_2$, and $r > R_2$. (*b*) What should the ratio of the charges q_1/q_2 and their relative signs be for the electric field to be zero for $r > R_2$? (*c*) Sketch the electric field lines for the situation in part (*b*) when q_1 is positive.

31 • A spherical shell of radius 6 cm carries a uniform surface charge density $\sigma = 9$ nC/m^2. (*a*) What is the total charge on the shell? Find the electric field at (*b*) $r = 2$ cm, (*c*) $r = 5.9$ cm, (*d*) $r = 6.1$ cm, and (*e*) $r = 10$ cm.

32 •• A sphere of radius 6 cm carries a uniform volume charge density $\rho = 450$ nC/m^3. (*a*) What is the total charge of the sphere? Find the electric field at (*b*) $r = 2$ cm, (*c*) $r = 5.9$ cm, (*d*) $r = 6.1$ cm, and (*e*) $r = 10$ cm. Compare your answers with Problem 31.

33 •• Consider two concentric conducting spheres (Figure 23-34). The outer sphere is hollow and initially has a charge $-7Q$ deposited on it. The inner sphere is solid and has a charge $+2Q$ on it. (*a*) How is the charge distributed on the outer sphere? That is, how much charge is on the outer surface and how much charge is on the inner surface? (*b*) Suppose a wire is connected between the inner and outer spheres. After electrostatic equilibrium is established, how much total charge is on the outside sphere? How much charge is on the outer surface of the outside sphere and how much is on the inner surface? Does the electric field at the surface of the inside sphere change when the wire is connected? If so, how? (*c*) Suppose we return to the original conditions in (*a*), with $+2Q$ on the inner sphere and $-7Q$ on the outer. We now connect the outer sphere to ground with a wire and then disconnect it. How much total charge will be on the outer sphere? How much charge will be on the inner surface of the outer sphere and how much will be on the outer surface?

Figure 23-34 Problem 33

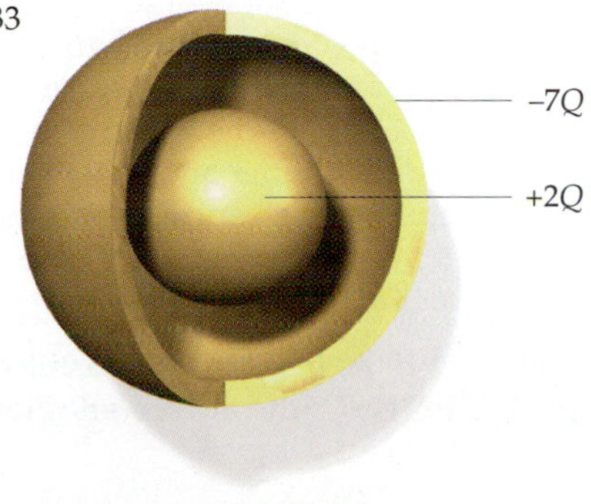

34 •• A nonconducting sphere of radius $R = 0.1$ m carries a uniform volume charge of charge density $\rho = 2.0$ nC/m^3. The magnitude of the electric field at $r = 2R$ is 1883 N/C. Find the magnitude of the electric field at $r = 0.5R$.

35 •• A nonconducting sphere of radius R carries a volume charge density that is proportional to the distance from the center: $\rho = Ar$ for $r \leq R$, where A is a constant; $\rho = 0$ for $r > R$. (*a*) Find the total charge on the sphere by summing the charges on shells of thickness dr and volume $4\pi r^2\,dr$. (*b*) Find the electric field E_r both inside and outside the charge distribution, and sketch E_r versus r.

36 •• Repeat Problem 35 for a sphere with volume charge density $\rho = B/r$ for $r < R$; $\rho = 0$ for $r > R$.

37 •• Repeat Problem 35 for a sphere with volume charge density $\rho = C/r^2$ for $r < R$; $\rho = 0$ for $r > R$.

38 ••• The charge density in a region of space is spherically symmetric and is given by $\rho(r) = Ce^{-r/a}$ when $r < R$ and $\rho = 0$ when $r > R$. Find the electric field as a function of r.

39 ••• A thick, nonconducting spherical shell of inner radius a and outer radius b has a uniform volume charge density ρ. Find (*a*) the total charge and (*b*) the electric field everywhere.

40 ••• A point charge of +5 nC is located at the origin. This charge is surrounded by a spherically symmetric negative charge distribution with volume density $\rho(r) = Ce^{-r/a}$. (*a*) Find the constant C in terms of a if the total charge of the system is zero? (*b*) What is the electric field at $r = a$?

41 ••• A nonconducting solid sphere of radius a with its center at the origin has a spherical cavity of radius b with its center at the point $x = b$, $y = 0$ as shown in Figure 23-35. The sphere has a uniform volume charge density ρ. Show that the electric field in the cavity is uniform and is given by $E_y = 0$, $E_x = \rho b/3\epsilon_0$. (*Hint:* Replace the cavity with spheres of equal positive and negative charge densities.)

Figure 23-35 Problem 41

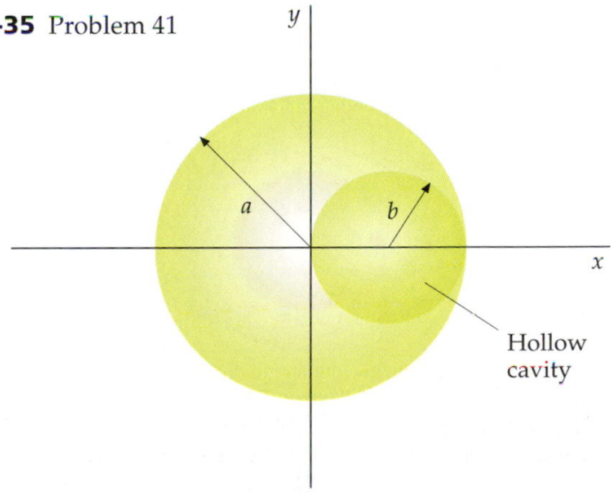

Cylindrical Charge Distributions

42 •• Show that the electric field due to an infinitely long, uniformly charged cylindrical shell of radius R carrying a surface charge density σ is given by

$$E_r = 0, \qquad r < R \qquad \text{23-28a}$$

$$E_r = \frac{\sigma R}{\epsilon_0 r} = \frac{\lambda}{2\pi\epsilon_0 r}, \qquad r > R \qquad \text{23-28b}$$

where $\lambda = 2\pi R\sigma$ is the charge per unit length on the shell.

43 •• A cylindrical shell of length 200 m and radius 6 cm carries a uniform surface charge density of $\sigma = 9\ \text{nC/m}^2$. (*a*) What is the total charge on the shell? Find the electric field at (*b*) $r = 2$ cm, (*c*) $r = 5.9$ cm, (*d*) $r = 6.1$ cm, and (*e*) $r = 10$ cm. (Use the results of Problem 42.)

44 •• An infinitely long nonconducting cylinder of radius R carries a uniform volume charge density of $\rho(r) = \rho_0$. Show that the electric field is given by

$$E_r = \frac{\rho R^2}{2\epsilon_0 r} = \frac{1}{2\pi\epsilon_0}\frac{\lambda}{r}, \qquad r > R \qquad \text{23-29a}$$

$$E_r = \frac{\rho}{2\epsilon_0} r = \frac{\lambda}{2\pi\epsilon_0 R^2} r, \qquad r < R \qquad \text{23-29b}$$

where $\lambda = \rho\pi R^2$ is the charge per unit length.

45 •• A cylinder of length 200 m and radius 6 cm carries a uniform volume charge density of $\rho = 300\ \text{nC/m}^3$. (*a*) What is the total charge of the cylinder? Use the formulas given in Problem 44 to calculate the electric field at a point equidistant from the ends at (*b*) $r = 2$ cm, (*c*) $r = 5.9$ cm, (*d*) $r = 6.1$ cm, and (*e*) $r = 10$ cm. Compare your results with those in Problem 43.

46 •• Consider two infinitely long, concentric cylindrical shells. The inner shell has a radius R_1 and carries a uniform surface charge density of σ_1, and the outer shell has a radius R_2 and carries a uniform surface charge density of σ_2. (*a*) Use Gauss's law to find the electric field in the regions $r < R_1$, $R_1 < r < R_2$, and $r > R_2$. (*b*) What is the ratio of the surface charge densities σ_2/σ_1 and their relative signs if the electric field is zero at $r > R_2$? What would the electric field between the shells be in this case? (*c*) Sketch the electric field lines for the situation in (*b*) if σ_1 is positive.

47 •• Figure 23-36 shows a portion of an infinitely long, concentric cable in cross section. The inner conductor carries a charge of 6 nC/m; the outer conductor is uncharged. (*a*) Find the electric field for all values of r, where r is the distance from the axis of the cylindrical system. (*b*) What are the surface charge densities on the inside and the outside surfaces of the outer conductor?

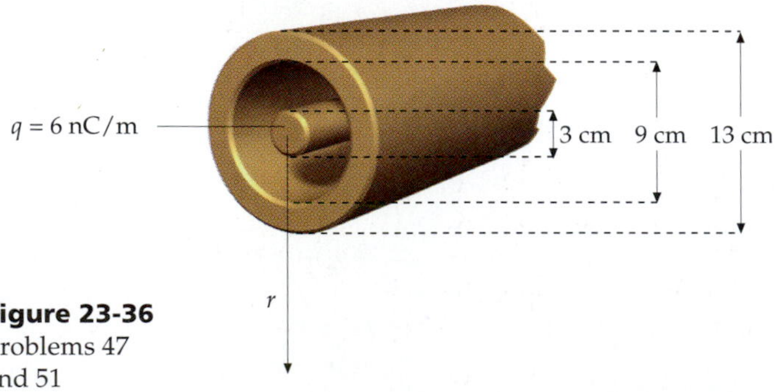

Figure 23-36
Problems 47 and 51

48 •• Repeat Problem 44 for a cylinder with volume charge density (a) $\rho(r) = ar$ and (b) $\rho = Cr^2$.

49 •• Repeat Problem 44 with $\rho = C/r$.

50 ••• An infinitely long, thick, nonconducting cylindrical shell of inner radius a and outer radius b has a uniform volume charge density ρ. Find the electric field everywhere.

51 ••• Suppose that the inner cylinder of Figure 23-36 is made of nonconducting material and carries a volume charge distribution given by $\rho(r) = C/r$, where $C = 200\ \text{nC/m}^2$. The outer cylinder is metallic. (*a*) Find the charge per meter carried by the inner cylinder. (*b*) Calculate the electric field for all values of r.

Charge and Field at Conductor Surfaces

52 • A penny is in an external electric field of magnitude 1.6 kN/C directed perpendicular to its faces. (*a*) Find the charge density on each face of the penny, assuming the faces are planes. (*b*) If the radius of the penny is 1 cm, find the total charge on one face.

53 • An uncharged metal slab has square faces with 12-cm sides. It is placed in an external electric field that is perpendicular to its faces. The total charge induced on one of the faces is 1.2 nC. What is the magnitude of the electric field?

54 • A charge of 6 nC is placed uniformly on a square sheet of nonconducting material of side 20 cm in the yz plane. (*a*) What is the surface charge density σ? (*b*) What is the magnitude of the electric field just to the right and just to the left of the sheet? (*c*) The same charge is placed on a square conducting slab of side 20 cm and thickness 1 mm. What is the surface charge density σ? (Assume that the charge distributes itself uniformly on the large square surfaces.) (*d*) What is the magnitude of the electric field just to the right and just to the left of each face of the slab?

55 • A spherical conducting shell with zero net charge has an inner radius a and an outer radius b. A point charge q is placed at the center of the shell. (*a*) Use Gauss's law and the properties of conductors in equilibrium to find the electric field in the regions $r < a$, $a < r < b$, and $b < r$. (*b*) Draw the electric field lines for this situation. (*c*) Find the charge density on the inner surface ($r = a$) and on the outer surface ($r = b$) of the shell.

56 •• The electric field just above the surface of the earth has been measured to be 150 N/C downward. What total charge on the earth is implied by this measurement?

57 •• A positive point charge of magnitude 2.5 μC is at the center of an uncharged spherical conducting shell of inner radius 60 cm and outer radius 90 cm. (*a*) Find the charge densities on the inner and outer surfaces of the shell and the total charge on each surface. (*b*) Find the electric field everywhere. (*c*) Repeat (*a*) and (*b*) with a net charge of +3.5 μC placed on the shell.

58 •• If the magnitude of an electric field in air is as great as 3×10^6 N/C, the air becomes ionized and begins to conduct electricity. This phenomenon is called dieletric breakdown. A charge of 18 μC is to be placed on a conducting sphere. What is the minimum radius of a sphere that can hold this charge without breakdown?

59 •• A square conducting slab with 5-m sides carries a net charge of 80 μC. (*a*) Find the charge density on each face of the slab and the electric field just outside one face of the slab. (*b*) The slab is placed to the right of an infinite charged nonconducting plane with charge density 2.0 μC/m^2 so that the faces of the slab are parallel to the plane. Find the electric field on each side of the slab far from its edges and the charge density on each face.

60 •• Imagine that a small hole has been punched through the wall of a thin, uniformly charged spherical shell whose surface charge density is σ. Find the electric field near the center of the hole.

General Problems

61 • True or false:

(*a*) If there is no charge in a region of space, the electric field on a surface surrounding the region must be zero everywhere.
(*b*) The electric field inside a uniformly charged spherical shell is zero.
(*c*) In electrostatic equilibrium, the electric field inside a conductor is zero.
(*d*) If the net charge on a conductor is zero, the charge density must be zero at every point on the surface of the conductor.

62 • If the electric field E is zero everywhere on a closed surface, is the net flux through the surface necessarily zero? What, then, is the net charge inside the surface?

63 • A point charge $-Q$ is at the center of a spherical conducting shell of inner radius R_1 and outer radius R_2 as shown in Figure 23-37. The charge on the inner surface of the shell is

(*a*) $+Q$.
(*b*) zero.
(*c*) $-Q$.
(*d*) dependent on the total charge carried by the shell.

Figure 23-37 Problems 63–67

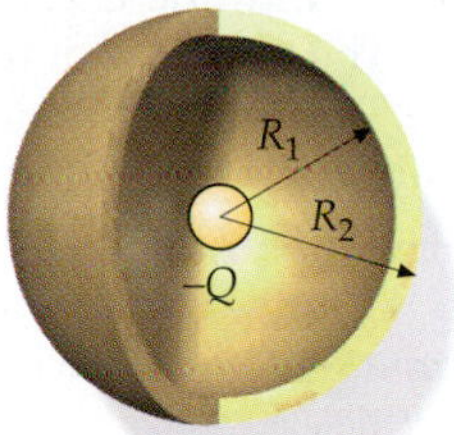

64 • For the configuration of Figure 23-37, the charge on the outer surface of the shell is

(*a*) $+Q$.
(*b*) zero.
(*c*) $-Q$.
(*d*) dependent on the total charge carried by the shell.

65 •• Suppose that the total charge on the conducting shell of Figure 23-37 is zero. It follows that the electric field for $r < R_1$ and $r > R_2$ points

(*a*) away from the center of the shell in both regions.
(*b*) toward the center of the shell in both regions.
(*c*) toward the center of the shell for $r < R_1$ and is zero for $r > R_2$.
(*d*) away from the center of the shell for $r < R_1$ and is zero for $> R_2$.

66 •• If the conducting shell in Figure 23-37 is grounded, which of the following statements is then correct?

(*a*) The charge on the inner surface of the shell is $+Q$ and that on the outer surface is $-Q$.
(*b*) The charge on the inner surface of the shell is $+Q$ and that on the outer surface is zero.
(*c*) The charge on both surfaces of the shell is $+Q$.
(*d*) The charge on both surfaces of the shell is zero.

67 •• For the configuration described in Problem 66, in which the conducting shell is grounded, the electric field for $r < R_1$ and $r > R_2$ points

(*a*) away from the center of the shell in both regions.
(*b*) toward the center of the shell in both regions.
(*c*) toward the center of the shell for $r < R_1$ and is zero for $r > R_2$.
(*d*) toward the center of the shell for $r < R_1$ and is zero for $r > R_1$.

68 •• If the net flux through a closed surface is zero, does it follow that the electric field E is zero everywhere on the surface? Does it follow that the net charge inside the surface is zero?

69 •• Equation 23-8 for the electric field on the perpendicular bisector of a finite line charge is different from Equation 23-9 for the electric field near an infinite line charge, yet Gauss's law would seem to give the same result for these two cases. Explain.

70 •• True or false: The electric field is discontinuous at all points at which the charge density is discontinuous.

71 •• Consider the three concentric metal spheres shown in Figure 23-38. Sphere I is solid, with radius R_1. Sphere II is hollow, with inner radius R_2 and outer radius R_3. Sphere III is hollow, with inner radius R_4 and outer radius R_5. Initially, all three spheres have zero excess charge. Then a negative charge $-Q_0$ is placed on sphere I and a positive charge $+Q_0$ is placed on sphere III. (*a*) After the charges have reached equilibrium, will the electric field in the space between spheres I and II point *toward* the center, *away* from the center, or neither? (*b*) How much charge will be on the inner surface of sphere II? Give the correct sign. (*c*) How much charge will be on the outer surface of sphere II? (*d*) How much charge will be on the inner surface of sphere III? (*e*) How much charge will be on the outer surface of sphere III? (*g*) Plot E versus r.

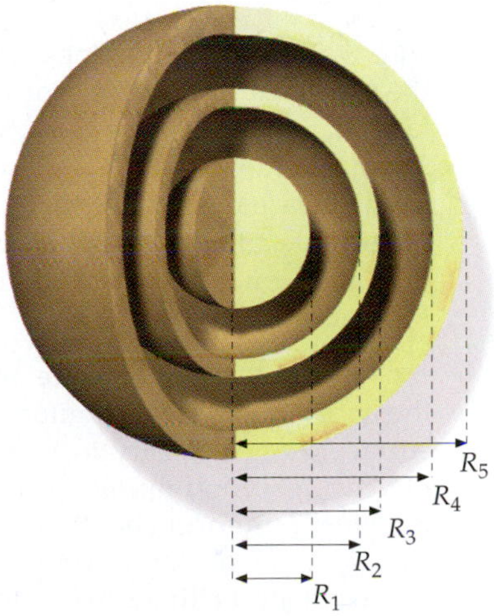

Figure 23-38 Problem 71

72 •• An early model of the hydrogen atom considered the atom to consist of a proton, which is a uniform charged sphere of radius R, with an electron in an orbit of radius r_0 *inside* the proton as shown in Figure 23-39. (*a*) Use Gauss's law to obtain the magnitude of E (the field due to the proton) at the position of the electron. Give your answer in terms of e (the charge on a proton), r_0, and R. (*b*) Find the frequency of revolution f in terms of r_0 and the velocity of the electron v. (*c*) What is the force on the electron in terms of m,v and r_0? (*d*) What is the frequency f in terms of m, e, R, ϵ_0, and r_0? (Each of your answers need not include all of the specified quantities.)

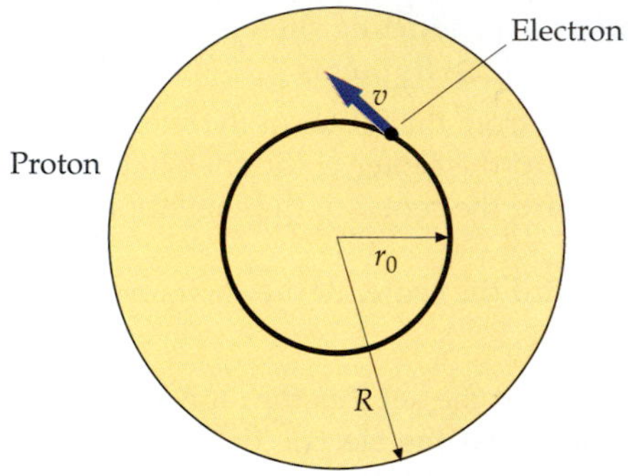

Figure 23-39 Problem 72

73 •• A nonuniform surface charge lies in the yz plane. At the origin, the surface charge density is $\sigma = 3.10\ \mu\text{C/m}^2$. Other charged objects are present as well. Just to the right of the origin, the x component of the electric field is $E_x = 4.65 \times 10^5$ N/C. What is E_x just to the left of the origin?

74 •• An infinite line charge of uniform linear charge density $\lambda = -1.5\ \mu\text{C/m}$ lies parallel to the y axis at $x = -2$ m. A point charge of 1.3 μC is located at $x = 1$ m, $y = 2$ m. Find the electric field at $x = 2$ m, $y = 1.5$ m.

75 •• Two infinite planes of charge lie parallel to each other and to the yz plane. One is at $x = -2$ m and has a surface charge density of $\sigma = -3.5\ \mu\text{C/m}^2$. The other is at $x = 2$ m and has a surface charge density of $\sigma = 6.0\ \mu\text{C/m}^2$. Find the electric field for (*a*) $x < -2$ m; (*b*) $-2\text{ m} < x < 2$ m; and (*c*) $x > 2$ m.

76 •• An infinitely long cylindrical shell is coaxial with the y axis and has a radius of 15 cm. It carries a uniform surface charge density $\sigma = 6\ \mu\text{C/m}^2$. A spherical shell of radius 25 cm is centered on the x axis at $x = 50$ cm and carries a uniform surface charge density $\sigma = -12\ \mu\text{C/m}^2$. Calculate the magnitude and direction of the electric field at (*a*) the origin; (*b*) $x = 20$ cm, $y = 10$ cm; and (*c*) $x = 50$ cm, $y = 20$ cm. (See Problem 42.)

77 •• An infinite plane in the xz plane carries a uniform surface charge density $\sigma_1 = 65$ nC/m^2. A second infinite plane carrying a uniform charge density $\sigma_2 = 45$ nC/m^2 intersects the xz plane at the z axis and makes an angle of 30° with the xz plane as shown in Figure 23-40. Find the electric field in the xy plane at (*a*) $x = 6$ m, $y = 2$ m and (*b*) $x = 6$ m, $y = 5$ m.

Figure 23-40 Problem 77

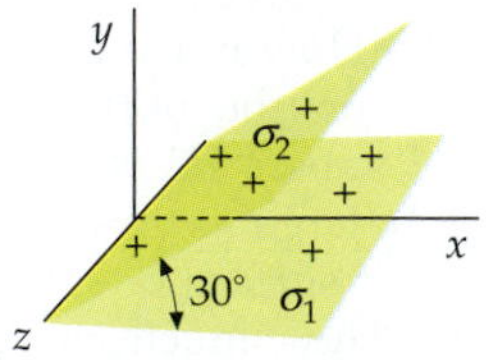

78 •• A ring of radius R carries a uniform, positive, linear charge density λ. Figure 23-41 shows a point P in the plane of the ring but not at the center. Consider the two elements of the ring of lengths s_1 and s_2 shown in the figure at distances r_1 and r_2, from point P. (*a*) What is the ratio of the charges of these elements? Which produces the greater field at point P? (*b*) What is the direction of the field at point P due to each element? What is the direction of the total electric field at point P? (*c*) Suppose that the electric field due to a point charge varied as $1/r$ rather than $1/r^2$. What would the electric field be at point P due to the elements shown? (*d*) How would your answers to parts (*a*), (*b*), and (*c*) differ if point P were inside a spherical shell of uniform charge and the elements were of areas s_1 and s_2?

Figure 23-41 Problem 78

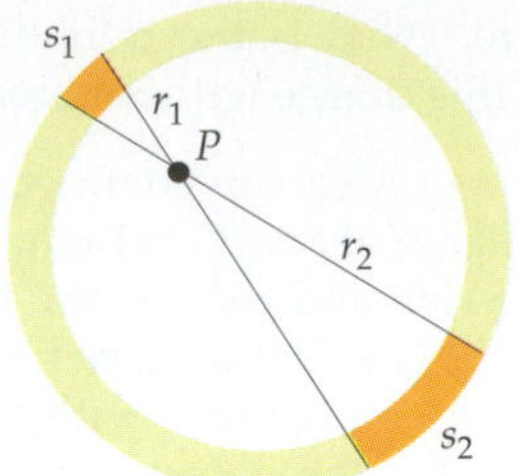

79 •• A ring of radius R that lies in the horizontal (xy) plane carries a charge Q uniformly distributed over its length. A mass m carries a charge q whose sign is opposite that of Q. (*a*) What is the minimum value of $|q|/m$ such that the mass will be in equilibrium under the action of gravity and the electrostatic force on the charge q? (*b*) If $|q|/m$ is twice what is calculated in (*a*), where will the mass be when it is in equilibrium?

80 •• A long, thin, nonconducting plastic rod is bent into a loop with radius R. Between the ends of the rods, a small gap of length l ($l \ll R$) remains. A charge Q is equally distributed on the rod. (*a*) Indicate the direction of the electric field at the center of the loop. (*b*) Find the magnitude of the electric field at the center of the loop.

81 •• A rod of length L lies perpendicular to an infinitely long, uniform line charge of charge density λ C/m (Figure 23-42). The near end of the rod is a distance d above the line charge. The rod carries a total charge Q uniformly distributed along its length. Find the force that the infinitely long line charge exerts on the rod.

Figure 23-42 Problem 81

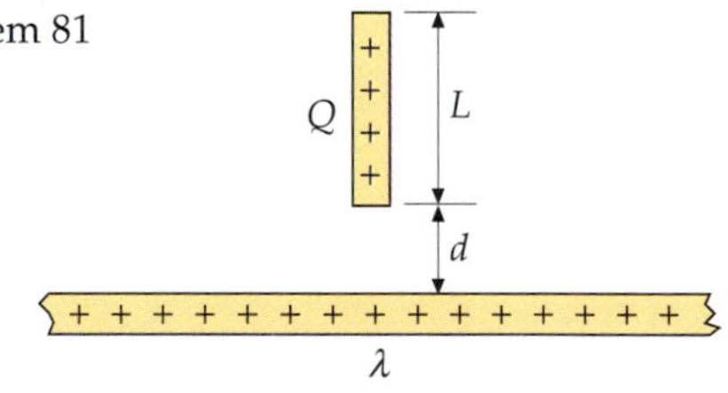

82 •• A nonconducting sphere 1.2 m in diameter with its center on the x axis at $x = 4$ m carries a uniform volume charge of density $\rho = 5\ \mu\text{C/m}^3$. Surrounding the sphere is a spherical shell with a diameter of 2.4 m and a uniform surface charge density $\sigma = -1.5\ \mu\text{C/m}^2$. Calculate the magnitude and direction of the electric field at (*a*) $x = 4.5$ m, $y = 0$; (*b*) $x = 4.0$ m, $y = 1.1$ m; and (*c*) $x = 2.0$ m, $y = 3.0$ m.

83 •• An infinite plane of charge with surface charge density $\sigma_1 = 3\ \mu\text{C/m}^2$ is parallel to the xz plane at $y = -0.6$

m. A second infinite plane of charge with surface charge density $\sigma_2 = -2\ \mu C/m^2$ is parallel to the yz plane at $x = 1$ m. A sphere of radius 1 m with its center in the xy plane at the intersection of the two charged planes ($x = 1$ m, $y = -0.6$ m) has a surface charge density $\sigma_3 = -3\ \mu C/m^2$. Find the magnitude and direction of the electric field on the x axis at (*a*) $x = 0.4$ m and (*b*) $x = 2.5$ m.

84 •• An infinite plane lies parallel to the yz plane at $x = 2$ m and carries a uniform surface charge density $\sigma = 2\ \mu C/m^2$. An infinite line charge of uniform linear charge density $\lambda = 4\ \mu C/m$ passes through the origin at an angle of 45° with the x axis in the xy plane. A sphere of volume charge density $\rho = -6\ \mu C/m^3$ and radius 0.8 m is centered on the x axis at $x = 1$ m. Calculate the magnitude and direction of the electric field in the xy plane at $x = 1.5$ m, $y = 0.5$ m.

85 •• An infinite line charge λ is located along the z axis. A mass m that carries a charge q whose sign is opposite to that of λ is in a circular orbit in the xy plane about the line charge. Obtain an expression for the period of the orbit in terms of m, q, R, and λ, where R is the radius of the orbit.

86 •• A ring of radius R that lies in the yz plane carries a positive charge Q uniformly distributed over its length. A particle of mass m that carries a negative charge of magnitude q is at the center of the ring. (*a*) Show that if $x \ll R$, the electric field along the axis of the ring is proportional to x. (*b*) Find the force on the mass m as a function of x. (*c*) Show that if m is given a small displacement in the x direction, it will perform simple harmonic motion. Calculate the period of that motion.

87 •• When the charges Q and q of Problem 86 are 5 μC and $-5\ \mu$C, respectively, and the radius of the ring is 8.0 cm, the mass m oscillates about its equilibrium position with an angular frequency of 21 rad/s. Find the angular frequency of oscillation of the mass if the radius of the ring is doubled to 16 cm and all other parameters remain unchanged.

88 •• Given the initial conditions of Problem 87, find the angular frequency of oscillation of the mass if the radius of the ring is doubled to 16 cm while keeping the linear charge density on the ring constant.

89 •• A nonconducting cylinder of radius 1.2 m and length 2.0 m carries a charge of 50 μC uniformly distributed throughout the cylinder. Find the electric field *on the cylinder axis* at a distance of (*a*) 0.5 m, (*b*) 2.0 m, and (*c*) 20 m from the center of the cylinder.

90 •• A uniform line charge of density λ lies on the x axis between $x = 0$ and $x = L$. Its total charge is $Q = 8$ nC. The electric field at $x = 2L$ is 600 N/C $\hat{i}$. Find the electric field at $x = 3L$.

91 •• Find the linear charge density λ (in C/m) of the line charge of Problem 90.

92 ••• A uniformly charged sphere of radius R is centered at the origin with a charge of Q. Find the force on a uniformly charged line oriented radially having a total charge q with its ends at $r = R$ and $r = R + d$.

93 ••• Two equal uniform line charges of length L lie on the x axis a distance d apart as shown in Figure 23-43. (*a*) What is the force that one line charge exerts on the other line charge? (*b*) Show that when $d \gg L$, the force tends toward the expected result of $k(\lambda L)^2/d^2$.

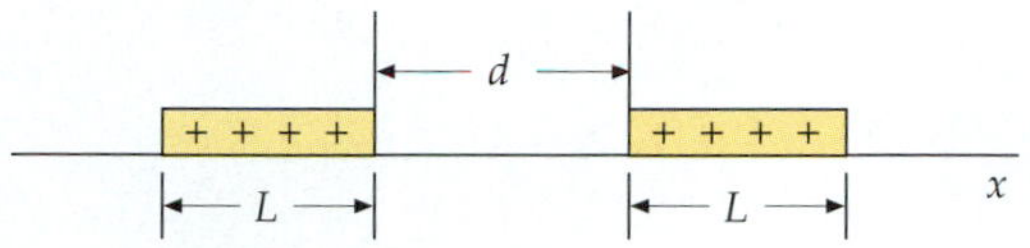

Figure 23-43 Problem 93

94 ••• A dipole $\vec{p}$ is located at a distance r from an infinitely long line charge with a uniform linear charge density λ. Assume that the dipole is aligned with the field due to the line charge. Determine the force that acts on the dipole.

95 ••• Suppose that the charge on the rod in Problem 81 is given by $\lambda(y) = ay^2$, where y is the distance from the midpoint of the rod, and that the total charge on the rod is Q. (*a*) Determine the constant a. (*b*) Find the force $d\vec{F}$ that acts on an element of charge $\lambda(y)\ dy$. (*c*) Integrate the force obtained in part (*b*) between $-L/2$ and $L/2$ to obtain the total force that acts on the rod.

96 ••• Repeat Problem 95 with the charge on the rod being $\lambda(y) = by$, where y is measured from the midpoint of the rod with the positive y direction up.

CHAPTER 24

Electric Potential

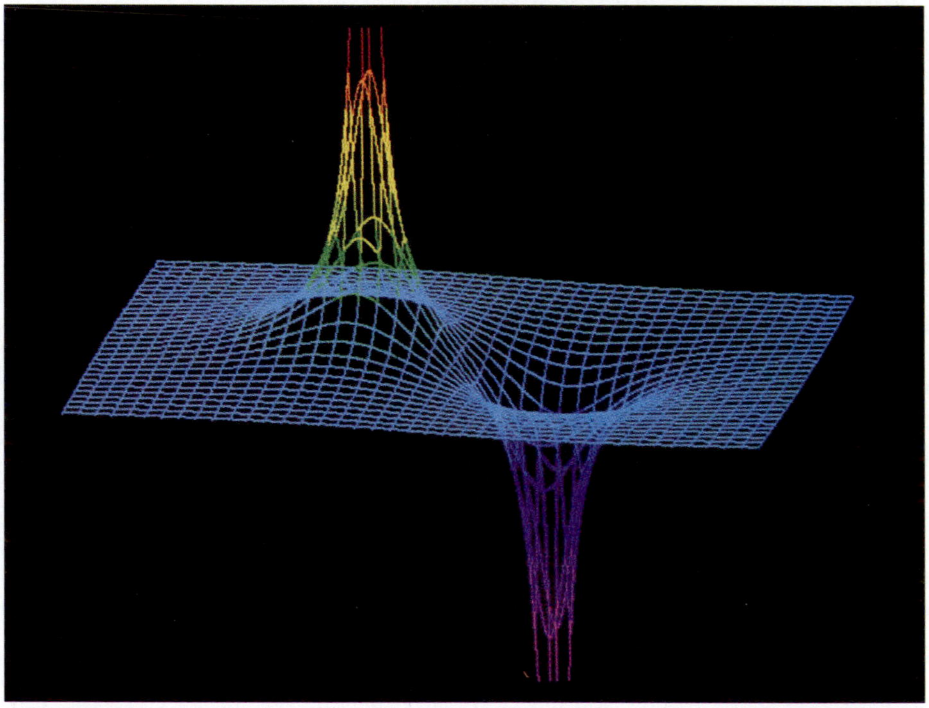

The electrostatic potential in the plane of an electric dipole. The potential due to each charge is proportional to the charge and inversely proportional to the distance from the charge.

The electric force between two charges is directed along the line of the charges and depends on the inverse square of their separation, the same as the gravitational force between two masses. Like the gravitational force, the electric force is conservative, so there is a potential-energy function U associated with it. If we place a test charge q_0 in an electric field, its potential energy is proportional to q_0. The potential energy per unit charge is a function of the position in space of the charge and is called the electric potential.

24-1 Potential Difference

In general, when a conservative force $\vec{F}$ undergoes a displacement $d\vec{\ell}$, the change in the potential-energy function dU is given by (Equation 6-21*b*)

$$dU = -\vec{F}\cdot d\vec{\ell}$$

The force exerted by an electric field $\vec{E}$ on a point charge q_0 is

$$\vec{F} = q_0\vec{E}$$

Thus, when a charge undergoes a displacement $d\vec{\ell}$ in an electric field $\vec{E}$, the change in the electrostatic potential energy is

$$dU = -q_0\vec{E}\cdot d\vec{\ell} \tag{24-1}$$

The potential-energy change is proportional to the charge q_0. The potential-energy change *per unit charge* is called the **potential difference** dV:

$$dV = \frac{dU}{q_0} = -\vec{E}\cdot d\vec{\ell} \tag{24-2a}$$

Definition—Potential difference

For a finite displacement from point a to point b, the change in potential is

$$\Delta V = V_b - V_a = \frac{\Delta U}{q_0} = -\int_a^b \vec{E}\cdot d\vec{\ell} \tag{24-2b}$$

Definition—Finite potential difference

The potential difference $V_b - V_a$ is the negative of the work done per unit charge by the electric field on a positive test charge when it moves from point a to point b. ΔV is also the positive work per charge that *you* must do against the electric field to move the charge from a to b.

The function V is called the **electric potential** or often just the **potential**. Like the electric field, the potential V is a function of position. Unlike the electric field, V is a scalar function, whereas $\vec{E}$ is a vector function. As with potential energy U, only *changes* in the potential V are important. We are free to choose the potential to be zero at any convenient point, just as we are when dealing with potential energy. If the electric potential and potential energy of a test charge are chosen to be zero at the same point, they are related by

$$U = q_0V \tag{24-3}$$

Relation between potential energy and potential

Continuity of *V*

In Chapter 23, we saw that the electric field is discontinuous by σ/ϵ_0 at a point where there is a surface charge density σ. The potential function, on the other hand, is continuous everywhere in space. We can see this from its definition. Consider an electric field in the x direction $\vec{E} = E_x\hat{i}$. The change in potential is given by Equation 24-2*a*.

$$dV = -\vec{E}\cdot d\vec{\ell} = -E_x\hat{i}\cdot(dx\,\hat{i} + dy\,\hat{j} + dz\,\hat{k})$$
$$= -E_x\,dx$$

Consider two nearby points x_1 and x_2. If V_1 is the potential at x_1 and V_2 is the potential at x_2, the potential difference can be written

$$\Delta V = (E_x)_{\text{av}}\,\Delta x = (E_x)_{\text{av}}\,(x_2 - x_1)$$

where $(E_x)_{\text{av}}$ is the average value of the electric field between the points. As x_2 approaches x_1, the potential difference ΔV approaches zero as long as $(E_x)_{\text{av}}$ is not infinite. The potential function V is thus continuous at any point not occupied by a point charge. Physically, if a test charge is moved a distance Δx, the work done by the field approaches zero as Δx approaches zero, as long as the electric field is not infinite.

Units

Since electric potential is the potential energy per unit charge, the SI unit for potential and potential difference is the joule per coulomb, called the **volt** (V):

$$1\ \text{V} = 1\ \text{J/C} \qquad \text{24-4}$$

The potential difference between two points (measured in volts) is sometimes called the **voltage**. In a 12-V car battery, the positive terminal has a potential 12 V higher than the negative terminal. If we attach an external circuit to the battery and one coulomb of charge is transferred from the positive terminal through the circuit to the negative terminal, the potential energy of the charge decreases by $Q\,\Delta V = (1\ \text{C})(12\ \text{V}) = 12\ \text{J}$.

We can see from Equation 24-2 that the dimensions of potential are also those of electric field times distance. Thus, the unit of the electric field is equal to one volt per meter:

$$1\ \text{N/C} = 1\ \text{V/m} \qquad \text{24-5}$$

In atomic and nuclear physics, we often have elementary particles with charges of magnitude e, such as electrons and protons, moving through potential differences of several to thousands or even millions of volts. Since energy has dimensions of electric charge times electric potential, a convenient unit of energy is the product of the electron charge e times a volt. This unit is called an **electron volt** (eV). The conversion between electron volts and joules is obtained by expressing the electronic charge in coulombs:

$$1\ \text{eV} = 1.6 \times 10^{-19}\ \text{C}\cdot\text{V} = 1.6 \times 10^{-19}\ \text{J} \qquad \text{24-6}$$

The electron volt

For example, an electron moving from the negative terminal to the positive terminal of a 12-V car battery gains potential energy of 12 eV.

Potential and Electric Field Lines

If we place a positive test charge q_0 in an electric field $\vec{E}$ and release it, it accelerates in the direction of $\vec{E}$. As the kinetic energy of the charge increases, its potential energy decreases. The charge therefore moves toward a region of lower potential energy, just as a mass falls toward a region of lower gravitational potential energy (Figure 24-1). Thus, as illustrated in Figure 24-2,

> Electric field lines point in the direction of decreasing electric potential.

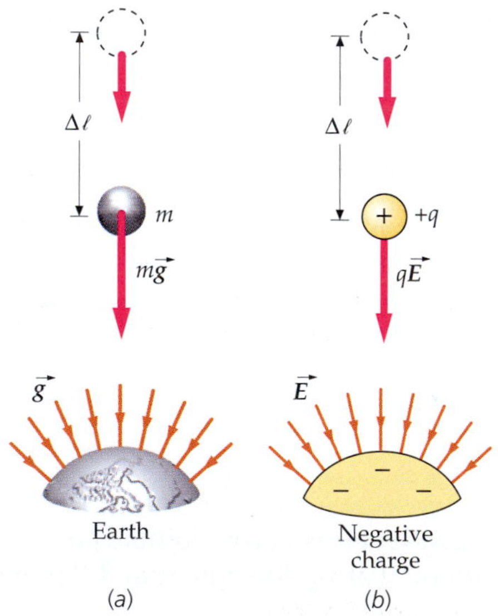

Figure 24-1 (*a*) The work done by the gravitational field on a mass decreases the gravitational potential energy. (*b*) The work done by the electric field on a positive charge $+q$ decreases the electrostatic potential energy.

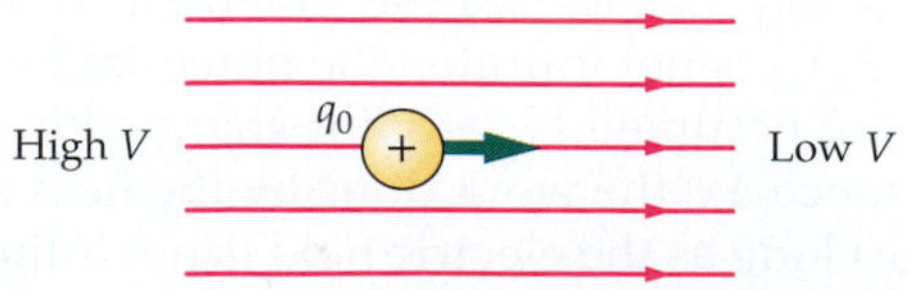

Figure 24-2 Electric field lines point in the direction of decreasing potential. When a positive test charge q_0 is placed in an electric field, it accelerates in the direction of the field. Its kinetic energy increases and its potential energy decreases.

Example 24-1

An electric field points in the positive x direction and has a constant magnitude of 10 N/C = 10 V/m. Find the potential as a function of x, assuming that $V = 0$ at $x = 0$.

1. By definition, the change in potential dV is related to the displacement $d\vec{\ell}$ and the electric field $\vec{E}$:

$$dV = -\vec{E}\cdot d\vec{\ell} = -(10\text{ V/m})\hat{i}\cdot(dx\,\hat{i} + dy\,\hat{j} + dz\,\hat{k}) = -(10\text{ V/m})\,dx$$

2. Integrate dV:

$$V = \int dV = \int -(10\text{ V/m})\,dx = -(10\text{ V/m})x + V_0$$

3. The constant of integration V_0 is found by setting $V = 0$ at $x = 0$:

$$V(0) = V_0 = 0$$

4. The potential is then:

$$V = -(10\text{ V/m})\,x$$

Remark The potential is zero at $x = 0$, and decreases by 10 V/m in the positive x direction.

Exercise Repeat this example for the electric field $\vec{E} = (10\text{ V/m}^2)x\hat{i}$. [*Answer* $V(x) = -(5\text{ V/m}^2)x^2$]

24-2 Potential due to a System of Point Charges

The electric potential at a distance r from a point charge q at the origin can be calculated from the electric field:

$$\vec{E} = \frac{kq}{r^2}\,\hat{r}$$

For an infinitesimal radial displacement $d\vec{\ell} = dr\,\hat{r}$, the change in potential is

$$dV = -\vec{E}\cdot d\vec{\ell} = -\frac{kq}{r^2}\,\hat{r}\cdot dr\hat{r} = -\frac{kq}{r^2}\,dr$$

Integrating, we obtain

$$V = +\frac{kq}{r} + V_0 \qquad \text{24-7}$$

Potential due to a point charge

where V_0 is a constant of integration.

If we define the potential to be zero at an infinite distance from the point charge ($r = \infty$), the constant V_0 is zero, and the potential at a distance r from the point charge is

$$V = \frac{kq}{r}, \qquad V = 0 \text{ at } r = \infty \qquad \text{24-8}$$

Coulomb potential

The potential given by Equation 24-8 is called the **Coulomb potential**. It is positive or negative depending on the sign of the charge q.

The potential energy U of a test charge q_0 placed a distance r from the point charge q is

$$U = q_0 V = \frac{kq_0 q}{r}, \qquad U = 0 \text{ at } r = \infty \qquad 24\text{-}9$$

Electrostatic potential energy of a two-charge system

This is the electrostatic potential energy of the two-charge system relative to $U = 0$ at infinite separation. If we release a test charge q_0 from rest at a distance r from q (and hold q fixed at the origin), the test charge will be accelerated outward (assuming that q has the same sign as q_0). Its kinetic energy at a great distance from q will be kq_0q/r. Alternatively, the work we must do against the electric field to bring a test charge q_0 from a great distance to a distance r from q is kq_0q/r (Figure 24-3).

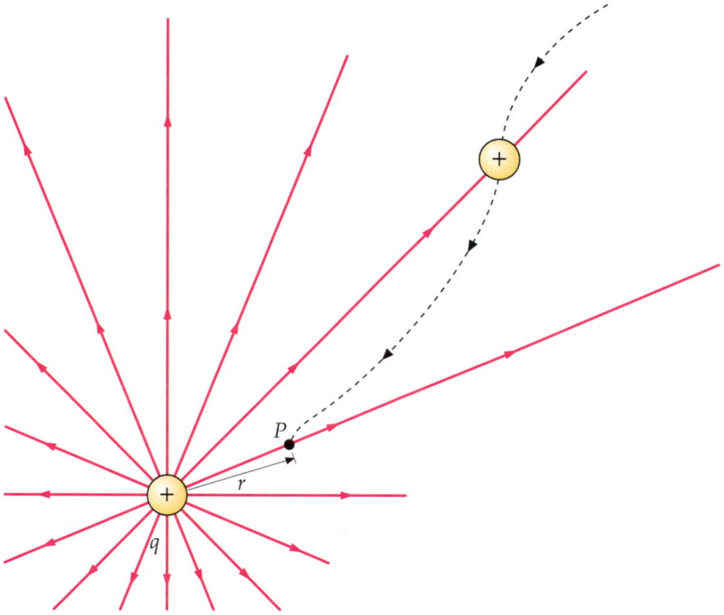

Figure 24-3 The work required to bring a test charge q_0 from infinity to a point P is kq_0q/r, where r is the distance from P to a charge q at the origin. The work per unit charge is kq/r, the electric potential at point P relative to zero potential at infinity. If the test charge is released from point P, the electric field does work kq_0q/r on the charge as the charge moves out to infinity.

Choosing the electrostatic potential energy of two charges to be zero at an infinite separation is analogous to the choice we made in Chapter 11 when we chose the gravitational potential energy of two point masses to be zero if the masses were very far apart.

Example 24-2

(*a*) What is the electric potential at a distance $r = 0.529 \times 10^{-10}$ m from a proton? (This is the average distance between the proton and electron in a hydrogen atom.) (*b*) What is the potential energy of the electron and the proton at this separation?

(*a*) Use $V = kq/r$ to calculate the potential V due to the proton:

$$V = \frac{kq}{r} = \frac{ke}{r} = \frac{(8.99 \times 10^9\ \text{N}\cdot\text{m}^2/\text{C}^2)(1.6 \times 10^{-19}\ \text{C})}{0.529 \times 10^{-10}\ \text{m}}$$

$$= 27.2\ \text{J/C} = 27.2\ \text{V}$$

(*b*) Use $U = q_0V$, with $q_0 = -e$ to calculate the electrostatic potential energy:

$$U = q_0V = (-e)(27.2\ \text{V}) = -27.2\ \text{eV}$$

Remarks If the electron were at rest at this distance from the proton, it would take 27.2 eV to remove it from the atom. However, the electron has kinetic energy equal to 13.6 eV, so its total energy in the atom is 13.6 eV − 27.2 eV = −13.6 eV. The energy needed to remove the electron from the atom is thus 13.6 eV. This energy is called the ionization energy.

Exercise What is the potential energy of the electron and proton in SI units? (*Answer* -4.35×10^{-18} J)

Example 24-3

In nuclear fission, a uranium-235 nucleus captures a neutron and splits apart into two lighter nuclei. Sometimes the two fission products are a barium nucleus (charge 56*e*) and a krypton nucleus (charge 36*e*). Assume that these nuclei are positive point charges separated by $r = 14.6 \times 10^{-15}$ m. Calculate the potential energy of this two-charge system in electron volts.

Picture the Problem The potential energy for two point charges separated by a distance r is $U = kq_1q_2/r$. To find this energy in electron volts we calculate the potential due to one of the charges kq_1/r in volts and multiply by the other charge.

1. Equation 24-9 gives the potential energy of the two charges:

$$U = \frac{kq_1q_2}{r} = \frac{k(56e)(36e)}{r}$$

2. Factor out e and substitute the given values:

$$U = \frac{k(56e)(36e)}{r} = e\,\frac{ke(56)(36)}{r}$$

$$= e\,\frac{(8.99 \times 10^9\ \mathrm{N \cdot m^2/C^2})(1.6 \times 10^{-19}\ \mathrm{C})(56)(36)}{14.6 \times 10^{-15}\ \mathrm{m}}$$

$$= e(1.99 \times 10^8\ \mathrm{V}) = 199\ \mathrm{MeV}$$

Remarks The separation distance r was chosen to be the sum of the radii of the two nuclei. After the fission, the two nuclei fly off because of their electrostatic repulsion. Their original potential energy of 199 MeV is converted into kinetic energy and thermal energy. Two or three neutrons are also released in the fission process. In a chain reaction, one or more of these neutrons produces a fission of another uranium nucleus. The average energy given off in chain reactions of this type is about 200 MeV per nucleus, as calculated in this example.

The potential at some point due to several point charges is the sum of the potentials due to each charge separately. (This follows from the superposition principle for the electric field. The work done by the net electric field is the sum of the work done by the electric fields due to each of the charges separately.) The potential due to a system of point charges q_i is thus given by

$$V = \sum_i \frac{kq_i}{r_i} \qquad \text{24-10}$$

Potential due to a system of point charges

where the sum is over all the charges, and r_i is the distance from the ith charge to the point P at which the potential is to be found.

Example 24-4

Two equal positive point charges of magnitude +5 nC are on the x axis. One is at the origin and the other is at $x = 8$ cm. Find the potential at (a) point P_1 on the x axis at $x = 4$ cm and (b) point P_2 on the y axis at $y = 6$ cm.

Picture the Problem The two positive point charges on the x axis are shown in Figure 24-4, and the potential is to be found at points P_1 and P_2.

Figure 24-4

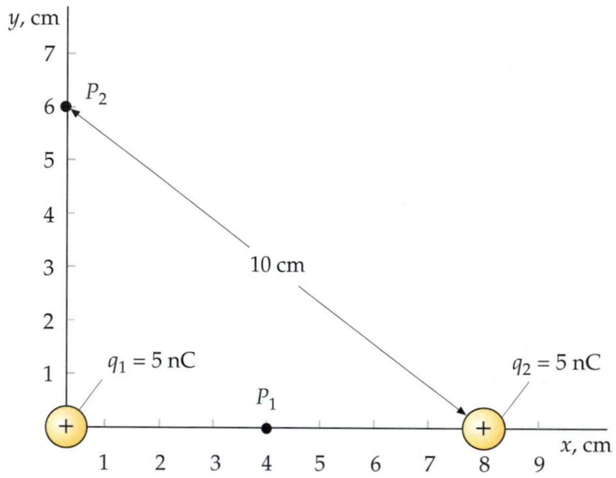

(a) 1. Use Equation 24-10 to write V as a function of the distances r_1 and r_2 to the charges:

$$V = \sum_i \frac{kq_i}{r_i} = \frac{kq_1}{r_1} + \frac{kq_2}{r_2}$$

2. Point P_1 is 4 cm from each charge, and the charges are equal.

$$r_1 = r_2 = r = 0.04 \text{ m}$$

$$q_1 = q_2 = q = 5 \times 10^{-9} \text{ C}$$

3. Use these to find the potential at point P_1:

$$V = \frac{kq_1}{r_1} + \frac{kq_2}{r_2} = \frac{2kq}{r}$$

$$= \frac{2 \times (8.99 \times 10^9 \text{ N·m}^2/\text{C}^2)(5 \times 10^{-9} \text{ C})}{0.04 \text{ m}}$$

$$= 2250 \text{ V}$$

(b) Point P_2 is 6 cm from one charge and 10 cm from the other. Use these to find the potential at point P_2:

$$V = \frac{(8.99 \times 10^9 \text{ N·m}^2/\text{C}^2)(5 \times 10^{-9} \text{ C})}{0.06 \text{ m}}$$

$$+ \frac{(8.99 \times 10^9 \text{ N·m}^2/\text{C}^2)(5 \times 10^{-9} \text{ C})}{0.10 \text{ m}}$$

$$= 749 \text{ V} + 450 \text{ V} \approx 1200 \text{ V}$$

Remarks Note that in (a), the electric field is zero at the point midway between the charges but the potential is not. It takes work to bring a test charge to this point from a long distance away, because the electric field is zero only at the final position.

Example 24-5

In Figure 24-5, a point charge q_1 is at the origin, and a second point charge q_2 is on the x axis at $x = a$. Find the potential everywhere on the x axis.

Picture the Problem The total potential is the sum of the potential due to each charge separately. The distance from q_1 to any point is $r_1 = |x|$ and the distance from q_2 to any point is $r_2 = |x - a|$.

Figure 24-5

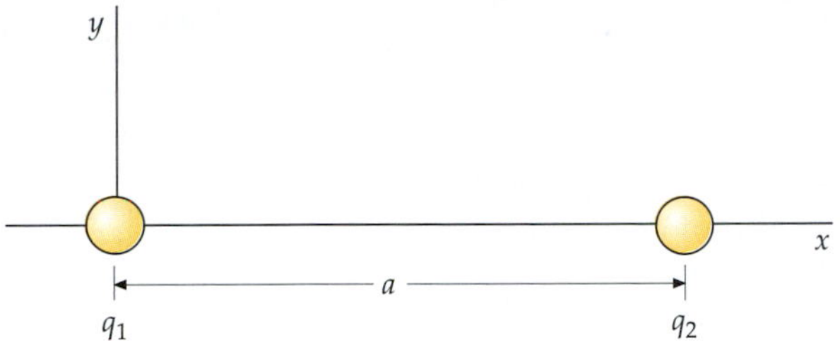

1. Write the potential as a function of the distances to the two charges:

$$V = \frac{kq_1}{r_1} + \frac{kq_2}{r_2} = \frac{kq_1}{|x|} + \frac{kq_2}{|x - a|}$$

2. To the right of both charges, $|x| = x$, and $|x - a| = x - a$:

$$V = \frac{kq_1}{x} + \frac{kq_2}{x - a}, \qquad x > a$$

Between the charges, $|x| = x$ and $|x - a| = a - x$:

$$V = \frac{kq_1}{x} + \frac{kq_2}{a - x}, \qquad 0 < x < a$$

To the left of both charges, $|x| = -x$ and $|x - a| = a - x$:

$$V = -\frac{kq_1}{x} + \frac{kq_2}{a - x}, \qquad x < 0$$

Remark Figure 24-6 shows V versus x for equal charges. The potential is always positive, and becomes infinite at each charge.

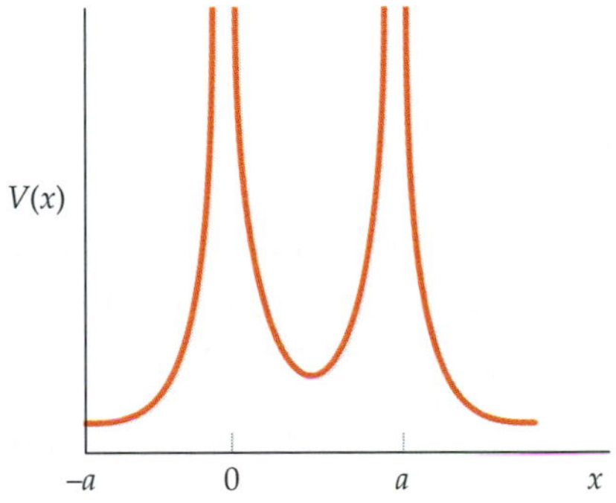

Figure 24-6

Example 24-6

An electric dipole consists of a positive charge $+q$ on the x axis at $x = +a$ and a negative charge $-q$ on the x axis at $x = -a$, as shown in Figure 24-7. Find the potential on the x axis for $x >> a$ in terms if the dipole moment $p = 2qa$.

Figure 24-7

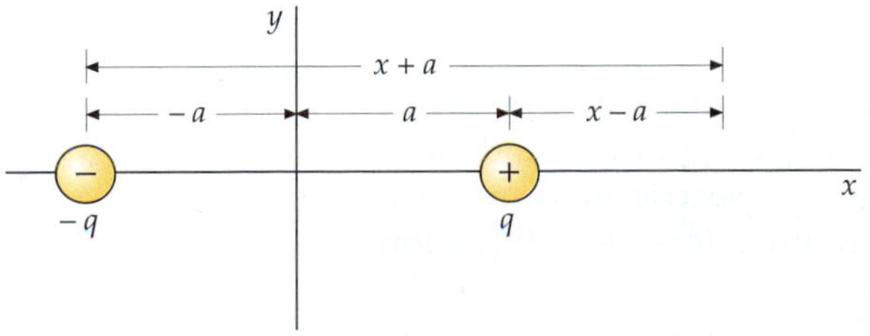

Picture the Problem The potential is the sum of the potential for each charge separately. For $x > a$, the distance to the positive charge is $x - a$ and the distance to the negative charge is $x + a$.

1. For $x > a$, the potential due to the two charges is:
$$V = \frac{kq}{x - a} + \frac{k(-q)}{x + a} = \frac{2kqa}{x^2 - a^2}$$

2. For $x >> a$, we can neglect a^2 compared with x^2 in the denominator. We then have:
$$V \approx \frac{2kqa}{x^2} = \frac{kp}{x^2}, \qquad x >> a \qquad \text{24-11}$$

Remarks Far from the dipole, the potential decreases as $1/r^2$, compared to $1/r$ for the potential of a point charge. A three-dimensional plot of the potential near the charges is shown on the opening page of this chapter.

24-3 Finding the Electric Field From the Potential

If we know the potential, we can use it to calculate the electric field. Consider a small displacement $d\vec{\ell}$ in an arbitrary electric field $\vec{E}$. The change in potential is

$$dV = -\vec{E}\cdot d\vec{\ell} = -E_\ell\, d\ell \qquad \text{24-12}$$

where E_ℓ is the component of $\vec{E}$ parallel to the displacement. Then

$$E_\ell = -\frac{dV}{d\ell} \qquad \text{24-13}$$

If the displacement $d\vec{\ell}$ is perpendicular to the electric field, the potential does not change. The greatest change in V occurs when the displacement $d\vec{\ell}$ is along $\vec{E}$. A vector that points in the direction of the greatest change in a scalar function and has a magnitude equal to the derivative of that function with respect to the distance in that direction is called the **gradient** of the function. The electric field $\vec{E}$ is the negative gradient of the potential V. The field lines point in the direction of the greatest decrease in the potential function.

If the potential depends only on x, there will be no change for displacements in the y or z direction, so $\vec{E}$ must be in the x direction. For a displacement in the x direction, $d\vec{\ell} = dx\hat{i}$, and Equation 24-12 becomes

$$dV(x) = -\vec{E}\cdot d\vec{\ell} = -\vec{E}\cdot dx\,\hat{i} = -E_x\, dx$$

Then

$$E_x = -\frac{dV(x)}{dx} \qquad \text{24-14}$$

Similarly, for a spherically symmetric charge distribution, the potential can be a function only of the radial distance r. Displacements perpendicular to the radial direction give no change in $V(r)$, so the electric field must be radial. A displacement in the radial direction is written $d\vec{\ell} = dr\,\hat{r}$. Equation 24-12 is then

$$dV(r) = -\vec{E}\cdot d\vec{\ell} = -\vec{E}\cdot dr\,\hat{r} = -E_r\, dr$$

and

$$E_r = -\frac{dV(r)}{dr} \qquad \text{24-15}$$

If we know either the potential or the electric field over some region of space, we can use one to calculate the other. The potential is often easier to calculate because it is a scalar function, whereas the electric field is a vector function. Note that we cannot calculate $\vec{E}$ if we know the potential V at just a single point—we must know V over a region of space to calculate $\vec{E}$.

Example 24-7

Find the electric field for the electric potential function $V(x)$ given by $V(x) = 100\text{ V} - (25\text{ V/m})x$.

This potential function depends only on x. The electric field is found from Equation 24-14:

$$\vec{E} = -\frac{dV}{dx}\hat{i} = +(25\text{ V/m})\hat{i}$$

Remarks This electric field is uniform and in the x direction. Note that the constant 100 V in the expression for $V(x)$ has no effect on the electric field. The electric field does not depend on the choice of zero for the potential function.

Exercise (*a*) At what point does V equal zero in this example? (*b*) Write the potential function corresponding to the same electric field with $V = 0$ at $x = 0$. [*Answers* (*a*) $x = 4$ m, (*b*) $V = -(25\text{ V/m})\,x$]

General Relation Between $\vec{E}$ and V

In vector notation, the gradient of V is written $\vec{grad}\,V$. Then

$$\vec{E} = -\vec{grad}\,V \qquad \text{24-16}$$

In general, the potential function can depend on x, y, and z. The rectangular components of the electric field are related to the partial derivatives of the potential with respect to x, y, or z, while the other variables are held constant. For example, the x component of the electric field is given by

$$E_x = -\frac{\partial V}{\partial x} \qquad \text{24-17a}$$

Similarly, the y and z components of the electric field are related to the potential by

$$E_y = -\frac{\partial V}{\partial y} \qquad \text{24-17b}$$

and

$$E_y = -\frac{\partial V}{\partial z} \qquad \text{24-17c}$$

Thus, Equation 24-16 in rectangular coordinates is

$$\vec{E} = -\vec{grad}\,V = -\left(\frac{\partial V}{\partial x}\hat{i} + \frac{\partial V}{\partial y}\hat{j} + \frac{\partial V}{\partial z}\hat{k}\right) \qquad \text{24-18}$$

24-4 Calculation of *V* for Continuous Charge Distributions

The potential due to a continuous distribution of charge can be calculated by choosing an element of charge dq, which we treat as a point charge, and changing the sum in Equation 24-10 to an integral:

$$V = \int \frac{k\,dq}{r} \qquad 24\text{-}19$$

Potential due to a continuous charge distribution

This equation assumes that $V = 0$ at an infinite distance from the charges, so we cannot use it when there is charge at infinity, as is the case for artificial charge distributions like an infinite line charge or an infinite plane charge.

V on the Axis of a Charged Ring

Figure 24-8 shows a uniformly charged ring of radius a and charge Q. The distance from an element of charge dq to the field point P on the axis of the ring is $r = \sqrt{x^2 + a^2}$. Since this distance is the same for all elements of charge on the ring, we can remove this term from the integral in Equation 24-19. The potential at point P due to the ring is thus

$$V = \int \frac{k\,dq}{r} = \int \frac{k\,dq}{\sqrt{x^2 + a^2}} = \frac{k}{\sqrt{x^2 + a^2}} \int dq$$

or

$$V = \frac{kQ}{\sqrt{x^2 + a^2}} \qquad 24\text{-}20$$

Potential on the axis of a uniformly charged ring

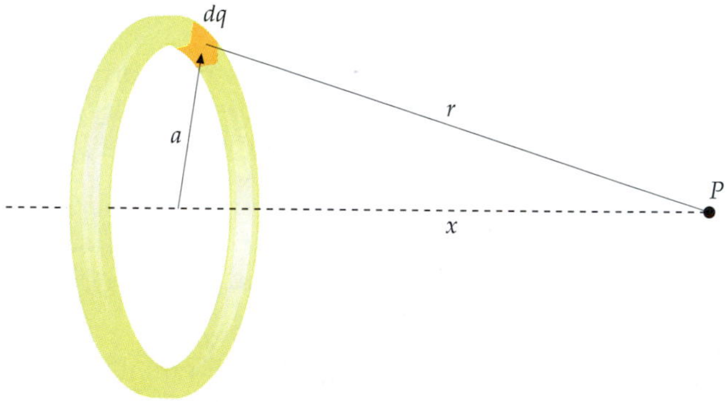

Figure 24-8 Geometry for the calculation of the electric potential at a point on the axis of a uniformly charged ring of radius a.

Note that when x is much greater than a, the potential approaches kQ/x, the same as for a point charge at the origin.

Example 24-8 *try it yourself*

A ring of radius 4 cm is in the yz plane with its center at the origin. The ring carries a uniform charge of 8 nC. A small particle of mass $m = 6$ mg $= 6 \times 10^{-6}$ kg and charge $q_0 = 5$ nC is placed at $x = 3$ cm and released. Find the speed of the particle when it is a great distance from the ring.

Picture the Problem As the particle moves along the x axis, its potential energy decreases and its kinetic energy increases. When the particle is very far away, its potential energy is zero, and by conservation of energy, its kinetic energy equals its original potential energy. The final speed is found from the final kinetic energy.

Cover the column to the right and try these on your own before looking at the answers.

Steps	*Answers*
1. Use $U = q_0V$ with V given by Equation 24-20 to calculate the initial potential energy of the point charge q_0 at a distance 3 cm from the center of the ring.	$U = q_0V = \frac{kQq_0}{\sqrt{x^2 + a^2}} = 7.19 \times 10^{-6}$ J

2. Use conservation of energy to write an equation for the kinetic energy of the particle when it is far from the ring.

$$U_f + K_f = U_i + K_i$$
$$0 + \tfrac{1}{2}mv^2 = 7.19 \times 10^6\ \text{J} + 0$$

3. Solve for the speed v.

$$v = 1.55\ \text{m/s}$$

Exercise What is the potential energy of the particle when it is at x = 9 cm? (*Answer* 3.65×10^{-6} J)

V on the Axis of a Uniformly Charged Disk

We can use our result for the potential on the axis of a ring charge to calculate the potential on the axis of a uniformly charged disk.

Example 24-9

Find the potential on the axis of a disk of radius R that carries a total charge Q distributed uniformly on its surface.

Picture the Problem We take the axis of the disk to be the x axis and we treat the disk as a set of ring charges. The ring of radius a and thickness da in Figure 24-9 has an area of $2\pi a\,da$ and its charge is $dq = \sigma\,dA = \sigma 2\pi a\,da$ where $\sigma = Q/\pi R^2$ is the surface charge density. The potential at point P is given by Equation 24-20. We then integrate from $a = 0$ to $a = R$ to find the total potential due to the disk.

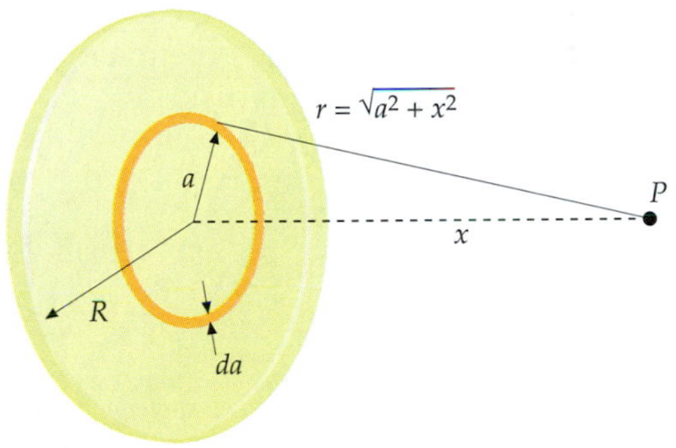

Figure 24-9

1. Write the potential dV at point P due to the charged ring of radius a:

$$dV = \frac{k\,dq}{(x^2 + a^2)^{1/2}} = \frac{k\sigma 2\pi a\,da}{(x^2 + a^2)^{1/2}}$$

2. Integrate from $a = 0$ to $a = R$:

$$V = \int_0^R \frac{k\sigma 2\pi a\,da}{(x^2 + a^2)^{1/2}} = k\sigma\pi \int_0^R (x^2 + a^2)^{-1/2} 2a\,da$$

3. The integral is of the form $\int u^n\,du$, with $u = x^2 + a^2$ and $n = -\frac{1}{2}$:

$$\int_0^R (x^2 + a^2)^{-1/2}\, 2a\,da = \left.\frac{(x^2 + a^2)^{+1/2}}{\frac{1}{2}}\right|_{a=0}^{a=R} = 2[(x^2 + R^2)^{1/2} - x]$$

4. Use this result to find V:

$$V = 2\pi k\sigma[(x^2 + R^2)^{1/2} - x]$$

Check the Result To find how the potential varies with x for $x >> R$, we write $V = 2\pi k\sigma x[(1 + R^2/x^2)^{1/2} - 1]$ and use the binomial expansion

$$\left(1 + \frac{R^2}{x^2}\right)^{1/2} \approx 1 + \frac{1}{2}\frac{R^2}{x^2} + \cdots$$

Then

$$V \approx 2\pi k\sigma x\left[1 + \frac{1}{2}\frac{R^2}{x^2} + \cdots - 1\right] = \frac{k(\sigma\pi R^2)}{x} = \frac{kQ}{x}$$

From Example 24-9 we see that the potential on the axis of a uniformly charged disk is

$$V = 2\pi k\sigma[(x^2 + R^2)^{1/2} - x] \tag{24-21}$$

Potential on the axis of a disk charge

Example 24-10

Calculate the electric field on the axis of (*a*) a uniformly charged ring and (*b*) a uniformly charged disk using the potential functions given above for these charge distributions.

Picture the Problem Since V depends only on x, the electric field has only an x component, given by $E_x = -dV/dx$.

(*a*) 1. Write Equation 24-20 for the potential on the axis of a uniformly charged ring:

$$V = \frac{kQ}{\sqrt{x^2 + a^2}} = kQ(x^2 + a^2)^{-1/2}$$

2. Compute $-dV/dx$ to find E_x:

$$E_x = -\frac{dV}{dx} = +\frac{1}{2}kQ(x^2 + a^2)^{-3/2}(2x) = \frac{kQx}{(x^2 + a^2)^{3/2}}$$

(*b*) 1. Write Equation 24-21 for the potential on the axis of a uniformly charged disk:

$$V = 2\pi k\sigma[(x^2 + R^2)^{1/2} - x]$$

2. Compute $-dV/dx$ to find E_x:

$$E_x = -\frac{dV}{dx} = -2\pi k\sigma\left[\frac{1}{2}(x^2 + a^2)^{-1/2}2x - 1\right]$$

$$= 2\pi k\sigma\left(1 - \frac{x}{\sqrt{x^2 + R^2}}\right)$$

Remarks The results for (*a*) and (*b*) are the same as Equations 23-10 and 23-11, which were calculated directly from Coulomb's law.

V due to an Infinite Plane of Charge

If we let R become very large, our disk approaches an infinite plane. As R approaches infinity the potential function (Equation 24-21) approaches infinity. However, we obtained Equation 24-21 from Equation 24-19, which assumes that $V = 0$ at infinity, so Equation 24-21 can't be used. For infinite charge distributions, we must choose $V = 0$ at some finite point rather than at infinity. For such cases, we first find the electric field $\vec{E}$ (by direct integration or from Gauss's law) and then calculate the potential from its definition $dV = -\vec{E}\cdot d\vec{\ell}$. For an infinite plane of charge of density σ in the yz plane, the electric field for positive x is given by

$$\vec{E} = \frac{\sigma}{2\epsilon_0}\hat{i} = 2\pi k\sigma\hat{i}$$

The potential is then

$$dV = -\vec{E}\cdot d\vec{\ell} = -(2\pi k\sigma\hat{i})\cdot(dx\hat{i} + dy\hat{j} + dz\hat{k}) = -2\pi k\sigma\, dx$$

Integrating, we obtain

$$V = V_0 - 2\pi k\sigma x$$

where the arbitrary constant V_0 is the potential at $x = 0$. Note that the potential decreases with distance from the plane and approaches $-\infty$ as x approaches $-\infty$ Therefore, we cannot choose the potential to be zero at $x = \infty$. For negative x, the electric field is

$$\vec{E} = -2\pi k\sigma\hat{i}$$

so

$$dV = -\vec{E}\cdot d\vec{\ell} = +2\pi k\sigma\, dx$$

and the potential is

$$V = V_0 + 2\pi k\sigma x$$

Since x is negative, the potential again decreases with distance from the plane and approaches $-\infty$ as x approaches $-\infty$. For either positive or negative x, the potential can be written

$$V = V_0 - 2\pi k\sigma|x| \qquad \text{24-22}$$

Potential near an infinite plane of charge

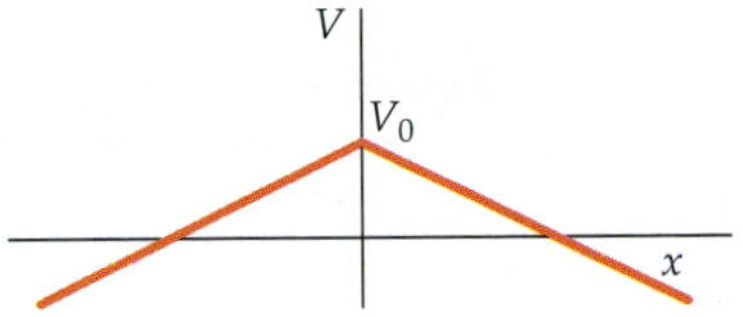

Figure 24-10 Plot of V versus x for an infinite plane of charge in the yz plane. Note that the potential is continuous at $x = 0$ even though the electric field is not.

Figure 24-10 shows this potential versus x.

Example 24-11

An infinite plane of charge density σ is in the yz plane at $x = 0$, and a point charge q is on the x axis at $x = a$ (Figure 24-11). Find the potential at some point P a distance r from the point charge for $x > 0$ (that is, to the right of the plane charge).

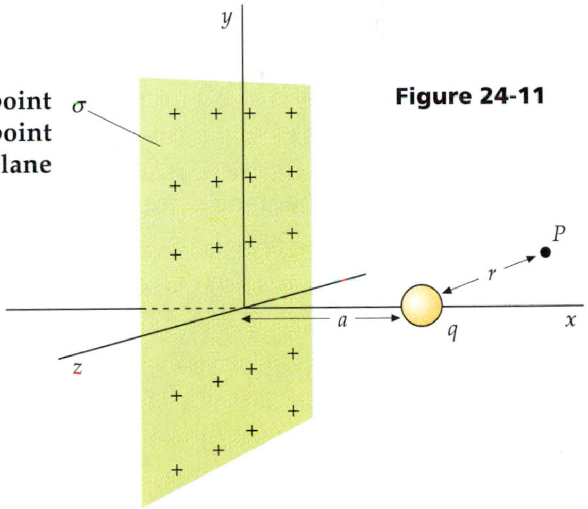

Figure 24-11

Picture the Problem The total potential is the sum of the potential due to the plane and the potential due to the point charge. Since we cannot choose V to be zero at $r = \infty$, we must include an arbitrary constant in our expression for $V(r)$ for a point charge. Since we have two constants, we will call one V_{01} and the other V_{02}. We are free to choose the zero of potential to be at any convenient finite point. For this example, we choose $V = 0$ at the origin.

1. The potential due to the plane is given by Equation 24-22 with $|x| = x$:

$$V_{plane} = V_{01} - 2\pi k\sigma x, \qquad V_{01} = \text{constant}$$

2. Equation 24-7 gives the potential due to a point charge:

$$V_{point} = \frac{kq}{r} + V_{02}, \qquad V_{02} = \text{constant}$$

3. Sum the above results to find the total potential V:

$$V = V_{plane} + V_{point} = V_{01} - 2\pi kx + \frac{kq}{r} + V_{02}$$

4. We choose $V = 0$ at the origin; that is, at $x = 0$ and $r = a$. This choice determines $V_{01} + V_{02}$:

$$V = 0 = V_{01} - 2\pi k\sigma(0) + \frac{kq}{a} + V_{02}$$

or

$$V_{01} + V_{02} = -\frac{kq}{a}$$

5. Substitute this for $V_{01} + V_{02}$ in the general expression for V:

$$V = -2\pi k\sigma x + \frac{kq}{r} - \frac{kq}{a}$$

Remarks In rectangular coordinates, $r = [(x-a)^2 + y^2 + z^2]^{1/2}$, thus

$$V = -2\pi k\sigma x + \frac{kq}{[(x-a)^2 + y^2 + z^2]^{1/2}} - \frac{kq}{a}$$

Exercise Find V for $x < 0$.

Answer

$$V = 2\pi k\sigma x + \frac{kq}{[(x-a)^2 + y^2 + z^2]^{1/2}} - \frac{kq}{a}$$

V Inside and Outside a Spherical Shell of Charge

We next find the potential due to a spherical shell of radius R with charge Q uniformly distributed on its surface. We are interested in the potential at all points inside and outside the shell. Since this shell is of finite extent, we could calculate the potential by direct integration of Equation 24-19, but this integration is somewhat difficult. Since the electric field for this charge distribution is easily obtained from Gauss's law, it is easiest to find the potential from the known electric field using $dV = -\vec{E}\cdot d\vec{\ell}$.

Outside the spherical shell, the electric field is radial and is the same as if all the charge were a point charge at the origin:

$$\vec{E} = \frac{kQ}{r^2}\hat{r}$$

The change in the potential for some displacement $d\vec{\ell} = dr\,\hat{r}$ outside the shell is then

$$dV = -\vec{E}\cdot d\vec{\ell} = -\frac{kQ}{r^2}\hat{r}\cdot dr\,\hat{r} = -\frac{kQ}{r^2}dr$$

Integrating, we obtain

$$V = \frac{kQ}{r} + V_0$$

where V_0 is the potential at $r = \infty$. Choosing the potential to be zero at $r = \infty$ gives

$$V = \frac{kQ}{r}, \qquad r > R$$

Inside the spherical shell, the electric field is zero. The change in potential for any displacement inside the shell is therefore also zero. Thus, the potential inside the shell must be constant. As r approaches R from outside the shell, the potential approaches kQ/R. Hence, the constant value of V inside must be kQ/R to make V continuous. Thus,

$$V = \begin{cases} \dfrac{kQ}{r}, & r \geq R \\[2ex] \dfrac{kQ}{R}, & r \leq R \end{cases} \qquad 24\text{-}23$$

Potential due to a spherical shell

This potential function is plotted in Figure 24-12.

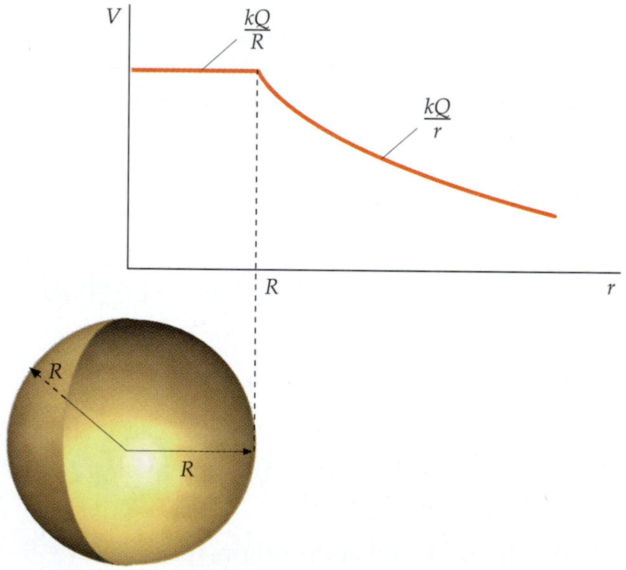

Figure 24-12 Electric potential of a uniformly charged spherical shell of radius R as a function of the distance r from the center of the shell. Inside the shell, the potential has the constant value kQ/R. Outside the shell, the potential is the same as that due to a point charge at the center of the sphere.

A common mistake is to think that the potential must be zero inside a spherical shell because the electric field is zero. But zero electric field merely implies that the potential does not change. Consider a spherical shell with a small hole so that we can move a test charge in and out of the shell. If we move the test charge from an infinite distance to the shell, the work per charge we must do is kQ/R. Inside the shell, there is no electric field, so it takes no work to move the test charge around inside the shell. The total amount of work per charge it takes to bring the test charge from infinity to any point inside the shell is just the work per charge it takes to bring it up to the shell radius R, which is kQ/R. The potential is therefore kQ/R everywhere inside the shell.

Exercise What is the potential of a spherical shell of radius 10 cm carrying a charge of 6 μC? (*Answer* 5.39×10^5 V = 539 kV)

Example 24-12 *try it yourself*

In one model, a proton is considered to be a spherical ball of charge of uniform volume charge density with radius R and total charge Q. The electric field inside the sphere is given by Equation 23-23*b*,

$$E_r = k\frac{Q}{R^3}r$$

Find the potential V both inside and outside the sphere.

Picture the Problem Outside the sphere, the charge looks like a point charge, so the potential is $V = kQ/r$. Inside the sphere, V can be found by integrating $dV = -\vec{E}\cdot d\vec{\ell}$. The constant of integration is found by requiring that V is continuous at $r = R$.

Cover the column to the right and try these on your own before looking at the answers.

Steps	*Answers*
1. Write down the potential V for $r \geq R$.	$V = \frac{kQ}{r}, \quad r \geq R$
2. For $r \leq R$, find dV from $dV = -\vec{E}\cdot d\vec{\ell}$	$dV = -\frac{kQ}{R^3}r\,dr, \quad r \leq R$
3. Find the indefinite integral of your expression in step 2. Be sure to include a constant of integration V_0.	$V = -\frac{kQ}{2R^3}r^2 + V_0$
4. Set your expression from step 3 equal to that from step 1 at $r = R$ and solve for the constant of integration V_0.	$V_0 = \frac{3kQ}{2R}$
5. Substitute your value for V_0 into your expression in step 3 to obtain $V(r)$ for $r \leq R$.	$V(r) = \frac{kQ}{2R}\left(3 - \frac{r^2}{R^2}\right), \quad r \leq R$

Check the Result Substituting $r = R$ in the result gives $V(R) = kQ/R$ as required. At $r = 0$, $V(0) = 3kQ/2R = 1.5\,kQ/R$, which is greater than $V(R)$, as it should be, because the electric field is in the positive radial direction for $r < R$, so work must be done to move a test charge from $r = R$ to $r = 0$.

Remarks Figure 24-13 shows $V(r)$ as a function of r. Note that both $V(r)$ and $E_r = -dV/dr$ are continuous everywhere.

Exercise What is $V(r)$ if we choose $V(R) = 0$? (*Answer* $V(r) = kQ/r - kQ/R$ for $r \geq R$; $V(r) = \frac{1}{2}(kQ/R)(1 - r^2/R^2)$ for $r \leq R$)

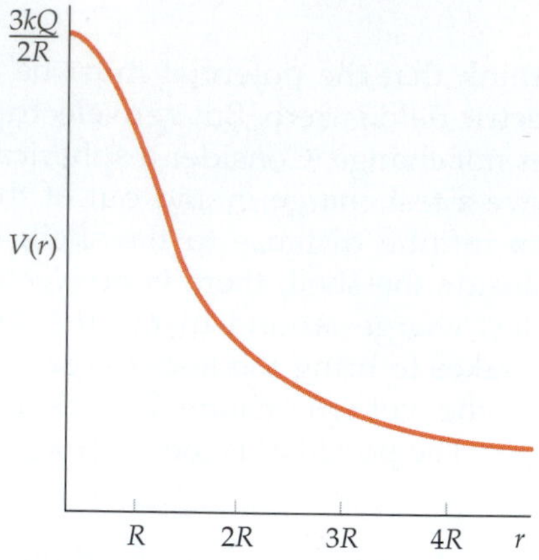

Figure 24-13

V due to an Infinite Line Charge

We will now calculate the potential due to a uniform infinite line charge. Let the charge per unit length be λ. Since this charge distribution extends to infinity, we find the potential from the electric field. In Chapter 23, we found that the electric field produced by an infinite line charge points away from the line (assuming λ to be positive) and is given by $E_r = 2k\lambda/r$. The change in potential for a displacement $d\vec{\ell}$ is then

$$dV = -\vec{E}\cdot d\vec{\ell} = -E_r\,dr = -\frac{2k\lambda}{r}\,dr$$

Integrating, we obtain

$$V = V_0 - 2k\lambda \ln r \qquad \text{24-24}$$

For a positive line charge, the electric field lines point away from the line and the potential decreases with increasing distance from the line charge. At large values of r, the potential decreases without limit. We therefore cannot choose for the potential to be zero at $r = \infty$. (Neither can we choose the potential to be zero at $r = 0$, because $\ln r$ approaches $-\infty$ as r approaches zero.) Instead, we choose V to be zero at some distance $r = a$. Substituting $r = a$ into Equation 24-24 and setting $V = 0$, we obtain

$$V = 0 = V_0 - 2k\lambda \ln a$$

or

$$V_0 = 2k\lambda \ln a$$

Then Equation 24-24 is

$$V = 2k\lambda \ln a - 2k\lambda \ln r$$

or

$$V = -2k\lambda \ln \frac{r}{a} \qquad \text{24-25}$$

Potential due to a line charge

24-5 Equipotential Surfaces

Since there is no electric field inside a conductor that is in static equilibrium, the change in potential as we move about inside the conductor is zero. The electric potential is thus the same throughout the conductor, that is, the conductor is an **equipotential volume** and its surface is an **equipotential surface**. Because the potential is constant on an equipotential surface, the change in V when a test charge is given a displacement $d\vec{\ell}$ parallel to the surface is $dV = -\vec{E} \cdot d\vec{\ell} = 0$. Then, since $\vec{E} \cdot d\vec{\ell}$ is zero, the electric field lines emanating from the equipotential surface must be perpendicular to the surface. Figures 24-14 and 24-15 show equipotential surfaces near a spherical conductor and a nonspherical conductor. Note that the field lines are everywhere perpendicular to the equipotential surfaces. If we move a short distance $d\ell$ along a field line from one equipotential surface to another, the potential changes by $dV = -\vec{E} \cdot d\vec{\ell} = -E\,d\ell$. Equipotential surfaces that have a fixed potential difference between them are more closely spaced where the electric field E is greater.

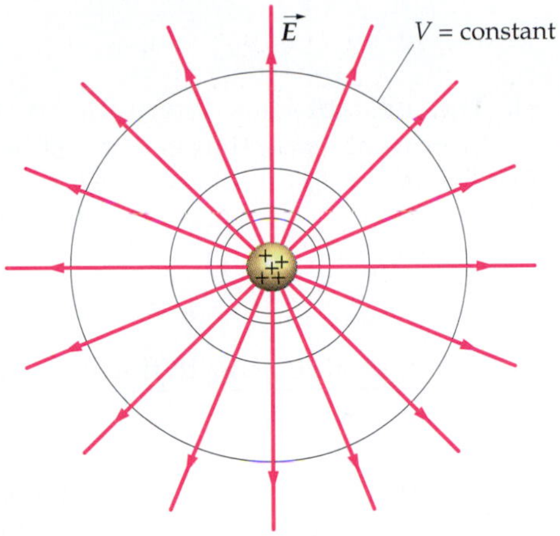

Figure 24-14 Equipotential surfaces and electric field lines outside a uniformly charged spherical conductor. The equipotential surfaces are spherical and the field lines are radial and perpendicular to the equipotential surfaces.

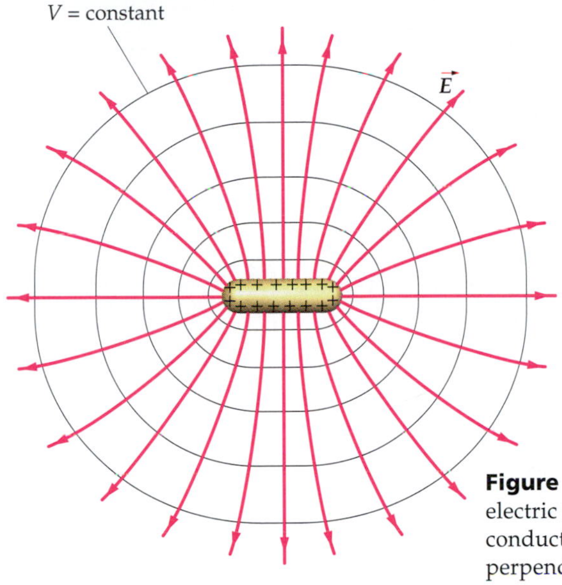

Figure 24-15 Equipotential surfaces and electric field lines outside a nonspherical conductor. Electric field lines are always perpendicular to equipotential surfaces.

Example 24-13

A hollow, uncharged spherical conductor has inner radius a and outer radius b. A positive point charge $+q$ is in the cavity at the center of the sphere. Find the potential $V(r)$ everywhere, assuming that $V = 0$ at $r = \infty$.

Picture the Problem We obtain the potential from $dV = -\vec{E} \cdot d\vec{\ell} = -E_r\,dr$. Inside the cavity, $E_r = kq/r^2$, so V is of the form $V = kq/r + V_0$. The conductor is an equipotential volume, so V is constant for $a \le r \le b$. The field lines inside the cavity must end on the inner surface of the cavity, so this surface has an induced charge of $-q$. Since the shell is uncharged, a positive charge $+q$ is on the outer surface. The three charges q at the center, $-q$ on the inner surface, and $+q$ on the outer surface produce a field $E_r = kq/r^2$ for $r > b$, so the potential for $r > b$ is $V = kq/r$.

1. Outside the shell, $V(r)$ is the same as that due to a point charge q at the origin. Choosing $V = 0$ at $r = \infty$, we have:

$$V(r) = \frac{kq}{r}, \qquad r \ge b$$

2. At $r = b$, the potential is kq/b. V remains at this constant value throughout the spherical shell:

$$V(r) = \frac{kq}{b}, \qquad a \le r \le b$$

3. Inside the cavity, V is the same as that due to a point charge q at the origin, but the arbitrary constant cannot be set equal to zero, because V has already been chosen to be zero at $r = \infty$:

$$V(r) = \frac{kq}{r} + V_0, \qquad r \le a$$

4. The constant V_0 is determined by the condition that V is continuous at $r = a$; that is, V must be kq/b at $r = a$:

$$V(a) = \frac{kq}{a} + V_0 = \frac{kq}{b}$$

or

$$V_0 = \frac{kq}{b} - \frac{kq}{a}$$

5. Use this value of V_0 to find $V(r)$ for $r \leq a$:

$$V(r) = \frac{kq}{r} + \frac{kq}{b} - \frac{kq}{a}, \qquad r \leq a$$

Remarks Figure 24-16 shows the electric potential as a function of the distance from the center of the cavity. Inside the conducting material, where $a \leq r \leq b$, the potential has the constant value kq/b. Outside the shell, the potential is the same as that of a point charge. Note that $V(r)$ is continuous everywhere. The electric field is discontinuous at the conductor surfaces, as reflected in the discontinuous slope of $V(r)$ at $r = a$ and $r = b$.

Figure 24-16

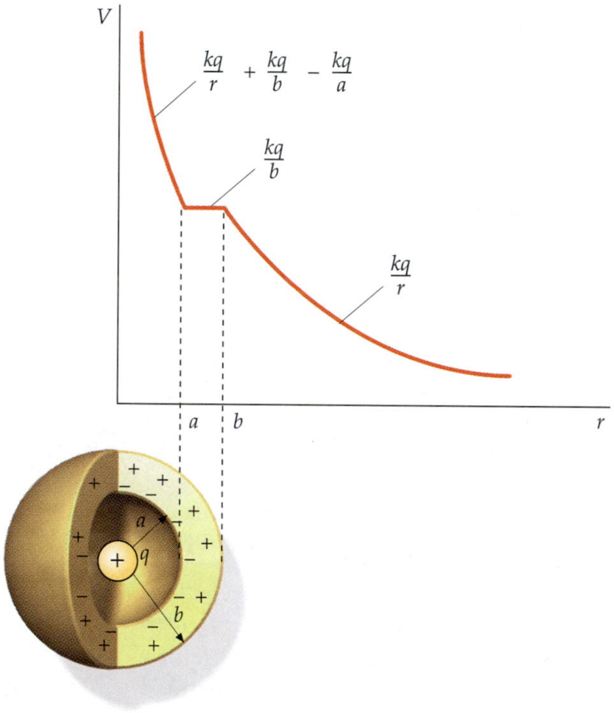

In general, two conductors that are separated in space will not be at the same potential. The potential difference between the conductors depends on their geometrical shapes, their separation, and the net charge on each. When two conductors are brought into contact, the charge on the conductors distributes itself so that electrostatic equilibrium is established and the electric field is zero inside both conductors. While in contact, the two conductors may be considered to be a single conductor with a single equipotential surface. If we put a spherical charged conductor in contact with a second spherical conductor that is uncharged, charge will flow to the neutral conductor until both conductors are at the same potential. If the conductors are identical, they share the original charge equally. If the conductors are now separated, each carries half the original charge, and both are at the same potential.

The Van de Graaff Generator

In Figure 24-17, a small conductor carrying a positive charge q is inside the cavity of a larger conductor. In equilibrium, the electric field is zero inside the conducting material of both conductors. The electric field lines that leave the positive charge q must end on the inner surface of the large conductor. This must occur no matter what the charge may be on the outside surface of the large conductor. Regardless of the charge on the large conductor, the small conductor in the cavity is at a greater potential because the electric field lines go from this conductor to the larger conductor. If the conductors are now connected, say, with a fine conducting wire, *all* the charge originally on the smaller conductor will flow to the larger one. When the connection is broken, there is no charge on the small conductor in the cavity, and there are no field lines between the conductors. The positive charge transferred from the smaller conductor resides completely on the outside surface of the larger conductor. If we put more positive charge on the small conductor in the cavity and again connect the conductors with a fine wire, all of the charge on the inner conductor will again flow to the outer conductor. This procedure can be repeated indefinitely. This method is used to produce large potentials in a device called the Van de Graaff generator, in which the charge is brought to the inner surface of a larger spherical conductor by a continuous charged belt (Figure 24-18). Work must be done by the motor driving the belt to bring the charge from the bottom to the top of the belt where the potential is very high. The greater the net charge on the outer conductor, the greater its potential, and the greater the electric field outside the conductor. A Van de Graaff accelerator is a device that uses the intense electric field produced by a Van de Graaff generator to accelerate positive particles, such as protons.

Figure 24-17 Small conductor carrying a positive charge inside a larger conductor.

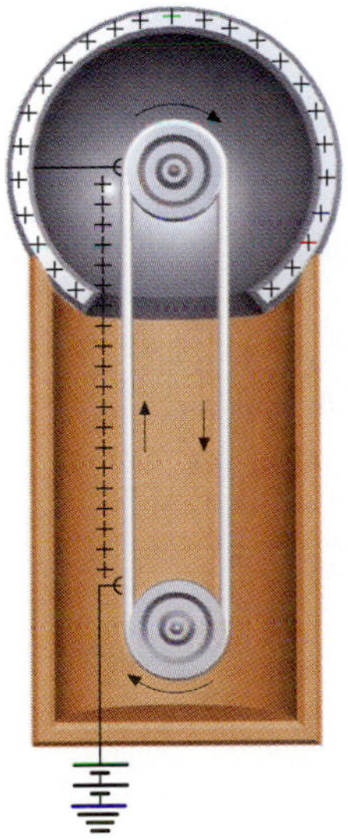

(*a*)

(*b*)

(*c*)

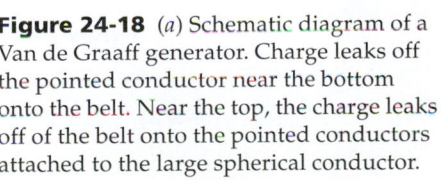

Figure 24-18 (*a*) Schematic diagram of a Van de Graaff generator. Charge leaks off the pointed conductor near the bottom onto the belt. Near the top, the charge leaks off of the belt onto the pointed conductors attached to the large spherical conductor. (*b*) This girl has been charged to a very high potential through contact with a demonstration Van de Graaff generator while standing on an insulating block. Her hair has acquired sufficient charge to show electrostatic repulsion. Care must be taken to acquire the charge gradually and to avoid rapid discharge to prevent a painful shock. (*c*) These large demonstration Van de Graaff generators in the Boston science museum are discharging to the grounded wire cage housing the operator.

Dielectric Breakdown

Many nonconducting materials become ionized in very high electric fields and become conductors. This phenomenon, called **dielectric breakdown,** occurs in air at an electric field strength of $E_{max} \approx 3 \times 10^6$ V/m = 3 MN/C. This limits the maximum potential that can be obtained in a Van de Graaff generator. The magnitude of the electric field for which dielectric breakdown occurs in a material is called the **dielectric strength** of that material. The dielectric strength of air is thus about 3 MV/m. The discharge through the conducting air resulting from dielectric breakdown is called **arc discharge.** The electric shock you receive when you touch a metal door knob after walking across a rug on a dry day is a familiar example of arc discharge. This occurs more often on dry days because moist air can conduct away charge as you acquire it. Lightning is an example of arc discharge on a large scale.

Example 24-14

A spherical conductor has a radius of 2 m. (*a*) What is the maximum charge that can be placed on the sphere before dielectric breakdown of the surrounding air occurs? (*b*) What is the maximum potential of the sphere?

Picture the Problem (*a*) We find the maximum charge by relating the charge to the electric field and setting the field equal to the dielectric strength of air, E_{max}. (*b*) The maximum potential is then found from the maximum charge calculated in (*a*).

(*a*) 1. The surface charge density on the conductor σ is related to the electric field just outside the conductor:

$$E = \frac{\sigma}{\epsilon_0}$$

2. Set this field equal to E_{max}:

$$E_{max} = 3 \times 10^6 \text{ N/C} = \frac{\sigma_{max}}{\epsilon_0}$$

3. The maximum charge Q is found from σ_{max}:

$$Q = 4\pi R^2 \sigma_{max} = 4\pi R^2(\epsilon_0 E_{max})$$
$$= 4\pi(2 \text{ m})^2(8.85 \times 10^{-12} \text{ C}^2/\text{N}\cdot\text{m}^2)(3 \times 10^6 \text{ N/C})$$
$$= 1.33 \times 10^{-3} \text{ C}$$

(*b*) Use this maximum charge to calculate the maximum potential of the sphere:

$$V_{max} = \frac{kQ_{max}}{R} = \frac{(8.99 \times 10^9 \text{ N}\cdot\text{m}^2/\text{C}^2)(1.33 \times 10^{-3} \text{ C})}{2 \text{ m}}$$
$$= 5.98 \times 10^6 \text{ V}$$

Example 24-15

Two charged spherical conductors of radius R_1 = 6 cm and R_2 = 2 cm are separated by a distance much greater than 6 cm and are connected by a conducting wire. A total charge Q = +80 nC is placed on one of the spheres. (*a*) What is the electric field near the surface of each sphere? (*b*) What is the electric potential of each sphere? (Assume that the charge on the connecting wire is negligible.)

Picture the Problem The total charge will be distributed with q_1 on sphere 1 and q_2 on sphere 2 such that the spheres will be at the same potential. We can use $V = kq/r$ for the potential of each sphere because they are far apart.

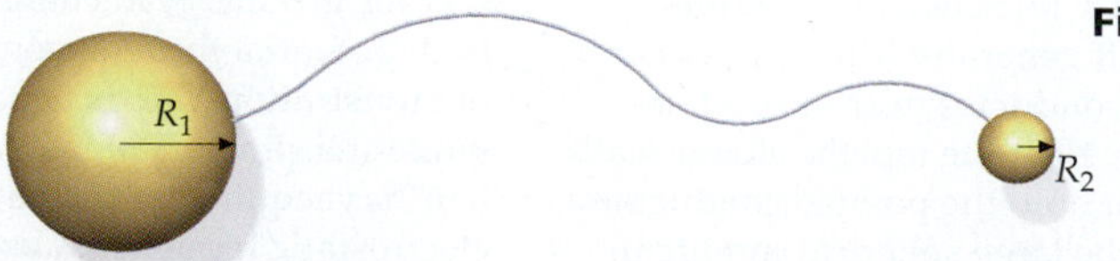

Figure 24-19

(*a*) 1. The electric field at the surface of each sphere is related to the charge on the sphere and its radius:

$$E_1 = \frac{kq_1}{R_1^2} \quad \text{and} \quad E_2 = \frac{kq_2}{R_2^2}$$

2. Conservation of charge gives us one relation between the charges q_1 and q_2:

$$q_1 + q_2 = Q = 80\text{ nC}$$

3. Equating the potential of the spheres gives us a second relation for the charges q_1 and q_2:

$$\frac{kq_1}{R_1} = \frac{kq_2}{R_2}$$

$$q_1 = \frac{R_1}{R_2}q_2 = \frac{6\text{ cm}}{2\text{ cm}}q_2 = 3q_2$$

4. Combine these results and solve for q_1 and q_2:

$$q_1 = 60\text{ nC}$$

$$q_2 = 20\text{ nC}$$

5. Use these results to calculate the electric fields at the surface of the spheres:

$$E_1 = \frac{kq_1}{R_1^2} = \frac{(8.99 \times 10^9\text{ N}\cdot\text{m}^2/\text{C}^2)(60 \times 10^{-9}\text{ C})}{(0.06\text{ m})^2} = 150\text{ kN/C}$$

$$E_2 = \frac{kq_2}{R_2^2} = \frac{(8.99 \times 10^9\text{ N}\cdot\text{m}^2/\text{C}^2)(20 \times 10^{-9}\text{ C})}{(0.02\text{ m})^2} = 450\text{ kN/C}$$

(*b*) Calculate the common potential from kq/R for either sphere:

$$V_1 = \frac{kq_1}{R_1} = \frac{(8.99 \times 10^9\text{ N}\cdot\text{m}^2/\text{C}^2)(60 \times 10^{-9}\text{ C})}{0.06\text{ m}} = 8.99\text{ kV}$$

Check the Result If we use sphere 2 to calculate V, we obtain $V_2 = kq_2/R_2 = (8.99 \times 10^9\text{ N}\cdot\text{m}^2/\text{C}^2)(20 \times 10^{-9}\text{ C})/0.02\text{ m} = 8.99 \times 10^3\text{ V}$. An additional check is available, since the electric field at the surface of each sphere is proportional to its charge density. Since the radius of sphere 1 is three times that of sphere 2, its surface area is 9 times that of sphere 2. And since it carries 3 times the charge, its charge density is $\frac{1}{3}$ that of sphere 2. Therefore, the field of sphere 1 should be $\frac{1}{3}$ that of sphere 2, which is what we found above.

When a charge is placed on a conductor of nonspherical shape, like that in Figure 24-20*a*, the surface of the conductor will be an equipotential surface, but the surface charge density and the electric field just outside the conductor will vary from point to point. Near a point where the radius of curvature is small, such as point A in the figure, the surface charge density and electric field will be large, whereas near a point where the radius of curvature is large, such as point B in the figure, the field and surface charge density will be small. We can understand this qualitatively by considering the ends of the conductor to be spheres of different radii. Let σ be the surface charge density.

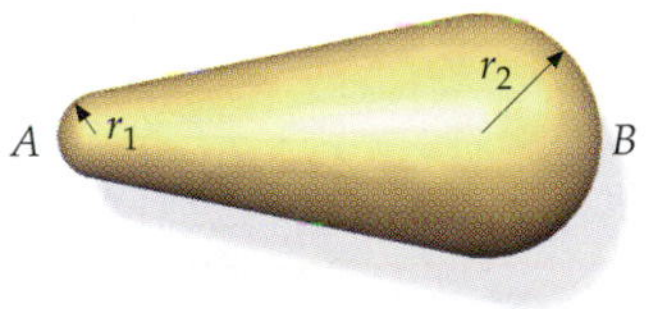

(*a*)

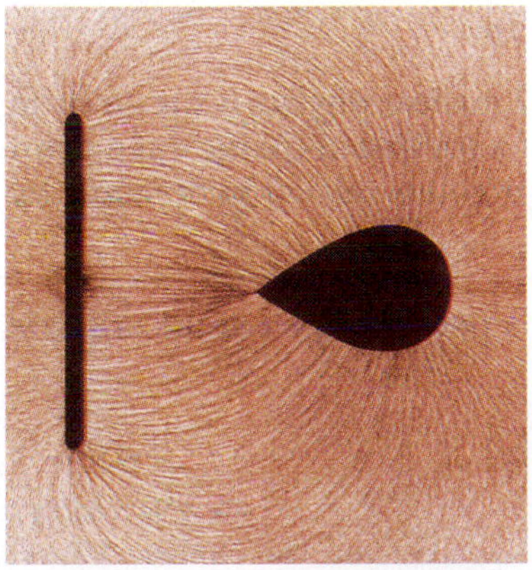

(*b*)

Figure 24-20 (*a*) A nonspherical conductor. If a charge is placed on such a conductor, it will produce an electric field that is stronger near point A, where the radius of curvature is small, than near point B, where the radius of curvature is large. (*b*) Electric field lines near a nonspherical conductor and plate carrying equal and opposite charges. The lines are shown by small bits of thread suspended in oil. Note that the electric field is strongest near points of small radius of curvature, such as at the ends of the plate and at the pointed left side of the conductor.

The potential of a sphere of radius r is

$$V = \frac{kq}{r} = \frac{1}{4\pi\epsilon_0}\frac{q}{r} \qquad \text{24-26}$$

Since the area of a sphere is $4\pi r^2$, the charge on a sphere is related to the charge density by $q = 4\pi r^2\sigma$. Substituting this expression for q into Equation 24-26 we have

$$V = \frac{1}{4\pi\epsilon_0}\frac{4\pi r^2\sigma}{r} = \frac{r\sigma}{\epsilon_0}$$

Solving for σ, we obtain

$$\sigma = \frac{\epsilon_0 V}{r} \qquad \text{24-27}$$

Since both "spheres" are at the same potential, the one with the smaller radius must have the greater surface charge density. And since $E_n = \sigma/\epsilon_0$, the electric field is greatest at points on the conductor where the radius of curvature is least.

For an arbitrarily shaped conductor, the potential at which dielectric breakdown occurs depends on the smallest radius of curvature of any part of the conductor. If the conductor has sharp points of very small radius of curvature, dielectric breakdown will occur at relatively low potentials. In the Van de Graaff generator (Figure 24-18*a*), the charge is transferred onto the belt by sharp-edged conductors near the bottom of the belt. The charge is removed from the belt by sharp-edged conductors near the top of the belt. Lightning rods at the top of a tall building draw the charge off a nearby cloud before the potential of the cloud can build up to a destructively large value.

Summary

1. Electric potential, which is defined as the electrostic potential energy per charge, is an important derived physical concept that is related to the electric field.
2. Because potential is a scalar quantity, it is often easier to calculate than the vector electric field. Once V is known, $\vec{E}$ can be calculated from V.

Topic	Remarks and Relevant Equations	
1. Potential Difference	The potential difference $V_b - V_a$ is defined as the negative of the work per unit charge done by the electric field when a test charge moves from point a to point b: $\Delta V = V_b - V_a = \frac{\Delta U}{q_0} = -\int_a^b \vec{E}\cdot d\vec{\ell}$	**24-2*b***
Potential difference for infinitesimal displacements	$dV = -\vec{E}\cdot d\vec{\ell}$	**24-2*a***
2. Electric Potential		
Potential due to a point charge	$V = \frac{kq}{r} + V_0 \qquad (V = V_0 \text{ at } r = \infty)$	**24-7**
Coulomb potential	$V = \frac{kq}{r} \qquad (V = 0 \text{ at } r = \infty)$	**24-8**

	Potential due to a system of point charges	$V = \sum_i \frac{kq_i}{r_i} \quad (V = 0 \text{ at } r = \infty)$	24-10
	Potential due to continuous charge distributions	$V = \int \frac{k\,dq}{r} \quad (V = 0 \text{ at } r = \infty)$ This expression can be used only if the charge distribution is contained in a finite volume so that the potential can be chosen to be zero at infinity.	24-19
	Potential and electric field lines	Electric field lines point in the direction of decreasing electric potential.	
	Continuity of electric potential	The potential function V is continuous everywhere in space.	
3.	**Finding the Electric Field from the Potential**	The electric field points in the direction of the greatest decrease in the potential. $E_\ell = -\frac{dV}{d\ell}$	24-13
	Gradient	A vector that points in the direction of the greatest change in a scalar function and has a magnitude equal to the derivative of that function with respect to the distance in that direction is called the gradient of the function. $\vec{E}$ is the negative gradient of V.	
	$\vec{E}$ in the x direction	$E_x = -\frac{dV(x)}{dx}$	24-14
	Radial electric field	$E_r = -\frac{dV(r)}{dr}$	24-15
4.	**General Relation Between $\vec{E}$ and V** (optional)	$\vec{E} = -\vec{grad}\, V = -\left(\frac{\partial V}{\partial x}\hat{i} + \frac{\partial V}{\partial y}\hat{j} + \frac{\partial V}{\partial z}\hat{k}\right)$	24-18
5.	**Units**		
	V and ΔV	The SI unit of potential and potential difference is the volt (V): $1\text{ V} = 1\text{ J/C}$	24-4
	Electric field	$1\text{ N/C} = 1\text{ V/m}$	24-5
	eV	The electron volt (eV) is the potential energy of a particle of charge e at a point where the potential is 1 volt: $1\text{ eV} = 1.6 \times 10^{-19}\text{ J}$	24-6
6.	**Potential Energy of Two Point Charges**	$U = q_0 V = \frac{kq_0 q}{r} \quad (U = 0 \quad \text{at} \quad r = \infty)$	24-9
7.	**Potential Functions**		
	On the axis of a uniformly charged ring	$V = \frac{kQ}{\sqrt{x^2 + a^2}}$	24-20
	On the axis of a uniformly charged disk	$V = 2\pi k\sigma[(x^2 + R^2)^{1/2} - x]$	24-21
	Near an infinite plane of charge	$V = V_0 - \frac{\sigma}{2\epsilon_0}\|x\| = V_0 - 2\pi k\sigma\|x\|$	24-22

For a spherical shell of charge	$V = \begin{cases} \dfrac{kQ}{r}, & r \geq R \\ \dfrac{kQ}{R}, & r \leq R \end{cases}$	**24-23**
For an infinite line charge	$V = -2k\lambda \ln \dfrac{r}{a}, \quad V = 0 \text{ at } r = a$	**24-25**
8. Charge on a Nonspherical Conductor	On a conductor of arbitrary shape, the surface charge density σ is greatest at points where the radius of curvature is smallest.	
9. Dielectric Breakdown	The amount of charge that can be placed on a conductor is limited by the fact that molecules of the surrounding medium become ionized in very high electric fields, causing the medium to become a conductor.	
Dielectric strength	The dielectric strength is the magnitude of the electric field at which dielectric breakdown occurs. The dielectric strength of air is $E_{max} \approx 3 \times 10^6 \text{ V/m} = 3 \text{ MV/m}$	

Problem-Solving Guide

1. Begin by drawing a neat diagram that includes the important features of the problem, including the location and value of charges in the system. It is often helpful to indicate equipotential surfaces and electric field lines.
2. The potential due to a system of point charges can be calculated from $V = \Sigma\, kq_i/r_i$.
3. The potential due to a continuous, finite system of charges can be calculated from $V = \int kdq/r$, or if the electric field is known, the potential can be found from $dV = -\vec{E}\cdot d\vec{\ell}$.

Summary of Worked Examples

Type of Calculation	Procedure and Relevant Examples	
1. Potential and Electric Field		
Calculate the change in electric potential V given the electric field $\vec{E}$.	Use $dV = -\vec{E}\cdot d\vec{\ell}$ and integrate.	**Examples 24-1, 24-12, and 24-13**
Determine $\vec{E}$ given the potential V.	Use $E_x = -dV(x)/dx$ for a potential that depends only on x or $E_r = -dV(r)/dr$ for a spherically symmetric charge distribution.	**Examples 24-7, 24-10**
Find the potential V for discrete charges:		
Point charge	Use $V = kq/r$.	**Example 24-2**
System of point charges	Use $V = \Sigma_i\, kq_i/r_i$.	**Examples 24-4, 24-5, 24-6**
Find V for a continuous charge distribution.	Integrate over the charge distribution ($V = \int kdq/r$), or calculate V from the electric field $V = -\int \vec{E}\cdot d\vec{\ell}$.	**Examples 24-9, 24-12, 24-13**
Find V for a system that includes point charges and continuous charge distributions.	Use $V_{point} = (kq/r) + V_i$ for each point charge where V_i is a different constant for each point charge that is determined by the choice of zero for V. Sum the potential due to the point charges and the potential due to the continuous charge distributions.	**Example 24-11**

2. Potential and Energy		
Find the potential energy of a two-charge system.	Use $U = kq_1q_2/r$.	**Examples 24-2, 24-3**
Find the change in kinetic energy ΔK as a charge moves through a region of electric potential.	Use conservation of energy, $\Delta K + \Delta U = 0$, where $\Delta U = q\,\Delta V$.	**Example 24-8**
3. Conductors		
Find the potential in problems involving conductors.	Use the fact that the surface of a conductor is an equipotential surface, and that V is constant everywhere within a conductor.	**Example 24-12**
Find the maximum charge on a conductor before dielectric breakdown.	Use $\sigma_{max} = \epsilon_0 E_{max}$, where E_{max} for air is equal to 3 MN/C.	**Example 24-14**
Charge sharing between conductors.	If two conductors are in contact, they must be at the same potential. This condition determines the ratio of charge on the conductors.	**Example 24-15**

Problems

In a few problems, you are given more data than you actually need; in a few other problems, you are required to supply data from your general knowledge, outside sources, or informed estimates.

Conceptual Problems

Problems from Optional and Exploring sections

- • Single-concept, single-step, relatively easy
- •• Intermediate-level, may require synthesis of concepts
- ••• Challenging, for advanced students

Potential and Potential Difference

1 • A uniform electric field of 2 kN/C is in the x direction. A positive point charge $Q = 3\ \mu$C is released from rest at the origin. (*a*) What is the potential difference $V(4\text{ m}) - V(0)$? (*b*) What is the change in the potential energy of the charge from $x = 0$ to $x = 4$ m? (*c*) What is the kinetic energy of the charge when it is at $x = 4$ m? (*d*) Find the potential $V(x)$ if $V(x)$ is chosen to be (*d*) zero at $x = 0$, (*e*) 4 kV at $x = 0$, and (*f*) zero at $x = 1$ m.

2 • An infinite plane of surface charge density $\sigma = +2.5\ \mu\text{C/m}^2$ is in the yz plane. (*a*) What is the magnitude of the electric field in newtons per coulomb? In volts per meter? What is the direction of $\vec{E}$ for positive values of x? (*b*) What is the potential difference $V_b - V_a$ when point b is at $x = 20$ cm and point a is at $x = 50$ cm? (*c*) How much work is required by an outside agent to move a test charge $q_0 = +1.5$ nC from point a to point b?

3 • Two large parallel conducting plates separated by 10 cm carry equal and opposite surface charge densities such that the electric field between them is uniform. The difference in potential between the plates is 500 V. An electron is released from rest at the negative plate. (*a*) What is the magnitude of the electric field between the plates? Is the positive or negative plate at the higher potential? (*b*) Find the work done by the electric field on the electron as the electron moves from the negative plate to the positive plate. Express your answer in both electron volts and joules. (*c*) What is the change in potential energy of the electron when it moves from the negative plate to the positive plate? What is its kinetic energy when it reaches the positive plate?

Potential and Potential Energy

4 • Explain the distinction between electric potential and electrostatic potential energy.

5 • A positive charge is released from rest in an electric field. Will it move toward a region of greater or smaller electric potential?

6 •• A lithium nucleus and an α particle are at rest. The lithium nucleus has a charge of $+3e$ and a mass of 7 u; the α particle has a charge of $+2e$ and a mass of 4 u. Which of the methods below would accelerate them both to the same kinetic energy?
(*a*) Accelerate them through the same electrical potential difference.
(*b*) Accelerate the α particle through potential V_1 and the lithium nucleus through $\frac{2}{3}V_1$.
(*c*) Accelerate the α particle through potential V_1 and the lithium nucleus through $\frac{7}{4}V_1$.
(*d*) Accelerate the α particle through potential V_1 and the lithium nucleus through $(2 \times 7)/(3 \times 4)V$.
(*e*) None of the above.

7 • A positive charge of magnitude 2 μC is at the origin. (*a*) What is the electric potential V at a point 4 m from the origin relative to $V = 0$ at infinity? (*b*) How much work must be done by an outside agent to bring a 3-μC charge from infinity to $r = 4$ m, assuming that the 2-μC charge is held fixed at the origin? (*c*) How much work must be done by an outside agent to bring the 2-μC charge from infinity to the origin if the 3-μC charge is first placed at $r = 4$ m and is then held fixed?

8 •• The distance between the K^+ and Cl^- ions in KCl is 2.80×10^{-10} m. Calculate the energy required to separate the two ions to an infinite distance apart, assuming them to be point charges initially at rest. Express your answer in eV.

9 •• Two identical masses m that carry equal charges q are separated by a distance d. Show that if both are released simultaneously their speeds when they are separated a great distance are $v/\sqrt{2}$, where v is the speed that one mass would have at a great distance from the other if it were released and the other held fixed.

10 •• Protons from a Van de Graaff accelerator are released from rest at a potential of 5 MV and travel through a vacuum to a region at zero potential. (*a*) Find the final speed of the 5-MeV protons. (*b*) Find the accelerating electric field if the same potential change occurred *uniformly* over a distance of 2.0 m.

11 •• An electron gun fires electrons at the screen of a television tube. The electrons start from rest and are accelerated through a potential difference of 30,000 V. What is the energy of the electrons when they hit the screen (*a*) in electron volts and (*b*) in joules? (*c*) What is the speed of impact of electrons with the screen of the picture tube?

12 •• (*a*) Derive an expression for the distance of closest approach of an α particle with kinetic energy E to a massive nucleus of charge Ze. Assume that the nucleus is fixed in space. (*b*) Find the distance of closest approach of a 5.0- and a 9.0-MeV α particle to a gold nucleus; the charge of the gold nucleus is $79e$. Neglect the recoil of the gold nucleus.

Systems of Point Charges

13 • Four 2-μC point charges are at the corners of a square of side 4 m. Find the potential at the center of the square (relative to zero potential at infinity) if (*a*) all the charges are positive, (*b*) three of the charges are positive and one is negative, and (*c*) two are positive and two are negative.

14 • Three point charges are on the x axis: q_1 is at the origin, q_2 is at $x = 3$ m, and q_3 is at $x = 6$ m. Find the potential at the point $x = 0$, $y = 3$ m if (*a*) $q_1 = q_2 = q_3 = 2\ \mu$C, (*b*) $q_1 = q_2 = 2\ \mu$C and $q_3 = -2\ \mu$C, and (*c*) $q_1 = q_3 = 2\ \mu$C and $q_2 = -2\ \mu$C.

15 • Points A, B, and C are at the corners of an equilateral triangle of side 3 m. Equal positive charges of 2 μC are at A and B. (*a*) What is the potential at point C? (*b*) How much work is required to bring a positive charge of 5 μC from infinity to point C if the other charges are held fixed? (*c*) Answer parts (*a*) and (*b*) if the charge at B is replaced by a charge of $-2\ \mu$C.

16 • A sphere with radius 60 cm has its center at the origin. Equal charges of 3 μC are placed at 60° intervals along the equator of the sphere. (*a*) What is the electric potential at the origin? (*b*) What is the electric potential at the north pole?

17 • Two point charges q and q' are separated by a distance a. At a point $a/3$ from q and along the line joining the two charges the potential is zero. Find the ratio q/q'.

18 •• Two positive charges $+q$ are on the x axis at $x = +a$ and $x = -a$. (*a*) Find the potential $V(x)$ as a function of x for points on the x axis. (*b*) Sketch $V(x)$ versus x. (*c*) What is the significance of the minimum on your curve?

19 •• A point charge of $+3e$ is at the origin and a second point charge of $-2e$ is on the x axis at $x = a$. (*a*) Sketch the potential function $V(x)$ versus x for all x. (*b*) At what point or points is $V(x)$ zero? (*c*) How much work is needed to bring a third charge $+e$ to the point $x = \frac{1}{2}a$ on the x axis?

Finding the Electric Field From the Potential

20 • If the electric potential is constant throughout a region of space, what can you say about the electric field in that region?

21 • If E is known at just one point, can V be found at that point?

22 • In what direction can you move relative to an electric field so that the electric potential does not change?

23 • A uniform electric field is in the negative x direction. Points a and b are on the x axis, a at $x = 2$ m and b at $x = 6$ m. (*a*) Is the potential difference $V_b - V_a$ positive or negative? (*b*) If the magnitude of $V_b - V_a$ is 10^5 V, what is the magnitude E of the electric field?

24 • The potential due to a particular charge distribution is measured at several points along the x axis as shown in Figure 24-21. For what value(s) in the range $0 < x < 10$ m is $E_x = 0$?

Figure 24-21 Problem 24

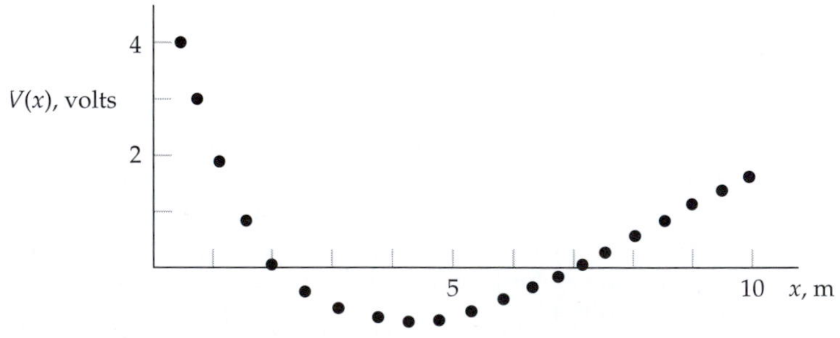

25 • A point charge $q = 3.00\ \mu$C is at the origin. (*a*) Find the potential V on the x axis at $x = 3.00$ m and at $x = 3.01$ m. (*b*) Does the potential increase or decrease as x increases? Compute $-\Delta V/\Delta x$, where ΔV is the change in potential from $x = 3.00$ m to $x = 3.01$ m and $\Delta x = 0.01$ m. (*c*) Find the electric field at $x = 3.00$ m, and compare its magnitude with $-\Delta V/\Delta x$ found in part (*b*). (*d*) Find the potential (to three significant figures) at the point $x = 3.00$ m, $y = 0.01$ m, and compare your result with the potential on the x axis at $x = 3.00$ m. Discuss the significance of this result.

26 • A charge of $+3.00\ \mu\text{C}$ is at the origin, and a charge of $-3.00\ \mu\text{C}$ is on the x axis at $x = 6.00$ m. (*a*) Find the potential on the x axis at $x = 3.00$ m. (*b*) Find the electric field on the x axis at $x = 3.00$ m. (*c*) Find the potential on the x axis at $x = 3.01$ m, and compute $-\Delta V/\Delta x$, where ΔV is the change in potential from $x = 3.00$ m to $x = 3.01$ m and $\Delta x = 0.01$ m. Compare your result with your answer to part (*b*).

27 • A uniform electric field is in the positive y direction. Points a and b are on the y-axis, a at $y = 2$ m and b at $y = 6$ m. (*a*) Is the potential difference $V_b - V_a$ positive or negative? (*b*) If the magnitude of $V_b - V_a$ is 2×10^4 V, what is the magnitude E of the electric field?

28 • In the following, V is in volts and x is in meters. Find E_x when (*a*) $V(x) = 2000 + 3000x$; (*b*) $V(x) = 4000 + 3000x$; (*c*) $V(x) = 2000 - 3000x$; and (*d*) $V(x) = -2000$, independent of x.

29 • The electric potential in some region of space is given by $V(x) = C_1 + C_2x^2$, where V is in volts, x is in meters, and C_1 and C_2 are positive constants. Find the electric field E in this region. In what direction is E?

30 •• A charge q is at $x = 0$ and a charge $-3q$ is at $x = 1$ m. (*a*) Find $V(x)$ for a general point on the x axis. (*b*) Find the points on the x axis where the potential is zero. (*c*) What is the electric field at these points? (*d*) Sketch $V(x)$ versus x.

31 •• An electric field is given by $E_x = 2.0x^3$ kN/C. Find the potential difference between the points on the x axis at $x = 1$ m and $x = 2$ m.

32 •• Three equal charges lie in the xy plane. Two are on the y axis at $y = -a$ and $y = +a$, and the third is on the x axis at $x = a$. (*a*) What is the potential $V(x)$ due to these charges at a point on the x axis? (*b*) Find E_x along the x axis from the potential function $V(x)$. Evaluate your answers to (*a*) and (*b*) at the origin and at $x = \infty$ to see if they yield the expected results.

General Relation Between $\vec{E}$ and V (optional)

33 ••• The electric potential in a region of space is given by $V = (2\ \text{V/m}^2)x^2 + (1\ \text{V/m}^2)yz$. Find the electric field at the point $x = 2$ m, $y = 1$ m, $z = 2$ m.

34 ••• A potential is given by

$$V(x, y, z) = \frac{kQ}{\sqrt{(x-a)^2 + y^2 + z^2}}$$

(*a*) Find the components E_x, E_y, and E_z of the electric field by differentiating this potential function. (*b*) What simple charge distribution might be responsible for this potential?

Calculating V for Continuous Charge Distributions

35 •• In the calculation of V at a point x on the axis of a ring of charge, does it matter whether the charge Q is uniformly distributed around the ring? Would either V or E_x be different if it were not?

36 •• (*a*) Sketch $V(x)$ versus x for the uniformly charged ring in the yz plane given by Equation 24-20. (*b*) At what point is $V(x)$ a maximum? (*c*) What is E_x at this point?

37 • A charge of $q = +10^{-8}$ C is uniformly distributed on a spherical shell of radius 12 cm. (*a*) What is the magnitude of the electric field just outside and just inside the shell? (*b*) What is the magnitude of the electric potential just outside and just inside the shell? (*c*) What is the electric potential at the center of the shell? What is the electric field at that point?

38 • A disk of radius 6.25 cm carries a uniform surface charge density $\sigma = 7.5\ \text{nC/m}^2$. Find the potential on the axis of the disk at a distance from the disk of (*a*) 0.5 cm, (*b*) 3.0 cm, and (*c*) 6.25 cm.

39 • An infinite line charge of linear charge density $\lambda = 1.5\ \mu\text{C/m}$ lies on the z axis. Find the potential at distances from the line charge of (*a*) 2.0 m, (*b*) 4.0 m, and (*c*) 12 m, assuming that $V = 0$ at 2.5 m.

40 •• Derive Equation 24-21 by integrating the electric field E_x along the axis of the disk. (See Equation 23-11.)

41 •• A rod of length L carries a charge Q uniformly distributed along its length. The rod lies along the y axis with its center at the origin. (*a*) Find the potential as a function of position along the x axis. (*b*) Show that the result obtained in (*a*) reduces to $V = kQ/x$ for $x \gg L$.

42 •• A disk of radius R carries a surface charge distribution of $\sigma = \sigma_0 R/r$. (*a*) Find the total charge on the disk. (*b*) Find the potential on the axis of the disk a distance x from its center.

43 •• Repeat Problem 42 if the surface charge density is $\sigma = \sigma_0 r^2/R^2$.

44 •• A rod of length L carries a charge Q uniformly distributed along its length. The rod lies along the y axis with one end at the origin. Find the potential as a function of position along the x axis.

45 •• A disk of radius R carries a charge density $+\sigma_0$ for $r < a$ and an equal but opposite charge density $-\sigma_0$ for $a < r < R$. The total charge carried by the disk is zero. (*a*) Find the potential a distance x along the axis of the disk. (*b*) Obtain an approximate expression for $V(x)$ when $x \gg R$.

46 •• Use the result obtained in Problem 45(*a*) to calculate the electric field along the axis of the disk. Then calculate the electric field by direct integration using Coulomb's law.

47 •• A rod of length L has a charge Q uniformly distributed along its length. The rod lies along the x axis with its center at the origin. (*a*) What is the electric potential as a function of position along the x axis for $x > L/2$? (*b*) Show that for $x \gg L/2$, your result reduces to that due to a point charge Q.

48 •• A conducting spherical shell of inner radius b and outer radius c is concentric with a small metal sphere of radius $a < b$. The metal sphere has a positive charge Q. The total charge on the conducting spherical shell is $-Q$. (*a*) What is the potential of the spherical shell? (*b*) What is the potential of the metal sphere?

49 •• Two very long, coaxial cylindrical shell conductors carry equal and opposite charges. The inner shell has radius a and charge $+q$; the other shell has radius b and charge $-q$. The length of each cylindrical shell is L. Find the potential difference between the shells.

50 •• A uniformly charged sphere has a potential on its surface of 450 V. At a radial distance of 20 cm from this surface, the potential is 150 V. What is the radius of the sphere, and what is the charge of the sphere?

51 •• Consider two infinite parallel planes of charge, one in the yz plane and the other at distance $x = a$. (*a*) Find the potential everywhere in space when $V = 0$ at $x = 0$ if the planes carry equal positive charge densities $+\sigma$. (*b*) Repeat the problem with charge densities equal and opposite, and the charge in the yz plane positive.

52 •• Show that for $x >> R$ the potential on the axis of a disk charge approaches kQ/x, where $Q = \sigma\pi R^2$ is the total charge on the disk. (*Hint:* Write $(x^2 + R^2)^{1/2} = x(1 + R^2/x^2)^{1/2}$ and use the binomial expression.)

53 •• In Example 24-12 you derived the expression

$$V(r) = \frac{kQ}{2R}\left(3 - \frac{r^2}{R^2}\right)$$

for the potential inside a solid sphere of constant charge density by first finding the electric field. In this problem you derive the same expression by direct integration. Consider a sphere of radius R containing a charge Q uniformly distributed. You wish to find V at some point $r < R$. (*a*) Find the charge q' inside a sphere of radius r and the potential V_1 at r due to this part of the charge. (*b*) Find the potential dV_2 at r due to the charge in a shell of radius r' and thickness dr' at $r' > r$. (*c*) Integrate your expression in (*b*) from $r' = r$ to $r' = R$ to find V_2. (*d*) Find the total potential V at r from $V = V_1 + V_2$.

54 ••• A nonconducting sphere of radius R has a volume charge density $\rho = \rho_0 r/R$, where ρ_0 is a constant. (*a*) Show that the total charge is $Q = \pi R^3\rho_0$. (*b*) Show that the total charge inside a sphere of radius $r < R$ is $q = Qr^4/R^4$. (*c*) Use Gauss's law to find the electric field E_r everywhere. (*d*) Use $dV = -E_r\, dr$ to find the potential V everywhere, assuming that $V = 0$ at $r = \infty$. (Remember that V is continuous at $r = R$.)

Equipotential Surfaces and Dielectric Breakdown

55 • Two charged metal spheres are connected by a wire, and sphere A is larger than sphere B (Figure 24-22). The magnitude of the electric potential of sphere A is

(*a*) greater than that at the surface of sphere B.
(*b*) less than that at the surface of sphere B.
(*c*) the same as that at the surface of sphere B.
(*d*) greater than or less than that at the surface of sphere B, depending on the radii of the spheres.
(*e*) greater than or less than that at the surface of sphere B, depending on the charge on the spheres.

Figure 24-22 Problem 55

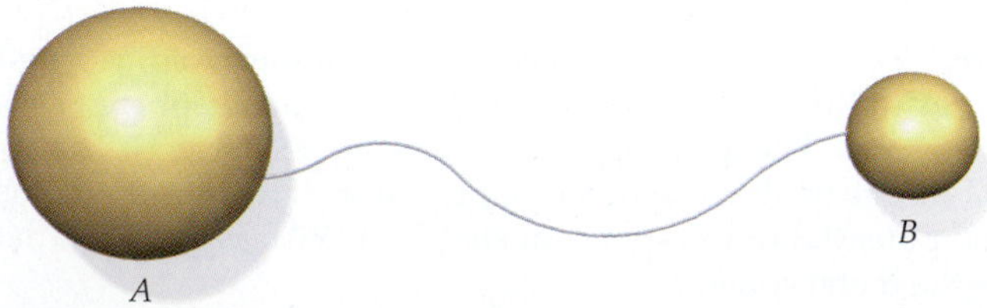

56 •• Figure 24-23 shows two parallel metal plates maintained at potentials of 0 and 60 V. Midway between the plates is a copper sphere. Sketch the equipotential surfaces and the electric field lines between the two plates.

Figure 24-23 Problem 56

57 •• Figure 24-24 shows a metal sphere carrying a charge $-Q$ and a point charge $+Q$. Sketch the electric field lines and equipotential surfaces in the vicinity of this charge system.

Figure 24-24 Problems 57 and 58

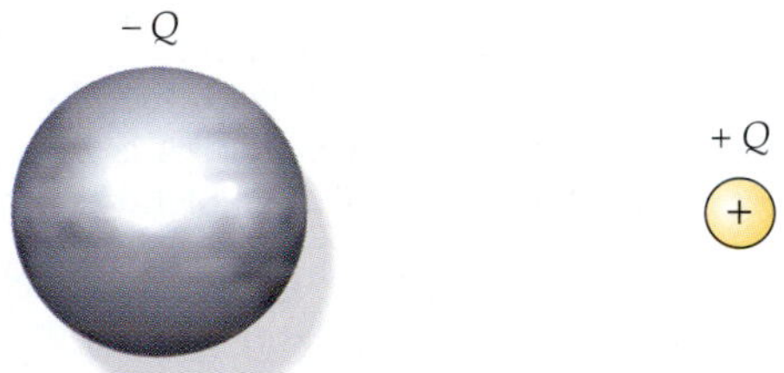

58 •• Repeat Problem 57 with the charge on the metal sphere changed to $+Q$.

59 •• Sketch the electric field lines and the equipotential surfaces both near and far from the conductor shown in Figure 24-20*a*, assuming that the conductor carries some charge Q.

60 •• Two equal positive charges are separated by a small distance. Sketch the electric field lines and the equipotential surfaces for this system.

61 • An infinite plane of charge has surface charge density 3.5 μC/m^2. How far apart are the equipotential surfaces whose potentials differ by 100 V?

62 • A point charge $q = +\frac{1}{9} \times 10^{-8}$ C is at the origin. Taking the potential to be zero at $r = \infty$, locate the equipotential surfaces at 20-V intervals from 20 to 100 V, and sketch them to scale. Are these surfaces equally spaced?

63 • (*a*) Find the maximum net charge that can be placed on a spherical conductor of radius 16 cm before dielectric breakdown of the air occurs. (*b*) What is the potential of the sphere when it carries this maximum charge?

64 • Find the greatest surface charge density σ_{max} that can exist on a conductor before dielectric breakdown of the air occurs.

65 •• Charge is placed on two conducting spheres that are very far apart and connected by a long thin wire (Figure 24-25). The larger sphere has a diameter twice that of the smaller. Which sphere has the largest electric field near its surface? By what factor is it larger than that at the surface of the other sphere?

Figure 24-25 Problem 65

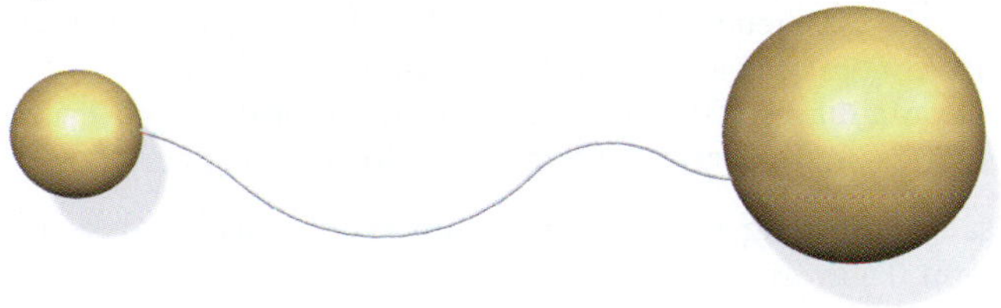

66 •• Charge is placed on two conducting spheres that are very far apart and connected by a long thin wire. The radius of the smaller sphere is 5 cm and that of the larger sphere is 12 cm. The electric field at the surface of the larger sphere is 200 kV/m. Find the surface charge density on each sphere.

67 •• Two concentric spherical shell conductors carry equal and opposite charges. The inner shell has radius a and charge $+q$; the outer shell has radius b and charge $-q$. Find the potential difference between the shells, $V_a - V_b$.

68 •• Two identical uncharged metal spheres connected by a wire are placed close by two similar conducting spheres with equal and opposite charges as shown in Figure 24-26. (*a*) Sketch the electric field lines between spheres 1 and 3 and between spheres 2 and 4. (*b*) What can be said about the potentials V_1, V_2, V_3, and V_4 of the spheres? (*c*) If spheres 3 and 4 are connected by a wire, prove that the final charge on each must be zero.

Figure 24-26 Problem 68

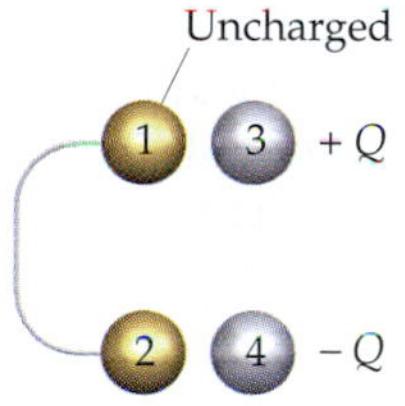

General Problems

69 • Two equal positive point charges $+Q$ are on the x axis. One is at $x = -a$ and the other is at $x = +a$. At the origin,

(*a*) $E = 0$ and $V = 0$.
(*b*) $E = 0$ and $V = 2kQ/a$.
(*c*) $\vec{E} = (2kQ^2/a^2)\,\hat{i}$ and $V = 0$.
(*d*) $\vec{E} = (2kQ^2/a^2)\,\hat{i}$ and $V = 2kQ/a$.
(*e*) none of the above is correct.

70 • The electrostatic potential is measured to be $V(x,y,z) = 4|x| + V_0$, where V_0 is a constant. The charge distribution responsible for this potential is

(*a*) a uniformly charged thread in the xy plane.
(*b*) a point charge at the origin.
(*c*) a uniformly charged sheet in the yz plane.
(*d*) a uniformly charged sphere of radius $1/\pi$ at the origin.

71 • Two point charges of equal magnitude but opposite sign are on the x axis; $+Q$ is at $x = -a$ and $-Q$ is at $x = +a$. At the origin,

(*a*) $E = 0$ and $V = 0$.
(*b*) $E = 0$ and $V = 2kQ/a$.
(*c*) $E = (2kQ^2/a^2)\,\hat{i}$ and $V = 0$.
(*d*) $E = (2kQ^2/a^2)\,\hat{i}$ and $V = 2kQ/a$.
(*e*) none of the above is correct.

72 •• True or false:

(*a*) If the electric field is zero in some region of space, the electric potential must also be zero in that region.
(*b*) If the electric potential is zero in some region of space, the electric field must also be zero in that region.
(*c*) If the electric potential is zero at a point, the electric field must also be zero at that point.
(*d*) Electric field lines always point toward regions of lower potential.
(*e*) The value of the electric potential can be chosen to be zero at any convenient point.
(*f*) In electrostatics, the surface of a conductor is an equipotential surface.
(*g*) Dielectric breakdown occurs in air when the potential is 3×10^6 V.

73 •• (*a*) V is constant on a conductor surface. Does this mean that σ is constant? (*b*) If E is constant on a conductor surface, does this mean that σ is constant? Does it mean that V is constant?

74 • An electric dipole has a positive charge of 4.8×10^{-19} C separated from a negative charge of the same magnitude by 6.4×10^{-10} m. What is the electric potential at a point 9.2×10^{-10} m from each of the two charges?

(*a*) 9.4 V
(*b*) Zero
(*c*) 4.2 V
(*d*) 5.1×10^9 V
(*e*) 1.7 V

75 • An electric field is given by $\vec{E} = ax\hat{i}$, where $\vec{E}$ is in newtons per coulomb, x is in meters, and a is a positive constant. (*a*) What are the SI units of a? (*b*) How much work is done by this field on a positive point charge q_0 when the charge moves from the origin to some point x? (*c*) Find the potential function $V(x)$ such that $V = 0$ at $x = 0$.

76 • Two positive charges $+q$ are on the y axis at $y = +a$ and $y = -a$. (*a*) Find the potential V for any point on the x axis. (*b*) Use your result in (*a*) to find the electric field at any point on the x axis.

77 • If a conducting sphere is to be charged to a potential of 10,000 V, what is the smallest possible radius of the sphere such that the electric field will not exceed the dielectric strength of air?

78 • An isolated aluminum sphere of radius 5.0 cm is at a potential of 400 V. How many electrons have been removed from the sphere to raise it to this potential?

79 • A point charge Q resides at the origin. A particle of mass $m = 0.002$ kg carries a charge of 4.0 μC. The particle is released from rest at $x = 1.5$ m. Its kinetic energy as it passes $x = 1.0$ m is 0.24 J. Find the charge Q.

80 • A conducting wedge is charged to a potential V with respect to a large conducting sheet (Figure 24-27). (*a*) Sketch the electric field lines and the equipotentials for this configuration. Where along the x axis is $|E|$ greatest? (*b*) An electron of mass m_e leaves the sheet with zero velocity. What is its speed v when it arrives at the wedge? (Ignore the effect of gravity.)

Figure 24-27 Problem 80

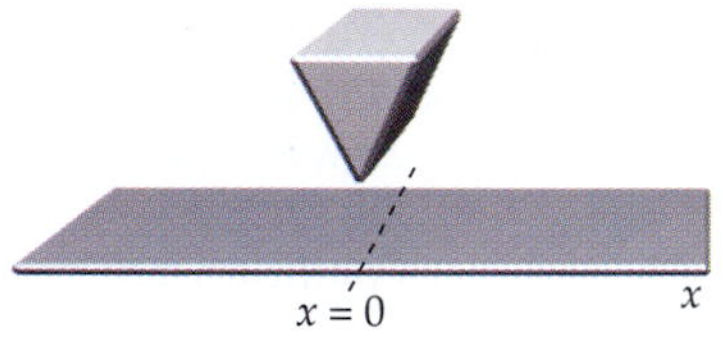

81 •• A Van de Graaff generator has a potential difference of 1.25 MV between the belt and the outer shell. Charge is supplied at the rate of 200 μC/s. What minimum power is needed to drive the moving belt?

82 •• A positive point charge $+Q$ is located at $x = -a$. (*a*) How much work is required to bring a second equal positive point charge $+Q$ from infinity to $x = +a$? (*b*) With the two equal positive point charges at $x = -a$ and $x = +a$, how much work is required to bring a third charge $-Q$ from infinity to the origin? (*c*) How much work is required to move the charge $-Q$ from the origin to the point $x = 2a$ along the semicircular path shown (Figure 24-28)?

Figure 24-28 Problem 82

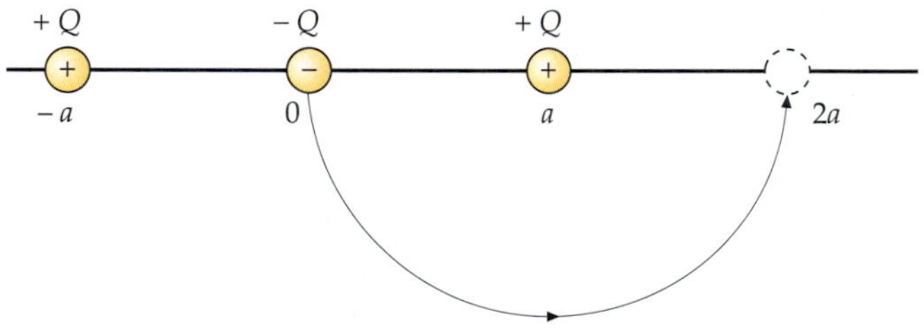

83 •• A charge of 2 nC is uniformly distributed around a ring of radius 10 cm that has its center at the origin and its axis along the x axis. A point charge of 1 nC is located at $x = 50$ cm. Find the work required to move the point charge to the origin. Give your answer in both joules and electron volts.

84 •• The centers of two metal spheres of radius 10 cm are 50 cm apart on the x axis. The spheres are initially neutral, but a charge Q is transferred from one sphere to the other, creating a potential difference between the spheres of 100 V. A proton is released from rest at the surface of the positively charged sphere and travels to the negatively charged sphere. At what speed does it strike the negatively charged sphere?

85 •• A spherical conductor of radius R_1 is charged to 20 kV. When it is connected by a long, fine wire to a second conducting sphere far away, its potential drops to 12 kV. What is the radius of the second sphere?

86 •• A uniformly charged ring of radius a and charge Q lies in the yz plane with its axis along the x axis. A point charge Q' is placed on the x axis at $x = 2a$. (*a*) Find the potential at any point on the x axis due to the total charge $Q + Q'$. (*b*) Find the electric field for any point on the x axis.

87 •• A metal sphere centered at the origin carries a surface charge of charge density $\sigma = 24.6$ nC/m^2. At $r = 2.0$ m, the potential is 500 V and the magnitude of the electric field is 250 V/m. Determine the radius of the metal sphere.

88 •• Along the axis of a uniformly charged disk, at a point 0.6 m from the center of the disk, the potential is 80 V and the magnitude of the electric field is 80 V/m; at a distance of 1.5 m, the potential is 40 V and the magnitude of the electric field is 23.5 V/m. Find the total charge residing on the disk.

89 •• When you touch a friend after walking across a rug on a dry day, you typically draw a spark of about 2 mm. Estimate the potential difference between you and your friend before the spark.

90 •• When ^{235}U captures a neutron, it fissions (splits) into two nuclei, in the process emitting several neutrons that can cause other uranium nuclei to fission. Assume that the fission products are two nuclei of equal charges of $+46e$ and that these nuclei are at rest just after fission and are separated by twice their radius, $2R \approx 1.3 \times 10^{-14}$ m. (*a*) Calculate the electrostatic potential energy of the fission fragments. This is approximately the energy released per fission. (*b*) About how many fissions per second are needed to produce 1 MW of power in a reactor?

91 •• A radioactive ^{210}Po nucleus emits an α particle of charge $+2e$ and energy 5.30 MeV. Assume that just after the α particle is formed and escapes from the nucleus, it is a distance R from the center of the daughter nucleus ^{206}Pb, which has a charge $+82e$. Calculate R by setting the electrostatic potential energy of the two particles at this separation equal to 5.30 MeV. (Neglect the size of the α particle.)

92 •• Two large, parallel, nonconducting planes carry equal and opposite charge densities of magnitude σ. The planes have area A and are separated by a distance d. (*a*) Find the potential difference between the planes. (*b*) A conducting slab having thickness a and area A, the same area as the planes, is inserted between the original two planes. The slab carries no net charge. Find the potential difference between the original two planes and sketch the electric field lines in the region between the original two planes.

93 •• A uniformly charged ring with a total charge of 100 μC and a radius of 0.1 m lies in the yz plane with its center at the origin. A meterstick has a point charge of 10 μC on the end marked 0 and a point charge of 20 μC on the end marked 100 cm. How much work does it take to bring the meterstick from a long distance away to a position along the x axis with the end marked 0 at $x = 0.2$ m and the other end at $x = 1.2$ m.

94 •• Three large conducting plates are parallel to one another with the outer plates connected by a wire. The inner plate is isolated and carries a charge density σ_t on the upper surface and σ_b on the lower surface, where $\sigma_t + \sigma_b = 12\ \mu C/m^2$. The inner plate is 1 mm from the top plate and 3 mm from the bottom plate. Find the surface charge densities σ_t and σ_b.

95 ••• A point charge q_1 is at the origin and a second point charge q_2 is on the x axis at $x = a$ as in Example 24-5. (*a*) Calculate the electric field everywhere on the x axis from the potential function given in that example. (*b*) Find the potential at a general point on the y axis. (*c*) Use your result from (*b*) to calculate the y component of the electric field on the y axis. Compare your result with that obtained directly from Coulomb's law.

96 ••• A particle of mass m carrying a positive charge q is constrained to move along the x axis. At $x = -L$ and $x = L$ are two ring charges of radius L (Figure 24-29). Each ring is centered on the x axis and lies in a plane perpendicular to it. Each carries a positive charge Q. (*a*) Obtain an expression for the potential due to the ring charges as a function of x for $-L < x < L$. (*b*) Show that in this region, $V(x)$ is a minimum at $x = 0$. (*c*) Show that for $x \ll L$, the potential is of the form $V(x) = V(0) + \alpha x^2$. (*d*) Derive an expression for the angular frequency of oscillation of the mass m if it is displaced slightly from the origin and released.

Figure 24-29 Problem 96

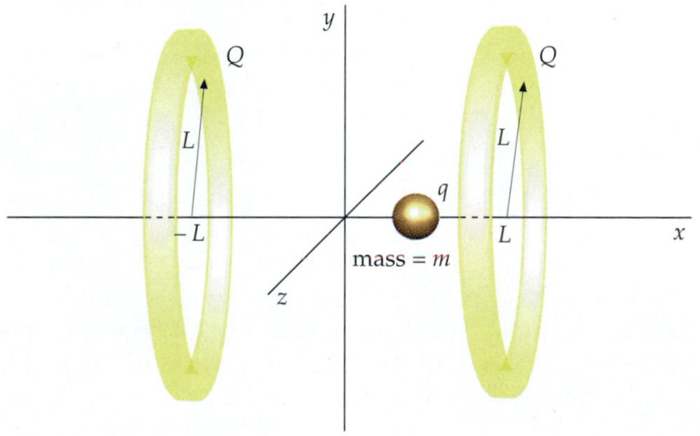

97 ••• Three concentric conducting spherical shells have radii a, b, and c such that $a < b < c$. Initially, the inner shell is uncharged, the middle shell has a positive charge Q, and the outer shell has a negative charge $-Q$. (*a*) Find the electric potential of the three shells. (*b*) If the inner and outer shells are now connected by a wire that is insulated as it passes through the middle shell, what is the electric potential of each of the three shells, and what is the final charge on each shell?

98 ••• Consider two concentric spherical metal shells of radii a and b, where $b > a$. The outer shell has a charge Q, but the inner shell is grounded. This means that the inner shell is at zero potential and that electric field lines leave the outer shell and go to infinity but other electric field lines leave the outer shell and end on the inner shell. Find the charge on the inner shell.

CHAPTER 25

Electrostatic Energy and Capacitance

A Leyden jar capacitor.

When we bring a point charge q from far away to a region where other charges are present, we must do work qV, where V is the potential at the final position due to the other charges in the vicinity. The work done is stored as electrostatic potential energy. The electrostatic potential energy of a system of charges is the total work needed to assemble the system.

When charge is placed on an isolated conductor, the potential of the conductor increases. The ratio of the charge to the potential is called the **capacitance** of the conductor. A useful device for storing charge and energy is the capacitor, which consists of two conductors, closely spaced but insulated from each other. When attached to a source of potential difference such as a battery, the conductors carry equal and opposite charges. The ratio of the magnitude of the charge on either conductor to the potential difference between the conductors is the capacitance of the capacitor. Capacitors have many uses. The flash attachment for your camera uses a capacitor to store the energy needed to provide the sudden flash of light. Capacitors are also used in the tuning circuits of devices such as radios, televisions, and cellular phones, allowing them to operate at specific frequencies.

The first capacitor was the Leyden jar, a glass container lined inside and out with gold foil. It was invented at the University of Leyden in the Netherlands by eighteenth-century experimenters who, while studying the effects of electric charges on people and animals, got the idea of trying to store a large amount of charge in a bottle of water. An experimenter held up a jar of water in one hand while charge was conducted to the water by a chain from a static electric generator. When he reached over to lift the chain out of the water with his other hand, he was knocked unconscious. Benjamin Franklin realized that the device for storing charge did not have to be jar-shaped and used foil-covered window glass, called Franklin panes. With several of these connected in parallel, he stored a large charge and attempted to kill a turkey with it. Instead, he knocked himself out. He later wrote, "I tried to kill a turkey but nearly succeeded in killing a goose."

25-1 Electrostatic Potential Energy

If we have a point charge q_1, the potential at a distance $r_{1,2}$ away is given by

$$V = \frac{kq_1}{r_{1,2}}$$

To bring up a second point charge q_2 from an infinite distance away to a distance $r_{1,2}$, we must do work:

$$W_2 = q_2 V = \frac{kq_2 q_1}{r_{1,2}}$$

To bring up a third charge, work must be done against the electric field produced by both q_1 and q_2. The work required to bring up a third charge q_3 to a distance $r_{1,3}$ from q_1 and a distance $r_{2,3}$ from q_2 is

$$W_3 = \frac{kq_3 q_1}{r_{1,3}} + \frac{kq_3 q_2}{r_{2,3}}$$

The total work required to assemble the three charges is the **electrostatic potential energy** U of the system of three point charges:

$$U = \frac{kq_2 q_1}{r_{1,2}} + \frac{kq_3 q_1}{r_{1,3}} + \frac{kq_3 q_2}{r_{2,3}} \qquad 25\text{-}1$$

This quantity of work is independent of the order in which the charges are brought to their final positions. In general,

> The electrostatic potential energy of a system of point charges is the work needed to bring the charges from an infinite separation to their final positions.

Electrostatic potential energy of a system

The first two terms on the right-hand side of Equation 25-1 can be written

$$\frac{kq_2 q_1}{r_{1,2}} + \frac{kq_3 q_1}{r_{1,3}} = q_1\left(\frac{kq_2}{r_{1,2}} + \frac{kq_3}{r_{1,3}}\right) = q_1 V_1$$

where V_1 is the potential due to charges q_2 and q_3. Similarly, the second and third terms represent the charge q_3 times the potential due to charges q_1 and q_2, and the first and third terms equal the charge q_2 times the potential due to charges q_1 and q_2. We can thus rewrite Equation 25-1 as

$$\begin{aligned} U &= \frac{kq_2 q_1}{r_{1,2}} + \frac{kq_3 q_1}{r_{1,3}} + \frac{kq_3 q_2}{r_{2,3}} \\ &= \frac{1}{2}\left(\frac{kq_2 q_1}{r_{1,2}} + \frac{kq_3 q_1}{r_{1,3}} + \frac{kq_3 q_2}{r_{2,3}} + \frac{kq_2 q_1}{r_{1,2}} + \frac{kq_3 q_1}{r_{1,3}} + \frac{kq_3 q_2}{r_{2,3}}\right) \\ &= \frac{1}{2}\left[q_1\left(\frac{kq_2}{r_{1,2}} + \frac{kq_3}{r_{1,3}}\right) + q_2\left(\frac{kq_3}{r_{2,3}} + \frac{kq_1}{r_{1,2}}\right) + q_3\left(\frac{kq_1}{r_{1,3}} + \frac{kq_2}{r_{2,3}}\right)\right] \end{aligned}$$

The electrostatic potential energy U of a system of n point charges is thus

$$U = \frac{1}{2}\sum_{i=1}^{n} q_i V_i \qquad 25\text{-}2$$

Electrostatic potential energy of a system of point charges

where V_i is the potential at the location of the ith charge due to all of the other charges.

Equation 25-2 also describes the electrostatic potential energy of a continuous charge distribution. Consider a spherical conductor of radius R. When the sphere carries a charge q, its potential relative to $V = 0$ at infinity is

$$V = \frac{kq}{R}$$

The work needed to bring an additional amount of charge dq from infinity to the conductor is $V\,dq$. This work equals the increase in the potential energy of the conductor:

$$dU = V\,dq = \frac{kq}{R}\,dq$$

The total potential energy U is the integral of dU as q increases from zero to its final value Q. Integrating from $q = 0$ to $q = Q$, we obtain

$$U = \frac{kQ^2}{2R} = \frac{1}{2}QV \qquad \text{25-3}$$

where $V = kQ/R$ is the potential on the surface of the charged sphere. Although we derived Equation 25-3 for a spherical conductor, it holds for any conductor. The potential of any conductor is proportional to its charge q, so we can write $V = \alpha q$, where α is some constant. Then the work needed to bring an additional charge dq from infinity to the conductor is $V\,dq = \alpha q\,dq$, and the total work needed to put a charge Q on the conductor is $\frac{1}{2}\alpha Q^2 = \frac{1}{2}QV$. If we have a set of n conductors with the ith conductor at potential V_i and carrying a charge Q_i, the electrostatic potential energy is

$$U = \frac{1}{2}\sum_{i=1}^{n} Q_i V_i \qquad \text{25-4}$$

Electrostatic potential energy of a system of conductors

Example 25-1

Points A, B, C, and D are at the corners of a square of side a, as shown in Figure 25-1. (*a*) Calculate the work required to place a positive charge q at each corner of the square by separately calculating the work required to bring each charge to its final position. (*b*) Show that Equation 25-2 gives the total work.

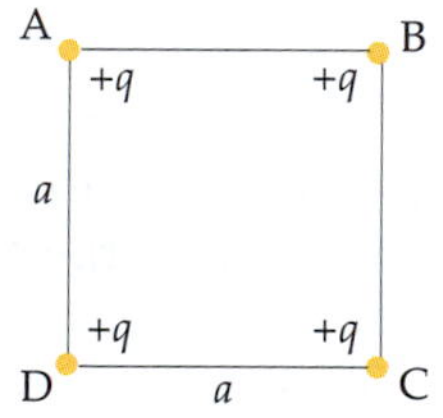

Figure 25-1

Picture the Problem No work is needed to place the first charge at point A since the potential there is zero when the other three charges are at infinity. As each additional charge is brought into place, work must be done because of the presence of the previous charges.

(*a*) 1. Place the first charge at A. To accomplish this step, the work W_A that is needed is zero:

$$W_A = 0$$

2. Bring the second charge to point B. The work required is $W_B = qV_A$, where V_A is the potential at B due to the first charge at A a distance a away:

$$W_B = qV_A = q\left(\frac{kq}{a}\right) = \frac{kq^2}{a}$$

3. $W_C = qV_C$, where V_C is the potential at C due to q at A a distance $\sqrt{2}\,a$ away, and q at B a distance a away:

$$W_C = qV_C = q\left(\frac{kq}{\sqrt{2}a} + \frac{kq}{a}\right) = \frac{kq^2}{\sqrt{2}a} + \frac{kq^2}{a}$$

4. Similar considerations give W_D, the work needed to bring the fourth charge to point D:

$$W_D = qV_D = q\left(\frac{kq}{a} + \frac{kq}{\sqrt{2}a} + \frac{kq}{a}\right) = \frac{2kq^2}{a} + \frac{kq^2}{\sqrt{2}a}$$

5. Summing the individual contributions gives the total work required to assemble the four charges:

$$W_{\text{total}} = W_A + W_B + W_C + W_D = \frac{4kq^2}{a} + \frac{2kq^2}{\sqrt{2}a}$$

$$= (4 + \sqrt{2})\frac{kq^2}{a}$$

(*b*) Calculate W_{total} from Equation 25-2. There are four identical terms, one from each charge:

$$W_{\text{total}} = U = \frac{1}{2}\sum_{i=1}^{4} q_i V_i = \frac{1}{2}\left[4 \times q\left(\frac{kq}{a} + \frac{kq}{a} + \frac{kq}{\sqrt{2}a}\right)\right]$$

$$= \frac{4kq^2}{a} + \frac{2kq^2}{\sqrt{2}a} = (4 + \sqrt{2})\frac{kq^2}{a}$$

Remark W_{total} is the total electrostatic energy of the charge distribution.

Exercise (*a*) How much work is required to bring a fifth positive charge q from infinity to the center of the square? (*b*) What is the total work required to assemble the five-charge system? (*Answers* (*a*) $4\sqrt{2}\,kq^2/a$, (*b*) $(4 + 5\sqrt{2})kq^2/a$)

25-2 Capacitance

The potential (relative to zero potential at infinity) of a single isolated conductor carrying a charge Q is proportional to the charge Q, and depends on the size and shape of the conductor. In general, the larger the conductor, the greater the amount of charge it can carry for a given potential. For example, the potential of a spherical conductor of radius R carrying a charge Q is

$$V = \frac{kQ}{R}$$

The ratio of charge Q to the potential V of an isolated conductor is called its capacitance C:

$$C = \frac{Q}{V} \qquad \text{25-5}$$

Definition—Capacitance

Capacitance is a measure of the capacity to store charge for a given potential difference. Since the potential is always proportional to the charge, this ratio does not depend on either Q or V, but only on the size and shape of the conductor. The capacitance of a spherical conductor is

$$C = \frac{Q}{V} = \frac{Q}{kQ/R} = \frac{R}{k} = 4\pi\epsilon_0 R \qquad \text{25-6}$$

The SI unit of capacitance is the coulomb per volt, which is called a **farad** (F) after the great English experimentalist Michael Faraday:

$$1\ \text{F} = 1\ \text{C/V} \qquad \text{25-7}$$

Since the farad is a rather large unit, submultiples such as the microfarad ($1\ \mu\text{F} = 10^{-6}$ F) or the picofarad ($1\ \text{pF} = 10^{-12}$ F) are often used. Since capacitance is in farads and R is in meters, we can see from Equation 25-6 that the SI unit for the permittivity of free space, ϵ_0, can also be written as a farad per meter:

$$\epsilon_0 = 8.85 \times 10^{-12}\ \text{F/m} = 8.85\ \text{pF/m} \qquad \text{25-8}$$

Exercise Find the radius of a spherical conductor that has a capacitance of 1 farad. (*Answer* 8.99×10^9 m, which is about 1400 times the radius of the earth)

We see from the above exercise that the farad is indeed a very large unit.

Exercise A sphere of capacitance C_1 carries a charge of 20 μC. If the charge is increased to 60 μC, what is the new capacitance C_2? (*Answer* $C_2 = C_1$. The capacitance does not depend on the charge. If the charge is tripled, the potential of the sphere will be tripled and the ratio Q/V, which depends only on the radius of the sphere, remains unchanged.)

Capacitors

A system of two conductors carrying equal but opposite charges is called a **capacitor.** A capacitor is usually charged by transferring a charge Q from one conductor to the other leaving one of the conductors with a charge $+Q$ and the other with a charge $-Q$. The capacitance of the device is defined to be Q/V, where Q is the magnitude of the charge on either conductor and V is the magnitude of the potential difference between the conductors.* To calculate the capacitance, we place equal and opposite charges on the conductors and then find the potential difference V by first finding the electric field $\vec{E}$ between them.

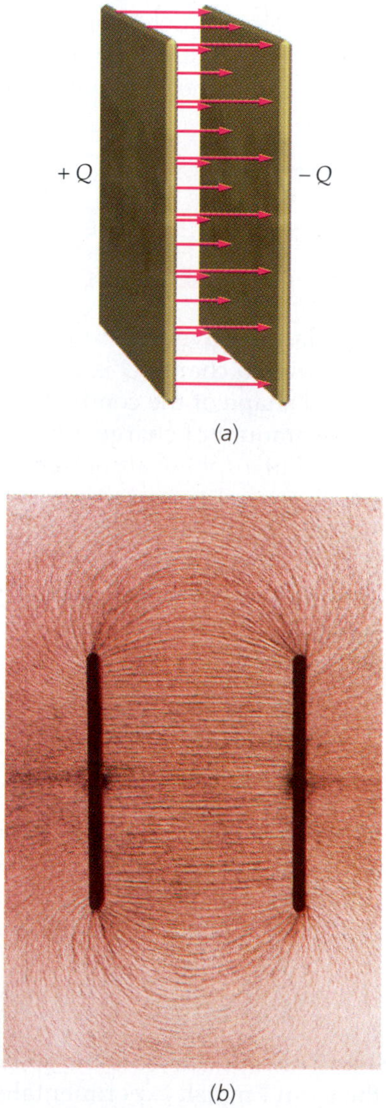

Figure 25-2 (*a*) Electric field lines between the plates of a parallel-plate capacitor. The lines are equally spaced between the plates, indicating that the field is uniform. (*b*) Electric field lines in a parallel-plate capacitor shown by small bits of thread suspended in oil.

Parallel-Plate Capacitors A common capacitor is the **parallel-plate capacitor,** which utilizes two parallel conducting plates. In practice, the plates may be thin metallic foils that are separated and insulated from one another by a thin plastic film. This "sandwich" is then rolled up, which allows for a large surface area in a relatively small space. Let A be the area of each plate and let d be the separation distance, which is small compared to the length and width of the plates. We place a charge $+Q$ on one plate and $-Q$ on the other. These charges attract each other and become uniformly distributed on the inside surfaces of the plates. Since the plates are close together, the electric field between them is approximately the same as the field between two equal and opposite infinite planes of charge. Each plate contributes a uniform field of magnitude $E = \sigma/2\epsilon_0$ (Equation 23-21) giving a total field $E = \sigma/\epsilon_0$, where $\sigma = Q/A$ is the magnitude of the charge per unit area on either plate. Since E is uniform between the plates (Figure 25-2), the potential difference between the plates equals the field times the plate separation d:

$$V = Ed = \frac{\sigma}{\epsilon_0}d = \frac{Qd}{\epsilon_0 A} \tag{25-9}$$

The capacitance of the parallel-plate capacitor is thus

$$C = \frac{Q}{V} = \frac{\epsilon_0 A}{d} \tag{25-10}$$

Capacitance of a parallel-plate capacitor

Note that since V is proportional to Q, the capacitance does not depend on either Q or V. For a parallel-plate capacitor, the capacitance is proportional to the area of the plates and is inversely proportional to the separation distance. In general, capacitance depends on the size, shape, and geometrical arrangement of the conductors and on the insulating medium between them.

* When we speak of the charge on a capacitor, we mean the magnitude of the charge on either conductor. The use of V rather than ΔV for the magnitude of the potential difference between the plates is standard and simplifies many of the equations relating to capacitance.

When a capacitor is connected to a battery,* as in Figure 25-3, charge is transferred from one conductor to the other until the potential difference between the conductors equals the potential difference across the battery terminals. The amount of charge transferred is $Q = CV$.

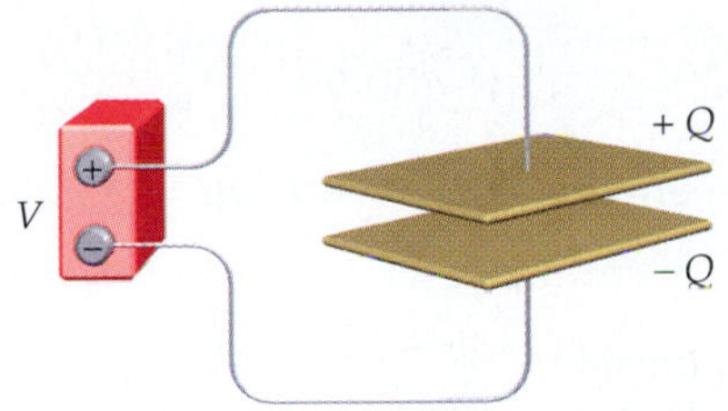

Figure 25-3 When the conductors in a capacitor are connected to the terminals of a battery, the battery transfers charge from one conductor to the other until the potential difference between the conductors equals that between the battery terminals. The amount of charge transferred is proportional to the potential difference.

*We will discuss batteries more fully in Chapter 26. Here, all we need to know is that a battery is a device that stores and supplies electrical energy and maintains a constant potential difference V between its terminals.

Example 25-2

A parallel-plate capacitor has square plates of side 10 cm separated by 1 mm. (*a*) Calculate the capacitance of this device. (*b*) If this capacitor is charged to 12 V, how much charge is transferred from one plate to another?

Picture the Problem The capacitance C is determined by the area and the separation of the plates. Once C is found, the charge for a given voltage V is found from the definition of capacitance $C = Q/V$.

(*a*) We find the capacitance using Equation 25-10:

$$C = \frac{\epsilon_0 A}{d} = \frac{(8.85\ \text{pF/m})(0.1\ \text{m})^2}{0.001\ \text{m}} = 88.5\ \text{pF}$$

(*b*) The charge transferred is found from the definition of capacitance:

$$Q = CV = (88.5\ \text{pF})(12\ \text{V}) = 1.06 \times 10^{-9}\ \text{C} = 1.06\ \text{nC}$$

Remarks Q is the magnitude of the charge on each plate of the capacitor. In this case, Q corresponds to roughly 6.6×10^9 electrons.

Exercise How large would the plates have to be for the capacitance to be 1 F? (*Answer* $A = 1.13 \times 10^8\ \text{m}^2$, which corresponds to a square 10.6 km on a side)

Cylindrical Capacitors A cylindrical capacitor consists of a small conducting cylinder or wire of radius r_1 and a larger, concentric cylindrical conducting shell of radius r_2. A coaxial cable, such as that used for cable television, can be thought of as a cylindrical capacitor. The capacitance per unit length of a coaxial cable is important in determining the transmission characteristics of the cable.

Example 25-3

Find an expression of the capacitance of a cylindrical capacitor consisting of two conductors both of length L. One cylinder has radius r_1 and the other is a coaxial cylindrical shell of inner radius r_2, with $r_1 < r_2 \ll L$ as shown in Figure 25-4.

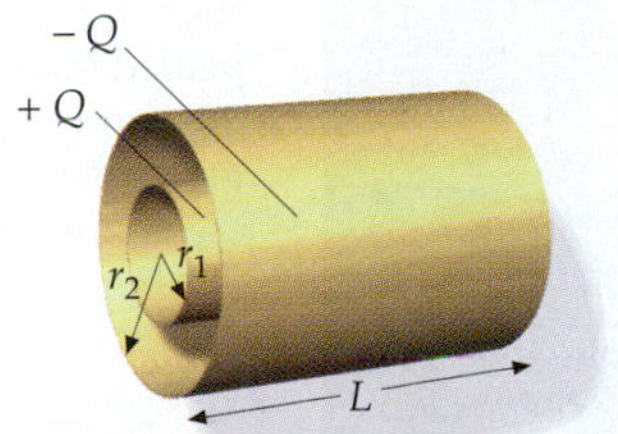

Figure 25-4

Picture the Problem We place charge $+Q$ on the inner conductor and $-Q$ on the outer conductor and calculate the potential difference $V = V_1 - V_2$ from the electric field between the conductors, which is found from Gauss's law. Since the electric field depends on r, we must integrate to find the potential difference.

1. The capacitance is defined as the ratio Q/V: $C = Q/V$

2. V is related to the electric field between the shells: $dV = -\vec{E}\cdot d\vec{\ell} = -E_r\,dr$

3. To find E_r we choose a cylindrical Gaussian surface of radius r between the conductors ($r_1 < r < r_2$). The area of the Gaussian surface is then $2\pi rL$. Gauss's law gives:

$$\oint_S E_n\,dA = \frac{1}{\epsilon_0}Q_{\text{inside}}$$

$$E_r 2\pi rL = \frac{Q}{\epsilon_0}$$

4. Solve for E_r:

$$E_r = \frac{1}{2\pi L\epsilon_0}\frac{Q}{r}$$

5. Integrate to find $V = V_1 - V_2$:

$$V = V_1 - V_2 = \int_{r_2}^{r_1} dV = \int_{r_2}^{r_1} -E_r\,dr$$

$$= \int_{r_1}^{r_2} +E_r\,dr = \frac{Q}{2\pi\epsilon_0 L}\int_{r_1}^{r_2}\frac{dr}{r} = \frac{Q}{2\pi\epsilon_0 L}\ln\frac{r_2}{r_1}$$

6. Substitute this result to find C:

$$C = \frac{Q}{V} = \frac{2\pi\epsilon_0 L}{\ln(r_2/r_1)}$$

Remarks The capacitance of a cylindrical capacitor is proportional to the length of the conductors.

Exercise How is the capacitance affected if the potential across a cylindrical capacitor is increased from 20 to 80 V? (*Answer* The capacitance of any capacitor does not depend on the potential. To increase V, you must increase the charge Q. The ratio Q/V depends only on the geometry of the capacitor.)

From Example 25-3 we see that the capacitance of a cylindrical capacitor is given by

$$C = \frac{2\pi\epsilon_0 L}{\ln(r_2/r_1)} \qquad 25\text{-}11$$

Capacitance of a cylindrical capacitor

A coaxial cable is a long cylindrical capacitor with a solid wire for the inner conductor and a braided-wire shield for the outer conductor. The outer rubber coating has been pealed back here to show the conductors and the white plastic insulator that separates the conductors.

Cutaway of a 200-μF capacitor used in an electronic strobe light.

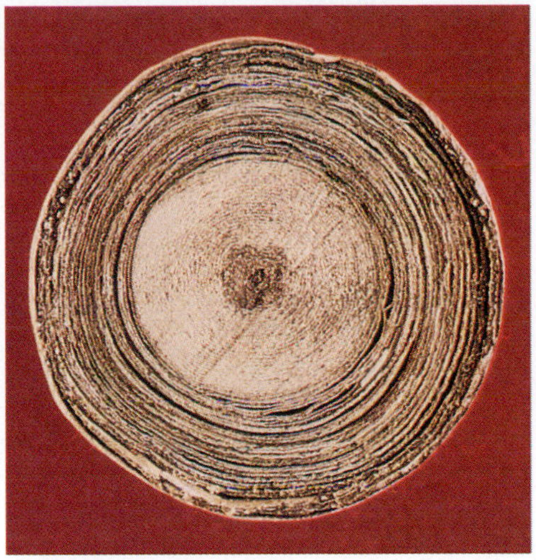

Cross section of a foil-wound capacitor.

A variable air-gap capacitor like those that were used in the tuning circuits of old radios. The semicircular plates rotate through the fixed plates, changing the amount of surface area between them, and hence the capacitance.

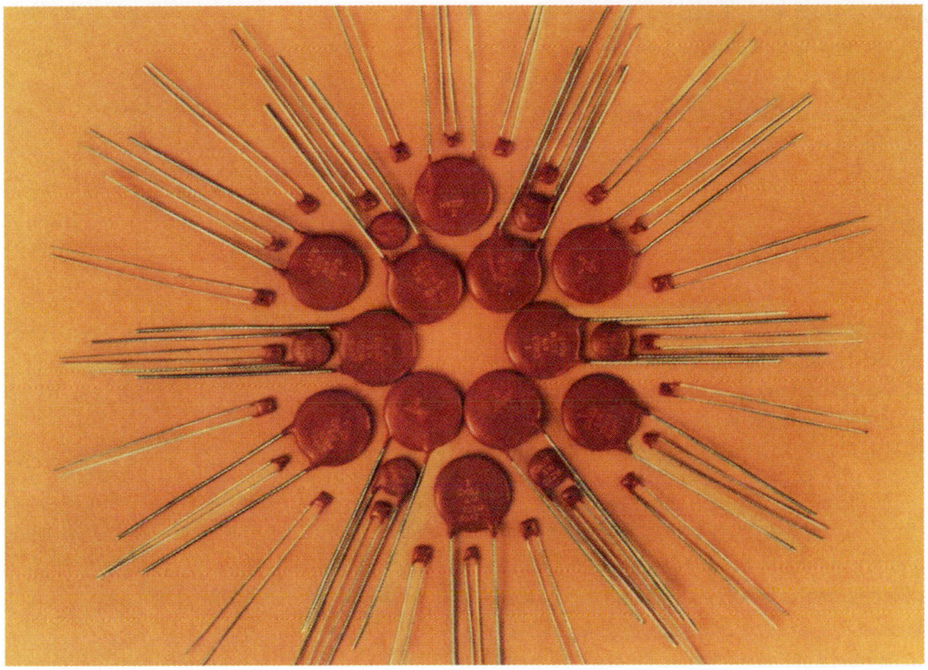

Ceramic capacitors for use in electronic circuits.

25-3 The Storage of Electrical Energy

When a capacitor is being charged, positive charge is transferred from the negatively charged conductor to the positively charged conductor. Work must therefore be done to charge a capacitor. Some of this work is stored as electrostatic potential energy.

Let q be the charge that has been transferred at some time during the charging process. The potential difference is then $V = q/C$. If a small amount of additional charge dq is now transferred from the negative conductor to the positive conductor through a potential increase of V (Figure 25-5), the potential energy of the charge is increased by

$$dU = V\,dq = \frac{q}{C}\,dq$$

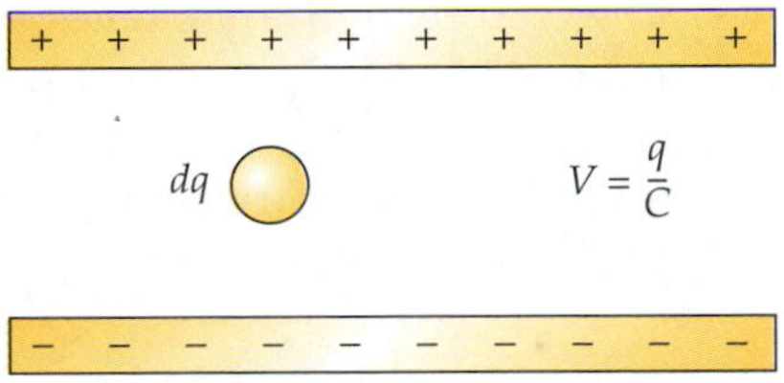

Figure 25-5 When a small amount of positive charge dq is moved from the negative conductor to the positive conductor, its potential energy is increased by $dU = V\,dq$, where V is the potential difference between the conductors.

The total increase in potential energy U is the integral of dU as q increases from zero to its final value Q (Figure 25-6):

$$U = \int dU = \int_0^Q \frac{q}{C}\,dq = \frac{1}{2}\frac{Q^2}{C}$$

This potential energy is the energy stored in the capacitor. Using $C = Q/V$, we can express this energy in a variety of ways:

$$U = \frac{1}{2}\frac{Q^2}{C} = \frac{1}{2}QV = \frac{1}{2}CV^2 \qquad \text{25-12}$$

Energy stored in a capacitor

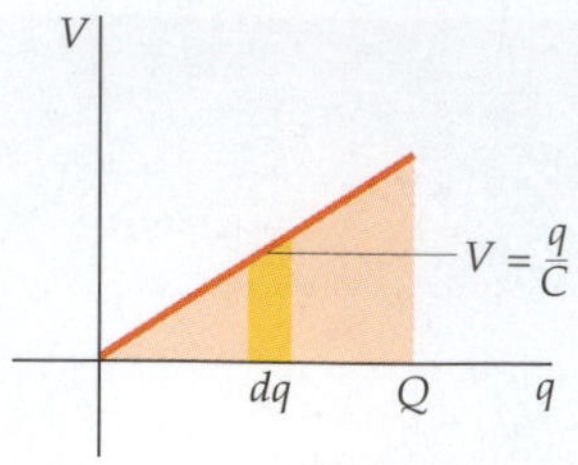

Figure 25-6 The work needed to charge a capacitor is the integral of $V\,dq$ from the original charge of $q = 0$ to the final charge of $q = Q$. This work is the area under the curve $\frac{1}{2}Q(Q/C)$.

Exercise A 15-μF capacitor is charged to 60 V. How much energy is stored in the capacitor? (*Answer* 0.027 J)

Exercise Obtain the expression for the electrostatic energy stored in a capacitor (Equation 25-12) from Equation 25-4 using $Q_1 = +Q$, $Q_2 = -Q$, and $V = V_1 - V_2$.

A capacitor is usually charged by connecting it to a battery, which maintains a potential difference V between its terminals. The work done by the battery in delivering a charge Q to the capacitor is QV, which is twice the energy stored in the capacitor. The additional work done by the battery is dissipated in heat in the connecting wires or radiation.*

*We will show in Section 26-6 that if the capacitor is connected to a battery by wires of some resistance R, half the energy supplied by the battery in charging the capacitor is dissipated as thermal energy in the wires.

Example 25-4

A parallel-plate capacitor with square plates 14 cm on a side and separated by 2.0 mm is connected to a battery and charged to 12 V. The battery is then disconnected from the capacitor and the plate separation is increased to 3.5 mm. (*a*) What is the charge on the capacitor? (*b*) How much energy was originally stored in the capacitor? (*c*) By how much is the energy increased when the plate separation is changed?

Picture the Problem (*a*) The charge on the capacitor can be calculated from the capacitance and then used to calculate the energy in (*b*). (*c*) Since the capacitor is removed from the battery, the charge remains constant as the plates are separated. The energy increase is found by using the charge and new potential to calculate the new energy, from which we subtract the original energy.

(*a*) 1. The charge on the capacitor equals the product of C and V: $\quad Q = CV$

2. Calculate the capacitance of the parallel-plate capacitor: $\quad C = \dfrac{\epsilon_0 A}{d} = \dfrac{(8.85\ \text{pF/m})(0.14\ \text{m})^2}{0.002\ \text{m}} = 86.7\ \text{pF}$

3. Substitute to calculate Q: $\quad Q = CV = (86.7\ \text{pF})(12\ \text{V}) = 1.04\ \text{nC}$

(*b*) Calculate the original energy stored: $\quad U = \frac{1}{2}QV = \frac{1}{2}(1.04\ \text{nC})(12\ \text{V}) = 6.24\ \text{nJ}$

(*c*) 1. After the capacitor is removed from the battery, the charge on the plates remains constant. The potential between the plates is the field E times the plate separation d: $\quad V = Ed$

2. Since E does not change, and the potential difference is 12 V when the plate separation is 2.0 mm, the potential difference when the separation is 3.5 mm is: $V' = (12\text{ V})\dfrac{3.5\text{ mm}}{2.0\text{ mm}} = 21\text{ V}$

3. At $d = 3.5$ mm, with $V' = 21$ V, the energy stored is: $U = \frac{1}{2}QV = \frac{1}{2}(1.04\text{ nC})(21\text{ V}) = 10.92\text{ nJ}$

4. Subtract the original energy to find the increase: $\Delta U = 10.92\text{ nJ} - 6.24\text{ nJ} = 4.68\text{ nJ}$

Remarks The additional energy calculated in part (*c*) comes from work done by the agent responsible for increasing the separation between the plates, which attract each other.

Figure 25-7 shows how this effect is used in a switch.

Exercise Find the original energy in the capacitor of this example directly from $U = \frac{1}{2}CV^2$. (*Answer* 6.24 nJ)

Exercise (*a*) Find the new capacitance C' in this example when separation of the plates is 3.5 mm. (*b*) Use your result to calculate the new energy from $U = Q^2/2C'$. (*Answers* (*a*) $C' = 49.6$ pF (*b*) $U = 10.92$ nJ)

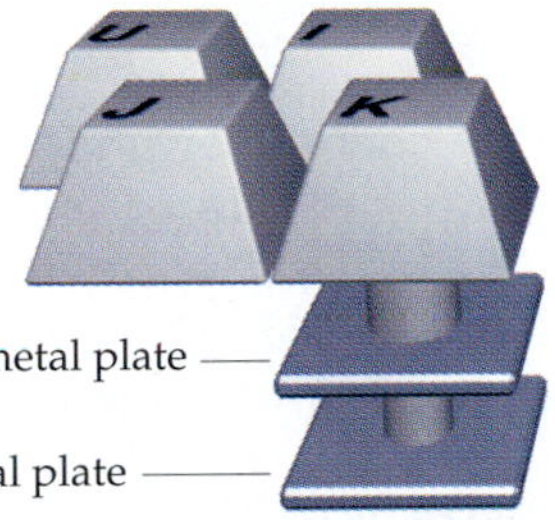

Figure 25-7 Capacitance switching in computer keyboards. A metal plate attached to each key acts as the top plate of a capacitor. Depressing the key decreases the separation between the top and bottom keys and increases the capacitance, which triggers the electronic circuitry of the computer to acknowledge the keystroke.

It is instructive to work part *c* of Example 25-4 in another way. The oppositely charged plates of a capacitor exert attractive forces on one another. Work must be done against these forces to increase the plate separation. Assume that the lower plate is held fixed and the upper plate is moved. The force on the upper plate is the charge on the plate Q times the electric field *due to the lower plate.* This field is half the total field between the plates, because the charge on the upper plate also contributes equally to the field. When the potential difference is 12 V and the separation is 2 mm, the total field between the plates is

$$E = \frac{V}{d} = \frac{12\text{ V}}{2\text{ mm}} = 6\text{ V/mm} = 6\text{ kV/m}$$

The force exerted on the upper plate by the bottom plate is thus

$$F = QE' = Q(\tfrac{1}{2}E) = (1.04\text{ nC})(3\text{ kV/m}) = 3.12\ \mu\text{N}$$

The work that must be done to move the upper plate a distance of $\Delta d = 1.5$ mm is then

$$W = F\,\Delta d = (3.12\ \mu\text{N})(1.5\text{ mm}) = 4.68\text{ nJ}$$

This work equals the increase in the energy stored.

Electrostatic Field Energy

In the process of charging a capacitor, an electric field is produced between the plates. The work required to charge the capacitor can be thought of as the work required to create the electric field. That is, we can think of the energy stored in a capacitor as energy stored in the electric field, called **electrostatic field energy**.

Consider a parallel-plate capacitor. We can relate the energy stored in the capacitor to the electric field E between the plates. The potential difference between the plates is related to the electric field by $V = Ed$, where d is the plate separation. The capacitance is given by $C = \epsilon_0 A/d$ (Equation 25-10). The energy stored is

$$U = \frac{1}{2}CV^2 = \frac{1}{2}\left(\frac{\epsilon_0 A}{d}\right)(Ed)^2 = \frac{1}{2}\epsilon_0 E^2(Ad)$$

The quantity Ad is the volume of the space between the plates of the capacitor containing the electric field. The energy per unit volume is called the **energy density** u_e. The energy density in an electric field E is thus

$$u_e = \frac{\text{energy}}{\text{volume}} = \frac{1}{2}\epsilon_0 E^2 \qquad 25\text{-}13$$

Energy density of an electrostatic field

Thus, the energy per unit volume of the electrostatic field is proportional to the square of the electric field. *Although we obtained Equation 25-13 by considering the electric field between the plates of a parallel-plate capacitor, the result applies to any electric field.* Whenever there is an electric field in space, the electrostatic energy per unit volume is given by Equation 25-13.

Exercise (*a*) Calculate the energy density u_e for Example 25-4 when the plate separation is 2.0 mm. (*b*) Show that the increase in energy in Example 25-4 is equal to u_e times the increase in volume (Δ vol) between the plates. (*Answers* (*a*) $u_e = \frac{1}{2}\epsilon_0 E^2 = 159.3\ \mu\text{J/m}^3$, (*b*) $\Delta\text{vol} = A\,\Delta d = 2.94 \times 10^{-5}\ \text{m}^3$, $u_e\,\Delta\text{vol} = 4.68$ nJ, in agreement with Example 25-4)

We can illustrate the generality of Equation 25-13 by calculating the electrostatic field energy of a spherical conductor of radius R that carries a charge Q. The electrostatic potential energy in terms of the charge Q and potential V is given by Equation 25-12:

$$U = \frac{kQ^2}{2R} = \frac{1}{2}QV \qquad 25\text{-}14$$

We now obtain the same result by considering the energy density of an electric field given by Equation 25-13. When the conductor carries a charge Q, the electric field is radial and is given by

$$E_r = 0, \qquad r < R \text{ (inside the conductor)}$$

$$E_r = \frac{kQ}{r^2}, \qquad r > R \text{ (outside the conductor)}$$

Since the electric field is spherically symmetric, we choose a spherical shell for our volume element. If the radius of the shell is r and its thickness is dr, the volume is $d\mathcal{V} = 4\pi r^2\,dr$ (Figure 25-8). The energy dU in this volume element is

$$dU = u_e\,d\mathcal{V} = \frac{1}{2}(\epsilon_0 E^2)4\pi r^2 dr$$

$$= \frac{1}{2}\epsilon_0\left(\frac{kQ}{r^2}\right)^2(4\pi r^2\,dr) = \frac{1}{2}(4\pi\epsilon_0 k^2)Q^2\frac{dr}{r^2} = \frac{1}{2}kQ^2\frac{dr}{r^2}$$

where we have used $4\pi\epsilon_0 = 1/k$. Since the electric field is zero for $r < R$, we obtain the total energy in the electric field by integrating from $r = R$ to $r = \infty$:

$$U = \int u_e\,d\mathcal{V} = \frac{1}{2}kQ^2\int_R^\infty \frac{dr}{r^2} = \frac{1}{2}k\frac{Q^2}{R} = \frac{1}{2}Q\left(\frac{kQ}{R}\right) = \frac{1}{2}QV \qquad 25\text{-}15$$

which is the same as Equation 25-12.

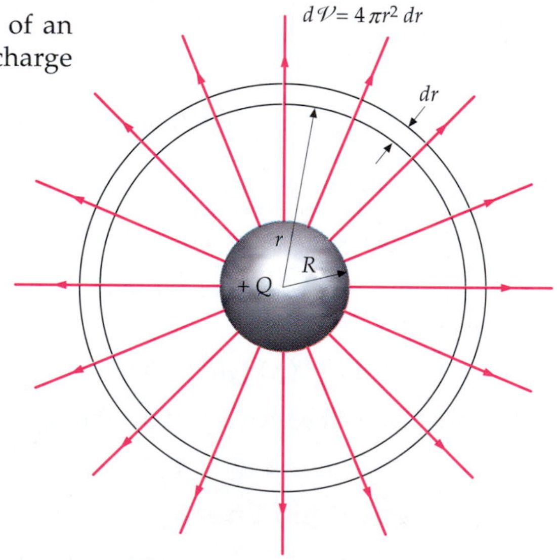

Figure 25-8 Geometry for the calculation of the electrostatic energy of a spherical conductor carrying a charge Q. The volume of the space between r and $r + dr$ is $d\mathcal{V} = 4\pi r^2\,dr$. The electrostatic field energy in this volume element is $u_e d\mathcal{V}$, where $u_e = \frac{1}{2}\epsilon_0 E^2$ is the energy density.

25-4 Combinations of Capacitors

Two or more capacitors are often used in combination. When two capacitors are connected as shown in Figure 25-9 such that the upper plates of the two capacitors are connected by a conducting wire and are therefore at a common potential V_a, and the lower plates are also connected together and are at a common potential V_b, the capacitors are said to be connected in **parallel.** (In electric circuits, a capacitor is indicated by the symbol ⊣⊢.) When capacitors are connected in parallel, the potential difference is the same across each capacitor.

In Figure 25-10, two capacitors are connected so that the magnitude of the charge on the two capacitors must be the same. These capacitors are said to be connected in **series.** The potential difference across the series combination is the sum of the potential differences across the individual capacitors.

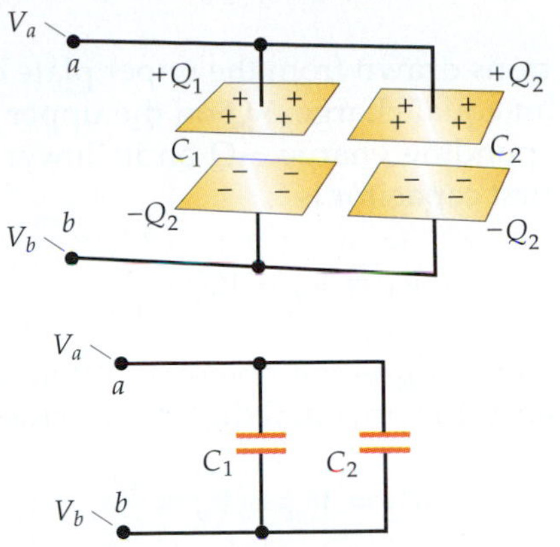

Figure 25-9 Two capacitors in parallel. The upper plates are connected together and are therefore at a common potential V_a; the lower plates are similarly connected together and therefore at a common potential V_b.

Parallel Capacitors

In Figure 25-9, assume that points a and b are connected to a battery or some other device that maintains a potential difference $V = V_a - V_b$ between the plates of each capacitor. If the capacitances are C_1 and C_2, the charges Q_1 and Q_2 stored on the plates are given by

$$Q_1 = C_1V$$

and

$$Q_2 = C_2V$$

The total charge stored is

$$Q = Q_1 + Q_2 = C_1V + C_2V = (C_1 + C_2)V$$

A combination of capacitors in a circuit can sometimes be replaced by a single capacitor that stores the same amount of charge for a given potential difference. The substitute capacitor is said to have an **equivalent capacitance**. The equivalent capacitance of two capacitors in parallel is the ratio of the total charge stored to the potential difference:

$$C_{eq} = \frac{Q}{V} = C_1 + C_2 \qquad \text{25-16}$$

Thus, for two capacitors in parallel, C_{eq} is the sum of the individual capacitances. When we add a second capacitor in parallel, we increase the capacitance. The conductor area is essentially increased, allowing more charge to be stored for the same potential difference.

The same reasoning can be extended to three or more capacitors connected in parallel, as in Figure 25-11:

$$C_{eq} = C_1 + C_2 + C_3 + \cdots \qquad \text{25-17}$$

Equivalent capacitance for capacitors in parallel

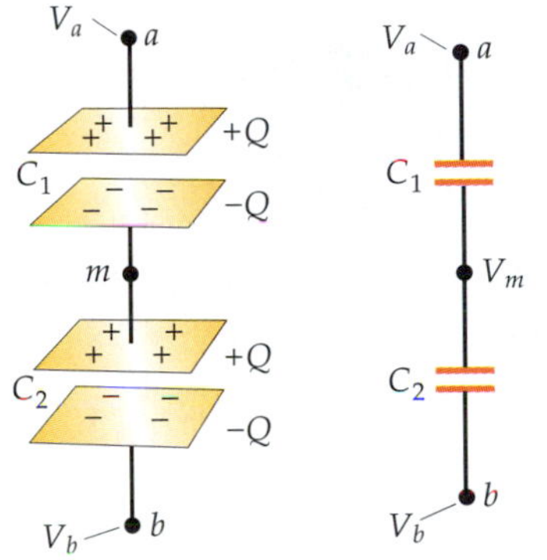

Figure 25-10 Two capacitors in series. The charge is the same on each capacitor.

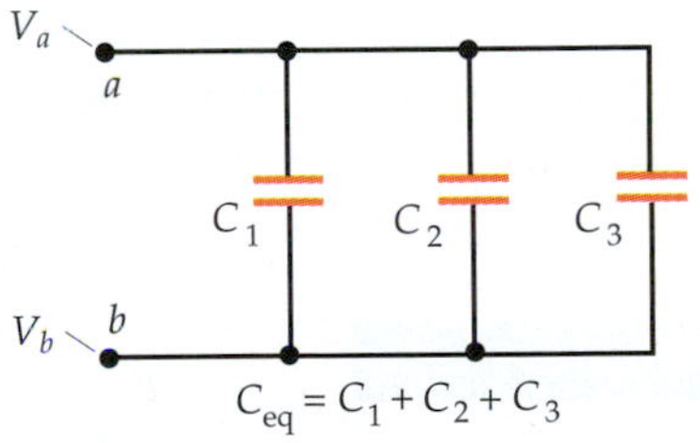

Figure 25-11 Three capacitors in parallel. The effect of adding a parallel capacitor to a circuit is an increase in the equivalent capacitance.

Series Capacitors

The capacitors in Figure 25-10 are connected in series. When points a and b are connected to the terminals of a battery, there is a potential difference $V = V_a - V_b$ across the two capacitors. If a charge $+Q$ is placed on the upper plate of the first capacitor, the electric field produced by that charge will induce an equal negative charge $-Q$ on its lower plate. This charge comes from electrons drawn from the upper plate of the second capacitor. Thus, there will be an equal charge $+Q$ on the upper plate of the second capacitor and a corresponding charge $-Q$ on its lower plate. The potential difference across the first capacitor is

$$V_1 = V_a - V_m = \frac{Q}{C_1}$$

where V_m is the potential of the adjacent plates and connecting wire. Similarly, the potential difference across the second capacitor is

$$V_2 = V_m - V_b = \frac{Q}{C_2}$$

The potential difference across the two capacitors in series is the sum of these potential differences:

$$V = V_a - V_b = V_1 + V_2 = \frac{Q}{C_1} + \frac{Q}{C_2} = Q\left(\frac{1}{C_1} + \frac{1}{C_2}\right) \qquad \text{25-18}$$

The equivalent capacitance of two capacitors in series is defined as

$$C_{eq} = \frac{Q}{V} \qquad \text{25-19}$$

Solving Equations 25-18 and 25-19 for V/Q gives

$$\frac{1}{C_{eq}} = \frac{1}{C_1} + \frac{1}{C_2} \qquad \text{25-20}$$

Equation 25-20 can be generalized to three or more capacitors connected in series:

$$\frac{1}{C_{eq}} = \frac{1}{C_1} + \frac{1}{C_2} + \frac{1}{C_3} + \cdots \qquad \text{25-21}$$

Equivalent capacitance for capacitors in series

Capacitor bank for storing energy to be used by the pulsed Nova laser at Lawrence Livermore Laboratories. The laser is used in fusion studies.

Exercise Two capacitors have capacitances of 20 and 30 μF. Find the equivalent capacitance if the capacitors are connected (*a*) in parallel and (*b*) in series. (*Answers* (*a*) 50 μF, (*b*) 12 μF)

Note that in the preceding exercise, the equivalent capacitance of the two capacitors in series is less than the capacitance of either capacitor. Adding a capacitor in series increases $1/C_{eq}$, which means the equivalent capacitance C_{eq} decreases.

Example 25-5

A 2-μF capacitor and a 4-μF capacitor are connected in series across an 18-V battery. Find the charge on the capacitors and the potential difference across each.

Picture the Problem Figure 25-12*a* shows the circuit in this Example and Figure 25-12*b* shows an equivalent capacitor that carries the same charge $Q = C_{eq}V$. After finding the charge, we can find the potential drop across each capacitor.

Figure 25-12

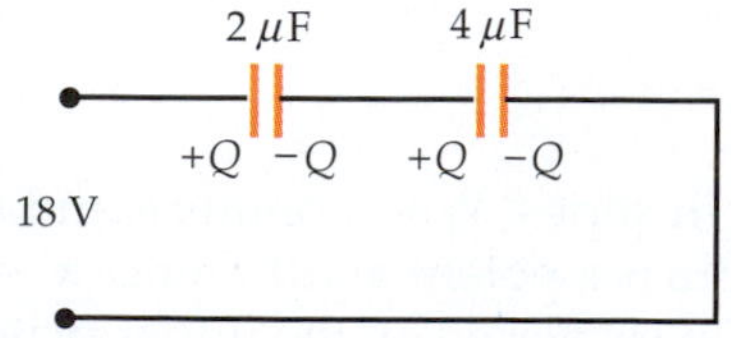

(*a*)

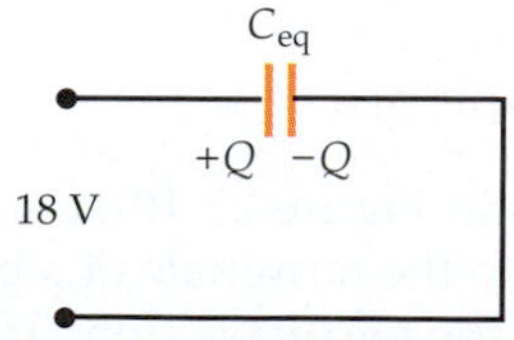

(*b*)

1. The charge on each capacitor equals the charge on the equivalent capacitor: $Q = C_{eq}V$

2. The equivalent capacitance of the series capacitors is found from:
$$\frac{1}{C_{eq}} = \frac{1}{C_1} + \frac{1}{C_2} = \frac{1}{2\mu F} + \frac{1}{4\mu F} = \frac{3}{4\mu F}$$
$$C_{eq} = \frac{4}{3}\,\mu F$$

3. Use this value to find the charge Q: $Q = C_{eq}V = \left(\frac{4}{3}\mu F\right)(18\text{ V}) = 24\ \mu C$

4. Use the result for Q to find the potential across the 2-μF capacitor: $V_1 = \frac{Q}{C_1} = \frac{24\ \mu C}{2\ \mu F} = 12\text{ V}$

5. Again use the result for Q to find the potential across the 4-μF capacitor: $V_2 = \frac{Q}{C_2} = \frac{24\ \mu C}{4\ \mu F} = 6\text{ V}$

Check the Result The sum of these potential differences is 18 V, as required.

Example 25-6 *try it yourself*

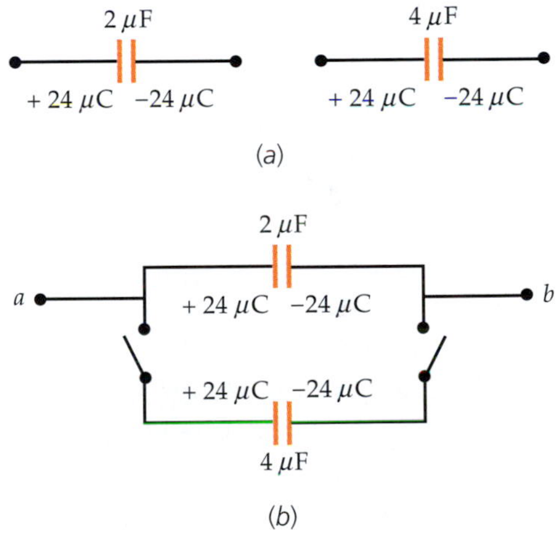

Figure 25-13

The two capacitors in Example 25-5 are removed from the battery and carefully disconnected from each other so that the charge on the plates is not disturbed (Figure 25-13*a*). They are then reconnected in a circuit containing open switches, positive plate to positive plate and negative plate to negative plate (Figure 25-13*b*). Find the potential difference across the capacitors and the charge on each capacitor when the switches are closed.

Picture the Problem Just after the two capacitors are disconnected from the battery, they carry equal charges of 24 μC. After switches S_1 and S_2 in the new circuit are closed, the capacitors are in parallel between points a and b. The potential across each of them is the same, and the equivalent capacitance of the system is $C_{eq} = C_1 + C_2$. The two positive plates form a single conductor with charge $Q = 48\ \mu C$, and the negative plates form a conductor with charge $-Q = -48\ \mu C$. Therefore, the potential difference is $V = Q/C_{eq}$, and the charges on the two capacitors are $Q_1 = C_1V$ and $Q_2 = C_2V$.

Cover the column to the right and try these on your own before looking at the answers.

Steps	***Answers***
1. Write V in terms of C_{eq} and the total charge Q.	$V = \frac{Q}{C_{eq}}$
2. Evaluate the equivalent capacitance.	$C_{eq} = 6\ \mu F$
3. Calculate V from C_{eq} and the charge $Q = 48\ \mu C$.	$V = 8\text{ V}$
4. Use V to find the charge on each capacitor.	$Q_1 = 16\ \mu C,\quad Q_2 = 32\ \mu C$

Check the Result Note that $Q_1 + Q_2 = Q = 48\ \mu C$, as required.

Exercise Find the energy stored in the capacitors before and after they are connected. (*Answer* $U_i = q^2/2C_1 + q^2/2C_2$, where $q = 24\ \mu C$. Thus, $U_i = 216\ \mu J$. $U_f = Q_1^2/2C_1 + Q_2^2/2C_2 = 192\ \mu J$. Note that $24\ \mu J$ is "lost" to thermal energy in the wires or radiated away.)

Example 25-7

(*a*) Find the equivalent capacitance of the network of three capacitors in Figure 25-14. (*b*) Find the charge on each capacitor and the voltage drop across it when the system is connected to a 6-V battery.

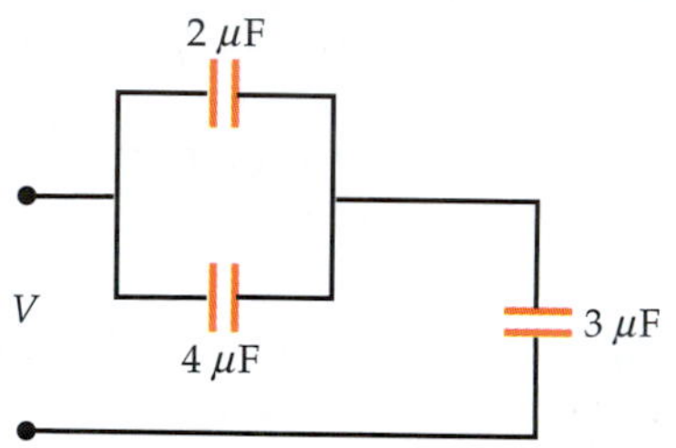

Figure 25-14

Picture the Problem (*a*) The 2- and 4-μF capacitors are connected in parallel, and the parallel combination is connected in series with the 3-μF capacitor. We first find the equivalent capacitance of the 2- and 4-μF capacitors (Figure 25-15*a*), then combine this equivalent capacitance with the 3-μF capacitor to reach a final equivalent capacitance (25-15*b*). (*b*) The charge on the 3-μF capacitor is the charge delivered by the battery $Q = C_{eq} V$ as shown in Figure 25-15*a*.

Figure 25-15

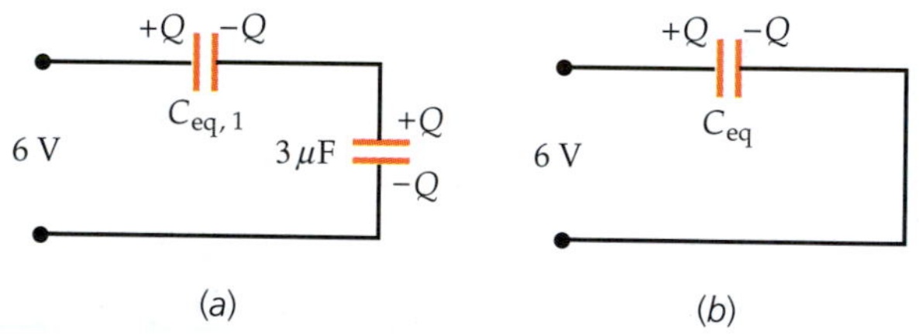

(*a*) 1. The equivalent capacitance of the two capacitors in parallel is the sum of the capacitances:

$$C_{eq,1} = C_1 + C_2 = 2\ \mu F + 4\ \mu F = 6\ \mu F$$

2. Find the equivalent capacitance of a 6-μF capacitor in series with a 3-μF capacitor:

$$\frac{1}{C_{eq}} = \frac{1}{C_1} + \frac{1}{C_2} = \frac{1}{6\ \mu F} + \frac{1}{3\ \mu F} = \frac{1}{2\ \mu F}$$

$$C_{eq} = 2\ \mu F$$

(*b*) 1. Calculate the charge Q delivered by the battery. This is also the charge on the 3-μF capacitor:

$$Q = C_{eq}V = (2\ \mu F)(6\ V) = 12\ \mu C$$

2. The potential drop across the 3-μF capacitor is Q/C:

$$V_3 = \frac{Q}{C_3} = \frac{12\ \mu C}{3\ \mu F} = 4\ V$$

3. The potential drop across the parallel combination $V_{2,4}$ is $Q/C_{eq,1}$:

$$V_{2,4} = \frac{Q}{C_{eq,1}} = \frac{12\ \mu C}{6\ \mu F} = 2\ V$$

4. The charge on each of the parallel capacitors is found from $Q_i = C_i V_{2,4}$, where $V_{2,4} = 2\ V$:

$$Q_2 = C_2 V_{2,4} = (2\ \mu F)(2\ V) = 4\ \mu C$$

$$Q_4 = C_4 V_{2,4} = (4\ \mu F)(2\ V) = 8\ \mu C$$

Check the Result The voltage drop across the parallel combination (2 V) plus that across the 3-μF capacitor (4 V) equals the voltage of the battery. Also, the sum of the charges on the parallel capacitors ($4\ \mu C + 8\ \mu C$) equals the total charge ($12\ \mu C$) on the 3-μF capacitor.

Exercise Find the energy stored in each capacitor. (*Answer* $U_2 = 4\ \mu J$, $U_3 = 24\ \mu J$, $U_4 = 8\ \mu J$. Note that $U_2 + U_3 + U_4 = 36\ \mu J = \frac{1}{2}QV = \frac{1}{2}C_{eq}V^2$.)

25-5 Dielectrics

A nonconducting material, such as air, glass, paper, or wood, is called a **dielectric**. When the space between the two conductors of a capacitor is occupied by a dielectric, the capacitance is increased by a factor κ that is characteristic of the dielectric, a fact discovered experimentally by Michael Faraday. The reason for this increase is that the electric field between the plates of a capacitor is weakened by the dielectric. Thus, for a given charge on the plates, the potential difference is reduced and the capacitance (Q/V) is increased.

If the original electric field between the plates of a capacitor without a dielectric is E_0, the field in a dielectric slab inserted between the plates is

$$E = \frac{E_0}{\kappa} \tag{25-22}$$

Electric field inside a dielectric

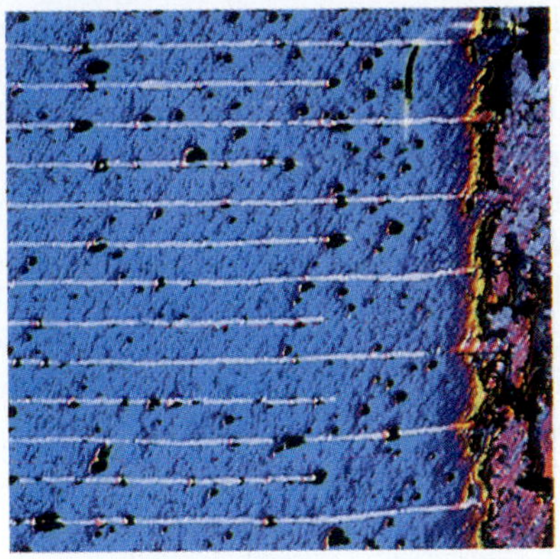

A cut section of a multilayer capacitor with a ceramic dielectric. The white lines are the edges of the conducting plates.

where κ is called the **dielectric constant.** For a parallel-plate capacitor of separation d, the potential difference between the plates is

$$V = Ed = \frac{E_0 d}{\kappa} = \frac{V_0}{\kappa}$$

where V is the potential difference with the dielectric and $V_0 = E_0 d$ is the original potential difference without the dielectric. The new capacitance is

$$C = \frac{Q}{V} = \frac{Q}{V_0/\kappa} = \kappa\frac{Q}{V_0}$$

or

$$C = \kappa C_0 \tag{25-23}$$

Effect of a dielectric on capacitance

where $C_0 = Q/V_0$ is the capacitance without the dielectric. The capacitance of a parallel-plate capacitor filled with a dielectric of constant κ is thus

$$C = \frac{\kappa\epsilon_0 A}{d} = \frac{\epsilon A}{d} \tag{25-24}$$

where

$$\epsilon = \kappa\epsilon_0 \tag{25-25}$$

is called the **permittivity** of the dielectric.

In the preceding discussion, we assumed that the charge on the plates of the capacitor did not change when the dielectric was inserted. This is true if the capacitor is charged and then removed from the charging source (the battery) before the insertion of the dielectric. If the dielectric is inserted while the battery is still connected, the battery supplies more charge to maintain the original potential difference. The total charge on the plates is then $Q = \kappa Q_0$. In either case, the capacitance (Q/V) is increased by the factor κ.

Exercise The 88.5-pF capacitor of Example 25-2 is filled with a dielectric of constant $\kappa = 2$. (*a*) Find the new capacitance. (*b*) Find the charge on the capacitor with the dielectric in place if the capacitor is attached to a 12-V battery. (*Answers* (*a*) 177 pF, (*b*) 2.12 nC)

Exercise The capacitor in the previous exercise is charged to 12 V without the dielectric and is then disconnected from the battery. The dielectric of constant $\kappa = 2$ is then inserted. Find the new values for (*a*) the charge Q, (*b*) the voltage V, and (*c*) the capacitance C. (*Answers* (*a*) $Q = 1.06$ nC, which is unchanged; (*b*) $V = 6$ V; (*c*) $C = 177$ pF)

Dielectrics not only increase the capacitance of a capacitor, they also provide a means for keeping parallel conducting plates apart, and they raise the potential difference at which dielectric breakdown occurs.* Consider a parallel-plate capacitor made from two sheets of metal foil that are separated by a thin plastic sheet. The plastic sheet allows the metal sheets to be very close together without actually being in electrical contact, and because the dielectric strength of plastic is greater than that of air, a greater potential difference can be attained before dielectric breakdown occurs. Table 25-1 lists the dielectric constants and dielectric strengths of some dielectrics. Note that for air, $\kappa \approx 1$, so for most situations we do not need to distinguish between air and a vacuum.

* Recall from Chapter 24 that for electric fields greater than about 3×10^6 V/m, air breaks down, that is, it becomes ionized and begins to conduct.

Table 25-1

Dielectric Constants and Strengths of Various Materials

Material	Dielectric Constant κ	Dielectric Strength, kV/mm
Air	1.00059	3
Bakelite	4.9	24
Glass (Pyrex)	5.6	14
Mica	5.4	10–100
Neoprene	6.9	12
Paper	3.7	16
Paraffin	2.1–2.5	10
Plexiglas	3.4	40
Polystyrene	2.55	24
Porcelain	7	5.7
Transformer oil	2.24	12

Example 25-8

A parallel-plate capacitor has square plates of side 10 cm and a separation of $d = 4$ mm. A dielectric slab of constant $\kappa = 2$ has the same area as the plates. (*a*) What is the capacitance without the dielectric? (*b*) What is the capacitance if the dielectric slab fills the space between the plates? (*c*) What is the capacitance if a dielectric slab of thickness 3 mm is inserted into the 4-mm gap?

Picture the Problem The capacitance without the dielectric, C_0, is found from the area and spacing of the plates (Figure 25-16*a*). When the capacitor is filled with a dielectric κ, (Figure 25-16*b*), the capacitance is $C = \kappa C_0$ (Equation (25-25). If the dielectric only partially fills the capacitor (Figure 25-16*c*), we calculate the potential difference V for a given charge Q, then apply the definition of capacitance, $C = Q/V$.

Figure 25-16

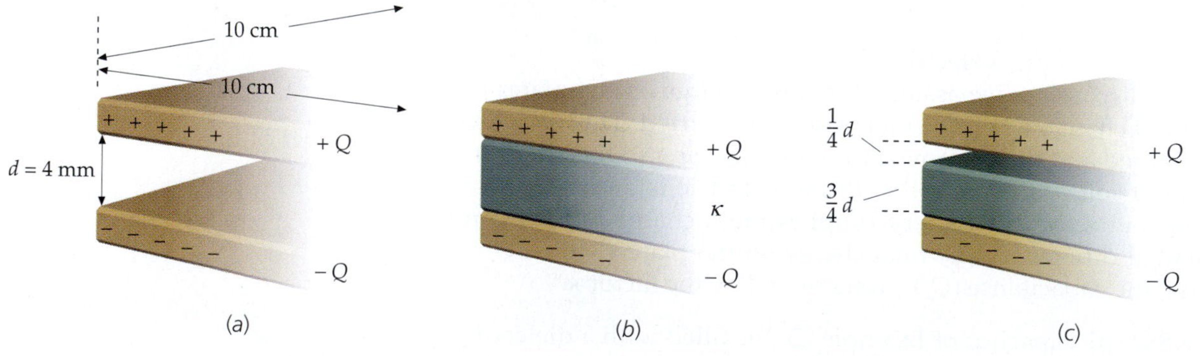

(*a*) If there is no dielectric, the capacitance C_0 is given by Equation 25-10:

$$C_0 = \frac{\epsilon_0 A}{d} = \frac{(8.85\text{ pF/m})(0.1\text{ m})^2}{0.004\text{ m}} = 22.1\text{ pF}$$

(*b*) When the capacitor is filled with a dielectric κ, its capacitance C is increased by the factor κ:

$$C = \kappa C_0 = (2)(22.1\text{ pF}) = 44.2\text{ pF}$$

(*c*) 1. The new capacitance is related to the original charge Q and the new potential difference V:

$$C = \frac{Q}{V}$$

2. The potential difference V between the plates is the sum of the potential difference for the gap plus the potential difference for the dielectric slab:

$$V = V_{\text{gap}} + V_{\text{slab}} = E_{\text{gap}}\left(\tfrac{1}{4}d\right) + E_{\text{slab}}\left(\tfrac{3}{4}d\right)$$

3. The field in the gap just outside the conductor is the original field E_0:

$$E_{\text{gap}} = E_0 = \frac{Q}{\epsilon_0 A}$$

4. The field in the dielectric slab is reduced by the factor κ:

$$E_{\text{slab}} = \frac{E_0}{\kappa}$$

5. Combining the previous two results yields V in terms of κ. Note that the original potential difference is $V_0 = E_0 d$:

$$V = E_0\left(\frac{1}{4}d\right) + \frac{E_0}{\kappa}\left(\frac{3}{4}d\right) = E_0 d\left(\frac{1}{4} + \frac{3}{4\kappa}\right) = V_0\left(\frac{\kappa + 3}{4\kappa}\right)$$

6. Using $C = Q/V$, we find the new capacitance in terms of the original capacitance, $C_0 = Q/V_0$:

$$C = \frac{Q}{V} = \frac{Q}{V_0(\kappa + 3)/4\kappa} = \frac{Q}{V_0}\left(\frac{4\kappa}{\kappa + 3}\right)$$

$$= C_0\left(\frac{4\kappa}{\kappa + 3}\right) = (22.1\text{ pF})\left(\frac{8}{5}\right) = 35.4\text{ pF}$$

Check the Result The absence of a dielectric corresponds to $\kappa = 1$. In this case, our result for the final step in (*c*) would reduce to $C = C_0$ as expected. Suppose that the dielectric slab were a conducting slab. In a conductor, $E = 0$, so according to Equation 25-22, κ for a conductor would equal infinity. For very large κ, the quantity $4\kappa/(\kappa + 3)$ is approximately 4, so the result for the final step in (*c*) approaches $4C_0$. A conducting slab simply extends the capacitor plate, hence the plate separation with the conducting dielectric in place would be $\frac{1}{4}d$. This means that C should be $4C_0$, as it is for very large κ.

Remarks Note that the results of this example are independent of the vertical position of the dielectric or conducting slab between the plates.

Energy Stored in the Presence of a Dielectric

The energy stored in a parallel-plate capacitor with dielectric is

$$U = \tfrac{1}{2}QV = \tfrac{1}{2}CV^2$$

We can express the capacitance C in terms of the area and separation of the plates, and the voltage difference V in terms of the electric field and plate separation, to obtain

$$U = \frac{1}{2}\left(\frac{\epsilon A}{d}\right)(Ed)^2 = \frac{1}{2}\epsilon E^2(Ad)$$

The quantity Ad is the volume between the plates containing the electric field. The energy per unit volume is thus

$$u_e = \tfrac{1}{2}\epsilon E^2 = \tfrac{1}{2}\kappa\epsilon_0 E^2 \qquad \text{25-26}$$

Example 25-9

Two parallel-plate capacitors, each having a capacitance of $C_1 = C_2 = 2\ \mu F$, are connected in parallel across a 12-V battery. Find (*a*) the charge on each capacitor and (*b*) the total energy stored in the capacitors.

The capacitors are then disconnected from the battery and a dielectric of constant $\kappa = 2.5$ is inserted between the plates of the capacitor C_2. After the dielectric is inserted, find (*c*) the potential difference across each capacitor, (*d*) the charge on each capacitor, and (*e*) the total energy stored in the capacitors.

Picture the Problem (*a* and *b*) The charge Q and total energy U can be found for each capacitor from its capacitance C and voltage V. (*c*) After the capacitors are removed from the battery, the total charge on them must remain the same. When the dielectric is inserted into one of the capacitors, its capacitance C_2 changes. The potential across the parallel combination can be found from the total charge and the equivalent capacitance.

(*a*) The charge on each capacitor is found from its capacitance C and voltage V:

$$Q = CV = (2\ \mu F)(12\ V) = 24\ \mu C$$

(*b*) 1. The energy stored in each capacitor is found from its capacitance C and voltage V:

$$U = \tfrac{1}{2}CV^2 = \tfrac{1}{2}(2\ \mu F)(12\ V)^2 = 144\ \mu J$$

2. The total energy is twice that stored in each capacitor:

$$U_{total} = 2U = 288\ \mu J$$

(*c*) 1. The potential across the parallel combination is related to the total charge Q_{total} and the equivalent capacitance C_{eq}:

$$V = \frac{Q_{total}}{C_{eq}}$$

2. The capacitance C_2 of the capacitor with the dielectric is increased by the factor κ. The equivalent capacitance is the sum of the capacitances:

$$C_{eq} = C_1 + C_2 = C_1 + \kappa C_1 = (2\ \mu F) + (2.5)(2\ \mu F) = 2\ \mu F + 5\ \mu F = 7\ \mu F$$

3. The total charge remains 48 μC. Substitute for Q_{total} and C_{eq} to calculate V:

$$V = \frac{Q_{total}}{C_{eq}} = \frac{48\ \mu C}{7\ \mu F} = 6.86\ V$$

(*d*) The charge on each capacitor is again derived from its capacitance and the voltage V:

$$Q_1 = C_1V = (2\ \mu F)(6.86\ V) = 13.7\ \mu C$$

$$Q_2 = C_2V = (5\ \mu F)(6.86\ V) = 34.3\ \mu C$$

(*e*) 1. The energy stored in each capacitor is found from its new capacitance C and new voltage V:

$$U_1 = \tfrac{1}{2}C_1V^2 = \tfrac{1}{2}(2\ \mu F)(6.86\ V)^2 = 47.1\ \mu J$$

$$U_2 = \tfrac{1}{2}C_2V^2 = \tfrac{1}{2}(5\ \mu F)(6.86\ V)^2 = 118\ \mu J$$

2. Add these energies to get the total energy stored:

$$U = U_1 + U_2 = 47.1\ \mu J + 118\ \mu J = 165\ \mu J$$

Check the Result When the dielectric is inserted into one of the capacitors, the field is weakened and the potential difference is lowered. Since the two capacitors are connected in parallel, charge must flow from the other capacitor so that the potential difference is the same across both capacitors. Note that the capacitor with the dielectric has the greater charge, and that when the charges calculated for each capacitor in (*d*) are added, $Q_1 + Q_2 = 13.7\ \mu C + 34.3\ \mu C = 48\ \mu C$, the result is the same as the original sum.

Remarks The total energy of 165 μJ is less than the original energy of 288 μJ. When the dielectric is inserted, it is pulled in and work is done on whatever was holding it. To remove the dielectric, work $W = 288\ \mu\text{J} - 165\ \mu\text{J} = 123\ \mu\text{J}$ must be done, and this work is stored as electrostatic potential energy.

Example 25-10 *try it yourself*

Find (*a*) the charge on each capacitor, and (*b*) the total energy stored in the capacitors of Example 25-9, if the dielectric is inserted into one of the capacitors while the battery is still connected.

Picture the Problem Since the battery is still connected, the potential difference across the capacitors remains 12 V. This condition determines the charge and energy stored in each capacitor. Let subscript 1 refer to the capacitor without the dielectric, and subscript 2 refer to the capacitor with the dielectric.

Cover the column to the right and try these on your own before looking at the answers.

Steps	*Answers*
(*a*) Calculate the charge on each capacitor from $Q = CV$ using the result that $C_1 = 2\ \mu\text{F}$ and $C_2 = 5\ \mu\text{F}$ as found in Example 25-9.	$Q_1 = C_1V = 24\ \mu\text{C}$ $Q_2 = C_2V = 60\ \mu\text{C}$
(*b*) 1. Calculate the energy stored in each capacitor from $U = \frac{1}{2}CV^2$. Check your results by using $U = \frac{1}{2}QV$.	$U_1 = 144\ \mu\text{J}$ $U_2 = 360\ \mu\text{J}$
2. Add your results for U_1 and U_2 to obtain the final energy.	$U_{\text{total}} = 504\ \mu\text{J}$

Remarks Note that Q_2 is two and a half times its value before the dielectric was inserted (since $\kappa = 2.5$). The battery supplies this additional charge in order to maintain a fixed potential difference. Because of the work done by the battery to supply this charge, the total energy of the system is higher with the dielectric in place (504 μJ) than without the dielectric (288 μJ).

25-6 Molecular View of a Dielectric

A dielectric weakens the electric field between the plates of a capacitor because the molecules in the dielectric produce an electric field in a direction opposite to the field produced by the plates. This electric field is due to the electric dipole moments of the molecules in the dielectric.

Although atoms and molecules are electrically neutral, they are affected by electric fields because they contain positive and negative charges that can respond to external fields. We can think of an atom as a very small, positively charged nucleus surrounded by a negatively charged electron cloud. In some atoms and molecules, the electron cloud is spherically symmetric, so its "center of charge" is at the center of the atom or molecule, coinciding with the positive charge. An atom or molecule like this has zero dipole moment and is said to be nonpolar. But in the presence of an external electric field, the positive and negative charge experience forces in opposite directions. The positive and negative charges then separate until the attractive force they exert on each other balances the forces due to the external electric field (Figure 25-17). The molecule is then said to be polarized and it behaves like an electric dipole.

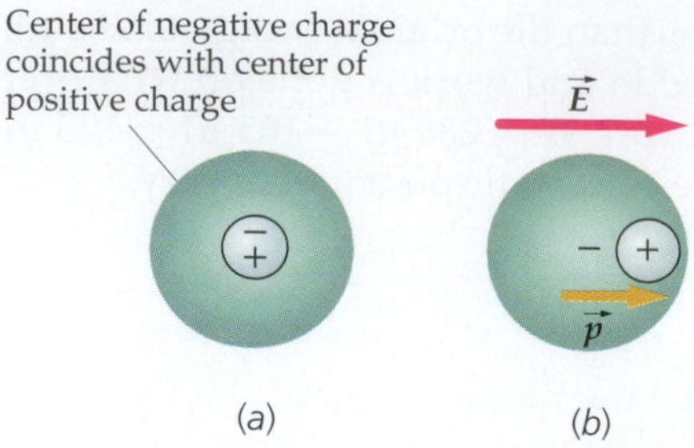

Figure 25-17 Schematic diagrams of the charge distributions of an atom or nonpolar molecule. (*a*) In the absence of an external electric field, the center of positive charge coincides with the center of negative charge. (*b*) In the presence of an external electric field, the centers of positive and negative charge are displaced, producing an induced dipole moment in the direction of the external field.

In some molecules (for example, HCl and H_2O), the centers of positive and negative charge do not coincide even in the absence of an external electric field. As we noted in Chapter 22, these polar molecules have a permanent electric dipole moment.

When a dielectric is placed in the field of a capacitor, its molecules are polarized such that there is a net dipole moment parallel to the field. If the molecules are polar, their dipole moments, originally oriented at random, tend to become aligned due to the torque exerted by the field.* If the molecules are nonpolar, the field induces dipole moments that are parallel to the field. In either case, the molecules in the dielectric are polarized in the direction of the external field (Figure 25-18).

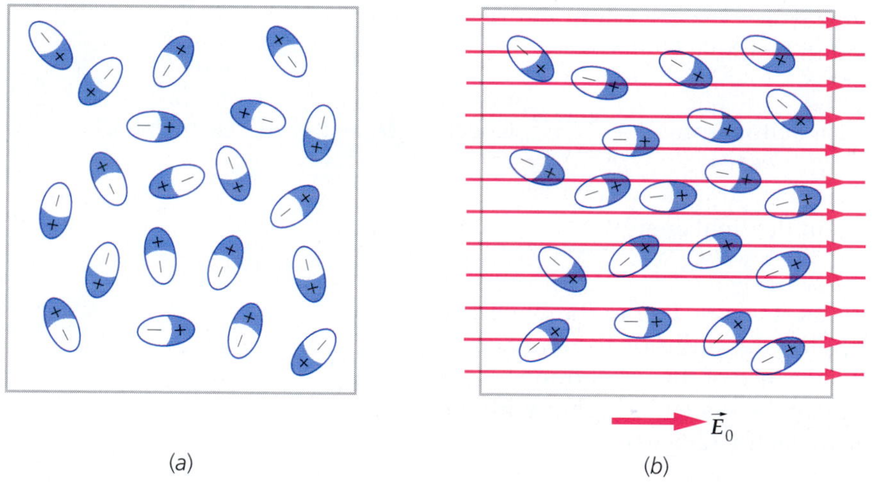

Figure 25-18 (*a*) The randomly oriented electric dipoles of a polar dielectric in the absence of an external electric field. (*b*) In the presence of an external electric field, the dipoles are partially aligned parallel to the field.

The net effect of the polarization of a homogeneous dielectric in a parallel-plate capacitor is the creation of a surface charge on the dielectric faces near the plates, as shown in Figure 25-19. The surface charge on the dielectric is called a **bound charge** because it is bound to the molecules of the dielectric and cannot move about like the free charge on the conducting capacitor plates. This bound charge produces an electric field opposite in direction to that produced by the free charge on the conductors. Thus, the net electric field between the plates is reduced, as illustrated in Figure 25-20.

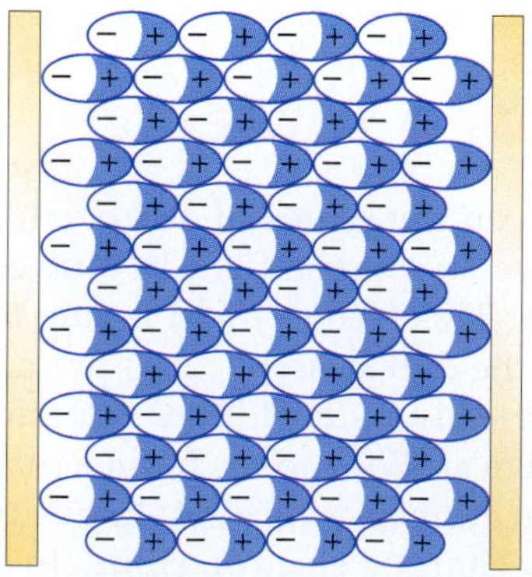

Figure 25-19 When a dielectric is placed between the plates of a capacitor, the electric field of the capacitor polarizes the molecules of the dielectric. The result is a bound charge on the surface of the dielectric that produces its own electric field; this field opposes the external field. The electric field between the plates is thus weakened by the dielectric.

* The degree of alignment depends on the external field and on the temperature. It is approximately proportional to pE/kT, where pE is the maximum energy of a dipole in a field E, and kT is the characteristic thermal energy.

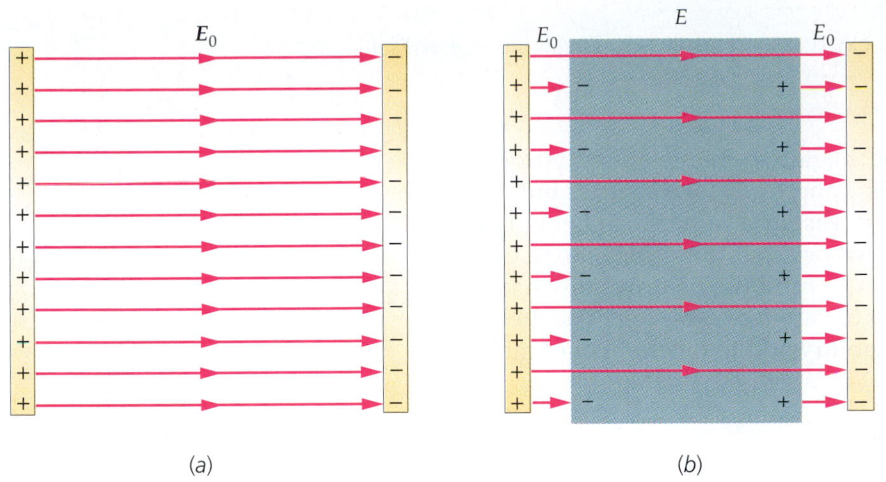

Figure 25-20 The electric field between the plates of a capacitor (*a*) with no dielectric and (*b*) with a dielectric. The surface charge on the dielectric weakens the original field between the plates.

Example 25-11

A hydrogen atom consists of a proton nucleus of charge $+e$ and an electron of charge $-e$. The charge distribution of the atom is spherically symmetric so the atom is nonpolar. Consider a model in which the hydrogen atom consists of a positive point charge $+e$ at the center of a uniformly charged spherical cloud of radius R and total charge $-e$. Show that when such an atom is placed in a uniform external electric field $\vec{E}$, the induced dipole moment is proportional to E, that is, $p = \alpha E$, where α is called the polarizability.

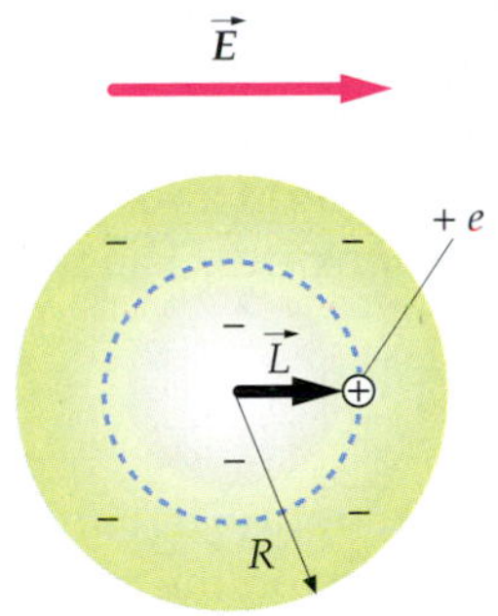

Figure 25-21

Picture the Problem In the external field, the positive charge is displaced from the center of the cloud by an amount L such that the force exerted by the field $e\vec{E}$ is balanced by the force exerted by the negative cloud $e\vec{E}'$, where $\vec{E}'$ is the field due to the cloud (Figure 25-21). We use Gauss's law to find E', and then we calculate the induced dipole moment $\vec{p} = e\vec{L}$.

1. Write the magnitude of the induced dipole moment in terms of e and L:

$$p = eL$$

2. We can find L by calculating the field E' due to the negatively charged cloud at a distance L from the center. We use Gauss's law to compute E'. Choose a spherical Gaussian surface of radius L concentric with the cloud. Then E' is constant on this surface:

$$\phi_{\text{net}} = \oint E_{\text{n}}\, dA = \frac{1}{\epsilon_0} Q_{\text{inside}}$$

$$E'(4\pi L^2) = \frac{1}{\epsilon_0} Q_{\text{inside}}$$

$$E' = \frac{1}{4\pi\epsilon_0 L^2} Q_{\text{inside}}$$

3. The charge inside the sphere of radius L equals the charge density times the volume:

$$Q_{\text{inside}} = \rho \frac{4}{3}\pi L^3 = \frac{-e}{\frac{4}{3}\pi R^3}\frac{4}{3}\pi L^3 = -e\frac{L^3}{R^3}$$

4. Substitute this value of Q_{inside} to calculate E':

$$E' = \frac{1}{4\pi\epsilon_0 L^2} Q_{\text{inside}} = \frac{1}{4\pi\epsilon_0 L^2}\left(-e\frac{L^3}{R^3}\right) = -e\frac{L}{4\pi\epsilon_0 R^3}$$

5. Solve for L:

$$L = \frac{-\,4\pi\epsilon_0 R^3 E'}{e}$$

6. E' is negative because it points inward on the Gaussian surface. At the positive charge, E' points to the left, so $E' = -E$:

$$E' = -E$$

7. Substitute these results for L and E' to express p in terms of the external field E:

$$p = eL = -4\pi\epsilon_0 R^3 E' = 4\pi\epsilon_0 R^3 E$$

Remarks The charge distribution of the negative charge in a hydrogen atom, obtained from quantum theory, is spherically symmetric, but the charge density decreases exponentially with distance rather than being uniform. Nevertheless, the above calculation shows that the dipole moment is indeed proportional to the external field $p = \alpha E$, and the polarizability α is of the order of $4\pi\epsilon_0 R^3$ where R is the radius of the atom or molecule. The dielectric constant κ can be related to the polarizability and to the number of molecules per unit volume.

Magnitude of the Bound Charge

The bound charge density σ_b on the surfaces of the dielectric is related to the dielectric constant κ and to the free charge density σ_f on the plates. Consider a dielectric slab between the plates of a parallel-plate capacitor as shown in Figure 25-22. If the dielectric is a very thin slab between plates that are close together, the electric field inside the dielectric slab due to the bound charge densities $+\sigma_b$ on the right and $-\sigma_b$ on the left is just the field due to two infinite-plane charge densities. The field E_b thus has the magnitude

$$E_b = \frac{\sigma_b}{\epsilon_0}$$

This field is directed to the left and subtracts from the electric field E_0 due to the free charge density on the capacitor plates, which has the magnitude

$$E_0 = \frac{\sigma_f}{\epsilon_0}$$

The magnitude of the net field $E = E_0/\kappa$ is the difference between these magnitudes:

$$E = E_0 - E_b = \frac{E_0}{\kappa}$$

or

$$E_b = E_0\left(1 - \frac{1}{\kappa}\right) = \frac{\kappa - 1}{\kappa}E_0$$

Writing σ_b/ϵ_0 for E_b and σ_f/ϵ_0 for E_0, we obtain

$$\sigma_b = \frac{\kappa - 1}{\kappa}\sigma_f = \left(1 - \frac{1}{\kappa}\right)\sigma_f \qquad 25\text{-}27$$

The bound charge density σ_b is always less than the free charge density σ_f on the capacitor plates, and it is zero if $\kappa = 1$, which is the case when there is no dielectric. For a conducting slab, $\kappa = \infty$ and $\sigma_b = \sigma_f$.

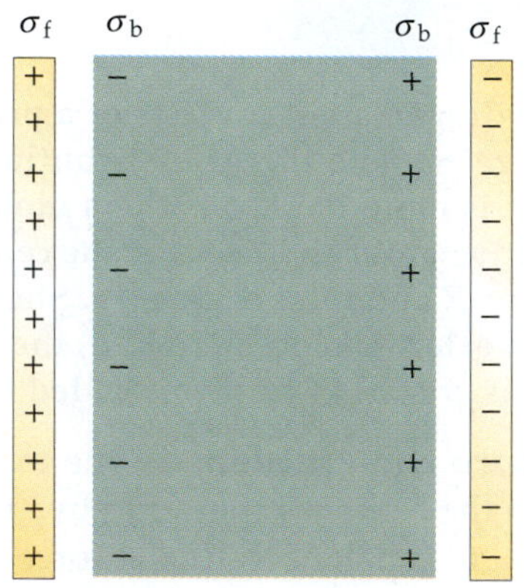

Figure 25-22 Parallel-plate capacitor with dielectric slab between the plates. If the plates are closely spaced, each of the surface charges can be considered an infinite plane charge. The electric field due to the free charge on the plates is directed to the right and has a magnitude $E_0 = \sigma_f/\epsilon_0$. That due to the bound charge is directed to the left and has a magnitude $E_b = \sigma_b/\epsilon_0$.

The Piezoelectric Effect

In certain crystals that contain polar molecules, such as quartz, tourmaline, and topaz, a mechanical stress applied to the crystal produces polarization of the molecules. This is known as the **piezoelectric effect**. The polarization of

optional

the stressed crystal causes a potential difference across the crystal, which can be used to produce an electric current. Piezoelectric crystals are used in transducers such as microphones, phonograph pickups, and vibration-sensing devices for converting mechanical strain into electrical signals. The converse piezoelectric effect, in which a voltage applied to such a crystal induces mechanical strain (deformation), is used in headphones and many other devices.

Because the natural frequency of vibration of quartz is in the range of radio frequencies, and because its resonance curve is very sharp,* it is used extensively to stabilize radio-frequency oscillators and to make accurate clocks.

* Resonance in ac circuits, which will be discussed in Chapter 31, is analogous to mechanical resonance, which was discussed in Chapter 14.

Summary

Topic	Remarks and Relevant Equations	
1. Electrostatic Potential Energy	The electrostatic potential energy of a system of charges is the work needed to bring the charges from an infinite separation to their final positions.	
Of point charges	$U = \frac{1}{2}\sum_{i=1}^{n} q_i V_i$	25-2
Of a conductor with charge Q at potential V	$U = \frac{1}{2}QV$	25-3
Of a system of conductors	$U = \frac{1}{2}\sum_{i=1}^{n} Q_i V_i$	25-4
Energy stored in a capacitor	$U = \frac{1}{2}\frac{Q^2}{C} = \frac{1}{2}QV = \frac{1}{2}CV^2$	25-12
Energy density in an electric field	$u_e = \frac{\text{energy}}{\text{volume}} = \frac{1}{2}\epsilon E^2 = \frac{1}{2}\kappa\epsilon_0 E^2$	25-26
2. Capacitor	A capacitor is a device for storing charge and energy. It consists of two conductors insulated from each other that carry equal and opposite charges.	
3. Capacitance	$C = \frac{Q}{V}$	25-5
Isolated conductor	Q is its total charge, V is its potential.	
Capacitor	Q is the magnitude of charge on either conductor, V is the potential difference.	
Isolated spherical conductor	$C = 4\pi\epsilon_0 R$	25-6

Parallel-plate capacitor	$C = \frac{\epsilon_0 A}{d}$	**25-10**
Cylindrical capacitor	$C = \frac{2\pi\epsilon_0 L}{\ln(r_2/r_1)}$	**25-11**

4. Equivalent Capacitance

Parallel capacitors	When capacitors are in parallel, the voltage is the same across each one, and the capacitances add: $C_{eq} = C_1 + C_2 + C_3 + \cdots$	**25-17**
Series capacitors	When capacitors are in series, the charge stored is the same on each one, and the voltage drops add: $\frac{1}{C_{eq}} = \frac{1}{C_1} + \frac{1}{C_2} + \frac{1}{C_3} + \cdots$	**25-21**

5. Dielectric

Macroscopic behavior	A nonconducting material is called a dielectric. When a dielectric is inserted between the plates of a capacitor, the electric field within the dielectric is weakened and the capacitance is thereby increased by the factor κ, the dielectric constant.	
Microscopic picture	The field in the dielectric of a capacitor is weakened because the dipole moments of the molecules (either preexisting or induced) tend to align with the field and thereby produce an electric field that opposes the external field. The aligned dipole moment of the dielectric is proportional to the external field.	
Electric field	$E = \frac{E_0}{\kappa}$	**25-22**
Capacitance	$C = \kappa C_0$	**25-23**
Permittivity ϵ	$\epsilon = \kappa\epsilon_0$	**25-25**
Uses of a dielectric	1. Increases capacitance 2. Increases dielectric strength 3. Physically separates conductors	

6. Piezoelectric Effect (optional) In certain crystals containing polar molecules, a mechanical stress polarizes the molecules, inducing a voltage across the crystal. Conversely, an applied voltage induces mechanical strain (deformation) in the crystal.

Problem-Solving Guide

1. Begin by drawing a neat diagram that includes the important features of the problem. Be sure to show the correct parallel or series connections between capacitors. When appropriate, indicate the charge on the plates of a capacitor and any dielectrics between the plates.
2. To find the capacitance of a capacitor, place a charge $+Q$ on one conductor and $-Q$ on the other. Then compute the electric field between the conductors and use it to find the potential difference V between the conductors. The capacitance is then $C = Q/V$.

Summary of Worked Examples

Type of Calculation	Procedure and Relevant Examples	
1. Energy		
Find the electrostatic potential energy of a system of point charges.	Use $U = \frac{1}{2}\Sigma\, q_i V_i$.	**Example 25-1**
2. Single Capacitors		
Find the capacitance of a parallel-plate capacitor.	Use $C = \epsilon_0 A/d$.	**Example 25-2**
Find the capacitance of any capacitor.	Place charges $+Q$ and $-Q$ on the two conductors and find the electric field E between them. Then calculate the potential difference by integrating $dV = -\vec{E}\, d\ell$ from one conductor to the other.	**Example 25-3**
Find the charge on a capacitor.	The charge is given by $Q = CV$.	**Example 25-2**
Find the energy stored in a capacitor.	Use $U = \frac{1}{2}Q^2/C = \frac{1}{2}QV = \frac{1}{2}CV^2$.	**Examples 25-4, 25-8, 25-6**
3. Combinations of Capacitors		
Calculate the equivalent capacitance.	Start by combining the smallest parallel or series units in the network, using $C_{eq} = C_1 + C_2 + \cdots$ (parallel) and $1/C_{eq} = 1/C_1 + 1/C_2 + \cdots$ (series). Keep replacing parallel and series units until the network is reduced to a single capacitor.	**Examples 25-5, 25-6, 25-7**
4. Capacitors with Dielectrics		
Determine the capacitance of a capacitor filled with a dielectric.	The capacitance is $C = \kappa C_0$, where C_0 is the capacitance without the dielectric.	**Examples 25-8, 25-9**
Calculate the capacitance of a capacitor partially filled a dielectric, or filled with more than one type of dielectric.	Find the potential difference V for a given charge Q, then use $C = Q/V$.	**Example 25-8**

Problems

Conceptual Problems

Problems from Optional and Exploring sections

In a few problems, you are given more data than you actually need; in a few other problems, you are required to supply data from your general knowledge, outside sources, or informed estimates.

• Single-concept, single-step, relatively easy

•• Intermediate-level, may require synthesis of concepts

••• Challenging, for advanced students

Electrostatic Potential Energy

1 • Three point charges are on the x axis: q_1 at the origin, q_2 at $x = 3$ m, and q_3 at $x = 6$ m. Find the electrostatic potential energy for (*a*) $q_1 = q_2 = q_3 = 2\ \mu$C, (*b*) $q_1 = q_2 = 2\ \mu$C and $q_3 = -2\ \mu$C, and (*c*) $q_1 = q_3 = 2\ \mu$C and $q_2 = -2\ \mu$C.

2 • Point charges q_1, q_2, and q_3 are at the corners of an equilateral triangle of side 2.5 m. Find the electrostatic potential energy of this charge distribution if (*a*) $q_1 = q_2 = q_3 = 4.2\ \mu$C, (*b*) $q_1 = q_2 = 4.2\ \mu$C and $q_3 = -4.2\ \mu$C, (*c*) $q_1 = q_2 = -4.2\ \mu$C and $q_3 = +4.2\ \mu$C.

3 • What is the electrostatic potential energy of an isolated spherical conductor of radius 10 cm that is charged to 2 kV?

4 •• Four point charges of magnitude 2 μC are at the corners of a square of side 4 m. Find the electrostatic potential energy if (*a*) all the charges are negative, (*b*) three of the charges are positive and one is negative, and (*c*) two are positive and two are negative.

5 •• Four charges are at the corners of a square centered at the origin as follows: q at $(-a, +a)$; $2q$ at (a, a); $-3q$ at

$(a, -a)$; and $6q$ at $(-a, -a)$. A fifth charge $+q$ is placed at the origin and released from rest. Find its speed when it is a great distance from the origin.

6 •• Four identical particles each with charge Q are at the corners of a square of side L. The particles are released one at a time proceeding clockwise around the square. Each particle is allowed to reach its final speed a long distance from the square before the next particle is released. What is the final kinetic energy of (*a*) the first particle released, (*b*) the second particle released, (*c*) the third particle released, and (*d*) the fourth particle released? (*e*) What is the final kinetic energy of each particle if the four particles are released simultaneously? Does this depend on whether or not the particles have identical masses?

Capacitance

7 • If the voltage across a parallel-plate capacitor is doubled, its capacitance

(*a*) doubles.
(*b*) drops by half.
(*c*) remains the same.

8 • If the charge on an isolated spherical conductor is doubled, its capacitance

(*a*) doubles.
(*b*) drops by half.
(*c*) remains the same.

9 • An isolated spherical conductor of radius 10 cm is charged to 2 kV. (*a*) How much charge is on the conductor? (*b*) What is the capacitance of the sphere? (*c*) How does the capacitance change if the sphere is charged to 6 kV?

10 • A capacitor has a charge of 30 μC. The potential difference between the conductors is 400 V. What is the capacitance?

11 • (*a*) If a parallel-plate capacitor has a 0.15-mm separation, what must its area be for it to have a capacitance of 1 F? (*b*) If the plates are square, what is the length of their sides?

Storage of Electrical Energy

12 • True or false: The electrostatic energy per unit volume at some point is proportional to the square of the electric field at that point.

13 • If the potential difference of a parallel-plate capacitor is doubled by changing the plate separation without changing the charge, by what factor does its stored electric energy change?

14 •• Half the charge is removed from a capacitor without changing its capacitance. What fraction of its stored energy is removed along with the charge?

15 •• A parallel-plate air capacitor is connected to a constant-voltage battery. If the separation between the capacitor plates is doubled while the capacitor remains connected to the battery, the energy stored in the capacitor

(*a*) quadruples.
(*b*) doubles.
(*c*) remains unchanged.
(*d*) drops to half its previous value.
(*e*) drops to one-fourth its previous value.

16 •• If the capacitor of Problem 15 is disconnected from the battery before the separation between the plates is doubled, the energy stored in the capacitor upon separation of the plates

(*a*) quadruples.
(*b*) doubles.
(*c*) remains unchanged.
(*d*) drops to half its previous value.
(*e*) drops to one-fourth its previous value.

17 • (*a*) A 3-μF capacitor is charged to 100 V. How much energy is stored in the capacitor? (*b*) How much additional energy is required to charge the capacitor from 100 to 200 V?

18 • A 10-μF capacitor is charged to $Q = 4\ \mu$C. (*a*) How much energy is stored in the capacitor? (*b*) If half the charge is removed, how much energy remains?

19 • (*a*) Find the energy stored in a 20-pF capacitor when it is charged to 5 μC. (*b*) How much additional energy is required to increase the charge from 5 to 10 μC?

20 • Find the energy per unit volume in an electric field that is equal to 3 MV/m, the dielectric strength of air.

21 • A parallel-plate capacitor with a plate area of 2 m^2 and a separation of 1.0 mm is charged to 100 V. (*a*) What is the electric field between the plates? (*b*) What is the energy per unit volume in the space between the plates? (*c*) Find the total energy by multiplying your answer to part (*b*) by the total volume between the plates. (*d*) Find the capacitance C. (*e*) Calculate the total energy from $U = \frac{1}{2}CV^2$, and compare your answer with your result for part (*c*).

22 •• Energy prospectors from a distant planet are inspecting the earth to decide if its electrical energy resources are worth stealing. Measurements reveal that earth's electric field extends upward for 1000 m and has an average magnitude of 200 V/m. Estimate the electrical energy stored in the atmosphere. (*Hint:* You may treat the atmosphere as a flat slab with an area equal to the surface area of the earth. Why?)

23 •• A parallel-plate capacitor with plates of area 500 cm^2 is charged to a potential difference V and is then disconnected from the voltage source. When the plates are moved 0.4 cm farther apart, the voltage between the plates increases by 100 V. (*a*) What is the charge Q on the positive plate of the capacitor? (*b*) How much does the energy stored in the capacitor increase due to the movement of the plates?

24 ••• A ball of charge of radius R has a uniform charge density ρ and a total charge $Q = \frac{4}{3}\pi R^3 \rho$. (*a*) Find the electrostatic energy density at a distance r from the center of the ball for $r < R$ and for $r > R$. (*b*) Find the energy in a spherical shell of volume $4\pi r^2\, dr$ for both $r < R$ and $r > R$. (*c*) Compute the total electrostatic energy by integrating your expressions from part (*b*), and show that your result can be written $U = \frac{3}{5}kQ^2/R$. Explain why this result is greater than that for a spherical conductor of radius R carrying a total charge Q.

Combinations of Capacitors

25 • True or false:

(*a*) The equivalent capacitance of two capacitors in parallel equals the sum of the individual capacitances.
(*b*) The equivalent capacitance of two capacitors in series is less than the capacitance of either capacitor alone.

26 •• Two initially uncharged capacitors of capacitance C_0 and $2C_0$, respectively, are connected in series across a battery. Which of the following is true?

(*a*) The capacitor $2C_0$ carries twice the charge of the other capacitor.
(*b*) The voltage across each capacitor is the same.
(*c*) The energy stored by each capacitor is the same.
(*d*) None of the above statements is correct.

27 • (*a*) How many 1.0-μF capacitors connected in parallel would it take to store a total charge of 1 mC with a potential difference of 10 V across each capacitor? (*b*) What would be the potential difference across the combination? (*c*) If the number of 1.0-μF capacitors found in part (*a*) is connected in series and the potential difference across each is 10 V, find the charge on each and the potential difference across the combination.

28 • A 3.0-μF capacitor and a 6.0-μF capacitor are connected in series, and the combination is connected in parallel with an 8.0-μF capacitor. What is the equivalent capacitance of this combination?

29 • Three capacitors are connected in a triangular network as shown in Figure 25-23. Find the equivalent capacitance across terminals *a* and *c*.

Figure 25-23 Problem 29

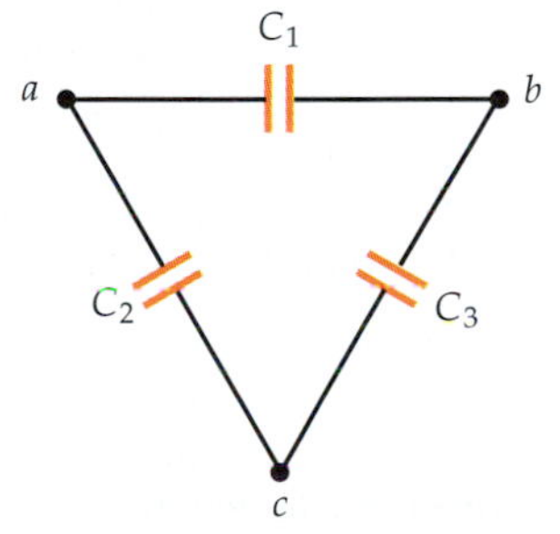

30 • A 10.0-μF capacitor and a 20.0-μF capacitor are connected in parallel across a 6.0-V battery. (*a*) What is the equivalent capacitance of this combination? (*b*) What is the potential difference across each capacitor? (*c*) Find the charge on each capacitor.

31 •• A 10.0-μF capacitor is connected in series with a 20.0-μF capacitor across a 6.0-V battery. (*a*) Find the charge on each capacitor. (*b*) Find the potential difference across each capacitor.

32 •• Three identical capacitors are connected so that their maximum equivalent capacitance is 15 μF. (*a*) Describe how the capacitors are combined. (*b*) There are three other ways to combine all three capacitors in a circuit. What are the equivalent capacitances for each arrangement?

33 •• For the circuit shown in Figure 25-24, find (*a*) the total equivalent capacitance between the terminals, (*b*) the charge stored on each capacitor, and (*c*) the total stored energy.

Figure 25-24
Problem 33

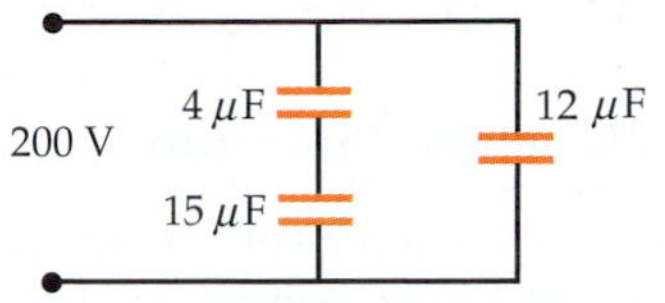

34 •• (*a*) Show that the equivalent capacitance of two capacitors in series can be written

$$C_{eq} = \frac{C_1C_2}{C_1 + C_2}$$

(*b*) Use this expression to show that $C_{eq} < C_1$ and $C_{eq} < C_2$. (*c*) Show that the correct expression for the equivalent capacitance of three capacitors in series is

$$C_{eq} = \frac{C_1C_2C_3}{C_1C_2 + C_2C_3 + C_1C_3}$$

35 •• For the circuit shown in Figure 25-25, find (*a*) the total equivalent capacitance between the terminals, (*b*) the charge stored on each capacitor, and (*c*) the total stored energy.

Figure 25-25
Problem 35

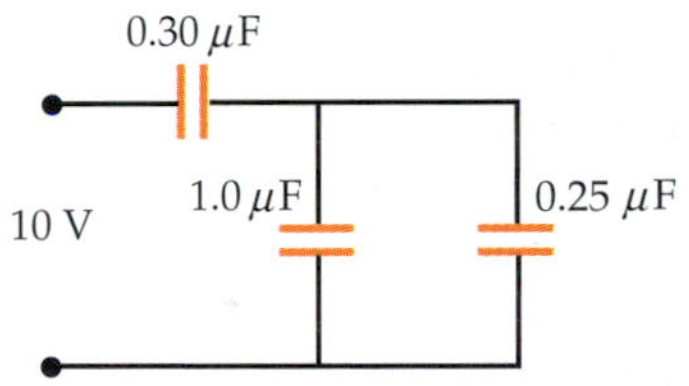

36 •• Five identical capacitors of capacitance C_0 are connected in a bridge network as shown in Figure 25-26. (*a*) What is the equivalent capacitance between points *a* and *b*? (*b*) Find the equivalent capacitance if the capacitance between *a* and *b* is changed to $10C_0$.

Figure 25-26 Problem 36

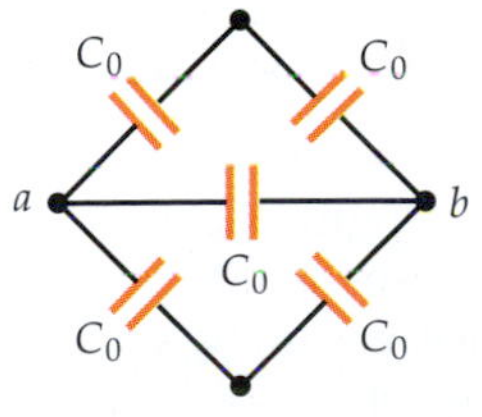

37 •• In Figure 25-27, $C_1 = 2\ \mu$F, $C_2 = 6\ \mu$F, and $C_3 = 3.5\ \mu$F. (*a*) Find the equivalent capacitance of this combination. (*b*) If the breakdown voltages of the individual capacitors are $V_1 = 100$ V, $V_2 = 50$ V, and $V_3 = 400$ V, what maximum voltage can be placed across points *a* and *b*?

Figure 25-27 Problem 37

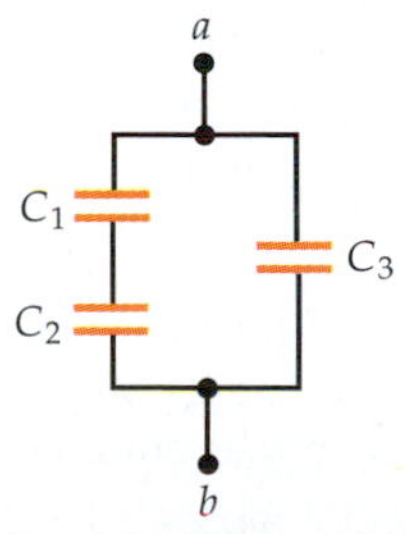

38 •• Design a network of capacitors that has a capacitance of 2 μF and breakdown voltage of 400 V using only

2-μF capacitors that have individual breakdown voltages of 100 V.

39 •• Find all the different possible equivalent capacitances that can be obtained using a 1.0-, a 2.0-, and a 4.0-μF capacitor in any combination that includes all three or any two of the capacitors.

Parallel-Plate Capacitors

40 • A parallel-plate capacitor has a capacitance of 2.0 μF and a plate separation of 1.6 mm. (*a*) What is the maximum potential difference between the plates such that dielectric breakdown of the air between the plates does not occur? (Use $E_{max} = 3$ MV/m.) (*b*) How much charge is stored at this maximum potential difference?

41 • An electric field of 2×10^4 V/m exists between the plates of a circular parallel-plate capacitor that has a plate separation of 2 mm. (*a*) What is the voltage across the capacitor? (*b*) What plate radius is required if the stored charge is 10 μC?

42 •• A parallel-plate, air-gap capacitor has a capacitance of 0.14 μF. The plates are 0.5 mm apart. (*a*) What is the area of each plate? (*b*) What is the potential difference if the capacitor is charged to 3.2 μC? (*c*) What is the stored energy? (*d*) How much charge can the capacitor carry before dielectric breakdown of the air between the plates occurs?

43 •• Design a 0.1-μF parallel-plate capacitor with air between the plates that can be charged to a maximum potential difference of 1000 V. (*a*) What is the minimum possible separation between the plates? (*b*) What minimum area must the plates of the capacitor have?

Cylindrical Capacitors

44 • A coaxial communications cable connecting two cities has an inner radius of 0.8 mm and an outer radius of 6 mm. Its length is 8×10^5 m (about 500 mi). Treat this cable as a cylindrical capacitor and calculate its capacitance.

45 • A Geiger tube consists of a wire of radius 0.2 mm and length 12 cm and a coaxial cylindrical shell conductor of the same length and a radius of 1.5 cm. (*a*) Find the capacitance, assuming that the gas in the tube has a dielectric constant of 1. (*b*) Find the charge per unit length on the wire when the potential difference between the wire and shell is 1.2 kV.

46 •• A cylindrical capacitor consists of a long wire of radius R_1 and length L with a charge $+Q$ and a concentric outer cylindrical shell of radius R_2, length L, and charge $-Q$. (*a*) Find the electric field and energy density at any point in space. (*b*) How much energy resides in a cylindrical shell between the conductors of radius r, thickness dr, and volume $2\pi rL\,dr$? (*c*) Integrate your expression from part (*b*) to find the total energy stored in the capacitor, and compare your result with that obtained using $U = \frac{1}{2}CV^2$.

47 ••• Three concentric thin conducting cylindrical shells have radii of 0.2, 0.5, and 0.8 cm. The space between the shells is filled with air. The innermost and outermost cylinders are connected together. Find the capacitance per unit length of this system.

Spherical Capacitors

48 •• A spherical capacitor consists of two thin concentric spherical shells of radii R_1 and R_2. (*a*) Show that the capacitance is given by $C = 4\pi\epsilon_0 R_1R_2/(R_2 - R_1)$. (*b*) Show that when the radii of the shells are nearly equal, the capacitance is given approximately by the expression for the capacitance of a parallel-plate capacitor, $C = \epsilon_0 A/d$, where A is the area of the sphere and $d = R_2 - R_1$.

49 •• A spherical capacitor has an inner sphere of radius R_1 with a charge of $+Q$ and an outer concentric spherical shell of radius R_2 with a charge of $-Q$. (*a*) Find the electric field and the energy density at any point in space. (*b*) Calculate the energy in the electrostatic field in a spherical shell of radius r, thickness dr, and volume $4\pi r^2\,dr$ between the conductors? (*c*) Integrate your expression from part (*b*) to find the total energy stored in the capacitor, and compare your result with that obtained using $U = \frac{1}{2}QV$.

50 ••• A spherical shell of radius R carries a charge Q distributed uniformly over its surface. Find the radius r of the sphere such that half the total electrostatic field energy of the system is contained within that sphere.

51 ••• Repeat Problem 50 if the charge Q resides not on a spherical shell but is distributed uniformly throughout a spherical volume of radius R. (See Problem 24.)

Disconnected and Reconnected Capacitors

52 •• A 2.0-μF capacitor is charged to a potential difference of 12.0 V. The wires connecting the capacitor to the battery are then disconnected from the battery and connected across a second, initially uncharged, capacitor. The potential difference across the 2.0-μF capacitor then drops to 4 V. What is the capacitance of the second capacitor?

53 •• A 100-pF capacitor and a 400-pF capacitor are both charged to 2.0 kV. They are then disconnected from the voltage source and are connected together, positive plate to positive plate and negative plate to negative plate. (*a*) Find the resulting potential difference across each capacitor. (*b*) Find the energy lost when the connections are made.

54 •• Two capacitors $C_1 = 4$ μF and $C_2 = 12$ μF are connected in series across a 12-V battery. They are carefully disconnected so that they are not discharged and are reconnected to each other with positive plate to positive plate and negative plate to negative plate. (*a*) Find the potential difference across each capacitor after they are connected. (*b*) Find the initial and final energy stored in the capacitors.

55 •• A 1.2-μF capacitor is charged to 30 V. After charging, the capacitor is disconnected from the voltage source and is connected to another uncharged capacitor. The final voltage is 10 V. (*a*) What is the capacitance of the other capacitor? (*b*) How much energy was lost when the connection was made?

56 •• Work Problem 53 if the capacitors are connected positive plate to negative plate after they have been charged to 2.0 kV.

57 •• Work Problem 54 if the two capacitors are first connected in parallel across the 12-V battery and are then connected, with the positive plate of each capacitor connected to the negative plate of the other.

58 •• A 20-pF capacitor is charged to 3.0 kV and then removed from the battery and connected to an uncharged 50-pF capacitor. (*a*) What is the new charge on each capacitor? (*b*) Find the initial energy stored in the 20-pF capacitor and the final energy stored in the two capacitors. Is electrostatic potential energy gained or lost when the two capacitors are connected?

59 ••• A parallel combination of three capacitors, $C_1 = 2\ \mu F$, $C_2 = 4\ \mu F$, and $C_3 = 6\ \mu F$, is charged with a 200-V source. The capacitors are then disconnected from both the voltage source and each other and are reconnected positive plates to negative plates as shown in Figure 25-28. (*a*) What is the voltage across each capacitor with switches S_1 and S_2 closed but switch S_3 open? (*b*) After switch S_3 is closed, what is the final charge on each capacitor? (*c*) Give the voltage across each capacitor after switch S_3 is closed.

Figure 25-28 Problem 59

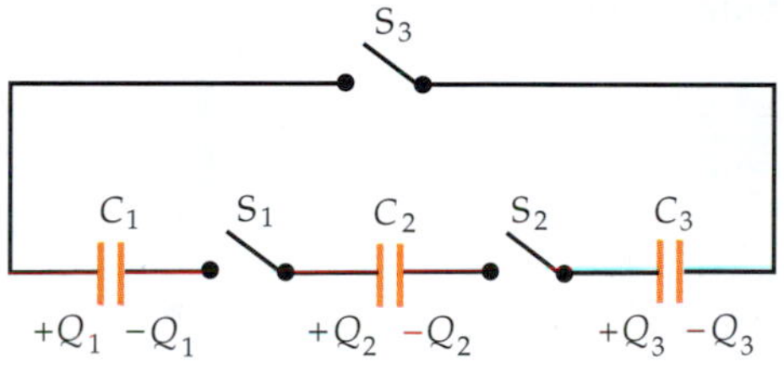

Dielectrics

60 • True or false: A dielectric inserted into a capacitor increases the capacitance.

61 • A parallel-plate capacitor is made by placing polyethylene ($\kappa = 2.3$) between two sheets of aluminum foil. The area of each sheet is 400 cm^2, and the thickness of the polyethylene is 0.3 mm. Find the capacitance.

62 •• Suppose the Geiger tube of Problem 45 is filled with a gas of dielectric constant $\kappa = 1.8$ and breakdown field of 2×10^6 V/m. (*a*) What is the maximum potential difference that can be maintained between the wire and shell? (*b*) What is the charge per unit length on the wire?

63 •• Repeat Problem 49 with the space between the two spherical shells filled with a dielectric of dielectric constant κ.

64 •• A certain dielectric with a dielectric constant $\kappa = 24$ can withstand an electric field of 4×10^7 V/m. Suppose we want to use this dielectric to construct a 0.1-μF capacitor that can withstand a potential difference of 2000 V. (*a*) What is the minimum plate separation? (*b*) What must the area of the plates be?

65 •• A parallel-plate capacitor has plates separated by a distance *s*. The space between the plates is filled with two dielectrics, one of thickness $\frac{1}{4}s$ and dielectric constant κ_1, the other with thickness $\frac{3}{4}s$ and dielectric constant κ_2. Find the capacitance of this capacitor in terms of C_0, the capacitance with no dielectrics.

66 •• A parallel-plate capacitor with no dielectric has a capacitance C_0. If the separation between the plates is *d*, and a slab with dielectric constant κ and thickness $t < d$ is placed in the capacitor, find the new capacitance.

67 •• The membrane of the axon of a nerve cell is a thin cylindrical shell of radius $r = 10^{-5}$ m, length $L = 0.1$ m, and thickness $d = 10^{-8}$ m. The membrane has a positive charge on one side and a negative charge on the other, and acts as a parallel-plate capacitor of area $A = 2\pi rL$ and separation *d*. Its dielectric constant is about $\kappa = 3$. (*a*) Find the capacitance of the membrane. If the potential difference across the membrane is 70 mV, find (*b*) the charge on each side of the membrane, and (*c*) the electric field through the membrane.

68 •• What is the dielectric constant of a dielectric on which the induced bound charge density is (*a*) 80% of the free charge density on the plates of a capacitor filled by the dielectric, (*b*) 20% of the free charge density, and (*c*) 98% of the free charge density?

69 •• Two parallel plates have charges *Q* and $-Q$. When the space between the plates is devoid of matter, the electric field is 2.5×10^5 V/m. When the space is filled with a certain dielectric, the field is reduced to 1.2×10^5 V/m. (*a*) What is the dielectric constant of the dielectric? (*b*) If $Q = 10$ nC, what is the area of the plates? (*c*) What is the total induced charge on either face of the dielectric?

70 •• Find the capacitance of the parallel-plate capacitor shown in Figure 25-29.

Figure 25-29 Problem 70

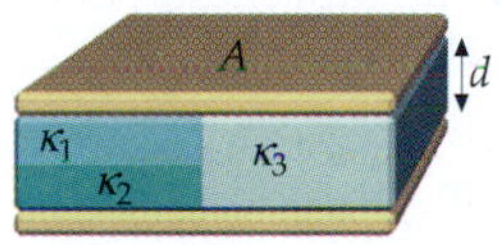

71 •• A parallel-plate capacitor has plates of area 600 cm^2 and a separation of 4 mm. The capacitor is charged to 100 V and is then disconnected from the battery. (*a*) Find the electric field E_0 and the electrostatic energy *U*. A dielectric of constant $\kappa = 4$ is then inserted, completely filling the space between the plates. Find (*b*) the new electric field *E*, (*c*) the potential difference *V*, and (*d*) the new electrostatic energy.

72 ••• A parallel-plate capacitor is constructed using a dielectric whose constant varies with position. The plates have area *A*. The bottom plate is at $y = 0$ and the top plate is at $y = y_0$. The dielectric constant is given as a function of *y* according to $\kappa = 1 + (3/y_0)\,y$. (*a*) What is the capacitance? (*b*) Find σ_b/σ_f on the surfaces of the dielectric. (*c*) Use Gauss's law to find the induced volume charge density $\rho(y)$ within this dielectric. (*d*) Integrate the expression for the volume charge density found in (*c*) over the dielectric, and show that the

total induced bound charge, including that on the surfaces, is zero.

General Problems

73 • True or false:

(*a*) The capacitance of a capacitor is defined as the total amount of charge it can hold.
(*b*) The capacitance of a parallel-plate capacitor depends on the voltage difference between the plates.
(*c*) The capacitance of a parallel-plate capacitor is proportional to the charge on its plates.

74 •• Two identical capacitors are connected in series to a 100-V battery. When only one capacitor is connected to this battery the energy stored is U_0. What is the total energy stored in the two capacitors when the series combination is connected to the battery?

(*a*) $4U_0$
(*b*) $2U_0$
(*c*) U_0
(*d*) $U_0/2$
(*e*) $U_0/4$

75 • Three capacitors have capacitances of 2.0, 4.0, and 8.0 μF. Find the equivalent capacitance if (*a*) the capacitors are connected in parallel and (*b*) they are connected in series.

76 • A 1.0-μF capacitor is connected in parallel with a 2.0-μF capacitor, and the combination is connected in series with a 6.0-μF capacitor. What is the equivalent capacitance of this combination?

77 • The voltage across a parallel plate capacitor with plate separation 0.5 mm is 1200 V. The capacitor is disconnected from the voltage source and the separation between the plates is increased until the energy stored in the capacitor has been doubled. Determine the final separation between the plates.

78 •• Determine the capacitance of each of the networks shown in Figure 25-30.

Figure 25-30 Problem 78

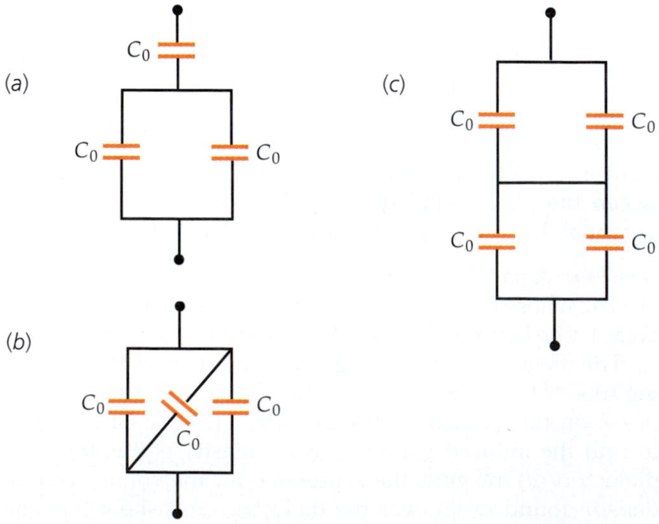

79 •• Figure 25-31 shows four capacitors connected in the arrangement known as a capacitance bridge. The capacitors are initially uncharged. What must be the relation between the four capacitances so that the potential between points *c* and *d* is zero when a voltage *V* is applied between points *a* and *b*?

Figure 25-31 Problem 79

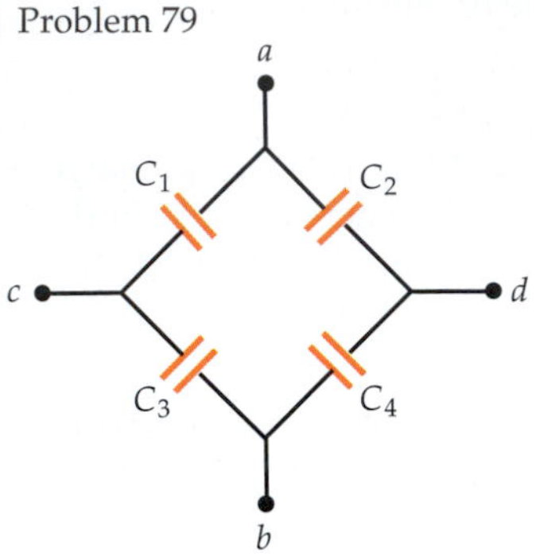

80 •• The plate areas and plate separations of the two parallel-plate capacitors shown in Figure 25-32 are identical. Half the region between the plates of capacitor C_1 is filled with a dielectric of dielectric constant κ. What fraction of the volume of capacitor C_2 should be filled with the same dielectric material so that the two capacitors have the same capacitance?

Figure 25-32 Problems 80 and 81

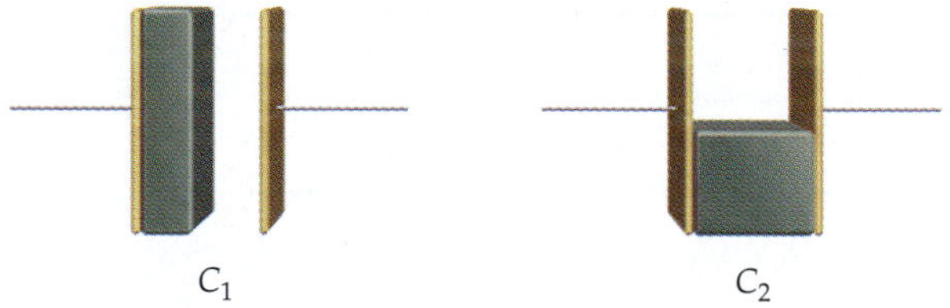

81 •• Repeat Problem 80 if the region filled with dielectric of capacitor (a) is two-thirds of the volume between the plates.

82 •• Two conducting spheres of radius *R* are separated by a distance large compared to their size. One initially has charge *Q* and the other is uncharged. A thin wire is then connected between them. What fraction of the initial energy is dissipated?

83 •• A parallel-plate capacitor of area *A* and separation *d* is charged to a potential difference *V* and then disconnected from the charging source. The plates are then pulled apart until the separation is 2*d*. Find expressions in terms of *A*, *d*, and *V* for (*a*) the new capacitance, (*b*) the new potential difference, and (c) the new stored energy. (*d*) How much work was required to change the plate separation from *d* to 2*d*?

84 •• A parallel-plate capacitor has capacitance C_0 with no dielectric. It is then filled with dielectric of constant κ. When a second capacitor of capacitance C' is connected in series with the first one, the capacitance of the series combination is C_0. Find C'.

85 •• Estimate the capacitance of a typical hot-air balloon.

86 •• A Leyden jar, the earliest type of capacitor, is a glass jar coated inside and out with metal foil. Suppose that a Leyden jar is a cylinder 40 cm high with 2.0-mm-thick walls and an inner diameter of 8 cm. Ignore any field fringing. (*a*) Find the capacitance of this Leyden jar if the dielectric constant κ of the glass is 5.0. (*b*) If the dielectric strength of the glass is 15 MV/m, what maximum charge can the Leyden jar carry without undergoing dielectric breakdown? (*Hint:* Treat the device as a parallel-plate capacitor.)

87 •• A parallel-plate capacitor is constructed from a layer of silicon dioxide of thickness 5×10^{-6} m between two conducting films. The dielectric constant of silicon dioxide is 3.8 and its dielectric strength is 8×10^6 V/m. (*a*) What voltage can be applied across this capacitor without dielectric breakdown? (*b*) What should the surface area of the layer of silicon dioxide be for a 10-pF capacitor? (*c*) Estimate the number of these capacitors that can fit into a square 1 cm by 1 cm.

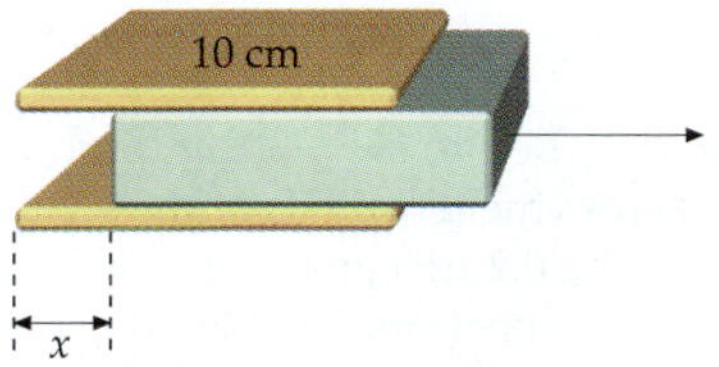

Figure 25-33 Problems 88 and 89

88 •• A parallel-plate capacitor has rectangular plates of length $L = 10$ cm and width $W = 4$ cm (Figure 25-33). The region between the plates is filled with a dielectric slab of dielectric constant $\kappa = 4$ which can slide along the length of the capacitor. Initially, the slab completely fills the rectangular region, and the capacitor holds a charge of 0.2 μC. How far should the dielectric slab be pulled so that the stored energy is double its initial value?

89 •• Suppose the capacitor of Problem 88 is connected to a constant voltage source of 20 V. How far should the dielectric slab be pulled so that the stored energy is reduced to half its initial value?

90 •• A parallel combination of two identical 2-μF parallel-plate capacitors is connected to a 100-V battery. The battery is then removed and the separation between the plates of one of the capacitors is doubled. Find the charge on each of the capacitors.

91 •• A parallel-plate capacitor has a capacitance C_0 and a plate separation d. Two dielectric slabs of constants κ_1 and κ_2, each of thickness $\frac{1}{2}d$ and having the same area as the plates, are inserted between the plates as shown in Figure 25-34. When the charge on the plates is Q, find (*a*) the electric field in each dielectric and (*b*) the potential difference between the plates. (*c*) Show that the new capacitance is given by $C = 2\kappa_1\kappa_2/(\kappa_1 + \kappa_2)\, C_0$. (*d*) Show that this system can be considered to be a series combination of two capacitors of thickness $\frac{1}{2}d$ filled with dielectrics of constant κ_1 and κ_2.

Figure 25-34 Problem 91

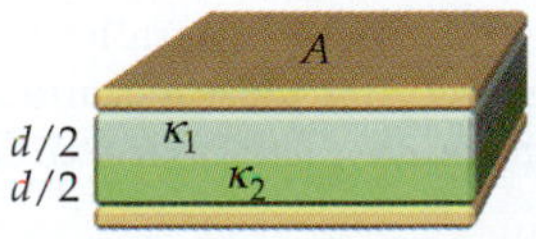

92 •• A parallel-plate capacitor has a plate area A and a separation d. A metal slab of thickness t and area A is inserted between the plates. (*a*) Show that the capacitance is given by $C = \epsilon_0 A/(d - t)$, regardless of where the metal slab is placed. (*b*) Show that this arrangement can be considered to be a capacitor of separation a in series with one of separation b, where $a + b + t = d$.

93 •• A parallel-plate capacitor is filled with two dielectrics of equal size as shown in Figure 25-35. (*a*) Show that this system can be considered to be two capacitors of area $\frac{1}{2}A$ connected in parallel. (*b*) Show that the capacitance is increased by the factor $(\kappa_1 + \kappa_2)/2$.

Figure 25-35 Problem 93

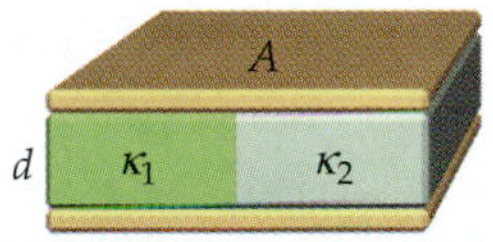

94 •• A parallel-plate capacitor of plate area A and separation x is given a charge Q and is then removed from the charging source. (*a*) Find the stored electrostatic energy as a function of x. (*b*) Find the increase in energy dU due to an increase in plate separation dx from $dU = (dU/dx)\, dx$. (*c*) If F is the force exerted by one plate on the other, the work needed to move one plate a distance dx is $F\, dx = dU$. Show that $F = Q^2/2\epsilon_0 A$. (*d*) Show that the force in part (*c*) equals $\frac{1}{2}EQ$, where Q is the charge on one plate and E is the electric field between the plates. Discuss the reason for the factor $\frac{1}{2}$ in this result.

95 •• A rectangular parallel-plate capacitor of length a and width b has a dielectric of width b partially inserted a distance x between the plates, as shown in Figure 25-36. (*a*) Find the capacitance as a function of x. Neglect edge effects. (*b*) Show that your answer gives the expected results for $x = 0$ and $x = a$.

Figure 25-36 Problems 95, 98, and 99

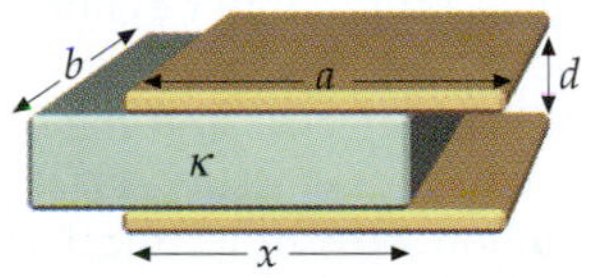

96 •• Two identical, 4-μF parallel-plate capacitors are connected in series across a 24-V battery. (*a*) What is the charge on each capacitor? (*b*) What is the total stored energy of the capacitors? A dielectric having a dielectric constant of 4.2 is inserted between the plates of one of the capacitors while the battery is still connected. (*c*) After the dielectric is inserted, what is the charge on each capacitor? (*d*) What is the potential difference across each capacitor? (*e*) What is the total stored energy of the capacitors?

97 •• A parallel-plate capacitor has a plate area of 1.0 m^2 and a plate separation distance of 0.5 cm. Completely filling the space between the conducting plates is a glass plate having a dielectric constant of 5.0. The capacitor is charged to a potential difference of 12.0 V and is then removed from its charging source. How much work is required to pull the glass plate out of the capacitor?

98 •• The capacitor shown in Figure 25-36 carries charges of $+Q$ and $-Q$ on its plates. (*a*) Find the stored energy as a function x. (*b*) Use the result of (*a*) to determine the force that acts on the dielectric slab.

99 •• The capacitor shown in Figure 25-36 is connected to a constant voltage source V. (*a*) Find the stored energy as a function of x. (*b*) Use the result of (*a*) to determine the force that acts on the dielectric slab.

100 •• A capacitor carries a charge of 15 μC when the potential between its plates is V. When the charge on the capacitor is increased to 18 μC, the potential between the plates increases by 6 V. Find the capacitance of the capacitor and the initial and final voltages.

101 ••• You are asked to construct a parallel-plate, air-gap capacitor that will store 100 kJ of energy. (*a*) What minimum volume is required between the plates of the capacitor? (*b*) Suppose you have developed a dielectric that can withstand 3×10^8 V/m and has a dielectric constant of 5. What volume of this dielectric between the plates of the capacitor is required for it to be able to store 100 kJ of energy?

102 ••• Consider two parallel-plate capacitors, C_1 and C_2, that are connected in parallel. The capacitors are identical except that C_2 has a dielectric inserted between its plates. A voltage source of 200 V is connected across the capacitors to charge them and is then disconnected. (*a*) What is the charge on each capacitor? (*b*) What is the total stored energy of the capacitors? (*c*) The dielectric is removed from C_2. What is the final stored energy of the capacitors? (*d*) What is the final voltage across the two capacitors?

103 ••• A capacitor is constructed of two concentric cylinders of radii a and b ($b > a$) having a length $L >> b$. A charge of $+Q$ is on the inner cylinder and a charge of $-Q$ is on the outer cylinder. The region between the two cylinders is filled with a dielectric having a dielectric constant κ. (*a*) Find the potential difference between the cylinders. (*b*) Find the density of the free charge σ_f on the inner cylinder and the outer cylinder. (*c*) Find the bound charge density σ_b on the inner cylindrical surface of the dielectric and the outer surface. (*d*) Find the total stored electrostatic energy. (*e*) If the dielectric will move without friction, how much mechanical work is required to remove the dielectric cylindrical shell?

104 ••• Two parallel-plate capacitors have the same separation and plate area. The capacitance of each is initially 10 μF. When a dielectric is inserted such that it completely fills the space between the plates of one of the capacitors, the capacitance of that capacitor increases to 35 μF. The 35- and 10-μF capacitors are connected in parallel and are charged to a potential difference of 100 V. The voltage source is then disconnected. (*a*) What is the stored energy of this system? (*b*) What are the charges on the two capacitors? (*c*) The dielectric is removed from the capacitor. What are the new charges on the plates of the capacitors? (*d*) What is the final stored energy of the system?

105 ••• A spherical weather balloon made of aluminized Mylar and filled with helium at atmospheric pressure can lift a payload of 0.2 kg. Determine the capacitance of the balloon. (Neglect the mass of the Mylar.)

106 ••• The two capacitors shown in Figure 25-37 have capacitances $C_1 = 0.4$ μF and $C_2 = 1.2$ μF. The voltages across the two capacitors are V_1 and V_2, respectively, and the total stored energy in the two capacitors is 1.14 mJ. If terminals b and c are connected together, the voltage $V_a - V_d = 80$ V; if terminal a is connected to b, and c is connected to d, the voltage $V_a - V_d = 20$ V. Find the initial voltages V_1 and V_2.

Figure 25-37 Problem 106

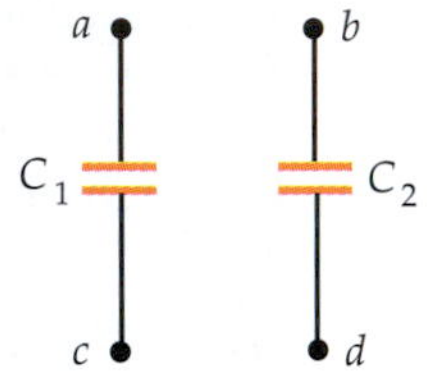

107 ••• Before switch S in Figure 25-38 is closed, the voltage across the terminals of the switch is 120 V and the voltage across the 0.2 μF capacitor is 40 V. The total energy stored in the two capacitors is 1440 μJ. After closing the switch, the voltage across each capacitor is 80 V, and the energy stored by the two capacitors has dropped to 960 μJ. Determine the capacitance of C_2 and the charge on that capacitor before the switch was closed.

Figure 25-38 Problem 107

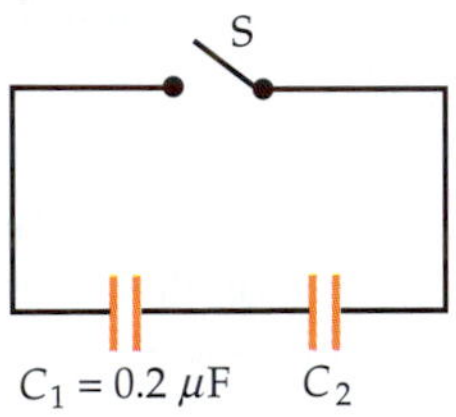

108 ••• A parallel-plate capacitor of area A and separation d is charged to a potential difference V and is then removed from the charging source. A dielectric slab of constant $\kappa = 2$, thickness d, and area $\frac{1}{2}A$ is inserted as shown in Figure 25-39. Let σ_1 be the free charge density at the conductor–dielectric surface and σ_2 be the free charge density at the conductor–air surface. (*a*) Why must the electric field have the same value inside the dielectric as in the free space between the plates? (*b*) Show that $\sigma_1 = 2\sigma_2$. (*c*) Show that the new capacitance is $3\epsilon_0 A/2d$ and that the new potential difference is $\frac{2}{3}V$.

Figure 25-39 Problem 108

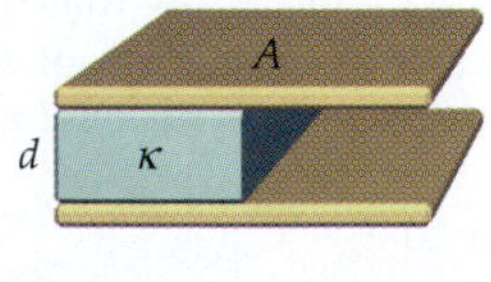

109 ••• Two identical, 10-μF parallel-plate capacitors are given equal charges of 100 μC each and are then removed from the charging source. The charged capacitors are connected by a wire between their positive plates and another wire between their negative plates. (*a*) What is the stored energy of the system? A dielectric having a dielectric constant of 3.2 is inserted between the plates of one of the capacitors such that it completely fills the region between the plates. (*b*) What is the final charge on each capacitor? (*c*) What is the final stored energy of the system?

110 ••• A capacitor has rectangular plates of length a and width b. The top plate is inclined at a small angle as shown in Figure 25-40. The plate separation varies from $d = y_0$ at the left to $d = 2y_0$ at the right, where y_0 is much less than a or b. Calculate the capacitance using strips of width dx and length b to approximate differential capacitors of area $b\,dx$ and separation $d = y_0 + (y_0/a)x$ that are connected in parallel.

Figure 25-40 Problem 110

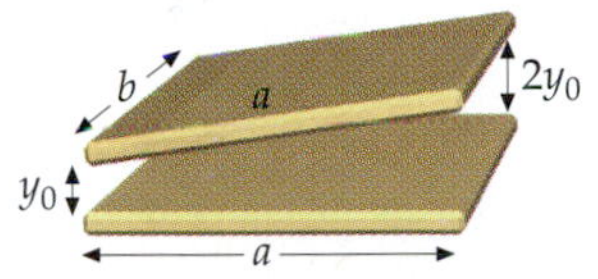

111 ••• Not all dielectrics that separate the plates of a capacitor are rigid. For example, the membrane of a nerve axon is a bilipid layer that has a finite compressibility. Consider a parallel-plate capacitor whose plate separation is maintained by a dielectric of dielectric constant $\kappa = 3.0$ and thickness $d = 0.2$ mm when the potential across the capacitor is zero. The dielectric, which has a dielectric strength of 40 kV/mm, is highly compressible, with a Young's modulus for compressive stress of $5 \times 10^6\ \mathrm{N/m^2}$. The capacitance of the capacitor in the limit $V \rightarrow 0$ is C_0. (*a*) Derive an expression for the capacitance as a function of voltage across the capacitor. (*b*) What is the maximum voltage that can be applied to the capacitor? (Assume that κ does not change under compression.) (*c*) What fraction of the total energy of the capacitor is electrostatic field energy and what fraction is mechanical stress energy stored in the compressed dielectric when the voltage across the capacitor is just below the breakdown voltage?

112 ••• A conducting sphere of radius R_1 is given a free charge Q. The sphere is surrounded by an uncharged concentric spherical dielectric shell having an inner radius R_1, an outer radius R_2, and a dielectric constant κ. The system is far removed from other objects. (*a*) Find the electric field everywhere in space. (*b*) What is the potential of the conducting sphere relative to $V = 0$ at infinity? (*c*) Find the total electrostatic potential energy of the system.

113 ••• A variable air capacitor like the one shown in the photograph on page 759 has a capacitance that changes between 0.02 and 0.12 μF as the shaft is rotated through an angle of 180°. A voltage of 100 V is maintained between the capacitor plates. Initially the capacitor is in its minimum capacitance position. (*a*) How much work must be done to rotate the shaft to the maximum capacitance position? (*b*) The shapes of the plates are designed so that the capacitance is a linear function of rotation angle. How much torque must be applied to rotate the capacitor to hold it in the position corresponding to $C = 0.07\ \mu$F?

114 ••• Repeat Problem 113 with a voltage of 100 V applied and then disconnected when the capacitor is fully charged.

CHAPTER 26

Electric Current and Direct-Current Circuits

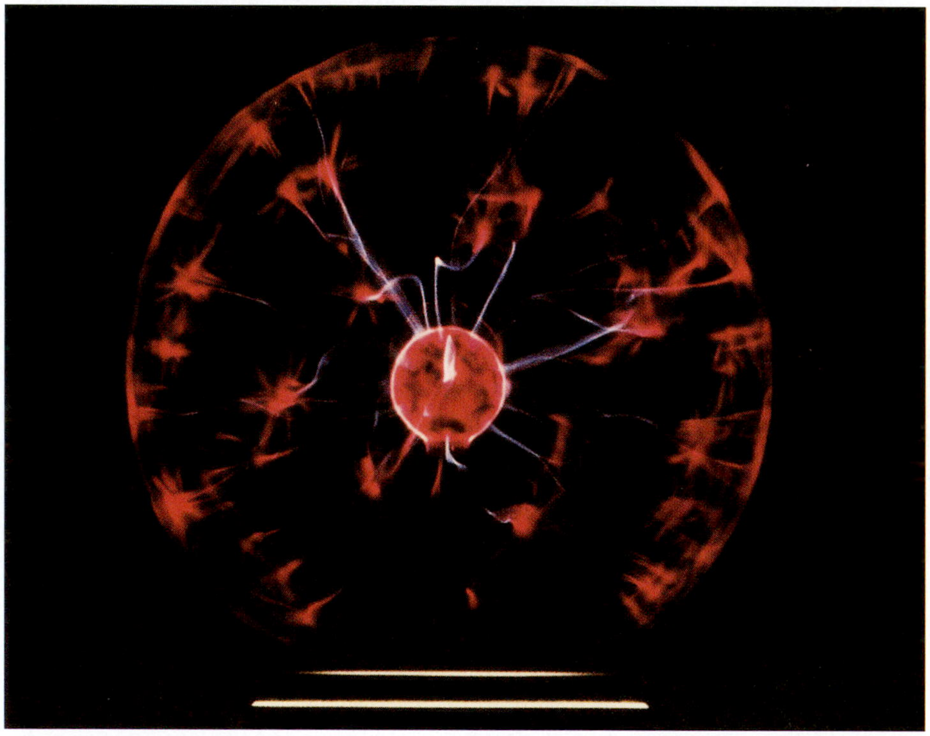

The plasma globe Eye of the Storm. In plasma globes, voltages ranging from 3 to 8 kV, at frequencies between 20 and 50 kHz, are applied between the sphere at the center of the globe and the outer shell, which is at ground. The region in between is filled with a mixture of inert gases. The applied voltage partially ionizes the gases, creating plasma filaments that conduct current. The colored displays are paths along which ionization is occurring.

When we turn on a light, we connect the wire filament in the light bulb across a potential difference that causes electric charge to flow through the wire, much the way a pressure difference in a garden hose causes water to flow through the hose. The flow of electric charge constitutes an electric current. Usually we think of currents as being in conducting wires, but the electron beam in a video monitor and a beam of charged ions from a particle accelerator also constitute electric currents.

When the direction of the current in a circuit element does not vary, the circuit is called a direct current (dc) circuit. Direct currents are usually produced by batteries connected to resistors and capacitors. In Chapter 31 we discuss alternating current (ac) circuits, in which the direction of the current alternates.

When a switch is thrown to turn on a circuit, an electric field propagates along the elements of the circuit at nearly the speed of light. Many complicated changes take place as the current builds up and charge accumulates at various points in the circuit, but an equilibrium or steady state is quickly established. The time for equilibrium to be established depends on the conductivity of the elements in the circuit, but is practically instantaneous for most purposes. In equilibrium, charge no longer accumulates at points along the circuit and the current is steady. (For circuits containing capacitors, the current may increase or decrease slowly, but appreciable changes occur only over times much longer than the time needed to reach the steady state.)

26-1 Current and the Motion of Charges

Electric **current** is defined as the rate of flow of electric charge through a cross-sectional area. Figure 26-1 shows a segment of a current-carrying wire in which charge carriers are moving. If ΔQ is the charge that flows through the cross-sectional area A in time Δt, the current I is

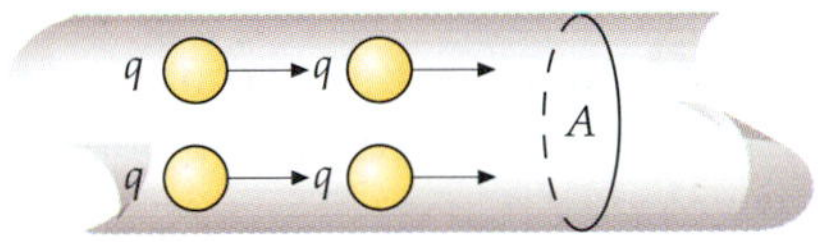

Figure 26-1 A segment of a current-carrying wire. If ΔQ is the amount of charge that flows through the cross-sectional area A in time Δt, the current is $I = \Delta Q/\Delta t$.

$$I = \frac{\Delta Q}{\Delta t} \qquad \text{26-1}$$

Definition—Electric current

The SI unit of current is the **ampere** (A):

$$1\ \text{A} = 1\ \text{C/s} \qquad \text{26-2}$$

By convention, the direction of current is considered to be the direction of flow of positive charge. This convention was established before it was known that free electrons are the particles that actually move in a conducting wire. Thus, electrons move in the direction *opposite* to the direction of the current. (In an accelerator that produces a proton beam, the direction of motion of the positively charged protons is in the direction of the current.)

In a conducting wire, the motion of negatively charged free electrons is quite complex. When there is no electric field in the wire, the free electrons move in random directions with relatively large speeds of the order of 10^6m/s.* Since the velocity vectors of the electrons are randomly oriented, the *average* velocity is zero. When an electric field is applied, a free electron experiences an acceleration due to the force $-e\vec{E}$, and acquires an additional velocity in the direction opposite the field. However, the kinetic energy acquired is quickly dissipated by collisions with the lattice ions in the wire. The electron is then accelerated again by the field. The net result of this repeated acceleration and dissipation of energy is that the electron has a small average velocity called its **drift velocity** opposite to the electric field.

The motion of the free electrons in a metal is similar to that of the molecules of a gas, such as air. In still air, the molecules move with large instantaneous velocities (due to their thermal energy) between collisions, but the average velocity is zero. When there is a breeze, the air molecules have a small drift velocity in the direction of the breeze superimposed on their much larger instantaneous velocities. Similarly, when there is no applied electric field, the "electron gas" in a metal has a zero average velocity, but when there is an applied electric field, the electron gas acquires a small drift velocity.

---

* The average energy of the free electrons in a metal is quite large even at very low temperatures. These electrons do not have the classical Maxwell–Boltzmann energy distribution and do not obey the classical equipartition theorem. We discuss the energy distribution of these electrons and calculate their average speed in Chapter 27.

Let n be the number of free charge-carrying particles per unit volume in a conducting wire of cross-sectional area A. We call n the **number density** of charge carriers. Assume that each particle carries a charge q and moves with a drift velocity v_d. In a time Δt, all the particles in the volume $Av_d\,\Delta t$, shown in Figure 26-2 as a shaded area, pass through the area element. The number of particles in this volume is $nAv_d\,\Delta t$, and the total charge is

$$\Delta Q = qnAv_d\,\Delta t$$

The current is thus

$$I = \frac{\Delta Q}{\Delta t} = qnAv_d \qquad 26\text{-}3$$

Relation between current and drift velocity

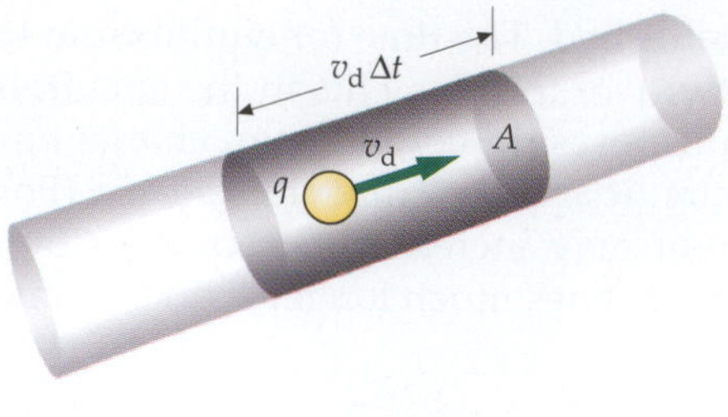

Figure 26-2 In time Δt, all the charges in the shaded volume pass through A. If there are n charge carriers per unit volume, each with charge q, the total charge in this volume is $\Delta Q = qnAv_d\,\Delta t$, where v_d is the drift velocity of the charge carriers.

Equation 26-3 can be used to find the current due to the flow of any type of charged particle, simply by substituting the velocity of the particle for the drift velocity v_d.

The number of charge carriers in a conductor can be measured by the Hall effect, which is discussed in Chapter 28. The result is that, in most metals, there is about one free electron per atom.

Example 26-1

A typical wire for laboratory experiments is made of copper and has a radius 0.815 mm. Calculate the drift velocity of electrons in such a wire carrying a current of 1 A, assuming one free electron per atom.

Picture the Problem Equation 26-3 relates the drift velocity to the number density of charge carriers, which equals the number density of copper atoms n_a. We can find n_a from the mass density of copper, its molecular mass, and Avogadro's number.

1. The drift velocity is related to the current and number density of charge carriers:

$$v_d = \frac{I}{nqA}$$

2. If there is one free electron per atom, the number density of free electrons equals the number density of atoms n_a:

$$n = n_a$$

3. The number density of atoms n_a is related to the mass density ρ_m. Avogadro's number N_A, and the molar mass M. For copper, $\rho = 8.93\ \text{g/cm}^3$ and $M = 63.5\ \text{g/mol}$:

$$n_a = \frac{\rho_m N_A}{M} = \frac{(8.93\ \text{g/cm}^3)(6.02\times10^{23}\ \text{atoms/mol})}{63.5\ \text{g/mol}}$$
$$= 8.47\times10^{22}\ \text{atoms/cm}^3 = 8.47\times10^{28}\ \text{atoms/m}^{-3}$$

4. The magnitude of the charge is e, and the area is related to the radius r of the wire:

$$q = e$$
$$A = \pi r^2$$

5. Substituting numerical values yields v_d:

$$v_d = \frac{I}{nqA} = \frac{I}{n_a e\pi r^2}$$
$$= \frac{1\ \text{C/s}}{(8.47\times10^{28}\ \text{m}^{-3})(1.6\times10^{-19}\ \text{C})\pi(0.000815\ \text{m})^2}$$
$$= 3.54\times10^{-5}\ \text{m/s}$$

Remark Typical drift velocities are of the order of a few hundredths of a millimeter per second, quite small by macroscopic standards.

Exercise How long would it take for an electron to drift from your car battery to the starter motor, a distance of about 1 m, if its drift velocity is 3.5×10^{-5} m/s? (*Answer* 7.9 h)

If electrons drift down a wire at such low speeds, why does an electric light come on instantly when the switch is thrown? An analogy with water in a hose may prove useful. When you turn on a faucet and water rushes into a long, empty hose, it takes quite a few seconds for the water to travel from the faucet to the nozzle. However, if the hose is already full of water, the water emerges from the nozzle almost instantaneously. Because of the water pressure at the faucet, the segment of water near the faucet pushes on the water immediately next to it, which pushes on the next segment of water and so on, until the last segment of water is pushed out the nozzle. This pressure wave moves down the hose at the speed of sound in water, and the water quickly reaches a steady flow rate.

Similarly, when a light is switched on, an electric field propagates down the wire with nearly the speed of light, and the free electrons throughout the wire acquire their drift velocity almost immediately. The charge that flows out of a segment of wire is replaced by an equal amount of charge that flows into the segment at the other end. Thus, charge starts moving through the filament almost immediately after the light switch is thrown. The transport of a significant amount of charge in a wire is accomplished not by a few charges moving rapidly down the wire, but by a very large number of charges slowly drifting down the wire.

Example 26-2

In a certain particle accelerator, a current of 0.5 mA is carried by a 5-MeV proton beam that has a radius of 1.5 mm. (*a*) Find the number density of protons in the beam. (*b*) If the beam hits a target, how many protons hit the target in 1 s?

(*a*)1. The number density is related to the current, charge, cross-sectional area, and speed:

$$n = \frac{I}{qAv}$$

2. We find the speed of the protons from their kinetic energy:

$$K = \tfrac{1}{2}mv^2 = 5 \text{ MeV}$$

$$= 5 \times 10^6 \text{ eV} \times \frac{1.6 \times 10^{-19} \text{ J}}{1 \text{ eV}}$$

$$= 8 \times 10^{-13} \text{ J}$$

3. Use $m = 1.67 \times 10^{-27}$ kg for the mass of a proton, and solve for the speed:

$$v = \sqrt{\frac{2K}{m}} = \sqrt{\frac{(2)(8 \times 10^{-13} \text{ J})}{1.67 \times 10^{-27} \text{ kg}}} = 3.10 \times 10^7 \text{ m/s}$$

4. Substitute to calculate n:

$$n = \frac{I}{qAv}$$

$$= \frac{0.5 \times 10^{-3} \text{ A}}{(1.6 \times 10^{-19} \text{ C/proton})\pi(1.5 \times 10^{-3} \text{ m})^2(3.10 \times 10^7 \text{ m/s})}$$

$$= 1.43 \times 10^{13} \text{ proton/m}^3$$

(*b*)1. The number of protons that hit the target in 1 s is related to the total charge ΔQ that hits in 1 s and the proton charge q:
$$N = \frac{\Delta Q}{q}$$

2. The charge ΔQ that strikes the target in some time Δt is the current times the time:
$$\Delta Q = I\,\Delta t = (0.5\text{ mA})(1\text{ s}) = 0.5\text{ mC}$$

3. The number of protons is then:
$$N = \frac{\Delta Q}{q} = \frac{0.5 \times 10^{-3}\text{ C}}{1.6 \times 10^{-19}\text{ C/proton}} = 3.13 \times 10^{15}\text{ protons}$$

Check the Result The number of protons hitting the target in time Δt is also the number in the volume $Av\,\Delta t$. Then $N = nAv\,\Delta t$. Substituting $n = (I/qAv)$ then gives $N = nAv\,\Delta t = (I/qAv)(Av)\Delta t = I\,\Delta t/q = \Delta Q/q$, which is what we used in part (*b*).

Remarks We were able to use the classical expression for kinetic energy in step 2 without taking relativity into consideration because the proton kinetic energy of 5 MeV is much less than the proton rest energy (about 931 MeV). The speed found, 3.1×10^7 m/s, is about one-tenth the speed of light.

26-2 Resistance and Ohm's Law

Current in a conductor is driven by an electric field $\vec{E}$ inside the conductor that exerts a force $q\vec{E}$ on the free charges. (In electrostatic equilibrium, the electric field must be zero inside a conductor, but when a conductor carries a current, it is no longer in electrostatic equilibrium and the free charge drifts down the conductor, driven by the electric field.) Since $\vec{E}$ is the direction of the force on a positive charge, it is in the direction of the current.

Figure 26-3 shows a wire segment of length ΔL and cross-sectional area A carrying a current I. Since the electric field points in the direction of decreasing potential, the potential at point a is greater than that at point b. Assuming that ΔL is small enough so that we may consider the electric field E to be constant across the segment, the potential difference V between points a and b is

$$V = V_a - V_b = E\,\Delta L \qquad \text{26-4}$$

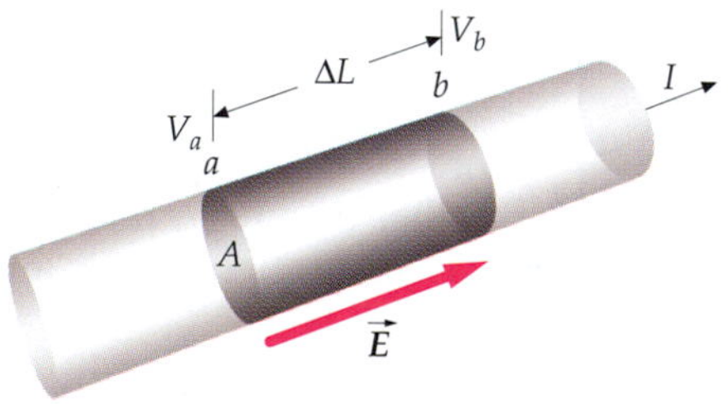

Figure 26-3 A segment of wire carrying a current I. The potential difference is related to the electric field by $V_a - V_b = E\,\Delta L$.

Again we use V rather than ΔV for the potential difference (which in this case is a potential *decrease*) to simplify the notation. The ratio of the potential drop to the current is called the **resistance** of the segment.

$$R = \frac{V}{I} \qquad \text{26-5}$$

Definition—Resistance

The SI unit of resistance, the volt per ampere, is called an ohm (Ω):

$$1\,\Omega = 1\text{ V/A} \qquad \text{26-6}$$

For many materials, the resistance does not depend on the voltage drop or the current. Such materials, which include most metals, are called **ohmic materials.** For ohmic materials, the potential drop across a segment is proportional to the current:

$$V = IR, \qquad R\text{ constant} \qquad \text{26-7}$$

Ohm's law

For **nonohmic materials,** the resistance depends on the current I, so V is not proportional to I. Figure 26-4 shows the potential difference V versus the current I for ohmic and nonohmic materials. For ohmic materials (Figure 26-4*a*), the relation is linear, but for nonohmic materials (Figure 26-4*b*), the relation is not linear. Ohm's law is not a fundamental law of nature, like Newton's laws or the laws of thermodynamics, but rather is an empirical description of a property shared by many materials.

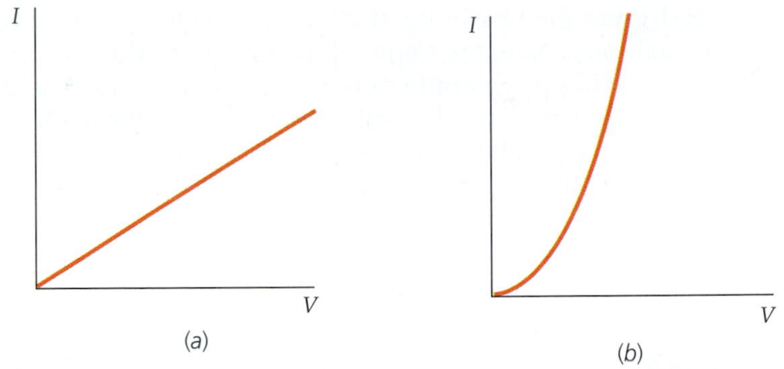

Figure 26-4 Plots of V versus I for (*a*) ohmic and (*b*) nonohmic materials. The resistance $R = V/I$ is independent of I for ohmic materials, as is indicated by the constant slope of the line in (*a*).

Exercise A wire of resistance 3 Ω carries a current of 1.5 A. What is the potential drop across the wire? (*Answer* 4.5 V)

The resistance of a conducting wire is found to be proportional to the length of the wire and inversely proportional to its cross-sectional area:

$$R = \rho\frac{L}{A} \qquad \text{26-8}$$

where the proportionality constant ρ is called the **resistivity** of the conducting material.* The unit of resistivity is the ohm-meter (Ω·m). Note that Equations 26-7 and 26-8 for electrical conduction and electrical resistance are of the same form as Equations 21-9 ($\Delta T = IR$) and 21-10 ($R = \Delta x/kA$) for thermal conduction and thermal resistance. For electrical equations, the potential difference V replaces the temperature difference ΔT and $1/\rho$ replaces the thermal conductivity k. Ohm was, in fact, led to his law by the similarity between the conduction of electricity and the conduction of heat.

Example 26-3

A Nichrome wire ($\rho = 10^{-6}$ Ω·m) has a radius of 0.65 mm. What length of wire is needed to obtain a resistance of 2.0 Ω?

Solve $R = \rho L/A$ (Equation 26-8) for L:

$$L = \frac{RA}{\rho} = \frac{(2\ \Omega)\pi(0.00065\ \text{m})^2}{10^{-6}\ \Omega\cdot\text{m}} = 2.65\ \text{m}$$

The resistivity of any given metal depends on the temperature. Figure 26-5 shows the temperature dependence of the resistivity of copper. This graph is nearly a straight line, which means that the resistivity varies nearly linearly with temperature.† In tables, the resistivity is usually given in terms of its value at 20°C, ρ_{20}, along with the **temperature coefficient of resistivity,** α, which is defined by

$$\rho = \rho_{20}[1 + \alpha\,(t_c - 20\ \text{C}°)] \qquad \text{26-9}$$

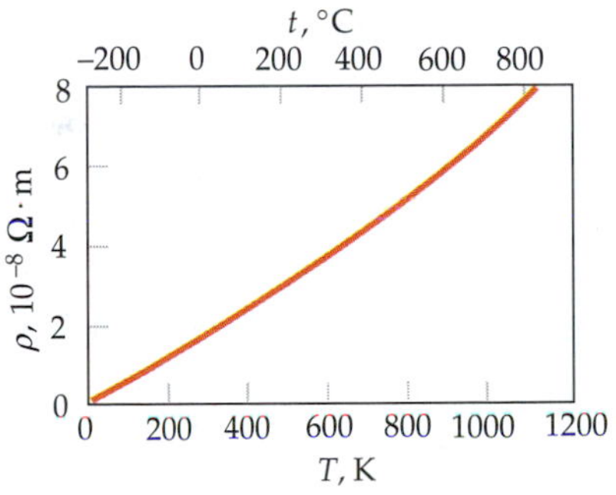

Figure 26-5 Plot of resistivity ρ versus temperature for copper. Since the Celsius and absolute temperatures differ only in the choice of zero, the resistivity has the same slope whether it is plotted against t or T.

* The symbol ρ used here for the resistivity was used in previous chapters for volume charge density. Care must be taken to distinguish which quantity ρ refers to. Usually this will be clear from the context.

† There is a breakdown in this linearity for all metals at very low temperatures that is not shown in Figure 26-5.

Table 26-1 gives the resistivity at 20°C and the temperature coefficient α for various materials. Note the tremendous range of values for ρ.

Electrical wires are manufactured in standard sizes. The diameter of the circular cross section is indicated by a *gauge number,* with higher numbers corresponding to smaller diameters, as can be seen from Table 26-2. Handbooks typically give R/L in ohms per centimeter or ohms per foot.

Table 26-1

Resistivities and Temperature Coefficients

Material	Resistivity ρ at 20°C, $\Omega\cdot$m	Temperature Coefficient α at 20°C, K^{-1}
Silver	1.6×10^{-8}	3.8×10^{-3}
Copper	1.7×10^{-8}	3.9×10^{-3}
Aluminum	2.8×10^{-8}	3.9×10^{-3}
Tungsten	5.5×10^{-8}	4.5×10^{-3}
Iron	10×10^{-8}	5.0×10^{-3}
Lead	22×10^{-8}	4.3×10^{-3}
Mercury	96×10^{-8}	0.9×10^{-3}
Nichrome	100×10^{-8}	0.4×10^{-3}
Carbon	3500×10^{-8}	-0.5×10^{-3}
Germanium	0.45	-4.8×10^{-2}
Silicon	640	-7.5×10^{-2}
Wood	10^{8}–10^{14}	
Glass	10^{10}–10^{14}	
Hard rubber	10^{13}–10^{16}	
Amber	5×10^{14}	
Sulfur	1×10^{15}	

Table 26-2

Wire Diameters and Cross-Sectional Areas for Commonly Used Copper Wires

Gauge Number	Diameter at 20°C, mm	Area, mm^2
4	5.189	21.15
6	4.115	13.30
8	3.264	8.366
10	2.588	5.261
12	2.053	3.309
14	1.628	2.081
16	1.291	1.309
18	1.024	0.8235
20	0.8118	0.5176
22	0.6438	0.3255

Example 26-4

Calculate the resistance per unit length of a 14-gauge copper wire.

1. From Equation 26-8, the resistance per unit length equals the resistivity per unit area:

$$\frac{R}{L} = \frac{\rho}{A}$$

2. Find the resistivity of copper from Table 26-1 and the area from Table 26-2:

$$\rho = 1.7 \times 10^{-8}\ \Omega\cdot\text{m}$$

$$A = 2.08\ \text{mm}^2$$

3. Use these values to find R/L:

$$\frac{R}{L} = \frac{\rho}{A} = \frac{1.7 \times 10^{-8}\ \Omega\cdot\text{m}}{2.08 \times 10^{-6}\ \text{m}^2} = 8.17 \times 10^{-3}\ \Omega/\text{m}$$

Remark 14-gauge copper wire is commonly used for low-current circuits. As this example shows, it has a very small resistance.

Carbon, which has a relatively high resistivity, is used in resistors found in electronic equipment. Resistors are often marked with colored stripes that indicate their resistance value.

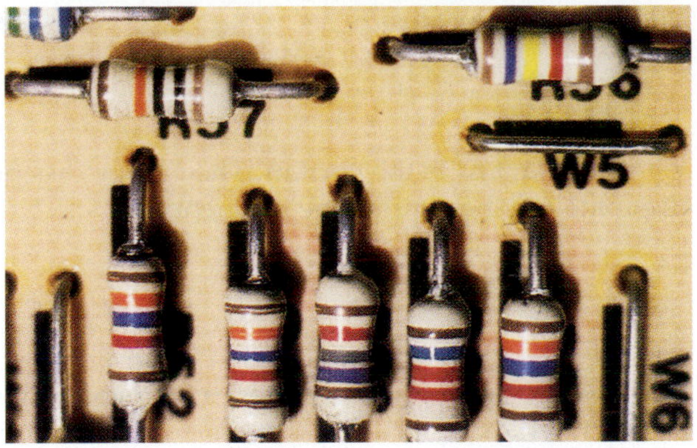

Color-coded carbon resistors on a circuit board.

Example 26-5

Assuming that an electric field E is uniform, find its magnitude in the 14-gauge copper wire of Example 26-4 when the wire is carrying a current of 1.3 A.

Picture the Problem We find the electric field as the voltage drop for a given length of wire, $E = V/L$. The voltage drop is found using Ohm's law, $V = IR$, and the resistance per length is given in Example 26-4.

1. The electric field equals the voltage drop per unit length: $E = \dfrac{V}{L}$

2. Write Ohm's law for the voltage drop: $V = IR$

3. Substitute this expression into the equation for E: $E = \dfrac{V}{L} = \dfrac{IR}{L} = I\dfrac{R}{L}$

4. Substitute the value of R/L found in Example 26-4 to calculate E: $E = I\dfrac{R}{L} = (1.3\ \text{A})(8.17 \times 10^{-3}\ \Omega/\text{m}) = 1.06 \times 10^{-2}\ \text{V/m}$

26-3 Energy in Electric Circuits

When there is an electric field in a conductor, the free electrons are accelerated for a short time, giving the "electron gas" increased kinetic energy; but this additional energy is quickly converted into thermal energy of the conductor by collisions between the electrons and the lattice ions of the conductor. The increase in thermal energy in a conductor is called **Joule heat.**

Consider the segment of wire of length ΔL and cross-sectional area A shown in Figure 26-6. In time Δt, charge ΔQ enters from the left at potential V_1 and the same amount of charge exits from the right at potential V_2. The effect is just as if the same charge ΔQ entered the segment at a high potential V_1 and left it at a low potential V_2. The change in potential energy of the charge passing though the segment is

$$\Delta U = \Delta Q(V_2 - V_1) = \Delta Q(-V) = -(\Delta Q)V$$

where $V = V_1 - V_2$ is the potential drop across the segment. The potential energy lost in this segment of the wire is thus

$$-\Delta U = (\Delta Q)V$$

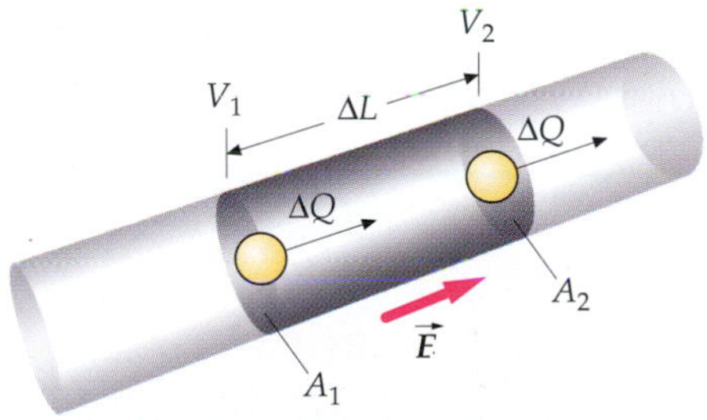

Figure 26-6 During a time Δt, an amount of charge ΔQ passes through area A_1, where the potential is V_1. During the same time interval, an equal amount of charge leaves the segment, passing through area A_2, where the potential is V_2. The effect is just as if the same charge ΔQ entered the segment at a high potential V_1 and left it at a low potential V_2, thereby losing potential energy in the segment.

The rate of energy loss is

$$-\frac{\Delta U}{\Delta t} = \frac{\Delta Q}{\Delta t}V = IV$$

where $I = \Delta Q/\Delta t$ is the current. The energy loss per unit time is the power P dissipated in the conducting segment:

$$P = VI \qquad \text{26-10}$$

Power dissipated in a conductor

If V is in volts and I is in amperes, the power is in watts. The power loss is the product of the decrease in potential energy per unit charge, V, and the charge flowing per unit time, I. Equation 26-10 applies to any device in a circuit. The power delivered to the device is the product of the potential drop and the current. In a conductor, this power goes into thermal energy in the conductor. Using $V = IR$, or $I = V/R$, we can write Equation 26-10 in other useful forms

$$P = VI = I^2R = \frac{V^2}{R} \qquad \text{26-11}$$

Power dissipated in a resistor

Example 26-6

A 12-Ω resistor carries a current of 3 A. Find the power dissipated in this resistor.

Picture the Problem Since we are given the current and the resistance, but not the potential drop, $P = I^2R$ is the most convenient equation to use. Alternatively, we could find the potential drop from $V = IR$, then use $P = IV$.

1. Compute I^2R: $P = I^2R = (3\text{ A})^2(12\ \Omega) = 108\text{ W}$

Check the Result The potential drop across the resistor is $V = IR = (3\text{ A})(12\ \Omega) = 36\text{ V}$. We can use this to find the power from $P = IV = (3\text{ A})(36\text{ V}) = 108\text{ W}$.

Exercise A wire of resistance 5 Ω carries a current of 3 A for 6 s. (*a*) How much power is put into the wire? (*b*) How much thermal energy is produced? (*Answers* (*a*) 45 W, (*b*) 270 J)

EMF and Batteries

To maintain a steady current in a conductor, we need a constant supply of electrical energy. A device that supplies electrical energy is called a **source of emf.** (The letters *emf* stand for *electromotive force*, a term that is now rarely used.) Examples of emf sources are a battery, which converts chemical energy into electrical energy, and a generator, which converts mechanical energy into electrical energy. A source of emf does work on the charge passing through it, raising the potential energy of the charge. The work per unit charge is called the **emf**, $\mathcal{E}$, of the source. The unit of emf is the volt, the same as the unit of potential difference. An **ideal battery** is a source of emf that maintains a constant potential difference between its two terminals, independent of the rate of flow of charge between them. The potential difference between the terminals of an ideal battery is equal in magnitude to the emf of the battery.

Figure 26-7 shows a simple circuit consisting of a resistance R connected to an ideal battery. In such diagrams, a battery is denoted by the symbol —|⊢. The resistance is indicated by the symbol -ΛΛΛ-. The straight lines indicate connecting wires of negligible resistance. The source of emf maintains a constant potential difference equal to $\mathcal{E}$ between points a and b, with point a being at the higher potential. There is negligible potential difference between points a and c or between points d and b because the connecting wire is assumed to have negligible resistance. The potential difference from c and d is therefore equal in magnitude to the emf $\mathcal{E}$, and the current in the resistor is given by $I = \mathcal{E}/R$. The direction of the current in this circuit is clockwise, as shown in the figure.

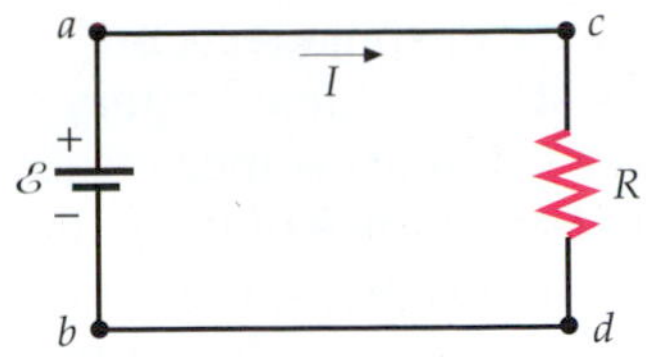

Figure 26-7 A simple circuit consisting of an ideal battery of emf $\mathcal{E}$, a resistance R, and connecting wires that are assumed to be without resistance.

Note that *inside* the source of emf, the charge flows from a region of low potential to a region of high potential, so it gains potential energy.* When charge ΔQ flows through the source of emf $\mathcal{E}$, its potential energy is increased by the amount $\Delta Q\,\mathcal{E}$. The charge then flows through the resistor, where this potential energy is converted into thermal energy. The rate at which energy is supplied by the source of emf is the power output:

$$P = \frac{\Delta Q\,\mathcal{E}}{\Delta t} = \mathcal{E}I \qquad \text{26-12}$$

In the simple circuit of Figure 26-7, the power put out by the source of emf equals that dissipated in the resistor.

A source of emf can be thought of as a charge pump that pumps the charge from a region of low electrical potential energy to a region of high electrical potential energy. Figure 26-8 shows a mechanical analog to the simple electric circuit just discussed.

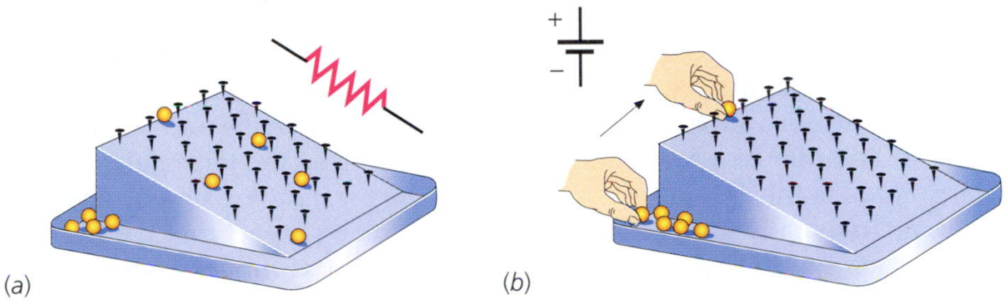

Figure 26-8 A mechanical analog of a simple circuit consisting of a resistance and source of emf. (*a*) The marbles start at some height h above the bottom and are accelerated between collisions with the nails by the gravitational field. The nails are analogous to the lattice ions in the resistor. During the collisions, the marbles transfer the kinetic energy they obtained between collisions to the nails. Because of the many collisions, the marbles have only a small, approximately constant, drift velocity toward the bottom. (*b*) When they reach the bottom, a child picks them up, lifts them to their original height h, and starts them again. The child, who does work mgh on each marble, is analogous to the source of emf. The energy source in this case is the internal chemical energy of the child.

In a **real battery**, the potential difference across the battery terminals, called the **terminal voltage**, is not simply equal to the emf of the battery. Consider the circuit consisting of a real battery and a resistor in Figure 26-9. If the current is varied by varying the resistance R and the terminal voltage is measured, the terminal voltage is found to decrease slightly as the current increases (Figure 26-10), just as if there were a small resistance within the battery.

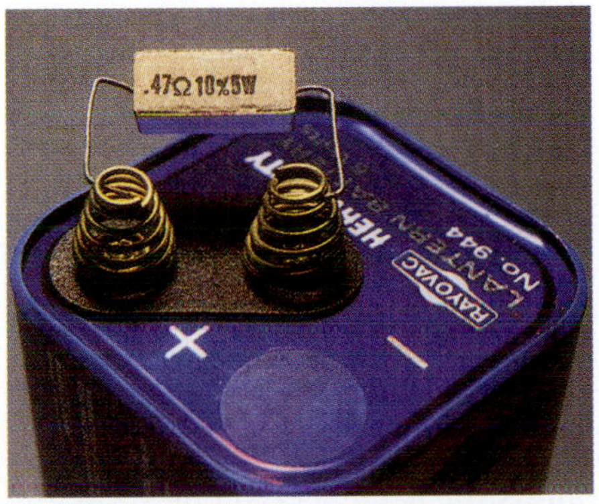

Figure 26-9 A simple circuit consisting of a real battery, a resistor, and connecting wires.

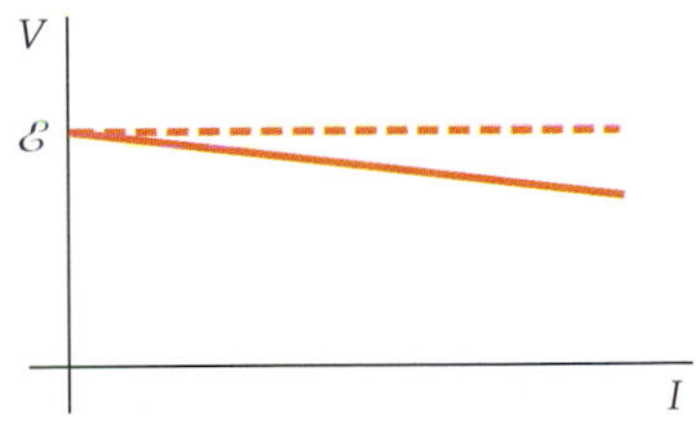

Figure 26-10 Terminal voltage V versus I for a real battery. The dashed line shows the terminal voltage of an ideal battery, which has the same magnitude as $\mathcal{E}$.

* When a battery is being charged by a generator or by another battery, the charge flows from a high-potential to a low-potential region within the battery being charged, thus losing electrostatic potential energy. The energy lost is converted to chemical energy and stored in the battery being charged.

Thus, we can consider a real battery to consist of an ideal battery of emf $\mathcal{E}$ plus a small resistance r, called the **internal resistance** of the battery.

The circuit diagram for a real battery and resistor is shown in Figure 26-11. If the current in the circuit is I, the potential at point a is related to that at point b by

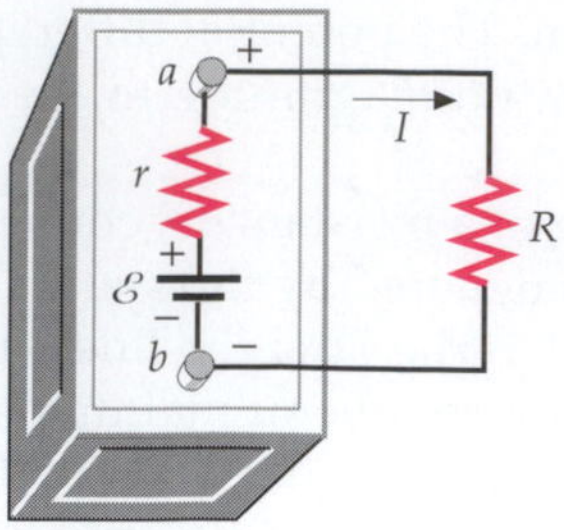

Figure 26-11 Circuit diagram for the circuit in Figure 26-9. A real battery can be represented by an ideal battery of emf $\mathcal{E}$ and a small resistance r.

$$V_a = V_b + \mathcal{E} - Ir$$

The terminal voltage is thus

$$V_a - V_b = \mathcal{E} - Ir \qquad \text{26-13}$$

The terminal voltage of the battery decreases linearly with current, as we saw in Figure 26-10. The potential drop across the resistor R is IR and is equal to the terminal voltage:

$$IR = V_a - V_b = \mathcal{E} - Ir$$

Solving for the current I, we obtain

$$I = \frac{\mathcal{E}}{R + r} \qquad \text{26-14}$$

The terminal voltage given by Equation 26-13 is less than the emf of the battery because of the potential drop across the internal resistance of the battery. Real batteries such as a good car battery usually have an internal resistance of the order of a few hundredths of an ohm, so the terminal voltage is nearly equal to the emf unless the current is very large. One sign of a bad battery is an unusually high internal resistance. If you suspect that your car battery is bad, checking the terminal voltage with a voltmeter, which draws very little current, is not sufficient. You need to check the terminal voltage while current is being drawn from the battery, such as while you are trying to start your car. Then the terminal voltage may drop considerably, indicating a high internal resistance and a bad battery.

Batteries are often rated in ampere-hours (A·h), which is the total charge they can deliver:

$$1\ \text{A·h} = 1\ \text{C/s}\ (3600\ \text{s}) = 3600\ \text{C}$$

The total energy stored in the battery is the total charge times the emf:

$$W = Q\mathcal{E} \qquad \text{26-15}$$

Example 26-7

An 11-Ω resistor is connected across a battery of emf 6 V and internal resistance 1 Ω. Find (*a*) the current, (*b*) the terminal voltage of the battery, (*c*) the power delivered by the emf source, (*d*) the power delivered to the external resistor, and (*e*) the power dissipated by the battery's internal resistance. (*f*) If the battery is rated at 150 A·h, how much energy does it store?

Picture the Problem The circuit diagram is the same as the one shown in Figure 26-11. We find the current from Equation 26-14 and then use it to find the terminal voltage and power delivered to the resistors.

(*a*) Equation 26-14 gives the current: $I = \dfrac{\mathcal{E}}{R + r} = \dfrac{6\text{ V}}{11\ \Omega + 1\ \Omega} = 0.5\text{ A}$

(*b*) Use the current to calculate the terminal voltage of the battery: $V_a - V_b = \mathcal{E} - Ir = 6\text{ V} - (0.5\text{ A})(1\ \Omega) = 5.5\text{ V}$

(*c*) The power delivered by the source of emf equals $\mathcal{E}I$: $P = \mathcal{E}I = (6\text{ V})(0.5\text{ A}) = 3\text{ W}$

(*d*) The power delivered to the external resistance equals I^2R: $I^2R = (0.5\text{ A})^2(11\ \Omega) = 2.75\text{ W}$

(*e*) The power dissipated in the internal resistance is I^2r: $I^2r = (0.5\text{ A})^2(1\ \Omega) = 0.25\text{ W}$

(*f*) The total energy stored is the emf times the total charge: $W = Q\mathcal{E} = 150\text{ A·h}\dfrac{3600\text{ C}}{1\text{ A·h}} \times 6\text{ V} = 3.24\text{ MJ}$

Remarks The value of the internal resistance is exaggerated in this example to simplify calculations. In other examples, we may simply ignore the internal resistance. Of the 3 W of power delivered by the battery, 2.75 W is dissipated as Joule heat in the resistor and 0.25 W is dissipated as Joule heat in the internal resistance of the battery.

Example 26-8 *try it yourself*

For a battery of given emf and internal resistance *r*, what value of external resistance *R* should be placed across the terminals to obtain the greatest Joule heating in *R*?

Picture the Problem The circuit diagram is the same as the one shown in Figure 26-11. The power input to R is I^2R, where $I = \mathcal{E}/(R + r)$. To find the maximum power, we compute dP/dR and set it equal to zero.

Cover the column to the right and try these on your own before looking at the answers.

Steps	*Answers*
1. Use Equation 26-14 to eliminate I from $P = I^2R$ so that P is written as a function of R and the constants $\mathcal{E}$ and r only.	$P = \dfrac{\mathcal{E}^2R}{(R + r)^2} = \mathcal{E}^2R(R + r)^{-2}$
2. Calculate the derivative dP/dR.	$\dfrac{dP}{dR} = \mathcal{E}^2(R + r)^{-2} - 2\mathcal{E}^2 R(R + r)^{-3}$
3. Set $dP/dR = 0$ and solve for R in terms of r.	$R = r$

Remarks The maximum value of P occurs when $R = r$, that is, when the load resistance equals the internal resistance. A similar result holds for alternating-current circuits. Choosing $R = r$ to maximize the power delivered to the load is known as *impedance matching*. A graph of P versus R is shown in Figure 26-12.

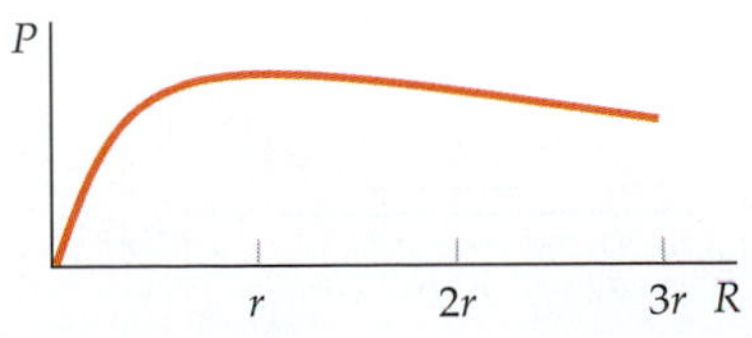

Figure 26-12

Exploring

Battery Technology

A battery is a set of chemical cells each of which consists of two metal electrodes immersed in a conducting solution called an electrolyte. Because of chemical reactions between the conductors and the electrolyte, one electrode, the cathode, becomes positively charged, and the other, the anode, becomes negatively charged. (*a*) In a dry cell, the electrolyte is a paste of ammonium chloride and other additives. The battery develops an emf of 1.5 V. (*b*) A 12-V storage battery consists of six cells that are rechargeable. Each cell has a lead dioxide cathode and a water solution of sulfuric acid for its electrolyte. (*c*) This giant battery, consisting of 200 cells, was built in 1870 in the basement of the Royal Institution, London, by Humphrey Davy.

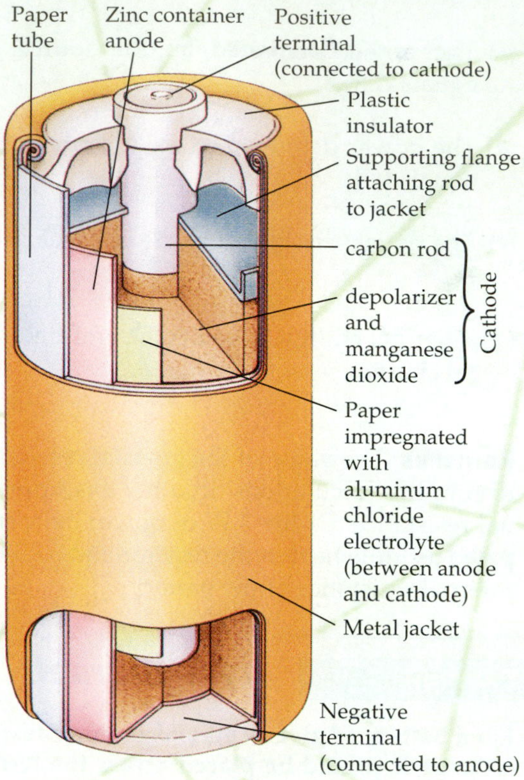

(*a*)

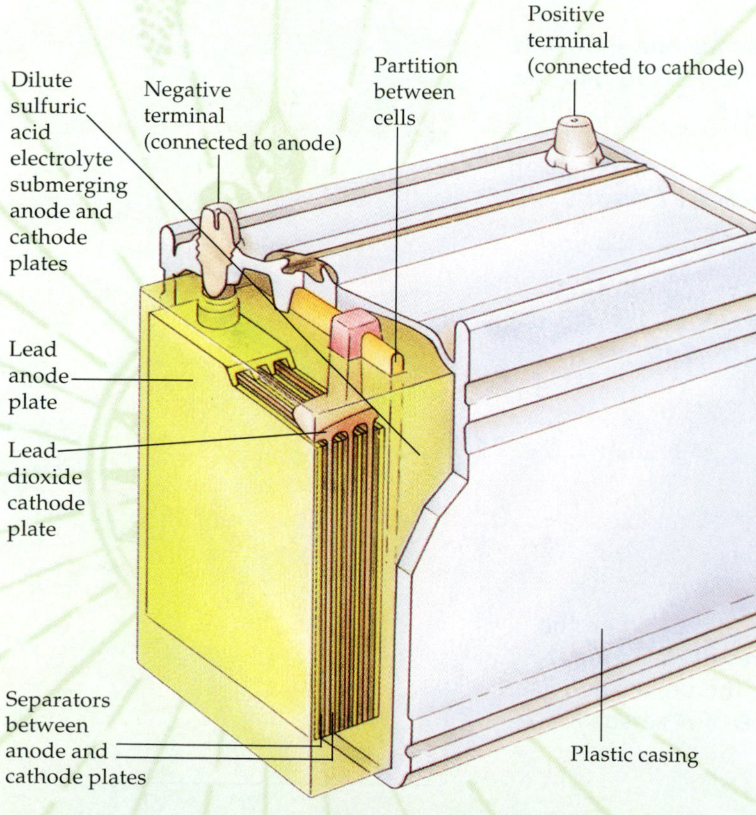

(*b*)

(*c*)

26-4 Combinations of Resistors

The analysis of a circuit can often be simplified by replacing two or more resistors by a single equivalent resistor that carries the same current with the same potential drop as the original resistors. The replacement of a set of resistors by an equivalent resistor is similar to the replacement of a set of capacitors by an equivalent capacitor, as discussed in Chapter 25.

Series Resistors

When two or more resistors are connected like R_1 and R_2 in Figure 26-13 so that they carry the same current I, the resistors are said to be connected in series. The potential drop across R_1 is IR_1 and that across R_2 is IR_2. The potential drop across the two resistors is the sum of the potential drops across the individual resistors:

$$V = IR_1 + IR_2 = I(R_1 + R_2) \qquad \text{26-16}$$

The single equivalent resistance R_{eq} that gives the same total potential drop V when carrying the same current I is found by setting V equal to IR_{eq} (Figure 26-13b). Then R_{eq} is given by

$$R_{eq} = R_1 + R_2$$

When there are more than two resistors in series, the equivalent resistance is

$$R_{eq} = R_1 + R_2 + R_3 + \cdots \qquad \text{26-17}$$

Equivalent resistance for resistors in series

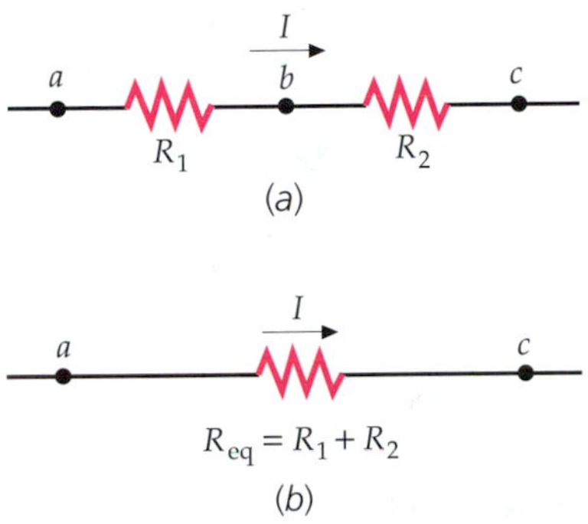

Figure 26-13 (a) Two resistors in series carry the same current. (b) The resistors in (a) can be replaced by a single equivalent resistance $R_{eq} = R_1 + R_2$ that gives the same total potential drop when carrying the same current as in (a).

Parallel Resistors

Two resistors that are connected as in Figure 26-14a, such that they have the same potential difference across them, are in parallel. Note that the resistors are connected at both ends by wires. Let I be the current from point a to point b. At point a the current splits into two parts, I_1 in resistor R_1 and I_2 in R_2. The total current is the sum of the individual currents:

$$I = I_1 + I_2 \qquad \text{26-18}$$

The potential drop across either resistor, $V = V_a - V_b$, is related to the currents by

$$V = I_1R_1 = I_2R_2 \qquad \text{26-19}$$

The equivalent resistance for parallel resistors is the resistance R_{eq} for which the same total current I produces the potential drop V (Figure 26-14b):

$$R_{eq} = \frac{V}{I}$$

Solving this equation for I and using $I = I_1 + I_2$, we have

$$I = \frac{V}{R_{eq}} = I_1 + I_2 = \frac{V}{R_1} + \frac{V}{R_2} \qquad \text{26-20}$$

where we have used Equation 26-19 for I_1 and I_2. The equivalent resistance for two resistors in parallel is therefore given by

$$\frac{1}{R_{eq}} = \frac{1}{R_1} + \frac{1}{R_2}$$

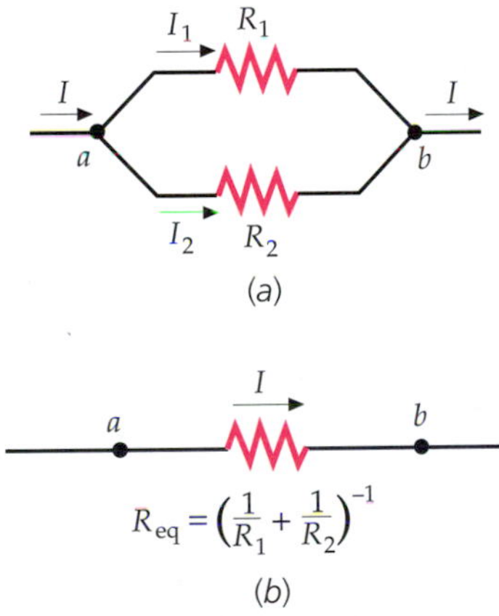

Figure 26-14 (a) Two resistors are in parallel when they are connected together at both ends so that the potential drop is the same across each. (b) The two resistors in (a) can be replaced by an equivalent resistance R_{eq} that is related to R_1 and R_2 by $1/R_{eq} = 1/R_1 + 1/R_2$.

This result can be generalized for combinations, such as that in Figure 26-15, in which three or more resistors are connected in parallel:

$$\frac{1}{R_{eq}} = \frac{1}{R_1} + \frac{1}{R_2} + \frac{1}{R_3} + \cdots \quad \text{26-21}$$

Equivalent resistance for resistors in parallel

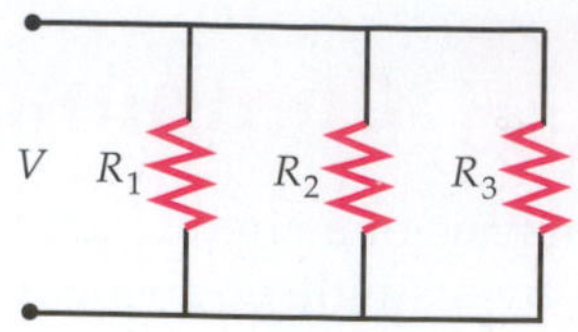

Figure 26-15 Three resistors in parallel.

Exercise A 2-Ω resistor and a 4-Ω resistor are connected (*a*) in series and (*b*) in parallel. Find the equivalent resistances for both cases. (*Answers* (*a*) 6 Ω, (*b*) 1.33 Ω)

Example 26-9

A potential difference of 12 V is applied across the parallel combination of 4- and 6-Ω resistors shown in Figure 26-16. Find (*a*) the equivalent resistance, (*b*) the total current, (*c*) the current in each resistor, (*d*) the power dissipated in each resistor, and (*e*) the power delivered by the 12-V battery.

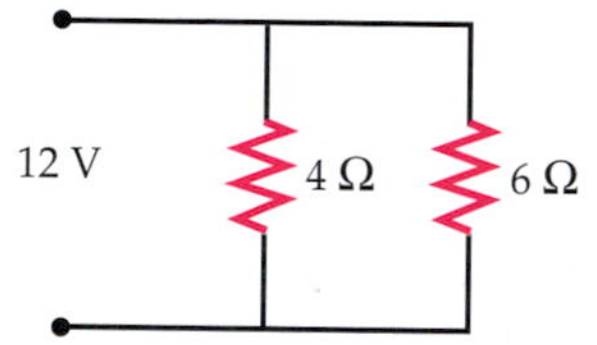

Figure 26-16

Picture the Problem Choose symbols and directions for the currents in Figure 26-17.

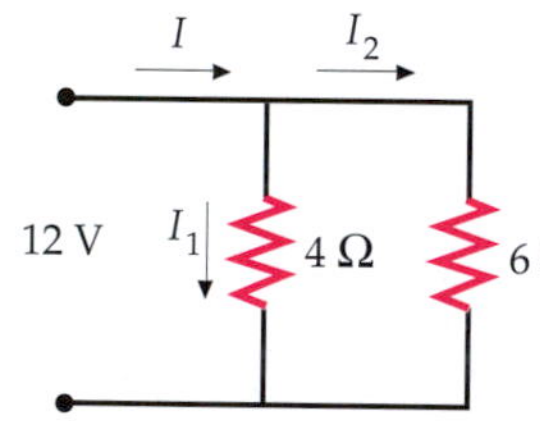

Figure 26-17

(*a*) Calculate the equivalent resistance:

$$\frac{1}{R_{eq}} = \frac{1}{4\ \Omega} + \frac{1}{6\ \Omega} = \frac{3}{12\ \Omega} + \frac{2}{12\ \Omega} = \frac{5}{12\ \Omega}$$

$$R_{eq} = \frac{12\ \Omega}{5} = 2.4\ \Omega$$

(*b*) The total current is the voltage drop divided by the equivalent resistance:

$$I = \frac{V}{R_{eq}} = \frac{12\ \text{V}}{2.4\ \Omega} = 5\ \text{A}$$

(*c*) We obtain the current in each resistor using Equation 26-19 and the fact that the potential drop is 12 V across each resistor:

$$V = IR$$

$$I_4 = \frac{12\ \text{V}}{4\ \Omega} = 3.0\ \text{A}$$

$$I_6 = \frac{12\ \text{V}}{6\ \Omega} = 2.0\ \text{A}$$

(*d*) Use these currents to find the power dissipated in each resistor:

$$P_4 = I_4^2 R = (3.0\ \text{A})^2(4\ \Omega) = 36\ \text{W}$$

$$P_6 = I_6^2 R = (2.0\ \text{A})^2(6\ \Omega) = 24\ \text{W}$$

(*e*) Use $P = VI$ to find the power delivered by the battery:

$$P = VI = (5\ \text{A})(12\ \text{V}) = 60\ \text{W}$$

Check the Result The power delivered by the battery equals the power dissipated in the two resistors $P = 60\ \text{W} = 36\ \text{W} + 24\ \text{W}$. In (*d*), we could have calculated the power dissipated in each resistor from $P_4 = VI_4 = (12\ \text{V})(3.0\ \text{A}) = 36\ \text{W}$ and $P_6 = VI_6 = (12\ \text{V})(2.0\ \text{A}) = 24\ \text{W}$.

Example 26-10 *try it yourself*

A 4-Ω resistor and a 6-Ω resistor are connected in series to a battery of emf 12 V and negligible internal resistance. Find (*a*) the equivalent resistance, (*b*) the current in the circuit, (*c*) the potential drop across each resistor, (*d*) the power dissipated in each resistor, and (*e*) the total power dissipated.

Cover the column to the right and try these on your own before looking at the answers.

Steps	*Answers*
(*a*)1. Draw a circuit diagram (Figure 26-18).	
2. Calculate R_{eq} for the two series resistors.	$R_{eq} = 10\ \Omega$
(*b*) Use $V = IR_{eq}$ to find the current.	$I = 1.2$ A
(*c*) Use Ohm's law to find the potential drop across each resistor.	4.8 V, 7.2 V
(*d*) Find the power dissipated in each resistor using $P = I^2R$. Check your result using $P = VI$ for each resistor.	5.76 W, 8.64 W
(*e*) Add your results from (*d*) to find the total power. Check your result using $P = VI$ and $P = I^2R_{eq}$.	14.4 W

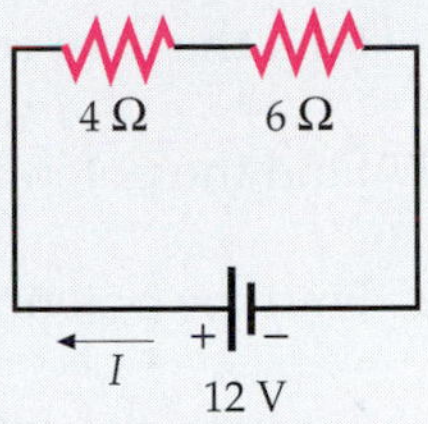

Figure 26-18

Remark Note that much less power is dissipated in the series circuit than in the corresponding parallel circuit of Example 26-9.

Note from Example 26-9 that the equivalent resistance of two parallel resistances is less than the resistance of either resistor alone. This is a general result. Suppose we have a single resistor R_1 carrying current I_1 with potential drop $V = I_1R_1$. If we add a second resistor in parallel, it will carry some additional current I_2 without affecting I_1. The equivalent resistance is $V/(I_1 + I_2)$, which is less than $R_1 = V/I_1$. Note also from Example 26-9 that the ratio of the currents in the two parallel resistors equals the inverse ratio of the resistances. This general result follows from Equation 26-19:

$$I_1R_1 = I_2R_2$$

$$\frac{I_1}{I_2} = \frac{R_2}{R_1} \quad \text{(parallel resistors)} \qquad 26\text{-}22$$

Example 26-11 *try it yourself*

For the circuit in Figure 26-19, find (*a*) the equivalent resistance of the circuit, (*b*) the total current in the source of emf, (*c*) the potential drop across each resistor, and (*d*) the current carried by each resistor.

Picture the Problem To find the equivalent resistance of the circuit, first replace the two parallel resistors by their equivalent resistance. Ohm's law can then be used to find the current and potential drops.

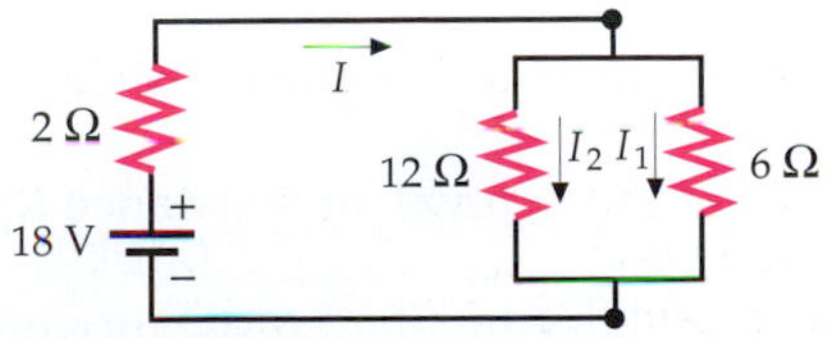

Figure 26-19

Cover the column to the right and try these on your own before looking at the answers.

Steps	Answers
(*a*)1. Find the equivalent resistance of the 6- and 12-Ω parallel combination.	$R_{eq} = 4\ \Omega$
2. Combine your result in step 1 with the 2-Ω resistor in series to find the total equivalent resistance of the circuit.	$R'_{eq} = 6\ \Omega$
(*b*) Find the total current using Ohm's law. This is the current in the 2-Ω resistor.	$I = 3$ A
(*c*)1. Find the potential drop across the 2-Ω resistor from $V_2 = IR$.	$V_2 = 6$ V
2. Find the potential drop across the parallel combination using $V_p = IR_{eq}$.	$V_p = 12$ V
(*d*) Find the current in the 6- and 12-Ω resistors from $I = V_p/R$.	$I_6 = 2$ A, $I_{12} = 1$ A

Check the Result The current in the 6-Ω resistor is twice that in the 12-Ω resistor, as we should expect. Also, these two currents sum to give I, the total current in the circuit, as they must. Finally, note that the potential drops across the 2-Ω resistor and the parallel combination sum to the emf of the battery; $V_2 + V_p = 6\text{ V} + 12\text{ V} = 18\text{ V}$.

Exercise Repeat this example with the 6-Ω resistor replaced by a wire of negligible resistance. (*Answers* (*a*) $R'_{eq} = 2\ \Omega$, (*b*) $I = 9$ A, (*c*) $V_2 = 18$ V, $V_0 = 0$, $V_{12} = 0$, (*d*) $I_2 = 9$ A, $I_0 = 9$ A, $I_{12} = 0$)

Exercise Repeat this example for the case in which the 6-Ω resistor is increased to infinity. (*Answers* (*a*) $R'_{eq} = 14\ \Omega$, (*b*) $I = 1.29$ A, (*c*) $V_2 = 2.57$ V, $V_\infty = 15.4$ V, $V_{12} = 15.4$ V, (*d*) $I_2 = 1.29$ A, $I_\infty = 0$, $I_{12} = 1.29$ A)

Example 26-12 *try it yourself*

Find the equivalent resistance between points *a* and *b* for the combination of resistors shown in Figure 26-20.

Picture the Problem You can analyze this complicated combination step by step. You first need to find the equivalent resistance R_{eq} of the 4- and 12-Ω parallel combination. Then combine your result with the 5-Ω resistance in series with the parallel combination. You are then left with two resistors in parallel.

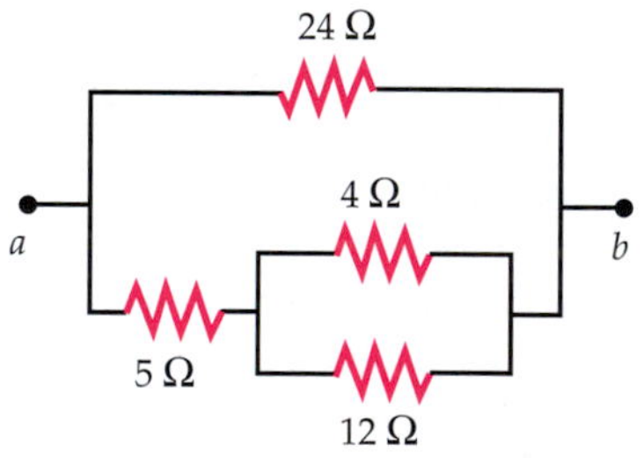

Figure 26-20

Cover the column to the right and try these on your own before looking at the answers.

Steps	Answers
1. Find the equivalent resistance R_{eq} of the 4- and 12-Ω resistors in parallel.	$R_{eq} = 3\ \Omega$
2. Find the equivalent resistance R'_{eq} of R_{eq} in series with the 5-Ω resistor.	$R'_{eq} = 8\ \Omega$
3. Find the equivalent resistance of R'_{eq} in parallel with the 24-Ω resistor.	$R''_{eq} = 6\ \Omega$

26-5 Kirchhoff's Rules

There are many simple circuits, such as that shown in Figure 26-21, that cannot be analyzed by merely replacing combinations of resistors by an equivalent resistance. The two resistors R_1 and R_2 in this circuit look as if they might be in parallel, but they are not. The potential drop is not the same across both resistors because of the presence of the emf source $\mathcal{E}_2$ in series with R_2. Nor are R_1 and R_2 in series, because they don't carry the same current.

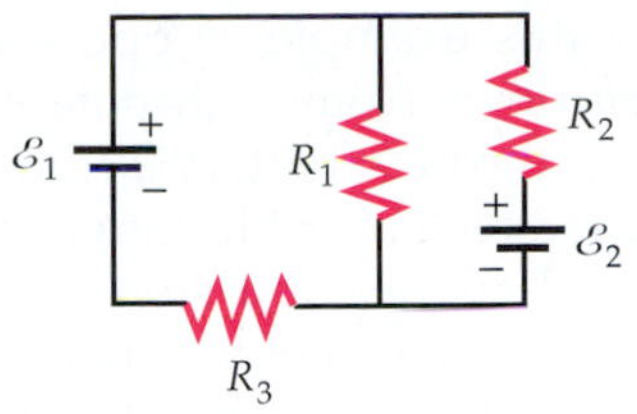

Figure 26-21 An example of a simple circuit that cannot be analyzed by replacing combinations of resistors in series or parallel with their equivalent resistances. The potential drops across R_1 and R_2 are not equal because of the emf source $\mathcal{E}_2$, so these resistors are not in parallel. (Parallel resistors would be connected together at both ends.) The resistors do not carry the same current, so they are not in series.

Two rules, called **Kirchhoff's rules**, apply to this and any other circuit:

1. When any closed-circuit loop is traversed, the algebraic sum of the changes in potential must equal zero.
2. At any junction point in a circuit where the current can divide, the sum of the currents into the junction must equal the sum of the currents out of the junction.

Kirchhoff's rules

Kirchhoff's first rule, called the **loop rule**, follows directly from the conservation of energy. If we have a charge q at some point where the potential is V, the potential energy of the charge is qV. As the charge traverses a loop in a circuit, it loses or gains energy as it passes through resistors, batteries, or other devices, but when it arrives back at its starting point, its energy must again be qV. That is, *the net change in the potential must be zero.*

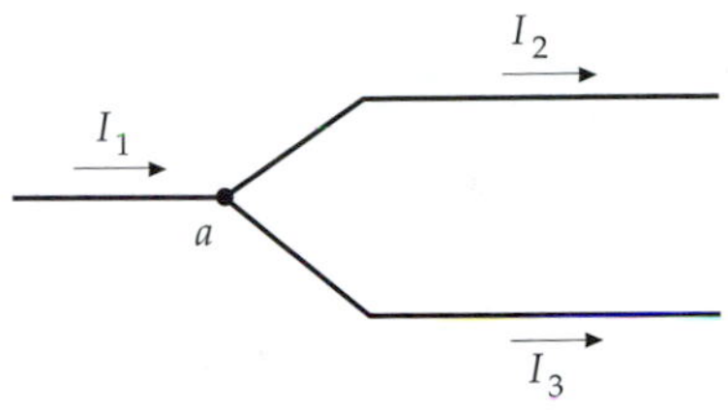

Figure 26-22 Illustration of Kirchhoff's junction rule. The current I_1 into point a equals the sum $I_2 + I_3$ of the currents out of point a.

Kirchhoff's second rule, called the **junction rule**, follows from the conservation of charge. Figure 26-22 shows the junction of three wires carrying currents I_1, I_2, and I_3. Since charge does not originate or accumulate at this point, the conservation of charge implies the junction rule, which for this case gives

$$I_1 = I_2 + I_3 \tag{26-23}$$

Single-Loop Circuits

As an example of using Kirchhoff's loop rule, consider the circuit shown in Figure 26-23 containing two batteries with internal resistances r_1 and r_2 and three external resistors. We wish to find the current in terms of the emfs.

Assume that I is clockwise, as indicated in Figure 26-23, and apply Kirchhoff's loop rule as we traverse the circuit in the assumed direction of the current, beginning at point a. The potential decreases and increases are given in the figure. Note that we encounter a potential drop as we traverse the source of emf between points c and d and a potential increase as we traverse the source of emf between f and g. Beginning at point a, we obtain from Kirchhoff's loop rule

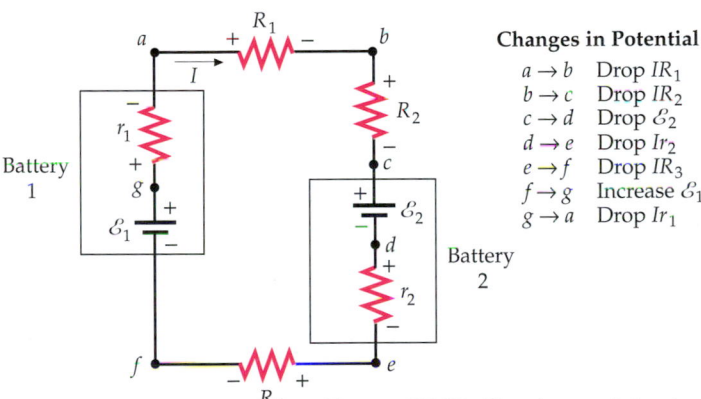

$$-IR_1 - IR_2 - \mathcal{E}_2 - Ir_2 - IR_3 + \mathcal{E}_1 - Ir_1 = 0$$

Solving for the current I, we obtain

$$I = \frac{\mathcal{E}_1 - \mathcal{E}_2}{R_1 + R_2 + R_3 + r_1 + r_2} \tag{26-24}$$

If $\mathcal{E}_2$ is greater than $\mathcal{E}_1$, we get a negative value for the current I, indicating that we have assumed the wrong direction for I.

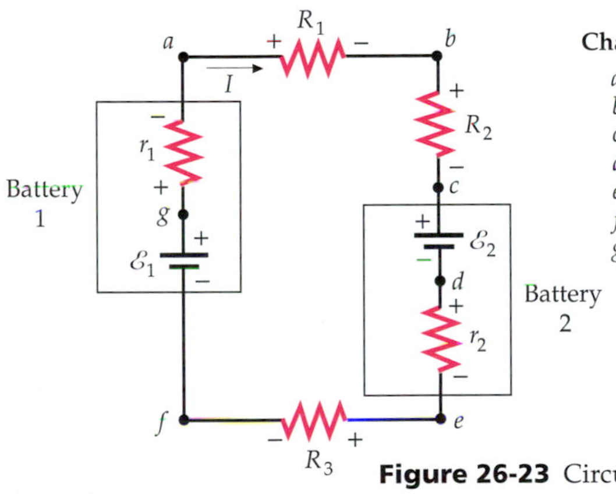

Changes in Potential

$a \to b$	Drop IR_1
$b \to c$	Drop IR_2
$c \to d$	Drop $\mathcal{E}_2$
$d \to e$	Drop Ir_2
$e \to f$	Drop IR_3
$f \to g$	Increase $\mathcal{E}_1$
$g \to a$	Drop Ir_1

Figure 26-23 Circuit containing two batteries and three external resistors. The plus and minus signs on the resistors are there to help us remember which side of each resistor is at the higher potential for the current direction we have assumed.

For this example, suppose that $\mathcal{E}_1$ is the greater emf. In battery 2, the charge flows from high potential to low potential. Therefore, a charge ΔQ moving through battery 2 from point c to point d loses energy $\mathcal{E}_2\,\Delta Q$. If battery 2 is a rechargeable battery, most of this electrical energy is converted into chemical energy and stored in the battery, which means that battery 2 is *charging*. (The rest of the energy is dissipated in the internal resistance.)

The analysis of a circuit is usually simplified if we choose one point to be at zero potential and then find the potentials of the other points relative to it. Since only potential differences are important, any point in a circuit can be chosen to have zero potential. In the following example, we choose point f in the figure to be at zero potential. This is indicated by the ground symbol ⏚ at point f.*

* As we saw in Section 18-2, the earth can be considered to be a very large conductor with a nearly unlimited supply of charge, which means that the potential of the earth remains essentially constant. In practice, electrical circuits are often grounded by connecting one point to the earth. The outside metal case of a washing machine, for example, is usually grounded by connecting it by a wire to a water pipe that is in contact with the earth. Since everything so grounded is at the same potential, it is convenient to designate this potential as zero.

Example 26-13

Suppose the elements in the circuit in Figure 26-23 have the values $\mathcal{E}_1 = 12$ V, $\mathcal{E}_2 = 4$ V, $r_1 = r_2 = 1\ \Omega$, $R_1 = R_2 = 5\ \Omega$, and $R_3 = 4\ \Omega$, as shown in Figure 26-24. (*a*) Find the potentials at points *a* through *g* in the figure, assuming that the potential at point *f* is zero. (*b*) Find the power input and output in the circuit.

Picture the Problem To find the potential differences, we first need to find the current I in the circuit. The voltage drop across each resistor is then IR. To discuss the energy balance, we calculate the power into or out of each element using Equations 26-10 and 26-11.

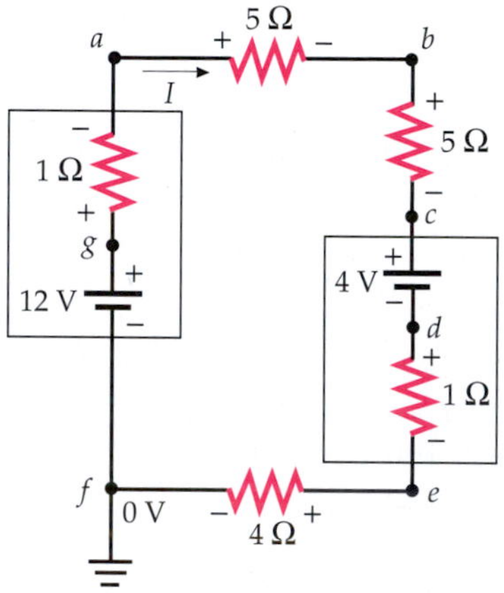

Figure 26-24

(*a*)1. The current I in the circuit is found using Equation 26-24:

$$I = \frac{12\text{ V} - 4\text{ V}}{5\ \Omega + 5\ \Omega + 4\ \Omega + 1\ \Omega + 1\ \Omega} = \frac{8\text{ V}}{16\ \Omega} = 0.5\text{ A}$$

2. We now find the potential at each labeled point in the circuit:

$$V_g = V_f + \mathcal{E}_1 = 0 + 12\text{ V} = 12\text{ V}$$
$$V_a = V_g - Ir_1 = 12\text{ V} - (0.5\text{ A})(1\ \Omega) = 11.5\text{ V}$$
$$V_b = V_a - IR_1 = 11.5\text{ V} - (0.5\text{ A})(5\ \Omega) = 9\text{ V}$$
$$V_c = V_b - IR_2 = 9\text{ V} - (0.5\text{ A})(5\ \Omega) = 6.5\text{ V}$$
$$V_d = V_c - \mathcal{E}_2 = 6.5\text{ V} - 4\text{ V} = 2.5\text{ V}$$
$$V_e = V_d - Ir_2 = 2.5\text{ V} - (0.5\text{ A})(1\ \Omega) = 2.0\text{ V}$$
$$V_f = V_e - IR_3 = 2.0\text{ V} - (0.5\text{ A})(4\ \Omega) = 0$$

(*b*)1. First, calculate the power delivered by the emf source $\mathcal{E}_1$:

$$P_{\mathcal{E}_1} = \mathcal{E}_1 I = (12\text{ V})(0.5\text{ A}) = 6\text{ W}$$

2. Part of this power is dissipated in the resistors, both internal and external:

$$P_R = I^2R_1 + I^2R_2 + I^2R_3 + I^2r_1 + I^2r_2$$
$$= (0.5\text{ A})^2(5\ \Omega + 5\ \Omega + 4\ \Omega + 1\ \Omega + 1\ \Omega) = 4.0\text{ W}$$

3. The remaining 2 W of power goes into charging battery 2:

$$P_{\mathcal{E}_2} = \mathcal{E}_2 I = (4\text{ V})(0.5\text{ A}) = 2\text{ W}$$

Remarks Figure 26-25 shows the potential at the labeled points of the circuit.

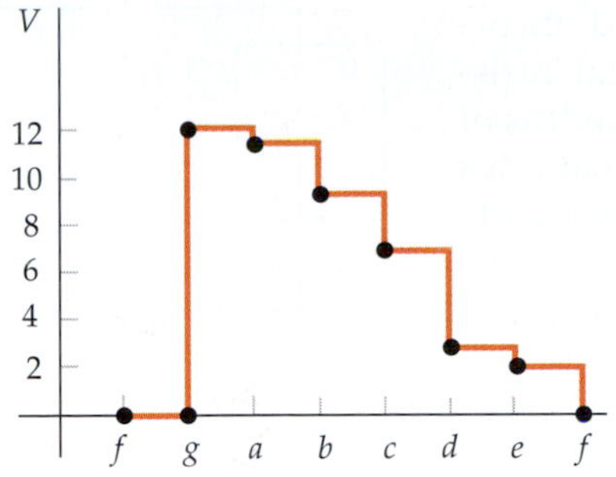

Figure 26-25

Note that the terminal voltage of the battery that is being charged in Example 26-13 is $V_c - V_e = 4.5$ V, which is greater than the emf of the battery. Because of its internal resistance, a battery is not completely reversible. If the same 4-V battery were to deliver 0.5 A to an external circuit, its terminal voltage would be 3.5 V (again assuming that its internal resistance is 1 Ω). If the internal resistance is very small, the terminal voltage of a battery is nearly equal to its emf, whether the battery is delivering current to an external circuit or is being charged. Some real batteries, such as those used in automobiles, are nearly reversible and can easily be recharged. Other types of batteries are not reversible. If you attempt to recharge one of these by driving current from its positive to its negative terminal, most, if not all, of the energy will go into heat rather than into the chemical energy of the battery, and the battery may explode.

Example 26-14

A good car battery is to be connected by jumper cables to a weak car battery to charge the weak one. (*a*) To which terminal of the weak battery should the positive terminal of the good battery be connected? (*b*) Assume that the good battery has an emf of $\mathcal{E}_1 = 12$ V and the weak battery has an emf of $\mathcal{E}_2 = 11$ V, that the internal resistances of the batteries are $r_1 = r_2 = 0.02\ \Omega$, and that the resistance of the jumper cables is $R = 0.01\ \Omega$. What will the charging current be? (*c*) What will the current be if the batteries are connected incorrectly?

(*a*) To charge the weak battery, we connect the terminals positive to positive and negative to negative to drive charge through the weak battery from the positive terminal to the negative terminal:

Figure 26-26

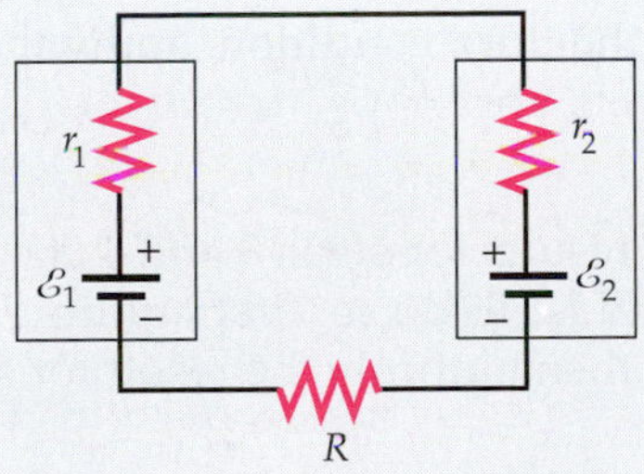

(*b*) Use Kirchhoff's loop rule to find the charging current:

$$I = \frac{\mathcal{E}_1 - \mathcal{E}_2}{R + r_1 + r_2} = \frac{12\text{ V} - 11\text{ V}}{0.05\ \Omega} = 20\text{ A}$$

(*c*) When the batteries are connected incorrectly, positive terminals to negative terminals, the emfs add:

$$I = \frac{\mathcal{E}_1 + \mathcal{E}_2}{R + r_1 + r_2} = \frac{12\text{ V} + 11\text{ V}}{0.05\ \Omega} = 460\text{ A}$$

Remarks If the batteries are connected incorrectly as shown in Figure 26-27, the total resistance of the circuit is of the order of hundredths of an ohm, the current is very large, and the batteries could explode in a shower of boiling battery acid.

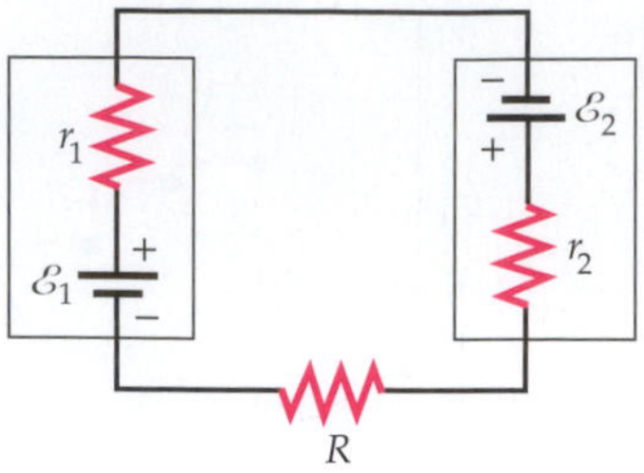

Figure 26-27 Two batteries connected incorrectly—dangerous!

Multiloop Circuits

To analyze circuits containing more than one loop, we need to use both of Kirchhoff's rules, with Kirchhoff's junction rule applied to points where the current splits into two or more parts.

Example 26-15

(*a*) Find the current in each part of the circuit shown in Figure 26-28. (*b*) Find the energy dissipated in the 4-Ω resistor in 3 s.

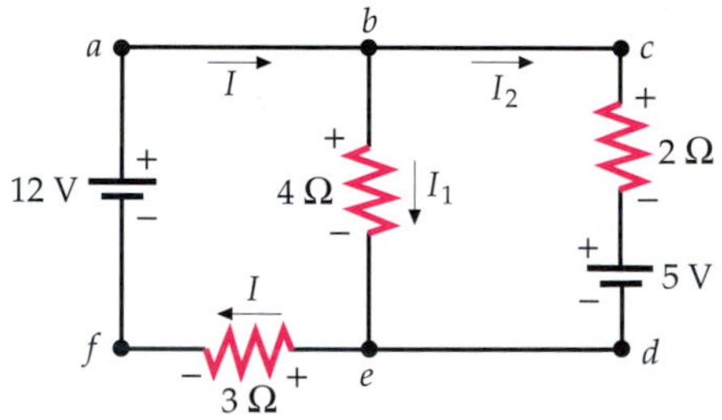

Figure 26-28

Picture the Problem There are three currents, I, I_1, and I_2, to be determined, so we need three conditions. One condition comes from applying the junction rule to point *b*. (We can also apply the junction rule to point *e*, the only other junction in the circuit, but it gives exactly the same information.) The other two conditions are obtained by applying the loop rule. There are three loops in the circuit: the two interior loops, *abefa* and *bcdeb*, and the exterior loop, *abcdefa*. We can use any two of these loops—the third will give redundant information.The direction of the current I_1 from *b* to *e* is not known before the circuit is analyzed. The plus and minus signs on the 4-Ω resistor are for the assumed direction of I_1 from *b* to *e*.

(*a*)1. Apply the junction rule to point *b*:

$$I = I_1 + I_2$$

2. Apply the loop rule to the outer loop, *abcdefa*:

$$12\text{ V} - (2\ \Omega)I_2 - 5\text{ V} - (3\ \Omega)(I_1 + I_2) = 0$$

3. Divide the above equation by 1 Ω, recalling that (1 V)/(1 Ω) = 1 A, then simplify:

$$7\text{ A} - 3I_1 - 5I_2 = 0$$

4. For the third condition, apply the loop rule to the loop on the left, *abefa*:

$$12\text{ V} - (4\ \Omega)I_1 - (3\ \Omega)(I_1 + I_2) = 0$$

$$12\text{ A} - 7I_1 - 3I_2 = 0$$

5. The results for steps 3 and 4 can be combined to solve for I_1 and I_2. To do so, first multiply the result for step 3 by 3, and then multiply the result for step 4 by 5:

$$21\text{ A} - 9I_1 - 15I_2 = 0$$

$$60\text{ A} - 35I_1 - 15I_2 = 0$$

6. Subtract the equations in step 5 to eliminate I_2, then solve for I_1:

$$39\text{ A} - 26I_1 = 0$$

$$I_1 = \frac{39\text{ A}}{26} = 1.5\text{ A}$$

7. Substitute I_1 in the results for step 3 or 4 to solve for I_2. Here, we choose step 3:

$$7\text{ A} - 3(1.5\text{ A}) - 5I_2 = 0$$

$$I_2 = \frac{2.5\text{ A}}{5} = 0.5\text{ A}$$

8. Finally, I_1 and I_2 determine I using the equation in step 1: $I = I_1 + I_2 = 1.5\text{ A} + 0.5\text{ A} = 2.0\text{ A}$

(*b*)1. The power dissipated in the 4-Ω resistor is found using $P = I_1^2R$: $P = I_1^2R = (1.5\text{ A})^2(4\ \Omega) = 9\text{ W}$

2. The total energy dissipated in a time t is $W = Pt$. In this case, $t = 3$ s: $W = Pt = (9\text{ W})(3\text{ s}) = 27\text{ J}$

Check the Result In Figure 26-29, we have chosen the potential to be zero at point f, and we have labeled the currents and the potentials at the other points. Note that $V_b - V_e = 6$ V and $V_e - V_f = 6$ V.

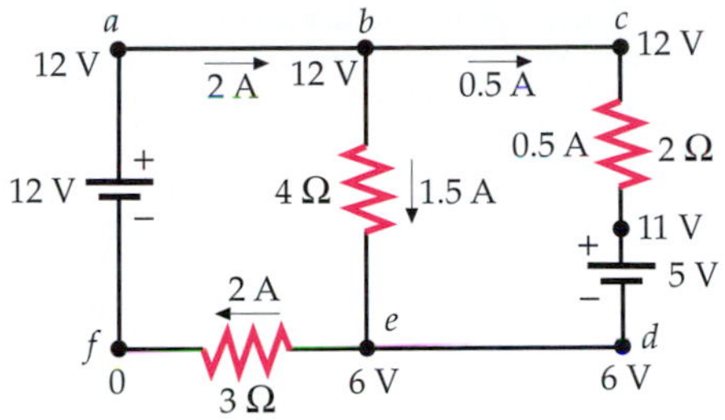

Figure 26-29

Remarks Applying the loop rule to the loop on the right, *bcde*, gives $-(2\ \Omega)I_2 - 5\text{ V} + (4\ \Omega)I_1 = 0$, or $-5\text{ A} + 4I_1 - 2I_2 = 0$. Note that this is just the result for step 3 minus the result for step 4 and hence contains no new information, as expected.

Exercise Find I_1 for the case in which the 3-Ω resistor approaches (*a*) zero resistance and (*b*) infinite resistance. (*Answers* (*a*) The potential drop across the 4-Ω resistor is 12 V; thus, $I_1 = 3$ A. (*b*) In this case, the loop on the left is an open circuit. Thus, $I_1 = (5\text{ V})/(2\ \Omega + 4\ \Omega) = 0.833$ A.)

Example 26-15 illustrates the general methods for the analysis of multiloop circuits:

1. Draw a sketch of the circuit.
2. Choose a direction for the current in each branch of the circuit, and label the currents in the circuit diagram. Add plus and minus signs to indicate the high- and low-potential sides of each resistor, capacitor, or source of emf.
3. Replace any combination of resistors in series or parallel with its equivalent resistance.
4. Apply the junction rule to each junction where the current divides.
5. Apply the loop rule to each loop until you obtain as many equations as unknowns.
6. Solve the equations to obtain the values of the unknowns.
7. Check your results by assigning a potential of zero to one point in the circuit and use the values of the currents found to determine the potentials at other points in the circuit.

General method for analyzing multiloop circuits

Example 26-16 *try it yourself*

(*a*) Find the current in each part of the circuit shown in Figure 26-30. Draw the circuit diagram with the correct magnitudes and directions for the current in each part. (*b*) Assign $V = 0$ to point *c* and then label the potential at each other point *a* through *f*.

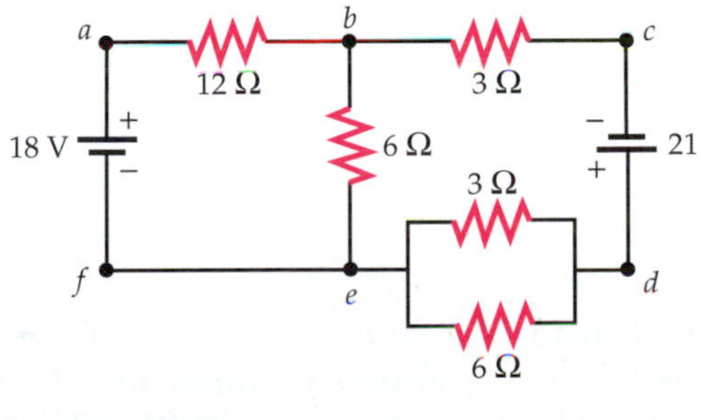

Figure 26-30

Picture the Problem First, replace the two parallel resistors by an equivalent resistance. Let I be the current through the 18-V battery, and I_1 be the current from *b* to *e*. The currents can then be found by applying the junction rule at points *b* and *c* and the loop rule to each loop.

Cover the column to the right and try these on your own before looking at the answers.

Steps	***Answers***
(*a*) 1. Find the equivalent resistance of the 3- and 6-Ω parallel resistors.	$R_{eq} = 2\ \Omega$
2. Apply the junction rule at points *b* and *e* and redraw the circuit diagram with the currents indicated (Figure 26-31).	**Figure 26-31** 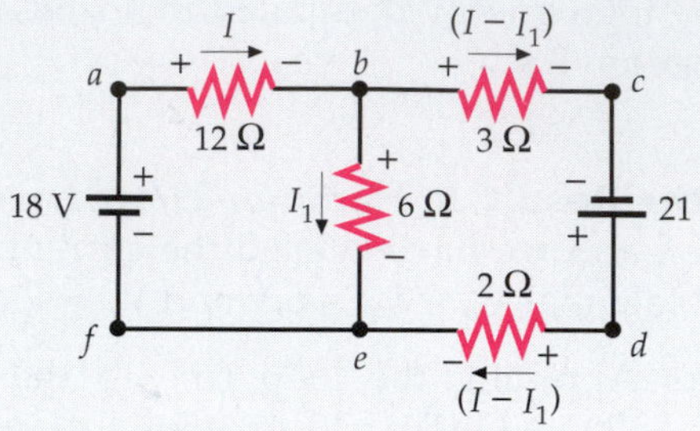
3. Apply Kirchhoff's loop rule to loop *abefa*.	$18\text{ V} - (12\ \Omega)I - (6\ \Omega)I_1 = 0$
4. Simplify your equation from step 3 to obtain an equation involving I and I_1.	$2I + I_1 = 3\text{ A}$
5. Apply Kirchhoff's loop rule to loop *bcdeb*.	$-(3\ \Omega)(I - I_1) + 21\text{ V} - (2\ \Omega)(I - I_1) + (6\ \Omega)I_1 = 0$
6. Simplify your equation in step 5.	$5I - 11I_1 = 21\text{ A}$
7. Solve your equations from steps 4 and 6 for I and I_1. One way to do this is to multiply the equation in step 4 by 11 and then add the equations to eliminate I_1.	$I = 2\text{ A},\quad I_1 = -1\text{ A}$
8. Find the current through the 21-V battery.	$I - I_1 = 3\text{ A}$
9. Use $V = (I - I_1)\,R_{eq}$ to find the potential drop across the parallel 3- and 6-Ω resistors.	$V = 6\text{ V}$
10. Use the result of step 9 to find the current in each of the parallel resistors.	$I_{3\Omega} = 2\text{ A},\ I_{6\Omega} = 1\text{ A}$
(*b*) Redraw Figure 26-31 showing the current through each part of the circuit (Figure 26-32). Begin with $V = 0$ at point *c* and calculate the potential at points *d*, *e*, *f*, *a*, and *b*.	**Figure 26-32** 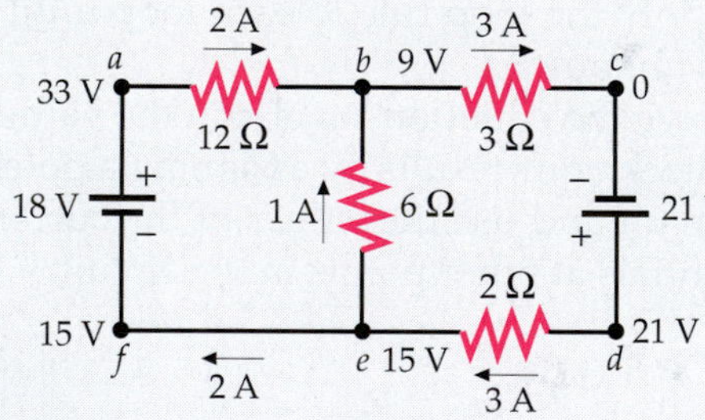

$$V_d = V_c + 21\text{ V} = 0 + 21\text{ V} = 21\text{ V}$$
$$V_e = V_d - (3\text{ A})(2\ \Omega) = 21\text{ V} - 6\text{ V} = 15\text{ V}$$
$$V_f = V_e = 15\text{ V}$$
$$V_a = V_f + 18\text{ V} = 15\text{ V} + 18\text{ V} = 33\text{ V}$$
$$V_b = V_a - (2\text{ A})(12\ \Omega) = 33\text{ V} - 24\text{ V} = 9\text{ V}$$

Check the Result From *b* to *c* the potential drops by $(3\text{ A})(3\ \Omega) = 9\text{ V}$, which gives $V_c = 0$, as assumed. From *e* to *b* the potential drops by $(1\text{ A})(6\ \Omega) = 6\text{ V}$ so $V_b = V_e - 6\text{ V} = 15\text{ V} - 6\text{ V} = 9\text{ V}$.

Ammeters, Voltmeters, and Ohmmeters

The devices that measure current, potential difference, and resistance are called **ammeters**, **voltmeters**, and **ohmmeters**, respectively. Often, all three of these meters are included in a single "multimeter" that can be switched from one use to another. You might use a voltmeter to measure the terminal voltage of your car battery and an ohmmeter to measure the resistance between two points in some electrical device at home (such as a toaster) when you suspect a short circuit or a broken wire.

To measure the current through a resistor in a simple circuit, we place an ammeter in series with the resistor, as shown in Figure 26-33, so that the ammeter and the resistor carry the same current. Since the ammeter has some resistance, the current in the circuit decreases slightly when the ammeter is inserted. Ideally, the ammeter should have a very small resistance so that the current to be measured is affected only slightly.

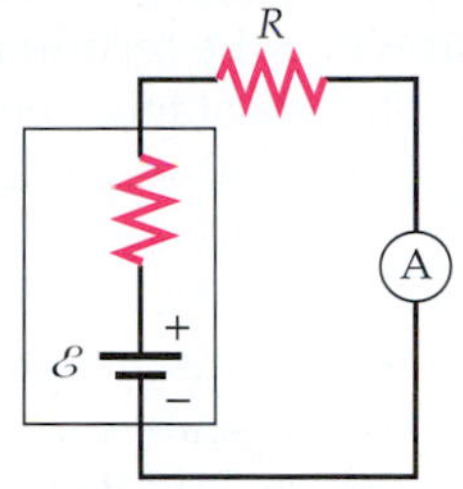

Figure 26-33 To measure the current in a resistor R, an ammeter Ⓐ is placed in series with the resistor so that it carries the same current as the resistor.

The potential difference across a resistor is measured by placing a voltmeter across the resistor in parallel with it, as shown in Figure 26-34, so that the potential drop across the voltmeter is the same as that across the resistor. The voltmeter reduces the resistance between points a and b, thus increasing the total current in the circuit and changing the potential drop across the resistor. A good voltmeter has a very large resistance so that its effect on the circuit is minimal.

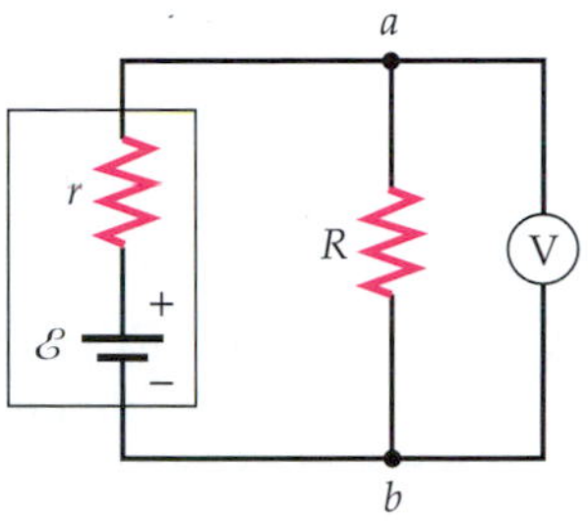

Figure 26-34 To measure the voltage drop across a resistor, a voltmeter Ⓥ is placed in parallel with the resistor so that the potential drops across the voltmeter and the resistor are the same.

The principal component of an ammeter and a voltmeter is a **galvanometer**, a device that detects small currents passing through it. The galvanometer is designed so that the scale reading is proportional to the current passing through. A typical galvanometer used in student laboratories consists of a coil of wire in the magnetic field of a permanent magnet. When the coil carries a current, the magnetic field exerts a torque on the coil causing it to rotate. A pointer attached to the coil indicates the reading on a scale. The coil itself contributes a small amount of resistance when the galvanometer is placed within a circuit. Many meters today have a digital readout rather than an indicator and a scale, but their basic operation is similar to that discussed here.

To construct an ammeter from a galvanometer, we place a small resistor called a **shunt resistor** in parallel with the galvanometer. The shunt resistance is usually much smaller than the resistance of the galvanometer so that most of the current is carried by the shunt resistor. The equivalent resistance of the ammeter is then approximately equal to the shunt resistance, and much smaller than the internal resistance of the galvanometer alone. To construct a voltmeter, we place a resistor with a large resistance in series with the galvanometer so that the equivalent resistance of the voltmeter is much larger than that of the galvanometer alone. Figure 26-35 illustrates the construction of an ammeter and voltmeter from a galvanometer. The resistance of the galvanometer R_g is shown separately in these schematic drawings, but it is actually part of the galvanometer.

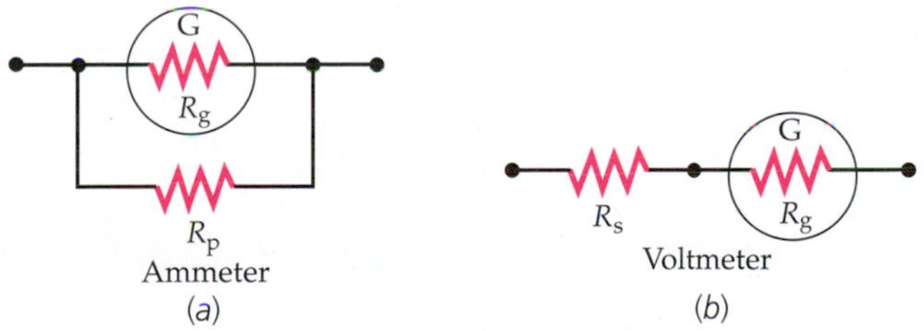

Figure 26-35 (*a*) An ammeter consists of a galvanometer Ⓖ whose resistance is R_g and a small parallel resistance R_p. (*b*) A voltmeter consists of a galvanometer Ⓖ and a large series resistance R_s.

A simple ohmmeter consists of a battery connected in series with a galvanometer and a resistor, as shown in Figure 26-36*a*. The resistance R_s is chosen such that when the terminals *a* and *b* are shorted (put in electrical contact, with negligible resistance between them), the current through the galvanometer gives a full-scale deflection. Thus, a full-scale deflection indicates no resistance between terminals *a* and *b*. A zero deflection indicates an infinite resistance between the terminals. When the terminals are connected across an unknown resistance *R*, the current through the galvanometer depends on *R*, so the scale can be calibrated to give a direct reading of *R*, as shown in Figure 26-36*b*. Because an ohmmeter sends a current through the resistance to be measured, some caution must be exercised when using this instrument. For example, you would not want to try to measure the resistance of a sensitive ammeter with an ohmmeter, because the current provided by the battery in the ohmmeter would probably damage the ammeter.

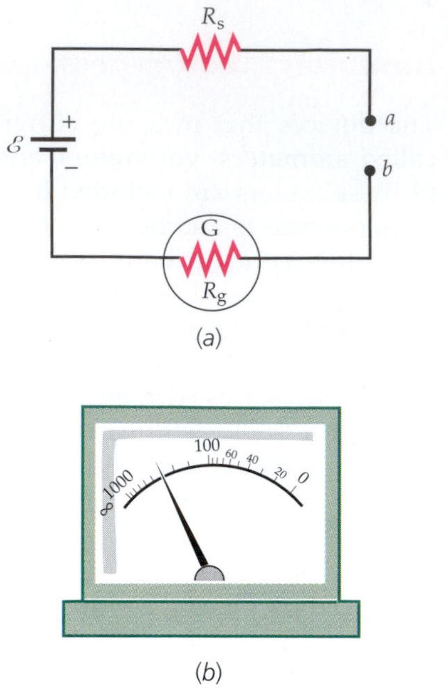

Figure 26-36 (*a*) An ohmmeter consists of a battery in series with a galvanometer and a resistor R_s, which is chosen such that the galvanometer gives full-scale deflection when points *a* and *b* are shorted. (*b*) When a resistor *R* is placed across *a* and *b*, the galvanometer needle deflects by an amount that depends on the value of *R*. The galvanometer scale is calibrated to give a readout in ohms.

26-6 *RC* Circuits

A circuit containing a resistor and capacitor is called an ***RC* circuit**. The current in an *RC* circuit flows in a single direction, as in all dc circuits, but the magnitude of the current varies with time. A practical example of an *RC* circuit is the circuit in the flash attachment of a camera. Before a flash photograph is taken, a battery in the flash attachment charges the capacitor through a resistor. When this is accomplished, the flash is ready. When the picture is taken, the capacitor discharges through the flash bulb. The capacitor is then recharged by the battery, and a short time later the flash is ready for another picture. Using Kirchhoff's rules, we can obtain equations for the charge *Q* and the current *I* as functions of time for both the charging and discharging of a capacitor through a resistor.

Discharging a Capacitor

Figure 26-37 shows a capacitor with initial charges of $+Q_0$ on the upper plate and $-Q_0$ on the lower plate. The capacitor is connected to a resistor *R* and a switch S, which is initially open. The potential difference across the capacitor is initially $V_0 = Q_0/C$, where *C* is the capacitance.

We close the switch at time $t = 0$. Since there is now a potential difference across the resistor, there must be a current in it. The initial current is

$$I_0 = \frac{V_0}{R} = \frac{Q_0}{RC} \qquad 26\text{-}25$$

The current is due to the flow of charge from the positive plate of the capacitor to the negative plate through the resistor. After a time, the charge on the capacitor is reduced. If we choose the direction of *I* to be clockwise, then the current equals the rate of *decrease* of that charge. If *Q* is the charge on the capacitor at any time, the current at that time is

$$I = -\frac{dQ}{dt} \qquad 26\text{-}26$$

Traversing the circuit in the direction of the current, we encounter a potential drop *IR* across the resistor and a potential increase *Q*/*C* across the capacitor. Thus, Kirchhoff's loop rule gives

$$\frac{Q}{C} - IR = 0 \qquad 26\text{-}27$$

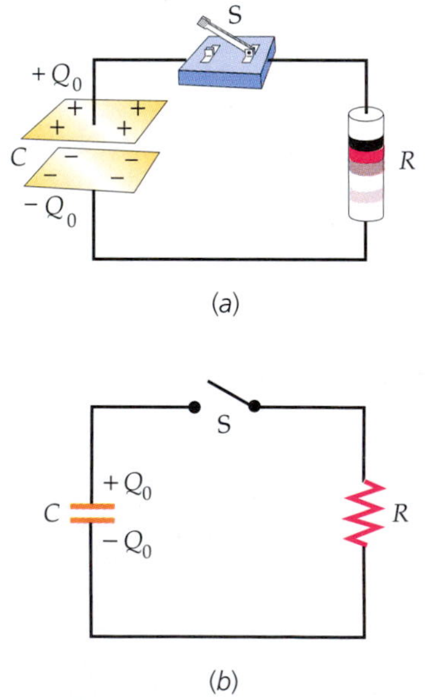

Figure 26-37 (*a*) A parallel-plate capacitor in series with a switch and a resistor *R*. (*b*) Circuit diagram for (*a*).

where both Q and I are functions of time and are related by Equation 26-26. Substituting $-dQ/dt$ for I in Equation 26-27, we have

$$\frac{Q}{C} + R\frac{dQ}{dt} = 0$$

or

$$\frac{dQ}{dt} = -\frac{1}{RC}Q \qquad 26\text{-}28$$

To solve this equation, we first separate the variables Q and t. Multiplying both sides of the equation by dt and dividing by Q, we obtain

$$\frac{dQ}{Q} = -\frac{dt}{RC} \qquad 26\text{-}29$$

Integrating from Q_0 at $t = 0$ to Q at time t gives

$$\ln\frac{Q}{Q_0} = -\frac{t}{RC}$$

Then

$$Q(t) = Q_0 e^{-t/RC} = Q_0 e^{-t/\tau} \qquad 26\text{-}30$$

where τ, called the **time constant**, is the time it takes for the charge to decrease to $1/e$ of its original value:

$$\tau = RC \qquad 26\text{-}31$$

Definition—Time constant

Figure 26-38 shows the charge on the capacitor in the circuit of Figure 26-37 as a function of time. After a time $t = \tau$, the charge is $Q = Q_0 e^{-1} = 0.37Q_0$, after a time $t = 2\tau$, the charge is $Q = Q_0 e^{-2} = 0.135Q_0$, and so forth. After a time equal to several time constants, the charge on the capacitor is negligible. This type of decrease, which is called an **exponential decrease**, is very common in nature. It occurs whenever the rate at which a quantity decreases is proportional to the quantity itself.*

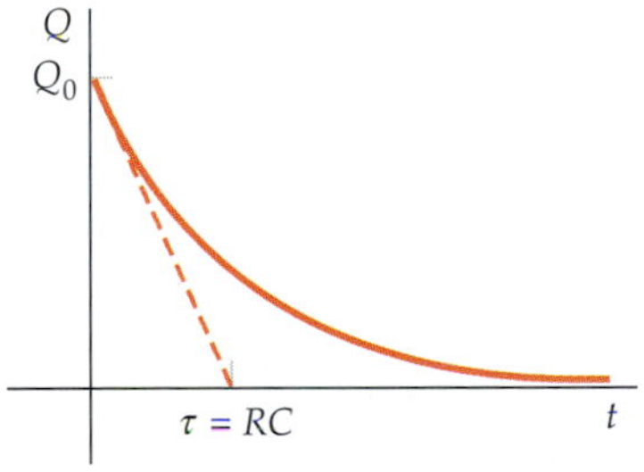

Figure 26-38 Plot of the charge on the capacitor versus time for the circuit in Figure 26-37 when the switch is closed at time $t = 0$. The time constant $\tau = RC$ is the time it takes for the charge to decrease to $e^{-1}Q_0$. The time constant is also the time it would take the capacitor to discharge fully if its discharge rate were constant, as indicated by the dashed line.

The decrease in the charge on a capacitor can be likened to the decrease in the amount of water in a bucket that has a small hole in the bottom. The rate at which the water flows out of the bucket is proportional to the pressure of the water, which is in turn proportional to the amount of water still in the bucket.

The current is obtained by differentiating Equation 26-30

$$I = -\frac{dQ}{dt} = \frac{Q_0}{RC}e^{-t/RC}$$

or

$$I = \frac{V_0}{R}e^{-t/RC} = I_0 e^{-t/\tau} \qquad 26\text{-}32$$

where $I_0 = Q_0/RC = V_0/R$ is the initial current. The current as a function of time is shown in Figure 26-39. The current also decreases exponentially with time and falls to $1/e$ of its initial value after a time $t = \tau = RC$.

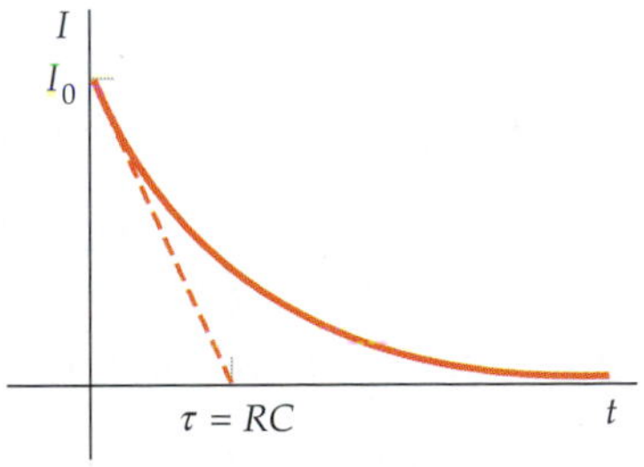

Figure 26-39 Plot of the current versus time for the circuit in Figure 26-37. The curve has the same shape as that in Figure 26-38. If the current decreased at a constant rate equal to its initial rate, it would reach zero after one time constant as indicated by the dashed line.

* We encountered exponential decreases in Chapter 14 when we studied the damped oscillator.

Example 26-17

A 4-μF capacitor is charged to 24 V and then connected across a 200-Ω resistor. Find (*a*) the initial charge on the capacitor, (*b*) the initial current through the 200-Ω resistor, (*c*) the time constant, and (*d*) the charge on the capacitor after 4 ms.

Picture the Problem The circuit diagram is the same as Figure 26-37.

(*a*) The initial charge is related to the capacitance and voltage:

$$Q_0 = CV = (4\ \mu\text{F})(24\ \text{V}) = 96\ \mu\text{C}$$

(*b*) The initial current is the initial voltage divided by the resistance:

$$I_0 = \frac{V_0}{R} = \frac{24\ \text{V}}{200\ \Omega} = 0.12\ \text{A}$$

(*c*) The time constant is RC:

$$\tau = RC = (200\ \Omega)(4\ \mu\text{F}) = 800\ \mu\text{s} = 0.8\ \text{ms}$$

(*d*) Substitute $t = 4$ ms into Equation 26-30 to find the charge on the capacitor at that time:

$$Q = Q_0 e^{-t/\tau} = (96\ \mu\text{C})e^{-(4\ \text{ms})/(0.8\ \text{ms})}$$
$$= (96\ \mu\text{C})e^{-5}$$
$$= 0.647\ \mu\text{C}$$

Exercise Find the current through the 200-Ω resistor at t = 4 ms. (*Answer* 0.809 mA)

Charging a Capacitor

Figure 26-40*a* shows a circuit for charging a capacitor. We will assume that the capacitor is initially uncharged. The switch, originally open, is closed at time $t = 0$. Charge immediately begins to flow through the resistor and onto the positive plate of the capacitor (Figure 26-40*b*). If the charge on the capacitor at some time is Q and the current in the circuit is I, Kirchhoff's loop rule gives

$$\mathcal{E} - V_R - V_C = 0$$

or

$$\mathcal{E} - IR - \frac{Q}{C} = 0 \qquad \text{26-33}$$

In this circuit, we've chosen the direction of I so that the current equals the rate at which the charge on the capacitor is *increasing*:

$$I = +\frac{dQ}{dt}$$

Substituting $+dQ/dt$ for I in Equation 26-33 gives

$$\mathcal{E} = R\frac{dQ}{dt} + \frac{Q}{C} \qquad \text{26-34}$$

At time $t = 0$, the charge on the capacitor is zero and the current is $I_0 = \mathcal{E}/R$. The charge then increases and the current decreases, as can be seen from Equation 26-33. The charge reaches a maximum value of $Q_f = C\mathcal{E}$ when the current I equals zero, as can also be seen from Equation 26-34.

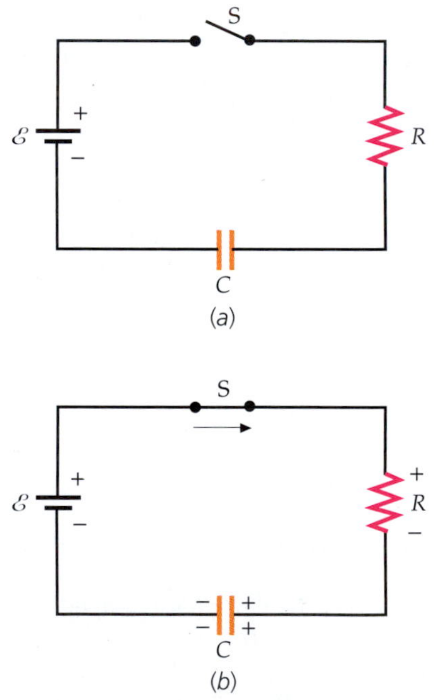

Figure 26-40 (*a*) Circuit for charging a capacitor to a potential difference $\mathcal{E}$. (*b*) After the switch is closed, there is a potential drop across the resistor and a charge on the capacitor.

Equation 26-34 can be solved in the same way as Equation 26-28. The details are left as a problem (see Problem 115). The result is

$$Q = C\mathcal{E}(1 - e^{-t/RC}) = Q_f(1 - e^{-t/\tau}) \quad \text{26-35}$$

where $Q_f = C\mathcal{E}$ is the final charge. The current is obtained from $I = dQ/dt$:

$$I = \frac{dQ}{dt} = -C\mathcal{E}e^{-t/RC}\left(\frac{-1}{RC}\right)$$

or

$$I = \frac{\mathcal{E}}{R}e^{-t/RC} = I_0e^{-t/\tau} \quad \text{26-36}$$

where the initial current in this case is $I_0 = \mathcal{E}/R$.

Figures 26-41 and 26-42 show the charge and the current as functions of time.

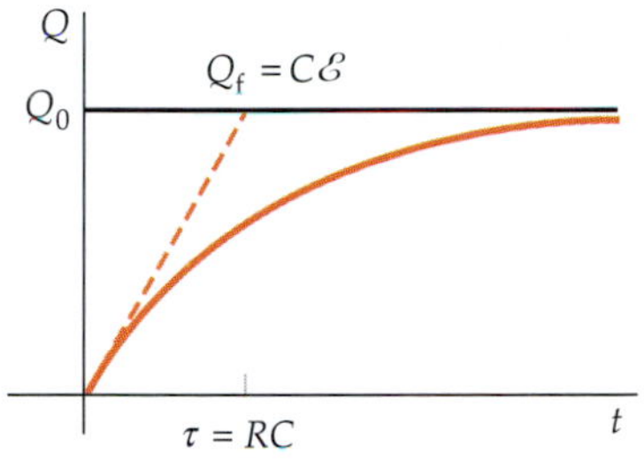

Figure 26-41 Plot of the charge on the capacitor versus time for the charging circuit of Figure 26-40 after the switch is closed at $t = 0$. After a time $t = \tau = RC$, the charge on the capacitor is $0.63C\mathcal{E}$, where $C\mathcal{E}$ is its final charge. If the charging rate were constant, the capacitor would be fully charged after a time $t = \tau$.

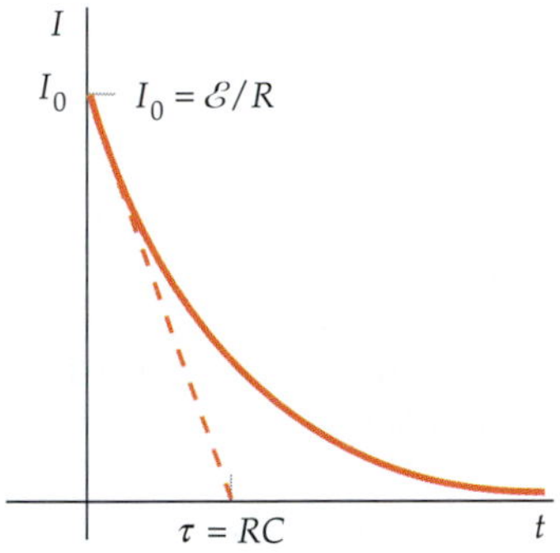

Figure 26-42 Plot of the current versus time for the charging circuit of Figure 26-40. The current is initially $\mathcal{E}/R$, and it decreases exponentially with time.

Exercise Show that Equation 26-35 does indeed satisfy Equation 26-34 by substituting $Q(t)$ and dQ/dt into Equation 26-34.

Exercise What fraction of the maximum charge is on the charging capacitor after a time $t = 2\tau$? (*Answer* 0.86)

Example 26-18 *try it yourself*

A 6-V battery of negligible internal resistance is used to charge a 2-μF capacitor through a 100-Ω resistor. Find (*a*) the initial current, (*b*) the final charge on the capacitor, and (*c*) the time required for the charge to reach 90% of its final value.

Cover the column to the right and try these on your own before looking at the answers.

Steps	***Answers***
(*a*) Find the initial current from $I_0 = \mathcal{E}/R$.	$I_0 = 0.06$ A
(*b*) Find the final charge from $Q = C\mathcal{E}$.	$Q_f = 12\ \mu C$
(*c*) 1. Set $Q = 0.9\ Q_f$ in Equation 26-35 and solve for $e^{+t/\tau}$.	$e^{t/\tau} = 10$
2. Take the natural logarithm of each side of your equation in the previous step.	$\ln(e^{t/\tau}) = t/\tau = \ln 10 = 2.3$
3. Solve for t.	$t = 2.3RC = 460\ \mu s$

Example 26-19

The capacitor in the circuit shown in Figure 26-43 is initially uncharged. Find the current through the battery (*a*) immediately after the switch is closed, and (*b*) a long time after the switch is closed.

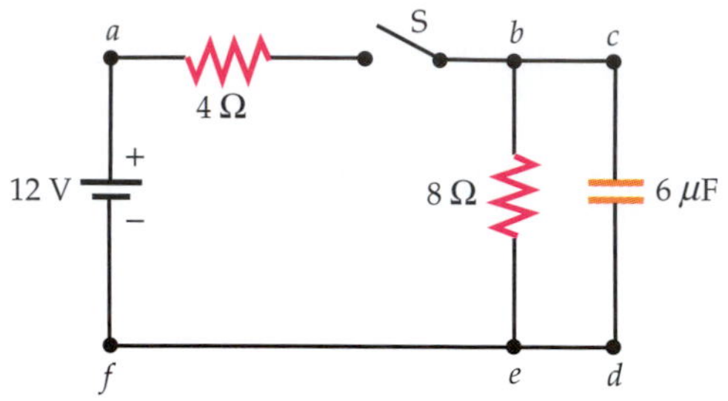

Figure 26-43

(*a*) Since the capacitor is initially uncharged, the potential is the same at points *d* and *c* just after the switch is closed. There is thus no initial current through the 8-Ω resistor between *b* and *e*. Apply the loop rule to the outer loop (*abcdefa*):

$$12\text{ V} - (4\ \Omega)I_0 = 0$$

$$I_0 = 3\text{ A}$$

(*b*) After a long time, the capacitor is fully charged, and no more charge flows onto or off of the plates. Apply the loop rule to the left loop (*abefa*):

$$12\text{ V} - (4\ \Omega)I_f - (8\ \Omega)I_f = 0$$

$$I_f = 1\text{ A}$$

Remarks The analysis of this circuit at the extreme times when the capacitor is either uncharged or fully charged is simple. When the capacitor is uncharged, it acts like a short circuit between points *c* and *d*, that is, the circuit is the same as the one shown in Figure 26-44*a*, where we have replaced the capacitor by a wire of zero resistance. When the capacitor is fully charged, it acts like an open circuit, as shown in Figure 26-44*b*.

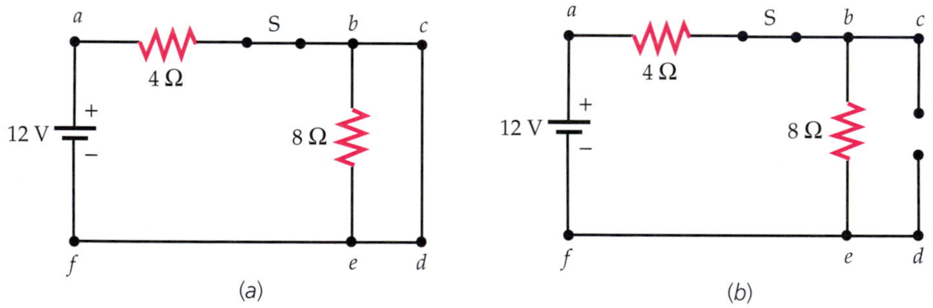

Figure 26-44

Energy Conservation in Charging a Capacitor

During the charging process, a total charge $Q_f = \mathcal{E}C$ flows through the battery. The battery therefore does work

$$W = Q_f\mathcal{E} = C\mathcal{E}^2$$

Half of this work is accounted for by the energy stored in the capacitor (see Equation 25-11):

$$U = \tfrac{1}{2}QV = \tfrac{1}{2}Q_f\mathcal{E} = \tfrac{1}{2}C\mathcal{E}^2$$

We now show that the other half of work done by the battery goes into Joule heat in the resistance of the circuit. The rate at which energy is put into the resistance R is

$$\frac{dW_R}{dt} = I^2R$$

Using Equation 26-36 for the current, we have

$$\frac{dW_R}{dt} = \left(\frac{\mathcal{E}}{R}e^{-t/RC}\right)^2 R = \frac{\mathcal{E}^2}{R}e^{-2t/RC}$$

We find the total Joule heat by integrating from $t = 0$ to $t = \infty$:

$$W_R = \int_0^\infty \frac{\mathcal{E}^2}{R}e^{-2t/RC}\,dt$$

The integration can be done by substituting $x = 2t/RC$. Then,

$$dt = \frac{RC}{2}dx$$

and

$$W_R = \frac{\mathcal{E}^2}{R}\frac{RC}{2}\int_0^\infty e^{-x}\,dx = \frac{\mathcal{E}^2C}{2}(-e^{-x})\Big|_0^\infty = \frac{\mathcal{E}^2C}{2}(-0 + 1)$$

The total Joule heat is thus

$$W_R = \frac{1}{2}\mathcal{E}^2C$$

This result is independent of the resistance R. Thus, when a capacitor is charged by a battery with a constant emf, half the energy provided by the battery is stored in the capacitor and half goes into thermal energy, independent of the resistance. The thermal energy includes the energy that goes into the internal resistance of the battery.

Summary

1. Ohm's law is an empirical law that holds only for certain materials.
2. Current, resistance, and emf are important *defined* quantities.
3. Kirchhoff's rules follow from the conservation of charge and the conservation of energy.

Topic	Remarks and Relevant Equations	
1. Electric Current	Electric current is the rate of flow of charge through a cross-sectional area.	
	$I = \frac{\Delta Q}{\Delta t} = qnAv_d$	**26-1, 26-3**
Drift velocity	In a conducting wire, electric current is the result of the slow drift of negatively charged electrons that are accelerated by an electric field in the wire and then collide with the lattice ions. Typical drift velocities of electrons in wires are of the order of a few millimeters per second.	

2.	**Resistance**		
	Definition	$R = \frac{V}{I}$	**26-5**
	Resistivity ρ	$R = \rho\frac{L}{A}$	**26-8**
3.	**Ohm's Law**	For ohmic materials, the resistance does not depend on the current or voltage drop: $V = IR \quad (R \text{ constant})$	**26-7**
4.	**Power**		
	Supplied to a device or segment	$P = VI$	**26-10**
	Dissipated in a resistor	$P = VI = I^2R = \frac{V^2}{R}$	**26-11**
5.	**emf**	A device that supplies energy to a circuit is called a source of emf.	
	Power supplied by an emf	$P = \mathcal{E}I$	**26-12**
6.	**Battery**		
	Ideal	An ideal battery is a source of emf that maintains a constant potential difference between its two terminals, independent of the rate of flow of charge between them.	
	Real	A real battery can be considered to be an ideal battery in series with a small resistance called its internal resistance.	
	Terminal voltage	$V_a - V_b = \mathcal{E} - Ir$	**26-13**
	Total energy	$W = Q\mathcal{E}$	**26-15**
7.	**Equivalent Resistance**		
	Series	$R_{eq} = R_1 + R_2 + R_3 + \cdots$	**26-17**
	Parallel	$\frac{1}{R_{eq}} = \frac{1}{R_1} + \frac{1}{R_2} + \frac{1}{R_3} + \cdots$	**26-21**
8.	**Kirchhoff's rules**	1. When any closed circuit loop is traversed, the algebraic sum of the changes in potential must equal zero. 2. At any junction in a circuit where the current can divide, the sum of the currents into the junction must equal the sum of the currents out of the junction.	
9.	**Measuring Devices**		
	Ammeter	An ammeter is a very low resistance device that is placed in series with a circuit element to measure the current in the element.	
	Voltmeter	A voltmeter is a very high resistance device that is placed in parallel with a circuit element to measure the voltage drop across the element.	

Ohmmeter	An ohmmeter is a device containing a battery that is used to measure the resistance of a circuit element placed across its terminals.	
10. Discharging a Capacitor		
Charge on the capacitor	$Q(t) = Q_0 e^{-t/RC} = Q_0 e^{-t/\tau}$	**26-30**
Current in the circuit	$I = \frac{V_0}{R} e^{-t/RC} = I_0 e^{-t/\tau}$	**26-32**
Time constant	$\tau = RC$	**26-31**
11. Charging a Capacitor		
Charge on the capacitor	$Q = C\mathcal{E}(1 - e^{-t/RC}) = Q_f(1 - e^{-t/\tau})$	**26-35**
Current in the circuit	$I = \frac{\mathcal{E}}{R} e^{-t/RC} = I_0 e^{-t/\tau}$	**26-36**

Problem-Solving Guide

To solve circuit problems

1. Draw a sketch of the circuit.
2. Choose a direction for the current in each branch of the circuit, and label the currents in the circuit diagram. Add plus and minus signs to indicate the high- and low-potential sides of each resistor, capacitor, and source of emf.
3. Replace any combination of resistors in series or parallel with its equivalent resistance.
4. Apply the junction rule to each junction where the current divides.
5. Apply the loop rule to several loops to obtain as many equations as unknowns.
6. Solve the equations to obtain the values of the unknowns.
7. Check your results by assigning a potential of zero to one point in the circuit and then using the values of the currents found to determine the potentials at other points in the circuit.

Summary of Worked Examples

Type of Calculation	Procedure and Relevant Examples
1. Resistance	
Find the resistance of a wire.	Use $R = \rho L/A$. **Examples 26-3, 26-4**
2. Energy	
Calculate the energy dissipated in a circuit.	Use $P = I^2R = V^2/R = VI$ for each resistor and $P = \mathcal{E}I$ for each emf. **Examples 26-6, 26-7, 26-8, 26-9, 26-10, 26-13, 26-15**
3. Equivalent Resistance	
Find the equivalent resistance.	Use $R_{eq} = R_1 + R_2 + \cdots$ (series) and $1/R_{eq} = 1/R_1 + 1/R_2 + \cdots$ (parallel). **Examples 26-6, 26-10, 26-11, 26-12**

4. Circuits Without Capacitors

Find the current in a single-loop circuit.	Apply Kirchhoff's loop rule.	**Examples 26-13, 26-14**
Find the currents in a multiloop circuit.	Apply Kirchhoff's loop and junction rules.	**Examples 26-15, 26-16**

5. *RC* Circuits

Find the charge on a capacitor.	Use $Q(t) = Q_0 e^{-t/RC}$ for a capacitor that is discharging, and $Q = C\mathcal{E}(1 - e^{-t/RC})$ for a capacitor that is charging.	**Examples 26-17, 26-18, 26-19**
Find the time constant.	Use $\tau = RC$.	**Examples 26-17, 26-18**
Find the initial and final current.	An uncharged capacitor is momentarily equivalent to a resistor of zero resistance. For a simple series circuit containing an uncharged capacitor, $I_i = \mathcal{E}/R$. A fully charged capacitor is equivalent to an open circuit. For a simple series circuit containing a fully charged capacitor, $I_f = 0$.	**Examples 26-17, 26-19**

Problems

Conceptual Problems

Problems from Optional and Exploring sections

In a few problems, you are given more data than you actually need; in a few other problems, you are required to supply data from your general knowledge, outside sources, or informed estimates.

• Single-concept, single-step, relatively easy
•• Intermediate-level, may require synthesis of concepts
••• Challenging, for advanced students

Current and Motion of Charges

1 • In our study of electrostatics, we concluded that there is no electric field within a conductor in electrostatic equilibrium. How is it that we can now discuss electric fields inside a conductor?

2 • A physics professor has assembled his class at the baggage-claim carousel of the local airport to demonstrate an analog of electrical current. "Think of each suitcase on the conveyor belt as a package of electrons carrying one coulomb of charge," he says. Counting and timing the suitcases reveals that the conveyor belt represents a wire carrying a constant 2-A current (constant as long as annoyed travelers could be kept away from their baggage by some of the huskier students). (*a*) How many suitcases will go by a given point in 5.0 min? (*b*) How many electrons does that represent?

3 • A 10-gauge copper wire carries a current of 20 A. Assuming one free electron per copper atom, calculate the drift velocity of the electrons.

4 • In a fluorescent tube of diameter 3.0 cm, 2.0×10^{18} electrons and 0.5×10^{18} positive ions (with a charge of $+e$) flow through a cross-sectional area each second. What is the current in the tube?

5 • In a certain electron beam, there are 5.0×10^6 electrons per cubic centimeter. Suppose the kinetic energy of each electron is 10.0 keV, and the beam is cylindrical, with a diameter of 1.00 mm. (*a*) What is the velocity of an electron in the beam? (*b*) Find the beam current.

6 •• A charge $+q$ moves in a circle of radius r with speed v. (*a*) Express the frequency f with which the charge passes a particular point in terms of r and v. (*b*) Show that the average current is qf and express it in terms of v and r.

7 •• A ring of radius a with a linear charge density λ rotates about its axis with angular velocity ω. Find an expression for the current.

8 •• A 10-gauge copper wire and a 14-gauge copper wire are welded together end to end. The wires carry a current of 15 A. If there is one free electron per copper atom in each wire, find the drift velocity of the electrons in each wire.

9 •• In a certain particle accelerator, a proton beam with a diameter of 2.0 mm constitutes a current of 1.0 mA. The kinetic energy of each proton is 20 MeV. The beam strikes a metal target and is absorbed by it. (*a*) What is the number n of protons per unit volume in the beam? (*b*) How many protons strike the target in 1.0 min? (*c*) If the target is initially uncharged, express the charge of the target as a function of time.

10 •• The current in a wire varies with time according to the relation $I = 20 + 3t^2$, where I is in amperes and t is in seconds. (*a*) How many coulombs are transported by the wire between $t = 0$ and $t = 10$ s? (*b*) What constant current would transport the same charge in the same time interval?

11 •• In a proton supercollider, the protons in a 5-mA beam move with nearly the speed of light. (*a*) How many protons are there per meter of the beam? (*b*) If the cross-sectional area of the beam is 10^{-6} m^2, what is the number density of protons?

Resistance and Ohm's Law

12 • Figure 26-8 illustrates a mechanical analog of a simple electric circuit. Devise another mechanical analog in which the current is represented by a flow of water instead of marbles.

13 • Two wires of the same material with the same length have different diameters. Wire A has twice the diameter of wire B. If the resistance of wire B is R, then what is the resistance of wire A?

(*a*) R
(*b*) $2R$
(*c*) $R/2$
(*d*) $4R$
(*e*) $R/4$

14 •• Discuss the difference between an emf and a potential difference.

15 •• Name several common sources of emf. What sort of energy is converted into electrical energy in each?

16 •• A metal bar is to be used as a resistor. Its dimensions are 2 by 4 by 10 units. To get the smallest resistance from this bar, one should attach leads to the opposite sides that have the dimensions of

(*a*) 2 by 4 units.
(*b*) 2 by 10 units.
(*c*) 4 by 10 units.
(*d*) All connections will give the same resistance.
(*e*) None of the above is correct.

17 •• Two cylindrical copper wires have the same mass. Wire A is twice as long as wire B. Their resistances are related by

(*a*) $R_A = 8R_B$.
(*b*) $R_A = 4R_B$.
(*c*) $R_A = 2R_B$.
(*d*) $R_A = R_B$.

18 • A 10-m-long wire of resistance 0.2 Ω carries a current of 5 A. (*a*) What is the potential difference across the wire? (*b*) What is the magnitude of the electric field in the wire?

19 • A potential difference of 100 V produces a current of 3 A in a certain resistor. (*a*) What is the resistance of the resistor? (*b*) What is the current when the potential difference is 25 V?

20 • A block of carbon is 3.0 cm long and has a square cross-sectional area with sides of 0.5 cm. A potential difference of 8.4 V is maintained across its length. (*a*) What is the resistance of the block? (*b*) What is the current in this resistor?

21 • A carbon rod with a radius of 0.1 mm is used to make a resistor. The resistivity of this material is 3.5×10^{-5} Ω·m. What length of the carbon rod will make a 10-Ω resistor?

22 • The third (current-carrying) rail of a subway track is made of steel and has a cross-sectional area of about 55 cm^2. What is the resistance of 10 km of this track? (Use ρ for iron.)

23 • What is the potential difference across one wire of a 30-m extension cord made of 16-gauge copper wire carrying a current of 5.0 A?

24 • How long is a 14-gauge copper wire that has a resistance of 2 Ω?

25 •• A cylinder of glass 1 cm long has a resistivity of 10^{12} Ω·m. How long would a copper wire of the same cross-sectional area need to be to have the same resistance as the glass cylinder?

26 •• An 80.0-m copper wire 1.0 mm in diameter is joined end to end with a 49.0-m iron wire of the same diameter. The current in each is 2.0 A. (*a*) Find the electric field in each wire. (*b*) Find the potential drop across each wire.

27 •• A copper wire and an iron wire with the same length and diameter carry the same current I. (*a*) Find the ratio of the potential drops across these wires. (*b*) In which wire is the electric field greater?

28 •• A variable resistance R is connected across a potential difference V that remains constant. When $R = R_1$, the current is 6.0 A. When R is increased to $R_2 = R_1 + 10.0$ Ω, the current drops to 2.0 A. Find (*a*) R_1 and (*b*) V.

29 •• A rubber tube 1 m long with an inside diameter of 4 mm is filled with a salt solution that has a resistivity of 10^{-3} Ω·m. Metal plugs form electrodes at the ends of the tube. (*a*) What is the resistance of the filled tube? (*b*) What is the resistance of the filled tube if it is uniformly stretched to a length of 2 m?

30 •• A wire of length 1 m has a resistance of 0.3 Ω. It is uniformly stretched to a length of 2 m. What is its new resistance?

31 •• Currents up to 30 A can be carried by 10-gauge copper wire. (*a*) What is the resistance of 100 m of 10-gauge copper wire? (*b*) What is the electric field in the wire when the current is 30 A? (*c*) How long does it take for an electron to travel 100 m in the wire when the current is 30 A?

32 •• A cube of copper has sides of 2.0 cm. If it is drawn out to form a 14-gauge wire, what will its resistance be?

33 ••• A semiconducting diode is a nonlinear device whose current I is related to the voltage V across the diode by $I = I_0(e^{eV/kT} - 1)$, where k is Boltzmann's constant, e is the magnitude of the charge on an electron, and T is the absolute temperature. If $I_0 = 10^{-9}$ A and $T = 293$ K, (*a*) what is the resistance of the diode for $V = 0.5$ V? (*b*) What is the resistance for $V = 0.6$ V?

34 ••• Find the resistance between the ends of the half ring shown in Figure 26-45. The resistivity of the material of the ring is ρ.

Figure 26-45 Problem 34

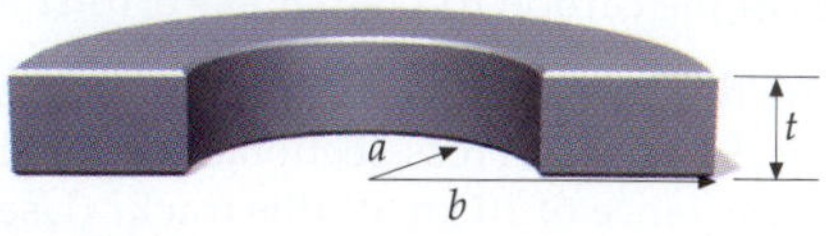

35 ••• The radius of a wire of length L increases linearly along its length according to $r = a + [(b - a)/L]\,x$, where x is the distance from the small end of radius a. What is the resistance of this wire in terms of its resistivity ρ, length L, radius a, and radius b?

36 ••• The space between two concentric spherical-shell conductors is filled with a material that has a resistivity of $10^9\ \Omega\cdot\text{m}$. If the inner shell has a radius of 1.5 cm and the outer shell has a radius of 5 cm, what is the resistance between the conductors? (*Hint:* Find the resistance of a spherical-shell element of the material of area $4\pi r^2$ and length dr, and integrate to find the total resistance of the set of shells in series.)

37 ••• The space between two metallic coaxial cylinders of length L and radii a and b is completely filled with a material having a resistivity ρ. (*a*) What is the resistance between the two cylinders? (See the hint in Problem 36.) (*b*) Find the current between the two cylinders if $\rho = 30\ \Omega\cdot\text{m}$, $a = 1.5$ cm, $b = 2.5$ cm, $L = 50$ cm, and a potential difference of 10 V is maintained between the two cylinders.

Temperature Dependence of Resistance

38 • A tungsten rod is 50 cm long and has a square cross-sectional area with sides of 1.0 mm. (*a*) What is its resistance at 20°C? (*b*) What is its resistance at 40°C?

39 • At what temperature will the resistance of a copper wire be 10% greater than it is at 20°C?

40 •• A toaster with a Nichrome heating element has a resistance of 80 Ω at 20°C and an initial current of 1.5 A. When the heating element reaches its final temperature, the current is 1.3 A. What is the final temperature of the heating element?

41 •• An electric space heater has a Nichrome heating element with a resistance of 8 Ω at 20°C. When 120 V are applied, the electric current heats the Nichrome wire to 1000°C. (*a*) What is the initial current drawn by the cold heating element? (*b*) What is the resistance of the heating element at 1000°C? (*c*) What is the operating wattage of this heater?

42 •• A 10-Ω Nichrome resistor is wired into an electronic circuit using copper leads (wires) of diameter 0.6 mm with a total length of 50 cm. (*a*) What additional resistance is due to the copper leads? (*b*) What percentage error in the total added resistance is produced by neglecting the resistance of the copper leads? (*c*) What change in temperature would produce a change in resistance of the Nichrome-wire equal to the resistance of the copper leads?

43 •• The filament of a certain lamp has a resistance that increases linearly with temperature. When a constant voltage is switched on, the initial current decreases until the filament reaches its steady-state temperature. The temperature coefficient of resistivity of the filament is $4 \times 10^{-3}\ \text{K}^{-1}$. The final current through the filament is one-eighth the initial current. What is the change in temperature of the filament?

44 ••• A wire of cross-sectional area A, length L_1, resistivity ρ_1, and temperature coefficient α_1 is connected end to end to a second wire of the same cross-sectional area, length L_2, resistivity ρ_2, and temperature coefficient α_2, so that the wires carry the same current. (*a*) Show that if $\rho_1 L_1 \alpha_1 + \rho_2 L_2 \alpha_2 = 0$, the total resistance R is independent of temperature for small temperature changes. (*b*) If one wire is made of carbon and the other is copper, find the ratio of their lengths for which R is approximately independent of temperature.

Energy in Electric Circuits

45 • A resistor carries a current I. The power dissipated in the resistor is P. What is the power dissipated if the same resistor carries current $3I$? (Assume no change in resistance.)

(*a*) P (*b*) $3P$ (*c*) $P/3$
(*d*) $9P$ (*e*) $P/9$

46 • The power dissipated in a resistor is P when the voltage drop across it is V. If the voltage drop is increased to 2 V (with no change in resistance), what is the power dissipated?

(*a*) P (*b*) $2P$ (*c*) $4P$
(*d*) $P/2$ (*e*) $P/4$

47 • A heater consists of a variable resistance connected across a constant voltage supply. To increase the heat output, should you decrease the resistance or increase it?

48 •• Two resistors dissipate the same amount of power. The potential drop across resistor A is twice that across resistor B. If the resistance of resistor B is R, what is the resistance of A?

(*a*) R (*b*) $2R$ (*c*) $R/2$
(*d*) $4R$ (*e*) $R/4$

49 • Find the power dissipated in a resistor connected across a constant potential difference of 120 V if its resistance is (*a*) 5 Ω and (*b*) 10 Ω.

50 • A 10,000-Ω carbon resistor used in electronic circuits is rated at 0.25 W. (*a*) What maximum current can this resistor carry? (*b*) What maximum voltage can be placed across this resistor?

51 • A 1-kW heater is designed to operate at 240 V. (*a*) What is its resistance, and what current does it draw? (*b*) What is the power dissipated in this resistor if it operates at 120 V? Assume that its resistance is constant.

52 • A battery has an emf of 12.0 V. How much work does it do in 5 s if it delivers a current of 3 A?

53 • A battery with 12-V emf has a terminal voltage of 11.4 V when it delivers a current of 20 A to the starter of a car. What is the internal resistance r of the battery?

54 • (*a*) How much power is delivered by the emf of the battery in Problem 53 when it delivers a current of 20 A? (*b*) How much of this power is delivered to the starter? (*c*) By how much does the chemical energy of the battery decrease when it delivers a current of 20 A to the starter for 3 min? (*d*) How much heat is developed in the battery when it delivers a current of 20 A for 3 min?

55 • A physics student runs a 1200-W electric heater constantly in her basement bedroom during the winter time. If electric energy costs 9 cents per kilowatt-hour, how much does this electric heating cost per 30-day month?

56 • A battery with an emf of 6 V and an internal resistance of 0.3 Ω is connected to a variable resistance R. Find the current and power delivered by the battery when R is (*a*) 0, (*b*) 5 Ω, (*c*) 10 Ω, and (*d*) infinite.

57 •• Staying up late to study, and having no stove to heat water, you use a 200-W heater from the lab to make coffee throughout the night. If 90% of the energy produced by the heater goes toward heating the water in your cup, (*a*) how long does it take to heat 0.25 kg of water from 15 to 100°C? (*b*) If you fall asleep while the water is heating, how long will it take to boil away after it reaches 100°C?

58 •• Suppose the bulb in a two-cell flashlight draws 4 W of power. The batteries go dead in 45 min and cost $7.99. (*a*) How many kilowatt-hours of energy can be supplied by the two batteries? (*b*) What is the cost per kilowatt-hour of energy if the batteries cannot be recharged? (*c*) If the batteries can be recharged at a cost of 9 cents per kilowatt-hour, what is the cost of recharging them?

59 •• A 12-V automobile battery with negligible internal resistance can deliver a total charge of 160 A·h. (*a*) What is the total stored energy in the battery? (*b*) How long could this battery provide 150 W to a pair of headlights?

60 •• A space heater in an old home draws a 12.5-A current. A pair of 12-gauge copper wires carries the current from the fuse box to the wall outlet, a distance of 30 m. The voltage at the fuse box is exactly 120 V. (*a*) What is the voltage delivered to the space heater? (*b*) If the fuse will blow at a current of 20 A, how many 60-W bulbs can be supplied by this line when the space heater is on? (Assume that the wires from the wall to the space heater and to the light fixtures have negligible resistance.)

61 •• A lightweight electric car is powered by ten 12-V batteries. At a speed of 80 km/h, the average frictional force is 1200 N. (*a*) What must be the power of the electric motor if the car is to travel at a speed of 80 km/h? (*b*) If each battery can deliver a total charge of 160 A·h before recharging, what is the total charge in coulombs that can be delivered by the 10 batteries before charging? (*c*) What is the total electrical energy delivered by the 10 batteries before recharging? (*d*) How far can the car travel at 80 km/h before the batteries must be recharged? (*e*) What is the cost per kilometer if the cost of recharging the batteries is 9 cents per kilowatt-hour?

62 ••• A 100-W heater is designed to operate with an emf of 120 V. (*a*) What is its resistance, and what current does it draw? (*b*) Show that if the potential difference across the heater changes by a small amount ΔV, the power changes by a small amount ΔP, where $\Delta P/P \approx 2\,\Delta V/V$. (*Hint:* Approximate the changes with differentials.) (*c*) Find the approximate power dissipated in the heater if the potential difference is decreased to 115 V.

Combinations of Resistors

63 • Two resistors are connected in parallel across a potential difference. The resistance of resistor A is twice that of resistor B. If the current carried by resistor A is I, then what is the current carried by B?

(*a*) I (*b*) $2I$ (*c*) $I/2$ (*d*) $4I$ (*e*) $I/4$

64 • Two resistors are connected in series across a potential difference. Resistor A has twice the resistance of resistor B. If the current carried by resistor A is I, then what is the current carried by B?

(*a*) I (*b*) $2I$ (*c*) $I/2$ (*d*) $4I$ (*e*) $I/4$

65 •• When two identical resistors are connected in series across the terminals of a battery, the power delivered by the battery is 20 W. If these resistors are connected in parallel across the terminals of the same battery, what is the power delivered by the battery?

(*a*) 5 W (*b*) 10 W (*c*) 20 W (*d*) 40 W (*e*) 80 W

66 • (*a*) Find the equivalent resistance between points *a* and *b* in Figure 26-46. (*b*) If the potential drop between *a* and *b* is 12 V, find the current in each resistor.

Figure 26-46 Problem 66

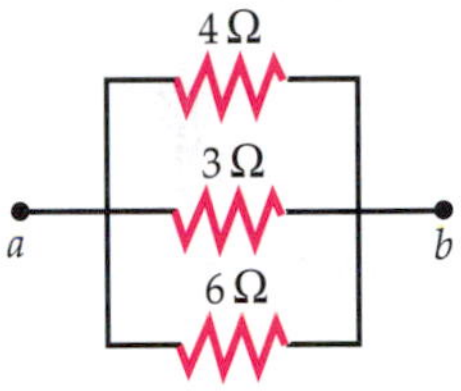

67 • Repeat Problem 66 for the resistor network shown in Figure 26-47.

Figure 26-47 Problem 67

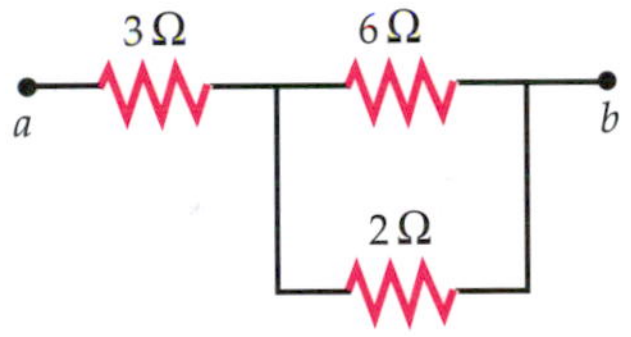

68 • Repeat Problem 66 for the resistor network shown in Figure 26-48.

Figure 26-48
Problems 68 and 69

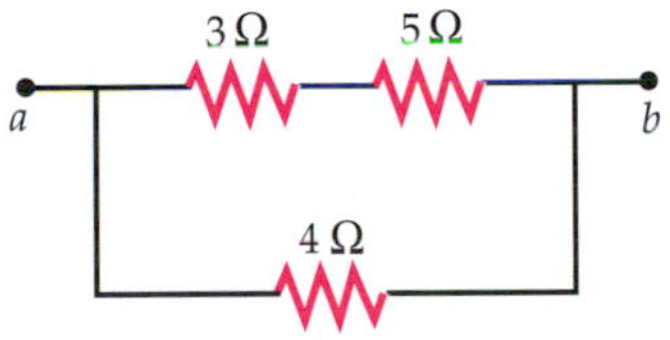

69 • In Figure 26-48, the current in the 4-Ω resistor is 4 A. (*a*) What is the potential drop between *a* and *b*? (*b*) What is the current in the 3-Ω resistor?

70 • (*a*) Show that the equivalent resistance between points *a* and *b* in Figure 26-49 is R. (*b*) What would be the effect of adding a resistance R between points *c* and *d*?

Figure 26-49 Problem 70

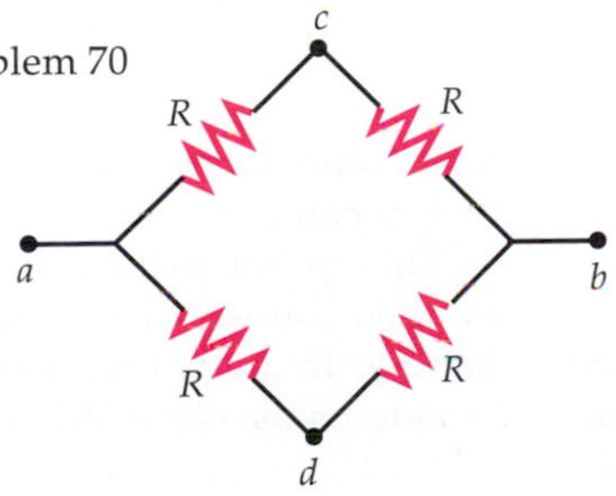

71 •• The battery in Figure 26-50 has negligible internal resistance. Find (*a*) the current in each resistor and (*b*) the power delivered by the battery.

Figure 26-50
Problem 71

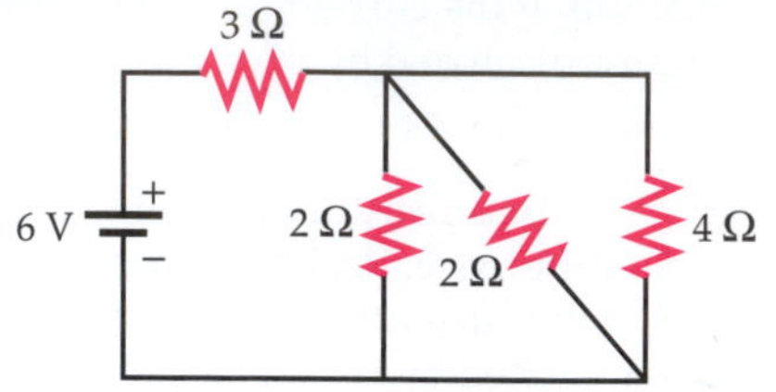

72 •• A battery has an emf $\mathcal{E}$ and an internal resistance r. When a 5.0-Ω resistor is connected across the terminals, the current is 0.5 A. When this resistor is replaced by an 11.0-Ω resistor, the current is 0.25 A. Find (*a*) the emf $\mathcal{E}$ and (*b*) the internal resistance r.

73 •• Consider the equivalent resistance of two resistors R_1 and R_2 connected in parallel as a function of the ratio $x = R_2/R_1$. (*a*) Show that $R_{\text{eq}} = R_1 x/(1 + x)$. (*b*) Sketch a plot of R_{eq} as a function of x.

74 •• Repeat Problem 66 for the resistor network shown in Figure 26-51.

Figure 26-51
Problem 74

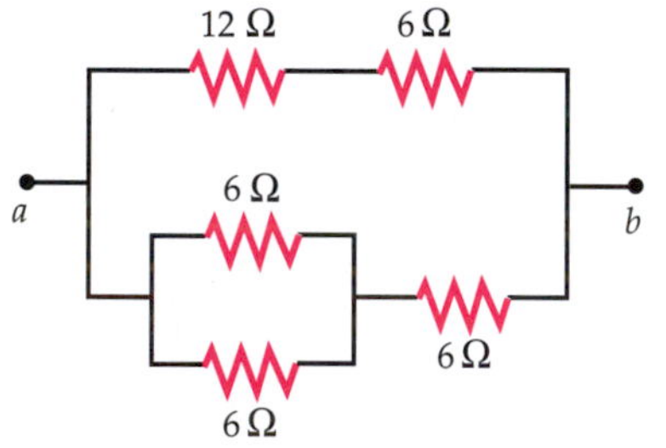

75 •• Repeat Problem 66 for the resistor network shown in Figure 26-52.

Figure 26-52
Problem 75

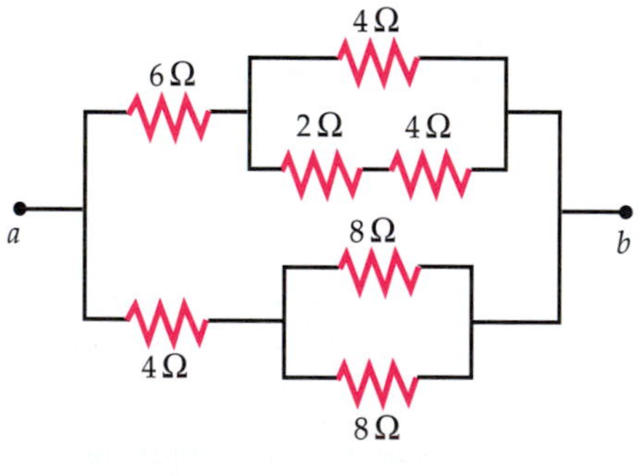

76 •• A length of wire has a resistance of 120 Ω. The wire is cut into N identical pieces which are then connected in parallel. The resistance of the parallel arrangement is 1.875 Ω. Find N.

77 •• A parallel combination of an 8-Ω resistor and an unknown resistor R is connected in series with a 16-Ω resistor and a battery. This circuit is then disassembled and the three resistors are then connected in series with each other and the same battery. In both arrangements, the current through the 8-Ω resistor is the same. What is the unknown resistance R?

78 •• For the resistance network shown in Figure 26-53, find (*a*) R_3 such that $R_{ab} = R_1$; (*b*) R_2 such that $R_{ab} = R_3$; and (*c*) R_1 such that $R_{ab} = R_1$.

Figure 26-53
Problems 78 and 79

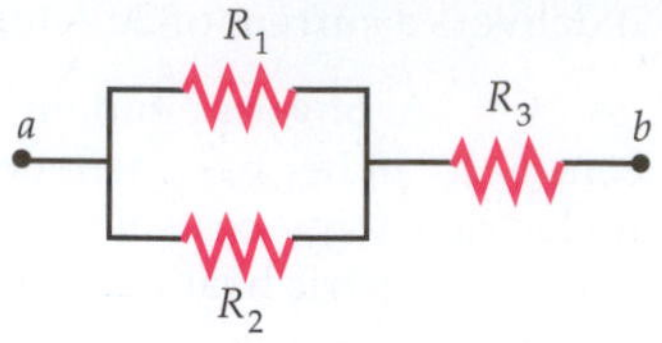

79 •• Check your results for Problem 78 using (*a*) $R_1 = 4$ Ω, $R_2 = 6$ Ω; (*b*) $R_1 = 4$ Ω, $R_3 = 3$ Ω; and (*c*) $R_2 = 6$ Ω, $R_3 = 3$ Ω.

80 ••• Nine 10-Ω resistors are connected as shown in Figure 26-54, and a potential difference of 20 V is applied between points a and b. (*a*) What is the equivalent resistance of this network? (*b*) Find the current in each of the nine resistors.

Figure 26-54
Problem 80

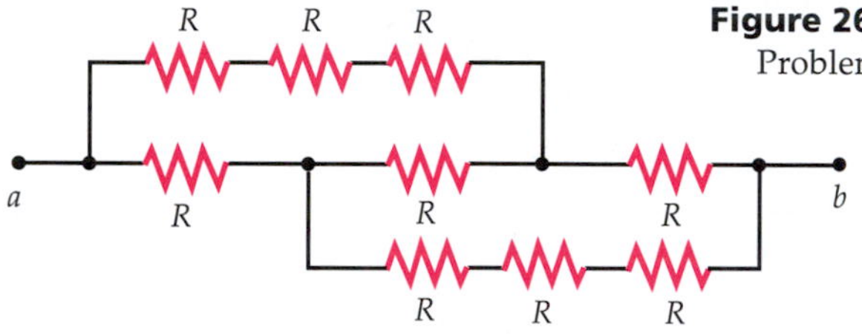

Kirchhoff's Rules

81 • Kirchoff's loop rule follows from

(*a*) conservation of charge.
(*b*) conservation of energy.
(*c*) Newton's laws.
(*d*) Coulomb's law.
(*e*) quantization of charge.

82 • In Figure 26-55, the emf is 6 V and $R = 0.5$ Ω. The rate of Joule heating in R is 8 W. (*a*) What is the current in the circuit? (*b*) What is the potential difference across R? (c) What is r?

Figure 26-55 Problem 82

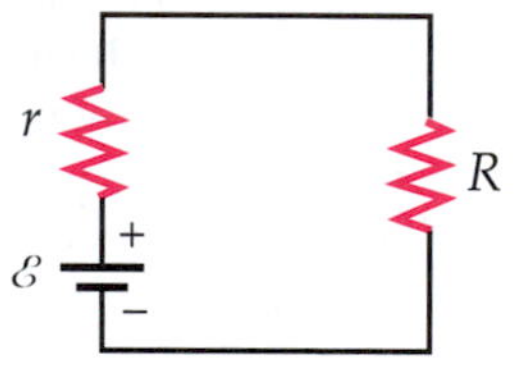

83 • For the circuit in Figure 26-56, find (*a*) the current, (*b*) the power delivered or absorbed by each emf, and (*c*) the rate of Joule heating in each resistor. (Assume that the batteries have negligible internal resistance.)

Figure 26-56 Problem 83

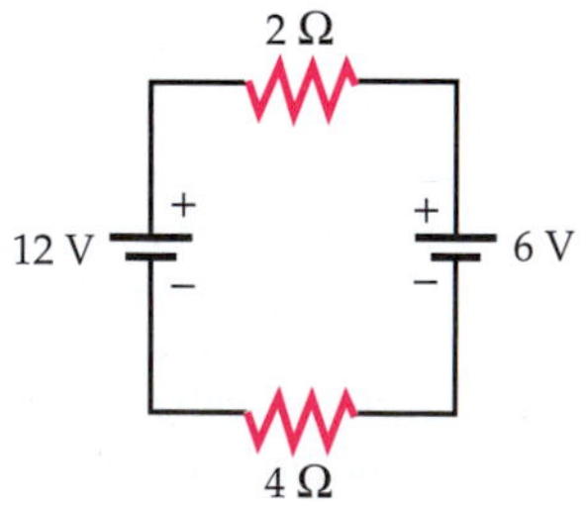

84 •• A sick car battery with an emf of 11.4 V and an internal resistance of 0.01 Ω is connected to a load of 2.0 Ω. To help the ailing battery, a second battery with an emf of 12.6 V and an internal resistance of 0.01 Ω is connected by jumper cables to the terminals of the first battery. (*a*) Draw a diagram of this circuit. (*b*) Find the current in each part of the circuit. (*c*) Find the power delivered by the second battery and discuss where this power goes, assuming that the emfs and internal resistances of both batteries remain constant.

85 •• In the circuit in Figure 26-57, the reading of the ammeter is the same with both switches open and both closed. Find the resistance R.

Figure 26-57 Problem 85

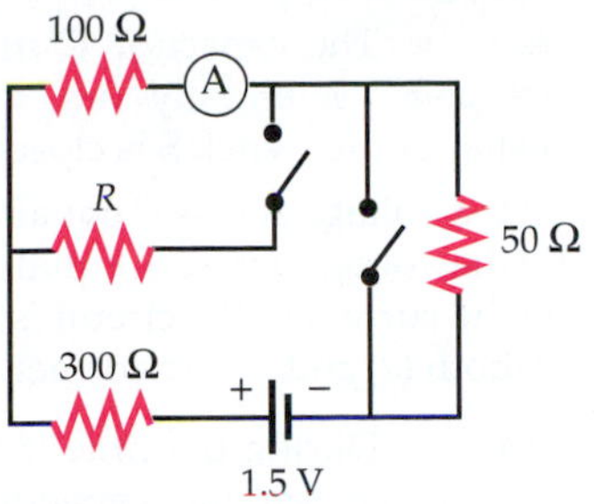

86 •• In the circuit in Figure 26-58, the batteries have negligible internal resistance, and the ammeter has negligible resistance. (*a*) Find the current through the ammeter. (*b*) Find the energy delivered by the 12-V battery in 3 s. (*c*) Find the total Joule heat produced in 3 s. (*d*) Account for the difference in your answers to parts (*b*) and (*c*).

Figure 26-58 Problem 86

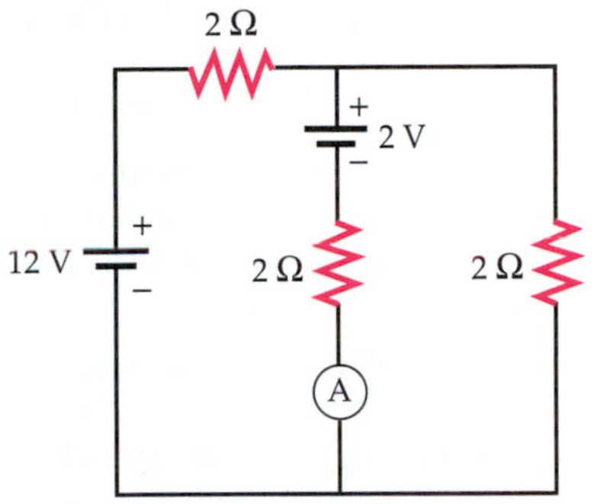

87 •• In the circuit in Figure 26-59, the batteries have negligible internal resistance. Find (*a*) the current in each resistor, (*b*) the potential difference between points *a* and *b*, and (*c*) the power supplied by each battery.

Figure 26-59 Problem 87

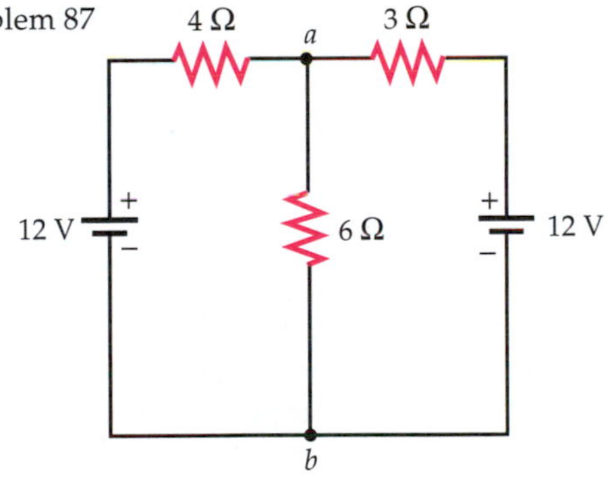

88 •• Repeat Problem 87 for the circuit in Figure 26-60.

Figure 26-60 Problem 88

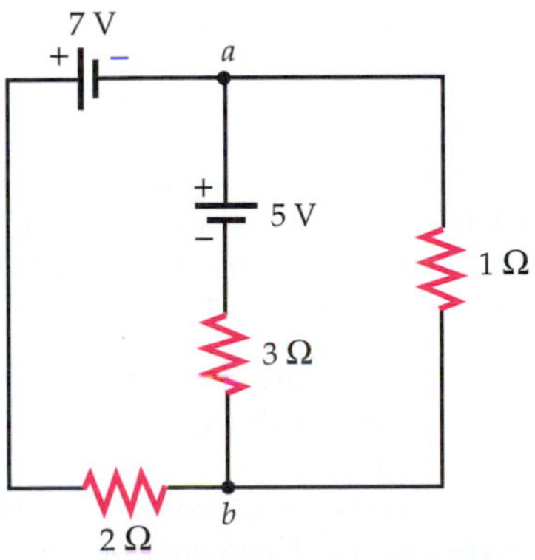

89 •• Two identical batteries, each with an emf $\mathcal{E}$ and an internal resistance r, can be connected across a resistance R either in series or in parallel. Is the power supplied to R greater when $R < r$ or when $R > r$?

90 •• For the circuit in Figure 26-61, find (*a*) the current in each resistor, (*b*) the power supplied by each emf, and (*c*) the power dissipated in each resistor.

Figure 26-61 Problem 90

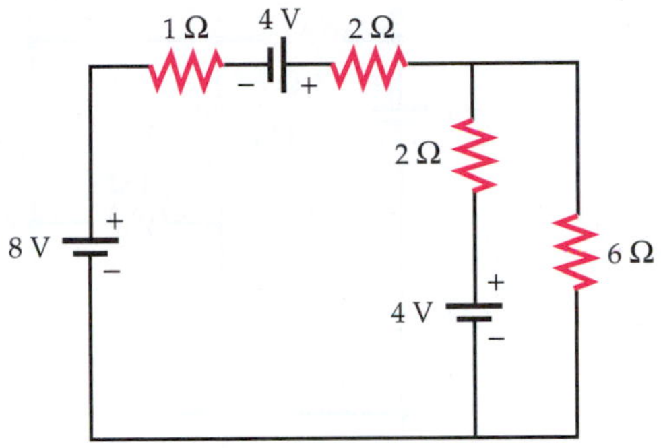

91 •• For the circuit in Figure 26-62, find the potential difference between points *a* and *b*.

Figure 26-62 Problem 91

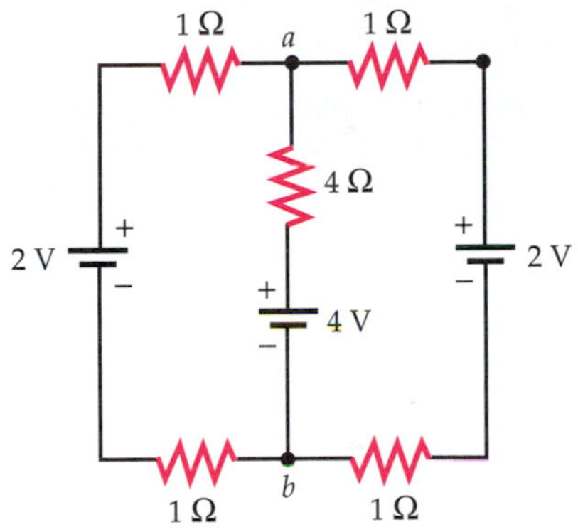

92 •• The battery in the circuit shown in Figure 26-63 has an internal resistance of 0.01 Ω. (*a*) An ammeter with a resistance of 0.01 Ω is inserted in series with the 0.74-Ω resistor at point *a*. What is the reading of the ammeter? (*b*) By what percentage is the current changed because of the ammeter? (*c*) The ammeter is removed and a voltmeter with a resistance of 1 kΩ is connected in parallel with the 0.74-Ω resistor from *a* to *b*. What is the reading of the voltmeter? (*d*) By what percentage is the voltage drop from *a* to *b* changed by the presence of the voltmeter?

Figure 26-63 Problem 92

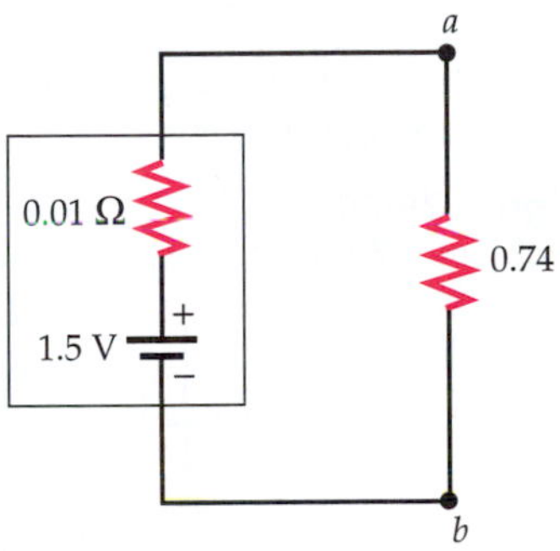

93 •• You have two batteries, one with $\mathcal{E} = 9.0$ V and $r = 0.8$ Ω and the other with $\mathcal{E} = 3.0$ V and $r = 0.4$ Ω. (*a*) Show how you would connect the batteries to give the largest current through a resistor R. Find the current for (*b*) $R = 0.2$ Ω, (*c*) $R = 0.6$ Ω, (*d*) $R = 1.0$ Ω, and (*e*) $R = 1.5$ Ω.

94 •• (*a*) Find the current in each part of the circuit shown in Figure 26-64. (*b*) Use your results from (*a*) to assign a potential at each indicated point assuming the potential at point *a* is zero.

Figure 26-64
Problem 94

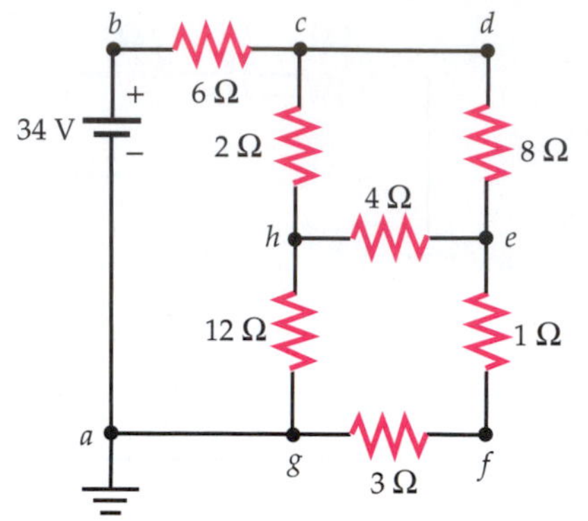

95 ••• In Problem 84, assume that the emf of the first battery increases at a constant rate of 0.2 V/h while the emf of the second battery and the internal resistances remain constant. (*a*) Find the current in each part of the circuit as a function of time. (*b*) Sketch a graph of the power delivered to the first battery as a function of time.

96 ••• (*a*) Find the current in each part of the circuit shown in Figure 26-65. (*b*) Use your results from (*a*) to assign a potential at each indicated point assuming the potential at point *a* is zero.

Figure 26-65
Problem 96

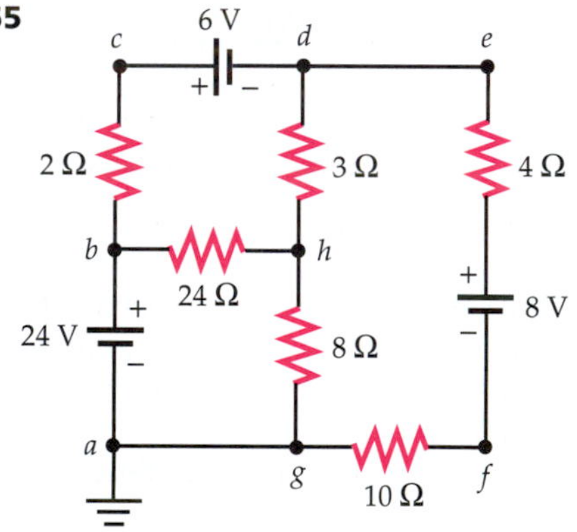

97 ••• Find the current in each resistor of the circuit shown in Figure 26-66.

Figure 26-66
Problems 97 and 98

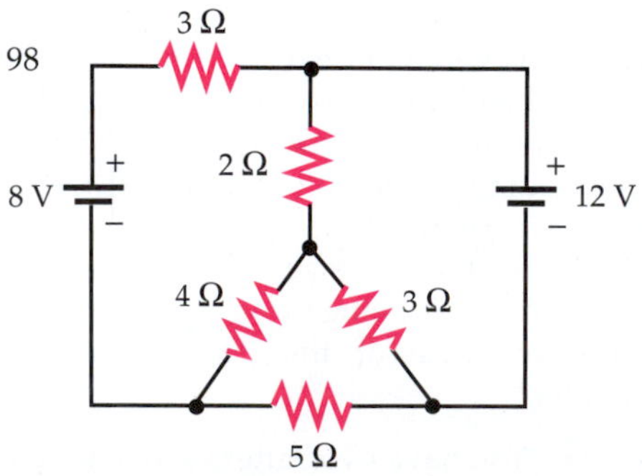

98 ••• Suppose that the emf of the left battery in Figure 26-66 is unknown but that the current delivered by the 12-V battery is known to be 0.6 A. Find the emf of the left battery and the current delivered by it.

RC Circuits

Figure 26-67 Problem 99

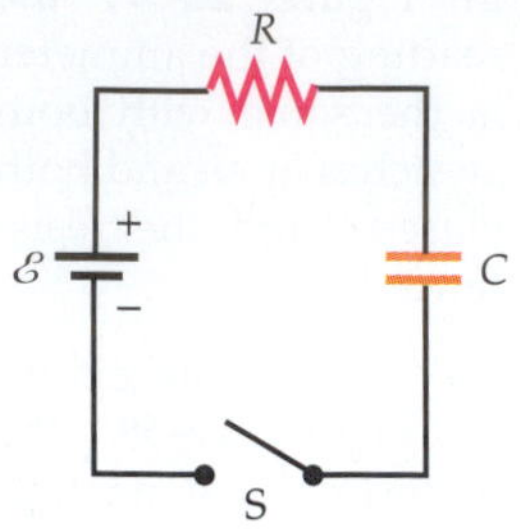

99 • The capacitor *C* in Figure 26-67 is initially uncharged. Just after the switch S is closed,

(*a*) the voltage across *C* equals ℰ.
(*b*) the voltage across *R* equals ℰ.
(*c*) the current in the circuit is zero.
(*d*) both (*a*) and (*c*) are correct.

100 •• During the time it takes to fully charge the capacitor of Figure 26-67,

(*a*) the energy supplied by the battery is $\frac{1}{2}C\mathcal{E}^2$.
(*b*) the energy dissipated in the resistor is $\frac{1}{2}C\mathcal{E}^2$.
(*c*) energy is dissipated in the resistor at a constant rate.
(*d*) the total charge flowing through the resistor is $\frac{1}{2}C\mathcal{E}$.

101 •• A battery is connected to a series combination of a switch, a resistor, and an initially uncharged capacitor. The switch is closed at $t = 0$. Which of the following statements is true?

(*a*) As the charge on the capacitor increases, the current increases.
(*b*) As the charge on the capacitor increases, the voltage drop across the resistor increases.
(*c*) As the charge on the capacitor increases, the current remains constant.
(*d*) As the charge on the capacitor increases, the voltage drop across the capacitor decreases.
(*e*) As the charge on the capacitor increases, the voltage drop across the resistor decreases.

102 •• A capacitor is discharging through a resistor. If it takes a time *T* for the charge on a capacitor to drop to half its initial value, how long does it take for the energy to drop to half its initial value?

103 •• A capacitor, resistor, and battery are connected in series. If *R* is doubled, how does this affect (*a*) the total energy stored, (*b*) the rate of energy storage, and (*c*) the time required to store $1/e$ of the final energy?

104 •• A capacitor, resistor, and battery are connected in series. If *C* is doubled, how does this affect (*a*) the total energy stored, (*b*) the rate of energy storage, and (*c*) the time required to store $1/e$ of the final energy?

105 • A 6-μF capacitor is charged to 100 V and is then connected across a 500-Ω resistor. (*a*) What is the initial charge on the capacitor? (*b*) What is the initial current just after the capacitor is connected to the resistor? (*c*) What is the time constant of this circuit? (*d*) How much charge is on the capacitor after 6 ms?

106 • (*a*) Find the initial energy stored in the capacitor of Problem 105. (*b*) Show that the energy stored in the capacitor is given by $U = U_0 e^{-2t/\tau}$, where U_0 is the initial energy and $\tau = RC$ is the time constant. (*c*) Sketch a plot of the energy *U* in the capacitor versus time *t*.

107 •• In the circuit of Figure 26-40, emf $\mathcal{E} = 50$ V and $C = 2.0$ μF; the capacitor is initially uncharged. At 4.0 s after switch S is closed, the voltage drop across the resistor is 20 V. Find the resistance of the resistor.

108 •• A 0.12-μF capacitor is given a charge Q_0. After 4 s, its charge is $\frac{1}{2}Q_0$. What is the effective resistance across this capacitor?

109 •• A 1.6-μF capacitor, initially uncharged, is connected in series with a 10-kΩ resistor and a 5.0-V battery of negligible internal resistance. (*a*) What is the charge on the capacitor after a very long time? (*b*) How long does it take the capacitor to reach 99% of its final charge?

110 •• Consider the circuit shown in Figure 26-68. From your knowledge of how capacitors behave in circuits, find (*a*) the initial current through the battery just after the switch is closed, (*b*) the steady-state current through the battery when the switch has been closed for a long time, and (*c*) the maximum voltage across the capacitor.

Figure 26-68 Problem 110

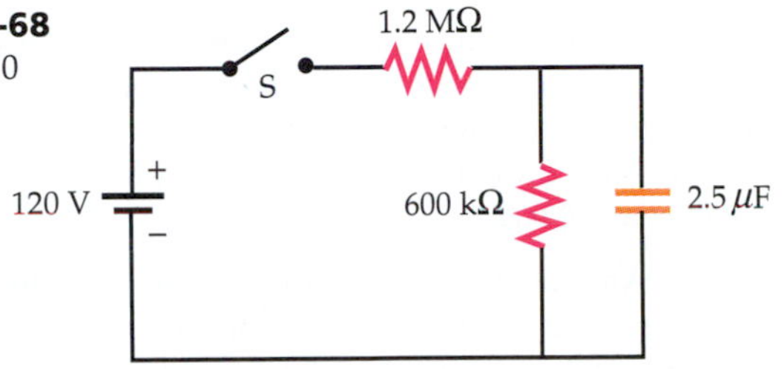

111 •• A 2-MΩ resistor is connected in series with a 1.5-μF capacitor and a 6.0-V battery of negligible internal resistance. The capacitor is initially uncharged. After a time $t = \tau = RC$, find (*a*) the charge on the capacitor, (*b*) the rate at which the charge is increasing, (*c*) the current, (*d*) the power supplied by the battery, (*e*) the power dissipated in the resistor, and (*f*) the rate at which the energy stored in the capacitor is increasing.

112 •• Repeat Problem 111 for the time $t = 2\tau$.

113 •• In the steady state, the charge on the 5-μF capacitor in the circuit in Figure 26-69 is 1000 μC. (*a*) Find the battery current. (*b*) Find the resistances R_1, R_2, and R_3.

Figure 26-69 Problem 113

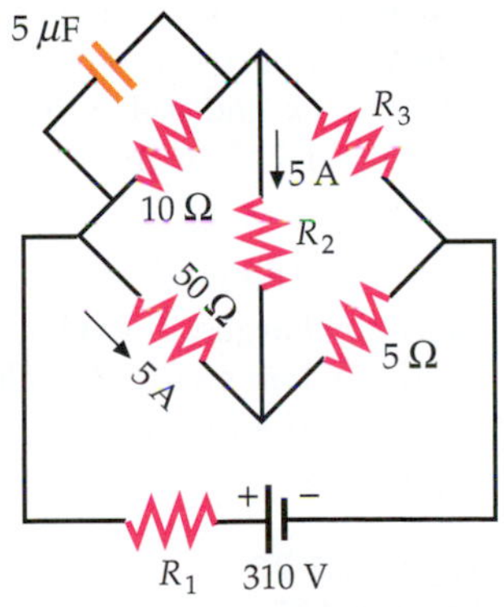

114 •• (*a*) What is the voltage across the capacitor in the circuit in Figure 26-70? (*b*) If the battery is disconnected, give the capacitor current as a function of time. (*c*) How long does it take the capacitor to discharge until the potential difference across it is 1 V?

Figure 26-70 Problem 114

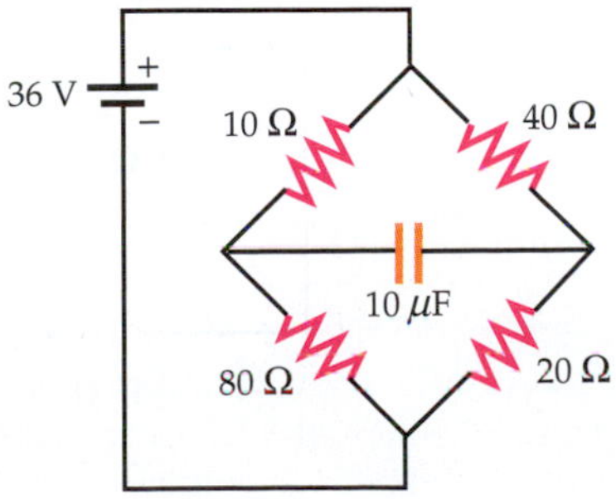

115 •• Show that Equation 26-34 can be written

$$\frac{dQ}{\mathcal{E}C - Q} = \frac{dt}{RC}$$

Integrate this equation to derive the solution given by Equation 26-35.

116 ••• A photojournalist's flash unit uses a 9.0-V battery pack to charge a 0.15-μF capacitor, which is then discharged through the flash lamp of 10.5-Ω resistance when a switch is closed. The minimum voltage necessary for the flash discharge is 7.0 V. The capacitor is charged through a 18-kΩ resistor. (*a*) How much time is required to charge the capacitor to the required 7.0 V? (*b*) How much energy is released when the lamp flashes? (*c*) How much energy is supplied by the battery during the charging cycle and what fraction of that energy is dissipated in the resistor?

117 ••• For the circuit in Figure 26-71, (*a*) what is the initial battery current immediately after switch S is closed? (*b*) What is the battery current a long time after switch S is closed? (*c*) What is the current in the 600-Ω resistor as a function of time?

Figure 26-71 Problem 117

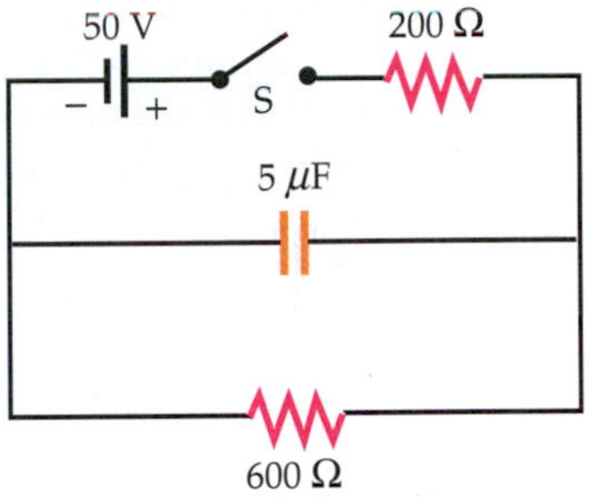

118 ••• For the circuit in Figure 26-72, (*a*) what is the initial battery current immediately after switch S is closed? (*b*) What is the battery current a long time after switch S is closed? (*c*) If the switch has been closed for a long time and is then opened, find the current through the 600-kΩ resistor as a function of time.

Figure 26-72 Problem 118

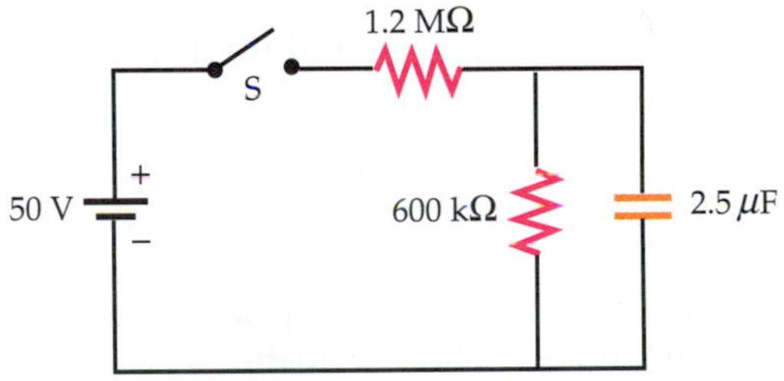

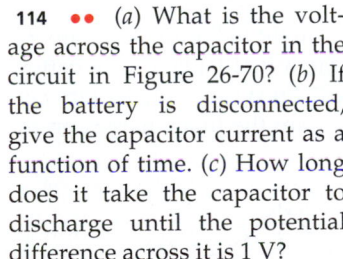

119 ••• In the circuit shown in Figure 26-73, the capacitor has a capacitance of 2.5 μF and the resistor a resistance of 0.5 MΩ. Before the switch is closed, the potential drop across the capacitor is 12 V, as shown. Switch S is closed at $t = 0$. (*a*) What is the current in R immediately after S is closed? (*b*) At what time t is the voltage across the capacitor 24 V?

Figure 26-73 Problems 119 and 120

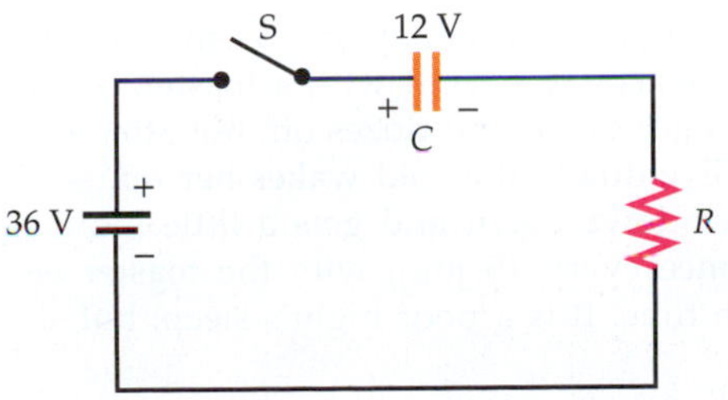

120 ••• Repeat Problem 119 if the capacitor is connected with reversed polarity.

General Problems

121 • A flash lamp is set off by the discharge of a capacitor that has been charged by a battery. Why not just connect the battery directly to the lamp?

122 • Which will produce more thermal energy when connected across an ideal battery, a small resistance or a large resistance?

123 • Do Kirchhoff's rules apply to circuits containing capacitors?

124 •• True or false:

(*a*) Ohm's law is $R = V/I$.
(*b*) Electrons drift in the direction of the current.
(*c*) A source of emf supplies power to an electrical circuit.
(*d*) When the potential drops by V in a segment of a circuit, the power supplied to that segment is IV.
(*e*) The equivalent resistance of two resistors in parallel is always less than the resistance of either resistor alone.
(*f*) The terminal voltage of a battery always equals its emf.
(*g*) The terminal voltage of a battery is always less than its emf.

125 •• In Figure 26-74, all three resistors are identical. The power dissipated is

(*a*) the same in R_1 as in the parallel combination of R_2 and R_3.
(*b*) the same in R_1 and R_2.
(*c*) greatest in R_1.
(*d*) smallest in R_1.

Figure 26-74 Problems 125 and 126

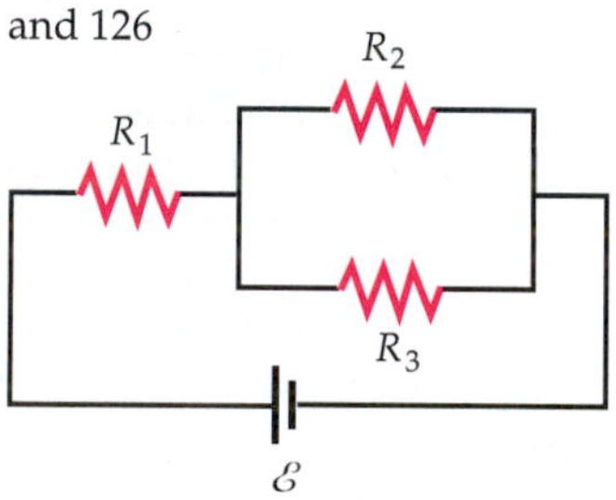

126 •• In Figure 26-74, $R_1 = 4\ \Omega$, $R_2 = 6\ \Omega$, and $R_3 = 12\ \Omega$. If we denote the currents through these resistors by I_1, I_2, and I_3, respectively, then

(*a*) $I_1 > I_2 > I_3$. (*b*) $I_2 = I_3$. (*c*) $I_3 > I_2$.
(*d*) none of the above is correct.

127 •• A 25-W light bulb is connected in series with a 100-W light bulb and a voltage V is placed across the combination. Which bulb is brighter? Explain.

128 • If the battery emf in Figure 26-74 is 24 V, then

(*a*) $I_2 = 4$ A. (*b*) $I_2 = 2$ A. (*c*) $I_2 = 1$ A.
(*d*) none of the above is correct.

129 • A 10.0-Ω resistor is rated as being capable of dissipating 5.0 W of power. (*a*) What maximum current can this resistor tolerate? (*b*) What voltage across this resistor will produce the maximum current?

130 • Margaret is economizing by turning off her space heater and warming herself with a toaster. She pushes the toaster plunger down and dozes off, but after 4 min it pops up again. Eventually the cold wakes her up, so she pushes the plunger down again and gets a little more sleep. This happens once every 15 min, with the toaster engaged for 4 min each time. It is a poor night's sleep, but she is determined to save money. Energy costs 9 cents per kilowatt-hour, and a 120-V source is used. (*a*) How much does it cost to operate an electric toaster for 4 min if its resistance is 11.0 Ω? (*b*) How much would it cost to operate a 5.0-Ω-heater connected across 120 V for 8 h?

131 • A 12-V car battery has an internal resistance of 0.4 Ω. (*a*) What is the current if the battery is shorted momentarily? (*b*) What is the terminal voltage when the battery delivers a current of 20 A to start the car?

132 •• The current drawn from a battery is 1.80 A when a 7.0-Ω resistor is connected across the battery terminals. If a second 12-Ω resistor is connected in parallel with the 7-Ω resistor, the battery delivers a current of 2.20 A. What are the emf and internal resistance of the battery?

133 •• A 16-gauge copper wire insulated with rubber can safely carry a maximum current of 6 A. (*a*) How great a potential difference can be applied across 40 m of this wire? (*b*) Find the electric field in the wire when it carries a current of 6 A. (*c*) Find the power dissipated in the wire when it carries a current of 6 A.

134 •• An automobile jumper cable 3 m long is constructed of multiple strands of copper wire that has an equivalent cross-sectional area of 10.0 mm^2. (*a*) What is the resistance of the jumper cable? (*b*) When the cable is used to start a car, it carries a current of 90 A. What is the potential drop that occurs across the jumper cable? (*c*) How much power is dissipated in the jumper cable?

135 •• A coil of Nichrome wire is to be used as the heating element in a water boiler that is required to generate 8.0 g of steam per second. The wire has a diameter of 1.80 mm and is connected to a 120-V power supply. Find the length of wire required.

136 •• A closed box has two metal terminals *a* and *b*. The inside of the box contains an unknown emf $\mathcal{E}$ in series with a resistance R. When a potential difference of 21 V is maintained between *a* and *b*, there is a current of 1 A between the terminals *a* and *b*. If this potential difference is reversed, a current of 2 A in the reverse direction is observed. Find $\mathcal{E}$ and R.

137 •• The capacitors in the circuit in Figure 26-75 are initially uncharged. (*a*) What is the initial value of the battery current when switch S is closed? (*b*) What is the battery current after a long time? (*c*) What are the final charges on the capacitors?

Figure 26-75 Problem 137

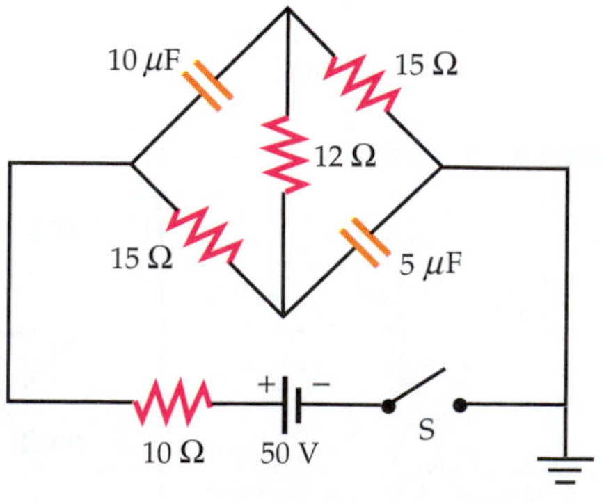

138 •• The circuit in Figure 26-76 is a slide-type *Wheatstone bridge.* It is used for determining an unknown resistance R_x in terms of the known resistances R_1, R_2, and R_0. The resistances R_1 and R_2 comprise a wire 1 m long. Point a is a sliding contact that is moved along the wire to vary these resistances. Resistance R_1 is proportional to the distance from the left end of the wire (labeled 0 cm) to point a, and R_2 is proportional to the distance from point a to the right end of the wire (labeled 100 cm). The sum of R_1 and R_2 remains constant. When points a and b are at the same potential, there is no current in the galvanometer and the bridge is said to be balanced. (Since the galvanometer is used to detect the absence of a current, it is called a *null detector.*) If the fixed resistance $R_0 = 200\ \Omega$, find the unknown resistance R_x if (*a*) the bridge balances at the 18-cm mark, (*b*) the bridge balances at the 60-cm mark, and (*c*) the bridge balances at the 95-cm mark.

Figure 26-76 Problems 138 and 139

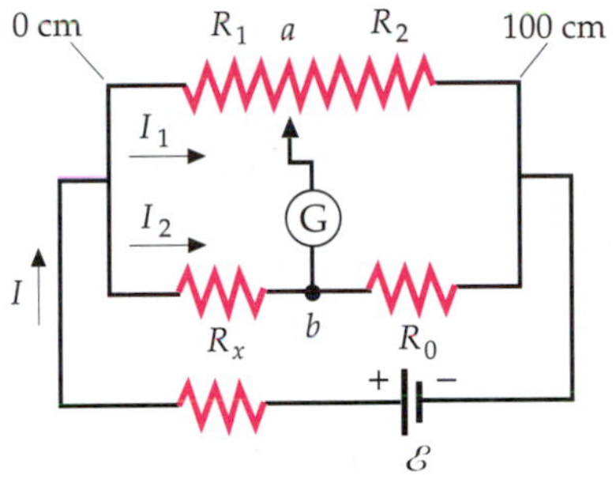

139 •• For the Wheatstone bridge of Problem 138, the bridge balances at the 98-cm mark when $R_0 = 200\ \Omega$. (*a*) What is the unknown resistance? (*b*) What effect would an error of 2 mm in the location of the balance point have on the measured value of the unknown resistance? (*c*) How should R_0 be changed so that the balance point for this unknown resistor will be nearer the 50-cm mark?

140 •• The wires in a house must be large enough in diameter so that they do not get hot enough to start a fire. Suppose a certain wire is to carry a current of 20 A, and it is determined that the Joule heating of the wire should not exceed 2 W/m. What diameter must a copper wire have to be safe for this current?

141 •• You are given n identical cells, each with emf $\mathcal{E}$ and internal resistance $r = 0.2\ \Omega$. When these cells are connected in parallel to form a battery, and a resistance R is connected to the battery terminal, the current through R is the same as when the cells are connected in series and R is attached to the terminals of that battery. Find the value of the resistor R.

142 •• A cyclotron produces a 3.50-μA proton beam of 60-MeV energy. The protons impinge and come to rest inside a 50-g copper target within the vacuum chamber. (*a*) Determine the number of protons that strike the target per second. (*b*) Find the energy deposited in the target per second. (*c*) How much time elapses before the target temperature rises 300°C? (Neglect cooling by radiation.)

143 •• Compact fluorescent light bulbs cost \$6 each and have an expected lifetime of 8000 h. These bulbs consume 20 W of power, but produce the illumination equivalent to 75-W incandescent bulbs. Incandescent bulbs cost about \$1.50 each and have an expected lifetime of 1200 h. If the average household has, on the average, six 75-W incandescent light bulbs on constantly, and if energy costs 11.5 cents per kilowatt-hour, how much money would a consumer save each year by installing the energy-efficient fluorescent light bulbs?

144 •• The space between the plates of a parallel-plate capacitor is filled with a dielectric of constant κ and resistivity ρ. (*a*) Show that the time constant for the decrease of charge on the plates is $\tau = \epsilon_0\kappa\rho$. (*b*) If the dielectric is mica, for which $\kappa = 5.0$ and $\rho = 9 \times 10^{13}\ \Omega{\cdot}\text{m}$, find the time it takes for the charge to decrease to $1/e^2 \approx 14\%$ of its initial value.

145 •• The belt of a Van de Graaff generator carries a surface charge density of 5 mC/m^2. The belt is 0.5 m wide and moves at 20 m/s. (*a*) What current does it carry? (*b*) If this charge is raised to a potential of 100 kV, what is the minimum power of the motor needed to drive the belt?

146 •• Conventional large electromagnets use water cooling to prevent excessive heating of the magnet coils. A large laboratory electromagnet draws 100 A when a voltage of 240 V is applied to the terminals of the energizing coils. To cool the coils, water at an initial temperature of 15°C is circulated through the coils. How many liters per second must pass through the coils if their temperature should not exceed 50°C?

147 ••• We show in Figure 26-77 the basis of the sweep circuit used in an oscilloscope. S is an electronic switch that closes whenever the potential across its terminals reaches a value V_c and opens when the potential has dropped to 0.2 V. The emf $\mathcal{E}$, much greater than V_c, charges the capacitor C through a resistor R_1. The resistor R_2 represents the small but finite resistance of the electronic switch. In a typical circuit, $\mathcal{E} = 800$ V, $V_c = 4.2$ V, $R_2 = 0.001\ \Omega$, $R_1 = 0.5\ \text{M}\Omega$ ($0.5 \times 10^6\ \Omega$), and $C = 0.02\ \mu$F. (*a*) What is the time constant for charging of the capacitor C? (*b*) Show that in the time required to bring the potential across S to the critical potential $V_c = 4.2$ V, the voltage across the capacitor increases almost linearly with time. (*Hint:* Use the expansion of the exponential for small values of exponent.) (*c*) What should be the value of R_1 so that C charges from 0.2 to 4.2 V in 0.1 s? (*d*) How much time elapses during the discharge of C through switch S? (*e*) At what rate is power dissipated in the resistor R_1 and in the switch resistance?

Figure 26-77 Problem 147

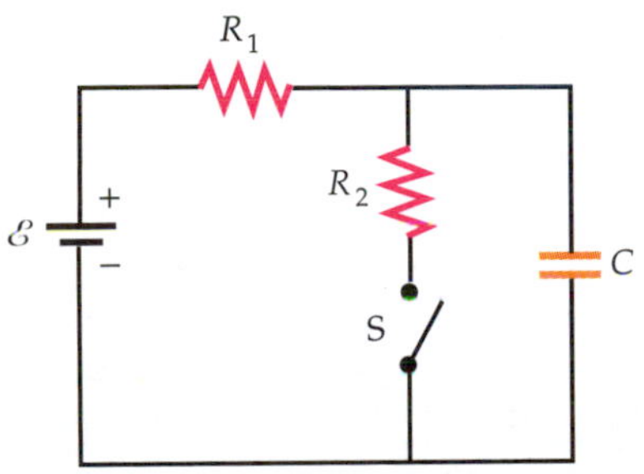

148 ••• In the circuit shown in Figure 26-78, $R_1 = 2.0\ \text{M}\Omega$, $R_2 = 5.0\ \text{M}\Omega$, and $C = 1.0\ \mu\text{F}$. At $t = 0$, switch S is closed, and at $t = 2.0$ s, switch S is opened. (*a*) Sketch the voltage across C and the current through R_2 between $t = 0$ and $t = 10$ s. (*b*) Find the voltage across the capacitor at $t = 2$ s and at $t = 8$ s.

Figure 26-78
Problem 148

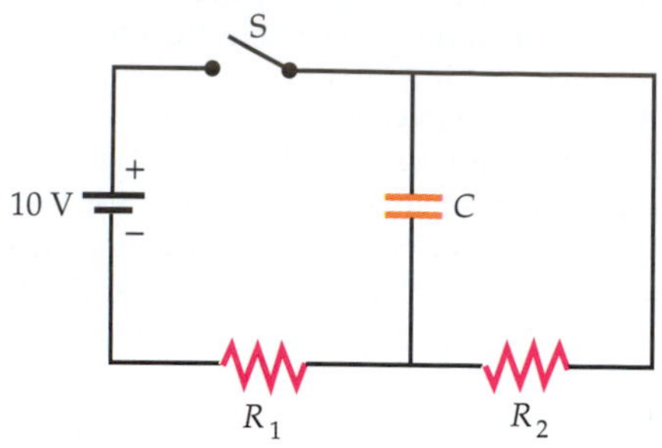

149 ••• If the capacitor in the circuit in Figure 26-70 is replaced by a 30-Ω resistor, what currents flow through the resistors?

150 ••• Two batteries with emfs $\mathcal{E}_1$ and $\mathcal{E}_2$ and internal resistances r_1 and r_2 are connected in parallel. Prove that if a resistor is connected in parallel with this combination, the optimal load resistance (the resistance at which maximum power is delivered) is $R = r_1r_2/(r_1 + r_2)$.

151 ••• Capacitors C_1 and C_2 are connected in parallel by a resistor and two switches as shown in Figure 26-79. Capacitor C_1 is initially charged to a voltage V_0, and capacitor C_2 is uncharged. The switches S are then closed. (*a*) What are the final charges on C_1 and C_2? (*b*) Compare the initial and final stored energies of the system. (*c*) What caused the decrease in the capacitor-stored energy?

Figure 26-79
Problems 151 and 152

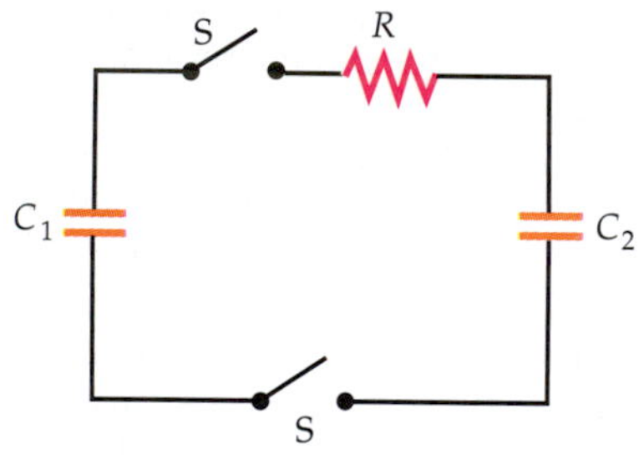

152 ••• (*a*) In Problem 151, find the current through R after the switches S are closed as a function of time. (*b*) Find the energy dissipated in the resistor as a function of time. (*c*) Find the total energy dissipated in the resistor and compare it with the loss of stored energy found in part (*b*) of Problem 151.

153 ••• In the circuit in Figure 26-80, the capacitors are initially uncharged. Switch S_2 is closed and then switch S_1 is closed. (*a*) What is the battery current immediately after S_1 is closed? (*b*) What is the battery current a long time after both switches are closed? (*c*) What is the final voltage across C_1? (*d*) What is the final voltage across C_2? (*e*) Switch S_2 is opened again after a long time. Give the current in the 150-Ω resistor as a function of time.

Figure 26-80 Problem 153

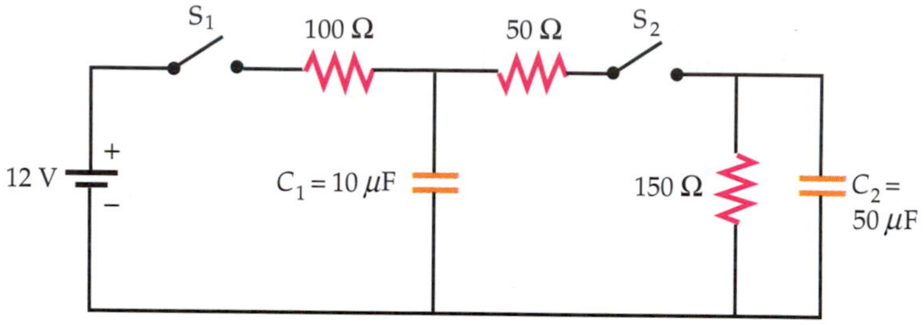

154 ••• In the *RC* circuit in Figure 26-40*a*, the capacitor is initially uncharged and the switch is closed at time $t = 0$. (*a*) What is the power supplied by the battery as a function of time? (*b*) What is the power dissipated in the resistor as a function of time? (*c*) What is the rate at which energy is stored in the capacitor as a function of time? Plot your answers to parts (*a*), (*b*), and (*c*) versus time on the same graph. (*d*) Find the maximum rate at which energy is stored in the capacitor as a function of the battery voltage e and the resistance R. At what time does this maximum occur?

155 ••• A linear accelerator produces a pulsed beam of electrons. The current is 1.6 A for the 0.1-μs duration of each pulse. (*a*) How many electrons are accelerated in each pulse? (*b*) What is the average current of the beam if there are 1000 pulses per second? (*c*) If each electron acquires an energy of 400 MeV, what is the average power output of the accelerator? (*d*) What is the peak power output? (*e*) What fraction of the time is the accelerator actually accelerating electrons? (This is called the *duty factor* of the accelerator.)

optional

CHAPTER 27

The Microscopic Theory of Electrical Conduction

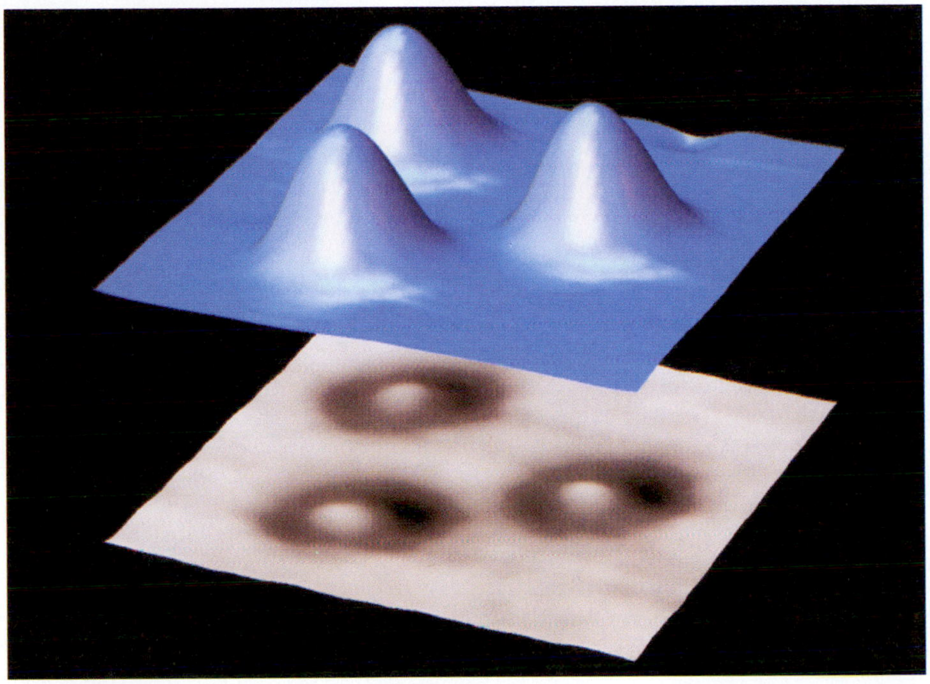

Simultaneously acquired topographic (top) and spectroscopic (bottom) images of three gadolinium atoms on top of a superconducting niobium surface. In the region near the gadolinium atoms the magnetic properties of these individual atoms break up Cooper electron pairs, thereby modifying the superconductivity of the niobium.

In earlier chapters we used Ohm's law to find the currents in various circuits while making no attempt to relate Ohm's law to the properties of metals. Here we introduce a microscopic model of electrical conduction to relate the resistivity of a metal to the properties of electrons and the lattice ions of the metal. This model is similar to the model of a gas that we discussed in Chapter 18 to relate the pressure exerted by a gas on the walls of its container to the average energy of the gas molecules, which in turn we could relate to the absolute temperature of the gas.

The first microscopic model of electric conduction was proposed by P. Drude in 1900 and developed by Hendrik A. Lorentz about 1909. This model successfully predicts Ohm's law and relates the resistivity of conductors to the mean speed v_{av} and the mean free path λ of the free electrons within the conductor. However, when v_{av} and λ are interpreted classically, there is a disagreement between the calculated and measured values of the resistivity, and a similar disagreement between the predicted and observed temperature dependence. Thus, the classical theory fails to adequately describe the resistivity of metals. Furthermore, the classical theory says nothing about the most striking property of solids, namely that some materials are conductors, others are insulators, and still others are semiconductors, materials whose resistivity falls between that of conductors and insulators.

When v_{av} and λ are interpreted using quantum theory, the magnitude and temperature dependence of the resistivity are correctly predicted. In addition, quantum theory allows us to determine if a material will be a conductor, insulator, or semiconductor.

27-1 A Microscopic Picture of Conduction

We consider a metal as a regular three-dimensional lattice of ions filling some volume V and containing a large number N of electrons that are free to move throughout the whole metal. Experimentally the number of free electrons in a metal is about 1 to 4 electrons per atom. In the absence of an electric field, the free electrons move about the metal randomly, much the way gas molecules move about in a container. We will often refer to these free electrons in a metal as an electron gas.

The current in a conducting wire segment is proportional to the voltage drop across the segment:

$$I = \frac{V}{R}$$

or

$$V = IR$$

The resistance R is proportional to the length L of the wire segment and inversely proportional to the cross-sectional area A:

$$R = \rho\frac{L}{A}$$

where ρ is the resistivity. According to Ohm's law, the resistance of a given wire segment is independent of the current in the segment and therefore independent of the voltage drop across the segment. For a uniform electric field E, the voltage across a segment of length L is $V = EL$. Then Ohm's law implies that the resistance, and therefore the resistivity, is independent of the applied electric field E.

Substituting $\rho L/A$ for R, and EL for V, we can write the current in terms of the electric field E and the resistivity. We have

$$I = \frac{V}{R} = \frac{EL}{\rho L/A} = \frac{1}{\rho}EA \qquad \text{27-1}$$

The current in a wire is related by Equation 26-3 to the number of electrons per unit volume $n_e = N/V$, the drift velocity v_d, the magnitude of the charge of an electron $q = e$, and the cross-sectional area A:

$$I = nqAv_d = n_e eAv_d \qquad \text{27-2}$$

Combining Equations 27-1 and 27-2 for the current I gives

$$n_e eAv_d = \frac{1}{\rho}EA$$

or

$$\rho = \frac{E}{n_e ev_d} \qquad \text{27-3}$$

If Ohm's law holds, ρ must be independent of E, so according to Equation 27-3, the drift velocity v_d must be proportional to E.

In the presence of an electric field, a free electron experiences a force of magnitude eE. If this were the only force acting, the electron would have an acceleration eE/m_e and its velocity would steadily increase. However, Equation 27-3 implies a steady-state situation with a constant drift velocity that is proportional to the field E. In the microscopic model, it is assumed that a free electron is accelerated for a short time and then makes a collision with a lattice ion. The velocity of the electron after the collision is completely unrelated to the drift velocity. The justification for this assumption is that the drift velocity is very small compared with the random thermal velocity.

Let τ be the average time since the last collision for an electron picked at random. Since the acceleration of each electron is eE/m_e, the drift velocity of the electrons is

$$v_d = \frac{eE}{m_e}\tau \qquad \text{27-4}$$

Using this result in Equation 27-3, we obtain

$$\rho = \frac{E}{n_e e(eE\tau/m_e)} = \frac{m_e}{n_e e^2 \tau} \qquad \text{27-5}$$

The time τ, called the **collision time**, is also the average time between collisions.* The average distance the electron travels between collisions is $v_{av}\tau$, which is called the **mean free path** λ:

$$\lambda = v_{av}\tau \qquad \text{27-6}$$

In terms of the mean free path and the mean speed, the resistivity is

$$\rho = \frac{m_e v_{av}}{n_e e^2 \lambda} \qquad \text{27-7}$$

Resistivity in terms of v_{av} and λ

According to Ohm's law, the resistivity is independent of the electric field E. Since m_e, n_e, and e^2 are constants, the only quantities that could possibly depend on E are the mean speed v_{av} and the mean free path λ. Let us examine these quantities to see if they can possibly depend on the applied field E.

Classical Interpretation of λ and v_{av}

Classically, at $T = 0$ all the free electrons in a conductor should have zero kinetic energy. As the conductor is heated, the lattice ions acquire an average kinetic energy of $\frac{3}{2}kT$, which is imparted to the electron gas by the collisions between the electrons and the ions. (This is a result of the equipartition theorem studied in Chapters 18 and 19.) The electron gas would then have a Maxwell–Boltzmann distribution just like a gas of molecules. In equilibrium, the electrons would be expected to have a mean kinetic energy of $\frac{3}{2}kT$, which at ordinary temperatures (~300 K) is about 0.04 eV. At $T = 300$ K, their root mean square (rms) speed,† which is slightly greater than the mean speed, is

$$v_{av} \approx v_{rms} = \sqrt{\frac{3kT}{m_e}} = \sqrt{\frac{3(1.38 \times 10^{-23}\,\text{J/K})(300\,\text{K})}{9.11 \times 10^{-31}\,\text{kg}}}$$

$$= 1.17 \times 10^5\ \text{m/s} \qquad \text{27-8}$$

* It is tempting but incorrect to think that if τ is the average time between collisions, the average time since its last collision is $\frac{1}{2}\tau$ rather than τ. If you find this confusing, you may take comfort in the fact that Drude used the incorrect result $\frac{1}{2}\tau$ in his original work.

† See Equation 18-23.

Note that this is about three billion times greater than the typical drift speed of 3.5×10^{-5} m/s calculated in Example 26-1. The very small drift velocity caused by the electric field therefore has essentially no effect on the very large mean speed of the electrons, so v_{av} in Equation 27-7 cannot depend on the electric field E.

The mean free path is related classically to the size of the lattice ions in the conductor and to the number of ions per unit volume. Consider one electron moving with speed v through a region of stationary ions, assumed to be hard spheres (Figure 27-1). The size of the electron is assumed to be negligible. The electron will collide with an ion if it comes within a distance r from the center of the ion, where r is the radius of the ion. In some time t_1, the electron moves a distance vt_1. If there is an ion whose center is in the cylindrical volume $\pi r^2 vt_1$, the electron will collide with it. The electron will then change directions and collide with another ion in time t_2 if the center of the ion is in the volume $\pi r^2 vt_2$. Thus, in the total time $t = t_1 + t_2 + \cdots$, the electron will collide with the number of ions whose centers are in the volume $\pi r^2 vt$. The number of ions in this volume is $n_{ion}\pi r^2 vt$, where n_{ion} is the number of ions per unit volume. The total path length divided by the number of collisions is the mean free path:

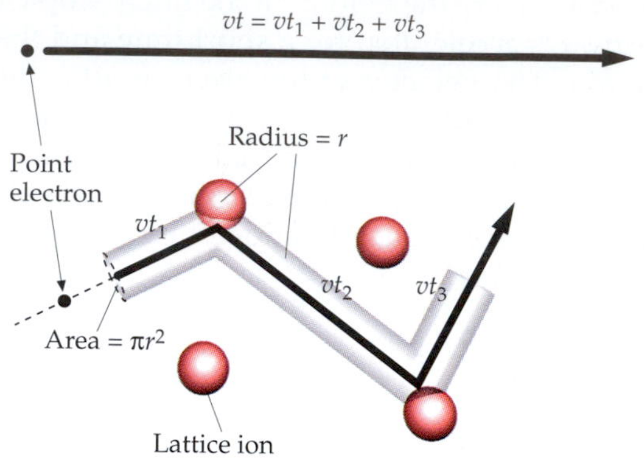

Figure 27-1 Model of an electron moving through the lattice ions of a conductor. The electron, which is considered to be a point, collides with an ion if it comes within a distance r of the center of the ion, where r is the radius of the ion. If the electron speed is v, it collides in time t with all the ions whose centers are in the volume $\pi r^2 vt$.

$$\lambda = \frac{vt}{n_{ion}\pi r^2 vt} = \frac{1}{n_{ion}\pi r^2} = \frac{1}{n_{ion}A} \qquad 27\text{-}9$$

where $A = \pi r^2$ is the cross-sectional area of a lattice ion.

Neither n_{ion} nor r depends on the electric field E, so λ also does not depend on E. Thus, according to the classical interpretation of v_{av} and λ, neither depend on E, so the resistivity ρ does not depend on E, in accordance with Ohm's law. However, the classical theory gives an incorrect temperature dependence for the resistivity. Since λ depends only on the number density of lattice ions and their radius, the only quantity in Equation 27-7 that depends on temperature in the classical theory is v_{av}, which is proportional to $\sqrt{T}$. But experimentally, ρ varies linearly with temperature. Furthermore, when ρ is calculated at $T = 300$ K using the Maxwell–Boltzmann distribution for v_{av} and Equation 27-9 for λ, the numerical result is about six times greater than the measured value.

The classical theory of conduction fails because electrons are not classical particles. The wave nature of the electrons must be considered. Because of the wave properties of electrons and the exclusion principle (to be discussed below), the energy distribution of the free electrons in a metal is not even approximately given by the Maxwell–Boltzmann distribution. Furthermore, the collision of an electron with a lattice ion is not similar to the collision of a baseball with a tree. Instead, it involves the scattering of electron waves by the lattice. To understand the quantum theory of conduction, we need a qualitative understanding of the energy distribution of free electrons in a metal. This will also help us understand the origin of contact potentials between two dissimilar metals in contact, and the contribution of free electrons to the heat capacity of metals.

27-2 The Fermi Electron Gas

We have used the term *electron gas* to describe the free electrons in a metal. Whereas the molecules in an ordinary gas such as air obey the classical Maxwell–Boltzmann energy distribution, the free electrons in a metal do not. Instead they obey a quantum energy distribution called the Fermi–Dirac distribution. Because the behavior of this electron gas is so different from a gas of molecules, the electron gas is often called a **Fermi electron gas**. The main features of a Fermi electron gas can be understood by considering an electron in a metal to be a particle in a box, a problem whose one-dimensional version we studied extensively in Chapter 17. We discuss the main features of a Fermi electron gas semiquantitatively in this section and leave the details of the Fermi–Dirac distribution to Section 27-6.

Energy Quantization in a Box

In Chapter 17 we found that the wavelength associated with an electron of momentum p is given by the de Broglie relation:

$$\lambda = \frac{h}{p} \tag{27-10}$$

where h is Planck's constant. When a particle is confined to a finite region of space such as a box, only certain wavelengths λ_n given by standing-wave conditions are allowed. For a one-dimensional box of length L, the standing-wave condition is

$$n\frac{\lambda_n}{2} = L \tag{27-11}$$

This results in the quantization of energy:

$$E_n = \frac{p_n^2}{2m} = \frac{(h/\lambda_n)^2}{2m} = \frac{h^2}{2m}\frac{1}{\lambda_n^2} = \frac{h^2}{2m}\frac{1}{(2L/n)^2}$$

or

$$E_n = n^2\frac{h^2}{8mL^2} \tag{27-12}$$

The wave function for the nth state is given by

$$\psi_n(x) = \sqrt{\frac{2}{L}}\sin\frac{n\pi x}{L} \tag{27-13}$$

The quantum number n characterizes the wave function for a particular state and the energy of that state. In three-dimensional problems, three quantum numbers arise, one associated with each dimension.

The Pauli Exclusion Principle

The distribution of electrons among the possible energy states is dominated by the exclusion principle, which was first enunciated by Wolfgang Pauli in 1925 to explain the electronic structure of atoms:

> No two electrons in an atom can be in the same quantum state; that is, they cannot have the same set of values for their quantum numbers.

Pauli exclusion principle

The exclusion principles applies to all "spin one-half" particles, which include electrons, protons, and neutrons.* These particles have a *spin* quantum number m_s which has two possible values, $+\frac{1}{2}$ and $-\frac{1}{2}$. The quantum state of

* Intrinsic spin is discussed briefly in Chapter 10, Section 5.

optional

a particle is characterized by the spin quantum number m_s, plus the quantum numbers associated with the spatial part of the wave function. Because the spin quantum numbers have just two possible values, the exclusion principle can be stated in terms of the spatial states:

> There can be at most two electrons with the same set of values for their *spatial* quantum numbers.

Exclusion principle in terms of spatial states

When there are more than two electrons in a system such as an atom or metal, only two can be in the lowest energy state. The third and fourth must go into the second-lowest state, and so on.

Particles that obey the exclusion principle are called **fermions**. Other particles such as α particles, deuterons, photons, and mesons do not obey the exclusion principle. These particles are called **bosons** and have either zero intrinsic spin, or integral spin quantum numbers.

Example 27-1

Compare the total energy of the ground state of 5 identical bosons of mass m in a one-dimensional box with that of 5 identical fermions of mass m in the same box.

Picture the Problem The ground state is the lowest possible energy state. The energy levels in a one-dimensional box are given by $E_n = n^2E_1$, where $E_1 = (h^2/8mL^2)$. The lowest energy for 5 bosons occurs when all the bosons are in the state $n = 1$ as shown in Figure 27-2*a*. For fermions, the lowest state occurs with two in state $n = 1$, two in $n = 2$, and one in $n = 3$ as shown in Figure 27-2*b*.

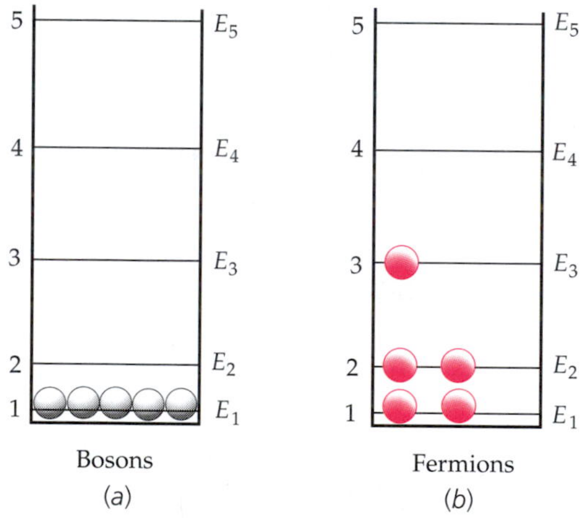

Figure 27-2

1. The energy of 5 bosons in the state $n = 1$ is: $E = 5E_1$

2. The energy of two fermions in the state $n = 1$, two in state $n = 2$, and one in state $n = 3$ is:
$$E = 2E_1 + 2E_2 + 1E_3 = 2E_1 + 2(2)^2E_1 + 1(3)^2E_1$$
$$= 2E_1 + 8E_1 + 9E_1 = 19E_1$$

Remark We see that the exclusion principle has a large effect on the energy of a multiple-particle system.

The Fermi Energy

When there are many electrons in a box, at $T = 0$ the electrons will occupy the lowest energy states consistent with the exclusion principle. If we have N electrons, we can put two electrons in the lowest energy level, two in the next lowest, and so on. The N electrons thus fill up the lowest $N/2$ energy levels

(Figure 27-3). The energy of the last filled (or half-filled) level at $T = 0$ is called the **Fermi energy** E_F. If the electrons moved in a one-dimensional box, the Fermi energy would be given by Equation 27-12 with $n = N/2$:

$$E_F = \left(\frac{N}{2}\right)^2 \frac{h^2}{8m_e L^2} = \frac{h^2}{32m_e}\left(\frac{N}{L}\right)^2 \qquad \text{27-14}$$

Fermi energy at $T = 0$ in one dimension

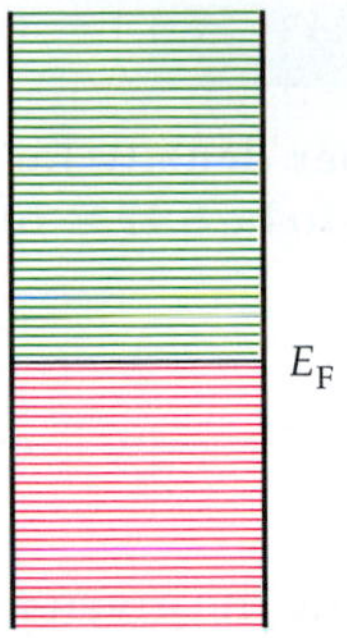

Figure 27-3 At $T = 0$ the electrons fill up the allowed energy states to the Fermi energy E_F. The levels are so closely spaced they can be assumed to be continuous.

In a one-dimensional box, the Fermi energy depends on the number of free electrons per unit length of the box.

Exercise Suppose there is an ion, and therefore a free electron, every 0.1 nm in a one-dimensional box. Calculate the Fermi energy. (*Hint:* Write Equation 27-14 as

$$E_F = \frac{(hc)^2}{32m_e c^2}\left(\frac{N}{L}\right)^2 = \frac{(1240\text{ eV}\cdot\text{nm})^2}{32(0.511\text{ MeV})}\left(\frac{N}{L}\right)^2$$

(*Answer* $E_F = 9.4$ eV)

In our model of conduction, the free electrons move in a *three-dimensional* box of volume V. The derivation of the Fermi energy in three dimensions is somewhat difficult, so we will just give the result. In three dimensions, the Fermi energy at $T = 0$ is given by

$$E_F = \frac{h^2}{8m_e}\left(\frac{3N}{\pi V}\right)^{2/3} \qquad \text{27-15}a$$

Fermi energy at $T = 0$ in three dimensions

The Fermi energy depends on the number of electrons per unit volume (the number density) N/V. Substituting numerical values for the constants gives

$$E_F = (0.365\text{ eV}\cdot\text{nm}^2)\left(\frac{N}{V}\right)^{2/3} \qquad \text{27-15}b$$

Fermi energy at $T = 0$ in three dimensions

Table 27-1 lists the free-electron number densities and Fermi energies at $T = 0$ for several metals.

Table 27-1

Free-Electron Number Densities and Fermi Energies at $T = 0$ for Selected Elements

	Element	N/V, Electrons/cm^3	E_F, eV
Al	Aluminum	18.1×10^{22}	11.7
Ag	Silver	5.86×10^{22}	5.50
Au	Gold	5.90×10^{22}	5.53
Cu	Copper	8.47×10^{22}	7.04
Fe	Iron	17.0×10^{22}	11.2
K	Potassium	1.4×10^{22}	2.11
Li	Lithium	4.70×10^{22}	4.75
Mg	Magnesium	8.60×10^{22}	7.11
Mn	Manganese	16.5×10^{22}	11.0
Na	Sodium	2.65×10^{22}	3.24
Sn	Tin	14.8×10^{22}	10.2
Zn	Zinc	13.2×10^{22}	9.46

Example 27-2

The number density for electrons in copper was calculated in Example 26-1 and found to be $8.47 \times 10^{22}/\text{cm}^3$. Calculate the Fermi energy at $T = 0$ for copper.

1. The Fermi energy is given by Equation 27-15: $E_F = (0.365\ \text{eV}\cdot\text{nm}^2)\left(\frac{N}{V}\right)^{2/3}$

2. Substitute the given number density for copper: $E_F = (0.365\ \text{eV}\cdot\text{nm}^2)\,(84.7/\text{nm}^3)^{2/3}$ $= 7.04\ \text{eV}$

Remark Note that the Fermi energy is much greater than kT at ordinary temperatures. For example, at $T = 300$ K, kT is only about 0.026 eV.

Exercise Use Equation 27-15*b* to calculate the Fermi energy at $T = 0$ for gold, which has a number density of 5.90×10^{22}. (*Answer* 5.53 eV)

The average energy of a free electron can be calculated from the complete energy distribution of the electrons, which is discussed in Section 27-6. At $T = 0$, the average energy turns out to be

$$E_{av} = \tfrac{3}{5}E_F \tag{27-16}$$

Average energy of electrons in a Fermi gas at T = 0

For copper, E_{av} is about 4 eV. This average energy is huge compared with typical thermal energies of about $kT \approx 0.026$ eV at a normal temperature of $T = 300$ K. This result is very different from the classical, Maxwell–Boltzmann distribution result that at $T = 0$, $E = 0$, and that at some temperature T, E is of the order of kT.

The Fermi Factor at $T = 0$

The probability of an energy state being occupied is called the **Fermi factor**, $f(E)$. At $T = 0$ all the states below E_F are filled, whereas all those above this energy are empty, as shown in Figure 27-4. Thus, at $T = 0$ the Fermi factor is simply

$$f(E) = 1, \qquad E < E_F$$
$$f(E) = 0, \qquad E > E_F \tag{27-17}$$

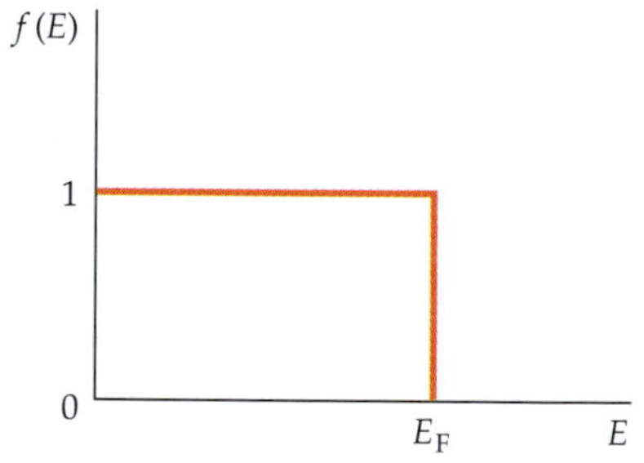

Figure 27-4 Fermi factor versus energy at $T = 0$.

The Fermi Factor for $T > 0$

At temperatures greater than $T = 0$, some electrons will occupy higher energy states because of thermal energy gained during collisions with the lattice. However, an electron cannot move to a higher or lower state unless it is unoccupied. Since the kinetic energy of the lattice ions is of the order of kT, electrons cannot gain much more energy than kT in collisions with the lattice ions. Therefore, only those electrons with energies within about kT of the Fermi energy can gain energy as the temperature is increased. At 300 K, kT is only 0.026 eV, so the exclusion principle prevents all but a very few electrons near the top of the energy distribution from gaining energy through random collisions with the lattice ions. Figure 27-5 shows the Fermi factor for some temperature T. Since for $T > 0$ there is no distinct energy that separates filled from unfilled levels, the definition of the Fermi energy must be slightly modified. At temperature T, the Fermi energy is defined to be that energy for which the probability of being occupied is $\frac{1}{2}$. For all but extremely high temperatures, the difference between the Fermi energy at temperature T and that at $T = 0$ is very small.

The **Fermi temperature** T_F is defined by

$$kT_F = E_F \tag{27-18}$$

For temperatures much lower than the Fermi temperature, the average energy of the lattice ions will be much less than the Fermi energy, and the electron energy distribution will not differ greatly from that at $T = 0$.

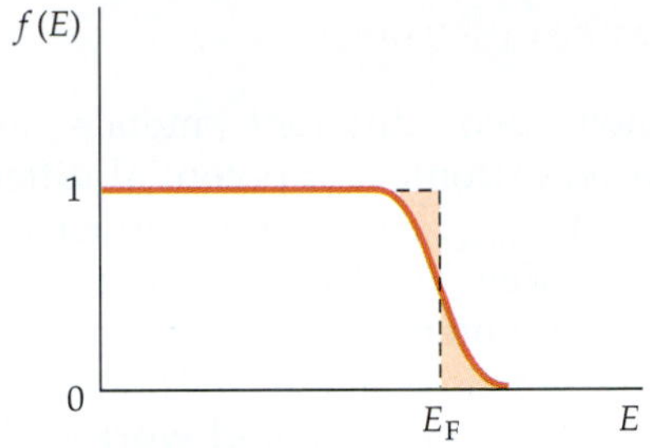

Figure 27-5 The Fermi factor for some temperature T. Some electrons with energies near the Fermi energy are excited, as indicated by the shaded regions. The Fermi energy is that value of E for which $f(E) = \frac{1}{2}$.

Example 27-3

Find the Fermi temperature for copper.

Use $E_F = 7.04$ eV and $k = 8.62 \times 10^{-5}$ eV/K in Equation 27-18:

$$T_F = \frac{E_F}{k} = \frac{7.04\text{ eV}}{8.62 \times 10^{-5}\text{ eV/K}} = 81{,}700\text{ K}$$

Remark We can see from this example that the Fermi temperature of copper is much greater than any temperature T for which copper remains a solid.

Because an electric field in a conductor accelerates all of the conduction electrons together, the exclusion principle does not prevent the free electrons in filled states from participating in conduction. Figure 27-6 shows the Fermi factor in one dimension versus *velocity* for an ordinary temperature. The factor is approximately 1 for speeds v_x in the range $-u_F < v_x < u_F$, where the Fermi speed u_F is related to the Fermi energy by $E_F = \frac{1}{2}mu_F^2$. Then

$$u_F = \sqrt{\frac{2E_F}{m_e}} \tag{27-19}$$

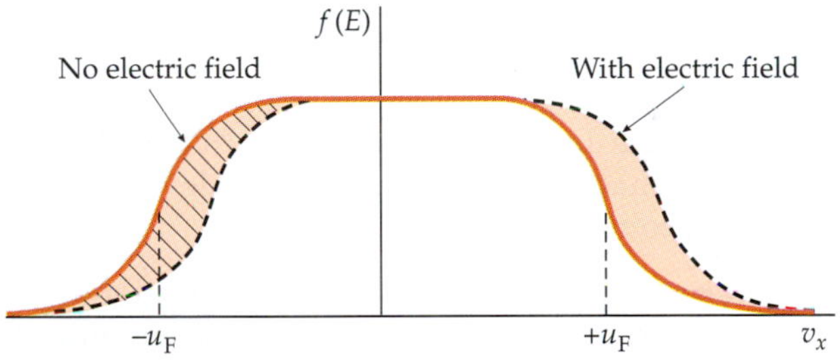

Figure 27-6 Fermi factor versus velocity in one dimension with no electric field (solid) and with an electric field in the $+x$ direction (dashed). The difference is greatly exaggerated.

Example 27-4

Calculate the Fermi speed for copper.

Use Equation 27-19 with $E_F = 7.04$ eV:

$$u_F = \sqrt{\frac{2(7.04\text{ eV})}{9.11 \times 10^{-31}\text{ kg}}\left(\frac{1.6 \times 10^{-19}\text{ J}}{1\text{ eV}}\right)} = 1.57 \times 10^6\text{ m/s}$$

The dashed curve in Figure 27-6 shows the Fermi factor after the electric field has been acting for some time t. Although all of the electrons have been shifted to higher velocities, the net effect is equivalent to shifting only the electrons near the Fermi energy.

optional

Contact Potential

When two different metals are placed in contact, a potential difference V_{contact} called the **contact potential** develops between them. The contact potential depends on the work functions of the two metals, ϕ_1 and ϕ_2 (we encountered work functions when the photoelectric effect was introduced in Chapter 17), and the Fermi energies of the two metals. When the metals are in contact, the total energy of the system is lowered if electrons near the boundary move from the metal with the higher Fermi energy into the metal with the lower Fermi energy until the Fermi energies of the two metals are the same, as shown in Figure 27-7. When equilibrium is established, the metal with the lower initial Fermi energy is negatively charged and the other is positively charged, so that between them there is a potential difference V_{contact} given by

$$V_{\text{contact}} = \frac{\phi_1 - \phi_2}{e} \qquad 27\text{-}20$$

Table 27-2 lists the work functions for several metals.

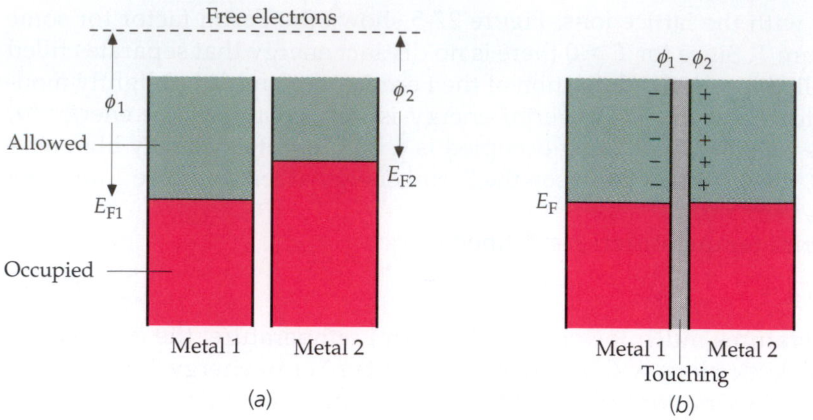

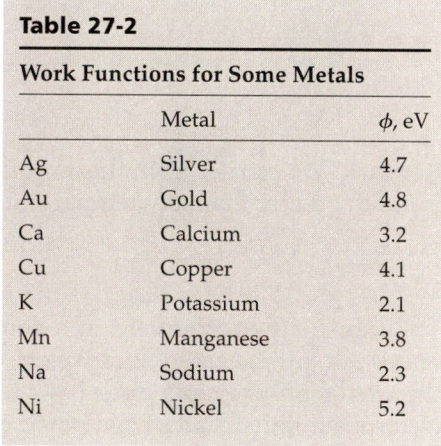

Figure 27-7 (*a*) Energy levels for two different metals with different Fermi energies and work functions. (*b*) When the metals are in contact, electrons flow from the metal that initially has the higher Fermi energy to the metal that initially has the lower Fermi energy until the Fermi energies are equal.

Table 27-2

Work Functions for Some Metals

	Metal	ϕ, eV
Ag	Silver	4.7
Au	Gold	4.8
Ca	Calcium	3.2
Cu	Copper	4.1
K	Potassium	2.1
Mn	Manganese	3.8
Na	Sodium	2.3
Ni	Nickel	5.2

Example 27-5

The threshold wavelength for the photoelectric effect is 271 nm for tungsten and 262 nm for silver. What is the contact potential developed when silver and tungsten are placed in contact?

Picture the Problem The contact potential is proportional to the difference in the work functions for the two metals. The work function ϕ can be found from the given threshold wavelengths using $\phi = hc/\lambda_t$ (Equation 17-4).

1. The contact potential is given by Equation 27-20: $\quad V_{\text{contact}} = \dfrac{\phi_1 - \phi_2}{e}$

2. The work function is related to the threshold wavelength: $\quad \phi = \dfrac{hc}{\lambda_t}$

3. Substitute $\lambda_t = 271$ nm for tungsten: $\quad \phi_W = \dfrac{hc}{\lambda_t} = \dfrac{1240\ \text{eV}\cdot\text{nm}}{271\ \text{nm}} = 4.58\ \text{eV}$

4. Substitute $\lambda_t = 262$ nm for silver: $\quad \phi_{Ag} = \dfrac{1240\ \text{eV}\cdot\text{nm}}{262\ \text{nm}} = 4.73\ \text{eV}$

5. The contact potential is thus: $\quad V_{\text{contact}} = \dfrac{\phi_{Ag} - \phi_W}{e} = 4.73\ \text{V} - 4.58\ \text{V}$

$$= 0.15\ \text{V}$$

Heat Capacity Due to Electrons in a Metal

The quantum-mechanical modification of the electron distribution in metals allows us to understand why the contribution of the electron gas to the heat capacity of a metal is much less that of the ions. According to the classical equipartition theorem, the energy of the lattice ions in n moles of a solid is $3nRT$, and thus the molar heat capacity is $C' = 3R$, where R is the universal gas constant (see Section 19-7). In a metal there is a free electron gas containing a number of electrons approximately equal to the number of lattice ions. If these electrons obey the classical equipartition theorem, they should have an energy of $\frac{3}{2}nRT$ and contribute an additional $\frac{3}{2}R$ to the molar heat capacity. But measured heat capacities of metals are just slightly greater than those of insulators. We can understand this because at some temperature T, only those electrons with energies near the Fermi energy can be excited by random collisions with the lattice ions. The number of these electrons is of the order of $(kT/E_F)N$, where N is the total number of electrons. The energy of these electrons is increased from that at $T = 0$ by an amount that is of the order of kT. So the total increase in thermal energy is of the order of $(kT/E_F)N \times kT$. We can thus express the energy of N electrons at temperature T as

$$E = NE_{av}(0) + \alpha N \frac{kT}{E_F} kT \qquad \text{27-21}$$

where α is some constant that we expect to be of the order of 1 if our reasoning is correct. The calculation of α is quite difficult. The result is $\alpha = \pi^2/4$. Using this result and writing E_F in terms of the Fermi temperature, $E_F = kT_F$, we obtain the following for the contribution of the electron gas to the heat capacity at constant volume:

$$C_v = \frac{dU}{dT} = 2\alpha Nk \frac{kT}{E_F} = \frac{\pi^2}{2} nR \frac{T}{T_F}$$

where we have written Nk in terms of the gas constant R ($Nk = nR$). The molar heat capacity at constant volume is then

$$C'_v = \frac{\pi^2}{2} R \frac{T}{T_F} \qquad \text{27-22}$$

We can see that because of the large value of T_F, the contribution of the electron gas is a small fraction of R at ordinary temperatures. Because $T_F =$ 81,700 K for copper, the molar heat capacity of the electron gas at $T = 300$ K is

$$C'_v = \frac{\pi^2}{2}\left(\frac{300\text{ K}}{81{,}700}\right) R \approx 0.02R$$

which is in good agreement with experiment.

27-3 Quantum Theory of Electrical Conduction

We can use Equation 27-7 for the resistivity if we use the Fermi speed u_F in place of v_{av}:

$$\rho = \frac{m_e u_F}{ne^2\lambda} \qquad \text{27-23}$$

We now have two problems. First, since the Fermi speed u_F is approximately independent of temperature, the resistivity given by Equation 27-23 is independent of temperature unless the mean free path depends on it. The second problem concerns magnitudes. As mentioned earlier, the classical expression for resistivity using v_{av} calculated from the Maxwell–Boltzmann distribution gives values that are about 6 times too large at $T = 300$ K. Since the Fermi speed u_F is about 16 times the Maxwell–Boltzmann value of v_{av}, the magnitude of ρ predicted by Equation 27-23 will be about 100 times greater than the experimentally determined value. The resolution of both of these problems lies in the calculation of the mean free path λ.

The Scattering of Electron Waves

In Equation 27-9 for the classical mean free path ($\lambda = 1/n_{ion}A$), the quantity $A = \pi r^2$ is the area of the lattice ion as seen by an electron. In the quantum calculation, the mean free path is related to the scattering of electron waves by the crystal lattice. Detailed calculations show that, for a *perfectly* ordered crystal, $\lambda = \infty$, that is, there is no scattering of the electron waves. The scattering of electron waves arises because of *imperfections* in the crystal lattice, which have nothing to do with the actual area of the lattice ions. According to the quantum theory of electron scattering, A depends merely on *deviations* of the lattice ions from a perfectly ordered array and not on the size of the ions. The most common causes of such deviations are thermal vibrations of the lattice ions or impurities.

We can use $\lambda = 1/n_{ion}A$ for the mean free path if we reinterpret the area A. Figure 27-8 compares the classical and quantum pictures of this area. In the quantum picture, the lattice ions are points that have no size but present an area $A = \pi r_0^2$, where r_0 is the amplitude of thermal vibrations. In Chapter 14 we saw that the energy of vibration in simple harmonic motion is proportional to the square of the amplitude, which is r_0^2. Thus, the effective area A is proportional to the energy of vibration of the lattice ions. From the equipartition theorem, we know that the average energy of vibration is proportional to kT.* Thus, A is proportional to T, and λ is proportional to $1/T$. Then the resistivity given by Equation 27-7 is proportional to T, in agreement with experiment.

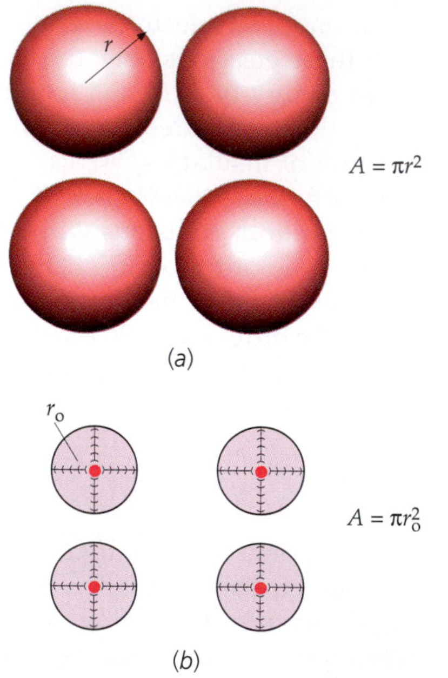

Figure 27-8 (*a*) Classical picture of the lattice ions as spherical balls of radius r that present an area πr^2 to the electrons. (*b*) Quantum-mechanical picture of the lattice ions as points that are vibrating in three dimensions. The area presented to the electrons is πr_0^2, where r_0 is the amplitude of oscillation of the ions.

The effective area A due to thermal vibrations can be calculated, and the results give values for the resistivity that are in agreement with experiment. At $T = 300$ K, for example, the effective area turns out to be about 100 times smaller than the actual area of a lattice ion. We see, therefore, that the free-electron model of metals gives a good account of electrical conduction if the classical mean speed v_{av} is replaced by the Fermi speed u_F and if the collisions between electrons and the lattice ions are interpreted in terms of the scattering of electron waves, for which only deviations from a perfectly ordered lattice are important.

The presence of impurities in a metal also causes deviations from perfect regularity in the crystal lattice. The effects of impurities on resistivity are approximately independent of temperature. The resistivity of a metal containing impurities can be written $\rho = \rho_t + \rho_i$, where ρ_t is the resistivity due to the thermal motion of the lattice ions and ρ_i is the resistivity due to impurities. Figure 27-9 shows typical resistance-versus-temperature curves for metals with impurities. As the temperature approaches zero, ρ_t approaches zero and the resistivity approaches the constant ρ_i due to impurities.

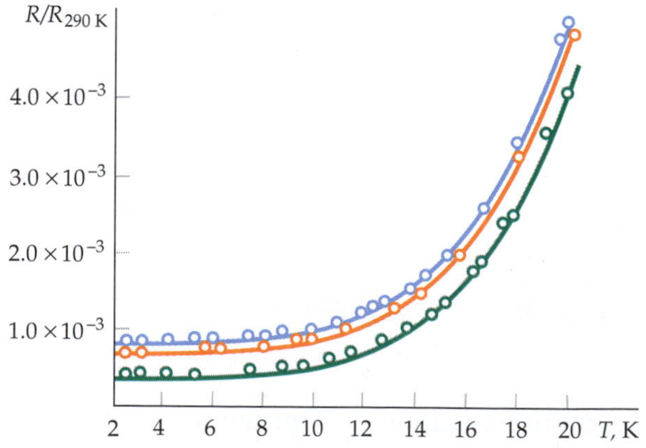

Figure 27-9 Relative resistance versus temperature for three samples of sodium. The three curves have the same temperature dependence but different magnitudes because of differing amounts of impurities in the samples.

* The equipartition theorem *does* hold for the lattice ions, which obey the Maxwell–Boltzmann energy distribution.

27-4 Band Theory of Solids

Resistivities vary enormously between insulators and conductors. For a typical insulator, such as quartz, $\rho \sim 10^{16}\ \Omega\cdot\text{m}$, whereas for a typical conductor, $\rho \sim 10^{-8}\ \Omega\cdot\text{m}$. The reason for this enormous variation is the variation in the number density of free electrons n_e. To understand this variation, we consider the effect of the lattice on the electron energy levels.

We begin by considering the energy levels of the individual atoms as they are brought together. The allowed energy levels in an isolated atom are often far apart. For example, in hydrogen, the lowest allowed energy ($E_1 = -13.6$ eV) is 10.2 eV below the next lowest ($E_2 = -13.6\ \text{eV})/4 = -3.4$ eV).* Let us consider two identical atoms and focus our attention on one particular energy level. When the atoms are far apart, the energy of a particular level is the same for each atom. As the atoms are brought closer together, the energy level for each atom changes because of the influence of the other atom. As a result, the level splits into two levels of slightly different energies for the two-atom system. If we bring three atoms close together, a particular energy level splits into three separate levels of slightly different energies. Figure 27-10 shows the energy splitting of two energy levels for six atoms as a function of the separation of the atoms.

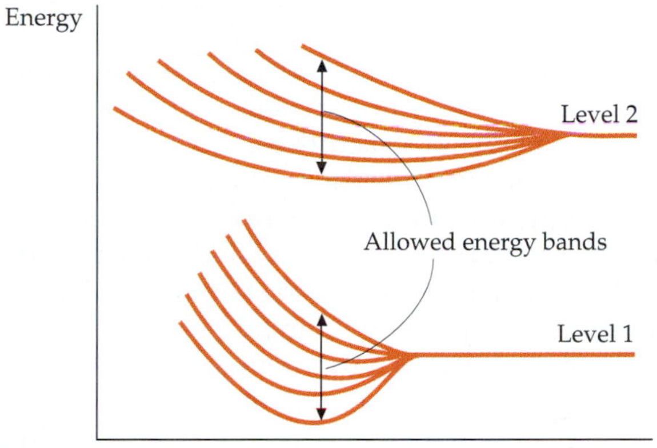

Figure 27-10 Energy splitting of two energy levels for six atoms as a function of the separation of the atoms. When there are many atoms, each level splits into a near-continuum of levels called a band.

If we have N identical atoms, a particular energy level in the isolated atom splits into N different, closely spaced energy levels when the atoms are close together. In a macroscopic solid, N is very large—of the order of 10^{23}—so each energy level splits into a very large number of levels called a **band**. The levels are spaced almost continuously within the band. There is a separate band of levels for each particular energy level of the isolated atom. The bands may be widely separated in energy, they may be close together, or they may even overlap, depending on the kind of atom and the type of bonding in the solid.

The lowest-energy bands, corresponding to the lowest energy levels of the atom in the lattice, are filled with electrons that are bound to the atom. The electrons that can take part in conduction occupy the higher energy bands. The highest energy band that contains electrons is called the **valence band**. The valence band may be completely filled with electrons or only partially filled, depending on the kind of atom and type of bonding in the solid.

We can now understand why some solids are conductors and others are insulators. If the valence band is only partially full, there are many available empty energy states in the band, and the electrons in the band can easily be raised to a higher energy state by an electric field. Accordingly, this material is a good conductor. If the valence band is full and there is a large energy gap between it and the next available band, a typical applied electric field will be too weak to excite an electron from the upper energy levels of the filled band across the large gap into the energy levels of the empty band, so the material is an insulator. The lowest band in which there are unoccupied states is called the **conduction band**. In a conductor, the valence band is only partially filled, so the valence band is also the conduction band. An energy gap between allowed bands is called a **forbidden energy band**.

The band structure for a conductor such as copper is shown in Figure

* The energy levels in hydrogen are discussed in Chapter 37.

27-11*a*. The lower bands (not shown) are filled with the inner electrons of the atoms. The valence band is only about half full. When an electric field is established in the conductor, the electrons in the conduction band are accelerated, which means that their energy is increased. This is consistent with the Pauli exclusion principle because there are many empty energy states just above those occupied by electrons in this band. These electrons are thus the conduction electrons.

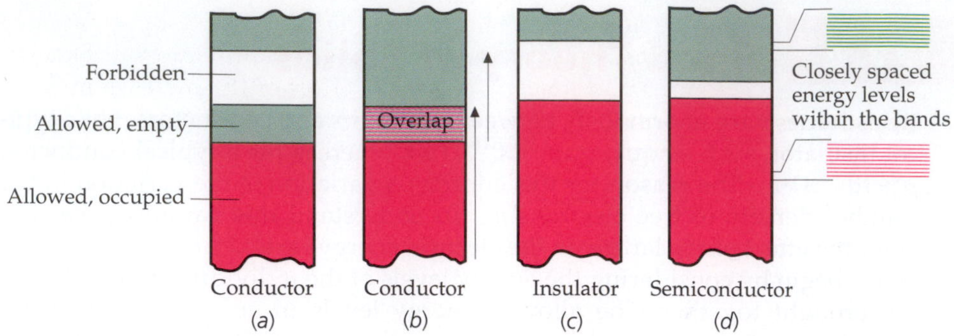

Figure 27-11 Four possible band structures for a solid. (*a*) A typical conductor. The valence band is only partially full, so electrons can be easily excited to nearby energy states. (*b*) A conductor in which the allowed energy bands overlap. (*c*) A typical insulator. There is a forbidden band with a large energy gap between the filled valence band and the conduction band. (*d*) A semiconductor. The energy gap between the filled valence band and the conduction band is very small, so some electrons are excited to the conduction band at normal temperatures, leaving holes in the valence band.

Figure 27-11*b* shows the band structure for magnesium, which is also a conductor. In this case, the highest occupied band is full, but there is an empty band above it that overlaps it. The two bands thus form a combined valence–conduction band that is only partially filled.

Figure 27-11*c* shows the band structure for a typical insulator. At $T = 0$ K, the valence band is completely full. The next energy band containing empty energy states, the conduction band, is separated from the valence band by a large energy gap. At $T = 0$, the conduction band is empty. At ordinary temperatures, a few electrons can be excited to states in this band, but most cannot be because the energy gap is large compared with the energy an electron might obtain by thermal excitation. Very few electrons can be thermally excited to the nearly empty conduction band, even at fairly high temperatures. When an electric field of ordinary magnitude is established in the solid, electrons cannot be accelerated because there are no empty energy states at nearby energies. We describe this by saying that there are no free electrons. The small conductivity that is observed is due to the very few electrons that are thermally excited into the nearly empty conduction band. When an electric field applied to an insulator is sufficiently strong to cause an electron to be excited across the energy gap to the empty band, dielectric breakdown occurs.

In some materials, the energy gap between the filled valence band and the empty conduction band is very small, as shown in Figure 27-11*d*. At $T = 0$, there are no electrons in the conduction band and the material is an insulator. However, at ordinary temperatures, there are an appreciable number of electrons in the conduction band due to thermal excitation. Such a material is called an **intrinsic semiconductor**. For typical semiconductors such as silicon and germanium, the energy gap is only about 1 eV. In the presence of an electric field, the electrons in the conduction band can be accelerated because there are empty states nearby. Also, for each electron in the conduction band there is a vacancy, or hole, in the nearly filled valence band. In the presence of an electric field, electrons in this band can also be excited to a vacant energy level. This contributes to the electric current and is most easily described as the motion of a hole in the direction of the field and opposite to the motion of the electrons. The hole thus acts like a positive charge. To visualize the conduction of holes, think of a two-lane, one-way road with one lane full of parked cars and the other empty. If a car moves out of the filled lane into the empty lane, it can move ahead freely. As the other cars move up to occupy the space left, the empty space propagates backward in the direction opposite the motion of the cars. Both the forward motion of the car in the nearly empty lane and the backward propagation of the empty space contribute to a net forward propagation of the cars.

An interesting characteristic of semiconductors is that the resistivity of the material decreases as the temperature increases, which is contrary to the case for normal conductors. The reason is that as the temperature increases, the number of free electrons increases because there are more electrons in the conduction band. The number of holes in the valence band also increases, of course. In semiconductors, the effect of the increase in the number of charge carriers, both electrons and holes, exceeds the effect of the increase in resistivity due to the increased scattering of the electrons by the lattice ions due to thermal vibrations. Semiconductors therefore have a negative temperature coefficient of resistivity.

27-5 Superconductivity

There are some materials for which the resistivity suddenly drops to zero below a certain temperature T_c, called the **critical temperature**. This amazing phenomenon, called **superconductivity**, was discovered in 1911 by the Dutch physicist H. Kamerlingh Onnes, who had developed a technique for liquefying helium (boiling point 4.2 K) and was putting his technique to work exploring the properties of materials at extremely low temperatures. Figure 27-12 shows his plot of the resistance of mercury versus temperature. The critical temperature for mercury is the same as the boiling point of helium, 4.2 K. Critical temperatures for other superconducting elements range from less than 0.1 K for hafnium and iridium to 9.2 K for niobium. The temperature range for superconductors goes much higher for a number of metallic compounds. For example, the superconducting alloy Nb_3Ge, discovered in 1973, has a critical temperature of 23.2 K, which was the highest known until 1986, when the discoveries of Bednorz and Müller launched the era of high-temperature superconductors, now defined as materials that exhibit superconductivity at temperatures above 77 K. To date (mid 1998), the highest temperature at which superconductivity has been demonstrated, using compounds containing mercury, is 125 K at atmospheric pressure and 164 K at high pressure.

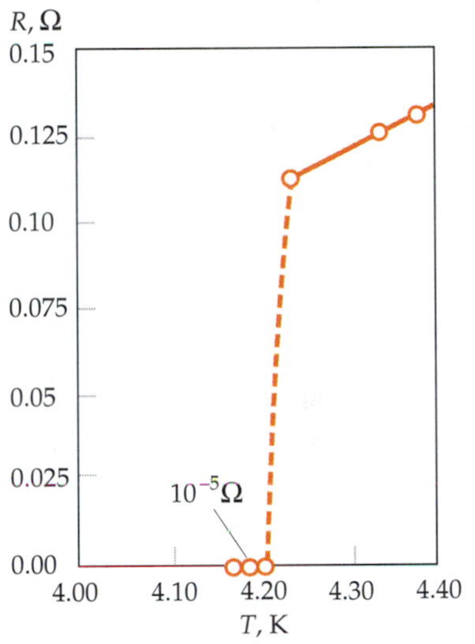

Figure 27-12 Plot by Kamerlingh Onnes of the resistance of mercury versus temperature, showing the sudden decrease at the critical temperature of $T = 4.2$ K.

The resistivity of a superconductor is zero. There can be a current in a superconductor even when the electric field in the superconductor is zero. Indeed, in superconducting rings in which there was no electric field, steady currents have been observed to persist for years without apparent loss. Despite the cost and inconvenience of refrigeration with expensive liquid helium, many superconducting magnets have been built using superconducting materials, because such magnets require no power expenditure to maintain the large current needed to produce a large magnetic field.

The discovery of high-temperature superconductors has revolutionized the study of superconductivity because relatively inexpensive liquid nitrogen, which boils at 77 K, can be used for a coolant. However, many problems, such as brittleness and the toxicity of the materials, make these new superconductors difficult to use. The search continues for new materials that will superconduct at even higher temperatures.

The BCS Theory

It had been recognized for some time that superconductivity is due to a collective action of the conducting electrons. In 1957, John Bardeen, Leon Cooper, and Robert Schrieffer published a successful theory of superconductivity now known by the initials of the inventors as the **BCS theory**. According to this theory, the electrons in a superconductor are coupled in pairs at low temperatures. The coupling comes about because of the interaction between electrons and the crystal lattice. One electron interacts with the lattice and perturbs it. The perturbed lattice interacts with another electron in such a way that there is an attraction between the two electrons that at low temperatures can exceed the Coulomb repulsion between them. The electrons form a bound state called a **Cooper pair**. The electrons in a Cooper pair have equal and opposite spins, so they form a system with zero spin. Each Cooper pair acts as a *single particle* with zero spin—in other words, as a boson. Bosons do not obey the exclusion principle. Any number of Cooper pairs may be in the same quantum state with the same energy. In the ground state of a superconductor (at $T = 0$), all the electrons are in Cooper pairs and all the Cooper pairs are in the same energy state. In the superconducting state, the Cooper pairs are correlated so that they all act together. An electric current can be produced in a superconductor because all of the electrons in this collective state move together. But energy cannot be dissipated by individual collisions of electron and lattice ions unless the temperature is high enough to break the binding of the Cooper pairs. The energy needed to break up a Cooper pair is similar to that needed to break up a molecule into its constituent atoms. This energy is called the **superconducting energy gap** E_g. In the BCS theory, this energy at absolute zero is related to the critical temperature by

$$E_g = 3.5kT_c \qquad \text{27-24}$$

The energy gap can be determined by measuring the current across a junction between a normal metal and a superconductor as a function of voltage. Consider two metals separated by a layer of insulating material, such as aluminum oxide, that is only a few nanometers thick. The insulating material between the metals forms a barrier that prevents most electrons from traversing the junction. But, as mentioned in Chapter 15, waves can tunnel through a barrier if the barrier is not too thick even if the energy of the wave is less than that of the barrier (see Figures 15-21 and 15-22).

When the materials on either side of the gap are normal nonsuperconducting metals, the current resulting from the tunneling of electrons through the insulating layer obeys Ohm's law for low applied voltages (Figure 27-13*a*). When one of the metals is a normal metal and the other is a superconductor, there is no current (at absolute zero) unless the applied voltage V is greater than a critical voltage $V_c = E_g/2e$, where E_g is the superconductor energy gap. Figure 27-13*b* shows the plot of current versus voltage for this situation. The current jumps abruptly when the energy $2eV_c$ absorbed by a Cooper pair is great enough to break up the pair. (The small current visible in Figure 27-13*b* before the critical voltage is reached is present because at any temperature above absolute zero some of the electrons in the superconductor are thermally excited above the energy gap and are therefore not paired.) The superconducting energy gap can thus be accurately measured by measuring the critical voltage V_c.

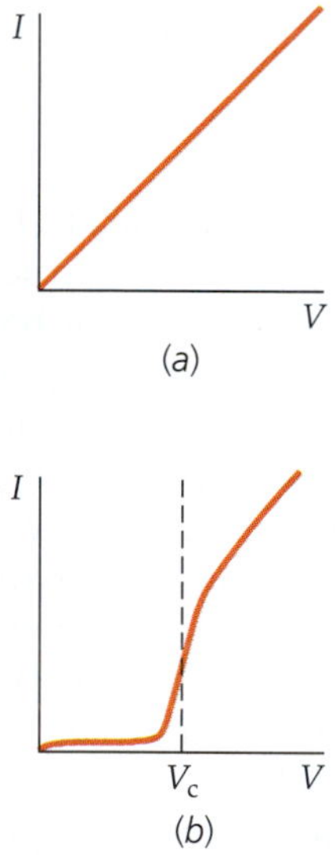

Figure 27-13 Tunneling current versus voltage for a junction of two metals separated by a thin oxide layer. (*a*) When both metals are normal metals, the current is proportional to the voltage, as predicted by Ohm's law. (*b*) When one metal is a normal metal and one is a superconductor, the current is approximately zero until the applied voltage exceeds the critical voltage $V_c = E_g/2e$.

Example 27-6

Calculate the superconducting energy gap for mercury (T_c = 4.2 K) predicted by the BCS theory.

1. The BCS prediction for the energy gap is: $E_g = 3.5kT_c$

2. Substitute $T_c = 4.2$ K:

$$E_g = 3.5kT_c$$
$$= 3.5(1.38 \times 10^{-23}\ \text{J/K})(4.2\ \text{K})\frac{1\ \text{eV}}{1.6 \times 10^{-19}\ \text{J}}$$
$$= 1.27 \times 10^{-3}\ \text{eV}$$

Note that the energy gap for a typical superconductor is much smaller than the energy gap for a typical semiconductor, which is of the order of 1 eV. As the temperature is increased from $T = 0$, some of the Cooper pairs are broken. Then there are fewer pairs available for each pair to interact with, and the energy gap is reduced until at $T = T_c$ the energy gap is zero (Figure 27-14).

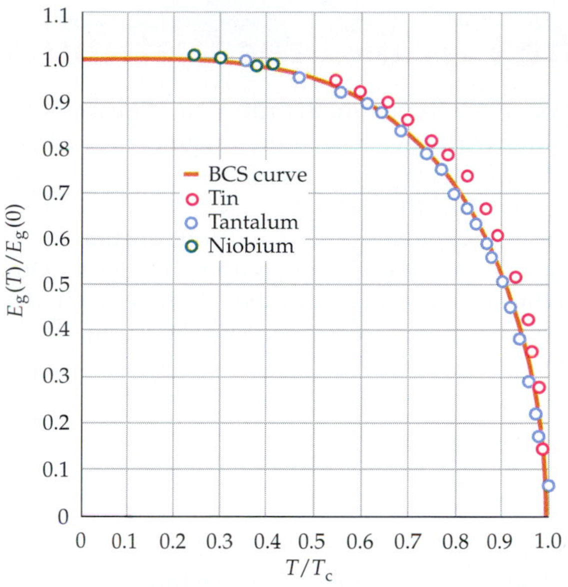

Figure 27-14 Ratio of the energy gap at temperature T to that at $T = 0$ as a function of the relative temperature T/T_c. The solid curve is that predicted by the BCS theory.

The Josephson Effect

When two superconductors are separated by a thin insulating barrier (for example, a layer of aluminum oxide a few nanometers thick), the junction is called a **Josephson junction**, based on the prediction in 1962 by Brian Josephson that Cooper pairs could tunnel across such a junction from one superconductor to the other with no resistance. The tunneling of Cooper pairs constitutes a current, which is observed even when there is no voltage applied across the junction. The current depends on the difference in phase of the wave functions that describe the Cooper pairs. Let ϕ_1 be the phase constant for the wave function of a Cooper pair in one superconductor. All the Cooper pairs in a superconductor act coherently, so they all have the same phase constant. If ϕ_2 is the phase constant for the Cooper pairs in the second superconductor, the current across the junction is given by

$$I = I_{max} \sin(\phi_2 - \phi_1) \qquad \text{27-25}$$

where I_{max} is the maximum current, which depends on the thickness of the barrier. This result has been observed experimentally and is known as the **dc Josephson effect**.

Josephson also predicted that if a dc voltage V were applied across a Josephson junction, there would be a current that alternates with frequency f given by

$$f = \frac{2eV}{h} \qquad \text{27-26}$$

This result, known as the **ac Josephson effect**, has been observed experimentally, and careful measurement of the frequency allows a precise determination of the ratio e/h. Because frequency can be measured very accurately, the ac Josephson effect is also used to establish precise voltage standards. The inverse effect, in which the application of an alternating voltage across a Josephson junction results in a dc current, has also been observed.

Example 27-7

Using $e = 1.602 \times 10^{-19}$ C and $h = 6.626 \times 10^{-34}$ J·s, calculate the frequency of the Josephson current if the applied voltage is 1 μV.

Substitute the given values into Equation 27-25 to calculate f:

$$f = \frac{2eV}{h} = \frac{2(1.602 \times 10^{-19}\text{ C})(10^{-6}\text{ V})}{6.626 \times 10^{-34}\text{ J}\cdot\text{s}}$$

$$= 4.835 \times 10^{8}\text{ Hz} = 483.5\text{ MHz}$$

optional

27-6 The Fermi–Dirac Distribution*

*This material is somewhat complicated and may be skipped over on a first reading.

The classical Maxwell–Boltzmann distribution (Equation 18-39) gives the number of molecules with energy E in the range between E and $E + dE$. It is the product of $g(E)\,dE$ where $g(E)$ is the density of states (number of energy states in the range dE) and the Boltzmann factor $e^{-E/kT}$, which is the probability of a state being occupied. The distribution function for free electrons in a metal is called the **Fermi–Dirac distribution**. The Fermi–Dirac distribution can be written in the same form as the Maxwell–Boltzmann distribution with the density of states calculated from quantum theory and the Boltzmann factor replaced by the Fermi factor. Let $n(E)\,dE$ be the number of electrons with energies between E and $E + dE$. This number is written

$$n(E)\,dE = g(E)\,dE\,f(E) \qquad 27\text{-}27$$

Energy distribution function

where $g(E)\,dE$ is the number of states between E and $E + dE$, $g(E)$ is called the density of states, and $f(E)$ is the probability of a state being occupied, which is the Fermi factor. The density of states in three dimensions is somewhat difficult to calculate, so we just give the result. For electrons in a metal of volume V, the density of states is

$$g(E) = \frac{8\pi\sqrt{2}m_e^{3/2}V}{h^3}E^{1/2} \qquad 27\text{-}28$$

Density of states

As in the classical Maxwell–Boltzmann distribution, the density of states is proportional to $E^{1/2}$.

At $T = 0$, the Fermi factor is given by Equation 27-17:

$$f(E) = 1, \qquad E < E_F$$

$$f(E) = 0, \qquad E > E_F$$

The integral of $n(E)\,dE$ over all energies gives the total number of electrons N. We can derive Equation 27-15*a* for the Fermi energy at $T = 0$ by integrating $n(E)\,dE$ from $E = 0$ to $E = \infty$. We obtain

$$N = \int_0^\infty n(E)\,dE = \int_0^{E_F} g(E)\,dE = \frac{8\pi\sqrt{2}m_e^{3/2}V}{h^3}\int_0^{E_F} E^{1/2}\,dE = \frac{16\pi\sqrt{2}m_e^{3/2}V}{3h^3}E_F^{3/2}$$

Note that at $T = 0$, $n(E)$ is zero for $E > E_F$ so we had to integrate only from $E = 0$ to $E = E_F$. Solving for E_F gives the Fermi energy at $T = 0$:

$$E_F = \frac{h^2}{8m_e}\left(\frac{3N}{\pi V}\right)^{2/3} \qquad 27\text{-}29$$

which is Equation 27-15*a*. In terms of the Fermi energy, the density of states (Equation 27-28) is

$$g(E) = \frac{8\pi\sqrt{2}m_e^{3/2}V}{h^3}E^{1/2} = \frac{3N}{2}E_F^{-3/2}E^{1/2} \qquad 27\text{-}30$$

Density of states in terms of E_F

The average energy at $T = 0$ is calculated from

$$E_{av} = \frac{\int_0^{E_F} Eg(E)\,dE}{\int_0^{E_F} g(E)\,dE} = \frac{1}{N}\int_0^{E_F} Eg(E)\,dE$$

where $N = \int_0^{E_F} g(E)\,dE$ is the total number of electrons. Performing the integration, we obtain Equation 27-16

$$E_{av} = \tfrac{3}{5}E_F \qquad 27\text{-}31$$

Average energy at $T = 0$

At $T > 0$, the Fermi factor is more complicated. It can be shown to be given by

$$f(E) = \frac{1}{e^{(E - E_F)/kT} + 1} \qquad \text{27-32}$$

Fermi factor

We can see from this equation that for E greater than E_F, $e^{(E-E_F)/kT}$ becomes very large as T approaches zero, so at $T = 0$, the Fermi factor is zero for $E > E_F$. On the other hand, for E less than E_F, $e^{(E-E_F)/kT}$ approaches 0 as T approaches zero, so at $T = 0$, $f(E) = 1$ for $E < E_F$. Thus the Fermi factor given by Equation 27-32 holds for all temperatures. Note also that for any nonzero value of T, $f(E) = \frac{1}{2}$ at $E = E_F$.

The complete Fermi–Dirac distribution function is thus

$$n(E)\,dE = \frac{8\pi\sqrt{2}m_e^{3/2}V}{h^3}E^{1/2}\frac{1}{e^{(E-E_F)/kT}+1}\,dE \qquad \text{27-33}$$

Fermi–Dirac distribution

We can see that for those few electrons with energies much greater than the Fermi energy, the Fermi factor approaches $1/e^{(E-E_F)/kT} = e^{(E_F-E)/kT} = e^{E_F}e^{-E/kT}$, which is proportional to $e^{-E/kT}$. Thus, the high-energy tail of the Fermi–Dirac energy distribution decreases as $e^{-E/kT}$, just like the classical Maxwell–Boltzmann energy distribution. The reason is that in this high-energy region, there are many unoccupied energy states and few electrons, so the Pauli exclusion principle is not important, and the distribution approaches the classical distribution. This result has practical importance because it applies to the conduction electrons in semiconductors.

Example 27-8

At what energy is the Fermi factor equal to 0.1 for copper at $T = 300$ K?

Picture the Problem We set $f(E) = 0.1$ in Equation 27-32 using $T = 300$ K and $E_F = 7.04$ eV from Table 27-1 and solve for E.

1. Solve Equation 27-32 for $e^{(E-E_F)/kT}$:

$$f(E) = \frac{1}{e^{(E-E_F)/kT}+1} = 0.1$$

$$e^{(E-E_F)/kT} = \frac{1}{f(E)} - 1 = \frac{1}{0.1} - 1 = 9$$

2. Take the natural logarithm of both sides:

$$\frac{E - E_F}{kT} = \ln 9 = 2.20$$

3. Solve for $E - E_F$:

$$E - E_F = 2.20kT = 2.20(8.62 \times 10^{-5}\ \text{eV/K})(300\ \text{K})$$

$$= 2.20(0.0259\ \text{eV}) = 0.0570\ \text{eV}$$

4. Solve for E using $E_F = 7.04$ eV from Table 27-1:

$$E = 7.04\ \text{eV} + 0.0570\ \text{eV} = 7.10\ \text{eV}$$

Remark The Fermi factor drops from about 1 to 0.1 at just 0.06 eV above the Fermi energy of about 7 eV.

optional

Example 27-9

Find the probability that an energy state in copper 0.1 eV above the Fermi energy is occupied at $T = 300$ K.

Picture the Problem The probability is the Fermi factor given in Equation 27-32, with $E_F = 7.04$ eV, and $E = 7.14$ eV.

1. The probability of a state being occupied equals the Fermi factor:

$$P = f(E) = \frac{1}{e^{(E-E_F)/kT} + 1}$$

2. Calculate the dimensionless exponent in the Fermi factor:

$$\frac{E - E_F}{kT} = \frac{7.14\text{ eV} - 7.04\text{ eV}}{(8.62 \times 10^{-5}\text{ eV/K})(300\text{ K})} = 3.87$$

3. Use this result to calculate the Fermi factor:

$$f = \frac{1}{e^{(E-E_F)/kT} + 1} = \frac{1}{e^{3.87} + 1}$$

$$= \frac{1}{48 + 1} = 0.0204 = 2.04\%$$

Remark The probability of an electron having an energy of 0.1 eV above the Fermi energy at 300 K is only about 2%.

Example 27-10 ***try it yourself***

Find the probability that an energy state in copper 0.1 eV *below* the Fermi energy is occupied at $T = 300$ K.

Picture the Problem The probability is the Fermi factor given in Equation 27-32, with $E_F = 7.04$ eV, and $E = 6.94$ eV.

Cover the column to the right and try these on your own before looking at the answers.

Steps	***Answers***
1. Write the Fermi factor.	$f(E) = \frac{1}{e^{(E-E_F)kT} + 1}$
2. Calculate the dimensionless exponent in the Fermi factor.	$\frac{E - E_F}{kT} = \frac{6.94\text{ eV} - 7.04\text{ eV}}{(8.62 \times 10^{-5}\text{ eV/K})(300\text{ K})} = -3.87$
3. Use your result in step 2 to calculate the Fermi factor.	$f = \frac{1}{e^{(E-E_F)/kT} + 1} = \frac{1}{e^{-3.87} + 1}$ $= \frac{1}{0.021 + 1} = 0.979 \approx 98\%$

Remark The probability of an electron having an energy of 0.1 eV *below* the Fermi energy at 300 K is about 98%.

Exercise What is the probability of an energy state 0.1 eV below the Fermi energy being unoccupied at 300 K? (*Answer* $1 - 0.98 = 0.02$ or 2%. This is the probability of there being a hole at this energy.)

Summary

Topic	Remarks and Relevant Equations	
1. Microscopic Picture of Conduction		
Resistivity	$\rho = \frac{m_e v_{av}}{n_e e^2 \lambda}$ where v_{av} is the average speed of the electrons and λ is their mean free path between collisions with the lattice ions.	27-7
Mean free path	$\lambda = \frac{1}{n_{ion}\pi r^2} = \frac{1}{n_{ion}A}$ where n_{ion} is the number of lattice ions per unit volume, r is their effective radius, and A is their effective cross-sectional area.	27-9
2. Classical Interpretation of v_{av} and λ	v_{av} is determined from the Maxwell–Boltzmann distribution, and r is the actual radius of a lattice ion.	
3. Quantum Interpretation of v_{av} and λ	v_{av} is determined from the Fermi–Dirac distribution and is approximately constant independent of temperature. The mean free path is determined from the scattering of electron waves, which occurs only because of deviations from a perfectly ordered array. The radius r is the amplitude of vibration of the lattice ion, which is proportional to $\sqrt{T}$, so A is proportional to T.	
4. Fermi Electron Gas		
Fermi energy E_F at $T = 0$	E_F is the energy of the highest filled (or half-filled) energy state.	
E_F at $T > 0$	E_F is the energy at which the probability of being occupied is $\frac{1}{2}$.	
Approximate magnitude of E_F	E_F is about 5 to 10 eV for most metals.	
Dependence of E_F on the number density of free electrons	$E_F = \frac{h^2}{8m_e}\left(\frac{3N}{\pi V}\right)^{2/3} = (0.365\ \text{eV}\cdot\text{nm}^2)\left(\frac{N}{V}\right)^{2/3}$	27-15*a,b*
Average energy at $T = 0$	$E_{av} = \frac{3}{5}E_F$	27-16
Fermi factor at $T = 0$	The Fermi factor $f(E)$ is the probability of a state being occupied $f(E) = 1, \quad E < E_F$ $f(E) = 0, \quad E > E_F$	27-17
Fermi temperature	$T_F = \frac{E_F}{k}$	27-18
Fermi speed	$u_F = \sqrt{\frac{2E_F}{m_e}}$	27-19
Contact potential	When two different metals are placed in contact, electrons flow from the metal with the higher Fermi energy to the one with the lower Fermi energy until the Fermi energies of the two metals are equal. In equilibrium, there is a potential difference be-	

optional

tween the metals that is equal to the difference in the work function of the two metals divided by the electronic charge e:

$$V_{\text{contact}} = \frac{\phi_1 - \phi_2}{e} \tag{27-20}$$

Heat capacity due to electrons	$C'_v = \frac{\pi^2}{2} R \frac{T}{T_F}$	**27-22**

5. Band Theory of Solids

When many atoms are brought together to form a solid, the individual energy levels are split into bands of allowed energies. The splitting depends on the type of bonding and the lattice separation. The highest-energy band containing electrons is called the valence band. In a conductor, the valence band is only partially full, so there are many available states for excited electrons. In an insulator, the valence band is completely full and there is a large energy gap between it and the next allowed band, the conduction band. In a semiconductor, the energy gap between the filled valence band and the empty conduction band is small; so at ordinary temperatures, an appreciable number of electrons are thermally excited into the conduction band.

6. Superconductivity

In a superconductor, the resistance drops suddenly to zero below a critical temperature T_c. Superconductors with critical temperatures as high as 125 K have been discovered.

BCS theory

Superconductivity is described by a theory of quantum mechanics called the BCS theory in which the free electrons form Cooper pairs. The energy needed to break up a Cooper pair is called the energy gap E_g. When all the electrons are paired, individual electrons cannot be scattered by a lattice ion, so the resistance is zero.

Tunneling

When a normal conductor is separated from a superconductor by a thin layer of oxide, electrons can tunnel through the energy barrier if the applied voltage across the layer is $E_g/2e$, where E_g is the energy needed to break up a Cooper pair. The energy gap E_g can be determined by a measurement of the tunneling current versus the applied voltage.

Josephson junction

A system of two superconductors separated by a thin layer of insulating material is called a Josephson junction.

dc Josephson effect

A dc current is observed to tunnel through a Josephson junction even in the absence of voltage across the junction.

ac Josephson effect

When a dc voltage V is applied across a Josephson junction, an ac current is observed with a frequency

$$f = \frac{2eV}{h} \tag{27-26}$$

Measurement of the frequency of this current allows a precise determination of the ratio e/h.

7. Fermi–Dirac Distribution

The number of electrons with energies between E and $E + dE$ is given by

$$n(E)\,dE = g(E)\,dE\,f(E) \tag{27-27}$$

where $g(E)$ is the density of states and $f(E)$ is the Fermi factor.

Density of states	$g(E) = \frac{8\pi\sqrt{2}m_e^{3/2}V}{h^3} E^{1/2}$	**27-28**
Fermi factor at temperature T	$f(E) = \frac{1}{e^{(E-E_F)/kT} + 1}$	**27-32**

Problem-Solving Guide

Summary of Worked Examples

Type of Calculation	Procedure and Relevant Examples	
1. The Fermi Electron Gas		
Calculate the Fermi energy from the number density of free electrons.	Use $E_F = (0.365 \text{ eV}\cdot\text{nm}^2)(N/V)^{2/3}$.	**Example 27-2**
Calculate the Fermi temperature.	Use $kT_F = E_F$ and Table 27-1 for E_F.	**Example 27-3**
Calculate the Fermi speed.	Use $u_F = \sqrt{2E_F/m_e}$ and Table 27-1 for E_F.	**Example 27-4**
Calculate the contact potential developed when two metals are placed in contact.	Find the work functions from the photoelectric thresholds and use $V_{\text{contact}} = (\phi_1 - \phi_2)/e$.	**Example 27-5**
2. Superconductors		
Find the superconducting gap from the critical temperature.	Use $E_g = 3.5kT_c$.	**Example 27-6**
Find the Josephson frequency for a given applied voltage.	Use $f = 2eV/h$.	**Example 27-7**
3. The Fermi–Dirac Distribution		
Find the energy for a given value of $f(E)$ at a given temperature.	Use $f(E) = 1/(e^{(E-E_F)/kT} + 1)$ and solve for E.	**Example 27-8**
Find the probability that a state of a given energy will be occupied.	Calculate the Fermi factor for the given energy.	**Examples 27-9, 27-10**

Problems

Conceptual Problems

Problems from Optional and Exploring sections

- • Single-concept, single-step, relatively easy
- •• Intermediate-level, may require synthesis of concepts
- ••• Challenging, for advanced students

In a few problems, you are given more data than you actually need; in a few other problems, you are required to supply data from your general knowledge, outside sources, or informed estimates.

Use Table 27-1 for the Fermi energies and electron number densities when needed.

Microscopic Picture of Conduction

1 • In the classical model of conduction, the electron loses energy on average in a collision because it loses the drift velocity it had picked up since the last collision. Where does this energy appear?

2 • A measure of the density of the free-electron gas in a metal is the distance r_s, which is defined as the radius of the sphere whose volume equals the volume per conduction electron. (*a*) Show that $r_s = (3/4\pi n)^{1/3}$, where n is the free-electron number density. (*b*) Calculate r_s for copper in nanometers.

3 • (*a*) Given a mean free path $\lambda = 0.4$ nm and a mean speed $v_{\text{av}} = 1.17 \times 10^5$ m/s for the current flow in copper at a temperature of 300 K, calculate the classical value for the resistivity ρ of copper. (*b*) The classical model suggests that the mean free path is temperature independent and that v_{av} depends on temperature. From this model, what would ρ be at 100 K?

The Fermi Electron Gas

4 • Calculate the number density of free electrons in (*a*) Ag ($\rho = 10.5$ g/cm^3) and (*b*) Au ($\rho = 19.3$ g/cm^3), assuming one free electron per atom, and compare your results with the values listed in Table 27-1.

5 • The density of aluminum is 2.7 g/cm^3. How many free electrons are present per aluminum atom?

6 • The density of tin is 7.3 g/cm^3. How many free electrons are present per tin atom?

7 • Calculate the Fermi temperature for (*a*) Al, (*b*) K, and (*c*) Sn.

8 • What is the speed of a conduction electron whose energy is equal to the Fermi energy E_F for (*a*) Na, (*b*) Au, and (*c*) Sn?

9 • Calculate the Fermi energy for (*a*) Al, (*b*) K, and (*c*) Sn using the number densities given in Table 27-1.

10 • Find the average energy of the conduction electrons at $T = 0$ in (*a*) copper and (*b*) lithium.

11 • Calculate (*a*) the Fermi temperature and (*b*) the Fermi energy at $T = 0$ for iron.

12 •• The pressure of an ideal gas is related to the average energy of the gas particles by $PV = \frac{2}{3}NE_{av}$, where N is the number of particles and E_{av} is the average energy. Use this to calculate the pressure of the Fermi electron gas in copper in newtons per square meter, and compare your result with atmospheric pressure, which is about 10^5 N/m^2. (*Note:* The units are most easily handled by using the conversion factors $1\ N/m^2 = 1\ J/m^3$ and $1\ eV = 1.6 \times 10^{-19}$ J.)

13 •• The bulk modulus B of a material can be defined by

$$B = -V\frac{\partial P}{\partial V}$$

(*a*) Use the ideal-gas relation $PV = \frac{2}{3}NE_{av}$ and Equations 27-15 and 27-16 to show that

$$P = \frac{2NE_F}{5V} = CV^{-5/3}$$

where C is a constant independent of V. (*b*) Show that the bulk modulus of the Fermi electron gas is therefore

$$B = \frac{5}{3}P = \frac{2NE_F}{3V}$$

(*c*) Compute the bulk modulus in newtons per square meter for the Fermi electron gas in copper and compare your result with the measured value of 140×10^9 N/m^2.

Contact Potential

14 • Thomas refuses to believe that a potential difference can be created simply by bringing two different metals into contact with each other. John talks him into making a small wager, and is about to cash in. (*a*) Which two metals from Table 27-2 would demonstrate his point most effectively? (*b*) What is the value of that contact potential?

15 • (*a*) In Problem 14, which choices of different metals would make the least impressive demonstration? (*b*) What is the value of that contact potential?

16 • Calculate the contact potential between (*a*) Ag and Cu, (*b*) Ag and Ni, and (*c*) Ca and Cu.

Quantum Theory of Electrical Conduction

17 • When the temperature of pure copper is lowered from 300 K to 4 K, its resistivity drops by a much greater factor than that of brass when it is cooled the same way. Why?

18 • The resistivities of Na, Au, and Sn at $T = 273$ K are 4.2 $\mu\Omega\cdot$cm, 2.04 $\mu\Omega\cdot$cm, and 10.6 $\mu\Omega\cdot$cm, respectively. Use these values and the Fermi speeds calculated in Problem 8 to find the mean free paths λ for the conduction electrons in these elements.

19 •• The resistivity of pure copper is increased by about 1×10^{-8} $\Omega\cdot$m by the addition of 1% (by number of atoms) of an impurity throughout the metal. The mean free path depends on both the impurity and the oscillations of the lattice ions according to the equation

$$\frac{1}{\lambda} = \frac{1}{\lambda_t} + \frac{1}{\lambda_i}$$

(*a*) Estimate λ_i from data given in Table 27-1. (*b*) If r is the effective radius of an impurity lattice ion seen by an electron, the scattering cross section is πr^2. Estimate this area using the fact that r is related to λ_i by Equation 27-9.

Band Theory of Solids

20 • A metal is a good conductor because the valence energy band for electrons is

(*a*) completely full.
(*b*) full, but there is only a small gap to a higher empty band.
(*c*) partly full.
(*d*) empty.
(*e*) None of these is correct.

21 • Insulators are poor conductors of electricity because

(*a*) there is a small energy gap between the valence band and the next higher band where electrons can exist.
(*b*) there is a large energy gap between the full valence band and the next higher band where electrons can exist.
(*c*) the valence band has a few vacancies for electrons.
(*d*) the valence band is only partly full.
(*e*) None of these is correct.

22 • You are an electron sitting at the top of the valence band in a silicon atom, longing to jump across the 1.14-eV energy gap that separates you from the bottom of the conduction band and all of the adventures that it may contain. What you need, of course, is a photon. What is the maximum photon wavelength that will get you across the gap?

23 • Work Problem 22 for germanium, for which the energy gap is 0.74 eV.

24 • Work Problem 22 for diamond, for which the energy gap is 7.0 eV.

25 •• A photon of wavelength 3.35 μm has just enough energy to raise an electron from the valence band to the conduction band in a lead sulfide crystal. (*a*) Find the energy gap between these bands in lead sulfide. (*b*) Find the temperature T for which kT equals this energy gap.

BCS Theory of Superconductivity

26 • (*a*) Use Equation 27-24 to calculate the superconducting energy gap for tin and compare your result with the measured value of 6×10^{-4} eV. (*b*) Use the measured value to calculate the wavelength of a photon having sufficient energy to break up Cooper pairs in tin ($T_c = 3.72$ K) at $T = 0$.

27 • Repeat Problem 26 for lead ($T_c = 7.19$ K), which has a measured energy gap of 2.73×10^{-3} eV.

The Fermi–Dirac Distribution

28 •• The number of electrons in the conduction band of an insulator or intrinsic semiconductor is governed chiefly by the Fermi factor. Since the valence band in these materials is nearly filled and the conduction band is nearly empty, the Fermi energy E_F is generally midway between the top of the valence band and bottom of the conduction band, i.e., at $E_g/2$, where E_g is the band gap between the two bands and the energy is measured from the top of the valence band. (*a*) In silicon, $E_g \approx 1.0$ eV. Show that in this case the Fermi factor for electrons at the bottom of the conduction band is given by $\exp(-E_g/2kT)$ and evaluate this factor. Discuss the significance of this result if there are 10^{22} valence electrons per cubic centimeter and the probability of finding an electron in the conduction band is given by the Fermi factor. (*b*) Repeat the calculation in (*a*) for an insulator with a band gap of 6.0 eV.

29 •• Show that at $E = E_F$, the Fermi factor is $F = 0.5$.

30 •• What is the difference between the energies at which the Fermi factor is 0.9 and 0.1 at 300 K in (*a*) copper, (*b*) potassium, and (*c*) aluminum.

31 •• What is the probability that a conduction electron in silver will have a kinetic energy of 4.9 eV at $T = 300$ K?

32 •• Show that $g(E) = (3N/2)E_F^{-3/2}\, E^{1/2}$ (Equation 27-30) follows from Equation 27-28 for $g(E)$, and Equation 27-15a for E_F.

33 •• Carry out the integration $E_{av} = (1/N)\int_0^{E_F} Eg(E)\, dE$ to show that the average energy at $T = 0$ is $\frac{3}{5}E_F$.

34 •• The density of the electron states in a metal can be written $g(E) = AE^{1/2}$, where A is a constant and E is measured from the bottom of the conduction band. (*a*) Show that the total number of states is $\frac{2}{3}AE_F^{3/2}$. (*b*) Approximately what fraction of the conduction electrons are within kT of the Fermi energy? (*c*) Evaluate this fraction for copper at $T =$ 300 K.

35 •• What is the probability that a conduction electron in silver will have a kinetic energy of 5.49 eV at $T = 300$ K?

36 •• Use the density-of-states function, Equation 27-28, to estimate the fraction of the conduction electrons in copper that can absorb energy from collisions with the vibrating lattice ions at (*a*) 77 K and (*b*) 300 K.

37 •• In an intrinsic semiconductor, the Fermi energy is about midway between the top of the valence band and the bottom of the conduction band. In germanium, the forbidden energy band has a width of 0.7 eV. Show that at room temperature the distribution function of electrons in the conduction band is given by the Maxwell–Boltzmann distribution function.

38 ••• (*a*) Show that for $E \geq 0$, the Fermi factor may be written as

$$f(E) = \frac{1}{Ce^{E/kT} + 1}$$

(*b*) Show that if $C >> e^{-E/kT}$, $f(E) = Ae^{-E/kT} << 1$; in other words, show that the Fermi factor is a constant times the classical Boltzmann factor if $A << 1$. (*c*) Use $\int n(E)\, dE = N$ and Equation 27-28 to determine the constant A. (*d*) Using the result obtained in part (*c*), show that the classical approximation is applicable when the electron concentration is very small and/or the temperature is very high. (*e*) Most semiconductors have impurities added in a process called doping, which increases the free electron concentration so that it is about $10^{17}/\text{cm}^3$ at room temperature. Show that for these systems, the classical distribution function is applicable.

39 ••• Show that the condition for the applicability of the classical distribution function for an electron gas ($A << 1$ in Problem 38) is equivalent to the requirement that the average separation between electrons is much greater than their de Broglie wavelength.

40 ••• The root-mean-square (rms) value of a variable is obtained by calculating the average value of the square of that variable and then taking the square root of the result. Use this procedure to determine the rms energy of a Fermi distribution. Express your result in terms of E_F and compare it to the average energy. Why do E_{av} and E_{rms} differ?

41 ••• When a star with a mass of about twice that of the sun exhausts its nuclear fuel, it collapses to a neutron star, a dense sphere of neutrons of about 10 km diameter. Neutrons are spin-$\frac{1}{2}$ particles and, like electrons, are subject to the exclusion principle. (*a*) Determine the neutron density of such a neutron star. (*b*) Find the Fermi energy of the neutron distribution.

General Problems

42 • True or false:

(*a*) Solids that are good electrical conductors are usually good heat conductors.
(*b*) The classical free-electron theory adequately explains the heat capacity of metals.
(*c*) At $T = 0$, the Fermi factor is either 1 or 0.
(*d*) The Fermi energy is the average energy of an electron in a solid.
(*e*) The contact potential between two metals is proportional to the difference in the work functions of the two metals.
(*f*) At $T = 0$, an intrinsic semiconductor is an insulator.
(*g*) Semiconductors conduct current in one direction only.

43 • How does the change in the resistivity of copper compare with that of silicon when the temperature increases?

44 • The density of potassium is 0.851 g/cm^3. How many free electrons are there per potassium atom?

45 • Calculate the number density of free electrons for (*a*) Mg ($\rho = 1.74\ \text{g/cm}^3$) and (*b*) Zn ($\rho = 7.1\ \text{g/cm}^3$), assuming two free electrons per atom, and compare your results with the values listed in Table 27-1.

46 •• Estimate the fraction of free electrons in copper that are in excited states above the Fermi energy at (*a*) room temperature of 300 K and (*b*) 1000 K.

47 ••• A 2-cm^2 wafer of pure silicon is irradiated with light having a wavelength of 775 nm. The intensity of the light beam is 4.0 W/m^2 and every photon that strikes the sample is absorbed and creates an electron–hole pair. (*a*) How many electron–hole pairs are produced in one second? (*b*) If the number of electron–hole pairs in the sample is 6.25×10^{11} in the steady state, at what rate do the electron–hole pairs recombine? (*c*) If every recombination event results in the radiation of one photon, at what rate is energy radiated by the sample?

CHAPTER 28

The Magnetic Field

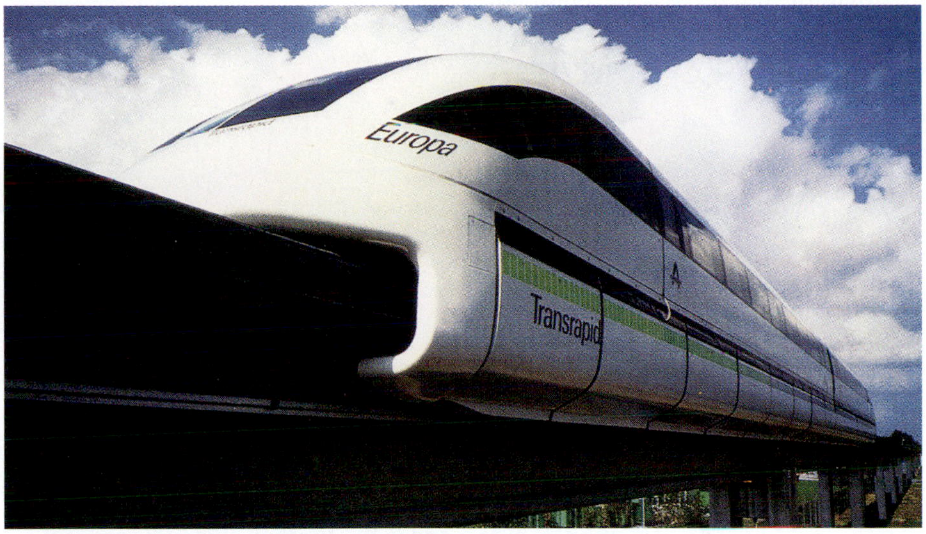

The superspeed Maglev System Transrapid train, using magnetic levitation, guidance, and propulsion, has achieved a peak speed greater than 500 km/h.

More than 2000 years ago, the Greeks were aware that a certain type of stone (now called magnetite) attracts pieces of iron, and there are written references to the use of magnets for navigation dating from the twelfth century.

In 1269, Pierre de Maricourt discovered that a needle laid at various positions on a spherical natural magnet orients itself along lines that pass through points at opposite ends of the sphere. He called these points the poles of the magnet. Subsequently, many experimenters noted that every magnet of whatever shape has two poles, designated the north and south poles, where the force exerted by the magnet is strongest. It was also noted that the like poles of two magnets repel each other and the unlike poles attract each other.

In 1600, William Gilbert discovered that the earth itself is a natural magnet with magnetic poles near the north and south geographic poles. Since the north pole of a compass needle points toward the south pole of a given magnet, what we call the north pole of the earth is actually a south magnetic pole, as illustrated in Figure 28-1.

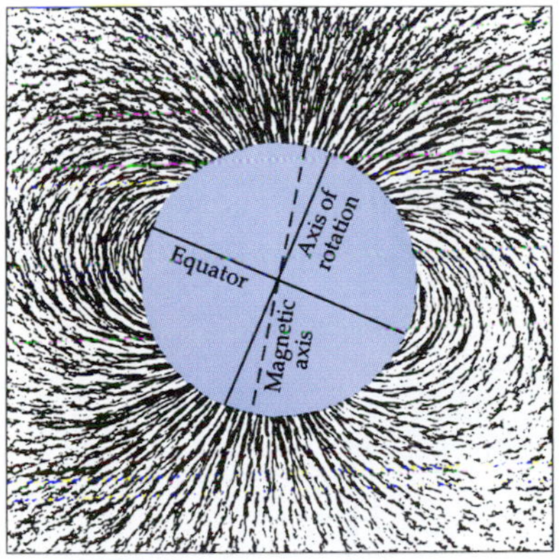

Figure 28-1 Magnetic field lines of the earth indicated by iron filings around a uniformly magnetized sphere. The field lines exit from the north magnetic pole, which is near the south geographic pole, and enter the south magnetic pole, which is near the north geographic pole.

Although electric charges and magnetic poles are similar in many respects, there is an important difference: Magnetic poles always occur in pairs. When a magnet is broken in half, equal and opposite poles appear at either side of the break point. The result is two magnets, each with a north and south pole. There has long been speculation as to the existence of an isolated magnetic pole, and in recent years considerable experimental effort has been made to find such an object. Thus far, there is no conclusive evidence that an isolated magnetic pole exists.

In this chapter, we consider the effects of a given magnetic field on moving charges and on wires carrying currents. The sources of magnetic fields are discussed in the next chapter.

28-1 The Force Exerted by a Magnetic Field

The existence of a magnetic field $\vec{B}$ at some point in space can be demonstrated with a compass needle. If there is a magnetic field, the needle will align itself in the direction of the field.

Experimentally it is observed that, when a charge q has velocity $\vec{v}$ in a magnetic field, there is a force on it that is proportional to q and to v, and to the sine of the angle between $\vec{v}$ and $\vec{B}$. Surprisingly, the force is perpendicular to both the velocity and the field. These experimental results can be summarized as follows. When a charge q moves with velocity $\vec{v}$ in a magnetic field $\vec{B}$, the magnetic force $\vec{F}$ on the charge is

$$\vec{F} = q\vec{v} \times \vec{B} \qquad 28\text{-}1$$

Magnetic force on a moving charge

Since $\vec{F}$ is perpendicular to both $\vec{v}$ and $\vec{B}$, it is perpendicular to the plane defined by these two vectors. The direction of $\vec{F}$ is given by the right-hand rule as $\vec{v}$ is rotated into $\vec{B}$, as illustrated in Figure 28-2.

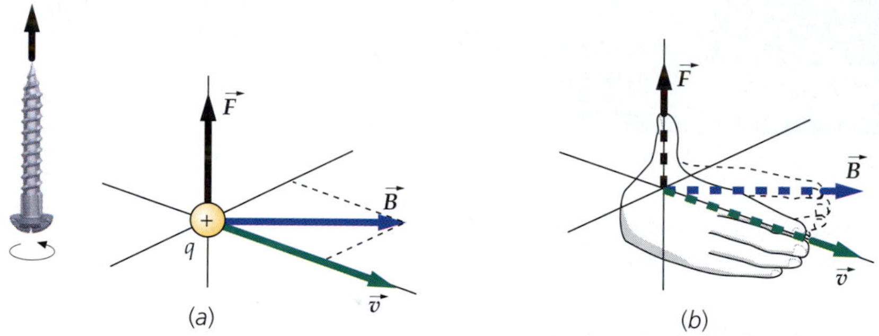

Figure 28-2 Right-hand rule for determining the direction of a force exerted on a charge moving in a magnetic field. (*a*) The force is perpendicular to both $\vec{v}$ and $\vec{B}$ and in the direction of the advance of a right-hand-threaded screw if turned in the same direction as to rotate $\vec{v}$ into $\vec{B}$. (*b*) If the fingers of the right hand are in the direction of $\vec{v}$ such that they can be curled into $\vec{B}$, the thumb points in the direction of $\vec{F}$.

Examples of the direction of the forces exerted on moving charges when the magnetic field vector $\vec{B}$ is in the vertical direction are given in Figure 28-3. Note that the direction of any particular magnetic field $\vec{B}$ can be found experimentally by measuring $\vec{F}$ and $\vec{v}$ for several velocities in different directions and then applying Equation 28-1.

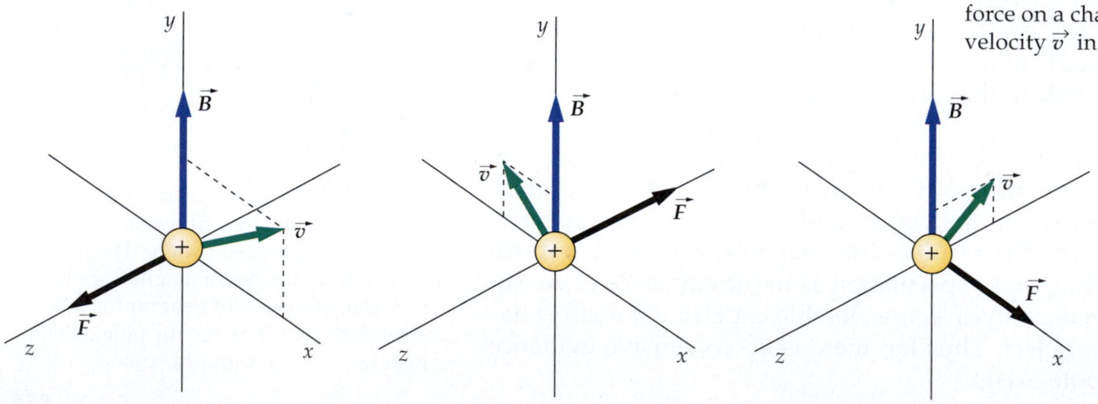

Figure 28-3 Direction of the magnetic force on a charged particle moving with velocity $\vec{v}$ in a magnetic field $\vec{B}$.

Equation 28-1 defines the **magnetic field** $\vec{B}$ in terms of the force exerted on a moving charge. The SI unit of magnetic field is the **tesla** (T). A charge of one coulomb moving with a velocity of one meter per second perpendicular to a magnetic field of one tesla experiences a force of one newton:

$$1\,\text{T} = 1\frac{\text{N/C}}{\text{m/s}} = 1\,\text{N/A}\cdot\text{m} \qquad \text{28-2}$$

This unit is rather large. The magnetic field of the earth has a magnitude of less than 10^{-4} T. The magnetic fields near powerful permanent magnets are about 0.1 to 0.5 T, and powerful laboratory and industrial electromagnets produce fields of 1 to 2 T. Fields greater than 10 T are quite difficult to produce because the resulting magnetic forces will tear the magnets apart or crush them. A commonly used unit, derived from the cgs system, is the **gauss** (G), which is related to the tesla as follows:

$$1\,\text{G} = 10^{-4}\,\text{T} \qquad \text{28-3}$$

Definition—Gauss

Since magnetic fields are often given in gauss, which is not an SI unit, remember to convert from gauss to teslas when making calculations.

Example 28-1

The magnetic field of the earth is measured at a point on the surface to have a magnitude of 0.6 G and is directed downward and northward, making an angle of about 70° with the horizontal, as shown in Figure 28-4. (The earth's magnetic field varies from place to place. These data are approximately correct for the central United States.) A proton of charge $q = 1.6 \times 10^{-19}$ C is moving horizontally in the northward direction with speed $v = 10$ Mm/s $= 10^7$ m/s. Calculate the magnetic force on the proton (*a*) using $F = qvB \sin\theta$ and (*b*) by expressing $\vec{v}$ and $\vec{B}$ in terms of the unit vectors $\hat{i}, \hat{j}, \hat{k}$, and computing $\vec{F} = q\vec{v} \times \vec{B}$.

Figure 28-4

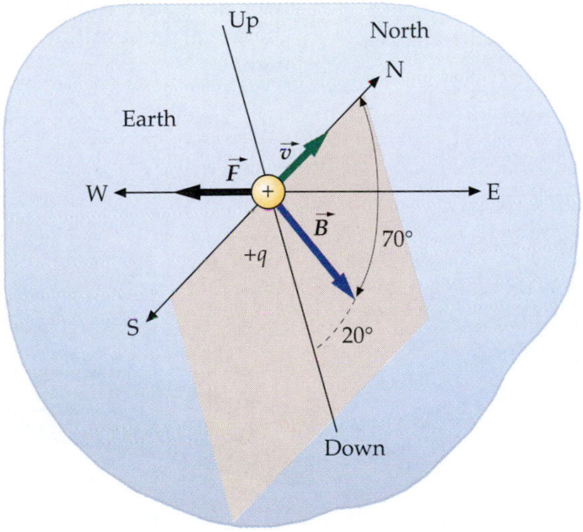

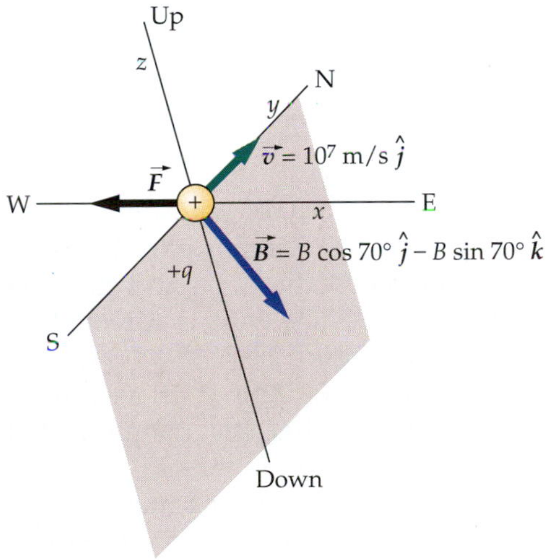

Figure 28-5

Picture the Problem Let the x and y directions be east and north, respectively, and let the z direction be upward (Figure 28-5). The velocity vector is then in the y direction.

(*a*) Calculate $F = qvB \sin\theta$ using $\theta = 70°$. From Figure 28-4 we see that the direction of the force is westward:

$$F = qvB \sin 70°$$
$$= (1.6 \times 10^{-19}\,\text{C})(10^7\,\text{m/s})(0.6 \times 10^{-4}\,\text{T})(0.94)$$
$$= 9.02 \times 10^{-17}\,\text{N}$$

(*b*)1. The magnetic force is the vector product of $q\vec{v}$ and $\vec{B}$: $\vec{F} = q\vec{v} \times \vec{B}$

2. Express $\vec{v}$ and $\vec{B}$ in terms of their components:

$$\vec{v} = v_y\hat{j}$$
$$\vec{B} = B_y\hat{j} + B_z\hat{k}$$

3. Write $\vec{F} = q\vec{v} \times \vec{B}$ in terms of these components:

$$\vec{F} = q\vec{v} \times \vec{B} = q(v_y\hat{j}) \times (B_y\hat{j} + B_z\hat{k})$$
$$= qv_yB_y(\hat{j} \times \hat{j}) + qv_yB_z(\hat{j} \times \hat{k}) = qv_yB_z\hat{i}$$

4. Evaluate $\vec{F}$:

$$\vec{F} = qv_y(-B\sin 70°)\hat{i}$$
$$= -(1.6 \times 10^{-19}\text{ C})(10^7\text{ m/s})(0.6 \times 10^{-4}\text{ T})(0.94)$$
$$= -9.02 \times 10^{-17}\text{ N}\,\hat{i}$$

Remark Note that the direction of $-\hat{i}$ is westward, so the force is directed westward as shown in Figure 28-5.

Exercise Find the force on a proton moving with velocity $\vec{v} = 4 \times 10^6\text{ m/s}\,\hat{i}$ in a magnetic field $\vec{B} = 2.0\text{ T}\,\hat{k}$. (*Answer* $-1.28 \times 10^{-12}\text{ N}\,\hat{j}$)

When a wire carries a current in a magnetic field, there is a force on the wire that is equal to the sum of the magnetic forces on the charged particles whose motion produces the current. Figure 28-6 shows a short segment of wire of cross-sectional area A and length ℓ carrying a current I. If the wire is in a magnetic field $\vec{B}$, the magnetic force on each charge is $q\vec{v}_d \times \vec{B}$, where $\vec{v}_d$ is the drift velocity of the charge carriers, which is the same as their average velocity. The number of charges in the wire segment is the number n per unit volume times the volume $A\ell$. Thus, the total force on the wire segment is

$$\vec{F} = (q\vec{v}_d \times \vec{B})nA\ell$$

From Equation 26-3, the current in the wire is

$$I = nqv_dA$$

Hence the force can be written

$$\vec{F} = I\vec{\ell} \times \vec{B} \qquad 28\text{-}4$$

Magnetic force on a segment of current-carrying wire

where $\vec{\ell}$ is a vector whose magnitude is the length of the wire and whose direction is parallel to the current. For the current in the positive x direction and the magnetic field in the xy plane, the force on the wire is directed along the positive z axis, as shown in Figure 28-7.

In Equation 28-4 it is assumed that the wire segment is straight and that the magnetic field does not vary over its length. The equation can be generalized for an arbitrarily shaped wire in any magnetic field. If we choose a very small wire segment $d\vec{\ell}$ and write the force on this segment as $d\vec{F}$, we have

$$d\vec{F} = I\,d\vec{\ell} \times \vec{B} \qquad 28\text{-}5$$

Magnetic force on a current element

where $\vec{B}$ is the magnetic field vector at the segment. The quantity $I\,d\vec{\ell}$ is called a **current element**. We find the total force on a current-carrying wire

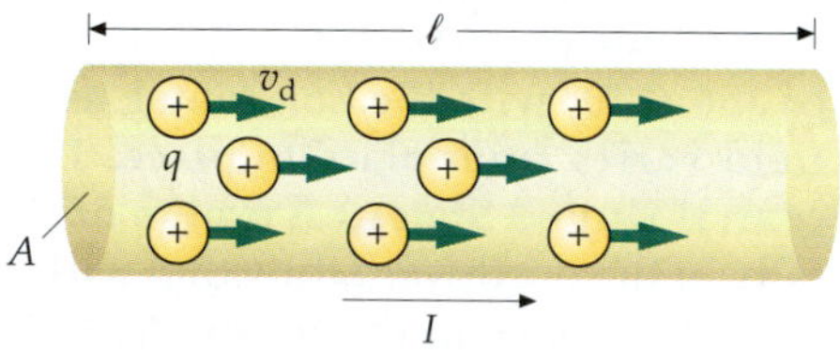

Figure 28-6 Wire segment of length ℓ carrying current I. If the wire is in a magnetic field there will be a force on each charge carrier resulting in a force on the wire.

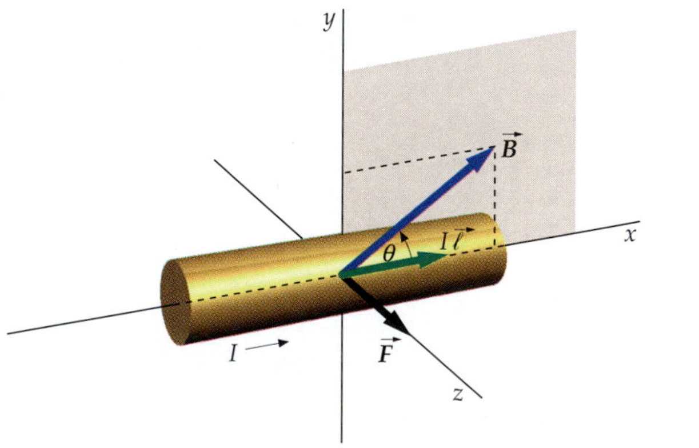

Figure 28-7 Magnetic force on a current-carrying segment of wire in a magnetic field. The current is in the x direction, and the magnetic field is in the xy plane and makes an angle θ with the x axis. The force $\vec{F}$ is in the z direction perpendicular to both $\vec{B}$ and $\vec{\ell}$ and has magnitude $I\ell B\sin\theta$.

by summing (integrating) the forces due to all the current elements in the wire. Equation 28-5 is the same as Equation 28-1 with the current element $I\,d\vec{\ell}$ replacing $q\vec{v}$.

Just as the electric field $\vec{E}$ can be represented by electric field lines, the magnetic field $\vec{B}$ can be represented by **magnetic field lines.** In both cases, the direction of the field is indicated by the direction of the field lines and the magnitude of the field is indicated by their density. There are, however, two important differences between electric field lines and magnetic field lines:

1. Electric field lines are in the direction of the electric force on a positive charge, but the magnetic field lines are perpendicular to the magnetic force on a moving charge.
2. Electric field lines begin on positive charges and end on negative charges; magnetic field lines form closed loops. Since isolated magnetic poles apparently do not exist, there are no points in space where magnetic field lines begin or end.

Figure 28-8 shows the magnetic field lines both inside and outside a bar magnet.

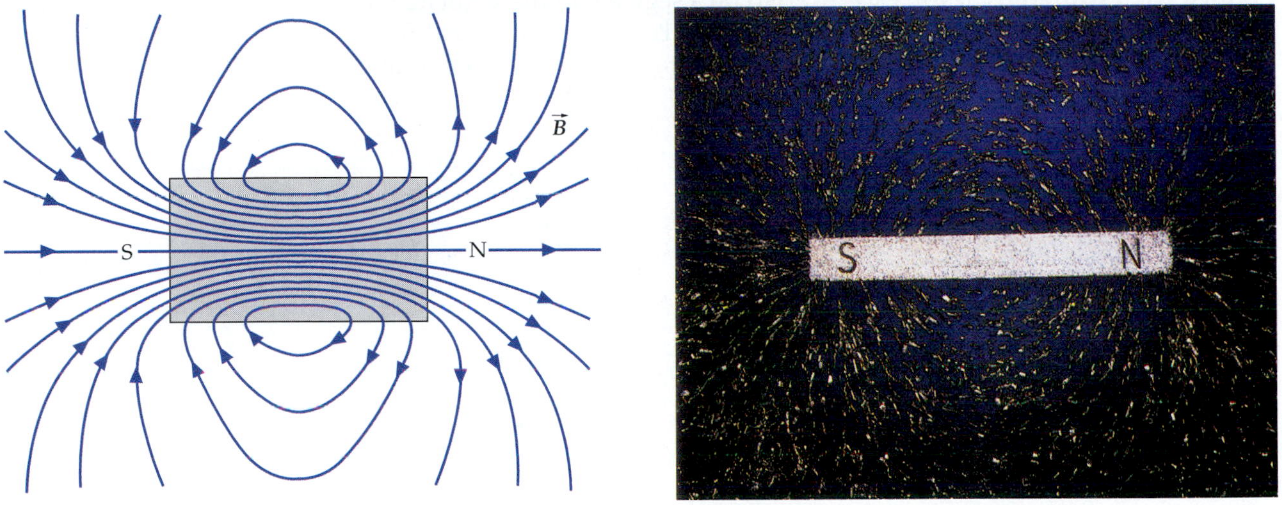

(*a*) (*b*)

Figure 28-8 (*a*) Magnetic field lines inside and outside a bar magnet. The lines emerge from the north pole and enter the south pole, but they have no beginning or end. Instead, they form closed loops. (*b*) Magnetic field lines outside a bar magnet as indicated by iron filings.

Example 28-2

A wire segment 3 mm long carries a current of 3 A in the x direction. It lies in a magnetic field of magnitude 0.02 T that is in the xy plane and makes an angle of 30° with the x axis, as shown in Figure 28-9. What is the magnetic force exerted on the wire segment?

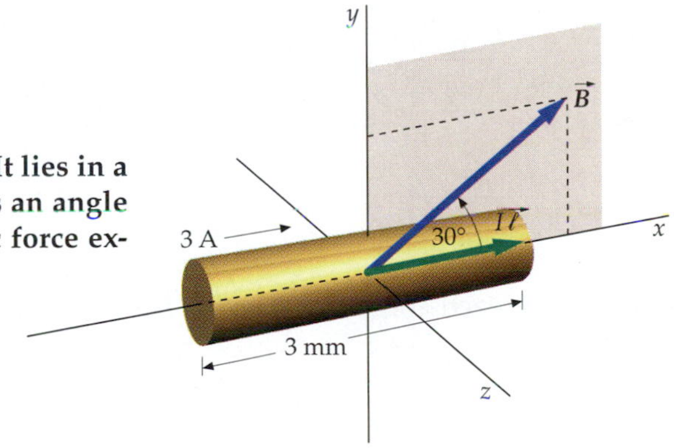

Figure 28-9

1. The magnetic force is in the direction of $\vec{\ell} \times \vec{B}$, which we see from Figure 28-9 is in the z direction. The magnetic force is given by Equation 28-4:

$$\vec{F} = I\vec{\ell} \times \vec{B} = I\ell B \sin 30^\circ\, \hat{k}$$
$$= (3.0\text{ A})(0.003\text{ m})(0.02\text{ T})(\sin 30^\circ)\hat{k}$$
$$= 9 \times 10^{-5}\text{ N}\,\hat{k}$$

Example 28-3

A wire bent into a semicircular loop of radius R lies in the xy plane. It carries a current I from point a to point b, as shown in Figure 28-10. There is a uniform magnetic field $\vec{B} = B\hat{k}$ perpendicular to the plane of the loop. Find the force acting on the semicircular loop part of the wire.

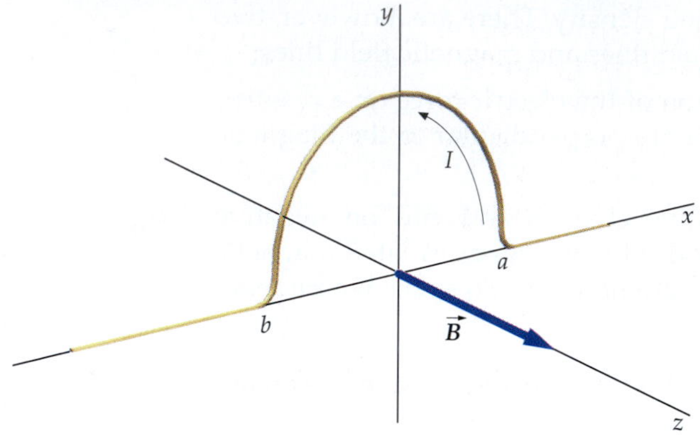

Figure 28-10

Picture the Problem The force $d\vec{F}$ exerted on a segment of the semicircular wire lies in the xy plane, as shown in Figure 28-11. We find the total force by expressing the x and y components of $d\vec{F}$ in terms of θ and integrating them separately from $\theta = 0$ to $\theta = \pi$.

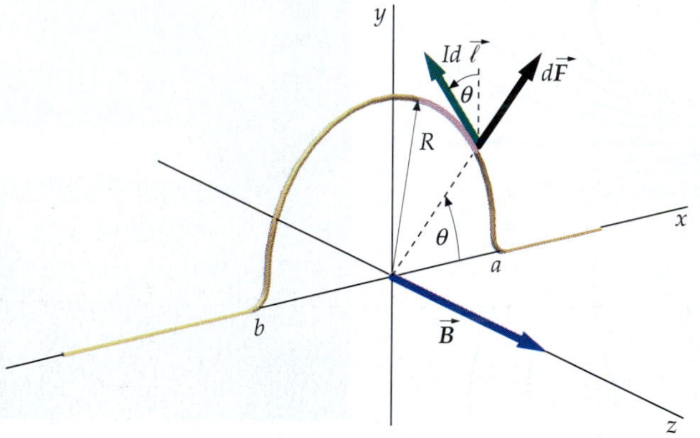

Figure 28-11

1. Write the force $d\vec{F}$ on a current element $d\vec{\ell}$:

$$d\vec{F} = I\,d\vec{\ell} \times \vec{B}$$

2. Express $d\vec{\ell}$ in terms of the unit vectors $\hat{i}$ and $\hat{j}$:

$$d\vec{\ell} = -d\ell \sin\theta\,\hat{i} + d\ell\cos\theta\,\hat{j}$$

3. Compute $I\,d\vec{\ell} \times \vec{B}$ using $d\ell = R\,d\theta$, and $\vec{B} = B\hat{k}$:

$$d\vec{F} = I\,d\vec{\ell} \times \vec{B}$$
$$= (-IR\sin\theta\,d\theta\,\hat{i} + IR\cos\theta\,d\theta\,\hat{j}) \times B\hat{k}$$
$$= IRB\sin\theta\,d\theta\,\hat{j} + IRB\cos\theta\,d\theta\,\hat{i}$$

4. Integrate each component of $d\vec{F}$ from $\theta = 0$ to $\theta = \pi$:

$$\vec{F} = IRB\hat{i}\int_0^{\pi}\cos\theta\,d\theta + IRB\hat{j}\int_0^{\pi}\sin\theta\,d\theta$$
$$\vec{F} = IRB\hat{i}(0) + IRB\hat{j}(2) = 2IRB\hat{j}$$

Check the Result The result that the x component of $\vec{F}$ is zero can be seen from symmetry. For the right half of the loop, $d\vec{F}$ points to the right; for the left half, $d\vec{F}$ points to the left.

Remark The net force on the semicircular wire is the same as if the semicircle were replaced by a straight-line segment of length $2R$ connecting points a and b. (This is a general result, as shown in Problem 16.)

28-2 Motion of a Point Charge in a Magnetic Field

The magnetic force on a charged particle moving through a magnetic field is always perpendicular to the velocity of the particle. The magnetic force thus changes the direction of the velocity but not its magnitude. *Therefore, magnetic fields do no work on particles and do not change their kinetic energy.*

In the special case where the velocity of a particle is perpendicular to a uniform magnetic field, as shown in Figure 28-12, the particle moves in a circular orbit. The magnetic force provides the centripetal force necessary for the centripetal acceleration v^2/r in circular motion. We can use Newton's second law to relate the radius of the circle to the magnetic field and the speed of the particle. If the velocity is $\vec{v}$, the magnitude of the net force is qvB, since $\vec{v}$ and $\vec{B}$ are perpendicular. Newton's second law gives

$$F = ma = m\frac{v^2}{r}$$

$$qvB = \frac{mv^2}{r}$$

or

$$r = \frac{mv}{qB} \qquad 28\text{-}6$$

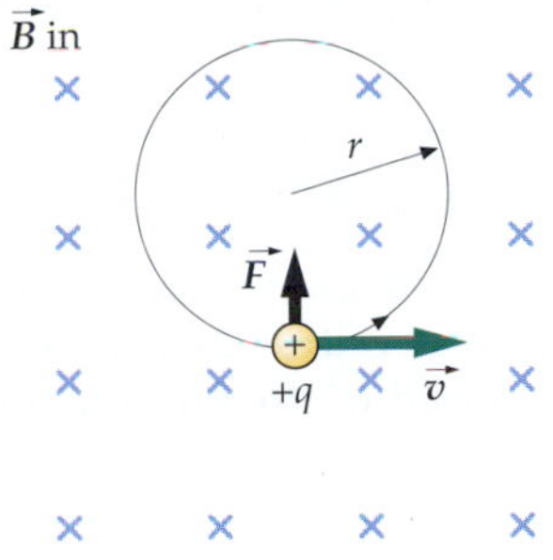

Figure 28-12 Charged particle moving in a plane perpendicular to a uniform magnetic field. The magnetic field is into the page as indicated by the crosses. (A field out of the plane of the page would be indicated by dots). The magnetic force is perpendicular to the velocity of the particle, causing it to move in a circular orbit.

The period of the circular motion is the time it takes the particle to travel once around the circumference of the circle. The period is related to the speed by

$$T = \frac{2\pi r}{v}$$

Substituting in $r = mv/qB$ from Equation 28-6, we obtain the period of the particle's circular motion, called the **cyclotron period**:

$$T = \frac{2\pi(mv/qB)}{v} = \frac{2\pi m}{qB} \qquad 28\text{-}7$$

Cyclotron period

The frequency of the circular motion, called the **cyclotron frequency,** is the reciprocal of the period:

$$f = \frac{1}{T} = \frac{qB}{2\pi m} \qquad 28\text{-}8$$

Cyclotron frequency

(*a*)

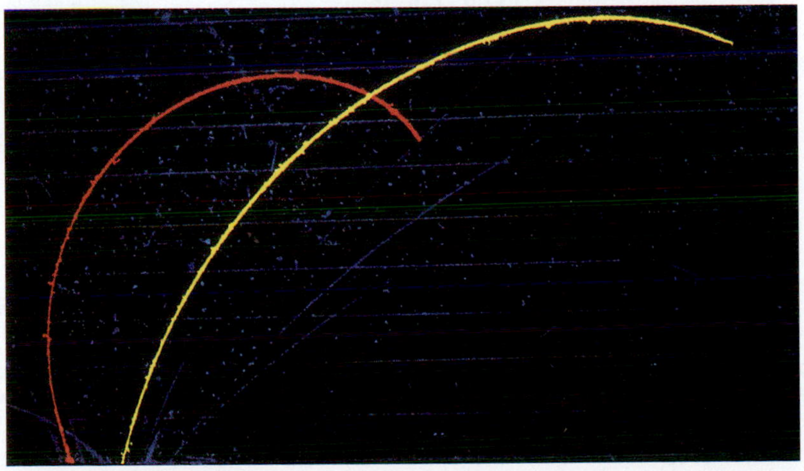

(*b*)

(*a*) Circular path of electrons moving in the magnetic field produced by two large coils. The electrons ionize the gas in the tube, causing it to give off a bluish glow that indicates the path of the beam. (*b*) False-color photograph showing tracks of a 1.6-MeV proton (red) and a 7-MeV α particle (yellow) in a cloud chamber. The radius of curvature is proportional to the momentum and inversely proportional to the charge of the particle. For these energies, the momentum of the α particle, which has twice the charge of the proton, is about four times that of the proton and so its radius of curvature is greater.

Note that the period and the frequency given by Equations 28-7 and 28-8 depend on the charge-to-mass ratio q/m but are independent of the radius r or velocity v. Two important applications of the circular motion of charged particles in a uniform magnetic field, the mass spectrometer and the cyclotron, are discussed later in this section.

Example 28-4

A proton of mass $m = 1.67 \times 10^{-27}$ kg and charge $q = e = 1.6 \times 10^{-19}$ C moves in a circle of radius 21 cm perpendicular to a magnetic field $B = 4000$ G. Find (*a*) the period of the motion and (*b*) the speed of the proton.

(*a*) Calculate the period T from Equation 28-7 with $B = 0.4$ T:

$$T = \frac{2\pi m}{qB} = \frac{2\pi(1.67 \times 10^{-27}\ \text{kg})}{(1.6 \times 10^{-19}\ \text{C})(0.4\ \text{T})} = 1.64 \times 10^{-7}\ \text{s}$$

(*b*) Calculate the speed v from Equation 28-6:

$$v = \frac{rqB}{m} = \frac{(0.21\ \text{m})(1.6 \times 10^{-19}\ \text{C})(0.4\ \text{T})}{1.67 \times 10^{-27}\ \text{kg}}$$

$$= 8.05 \times 10^{6}\ \text{m/s}$$

Remark The radius of the circular motion is proportional to the speed, but the period is independent of both the speed and radius.

Check the Result Note that the product of the speed v and the period T equals the circumference of the circle $2\pi r$ as expected: $vT = (8.05 \times 10^6\ \text{m/s})(1.64 \times 10^{-7}\ \text{s}) = 1.32$ m; $2\pi r = 2\pi(0.21\ \text{m}) = 1.32$ m.

Suppose that a charged particle enters a uniform magnetic field with a velocity that is not perpendicular to $\vec{B}$. We can resolve the velocity of the particle into components $v_{\parallel}$ parallel to $\vec{B}$ and $v_{\perp}$ perpendicular to $\vec{B}$. The motion due to the perpendicular component is the same as that just discussed. The component of the velocity parallel to $\vec{B}$ is not affected by the magnetic field and therefore remains constant. The path of the particle is thus a helix, as shown in Figure 28-13.

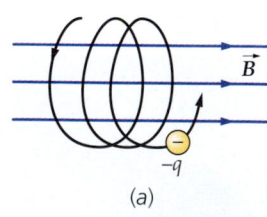

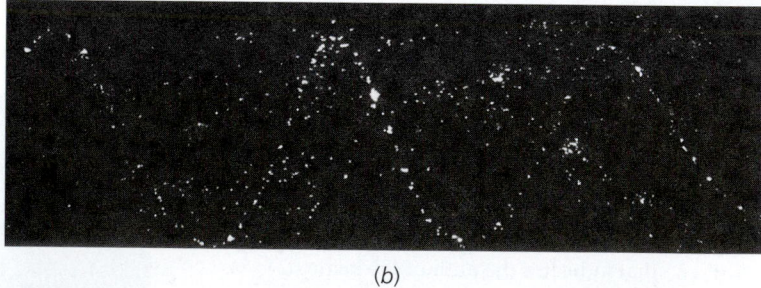

Figure 28-13 (*a*) When a particle has a velocity component parallel to a magnetic field as well as one perpendicular to the field, it moves in a helical path around the field lines. (*b*) Cloud-chamber photograph of the helical path of an electron moving in a magnetic field. The path of the electron is made visible by the condensation of water droplets in the cloud chamber.

The motion of charged particles in nonuniform magnetic fields can be quite complex. Figure 28-14 shows a **magnetic bottle,** an interesting magnetic field configuration in which the field is weak at the center and strong at both ends. A detailed analysis of the motion of a charged particle in such a field shows that the particle spirals around the field lines and becomes trapped, oscillating back and forth between points P_1 and P_2 in the figure. Such magnetic field configurations are used to confine dense beams of

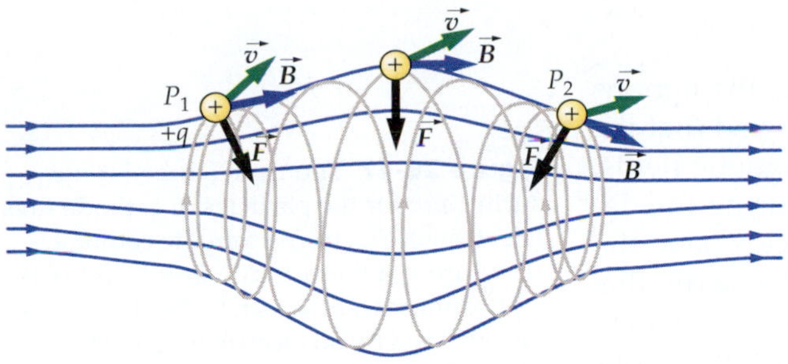

Figure 28-14 Magnetic bottle. When a charged particle moves in such a field, which is strong at both ends and weak in the middle, the particle becomes trapped and moves back and forth spiraling around the field lines.

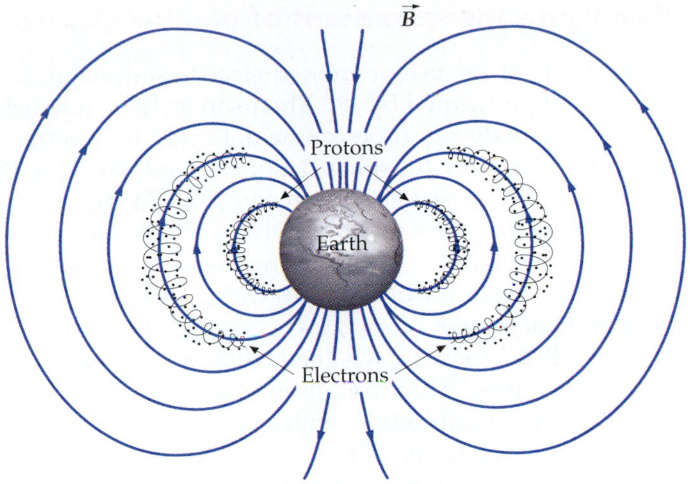

Figure 28-15 Van Allen belts. Protons (inner belts) and electrons (outer belts) are trapped in the earth's magnetic field and spiral around the field lines between the north and south poles.

charged particles, called *plasmas,* in nuclear fusion research. A similar phenomenon is the oscillation of ions back and forth between the earth's magnetic poles in the Van Allen belts (Figure 28-15).

The Velocity Selector

The magnetic force on a charged particle moving in a uniform magnetic field can be balanced by an electric force if the magnitudes and directions of the magnetic and electric fields are properly chosen. Since the electric force is in the direction of the electric field (for positive particles) and the magnetic force is perpendicular to the magnetic field, the electric and magnetic fields in the region through which the particle is moving must be perpendicular to each other if the forces are to balance. Such a region is said to have **crossed fields**.

Figure 28-16 shows a region of space between the plates of a capacitor where there is an electric field and a perpendicular magnetic field (produced by a magnet with poles above and below the paper). Consider a particle of charge q entering this space from the left. The net force on the particle is

$$\vec{F} = q\vec{E} + q\vec{v} \times \vec{B}$$

If q is positive, the electric force of magnitude qE is down and the magnetic force of magnitude qvB is up. If the charge is negative, each of these forces is reversed. The two forces balance if $qE = qvB$ or

$$v = \frac{E}{B} \tag{28-9}$$

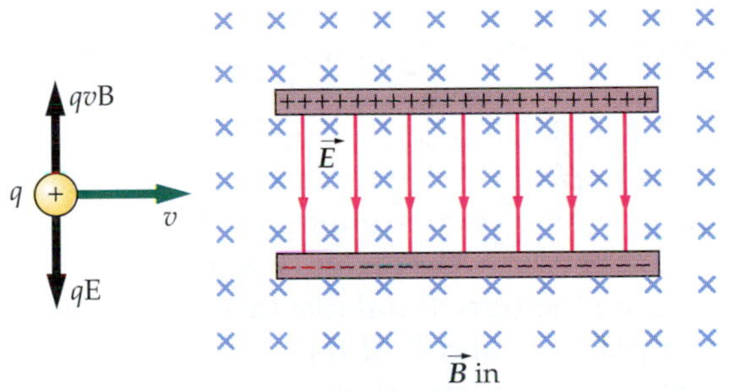

Figure 28-16 Crossed electric and magnetic fields. When a positive particle moves to the right, it experiences a downward electric force and an upward magnetic force. These forces balance if the speed of the particle is related to the field strengths by $vB = E$.

For given magnitudes of the electric and magnetic fields, the forces balance only for particles with the speed given by Equation 28-9. Any particle with this speed, regardless of its mass or charge, will traverse the space undeflected. A particle with a greater speed will be deflected in the direction of the magnetic force, and one with less speed will be deflected in the direction of the electric force. This arrangement of fields is often used as a **velocity selector**, a device that allows only particles with speed given by Equation 28-9 to pass.

Exercise A proton is moving in the x direction in a region of crossed fields where $\vec{E} = 2 \times 10^5\,\mathrm{N/C}\,\hat{k}$ and $\vec{B} = -3000\,\mathrm{G}\,\hat{j}$. (*a*) What is the speed of the proton if it is not deflected? (*b*) If the proton moves with twice this speed, in which direction will it be deflected? (*Answers* (*a*) 667 km/s (*b*) in the negative z direction)

optional

Thomson's Measurement of *q*/*m* for Electrons

An example of the use of crossed electric and magnetic fields is the famous experiment performed by J. J. Thomson in 1897 in which he showed that the rays of a cathode-ray tube can be deflected by electric and magnetic fields, indicating that they must consist of charged particles. By measuring the deflections of these particles Thomson showed that all the particles have the same charge-to-mass ratio q/m. He also showed that particles with this charge-to-mass ratio can be obtained using any material for a source, which means that these particles, now called electrons, are a fundamental constituent of all matter.

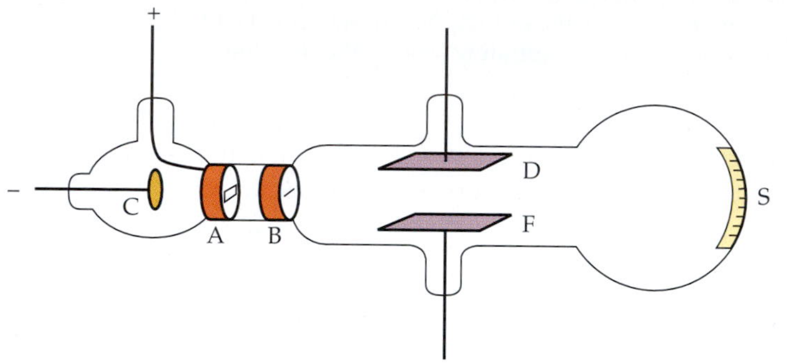

Figure 28-17 Thomson's tube for measuring q/m for the particles of cathode rays (electrons). Electrons from the cathode C pass through the slits at A and B and strike a phosphorescent screen S. The beam can be deflected by an electric field between plates D and F or by a magnetic field (not shown).

Figure 28-17 shows a schematic diagram of the cathode-ray tube Thomson used. Electrons are emitted from the cathode C, which is at a negative potential relative to the slits A and B. An electric field in the direction from A to C accelerates the electrons, and they pass through slits A and B into a field-free region. The electrons then enter the electric field between the capacitor plates D and F that is perpendicular to the velocity of the electrons. This field accelerates them vertically for the short time they are between the plates. The electrons are deflected and strike the phosphorescent screen S at the far right side of the tube at some deflection Δy from the point at which they strike when there is no field between the plates. The screen glows where the electrons strike it, indicating the location of the beam. The initial speed of the electrons v_0 is determined by introducing a magnetic field $\vec{B}$ between the plates in a direction that is perpendicular to both the electric field and the initial velocity of the electrons. The magnitude of $\vec{B}$ is adjusted until the beam is not deflected. The speed is then found from Equation 28-9.

With the magnetic field turned off, the beam is deflected by an amount Δy, which consists of two parts: the deflection Δy_1, which occurs while the electrons are between the plates, and the deflection Δy_2, which occurs after the electrons leave the region between the plates (Figure 28-18).

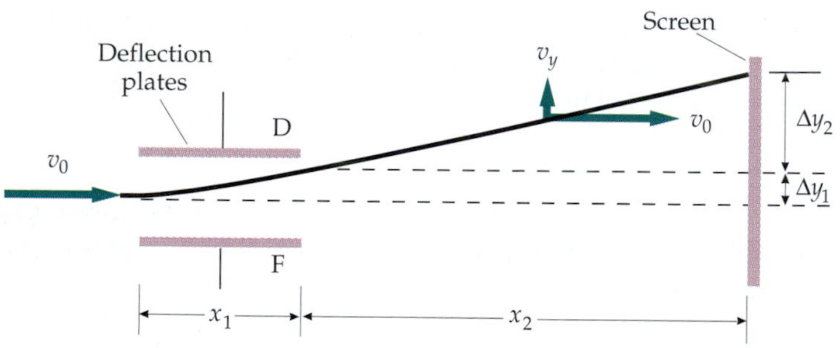

Figure 28-18 The total deflection of the beam in the J.J. Thomson experiments consists of the deflection y_1 while the electrons are between the plates plus the deflection y_2 that occurs in the field-free region between the plates and the screen.

Let x_1 be the horizontal distance across the deflection plates D and F. If the electron is moving horizontally with speed v_0 when it enters the plates, the time spent between the plates is $t_1 = x_1/v_0$, and the vertical velocity when it leaves the plates is

$$v_y = at_1 = \frac{qE}{m}t_1 = \frac{qE}{m}\frac{x_1}{v_0}$$

where E is the electric field between the plates. The deflection in this region is

$$\Delta y_1 = \frac{1}{2}at_1^2 = \frac{1}{2}\frac{qE}{m}\left(\frac{x_1}{v_0}\right)^2$$

The electron then travels an additional horizontal distance x_2 in the field-free region from the deflection plates to the screen. Since the velocity of the electron is constant in this region, the time to reach the screen is $t_2 = x_2/v_0$, and the additional vertical deflection is

$$\Delta y_2 = v_y t_2 = \frac{qE}{m}\frac{x_1}{v_0}\frac{x_2}{v_0}$$

The total deflection at the screen is therefore

$$\Delta y = \Delta y_1 + \Delta y_2 = \frac{1}{2}\frac{qE}{m}\left(\frac{x_1}{v_0}\right)^2 + \frac{qE}{m}\frac{x_1 x_2}{v_0^2} \qquad \text{28-10}$$

The measured deflection Δy can be used to determine the charge-to-mass ratio, q/m, from Equation 28-10.

Example 28-5

Electrons pass undeflected through the plates of Thomson's apparatus when the electric field is 3000 V/m and there is a crossed magnetic field of 1.40 G. If the plates are 4 cm long and the ends of the plates are 30 cm from the screen, find the deflection on the screen when the magnetic field is turned off.

Picture the Problem The mass and charge of the electron are known: $m = 9.11 \times 10^{-31}$ kg and $q = e = 1.6 \times 10^{-19}$ C. The speed of the electron can be found from the ratio of the magnetic and electric fields.

1. The total deflection of the electron is given by Equation 28-10:

$$\Delta y = \Delta y_1 + \Delta y_2 = \frac{1}{2}\frac{qE}{m}\left(\frac{x_1}{v_0}\right)^2 + \frac{qE}{m}\frac{x_1 x_2}{v_0^2}$$

2. The speed v_0 equals E/B:

$$v_0 = \frac{E}{B} = \frac{3000\ \text{V/m}}{1.40 \times 10^{-4}\ \text{T}} = 2.14 \times 10^7\ \text{m/s}$$

3. Substitute this value for v_0, the given value of E, and the known values for m and q to find Δy:

$$\Delta y = \frac{1}{2}\frac{(1.6 \times 10^{-19}\ \text{C})(3000\ \text{V/m})}{9.11 \times 10^{-31}\ \text{kg}}\left(\frac{0.04\ \text{m}}{2.14 \times 10^7\ \text{m/s}}\right)^2$$

$$+ \frac{(1.6 \times 10^{-19}\ \text{C})(3000\ \text{V/m})}{9.11 \times 10^{-31}\ \text{kg}}\frac{(0.04\ \text{m})(0.30\ \text{m})}{(2.14 \times 10^7\ \text{m/s})^2}$$

$$= 9.20 \times 10^{-4}\ \text{m} + 1.38 \times 10^{-2}\ \text{m}$$

$$= 0.92\ \text{mm} + 13.8\ \text{mm} = 14.7\ \text{mm}$$

The Mass Spectrometer

The **mass spectrometer,** first designed by Francis William Aston in 1919, was developed as a means of measuring the masses of isotopes. Such measurements are an important way of determining both the presence of isotopes and their abundance in nature. For example, natural magnesium has been found to consist of 78.7% ^{24}Mg, 10.1% ^{25}Mg, and 11.2% ^{26}Mg. These isotopes have masses in the approximate ratio 24:25:26.

Figure 28-19 shows a simple schematic drawing of a mass spectrometer. Ions from an ion source are accelerated by an electric field and enter a uniform magnetic field. If the ions start from rest and move through a potential drop ΔV, their kinetic energy when they enter the magnetic field equals their loss in potential energy, $q\,\Delta V$:

$$\tfrac{1}{2}mv^2 = q\,\Delta V \qquad \text{28-11}$$

The ions move in a semicircle of radius r given by Equation 28-6, $r = mv/qB$, and strike a photographic plate at point P_2, a distance $2r$ from the point P_1 where they entered the magnetic field.

The speed v can be eliminated from Equations 28-6 and 28-11 to find m/q in terms of the known quantities ΔV, B, and r. We first solve Equation 28-6 for v

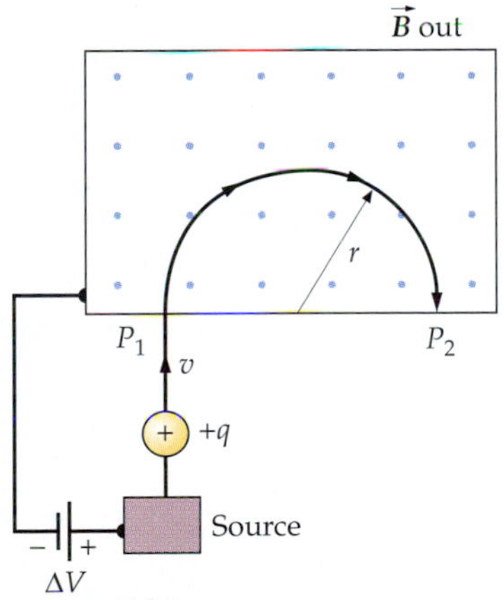

Figure 28-19 Schematic drawing of a mass spectrometer. Ions from an ion source are accelerated through a potential difference ΔV and enter a uniform magnetic field. The magnetic field is out of the plane of the page as indicated by the dots. The ions are bent into a circular arc and emerge at P_2. The radius of the circle varies with the mass of the ion.

optional

and square each term, which gives

$$v^2 = \frac{r^2q^2B^2}{m^2}$$

Substituting this expression for v^2 into Equation 28-11, we obtain

$$\frac{1}{2}m\left(\frac{r^2q^2B^2}{m^2}\right) = q\,\Delta V$$

Simplifying this equation and solving it for m/q, we obtain

$$\frac{m}{q} = \frac{B^2r^2}{2\,\Delta V} \qquad \text{28-12}$$

In Aston's original mass spectrometer, mass differences could be measured to a precision of about 1 part in 10,000. The precision has been improved by introducing a velocity selector between the ion source and the magnet, which increases the degree of accuracy with which the velocities of the incoming ions can be determined.

Example 28-6

A ^{58}Ni ion of charge +*e* and mass 9.62 × 10^{-26} kg is accelerated through a potential difference of 3 kV and deflected in a magnetic field of 0.12 T. (*a*) Find the radius of curvature of the orbit of the ion. (*b*) Find the difference in the radii of curvature of ^{58}Ni ions and ^{60}Ni ions. (Assume that the mass ratio is 58/60.)

Picture the Problem The radius of curvature r can be found using Equation 28-12. Using the mass dependence of r, we can find the radius for ^{60}Ni ions from the radius for ^{58}Ni ions, and then take the difference.

(*a*) Solve Equation 28-12 for r:

$$r = \sqrt{\frac{2m\,\Delta V}{qB^2}} = \left[\frac{2(9.62\times10^{-26}\text{ kg})(3000\text{ V})}{(1.6\times10^{-19}\text{ C})(0.12\text{ T})^2}\right]^{1/2}$$

$$= 0.501\text{ m}$$

(*b*)1. Let r_1 and r_2 be the radius of the orbit of the ^{58}Ni ion and the ^{60}Ni ion, respectively. Use the result in (*a*) to find the ratio of r_2 to r_1:

$$\frac{r_2}{r_1} = \sqrt{\frac{m_2}{m_1}} = \sqrt{\frac{60}{58}} = 1.017$$

2. Use the result of the previous step to calculate r_2 for ^{60}Ni:

$$r_2 = 1.017r_1 = (1.017)(0.501\text{ m}) = 0.510\text{ m}$$

3. The difference in orbital radii is $r_2 - r_1$:

$$r_2 - r_1 = 0.510\text{ m} - 0.501\text{ m} = 0.009\text{ m} = 9\text{ mm}$$

The Cyclotron

The cyclotron was invented by E. O. Lawrence and M. S. Livingston in 1934 to accelerate particles such as protons or deuterons to high kinetic energies.* The high-energy particles are used to bombard atomic nuclei, causing nuclear reactions that are then studied to obtain information about the nucleus. High-energy protons and deuterons are also used to produce radioactive materials and for medical purposes.

Figure 28-20 is a schematic drawing of a cyclotron. The particles move in two semicircular metal containers called *dees,* after their shape. The dees are housed in a vacuum chamber that is in a uniform magnetic field provided by

* A deuteron is the nucleus of heavy hydrogen, ^{2}H, which consists of a proton and neutron tightly bound together.

an electromagnet. The region in which the particles move must be evacuated so that the particles will not be scattered in collisions with air molecules and lose energy. A potential difference ΔV, which alternates in time with a period T, is maintained between the dees. The period is chosen to be the cyclotron period $T = 2\pi m/qB$ (Equation 28-7). The potential difference creates an electric field across the gap between the dees. At the same time, there is no electric field within each dee because the metal dees are shielded.

Positively charged particles are initially injected into dee_1 with a small velocity from an ion source S near the center of the dees. They move in a semicircle in dee_1 and arrive at the gap between dee_1 and dee_2 after a time $\frac{1}{2}T$. The potential is adjusted so that dee_1 is at a higher potential than dee_2 when the particles arrive at the gap between them. Each particle is therefore accelerated across the gap by the electric field and gains kinetic energy equal to $q\,\Delta V$.

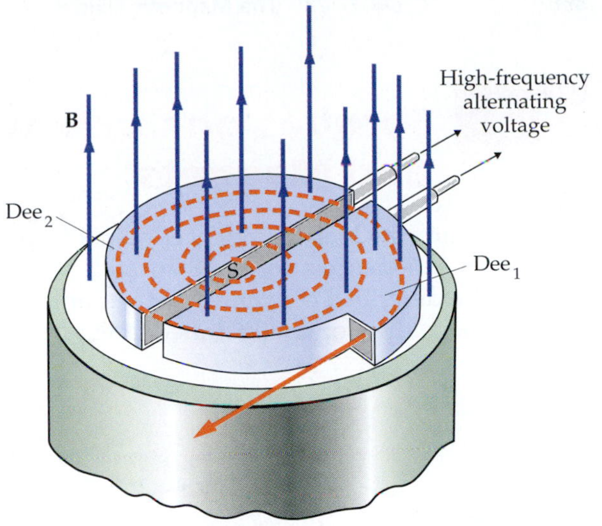

Figure 28-20 Schematic drawing of a cyclotron. The upper pole face of the magnet has been omitted. Charged particles such as protons are accelerated from a source at the center by the potential difference across the gap between the dees. When they arrive at the gap again, the potential difference has changed sign so they are again accelerated across the gap, and move in a larger circle. The potential difference across the gap alternates with the cyclotron frequency of the particle, which is independent of the radius of the circle.

Because it now has more kinetic energy, the particle moves in a semicircle of larger radius in dee_2. It arrives at the gap again after a time $\frac{1}{2}T$, because the period is independent of the particle's speed. By this time, the potential between the dees has been reversed so that dee_2 is now at the higher potential. Once more the particle is accelerated across the gap and gains additional kinetic energy equal to $q\,\Delta V$. Each time the particle arrives at the gap, it is accelerated and gains kinetic energy equal to $q\,\Delta V$. Thus, it moves in larger and larger semicircular orbits until it eventually leaves the magnetic field. In the typical cyclotron, each particle may make 50 to 100 revolutions and exit with energies of up to several hundred mega-electron-volts.

The kinetic energy of a particle leaving a cyclotron can be calculated by setting r in Equation 28-6 equal to the maximum radius of the dees and solving the equation for v:

$$r = \frac{mv}{qB}, \qquad v = \frac{qBr}{m}$$

Then, $$K = \frac{1}{2}mv^2 = \frac{1}{2}\left(\frac{q^2B^2}{m}\right)r^2 \qquad \text{28-13}$$

Example 28-7

A cyclotron for accelerating protons has a magnetic field of 1.5 T and a maximum radius of 0.5 m. (*a*) What is the cyclotron frequency? (*b*) What is the kinetic energy of the protons when they emerge?

(*a*) The cyclotron frequency is given by Equation 28-8:

$$f = \frac{qB}{2\pi m} = \frac{(1.6 \times 10^{-19}\ \text{C})(1.5\ \text{T})}{2\pi(1.67 \times 10^{-27}\ \text{kg})} = 2.29 \times 10^{7}\ \text{Hz}$$
$$= 22.9\ \text{MHz}$$

(*b*)1. The kinetic energy of the emerging protons is given by Equation 28-13:

$$K = \frac{1}{2}\left[\frac{(1.6 \times 10^{-19}\ \text{C})^2\,(1.5\ \text{T})^2}{1.67 \times 10^{-27}\ \text{kg}}\right](0.5\ \text{m})^2 = 4.31 \times 10^{-12}\ \text{J}$$

2. The energies of protons and other elementary particles are usually expressed in electron volts. Use 1 eV = 1.6×10^{-19} J to convert to eV:

$$K = 4.31 \times 10^{-12}\ \text{J} \times \frac{1\ \text{eV}}{1.6 \times 10^{-19}\ \text{J}} = 26.9\ \text{MeV}$$

optional

28-3 Torques on Current Loops and Magnets

A current-carrying loop experiences no net force in a uniform magnetic field, but it does experience a torque that tends to twist it. The orientation of the loop can be described conveniently by a unit vector $\hat{n}$ that is perpendicular to the plane of the loop as illustrated in Figure 28-21.

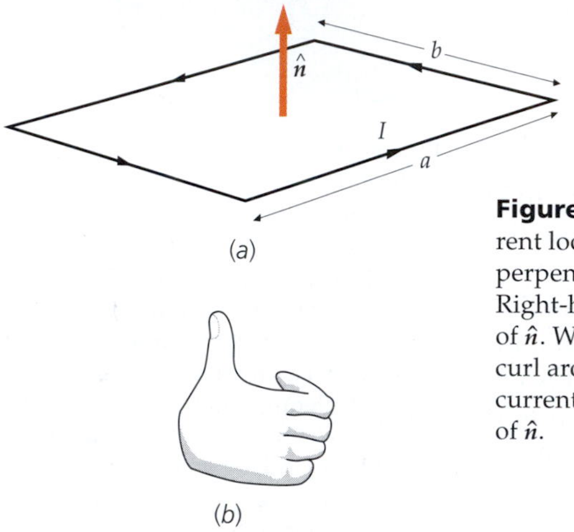

Figure 28-21 (*a*) The orientation of a current loop is described by the unit vector $\hat{n}$ perpendicular to the plane of the loop. (*b*) Right-hand rule for determining the sense of $\hat{n}$. When the fingers of the right hand curl around the loop in the direction of the current, the thumb points in the direction of $\hat{n}$.

Figure 28-22 shows the forces exerted by a uniform magnetic field on a rectangular loop whose normal unit vector $\hat{n}$ makes an angle θ with the magnetic field $\vec{B}$. The net force on the loop is zero. The forces F_1 and F_2 have the magnitude

$$F_1 = F_2 = IaB$$

These forces form a couple so the torque is the same about any point. Point P in Figure 28-22 is a convenient point about which to compute the torque. The magnitude of the torque is

$$\tau = F_2 b \sin\theta = IaBb \sin\theta = IAB \sin\theta$$

where $A = ab$ is the area of the loop. For a loop with N turns, the torque has the magnitude

$$\tau = NIAB \sin\theta$$

This torque tends to twist the loop so that its plane is perpendicular to $\vec{B}$ (i.e., so that $\vec{n}$ is in the same direction as $\vec{B}$).

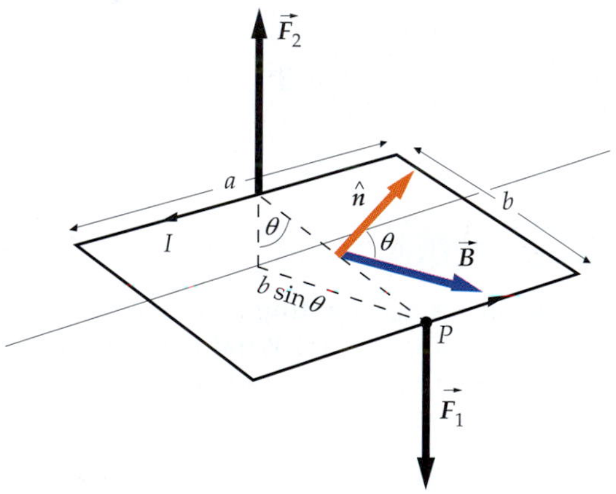

Figure 28-22 Rectangular current loop whose unit normal $\hat{n}$ makes an angle θ with a uniform magnetic field **B**. The torque on the loop has magnitude $IAB \sin\theta$ and is in the direction such that $\hat{n}$ tends to rotate into **B**.

The torque can be written conveniently in terms of the **magnetic dipole moment** μ (also referred to simply as the **magnetic moment**) of the current loop, which is defined as

$$\vec{\mu} = NIA\hat{n} \qquad \text{28-14}$$

Magnetic dipole moment of a current loop

The SI unit of magnetic moment is the ampere-meter^2 ($A \cdot m^2$). In terms of the magnetic dipole moment, the torque on the current loop is given by

$$\vec{\tau} = \vec{\mu} \times \vec{B} \qquad \text{28-15}$$

Torque on a current loop

Equation 28-15, which we have derived for a rectangular loop, holds in general for a flat loop of any shape. The torque on any loop is the cross product of the magnetic moment $\vec{\mu}$ of the loop and the magnetic field $\vec{B}$, where the magnetic moment is defined to be a vector that is perpendicular to the area of the loop (Figure 28-23) and has magnitude equal to NIA. Comparing Equation 28-15 with Equation 22-11 for the torque on an electric dipole, we see that a current loop in a magnetic field acts like an electric dipole in an electric field.

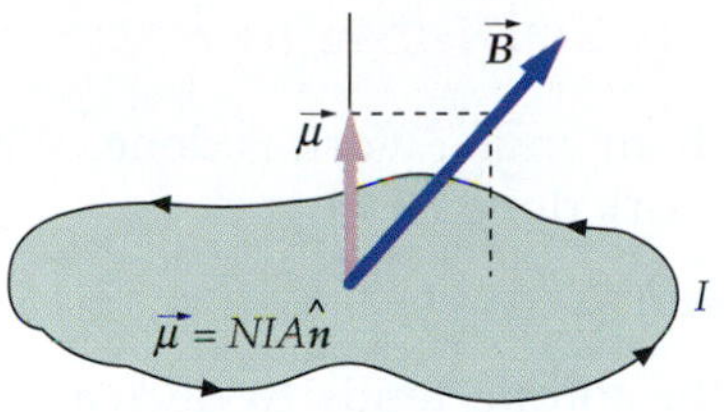

Figure 28-23 A current loop of arbitrary shape is described by its magnetic moment $\vec{\mu} = NIA\hat{n}$. In a magnetic field $\vec{B}$, it experiences a torque $\vec{\mu} \times \vec{B}$.

Example 28-8

A circular loop of radius 2 cm has 10 turns of wire and carries a current of 3 A. The axis of the loop makes an angle of 30° with a magnetic field of 8000 G. Find the magnitude of the torque on the loop.

1. The magnitude of the torque is given by Equation 28-15:
   $\tau = |\vec{\mu} \times \vec{B}| = \mu B \sin\theta = \mu B \sin 30°$
2. Calculate the magnitude of the magnetic moment of the loop:
   $\mu = NIA = (10)(3\text{ A})\pi(0.02\text{ m})^2 = 3.77 \times 10^{-2}\text{ A}\cdot\text{m}^2$
3. Substitute this value of μ and the given value of B to calculate τ:
   $\tau = \mu B \sin\theta = (3.77 \times 10^{-2}\text{ A}\cdot\text{m}^2)(0.8\text{ T})(\sin 30°)$
   $= 1.51 \times 10^{-2}\text{ N}\cdot\text{m}$

Example 28-9 *try it yourself*

A circular wire loop of radius R, mass m, and current I lies on a rough surface (Figure 28-24). There is a horizontal magnetic field $\vec{B}$. How large can the current I be before one edge of the loop will lift off the surface?

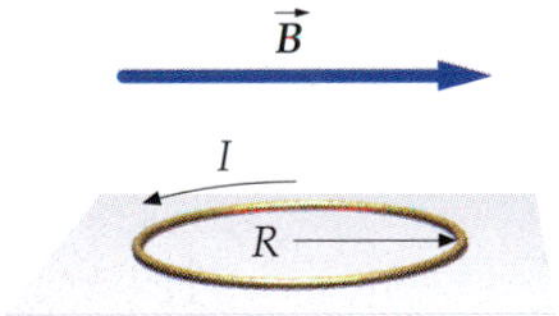

Figure 28-24

Picture the Problem The loop will start to lift off when the magnetic torque equals the gravitational torque (Figure 28-25).

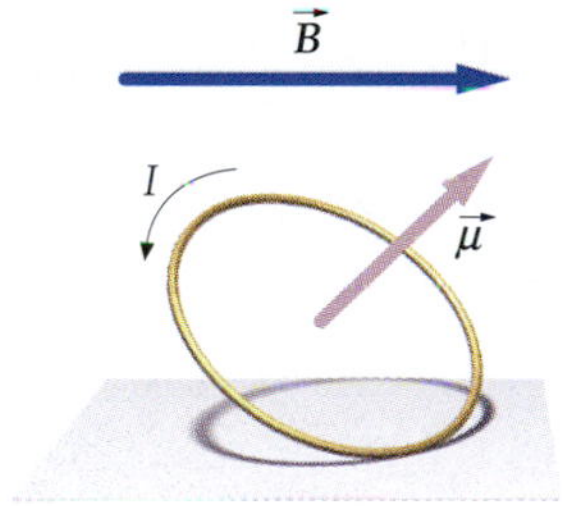

Figure 28-25

Cover the column to the right and try these on your own before looking at the answers.

Steps	***Answers***
1. Find the magnetic torque acting on the loop.	$\tau_m = \mu B = I\pi R^2 B$
2. Find the gravitational torque exerted on the loop.	$\tau_g = mgR$
3. Equate the torques and solve for I.	$I = \dfrac{mg}{\pi RB}$

Potential Energy of a Magnetic Dipole in a Magnetic Field

When a torque is exerted through an angle, work is done. When a dipole is rotated through an angle $d\theta$, the work done is

$$dW = -\tau\, d\theta = -\mu B \sin\theta\, d\theta$$

The minus sign arises because the torque tends to decrease θ. Setting this work equal to the decrease in potential energy, we have

$$dU = -dW = +\mu B \sin\theta\, d\theta$$

Integrating, we obtain

$$U = -\mu B \cos\theta + U_0$$

If we choose the potential energy to be zero when $\theta = 90°$, then $U_0 = 0$ and the potential energy of the dipole is

$$U = -\mu B \cos\theta = -\vec{\mu}\cdot\vec{B} \qquad \text{28-16}$$

Equation 28-16 gives the potential energy of a magnetic dipole at an angle θ to a magnetic field.

Example 28-10

A square 12-turn coil with sides of length 40 cm carries a current of 3 A. It lies in the xy plane as shown in a uniform magnetic field $\vec{B} = 0.3\,\mathrm{T}\hat{i} + 0.4\,\mathrm{T}\hat{k}$. Find (*a*) the magnetic moment of the coil and (*b*) the torque exerted on the coil. (c) Find the potential energy of the coil.

Picture the Problem From Figure 28-26 we see that the magnetic moment of the loop is in the positive z direction.

Figure 28-26

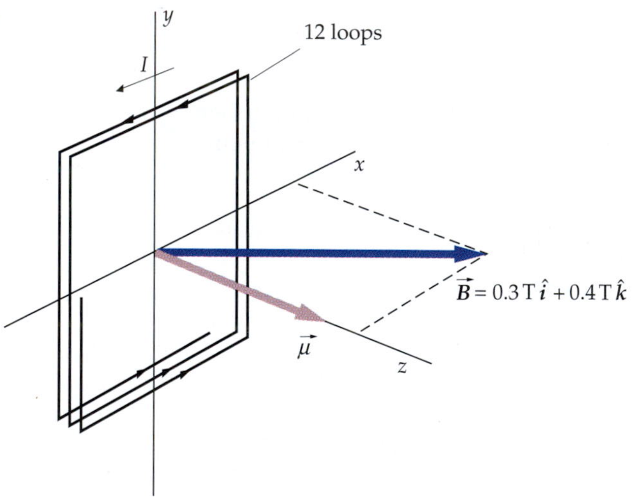

(*a*) Calculate the magnetic moment of the loop:

$$\vec{\mu} = NIA\hat{k} = (12)(3\text{ A})(0.40\text{ m})^2\hat{k} = 5.76\ \mathrm{A\cdot m^2}\,\hat{k}$$

(*b*) The torque on the current loop is given by Equation 28-15:

$$\vec{\tau} = \vec{\mu}\times\vec{B} = (5.76\ \mathrm{A\cdot m^2}\,\hat{k})\times(0.3\,\mathrm{T}\,\hat{i} + 0.4\,\mathrm{T}\,\hat{k}) = 1.73\ \mathrm{N\cdot m}\,\hat{j}$$

(*c*) The potential energy is the negative dot product of $\vec{\mu}$ and $\vec{B}$:

$$U = -\vec{\mu}\cdot\vec{B} = -(5.76\ \mathrm{A\cdot m^2}\,\hat{k})\cdot(0.3\,\mathrm{T}\,\hat{i} + 0.4\,\mathrm{T}\,\hat{k}) = -2.30\ \mathrm{J}$$

Remarks We have used $\hat{k} \times \hat{k} = 0$ and $\hat{k} \times \hat{i} = \hat{j}, \hat{k} \cdot \hat{i} = 0$ and $\hat{k} \cdot \hat{k} = 1$. The torque is in the y direction.

Exercise Calculate U if the coil rotates so that $\vec{\mu}$ is aligned with $\vec{B}$. (Answer $U = -\mu B = -(5.76\ \text{A}\cdot\text{m}^2)(0.5\ \text{T}) = -2.88\ \text{J}$. Note that this potential energy is lower than that found in the example. The torque tends to rotate the loop toward a position of lower potential energy.)

When a small permanent magnet such as a compass needle is placed in a magnetic field $\vec{B}$, the field exerts a torque on the magnet that tends to rotate the magnet so that it lines up with the field. This effect also occurs with previously unmagnetized iron filings, which become magnetized in the presence of a $\vec{B}$ field. The bar magnet is characterized by a magnetic moment $\vec{\mu}$ that points from the south pole to the north pole. A small bar magnet thus behaves like a current loop. This is not a coincidence. The origin of the magnetic moment of a bar magnet is, in fact, microscopic current loops that result from the motion of electrons in the atoms of the magnet.

Example 28-11

A nonconducting disk of mass M and radius R has a surface charge density σ and rotates with angular velocity ω about its axis. Find the magnetic moment of the rotating disk.

Picture the Problem We find the magnetic moment of a circular element of radius r and width dr and integrate (Figure 28-27). The charge on the element is $dq = \sigma\ dA = \sigma 2\pi r\ dr$. If the charge is positive, the magnetic moment is in the direction of $\vec{\omega}$, so we need only calculate the magnitude.

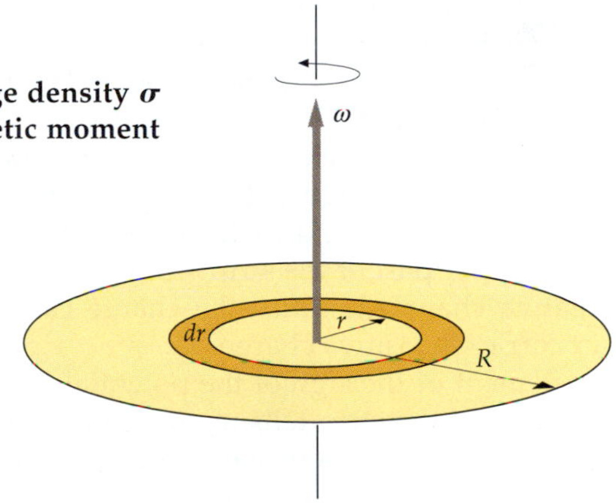

Figure 28-27

1. The magnetic moment of the strip shown is the current times the area of the loop: $d\mu = (dI)A = (dI)\pi r^2$

2. The current in the strip is the total charge on the strip times its frequency of rotation $f = \omega/2\pi$: $dI = (dq)f = (\sigma\, dA)\dfrac{\omega}{2\pi} = (\sigma\, 2\pi r\, dr)\dfrac{\omega}{2\pi} = \sigma\omega r\, dr$

3. Substitute to obtain the magnetic moment of the strip $d\mu$ in terms of r and dr: $d\mu = (dI)\pi r^2 = (\sigma\omega r\, dr)\pi r^2 = \pi\sigma\omega r^3\, dr$

4. Integrate from $r = 0$ to $r = R$: $\mu = \int d\mu = \int_0^R \pi\sigma\omega r^3\, dr = \dfrac{1}{4}\pi\sigma\omega R^4$

5. Use the fact that $\vec{\mu}$ is parallel to $\vec{\omega}$ if σ is positive to write the magnetic moment as a vector: $\vec{\mu} = \frac{1}{4}\pi\sigma R^4\vec{\omega}$

Remarks In terms of the total charge $Q = \sigma\pi R^2$, the magnetic moment is $\vec{\mu} = \frac{1}{4}QR^2\vec{\omega}$. The angular momentum of the disk is $\vec{L} = (\frac{1}{2}MR^2)\vec{\omega}$, so the magnetic moment can be written $\vec{\mu} = (Q/2M)\vec{L}$, which is a general result. (See Problem 53.)

28-4 The Hall Effect

As we have seen, charges moving in a magnetic field experience a force perpendicular to their motion. When these charges are traveling in a conducting wire, they will be pushed to one side of the wire. This results in a separation of charge in the wire called the **Hall effect**. This phenomenon allows us to determine the sign of the charge on the charge carriers and the number of charge carriers per unit volume n in a conductor. It also provides a convenient method for measuring magnetic fields.

Figure 28-28 shows two conducting strips, each of which carries a current I to the right because the left sides of the strips are connected to the positive terminal of a battery and the right sides are connected to the negative terminal. The strips are in a magnetic field that is directed into the paper. Let us assume for the moment that the current in the strip consists of positively charged particles moving to the right as shown in Figure 28-28*a*. The magnetic force on these particles is $q\vec{v}_d \times \vec{B}$ (where $\vec{v}_d$ is the drift velocity of the charge carriers). This force is directed upward. The positive particles therefore move up to the top of the strip, leaving the bottom of the strip with an excess negative charge. This separation of charge produces an electrostatic field in the strip that opposes the magnetic force on the charge carriers. When the electrostatic and magnetic forces balance, the charge carriers no longer move upward. In this equilibrium situation, the upper part of the strip is positively charged, so it is at a greater potential than the negatively charged lower part. If the current consists of negatively charged particles, as shown in Figure 28-28*b*, the charge carriers in the strip must move to the left (since the current is still to the right). The magnetic force $q\vec{v}_d \times \vec{B}$ is again up because the signs of both q and $\vec{v}_d$ have been changed. Again the carriers are forced to the upper part of the strip, but the upper part of the strip now carries a negative charge (because the charge carriers are negative) and the lower part carries a positive charge.

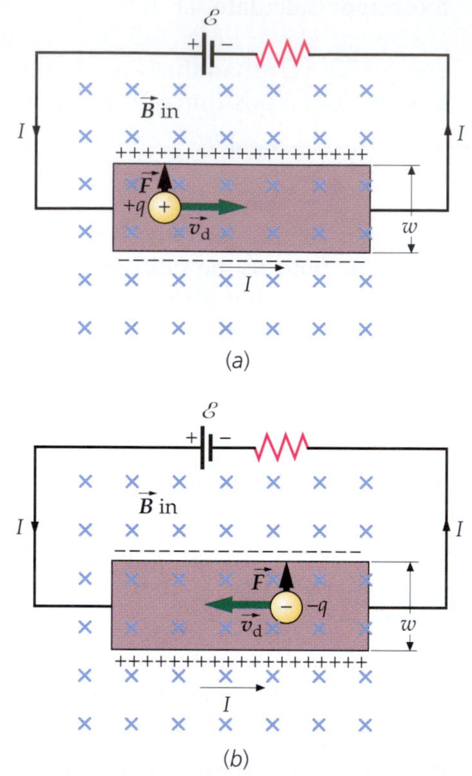

Figure 28-28 The Hall effect. The magnetic field is directed into the plane of the page as indicated by the crosses. The magnetic force on a charged particle is upward for a current to the right whether the current is due to (*a*) positive particles moving to the right or (*b*) negative particles moving to the left.

A measurement of the sign of the potential difference between the upper and lower parts of the strip tells us the sign of the charge carriers. In semiconductors, the charge carriers may be negative electrons or positive holes. A measurement of the sign of the potential difference tells us which are dominant for a particular semiconductor. For a normal metallic conductor, we find that the upper part of the strip in Figure 28-28 is at a lower potential than the lower part—which means that the upper part must carry a negative charge. Thus, Figure 28-28*b* is the correct illustration of the current in a normal conductor. It was this type of experiment that led to the discovery that the charge carriers in metallic conductors are negative.

The potential difference between the top and bottom of the strip is called the **Hall voltage**. We can calculate the magnitude of the Hall voltage in terms of the drift velocity. The magnitude of the magnetic force on the charge carriers in the strip is qv_dB. This magnetic force is balanced by the electrostatic force of magnitude qE, where E is the electric field due to the charge separation. Thus, we have $E = v_dB$. If the width of the strip is w, the potential difference is Ew. The Hall voltage is therefore

$$V_H = Ew = v_dBw \qquad \text{28-17}$$

Exercise A conducting strip of width $w = 2.0$ cm is placed in a magnetic field of 8000 G. The Hall voltage is measured to be 0.64 μV. Calculate the drift velocity of the electrons. (*Answer* 4.0×10^{-5} m/s)

Since the drift velocity for ordinary currents is very small, we can see from Equation 28-17 that the Hall voltage is very small for ordinary-sized strips and magnetic fields. From measurements of the Hall voltage for a strip of a given size, we can determine the number of charge carriers per unit volume in the strip. The current is given by Equation 25-3:

$$I = nqv_dA$$

where A is the cross-sectional area of the strip. For a strip of width w and thickness t, the cross-sectional area is $A = wt$. Since the charge carriers are electrons, the quantity q is the charge on one electron e. The number density of charge carriers n is thus given by

$$n = \frac{I}{Aqv_d} = \frac{I}{wtev_d} \qquad \text{28-18}$$

Substituting $v_dw = V_H/B$ from Equation 28-17, we have

$$n = \frac{IB}{teV_H} \qquad \text{28-19}$$

Example 28-12

A silver slab of thickness 1 mm and width 1.5 cm carries a current of 2.5 A in a region in which there is a magnetic field of magnitude 1.25 T perpendicular to the slab. The Hall voltage is measured to be 0.334 μV. (*a*) Calculate the number density of the charge carriers. (*b*) Compare your answer in part (*a*) to the number density of atoms in silver, which has a mass density of $\rho = 10.5\ \text{g/cm}^3$ and a molar mass of $M = 107.9$ g/mol.

(*a*) Substitute numerical values into Equation 28-19 to find n:

$$n = \frac{IB}{teV_H} = \frac{(2.5\ \text{A})(1.25\ \text{T})}{(0.001\ \text{m})(1.6 \times 10^{-19}\ \text{C})(3.34 \times 10^{-7}\ \text{V})}$$

$$= 5.85 \times 10^{28}\ \text{electrons/m}^3$$

(*b*) The number of atoms per unit volume is $\rho N_A/M$:

$$n_a = \rho\frac{N_A}{M} = (10.5\ \text{g/cm}^3)\frac{6.02 \times 10^{23}\ \text{atoms/mol}}{107.9\ \text{g/mol}}$$

$$= 5.86 \times 10^{22}\ \text{atoms/cm}^3 = 5.86 \times 10^{28}\ \text{atoms/m}^3$$

Remark These results indicate that the number of charge carriers in silver is very nearly one per atom.

The Hall voltage provides a convenient method for measuring magnetic fields. If we rearrange Equation 28-19, we can write for the Hall voltage

$$V_H = \frac{I}{nte}B \qquad \text{28-20}$$

A given strip can be calibrated by measuring the Hall voltage for a given current in a known magnetic field. The strip can then be used to measure an unknown magnetic field B by measuring the Hall voltage for a given current.

The Quantum Hall Effect

According to Equation 28-20, the Hall voltage should increase linearly with magnetic field B for a given current in a given slab. In 1980, while studying the Hall effect in semiconductors at very low temperatures and very large magnetic fields, the German physicist Klaus von Klitzing discovered that a

optional

plot of V_H versus B resulted in a series of plateaus, as shown in Figure 28-29, rather than a straight line. That is, the Hall voltage is quantized. For this discovery of the quantum Hall effect, von Klitzing won the Nobel Prize in physics in 1985.

In the theory of the quantum Hall effect, the Hall resistance, defined as $R_H = V_H/I$, can take on only the values

$$R_H = \frac{V_H}{I} = \frac{R_K}{n}, \qquad n = 1, 2, 3, \ldots \qquad \text{28-21}$$

where n is an integer, and R_K, called the **von Klitzing constant**, is related to the fundamental electronic charge e and Planck's constant h by

$$R_K = \frac{h}{e^2} \qquad \text{28-22}$$

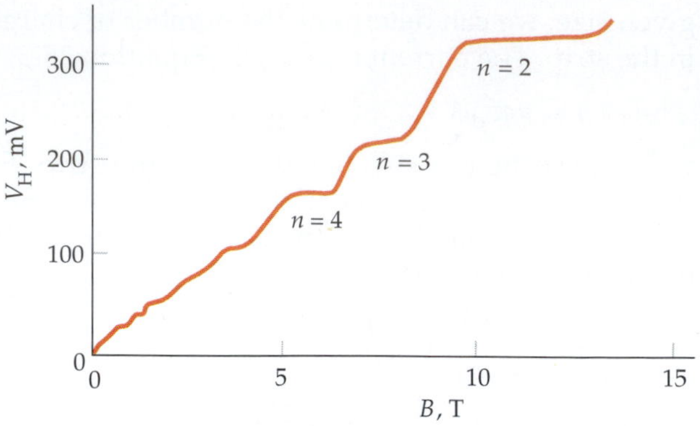

Figure 28-29 A plot of the Hall voltage versus applied magnetic field shows plateaus, indicating that the Hall voltage is quantized. These data were taken at a temperature of 1.39 K with the current I held fixed at 25.52 μA.

Because the von Klitzing constant can be measured to an accuracy of a few parts per billion, the quantum Hall effect is now used to define a standard of resistance. As of January 1990, the ohm is now defined so that R_K has the value 25,812.807 Ω exactly.

Recent experiments have shown that under certain special conditions, the Hall resistance is given by Equation 28-22 with the integer n replaced by a series of rational fractions. At present, the theory of this fractional quantum Hall effect is incomplete.

Summary

1. The magnetic field describes the condition in space in which moving charges experience a force perpendicular to their velocity.
2. The magnetic force is part of the electromagnetic force, one of the four fundamental forces of nature.
3. The magnitude and direction of a magnetic field $\vec{B}$ are defined by the force $\vec{F} = q\vec{v} \times \vec{B}$ exerted on moving charges.

Topic	Remarks and Relevant Equations	
1. Magnetic Force		
On a moving charge	$\vec{F} = q\vec{v} \times \vec{B}$	**28-1**
On a current element	$d\vec{F} = I\, d\vec{\ell} \times \vec{B}$	**28-5**
Unit of the magnetic field	The SI unit of magnetic fields is the tesla (T). A commonly used unit is the gauss (G), which is related to the tesla by $1\ \text{T} = 10^4\ \text{G}$	**28-3**
2. Motion of Point Charges	A particle of mass m and charge q moving with speed v in a plane perpendicular to a magnetic field moves in a circular orbit. The period and frequency of this circular motion are independent of the radius of the orbit and of the speed of the particle.	
Radius of circular orbit	$r = \frac{mv}{qB}$	**28-6**

Cyclotron period	$T = \dfrac{2\pi m}{qB}$	28-7
Cyclotron frequency	$f = \dfrac{1}{T} = \dfrac{qB}{2\pi m}$	28-8
Velocity selector (optional)	A velocity selector consists of crossed electric and magnetic fields such that the electric and magnetic forces balance for a particle whose speed is given by $v = \dfrac{E}{B}$	28-9
Measurement of q/m (optional)	The deflection of a charged particle in an electric field depends on the speed of the particle and is proportional to the charge-to-mass ratio q/m of the particle. J.J. Thomson used crossed electric and magnetic fields to measure the speed of cathode rays and then measured q/m for these particles by deflecting them in an electric field. He showed that all cathode rays consist of particles that all have the same charge-to-mass ratio. These particles are now called electrons.	
Mass spectrometer (optional)	The mass-to-charge ratio of an ion of known speed can be determined by measuring the radius of the circular path taken by the ion in a known magnetic field.	

3. Current Loops

Magnetic moment	$\vec{\mu} = NIA\hat{n}$	28-14
Torque	$\vec{\tau} = \vec{\mu} \times \vec{B}$	28-15
Potential energy	$U = -\mu B \cos\theta = -\vec{\mu} \cdot \vec{B}$	28-16
Net force	The net force on a current loop in a uniform magnetic field is zero.	

4. The Hall Effect

When a conducting strip carrying a current is placed in a magnetic field, the magnetic force on the charge carriers causes a separation of charge called the Hall effect. This results in a voltage V_H, called the Hall voltage. The sign of the charge carriers can be determined from a measurement of the sign of the Hall voltage, and the number of carriers per unit volume can be determined from the magnitude of V_H.

Hall voltage	$V_H = v_d B w = \dfrac{I}{nte} B$	28-17, 28-20
Quantum Hall effect (optional)	Measurements at very low temperatures in very large magnetic fields indicate that the Hall resistance $R_H = V_H/I$ is quantized and can take on only the values given by $R_H = \dfrac{V_H}{I} = \dfrac{R_K}{n}, \quad n = 1, 2, 3, \ldots$	28-21
von Klitzing constant (optional)	$R_K = \dfrac{h}{e^2} \approx 25{,}813\ \Omega$	28-22

Problem-Solving Guide

1. Begin by drawing a neat diagram that includes the important features of the problem.
2. The vector nature of the magnetic force is clarified with a sketch showing $\vec{v}$ and $\vec{B}$. Your sketch will help greatly in applying the right-hand rule.
3. For problems involving current-carrying wires, show the direction of I and again apply the right-hand rule.

Summary of Worked Examples

Type of Calculation	Procedure and Relevant Examples	
1. Magnetic Force		
Find the magnetic force on a charged particle.	Use $\vec{F} = q\vec{v} \times \vec{B}$.	**Example 28-1**
Find the magnetic force on a current-carrying wire.	Use $\vec{F} = I\vec{\ell} \times \vec{B}$.	**Example 28-2**
2. Motion of Particles		
Determine the radius and frequency of motion for a charged particle in a magnetic field.	Applying Newton's second law—equating the magnetic force and the centripetal force—yields the relations $r = mv/qB$ and $f = 1/T = qB/2\pi m$.	**Examples 28-3, 28-5, 28-6**
3. Magnetic Moments		
Find the magnetic moment of a loop.	The magnitude of the magnetic moment is $\mu = NIA$. The direction is found using the right-hand rule.	**Examples 28-8, 28-9, 28-10**
Find the torque acting on a current loop.	Use $\vec{\tau} = \vec{\mu} \times \vec{B}$.	**Examples 28-8, 28-9, 28-10**
4. The Hall Effect		
Calculate the density of charge carriers in a conductor.	Use $n = \dfrac{IB}{teV_H}$	**Example 28-12**

Problems

Conceptual Problems
Problems from Optional and Exploring sections

In a few problems, you are given more data than you actually need; in a few other problems, you are required to supply data from your general knowledge, outside sources, or informed estimates.

• Single-concept, single-step, relatively easy
•• Intermediate-level, may require synthesis of concepts
••• Challenging, for advanced students

Force Exerted by a Magnetic Field

1 • When a cathode-ray tube is placed horizontally in a magnetic field that is directed vertically upward, the electrons emitted from the cathode follow one of the dashed paths to the face of the tube in Figure 28-30. The correct path is ________.

(*a*) 1 (*b*) 2 (*c*) 3
(*d*) 4 (*e*) 5

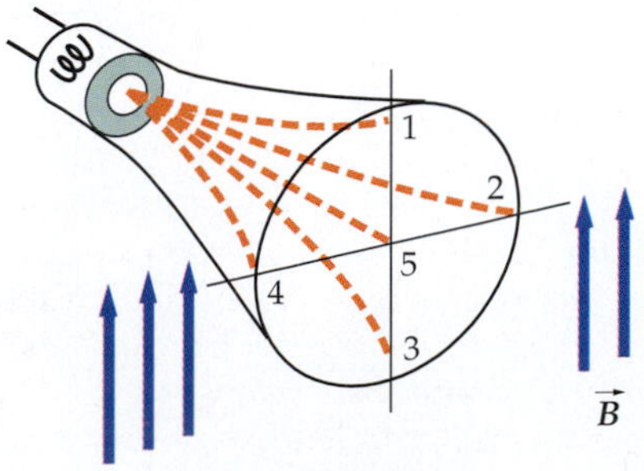

Figure 28-30 Problem 1

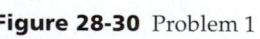

2 • Why not define $\vec{B}$ to be in the direction of $\vec{F}$, as we do for $\vec{E}$?

3 • Find the magnetic force on a proton moving with velocity 4.46 Mm/s in the positive x direction in a magnetic field of 1.75 T in the positive z direction.

4 • A charge $q = -3.64$ nC moves with a velocity of 2.75×10^6 m/s $\hat{i}$. Find the force on the charge if the magnetic field is (*a*) $\vec{B} = 0.38$ T $\hat{j}$, (*b*) $\vec{B} = 0.75$ T $\hat{i} + 0.75$ T $\hat{j}$, (*c*) $\vec{B} = 0.65$ T $\hat{i}$, (*d*) $\vec{B} = 0.75$ T $\hat{i} + 0.75$ T $\hat{k}$.

5 • A uniform magnetic field of magnitude 1.48 T is in the positive z direction. Find the force exerted by the field on a proton if the proton's velocity is (*a*) $\vec{v} = 2.7$ Mm/s $\hat{i}$, (*b*) $\vec{v} = 3.7$ Mm/s $\hat{j}$, (*c*) $\vec{v} = 6.8$ Mm/s $\hat{k}$, and (*d*) $\vec{v} = 4.0$ Mm/s $\hat{i} + 3.0$ Mm/s $\hat{j}$.

6 • An electron moves with a velocity of 2.75 Mm/s in the xy plane at an angle of 60° to the x axis and 30° to the y axis. A magnetic field of 0.85 T is in the positive y direction. Find the force on the electron.

7 • A straight wire segment 2 m long makes an angle of 30° with a uniform magnetic field of 0.37 T. Find the magnitude of the force on the wire if it carries a current of 2.6 A.

8 • A straight wire segment $I\vec{\ell} = (2.7\text{ A})(3\text{ cm}\ \hat{i} + 4\text{ cm}\ \hat{j})$ is in a uniform magnetic field $\vec{B} = 1.3$ T $\hat{i}$. Find the force on the wire.

9 • What is the force (magnitude and direction) on an electron with velocity $\vec{v} = (2\hat{i} - 3\hat{j}) \times 10^6$ m/s in a magnetic field $\vec{B} = (0.8\hat{i} + 0.6\hat{j} - 0.4\hat{k})$ T?

10 •• The wire segment in Figure 28-31 carries a current of 1.8 A from a to b. There is a magnetic field $\vec{B} = 1.2$ T $\hat{k}$. Find the total force on the wire and show that it is the same as if the wire were a straight segment from a to b.

Figure 28-31 Problem 10

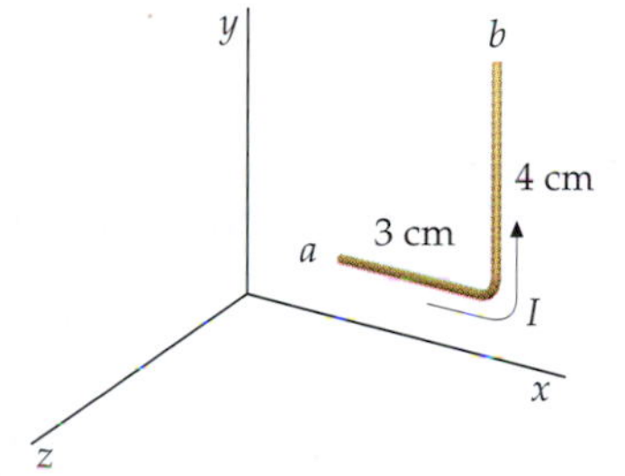

11 •• A straight, stiff, horizontal wire of length 25 cm and mass 50 g is connected to a source of emf by light, flexible leads. A magnetic field of 1.33 T is horizontal and perpendicular to the wire. Find the current necessary to float the wire, that is, the current such that the magnetic force balances the weight of the wire.

12 •• A simple gaussmeter for measuring horizontal magnetic fields consists of a stiff 50-cm wire that hangs from a conducting pivot so that its free end makes contact with a pool of mercury in a dish below. The mercury provides an electrical contact without constraining the movement of the wire. The wire has a mass of 5 g and conducts a current downward. (*a*) What is the equilibrium angular displacement of the wire from vertical if the horizontal magnetic field is 0.04 T and the current is 0.20 A? (*b*) If the current is 20 A and a displacement from vertical of 0.5 mm can be detected for the free end, what is the horizontal magnetic field sensitivity of this gaussmeter?

13 •• A current-carrying wire is bent into a semicircular loop of radius R that lies in the xy plane. There is a uniform magnetic field $\vec{B} = B\hat{k}$ perpendicular to the plane of the loop (Figure 28-32). Show that the force acting on the loop is $\vec{F} = 2IRB\hat{j}$.

Figure 28-32 Problem 13

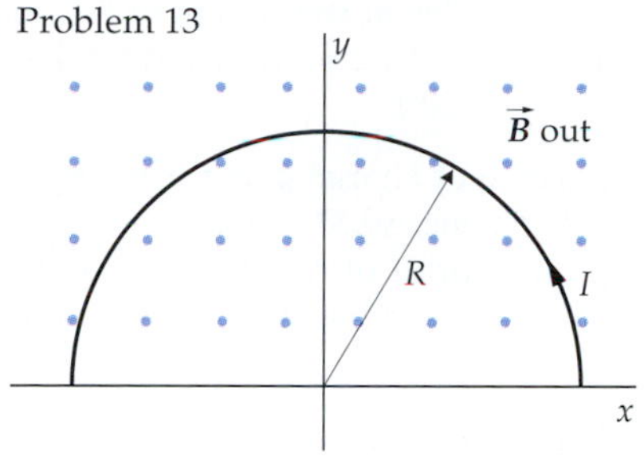

14 •• A 10-cm length of wire carries a current of 4.0 A in the positive z direction. The force on this wire due to a magnetic field $\vec{B}$ is $\vec{F} = (-0.2\hat{i} + 0.2\hat{j})$ N. If this wire is rotated so that the current flows in the positive x direction, the force on the wire is $\vec{F} = 0.2\hat{k}$ N. Find the magnetic field $\vec{B}$.

15 •• A 10-cm length of wire carries a current of 2.0 A in the positive x direction. The force on this wire due to the presence of a magnetic field $\vec{B}$ is $\vec{F} = (3.0\hat{j} + 2.0\hat{k})$ N. If this wire is now rotated so that the current flows in the positive y direction, the force on the wire is $\vec{F} = (-3.0\hat{i} - 2.0\hat{k})$ N. Determine the magnetic field $\vec{B}$.

16 ••• A wire bent in some arbitrary shape carries a current I in a uniform magnetic field $\vec{B}$. Show explicitly that the total force on the part of the wire from some point a to some point b is $\vec{F} = I\vec{\ell} \times \vec{B}$, where $\vec{\ell}$ is the vector from a to b.

Motion of a Point Charge in a Magnetic Field

17 • True or false: The magnetic force does not accelerate a particle because the force is perpendicular to the velocity of the particle.

18 • A moving charged particle enters a region in which it is suddenly deflected perpendicular to its motion. How can you tell if the deflection was caused by a magnetic field or an electric field?

19 • A proton moves in a circular orbit of radius 65 cm perpendicular to a uniform magnetic field of magnitude 0.75 T. (*a*) What is the period for this motion? (*b*) Find the speed of the proton. (*c*) Find the kinetic energy of the proton.

20 • An electron of kinetic energy 45 keV moves in a circular orbit perpendicular to a magnetic field of 0.325 T. (*a*) Find the radius of the orbit. (*b*) Find the frequency and period of the motion.

21 • An electron from the sun with a speed of 1×10^7 m/s enters the earth's magnetic field high above the equator where the magnetic field is 4×10^{-7} T. The electron moves nearly in a circle except for a small drift along the direction of the earth's magnetic field that will take it toward the north pole. (*a*) What is the radius of the circular motion? (*b*) What is the radius of the circular motion near the north pole where the magnetic field is 2×10^{-5} T?

22 •• Protons and deuterons (each with charge $+e$) and alpha particles (with charge $+2e$) of the same kinetic energy enter a uniform magnetic field $\vec{B}$ that is perpendicular to their velocities. Let r_p, r_d, and r_α be the radii of their circular orbits. Find the ratios r_d/r_p and r_α/r_p. Assume that $m_\alpha = 2m_d = 4m_p$.

23 •• A proton and an alpha particle move in a uniform magnetic field in circles of the same radii. Compare (*a*) their velocities, (*b*) their kinetic energies, and (*c*) their angular momenta. (See Problem 22.)

24 •• A particle of charge q and mass m has momentum $p = mv$ and kinetic energy $K = \frac{1}{2}mv^2 = p^2/2m$. If the particle moves in a circular orbit of radius r perpendicular to a uniform magnetic field B, show that (*a*) $p = Bqr$ and (*b*) $K = B^2q^2r^2/2m$.

25 •• A beam of particles with velocity $\vec{v}$ enters a region of uniform magnetic field $\vec{B}$ that makes a small angle θ with $\vec{v}$. Show that after a particle moves a distance $2\pi(m/qB)v \cos \theta$ measured along the direction of $\vec{B}$, the velocity of the particle is in the same direction as it was when it entered the field.

26 •• A proton with velocity $v = 10^7$ m/s enters a region of uniform magnetic field $B = 0.8$ T, which is into the page, as shown in Figure 28-33. The angle $\theta = 60°$. Find the angle ϕ and the distance d.

Figure 28-33
Problems 26 and 27

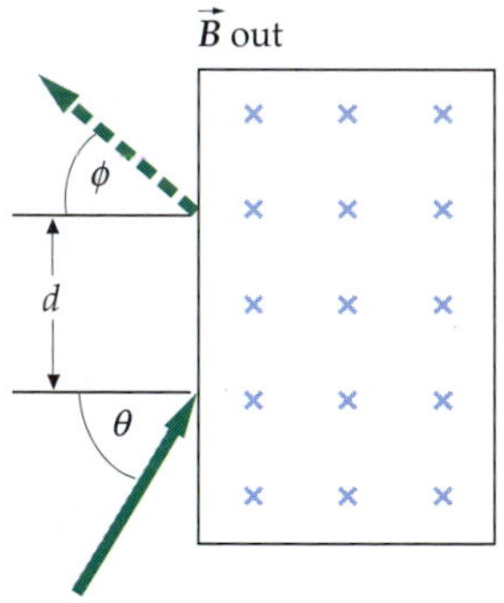

27 •• Suppose that in Figure 28-33 $B = 0.6$ T, the distance $d = 0.4$ m, and $\theta = 24°$. Find the speed v and the angle ϕ if the particles are (*a*) protons and (*b*) deuterons.

Velocity Selectors (optional)

28 • A beam of positively charged particles passes undeflected from left to right through a velocity selector in which the electric field is up. The beam is then reversed so that it travels from right to left. Will the beam now be deflected in the velocity selector? If so, in which direction?

29 • A velocity selector has a magnetic field of magnitude 0.28 T perpendicular to an electric field of magnitude 0.46 MV/m. (*a*) What must the speed of a particle be for it to pass through undeflected? What energy must (*b*) protons and (*c*) electrons have to pass through undeflected?

30 • A beam of protons moves along the x axis in the positive x direction with a speed of 12.4 km/s through a region of crossed fields balanced for zero deflection. (*a*) If there is a magnetic field of magnitude 0.85 T in the positive y direction, find the magnitude and direction of the electric field. (*b*) Would electrons of the same velocity be deflected by these fields? If so, in what direction?

Measuring q/m (optional)

31 •• The plates of a Thomson q/m apparatus are 6.0 cm long and are separated by 1.2 cm. The end of the plates is 30.0 cm from the tube screen. The kinetic energy of the electrons is 2.8 keV. (*a*) If a potential of 25.0 V is applied across the deflection plates, by how much will the beam deflect? (*b*) Find the magnitude of the crossed magnetic field that will allow the beam to pass through undeflected.

Mass Spectrometer (optional)

32 •• Chlorine has two stable isotopes, ^{35}Cl and ^{37}Cl, whose natural abundances are about 76% and 24%, respectively. Singly ionized chlorine gas is to be separated into its isotopic components using a mass spectrometer. The magnetic field in the spectrometer is 1.2 T. What is the minimum value of the potential through which these ions must be accelerated so that the separation between them is 1.4 cm?

33 •• A singly ionized ^{24}Mg ion (mass 3.983×10^{-26} kg) is accelerated through a 2.5-kV potential difference and deflected in a magnetic field of 557 G in a mass spectrometer. (*a*) Find the radius of curvature of the orbit for the ion. (*b*) What is the difference in radius for ^{26}Mg and ^{24}Mg ions? (Assume that their mass ratio is 26/24.)

34 •• A beam of ^{6}Li and ^{7}Li ions passes through a velocity selector and enters a magnetic spectrometer. If the diameter of the orbit of the ^{6}Li ions is 15 cm, what is the diameter of that for ^{7}Li ions?

35 •• In Example 28-6, determine the time required for a ^{58}Ni ion and a ^{60}Ni ion to complete the semicircular path.

36 •• Before entering a mass spectrometer, ions pass through a velocity selector consisting of parallel plates separated by 2.0 mm and having a potential difference of 160 V. The magnetic field between the plates is 0.42 T. The magnetic field in the mass spectrometer is 1.2 T. Find (*a*) the speed of the ions entering the mass spectrometer and (*b*) the difference in the diameters of the orbits of singly ionized ^{238}U and ^{235}U. (The mass of a ^{235}U ion is 3.903×10^{-25} kg.)

The Cyclotron (optional)

37 •• A cyclotron for accelerating protons has a magnetic field of 1.4 T and a radius of 0.7 m. (*a*) What is the cyclotron frequency? (*b*) Find the maximum energy of the protons when they emerge. (*c*) How will your answers change if deuterons, which have the same charge but twice the mass, are used instead of protons?

38 •• A certain cyclotron with magnetic field of 1.8 T is designed to accelerate protons to 25 MeV. (*a*) What is the cyclotron frequency? (*b*) What must the minimum radius of the magnet be to achieve a 25-MeV emergence energy? (*c*) If the alternating potential applied to the dees has a maximum value of 50 kV, how many revolutions must the protons make before emerging with an energy of 25 MeV?

39 •• Show that the cyclotron frequencies of deuterons and alpha particles are the same and are half that of a proton in the same magnetic field. (See Problem 22.)

40 •• Show that the radius of the orbit of a charged particle in a cyclotron is proportional to the square root of the number of orbits completed.

Torques on Current Loops and Magnets

41 • What orientation of a current loop gives maximum torque?

42 • A small circular coil of 20 turns of wire lies in a uniform magnetic field of 0.5 T such that the normal to the plane of the coil makes an angle of 60° with the direction of $\vec{B}$. The radius of the coil is 4 cm, and it carries a current of 3 A. (*a*) What is the magnitude of the magnetic moment of the coil? (*b*) What is the magnitude of the torque exerted on the coil?

43 • What is the maximum torque on a 400-turn circular coil of radius 0.75 cm that carries a current of 1.6 mA and resides in a uniform magnetic field of 0.25 T?

44 • A current-carrying wire is bent into the shape of a square of sides $L = 6$ cm and is placed in the xy plane. It carries a current $I = 2.5$ A. What is the torque on the wire if there is a uniform magnetic field of 0.3 T (*a*) in the z direction, and (*b*) in the x direction?

45 • Repeat Problem 44 if the wire is bent into an equilateral triangle of sides 8 cm.

46 •• A rigid, circular loop of radius R and mass M carries a current I and lies in the xy plane on a rough, flat table. There is a horizontal magnetic field of magnitude B. What is the minimum value of B such that one edge of the loop will lift off the table?

47 •• A rectangular, 50-turn coil has sides 6.0 and 8.0 cm long and carries a current of 1.75 A. It is oriented as shown in Figure 28-34 and pivoted about the z axis. (*a*) If the wire in the xy plane makes an angle $\theta = 37°$ with the y axis as shown, what angle does the unit normal $\hat{n}$ make with the x axis? (*b*) Write an expression for $\hat{n}$ in terms of the unit vectors $\hat{i}$ and $\hat{j}$. (*c*) What is the magnetic moment of the coil? (*d*) Find the torque on the coil when there is a uniform magnetic field $\vec{B} = 1.5\text{ T}\,\hat{j}$. (*e*) Find the potential energy of the coil in this field.

Figure 28-34 Problems 47 and 48

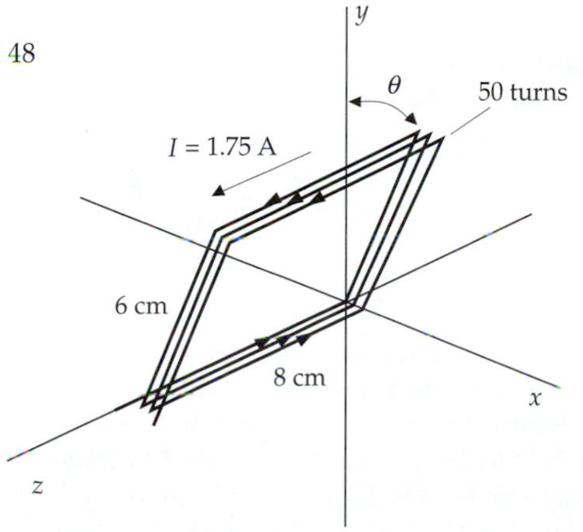

48 •• The coil in Problem 47 is pivoted about the z axis and held at various positions in a uniform magnetic field $\vec{B} = 2.0\text{ T}\,\hat{j}$. Sketch the position of the coil and find the torque exerted when the unit normal is (*a*) $\hat{n} = \hat{i}$, (*b*) $\hat{n} = \hat{j}$, (*c*) $\hat{n} = -\hat{j}$, and (*d*) $\hat{n} = (\hat{i} + \hat{j})/\sqrt{2}$.

Magnetic Moments

49 • The SI unit for the magnetic moment of a current loop is A·m^2. Use this to show that 1 T = 1 N/A·m.

50 •• A small magnet of length 6.8 cm is placed at an angle of 60° to the direction of a uniform magnetic field of magnitude 0.04 T. The observed torque has a magnitude of 0.10 N·m. Find the magnetic moment of the magnet.

51 •• A wire loop consists of two semicircles connected by straight segments (Figure 28-35). The inner and outer radii are 0.3 and 0.5 m, respectively. A current of 1.5 A flows in this loop with the current in the outer semicircle in the clockwise direction. What is the magnetic moment of this current loop?

Figure 28-35 Problem 51

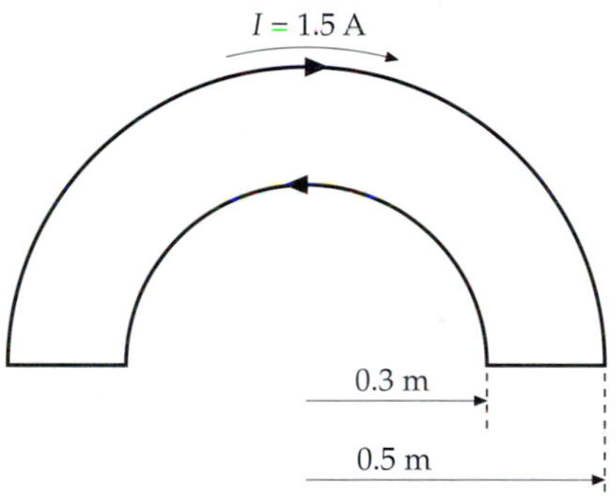

52 •• A wire of length L is wound into a circular coil of N loops. Show that when this coil carries a current I, its magnetic moment has the magnitude $IL^2/4\pi N$.

53 •• A particle of charge q and mass m moves in a circle of radius r and with angular velocity ω. (*a*) Show that the average current is $I = q\omega/2\pi$ and that the magnetic moment has the magnitude $\mu = \frac{1}{2}q\omega r^2$. (*b*) Show that the angular momentum of this particle has the magnitude $L = mr^2\omega$ and that the magnetic moment and angular momentum vectors are related by $\vec{\mu} = (q/2m)\vec{L}$.

54 ••• A single loop of wire is placed around the circumference of a rectangular piece of cardboard whose length and width are 70 and 20 cm, respectively. The cardboard is now folded along a line perpendicular to its length and midway between the two ends so that the two planes formed by the folded cardboard make an angle of 90°. If the wire loop carries a current of 0.2 A, what is the magnitude of the magnetic moment of this system?

55 ••• Repeat Problem 54 if the line along which the cardboard is folded is 40 cm from one end.

56 ••• A hollow cylinder has length L and inner and outer radii R_i and R_o, respectively (Figure 28-36). The cylinder carries a uniform charge density ρ. Derive an expression for the magnetic moment as a function of ω, the angular velocity of rotation of the cylinder about its axis.

Figure 28-36 Problem 56

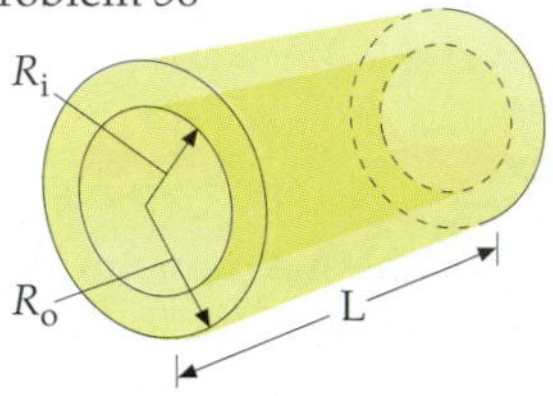

57 ••• A nonconducting rod of mass M and length ℓ has a uniform charge per unit length λ and rotates with angular velocity $\vec{\omega}$ about an axis through one end and perpendicular to the rod. (*a*) Consider a small segment of the rod of length dx and charge $dq = \lambda\, dx$ at a distance x from the pivot (Figure 28-37). Show that the magnetic moment of this segment is $\frac{1}{2}\lambda\omega x^2\, dx$. (*b*) Integrate your result to show that the total magnetic moment of the rod is $\mu = \frac{1}{6}\lambda\omega\, \ell^3$. (*c*) Show that the magnetic moment $\vec{\mu}$ and angular momentum $\vec{L}$ are related by $\vec{\mu} = (Q/2M)\vec{L}$, where Q is the total charge on the rod.

Figure 28-37 Problem 57

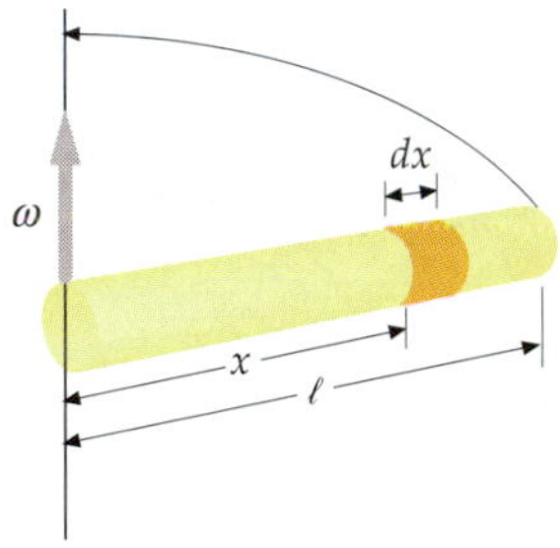

58 ••• A nonuniform, nonconducting disk of mass M, radius R, and total charge Q has a surface charge density $\sigma = \sigma_0 r/R$ and a mass per unit area $\sigma_m = (M/Q)\sigma$. The disk rotates with angular velocity ω about its axis. (*a*) Show that the magnetic moment of the disk has a magnitude $\mu = \frac{1}{5}\pi\omega\sigma_0 R^4 = \frac{3}{10}Q\omega R^2$. (*b*) Show that the magnetic moment $\vec{\mu}$ and angular momentum $\vec{L}$ are related by $\vec{\mu} = (Q/2M)\vec{L}$.

59 ••• A spherical shell of radius R carries a surface charge density σ. The sphere rotates about its diameter with angular velocity ω. Find the magnetic moment of the rotating sphere.

60 ••• A solid sphere of radius R carries a uniform volume charge density ρ. The sphere rotates about its diameter with angular velocity ω. Find the magnetic moment of this rotating sphere.

61 ••• A solid cylinder of radius R and length L carries a uniform charge density $+\rho$ between $r = 0$ and $r = R_s$ and an equal charge density of opposite sign, $-\rho$, between $r = R_s$ and $r = R$. What must be the radius R_s so that on rotation of the cylinder about its axis the magnetic moment is zero?

62 ••• A solid cylinder of radius R and length L carries a uniform charge density $\rho = -\rho_0$ between $r = 0$ and $r = \frac{1}{2}R$ and a positive charge density of equal magnitude, $+\rho_0$, between $r = \frac{1}{2}R$ and $r = R$ (Figure 28-38). The cylinder rotates about its axis with angular velocity $\vec{\omega}$. Derive an expression for the magnetic moment of the cylinder.

Figure 28-38 Problem 62

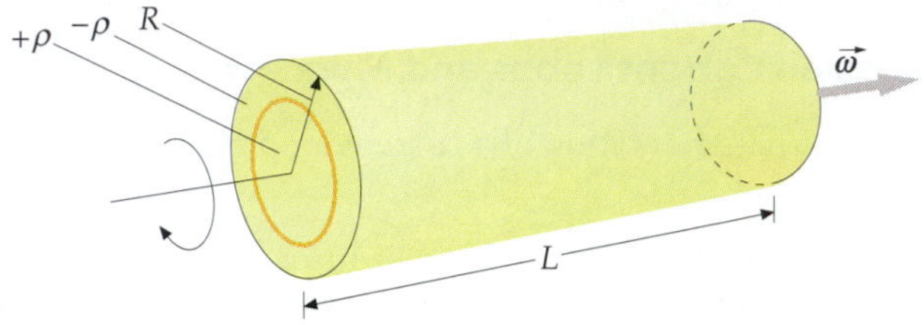

63 ••• A cylindrical shell of length L with inner radius R_i and outer radius R_o carries a uniform charge density, $+\rho_0$, between R_i and radius R_s and an equal charge density of opposite sign, $-\rho_0$, between R_s and R_o. The cylinder rotates about its axis with angular velocity $\vec{\omega}$. Derive an expression for the magnetic moment of this cylinder.

64 ••• A solid sphere of radius R carries a uniform charge density, $+\rho_0$, between $r = 0$ and $r = R_s$ and an equal charge density of opposite sign, $-\rho_0$, between $r = R_s$ and $r = R$. The sphere rotates about its diameter with angular velocity ω. Find R_s such that magnetic moment of the sphere is zero. What is the net charge carried by the sphere?

65 ••• A solid sphere of radius R carries a uniform charge density, $+\rho_0$, between $r = 0$ and $r = \frac{1}{2}R$ and an equal charge density of opposite sign, $-\rho_0$, between $r = \frac{1}{2}R$ and $r = R$. The sphere rotates about its diameter with angular velocity ω. Derive an expression for the magnetic moment of this rotating sphere.

The Hall Effect

66 • A metal strip 2.0 cm wide and 0.1 cm thick carries a current of 20 A in a uniform magnetic field of 2.0 T, as shown in Figure 28-39. The Hall voltage is measured to be 4.27 μV. (*a*) Calculate the drift velocity of the electrons in the strip. (*b*) Find the number density of the charge carriers in the strip. (*c*) Is point *a* or *b* at the higher potential?

Figure 28-39
Problems 66 and 67

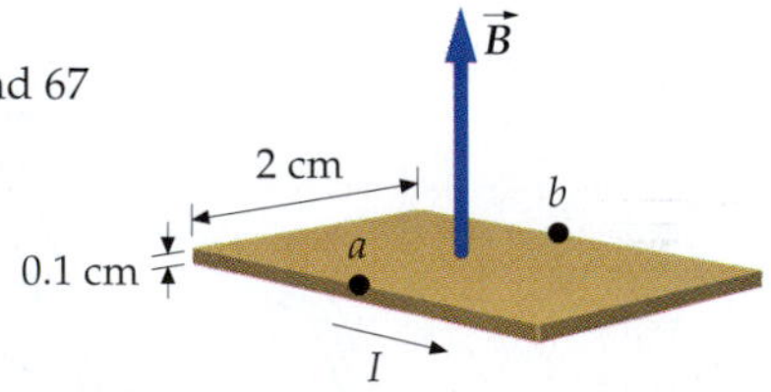

67 •• The number density of free electrons in copper is 8.47×10^{22} electrons per cubic centimeter. If the metal strip in Figure 28-39 is copper and the current is 10 A, find (*a*) the drift velocity v_d and (*b*) the Hall voltage. (Assume that the magnetic field is 2.0 T.)

68 •• A copper strip ($n = 8.47 \times 10^{22}$ electrons per cubic centimeter) 2 cm wide and 0.1 cm thick is used to measure the magnitudes of unknown magnetic fields that are perpendicular to the strip. Find the magnitude of B when $I = 20$ A and the Hall voltage is (*a*) 2.00 μV, (*b*) 5.25 μV, and (*c*) 8.00 μV.

69 •• Because blood contains charged ions, moving blood develops a Hall voltage across the diameter of an artery. A large artery with a diameter of 0.85 cm has a flow speed of 0.6 m/s. If a section of this artery is in a magnetic field of 0.2 T, what is the potential difference across the diameter of the artery?

70 •• The Hall coefficient R is defined as $R = E_y/J_xB_z$, where J_x is the current per unit area in the x direction in the slab, B_z is the magnetic field in the z direction, and E_y is the resulting Hall field in the y direction. Show that the Hall coefficient is $1/nq$, where q is the charge of the charge carriers, -1.6×10^{-19} C if they are electrons. (The Hall coefficients of monovalent metals, such as copper, silver, and sodium, are therefore negative.)

71 •• Aluminum has a density of 2.7×10^3 kg/m^3 and a molar mass of 27 g/mol. The Hall coefficient of aluminum is $R = -0.3 \times 10^{-10}$ m^3/C. (See Problem 70 for the definition of R.) Find the number of conduction electrons per aluminum atom.

72 •• Magnesium is a divalent metal. Its density is 1.74×10^3 kg/m^3 and its molar mass is 24.3 g/mol. Assuming that each magnesium atom contributes two conduction electrons, what should be the Hall coefficient of magnesium? How does your result compare to the measured value of -0.94×10^{-10} m^3/C?

General Problems

73 • True or false:

(*a*) The magnetic force on a moving charged particle is always perpendicular to the velocity of the particle.
(*b*) The torque on a magnet tends to align the magnetic moment in the direction of the magnetic field.
(*c*) A current loop in a uniform magnetic field behaves like a small magnet.
(*d*) The period of a particle moving in a circle in a magnetic field is proportional to the radius of the circle.
(*e*) The drift velocity of electrons in a wire can be determined from the Hall effect.

74 • Show that the force on a current element is the same in direction and magnitude regardless of whether positive charges, negative charges, or a mixture of positive and negative charges create the current.

75 • A proton with a charge $+e$ is moving with a speed v at 50° to the direction of a magnetic field $\vec{B}$. The component of the resulting force on the proton in the direction of $\vec{B}$ is

(*a*) $evB \sin 50° \cos 50°$.
(*b*) $evB \cos 50°$.
(*c*) zero.
(*d*) $evB \sin 50°$.
(*e*) none of these.

76 • If the magnetic field vector is directed toward the north and a positively charged particle is moving toward the east, what is the direction of the magnetic force on the particle?

77 • A positively charged particle is moving northward in a magnetic field. The magnetic force on the particle is toward the northeast. What is the direction of the magnetic field?

(*a*) Up
(*b*) West
(*c*) South
(*d*) Down
(*e*) This situation cannot exist.

78 • A ^{7}Li nucleus with a charge of $+3e$ and a mass of 7 u (1 u = 1.66×10^{-27} kg) and a proton with charge $+e$ and mass 1 u are both moving in a plane perpendicular to a magnetic field $\vec{B}$. The two particles have the same momentum. The ratio of the radius of curvature of the path of the proton R_p to that of the ^{7}Li nucleus, R_{Li} is

(*a*) $R_p/R_{Li} = 3$.
(*b*) $R_p/R_{Li} = 1/3$.
(*c*) $R_p/R_{Li} = 1/7$.
(*d*) $R_p/R_{Li} = 3/7$.
(*e*) none of these.

79 • An electron moving with velocity v to the right enters a region of uniform magnetic field that points out of the paper. After the electron enters this region, it will be

(*a*) deflected out of the plane of the paper.
(*b*) deflected into the plane of the paper.
(*c*) deflected upward.
(*d*) deflected downward.
(*e*) undeviated in its motion.

80 • How are magnetic field lines similar to electric field lines? How are they different?

81 • A long wire parallel to the x axis carries a current of 6.5 A in the positive x direction. There is a uniform magnetic field $\vec{B} = 1.35\text{ T}\,\hat{j}$. Find the force per unit length on the wire.

82 • An alpha particle (charge $+2e$) travels in a circular path of radius 0.5 m in a magnetic field of 1.0 T. Find (*a*) the period, (*b*) the speed, and (*c*) the kinetic energy (in electron volts) of the alpha particle. Take $m = 6.65 \times 10^{-27}$ kg for the mass of the alpha particle.

83 • If a current I in a given wire and a magnetic field $\vec{B}$ are known, the force $\vec{F}$ on the current is uniquely determined. Show that knowing $\vec{F}$ and I does not provide complete knowledge of $\vec{B}$.

84 •• The pole strength q_m of a bar magnet is defined by $q_m = |\vec{\mu}|/L$, where L is the length of the magnet. Show that the torque exerted on a bar magnet in a uniform magnetic field $\vec{B}$ is the same as if a force $+q_m\vec{B}$ is exerted on the north pole and a force $-q_m\vec{B}$ is exerted on the south pole.

85 •• A particle of mass m and charge q enters a region where there is a uniform magnetic field B along the x axis. The initial velocity of the particle is $\vec{v} = v_{0x}\hat{i} + v_{0y}\hat{j}$ so the particle moves in a helix. (*a*) Show that the radius of the helix is $r = mv_{0y}/qB$. (*b*) Show that the particle takes a time $t = 2\pi m/qB$ to make one orbit around the helix.

86 •• A metal crossbar of mass M rides on a pair of long, horizontal conducting rails separated by a distance ℓ and connected to a device that supplies constant current I to the circuit, as shown in Figure 28-40. A uniform magnetic field B is established as shown. (*a*) If there is no friction and the bar starts from rest at $t = 0$, show that at time t the bar has velocity $v = (BI\ell/M)t$. (*b*) In which direction will the bar move? (*c*) If the coefficient of static friction is μ_s, find the minimum field B necessary to start the bar moving.

Figure 28-40 Problems 86 and 87

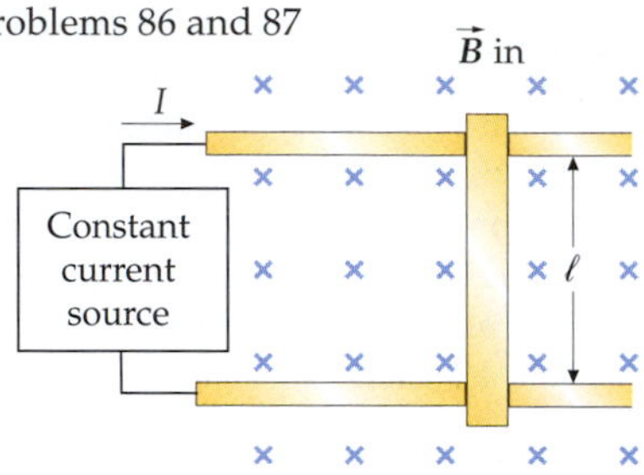

87 •• Assume that the rails in Figure 28-40 are frictionless but tilted upward so that they make an angle θ with the horizontal. (*a*) What vertical magnetic field B is needed to keep the bar from sliding down the rails? (*b*) What is the acceleration of the bar if B has twice the value found in part (*a*)?

88 •• A long, narrow bar magnet that has magnetic moment $\vec{\mu}$ parallel to its long axis is suspended at its center as a frictionless compass needle. When placed in a magnetic field $\vec{B}$, the needle lines up with the field. If it is displaced by a small angle θ, show that the needle will oscillate about its equilibrium position with frequency $f = \frac{1}{2\pi}\sqrt{\mu B/I}$, where I is the moment of inertia about the point of suspension.

89 •• A conducting wire is parallel to the y axis. It moves in the positive x direction with a speed of 20 m/s in a magnetic field $\vec{B} = 0.5\text{ T}\,\hat{k}$. (*a*) What are the magnitude and direction of the magnetic force on an electron in the conductor? (*b*) Because of this magnetic force, electrons move to one end of the wire leaving the other end positively charged, until the electric field due to this charge separation exerts a force on the electrons that balances the magnetic force. Find the magnitude and direction of this electric field in the steady state. (*c*) Suppose the moving wire is 2 m long. What is the potential difference between its two ends due to this electric field?

90 ••• The rectangular frame in Figure 28-41 is free to rotate about the axis A–A on the horizontal shaft. The frame is 10 cm long and 6 cm wide and the rods that make up the frame have a mass per unit length of 20 g/cm. A uniform magnetic field $B = 0.2$ T is directed as shown. A current may be sent around the frame by means of the wires attached at the top. (*a*) If no current passes through the frame, what is the period of this physical pendulum for small oscillations? (*b*) If a current of 8.0 A passes through the frame in the direction indicated by the arrow, what is then the period of this physical pendulum? (*c*) Suppose the direction of the current is opposite to that shown. The frame is displaced from the vertical by some angle θ. What must be the magnitude of the current so that this frame will be in equilibrium?

Figure 28-41 Problem 90

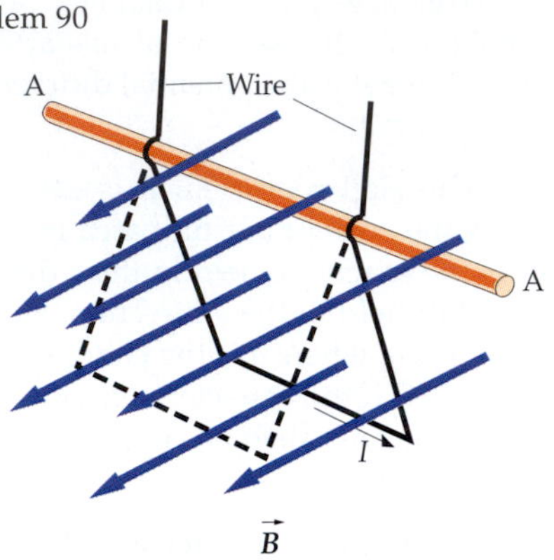

91 ••• A stiff, straight, horizontal wire of length 25 cm and mass 20 g is supported by electrical contacts at its ends, but is otherwise free to move vertically upward. The wire is in a uniform, horizontal magnetic field of magnitude 0.4 T perpendicular to the wire. A switch connecting the wire to a battery is closed and the wire is shot upward, rising to a maximum height h. The battery delivers a total charge of 2 C during the short time it makes contact with the wire. Find the height h.

92 ••• A solid sphere of radius R carries a charge density $-\rho_0$ in the region $r = 0$ to $r = R_s$ and an equal charge density of opposite sign, $+\rho_0$, between $r = R_s$ and $r = R$. The net charge carried by the sphere is zero. (*a*) What must be the ratio R/R_s? (*b*) If this sphere rotates with angular velocity ω about its diameter, what is its magnetic moment?

93 ••• A circular loop of wire with mass M carries a current I in a uniform magnetic field. It is initially in equilibrium with its magnetic moment vector aligned with the magnetic field. The loop is given a small twist about a diameter and then released. What is the period of the motion? (Assume that the only torque exerted on the loop is due to the magnetic field.)

94 ••• A small bar magnet has a magnetic moment $\vec{\mu}$ that makes an angle θ with the x axis and lies in a nonuniform magnetic field given by $\vec{B} = B_x(x)\hat{i} + B_y(y)\hat{j}$. Use $F_x = -dU/dx$ and $F_y = -dU/dy$ to show that there is a net force on the magnet that is given by

$$\vec{F} \approx \mu_x \frac{\partial B_x}{\partial x}\hat{i} + \mu_y \frac{\partial B_y}{\partial_y}\hat{j}$$

CHAPTER 29

Sources of the Magnetic Field

These coils at the Kettering Magnetics Laboratory at Oakland University are called Helmholtz coils. They are used to cancel the earth's magnetic field and to provide a uniform magnetic field in a small region of space for studying the magnetic properties of matter.

The earliest known sources of magnetism were permanent magnets. One month after Oersted announced his discovery that a compass needle is deflected by an electric current, Jean Baptiste Biot and Felix Savart announced the results of their measurements of the force on a magnet near a long, current-carrying wire and analyzed these results in terms of the magnetic field produced by each element of the current. André-Marie Ampère extended these experiments and showed that current elements also experience a force in the presence of a magnetic field and that two currents exert forces on each other.

We begin by considering the magnetic field produced by a single moving charge and by the moving charges in a current element. We then calculate the magnetic fields produced by some common current configurations, such as a straight wire segment, a long, straight wire, a current loop, and a solenoid.

Next we discuss Ampère's law, which relates the line integral of the magnetic field around a closed loop to the total current that passes through the loop. Finally, we consider the magnetic properties of matter.

29-1 The Magnetic Field of Moving Point Charges

When a point charge q moves with velocity $\vec{v}$, it produces a magnetic field $\vec{B}$ in space given by

$$\vec{B} = \frac{\mu_0}{4\pi}\frac{q\vec{v} \times \hat{r}}{r^2} \qquad \text{29-1}$$

Magnetic field of a moving charge

where $\hat{r}$ is a unit vector that points from the charge q to the field point P (Figure 29-1) and μ_0 is a constant of proportionality called the **permeability of free space***, which has the value

$$\mu_0 = 4\pi \times 10^{-7}\,\mathrm{T{\cdot}m/A} = 4\pi \times 10^{-7}\,\mathrm{N/A^2} \qquad \text{29-2}$$

The units of μ_0 are such that B is in teslas when q is in coulombs, v is in meters per second, and r is in meters. The unit $\mathrm{N/A^2}$ comes from the fact that $1\,\mathrm{T} = 1\,\mathrm{N/A{\cdot}m}$. The constant $1/4\pi$ is arbitrarily included in Equation 29-1 so that the factor 4π will not appear in Ampere's law (Equation 29-15), which we will study in Section 29-3.

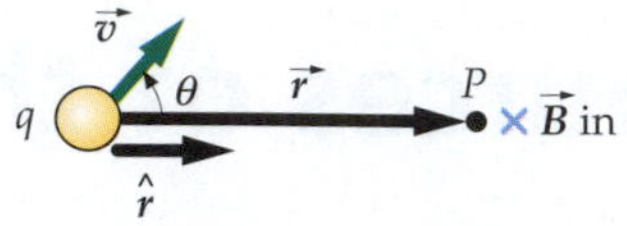

Figure 29-1 A point charge q moving with velocity $\vec{v}$ produces a magnetic field $\vec{B}$ at a field point P that is in the direction $\vec{v} \times \hat{r}$, where $\hat{r}$ is the unit vector pointing from the charge to the field point. The field varies inversely as the square of the distance from the charge to the field point and is proportional to the sine of the angle between $\vec{v}$ and $\hat{r}$. (The blue x at the field point indicates that the direction of the field is into the page.)

* Some care must be taken not to confuse the constant μ_0 with the magnetic moment μ.

Example 29-1

A point charge of magnitude $q = 4.5$ nC is moving with speed $v = 3.6 \times 10^7$ m/s parallel to the x axis along the line $y = 3$ m. Find the magnetic field at the origin produced by this charge when the charge is at the point $x = -4$ m, $y = 3$ m, as shown in Figure 29-2.

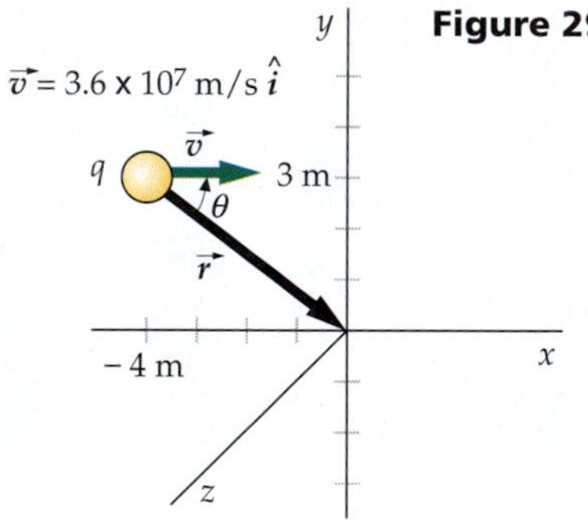

Figure 29-2

1. The magnetic field is given by Equation 29-1: $\vec{B} = \dfrac{\mu_0}{4\pi}\dfrac{q\vec{v} \times \hat{r}}{r^2}$, with $\vec{v} = v\hat{i}$

2. Find $\vec{r}$ and r from Figure 29-2, and write $\hat{r}$ in terms of $\hat{i}$ and $\hat{j}$:
$\vec{r} = 4\,\mathrm{m}\,\hat{i} - 3\,\mathrm{m}\,\hat{j}$
$r = \sqrt{4^2 + 3^2}\,\mathrm{m} = 5\,\mathrm{m}$
$\hat{r} = \dfrac{\vec{r}}{r} = \dfrac{4\,\mathrm{m}\,\hat{i} - 3\,\mathrm{m}\,\hat{j}}{5\,\mathrm{m}} = 0.8i - 0.6\hat{j}$

3. Evaluate the cross product: $\vec{v} \times \hat{r} = (v\hat{i}) \times (0.8\hat{i} - 0.6\hat{j}) = -0.6v\hat{k}$

4. Substitute the above results in Equation 29-1 to obtain $\vec{B}$:

$$\vec{B} = \frac{\mu_0}{4\pi}\frac{q\vec{v} \times \hat{r}}{r^2} = \frac{\mu_0}{4\pi}\frac{q(-0.6v\hat{k})}{r^2}$$

$$= -(10^{-7}\,\mathrm{T\cdot m/A})\frac{(4.5 \times 10^{-9}\,\mathrm{C})(0.6)(3.6 \times 10^{7}\,\mathrm{m/s})}{(5\,\mathrm{m})^2}\hat{k}$$

$$= -3.89 \times 10^{-10}\,\mathrm{T}\,\hat{k}$$

Remarks: It is also possible to obtain $\vec{B}$ without finding an explicit expression for the unit vector $\hat{r}$. From Figure 29-2 we note that $\vec{v} \times \hat{r}$ is in the negative z direction. In addition, the magnitude of $\vec{v} \times \hat{r}$ is $v \sin\theta$, where $\sin\theta = 3\,\mathrm{m}/5\,\mathrm{m} = 0.6$. Combining these results, we have $\vec{v} \times \hat{r} = v\sin\theta\,(-\hat{k}) = -v(0.6)\hat{k}$, in agreement with our result in step 2. Finally, this example shows that the magnetic field due to a moving charge is quite small. For comparison, the earth's magnetic field near its surface has a magnitude of about 10^{-4} T.

Exercise Find the magnetic field on the y axis at $y = 3$ m and at $y = 6$ m. (*Answers* $\vec{B} = 0$, $\vec{B} = 3.89 \times 10^{-10}\,\mathrm{T}\,\hat{k}$)

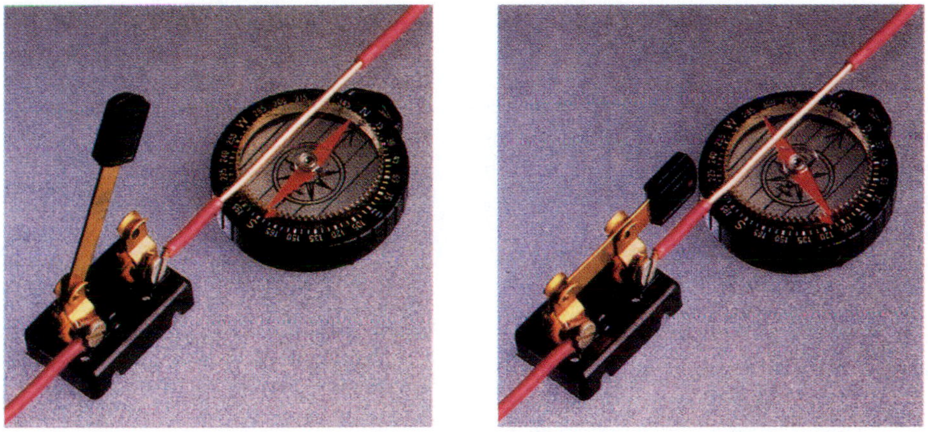

(*a*) (*b*)

Oersted's experiment. (*a*) With no current in the wire, the compass needle points north. (*b*) When the wire carries a current, the needle is deflected in the direction of the resultant magnetic field. The current in the wire is directed upward from left to right. The insulation has been stripped from the wire to improve the contrast of the photograph.

29-2 The Magnetic Field of Currents: The Biot–Savart Law

In the previous chapter we extended our discussion of forces on point charges to forces on current elements by replacing $q\vec{v}$ with the current element $I\,d\vec{\ell}$. We do the same for the magnetic field produced by a current element. The magnetic field $d\vec{B}$ produced by a current element $I\,d\vec{\ell}$ is given by Equation 29-1 with $q\vec{v}$ replaced by $I\,d\vec{\ell}$:

$$d\vec{B} = \frac{\mu_0}{4\pi}\frac{I\,d\vec{\ell} \times \hat{r}}{r^2} \qquad 29\text{-}3$$

Biot–Savart law

Equation 29-3, known as the **Biot–Savart law,** was also deduced by Ampère. The Biot–Savart law and Equation 29-1 are analogous to Coulomb's law for the electric field of a point charge. The source of the magnetic field is a moving charge $q\vec{v}$ or a current element $I\,d\vec{\ell}$, just as the charge q is the source of the electrostatic field. The magnetic field decreases with the square of the distance from the moving charge or current element, just as the electric field decreases with the square of the distance from a point charge. However, the directional aspects of the electric and magnetic fields are quite different. Whereas the electric field points in the radial direction $\hat{r}$ from the point charge to the field point (for a positive charge), the magnetic field is perpendicular both to $\hat{r}$ and to the direction of motion $\vec{v}$ of the charges, which is along the direction of the current element. At a point along the line of a current element, such as point P_2 in Figure 29-3, the magnetic field due to that element is zero, because $I\,d\vec{\ell}$ is parallel to the vector $\hat{r}$.

The magnetic field due to the total current in a circuit can be calculated by using the Biot–Savart law to find the field due to each current element and then summing (integrating) over all the current elements in the circuit. This calculation is difficult for all but the simplest circuit geometries.

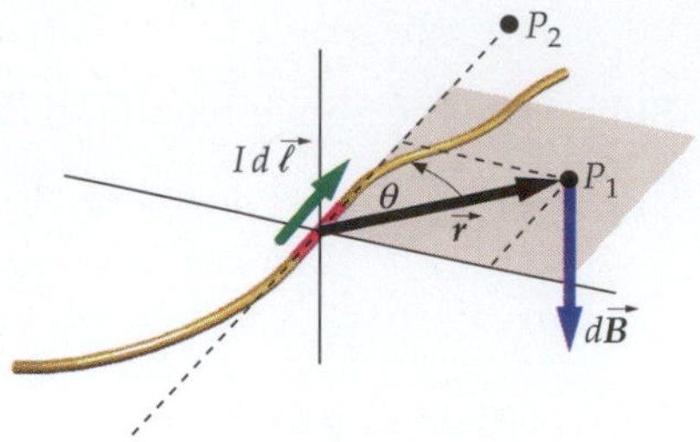

Figure 29-3 The current element $I\,d\vec{\ell}$ produces a magnetic field at point P_1 that is perpendicular to both $I\,d\vec{\ell}$ and $\hat{r}$. It produces no magnetic field at point P_2, which is along the line of $I\,d\vec{\ell}$.

$\vec{B}$ due to a Current Loop

Figure 29-4 shows a current element $I\,d\vec{\ell}$ of a current loop of radius R and the unit vector $\hat{r}$ that is directed from the element to the center of the loop. The magnetic field at the center of the loop due to this element is directed along the axis of the loop, and its magnitude is given by

$$dB = \frac{\mu_0}{4\pi}\frac{I\,d\ell \sin\theta}{R^2}$$

where θ is the angle between $I\,d\vec{\ell}$ and $\hat{r}$, which is 90° for each current element, so $\sin\theta = 1$. The magnetic field due to the entire current is found by integrating over all the current elements in the loop. Since R is the same for all elements, we obtain

$$B = \int dB = \frac{\mu_0}{4\pi}\frac{I}{R^2}\oint d\ell$$

The integral of $d\ell$ around the complete loop gives the total length $2\pi R$, the circumference of the loop. The magnetic field due to the entire loop is thus

$$B = \frac{\mu_0}{4\pi}\frac{I2\pi R}{R^2} = \frac{\mu_0 I}{2R} \qquad 29\text{-}4$$

B at the center of a current loop

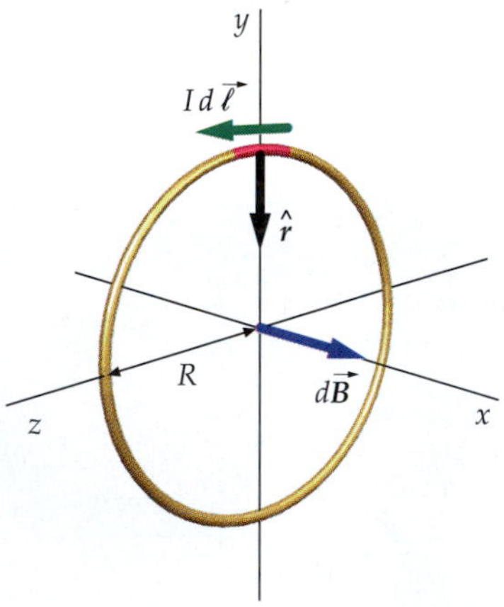

Figure 29-4 Current element for calculating the magnetic field at the center of a circular current loop. Each element produces a magnetic field that is directed along the axis of the loop.

Exercise Find the current in a circular loop of radius 8 cm that will give a magnetic field of 2 G at the center of the loop. (*Answer* 25.5 A)

Figure 29-5 shows the geometry for calculating the magnetic field at a point on the axis of a circular current loop a distance x from its center. We first consider the current element at the top of the loop. Here, as everywhere around the loop, $I\,d\vec{\ell}$ is tangent to the loop and perpendicular to the vector $\hat{r}$ from the current element to the field point P. The magnetic field $d\vec{B}$ due to this element is in the direction shown in the figure, perpendicular to $\vec{r}$ and also perpendicular to $I\,d\vec{\ell}$. The magnitude of $d\vec{B}$ is

$$|d\vec{B}| = \frac{\mu_0}{4\pi}\frac{I|d\vec{\ell} \times \hat{r}|}{r^2} = \frac{\mu_0}{4\pi}\frac{I\,d\ell}{x^2 + R^2}$$

where we have used the facts that $r^2 = x^2 + R^2$ and that $d\vec{\ell}$ and $\hat{r}$ are perpendicular, so $|d\vec{\ell} \times \hat{r}| = d\ell$.

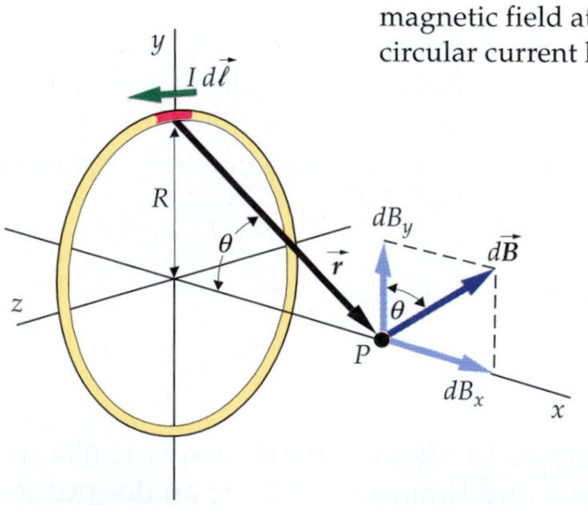

Figure 29-5 Geometry for calculating the magnetic field at a point on the axis of a circular current loop.

When we sum around all the current elements in the loop, the components of $d\vec{B}$ perpendicular to the axis of the loop, such as dB_y in Figure 29-5, sum to zero, leaving only the components dB_x that are parallel to the axis. We thus compute only the x component of the field. From Figure 29-5, we have

$$dB_x = dB \sin\theta = dB\left(\frac{R}{\sqrt{x^2+R^2}}\right) = \frac{\mu_0}{4\pi}\frac{I\,d\ell}{x^2+R^2}\frac{R}{\sqrt{x^2+R^2}}$$

To find the field due to the entire loop of current, we integrate dB_x around the loop:

$$B_x = \oint dB_x = \oint \frac{\mu_0}{4\pi}\frac{IR}{(x^2+R^2)^{3/2}}\,d\ell$$

Since neither x nor R varies as we sum over the elements in the loop, we can remove these quantities from the integral. Then,

$$B_x = \frac{\mu_0 IR}{4\pi(x^2+R^2)^{3/2}}\oint d\ell$$

The integral of $d\ell$ around the loop gives $2\pi R$. Thus,

$$B_x = \frac{\mu_0}{4\pi}\frac{IR(2\pi R)}{(x^2+R^2)^{3/2}} = \frac{\mu_0}{4\pi}\frac{2\pi R^2 I}{(x^2+R^2)^{3/2}} \qquad \text{29-5}$$

B on the axis of a current loop

Exercise Show that Equation 29-5 reduces to $B_x = \frac{\mu_0 I}{2R}$ (Equation 29-4) at the center of the loop.

At great distances from the loop, x is much greater than R, so $(x^2 + R^2)^{3/2} \approx (x^2)^{3/2} = |x|^3$. Then,

$$B_x = \frac{\mu_0}{4\pi}\frac{2I\pi R^2}{|x^3|}$$

or

$$B_x = \frac{\mu_0}{4\pi}\frac{2\mu}{|x^3|} \qquad \text{29-6}$$

Magnetic-dipole field on the axis of the dipole

where $\mu = I\pi R^2$ is the magnitude of the magnetic moment of the loop. Note the similarity of this expression and the electric field on the axis of an electric dipole of moment p (Equation 22-10):

$$E_x = \frac{1}{4\pi\epsilon_0}\frac{2p}{|x^3|}$$

Although it has not been demonstrated, our result that a current loop produces a magnetic dipole field far away holds in general for any point whether it is on or off of the axis of the loop. Thus, a current loop behaves as a magnetic dipole both in that it experiences a torque $\vec{\mu} \times \vec{B}$ when placed in an external magnetic field (as was shown in Chapter 28) and in that it produces a magnetic dipole field at a great distance from it. Figure 29-6 shows the magnetic field lines for a current loop.

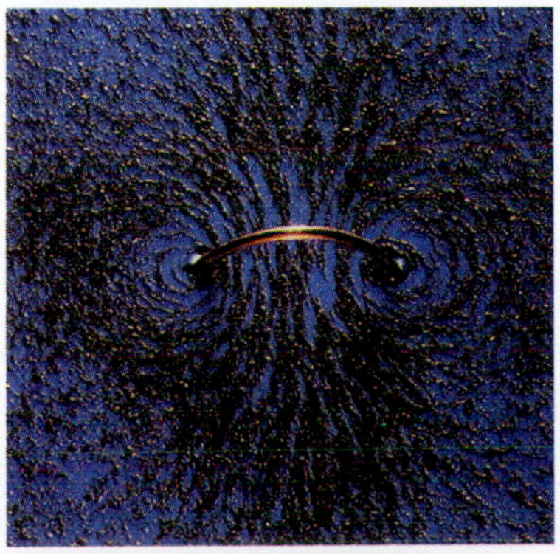

Figure 29-6 The magnetic field lines of a circular current loop indicated by iron filings.

Example 29-2

A circular loop of radius 5.0 cm has 12 turns and lies in the yz plane. It carries a current of 4 A in the direction such that the magnetic moment of the loop is along the x axis. Find the magnetic field on the x axis at (*a*) $x = 0$, (*b*) $x = 15$ cm, and (*c*) $x = 3$ m.

Picture the Problem The magnetic field due to a loop with N turns is N times that due to a single turn. (*a*) At $x = 0$ (center of the loops) $B = \mu_0 NI/2R$ (from Equation 29-4). Equation 29-5 gives the magnetic field on axis due to the current in a single turn. Far from the loop, as in part (*c*), the field can be found using Equation 29-6. In this case, since we have N loops, the magnetic moment is $\mu = NI\pi R^2$.

(*a*) B_x at the center is N times that given by Equation 29-4 for a single loop:

$$B_x = \frac{\mu_0 NI}{2R} = \frac{\mu_0}{4\pi}\frac{2\pi NI}{R} = (10^{-7}\,\mathrm{T\cdot m/A})\frac{2\pi(12)(4\,\mathrm{A})}{0.05\,\mathrm{m}}$$

$$= 6.03 \times 10^{-4}\,\mathrm{T}$$

(*b*) B_x on the axis is N times that given by Equation 29-5:

$$B_x = \frac{\mu_0}{4\pi}\frac{2\pi R^2 NI}{(x^2 + R^2)^{3/2}}$$

$$= (10^{-7}\,\mathrm{T\cdot m/A})\frac{2\pi(0.05\,\mathrm{m})^2(12)(4\,\mathrm{A})}{[(0.15\,\mathrm{m})^2 + (0.05\,\mathrm{m})^2]^{3/2}}$$

$$= 1.91 \times 10^{-5}\,\mathrm{T}$$

(*c*) 1. Since 3 m is much greater than the radius $R = 0.05$ m, we can use Equation 29-6 for the magnetic field far from the loop:

$$B_x = \frac{\mu_0}{4\pi}\frac{2\mu}{x^3}$$

2. The magnitude of the magnetic moment of the loop is NIA:

$$\mu = NI\pi R^2 = (12)(4\,\mathrm{A})\pi(0.05\,\mathrm{m})^2 = 0.377\,\mathrm{A\cdot m^2}$$

3. Substitute μ and $x = 3$ m into B_x in step 1:

$$B_x = \frac{\mu_0}{4\pi}\frac{2\mu}{x^3} = (10^{-7}\,\mathrm{T\cdot m/A})\frac{2(0.377\,\mathrm{A\cdot m^2})}{(3\,\mathrm{m})^3}$$

$$= 2.79 \times 10^{-9}\,\mathrm{T}$$

Remarks: Note in step 1 that the field produced by a current loop is typically much larger than the field due to a single moving charge (see Example 29-1). Since $x = 60R$ in (*c*), we used an approximation that is valid for $x \gg R$.

Exercise Find the magnetic field on the x axis at $x = -15$ cm. (*Answer* From Equation 29-5 we see that B_x is symmetric in x, thus $B_x = 1.91 \times 10^{-5}$ T.)

Example 29-3 *try it yourself*

A small bar magnet of magnetic moment $\mu = 0.03\,\mathrm{A\cdot m^2}$ is placed at the center of the loop of Example 29-2 so that its magnetic moment lies in the xy plane and makes an angle of 30° with the x axis. Neglecting any variation in B over the region of the magnet, find the torque on the magnet.

Picture the Problem The torque on a magnetic moment is given by $\vec{\tau} = \vec{\mu} \times \vec{B}$. Since $\vec{B}$ is in the x direction, you can see from Figure 29-7 that $\vec{\mu} \times \vec{B}$ is in the negative z direction.

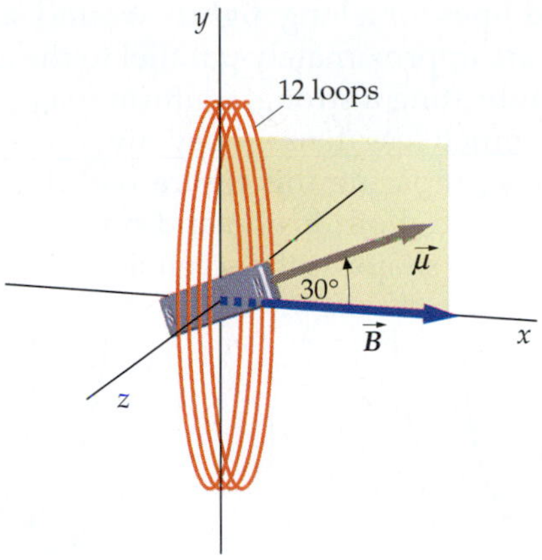

Figure 29-7

Cover the column to the right and try these on your own before looking at the answers.

Steps	***Answers***
1. Compute the magnitude of the torque from $\vec{\tau} = \vec{\mu} \times \vec{B}$.	$\tau = 9.04 \times 10^{-6}\,\mathrm{N\cdot m}$
2. Indicate the direction with a unit vector.	$\vec{\tau} = -(9.04 \times 10^{-6}\,\mathrm{N\cdot m})\hat{k}$

$\vec{B}$ due to a Current in a Solenoid

A **solenoid** is a wire tightly wound into a helix of closely spaced turns, as illustrated in Figure 29-8. It is used to produce a strong, uniform magnetic field in the region surrounded by its loops. Its role in magnetism is analogous to that of the parallel-plate capacitor, which produces a strong, uniform electric field between its plates. The magnetic field of a solenoid is essentially that of a set of N identical current loops placed side by side. Figure 29-9 shows the magnetic field lines for two such loops.

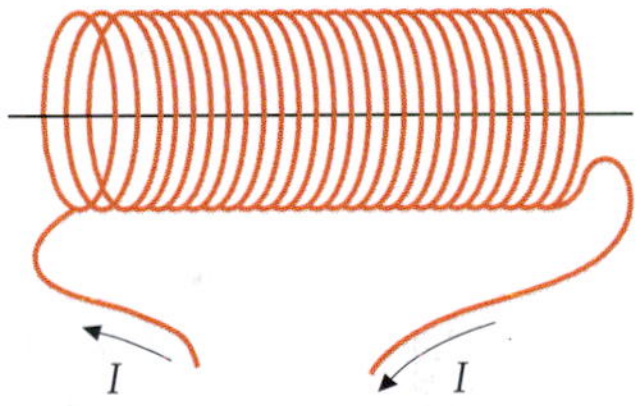

Figure 29-8 A tightly wound solenoid can be considered a set of circular current loops placed side by side that carry the same current. It produces a uniform magnetic field inside the loops.

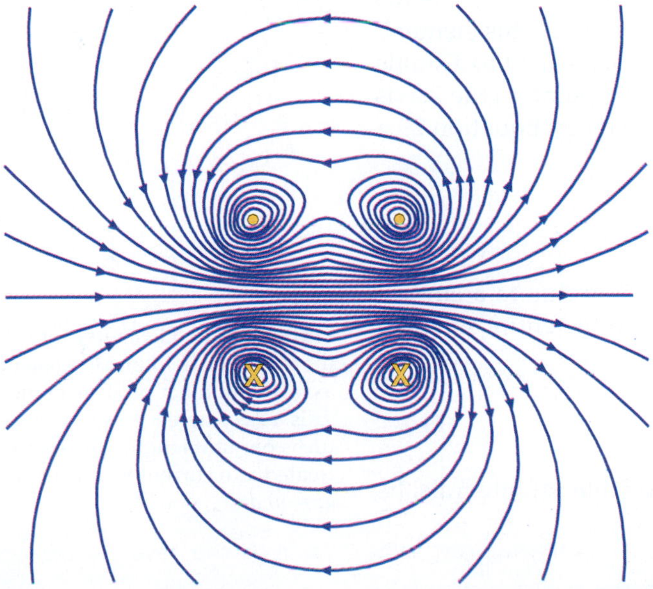

Figure 29-9 Magnetic field lines due to two loops carrying the same current in the same sense. The points where the loops intersect the plane of the page are marked by an X where the current enters and a dot where the current emerges. In the region between the loops, the magnetic fields of the individual loops add so the resultant field is strong. In the regions away from the loops, the resultant field is weak.

Figure 29-10 shows the magnetic field lines for a long, tightly wound solenoid. Inside the solenoid, the field lines are approximately parallel to the axis and are closely and uniformly spaced, indicating a strong, uniform magnetic field. Outside the solenoid, the lines are much less dense. They diverge from one end and converge at the other end. Comparing this figure with Figure 28-8, we see that the field lines of a solenoid, both inside and outside, are identical to those of a bar magnet of the same shape as the solenoid.

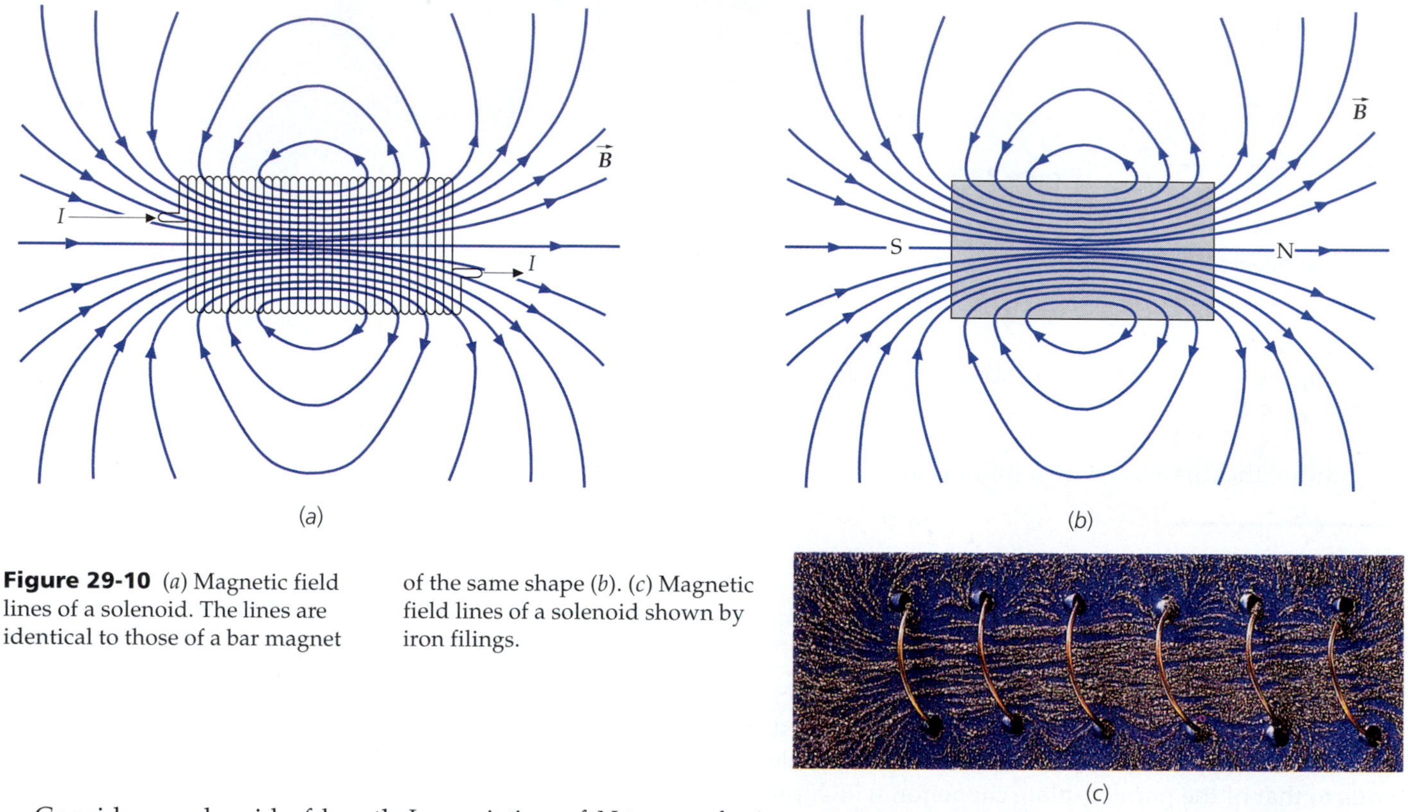

Figure 29-10 (*a*) Magnetic field lines of a solenoid. The lines are identical to those of a bar magnet of the same shape (*b*). (*c*) Magnetic field lines of a solenoid shown by iron filings.

Consider a solenoid of length L consisting of N turns of wire carrying a current I. We choose the axis of the solenoid to be along the x axis, with the left end at $x = -a$ and the right end at $x = +b$ as shown in Figure 29-11. We will calculate the magnetic field at the origin. The figure shows an element of the solenoid of length dx at a distance x from the origin. If $n = N/L$ is the number of turns per unit length, there are $n\,dx$ turns of wire in this element, with each turn carrying a current I. The element is thus equivalent to a single loop carrying a current $di = nI\,dx$. The magnetic field at a point on the x axis due to a loop at the origin carrying a current $nI\,dx$ is given by Equation 29-5 with I replaced by $nI\,dx$:

$$dB_x = \frac{\mu_0}{4\pi}\frac{2\pi R^2 nI\,dx}{(x^2 + R^2)^{3/2}}$$

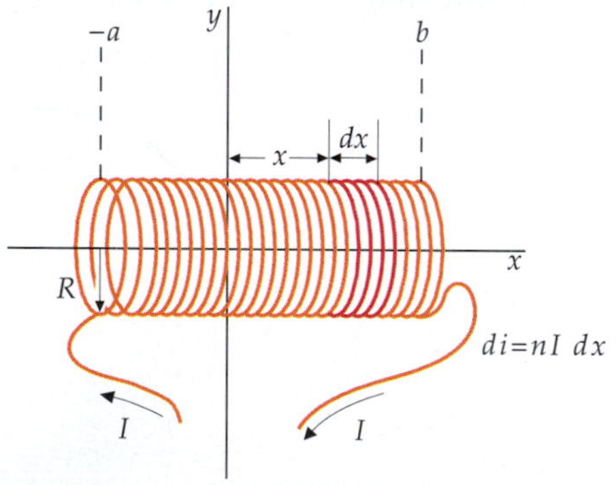

Figure 29-11 Geometry for calculating the magnetic field inside a solenoid on its axis. The number of turns in the element dx is $n\,dx$, where $n = N/\ell$ is the number of turns per unit length. The element dx is treated as a current loop carrying a current $di = nI\,dx$.

This expression also gives the magnetic field at the origin due to a current loop at x. We find the magnetic field at the origin due to the entire solenoid by integrating this expression from $x = -a$ to $x = b$:

$$B_x = \frac{\mu_0}{4\pi} 2\pi R^2 nI \int_{-a}^{b} \frac{dx}{(x^2 + R^2)^{3/2}} \qquad \text{29-7}$$

The integral in Equation 29-7 can be found in standard tables of integrals. Its value is

$$\int_{-a}^{b} \frac{dx}{(x^2 + R^2)^{3/2}} = \left.\frac{x}{R^2\sqrt{x^2 + R^2}}\right|_{-a}^{b} = \frac{b}{R^2\sqrt{b^2 + R^2}} + \frac{a}{R^2\sqrt{a^2 + R^2}}$$

Substituting this into Equation 29-7, we obtain

$$B_x = \frac{1}{2}\mu_0 nI\left(\frac{b}{\sqrt{b^2 + R^2}} + \frac{a}{\sqrt{a^2 + R^2}}\right) \tag{29-8}$$

For a long solenoid for which a and b are much larger than R, the two terms in the parentheses each tend toward 1. For this approximation, the magnetic field is

$$B_x = \mu_0 nI \tag{29-9}$$

B inside a long solenoid

If the origin is at one end of the solenoid, either a or b is zero. Then, if the other end is far away compared with the radius, one of the terms in the parentheses of Equation 29-9 is zero and the other is 1, so $B \approx \frac{1}{2}\mu_0 nI$. Thus, the magnitude of $\vec{B}$ at a point near either end of a long solenoid is about half that at points within the solenoid away from the ends. Figure 29-12 gives a plot of the magnetic field on the axis of a solenoid versus position (with the origin at the center of the solenoid). The approximation that the field is constant independent of the position along the axis is quite good except for very near the ends.

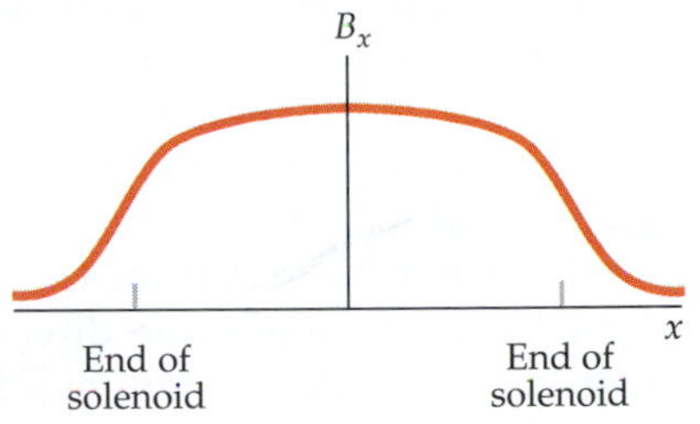

Figure 29-12 Graph of the magnetic field on the axis inside a solenoid versus the position x on the axis. The field inside the solenoid is nearly constant except near the ends.

Example 29-4

Find the magnetic field at the center of a solenoid of length 20 cm, radius 1.4 cm, and 600 turns that carries a current of 4 A.

1. We will calculate the field exactly using Equation 29-8.

$$B_x = \frac{1}{2}\mu_0 nI\left(\frac{b}{\sqrt{b^2 + R^2}} + \frac{a}{\sqrt{a^2 + R^2}}\right)$$

2. For a point at the center of the solenoid, $a = b = 10$ cm. Thus, each term in the parentheses in Equation 29-8 has the value:

$$\frac{a}{\sqrt{a^2 + R^2}} = \frac{b}{\sqrt{b^2 + R^2}} = \frac{10\text{ cm}}{\sqrt{(10\text{ cm})^2 + (1.4\text{ cm})^2}} = 0.990$$

3. Substitute these results into B_x in step 1:

$$B_x = \tfrac{1}{2}\mu_0 nI(0.990 + 0.990)$$
$$= (0.5)(4\pi \times 10^{-7}\text{ T}\cdot\text{m/A})(600\text{ turns}/0.2\text{ m})(4\text{ A})(0.990 + 0.990)$$
$$= 1.50 \times 10^{-2}\text{ T}$$

Remarks: Note that the approximation obtained using Equation 29-9 amounts to replacing 0.99 by 1.00, which differs by only 1%. Note also that the magnitude of the magnetic field inside this solenoid is fairly large—about 250 times the magnetic field of the earth.

Exercise: Calculate B_x using the long-solenoid approximation. (*Answer* 1.51×10^{-2} T)

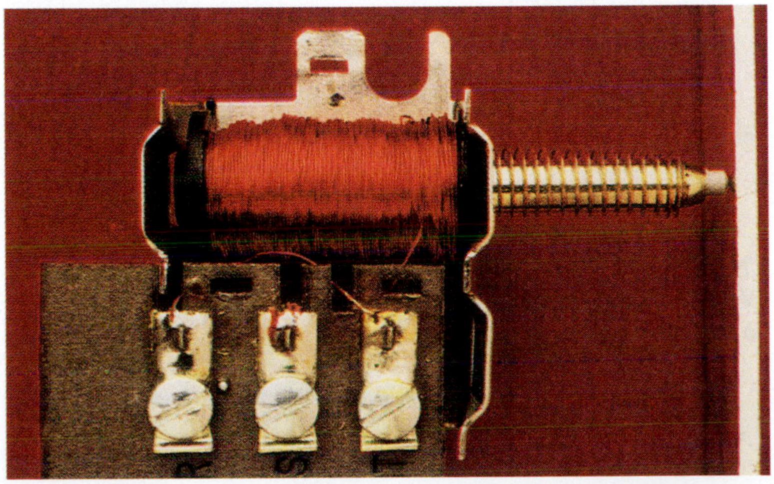

A cross section of a doorbell. When the outer solenoid is energized, its magnetic field causes the inner plunger to strike the bell.

$\vec{B}$ due to a Current in a Straight Wire

Figure 29-13 shows the geometry for calculating the magnetic field $\vec{B}$ at a point P due to the current in the straight wire segment shown. We choose the x axis to be along the wire and point P to be on the y axis. Because of the symmetry in this problem, any direction perpendicular to the wire could be chosen for the y axis.

A typical current element $I\,d\vec{\ell}$ at a distance x from the origin is shown. The vector $\vec{r}$ points from the element to the field point P. The direction of the magnetic field at P due to this element is the direction of $I\,d\vec{\ell} \times \vec{r}$, which is out of the paper. Note that the magnetic fields due to all the current elements of the wire are in this same direction. Thus, we need to compute only the magnitude of the field. The field due to the current element shown has the magnitude (Equation 29-3)

$$dB = \frac{\mu_0}{4\pi}\frac{I\,dx}{r^2}\sin\phi$$

It is more convenient to write this in terms of θ rather than ϕ:

$$dB = \frac{\mu_0}{4\pi}\frac{I\,dx}{r^2}\cos\theta \qquad \text{29-10}$$

To sum over all the current elements, we need to relate the variables θ, r, and x. It turns out to be easiest to express x and r in terms of θ. We have

$$x = y\tan\theta$$

Then,

$$dx = y\sec^2\theta\,d\theta = y\frac{r^2}{y^2}\,d\theta = \frac{r^2}{y}\,d\theta$$

where we have used $\sec\theta = r/y$. Substituting this expression for dx into Equation 29-10, we obtain

$$dB = \frac{\mu_0}{4\pi}\frac{I}{r^2}\frac{r^2\,d\theta}{y}\cos\theta = \frac{\mu_0}{4\pi}\frac{I}{y}\cos\theta\,d\theta$$

Let us first calculate the contribution from the current elements to the right of the point $x = 0$. We sum over these elements by integrating from $\theta = 0$ to $\theta = \theta_1$, where θ_1 is the angle between the line perpendicular to the wire and the line from P to the right end of the wire, as shown in Figure 29-13*b*. For this contribution, we have

$$B_1 = \int_0^{\theta_1}\frac{\mu_0}{4\pi}\frac{I}{y}\cos\theta\,d\theta$$

$$= \frac{\mu_0}{4\pi}\frac{I}{y}\int_0^{\theta_1}\cos\theta\,d\theta = \frac{\mu_0}{4\pi}\frac{I}{y}\sin\theta_1$$

Similarly, the contribution from elements to the left of $x = 0$ is

$$B_2 = \frac{\mu_0}{4\pi}\frac{I}{y}\sin\theta_2$$

The total magnetic field due to the wire segment is the sum of B_1 and B_2. Writing R instead of y for the perpendicular distance from the wire segment to the field point, we obtain

$$B = \frac{\mu_0}{4\pi}\frac{I}{R}(\sin\theta_1 + \sin\theta_2) \qquad \text{29-11}$$

B due to a straight wire segment

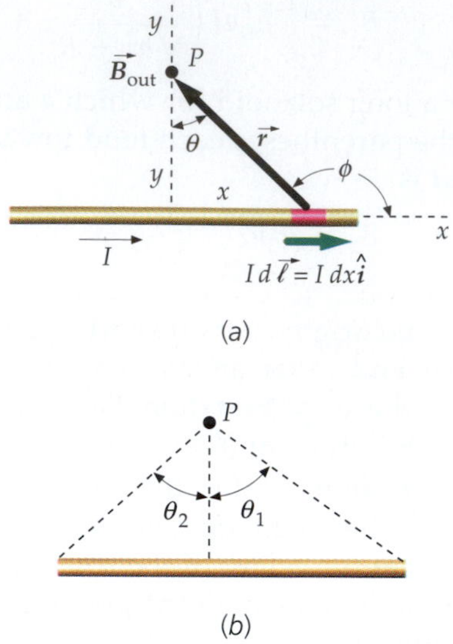

Figure 29-13 (*a*) Geometry for calculating the magnetic field at point P due to a straight current segment. Each element of the segment contributes to the total magnetic field at point P, which is directed out of the paper. (*b*) The result is expressed in terms of the angles θ_1 and θ_2.

This result gives the magnetic field due to any wire segment in terms of the perpendicular distance R and the angles subtended at the field point by the ends of the wire. If the wire is very long, these angles are nearly 90°. The result for a very long wire is obtained from Equation 29-11 by setting $\theta_1 = \theta_2 = 90°$:

$$B = \frac{\mu_0}{4\pi}\frac{2I}{R} \tag{29-12}$$

B due to a long, straight wire

At any point in space, the magnetic field lines of a long, straight, current-carrying wire are tangent to a circle of radius R about the wire, where R is the perpendicular distance from the wire to the field point. The direction of $\vec{B}$ can be determined by applying the right-hand rule as shown in Figure 29-14*a*. The magnetic field lines thus encircle the wire as shown in Figure 29-14*b*.

The result expressed by Equation 29-12 was found experimentally by Biot and Savart in 1820. From an analysis of it, they were able to discover the expression given in Equation 29-3 for the magnetic field due to a current element.

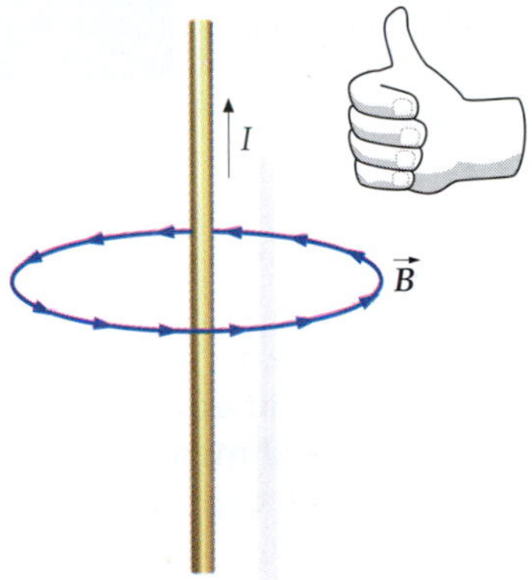

(*a*)

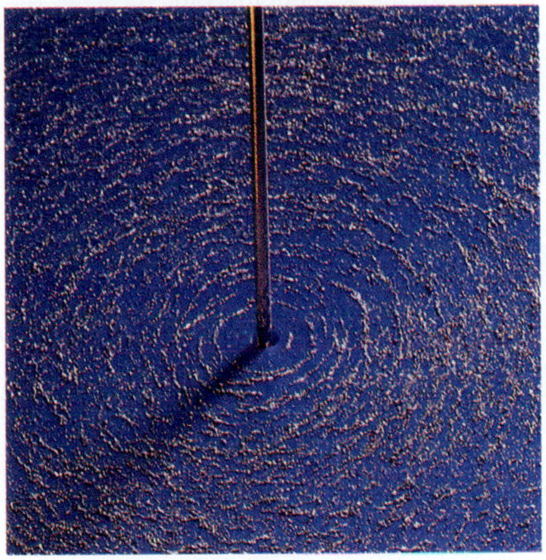

(*b*)

Figure 29-14 (*a*) Right-hand rule for determining the direction of the magnetic field due to a long, straight, current-carrying wire. The magnetic field lines encircle the wire in the direction of the fingers of the right hand when the thumb points in the direction of the current. (*b*) Magnetic field lines due to a long wire indicated by iron filings.

Example 29-5

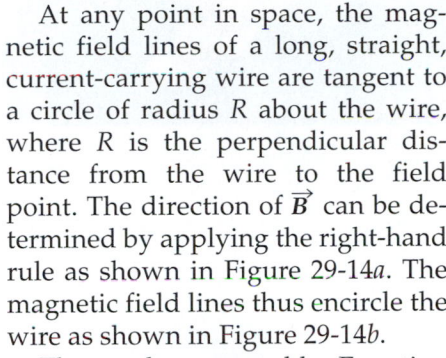

Find the magnetic field at the center of a square current loop of side L = 50 cm carrying a current of 1.5 A.

Picture the Problem The magnetic field at the center of the loop is the sum of the contributions from each of the four sides of the loop. From Figure 29-15 we can see that each side of the loop produces a field of equal magnitude pointing out of the page. Thus, we use Equation 29-11 for a given side, then multiply by 4 for the total field.

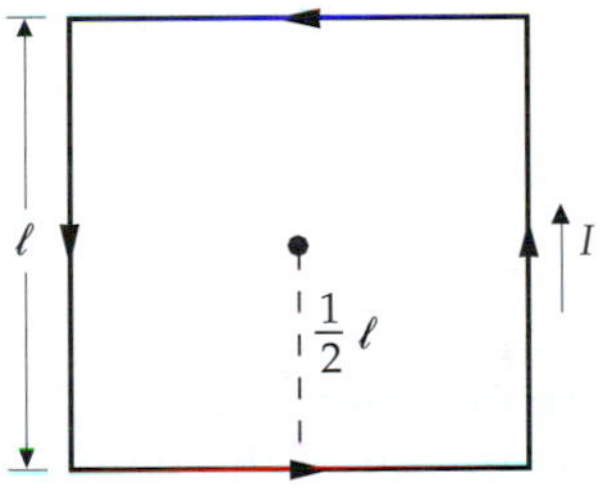

Figure 29-15

1. The total field is 4 times the field B_s due to a side: $B = 4B_s$

2. Calculate the magnetic field B_s due to a given side of the loop. Note from the figure that $R = \frac{1}{2}L$ and $\theta_1 = \theta_2 = 45°$:

$$B_s = \frac{\mu_0}{4\pi}\frac{I}{\frac{1}{2}L}(\sin 45° + \sin 45°)$$

$$= (10^{-7}\,\mathrm{T{\cdot}m/A})\frac{1.5\ \mathrm{A}}{0.25\ \mathrm{m}}(2\sin 45°) = 8.49 \times 10^{-7}\ \mathrm{T}$$

3. Multiply this value by 4 to find the total field: $B = 4B_s = 4(8.49 \times 10^{-7}\ \mathrm{T}) = 3.39 \times 10^{-6}\ \mathrm{T}$

Exercise Compare the magnetic field at the center of a circular current loop of radius R with that at the center of a square current loop of side $L = 2R$. Which is larger? (*Answer* B is larger for the circle by about 10%)

Exercise Find the magnetic field at a distance of 20 cm from a long, straight wire carrying a current of 5 A. (*Answer* $B = 5.00 \times 10^{-6}$ T)

We note from the exercise above that the magnetic field near a wire carrying a current of ordinary size is small. At 20 cm from a long, straight wire carrying 5 A, it is only about 10% of the magnetic field due to the earth.

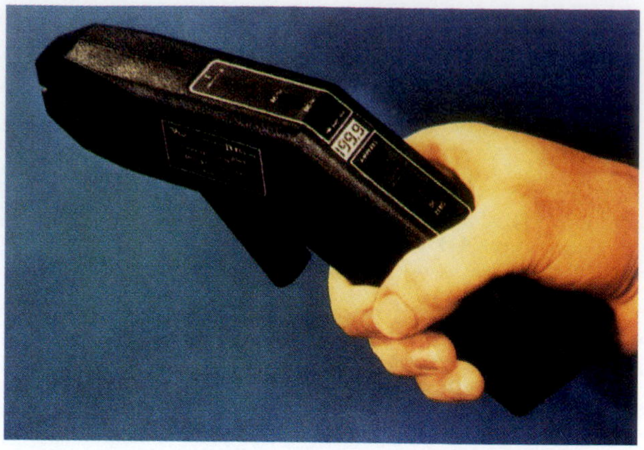

A current gun for measuring electric current. The jaws of the current gun clamp around a current-carrying wire without touching the wire. The magnetic field produced by the wire is measured with a Hall-effect device mounted in the current gun. The Hall-effect device puts out a voltage proportional to the magnetic field, which in turn is proportional to the current in the wire.

Example 29-6

A long, straight wire carrying a current of 1.7 A in the positive z direction lies along the line $x = -3$ cm, $y = 0$. A similar wire carrying a current of 1.7 A in the positive z direction lies along the line $x = +3$ cm, $y = 0$ as shown in Figure 29-16. Find the magnetic field at a point P on the y axis at $y = 6$ cm.

Figure 29-16

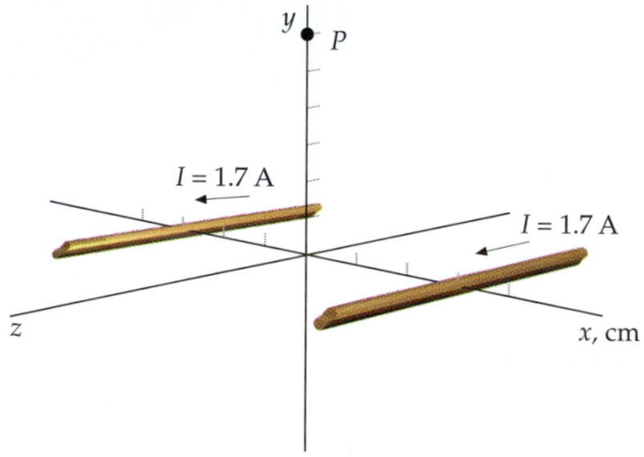

Picture the Problem The magnetic field at point P is the vector sum of the field $\vec{B}_L$ due to the wire on the left in Figure 29-17, and the field $\vec{B}_R$ due to the wire on the right. Since each wire carries the same current, and is the same distance from P, the magnitudes $\vec{B}_L$ and $\vec{B}_R$ are equal. $\vec{B}_L$ is perpendicular to the radius from the left wire to point P, and $\vec{B}_R$ is perpendicular to the radius from the right wire to the point P.

Figure 29-17

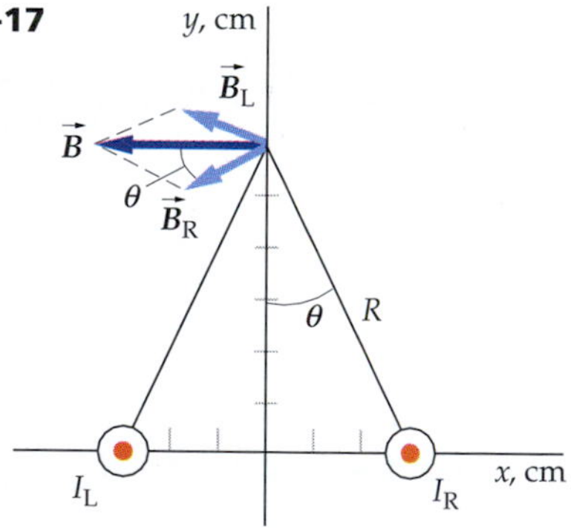

1. The field at P is the vector sum of the fields $\vec{B}_L$ and $\vec{B}_R$: $\quad \vec{B} = \vec{B}_L + \vec{B}_R$

2. From Figure 29-17 we see that the resultant magnetic field is in the negative x direction and has the magnitude $2B_L \cos\theta$: $\quad \vec{B} = -2B_L \cos\theta\,\hat{i}$

3. The magnitudes of $\vec{B}_L$ and $\vec{B}_R$ are given by Equation 29-12: $\quad B_L = B_R = \dfrac{\mu_0}{4\pi}\dfrac{2I}{R}$

4. R is the distance from each wire to the point P. We find R from the figure and substitute into the expression for B_L and B_R:

$$R = \sqrt{(3\text{ cm})^2 + (6\text{ cm})^2} = 6.71\text{ cm}$$

$$B_L = B_R = (10^{-7}\,\text{T}\cdot\text{m/A})\frac{2(1.7\text{ A})}{0.0671\text{ m}} = 5.07 \times 10^{-6}\text{ T}$$

5. We obtain $\cos\theta$ from the figure:

$$\cos\theta = \frac{6\text{ cm}}{R} = \frac{6\text{ cm}}{6.71\text{ cm}} = 0.894$$

6. Substitute the values of $\cos\theta$ and B_L into the equation in step 2 for $\vec{B}$:

$$\vec{B} = -2(5.07 \times 10^{-6}\text{ T})(0.894)\hat{i} = -9.07 \times 10^{-6}\text{ T}\,\hat{i}$$

Exercise Find $\vec{B}$ at the origin. (*Answer* 0)

Exercise Find $\vec{B}$ at the origin assuming that I_R goes *into* the page. (*Answer* $\vec{B} = 2.27 \times 10^{-5}\text{ T}\,\hat{j}$)

Definition of the Ampere

We can use Equation 29-12 for the magnetic field due to a long, straight, current-carrying wire and $d\vec{F} = I\,d\vec{\ell} \times \vec{B}$ (Equation 28-5) for the force exerted by a magnetic field on a segment of a current-carrying wire to find the force exerted by one long, straight current on another. Figure 29-18 shows two long, parallel wires carrying currents in the same direction. We consider the force on a segment $d\vec{\ell}_2$ carrying current I_2 as shown. The magnetic field $\vec{B}_1$ at this segment due to current I_1 is perpendicular to the segment $I_2\,d\vec{\ell}_2$ as shown. This is true for all current elements along the wire. The magnetic force $d\vec{F}_2$ on current segment $I_2\,d\vec{\ell}_2$ is directed toward current I_1. Similarly, a current segment $I_1\,d\vec{\ell}_1$ will experience a magnetic force directed toward current I_2 due to a magnetic field arising from current I_2. Thus, two parallel currents attract each other. If one of the currents is reversed, the force will be reversed, so two antiparallel currents will repel each other. The attraction or repulsion of parallel or antiparallel currents was discovered experimentally by Ampère one week after he heard of Oersted's discovery of the effect of a current on a compass needle.

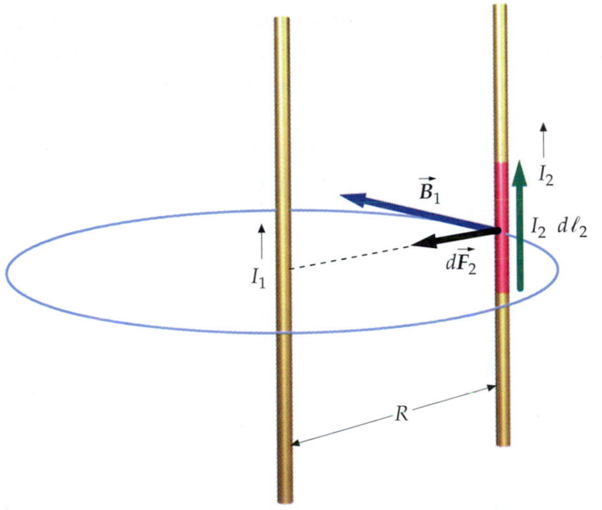

Figure 29-18 Two long, straight wires carrying parallel currents. The magnetic field $\vec{B}_1$ due to current I_1 is perpendicular to current I_2. The force on current I_2 is toward current I_1. There is an equal and opposite force exerted by current I_2 on I_1. The currents thus attract each other.

The magnitude of the magnetic force on the segment $I_2\,d\vec{\ell}_2$ is

$$dF_2 = |I\,d\vec{\ell}_2 \times \vec{B}_1|$$

Since the magnetic field at segment $I_2\,d\vec{\ell}_2$ is perpendicular to the current segment, we have

$$dF_2 = I_2\,d\ell_2 B_1$$

If the distance R between the wires is much less than their length, the field at $I_2\,d\vec{\ell}_2$ due to current I_1 will approximate the field due to an infinitely long, current-carrying wire, which is given by Equation 29-12. The magnitude of the force on the segment $I_2\,d\vec{\ell}_2$ is therefore

$$dF_2 = I_2\,d\ell_2\frac{\mu_0 I_1}{2\pi R}$$

The force per unit length is

$$\frac{dF_2}{d\ell_2} = I_2\frac{\mu_0 I_1}{2\pi R} = 2\frac{\mu_0}{4\pi}\frac{I_1 I_2}{R}$$ 29-13

In Chapter 18, the coulomb was defined in terms of the ampere, but the definition of the ampere was deferred. The ampere is defined as follows:

> The ampere is that constant current which, if maintained in two straight, parallel conductors of infinite length, of negligible circular cross section, and placed one meter apart in a vacuum, would produce between these conductors a force equal to 2×10^{-7} newtons per meter of length.

Definition—Ampere

This definition of the ampere makes the permeability of free space μ_0 equal to exactly $4\pi \times 10^{-7}\ \mathrm{N/A^2}$. It also allows the unit of current (and therefore the unit of electric charge) to be determined by a mechanical measurement. In practice, currents much closer together than 1 m are used so that the force can be measured accurately with long but finite wires.

Figure 29-19 shows a **current balance,** a device that can be used to calibrate an ammeter from the definition of the ampere. The upper conductor, directly above the lower conductor, is free to rotate about knife edge contacts and is balanced so that the wires (or conducting rods) are a small distance apart. The conductors are connected in series to carry the same current but in opposite directions so that they will repel each other. Weights are placed on the upper conductor until it balances again at the original separation. The force of repulsion is thus determined by measuring the total weight needed to balance the upper conductor.

(*a*)

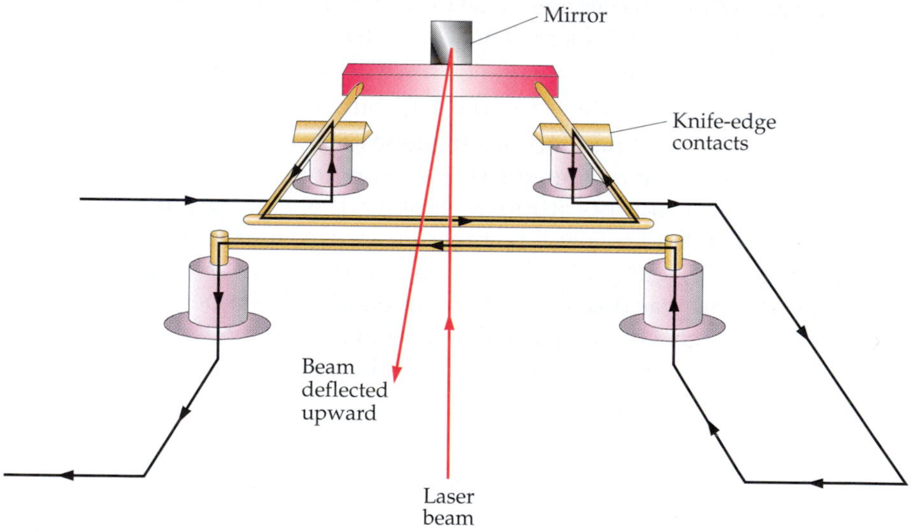

(*b*)

Figure 29-19 (*a*) Current balance used in an elementary physics laboratory to calibrate an ammeter. (*b*) A schematic diagram of the current balance in (*a*). The two parallel rods in front carry equal but oppositely directed currents and therefore repel each other. The force of repulsion is balanced by weights placed on the upper rod, which is part of a rectangle that is balanced on knife edges at the back. The mirror on top is used to reflect a beam of laser light for accurately determining the position of the upper rod.

Example 29-7 *try it yourself*

Two straight rods 50 cm long and 1.5 mm apart in a current balance carry currents of 15 A each in opposite directions. What mass must be placed on the upper rod to balance the magnetic force of repulsion?

Picture the Problem Equation 29-13 gives the magnitude of the magnetic force per unit length exerted by the lower rod on the upper rod. Find this force for a rod of length L and set it equal to the weight mg.

Cover the column to the right and try these on your own before looking at the answers.

Steps	**Answers**
1. Set the weight mg equal to the magnetic force of repulsion of the rods.	$mg = 2\left(\frac{\mu_0}{4\pi}\right)\frac{I_1 I_2}{R}L$
2. Solve for the mass m.	$m = 1.53 \times 10^{-3}\ \text{kg} = 1.53\ \text{g}$

Remark Since only 1.53 g are required to balance the system, we see that the magnetic force between two current-carrying wires is relatively small, even for currents as large as 15 A separated by only 1.5 mm.

29-3 Gauss's Law for Magnetism

The magnetic field lines as shown in Figures 29-6, 29-9, and 29-10 differ from electric field lines in that the lines of $\vec{B}$ form closed curves, whereas lines of $\vec{E}$ begin and end on electric charges. The magnetic equivalent of an electric charge is a magnetic pole, such as appears to be at the ends of a bar magnet. However, although the magnetic field lines appear to diverge from the north pole outside a bar magnet (Figure 29-10*b*), the lines inside the magnet point toward the pole. These lines enter the south pole of the magnet from the outside, but on the inside of the magnet the lines leave the south pole. If one end of a bar magnet is enclosed by a surface, the number of magnetic field lines that leave the surface is exactly equal to the number that enter the surface. That is, the net flux of the field through any closed surface is zero.*

$$\phi_{\text{m,net}} = \oint_S B_n\,dA = 0 \qquad \text{29-14}$$

Gauss's law for magnetism

where the definition of the magnetic flux ϕ_{m} is exactly analogous to the electric flux with $\vec{B}$ replacing $\vec{E}$. This result is called Gauss's law for magnetism. It is the mathematical statement that there are no points in space from which magnetic field lines diverge, or to which they converge. That is, isolated magnetic poles do not exist. The fundamental unit of magnetism is the magnetic dipole. Figure 29-20 compares the lines of $\vec{B}$ for a magnetic dipole with the lines of $\vec{E}$ for an electric dipole. Note that far from the dipoles the lines are identical. But inside the dipole, the lines of $\vec{E}$ are opposite the lines of $\vec{B}$. The lines of $\vec{E}$ diverge from the positive charge and converge to the negative charge, whereas the lines of $\vec{B}$ are continuous loops.

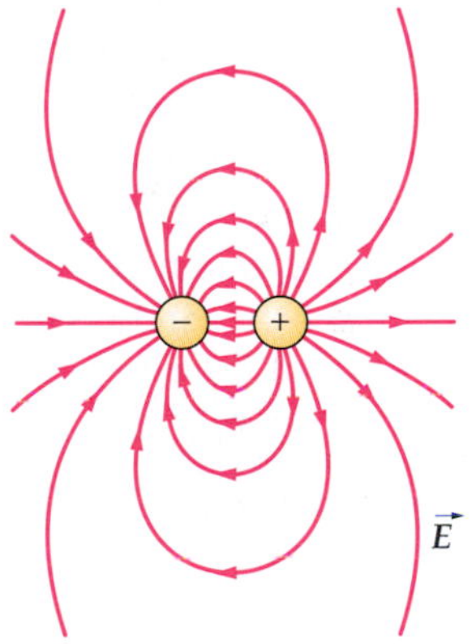

(*a*)

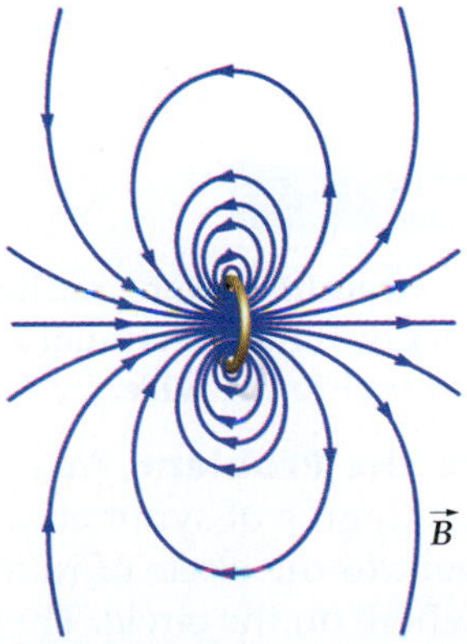

(*b*)

Figure 29-20 (*a*) Electric field lines of an electric dipole. (*b*) Magnetic field lines of a magnetic dipole. Far from the dipoles, the field lines are identical. In the region between the charges in (*a*), the electric field is opposite the dipole moment, whereas inside the loop in (*b*), the magnetic field is parallel to the dipole moment.

* Recall that the net flux of the electric field is a measure of the net number of lines that leave a closed surface and is equal to $Q_{\text{inside}}/\epsilon_0$.

29-4 Ampère's Law

In Chapter 23 we found that for highly symmetric charge distributions, we could calculate the electric field more easily using Gauss's law than Coulomb's law. A similar situation exists in magnetism. Ampère's law, which relates the tangential component of $\vec{B}$ summed around a closed curve C to the current I_C that passes through the curve can be used to obtain an expression for the magnetic field in situations that have a high degree of symmetry. In mathematical form, **Ampère's law** is

$$\oint_C \vec{B}\cdot d\vec{\ell} = \mu_0 I_C, \qquad C \text{ is any closed curve} \qquad 29\text{-}15$$

Ampère's law

where I_C is the net current that penetrates the area bounded by the curve C. Ampère's law holds for any curve C as long as the currents are continuous, that is, they do not begin or end at any finite point. It is useful in calculating the magnetic field $\vec{B}$ in situations that have a high degree of symmetry so that the line integral $\oint_C \vec{B}\cdot d\vec{\ell}$ can be written as the product of B and some distance. Ampere's law and Gauss's law are both of considerable theoretical importance, and both hold whether or not there is symmetry, but if there is no symmetry, neither is useful in calculating electric or magnetic fields.

The simplest application of Ampère's law is to find the magnetic field of an infinitely long, straight, current-carrying wire. Figure 29-21 shows a circular curve around a point on a long wire with its center at the wire. If we assume that we are far from the ends of the wire, we can use symmetry to rule out the possibility of any component of $\vec{B}$ parallel to the wire. We may then assume that the magnetic field is tangent to this circle and has the same magnitude B at any point on the circle. Ampère's law then gives

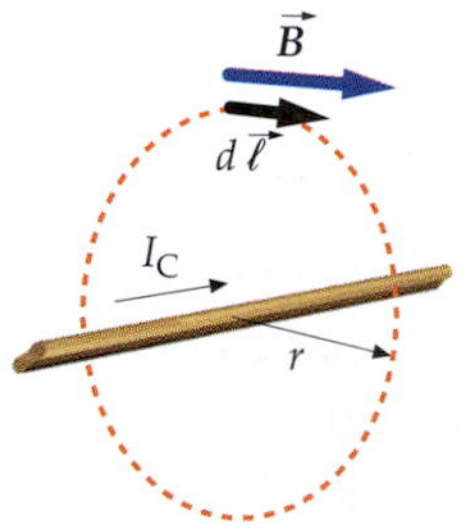

Figure 29-21 Geometry for calculating the magnetic field of a long, straight, current-carrying wire using Ampère's law. On a circle around the wire, the magnetic field is constant and tangent to the circle.

$$\oint_C \vec{B}\cdot d\vec{\ell} = B\oint_C d\ell = \mu_0 I_C$$

where we have taken B out of the integral because it has the same value everywhere on the circle. The integral of $d\ell$ around the circle equals $2\pi r$, the circumference of the circle. The current I_C is the current I in the wire. We thus obtain

$$B(2\pi r) = \mu_0 I$$

$$B = \frac{\mu_0 I}{2\pi R} = \frac{\mu_0}{4\pi}\frac{2I}{R}$$

which is Equation 29-12.

Example 29-8

A long, straight wire of radius R carries a current I that is uniformly distributed over the cross-sectional area of the wire. Find the magnetic field both outside and inside the wire.

Picture the Problem We can use Ampère's law to calculate $\vec{B}$ because of the high degree of symmetry. At a distance r (Figure 29-22), we know that $\vec{B}$ is tangent to the circle of radius r about the wire and constant in magnitude everywhere on the circle. The current through C depends on whether r is less than or greater than the radius of the wire a.

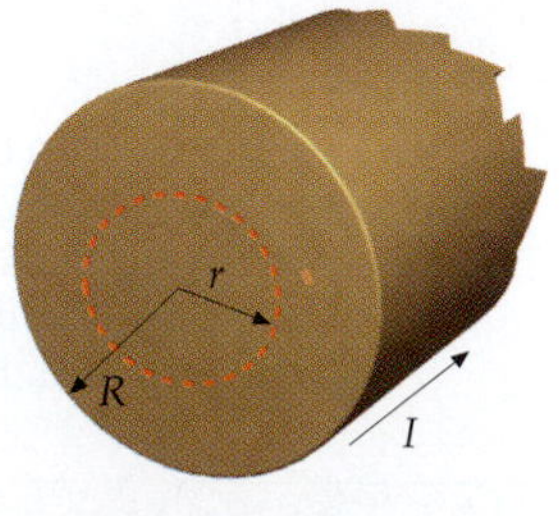

Figure 29-22

1. Apply Ampère's law to a circle of radius r:

$$\oint_C \vec{B}\cdot d\vec{\ell} = B\oint_C d\ell = B2\pi r = \mu_0 I_C$$

$$B = \frac{\mu_0}{2\pi}\frac{I_C}{r}$$

2. Ouside the wire, $r > R$, and the total current passes through the curve C:

$$I_C = I$$

$$B = \frac{\mu_0}{2\pi}\frac{I}{r}$$

3. Inside the wire, $r < R$ and the current passing through C is $(\pi r^2/\pi R^2)$ times the total current I:

$$I_C = \frac{\pi r^2}{\pi R^2}I = \frac{r^2}{R^2}I$$

$$B = \frac{\mu_0}{2\pi}\frac{I_C}{r} = \frac{\mu_0}{2\pi}\frac{(r^2/R^2)I}{r} = \frac{\mu_0}{2\pi R^2}Ir$$

Remark Inside the wire, the field increases with distance from the center of the wire. Figure 29-23 shows the graph of B versus r for this example.

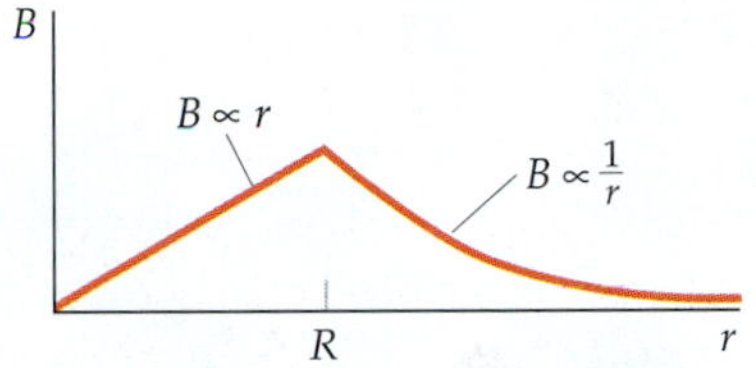

Figure 29-23

We see from Example 29-8 that the magnetic field due to a current uniformly distributed over a wire of radius R is given by

$$B = \frac{\mu_0}{2\pi R^2}Ir, \qquad r \le R$$

$$B = \frac{\mu_0}{2\pi}\frac{I}{r}, \qquad r \ge R \qquad \text{29-16}$$

For our next application of Ampère's law, we calculate the magnetic field of a tightly wound **toroid,** which consists of loops of wire wound around a doughnut-shaped form as shown in Figure 29-24. There are N turns of wire, each carrying a current I. To calculate B, we evaluate the line integral $\oint_C \vec{B}\cdot d\vec{\ell}$ around a circle of radius r centered in the middle of the toroid. By symmetry, $\vec{B}$ is tangent to this circle and constant in magnitude at every point on the circle. Then,

$$\oint_C \vec{B}\cdot d\vec{\ell} = B2\pi r = \mu_0 I_C$$

Let a and b be the inner and outer radii of the toroid, respectively. The total current through the circle of radius r for $a < r < b$ is NI. Ampère's law then gives

$$\oint_C \vec{B}\cdot d\vec{\ell} = B2\pi r = \mu_0 I_C = \mu_0 NI$$

or

$$B = \frac{\mu_0 NI}{2\pi r}, \qquad a < r < b \qquad \text{29-17}$$

B inside a tightly wound toroid

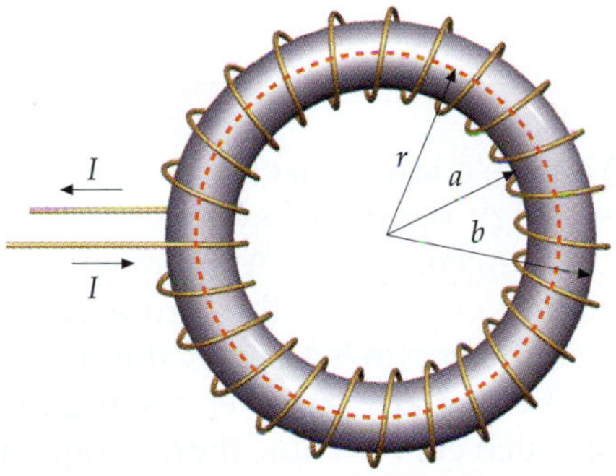

Figure 29-24 A toroid consists of loops of wire wound around a doughnut-shaped form. The magnetic field at any distance r can be found by applying Ampère's law to the circle of radius r.

If r is less than a, there is no current through the circle of radius r. If r is greater than b, the total current through r is zero because for each current I

into the page at the inner surface of the toroid in Figure 29-24, there is an equal current I out of the page at the outer surface. Thus, the magnetic field is zero for both $r < a$ and $r > b$:

$$B = 0, \qquad r < a \quad \text{or} \quad r > b$$

The magnetic field inside the toroid is not uniform but decreases with r. However, if the diameter of the loops of the toroid, $b - a$, is much less than the radius of the doughnut, the variation in r from $r = a$ to $r = b$ is small, and B is approximately uniform, as it is in a solenoid.

(*a*)

(*b*)

(*a*) The Tokamak fusion test reactor is a large toroid that produces a magnetic field for confining charged particles. Coils containing over 10 km of water-cooled copper wire carry a pulsed current, which has a peak value of 73,000 A and produces a magnetic field of 5.2 T for about 3 s. (*b*) Inspection of the assembly of the Tokamak from inside the toroid.

Limitations of Ampère's Law

Ampère's law is useful for calculating the magnetic field only when there is a high degree of symmetry. Consider the current loop shown in Figure 29-25. According to Ampère's law, the line integral $\oint_C \vec{B} \cdot d\vec{\ell}$ around a curve such as curve C in the figure equals μ_0 times the current I in the loop. Although Ampère's law is valid for this curve, the magnetic field $\vec{B}$ is not constant along any curve encircling the current, nor is it everywhere tangent to any such curve. Thus, there is not enough symmetry in this situation to allow us to calculate $\vec{B}$ using Ampère's law.

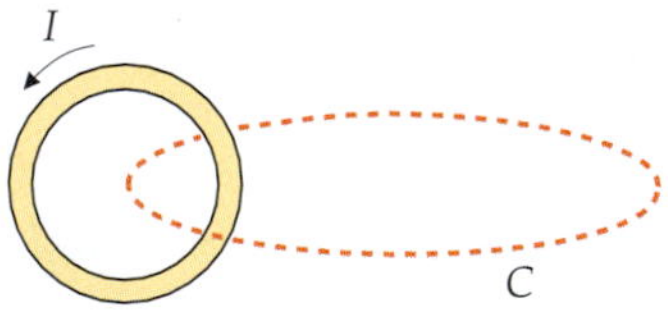

Figure 29-25 Ampère's law holds for the curve C encircling the current in the circular loop, but it is not useful for finding $\vec{B}$, because $\vec{B}$ is neither constant along the curve nor tangent to it.

Figure 29-26 shows a finite current segment of length ℓ. We wish to find the magnetic field at point P, which is equidistant from the ends of the segment and at a distance r from the center of the segment. A direct application of Ampère's law gives

$$B = \frac{\mu_0}{2\pi}\frac{I}{r}$$

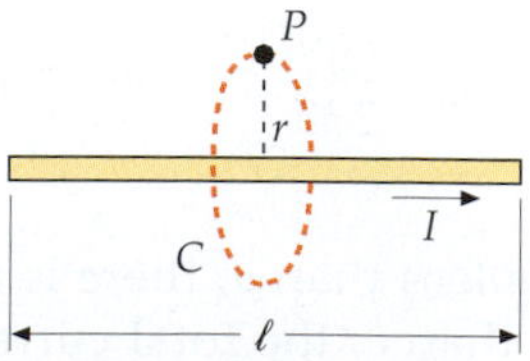

Figure 29-26 The application of Ampère's law to find the magnetic field on the bisector of a finite current segment gives an incorrect result.

This result is the same as for an infinitely long wire, since the same symmetry arguments apply. It does not agree with the result obtained from the Biot–Savart law, which depends on the length of the current segment and which agrees with experiment. If the current segment is just one part of a continuous circuit carrying a current, as shown in Figure 29-27, Ampère's law for curve C is valid, but it cannot be used to find the magnetic field at point P because there is no symmetry.

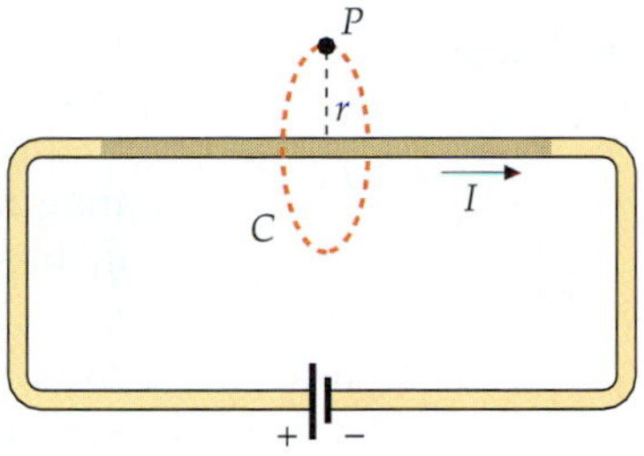

Figure 29-27 If the current segment in Figure 29-25 is part of a complete circuit, Ampère's law for the curve C is valid, but there is not enough symmetry to use it to find the magnetic field at point P.

In Figure 29-28, the current in the segment arises from a small spherical conductor with initial charge $+Q$ at the left of the segment and another one at the right with charge $-Q$. When they are connected, a current $I = -dQ/dt$ exists in the segment for a short time, until the spheres are uncharged. For this case, we *do* have the symmetry needed to assume that $\vec{B}$ is tangential to the curve and constant in magnitude along the curve. For a situation like this, in which the current is discontinuous in space, Ampère's law is not valid. In Chapter 32, we will see how Maxwell was able to modify Ampère's law so that it holds for all currents. When Maxwell's generalized form of Ampère's law is used to calculate the magnetic field for a current segment, such as that shown in Figure 29-28, the result agrees with that found from the Biot–Savart law.

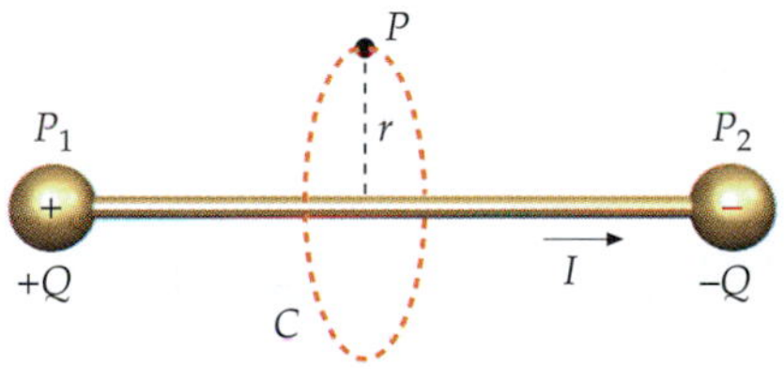

Figure 29-28 If the current segment in Figure 29-26 is due to a momentary flow of charge from a small conductor on the left to the one at the right, there is enough symmetry to use Ampère's law to compute the magnetic field at P, but Ampère's law is not valid because the current is not continuous in space.

29-5 Magnetism in Matter

Atoms have magnetic dipole moments due to the motion of their electrons and due to the intrinsic magnetic dipole moment associated with the spin of the electrons. Unlike the situation with electric dipoles, the alignment of magnetic dipoles parallel to an external magnetic field tends to *increase* the field. We can see this difference by comparing the electric field lines of an electric dipole with the magnetic field lines of a magnetic dipole, such as a small current loop, as in Figure 29-20. Far from the dipoles, the field lines are identical. However, between the charges of the electric dipole, the electric field lines are opposite the direction of the dipole moment, whereas inside the current loop, the magnetic field lines are parallel to the magnetic dipole moment. Thus, inside a magnetically polarized material, the magnetic dipoles create a magnetic field that is *parallel* to the magnetic-dipole-moment vectors.

Exploring

The Magnetic Force and Conservation of Momentum

The magnetic force exerted by one moving charge on another is found by combining Equation 28-1 for the force on a moving charge in a magnetic field and Equation 29-1 for the magnetic field of a charge. The force $\vec{F}_{1,2}$ exerted by a charge q_1 moving with velocity v_1 on a charge q_2 moving with velocity v_2 is given by

$$\vec{F}_{1,2} = q_2\vec{v}_2 \times \vec{B}_1$$

$$= q_2\vec{v}_2 \times \left(\frac{\mu_0}{4\pi}\frac{q_1\vec{v}_1 \times \hat{r}_{1,2}}{r_{1,2}^2}\right) \qquad 1a$$

where $\vec{B}_1$ is the magnetic field at the position of charge q_2 due to charge q_1, and $\hat{r}_{1,2}$ is the unit vector pointing from q_1 to q_2. Similarly, the force $\vec{F}_{2,1}$ exerted by a charge q_2 moving with velocity v_2 on a charge q_1 moving with velocity v_1 is given by

$$\vec{F}_{2,1} = q_1\vec{v}_1 \times \vec{B}_2$$

$$= q_1\vec{v}_1 \times \left(\frac{\mu_0}{4\pi}\frac{q_2\vec{v}_2 \times \hat{r}_{2,1}}{r_{2,1}^2}\right) \qquad 1b$$

These relations are remarkable in that the force exerted by charge q_1 on charge q_2 is generally not equal and opposite to that exerted by charge q_2 on charge q_1. That is, these forces do not obey Newton's third law, as can be demonstrated by considering the special case illustrated in Figure 1. Here, the magnetic field $\vec{B}_1$ due to charge q_1 at charge q_2 is in the negative z direction, and the force on q_2 is to the left in the negative x direction. However, the magnetic field $\vec{B}_2$ due to q_2 at q_1 is zero because q_1 lies along the line of motion of q_2. Thus, there is no magnetic force exerted by q_2 on q_1. There is thus a net force $\vec{F}_{1,2}$ acting on the two-charge system. The system will accelerate in the direction of this force, and linear momentum will not be conserved.

This apparent violation of the law of conservation of linear momentum results from our treating the force exerted by one charge on another as an action-at-a-distance force and neglecting the momentum carried by the electric and magnetic fields of the moving charges. We saw in Chapter 22 that there is energy associated with an electric field, and we will see later that there is also energy associated with a magnetic field. Advanced treatments of the electric and magnetic fields of moving charges show that there is also momentum associated with these fields. When the charges move, as in Figure 1, the linear momentum produced when the system accelerates to the left is balanced by momentum in the opposite direction carried by the fields. Thus, when we include the momentum of the fields, the total momentum of the system is conserved.

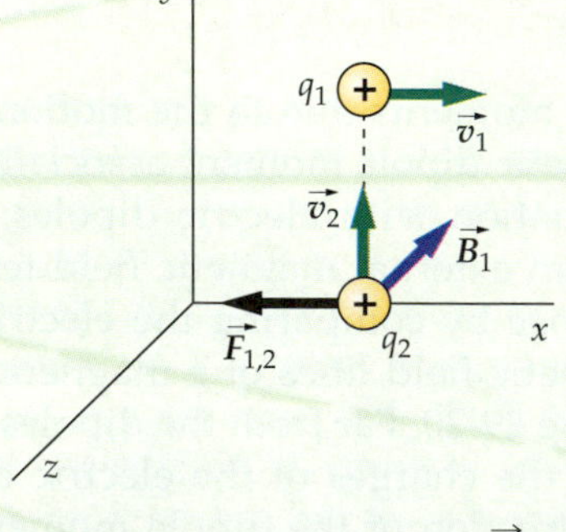

Figure 1 The forces exerted by moving charges on each other are not equal and opposite. The magnetic field $\vec{B}_1$ at charge q_2 due to charge q_1 is in the negative z direction, so it exerts a force $\vec{F}_{1,2}$ on q_2 to the left in the negative x direction. However, $\vec{B}_2$ at charge q_1 due to charge q_2 is zero, so there is no force on q_1.

Angular momentum is also carried by the electromagnetic fields produced by moving charges. Consider Figure 2, which shows a point charge q_1, $\vec{R} = x\hat{i} + y\hat{j}$, moving parallel to the x axis with velocity $\vec{v}_1 = v_1\hat{i}$, and a second point charge q_2 at the origin and moving along the x axis with velocity $\vec{v}_2 = v_2\hat{i}$. Let us calculate the magnetic force exerted by each charge on the other, assuming both charges to be positive.

We first find the force on charge q_1. We note that the vector $\vec{r}_{2,1}$ from charge q_2 to charge q_1 is just $\vec{R}$. Using $\hat{r}_{2,1} = \vec{R}/R$ we have

$$\frac{\vec{v}_2 \times \hat{r}_{2,1}}{r_{2,1}^2} = \frac{\vec{v}_2 \times \vec{R}}{R^3} = \frac{v_2\hat{i} \times (x\hat{i} + y\hat{j})}{R^3} = \frac{yv_2}{R^3}\hat{k}$$

so the magnetic field produced by charge q_2 at the position of charge q_1 is

$$\vec{B}_2 = \frac{\mu_0}{4\pi}\frac{q_2\vec{v}_2 \times \hat{r}_{2,1}}{r_{2,1}^2} = \frac{\mu_0}{4\pi}\frac{q_2 y v_2}{R^3}\hat{k}$$

The magnetic force exerted by charge q_2 on charge q_1 is then

$$\vec{F}_{2,1} = q_1\vec{v}_1 \times \vec{B}_2 = q_1 v_1\hat{i} \times \frac{\mu_0}{4\pi}\frac{q_2 y v_2}{R^3}\hat{k}$$

$$= -\frac{\mu_0}{4\pi}\frac{q_1 q_2 v_1 v_2 y}{R^3}\hat{j}$$

This force is downward and parallel to the y axis, as shown.

To find the magnetic force exerted by charge q_1 on charge q_2, we note that the vector $\vec{r}_{1,2}$ from charge q_1 to charge q_2 is $-\vec{R}$. Then the magnetic field produced by charge q_1 at the position of charge q_2 is

$$\vec{B}_1 = \frac{\mu_0}{4\pi}\frac{q_1\vec{v}_1 \times \hat{r}_{1,2}}{r_{1,2}^2}$$

$$= \frac{\mu_0}{4\pi}\frac{q_1\vec{v}_1 \times (-\vec{R})}{R^3} = -\frac{\mu_0}{4\pi}\frac{q_1 y v_1}{R^3}\hat{k}$$

The magnetic force exerted by charge q_1 on charge q_2 is then

$$\vec{F}_{1,2} = q_2\vec{v}_2 \times \vec{B}_1 = q_2 v_2\hat{i} \times \left(-\frac{\mu_0}{4\pi}\frac{q_1 y v_1}{R^3}\hat{k}\right)$$

$$= +\frac{\mu_0}{4\pi}\frac{q_1 q_2 v_1 v_2 y}{R^3}\hat{j}$$

In this case, the forces are equal and opposite, as shown in Figure 2*b*, but they are not along the line joining the two particles. The magnetic forces thus exert a torque on the two-particle system. Here, the apparent lack of conservation of angular momentum implied by the existence of this torque is resolved by the consideration of the angular momentum carried by the electromagnetic field.

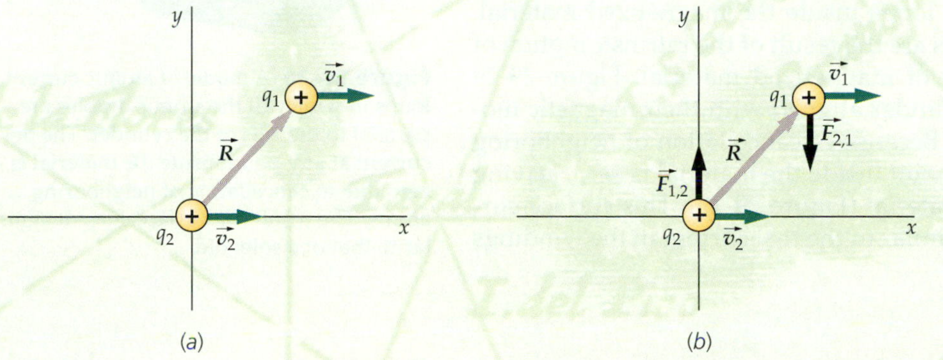

Figure 2 (*a*) Two charges moving in parallel directions. (*b*) The magnetic forces exerted by the charges on each other are equal and opposite, but they are not along the line joining the charges.

Materials fall into three categories—**paramagnetic**, **diamagnetic**, and **ferromagnetic**—according to the behavior of their magnetic moments in an external magnetic field. Paramagnetism arises from the partial alignment of the electron spins (in metals) or of atomic or molecular magnetic moments by an applied magnetic field in the direction of the field. In paramagnetic materials, the magnetic dipoles do not interact strongly with each other and are normally randomly oriented. In the presence of an external magnetic field, the dipoles are partially aligned in the direction of the field, thereby increasing the field. However, in external magnetic fields of ordinary strength at ordinary temperatures, only a very small fraction of the molecules are aligned because thermal motion tends to randomize their orientation. The increase in the total magnetic field is therefore very small. Ferromagnetism is much more complicated. Because of a strong interaction between neighboring magnetic dipoles, a high degree of alignment occurs even in weak external magnetic fields, causing a very large increase in the total field. Even when there is no external magnetic field, a ferromagnetic material may have its magnetic dipoles aligned, as in permanent magnets. Diamagnetism arises from the orbital magnetic dipole moments induced by an applied magnetic field. These magnetic moments are opposite the direction of the applied magnetic field so they decrease the total magnetic field B. This effect actually occurs in all materials, but because the induced magnetic moments are very small compared to the permanent magnetic moments, diamagnetism is masked by paramagnetic or ferromagnetic effects. Diamagnetism is thus observed only in materials that have no permanent magnetic moments.

Magnetization and Magnetic Susceptibility

When some material is placed in a strong magnetic field, such as that of a solenoid, the magnetic field of the solenoid tends to align the magnetic dipole moments (either permanent or induced) inside the material, and the material is said to be magnetized. We describe a magnetized material by its **magnetization $\vec{M}$**, which is defined as the net magnetic dipole moment per unit volume of the material:

$$\vec{M} = \frac{d\vec{\mu}}{dV} \quad \text{29-18}$$

Long before we had any understanding of atomic or molecular structure, Ampère proposed a model of magnetism in which the magnetization of materials is due to microscopic current loops inside the magnetized material. We now know that these current loops are the result of the intrinsic motion of atomic charges. Consider a cylinder of magnetized material. Figure 29-29 shows atomic current loops in the cylinder aligned with their magnetic moments along the axis of the cylinder. Because of cancellation of neighboring current loops, the net current at any point inside the material is zero, leaving a net current on the surface of the material (Figure 29-30). This surface current, called an **amperian current,** is similar to the real current in the windings of the solenoid.

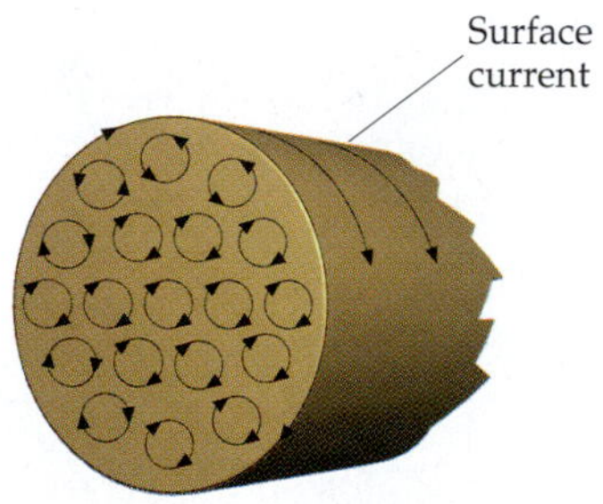

Figure 29-29 A model of atomic current loops in which all the atomic dipoles are parallel to the axis of the cylinder. The net current at any point inside the material is zero due to cancellation of neighboring atoms. The result is a surface current similar to that of a solenoid.

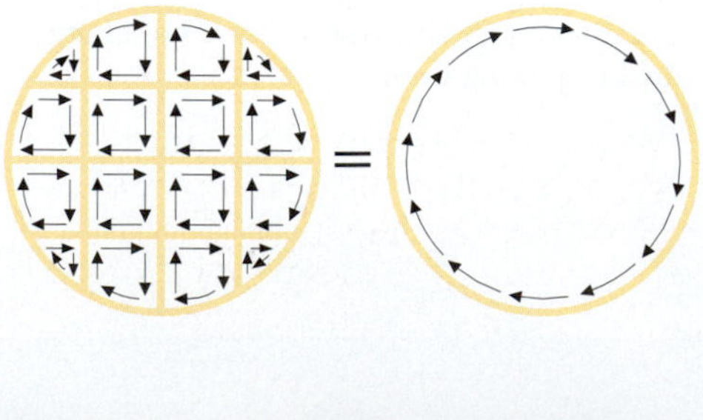

Figure 29-30 The currents in the adjacent current loops in the interior of a uniformly magnetized material cancel, leaving only a surface current. Cancellation occurs at every interior point independent of the shape of the loops.

Figure 29-31 shows a small disk of cross-sectional area A, length $d\ell$, and volume $dV = A\,d\ell$. Let di be the amperian current on the surface of the disk. The magnitude of the magnetic dipole moment of the disk is the same as that of a current loop of area A carrying a current di:

$$d\mu = A\,di$$

The magnetization of the disk is the magnetic moment per unit volume:

$$M = \frac{d\mu}{dV} = \frac{A\,di}{A\,d\ell} = \frac{di}{d\ell} \qquad \text{29-19}$$

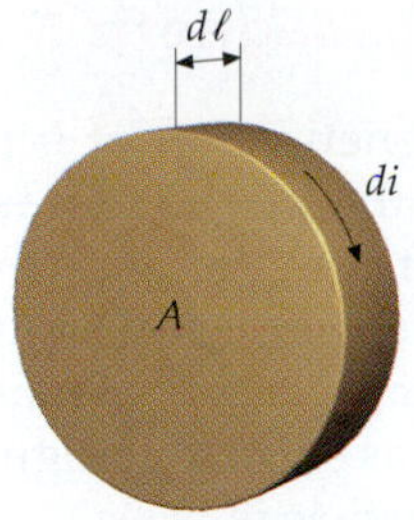

Figure 29-31 Disk element for relating the magnetization M to the surface current per unit length.

Thus, the magnitude of the magnetization vector is the amperian current per unit length along the surface of the magnetized material. We see from this result that the units of M are amperes per meter.

Consider a cylinder that has a uniform magnetization $\vec{M}$ parallel to its axis. The effect of the magnetization is the same as if the cylinder carried a surface current per unit length of magnitude M. This current is similar to the current carried by a tightly wound solenoid. For a solenoid, the current per unit length is nI, where n is the number of turns per unit length and I is the current in each turn. The magnitude of the magnetic field B_m inside the cylinder and far from its ends is thus given by Equation 29-9 for a solenoid with nI replaced by M:

$$B_m = \mu_0 M \qquad \text{29-20}$$

Suppose we place a cylinder of magnetic material inside a long solenoid with n turns per unit length that carries a current I. The applied field of the solenoid $\vec{B}_{app}$ ($B_{app} = \mu_0 nI$) magnetizes the material so that it has a magnetization $\vec{M}$. The resultant magnetic field at a point inside the solenoid and far from its ends due to the current in the solenoid plus the magnetized material is

$$\vec{B} = \vec{B}_{app} + \mu_0\vec{M} \qquad \text{29-21}$$

For paramagnetic and ferromagnetic materials, $\vec{M}$ is in the same direction as $\vec{B}_{app}$; for diamagnetic materials, $\vec{M}$ is opposite to $\vec{B}_{app}$. For paramagnetic and diamagnetic materials, the magnetization is found to be proportional to the applied magnetic field that produces the alignment of the magnetic dipoles in the material. We can thus write

$$\vec{M} = \chi_m \frac{\vec{B}_{app}}{\mu_0} \qquad \text{29-22}$$

where χ_m is a dimensionless number called the **magnetic susceptibility.** Equation 29-21 is then

$$\vec{B} = \vec{B}_{app} + \mu_0\vec{M} = \vec{B}_{app}(1 + \chi_m) = K_m\vec{B}_{app} \qquad \text{29-23}$$

where

$$K_m = 1 + \chi_m \qquad \text{29-24}$$

is called the **relative permeability** of the material. For paramagnetic materials, χ_m is a small positive number that depends on temperature. For diamagnetic materials, it is a small negative number independent of temperature. Table 29-1 lists the magnetic susceptibility of various paramagnetic and diamagnetic materials. We see that the magnetic susceptibility for the solids listed is of the order of 10^{-5}, and $K_m \approx 1$.

The magnetization of ferromagnetic materials, which we discuss shortly, is much more complicated. The relative permeability K_m defined as the ratio B/B_{app} is not constant, and has maximum values ranging from 5000 to 100,000. In the case of permanent magnets, K_m is not even defined, since such materials exhibit magnetization even in the absence of an applied field.

Table 29-1

Magnetic Susceptibility of Various Materials at 20°C

Material	χ_m
Aluminum	2.3×10^{-5}
Bismuth	-1.66×10^{-5}
Copper	-0.98×10^{-5}
Diamond	-2.2×10^{-5}
Gold	-3.6×10^{-5}
Magnesium	1.2×10^{-5}
Mercury	-3.2×10^{-5}
Silver	-2.6×10^{-5}
Sodium	-0.24×10^{-5}
Titanium	7.06×10^{-5}
Tungsten	6.8×10^{-5}
Hydrogen (1 atm)	-9.9×10^{-9}
Carbon dioxide (1 atm)	-2.3×10^{-9}
Nitrogen (1 atm)	-5.0×10^{-9}
Oxygen (1 atm)	2090×10^{-9}

Atomic Magnetic Moments

The magnetization of a paramagnetic or ferromagnetic material can be related to the permanent magnetic moments of the individual atoms or electrons of the material. The orbital magnetic moment of an atomic electron can be derived semiclassically, even though it is quantum mechanical in origin. Consider a particle of mass m and charge q moving with speed v in a circle of radius r as shown in Figure 29-32. The magnitude of the angular momentum of the particle is

$$L = mvr \qquad 29\text{-}25$$

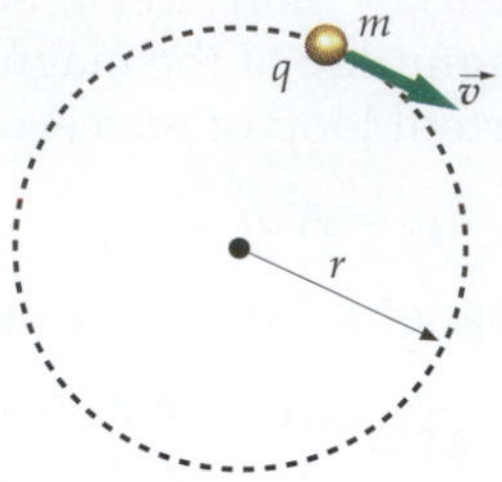

Figure 29-32 Particle of charge q and mass m moving in a circle of radius r. The angular momentum is into the paper and has a magnitude mvr, and the magnetic moment is into the paper (if q is positive) and has a magnitude $\frac{1}{2}qvr$.

The magnitude of the magnetic moment is the product of the current and the area of the circle:

$$\mu = IA = I\pi r^2$$

If T is the time for the charge to complete one revolution, the current (charge passing a point per unit time) is q/T. Since the period T is the distance $2\pi r$ divided by the velocity v, the current is

$$I = \frac{q}{T} = \frac{qv}{2\pi r}$$

The magnetic moment is then

$$\mu = IA = \frac{qv}{2\pi r}\pi r^2 = \frac{1}{2}qvr \qquad 29\text{-}26$$

Using $vr = L/m$ from Equation 29-25, we have for the magnetic moment

$$\mu = \frac{q}{2m}L$$

If the charge q is positive, the angular momentum and magnetic moment are in the same direction. We can therefore write

$$\vec{\mu} = \frac{q}{2m}\vec{L} \qquad 29\text{-}27$$

Classical relation between magnetic moment and angular momentum

Equation 29-27 is the general classical relation between magnetic moment and angular momentum. It also holds in the quantum theory of the atom for orbital angular momentum, but not for the intrinsic spin angular momentum of the electron. For electron spin, the magnetic moment is twice that predicted by this equation.* The extra factor of 2 is a result from quantum theory that has no analog in classical mechanics.

Since angular momentum is quantized, the magnetic moment of an atom is also quantized. The quantum of angular momentum is $\hbar = h/2\pi$, where h is Planck's constant, so we express the magnetic moment in terms of $\vec{L}/\hbar$

$$\vec{\mu} = \frac{q\hbar}{2m}\frac{\vec{L}}{\hbar}$$

For an electron, $m = m_e$ and $q = -e$, so the magnetic moment of the electron due to its orbital motion is

$$\vec{\mu}_\ell = -\frac{e\hbar}{2m_e}\frac{\vec{L}}{\hbar} = -\mu_B\frac{\vec{L}}{\hbar} \qquad 29\text{-}28$$

Magnetic moment due to orbital motion of an electron

* Precise measurements indicate that the magnetic moment of the electron due to its spin is 2.00232 times that predicted by Equation 29-25. This result, and the phenomenon of electron spin itself, was predicted in 1927 by P. Dirac, who combined special relativity and quantum mechanics into a relativistic wave equation called the Dirac equation. The fact that the intrinsic magnetic moment of the electron is approximately twice what we would expect makes it clear that the simple model of the electron as a spinning ball is not to be taken literally.

where

$$\mu_B = \frac{e\hbar}{2m_e} = 9.27 \times 10^{-24}\ \text{A}\cdot\text{m}^2 = 9.27 \times 10^{-24}\ \text{J/T} \qquad \text{29-29}$$
$$= 5.79 \times 10^{-5}\ \text{eV/T}$$

Bohr magneton

is the quantum unit of magnetic moment called a **Bohr magneton.** The magnetic moment of an electron due to its intrinsic spin angular momentum $\vec{S}$ is

$$\vec{\mu}_s = -2 \times \frac{e\hbar}{2m_e}\frac{\vec{S}}{\hbar} = -2\mu_B \frac{\vec{S}}{\hbar} \qquad \text{29-30}$$

Magnetic moment due to electron spin

Although the calculation of the magnetic moment of any atom is a complicated problem in quantum theory, the result for all electrons, according to both theory and experiment, is that the magnetic moment is of the order of a few Bohr magnetons. For atoms with zero net angular momentum, the net magnetic moment is zero.*

If all the atoms or molecules in some material have their magnetic moments aligned, the magnetic moment per unit volume of the material is the product of the number of molecules per unit volume n and the magnetic moment μ of each molecule. For this extreme case, the **saturation magnetization** M_s is

$$M_s = n\mu \qquad \text{29-31}$$

The number of molecules per unit volume can be found from the molecular mass $\mathscr{M}$, the density ρ of the material, and Avogadro's number N_A:

$$n = \frac{N_A\ \text{(atoms/mol)}}{\mathscr{M}\text{(kg/mol)}}\rho(\text{kg/m}^3) \qquad \text{29-32}$$

*The shell structure of atoms is discussed in Chapter 37.

Example 29-9

Find the saturation magnetization and the magnetic field it produces for iron, assuming that each iron atom has a magnetic moment of 1 Bohr magneton.

Picture the Problem We find the number of molecules per unit volume from the density of iron, $\rho = 7.9 \times 10^3\ \text{kg/m}^3$, and its molecular mass $\mathscr{M} = 55.8 \times 10^{-3}\ \text{kg/mol}$.

1. The saturation magnetic field is the product of the number of molecules per unit volume and the magnetic moment of each molecule:

$$M_s = n\mu$$

2. Calculate the number of molecules per unit volume from Avogadro's number, the molecular mass, and the density:

$$n = \frac{N_A}{\mathscr{M}}\rho = \frac{6.02 \times 10^{23}\ \text{atoms/mol}}{55.8 \times 10^{-3}\ \text{kg/mol}}(7.9 \times 10^3\ \text{kg/m}^3)$$
$$= 8.52 \times 10^{28}\ \text{atoms/m}^3$$

3. Substitute this result and $\mu = 1$ Bohr magneton to calculate the saturation magnetization:

$$M_s = n\mu$$
$$= (8.52 \times 10^{28}\ \text{atoms/m}^3)(9.27 \times 10^{-24}\ \text{A}\cdot\text{m}^2)$$
$$= 7.90 \times 10^5\ \text{A/m}$$

4. The magnetic field on the axis inside a long iron cylinder resulting from this maximum magnetization is given by $B = \mu_0 M_s$:

$$B = \mu_0 M_s$$
$$= (4\pi \times 10^{-7}\ \text{T}\cdot\text{m/A})(7.90 \times 10^5\ \text{A/m})$$
$$= 0.993\ \text{T} \approx 1\ \text{T}$$

Remarks The measured saturation magnetic field of annealed iron is about 2.16 T, indicating that the magnetic moment of an iron atom is slightly greater than 2 Bohr magnetons. This magnetic moment is due mainly to the spins of two unpaired electrons in the iron atom.

optional

Paramagnetism

Paramagnetism occurs in materials whose atoms have permanent magnetic moments that interact with each other only very weakly, resulting in a very small, positive magnetic susceptibility χ_m. When there is no external magnetic field, these magnetic moments are randomly oriented. In the presence of an external magnetic field, they tend to line up parallel to the field, but this is counteracted by the tendency for the magnetic moments to be randomly oriented due to thermal motion. The fraction of the moments that line up with the field depends on the strength of the field and on the temperature. This fraction is usually small because the energy of a magnetic moment in an external magnetic field is typically much smaller than the thermal energy of an atom of the material, which is of the order of kT, where k is Boltzmann's constant and T is the absolute temperature.

The potential energy of a magnetic dipole of moment $\vec{\mu}$ in an external magnetic field $\vec{B}$ is given by Equation 28-16:

$$U = -\mu B \cos\theta = -\vec{\mu}\cdot\vec{B}$$

The potential energy when the moment is parallel with the field ($\theta = 0$) is thus lower than when it is antiparallel ($\theta = 180°$) by the amount $2\mu B$. For a typical magnetic moment of 1 Bohr magneton and a typical strong magnetic field of 1 T, the difference in potential energy is

$$\Delta U = 2\mu_B B = 2(5.79 \times 10^{-5}\ \text{eV/T})(1\ \text{T}) = 1.16 \times 10^{-4}\ \text{eV}$$

At a normal temperature of $T = 300$ K, the typical thermal energy kT is

$$kT = (8.62 \times 10^{-5}\ \text{eV/K})(300\ \text{K}) = 2.59 \times 10^{-2}\ \text{eV}$$

which is about 200 times greater than $2\mu_B B$. Thus, even in a very strong magnetic field of 1 T, most of the magnetic moments will be randomly oriented because of thermal motions unless the temperature is very low.

Figure 29-33 shows a plot of the magnetization M versus an applied external magnetic field B_{app} at a given temperature. In very strong fields, nearly all the magnetic moments are aligned with the field and $M \approx M_s$. (For magnetic fields attainable in the laboratory, this can occur only for very low temperatures.) When $B_{app} = 0$, $M = 0$, indicating that the orientation of the moments is completely random. In weak fields, the magnetization is approximately proportional to the applied field, as indicated by the orange dashed line in the figure. In this region, the magnetization is given by

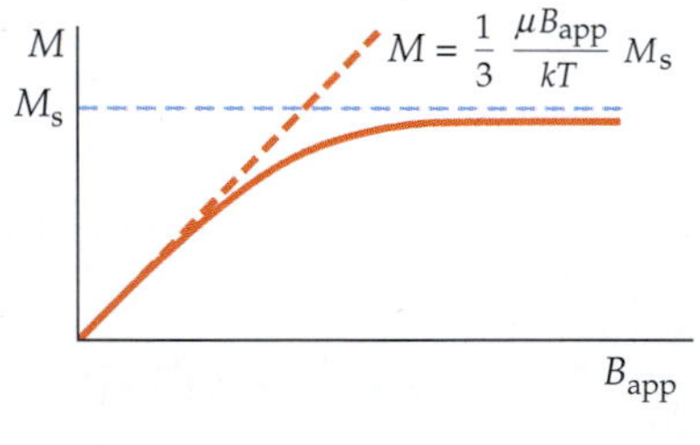

Figure 29-33 Plot of magnetization M versus applied field B_{app}. In very strong fields, the magnetization approaches the saturation value M_s. This can be achieved only at very low temperatures. In weak fields, the magnetization is approximately proportional to B_{app}, a result known as Curie's law.

$$M = \frac{1}{3}\frac{\mu B_{app}}{kT} M_s \qquad 29\text{-}33$$

Curie's law

Note that $(\mu B_{\text{app}}/kT)$ is the ratio of the maximum energy of a dipole in the magnetic field to the characteristic thermal energy. The result that the magnetization varies inversely with the absolute temperature was discovered experimentally by Pierre Curie and is known as **Curie's law.**

Liquid oxygen, which is paramagnetic, is attracted by the magnetic field of a permanent magnet. A net force is exerted on the magnetic dipoles because the magnetic field is not uniform.

Example 29-10

If $\mu = \mu_B$, at what temperature will the magnetization be 1% of the saturation magnetization in an applied magnetic field of 1 T?

1. Substitute $M = 0.01\, M_s$ into Curie's law:

$$M = \frac{1}{3}\frac{\mu B_{\text{app}}}{kT}M_s = 0.01M_s$$

2. Solve for T:

$$T = \frac{\mu B_{\text{app}}}{0.03k} = \frac{(5.79 \times 10^{-5}\text{ eV/T})(1\text{ T})}{(0.03)(8.62 \times 10^{-5}\text{ eV/K})} = 22.4\text{ K}$$

Remark From this example, we see that even in a strong applied magnetic field of 1 T, the magnetization is less than 1% of saturation at temperatures above 22.4 K.

Exercise If $\mu = \mu_B$, what fraction of the saturation magnetization is M at 300 K for an external magnetic field of 15,000 G? (*Answer* $M/M_s = 1.12 \times 10^{-3}$)

Ferromagnetism

Ferromagnetism occurs in pure iron, cobalt, and nickel, and in alloys of these metals with each other. It also occurs in gadolinium, dysprosium, and a few compounds. Ferromagnetism arises from a strong interaction between the electrons in a partially full band in a metal or between the localized electrons that form magnetic moments on neighboring atoms or molecules. This interaction, called the **exchange interaction**, lowers the energy of a pair of electrons with parallel spins.

Ferromagnetic materials have very large, positive values of magnetic susceptibility χ_m (as measured under conditions described below). In these substances, a small external magnetic field can produce a very large degree of alignment of the atomic magnetic dipole moments. In some cases, the align-

A chunk of magnetite (lodestone) attracts the needle of a compass.

optional

ment can persist even when the external magnetizing field is removed. This occurs because the magnetic dipole moments exert strong forces on their neighbors so that over a small region of space the moments are aligned with each other even when there is no external field. The region of space over which the magnetic dipole moments are aligned is called a **magnetic domain.** The size of a domain is usually microscopic. Within the domain, all the magnetic moments are aligned, but the direction of alignment varies from domain to domain so that the net magnetic moment of a macroscopic piece of ferromagnetic material is zero in the normal state. Figure 29-34 illustrates this situation. The dipole forces that produce this alignment are predicted by quantum theory but cannot be explained with classical physics. At temperatures above a critical temperature, called the **Curie temperature,** thermal agitation is great enough to break up this alignment, and ferromagnetic materials become paramagnetic.

(*a*)

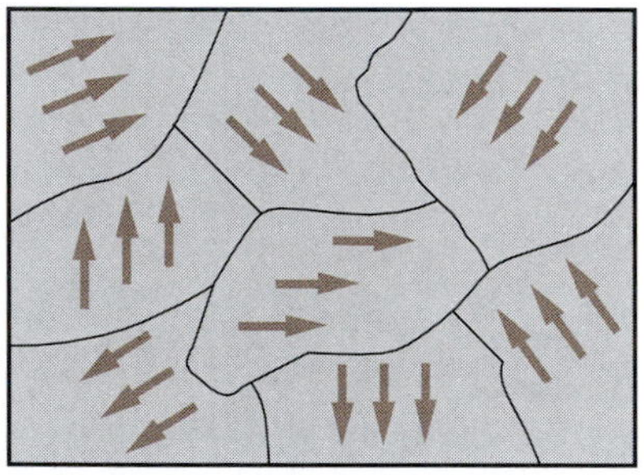

(*b*)

Figure 29-34 (*a*) Schematic illustration of ferromagnetic domains. Within a domain, the magnetic dipoles are aligned, but the direction of alignment varies from domain to domain so that the net magnetic moment is zero. A small external magnetic field may cause the enlargement of those domains that are aligned parallel to the field, or it may cause the alignment within a domain to rotate. In either case, the result is a net magnetic moment parallel to the field. (*b*) Magnetic domains on the surface of an Fe–3% Si crystal observed using a scanning electron microscope with polarization analysis. The four colors indicate four possible domain orientations.

When an external magnetic field is applied, the boundaries of the domains may shift or the direction of alignment within a domain may change so that there is a net macroscopic magnetic moment in the direction of the applied field. Since the degree of alignment is large for even a small external field, the magnetic field produced in the material by the dipoles is often much greater than the external field.

Let us consider what happens when we magnetize a long iron rod by placing it inside a solenoid and gradually increase the current in the solenoid windings. We assume that the rod and the solenoid are long enough to permit us to neglect end effects. Since the induced magnetic moments are in the same direction as the applied field, $\vec{B}_{\text{app}}$ and $\vec{M}$ are parallel. Then

$$B = B_{\text{app}} + \mu_0 M = \mu_0 n I + \mu_0 M \qquad \text{29-34}$$

In ferromagnetic materials, the magnetic field $\mu_0 M$ due to the magnetic moments is often greater than the magnetizing field B_{app} by a factor of several thousand.

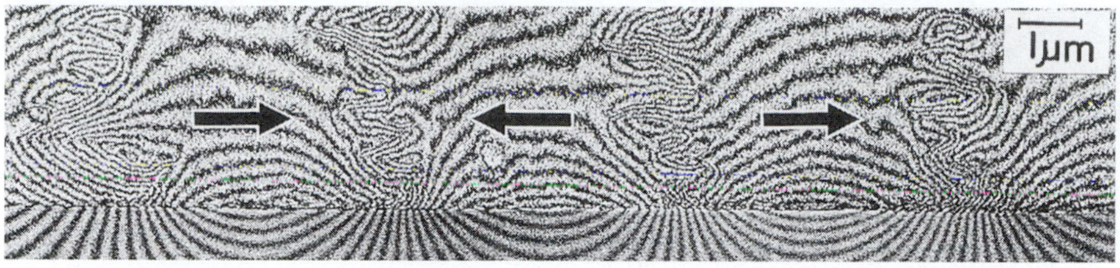

(*a*)

(*b*)

(*a*) Magnetic field lines on a cobalt magnetic recording tape. The solid arrows indicate the encoded magnetic bits. (*b*) Cross section of a magnetic tape recording head. Current from an audio amplifier is sent to wires around a magnetic core in the recording head where it produces a magnetic field. When the tape passes over a gap in the core of the recording head, the fringing magnetic field encodes information on the tape.

Figure 29-35 shows a plot of B versus the magnetizing field B_{app}. As the current is gradually increased from zero, B increases from zero along the part of the curve from the origin O to point P_1. The flattening of this curve near point P_1 indicates that the magnetization M is approaching its saturation value M_s, at which all the atomic magnetic moments are aligned. Above saturation, B increases only because the magnetizing field $B_{app} = \mu_0 nI$ increases. When B_{app} is gradually decreased from point P_1, there is not a corresponding decrease in the magnetization. The shift of the domains in a ferromagnetic material is not completely reversible, and some magnetization remains even when B_{app} is reduced to zero, as indicated in the figure. This effect is called **hysteresis,** from the Greek word *hysteros* meaning later or behind, and the curve in Figure 29-35 is called a **hysteresis curve.** The value of the magnetic field at point r when B_{app} is zero is called the **remnant field** B_r. At this point, the iron rod is a permanent magnet. If the current in the solenoid is now reversed so that B_{app} is in the opposite direction, the magnetic field B is gradually brought to zero at point c. The remaining part of the hysteresis curve is obtained by further increasing the current in the opposite direction until point P_2 is reached, which corresponds to saturation in the opposite direction, and then decreasing the current to zero at point P_3 and increasing it again in its original direction.

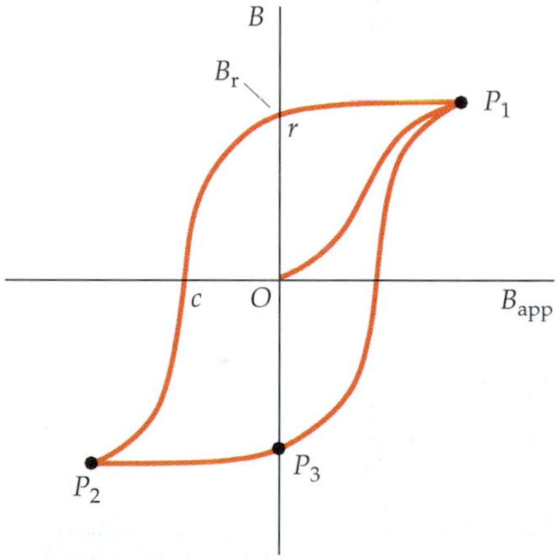

Figure 29-35 Plot of B versus the applied field B_{app}. The outer curve is called a hysteresis curve. The field B_r is called the remnant field. It remains when the applied field returns to zero.

Since the magnetization M depends on the previous history of the material, and since it can have a large value even when the applied field is zero, it is not simply related to the applied field B_{app}. However, if we confined ourselves to that part of the magnetization curve from the origin to point P_1 in Figure 29-35, $\vec{M}$ and $\vec{B}_{app}$ are parallel and M is zero when B_{app} is zero. We can then define the magnetic susceptibility as in Equation 29-22,

$$M = \chi_m \frac{B_{app}}{\mu_0}$$

and

$$B = B_{app} + \mu_0 M = B_{app}(1 + \chi_m) = K_m \mu_0 nI = \mu nI \qquad \text{29-35}$$

where

$$\mu = (1 + \chi_m)\mu_0 = K_m \mu_0 \qquad \text{29-36}$$

is called the **permeability** of the material. (For paramagnetic and diamagnetic materials, χ_m is much less than 1 so the permeability μ and the permeability of free space μ_0 are very nearly equal.)

Since B does not vary linearly with B_{app}, as can be seen from Figure 29-35, the relative permeability is not constant. The maximum value of K_m occurs at

optional

Table 29-2

Maximum Values of $\mu_0 M$ and K_m for Some Ferromagnetic Materials

Material	$\mu_0 M_s$, T	K_m
Iron (annealed)	2.16	5,500
Iron-silicon (96% Fe, 4% Si)	1.95	7,000
Permalloy (55% Fe, 45% Ni)	1.60	25,000
Mu-metal (77% Ni, 16% Fe, 5% Cu, 2% Cr)	0.65	100,000

a magnetization that is considerably less than the saturation magnetization. Table 29-2 lists the saturation magnetic field $\mu_0 M_s$ and the maximum values of K_m for some ferromagnetic materials. Note that the maximum values of K_m are much greater than 1.

The area enclosed by the hysteresis curve is proportional to the energy dissipated as heat in the irreversible process of magnetizing and demagnetizing. If the hysteresis effect is small, so that the area inside the curve is small, indicating a small energy loss, the material is called **magnetically soft.** Soft iron is an example. The hysteresis curve for a magnetically soft material is shown in Figure 29-36. Here the remnant field B_r is nearly zero, and the energy loss per cycle is small. Magnetically soft materials are used for transformer cores to allow the magnetic field B to change without incurring large energy losses as the field alternates. On the other hand, a large remnant field is desirable in a permanent magnet. **Magnetically hard** materials, such as carbon steel and the alloy Alnico 5, are used for permanent magnets.

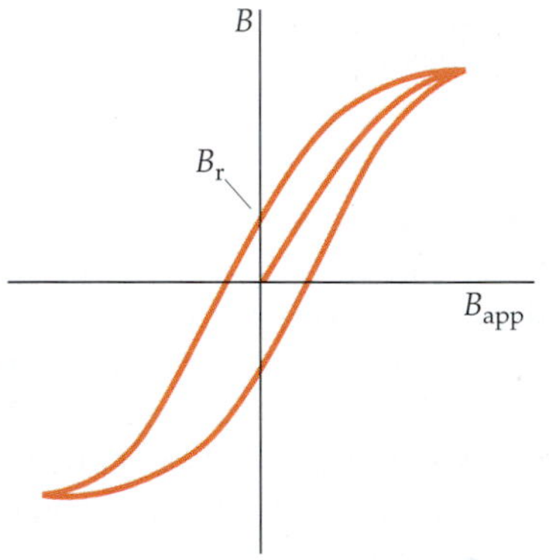

Figure 29-36 Hysteresis curve for a magnetically soft material. The remnant field is very small compared with that for a magnetically hard material such as that in Figure 29-35.

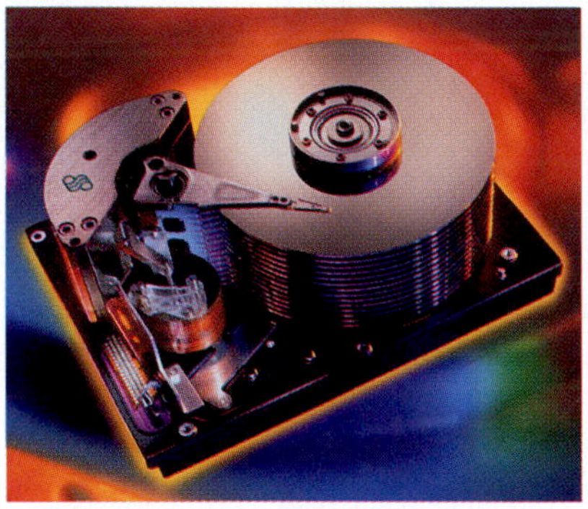

(a)

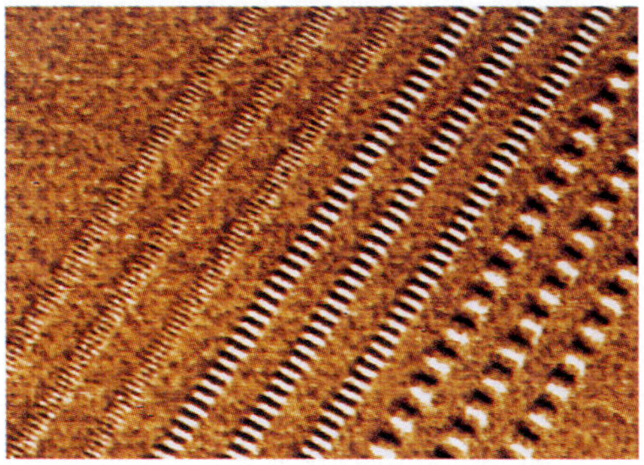

(b)

(*a*) An extremely high-capacity hard-disk drive for magnetic storage of information, capable of storing over 47 gigabytes of information. (*b*) A magnetic test pattern on a hard disk, magnified 2400 times. The light and dark regions correspond to oppositely directed magnetic fields. The smooth region just outside the pattern is a region of the disk that has been erased just prior to writing.

Example 29-11

A long solenoid with 12 turns per centimeter has a core of annealed iron. When the current is 0.50 A, the magnetic field inside the iron core is 1.36 T. Find (*a*) the applied field B_{app}, (*b*) the relative permeability K_m, and (*c*) the magnetization M.

Picture the Problem The applied field is just that of a long solenoid given by $B_{app} = \mu_0 nI$. Since the total magnetic field is given, we can find the relative permeability from its definition ($K_m = B/B_{app}$) and we can find M from $B = B_{app} + \mu_0 M$.

optional

(*a*) The applied field is given by Equation 29-10:

$$B_{app} = \mu_0 nI$$
$$= (4\pi \times 10^{-7}\ \text{T·m/A})(1200\ \text{turns/m})(0.50\ \text{A})$$
$$= 7.54 \times 10^{-4}\ \text{T}$$

(*b*) The relative permeability is the ratio of B to B_{app}:

$$K_m = \frac{B}{B_{app}} = \frac{1.36\ \text{T}}{7.54 \times 10^{-4}\ \text{T}} = 1.80 \times 10^3 = 1800$$

(*c*) The magnetization M is found from Equation 29-34:

$$\mu_0 M = B - B_{app}$$
$$= 1.36\ \text{T} - 7.54 \times 10^{-4}\ \text{T} \approx B = 1.36\ \text{T}$$
$$M = \frac{B}{\mu_0} = \frac{1.36\ \text{T}}{4\pi \times 10^{-7}\ \text{T·m/A}} = 1.08 \times 10^6\ \text{A/m}$$

Remarks The applied magnetic field of 7.54×10^{-4} T is a negligible fraction of the total field of 1.36 T. Note that the value for K_m of 1800 is considerably smaller than the maximum value of 5500 in Table 29-2. Note also that the susceptibility $\chi_m = K_m - 1 \approx K_m$ to the three-place accuracy with which we calculated K_m.

Diamagnetism

Diamagnetic materials are those having very small, negative values of magnetic susceptibility χ_m. Diamagnetism was discovered by Faraday in 1846 when he found that a piece of bismuth is repelled by either pole of a magnet, indicating that the external field of the magnet induces a magnetic moment in bismuth in the direction opposite the field.

We can understand this effect qualitatively from Figure 29-37, which shows two positive charges moving in circular orbits with the same speed but in opposite directions. Their magnetic moments are in opposite directions and therefore cancel.* In the presence of an external magnetic field $\vec{B}$ directed into the paper, the charges experience an extra force $q\vec{v} \times \vec{B}$, which is along the radial direction. For the charge on the left, this extra force is inward, increasing the centripetal force. If the charge is to remain in the same circular orbit, it must speed up so that mv^2/r equals the total centripetal force.† Its magnetic moment, which is outward, is thus increased. For the charge on the right, the additional force is outward, so the particle must slow down to maintain its circular orbit. Its magnetic moment, which is inward, is decreased. In each case, the *change* in the magnetic moment of the charges in the direction out of the page, opposite that of the external applied field. Since the permanent magnetic moments of the two charges are equal and oppositely directed, they add to zero, leaving only the induced magnetic moments, which are both opposite the direction of the applied magnetic field.

A material will be diamagnetic if its atoms have zero net angular momentum and therefore no permanent magnetic moment. (The net angular momentum of an atom depends on the electronic structure of the atom, a subject we study in Chapter 37.) The induced magnetic moments that cause diamag-

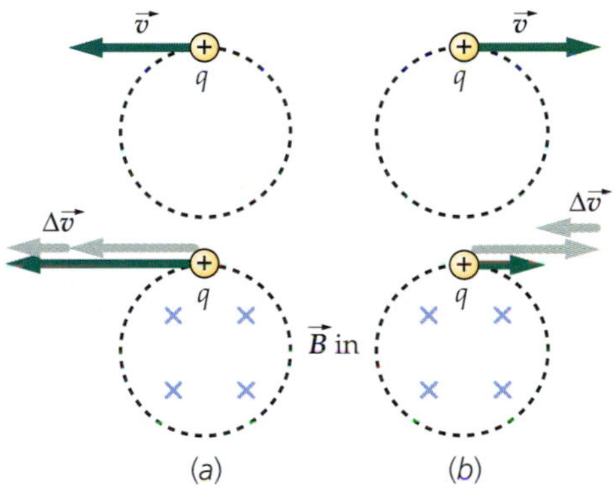

Figure 29-37 (*a*) A positive charge moving counterclockwise in a circle has its magnetic moment outward. When an external, inward magnetic field is turned on, the magnetic force increases the centripetal force so the speed of the particle must increase. The change in the magnetic moment is outward. (*b*) A positive charge moving clockwise in a circle has its magnetic moment inward. When an external, inward magnetic field is turned on, the magnetic force decreases the centripetal force so the speed of the particle must decrease. As in (*a*), the change in the magnetic moment is outward.

* It is simpler to consider positive charges even though it is the negatively charged electrons that provide the magnetic moments in matter.

† The electron speeds up because of an electric field induced by the changing magnetic field, an effect called induction, which we discuss in Chapter 30.

optional

netism have magnitudes of the order of 10^{-5} Bohr magnetons. Since this is much smaller than the permanent magnetic moments of the atoms of paramagnetic or ferromagnetic materials, the diamagnetic effect in these atoms is masked by the alignment of their permanent magnetic moments. However, since this alignment decreases with temperature, all materials are theoretically diamagnetic at sufficiently high temperatures.

When a superconductor is placed in an external magnetic field, electric currents are induced on its surface so that the net magnetic field in the superconductor is zero. Consider a superconducting rod inside a solenoid of n turns per unit length. When the solenoid is connected to a source of emf so that it carries a current I, the magnetic field due to the solenoid is $\mu_0 nI$. A surface current of $-nI$ per unit length is induced on the superconducting rod that cancels out the field due to the solenoid so that the net field inside the superconductor is zero. From Equation 29-23,

$$\vec{B} = \vec{B}_{\text{app}}(1 + \chi_{\text{m}}) = 0$$

so

$$\chi_{\text{m}} = -1$$

A superconductor is thus a perfect diamagnet with a magnetic susceptibility of -1.

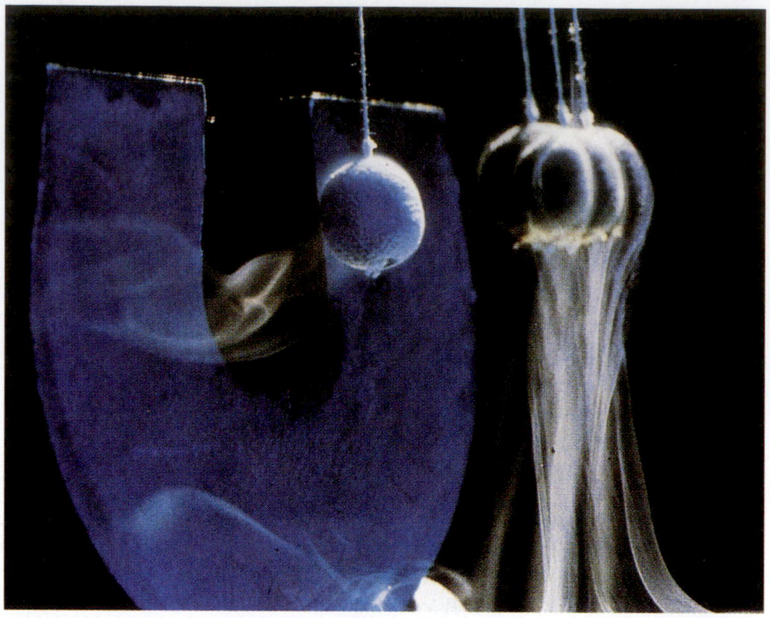

A superconductor is a perfect diamagnet. Here, the superconducting pendulum bob is repelled by the permanent magnet.

Summary

1. Magnetic fields arise from moving charges, and therefore from currents.
2. The Biot–Savart law describes the magnetic field produced by a current element.
3. Ampere's law relates the line integral of the magnetic field along some curve to the current that passes through any area bounded by the curve.
4. The magnetization vector $\vec{M}$ describes the magnetic moment per unit volume of matter.
5. The classical relation $\vec{\mu} = (q/2m)\vec{L}$ is derived from the definitions of angular momentum and magnetic moment.
6. The Bohr magneton is a convenient unit for atomic and nuclear magnetic moments.

Topic	Remarks and Relevant Equations	
1. Magnetic Field $\vec{B}$		
Due to a moving point charge	$\vec{B} = \dfrac{\mu_0}{4\pi}\dfrac{q\vec{v} \times \hat{r}}{r^2}$	29-1
	where $\hat{r}$ is a unit vector that points from the charge to the field point, and μ_0 is the permeability of free space:	
	$\mu_0 = 4\pi \times 10^{-7}\,\text{T·m/A} = 4\pi \times 10^{-7}\,\text{N/A}^2$	29-2

Due to a current element (Biot–Savart law)

$$d\vec{B} = \frac{\mu_0}{4\pi}\frac{I\,d\vec{\ell} \times \hat{r}}{r^2}$$ **29-3**

On the axis of a current loop

$$B_x = \frac{\mu_0}{4\pi}\frac{2\pi R^2 I}{(x^2 + R^2)^{3/2}}$$ **29-5**

On the axis of a current loop far from the loop

$$B_x = \frac{\mu_0}{4\pi}\frac{2\mu}{|x^3|}$$

where μ is the magnetic moment. **29-6**

Inside a solenoid, far from the ends

$$B = \mu_0 n I$$

where n is the number of turns per unit length. **29-9**

Due to a straight, current-carrying wire

$$B = \frac{\mu_0}{4\pi}\frac{I}{R}(\sin\theta_1 + \sin\theta_2)$$ **29-11**

where R is the perpendicular distance to the wire and θ_1 and θ_2 are the angles subtended at the field point by the ends of the wire.

Due to an infinitely long, straight wire

$$B = \frac{\mu_0}{4\pi}\frac{2I}{R}$$ **29-12**

The direction of $\vec{B}$ is such that the lines of $\vec{B}$ encircle the wire in the direction of the fingers of the right hand if the thumb points in the direction of the current.

Inside a toroid

$$B = \frac{\mu_0 N I}{2\pi r}, \qquad a < r < b$$ **29-17**

2. Magnetic Field Lines

The magnetic field is indicated by lines parallel to $\vec{B}$ at any point whose density is proportional to the magnitude of $\vec{B}$. Magnetic lines do not begin or end at any point in space. Instead, they form continuous loops.

3. Gauss's Law for Magnetism

$$\phi_{m,net} = \oint_S B_n\,dA = 0$$ **29-14**

4. Magnetic Poles

Magnetic poles always occur in pairs. Isolated magnetic poles have not been found.

5. Ampère's Law

$$\oint_C \vec{B}\cdot d\vec{\ell} = \mu_0 I_C$$

where C is any closed curve. **29-15**

Validity of Ampère's law

Ampère's law is valid only if the currents are continuous. It can be used to derive expressions for the magnetic field for situations with a high degree of symmetry, such as a long, straight, current-carrying wire, or a long, tightly wound solenoid.

6. Magnetic Matter

Matter can be classified as either paramagnetic, ferromagnetic, or diamagnetic.

Magnetization

A magnetized material is described by its magnetization vector $\vec{M}$, which is defined to be the net magnetic dipole moment per unit volume of the material:

$$\vec{M} = \frac{d\vec{\mu}}{dV}$$ **29-18**

The magnetic field due to a uniformly magnetized cylinder is the same as if the cylinder carried a current per unit length of magnitude M on its surface. This current, which is due to the intrinsic motion of the atomic charges in the cylinder, is called an amperian current.

7. $\vec{B}$ in Magnetic Materials

$$\vec{B} = \vec{B}_{\text{app}} + \mu_0 \vec{M} \qquad \textbf{29-21}$$

Magnetic susceptibility χ_m

$$\vec{M} = \chi_m \frac{\vec{B}_{\text{app}}}{\mu_0} \qquad \textbf{29-22}$$

For paramagnetic materials, χ_m is a small positive number that depends on temperature. For diamagnetic materials (other than superconductors), it is a small negative constant independent of temperature. For superconductors, $\chi_m = -1$. For ferromagnetic materials, the magnetization depends not only on the magnetizing current but also on the past history of the material.

Relative permeability

$$\vec{B} = K_m \vec{B}_{\text{app}} \qquad \textbf{29-23}$$

where

$$K_m = 1 + \chi_m \qquad \textbf{29-24}$$

8. Atomic Magnetic Moments

$$\vec{\mu} = \frac{q}{2m}\vec{L} \quad \text{(classical)} \qquad \textbf{29-27}$$

where $\vec{L}$ is the angular momentum of the particle.

Orbital

$$\vec{\mu}_\ell = -\frac{e\hbar}{2m_e}\frac{\vec{L}}{\hbar} = -\mu_B \frac{\vec{L}}{\hbar} \qquad \textbf{29-28}$$

Spin

$$\vec{\mu}_s = -2 \times \frac{e\hbar}{2m_e}\frac{\vec{S}}{\hbar} = -2\mu_B \frac{\vec{S}}{\hbar} \qquad \textbf{29-30}$$

Bohr magneton

$$\mu_B = \frac{e\hbar}{2m_e} = 9.27 \times 10^{-24}\ \text{A}\cdot\text{m}^2$$

$$= 9.27 \times 10^{-24}\ \text{J/T} = 5.79 \times 10^{-5}\ \text{eV/T} \qquad \textbf{29-29}$$

where

$$\hbar = \frac{h}{2\pi} = 1.05 \times 10^{-34}\ \text{J}\cdot\text{s}$$

and $h = 6.67 \times 10^{-34}$ J·s is Planck's constant.

9. Paramagnetism (optional)

Paramagnetic materials have permanent atomic magnetic moments that have random directions in the absence of an external magnetic field. In an external field, some of these dipoles are aligned, producing a small contribution to the total field that adds to the external field. The degree of alignment is small except in very strong fields and at very low temperatures. At ordinary temperatures, thermal motion tends to maintain the random directions of the magnetic moments.

Curie's law

In weak fields, the magnetization is approximately proportional to the applied field and inversely proportional to the absolute temperature.

$$M = \frac{1}{3}\frac{\mu B_{\text{app}}}{kT} M_s \qquad \textbf{29-33}$$

10. Ferromagnetism (optional)

Ferromagnetic materials have small regions of space called magnetic domains in which the permanent atomic magnetic moments are aligned. When the material is unmagnetized, the direction of alignment in one domain is independent of that in another so that no net magnetic field is produced. When the material is magnetized, the domains of a ferromagnetic material are aligned, producing a very strong contribution to the magnetic field. This alignment can persist even when the external field is removed, thus leading to permanent magnetism.

11. Diamagnetism (optional)	Diamagnetic materials are those in which the magnetic moments of all electrons in each atom cancel, leaving each atom with zero magnetic moment in the absence of an external field. In an external field, a very small magnetic moment is induced that tends to weaken the field. This effect is independent of temperature. Superconductors are diamagnetic with susceptibility equal to -1.

Problem-Solving Guide

1. Begin by drawing a neat diagram that includes the important features of the problem. The vector nature of the magnetic field can be clarified with a diagram showing $\vec{v}$ and $\hat{r}$ (for moving point charges) or showing $d\vec{\ell}$ and $\hat{r}$ (for a current element). Indicate the direction of $\vec{B}$.
2. Ampère's law can be used to calculate B in problems with cylindrical symmetry such as a very long, current-carrying wire.

Summary of Worked Examples

Type of Calculation	Procedure and Relevant Examples
1. $\vec{B}$ due to Moving Charges	
Find the magnetic field produced by a moving point charge.	Use $\vec{B} = \frac{\mu_0}{4\pi}\frac{q\vec{v} \times \hat{r}}{r^2}$ where $\hat{r}$ points from the charge to the field point. **Example 29-1**
2. $\vec{B}$ due to Currents	
Find the magnetic field on the axis of a current loop.	At a distance x from the loop, B_x is given by $B_x = \frac{\mu_0}{4\pi}\frac{2\pi R^2 I}{(x^2 + R^2)^{3/2}}$ At the center of the loop ($x = 0$) $B_x = \mu_0 I/2R$. At great distances ($x >> R$), $B_x = \frac{\mu_0}{4\pi}\frac{2\mu}{x^3}$ where $\mu = NI\pi R^2$. **Example 29-2**
Find the magnetic field inside a solenoid.	The general result is $B_x = \frac{1}{2}\mu_0 nI\left(\frac{b}{\sqrt{b^2 + R^2}} + \frac{a}{\sqrt{a^2 + R^2}}\right)$ where a and b are the distances from the field point to either end of the solenoid. For a long solenoid, $B = \mu_0 nI$. **Example 29-4**
Find the magnetic field due to a current in a straight wire.	Use $B = \frac{\mu_0}{4\pi}\frac{I}{R}(\sin\theta_1 + \sin\theta_2)$ where θ_1 and θ_2 are the angles from the field point to either end of the wire. For a long wire, $B = \frac{\mu_0 I}{2\pi R} = \frac{\mu_0}{4\pi}\frac{2I}{R}$ **Examples 29-5, 29-6, 29-7**

	Find the force exerted by one wire on another.	Calculate $\vec{B}$ at one wire due to the current in the other wire. Then use $$d\vec{F} = I\,d\vec{\ell} \times \vec{B} \quad \text{or} \quad \frac{F}{d\ell} = 2\frac{\mu_0}{4\pi}\frac{I_1 I_2}{R}$$ for the force between two parallel current-carrying wires. **Example 29-7**
3.	**Torques on Magnets**	The torque exerted on a magnet in a field $\vec{B}$ is given by $$\vec{\tau} = \vec{\mu} \times \vec{B}$$ **Example 29-3**
4.	**Ampère's Law**	
	Use Ampère's law to find the magnetic field.	Find a curve C such that $\vec{B}$ is tangential to the curve and constant along it. Then use $\oint_C \vec{B}\cdot d\vec{\ell} = \mu_0 I_C$ and remove B from the integral. **Example 29-8**
5.	**$\vec{B}$ in Matter**	
	Find the saturation magnetic field in some material.	Multiply the number of molecules per unit volume by the magnetic moment of each molecule to obtain the magnetization M. Then use $B_m = \mu_0 M$. **Example 29-9**
	Find the relative permeability and magnetization of a ferrromagnetic material given the total field (optional).	Calculate the applied field B_{app} and use $K_m = B/B_{app}$. Then use $\mu_0 M = B - B_{app}$. **Example 29-11**

Problems

Conceptual Problems

Problems from Optional and Exploring sections

In a few problems, you are given more data than you actually need; in a few other problems, you are required to supply data from your general knowledge, outside sources, or informed estimates.

- • Single-concept, single-step, relatively easy
- •• Intermediate-level, may require synthesis of concepts
- ••• Challenging, for advanced students

Moving Point Charges

1 • Compare the directions of the electric and magnetic forces between two positive charges, which move along parallel paths (*a*) in the same direction, and (*b*) in opposite directions.

2 • At time $t = 0$, a particle with charge $q = 12\ \mu\text{C}$ is located at $x = 0$, $y = 2$ m; its velocity at that time is $\vec{v} = 30\ \text{m/s}\,\hat{i}$. Find the magnetic field at (*a*) the origin; (*b*) $x = 0$, $y = 1$ m; (*c*) $x = 0$, $y = 3$ m; and (*d*) $x = 0$, $y = 4$ m.

3 • For the particle in Problem 2, find the magnetic field at (*a*) $x = 1$ m, $y = 3$ m; (*b*) $x = 2$ m, $y = 2$ m; and (*c*) $x = 2$ m, $y = 3$ m.

4 • A proton (charge $+e$) traveling with a velocity of $\vec{v} = 1 \times 10^4\ \text{m/s}\,\hat{i} + 2 \times 10^4\ \text{m/s}\,\hat{j}$ is located at $x = 3$ m, $y = 4$ m at some time t. Find the magnetic field at the following positions: (*a*) $x = 2$ m, $y = 2$ m; (*b*) $x = 6$ m, $y = 4$ m; and (*c*) $x = 3$ m, $y = 6$ m.

5 • An electron orbits a proton at a radius of 5.29×10^{-11} m. What is the magnetic field at the proton due to the orbital motion of the electron?

6 •• Two equal charges q located at $(0, 0, 0)$ and $(0, b, 0)$ at time zero are moving with speed v in the positive x direction ($v \ll c$). Find the ratio of the magnitudes of the magnetic and electrostatic force on each.

The Biot–Savart Law

7 • The Biot–Savart law is similar to Coulomb's law in that both

(*a*) are inverse square laws.
(*b*) deal with forces on charged particles.
(*c*) deal with excess charges.
(*d*) include the permeability of free space.
(*e*) are not electrical in nature.

8 • A small current element $I\,d\vec{\ell}$, with $d\vec{\ell} = 2\ \text{mm}\,\hat{k}$ and $I = 2$ A, is centered at the origin. Find the magnetic field $d\vec{B}$ at the following points: (*a*) on the x axis at $x = 3$ m, (*b*) on the x axis at $x = -6$ m, (*c*) on the z axis at $z = 3$ m, and (*d*) on the y axis at $y = 3$ m.

9 • For the current element in Problem 8, find the magnitude and direction of $d\vec{B}$ at $x = 0$, $y = 3$ m, $z = 4$ m.

10 • For the current element in Problem 8, find the magnitude of $d\vec{B}$ and indicate its direction on a diagram at (*a*) $x = 2$ m, $y = 4$ m, $z = 0$ and (*b*) $x = 2$ m, $y = 0$, $z = 4$ m.

Current Loops

11 • Is $\vec{B}$ uniform everywhere within a current loop? Explain.

12 • A single loop of wire of radius 3 cm carries a current of 2.6 A. What is the magnitude of B on the axis of the loop at (*a*) the center of the loop, (*b*) 1 cm from the center, (*c*) 2 cm from the center, and (*d*) 35 cm from the center?

13 • A single-turn, circular loop of radius 10.0 cm is to produce a field at its center that will just cancel the earth's magnetic field at the equator, which is 0.7 G directed north. Find the current in the loop and make a sketch showing the orientation of the loop and the current.

14 •• For the loop of wire in Problem 13, at what point along the axis of the loop is the magnetic field (*a*) 10% of the field at the center, (*b*) 1% of the field at the center, and (*c*) 0.1% of the field at the center?

15 •• A single-turn circular loop of radius 8.5 cm is to produce a field at its center that will just cancel the earth's field of magnitude 0.7 G directed at 70° below the horizontal north direction. Find the current in the loop and make a sketch showing the orientation of the loop and the current.

16 •• A circular current loop of radius R carrying a current I is centered at the origin with its axis along the x axis. Its current is such that it produces a magnetic field in the positive x direction. (*a*) Sketch a graph of B_x versus x for points on the x axis. Include both positive and negative values of x. Compare this graph with that for E_x due to a charged ring of the same size. (*b*) A second, identical current loop, carrying an equal current in the same sense, is in a plane parallel to the yz plane with its center at $x = d$. Sketch graphs of the magnetic field on the x axis due to each loop separately and the resultant field due to the two loops. Show from your sketch that dB_x/dx is zero midway between the two loops.

17 •• Two coils that are separated by a distance equal to their radius and that carry equal currents such that their axial fields add are called Helmholtz coils. A feature of Helmholtz coils is that the resultant magnetic field between the coils is very uniform. Let $R = 10$ cm, $I = 20$ A, and $N = 300$ turns for each coil. Place one coil in the yz plane with its center at the origin and the other in a parallel plane at $x = 10$ cm. (*a*) Calculate the resultant field B_x at $x = 5$ cm, $x = 7$ cm, $x = 9$ cm, and $x = 11$ cm. (*b*) Use your results and the fact that B_x is symmetric about the midpoint of the coils to sketch B_x versus x. (See also Problem 18.)

18 ••• Two Helmholtz coils with radii R have their axes along the x axis (see Problem 17). One coil is in the yz plane and the other is in a parallel plane at $x = R$. Show that at the midpoint of the coils ($x = \frac{1}{2}R$), $dB_x/dx = 0$, $d^2B_x/dx^2 = 0$, and $d^3B_x/dx^3 = 0$. This shows that the magnetic field at points near the midpoint is approximately equal to that at the midpoint.

Straight-Line Current Segments

19 • Two wires lie in the plane of the paper and carry equal currents in opposite directions, as shown in Figure 29-38. At a point midway between the wires, the magnetic field is

(*a*) zero.
(*b*) into the page.
(*c*) out of the page.
(*d*) toward the top or bottom of the page.
(*e*) toward one of the two wires.

Figure 29-38 Problem 19

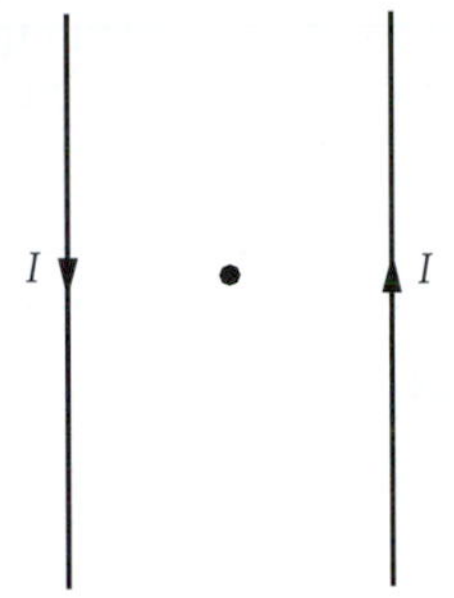

20 • Two parallel wires carry currents I_1 and $I_2 = 2I_1$ in the same direction. The forces F_1 and F_2 on the wires are related by

(*a*) $F_1 = F_2$. (*b*) $F_1 = 2F_2$. (*c*) $2F_1 = F_2$.
(*d*) $F_1 = 4F_2$. (*e*) $4F_1 = F_2$.

21 • A wire carries an electrical current straight up. What is the direction of the magnetic field due to the wire a distance of 2 m north of the wire?

(*a*) North (*b*) East (*c*) West
(*d*) South (*e*) Upward

22 • Two current-carrying wires are perpendicular to each other. The current in one flows vertically upward and the current in the other flows horizontally toward the east. The horizontal wire is one meter south of the vertical wire. What is the direction of the net magnetic force on the horizontal wire?

(*a*) North (*b*) East (*c*) West (*d*) South
(*e*) There is no net magnetic force on the horizontal wire.

23 • A long, straight wire carries a current of 10 A. Find the magnitude of B at (*a*) 10 cm, (*b*) 50 cm, and (*c*) 2 m from the center of the wire.

Problems 24 to 29 refer to Figure 29-39, which shows two long, straight wires in the xy plane and parallel to the x axis. One wire is at $y = -6$ cm and the other is at $y = +6$ cm. The current in each wire is 20 A.

Figure 29-39 Problems 24–29

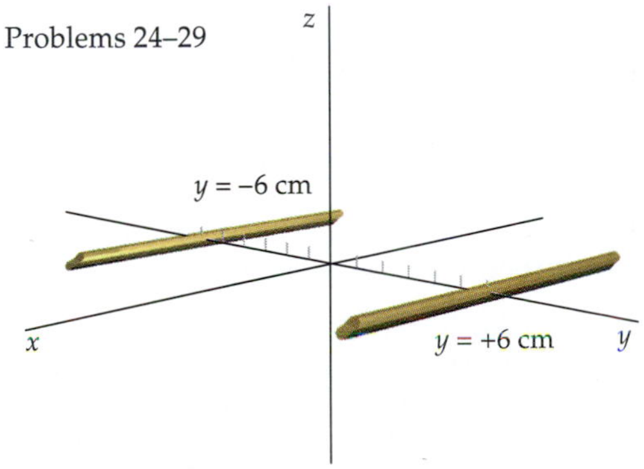

24 • If the currents in Figure 29-39 are in the negative x direction, find $\vec{B}$ at the points on the y axis at (*a*) $y = -3$ cm, (*b*) $y = 0$, (*c*) $y = +3$ cm, and (*d*) $y = +9$ cm.

25 • Sketch B_z versus y for points on the y axis when both currents are in the negative x direction.

26 • Find $\vec{B}$ at points on the y axis as in Problem 24 when the current in the wire at $y = -6$ cm is in the negative x direction and the current in the wire at $y = +6$ cm is in the positive x direction.

27 • Sketch B_z versus y for points on the y axis when the directions of the currents are opposite to those in Problem 26.

28 • Find $\vec{B}$ on the z axis at $z = +8$ cm if (*a*) the currents are parallel, as in Problem 24 and (*b*) the currents are antiparallel, as in Problem 26.

29 • Find the magnitude of the force per unit length exerted by one wire on the other.

30 • Two long, straight, parallel wires 8.6 cm apart carry currents of equal magnitude I. They repel each other with a force per unit length of 3.6 nN/m. (*a*) Are the currents parallel or antiparallel? (*b*) Find I.

31 •• The current in the wire of Figure 29-40 is 8.0 A. Find B at point P due to each wire segment and sum to find the resultant B.

Figure 29-40 Problem 31

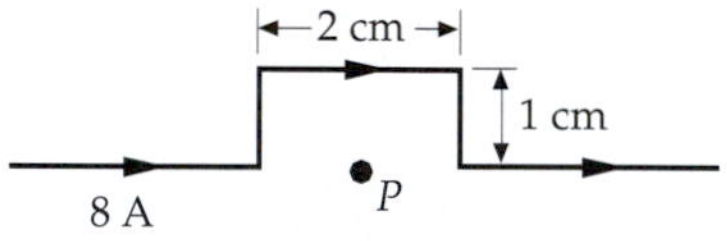

32 •• A wire of length 16 cm is suspended by flexible leads above a long, straight wire. Equal but opposite currents are established in the wires such that the 16-cm wire floats 1.5 mm above the long wire with no tension in its suspension leads. If the mass of the 16-cm wire is 14 g, what is the current?

33 •• Three long, parallel, straight wires pass through the corners of an equilateral triangle of sides 10 cm as shown in Figure 29-41, where a dot means that the current is out of the paper and a cross means that it is into the paper. If each current is 15.0 A, find (*a*) the force per unit length on the upper wire, and (*b*) the magnetic field B at the upper wire due to the two lower wires.

Figure 29-41 Problems 33 and 34

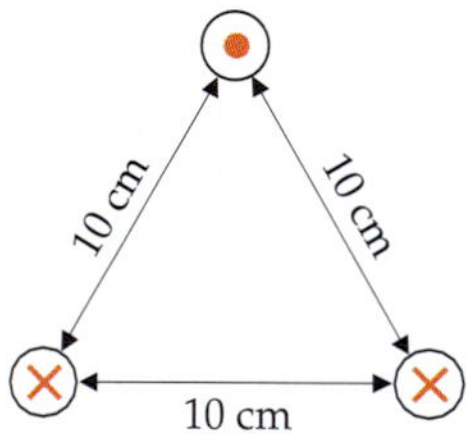

34 •• Work Problem 33 with the current in the lower right corner of Figure 29-41 reversed.

35 •• An infinitely long, insulated wire lies along the x axis and carries current I in the positive x direction. A second infinitely long, insulated wire lies along the y axis and carries current I in the positive y direction. Where in the xy plane is the resultant magnetic field zero?

36 •• An infinitely long wire lies along the z axis and carries a current of 20 A in the positive z direction. A second infinitely long wire is parallel to the z axis at $x = 10$ cm. (*a*) Find the current in the second wire if the magnetic field at $x =$ 2 cm is zero. (*b*) What is the magnetic field at $x = 5$ cm?

37 •• Three very long, parallel wires are at the corners of a square, as shown in Figure 29-42. They each carry a current of magnitude I. Find the magnetic field B at the unoccupied corner of the square when (*a*) all the currents are into the paper, (*b*) I_1 and I_3 are in and I_2 is out, and (*c*) I_1 and I_2 are in and I_3 is out.

Figure 29-42 Problem 37

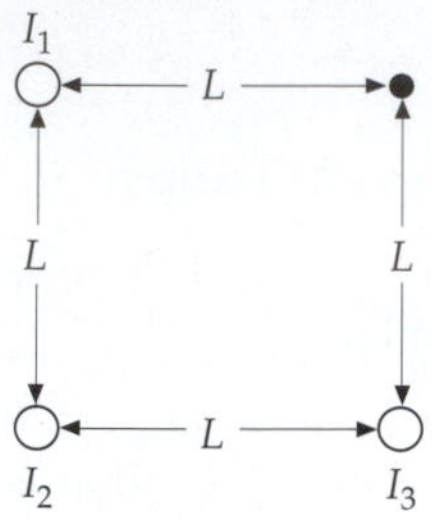

38 •• Four long, straight, parallel wires each carry current I. In a plane perpendicular to the wires, the wires are at the corners of a square of side a. Find the force per unit length on one of the wires if (*a*) all the currents are in the same direction, and (*b*) the currents in the wires at adjacent corners are oppositely directed.

39 •• An infinitely long, nonconducting cylinder of radius R lies along the z axis. Five long, conducting wires are parallel to the cylinder and spaced equally on the upper half of its surface. Each wire carries a current I in the positive z direction. Find the magnetic field on the z axis.

Solenoids

40 • A solenoid with length 30 cm, radius 1.2 cm, and 300 turns carries a current of 2.6 A. Find B on the axis of the solenoid (*a*) at the center, (*b*) inside the solenoid at a point 10 cm from one end, and (*c*) at one end.

41 • A solenoid 2.7 m long has a radius of 0.85 cm and 600 turns. It carries a current I of 2.5 A. What is the approximate magnetic field B on the axis of the solenoid?

42 ••• A solenoid has n turns per unit length and radius R and carries a current I. Its axis is along the x axis with one end at $x = -\frac{1}{2}\ell$ and the other end at $x = +\frac{1}{2}\ell$, where ℓ is the total length of the solenoid. Show that the magnetic field B at a point on the axis outside the solenoid is given by

$$B = \tfrac{1}{2}\mu_0 nI(\cos\theta_1 - \cos\theta_2) \qquad \textbf{29-35}$$

where

$$\cos\theta_1 = \frac{x + \frac{1}{2}\ell}{[R^2 + (x + \frac{1}{2}\ell)^2]^{1/2}}$$

and

$$\cos\theta_2 = \frac{x - \frac{1}{2}\ell}{[R^2 + (x - \frac{1}{2}\ell)^2]^{1/2}}$$

43 ••• In Problem 42, a formula for the magnetic field along the axis of a solenoid is given. For $x >> \ell$ and $\ell > R$, the angles θ_1 and θ_2 in Equation 29-35 are very small, so the small-angle approximation $\cos\theta \approx 1 - \theta^2/2$ is valid. (*a*) Draw a diagram and show that

$$\theta_1 \approx \frac{R}{x + \frac{1}{2}\ell} \quad \text{and} \quad \theta_2 \approx \frac{R}{x - \frac{1}{2}\ell}$$

(*b*) Show that the magnetic field at a point far from either end of the solenoid can be written

$$B = \frac{\mu_0}{4\pi}\left(\frac{q_m}{r_1^2} - \frac{q_m}{r_2^2}\right) \qquad \text{29-37}$$

where $r_1 = x - \frac{1}{2}\ell$ is the distance to the near end of the solenoid, $r_2 = x + \frac{1}{2}\ell$ is the distance to the far end, and $q_m = nI\pi R^2 = \mu/\ell$, where $\mu = NI\pi R^2$ is the magnetic moment of the solenoid.

44 ••• In this problem, you will derive Equation 29-37 by another method. Consider a long, tightly wound solenoid of length ℓ and radius $R \ll \ell$ lying along the axis with its center at the origin. It has N turns and carries a current I. Consider an element of the solenoid of length dx. (*a*) What is the magnetic moment of this element? (*b*) Show that the magnetic field dB due to this element at a point on the x axis x_0 far from the element is given by

$$dB = \frac{\mu_0}{2\pi} nIA \frac{dx}{x'^3}$$

where $A = \pi R^2$ and $x' = x_0 - x$ is the distance from the element to the field point. (*c*) Integrate this expression from $x = -\frac{1}{2}\ell$ to $x = +\frac{1}{2}\ell$ to obtain Equation 29-37.

Ampère's Law

45 • Ampère's law is valid

(*a*) when there is a high degree of symmetry.
(*b*) when there is no symmetry.
(*c*) when the current is constant.
(*d*) when the magnetic field is constant.
(*e*) in all of these situations if the current is continuous.

46 • A long, straight, thin-walled, cylindrical shell of radius R carries a current I. Find B inside and outside the cylinder.

47 • In Figure 29-43, one current is 8 A into the paper, the other current is 8 A out of the paper, and each curve is a circular path. (*a*) Find $\oint_C \vec{B} \cdot d\vec{\ell}$ for each path indicated. (*b*) Which path, if any, can be used to find B at some point due to these currents?

Figure 29-43 Problem 47

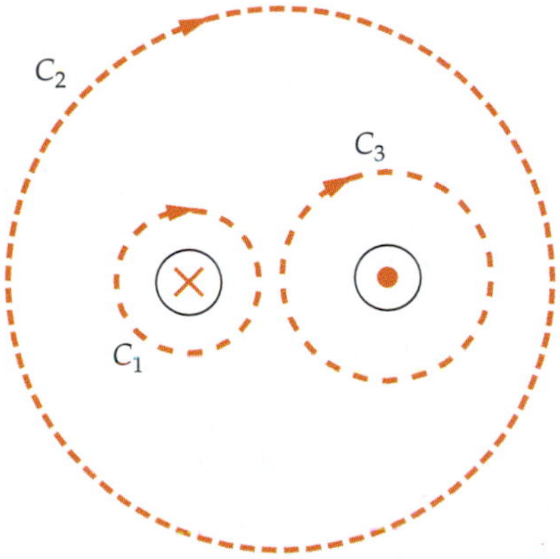

48 • A very long, coaxial cable consists of an inner wire and a concentric outer cylindrical conducting shell of radius R. At one end, the wire is connected to the shell. At the other end, the wire and shell are connected to opposite terminals of a battery, so there is a current down the wire and back up the shell. Assume that the cable is straight. Find B (*a*) at points between the wire and the shell far from the ends, and (*b*) outside the cable.

49 •• A wire of radius 0.5 cm carries a current of 100 A that is uniformly distributed over its cross-sectional area. Find B (*a*) 0.1 cm from the center of the wire, (*b*) at the surface of the wire, and (*c*) at a point outside the wire 0.2 cm from the surface of the wire. (*d*) Sketch a graph of B versus the distance from the center of the wire.

50 •• Show that a uniform magnetic field with no fringing field, such as that shown in Figure 29-44, is impossible because it violates Ampère's law. Do this by applying Ampère's law to the rectangular curve shown by the dashed lines.

Figure 29-44 Problem 50

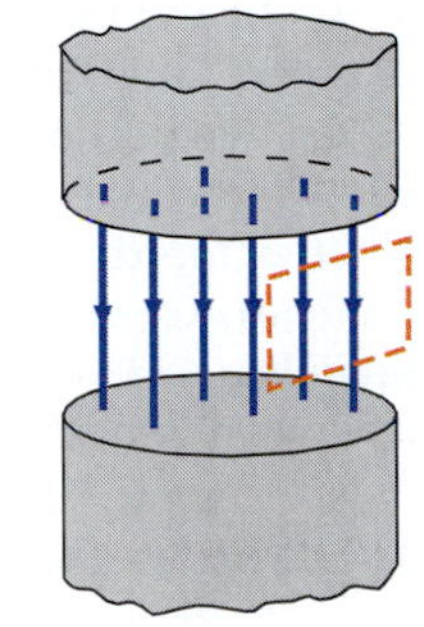

51 •• A coaxial cable consists of a solid inner cylindrical conductor of radius 1.00 mm and an outer cylindrical shell conductor of inner radius 2.00 mm and outer radius 3.00 mm. There is a current of 18 A down the inner wire and an equal return current in the outer conductor. The currents are uniform over the cross section of each conductor. Find the numerical value of $\oint_C \vec{B} \cdot d\vec{\ell}$ for a closed circular path (centered on the axis of the cable and in a plane perpendicular to the axis) which has a radius r for (*a*) $r = 1.50$ mm, (*b*) $r = 2.50$ mm, and (*c*) $r = 3.50$ mm.

52 •• An infinitely long, thick, cylindrical shell of inner radius a and outer radius b carries a current I uniformly distributed across a cross section of the shell. Find the magnetic field for (*a*) $r < a$, (*b*) $a < r < b$, and (*c*) $r > b$.

53 •• Figure 29-45 shows a solenoid carrying a current I with n turns per unit length. Apply Ampère's law to the rectangular curve shown to derive an expression for B assuming that it is uniform inside the solenoid and zero outside it.

Figure 29-45 Problem 53

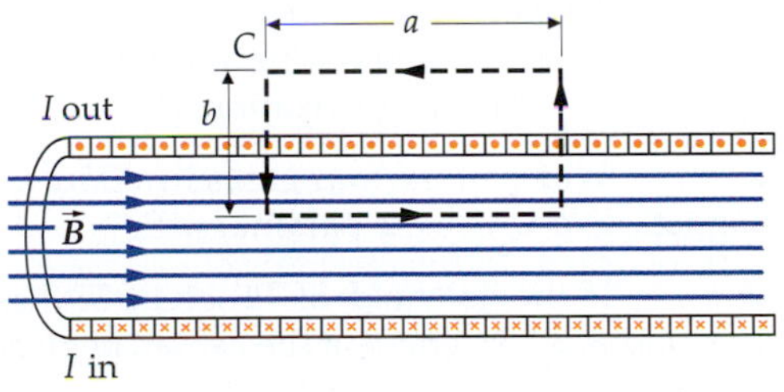

54 •• A tightly wound toroid of inner radius 1 cm and outer radius 2 cm has 1000 turns of wire and carries a current of 1.5 A. (*a*) What is the magnetic field at a distance of 1.1 cm from the center? (*b*) What is the field 1.5 cm from the center?

55 •• The xz plane contains an infinite sheet of current in the positive z direction. The current per unit length (along

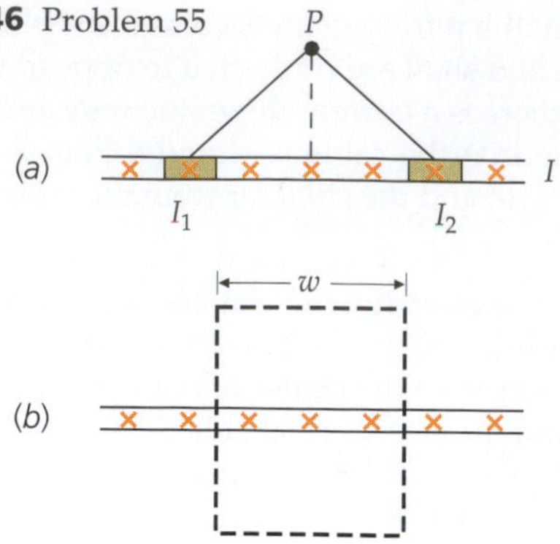

Figure 29-46 Problem 55

the x direction) is λ. Figure 29-46a shows a point P above the sheet ($y > 0$) and two portions of the current sheet labeled I_1 and I_2. (a) What is the direction of the magnetic field $\vec{B}$ at P due to the two portions of the current shown? (b) What is the direction of the magnetic field $\vec{B}$ at P due to the entire sheet? (c) What is the direction of $\vec{B}$ at a point below the sheet ($y < 0$)? (d) Apply Ampère's law to the rectangular curve shown in Figure 29-46b to show that the magnetic field at any point above the sheet is given by $\vec{B} = -\frac{1}{2}\mu_0\lambda\hat{i}$.

Magnetization and Magnetic Susceptibility

56 • True or false:

(a) Diamagnetism is the result of induced magnetic dipole moments.
(b) Paramagnetism is the result of the partial alignment of permanent magnetic dipole moments.

57 • If the magnetic susceptibility is positive,

(a) paramagnetic effects or ferromagnetic effects must be greater than diamagnetic effects.
(b) diamagnetic effects must be greater than paramagnetic effects.
(c) diamagnetic effects must be greater than ferromagnetic effects.
(d) ferromagnetic effects must be greater than paramagnetic effects.
(e) paramagnetic effects must be greater than ferromagnetic effects.

58 • A tightly wound solenoid 20 cm long has 400 turns and carries a current of 4 A such that its axial field is in the z direction. Neglecting end effects, find B and B_{app} at the center when (a) there is no core in the solenoid, and (b) there is an iron core with a magnetization $M = 1.2 \times 10^6$ A/m.

59 • Which of the four gases listed in Table 29-1 are diamagnetic and which are paramagnetic?

60 • If the solenoid of Problem 58 has an aluminum core, find B_{app}, M, and B at the center, neglecting end effects.

61 • Repeat Problem 60 for a tungsten core.

62 • A long solenoid is wound around a tungsten core and carries a current. (a) If the core is removed while the current is held constant, does the magnetic field inside the solenoid decrease or increase? (b) By what percentage?

63 • When a sample of liquid is inserted into a solenoid carrying a constant current, the magnetic field inside the solenoid decreases by 0.004%. What is the magnetic susceptibility of the liquid?

64 • A long solenoid carrying a current of 10 A has 50 turns/cm. What is the magnetic field in the interior of the solenoid when the interior is (a) a vacuum, (b) filled with aluminum, and (c) filled with silver?

65 •• An engineer intends to fill a solenoid with a mixture of oxygen and nitrogen at room temperature and 1 atmosphere pressure such that K_m is exactly 1. Assume that the magnetic dipole moments of the gas molecules are all aligned and that the susceptibility of a gas is proportional to the number density of its molecules. What should the ratio of the number densities of oxygen to nitrogen molecules be so that $K_m = 1$?

66 •• A cylinder of magnetic material is placed in a long solenoid of n turns per unit length and current I. Table 29-3 gives the magnetic field B versus nI. Use these values to plot B versus B_{app} and K_m versus nI.

Table 29-3

nI, A/m	0	50	100	150	200	500	1000	10,000
B, T	0	0.04	0.67	1.00	1.2	1.4	1.6	1.7

67 •• A small magnetic sample is in the form of a disk having a radius of 1.4 cm, a thickness of 0.3 cm, and a uniform magnetization along its axis throughout its volume. The magnetic moment of the sample is 1.5×10^{-2} A·m². (a) What is the magnetization $\vec{M}$ of the sample? (b) If this magnetization is due to the alignment of N electrons each with a magnetic moment of 1 μ_B, what is N? (c) If the magnetization is along the axis of the disk, what is the magnitude of the amperian surface current?

68 •• The magnetic moment of the earth is about 9×10^{22} A·m². (a) If the magnetization of the earth's core were 1.5×10^9 A/m, what is the core volume? (b) What is the radius of such a core if it were spherical and centered with the earth?

Atomic Magnetic Moments

69 •• Nickel has a density of 8.7 g/cm³ and molecular mass of 58.7 g/mol. Its saturation magnetization is given by $\mu_0 M_s = 0.61$ T. Calculate the magnetic moment of a nickel atom in Bohr magnetons.

70 •• Repeat Problem 69 for cobalt, which has a density of 8.9 g/cm³, a molecular mass of 58.9 g/mol, and a saturation magnetization given by $\mu_0 M_s = 1.79$ T.

Paramagnetism (optional)

71 • Show that Curie's law predicts that the magnetic susceptibility of a paramagnetic substance is $\chi_m = \mu\mu_0 M_s/3kT$.

72 •• In a simple model of paramagnetism, we can consider that some fraction f of the molecules have their magnetic moments aligned with the external magnetic field and that the rest of the molecules are randomly oriented and so do not contribute to the magnetic field. (*a*) Use this model and Curie's law to show that at temperature T and external magnetic field B the fraction of aligned molecules is $f = \mu B/3kT$. (*b*) Calculate this fraction for $T = 300$ K, $B = 1$ T, assuming μ to be 1 Bohr magneton.

73 •• Assume that the magnetic moment of an aluminum atom is 1 Bohr magneton. The density of aluminum is 2.7 g/cm^3, and its molecular mass is 27 g/mol. (*a*) Calculate M_s and $\mu_0 M_s$ for aluminum. (*b*) Use the results of Problem 71 to calculate χ_m at $T = 300$ K. (*c*) Explain why the result for part (*b*) is larger than the value listed in Table 29-1.

74 •• A toroid with N turns carrying a current I has mean radius R and cross-sectional radius r, where $r \ll R$ (Figure 29-47). When the toroid is filled with material, it is called a *Rowland ring*. Find B_{app} and B in such a ring, assuming a magnetization $\vec{M}$ everywhere parallel to $\vec{B}_{app}$.

Figure 29-47 Problem 74

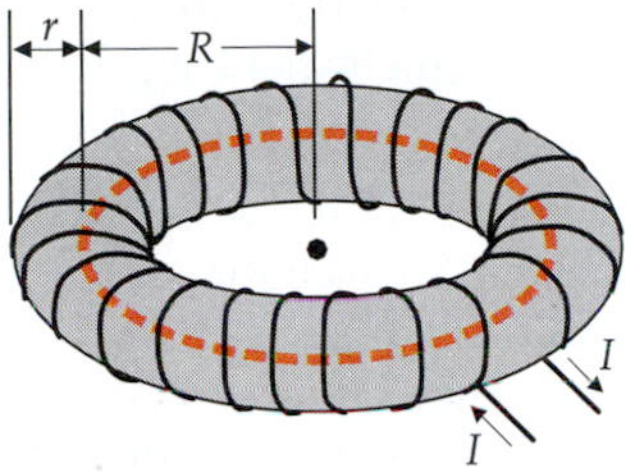

75 •• A toroid is filled with liquid oxygen that has a susceptibility of 4×10^{-3}. The toroid has 2000 turns and carries a current of 15 A. Its mean radius is 20 cm, and the radius of its cross section is 0.8 cm. (*a*) What is the magnetization M? (*b*) What is the magnetic field B? (*c*) What is the percentage increase in B produced by the liquid oxygen?

76 •• A toroid has an average radius of 14 cm and a cross-sectional area of 3 cm^2. It is wound with fine wire, 60 turns/cm measured along its mean circumference, and the wire carries a current of 4 A. The core is filled with a paramagnetic material of magnetic susceptibility 2.9×10^{-4}. (*a*) What is the magnitude of the magnetic field within the substance? (*b*) What is the magnitude of the magnetization? (*c*) What would the magnitude of the magnetic field be if there were no paramagnetic core present?

Ferromagnetism (optional)

77 • For annealed iron, the relative permeability K_m has its maximum value of about 5500 at $B_{app} = 1.57 \times 10^{-4}$ T. Find M and B when K_m is maximum.

78 •• The saturation magnetization for annealed iron occurs when $B_{app} = 0.201$ T. Find the permeability μ and the relative permeability K_m of annealed iron at saturation. (See Table 29-2.)

79 •• The coercive force is defined to be the applied magnetic field needed to bring B back to zero along the hysteresis curve (point c in Figure 29-35). For a certain permanent bar magnet, the coercive force $B_{app} = 5.53 \times 10^{-2}$ T. The bar magnet is to be demagnetized by placing it inside a 15-cm-long solenoid with 600 turns. What minimum current is needed in the solenoid to demagnetize the magnet?

80 •• A long solenoid with 50 turns/cm carries a current of 2 A. The solenoid is filled with iron, and B is measured to be 1.72 T. (*a*) Neglecting end effects, what is B_{app}? (*b*) What is M? (*c*) What is the relative permeability K_m?

81 •• When the current in Problem 80 is 0.2 A, the magnetic field is measured to be 1.58 T. (*a*) Neglecting end effects, what is B_{app}? (*b*) What is M? (*c*) What is the relative permeability K_m?

82 •• A long, iron-core solenoid with 2000 turns/m carries a current of 20 mA. At this current, the relative permeability of the iron core is 1200. (*a*) What is the magnetic field within the solenoid? (*b*) With the iron core removed, what current will produce the same field within the solenoid?

83 •• Two long, straight wires 4.0 cm apart are embedded in a uniform insulator having relative permeability of $K_m = 120$. The wires carry 40 A in opposite directions. (*a*) What is the magnetic field at the midpoint of the plane of the wires? (*b*) What is the force per unit length on the wires?

84 •• The toroid of Problem 75 has its core filled with iron. When the current is 10 A, the magnetic field in the toroid is 1.8 T. (*a*) What is the magnetization M? (*b*) Find the values for K_m, μ, and χ_m for the iron sample.

85 •• Find the magnetic field in the toroid of Problem 76 if the current in the wire is 0.2 A and soft iron, having a relative permeability of 500, is substituted for the paramagnetic core?

86 •• A long, straight wire with a radius of 1.0 mm is coated with an insulating ferromagnetic material that has a thickness of 3.0 mm and a relative magnetic permeability of $K_m = 400$. The coated wire is in air and the wire itself is nonmagnetic. The wire carries a current of 40 A. (*a*) Find the magnetic field inside the wire as a function of radius r. (*b*) Find the magnetic field inside the ferromagnetic material as a function of radius r. (*c*) Find the magnetic field outside the ferromagnetic material as a function of r. (*d*) What must the magnitudes and directions of the amperian currents be on the surfaces of the ferromagnetic material to account for the magnetic fields observed?

General Problems

87 • True or false:

(*a*) The magnetic field due to a current element is parallel to the current element.
(*b*) The magnetic field due to a current element varies inversely with the square of the distance from the element.
(*c*) The magnetic field due to a long wire varies inversely with the square of the distance from the wire.
(*d*) Ampère's law is valid only if there is a high degree of symmetry.
(*e*) Ampère's law is valid only for continuous currents.

88 • Can a particle have angular momentum and not have a magnetic moment?

89 • Can a particle have a magnetic moment and not have angular momentum?

90 • A circular loop of wire carries a current I. Is there angular momentum associated with the magnetic moment of the loop? If so, why is it not noticed?

91 • A hollow tube carries a current. Inside the tube, $\vec{B} = 0$. Why is this the case, since $\vec{B}$ is strong inside a solenoid?

92 • When a current is passed through the wire in Figure 29-48, will it tend to bunch up or form a circle?

Figure 29-48 Problem 92

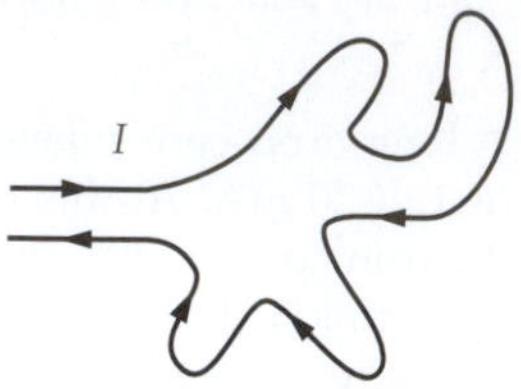

93 • Find the magnetic field at point P in Figure 29-49.

Figure 29-49 Problem 93

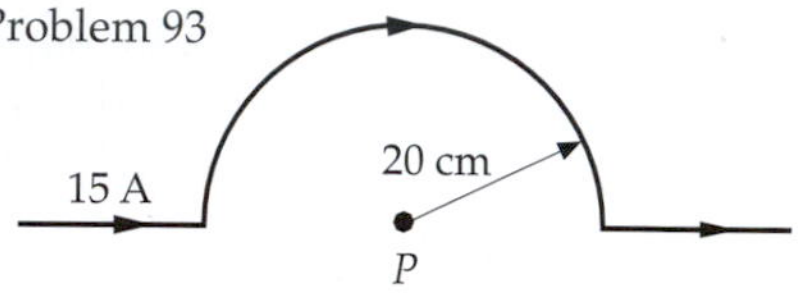

94 • In Figure 29-50, find the magnetic field at point P, which is at the common center of the two semicircular arcs.

Figure 29-50 Problem 94

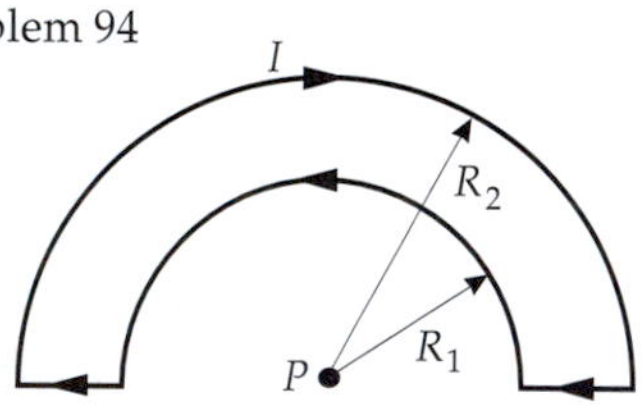

95 •• A wire of length ℓ is wound into a circular coil of N loops and carries a current I. Show that the magnetic field at the center of the coil is given by $B = \mu_0 \pi N^2 I/\ell$.

96 •• A very long wire carrying a current I is bent into the shape shown in Figure 29-51. Find the magnetic field at point P.

Figure 29-51 Problem 96

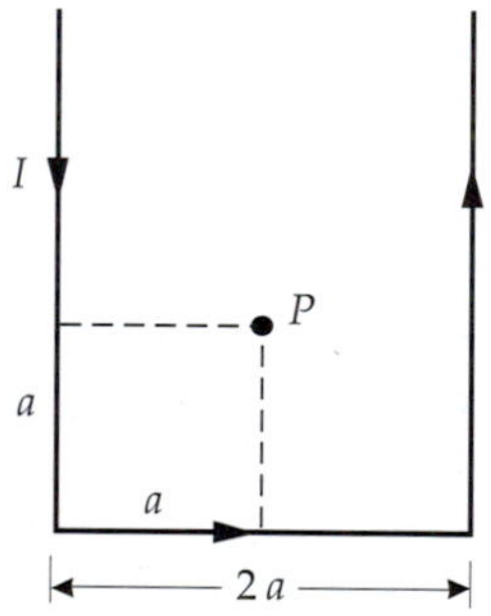

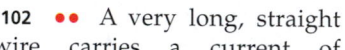

97 •• A loop of wire of length ℓ carries a current I. Compare the magnetic fields at the center of the loop when it is (*a*) a circle, (*b*) a square, and (*c*) an equilateral triangle. Which field is largest?

98 •• A power cable carrying 50.0 A is 2.0 m below the earth's surface, but its direction and precise position are unknown. Show how you could locate the cable using a compass. Assume that you are at the equator, where the earth's magnetic field is 0.7 G north.

99 •• A long, straight wire carries a current of 20 A as shown in Figure 29-52. A rectangular coil with two sides parallel to the straight wire has sides 5 cm and 10 cm with the near side a distance 2 cm from the wire. The coil carries a current of 5 A. (*a*) Find the force on each segment of the rectangular coil due to the current in the long, straight wire. (*b*) What is the net force on the coil?

Figure 29-52 Problem 99

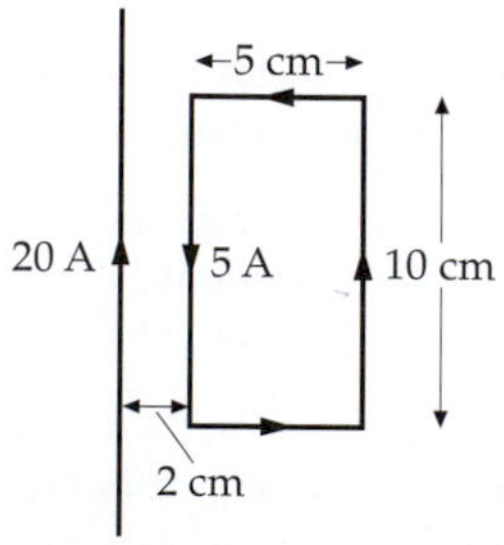

100 •• The closed loop shown in Figure 29-53 carries a current of 8.0 A in the counterclockwise direction. The radius of the outer arc is 60 cm, that of the inner arc is 40 cm. Find the magnetic field at point P.

Figure 29-53 Problem 100

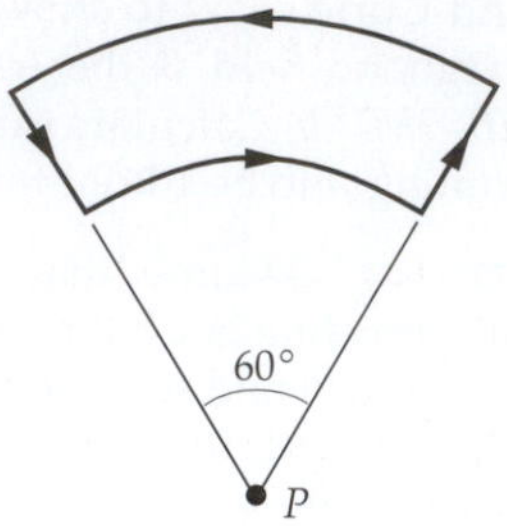

101 •• A closed circuit consists of two semicircles of radii 40 and 20 cm that are connected by straight segments as shown in Figure 29-54. A current of 3.0 A flows around this circuit in the clockwise direction. Find the magnetic field at point P.

Figure 29-54 Problem 101

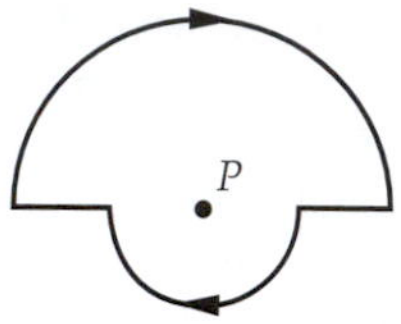

102 •• A very long, straight wire carries a current of 20.0 A. An electron 1.0 cm from the center of the wire is moving with a speed of 5.0×10^6 m/s. Find the force on the electron when it moves (*a*) directly away from the wire, (*b*) parallel to the wire in the direction of the current, and (*c*) perpendicular to the wire and tangent to a circle around the wire.

103 •• A current I is uniformly distributed over the cross section of a long, straight wire of radius 1.40 mm. At the surface of the wire, the magnitude of the magnetic field is $B = 2.46$ mT. Find the magnitude of the magnetic field at (*a*) 2.10 mm from the axis and (*b*) 0.60 mm from the axis. (*c*) Find the current I.

104 •• A large, 50-turn circular coil of radius 10.0 cm carries a current of 4.0 A. At the center of the large coil is a small 20-turn coil of radius 0.5 cm carrying a current of 1.0 A. The planes of the two coils are perpendicular. Find the torque exerted by the large coil on the small coil. (Neglect any variation in B due to the large coil over the region occupied by the small coil.)

105 •• Figure 29-55 shows a bar magnet suspended by a thin wire that provides a restoring torque $-\kappa\theta$. The magnet is 16 cm long, has a mass of 0.8 kg, a dipole moment of $\mu = 0.12$ A·m^2, and it is located in a region where a uniform magnetic field B can be established. When the external magnetic field is 0.2 T and the magnet is given a small angular displacement $\Delta\theta$, the bar magnet oscillates about its equilibrium position with a period of 0.500 s. Determine the constant κ and the period of this torsional pendulum when $B = 0$.

Figure 29-55 Problem 105

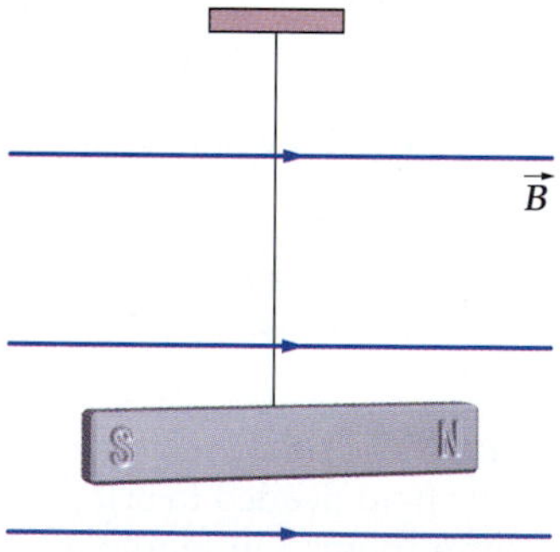

106 •• A long, narrow bar magnet that has magnetic moment $\vec{\mu}$ parallel to its long axis is suspended at its center as a frictionless compass needle. When placed in a magnetic field $\vec{B}$, the needle lines up with the field. If it is displaced by a small angle θ, show that the needle will oscillate about its equilibrium position with frequency $f = (1/2\pi)\sqrt{\mu B/I}$, where I is the moment of inertia about the point of suspension.

107 •• A small bar magnet of mass 0.1 kg, length 1 cm, and magnetic moment $\mu = 0.04\ \text{A·m}^2$ is located at the center of a 100-turn loop of 0.2 m diameter. The loop carries a current of 5.0 A. At equilibrium, the bar magnet is aligned with the field due to the current loop. The bar magnet is given a displacement along the axis of the loop and released. Show that if the displacement is small, the bar magnet executes simple harmonic motion, and find the period of this motion.

108 •• Suppose the needle in Problem 106 is a uniformly magnetized iron rod that is 8 cm long and has a cross-sectional area of 3 mm^2. Assume that the magnetic dipole moment for each iron atom is $2.2\mu_B$ and that all the iron atoms have their dipole moments aligned. Calculate the frequency of small oscillations about the equilibrium position when the magnetic field is 0.5 G.

109 •• The needle of a magnetic compass has a length of 3 cm, a radius of 0.85 mm, and a density of $7.96 \times 10^3\ \text{kg/m}^3$. It is free to rotate in a horizontal plane, where the horizontal component of the earth's magnetic field is 0.6 G. When disturbed slightly, the compass executes simple harmonic motion about its midpoint with a frequency of 1.4 Hz. (*a*) What is the magnetic dipole moment of the needle? (*b*) What is the magnetization M? (*c*) What is the amperian current on the surface of the needle? (See Problem 106.)

110 •• An iron bar of length 1.4 m has a diameter of 2 cm and a uniform magnetization of 1.72×10^6 A/m directed along the bar's length. The bar is stationary in space and is suddenly demagnetized so that its magnetization disappears. What is the rotational angular velocity of the bar if its angular momentum is conserved? (Assume that Equation 29-27 holds where m is the mass of an electron and $q = -e$.)

111 •• The magnetic dipole moment of an iron atom is $2.219\mu_B$. (*a*) If all the atoms in an iron bar of length 20 cm and cross-sectional area 2 cm^2 have their dipole moments aligned, what is the dipole moment of the bar? (*b*) What torque must be supplied to hold the iron bar perpendicular to a magnetic field of 0.25 T?

112 •• A relatively inexpensive ammeter called a *tangent galvanometer* can be made using the earth's field. A plane circular coil of N turns and radius R is oriented such that the field B_c it produces in the center of the coil is either east or west. A compass is placed at the center of the coil. When there is no current in the coil, the compass needle points north. When there is a current I, the compass needle points in the direction of the resultant magnetic field $\vec{B}$ at an angle θ to the north. Show that the current I is related to θ and the horizontal component of the earth's field B_e by

$$I = \frac{2RB_e}{\mu_0 N}\tan\theta$$

113 •• An infinitely long, straight wire is bent as shown in Figure 29-56. The circular portion has a radius of 10 cm with its center a distance r from the straight part. Find r such that the magnetic field at the center of the circular portion is zero.

Figure 29-56 Problem 113

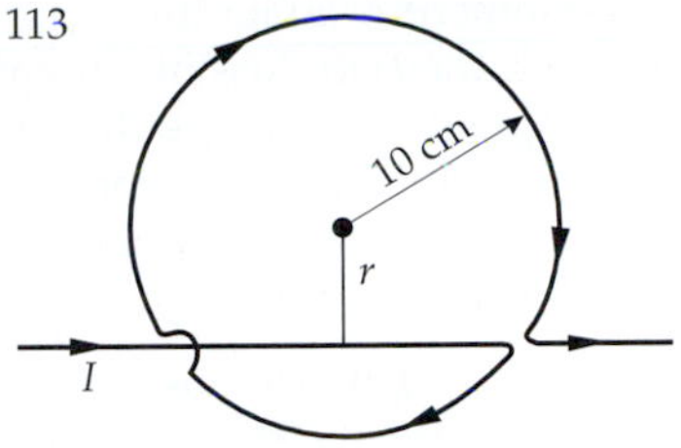

114 •• (*a*) Find the magnetic field at point P for the wire carrying current I shown in Figure 29-57. (*b*) Use your result from (*a*) to find the field at the center of a polygon of N sides. Show that when N is very large, your result approaches that for the magnetic field at the center of a circle.

Figure 29-57 Problem 114

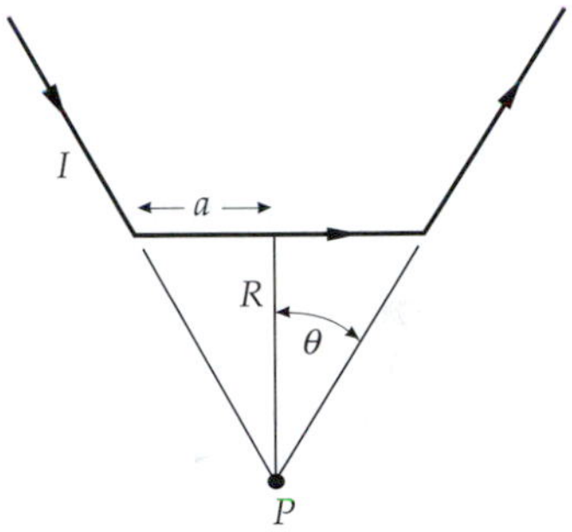

115 •• The current in a long, cylindrical conductor of radius $R = 10$ cm varies with distance from the axis of the cylinder according to the relation $I(r) = (50\ \text{A/m})r$. Find the magnetic field at (a) $r = 5$ cm (b) at $r = 10$ cm, and (c) $r = 20$ cm.

116 •• Figure 29-58 shows a square loop, 20 cm per side, in the xy plane with its center at the origin. The loop carries a current of 5 A. Above it at $y = 0$, $z = 10$ cm is an infinitely long wire parallel to the x axis carrying a current of 10 A. (*a*) Find the torque on the loop. (*b*) Find the net force on the loop.

Figure 29-58 Problem 116

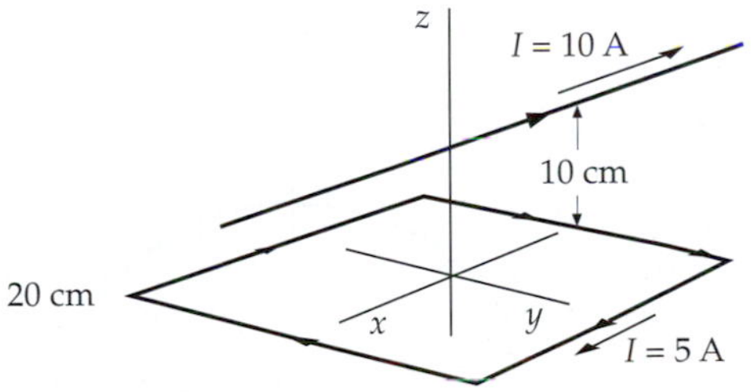

117 •• In the Bohr model of the hydrogen atom, an electron in the ground state orbits a proton at a radius of 5.29×10^{-11} m. In a reference frame in which the orbiting electron is at rest, the proton circulates about the electron at a radius of 5.29×10^{-11} m with the same angular velocity as that of the electron in the reference frame in which the proton is at rest.

Consequently, in the rest frame of the electron, the magnetic field due to the motion of the proton has the same magnitude as that calculated in Problem 5. The electron has an intrinsic magnetic moment of magnitude μ_B. Find the energy difference between the two possible orientations of the electron's intrinsic magnetic moment, either parallel or antiparallel to the magnetic field due to the apparent motion of the proton. (This energy difference is readily observed spectroscopically and is known as the *fine structure splitting*.)

118 •• The proton also has an intrinsic magnetic moment whose magnitude is $1.52 \times 10^{-3}\mu_B$. The orientation of the proton's magnetic moment is quantized; it can only be parallel or antiparallel to the magnetic field at the location of the proton. Using the result of Problem 117, determine the energy difference of the hydrogen atom in the Bohr model for the two possible orientations of the proton's magnetic moment; neglect the magnetic field at the proton due to the electron's intrinsic magnetic moment. (This energy difference is referred to as *hyperfine structure splitting*.)

119 •• In the calculation of the hyperfine structure splitting in Problem 117, you neglected the magnetic field at the proton's position due to the intrinsic magnetic moment of the electron. Calculate the magnetic field due to the intrinsic magnetic moment of the electron at a distance of 5.29×10^{-11} m and compare its magnitude at the location of the proton with that due to the orbital motion of the electron. (*Hint*: Assume the electron spin is perpendicular to the plane of its orbit, and show that the magnitude of the magnetic field at a great distance from a magnetic dipole on a line perpendicular to the dipole is given by $B = (\mu_0/4\pi)\mu/r^3$.)

120 ••• A disk of radius R carries a fixed charge density σ and rotates with angular velocity ω. (*a*) Consider a circular strip of radius r and width dr with charge dq. Show that the current produced by this strip $dI = (\omega/2\pi)\, dq = \omega\sigma r\, dr$. (*b*) Use your result from part (*a*) to show that the magnetic field at the center of the disk is $B = \frac{1}{2}\mu_0\sigma\omega R$. (*c*) Use your result from part (*a*) to find the magnetic field at a point on the axis of the disk a distance x from the center.

121 ••• A very long, straight conductor with a circular cross section of radius R carries a current I. Inside the conductor, there is a cylindrical hole of radius a whose axis is parallel to the axis of the conductor a distance b from it (Figure 29-59). Let the z axis be the axis of the conductor, and let the axis of the hole be at $x = b$. Find the magnetic field $\vec{B}$ at the point (*a*) on the x axis at $x = 2R$, and (*b*) on the y axis at $y = 2R$. (*Hint*: Consider a uniform current distribution throughout the cylinder of radius R plus a current in the opposite direction in the hole.)

Figure 29-59 Problem 121

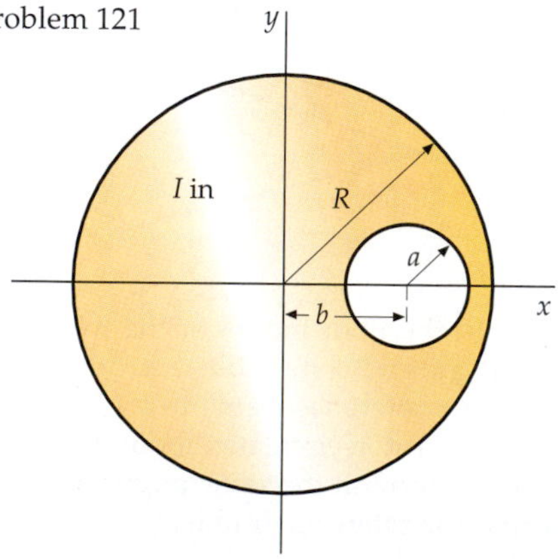

122 ••• For the cylinder with the hole in Problem 121, show that the magnetic field inside the hole is uniform, and find its magnitude and direction.

123 ••• A square loop of side ℓ lies in the yz plane with its center at the origin. It carries a current I. Find the magnetic field B at any point on the x axis and show from your expression that for x much larger than ℓ,

$$B \approx \frac{\mu_0}{4\pi}\frac{2\mu}{x^3}$$

where $\mu = I\ell^2$ is the magnetic moment of the loop.

124 ••• A circular loop carrying current I lies in the yz plane with its axis along the x axis. (*a*) Evaluate the line integral $\oint_C \vec{B}\cdot d\vec{\ell}$ along the axis of the loop from $x = -\ell_1$ to $x = +\ell_1$. (*b*) Show that when $\ell_1 \to \infty$, the line integral approaches $\mu_0 I$. This result can be related to Ampère's law by closing the curve of integration with a semicircle of radius ℓ on which $B \approx 0$ for ℓ very large.

125 ••• The current in a long cylindrical conductor of radius R is given by $I(r) = I_0(1 - e^{r/a})$. Derive expressions for the magnetic field for $r < R$ and for $r > R$.

126 ••• In Example 29-8 we calculated the magnetic field inside and outside a wire of radius R carrying a uniform current I. Consider a filament of current at a distance r from the center of the wire. Show that this filament experiences a force directed toward the center of the wire and that, therefore, the current distribution cannot be truly uniform. This so-called *pinch effect* depends on the magnitude of the current, but is generally negligibly small. Consider a copper wire of 2.0 cm diameter that carries a nominally uniform current of 400 A. Calculate the force per meter exerted on a 0.1-mm-thick annular region of current at the periphery of the wire. Assume that the current is carried by the conduction electrons. How far toward the center of the wire would this annular current region have to move so that the electrostatic force between the electrons and the fixed positive ions at the periphery just balances the magnetic force?

CHAPTER 30

Magnetic Induction

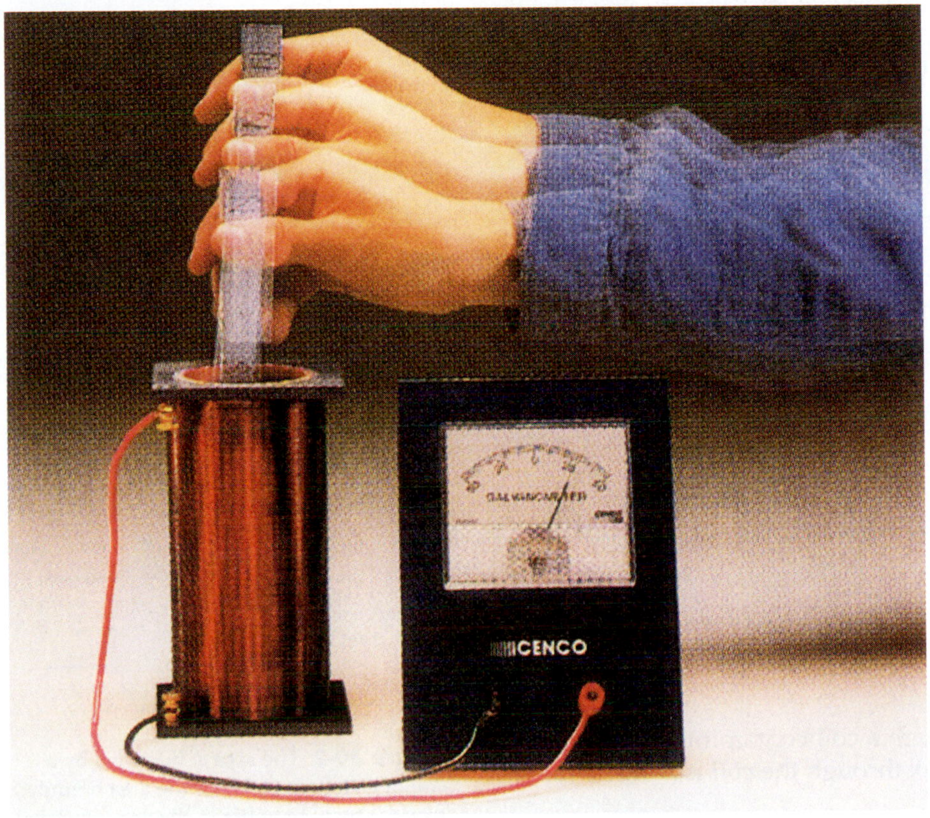

Demonstration of induced emf. When the magnet is moving toward or away from the coil, an emf is induced in the coil, as shown by the galvanometer's deflection. No deflection is observed when the magnet is stationary.

In the early 1830s, Michael Faraday in England and Joseph Henry in America independently discovered that a *changing* magnetic field induces a current in a wire. The emfs and currents caused by changing magnetic fields are called **induced emfs** and **induced currents.** The process itself is referred to as **magnetic induction.**

When you pull the plug of an electric cord from its socket, you sometimes observe a small spark. Before the cord is disconnected, it carries a current, which produces a magnetic field encircling the current. When the cord is disconnected, the current abruptly ceases and the magnetic field around it collapses. The changing magnetic field produces an emf that tries to maintain the original current, resulting in a spark across the plug. Once the magnetic field reaches zero it is no longer changing, and the emf is zero.

Changing magnetic fields can result from changing currents or from moving magnets. The chapter-opening photo illustrates a simple classroom demonstration of induced currents. The ends of a coil are attached to a galvanometer and a strong magnet is moved toward or away from the coil. The momentary deflection shown by the galvanometer *during* the motion indicates that there is an induced electric current in the coil–galvanometer circuit. A current is also induced if the coil is moved toward or away from the magnet, or if the coil is rotated in a fixed magnetic field. A coil rotating in a magnetic field is the basic element of a generator, which converts mechanical energy into electrical energy.

All of the various methods of magnetic induction can be summarized by a single relation known as Faraday's law, which relates the induced emf in a circuit to the change in magnetic flux through the circuit.

30-1 Magnetic Flux

The flux of a magnetic field through a surface is defined similarly to the flux of an electric field (Section 22-2). Let dA be an element of area on the surface and $\hat{n}$ be the unit vector perpendicular to the element (Figure 30-1). The magnetic flux $\phi_{\rm m}$ is then defined to be

$$\phi_{\rm m} = \int_S \vec{B}\cdot\hat{n}\, dA = \int_S B_n\, dA \qquad 30\text{-}1$$

Definition—Magnetic flux

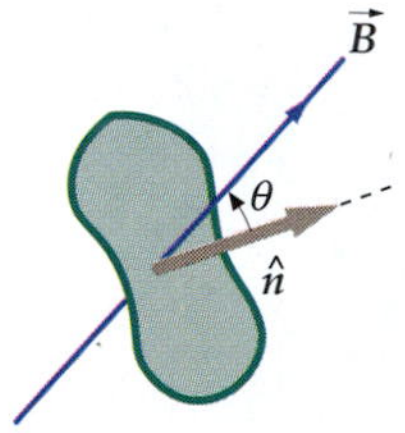

Figure 30-1 When $\vec{B}$ makes an angle θ with the normal to the area of a loop, the flux through the loop is $B \cos\theta\, A$.

The unit of magnetic flux is that of magnetic field times area, tesla-meter squared, which is called a **weber** (Wb):

$$1\text{ Wb} = 1\text{ T}\cdot\text{m}^2 \qquad 30\text{-}2$$

Since B is proportional to the number of field lines per unit area, the magnetic flux is proportional to the number of lines through the area.

Exercise Show that a weber per second is a volt.

If the surface is a plane with area A, and $\vec{B}$ is constant in magnitude and direction over the surface and makes an angle θ with the unit normal vector, the flux is

$$\phi_{\rm m} = BA\cos\theta$$

We are often interested in the flux through a coil containing several turns of wire. If the coil contains N turns, the flux through the coil is N times the flux through each turn (Figure 30-2):

$$\phi_{\rm m} = NBA\cos\theta \qquad 30\text{-}3$$

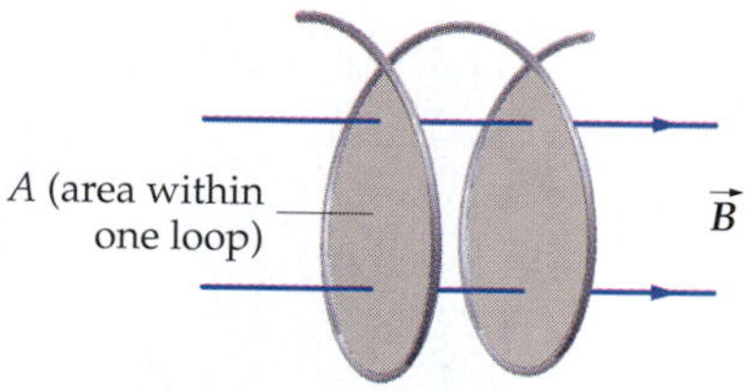

Figure 30-2 The area A bounded by a coil of two turns is twice the area bounded by each turn. In general, the area bounded by a coil of N turns is N times that of each turn.

Example 30-1

Find the magnetic flux through a solenoid that is 40 cm long, has a radius of 2.5 cm, has 600 turns, and carries a current of 7.5 A.

Picture the Problem The magnetic field $\vec{B}$ inside the solenoid is uniform and along the axis of the solenoid. It is therefore perpendicular to the plane of the coils. We thus need to find B inside the solenoid and then multiply B by NA.

1. The magnetic flux is the product of the number of turns, the magnetic field, and the area of the coils: $\phi_m = NBA$

2. The magnetic field inside the solenoid is given by $B = \mu_0 nI$, where $n = N/\ell$ is the number of turns per unit length: $\phi_m = N\mu_0(N/\ell)IA = \mu_0 N^2 IA/\ell$

3. Express the area of the coils in terms of its radius: $A = \pi r^2$

4. Substitute the given values to calcuate the flux:
$$\phi_m = \mu_0 N^2 IA/\ell$$
$$= (4\pi \times 10^{-7}\ \mathrm{T{\cdot}m/A})(600\ \mathrm{turns})^2(7.5\ \mathrm{A})\pi(0.025\ \mathrm{m})^2/(0.40\ \mathrm{m})$$
$$= 1.66 \times 10^{-2}\ \mathrm{Wb}$$

Remark Note that since $\phi_m = NBA$ and B is proportional to the number of turns N, the flux is proportional to N^2.

30-2 Induced emf and Faraday's Law

Experiments by Faraday, Henry, and others showed that if the magnetic flux through an area bounded by a circuit is changed by any means, an emf equal in magnitude to the rate of change of the flux is induced in the circuit. We usually detect the emf by observing a current in the circuit, but it is present even when the circuit is incomplete (not closed) and there is no current. Previously we considered emfs that were localized in a specific part of the circuit, such as between the terminals of the battery. However, induced emfs can be considered to be distributed throughout the circuit.

The magnetic flux through a circuit can be changed in many different ways. The current producing the magnetic field may be increased or decreased, permanent magnets may be moved toward the circuit or away from it, the circuit itself may be moved toward or away from the source of the flux, the orientation of the circuit may be changed, or the area of the circuit in a fixed magnetic field may be increased or decreased. In every case, an emf is induced in the circuit that is equal in magnitude to the rate of change of the magnetic flux.

Figure 30-3 shows a single loop of wire in a magnetic field. If the flux through the loop is changing, an emf is induced in the loop. Since emf is the work done per unit charge, there must be a force exerted on the charge associated with the emf. The force per unit charge is the electric field $\vec{E}$, which in this case is induced by the changing flux. The line integral of the electric field around a complete circuit equals the work done per unit charge, which, by definition, is the emf in the circuit:

$$\mathcal{E} = \oint_C \vec{E}\cdot d\vec{\ell} \qquad 30\text{-}4$$

Definition—emf

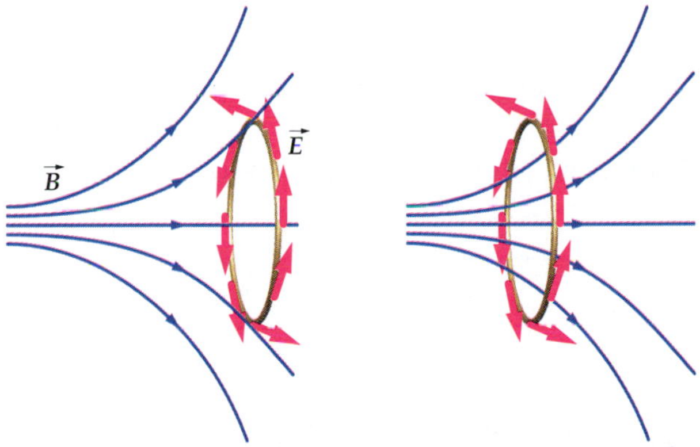

Figure 30-3 When the magnetic flux through the wire loop is changing, an emf is induced in the loop. The emf is distributed throughout the loop and is equivalent to a nonconservative electric field $\vec{E}$ parallel to the wire. In this figure, the direction of $\vec{E}$ corresponds to the case in which the flux through the loop is increasing.

The electric fields that we studied previously resulted from static electric charges. These fields are conservative, meaning that the work done by the electrostatic field around a closed curve is zero. The electric field resulting from changing magnetic flux is not conservative. Its line integral around a closed curve equals the induced emf, which equals the rate of change of the magnetic flux:

$$\mathcal{E} = \oint_C \vec{E} \cdot d\vec{\ell} = -\frac{d\phi_m}{dt} \qquad 30\text{-}5$$

Faraday's law

This result is known as **Faraday's law.** The negative sign in Faraday's law has to do with the direction of the induced emf, which we will discuss shortly.

Example 30-2

A uniform magnetic field makes an angle of 30° with the axis of a circular coil of 300 turns and a radius of 4 cm. The field changes at a rate of 85 T/s. Find the magnitude of the induced emf in the coil.

Picture the Problem The induced emf equals N times the rate of change of the flux through each turn. Since B is uniform, the flux through each turn is simply $\phi_m = BA \cos\theta$, where $A = \pi r^2$ is the area of the coil.

1. The magnitude of the induced emf is given by Faraday's law: $|\mathcal{E}| = \frac{d\phi_m}{dt}$

2. For a uniform field, the flux is: $\phi_m = NBA\cos\theta$

3. Substitute this expression for ϕ_m and calculate $|\mathcal{E}|$: $|\mathcal{E}| = \frac{d\phi_m}{dt} = \frac{d}{dt}(NBA\cos\theta) = NA\cos\theta\frac{dB}{dt}$

$$= (300)(3.14)(0.04\text{ m})^2 \cos 30° (85\text{ T/s}) = 111\text{ V}$$

Exercise If the resistance of the coil is 200 Ω, what is the induced current? (*Answer* 0.555 A)

Example 30-3 *try it yourself*

An 80-turn coil has a radius of 5.0 cm and a resistance of 30 Ω. At what rate must a perpendicular magnetic field change to produce a current of 4.0 A in the coil?

Picture the Problem The rate of change of the magnetic field is related to the rate of change of the flux, which is related to the induced emf by Faraday's law. The emf in the coil equals IR.

Cover the column to the right and try these on your own before looking at the answers.

Steps	**Answers**
1. Write the magnetic flux in terms of B, N, and the radius r, and solve for B.	$\phi_m = N\pi r^2 B$ $B = \dfrac{\phi_m}{N\pi r^2}$
2. Take the time derivative of B.	$\dfrac{dB}{dt} = \dfrac{1}{N\pi r^2}\dfrac{d\phi_m}{dt}$
3. Use Faraday's law to relate the rate of change of the flux to the emf.	$\dfrac{d\phi_m}{dt} = \mathcal{E}$
4. Calculate the emf in the coil from the current and resistance of the coil.	$\mathcal{E} = IR = 120\text{ V}$
5. Substitute numerical values of E, N, and r to calculate dB/dt.	$\dfrac{dB}{dt} = \dfrac{1}{N\pi r^2}\dfrac{d\phi_m}{dt} = 191\text{ T/s}$

Example 30-4

A magnetic field $\vec{B}$ is perpendicular to the plane of the page and uniform in a circular region of radius R as shown in Figure 30-4. Outside of the circular region, $\vec{B}$ decreases to 0. The rate of change of the magnitude of $\vec{B}$ is dB/dt. What is the magnitude of the induced electric field in the plane of the page (*a*) at a distance $r < R$ from the center of the circular region, and (*b*) at a distance $r > R$, where $B = 0$.

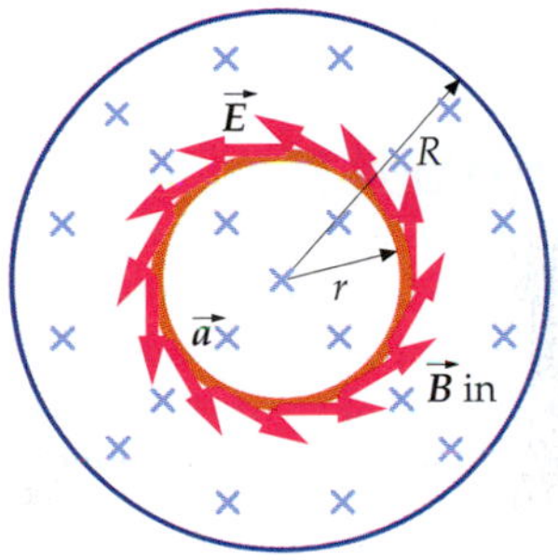

Figure 30-4

Picture the Problem The magnetic field $\vec{B}$ is into the page and uniform over a circular region of radius R. When B changes, the magnetic flux changes and an emf $\mathcal{E} = \oint_C \vec{E}\cdot d\vec{\ell}$ is induced around any curve enclosing the flux. The induced electric field is found by applying Faraday's law. Since we are interested only in magnitudes, we neglect the minus sign and use $\oint_C \vec{E}\cdot d\vec{\ell} = d\phi_m/dt$. To take advantage of the system's symmetry we choose a circular curve of radius r to compute the line integral. By symmetry, $\vec{E}$ is tangent to this curve and has the same magnitude at any point on it. We then calculate the magnetic flux ϕ_m and take its time derivative. Setting the integral and the time derivative equal yields an expression for E.

(*a*) 1. E is found from the line integral for a circle of radius $r < R$. $\vec{E}$ is tangent to the circle and has a constant magnitude:	$\oint_C \vec{E}\cdot d\vec{\ell} = E(2\pi r)$
2. The line integral is calculated from Faraday's law:	$\oint_C \vec{E}\cdot d\vec{\ell} = \dfrac{d\phi_m}{dt}$
3. For $r < R$, B is constant over the circle. Since $\vec{B}$ is perpendicular to the plane of the circle, the flux is simply BA:	$\phi_m = BA = B\pi r^2$
4. Calculate the time derivative of ϕ_m:	$\dfrac{d\phi_m}{dt} = \pi r^2\dfrac{dB}{dt}$

5. Substitute these results for $\oint_C \vec{E}\cdot d\vec{\ell}$ and solve for E:

$$2\pi r E = \pi r^2 \frac{dB}{dt}$$

$$E = \frac{r}{2}\frac{dB}{dt}, \quad r < R$$

(*b*)1. For a circle of radius $r > R$, where the magnetic field is zero, the line integral is the same as before:

$$\oint_C \vec{E}\cdot d\vec{\ell} = E(2\pi r)$$

2. Since $B = 0$ for $r > R$, the magnetic flux is $\pi R^2 B$:

$$\phi_m = \pi R^2 B$$

3. Apply Faraday's law to find E:

$$2\pi r E = \pi R^2 \frac{dB}{dt}$$

$$E = \frac{R^2}{2r}\frac{dB}{dt}, \quad r > R$$

Remarks Note that the electric field in this example is produced by a changing magnetic field rather than by electric charges. If charges had caused the field, $\vec{E}$ would have to start on positive charges and end on negative charges. Since charges are not present, however, $\vec{E}$ forms circles that have no beginning and no end. Note also that the emf exists in any closed curve bounding the area through which the magnetic flux is changing whether or not there is a wire or circuit along the curve.

Example 30-5

A small coil of *N* turns has its plane perpendicular to a uniform magnetic field $\vec{B}$ as shown in Figure 30-5. The coil is connected to a current integrator Ⓒ, a device designed to measure the total charge passing through it. Find the charge passing through the coil if the coil is rotated through 180° about its diameter.

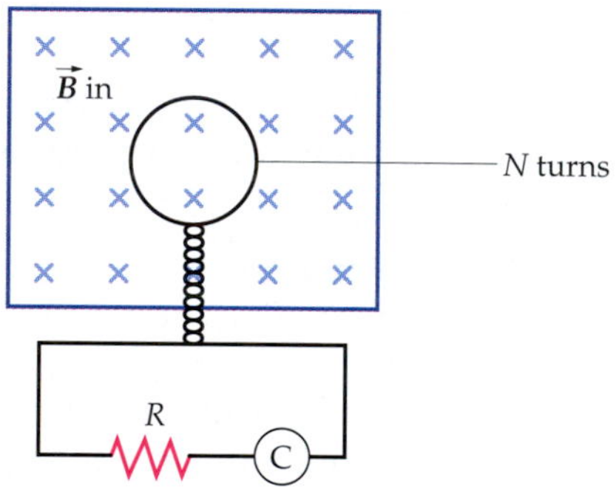

Figure 30-5

Picture the Problem When the coil in Figure 30-5 is rotated, the magnetic flux through it changes, causing an induced emf, $\mathcal{E}$. The emf in turn causes a current $I = \mathcal{E}/R$, where R is the total resistance of the circuit. Since $I = dQ/dt$, we can find the charge passing through the coil by integrating I; that is, $Q = \int dQ = \int I\,dt$.

1. The total charge is the integral of the current:

$$Q = \int dQ = \int I\,dt$$

2. The current is related to the emf by Ohm's law:

$$I = \frac{\mathcal{E}}{R}$$

3. The magnitude of the emf $\mathcal{E}$ is given by Faraday's law:

$$|\mathcal{E}| = \frac{d\phi_m}{dt}$$

4. Substitute these results to find the charge:

$$Q = \int I\,dt = \int \frac{\mathcal{E}}{R}\,dt = \frac{1}{R}\int \frac{d\phi_m}{dt}\,dt = \frac{1}{R}\int d\phi_m$$

$$= \frac{\Delta\phi_m}{R}$$

5. The initial flux through the coil is $\phi_{mi} = NBA$. When the coil is flipped, the flux reverses. Thus, $\phi_{mf} = -NBA$. This gives us $\Delta\phi_m$:
   $\Delta\phi_m = \phi_{mf} - \phi_{mi} = (NBA) - (-NBA) = 2NBA$

6. Combining the previous two results yields Q:
   $$Q = \frac{2NBA}{R}$$

Remarks Note that the charge Q is independent of the time involved in rotating the coil—all that matters is the change in magnetic flux. A coil used in this way is called a *flip coil*. It is used to measure magnetic fields. For example, if the current integrator Ⓒ measures a total charge Q passing through the coil when it is flipped, the magnetic field can be found from $B = RQ/2NA$.

Exercise A flip coil of 40 turns has a radius of 3 cm and a resistance of 16 Ω. If the coil is turned through 180° in a magnetic field of 5000 G, how much charge passes through it? (*Answer* 7.07 mC)

30-3 Lenz's Law

The negative sign in Faraday's law has to do with the direction of the induced emf, which can be found from a general physical principle known as **Lenz's law:**

> The induced emf and induced current are in such a direction as to oppose the change that produces them.

Lenz's law

Note that we didn't specify just what kind of change causes the induced emf and current. We purposefully left the statement vague to cover a variety of conditions, which we will now illustrate.

Figure 30-6 shows a bar magnet moving toward a loop that has a resistance R. Since $\vec{B}$ from the bar magnet is to the right, out of the north pole of the magnet, the movement of the magnet toward the loop tends to increase the flux through the loop to the right. (The magnetic field at the loop is stronger when the magnet is closer.) The induced current in the loop produces a magnetic field of its own. This induced current is in the direction shown, so the magnetic flux it produces is opposite that of the magnet. The induced magnetic field tends to *decrease* the flux through the loop. If the magnet were moved away from the loop, which would decrease the flux through the loop due to the magnet, the induced current would be in the opposite direction from that in Figure 30-6. In that case, the current would produce a magnetic field to the right, which would tend to increase the flux through the loop. As we might expect, moving the loop toward or away from the magnet has the same effect as moving the magnet. Only the relative motion is important.

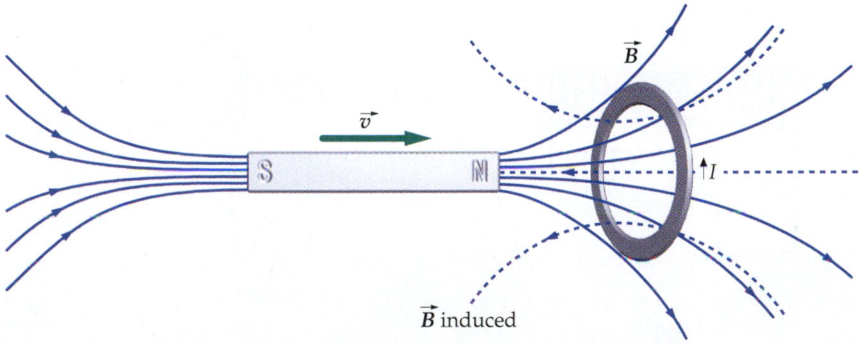

Figure 30-6 When the bar magnet is moving toward the loop, the emf induced in the loop produces a current in the direction shown. The magnetic field due to the induced current in the loop (indicated by the dashed lines) produces a flux that opposes the increase in flux through the loop due to the motion of the magnet.

Figure 30-7 shows the induced magnetic moment of the current loop when the magnet is moving toward it as in Figure 30-6. The loop acts like a small magnet with its north pole to the left and its south pole to the right. Since opposite poles attract and like poles repel, the induced magnetic moment of the loop repels the bar magnet, that is, it opposes its motion toward the loop. Thus, we can express Lenz's law in terms of forces rather than flux. If the bar magnet is moved toward the loop, the induced current must produce a magnetic moment to oppose this change.

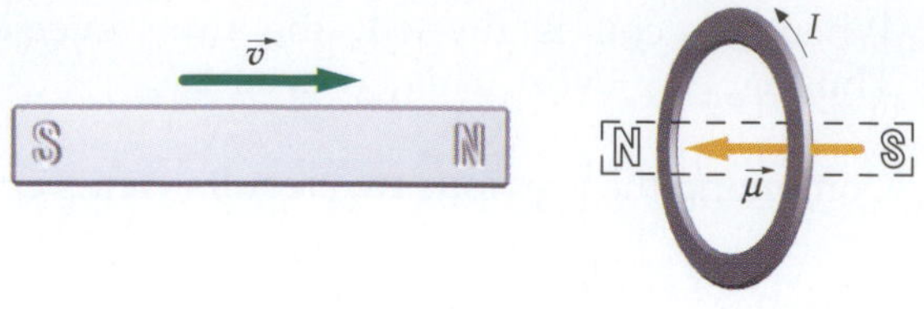

Figure 30-7 The magnetic moment of the loop $\vec{\mu}$ (shown in outline as if it were a magnet) due to the induced current is such as to oppose the motion of the bar magnet. Here the bar magnet is moving toward the loop so the induced magnetic moment repels the bar magnet.

Lenz's law is required by the law of conservation of energy. If the current in the loop in Figure 30-7 were opposite the direction shown, the induced magnetic moment of the loop would attract the magnet when it is moving toward the loop and cause it to accelerate toward the loop. If we begin with the magnet a great distance from the loop and give it a very slight push toward the loop, the force due to the induced current would be toward the loop, which would increase the velocity of the magnet. As the speed of the magnet increases, the rate of change of the flux would increase, thereby increasing the induced current. This would further increase the force on the magnet. Hence, the kinetic energy of the magnet and the rate at which Joule heat is produced in the loop (I^2R) would both increase with no source of energy. This would violate the law of conservation of energy.

In Figure 30-8, the bar magnet is at rest, and the loop is moving away from it. The induced current and magnetic moment are shown in the figure. In this case, the magnetic moment of the loop attracts the bar magnet, thus opposing the motion of the loop as required by Lenz's law.

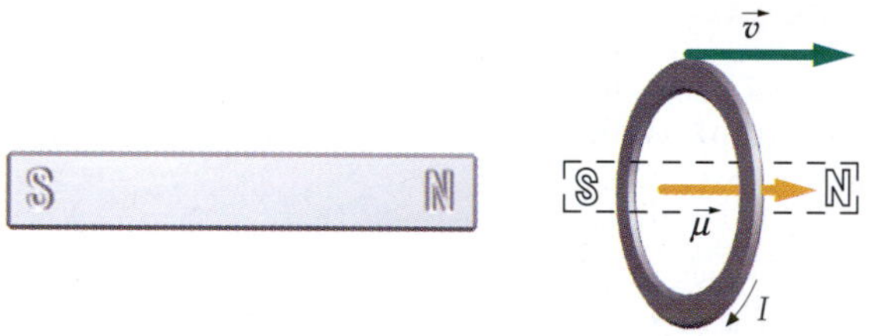

Figure 30-8 When the loop is moving away from the stationary bar magnet, the induced magnetic moment in the loop attracts the bar magnet, again opposing the relative motion.

In Figure 30-9, when the current in circuit 1 is changing, there is a changing flux through circuit 2. Suppose that the switch S in circuit 1 is initially open so that there is no current in the circuit (Figure 30-9*a*). When we close the switch (Figure 30-9*b*), the current in circuit 1 does not reach its steady value $\mathcal{E}_1/R_1$ instantaneously but takes some time to change from zero to this value. During this time, while the current is increasing, the flux through circuit 2 is changing, and there is an induced current in that circuit in the direction shown. When the current in circuit 1 reaches its steady value, the flux through circuit 2 is no longer changing, so there is no induced current in circuit 2. An induced current in circuit 2 in the opposite direction appears momentarily when the switch in circuit 1 is opened (Figure 30-9*c*) and the current is decreasing to zero. It is important to understand that there is an induced emf *only while the flux is changing*. The emf does not depend on the magnitude of the flux, only on its rate of change. If there is a large, steady flux through a circuit, there is no induced emf.

Figure 30-9 (*a*) Two adjacent circuits. (*b*) Just after the switch is closed, I_1 is increasing in the direction shown. The changing flux in circuit 2 induces the current I_2. The flux due to I_2 opposes the increase in flux due to I_1. (*c*) As the switch is opened, I_1 decreases and B decreases. The induced current I_2 then tends to maintain the flux in the circuit, opposing the change.

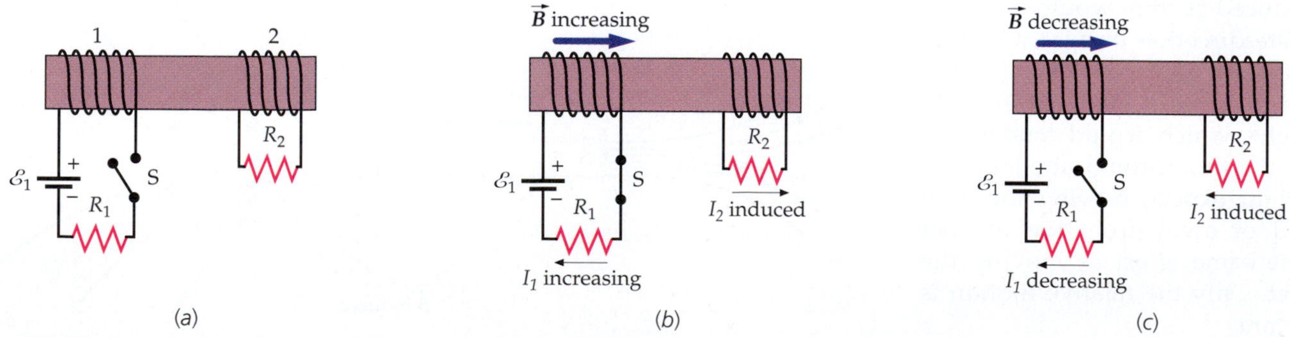

For our next example, we consider the single, isolated circuit shown in Figure 30-10. When there is a current in the circuit, there is a magnetic flux through the coil due to its own current. When the current is changing, the flux in the coil is changing and there is an induced emf in the circuit. This *self-induced* emf opposes the change in the current. It is therefore called a **back emf**. Because of this self-induced emf, the current in a circuit cannot jump instantaneously from zero to some finite value or from some finite value to zero. Henry first noticed this effect when he was experimenting with a circuit consisting of many turns of a wire like that in Figure 30-10. This arrangement gives a large flux through the circuit for even a small current. Henry noticed a spark across the switch when he tried to break the circuit. Such a spark is due to the large induced emf that occurs when the current varies rapidly, as during the opening of the switch. In this case, the induced emf is directed so as to maintain the original current. The large induced emf produces a large voltage drop across the switch as it is opened. The electric field between the contacts of the switch is large enough to tear electrons from surrounding air molecules, causing dielectric breakdown. When the molecules in the air dielectric are ionized, the air conducts electric current in the form of a spark.

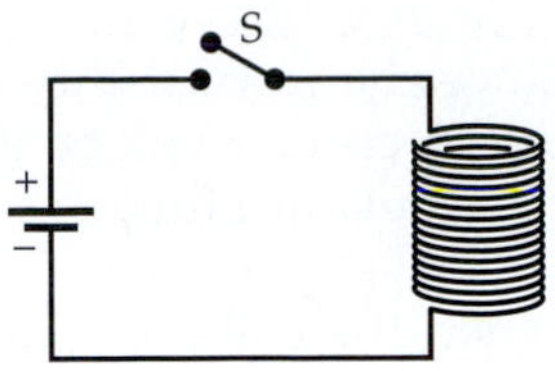

Figure 30-10 The coil with many turns of wire gives a large flux for a given current in the circuit. When the current changes, there is a large emf induced in the coil opposing the change.

Example 30-6

A rectangular coil of 80 turns, 20 cm wide and 30 cm long, is located in a magnetic field B = 0.8 T directed into the page (Figure 30-11), with only half of the coil in the region of the magnetic field. The resistance of the coil is 30 Ω. Find the magnitude and direction of the induced current if the coil is moved with a speed of 2 m/s (*a*) to the right, (*b*) up, and (*c*) down.

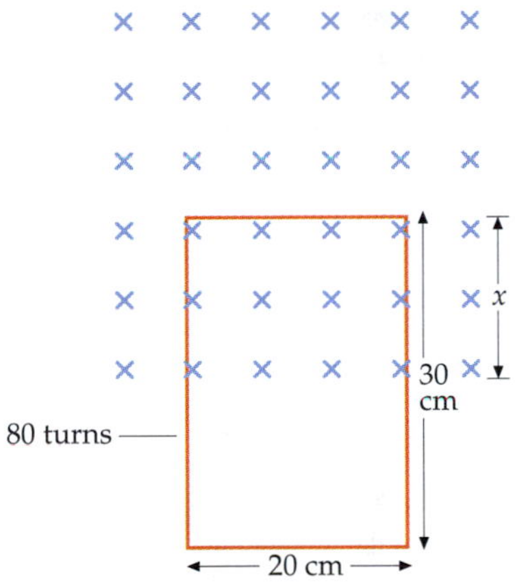

Figure 30-11

Picture the Problem The current equals the induced emf divided by the resistance. We can calculate the emf induced in the circuit as the coil moves by calculating the rate of change of the flux through the coil. The flux is proportional to the distance x. The direction of the current is found from Lenz's law.

(*a*)1. The magnitude of the induced current equals the emf divided by the resistance:

$$I = \frac{|\mathcal{E}|}{R}$$

2. The magnitude of the induced emf is given by Faraday's law:

$$|\mathcal{E}| = \frac{d\phi_m}{dt}$$

3. When the coil is moving to the right (or to the left), the flux does not change (until the coil leaves the region of magnetic field). The current is therefore zero:

$$|\mathcal{E}| = \frac{d\phi_m}{dt} = 0$$

$$I = 0$$

(*b*)1. The flux is the product of B and the area, which is given by (20 cm)x:

$$\phi_m = NB(20\text{ cm})x$$

2. Compute the rate of change of the flux when the coil is moving up:

$$\frac{d\phi_m}{dt} = NB(20\text{ cm})\frac{dx}{dt} = (80)(0.8\text{ T})(0.20\text{ m})(2\text{ m/s}) = 25.6\text{ V}$$

3. Calculate the magnitude of the current:

$$I = |\mathcal{E}|/R = \frac{25.6\text{ V}}{30\ \Omega} = 0.853\text{ A}$$

4. Since the inward flux is increasing, the induced current will be in the sense as to produce outward flux: The current is counterclockwise.

(*c*) When the coil moves downward at 2 m/s, the current has the same magnitude as when it moves upward, but is oppositely directed: $I = 0.853$ A clockwise

Remarks As the coil moves downward, the inward flux decreases so the induced current is clockwise so as to produce a flux inward.

optional

30-4 Motional emf

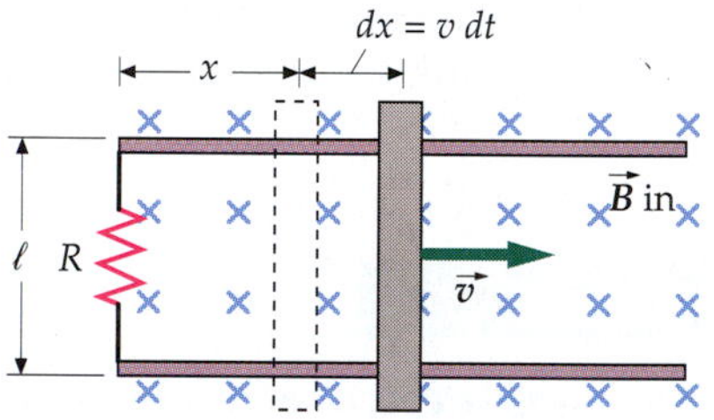

Figure 30-12 A conducting rod sliding on conducting rails in a magnetic field. As the rod moves to the right, the area of the circuit increases so the magnetic flux through the circuit into the paper increases. An emf of magnitude $B\ell v$ is induced in the circuit, producing a counterclockwise current that produces flux out of the paper opposing the charge.

Figure 30-12 shows a conducting rod sliding to the right along conducting rails that are connected by a resistor. A uniform magnetic field $\vec{B}$ is directed into the paper. Since the area of the circuit increases as the rod moves to the right, the magnetic flux through the circuit is increasing. An emf is therefore induced in the circuit. Let ℓ be the separation of the rails and x be the distance from the left end of the rails to the rod at some time. The area enclosed by the circuit is then ℓx, and the magnetic flux through the circuit at this time is

$$\phi_{\mathrm{m}} = BA = B\ell x$$

When the rod moves through a distance dx, the area enclosed by the circuit changes by $dA = \ell\, dx$ and the flux changes by $d\phi_{\mathrm{m}} = B\ell\, dx$. The rate of change of the flux is

$$\frac{d\phi_{\mathrm{m}}}{dt} = B\ell\frac{dx}{dt} = B\ell v$$

where $v = dx/dt$ is the speed of the rod. The magnitude of the emf induced in this circuit is therefore

$$|\mathcal{E}| = \frac{d\phi_{\mathrm{m}}}{dt} = B\ell v$$

The direction of the emf in this case is such as to produce a current in the counterclockwise sense. The flux produced by this induced current is out of the paper, opposing the increase in flux due to the motion of the rod. Because of the induced current, which is upward in the rod, there is a magnetic force on the rod of magnitude $I\ell B$. The direction of this force, obtained from the right-hand rule, is to the left, opposing the motion of the rod. If the rod is given some initial velocity $\vec{v}$ to the right and is then released, the force due to the induced current slows the rod until it stops. To maintain the motion of the rod, an external force must be exerted on it to the right.

The emf induced in a conductor moving through a magnetic field is called **motional emf.** More generally,

> Motional emf is any emf induced by the relative motion of a magnetic field and a current path.

Motional emf defined

Motional emf is induced in a conducting rod or wire moving in a magnetic field even when there is no complete circuit and thus no current.

Figure 30-13 shows an electron in a conducting rod that is moving through a uniform magnetic field directed into the paper. Because the electron is moving horizontally with the rod, there is a magnetic force on the electron that has a downward component of magnitude qvB. Because of this magnetic force, free electrons in the rod move downward, producing a net negative charge at the bottom and leaving a net positive charge at the top. The electrons continue to move down until the electric field produced by the separated charges exerts an upward force of magnitude qE on the electrons that balances the magnetic force qvB. In equilibrium, the electric field in the rod is thus

$$E = vB$$

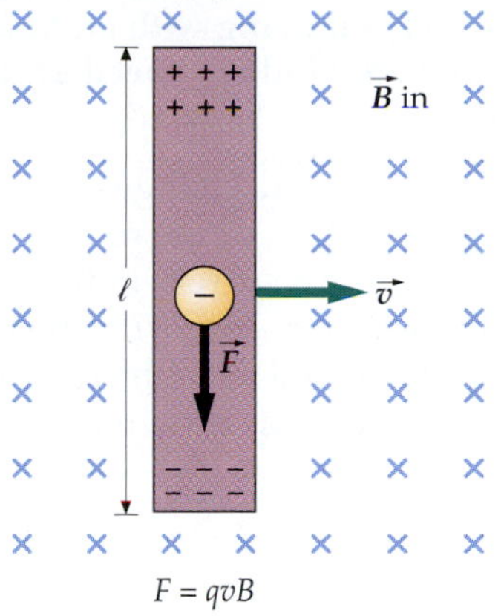

Figure 30-13 An electron in a conducting rod that is moving through a magnetic field experiences a magnetic force that has a downward component. Electrons move to the bottom of the rod, leaving the top of the rod positive. The charge separation produces an electric field of magnitude $E = vB$. The potential at the top of the rod is greater than that at the bottom by $E\ell = vB\ell$.

The potential difference across the rod is

$$\Delta V = E\ell = vB\ell$$

This potential difference equals the magnitude of the induced emf, that is, the motional emf:

$$|\mathcal{E}| = vB\ell \qquad 30\text{-}6$$

Motional emf

Exercise A rod 40 cm long moves at 12 m/s in a plane perpendicular to a magnetic field of 3000 G. Its velocity is perpendicular to its length. Find the emf induced in the rod. (*Answer* 1.44 V)

Example 30-7 *try it yourself*

In Figure 30-12, let $B = 0.6$ T, $v = 8$ m/s, $\ell = 15$ cm, and $R = 25\ \Omega$, and assume that the resistance of the rods and rails is negligible. Find (*a*) the induced emf in the circuit, (*b*) the current in the circuit, (*c*) the force needed to move the rod with constant velocity, and (*d*) the power dissipated in the resistor.

Cover the column to the right and try these on your own before looking at the answers.

Steps	***Answers***
(*a*) Calculate the induced emf from Equation 30-6.	$\mathcal{E} = Bv\ell = 0.72$ V
(*b*) Find the current from Ohm's law.	$I = \dfrac{\mathcal{E}}{R} = 28.8$ mA
(*c*) The force needed to move the rod with constant velocity is equal and opposite to the force exerted by the magnetic field on the rod, which has the magnitude $IB\ell$. Calculate the magnitude of this force.	$F = IB\ell = 2.59$ mN
(*d*)1. Find the power dissipated in the resistor.	$P = I^2R = 20.7$ mW
2. Check your answer for the previous step by computing the power input of the force from $P = Fv$.	$P = Fv = 20.7$ mW

optional

Example 30-8

A rod of mass m slides on frictionless, conducting rails in a region of constant magnetic field B. At time $t = 0$, the rod is moving with an initial speed v_0 and the external force acting on it is removed. Find the speed of the rod as a function of time.

Picture the Problem The speed of the rod changes because a magnetic force acts on the induced current. The motion of the rod through a magnetic field induces an emf $\mathcal{E} = B\ell v$ and therefore a current in the rod, $I = \mathcal{E}/R$. This causes a magnetic force to act on the rod, $F = IB\ell$. With the force known, we apply Newton's second law to find the speed as a function of time. Take the direction of the initial velocity to be positive, as shown in Figure 30-14.

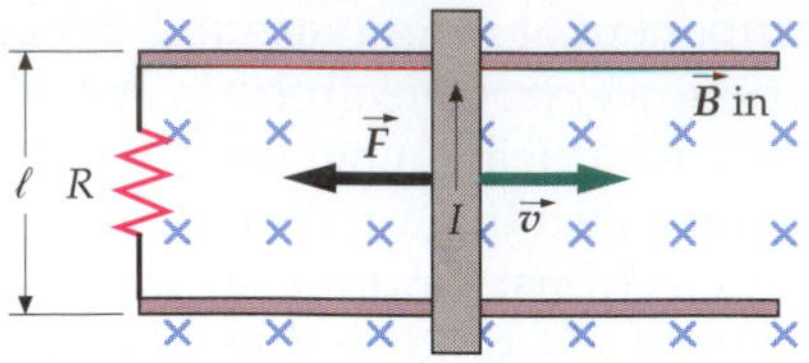

Figure 30-14

1. Apply Newton's second law to the rod: $F = ma = m\dfrac{dv}{dt}$

2. The force exerted on the rod is the magnetic force, which is proportional to the current: $F = IB\ell$

3. The current equals the motional emf divided by the resistance of the rod: $I = \dfrac{\mathcal{E}}{R} = \dfrac{B\ell v}{R}$

4. With these results we find the magnitude of the magnetic force exerted on the rod: $F = IB\ell = \dfrac{B\ell v}{R}B\ell = \dfrac{B^2\ell^2 v}{R}$

5. The force is opposite to the direction of motion, as shown in Figure 30-14, hence it is negative. Newton's second law then gives: $-\dfrac{B^2\ell^2 v}{R} = m\dfrac{dv}{dt}$

6. Separate variables, then integrate the speed from v_0 to v and integrate the time from 0 to t:

$$\frac{dv}{v} = -\frac{B^2\ell^2}{mR}dt$$

$$\int_{v_0}^{v}\frac{dv}{v} = -\frac{B^2\ell^2}{mR}\int_0^t dt$$

$$\ln\frac{v}{v_0} = -\frac{B^2\ell^2}{mR}t$$

7. Solve for v: $v = v_0 e^{-(B^2\ell^2/mR)t}$

Remarks If the force were constant, the rod's speed would decrease linearly with time. However, because the force is proportional to the rod's speed, as found in step 4, the force is large initially but decreases as the speed decreases. In principle, the rod never stops moving. Even so, it travels only a finite distance. (See Problem 37.)

30-5 Eddy Currents

In the examples we have discussed, the currents produced by a changing flux were set up in definite circuits. Often a changing flux sets up circulating currents, called *eddy currents*, in a piece of bulk metal like the core of a transformer. The heat produced by such current constitutes a power loss in the transformer. Consider a conducting slab between the pole faces of an electromagnet (Figure 30-15). If the magnetic field B between the pole faces is changing with time (as it will if the current in the magnet windings is alternating current), the flux through any closed loop in the slab such as through the curve C indicated in the figure will be changing. Since path C is in a conductor, there will be a current along the path. At the right in Figure 30-15 we have indicated just one of the many closed paths that will contain currents if B varies.

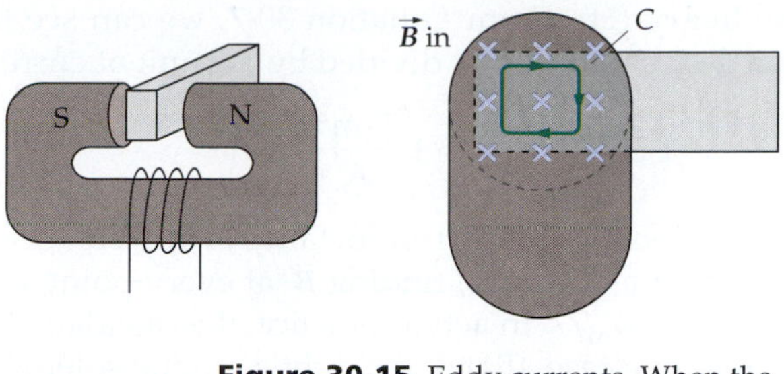

Figure 30-15 Eddy currents. When the magnetic field through a metal slab is changing, an emf is induced in any closed loop in the metal such as loop C. The induced emf causes a current in the loop.

The existence of eddy currents can be demonstrated by pulling a copper or aluminum sheet between the poles of a strong permanent magnet (Figure 30-16). Part of the area enclosed by curve C in this figure is in the magnetic field, and part is outside the field. As the sheet is pulled to the right, the flux through this curve decreases (assuming that the flux into the paper is positive). According to Faraday's law and Lenz's law, a clockwise current will be induced around this curve. Since this current is directed upward in the region between the pole faces, the magnetic field exerts a force on the current to the left, opposing motion of the sheet. You can feel this force on the sheet if you try to pull a conducting sheet suddenly through a strong magnetic field.

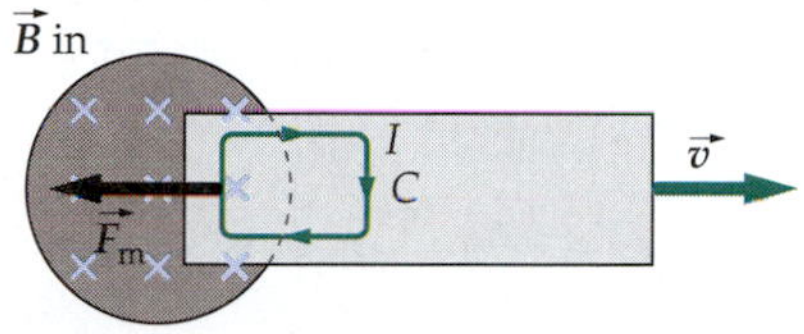

Figure 30-16 Demonstration of eddy currents. When the metal sheet is pulled to the right, there is a magnetic force to the left on the induced current opposing the motion.

Eddy currents are usually unwanted because power is lost in the form of heat generated by the current, and the heat itself must be dissipated. The power loss can be reduced by increasing the resistance of the possible paths for the eddy currents, as shown in Figure 30-17*a*. Here the conducting slab is laminated, that is, made up of small strips glued together. Because insulating glue separates the strips, the eddy currents are essentially confined to the strips. The large eddy-current loops are broken up, and the power loss is greatly reduced. Similarly if the sheet has cuts in it, as in Figure 30-17*b*, the eddy currents are lessened and the magnetic force is greatly reduced.

Eddy currents are not always undesirable. For example, they are often used to damp unwanted oscillations. With no damping present, sensitive mechanical balance scales used to weigh small masses might oscillate back and forth around their equilibrium reading many times. Such scales are usually designed so that a small piece of metal moves between the poles of a magnet as the scales oscillate. The resulting eddy currents dampen the oscillations so that equilibrium is quickly reached. Eddy currents also have a role in the magnetic braking systems of some rapid transit cars. A large electromagnet is positioned in the vehicle over the rails. When the magnet is energized by a current in its windings, eddy currents are induced in the rails by the motion of the magnet, and the magnetic forces provide a drag force on the car that stops it.

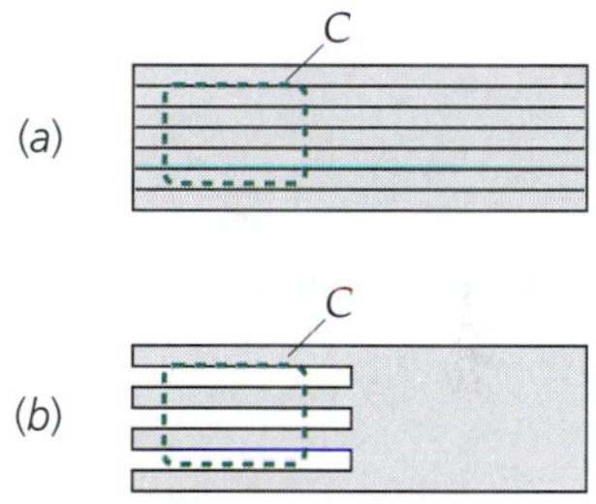

Figure 30-17 Eddy currents in a metal slab can be reduced by disrupting the conduction paths in the slab. (*a*) If the slab is constructed from strips of metal glued together, the insulating glue between the slabs increases the resistance of the closed loop C. (*b*) Slots cut into the metal slab also reduce the eddy current.

30-6 Inductance

Self-inductance

The magnetic flux through a circuit is related to the current in that circuit and the currents in other, nearby circuits.* Consider a coil carrying a current I. The current produces a magnetic field B that varies from point to point, but B is proportional to I at every point. The magnetic flux through the coil is therefore also proportional to I:

$$\phi_m = LI \qquad 30\text{-}7$$

Definition—Self-inductance

where L is a constant called the **self-inductance** of the coil. The self-inductance depends on the geometric shape of the coil. The SI unit of inductance is

*We will assume that there are no permanent magnets around.

the **henry** (H). From Equation 30-7, we can see that the unit of inductance equals the unit of flux divided by the unit of current:

$$1\text{ H} = 1\frac{\text{Wb}}{\text{A}} = 1\frac{\text{T}\cdot\text{m}^2}{\text{A}}$$

In principle, the self-inductance of any coil or circuit can be calculated by assuming a current I, finding $\vec{B}$ at every point, calculating the flux ϕ_m, and using $L = \phi_m/I$. In actual practice, the calculation is very difficult. However, there is one case, that of the tightly wound solenoid, for which the self-inductance can be calculated directly. The magnetic flux through a solenoid of length ℓ and N turns carrying a current I was calculated in Example 30-1:

$$\phi_m = \frac{\mu_0 N^2 IA}{\ell} = \mu_0 n^2 IA\ell \qquad \text{30-8}$$

where $n = N/\ell$ is the number of turns per unit length. As expected, the flux is proportional to the current I. The proportionality constant is the self-inductance:

$$L = \frac{\phi_m}{I} = \mu_0 n^2 A\ell \qquad \text{30-9}$$

Self-inductance of a solenoid

The self-inductance of a solenoid is proportional to the square of the number of turns per unit length n and to the volume $A\ell$. Thus, like capacitance, self-inductance depends only on geometric factors. From the dimensions of Equation 30-9, we can see that μ_0 can be expressed in henrys per meter:

$$\mu_0 = 4\pi \times 10^{-7}\text{ H/m}$$

Example 30-9

Find the self-inductance of a solenoid of length 10 cm, area 5 cm², and 100 turns.

Picture the Problem We can calculate the self-inductance in henrys from Equation 30-9 if we put all the quantities in SI units.

1. L is given by Equation 30-9:

$$L = \mu_0 n^2 A\ell$$

2. Convert the given quantities to SI units:

$$\ell = 10\text{ cm} = 0.1\text{ m}$$

$$A = 5\text{ cm}^2 = 5 \times 10^{-4}\text{ m}^2$$

$$n = N/\ell = (100\text{ turns})/(0.1\text{ m}) = 1000\text{ turns/m}$$

$$\mu_0 = 4\pi \times 10^{-7}\text{ H/m}$$

3. Substitute the given quantities:

$$L = \mu_0 n^2 A\ell$$

$$= (4\pi \times 10^{-7}\text{ H/m})(10^3\text{ turns/m})^2(5 \times 10^{-4}\text{ m}^2)(0.1\text{ m})$$

$$= 6.28 \times 10^{-5}\text{ H}$$

When the current in a circuit is changing, the magnetic flux due to the current is also changing, so an emf is induced in the circuit. Since the self-inductance of a circuit is constant, the change in flux is related to the change in current by

$$\frac{d\phi_m}{dt} = \frac{d(LI)}{dt} = L\frac{dI}{dt}$$

According to Faraday's law, we have

$$\mathcal{E} = -\frac{d\phi_m}{dt} = -L\frac{dI}{dt} \qquad \text{30-10}$$

Thus, the self-induced emf is proportional to the rate of change of the current. A coil or solenoid with many turns has a large self-inductance and is called an **inductor**. In circuits it is denoted by the symbol 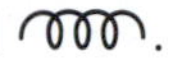. We can often neglect the self-inductance of the rest of the circuit compared with that of an inductor.

Exercise At what rate must the current in the solenoid of Example 30-9 change to induce an emf of 20 V? (*Answer* 3.18×10^5 A/s)

Mutual Inductance

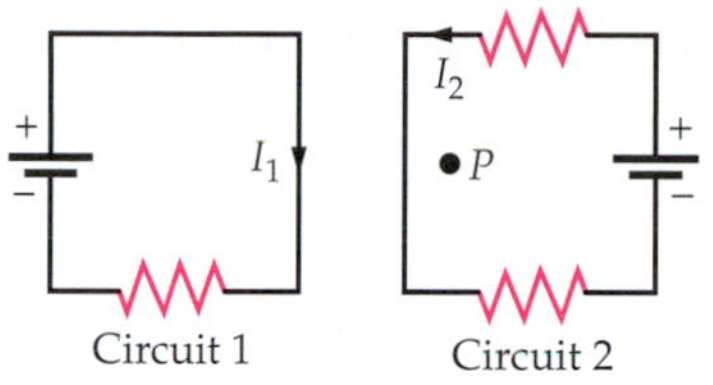

Figure 30-18 Two adjacent circuits. The magnetic field at point P is partly due to current I_1 and partly due to I_2. The flux through either circuit is the sum of two terms, one proportional to I_1 and the other to I_2.

When two or more circuits are close to each other, as in Figure 30-18, the magnetic flux through one circuit depends not only on the current in that circuit but also on the current in the nearby circuits. Let I_1 be the current in circuit 1 on the left in Figure 30-18, and let I_2 be the current in circuit 2 on the right. The magnetic field at some point P is due partly to I_1 and partly to I_2. The contribution to the overall field from each circuit is proportional to the current in each circuit. We can therefore write the flux through circuit 2, ϕ_{m2}, as the sum of two parts, one proportional to the current I_1 and the other proportional to the current I_2:

$$\phi_{m2} = L_2 I_2 + M_{2,1} I_1 \qquad \text{30-11}a$$

Definition—Mutual inductance

where L_2 is the self-inductance of circuit 2 and $M_{2,1}$ is called the **mutual inductance** of the two circuits. The mutual inductance depends on the geometrical arrangement of the two circuits. For instance, if the circuits are far apart, the flux through circuit 2 due to the current I_1 will be small and the mutual inductance will be small. An equation similar to Equation 30-11*a* can be written for the flux through circuit 1:

$$\phi_{m1} = L_1 I_1 + M_{1,2} I_2 \qquad \text{30-11}b$$

where L_1 is the self-inductance of circuit 1.

We can calculate the mutual inductance for two tightly wound concentric solenoids like the ones shown in Figure 30-19. Let ℓ be the length of both solenoids, and let the inner solenoid have N_1 turns and radius r_1 and the outer solenoid have N_2 turns and radius r_2. We will first calculate the mutual inductance $M_{2,1}$ by assuming that the inner solenoid carries a current I_1 and finding the magnetic flux ϕ_{m2} due to this current through the outer solenoid.

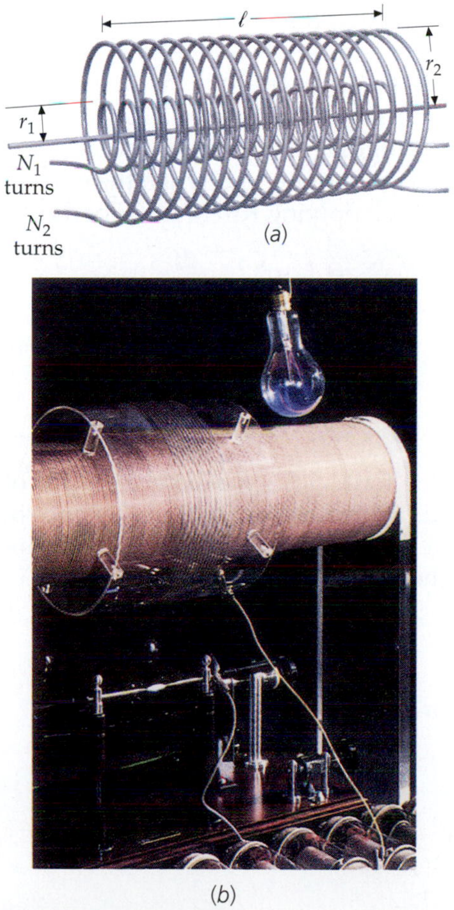

Figure 30-19 (*a*) A long, narrow solenoid inside a second solenoid of the same length. A current in either solenoid produces magnetic flux in the other. (*b*) Tesla coil illustrating the geometry of the wires in part (*a*). Such a device functions as a transformer (Chapter 31). Here, low-voltage alternating current in the outer winding is transformed into a higher-voltage alternating current in the inner winding. Induced alternating voltage from the changing fields is great enough to light the bulb above the coil.

The magnetic field due to the current in the inner solenoid is constant in the space within the solenoid and has magnitude

$$B_1 = \mu_0(N_1/\ell)I_1 = \mu_0 n_1 I_1 \qquad \text{30-12}$$

The flux through the outer solenoid due to this magnetic field is therefore

$$\phi_{m2} = N_2 B_1(\pi r_1^2) = n_2 \ell B_1(\pi r_1^2) = \mu_0 n_2 n_1 \,\ell(\pi r_1^2) I_1$$

Note that the area used to compute the flux through the outer solenoid is not the area of that solenoid, πr_2^2, but rather is the area of the inner solenoid, πr_1^2, because the magnetic field due to the inner solenoid is zero outside the inner solenoid. The mutual inductance $M_{1,2}$ is thus

$$M_{2,1} = \frac{\phi_{m2}}{I_1} = \mu_0 n_2 n_1 \ell \pi r_1^2 \qquad \text{30-13}$$

Exercise Calculate the mutual inductance $M_{1,2}$ of the concentric solenoids of Figure 30-19 by finding the flux through the inner solenoid due to a current I_2 in the outer solenoid. (*Answer* $M_{1,2} = M_{2,1} = \mu_0 n_2 n_1 \ell \pi r_1^2$)

Note from the exercise above that $M_{1,2} = M_{2,1}$. It can be shown that this is a general result. We will therefore drop the subscripts for mutual inductance and simply write M.

30-7 Magnetic Energy

An inductor stores magnetic energy, just as a capacitor stores electrical energy. Consider the circuit consisting of an inductance L and a resistance R in series with a battery of emf $\mathcal{E}_0$ and a switch S shown in Figure 30-20. We assume that R and L are the resistance and inductance of the entire circuit. The switch is initially open, so there is no current in the circuit. A short time after the switch is closed, there is a current I in the circuit, a potential drop IR across the resistor, and a back emf of magnitude $L\,dI/dt$ in the inductor. In the circuit diagram, we put plus and minus signs on the inductor to indicate the direction of the emf when the current is increasing, that is, when dI/dt is positive. Applying Kirchhoff's loop rule to this circuit gives

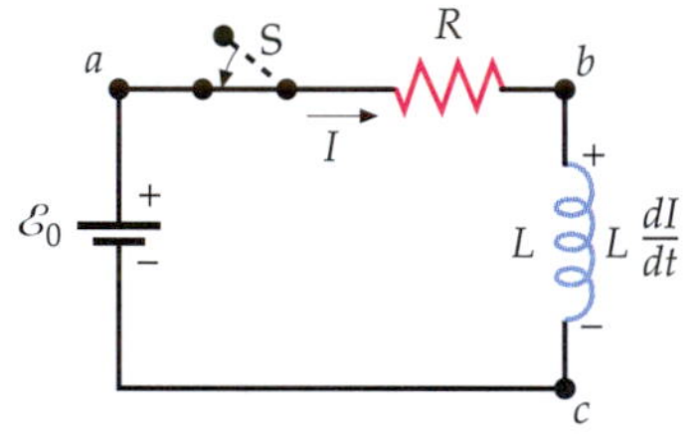

Figure 30-20 Just after the switch S is closed in this circuit, the current begins to increase and a back emf of magnitude $L\,dI/dt$ is generated in the inductor. The potential drop across the resistor IR plus the potential drop across the inductor equals the emf of the battery.

$$\mathcal{E}_0 - IR - L\frac{dI}{dt} = 0 \qquad \text{30-14}$$

If we multiply each term by the current I and rearrange, we obtain

$$\mathcal{E}_0 I = I^2 R + LI\frac{dI}{dt} \qquad \text{30-15}$$

The term $\mathcal{E}_0 I$ is the power output of the battery. The term I^2R is the power dissipated as heat in the resistance of the circuit. The term $LI\,dI/dt$ is the rate at which energy is put into the inductor. If U_m is the energy in the inductor, then

$$\frac{dU_m}{dt} = LI\frac{dI}{dt}$$

or

$$dU_m = LI\,dI$$

Integrating this equation from time $t = 0$, when the current is zero, to $t = \infty$, when the current has reached its final value I_f, we obtain

$$U_m = \int dU_m = \int_0^{I_f} LI\,dI = \frac{1}{2}LI_f^2$$

The energy stored in an inductor carrying a current I is thus given by

$$U_m = \tfrac{1}{2}LI^2 \qquad \text{30-16}$$

Energy stored in an inductor

When a current is produced in an inductor, a magnetic field is created in the space within the inductor coil. We can think of the energy stored in an inductor as energy stored in the magnetic field. For the special case of a solenoid, the magnetic field is related to the current I and the number of turns per unit length n by

$$B = \mu_0 n I$$

and the self-inductance is given by Equation 30-9:

$$L = \mu_0 n^2 A\ell$$

where A is the cross-sectional area and ℓ is the length. Substituting $B/\mu_0 n$ for I and $\mu_0 n^2 A\ell$ for L in Equation 30-16, we obtain

$$U_m = \frac{1}{2}LI^2 = \frac{1}{2}\mu_0 n^2 A\ell\left(\frac{B}{\mu_0 n}\right)^2 = \frac{B^2}{2\mu_0}A\ell$$

The quantity $A\ell$ is the volume of the space within the solenoid containing the magnetic field. The energy per unit volume is the **magnetic energy density** u_m:

$$u_m = \frac{B^2}{2\mu_0} = \tfrac{1}{2}\left(\frac{1}{\mu_0}\right)B^2 \qquad \text{30-17}$$

Magnetic energy density

Although we derived this by considering the special case of the magnetic field in a solenoid, it is a general result. Whenever there is a magnetic field in space, the magnetic energy per unit volume is given by Equation 30-17. Note the similarity to the energy density in an electric field (Equation 25-13):

$$u_e = \tfrac{1}{2}\epsilon_0 E^2 \qquad \text{30-18}$$

Example 30-10

A certain region of space contains a magnetic field of 200 G and an electric field of 2.5×10^6 N/C. Find (*a*) the total energy density and (*b*) the energy in a cubical box of side $\ell = 12$ cm.

Picture the Problem The total energy density u is the sum of the electrical and magnetic energy densities, $u = u_e + u_m$. The energy in a volume $\mathcal{V}$ is given by $U = u\mathcal{V}$.

(*a*)1. Calculate the electrical energy density:

$$u_e = \tfrac{1}{2}\epsilon_0 E^2 = \tfrac{1}{2}(8.85\times10^{-12}\,\mathrm{C^2/N\cdot m^2})(2.5\times10^6\,\mathrm{N/C})^2 = 27.7\,\mathrm{J/m^3}$$

2. Calculate the magnetic energy density:

$$u_m = \frac{1}{2}\frac{B^2}{\mu_0} = \frac{1}{2}\frac{(0.02\,\mathrm{T})^2}{(4\pi\times10^{-7}\,\mathrm{N/A^2})} = 159\,\mathrm{J/m^3}$$

3. The total energy density is the sum of the above two contributions:

$$u = u_e + u_m = 27.7\,\mathrm{J/m^3} + 159\,\mathrm{J/m^3} = 187\,\mathrm{J/m^3}$$

(*b*) The total energy in the box is $U = u\mathcal{V}$, where $\mathcal{V} = \ell^3$ is the volume of the box:

$$U = u\mathcal{V} = u\ell^3 = (187\,\mathrm{J/m^3})(0.12\,\mathrm{m})^3 = 0.323\,\mathrm{J}$$

optional

30-8 *RL* Circuits

A circuit containing a resistor and an inductor such as that shown in Figure 30-20 is called an ***RL* circuit.** Since all circuits have resistance and self-inductance, the analysis of an *RL* circuit can be applied to some extent to all circuits.* For the circuit in Figure 30-20, application of Kirchhoff's loop rule gave us (Equation 30-14):

$$\mathcal{E}_0 - IR - L\frac{dI}{dt} = 0$$

Let's look at some general features of the current before we solve this equation. Just after we close the switch in the circuit, the current is zero so IR is zero, and the back emf $L\,dI/dt$ equals the emf of the battery, $\mathcal{E}_0$. Setting $I = 0$ in Equation 30-14, we get

$$\left(\frac{dI}{dt}\right)_0 = \frac{\mathcal{E}_0}{L} \qquad 30\text{-}19$$

As the current increases, IR increases, and dI/dt decreases. Note that the current cannot jump suddenly from zero to some finite value as it would if there were no inductance. When there is some inductance ($L \neq 0$), dI/dt is finite and therefore the current must be continuous in time. After a short time, the current has reached a positive value I, and the rate of change of the current is

$$\frac{dI}{dt} = \frac{\mathcal{E}_0}{L} - \frac{IR}{L}$$

At this time the current is still increasing, but its rate of increase is less than at $t = 0$. The final value of the current can be obtained by setting dI/dt equal to zero:

$$I_f = \frac{\mathcal{E}_0}{R} \qquad 30\text{-}20$$

Figure 30-21 shows the current in this circuit as a function of time. This figure is similar to that for the charge on a capacitor when the capacitor is charged in an *RC* circuit (Figure 26-41).

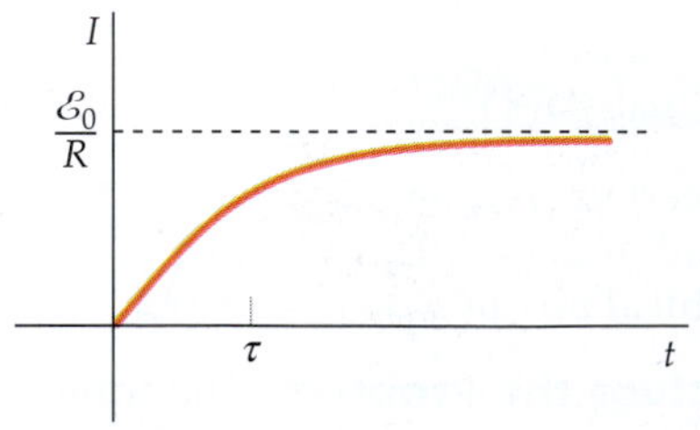

Figure 30-21 Current versus time in an *LR* circuit. At a time $t = \tau = L/R$, the current is at 63% of its maximum value $\mathcal{E}_0/R$.

* All circuits also have some capacitance between parts of the circuits at different potentials. We will consider the effects of capacitance in Chapter 31 when we study ac circuits. Here we will neglect capacitance to simplify the analysis and to focus on the effects of inductance.

Equation 30-14 is of the same form as Equation 26-34 for the charging of a capacitor and can be solved in the same way. The result is

$$I = \frac{\mathcal{E}_0}{R}(1 - e^{-Rt/L}) = \frac{\mathcal{E}_0}{R}(1 - e^{-t/\tau}) = I_f(1 - e^{-t/\tau}) \qquad \text{30-21}$$

where

$$\tau = \frac{L}{R} \qquad \text{30-22}$$

is the **time constant** of the circuit. The larger the self-inductance L or the smaller the resistance R, the longer it takes for the current to build up.

Example 30-11

A coil of self-inductance 5.0 mH and a resistance of 15.0 Ω is placed across the terminals of a 12-V battery of negligible internal resistance. (*a*) What is the final current? (*b*) What is the current after 100 μs?

Picture the Problem The final current is $I_f = \mathcal{E}_0/R$ as given in Equation 30-20. The current as a function of time is given by Equation 30-21, $I = I_f(1 - e^{-t/\tau})$, where $\tau = L/R$.

(*a*) Use Equation 30-20 to find the final current, I_f:

$$I_f = \frac{\mathcal{E}_0}{R} = \frac{12\text{ V}}{15\ \Omega} = 0.800\text{ V}$$

(*b*)1. Use Equation 30-21 to write the current I at any time t:

$$I = I_f(1 - e^{-t/\tau})$$

2. Calculate the time constant τ:

$$\tau = \frac{L}{R} = \frac{5 \times 10^{-3}\text{ H}}{15\ \Omega} = 333\ \mu\text{s}$$

3. Use this result for τ and calculate I for $t = 300\ \mu$s:

$$I = I_f(1 - e^{-t/\tau}) = (0.800\text{ A})(1 - e^{-100/333})$$
$$= (0.800\text{ A})(1 - 0.741) = 0.207\text{ A}$$

Exercise How much energy is stored in this inductor when the final current has been attained? (*Answer* $U_m = \frac{1}{2}LI_f^2 = 1.6 \times 10^{-3}$ J)

In Figure 30-22, the circuit has an additional switch that allows us to remove the battery, and an additional resistor R_1 to protect the battery so that it is not shorted when both switches are momentarily closed. When S_2 is open and S_1 is closed, the current builds up in the circuit just as discussed, except that the total resistance is now $R_1 + R$ and the final current is $\mathcal{E}_0/(R + R_1)$. Suppose that S_1 has been closed for a long time, so that the current is approximately steady at its final value, which we will call I_0. At time $t = 0$ we close switch S_2 and open switch S_1 (to remove the battery from consideration completely). We now have a circuit with just a resistor and an inductor (loop *abcda*) carrying an initial current I_0. Applying Kirchhoff's loop rule to this circuit gives

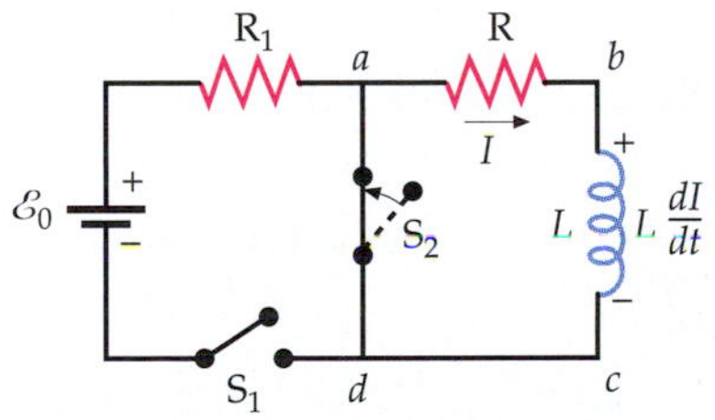

Figure 30-22 An *RL* circuit with two switches so that the battery can be removed from the circuit. After the current in the inductor reaches its maximum value with S_1 closed, S_2 is closed and S_1 is opened.

$$-IR - L\frac{dI}{dt} = 0$$

or

$$\frac{dI}{dt} = -\frac{R}{L}I \qquad \text{30-23}$$

Equation 30-23 is of the same form as Equation 26-28 for the discharge of a capacitor. It can be solved by direct integration. We will omit the details and merely state the solution. The current I is given by

$$I = I_0 e^{-Rt/L} = I_0 e^{-t/\tau} \qquad \text{30-24}$$

where $\tau = L/R$ is the time constant. Figure 30-23 shows the current as a function of time.

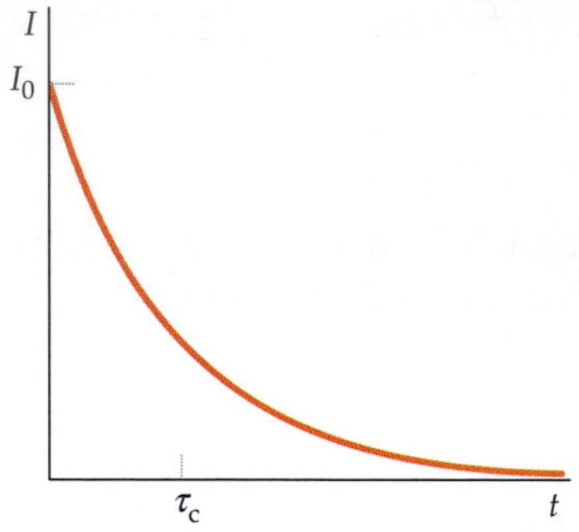

Figure 30-23 Current versus time for the circuit in Figure 30-22. The current decreases exponentially with time.

Exercise What is the time constant of a circuit of resistance 85 Ω and inductance 6 mH? (*Answer* 70.6 μs)

Example 30-12

Find the total heat produced in the resistor R in Figure 30-22 when the current in the inductor decreases from its initial value of I_0 to 0.

Picture the Problem The rate of heat production I^2R varies with time so we must integrate.

1. The rate of heat production is I^2R: $$P = \frac{dW}{dt} = I^2R$$

2. The total energy dissipated as heat in the resistor is the integral of dW from $t = 0$ to $t = \infty$: $$W = \int_0^\infty I^2R\,dt$$

3. The current I is given by Equation 30-24: $$I = I_0 e^{-Rt/L}$$

4. Substitute this current into the integral: $$W = \int_0^\infty I^2R\,dt = \int_0^\infty I_0^2 e^{-2Rt/L} R\,dt$$

5. The integration can be done by substituting $x = 2Rt/L$: $$x = 2Rt/L$$ $$dt = \frac{L}{2R}\,dx$$ $$W = I_0^2 R \frac{L}{2R} \int_0^\infty e^{-x}\,dx$$

6. The integral in step 5 is 1. Then: $$W = \tfrac{1}{2} L I_0^2$$

Remark The total heat produced equals the energy $\frac{1}{2}LI_0^2$ originally stored in the inductor.

Example 30-13

For the circuit shown in Figure 30-24, find the currents I_1, I_2, and I_3 (*a*) immediately after switch S is closed and (*b*) a long time after switch S has been closed. After the switch has been closed for a long time, it is opened. Find the three currents (*c*) immediately after switch S is opened and (*d*) a long time after switch S was opened.

Figure 30-24

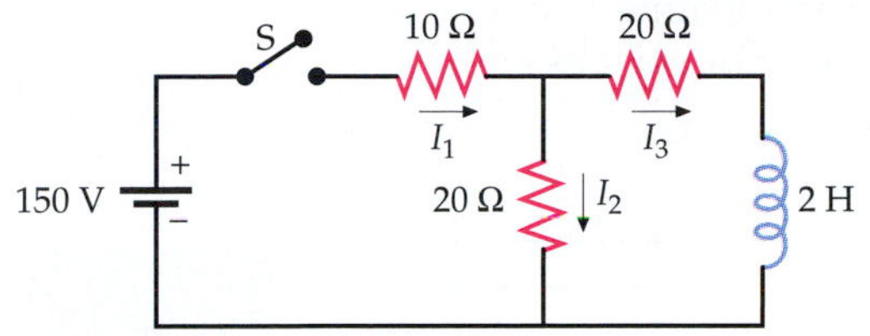

Picture the Problem (*a*) We simplify our calculations by using the fact that the current in an inductor cannot change abruptly. Thus, the current in the inductor must be zero just after the switch is closed, because it is zero before. (*b*) When the current reaches its final value, dI/dt equals zero, so there is no potential drop across the inductor. The inductor thus acts like a short circuit, that is, like a wire with zero resistance. (*c*) Immediately after the switch is opened, the current in the inductor is the same as it was before. (*d*) A long time after the switch is opened, all the currents must be zero.

(*a*) The current through the inductor is zero, just as it was before the switch was closed. The current in the left loop equals the emf divided by the equivalent resistance of the two resistors in series:

$$I_1 = I_2 = \frac{150\text{ V}}{10\ \Omega + 20\ \Omega} = 5\text{ A}$$

$$I_3 = 0$$

(*b*)1. After a long time, the current is steady and the inductor acts like a short circuit, so we have two 20-Ω resistors in parallel, with the combination in series with the 10-Ω resistor. Redraw the circuit (Figure 30-25) and find the equivalent resistance of the parallel resistors:

$$\frac{1}{R_{eq}} = \frac{1}{20} + \frac{1}{20} = \frac{2}{20} = \frac{1}{10}$$

$$R_{eq} = 10\ \Omega$$

Figure 30-25

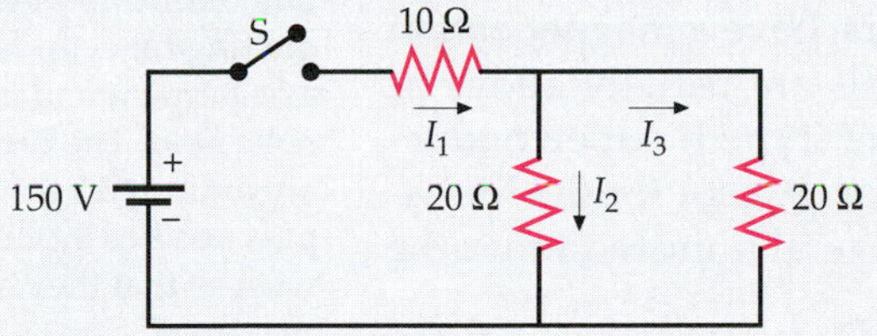

2. Find the current I_1:

$$I_1 = \frac{150\text{ V}}{10\ \Omega + 10\ \Omega} = 7.5\text{ A}$$

3. Find I_2 and I_3 using the fact that the current in the parallel 20-Ω resistors must be equal:

$$I_2 = I_3 = 3.75\text{ A}$$

(*c*) When the switch is reopened, I_1 must be zero, and the current in the inductor I_3 momentarily remains the same at 3.75 A. Then I_2 must equal $-I_3$:

$$I_1 = 0$$

$$I_3 = 3.75\text{ A}$$

$$I_2 = -I_3 = -3.75\text{ A}$$

(*d*) A long time after the switch is opened, all the currents must be zero.

$$I_1 = I_2 = I_3 = 0$$

30-9 Magnetic Properties of Superconductors

Superconductors have resistivities of zero below a critical temperature T_c, which varies from material to material. In the presence of a magnetic field B, the critical temperature is lower than it is when there is no field. As the magnetic field increases, the critical temperature decreases. If the magnetic field is greater than some critical field B_c, superconductivity does not exist at any temperature.

Meissner Effect

When a superconductor is cooled below the critical temperature in an external magnetic field, the magnetic field lines are expelled from the superconductor so the magnetic field inside the superconductor is zero (Figure 30-26). This effect was discovered by Meissner and Ochsenfeld in 1933 and is now known as the **Meissner effect.** The mechanism by which the magnetic field lines are expelled is an induced superconducting current on the surface of the superconductor. The magnetic levitation shown on page 656 results from the repulsion between the permanent magnet producing the external field and the magnetic field produced by the currents induced in the superconductor. Only certain superconductors called **type I superconductors** exhibit the complete Meissner effect. Figure 30-27*a* shows a plot of the magnetization M times μ_0 versus the applied magnetic field B_{app} for a type I superconductor. For a magnetic field less than the critical field B_c, the magnetic field $\mu_0 M$ induced in the superconductor is equal and opposite to the external magnetic field. The values of B_c for type I superconductors are always too small for such materials to be useful in the coils of a superconducting magnet.

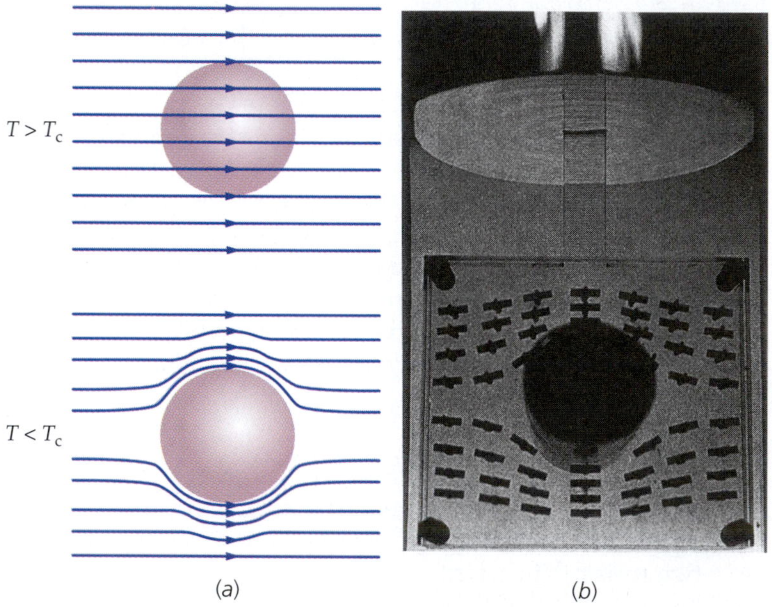

Figure 30-26 (*a*) The Meissner effect in a superconducting sphere cooled in a constant applied magnetic field. As the temperature drops below the critical temperature T_c, the magnetic field lines are expelled from the sphere. (*b*) Demonstration of the Meissner effect. A superconducting tin cylinder is situated with its axis perpendicular to a horizontal magnetic field. The directions of the field lines are indicated by weakly magnetized compass needles mounted in a Lucite sandwich so that they are free to turn.

Other materials, known as **type II superconductors,** have a magnetization curve similar to that in Figure 30-27*b*. Such materials are usually alloys or metals that have large resistivities in the normal state. Type II superconductors exhibit the electrical properties of superconductors except for the Meissner effect up to the critical field B_{c2}, which may be several hundred times the

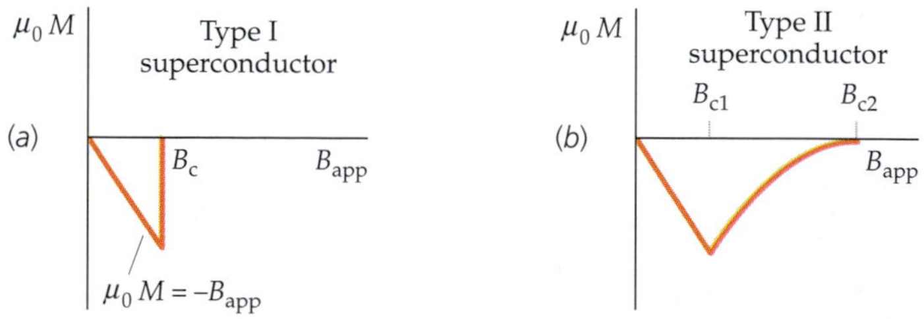

Figure 30-27 Plots of μ_0 times the magnetization M versus applied magnetic field for type I and type II superconductors. (*a*) In a type I superconductor, the resultant magnetic field is zero below a critical applied field B_c because the field due to induced currents on the surface of the superconductor exactly cancels the applied field. Above the critical field, the material is a normal conductor and the magnetization is too small to be seen on this scale. (*b*) In a type II superconductor, the magnetic field starts to penetrate the superconductor at a field B_{c1}, but the material remains superconducting up to a field B_{c2}, after which it becomes a normal conductor.

typical values of critical fields for type I superconductors. For example, the alloy Nb_3Ge has a critical field $B_{c2} = 34$ T. Such materials can be used for high-field superconducting magnets. Below the critical field B_{c1}, the behavior of a type II superconductor is the same as that of a type I superconductor. In the region between fields B_{c1} and B_{c2}, the superconductor is said to be in a vortex state.

Flux Quantization

Consider a superconducting ring of area A carrying a current. There can be a magnetic flux $\phi_m = B_n A$ through the ring due to the current in the ring and due also perhaps to other currents external to the ring. According to Faraday's law, if the flux changes, an emf will be induced in the ring that is proportional to the rate of change of the flux. But there can be no emf in a superconducting ring because it is resistanceless. The flux through the ring is thus frozen and cannot change.

Another effect, which results from the quantum-mechanical treatment of superconductivity, is that the total flux through the loop is quantized and is given by

$$\phi_m = n\frac{h}{2e}, \qquad n = 1, 2, 3, \ldots \tag{30-25}$$

The smallest unit of flux, called a **fluxon,** is

$$\phi_0 = \frac{h}{2e} = 2.0678 \times 10^{-15}\ \mathrm{T\cdot m^2} \tag{30-26}$$

Summary

1. Faraday's law and Lenz' law are fundamental laws of physics.
2. Inductance is a property of a circuit element that relates the flux through the element to the current.

Topic	Remarks and Relevant Equations	
1. Magnetic Flux ϕ_m		
General definition	$\phi_m = \int_S \vec{B}\cdot\hat{n}\, dA$	30-1
Constant field, coil of n turns	$\phi_m = NBA\cos\theta$	30-3
Units	$1\ \mathrm{Wb} = 1\ \mathrm{T\cdot m^2}$	30-2
Due to current in a circuit	$\phi_m = LI$	30-7
Due to current in two circuits	$\phi_{m1} = L_1I_1 + MI_2 \qquad \phi_{m2} = L_2I_2 + MI_1$	30-11
Quantization (optional)	$\phi_m = n\frac{h}{2e}, \qquad n = 1, 2, 3, \ldots$	30-25
Fluxon (optional)	$\phi_0 = \frac{h}{2e} = 2.0678 \times 10^{-15}\ \mathrm{T\cdot m^2}$	30-26

2.	**emf**		
	Definition	$\mathcal{E} = \oint_C \vec{E} \cdot d\vec{\ell}$	**30-4**
	Motional (optional)	$\lvert\mathcal{E}\rvert = vB\ell$	**30-6**
	Self-induced (Back emf)	$\mathcal{E} = -L\dfrac{dI}{dt}$	**30-10**
3.	**Faraday's Law**	$\mathcal{E} = \oint_C \vec{E} \cdot d\vec{\ell} = -\dfrac{d\phi_m}{dt}$	**30-5**
4.	**Lenz's Law**	The induced emf and induced current are in such a direction as to oppose the change that produces them.	
5.	**Inductance**		
	Self-inductance	$\phi_m = LI$	**30-7**
	Self-inductance of a solenoid	$L = \mu_0 n^2 A\ell$	**30-9**
	Mutual inductance	$\phi_{m1} = MI_2 \qquad \phi_{m2} = MI_1$	
	Units	$1\,\text{H} = 1\,\dfrac{\text{Wb}}{\text{A}} = 1\,\dfrac{\text{T}\cdot\text{m}^2}{\text{A}}$ $\mu_0 = 4\pi \times 10^{-7}\,\text{H/m}$	
6.	**Magnetic Energy**		
	Energy in an inductor	$U_m = \frac{1}{2}LI^2$	**30-16**
	Energy density in a magnetic field	$u_m = \dfrac{B^2}{2\mu_0}$	**30-17**
7.	***RL* Circuits** (optional)	In an *RL* circuit, which consists of a resistance *R*, an inductance *L*, and a battery of emf $\mathcal{E}_0$ in series, the current does not reach its maximum value *I* instantaneously but rather takes some time to build up. If the current is initially zero, its value at some later time *t* is given by	
	Current	$I = \dfrac{\mathcal{E}_0}{R}(1 - e^{-Rt/L}) = \dfrac{\mathcal{E}_0}{R}(1 - e^{-t/\tau}) = I_f(1 - e^{-t/\tau})$	**30-21**
	Time constant τ	$\tau = \dfrac{L}{R}$	**30-22**

Problem-Solving Guide

Begin by drawing a neat diagram that includes the important features of the problem. Show the direction of $\vec{B}$ in order to correctly calculate the magnetic flux. When *B* changes with time, it is helpful to include in your drawing the direction of the induced electric field $\vec{E}$. In circuit problems, be sure to show the direction of the current *I* and to show which side of the inductor is at higher potential.

Summary of Worked Examples

Type of Calculation	Procedure and Relevant Examples
1. Flux	
Find the magnetic flux through a coil of N turns.	Use $\phi_m = NBA \cos\theta$ if B is uniform, or $\phi_m = \int_S NB_n\,dA$ if B varies. **Examples 30-1, 30-2**
2. Induced emf	
Find the induced electric field $\vec{E}$, or the induced emf $\mathcal{E}$ due to a changing magnetic flux.	Use Faraday's law $\mathcal{E} = \oint_C \vec{E}\cdot d\vec{\ell} = -\frac{d\phi_m}{dt}$ **Examples 30-2, 30-3, 30-4, 30-5, 30-6**
Find the emf produced by motion through a magnetic field.	The magnitude of the motional emf is $\mathcal{E} = vB\ell$. The direction is given by Lenz's law. **Examples 30-7, 30-8**
3. Inductance	
Calculate the inductance of a solenoid.	Use $L = \mu_0 n^2 A\ell$. **Example 30-9**
4. Magnetic Energy	
Calculate the magnetic energy and the magnetic energy density.	The magnetic energy is $U_m = \frac{1}{2}LI^2$ and the magnetic energy density is $u_m = \frac{B^2}{2\mu_0}$. **Example 30-10**
5. *RL* Circuits	
Find the current in a simple *RL* circuit.	Use $I = I_f(1 - e^{-t/\tau})$ for a switch closed at $t = 0$ and $I = I_0 e^{-t/\tau}$ for a switch opened at $t = 0$. In each case, $\tau = L/R$. **Example 30-11**
Determine the current in a circuit containing inductors immediately after a switch is opened or closed, or long after a switch is closed.	Immediately after a switch is opened or closed the current in an inductor remains the same. Long after a switch is closed, an inductor acts like a wire with no resistance. **Example 30-11**

Problems

Conceptual Problems

Problems from Optional and Exploring sections

In a few problems, you are given more data than you actually need; in a few other problems, you are required to supply data from your general knowledge, outside sources, or informed estimates.

- • Single-concept, single-step, relatively easy
- •• Intermediate-level, may require synthesis of concepts
- ••• Challenging, for advanced students

Magnetic Flux

1 • A uniform magnetic field of magnitude 2000 G is parallel to the x axis. A square coil of side 5 cm has a single turn and makes an angle θ with the z axis as shown in Figure 30-28. Find the magnetic flux through the coil when (*a*) $\theta = 0°$, (*b*) $\theta = 30°$, (*c*) $\theta = 60°$, and (*d*) $\theta = 90°$.

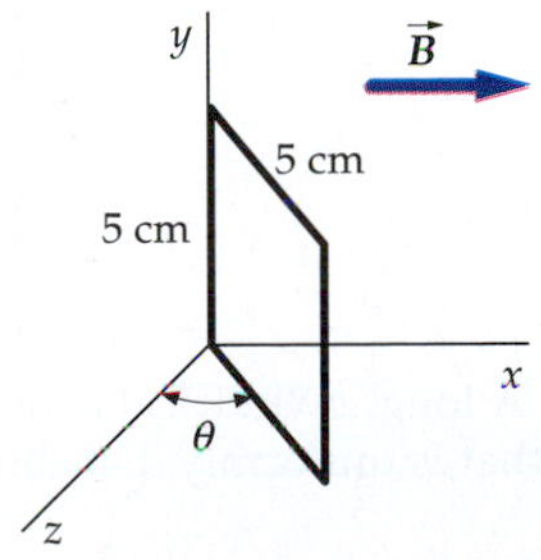

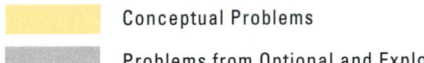

Figure 30-28 Problem 1

2 • A circular coil has 25 turns and a radius of 5 cm. It is at the equator, where the earth's magnetic field is 0.7 G north. Find the magnetic flux through the coil when its plane is (*a*) horizontal, (*b*) vertical with its axis pointing north, (*c*) vertical with its axis pointing east, and (*d*) vertical with its axis making an angle of 30° with north.

3 • A magnetic field of 1.2 T is perpendicular to a square coil of 14 turns. The length of each side of the coil is 5 cm. (*a*) Find the magnetic flux through the coil. (*b*) Find the magnetic flux through the coil if the magnetic field makes an angle of 60° with the normal to the plane of the coil.

4 • A circular coil of radius 3.0 cm has its plane perpendicular to a magnetic field of 400 G. (*a*) What is the magnetic flux through the coil if the coil has 75 turns? (*b*) How many turns must the coil have for the flux to be 0.015 Wb?

5 • A uniform magnetic field $\vec{B}$ is perpendicular to the base of a hemisphere of radius R. Calculate the magnetic flux through the spherical surface of the hemisphere.

6 •• Find the magnetic flux through a solenoid of length 25 cm, radius 1 cm, and 400 turns that carries a current of 3 A.

7 •• Work Problem 6 for an 800-turn solenoid of length 30 cm, and radius 2 cm, carrying a current of 2 A.

8 •• A circular coil of 15 turns of radius 4 cm is in a uniform magnetic field of 4000 G in the positive x direction. Find the flux through the coil when the unit vector perpendicular to the plane of the coil is (*a*) $\hat{n} = \hat{i}$, (*b*) $\hat{n} = \hat{j}$, (*c*) $\hat{n} = (\hat{i} + \hat{j})/\sqrt{2}$, (*d*) $\hat{n} = \hat{k}$, and (*e*) $\hat{n} = 0.6\hat{i} + 0.8\hat{j}$.

9 •• A solenoid has n turns per unit length, radius R_1, and carries a current I. (*a*) A large circular loop of radius $R_2 > R_1$ and N turns encircles the solenoid at a point far away from the ends of the solenoid. Find the magnetic flux through the loop. (*b*) A small circular loop of N turns and radius $R_3 < R_1$ is completely inside the solenoid, far from its ends, with its axis parallel to that of the solenoid. Find the magnetic flux through this small loop.

10 •• A long, straight wire carries a current I. A rectangular loop with two sides parallel to the straight wire has sides a and b with its near side a distance d from the straight wire, as shown in Figure 30-29. (*a*) Compute the magnetic flux through the rectangular loop. (*Hint:* Calculate the flux through a strip of area $dA = b\,dx$ and integrate from $x = d$ to $x = d + a$.) (*b*) Evaluate your answer for $a = 5$ cm, $b = 10$ cm, $d = 2$ cm, and $I = 20$ A.

Figure 30-29
Problems 10 and 46

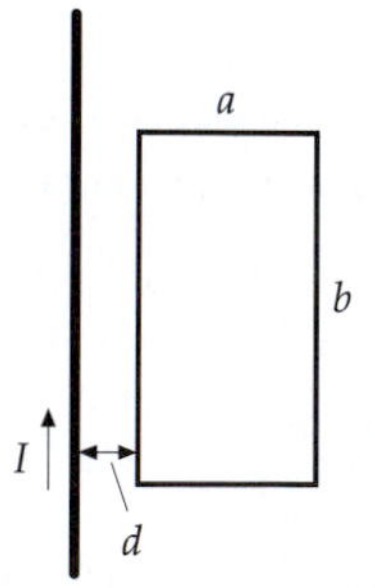

11 ••• A long, cylindrical conductor of radius R carries a current I that is uniformly distributed over its cross-sectional area. Find the magnetic flux per unit length through the area indicated in Figure 30-30.

Figure 30-30 Problem 11

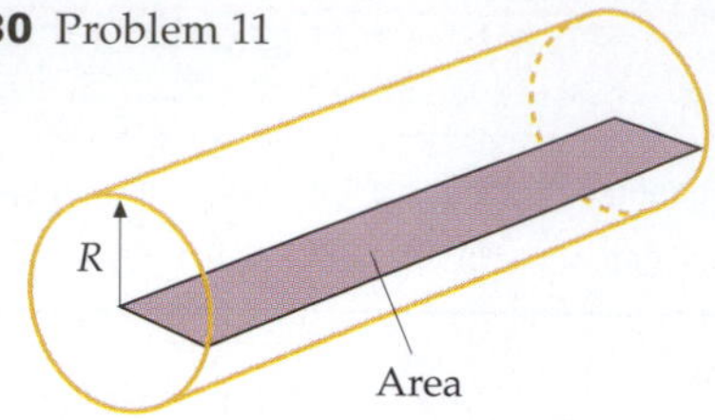

12 ••• A rectangular coil in the plane of the page has dimensions a and b. A long wire that carries a current I is placed directly above the coil (Figure 30-31). (*a*) Obtain an expression for the magnetic flux through the coil as a function of x for $0 \le x \le 2b$. (*b*) For what value of x is flux through the coil a maximum? For what value of x is the flux a minimum?

Figure 30-31 Problem 12

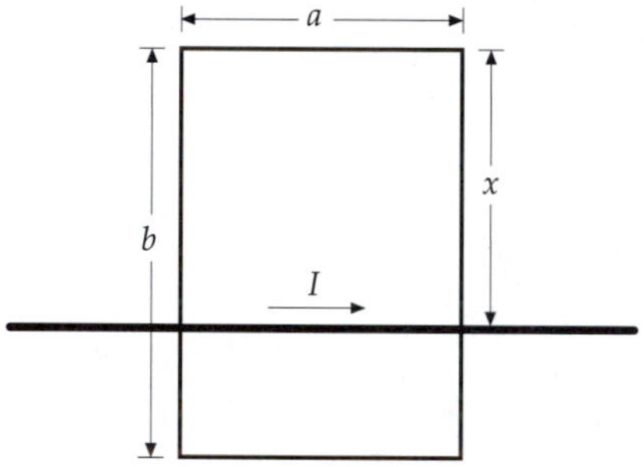

Induced emf and Faraday's Law

13 • A conducting loop lies in the plane of this page and carries a clockwise induced current. Which of the following statements could be true?

(*a*) A constant magnetic field is directed into the page.
(*b*) A constant magnetic field is directed out of the page.
(*c*) An increasing magnetic field is directed into the page.
(*d*) A decreasing magnetic field is directed into the page.
(*e*) A decreasing magnetic field is directed out of the page.

14 • A uniform magnetic field $\vec{B}$ is established perpendicular to the plane of a loop of radius 5.0 cm, resistance 0.4 Ω, and negligible self-inductance. The magnitude of $\vec{B}$ is increasing at a rate of 40 mT/s. Find (*a*) the induced emf in the loop, (*b*) the induced current in the loop, and (*c*) the rate of Joule heating in the loop.

15 • The flux through a loop is given by $\phi_m = (t^2 - 4t) \times 10^{-1}$ Wb, where t is in seconds. (*a*) Find the induced emf $\mathcal{E}$ as a function of time. (*b*) Find both ϕ_m and $\mathcal{E}$ at $t = 0$, $t = 2$ s, $t = 4$ s, and $t = 6$ s.

16 • (*a*) For the flux given in Problem 15, sketch graphs of ϕ_m and $\mathcal{E}$ versus t. (*b*) At what time is the flux minimum? What is the emf at this time? (*c*) At what times is the flux zero? What is the emf at these times?

17 • The magnetic field in Problem 4 is steadily reduced to zero in 0.8 s. What is the magnitude of the emf induced in the coil of part (*b*)?

18 • A solenoid of length 25 cm and radius 0.8 cm with 400 turns is in an external magnetic field of 600 G that makes an angle of 50° with the axis of the solenoid. (*a*) Find the magnetic flux through the solenoid. (*b*) Find the magnitude of the emf induced in the solenoid if the external magnetic field is reduced to zero in 1.4 s.

19 •• A 100-turn circular coil has a diameter of 2.0 cm and resistance of 50 Ω. The plane of the coil is perpendicular to a uniform magnetic field of magnitude 1.0 T. The direction of the field is suddenly reversed. (*a*) Find the total charge that passes through the coil. If the reversal takes 0.1 s, find (*b*) the average current in the coil and (*c*) the average emf in the coil.

20 •• At the equator, a 1000-turn coil with a cross-sectional area of 300 cm^2 and a resistance of 15.0 Ω is aligned with its plane perpendicular to the earth's magnetic field of 0.7 G. If the coil is flipped over, how much charge flows through it?

21 •• A circular coil of 300 turns and radius 5.0 cm is connected to a current integrator. The total resistance of the circuit is 20 Ω. The plane of the coil is originally aligned perpendicular to the earth's magnetic field at some point. When the coil is rotated through 90°, the charge that passes through the current integrator is measured to be 9.4 μC. Calculate the magnitude of the earth's magnetic field at that point.

22 •• An elastic circular conducting loop is expanding at a constant rate so that its radius is given by $R = R_0 + vt$. The loop is in a region of constant magnetic field perpendicular to the loop. What is the emf generated in the expanding loop? Neglect possible effects of self-inductance.

23 •• The wire in Problem 12 is placed at $x = b/4$. (*a*) Obtain an expression for the emf induced in the coil if the current varies with time according to $I = 2t$. (*b*) If $a = 1.5$ m and $b = 2.5$ m, what should be the resistance of the coil so that the induced current is 0.1 A? What is the direction of this current?

24 •• Repeat Problem 23 if the wire is placed at $x = b/3$.

Lenz's Law

25 • Give the direction of the induced current in the circuit on the right in Figure 30-32 when the resistance in the circuit on the left is suddenly (*a*) increased and (*b*) decreased.

Figure 30-32 Problem 25

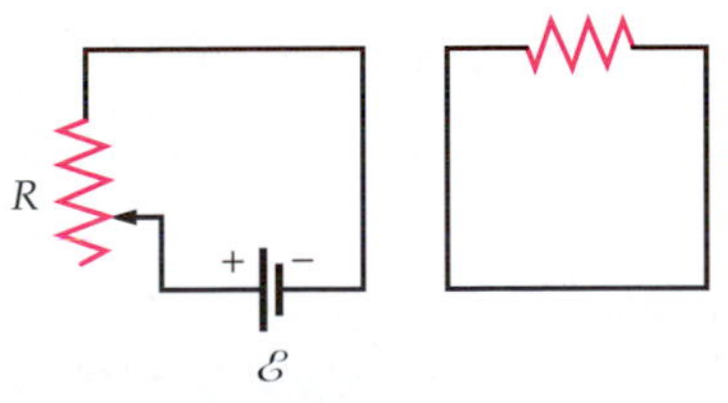

26 •• The two circular loops in Figure 30-33 have their planes parallel to each other. As viewed from A toward B, there is a counterclockwise current in loop A. Give the direction of the current in loop B and state whether the loops attract or repel each other if the current in loop A is (*a*) increasing and (*b*) decreasing.

Figure 30-33 Problem 26

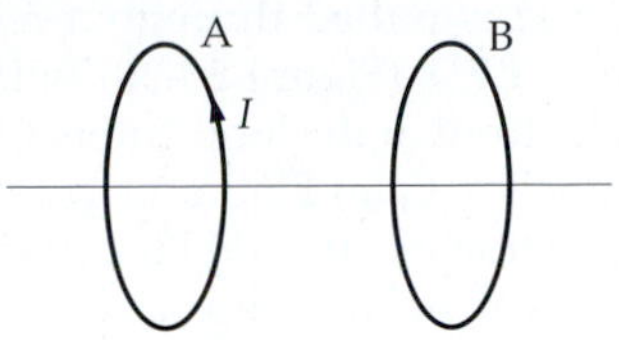

27 •• A bar magnet moves with constant velocity along the axis of a loop as shown in Figure 30-34. (*a*) Make a qualitative graph of the flux ϕ_m through the loop as a function of time. Indicate the time t_1 when the magnet is halfway through the loop. (*b*) Sketch a graph of the current I in the loop versus time, choosing I to be positive when it is counterclockwise as viewed from the left.

Figure 30-34 Problem 27

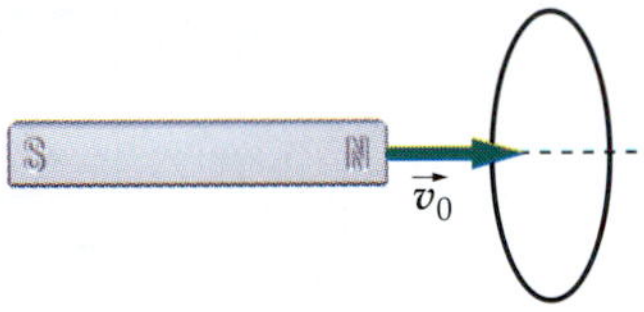

28 •• A bar magnet is mounted on the end of a coiled spring in such a way that it moves with simple harmonic motion along the axis of a loop as shown in Figure 30-35. (*a*) Make a qualitative graph of the flux ϕ_m through the loop as a function of time. Indicate the time t_1 when the magnet is halfway through the loop. (*b*) Sketch the current I in the loop versus time, choosing I to be positive when it is counterclockwise as viewed from above.

Figure 30-35 Problem 28

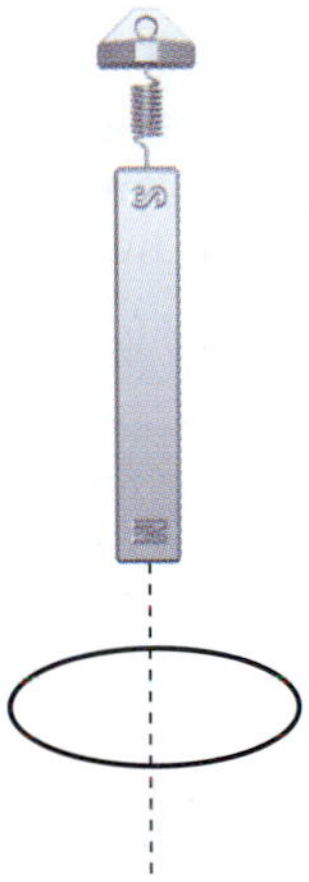

Motional emf (optional)

29 • A rod 30 cm long moves at 8 m/s in a plane perpendicular to a magnetic field of 500 G. The velocity of the rod is perpendicular to its length. Find (*a*) the magnetic force on an electron in the rod, (*b*) the electrostatic field $\vec{E}$ in the rod, and (*c*) the potential difference V between the ends of the rod.

30 • Find the speed of the rod in Problem 29 if the potential difference between the ends is 6 V.

31 • In Figure 30-14, let B be 0.8 T, $v = 10.0$ m/s, $\ell = 20$ cm, and $R = 2$ Ω. Find (*a*) the induced emf in the circuit, (*b*) the current in the circuit, and (*c*) the force needed to move the rod with constant velocity assuming negligible friction. Find (*d*) the power input by the force found in part (*c*), and (*e*) the rate of Joule heat production I^2R.

32 • Work Problem 31 for $B = 1.5$ T, $v = 6$ m/s, $\ell = 40$ cm, and $R = 1.2$ Ω.

33 •• A 10-cm by 5-cm rectangular loop with resistance 2.5 Ω is pulled through a region of uniform magnetic field $B = 1.7$ T (Figure 30-36) with constant speed $v = 2.4$ cm/s. The front of the loop enters the region of the magnetic field at time $t = 0$. (*a*) Find and graph the flux through the loop as a function of time. (*b*) Find and graph the induced emf and the current in the loop as functions of time. Neglect any self-inductance of the loop and extend your graphs from $t = 0$ to $t = 16$ s.

Figure 30-36 Problem 33

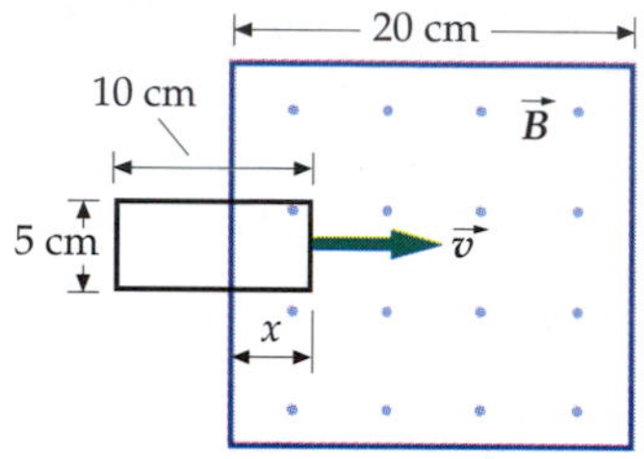

34 •• A uniform magnetic field of magnitude 1.2 T is in the z direction. A conducting rod of length 15 cm lies parallel to the y axis and oscillates in the x direction with displacement given by $x = (2\text{ cm})\cos 120\pi t$. What is the emf induced in the rod?

35 •• In Figure 30-37, the rod has a resistance R and the rails are horizontal and have negligible resistance. A battery of emf $\mathcal{E}$ and negligible internal resistance is connected between points a and b such that the current in the rod is downward. The rod is placed at rest at $t = 0$. (*a*) Find the force on the rod as a function of the speed v and write Newton's second law for the rod when it has speed v. (*b*) Show that the rod will approach a terminal speed and find an expression for it. (*c*) What is the current when the rod moves at its terminal speed?

Figure 30-37
Problems 35 and 38

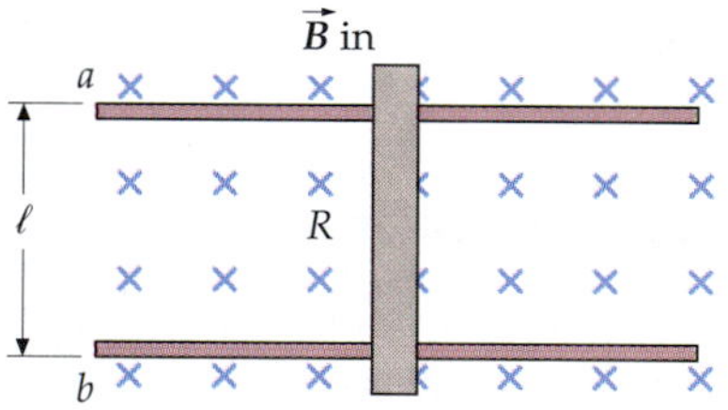

36 •• In Example 30-8, find the total energy dissipated in the resistance and show that it is equal to $\frac{1}{2}mv_0^2$.

37 •• Find the total distance traveled by the rod in Example 30-8.

38 •• In Figure 30-37, the rod has a resistance R and the rails have negligible resistance. A capacitor with charge Q_0 and capacitance C is connected between points a and b such that the current in the rod is downward. The rod is placed at rest at $t = 0$. (*a*) Write the equation of motion for the rod on the rails. (*b*) Show that the terminal speed of the rod down the rails is related to the final charge on the capacitor.

39 •• In Figure 30-38, a conducting rod of mass m and negligible resistance is free to slide without friction along two parallel rails of negligible resistance separated by a distance ℓ and connected by a resistance R. The rails are attached to a long, inclined plane that makes an angle θ with the horizontal. There is a magnetic field B directed upward. (*a*) Show that there is a retarding force directed up the incline given by $F = (B^2\ell^2 v\cos^2\theta)/R$. (*b*) Show that the terminal speed of the rod is $v_t = (mgR\sin\theta)/(B^2\ell^2\cos^2\theta)$.

Figure 30-38 Problems 39, 42, and 43

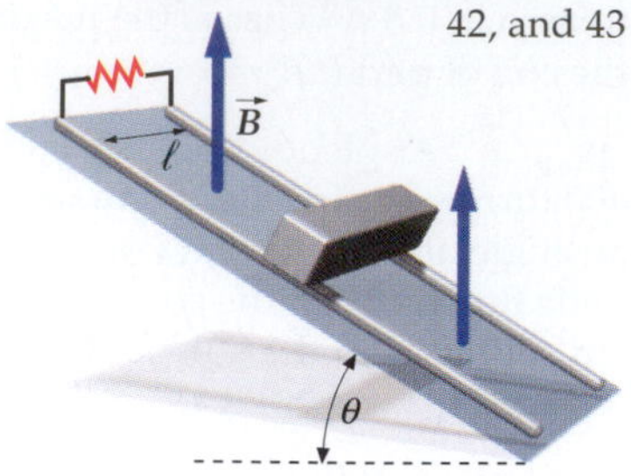

40 •• A simple pendulum has a wire of length ℓ supporting a metal ball of mass m. The wire has negligible mass and moves in a uniform horizontal magnetic field B. This pendulum executes simple harmonic motion having angular amplitude θ_0. What is the emf generated along the wire?

41 •• A wire lies along the z axis and carries current $I = 20$ A in the positive z direction. A small conducting sphere of radius $R = 2$ cm is initially at rest on the y axis at a distance $h = 45$ m above the wire. The sphere is dropped at time $t = 0$. (*a*) What is the electric field at the center of the sphere at $t = 3$ s? Assume that the only magnetic field is that produced by the wire. (*b*) What is the voltage across the sphere at $t = 3$ s?

42 •• In Figure 30-38, let $\theta = 30°$, $m = 0.4$ kg, $\ell = 15$ m, and $R = 2.0$ Ω. The rod starts from rest at the top of the inclined plane at $t = 0$. The rails have negligible resistance. There is a constant, vertically directed magnetic field of magnitude $B = 1.2$ T. (*a*) Find the emf induced in the rod as a function of its velocity down the rails. (*b*) Write Newton's law of motion for the rod; show that the rod will approach a terminal speed and determine its value.

43 •• When the rod of Problem 42 moves at its terminal speed, what is the power dissipated in the resistor? What is the rate of change of the potential energy of the rod?

44 ••• A solid conducting cylinder of radius 0.1 m and mass of 4 kg rests on horizontal conducting rails (Figure 30-39). The rails, separated by a distance $a = 0.4$ m, have a rough surface so the cylinder rolls rather than slides. A 12-V battery is connected to the rails as shown. The only significant resistance in the circuit is the contact resistance of 6 Ω between the cylinder and rails. The system is in a uniform vertical magnetic field. The cylinder is initially at rest next to the battery. (*a*) What must be the magnitude and direction of $\vec{B}$ so that the cylinder has an initial acceleration of 0.1 m/s^2 to the right? (*b*) Find the force on the cylinder as a function of its speed v. (*c*) Find the terminal velocity of the cylinder. (*d*) What is the kinetic energy of the cylinder when it has reached its terminal velocity? (Neglect the magnetic field due to the current in the battery–rails–cylinder loop and assume that the current density in the cylinder is uniform.)

Figure 30-39 Problem 44

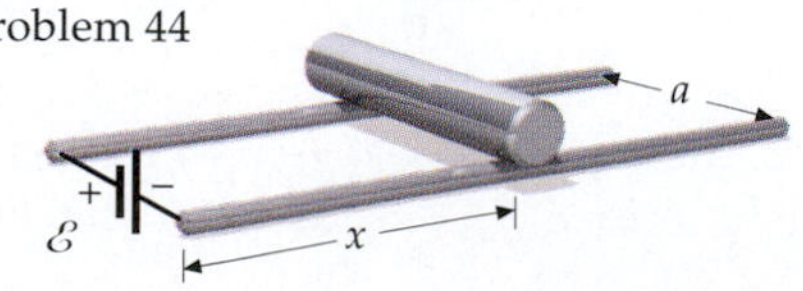

45 ••• A rod of length ℓ is perpendicular to a long wire carrying current I, as shown in Figure 30-40. The near end of the rod is a distance d away from the wire. The rod moves with a speed v in the direction of the current I. (*a*) Show that the potential difference between the ends of the rod is given by

$$V = \frac{\mu_0 I}{2\pi} v \ln \frac{d + \ell}{d}$$

(*b*) Use Faraday's law to obtain this result by considering the flux through a rectangular area $A = \ell v t$ swept out by the rod.

Figure 30-40 Problem 45

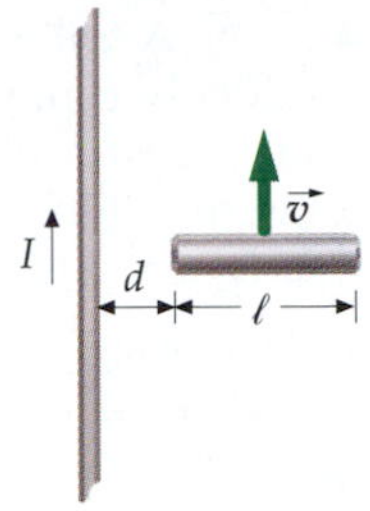

46 ••• The loop in Problem 10 moves away from the wire with a constant speed v. At time $t = 0$, the left side of the loop is a distance d from the long straight wire. (*a*) Compute the emf in the loop by computing the motional emf in each segment of the loop that is parallel to the long wire. Explain why you can neglect the emf in the segments that are perpendicular to the wire. (*b*) Compute the emf in the loop by first computing the flux through the loop as a function of time and then using $\mathcal{E} = -d\phi_m/dt$ and compare your answer with that obtained in part (*a*).

47 ••• A conducting rod of length ℓ rotates at constant angular velocity about one end, in a plane perpendicular to a uniform magnetic field B (Figure 30-41). (*a*) Show that the magnetic force on a charge q at a distance r from the pivot is $Bqr\omega$. (*b*) Show that the potential difference between the ends of the rod is $V = \frac{1}{2}B\omega\ell^2$. (*c*) Draw any radial line in the plane from which to measure $\theta = \omega t$. Show that the area of the pie-shaped region between the reference line and the rod is $A = \frac{1}{2}\ell^2\theta$. Compute the flux through this area, and show that $\mathcal{E} = \frac{1}{2}B\omega\ell^2$ follows when Faraday's law is applied to this area.

Figure 30-41 Problem 47

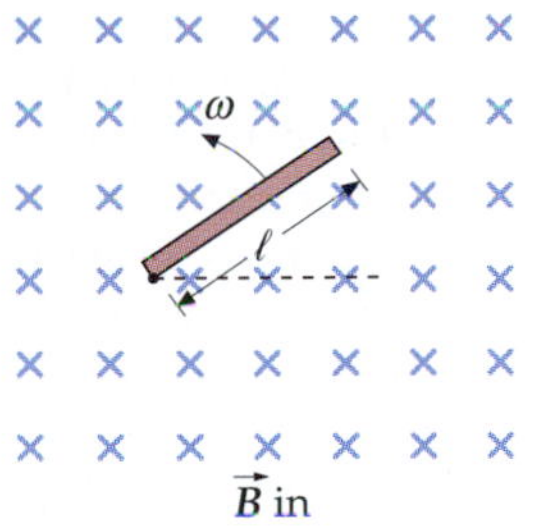

Inductance

48 • How would the self-inductance of a solenoid be changed if (*a*) the same length of wire were wound onto a cylinder of the same diameter but twice the length; (*b*) twice as much wire were wound onto the same cylinder; and (*c*) the same length of wire were wound onto a cylinder of the same length but twice the diameter?

49 • A coil with a self-inductance of 8.0 H carries a current of 3 A that is changing at a rate of 200 A/s. Find (*a*) the magnetic flux through the coil and (*b*) the induced emf in the coil.

50 • A coil with self-inductance L carries a current I, given by $I = I_0 \sin 2\pi ft$. Find and graph the flux ϕ_m and the self-induced emf as functions of time.

51 •• A solenoid has a length of 25 cm, a radius of 1 cm, and 400 turns, and carries a 3-A current. Find (*a*) B on the axis at the center of the solenoid; (*b*) the flux through the solenoid, assuming B to be uniform; (*c*) the self-inductance of the solenoid; and (*d*) the induced emf in the solenoid when the current changes at 150 A/s.

52 •• Two solenoids of radii 2 cm and 5 cm are coaxial. They are each 25 cm long and have 300 and 1000 turns, respectively. Find their mutual inductance.

53 •• A long, insulated wire with a resistance of 18 Ω/m is to be used to construct a resistor. First, the wire is bent in half, and then the doubled wire is wound in a cylindrical form as shown in Figure 30-42. The diameter of the cylindrical form is 2 cm, its length is 25 cm, and the total length of wire is 9 m. Find the resistance and inductance of this wire-wound resistor.

Figure 30-42 Problem 53

54 •• Figure 30-43 shows two long solenoids each with 2000 turns of wire. The outer solenoid is 20 cm long and has a diameter of 2 cm. The inner solenoid is 10 cm long and has a diameter of 1 cm. Find the effective inductance of this arrangement.

Figure 30-43 Problem 54

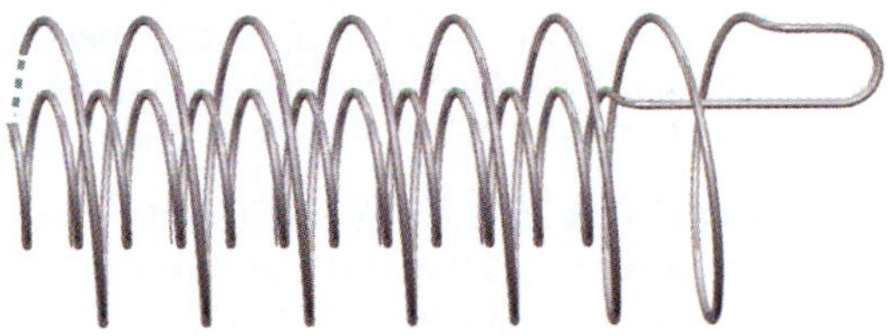

55 ••• In Figure 30-44, circuit 2 has a total resistance of 300 Ω. A total charge of 2×10^{-4} C flows through the galvanometer in circuit 2 when switch S in circuit 1 is closed. After a long time, the current in circuit 1 is 5 A. What is the mutual inductance between the two coils?

Figure 30-44 Problem 55

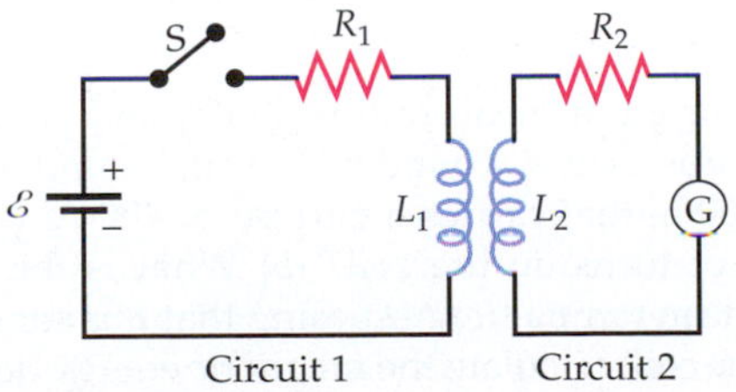

56 ••• Show that the inductance of a toroid of rectangular cross section as shown in Figure 30-45 is given by

$$L = \frac{\mu_0 N^2 H \ln(b/a)}{2\pi}$$

where N is the total number of turns, a is the inside radius, b is the outside radius, and H is the height of the toroid.

Figure 30-45
Problem 56

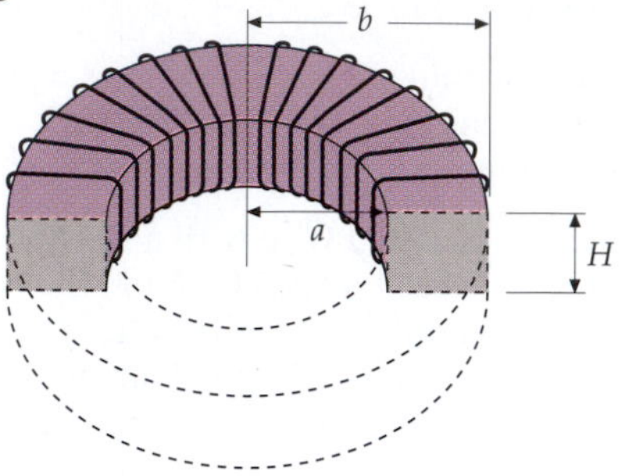

Magnetic Energy

57 • If the current through an inductor were doubled, the energy stored in the inductor would be

(*a*) the same.
(*b*) doubled.
(*c*) quadrupled.
(*d*) halved.
(*e*) quartered.

58 • A coil with a self-inductance of 2.0 H and a resistance of 12.0 Ω is connected across a 24-V battery of negligible internal resistance. (*a*) What is the final current? (*b*) How much energy is stored in the inductor when the final current is attained?

59 • Find (*a*) the magnetic energy, (*b*) the electric energy, and (*c*) the total energy in a volume of 1.0 m^3 in which there is an electric field of 10^4 V/m and a magnetic field of 5000 G.

60 •• In a plane electromagnetic wave such as a light wave, the magnitudes of the electric and magnetic fields are related by $E = cB$, where $c = 1/\sqrt{\epsilon_0 \mu_0}$ is the speed of light. Show that in this case the electric and the magnetic energy densities are equal.

61 •• A solenoid of 2000 turns, area 4 cm^2, and length 30 cm carries a current of 4.0 A. (*a*) Calculate the magnetic energy stored in the solenoid from $\frac{1}{2}LI^2$. (*b*) Divide your answer in part (*a*) by the volume of the solenoid to find the magnetic energy per unit volume in the solenoid. (*c*) Find B in the solenoid. (*d*) Compute the magnetic energy density from $u_m = B^2/2\mu_0$, and compare your answer with your result for part (*b*).

62 •• A long, cylindrical wire of radius $a = 2$ cm carries current $I = 80$ A uniformly distributed over its cross-sectional area. Find the magnetic energy per unit length within the wire.

63 •• A toroid of mean radius 25 cm and circular cross section of radius 2 cm is wound with a superconducting wire of length 1000 m that carries a current of 400 A. (*a*) What is the number of turns on the coil? (*b*) What is the magnetic field at the mean radius? (*c*) Assuming that B is constant over the area of the coil, calculate the magnetic energy density and the total energy stored in the toroid.

RL Circuits

64 • A coil of resistance 8.0 Ω and self-inductance 4.0 H is suddenly connected across a constant potential difference of 100 V. Let $t = 0$ be the time of connection, at which the current is zero. Find the current I and its rate of change dI/dt at times (*a*) $t = 0$, (*b*) $t = 0.1$ s, (*c*) $t = 0.5$ s, and (*d*) $t = 1.0$ s.

65 • The current in a coil with a self-inductance of 1 mH is 2.0 A at $t = 0$, when the coil is shorted through a resistor. The total resistance of the coil plus the resistor is 10.0 Ω. Find the current after (*a*) 0.5 ms and (*b*) 10 ms.

66 •• In the circuit of Figure 30-20, let $\mathcal{E}_0 = 12.0$ V, $R = 3.0$ Ω, and $L = 0.6$ H. The switch is closed at time $t = 0$. At time $t = 0.5$ s, find (*a*) the rate at which the battery supplies power, (*b*) the rate of Joule heating, and (*c*) the rate at which energy is being stored in the inductor.

67 •• Do Problem 66 for the times $t = 1$ s and $t = 100$ s.

68 •• The current in an *RL* circuit is zero at time $t = 0$ and increases to half its final value in 4.0 s. (*a*) What is the time constant of this circuit? (*b*) If the total resistance is 5 Ω, what is the self-inductance?

69 •• How many time constants must elapse before the current in an *RL* circuit that is initially zero reaches (*a*) 90%, (*b*) 99%, and (*c*) 99.9% of its final value?

70 •• A coil with inductance 4 mH and resistance 150 Ω is connected across a battery of emf 12 V and negligible internal resistance. (*a*) What is the initial rate of increase of the current? (*b*) What is the rate of increase when the current is half its final value? (*c*) What is the final current? (*d*) How long does it take for the current to reach 99% of its final value?

71 •• A large electromagnet has an inductance of 50 H and a resistance of 8.0 Ω. It is connected to a dc power source of 250 V. Find the time for the current to reach (*a*) 10 A and (*b*) 30 A.

72 •• Given the circuit shown in Figure 30-46, assume that switch S has been closed for a long time so that steady currents exist in the circuit and that the inductor L is made of superconducting wire so that its resistance may be considered to be zero. (*a*) Find the battery current, the current in the 100-Ω resistor, and the current through the inductor. (*b*) Find the initial voltage across the inductor when switch S is opened. (*c*) Give the current in the inductor as a function of time measured from the instant of opening switch S.

Figure 30-46
Problem 72

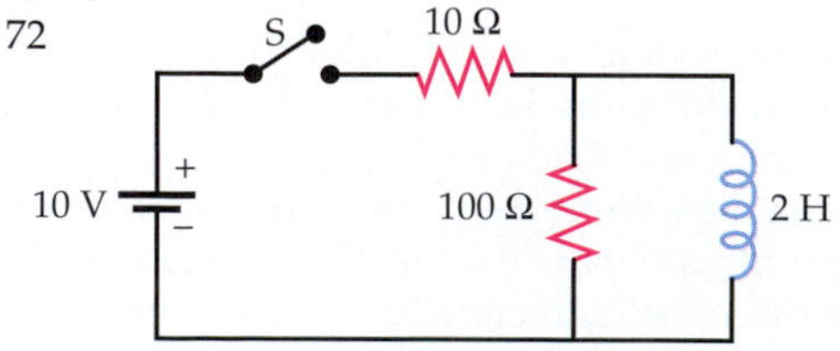

73 •• Compute the initial slope dI/dt at $t = 0$ from Equation 30-24, and show that if the current decreased steadily at this rate, it would be zero after one time constant.

74 •• An inductance L and resistance R are connected in series with a battery as in Figure 30-22. A long time after switch S_1 is closed, the current is 2.5 A. When the battery is switched out of the circuit by opening switch S_1 and closing S_2, the current drops to 1.5 A in 45 ms. (*a*) What is the time constant for this circuit? (*b*) If $R = 0.4\ \Omega$, what is L?

75 •• When the current in a certain coil is 5.0 A and is increasing at the rate of 10.0 A/s, the potential difference across the coil is 140 V. When the current is 5.0 A and is decreasing at the rate of 10 A/s, the potential difference is 60 V. Find the resistance and self-inductance of the coil.

76 •• For the circuit of Figure 30-47, (*a*) find the rate of change of the current in each inductor and in the resistor just after the switch is closed. (*b*) What is the final current? (Use the result from Problem 92.)

Figure 30-47 Problem 76

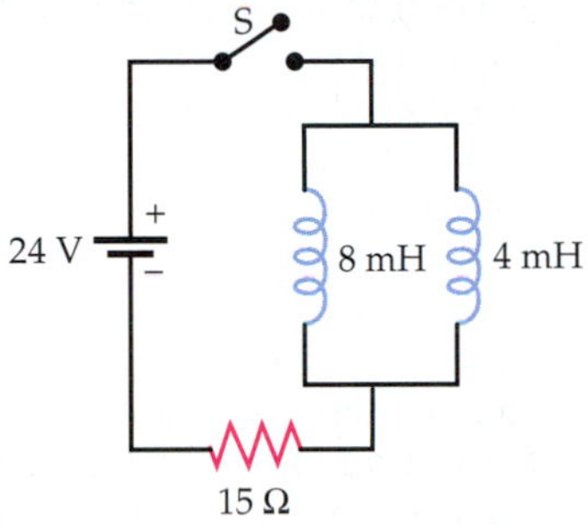

77 •• For the circuit of Example 30-11, find the time at which the power dissipation in the resistor equals the rate at which magnetic energy is stored in the inductor.

78 ••• In the circuit of Figure 30-21, let $\mathcal{E}_0 = 12.0$ V, $R = 3.0\ \Omega$, and $L = 0.6$ H. The switch is closed at time $t = 0$. From time $t = 0$ to $t = \tau$, find (*a*) the total energy that has been supplied by the battery, (*b*) the total energy that has been dissipated in the resistor, and (*c*) the energy that has been stored in the inductor. (*Hint:* Find the rates as functions of time and integrate from $t = 0$ to $t = \tau = L/R$.)

General Problems

79 • Two identical bar magnets are dropped from equal heights. Magnet A is dropped from above bare earth, whereas magnet B is dropped from above a metal plate. Which magnet strikes first?

(*a*) Magnet A
(*b*) Magnet B
(*c*) Both strike at the same time.
(*d*) Whichever has the N pole toward the ground.
(*e*) Whichever has the S pole toward the ground.

80 • True or false:

(*a*) The induced emf in a circuit is proportional to the magnetic flux through the circuit.
(*b*) There can be an induced emf at an instant when the flux through the circuit is zero.
(*c*) Lenz's law is related to the conservation of energy.
(*d*) The inductance of a solenoid is proportional to the rate of change of the current in it.
(*e*) The magnetic energy density at some point in space is proportional to the square of the magnetic field at that point.

81 • A bar magnet is dropped inside a long vertical tube. If the tube is made of metal, the magnet quickly approaches a terminal speed, but if the tube is made of cardboard, it does not. Explain.

82 • A circular coil of radius 3.0 cm has 6 turns. A magnetic field $B = 5000$ G is perpendicular to the coil. (*a*) Find the magnetic flux through the coil. (*b*) Find the magnetic flux through the coil if the coil makes an angle of 20° with the magnetic field.

83 • The magnetic field in Problem 82 is steadily reduced to zero in 1.2 s. Find the emf induced in the coil when (*a*) the magnetic field is perpendicular to the coil and (*b*) the magnetic field makes an angle of 20° with the normal to the coil.

84 • A 100-turn coil has a radius of 4.0 cm and a resistance of 25 Ω. At what rate must a perpendicular magnetic field change to produce a current of 4.0 A in the coil?

85 •• Figure 30-48 shows an ac generator. It consists of a rectangular loop of dimensions a and b with N turns connected to slip rings. The loop rotates with an angular velocity ω in a uniform magnetic field $\vec{B}$. (*a*) Show that the potential difference between the two slip rings is $\mathcal{E} = NBab\omega \sin \omega t$. (*b*) If $a = 1.0$ cm, $b = 2.0$ cm, $N = 1000$, and $B = 2$ T, at what angular frequency ω must the coil rotate to generate an emf whose maximum value is 110 V?

Figure 30-48 Problems 85 and 86

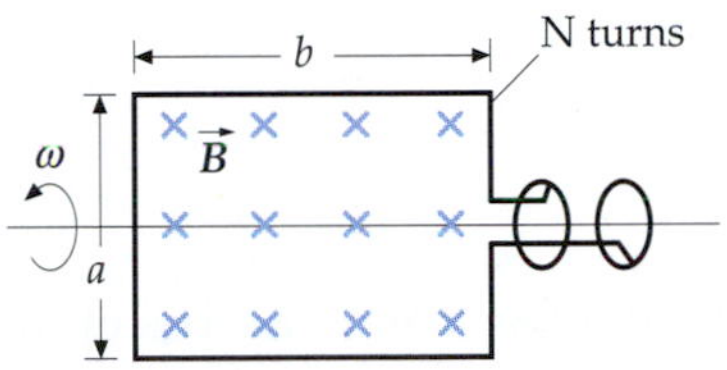

86 •• Prior to about 1960, magnetic field strength was measured by means of a rotating coil gaussmeter. This device used a small loop of many turns rotating on an axis perpendicular to the magnetic field at fairly high speed and connected to an ac voltmeter by means of slip rings like those shown in Figure 30-48. The sensing coil for a rotating coil gaussmeter has 400 turns and an area of 1.4 cm^2. The coil rotates at 180 rpm. If the magnetic field strength is 0.45 T, find the maximum induced emf in the coil and the orientation of the coil relative to the field for which this maximum induced emf occurs.

87 •• Show that the effective inductance for two inductors L_1 and L_2 connected in series such that none of the flux from either passes through the other is given by $L_{eff} = L_1 + L_2$.

88 •• The rectangular coil in Figure 30-49 has 80 turns, is 25 cm wide and 30 cm long, and is located in a magnetic field $B = 1.4$ T directed out of the page as shown, with only half of the coil in the region of the magnetic field. The resistance of the coil is 24 Ω. Find the magnitude and direction of the induced current if the coil is moved with a speed of 2 m/s (*a*) to the right, (*b*) up, (*c*) to the left, and (*d*) down.

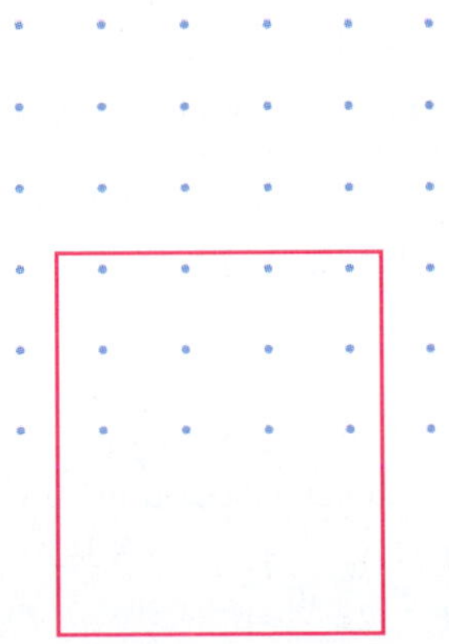

Figure 30-49 Problems 88–90

89 •• Suppose the coil of Problem 88 is rotated about its vertical centerline at constant angular velocity of 2 rad/s. Find the induced current as a function of time.

90 •• Suppose the coil of Problem 88 is rotated about its horizontal centerline at constant angular velocity of 2 rad/s. Find the induced current as a function of time.

91 •• Show that if the flux through each turn of an N-turn coil of resistance R changes from ϕ_{m1} to ϕ_{m2} the total charge passing through the coil is given by $Q = N(\phi_{m1} - \phi_{m2})/R$.

92 •• Show that the effective inductance for two inductors L_1 and L_2 connected in parallel such that none of the flux from either passes through the other is given by

$$\frac{1}{L_{eff}} = \frac{1}{L_1} + \frac{1}{L_2}$$

93 •• A long solenoid has n turns per unit length and carries a current given by $I = I_0 \sin \omega t$. The solenoid has a circular cross section of radius R. Find the induced electric field at a radius r from the axis of the solenoid for (*a*) $r < R$ and (*b*) $r > R$.

94 ••• A thin-walled hollow wire of radius a lies with its axis along the z axis and carries current I in the positive z direction. A second identical wire is parallel to the first with its axis along the line $x = d$. The second wire carries current I in the negative z direction. (*a*) Find the magnetic flux per unit length through the space in the xz plane between the wires. (*b*) If the far ends of the wires are connected together so that the parallel wires form two sides of a loop, find the self-inductance per unit length of the loop.

95 ••• A coaxial cable consists of two very thin-walled conducting cylinders of radii r_1 and r_2 (Figure 30-50). Current I goes in one direction down the inner cylinder and in the opposite direction in the outer cylinder. (*a*) Use Ampère's law to find B. Show that $B = 0$ except in the region between the conductors. (*b*) Show that the magnetic energy density in the region between the cylinders is

$$u_m = \frac{\mu_0 I^2}{8\pi^2 r^2}$$

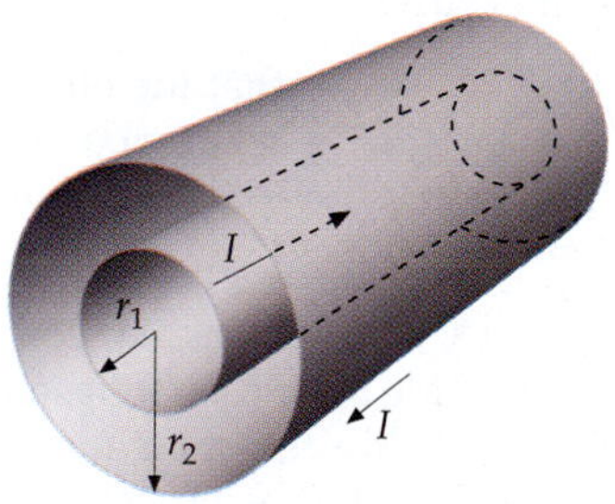

Figure 30-50
Problems 95 and 96

(*c*) Find the magnetic energy in a cylindrical shell volume element of length ℓ and volume $dV = \ell 2\pi r\, dr$, and integrate your result to show that the total magnetic energy in the volume of length ℓ is

$$U_m = \frac{\mu_0}{4\pi} I^2 \ell \ln \frac{r_2}{r_1}$$

(*d*) Use the result in part (*c*) and $U_m = \frac{1}{2}LI^2$ to show that the self-inductance per unit length is

$$\frac{L}{\ell} = \frac{\mu_0}{2\pi} \ln \frac{r_2}{r_1}$$

96 ••• In Figure 30-50, compute the flux through a rectangular area of sides ℓ and $r_2 - r_1$ between the conductors. Show that the self-inductance per unit length can be found from $\phi_m = LI$ (see part (*d*) of Problem 95).

97 ••• Figure 30-51 shows a rectangular loop of wire, 0.30 m wide and 1.50 m long, in the vertical plane and perpendicular to a uniform magnetic field $B = 0.40$ T, directed inward as shown. The portion of the loop not in the magnetic field is 0.10 m long. The resistance of the loop is 0.20 Ω and its mass is 0.50 kg. The loop is released from rest at $t = 0$. (*a*) What is the magnitude and direction of the induced current when the loop has a downward velocity v? (*b*) What is the force that acts on the loop as a result of this current? (*c*) What is the net force acting on the loop? (*d*) Write the equation of motion of the loop. (*e*) Obtain an expression for the velocity of the loop as a function of time. (*f*) Integrate the expression obtained in part (*e*) to find the displacement y as a function of time. (*g*) From the result obtained in part (*f*) find t for $y = 1.40$ m, i.e., the time when the loop leaves the region of magnetic field. (*h*) Find the velocity of the loop at that instant. (*i*) What would be the velocity of the loop after it has dropped 1.40 m if $B = 0$?

Figure 30-51
Problems 97 and 98

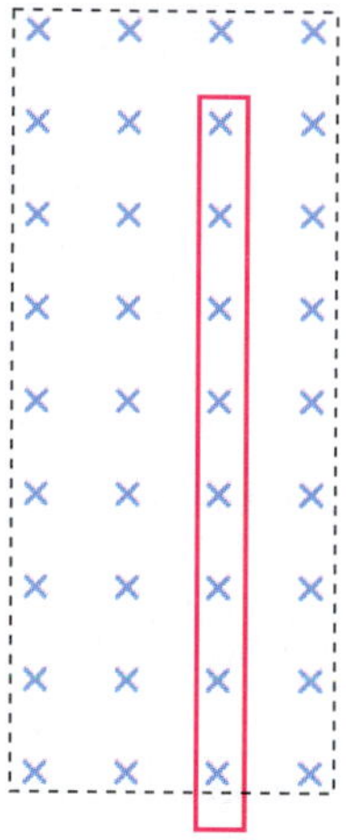

98 ••• The loop of Problem 97 is attached to a plastic spring of spring constant κ (Figure 30-52). (*a*) When $B = 0$, the period of small-amplitude vertical oscillations of the mass–spring system is 0.8 s. Find the spring constant κ. (*b*) When $B \neq 0$, a current is induced in the loop as a result of its up and down motion. Obtain an expression for the induced current as a function of time when $B = 0.40$ T. (*c*) Show that the induced current acts as a damping mechanism. (*d*) Determine the value of the magnetic field for which the Q of the mass–spring system is 100.

Figure 30-52
Problems 98 and 99

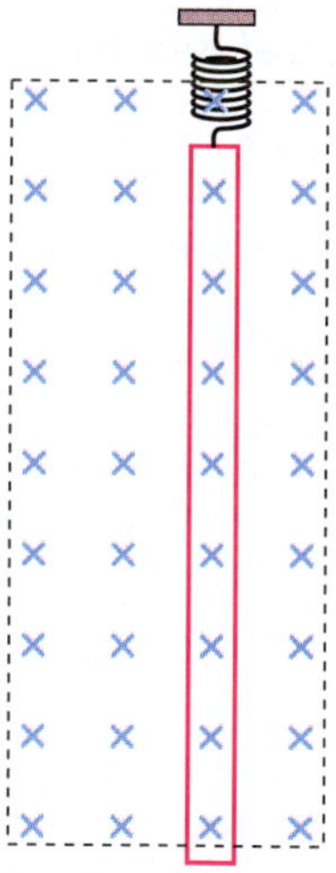

99 ••• Show that the effective inductance of two inductors, L_1 and L_2, connected in series and in close proximity, is $L_{eff} = L_1 + L_2 \pm 2M$. When should the plus sign be used in this expression, and when should the minus sign be used?

CHAPTER 31

Alternating-Current Circuits

Most high-voltage long-distance power transmission uses alternating current because converting ac to dc at the sending end (such as a hydroelectric power plant) and from dc back to ac at the consuming end has been expensive. However, recent advances in technology have revived interest in high-voltage direct current (HVDC). Shown here is a dc-to-ac conversion station near Boston, linked by HVDC lines to a 2000-MW hydroelectric generator unit in James Bay, Quebec. This particular station is used to convert direct current to alternating current.

More than 99% of the electrical energy used today is produced by electrical generators in the form of alternating current, which has a great advantage over direct current in that electrical energy can be transported over long distances at very high voltages and low currents to reduce energy losses due to Joule heat. It can then be transformed, with almost no energy loss, to lower and safer voltages and correspondingly higher currents for everyday use. The transformer that accomplishes this change in voltage and current works on the basis of magnetic induction. In North America, power is delivered by a sinusoidal current of frequency 60 Hz. Devices such as radios, television sets, and microwave ovens detect or generate alternating currents of much greater frequencies.

Alternating current is easily produced by magnetic induction in an ac generator, which is designed to put out a sinusoidal emf. We will see that when the generator output is sinusoidal, the current in an inductor, a capacitor, or a resistor is also sinusoidal, though it is generally not in phase with the generator's emf. When the emf and current are both sinusoidal, their maximum values can be easily related. The study of sinusoidal currents is important because even currents that are not sinusoidal can be analyzed in terms of sinusoidal components using Fourier analysis.

31-1 ac Generators

Figure 31-1 shows a simple **generator** consisting of a coil of area A and N turns rotating in a uniform magnetic field. The ends of the coil are connected to rings called slip rings that rotate with the coil. They make electrical contact through stationary conducting brushes in contact with the rings.

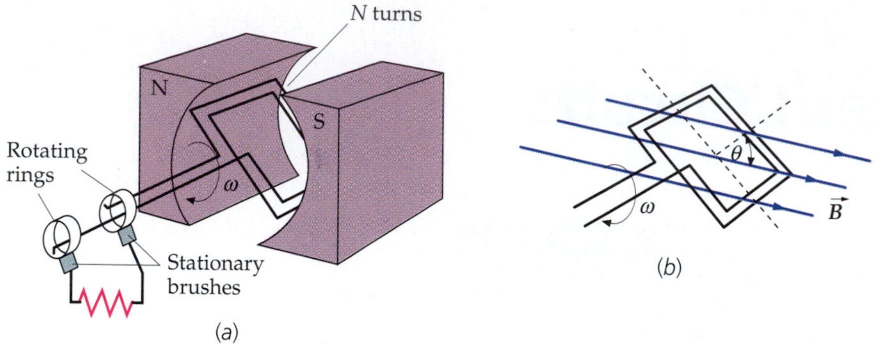

Figure 31-1 (*a*) An ac generator. A coil rotating with constant angular frequency ω in a magnetic field $\vec{B}$ generates a sinusoidal emf. Energy from a waterfall or a steam turbine is used to rotate the coil to produce electrical energy. The emf is supplied to an external circuit by the brushes in contact with the rings. (*b*) At this instant, the normal to the plane of the coil makes an angle θ with the magnetic field and the flux is $BA\cos\theta$.

When the line perpendicular to the plane of the coil makes an angle θ with a uniform magnetic field $\vec{B}$, as shown in the figure, the magnetic flux through the coil is

$$\phi_m = NBA\cos\theta \qquad \text{31-1}$$

When the coil is mechanically rotated, the flux through it will change, and an emf will be induced. If ω is the angular velocity of rotation and the initial angle is δ, the angle at some later time t is given by

$$\theta = \omega t + \delta$$

Then

$$\phi_m = NBA\cos(\omega t + \delta) = NBA\cos(2\pi f t + \delta)$$

The emf in the coil will then be

$$\mathcal{E} = -\frac{d\phi_m}{dt} = -NBA\frac{d}{dt}\cos(\omega t + \delta) = +NBA\omega\sin(\omega t + \delta) \qquad \text{31-2}$$

or

$$\mathcal{E} = \mathcal{E}_{max}\sin(\omega t + \delta) \qquad \text{31-3}$$

where

$$\mathcal{E}_{max} = NBA\omega \qquad \text{31-4}$$

We can thus produce a sinusoidal emf in a coil by rotating it with constant angular velocity in a magnetic field. Although practical generators are considerably more complicated, they work on the same principle that an alternating emf is produced in a coil rotating in a magnetic field, and they are designed so that the emf produced is sinusoidal. In circuit diagrams, an ac generator is represented by the symbol ⊖.

The same coil in a magnetic field that can be used to generate an alternating emf can also be used as an ac **motor.** Instead of mechanically rotating the coil to generate an emf, we apply an alternating current to the coil from another ac generator. The torque due to the magnetic force on the wire rotates the coil. As the coil rotates in the magnetic field, a back emf is generated that tends to counter the emf that supplies the current. When the motor is first turned on, there is no back emf and the current is very large, being limited only by the resistance in the circuit. As the motor begins to rotate, the back emf increases and the current decreases.

(a) (b)

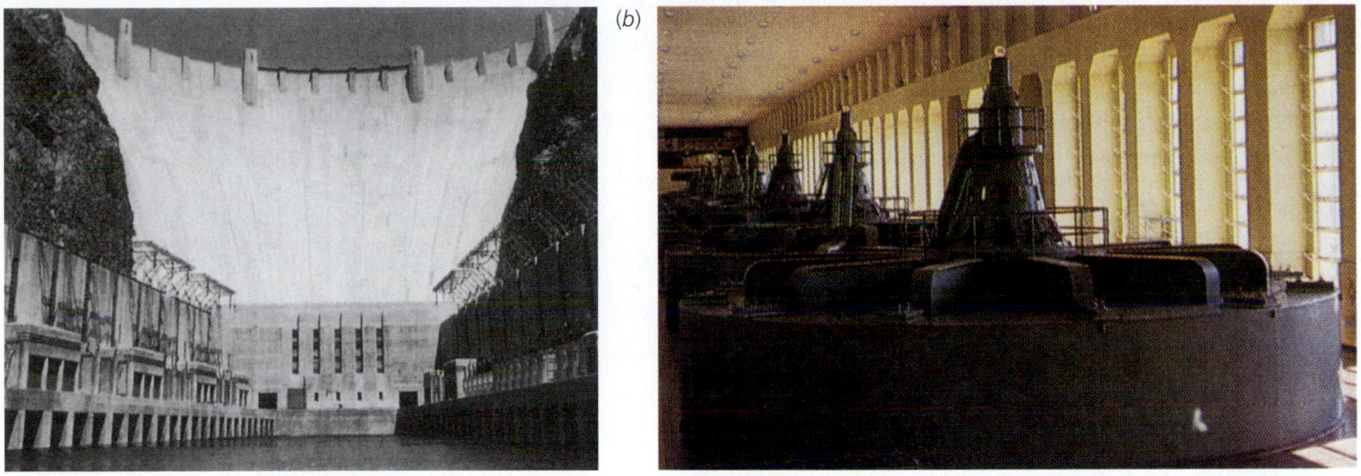

(a) River-level view of Hoover Dam with the Nevada wing of its power plant on the left and the Arizona wing on the right. The mechanical energy of falling water drives turbines (b) for the generation of electricity.

(c) Schematic drawing of Hoover Dam showing the intake towers and pipes (penstocks) that carry the water to the generators below.

(c)

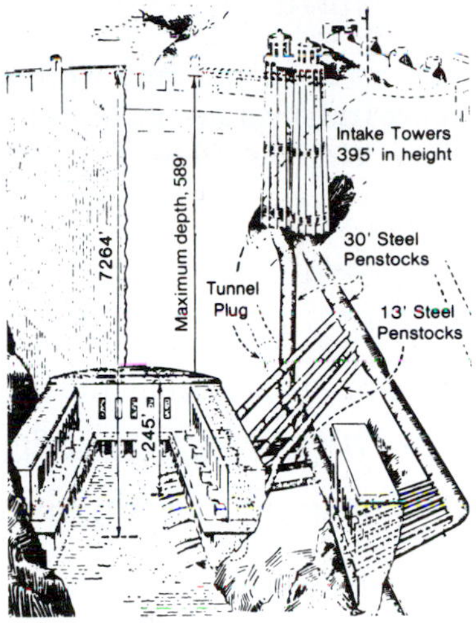

Exercise A 250-turn coil has an area of 3 cm^2. If it rotates in a magnetic field of 0.4 T at 60 Hz, what is $\mathcal{E}_{max}$? (*Answer* $\mathcal{E}_{max} = 11.3$ V)

31-2 Alternating Current in a Resistor

Figure 31-2 shows a simple ac circuit consisting of a generator and a resistor. The plus and minus signs around the generator symbol indicate the sides of higher and lower potential, and the same signs around the resistor indicate the assumed direction of the current. The voltage drop across the resistor V_R is given by

$$V_R = V_+ - V_- = IR \qquad \text{31-5}$$

If $\mathcal{E}$ is the emf supplied by the generator, applying Kirchhoff's loop rule to this circuit gives

$$\mathcal{E} - V_R = 0$$

If the generator produces an emf given by Equation 31-3, we have

$$\mathcal{E}_{max} \sin(\omega t + \delta) - IR = 0$$

In this equation, the phase constant δ is arbitrary. It is convenient to choose $\delta = \pi/2$ so that $\mathcal{E} = \mathcal{E}_{max} \sin(\omega t + \delta) = \mathcal{E}_{max} \cos \omega t$. Then

$$\mathcal{E}_{max} \cos \omega t - IR = 0 \qquad \text{31-6}$$

The current in the resistor is

$$I = \frac{\mathcal{E}_{max}}{R} \cos \omega t = I_{max} \cos \omega t \qquad \text{31-7}$$

where

$$I_{max} = \frac{\mathcal{E}_{max}}{R} \qquad \text{31-8}$$

Note that the current through the resistor is in phase with the voltage across the resistor.

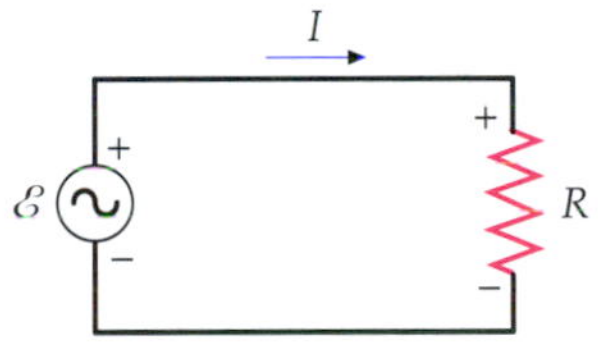

Figure 31-2 An ac generator in series with a resistor R.

The power dissipated in the resistor varies with time. Its instantaneous value is

$$P = I^2R = (I_{max} \cos \omega t)^2 R = I_{max}^2 R \cos^2 \omega t \tag{31-9}$$

Figure 31-3 shows the power as a function of time. It varies from zero to its maximum value $I_{max}^2 R$ as shown. We are usually interested in the average power over one or more cycles:

$$P_{av} = (I^2R)_{av} = I_{max}^2 R(\cos^2 \omega t)_{av}$$

The average value of $\cos^2 \omega t$ over one or more periods is $\frac{1}{2}$.* The average power dissipated in the resistor is thus

$$P_{av} = (I^2R)_{av} = \tfrac{1}{2} I_{max}^2 R \tag{31-10}$$

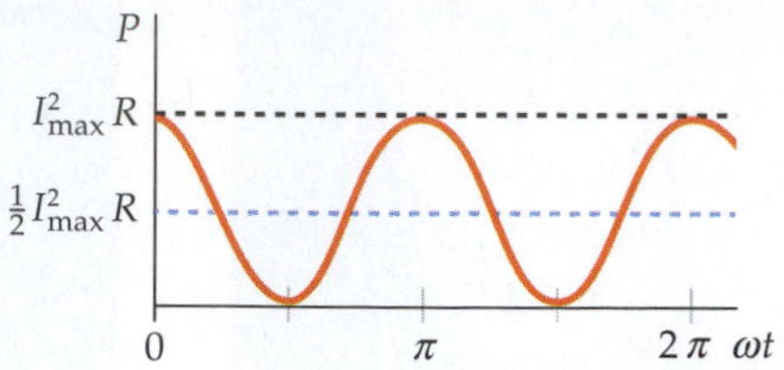

Figure 31-3 Plot of the power dissipated in the resistor in Figure 31-2 versus time. The power varies from zero to a maximum value $I_{max}^2 R$. The average power is half the maximum power.

rms Values

Most ac ammeters and voltmeters are designed to measure **root-mean-square (rms) values** of current and voltage rather than the maximum or peak values. The **rms** value of a current I_{rms} is defined by

$$I_{rms} = \sqrt{(I^2)_{av}} \tag{31-11}$$

Definition—rms current

For a sinusoidal current, the average value of I^2 is

$$(I^2)_{av} = [(I_{max} \cos \omega t)^2]_{av} = \tfrac{1}{2} I_{max}^2$$

Substituting $\frac{1}{2} I_{max}^2$ for $(I^2)_{av}$ in Equation 31-11, we obtain

$$I_{rms} = \frac{1}{\sqrt{2}} I_{max} \tag{31-12}$$

rms value related to maximum value

The rms value of any quantity that varies sinusoidally equals the maximum value of that quantity divided by $\sqrt{2}$.

Substituting I_{rms}^2 for $\frac{1}{2} I_{max}^2$ in Equation 31-10, we obtain for the average power dissipated in the resistor

$$P_{av} = I_{rms}^2 R \tag{31-13}$$

The rms current equals the steady dc current that would produce the same Joule heating as the actual ac current.

For the simple circuit in Figure 31-2, the average power delivered by the generator is equal to that dissipated in the resistor:

$$P_{av} = (\mathcal{E} I)_{av} = [(\mathcal{E}_{max} \cos \omega t)(I_{max} \cos \omega t)]_{av}$$

$$= \mathcal{E}_{max} I_{max} (\cos^2 \omega t)_{av}$$

or

$$P_{av} = \tfrac{1}{2} \mathcal{E}_{max} I_{max}$$

* This can be seen from the identity $\cos^2 \omega t + \sin^2 \omega t = 1$. A plot of $\sin^2 \omega t$ looks the same as one of $\cos^2 \omega t$ except that it is shifted by 90°. Both have the same average value over one or more periods, and since their sum is 1, the average value of each must be $\frac{1}{2}$.

Using $I_{rms} = I_{max}/\sqrt{2}$ and $\mathcal{E}_{rms} = \mathcal{E}_{max}/\sqrt{2}$, this can be written

$$P_{av} = \mathcal{E}_{rms}I_{rms} \tag{31-14}$$

Average power delivered by a generator

The rms current is related to the rms emf in the same way that the maximum current is related to the maximum emf. We can see this by dividing each side of Equation 31-8 by $\sqrt{2}$ and using $I_{rms} = I_{max}/\sqrt{2}$ and $\mathcal{E}_{rms} = \mathcal{E}_{max}/\sqrt{2}$:

$$I_{rms} = \frac{\mathcal{E}_{rms}}{R} \tag{31-15}$$

Equations 31-13, 31-14, and 31-15 are of the same form as the corresponding equations for direct-current circuits with I replaced by I_{rms} and $\mathcal{E}$ replaced by $\mathcal{E}_{rms}$. We can therefore calculate the power input and the heat generated using the same equations that we used for direct current if we use rms values for the current and emf.

In the circuit of Figure 31-2, which contains only a generator and a resistor, the voltage drop across the resistor equals the voltage of the generator. In more complicated circuits containing several elements, the voltage drop across a resistor is usually not equal to the generator voltage. It is useful, therefore, to write Equation 31-15 in terms of the voltage drop across the resistor $V_{R,rms}$:

$$I_{rms} = \frac{V_{R,rms}}{R} \tag{31-16}$$

Exercise A 12-Ω resistor is connected across a sinusoidal emf that has a peak value of 48 V. Find (*a*) the rms current, (*b*) the average power, and (*c*) the maximum power. (*Answers* (*a*) 2.83 A, (*b*) 96 W, (*c*) 192 W)

The ac power supplied by power companies to dwellings in the United States has a frequency of 60 Hz and a voltage of 120 V_{rms}.* If you plug in a 1600-W heater, it will draw a current of

$$I_{rms} = \frac{P_{av}}{\mathcal{E}_{rms}} = \frac{1600\text{ W}}{120\text{ V}} = 13.3\text{ A}$$

The voltage across the outlets is maintained at 120 V, independent of the current drawn. Thus, all appliances plugged into the outlets of a single circuit are essentially in parallel. If you plug a 500-W toaster into another outlet of the same circuit, it will draw a current of 500 W/120 V = 4.17 A and the total current through the circuit will be 17.5 A. Currents greater than about 20 A may overheat household wiring and create a fire hazard. Each circuit is therefore equipped with a circuit breaker (or a fuse in older houses) that trips (or blows) when the current exceeds 20 A. The maximum power load that can be handled by a circuit with a 20-A circuit breaker is

$$P_{av} = \mathcal{E}_{rms}I_{rms} = (120\text{ V})(20\text{ A}) = 2.4\text{ kW}$$

Since most modern houses require considerably more than 2.4 kW of power, several circuits are supplied, each with its own circuit breaker and each having several outlets.

---

* For some high-power appliances, such as an electric clothes dryer or an oven, separate lines carrying power at 240 V are often required. For a given power requirement, only half as much current is required at 240 V as at 120 V, but a shock at 240 V is much more likely to be fatal than one at 120 V.

Example 31-1

Find (*a*) the average current and (*b*) the rms current for the sawtooth waveform shown in Figure 31-4. In the region $0 < t < T$, the current is given by $I = (I_0/T)t$.

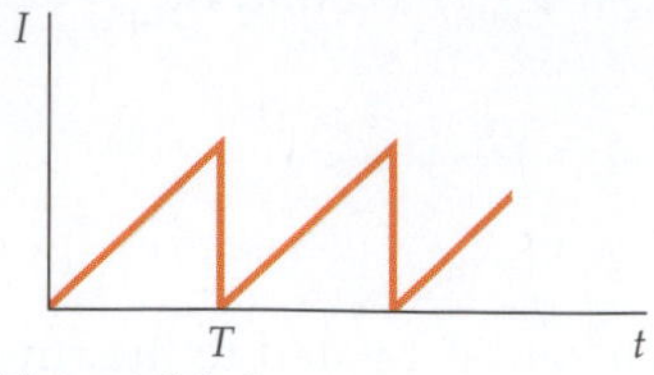

Figure 31-4

Picture the Problem The average of any quantity over a time interval T is the integral of the quantity over the interval divided by T. We use this to find both the average current, I_{av}, and the average of the current squared, $(I^2)_{av}$.

(*a*) Calculate I_{av} by integrating I from $t = 0$ to $t = T$ and dividing by T:

$$I_{av} = \frac{1}{T}\int_0^T I\,dt = \frac{1}{T}\int_0^T \frac{I_0}{T}t\,dt = \frac{I_0}{T^2}\frac{T^2}{2} = \frac{1}{2}I_0$$

(*b*)1. Find $(I^2)_{av}$ by integrating I^2:

$$(I^2)_{av} = \frac{1}{T}\int_0^T I^2\,dt = \frac{1}{T}\int_0^T \left(\frac{I_0}{T}\right)^2 t^2\,dt = \frac{I_0^2}{T^3}\frac{T^3}{3} = \frac{1}{3}I_0^2$$

2. The rms current is the square root of the above result:

$$I_{rms} = \sqrt{(I^2)_{av}} = \frac{I_0}{\sqrt{3}}$$

31-3 Alternating Current in Inductors and Capacitors

Alternating current behaves differently than direct current in inductors and capacitors. When a capacitor becomes fully charged in a dc circuit, it stops the current, that is, it acts like an open circuit. But if the current alternates, charge continually flows onto or off of the plates of the capacitor. We will see that at high frequencies a capacitor hardly impedes the current at all, that is, it acts like a short circuit. Conversely, an inductor coil usually has a very small resistance and is essentially a short circuit for direct current, but when the current is changing, a back emf is generated in an inductor that is proportional to dI/dt. At high frequencies the back emf is large and the inductor acts like an open circuit.

Inductors in ac Circuits

Figure 31-5 shows an inductor coil in series with an ac generator. When the current increases in the inductor, a back emf of magnitude $L\,dI/dt$ is generated due to the changing flux. Usually this back emf is much greater than the drop IR due to the resistance of the coil, so we normally neglect the resistance of the coil. The voltage drop across the inductor V_L is then given by

$$V_L = V_+ - V_- = L\frac{dI}{dt} \qquad \text{31-17}$$

Applying Kirchhoff's loop rule to this circuit gives

$$\mathcal{E} - V_L = 0$$

or

$$V_L = \mathcal{E} = \mathcal{E}_{max}\cos\omega t$$

$$L\frac{dI}{dt} = \mathcal{E}_{max}\cos\omega t \qquad \text{31-18}$$

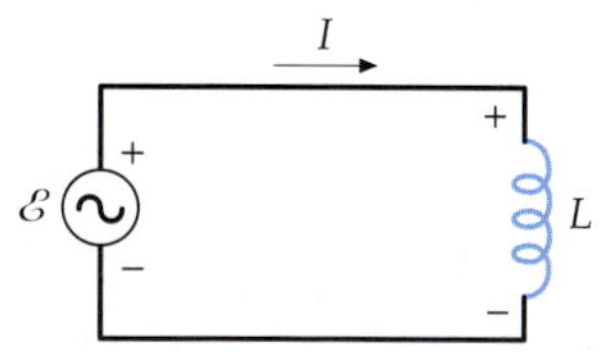

Figure 31-5 An ac generator in series with an inductor L. Plus and minus signs have been placed on the inductor to indicate the direction of the potential drop when dI/dt is positive for the assumed direction of the current. Note that for a positive value of dI/dt, the point at which the current enters the inductor is at a higher potential than the point at which the current leaves.

Multiplying both sides by dt and dividing by L, we obtain

$$dI = \frac{\mathcal{E}_{max}}{L} \cos \omega t \, dt \tag{31-19}$$

We solve for the current I by integrating both sides of the equation:

$$I = \frac{\mathcal{E}_{max}}{L} \int \cos \omega t \, dt = \frac{\mathcal{E}_{max}}{\omega L} \sin \omega t + C \tag{31-20}$$

where the constant of integration C is the dc component of the current. Setting the dc component of the current to be zero, we have

$$I = \frac{\mathcal{E}_{max}}{\omega L} \sin \omega t = I_{max} \sin \omega t \tag{31-21}$$

where

$$I_{max} = \frac{\mathcal{E}_{max}}{\omega L} \tag{31-22}$$

The current $I = I_{max} \sin \omega t$ is 90° out of phase with the voltage across the inductor $V_L = \mathcal{E}_{max} \cos \omega t$. From Figure 31-6, which shows I and V_L as functions of time, we can see that the maximum value of the voltage occurs 90° or one-fourth period before the corresponding maximum value of the current. The voltage drop across an inductor is said to *lead the current by 90°*. We can understand this physically. When I is zero but increasing, dI/dt is maximum, so the back emf induced in the inductor is at its maximum. One-quarter cycle later, I is maximum. At this time, dI/dt is zero, so V_L is zero. Using the trigonometric identity $\sin \omega t = \cos (\omega t - \pi/2)$, Equation 31-21 for the current can be written

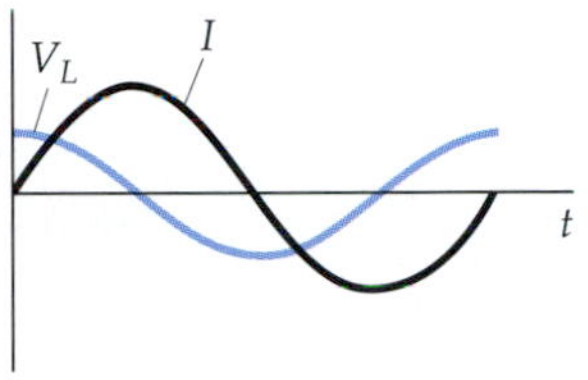

Figure 31-6 Current and voltage across the inductor in Figure 31-5 as functions of time. The maximum voltage occurs one-fourth period before the maximum current. Thus, the voltage is said to lead the current by one-fourth period or 90°.

$$I = I_{max} \cos (\omega t - \pi/2) \tag{31-23}$$

The relation between the maximum current and the maximum voltage (or between the rms current and rms voltage) for an inductor can be written in a form similar to Equation 31-15 for a resistor. From Equation 31-22, we have

$$I_{max} = \frac{\mathcal{E}_{max}}{\omega L} = \frac{\mathcal{E}_{max}}{X_L} \tag{31-24}$$

where

$$X_L = \omega L \tag{31-25}$$

Definition—Inductive reactance

is called the **inductive reactance.** Since $I_{rms} = I_{max}/\sqrt{2}$ and $\mathcal{E}_{rms} = \mathcal{E}_{max}/\sqrt{2}$ the rms current is given by

$$I_{rms} = \frac{\mathcal{E}_{rms}}{X_L} \tag{31-26}$$

Like resistance, inductive reactance has units of ohms. As we can see from Equation 31-26, the larger the reactance for a given emf, the smaller the current. Unlike resistance, the inductive reactance depends on the frequency of the current—the greater the frequency, the greater the reactance.

The instantaneous power input to the inductor from the generator is

$$P = \mathcal{E}I = (\mathcal{E}_{max} \cos \omega t)(I_{max} \sin \omega t) = \mathcal{E}_{max} I_{max} \cos \omega t \sin \omega t$$

The average power into the inductor is zero. We can see this by using

$$\cos \omega t \sin \omega t = \tfrac{1}{2} \sin 2\,\omega t$$

The value of this term oscillates twice during each cycle and is negative as often as it is positive. Thus, on the average, no energy is dissipated in an inductor. (This is true only if the resistance of the inductor can be neglected.)

Example 31-2

A 40-mH inductor is placed across an ac generator that has a maximum emf of 120 V. Find the inductive reactance and the maximum current when the frequency is (*a*) 60 Hz and (*b*) 2000 Hz.

Picture the Problem We calculate the inductive reactance at each frequency and use Equation 31-24 to find the maximum current.

(*a*)1. The maximum current equals the maximum emf divided by the inductive reactance:

$$I_{max} = \frac{\mathcal{E}_{max}}{X_L}$$

2. Compute the inductive reactance at 60 Hz:

$$X_{L1} = \omega_1 L = 2\pi f_1 L = (2\pi)(60 \text{ Hz})(40 \times 10^{-3} \text{ H}) = 15.1\ \Omega$$

3. Use this value of X_L to compute the maximum current at 60 Hz:

$$I_{1,max} = \frac{120 \text{ V}}{15.1\ \Omega} = 7.95 \text{ A}$$

(*b*)1. Compute the inductive reactance at 2000 Hz:

$$X_{L2} = \omega_2 L = 2\pi f_2 L = (2\pi)(2000 \text{ Hz})(40 \times 10^{-3} \text{ H}) = 503\ \Omega$$

2. Use this value of X_L to compute the maximum current at 2000 Hz:

$$I_{2,max} = \frac{120 \text{ V}}{503\ \Omega} = 0.239 \text{ A}$$

Capacitors in ac Circuits

When a capacitor is connected across the terminals of a generator (Figure 31-7), the voltage drop across the capacitor is

$$V_C = V_+ - V_- = \frac{Q}{C} \qquad \text{31-27}$$

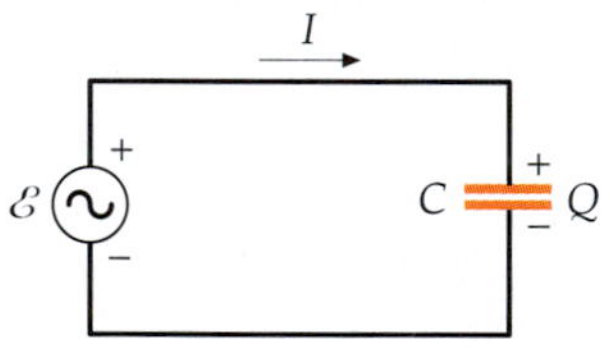

Figure 31-7 An ac generator in series with a capacitor *C*. Again plus and minus signs on the capacitor plates indicate a positive charge on the plate where the current enters and a negative charge on the plate where the current leaves. The current is related to the charge by $I = dQ/dt$.

From Kirchhoff's loop rule, we have

$$\mathcal{E} - V_C = 0$$

or

$$V_C = \mathcal{E} = \mathcal{E}_{max} \cos \omega t = \frac{Q}{C}$$

Thus

$$Q = \mathcal{E}_{max} C \cos \omega t$$

The current is

$$I = \frac{dQ}{dt} = -\omega \mathcal{E}_{max} C \sin \omega t = -I_{max} \sin \omega t$$

where

$$I_{max} = \omega \mathcal{E}_{max} C \qquad \text{31-28}$$

Using the trigonometric identity $\sin \omega t = -\cos(\omega t + \pi/2)$, we obtain

$$I = -\omega C \mathcal{E}_{max} \sin \omega t = I_{max} \cos(\omega t + \pi/2) \qquad \text{31-29}$$

As with the inductor, the current $I = I_{max} \cos(\omega t + \pi/2)$ is not in phase with the voltage drop across the capacitor, $V_C = \mathcal{E}_{max} \cos \omega t$. From Figure 31-8, we see that the maximum value of the voltage occurs 90° or one-fourth period *after* the maximum value of the current. Thus, *the voltage drop across a capacitor lags the current by 90°*. Again, we can understand this physically. The maximum rate of charge buildup $dQ/dt = I$ occurs when the charge Q is zero and therefore when V_C is zero ($\omega t = 3\pi/2$ in Figure 31-7). As the charge on the capacitor plate increases, the current decreases until the charge is a maximum (so that V_C is a maximum) and the current is zero ($\omega t = 2\pi$ or $\omega t = 0$ in Figure 31-7). The current then becomes negative as the charge flows back in the opposite direction, off the capacitor.

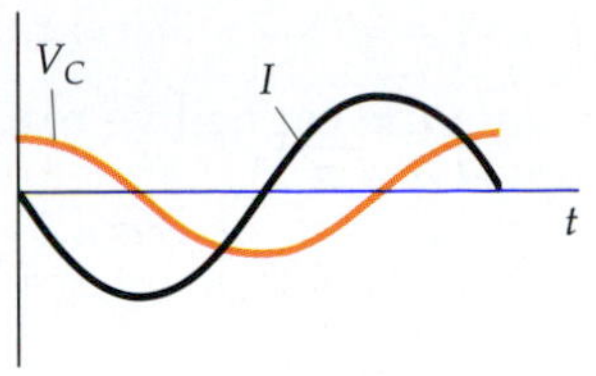

Figure 31-8 Current and voltage across the capacitor in Figure 31-7 versus time. The maximum voltage occurs one-fourth period after the maximum current. Thus, the voltage is said to lag the current by 90°.

Again, we can relate the current to the emf in a form similar to Equation 31-8 for a resistor. From Equation 31-28, we have

$$I_{max} = \omega C \mathcal{E}_{max} = \frac{\mathcal{E}_{max}}{1/\omega C} = \frac{\mathcal{E}_{max}}{X_C}$$

and, similarly,

$$I_{rms} = \frac{\mathcal{E}_{rms}}{X_C} \qquad \text{31-30}$$

where

$$X_C = \frac{1}{\omega C} \qquad \text{31-31}$$

Definition—Capacitive reactance

is called the **capacitive reactance** of the circuit. Like resistance and inductive reactance, capacitive reactance has units of ohms, and like inductive reactance, capacitive reactance depends on the frequency of the current. In this case, the greater the frequency, the smaller the reactance. The average power input to a capacitor from an ac generator is zero, as it is for an inductor. This is because the emf is proportional to cos ωt and the current is proportional to sin ωt and $(\cos \omega t \sin \omega t)_{av} = 0$. Thus, like inductors with no resistance, capacitors dissipate no energy.

Since charge cannot pass across the space between the plates of a capacitor, it may seem strange that there is a continuing alternating current in the circuit of Figure 31-7. Consider an initially uncharged capacitor across a source of emf, with the upper plate attached to the positive terminal. Initially, positive charge flows to the upper plate and away from the lower plate. The effect is the same as if the charge actually flows *across* the space between the plates. If the source of emf is an ac generator, the charge on one plate and the potential difference changes sign every half-period. If we double the frequency, we double the amount of charge that flows onto and off of the plate in a given time, thus we double the current. Hence, the greater the frequency, the less the capacitor impedes the flow of charge.

Example 31-3

A 20-μF capacitor is placed across a generator that has a maximum emf of 100 V. Find the capacitive reactance and the maximum current when the frequency is 60 Hz and when it is 5000 Hz.

Picture the Problem The capacitive reactance is $X_C = 1/\omega C$ and the maximum current is $I_{max} = \mathcal{E}_{max}/X_C$.

1. Calculate the capacitive reactance at 60 Hz and at 5000 Hz:

$$X_{C1} = \frac{1}{\omega_1 C} = \frac{1}{2\pi f_1 C} = \frac{1}{2\pi(60\text{ Hz})(20\times 10^{-6}\text{ F})} = 133\ \Omega$$

$$X_{C2} = \frac{1}{\omega_2 C} = \frac{1}{2\pi f_2 C} = \frac{1}{2\pi(5000\text{ Hz})(20\times 10^{-6}\text{ F})} = 1.59\ \Omega$$

2. Use these values of X_C to find the maximum currents:

$$I_{1,\text{max}} = \frac{\mathcal{E}_{\text{max}}}{X_{C1}} = \frac{100\text{ V}}{133\ \Omega} = 0.752\text{ A}$$

$$I_{2,\text{max}} = \frac{\mathcal{E}_{\text{max}}}{X_{C2}} = \frac{100\text{ V}}{1.59\ \Omega} = 62.9\text{ A}$$

Remark Note that the current increases with frequency, as expected.

In the circuits of Figures 31-5 and 31-7, which contain only a generator and an inductor or capacitor, the voltage drop across the inductor or capacitor equals the voltage of the generator. In more complicated circuits containing three or more elements, the voltage drop across each element is usually not equal to the generator voltage. It is useful, therefore, to write Equations 31-26 and 31-30 in terms of the voltage drops across the inductor and capacitor, respectively. If $V_{L,\text{rms}}$ is the rms voltage drop across an inductor, the rms current in the inductor is given by

$$I_{\text{rms}} = \frac{V_{L,\text{rms}}}{\omega L} = \frac{V_{L,\text{rms}}}{X_L} \qquad \text{31-32}$$

The voltage drop across the inductor leads the current by 90°. Similarly, if $V_{C,\text{rms}}$ is the rms voltage across a capacitor, the rms current in the capacitor is given by

$$I_{\text{rms}} = \frac{V_{C,\text{rms}}}{1/\omega C} = \frac{V_{C,\text{rms}}}{X_C} \qquad \text{31-33}$$

The voltage drop across the capacitor lags the current by 90°. Equations 31-32 and 31-33 can also be written in terms of the maximum voltages and maximum currents.

optional

31-4 Phasors

The phase relations between the current and the voltage drop in a capacitor or inductor can be represented by two-dimensional vectors called **phasors.** In Figure 31-9, the voltage across a resistor V_R is represented by a vector $\vec{V}_R$ that has magnitude $I_{\text{max}}R$ and makes an angle θ with the x axis. This voltage is in phase with the current. In general, a steady-state current in an ac circuit varies with time as

$$I = I_{\text{max}} \cos\theta = I_{\text{max}} \cos(\omega t - \delta) \qquad \text{31-34}$$

where ω is the angular frequency and δ is some phase constant. The voltage drop across a resistor is then given by

$$V_R = IR = I_{\text{max}}R \cos(\omega t - \delta) \qquad \text{31-35}$$

The instantaneous value of the voltage drop across a resistor is thus equal to the x component of the phasor vector $\vec{V}_R$, which rotates counterclockwise

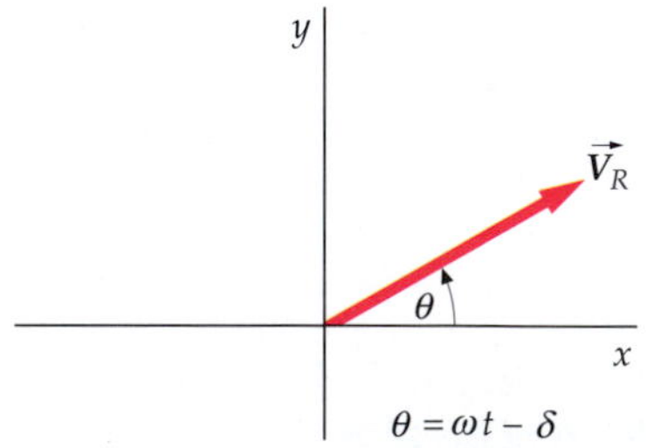

Figure 31-9 The voltage across a resistor can be represented by a vector $\vec{V}_R$, called a phasor, that has magnitude $I_{\text{max}}R$ and makes an angle $\theta = \omega t - \delta$ with the x axis. The phasor rotates with an angular frequency ω. The voltage $V_R = IR$ is the x component of $\vec{V}_R$.

with an angular frequency ω. The current I may be written as the x component of a phasor $\vec{I}$ having the same orientation as $\vec{V}_R$.

When several components are connected together in a series circuit, their voltages add. When they are connected in parallel, their currents add. Adding sines or cosines of different amplitudes and phases algebraically is awkward. It is much easier to do this by vector addition.

Phasors are used as follows. Any ac voltage or current is written in the form $A\cos(\omega t - \delta)$, which in turn is treated as the x component A_x of a phasor $\vec{A}$ that makes an angle $(\omega t - \delta)$ with the x axis. Instead of adding two voltages or currents algebraically as $A\cos(\omega t - \delta_1) + B\cos(\omega t - \delta_2)$, we represent these quantities as phasors $\vec{A}$ and $\vec{B}$ and find the phasor sum $\vec{C} = \vec{A} + \vec{B}$ geometrically. The resultant voltage or current is then the x component of the resultant phasor, $C_x = A_x + B_x$. The geometric representation conveniently shows the relative amplitudes and phases of the phasors.

Consider a circuit containing an inductor L, a capacitor C, and a resistor R all connected in series. They all carry the same current, which is represented as the x component of the current phasor $\vec{I}$. The voltage across the inductor V_L is represented by a phasor $\vec{V}_L$ that has magnitude $I_{max}X_L$ and leads the current phasor $\vec{I}$ by 90°. Similarly, the voltage across the capacitor V_C is represented by a phasor $\vec{V}_C$ that has magnitude $I_{max}X_C$ and lags the current by 90°. Figure 31-10 shows the three phasors $\vec{V}_R$, $\vec{V}_L$, and $\vec{V}_C$. As time goes on, the three phasors rotate counterclockwise with an angular frequency ω, so the relative positions of the vectors do not change. At any time, the instantaneous value of the voltage drop across any of these elements equals the x component of the corresponding phasor.

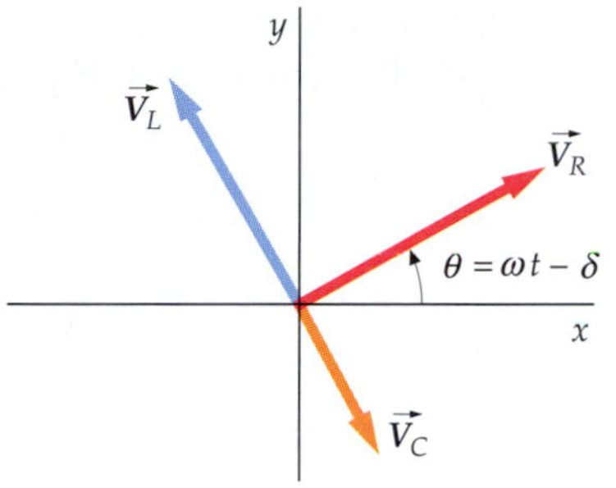

Figure 31-10 Phasor representations of the voltages V_R, V_L, and V_C. Each vector rotates in the counterclockwise direction with an angular frequency ω. At any instant, the voltage across an element equals the x component of the corresponding phasor, and the sum of the voltages equals the x component of the vector sum $\vec{V}_R + \vec{V}_L + \vec{V}_C$.

31-5 *LC* and *RLC* Circuits Without a Generator

Figure 31-11 shows a simple circuit with inductance and capacitance but no resistance. Such a circuit is called an ***LC* circuit**. We assume that the capacitor carries an initial charge Q_0 and that the switch is initially open. After the switch is closed at $t = 0$, the charge begins to flow through the inductor. In the figure, the signs of Q on the capacitor and the direction of the current I have been chosen such that

$$I = \frac{dQ}{dt}$$

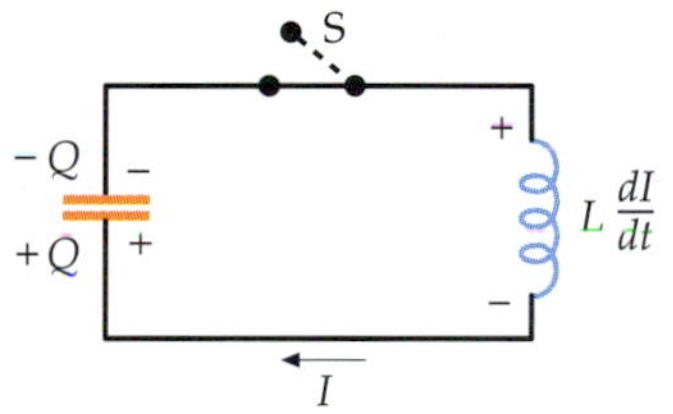

Figure 31-11 An *LC* circuit. When the switch is closed, the initially charged capacitor discharges through the inductor, producing a back emf.

Applying Kirchhoff's loop rule to the circuit for the assumed signs of Q and I, we have

$$L\frac{dI}{dt} + \frac{Q}{C} = 0 \qquad \text{31-36}$$

Substituting dQ/dt for I gives

$$L\frac{d^2Q}{dt^2} + \frac{Q}{C} = 0 \qquad \text{31-37}$$

This is of the same form as the equation for the acceleration of a mass on a spring:

$$m\frac{d^2x}{dt^2} + kx = 0$$

optional

The behavior of an LC circuit is thus analogous to that of a mass on a spring, with L analogous to the mass m, Q analogous to the position x, and $1/C$ analogous to the spring constant k. Also, the current I is analogous to the velocity v, since $v = dx/dt$ and $I = dQ/dt$. In mechanics, the mass of an object describes the inertia of the object. The greater the mass, the more difficult it is to change the velocity of the object. Similarly, the inductance L can be thought of as the inertia of an ac circuit. The greater the inductance, the more difficult it is to change the current I.

If we divide each term in Equation 31-37 by L and rearrange, we obtain

$$\frac{d^2Q}{dt^2} = -\frac{1}{LC}Q \qquad \text{31-38}$$

which is analogous to

$$\frac{d^2x}{dt^2} = -\frac{k}{m}x = -\omega^2 x \qquad \text{31-39}$$

where $\omega^2 = k/m$. In Chapter 14, we found that we could write the solution of Equation 31-39 for simple harmonic motion in the form

$$x = A\cos(\omega t - \delta)$$

where $\omega = \sqrt{k/m}$ is the angular frequency, A is the amplitude, and δ is the phase constant, which depends on the initial conditions. We can put Equation 31-38 in this same form by writing ω^2 for $1/LC$. Then

$$\frac{d^2Q}{dt^2} = -\omega^2 Q \qquad \text{31-40}$$

$$\omega = \frac{1}{\sqrt{LC}} \qquad \text{31-41}$$

The solution of Equation 31-40 is

$$Q = A\cos(\omega t - \delta)$$

The current is found by differentiating:

$$I = \frac{dQ}{dt} = -\omega A \sin(\omega t - \delta)$$

If we choose our initial conditions to be $Q = Q_0$ and $I = 0$ at $t = 0$, the phase constant δ is zero and $A = Q_0$. Our solutions are then

$$Q = Q_0 \cos \omega t \qquad \text{31-42}$$

and

$$I = -\omega Q_0 \sin \omega t = -I_{max} \sin \omega t \qquad \text{31-43}$$

where $I_{max} = \omega Q_0$.

Figure 31-12 shows graphs of Q and I versus time. The charge oscillates between the values $+Q_0$ and $-Q_0$ with angular frequency $\omega = \sqrt{1/LC}$. The current oscillates between $+\omega Q_0$ and $-\omega Q_0$ with the same frequency and is 90° out of phase with the charge. The current is maximum when the charge is zero and zero when the charge is maximum.

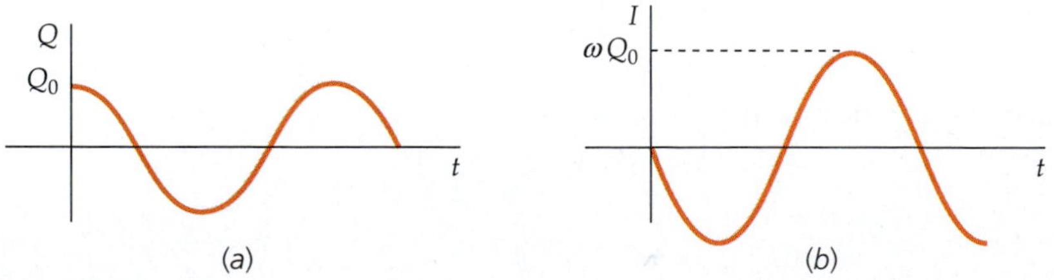

Figure 31-12 Graphs of (*a*) Q versus t and (*b*) I versus t for the LC circuit of Figure 31-11.

In our study of the oscillations of a mass on a spring, we found that the total energy is constant, and that it oscillates between potential and kinetic energy. In our *LC* circuit, we also have two kinds of energy, electric energy and magnetic energy. The electric energy stored in the capacitor is

$$U_e = \frac{1}{2}QV_C = \frac{1}{2}\frac{Q^2}{C}$$

Substituting $Q_0 \cos \omega t$ for Q, we have for the electric energy

$$U_e = \frac{1}{2}\frac{Q_0^2}{C}\cos^2 \omega t \qquad \text{31-44}$$

The electric energy oscillates between its maximum value $Q_0^2/2C$ and zero. The magnetic energy stored in the inductor is

$$U_m = \frac{1}{2}LI^2 \qquad \text{31-45}$$

Substituting $I = -\omega Q_0 \sin \omega t$ (Equation 31-43), we get

$$U_m = \frac{1}{2}L\omega^2 Q_0^2 \sin^2 \omega t = \frac{1}{2}\frac{Q_0^2}{C}\sin^2 \omega t \qquad \text{31-46}$$

where we have used $\omega^2 = 1/LC$. The magnetic energy also oscillates between its maximum value of $Q_0^2/2C$ and zero. The sum of the electrostatic and magnetic energies is the total energy, which is constant in time:

$$U_{total} = U_e + U_m = \frac{1}{2}\frac{Q_0^2}{C}\cos^2 \omega t + \frac{1}{2}\frac{Q_0^2}{C}\sin^2 \omega t = \frac{1}{2}\frac{Q_0^2}{C}$$

This equals the energy initially stored on the capacitor.

Example 31-4

A 2-μF capacitor is charged to 20 V and is then connected across a 6-μH inductor. (*a*) What is the frequency of oscillation? (*b*) What is the maximum value of the current?

(*a*) The frequency of oscillation depends only on the values of the capacitance and inductance:

$$f = \frac{\omega}{2\pi} = \frac{1}{2\pi\sqrt{LC}} = \frac{1}{2\pi\sqrt{(6 \times 10^{-6}\text{ H})(2 \times 10^{-6}\text{ F})}}$$
$$= 4.59 \times 10^4 \text{ Hz}$$

(*b*)1. The maximum value of the current is related to the maximum value of the charge:

$$I_{max} = \omega Q_0 = \frac{Q_0}{\sqrt{LC}}$$

2. Find the initial charge on the capacitor from the initial voltage:

$$Q_0 = CV_0 = (2\ \mu\text{F})(20\text{ V}) = 40\ \mu\text{C}$$

3. Use the value of Q_0 to calculate I_{max}:

$$I_{max} = \frac{40\ \mu\text{C}}{\sqrt{(6\ \mu\text{H})(2\ \mu\text{F})}} = 11.5\text{ A}$$

Exercise A 5-μF capacitor is charged and is then discharged through an inductor. What should the value of the inductance be so that the current oscillates with frequency 8 kHz? (*Answer* 79.2 μH)

If we include a resistor in series with the capacitor and inductor as in Figure 31-13, we have an ***RLC* circuit**. Kirchhoff's loop rule gives

$$L\frac{dI}{dt} + \frac{Q}{C} + IR = 0 \qquad \text{31-47}a$$

or

$$L\frac{d^2Q}{dt^2} + \frac{Q}{C} + R\frac{dQ}{dt} = 0 \qquad \text{31-47}b$$

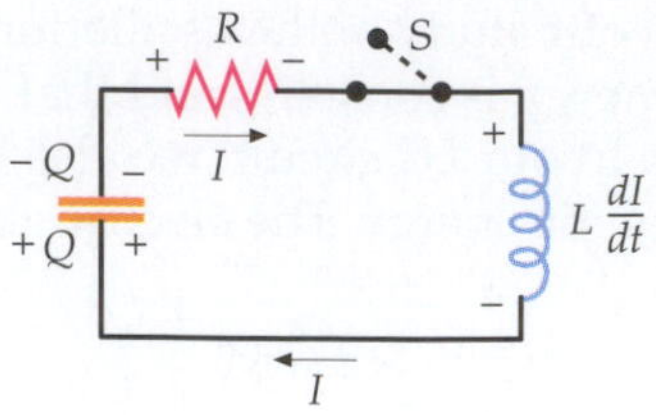

Figure 31-13 An *RLC* circuit.

where we have used $I = dQ/dt$ as before. Equations 31-47*a* and *b* are analogous to the equation for a damped harmonic oscillator (see Equation 14-32):

$$m\frac{d^2x}{dt^2} + kx + b\frac{dx}{dt} = 0$$

The first term, $L\,dI/dt = L\,d^2Q/dt^2$, is analogous to the mass times the acceleration, $m\,dv/dt = m\,d^2x/dt^2$; the second term, Q/C, is analogous to the restoring force kx; and the third term, $IR = R\,dQ/dt$, is analogous to the damping term, $bv = b\,dx/dt$. In the oscillation of a mass on a spring, the damping constant b leads to a dissipation of mechanical energy as heat. In an *RLC* circuit, the resistance R is analogous to the damping constant b and leads to a dissipation of electrical energy as Joule heat.

If the resistance is small, the charge and current oscillate with (angular) frequency* that is very nearly equal to $\omega_0 = 1/\sqrt{LC}$, which is called the natural frequency of the circuit, but the oscillations are damped. We can understand this qualitatively from energy considerations. If we multiply each term in Equation 31-47*a* by the current I, we obtain

$$IL\frac{dI}{dt} + I\frac{Q}{C} + I^2R = 0 \qquad \text{31-48}$$

The first term in this equation is the current times the voltage across the inductor. This is the rate at which energy is put into the inductor or taken out of it; that is, it is the rate of change of the magnetic energy, $d(\frac{1}{2}LI^2)/dt$, which is positive or negative depending on whether I and dI/dt have the same sign or different signs. Similarly, the second term is the current times the voltage across the capacitor. This is the rate of change of the energy of the capacitor, which may be positive or negative. The last term, I^2R, is the rate at which energy is dissipated in the resistor as Joule heat and is always positive. The sum of the electric and magnetic energies is not constant for this circuit because energy is continually dissipated in the resistor. Figure 31-14 shows graphs of Q versus t and I versus t for a small resistance R. If we increase R, the oscillations become more heavily damped until a critical value of R is reached for which there is not even one oscillation. Figure 31-15 shows Q versus t when the value of R is greater than the critical damping value.

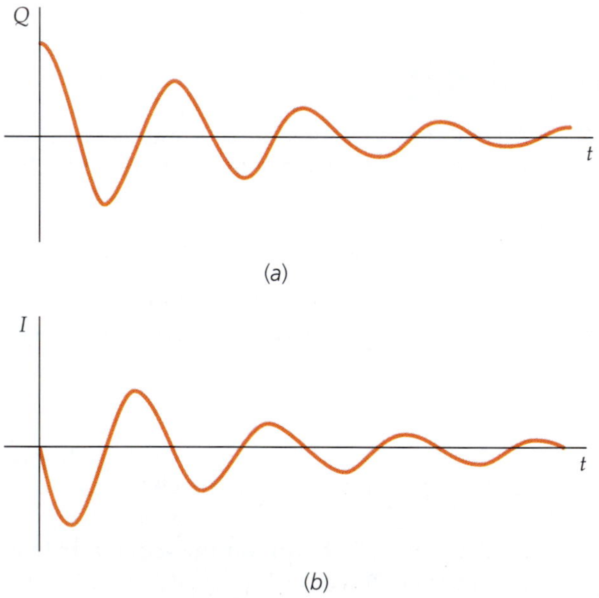

Figure 31-14 Graphs of (*a*) Q versus t and (*b*) I versus t for the *RLC* circuit of Figure 31-13 when R is small enough so that the oscillations are underdamped.

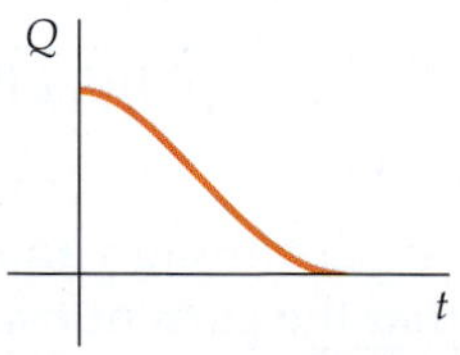

Figure 31-15 Graph of Q versus t for the *RLC* circuit of Figure 31-14 when R is so large that the oscillations are overdamped.

* As we did in Chapter 14 when we discussed mechanical oscillations, we usually omit the word *angular* when the omission will not cause confusion.

31-6 *RLC* Circuits With a Generator

Series *RLC* Circuit

Figure 31-16 shows a series *RLC* circuit with a generator. If the generator emf varies with time as $\mathcal{E} = \mathcal{E}_{max} \cos \omega t$, Kirchhoff's loop rule gives

$$\mathcal{E}_{max} \cos \omega t - L\frac{dI}{dt} - \frac{Q}{C} - IR = 0$$

Using $I = dQ/dt$ and rearranging, we obtain

$$L\frac{d^2Q}{dt^2} + R\frac{dQ}{dt} + \frac{Q}{C} = \mathcal{E}_{max} \cos \omega t \qquad \text{31-49}$$

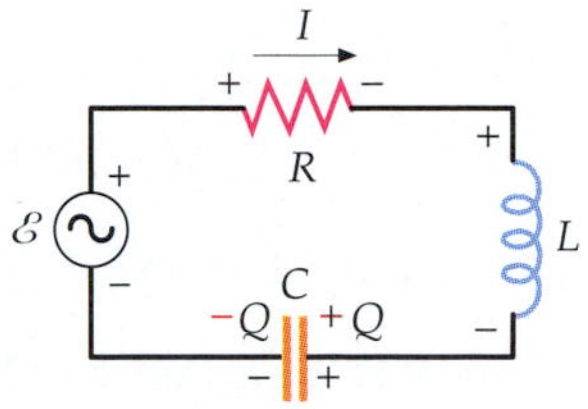

Figure 31-16 A series *RLC* circuit with an ac generator.

This equation is analogous to Equation 14-47 for the forced oscillation of a mass on a spring*:

$$m\frac{d^2x}{dt^2} + b\frac{dx}{dt} + m\omega_0^2 x = F_0 \cos \omega t$$

We will discuss the solution of Equation 31-49 qualitatively as we did with Equation 14-47 for the forced oscillator. The current in the circuit consists of a transient current that depends on the initial conditions (such as the initial phase of the generator and the initial charge on the capacitor) and a steady-state current that does not. We will ignore the transient current, which decreases exponentially with time and is eventually negligible, and concentrate on the steady-state current. The steady-state current obtained by solving Equation 31-49 is

$$I = I_{max} \cos (\omega t - \delta) \qquad \text{31-50}$$

where the phase angle δ is given by

$$\tan \delta = \frac{X_L - X_C}{R} \qquad \text{31-51}$$

The maximum current is

$$I_{max} = \frac{\mathcal{E}_{max}}{\sqrt{R^2 + (X_L - X_C)^2}} = \frac{\mathcal{E}_{max}}{Z} \qquad \text{31-52}$$

Current in a series RLC circuit

where

$$Z = \sqrt{R^2 + (X_L - X_C)^2} \qquad \text{31-53}$$

Impedance of a series RLC circuit

The quantity $X_L - X_C$ is called the **total reactance,** and Z is called the **impedance.** Combining these results, we have

$$I = \frac{\mathcal{E}_{max}}{Z} \cos (\omega t - \delta) \qquad \text{31-54}$$

Figure 31-17 Phase relations among voltages in a series *RLC* circuit. The voltage across the resistor is in phase with the current. The voltage across the inductor V_L leads the current by 90°. The voltage across the capacitor lags the current by 90°. The sum of the vectors representing these voltages gives a vector at an angle δ with the current representing the applied emf. For the case shown here, V_L is greater than V_C and the current lags the emf by δ.

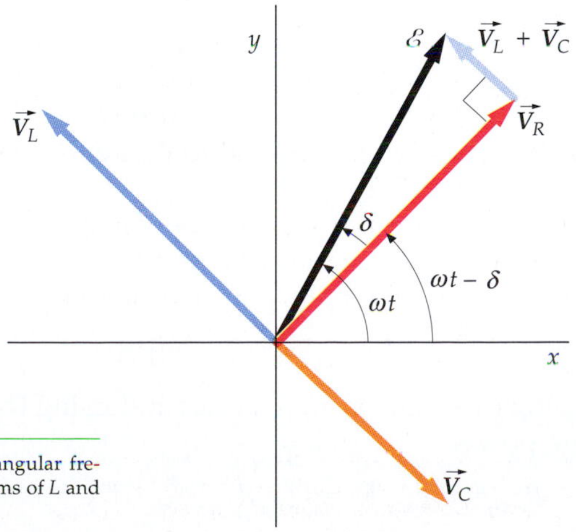

Equation 31-54 can also be obtained from a simple diagram using the phasor representations. Figure 31-17 shows the phasors representing the voltage drops across the resistance, the inductance, and the capacitance. The x component of each of these vectors equals the instantaneous voltage drop across the corresponding element. Since the sum of the x components equals the x component of the sum of the vectors, the sum of the x components equals the sum of the voltage drops across these elements, which by Kirchhoff's loop rule equals the instantaneous emf.

* In Equation 14-53 the force constant k was written in terms of the mass m and the natural angular frequency ω_0 using $k = m\omega_0^2$. The capacitance in Equation 31-49 could be similarly written in terms of L and the natural angular frequency using $1/C = L\omega_0^2$.

optional

If we represent the applied emf, $\mathcal{E} = \mathcal{E}_{max} \cos \omega t$, as a phasor $\vec{\mathcal{E}}$ that has the magnitude $\mathcal{E}_{max}$, we have

$$\vec{\mathcal{E}} = \vec{V}_R + \vec{V}_L + \vec{V}_C \quad \text{31-55}$$

In terms of the magnitudes,

$$\mathcal{E} = |\vec{V}_R + \vec{V}_L + \vec{V}_C| = \sqrt{V_{R,max}^2 + (V_{L,max} - V_{C,max})^2}$$

But $V_R = I_{max}R$, $V_L = I_{max}X_L$, and $V_C = I_{max}X_C$. Thus,

$$\mathcal{E}_{max} = I_{max}\sqrt{R^2 + (X_L - X_C)^2} = I_{max}Z$$

The phasor $\vec{\mathcal{E}}$ makes an angle δ with $\vec{V}_R$ as shown in Figure 31-17. From the figure, we can see that

$$\tan \delta = \frac{|\vec{V}_L + \vec{V}_C|}{|\vec{V}_R|} = \frac{I_{max}X_L - I_{max}X_C}{I_{max}R} = \frac{X_L - X_C}{R}$$

in agreement with Equation 31-51. Since $\vec{\mathcal{E}}$ makes an angle ωt with the x axis, $\vec{V}_R$ makes an angle $\omega t - \delta$ with the x axis. This voltage is in phase with the current, which is therefore given by

$$I = I_{max} \cos(\omega t - \delta) = \frac{\mathcal{E}_{max}}{Z} \cos(\omega t - \delta)$$

This is Equation 31-54. The relation between the impedance Z and the resistance R and the total reactance $X_L - X_C$ can be remembered using the right triangle shown in Figure 31-18.

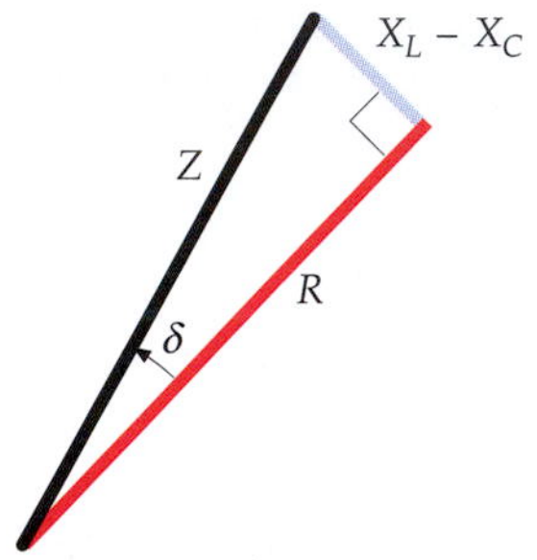

Figure 31-18 Triangle relating capacitive and inductive reactance, resistance, impedance, and the phase angle in an *RLC* circuit.

Resonance

When X_L and X_C are equal, the total reactance is zero, and the impedance Z has its smallest value R. Then I_{max} has its greatest value and the phase angle δ is zero, which means that the current is in phase with the applied emf. The value of ω for which X_L and X_C are equal is obtained from

$$X_L = X_C$$

$$\omega L = \frac{1}{\omega C}$$

or

$$\omega = \frac{1}{\sqrt{LC}} = \omega_0$$

When the generator frequency ω equals the natural frequency ω_0, the impedance is smallest, I_{max} is greatest, and the circuit is said to be at **resonance**. The natural frequency ω_0 is therefore also called the **resonance frequency**. This resonance condition in a driven *RLC* circuit is similar to that in a driven simple harmonic oscillator.

Since neither an inductor nor a capacitor dissipates energy, the average power delivered to a series *RLC* circuit is the average power supplied to the resistor. The instantaneous power supplied to the resistor is

$$P = I^2R = [I_{max} \cos(\omega t - \delta)]^2R$$

Averaging over one or more cycles and using $(\cos^2 \theta)_{av} = \frac{1}{2}$, we obtain for the average power

$$P_{av} = \tfrac{1}{2}I_{max}^2R = I_{rms}^2R \quad \text{31-56}$$

Using $R/Z = \cos\delta$ from Figure 31-18 and $I = \mathcal{E}/Z$ this can be written

$$P_{\rm av} = \tfrac{1}{2}\mathcal{E}_{\max} I_{\max} \cos\delta = \mathcal{E}_{\rm rms} I_{\rm rms} \cos\delta \qquad \text{31-57}$$

The quantity $\cos\delta$ is called the **power factor** of the *RLC* circuit. At resonance, δ is zero, and the power factor is 1.

The power can also be expressed as a function of the angular frequency ω. Using $I_{\rm rms} = \mathcal{E}_{\rm rms}/Z$, Equation 31-56 becomes

$$P_{\rm av} = I_{\rm rms}^2 R = \mathcal{E}_{\rm rms}^2 \frac{R}{Z^2}$$

From the definition of impedance Z, we have

$$Z^2 = (X_L - X_C)^2 + R^2 = \left(\omega L - \frac{1}{\omega C}\right)^2 + R^2$$

$$= \frac{L^2}{\omega^2}\left(\omega^2 - \frac{1}{LC}\right)^2 + R^2$$

$$= \frac{L^2}{\omega^2}(\omega^2 - \omega_0^2)^2 + R^2$$

where we have used $\omega_0^2 = 1/LC$. Using this expression for Z^2, we obtain the average power as a function of ω:

$$P_{\rm av} = \frac{\mathcal{E}_{\rm rms}^2 R\omega^2}{L^2(\omega^2 - \omega_0^2)^2 + \omega^2 R^2} \qquad \text{31-58}$$

Figure 31-19 shows the average power supplied by the generator to the circuit as a function of generator frequency for two different values of the resistance R. These curves, called **resonance curves,** are the same as the power-versus-frequency curves for a driven damped oscillator (see Section 14-5). The average power is maximum when the generator frequency equals the resonance frequency. When the resistance is small, the resonance curve is narrow; when it is large, the curve is broad. A resonance curve can be characterized by the **resonance width** $\Delta\omega$. As shown in Figure 31-19, the resonance width is the frequency difference between the two points on the curve where the power is half its maximum value. When the width is small compared with the resonance frequency, the resonance is sharp, that is, the resonance curve is narrow.

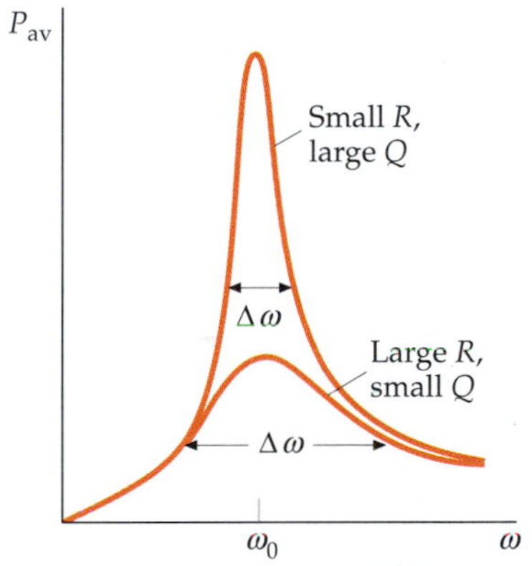

Figure 31-19 Plot of average power versus frequency for a series *RLC* circuit. The power is maximum when the frequency of the generator ω equals the natural frequency of the circuit $\omega_0 = 1/\sqrt{LC}$. If the resistance is small, the Q factor is large and the resonance is sharp. The resonance width $\Delta\omega$ of the curves is measured between points where the power is half its maximum value.

In Chapter 14, the Q factor for a mechanical oscillator was defined as $Q = \omega_0 m/b$ (Equation 14-39) where m is the mass and b is the damping constant. We then saw that $Q = 2\pi E/|\Delta E|$ (Equation 14-41), where E is the total energy of the system and ΔE is the energy lost in one cycle. The **Q factor** for an *RLC* circuit can be defined in a similar way. Since L is analogous to the mass m and R is analogous to the damping constant b, the Q factor for an *RLC* circuit is given by

$$Q = \frac{2\pi E}{|\Delta E|} = \frac{\omega_0 L}{R} \qquad \text{31-59}$$

When the resonance curve is reasonably narrow (that is, when Q is greater than about 2 or 3), the Q factor can be approximated by

$$Q \approx \frac{\omega_0}{\Delta\omega} = \frac{f_0}{\Delta f} \qquad \text{31-60}$$

Q factor for an RLC circuit

Resonance circuits are used in radio receivers, where the resonance frequency of the circuit is varied by varying the capacitance. Resonance occurs when the natural frequency of the circuit equals one of the frequencies of the radio waves picked up at the antenna. At resonance, there is a relatively large

optional

current in the antenna circuit. If the Q factor of the circuit is sufficiently high, currents due to other station frequencies off resonance will be negligible compared with those due to the station frequency to which the circuit is tuned.

Example 31-5

A series RLC circuit with $L = 2$ H, $C = 2\ \mu$F, and $R = 20\ \Omega$ is driven by a generator with a maximum emf of 100 V and a variable frequency. Find (*a*) the resonance frequency f_0, (*b*) the Q value, (*c*) the width of the resonance Δf, and (*d*) the maximum current at resonance.

Picture the Problem The resonance frequency is found from $\omega_0 = 1/\sqrt{LC}$ and the Q value is found from $Q = \omega_0 L/R$.

(*a*) The resonance frequency is $f_0 = \omega_0/2\pi$:

$$f_0 = \frac{\omega_0}{2\pi} = \frac{1}{2\pi\sqrt{LC}}$$

$$= \frac{1}{2\pi\sqrt{(2\text{ H})(2\times 10^{-6}\text{ F})}} = 79.6\text{ Hz}$$

(*b*) Use this result to calculate Q:

$$Q = \frac{\omega_0 L}{R} = \frac{2\pi(79.6\text{ Hz})(2\text{ H})}{20\ \Omega} = 50$$

(*c*) Use the value of Q to find the width of the resonance Δf:

$$\Delta f = \frac{f_0}{Q} = \frac{79.6\text{ Hz}}{50} = 1.59\text{ Hz}$$

(*d*) At resonance the impedance is just R and I_{max} is $\mathcal{E}/R$:

$$I_{max} = \frac{\mathcal{E}_{max}}{R} = \frac{100\text{ V}}{20\ \Omega} = 5\text{ A}$$

Remark The width is only 1.59 Hz, which is small compared with the resonance frequency of 79.6 Hz, so the resonance peak is quite sharp.

Example 31-6 *try it yourself*

If the generator in Example 31-5 has a frequency of 60 Hz, find (*a*) the maximum current I_{max}, (*b*) the phase angle δ, (*c*) the power factor, and (*d*) the average power delivered.

Picture the Problem The maximum current is the maximum emf divided by the total impedance of the circuit. The phase angle δ is found from $\tan\delta = (X_L - X_C)/R$. You can use either Equation 31-56 or 31-57 to find the average power delivered.

Cover the column to the right and try these on your own before looking at the answers.

Steps	***Answers***
(*a*) 1. Write the maximum current in terms of the maximum emf and the impedance.	$I_{max} = \mathcal{E}_{max}/Z$
2. Calculate the capacitive and inductive reactances and the total reactance.	$X_C = 1326\ \Omega,\quad X_L = 754\ \Omega$ $X_L - X_C = -572\ \Omega$

3. Calculate the total impedance Z. — $Z \approx 572\ \Omega$

4. Use the results of steps 2 and 3 to calculate I_{max}. — $I_{max} = 0.175$ A

(*b*) Use the results of steps 2 and 3 above to calculate tan δ. Then find δ. — $\tan \delta = \dfrac{X_L - X_C}{R} = -28.6, \quad \delta = -88.0°$

(*c*) Use your value of δ to compute the power factor. — $\cos \delta = 0.0349$

(*d*) Calculate the average power delivered from Equation 31-56, and check your result using the power factor found in (*c*). — $P_{av} = \frac{1}{2} I_{max}^2 R = 0.306$ W; $P_{av} = \frac{1}{2}\mathcal{E}_{max} I_{max} \cos \delta = 0.305$ W

Remarks The slight discrepancy in the two calculations of power in (*d*) is due to rounding error in the current.

The generator frequency of 60 Hz is well below the resonance frequency of 79.6 Hz. (Recall that the width as calculated in Example 31-5 is only 1.59 Hz.) As a result, the total reactance is much greater in magnitude than the resistance. This is always the case far from resonance. Similarly, the maximum current of 0.175 A is much less than I_{max} at resonance, which was found to be 5 A. Finally, we see from Figure 31-17 that a negative phase angle means that the current leads the generator voltage.

Example 31-7 *try it yourself*

Find the maximum voltage across the resistor, the inductor, and the capacitor at resonance for the circuit in Example 31-5.

Picture the Problem The maximum voltage across the resistor is I_{max} times R. Similarly, the maximum voltage across the inductor or capacitor is I_{max} times the appropriate reactance. We found $I_{max} = 5$ A and $f_0 = 79.6$ Hz in Example 31-5.

Cover the column to the right and try these on your own before looking at the answers.

Steps	***Answers***
1. Calculate $V_{R,max} = I_{max}R$.	$V_{R,max} = I_{max}R = 100$ V
2. Express $V_{L,max}$ in terms of I_{max} and X_L.	$V_{L,max} = I_{max}X_L$
3. Calculate X_L at resonance.	$X_L = \omega_0 L = 2\pi f_0 L = 1000\ \Omega$
4. Use your result in step 3 to calculate $V_{L,max}$.	$V_{L,max} = I_{max}X_L = 5000$ V
5. Express $V_{C,max}$ in terms of I_{max} and X_C.	$V_{C,max} = I_{max}X_C$
6. Calculate X_C at resonance.	$X_C = \dfrac{1}{\omega_0 C} = 1000\ \Omega$
7. Use your result in step 6 to calculate $V_{C,max}$.	$V_{C,max} = I_{max}X_C = 5000$ V

Remarks The inductive and capacitive reactances are equal, as we would expect since we found the resonance frequency by setting them equal. The phasor diagram for the voltages across the resistor, capacitor, and inductor is shown in Figure 31-20. The maximum voltage across the resistor is a relatively safe 100 V, equal to the maximum emf of the generator. However, the maximum voltages across the inductor and the capacitor are a dangerously high 5000 V. These voltages are 180° out of phase. At resonance, the voltage across the inductor at any instant is the negative of that across the capacitor, so they always sum to zero, leaving the voltage across the resistor equal to the emf in the circuit.

Figure 31-20

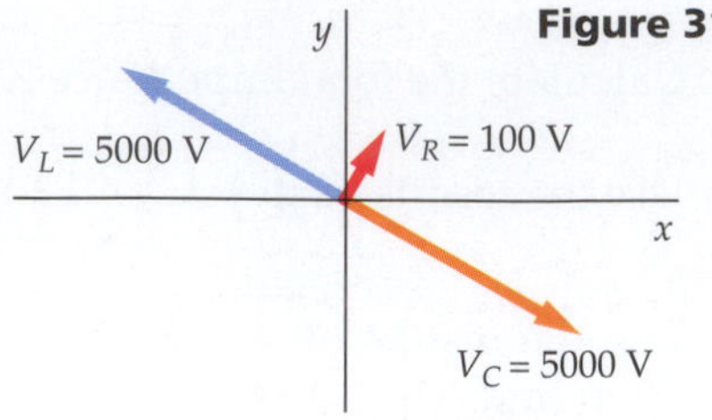

Example 31-8

A resistor R and capacitor C are in series with a generator as shown in Figure 31-21. The generator voltage is given by $V_{in} = V_0 \cos \omega t$. Find the rms voltage across the capacitor $V_{out,rms}$ as a function of frequency ω.

Figure 31-21

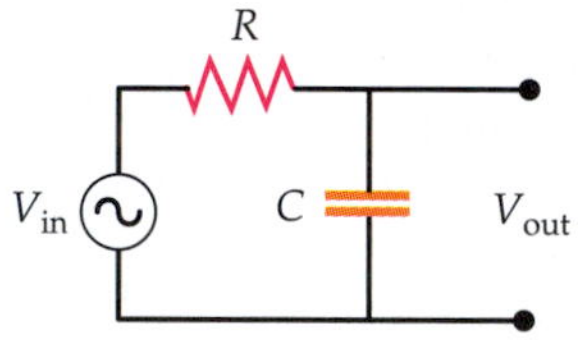

Picture the Problem The rms voltage across the capacitor is related to the rms current and the capacitive reactance. The rms current is found from the input voltage and the impedance.

1. The voltage across the capacitor is I_{rms} times X_C:
$$V_{out\ rms} = I_{rms} X_C$$

2. The rms current depends on the input voltage and the impedance:
$$I_{rms} = \frac{V_{in\ rms}}{Z}$$

3. In this circuit, only R and X_C contribute to the total impedance:
$$Z = \sqrt{R^2 + X_C^2}$$

4. Substitute these values and $X_C = 1/\omega C$ to find the output voltage:
$$V_{out\ rms} = I_{rms} X_C = \frac{V_{in\ rms} X_C}{\sqrt{R^2 + X_C^2}}$$
$$= \frac{V_{in\ rms}(1/\omega C)}{\sqrt{R^2 + (1/\omega C)^2}} = \frac{V_{in\ rms}}{\sqrt{\omega^2 C^2 R^2 + 1}}$$

Remarks This circuit is called an *RC low-pass filter*, since it transmits low frequencies with greater amplitude than high frequencies. In fact, the output voltage equals the input voltage in the limit that $\omega \to 0$, but approaches zero for $\omega \to \infty$, as shown in the graph of the ratio of output voltage to input voltage in Figure 31-22.

Exercise Find the output voltage for this circuit if the capacitor is replaced by an inductor L. (*Answer* $V_{out\ rms} = V_{in\ rms}(\omega L)/\sqrt{R^2 + (\omega L)^2}$. This circuit is a *high-pass filter*.)

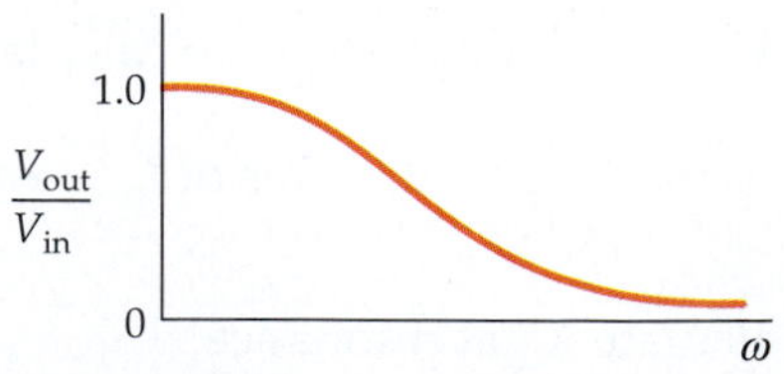

Figure 31-22

A shipboard radio, circa 1920. Exposed at the operator's left are the inductance coils and capacitor plates of the tuning circuit.

Parallel *RLC* Circuit

Figure 31-23 shows a resistor R, capacitor C, and an inductor L connected in parallel across an ac generator. The total current I from the generator divides into three currents, the current I_R in the resistor, the current I_C in the capacitor, and the current I_L in the inductor. The instantaneous voltage V is the same across each element. The current in the resistor is in phase with the voltage and has magnitude V/R. Since the voltage drop across an inductor *leads* the current in the inductor by 90°, I_L *lags* the voltage by 90° and has magnitude V/X_L. Similarly, the I_C leads the voltage by 90° and has magnitude V/X_C. These currents are represented by phasors in Figure 31-24. The total current I is the x component of the vector sum of the individual currents as shown in the figure. The magnitude of the total current is

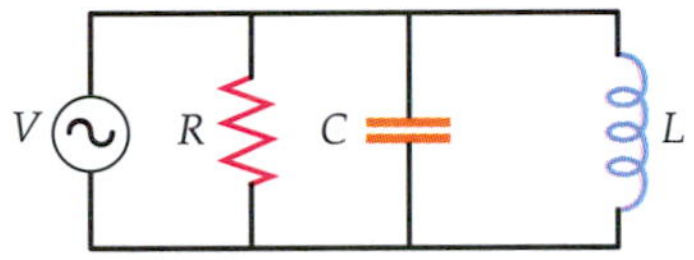

Figure 31-23 A parallel *RLC* circuit.

$$I = \sqrt{I_R^2 + (I_L - I_C)^2} = \sqrt{\left(\frac{V}{R}\right)^2 + \left(\frac{V}{X_L} - \frac{V}{X_C}\right)^2} = \frac{V}{Z} \qquad \text{31-61}$$

where the impedance Z is related to the resistance and the capacitive and inductive reactances by

$$\frac{1}{Z} = \sqrt{\left(\frac{1}{R}\right)^2 + \left(\frac{1}{X_L} - \frac{1}{X_C}\right)^2} \qquad \text{31-62}$$

At resonance, the generator frequency ω equals the natural frequency $\omega_0 = 1/\sqrt{LC}$, and the inductive and capacitive reactances are equal. Then from Equation 31-62 we see that $1/Z$ has its minimum value $1/R$, so the impedance Z is maximum and the total current is minimum. We can understand this if we note that at resonance $X_C = X_L$, and the currents in the inductor and capacitor are equal but 180° out of phase, so the total current is just the current in the resistor.

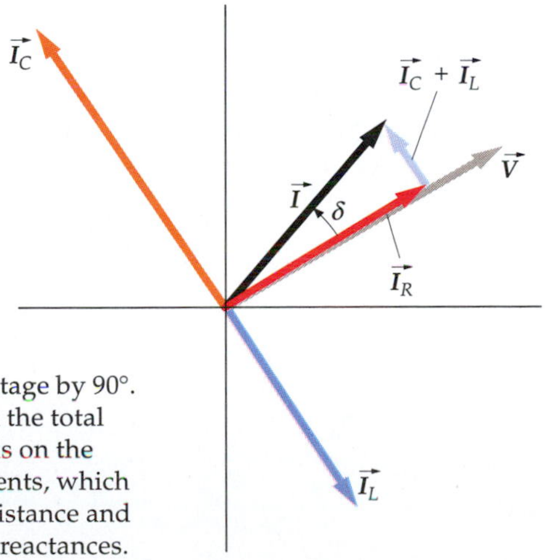

Figure 31-24 Phasor diagram for the currents in the parallel *RLC* circuit of Figure 31-23. The voltage is the same across each element. The current in the resistor is in phase with the voltage. The current in the capacitor leads the voltage by 90° and that in the inductor lags the voltage by 90°. The phase difference δ between the total current and the voltage depends on the relative magnitudes of the currents, which depend on the values of the resistance and of the capacitive and inductive reactances.

Exploring

Solving *RLC* Circuits Using Complex Numbers

The methods discussed in the essay "Exploring . . . Using Complex Numbers to Solve the Oscillator Equations" in Chapter 14 for solving the driven oscillator equations apply equally well to the solution of *RLC* circuits. Equation 31-49 is

$$L\frac{d^2Q}{dt^2} + R\frac{dQ}{dt} + \frac{Q}{C} = \mathcal{E}_{max}\cos\omega t$$

Since we are interested in the current, we write $I = dQ/dt$ and $Q = \int I\,dt$. Then

$$L\frac{dI}{dt} + RI + \frac{1}{C}\int I\,dt = \mathcal{E}_{max}\cos\omega t \qquad 1$$

We note that the right side of this equation is the real part of $\mathcal{E}_{max}e^{i\omega t}$. To solve Equation 1 we then find the complex function z that satisfies

$$L\frac{dz}{dt} + Rz + \frac{1}{C}\int z\,dt = \mathcal{E}_{max}e^{i\omega t} \qquad 2$$

The real part of our solution will be the current I. As we did in Chapter 14, we try

$$z = I_0 e^{i\omega t} \qquad 3$$

Then $dz/dt = i\omega z$, and $\int z\,dt = (1/i\omega)z$. Substituting these results into Equation 2 gives

$$i\omega Lz + Rz + \frac{1}{i\omega C}z = \mathcal{E}_{max}e^{i\omega t} = \frac{\mathcal{E}_{max}}{I_0}z \qquad 4$$

Divide each term by z and write $\omega L = X_L$ and $1/\omega C = X_C$. Then, using $X_C/i = -iX_C$, we obtain

$$i(X_L - X_C) + R = \frac{\mathcal{E}_{max}}{I_0} \qquad 5$$

Solve for I_0

$$I_0 = \frac{\mathcal{E}_{max}}{i(X_L - X_C) + R} \qquad 6$$

Again, we put the denominator of Equation 6 in polar form (Figure 1):

$$i(X_L - X_C) + R = \sqrt{(X_L - X_C)^2 + R^2}\,e^{i\delta}$$
$$= Ze^{i\delta} \qquad 7$$

where $Z = \sqrt{(X_L - X_C)^2 + R^2}$ is the impedance and

$$\tan\delta = \frac{X_L - X_C}{R} \qquad 8$$

Then

$$I_0 = \frac{\mathcal{E}_{max}}{Ze^{i\delta}} = \frac{\mathcal{E}_{max}}{Z}e^{-i\delta}$$

and

$$z = I_0e^{i\omega t} = \frac{\mathcal{E}_{max}}{Z}e^{-i\delta}e^{i\omega t} = \frac{\mathcal{E}_{max}}{Z}e^{i(\omega t - \delta)}$$

The current is thus given by

$$I = \mathrm{Re}(z) = \frac{\mathcal{E}_{max}}{Z}\cos(\omega t - \delta)$$

which is the same as Equation 31-54.

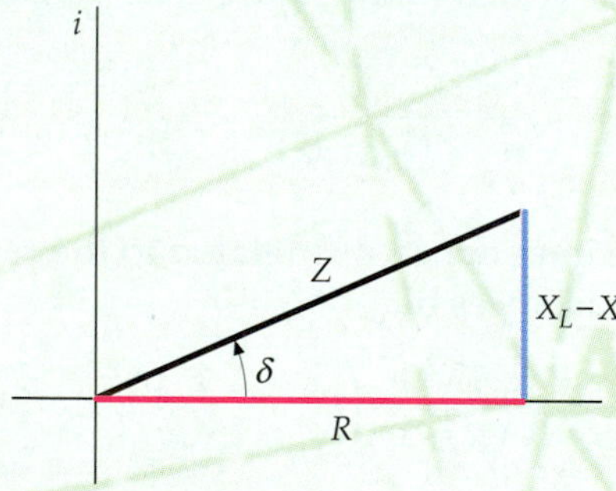

Figure 1 Polar representation of $i(X_L - X_C) + R$. The magnitude is $Z = \sqrt{(X_L - X_C)^2 + R^2}$ and the phase angle is δ.

Complex Impedances

The complex number technique described above can be extended into a generalized form of Ohm's

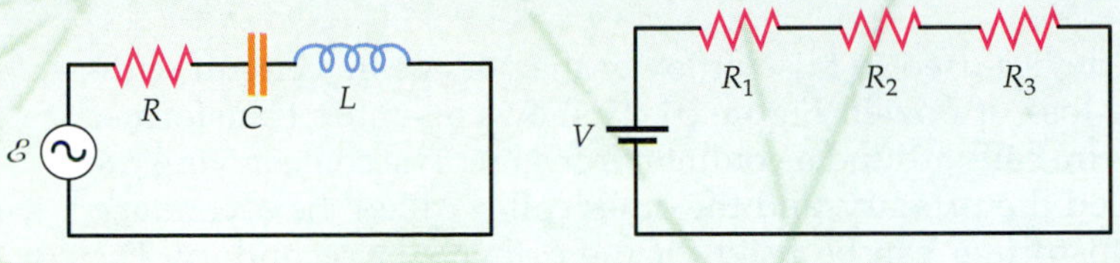

Figure 2 (*a*) *RLC* series circuit. (*b*) Analogous dc circuit.

law that makes ac problems look like dc problems. Consider the simple dc circuit shown next to our series *RLC* circuit in Figure 2. In the dc circuit, the applied voltage equals the sum of the voltage drops:

$$V = I(R_1 + R_2 + R_3)$$

If we multiply each term in Equation 4 by I_0/z, we obtain

$$\mathscr{E}_{\text{max}} = I_0\left(R + \frac{1}{i\omega C} + i\omega L\right) \qquad 9$$

The quantities R, $1/i\omega C$, and $i\omega L$ are the complex impedances for these circuit elements. Just as the resistance is the ratio of the voltage to current for a resistor, the complex impedance is the ratio of the complex voltage to the complex current for a circuit element. We illustrate the use of complex impedance by considering a circuit with an inductor and capacitor in parallel, and with the combination in series with a resistor, as shown in Figure 3*a*. The analogous dc circuit is shown in Figure 3*b*.

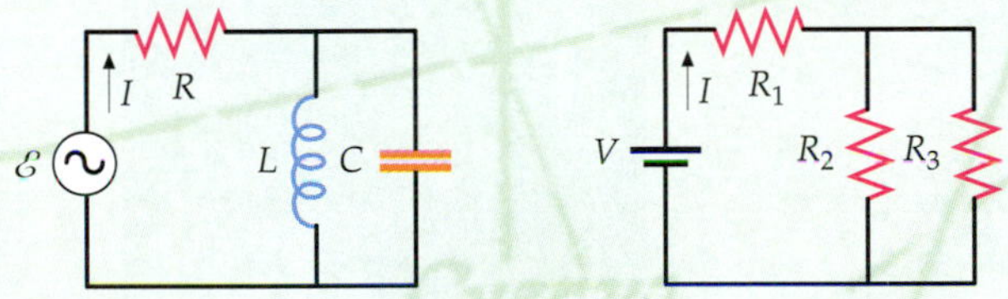

Figure 3 (*a*) An ac circuit with inductance and capacitance in parallel, and with the combination in series with a resistance. (*b*) An analogous dc circuit.

To find the current in the dc circuit, we first find the total resistance by adding R_1 to the resistance of the parallel combination of R_2 and R_3. The result is

$$R_{\text{eq}} = R_1 + \frac{R_2R_3}{R_2 + R_3}$$

The total impedance of the ac circuit is found in the same way.

$$Z_{\text{eq}} = Z_1 + \frac{Z_2Z_3}{Z_2 + Z_3} = R + \frac{(i\omega L)(1/i\omega C)}{i\omega L + 1/i\omega C}$$

$$= R + i\frac{L/C}{(1/\omega C) - \omega L} = R + i\frac{L/C}{X_C - X_L} \qquad 10$$

In polar form, the complex equivalent impedance is

$$Z_{\text{eq}} = \sqrt{R^2 + \left(\frac{L/C}{X_C - X_L}\right)^2}\, e^{i\delta} \qquad 11$$

where

$$\tan\delta = \frac{L/C}{R(X_C - X_L)} \qquad 12$$

If the emf is given by $\mathscr{E} = \mathscr{E}_{\text{max}} \cos \omega t = \text{Re}\,(\mathscr{E}_{\text{max}}e^{i\omega t})$, the current is given by

$$I = \text{Re}\left(\frac{\mathscr{E}_{\text{max}}e^{i\omega t}}{Z_{\text{eq}}}\right) = I_{\text{max}} \cos(\omega t - \delta) \qquad 13$$

where

$$I_{\text{max}} = \frac{\mathscr{E}_{\text{max}}}{|Z_{\text{eq}}|} = \frac{\mathscr{E}_{\text{max}}}{\sqrt{R^2 + \left(\frac{L/C}{X_C - X_L}\right)^2}} \qquad 14$$

and the phase angle δ is given by Equation 12. Note that the current goes to zero when $X_C = X_L$, that is, at the frequency $\omega = 1/\sqrt{LC}$ for which $\omega L = 1/\omega C$. At this frequency, the parallel combination is equivalent to an *open* circuit rather than a *short* circuit in the series combination.

31-7 The Transformer

A transformer is a device used to raise or lower the voltage in a circuit without an appreciable loss of power. Figure 31-25 shows a simple transformer consisting of two wire coils around a common iron core. The coil carrying the input power is called the primary, and the other coil is called the secondary. Either coil of a transformer can be used for the primary or secondary. The transformer operates on the principle that an alternating current in one circuit induces an alternating emf in a nearby circuit due to the mutual inductance of the two circuits. The iron core increases the magnetic field for a given current and guides it so that nearly all the magnetic flux through one coil goes through the other coil. If no power were lost, the product of the voltage and the current in the secondary circuit would equal the product of the voltage and the current in the primary circuit. Thus, if the voltage is raised, the current is lowered, and vice versa. Power losses arise because of the Joule heating in the small resistances in both coils, or in current loops within the core* and from hysteresis in the iron cores. We will neglect these losses and consider an ideal transformer of 100% efficiency, for which all of the power supplied to the primary coil appears in the secondary coil. Actual transformers are often 90 to 95% efficient.

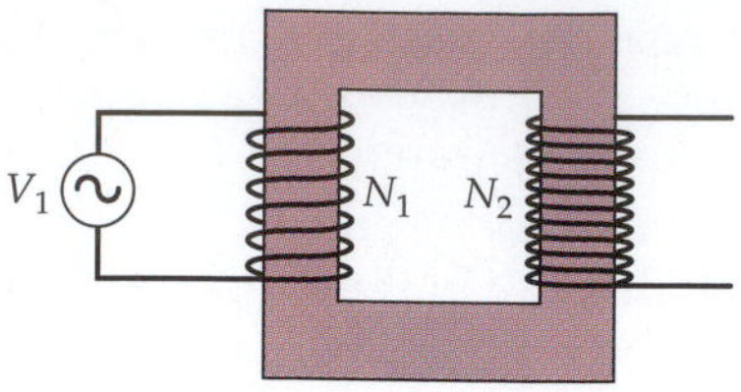

Figure 31-25 Transformer with N_1 turns in the primary and N_2 turns in the secondary.

Consider a transformer with an emf V_1 across the primary of N_1 turns; the secondary coil of N_2 turns is an open circuit. Because of the iron core, there is a large flux through each coil even when the magnetizing current I_m in the primary circuit is very small. We can ignore the resistances of the coils, which are negligible in comparison with their inductive reactances. The primary is then a simple circuit consisting of an ac generator and a pure inductance like that discussed in Section 31-2. The (magnetizing) current and the voltage in the primary are out of phase by 90°, and the average power dissipated in the primary coil is zero. If ϕ_{turn} is the magnetic flux in one turn of the primary coil, the voltage drop across the primary coil is $V_{L1} = N_1\, d\phi_{turn}/dt$. Applying Kirchhoff's loop rule to the primary circuit then gives

$$V_1 - N_1\frac{d\phi_{turn}}{dt} = 0$$

or

$$V_1 = N_1\frac{d\phi_{turn}}{dt} \qquad \text{31-63}$$

If there is no flux leakage out of the iron core, the flux through each turn is the same for both coils. Thus, the total flux through the secondary coil is $N_2\,\phi_{turn}$, and the voltage across the secondary coil is

$$V_2 = N_2\frac{d\phi_{turn}}{dt} \qquad \text{31-64}$$

Comparing equations 31-63 and 31-64, we can see that

$$V_2 = \frac{N_2}{N_1}V_1 \qquad \text{31-65}$$

If N_2 is greater than N_1, the voltage in the secondary coil is greater than that in the primary coil, and the transformer is called a step-up transformer. If N_2 is less than N_1, the voltage in the secondary coil is less than that in the primary coil, and the transformer is called a step-down transformer.

(a)

(b)

(*a*) A suburban power substation where transformers step down voltage from high-voltage transmission lines. (*b*) A power box with transformer for stepping down voltage for distribution to homes.

---

* The induced currents, called eddy currents, can be greatly reduced by using a core of laminated metal to break up current paths.

When we put a resistance R, called a load resistance, across the secondary coil, there will then be a current I_2 in the secondary circuit that is in phase with the voltage V_2 across the resistance. This current sets up an additional flux ϕ'_{turn} through each turn that is proportional to N_2I_2. This flux opposes the original flux set up by the original magnetizing current I_{m} in the primary. However, the voltage across the primary coil is determined by the generator emf, which is unaffected by the secondary circuit. According to Equation 31-64, the flux in the iron core must change at the original rate; that is, the total flux in the iron core must be the same as when there is no load across the secondary. The primary coil thus draws an additional current I_1 to maintain the original flux ϕ_{turn}. The flux through each turn produced by this additional current is proportional to N_1I_1. Since this flux equals $-\phi'_{\text{turn}}$, the additional current I_1 in the primary is related to the current I_2 in the secondary by

$$N_1I_1 = -N_2I_2 \qquad \text{31-66}$$

These currents are 180° out of phase and produce counteracting fluxes. Since I_2 is in phase with V_2, the additional current I_1 is in phase with the applied emf. The power input from the generator is $V_{1,\text{rms}}\, I_{1,\text{rms}}$, and the power output is $V_{2,\text{rms}}I_{2,\text{rms}}$. (The magnetizing current does not contribute to the power input because it is 90° out of phase with the generator voltage.) If there are no losses,

$$V_{1,\text{rms}}I_{1,\text{rms}} = V_{2,\text{rms}}I_{2,\text{rms}} \qquad \text{31-67}$$

In most cases the additional current in the primary I_1 is much greater than the original magnetizing current I_{m} that is drawn from the generator when there is no load. This can be demonstrated by putting a light bulb in series with the primary coil. The bulb is much brighter when there is a load across the secondary than when the secondary circuit is open. If I_{m} can be neglected, Equation 31-67 relates the total currents in the primary and secondary circuits.

Example 31-9

A doorbell requires 0.4 A at 6 V. It is connected to a transformer whose primary, containing 2000 turns, is connected to a 120-V ac line. (*a*) How many turns should there be in the secondary? (*b*) What is the current in the primary?

Picture the Problem We can find the number of turns from the turns ratio, which equals the voltage ratio. The primary current can be found by equating the power out to the power in.

(*a*)1. The turns ratio can be obtained from Equation 31-65:

$$\frac{N_2}{N_1} = \frac{V_2}{V_1} = \frac{6\text{ V}}{120\text{ V}} = \frac{1}{20}$$

2. Solve for the number of turns in the secondary, N_2:

$$N_2 = \frac{1}{20}(2000\text{ turns}) = 100\text{ turns}$$

(*b*)1. Since we are assuming 100% efficiency in power transmission, the input and output currents are related by Equation 31-66:

$$V_2I_2 = V_1I_1$$

2. Solve for the current in the primary, I_1:

$$I_1 = \frac{V_2}{V_1}I_2 = \frac{6\text{ V}}{120\text{ V}}(0.4\text{ A}) = 0.02\text{ A}$$

optional

An important use of transformers is in the transport of electrical power. To minimize the I^2R heat loss in transmission lines, it is economical to use a high voltage and a low current. On the other hand, safety and other considerations require that power be delivered to consumers at lower voltages and therefore with higher currents. Suppose, for example, that each person in a city with a population of 50,000 uses 1.2 kW of electric power. (The per capita consumption of power in the United States is actually somewhat higher than this). At 120 V, the current required for each person would be

$$I = \frac{1200 \text{ W}}{120 \text{ V}} = 10 \text{ A}$$

The total current for 50,000 people would then be 500,000 A. The transport of such a current from a power-plant generator to a city many kilometers away would require conductors of enormous size, and the I^2R power loss would be substantial. Rather than transmit the power at 120 V, step-up transformers are used at the power plant to step up the voltage to some very large value, such as 600,000 V. For this voltage, the current needed is only

$$I = \frac{120 \text{ V}}{600{,}000 \text{ V}}(500{,}000 \text{ A}) = 100 \text{ A}$$

To reduce the voltage to a safer level for transport within a city, power stations are located just outside the city to step down the voltage to a safer value, such as 10,000 V. Transformers in boxes attached to the power poles outside each house again step down the voltage to 120 V (or 240 V) for distribution to the house. It is because of the ease of stepping the voltage up or down with transformers that alternating current rather than direct current is in common use.

Example 31-10

A transmission line has a resistance of 0.02 Ω/km. Calculate the I^2R power loss if 200 kW of power is transmitted from a power generator to a city 10 km away at (*a*) 240 V and (*b*) 4.4 kV.

Picture the Problem First, note that the total resistance of 10 km of wire is $R = (0.02\ \Omega/\text{km})(10 \text{ km}) = 0.2\ \Omega$. In each case, begin by finding the current needed to transmit 200 kW using $P = IV$, then find the power loss using I^2R.

(*a*) 1. Find the current needed to transmit 200 kW of power at 240 V:

$$I = \frac{P}{V} = \frac{200 \text{ kW}}{240 \text{ V}} = 833 \text{ A}$$

2. Calculate the power loss:

$$I^2R = (833 \text{ A})^2(0.2\ \Omega) = 139{,}000 \text{ W}$$

(*b*) 1. Now find the current needed to transmit 200 kW of power at 4.4 kV:

$$I = \frac{P}{V} = \frac{200 \text{ kW}}{4.4 \text{ kV}} = 45.5 \text{ A}$$

2. Calculate the power loss:

$$I^2R = (45.5 \text{ A})^2(0.2\ \Omega) = 414 \text{ W}$$

Remark Note that with a transmission voltage of 240 V almost 70% of the power is wasted through heat loss, but with transmission at 4.4 kV only about 0.2% is lost. This illustrates the advantage of high-voltage power transmission.

Exploring

Electric Motors

John Dentler
United States Naval Academy

The wide variety of applications for electric motors requires many different designs. Electric-clock motors must operate at a precise speed. Automobile starters must deliver a tremendous torque from a standstill. A hand-held hair dryer must be lightweight and operate at several different speeds. Engineers design motors for various applications using models derived from the physical principles discussed in this text. These models are equations that predict a motor's performance for a set of specific applications or loads.

The rotating motor (Figure 1) has current-carrying conductors that react with an external field. The field, called the *stator field,* is created and controlled by the coil of wire visible at the bottom of Figure 1. Flux from the wire coil passes through the core, creating a north pole on the left and a south pole on the right of the rotating element. The rotating assembly, called the *armature,* consists of an iron cylinder with eight slots that contain conductors. If current can be driven through these conductors in the direction shown (front to back near the south pole and back to front near the north pole), then a net clockwise torque (down on the south-pole side and up on the north-pole side) will be developed to turn the armature.

Constructing a device that maintains the proper current direction in each conductor as the armature turns is a complicated task. Such devices are called *commutator brush assemblies.* The commutator brush assembly shown in Figure 2 consists of four segments that protrude along the motor shaft and two brushes that conduct current from a source to the segments. Each segment is connected to two conductors, which run through the slots of the rotating assembly. The conductors are interconnected through wires in the rear of the rotating assembly and by the commutator segments in the front of the assembly. This method of connection results in two parallel paths between the brushes; thus all of the conductors are used all of the time.

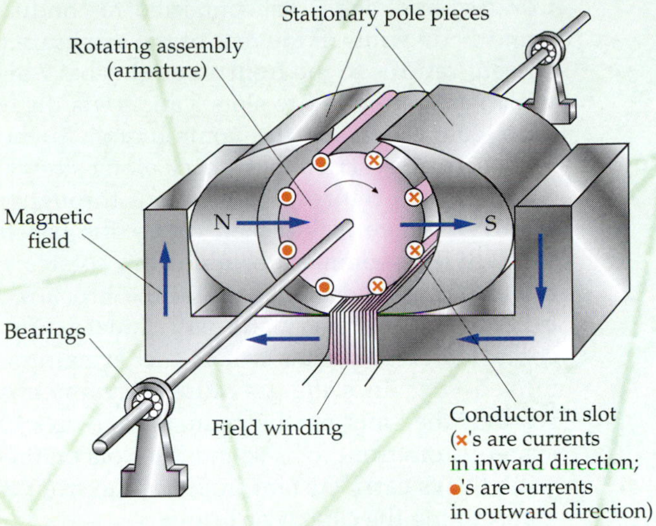

Figure 1 A simple rotating electric motor.

In the commutator shown in Figure 2, current delivered from the brush on the right follows one of the two parallel paths through the armature. The conductors in slots 2 and 5 both carry the current from the front to the back of the armature.

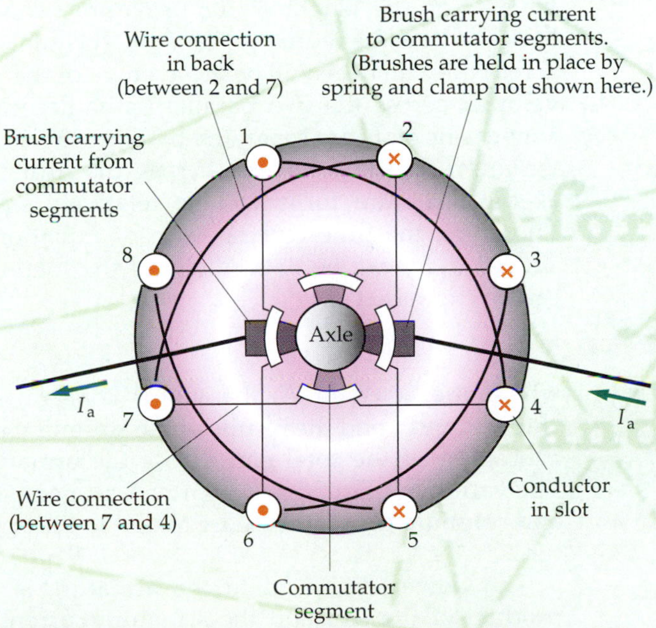

Figure 2 A commutator brush assembly for the motor in Figure 1. A commutator, in its most general sense, is a switching device. The device shown switches current direction through the armature to maintain the clockwise rotation.

Conductors 2 and 5 are connected to conductors 7 and 8 via wires in the rear of the armature. The current returns to the front through slots 7 and 8, which are connected to slots 3 and 4 via the common connection on the commutator segments. The current is carried to the back along slots 3 and 4 and is then returned to the front through slots 1 and 6, where it is picked off by the brush on the left. The commutator assembly rotates with the armature and moves under the brushes. The brushes are stationary and will contact different commutator segments when the armature has moved 90°. Since the armature is symmetrically wound, the slots on the right will always carry current from front to back and the slots on the left will always carry current from back to front, thus maintaining the clockwise torque.

The total torque turning the motor is the sum of the torques exerted by the conductors in each slot. In any position there are four armature conductors acting on the right and four on the left; therefore, the torque is approximately constant. Similarly, the total emf developed between the brushes is the sum of the emfs on each conductor. At any position there are two parallel paths, each consisting of four conductors. From Section 30-4, the emf developed across a single length of wire can be shown to be

$$\mathcal{E} = B\ell r\omega \sin(\omega t + \delta) \qquad 1$$

where ℓ is the length of the armature (front to back) and r is the radius of the rotor. The total emf across the armature will be the average of that developed across the two parallel paths described above. The slots are separated by only 45°, so the variation in the total emf across the armature as the armature turns will be relatively small. Therefore, the time-varying term of Equation 1, $\sin(\omega t + \delta)$, can be discarded and the total emf can be expressed as

$$\mathcal{E}_{\text{total}} = BK\omega \qquad 2$$

where the constant K, the motor constant, includes ℓ and r and the results of the summing and averaging of the total emf across the armature. The validity of Equation 2 improves as more slots and commutator segments are added to the armature.

The power delivered to the armature is the product of the emf and the armature current I_a. For a rotating motor, the load is a torque τ applied to the shaft opposing the direction of rotation. The mechanical power delivered to the load is the product of the torque and the angular velocity. At equilibrium, the driving torque from the motor is equal and opposite to the load torque. Thus,

$$P = \mathcal{E}I_a = \tau\omega \qquad 3$$

Substituting $BK\omega$ for the emf from Equation 7, we obtain

$$P = BKI_a\omega = \tau\omega \qquad 4$$

We can represent the armature by a simple voltage source with an external resistance R_a. The field winding connections of the coil shown at the bottom of Figure 1 can be connected either in series or in parallel (shunt) with the armature. These two methods of connection yield motors with extremely different characteristics.

Parallel (Shunt) Connection

Figure 3 shows the circuit for the shunt or parallel field connection. A variable resistance controls the field and thereby the speed of the motor. Applying Kirchhoff's loop rule to this circuit yields

$$V - I_aR_a - BK\omega = 0 \qquad 5$$

which can be rearranged to express the rotational speed ω in terms of the armature current I_a:

$$\omega = -\frac{R_a}{BK}I_a + \frac{V}{BK} \qquad 6$$

If we substitute τ/BK for the current from Equation 4, the rotational speed is

$$\omega = -\frac{R_a}{(BK)^2}\tau + \frac{V}{BK} \qquad 7$$

Equation 7 is a linear equation relating the rotational speed to the load. The speed can be controlled either by varying the voltage V or, more commonly, by varying the current into the coil by varying the resistance.

At high armature currents, the armature core saturates, the voltage drop due to the armature inductance becomes significant, and the relationship between the torque and speed becomes nonlinear. However, for normal loads, Equation 7

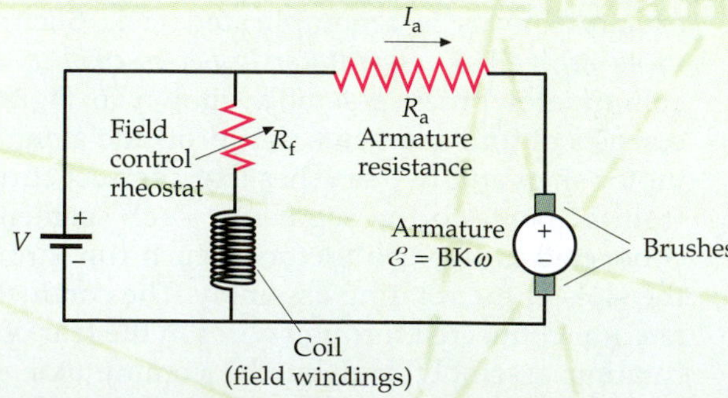

Figure 3 Circuit for a typical dc shunt motor.

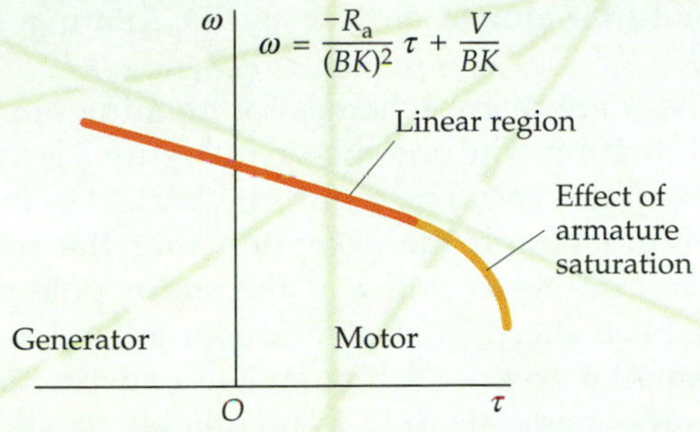

Figure 4 Graph of torque versus rotational speed showing the effect of armature saturation on the performance characteristic of a typical dc shunt motor.

accurately describes the motor's operation. Figure 4 shows how the *performance characteristic* of the motor, or its speed versus torque, is affected by armature saturation.

Series Connection

In the motor circuit of Figure 5, the coil is connected in series with the armature, so the field strength is a function of the armature current.

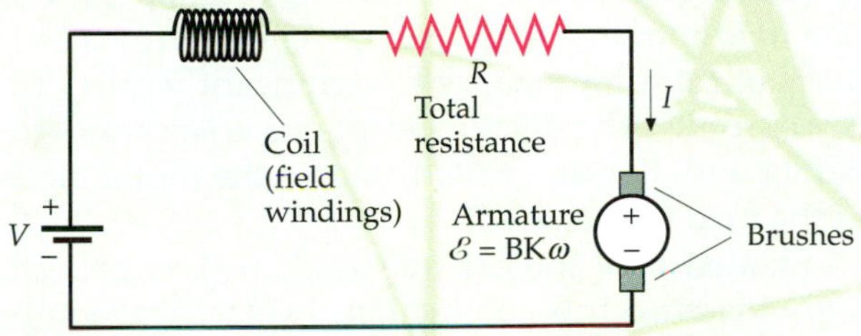

Figure 5 Circuit for a typical dc series motor.

If the armature current is small and the field does not saturate, the product of the field strength and the motor constant K can be expressed as a linear function of the armature current:

$$BK = CI$$

where C is some constant. Substituting this value of BK in the expressions for the armature emf, power, and torque yields

$$\mathcal{E} = CI\omega P = CI^2\omega \quad \text{and} \quad \tau = CI^2$$

The Kirchhoff loop rule then gives

$$V - IR - CI\omega = 0$$

where R represents the total resistance of the coil and armature, and I represents the only current in the circuit. This gives the following speed-versus-current equation:

$$\omega = \frac{V}{CI} - \frac{R}{C}$$

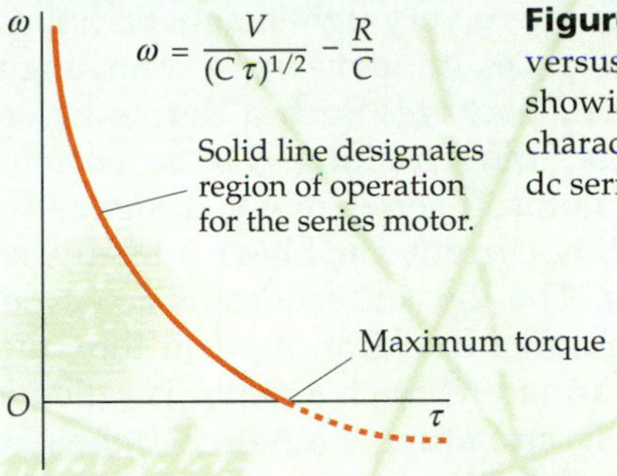

Figure 6 Graph of torque versus rotational speed showing the performance characteristic of the typical dc series motor.

Substituting $\sqrt{\tau/C}$ for I, we obtain the equation for the speed versus torque, which is the performance characteristic for the series motor:

$$\omega = \frac{V}{(C\tau)^{1/2}} - \frac{R}{C}$$

Figure 6 shows the performance characteristic for the series motor. Comparing this performance characteristic with that for the shunt motor reveals striking differences. At low torques, the series motor runs very fast, almost without limit (the only load is the friction of its bearings and the air around the rotor) whereas the shunt motor is regulated to run close to the speed V/BK. At high torques, the speed of the shunt motor tapers off and the motor stalls, but the series motor delivers its greatest torque when the motor is stopped. A series motor is therefore the best choice to start a car engine, which requires a high torque at $\omega = 0$. On the other hand, a shunt motor is the better choice to drive a speed-sensitive load, like a tape recorder.

With only minor modifications, the principles of dc motor construction and operation also apply to ac motors. The torque of a series motor is proportional to I^2 and is thus independent of current direction. This is because the same current is in both the stationary field and rotating armature. With this cursory examination, it might be concluded that any series dc motor would run on alternating current. However, an assumption made to simplify the analysis of the dc motor was that inductance could be ignored. Inductance cannot be ignored when driving a motor from an ac source. Inductance has two effects: (1) It acts as a throttle limiting the amount of ac current for a given input voltage, and (2) it changes the phase relationship of the current and voltage.

A dc shunt motor typically has field windings with high resistance and an armature with high inductance. Applying alternating current to such a motor would create a phase difference between the field and the armature currents resulting in unsatisfactory performance.

A series motor has a very tight magnetic circuit with close tolerances to develop a very high torque in a small package. Such a device has a high inductance, thus limiting the ac current drawn by the motor. A series motor designed to run on alternating current must have a relatively low inductance. The low inductance is achieved by limiting the amount of iron used in the pole pieces and the armature. Such a motor is called a *universal motor.* By its nature, it is both lightweight and limited to driving devices with relatively light loads such as vacuum cleaners, food blenders, hair dryers, and sewing machines. Its performance characteristic is similar to that for the dc series motor shown in Figure 6.

The most common ac motor is the *induction motor.* This motor has a rotating assembly like the one shown in Figure 1, but unlike the dc motor, the commutator and interconnecting wires are replaced with shorting plates, connecting all the slotted conductors, mounted on the front and back. The challenge becomes how to make the shorted rotor rotate. The solution is to make the field from the stator appear to rotate. If the field rotates, there will be a relative velocity between the rotor and the stator field. An emf develops across the shorted rotor, driving current through the conductors in the slots. The rotating stator field produces a torque on the induced current in the rotor. The rotor moves such as to minimize the relative motion between it and the field. Thus, the rotor turns almost as fast as the rotating stator field.

There are many schemes for creating apparent field rotation. The one shown in Figure 7 is known as the *shaded pole.* The motor is identical to the one in Figure 1 with the exception that the rotor is shorted on either end and the stator pole pieces have been sliced, with a conducting band around the small pieces of each pole. This construction allows the magnetic field to be quickly established through the faces of the large pole pieces and to be delayed through the small faces by the inductance of the conducting band. The phase delay between the field through the large pole faces and the field through the small pole faces creates the appearance of a rotating field.

The performance characteristic of a typical induction motor is shown in Figure 8. Normal operation is close to the speed of field rotation. If the motor in Figure 7 were connected to a 60-Hz source, the rotational speed would be somewhat less than 60 rev/s. The maximum torque shown on the performance characteristic curve occurs where the difference between the rotor speed and the field rotation speed is large enough for the effects of rotor inductance to significantly delay rotor currents. The delayed rotor currents cannot interact with the stator field, and so the motor stalls if the load is increased.

Shaded-pole motors are used in devices with light loads such as cooling fans in electrical equipment. More complex schemes for creating field rotation are used in the induction motors for refrigerators and air conditioners. Large industrial induction motors use three-phase electricity to rotate the field.

Figure 7 A shaded-pole induction motor.

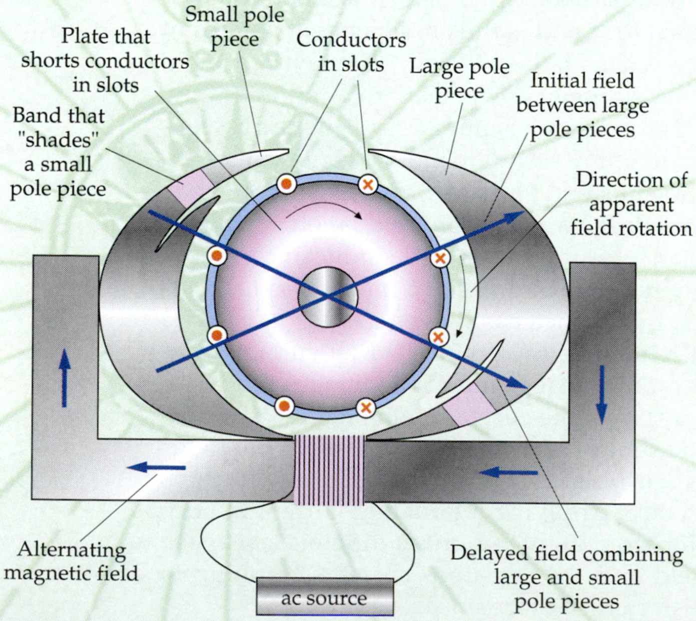

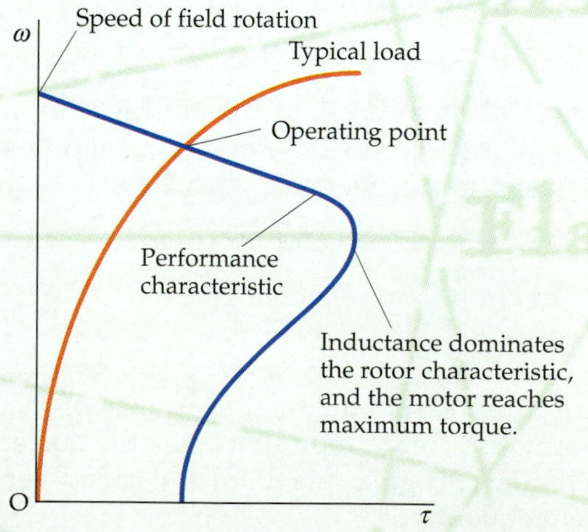

Figure 8 Graph of torque versus speed showing the performance characteristic of a typical induction motor. The load line indicated is typical for a centrifugal pump.

Summary

1. Reactance is a frequency-dependent property of capacitors and inductors that is analogous to the resistance of a resistor.
2. Impedance is a frequency-dependent property of an ac circuit or circuit loop that is analogous to the resistance in a dc circuit.
3. Phasors are two-dimensional vectors that allow us to picture the phase relations in a circuit.
4. Resonance occurs when the frequency of the generator equals the natural frequency of the oscillating circuit.

Topic	Remarks and Relevant Equations	
1. ac Generator	An ac generator is a device for transforming mechanical energy into electrical energy. This transformation is accomplished by using the mechanical energy to rotate a conducting coil in a magnetic field.	
emf generated	$\mathcal{E} = \mathcal{E}_{max} \sin(\omega t + \delta) = NBA\omega \sin(\omega t + \delta)$	31-3, 31-4
2. Current		
rms current	$I_{rms} = \sqrt{(I^2)_{av}}$	31-11
rms current and maximum current	$I_{rms} = \dfrac{I_{max}}{\sqrt{2}}$	31-12
In a resistor	$I_{rms} = \dfrac{V_{R,rms}}{R}$, voltage and current in phase	31-16
In an inductor	$I_{rms} = \dfrac{V_{L,rms}}{\omega L} = \dfrac{V_{L,rms}}{X_L}$, voltage leads current by 90°	31-32
In a capacitor	$I_{rms} = \dfrac{V_{C,rms}}{1/\omega C} = \dfrac{V_{C,rms}}{X_C}$, voltage lags current by 90°	31-33
3. Reactance		
Inductive reactance	$X_L = \omega L$	31-25
Capacitive reactance	$X_C = \dfrac{1}{\omega C}$	31-31
4. Average Power Dissipation		
In a resistor	$P_{av} = \frac{1}{2}\mathcal{E}_{max} I_{max} = \mathcal{E}_{rms} I_{rms} = I^2_{rms} R$	31-13, 31-14, 31-56
In an inductor	$P_{av} = 0$	
In a capacitor	$P_{av} = 0$	
5. Phasors	Phasors are two-dimensional vectors that represent the current $\vec{I}$, the voltage across a resistor $\vec{V}_R$, the voltage across a capacitor $\vec{V}_C$, and the voltage across an inductor $\vec{V}_L$ in	

an ac circuit. These phasors rotate in the counterclockwise direction with an angular frequency ω that is equal to the angular frequency of the current. $\vec{V}_R$ is in phase with the current, $\vec{V}_L$ leads the current by 90°, and $\vec{V}_C$ lags the current by 90°. The x component of each phasor equals the magnitude of the current or the corresponding voltage drop at any instant.

6. *LC* and *RLC* Series Circuit

If a capacitor is discharged through an inductor, the charge and voltage on the capacitor oscillate with angular frequency

$$\omega = \frac{1}{\sqrt{LC}} \qquad \textbf{31-41}$$

The current in the inductor oscillates with the same frequency, but it is out of phase with the charge by 90°. The energy oscillates between electric energy in the capacitor and magnetic energy in the inductor. If the circuit also has resistance, the oscillations are damped because energy is dissipated in the resistor.

7. Series *RLC* Circuit Driven by a Generator of Frequency ω

Current

$$I = \frac{\mathcal{E}_{\rm max}}{Z}\cos(\omega t - \delta) \qquad \textbf{31-54}$$

Impedance Z

$$Z = \sqrt{R^2 + (X_L - X_C)^2} \qquad \textbf{31-53}$$

Phase angle δ

$$\tan\delta = \frac{X_L - X_C}{R} \qquad \textbf{31-51}$$

Average power

$$P_{\rm av} = \tfrac{1}{2}I_{\rm max}^2 R = I_{\rm rms}^2 R = \mathcal{E}_{\rm rms} I_{\rm rms}\cos\delta$$

$$= \frac{\mathcal{E}_{\rm rms}^2 R\omega^2}{L^2(\omega^2 - \omega_0^2)^2 + \omega^2 R^2} \qquad \textbf{31-56, 31-57, 31-58}$$

Power factor

The quantity $\cos\delta$ in Equation 31-57 is called the power factor of the *RLC* circuit. At resonance, δ is zero, and the power factor is 1. Then

$$(P_{\rm av})_{\rm max} = \mathcal{E}_{\rm rms} I_{\rm rms}$$

Resonance

When the current is maximum, the circuit is said to be at resonance. The conditions for resonance are

$$X_L = X_C, \quad Z = R$$

$$\omega = \omega_0 = \frac{1}{\sqrt{LC}}$$

$$\delta = 0$$

8. *Q* Factor

The sharpness of the resonance curve is described by the Q factor,

$$Q = \frac{\omega_0 L}{R} \qquad \textbf{31-59}$$

When the resonance curve is reasonably narrow, the Q factor can be approximated by

$$Q \approx \frac{\omega_0}{\Delta\omega} = \frac{f_0}{\Delta f} \qquad \textbf{31-60}$$

9. Transformers

A transformer is a device used to raise or lower the voltage in a circuit without an appreciable loss in power. For a transformer with N_1 turns in the primary and N_2 turns in the secondary, the voltage across the secondary coil is related to the generator emf across the primary coil by

$$V_2 = \frac{N_2}{N_1} V_1 \qquad \textbf{31-65}$$

If there are no losses,

$$V_{1,\text{rms}} I_{1,\text{rms}} = V_{2,\text{rms}} I_{2,\text{rms}} \qquad \textbf{31-67}$$

Problem-Solving Guide

1. Draw a neat sketch of the appropriate circuit, showing all circuit elements—capacitors, resistors, inductors, and generators. Indicate a direction for the current I and label the appropriate high- and low-voltage sides of the elements.
2. When using phasor analysis, draw vectors for $\vec{V}_R$, $\vec{V}_L$, and $\vec{V}_C$, being careful to draw $\vec{V}_C$ at 90° clockwise from $\vec{V}_R$, and $\vec{V}_L$ at 90° counterclockwise from $\vec{V}_R$.

Summary of Worked Examples

Type of Calculation	Procedure and Relevant Examples
1. rms Current	
Find the average and rms current for a given time interval.	The average current is $I_{\text{av}} = \frac{1}{T}\int_0^T I\,dt$. The rms current is $I_{\text{rms}} = \sqrt{(I^2)_{\text{av}}}$ where $(I^2)_{\text{av}} = \frac{1}{T}\int_0^T I^2\,dt$ **Example 31-1**
2. Reactance	
Find the reactance of inductors and capacitors.	Use $X_L = \omega L$ and $X_C = 1/\omega C$. **Examples 31-2, 31-3**
3. *LC* and *RLC* Circuits	
Determine the frequency and maximum current in an *LC* circuit.	The frequency of oscillation is $\omega = 1/\sqrt{LC}$ and the maximum current is $I_{\max} = \omega Q_0$, where Q_0 is the initial charge on the capacitor. **Example 31-4**
Analyze an *RLC* circuit with a generator.	Use $\omega_0 = 1/\sqrt{LC}$, $I_{\max} = \mathcal{E}_{\max}/Z$, $Z = \sqrt{R^2 + (X_L - X_C)^2}$, $\tan\delta = (X_L - X_C)/R$, and $Q = \omega_0 L/R$. **Examples 31-5, 31-6, 31-7, 31-8, 31-9**
4. Transformers	
Relate the currents in the primary and secondary coils of a transformer.	The currents are related by $N_1 I_1 = -N_2 I_2$, where subscripts 1 and 2 refer to the primary and secondary coils, respectively. **Example 31-10**

Problems

Conceptual Problems

Problems from Optional and Exploring sections

In a few problems, you are given more data than you actually need; in a few other problems, you are required to supply data from your general knowledge, outside sources, or informed estimates.

- • Single-concept, single-step, relatively easy
- •• Intermediate-level, may require synthesis of concepts
- ••• Challenging, for advanced students

Generators

1 • A 200-turn coil has an area of 4 cm^2 and rotates in a magnetic field of 0.5 T. (*a*) What frequency will generate a maximum emf of 10 V? (*b*) If the coil rotates at 60 Hz, what is the maximum emf?

2 • In what magnetic field must the coil of Problem 1 be rotating to generate a maximum emf of 10 V at 60 Hz?

3 • A 2-cm by 1.5-cm rectangular coil has 300 turns and rotates in a magnetic field of 4000 G. (*a*) What is the maximum emf generated when the coil rotates at 60 Hz? (*b*) What must its frequency be to generate a maximum emf of 110 V?

4 • The coil of Problem 3 rotates at 60 Hz in a magnetic field B. What value of B will generate a maximum emf of 24 V?

Alternating Current in a Resistor

5 • As the frequency in the simple ac circuit in Figure 31-26 increases, the rms current through the resistor

(*a*) increases.
(*b*) does not change.
(*c*) may increase or decrease depending on the magnitude of the original frequency.
(*d*) may increase or decrease depending on the magnitude of the resistance.
(*e*) decreases.

Figure 31-26 Problem 5

6 • If the rms voltage in an ac circuit is doubled, the peak voltage is

(*a*) increased by a factor of 2.
(*b*) decreased by a factor of 2.
(*c*) increased by a factor of $\sqrt{2}$.
(*d*) decreased by a factor of $\sqrt{2}$.
(*e*) not changed.

7 • A 100-W light bulb is plugged into a standard 120-V (rms) outlet. Find (*a*) I_{rms}, (*b*) I_{max}, and (*c*) the maximum power.

8 • A 3-Ω resistor is placed across a generator having a frequency of 60 Hz and a maximum emf of 12.0 V. (*a*) What is the angular frequency ω of the current? (*b*) Find I_{max} and I_{rms}. What is (*c*) the maximum power into the resistor, (*d*) the minimum power, and (*e*) the average power?

9 • A circuit breaker is rated for a current of 15 A rms at a voltage of 120 V rms. (*a*) What is the largest value of I_{max} that the breaker can carry? (*b*) What average power can be supplied by this circuit?

Alternating Current in Inductors and Capacitors

10 • If the frequency in the circuit shown in Figure 31-27 is doubled, the inductance of the inductor will

(*a*) increase by a factor of 2.
(*b*) not change.
(*c*) decrease by a factor of 2.
(*d*) increase by a factor of 4.
(*e*) decrease by a factor of 4.

Figure 31-27 Problems 10 and 11

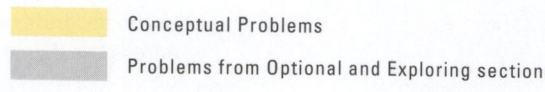

11 • If the frequency in the circuit shown in Figure 31-27 is doubled, the inductive reactance of the inductor will

(*a*) increase by a factor of 2.
(*b*) not change.
(*c*) decrease by a factor of 2.
(*d*) increase by a factor of 4.
(*e*) decrease by a factor of 4.

12 • If the frequency in the circuit in Figure 31-28 is doubled, the capacitative reactance of the circuit will

(*a*) increase by a factor of 2.
(*b*) not change.
(*c*) decrease by a factor of 2.
(*d*) increase by a factor of 4.
(*e*) decrease by a factor of 4.

Figure 31-28 Problem 12

13 • In a circuit consisting of a generator and an inductor, are there any times when the inductor absorbs power from the generator? Are there any times when the inductor supplies power to the generator?

14 • In a circuit consisting of a generator and a capacitor, are there any times when the capacitor absorbs power from the generator? Are there any times when the capacitor supplies power to the generator?

15 • What is the reactance of a 1.0-mH inductor at (*a*) 60 Hz, (*b*) 600 Hz, and (*c*) 6 kHz?

16 • An inductor has a reactance of 100 Ω at 80 Hz. (*a*) What is its inductance? (*b*) What is its reactance at 160 Hz?

17 • At what frequency would the reactance of a 10.0-μF capacitor equal that of a 1.0-mH inductor?

18 • What is the reactance of a 1.0-nF capacitor at (*a*) 60 Hz, (*b*) 6 kHz, and (*c*) 6 MHz?

19 • An emf of 10.0 V maximum and frequency 20 Hz is applied to a 20-μF capacitor. Find (*a*) I_{max} and (*b*) I_{rms}.

20 • At what frequency is the reactance of a 10-μF capacitor (*a*) 1 Ω, (*b*) 100 Ω, and (*c*) 0.01 Ω?

Phasors (optional)

21 • Draw the resultant phasor diagram for a series *RLC* circuit when $V_L < V_C$. Show on your diagram that the emf will lag the current by the phase angle δ given by

$$\tan\delta = \frac{V_C - V_L}{V_R}$$

22 •• Two ac voltage sources are connected in series with a resistor $R = 25\ \Omega$. One source is given by $V_1 = (5.0\ \text{V})\cos(\omega t - \alpha)$, and the other source is $V_2 = (5.0\ \text{V})\cos(\omega t + \alpha)$, with $\alpha = \pi/6$. (*a*) Find the current in *R* using a trigonometric identity for the sum of two cosines. (*b*) Use phasor diagrams to find the current in *R*. (*c*) Find the current in *R* if $\alpha = \pi/4$ and the amplitude of V_2 is increased from 5.0 to 7.0 V.

LC and *RLC* Circuits Without a Generator (optional)

23 • The SI units of inductance times capacitance are

(*a*) seconds squared.
(*b*) hertz.
(*c*) volts.
(*d*) amperes.
(*e*) ohms.

24 •• Making *LC* circuits with oscillation frequencies of thousands of hertz or more is easy, but making *LC* circuits that have small frequencies is difficult. Why?

25 • Show from the definitions of the henry and the farad that $1/\sqrt{LC}$ has the unit s^{-1}.

26 • (*a*) What is the period of oscillation of an *LC* circuit consisting of a 2-mH coil and a 20-μF capacitor? (*b*) What inductance is needed with an 80-μF capacitor to construct an *LC* circuit that oscillates with a frequency of 60 Hz?

27 •• An *LC* circuit has capacitance C_1 and inductance L_1. A second circuit has $C_2 = \frac{1}{2}C_1$ and $L_2 = 2L_1$, and a third circuit has $C_3 = 2C_1$ and $L_3 = \frac{1}{2}L_1$. (*a*) Show that each circuit oscillates with the same frequency. (*b*) In which circuit would the maximum current be greatest if the capacitor in each were charged to the same potential *V*?

28 •• A 5-μF capacitor is charged to 30 V and is then connected across a 10-mH inductor. (*a*) How much energy is stored in the system? (*b*) What is the frequency of oscillation of the circuit? (*c*) What is the maximum current in the circuit?

RL Circuits With a Generator (optional)

29 • A coil can be considered to be a resistance and an inductance in series. Assume that $R = 100\ \Omega$ and $L = 0.4$ H. The coil is connected across a 120-V-rms, 60-Hz line. Find (*a*) the power factor, (*b*) the rms current, and (*c*) the average power supplied.

30 •• A resistance *R* and a 1.4-H inductance are in series across a 60-Hz ac voltage. The voltage across the resistor is 30 V and the voltage across the inductor is 40 V. (*a*) What is the resistance *R*? (*b*) What is the ac input voltage?

31 •• A coil has a dc resistance of 80 Ω and an impedance of 200 Ω at a frequency of 1 kHz. One may neglect the wiring capacitance of the coil at this frequency. What is the inductance of the coil?

32 •• A single transmission line carries two voltage signals given by $V_1 = (10\ \text{V})\cos 100t$ and $V_2 = (10\ \text{V})\cos 10{,}000t$, where *t* is in seconds. A series inductor of 1 H and a shunting resistor of 1 kΩ is inserted into the transmission line as indicated in Figure 31-29. (*a*) What is the voltage signal observed at the output side of the transmission line? (*b*) What is the ratio of the low-frequency amplitude to the high-frequency amplitude?

Figure 31-29 Problem 32

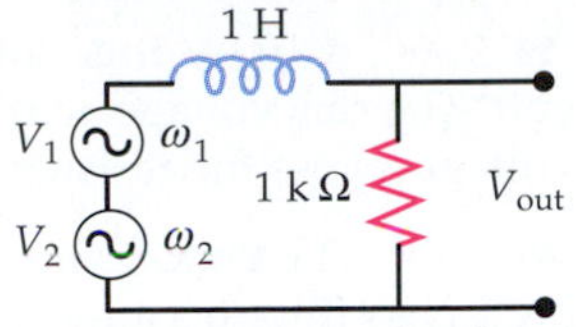

33 •• A coil with resistance and inductance is connected to a 120-V-rms, 60-Hz line. The average power supplied to the coil is 60 W, and the rms current is 1.5 A. Find (*a*) the power factor, (*b*) the resistance of the coil, and (*c*) the inductance of the coil. (*d*) Does the current lag or lead the voltage? What is the phase angle δ?

34 •• A 36-mH inductor with a resistance of 40 Ω is connected to a source whose voltage is $\mathcal{E} = (345\ \text{V})\cos(150\pi t)$, where *t* is in seconds. Determine the maximum current in the circuit, the maximum and rms voltages across the inductor, the average power dissipation, and the maximum and average energy stored in the magnetic field of the inductor.

35 •• A coil of resistance *R*, inductance *L*, and negligible capacitance has a power factor of 0.866 at a frequency of 60 Hz. What is the power factor for a frequency of 240 Hz?

36 •• A resistor and an inductor are connected in parallel across an emf $\mathcal{E} = \mathcal{E}_{max}\cos\omega t$ as shown in Figure 31-30. Show that (*a*) the current in the resistor is $I_R = (\mathcal{E}_{max}/R)\cos\omega t$, (*b*) the current in the inductor is $I_L = (\mathcal{E}_{max}/X_L)\cos(\omega t - 90°)$, and (*c*) $I = I_R + I_L = I_{max}\cos(\omega t - \delta)$, where $\tan\delta = R/X_L$ and $I_{max} = \mathcal{E}_{max}/Z$ with $Z^{-2} = R^{-2} + X_L^{-2}$.

Figure 31-30 Problem 36

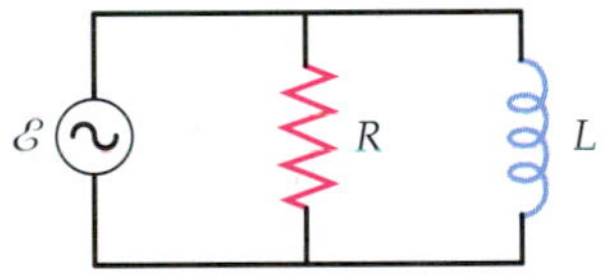

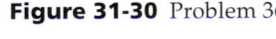

37 •• Figure 31-31 shows a load resistor $R_L = 20\ \Omega$ connected to a high-pass filter consisting of an inductor $L = 3.2$ mH and a resistor $R = 4\ \Omega$. The input voltage is $\mathcal{E} = (100\ \text{V})\cos(2\pi ft)$. Find the rms currents in *R*, *L*, and R_L if (*a*) $f = 500$ Hz and (*b*) $f = 2000$ Hz. (*c*) What fraction of the total power delivered by the voltage source is dissipated in the load resistor if the frequency is 500 Hz and if the frequency is 2000 Hz?

Figure 31-31
Problems 37 and 96

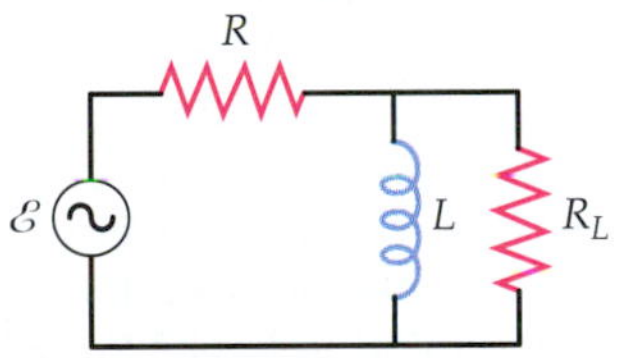

38 •• An ac source $\mathcal{E}_1 = (20\ \text{V})\cos(2\pi ft)$ in series with a battery $\mathcal{E}_2 = 16$ V is connected to a circuit consisting of resistors $R_1 = 10\ \Omega$ and $R_2 = 8\ \Omega$ and an inductor $L = 6$ mH (Figure 31-32). Find the power dissipated in R_1 and R_2 if (*a*) $f = 100$ Hz, (*b*) $f = 200$ Hz, and (*c*) $f = 800$ Hz.

Figure 31-32 Problem 38

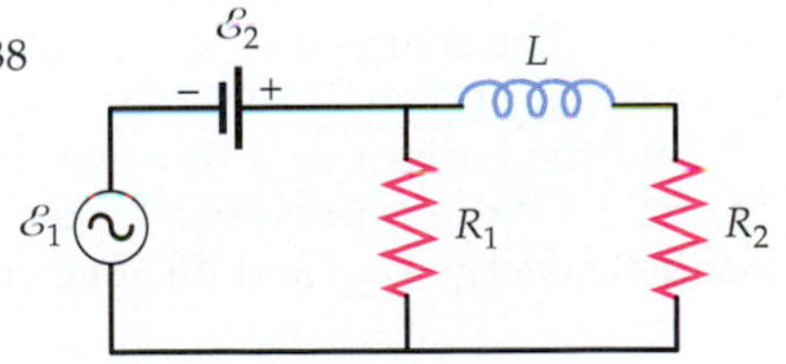

RC and *RL* Circuits With a Generator (optional)

39 •• A 100-V-rms voltage is applied to a series *RC* circuit. The rms voltage across the capacitor is 80 V. What is the voltage across the resistor?

40 •• The circuit shown in Figure 31-33 is called an *RC* high-pass filter because high input frequencies are transmitted with greater amplitude than low input frequencies. (*a*) If the input voltage is $V_{in} = V_0 \cos \omega t$, show that the output voltage is

$$V_{out} = \frac{V_0}{\sqrt{(1/\omega RC)^2 + 1}}$$

(*b*) At what angular frequency is the output voltage half the input voltage? (*c*) Sketch a graph of V_{out}/V_0 as a function of ω.

Figure 31-33 Problem 40

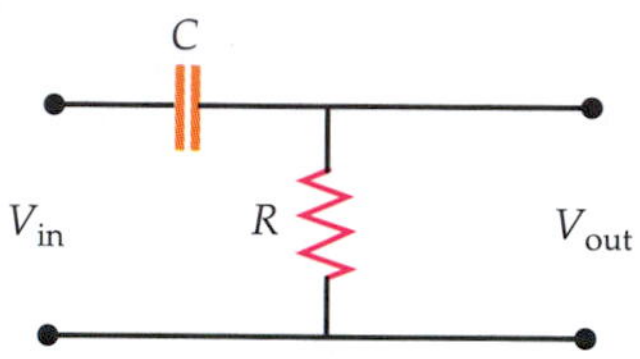

41 •• A coil draws 15 A when connected to a 220-V 60-Hz ac line. When it is in series with a 4-Ω resistor and the combination is connected to a 100-V battery, the battery current after a long time is observed to be 10 A. (*a*) What is the resistance in the coil? (*b*) What is the inductance of the coil?

42 •• Figure 31-34 shows a load resistor $R_L = 20\ \Omega$ connected to a low-pass filter consisting of a capacitor $C = 8\ \mu F$ and resistor $R = 4\ \Omega$. The input voltage is $\mathcal{E} = (100\ V) \cos (2\pi ft)$. Find the rms currents in *R*, *C*, and R_L if (*a*) $f = 500$ Hz and (*b*) $f = 2000$ Hz. (*c*) What fraction of the total power delivered by the voltage source is dissipated in the load resistor if the frequency is 500 Hz and if the frequency is 2000 Hz?

Figure 31-34 Problems 42 and 97

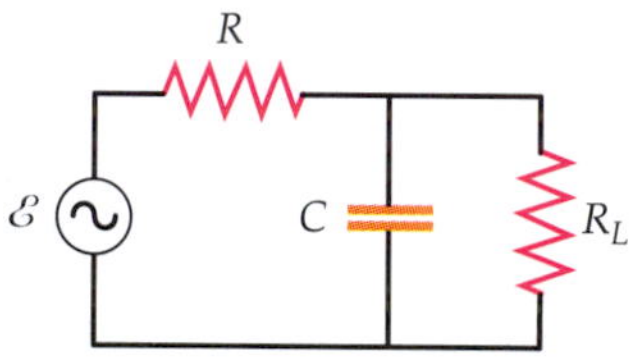

LC Circuits With a Generator (optional)

43 •• The generator voltage in Figure 31-35 is given by $\mathcal{E} = (100\ V) \cos (2\pi ft)$. (*a*) For each branch, what is the amplitude of the current and what is its phase relative to the applied voltage? (*b*) What is the angular frequency ω such that the current in the generator vanishes? (*c*) At this resonance, what is the current in the inductor? What is the current in the capacitor? (*d*) Draw a phasor diagram showing the general relationships between the applied voltage, the generator current, the capacitor current, and the inductor current for the case where the inductive reactance is larger than the capacitive reactance.

Figure 31-35 Problem 43

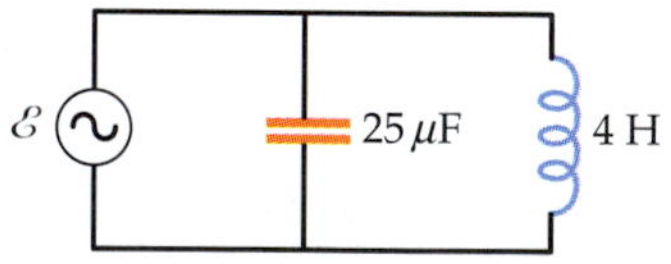

44 •• The charge on the capacitor of a series *LC* circuit is given by $Q = (15\ \mu C) \cos (1250t + \pi/4)$ where *t* is in seconds. (*a*) Find the current as a function of time. (*b*) Find *C* if $L = 28$ mH. (*c*) Write expressions for the electrical energy U_e, the magnetic energy U_m, and the total energy *U*.

45 ••• One method for measuring the compressibility of a dielectric material uses an *LC* circuit with a parallel-plate capacitor. The dielectric is inserted between the plates and the change in resonance frequency is determined as the capacitor plates are subjected to a compressive stress. In such an arrangement, the resonance frequency is 120 MHz when a dielectric of thickness 0.1 cm and dielectric constant $\kappa = 6.8$ is placed between the capacitor plates. Under a compressive stress of 800 atm, the resonance frequency decreases to 116 MHz. Find Young's modulus of the dielectric material.

46 ••• Figure 31-36 shows an inductance *L* and a parallel plate capacitor of width $w = 20$ cm and thickness 0.2 cm. A dielectric with dielectric constant $\kappa = 4.8$ that can completely fill the space between the capacitor plates can be slid between the plates. The inductor has an inductance $L = 2$ mH. When half the dielectric is between the capacitor plates, i.e, when $x = \frac{1}{2}w$, the resonant frequency of this *LC* combination is 90 MHz. (*a*) What is the capacitance of the capacitor without the dielectric? (*b*) Find the resonance frequency as a function of *x*.

Figure 31-36 Problem 46

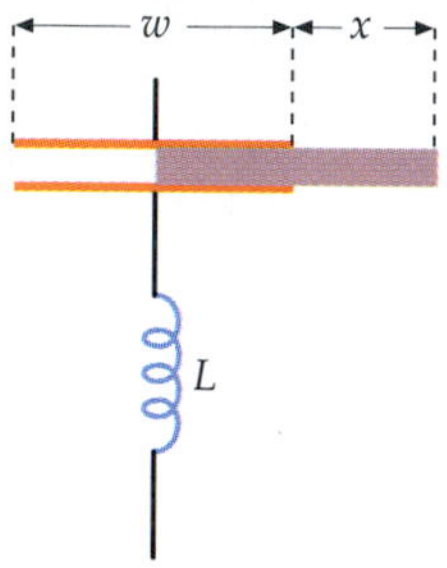

RLC Circuits With a Generator (optional)

47 • True or false:

(*a*) An *RLC* circuit with a high *Q* factor has a narrow resonance curve.

(*b*) At resonance, the impedance of an *RLC* circuit equals the resistance *R*.

(*c*) At resonance, the current and generator voltage are in phase.

48 • Does the power factor depend on the frequency?

49 • Are there any disadvantages to having a radio tuning circuit with an extremely large *Q* factor?

50 • What is the power factor for a circuit that has inductance and capacitance but no resistance?

51 • A series *RLC* circuit in a radio receiver is tuned by a variable capacitor so that it can resonate at frequencies from 500 to 1600 kHz. If $L = 1.0\ \mu H$, find the range of capacitances necessary to cover this range of frequencies.

52 • (*a*) Find the power factor for the circuit in Example 31-5 when $\omega = 400$ rad/s. (*b*) At what angular frequency is the power factor 0.5?

53 • An ac generator with a maximum emf of 20 V is connected in series with a 20-μF capacitor and an 80-Ω resistor. There is no inductance in the circuit. Find (*a*) the power factor, (*b*) the rms current, and (*c*) the average power if the angular frequency of the generator is 400 rad/s.

54 •• Show that the formula $P_{av} = R\mathcal{E}_{rms}^2/Z^2$ gives the correct result for a circuit containing only a generator and (*a*) a resistor, (*b*) a capacitor, and (*c*) an inductor.

55 •• A series *RLC* circuit with $L = 10$ mH, $C = 2\ \mu$F, and $R = 5\ \Omega$ is driven by a generator with a maximum emf of 100 V and a variable angular frequency ω. Find (*a*) the resonant frequency ω_0 and (*b*) I_{rms} at resonance. When $\omega = 8000$ rad/s, find (*c*) X_C and X_L, (*d*) Z and I_{rms}, and (*e*) the phase angle δ.

56 •• For the circuit in Problem 55, let the generator frequency be $f = \omega/2\pi = 1$ kHz. Find (*a*) the resonance frequency $f_0 = \omega_0/2\pi$, (*b*) X_C and X_L, (*c*) the total impedance Z and I_{rms}, and (*d*) the phase angle δ.

57 •• Find the power factor and the phase angle δ for the circuit in Problem 55 when the generator frequency is (*a*) 900 Hz, (*b*) 1.1 kHz, and (*c*) 1.3 kHz.

58 •• Find (*a*) the Q factor and (*b*) the resonance width for the circuit in Problem 55. (*c*) What is the power factor when $\omega = 8000$ rad/s?

59 •• FM radio stations have carrier frequencies that are separated by 0.20 MHz. When the radio is tuned to a station, such as 100.1 MHz, the resonance width of the receiver circuit should be much smaller than 0.2 MHz so that adjacent stations are not received. If $f_0 = 100.1$ MHz and $\Delta f = 0.05$ MHz, what is the Q factor for the circuit?

60 •• A coil is connected to a 60-Hz, 100-V ac generator. At this frequency the coil has an impedance of 10 Ω and a reactance of 8 Ω. (*a*) What is the current in the coil? (*b*) What is the phase angle between the current and the applied voltage? (*c*) What series capacitance is required so that the current and voltage are in phase? (*d*) What then is the voltage measured across the capacitor?

61 •• A 0.25-H inductor and a capacitor *C* are connected in series with a 60-Hz ac generator. An ac voltmeter is used to measure the rms voltages across the inductor and capacitor separately. The rms voltage across the capacitor is 75 V and that across the inductor is 50 V. (*a*) Find the capacitance *C* and the rms current in the circuit. (*b*) What would be the measured rms voltage across both the capacitor and inductor together?

62 •• (*a*) Show that Equation 31-51 can be written as

$$\tan\delta = \frac{L(\omega^2 - \omega_0^2)}{\omega R}$$

Find δ approximately at (*b*) very low frequencies and (*c*) very high frequencies.

63 •• (*a*) Show that in a series *RC* circuit with no inductance, the power factor is given by

$$\cos\delta = \frac{RC\omega}{\sqrt{1 + (RC\omega)^2}}$$

(*b*) Sketch a graph of the power factor versus ω.

64 •• In the circuit in Figure 31-37, the ac generator produces an rms voltage of 115 V when operated at 60 Hz. What is the rms voltage across points (*a*) *AB*, (*b*) *BC*, (*c*) *CD*, (*d*) *AC*, and (*e*) *BD*?

Figure 31-37 Problem 64

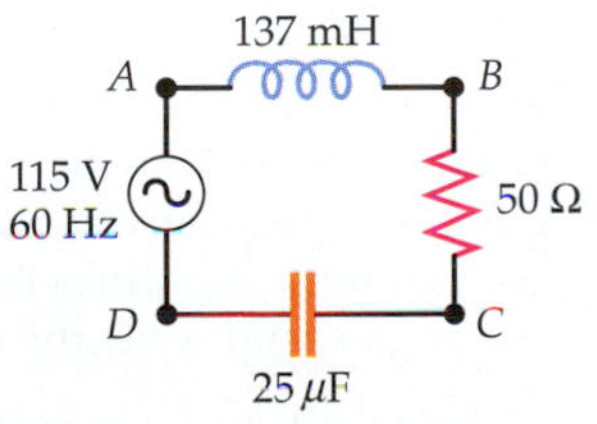

65 •• A variable-frequency ac generator is connected to a series *RLC* circuit for which $R = 1$ kΩ, $L = 50$ mH, and $C = 2.5\ \mu$F. (*a*) What is the resonance frequency of the circuit? (*b*) What is the Q value? (*c*) At what frequencies is the value of the average power delivered by the generator half of its maximum value?

66 •• An experimental physicist wishes to design a series *RLC* circuit with a Q value of 10 and a resonance frequency of 33 kHz. She has a 45-mH inductor with negligible resistance. What values for the resistance *R* and capacitance *C* should she use?

67 •• When an *RLC* series circuit is connected to a 120-V-rms, 60-Hz line, the current is $I_{rms} = 11.0$ A and the current leads the voltage by 45°. (*a*) Find the power supplied to the circuit. (*b*) What is the resistance? (*c*) If the inductance $L = 0.05$ H, find the capacitance *C*. (*d*) What capacitance or inductance should you add to make the power factor 1?

68 •• A series *RLC* circuit is driven at a frequency of 500 Hz. The phase angle between the applied voltage and current is determined from an oscilloscope measurement to be $\delta = 75°$. If the total resistance is known to be 35 Ω and the inductance is 0.15 H, what is the capacitance of the circuit?

69 •• A series *RLC* circuit with $R = 400\ \Omega$, $L = 0.35$ H, and $C = 5\ \mu$F is driven by a generator of variable frequency f. (*a*) What is the resonance frequency f_0? Find f and f/f_0 when the phase angle δ is (*b*) 60°, and (*c*) −60°.

70 •• Sketch the impedance *Z* versus ω for (*a*) a series *LR* circuit, (*b*) a series *RC* circuit, and (*c*) a series *RLC* circuit.

71 •• Given the circuit shown in Figure 31-38, (*a*) find the power loss in the inductor. (*b*) Find the resistance *r* of the inductor. (*c*) Find the inductance *L*.

Figure 31-38 Problem 71

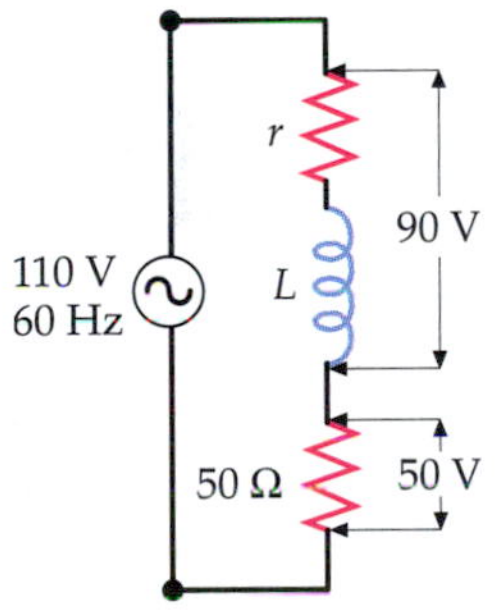

72 •• Show that Equation 31-52 can be written as

$$I_{max} = \frac{\omega \mathcal{E}_{max}}{\sqrt{L^2(\omega^2 - \omega_0^2)^2 + \omega^2 R^2}}$$

73 •• In a series *RLC* circuit, $X_C = 16\ \Omega$ and $X_L = 4\ \Omega$ at some frequency. The resonance frequency is $\omega_0 = 10^4$ rad/s. (*a*) Find *L* and *C*. If $R = 5\ \Omega$ and $\mathcal{E}_{max} = 26$ V, find (*b*) the Q factor and (*c*) the maximum current.

74 •• In a series *RLC* circuit connected to an ac generator whose maximum emf is 200 V, the resistance is 60 Ω and the capacitance is 8.0 μF. The inductance can be varied from 8.0 mH to 40.0 mH by the insertion of an iron core in the solenoid. The angular frequency of the generator is 2500 rad/s. If the capacitor voltage is not to exceed 150 V, find (*a*) the maximum current and (*b*) the range of inductance that is safe to use.

75 •• A certain electrical device draws 10 A rms and has an average power of 720 W when connected to a 120-V-rms, 60-Hz power line. (*a*) What is the impedance of the device? (*b*) What series combination of resistance and reactance is this device equivalent to? (*c*) If the current leads the emf, is the reactance inductive or capacitive?

76 •• A method for measuring inductance is to connect the inductor in series with a known capacitance, a known resistance, an ac ammeter, and a variable-frequency signal generator. The frequency of the signal generator is varied and the emf is kept constant until the current is maximum. (*a*) If $C = 10\ \mu\text{F}$, $\mathcal{E}_{\text{max}} = 10$ V, $R = 100\ \Omega$, and I is maximum at $\omega = 5000$ rad/s, what is L? (*b*) What is I_{max}?

77 •• A resistor and a capacitor are connected in parallel across a sinusoidal emf $\mathcal{E} = \mathcal{E}_{\text{max}} \cos \omega t$ as shown in Figure 31-39. (*a*) Show that the current in the resistor is $I_R = (\mathcal{E}_{\text{max}}/R) \cos \omega t$. (*b*) Show that the current in the capacitor branch is $I_C = (\mathcal{E}_{\text{max}}/X_C) \cos (\omega t + 90°)$. (*c*) Show that the total current is given by $I = I_R + I_C = I_{\text{max}} \cos (\omega t + \delta)$, where $\tan \delta = R/X_C$ and $I_{\text{max}} = \mathcal{E}_{\text{max}}/Z$ with $Z^{-2} = R^{-2} + X_C^{-2}$.

Figure 31-39 Problem 77

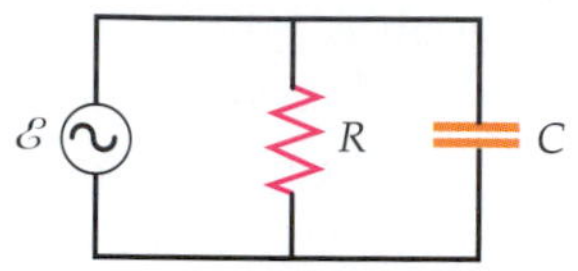

78 •• The impedances of motors, transformers, and electromagnets have inductive reactance. Suppose that the phase angle of the total impedance of a large industrial plant is 25° when the plant is under full operation and using 2.3 MW of power. The power is supplied to the plant from a substation 4.5 km from the plant; the 60 Hz rms line voltage at the plant is 40,000 V. The resistance of the transmission line from the substation to the plant is 5.2 Ω. The cost per kilowatt-hour is 0.07 dollars. The plant pays only for the actual energy used. (*a*) What are the resistance and inductive reactance of the plant's total load? (*b*) What is the current in the power lines and what must be the rms voltage at the substation to maintain the voltage at the plant at 40,000 V? (*c*) How much power is lost in transmission? (*d*) Suppose that the phase angle of the plant's impedance were reduced to 18° by adding a bank of capacitors in series with the load. How much money would be saved by the electric utility during one month of operation, assuming the plant operates at full capacity for 16 h each day? (*e*) What must be the capacitance of this bank of capacitors?

79 •• In the circuit shown in Figure 31-40, $R = 10\ \Omega$, $R_L = 30\ \Omega$, $L = 150$ mH, and $C = 8\ \mu\text{F}$; the frequency of the ac source is 10 Hz and its amplitude is 100 V. (*a*) Using phasor diagrams, determine the impedance of the circuit when switch S is closed. (*b*) Determine the impedance of the circuit when switch S is open. (*c*) What are the voltages across the load resistor R_L when switch S is closed and when it is open? (*d*) Repeat parts (*a*), (*b*), and (*c*) with the frequency of the source changed to 1000 Hz. (*e*) Which arrangement is a better low-pass filter, S open or S closed?

Figure 31-40 Problem 79

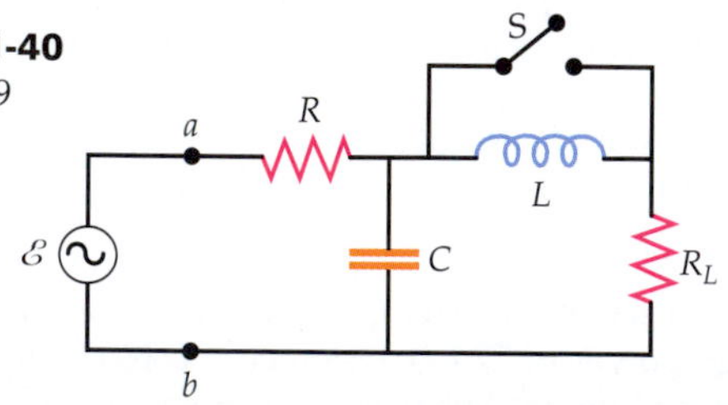

80 •• In the circuit shown in Figure 31-41, $R_1 = 2\ \Omega$, $R_2 = 4\ \Omega$, $L = 12$ mH, $C = 30\ \mu\text{F}$, and $\mathcal{E} = (40\text{ V}) \cos (\omega t)$. (*a*) Find the resonance frequency. (*b*) At the resonance frequency, what are the rms currents in each resistor and the rms current supplied by the source emf?

Figure 31-41 Problems 80, 98, and 99

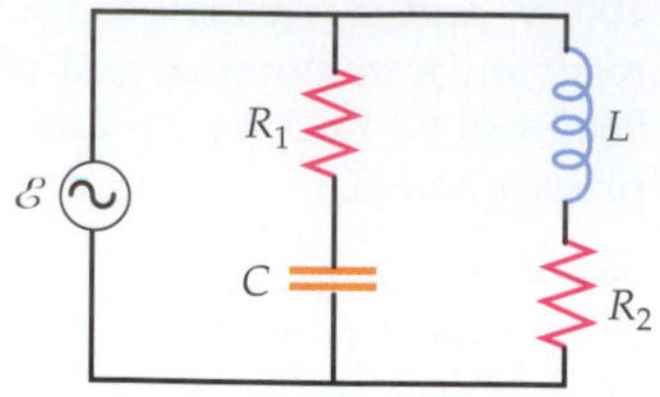

81 •• For the circuit in Figure 31-23, derive an expression for the Q of the circuit assuming the resonance is sharp.

82 •• For the circuit in Figure 31-23, $L = 4$ mH. (*a*) What capacitance C will result in a resonance frequency of 4 kHz? (*b*) When C has the value found in (*a*), what should be the resistance R so that the Q of the circuit is 8?

83 •• If the capacitance of C in Problem 82 is reduced to half the value found in Problem 82, what then are the resonance frequency and the Q of the circuit? What should be the resistance R to give $Q = 8$?

84 •• A series circuit consists of a 4.0-nF capacitor, a 36-mH inductor, and a 100-Ω resistor. The circuit is connected to a 20-V ac source whose frequency can be varied over a wide range. (*a*) Find the resonance frequency f_0 of the circuit. (*b*) At resonance, what is the rms current in the circuit and what are the rms voltages across the inductor and capacitor? (*c*) What is the rms current and what are the rms voltages across the inductor and capacitor at $f = f_0 + \frac{1}{2}\Delta f$, where Δf is the width of the resonance?

85 •• Repeat Problem 84 with the 100-Ω resistor replaced by a 40-Ω resistor.

86 ••• In the parallel circuit shown in Figure 31-42, $V_{\text{max}} = 110$ V. (*a*) What is the impedance of each branch? (*b*) For each branch, what is the current amplitude and its phase relative to the applied voltage? (*c*) Give the current phasor diagram, and use it to find the total current and its phase relative to the applied voltage.

Figure 31-42 Problem 86

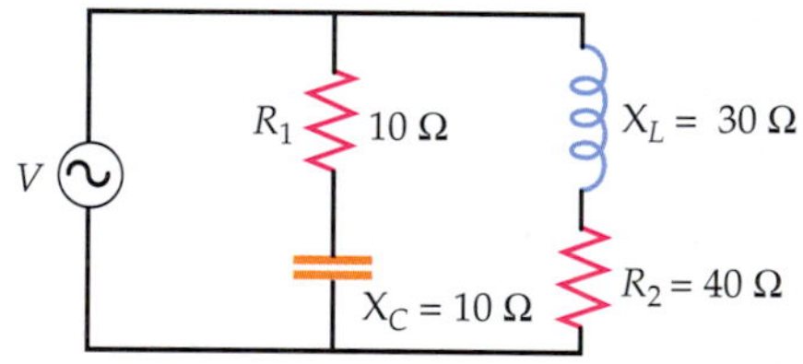

87 ••• (*a*) Show that Equation 31-51 can be written as

$$\tan \delta = \frac{Q(\omega^2 - \omega_0^2)}{\omega\omega_0}$$

(*b*) Show that near resonance

$$\tan \delta \approx \frac{2Q(\omega - \omega_0)}{\omega}$$

(*c*) Sketch a plot of δ versus x, where $x = \omega/\omega_0$, for a circuit with high Q and for one with low Q.

88 ••• Show by direct substitution that the current given by Equation 31-50 with δ and I_{max} given by Equations 31-51 and 31-52, respectively, satisfies Equation 31-49. (*Hint:* Use trigonometric identities for the sine and cosine of the sum of two angles, and write the equation in the form

$$A \sin \omega t + B \cos \omega t = 0$$

Since this equation must hold for all times, $A = 0$ and $B = 0$.)

89 ••• An ac generator is in series with a capacitor and an inductor in a circuit with negligible resistance. (*a*) Show that the charge on the capacitor obeys the equation

$$L\frac{d^2Q}{dt^2} + \frac{Q}{C} = \mathcal{E}_{max}\cos\omega t$$

(*b*) Show by direct substitution that this equation is satisfied by $Q = Q_{max}\cos wt$ if

$$Q_{max} = -\frac{\mathcal{E}_{max}}{L(\omega^2 - \omega_0^2)}$$

(*c*) Show that the current can be written as $I = I_{max}\cos(\omega t - \delta)$, where

$$I_{max} = \frac{\omega\mathcal{E}_{max}}{L|\omega^2 - \omega_0^2|} = \frac{\mathcal{E}_{max}}{|X_L - X_C|}$$

and $\delta = -90°$ for $\omega < \omega_0$ and $\delta = 90°$ for $\omega > \omega_0$.

90 ••• Figure 31-19 shows a plot of average power P_{av} versus generator frequency ω for an *RLC* circuit with a generator. The average power P_{av} is given by Equation 31-58. The "full width at half-maximum" $\Delta\omega$ is the width of the resonance curve between the two points where P_{av} is one-half its maximum value. Show that, for a sharply peaked resonance, $\Delta\omega \approx R/L$ and, hence, that $Q \approx \omega_0/\Delta\omega$ in this case (Equation 31-60). [*Hint:* At resonance, the denominator of the expression on the right of Equation 31-58 is ω^2R^2. The half-power points will occur when the denominator is twice the value near resonance, that is, when $L^2(\omega^2 - \omega_0^2)^2 = \omega^2R^2 \approx \omega_0^2R^2$. Let ω_1 and ω_2 be the solutions of this equation. For a sharply peaked resonance, $\omega_1 \approx \omega_0$ and $\omega_2 \approx \omega_0$. Then, using the fact that $\omega + \omega_0 \approx 2\omega_0$, one finds that $\Delta\omega = \omega_2 - \omega_1 \approx R/L$.]

91 ••• Show by direct substitution that Equation 31-47*b* is satisfied by $Q = Q_0e^{-Rt/2L}\cos\omega' t$ where $\omega' = \sqrt{(1/LC) - (R/2L)^2}$ and Q_0 is the charge on the capacitor at $t = 0$.

92 ••• (*a*) Compute the current $I = dQ/dt$ from the solution of Equation 31-47*b* given in Problem 91, and show that

$$I = -I_0\left(\sin\omega' t + \frac{R}{2L\omega'}\cos\omega' t\right)e^{-Rt/2L}$$

where $I_0 = \omega' Q_0$. (*b*) Show that this can be written

$$I = -\frac{I_0}{\cos\delta}(\cos\delta\sin\omega' t + \sin\delta\cos\omega' t)e^{-Rt/2L}$$

$$= -\frac{I_0}{\cos\delta}\sin(\omega' t + \delta)e^{-Rt/2L}$$

where $\tan\delta = R/2L\omega'$. When $R/2L\omega'$ is small, $\cos\delta \approx 1$, and $I \approx I_0\sin(\omega' t + \delta)e^{-Rt/2L}$.

93 ••• One method for measuring the magnetic susceptibility of a sample uses an *LC* circuit consisting of an air-core solenoid and a capacitor. The resonant frequency of the circuit without the sample is determined and then measured again with the sample inserted in the solenoid. Suppose the solenoid is 4.0 cm long, 0.3 cm in diameter, and has 400 turns of fine wire. Assume that the sample that is inserted in the solenoid is also 4.0 cm long and fills the air space. Neglect end effects. (In practice, a test sample of known susceptibility of the same shape as the unknown is used to calibrate the instrument.) (*a*) What is the inductance of the empty solenoid? (*b*) What should be the capacitance of the capacitor so that the resonance frequency of the circuit without a sample is 6.0000 MHz? (*c*) When a sample is inserted in the solenoid, the resonance frequency drops to 5.9989 MHz. Determine the sample's susceptibility.

94 ••• A concentric cable of cylindrical cross section has an inner conductor of 0.4 cm diameter and an outer conductor of 2.0 cm diameter. Air fills the space between the conductors. (*a*) Find the resonance frequency of a one-meter length of this conductor. (*b*) What length of conductor will result in a resonance frequency of 18 GHz?

95 ••• Repeat Problem 94 if the inner and outer conductors of the cable are separated by a dielectric of dielectric constant $\kappa = 5.8$.

96 ••• At what frequency will the voltage across the load resistor of Problem 37 be half the source voltage?

97 ••• At what frequency will the voltage across the load resistor of Problem 42 be half the source voltage?

98 ••• (*a*) Find the angular frequency ω for the circuit in Problem 80 such that the magnitude of the reactance of the two parallel branches are equal. (*b*) At that frequency, what is the power dissipation in each of the two resistors?

99 ••• (*a*) For the circuit of Problem 80, find the angular frequency ω for which the power dissipation in the two resistors is the same. (*b*) At that angular frequency, what is the reactance of each of the two parallel branches? (*c*) Draw a phasor diagram showing the current through each of the two parallel branches. (*d*) What is the impedance of the circuit?

The Transformer (optional)

100 • A transformer is used to change

(*a*) capacitance. (*b*) frequency. (*c*) voltage.
(*d*) power. (*e*) none of these.

101 • True or false: If a transformer increases the current, it must decrease the voltage.

102 •• An ideal transformer has N_1 turns on its primary and N_2 turns on its secondary. The power dissipated in a load resistance R connected across the secondary is P_2 when the primary voltage is V_1. The current in the primary windings is then

(*a*) P_2/V_1. (*b*) $(N_1/N_2)(P_2/V_1)$.
(*c*) $(N_2/N_1)(P_2/V_1)$. (*d*) $(N_2/N_1)^2(P_2/V_1)$.

103 • An ac voltage of 24 V is required for a device whose impedance is 12 Ω. (*a*) What should the turn ratio of a transformer be so the device can be operated from a 120-V line? (*b*) Suppose the transformer is accidentally connected reversed, i.e., with the secondary winding across the 120-V line and the 12-Ω load across the primary. How much current will then flow in the primary winding?

104 • A transformer has 400 turns in the primary and 8 turns in the secondary. (*a*) Is this a step-up or step-down transformer? (*b*) If the primary is connected across 120 V rms, what is the open-circuit voltage across the secondary? (*c*) If the primary current is 0.1 A, what is the secondary current, assuming negligible magnetization current and no power loss?

105 • The primary of a step-down transformer has 250 turns and is connected to a 120-V-rms line. The secondary is to supply 20 A at 9 V. Find (*a*) the current in the primary and (*b*) the number of turns in the secondary, assuming 100% efficiency.

106 • A transformer has 500 turns in its primary, which is connected to 120 V rms. Its secondary coil is tapped at three places to give outputs of 2.5, 7.5, and 9 V. How many turns are needed for each part of the secondary coil?

107 • The distribution circuit of a residential power line is operated at 2000 V rms. This voltage must be reduced to 240 V rms for use within the residences. If the secondary side of the transformer has 400 turns, how many turns are in the primary?

108 •• An audio oscillator (ac source) with an internal resistance of 2000 Ω and an open-circuit rms output voltage of 12 V is to be used to drive a loudspeaker with a resistance of 8 Ω. What should be the ratio of primary to secondary turns of a transformer so that maximum power is transferred to the speaker? Suppose a second identical speaker is connected in parallel with the first speaker. How much power is then supplied to the two speakers combined?

109 •• One use of a transformer is for *impedance matching.* For example, the output impedance of a stereo amplifier is matched to the impedance of a speaker by a transformer. In Equation 31-67, the currents I_1 and I_2 can be related to the impedance Z in the secondary since $I_2 = V_2/Z$. Using Equations 31-65 and 31-66, show that

$$I_1 = \frac{\mathcal{E}}{(N_1/N_2)^2 Z}$$

and, therefore, $Z_{\text{eff}} = (N_1/N_2)^2 Z$.

General Problems

110 • True or false:

(*a*) Alternating current in a resistance dissipates no power because the current is negative as often as it is positive.
(*b*) At very high frequencies, a capacitor acts like a short circuit.

111 • A 5.0-kW electric clothes dryer runs on 240 V rms. Find (*a*) I_{rms} and (*b*) I_{max}. (*c*) Find the same quantities for a dryer of the same power that operates at 120 V rms.

112 • Find the reactance of a 10.0-μF capacitor at (*a*) 60 Hz, (*b*) 6 kHz, and (*c*) 6 MHz.

113 • Sketch a graph of X_L versus f for L = 3 mH.

114 • Sketch a graph of X_C versus f for C = 100 μF.

115 •• A resistance R carries a current I = (5.0 A) sin $120\pi t$ + (7.0 A) sin $240\pi t$. (*a*) What is the rms current? (*b*) If the resistance R is 12 Ω, what is the power dissipated in the resistor? (*c*) What is the rms voltage across the resistor?

116 •• Figure 31-43 shows the voltage V versus time t for a "square-wave" voltage. If V_0 = 12 V, (*a*) what is the rms voltage of this waveform? (*b*) If this alternating waveform is rectified by eliminating the negative voltages so that only the positive voltages remain, what now is the rms voltage of the rectified waveform?

Figure 31-43 Problem 116

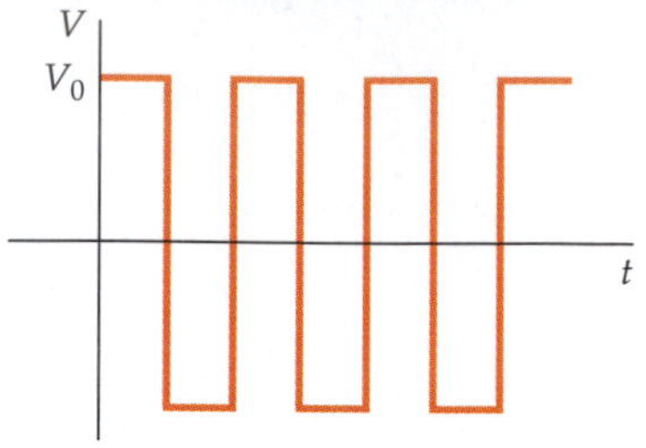

117 •• A pulsed current has a constant value of 15 A for the first 0.1 s of each second and is then 0 for the next 0.9 s of each second. (*a*) What is the rms value for this current waveform? (*b*) Each current pulse is generated by a voltage pulse of maximum value 100 V. What is the average power delivered by the pulse generator?

118 •• A circuit consists of two capacitors, a 24-V battery, and an ac voltage connected as shown in Figure 31-44. The ac voltage is given by $\mathcal{E}$ = (20 V) cos ($120\pi t$) where t is in seconds. (*a*) Find the charge on each capacitor as a function of time. Assume that transient effects have had sufficient time to decay. (*b*) What is the steady-state current? (*c*) What is the maximum energy stored in the capacitors? (*d*) What is the minimum energy stored in the capacitors?

Figure 31-44
Problem 118

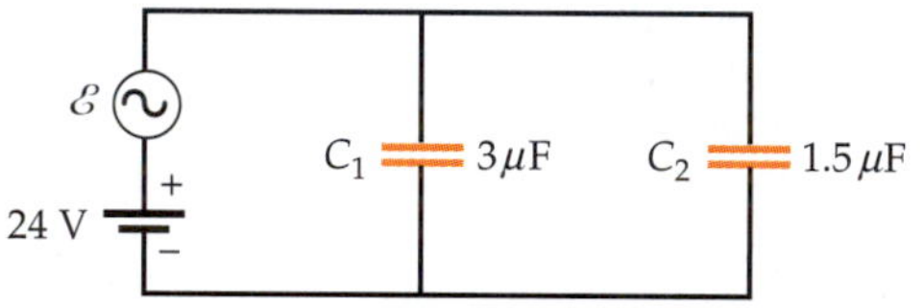

119 •• What are the average and rms values of current for the two current waveforms shown in Figure 31-45?

Figure 31-45
Problem 119

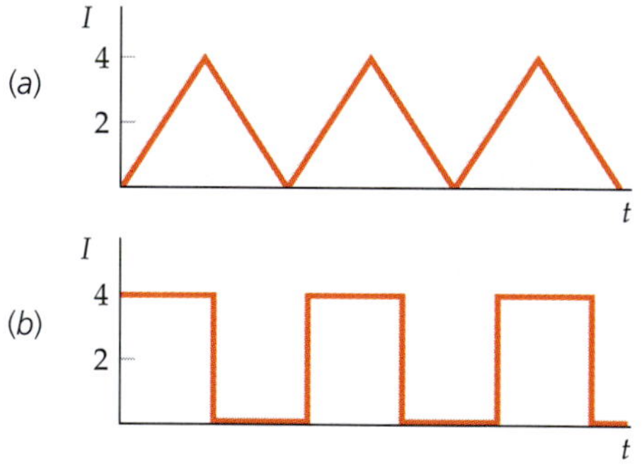

120 •• In the circuit shown in Figure 31-46, $\mathcal{E}_1$ = (20 V) cos ($2\pi ft$), f = 180 Hz, $\mathcal{E}_2$ = 18 V, and R = 36 Ω. Find the maximum, minimum, average, and rms values of the current through the resistor.

Figure 31-46
Problems 120–122

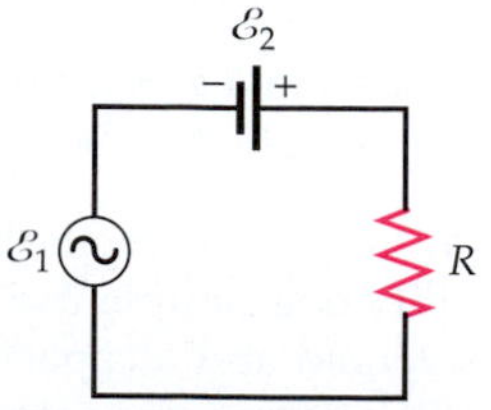

121 •• Repeat Problem 120 if the resistor R is replaced by a 2-μF capacitor.

122 •• Repeat Problem 120 if the resistor R is replaced by a 12-mH inductor.

CHAPTER 32

Maxwell's Equations and Electromagnetic Waves

A multiple-exposure view showing the 26-m tracking antenna at Wallops Station, Virginia, and a total solar eclipse. Electromagnetic radiation at radio wavelengths, like that at optical wavelengths, is not readily absorbed by the earth's atmosphere, making it a viable means of communication between distant points on the ground or between ground and space.

Maxwell's equations, first proposed by the great Scottish physicist James Clerk Maxwell, relate the electric and magnetic field vectors $\vec{E}$ and $\vec{B}$ to their sources, which are electric charges, currents, and changing fields. These equations summarize the experimental laws of electricity and magnetism—the laws of Coulomb, Gauss, Biot–Savart, Ampère, and Faraday. These experimental laws hold in general except for Ampère's law, which does not apply to discontinuous currents such as those that occur when charging or discharging a capacitor. Maxwell was able to generalize Ampère's law with the invention of the displacement current (Section 32-1). He was then able to show that the generalized laws of electricity and magnetism imply the existence of electromagnetic waves.

Maxwell's equations play a role in classical electromagnetism analogous to that of Newton's laws in classical mechanics. In principle, all problems in classical electricity and magnetism can be solved using Maxwell's equations, just as all problems in classical mechanics can be solved using Newton's laws. Maxwell's equations are considerably more complicated than Newton's laws, however, and their application to most problems involves mathematics beyond the scope of this book. Nevertheless, Maxwell's equations are of great theoretical importance. For example, Maxwell showed that these equations can be combined to yield a wave equation for the electric and magnetic field vectors $\vec{E}$ and $\vec{B}$. Such **electromagnetic waves** are caused by accelerating charges, for example, the charges in an alternating current in an antenna. They were first produced in the laboratory by Heinrich Hertz in 1887. Maxwell showed that the speed of electromagnetic waves in free space should be

$$c = \frac{1}{\sqrt{\mu_0 \epsilon_0}} \tag{32-1}$$

where ϵ_0, the permittivity of free space, is the constant appearing in Coulomb's and Gauss's laws and μ_0, the permeability of free space, is the constant appearing in the Biot–Savart law and Ampère's law. When the measured value of ϵ_0 and the defined value of μ_0 are put into Equation 32-1, the speed of electromagnetic waves is found to be about 3×10^8 m/s, the same as the measured speed of light. Maxwell noted this "coincidence" with great excitement and correctly surmised that light itself is an electromagnetic wave.

32-1 Maxwell's Displacement Current

Ampère's law (Equation 29-15) relates the line integral of the magnetic field around some closed curve C to the current that passes through any area bounded by that curve:

$$\oint_C \vec{B} \cdot d\vec{\ell} = \mu_0 I, \qquad \text{for any closed curve } C \tag{32-2}$$

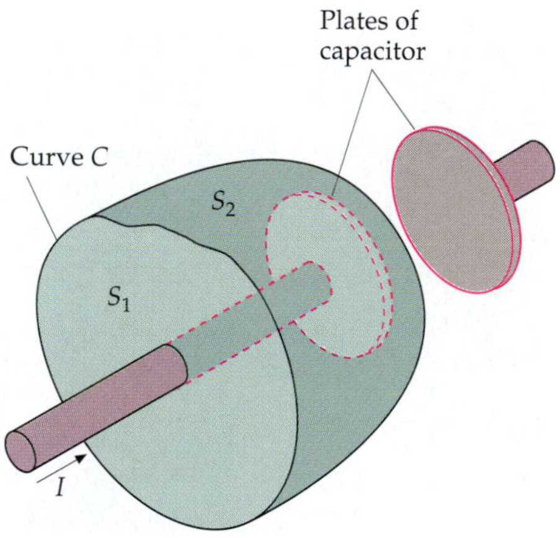

Figure 32-1 Two surfaces S_1 and S_2 bounded by the same curve C. The current I passes through surface S_1 but not S_2. Ampère's law, which relates the line integral of the magnetic field $\vec{B}$ around the curve C to the total current passing through any surface bounded by C, is not valid when the current is not continuous, as when it stops at the capacitor plate here.

Maxwell recognized a flaw in Ampère's law. Figure 32-1 shows two different surfaces bounded by the same curve C, which encircles a wire carrying current into a capacitor plate. The current through surface S_1 is I, but there is no current through surface S_2 because the charge stops on the capacitor plate. There is thus ambiguity in the phrase "the current through any surface bounded by the curve." Such a problem always arises when the current is discontinuous.

Maxwell showed that the law can be generalized to include all situations if the current I in the equation is replaced by the sum of the conduction current I and another term I_d, called **Maxwell's displacement current,** defined as

$$I_d = \epsilon_0 \frac{d\phi_e}{dt} \tag{32-3}$$

Definition—Displacement current

where ϕ_e is the flux of the electric field through the same surface bounded by the curve C. The generalized form of Ampère's law is then

$$\oint_C \vec{B} \cdot d\vec{\ell} = \mu_0(I + I_d) = \mu_0 I + \mu_0 \epsilon_0 \frac{d\phi_e}{dt} \tag{32-4}$$

Generalized form of Ampère's law

We can understand this generalization by considering Figure 32-1 again. Let us call the sum $I + I_d$ the generalized current. According to the argument just stated, the same generalized current must cross any area bounded by the curve C. Thus, there can be no net generalized current into or out of the closed volume. If there is a net true current I into the volume, there must be an equal net displacement current I_d out of the volume. In the volume in the figure, there is a net conduction current I into the volume that increases the charge within the volume:

$$I = \frac{dQ}{dt}$$

The flux of the electric field out of the volume is related to the charge by Gauss's law:

$$\phi_{net} = \oint_S E_n \, dA = \frac{1}{\epsilon_0} Q_{inside}$$

The rate of increase of the charge is thus proportional to the rate of increase of the net flux out of the volume:

$$\frac{dQ}{dt} = \epsilon_0 \frac{d\phi_{e,net}}{dt} = I_d$$

Thus, the net conduction current into the volume equals the net displacement current out of the volume. The generalized current is always continuous.

It is interesting to compare Equation 32-4 to Faraday's law (Equation 30-5):

$$\mathcal{E} = \oint_C \vec{E} \cdot d\vec{\ell} = -\frac{d\phi_m}{dt} \qquad \text{32-5}$$

According to Faraday's law, a changing magnetic flux produces an electric field whose line integral around a closed curve is proportional to the rate of change of magnetic flux through the curve. Maxwell's modification of Ampère's law shows that a changing electric flux produces a magnetic field whose line integral around a curve is proportional to the rate of change of the electric flux. We thus have the interesting reciprocal result that a changing magnetic field produces an electric field (Faraday's law) and a changing electric field produces a magnetic field (generalized form of Ampère's law). Note that there is no magnetic analog of a conduction current I. This is because the magnetic monopole, the magnetic analog of an electric charge, does not exist.

Example 32-1

A parallel-plate capacitor has closely spaced circular plates of radius R. Charge is flowing onto the positive plate and off of the negative plate at the rate $I = dQ/dt = 2.5$ A. Compute the displacement current between the plates.

Picture the Problem The displacement current is $I_d = \epsilon_0 \, d\phi_{e/dt}$, where ϕ_e is the electric flux between the plates. Since the parallel plates are closely spaced, we can consider the electric field to be uniform and perpendicular to the plates within the capacitor, and zero outside the capacitor. Thus, the electric flux is simply $\phi_e = EA$, where E is the electric field between the plates and A is the area of the plates.

1. The displacement current is found by taking the time derivative of the electric flux: $\quad I_d = \epsilon_0 \dfrac{d\phi_e}{dt}$

2. The flux equals the electric field times the area: $\quad \phi_e = EA$

3. The electric field is proportional to the charge density on the plates, which we treat as infinite sheets:

$$E = \frac{\sigma}{\epsilon_0} = \frac{Q/A}{\epsilon_0}$$

4. Substitute these results to calculate I_d:

$$I_d = \epsilon_0 \frac{d(EA)}{dt} = \epsilon_0 A \frac{d}{dt}\left(\frac{Q}{A\epsilon_0}\right)$$

$$= \frac{dQ}{dt} = 2.5 \text{ A}$$

Remark Note that the displacement current across the gap of the capacitor is equal to the conduction current in the wires leading in and out of the capacitor.

Example 32-2

The circular plates in Example 32-1 have a radius of $R = 3.0$ cm. Find the magnetic field at a point between the plates at a distance $r = 2.0$ cm from the axis of the plates when the current into the positive plate is 2.5 A.

Picture the Problem We find B from the generalized form of Ampère's law (Equation 32-4). We chose a circular path of radius $r = 2.0$ cm about the center line joining the plates as shown in Figure 32-2 to compute $\oint \vec{B}\cdot d\vec{\ell}$. By symmetry, $\vec{B}$ is tangent to this circle and has the same magnitude everywhere on it.

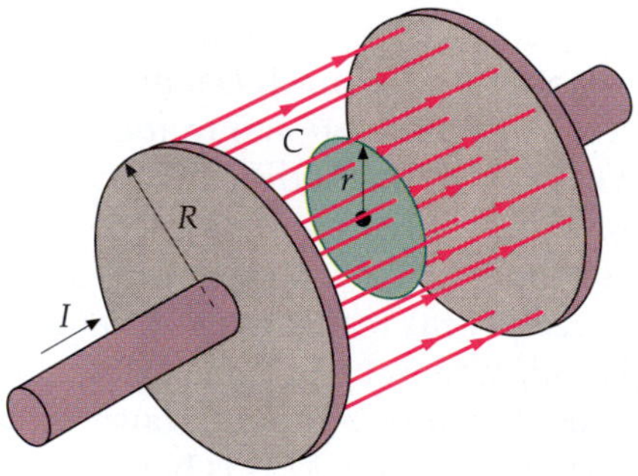

Figure 32-2

1. We find B from the generalized form of Ampère's law:

$$\oint \vec{B}\cdot d\vec{\ell} = \mu_0 I + \mu_0\epsilon_0 \frac{d\phi_e}{dt}$$

2. By symmetry the line integral is merely B times the circumference of the circle:

$$\oint \vec{B}\cdot d\vec{\ell} = B(2\pi r)$$

3. Since there is no conduction current between the plates of the capacitor, the generalized current is just the displacement current:

$$\oint \vec{B}\cdot d\vec{\ell} = B(2\pi r) = \mu_0\epsilon_0 \frac{d\phi_e}{dt}$$

4. The electric flux equals the product of the constant field E and the area bounded by the curve:

$$\phi_e = \pi r^2 E = \pi r^2 \frac{\sigma}{\epsilon_0} = \pi r^2 \frac{Q}{\epsilon_0 \pi R^2} = \frac{r^2 Q}{\epsilon_0 R^2}$$

5. Substitute these results into step 1 and solve for B:

$$B(2\pi r) = \mu_0\epsilon_0 \frac{d\phi_e}{dt} = \mu_0\epsilon_0 \frac{d}{dt}\left(\frac{r^2 Q}{\epsilon_0 R^2}\right) = \mu_0 \frac{r^2}{R^2}\frac{dQ}{dt}$$

$$B = \frac{\mu_0}{2\pi}\frac{r}{R^2}\frac{dQ}{dt} = \frac{\mu_0}{2\pi}\frac{r}{R^2} I$$

$$= (2 \times 10^{-7} \text{ T}\cdot\text{m/A})\left(\frac{0.02 \text{ m}}{(0.03 \text{ m})^2}\right)(2.5 \text{ A}) = 1.11 \times 10^{-5} \text{ T}$$

32-2 Maxwell's Equations

Maxwell's equations are

$$\oint_S E_n \, dA = \frac{1}{\epsilon_0} Q_{\text{inside}} \qquad \text{32-6}a$$

$$\oint_S B_n \, dA = 0 \qquad \text{32-6}b$$

$$\oint_C \vec{E} \cdot d\vec{\ell} = -\frac{d}{dt}\int_S B_n \, dA \qquad \text{32-6}c$$

$$\oint_C \vec{B} \cdot d\vec{\ell} = \mu_0 I + \mu_0\epsilon_0 \frac{d}{dt}\int_S E_n \, dA \qquad \text{32-6}d$$

Maxwell's equations

Equation 32-6*a* is Gauss's law; it states that the flux of the electric field through any closed surface equals $1/\epsilon_0$ times the net charge inside the surface. As discussed in Chapter 23, Gauss's law implies that the electric field due to a point charge varies inversely as the square of the distance from the charge. This law describes how electric field lines diverge from a positive charge and converge on a negative charge. Its experimental basis is Coulomb's law.

Equation 32-6*b*, sometimes called Gauss's law for magnetism, states that the flux of the magnetic field vector $\vec{B}$ is zero through any closed surface. This equation describes the experimental observation that magnetic field lines do not diverge from any point in space or converge on any point; that is, it implies that isolated magnetic poles do not exist.

Equation 32-6*c* is Faraday's law; it states that the integral of the electric field around any closed curve *C*, which is the emf, equals the (negative) rate of change of the magnetic flux through any surface *S* bounded by the curve. (*S* is not a closed surface, so the magnetic flux through *S* is not necessarily zero.) Faraday's law describes how electric field lines encircle any area through which the magnetic flux is changing, and it relates the electric field vector $\vec{E}$ to the rate of change of the magnetic field vector $\vec{B}$.

Equation 32-6*d*, Ampère's law modified to include Maxwell's displacement current, states that the line integral of the magnetic field $\vec{B}$ around any closed curve *C* equals μ_0 times the current through any surface bounded by the curve plus $\mu_0\epsilon_0$ times the rate of change of the electric flux through the surface. This law describes how the magnetic field lines encircle an area through which a current is passing or through which the electric flux is changing.

In Section 32-4 we show how wave equations for both the electric field $\vec{E}$ and the magnetic field $\vec{B}$ can be derived from Maxwell's equations.

32-3 Electromagnetic Waves

Figure 32-3 shows the electric and magnetic field vectors of an electromagnetic wave. The electric and magnetic fields are perpendicular to each other and perpendicular to the direction of propagation of the wave. Electromagnetic waves are thus transverse waves. The magnitudes of $\vec{E}$ and $\vec{B}$ are in phase and are related by

$$E = cB \qquad \text{32-7}$$

where $c = 1/\sqrt{\mu_0\epsilon_0}$ is the speed of

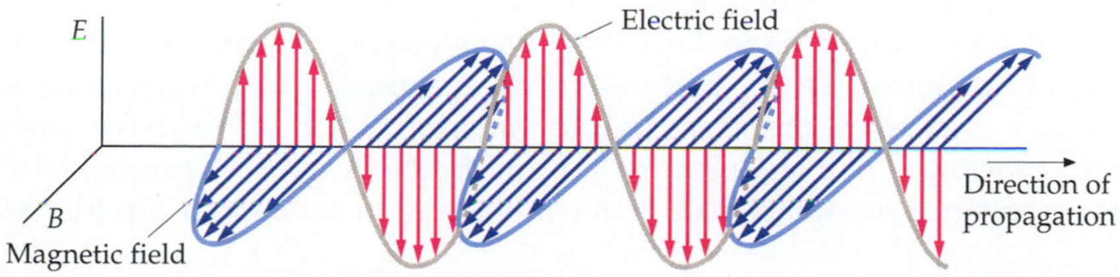

Figure 32-3 The electric and magnetic field vectors in an electromagnetic wave. The fields are in phase, perpendicular to each other, and perpendicular to the direction of propagation of the wave.

the wave. In general, the direction of propagation of an electromagnetic wave is the direction of the cross product $\vec{E} \times \vec{B}$.

The Electromagnetic Spectrum

The various types of electromagnetic waves—light, radio waves, X rays, gamma rays, microwaves, and others—differ only in wavelength and frequency, which are related to the speed c in the usual way, $f = c/\lambda$. Table 32-1 gives the **electromagnetic spectrum** and the names usually associated with the various frequency and wavelength ranges. These ranges are often not well defined and sometimes overlap. For example, electromagnetic waves with wavelengths of about 0.1 nm are usually called X rays, but if they originate from nuclear radioactivity, they are called gamma rays.

Table 32-1

The Electromagnetic Spectrum

Frequency, Hz | Wavelength, m

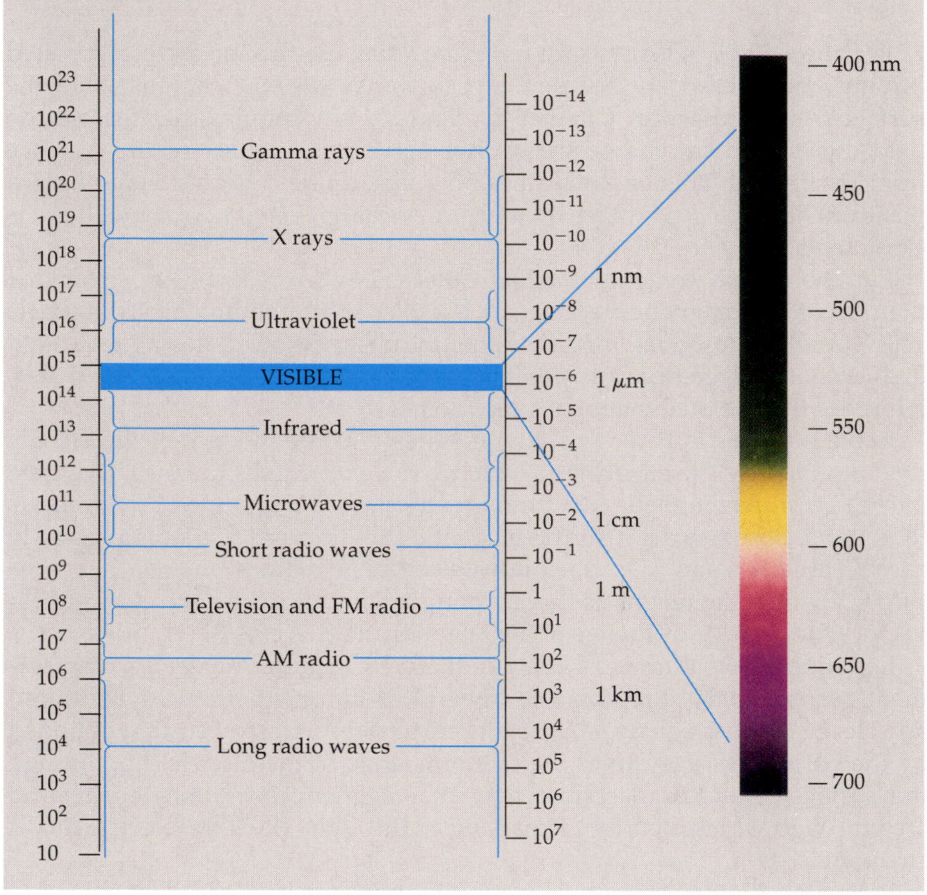

The human eye is sensitive to electromagnetic radiation with wavelengths from about 400 to 700 nm, the range called **visible light**. The shortest wavelengths in the visible spectrum correspond to violet light and the longest to red light, with all the colors of the rainbow falling between these extremes. Electromagnetic waves with wavelengths just beyond the visible spectrum on the short-wavelength side are called **ultraviolet rays**, and those with wavelengths just beyond the visible spectrum on the long-wavelength side are called **infrared waves**. Heat radiation given off by bodies at ordinary temperatures is in the infrared region of the electromagnetic spectrum. There are no limits on the wavelengths of electromagnetic radiation; that is, all wavelengths (or frequencies) are theoretically possible.

The differences in wavelengths of the various kinds of electromagnetic waves have important physical consequences. As we know, the behavior of waves depends strongly on the relative sizes of the wavelengths and the physical objects or apertures the waves encounter. Since the wavelengths of light are in the rather narrow range from about 400 to 700 nm, they are much smaller than most obstacles, so the ray approximation (introduced in Section 15-4) is often valid. The wavelength and frequency are also important in determining the kinds of interactions between electromagnetic waves and matter. X rays, for example, have very short wavelengths and high frequencies. They easily penetrate many materials that are opaque to lower-frequency light waves, which are absorbed by the materials. Microwaves have wavelengths of the order of a few centimeters and frequencies that are close to the natural resonance frequencies of water molecules in solids and liquids. Microwaves are therefore readily absorbed by the water molecules in foods, which is the mechanism by which food is heated in microwave ovens.

Production of Electromagnetic Waves

Electromagnetic waves are produced when free electric charges accelerate, or when electrons bound to atoms and molecules make transitions to lower energy states. Radio waves, which have frequencies from about 550 to 1600 kHz for AM and from about 88 to 108 MHz for FM, are produced by macroscopic electric currents oscillating in radio antennas. The frequency of the emitted waves equals the frequency of oscillation of the charges.

A continuous spectrum of X rays is produced by the deceleration of electrons when they crash into a metal target. The radiation produced is called **bremsstrahlung** (German for braking radiation). Accompanying the broad, continuous bremsstrahlung spectrum is a discrete spectrum of X-ray lines produced by transitions of inner electrons in the atoms of the target material.

Synchrotron radiation arises from the circular orbital motion of charged particles (usually electrons or positrons) in nuclear accelerators called synchrotrons. Originally considered a nuisance by accelerator scientists, synchrotron radiation X rays are now produced and used as a medical diagnostic tool because of the ease of manipulating the beams with reflection and diffraction optics. Synchrotron radiation is also emitted by charged particles trapped in magnetic fields associated with stars and galaxies. It is believed that most low-frequency radio waves reaching the earth from outer space originate as synchrotron radiation.

Heat is radiated by the thermally excited molecular charges. The spectrum of heat radiation is the blackbody radiation spectrum discussed in Section 21-4.

Light waves, which have frequencies of the order of 10^{14} Hz, are generally produced by transitions of bound atomic charges. We discuss sources of light waves in Chapter 33.

Electric-Dipole Radiation

Figure 32-4 is a schematic drawing of an electric-dipole radio antenna consisting of two conducting rods along a line fed by an alternating-current generator. At time $t = 0$ (Figure 32-4*a*), the ends of the rods are charged, and there is an electric field near the rod parallel to the rod. There is also a magnetic field, not shown, encircling the rods due to the current in the rods. These fields move out away from the rods with the speed of light. After one-fourth period, at $t = T/4$ (Figure 32-4*b*), the rods are uncharged, and the electric field near the rod is zero. At $t = T/2$ (Figure 32-4*c*), the rods are again charged, but the charges are opposite those at $t = 0$. The electric and magnetic fields at a great distance from the antenna are quite different from the fields near the antenna. Far from the antenna, the electric and magnetic fields oscillate in phase with simple harmonic

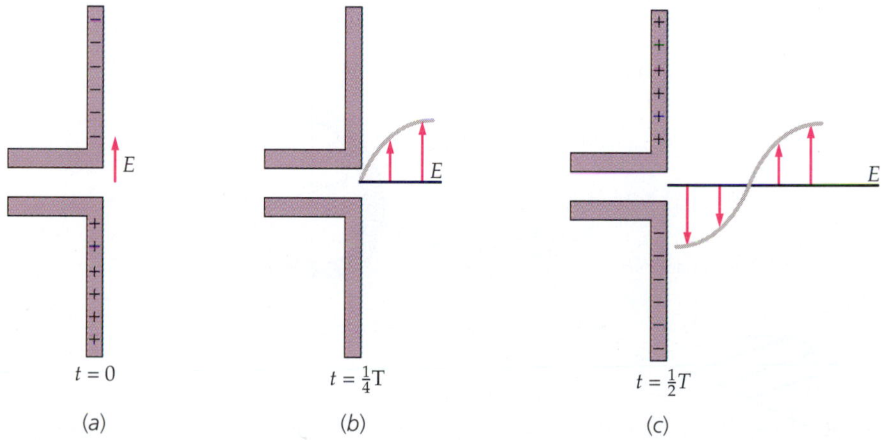

Figure 32-4 An electric-dipole antenna for radiating electromagnetic waves. Alternating current is supplied to the antenna by a generator (not shown). The electric field due to the charges in the antenna propagates outward at the speed of light. There is also a propagating magnetic field (not shown) perpendicular to the paper due to the current in the antenna.

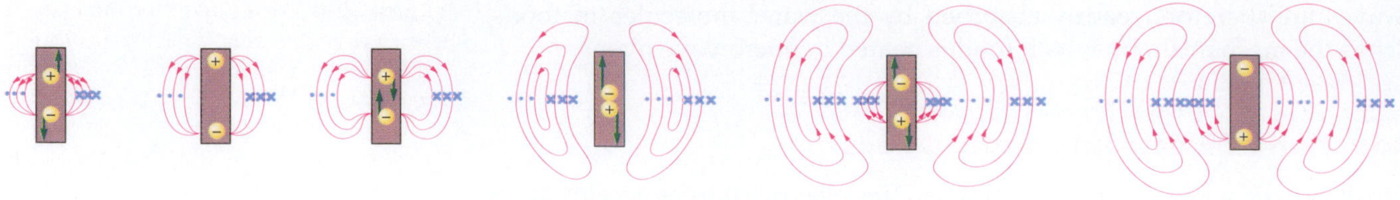

motion, perpendicular to each other and to the direction of propagation of the wave. Figure 32-5 shows the electric and magnetic fields far from an electric-dipole antenna.

Electromagnetic waves of radio or television frequencies can be detected by a dipole antenna placed parallel to the electric field so that it induces an alternating current in the antenna (Figure 32-6). They can also be detected by a loop antenna placed perpendicular to the magnetic field so that the changing magnetic flux through the loop induces a current in the loop (Figure 32-7). Electromagnetic waves of frequency in the visible light range are detected by the eye or by photographic film, both of which are mainly sensitive to the electric field.

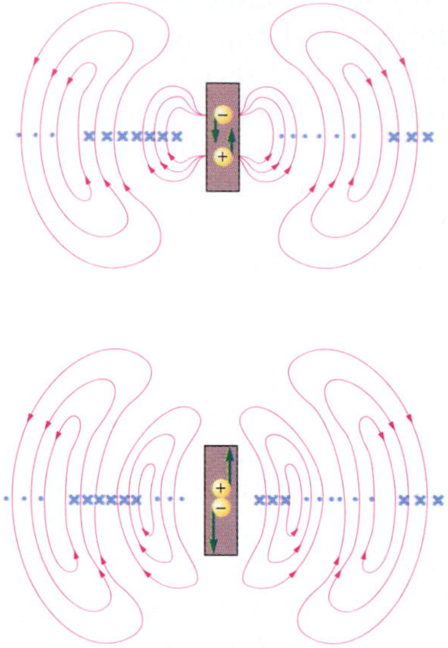

Figure 32-5 Electric and magnetic field lines produced by an oscillating electric dipole.

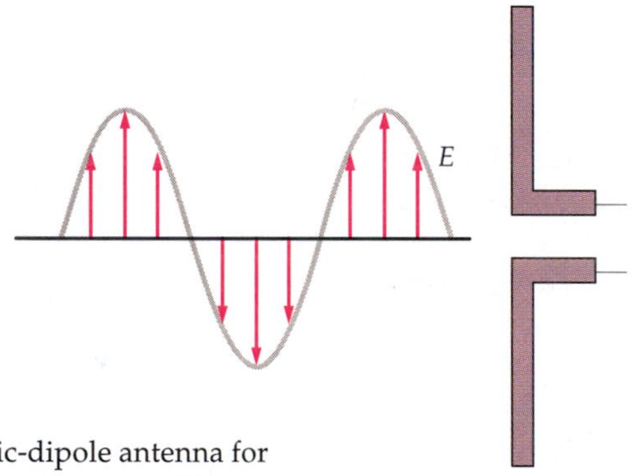

Figure 32-6 Electric-dipole antenna for detecting electromagnetic waves. The alternating electric field of the wave produces an alternating current in the antenna.

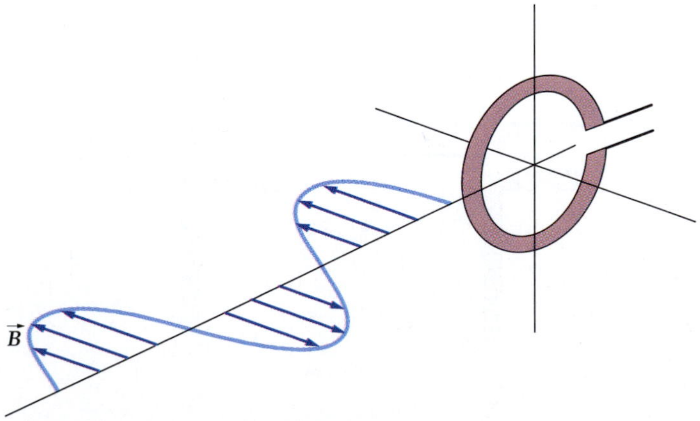

Figure 32-7 A loop antenna for detecting electromagnetic radiation. The alternating magnetic flux in the loop due to the magnetic field of the radiation induces an alternating current in the loop.

The radiation from a dipole antenna such as that in Figure 32-4 is called electric-dipole radiation. Many electromagnetic waves exhibit the characteristics of electric-dipole radiation. An important feature of this type of radiation is that the intensity of the electromagnetic waves radiated by a dipole antenna is zero along the axis of the antenna and maximum in the directions perpendicular to the axis. If the dipole is in the y direction with its center at the origin as in Figure 32-8, the intensity is zero along the y axis and maximum in the xz plane. In the direction of a line making an angle θ with the y axis, the intensity is proportional to $\sin^2 \theta$.

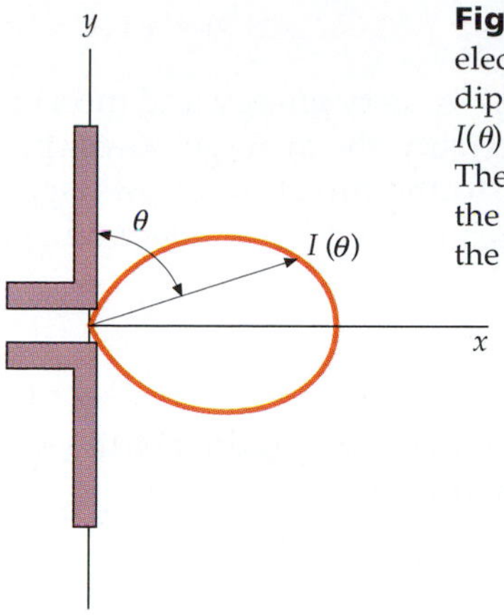

Figure 32-8 Polar plot of the intensity of electromagnetic radiation from an electric-dipole antenna versus angle. The intensity $I(\theta)$ is proportional to the length of the arrow. The intensity is maximum perpendicular to the antenna at $\theta = 90°$ and minimum along the antenna at $\theta = 0°$ or $\theta = 180°$.

Example 32-3

An antenna consisting of a single loop of wire of radius 10 cm is used to detect electromagnetic waves for which $E_{rms} = 0.15$ V/m. Find the rms emf induced in the loop if the wave frequency is (*a*) 600 kHz and (*b*) 600 MHz.

Picture the Problem The induced emf in the coil is related to the rate of change of the magnetic flux by Faraday's law.

(*a*)1. Faraday's law relates the magnitude of the emf to the rate of change of the magnetic flux:

$$|\mathcal{E}| = \frac{d\phi_m}{dt} = \pi r^2 \frac{dB}{dt}$$

$$\mathcal{E}_{rms} = \pi r^2 \left(\frac{dB}{dt}\right)_{rms}$$

2. Compute dB_{rms}/dt from a sinusoidal B:

$$B = B_0 \sin (kx - \omega t)$$

$$\frac{dB}{dt} = -\omega B_0 \cos (kx - \omega t)$$

$$\left(\frac{dB}{dt}\right)_{rms} = \omega B_{rms}$$

3. Relate B_{rms} to E_{rms}:

$$B_{rms} = \frac{E_{rms}}{c}$$

$$\left(\frac{dB}{dt}\right)_{rms} = \frac{\omega E_{rms}}{c} = \frac{2\pi f}{c} E_{rms}$$

4. Calculate $\mathcal{E}_{rms}$ at $f = 600$ Hz:

$$\mathcal{E}_{rms} = \pi r^2 \left(\frac{dB}{dt}\right)_{rms} = \pi r^2 \frac{2\pi f}{c} E_{rms}$$

$$= \pi(0.1\text{ m})^2\, 2\pi(6 \times 10^5\text{ Hz})(0.15\text{ V/m})/(3 \times 10^8\text{ m/s})$$

$$= 5.92 \times 10^{-5}\text{ V}$$

(*b*) The induced emf is proportional to the frequency, so at 600 MHz it will be 1000 times greater than at 600 kHz:

$$\mathcal{E}_{rms} = (10^3)(5.92 \times 10^{-5}\text{ V}) = 0.0592\text{ V}$$

Energy and Momentum in an Electromagnetic Wave

Like all waves, electromagnetic waves carry energy and momentum. The energy carried is described by the intensity, the average power per unit area incident on a surface perpendicular to the direction of propagation. The momentum per unit time per unit area carried by an electromagnetic wave is called the **radiation pressure**.

Intensity The intensity of a wave equals the product of the wave speed and the average energy density u_{av} (Section 15-3).

$$I = u_{av}\, c$$

The total energy density in the wave u is the sum of the electric and magnetic energy densities. The electric energy density u_e (Equation 25-13) and magnetic energy density u_m (Equation 30-19) are given by

$$u_e = \frac{1}{2}\epsilon_0 E^2 \quad \text{and} \quad u_m = \frac{B^2}{2\mu_0}$$

In an electromagnetic wave in free space, E equals cB, so we can express the magnetic energy density in terms of the electric field:

$$u_m = \frac{B^2}{2\mu_0} = \frac{(E/c)^2}{2\mu_0} = \frac{E^2}{2\mu_0 c^2} = \frac{1}{2}\epsilon_0 E^2$$

where we have used $c^2 = 1/\epsilon_0\mu_0$. Thus, the electric and magnetic energy densities are equal. Using $E = cB$, we may express the total energy density in several useful ways:

$$u = u_e + u_m = \epsilon_0 E^2 = \frac{B^2}{\mu_0} = \frac{EB}{\mu_0 c} \qquad 32\text{-}8$$

Energy density in an electromagnetic wave

To compute the *average* energy density, we replace the instantaneous fields E and B by their rms values $E_{rms} = E_0/\sqrt{2}$ and $B_{rms} = B_0/\sqrt{2}$, where E_0 and B_0 are the maximum values of the fields. The intensity is then

$$I = u_{av}c = \frac{E_{rms}B_{rms}}{\mu_0} = \frac{1}{2}\frac{E_0 B_0}{\mu_0} = |\vec{S}|_{av} \qquad 32\text{-}9$$

Intensity of an electromagnetic wave

where the vector

$$\vec{S} = \frac{\vec{E} \times \vec{B}}{\mu_0} \qquad 32\text{-}10$$

Definition—Poynting vector

is called the **Poynting vector** after its discoverer, Sir John Poynting. The average magnitude of $\vec{S}$ is the intensity of the wave, and the direction of $\vec{S}$ is the direction of propagation of the wave.

Radiation Pressure We now show by a simple example that an electromagnetic wave carries momentum. Consider a wave moving along the x axis that is incident on a stationary charge as shown in Figure 32-9. For simplicity, we assume that $\vec{E}$ is in the y direction and $\vec{B}$ is in the z direction, and we neglect

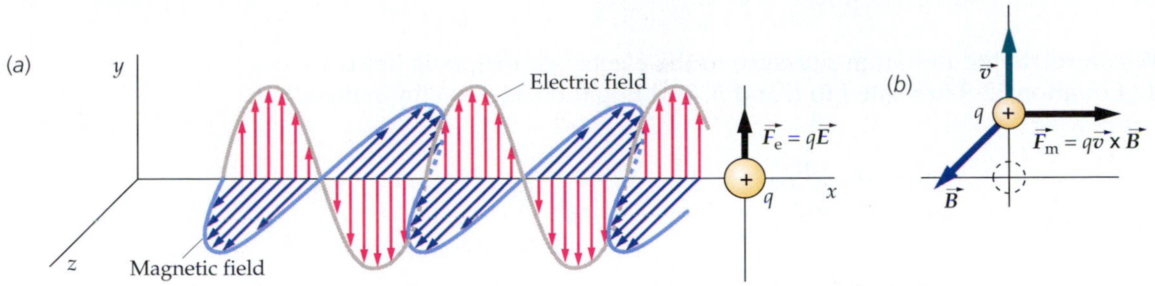

Figure 32-9 An electromagnetic wave incident on a point charge that is initially at rest on the x axis. (a) The electric force $q\vec{E}$ accelerates the charge in the upward direction. (b) When the velocity of the charge is $\vec{v}$ upward, the magnetic force $q\vec{v} \times \vec{B}$ accelerates the charge in the direction of the wave.

the time dependence of the fields. The particle experiences a force $q\vec{E}$ in the y direction and is thus accelerated by the electric field. At any time t, the velocity in the y direction is

$$v_y = at = \frac{qE}{m}t$$

After a short time t_1, the charge has acquired kinetic energy equal to

$$K = \frac{1}{2}mv_y^2 = \frac{1}{2}\frac{mq^2E^2t_1^2}{m^2} = \frac{1}{2}\frac{q^2E^2}{m}t_1^2 \qquad \text{32-11}$$

When the charge is moving in the y direction, it experiences a magnetic force

$$\vec{F}_m = q\vec{v} \times \vec{B} = qv_y\hat{j} \times B\hat{k} = qv_yB\hat{i} = \frac{q^2EB}{m}t\hat{i}$$

Note that this force is in the direction of propagation of the wave. Using $F_x = dp_x/dt$, we find for the momentum p_x transferred by the wave to the particle in time t_1:

$$p_x = \int_0^{t_1} F_x\,dt = \int_0^{t_1} \frac{q^2EB}{m}t\,dt = \frac{1}{2}\frac{q^2EB}{m}t_1^2$$

If we use $B = E/c$, this becomes

$$p_x = \frac{1}{c}\left(\frac{1}{2}\frac{q^2E^2}{m}t_1^2\right) \qquad \text{32-12}$$

Comparing Equations 32-11 and 32-12, we see that the momentum acquired by the charge in the direction of the wave is $1/c$ times the energy. Although our simple calculation was not rigorous, the results are correct. The magnitude of the momentum carried by an electromagnetic wave is $1/c$ times the energy carried by the wave:

$$p = \frac{U}{c} \qquad \text{32-13}$$

Momentum and energy in an electromagnetic wave

Since the intensity is the energy per unit time per unit area, the intensity divided by c is the momentum carried by the wave per unit time per unit area. The momentum carried per unit time is a force. The intensity divided by c is thus a force per unit area, which is a pressure. This pressure is the radiation pressure P_r:

$$P_r = \frac{I}{c} \qquad \text{32-14}$$

Radiation pressure and intensity

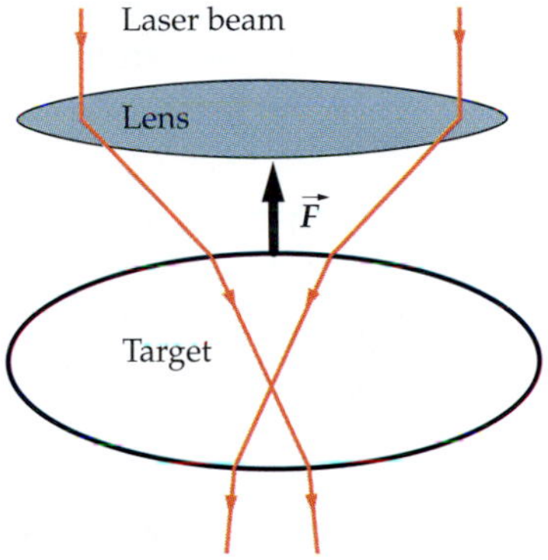

"Laser tweezers" make use of the momentum carried by electromagnetic waves to manipulate targets on a molecular scale. The two rays shown are refracted as they pass through a transparent target, such as a biological cell, or on an even smaller scale, a tiny transparent bead attached to a large molecule within a cell. At each refraction, the rays are bent downward, which increases the downward component of momentum of the rays. The target thus exerts a downward force on the laser beams, and the laser beams exert an upward force on the target, which pulls the target toward the laser source. The force is typically of the order of piconewtons. Laser tweezers have been used to accomplish such astonishing feats as stretching out coiled DNA.

We can relate the radiation pressure to the electric or magnetic fields by using Equation 32-9 to relate I to E and B, and Equation 32-7 to eliminate either E or B:

$$P_r = \frac{I}{c} = \frac{E_0 B_0}{2\mu_0 c} = \frac{E_{rms} B_{rms}}{\mu_0 c} = \frac{E_0^2}{2\mu_0 c^2} = \frac{B_0^2}{2\mu_0} \qquad 32\text{-}15$$

Radiation pressure in terms of E and B

Consider an electromagnetic wave incident normally on some surface. If the surface absorbs energy U from the electromagnetic wave, it also absorbs momentum p given by Equation 32-14, and the pressure exerted on the surface equals the radiation pressure. If the wave is reflected, the momentum transferred is $2p$ because the wave now carries momentum in the opposite direction. The pressure exerted on the surface by the wave is then twice the radiation pressure.

Example 32-4

A light bulb emits spherical electromagnetic waves uniformly in all directions. Find (*a*) the intensity, (*b*) the radiation pressure, and (*c*) the electric and magnetic fields at a distance of 3 m from the bulb, assuming that 50 W of electromagnetic radiation is emitted.

Picture the Problem At a distance r from the bulb, the energy is spread uniformly over an area $4\pi r^2$. The intensity is the power divided by the area. The radiation pressure can then be found from $P_r = I/c$.

(*a*)1. Divide the power output by the area to find the intensity:

$$I = \frac{50\text{ W}}{4\pi r^2}$$

2. Substitute $r = 3$ m:

$$I = \frac{50\text{ W}}{4\pi(3\text{ m})^2} = 0.442\text{ W/m}^2$$

(*b*) The radiation pressure is the intensity divided by the speed of light:

$$P_r = \frac{I}{c} = \frac{0.442\text{ W/m}^2}{3\times10^8\text{ m/s}} = 1.47\times10^{-9}\text{ Pa}$$

(*c*)1. B_0 is related to P_r by Equation 32-15:

$$B_0 = (2\mu_0 P_r)^{1/2} = [2(4\pi\times10^{-7}\text{ T}\cdot\text{m/A})(1.47\times10^{-9}\text{ Pa})]^{1/2}$$
$$= 6.08\times10^{-8}\text{ T}$$

2. The maximum value of the electric field E_0 is c times B_0:

$$E_0 = cB_0 = (3\times10^8\text{ m/s})(6.08\times10^{-8}\text{ T})$$
$$= 18.2\text{ V/m}$$

3. The electric and magnetic fields at that point are of the form:

$$E = E_0 \sin\omega t \text{ and}$$
$$B = B_0 \sin\omega t$$

with $E_0 = 18.2$ V/m and $B_0 = 6.08\times10^{-8}$ T.

Remark Note that the pressure calculated in (*b*) is very small compared with atmospheric pressure, which is of the order of 10^5 Pa.

Example 32-5

You are stranded in space a distance of 20 m from your spaceship. You carry a 1-kW laser. If your total mass, including your space suit and laser, is 95 kg, how long will it take you to reach the ship if you point the laser directly away from it?

Picture the Problem The laser emits light, which carries with it momentum. By momentum conservation, you are given an equal and opposite momentum toward the ship. The momentum carried by light is $p = U/c$, where U is the energy of the light. If the power of the laser is $P = dU/dt$, then the rate of change of momentum produced by the laser is $dp/dt = (dU/dt)/c = P/c$. This is the force exerted on you, which is constant.

1. The time taken is related to the distance and acceleration: $x = \frac{1}{2}at^2; \quad t = \sqrt{\frac{2x}{a}}$

2. Your acceleration is the force divided by your mass: $a = \frac{F}{m} = \frac{P/c}{m} = \frac{1000\text{ W}}{(95\text{ kg})(3 \times 10^8\text{ m/s})} = 3.51 \times 10^{-8}\text{ m/s}^2$

3. Use this acceleration to calculate the time t: $t = \sqrt{\frac{2x}{a}} = \sqrt{\frac{2(20\text{ m})}{3.51 \times 10^{-8}\text{ m/s}^2}} = 3.38 \times 10^4\text{ s} = 9.38\text{ h}$

Remarks Note that the acceleration found here is extremely small—only about one-billionth the acceleration of gravity. Your speed when you reach the ship would be $v = at = 1.19$ mm/s, which is practically imperceptible.

32-4 The Wave Equation for Electromagnetic Waves

optional

In Section 15-1, we saw that waves on a string obey a partial differential equation called the **wave equation**:

$$\frac{\partial^2 y(x, t)}{\partial x^2} = \frac{1}{v^2}\frac{\partial^2 y(x, t)}{\partial t^2} \qquad 32\text{-}16$$

where $y(x, t)$ is the wave function, which for string waves is the displacement of the string.* The velocity of the wave is given by $v = \sqrt{F/\mu}$, where F is the tension and μ is the linear mass density. The solutions to this equation are harmonic wave functions of the form

$$y(x, t) = y_0 \sin (kx - \omega t)$$

where $k = 2\pi/\lambda$ is the wave number and $\omega = 2\pi f$ is the angular frequency.

Maxwell's equations imply that both $\vec{E}$ and $\vec{B}$ obey wave equations similar to Equation 32-16. We consider only free space, in which there are no charges or currents, and we assume that the electric and magnetic fields $\vec{E}$ and $\vec{B}$ are functions of time and one space coordinate only, which we will take to be the x coordinate. Such a wave is called a **plane wave,** because field

* The derivatives are partial derivatives because the wave function depends on both x and t.

optional

quantities are constant across any plane perpendicular to the x axis. For a plane electromagnetic wave traveling parallel to the x axis, the x components of the fields are zero, so the vectors $\vec{E}$ and $\vec{B}$ are perpendicular to the x axis and each obeys the wave equation:

$$\frac{\partial^2 \vec{E}}{\partial x^2} = \frac{1}{c^2}\frac{\partial^2 \vec{E}}{\partial t^2} \qquad 32\text{-}17a$$

Wave equation for $\vec{E}$

$$\frac{\partial^2 \vec{B}}{\partial x^2} = \frac{1}{c^2}\frac{\partial^2 \vec{B}}{\partial t^2} \qquad 32\text{-}17b$$

Wave equation for $\vec{B}$

where $c = 1/\sqrt{\mu_0 \epsilon_0}$ is the speed of the waves.

Derivation of the Wave Equation

We can relate the space derivative of one of the field vectors to the time derivative of the other by applying Equations 32-6*c* and 32-6*d* to appropriately chosen curves in space. We first relate the space derivative of E_y to the time derivative of B_z by applying Equation 32-6*c* (which is Faraday's law) to the rectangular curve of sides Δx and Δy lying in the xy plane (Figure 32-10). If Δx and Δy are very small, the line integral of $\vec{E}$ around this curve is approximately

$$\oint \vec{E}\cdot d\vec{\ell} = E_y(x_2)\Delta y - E_y(x_1)\Delta y$$

where $E_y(x_1)$ is the value of E_y at the point x_1 and $E_y(x_2)$ is the value of E_y at the point x_2. The contributions of the type $E_x\,\Delta x$ from the top and bottom of this curve are zero because $E_x = 0$. Since Δx is very small, we can approximate the difference in E_y at the points x_1 and x_2 by

$$E_y(x_2) - E_y(x_1) = \Delta E \approx \frac{\partial E_y}{\partial x}\Delta x$$

Then

$$\oint \vec{E}\cdot d\vec{\ell} \approx \frac{\partial E_y}{\partial x}\Delta x\,\Delta y$$

The flux of the magnetic field through this curve is approximately

$$\int_S B_n\,dA = B_z\,\Delta x\,\Delta y$$

Faraday's law then gives

$$\frac{\partial E_y}{\partial x}\Delta x\,\Delta y = -\frac{\partial B_z}{\partial t}\Delta x\,\Delta y$$

or

$$\frac{\partial E_y}{\partial x} = -\frac{\partial B_z}{\partial t} \qquad 32\text{-}18$$

Equation 32-18 implies that if there is a component of the electric field E_y that depends on x, there must be a component of the magnetic field B_z that

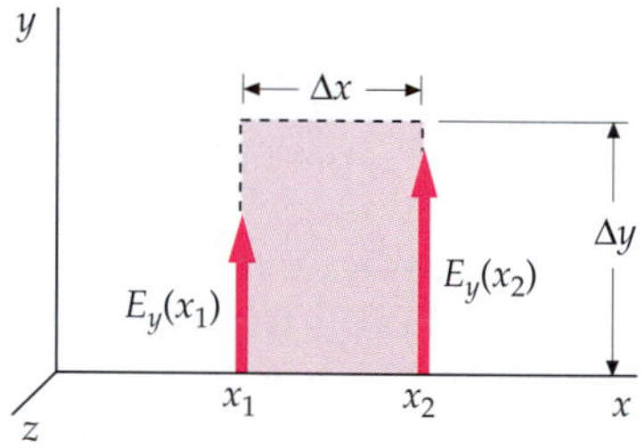

Figure 32-10 A rectangular curve in the xy plane for the derivation of Equation 32-18.

depends on time or, conversely, that if there is a component of the magnetic field B_z that depends on time, there must be a component of the electric field E_y that depends on x. We can get a similar equation relating the space derivative of the magnetic field B_z to the time derivative of the electric field E_y by applying Equation 32-6d to the curve of sides Δx and Δz in the xz plane shown in Figure 32-11.

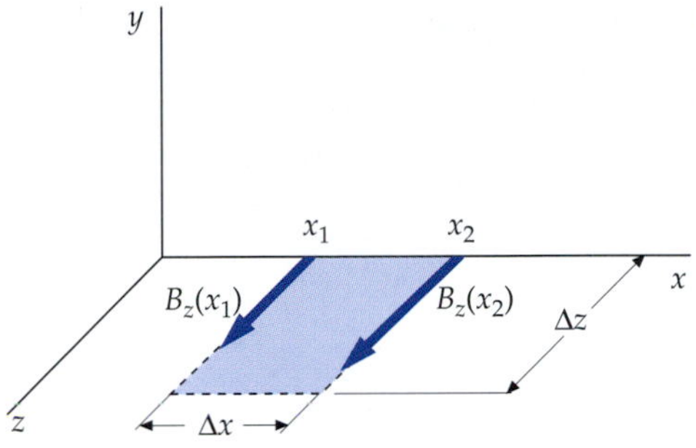

Figure 32-11 A rectangular curve in the xz plane for the derivation of Equation 32-19.

For the case of no conduction currents, Equation 32-6d is

$$\oint \vec{B}\cdot d\vec{\ell} = \mu_0\epsilon_0 \frac{d}{dt}\int_S E_n\, dA$$

The details of this calculation are similar to those for Equation 32-18. The result is

$$\frac{\partial B_z}{\partial x} = -\mu_0\epsilon_0 \frac{\partial E_y}{\partial t} \qquad \text{32-19}$$

We can eliminate either B_z or E_y from Equations 32-18 and 32-19 by differentiating either equation with respect to x or t. If we differentiate both sides of Equation 32-18 with respect to x, we obtain

$$\frac{\partial}{\partial x}\left(\frac{\partial E_y}{\partial x}\right) = -\frac{\partial}{\partial x}\left(\frac{\partial B_z}{\partial t}\right)$$

or

$$\frac{\partial^2 E_y}{\partial x^2} = -\frac{\partial}{\partial t}\left(\frac{\partial B_z}{\partial x}\right)$$

where the order of the time and space derivatives on the right side have been interchanged. We now use Equation 32-19 for $\partial B_z/\partial x$:

$$\frac{\partial^2 E_y}{\partial x^2} = -\frac{\partial}{\partial t}\left(-\mu_0\epsilon_0 \frac{\partial E_y}{\partial t}\right)$$

which yields the wave equation

$$\frac{\partial^2 E_y}{\partial x^2} = \mu_0\epsilon_0 \frac{\partial^2 E_y}{\partial t^2} \qquad \text{32-20}$$

Comparing this equation with Equation 32-16, we see that E_y obeys a wave equation for waves with speed $c = 1/\sqrt{\mu_0\epsilon_0}$, which is Equation 32-1.

If we had instead chosen to eliminate E_y from Equations 32-18 and 32-19 (by differentiating Equation 32-18 with respect to t, for example), we would have obtained an equation identical to Equation 32-20 except with B_z replacing E_y. We can thus see that both the electric field E_y and the magnetic field B_z obey a wave equation for waves traveling with the velocity $1/\sqrt{\mu_0\epsilon_0}$, which is the velocity of light.

optional

By following the same line of reasoning as used above, and applying Equation 32-6*c* (Faraday's law) to the curve in the *xz* plane (Figure 32-11), we would obtain

$$\frac{\partial E_z}{\partial x} = \frac{\partial B_y}{\partial t} \qquad \text{32-21}$$

Similarly, the application of Equation 32-6*d* to the curve in the *xy* plane (Figure 32-10) gives

$$\frac{\partial B_y}{\partial x} = \mu_0 \epsilon_0 \frac{\partial E_z}{\partial t} \qquad \text{32-22}$$

We can use these results to show that, for a wave propagating in the x direction, the components E_z and B_y also obey the wave equation.

Consider the harmonic wave function of the form

$$E_y = E_{y0} \sin (kx - \omega t) \qquad \text{32-23}$$

If we substitute this solution into Equation 32-18, we can see that the magnetic field B_z is in phase with the electric field E_y. We have

$$\frac{\partial B_z}{\partial t} = -\frac{\partial E_y}{\partial x} = -kE_{y0} \cos (kx - \omega t)$$

Solving for B_z gives

$$B_z = \frac{k}{\omega} E_{y0} \sin (kx - \omega t) = B_{z0} \sin (kx - \omega t) \qquad \text{32-24}$$

where

$$B_{z0} = \frac{k}{\omega} E_{y0} = \frac{E_{y0}}{c}$$

and $c = \omega/k$ is the velocity of the wave. (We have omitted the arbitrary constant of integration because it plays no part in the wave that we are interested in.) Since the electric and magnetic fields oscillate in phase with the same frequency, we have the general result that the magnitude of the electric field is c times the magnitude of the magnetic field for an electromagnetic wave:

$$E = cB$$

which is Equation 32-7.

We see that Maxwell's equations imply wave equations 32-17*a* and 32-17*b* for the electric and magnetic fields; and that if E_y varies harmonically, as in Equation 32-23, the magnetic field B_z is in phase with E_y and has an amplitude related to the amplitude of E_y by $B_z = E_y/c$. The electric and magnetic fields are perpendicular to each other and to the direction of the wave propagation, as shown in Figure 32-3.

Example 32-6

The electric field vector of an electromagnetic wave is given by $\vec{E}(x, t) = E_0 \sin (kx - \omega t)\hat{j} + E_0 \cos (kx - \omega t)\hat{k}$. (*a*) Find the corresponding magnetic field. (*b*) Compute $\vec{E} \cdot \vec{B}$ and $\vec{E} \times \vec{B}$.

Picture the Problem We find B_y using either Equation 32-21 or 32-22, and B_z using Equation 32-18 or 32-19. The products $\vec{E}\cdot\vec{B}$ and $\vec{E}\times\vec{B}$ are found using standard vector operations.

(*a*)1. Use Equation 32-21 to obtain a relation for B_y:

$$\frac{\partial B_y}{\partial t} = \frac{\partial E_z}{\partial x}$$

$$= \frac{\partial}{\partial x}[E_0 \cos (kx - \omega t)] = -kE_0 \sin (kx - \omega t)$$

2. Integrate the above result (neglecting the arbitrary constant of integration) to find B_y. Let $B_0 = kE_0/\omega = E_0/c$:

$$B_y = [kE_0 \cos (kx - \omega t)]\left(\frac{-1}{\omega}\right) = -B_0 \cos (kx - \omega t)$$

3. Equation 32-18 gives us a relation for B_z:

$$\frac{\partial B_z}{\partial t} = -\frac{\partial E_y}{\partial x}$$

$$= -\frac{\partial}{\partial x}[E_0 \sin (kx - \omega t)] = -kE_0 \cos (kx - \omega t)$$

4. Integrating and using $B_0 = kE_0/\omega = E_0/c$, we find B_z:

$$B_z = [-kE_0 \sin (kx - \omega t)]\left(\frac{-1}{\omega}\right) = B_0 \sin (kx - \omega t)$$

(*b*)1. Let $\theta = kx - \omega t$ to simplify the notation and calculate $\vec{E}\cdot\vec{B}$:

$$\begin{aligned}\vec{E}\cdot\vec{B} &= [E_0 \sin \theta\hat{j} + E_0 \cos \theta\,\hat{k}]\cdot[-B_0 \cos \theta\,\hat{j} + B_0 \sin \theta\,\hat{k}]\\ &= -E_0B_0 \sin \theta \cos \theta\,\hat{j}\cdot\hat{j} + E_0B_0 \sin^2 \theta\,\hat{j}\cdot\hat{k}\\ &\quad -E_0B_0 \cos^2 \theta\,\hat{k}\cdot\hat{j} + E_0B_0 \cos \theta \sin \theta\,\hat{k}\cdot\hat{k}\\ &= -E_0B_0 \sin \theta \cos \theta + 0 - 0 + E_0B_0 \cos \theta \sin \theta = 0\end{aligned}$$

2. Calculate $\vec{E}\times\vec{B}$:

$$\begin{aligned}\vec{E}\times\vec{B} &= [E_0 \sin \theta\,\hat{j} + E_0 \cos \theta\,\hat{k}] \times [-B_0 \cos \theta\,\hat{j} + B_0 \sin \theta\,\hat{k}]\\ &= E_0B_0 \sin^2 \theta\,\hat{j} \times \hat{k} + (-E_0B_0 \cos^2 \theta\,\hat{k} \times \hat{j})\\ &= E_0B_0 \sin^2 \theta\,\hat{i} + E_0B_0 \cos^2 \theta\,\hat{i} = E_0B_0\hat{i}\end{aligned}$$

Remarks We see that $\vec{E}$ and $\vec{B}$ are perpendicular to one another, and that $\vec{E}\times\vec{B}$ is in the direction of propagation of the wave. This type of electromagnetic wave is said to be *circularly polarized*. At a fixed value of x, both $\vec{E}$ and $\vec{B}$ rotate in a circle in a plane perpendicular to x with angular frequency ω. The fields $\vec{E}$ and $\vec{B}$ are constant in magnitude, as can be seen by noting that $\vec{E}\cdot\vec{E}$ and $\vec{B}\cdot\vec{B}$ are constant.

Exercise Calculate $\vec{E}\cdot\vec{E}$ and $\vec{B}\cdot\vec{B}$. [*Answers* $\vec{E}\cdot\vec{E} = E_y^2 + E_z^2 = E_0^2 \sin^2 (kx - \omega t) + E_0^2 \cos^2 (kx - \omega t) = E_0^2$ and $\vec{B}\cdot\vec{B} = B_y^2 + B_z^2 = B_0^2 \cos^2 (kx - \omega t) + B_0^2 \sin^2 (kx - \omega t) = B_0^2$]

Exploring

James Clerk Maxwell (1831–1879)

C. W. F. Everitt
Stanford University

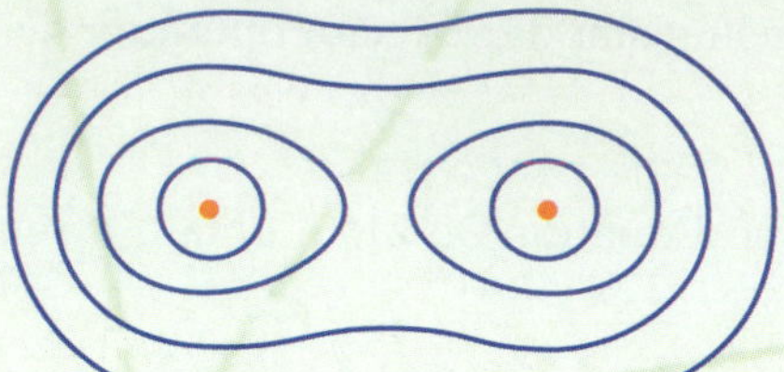

(*a*)

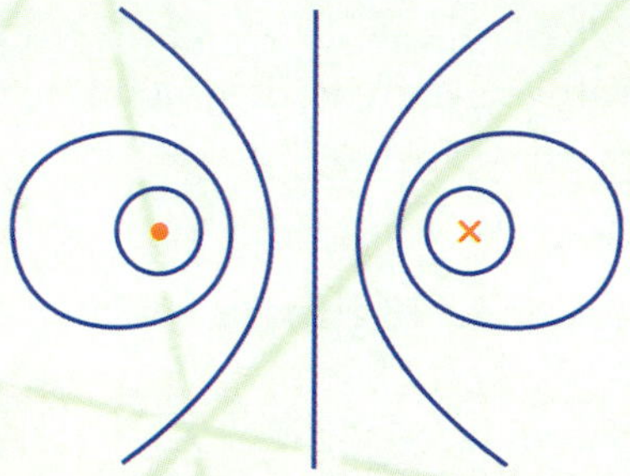

(*b*)

Figure 1 Faraday's explanation of forces between current-carrying wires. The two diagrams show the lines of force observed when currents are flowing in parallel wires. Faraday assumed that the lines of force tend to shorten and repel each other sideways. (*a*) For wires with currents flowing in the same direction, the lines of force pull the two wires together. (*b*) For wires with currents flowing in opposite directions, the lines of force push the wires apart.

One day in 1877 a Scottish undergraduate named Donald MacAlister wrote home from Cambridge University that he had just had dinner with a professor who was "one of the best of our men, and a thorough old Scotch laird in ways and speech." James Clerk Maxwell was wealthy, an expert swimmer and horseman, proprietor of an estate of 2000 acres in Scotland—and a scientist whose writings remain astonishingly up to date. The greatest mathematical physicist since Newton, he created the electromagnetic theory of light, predicted the existence of radio waves, wrote the first significant paper on control theory, and was joint inventor with Ludwig Boltzmann of statistical mechanics. He also performed with his wife's aid a brilliant series of experiments on color vision and took the first color photograph. In the 2 years before his death due to cancer in 1879, at the age of 48, he would lay the foundations of another new subject that was to reach fruition in the twentieth century, rarefied gas dynamics.

Maxwell's electromagnetic theory of light was rooted in the work of two men, Michael Faraday and William Thomson. Faraday's invention of the electric motor and his research on electromagnetic induction, electrochemistry, dielectric and diamagnetic action, and magneto-optical rotation made him in Maxwell's words "the nucleus of everything electric since 1830." His contributions to theory lay in his progressively advancing ideas about lines of electric and magnetic force, in particular the geometrical relations governing electromagnetic phenomena and the idea that magnetic forces might be accounted for not by direct attractions and repulsions between elements of current but by attributing to lines of force the property of shortening themselves and repelling each other sideways (Figure 1). Thomson's role was to relate lines of force to existing theories in electrostatics and magnetostatics, to invent a number of highly ingenious analytical techniques for solving electrical problems, and to emphasize the cardinal importance of energy principles in electromagnetism. Maxwell then introduced a series of new concepts: the *electrotonic function* (vector potential), the energy density of the field, and the displacement current; he organized the subject into a coherent structure and in 1861 made the momentous discovery of the equivalence between light and electromagnetic waves.

The first part of Maxwell's paper "On Faraday's Lines of Force" (1855–1856) developed an analogy, due in essence to Thomson, between lines of electric and magnetic force and streamlines in a moving incompressible fluid. Maxwell applied this to interpret many of Faraday's observations, prefacing his paper with a luminous discussion of the significance of analogies in physics.

Next, still building on Faraday and Thomson, Maxwell extended the discussion to electromagnetism. He formulated a group of equations summarizing the relations of the electric and magnetic fields to the charges and currents producing them—the beginnings of what we now call Maxwell's equations. They described the phenomena with elegant precision from a point of view completely different from the then-popular action-at-a-distance theories of André-Marie Ampère and Wilhelm Weber.

After such a brilliant start one might have expected a rush of papers following up the new ideas. But other physicists ignored them, and Maxwell had the habit of investigating different subjects in turn, often with long intervals between successive papers in the same field. Six years elapsed before the appearance of his next paper, "On Physical Lines of Force," published in four parts in 1861–1862. During the interval Maxwell made brilliant contributions to three distinct subjects before returning to electromagnetism: color vision, the theory of Saturn's rings, and the kinetic theory of gases.

"On Physical Lines of Force" contained Maxwell's extraordinary molecular-vortex model of the electromagnetic field. To account for the pattern of stresses associated with lines of force by Faraday, Maxwell investigated the properties of a medium occupying all space in which tiny molecular vortices rotate with their axes parallel to the lines of force. The closer together the lines are, the faster the rotation of the vortices. In a medium of this kind the lines of force do tend to shorten themselves and repel each other sideways, yielding the right forces between currents and magnets. The question is, what makes the vortices rotate? Here Maxwell put forward an idea as ingenious as it was weird. He postulated that an electric current consists in the motion of tiny particles that mesh like gear wheels with the vortices, and that the medium is filled with similar particles between the vortices. Figure 2 gives the picture. Maxwell remarks:

> I do not bring [this hypothesis] forward as a mode of connexion existing in nature . . . [but] I venture to say that anyone who understands [its] provisional and temporary character . . . will find himself helped rather than hindered by it in his search for the true interpretation of [electromagnetic] phenomena.

The question then was how to fit electrostatic phenomena into the model. Maxwell made the medium an elastic one. Thus, magnetic forces were accounted for by rotations in the medium, and electric forces by its elastic distortion. Any elastic medium will transmit waves. In Maxwell's medium the velocity of the waves turned out to be related to the ratio of electric to magnetic forces. Putting in numbers from an experiment of 1856 by G. Kohlrausch and W. Weber, Maxwell found to his astonishment that the propagation velocity was equal to the velocity of light. With excitement he wrote, "we can scarcely avoid the inference that *light consists in the transverse undulation of the same medium which is the cause of electric and magnetic phenomena.*"

Having made the great discovery, Maxwell promptly jettisoned his model. Instead of attempting a more refined mechanical explanation of the phenomena, he formulated a system of electromagnetic equations from which he deduced that waves of electric and magnetic force would propagate through space with the velocity of light. That is why his is called an *electromagnetic* theory of light, in contrast to the theories of the mechanical ether that preceded it. The theory appeared in two papers of 1865 and 1868, and in its most general form in the great *Treatise on Electricity and Magnetism,* published in 1873. This was a

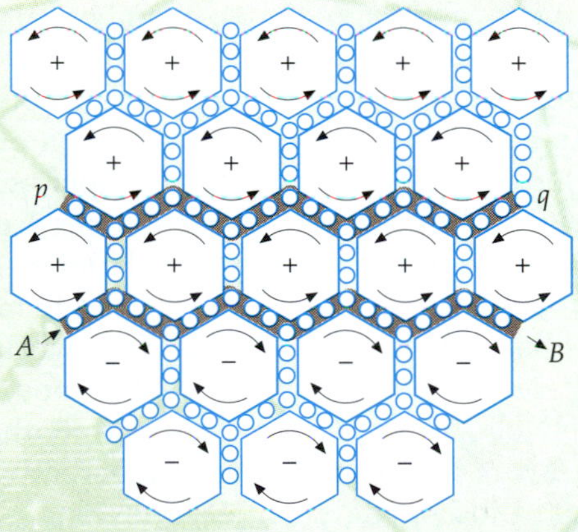

Figure 2 Maxwell's vortex model of the magnetic field. The rotating vortices represent lines of magnetic force. They mesh with small particles that act like gear wheels. In free space the particles are restrained from moving, except for a small elastic reaction (the displacement current), but in a conducting wire they are free to move. Their motion constitutes an electric current, which in turn sets the vortices in rotation, creating the magnetic field around the wire. *A* and *B* represent current through a wire, and *p* and *q* represent an induced current in an adjacent wire. (Redrawn from *The Scientific Papers of James Clerk Maxwell,* Vol. I, Figure 2 after p. 488.)

work of such scope that Robert Andrews Millikan, author of the famous oil-drop experiment to measure the charge on the electron, ranked it with Newton's *Principia* in considering them the two most influential books in the history of physics, "the one creating our modern mechanical world and the other our modern electrical world."

Equally profound were Maxwell's contributions to statistical and molecular physics. They began with a paper in 1859 on the kinetic theory of gases, in which Maxwell introduced the velocity distribution function and enunciated the *equipartition theorem* (Section 18-5), which in its original form stated that the average translational and rotational energies of large numbers of colliding molecules, whether of the same or different species, are equal. One result was Maxwell's estimate of the mean free path of a gas molecule, which Loschmidt in 1865 applied to make the first serious estimates of the diameters of molecules. Later, Maxwell developed the general theory of transport phenomena, from which the Boltzmann equation is derived; invented the concept of ensemble averaging; created rarefied-gas dynamics; and conceived that "very small BUT lively being" the Maxwell demon.

The demon, so named by Kelvin, is one of the earliest examples in physics of a "thought experiment." Maxwell imagined two chambers of gas, A and B, separated by a wall in which there was a trap door guarded by a tiny being with eyesight so acute that it could discern the motion of individual molecules. By opening the door when a fast molecule approached from chamber A or as a slow molecule approached from chamber B, the demon could redistribute the velocities to make B hotter than A without doing any work and thereby defeat the second law of thermodynamics. Maxwell's point was to demonstrate that the second law of thermodynamics is inherently a statistical law and not a dynamical one.

The work by Maxwell and Boltzmann on statistical mechanics had profound implications for modern physics. Brilliant as its successes were, the failures were—as Maxwell saw—in some ways even more striking. The equipartition theorem gave an answer for the ratio of the specific heats of gases that disagreed with experiment, while some of Boltzmann's theorems "proved too much" because they would apply to the properties of solids and liquids as well as gases. These questions remained shrouded in mystery until the emergence in 1900 of Planck's quantum hypothesis. Writing about them in 1877, Maxwell confessed his bewilderment and stated that nothing remained but to adopt the attitude of "thoroughly conscious ignorance that is the prelude to every real advance in knowledge."

Maxwell was an unusually sensitive man, with strong religious feeling and a fascinating and astonishing sense of humor. Many of his letters reveal a delightfully sly irony. He also had some talent for writing poetry, usually light, but occasionally touching a deeper note. The last stanza of one poem to his wife, written in 1867, was

All powers of mind, all force of will
  May lie in dust when we are dead,
But love is ours, and shall be still
  When earth and seas are fled.

James Clerk Maxwell (1831–1879) with his wife, Katherine Mary, and their dog.

Summary

1. Maxwell's equations summarize the fundamental laws of physics that govern electricity and magnetism.
2. Electromagnetic waves include light, radio and television waves, X rays, gamma rays, microwaves, and others.

Topic	Remarks and Relevant Equations	
1. Maxwell's Displacement Current	Ampère's law can be generalized to apply to discontinuous currents if the conduction current I is replaced by $I + I_d$, where I_d is Maxwell's displacement current: $$I_d = \epsilon_0 \frac{d\phi_e}{dt}$$	32-3
Generalized form of Ampère's law	$$\oint_C \vec{B} \cdot d\vec{\ell} = \mu_0(I + I_d) = \mu_0 I + \mu_0 \epsilon_0 \frac{d\phi_e}{dt}$$	32-4
2. Maxwell's Equations	The laws of electricity and magnetism are summarized by Maxwell's equations.	
Gauss's law	$$\oint_S E_n \, dA = \frac{1}{\epsilon_0} Q_{inside}$$	32-6*a*
Gauss's law for magnetism (isolated magnetic poles do not exist)	$$\oint_S B_n \, dA = 0$$	32-6*b*
Faraday's law	$$\oint_C \vec{E} \cdot d\vec{\ell} = -\frac{d}{dt} \int_S B_n \, dA$$	32-6*c*
Ampère's law modified	$$\oint_C \vec{B} \cdot d\vec{\ell} = \mu_0 I + \mu_0 \epsilon_0 \frac{d}{dt} \int_S E_n \, dA$$	32-6*d*
3. Electromagnetic Waves	In an electromagnetic wave, the electric and magnetic field vectors are perpendicular to each other and to the direction of propagation. Their magnitudes are related by $$E = cB$$	32-7
Wave speed	$$c = \frac{1}{\sqrt{\mu_0 \epsilon_0}} \approx 3 \times 10^8 \text{ m/s}$$	32-1
Electromagnetic spectrum	The various types of electromagnetic waves—light, radio waves, X rays, gamma rays, microwaves, and others—differ only in wavelength and frequency. The human eye is sensitive to the range from about 400 to 700 nm.	
Electric-dipole radiation	Electromagnetic waves are produced when electric charges accelerate. Oscillating charges in an electric-dipole antenna radiate electromagnetic waves with an intensity that is maximum in directions perpendicular to the antenna and zero along the axis of the antenna. Perpendicular to the antenna and far away from it, the electric field of the electromagnetic wave is parallel to the antenna.	
Energy density in an electromagnetic wave	$$u = u_e + u_m = \epsilon_0 E^2 = \frac{B^2}{\mu_0} = \frac{EB}{\mu_0 c}$$	32-8
Intensity of an electromagnetic wave	$$I = u_{av} c = \frac{E_{rms} B_{rms}}{\mu_0} = \frac{1}{2}\frac{E_0 B_0}{\mu_0} = \lvert \vec{S} \rvert_{av}$$	32-9

Poynting vector	$\vec{S} = \dfrac{\vec{E} \times \vec{B}}{\mu_0}$	**32-10**
Momentum in an electromagnetic wave	$p = \dfrac{U}{c}$	**32-13**
Radiation pressure	$P_r = \dfrac{I}{c}$	**32-14**

4. Wave Equation (optional)

Maxwell's equations imply that the electric and magnetic field vectors in free space obey a wave equation

$$\frac{\partial^2 \vec{E}}{\partial x^2} = \frac{1}{c^2}\frac{\partial^2 \vec{E}}{\partial t^2} \qquad \textbf{32-17}a$$

$$\frac{\partial^2 \vec{B}}{\partial x^2} = \frac{1}{c^2}\frac{\partial^2 \vec{B}}{\partial t^2} \qquad \textbf{32-17}b$$

Problem-Solving Guide

Begin by drawing a neat diagram that includes the important features of the problem. In problems involving Maxwell's displacement current, it is useful to draw the surface through which the electric flux is to be calculated and the bounding curve C.

Summary of Worked Examples

Type of Calculation	Procedure and Relevant Examples
1. Displacement Current	
Find the displacement current due to a time-dependent electric field.	The displacement current I_d is given by Equation 32-3, $I_d = \epsilon_0\, d\phi_e/dt$, where ϕ_e is the electric flux. **Example 32-1**
Find the magnetic field due to a displacement current.	The magnetic field can be calculated from the generalized form of Ampère's law, $\oint_C \vec{B} \cdot d\vec{\ell} = \mu_0(I + I_d)$, where I is the conduction current and I_d is the displacement current. **Example 32-2**
2. Electromagnetic Waves	
Calculate the rms-induced emf in a loop antenna due to electromagnetic waves of frequency f.	Use $\mathcal{E}_{rms} = \pi r^2 (dB/dt)_{rms}$ and $B = E/c$. **Example 32-3**
Find the intensity, momentum, and radiation pressure associated with an electromagnetic wave.	Use $I = E_0 B_0/2\mu_0$, $p = U/c$, and $P_r = I/c$. **Examples 32-4, 32-5**

Problems

Conceptual Problems

Problems from Optional and Exploring sections

In a few problems, you are given more data than you actually need; in a few other problems, you are required to supply data from your general knowledge, outside sources, or informed estimates.

• Single-concept, single-step, relatively easy
•• Intermediate-level, may require synthesis of concepts
••• Challenging, for advanced students

Maxwell's Displacement Current

1 • A parallel-plate capacitor in air has circular plates of radius 2.3 cm separated by 1.1 mm. Charge is flowing onto the upper plate and off the lower plate at a rate of 5 A. (*a*) Find the time rate of change of the electric field between the plates. (*b*) Compute the displacement current between the plates and show that it equals 5 A.

2 • In a region of space, the electric field varies according to $E = (0.05\ \mathrm{N/C}) \sin 2000t$, where t is in seconds. Find the maximum displacement current through a 1-m^2 area perpendicular to $\vec{E}$.

3 •• For Problem 1, show that at a distance r from the axis of the plates the magnetic field between the plates is given by $B = (1.89 \times 10^{-3}\ \mathrm{T/m})r$ if r is less than the radius of the plates.

4 •• (*a*) Show that for a parallel-plate capacitor the displacement current is given by $I_d = C\, dV/dt$, where C is the capacitance and V the voltage across the capacitor. (*b*) A parallel plate capacitor $C = 5$ nF is connected to an emf $\mathcal{E} = \mathcal{E}_0 \cos \omega t$, where $\mathcal{E}_0 = 3$ V and $\omega = 500\pi$. Find the displacement current between the plates as a function of time. Neglect any resistance in the circuit.

5 •• Current of 10 A flows into a capacitor having plates with areas of 0.5 m^2. (*a*) What is the displacement current between the plates? (*b*) What is dE/dt between the plates for this current? (*c*) What is the line integral of $\vec{B} \cdot d\vec{\ell}$ around a circle of radius 10 cm that lies within and parallel to the plates?

6 •• A parallel-plate capacitor with circular plates is given a charge Q_0. Between the plates is a leaky dielectric having a dielectric constant of κ and a resistivity ρ. (*a*) Find the conduction current between the plates as a function of time. (*b*) Find the displacement current between the plates as a function of time. What is the total (conduction plus displacement) current? (*c*) Find the magnetic field produced between the plates by the leakage discharge current as a function of time. (*d*) Find the magnetic field between the plates produced by the displacement current as a function of time. (*e*) What is the total magnetic field between the plates during discharge of the capacitor?

7 •• The leaky capacitor of Problem 6 is charged such that the voltage across the capacitor is given by $V(t) = (0.01\ \mathrm{V/s})t$. (*a*) Find the conduction current as a function of time. (*b*) Find the displacement current. (*c*) Find the time for which the displacement current is equal to the conduction current.

8 •• The space between the plates of a capacitor is filled with a material of resistivity $\rho = 10^4\ \Omega \cdot \mathrm{m}$ and dielectric constant $\kappa = 2.5$. The parallel plates are circular with a radius of 20 cm and are separated by 1 mm. The voltage across the plates is given by $V_0 \cos \omega t$, with $V_0 = 40$ V and $\omega = 120\pi$ rad/s. (*a*) What is the displacement current density? (*b*) What is the conduction current between the plates? (*c*) At what angular frequency is the total current 45° out of phase with the applied voltage?

9 ••• In this problem, you are to show that the generalized form of Ampère's law (Equation 32-4) and the Biot–Savart law give the same result in a situation in which they both can be used. Figure 32-12 shows two charges $+Q$ and $-Q$ on the x axis at $x = -a$ and $x = +a$, with a current $I = -dQ/dt$ along the line between them. Point P is on the y axis at $y = R$.

Figure 32-12 Problem 9

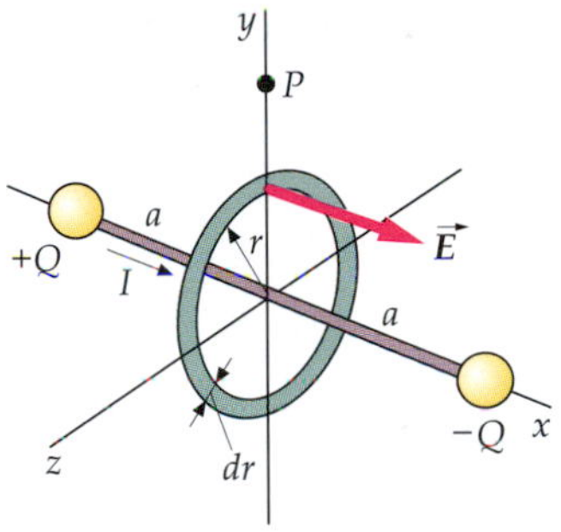

(*a*) Use the Biot–Savart law to show that the magnitude of B at point P is

$$B = \frac{\mu_0 I a}{2\pi R} \frac{1}{\sqrt{R^2 + a^2}}$$

(*b*) Consider a circular strip of radius r and width dr in the yz plane with its center at the origin. Show that the flux of the electric field through this strip is

$$E_x\, dA = \frac{Q}{\epsilon_0} a(r^2 + a^2)^{-3/2}\, r\, dr$$

(*c*) Use your result for part (*b*) to find the total flux ϕ_e through a circular area of radius R. Show that

$$\epsilon_0 \phi_e = Q\left(1 - \frac{a}{\sqrt{a^2 + R^2}}\right)$$

(*d*) Find the displacement current I_d, and show that

$$I + I_d = I \frac{a}{\sqrt{a^2 + R^2}}$$

(*e*) Then show that Equation 32-4 gives the same result for B as that found in part (*a*).

Maxwell's Equations

10 •• Theorists have speculated on the possible existence of magnetic monopoles, and there have been several, as yet unsuccessful, experimental searches for such monopoles. Suppose magnetic monopoles were found and that the magnetic field at a distance r from a monopole of strength q_m is given by $B = (\mu_0/4\pi)q_m/r^2$. How would Maxwell's equations have to be modified to be consistent with such a discovery?

11 •• Show that the normal component of the magnetic field $\vec{B}$ is continuous across a surface. Do this by applying Gauss's law for $\vec{B}$ ($\int B_n \, dA = 0$) to a pillbox Gaussian surface that has a face on each side of the surface.

The Electromagnetic Spectrum

12 • Which waves have greater frequencies, light waves or X rays?

13 • Are the frequencies of ultraviolet radiation greater or less than those of infrared radiation?

14 • What kind of waves have wavelengths of the order of a few meters?

15 • Find the wavelength for (*a*) a typical AM radio wave with a frequency of 1000 kHz and (*b*) a typical FM radio wave of 100 MHz.

16 • What is the frequency of a 3-cm microwave?

17 • What is the frequency of an X ray with a wavelength of 0.1 nm?

Electric-Dipole Radiation

18 • The detection of radio waves can be accomplished with either a dipole antenna or a loop antenna. The dipole antenna detects the (pick one) [*electric*] [*magnetic*] field of the wave, and the loop antenna detects the [*electric*] [*magnetic*] field of the wave.

19 • A transmitter uses a loop antenna with the loop in the horizontal plane. What should be the orientation of a dipole antenna at the receiver for optimum signal reception?

20 •• The intensity of radiation from an electric dipole is proportional to $(\sin^2 \theta)/r^2$, where θ is the angle between the electric dipole moment and the position vector $\vec{r}$. A radiating electric dipole lies along the z axis (its dipole moment is in the z direction). Let I_1 be the intensity of the radiation at a distance $r = 10$ m and at angle $\theta = 90°$. Find the intensity (in terms of I_1) at (*a*) $r = 30$ m, $\theta = 90°$; (*b*) $r = 10$ m, $\theta = 45°$; and (*c*) $r = 20$ m, $\theta = 30°$.

21 •• (*a*) For the situation described in Problem 20, at what angle is the intensity at $r = 5$ m equal to I_1? (*b*) At what distance is the intensity equal to I_1 at $\theta = 45°$?

22 •• The transmitting antenna of a radio station is a dipole located atop a mountain 2000 m above sea level. The intensity of the signal on a nearby mountain 4 km distant and also 2000 m above sea level is 4×10^{-12} W/m². What is the intensity of the signal at sea level and 1.5 km from the transmitter? (See Problem 20.)

23 ••• A radio station that uses a vertical dipole antenna broadcasts at a frequency of 1.20 MHz with total power output of 500 kW. The radiation pattern is as shown in Figure 32-8, i.e., the intensity of the signal varies as $\sin^2 \theta$, where θ is the angle between the direction of propagation and the vertical, and is independent of azimuthal angle. Calculate the intensity of the signal at a horizontal distance of 120 km from the station. What is the intensity at that point as measured in photons per square centimeter per second?

24 ••• At a distance of 30 km from a radio station broadcasting at a frequency of 0.8 MHz, the intensity of the electromagnetic wave is 2×10^{-13} W/m². The transmitting antenna is a vertical dipole. What is the total power radiated by the station?

25 ••• A small private plane approaching an airport is flying at an altitude of 2500 m above ground. The airport's flight control system transmits 100 W at 24 MHz, using a vertical dipole antenna. What is the intensity of the signal at the plane's receiving antenna when the plane's position on a map is 4 km from the airport?

Energy and Momentum in an Electromagnetic Wave

26 • An electromagnetic wave has an intensity of 100 W/m². Find (*a*) the radiation pressure P_r, (*b*) E_{rms}, and (*c*) B_{rms}.

27 • The amplitude of an electromagnetic wave is $E_0 = 400$ V/m. Find (*a*) E_{rms}, (*b*) B_{rms}, (*c*) the intensity I, and (*d*) the radiation pressure P_r.

28 • The rms value of the electric field in an electromagnetic wave is $E_{rms} = 400$ V/m. (*a*) Find B_{rms}, (*b*) the average energy density, and (*c*) the intensity.

29 • Show that the units of $E = cB$ are consistent; that is, show that when B is in teslas and c is in meters per second, the units of cB are volts per meter or newtons per coulomb.

30 • The root-mean-square value of the magnitude of the magnetic field in an electromagnetic wave is $B_{rms} = 0.245\ \mu$T. Find (*a*) E_{rms}, (*b*) the average energy density, and (*c*) the intensity.

31 •• (*a*) An electromagnetic wave of intensity 200 W/m² is incident normally on a rectangular black card with sides of 20 and 30 cm that absorbs all the radiation. Find the force exerted on the card by the radiation. (*b*) Find the force exerted by the same wave if the card reflects all the radiation incident on it.

32 •• Find the force exerted by the electromagnetic wave on the reflecting card in part (*b*) of Problem 31 if the radiation is incident at an angle of 30° to the normal.

33 •• An AM radio station radiates an isotropic sinusoidal wave with an average power of 50 kW. What are the amplitudes of E_{max} and B_{max} at a distance of (*a*) 500 m, (*b*) 5 km, and (*c*) 50 km?

34 •• The intensity of sunlight striking the earth's upper atmosphere (called the solar constant) is 1.35 kW/m^2. (*a*) Find E_{rms} and B_{rms} due to the sun at the upper atmosphere of the earth. (*b*) Find the average power output of the sun. (*c*) Find the intensity and the radiation pressure at the surface of the sun.

35 •• A demonstration laser has an average output power of 0.9 mW and a beam diameter of 1.2 mm. What is the force exerted by the laser beam on (*a*) a 100% absorbing black surface? (*b*) a 100% reflecting surface?

36 •• A laser beam has a diameter of 1.0 mm and average power of 1.5 mW. Find (*a*) the intensity of the beam, (*b*) E_{rms}, (*c*) B_{rms}, and (*d*) the radiation pressure.

37 •• Instead of sending power by a 750-kV, 1000-A transmission line, one desires to beam this energy via an electromagnetic wave. The beam has a uniform intensity within a cross-sectional area of 50 m^2. What are the rms values of the electric and the magnetic fields?

38 •• A laser pulse has an energy of 20 J and a beam radius of 2 mm. The pulse duration is 10 ns and the energy density is constant within the pulse. (*a*) What is the spatial length of the pulse? (*b*) What is the energy density within the pulse? (*c*) Find the electric and magnetic amplitudes of the laser pulse.

39 •• The electric field of an electromagnetic wave oscillates in the y direction and the Poynting vector is given by

$$\vec{S}(x,t) = (100\ \mathrm{W/m^2})\cos^2[10x - (3 \times 10^9)t]\hat{i}$$

where x is in meters and t is in seconds. (*a*) What is the direction of propagation of the wave? (*b*) Find the wavelength and the frequency. (*c*) Find the electric and magnetic fields.

40 •• A pulsed laser fires a 1000-MW pulse of 200-ns duration at a small object of mass 10 mg suspended by a fine fiber 4 cm long. If the radiation is completely absorbed without other effects, what is the maximum angle of deflection of this pendulum?

41 •• A 10- by 15-cm card has a mass of 2 g and is perfectly reflecting. The card hangs in a vertical plane and is free to rotate about a horizontal axis through the top edge. The card is illuminated uniformly by an intense light that causes the card to make an angle of 1° with the vertical. Find the intensity of the light.

42 •• A valuable 0.08-kg gem and a 105-kg spaceperson are separated by 95 m. Both objects are initially at rest. The spaceperson has a 1.5-kW laser that can be used as a photon rocket motor to propel the person toward the diamond. How long would it take the spaceperson to move 95 m using the laser rocket propulsion?

43 •• It has been suggested that spacecraft could be propelled by the radiation pressure from the sun. What must be the surface mass density (kg/m^2) of a perfectly reflecting sheet so that at a distance of one astronomical unit the force due to radiation pressure is twice that due to the gravitational attraction between the reflecting sheet and the sun? (*Note:* One astronomical unit is the average radius of the earth's orbit.) How will the ratio of radiation force to gravitational force change as the reflecting sheet accelerates away from the sun?

44 •• Suppose a mass of 50 kg is attached to a perfectly reflecting sheet whose surface mass density is that obtained in Problem 43. What must be the surface area of the sheet so that at a distance of one astronomical unit the acceleration of the system away from the sun is 0.4 mm/s^2? How does the acceleration vary with distance from the sun?

Blackbody Radiation

A blackbody is an object that is a perfect absorber; that is, it absorbs all radiation incident on it. It is also a perfect radiator. The power radiated by a blackbody of area A at temperature T is given by the Stefan–Boltzmann law (Equation 21-17 with e = 1),

$$P_r = \sigma A T^4$$

where $\sigma = 5.6703 \times 10^{-8}\ W/m^2 \cdot K^4$.

45 •• A very long wire of radius 4 mm is heated to 1000 K. The surface of the wire is an ideal blackbody radiator. (*a*) What is the total power radiated per unit length? Find (*b*) the magnitude of the Poynting vector S, (*c*) E_{rms}, and (*d*) B_{rms} at a distance of 25 cm from the wire.

46 •• A blackbody sphere of radius R is a distance 2×10^{11} m from the sun. The effective area of the body for absorption of energy from the sun is πR^2, but the area for radiation by the object is $4\pi R^2$. The power output of the sun is 3.83×10^{26} W. What is the temperature of the sphere?

47 •• (*a*) If the earth were an ideal blackbody with infinite thermal conductivity and no atmosphere, what would be the temperature of the earth? (*b*) If 40% of the incident sun's energy were reflected, what then would be the temperature of the earth? (See Problem 46.)

The Wave Equation for Electromagnetic Waves (optional)

48 • Show by direct substitution that Equation 32-17*a* is satisfied by the wave function

$$E_y = E_0 \sin(kx - \omega t) = E_0 \sin k(x - ct)$$

where $c = \omega/k$.

49 • Use the known values of μ_0 and ϵ_0 in SI units to compute $c = 1/\sqrt{\epsilon_0\mu_0}$ and show that it is approximately 3×10^8 m/s.

50 ••• (*a*) Using arguments similar to those given in the text, show that for a plane wave, in which E and B are independent of y and z,

$$\frac{\partial E_z}{\partial x} = \frac{\partial B_y}{\partial t}$$

and

$$\frac{\partial B_y}{\partial x} = \mu_0\epsilon_0\frac{\partial E_z}{\partial t}$$

(*b*) Show that E_z and B_y also satisfy the wave equation.

General Problems

51 • True or false:

(*a*) Maxwell's equations apply only to fields that are constant over time.

(*b*) The wave equation can be derived from Maxwell's equations.

(*c*) Electromagnetic waves are transverse waves.

(*d*) In an electromagnetic wave in free space, the electric and magnetic fields are in phase.

(*e*) In an electromagnetic wave in free space, the electric and magnetic field vectors $\vec{E}$ and $\vec{B}$ are equal in magnitude.

(*f*) In an electromagnetic wave in free space, the electric and magnetic energy densities are equal.

52 • (*a*) Show that if E is in volts per meter and B is in teslas, the units of the Poynting vector $\vec{S} = \vec{E} \times \vec{B}/\mu_0$ are watts per square meter. (*b*) Show that if the intensity I is in watts per square meter, the units of radiation pressure $P_r = I/c$ are newtons per square meter.

53 •• A loop antenna that may be rotated about a vertical axis is used to locate an unlicensed amateur radio transmitter. If the output of the receiver is proportional to the intensity of the received signal, how does the output of the receiver vary with the orientation of the loop antenna?

54 •• An electromagnetic wave has a frequency of 100 MHz and is traveling in a vacuum. The magnetic field is given by $\vec{B}(z,t) = (10^{-8}\text{ T})\cos(kz - \omega t)\hat{i}$. (*a*) Find the wavelength, and the direction of propagation of this wave. (*b*) Find the electric vector $\vec{E}(z, t)$. (*c*) Give Poynting's vector, and find the intensity of this wave.

55 •• A circular loop of wire can be used to detect electromagnetic waves. Suppose a 100-MHz FM station radiates 50 kW uniformly in all directions. What is the maximum rms voltage induced in a loop of radius 30 cm at a distance of 10^5 m from the station?

56 •• The electric field from a radio station some distance from the transmitter is given by $E = (10^{-4}\text{ N/C}) \times \cos 10^6 t$, where t is in seconds. (*a*) What voltage is picked up on a 50-cm wire oriented along the electric field direction? (*b*) What voltage can be induced in a loop of radius 20 cm?

57 •• A circular capacitor of radius a has a thin wire of resistance R connecting the centers of the two plates. A voltage $V_0 \sin \omega t$ is applied between the plates. (*a*) What is the current drawn by this capacitor? (*b*) What is the magnetic field as a function of radial distance r from the centerline within the plates of this capacitor? (*c*) What is the phase angle between current and applied voltage?

58 •• A 20-kW beam of radiation is incident normally on a surface that reflects half of the radiation. What is the force on this surface?

59 •• Show that the relation between the momentum carried by an electromagnetic wave and the energy, Equation 32-13, can also be derived using the Einstein–Planck relation, $E = hf$; the de Broglie equation, $p = h/\lambda$; and $c = f\lambda$.

60 •• The electric fields of two harmonic waves of angular frequency ω_1 and ω_2 are given by $\vec{E}_1 = E_{1,0}\cos(k_1 x - \omega_1 t)\hat{j}$ and $\vec{E}_2 = E_{2,0}\cos(k_2 x - \omega_2 t + \delta)\hat{j}$. Find (*a*) the instantaneous Poynting vector for the resultant wave motion and (*b*) the time-average Poynting vector. If $\vec{E}_2 = E_{2,0}\cos(k_2 x + \omega_2 t + \delta)\hat{j}$, find (*c*) the instantaneous Poynting vector for the resultant wave motion and (*d*) the time-average Poynting vector.

61 •• At the surface of the earth, there is an approximate average solar flux of 0.75 kW/m^2. A family wishes to construct a solar energy conversion system to power their home. If the conversion system is 30% efficient and the family needs a maximum of 25 kW, what effective surface area is needed for perfectly absorbing collectors?

62 •• Suppose one has an excellent radio capable of detecting a signal as weak as 10^{-14} W/m^2. This radio has a 2000-turn coil antenna having a radius of 1 cm wound on an iron core that increases the magnetic field by a factor of 200. The radio frequency is 140 kHz. (*a*) What is the amplitude of the magnetic field in this wave? (*b*) What is the emf induced in the antenna? (*c*) What would be the emf induced in a 2-m wire oriented in the direction of the electric field?

63 •• A 654-nm laser whose beam diameter is 0.4 mm points upward. A small, perfectly reflecting spherical particle having a diameter of 10 μm and a density of 0.2 g/cm^3 is supported against gravity by the radiation pressure from the laser beam. Determine the power output of this laser.

The following two problems do not concern waves, but illustrate the use of the Poynting vector to describe the flow of electromagnetic energy.

64 ••• A long, cylindrical conductor of length L, radius a, and resistivity ρ carries a steady current I that is uniformly distributed over its cross-sectional area. (*a*) Use Ohm's law to relate the electric field E in the conductor to I, ρ, and a. (*b*) Find the magnetic field B just outside the conductor. (*c*) Use the results for parts (*a*) and (*b*) to compute the Poynting vector $\vec{S} = \vec{E} \times \vec{B}/\mu_0$ at $r = a$ (the edge of the conductor). In what direction is $\vec{S}$? (*d*) Find the flux $\oint S_n\, dA$ through the surface of the conductor into the conductor, and show that the rate of energy flow into the conductor equals I^2R, where R is the resistance. (Here S_n is the *inward* component of $\vec{S}$ perpendicular to the surface of the conductor.)

65 ••• A long solenoid of n turns per unit length has a current that slowly increases with time. The solenoid has radius R, and the current in the windings has the form $I(t) = at$. (*a*) Find the induced electric field at a distance $r < R$ from the solenoid axis. (*b*) Find the magnitude and direction of the Poynting vector $\vec{S}$ at the cylindrical surface $r = R$ just inside the solenoid windings. (*c*) Calculate the flux $\oint S_n\, dA$ into the solenoid, and show that it equals the rate of increase of the magnetic energy inside the solenoid. (Here S_n is the *inward* component of $\vec{S}$ perpendicular to the surface of the solenoid.)

66 ••• Small particles might be blown out of solar systems by the radiation pressure of sunlight. Assume that the particles are spherical with a radius r and a density of 1 g/cm^3 and that they absorb all the radiation in a cross-sectional area of πr^2. They are a distance R from the sun, which has a power output of 3.83×10^{26} W. What is the radius r for which the radiation force of repulsion just balances the gravitational force of attraction to the sun?

67 ••• Some science fiction writers have used solar sails to propel interstellar spaceships. Imagine a giant sail erected on a spacecraft subjected to the solar radiation pressure. (*a*) Show that the spacecraft's acceleration is given by

$$a = \frac{P_S A}{4\pi r^2 cm}$$

where P_s is the power output of the sun and is equal to 3.8×10^{26} W, A is the surface area of the sail, m is the total mass of the spacecraft, r is the distance from the sun, and c is the speed of light. (*b*) Show that the velocity of the spacecraft at a distance r from the sun is found from

$$v^2 = v_0^2 + \left(\frac{P_S A}{2\pi mc}\right)\left(\frac{1}{r_0} - \frac{1}{r}\right)$$

where v_0 is the initial velocity at r_0. (*c*) Compare the relative accelerations due to the radiation pressure and the gravitational force. Use reasonable values for A and m. Will such a system work?

68 ••• Novelty stores sell a device called a radiometer (Figure 32-13), in which a balanced vane spins rapidly. A card is mounted on each arm of the vane. One side of each card is white and the other is black. Assume that the mass of each card is 2 g, that the light-collecting area for each card is 1 cm^2, and that each arm of the vane has a length of 2 cm. (*a*) If a 100-W light bulb produces 50 W of electromagnetic energy and the bulb is 50 cm from the radiometer, find the maximum angular acceleration of the vane. (Estimate the moment of inertia of the vane by assuming that all the mass of each card is at the end of the arms.) (*b*) How long will it take for the vane to accelerate to 10 rev/min if it starts from rest and is subject to the maximum angular acceleration at all times? (*c*) Can the radiation pressure account for the rapid motion of the radiometer?*

69 ••• When an electromagnetic wave is reflected at normal incidence on a perfectly conducting surface, the electric field vector of the reflected wave at the reflecting surface is the negative of that of the incident wave. (*a*) Explain why this should be. (*b*) Show that the superposition of incident and reflected waves result in a standing wave. (*c*) What is the relationship between the magnetic field vector of the incident and reflected waves at the reflecting surface?

70 ••• An intense point source of light radiates 1 MW isotropically. The source is located 1.0 m above an infinite perfectly reflecting plane. Determine the force that acts on the plane.

Figure 32-13 Problem 68

* The radiometer actually spins in the opposite direction from what would be expected if the force were due to radiation pressure. The reason is that the air near the black side is warmer than that near the white side, so the air molecules hitting the black side have greater energy than those hitting the white side.

PART V

light

Stress patterns around a crack in a sheet of transparent plastic are revealed by polarized light. The stress is perpendicular to the crack. Two smaller cracks have propagated from the lower end of the large one, creating additional patterns of stress. Smaller circular flaws surround the upper end of the large crack.

CHAPTER 33

Properties of Light

A bright primary rainbow and the fainter secondary rainbow in a sheet of rain over Lake Michigan. The primary bow is formed by light rays that enter spherical drops of water and are reflected once internally before leaving the drops. The secondary bow results from rays that experience two internal reflections before leaving the drops.

The human eye is sensitive to electromagnetic radiation with wavelengths from about 400 to 700 nm. The shortest wavelengths in the visible spectrum correspond to violet light and the longest to red light. The perceived colors of light are the result of the physiological and psychological response of the eye–brain sensing system to the different frequencies of visible light. Although the correspondence between perceived color and frequency is quite good, there are many interesting deviations. For example, a mixture of red light and green light is perceived by the eye–brain sensing system as yellow even in the absence of light in the yellow region of the spectrum. In this chapter, we study how light is produced, how its speed is measured, and how it is scattered, reflected, refracted, and polarized.

33-1 Wave–Particle Duality

The wave nature of light was first demonstrated by Thomas Young, who observed the interference pattern of two coherent light sources produced by illuminating a pair of narrow, parallel slits with a single source.* The wave

* The wave–particle duality of light and electrons is discussed in detail in Chapter 17. General wave properties such as propagation, reflection, refraction, interference, and coherence are discussed in Chapters 15 and 16.

theory of light culminated in 1860 with Maxwell's prediction of electromagnetic waves. The particle nature of light was first proposed by Albert Einstein in 1905 in his explanation of the photoelectric effect. A particle of light called a **photon** has energy E that is related to the frequency f and wavelength λ of the light wave by the Einstein equation

$$E = hf = \frac{hc}{\lambda} \qquad \text{33-1}$$

Einstein equation for photon energy

where c is the speed of light and h is Planck's constant:

$$h = 6.626 \times 10^{-34}\ \text{J}\cdot\text{s} = 4.136 \times 10^{-15}\ \text{eV}\cdot\text{s}$$

Since energies are often given in electron volts and wavelengths are given in nanometers, it is convenient to express the combination hc in eV·nm. We have

$$hc = (4.14 \times 10^{-15}\ \text{eV}\cdot\text{s})(3 \times 10^{8}\ \text{m/s}) = 1.24 \times 10^{-6}\ \text{eV}\cdot\text{m}$$

or

$$hc = 1240\ \text{eV}\cdot\text{nm} \qquad \text{33-2}$$

The propagation of light is governed by its wave properties, whereas the exchange of energy between light with matter is governed by its particle properties. This wave–particle duality is a general property of nature. For example, electrons (and other so-called "particles") also propagate as waves and exchange energy as particles.

33-2 Light Spectra

Newton was the first to recognize that white light is a mixture of light of all colors of approximately equal intensity. He demonstrated this by letting sunlight fall on a glass prism and observing the spectrum of refracted light (Figure 33-1). Because the angle of refraction of a glass prism depends slightly on wavelength, the refracted beam is spread out in space into its component col-

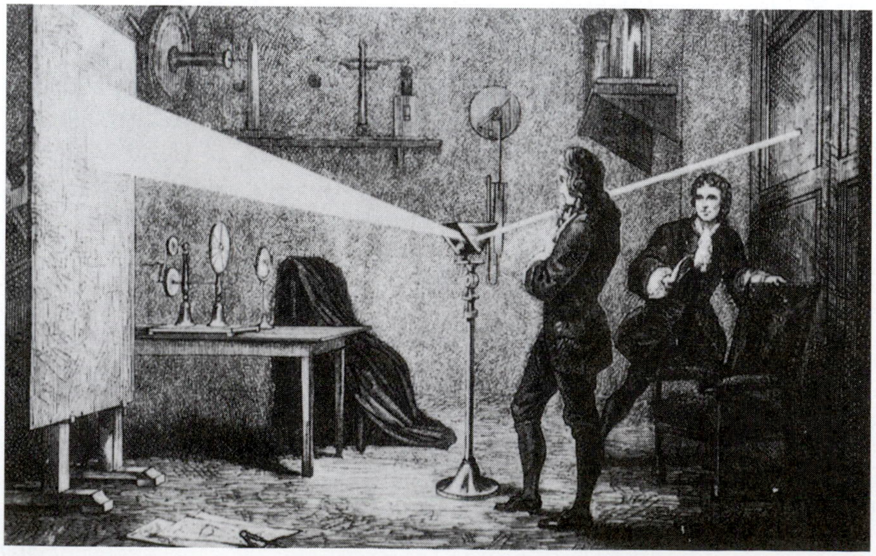

Figure 33-1 Newton demonstrating the spectrum of sunlight with a glass prism.

ors or wavelengths, like a rainbow. Figure 33-2 shows a spectroscope, which is a device for analyzing the spectra of a light source. Light from the source passes through a narrow slit, traverses a lens to make the beam parallel, and falls on a glass prism. The refracted beam is viewed with a telescope, which is mounted on a rotating platform so that the angle of the refracted beam, which depends on the wavelength, can be measured. The spectrum of the light source can thus be analyzed in terms of its component wavelengths. The spectrum of sunlight contains a continuous range of wavelengths and is therefore called a **continuous spectrum**. The light emitted by the atoms in low-pressure gases contains only a discrete set of wavelengths. Each wavelength emitted by the source produces a separate image of the collimating slit in the spectroscope. Such a spectrum is called a **line spectrum**. The continuous visible spectrum and the line spectra from several elements are shown in the photograph below.

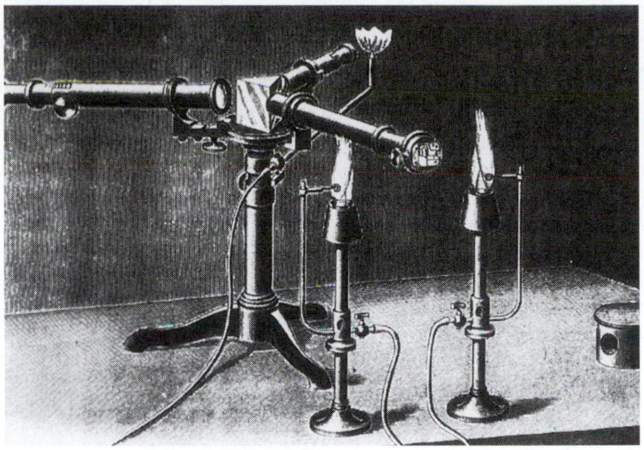

Figure 33-2 A late nineteenth-century spectroscope belonging to Gustave Kirchhoff. Modern student spectroscopes usually share the same general design.

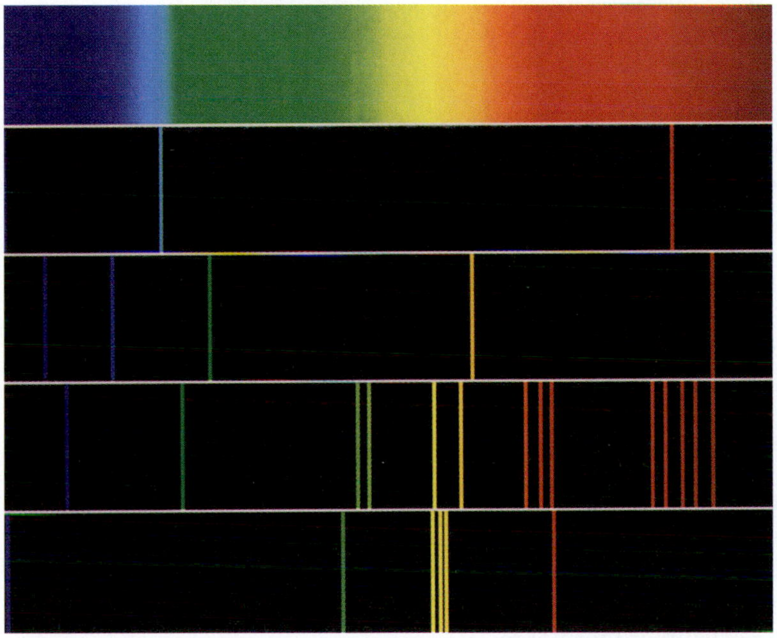

The continuous visible spectrum (*top*) and the line spectra of *(from top to bottom)* hydrogen, helium, barium, and mercury.

33-3 Sources of Light

Line Spectra The most common sources of visible light are transitions of the outer electrons in atoms. Normally an atom is in its ground state with its electrons at their lowest allowed energy levels, consistent with the exclusion principle. (The exclusion principle, which was first enunciated by Wolfgang Pauli in 1925 to explain the electronic structure of atoms, states that no two electrons in an atom can be in the same quantum state.) The lowest energy electrons are closest to the nucleus and are tightly bound, forming a stable inner core. The one or two electrons in the highest energy states are much farther from the nucleus and are relatively easily excited to vacant higher energy states. These outer electrons are responsible for the energy changes in the atom that result in the emission or absorption of visible light.

When an atom collides with another atom or with a free electron, or when it absorbs electromagnetic energy, the outer electrons can be excited to higher energy states. After a time of about 10^{-8} s, these outer electrons spontaneously make transitions to lower energy states with the emission of a photon. This process, called **spontaneous emission**, is random; the photons emitted from two different atoms are not correlated. The emitted light is thus incoherent. By conservation of energy, the energy of an emitted photon is the energy difference ΔE between the initial and final state. The frequency of the light wave is related to the energy by the Einstein equation, $\Delta E = hf$. The wavelength of the emitted light is then

$$\lambda = \frac{c}{f} = \frac{hc}{hf} = \frac{hc}{\Delta E} \qquad \text{33-3}$$

The photon energies corresponding to shortest (400 nm) and longest (700 nm) wavelengths in the visible spectrum are

$$E_{400\text{nm}} = \frac{hc}{\lambda} = \frac{1240 \text{ eV}\cdot\text{nm}}{400 \text{ nm}} = 3.1 \text{ eV} \qquad \text{33-4}a$$

and

$$E_{700\text{nm}} = \frac{hc}{\lambda} = \frac{1240 \text{ eV}\cdot\text{nm}}{700 \text{ nm}} = 1.77 \text{ eV} \qquad \text{33-4}b$$

Because the energy levels in atoms form a discrete set, the emission spectrum of light from single atoms or atoms in low-pressure gases consists of a set of sharp discrete lines that are characteristic of the element. These narrow lines are broadened somewhat by Doppler shifts due to the motion of the atom relative to the observer and by collisions with other atoms, but, generally, if the gas density is low enough, the lines are narrow and well separated from one another. The study of the line spectra of hydrogen and other atoms led to the first understanding of the energy levels of atoms.

Continuous Spectra When atoms are close together and interact strongly, as in liquids and solids, the energy levels of the individual atoms are spread out into energy bands, resulting in essentially continuous bands of energy levels. When the bands overlap, as they often do, the result is a continuous spectrum of possible energies and a continuous emission spectrum. In an incandescent material such as a hot metal filament, electrons are randomly accelerated by frequent collisions, resulting in a broad spectrum of thermal radiation. The rate at which an object radiates thermal energy is proportional to the fourth power of its absolute temperature.* The radiation emitted by an object at temperatures below about 600°C is concentrated in the infrared and is not visible. As an object is heated, the energy radiated extends to shorter and shorter wavelengths. Between about 600 and 700°C, enough of the radiated energy is in the visible spectrum for the object to glow a dull red. At higher and higher temperatures, the object becomes bright red and then white. The wavelength at which the power is a maximum varies inversely with the temperature, a result known as Wien's displacement law. The surface of the sun at $T = 6000$ K emits a continuous spectrum of approximately constant intensity over the visible range of wavelengths.

* This is known as the Stefan–Boltzmann law. This and other properties of thermal radiation such as Wien's displacement law were discussed more fully in Section 21-4.

Absorption, Scattering, and Stimulated Emission

Radiation is emitted when an atom makes a transition from an excited state to a state of lower energy; radiation is absorbed when an atom makes a transition from a lower state to a higher state. When atoms are irradiated with a continuous spectrum of radiation, the transmitted spectrum shows dark lines corresponding to absorption of light at discrete wavelengths. The absorption spectra of atoms were the first line spectra observed. Since atoms and molecules at normal temperatures are in either their ground states or low-lying excited states, absorption spectra are usually simpler than emission spectra.

Figure 33-3 illustrates several interesting phenomena that can occur when a photon is incident on an atom. In Figure 33-3*a*, the energy of the incoming photon is too small to excite the atom to an excited state, so the atom remains in its ground state and the photon is said to be scattered. Since the incoming and outgoing or scattered photons have the same energy, the scattering is said to be elastic. If the wavelength of the incident light is large compared with the size of the atom, the scattering can be described in terms of classical electromagnetic theory and is called **Rayleigh scattering** after Lord Rayleigh, who worked out the theory in 1871. The probability of Rayleigh scattering varies as $1/\lambda^4$. This means that blue light is scattered much more readily than red light, which accounts for the bluish color of the sky. The removal of blue light by Rayleigh scattering also accounts for the reddish color of the transmitted light seen in sunsets.

Figure 33-3*b* shows **inelastic scattering,** which occurs when the incident photon has enough energy to cause the atom to make a transition to an excited state. The energy of the scattered photon hf' is less than that of the incident photon hf by ΔE, the difference between the energy of the ground state and the energy of the excited state. Inelastic scattering of light from molecules was first observed by the Indian physicist C. V. Raman and is often referred to as **Raman scattering.**

In Figure 33-3*c*, the energy of the incident photon is just equal to the difference in energy between the ground state and the first excited state of the atom. The atom makes a transition to its first excited state and then after a short delay decays by spontaneous emission back to the ground state with the emission of a photon whose energy is equal to that of the incident photon. The phase of the emitted photon is not correlated with the phase of the incident photon. This multistep process is called **resonance absorption.**

In Figure 33-3*d*, the energy of the incident photon is great enough to excite the atom to one of its higher excited states. The atom then loses its energy by spontaneous emission as it makes one or more transitions to lower energy states. A common example occurs when the atom is excited by ultraviolet light and emits visible light as it returns to its ground state. This process is called **fluorescence.** Since the lifetime of a typical excited atomic energy state is of the order of 10^{-8} s, this process appears to occur instantaneously. However, some excited states have much longer lifetimes—of the order of milliseconds or occasionally seconds or even minutes. Such a state is called a **metastable state. Phosphorescent materials** have very long-lived metastable states, and so emit light long after the original excitation.

Figure 33-3*e* illustrates the photoelectric effect, in which the absorption of the photon ionizes the atom by causing the emis-

Figure 33-3 Photon–atom interactions.

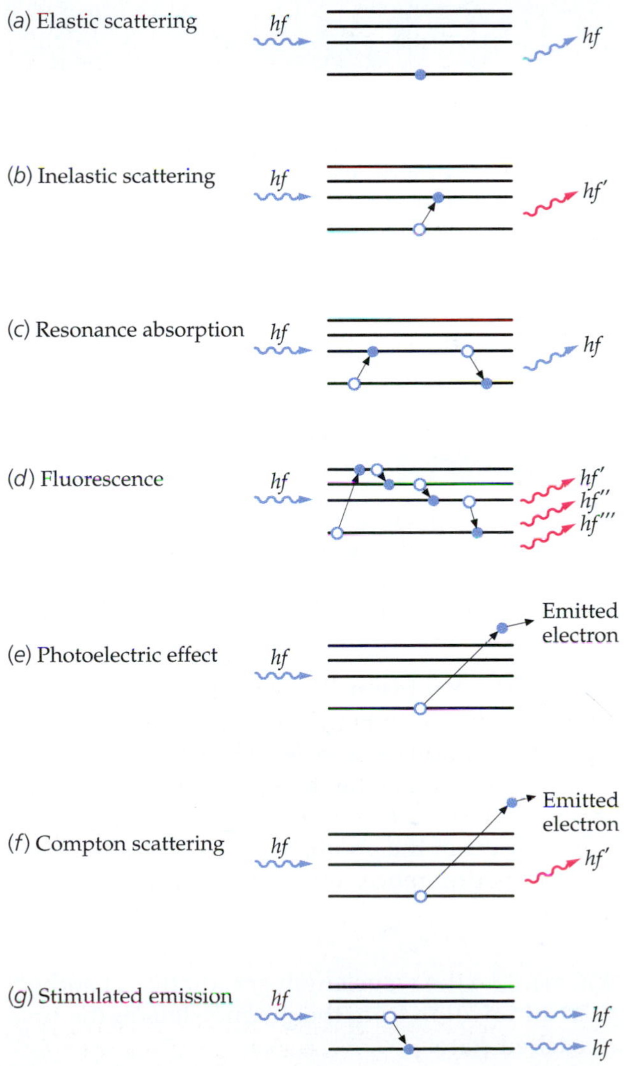

sion of an electron. Figure 33-3*f* illustrates Compton scattering, which occurs if the energy of the incident photon is much greater than the ionization energy. Note that in Compton scattering, a photon is emitted, whereas in the photoelectric effect, the photon is absorbed with none emitted.

Figure 33-3*g* illustrates **stimulated emission.** This process occurs if the atom or molecule is initially in an excited state of energy E_2, and the energy of the incident photon is equal to $E_2 - E_1$, where E_1 is the energy of a lower state or the ground state. In this case, the oscillating electromagnetic field associated with the incident photon stimulates the excited atom or molecule, which then emits a photon in the same direction as the incident photon and in phase with it. In stimulated emission, the phase of the light emitted from one atom is related to that emitted by every other atom, so the resulting light is coherent. As a result, interference of the light from different atoms can be observed.

(*a*)

(*b*)

A collection of minerals in (*a*) daylight and (*b*) ultraviolet light (sometimes called "black light"). Identified by number in the schematic (*c*), they are 1, powellite; 2, willemite; 3, scheelite; 4, calcite; 5, calcite and willemite composite; 6, optical calcite; 7, willemite; and 8, opal. The change in color is due to the minerals fluorescing under the ultraviolet light. In optical calcite, both fluorescence and phosphorescence occur.

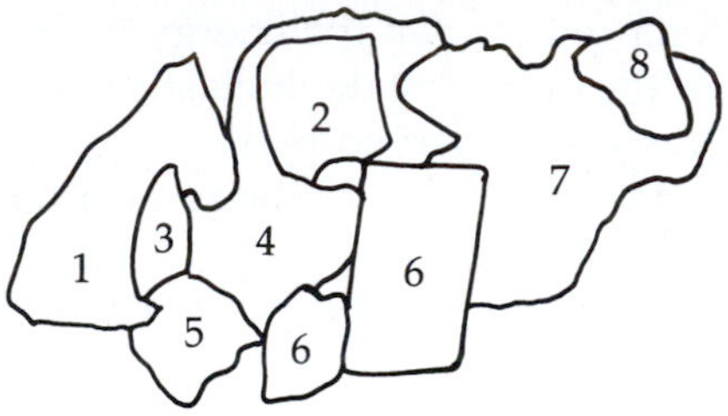

(*c*)

Example 33-1

The first excited state of potassium is E_2 = 1.62 eV above the ground state E_1, which we take to be zero. Potassium also has energy levels at E_3 = 2.61 eV and E_4 = 3.07 eV. (*a*) What is the maximum wavelength of radiation that can be absorbed by potassium in its ground state? Calculate the wavelength when the atom makes a transition from (*b*) E_4 to the ground state, and (*c*) from E_4 to E_3.

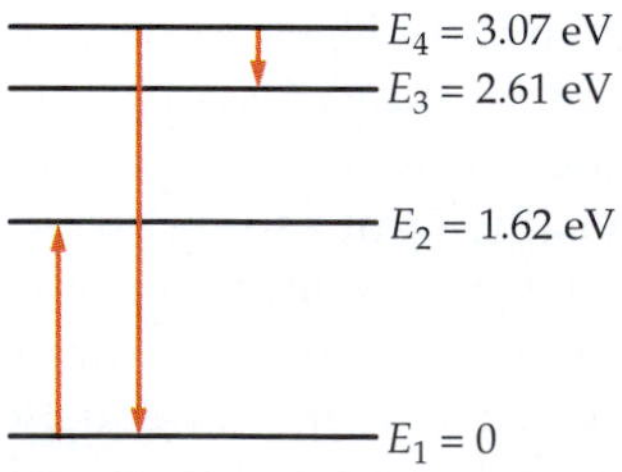

Figure 33-4

Picture the Problem The ground state E_1 and the first three excited energy levels are shown in Figure 33-4. (*a*) Since the wavelength is related to the energy of a photon by $\lambda = hc/\Delta E$, longer wavelengths correspond to smaller energy differences. The smallest energy difference for a transition originating at the ground state is from the ground state to the first excited state. (*b*) The wavelengths of the photons given off when the atom de-excites are related to the energy differences by $\lambda = hc/\Delta E$.

(*a*) Calculate the wavelength of radiation absorbed in a transition from the ground state to the first excited state:

$$\lambda = \frac{hc}{\Delta E} = \frac{1240\ \text{eV}\cdot\text{nm}}{1.62\ \text{eV} - 0} = 765\ \text{nm}$$

(*b*) For the transition from E_4 to the ground state, the photon energy is $E_4 - E_1 = E_4$. Calculate the wavelength of radiation emitted in this transition:

$$\lambda = \frac{hc}{\Delta E} = \frac{hc}{E_4 - E_1} = \frac{1240 \text{ eV}\cdot\text{nm}}{3.07 \text{ eV} - 0} = 404 \text{ nm}$$

(*c*) For the transition from E_4 to E_3, the photon energy is $E_4 - E_3$. Calculate the wavelength of radiation emitted in this transition:

$$\lambda = \frac{hc}{\Delta E} = \frac{hc}{E_4 - E_3} = \frac{1240 \text{ eV}\cdot\text{nm}}{3.07 \text{ eV} - 2.61 \text{ eV}} = 2700 \text{ nm}$$

Remarks The wavelength of radiation emitted in the transition from E_2 to the ground state E_1 is 765 nm, the same as that for radiation absorbed in the transition from the ground state to E_2. This transition and the one from E_4 to the ground state both result in photons in the visible spectrum.

Lasers

The laser (*l*ight *a*mplification by *s*timulated *e*mission of *r*adiation) is a device that produces a strong beam of coherent photons by stimulated emission. Consider a system consisting of atoms that have a ground state of energy E_1 and an excited metastable state of energy E_2. If these atoms are irradiated by photons of energy $E_2 - E_1$, those atoms in the ground state can absorb a photon and make the transition to state E_2, whereas those atoms already in the excited state may be stimulated to decay back to the ground state. The relative probabilities of absorption and stimulated emission were first worked out by Einstein, who showed them to be equal. Ordinarily, nearly all the atoms of the system at normal temperature will initially be in the ground state, so absorption will be the main effect. To produce more stimulated-emission transitions than absorption transitions, we must arrange to have more atoms in the excited state than in the ground state. This condition, called population inversion, can be achieved by a method called optical pumping in which atoms are "pumped" up to levels of energy greater than E_2 by the absorption of an intense auxiliary radiation. The atoms then decay down to state E_2 either by spontaneous emission or by nonradiative transitions such as those due to collisions.

Figure 33-5 shows a schematic diagram of the first laser, a ruby laser built by Theodore Maiman in 1960. It consists of a ruby rod a few centimeters long surrounded by a helical gaseous flashtube that emits a broad spectrum of light. The ends of the ruby rod are flat and perpendicular to the axis of the rod. Ruby is a transparent crystal of Al_2O_3 with a small amount (about 0.05%) of chromium. It appears red because the chromium ions (Cr^{3+}) have strong absorption bands in the blue and green regions of the visible spectrum, as shown in Figure 33-6. The energy levels of chromium that are important for the operation of a ruby laser are shown in Figure 33-7. When the

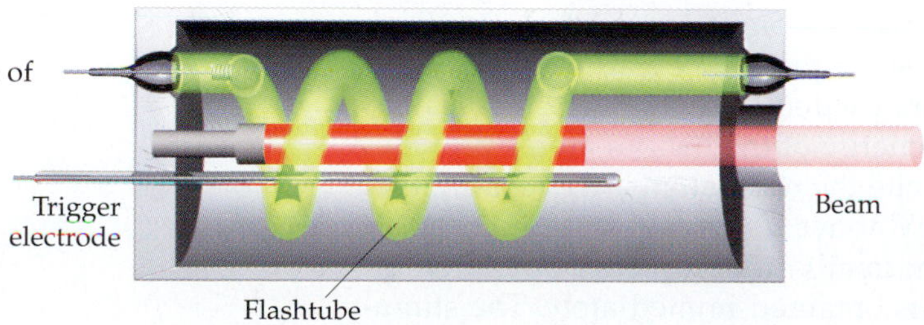

Figure 33-5 Schematic diagram of the first ruby laser.

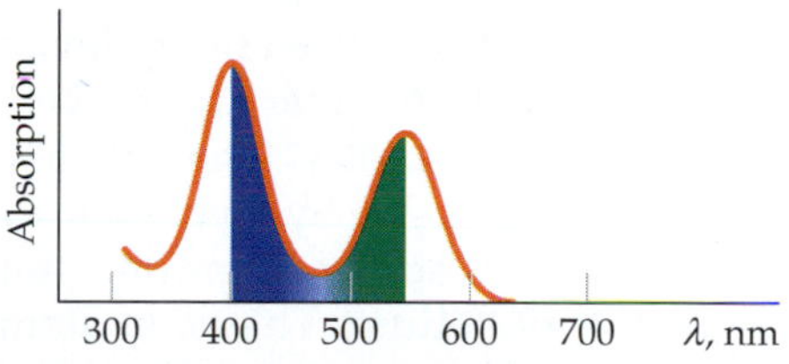

Figure 33-6 Absorption versus wavelength for Cr^{3+} in ruby. Ruby appears red because of the strong absorption of green and blue light by the chromium ions.

flashtube is fired, there is an intense burst of light lasting a few milliseconds. Absorption excites many of the chromium ions to the bands of energy levels indicated by the shading in Figure 33-7. The chromium ions then relax, giving up their energy to the crystal in nonradiative transitions as they drop down to a pair of metastable states labeled E_2 in the figure. These metastable states are about 1.79 eV above the ground state. If the flash is intense enough, more atoms will make the transition to the states E_2 than remain in the ground state. As a result, the populations of the ground state and the metastable states become inverted. When some of the atoms in the states E_2 decay to the ground state by spontaneous emission, they emit photons of energy 1.79 eV and wavelength 694.3 nm. Some of these photons stimulate other excited atoms to emit photons of the same energy and wavelength.

In the ruby laser, both ends of the crystal are silvered such that one end is almost totally reflecting and the other end is only partially reflecting (about 99%). When photons traveling parallel to the axis of the crystal strike the silvered ends, all are reflected from the back face and most are reflected from the front face with a few photons escaping through the partially silvered front face. During each pass through the crystal, the photons stimulate more and more atoms so that the photon beam builds up and an intense beam is emitted (Figure 33-8). Modern ruby lasers generate intense light beams with energies ranging from 50 to 100 J in pulses lasting a few milliseconds. The beam can have a diameter as small as 1 mm and an angular divergence as small as 0.25 milliradian to about 7 milliradians.

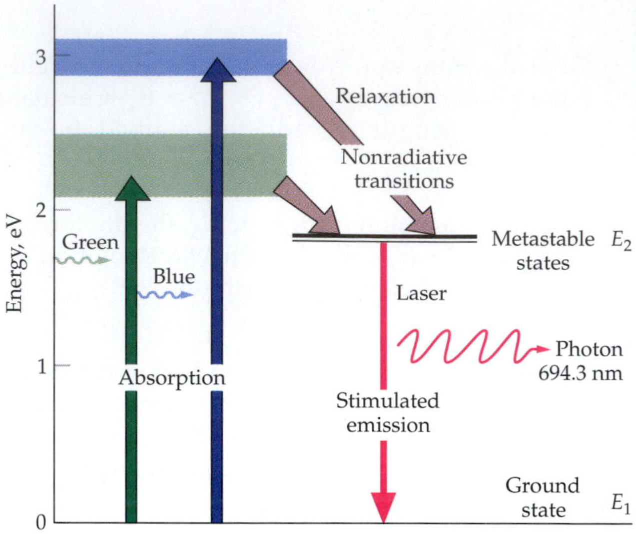

Figure 33-7 Energy levels in a ruby laser. To make the population of the metastable states greater than that of the ground state, the ruby crystal is subjected to intense radiation that contains energy in the green and blue wavelengths. This excites atoms from the ground state to the bands of energy levels indicated by the shading, from which they decay to the metastable states by nonradiative transitions.

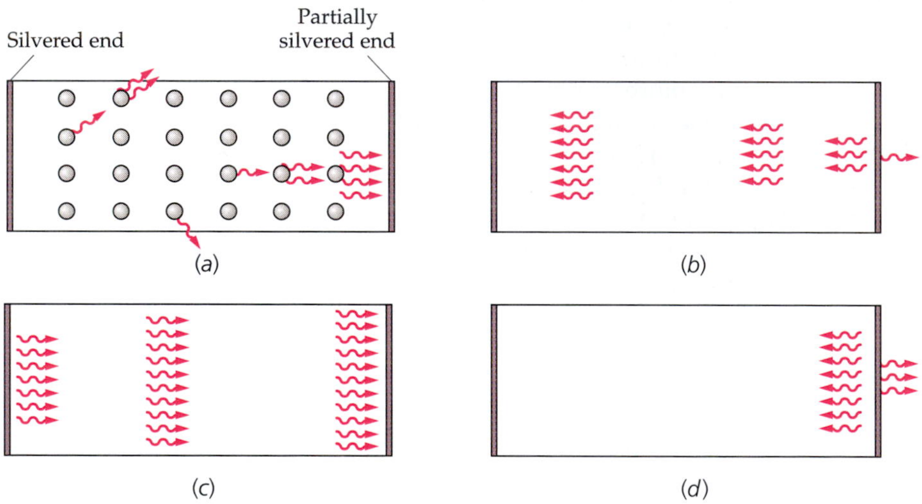

Figure 33-8 Buildup of photon beam in a laser. (*a*) When irradiated, some atoms spontaneously emit photons, some of which travel to the right and stimulate other atoms to emit photons parallel to the axis of the crystal. (*b*) Of the four photons that strike the right face, one is transmitted and three are reflected. As the reflected photons traverse the laser crystal, they stimulate other atoms to emit photons and the beam builds up. By the time the beam reaches the right face again (*c*), it comprises many photons. (*d*) Some of these photons are transmitted, the rest are reflected.

Population inversion is achieved somewhat differently in the continuous helium–neon laser. The energy levels of helium and neon that are important for operation of the laser are shown in Figure 33-9. Helium has an excited energy state $E_{2,\mathrm{He}}$ that is 20.61 eV above its ground state. Helium atoms are excited to state $E_{2,\mathrm{He}}$ by an electric discharge. Neon has an excited state $E_{3,\mathrm{Ne}}$ that is 20.66 eV above its ground state. This is just 0.05 eV above the first excited state of helium. The neon atoms are excited to state $E_{3,\mathrm{Ne}}$ by collisions with excited helium atoms. The kinetic energy of the helium atoms provides the extra 0.05 eV of energy needed to excite the neon atoms. There is another excited state of neon $E_{2,\mathrm{Ne}}$ that is 18.70 eV above its ground state and 1.96 eV below state $E_{3,\mathrm{Ne}}$. Since state $E_{2,\mathrm{Ne}}$ is normally unoccupied, population inversion between states $E_{3,\mathrm{Ne}}$ and $E_{2,\mathrm{Ne}}$ is obtained immediately. The stimu-

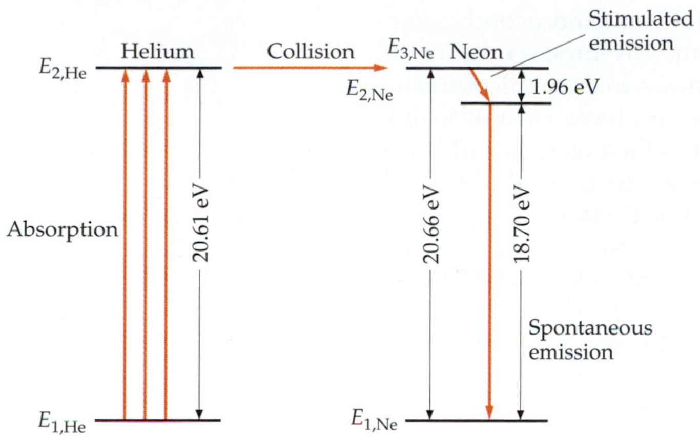

Figure 33-9 Energy levels of helium and neon that are important for the helium–neon laser. The helium atoms are excited by electrical discharge to an energy state 20.61 eV above the ground state. They collide with neon atoms, exciting some neon atoms to an energy state 20.66 eV above the ground state. Population inversion is thus achieved between this level and one 1.96 eV below it. The spontaneous emission of photons of energy 1.96 eV stimulates other atoms in the upper state to emit photons of energy 1.96 eV.

lated emission that occurs between these states results in photons of energy 1.96 eV and wavelength 632.8 nm, which produces a bright red light. After stimulated emission, the atoms in state $E_{2,\mathrm{Ne}}$ decay to the ground state by spontaneous emission.

Note that there are four energy levels involved in the helium–neon laser, whereas the ruby laser involved only three levels. In a three-level laser, population inversion is difficult to achieve because more than half the atoms in the ground state must be excited. In a four-level laser, population inversion is easily achieved because the state after stimulated emission is not the ground state but an excited state that is normally unpopulated.

Figure 33-10 shows a schematic diagram of a helium–neon laser commonly used for physics demonstrations. It consists of a gas tube containing 15% helium gas and 85% neon gas. A totally reflecting flat mirror is mounted at one end of the gas tube and a partially reflecting concave mirror is placed at the other end. The concave mirror focuses parallel light at the flat mirror and also acts as a lens that transmits part of the light so that it emerges as a parallel beam.

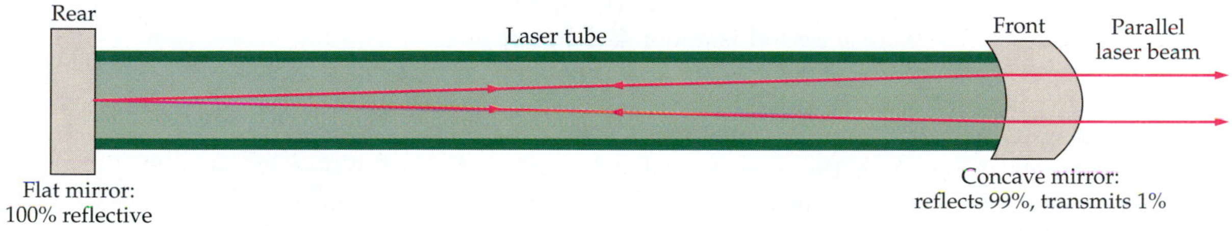

Figure 33-10 Schematic drawing of a helium–neon laser. The use of a concave mirror rather than a second plane mirror makes the alignment of the mirrors less critical than it is for the ruby laser. The concave mirror on the right also serves as a lens that focuses the emitted light into a parallel beam.

A laser beam is coherent, very narrow, and intense. Its coherence makes the laser beam useful in the production of holograms, which we discuss in Chapter 35. The precise direction and small angular spread of the beam make it useful as a surgical tool for destroying cancer cells or reattaching a detached retina. Lasers are also used by surveyors for precise alignment over large distances. Distances can be accurately measured by reflecting a laser pulse from a mirror and measuring the time the pulse takes to travel to the mirror and back. The distance to the moon has been measured to within a few centimeters using a mirror placed on the moon for that purpose. Laser beams are also used in fusion research. An intense laser pulse is focused on tiny pellets of deuterium–tritium in a combustion chamber. The beam heats the pellets to temperatures of the order of 10^8 K in a very short time, causing the deuterium and tritium to fuse and release energy.

Laser technology is advancing so fast that it is possible to mention only a few of the recent developments. In addition to the ruby laser, there are many other solid-state lasers with output wavelengths ranging from about 170 nm to about 3900 nm. Lasers that generate more than 1 kW of continuous power have been constructed. Pulsed lasers can now deliver nanosecond pulses of power exceeding 10^{14} W. Various gas lasers can now produce beams of wavelengths ranging from the far infrared to the ultraviolet. Semiconductor lasers (also known as diode lasers or junction lasers) have shrunk in just 10 years from the size of a pinhead to mere billionths of a meter. Liquid lasers using chemical dyes can be tuned over a range of wavelengths (about 70 nm for continuous lasers and more than 170 nm for pulsed lasers). A relatively new laser, the free-electron laser, extracts light energy from a beam of free electrons moving through a spatially varying magnetic field. The free-electron laser has the potential for very high power and high efficiency and can be tuned over a large range of wavelengths. There appears to be no limit to the variety and uses of modern lasers.

33-4 The Speed of Light

The first effort to measure the speed of light was made by Galileo. He and a partner stood on hilltops about three kilometers apart, each with a lantern and a shutter to cover it. Galileo proposed to measure the time it took for light to travel back and forth between the experimenters. First, one would uncover his lantern, and when the other saw the light, he would uncover his. The time between the first partner's uncovering his lantern and his seeing the light from the other lantern would be the time it took for light to travel back and forth between the experimenters. Though this method is sound in principle, the speed of light is so great that the time interval to be measured is much smaller than fluctuations in human response time, so Galileo was unable to obtain any value for the speed of light.

The first indication of the true magnitude of the speed of light came from astronomical observations of the period of Io, one of the moons of Jupiter. This period is determined by measuring the time between eclipses (when the moon Io disappears behind Jupiter). The eclipse period is about 42.5 h, but measurements made when the earth is moving away from Jupiter along path ABC in Figure 33-11 give a greater time for this period than do measurements made when the earth is moving toward Jupiter along path CDA in the figure. Since these measurements differ from the average value by only about 15 s, the discrepancies were difficult to measure accurately. In 1675, the astronomer Ole Römer attributed these discrepancies to the fact that the speed of light is not infinite. During the 42.5 h between eclipses of Jupiter's moon, the distance between the earth and Jupiter changes, making the path for the light longer or shorter. Römer devised the following method for measuring the cumulative effect of these discrepancies. Because Jupiter moves much more slowly than earth, we can neglect its motion. When the earth is at point A, nearest to Jupiter, the distance between the earth and Jupiter is changing negligibly. The period of Io's eclipse is measured, providing the time between the beginnings of successive eclipses. Based on this measurement, the number of eclipses in 6 months is computed, and the time when an eclipse should begin a half-year later when the earth is at point C is predicted. When the earth is actually at C, the observed beginning of the eclipse is about 16.6 min later than predicted. This is the time it takes light to travel a distance equal to the diameter of the earth's orbit.

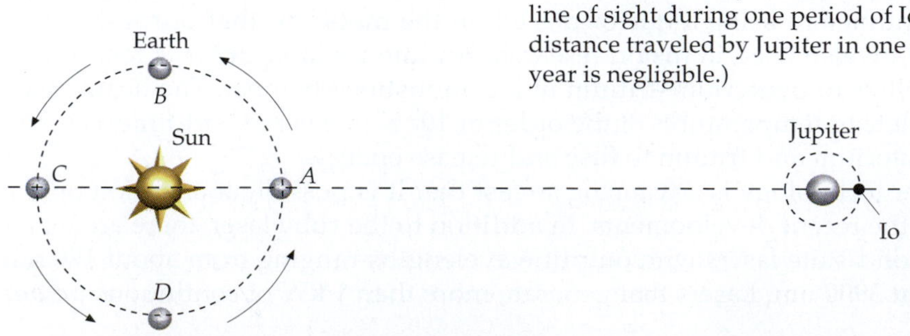

Figure 33-11 Römer's method of measuring the speed of light. The time between eclipses of Jupiter's moon Io appears to be greater when the earth is moving along path ABC than when it is moving along path CDA. The difference is due to the time it takes light to travel the distance traveled by the earth along the line of sight during one period of Io. (The distance traveled by Jupiter in one earth year is negligible.)

(*a*)

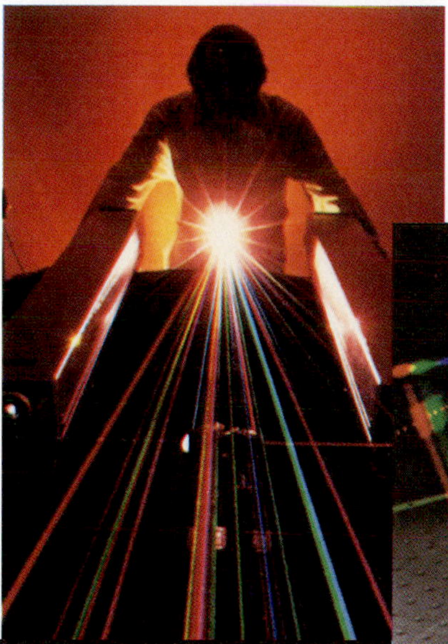

(*b*)

(*c*)

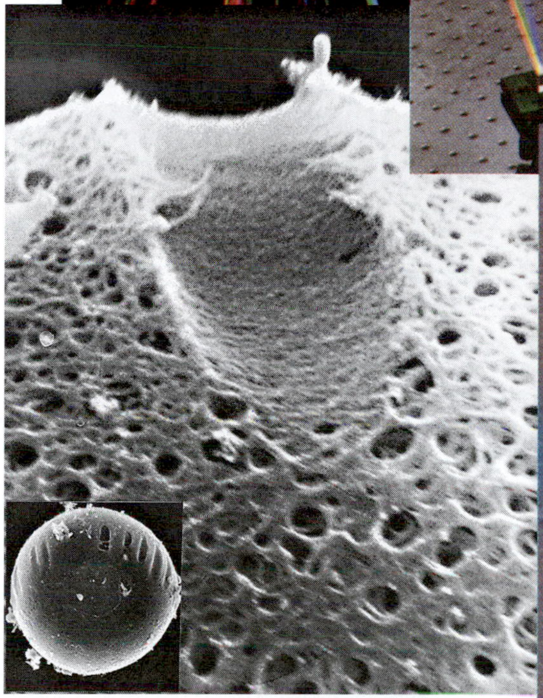

(*d*)

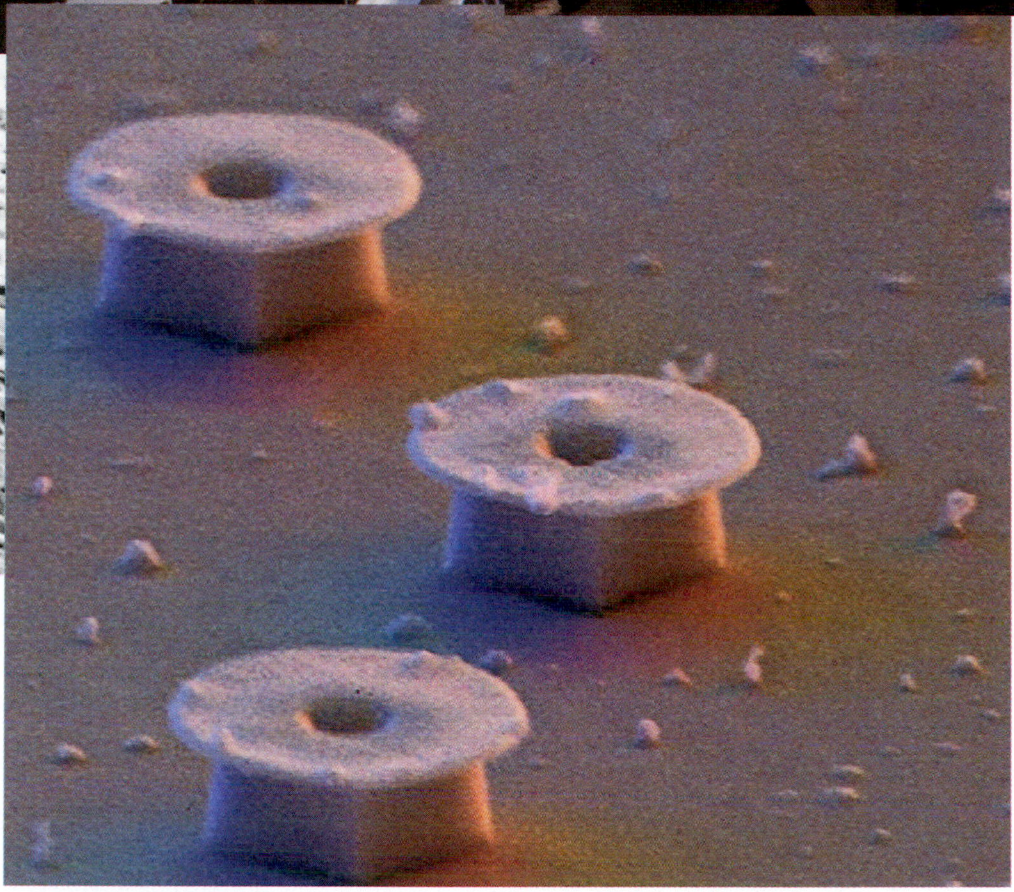

(*e*)

(*a*) Beams from a krypton and an argon laser, split into their component wavelengths. In these gas lasers, krypton and argon atoms have been stripped of multiple electrons, forming positive ions. The light-emitting energy transitions occur when excited electrons in the ions decay from one upper energy level to another. Here, several energy transitions are occurring at once, each corresponding to emitted light of a different wavelength.
(*b*) A femtosecond pulsed laser. By a technique known as "modelocking," different excited modes within a laser's cavity can be made to interfere with one another and create a series of ultrashort pulses, picoseconds long, that correspond to the time it takes light to bounce back and forth once within the cavity. Ultrashort pulses have been used as probes to study the behavior of molecules during chemical reactions.
(*c*) A carbon dioxide laser takes just two minutes to cut out a steel saw blade.
(*d*) A groove etched in the zona pellucida (protective outer covering) of a mouse egg by a "laser scissor" facilitates implantation. This technique has already been applied in human fertility therapies. Several effects contribute to the ability of the finely focused laser to cut on such a delicate scale—photon absorption may heat the target, break molecular bonds, or drive chemical reactions.
(*e*) The so-called nanolasers shown are semiconductor disks mere microns in diameter and fractions of a micron in width. These tiny lasers work like their larger counterparts, generating and trapping photons until stimulated emission creates enough intensity for the photons to break out from the perimeters of the disks. Exploiting quantum effects that prevail on this microscopic scale, nanolasers promise great efficiency and are being explored as ultrafast, low-energy switching devices.

Example 33-2

The diameter of the earth's orbit is 3.00×10^{11} m. If light takes 16.6 min to travel this distance, what is the speed of light in meters per second?

Picture the Problem We divide the distance traveled by the time and convert the time from minutes to seconds.

1. The speed is the distance divided by the time:

$$c = \frac{\Delta x}{\Delta t} = \frac{3.00 \times 10^{11}\ \text{m}}{16.6\ \text{min}} \times \frac{1\ \text{min}}{60\ \text{s}} = 3.01 \times 10^{8}\ \text{m/s}$$

Remark Römer obtained a considerably smaller value for c because he used 22 min for Δt.

The first nonastronomical measurement of the speed of light was made by the French physicist Fizeau in 1849. On a hill in Paris, Fizeau placed a light source and a system of lenses arranged such that the light reflected from a semitransparent mirror was focused on a gap in a toothed wheel as shown in Figure 33-12. On a distant hill (about 8.63 km away), he placed a mirror to reflect the light back to be viewed by an observer as shown. The toothed wheel was rotated, and the speed of rotation was varied. At low speeds of rotation, no light was visible because the light that passed through a gap in the rotating wheel and was reflected back by the mirror was obstructed by the next tooth of the wheel. The speed of rotation was then increased. The light suddenly became visible when the rotation speed was such that the reflected light passed through the next gap in the wheel. The time for the wheel to rotate through the angle between successive gaps equals the time for the light to make the round trip to the distant mirror and back.

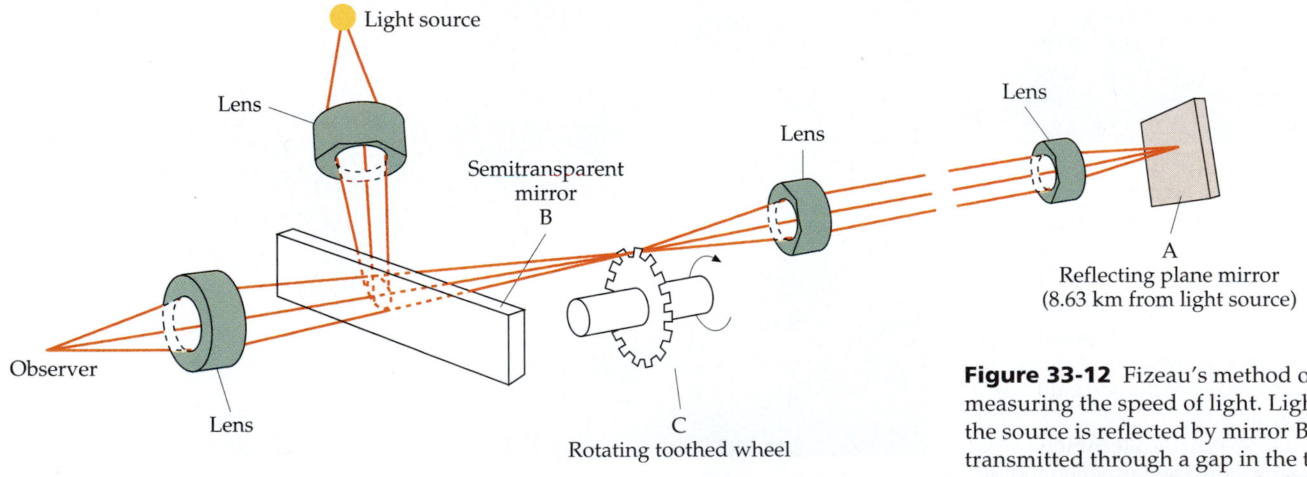

Figure 33-12 Fizeau's method of measuring the speed of light. Light from the source is reflected by mirror B and is transmitted through a gap in the toothed wheel to mirror A. The speed of light is determined by measuring the angular speed of the wheel that will permit the reflected light to pass through the next gap in the toothed wheel so that an image of the source is observed.

Fizeau's method was improved upon by Foucault, who replaced the toothed wheel with an eight-sided rotating mirror as shown in Figure 33-13. Light strikes one face of the mirror, is reflected to a distant fixed mirror, and is then reflected back to another face of the rotating mirror and then to an observing telescope. When the mirror rotates through one-eighth of a turn during the time it takes for the light to travel to the fixed mirror and back (or $n/8$ turns, where n is an integer) another face of the mirror is in the right position for the reflected light to enter the telescope. In about 1850, Foucault measured the speed of light in air and in water and showed that it is less in water. Using essentially the same method, the American physicist A. A. Michelson made precise measurements of the speed of light from 1880 to 1930.

Another method of determining the speed of light involves the measurement of the electrical constants ϵ_0 and μ_0 to determine c from $c = 1/\sqrt{\epsilon_0\mu_0}$.

The various methods we have discussed for measuring the speed of light are all in general agreement. Today, the speed of light is defined to be exactly

$$c = 299{,}792{,}458 \text{ m/s} \qquad 33\text{-}5$$

and the standard unit of length, the meter, is defined in terms of this speed. A measurement of the speed of light is therefore now a measurement of the size of the meter, which is the distance light travels in 1/299,792,458 s. The value 3×10^8 m/s for the speed of light is accurate enough for nearly all calculations. The speed of radio waves and all other electromagnetic waves (in a vacuum) is the same as the speed of light.

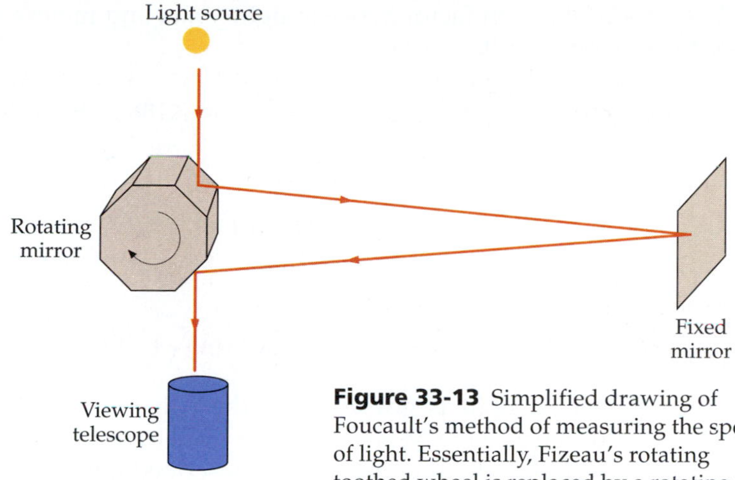

Figure 33-13 Simplified drawing of Foucault's method of measuring the speed of light. Essentially, Fizeau's rotating toothed wheel is replaced by a rotating octagonal mirror. When the mirror makes an eighth of a revolution during the time it takes for the light to travel to the fixed mirror and back, the next face of the mirror is in the proper position to reflect the light into the telescope.

Example 33-3

In Fizeau's experiment, the wheel had 720 teeth, and light was observed when the wheel rotated at 25.2 revolutions per second. If the distance from the wheel to the distant mirror was 8.63 km, what was Fizeau's value for the speed of light?

Picture the Problem The time taken for the light to travel from the wheel to the mirror and back is the time for the wheel to rotate from one tooth to the next, which is the time for the wheel to make $\frac{1}{720}$ revolution.

1. The speed is the distance divided by the time: $c = \dfrac{\Delta x}{\Delta t}$

2. The distance is twice the distance from the wheel to the mirror: $\Delta x = 2 \times 8.63 \text{ km} = 17.3 \text{ km}$

3. Calculate the time to make $\frac{1}{720}$ revolution: $\Delta t = \dfrac{1 \text{ s}}{25.2 \text{ rev}}\left(\dfrac{1}{720}\text{ rev}\right) = 5.51 \times 10^{-5} \text{ s}$

4. Substitute these values to calculate c: $c = \dfrac{\Delta x}{\Delta t} = \dfrac{17.3 \times 10^3 \text{ m}}{5.51 \times 10^{-5} \text{ s}} = 3.14 \times 10^8 \text{ m/s}$

Remark This result is about 5% too high.

Exercise Space travelers on the moon use electromagnetic waves to communicate with the space control center on earth. Use $c = 3 \times 10^8$ m/s to calculate the time delay for their signal to reach the earth, which is 3.84×10^8 m away. (*Answer* 1.28 s each way)

Large distances are often given in terms of the distance traveled by light in a given time. For example, the distance to the sun is 8.33 light-minutes, written 8.33 c-min. A light-year is the distance light travels in one year. We can easily find a conversion factor between light-years and meters. The number of seconds in one year is

$$1\text{ y} = 1\text{ y} \times \frac{365.24\text{ d}}{1\text{ y}} \times \frac{24\text{ h}}{1\text{ d}} \times \frac{3600\text{ s}}{1\text{ h}} = 3.156 \times 10^7\text{ s}$$

The number of meters in one light-year is thus

$$1\ c\text{-year} = (2.998 \times 10^8\text{ m/s})(3.156 \times 10^7\text{ s}) = 9.46 \times 10^{15}\text{ m} \qquad 33\text{-}6$$

33-5 The Propagation of Light

The propagation of light is governed by the wave equation discussed in Chapter 32. But long before Maxwell's theory of electromagnetic waves, the propagation of light and other waves was described empirically by two interesting and very different principles attributed to the Dutch physicist Christian Huygens (1629–1695) and the French mathematician Pierre de Fermat (1601–1665).

Huygens' Principle

Figure 33-14 shows a portion of a spherical wavefront emanating from a point source. The wavefront is the locus of points of constant phase. If the radius of the wavefront is r at time t, its radius at time $t + \Delta t$ is $r + c\,\Delta t$, where c is the speed of the wave. However, if a part of the wave is blocked by some obstacle or if the wave passes through a different medium, as in Figure 33-15, the determination of the new wavefront at time $t + \Delta t$ is much more difficult.

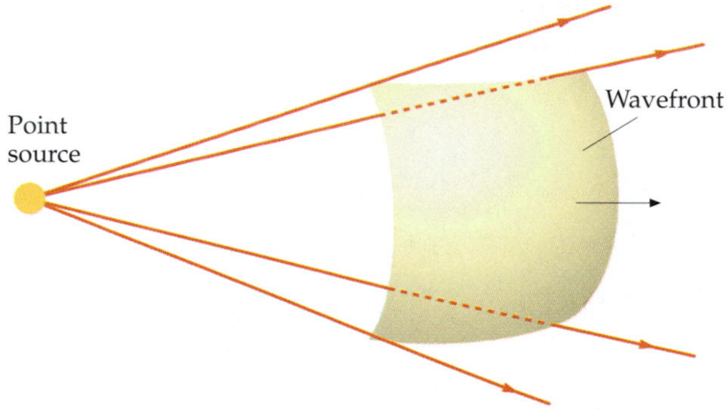

Figure 33-14 Spherical wavefront from a point source.

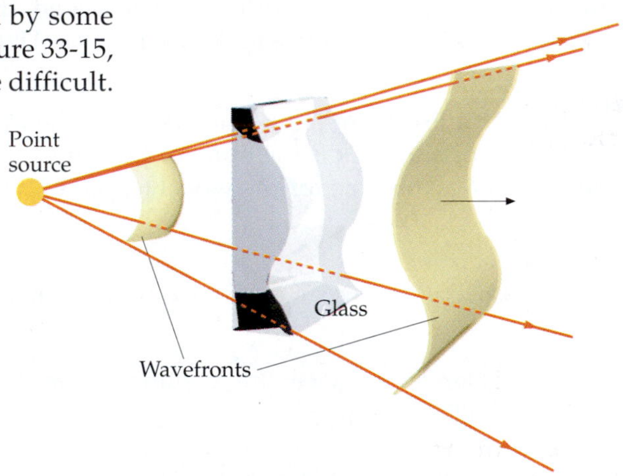

Figure 33-15 Wavefront from a point source before and after passing through an irregularly shaped piece of glass.

The propagation of any wave through space can be described using a geometric method discovered by Huygens about 1678, which is now known as **Huygens' principle** or **Huygens' construction:**

> Each point on a primary wavefront serves as the source of spherical secondary wavelets that advance with a speed and frequency equal to those of the primary wave. The primary wavefront at some later time is the envelope of these wavelets.

Huygens' principle

Figure 33-16 shows the application of Huygens' principle to the propagation of a plane wave and of a spherical wave. Of course, if each point on a wavefront were really a point source, there would be waves in the backward direction as well. Huygens ignored these back waves.

Huygens' principle was later modified by Fresnel so that the new wavefront was calculated from the old wavefront by superposition of the wavelets considering their relative amplitudes and phases. Kirchhoff later showed that the Huygens–Fresnel principle was a consequence of the wave equation, thus putting it on a firm mathematical basis. Kirchhoff showed that the intensity of the wavelets depends on the angle and is zero in the backward direction.

We will use Huygens' principle to derive the laws of reflection and refraction in Section 33-8. In Chapter 35, we apply Huygens' principle with Fresnel's modification to calculate the diffraction pattern of a single slit. Because the wavelength of light is so small, we can often use the ray approximation to describe its propagation.

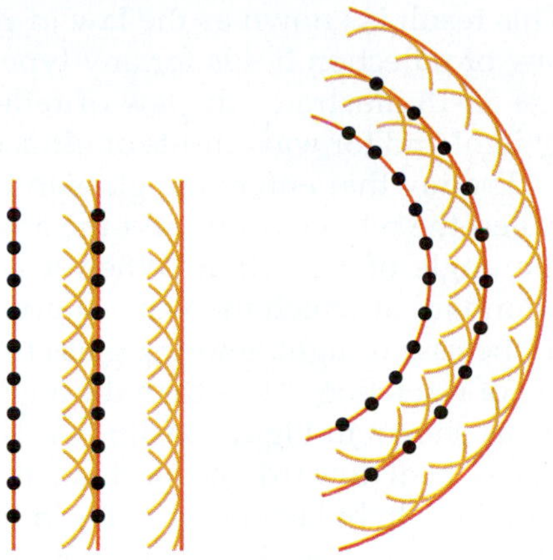

Figure 33-16 Huygens' construction for the propagation to the right of (*a*) a plane wave and (*b*) an outgoing spherical, or circular, wave.

Fermat's Principle

The propagation of light can also be described by Fermat's principle:

> The path taken by light in traveling from one point to another is such that the time of travel is a minimum.*

Fermat's principle

In Section 33-8 we will use Fermat's principle to derive the laws of reflection and refraction.

33-6 Reflection and Refraction

The speed of light in a transparent medium such as air, water, or glass is less than the speed $c = 3 \times 10^8$ m/s in vacuum. A transparent medium is characterized by the **index of refraction** n, which is defined as the ratio of the speed of light in a vacuum, c, to the speed in the medium, v:

$$n = \frac{c}{v} \qquad \text{33-7}$$

Definition—Index of refraction

For water, $n = 1.33$, whereas for glass n ranges from about 1.5 to 1.66 depending on the type of glass. Diamond has a very high index of refraction of about 2.4. The index of refraction of air is about 1.0003 so for most purposes we can assume the speed of light in air is the same as in vacuum.

When a beam of light strikes a boundary surface separating two different media, such as an air–glass interface, part of the light energy is reflected and part enters the second medium. If the incident light is not perpendicular to the surface, then the transmitted light is not parallel to the incident light. The change in direction of the transmitted ray is called **refraction.** Figure 33-17 shows a light ray striking a smooth air–glass interface. The angle θ_1 between the incident ray and the normal (the line perpendicular to the surface) is called the **angle of incidence,** and the plane defined by these two lines is called the **plane of incidence.** The reflected ray lies in the plane of incidence and makes an angle θ_1' with the normal that is equal to the angle of incidence as shown in the figure:

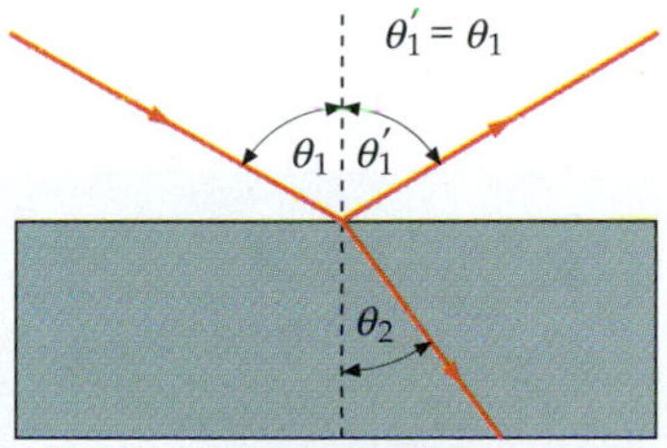

Figure 33-17 The angle of reflection θ_1' equals the angle of incidence θ_1. The angle of refraction θ_2 is less than the angle of incidence if the light speed in the second medium is less than that in the incident medium.

$$\theta_1' = \theta_1 \qquad \text{33-8}$$

Law of reflection

* A more complete and general statement is that the time of travel is stationary with respect to variations in path; that is, if t is expressed in terms of some parameter x, the path taken will be such that $dt/dx = 0$. The important characteristic of a stationary path is that the time taken along nearby paths will be approximately the same as that along the true path.

This result is known as the **law of reflection.** The law of reflection holds for any type of wave. Figure 33-18 illustrates the law of reflection for rays of light and for wavefronts of ultrasonic waves.

The ray that enters the glass in Figure 33-17 is called the refracted ray, and the angle θ_2 is called the angle of refraction. When a wave crosses a boundary at which the wave speed is reduced, as in the case of light entering glass from air, the angle of refraction is less than the angle of incidence θ_1, as shown in Figure 33-17; that is, the refracted ray is bent toward the normal. If, on the other hand, the light beam originates in the glass and is refracted into the air, then the refracted ray is bent away from the normal.

The angle of refraction θ_2 depends on the angle of incidence and on the relative speed of light waves in the two mediums. If v_1 is the wave speed in the incident medium and v_2 is the wave speed in the transmission medium, the angles of incidence and refraction are related by

$$\frac{1}{v_1}\sin\theta_1 = \frac{1}{v_2}\sin\theta_2 \qquad \text{33-9}a$$

Equation 33-9*a* holds for the refraction of any kind of wave incident on a boundary interface separating two media.

In terms of the indexes of refraction of the two media n_1 and n_2, Equation 33-9*a* is

$$n_1 \sin\theta_1 = n_2 \sin\theta_2 \qquad \text{33-9}b$$

Snell's law of refraction

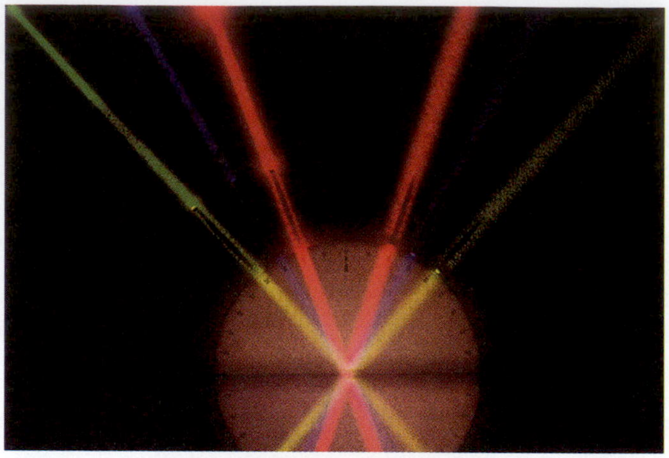

(*a*)

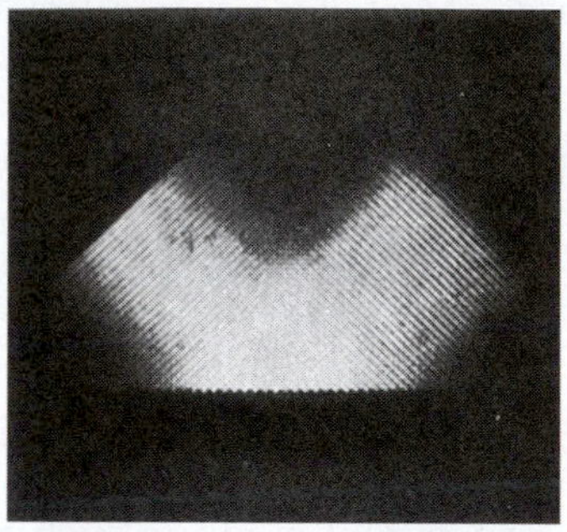

(*b*)

Figure 33-18 (*a*) Light rays reflecting from an air–glass interface showing equal angles of incidence and reflection. (*b*) Ultrasonic plane waves in water reflecting from a steel plate.

This result was discovered experimentally in 1621 by the Dutch scientist Willebrod Snell and is known as **Snell's law** or the **law of refraction.** It was independently discovered a few years later by René Descartes.

Reflection and refraction of a beam of light incident on a glass slab.

Physical Mechanisms for Reflection and Refraction

The physical mechanism of the reflection and refraction of light can be understood in terms of the absorption and reradiation of the light by the atoms in the reflecting or refracting medium. When light traveling in air strikes a glass surface, the atoms in the glass absorb the light and reradiate it at the same frequency in all directions. The waves radiated backward by the glass atoms interfere constructively at an angle equal to the angle of incidence to produce the reflected wave.

The transmitted wave is the result of the interference of the incident wave and the wave produced by the absorption and reradiation of light energy by the atoms in the medium. For light entering glass from air, there is a phase lag between the reradiated wave and the incident wave. There is therefore also a phase lag between the resultant wave and the incident wave. This phase lag means that the position of a wave crest of the transmitted wave is retarded relative to the position of a wave crest of the incident wave in the medium. As a result, the transmitted wave does not travel as far in a given time as the original incident wave; that is, the velocity of the transmitted wave is less than that of the incident wave. The index of refraction is therefore greater than 1. The frequency of the light in the second medium is the same as that of the incident light—the atoms absorb and reradiate the light at the same frequency—but the wave speed is different, so the wavelength of the transmitted light is different from that of the incident light. If λ is the wavelength of light in a vacuum, the wavelength λ' in a medium of index of refraction n is

$$\lambda' = \frac{v}{f} = \frac{c/n}{f} = \frac{\lambda}{n} \qquad 33\text{-}10$$

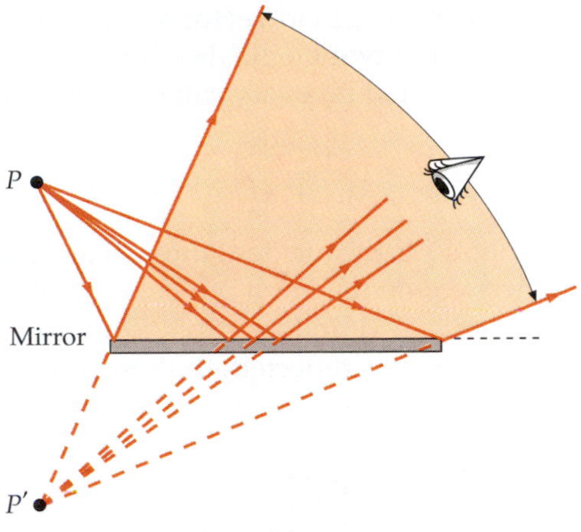

(a)

(b)

Figure 33-19 (*a*) Specular reflection from a smooth surface. (*b*) Specular reflection of trees from water.

Specular and Diffuse Reflection

Figure 33-19*a* shows a bundle of light rays from a point source P that are reflected from a flat surface. After reflection, the rays diverge exactly as if they came from a point P' behind the surface. (This point is called the *image point*. We will study the formation of images by reflecting and refracting surfaces in the next chapter.) When these rays enter the eye, they cannot be distinguished from rays actually diverging from a source at P'.

Reflection from a smooth surface is called **specular reflection.** It differs from **diffuse reflection,** which is illustrated in Figure 33-20. Here, because the surface is rough, the rays from a point reflect in random directions and do not diverge from any point, so there is no image. The reflection of light from the page of this book is diffuse reflection. The glass used in picture frames is sometimes ground slightly to give diffuse reflection and thereby cut down on glare from the light used to illuminate the picture. Diffuse reflection from the surface of the road allows you to see the road when you are driving at night because some of the light from your headlights reflects back toward you.

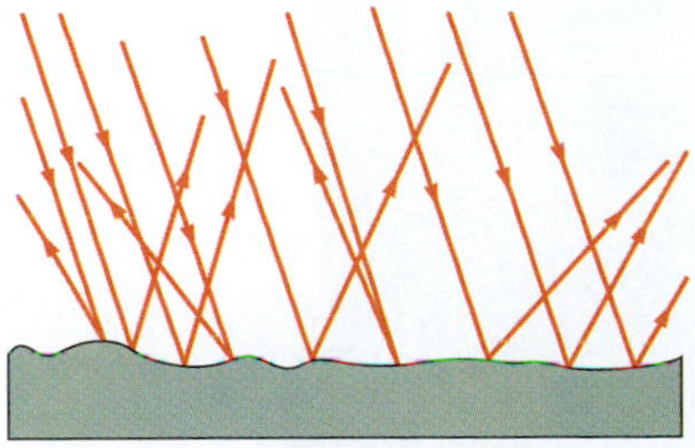
(a)

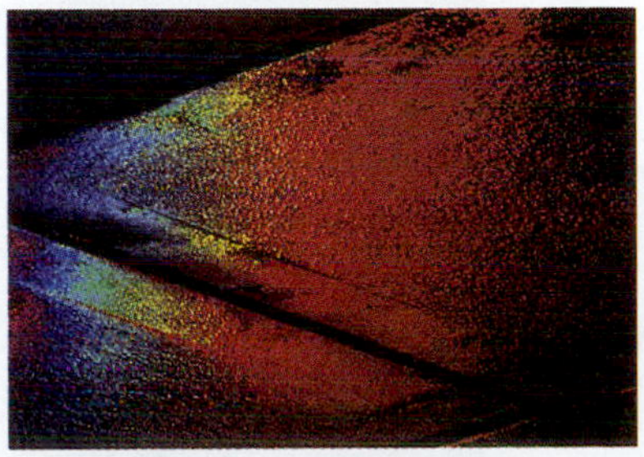
(b)

Figure 33-20 (*a*) Diffuse reflection from a rough surface. (*b*) Diffuse reflection of colored lights from a sidewalk.

Relative Intensity of Reflected and Transmitted Light

The fraction of light energy reflected at a boundary such as an air–glass interface depends in a complicated way on the angle of incidence, the orientation of the electric field vector associated with the wave, and the indices of refraction of the two media. For the special case of normal incidence ($\theta_1 = \theta_1' = 0$), the reflected intensity can be shown to be

$$I = \left(\frac{n_1 - n_2}{n_1 + n_2}\right)^2 I_0 \qquad \text{33-11}$$

where I_0 is the incident intensity and n_1 and n_2 are the indexes of refraction of the two media. For a typical case of reflection from an air–glass interface for which $n_1 = 1$ and $n_2 = 1.5$, Equation 33-11 gives $I = I_0/25$. Only about 4% of the energy is reflected; the rest is transmitted.

Example 33-4

Light traveling in air enters water with an angle of incidence of 45°. If the index of refraction of water is 1.33, what is the angle of refraction?

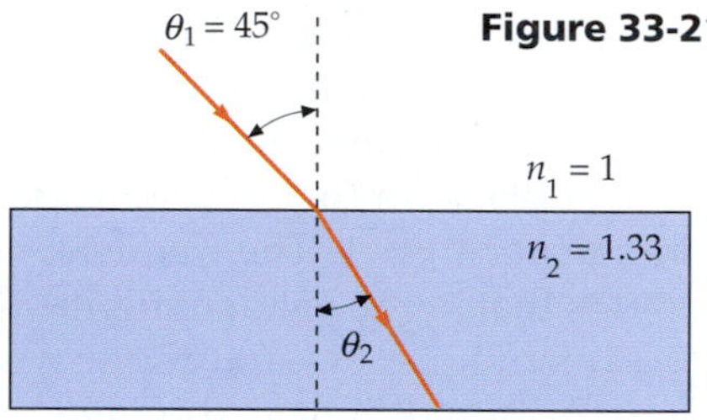

Figure 33-21

Picture the Problem The angle of refraction is found using Snell's law. Let subscripts 1 and 2 refer to the air and water, respectively. Then $n_1 = 1$, $\theta_1 = 45°$, $n_2 = 1.33$, and θ_2 is the angle of refraction (Figure 33-21).

1. Use Snell's law to solve for sin θ_2, the sine of the angle of refraction:

$$n_1 \sin\theta_1 = n_2 \sin\theta_2$$

$$\sin\theta_2 = \left(\frac{n_1}{n_2}\right)\sin\theta_1 = \left(\frac{1}{1.33}\right)\sin 45° = \frac{0.707}{1.33} = 0.532$$

2. Find the angle whose sine is 0.532:

$$\theta_2 = \sin^{-1} 0.532 = 32.1°$$

Remark Note that light is bent closer to the normal in the medium with the larger index of refraction.

Total Internal Reflection

Figure 33-22 shows a point source in glass with rays striking the glass–air interface at various angles. All the rays not perpendicular to the interface are bent away from the normal. As the angle of incidence is increased, the angle of refraction increases until a critical angle of incidence θ_c is reached for

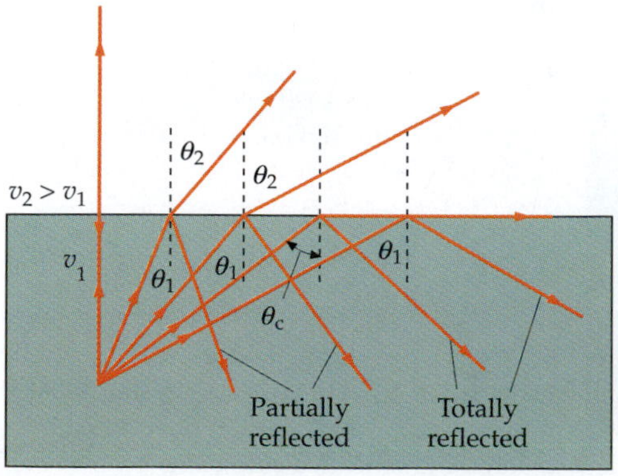

Figure 33-22 (*a*) Total internal reflection. As the angle of incidence is increased, the angle of refraction is increased until, at a critical angle of incidence θ_c, the angle of refraction is 90°. For angles of incidence greater than the critical angle, there is no refracted ray. (*b*) Photograph of refraction and total internal reflection from a water–air interface.

which the angle of refraction is 90°. For incident angles greater than this critical angle, there is no refracted ray. All the energy is reflected. This phenomenon is called **total internal reflection.** The critical angle can be found in terms of the indexes of refraction of the two media by solving Equation 33-9*b* for $\sin \theta_1$ and setting θ_2 equal to 90°:

$$\sin \theta_c = \frac{n_2}{n_1} \sin 90° = \frac{n_2}{n_1} \qquad \text{33-12}$$

Critical angle for total internal reflection

Note that total internal reflection occurs only when the light is originally in the medium with the higher index of refraction. Mathematically, if n_2 is greater than n_1, Snell's law cannot be satisfied because there is no real angle whose sine is greater than 1.

Example 33-5 *try it yourself*

A particular glass has an index of refraction of $n = 1.50$. What is the critical angle for total internal reflection for light leaving this glass and entering air, for which $n = 1.00$?

Cover the column to the right and try these on your own before looking at the answers.

Steps	*Answers*
1. Calculate the sine of the critical angle from Equation 33-12.	$\sin \theta_c = 0.667$
2. Use your result to find the angle.	$\theta_c = 41.8°$

Figure 33-23*a* shows light incident normally on one of the short sides of a 45–45–90° glass prism. If the index of refraction of the prism is 1.5, the critical angle for total internal reflection is 41.8°, as you found in Example 33-5. Since the angle of incidence of the ray on the glass–air interface is 45°, the light will be totally reflected and will exit perpendicular to the other face of the prism as shown. In Figure 33-23*b*, the light is incident perpendicular to the hypotenuse of the prism and is totally reflected twice such that it emerges at 180° to its original direction. Prisms are used to change the directions of light rays. In binoculars, four prisms are used to reinvert the image that was inverted by the binocular lens. Diamonds have a very high index of refraction ($n \approx 2.4$), so nearly all the light that enters a diamond is eventually reflected back out, giving the diamond its sparkle.

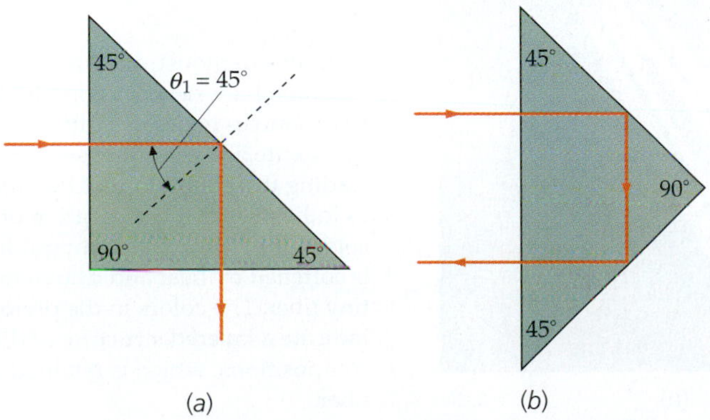

Figure 33-23 (*a*) Light entering through one of the short sides of a 45–45–90° glass prism is totally reflected. (*b*) Light entering through the long side of the prism is totally reflected twice.

Fiber Optics An interesting application of total internal reflection is the transmission of a beam of light down a long, narrow, transparent glass fiber (Figure 33-24*a*). If the beam begins approximately parallel to the axis of the fiber, it will strike the walls of the fiber at angles greater than the critical angle (if the bends in the fiber are not too sharp) and no light energy will be lost through the walls of the fiber. A bundle of such fibers can be used for imaging, as illustrated in Figure 33-24*b*. Fiber optics has many applications in medicine and in communications. In medicine, light is transmitted along tiny fibers to visually probe various internal organs without surgery. In communications, the rate at which information can be transmitted is related to the signal frequency. A transmission system using light of frequencies of the order of 10^{14} Hz can transmit information at a much greater rate than one using radio waves, which have frequencies of the order of 10^{6} Hz. In telecommunication systems, a single glass fiber the thickness of a human hair can transmit audio or video information equivalent to 32,000 voices speaking simultaneously.

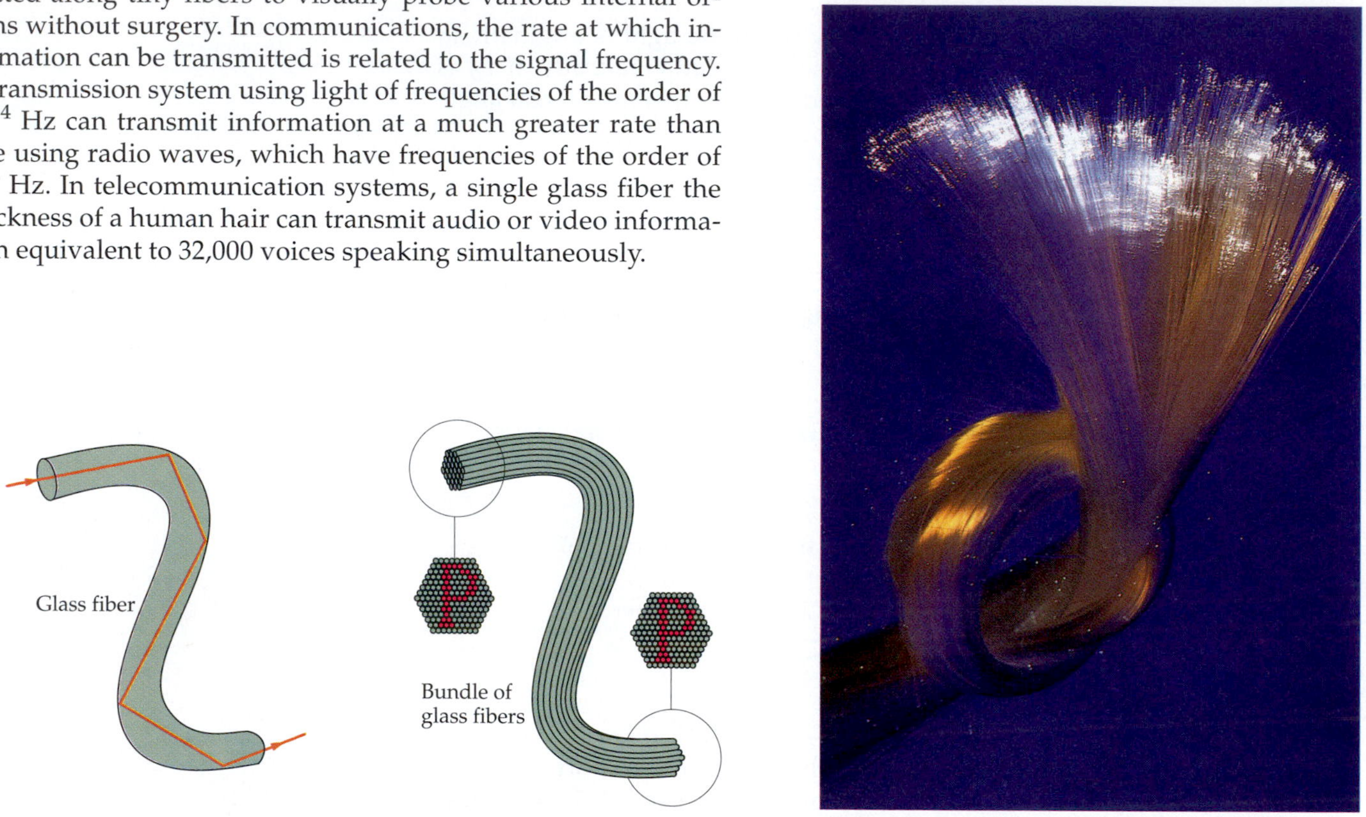

(*a*) (*b*) (*c*)

Figure 33-24 (*a*) A light pipe. Light inside the pipe is always incident at an angle greater than the critical angle, so no light escapes the pipe by refraction. (*b*) Light from the object is transported by a bundle of glass fibers to form an image of the object at the other end of the pipe. (*c*) Light emerging from a bundle of glass fibers.

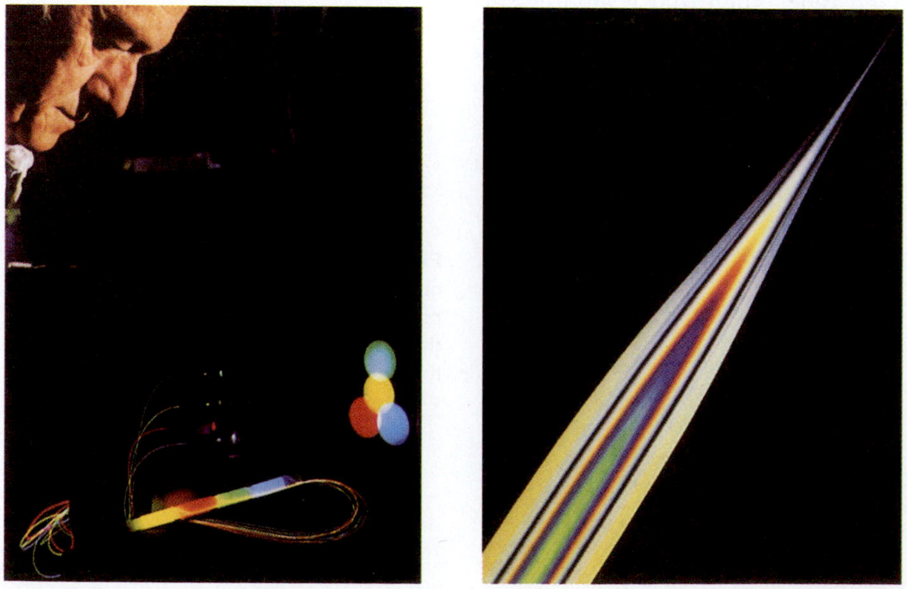

(*a*) (*b*)

(*a*) In this demonstration at the Naval Research Laboratory, a combination of laser sources generates different colors that excite adjacent fiber sensor elements, leading to a separation of the information as indicated by the separation of the colors. (*b*) The tip of a light guide preform is softened by heat and drawn into a long, tiny fiber. The colors in the preform indicate a layered structure of differing compositions, which is retained in the fiber.

Mirages

When the index of refraction of a medium changes gradually, the refraction is continuous, leading to a gradual bending of the light. An interesting example of this is the formation of a mirage. On a hot day, there is often a layer of air near the ground that is warmer and therefore less dense than the air just above it. The speed of light is slightly greater in this less dense layer, so a light beam passing from the cooler layer into the warmer layer is bent. Figure 33-25*a* shows the light from a tree when the surrounding air is all at the same temperature. The wavefronts are spherical, and the rays are straight lines. In Figure 33-25*b*, the air near the ground is warmer, resulting in a greater speed of light there. The portions of the wavefronts near the ground travel faster and get ahead of the higher portions, creating a nonspherical wavefront and causing a curving of the rays. Thus, the ray shown initially heading for the ground is bent upward. As a result, the viewer sees an image of the tree looking as if it were reflected off a water surface on the ground. When driving on a very hot day, you may have noticed apparent wet spots on the highway that disappear when you get to them. This is due to the refraction of light from a hot air layer near the pavement.

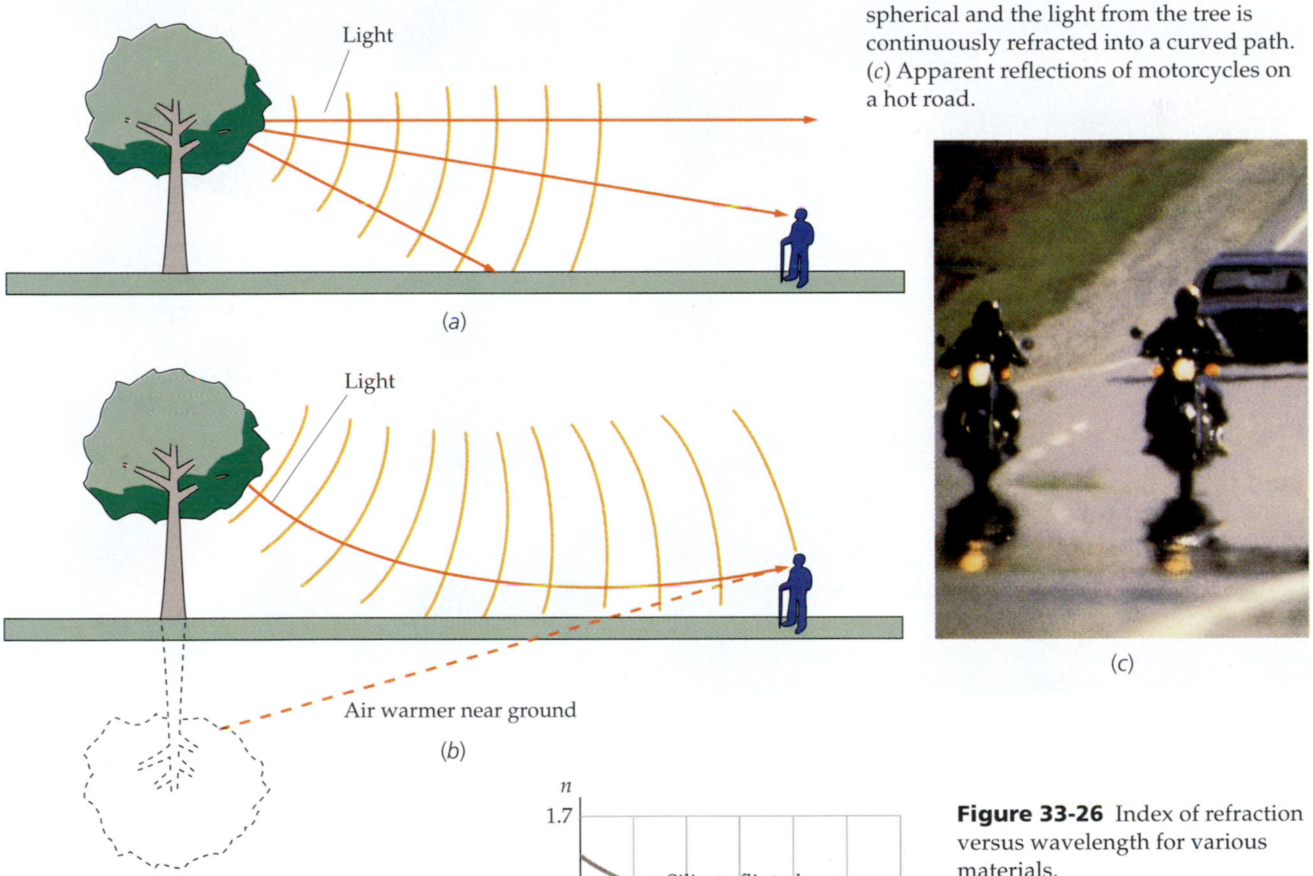

Figure 33-25 A mirage. (*a*) When the air is at a uniform temperature, the wavefronts of the light from the tree are spherical. (*b*) When the air near the ground is warmer, the wavefronts are not spherical and the light from the tree is continuously refracted into a curved path. (*c*) Apparent reflections of motorcycles on a hot road.

Dispersion

The index of refraction of a material has a slight dependence on wavelength. For many materials, n decreases slightly as the wavelength increases, as shown in Figure 33-26. The dependence of the index of refraction on wavelength (and therefore on frequency) is called **dispersion.** When a beam of white light is incident at some angle on the surface of a glass prism, the angle of refraction (which is measured relative to the normal) for the

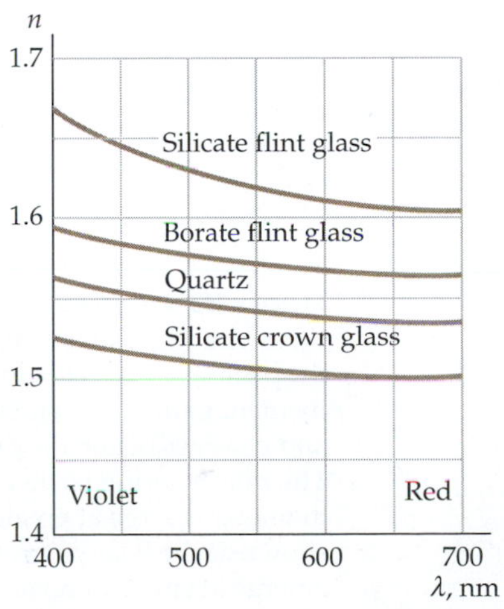

Figure 33-26 Index of refraction versus wavelength for various materials.

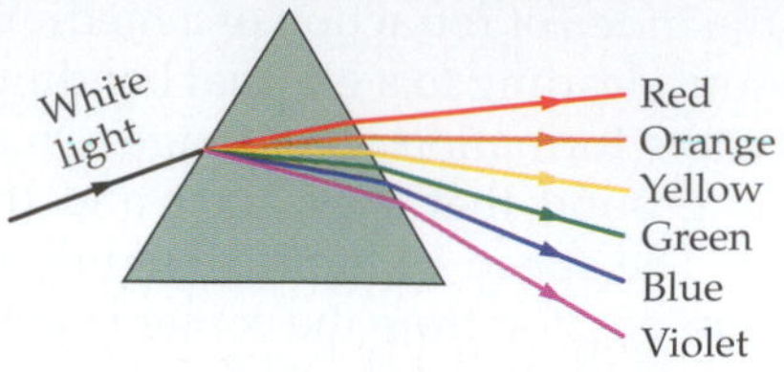

Figure 33-27 A beam of white light incident on a glass prism is dispersed into its component colors. The index of refraction decreases as the wavelength increases so that the longer wavelengths (red) are bent less than the shorter wavelengths (blue).

shorter wavelengths is slightly smaller than that for the longer wavelengths. The light of shorter wavelength (toward the violet end of the spectrum) is therefore bent more toward the normal than that of longer wavelength. The beam of white light is thus spread out or dispersed into its component colors or wavelengths (Figure 33-27).

Rainbows The rainbow is a familiar example of dispersion, in this case the dispersion of sunlight. Figure 33-28 is a diagram originally drawn by Descartes showing parallel rays of light from the sun entering a spherical water drop. First, the rays are refracted as they enter the drop. They are then reflected from the water–air interface on the other side of the drop and finally are refracted again as they leave the drop.

From Figure 33-28 we can see that the angle made by the emerging rays and the diameter (ray 1) reaches a maximum around ray 7 and then decreases. The concentration of rays emerging at approximately the maximum angle gives rise to the rainbow. By construction using the law of refraction, Descartes showed that the maximum angle is about 42°. To observe a rain-

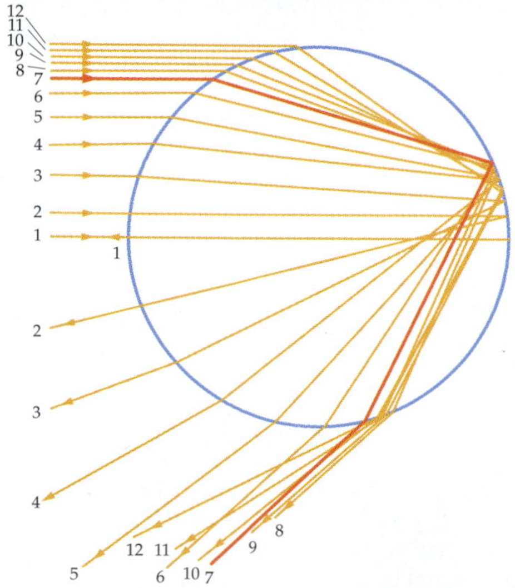

Figure 33-28 Descartes' construction of parallel rays of light entering a spherical water drop. Ray 1 enters the drop along a diameter and is reflected back along its incident path. Ray 2 enters slightly above the diameter and emerges below the diameter at a small angle with it. The rays entering farther and farther away from the diameter emerge at greater and greater angles up to ray 7, shown as the heavy line. Rays entering above ray 7 emerge at smaller and smaller angles with the diameter.

bow, we must therefore look at the water drops at an angle of 42° relative to the line back to the sun, as shown in Figure 33-29. The angular radius of the rainbow is therefore 42°.

The separation of the colors in the rainbow results from the fact that the index of refraction of water depends slightly on the wavelength of light. The angular radius of the bow will therefore depend slightly on the wavelength of the light. The observed rainbow is made up of light rays from many different droplets of water (Figure 33-30). The color seen at a particular angular radius corresponds to the wavelength of light that allows the light to reach the eye from the droplets at that angular radius. Because n_{water} is smaller for red light than for blue light, the red part of the rainbow is at a slightly greater angular radius than the blue part of the rainbow, so red is at the outer side of the rainbow.

When a light ray strikes a surface separating water and air, part of the light is reflected and part is refracted. A secondary rainbow results from the light rays that are reflected twice within a droplet (Figure 33-31). The secondary bow has an angular radius of 51°, and its color sequence is the reverse of that of the primary bow; that is, the violet is on the outside in the secondary bow. Because of the small fraction of light reflected from a water–air interface, the secondary bow is considerably fainter than the primary bow.

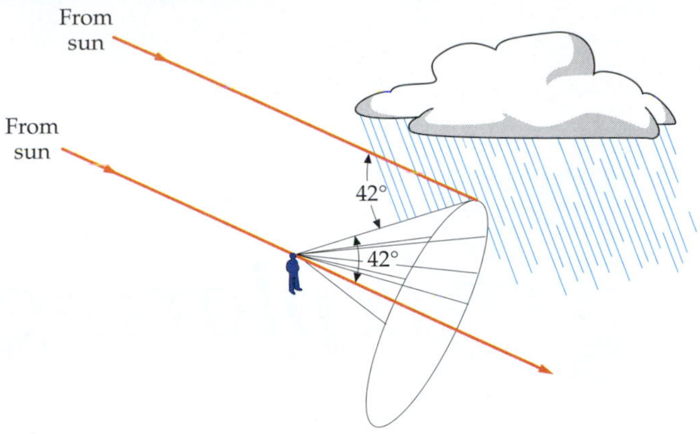

Figure 33-29 A rainbow is viewed at an angle of 42° from the line to the sun, as predicted by Descartes' construction in Figure 33-28.

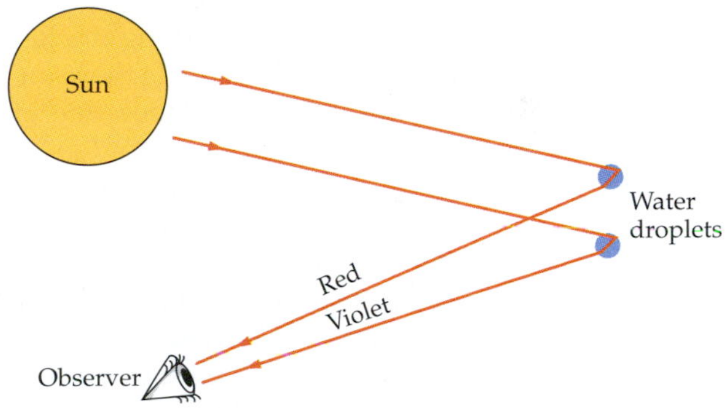

Figure 33-30 The rainbow results from light from many different water droplets.

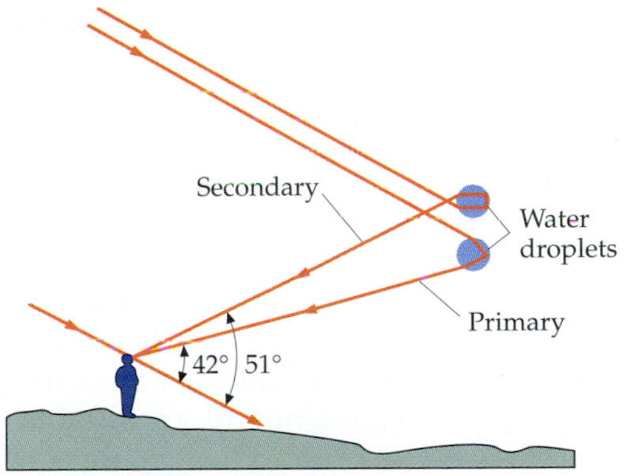

Figure 33-31 The secondary rainbow results from light rays that are reflected twice within a water droplet.

(*a*)

(*b*)

(*a*) This 22° halo around the sun results from reflection and refraction from hexagonal ice crystals that are randomly oriented in the atmosphere. (*b*) When the ice crystals are not randomly oriented but are falling with their flat bases horizontal, only parts of the halo on each side of the sun, called "sun dogs," are seen.

*e*xploring

Beyond the (Visible) Rainbow*

Robert Greenler
University of Wisconsin, Milwaukee

One day while I was sitting at my desk woolgathering rather than addressing the task at hand, the question occurred to me: Is there an infrared rainbow in the sky? How does one explore such a question? Here is the process I went through. For there to be an infrared rainbow, a number of conditions must be met. First, the source of light must emit infrared radiation (the sun emits light over the entire electromagnetic spectrum from X rays to radio waves). Second, the infrared radiation must pass through the earth's atmosphere (water vapor and carbon dioxide in the atmosphere absorb some infrared wavelengths, but others pass through unimpeded). The rainbow is caused by light rays that enter a droplet of water and are reflected internally before emerging from the drop. For there to be an infrared rainbow, the third requirement is that the infrared rays would have to pass through a water droplet. This is a serious consideration. Just because a droplet of water appears transparent in visible light, we cannot assume that it is transparent to infrared "light"; indeed, liquid water does absorb over a broad range of infrared wavelengths. However, the measured spectral transmittance of water shows that water drops should be quite transparent from the visible region out to an infrared wavelength of about 1300 nm. Finally, after emerging from the raindrop, the infrared rays that have survived all these losses must again pass through air to the (unseeing) eye of the would-be observer.

The Search

This line of reasoning produced a tentative answer to the question that prompted the speculation: Yes, there should be an infrared rainbow in the sky and it should lie in a band just outside of the red of the visible rainbow.

I decided to try to photograph this invisible bow using film that is sensitive to a portion of the infrared spectrum. Figure 1 shows the curve of the sensitivity of the film. The figure also shows a curve of the sensitivity of the human eye, as a way of defining the limits of the visible spectral region (extending from about 400 nm at the violet end of the spectrum to 700 nm at the red end). Note that the infrared film has a sensitivity extending out to about 930 nm. Because the film is sensitive not only to the infrared but to the entire visible region (it is *very* sensitive to blue light), we used a filter that is opaque to visible light and transmits only wavelengths longer than about 800 nm. As can be seen from Figure 1, this combination of film and filter will permit the recording of only those wavelengths in a band between 800 and 930 nm, well removed from the visible spectral region.

The Capture

Anyone who has tried to photograph rainbows knows that they usually occur when a camera is not at hand and fade just before one is located. I decided to first try an easier subject—the rainbow in a water spray. Figure 2 shows one of the first photographic results. A garden hose with many holes was wrapped back and forth across a board resting on top of the ladder. And in the spray of the hose—the infrared rainbow! You can also see the fainter, secondary rainbow outside the

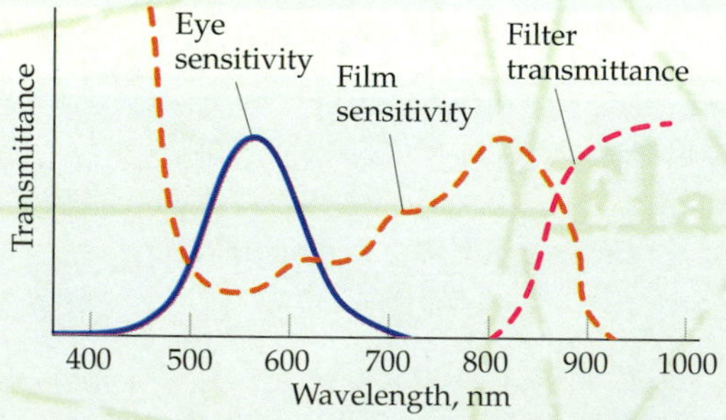

Figure 1 The infrared film has a sensitivity extending through the visible and into the near infrared range. The filter is opaque to visible light but transmits in the infrared for wavelengths longer than 800 nm. The combination of film and filter records an image with wavelengths between 800 and 930 nm, well outside the visible spectrum.

* This essay is adapted from an article that first appeared in *Optic News*, published by the Optical Society of America, in November 1988. Robert Greenler has been professor of physics at the University of Wisconsin–Milwaukee since 1962, where he has been instrumental in the development of the Laboratory for Surface Studies at Milwaukee. His research interests include the study of the structure of molecules adsorbed on solid surfaces, optical effects of the sky, and the understanding of the iridescent colors seen in many biological organisms.

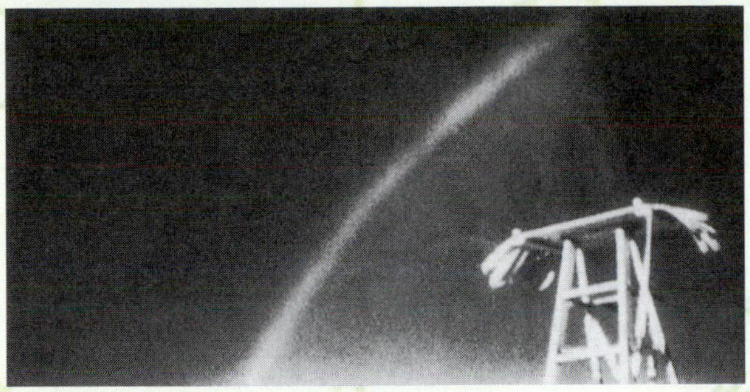

Figure 2 An infrared rainbow photographed in the spray of a garden hose. The fainter secondary bow is shown outside (to the left of) the primary bow. The fringes seen inside the primary bow are caused by interference effects.

brighter, primary bow. This corresponds to the secondary bow seen rather commonly in visible light and it results from rays that enter a water droplet and experience two internal reflections before leaving the drop (see Figure 33-31).

There is another interesting feature in this infrared photograph: immediately *inside* (to the right side of) the bright primary bow there is another bright band—or perhaps two bands. Such fringes, sometimes seen inside a visible bow, are called supernumerary bows and result from the interference of light waves (see Chapter 35).

Close inspection of the negative of Figure 2 reveals yet another feature, which is difficult to reproduce in a printed picture. There is a faint fringe just *outside* the secondary bow. A process similar to the one that produces supernumerary bows inside the primary should, in theory, produce a similar set of fringes outside the secondary. I have never seen any of those fringes associated with any rainbow or with any rainbow photograph, but they are visible in the original of this photograph. They are even visible when I project the slide onto a screen. For the first attempt, that was quite an exciting collection of effects.

Other Effects in the Infrared Photos

Some other features of these infrared photographs are worth considering. If the only radiation that produced these photographic images is invisible, infrared radiation, is it surprising that we can see the ladder, trees, and grass? One should not be too surprised. These objects absorb some wavelengths and reflect or scatter others. Objects that absorb the infrared appear dark in the photos and those that scatter it strongly appear bright. To make it clear just what these photos show, we need to understand the difference between reflected (or scattered) radiation and emitted radiation.

Normally when you look at objects in your landscape, you see them only by the light they scatter. However, if the temperature of an object is high enough, it emits light. If it is very hot—you might call it "white hot"—it emits a broad spectrum of wavelengths with the peak of the emission curve in the visible spectrum. If the object cools down a bit, the peak in the emission curve moves to longer wavelengths. The result is that there is more red light than blue light being emitted, and the appropriate description for its temperature is "red hot." At a lower temperature, you might see a dull red glow. At this point, the peak of the emission curve is in the infrared with just a small amount of emission in the red end of the visible region. At a slightly lower temperature, the object appears dark; the emission peak has moved further into the infrared, so no visible radiation can be seen. If the object cools to where it is warm to the touch, its emission peak is far out in the infrared—perhaps at 10,000 nm—and it is emitting almost nothing in the visible or in the near infrared region to which photographic films are sensitive.

If, however, you could produce a picture with 10,000-nm radiation, objects slightly warmer than their surroundings would appear to be bright—they would be glowing with emitted infrared radiation. There *are* ways to produce such pictures; they are used to show sources of heat loss in homes or to record relatively warm spots on a human body (thermograms) that may indicate the site of some physical disorder. These pictures are usually described as infrared pictures, but they are quite different from the photographs taken with infrared-sensitive film. This film is sensitive only to the near infrared, but the "heat pictures" result from emitted radiation in the far infrared. So the infrared photographs shown here show only the infrared radiation from the sun that is scattered by the leaves or ladder or transformed by raindrop spheres into an invisible rainbow.

Another interesting feature of the photographs is the darkness of the clear sky background. We see light in the clear, clean sky, away from the sun, due to scattering from the molecules of gases in the air. These small scattering particles (much smaller than the wavelength of the light) scatter the shorter waves more effectively than the longer waves. Thus, more blue light is scattered than red light, giving the sky its blue color. This same effect, which makes the sky darker in red light than in blue, makes it even darker in the infrared light sampled by these photographs.

optional

Calculating the Angular Radius of the Rainbow We can calculate the angular radius of the rainbow from the laws of reflection and refraction. Figure 33-32 shows a ray of light incident on a spherical water droplet at point A. The angle of refraction θ_2 is related to the angle of incidence θ_1 by Snell's law:

$$n_{\text{air}} \sin \theta_1 = n_{\text{water}} \sin \theta_2 \qquad \text{33-13}$$

Point P in Figure 33-32 is the intersection of the line of the incident ray and the line of the emerging ray. The angle ϕ_d is called the angle of deviation of the ray. The angle 2β is the angular radius of the rainbow. 2β is related to ϕ_d by:

$$\phi_d + 2\beta = \pi \qquad \text{33-14}$$

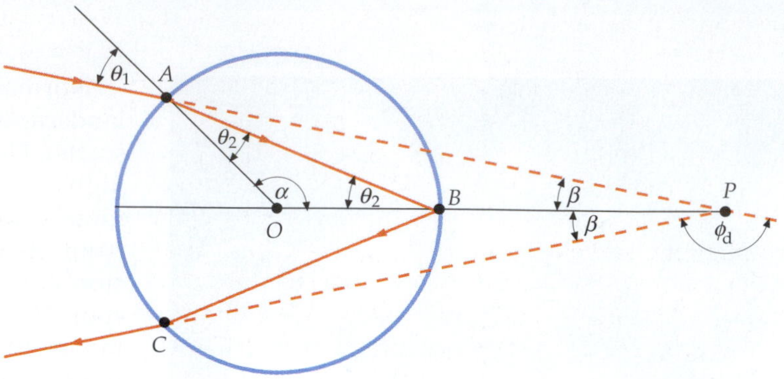

Figure 33-32 Light ray incident on a spherical water drop. The refracted ray strikes the back of the water droplet at point B. It makes an angle θ_2 with the radial line OB and is reflected at an equal angle. The ray is refracted again at point C, where it leaves the droplet.

We wish to relate the angle of deviation ϕ_d to the angle of incidence θ_1. From the triangle AOB, we have

$$2\theta_2 + \alpha = \pi \qquad \text{33-15}$$

Similarly, from the triangle AOP, we have

$$\theta_1 + \beta + \alpha = \pi \qquad \text{33-16}$$

Eliminating α from Equations 33-15 and 33-16 and solving for β gives

$$\beta = \pi - \theta_1 - \alpha = \pi - \theta_1 - (\pi - 2\theta_2) = 2\theta_2 - \theta_1$$

Substituting this value for β into Equation 33-14 gives the angle of deviation:

$$\phi_d = \pi - 2\beta = \pi - 4\theta_2 + 2\theta_1 \qquad \text{33-17}$$

Equation 33-17 can be combined with Snell's law to eliminate θ_2 and give the angle of deviation ϕ_d in terms of the angle of incidence θ_1:

$$\phi_d = \pi + 2\theta_1 - 4 \arcsin\left(\frac{n_{\text{air}} \sin \theta_1}{n_{\text{water}}}\right) \qquad \text{33-18}$$

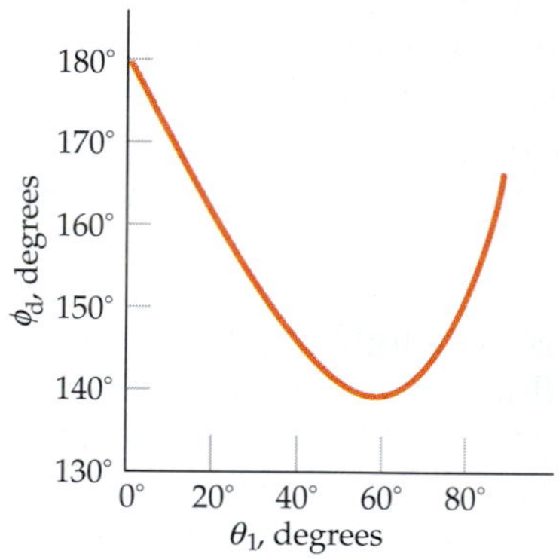

Figure 33-33 Plot of the angle of deviation ϕ_d as a function of incident angle θ_1. The angle of deviation has its minimum value of 138° when the angle of incidence is 60°. Since $d\phi_d/d\theta_1 = 0$ at minimum deviation, the deviation of rays with incident angles slightly less or slightly greater than 60° will be approximately the same.

Figure 33-33 shows a plot of ϕ_d versus θ_1. The angle of deviation ϕ_d has its minimum value when $\theta_1 = 60°$. At this angle of incidence, the angle of deviation is $\phi_d = 138°$. This angle is called the **angle of minimum deviation.** At incident angles that are slightly greater or slightly smaller than 60°, the angle of deviation is approximately the same. Therefore, the light reflected by the water droplet will be concentrated near the angle of minimum deviation. The angular radius of the rainbow is thus

$$2\beta = \pi - \phi_d = 180° - 138° = 42° \qquad \text{33-19}$$

33-7 Polarization

In a transverse wave, the vibration is perpendicular to the direction of propagation of the wave. For example, in a light wave traveling in the z direction, the electric field is perpendicular to the z direction. (The magnetic field of a light wave is also perpendicular to the z direction.) If the vibration remains parallel to a fixed line in space, the wave is said to be **linearly polarized.** We can visualize polarization most easily by considering mechanical waves on a string. If one end of a string held horizontal is moved up and down, the resulting waves on the string are linearly polarized with each element of the string vibrating in the vertical direction. Similarly, if one end is moved along a horizontal line, the displacements of the string are linearly polarized in the horizontal direction. If one end of the string is moved with constant speed in a circle, the resulting wave is said to be **circularly polarized.** In this case, each element of the string moves in a circle. Unpolarized waves can be produced by moving the end of the string vertically and horizontally in a random way. Then, if the string itself is in the z direction, the vibrations will have both x and y components that vary randomly.

Most waves produced by a single source are polarized. For example, electromagnetic waves produced by a single dipole antenna are linearly polarized with the electric field vector parallel to the antenna. Waves produced by many sources are usually unpolarized. A typical light source, for example, contains millions of atoms acting independently. The electric field for such a wave can be resolved into x and y components that vary randomly because there is no correlation between the individual atoms producing the light.

The polarization of electromagnetic waves can be demonstrated with microwaves, which have wavelengths on the order of centimeters. In a typical microwave generator, polarized waves are radiated by a dipole antenna. In Figure 33-34, the dipole antenna is vertical, so the electric field vector $\vec{E}$ of the radiated waves is vertical. An absorber can be made of a screen of parallel straight wires. When the wires are vertical, as in Figure 33-34*a*, the electric field parallel to the wires sets up currents in the wires and energy is absorbed. When the wires are horizontal and therefore perpendicular to $\vec{E}$, as in Figure 33-34*b*, no currents are set up and the waves are transmitted.

There are four phenomena that produce polarized electromagnetic waves from unpolarized waves: (1) absorption, (2) reflection, (3) scattering, and (4) birefringence (also called double refraction), each of which is examined in the upcoming sections.

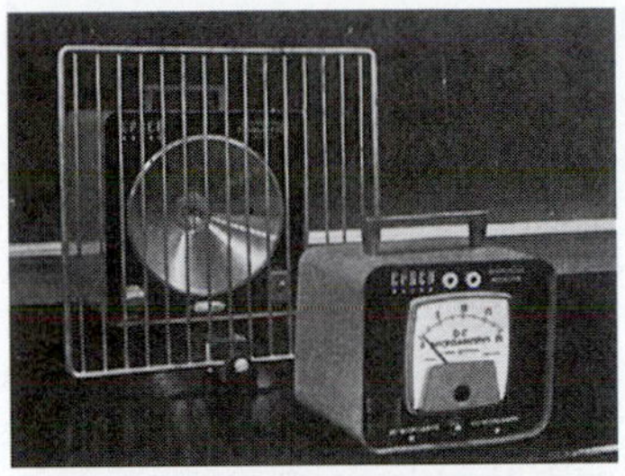

(*a*)

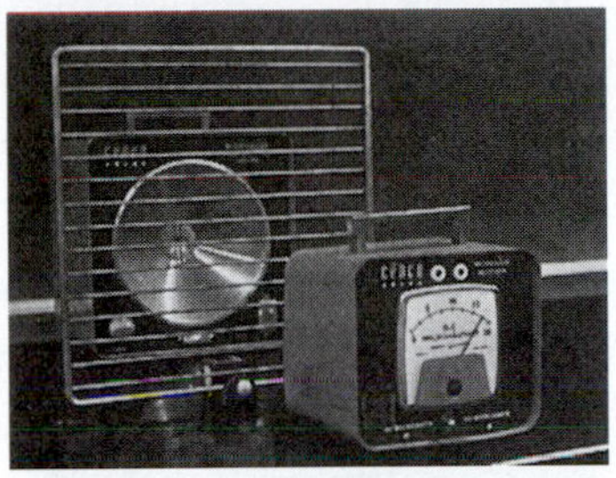

(*b*)

Figure 33-34 Demonstration showing the polarization of microwaves. The electric field of the microwaves is vertical, parallel to the vertical dipole antenna. (*a*) When the metal wires of the absorber are vertical, electric currents are set up in the wires and energy is absorbed, as indicated by the low reading on the microwave detector. (*b*) When the wires are horizontal, no currents are set up, and the microwaves are transmitted, as indicated by the high reading on the detector.

Polarization by Absorption

Several naturally occurring crystals, when cut into appropriate shapes, absorb and transmit light differently depending on the polarization of the light. These crystals can be used to produce linearly polarized light. In 1938, E. H. Land invented a simple commercial polarizing film called Polaroid. This material contains long-chain hydrocarbon molecules that are aligned when the sheet is stretched in one direction during the manufacturing process. These chains become conducting at optical frequencies when the sheet is dipped in a solution containing iodine. When light is incident with its electric field vector parallel to the chains, electric currents are set up along the chains, and the light energy is absorbed, just as the microwaves are absorbed by the wires in Figure 33-34. If the electric field is perpendicular to the chains, the light is transmitted. The direction perpendicular to the chains is called the **transmission axis.** We will make the simplifying assumption that all the light is transmitted when the electric field is parallel to the transmission axis and all is absorbed when it is perpendicular to the transmission axis.

Consider an unpolarized light beam traveling in the z direction incident on a polarizing film with its transmission axis in the x direction. On the average, half of the incident light has its electric field in the y direction and half has it in the x direction. Thus, half the intensity is transmitted, and the transmitted light is linearly polarized with its electric field in the x direction.

Suppose we have a second polarizing film whose transmission axis makes an angle θ with that of the first, as shown in Figure 33-35. If $\vec{E}$ is the electric field between the films, its component along the direction of the transmission axis of the sec-

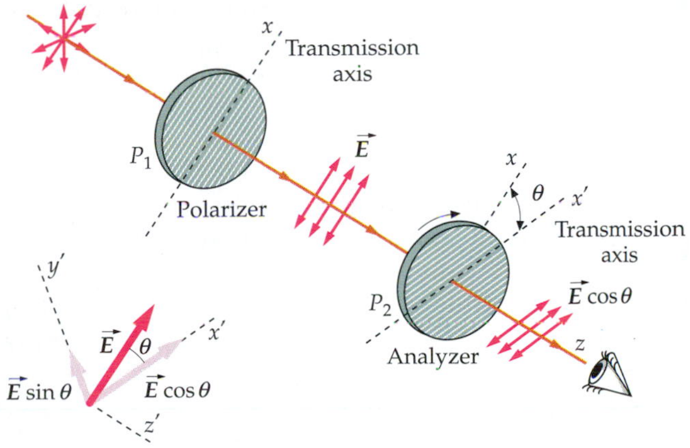

Figure 33-35 Two polarizing films with their transmission axes making an angle θ with each other. Only the component $E\cos\theta$ is transmitted through the second film. If the intensity between the films is I_0, that transmitted by both films is $I_0\cos^2\theta$.

ond film is $E \cos \theta$. Since the intensity of light is proportional to E^2, the intensity of light transmitted by both films will be given by

$$I = I_0 \cos^2 \theta \tag{33-20}$$

where I_0 is the intensity incident on the second film and is, of course, half the intensity incident on the first film. When two polarizing elements are placed in succession in a beam of light as described here, the first is called the **polarizer** and the second is called the **analyzer.** If the polarizer and analyzer are crossed, that is, if their transmission axes are perpendicular to each other, no light gets through. Equation 33-20 is known as **Malus's law** after its discoverer, E. L. Malus (1775–1812). It applies to any two polarizing elements whose transmission axes make an angle θ with each other.

(*a*) (*b*)

(*a*) Cross polarizers block out all of the light. (*b*) In a liquid crystal display, the crystal is between crossed polarizers. Light incident on the crystal is transmitted because the crystal rotates the direction of polarization of the light 90°. The light is reflected back out through the crystal by a mirror behind the crystal, and a uniform background is seen. When a voltage is applied across a small segment of the crystal, the polarization is not rotated, so no light is transmitted and the segment appears black.

Example 33-6

Unpolarized light of intensity 3.0 W/m^2 is incident on two polarizing films whose transmission axes make an angle of 60° (Figure 33-36). What is the intensity of light transmitted by the second film?

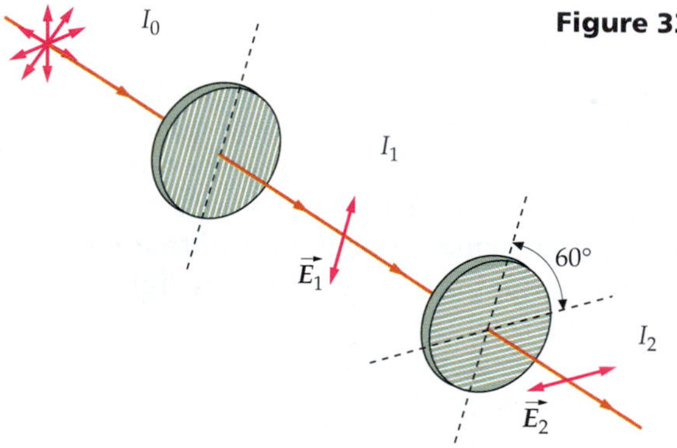

Figure 33-36

Picture the Problem Since the incident light is unpolarized, half the incident intensity is transmitted by the first polarizing film. The second film reduces the intensity by a factor of $\cos^2 \theta$, with $\theta = 60°$.

1. The intensity transmitted by the second film is related to the incident intensity I_1 by Equation 33-18: $I_2 = I_1 \cos^2 \theta$

2. The intensity incident on the second film I_1 is half the intensity of unpolarized light incident on the first film: $I_1 = \frac{1}{2} I_0$

3. Combine these results and substitute the given data:
$$I_2 = \tfrac{1}{2} I_0 \cos^2 60° = \tfrac{1}{2}(3.0\ \text{W/m}^2)(0.500)^2 = 0.375\ \text{W/m}^2$$

Remarks Half the intensity passes through the first film no matter what its orientation is. Note that the second film rotates the direction of polarization.

Polarization by Reflection

When unpolarized light is reflected from a plane surface boundary between two transparent media, such as air and glass or air and water, the reflected light is partially polarized. The degree of polarization depends on the angle between the incident ray and the normal to the surface (called the angle of incidence) and on the ratio of the wave speeds in the two media. For a certain angle of incidence called the polarizing angle θ_p, the reflected light is completely polarized. At the polarizing angle the reflected and refracted rays are perpendicular to each other. This result was discovered experimentally by Sir David Brewster in 1812.

Figure 33-37 shows light incident at the polarizing angle θ_p for which the reflected light is completely polarized. The electric field of the incident light can be resolved into components parallel and perpendicular to the plane of incidence. The reflected light is completely polarized with its electric field perpendicular to the plane of incidence. We can relate the polarizing angle to the indexes of refraction of the media using Snell's law. If n_1 is the index of refraction of the first medium and n_2 is that of the second medium, Snell's law gives

$$n_1 \sin \theta_p = n_2 \sin \theta_2$$

where θ_2 is the angle of refraction. From Figure 33-37, we can see that the sum of the angle of reflection and the angle of refraction is 90°. Since the angle of reflection equals the angle of incidence, we have

$$\theta_2 = 90° - \theta_p$$

Then

$$n_1 \sin \theta_p = n_2 \sin(90° - \theta_p) = n_2 \cos \theta_p$$

or

$$\tan \theta_p = \frac{n_2}{n_1} \qquad \text{33-21}$$

Brewster's law

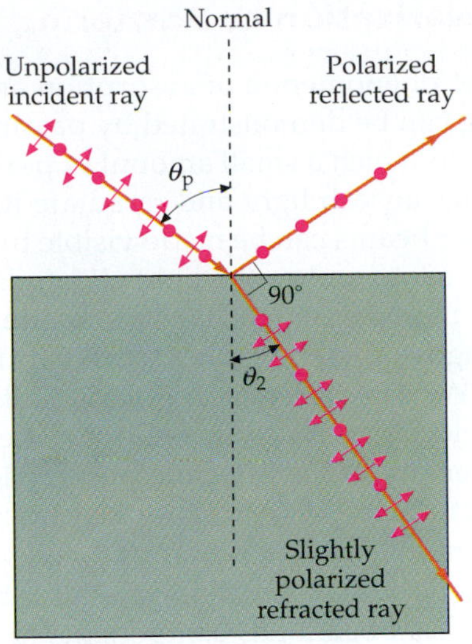

Figure 33-37 Polarization by reflection. The incident wave is unpolarized and has components of the electric field parallel to the plane of incidence (arrows) and components perpendicular to this plane (dots). For incidence at the polarizing angle, the reflected wave is completely polarized, with its electric field perpendicular to the plane of incidence.

Equation 33-21 is known as **Brewster's law.**

Although the reflected light is completely polarized for this angle of incidence, the transmitted light is only partially polarized because only a small fraction of the incident light is reflected. If the incident light itself is polarized with $\vec{E}$ in the plane of incidence, there is no reflected light when the angle of incidence is θ_p. We can understand this qualitatively from Figure 33-38. If we consider the molecules in the second medium to be oscillating parallel to the electric field of the refracted ray, there can be no reflected ray because no energy is radiated along the line of oscillation.

Because of the polarization of reflected light, sunglasses made of polarizing material can be very effective in cutting out glare. If light is reflected from a horizontal surface such as a lake or snow on the ground, the plane of incidence will be vertical and the electric field of the reflected light will be predominantly horizontal. Polarized sunglasses with a vertical transmission axis will then reduce glare by absorbing much of the reflected light. If you have polarized sunglasses, you can observe this effect by looking through the glasses at reflected light and then rotating the glasses 90°; much more of the light will be transmitted.

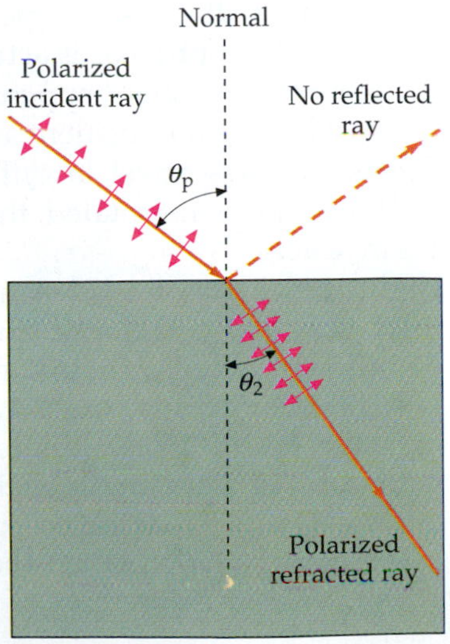

Figure 33-38 Polarized light incident at the polarizing angle. When the light is polarized with $\vec{E}$ in the plane of incidence, there is no reflected ray.

optional

Polarization by Scattering

The phenomenon of absorption and reradiation is called **scattering.** Scattering can be demonstrated by passing a light beam through a container of water to which a small amount of powdered milk has been added. The milk particles absorb light and reradiate it, making the light beam visible. Similarly, laser beams can be made visible by introducing chalk or smoke particles into the air to scatter the light. A familiar example of light scattering is that from air molecules, which tend to scatter short wavelengths more than long wavelengths, thereby giving the sky its blue color.

We can understand polarization by scattering if we think of an absorbing molecule as an electric-dipole antenna that radiates waves with a maximum intensity in the direction perpendicular to the antenna with the electric field vector parallel to the antenna and zero intensity in the direction along the antenna. Figure 33-39 shows a beam of unpolarized light that initially travels along the z axis, striking a scattering center at the origin. The electric field in the light beam has components in both the x and y directions perpendicular to the direction of motion of the light beam. These fields set up oscillations of the scattering center in both the x and y directions, but there is no oscillation in the z direction. The oscillation of the scattering center in the x direction produces light along the y axis but not along the x axis, which is along the line of oscillation. The light radiated along the y axis is thus polarized in the x direction. Similarly, the light radiated along the x axis is polarized in the y direction. This can be seen easily by examining the scattered light with a piece of polarizing film.

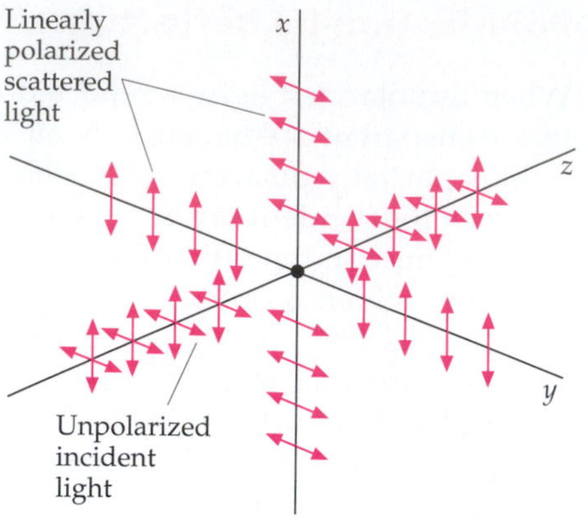

Figure 33-39 Polarization by scattering. Unpolarized light propagating in the z direction is incident on a scattering center at the origin. The light scattered in the x direction is polarized in the y direction and that scattered in the y direction is polarized in the x direction.

Polarization by Birefringence

Birefringence is a complicated phenomenon that occurs in calcite and other noncubic crystals and in some stressed plastics such as cellophane. Most materials are **isotropic**, that is, the speed of light passing through the material is the same in all directions. Because of their atomic structure, birefringent materials are **anisotropic.** The speed of light depends on its direction of propagation through the material. When a light ray is incident on such materials, it may be separated into two rays called the *ordinary ray* and the *extraordinary ray.* These rays are polarized in mutually perpendicular directions, and they travel with different speeds. Depending on the relative orientation of the material and the incident light, the rays may also travel in different directions.

There is one particular direction in a birefringent material in which both rays propagate with the same speed. This direction is called the **optic axis** of the material. (The optic axis is actually a *direction* rather than a line in the material.) Nothing unusual happens when light travels along the optic axis. However, when light is incident at an angle to the optic axis, as shown in Figure 33-40, the rays travel in different directions and emerge separated in space. If the material is rotated, the extraordinary ray (the e ray in the figure) rotates in space.

If light is incident on a birefringent plate perpendicular to its crystal face and perpendicular to the optic axis, the two rays travel in the same direction but at different speeds. The number of wavelengths in the two rays in the plate is different because the wavelengths ($\lambda = v/f$) of the rays differ. The rays emerge with a phase difference that depends on the thickness of the plate and on the wavelength of the incident light. In a **quarter-wave plate,** the thickness is such that there is a 90° phase difference between the waves of a particular wavelength when they emerge. In a **half-wave plate,** the rays emerge with a phase difference of 180°.

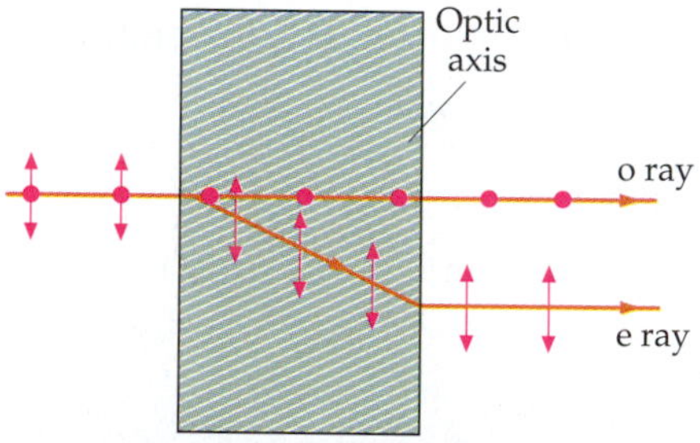

(*a*)

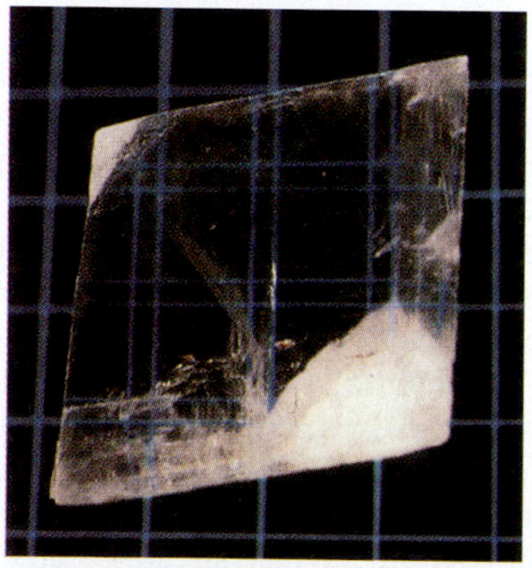

(*b*)

Figure 33-40 (*a*) A narrow beam of light incident on a birefringent crystal such as calcite is split into two beams, called the ordinary ray (o ray) and the extraordinary ray (e ray), that have mutually perpendicular polarizations. If the crystal is rotated, the extraordinary ray rotates in space. (*b*) A double image of the cross hatching is produced by this birefringent crystal of calcium carbonate.

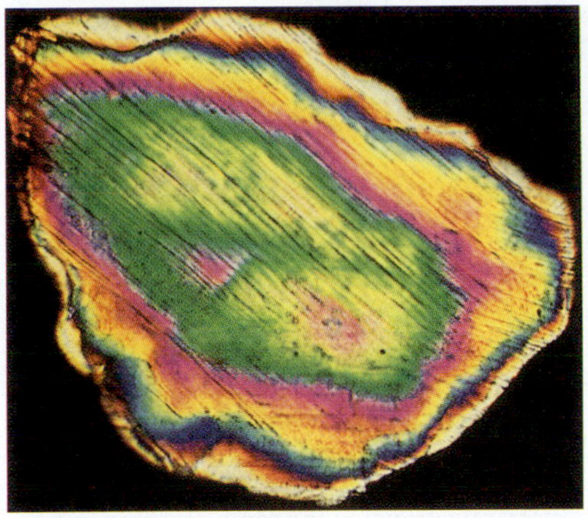

(a)

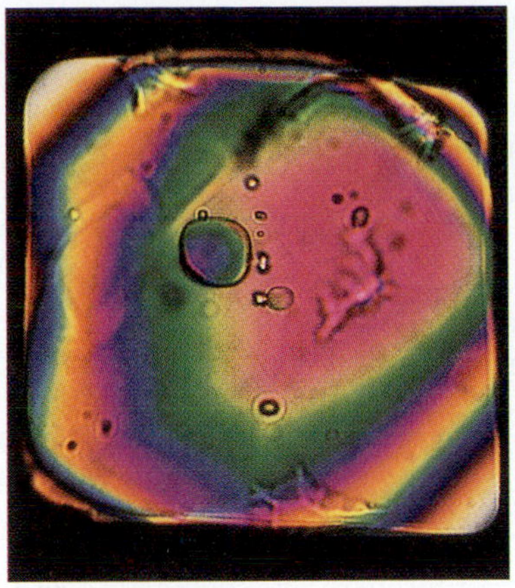

(b)

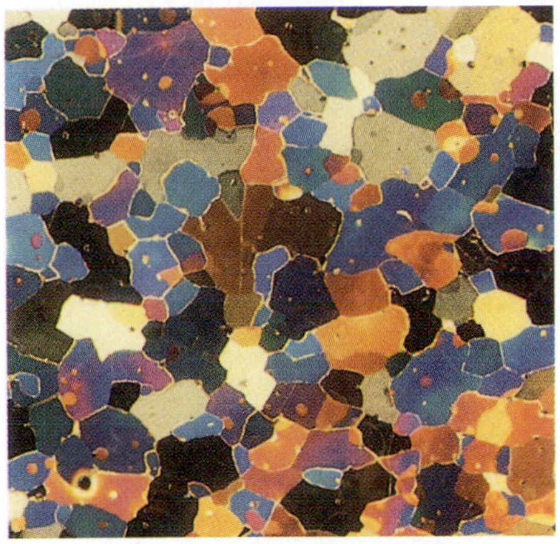

(c)

(d)

(e)

When the transmission axes of two polarizing films are perpendicular, the polarizers are said to be crossed and no light is transmitted. However, many materials are birefringent or become so under stress. Such materials rotate the direction of polarization of the light so that light of a particular wavelength is transmitted through both polarizers. When a birefringent material is viewed between crossed polarizers, information about its internal structure is revealed. (*a*) A shocked quartz grain from the site of a meteorite crater. The layered structure, evidenced by the parallel lines, arises from the shock of the impact of the meteor. (*b*) A grain of quartz typically found in silicic volcanic rocks. No shock lines are seen. (*c*) Thin sections of ice core from the antarctic ice sheet reveal bubbles of trapped CO_2, which appear amber-colored. This sample was taken from a depth of 194 m, corresponding to air trapped 1600 years ago, whereas that in (*d*) is from a depth of 56 m, corresponding to air trapped 450 years ago. Ice core measurements have replaced the less reliable technique of analyzing carbon in tree rings to compare current atmospheric CO_2 levels with those of the recent past. (*e*) Robert Mark of the Princeton School of Architecture examines the stress patterns in a plastic model of the nave structure of Chartres Cathedral.

Suppose that the incident light is linearly polarized such that the electric field vector is at 45° to the optic axis, as illustrated in Figure 33-41. The ordinary and extraordinary rays start out in phase and have equal amplitudes. With a quarter-wave plate, the waves emerge with a phase difference of 90°, so the resultant electric field has components $E_x = E_0 \sin \omega t$ and $E_y = E_0 \sin(\omega t + 90°) = E_0 \cos \omega t$. The electric field vector thus rotates in a circle and the wave is circularly polarized.

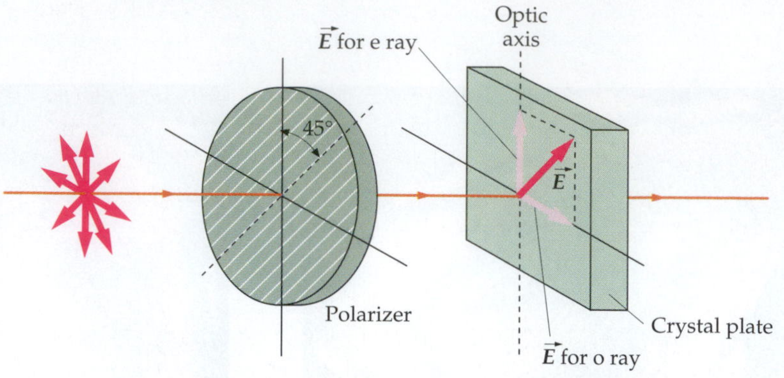

Figure 33-41 Polarized light emerging from the polarizer is incident on a birefringent crystal such that the electric field vector makes a 45° angle with the optic axis, which is perpendicular to the light beam. The ordinary and extraordinary rays travel in the same direction but at different speeds. The polarization of the emerging light depends on the thickness of the crystal and the wavelength of the light.

With a half-wave plate, the waves emerge with a phase difference of 180°, so the resultant electric field is linearly polarized with components $E_x = E_0 \sin \omega t$ and $E_y = E_0 \sin(\omega t + 180°) = -E_0 \sin \omega t$. The net effect is that the direction of polarization of the wave is rotated by 90° relative to that of the incident light, as shown in Figure 33-42.

Interesting and beautiful patterns, like those on page 1026, can be observed by placing birefringent materials such as cellophane or stressed plastic between two polarizing sheets with their transmission axes perpendicular to each other. Ordinarily, no light is transmitted through crossed polarizing sheets. However, if we place a birefringent material between the crossed Polaroids, the material acts as a half-wave plate for light of a certain color depending on the material's thickness. The direction of polarization is rotated and some light gets through both films. Various glasses and plastics become birefringent when under stress. The stress patterns can be observed when the material is placed between crossed polarizing sheets.

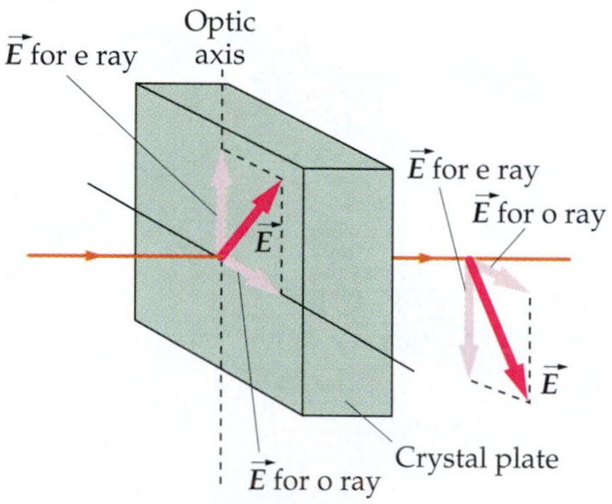

Figure 33-42 When the birefringent crystal in Figure 33-41 is a half-wave plate, the direction of polarization of the emerging light is rotated by 90°.

33-8 Derivation of the Laws of Reflection and Refraction

The laws of reflection and refraction can be derived from either Huygens' principle or Fermat's principle.

Huygens' Principle

Reflection Figure 33-43 shows a plane wavefront AA' striking a mirror at point A. As can be seen from the figure, the angle ϕ_1 between the wavefront and the mirror is the same as the angle of incidence θ_1, which is the angle between the perpendicular to the mirror and the rays that are perpendicular to the wavefronts. According to Huygens' principle, each point on a given wavefront can be considered to be a point source of secondary wavelets. The position of the wavefront after a time t is found by constructing wavelets of radius ct with their centers on the wavefront AA'. Wavelets that do not strike the mirror form the portion of the new wavefront BB'. Wavelets that do strike the mirror are reflected and form the portion of the new wavefront BB''. By a similar construction, the wavefront $C''C$ is obtained from the Huygens' wavelets originating on the wavefront $B''B$. Figure 33-44 is an enlargement of a portion of Figure 33-43 showing AP, which is part of the original wavefront. During the time t, the wavelet from point P reaches the mirror at point B, and the wavelet from point A reaches point B''. The reflected wave BB'' makes an angle ϕ_1' with the mirror that

Figure 33-43 Plane wave reflected at a plane mirror. The angle θ_1 between the incident ray and the normal to the mirror is the angle of incidence. It is equal to the angle ϕ_1 between the incident wavefront and the mirror.

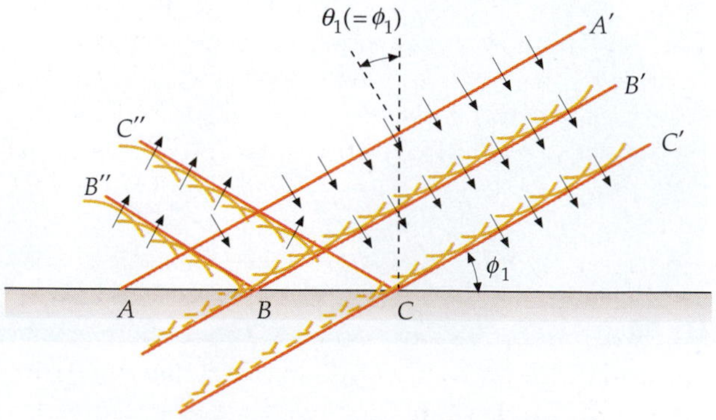

is equal to the angle of reflection θ_1' between the reflected ray and the normal to the mirror. The triangles ABP and BAB'' are both right triangles with a common side AB and equal sides $AB'' = BP = ct$. Hence, these triangles are congruent, and the angles ϕ_1 and ϕ_1' are equal, implying that the angle of reflection θ_1' equals the angle of incidence θ_1.

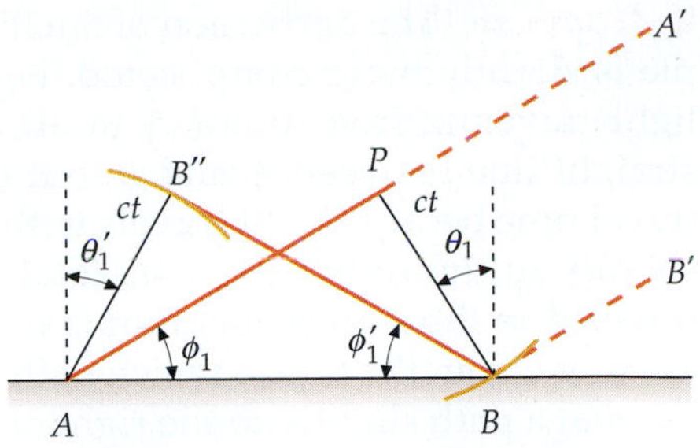

Figure 33-44 Geometry of Huygen's construction for the calculation of the law of reflection. The wavefront AP initially strikes the mirror at point A. After a time t, the Huygens' wavelet from P strikes the mirror at point B, and the one from A reaches point B'.

Refraction Figure 33-45 shows a plane wave incident on an air–glass interface. We apply Huygens' construction to find the wavefront of the transmitted wave. Line AP indicates a portion of the wavefront in medium 1 that strikes the glass surface at an angle ϕ_1. In time t the wavelet from P travels the distance $v_1 t$ and reaches the point B on the line AB separating the two media, while the wavelet from point A travels a shorter distance $v_2 t$ into the second medium. The new wavefront BB' is not parallel to the original wavefront AP because the speeds v_1 and v_2 are different. From the triangle APB,

$$\sin \phi_1 = \frac{v_1 t}{AB}$$

or

$$AB = \frac{v_1 t}{\sin \phi_1} = \frac{v_1 t}{\sin \theta_1}$$

using the fact that the angle ϕ_1 equals the angle of incidence θ_1. Similarly, from triangle $AB'B$,

$$\sin \phi_2 = \frac{v_2 t}{AB}$$

or

$$AB = \frac{v_2 t}{\sin \phi_2} = \frac{v_2 t}{\sin \theta_2}$$

where $\theta_2 = \phi_2$ is the angle of refraction. Equating the two values for AB, we obtain

$$\frac{\sin \theta_1}{v_1} = \frac{\sin \theta_2}{v_2} \qquad \text{33-22}$$

Substituting $v_1 = c/n_1$ and $v_2 = c/n_2$ in this equation and multiplying by c, we obtain $n_1 \sin \theta_1 = n_2 \sin \theta_2$, which is Snell's law.

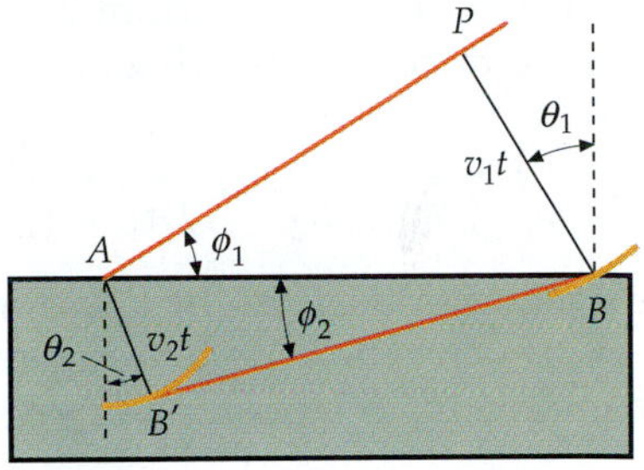

Figure 33-45 Application of Huygens' principle to the refraction of plane waves at the surface separating a medium in which the wave speed is v_1 from a medium in which the wave speed v_2 is less than v_1. The angle of refraction in this case is less than the angle of incidence.

Fermat's Principle

Reflection Figure 33-46 shows two paths in which light leaves point A, strikes the plane surface, which we can consider to be a mirror, and travels to point B. The problem for the application of Fermat's principle to reflection can be stated as follows: At what point P in the figure must the light strike the mirror so that it will travel from point A to point B in the least time? Since the light is traveling in the same medium for this problem, the time will be minimum when the distance is minimum. In Figure 33-46 the distance APB is the same as the distance $A'PB$, where A' is the image of the source A. Point A' lies along the perpendicular from A to the mirror and is equidistant behind the mirror. As we vary point P, the distance $A'PB$ is least when the points A', P, and B lie on a straight line. We can see from the figure that this occurs when the angle of incidence equals the angle of reflection.

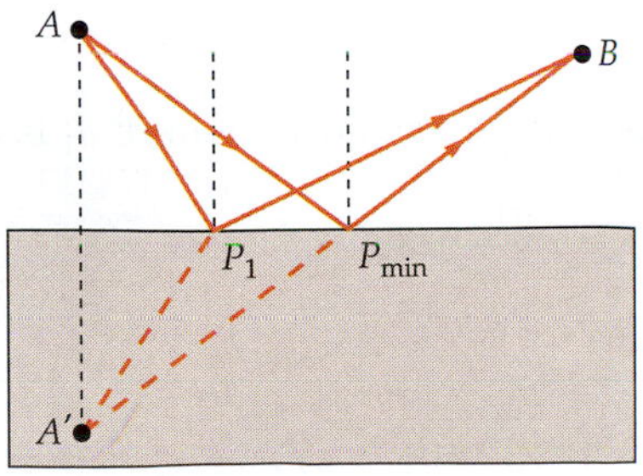

Figure 33-46 Geometry for deriving the law of reflection from Fermat's principle. The time it takes for the light to travel from point A to point B is a minimum when the light strikes the surface at point P.

optional

Refraction The derivation of Snell's law of refraction from Fermat's principle is slightly more complicated. Figure 33-47 shows the possible paths for light traveling from point A in air to point B in glass. Point P_1 is on the straight line between A and B, but this path is not the one for the shortest travel time because light travels with a smaller speed in the glass. If we move slightly to the right of P_1, the total path length is greater, but the distance traveled in the slower medium is less than for the path through P_1. It is not apparent from the figure which path is that of least time, but it is not surprising that a path slightly to the right of the straight-line path takes less time because the time gained by traveling a shorter distance in the glass more than compensates for the time lost traveling a longer distance in the air. As we move the point of intersection of the possible path to the right of point P_1, the total time of travel from A to B decreases until we reach a minimum at point P_{min}. Beyond this point, the time saved by traveling a shorter distance in the glass does not compensate for the greater time required for the greater distance traveled in the air.

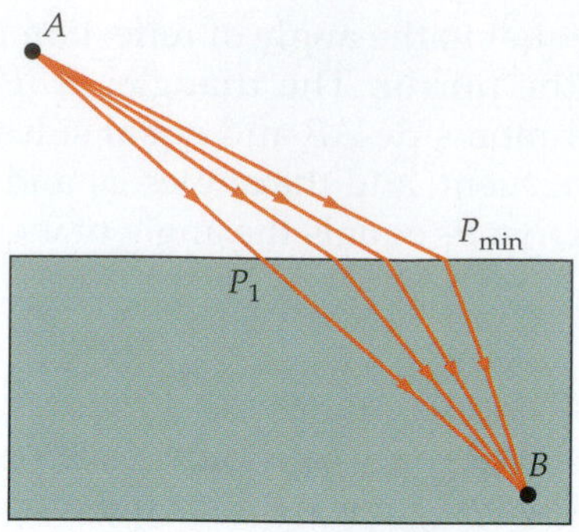

Figure 33-47 Geometry for deriving Snell's law from Fermat's principle. The point P_{min} is the point at which light must strike the glass in order that the travel time from A to B is a minimum.

Figure 33-48 shows the geometry for finding the path of least time. If L_1 is the distance traveled in medium 1 with index of refraction n_1, and L_2 is the distance traveled in medium 2 with index of refraction n_2, the time for light to travel the total path AB is

$$t = \frac{L_1}{v_1} + \frac{L_2}{v_2} = \frac{L_1}{c/n_1} + \frac{L_2}{c/n_2} = \frac{n_1 L_1}{c} + \frac{n_2 L_2}{c} \qquad \text{33-23}$$

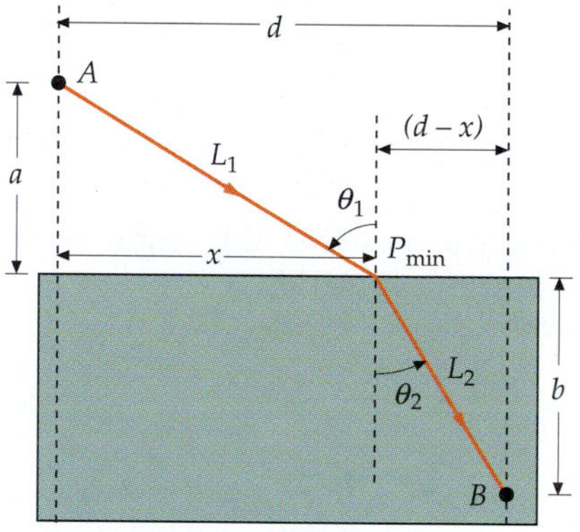

Figure 33-48 Geometry for calculating the minimum time in the derivation of Snell's law from Fermat's principle.

We wish to find the point P_{min} for which this time is a minimum. We do this by expressing the time in terms of a single parameter x, as shown in the figure, indicating the position of point P_{min}. In terms of the distance x,

$$L_1^2 = a^2 + x^2 \qquad \text{and} \qquad L_2^2 = b^2 + (d - x)^2 \qquad \text{33-24}$$

Figure 33-49 shows the time t as a function of x. At the value of x for which the time is a minimum, the slope of the graph of t versus x is zero:

$$\frac{dt}{dx} = 0$$

Differentiating each term in Equation 33-23 with respect to x and setting the result equal to zero, we obtain

$$\frac{dt}{dx} = \frac{1}{c}\left(n_1 \frac{dL_1}{dx} + n_2 \frac{dL_2}{dx}\right) = 0 \qquad \text{33-25}$$

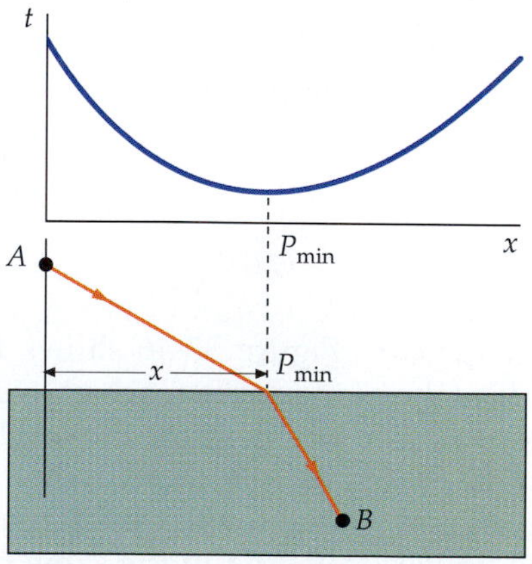

Figure 33-49 Graph of the time it takes for light to travel from A to B versus x, measured along the refracting surface. The time is a minimum at the point at which the angles of incidence and refraction obey Snell's law.

We can compute these derivatives from Equations 33-24. We have

$$2L_1 \frac{dL_1}{dx} = 2x \qquad \text{or} \qquad \frac{dL_1}{dx} = \frac{x}{L_1}$$

But x/L_1 is just $\sin\theta_1$, where θ_1 is the angle of incidence. Thus,

$$\frac{dL_1}{dx} = \sin\theta_1 \qquad \text{33-26}$$

Similarly,

$$2L_2 \frac{dL_2}{dx} = 2(d - x)(-1)$$

or

$$\frac{dL_2}{dx} = -\frac{d - x}{L_2} = -\sin\theta_2 \qquad \text{33-27}$$

where θ_2 is the angle of refraction. From Equation 33-25,

$$n_1\frac{dL_1}{dx} + n_2\frac{dL_2}{dx} = 0 \qquad \text{33-28}$$

Substituting the results of Equations 33-26 and 33-27 for dL_1/dx and dL_2/dx gives

$$n_1 \sin \theta_1 + n_2(-\sin \theta_2) = 0$$

or

$$n_1 \sin \theta_1 = n_2 \sin \theta_2$$

which is Snell's law.

Summary

Topic	Remarks and Relevant Equations	
1. Visible Light	The human eye is sensitive to electromagnetic radiation with wavelengths from about 400 nm (violet) to about 700 nm (red). The photon energies range from about 1.8 eV to 3.1 eV. A uniform mixture of wavelengths, such as that emitted by the sun, appears white to our eyes.	
2. Wave–Particle Duality	Light propagates like a wave, but interacts with matter like a particle.	
Photon energy	$E = hf = \frac{hc}{\lambda}$	33-1
Planck's constant	$h = 6.626 \times 10^{-34}\ \mathrm{J \cdot s} = 4.136 \times 10^{-15}\ \mathrm{eV \cdot s}$	
hc	$hc = 1240\ \mathrm{eV \cdot nm}$	33-2
3. Emission of Light	Light is emitted when an outer atomic electron makes a transition from an excited state to a state of lower energy.	
Line spectra	Atoms in dilute gases emit a discrete set of wavelengths called a line spectra. The photon energy $E = hf = hc/\lambda$ equals the difference in energy of the initial and final states of the atom.	
Continuous spectra	Atoms in high-density gases, liquids, or solids have continuous bands of energy levels so they emit a continuous spectrum of light. Thermal radiation is visible if the temperature of the emitting object is above about 600°C.	
Spontaneous emission	An atom in an excited state will spontaneously make a transition to a lower state with the emission of a photon. This process is random with a characteristic lifetime of about 10^{-8} s. The photons from two or more atoms are not correlated, so the light is incoherent.	
Stimulated emission	Stimulated emission occurs if an atom is initially in an excited state and a photon of energy equal to the energy difference between that state and a lower state is incident on the atom. The oscillating electromagnetic field of the incident photon stimulates the excited atom to emit another photon in the same direction and in phase with the incident photon. The emitted light is coherent.	

4. Lasers	A laser produces an intense, coherent, and narrow beam of photons as the result of stimulated emission. The operation of a laser depends on population inversion, in which there are more atoms in an excited state than in the ground state or a lower state.
5. Speed of Light	The SI unit of length, the meter, is defined so that the speed of light in vacuum is exactly $$c = 299{,}792{,}457 \text{ m/s} \qquad \textbf{33-5}$$
c in a medium	$$v = \frac{c}{n} \qquad \textbf{33-7}$$ where n is the index of refraction
6. Huygens' Principle	Each point on a primary wavefront serves as the source of spherical secondary wavelets that advance with a speed and frequency equal to that of the primary wave. The primary wavefront at some later time is the envelope of these wavelets.
7. Reflection and Refraction	When light is incident on a surface separating two media in which the speed of light differs, part of the light energy is transmitted and part is reflected.
Law of reflection	The reflected ray lies in the plane of incidence and makes an angle θ_1' with the normal that is equal to the angle of incidence $$\theta_1' = \theta_1 \qquad \textbf{33-8}$$
Reflected intensity, normal incidence	$$I = \left(\frac{n_1 - n_2}{n_1 + n_2}\right)^2 I_0 \qquad \textbf{33-11}$$
Index of refraction	$$n = \frac{c}{v} \qquad \textbf{33-7}$$
Law of refraction (Snell's law)	$$n_1 \sin\theta_1 = n_2 \sin\theta_2 \qquad \textbf{33-9}\textbf{\textit{b}}$$
Total internal reflection	When light is traveling in a medium with an index of refraction n_1 and is incident on the boundary of a second medium with a lower index of refraction $n_2 < n_1$, the light is totally reflected if the angle of incidence is greater than the critical angle θ_c given by
Critical angle	$$\sin\theta_c = \frac{n_2}{n_1} \qquad \textbf{33-12}$$
Dispersion	The speed of light in a medium and therefore the index of refraction of that medium depends on the wavelength of light. Because of dispersion, a beam of white light incident on a refracting prism is dispersed into its component colors. Similarly, the reflection and refraction of sunlight by raindrops produces a rainbow.
8. Polarization	Transverse waves can be polarized. The four phenomena that produce polarized electromagnetic waves from unpolarized waves are (1) absorption, (2) scattering, (3) reflection, and (4) birefringence.
Malus's Law	When two polarizers have their transmission axes at an angle θ, the intensity transmitted by the second polarizer is reduced by the factor $\cos^2\theta$: $$I = I_0 \cos^2\theta \qquad \textbf{33-20}$$

Problem-Solving Guide

Ray diagrams are particularly useful in problems involving reflection and refraction. Be sure to label the appropriate angles and indicate the index of refraction.

Summary of Worked Examples

Type of Calculation	Procedure and Relevant Examples
1. Absorption and Emission of Light	
Find the wavelength of light absorbed or emitted.	Use $\lambda = hc/\Delta E$, where ΔE is the difference in energy between the initial and final states of the atom. **Example 33-1**
2. Speed of Light	
Find c given the distance and time.	Use $c = \Delta x/\Delta t$. **Example 33-2**
Find c from Fizeau's or Foucault's experiment.	Use $c = \Delta x/\Delta t$ and calculate the time Δt for the wheel or mirror to rotate from one gap or from one mirror to the next. **Example 33-3**
3. Reflection	
Find the angle of reflection.	The angle of reflection is always equal to the angle of incidence, $\theta_1' = \theta_1$.
4. Refraction	
Find the angle of refraction.	Use Snell's law, $n_1 \sin\theta_1 = n_2 \sin\theta_2$. **Example 33-4**
Determine the critical angle for total internal reflection.	The critical angle is given by $\sin\theta_c = n_2/n_1$. **Example 33-5**
5. Polarization	
Find the intensity of light transmitted through a polarizer.	If the incident light is unpolarized, half the intensity will be transmitted through a polarizing film. If the incident light is polarized, the transmitted intensity will be $I_0 \cos^2\theta$. **Example 33-6**

Problems

Conceptual Problems

Problems from Optional and Exploring sections

In a few problems, you are given more data than you actually need; in a few other problems, you are required to supply data from your general knowledge, outside sources, or informed estimates.

• Single-concept, single-step, relatively easy

•• Intermediate-level, may require synthesis of concepts

••• Challenging, for advanced students

Use n = 1.33 for water and n = 1.5 for glass unless otherwise specified.

Light Sources

1 •• Why is helium needed in a helium–neon laser? Why not just use neon?

2 •• When a beam of visible white light passes through a gas of atomic hydrogen and is viewed with a spectroscope, dark lines are observed at the wavelengths of the emission series. The atoms that participate in the resonance absorption then emit this same wavelength light as they return to the ground state. Explain why the observed spectrum nevertheless exhibits pronounced dark lines.

3 • A pulse from a ruby laser has an average power of 10 MW and lasts 1.5 ns. (*a*) What is the total energy of the pulse? (*b*) How many photons are emitted in this pulse?

4 • A helium–neon laser emits light of wavelength 632.8 nm and has a power output of 4 mW. How many photons are emitted per second by this laser?

5 • The first excited state of an atom of a gas is 2.85 eV above the ground state. (*a*) What is the wavelength of radiation for resonance absorption? (*b*) If the gas is irradiated with monochromatic light of 320 nm wavelength, what is the wavelength of the Raman scattered light?

6 •• A gas is irradiated with monochromatic ultraviolet light of 368 nm wavelength. Scattered light of the same wavelength and of 658 nm wavelength is observed. Assuming that the gas atoms were in their ground state prior to irradiation, find the energy difference between the ground state and the atomic state excited by the irradiation.

7 •• Sodium has excited states 2.11 eV, 3.2 eV, and 4.35 eV above the ground state. (*a*) What is the maximum wavelength of radiation that will result in resonance fluorescence? What is the wavelength of the fluorescent radiation? (*b*) What wavelength will result in excitation of the state 4.35 eV above the ground state? If that state is excited, what are the possible wavelengths of resonance fluorescence that might be observed?

8 •• Singly ionized helium is a hydrogen-like atom with a nuclear charge of $2e$. Its energy levels are given by $E_n = -4E_0/n^2$, where $E_0 = 13.6$ eV. If a beam of visible white light is sent through a gas of singly ionized helium, at what wavelengths will dark lines be found in the spectrum of the transmitted radiation?

The Speed of Light

9 • Estimate the time required for light to make the round trip in Galileo's experiment to determine the speed of light.

10 • Mission Control sends a brief wake-up call to astronauts in a far away spaceship. Five seconds after the call is sent, Mission Control can hear the groans of the astronauts. How far away (at most) from the earth is the spaceship?

(*a*) 7.5×10^8 m
(*b*) 15×10^8 m
(*c*) 30×10^8 m
(*d*) 45×10^8 m
(*e*) The spaceship is on the moon.

11 • The spiral galaxy in the Andromeda constellation is about 2×10^{19} km away from us. How many light-years is this?

12 • On a spacecraft sent to Mars to take pictures, the camera is triggered by radio waves, which like all electromagnetic waves travel with the speed of light. What is the time delay between sending the signal from the earth and receiving it on Mars? (Take the distance to Mars to be 9.7×10^{10} m.)

13 • The distance from a point on the surface of the earth to one on the surface of the moon is measured by aiming a laser light beam at a reflector on the surface of the moon and measuring the time required for the light to make a round trip. The uncertainty in the measured distance Δx is related to the uncertainty in the time Δt by $\Delta x = c\,\Delta t$. If the time intervals can be measured to ± 1.0 ns, find the uncertainty of the distance in meters.

14 •• In Galileo's attempt to determine the speed of light, he and his assistant were located on hilltops about 3 km apart. Galileo flashed a light and received a return flash from his assistant. (*a*) If his assistant had an instant reaction, what time difference would Galileo need to be able to measure for this method to be successful? (*b*) How does this time compare with human reaction time, which is about 0.2 s?

Reflection and Refraction

15 • How does a thin layer of water on the road affect the light you see reflected off the road from your own headlights? How does it affect the light you see reflected from the headlights of an oncoming car?

16 • A ray of light passes from air into water, striking the surface of the water with an angle of incidence of 45°. Which of the following four quantities change as the light enters the water: (1) wavelength, (2) frequency, (3) speed of propagation, (4) direction of propagation?

(*a*) 1 and 2 only
(*b*) 2, 3, and 4 only
(*c*) 1, 3, and 4 only
(*d*) 3 and 4 only
(*e*) 1, 2, 3, and 4

17 •• The density of the atmosphere decreases with height, as does the index of refraction. Explain how one can see the sun after it has set. Why does the setting sun appear flattened?

18 • Calculate the fraction of light energy reflected from an air–water interface at normal incidence.

19 • Find the angle of refraction of a beam of light in air that hits a water surface at an angle of incidence of (*a*) 20°, (*b*) 30°, (*c*) 45°, and (*d*) 60°. Show these rays on a diagram.

20 • Repeat Problem 18 for a beam of light initially in water that is incident on a water–air interface.

21 • Find the speed of light in water and in glass.

22 • The index of refraction for silicate flint glass is 1.66 for light with a wavelength of 400 nm and 1.61 for light with a wavelength of 700 nm. Find the angles of refraction for light of these wavelengths that is incident on this glass at an angle of 45°.

23 •• A slab of glass with an index of refraction of 1.5 is submerged in water with an index of refraction of 1.33. Light in the water is incident on the glass. Find the angle of refraction if the angle of incidence is (*a*) 60°, (*b*) 45°, and (*c*) 30°.

24 •• Repeat Problem 23 for a beam of light initially in the glass that is incident on the glass–water interface at the same angles.

25 •• Light is incident normally on a slab of glass with an index of refraction $n = 1.5$. Reflection occurs at both surfaces of the slab. About what percentage of the incident light energy is transmitted by the slab?

26 •• This problem is a refraction analogy. A band is marching down a football field with a constant speed v_1. About midfield, the band comes to a section of muddy ground that has a sharp boundary making an angle of 30° with the 50-yd line as shown in Figure 33-50. In the mud, the marchers move with speed $v_2 = \frac{1}{2}v_1$. Diagram how each line of marchers is bent as it encounters the muddy section of the field so that the band is eventually marching in a different direction. Indicate the original direction by a ray and the final direction by a second ray, and find the angles between the rays and the line perpendicular to the boundary. Is their direction of motion bent toward the perpendicular to the boundary or away from it?

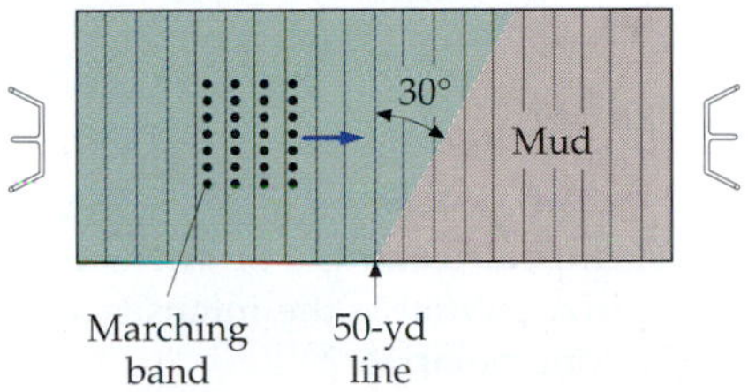

Figure 33-50 Problem 26

27 •• A point source of light is 5 cm above a plane reflecting surface (such as a mirror). Draw a ray from the source that strikes the surface at an angle of incidence of 45° and two more rays that strike the surface at angles slightly less than 45°, and draw the reflected ray for each. The reflected rays appear to diverge from a point called the image of the light source. Draw dotted lines extending the reflected rays back until they meet at a point behind the surface to locate the image point.

28 •• In Figure 33-51, light is initially in a medium (such as air) of index of refraction n_1. It is incident at angle θ_1 on the surface of a liquid (such as water) of index of refraction n_2. The light passes through the layer of water and enters glass of index of refraction n_3. If θ_3 is the angle of refraction in the glass, show that $n_1 \sin\theta_1 = n_3 \sin\theta_3$. That is, show that the second medium can be neglected when finding the angle of refraction in the third medium.

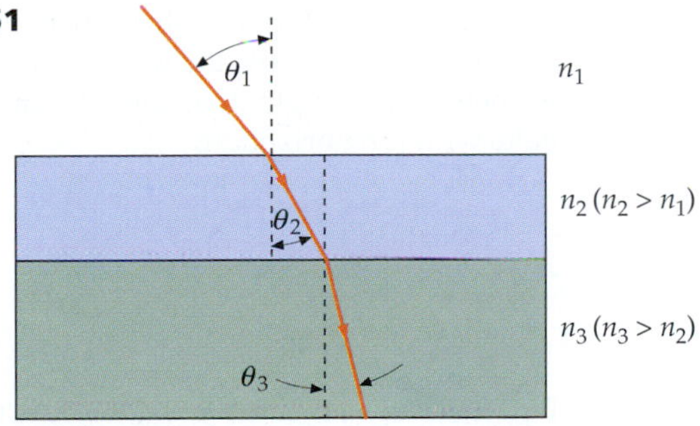

Figure 33-51 Problem 28

29 ••• Figure 33-52 shows a beam of light incident on a glass plate of thickness d and index of refraction n. (*a*) Find the angle of incidence such that the perpendicular separation between the ray reflected from the top surface and that reflected from the bottom surface and exiting the top surface is a maximum. (*b*) What is this angle of incidence if the index of refraction of the glass is 1.60? What is the separation of the two beams if the thickness of the glass plate is 4.0 cm?

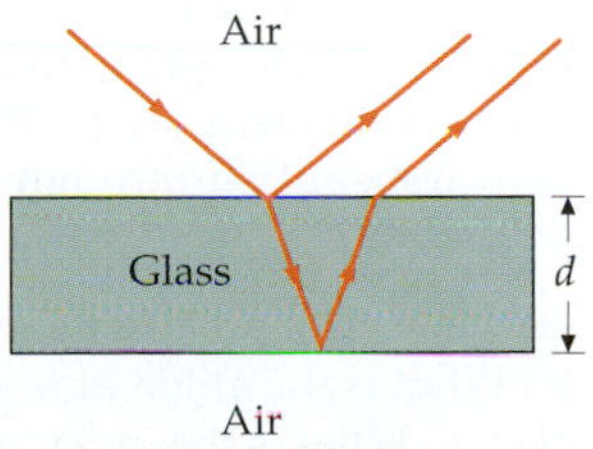

Figure 33-52 Problem 29

30 ••• Consider the situation shown in Figure 33-53. The index of refraction of the glass plate is n. Find the angle of incidence such that the perpendicular separation between the two beams emerging from the top surface is the same as the perpendicular displacement of the beam emerging from the bottom surface from the incident beam.

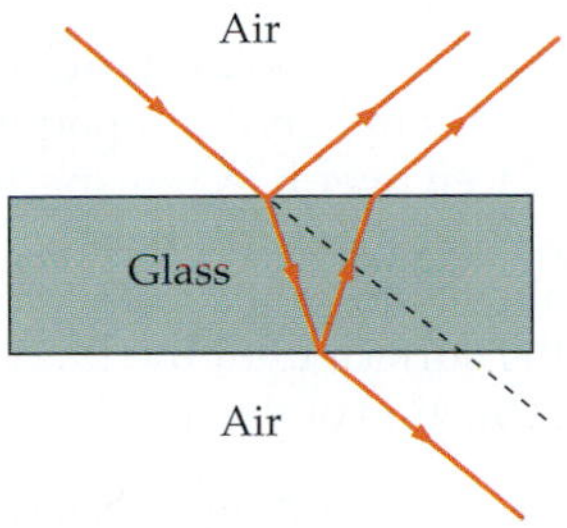

Figure 33-53 Problem 30

Deriving the Laws of Reflection and Refraction (optional)

31 • A physics student playing pocket billiards wants to strike her cue ball so that it hits a cushion and then hits the eight ball squarely. She chooses several points on the cushion and for each point measures the distance from it to the cue ball and to the eight ball. She aims at the point for which the sum of these distances is least. (*a*) Will her cue ball hit the eight ball? (*b*) How is her method related to Fermat's principle?

32 • A swimmer at S in Figure 33-54 develops a leg cramp while swimming near the shore of a calm lake and calls for help. A lifeguard at L hears the call. The lifeguard can run 9 m/s and swim 3 m/s. He knows physics and chooses a path that will take the least time to reach the swimmer. Which of the paths shown in Figure 33-54 does he take?

Figure 33-54 Problem 32

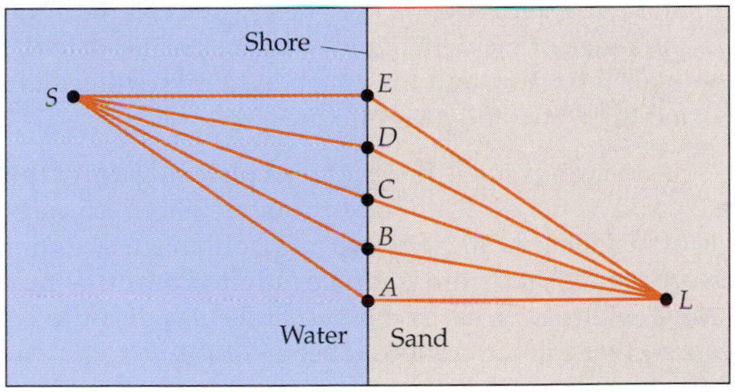

Total Internal Reflection

33 • What is the critical angle for total internal reflection for light traveling initially in water that is incident on a water–air interface?

34 •• A glass surface ($n = 1.50$) has a layer of water ($n = 1.33$) on it. Light in the glass is incident on the glass–water interface. Find the critical angle for total internal reflection.

35 •• A point source of light is located 5 m below the surface of a large pool of water. Find the area of the largest circle on the pool's surface through which light coming directly from the source can emerge.

36 •• Light is incident normally on the largest face of an isosceles-right-triangle prism. What is the speed of light in this prism if the prism is just barely able to produce total internal reflection?

37 •• A point source of light is located at the bottom of a steel tank, and an opaque circular card of radius 6.0 cm is placed over it. A transparent fluid is gently added to the tank so that the card floats on the surface with its center directly above the light source. No light is seen by an observer above the surface until the fluid is 5 cm deep. What is the index of refraction of the fluid?

38 •• A grain of sand is directly below the center of the base of a cube of transparent material. The grain of sand is visible when viewed through the top surface but cannot be seen when looking into any of the sides of the cube. What is the minimum index of refraction of the material of which the cube is made?

39 ••• Light is incident normally upon one face of a prism of glass with an index of refraction n (Figure 33-55). The light is totally reflected at the right side. (*a*) What is the minimum value n can have? (*b*) When the prism is immersed in a liquid whose index of refraction is 1.15, there is still total reflection, but when it is immersed in water, whose index of refraction is 1.33, there is no longer total reflection. Use this information to establish limits for possible values of n.

Figure 33-55 Problem 39

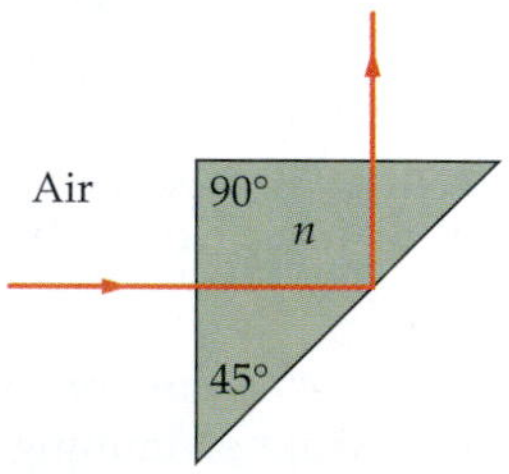

40 ••• Investigate how a thin film of water on a glass surface affects the critical angle for total reflection. Take $n = 1.5$ for glass and $n = 1.33$ for water. (*a*) What is the critical angle for total internal reflection at the glass–water interface? (*b*) Is there any range of incident angles that are greater than θ_c for glass-to-air refraction and for which light rays will leave the glass and the water and pass into the air?

41 ••• A laser beam is incident on a plate of glass of thickness 3 cm. The glass has an index of refraction of 1.5 and the angle of incidence is 40°. The top and bottom surfaces of the glass are parallel and both produce reflected beams of nearly the same intensity. What is the perpendicular distance d between the two adjacent reflected beams?

42 ••• Figure 33-56 shows a glass prism of index of refraction $n = 1.52$ in the shape of an isosceles triangle with base angles of 45°. (*a*) Find the maximum angle of incidence of the beam incident on the side face so that it suffers total internal reflection at the base. (*b*) What is the maximum value of the index of refraction of the prism so that the light beam will suffer total internal reflection at the base whatever the angle of incidence?

Figure 33-56 Problem 42

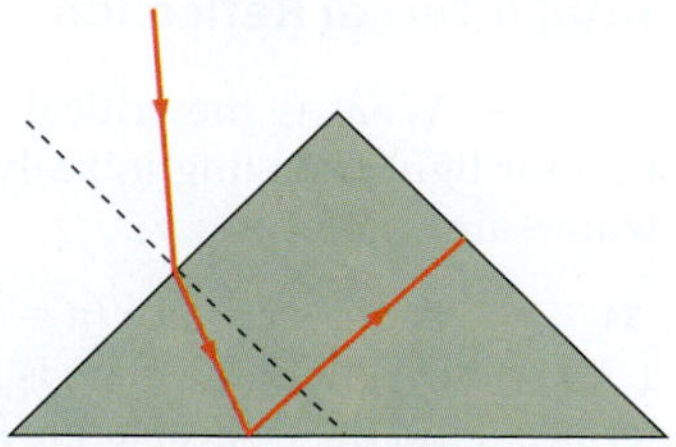

Dispersion

43 •• A beam of light strikes the plane surface of silicate flint glass at an angle of incidence of 45°. The index of refraction of the glass varies with wavelength as shown in the graph in Figure 33-26. How much smaller is the angle of refraction for violet light of wavelength 400 nm than that for red light of wavelength 700 nm?

44 •• Repeat Problem 43 for quartz.

Polarization

45 • Two polarizers have their transmission axes at an angle θ. Unpolarized light of intensity I is incident upon the first polarizer. What is the intensity of the light transmitted by the second polarizer?

(*a*) $I \cos^2 \theta$
(*b*) $(I \cos^2 \theta)/2$
(*c*) $(I \cos^2 \theta)/4$
(*d*) $I \cos \theta$
(*e*) $(I \cos \theta)/4$
(*f*) None of the above.

46 • Which of the following is *not* a phenomenon whereby polarized light can be produced from unpolarized light?

(*a*) absorption
(*b*) reflection
(*c*) birefringence
(*d*) diffraction
(*e*) scattering

47 • What is the polarizing angle for (*a*) water with $n = 1.33$ and (*b*) glass with $n = 1.5$?

48 • Light known to be polarized in the horizontal direction is incident on a polarizing sheet. It is observed that only 15.0% of the intensity of the incident light is transmitted through the sheet. What angle does the transmission axis of the sheet make with the horizontal?

(*a*) 8.6°
(*b*) 21°
(*c*) 23°
(*d*) 67°
(*e*) 81°

49 • Two polarizing sheets have their transmission axes crossed so that no light gets through. A third sheet is inserted between the first two so that its transmission axis makes an angle θ with that of the first sheet. Unpolarized light of intensity I_0 is incident on the first sheet. Find the intensity of the light transmitted through all three sheets if (*a*) $\theta = 45°$ and (*b*) $\theta = 30°$.

50 •• The polarizing angle for a certain substance is 60°. (*a*) What is the angle of refraction of light incident at this angle? (*b*) What is the index of refraction of this substance?

51 •• Two polarizing sheets have their transmission axes crossed and a third sheet is inserted so that its transmission axis makes an angle θ with that of the first sheet as in Problem 49. Find the intensity of the transmitted light as a function of θ. Show that the intensity transmitted through all three sheets is maximum when $\theta = 45°$.

52 •• If the middle polarizing sheet in Problem 51 is rotating at an angular velocity ω about an axis parallel to the light beam, find the intensity transmitted through all three sheets as a function of time. Assume that $\theta = 0$ at time $t = 0$.

53 •• A stack of $N + 1$ ideal polarizing sheets is arranged with each sheet rotated by an angle of $\pi/2N$ rad with respect to the preceding sheet. A plane linearly polarized light wave of intensity I_0 is incident normally on the stack. The incident light is polarized along the transmission axis of the first sheet and therefore perpendicular to the transmission axis of the last sheet in the stack. (*a*) What is the transmitted intensity through the stack? (*b*) For 3 sheets ($N = 2$), what is the transmitted intensity? (*c*) For 101 sheets, what is the transmitted intensity? (*d*) What is the direction of polarization of the transmitted beam in each case?

54 •• Show that a linearly polarized wave can be thought of as a superposition of a right and a left circularly polarized wave.

55 •• Suppose that in Problem 49 the middle sheet is replaced by two polarizing sheets. If the angles between the directions of polarization of adjacent sheets is 30°, what is the intensity of the transmitted light? How does this compare with the intensity obtained in Problem 49*a*?

56 •• In a circularly polarized wave, the magnitude of the electric field is constant. If the wave propagates along the z axis, the angle between $\vec{E}$ and the x axis changes by 2π radians over one wavelength. Write expressions for the electric and magnetic fields of a circularly polarized wave of angular frequency ω propagating in vacuum in the positive z direction.

57 •• Show that the electric field of a circularly polarized wave propagating in the x direction can be expressed by

$$\vec{E} = E_0 \sin(kx - \omega t)\,\hat{j} + E_0 \cos(kx - \omega t)\,\hat{k}$$

58 •• For the wave whose electric field is given by the expression in Problem 57, what is the corresponding expression for the magnetic field $\vec{B}$?

59 •• Find expressions for the electric field $\vec{E}$ and magnetic field $\vec{B}$ for a circularly polarized wave propagating in the negative z direction. (See Problems 57 and 58.)

60 •• A circularly polarized wave is said to be *right circularly polarized* if the electric and magnetic fields rotate clockwise when viewed along the direction of propagation and *left circularly polarized* if the field rotate counterclockwise. What is the sense of the circular polarization for the wave described by the expression in Problem 57? What would be the corresponding expression for a circularly polarized wave of the opposite sense?

61 •• Vertically polarized light of intensity I_0 is incident on a stack of N ideal polarizing sheets whose angles with respect to the vertical are $\theta_n = n\pi/2N$. Determine the direction of polarization of the transmitted light and its intensity. Show that as $N \to \infty$ the direction of polarization is rotated without loss of intensity.

General Problems

62 • True or false:

(*a*) Light and radio waves travel with the same speed through a vacuum.
(*b*) Most of the light incident normally on an air–glass interface is reflected.
(*c*) The angle of refraction of light is always less than the angle of incidence.
(*d*) The index of refraction of water is the same for all wavelengths in the visible spectrum.
(*e*) Longitudinal waves cannot be polarized.

63 •• Of the following statements about the speeds of the various colors of light in glass, which are true?

(*a*) All colors of light have the same speed in glass.
(*b*) Violet has the highest speed, red the lowest.
(*c*) Red has the highest speed, violet the lowest.
(*d*) Green has the highest speed, red and violet the lowest.
(*e*) Red and violet have the highest speed, green the lowest.

64 •• It is a common experience that on a calm, sunny day one can hear voices of persons in a boat over great distances. Explain this phenomenon, keeping in mind that sound is reflected from the surface of the water and that the temperature of the air just above the water's surface is usually less than that at a height of 10 or 20 m above the water.

65 • A beam of monochromatic red light with a wavelength of 700 nm in air travels in water. (*a*) What is the wavelength in water? (*b*) Does a swimmer underwater observe the same color or a different color for this light?

66 • As the speed of computer operations increases, computer architecture acquires greater importance; the time required to transfer a signal between the central processing unit (CPU) and memory can be a limiting factor in determining the time required for computation. What is the maximum separation between a memory chip and the CPU to allow transfer information between these units in less than 0.5 ns?

67 •• The critical angle for total internal reflection for a substance is 45°. What is the polarizing angle for this substance?

68 •• Figure 33-57 shows two plane mirrors that make an angle θ with each other. Show that the angle between the incident and reflected rays is 2θ.

Figure 33-57 Problem 68

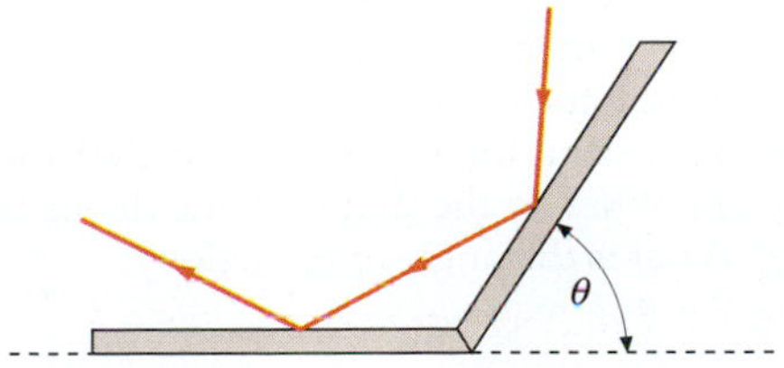

69 •• A silver coin sits on the bottom of a swimming pool that is 4 m deep. A beam of light reflected from the coin emerges from the pool making an angle of 20° with respect to the water's surface and enters the eye of an observer. Draw a ray from the coin to the eye of the observer. Extend this ray, which goes from the water–air interface to the eye, straight back until it intersects with the vertical line drawn through the coin. What is the apparent depth of the swimming pool to this observer?

70 •• Two affluent students decide to improve on Galileo's experiment to measure the speed of light. One student goes to London and calls the other in New York on the telephone. The telephone signals are transmitted by reflecting electromagnetic waves from a satellite that is 37.9 Mm above the earth's surface. If the distance between London and New York is neglected, the distance traveled is twice this distance. One student claps his hands, and when the other student hears the sound over the phone, she claps her hands. The first student measures the time between his clap and his hearing the second one. Calculate this time lapse, neglecting the students' response times. Do you think this experiment would be successful? What improvements for measuring this time interval would you suggest? (Time delays in the electronic circuits that are greater than those due to the light traveling to the satellite and back make this experiment not feasible.)

71 •• Fishermen always insist on silence because noise on shore will scare fish away. Suppose a fisherman cast a baited hook 20 m from the shore of a calm lake to a point where the depth is 15 m. Show that noise on shore cannot possibly be sensed by fish at that point. *Note:* The speed of sound in air is 330 m/s; the speed of sound in water is 1450 m/s.

72 •• A swimmer at the bottom of a pool 3 m deep looks up and sees a circle of light. If the index of refraction of the water in the pool is 1.33, find the radius of the circle.

73 •• Show that when a mirror is rotated through an angle θ, the reflected beam of light is rotated through 2θ.

74 •• Use Figure 33-26 to calculate the critical angles for total internal reflection for light initially in silicate flint glass that is incident on a glass–air interface if the light is (*a*) violet light of wavelength 400 nm, and (*b*) red light of wavelength 700 nm.

75 •• Show that for normally incident light, the intensity transmitted through a glass slab with an index of refraction of n is approximately given by

$$I_T = I_0\left[\frac{4n}{(n+1)^2}\right]^2$$

76 •• A ray of light begins at the point $x = -2$ m, $y = 2$ m, strikes a mirror in the xz plane at some point x, and reflects through the point $x = 2$ m, $y = 6$ m. (*a*) Find the value of x that makes the total distance traveled by the ray a minimum. (*b*) What is the angle of incidence on the reflecting plane? What is the angle of reflection?

77 •• Light passes symmetrically through a prism having an apex angle of α as shown in Figure 33-58. (*a*) Show that the angle of deviation δ is given by

$$\sin\frac{\alpha+\delta}{2} = n\sin\frac{\alpha}{2}$$

(*b*) If the refractive index for red light is 1.48 and that for violet light is 1.52, what is the angular separation of visible light for a prism with an apex angle of 60°?

Figure 33-58 Problems 77, 88, and 89

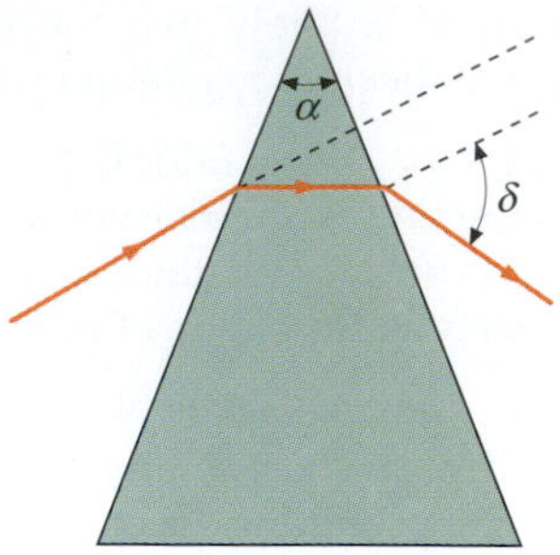

78 •• (*a*) For a light ray inside a transparent medium having a planar interface with a vacuum, show that the polarizing angle and the critical angle for internal reflection satisfy $\tan\theta_p = \sin\theta_c$. (*b*) Which angle is larger?

79 •• Light is incident from air on a transparent substance at an angle of 58.0° with the normal. The reflected and refracted rays are observed to be mutually perpendicular. (*a*) What is the index of refraction of the transparent substance? (*b*) What is the critical angle for total internal reflection in this substance?

80 •• A light ray in dense flint glass with an index of refraction of 1.655 is incident on the glass surface. An unknown liquid condenses on the surface of the glass. Total internal reflection on the glass–liquid interface occurs for an angle of incidence on the glass–liquid interface of 53.7°. (*a*) What is the refractive index of the unknown liquid? (*b*) If the liquid is removed, what is the angle of incidence for total internal reflection? (*c*) For the angle of incidence found in part (*b*), what is the angle of refraction of the ray into the liquid film? Does a ray emerge from the liquid film into the air above? Assume that the glass and liquid have perfect planar surfaces.

81 •• Given that the index of refraction for red light in water is 1.3318 and that the index of refraction for blue light in water is 1.3435, find the angular separation of these colors in the primary rainbow. (Use the equation given in Problem 86.)

82 •• A ray of light falls on a rectangular glass block ($n = 1.5$) that is almost completely submerged in water ($n = 1.33$) as shown in Figure 33-59. (*a*) Find the angle θ for which total internal reflection just occurs at point P. (*b*) Would total internal reflection occur at point P for the value of θ found in part (*a*) if the water were removed? Explain.

Figure 33-59 Problem 82

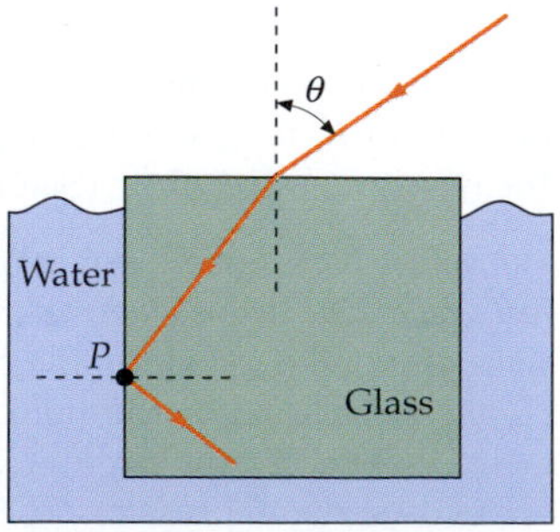

83 •• (*a*) Use the result for Problem 75 to find the ratio of the transmitted intensity to the incident intensity through N

parallel slabs of glass for light of normal incidence. (*b*) Find this ratio for three slabs of glass with $n = 1.5$. (*c*) How many slabs of glass with $n = 1.5$ will reduce the intensity to 10% of the incident intensity?

84 •• Light is incident on a slab of transparent material at an angle θ_1 as shown in Figure 33-60. The slab has a thickness t and an index of refraction n. Show that

$$n = \frac{\sin \theta_1}{\sin[\arctan(d/t)]}$$

where d is the distance shown in the figure and $\arctan(d/t)$ is the angle whose tangent is d/t.

Figure 33-60 Problem 84

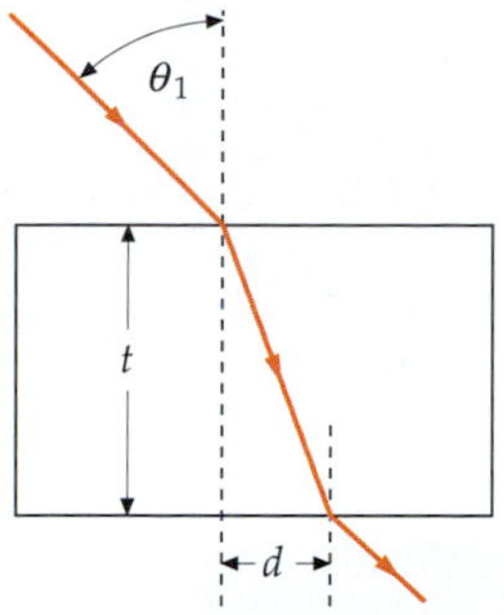

85 •• Suppose rain falls vertically from a stationary cloud 10,000 m above a confused marathoner running in a circle with constant speed of 4 m/s. The rain has a terminal speed of 9 m/s. (*a*) What is the angle that the rain appears to make with the vertical to the marathoner? (*b*) What is the apparent motion of the cloud as observed by the marathoner? (*c*) A star on the axis of the earth's orbit appears to have a circular orbit of angular diameter of 41.2 seconds of arc. How is this angle related to the earth's speed in its orbit and the velocity of photons received from this distant star? (*d*) What is the speed of light as determined from the data in part (*c*)?

86 ••• Equation 33-18 gives the relation between the angle of deviation ϕ_d of a light ray incident on a spherical drop of water in terms of the incident angle θ_1 and the index of refraction of water. (*a*) Assume that $n_{air} = 1$, and differentiate ϕ_d with respect to θ_1. [*Hint:* If $y = \arcsin x$, then $dy/dx = (1 - x^2)^{-1/2}$.] (*b*) Set $d\phi_d/d\theta_1 = 0$ and show that the angle of incidence θ_{1m} for minimum deviation is given by

$$\cos \theta_{1m} = \sqrt{\frac{n^2 - 1}{3}}$$

and find θ_{1m} for water, where the index of refraction for water is 1.33.

87 ••• (*a*) Show that a light ray transmitted through a glass slab emerges parallel to the incident ray but displaced from it. (*b*) For an incident angle of 60°, glass of index of refraction $n = 1.5$, and a slab of thickness 10 cm, find the displacement measured perpendicularly from the incident ray.

88 ••• Show that if the apex angle α of the prism of Problem 77 is small, the angle of deviation δ is given by $\delta = (n - 1)\alpha$, independent of the angle of incidence.

89 ••• Show that the angle of deviation δ is a minimum if the angle of incidence is such that the ray passes through the prism symmetrically as shown in Figure 33-58.

CHAPTER 34

Optical Images

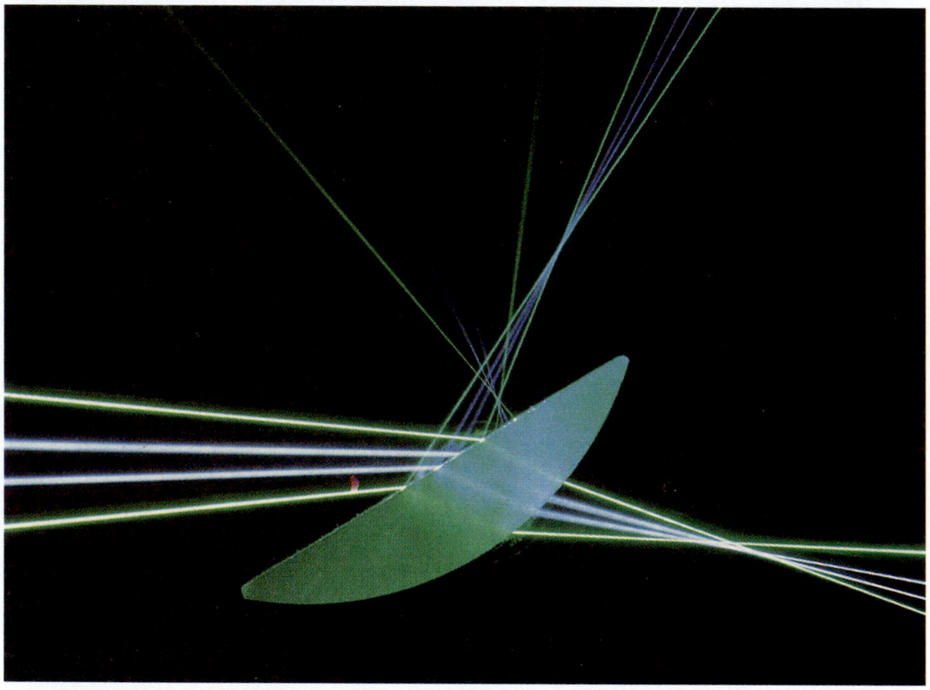

The focusing of rays by reflection and refraction is illustrated by these laser beams incident on a glass lens.

Because the wavelength of light is very small compared with most obstacles and openings, diffraction—the bending of waves around corners—is often negligible, and the ray approximation, in which waves are considered to propagate in straight lines, is valid. In this chapter we apply the laws of reflection and refraction to the formation of images by mirrors and lenses.

34-1 Mirrors

Plane Mirrors

Figure 34-1 shows a bundle of light rays emanating from a point source P and reflected from a plane mirror. After reflection, the rays diverge exactly as if they came from a point P' behind the plane of the mirror. The point P' is called the **image** of the **object** P. When these reflected rays enter the eye, they cannot be distinguished from rays diverging from a source at P' with no mirror present. This image is called a **virtual image** because the light does not actually emanate from it. The image point P' lies on the line through the ob-

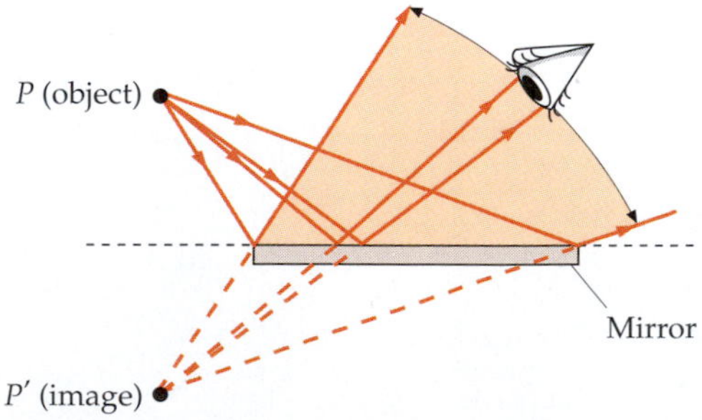

Figure 34-1 Image formed by a plane mirror. The rays from point P that strike the mirror and enter the eye appear to come from the image point P' behind the mirror. The image can be seen by the eye placed anywhere in the shaded region.

ject P perpendicular to the plane of the mirror at a distance behind the plane equal to that from the plane to the object, as shown in Figure 34-1. The image can be seen by an eye anywhere in the shaded region indicated, in which a straight line from the image to the eye passes through the mirror. The object need not be directly in front of the mirror. As long as the object is not behind the plane of the mirror, there is some position at which the eye can be placed to view the image.

If you hold up your right hand and look in the mirror, the image you see is the same size, but it looks like a left hand (Figure 34-2). This right-to-left reversal is a result of **depth inversion**—the hand is transformed from a right hand to a left hand because the front and back of the hand are reversed by the mirror. Depth inversion is also illustrated in Figure 34-3. Figure 34-4 shows the image of a simple rectangular coordinate system. The mirror transforms a right-handed coordinate system for which $\hat{i} \times \hat{j} = \hat{k}$, into a left-handed coordinate system for which $\hat{i} \times \hat{j} = -\hat{k}$.

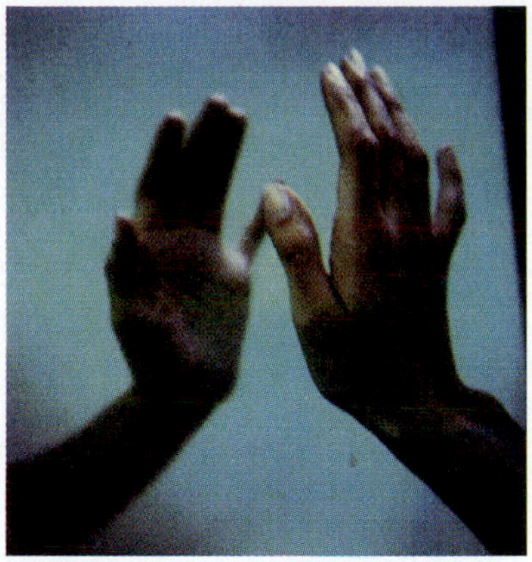

Figure 34-2 The image of a right hand in a plane mirror is a left hand. This right-to-left reversal is a result of depth inversion.

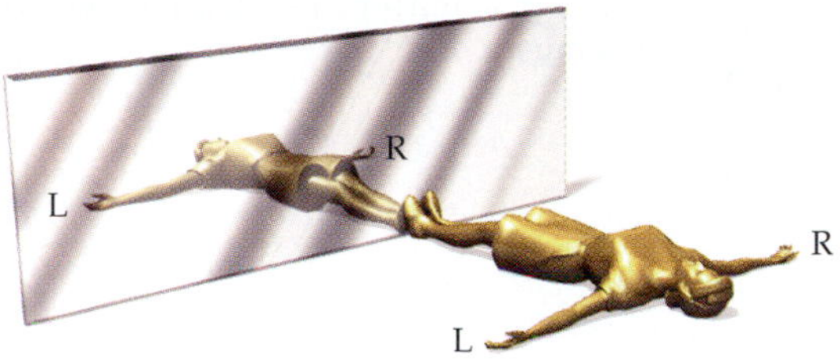

Figure 34-3 A person lying down with her feet against the mirror. The image is depth inverted.

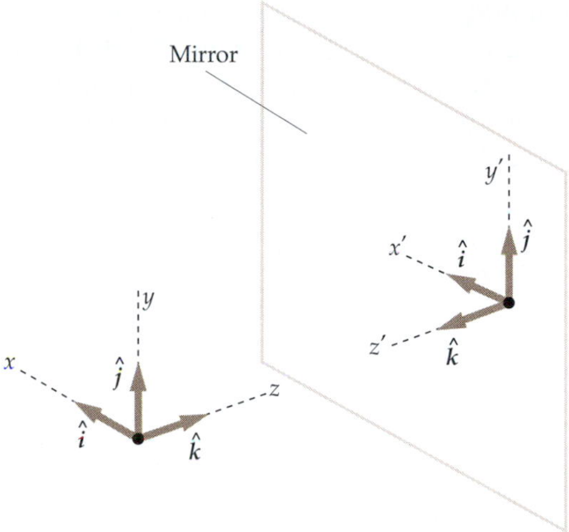

Figure 34-4 Image of a rectangular coordinate system in a plane mirror. The arrow along the z axis is reversed in the image. The image of the original right-handed coordinate system, for which $\hat{i} \times \hat{j} = \hat{k}$, is a left-handed coordinate system, for which $\hat{i} \times \hat{j} = -\hat{k}$.

Figure 34-5 shows an arrow of height y standing parallel to a plane mirror a distance s from it. We can locate the image of the arrowhead (and of any other point on the arrow) by drawing two rays. One ray, drawn perpendicular to the mirror, hits the mirror at point A and is reflected back onto itself. The other ray, making an angle θ with the normal to the mirror, is reflected, making an equal angle θ with the x axis. The extension of these two rays back behind the mirror locates the image of the arrowhead, as shown by the dashed lines in the figure. We can see from this figure that the image is the same distance behind the mirror as the object is in front of the mirror, and that the image is erect (points in the same direction as the object) and is the same size as the object.

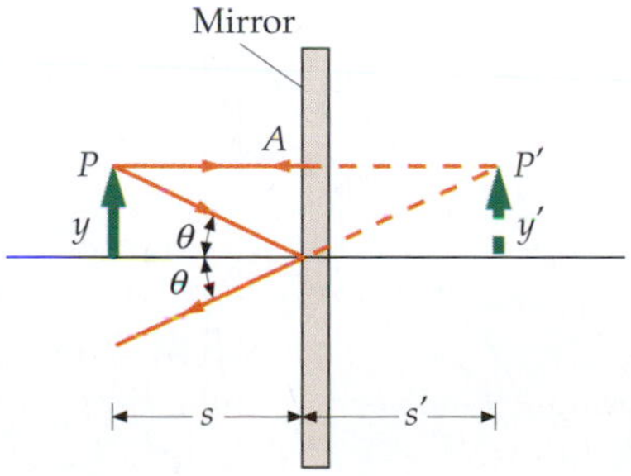

Figure 34-5 Ray diagram for locating the image of an arrow in a plane mirror.

The formation of multiple images by two plane mirrors making an angle with each other is illustrated in Figure 34-6. We frequently see this phenomenon in clothing stores that provide adjacent mirrors. Light reflected from mirror 1 strikes mirror 2 just as if it came from the image point P_1. The image P_1' is the object for mirror 2. Its image is at point $P_{1,2}''$. This image will be formed whenever the image point P_1 is in front of the plane of mirror 2. The image at point P_2' is due to rays from the object that reflect directly from mirror 2. Since P_2' is behind the plane of mirror 1, it cannot serve as an object point for a further image in mirror 1. The number of multiple images formed by two mirrors depends on the angle between the mirrors and the position of the object.

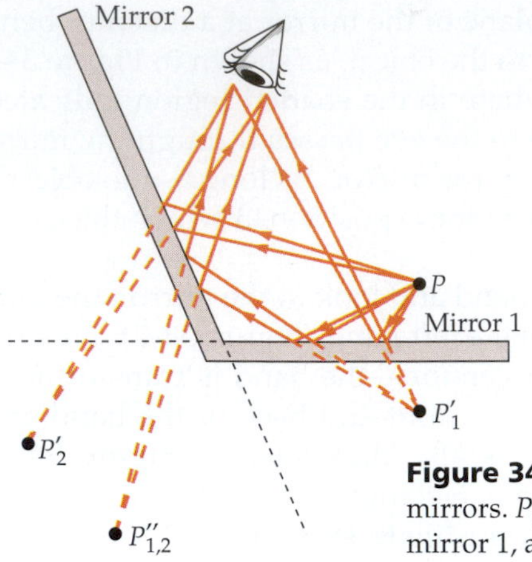

Figure 34-6 Images formed by two plane mirrors. P_1' is the image of the object P in mirror 1, and P_2' is the image of the object in mirror 2. Point $P_{1,2}''$ is the image of P_1' in mirror 2 seen when light rays from the object reflect first from mirror 1 and then from mirror 2. The image P_2' does not have an image in mirror 1 because it is behind that mirror.

Figure 34-7 shows two mirrors at right angles to each other. Rays from the object to the eye that strike mirror 1 and then mirror 2 are shown in Figure 34-7*a*. In this case, the image point $P_{1,2}''$ is the same as that for rays that strike mirror 2 first and then mirror 1, as can be seen from Figure 34-7*b*. If you stand in front of two vertical mirrors that are perpendicular to each other, such as in the corner of a room, the image you see is the same as that seen by others who are facing you because depth inversion occurs twice, once in each mirror.

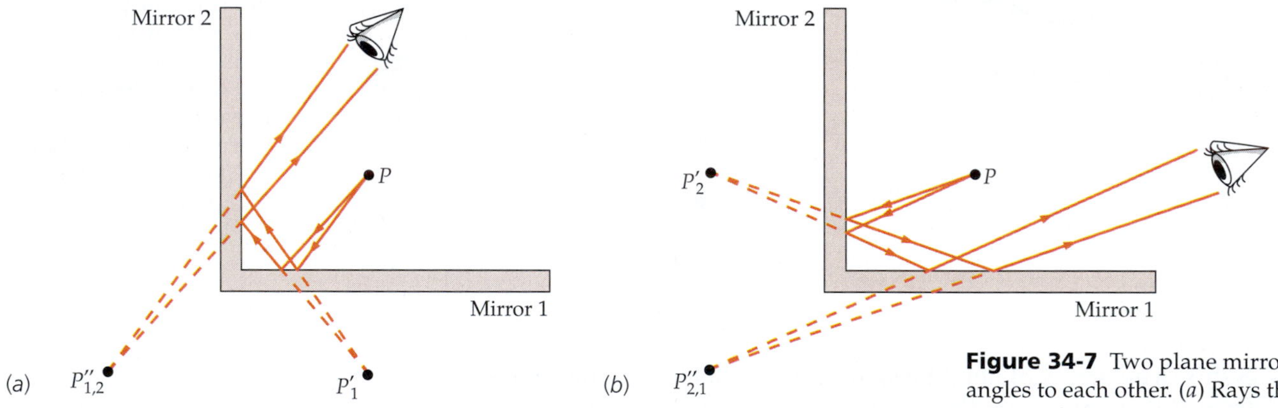

Figure 34-7 Two plane mirrors at right angles to each other. (*a*) Rays that strike mirror 1 first and then mirror 2. The image of P_1' in mirror 2 is $P_{1,2}''$. (*b*) Rays that strike mirror 2 first and then mirror 1. The image of P_2' in mirror 1 is $P_{2,1}''$, which coincides with $P_{1,2}''$ for perpendicular mirrors.

Figure 34-8 illustrates the fact that a horizontal ray reflected from two perpendicular vertical mirrors is exactly reversed in direction no matter what angle the ray makes with the mirrors. If three mirrors are placed perpendicular to each other like the sides of an inside corner of a box, any ray incident on any of the mirrors from any direction is exactly reversed. A set of mirrors of this type was placed on the moon facing the earth. A laser beam from earth directed at the mirrors is reflected back to the same place on the earth. Such a beam has been used to measure the distance to the mirrors to within a few centimeters by measuring the time it takes for the light to reach the mirrors and return.

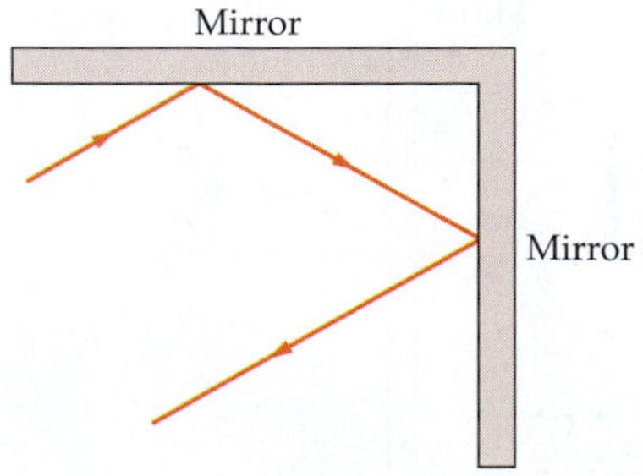

Figure 34-8 A ray striking one of two perpendicular plane mirrors is reflected from the second mirror in the direction opposite the original direction for any angle of incidence.

Spherical Mirrors

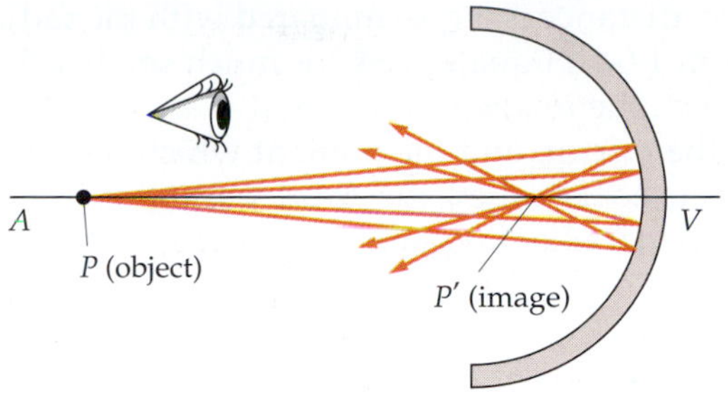

Figure 34-9 Rays from a point object P on the axis AV of a concave spherical mirror form an image at P'. The image is sharp if the rays strike the mirror near the axis.

Figure 34-9 shows a bundle of rays from a point source P on the axis of a concave spherical mirror reflecting from the mirror and converging at point P'. The rays then diverge from this point just as if there were an object at that point. This image is called a **real image,** because light actually does emanate from the image point. The image can be seen by an eye at the left of the image looking into the mirror. It could also be observed on a ground-glass viewing screen or photographic film placed at the image point. A virtual image, such as that formed by a plane mirror as discussed in the previous section, cannot be observed on a screen at the image point because there is no light there. Despite this distinction between real and virtual images, the eye makes no distinction between them. The light rays diverging from a real image and those appearing to diverge from a virtual image are the same to the eye.

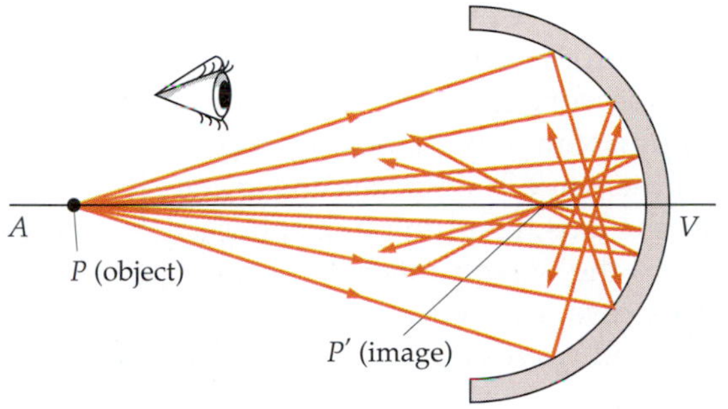

Figure 34-10 Spherical aberration of a mirror. Nonparaxial rays that strike the mirror at points far from the axis AV are not reflected through the image point P'. These rays blur the image.

From Figure 34-10, we can see that only rays that strike the spherical mirror at points near the axis AV are reflected through the image point. Such rays are called **paraxial rays.** Rays that strike the mirror at points far from the axis, called *nonparaxial rays,* converge to different points near the image point. Such rays cause the image to appear blurred, an effect called **spherical aberration.** The image can be sharpened by blocking off all but the central part of the mirror so that nonparaxial rays do not strike it. The image is then sharper, but its brightness is reduced because less light is reflected to the image point.

Figure 34-11 shows a ray from an object point P reflecting off the mirror and passing through the image point P'. Point C is the center of curvature of the mirror. The incident and reflected rays make equal angles with the radial line CA, which is perpendicular to the surface of the mirror.

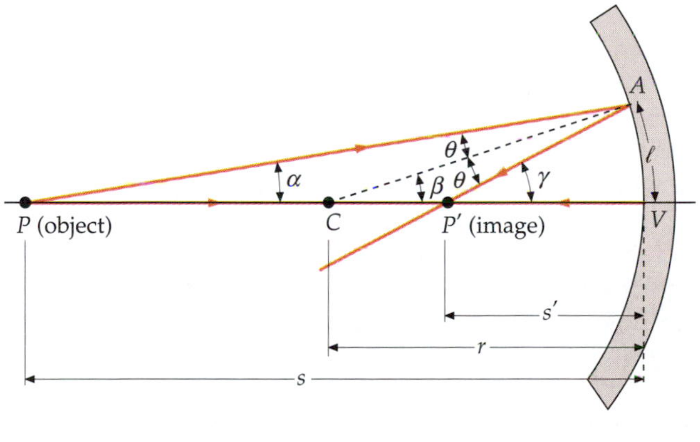

Figure 34-11 Geometry for calculating the image distance s' from the object distance s and the radius of curvature r. The angle β is an exterior angle to the triangle PAC, therefore $\beta = \alpha + \theta$. Similarly, from the triangle PAP', $\gamma = \alpha + 2\theta$. Eliminating θ from these equations gives $2\beta = \alpha + \gamma$. Equation 34-1 follows directly if we assume the following small-angle approximations: $\alpha \approx \ell/s$, $\beta \approx \ell/r$, and $\gamma \approx \ell/s'$

The image distance s' from the vertex of the mirror V to P' can be related to the object distance s from the vertex V to point P and the radius of curvature r of the mirror by elementary geometry. The result is

$$\frac{1}{s} + \frac{1}{s'} = \frac{2}{r} \qquad \text{34-1}$$

The derivation of this equation assumes that the rays are paraxial.

When the object distance is large compared with the radius of curvature of the mirror, the term $1/s$ in Equation 34-1 is much smaller than $2/r$ and can be neglected. For $s = \infty$, the image distance is $s' = \frac{1}{2}r$. This distance is called the **focal length** f of the mirror, and the point at which parallel rays incident on the mirror are focused is called the **focal point** F as illustrated in Figure 34-12*a*. (Again, only paraxial rays are focused at a single point.)

$$f = \frac{1}{2}r \qquad \text{34-2}$$

Focal length for a mirror

The focal length of a spherical mirror is half the radius of curvature. In terms of the focal length f, Equation 34-1 is

$$\frac{1}{s} + \frac{1}{s'} = \frac{1}{f} \qquad \text{34-3}$$

Mirror equation

Equation 34-3 is called the **mirror equation**.

When an object is very far from the mirror, the rays are parallel, and the wavefronts are approximately planes (Figure 34-12*b*). In Figure 34-12*b*, note how the edges of the wavefront hit the concave mirror surface before the central portion near the axis, resulting in a spherical wavefront upon reflection. Figure 34-13 shows the wavefronts and rays for plane waves striking a convex mirror. In this case, the central part of the wavefront strikes the mirror first, and the reflected waves appear to come from the focal point behind the mirror.

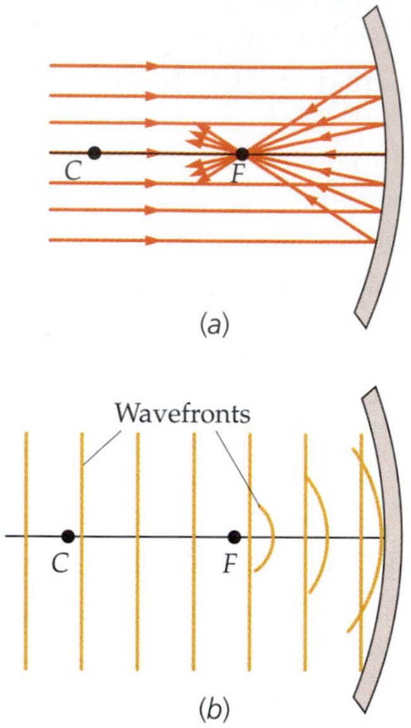

Figure 34-12 (*a*) Parallel rays strike a concave mirror and are reflected through the focal point F at a distance $r/2$. (*b*) The incoming wavefronts are plane waves; upon reflection, they become spherical waves that converge at the focal point.

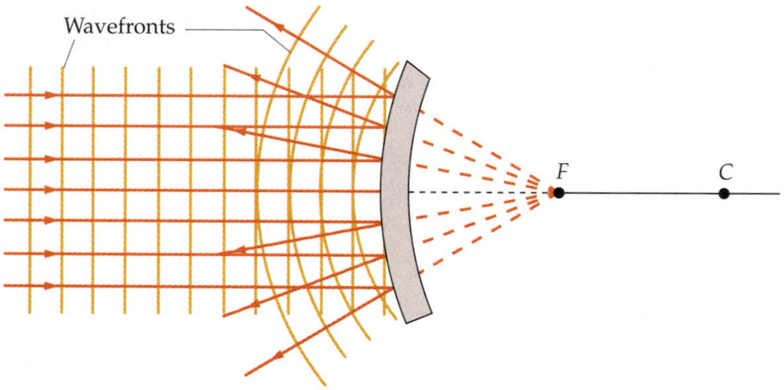

Figure 34-13 Reflection of plane waves from a convex mirror. The outgoing wavefronts are spherical as if emanating from the focal point F behind the mirror. The rays are perpendicular to the wavefronts, and appear to diverge from F.

Figure 34-14 illustrates a property of waves called **reversibility.** If we reverse the direction of a reflected ray, the law of reflection assures us that the reflected ray will be along the original incoming ray but in the opposite direction. (Reversibility holds also for refracted rays, which are discussed in later sections.) Thus, if we have a real image of an object formed by a reflecting (or refracting) surface, we can place an object at the image point and a new image will be formed at the position of the original object.

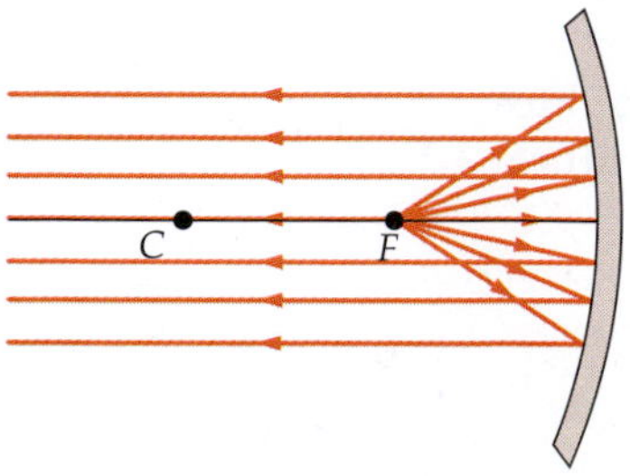

Figure 34-14 Reversibility. Rays diverging from a point source at the focal point of a concave mirror are reflected from the mirror as parallel rays. The rays are the same as in Figure 34-12*a* but in the reverse direction.

Example 34-1

An object is 12 cm from a concave mirror with a radius of curvature of 6 cm. Find (*a*) the focal length of the mirror and (*b*) the image distance.

Picture the Problem The focal length of a spherical mirror is half the radius of curvature. Once the focal length is known, the image distance can be found using the mirror equation (Equation 34-3).

(*a*) The focal length is half the radius of curvature: $f = \frac{1}{2}r = \frac{1}{2}(6\text{ cm}) = 3\text{ cm}$

(*b*)1. Use the mirror equation to find a relation for the image distance s': $\frac{1}{s} + \frac{1}{s'} = \frac{1}{f}$ or $\frac{1}{12\text{ cm}} + \frac{1}{s'} = \frac{1}{3\text{ cm}}$

2. Solve for s': $\frac{1}{s'} = \frac{4}{12\text{ cm}} - \frac{1}{12\text{ cm}} = \frac{3}{12\text{ cm}}$

$s' = 4\text{ cm}$

Exercise A concave mirror has a focal length of 4 cm. (*a*) What is its radius of curvature? (*b*) Find the image distance for an object 2 cm from the mirror. (*Answers* (*a*) 8 cm; (*b*) $s' = -4$ cm)

Ray Diagrams for Mirrors

A useful method to locate images is by geometric construction of a **ray diagram**, as illustrated in Figure 34-15, where the object is a human figure perpendicular to the axis a distance s from the mirror. By the judicious choice of rays from the head of the figure, we can quickly locate the image. There are three **principal rays** that are convenient to use:

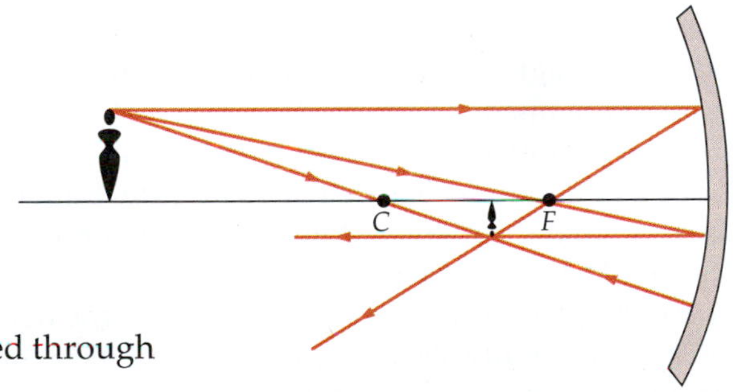

Figure 34-15 Ray diagram for the location of the image by geometric construction.

1. The **parallel ray,** drawn parallel to the axis. This ray is reflected through the focal point.
2. The **focal ray,** drawn through the focal point. This ray is reflected parallel to the axis.
3. The **radial ray,** drawn through the center of curvature. This ray strikes the mirror perpendicular to its surface and is thus reflected back on itself.

Principal rays for a mirror

These rays are shown in Figure 34-15. The intersection of any two rays locates the image point of the head. The third ray can be used to provide a check. Ray diagrams are easier to draw if the mirror is replaced by a straight line that extends as far as necessary to intercept the rays, as shown in Figure 34-16. Note that the image in this case is inverted and smaller than the object.

When the object is between the mirror and its focal point, the rays reflected from the mirror do not converge but appear to diverge from a point

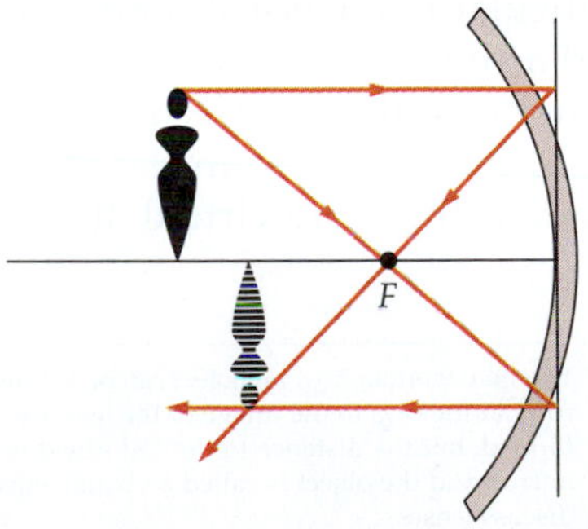

Figure 34-16 Ray diagrams are easier to construct if the curved surface is replaced by a plane.

behind the mirror, as illustrated in Figure 34-17. In this case, the image is virtual and erect (*erect* meaning not inverted relative to the object). For an object between the mirror and the focal point, s is less than $\frac{1}{2}r$, so the image distance s' calculated from Equation 34-1 turns out to be negative. We can apply Equations 34-1, 34-2, and 34-3 to this case and to convex mirrors if we adopt a convenient sign convention. Whether the mirror is convex or concave, real images can be formed only in front of the mirror, that is, on the same side of the mirror as the object. Virtual images are formed behind the mirror where there are no actual light rays. Our sign convention is as follows:

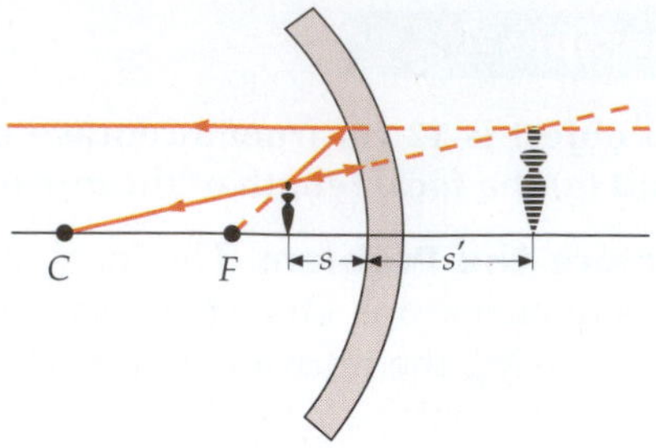

Figure 34-17 A virtual image is formed by a concave mirror when the object is inside the focal point. Here the image is located by the radial ray, which is reflected back on itself, and the focal ray, which is reflected parallel to the axis. These rays appear to diverge from a point behind the mirror found by extending them.

s	+ if the object is in front of the mirror (real object) − if the object is behind the mirror (virtual object)*
s'	+ if the image is in front of the mirror (real image) − if the image is behind the mirror (virtual image)
r, f	+ if the center of curvature is in front of the mirror (concave mirror) − if the center of curvature is behind the mirror (convex mirror)

Sign conventions for reflection

With these sign conventions, Equations 34-1, 34-2, and 34-3 can be used for all situations with any type of mirror.

The ratio of the image size to the object size is defined as the **lateral magnification** of the image. From Figure 34-18 we see that the lateral magnification is

$$m = \frac{y'}{y} = -\frac{s'}{s} \qquad \text{34-4}$$

Lateral magnification

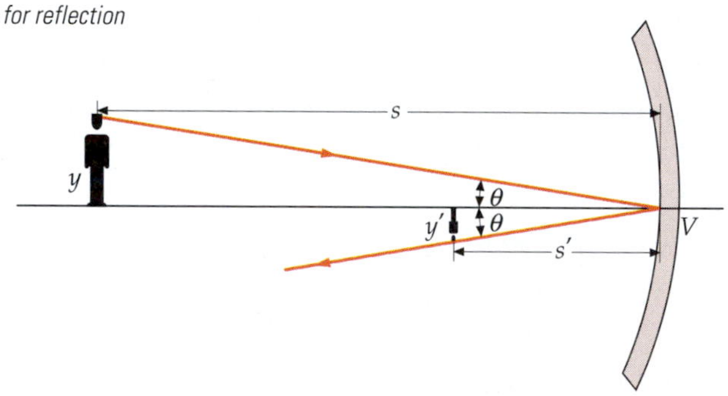

Figure 34-18 Geometry for finding the lateral magnification. A ray from the top of the object to the mirror vertex makes an angle θ with the axis. The reflected ray to the top of the image makes an equal angle with the axis. A comparison of the triangles formed by these rays, the axis, and the object and image shows that the lateral magnification y'/y equals the ratio of the distances s'/s. The minus sign in Equation 34-4 results from the fact that y'/y is negative when s and s' are both positive.

A negative magnification, which occurs when both s and s' are positive, indicates that the image is inverted.

For plane mirrors, the radius of curvature is infinite. The focal length given by Equation 34-2 is then also infinite. Equation 34-3 then gives $s' = -s$, indicating that the image is behind the mirror at a distance equal to the object distance. The magnification given by Equation 34-4 is then +1, indicating that the image is erect and the same size as the object.

Although the preceding equations coupled with our sign conventions are relatively easy to use, we often need to know only the approximate location and magnification of the image and whether it is real or virtual, and erect or inverted. This knowledge is usually easiest to obtain by constructing a ray diagram. It is always a good idea to use both the graphical method and the algebraic method to locate an image so that one method serves as a check on the results of the other.

Convex Mirrors Figure 34-19 shows a ray diagram for an object in front of a convex mirror. The central ray heading toward the center of curvature C is perpendicular to the mirror and is reflected back on itself. The parallel ray is reflected as if it came from the focal point F behind the mirror. The focal ray (not shown) would be drawn toward the focal point and would be reflected parallel to the axis. We can see from the figure that the image is behind the mirror and is therefore virtual. It is also erect and smaller than the object.

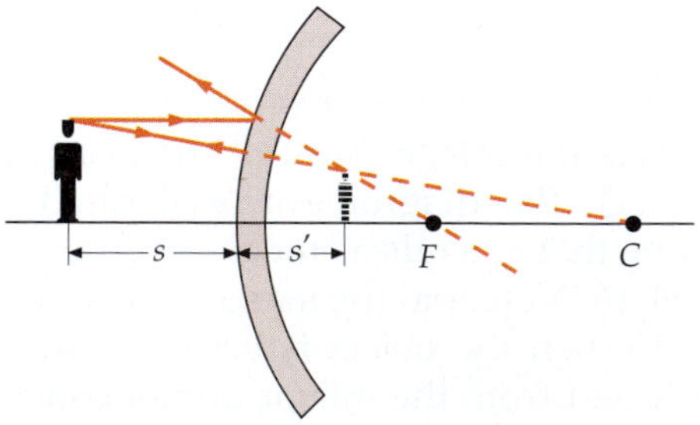

Figure 34-19 Ray diagram for a convex mirror.

* You may wonder how an object can be behind a mirror. This occurs when there is a lens in front of the mirror and the rays to the image of the lens are intercepted by the mirror. The image of the lens is then never formed, but the distance to the unformed image behind the mirror is taken as the object distance for the mirror, and the object is called a virtual object. We will discuss examples of this in Section 34-2 when we discuss lenses.

Example 34-2

An object 2 cm high is 10 cm from a convex mirror with a radius of curvature of 10 cm. (*a*) Locate the image and (*b*) find its height.

Picture the Problem The ray diagram for this problem is the same as Figure 34-19. From this we see that the image is erect, virtual, and smaller than the object. To find the exact location and size, we use the mirror equation with $s = 10$ cm and $r = -10$ cm.

(*a*) 1. The image distance s' is related to the object distance s and the focal length f by the mirror equation:

$$\frac{1}{s} + \frac{1}{s'} = \frac{1}{f}$$

2. Calculate the focal length of the mirror:

$$f = \frac{1}{2}r = \frac{1}{2}(-10\text{ cm}) = -5\text{ cm}$$

3. Substitute $s = 10$ cm and $f = -5$ cm into the mirror equation to find the image distance:

$$\frac{1}{10\text{ cm}} + \frac{1}{s'} = \frac{1}{f} = \frac{1}{-5\text{ cm}}$$

4. Solve for s':

$$\frac{1}{s'} = -\frac{2}{10\text{ cm}} - \frac{1}{10\text{ cm}} = -\frac{3}{10\text{ cm}}$$

$$s' = -3.33\text{ cm}$$

(*b*) 1. The height of the image is m times the height of the object:

$$y' = my$$

2. Calculate the magnification m:

$$m = -\frac{s'}{s} = -\frac{-3.33\text{ cm}}{10\text{ cm}} = +0.333$$

3. Use m to find the height of the image:

$$y' = my = (0.333)(2\text{ cm}) = 0.666\text{ cm}$$

Remarks The image distance is negative, indicating a virtual image behind the mirror. The magnification is positive, indicating that the image is erect.

Exercise Find the image distance and magnification for an object 5 cm away from the mirror in Example 34-2 and draw a ray diagram. (*Answers* $s' = -2.5$ cm, $m = +0.5$, the image is erect, virtual, and reduced in size.)

(*a*)

(*b*)

(*a*) A convex mirror resting on paper with equally spaced parallel stripes. Note the large number of lines imaged in a small space and the reduction in size and distortion in shape of the image. (*b*) A convex mirror is used for security in a store.

34-2 Lenses

Images Formed by Refraction

Figure 34-20 illustrates the formation of an image by refraction at a spherical surface separating two media with indexes of refraction n_1 and n_2. In this figure, n_2 is greater than n_1, so the waves travel more slowly in the second medium.

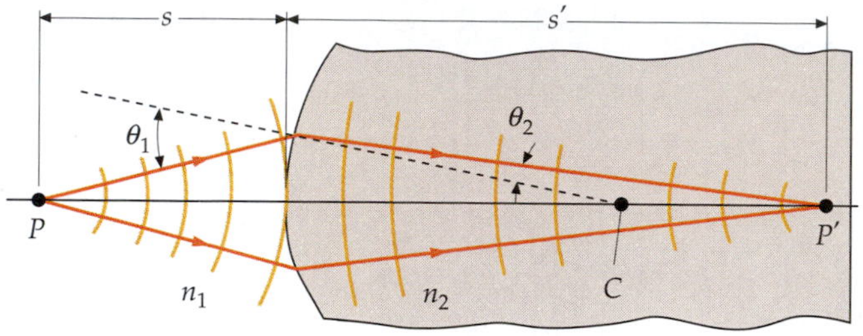

Figure 34-20 Image formed by refraction at a spherical surface between two media where the waves move slower in the second medium.

Again, only paraxial rays converge to one point. An equation relating the image distance to the object distance, the radius of curvature, and the indexes of refraction can be derived by applying Snell's law of refraction to these rays and using small-angle approximations. The geometry is shown in Figure 34-21. The result is

$$\frac{n_1}{s} + \frac{n_2}{s'} = \frac{n_2 - n_1}{r} \tag{34-5}$$

Refraction at a single surface

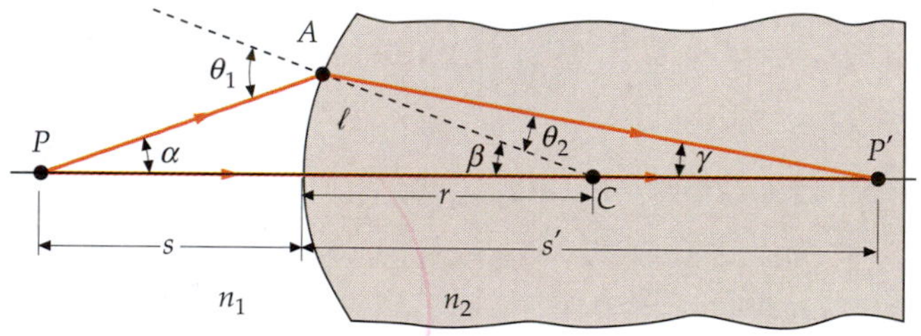

Figure 34-21 Geometry for relating the image position to the object position for refraction at a single spherical surface. The angles θ_1 and θ_2 are related by Snell's law: $n_1 \sin\theta_1 = n_2 \sin\theta_2$. The small-angle approximation $\sin\theta = \theta$ gives $n_1\theta_1 = n_2\theta_2$. From triangle ACP', we have $\beta = \theta_2 + \gamma = (n_1/n_2)\theta_1 + \gamma$. We can obtain another relation for θ_1 from triangle PAC: $\theta_1 = \alpha + \beta$. Eliminate θ_1 from these two equations: $n_1\alpha + n_1\beta + n_2\gamma = n_2\beta$. Simplify: $n_1\alpha + n_2\gamma = (n_2 - n_1)\beta$. Using the small-angle approximations $\alpha \approx \ell/s$, $\beta \approx \ell/r$, and $\gamma \approx \ell/s'$ gives Equation 34-5.

In refraction, real images are formed in back of the surface, which we will call the transmission side, whereas virtual images occur on the incident side in front of the surface. The sign conventions we use for refraction are similar to those for reflection:

s	+ (real object) for objects in front of the surface (incident side) − (virtual object) for objects in back of the surface (transmission side)
s'	+ (real image) for images in back of the surface (transmission side) − (virtual image) for images in front of the surface (incident side)
r, f	+ if the center of curvature is on the transmission side − if the center of curvature is on the incident side

Sign conventions for refraction

If we compare these sign conventions with those for reflection, we see that s' is positive and the image is real when the image is on the side of the surface traversed by the reflected or refracted light. For reflection, this side is in front of the mirror, whereas for refraction, it is behind the refracting surface. Similarly, r and f are positive when the center of curvature is on the side traversed by the reflected or refracted light.

Example 34-3 *try it yourself*

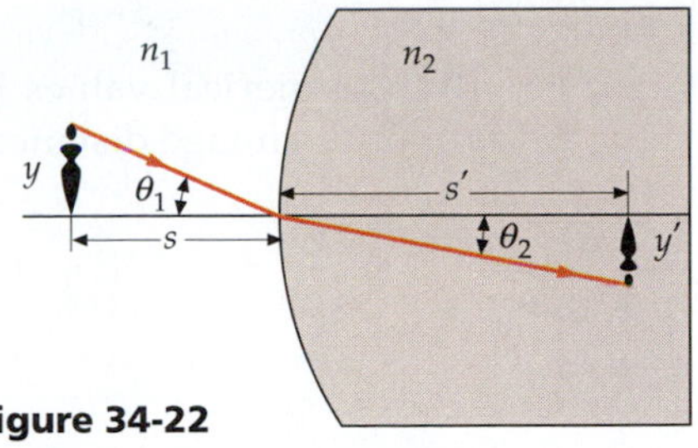

Figure 34-22

Derive an expression for the magnification $m = y'/y$ of an image formed by a refracting surface.

Picture the Problem The magnification is the ratio of y' to y. These heights are related to the tangents of the angles θ_1 and θ_2, as shown in Figure 34-22. The angles are related by Snell's law. For paraxial rays, you can approximate $\tan\theta \approx \sin\theta \approx \theta$.

Cover the column to the right and try these on your own before looking at the answers.

Steps	*Answers*
1. Write expressions for $\tan\theta_1$ and $\tan\theta_2$ in terms of the heights y and $-y'$, and the object and image distances s and s'. (Since y' is negative, use $-y'$ so that $\tan\theta_2$ is positive.)	$\tan\theta_1 = \frac{y}{s}$; $\tan\theta_2 = \frac{-y'}{s'}$
2. Apply the small angle approximation $\tan\theta \approx \theta$ to your expressions.	$\theta_1 = \frac{y}{s}$; $\theta_2 = \frac{-y'}{s'}$
3. Write Snell's law relating the angles θ_1 and θ_2 using the small-angle approximation $\sin\theta \approx \theta$.	$n_1\theta_1 = n_2\theta_2$
4. Substitute the expressions for θ_1 and θ_2 found in step 2.	$n_1\left(\frac{y}{s}\right) = n_2\left(\frac{-y'}{s'}\right)$
5. Solve for the magnification $m = y'/y$.	$m = \frac{y'}{y} = -\frac{n_1 s'}{n_2 s}$

We see from Example 34-3 that the magnification due to refraction at a spherical surface is

$$m = \frac{y'}{y} = -\frac{n_1 s'}{n_2 s} \tag{34-6}$$

Example 34-4

A fish is in a spherical bowl of water with an index of refraction of 1.33. The radius of the bowl is 15 cm. The fish looks through the bowl and sees a cat sitting on the table with its nose 10 cm from the bowl (Figure 34-23). Find (*a*) the image distance for the cat's nose, and (*b*) the magnification of the nose. Neglect any effect of the bowl's thin glass wall.

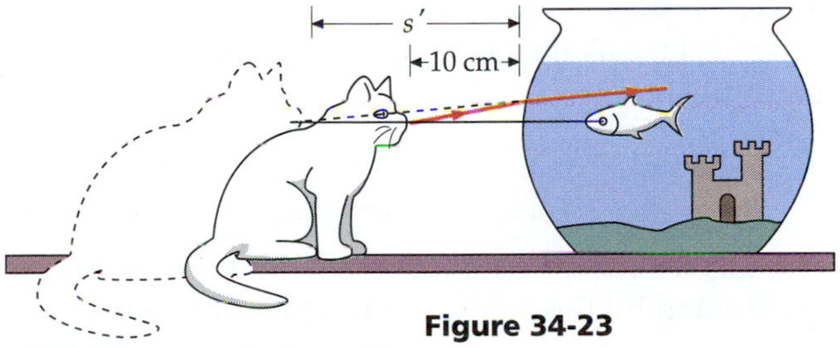

Figure 34-23

Picture the Problem We find the image distance s' using Equation 34-5 and the magnification using Equation 34-6. Since we are interested in light that goes from the cat's nose to the fish, it follows that air is the incident side and water is the transmission side. With these identifications, we have $n_1 = 1$, $n_2 = 1.33$, $s = +10$ cm (real object), and $r = +15$ cm (center of curvature on the transmission side).

(*a*)1. Substitute numerical values into Equation 34-5 to find a relation for the image distance s':

$$\frac{n_1}{s} + \frac{n_2}{s'} = \frac{n_2 - n_1}{r}$$

$$\frac{1}{10\text{ cm}} + \frac{1.33}{s'} = \frac{1.33 - 1}{15\text{ cm}}$$

$$\frac{1.33}{s'} = \frac{0.33}{15\text{ cm}} - \frac{1}{10\text{ cm}}$$

2. Solve for s':

$$s' = -17.1\text{ cm}$$

(*b*) Substitute numerical values into Equation 34-6 to find the magnification m:

$$m = -\frac{n_1 s'}{n_2 s} = -\frac{(1)(-17.1\text{ cm})}{(1.33)(10\text{ cm})} = 1.29$$

Remarks Since s' is negative, the image is virtual; that is, the image is on the incident side of the bowl, as shown in Figure 34-23. The cat appears to be slightly farther away ($|s'| > s$), but larger ($|m| > 1$). The fact that m is positive indicates that the image is upright.

Exercise If the fish is 7.5 cm from the surface of the bowl, find (*a*) the location of the fish and (*b*) its magnification, as seen by the cat. (*Answers* $n_1 = 1.33$, $n_2 = 1$, $s = 7.5$ cm, $r = -15$ cm, thus (*a*) $s' = -6.44$ cm, and (*b*) $m = 1.14$. The fish appears slightly closer and larger.)

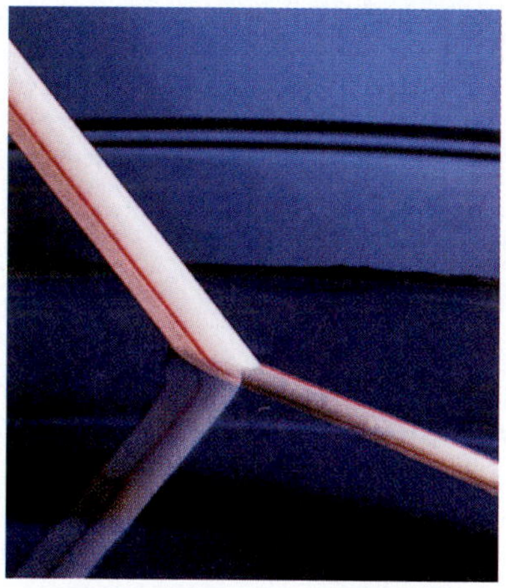

Because of refraction, the apparent depth of the submerged portion of the straw is less than the real depth. Consequently, the straw appears to be bent. A reflected image of the straw is also seen.

We can use Equation 34-5 to find the **apparent depth** of an object under water when it is viewed from directly overhead. For this case, the surface is a plane surface, so the radius of curvature is infinite. The image and object distances are related by

$$\frac{n_1}{s} + \frac{n_2}{s'} = 0$$

where n_1 is the index of refraction of the first medium (water) and n_2 is that of the second medium (air). The apparent depth is therefore

$$s' = -\frac{n_2}{n_1}s \qquad 34\text{-}7$$

The negative sign indicates that the image is virtual and on the same side of the refracting surface as the object, as shown in the ray diagram in Figure 34-24. The magnification is

$$m = -\frac{n_1 s'}{n_2 s} = +1$$

Since $n_2 = 1$ for air, we see from Equation 34-7 that the apparent depth equals the real depth divided by the index of refraction of water.

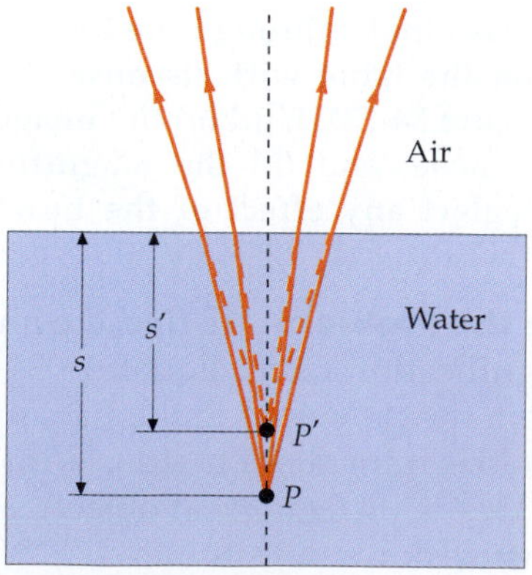

Figure 34-24 Ray diagram for the image of an object in water as viewed from directly overhead. The depth of the image is less than the depth of the object.

Example 34-5

Find the apparent depth of a fish resting 1 m below the surface of water ($n = \frac{4}{3}$).

Picture the Problem Since the light originates in the water, use $n_1 = \frac{4}{3}$, and $n_2 = 1$.

Calculate s' from Equation 34-7 using the given values for n_1, n_2, and s:

$$s' = \frac{1}{4/3}(1\text{ m}) = \frac{3}{4}(1\text{ m}) = 0.75\text{ m}$$

Remarks The apparent depth is three-fourths the actual depth, so the fish appears to be 75 cm below the surface. Note that this result holds only when the object is viewed from directly overhead so that the rays are paraxial.

Thin Lenses

The most important application of Equation 34-5 for refraction at a single surface is finding the position of the image formed by a lens. This is done by considering the refraction at each surface of the lens separately to derive an equation relating the image distance to the object distance, the radius of curvature of each surface of the lens, and the index of refraction of the lens.

We will consider a very thin lens of index of refraction n with air on both sides. Let the radii of curvature of the surfaces of the lens be r_1 and r_2. If an object is at a distance s from the first surface (and therefore from the lens), the distance s_1' of the image due to refraction at the first surface can be found using Equation 34-5:

$$\frac{1}{s} + \frac{n}{s_1'} = \frac{n-1}{r_1} \qquad \text{34-8}$$

This image is not formed because the light is again refracted at the second surface. Figure 34-25 shows the case when the image distance s_1' for the first surface is negative, indicating a virtual image to the left of the surface. Rays in the glass refracted from the first surface diverge as if they came from the image point P_1'. They strike the second surface at the same angles as if there were an object at this image point. The image for the first surface therefore becomes the object for the second surface. Since the lens is of negligible thickness, the object distance is equal in magnitude to s_1', but since object distances in front of the surface are positive whereas image distances are negative there, the object distance for the second surface is $s_2' = -s_1'$.* We now write Equation 34-5 for the second surface with $n_1 = n$, $n_2 = 1$, and $s = -s_1'$. The image distance for the second surface is the final image distance s' for the lens:

$$\frac{n}{-s_1'} + \frac{1}{s'} = \frac{1-n}{r_2} \qquad \text{34-9}$$

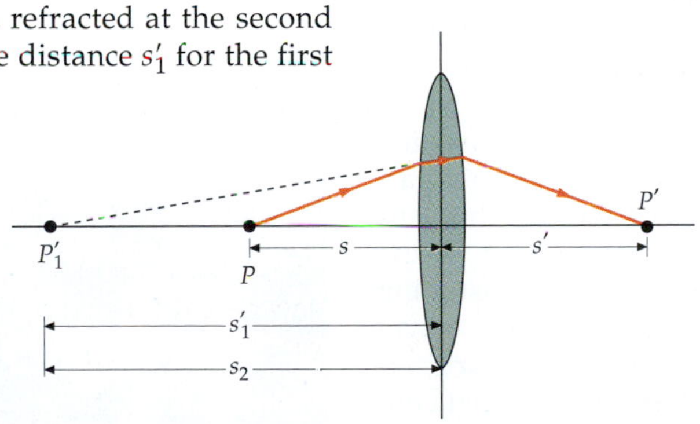

Figure 34-25 Refraction occurs at both surfaces of a lens. Here, the refraction at the first surface leads to a virtual image at P_1'. The rays strike the second surface as if they came from P_1'. Since image distances are negative when the image is on the incident side of the surface whereas object distances are positive for objects there, $s_2 = -s_1'$ is the object distance for the second surface of the lens.

* If s_1' were positive, the rays would converge as they strike the second surface. The object for the second surface would then be to the right of the surface. This object would be a virtual object. Again, $s_2 = -s_1'$.

We can eliminate the image distance for the first surface s_1' by adding Equations 34-8 and 34-9. We obtain

$$\frac{1}{s} + \frac{1}{s'} = (n - 1)\left(\frac{1}{r_1} - \frac{1}{r_2}\right) \qquad \text{34-10}$$

Equation 34-10 gives the image distance s' in terms of the object distance s and the properties of the thin lens—r_1, r_2, and the index of refraction n. As with mirrors, the focal length f of a thin lens is defined as the image distance when the object distance is infinite. Setting s equal to infinity and writing f for the image distance s', we obtain

$$\frac{1}{f} = (n - 1)\left(\frac{1}{r_1} - \frac{1}{r_2}\right) \qquad \text{34-11}$$

Lens-maker's equation

Equation 34-11 is called the **lens-maker's equation;** it gives the focal length of a thin lens in terms of the properties of the lens. Substituting $1/f$ for the right side of Equation 34-10, we obtain

$$\frac{1}{s} + \frac{1}{s'} = \frac{1}{f} \qquad \text{34-12}$$

Thin-lens equation

This **thin-lens equation** is the same as the mirror equation (Equation 34-3). Recall, however, that the sign conventions for refraction are somewhat different from those for reflection. For lenses, the image distance s' is positive when the image is on the transmission side of the lens, that is, when it is on the side opposite the side upon which light is incident. The sign convention for r in Equation 34-11 is the same as that for refraction at a single surface. The radius is positive if the center of curvature is on the transmission side of the lens and negative if it is on the incident side.

Figure 34-26*a* shows the wavefronts of plane waves incident on a double convex lens. The central part of the wavefront strikes the lens first. Since the wave speed in the lens is less than that in air (assuming $n > 1$), the central part of the wavefront lags behind the outer parts, resulting in a spherical wave that converges at the focal point F'. The rays for this situation are shown in Figure 34-26*b*. Such a lens is called a **converging lens.** Since its focal length as calculated from Equation 34-11 is positive, it is also called a **positive lens.** Any lens that is thicker in the middle than at the edges is a converging lens (providing that the index of refraction of the lens is greater than that of the surrounding medium). Figure 34-27 shows the

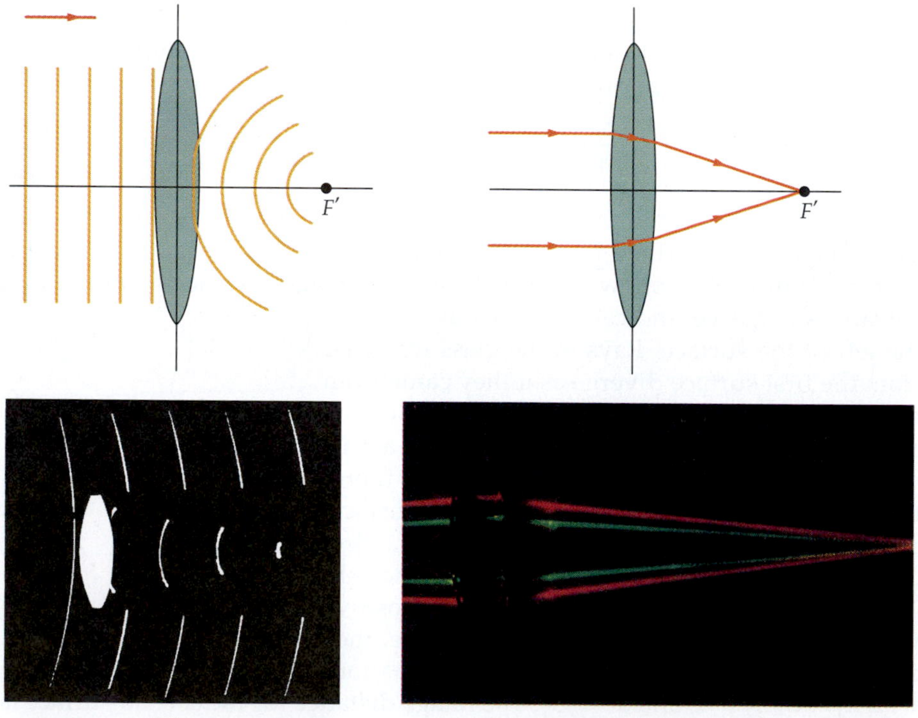

(*a*) (*b*)

Figure 34-26 (*a*) *Top:* Wavefronts for plane waves striking a converging lens. The central part of the wavefront is retarded more by the lens than the outer part, resulting in a spherical wave that converges at the focal point F'. *Bottom:* Wavefronts passing through a lens, shown by a photographic technique called *light-in-flight-recording* that uses a pulsed laser to make a hologram of the wavefronts of light. (*b*) *Top:* Rays for plane waves striking a converging lens. The rays are bent at each surface and converge at the focal point. *Bottom:* Photograph of rays focused by a converging lens.

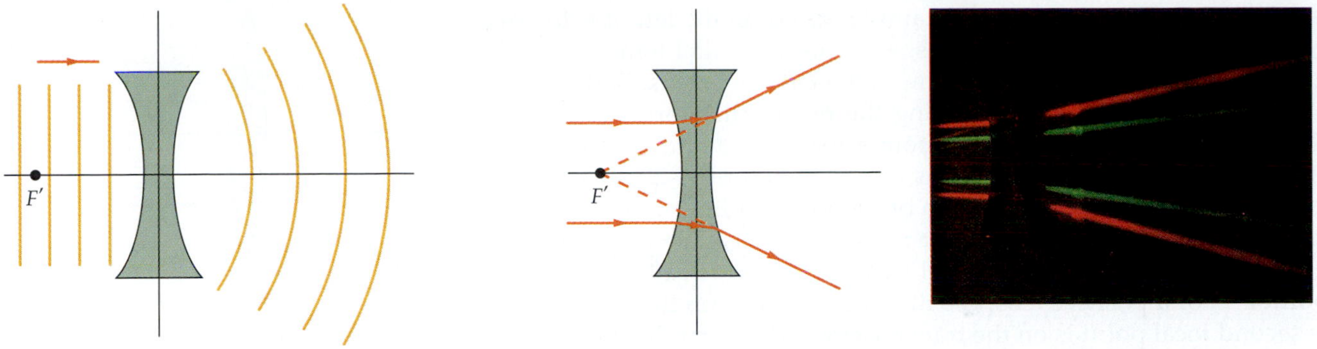

Figure 34-27 (*a*) Wavefronts for plane waves striking a diverging lens. Here, the outer parts of the wavefronts are retarded more than the central part, resulting in a spherical wave that diverges as it moves out as if it came from the focal point F' in front of the lens. (*b*) *Left:* Rays for plane waves striking the same diverging lens. The rays are bent outward and diverge as if they came from the focal point F'. *Right:* Photograph of rays passing through a diverging lens.

wavefronts and rays for plane waves incident on a double concave lens. In this case, the outer part of the wavefronts lag behind the central parts, resulting in outgoing spherical waves that diverge from a focal point on the incident side of the lens. The focal length of this lens is negative. Any lens (with index of refraction greater than that of the surrounding medium) that is thinner in the middle than at the edges is a **diverging,** or **negative, lens.**

Example 34-6

A double convex thin glass lens with index of refraction $n = 1.5$ has radii of curvature of magnitude 10 cm and 15 cm as shown in Figure 34-28. Find its focal length.

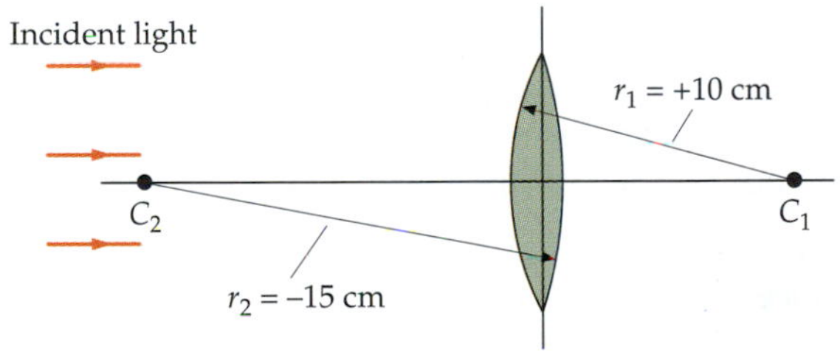

Figure 34-28

Picture the Problem We can find the focal length using the lens-maker's equation (Equation 34-11). Here, light is incident on the surface with the smaller radius of curvature. The center of curvature of this surface, C_1, is on the transmission side of the lens, thus $r_1 = +10$ cm. For the second surface, the center of curvature, C_2, is on the incident side, hence $r_2 = -15$ cm.

Numerical substitution in Equation 34-11 yields the focal length f:

$$\frac{1}{f} = (1.5 - 1)\left(\frac{1}{10\text{ cm}} - \frac{1}{-15\text{ cm}}\right) = 0.5\left(\frac{1}{6\text{ cm}}\right)$$

$$f = 12\text{ cm}$$

Remark Note that both surfaces tend to converge the light rays, hence they both make a positive contribution to the focal length of the lens.

Exercise A double convex thin lens has an index of refraction $n = 1.6$ and radii of curvature of equal magnitude. If its focal length is 15 cm, what is the magnitude of the radius of curvature of each surface? (*Answer* 18 cm)

Exercise Show that if you reverse the direction of the incoming light so that it is incident on the surface with the greater radius of curvature, you get the same result for the focal length.

If parallel light strikes the lens of Example 34-6 from the left, it is focused at a point 12 cm to the right of the lens, whereas if parallel light strikes the lens from the right, it is focused at 12 cm to the left of the lens. Both of these points are focal points of the lens. Using the reversibility property of light rays, we can see that light diverging from a focal point and striking a lens will leave the lens as a parallel beam, as shown in Figure 34-29. In a particular lens problem in which the direction of the incident light is specified, the object point for which light emerges as a parallel beam is called the **first focal point** F and the point at which parallel light is focused is called the **second focal point** F'. For a positive lens, the first focal point is on the incident side and the second focal point is on the transmission side. If parallel light is incident on the lens at a small angle with the axis, as in Figure 34-30, it is focused at a point in the **focal plane** a distance f from the lens.

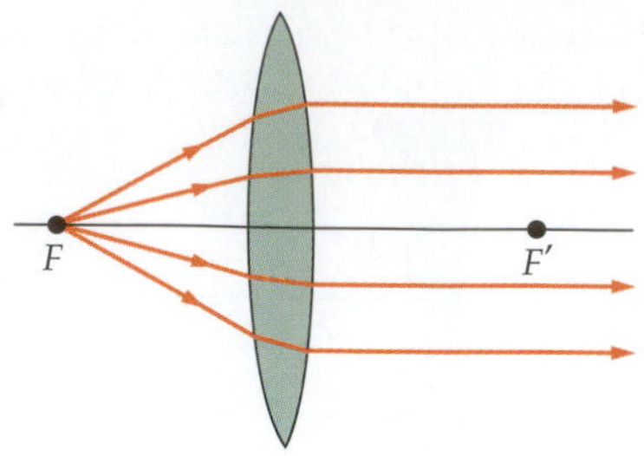

Figure 34-29 Light rays diverging from the focal point of a positive lens emerge parallel to the axis.

The reciprocal of the focal length is called the **power of a lens.** When the focal length is expressed in meters, the power is given in reciprocal meters called **diopters** (D):

$$P = \frac{1}{f} \text{ diopters} \tag{34-13}$$

The power of a lens measures its ability to focus parallel light at a short distance from the lens. The shorter the focal length, the greater the power. For example, a lens with a focal length of 25 cm = 0.25 m has a power of 4.0 D. A lens with a focal length of 10 cm = 0.10 m has a power of 10 D. Since the focal length of a diverging lens is negative, its power is also negative.

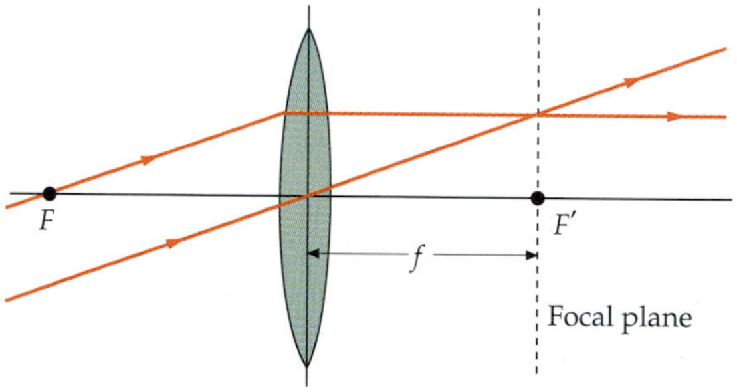

Figure 34-30 Parallel rays incident on the lens at an angle to its axis are focused at a point in the focal plane of the lens.

Example 34-7

A double concave lens has an index of refraction of 1.5 and radii of curvature of magnitude 10 cm and 15 cm. Find (*a*) its focal length (*b*) its power.

Picture the Problem For the orientation of the lens relative to the incident light shown in Figure 34-31, the radius of curvature of the first surface is $r_1 = -15$ cm and that of the second surface is $r_2 = +10$ cm.

Figure 34-31

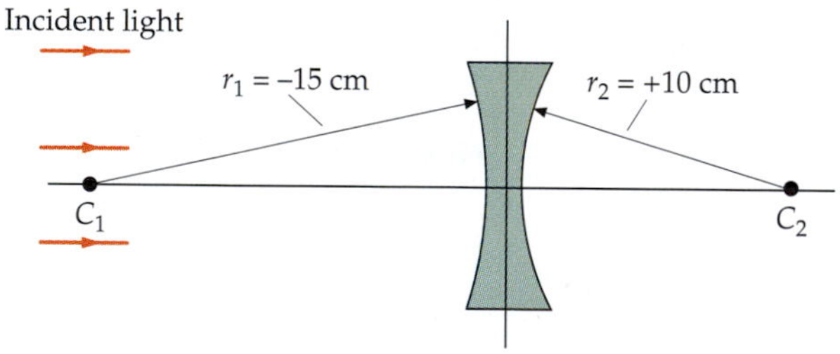

(*a*) Calculate f from the lens-maker's equation using the given value of n and the values of r_1 and r_2 for the orientation shown:

$$\frac{1}{f} = (1.5 - 1.0)\left(\frac{1}{-15\text{ cm}} - \frac{1}{+10\text{ cm}}\right)$$

$$f = -12\text{ cm}$$

(*b*) The power is the reciprocal of the focal length expressed in meters:

$$P = \frac{1}{f} = \frac{1}{-0.12\text{ m}} = -8.33\text{ D}$$

Remark We obtain the same result no matter which surface the light strikes first.

In laboratory experiments involving lenses, it is usually much easier to measure the focal length than to calculate it from the radii of curvature of the surfaces.

The weight and bulk of a large-diameter lens can be reduced by constructing the lens from annular segments at different angles such that light from a point is refracted by the segments into a parallel beam. Such an arrangement is called a Fresnel lens. Several Fresnel lenses are used in this lighthouse to produce intense parallel beams of light from a source at the focal point of the lenses.

Ray Diagrams for Lenses

As with images formed by mirrors, it is convenient to locate the images of lenses by graphical methods. Figure 34-32 illustrates the graphical method for a converging lens. For the sake of simplicity, we consider the rays to bend at the plane through the center of the lens. For a positive lens, the three principal rays are

1. The **parallel ray,** drawn parallel to the axis. This ray is bent through the second focal point of the lens.
2. The **central ray,** drawn through the center (the vertex) of the lens. This ray is undeflected. (The faces of the lens are parallel at this point, so the ray emerges in the same direction but displaced slightly. Since the lens is thin, the displacement is negligible.)
3. The **focal ray,** drawn through the first focal point. This ray emerges parallel to the axis.

Principal rays for a positive lens

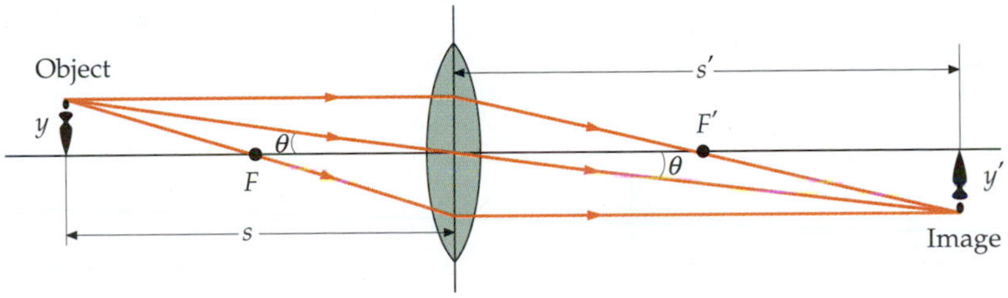

Figure 34-32 Ray diagram for a thin converging lens. For the sake of simplicity, we assume that all the bending of light takes place at the central plane. The ray through the center is undeflected because the lens surfaces there are parallel and close together.

These three rays converge to the image point, as shown in Figure 34-32. In this case, the image is real and inverted. From the figure, we have $\tan\theta = y/s = -y'/s'$. The lateral magnification is then

$$m = \frac{y'}{y} = -\frac{s'}{s} \qquad 34\text{-}14$$

This expression is the same as that for mirrors. Again, a negative magnification indicates that the image is inverted.

The principal rays for a negative, or diverging, lens are

1. The **parallel ray,** drawn parallel to the axis. This ray diverges from the lens as if it came from the second focal point.
2. The **central ray,** drawn through the center (the vertex) of the lens. This ray is undeflected.
3. The **focal ray,** drawn toward the first focal point. This ray emerges parallel to the axis.

Principal rays for a negative lens

The ray diagram for a diverging lens is shown in Figure 34-33.

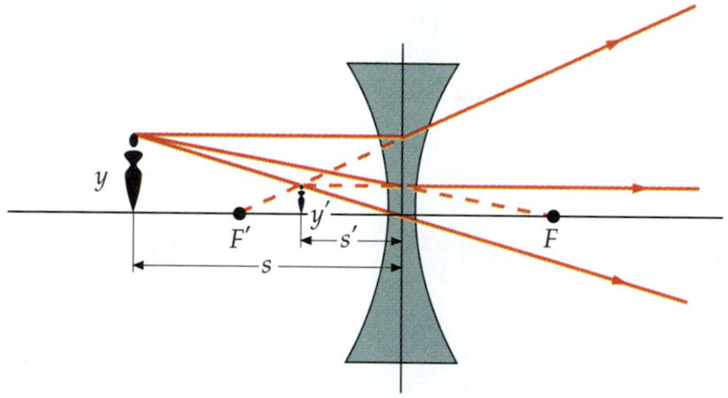

Figure 34-33 Ray diagram for a diverging lens. The parallel ray is bent away from the axis as if it came from the second focal point F'. The ray toward the first focal point F emerges parallel to the axis.

Example 34-8

An object 1.2 cm high is placed 4 cm from the double convex lens of Example 34-6. Locate the image both graphically and algebraically, state whether it is real or virtual, and find its height.

1. Draw the parallel ray. This ray leaves the object parallel to the axis, then is bent by the lens to pass through the second focal point, F' (Figure 34-34):

Figure 34-34

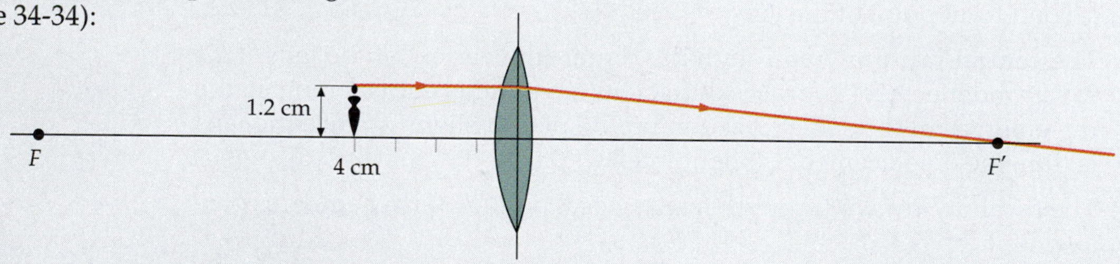

2. Draw the central ray, which passes undeflected through the center of the lens. Since the two rays are diverging on the transmission side, we extend them back to the incident side to find the image (Figure 34-35):

Figure 34-35

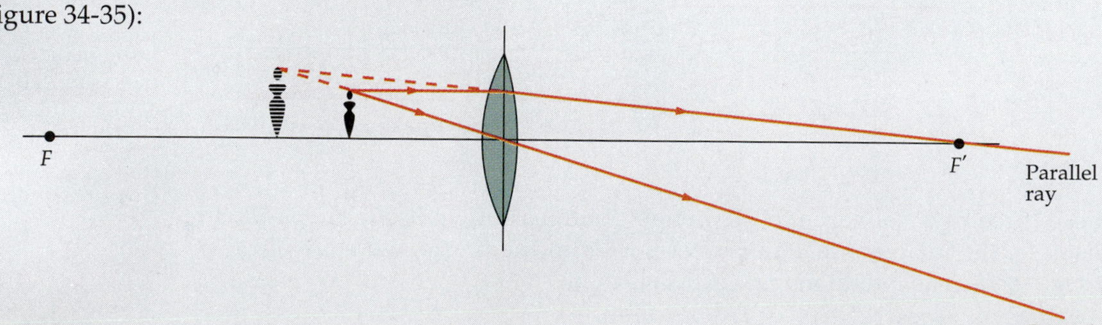

3. As a check, we also draw the focal ray. This ray leaves the object on a line passing through the first focal point, then emerges parallel to the axis. Note that the image is virtual, erect, and enlarged (Figure 34-36):

Figure 34-36

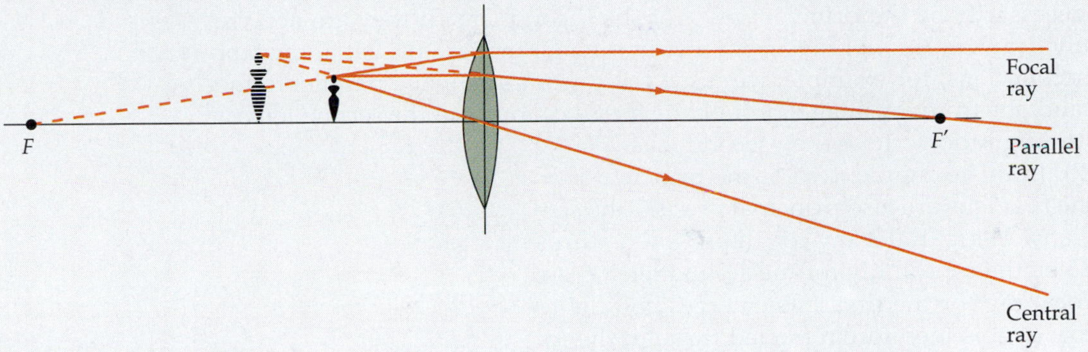

4. We now verify the results of the ray diagram algebraically. First, find the image distance using Equation 34-12:

$$\frac{1}{4\text{ cm}} + \frac{1}{s'} = \frac{1}{12\text{ cm}}$$

$$\frac{1}{s'} = \frac{1}{12\text{ cm}} - \frac{1}{4\text{ cm}} = -\frac{1}{6\text{ cm}}$$

$$s' = -6\text{ cm}$$

5. The height of the image is found from the height of the object and the magnification:

$$h' = mh$$

6. The magnification m is given by Equation 34-14:

$$m = -\frac{s'}{s} = -\frac{-6\text{ cm}}{4\text{ cm}} = +1.5$$

7. Using this result we find the height of the image, h':

$$h' = mh = (1.5)(1.2\text{ cm}) = 1.8\text{ cm}$$

Remarks Note the agreement between the algebraic and ray-diagram results. Algebraically, we find that the image is 6 cm from the lens on the incident side (since $s' < 0$); that is, the image is 2 cm to the left of the object. Since $m > 0$, it follows that the image is erect, and because $m > 1$, the image is enlarged. It is good practice to solve lens problems both ways and compare the results.

Exercise An object is placed 15 cm from a double convex lens of focal length 10 cm. Find the image distance and the magnification. Draw a ray diagram. Is the image real or virtual? Erect or inverted? (Answers $s' = 30$ cm, $m = -2$, real, inverted)

Exercise Work the previous exercise for an object placed 5 cm from a lens with a focal length of 10 cm. (Answers $s' = -10$ cm, $m = 2$, virtual, erect)

Combinations of Lenses

If we have two or more thin lenses, we can find the final image produced by the system by finding the image distance for the first lens and using it along with the distance between lenses to find the object distance for the second lens. That is, we consider each image, whether it is real or virtual and whether it is formed or not, as the object for the next lens.

Example 34-9

A second lens of focal length +6 cm is placed 12 cm to the right of the lens in Example 34-8. Locate the final image.

Picture the Problem The rays used to locate the image of the first lens will not necessarily be the principal rays for the second lens. If they are not, we merely draw additional rays from the first image that are principal rays for the second lens; for example, we draw a ray from the image parallel to the axis and one from the image through the first focal point of the second lens or one through the vertex of the second lens (Figure 34-37). In this example, two of the principal rays for the first lens are also principal rays for the second lens. The parallel ray for the first lens turns out to be the central ray for the second lens. Also, the focal ray for the first lens emerges parallel to the axis and is therefore refracted through the focal point of the second lens. (In the figure, we have extended the central ray for the first lens so that it passes through the image found from the other two rays.)

Algebraically, we use $s_2 = 18$ cm because the first image is 6 cm to the left of the first lens and therefore 18 cm to the left of the second lens.

Figure 34-37

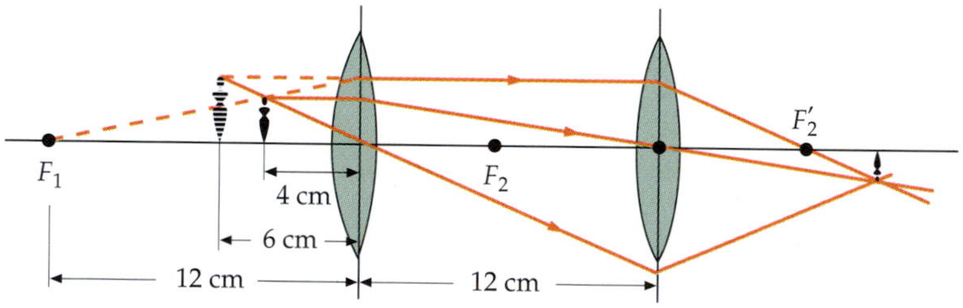

Use $s_2 = 18$ cm and $f = 6$ cm to calculate s_2':

$$\frac{1}{s_2} + \frac{1}{s_2'} = \frac{1}{f_2}$$

$$\frac{1}{18\text{ cm}} + \frac{1}{s_2'} = \frac{1}{6\text{ cm}}$$

$$s_2' = 9\text{ cm}$$

Example 34-10 *try it yourself*

Two lenses, each of focal length 10 cm, are 15 cm apart. Find the final image of an object 15 cm from one of the lenses.

Picture the Problem Use a ray diagram to find the approximate location of the image formed by lens 1. When these rays strike lens 2 they are further refracted, leading to the final image. Precise results are obtained algebraically using the thin lens equation for both lens 1 and lens 2.

Cover the column to the right and try these on your own before looking at the answers.

Steps	*Answers*
1. Draw the parallel (a), central (b), and focal (c) rays for lens 1 (Figure 34-38). If lens 2 did not alter these rays, they would form an image at I_1.	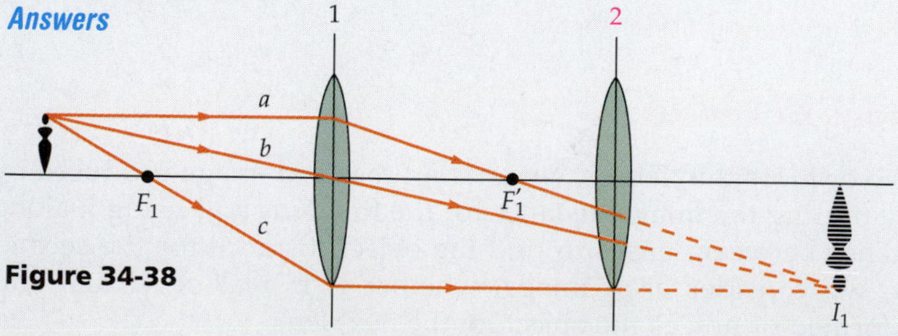

Figure 34-38

2. Note that the focal ray (c) strikes lens 2 parallel to the axis, hence it is refracted through the second focal point of lens 2. To find the final image, include one additional ray (d) that passes through the first focal point of lens 2. It would meet the other rays at I_1, but lens 2 refracts it parallel to the axis. The intersection of (c) and (d) gives the image location (Figure 34-39).

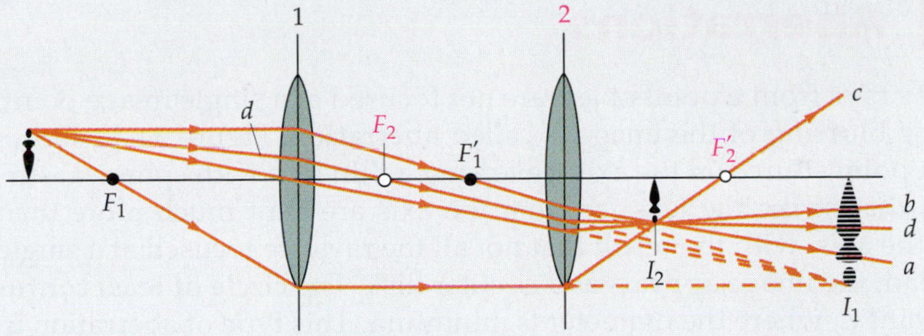

Figure 34-39

3. To proceed algebraically, use the thin-lens equation to find the image distance s_1' produced by lens 1. — $s_1' = 30$ cm

4. For lens 2, the image, I_1 is 15 cm from the lens on the transmission side, hence $s_2 = -15$ cm. Use this to find the final image distance s_2'. — $s_2' = 6$ cm

Remark From the ray diagram we see that the final image is real, inverted, and slightly reduced.

When two thin lenses of focal lengths f_1 and f_2 are placed together, the equivalent focal length of the combination f is given by

$$\frac{1}{f} = \frac{1}{f_1} + \frac{1}{f_2} \qquad \text{34-15}$$

The power of two lenses in contact is given by

$$P = P_1 + P_2 \qquad \text{34-16}$$

Example 34-11 *try it yourself*

For two lenses close together, derive the relation

$$\frac{1}{f} = \frac{1}{f_1} + \frac{1}{f_2}$$

Picture the Problem Apply the thin-lens equation to each lens using the fact that the distance between the lenses is zero so the object distance for the second lens is the negative of the image distance for the first lens.

Cover the column to the right and try these on your own before looking at the answers.

Steps	***Answers***
1. Write the thin-lens equation for lens 1.	$\frac{1}{s} + \frac{1}{s_1'} = \frac{1}{f_1}$
2. Using $s_2 = -s_1'$, write the thin-lens equation for lens 2.	$\frac{1}{-s_1'} + \frac{1}{s'} = \frac{1}{f_2}$
3. Add your two resulting equations to eliminate s_1'.	$\frac{1}{s} + \frac{1}{s'} = \frac{1}{f_1} + \frac{1}{f_2} = \frac{1}{f}$

optional

34-3 Aberrations

When all the rays from a point object are not focused at a single image point, the resulting blurring of the image is called **aberration.** Figure 34-40 shows rays from a point source on the axis traversing a thin lens with spherical surfaces. Rays that strike the lens far from the axis are bent much more than those near the axis, with the result that not all the rays are focused at a single point. Instead, the image appears as a circular disk. The **circle of least confusion** is at point C, where the diameter is minimum. This type of aberration in a lens is called **spherical aberration;** it is the same as the spherical aberration of mirrors discussed in Section 34-2. Similar but more complicated aberrations called *coma* (for the comet-shaped image) and *astigmatism* occur when objects are off axis. The aberration in the shape of the image of an extended object due to the fact that the magnification depends on the distance of the object point from the axis is called **distortion.** We will not discuss these aberrations further except to point out that they do not arise from any defect in the lens or mirror but instead result from the application of the laws of refraction and reflection to spherical surfaces. They are not evident in our simple equations because we used small-angle approximations in the derivation of these equations.

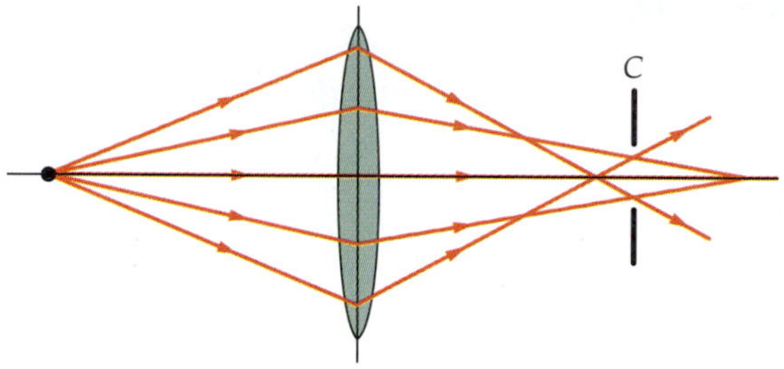

Figure 34-40 Spherical aberration in a lens. Rays from a point object on the axis are not focused at a point. Spherical aberration can be reduced by blocking off the outer parts of the lens, but this also reduces the amount of light reaching the image.

Some aberrations can be eliminated or partially corrected by using nonspherical surfaces for mirrors or lenses, but nonspherical surfaces are usually much more difficult and costly to produce than spherical surfaces. One example of a nonspherical reflecting surface is the parabolic mirror illustrated in Figure 34-41. Rays that are parallel to the axis of a parabolic surface are reflected and focused at a common point no matter how far they are from the axis. Parabolic reflecting surfaces are sometimes used in large astronomical telescopes, which need a large reflecting surface to gather as much light as possible to make the image as intense as possible (reflecting telescopes are described in the upcoming optional Section 34-4). Satellite dishes use parabolic surfaces to focus microwaves from communications satellites. A parabolic surface can also be used in a searchlight to produce a parallel beam of light from a small source placed at the focal point of the surface.

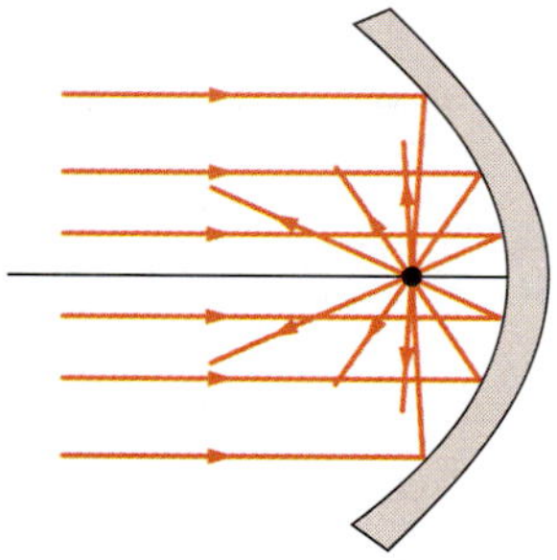

Figure 34-41 A parabolic mirror focuses all rays parallel to the axis to a single point with no spherical aberration.

An important aberration found with lenses but not with mirrors is **chromatic aberration,** which is due to variations in the index of refraction with wavelength. From Equation 34-11, we can see that the focal length of a lens depends on its index of refraction and is therefore different for different wavelengths. Since n is slightly greater for blue light than for red light, the focal length for blue light will be shorter than that for red light. Because chromatic aberration does not occur for mirrors, many large telescopes use a large mirror instead of the large, light-gathering (objective) lens.

Chromatic and other aberrations can be partially corrected by using combinations of lenses instead of a single lens. For example, a positive lens and a negative lens of greater focal length can be used together to produce a converging lens system that has much less chromatic aberration than a single lens of the same focal length. The lens of a good camera typically contains six elements to correct for the various aberrations that are present.

34-4 Optical Instruments

The Eye

The optical system of prime importance is the eye, shown in Figure 34-42. Light enters the eye through a variable aperture, the pupil, and is focused by the cornea–lens system on the retina, a film of nerve fibers covering the back surface. The retina contains tiny sensing structures called *rods* and *cones*, which receive the image and transmit the information along the optic nerve to the brain. The shape of the crystalline lens can be altered slightly by the action of the ciliary muscle. When the eye is focused on an object far away, the muscle is relaxed and the cornea–lens system has its maximum focal length, about 2.5 cm, the distance from the cornea to the retina. When the object is brought closer to the eye, the ciliary muscle increases the curvature of the lens slightly, thereby decreasing its focal length, so that the image is again focused on the retina. This process is called *accommodation*. If the object is too close to the eye, the lens cannot focus the light on the retina and the image is blurred. The closest point for which the lens can focus the image on the retina is called the **near point**. The distance from the eye to the near point varies greatly from one person to another and changes with age. At 10 years, the near point may be as close as 7 cm, whereas at 60 years it may recede to 200 cm because of the loss of flexibility of the lens. The standard value taken for the near point is 25 cm.

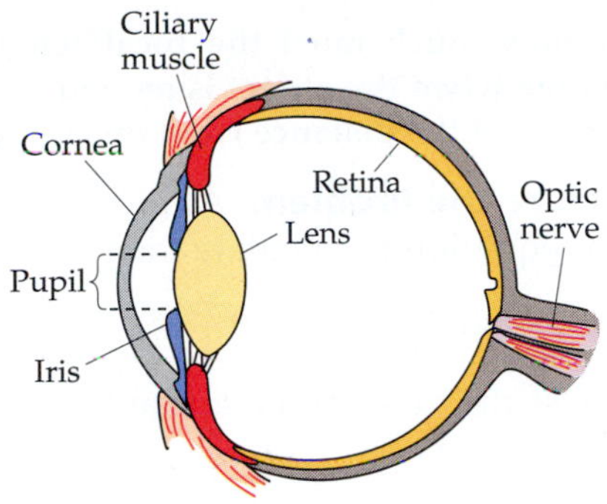

Figure 34-42 The human eye. The amount of light entering the eye is controlled by the iris, which regulates the size of the pupil. The lens thickness is controlled by the ciliary muscle. The cornea and lens together focus the image on the retina, which contains about 125 million receptors called rods and cones and about 1 million optic-nerve fibers.

If the eye underconverges, resulting in the images being focused behind the retina, the person is said to be farsighted. A farsighted person can see distant objects where little convergence is required, but has trouble seeing close objects. Farsightedness is corrected with a converging (positive) lens (Figure 34-43).

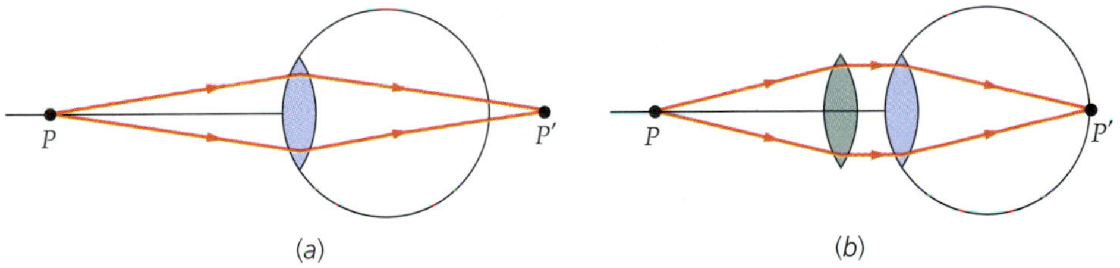

Figure 34-43 (*a*) A farsighted eye focuses rays from a nearby object to a point behind the retina. (*b*) A converging lens corrects this defect by bringing the image onto the retina. These diagrams and those following are drawn as if all the focusing of the eye is done at the lens, whereas in fact the lens and cornea system act more like a spherical refracting surface than a thin lens.

On the other hand, the eye of a nearsighted person overconverges and focuses light from distant objects in front of the retina. A nearsighted person can see nearby objects for which the widely diverging incident rays can be focused on the retina, but has trouble seeing distant objects. Nearsightedness is corrected with a diverging (negative) lens (Figure 34-44).

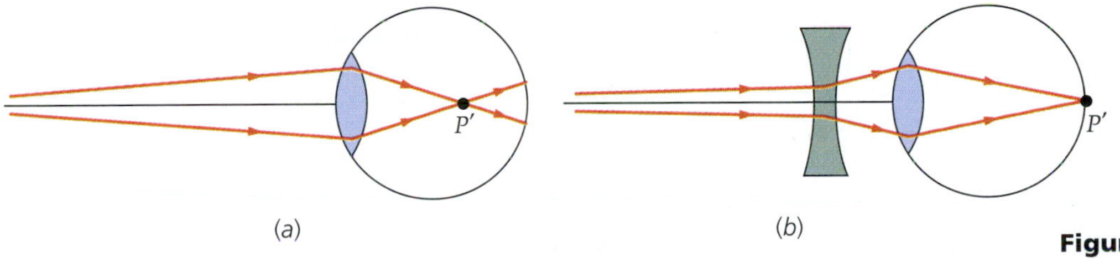

Figure 34-44 (*a*) A nearsighted eye focuses rays from a distant object to a point in front of the retina. (*b*) A diverging lens corrects this defect.

Another common defect of vision is astigmatism, which is caused by the cornea being not quite spherical but having a different curvature in one plane than in another. This results in a blurring of the image of a point object into a short line. Astigmatism is corrected by glasses using lenses of cylindrical rather than spherical shape.

optional

Example 34-12

By how much must the focal length of the cornea–lens system of the eye change when the object is moved from infinity to the near point at 25 cm? Assume that the distance from the cornea to the retina is 2.5 cm.

Picture the Problem At infinity the focal length is 2.5 cm. We use the thin-lens equation to calculate the focal length when $s = 25$ cm and $s' = 2.5$ cm.

1. Use the thin-lens equation to calculate f:

$$\frac{1}{f} = \frac{1}{25\text{ cm}} + \frac{1}{2.5\text{ cm}} = \frac{1}{25\text{ cm}} + \frac{10}{25\text{ cm}} = \frac{11}{25\text{ cm}}$$

$$f = 2.27\text{ cm}$$

2. Subtract the original focal length of 2.5 cm to find the change:

$$\Delta f = 2.27\text{ cm} - 2.5\text{ cm} = -0.23\text{ cm}$$

Remarks In terms of the power of the cornea–lens system, when the focal length is 2.5 cm = 0.025 m for distant objects, the power is $P = 1/f = 40$ D. When the focal length is 2.27 cm, the power is 44 D.

Exercise Find the change in the focal length of the eye when an object originally at 4 m is brought to 40 cm from the eye. (Assume that the distance from the cornea to the retina is 2.5 cm.) (*Answer* -0.13 cm)

The apparent size of an object is determined by the size of the image on the retina. The larger the image on the retina, the greater the number of rods and cones activated. From Figure 34-45 we see that the size of the image on the retina is greater when the object is close than it is when the object is far away. The apparent size of an object is thus greater when it is closer to the eye. The image size is proportional to the angle θ subtended by the object at the eye. For Figure 34-45,

$$\theta = \frac{y'}{2.5\text{ cm}} \qquad 34\text{-}17$$

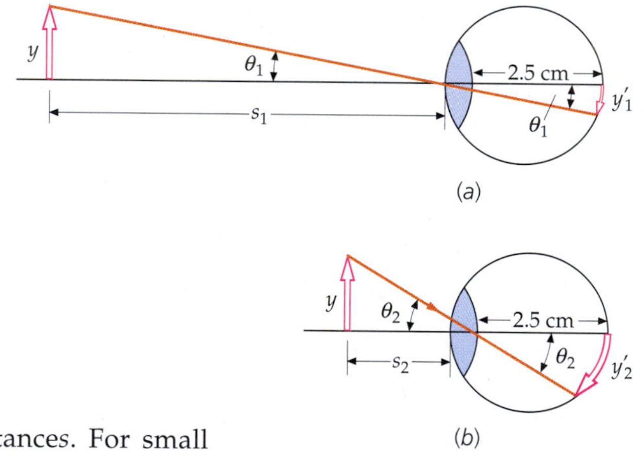

Figure 34-45 (*a*) A distant object of height y looks small because the image on the retina is small. (*b*) When the same object is closer, it looks larger because the image on the retina is larger. The angle subtended is $\theta = y'/(2.5\text{ cm})$.

The angle θ is related to the object size y and object distances. For small angles,

$$\theta \approx \tan\theta = \frac{y}{s} \qquad 34\text{-}18$$

Combining Equations 34-17 and 34-18 gives

$$y' = (2.5\text{ cm})\,\theta \approx (2.5\text{ cm})\,\frac{y}{s} \qquad 34\text{-}19$$

The size of the image on the retina is proportional to the size of the object and inversely proportional to the distance between the object and the eye. Since the near point is the closest point to the eye for which a sharp image can be formed on the retina, the distance to the near point is called the *distance of most distinct vision.*

optional

Example 34-13

The near point of a person's eye is 75 cm. What power reading glasses should be used to bring the near point to 25 cm? Assume that the lens of the glasses is in contact with the lens of the eye.

Picture the Problem Figure 34-46 shows a diagram of an object 25 cm from a converging lens that produces a virtual, erect image at $s' = -75$ cm.

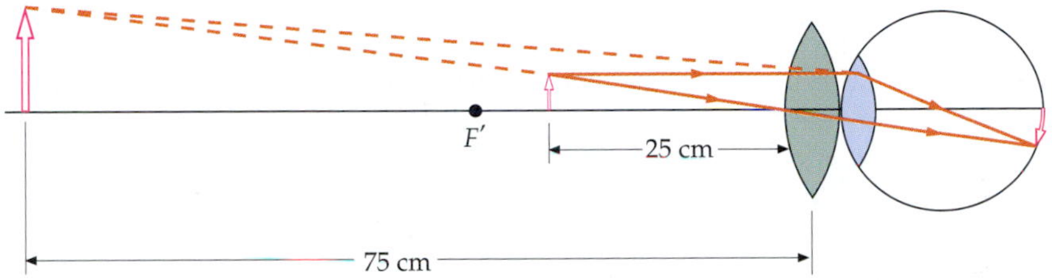

Figure 34-46

Use the thin-lens equation with $s = 25$ cm and $s' = -75$ cm to calculate the power, $1/f$:

$$\frac{1}{25\text{ cm}} + \frac{1}{-75\text{ cm}} = \frac{1}{f}$$

$$\frac{1}{f} = \frac{2}{75\text{ cm}} = \frac{1}{0.375\text{ m}} = 2.67\text{ diopter}$$

Remarks If your near point is 75 cm, you are farsighted. To read a book you must hold it about 75 cm from your eye to be able to focus on the print. The image of the print on your retina is then very small. A converging lens of the type found in reading glasses allows you to bring the book closer to the eye, which makes the retinal image of the print larger. In this example we assumed that the lens was in contact with the eye. For reading glasses, which are perched a short distance in front of the eye, the results are slightly different.

Exercise Calculate the power of the lens of the eye for which the near point is 75 cm, and calculate the combined power of the two lenses in contact. Compare this with the power of a lens for which $s' = 2.5$ cm when $s = 25$ cm. (*Answers* $P_{\text{eye}} = 41.33$ D; $P_{\text{c}} = 41.33\text{ D} + 2.67\text{ D} = 44$ D; $P = 44$ D)

The Simple Magnifier

We saw in Example 34-13 that the apparent size of an object can be increased by using a converging lens to allow the object to be brought closer to the eye, thus increasing the size of the image on the retina. Such a converging lens is called a simple magnifier. In Figure 34-47*a*, a small object of height y is at the near point of the eye at a distance x_{np}. The angle subtended, θ_{o}, is given approximately by

$$\theta_{\text{o}} = \frac{y}{x_{\text{np}}}$$

In Figure 34-47*b*, a converging lens of focal length f, smaller than x_{np}, is placed in front of the eye, and the object is placed at the focal point of

Figure 34-47 (*a*) An object at the near point subtends an angle θ_{o} at the eye. (*b*) When the object is at the focal point of the converging lens, the rays emerge from the lens parallel and enter the eye as if they came from an object a very large distance away. The image can thus be viewed at infinity by the relaxed eye. When f is less than the near point, the converging lens allows the object to be brought closer to the eye, increasing the angle subtended by the object to θ, thereby increasing the size of the image on the retina.

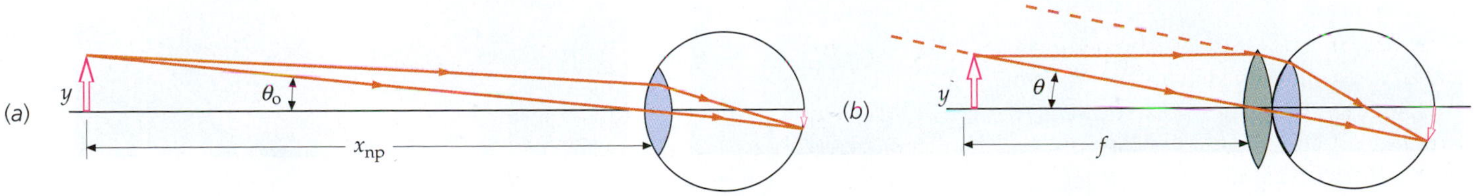

the lens. The rays emerge from the lens parallel, indicating that the image is at an infinite distance in front of the lens. The parallel rays are focused by the relaxed eye on the retina. Assuming that the lens is in contact with the eye, the angle subtended by the object is now approximately

$$\theta = \frac{y}{f}$$

The ratio θ/θ_o is called the *angular magnification* or *magnifying power* M of the lens:

$$M = \frac{\theta}{\theta_\mathrm{o}} = \frac{x_\mathrm{np}}{f} \qquad \text{34-20}$$

Simple magnifiers are used as eyepieces (called oculars) in microscopes and telescopes to view the image formed by another lens or lens system. To correct aberrations, combinations of lenses that result in a short positive focal length may be used in place of a single lens, but the principle of the simple magnifier is the same.

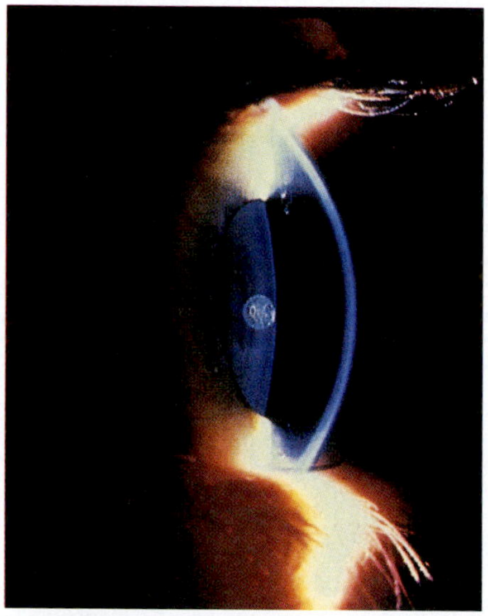

(*a*)

(*b*)

(*a*) The human eye in profile. (*b*) The lens of the eye is kept in place by the ciliary muscle (shown here in the upper left), which rings the lens. When the ciliary muscle contracts, the lens tends to bulge. The greater lens curvature enables the eye to focus on nearby objects. (*c*) Some of the 120 million rods and 7 million cones in the eye, magnified approximately 5000 times. The rods (the more slender of the two) are more sensitive in dim light, whereas the cones are more sensitive to color. The rods and cones form the bottom layer of the retina and are covered by nerve cells, blood vessels, and supporting cells. Most of the light entering the eye is reflected or absorbed before reaching the rods and cones. The light that does reach them triggers electrical impulses along nerve fibers that ultimately reach the brain. (*d*) A neural net used in the vision system of certain robots. Loosely modeled on the human eye, it contains 1920 sensors.

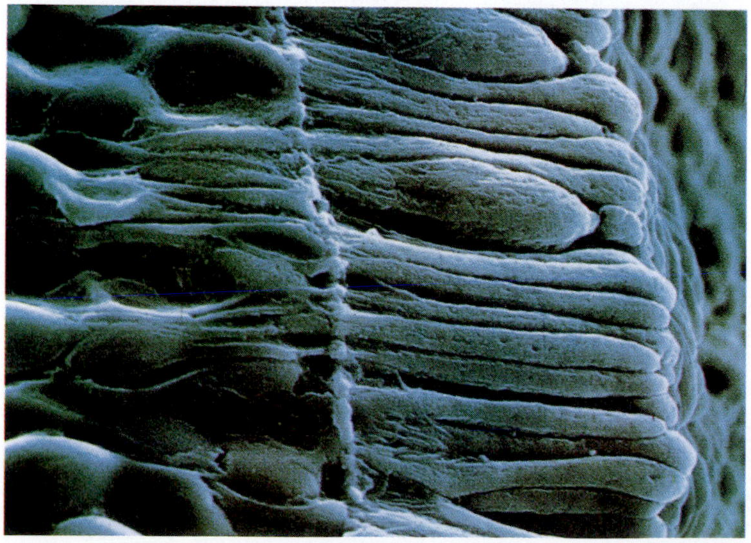

(*c*)

(*d*)

Example 34-14 *try it yourself*

A person with a near point of 25 cm uses a 40-D lens as a simple magnifier. What angular magnification is obtained?

Picture the Problem The angular magnification is found from the focal length f (Equation 34-20), which is the reciprocal of the power.

Cover the column to the right and try these on your own before looking at the answers.

Steps	*Answers*
1. Calculate the focal length of the lens.	$f = 2.5$ cm
2. Use your result in Equation 34-20 to calculate the angular magnification.	$M = 10$

Remark The object looks 10 times larger because it can be placed at 2.5 cm rather than at 25 cm from the eye, thus increasing the image on the retina tenfold.

Exercise What is the magnification in this example if the near point of the person is 30 cm rather than 25 cm? (*Answer* $M = 12$)

The Compound Microscope

The compound microscope (Figure 34-48) is used to look at very small objects at short distances. In its simplest form, it consists of two converging lenses. The lens nearest the object, called the **objective,** forms a real image of the object. This image is enlarged and inverted. The lens nearest the eye, called the **eyepiece** or **ocular,** is used as a simple magnifier to view the image formed by the objective. The eyepiece is placed such that the image formed by the objective falls at the first focal point of the eyepiece. The light thus emerges from the eyepiece as a parallel beam as if it were coming from a point a great distance in front of the lens. (This is commonly called "viewing the image at infinity.")

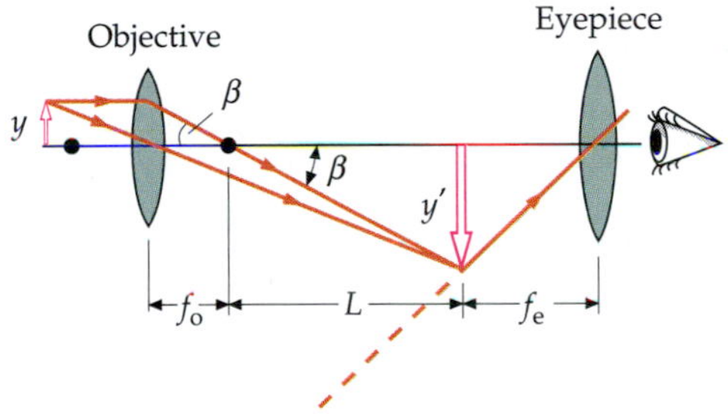

Figure 34-48 Schematic diagram of a compound microscope consisting of two positive lenses, the objective of focal length f_o and the ocular, or eyepiece, of focal length f_e. The real image of the object formed by the objective is viewed by the eyepiece, which acts as a simple magnifier. The final image is at infinity.

The distance between the second focal point of the objective and the first focal point of the eyepiece is called the **tube length** L. It is typically fixed at about 16 cm. The object is placed just outside the first focal point of the objective so that an enlarged image is formed at the first focal point of the eyepiece a distance $L + f_o$ from the objective, where f_o is the focal length of the objective. From Figure 34-48, $\tan \beta = y/f_o = -y'/L$. The lateral magnification of the objective is therefore

$$m_o = \frac{y'}{y} = -\frac{L}{f_o} \qquad \text{34-21}$$

The angular magnification of the eyepiece is

$$M_e = \frac{x_{np}}{f_e}$$

where x_{np} is the near point of the viewer (the nearest point at which the viewer can focus), and f_e is the focal length of the eyepiece. The magnifying power of the compound microscope is the product of the lateral magnification of the objective and the angular magnification of the eyepiece:

$$M = m_o M_e = -\frac{L}{f_o}\frac{x_{np}}{f_e} \qquad \text{34-22}$$

Magnifying power of a microscope

optional

Example 34-15

A microscope has an objective lens of focal length 1.2 cm and an eyepiece of focal length 2.0 cm separated by 20 cm. (*a*) Find the magnifying power if the near point of the viewer is 25 cm. (*b*) Where should the object be placed if the final image is to be viewed at infinity?

(*a*)1. The magnifying power is given by Equation 34-22:
$$M = -\frac{L}{f_o}\frac{x_{np}}{f_e}$$

2. The tube length L is the distance between the lenses minus the focal distances:
$$L = 20\text{ cm} - 2\text{ cm} - 1.2\text{ cm} = 16.8\text{ cm}$$

3. Substitute this value for L and the given values of x_{np}, f_o, and f_e to calculate M:
$$M = -\frac{L}{f_o}\frac{x_{np}}{f_e} = -\frac{16.8\text{ cm}}{1.2\text{ cm}}\frac{25\text{ cm}}{2\text{ cm}} = -175$$

(*b*)1. Calculate the object distance s in terms of the image distance for the objective s' and the focal length f_o:
$$\frac{1}{s} + \frac{1}{s'} = \frac{1}{f_o}$$

2. From Figure 34-48, the image distance for the image of the objective is $f_o + L$:
$$s' = f_o + L = 1.2\text{ cm} + 16.8\text{ cm} = 18\text{ cm}$$

3. Substitute to calculate s:
$$\frac{1}{s} + \frac{1}{18\text{ cm}} = \frac{1}{1.2\text{ cm}}$$
$$s = 1.29\text{ cm}$$

Remark The object should thus be placed at 1.29 cm from the objective or 0.09 cm outside its first focal point.

The Telescope

A telescope is used to view objects that are far away and often large. It works by creating an image of the object that is much closer than the object. The astronomical telescope, illustrated schematically in Figure 34-49, consists of two positive lenses—an objective lens that forms a real, inverted image and an eyepiece that is used as a simple magnifier to view that image. Because the object is very far away, the image of the objective lies at the focal point of the objective, and the image distance equals the focal length f_o. The image formed by the objective is much smaller than the object because the object distance is much larger than the focal length of the objective. For example, if we are looking at the moon, the image of the moon formed by the objective is much smaller than the moon itself. The purpose of the objective is not to magnify the object, but to produce an image that is close so it can be viewed by the eyepiece. The eyepiece is placed a distance f_e from the image, where f_e is the focal length of the eyepiece, so the final image can be viewed at infinity. Since this image is at the second focal point of the objective and at the first focal point of the ocular, the objective and ocular must be separated by the sum of the focal lengths of the objective and eyepiece, $f_o + f_e$.

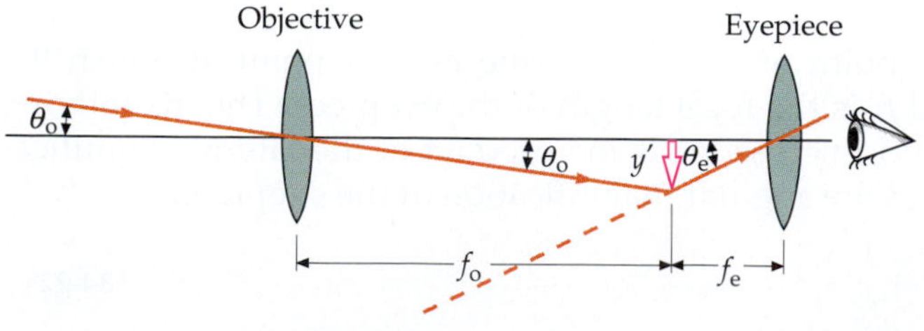

Figure 34-49 Schematic diagram of an astronomical telescope. The objective forms a real image of a distant object near its second focal point, which coincides with the first focal point of the eyepiece. The eyepiece serves as a simple magnifier to view the image.

The magnifying power of the telescope is the angular magnification θ_e/θ_o, where θ_e is the angle subtended by the final image as viewed through the eyepiece and θ_o is the angle subtended by the object when it is viewed directly by the unaided eye. The angle θ_o is the same as that subtended by the object at the objective shown in Figure 34-49. (The distance from a distant object, such as the moon, to the objective is essentially the same as the distance to the eye.) From this figure, we can see that

$$\tan \theta_o = -\frac{y'}{f_o} \approx \theta_o$$

where we have used the small-angle approximation $\tan \theta \approx \theta$ and have introduced a negative sign to make θ_o positive when y' is negative. The angle θ_e in the figure is that subtended by the final image:

$$\tan \theta_e = -\frac{y'}{f_e} \approx \theta_e$$

Since y' is negative, θ_e is negative, indicating that the image is inverted. The magnifying power of the telescope is then

$$M = \frac{\theta_e}{\theta_o} = -\frac{f_o}{f_e} \qquad \text{34-23}$$

Magnifying power of a telescope

From Equation 34-23, we can see that a large magnifying power is obtained with an objective of large focal length and an eyepiece of short focal length.

Exercise The world's largest refracting telescope is at the Yerkes Observatory of the University of Chicago at Williams Bay, Wisconsin. The objective has a diameter of 102 cm and a focal length of 19.5 m. The focal length of the eyepiece is 10 cm. What is its magnifying power? (*Answer* -195)

The main consideration with an astronomical telescope is not its magnifying power but its light-gathering power, which depends on the size of the objective. The larger the objective, the brighter the image. Very large lenses without aberrations are difficult to produce. In addition, there are mechanical problems in supporting very large, heavy lenses by their edges. A reflecting telescope (Figures 34-50 and 34-51) uses a concave mirror instead of a lens for its objective. This offers several advantages. For one, a mirror does not produce chromatic aberration. In addition, mechanical support is much simpler, since the mirror weighs far less than a lens of equivalent optical quality and can be supported over its entire back surface. In modern telescopes, the objective mirror consists of several dozen adaptive mirror segments that can be adjusted individually to correct for minute variations in gravitational stress when the telescope is tilted, and to compensate for thermal expansions and contractions and other changes caused by climate conditions.

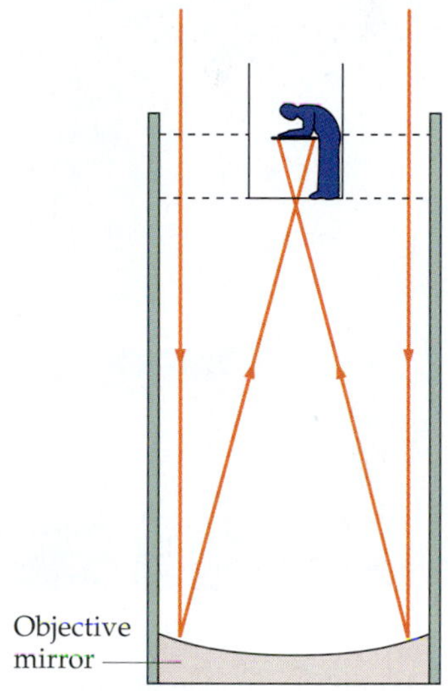

Figure 34-50 A reflecting telescope uses a mirror for its objective. Because the viewer compartment blocks off some of the incoming light, the arrangement shown here is used only in telescopes with very large objective mirrors.

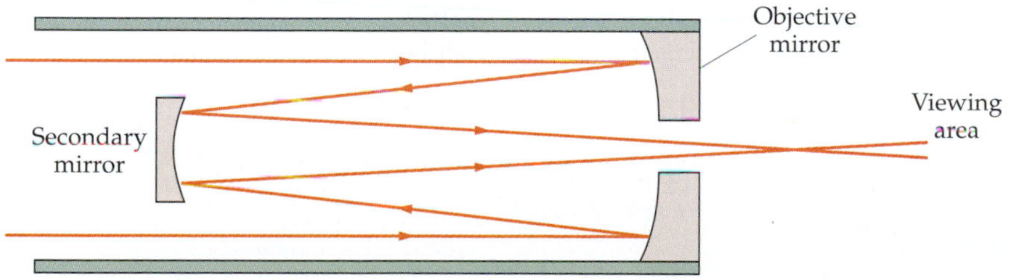

Figure 34-51 This reflecting telescope has a secondary mirror to redirect the light through a small hole in the objective mirror, thus providing more room for auxiliary instruments in the viewing area.

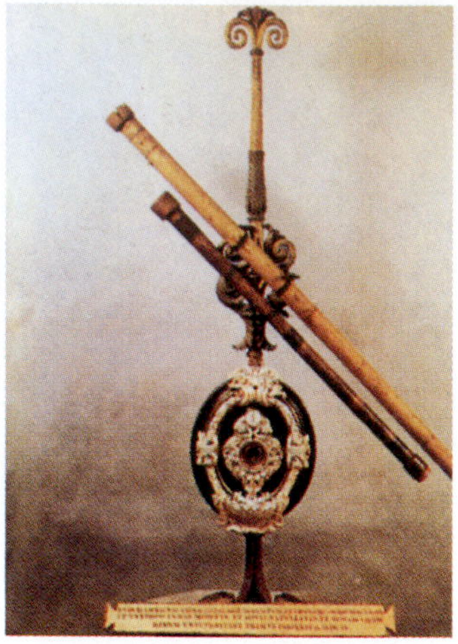

(*a*)

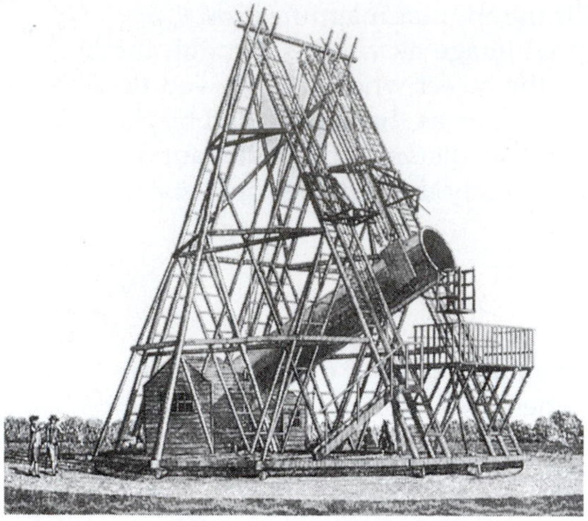

(*b*)

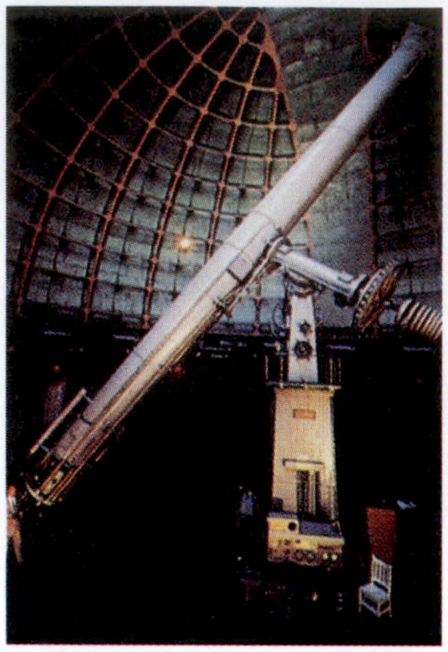

(*c*)

(*d*)

(*e*)

Astronomy at optical wavelengths began with Galileo approximately 400 years ago. In this century astronomers began to explore the electromagnetic spectrum at other wavelengths, beginning with radio astronomy in the 1940s, satellite-based x-ray astronomy in the early 1960s, and more recently, ultraviolet, infrared, and gamma-ray astronomy. (*a*) Galileo's seventeenth-century telescope, with which he discovered mountains on the moon, sunspots, Saturn's rings, and the bands and moons of Jupiter. (*b*) An engraving of the reflector telescope built in the 1780s and used by the great astronomer Hershel, who was the first to observe galaxies outside our own. (*c*) Because it is difficult to make large, flaw-free lenses, refractor telescopes like this 91.4-cm telescope at Lick Observatory have been superseded in light-gathering power by reflector telescopes. (*d*) The great astronomer Hubble, who discovered the apparent expansion of the universe, is shown seated in the observer's cage of the 5.08-m Hale reflecting telescope, which is large enough for the observer to sit at the prime focus itself. (*e*) This 10-m optical reflector at the Whipple Observatory in southern Arizona is the largest instrument designed exclusively for use in gamma-ray astronomy. High-energy gamma rays of unknown origin strike the upper atmosphere and create cascades of particles. Among these particles are high-energy electrons that emit Cerenkov radiation observable from the ground. According to one hypothesis, high-energy gamma rays are emitted as matter is accelerated toward ultradense rotating stars called pulsars.

optional

(*a*)

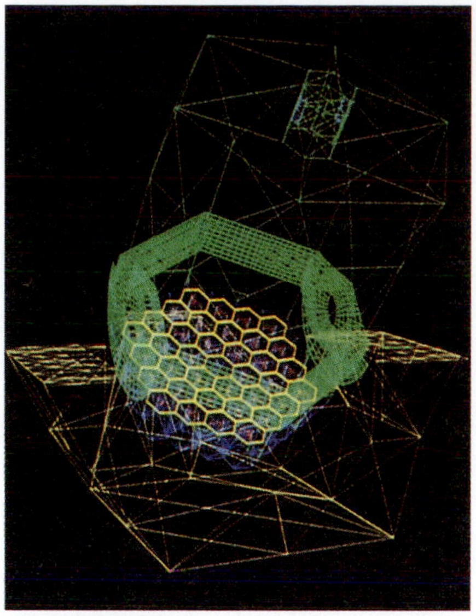

(*b*)

(*c*)

(*a*) The Keck Observatory, atop the inactive volcano of Mauna Kea, Hawaii, houses the world's largest optical telescope. The clear, dry air and lack of light pollution make the remote heights of Mauna Kea an ideal site for astronomical observations. (*b*) The Keck telescope is composed of 36 hexagonal mirror segments performing together as if they were a single mirror 10 m wide—roughly twice as large as the largest single-mirror telescope presently in operation.

(*c*) Beneath each Keck mirror is a system of computer-controlled sensors and motor-driven actuators that can continuously vary the mirror's shape. These variations, which are sensitive to within 100 nm, enable the system to compensate for variations in the alignments of the segments due to minute variations in gravitational stress when the telescope is tilted, thermal expansions and contractions, and fluctuations caused by gusts of wind on the mountain top.

The Hubble Space Telescope, high above the atmospheric turbulence that limits the ability of ground-based telescopes to resolve images at optical wavelengths.

Summary

Topic	Remarks and Relevant Equations	
1. Virtual and Real Images and Objects	An image is real if light actually converges at the image point and diverges (such as in front of a concave mirror, or behind a thin lens). It is virtual if light only appears to diverge from the image point (such as behind a plane mirror).	
Virtual object	Images due to reflection or refraction at one surface are often the objects for the next surface. If such an image is not formed because the light is interrupted by the surface, it is a virtual object.	
2. Mirrors		
Focal length	The focal length is the image distance when the object is at infinity so the incident light is parallel to the axis: $f = \frac{1}{2}r$	**34-2**
Equation for locating an image	$\frac{1}{s} + \frac{1}{s'} = \frac{1}{f}$	**34-3**
Lateral magnification	$m = \frac{y'}{y} = -\frac{s'}{s}$	**34-4**
Ray diagrams	Images can be located by a ray diagram using any two of three principle rays: 1. The parallel ray, drawn parallel to the axis, is reflected through the focal point. 2. The focal ray, drawn through the focal point, is reflected parallel to the axis. 3. The radial ray, drawn through the center of curvature, strikes the mirror perpendicular to its surface and is thus reflected back on itself.	
Sign conventions	s + if the object is in front of the mirror (real object) − if the object is behind the mirror (virtual object) s' + if the image is in front of the mirror (real image) − if the image is behind the mirror (virtual image) r, f + if the center of curvature is in front of the mirror (concave mirror) − if the center of curvature is behind the mirror (convex mirror)	
3. Images Formed by Refraction at a Spherical Surface		
Location	$\frac{n_1}{s} + \frac{n_2}{s'} = \frac{n_2 - n_1}{r}$	**34-5**
Magnification	$m = \frac{y'}{y} = -\frac{n_1 s'}{n_2 s}$	**34-6**
Sign convention	s + (real object) for objects in front of the surface (incident side) − (virtual object) for objects in back of the surface (transmission side) s' + (real image) for images in back of the surface (transmission side) − (virtual image) for images in front of the surface (incident side) r, f + if the center of curvature is on the transmission side − if the center of curvature is on the incident side	

4. Thin Lenses			
Focal length (lens-maker's equation)		$\frac{1}{f} = (n-1)\left(\frac{1}{r_1} - \frac{1}{r_2}\right)$ A positive lens ($f > 0$) is a converging lens (like a double convex lens). A negative lens ($f < 0$) is a diverging lens (like a double concave lens).	34-11
Power		$P = \frac{1}{f}$ diopters	34-13
Equation for locating image		$\frac{1}{s} + \frac{1}{s'} = \frac{1}{f}$	34-12
Magnification		$m = \frac{y'}{y} = -\frac{s'}{s}$	34-14
Ray diagrams		Images can be located by a ray diagram using any two of the three principal rays:	
	Positive lens	1. The parallel ray, drawn parallel to the axis, is bent through the second focal point of the lens. 2. The central ray, drawn through the center (the vertex) of the lens, is undeflected. 3. The focal ray, drawn through the first focal point, emerges parallel to the axis.	
	Negative lens	1. The parallel ray, drawn parallel to the axis, diverges from the lens as if it came from the second focal point. 2. The central ray, drawn through the center (the vertex) of the lens, is undeflected. 3. The focal ray, drawn toward the first focal point, emerges parallel to the axis.	
Sign convention		The sign conventions are the same as for refraction at a spherical surface.	
5. Aberrations (optional)		Blurring of the image of a single object point is called aberration. Spherical aberration results from the fact that a spherical surface focuses only paraxial rays (those that travel close to the axis) at a single point. Nonparaxial rays are focused at nearby points depending on the angle made with the axis. Spherical aberration can be reduced by reducing the size of the spherical surface, which also reduces the amount of light reaching the image. Chromatic aberration, which occurs with lenses but not mirrors, results from the variation in the index of refraction with wavelength. Lens aberrations are most commonly reduced by using a series of lens elements.	
6. The Eye (optional)		The cornea–lens system of the eye focuses light on the retina, where it is sensed by the rods and cones that send information along the optic nerve to the brain. When the eye is relaxed, the focal length of the cornea–lens system is about 2.5 cm, the distance to the retina. When objects are brought near the eye, the lens changes shape to decrease the overall focal length so that the image remains focused on the retina. The closest distance for which the image can be focused on the retina is called the near point, typically about 25 cm. The apparent size of an object depends on the size of the image on the retina. The closer the object, the larger the image on the retina and therefore the larger the apparent size of the object.	
7. Simple Magnifier (optional)		A simple magnifier consists of a lens with a positive focal length that is smaller than the near point.	
Angular magnification (or magnifying power)		$M = \frac{\theta}{\theta_o} = \frac{x_{np}}{f}$	34-20

8. Compound Microscope (optional)	The compound microscope is used to look at very small objects that are nearby. Its simplest form consists of two lenses, an objective and an ocular or eyepiece. The object to be viewed is placed just outside the focal point of the objective, which forms an enlarged image of the object at the focal point of the eyepiece. The eyepiece acts as a simple magnifier to view the final image.
Magnifying power	$M = m_o M_e = -\frac{L}{f_o}\frac{x_{np}}{f_e}$ **34-22** where L is the tube length, the distance between the second focal point of the objective and the first focal point of the eyepiece.
9. Telescope (optional)	The telescope is used to view objects far away. The objective of the telescope forms a real image that is much smaller than the object but much closer. The eyepiece is then used as a simple magnifier to view the image. A reflecting telescope uses a mirror for its objective.
Magnifying power	$M = \frac{\theta_e}{\theta_o} = -\frac{f_o}{f_e}$ **34-23**

Problem-Solving Guide

The approximate location, size, and orientation of an image are most easily determined by a ray diagram.

Summary of Worked Examples

Type of Calculation	Procedure and Relevant Examples
1. Mirrors	
Find the image distance and size.	Use $1/s + 1/s' = 1/f$, where $f = \frac{1}{2}r$ is the focal length of the mirror. Draw a ray diagram to check your result. To find the size, use $m = y'/y = -s'/s$. **Examples 34-1, 34-2**
2. Refracting Surfaces	
Find the image distance and size.	Use $n_1/s + n_2/s' = (n_2 - n_1)/r$ to find the distance. To find the size, use $m = y'/y = -n_1 s'/n_2 s$. **Example 34-4**
Find the apparent depth of an object.	The apparent depth equals the real depth divided by the index of refraction of the medium. **Example 34-5**
3. Thin Lenses	
Find the focal length.	Use $1/f = (n - 1)(1/r_1 - 1/r_2)$. Be sure to use the correct sign for each radius. **Examples 34-6, 34-7**
Find the power.	The power is the reciprocal of the focal length expressed in meters. **Example 34-7**
Find the image distance and size.	Use $1/s + 1/s' = 1/f$ to locate the image and $m = y'/y = -s'/s$ to find the size. The image is real if s' is positive, virtual if s' is negative. Check your result with a ray diagram. **Example 34-8**
4. Lens Combinations	To find the final image, first find the image distance for the first lens, then use it to find the object distance for the second lens. **Examples 34-9, 34-10**

5. Optical Instruments (optional)

Find the change in focal length of the eye lens.	The image distance is fixed at $s' = 2.5$ cm, the distance from the lens to the retina. Use the thin-lens equation to find f for a given s.	**Example 34-12**
Find the power of reading glasses needed.	Use the same procedure as for lens combinations. If the glasses are right next to the eye, the object distance for the eye lens will be the negative of the image distance for the glasses lens.	**Example 34-13**
Find the magnifying power of a simple magnifier.	Use $M = x_{np}/f$, where x_{np} is the near point of the eye.	**Example 34-14**
Find the magnifying power of a compound microscope.	Use $M = -f_o/f_e$.	**Example 34-15**

Problems

Conceptual Problems

Problems from Optional and Exploring sections

In a few problems, you are given more data than you actually need; in a few other problems, you are required to supply data from your general knowledge, outside sources, or informed estimates.

• Single-concept, single-step, relatively easy

•• Intermediate-level, may require synthesis of concepts

••• Challenging, for advanced students

Use $n = 1.33$ for the index of refraction of water unless otherwise specified.

Plane Mirrors

1 • Can a virtual image be photographed?

2 • Suppose each axis of a coordinate system like the one in Figure 34-4 is painted a different color. One photograph is taken of the coordinate system and another is taken of its image in a plane mirror. Is it possible to tell that one of the photographs is of a mirror image rather than both being photographs of the real coordinate system from different angles?

3 • The image of the point object P in Figure 34-52 is viewed by an eye as shown. Draw a bundle of rays from the object that reflect from the mirror and enter the eye. For this object position and mirror, indicate the region of space in which the eye can see the image.

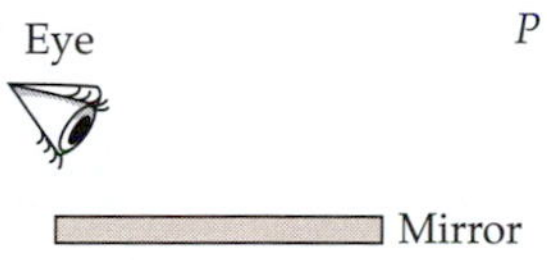

Figure 34-52 Problem 3

4 • A person 1.62 m tall wants to be able to see her full image in a plane mirror. (*a*) What must be the minimum height of the mirror? (*b*) How far above the floor should it be placed, assuming that the top of the person's head is 15 cm above her eye level? Draw a ray diagram.

5 • Two plane mirrors make an angle of 90°. Show by considering various object positions that there are three images for any position of an object. Draw appropriate bundles of rays from the object to the eye for viewing each image.

6 • (*a*) Two plane mirrors make an angle of 60° with each other. Show on a sketch the location of all the images formed of a point object on the bisector of the angle between the mirrors. (*b*) Repeat for an angle of 120°.

7 •• When two plane mirrors are parallel, such as on opposite walls in a barber shop, multiple images arise because each image in one mirror serves as an object for the other mirror. A point object is placed between parallel mirrors separated by 30 cm. The object is 10 cm in front of the left mirror and 20 cm in front of the right mirror. (*a*) Find the distance from the left mirror to the first four images in that mirror. (*b*) Find the distance from the right mirror to the first four images in that mirror.

Spherical Mirrors

8 •• True or False

(*a*) The virtual image formed by a concave mirror is always smaller than the object.
(*b*) A concave mirror always forms a virtual image.
(*c*) A convex mirror never forms a real image of a real object.
(*d*) A concave mirror never forms an enlarged real image of an object.

9 •• Under what condition will a concave mirror produce an erect image? A virtual image? An image smaller than the object? An image larger than the object?

10 •• Answer Problem 9 for a convex mirror.

11 •• Convex mirrors are often used for rear-view mirrors on cars and trucks to give a wide-angle view. Below the mirror is written, "Warning, objects are closer than they appear." Yet according to a ray diagram such as Figure 34-19, the image distance for distant objects is much smaller than the object distance. Why then do they appear more distant?

12 •• As an object is moved from a great distance toward the focal point of a concave mirror, the image moves from

(*a*) a great distance toward the focal point and is always real.
(*b*) the focal point to a great distance from the mirror and is always real.
(*c*) the focal point toward the center of curvature of the mirror and is always real.
(*d*) the focal point to a great distance from the mirror and changes from a real to a virtual image.

13 • A concave spherical mirror has a radius of curvature of 40 cm. Draw ray diagrams to locate the image (if one is formed) for an object at a distance of (*a*) 100 cm, (*b*) 40 cm, (*c*) 20 cm, and (*d*) 10 cm from the mirror. For each case, state whether the image is real or virtual; erect or inverted; and enlarged, reduced, or the same size as the object.

14 • Use the mirror equation to locate and describe the images for the object distances and mirror of Problem 13.

15 • Repeat Problem 13 for a convex mirror with the same radius of curvature.

16 • Use the mirror equation to locate and describe the images for the object distances and convex mirror of Problem 15.

17 • Show that a convex mirror cannot form a real image of a real object, no matter where the object is placed, by showing that s' is always negative for a positive s.

18 • A dentist wants a small mirror that will produce an upright image with a magnification of 5.5 when the mirror is located 2.1 cm from a tooth. (*a*) What should the radius of curvature of the mirror be? (*b*) Should it be concave or convex?

19 •• Convex mirrors are used in stores to provide a wide angle of surveillance for a reasonable mirror size. The mirror shown in Figure 34-53 allows a clerk 5 m away from the mirror to survey the entire store. It has a radius of curvature of 1.2 m. (*a*) If a customer is 10 m from the mirror, how far from the mirror surface is his image? (*b*) Is the image in front of or behind the mirror? (*c*) If the customer is 2 m tall, how high is his image?

Figure 34-53 Problem 19

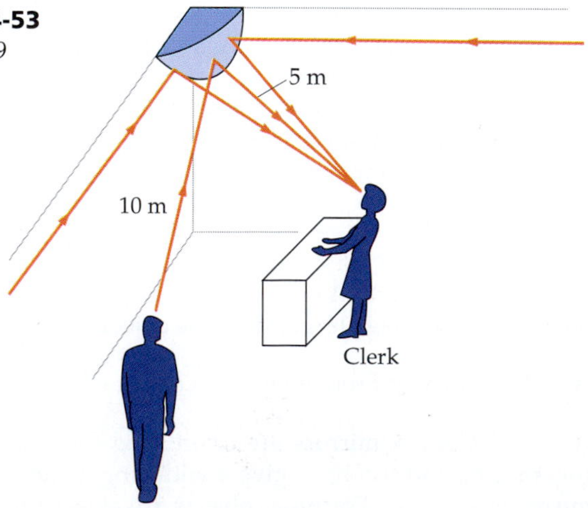

20 •• A certain telescope uses a concave spherical mirror with a radius of curvature of 8 m. Find the location and diameter of the image of the moon formed by this mirror. The moon has a diameter of 3.5×10^6 m and is 3.8×10^8 m from the earth.

21 •• A concave spherical mirror has a radius of curvature of 6.0 cm. A point object is on the axis 9 cm from the mirror. Construct a precise ray diagram showing rays from the object that make angles of 5°, 10°, 30°, and 60° with the axis, strike the mirror, and are reflected back across the axis. (Use a compass to draw the mirror, and use a protractor to measure the angles needed to find the reflected rays.) What is the spread δx of the points where these rays cross the axis?

22 •• A concave mirror has a radius of curvature 6.0 cm. Draw rays parallel to the axis at 0.5, 1.0, 2.0, and 4.0 cm above the axis and find the points at which the reflected rays cross the axis. (Use a compass to draw the mirror and a protractor to find the angle of reflection for each ray.) (*a*) What is the spread δx of the points where these rays cross the axis? (*b*) By what percentage could this spread be reduced if the edge of the mirror were blocked off so that parallel rays more than 2.0 cm from the axis could not strike the mirror?

23 •• An object placed 8 cm from a concave spherical mirror produces a virtual image 10 cm behind the mirror. (*a*) If the object is moved back to 25 cm from the mirror, where is the image located? (*b*) Is it real or virtual?

24 •• An object located 100 cm from a concave mirror forms a real image 75 cm from the mirror. The mirror is then turned around so that its convex side faces the object. The mirror is moved so that the image is now 35 cm behind the mirror. How far was the mirror moved? Was it moved toward or away from the object?

25 •• Parallel light from a distant object strikes the large mirror in Figure 34-54 ($r = 5$ m) and is reflected by the small mirror that is 2 m from the large mirror. The small mirror is actually spherical, not planar as shown. The light is focused at the vertex of the large mirror. (*a*) What is the radius of curvature of the small mirror? (*b*) Is it convex or concave?

Figure 34-54 Problem 25

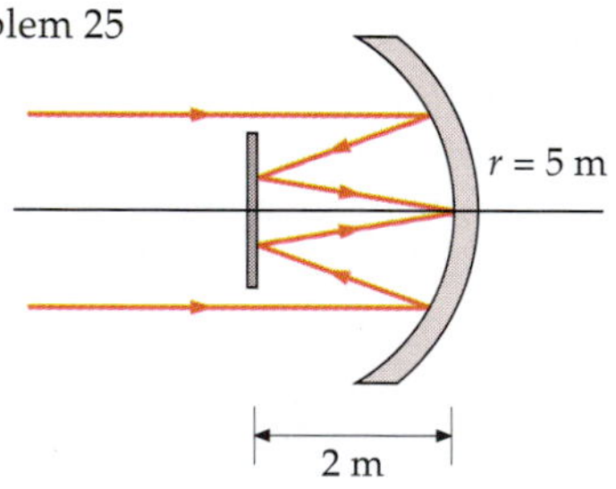

26 •• A woman uses a concave makeup mirror with a radius of curvature of 1.5 m. How far from the mirror should her face be for the image to be 80 cm from her face?

Images Formed by Refraction

27 • A bird above the water is viewed by a scuba diver submerged beneath the water's surface directly below the bird. Does the bird appear to the diver to be closer to or farther from the surface than it actually is?

28 • A sheet of paper with writing on it is protected by a thick glass plate having an index of refraction of 1.5. If the plate is 2 cm thick, at what distance beneath the top of the plate does the writing appear when it is viewed from directly overhead?

29 • A fish is 10 cm from the front surface of a fish bowl of radius 20 cm. (*a*) Where does the fish appear to be to someone in air viewing it from in front of the bowl? (*b*) Where does the fish appear to be when it is 30 cm from the front surface of the bowl?

30 •• A very long glass rod of 2-cm diameter has one end ground to a convex spherical surface of radius 5 cm. Its index of refraction is 1.5. (*a*) A point object in air is on the axis of the rod 20 cm from the surface. Find the image and state whether it is real or virtual. Repeat for (*b*) an object 5 cm from the surface and (*c*) an object very far from the surface. Draw a ray diagram for each case.

31 •• At what distance from the rod of Problem 30 should the object be placed so that the light rays in the rod are parallel? Draw a ray diagram for this situation.

32 •• Repeat Problem 30 for a glass rod with a concave hemispherical surface of radius −5 cm.

33 •• Repeat Problem 30 when the glass rod and objects are immersed in water.

34 •• Repeat Problem 30 for a glass rod with a concave hemispherical surface of radius −5 cm when it and the objects are immersed in water.

35 •• A glass rod 96 cm long with an index of refraction of 1.6 has its ends ground to convex spherical surfaces of radii 8 and 16 cm. A point object is in air on the axis of the rod 20 cm from the end with the 8-cm radius. (*a*) Find the image distance due to refraction at the first surface. (*b*) Find the final image due to refraction at both surfaces. (*c*) Is the final image real or virtual?

36 •• Repeat Problem 35 for a point object in air on the axis of the rod 20 cm from the end with the 16-cm radius.

Thin Lenses

37 • Under what conditions will the focal length of a thin lens be positive? Negative?

38 • The focal length of a simple lens is different for different colors of light. Why?

39 •• An object is placed 40 cm from a lens of focal length −10 cm. The image is

(*a*) real, inverted, and diminished.
(*b*) real, inverted, and enlarged.
(*c*) virtual, inverted, and diminished.
(*d*) virtual, upright, and diminished.
(*e*) virtual, upright, and enlarged.

40 •• If a real object is placed just inside the focal point of a converging lens, the image is

(*a*) real, inverted, and enlarged.
(*b*) virtual, erect, and diminished.
(*c*) virtual, erect, and enlarged.
(*d*) real, inverted, and diminished.

41 • The following thin lenses are made of glass with an index of refraction of 1.5. Make a sketch of each lens, and find its focal length in air: (*a*) double convex, $r_1 = 10$ cm and $r_2 = -21$ cm; (*b*) plano-convex, $r_1 = \infty$ and $r_2 = -10$ cm; (*c*) double concave, $r_1 = -10$ cm and $r_2 = +10$ cm; (*d*) plano-concave, $r_1 = \infty$ and $r_2 = +20$ cm.

42 • Glass with an index of refraction of 1.6 is used to make a thin lens that has radii of equal magnitude. Find the radii of curvature and make a sketch of the lens if the focal length in air is (*a*) +5 cm and (*b*) −5 cm.

43 • Find the focal length of a glass lens of index of refraction 1.62 that has a concave surface with radius of magnitude 100 cm and a convex surface with a radius of magnitude 40 cm.

44 • A double-concave lens of index of refraction 1.45 has radii of magnitudes 30 and 25 cm. An object is located 80 cm to the left of the lens. Find (*a*) the focal length of the lens, (*b*) the location of the image, and (*c*) the magnification of the image. (*d*) Is the image real or virtual? Upright or inverted?

45 • The following thin lenses are made of glass of index of refraction 1.6. Make a sketch of each lens, and find its focal length in air: (*a*) $r_1 = 20$ cm, $r_2 = 10$ cm; (*b*) $r_1 = 10$ cm, $r_2 = 20$ cm; (*c*) $r_1 = -10$ cm, $r_2 = -20$ cm.

46 • For the following object distances and focal lengths of thin lenses in air, find the image distance and the magnification and state whether the image is real or virtual and erect or inverted: (*a*) $s = 40$ cm, $f = 20$ cm; (*b*) $s = 10$ cm, $f = 20$ cm; (*c*) $s = 40$ cm, $f = -30$ cm; (*d*) $s = 10$ cm, $f = -30$ cm.

47 • An object 3.0 cm high is placed 20 cm in front of a thin lens of power 20 D. Draw a precise ray diagram to find the position and size of the image and check your results using the thin-lens equation.

48 • Repeat Problem 47 for an object 1.0 cm high placed 10 cm in front of a thin lens of power 20 D.

49 • Repeat Problem 47 for an object 1.0 cm high placed 10 cm in front of a thin lens whose power is −20 D.

50 •• (*a*) What is meant by a negative object distance? How can it occur? Find the image distance and magnification and state whether the image is virtual or real and erect or inverted for a thin lens in air when (*b*) $s = -20$ cm, $f = +20$ cm and (*c*) $s = -10$ cm, $f = -30$ cm. Draw a ray diagram for each of these cases.

51 •• Two converging lenses, each of focal length 10 cm, are separated by 35 cm. An object is 20 cm to the left of the first lens. (*a*) Find the position of the final image using both a ray diagram and the thin-lens equation. (*b*) Is the image real or virtual? Erect or inverted? (*c*) What is the overall lateral magnification of the image?

52 •• Work Problem 51 for a second lens that is a diverging lens of focal length −15 cm.

53 •• A thin lens of index of refraction 1.5 has one convex side with a radius of magnitude 20 cm. When an object 1 cm in height is placed 50 cm from this lens, an upright image 2.15 cm in height is formed. (*a*) Calculate the radius of the

second side of the lens. Is it concave or convex? (*b*) Draw a sketch of the lens.

54 •• (*a*) Show that to obtain a magnification of magnitude m with a converging thin lens of focal length f, the object distance must be given by $s = (m - 1)f/m$. (*b*) A camera lens with 50-mm focal length is used to take a picture of a person 1.75 m tall. How far from the camera should the person stand so that the image size is 24 mm?

55 •• An object is 15 cm in front of a positive lens of focal length 15 cm. A second positive lens of focal length 15 cm is 20 cm from the first lens. Find the final image and draw a ray diagram.

56 •• Work Problem 55 for a second lens with a focal length of −15 cm.

57 ••• In a convenient form of the thin-lens equation used by Newton, the object and image distances are measured from the focal points. Show that if $x = s - f$ and $x' = s' - f$, the thin-lens equation can be written as $xx' = f^2$, and the lateral magnification is given by $m = -x'/f = -f/x$. Indicate x and x' on a sketch of a lens.

58 ••• An object is placed 2.4 m from a screen, and a lens of focal length f is placed between the object and the screen so that a real image of the object is formed on the screen. When the lens is moved 1.2 m toward the screen, another real image of the object is formed on the screen. (*a*) Where was the lens located before it was moved? (*b*) What is the focal length of the lens?

59 ••• An object is 17.5 cm to the left of a lens of focal length 8.5 cm. A second lens of focal length −30 cm is 5 cm to the right of the first lens. (*a*) Find the distance between the object and the final image formed by the second lens. (*b*) What is the overall magnification? (*c*) Is the final image real or virtual? Upright or inverted?

Aberrations (optional)

60 • Chromatic aberration is a common defect of

(*a*) concave and convex lenses.
(*b*) concave lenses only.
(*c*) concave and convex mirrors.
(*d*) all lenses and mirrors.

61 • True or false:

(*a*) Aberrations occur only for real images.
(*b*) Chromatic aberration does not occur with mirrors.

62 • A double-convex lens of radii $r_1 = +10$ cm and $r_2 = -10$ cm is made from glass with indexes of refraction of 1.53 for blue light and 1.47 for red light. Find the focal length of this lens for (*a*) red light and (*b*) blue light.

The Eye (optional)

In the following problems, take the distance from the cornea–lens system of the eye to the retina to be 2.5 cm.

63 • If an object is placed 25 cm from the eye of a farsighted person who does not wear corrective lenses, a sharp image is formed

(*a*) behind the retina, and the corrective lens should be convex.
(*b*) behind the retina, and the corrective lens should be concave.
(*c*) in front of the retina, and the corrective lens should be convex.
(*d*) in front of the retina, and the corrective lens should be concave.

64 •• Myopic (nearsighted) persons sometimes claim to see better under water without corrective lenses. Why?

(*a*) The accommodation of the eye's lens is better under water.
(*b*) Refraction at the water–cornea interface is less than at the air–cornea interface.
(*c*) Refraction at the water–cornea interface is greater than at the air–cornea interface.
(*d*) No reason; the effect is only an illusion and not really true.

65 •• A nearsighted person who wears corrective lenses would like to examine an object at close distance. Identify the correct statement.

(*a*) The corrective lenses give an enlarged image and should be worn while examining the object.
(*b*) The corrective lenses give a reduced image of the object and should be removed.
(*c*) The corrective lenses result in a magnification of unity; it does not matter whether they are worn or removed.

66 • Suppose the eye were designed like a camera with a lens of fixed focal length $f = 2.5$ cm that could move toward or away from the retina. Approximately how far would the lens have to move to focus the image of an object 25 cm from the eye onto the retina? (*Hint:* Find the distance from the retina to the image behind it for an object at 25 cm.)

67 • Find the change in the focal length of the eye when an object originally at 3 m is brought to 30 cm from the eye.

68 • Find (*a*) the focal length and (*b*) the power of a lens that will produce an image at 80 cm from the eye of a book that is 30 cm from the eye.

69 • A farsighted person requires lenses with a power of 1.75 D to read comfortably from a book that is 25 cm from the eye. What is that person's near point without the lenses?

70. • If two point objects close together are to be seen as two distinct objects, the images must fall on the retina on two different cones that are not adjacent. That is, there must be an unactivated cone between them. The separation of the cones is about 1 μm. (*a*) What is the smallest angle the two points can subtend? (See Figure 34-55.) (*b*) How close can two points be if they are 20 m from the eye?

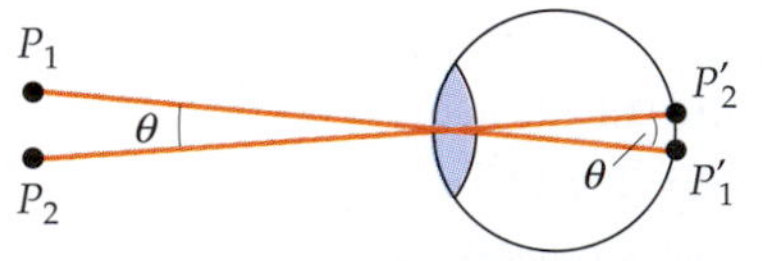

Figure 34-55 Problem 70. The two points will look like two separate points only if their images fall on two different, nonadjacent cones of the retina.

71 •• A person with a near point of 80 cm needs to read from a computer screen that is 45 cm from her eye. (*a*) Find the focal length of the lenses in reading glasses that will produce an image of the screen at 80 cm from her eye. (*b*) What is the power of the lenses?

72 •• A nearsighted person cannot focus clearly on objects more distant than 225 cm from her eye. What power lenses are required for her to see distant objects clearly?

73 •• Since the index of refraction of the lens of the eye is not very different from that of the surrounding material, most of the refraction takes place at the cornea, where n changes abruptly from 1.0 in air to about 1.4. Assuming the cornea to be a homogeneous sphere with an index of refraction of 1.4, calculate its radius if it focuses parallel light on the retina a distance 2.5 cm away. Do you expect your result to be larger or smaller than the actual radius of the cornea?

74 •• The near point of a certain person is 80 cm. Reading glasses are prescribed so that he can read a book at 25 cm from his eye. The glasses are 2 cm from the eye. What diopter lens should be used in the glasses?

75 ••• At age 45, a person is fitted for reading glasses of power 2.1 D in order to read at 25 cm. By the time she reaches 55, she discovers herself holding her newspaper at a distance of 40 cm in order to see it clearly with her glasses on. (*a*) Where was her near point at age 45? (*b*) Where is her near point at age 55? (*c*) What power is now required for the lenses of her reading glasses so that she can again read at 25 cm? (Assume that the glasses are 2.2 cm from her eyes.)

76 ••• An aging physics professor discovers that he can see objects clearly only between 0.75 m and 2.5 m so he decides he needs bifocals. The upper part of the lens allows him to see objects clearly at infinity, and the lower part allows him to see objects clearly at 25 cm. Assume that the lens is 2 cm from his eye. (*a*) Calculate the power of the lens required for the upper part of his bifocals. (*b*) Calculate the power of the lens required for the lower part of his bifocals. (*c*) Is there a range of distance over which he cannot see objects clearly no matter which part of the bifocals he looks through? If so, what is that range?

The Simple Magnifier (optional)

77 • A person with a near-point distance of 30 cm uses a simple magnifier of power 20 D. What is the magnification obtained if the final image is at infinity?

78 • A person with a near-point distance of 25 cm wishes to obtain a magnifying power of 5 with a simple magnifier. What should be the focal length of the lens used?

79 • What is the magnifying power of a lens of focal length 7 cm when the image is viewed at infinity by a person whose near point is at 35 cm?

80 •• A lens of focal length 6 cm is used as a simple magnifier with the image at infinity by one person whose near point is 25 cm and by another whose near point is 40 cm. What is the effective magnifying power of the lens for each person? Compare the size of the image on the retina when each looks at the same object with the magnifier.

81 •• A botanist examines a leaf using a convex lens of power 12 D as a simple magnifier. What is the expected angular magnification if (*a*) the final image is at infinity, and (*b*) the final image is at 25 cm?

82 •• (*a*) Show that if the final image of a simple magnifier is to be at the near point of the eye rather than at infinity, the angular magnification is given by

$$M = \frac{x_{\rm np}}{f} + 1$$

(*b*) Find the magnification of a 20-D lens for a person with a near point of 30 cm if the final image is at the near point. Draw a ray diagram for this situation.

83 •• Show that when the image of a simple magnifier is viewed at the near point, the lateral and angular magnification of the magnifier are equal.

Microscopes (optional)

84 •• A microscope objective has a focal length of 0.5 cm. It forms an image at 16 cm from its second focal point. What is the magnifying power for a person whose near point is at 25 cm if the focal length of the eyepiece is 3 cm?

85 •• A microscope has an objective of focal length 16 mm and an eyepiece that gives an angular magnification of 5 for a person whose near point is 25 cm. The tube length is 18 cm. (*a*) What is the lateral magnification of the objective? (*b*) What is the magnifying power of the microscope?

86 •• A crude symmetric hand-held microscope consists of two converging 20-D lenses fastened in the ends of a tube 30 cm long. (*a*) What is the "tube length" of this microscope? (*b*) What is the lateral magnification of the objective? (*c*) What is the magnifying power of the microscope? (*d*) How far from the objective should the object be placed?

87 •• Repeat Problem 86 for the same two lenses separated by 40 cm.

88 •• A compound microscope has an object with a power of 45 D and an eyepiece with a power of 80 D. The lenses are separated by 28 cm. Assuming that the final image is formed 25 cm from the eye, what is the magnifying power?

89 ••• A microscope has a magnifying power of 600 and an eyepiece of angular magnification of 15. The objective lens is 22 cm from the eyepiece. Without making any approximations, calculate (*a*) the focal length of the eyepiece, (*b*) the location of the object such that it is in focus for a normal relaxed eye, and (*c*) the focal length of the objective lens.

Telescopes (optional)

90 • A simple telescope has an objective with a focal length of 100 cm and an eyepiece of focal length 5 cm. It is used to look at the moon, which subtends an angle of about 0.009 rad. (*a*) What is the diameter of the image formed by the objective? (*b*) What angle is subtended by the final image at infinity? (*c*) What is the magnifying power of the telescope?

91 • The objective lens of the refracting telescope at the Yerkes Observatory has a focal length of 19.5 m. When it is used to look at the moon, which subtends an angle of about 0.009 rad, what is the diameter of the image of the moon formed by the objective?

92 •• The 200-in (5.1-m) mirror of the reflecting telescope at Mt. Palomar has a focal length of 1.68 m. (*a*) By what factor is the light-gathering power increased over the 40-in (1.016-m) diameter refracting lens of the Yerkes Observatory telescope? (*b*) If the focal length of the eyepiece is 1.25 cm, what is the magnifying power of this telescope?

93 •• An astronomical telescope has a magnifying power of 7. The two lenses are 32 cm apart. Find the focal length of each lens.

94 •• A disadvantage of the astronomical telescope for terrestrial use (for example, at a football game) is that the image is inverted. A Galilean telescope uses a converging lens as its objective, but a diverging lens as its eyepiece. The image formed by the objective is behind the eyepiece at its focal point so that the final image is virtual, erect, and at infinity. (*a*) Show that the magnifying power is $M = -f_o/f_e$, where f_o is the focal length of the objective and f_e is that of the eyepiece (which is negative). (*b*) Draw a ray diagram to show that the final image is indeed virtual, erect, and at infinity.

95 •• A Galilean telescope (see Problem 94) is designed so that the final image is at the near point, which is 25 cm (rather than at infinity). The focal length of the objective is 100 cm and that of the eyepiece is −5 cm. (*a*) If the object distance is 30 m, where is the image of the objective? (*b*) What is the object distance for the eyepiece so that the final image is at the near point? (*c*) How far apart are the lenses? (*d*) If the object height is 1.5 m, what is the height of the final image? What is the angular magnification?

96 ••• A hunter lost in the mountains tries to make a telescope from two lenses of power 2.0 and 6.5 D, and a cardboard tube. (*a*) What is the maximum possible magnifying power? (*b*) How long must the tube be? (*c*) Which lens should be used as the eyepiece? Why?

97 ••• If you look into the wrong end of a telescope, that is, into the objective, you will see distant objects reduced in size. For a refracting telescope with an objective of focal length 2.25 m and an eyepiece of focal length 1.5 cm, by what factor is the angular size of the object reduced?

General Problems

98 • The image of a real object formed by a convex mirror

(*a*) is always real and inverted.
(*b*) is always virtual and enlarged.
(*c*) may be real.
(*d*) is always virtual and diminished.

99 • The glass of a converging lens has an index of refraction of 1.6. When the lens is in air, its focal length is 30 cm. If immersed in water, its focal length will be

(*a*) greater than 30 cm. (*b*) less than 30 cm.
(*c*) the same as before, 30 cm. (*d*) negative.

100 •• True or false:

(*a*) A virtual image cannot be displayed on a screen.
(*b*) A negative image distance implies that the image is virtual.
(*c*) All rays parallel to the axis of a spherical mirror are reflected through a single point.
(*d*) A diverging lens cannot form a real image from a real object.
(*e*) The image distance for a positive lens is always positive.

101 • Show that a diverging lens can never form a real image from a real object. (*Hint:* Show that s' is always negative.)

102 • A camera uses a positive lens to focus light from an object onto a film. Unlike the eye, the camera lens has a fixed focal length, but the lens itself can be moved slightly to vary the image distance to the image on the film. A telephoto lens has a focal length of 200 mm. By how much must it move to change from focusing on an object at infinity to one at a distance of 30 m?

103 • A wide-angle lens of a camera has a focal length of 28 mm. By how much must it move to change from focusing on an object at infinity to one at a distance of 5 m? (See Problem 102.)

104 • A converging lens made of polystyrene (index of refraction, 1.59) has a focal length of 50 cm. One surface is convex with radius of magnitude 50 cm. Find the radius of the second surface. Is it convex or concave?

105 • A thin converging lens of focal length 10 cm is used to obtain an image that is twice as large as a small object. Find the object and image distances if (*a*) the image is to be erect and (*b*) the image is to be inverted. Draw a ray diagram for each case.

106 •• A scuba diver wears a diving mask with a face plate that bulges outward with a radius of curvature of 0.5 m. There is thus a convex spherical surface between the water and the air in the mask. A fish is 2.5 m in front of the diving mask. (*a*) Where does the fish appear to be? (*b*) What is the magnification of the image of the fish?

107 •• You wish to see an image of your face for applying makeup or shaving. If you want the image to be upright, virtual, and magnified 1.5 times when your face is 30 cm from the mirror, what kind of mirror should you use, convex or concave, and what should its focal length be?

108 •• A small object is 20 cm from a thin positive lens of focal length 10 cm. To the right of the lens is a plane mirror that crosses the axis at the second focal point of the lens and is tilted so that the reflected rays do not go back through the lens (Figure 34-56). (*a*) Find the position of the final image. (*b*) Is this image real or virtual? (*c*) Sketch a ray diagram showing the final image.

Figure 34-56
Problem 108

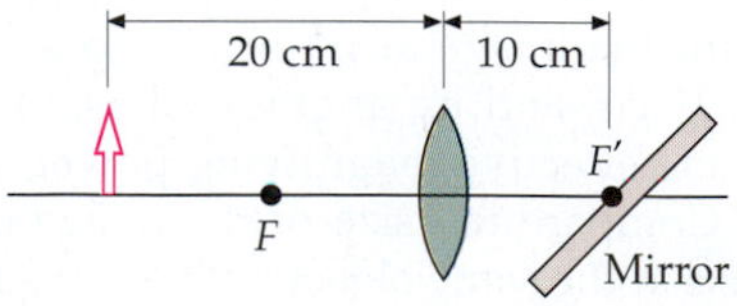

109 •• A 35-mm camera has a picture size of 24 mm by 36 mm. It is used to take a picture of a person 175 cm tall so that the image just fills the height (24 mm) of the film. How far should the person stand from the camera if the focal length of the lens is 50 mm?

110 •• A 35-mm camera with interchangeable lenses is used to take a picture of a hawk that has a wing span of 2 m. The hawk is 30 m away. What would be the ideal focal length of the lens used so that the image of the wings just fills the width of the film, which is 36 mm?

111 •• An object is placed 12 cm to the left of a lens of focal length 10 cm. A second lens of focal length 12.5 cm is placed 20 cm to the right of the first lens. (*a*) Find the position of the final image. (*b*) What is the magnification of the image? (*c*) Sketch a ray diagram showing the final image.

112 •• (*a*) Show that if f is the focal length of a thin lens in air, its focal length in water is

$$f' = \frac{n_W(n-1)}{n-n_W} f$$

where n_W is the index of refraction of water and n is that of the lens. (*b*) Calculate the focal length in air and in water of a double-concave lens of index of refraction $n = 1.5$ that has radii of magnitudes 30 and 35 cm.

113 •• (*a*) Find the focal length of a *thick* double-convex lens with an index of refraction of 1.5, a thickness of 4 cm, and radii of +20 and −20 cm. (*b*) Find the focal length of this lens in water.

114 •• A 2-cm-thick layer of water ($n = 1.33$) floats on top of a 4-cm-thick layer of carbon tetrachloride ($n = 1.46$) in a tank. How far below the top surface of the water does the bottom of the tank appear to be to an observer looking from above at normal incidence?

115 •• While sitting in your car, you see a jogger in your side mirror, which is convex with a radius of curvature of magnitude 2 m. The jogger is 5 m from the mirror and is approaching at 3.5 m/s. How fast does the jogger appear to be running when viewed in the mirror?

116 •• In the seventeenth century, Antonie van Leeuwenhoek, the first great microscopist, used simple spherical lenses made first of water droplets and then of glass for his first instruments. He made staggering discoveries with these simple lenses. Consider a glass sphere of radius 2.0 mm with an index of refraction of 1.50. Find the focal length of this lens. *Hint:* Use the equation for refraction at a single spherical surface to find the image distance for an infinite object distance for the first surface. Then use this image point as the object point for the second surface.

117 ••• An object is 15 cm to the left of a thin convex lens of focal length 10 cm. A concave mirror of radius 10 cm is 25 cm to the right of the lens. (*a*) Find the position of the final image formed by the mirror and lens. (*b*) Is the image real or virtual? Erect or inverted? (*c*) Show on a diagram where your eye must be to see this image.

118 ••• Find the final image for the situation in Problem 108 when the mirror is not tilted. Assume that the image is viewed by an eye to the left of the object looking through the lens into the mirror.

119 ••• When a bright light source is placed 30 cm in front of a lens, there is an erect image 7.5 cm from the lens. There is also a faint inverted image 6 cm in front of the lens due to reflection from the front surface of the lens. When the lens is turned around, this weaker, inverted image is 10 cm in front of the lens. Find the index of refraction of the lens.

120 ••• A horizontal concave mirror with radius of curvature of 50 cm holds a layer of water with an index of refraction of 1.33 and a maximum depth of 1 cm. At what height above the mirror must an object be placed so that its image is at the same position as the object?

121 ••• A lens with one concave side with a radius of magnitude 17 cm and one convex side with a radius of magnitude 8 cm has a focal length in air of 27.5 cm. When placed in a liquid with an unknown index of refraction, the focal length increases to 109 cm. What is the index of refraction of the liquid?

122 ••• A glass ball of radius 10 cm has an index of refraction of 1.5. The back half of the ball is silvered so that it acts as a concave mirror (Figure 34-57). Find the position of the final image seen by an eye to the left of the object and ball for an object at (*a*) 30 cm and (*b*) 20 cm to the left of the front surface of the ball.

Figure 34-57 Problem 122

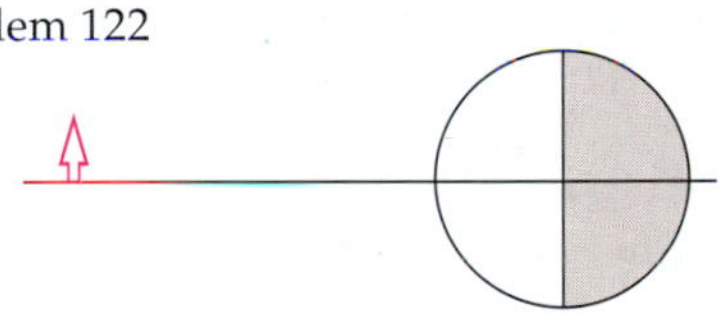

123 ••• (*a*) Show that a small change dn in the index of refraction of a lens material produces a small change in the focal length df given approximately by $df/f = -dn/(n-1)$. (*b*) Use this result to find the focal length of a thin lens for blue light, for which $n = 1.53$, if the focal length for red light, for which $n = 1.47$, is 20 cm.

124 ••• The lateral magnification of a spherical mirror or a thin lens is given by $m = -s'/s$. Show that for objects of small horizontal extent, the longitudinal magnification is approximately $-m^2$. (*Hint:* Show that $ds'/ds = s'^2/s^2$.)

CHAPTER 35

Interference and Diffraction

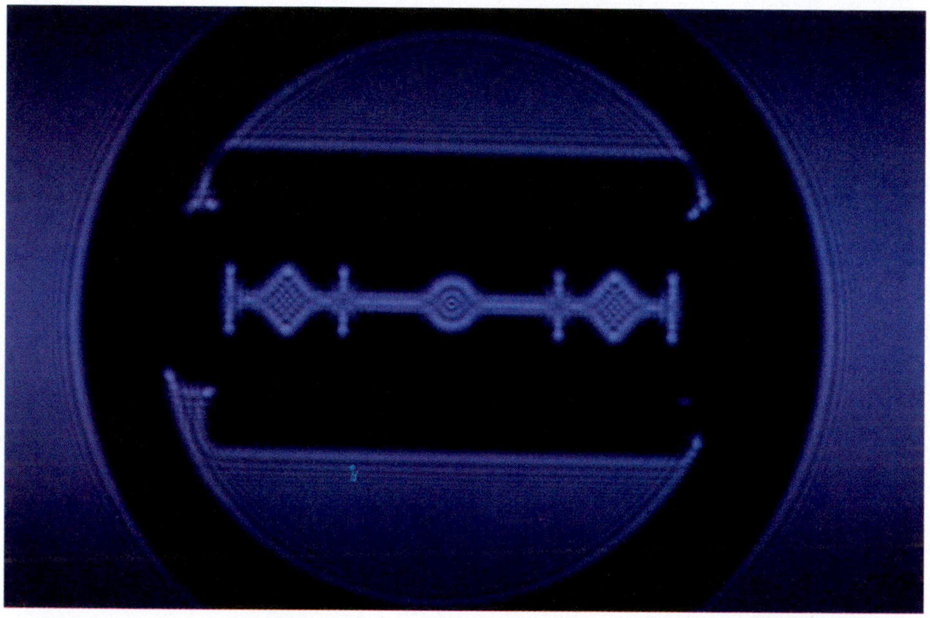

The diffraction of light incident on a razor blade.

Interference and diffraction are the important phenomena that distinguish waves from particles.* Interference is the combining by superposition of two or more waves that meet at one point in space. Diffraction is the bending of waves around corners that occurs when a portion of a wavefront is cut off by a barrier or obstacle. The pattern of the resulting wave can be calculated by treating each point on the original wavefront as a point source according to Huygens' principle and calculating the interference pattern resulting from these sources.

35-1 Phase Difference and Coherence

When two harmonic waves of the same frequency and wavelength but differing in phase combine, the resultant wave is a harmonic wave whose amplitude depends on the phase difference. If the phase difference is zero or an integer times 360°, the waves are in phase and interfere constructively. The resultant amplitude equals the sum of the individual amplitudes, and the intensity (which is proportional to the square of the amplitude) is maximum. If the phase difference is 180° or any odd integer times 180°, the waves are out of phase and interfere destructively. The resultant amplitude is then the dif-

*Before you study this chapter, you should review Chapters 15, 16, and 17, where the general topics of interference and diffraction of waves are discussed.

ference between the individual amplitudes, and the intensity is a minimum. If the amplitudes are equal, the maximum intensity is four times that of either source and the minimum intensity is zero.

A phase difference between two waves is often the result of a difference in path length. A path difference of one wavelength produces a phase difference of 360°, which is equivalent to no phase difference at all. A path difference of one-half wavelength produces a 180° phase difference. In general, a path difference of Δr contributes a phase difference δ given by

$$\delta = \frac{\Delta r}{\lambda} 2\pi = \frac{\Delta r}{\lambda} 360° \qquad \text{35-1}$$

Example 35-1

(*a*) What is the minimum path difference that will produce a phase difference of 180° for light of wavelength 800 nm? (*b*) What phase difference will that path difference produce in light of wavelength 700 nm?

Picture the Problem For both questions we use the relationship between phase difference and path difference given in Equation 35-1.

(*a*) Set $\delta = 180°$ in Equation 35-1 and solve for Δr:

$$\delta = \frac{\Delta r}{\lambda} 360° = 180°$$

$$\Delta r = \frac{\lambda \delta}{360°} = \frac{1}{2}\lambda = \frac{1}{2}(800 \text{ nm}) = 400 \text{ nm}$$

(*b*) Set $\lambda = 700$ nm, $\Delta r = 400$ nm, and solve for δ:

$$\delta = \frac{\Delta r}{\lambda} 360° = \frac{400 \text{ nm}}{700 \text{ nm}} 360° = 206° = 3.59 \text{ rad}$$

Another cause of phase difference is the 180° phase change a wave sometimes undergoes upon reflection from a boundary surface. This phase change is analogous to the inversion of a pulse on a string when it reflects from a point where the density suddenly increases, such as when a light string is attached to a heavier string or rope. The inversion of the reflected pulse is equivalent to a phase change of 180° for a sinusoidal wave, which can be thought of as a series of pulses. When light traveling in air strikes the surface of a medium in which light travels more slowly, such as glass or water, there is a 180° phase change in the reflected light. When light is originally traveling in glass or water, there is no phase change in the light reflected from the glass–air or water–air interface. This is analogous to the reflection without inversion of a pulse on a heavy string at a point where the heavy string is attached to a lighter string.

As we saw in Chapter 16, interference of waves from two sources is not observed unless the sources are coherent. Because a light beam is usually the result of millions of atoms radiating independently, the phase difference between the waves from such sources fluctuates randomly many times per second, so two light sources are usually not coherent. Coherence in optics is often achieved by splitting the light beam from a single source into two or more beams that can then be combined to produce an interference pattern. The light beam can be split by reflecting the light from the two closely spaced surfaces of a thin film (Section 35-2), by diffracting the beam through two small openings or slits in an opaque barrier (Section 35-3), or by using a single point source and its image in a plane mirror for the two sources (Section 35-3). Today, lasers are the most important sources of coherent light in the laboratory.

35-2 Interference in Thin Films

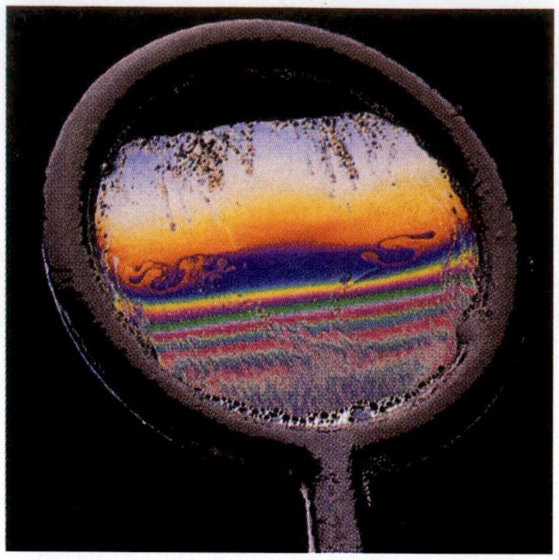

Interference of light from the front and back surface of a thin soap film. At the top where the film is very thin, the rays from the front surface of the film (which undergo a 180° phase change) and the rays from the back surface of the film (which do not change phase) interfere destructively and the film appears dark. At other parts of the film, the interference is destructive or constructive depending on the wavelength and on the thickness of the film.

You have probably noticed the colored bands in a soap bubble or in the film on the surface of oily water. These bands are due to the interference of light reflected from the top and bottom surfaces of the film. The different colors arise because of variations in the thickness of the film, causing interference for different wavelengths at different points.

Consider a thin film of water (such as a small section of a soap bubble) of uniform thickness viewed at small angles with the normal as shown in Figure 35-1. Part of the light is reflected from the upper, air–water interface where it undergoes a 180° phase change. Some of the light enters the film and is partially reflected by the bottom water–air interface. There is no phase change in this reflected light. If the light is nearly perpendicular to the surfaces, both the ray reflected from the top surface and the one reflected from the bottom surface can enter the eye at point P in the figure. The path difference between these two rays is $2t$, where t is the thickness of the film. This path difference produces a phase difference of $(2t/\lambda')360°$, where $\lambda' = \lambda/n$ is the wavelength of the light in the film, and n is the index of refraction of the film. The total phase difference between these two rays is thus 180° plus that due to the path difference. Destructive interference occurs when the path difference $2t$ is zero or a whole number of wavelengths λ' (in the film). Constructive interference occurs when the path difference is an odd number of half-wavelengths.

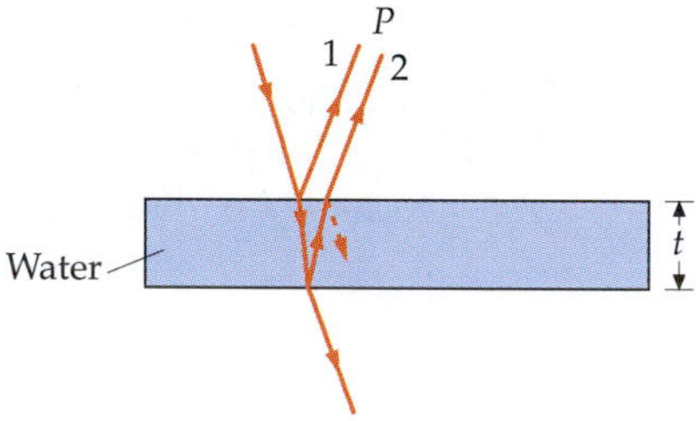

Figure 35-1 Light rays reflected from the top and bottom surfaces of a thin film are coherent because both rays come from the same source. If the light is incident nearly normally, the two reflected rays will be very close to each other and will produce interference.

When a thin water film lies on a glass surface as in Figure 35-2, the ray that reflects from the lower, water–glass interface also undergoes a 180° phase change because the index of refraction of glass (about 1.5) is greater than that of water (about 1.33). Thus, both the rays shown in the figure have undergone a 180° phase change upon reflection. The phase difference between these rays is due solely to the path difference and is given by $\delta = (2t/\lambda')360°$.

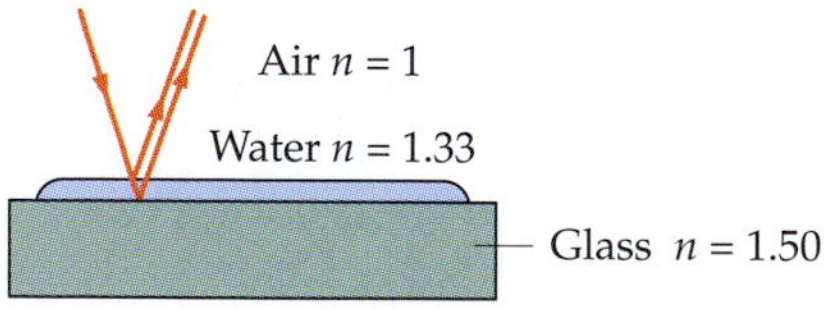

Figure 35-2 Interference of light reflected from a thin film of water resting on a glass surface. In this case, both rays undergo a change in phase of 180° upon reflection.

When a thin film of varying thickness is viewed with monochromatic light, such as the yellow light from a sodium lamp, alternating bright and dark bands or lines called **fringes** are observed. The distance between a bright fringe and a dark fringe is that distance over which the film's thickness changes such that the path difference $2t$ is $\lambda'/2$. Figure 35-3a shows the interference pattern observed when light is reflected from an air film between a spherical glass surface and a plane glass surface in contact. These circular interference fringes are known as **Newton's rings.** Typical rays re-

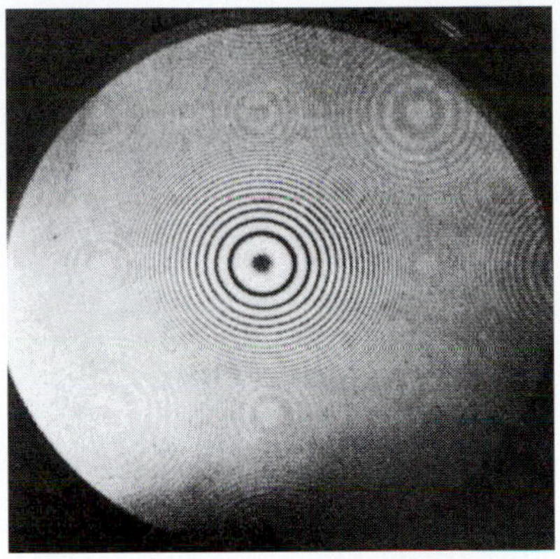

(a)

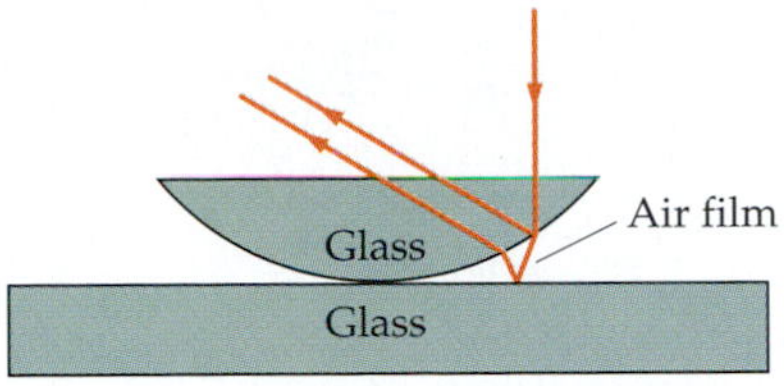

(b)

Figure 35-3 (*a*) Newton's rings observed with light reflected from a thin film of air between a plane glass surface and a spherical glass surface. At the center, the thickness of the air film is negligible and the interference is destructive because of the phase change of one of the rays. (*b*) Glass surfaces for the observation of Newton's rings shown in part (*a*). The thin film in this case is the film of air between the glass surfaces.

flected at the top and bottom of the air film are shown in Figure 35-3*b*. Near the point of contact of the surfaces, where the path difference between the ray reflected from the upper glass–air interface and the ray reflected from the lower air–glass interface is essentially zero or is at least small compared with the wavelength of light, the interference is perfectly destructive because of the 180° phase shift of the ray reflected from the lower air–glass interface. This central region in Figure 35-3*a* is therefore dark. The first bright fringe occurs at the radius at which the path difference is $\lambda/2$, which contributes a phase difference of 180°. This adds to the phase shift due to reflection to produce a total phase difference of 360°, which is equivalent to a zero phase difference. The second dark region occurs at the radius at which the path difference is λ, and so on.

Example 35-2

A wedge-shaped film of air is made by placing a small slip of paper between the edges of two flat pieces of glass as shown in Figure 35-4. Light of wavelength 500 nm is incident normally on the glass, and interference fringes are observed by reflection. If the angle θ made by the plates is 3×10^{-4} rad, how many interference fringes per centimeter are observed?

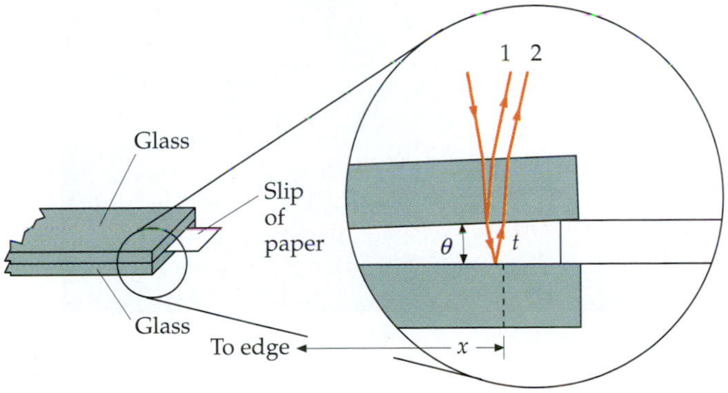

Figure 35-4 The angle θ, which is less than 0.02°, is exaggerated. The incoming and outgoing rays are essentially perpendicular to all air–glass interfaces.

Picture the Problem We find the number of fringes per centimeter by finding the horizontal distance x to the mth fringe and solving for m/x. Because the ray reflected from the bottom plate undergoes a 180° phase shift, the point of contact (where the path difference is zero) will be dark. The first dark fringe after this point occus when $2t = \lambda'$, where $\lambda' = \lambda$ is the wavelength in the air film, and t is the plate separation at x as shown in Figure 35-4. Since the angle q is small, we can use the small-angle approximation $\theta \approx t/x$.

1. The mth fringe occurs when the path difference $2t$ equals m wavelengths:

$$2t = m\lambda' = m\lambda$$

$$m = \frac{2t}{\lambda}$$

2. The thickness t is related to the angle θ:

$$\theta = \frac{t}{x}$$

3. Substitute $t = x\theta$ into the equation for m:

$$m = \frac{2x\theta}{\lambda}$$

4. Calculate m/x:

$$\frac{m}{x} = \frac{2\theta}{\lambda} = \frac{2(3 \times 10^{-4})}{5 \times 10^{-7}\ \text{m}} = 1200\ \text{m}^{-1} = 12\ \text{cm}^{-1}$$

Remarks We therefore observe 12 dark fringes per centimeter. In practice, the number of fringes per centimeter, which is easy to count, can be used to determine the angle. Note that if the angle of the wedge is increased, the fringes become more closely spaced.

Exercise How many fringes per centimeter are observed if light of wavelength 650 nm is used? (*Answer* $9.2\ \text{cm}^{-1}$)

Figure 35-5*a* shows interference fringes produced by a wedge-shaped air film between two flat glass plates as in Example 35-2. Plates that produce straight fringes such as those in Figure 35-5*a* are said to be **optically flat.** A similar wedge-shaped air film formed by two ordinary glass plates yields the irregular fringe pattern in Figure 35-5*b*, which indicates that these plates are not optically flat.

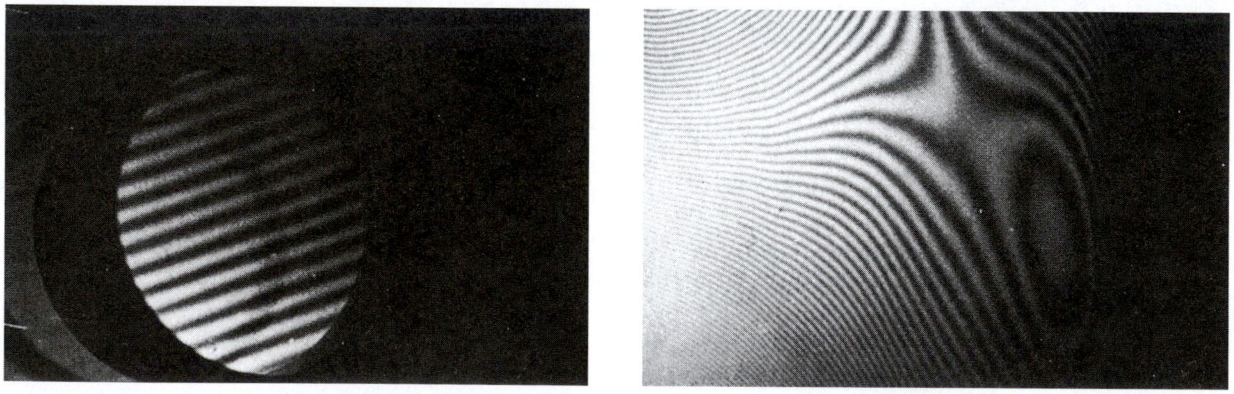

(*a*) (*b*)

Figure 35-5 (*a*) Straight-line fringes from a wedge-shaped film of air like that in Figure 35-4. The straightness of the fringes indicates that the glass plates are optically flat. (*b*) Fringes from a wedge-shaped film of air between glass plates that are not optically flat.

One application of interference effects in thin films is in nonreflecting lenses, which are made by covering a lens with a thin film of a material that has an index of refraction of about 1.22, which is between that of glass and air. Then the intensities of the light reflected from the top and bottom surfaces of the film are approximately equal, and since both rays undergo a 180° phase change, there is no phase difference between the rays due to reflection. The thickness of the film is chosen to be $\lambda'/4 = \lambda/4n$ where λ is in the middle of the visible spectrum, so that there is a phase change of 180° due to the path difference of $\lambda'/2$. Reflection from the coated surface is thus minimized.

35-3 The Two-Slit Interference Pattern

Interference patterns of light from two or more sources can be observed only if the sources are coherent. The interference in thin films discussed previously can be observed because the two beams come from the same light source but are separated by reflection. In Young's famous experiment, in which he demonstrated the wave nature of light, two coherent light sources are produced by illuminating two very narrow parallel slits with a single light source. We saw in Chapter 15 that when a wave encounters a barrier with a very small opening, the opening acts as a point source of waves (Figure 35-6). In Young's experiment, diffraction causes each slit to act as a line source (which is equivalent to a point source in two dimensions). The interference pattern is observed on a screen far from the slits (Figure 35-7*a*). At very large distances from the slits, the lines from the two slits to some point P on the screen are approximately parallel, and the path difference is approximately $d \sin\theta$, where d is the separation of the slits as shown in Figure 35-7*b*. We thus have interference maxima at an angle given by

$$d \sin\theta = m\lambda, \qquad m = 0, 1, 2, \ldots \qquad \text{35-2}$$

Two-slit interference maxima

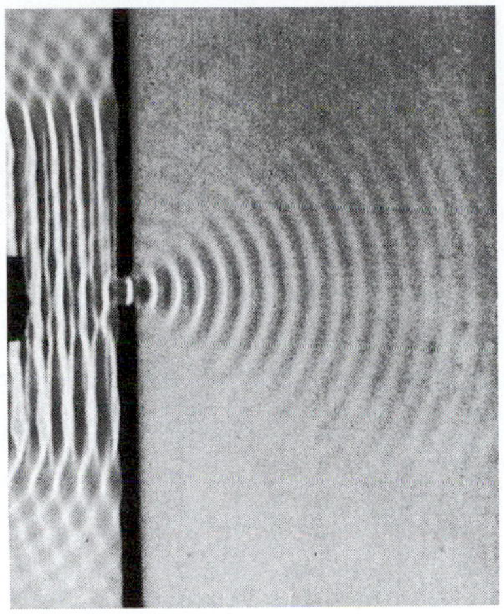

Figure 35-6 Plane water waves in a ripple tank encountering a barrier with a small opening. The waves to the right of the barrier are circular waves that are concentric about the opening just as if there were a point source at the opening.

where m is called the **order number**. The interference minima occur at

$$d \sin\theta = (m + \tfrac{1}{2})\lambda, \qquad m = 0, 1, 2, \ldots \qquad \text{35-3}$$

Two-slit interference minima

The phase difference δ at a point P is

$$\delta = \frac{2\pi}{\lambda} d \sin\theta \qquad \text{35-4}$$

We can relate the distance y_m measured along the screen from the central point to the mth bright fringe (see Figure 35-7*b*) to the distance L from the slits to the screen:

$$\tan\theta = \frac{y_m}{L}$$

For small angles (which is nearly always the case), $\tan\theta \approx \sin\theta$. Substituting y_m/L for $\sin\theta$ in Equation 35-2 and solving for y_m gives

$$y_m = m\frac{\lambda L}{d} \qquad \text{35-5}$$

Distance on screen to the mth bright fringe

From this result we see that the fringes are equally spaced on the screen.

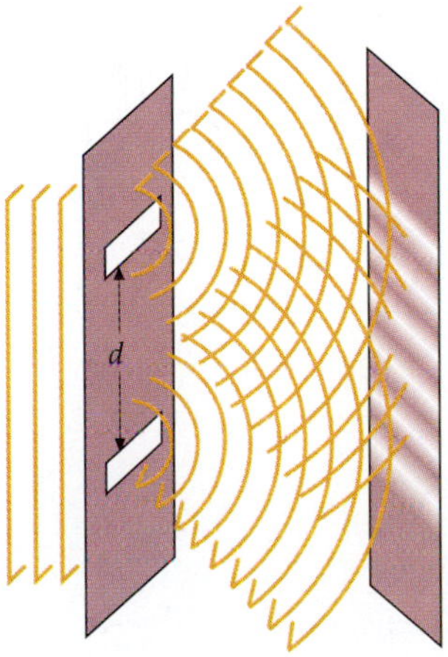

(*a*)

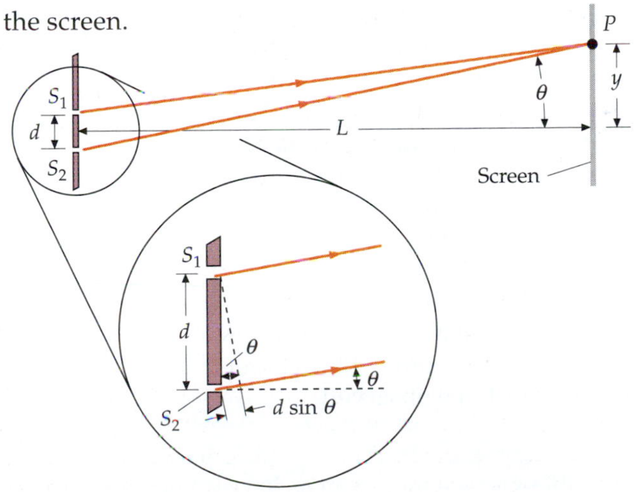

(*b*)

Figure 35-7 (*a*) Two slits act as coherent sources of light for the observation of interference in Young's experiment. Cylindrical waves from the slits overlap and produce an interference pattern on a screen. (*b*) Geometry for relating the distance y measured along the screen to L and θ. When the screen is very far away compared with the slit separation, the rays from the slits to a point on the screen are approximately parallel, and the path difference between the two rays is $d \sin\theta$.

Calculation of Intensity

To calculate the intensity of the light on the screen at a general point P, we need to add two harmonic wave functions that differ in phase.* The wave functions for electromagnetic waves are the electric field vectors. Let E_1 be the electric field at some point P on the screen due to the waves from slit 1, and let E_2 be the electric field at that point due to waves from slit 2. Since the angles of interest are small, we can assume that these fields are parallel and we therefore consider only their magnitudes. Both electric fields oscillate with the same frequency (they result from a single source that illuminates both slits) and they have the same amplitude. (The path difference is only of the order of a few wavelengths of light at most.) They have a phase difference δ given by Equation 35-4. If we represent these wave functions by

$$E_1 = A_0 \sin \omega t$$

and

$$E_2 = A_0 \sin (\omega t + \delta)$$

the resultant wave function is

$$E = E_1 + E_2 = A_0 \sin \omega t + A_0 \sin (\omega t + \delta)$$

$$= 2A_0 \cos \tfrac{1}{2}\delta \sin (\omega t + \tfrac{1}{2}\delta) \qquad \text{35-6}$$

where we used

$$\sin \alpha + \sin \beta = 2 \cos \tfrac{1}{2}(\alpha - \beta) \sin \tfrac{1}{2}(\alpha + \beta) \qquad \text{35-7}$$

The amplitude of the resultant wave is thus $2A_0 \cos \frac{1}{2}\delta$. It has its maximum value of $2A_0$ when the waves are in phase and is zero when they are 180° out of phase. Since the intensity is proportional to the square of the amplitude, the intensity at any point P is

$$I = 4I_0 \cos^2 \tfrac{1}{2}\delta \qquad \text{35-8}$$

Intensity in terms of phase difference

where I_0 is the intensity of the light on the screen from either slit separately. The phase angle δ is related to the position on the screen by Equation 35-4.

Figure 35-8*a* shows the intensity pattern as seen on a screen. A graph of the intensity as a function of $\sin \theta$ is shown in Figure 35-8*b*. For small θ, this is equivalent to a plot of intensity versus y since $y \approx L \sin \theta$. The intensity I_0 is that from each slit separately. The dashed line in Figure 35-8*b* shows the average intensity $2I_0$, which is the result of averaging over many interference maxima and minima. This is the intensity that would arise from the two sources if they acted independently without interference, that is, if they were not coherent. Then total phase difference between them would fluctuate randomly so that only the average intensity would be observed.

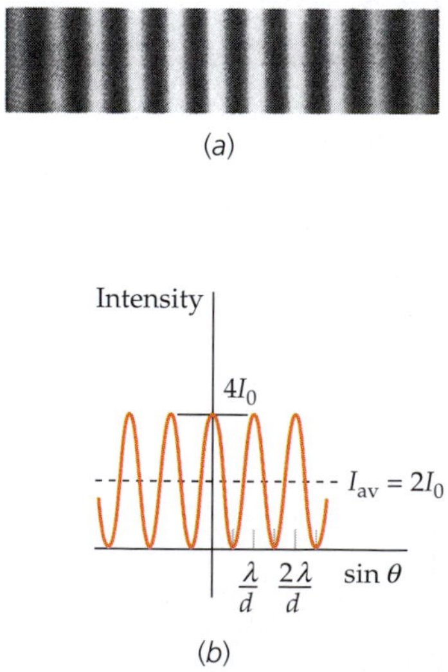

Figure 35-8 (*a*) The interference pattern observed on a screen far away from the two slits of Figure 35-7. (*b*) Plot of intensity versus $\sin \theta$. The maximum intensity is $4I_0$, where I_0 is the intensity due to each slit separately. The average intensity (dashed line) is $2I_0$.

* We did this in Chapter 16 where we discussed the general superposition of two waves.

Figure 35-9 shows another method of producing the two-slit interference pattern, an arrangement known as **Lloyd's mirror.** A single slit is placed at a distance $\frac{1}{2}d$ above the plane of a mirror. Light striking the screen directly from the source interferes with that reflected from the mirror. The reflected light can be considered to come from the virtual image of the slit formed by the mirror. Because of the 180° change in phase upon reflection at the mirror, the interference pattern is that of two coherent line sources that differ in phase by 180°. The pattern is the same as that shown in Figure 35-8 for two slits except that the maxima and minima are interchanged. The central fringe just above the mirror at a point equidistant from the sources is dark. Constructive interference occurs at points for which the path difference is a half-wavelength or any odd number of half-wavelengths. At these points, the 180° phase difference due to the path difference combines with the 180° phase difference of the sources to produce constructive interference.

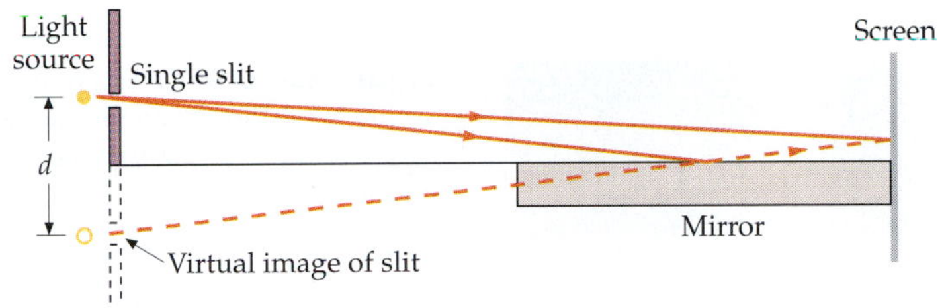

Figure 35-9 Lloyd's mirror for producing a two-slit interference pattern. The two sources (the slit and its image) are coherent and are 180° out of phase. The central interference band at the point equidistant from the sources is dark.

Example 35-3 *try it yourself*

Two narrow slits separated by 1.5 mm are illuminated by yellow light of wavelength 589 nm from a sodium lamp. Find the spacing of the fringes observed on a screen 3 m away.

Picture the Problem The distance y_m measured along the screen to the mth bright fringe is given by Equation 35-5, with $L = 3$ m, $d = 1.5$ mm, and $\lambda = 589$ nm. The spacing of the fringes is y_m/m, the distance divided by the number of fringes.

Cover the column to the right and try these on your own before looking at the answers.

Steps	***Answers***
1. Solve Equation 35-5 for y_m/m.	$\frac{y_m}{m} = \lambda\frac{L}{d}$
2. Substitute the given values.	$\frac{y_m}{m} = 1.18\text{ mm}$

Exercise A point source of light ($\lambda = 589$ nm) is placed 0.4 mm above a mirror. Interference fringes are observed on a screen 6 m away. Find the spacing of the fringes. (*Answer* 4.42 mm)

35-4 Diffraction Pattern of a Single Slit

In our discussion of the interference patterns produced by two or more slits, we assumed that the slits were very narrow so that we could consider them to be line sources of cylindrical waves, which in our two-dimensional diagrams are point sources of circular waves. We could therefore assume that the intensity due to one slit acting alone was the same (I_0) at any point P on the screen independent of the angle θ made between the ray to point P and the normal line between the slit and the screen. When the slit is not narrow, the intensity on a screen far away is not independent of angle but decreases as the angle increases. Consider a slit of width a. Figure 35-10 shows the intensity pattern on a screen far away from the slit of width a as a function of $\sin\theta$. We can see that the intensity is maximum in the forward direction ($\sin\theta = 0$) and decreases to zero at an angle that depends on the slit width a and the wavelength λ.

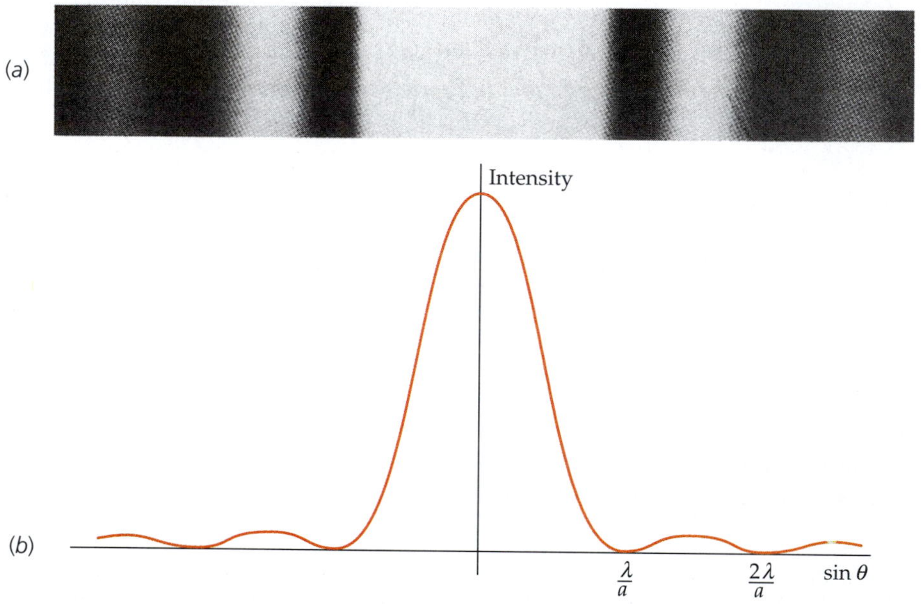

Figure 35-10 (*a*) Diffraction pattern of a single slit as observed on a screen far away. (*b*) Plot of intensity versus $\sin\theta$ for the pattern in (*a*).

Most of the light intensity is concentrated in the broad **central diffraction maximum,** though there are minor secondary maxima bands on either side of the central maximum. The first zeroes in the intensity occur at angles given by

$$\sin\theta = \frac{\lambda}{a} \qquad 35\text{-}9$$

Note that for a given wavelength λ, the width of the central maximum varies inversely with the width of the slit. If we *increase* the slit width a, the angle θ at which the intensity first becomes zero *decreases,* giving a more narrow central diffraction maximum. Conversely, if we *decrease* the slit width, the angle of the first zero *increases,* giving a wider central diffraction maximum. When a is very small, there are no points of zero intensity in the pattern, and the slit acts as a line source (a point source in two dimensions), radiating light energy essentially equally in all directions.

Multiplying both sides of Equation 35-9 by a gives

$$a \sin\theta = \lambda \qquad 35\text{-}10$$

The quantity $a \sin\theta$ is the path difference between a light ray leaving the top of the slit and one leaving the bottom of the slit. We see that the first diffraction *minimum* occurs when these two rays are in phase, that is, when their path difference equals 1 wavelength. We can understand this result by considering each point on a wavefront to be a point source of light in accordance with Huygens' principle. In Figure 35-11 we have placed a line of dots on the wavefront at the slit to represent these point sources schematically. Suppose, for example, that we have 100 such dots and that we look at an angle θ for which $a \sin\theta = \lambda$, that is, the angle for which the waves from the top and bottom of the slit are in phase. Let us consider the slit to be divided into two regions, with the first 50 sources in the first, upper region and sources 51 through 100 in the second, lower region. When the path difference between the top and bottom of the slit equals one wavelength, the path difference between source 1 (the first source in the upper region) and source 51 (the first source in the lower region) is $\frac{1}{2}\lambda$. The waves from these two sources will be out of phase by 180° and will thus cancel. Similarly, waves from the second source in each region (source 2 and source 52) will cancel. Continuing this argument, we can see that the waves from each pair of sources separated by $a/2$ will cancel. Thus, there will be no light energy at this angle. We can extend this argument to the second and third minima in the diffraction pattern of Figure 35-10. At an angle such that $a \sin\theta = 2\lambda$, we can divide the slit into four regions, two for the top half and two for the bottom half. Using this same argument, the light intensity from the top half is zero because of the cancellation of pairs of sources, and, similarly, the light intensity from the bottom half is zero. The general expression for the points of zero intensity in the diffraction pattern of a single slit is thus

$$a \sin\theta = m\lambda, \qquad m = 1, 2, 3, \ldots \tag{35-11}$$

Points of zero intensity for a single-slit diffraction pattern

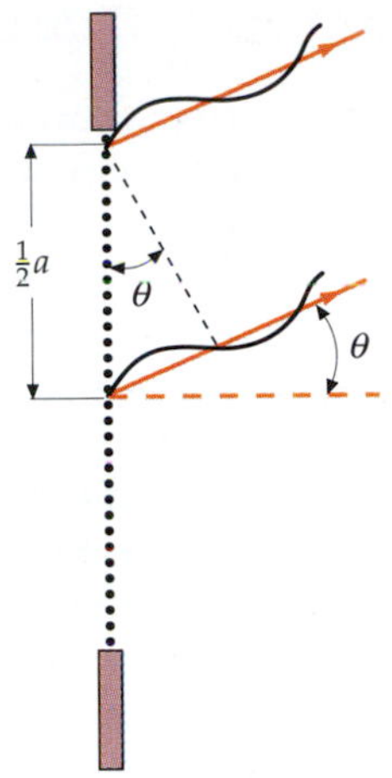

Figure 35-11 A single slit is represented by a large number of point sources of equal amplitude. At the first diffraction minimum of a single slit, the waves from the source near the top and those from the source just below the middle of the slit are 180° out of phase and cancel, as do all other pairs of sources.

Usually, we are just interested in the first occurrence of a minimum in the light intensity because nearly all of the light energy is contained in the central diffraction maximum.

In Figure 35-12, the distance y from the central maximum to the first diffraction minimum is related to the angle θ and the distance L from the slit to the screen by

$$\tan\theta = \frac{y}{L}$$

Since this angle is very small, $\tan\theta \approx \sin\theta$. Then, according to Equation 35-11, we have $\sin\theta = \lambda/a \approx y/L$, or

$$y = \frac{L\lambda}{a} \tag{35-12}$$

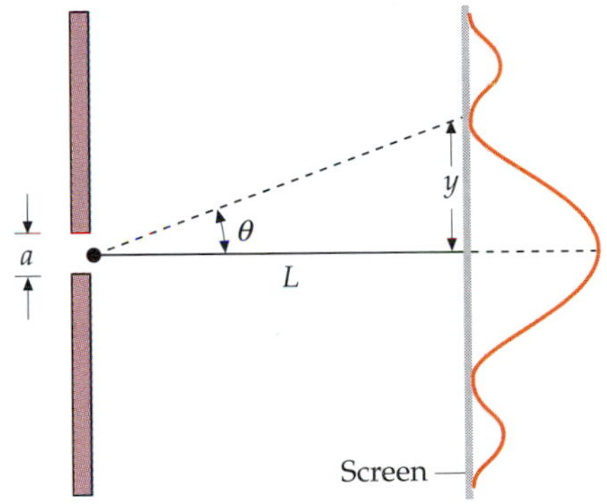

Figure 35-12 The distance y measured along the screen from the central maximum to the first diffraction minimum is related to the angle θ by $\tan\theta = y/L$, where L is the distance to the screen. Since the angle is very small, $\tan\theta \approx \sin\theta$. Then $y = L\tan\theta \approx L\sin\theta = L\lambda/a$.

Example 35-4

In a lecture demonstration of single-slit diffraction, a laser beam of wavelength 700 nm passes through a vertical slit 0.2 mm wide and hits a screen 6 m away. Find the width of the central diffraction maximum on the screen, that is, the distance between the first minimum on the left and the first minimum on the right of the central maximum.

Picture the Problem Referring to Figure 35-12, the width of the central diffraction maximum is $2y$.

Substitute the given data to calculate $2y$ from Equation 35-12:

$$2y = \frac{2L\lambda}{a} = \frac{2(6\text{ m})(700 \times 10^{-9}\text{ m})}{0.0002\text{ m}} = 4.2 \times 10^{-2}\text{ m} = 4.2\text{ cm}$$

Interference–Diffraction Pattern of Two Slits

When there are two or more slits, the intensity pattern on a screen far away is a combination of the single-slit diffraction pattern and the multiple-slit interference pattern we have studied. Figure 35-13 shows the intensity pattern on a screen far from two slits whose separation d is 10 times the width a of each slit. The pattern is the same as the two-slit pattern with very narrow slits (Figure 35-10) except that it is modulated by the single-slit diffraction pattern; that is, the intensity due to each slit separately is now not constant but decreases with angle as shown in Figure 35-13b.

Note that in Figure 35-13 the central diffraction maximum contains 19 interference maxima—the central interference maximum and 9 maxima on either side. The tenth interference maximum on either side of the central one is at the angle θ given by $\sin\theta = 10\lambda/d = \lambda/a$ since $d = 10a$. This coincides with the position of the first diffraction minimum, so this interference maximum is not seen. At these points, the light from the two slits would be in phase and would interfere constructively, but there is no light from either slit because the points are diffraction minima. We can see in general that if $m = d/a$, the mth interference maximum will fall at the first diffraction minimum. Since the mth fringe is not seen, there will be $m - 1$ fringes on each side of the central fringe for a total of N fringes in the central maximum, where N is given by

$$N = 2(m - 1) + 1 = 2m - 1 \qquad 35\text{-}13$$

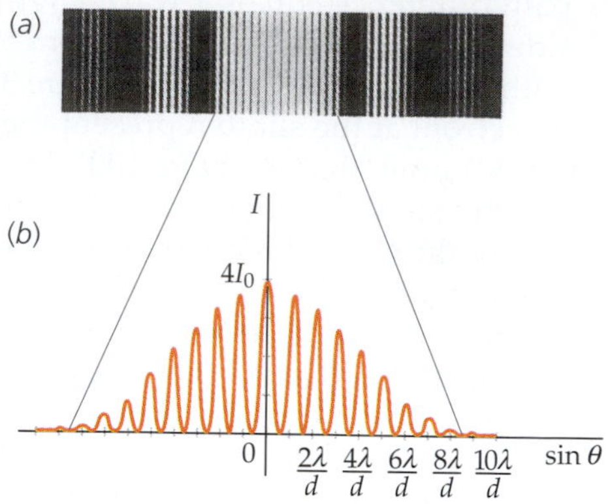

Figure 35-13 (a) Interference–diffraction pattern for two slits whose separation d is equal to 10 times their width a. The tenth interference maximum on either side of the central interference maximum is missing because it falls at the first diffraction minimum. (b) Plot of intensity versus $\sin\theta$ for the central band of the pattern in (a).

Example 35-5

Two slits of width $a = 0.015$ mm are separated by a distance $d = 0.06$ mm and illuminated by light of wavelength $\lambda = 650$ nm. How many bright fringes are seen in the central diffraction maximum?

Picture the Problem We need to find the value of m for which the mth interference maximum coincides with the first diffraction minimum. Then there will be $N = 2m - 1$ fringes in the central maximum.

1. Find the angle θ_1 of the first diffraction minimum: $\sin\theta_1 = \dfrac{\lambda}{a}$ (first diffraction minimum)

2. Find the angle θ_m of the mth interference maxima: $\sin\theta_m = \dfrac{m\lambda}{d}$ (mth interference maxima)

3. Set these angles equal and solve for m: $\dfrac{m\lambda}{d} = \dfrac{\lambda}{a}$

$$m = \frac{d}{a} = \frac{0.06\text{ mm}}{0.015\text{ mm}} = 4$$

4. Use $N = 2m - 1$ to find the number of bright fringes in the central maximum: $N = 2m - 1 = 2(4) - 1 = 7$ fringes

Remarks The first diffraction minimum coincides with the fourth bright fringe. Therefore, there are 3 bright fringes visible on either side of the central diffraction maximum. These six maxima, plus the central interference maximum, combine for a total of 7 bright fringes in the central diffraction maximum.

35-5 Using Phasors to Add Harmonic Waves

To calculate the interference pattern produced by three, four, or more coherent light sources and to calculate the diffraction pattern of a single slit, we need to combine several harmonic waves of the same frequency that differ in phase. A simple geometric interpretation of harmonic wave functions leads to a method of adding harmonic waves of the same frequency by geometric construction.

Let the wave functions for two waves at some point be $E_1 = A_1 \sin \alpha$ and $E_2 = A_2 \sin (\alpha + \delta)$, where $\alpha = \omega t$. Our problem is then to find the sum

$$E_1 + E_2 = A_1 \sin \alpha + A_2 \sin (\alpha + \delta)$$

We can represent each wave function by a two-dimensional vector as shown in Figure 35-14. The geometric method of addition is based on the fact that the y (or x) component of the resultant of two vectors equals the sum of the y (or x) components of the vectors as illustrated in the figure. The wave function y_1 is represented by the vector $\vec{A}_1$. As the time varies, this vector rotates in the xy plane with angular frequency ω. Such a vector is called a **phasor.** (We encountered phasors in our study of ac circuits in Section 31-4.) The wave function E_2 is the y component of a phasor of magnitude A_2 that makes an angle $\alpha + \delta$ with the x axis. By the laws of vector addition, the sum of these components equals the y component of the resultant phasor $\vec{A}$, as shown in Figure 35-14. The y component of the resultant phasor, $A \sin (\alpha + \delta')$, is a harmonic wave function that is the sum of the two original wave functions:

$$A_1 \sin \alpha + A_2 \sin (\alpha + \delta) = A \sin (\alpha + \delta') \qquad 35\text{-}14$$

where A (the amplitude of the resultant wave) and δ' (the phase of the resultant wave relative to the first wave) are found by adding the phasors representing the waves. As time varies, α varies. The phasors representing the two wave functions and the resultant phasor representing the resultant wave function rotate in space, but their relative positions do not change because they all rotate with the same angular velocity ω.

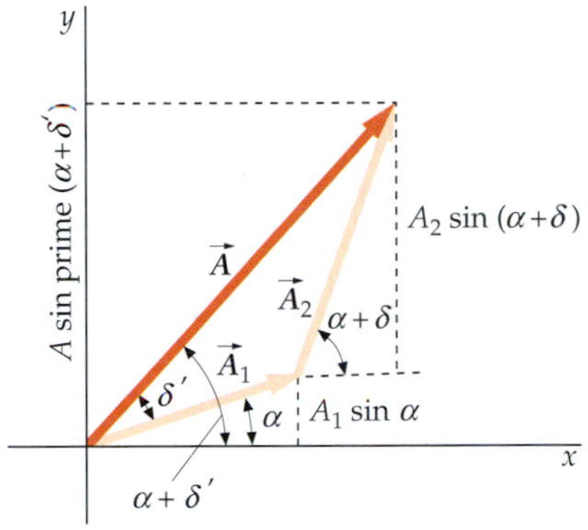

Figure 35-14 Phasor representation of wave functions.

Example 35-6 *try it yourself*

Use the phasor method of addition to derive Equation 35-14 for the superposition of two waves of the same amplitude.

Picture the Problem Represent the waves $y_1 = A_0 \sin \alpha$ and $y_2 = A_0 \sin (\alpha + \delta)$ by phasors of length A_0 making an angle δ with one another. The resultant wave $y_r = A \sin (\alpha + \delta')$ is represented by the sum of these vectors, which form an isosceles triangle as shown in Figure 35-15.

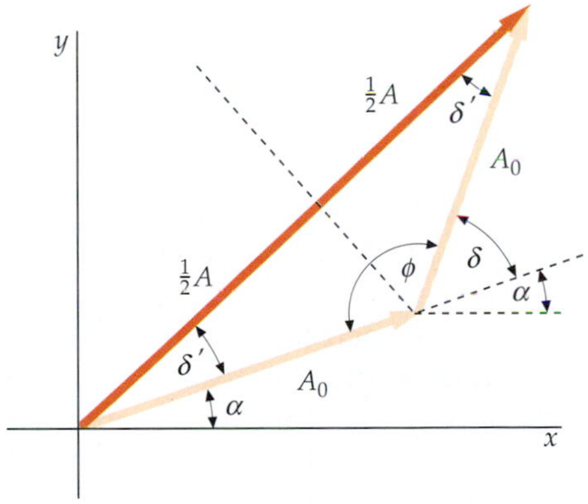

Figure 35-15

Cover the column to the right and try these on your own before looking at the answers.

Steps	Answers
1. Find the phase angle δ' in terms of ϕ from the fact that the three angles in the triangle must sum to 180°:	$\delta' + \delta' + \phi = 180°$
2. Relate ϕ to δ:	$\delta + \phi = 180°$

3. Eliminate ϕ and solve for δ'. $\qquad \delta' = \frac{1}{2}\delta$

4. Write $\cos \delta'$ in terms of A and A_0. $\qquad \cos \delta' = \dfrac{\frac{1}{2}A}{A_0}$

5. Solve for A in terms of δ. $\qquad A = 2A_0 \cos \delta' = 2A_0 \cos \frac{1}{2}\delta$

6. Use your results for A and δ' to write the resultant wave function. $\qquad y_r = A \sin (\alpha + \delta') = (2A_0 \cos \frac{1}{2}\delta) \sin (\alpha + \frac{1}{2}\delta)$

Exercise Find the resultant of the two waves $E_1 = 4 \sin (\omega t)$ and $E_2 = 3 \sin (\omega t + 90°)$ (*Answer* $E_1 + E_2 = 5 \sin (\omega t + 37°)$)

The Interference Pattern of Three or More Equally Spaced Sources

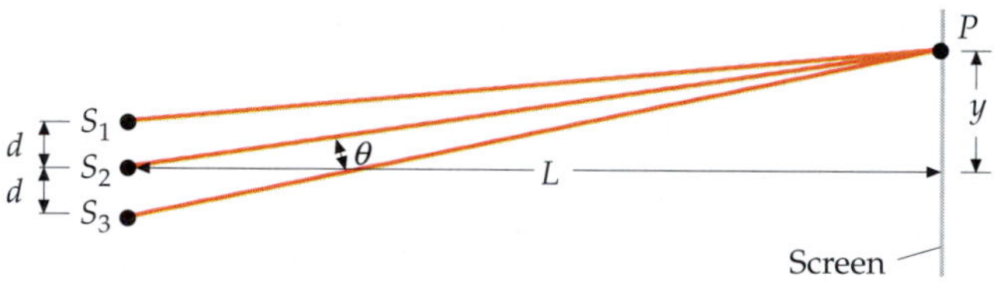

Figure 35-16 Geometry for calculating the intensity pattern far from three equally spaced sources that are in phase.

We can apply the phasor method of addition to calculate the interference pattern of three or more equally spaced, coherent sources in phase. We are most interested in the interference maxima and minima. Figure 35-16 illustrates the case of three sources. The geometry is the same as for two sources. At a great distance from the sources, the rays from the sources to a point P on the screen are approximately parallel. The path difference between the first and second source is then $d \sin \theta$, as before, and that between the first and third source is $2d \sin \theta$. The wave at point P is the sum of three waves. Let $\alpha = \omega t$ be the phase of the first wave at point P. We thus have the problem of adding three waves of the form

$$E_1 = A_0 \sin \alpha$$
$$E_2 = A_0 \sin (\alpha + \delta) \qquad \text{35-15}$$
$$E_3 = A_0 \sin (\alpha + 2\delta)$$

where

$$\delta = \frac{2\pi}{\lambda} d \sin \theta \approx \frac{2\pi}{\lambda}\frac{yd}{L} \qquad \text{35-16}$$

as in the two-slit problem.

At $\theta = 0$, $\delta = 0$, so all the waves are in phase. The amplitude of the resultant wave is 3 times that of each individual wave and the intensity is 9 times that due to each source acting separately. As the angle δ increases from $\theta = 0$, the phase angle δ increases and the intensity decreases. The position $\theta = 0$ is thus a position of maximum intensity.

Figure 35-17 shows the phasor addition of three waves for a phase angle δ of about 30°. (This corresponds to a point P on the screen for which θ is given by $\sin \theta = \lambda\delta/2\pi d = \lambda/12d$.) The resultant amplitude A is considerably less than 3 times that of each source. As the phase angle δ increases, the resultant amplitude decreases until the amplitude is zero at $\delta = 120°$. For this phase difference, the three phasors form an equilateral triangle (Figure 35-18). This first interference minimum for three sources occurs at a smaller phase angle (and therefore at a smaller space angle θ) than it does for only two sources (for which the first minimum occurs at $\delta = 180°$). As δ increases from 120°, the resultant amplitude increases, reaching a secondary maximum near $\delta =$

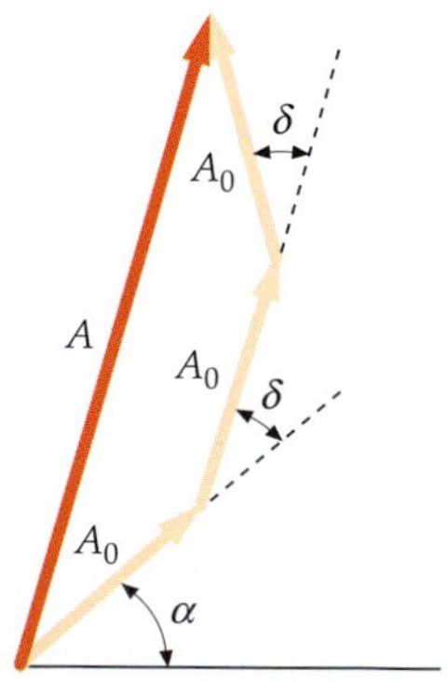

Figure 35-17 Phasor diagram for determining the resultant amplitude A due to three waves, each of amplitude A_0, that have phase differences of δ and 2δ due to path differences of $d \sin \theta$ and $2d \sin \theta$. The angle $\alpha = \omega t$ varies with time, but this does not affect the calculation of A.

180°. At the phase angle $\delta = 180°$, the amplitude is the same as that from a single source since the waves from the first two sources cancel each other, leaving only the third. The intensity of the secondary maximum is one-ninth that of the maximum at $\theta = 0$. As δ increases beyond 180°, the amplitude again decreases and is zero at $\delta = 180° + 60° = 240°$. For δ greater than 240°, the amplitude increases and is again 3 times that of each source when $\delta = 360°$. This phase angle corresponds to a path difference of 1 wavelength for the waves from the first two sources and 2 wavelengths for the waves from the first and third sources. Hence, the three waves are in phase at this point. The largest maxima, called the principal maxima, are at the same positions as for just two sources, which are those points corresponding to the angles θ given by

$$d \sin \theta = m\lambda, \qquad m = 0, 1, 2, \ldots \tag{35-17}$$

These maxima are stronger and narrower than those for two sources. They occur at points for which the path difference between adjacent sources is zero or an integral number of wavelengths.

These results can be generalized to more than three sources. For four equally spaced sources that are in phase, the principal interference maxima are again given by Equation 35-17, but these maxima are even more intense, they are narrower, and there are two small secondary maxima between each pair of principal maxima. At $\theta = 0$, the intensity is 16 times that due to a single source. The first interference minimum occurs when δ is 90°, as can be seen from the phasor diagram of Figure 35-19. The first secondary maximum is near $\delta = 120°$, where the waves from three of the sources cancel, leaving only the wave from the fourth source. The intensity of the secondary maximum is approximately one-sixteenth that of the central maximum. There is another minimum at $\delta = 180°$, another secondary maximum near $\delta = 240°$, and another minimum at $\delta = 270°$ before the next principal maximum at $\delta = 360°$.

Figure 35-20 shows the intensity patterns for two, three, and four equally spaced sources. Figure 35-21 shows a graph of I/I_0, where I_0 is the intensity due to each source acting separately. For three sources, there is a very small secondary maximum between each pair of principal maxima, and the principal maxima are sharper and more intense than those due to just two sources. For four sources, there are two small secondary maxima between each pair of principal maxima, and the principal maxima are even more narrow and intense.

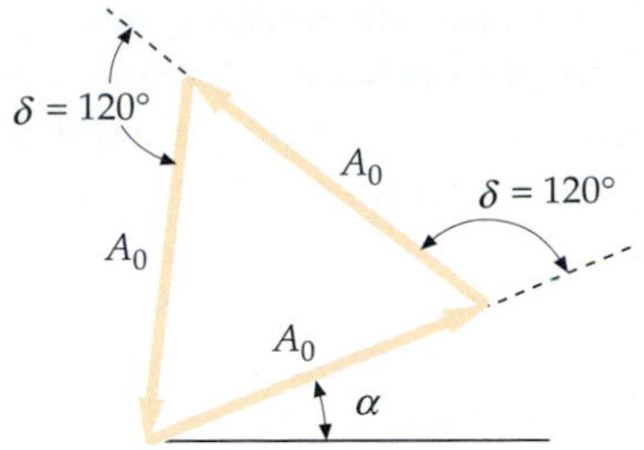

Figure 35-18 The resultant amplitude for the waves from three sources is zero when δ is 120°. This interference minimum occurs at a smaller angle θ than does the first minimum for two sources, which occurs when δ is 180°.

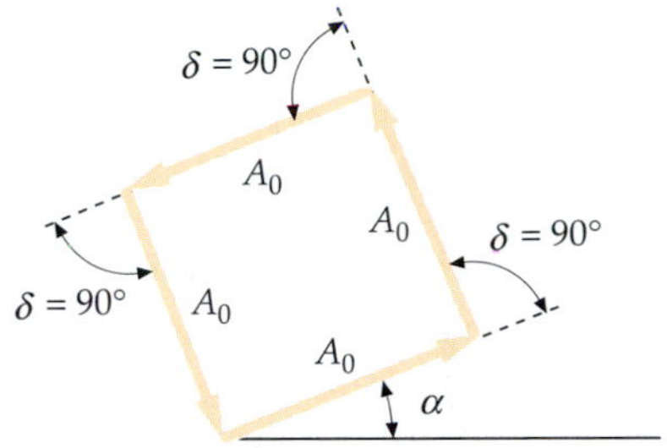

Figure 35-19 Phasor diagram for the first minimum for four equally spaced in-phase sources. The amplitude is zero when the phase difference of the waves from adjacent sources is 90°.

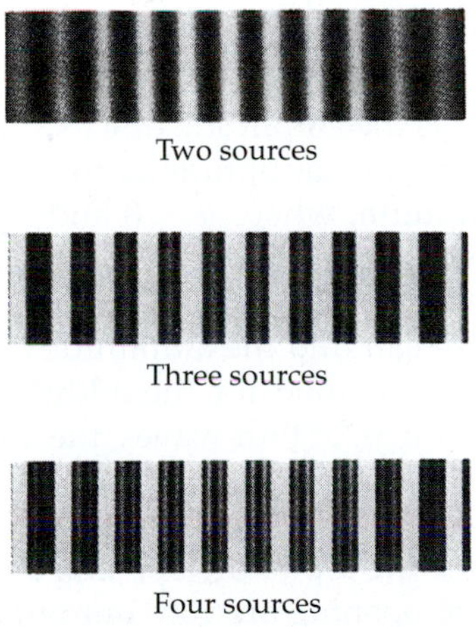

Figure 35-20 Intensity patterns for two, three, and four equally spaced coherent sources. There is a secondary maximum between each pair of principal maxima for three sources, and two secondary maxima for four sources.

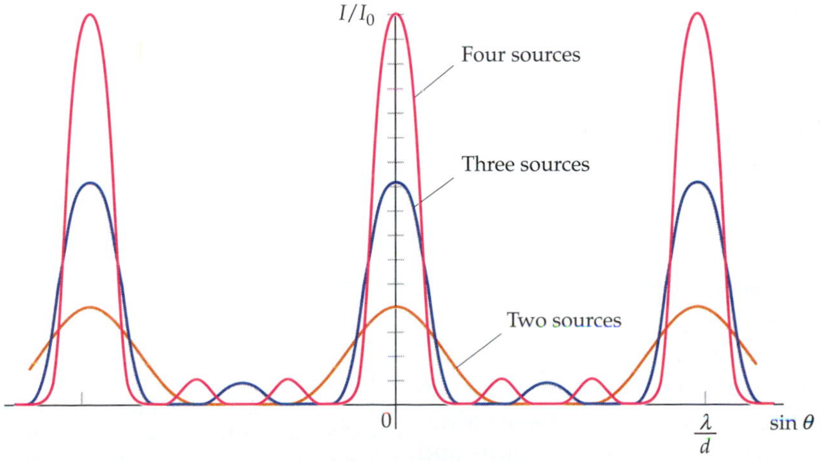

Figure 35-21 Plot of relative intensity versus $\sin \theta$ for two, three, and four equally spaced coherent sources.

From this discussion, we can see that as we increase the number of sources, the intensity becomes more and more concentrated in the principal maxima given by Equation 35-17, and these maxima become narrower. For N sources, the intensity of the principal maxima is N^2 times that due to a single source. The first minimum occurs at a phase angle of $\delta = 360°/N$, for which the N phasors form a closed polygon of N sides. There are $N - 2$ secondary maxima between each pair of principal maxima. These secondary maxima are very weak compared with the principal maxima. As the number of sources is increased, the principal maxima become sharper and more intense, and the intensities of the secondary maxima become negligible compared to those of the principal maxima.

Calculating the Single-Slit Diffraction Pattern

We now use the phasor method of addition of harmonic waves to calculate the intensity pattern shown in Figure 35-10. We assume that the slit of width a is divided into N equal intervals and that there is a point source of waves at the midpoint of each interval (Figure 35-22). If d is the distance between two adjacent sources and a is the width of the opening, we have $d = a/N$. Since the screen on which we are calculating the intensity is far from the sources, the rays from the sources to a point P on the screen are approximately parallel. The path difference between any two adjacent sources is $d \sin \theta$, and the phase difference is

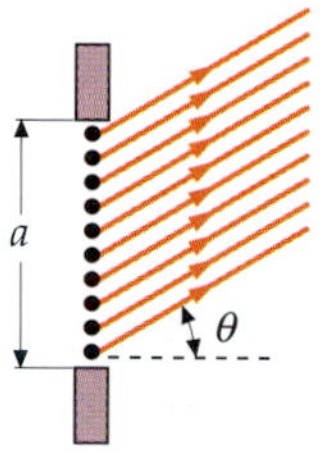

Figure 35-22 Diagram for calculating the diffraction pattern far away from a narrow slit. The slit width a is assumed to contain a large number of in-phase point sources separated by a distance d. The rays from these sources to a point far away are approximately parallel. The path difference for the waves from adjacent sources is $d \sin \theta$.

$$\delta = \frac{2\pi}{\lambda} d \sin \theta$$

If A_0 is the amplitude due to a single source, the amplitude at the central maximum, where $\theta = 0$ and all the waves are in phase, is $A_{max} = NA_0$ (Figure 35-23).

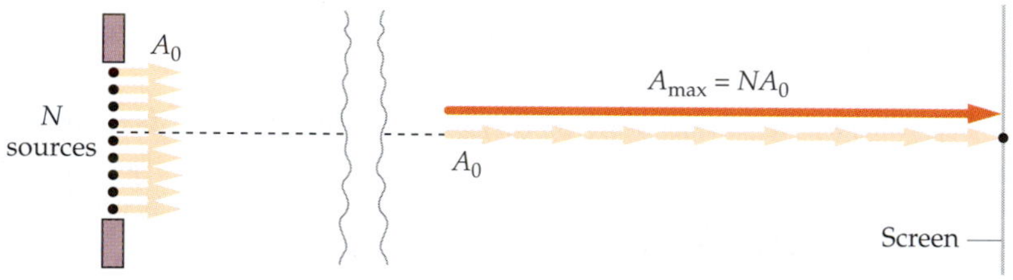

Figure 35-23 A single slit is represented by N sources, each of amplitude A_0. At the central maximum point at $\theta = 0$, the waves from the sources add in phase, giving a resultant amplitude $A_{max} = NA_0$.

We can find the amplitude at some other point at an angle θ by using the phasor method for the addition of harmonic waves. As in the addition of two, three, or four waves, the intensity is zero at any point where the phasors representing the waves form a closed polygon. In this case the polygon has N sides (Figure 35-24). At the first minimum, the wave from the first source just below the top of the opening and that from the source just below the middle of the opening are 180° out of phase. In this case, the waves from the source near the top of the opening differ from those from the bottom by nearly 360°. (The phase difference is, in fact, $360° - 360°/N$.) Thus, if the number of sources is very large, we get complete cancellation when the waves from the first and last sources are out of phase by 360°, corresponding to a path difference of 1 wavelength, in agreement with Equation 35-11.

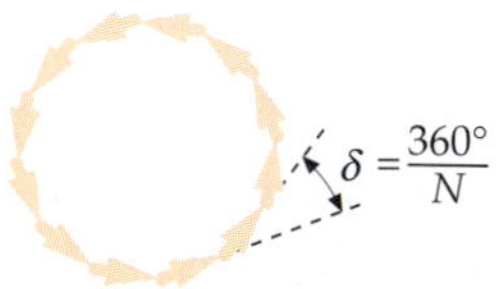

Figure 35-24 Phasor diagram for calculating the first minimum in the single-slit diffraction pattern. When the waves from the N sources completely cancel, the N phasors form a closed polygon. The phase difference between the waves from adjacent sources is then $\delta = 360°/N$. When N is very large, the waves from the first and last sources are approximately in phase.

We will now calculate the amplitude at a general point at which the waves from two adjacent sources differ in phase by δ. Figure 35-25 shows the phasor diagram for the addition of N waves where the subsequent waves differ in phase from the first wave by $\delta, 2\delta, \ldots, (N-1)\delta$. When N is very large and δ is very small, the phasor diagram approximates the arc of a circle. The resultant amplitude A is the length of the chord of this arc. We will calculate this resultant amplitude in terms of the phase difference ϕ between the first wave and the last wave. From Figure 35-25, we have

$$\sin \tfrac{1}{2}\phi = \frac{A/2}{r}$$

or

$$A = 2r \sin \tfrac{1}{2}\phi \qquad \text{35-18}$$

where r is the radius of the arc. Since the length of the arc is $A_{\max} = NA_0$ and the angle subtended is ϕ, we have

$$\phi = \frac{A_{\max}}{r} \qquad \text{35-19}$$

or

$$r = \frac{A_{\max}}{\phi}$$

Substituting this into Equation 35-18 gives

$$A = \frac{2A_{\max}}{\phi}\sin \tfrac{1}{2}\phi = A_{\max}\frac{\sin \frac{1}{2}\phi}{\frac{1}{2}\phi}$$

Since the amplitude at the center of the central maximum ($\theta = 0$) is $A_{\max}$, the ratio of the intensity at any other point to that at the center of the central maximum is

$$\frac{I}{I_0} = \frac{A^2}{A_{\max}^2} = \left(\frac{\sin \frac{1}{2}\phi}{\frac{1}{2}\phi}\right)^2$$

or

$$I = I_0\left(\frac{\sin \frac{1}{2}\phi}{\frac{1}{2}\phi}\right)^2 \qquad \text{35-20}$$

Intensity for a single-slit diffraction pattern

The phase difference ϕ between the first and last waves is $2\pi/\lambda$ times the path difference $a \sin \theta$ between the top and bottom of the opening:

$$\phi = \frac{2\pi}{\lambda} a \sin \theta \qquad \text{35-21}$$

Equations 35-20 and 35-21 describe the intensity pattern shown in Figure 35-10. The first minimum occurs at $a \sin \theta = \lambda$, the point where the waves from the top and bottom of the opening have a path difference of λ and are in phase. The second minimum occurs at $a \sin \theta = 2\lambda$, where the waves from the top and bottom of the opening have a path difference of 2λ.

There is a secondary maximum approximately midway between the first and second minima at $a \sin \theta \approx \frac{3}{2}\lambda$. Figure 35-26 shows the phasor diagram for determining the approximate intensity of this secondary maximum. The phase difference between the first and last waves is approximately $360° + 180°$. The phasors thus complete $1\frac{1}{2}$ circles. The resultant amplitude is the diameter of a circle with a circumference that is two-thirds the total length $A_{\max}$. If $C = \frac{2}{3}A_{\max}$ is the circumference, the diameter A is

$$A = \frac{C}{\pi} = \frac{\frac{2}{3}A_{\max}}{\pi} = \frac{2}{3\pi}A_{\max}$$

and

$$A^2 = \frac{4}{9\pi^2}A_{\max}^2$$

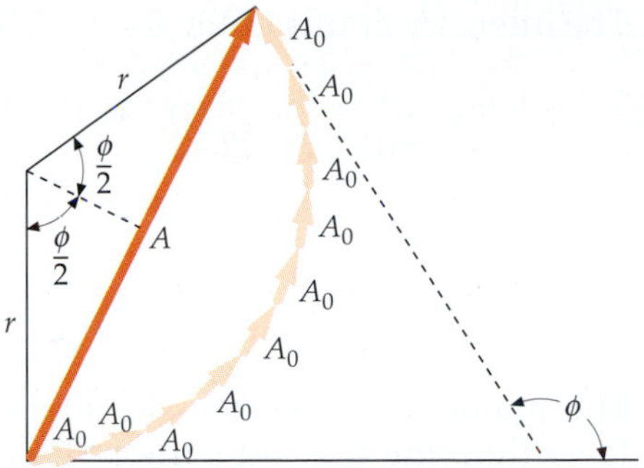

Figure 35-25 Phasor diagram for calculating the resultant amplitude due to the waves from N sources in terms of the phase difference ϕ between the wave from the first source just below the top of the slit and that from the last source just above the bottom of the slit. When N is very large, the resultant amplitude A is the chord of a circular arc of length $NA_0 = A_{\max}$.

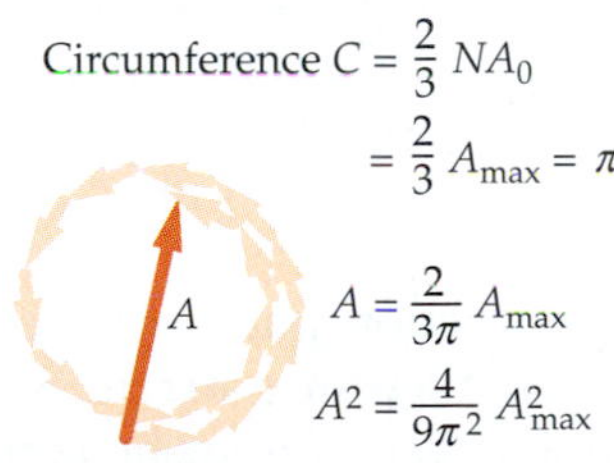

Figure 35-26 Phasor diagram for calculating the approximate amplitude of the first secondary maximum of the single-slit diffraction pattern. This secondary maximum occurs near the midpoint between the first and second minima when the N phasors complete $1\frac{1}{2}$ circles.

optional

The intensity at this point is

$$I = \frac{4}{9\pi^2} I_0 = \frac{1}{22.2} I_0 \qquad \text{35-22}$$

Calculating the Interference–Diffraction Pattern of Two Slits

The intensity of the two-slit interference–diffraction pattern can be calculated from the two-slit pattern (Equation 35-8) with the intensity of each slit (I_0 in that equation) replaced by the diffraction-pattern intensity due to each slit, I, given by Equation 35-20. The intensity for the two-slit interference–diffraction pattern is thus

$$I = 4I_0\left(\frac{\sin \frac{1}{2}\phi}{\frac{1}{2}\phi}\right)^2 \cos^2 \tfrac{1}{2}\delta \qquad \text{35-23}$$

Interference–diffraction intensity for two slits

where ϕ is the difference in phase between rays from the top and bottom of each slit, which is related to the width of each slit by

$$\phi = \frac{2\pi}{\lambda} a \sin \theta$$

and δ is the difference in phase between rays from the centers of two adjacent slits, which is related to the slit separation by

$$\delta = \frac{2\pi}{\lambda} d \sin \theta$$

In Equation 35-23, the intensity I_0 is the intensity at $\theta = 0$ due to one slit alone.

35-6 Fraunhofer and Fresnel Diffraction

Diffraction patterns like the single-slit pattern in Figure 35-10 that are observed at points for which the rays from an aperture or obstacle are nearly parallel are called **Fraunhofer diffraction patterns.** Fraunhofer patterns can be observed at great distances from the obstacle or aperture so that the rays reaching any point are approximately parallel, or they can be observed using a lens to focus parallel rays on a viewing screen placed in the focal plane of the lens.

The diffraction pattern observed near an aperture or obstacle is called a **Fresnel diffraction pattern.** Because the rays from an aperture or obstacle close to a screen cannot be considered parallel, Fresnel diffraction is much more difficult to analyze. Figure 35-27 illustrates the difference between the Fresnel and Fraunhofer patterns for a single slit.*

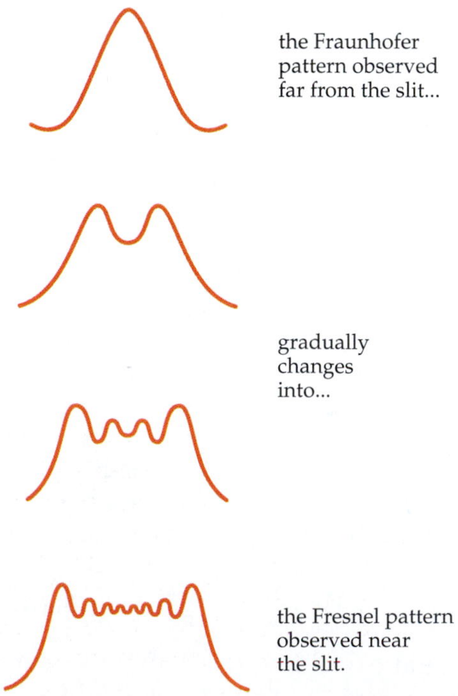
As the screen is moved closer,
the Fraunhofer pattern observed far from the slit...
gradually changes into...
the Fresnel pattern observed near the slit.

Figure 35-27 Diffraction patterns for a single slit at various screen distances.

* See Richard E. Haskel, "A Simple Experiment on Fresnel Diffraction," *American Journal of Physics*, vol. 38, 1970, p. 1039.

Figure 35-28*a* shows the Fresnel diffraction pattern of an opaque disk. Note the bright spot at the center of the pattern caused by the constructive interference of the light waves diffracted from the edge of the disk. This pattern is of some historical interest. In an attempt to discredit Fresnel's wave theory of light, Poisson pointed out that it predicted a bright spot at the center of the shadow, which he assumed was a ridiculous contradiction of fact. However, Fresnel immediately demonstrated experimentally that such a spot does, in fact, exist. This demonstration convinced many doubters of the validity of the wave theory of light. The Fresnel diffraction pattern of a circular aperture is shown in Figure 35-28*b*. Comparing this with the pattern of the opaque disk in Figure 35-28*a*, we can see that the two patterns are complements of each other.

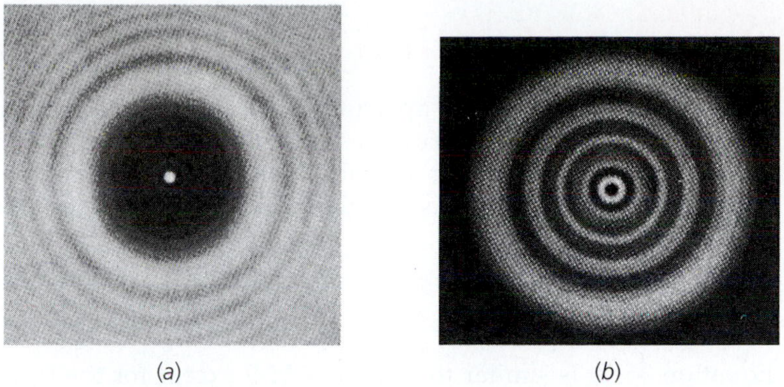

Figure 35-28 (*a*) Fresnel diffraction pattern of an opaque disk. At the center of the shadow, the light waves diffracted from the edge of the disk are in phase and produce a bright spot called the *Poisson spot.* (*b*) Fresnel diffraction pattern of a circular aperture. Compare this with part (*a*).

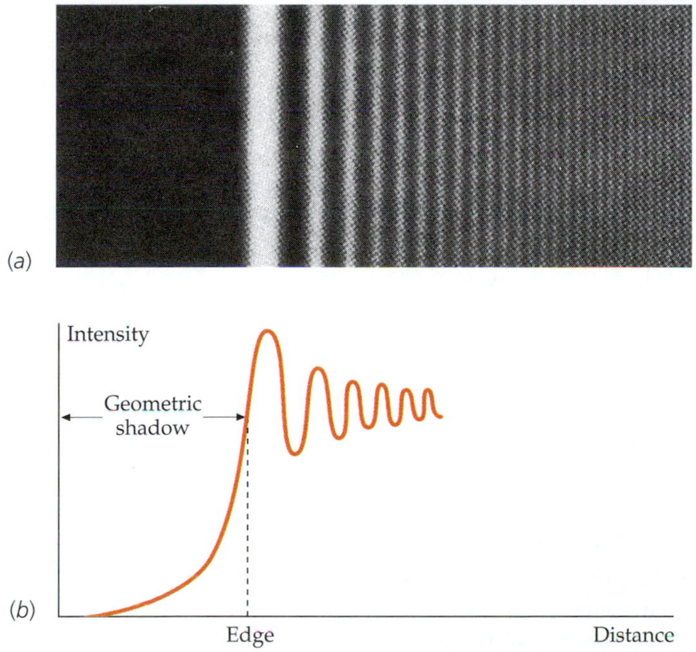

Figure 35-29 (*a*) Fresnel diffraction of a straightedge. (*b*) Intensity versus distance along a line perpendicular to the edge.

Figure 35-29*a* shows the Fresnel diffraction pattern of a straight edge illuminated by light from a point source. A graph of the intensity versus distance (measured along a line perpendicular to the edge) is shown in Figure 35-29*b*. The light intensity does not fall abruptly to zero in the geometric shadow, but it decreases rapidly and is negligible within a few wavelengths of the edge. The Fresnel diffraction pattern of a rectangular aperture is shown in Figure 35-30. These patterns cannot be seen with broad light sources like an ordinary light bulb because the dark fringes of the pattern produced by light from one point on the source overlap the bright fringes of the pattern produced by light from another point.

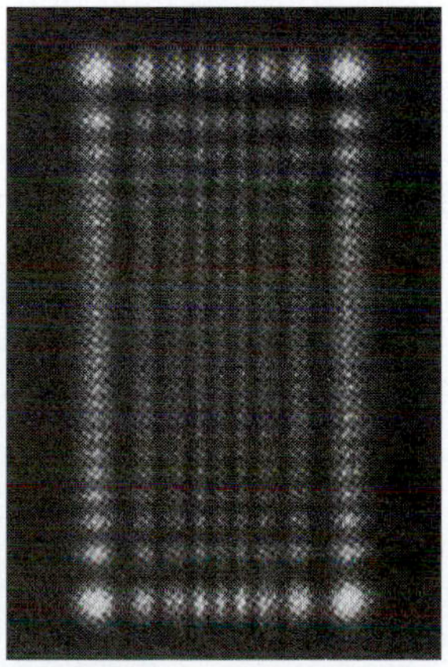

Figure 35-30 Fresnel diffraction of a rectangular aperture.

35-7 Diffraction and Resolution

Figure 35-31 Fraunhofer diffraction pattern of a circular aperture.

Diffraction due to a circular aperture has important implications for the resolution of many optical instruments. Figure 35-31 shows the Fraunhofer diffraction pattern of a circular aperture. The angle θ subtended by the first diffraction minimum is related to the wavelength and the diameter of the opening D by

$$\sin\theta = 1.22\frac{\lambda}{D} \qquad 35\text{-}24$$

Equation 35-24 is similar to Equation 35-9 except for the factor 1.22, which arises from the mathematical analysis, which is similar to that for a single slit but more complicated because of the circular geometry. In many applications, the angle θ is small, so $\sin\theta$ can be replaced by θ. The first diffraction minimum is then at an angle θ given by

$$\theta \approx 1.22\frac{\lambda}{D} \qquad 35\text{-}25$$

Figure 35-32 shows two point sources that subtend an angle α at a circular aperture far from the sources. The intensities of the Fraunhofer diffraction pattern are also indicated in this figure. If α is much greater than $1.22\lambda/D$, the sources will be seen as two sources. However, as α is decreased, the overlap of the diffraction patterns increases, and it becomes difficult to distinguish the two sources from one source. At the critical angular separation, α_c, given by

$$\alpha_c = 1.22\frac{\lambda}{D} \qquad 35\text{-}26$$

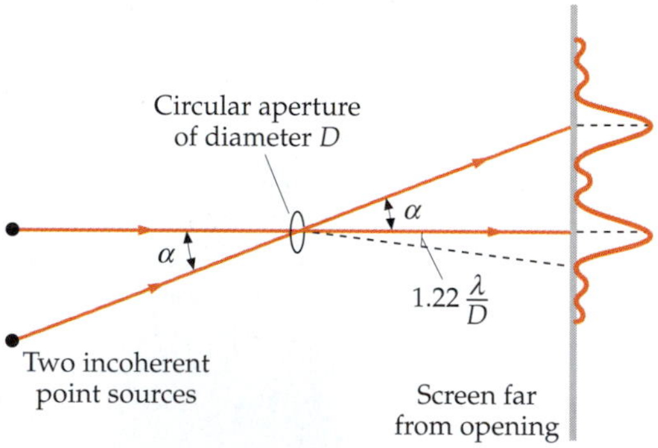

Figure 35-32 Two distant sources that subtend an angle α. If α is much greater than $1.22\lambda/D$, where λ is the wavelength of light and D is the diameter of the aperture, the diffraction patterns have little overlap and the sources are easily seen as two sources. If α is not much greater than $1.22\lambda/D$, the overlap of the diffraction patterns makes it difficult to distinguish two sources from one.

the first minimum of the diffraction pattern of one source falls on the central maximum of the other source. These objects are said to be just resolved by **Rayleigh's criterion for resolution.** Figure 35-33 shows the diffraction patterns for two sources when α is greater than the critical angle for resolution and when α is just equal to the critical angle for resolution.

Equation 35-26 has many applications. The *resolving power* of an optical instrument such as a microscope or telescope is the ability of the instrument to resolve two objects that are close together. The images of the objects tend to overlap because of diffraction at the entrance aperture of the instrument. We can see from Equation 35-26 that the resolving power can be increased either by increasing the diameter D of the lens (or mirror) or by decreasing the wavelength λ. Astronomical telescopes use large objective lenses or mirrors to increase their resolution as well as to increase their light-gathering power. In a microscope, a film of transparent oil with index of refraction of about 1.55 is sometimes used under the objective to decrease the wavelength of the light ($\lambda' = \lambda/n$). The wavelength can be reduced further by using ultraviolet light and photographic film; however, ordinary glass is opaque to ultraviolet light, so the lenses in an ultraviolet microscope must be made from quartz or fluorite. To obtain very high resolutions, electron microscopes are used—microscopes that

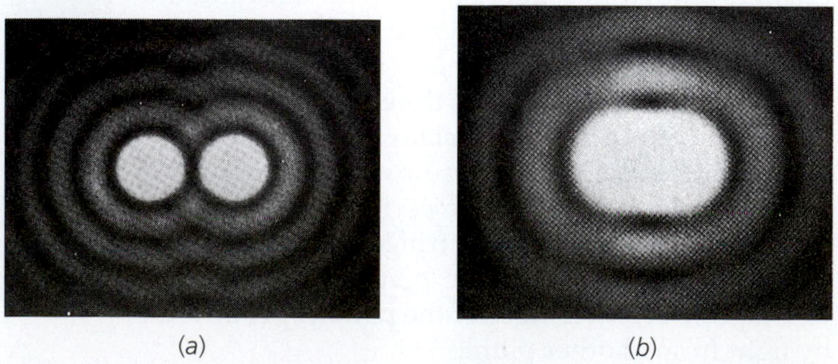

Figure 35-33 Diffraction patterns for a circular aperture and two incoherent point sources when (*a*) α is much greater than $1.22\lambda/D$ and (*b*) when α is at the limit of resolution, $\alpha_c = 1.22\lambda/D$.

use electrons rather than light. The wavelengths of electrons vary inversely with the square root of their kinetic energy and can be made as small as desired.*

*The wave properties of electrons are discussed in Chapter 17.

Example 35-7

(*a*) What minimum angular separation must two point objects have if they are to be just resolved by the eye? (*b*) How far apart must they be if they are 100 m away? Assume that the diameter of the pupil of the eye is 5 mm and that the wavelength of the light is 600 nm.

Picture the Problem (*a*) The critical angle is calculated from $\alpha_c = 1.22\ \lambda/D$. (*b*) If the objects are separated by a distance y and are 100 m away, they will be just resolved if $\tan \alpha_c = y/(100\text{ m})$.

(*a*) Use $\alpha_c = 1.22\lambda/D$ with $D = 5$ mm and $\lambda = 600$ nm:

$$\alpha_c = 1.22\,\frac{6 \times 10^{-7}\text{ m}}{5 \times 10^{-3}\text{ m}} = 1.46 \times 10^{-4}\text{ rad}$$

(*b*) Set $\tan \alpha_c = y/100$ m and solve for y:

$$\tan \alpha_c = \frac{y}{100\text{ m}}$$

$$y = (100\text{ m})\tan \alpha_c \approx (100\text{ m})\alpha_c$$

$$= (100\text{ m})(1.46 \times 10^{-4}) = 1.46 \times 10^{-2}\text{ m} = 1.46\text{ cm}$$

Exercise Two objects are 4 cm apart. How far away from them can you be and still resolve them with your eye if $\lambda = 600$ nm and the diameter of the pupil of your eye is 5 mm? (*Answer* 274 m)

It is instructive to compare the limitation on resolution of the eye due to diffraction as seen in Example 35-7 with that due to the separation of the receptors (cones) on the retina. To be seen as two distinct objects, the images of the objects must fall on the retina on two nonadjacent cones. (See Problem 70 in Chapter 34.) Because the retina is about 2.5 cm from the eye lens, the distance y on the retina corresponding to an angular separation of 1.5×10^{-4} rad is found from

$$\alpha_c = 1.5 \times 10^{-4}\text{ rad} = \frac{y}{2.5\text{ cm}}$$

or

$$y = 3.75 \times 10^{-4}\text{ cm} = 3.75 \times 10^{-6}\text{ m} = 3.75\ \mu\text{m}$$

The actual separation of the cones in the fovea centralis, where the cones are the most tightly packed, is about 1 μm. Outside this region, they are about 3 to 5 μm apart.

35-8 Diffraction Gratings

optional

A useful tool for measuring the wavelength of light is the **diffraction grating,** which consists of a large number of equally spaced lines or slits on a flat surface. Such a grating can be made by cutting parallel, equally spaced grooves on a glass or metal plate with a precision ruling machine. With a reflection grating, light is reflected from the ridges between the lines. Phonograph records and compact disks exhibit some of the properties of reflection gratings. In a transmission grating, the light passes through the clear gaps between the rulings. Inexpensive plastic gratings with 10,000 or more slits per centimeter are common items. The spacing of the slits in a grating with 10,000 slits per centimeter is $d = (1 \text{ cm})/10{,}000 = 10^{-4}$ cm.

Compact disks act as reflection gratings.

Consider a plane light wave incident normally on a transmission grating (Figure 35-34). Assume that the width of each slit is very small so that it produces a widely diffracted beam. The interference pattern produced on a screen a large distance from the grating is that due to a large number of equally spaced light sources. Suppose we have N slits with separation d between adjacent slits. At $\theta = 0$, the light from each slit is in phase with that from all the other slits so the amplitude of the wave is NA_0, where A_0 is the amplitude from each slit, and the intensity is N^2I_0, where I_0 is the intensity due to each slit. At an angle θ such that $d \sin\theta = \lambda$, the path difference between any two successive slits is λ, so again the light from each slit is in phase with that from all the other slits and the intensity is N^2I_0. The interference maxima are thus at angles θ given by

$$d \sin\theta = m\lambda, \qquad m = 0, 1, 2, \ldots \qquad \text{35-27}$$

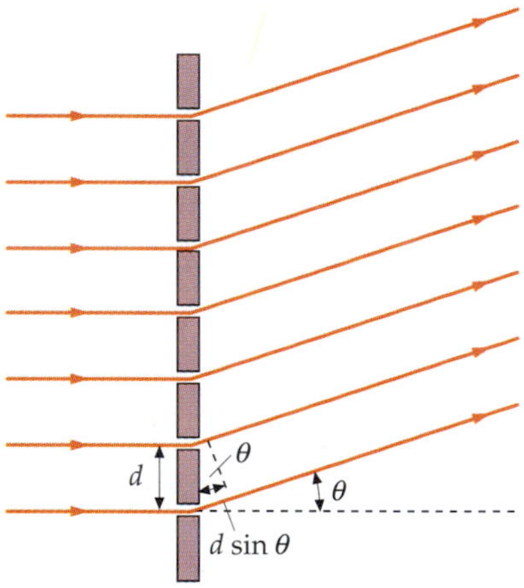

Figure 35-34 Light incident normally on a diffraction grating. At an angle θ, the path difference between rays from adjacent slits is $d \sin\theta$.

The position of an interference maximum does not depend on the number of sources, but the more sources there are, the sharper and more intense the maximum will be.

To see that the interference maxima will be sharper when there are many slits, consider the case of N slits. The distance from the first slit to the Nth slit is $(N-1)d \approx Nd$. When the path difference for the light from the first slit and that from the Nth slit is λ, the resulting intensity will be zero. (We saw this in our discussion of single-slit diffraction.) Since the first and Nth slits are separated by approximately Nd, the intensity will be zero at angle θ_{min} given by

$$Nd \sin\theta_{min} = \lambda$$

$$\sin\theta_{min} \approx \theta_{min} = \frac{\lambda}{Nd}$$

The width of the interference maximum $2\theta_{min}$ is thus proportional to $1/N$. Thus, the greater the number of slits N, the sharper the maximum. Since the intensity in the maximum is proportional to N^2I_0, the amount of light in the maximum is proportional to NI_0.

Figure 35-35*a* shows a student spectroscope that uses a diffraction grating to analyze light. In student laboratories, the light source is typically a glass tube containing atoms of a gas, for example, helium or sodium vapor, that are excited by a bombardment of electrons accelerated by high voltage across the tube. The light emitted by such a source contains only certain wavelengths that are characteristic of the atoms in the source. Light from the source passes through a narrow collimating slit and is made parallel by a lens. Parallel light from the lens is incident on the grating. Instead of falling on a screen a large distance away, the parallel light from the grating is focused by a telescope and viewed by the eye. The telescope is mounted on a rotating platform that has been calibrated so that the angle θ can be measured. In the forward direction ($\theta = 0$), the central maximum for all wavelengths is seen. If light of a particular wavelength λ is emitted by the source, the first interference maximum is seen at the angle θ given by Equation 35-27

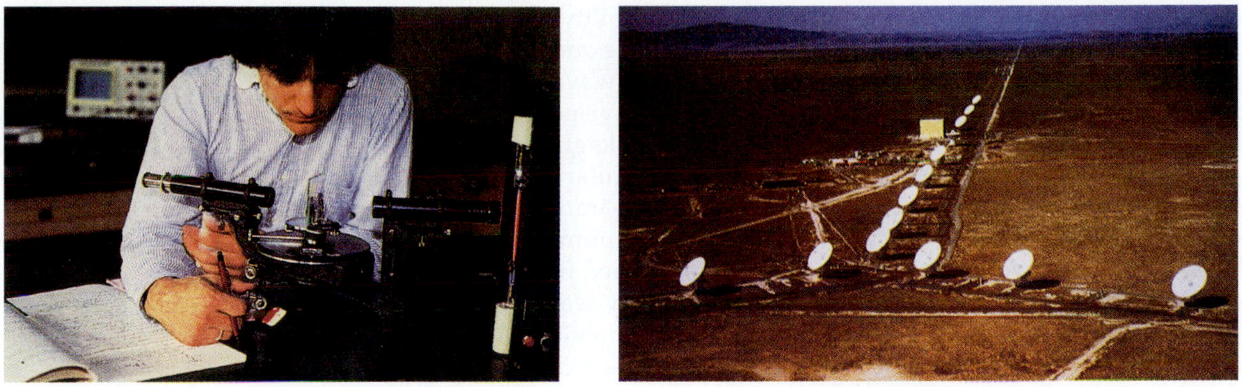

(*a*) (*b*)

Figure 35-35 (*a*) Typical student spectroscope. Light from a collimating slit near the source is made parallel by a lens and falls on a grating. The diffracted light is viewed with a telescope at an angle that can be accurately measured. (*b*) Aerial view of the very large array (VLA) radio telescope in New Mexico. Radio signals from distant galaxies add constructively when Equation 35-27 is satisfied, where d is the distance between two adjacent telescopes.

with $m = 1$. Each wavelength emitted by the source produces a separate image of the collimating slit in the spectroscope called a **spectral line.** The set of lines corresponding to $m = 1$ is called the **first-order spectrum.** The **second-order spectrum** corresponds to $m = 2$ for each wavelength. Higher orders may be seen if the angle θ given by Equation 35-27 is less than 90°. Depending on the wavelengths, the orders may be mixed; that is, the third-order line for one wavelength may occur before the second-order line for another wavelength. If the spacing of the slits in the grating is known, the wavelengths emitted by the source can be determined by measuring the angle θ.

Example 35-8

Sodium light is incident on a diffraction grating with 12,000 lines per centimeter. At what angles will the two yellow lines of wavelengths 589.00 nm and 589.59 nm be seen in the first order?

Picture the Problem Apply $d \sin\theta = m\lambda$ to each wavelength, with $m = 1$ and $d = 1\text{ cm}/12{,}000$.

1. The angle θ is given by $d \sin\theta = m\lambda$, with $m = 1$:

$$\sin\theta = \frac{m\lambda}{d} = \frac{\lambda}{d}$$

2. Calculate θ for $\lambda = 589$ nm:

$$\sin\theta = \frac{\lambda}{d} = \frac{589 \times 10^{-9}\text{ m}}{(1\text{ cm}/12{,}000)} = 0.7068$$

$$\theta = \sin^{-1} 0.7068 = 44.98°$$

3. Repeat the calculation for $\lambda = 589.59$ nm:

$$\sin\theta = \frac{\lambda}{d} = \frac{589.59 \times 10^{-9}\text{ m}}{1\text{ cm}/12{,}000} = 0.7075$$

$$\theta = \sin^{-1} 0.7075 = 45.03°$$

Remark Note that light of greater wavelength is diffracted through greater angles.

Exercise Find the angles for the two yellow lines if the grating has 15,000 lines per centimeter. (*Answers* 62.07° and 62.18°)

An important feature of a spectroscope is its ability to resolve spectral lines of two nearly equal wavelengths λ_1 and λ_2. For example, the two prominent yellow lines in the spectrum of sodium have wavelengths 589.00 and 589.59 nm. These can be seen as two separate wavelengths if their interference maxima do not overlap. According to Rayleigh's criterion for resolution, these wavelengths are resolved if the angular separation of their interference maxima is greater than the angular separation between one interference maximum and the first interference minimum on either side of it. The **resolving power** of a diffraction grating is defined to be $\lambda/|\Delta\lambda|$, where $|\Delta\lambda|$ is the smallest difference between two nearby wavelengths, each approximately equal to λ, that may be resolved. The resolving power is proportional to the number of slits illuminated because the more slits illuminated, the sharper the interference maxima. The resolving power R can be shown to be

$$R = \frac{\lambda}{|\Delta\lambda|} = mN \qquad \text{35-28}$$

where N is the number of slits and m is the order number (see Problem 73). We can see from Equation 35-28 that to resolve the two yellow lines in the sodium spectrum the resolving power must be

$$R = \frac{589.00 \text{ nm}}{589.59 - 589.00 \text{ nm}} = 998$$

Thus, to resolve the two yellow sodium lines in the first order ($m = 1$), we need a grating containing about 1000 slits in the area illuminated by the light.

Holograms

An interesting application of diffraction gratings is the production of a three-dimensional photograph called a **hologram** (Figure 35-36). In an ordinary photograph, the intensity of reflected light from an object is recorded on a film. When the film is viewed by transmitted light, a two-dimensional image is produced. In a hologram, a beam from a laser is split into two beams, a reference beam and an object beam. The object beam reflects from the object to be photographed and the interference pattern between it and the reference beam is recorded on a photographic film. This can be done because the laser

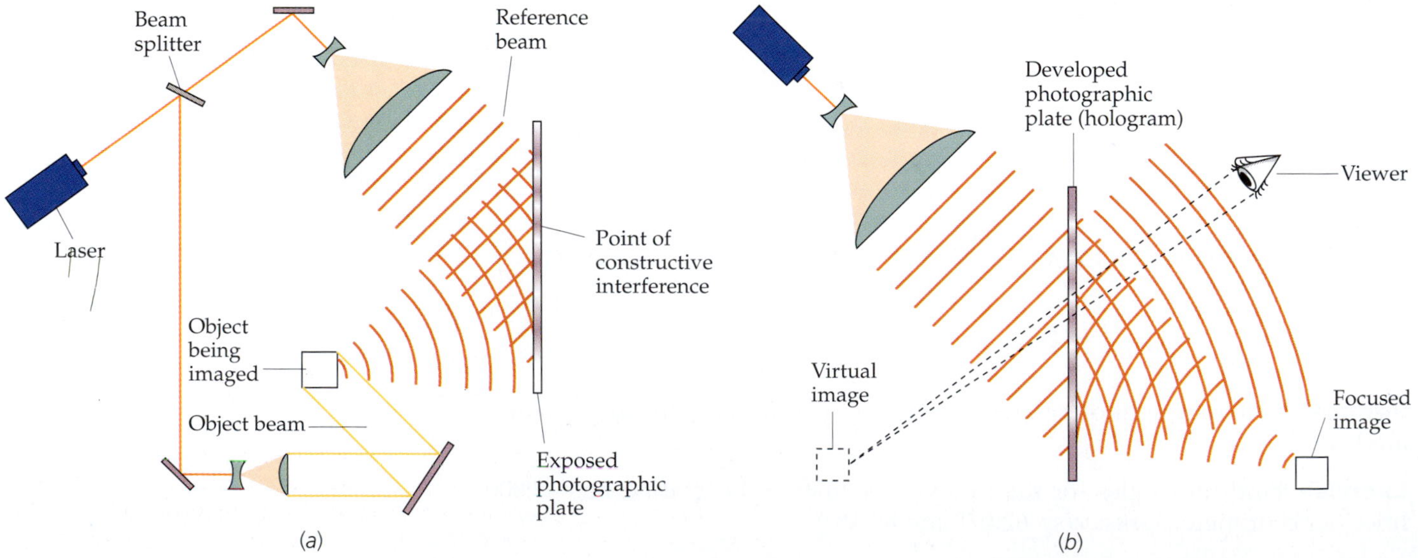

Figure 35-36 (*a*) The production of a hologram. The interference pattern produced by the reference beam and object beam is recorded on a photographic film. (*b*) When the film is developed and illuminated by coherent laser light, a three-dimensional image is seen.

beam is coherent so that the relative phase difference between the reference beam and object beam can be kept constant during the exposure. The interference fringes on the film act as a diffraction grating. When the film is illuminated with a laser, a three-dimensional replica of the object is produced.

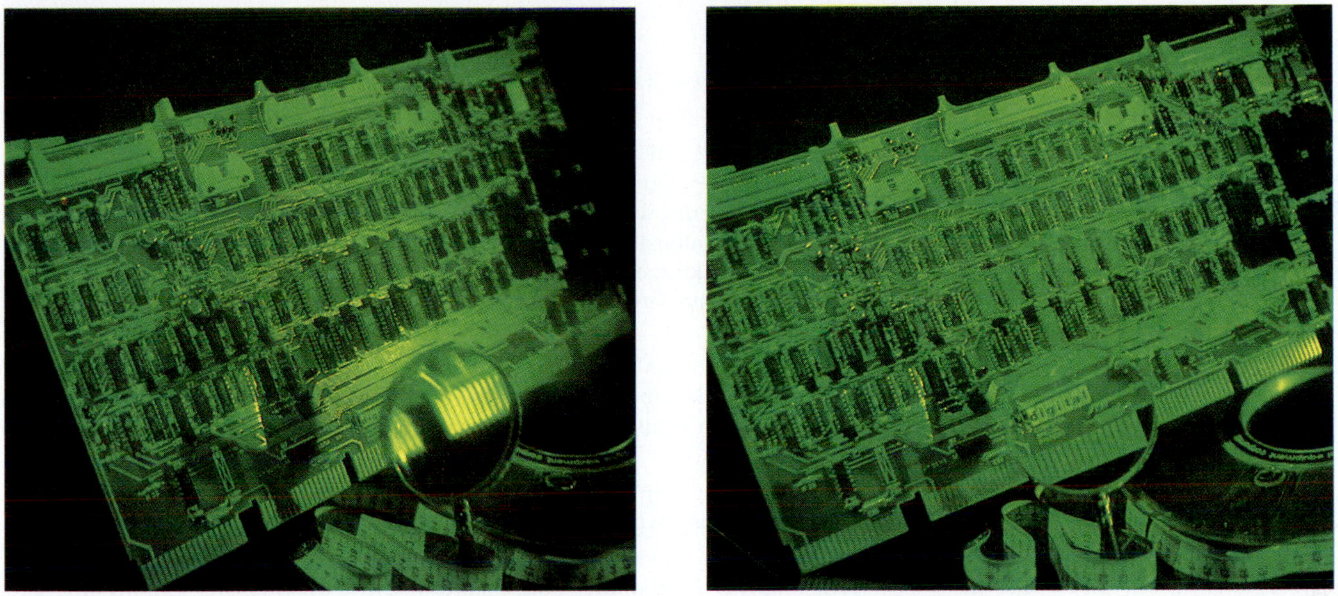

A hologram viewed from two different angles. Note that different parts of the circuit board appear behind the front magnifying lens.

Holograms that you see on credit cards or postage stamps, called rainbow holograms, are more complicated. A horizontal strip of the original hologram is used to make a second hologram. The three-dimensional image can be seen as the viewer moves from side to side, but if viewed with laser light, the image disappears when the viewer's eyes move above or below the slit image. When viewed with white light, the image is seen in different colors as the viewer moves in the vertical direction.

Summary

Topic	Remarks and Relevant Equations	
1. Interference	Two light rays interfere constructively if their phase difference is zero or an integer times 360°. They interfere destructively if their phase difference is 180° or an odd integer times 180°.	
Phase difference due to path difference	$\delta = \frac{\Delta r}{\lambda} 2\pi = \frac{\Delta r}{\lambda} 360°$	**35-1**
Phase difference due to reflection	A phase difference of 180° is introduced when a light wave is reflected from a boundary between two media for which the wave speed is greater in the original medium, such as the one between air and glass.	
Thin films	The interference of light rays reflected from the top and bottom surfaces of a thin film produces colored bands or fringes, commonly observed in soap films or oil films. The difference in phase between the two rays results from the path difference of twice	

the thickness of the film plus any phase change due to reflection of one or both of the rays.

Two slits	The path difference at an angle θ on a screen far away from two narrow slits separated by a distance d is $d \sin \theta$. If the intensity due to each slit separately is I_0, the intensity at points of constructive interference is $4I_0$ and that at points of destructive interference is 0.
Interference maxima	$d \sin \theta = m\lambda, \quad m = 0, 1, 2, \ldots$ **35-2**
Interference minima	$d \sin \theta = (m + \frac{1}{2})\lambda, \quad m = 0, 1, 2, \ldots$ **35-3**
2. Diffraction	Diffraction occurs whenever a portion of a wavefront is limited by an obstacle or aperture. The intensity of light at any point in space can be computed using Huygens' principle by taking each point on the wavefront to be a point source and computing the resulting interference pattern.
Fraunhofer patterns	Fraunhofer patterns are observed at great distances from the obstacle or aperture so that the rays reaching any point are approximately parallel, or they can be observed using a lens to focus parallel rays on a viewing screen placed in the focal plane of the lens.
Fresnel patterns	Fresnel patterns are observed at points close to the source.
Single slit	When light is incident on a single slit of width a, the intensity pattern on a screen far away shows a broad central diffraction maximum that decreases to zero at an angle θ given by $\sin \theta = \frac{\lambda}{a}$ **35-9** The width of the central maximum is inversely proportional to the width of the slit. Other zeros in the single-slit diffraction pattern occur at angles given by $a \sin \theta = m\lambda, \quad m = 1, 2, 3, \ldots$ **35-11** On each side of the central maximum are secondary maxima of much smaller intensity.
Two slits	The interference–diffraction pattern of two slits of finite size is the two-slit interference pattern modulated by the single-slit diffraction pattern. If $d/a = m$, the mth interference maximum coincides with the first diffraction minimum and will not be seen. Then there will be $2m - 1$ interference maxima within the central diffraction maximum.
Resolution of two sources	When light from two point sources that are close together passes through an aperture, the diffraction patterns of the sources may overlap. If the overlap is too great, the two sources cannot be resolved as two separate sources. When the central diffraction maximum of one source falls at the diffraction minimum of the other source, the two sources are said to be just resolved by Rayleigh's criterion for resolution. For a circular aperture of diameter D, the critical angular separation of two sources for resolution by Rayleigh's criterion is
Rayleigh's criterion	$\alpha_c = 1.22 \frac{\lambda}{D}$ **35-26**
Gratings (optional)	A diffraction grating consisting of a large number of closely spaced lines or slits is used to measure the wavelength of light emitted by a source. The positions of the mth

order interference maxima from a grating are at angles given by

$$d \sin \theta = m\lambda, \qquad m = 0, 1, 2, \ldots$$

The resolving power of a grating is

$$R = \frac{\lambda}{|\Delta\lambda|} = mN \qquad \text{35-27}$$

where N is the number of slits of the grating that are illuminated and m is the order number.

3. **Phasors** (optional)	Two or more harmonic waves can be added by representing each wave as a vector called a phasor. The phase difference between the waves is represented as the angle between the phasors.

Problem-Solving Guide

Summary of Worked Examples

Type of Calculation	Procedure and Relevant Examples	
1. Interference		
Relate phase difference to path difference.	Use $\delta = \frac{\Delta r}{\lambda} 2\pi = \frac{\Delta r}{\lambda} 360°$.	**Example 35-1**
Calculate the number of interference fringes in a thin film.	Find the distance x to the mth fringe. Then compute m/x.	**Example 35-2**
Locate fringes in a two-slit interference pattern.	The mth maximimum is at $y_m = m\lambda L/d$.	**Example 35-3**
2. Diffraction		
Locate the zero-intensity points for single-slit diffraction.	Use $a \sin \theta = m\lambda$, where $m = 1, 2, 3, \ldots$.	**Example 35-4**
Find the number of interference maxima in the central diffraction maximum for the two-slit pattern.	Use $N = 2m - 1$, where $m = d/a$.	**Example 35-5**
Find the critical angle for resolution.	Use $\alpha_c = 1.22\ \lambda/D$.	**Example 35-7**
Locate interference maxima produced by a diffraction grating.	Use $d \sin \theta = m\lambda$, where $m = 0, 1, 2, \ldots$.	**Example 35-8**
3. Phasors (optional)		
Find the resultant of the sum of two or more harmonic waves.	Use phasors to represent the waves and add vectorily.	**Example 35-6**

Problems

Conceptual Problems

Problems from Optional and Exploring sections

In a few problems, you are given more data than you actually need; in a few other problems, you are required to supply data from your general knowledge, outside sources, or informed estimates.

• Single-concept, single-step, relatively easy

•• Intermediate-level, may require synthesis of concepts

••• Challenging, for advanced students

Phase Difference and Coherence

1 • When destructive interference occurs, what happens to the energy in the light waves?

2 • Which of the following pairs of light sources are coherent: (*a*) two candles; (*b*) one point source and its image in a plane mirror; (*c*) two pinholes uniformly illuminated by the same point source; (*d*) two headlights of a car; (*e*) two images of a point source due to reflection from the front and back surfaces of a soap film.

3 • (*a*) What minimum path difference is needed to introduce a phase shift of 180° in light of wavelength 600 nm? (*b*) What phase shift will that path difference introduce in light of wavelength 800 nm?

4 • Light of wavelength 500 nm is incident normally on a film of water 10^{-4} cm thick. The index of refraction of water is 1.33. (*a*) What is the wavelength of the light in the water? (*b*) How many wavelengths are contained in the distance $2t$, where t is the thickness of the film? (*c*) What is the phase difference between the wave reflected from the top of the air–water interface and the one reflected from the bottom of the water–air interface after it has traveled this distance?

5 •• Two coherent microwave sources that produce waves of wavelength 1.5 cm are in the xy plane, one on the y axis at $y = 15$ cm and the other at $x = 3$ cm, $y = 14$ cm. If the sources are in phase, find the difference in phase between the two waves from these sources at the origin.

Interference in Thin Films

6 • The spacing between Newton's rings decreases rapidly as the diameter of the rings increases. Explain qualitatively why this occurs.

7 •• If the angle of a wedge-shaped air film such as that in Example 35-2 is too large, fringes are not observed. Why?

8 •• Why must a film used to observe interference colors be thin?

9 • A loop of wire is dipped in soapy water and held so that the soap film is vertical. (*a*) Viewed by reflection with white light, the top of the film appears black. Explain why. (*b*) Below the black region are colored bands. Is the first band red or violet? (*c*) Describe the appearance of the film when it is viewed by *transmitted* light.

10 • A wedge-shaped film of air is made by placing a small slip of paper between the edges of two flat plates of glass. Light of wavelength 700 nm is incident normally on the glass plates, and interference bands are observed by reflection. (*a*) Is the first band near the point of contact of the plates dark or bright? Why? (*b*) If there are five dark bands per centimeter, what is the angle of the wedge?

11 •• The diameters of fine wires can be accurately measured using interference patterns. Two optically flat pieces of glass of length L are arranged with the wire between them as shown in Figure 35-37. The setup is illuminated by monochromatic light, and the resulting interference fringes are detected. Suppose $L = 20$ cm and yellow sodium light ($\lambda \approx 590$ nm) is used for illumination. If 19 bright fringes are seen along this 20-cm distance, what are the limits on the diameter of the wire? *Hint:* The nineteenth fringe might not be right at the end, but you do not see a twentieth fringe at all.

Figure 35-37 Problem 11

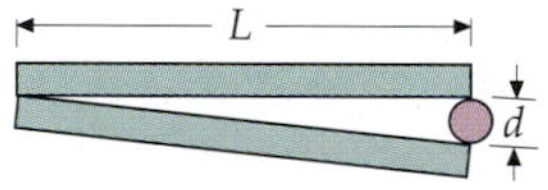

12 •• Light of wavelength 600 nm is used to illuminate normally two glass plates 22 cm in length that touch at one end and are separated at the other end by a wire of radius 0.025 mm. How many bright fringes appear along the total length of the plates?

13 •• A thin film having an index of refraction of 1.5 is surrounded by air. It is illuminated normally by white light and is viewed by reflection. Analysis of the resulting reflected light shows that the wavelengths 360, 450, and 602 nm are the only missing wavelengths in or near the visible portion of the spectrum. That is, for these wavelengths, there is destructive interference. (*a*) What is the thickness of the film? (*b*) What visible wavelengths are brightest in the reflected interference pattern? (*c*) If this film were resting on glass with an index of refraction of 1.6, what wavelengths in the visible spectrum would be missing from the reflected light?

14 •• A drop of oil ($n = 1.22$) floats on water ($n = 1.33$). When reflected light is observed from above as shown in Figure 35-38, what is the thickness of the drop at the point where the second red fringe, counting from the edge of the drop, is observed? Assume red light has a wavelength of 650 nm.

Figure 35-38 Problem 14

15 •• A film of oil of index of refraction $n = 1.45$ rests on an optically flat piece of glass of index of refraction $n = 1.6$. When illuminated with white light at normal incidence, light of wavelengths 690 and 460 nm is predominant in the reflected light. Determine the thickness of the oil film.

16 •• A film of oil of index of refraction $n = 1.45$ floats on water ($n = 1.33$). When illuminated with white light at normal incidence, light of wavelengths 700 nm and 500 nm is predominant in the reflected light. Determine the thickness of the oil film.

Newton's Rings

17 •• A Newton's-ring apparatus consists of a glass lens with radius of curvature R that rests on a flat glass plate as shown in Figure 35-39. The thin film is air of variable thickness. The pattern is viewed by reflected light. (*a*) Show that for a thickness t the condition for a bright (constructive) interference ring is

Figure 35-39 Problem 17

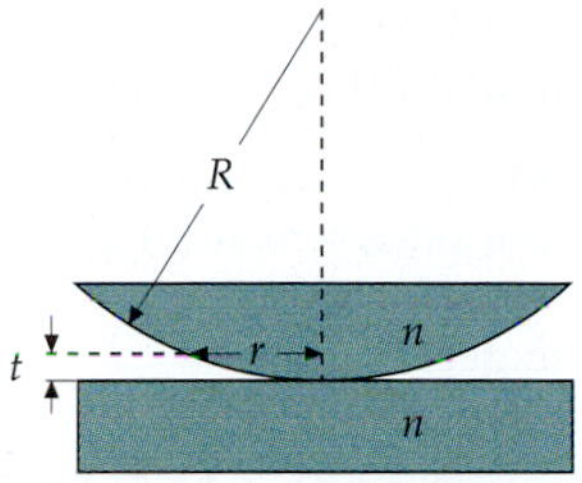

$$t = \left(m + \frac{1}{2}\right)\frac{\lambda}{2}, \quad m = 0, 1, 2, \ldots$$

(*b*) Apply the Pythagorean theorem to the triangle of sides r, $R - t$, and hypotenuse R to show that for $t \ll R$, the radius of a fringe is related to t by

$$r = \sqrt{2tR} \qquad 35\text{-}29$$

(*c*) How would the transmitted pattern look in comparison with the reflected one? (*d*) Use $R = 10$ m and a diameter of 4 cm for the lens. How many bright fringes would you see if the apparatus were illuminated by yellow sodium light ($\lambda \approx$ 590 nm) and were viewed by reflection? (*e*) What would be the diameter of the sixth bright fringe? (*f*) If the glass used in the apparatus has an index of refraction $n = 1.5$ and water ($n_w = 1.33$) is placed between the two pieces of glass, what change will take place in the bright fringes?

18 •• A plano-convex glass lens of radius of curvature 2.0 m rests on an optically flat glass plate. The arrangement is illuminated from above with monochromatic light of 520-nm wavelength. The indexes of refraction of the lens and plate are 1.6. Determine the radii of the first and second bright fringe in the reflected light. (Use Equation 35-29 from Problem 17 to relate r to t.)

19 •• Suppose that before the lens of Problem 18 is placed on the plate a film of oil of refractive index 1.82 is deposited on the plate. What will then be the radii of the first and second bright fringes? (Use Equation 35-29 from Problem 17 to relate r to t.)

Two-Slit Interference Pattern

20 • A double-slit interference experiment is set up in a chamber that can be evacuated. Using monochromatic light, an interference pattern is observed when the chamber is open to air. As the chamber is evacuated one will note that

(*a*) the interference fringes remain fixed.
(*b*) the interference fringes move closer together.
(*c*) the interference fringes move farther apart.
(*d*) the interference fringes disappear completely.

21 • Two narrow slits separated by 1 mm are illuminated by light of wavelength 600 nm, and the interference pattern is viewed on a screen 2 m away. Calculate the number of bright fringes per centimeter on the screen.

22 • Using a conventional two-slit apparatus with light of wavelength 589 nm, 28 bright fringes per centimeter are observed on a screen 3 m away. What is the slit separation?

23 • Light of wavelength 633 nm from a helium–neon laser is shone normally on a plane containing two slits. The first interference maximum is 82 cm from the central maximum on a screen 12 m away. (*a*) Find the separation of the slits. (*b*) How many interference maxima can be observed?

24 •• Two narrow slits are separated by a distance d. Their interference pattern is to be observed on a screen a large distance L away. (*a*) Calculate the spacing y of the maxima on the screen for light of wavelength 500 nm when $L =$ 1 m and $d = 1$ cm. (*b*) Would you expect to observe the interference of light on the screen for this situation? (*c*) How close together should the slits be placed for the maxima to be separated by 1 mm for this wavelength and screen distance?

25 •• Light is incident at an angle ϕ with the normal to a vertical plane containing two slits of separation d (Figure 35-40). Show that the interference maxima are located at angles θ given by $\sin\theta + \sin\phi = m\lambda/d$.

Figure 35-40 Problems 25 and 26

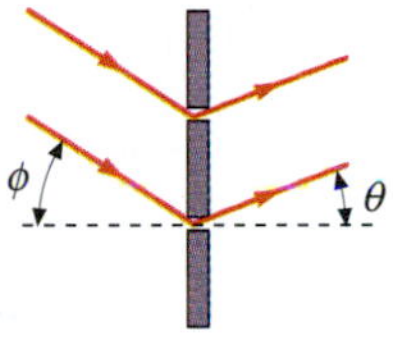

26 •• White light falls at an angle of 30° to the normal of a plane containing a pair of slits separated by 2.5 μm. What visible wavelengths give a bright interference maximum in the transmitted light in the direction normal to the plane? (See Problem 25.)

27 •• Laser light falls normally on three evenly spaced, very narrow slits. When one of the side slits is covered, the first-order maximum is at 0.60° from the normal. If the center slit is covered and the other two are open, find (*a*) the angle of the first-order maximum and (*b*) the order number of the maximum that now occurs at the same angle as the fourth-order maximum did before.

Diffraction Pattern of a Single Slit

28 • As the width of a slit producing a single-slit diffraction pattern is slowly and steadily reduced, how will the diffraction pattern change?

29 • Equation 35-2, $d \sin\theta = m\lambda$, and Equation 35-11, $a \sin\theta = m\lambda$, are sometimes confused. For each equation, define the symbols and explain the equation's application.

30 • Light of wavelength 600 nm is incident on a long, narrow slit. Find the angle of the first diffraction minimum if the width of the slit is (*a*) 1 mm, (*b*) 0.1 mm, and (*c*) 0.01 mm.

31 • The single-slit diffraction pattern of light is observed on a screen a large distance L from the slit. Note from Equation 35-12 that the width $2y$ of the central maximum varies inversely with the width a of the slit. Calculate the width $2y$ for $L = 2$ m, $\lambda = 500$ nm, and (*a*) $a = 0.1$ mm, (*b*) $a = 0.01$ mm, and (*c*) $a = 0.001$ mm.

32 • Plane microwaves are incident on a long, narrow metal slit of width 5 cm. The first diffraction minimum is observed at $\theta = 37°$. What is the wavelength of the microwaves?

33 •• For a ruby laser of wavelength 694 nm, the end of the ruby crystal is the aperture that determines the diameter of the light beam emitted. If the diameter is 2 cm and the laser is aimed at the moon, 380,000 km away, find the approximate diameter of the light beam when it reaches the moon, assuming the spread is due solely to diffraction.

Interference–Diffraction Pattern of Two Slits

34 • How many interference maxima will be contained in the central diffraction maximum in the diffraction–interference pattern of two slits if the separation d of the slits is 5 times their width a? How many will there be if $d = Na$ for any value of N?

35 •• A two-slit Fraunhofer interference–diffraction pattern is observed with light of wavelength 500 nm. The slits have a separation of 0.1 mm and a width of a. (*a*) Find the width a if the fifth interference maximum is at the same angle as the first diffraction minimum. (*b*) For this case, how many bright interference fringes will be seen in the central diffraction maximum?

36 •• A two-slit Fraunhofer interference–diffraction pattern is observed with light of wavelength 700 nm. The slits have widths of 0.01 mm and are separated by 0.2 mm. How many bright fringes will be seen in the central diffraction maximum?

37 •• Suppose that the *central* diffraction maximum for two slits contains 17 interference fringes for some wavelength of light. How many interference fringes would you expect in the first *secondary* diffraction maximum?

38 •• Light of wavelength 550 nm illuminates two slits of width 0.03 mm and separation 0.15 mm. (*a*) How many interference maxima fall within the full width of the central diffraction maximum? (*b*) What is the ratio of the intensity of the third interference maximum to the side of the centerline (not counting the center interference maximum) to the intensity of the center interference maximum?

Using Phasors to Add Harmonic Waves (optional)

39 • Find the resultant of the two waves $E_1 = 2 \sin \omega t$ and $E_2 = 3 \sin (\omega t + 270°)$.

40 • Find the resultant of the two waves $E_1 = 4 \sin \omega t$ and $E_2 = 3 \sin (\omega t + 60°)$.

41 •• At the second secondary maximum of the diffraction pattern of a single slit, the phase difference between the waves from the top and bottom of the slit is approximately 5π. The phasors used to calculate the amplitude at this point complete 2.5 circles. If I_0 is the intensity at the central maximum, find the intensity I at this second secondary maximum.

42 •• (*a*) Show that the positions of the interference minima on a screen a large distance L away from three equally spaced sources (spacing d, with $d >> \lambda$) are given approximately by

$$y = \frac{n\lambda L}{3d}, \quad \text{where } n = 1, 2, 4, 5, 7, 8, 10, \ldots$$

that is, n is not a multiple of 3. (*b*) For $L = 1$ m, $\lambda = 5 \times 10^{-7}$ m, and $d = 0.1$ mm, calculate the width of the principal interference maxima (the distance between successive minima) for three sources.

43 •• (*a*) Show that the positions of the interference minima on a screen a large distance L away from four equally spaced sources (spacing d, with $d >> \lambda$) are given approximately by

$$y = \frac{n\lambda L}{4d}, \quad \text{where } n = 1, 2, 3, 5, 6, 7, 9, 10, \ldots$$

that is, n is not a multiple of 4. (*b*) For $L = 2$ m, $\lambda = 6 \times 10^{-7}$ m, and $d = 0.1$ mm, calculate the width of the principal interference maxima (the distance between successive minima) for four sources. Compare this width with that for two sources with the same spacing.

44 •• Light of wavelength 480 nm falls normally on four slits. Each slit is 2 μm wide and is separated from the next by 6 μm. (*a*) Find the angle from the center to the first point of zero intensity of the single-slit diffraction pattern on a distant screen. (*b*) Find the angles of any bright interference maxima that lie inside the central diffraction maximum. (*c*) Find the angular spread between the central interference maximum and the first interference minimum on either side of it. (*d*) Sketch the intensity as a function of angle.

45 ••• Three slits, each separated from its neighbor by 0.06 mm, are illuminated by a coherent light source of wavelength 550 nm. The slits are extremely narrow. A screen is located 2.5 m from the slits. The intensity on the centerline is 0.05 W/m^2. Consider a location 1.72 cm from the centerline. (*a*) Draw the phasors, according to the phasor model for the addition of harmonic waves, appropriate for this location. (*b*) From the phasor diagram, calculate the intensity of light at this location.

46 ••• Four coherent sources are located on the y axis at $+3\lambda/4$, $+\lambda/4$, $-\lambda/4$, and $-3\lambda/4$. They emit waves of wavelength λ and intensity I_0. (*a*) Calculate the net intensity I as a function of the angle θ measured from the $+x$ axis. (*b*) Make a polar plot of $I(\theta)$.

47 ••• For single-slit diffraction, calculate the first three values of ϕ (the total phase difference between rays from each edge of the slit) that produce subsidiary maxima by (*a*) using the phasor model and (*b*) setting $dI/d\phi = 0$, where I is given by Equation 35-20.

Diffraction and Resolution

48 • Light of wavelength 700 nm is incident on a pinhole of diameter 0.1 mm. (*a*) What is the angle between the central maximum and the first diffraction minimum for a Fraunhofer diffraction pattern? (*b*) What is the distance between the central maximum and the first diffraction minimum on a screen 8 m away?

49 • Two sources of light of wavelength 700 nm are 10 m away from the pinhole of Problem 48. How far apart must the sources be for their diffraction patterns to be resolved by Rayleigh's criterion?

50 • Two sources of light of wavelength 700 nm are separated by a horizontal distance x. They are 5 m from a vertical slit of width 0.5 mm. What is the least value of x for which the diffraction pattern of the sources can be resolved by Rayleigh's criterion?

51 • The headlights on a small car are separated by 112 cm. At what maximum distance could you resolve them if the diameter of your pupil is 5 mm and the effective wavelength of the light is 550 nm?

52 • You are told not to shoot until you see the whites of their eyes. If their eyes are separated by 6.5 cm and the diameter of your pupil is 5 mm, at what distance can you resolve the two eyes using light of wavelength 550 nm?

53 •• (*a*) How far apart must two objects be on the moon to be resolved by the eye? Take the diameter of the pupil of the eye to be 5 mm, the wavelength of the light to be 600 nm, and the distance to the moon to be 380,000 km. (*b*) How far apart must the objects on the moon be to be resolved by a telescope that has a mirror of diameter 5 m?

54 •• The ceiling of your lecture hall is probably covered with acoustic tile, which has small holes separated by about 6 mm. (*a*) Using light with a wavelength of 500 nm, how far could you be from this tile and still resolve these holes? The diameter of the pupil of your eye is about 5 mm. (*b*) Could you resolve these holes better with red light or with violet light?

55 •• The telescope on Mount Palomar has a diameter of 200 inches. Suppose a double star were 4 lightyears away. Under ideal conditions, what must be the minimum separation of the two stars for their images to be resolved using light of wavelength 550 nm?

56 •• The star Mizar in Ursa Major is a binary system of stars of nearly equal magnitudes. The angular separation between the two stars is 14 seconds of arc. What is the minimum diameter of the pupil that allows resolution of the two stars using light of wavelength 550 nm?

Diffraction Gratings (optional)

57 • When a diffraction grating is illuminated with white light, the first-order maximum of green light

(*a*) is closer to the central maximum than that of red light.
(*b*) is closer to the central maximum than that of blue light.
(*c*) overlaps the second order maximum of red light.
(*d*) overlaps the second order maximum of blue light.

58 • A diffraction grating with 2000 slits per centimeter is used to measure the wavelengths emitted by hydrogen gas. At what angles θ in the first-order spectrum would you expect to find the two violet lines of wavelengths 434 and 410 nm?

59 • With the grating used in Problem 58, two other lines in the first-order hydrogen spectrum are found at angles $\theta_1 = 9.72 \times 10^{-2}$ rad and $\theta_2 = 1.32 \times 10^{-1}$ rad. Find the wavelengths of these lines.

60 • Repeat Problem 58 for a diffraction grating with 15,000 slits per centimeter.

61 • What is the longest wavelength that can be observed in the fifth-order spectrum using a diffraction grating with 4000 slits per centimeter?

62 •• A diffraction grating of 2000 slits per centimeter is used to analyze the spectrum of mercury. (*a*) Find the angular separation in the first-order spectrum of the two lines of wavelengths 579.0 and 577.0 nm. (*b*) How wide must the beam on the grating be for these lines to be resolved?

63 •• A diffraction grating with 4800 lines per centimeter is illuminated at normal incidence with white light (wavelength range 400 to 700 nm). For how many orders can one observe the complete spectrum in the transmitted light? Do any of these orders overlap? If so, describe the overlapping regions.

64 •• A square diffraction grating with an area of 25 cm^2 has a resolution of 22,000 in the fourth order. At what angle should you look to see a wavelength of 510 nm in the fourth order?

65 •• Sodium light of wavelength 589 nm falls normally on a 2-cm-square diffraction grating ruled with 4000 lines per centimeter. The Fraunhofer diffraction pattern is projected onto a screen at 1.5 m by a lens of focal length 1.5 m placed immediately in front of the grating. Find (*a*) the positions of the first two intensity maxima on one side of the central maximum, (*b*) the width of the central maximum, and (*c*) the resolution in the first order.

66 •• The spectrum of neon is exceptionally rich in the visible region. Among the many lines are two at wavelengths of 519.313 and 519.322 nm. If light from a neon discharge tube is normally incident on a transmission grating with 8400 lines per centimeter and the spectrum is observed in second order, what must be the width of the grating that is illuminated so that these two lines can be resolved?

67 •• Mercury has several stable isotopes, among them ^{198}Hg and ^{202}Hg. The strong spectral line of mercury at about 546.07 nm is a composite of spectral lines from the various mercury isotopes. The wavelengths of this line for ^{198}Hg and ^{202}Hg are 546.07532 and 546.07355 nm, respectively. What must be the resolving power of a grating capable of resolving these two isotopic lines in the third-order spectrum? If the grating is illuminated over a 2-cm-wide region, what must be the number of lines per centimeter of the grating?

68 •• A transmission grating is used to study the spectral region extending from 480 to 500 nm. The angular spread of this region is 12° in third order. (*a*) Find the number of lines

this region is 12° in third order. (*a*) Find the number of lines per centimeter. (*b*) How many orders are visible?

69 •• White light is incident normally on a transmission grating and the spectrum is observed on a screen 8.0 m from the grating. In the second-order spectrum, the separation between light of 520- and 590-nm wavelengths is 8.4 cm. (*a*) Determine the number of lines per centimeter of the grating. (*b*) What is the separation between these two wavelengths in the first-order and third-order spectra?

70 ••• A diffraction grating has n lines per meter. Show that the angular separation of two lines of wavelengths λ and $\lambda + \Delta\lambda$ meters is approximately

$$\Delta\theta = \frac{\Delta\lambda}{\sqrt{(1/nm)^2 - \lambda^2}}$$

71 ••• When assessing a diffraction grating, we are interested not only in its resolving power R, which is the ability of the grating to separate two close wavelengths, but also in the dispersion D of the grating. This is defined by $D = \Delta\theta_m/\Delta\lambda$ in the mth order. (*a*) Show that D can be written

$$D = \frac{m}{\sqrt{d^2 - m^2\lambda^2}}$$

where d is the slit spacing. (*b*) If a diffraction grating with 2000 slits per centimeter is to resolve the two yellow sodium lines in the second order (wavelengths 589.0 and 589.6 nm), how many slits must be illuminated by the beam? (*c*) What would the separation be between these resolved yellow lines if the pattern were viewed on a screen 4 m from the grating?

72 ••• For a diffraction grating in which all the surfaces are normal to the incident radiation, most of the energy goes into the zeroth order, which is useless from a spectroscopic point of view since in zeroth order all the wavelengths are at 0°. Therefore, modern gratings have shaped, or *blazed*, grooves as shown in Figure 35-41. This shifts the specular reflection, which contains most of the energy, from the zeroth order to some higher order. (*a*) Calculate the blaze angle ϕ in terms of a (the groove separation), λ (the wavelength), and m (the order in which specular reflection is to occur). (*b*) Calculate the proper blaze angle for the specular reflection to occur in the second order for light of wavelength 450 nm incident on a grating with 10,000 lines per centimeter.

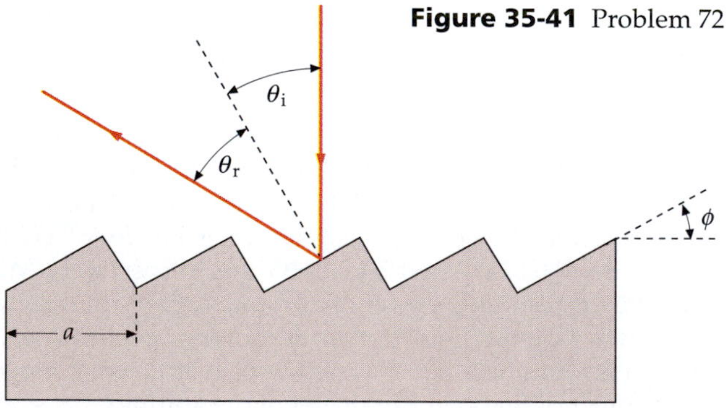

Figure 35-41 Problem 72

73 ••• In this problem you will derive Equation 35-28 for the resolving power of a diffraction grating containing N slits separated by a distance d. To do this you will calculate the angular separation between the maximum and minimum for some wavelength λ and set it equal to the angular separation of the mth-order maximum for two nearby wavelengths. (*a*) Show that the phase difference ϕ between the light from two adjacent slits is given by

$$\phi = \frac{2\pi d}{\lambda}\sin\theta$$

(*b*) Differentiate this expression to show that a small change in angle $d\theta$ results in a change in phase of $d\phi$ given by

$$d\phi = \frac{2\pi d}{\lambda}\cos\theta\, d\theta$$

(*c*) For N slits, the angular separation between an interference maximum and interference minimum corresponds to a phase change of $d\phi = 2\pi/N$. Use this to show that the angular separation $d\theta$ between the maximum and minimum for some wavelength λ is given by

$$d\theta = \frac{\lambda}{Nd\cos\theta} \qquad \text{35-30}$$

(*d*) The angle of the mth-order interference maximum for wavelength λ is given by Equation 35-27. Compute the differential of each side of this equation to show that angular separation of the mth-order maximum for two nearly equal wavelengths differing by $d\lambda$ is given by

$$d\theta \approx \frac{m\, d\lambda}{d\cos\theta} \qquad \text{35-31}$$

(*e*) According to Rayleigh's criterion, two wavelengths will be resolved in the mth order if the angular separation of the wavelengths given by Equation 35-31 equals the angular separation of the interference maximum and interference minimum given by Equation 35-30. Use this to derive Equation 35-28 for the resolving power of a grating.

General Problems

74 • True or false:

(*a*) When waves interfere destructively, the energy is converted into heat energy.
(*b*) Interference is observed only for waves from coherent sources.
(*c*) In the Fraunhofer diffraction pattern for a single slit, the narrower the slit, the wider the central maximum of the diffraction pattern.
(*d*) A circular aperture can produce both a Fraunhofer and a Fresnel diffraction pattern.
(*e*) The ability to resolve two point sources depends on the wavelength of the light.

75 • In a lecture demonstration, laser light is used to illuminate two slits separated by 0.5 mm, and the interference pattern is observed on a screen 5 m away. The distance on the screen from the centerline to the thirty-seventh bright fringe is 25.7 cm. What is the wavelength of the light?

76 • A long, narrow, horizontal slit lies 1 mm above a plane mirror, which is in the horizontal plane. The interference pattern produced by the slit and its image is viewed on a screen 1 m from the slit. The wavelength of the light is 600 nm. (*a*) Find the distance from the mirror to the first maximum. (*b*) How many dark bands per centimeter are seen on the screen?

77 • In a lecture demonstration, a laser beam of wavelength 700 nm passes through a vertical slit 0.5 mm wide and hits a screen 6 m away. Find the horizontal length of the principal diffraction maximum on the screen; that is, find the distance between the first minimum on the left and the first minimum on the right of the central maximum.

78 • What minimum aperture, in millimeters, is required for opera glasses (binoculars) if an observer is to be able to distinguish the soprano's individual eyelashes (separated by 0.5 mm) at an observation distance of 25 m? Assume the effective wavelength of the light to be 550 nm.

79 • The diameter of the aperture of the radio telescope at Arecibo, Puerto Rico, is 300 m. What is the resolving power of the telescope when tuned to detect microwaves of 3.2 cm wavelength?

80 •• A thin layer of a transparent material with an index of refraction of 1.30 is used as a nonreflective coating on the surface of glass with an index of refraction of 1.50. What should the thickness of the material be for it to be nonreflecting for light of wavelength 600 nm?

81 •• A *Fabry–Perot interferometer* consists of two parallel, half-silvered mirrors separated by a small distance a. Show that when light is incident on the interferometer with an angle of incidence θ, the transmitted light will have maximum intensity when $a = (m\lambda/2)\cos\theta$.

82 •• A mica sheet 1.2 μm thick is suspended in air. In reflected light, there are gaps in the visible spectrum at 421, 474, 542, and 633 nm. Find the index of refraction of the mica sheet.

83 •• A camera lens is made of glass with an index of refraction of 1.6. This lens is coated with a magnesium fluoride film ($n = 1.38$) to enhance its light transmission. This film is to produce zero reflection for light of wavelength 540 nm. Treat the lens surface as a flat plane and the film as a uniformly thick flat film. (*a*) How thick must the film be to accomplish its objective in the first order? (*b*) Would there be destructive interference for any other visible wavelengths? (*c*) By what factor would the reflection for light of wavelengths 400 and 700 nm be reduced by this film? Neglect the variation in the reflected light amplitudes from the two surfaces.

84 •• In a pinhole camera, the image is fuzzy because of geometry (rays arrive at the film through different parts of the pinhole) and because of diffraction. As the pinhole is made smaller, the fuzziness due to geometry is reduced, but the fuzziness due to diffraction is increased. The optimum size of the pinhole for the sharpest possible image occurs when the spread due to diffraction equals that due to the geometric effects of the pinhole. Estimate the optimum size of the pinhole if the distance from it to the film is 10 cm and the wavelength of the light is 550 nm.

85 •• The Impressionist painter Georges Seurat used a technique called "pointillism," in which his paintings are composed of small, closely spaced dots of pure color, each about 2 mm in diameter. The illusion of the colors blending together smoothly is produced in the eye of the viewer by diffraction effects. Calculate the minimum viewing distance for this effect to work properly. Use the wavelength of visible light that requires the *greatest* distance, so that you're sure the effect will work for *all* visible wavelengths. Assume that the pupil of the eye has a diameter of 3 mm.

86 ••• A *Jamin refractometer* is a device for measuring or comparing the indexes of refraction of fluids. A beam of monochromatic light is split into two parts, each of which is directed along the axis of a separate cylindrical tube before being recombined into a single beam that is viewed through a telescope. Suppose that each tube is 0.4 m long and that sodium light of wavelength 589 nm is used. Both tubes are initially evacuated, and constructive interference is observed in the center of the field of view. As air is slowly allowed to enter one of the tubes, the central field of view changes to dark and back to bright a total of 198 times. (*a*) What is the index of refraction of air? (*b*) If the fringes can be counted to ±0.25 fringe, where one fringe is equivalent to one complete cycle of intensity variation at the center of the field of view, to what accuracy can the index of refraction of air be determined by this experiment?

87 ••• Light of wavelength λ is diffracted through a single slit of width a, and the resulting pattern is viewed on a screen a long distance L away from the slit. (*a*) Show that the width of the central maximum on the screen is approximately $2L\lambda/a$. (*b*) If a slit of width $2L\lambda/a$ is cut in the screen and is illuminated, show that the width of its central diffraction maximum at the same distance L is a to the same approximation.

88 ••• Television viewers in rural areas often find that the picture flickers (fades in and out) as an airplane flies across the sky in the vicinity. The flickering arises from the interference between the signal directly from the transmitter and that reflected to the antenna from the airplane. Suppose the receiver is 36 km from the transmitter broadcasting at a frequency of 86.0 MHz and an airplane is flying at a height of about 600 m above the receiver toward the transmitter. The rate of oscillation of the picture's intensity is 4 Hz. (*a*) Determine the speed of the plane. (*b*) If the picture's intensity is a maximum when the plane is directly overhead, what is the exact height of the plane above the receiving antenna?

89 ••• For the situation described in Problem 88, show that the rate of oscillation of the picture's intensity is a minimum when the airplane is directly above the midpoint between the transmitter and receiving antenna.

90 ••• A double-slit experiment uses a helium–neon laser with a wavelength of 633 nm and a slit separation of 0.12 mm. When a thin sheet of plastic is placed in front of one of the slits, the interference pattern shifts by 5.5 fringes. When the experiment is repeated under water, the shift is 3.5 fringes. Calculate (*a*) the thickness of the plastic sheet and (*b*) the index of refraction of the plastic sheet.

91 ••• Two coherent sources are located on the y axis at $+\lambda/4$ and $-\lambda/4$. They emit waves of wavelength λ and intensity I_0. (*a*) Calculate the net intensity I as a function of the angle θ measured from the $+x$ axis. (*b*) Make a polar plot of $I(\theta)$.

PART VI

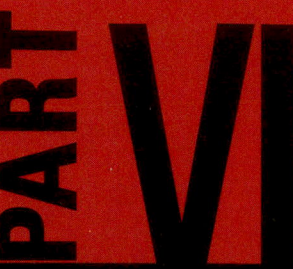

The surface of a crystal of silicon, an important ingredient in many semiconductor devices, is shown in this scanning tunneling micrograph at a magnification of about 1,000,000,000. Individual atoms are seen as hills. A computer plots digital information collected by the microscope and assigns false colors to accentuate the crystalline structure.

modern physics: quantum mechanics, relativity, and the structure of matter

CHAPTER 36

Applications of the Schrödinger Equation

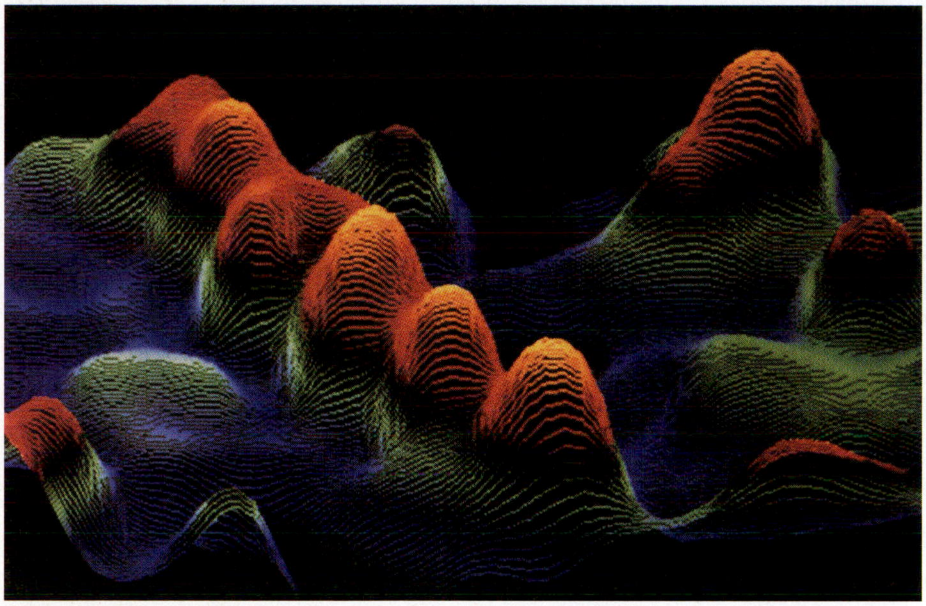

False-color scanning tunneling micrograph of a segment of a DNA molecule magnified about one and a half million times. A sample of double-stranded DNA was dissolved in a salt solution and deposited on graphite prior to being imaged in air by the microscope. The row of orange-yellow peaks corresponds to the ridges of the DNA double helix.

In Chapter 17 we found that electrons and other "particles" have wave properties and are described by a wave function $\Psi(x, t)$. The probability of finding the particle in some region of space is proportional to the square of the wave function. We mentioned that the wave function is a solution of the Schrödinger equation, and we discussed some solutions qualitatively without reference to the equation itself. In particular, we showed how the standing-wave conditions lead to quantization of energy for a particle confined to a one-dimensional box.

This chapter is a continuation of the material introduced in Chapter 17. We discuss the Schrödinger equation and apply it to the particle in the box problem and to several other situations in which a particle is confined to a region of space to illustrate how boundary conditions lead to energy quantization. We then show how the Schrödinger equation leads to barrier penetration, and discuss the extension of the Schrödinger equation to more than one dimension and to more than one particle. You should review the material in Chapter 17 before moving on to the following discussion.

36-1 The Schrödinger Equation

Like the classical wave equation (Equation 15-9*b*), the Schrödinger equation is a partial differential equation in space and time. Like Newton's laws of motion, it cannot be derived. Its validity, like that of Newton's laws, lies in its agreement with experiment. In one dimension, the Schrödinger equation is[†]

$$-\frac{\hbar^2}{2m}\frac{\partial^2\Psi(x,t)}{\partial x^2} + U\Psi(x,t) = i\hbar\frac{\partial\Psi(x,t)}{\partial t} \qquad 36\text{-}1$$

Time-dependent Schrödinger equation

where U is the potential-energy function. Equation 36-1 is called the **time-dependent Schrödinger equation.** Unlike the classical wave equation, it relates the second space derivative of the wave function to the *first* time derivative of the wave function, and it contains the imaginary number $i = \sqrt{-1}$. The wave functions that are solutions of this equation are not necessarily real. $\Psi(x, t)$ is not a measurable function like the classical wave functions for sound or string waves. The probability of finding a particle in some region of space dx is certainly real though, so we must modify slightly the equation for probability density given in Chapter 17 (Equation 17-14). We take for the probability of finding a particle in some region dx

$$P(x,t)\,dx = |\Psi(x,t)|^2\,dx = \Psi^*\,\Psi\,dx \qquad 36\text{-}2$$

where Ψ^*, the complex conjugate of Ψ, is obtained from Ψ by replacing i by $-i$ wherever it appears.[‡]

When the potential energy U does not depend on time (for example, when a particle's wave function corresponds to a standing wave), the time-dependent Schrödinger equation can be simplified by writing the wave function in the form

$$\Psi(x,t) = \psi(x)e^{-i\omega t} \qquad 36\text{-}3$$

The right side of Equation 36-1 is then

$$i\hbar\frac{\partial\Psi(x,t)}{\partial t} = i\hbar(-i\omega)\psi(x)e^{-i\omega t} = \hbar\omega\psi(x)e^{-i\omega t} = E\psi(x)e^{-i\omega t}$$

where $E = \hbar\omega$ is the energy of the particle. Substituting $\psi(x)e^{-i\omega t}$ into Equation 36-1 and canceling the common factor $e^{-i\omega t}$ we obtain an equation for $\psi(x)$ called the **time-independent Schrödinger equation**

$$-\frac{\hbar^2}{2m}\frac{d^2\psi(x)}{dx^2} + U(x)\psi(x) = E\psi(x) \qquad 36\text{-}4$$

Time-independent Schrödinger equation

where we have written U as $U(x)$ to emphasize the fact that there is no time dependence in the equation.

The calculation of the allowed energy levels in a system involves only the time-independent Schrödinger equation, whereas finding the probabilities of transition between these levels requires the solution of the time-dependent equation. In this book we will be concerned only with the time-independent Schrödinger equation.

The solution of Equation 36-4 depends on the form of the potential-energy function $U(x)$. When $U(x)$ is such that the particle is confined to some region of space, only certain discrete energies E_n give solutions ψ_n that can satisfy the normalization condition (Equation 17-15):

$$\int_{-\infty}^{\infty} |\psi|^2\,dx = 1$$

[†] Although we simply state the Schrödinger equation, Schrödinger himself had a vast knowledge of classical wave theory that led him to this equation.

[‡] Every complex number can be written in the form $z = a + bi$, where a and b are real numbers and $i = \sqrt{-1}$. The complex conjugate of z is $z^* = a - bi$, so $z^*z = (a + bi)(a - bi) = a^2 + b^2 = |z|^2$. Complex numbers are discussed more fully in Appendix D.

The complete time-dependent wave functions are then given, from Equation 36-3, by

$$\Psi_n(x, t) = \psi_n(x)e^{-i\omega_n t} = \psi_n(x)e^{-i(E_n/\hbar)t} \qquad 36\text{-}5$$

We will illustrate the use of the time-independent Schrödinger equation by solving it for the problem of a particle in a box. The potential energy for a one-dimensional box from $x = 0$ to $x = L$ is shown in Figure 36-1. It is called an **infinite square-well potential** and is described mathematically by

$$\begin{aligned} U(x) &= 0, \qquad 0 < x < L \\ U(x) &= \infty, \qquad x < 0 \quad \text{or} \quad x > L \end{aligned} \qquad 36\text{-}6$$

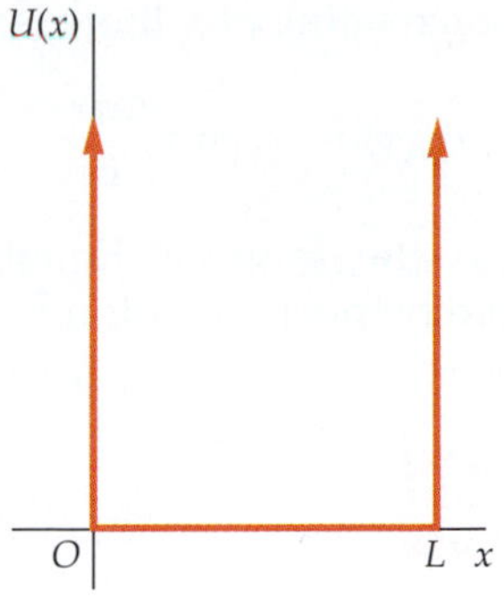

Figure 36-1 The infinite square-well potential energy. For $x < 0$ and $x > L$, the potential energy $U(x)$ is infinite. The particle is confined to the region in the well $0 < x < L$.

Inside the box, the potential energy is zero, whereas outside the box it is infinite. Since we require the particle to be in the box, we have $\psi(x) = 0$ everywhere outside the box. We then need to solve Schrödinger's equation inside the box subject to the condition that $\psi(x)$ must be zero at $x = 0$ and at $x = L$.

Inside the box, Schrödinger's equation is

$$-\frac{\hbar^2}{2m}\frac{d^2\psi(x)}{dx^2} = E\psi(x)$$

or

$$\frac{d^2\psi(x)}{dx^2} = -\frac{2mE}{\hbar^2}\psi(x) = -k^2\psi(x) \qquad 36\text{-}7$$

where

$$k^2 = \frac{2mE}{\hbar^2} \qquad 36\text{-}8$$

The general solution of Equation 36-7 can be written as

$$\psi(x) = A \sin kx + B \cos kx \qquad 36\text{-}9$$

where A and B are constants. At $x = 0$, we have

$$\psi(0) = A \sin(k0) + B \cos(k0) = 0 + B$$

The boundary condition $\psi(x) = 0$ at $x = 0$ thus gives $B = 0$, and Equation 36-9 becomes

$$\psi(x) = A \sin kx \qquad 36\text{-}10$$

The wave function is thus a sine wave with the wavelength λ related to the wave number k in the usual way, $\lambda = 2\pi/k$. The boundary condition $\psi(x) = 0$ at $x = L$ restricts the possible values of k and therefore the values of the wavelength λ, and (from Equation 36-8) the energy $E = \hbar^2k^2/2m$. We have

$$\psi(L) = A \sin kL = 0 \qquad 36\text{-}11$$

This condition is satisfied if kL is π or any integer times π, that is, if k is restricted to the values k_n given by

$$k_n = n\frac{\pi}{L}, \qquad n = 1, 2, 3, \ldots \qquad 36\text{-}12$$

Substituting this result into Equation 36-8 and solving for E gives us the allowed energy values:

$$E_n = \frac{\hbar^2 k_n^2}{2m} = \frac{\hbar^2}{2m}\left(n\frac{\pi}{L}\right)^2 = n^2\left(\frac{h^2}{8mL^2}\right) = n^2 E_1 \qquad 36\text{-}13$$

$$E_1 = \frac{h^2}{8mL^2} \qquad 36\text{-}14$$

where we have used $\hbar = h/2\pi$. Equation 36-14 is the same as Equation 17-19, which we obtained by fitting an integral number of half-wavelengths into the box.

For each value of n there is wave function $\psi_n(x)$ given by

$$\psi_n(x) = A_n \sin \frac{n\pi x}{L} \qquad \text{36-15}$$

which is the same as Equation 17-22 with the constant $A_n = \sqrt{2/L}$ determined by normalization.†

36-2 A Particle in a Finite Square Well

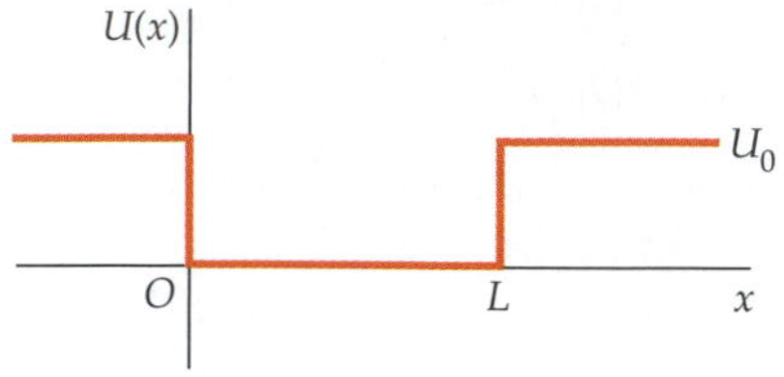

Figure 36-2 The finite square-well potential energy.

The quantization of energy that we found for a particle in an infinite square well is a result that follows from the general solution of the Schrödinger equation for any particle confined to some region of space. We will illustrate this by considering the qualitative behavior of the wave function for a slightly more general potential-energy function, the finite square well shown in Figure 36-2. This potential-energy function is described mathematically by

$$\begin{aligned} U(x) &= U_0, \quad x < 0 \\ U(x) &= 0, \quad 0 < x < L \\ U(x) &= U_0, \quad x > L \end{aligned} \qquad \text{36-16}$$

This potential-energy function is discontinuous at $x = 0$ and $x = L$, but it is finite everywhere. The solutions of the Schrödinger equation for this type of potential-energy function depend on whether the total energy E is greater or less than U_0. We will not discuss the case of $E > U_0$, except to remark that in that case the particle is not confined and any value of the energy is allowed; that is, there is no energy quantization. Here we assume that $E < U_0$.

Inside the well, $U(x) = 0$, and the time-independent Schrödinger equation is the same as for the infinite well (Equation 36-7):

$$\frac{d^2\psi(x)}{dx^2} = -\frac{2mE}{\hbar^2}\psi(x) = -k^2\psi(x)$$

where $U(x)$ is the potential energy and E is the total energy, with

$$k^2 = \frac{2mE}{\hbar^2}$$

The general solution is of the form

$$\psi(x) = A \sin kx + B \cos kx$$

In this case, $\psi(x)$ is not required to be zero at $x = 0$, so B is not zero. Outside the well, the time-independent Schrödinger equation is

$$\frac{d^2\psi(x)}{dx^2} = \frac{2m}{\hbar^2}(U_0 - E)\psi(x) = \alpha^2\psi(x) \qquad \text{36-17}$$

where

$$\alpha^2 = \frac{2m}{\hbar^2}(U_0 - E) > 0 \qquad \text{36-18}$$

The wave functions and allowed energies for the particle can be found by solving Equation 36-17 for $\psi(x)$ outside the well and then requiring that $\psi(x)$ and $d\psi(x)/dx$ be continuous at the boundaries $x = 0$ and $x = L$. The solution of Equation 36-17 is not difficult (for positive values of x, it is of the form $\psi(x) = Ce^{-\alpha x}$), but applying the boundary conditions involves much tedious

† See Equation 17-15.

algebra and is not important for our purpose. The important feature of Equation 36-17 is that the second derivative of $\psi(x)$, which is related to the curvature of the wave function, has the same sign as the wave function ψ. If ψ is positive, $d^2\psi/dx^2$ is also positive and the wave function curves away from the axis as shown in Figure 36-3*a*. Similarly, if ψ is negative, $d^2\psi/dx^2$ is negative and ψ again curves away from the axis as shown in Figure 36-3*b*. This behavior is very different from that inside the well, where ψ and $d^2\psi/dx^2$ have opposite signs so that ψ always curves toward the axis like a sine or cosine function. Because of this behavior outside the well, for most values of the energy E in Equation 36-17, $\psi(x)$ becomes infinite as x approaches $\pm\infty$; that is, most wave functions $\psi(x)$ are not well behaved outside the well. Though they satisfy the Schrödinger equation, such functions are not proper wave functions because they cannot be normalized. The solutions of the Schrödinger equation are well behaved (that is, approach 0 as $|x|$ becomes very large) only for certain values of the energy. These energy values are the allowed energies for the finite square well.

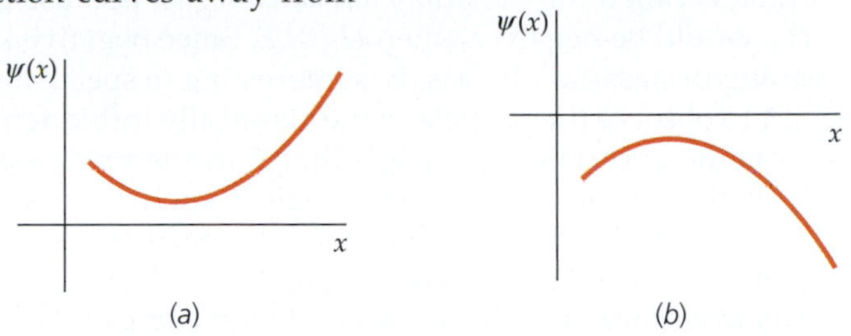

Figure 36-3 (*a*) A positive function with positive curvature. (*b*) A negative function with negative curvature.

Figure 36-4 shows a well-behaved wave function with a wavelength λ_1 inside the well corresponding to the ground-state energy. The behavior of the wave functions corresponding to nearby wavelengths and energies is also shown. Figure 36-5 shows the wave functions and probability distributions for the ground state and first two excited states. From this figure, we can see that the wavelengths inside the well are slightly longer than the corresponding wavelengths for the infinite well (Figure 17-14), so the corresponding energies are slightly less than those for the infinite well. Another feature of the finite-well problem is that there are only a finite number of allowed energies. For very small values of U_0, there is only one allowed energy.

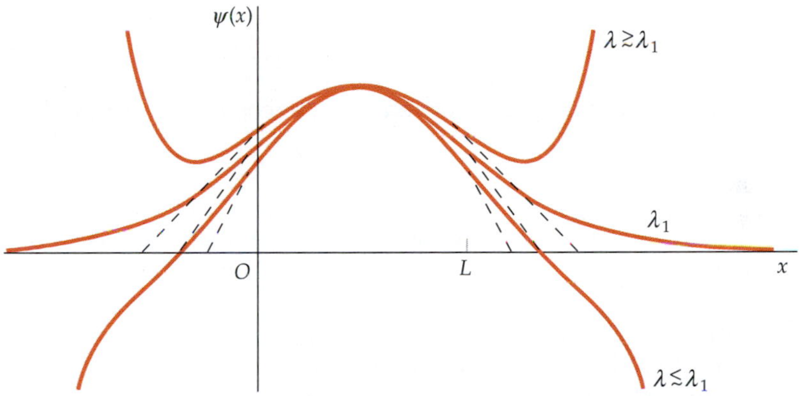

Figure 36-4 Functions satisfying the Schrödinger equation with wavelengths near the wavelength λ_1 corresponding to the ground-state energy $E_1 = \hbar^2/2m\lambda_1^2$ in the finite well. If λ is slightly greater than λ_1, the function approaches infinity, like the function in Figure 36-3*a*. At the critical wavelength λ_1, the function and its slope approach zero together. If λ is slightly less than λ_1, the function crosses the x axis while the slope is still negative. The slope then becomes more negative because its rate of change $d^2\psi/dx^2$ is now negative. This function approaches negative infinity as x approaches infinity.

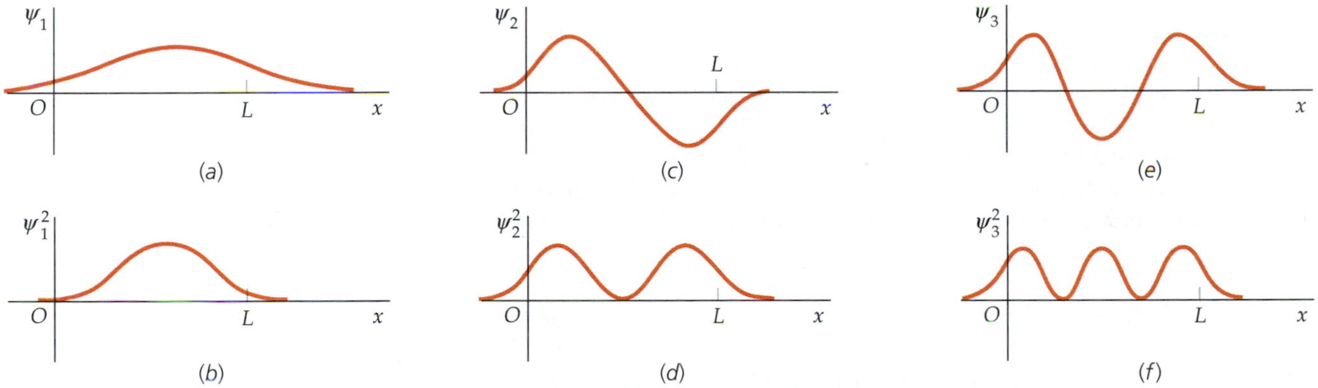

Figure 36-5 Graphs of the wave functions $\psi_n(x)$ and probability distributions $\psi_n^2(x)$ for $n = 1, 2,$ and 3 for the finite square well. Compare these graphs with those of Figure 17-14 for the infinite square well, where the wave functions are zero at $x = 0$ and $x = L$. The wavelengths here are slightly longer than the corresponding ones for the infinite well, so the allowed energies are somewhat smaller.

Note that the wave function penetrates beyond the edges of the well at $x = L$ and $x = 0$, indicating that there is some small probability of finding the particle in the region in which its total energy E is less than its potential energy U_0. This region is called the *classically forbidden region* because the kinetic energy, $E - U_0$, would be negative when $U_0 > E$. Since negative kinetic energy has no meaning in classical physics, it is interesting to speculate on the result of an attempt to observe the particle in the classically forbidden region. It can be shown from the uncertainty principle that if an attempt is made to localize the particle in the classically forbidden region, such a measurement introduces an uncertainty in the momentum of the particle corresponding to a minimum kinetic energy that is greater than $U_0 - E$. This is just great enough to prevent us from measuring a negative kinetic energy. The penetration of the wave function into a classically forbidden region does have important consequences in barrier penetration, which will be discussed in Section 36-4.

Much of our discussion of the finite-well problem applies to any problem in which $E > U(x)$ in some region and $E < U(x)$ outside that region, as we see in the next section.

36-3 The Harmonic Oscillator

The potential energy for a particle of mass m attached to a spring of force constant K is

$$U(x) = \tfrac{1}{2}Kx^2 = \tfrac{1}{2}m\omega_0^2 x^2 \qquad \text{36-19}$$

where $\omega_0 = \sqrt{K/m}$ is the natural frequency of the oscillator. Classically, the object oscillates between $x = +A$ and $x = -A$. Its total energy is $E = \tfrac{1}{2}m\omega_0^2 A^2$, which can have any positive value, or zero.

This potential-energy function, shown in Figure 36-6, applies to any system undergoing small oscillations about a position of stable equilibrium. For example, it could apply to the oscillations of the atoms of a diatomic molecule such as H_2 or HCl oscillating about their equilibrium separation. In the region $-A \le x \le A$ between the classical turning points, the total energy is greater than the potential energy and the Schrödinger equation can be written

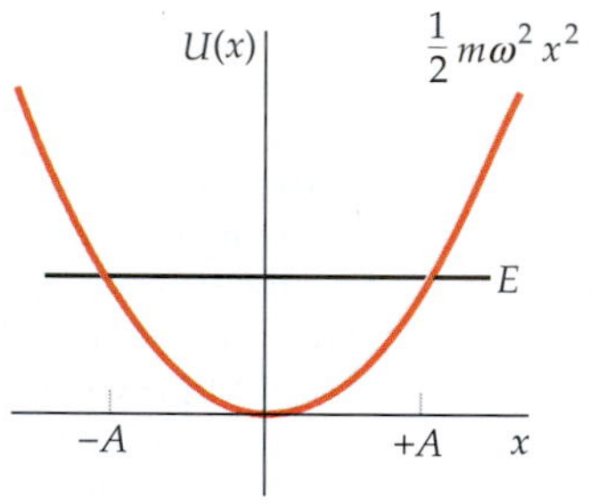

Figure 36-6 Harmonic oscillator potential.

$$\frac{d^2\psi(x)}{dx^2} = -k^2\psi(x) \qquad \text{36-20}$$

where $k^2 = 2m[E - U(x)]/\hbar^2$ now depends on x. The solutions of this equation are no longer simple sine or cosine functions because the wave number $k = 2\pi/\lambda$ now varies with x; but since $d^2\psi/dx^2$ and ψ have opposite signs, ψ will always curve toward the axis and the solutions will oscillate.

Outside the classical turning points ($|x| > A$), the potential energy is greater than the total energy and the Schrödinger equation is similar to Equation 36-17:

$$\frac{d^2\psi(x)}{dx^2} = +\alpha^2\psi(x) \qquad \text{36-21}$$

except that here $\alpha^2 = (2m/\hbar^2)[U(x) - E] > 0$ depends on x. For $|x| > A$, $d^2\psi/dx^2$ and ψ have the same sign, so ψ will curve away from the axis and there will be only certain values of E for which solutions exist that approach zero as x approaches infinity.

For the harmonic oscillator potential energy function, the Schrödinger equation is

$$-\frac{\hbar^2}{2m}\frac{d^2\psi(x)}{dx^2} + \tfrac{1}{2}m\omega_0^2x^2\psi(x) = E\psi(x) \qquad \text{36-22}$$

Wave Functions and Energy Levels

Rather than pursue a general solution to the Schrödinger equation for this system, we simply present the solution for the ground state and the first excited state.

The ground-state wave function $\psi_0(x)$ is found to be a Gaussian function centered at the origin:

$$\psi_0(x) = A_0 e^{-ax^2} \qquad \text{36-23}$$

where A_0 and a are constants. This function and the wave function for the first excited state are shown in Figure 36-7.

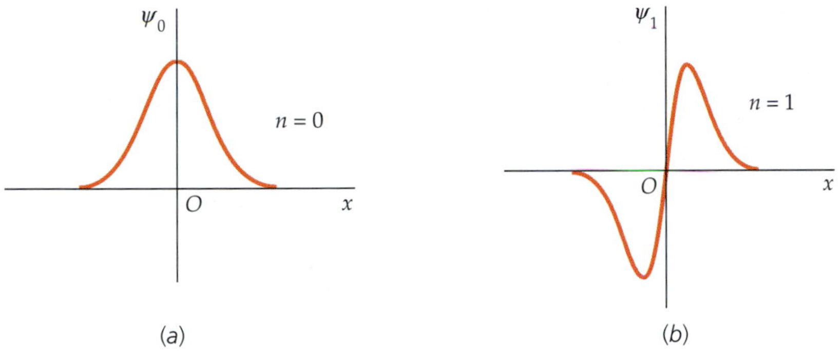

Figure 36-7 (*a*) Ground-state wave function for the harmonic oscillator potential. (*b*) Wave function for the first excited state of the harmonic oscillator potential.

Example 36-1

Verify that $\psi_0(x) = A_0 e^{-ax^2}$ is a solution of the Schrödinger equation for the harmonic oscillator.

Picture the Problem We take the first and second derivative of ψ with respect to x and substitute into Equation 36-22. Since this is the ground-state wave function, we write E_0 for the energy E.

1. Compute $d\psi_0/dx$: $\quad \dfrac{d\psi_0(x)}{dx} = -2axA_0e^{-ax^2}$

2. Compute $d^2\psi_0/dx^2$: $\quad \dfrac{d^2\psi_0(x)}{dx^2} = -2aA_0e^{-ax^2} + 4a^2x^2A_0e^{-ax^2}$

3. Substitute these derivatives into the Schrödinger equation:

$$-\frac{\hbar^2}{2m}(-2aA_0e^{-ax^2} + 4a^2x^2A_0e^{-ax^2}) + \frac{1}{2}m\omega_0^2x^2A_0e^{-ax^2} = E_0A_0e^{-ax^2}$$

4. Cancel the common factor $A_0e^{-ax^2}$: $\quad -\dfrac{\hbar^2}{2m}(-2a + 4a^2x^2) + \dfrac{1}{2}m\omega_0^2x^2 = E_0$

5. The equation in step 4 must hold for all x. Substitute $x = 0$: $\quad E_0 = \dfrac{\hbar^2 a}{m}$

6. Substitute this result into the equation in step 4: $-\frac{\hbar^2}{2m}(4a^2x^2) + \frac{1}{2}m\omega_0^2x^2 = 0$

7. The coefficients of the x^2 terms must equal zero: $-\frac{\hbar^2}{2m}(4a^2) + \frac{1}{2}m\omega_0^2 = 0$

8. Solve for a: $a^2 = \frac{m^2\omega_0^2}{4\hbar^2}$

$$a = \frac{m\omega_0}{2\hbar}$$

6. Substitute this result into the equation for E_0 in step 5: $E_0 = \frac{\hbar^2 a}{m} = \frac{1}{2}\hbar\omega_0$

Remark We have shown that the given function satisfies the Schrödinger equation for any value of A_0 as long as the energy E_0 is given by $E_0 = \frac{1}{2}\hbar\omega_0$. In general, when we have an equation such as $Ax^3 + Bx^2 + Cx + D = 0$, which must hold for *all values* of x, each coefficient must be zero. That is, $A = B = C = D = 0$.

We see from this example that the ground-state energy is given by

$$E_0 = \frac{\hbar^2 a}{m} = \frac{1}{2}\hbar\omega_0 \qquad 36\text{-}24$$

The first excited state has a node in the center of the potential well, just as with the particle in a box.† The wave function $\psi_1(x)$ is

$$\psi_1(x) = A_1xe^{-ax^2} \qquad 36\text{-}25$$

where $a = m\omega_0/2\hbar$, as in Example 36-1. This function is also shown in Figure 36-7. Substituting $\psi_1(x)$ into the Schrödinger equation as was done for $\psi_0(x)$ in Example 36-1 yields the energy of the first excited state,

$$E_1 = \frac{3}{2}\hbar\omega_0$$

In general, the energy of the nth excited state of the harmonic oscillator is

$$E_n = (n + \tfrac{1}{2})\hbar\omega_0 \qquad 36\text{-}26$$

as indicated in Figure 36-8. The fact that the energy levels are evenly spaced by the amount $\hbar\omega_0$ is a peculiarity of the harmonic oscillator potential. As we saw in Chapter 17, the energy levels for a particle in a box, or for the hydrogen atom, are not evenly spaced. The precise spacing of energy levels is closely tied to the particular form of the potential-energy function.

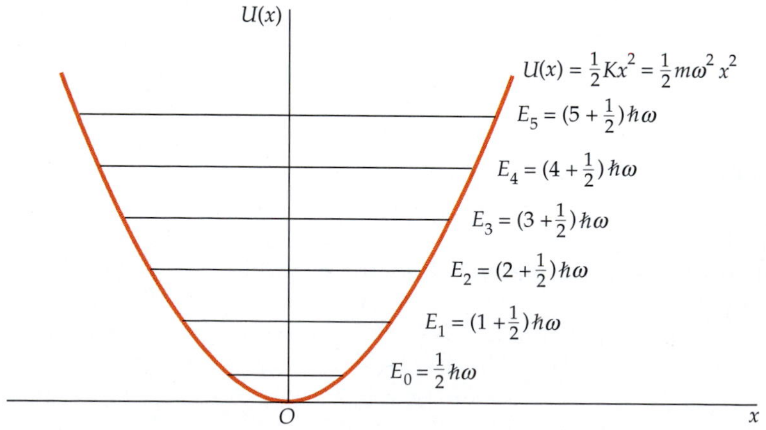

Figure 36-8 Energy levels in the harmonic oscillator potential.

† Each higher-energy state has one additional node in the wave function.

36-4 Reflection and Transmission of Electron Waves: Barrier Penetration

In Sections 36-2 and 36-3, we were concerned with bound-state problems in which the potential energy is larger than the total energy for large values of $|x|$. In this section, we consider some simple examples of unbound states for which E is greater than $U(x)$. For these problems, $d^2\psi/dx^2$ and ψ have opposite signs, so $\psi(x)$ curves toward the axis and does not become infinite at large values of $|x|$.

Step Potential

Consider a particle of energy E moving in a region in which the potential energy is the step function

$$U(x) = 0, \qquad x < 0$$

$$U(x) = U_0, \qquad x > 0$$

as shown in Figure 36-9. We are interested in what happens when a particle moving from left to right encounters the step.

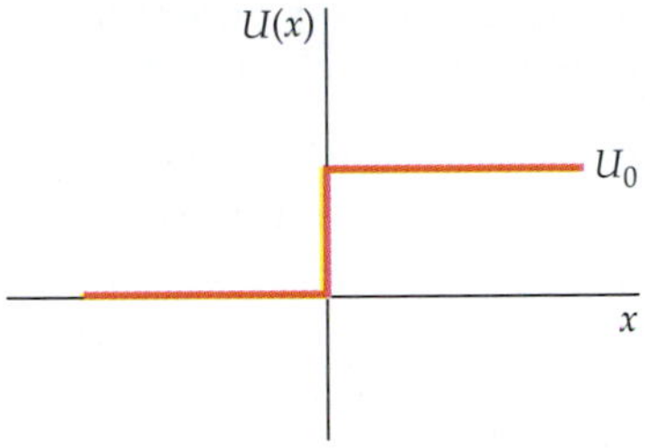

Figure 36-9 Step potential. A classical particle incident from the left, with total energy $E > U_0$, is always transmitted. The change in potential energy at $x = 0$ merely provides an impulsive force that reduces the speed of the particle. A wave incident from the left is partially transmitted and partially reflected because the wavelength changes abruptly at $x = 0$.

The classical answer is simple. To the left of the step, the particle moves with a speed $v = \sqrt{2E/m}$. At $x = 0$, an impulsive force acts on the particle. If the initial energy E is less than U_0, the particle will be turned around and will then move to the left at its original speed; that is, the particle will be reflected by the step. If E is greater than U_0, the particle will continue to move to the right but with reduced speed given by $v = \sqrt{2(E - U_0)/m}$. We can picture this classical problem as a ball rolling along a level surface and coming to a steep hill of height h given by $mgh = U_0$. If the initial kinetic energy of the ball is less than mgh, the ball will roll part way up the hill and then back down and to the left along the lower surface at its original speed. If E is greater than mgh, the ball will roll up the hill and proceed to the right at a lesser speed.

The quantum-mechanical result is similar when E is less than U_0. Figure 36-10 shows the wave function for the case $E < U_0$. The wave function does not go to zero at $x = 0$ but rather decays exponentially, like the wave function for the bound state in a finite square-well problem. The wave penetrates slightly into the classically forbidden region $x > 0$, but it is eventually completely reflected. This problem is somewhat similar to that of total internal reflection in optics.

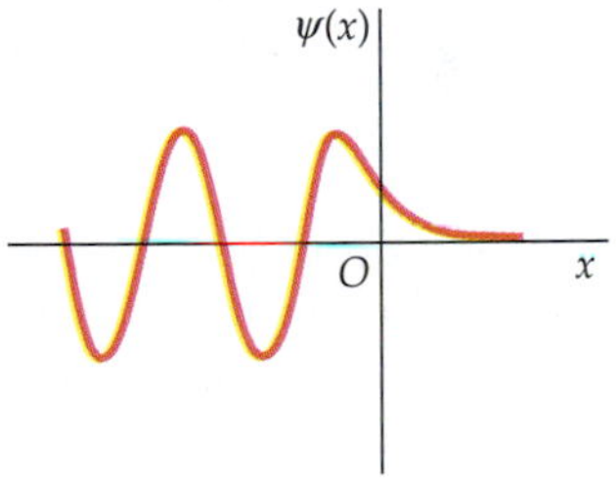

Figure 36-10 When the total energy E is less than U_0, the wave function penetrates slightly into the region $x > 0$. However, the probability of reflection for this case is 1, so no energy is transmitted.

For $E > U_0$, the quantum-mechanical result differs markedly from the classical result. At $x = 0$, the wavelength changes abruptly from $\lambda_1 = h/p_1 = h/\sqrt{2mE}$ to $\lambda_2 = h/p_2 = h/\sqrt{2m(E - U_0)}$. We know from our study of waves that when the wavelength changes suddenly, part of the wave is reflected and part is transmitted. Since the motion of an electron (or other particle) is governed by a wave equation, the electron will be sometimes transmitted and sometimes reflected. The probabilities of reflection and transmission can be calculated by solving the Schrödinger equation in each region of space and comparing the amplitudes of the transmitted and reflected waves with that of the incident wave. This calculation and its result are similar to finding the fraction of light reflected from an air–glass interface. If R is the probability of reflection, called the **reflection coefficient,** this calculation gives

$$R = \frac{(k_1 - k_2)^2}{(k_1 + k_2)^2} \qquad 36\text{-}27$$

where k_1 is the wave number for the incident wave and k_2 is that for the transmitted wave. This result is the same as that in optics for the reflection of light at normal incidence from the boundary between two media having different indexes of refraction n (Equation 33-11). The probability of transmission T, called the **transmission coefficient,** can be calculated from the reflection coefficient, since the probability of transmission plus the probability of reflection must equal 1:

$$T + R = 1 \qquad \text{36-28}$$

Example 36-2

A particle of energy E_0 traveling in a region in which the potential energy is zero is incident on a potential barrier of height $U_0 = 0.2E_0$. Find the probability that the particle will be reflected.

Picture the Problem We need to calculate the wave numbers k_1 and k_2 and use them to calculate the reflection coefficient R from Equation 36-27. The wave numbers are related to the kinetic energy K by $K = p^2/2m = \hbar^2k^2/2m$.

1. The probability of reflection is the reflection coefficient:

$$R = \frac{(k_1 - k_2)^2}{(k_1 + k_2)^2}$$

2. Calculate k_1 from the initial kinetic energy E_0:

$$\frac{\hbar^2 k_1^2}{2m} = E_0$$

$$k_1 = \sqrt{2mE_0/\hbar^2} = 1.41\sqrt{mE_0/\hbar^2}$$

3. Relate k_2 to the final kinetic energy K_2:

$$\frac{\hbar^2 k_2^2}{2m} = K_2 = E_0 - U_0 = E_0 - 0.2E_0 = 0.8E_0$$

4. Solve for k_2:

$$k_2 = \sqrt{2m(0.8E_0)/\hbar^2} = 1.26\sqrt{mE_0/\hbar^2}$$

5. Substitute these values to calculate R:

$$R = \frac{(k_1 - k_2)^2}{(k_1 + k_2)^2} = \frac{(1.41 - 1.26)^2}{(1.41 + 1.26)^2} = 0.00316$$

Remark The probability of reflection is only 0.3%. This probability is small because the barrier height reduces the kinetic energy by only 20%. Since k is proportional to the square root of the kinetic energy, the wave number and therefore the wave length is changed by only 10%.

Exercise Express the index of refraction n of light in terms of the wave number k, and show that Equation 33-11 for the reflection of light at normal incidence is the same as Equation 36-27.

In quantum mechanics, a localized particle is represented by a wave packet, which has a maximum at the most probable position of the particle. Figure 36-11 shows a wave packet representing a particle of energy E incident on a step potential of height U_0, which is less than E. After the encounter, there are two wave packets. The relative heights of the transmitted packet and reflected packet indicate the relative probabilities of transmission and reflection. For the situation shown here, E is much greater than U_0, and the probability of transmission is much greater than that of reflection.

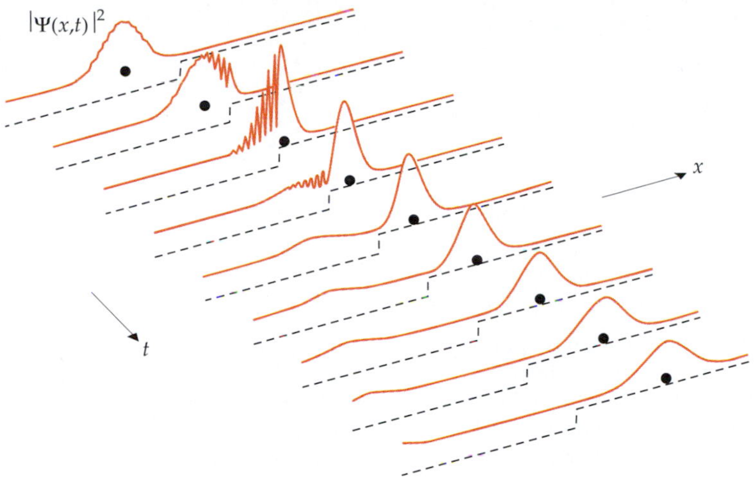

Figure 36-11 Time development of a one-dimensional wave packet representing a particle incident on a step potential for $E > U_0$. The position of a classical particle is indicated by the dot. Note that part of the packet is transmitted and part is reflected.

Barrier Penetration

Figure 36-12*a* shows a rectangular potential barrier of height U_0 and width a given by

$$U(x) = 0, \qquad x < 0$$
$$U(x) = U_0, \qquad 0 < x < a$$
$$U(x) = 0, \qquad x > a$$

We consider a particle of energy E, which is slightly less than U_0, that is incident on the barrier from the left. Classically, the particle would always be reflected. However, a wave incident from the left does not decrease immediately to zero at the barrier but will instead decay exponentially in the classically forbidden region $0 < x < a$. Upon reaching the far wall of the barrier ($x = a$), the wave function must join smoothly to a sinusoidal wave function to the right of the barrier as shown in Figure 36-12*b*. This implies that there is some probability of the particle (which is represented by the wave function) being found on the far side of the barrier even though, classically, it should never pass through the barrier. For the case in which the quantity $\alpha a = \sqrt{2ma^2(U_0 - E)/\hbar^2}$ is much greater than 1, the transmission coefficient is proportional to $e^{-2\alpha a}$:

$$T \propto e^{-2\alpha a} \qquad \text{36-29}$$

Transmission through a barrier

with $\alpha = \sqrt{2m(U_0 - E)/\hbar^2}$. The probability of penetration of the barrier thus decreases exponentially with the barrier thickness a and with the square root of the relative barrier height $(U_0 - E)$.

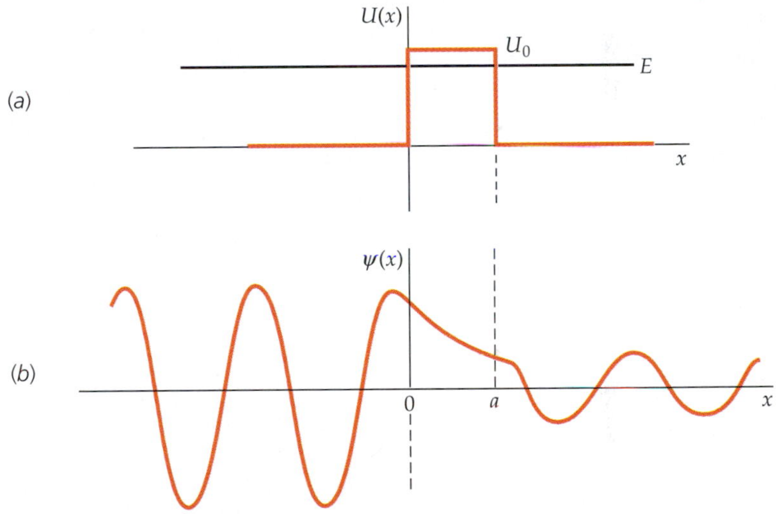

Figure 36-12 (*a*) Rectangular potential barrier. (*b*) Penetration of the barrier by a wave with total energy less than the barrier energy. Part of the wave is transmitted by the barrier even though, classically, the particle cannot enter the region $0 < x < a$ in which the potential energy is greater than the total energy.

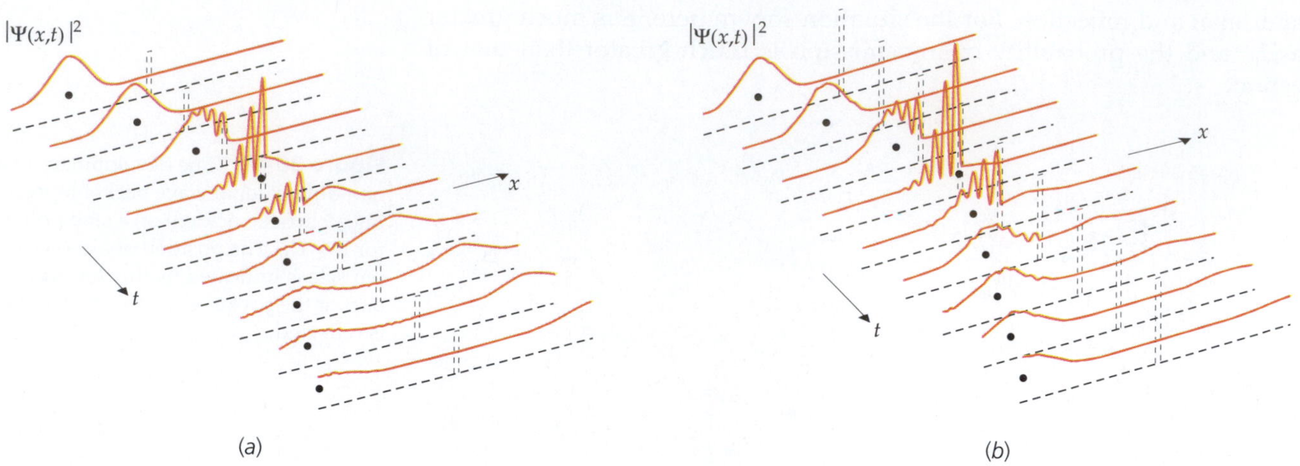

Figure 36-13 Barrier penetration. (*a*) A wave packet representing a particle incident on a barrier of height just slightly greater than the energy of the particle. For this particular choice of energies, the probability of transmission is approximately equal to the probability of reflection, as indicated by the relative sizes of the transmitted and reflected packets. (*b*) The same particle incident on a barrier of height much greater than the energy of the particle. A very small part of the packet tunnels through the barrier. In both drawings, the position of a classical particle is indicated by a dot.

Figure 36-13*a* shows a wave packet incident on a potential barrier of height U_0 that is considerably greater than the energy of the particle. The probability of penetration is very small, as indicated by the relative sizes of the reflected and transmitted packets. In Figure 36-13*b* the barrier is just slightly greater than the energy of the particle. In this case the probability of penetration is about the same as the probability of reflection. Figure 36-14 shows a particle incident on two potential barriers of height just slightly greater than the energy of the particle.

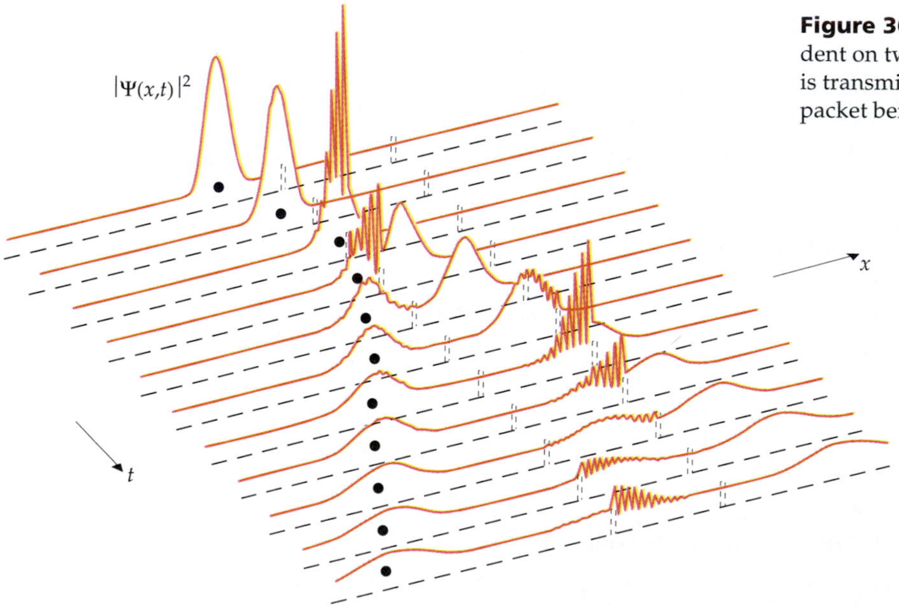

Figure 36-14 Wave packet representing a particle incident on two barriers. At each encounter, part of the packet is transmitted and part reflected, resulting in part of the packet being trapped between the barriers for some time.

As we have mentioned, the penetration of a barrier is not unique to quantum mechanics. When light is totally reflected from a glass–air interface, the light wave can penetrate the air barrier if a second piece of glass is brought within a few wavelengths of the first. This effect can be demonstrated with a laser beam and two 45° prisms (Figure 36-15). Similarly, water waves in a ripple tank can penetrate a gap of deep water (Figure 36-16).

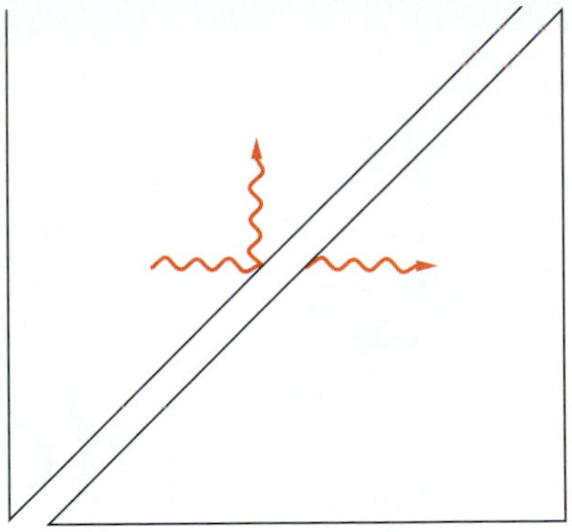

Figure 36-15 Penetration of an optical barrier. If the second prism is close enough to the first, part of the wave penetrates the air barrier even when the angle of incidence in the first prism is greater than the critical angle.

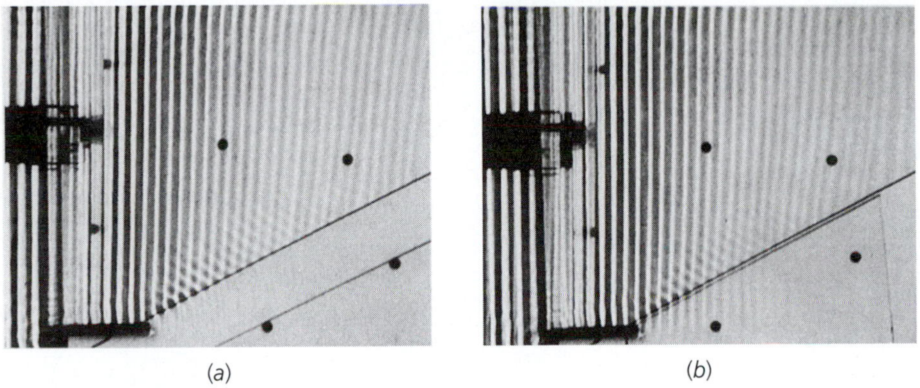

Figure 36-16 Penetration of a barrier by water waves in a ripple tank. In (*a*) the waves are totally reflected from a gap of deeper water. When the gap is very narrow, as in (*b*), a transmitted wave appears.

The theory of barrier penetration was used by George Gamow in 1928 to explain the enormous variation in the half-lives for α decay of radioactive nuclei. (Alpha particles are helium nuclei emitted from larger atoms in radioactive decay; they consist of two protons and two neutrons tightly bound together.) In general, the smaller the energy of the emitted α particle, the longer the half-life. The energies of α particles from natural radioactive sources range from about 4 to 7 MeV, whereas the half-lives range from about 10^{-5} second to 10^{10} years. Gamow represented a radioactive nucleus by a potential well containing an α particle as shown in Figure 36-17. Without knowing very much about the nuclear force that is exerted on the α particle within the nucleus, he represented it by a square well. Just outside the well, the α particle with its charge of $+2e$ is repelled by the nucleus with its charge $+Ze$, where Ze is the remaining nuclear charge. This force is represented by the Coulomb potential energy $+k(2e)(Ze)/r$. The energy E is the measured kinetic energy of the emitted α particle, because when it is far from the nucleus its potential energy is zero. After the α particle is formed inside the radioactive nucleus, it bounces back and forth inside the nucleus, hitting the barrier at the nuclear radius R. Each time it strikes the barrier, there is some small probability of its penetrating and appearing outside the nucleus. We can see from Figure 36-17 that a small increase in E reduces the relative height of the barrier $U - E$ and also its thickness. Because the probability of penetration is so sensitive to the barrier thickness and relative height, a small increase in E leads to a large increase in the probability of transmission and therefore to a shorter lifetime. Gamow was able to derive an expression for the half-life as a function of E that is in excellent agreement with experimental results.

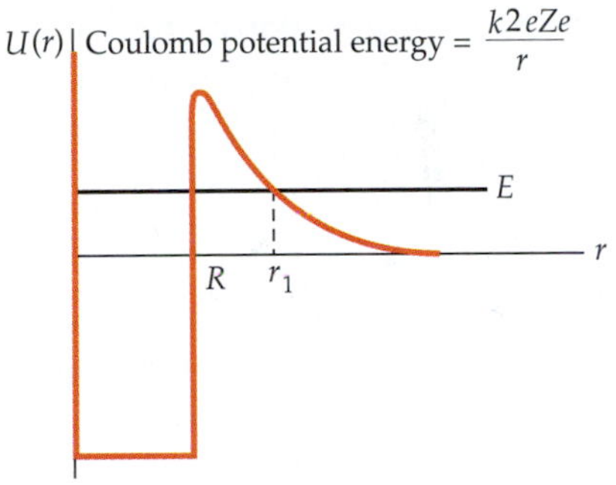

Figure 36-17 Model of a potential-energy function for an α particle in a radioactive nucleus. The strong attractive nuclear force when r is less than the nuclear radius R can be approximately described by the potential well shown. Outside the nucleus the nuclear force is negligible, and the potential is given by Coulomb's law, $U(r) = +k(2e)(Ze)/r$, where Ze is the nuclear charge and $2e$ is the charge of the α particle.

In the **scanning tunneling electron microscope** developed in the 1980s, a thin space between a material specimen and a tiny probe acts as a barrier to electrons bound in the specimen. A small voltage applied between the probe and specimen causes the electrons to *tunnel* through the vacuum separating the two surfaces if the surfaces are close enough together. The tunneling current is extremely sensitive to the size of the gap between the probe and specimen. If a constant tunneling current is maintained as the probe scans the specimen, the surface of the specimen can be mapped out by the motions of the probe. In this way, the surface features of a specimen can be measured with a resolution of the order of the size of an atom.

Exploring

The Scanning Tunneling Microscope

Ellen Williams
University of Maryland

Although all scientists believe that matter consists of atoms, the evidence that atoms exist is mostly indirect. Three methods are known for actually imaging individual atoms: transmission electron microscopy, field ion microscopy, and scanning tunneling microscopy. Of these, the last is the most recently developed and the most versatile. Gert Binnig and Heinrich Rohrer were awarded the 1986 Nobel Prize in Physics (which they shared with a third scientist) for developing this technique.

To understand scanning tunneling microscopy (STM), we first need to look at the behavior of electrons in metals. Classically, the most weakly bound electrons—those at the Fermi energy—can never leave the metal unless they are given the energy necessary to go over the potential barrier represented by their work function ϕ. Quantum mechanically, however, electrons near the Fermi energy can *tunnel* through the potential barrier. By placing two pieces of metal close to one another, as shown in Figure 1, a finite square-well barrier can be created. The probability that electrons at the Fermi energy will tunnel through the barrier is proportional to $e^{-\alpha a}$, where a is the distance separating the two pieces of metal, and α depends on the barrier height—in this case, the work function. This exponential dependence of the transmission probability on separation is what makes STM possible.

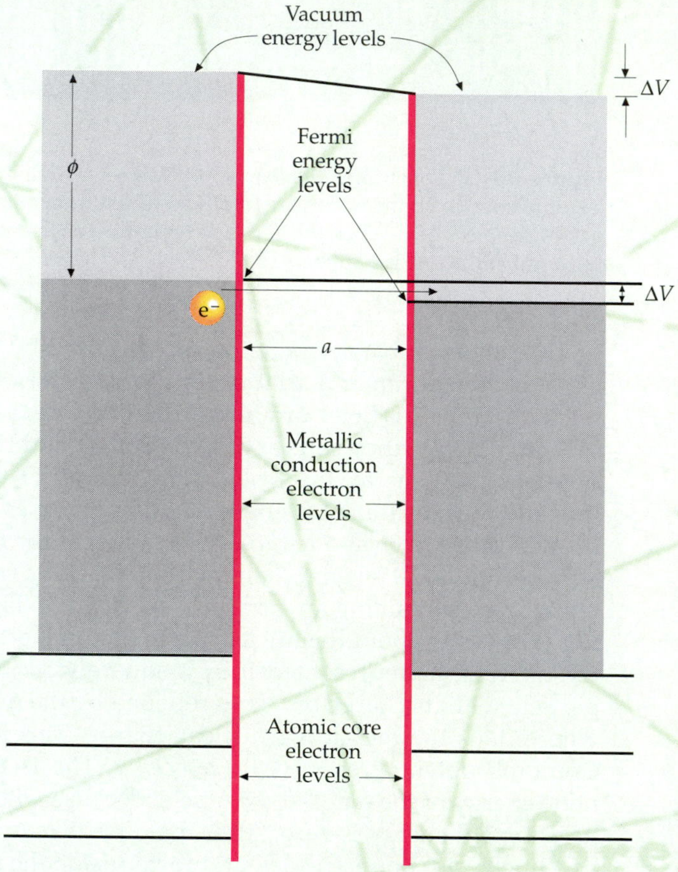

Figure 1 Energy-level diagram for the conduction electrons in two metals separated by a distance a. The electrons move as essentially free particles within the metal. The most weakly bound electrons have the highest energy (the Fermi energy E_F) and are held within the solid by a potential-energy barrier known as the work function ϕ. To induce a measurable electron tunneling current, a small voltage difference ΔV is applied between the two metals. This offsets their energy levels as shown, so that electrons can flow from the occupied states near E_F on the more negative metal (on the left in the figure) to the unoccupied states just above E_F on the more positive metal (on the right in the figure).

Ellen Williams earned her BS in chemistry at Michigan State University and her doctorate at the California Institute of Technology, studying the properties of thin layers of molecules on surfaces. Since then, she has been at the Department of Physics and Astronomy at the University of Maryland.

The mechanism of STM is illustrated in Figure 2. If a pointed metal probe is placed sufficiently close to a sample and a small voltage is applied between the probe and the sample, electron tunneling can occur. The net flow of electrons can be measured as a tunneling current proportional to the transmission probability. If we then scan the probe back and forth above the sample, any bumps on the sample surface will change the separation. Because of the exponential relationship between the separation and the transmission probability, changes as small as 0.01 nm result in measurable changes in the tunneling current. Measurement of the tunneling current while scanning thus generates a topographic map of the surface. In principle, it is possible to image individual atoms on the surface using STM.

Three formidable problems had to be overcome before individual atoms could be imaged. Vibration was an important challenge, because the separation between the sample and the probe must be very small—typically, only a few nanometers (comparable to the size of atoms). A minor perturbation can jam the probe into the sample, ruining the experiment. The most common problem is floor vibration, typically with an amplitude of about 1 μm—a thousand times larger than the allowable separation between tip and sample. Thus, very careful engineering is required.

The second problem is probe sharpness, which determines how small a structure can be imaged. Electrochemical etching can sharpen the end of a metal wire to a radius of about 1 μm (1000 nm). A probe with such a large surface area would allow tunneling to occur over too large a region of the sample surface. To resolve small features such as atoms, the probe must be comparable in size to the features. In trying to fabricate such a probe, we are rewarded for our technological shortcomings. Our most polished wires are rough on an atomic scale, having mini-tips much like the one illustrated in Figure 2. The end of such a mini-tip will present one or a few atoms close to the surface. The exponential dependence of transmission probability on separation then guarantees that tunneling can occur preferentially from the end of the mini-tip.

The third problem in STM is position control. How can the probe be moved around to an accuracy of less than 0.1 nm? The answer lies in a type of material known as piezoelectric ceramic, which expands and contracts when an external voltage is applied to electrodes on opposite faces. Typically, expansions are on the order of a few tenths of a nanometer per volt. A probe attached to a piece of piezoelectric ceramic can be moved with great precision.

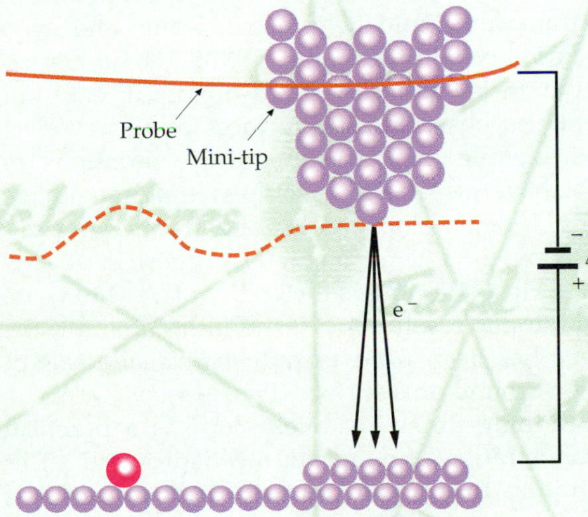

Figure 2 Schematic illustration of the path of a probe (dashed line) scanned across a surface while constant tunneling current is maintained. If the probe is very large (as illustrated by the solid line), tunneling occurs over a large area and atomic features cannot be resolved. However, if the probe has a mini-tip of atomic dimensions, then tunneling occurs into a small area, allowing very small features (even individual atoms) to be imaged.

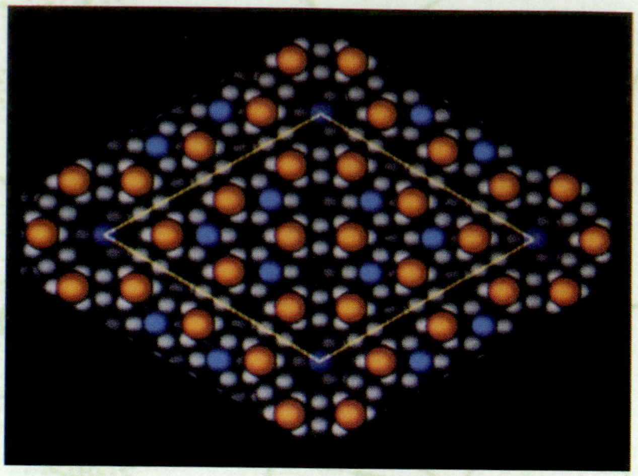

(a)

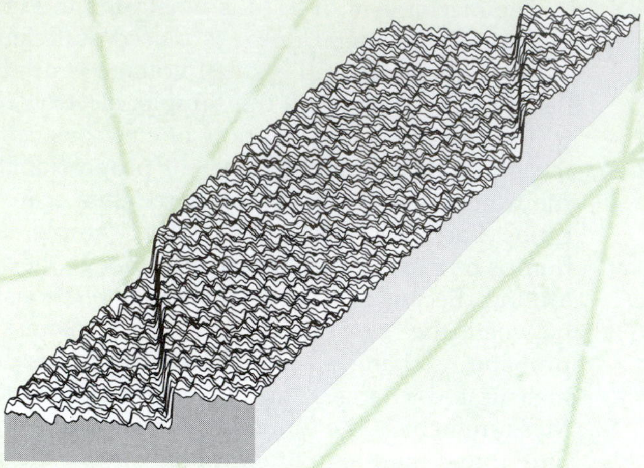

(b)

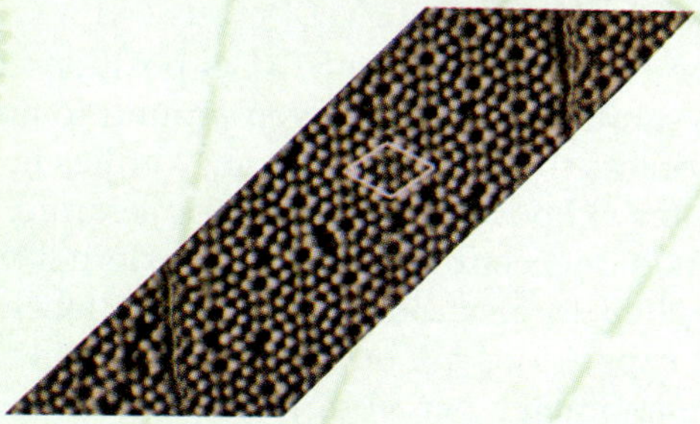

(c)

Figure 3 STM imaging of atoms and steps on a silicon surface. (*a*) Model of the atoms on a silicon surface. The red circles represent atoms that protrude highest above the surface, the blue circles are atoms in a lower layer, and the gray circles are atoms in a still lower layer. The white diamond shows the repeating unit of the structure. The length of each side of this unit cell is 2.7 nm. (*b*) Traces of height versus position (line scans) obtained by an STM over an area of approximately 10 by 35 nm. (*c*) The same data as presented in (*b*) in a gray-scale representation, allowing the imaging of the highest layer of atoms in the atomic model of (*a*) to be immediately recognized. The unit cell of the structure is indicated for comparison with (*a*).

The power of STM is illustrated in Figure 3, where an atomic model of a silicon surface is compared with STM images of the real surface. The data from an STM scan consist of values of surface height versus position and can be presented as a line scan (the dashed line in Figure 2). The line scan image of Figure 3*b* is easier to visualize if the data are represented by a gray scale (Figure 3*c*): Height is represented by intensity of color—ranging from white for the highest points to black for the lowest. Note the striking correspondence to the model in Figure 3*a*. The deep holes correspond to the positions of missing atoms in the model, and the bright spots are due to the atoms that protrude above the average surface plane. Two abrupt changes in height also appear in this image; these surface steps are important in practical processes such as crystal growth and microfabrication.

In addition to studies of atoms at surfaces, STM has a range of practical applications, in part because STM is quite insensitive to its microscopic environment. Tunneling microscopes operate in vacuum, air, liquid helium, oil, water, and even electrolytic solutions. Thus, STM is useful, for example, in imaging DNA in a biological environment and observing the surfaces of battery electrodes while they are operating. Variations of STM have also been developed that can image samples that are not conductors (atomic force microscopy) or image magnetic properties at surfaces. Most stunning is the ability of STM to write with atomic resolution. At the ultimate limit, the probe has been used to pull *individual atoms* of xenon around on a surface (Figure 4).

Scanning tunneling microscopy is a practical demonstration of quantum mechanics and an illustration that understanding basic concepts of physics can yield tremendous gains in advanced technology. It is also an object lesson in the long-term and often unforeseeable benefits that accrue from developing fundamental ideas.

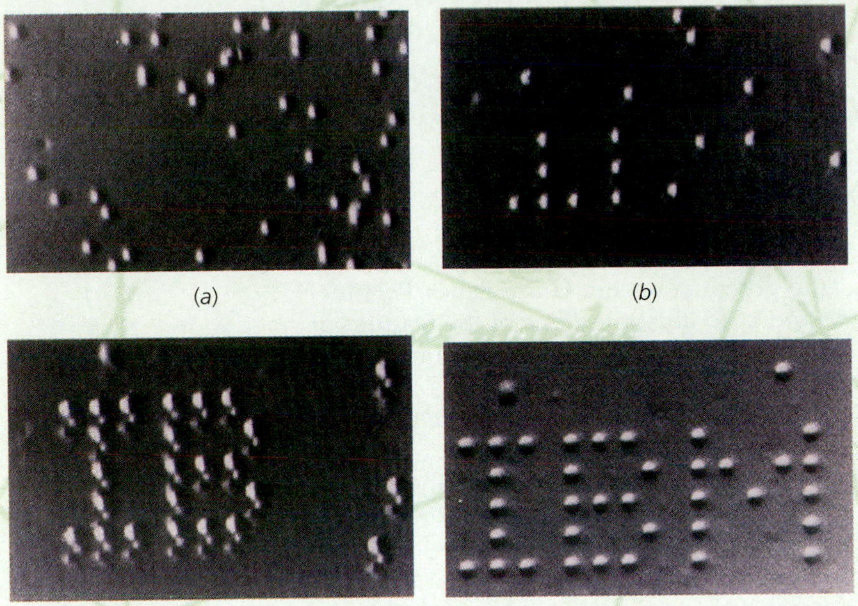

(a) (b)

(c) (d)

Figure 4 Sequence of STM images taken during the construction of a patterned array of xenon atoms on a nickel surface at a temperature of 4 K. Xenon atoms were allowed to stick randomly on the surface from the gas phase (upper left). The STM tip was then used to "nudge" the atoms one by one across the surface to spell out the name of the company that sponsored the development of STM.

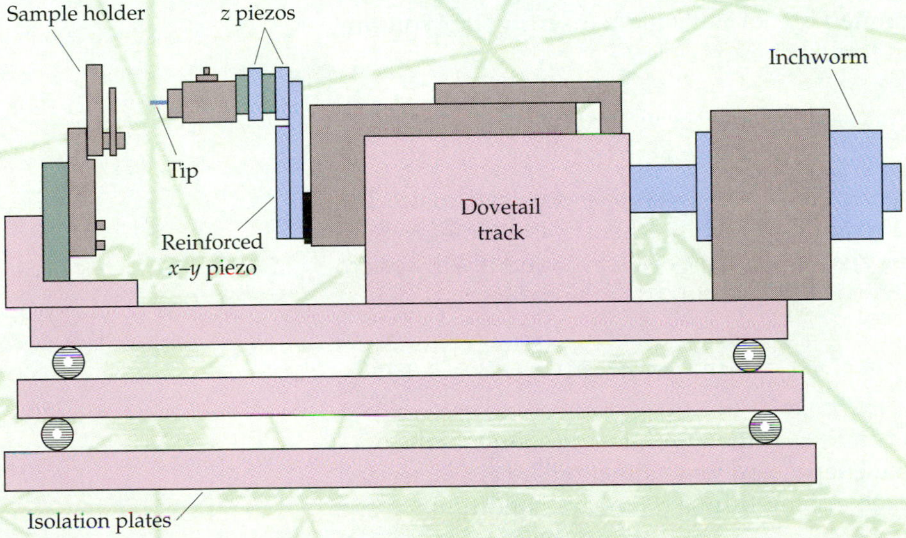

Figure 5 Schematic diagram of an STM. The sample holder is rigidly mounted to the top plate in a stack of isolation plates. The tip is fixed onto the x, y, z scanning piezos. To position the tip so tunneling can occur, these piezos are mounted on a heavy block that can slide on a dovetail track. The block is then pushed by an electronically controlled "walker" device called an inchworm, which can move forward and back in steps of 4 nm.

36-5 The Schrödinger Equation in Three Dimensions

The one-dimensional time-independent Schrödinger equation is easily extended to three dimensions. In rectangular coordinates, it is

$$-\frac{\hbar^2}{2m}\left(\frac{\partial^2\psi}{\partial x^2}+\frac{\partial^2\psi}{\partial y^2}+\frac{\partial^2\psi}{\partial z^2}\right)+U\psi=E\psi \qquad \text{36-30}$$

where the wave function ψ and the potential energy U are generally functions of all three coordinates, x, y, and z. To illustrate some of the features of problems in three dimensions, we consider a particle in a three-dimensional infinite square well given by $U(x, y, z) = 0$ for $0 < x < L$, $0 < y < L$, and $0 < z < L$. Outside this cubical region $U(x, y, z) = \infty$. For this problem, the wave function must be zero at the edges of the well.

There are standard methods in partial differential equations for solving Equation 36-30. We can guess the form of the solution from our knowledge of probability. For a one-dimensional box along the x axis, we have found the probability that a particle is in the region dx at x to be $A_1^2 \sin^2 k_1x\, dx$ (from Equation 36-10), where A_1 is a normalization constant and $k_1 = n\pi/L$ is the wave number. Similarly, for a box along the y axis, the probability of a particle being in a region dy at y is $A_2^2 \sin^2 k_2y\, dy$. The probability of two independent events occurring is the product of the probabilities of each event occurring.† So the probability of a particle being in region dx at x *and* in region dy at y is $A_1^2 \sin^2 k_1x\, dx\, A_2^2 \sin^2 k_2y\, dy = A_1^2 \sin^2 k_1x\, A_2^2 \sin^2 k_2y\, dx\, dy$. The probability of a particle being in the region dx, dy, and dz is $\psi(x, y, z)\, dx\, dy\, dz$, where $\psi(x, y, z)$ is the solution of Equation 36-30. This solution is of the form

$$\psi(x, y, z) = A \sin k_1x \sin k_2y \sin k_3z \qquad \text{36-31}$$

where the constant A is determined by normalization. Inserting this solution into Equation 36-30, we obtain for the energy

$$E = \frac{\hbar^2}{2m}(k_1^2 + k_2^2 + k_3^2)$$

which is equivalent to $E = (p_x^2 + p_y^2 + p_z^2)/2m$, with $p_x = \hbar k_1$, and so on. The wave function will be zero at $x = L$ if $k_1 = n_1\pi/L$, where n_1 is an integer. Similarly, the wave function will be zero at $y = L$ if $k_2 = n_2\pi/L$, and it will be zero at $z = L$ if $k_3 = n_3\pi/L$. The energy is thus quantized to the values

$$E_{n_1,n_2,n_3} = \frac{\hbar^2\pi^2}{2mL^2}(n_1^2 + n_2^2 + n_3^2) = E_1(n_1^2 + n_2^2 + n_3^2) \qquad \text{36-32}$$

where n_1, n_2, and n_3 are integers and E_1 is the ground-state energy of the one-dimensional well. Note that the energy and wave function are characterized by three quantum numbers, each arising from a boundary condition for one of the coordinates.

The lowest energy state (the ground state) for the cubical well occurs when $n_1 = n_2 = n_3 = 1$ and has the value

$$E_{1,1,1} = \frac{3\hbar^2\pi^2}{2mL^2} = 3E_1$$

† For example, if you throw two dice, the probability of the first die coming up 6 is 1/6 and the probability of the second die coming up an odd number is 1/2. The probability of the first die coming up 6 *and* the second die coming up an odd number is (1/6)(1/2) = 1/12.

The first excited energy level can be obtained in three different ways: $n_1 = 2$, $n_2 = n_3 = 1$; $n_2 = 2$, $n_1 = n_3 = 1$; or $n_3 = 2$, $n_1 = n_2 = 1$. Each has a different wave function. For example, the wave function for $n_1 = 2$ and $n_2 = n_3 = 1$ is

$$\psi_{2,1,1} = A \sin \frac{2\pi x}{L} \sin \frac{\pi y}{L} \sin \frac{\pi z}{L} \qquad \text{36-33}$$

There are thus three different quantum states as described by the three different wave functions corresponding to the same energy level. An energy level with which more than one wave function is associated is said to be **degenerate.** In this case, there is threefold degeneracy. Degeneracy is related to the spatial symmetry of the problem. If, for example, we consider a noncubic well where $U = 0$ for $0 < x < L_1$, $0 < y < L_2$, and $0 < z < L_3$, the boundary conditions at the edges would lead to the quantum conditions $k_1L_1 = n_1\pi$, $k_2L_2 = n_2\pi$, and $k_3L_3 = n_3\pi$, and the total energy would be

$$E_{n_1,n_2,n_3} = \frac{\hbar^2\pi^2}{2m}\left(\frac{n_1^2}{L_1^2} + \frac{n_2^2}{L_2^2} + \frac{n_3^2}{L_3^2}\right) \qquad \text{36-34}$$

These energy levels are not degenerate if L_1, L_2, and L_3 are not equal. Figure 36-18 shows the energy levels for the ground state and first two excited states for an infinite cubic well in which the excited states are degenerate and for a noncubic infinite well in which L_1, L_2, and L_3 are slightly different so that the excited levels are slightly split apart and the degeneracy is removed.

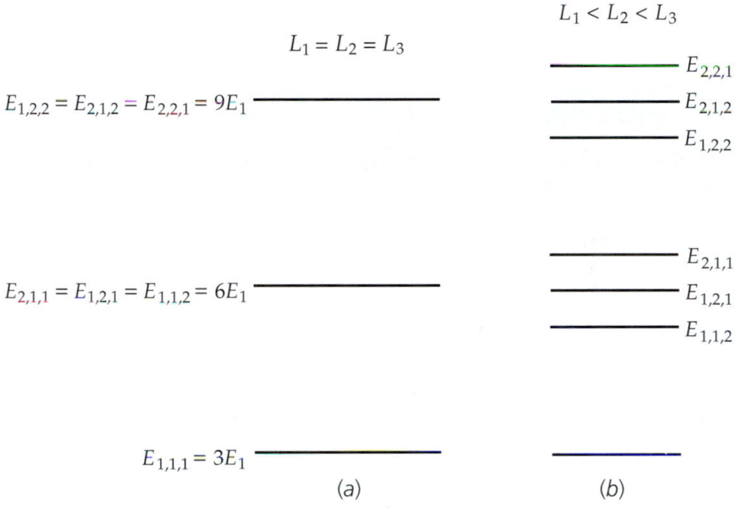

Figure 36-18 Energy-level diagrams for (*a*) a cubic infinite well and (*b*) a noncubic infinite well. In (*a*) the energy levels are degenerate; that is, there are two or more wave functions having the same energy. The degeneracy is removed when the symmetry of the potential is removed, as in (*b*).

Example 36-3

A particle is in a three-dimensional box with $L_3 = L_2 = 2L_1$. Give the quantum numbers n_1, n_2, n_3 that correspond to the lowest ten quantum states of this box.

Picture the Problem We can use Equation 36-34 to write the energies in terms of the ratios $L_2/L_1 = 2$ and $L_3/L_1 = 2$, then find by inspection the values of the quantum numbers that give the lowest energies.

1. The energy of a level is given by Equation 36-34:

$$E_{n_1,n_2,n_3} = \frac{\hbar^2\pi^2}{2m}\left(\frac{n_1^2}{L_1^2} + \frac{n_2^2}{L_2^2} + \frac{n_3^2}{L_3^2}\right)$$

2. Factor out $1/L_1^2$:

$$E_{n_1,n_2,n_3} = \frac{\hbar^2\pi^2}{2mL_1^2}\left(n_1^2 + n_2^2\frac{L_1^2}{L_2^2} + n_3^2\frac{L_1^2}{L_3^2}\right)$$
$$= E_1(n_1^2 + n_2^2/4 + n_3^2/4)$$

3. The lowest energy is $E_{1,1,1}$:

$$E_{1,1,1} = E_1\,(1^2 + 1^2/4 + 1^2/4) = 1.5E_1 \quad \text{(1st)}$$

4. The energy increases the least when we increase n_2 or n_3. Try various values of the quantum numbers:

$$E_{1,2,1} = E_{1,1,2} = E_1(1^2 + 2^2/4 + 1/4) = 2.25E_1 \quad \text{(2nd and 3rd)}$$
$$E_{1,2,2} = E_1(1^2 + 2^2/4 + 2^2/4) = 3.0E_1 \quad \text{(4th)}$$
$$E_{1,3,1} = E_{1,1,3} = E_1(1^2 + 3^2/4 + 1^2/4) = 3.50E_1 \quad \text{(5th and 6th)}$$
$$E_{1,3,2} = E_{1,2,3} = E_1(1^2 + 3^2/4 + 2^2/4) = 4.25E_1 \quad \text{(7th and 8th)}$$
$$E_{2,1,1} = E_1(2^2 + 1^2/4 + 1^2/4) = 4.5E_1 \quad \text{(9th)}$$
$$\left.\begin{aligned} E_{2,2,1} = E_{2,1,2} &= E_1(2^2 + 2^2/4 + 1^2/4) = 5.25E_1 \\ E_{1,4,1} = E_{1,1,4} &= E_1(1^2 + 4^2/4 + 1^2/4) = 5.25E_1 \end{aligned}\right\} \text{(10}^{\text{th}}\text{, 11}^{\text{th}}\text{, 12}^{\text{th}}\text{, and 13}^{\text{th}}\text{)}$$

Remarks Note the degeneracy of the levels.

Exercise Find the quantum numbers and energies of the next four levels in step 4. (Answer $E_{1,3,3} = 5.5E_1$, $E_{1,4,2} = E_{1,2,4} = E_{2,2,2} = 6.0E_1$)

Example 36-4 ***try it yourself***

Write the degenerate wave functions for the 4th and 5th excited states (levels 5 and 6) of the results in step 4 of Example 36-3

Picture the Problem Use Equation 36-33 with $k_i = n_i\pi/L$.

Cover the column to the right and try these on your own before looking at the answers.

Steps	***Answers***
1. Write the wave functions corresponding to the energies $E_{1,3,1}$ and $E_{1,1,3}$.	$\psi_{1,3,1} = A \sin\frac{\pi x}{L} \sin\frac{3\pi y}{L} \sin\frac{\pi z}{L}$ $\psi_{1,1,3} = A \sin\frac{\pi x}{L} \sin\frac{\pi y}{L} \sin\frac{3\pi z}{L}$

36-6 The Schrödinger Equation for Two Identical Particles

Our discussion of quantum mechanics has thus far been limited to situations in which a single particle moves in some force field characterized by a potential-energy function U. The most important physical problem of this type is the hydrogen atom, in which a single electron moves in the Coulomb potential of the proton nucleus. This problem is actually a two-body problem since the proton also moves in the field of the electron. However, the motion of the much more massive proton requires only a very small correction to the energy of the atom that is easily made in both classical and quantum mechanics. When we consider more complicated problems, such as the helium atom, we must apply quantum mechanics to two or more electrons moving in an external field. Such problems are complicated by the interaction of the electrons with each other and also by the fact that the electrons are identical.

The interaction of two electrons with each other is electromagnetic and is essentially the same as the classical interaction of two charged particles. The Schrödinger equation for an atom with two or more electrons cannot be solved exactly, so approximation methods must be used. This is not very different from the situation in classical problems with three or more particles. However, the complications arising from the identity of electrons are purely quantum mechanical and have no classical counterpart. They are due to the fact that it is impossible to keep track of which electron is which. Classically, identical particles can be identified by their positions, which can be determined with unlimited accuracy. This is impossible quantum mechanically because of the uncertainty principle. Figure 36-19 offers a schematic illustration of the problem.

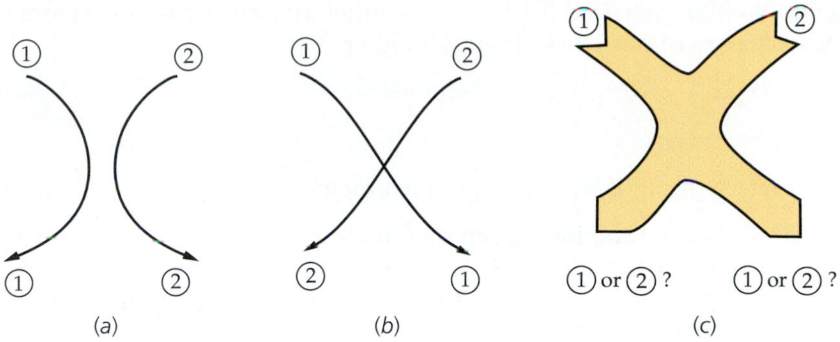

Figure 36-19 Two possible classical electron paths (*a* and *b*). If electrons were classical particles they could be distinguished by the path followed. But because of the quantum-mechanical wave properties of electrons, the paths are spread out, as indicated by the shaded region in (*c*). It is impossible to distinguish which electron is which after they separate.

The indistinguishability of identical particles has important consequences. For instance, consider the very simple case of two identical, noninteracting particles in a one-dimensional infinite square well. The time-independent Schrödinger equation for two particles, each of mass m, is

$$-\frac{\hbar^2}{2m}\frac{\partial^2\psi(x_1, x_2)}{\partial x_1^2} - \frac{\hbar^2}{2m}\frac{\partial^2\psi(x_1, x_2)}{\partial x_2^2} + U\psi(x_1, x_2) = E\psi(x_1, x_2) \qquad \text{36-35}$$

where x_1 and x_2 are the coordinates of the two particles. If the particles interact, the potential energy U contains terms with both x_1 and x_2 that cannot be separated into separate terms containing only x_1 or x_2. For example, the electrostatic repulsion of two electrons in one dimension is represented by the potential energy $ke^2/|x_2 - x_1|$. However, if the particles do not interact (as we are assuming here), we can write $U = U_1(x_1) + U_2(x_2)$. For the infinite square well, we need only solve the Schrödinger equation inside the well where $U = 0$, and require that the wave function be zero at the walls of the well. With $U = 0$, Equation 36-35 looks just like the expression for a particle in a two-dimensional well (Equation 36-30, with no z and with y replaced by x_2).

Solutions of this equation can be written in the form[†]

$$\psi_{n,m} = \psi_n(x_1)\psi_m(x_2) \qquad \text{36-36}$$

where ψ_n and ψ_m are the single-particle wave functions for a particle in an infinite well and n and m are the quantum numbers of particles 1 and 2, respectively. For example, for $n = 1$ and $m = 2$, the wave function is

$$\psi_{1,2} = A \sin\frac{\pi x_1}{L} \sin\frac{2\pi x_2}{L} \qquad \text{36-37}$$

The probability of finding particle 1 in dx_1 *and* particle 2 in dx_2 is $\psi^2_{n,m}(x_1, x_2)\, dx_1\, dx_2$, which is just the product of the separate probabilities $\psi^2_n(x_1)\, dx_1$ and $\psi^2_m(x_2)\, dx_2$. However, even though we have labeled the particles 1 and 2, we cannot distinguish which is in dx_1 and which is in dx_2 if they are identical. The mathematical descriptions of identical particles must be the same if we interchange the labels. The probability density $\psi^2(x_1, x_2)$ must therefore be the same as $\psi^2(x_2, x_1)$:

$$\psi^2(x_2, x_1) = \psi^2(x_1, x_2) \qquad \text{36-38}$$

[†] Again, this result can be obtained by solving Equation 36-35, but it also can be understood in terms of our knowledge of probability. The probability of electron 1 being in region dx_1 *and* electron 2 being in region dx_2 is the product of the individual probabilities.

Equation 36-38 is satisfied if $\psi(x_2, x_1)$ is either **symmetric** or **antisymmetric** on the exchange of particles—that is, if either

$$\psi(x_2, x_1) = \psi(x_1, x_2) \qquad \text{symmetric} \tag{36-39}$$

or

$$\psi(x_2, x_1) = -\psi(x_1, x_2) \qquad \text{antisymmetric} \tag{36-40}$$

Note that the wave functions given by Equations 36-36 and 36-37 are neither symmetric nor antisymmetric. If we interchange x_1 and x_2 in these wave functions, we get a different wave function, which implies that the particles can be distinguished.

We can find symmetric and antisymmetric wave functions that are solutions of the Schrödinger equation by adding or subtracting $\psi_{n,m}$ and $\psi_{m,n}$. Adding them, we obtain

$$\psi_S = A'[\psi_n(x_1)\psi_m(x_2) + \psi_n(x_2)\psi_m(x_1)] \quad \text{symmetric} \tag{36-41}$$

and subtracting them, we obtain

$$\psi_A = A'[\psi_n(x_1)\psi_m(x_2) - \psi_n(x_2)\psi_m(x_1)] \quad \text{antisymmetric} \tag{36-42}$$

For example, the symmetric and antisymmetric wave functions for the first excited state of two identical particles in an infinite square well would be

$$\psi_S = A'\left(\sin\frac{\pi x_1}{L}\sin\frac{2\pi x_2}{L} + \sin\frac{\pi x_2}{L}\sin\frac{2\pi x_1}{L}\right) \tag{36-43}$$

and

$$\psi_A = A'\left(\sin\frac{\pi x_1}{L}\sin\frac{2\pi x_2}{L} - \sin\frac{\pi x_2}{L}\sin\frac{2\pi x_1}{L}\right) \tag{36-44}$$

There is an important difference between antisymmetric and symmetric wave functions. If $n = m$, the antisymmetric wave function is identically zero for all values of x_1 and x_2 whereas the symmetric wave function is not. Thus, if the wave function describing two identical particles is antisymmetric, the quantum numbers n and m of two particles cannot be the same. This is an example of the **Pauli exclusion principle,** which was first stated by Wolfgang Pauli for electrons in an atom:

> No two electrons in an atom can have the same quantum numbers.

Pauli exclusion principle

It is found that electrons, protons, neutrons, and some other particles have antisymmetric wave functions and obey the Pauli exclusion principle. These particles are called **fermions.** Other particles, such as α particles, deuterons, photons, and mesons, have symmetric wave functions and do not obey the Pauli exclusion principle. These particles are called **bosons.**

Summary

1. The Schrödinger equation is a differential equation that relates the second spatial derivative of a wave function to its first time derivative. Wave functions that describe physical situations are solutions of this differential equation.
2. Because a wave function must be normalizable, it must be well behaved; that is, it must approach zero as x approaches infinity. For bound systems such as a particle in a

box, a simple harmonic oscillator, or an electron in an atom, this requirement leads to energy quantization.

3. The well-behaved wave functions for bound systems describe standing waves.

Topic	Remarks and Relevant Equations	
1. Time-Independent Schrödinger Equation	$-\frac{\hbar^2}{2m}\frac{d^2\psi(x)}{dx^2} + U(x)\psi(x) = E\psi(x)$	**36-4**
Allowable solutions	In addition to satisfying the Schrödinger equation, a wave function $\psi(x)$ must be continuous and (if U is not infinite) must have a continuous first derivative $d\psi/dx$. Because the probability of finding an electron somewhere must be 1, the wave function must obey the normalization condition $\int_{-\infty}^{\infty} \lvert\psi\rvert^2\,dx = 1$ This condition implies the boundary condition that ψ must approach 0 as x approaches $\pm\infty$. Such boundary conditions lead to the quantization of energy.	
2. Confined Particles	When the total energy E is greater than the potential energy $U(x)$ in some region (the classically allowed region) and less than $U(x)$ outside that region, the wave function oscillates within the classically allowed region and increases or decreases exponentially outside that region. The wave function approaches zero as x approaches ∞ only for certain values of the total energy E. The energy is thus quantized.	
Finite square well	In a finite well of height U_0, there are only a finite number of allowed energies, and these are slightly less than the corresponding energies in an infinite well.	
Simple harmonic oscillator	In the oscillator potential energy function $U(x) = \frac{1}{2}m\omega_0^2x^2$, the allowed energies are equally spaced and given by	
	$E_n = (n + \frac{1}{2})\hbar\omega_0$	**36-26**
	The ground-state wave function is given by	
	$\psi_0(x) = A_0 e^{-ax^2}$	**36-23**
	where A_0 is the normalization constant and $a = m\omega_0/2\hbar$.	
3. Reflection and Barrier Penetration	When the potential changes abruptly over a small distance, a particle may be reflected even though $E > U(x)$. A particle may penetrate a region in which $E < U(x)$. Reflection and penetration of electron waves are similar to those for other kinds of waves.	
4. Schrödinger Equation in Three Dimensions	The wave function for a particle in a three-dimensional box can be written $\psi(x, y, z) = \psi_1(x)\psi_2(y)\psi_3(z)$ where ψ_1, ψ_2, and ψ_3 are wave function for a one-dimensional box.	
Degeneracy	When more than one wave function is associated with the same energy level, the energy level is said to be degenerate. Degeneracy arises because of spatial symmetry.	
5. Schrödinger Equation for Two Identical Particles	A wave function that describes two identical particles must be either symmetric or antisymmetric when the coordinates of the particles are exchanged. Fermions, which include electrons, protons, and neutrons, are described by antisymmetric wave functions and obey the Pauli exclusion principle, which states that no two particles can have the same values for their quantum number. Bosons, which include α particles, deuterons, photons, and mesons, have symmetric wave functions and do not obey the Pauli exclusion principle.	

Problem-Solving Guide

Begin by drawing a neat diagram that includes the important features of the problem.

Summary of Worked Examples

Type of Calculation	Procedure and Relevant Examples
1. Harmonic Oscillator	
Verify that a given function satisfies the Schrödinger equation	Take the first and second derivative with respect to x and substitute into the Schrödinger equation. **Example 36-1**
2. Transmission and Reflection	
Calculate the probability of reflection or transmission when $E > U_0$.	Calculate the reflection coefficient given by Equation 36-27. For transmission use $R + T = 1$ **Example 36-2**
3. Schrödinger Equation in Three Dimensions	
Find the quantum numbers, energies, and wave functions of the lowest states in a three-dimensional square well.	Write the energies in terms of the quantum numbers and try the various combinations of the smallest quantum numbers. The wave functions are determined by the quantum numbers. **Examples 36-3, 36-4**

Problems

Conceptual Problems

Problems from Optional and Exploring sections

In a few problems, you are given more data than you actually need; in a few other problems, you are required to supply data from your general knowledge, outside sources, or informed estimates.

- • Single-concept, single-step, relatively easy
- •• Intermediate-level, may require synthesis of concepts
- ••• Challenging, for advanced students

A Particle in a Finite Square Well

1 • True or false: Boundary conditions on the wave function lead to energy quantization.

2 • Sketch (*a*) the wave function and (*b*) the probability distribution for the $n = 4$ state for the finite square-well potential.

3 • Sketch (*a*) the wave function and (*b*) the probability distribution for the $n = 5$ state for the finite square-well potential.

The Harmonic Oscillator

4 •• Show that the expectation value $<x> = \int x|\psi|^2\,dx$ is zero for both the ground and the first excited states of the harmonic oscillator.

5 •• Use the procedure of Example 36-1 to verify that the energy of the first excited state of the harmonic oscillator is $E_1 = \frac{3}{2}\hbar\omega_0$. (*Note:* Rather than solve for a again, use the result $a = m\omega_0/2\hbar$ obtained in Example 36-1.)

6 ••• Show that the normalization constant A_0 of Equation 36-23 is $A_0 = (2m\omega_0/h)^{1/4}$.

7 ••• Find the normalization constant A_1 for the wave function of the first excited state of the harmonic oscillator, Equation 36-25.

8 ••• Find the expectation value $<x^2> = \int x^2|\psi|^2\,dx$ for the ground state of the harmonic oscillator. Use it to show that the average potential energy equals half the total energy.

9 ••• Verify that $\psi_1(x) = A_1 x e^{-ax^2}$ is the wave function corresponding to the first excited state of a harmonic oscillator by substituting it into the time-independent Schrödinger equation and solving for a and E.

10 ••• Find the expectation value $<x^2> = \int x^2|\psi|^2\,dx$ for the first excited state of the harmonic oscillator.

11 ••• Classically, the average kinetic energy of the harmonic oscillator equals the average potential energy. We may assume that this is also true for the quantum mechanical harmonic oscillator. Use this condition to determine the expectation value of p^2 for the ground state of the harmonic oscillator.

12 ••• We know that for the classical harmonic oscillator, $p_{av} = 0$. It can be shown that for the quantum mechanical harmonic oscillator $\langle p \rangle = 0$. Use the results of Problems 4, 6, and 11 to determine the uncertainty product $\Delta x\, \Delta p$ for the ground state of the harmonic oscillator.

Reflection and Transmission of Electron Waves: Barrier Penetration

13 •• A free particle of mass m with wave number k_1 is traveling to the right. At $x = 0$, the potential jumps from zero to U_0 and remains at this value for positive x. (*a*) If the total energy is $E = \hbar^2 k_1^2/2m = 2U_0$, what is the wave number k_2 in the region $x > 0$? Express your answer in terms of k_1 and in terms of U_0. (*b*) Calculate the reflection coefficient R at the potential step. (*c*) What is the transmission coefficient T? (*d*) If one million particles with wave number k_1 are incident upon the potential step, how many particles are expected to continue along in the positive x direction? How does this compare with the classical prediction?

14 •• Suppose that the potential jumps from zero to $-U_0$ at $x = 0$ so that the free particle speeds up instead of slowing down. The wave number for the incident particle is again k_1, and the total energy is $2U_0$. (*a*) What is the wave number for the particle in the region of positive x? (*b*) Calculate the reflection coefficient R at $x = 0$. (*c*) What is the transmission coefficient T? (*d*) If one million particles with wave number k_1 are incident upon the potential step, how many particles are expected to continue along in the positive x direction? How does this compare with the classical prediction?

15 •• Work Problem 13 for the case in which the energy of the incident particle is $1.01U_0$ instead of $2U_0$.

16 •• A particle of energy E approaches a step barrier of height U. What should be the ratio E/U so that the reflection coefficient is $\frac{1}{2}$?

17 •• Use Equation 36-29 to calculate the order of magnitude of the probability that a proton will tunnel out of a nucleus in one collision with the nuclear barrier if it has energy 6 MeV below the top of the potential barrier and the barrier thickness is 10^{-15} m.

18 •• A 10-eV electron is incident on a potential barrier of height 25 eV and width of 1 nm. (*a*) Use Equation 36-29 to calculate the order of magnitude of the probability that the electron will tunnel through the barrier. (*b*) Repeat your calculation for a width of 0.1 nm.

The Schrödinger Equation in Three Dimensions

19 • A particle is confined to a three-dimensional box that has sides L_1, $L_2 = 2L_1$, and $L_3 = 3L_1$. Give the quantum numbers n_1, n_2, n_3 that correspond to the lowest ten quantum states of this box.

20 • Give the wave functions for the lowest ten quantum states of the particle in Problem 19.

21 • (*a*) Repeat Problem 19 for the case $L_2 = 2L_1$ and $L_3 = 4L_1$. (*b*) What quantum numbers correspond to degenerate energy levels?

22 • Give the wave functions for the lowest ten quantum states of the particle in Problem 21.

23 • A particle moves in a potential well given by $U(x, y, z) = 0$ for $-L/2 < x < L/2$, $0 < y < L$, and $0 < z < L$, and $U = \infty$ outside these ranges. (*a*) Write an expression for the ground-state wave function for this particle. (*b*) How do the allowed energies compare with those for a box having $U = 0$ for $0 < x < L$, rather than for $-L/2 < x < L/2$?

24 •• A particle moves freely in the two-dimensional region defined by $0 \le x \le L$ and $0 \le y \le L$. (*a*) Find the wave function satisfying Schrödinger's equation. (*b*) Find the corresponding energies. (*c*) Find the lowest two states that are degenerate. Give the quantum numbers for this case. (*d*) Find the lowest three states that have the same energy. Give the quantum numbers for the three states having the same energy.

25 •• What is the next energy level above those found in Problem 24*c* for a particle in a two-dimensional square box for which the degeneracy is greater than 2?

Identical Particles

26 • Show that Equation 36-37 satisfies Equation 36-35 with $U = 0$, and find the energy of this state.

27 • What is the ground-state energy of ten noninteracting bosons in a one-dimensional box of length L?

28 • What is the ground-state energy of ten noninteracting fermions, such as neutrons, in a one-dimensional box of length L? (Because the quantum number associated with spin can have two values, each spatial state can hold two neutrons.)

Orthogonality of Wave Functions

The integral of two functions over some space interval is somewhat analogous to the dot product of two vectors. If this integral is zero, the functions are said to be orthogonal, which is analogous to two vectors being perpendicular. The following problems illustrate the general principle that any two wave functions corresponding to different energy levels in the same potential are orthogonal.

29 •• Show that the ground-state wave function and that of the first excited state of the harmonic oscillator are orthogonal; i.e., show that $\int \psi_0(x)\psi_1(x)\,dx = 0$.

30 •• The wave function for the state $n = 2$ of the harmonic oscillator is $\psi_2(x) = A_2(2ax^2 - \frac{1}{2})e^{-ax^2}$, where A_2 is the normalization constant for this wave function. Show that the wave functions for the states $n = 1$ and $n = 2$ of the harmonic oscillator are orthogonal.

31 ••• For the wave functions $\psi_n(x) = \sqrt{2/L}\,\sin(n\pi x/L)$ corresponding to a particle in an infinite square well potential from 0 to L, show that $\int \psi_n(x)\psi_m(x)\,dx = 0$, that is, ψ_n and ψ_m are orthogonal.

General Problems

32 •• Consider a particle in a one-dimensional box of length L that is centered at the origin. (*a*) What are the values of $\psi_1(0)$ and $\psi_2(0)$? (*b*) What are the values of $<x>$ for the states $n = 1$ and $n = 2$? (*c*) Evaluate $<x^2>$ for the states $n = 1$ and $n = 2$. (See Problem 59 in Chapter 17.)

33 •• Eight identical noninteracting fermions (such as neutrons) are confined to a two-dimensional square box of side length L. Determine the energies of the three lowest states. (See Problem 24.)

34 •• A particle is confined to a two-dimensional box defined by the following boundary conditions: $U(x, y) = 0$ for $-L/2 \leq x \leq L/2$ and $-3L/2 \leq y \leq 3L/2$; and $U(x, y) = \infty$ elsewhere. (*a*) Determine the energies of the lowest three bound states. Are any of these states degenerate? (*b*) Identify the lowest doubly degenerate bound state by appropriate quantum numbers and determine its energy.

35 •• A particle moves in a potential given by $U(x) = A|x|$. Without attempting to solve the Schrödinger equation, sketch the wave function for (*a*) the ground-state energy of a particle inside this potential and (*b*) the first excited state for this potential.

36 ••• The classical probability distribution function for a particle in a one-dimensional box of length L is $P = 1/L$. (See Example 17-5.) (*a*) Show that the classical expectation value of x^2 for a particle in a one-dimensional box of length L centered at the origin (Problem 32) is $L^2/12$. (*b*) Find the quantum expectation value of x^2 for the nth state of a particle in the one-dimensional box of Problem 32 and show that it approaches the classical limit $L^2/12$ for $n >> 1$.

37 ••• Show that Equations 36-27 and 36-28 imply that the transmission coefficient for particles of energy E incident on a step barrier $U_0 < E$ is given by

$$T = \frac{4k_1k_2}{(k_1 + k_2)^2} = \frac{4r}{(1 + r)^2}$$

where $r = k_2/k_1$.

38 ••• (*a*) Show that for the case of a particle of energy E incident on a step barrier $U_0 < E$, the wave numbers k_1 and k_2 are related by

$$\frac{k_2}{k_1} = r = \sqrt{1 - \frac{U_0}{E}}$$

Use this and the results of Problem 37 to calculate the transmission coefficient T and the reflection coefficient R for the case (*b*) $E = 1.2U_0$, (*c*) $E = 2.0U_0$, and (*d*) $E = 10.0U_0$.

39 ••• Determine the normalization constant A_2 in Problem 30.

40 ••• Consider the time-independent one-dimensional Schrödinger equation when the potential function is symmetric about the origin, i.e., when $U(x) = U(-x)$. (*a*) Show that if $\psi(x)$ is a solution of the Schrödinger equation with energy E, then $\psi(-x)$ is also a solution with the same energy E, and that, therefore, $\psi(x)$ and $\psi(-x)$ can differ by only a multiplicative constant. (*b*) Write $\psi(x) = C\psi(-x)$, and show that $C = \pm 1$. Note that $C = +1$ means that $\psi(x)$ is an even function of x, and $C = -1$ means that $\psi(x)$ is an odd function of x.

41 ••• In this problem you will derive the ground-state energy of the harmonic oscillator using the precise form of the uncertainty principle, $\Delta x\, \Delta p \geq \hbar/2$, where Δx and Δp are defined to be the standard deviations $(\Delta x)^2 = [(x - x_{av})^2]_{av}$ and $(\Delta p)^2 = [(p - p_{av})^2]_{av}$ (see Equation 18-35). Proceed as follows:

1. Write the total classical energy in terms of the position x and momentum p using $U(x) = \frac{1}{2}m\omega^2x^2$ and $K = p^2/2m$.
2. Use the result of Equation 18-35 to write $(\Delta x)^2 = [(x - x_{av})^2]_{av} = (x^2)_{av} - x_{av}^2$ and $(\Delta p)^2 = [(p - p_{av})^2]_{av} = (p^2)_{av} - p_{av}^2$.
3. Use the symmetry of the potential energy function to argue that x_{av} and p_{av} must be zero, so that $(\Delta x)^2 = (x^2)_{av}$ and $(\Delta p)^2 = (p^2)_{av}$.
4. Assume that $\Delta p = \hbar/2\Delta x$ to eliminate $(p^2)_{av}$ from the average energy $E_{av} = (p^2)_{av}/2m + \frac{1}{2}m\omega^2(x^2)_{av}$ and write E_{av} as $E_{av} = \hbar^2/8mZ + \frac{1}{2}m\omega^2Z$, where $Z = (x^2)_{av}$.
5. Set $dE/dZ = 0$ to find the value of Z for which E is a minimum.
6. Show that the minimum energy is given by $(E_{av})_{min} = +\frac{1}{2}\hbar\omega$.

42 ••• A particle of mass m near the earth's surface at $z = 0$ can be described by the potential energy

$$U = mgz, \quad z > 0$$

$$U = \infty, \quad z < 0$$

For some positive value of total energy E, indicate the classically allowed region on a sketch of $U(z)$ versus z. Sketch also the classical kinetic energy versus z. The Schrödinger equation for this problem is quite difficult to solve. Using arguments similar to those in Section 36-2 about the curvature of the wave function as given by the Schrödinger equation, sketch your "guesses" for the shape of the wave function for the ground state and the first two excited states.

CHAPTER 37

Atoms

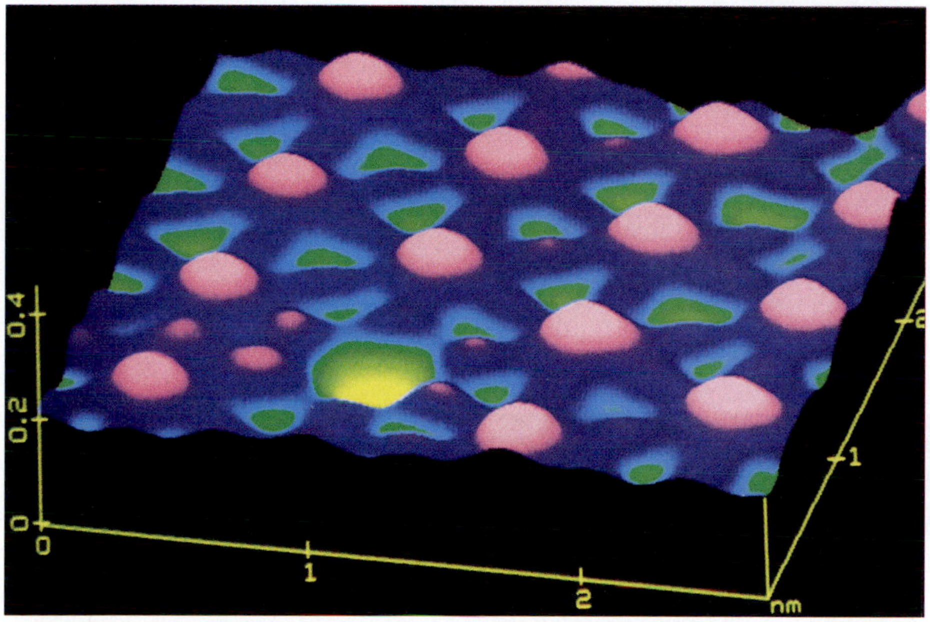

A scanning tunneling microscope image of iodine atoms (pink) adsorbed on platinum. The yellow pocket represents the gap where an iodine atom has been dislodged.

Slightly more than 100 different elements have been discovered. Each is characterized by an atom that contains a number of protons Z, an equal number of electrons, and a number of neutrons N. The number of protons Z is called the **atomic number.** The lightest atom, hydrogen (H), has $Z = 1$; the next lightest, helium (He), has $Z = 2$; the next lightest, lithium (Li), has $Z = 3$; and so forth. Nearly all the mass of the atom is concentrated in a tiny nucleus, which contains the protons and neutrons. The nuclear radius is typically about 1 to 10 fm (1 fm $= 10^{-15}$ m). The distance between the nucleus and the electrons is about 0.1 nm = 100,000 fm. This distance determines the "size" of the atom.

The chemical and physical properties of an element are determined by the number and arrangement of the electrons in the atom. Because each proton has a positive charge $+e$, the nucleus has a total positive charge $+Ze$. The electrons are negatively charged ($-e$), so they are attracted to the nucleus and repelled by each other. Since electrons and protons have equal but opposite charges and there are an equal number of electrons and protons in an atom, atoms are electrically neutral. Atoms that lose or gain one or more electrons are then electrically charged and are called *ions.*

We will begin our study of atoms by discussing the Bohr model, a semiclassical model developed by Niels Bohr in 1913 to explain the spectra emitted by hydrogen atoms. Although this "pre-quantum mechanics" model has many shortcomings, it provides a useful framework for the discussion of atomic phenomena. For example, even though we now know that the electron does not circle the nucleus in well-defined orbits as in the Bohr model but instead is described by a wave function that satisfies the Schrödinger equation, the probability distributions that follow from the full quantum theory do in fact have maxima at the positions of the Bohr orbits. After discussing the Bohr model, we will apply our knowledge of quantum mechanics from Chapter 36 to give a qualitative description of the hydrogen atom. We will then discuss the structure of other atoms and the periodic table of the elements.

37-1 The Nuclear Atom

Atomic Spectra

By the beginning of the twentieth century, a large body of data had been collected on the emission of light by atoms in a gas when they are excited by an electric discharge. Viewed through a spectroscope with a narrow-slit aperture, this light appears as a discrete set of lines of different colors or wavelengths; the spacing and intensities of the lines are characteristic of the element. The wavelengths of these spectral lines could be accurately determined, and much effort went into finding regularities in the spectra. Figure 37-1 shows line spectra for hydrogen and mercury.

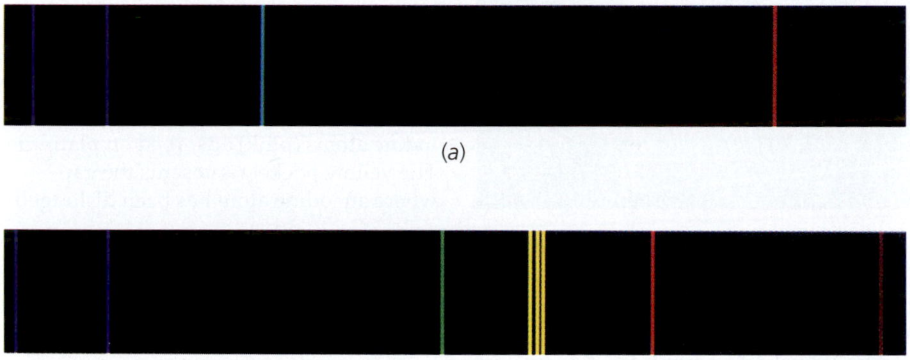

Figure 37-1 (*a*) Line spectrum of hydrogen; (*b*) line spectrum of mercury.

In 1884 a Swiss school teacher, Johann Balmer, found that the wavelengths of the lines in the visible spectrum of hydrogen can be represented by the formula

$$\lambda = (364.6\text{ nm})\frac{m^2}{m^2 - 4}, \qquad m = 3, 4, 5, \ldots \tag{37-1}$$

Balmer suggested that this might be a special case of a more general expression that would be applicable to the spectra of other elements. Such an expression, found by Johannes R. Rydberg and Walter Ritz and known as the **Rydberg–Ritz formula,** gives the reciprocal wavelength as

$$\frac{1}{\lambda} = R\left(\frac{1}{n_2^2} - \frac{1}{n_1^2}\right) \tag{37-2}$$

where n_1 and n_2 are integers with $n_1 > n_2$ and R is the **Rydberg constant,** which is the same for all spectral series of the same element and varies only slightly in a regular way from element to element. For hydrogen, R has the value

$$R_H = 1.096776 \times 10^7 \text{ m}^{-1}$$

The Rydberg–Ritz formula gives the wavelengths for all the lines in the spectra of hydrogen as well as alkali elements such as lithium and sodium.

Many attempts were made to construct a model of the atom that would yield these formulas for its radiation spectrum. The most popular model, due to J. J. Thomson, considered various arrangements of electrons embedded in some kind of fluid that contained most of the mass of the atom and had enough positive charge to make the atom electrically neutral. Thomson's model, called the "plum pudding" model, is illustrated in Figure 37-2. Since classical electromagnetic theory predicted that a charge oscillating with frequency f would radiate electromagnetic energy of that frequency, Thomson searched for configurations that were stable and had normal modes of vibration of frequencies equal to those of the spectrum of the atom. A difficulty of this model and all others was that according to classical physics, electric forces alone cannot produce stable equilibrium. Thomson was unsuccessful in finding a model that predicted the observed frequencies for any atom.

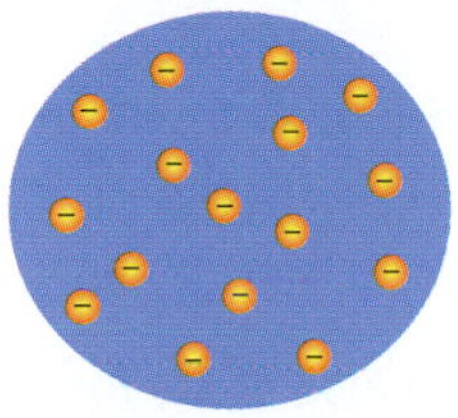

Figure 37-2 J. J. Thomson's plum pudding model of the atom. In this model, the negative electrons are embedded in a fluid of positive charge. For a given configuration in such a system, the resonance frequencies of oscillations of the electrons can be calculated. According to classical theory, the atom should radiate light of frequency equal to the frequency of oscillation of the electrons. Thomson could not find any configuration that would give frequencies in agreement with the measured frequencies of the spectrum of any atom.

The Thomson model was essentially ruled out by a set of experiments by H. W. Geiger and E. Marsden under the supervision of E. Rutherford about 1911, in which alpha particles from radioactive radium were scattered by atoms in a gold foil. Rutherford showed that the number of alpha particles scattered at large angles could not be accounted for by an atom in which the positive charge was distributed throughout the atomic size (known to be about 0.1 nm in diameter) but required that the positive charge and most of the mass of the atom be concentrated in a very small region, now called the nucleus, of diameter of the order of 10^{-6} nm = 1 fm.

37-2 The Bohr Model of the Hydrogen Atom

Niels Bohr, working in the Rutherford laboratory in 1913, proposed a model of the hydrogen atom that extended the work of Planck, Einstein, and Rutherford and successfully predicted the observed spectra. According to Bohr's model, the electron of the hydrogen atom moves under the influence of the Coulomb attraction to the positive nucleus according to classical mechanics, which predicts circular or elliptical orbits with the force center at one focus, as in the motion of the planets around the sun. For simplicity he chose a circular orbit as shown in Figure 37-3.

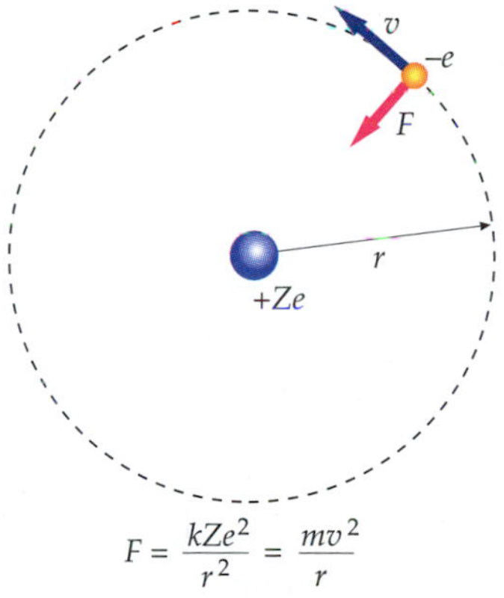

Figure 37-3 Electron of charge $-e$ traveling in a circular orbit of radius r around the nuclear charge $+Ze$. The attractive electrical force kZe^2/r^2 provides the centripetal force holding the electron in its orbit.

Energy in a Circular Orbit

Consider an electron of charge $-e$ moving in a circular orbit of radius r about a positive charge Ze such as the nucleus of a hydrogen atom ($Z = 1$) or of a singly ionized helium atom ($Z = 2$). The total energy of the electron can be related to the radius of the orbit. The potential energy of the electron of charge $-e$ at a distance r from a positive charge Ze is

$$U = \frac{kq_1q_2}{r} = \frac{k(Ze)(-e)}{r} = -\frac{kZe^2}{r} \qquad \text{37-3}$$

where k is the Coulomb constant. The kinetic energy K can be obtained as a function of r by using Newton's second law, $\Sigma\vec{F} = m\vec{a}$. Setting the Coulomb attractive force equal to the mass times the centripetal acceleration gives

$$\frac{kZe^2}{r^2} = m\frac{v^2}{r} \qquad \text{37-4a}$$

Then

$$K = \frac{1}{2}mv^2 = \frac{1}{2}\frac{kZe^2}{r} \qquad \text{37-4b}$$

The kinetic energy thus varies inversely with r like the potential energy. Note that the magnitude of the potential energy is twice that of the kinetic energy,

$$U = -2K \qquad \text{37-5}$$

This is a general result in $1/r^2$ force fields. It also holds for circular orbits in a gravitational field (see Example 11-6 in Section 11-3). The total energy is the sum of the kinetic energy and potential energy

$$E = K + U = \frac{1}{2}\frac{kZe^2}{r} - \frac{kZe^2}{r}$$

or

$$E = -\frac{1}{2}\frac{kZe^2}{r} \qquad \text{37-6}$$

Energy in a circular orbit for a $1/r^2$ force

Although mechanical stability is achieved because the Coulomb attractive force provides the centripetal force necessary for the electron to remain in orbit, classical *electromagnetic* theory says that such an atom would be unstable electrically because the electron must accelerate when moving in a circle and therefore radiate electromagnetic energy of frequency equal to that of its motion. According to the classical theory, such an atom would quickly collapse, the electron spiraling into the nucleus as it radiates away its energy.

Bohr's Postulates

Bohr "solved" the difficulty of the collapsing atom by *postulating* that only certain orbits, called stationary states, are allowed, and in these orbits the electron does not radiate. An atom radiates only when the electron makes a transition from one allowed orbit (stationary state) to another.

> The electron in the hydrogen atom can move only in certain nonradiating, circular orbits called stationary states.

Bohr's first postulate: nonradiating orbits

The second postulate relates the frequency of radiation to the energies of the stationary states. If E_i and E_f are the initial and final energies of the atom, the frequency of the emitted radiation during a transition is given by

$$f = \frac{E_i - E_f}{h} \qquad 37\text{-}7$$

Bohr's second postulate: photon frequency from energy conservation

where h is Planck's constant. This postulate is equivalent to the assumption of conservation of energy with the emission of a photon of energy hf. Combining Equations 37-6 and 37-7, we obtain for the frequency

$$f = \frac{E_1 - E_2}{h} = \frac{1}{2}\frac{kZe^2}{h}\left(\frac{1}{r_2} - \frac{1}{r_1}\right) \qquad 37\text{-}8$$

where r_1 and r_2 are the radii of the initial and final orbits.

To obtain the frequencies implied by the Rydberg–Ritz formula, $f = c/\lambda = cR(1/n_2^2 - 1/n_1^2)$, it is evident that the radii of stable orbits must be proportional to the squares of integers. Bohr searched for a quantum condition for the radii of the stable orbits that would yield this result. After much trial and error, he found that he could obtain it if he postulated that the angular momentum of the electron in a stable orbit equals an integer times $\hbar$ ("h bar," Planck's constant divided by 2π). Since the angular momentum of a circular orbit is just mvr, this postulate is

$$mvr = \frac{nh}{2\pi} = n\hbar, \qquad n = 1, 2, \ldots \qquad 37\text{-}9$$

Bohr's third postulate: quantized angular momentum

where $\hbar = h/2\pi = 1.055 \times 10^{-34}\ \text{J}\cdot\text{s} = 6.582 \times 10^{-16}\ \text{eV}\cdot\text{s}$.

Equation 37-9 relates the speed v to the radius r. Equation 37-4a from Newton's second law gives us another equation relating the speed to the radius:

$$\frac{kZe^2}{r^2} = m\frac{v^2}{r}$$

or

$$v^2 = \frac{kZe^2}{mr} \qquad 37\text{-}10$$

We can determine r by eliminating v between Equations 37-9 and 37-10. Solving Equation 37-9 for v and squaring gives

$$v^2 = n^2\frac{\hbar^2}{m^2r^2}$$

Comparing this result with Equation 37-10, we get

$$n^2\frac{\hbar^2}{m^2r^2} = \frac{kZe^2}{mr}$$

Solving for r, we get

$$r = n^2\frac{\hbar^2}{mkZe^2} = n^2\frac{a_0}{Z} \qquad 37\text{-}11$$

Radius of the Bohr orbits

where a_0 is called the **first Bohr radius**.

$$a_0 = \frac{\hbar^2}{mke^2} \approx 0.0529\ \text{nm} \qquad 37\text{-}12$$

First Bohr radius

Substituting the expressions for r in Equation 37-11 into Equation 37-8 for the frequency gives

$$f = \frac{1}{2}\frac{kZe^2}{h}\left(\frac{1}{r_2} - \frac{1}{r_1}\right) = Z^2\frac{mk^2e^4}{4\pi\hbar^3}\left(\frac{1}{n_2^2} - \frac{1}{n_1^2}\right) \qquad \text{37-13}$$

If we compare this expression with $Z = 1$ for $f = c/\lambda$ with the empirical Rydberg–Ritz formula (Equation 37-2), we obtain for the Rydberg constant

$$R = \frac{mk^2e^4}{4\pi c\hbar^3} \qquad \text{37-14}$$

Using the values of m, e, and $\hbar$ known in 1913, Bohr calculated R and found his result to agree (within the limits of the uncertainties of the constants) with the value obtained from spectroscopy.

Example 37-1

For waves in a circle, the standing wave condition is $n\lambda = 2\pi r$. Show that this condition for electron waves implies quantization of angular momentum.

1. Write the standing-wave condition: $n\lambda = 2\pi r$
2. Use the de Broglie relation to relate the momentum to λ: $p = \frac{h}{\lambda} = \frac{h}{2\pi r/n} = n\frac{h}{2\pi r} = n\frac{\hbar}{r}$
3. The angular momentum of an electron in a circular orbit is mvr: $L = mvr = pr = n\hbar$

Energy Levels

The total energy of the electron in the hydrogen atom is related to the radius of the circular orbit by Equation 37-6. If we substitute the quantized values of r as given by Equation 37-11, we obtain

$$E_n = -\frac{1}{2}\frac{kZe^2}{r} = -\frac{1}{2}\frac{kZe^2}{n^2a_0/Z} = -\frac{1}{2}\frac{kZe^2}{n^2\hbar^2/mkZe^2}$$

or

$$E_n = -\frac{mk^2e^4}{2\hbar^2}\frac{Z^2}{n^2} = -Z^2\frac{E_0}{n^2}, \qquad n = 1, 2, \ldots \qquad \text{37-15}$$

Energy levels

where

$$E_0 = \frac{mk^2e^4}{2\hbar^2} = \frac{1}{2}\frac{ke^2}{a_0} \approx 13.6 \text{ eV} \qquad \text{37-16}$$

The energies E_n with $Z = 1$ are the quantized allowed energies for the hydrogen atom.

Transitions between these allowed energies result in the emission or absorption of a photon whose frequency is given by $f = (E_i - F_f)/h$, and whose wavelength is

$$\lambda = \frac{c}{f} = \frac{hc}{E_i - E_f} \qquad \text{37-17}$$

As we found in Chapter 17, it is convenient to have the value of hc in electron-volt nanometers:

$$hc = 1240 \text{ eV}\cdot\text{nm} \tag{37-18}$$

Since the energies are quantized, the frequencies and wavelengths of the radiation emitted by the hydrogen atom are quantized in agreement with the observed line spectrum.

Figure 37-4 shows the energy-level diagram for hydrogen. The energy of the hydrogen atom in the ground state is $E_1 = -13.6$ eV. As n approaches infinity the energy approaches zero, the highest energy state. The process of removing an electron from an atom is called ionization, and the energy required to remove the electron is the **ionization energy**. The ionization energy of the hydrogen atom, which is also the binding energy of the atom, is 13.6 eV. A few transitions from a higher to a lower state are indicated in Figure 37-4. When Bohr published his model of the hydrogen atom, the Balmer series, corresponding to $n_2 = 2$ and $n_1 = 3, 4, 5, \ldots$, and the Paschen series, corresponding to $n_2 = 3$ and $n_1 = 4, 5, 6, \ldots$, were known. In 1916, T. Lyman found the series corresponding to $n_2 = 1$, and in 1922 and 1924, F. Brackett and H. A. Pfund, respectively, found series corresponding to $n_2 = 4$ and $n_2 = 5$. Only the Balmer series lies in the visible portion of the electromagnetic spectrum.

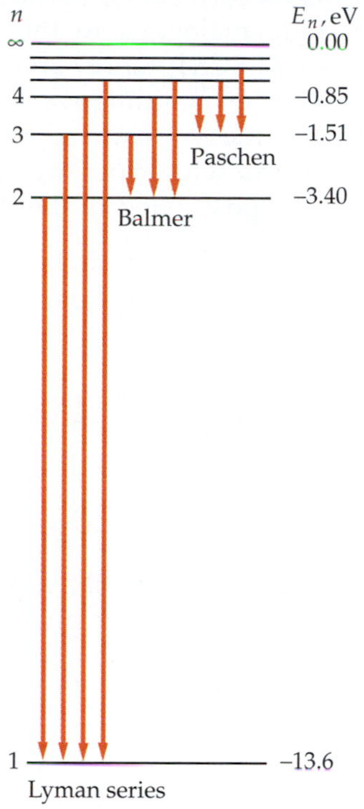

Figure 37-4 Energy-level diagram for hydrogen showing the first few transitions in each of the Lyman, Balmer, and Paschen series. The energies of the levels are given by Equation 37-15.

Example 37-2

Find (*a*) the energy and (*b*) wavelength of the line with the longest wavelength in the Lyman series.

Picture the Problem From Figure 37-4, we can see that the Lyman series corresponds to transitions ending at the ground-state energy, $E_f = E_1 = -13.6$ eV. Since λ varies inversely with energy, the transition with the longest wavelength is the transition with the lowest energy, which is that from the first excited state $n = 2$ to the ground state $n = 1$.

(*a*) The energy of the photon is the difference in the energies of the initial and final atomic state:

$$E = E_i - E_f$$
$$= E_2 - E_1 = \frac{-13.6 \text{ eV}}{2^2} - \frac{-13.6 \text{ eV}}{1^2}$$
$$= -3.40 \text{ eV} + 13.6 \text{ eV} = 10.2 \text{ eV}$$

(*b*) The wavelength of the photon is:

$$\lambda = \frac{hc}{E_2 - E_1} = \frac{1240 \text{ eV}\cdot\text{nm}}{10.2 \text{ eV}} = 121.6 \text{ nm}$$

Remark This photon is outside the visible spectrum, in the ultraviolet region. Since all the other lines in the Lyman series have even greater energies and shorter wavelengths, the Lyman series is completely in the ultraviolet region.

Exercise Find the shortest wavelength for a line in the Lyman series. (*Answer* 91.2 nm)

Despite its spectacular successes, the Bohr model of the hydrogen atom had many shortcomings. There was no justification for the postulates of stationary states, or for the quantization of angular momentum other than the fact that these postulates led to energy levels that agreed with spectroscopic data. Furthermore, attempts to apply the model to more complicated atoms had little success. The quantum-mechanical theory resolves these difficulties. The stationary states of the Bohr model correspond to the standing-wave solutions of the Schrödinger equation analogous to the standing electron waves for a particle in a box discussed in Chapters 17 and 36. Energy quantization is a direct consequence of the standing-wave solutions of the Schrödinger equation. For hydrogen these quantized energies agree with those obtained from the Bohr model and with experiment. The quantization of angular momentum that had to be postulated in the Bohr model is predicted by the quantum theory.

37-3 Quantum Theory of Atoms

The Schrödinger Equation in Spherical Coordinates

In quantum theory, the electron is described by its wave function ψ. The absolute square of the electron wave function $|\psi|^2$ gives the probability of finding the electron in some region of space. Boundary conditions on the wave function lead to the quantization of the wavelengths and frequencies and thereby to the quantization of the electron energy.

Consider a single electron of mass m moving in three dimensions in a region in which the potential energy is U. The time-independent Schrödinger equation for such a particle is given by Equation 36-30:

$$-\frac{\hbar^2}{2m}\left(\frac{\partial^2\psi}{\partial x^2}+\frac{\partial^2\psi}{\partial y^2}+\frac{\partial^2\psi}{\partial z^2}\right)+U\psi=E\psi \qquad \text{37-19}$$

For an isolated atom, the potential energy U depends only on the radial distance $r=\sqrt{x^2+y^2+z^2}$. The problem is then most conveniently treated using the spherical coordinates r, θ, and ϕ, which are related to the rectangular coordinates x, y, and z by

$$z = r\cos\theta$$

$$x = r\sin\theta\cos\phi \qquad \text{37-20}$$

$$y = r\sin\theta\sin\phi$$

These relations are shown in Figure 37-5. The transformation of the three-dimensional Schrödinger equation into spherical coordinates is straightforward but involves much tedious calculation, which we will omit. The result is

$$-\frac{\hbar^2}{2mr^2}\frac{\partial}{\partial r}\left(r^2\frac{\partial\psi}{\partial r}\right)-\frac{\hbar^2}{2mr^2}\left[\frac{1}{\sin\theta}\frac{\partial}{\partial\theta}\left(\sin\theta\frac{\partial\psi}{\partial\theta}\right)+\frac{1}{\sin^2\theta}\frac{\partial^2\psi}{\partial\phi^2}\right]+U(r)\psi=E\psi \qquad \text{37-21}$$

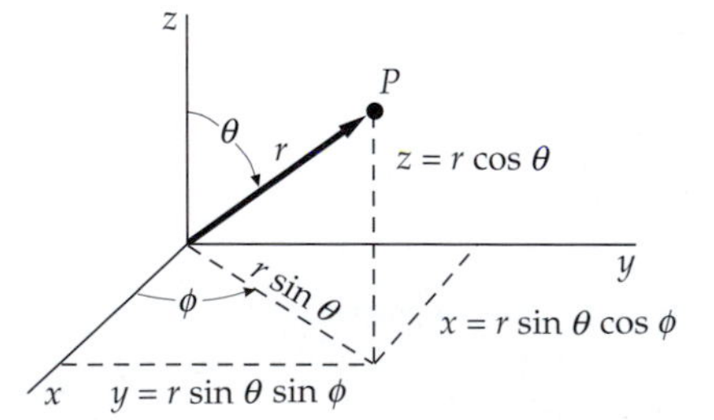

Figure 37-5 Geometric relations between spherical and rectangular coordinates.

Despite the formidable appearance of this equation, it was not difficult for Schrödinger to solve because it is similar to other partial differential equations in classical physics that had been thoroughly studied. We will not solve this equation but merely discuss qualitatively some of the interesting features of the wave functions that satisfy it.

The first step in the solution of a partial differential equation such as Equation 37-21 is to separate the variables by writing the wave function $\psi(r, \theta, \phi)$ as a product of functions of each single variable:

$$\psi(r, \theta, \phi) = R(r)f(\theta)g(\phi) \qquad 37\text{-}22$$

where R depends only on the radial coordinate r, f depends only on θ, and g depends only on ϕ. When this form of $\psi(r, \theta, \phi)$ is substituted into Equation 37-21, the partial differential equation can be transformed into three ordinary differential equations, one for $R(r)$, one for $f(\theta)$, and one for $g(\phi)$. The potential energy $U(r)$ appears only in the equation for $R(r)$, which is called the **radial equation.** The particular form of $U(r)$ given in Equation 37-19 therefore has no effect on the solutions of the equations for $f(\theta)$ and $g(\phi)$, and therefore has no effect on the angular dependence of the wave function $\psi(r, \theta, \phi)$. These solutions are applicable to any problem in which the potential energy depends only on r.

Quantum Numbers in Spherical Coordinates

In three dimensions, the requirement that the wave function be continuous and normalizable introduces three quantum numbers, one associated with each dimension. In spherical coordinates the quantum number associated with r is labeled n, that associated with θ is labeled ℓ, and that associated with ϕ is labeled m. The quantum numbers n_1, n_2, and n_3 that we found in Chapter 36 for a particle in a three-dimensional square well in rectangular coordinates x, y, and z were independent of one another, but the quantum numbers associated with wave functions in spherical coordinates are interdependent. The possible values of these quantum numbers are

$$\begin{aligned} n &= 1, 2, 3, \ldots \\ \ell &= 0, 1, 2, \ldots, n-1 \\ m &= -\ell, (-\ell + 1), \ldots 0, 1, 2, \ldots \ell \end{aligned} \qquad 37\text{-}23$$

Quantum numbers in spherical coordinates

That is, n can be any positive integer; ℓ can be 0 or any positive integer up to $n - 1$; and m can have $2\ell + 1$ possible values, ranging from $-\ell$ to $+\ell$ in integral steps.

The number n is called the **principal quantum number.** It is associated with the dependence of the wave function on the distance r and therefore with the probability of finding the electron at various distances from the nucleus. The quantum numbers ℓ and m are associated with the angular momentum of the electron and with the angular dependence of the electron wave function. The quantum number ℓ is called the **orbital quantum number.** The magnitude of the orbital angular momentum L of the electron is related to ℓ by

$$L = \sqrt{\ell(\ell + 1)}\,\hbar \qquad 37\text{-}24$$

The quantum number m is called the **magnetic quantum number.** It is related to the component of the angular momentum in some direction. Ordinarily, all directions are equivalent, but one particular direction can be specified by placing the atom in a magnetic field. If the z direction is chosen for the magnetic field, the z component of the angular momentum of the electron is given by the quantum condition

$$L_z = m\hbar \qquad 37\text{-}25$$

This quantum condition arises from the boundary condition on the coordinate ϕ that the probability of finding the electron at some angle ϕ_1 must be the same as that of finding it at angle $\phi_1 + 2\pi$ because these are the same points in space.

If we measure the angular momentum of the electron in units of $\hbar$, we see that the angular momentum is quantized to the value $\sqrt{\ell(\ell+1)}$ units and that its component along any direction can have only the $2\ell + 1$ values ranging from $-\ell$ to $+\ell$ units. Figure 37-6 shows a vector-model diagram illustrating the possible orientations of the angular-momentum vector for $\ell = 2$. Note that only specific values of θ are allowed; that is, the directions in space are quantized.

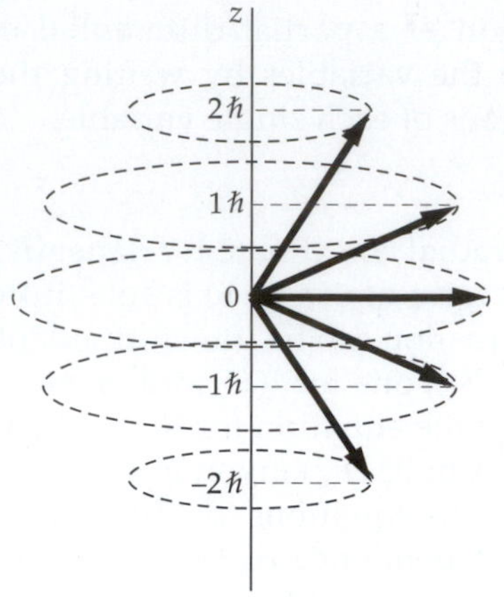

Figure 37-6 Vector-model diagram illustrating the possible values of the z component of the angular-momentum vector for the case $\ell = 2$. The magnitude of the angular momentum is $L = \hbar\sqrt{\ell(\ell+1)} = \hbar\sqrt{2(2+1)} = \hbar\sqrt{6}$.

Example 37-3

If the angular momentum is characterized by the quantum number $\ell = 2$, what are the possible values of L_z, and what is the smallest possible angle between $\vec{L}$ and the z axis?

Picture the Problem The possible orientations of $\vec{L}$ and an arbitrary z axis are shown in Figure 37-6.

1. Write the possible values of L_z:

$$L_z = m\hbar, \quad \text{where } m = -2, -1, 0, +1, \text{or } +2$$

2. Express the angle θ between $\vec{L}$ and the z axis in terms of L and L_z:

$$\cos\theta = \frac{L_z}{L} = \frac{m\hbar}{\sqrt{\ell(\ell+1)}\,\hbar} = \frac{m}{\sqrt{\ell(\ell+1)}}$$

3. The smallest angle occurs when $m = +\ell$ or $-\ell$. Calculate this angle for $\ell = 2$:

$$\cos\theta = \frac{2}{\sqrt{2(2+1)}} = \frac{2}{\sqrt{6}} = 0.816$$

$$\theta = 35.3^\circ$$

Remark We note the somewhat strange result that the angular-momentum vector cannot lie along the z axis.

Exercise An atom has an angular momentum characterized by the quantum number $\ell = 4$. What are the possible values of m? (*Answer* $-4, -3, -2, -1, 0, 1, 2, 3, 4$)

37-4 Quantum Theory of the Hydrogen Atom

We can treat the simplest atom, the hydrogen atom, as a stationary nucleus, the proton, that has a single particle, an electron, moving with kinetic energy $p^2/2m$ and potential energy $U(r)$ due to the electrostatic attraction of the proton*:

$$U(r) = -\frac{kZe^2}{r} \tag{37-26}$$

* We include the factor Z, which is 1 for hydrogen, so that we can apply our results to other one-electron atoms, such as ionized helium He^+, for which $Z = 2$.

For this potential energy, the Schrödinger equation can be solved exactly. In the lowest energy state, the ground state, the principal quantum number n has the value 1, ℓ is 0, and m is 0.

Energy Levels

The allowed energies of the hydrogen atom that result from the solution of the Schrödinger equation are

$$E_n = -\frac{mk^2e^4}{2\hbar^2}\frac{Z^2}{n^2} = -Z^2\frac{E_0}{n^2}, \qquad n = 1, 2, 3, \ldots \tag{37-27}$$

Energy levels for hydrogen

where

$$E_0 = \frac{mk^2e^4}{2\hbar^2} \approx 13.6 \text{ eV} \tag{37-28}$$

These energies are the same as in the Bohr model. Note that the energy is negative, indicating that the electron is bound to the nucleus (thus the term *bound state*), and that the energy depends only on the principal quantum number n. The fact that the energy does not depend on the orbital quantum number ℓ is a peculiarity of the inverse-square force and holds only for an inverse r potential such as Equation 37-26. For more complicated atoms having several electrons, the interaction of the electrons leads to a dependence of the energy on ℓ. In general, the lower the value of ℓ, the lower the energy for such atoms. Since there is usually no preferred direction in space, the energy for any atom does not ordinarily depend on the magnetic quantum number m, which is related to the z component of the angular momentum. The energy does depend on m if the atom is in a magnetic field.

Figure 37-7 shows an energy-level diagram for hydrogen. This diagram is similar to Figure 37-4, except that the states with the same value of n but different values of ℓ are shown separately. These states (called *terms*) are referred to by giving the value of n along with a code letter: S for $\ell = 0$, P for $\ell = 1$, D for $\ell = 2$, and F for $\ell = 3$.* When an atom makes a transition from one allowed energy state to another, electromagnetic radiation in the form of a photon is emitted or absorbed. Such transitions result in spectral lines that are characteristic of the atom. The transitions obey the **selection rules**

$$\Delta m = 0 \quad \text{or} \quad \pm 1$$
$$\Delta \ell = \pm 1 \tag{37-29}$$

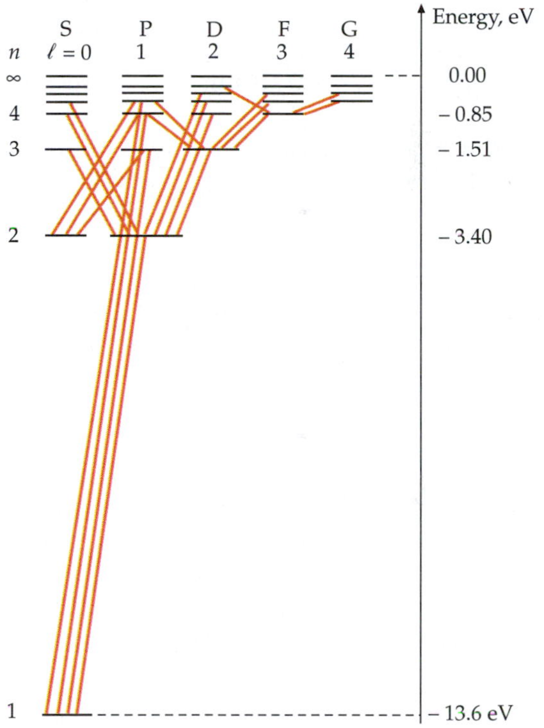

Figure 37-7 Energy-level diagram for hydrogen. The diagonal lines show transitions that involve emission or absorption of radiation and obey the selection rule $\Delta\ell = \pm 1$. States with the same value of n but different values of ℓ have the same energy $-E_0/n^2$, where $E_0 = 13.6$ eV as in the Bohr model.

These selection rules are related to the conservation of angular momentum and to the fact that the photon itself has an intrinsic angular momentum that has a maximum component along any axis of $1\hbar$. The wavelengths of the spectral lines emitted by hydrogen (and by other atoms) are related to the energy levels by

$$hf = \frac{hc}{\lambda} = E_i - E_f \tag{37-30}$$

where E_i and E_f are the energies of the initial and final states.

* These code letters are remnants of spectroscopists' descriptions of various spectral lines as *sharp, principal, diffuse,* and *fundamental*. For values greater than 3, the letters follow alphabetically; thus, G is used for $\ell = 4$, and so forth.

Wave Functions and Probability Densities

The solutions of the Schrödinger equation in spherical coordinates are characterized by the quantum numbers n, ℓ, and m, and are written $\psi_{n\ell m}$. For any given value of n, there are n possible values of ℓ ($\ell = 0, 1, \ldots, n - 1$), and for each value of ℓ, there are $2\ell + 1$ possible values of m. For hydrogen, the energy depends only on n, so there are generally many different wave functions that correspond to the same energy (except at the lowest energy level, for which $n = 1$ and therefore ℓ and m must be 0). These energy levels are therefore degenerate (see Section 36-5). The origins of this degeneracy are the $1/r$ dependence of the potential energy and the fact that, in the absence of any external fields, there is no preferred direction in space.

The Ground State In the lowest energy state, the ground state, the principal quantum number n has the value 1, ℓ is 0, and m is 0. The energy is -13.6 eV, and the angular momentum is zero. Note that this differs from the Bohr model in which the angular momentum in the ground state is $1\hbar$. The wave function for the ground state is

$$\psi_{1,0,0} = C_{1,0,0}e^{-Zr/a_0} \qquad 37\text{-}31$$

where

$$a_0 = \frac{\hbar^2}{mke^2} = 0.0529 \text{ nm}$$

is the first Bohr radius and $C_{1,0,0}$ is a constant that is determined by normalization. In three dimensions, the normalization condition is

$$\int |\psi|^2 \, dV = 1$$

where dV is a volume element and the integration is performed over all space. In spherical coordinates, the volume element (Figure 37-8) is

$$dV = (r \sin\theta \, d\phi)(r \, d\theta) \, dr = r^2 \sin\theta \, dr \, d\theta \, d\phi$$

Figure 37-8 Volume element in spherical coordinates.

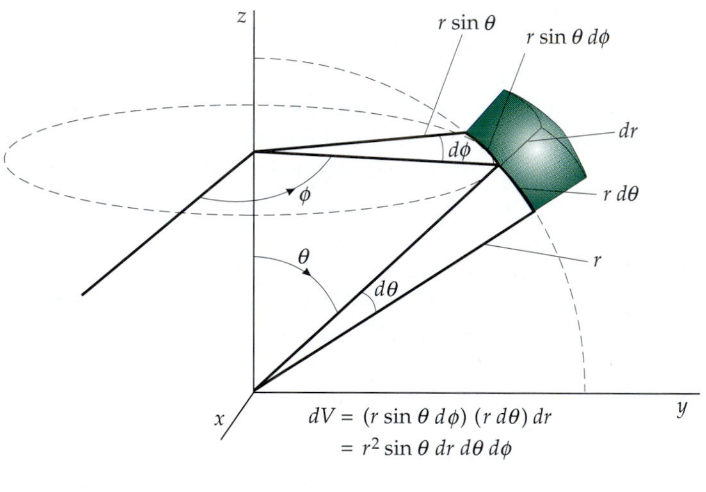

We integrate over all space by integrating over ϕ from $\phi = 0$ to $\phi = 2\pi$, over θ from $\theta = 0$ to $\theta = \pi$, and over r from $r = 0$ to $r = \infty$. The normalization condition is thus

$$\int |\psi|^2 \, dV = \int_0^\infty \int_0^\pi \int_0^{2\pi} |\psi|^2 \, r^2 \sin\theta \, dr \, d\theta \, d\phi$$

$$= \int_0^{2\pi} d\phi \int_0^\pi \sin\theta \, d\theta \int_0^\infty |\psi|^2 \, r^2 \, dr$$

$$= \int_0^{2\pi} d\phi \int_0^\pi \sin\theta \, d\theta \int_0^\infty C_{1,0,0}^2 e^{-2Zr/a_0} \, r^2 \, dr = 1$$

Since there is no θ or ϕ dependence in $\psi_{1,0,0}$, the integration over the angles gives 4π. From a table of integrals, we obtain

$$\int_0^\theta e^{-2Zr/a_0} \, r^2 \, dr = \frac{a_0^3}{4Z^3}$$

Then

$$4\pi C_{1,0,0}^2 \left(\frac{a_0^3}{4Z^3}\right) = 1$$

and

$$C_{1,0,0} = \frac{1}{\sqrt{\pi}}\left(\frac{Z}{a_0}\right)^{3/2} \qquad 37\text{-}32$$

The normalized ground-state wave function is thus

$$\psi_{1,0,0} = \frac{1}{\sqrt{\pi}}\left(\frac{Z}{a_0}\right)^{3/2} e^{-Zr/a_0} \qquad 37\text{-}33$$

The probability of finding the electron in a volume dV is $|\psi|^2\, dV$. The probability density $|\psi|^2$ is illustrated in Figure 37-9. Note that this probability density is spherically symmetric; that is, it depends only on r and not on θ or ϕ. The probability density is maximum at the origin.

We are more often interested in the probability of finding the electron at some radial distance r between r and $r + dr$. This radial probability $P(r)\, dr$ is the probability density $|\psi|^2$ times the volume of the spherical shell of thickness dr, which is $dV = 4\pi r^2\, dr$. The probability of finding the electron in the range from r to $r + dr$ is thus $P(r)\, dr = |\psi|^2 4\pi r^2\, dr$, and the **radial probability density** is

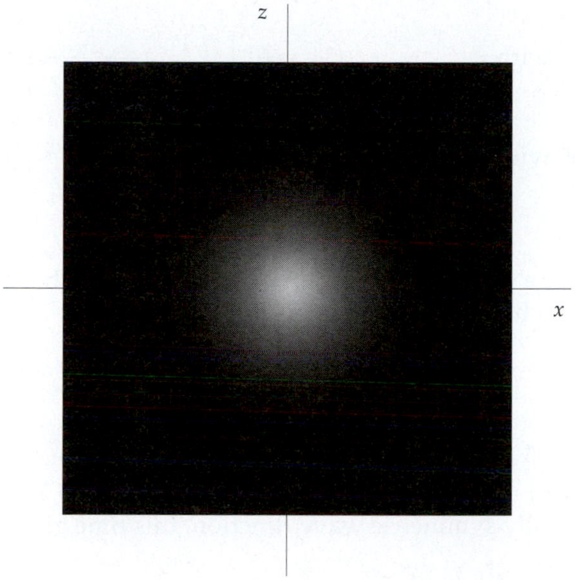

Figure 37-9 Computer-generated picture of the probability density $|\psi|^2$ for the ground state of hydrogen. The quantity $e|\psi|^2$ can be thought of as the electron charge density in the atom. The density is spherically symmetric, is greatest at the origin, and decreases exponentially with r.

$$P(r) = 4\pi r^2 |\psi|^2 \qquad 37\text{-}34$$

Radial probability density

For the hydrogen atom in the ground state, the radial probability density is

$$P(r) = 4\pi r^2|\psi|^2 = 4\pi C_{1,0,0}^2 r^2 e^{-2Zr/a_0} = 4\left(\frac{Z}{a_0}\right)^3 r^2 e^{-2Zr/a_0} \qquad 37\text{-}35$$

Figure 37-10 shows the radial probability density $P(r)$ as a function of r. The maximum value of $P(r)$ occurs at $r = a_0/Z$, which for $Z = 1$ is the first Bohr radius. In contrast to the Bohr model, in which the electron stays in a well-defined orbit at $r = a_0$, we see that it is possible for the electron to be found at any distance from the nucleus. However, the most probable distance is a_0 (assuming $Z = 1$), and the chance of finding the electron at a much different distance is small. It is often useful to think of the electron in an atom as a charged cloud of charge density $\rho = e|\psi|^2$, but we should remember that when it interacts with matter, an electron is always observed as a single charge.

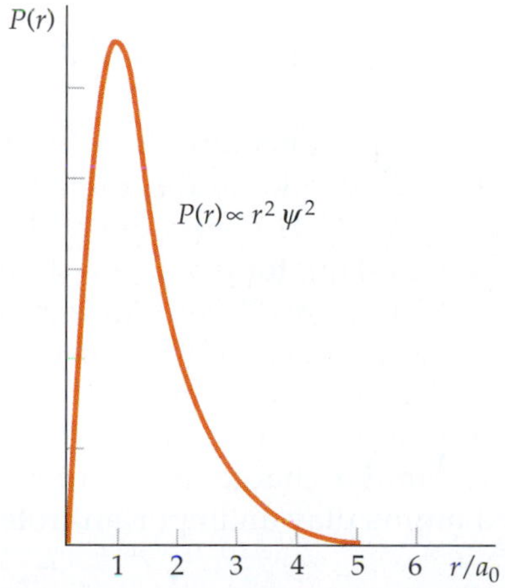

Figure 37-10 Radial probability density $P(r)$ versus r/a_0 for the ground state of the hydrogen atom. $P(r)$ is proportional to $r^2\psi^2$. The value of r for which $P(r)$ is maximum is the most probable distance $r = a_0$.

Example 37-4

Find the probability of finding the electron in the range $\Delta r = 0.06a_0$ at (*a*) $r = a_0$ and (*b*) $r = 2a_0$ for the ground state of the hydrogen atom.

Picture the Problem Because the range Δr is so small, the variation in the radial probability density $P(r)$ can be neglected. The probability of finding the electron in some small range Δr is then $P(r)\,\Delta r$:

(*a*) Use Equation 37-35 with $Z = 1$ and $r = a_0$:

$$P(r)\,\Delta r = \left[4\left(\frac{1}{a_0}\right)^3 r^2 e^{-2r/a_0}\right]\Delta r = \left[4\left(\frac{1}{a_0}\right)^3 a_0^2 e^{-2}\right](0.06a_0) = 0.0325$$

(*b*) Use Equation 37-35 with $Z = 1$ and $r = 2a_0$:

$$P(r)\,\Delta r = \left[4\left(\frac{1}{a_0}\right)^3 r^2 e^{-2r/a_0}\right]\Delta r = \left[4\left(\frac{1}{a_0}\right)^3 4a_0^2 e^{-4}\right](0.06a_0) = 0.0176$$

Remark There is about a 3% chance of finding the electron in this range at $r = a_0$, but at $r = 2a_0$ the chance is only about 0.2%.

The First Excited State

In the first excited state, $n = 2$ and ℓ can be either 0 or 1. For $\ell = 0$, $m = 0$, and we again have a spherically symmetric wave function, this time given by

$$\psi_{2,0,0} = C_{2,0,0}\left(2 - \frac{Zr}{a_0}\right)e^{-Zr/2a_0} \qquad \text{37-36}$$

For $\ell = 1$, m can be $+1$, 0, or -1. The corresponding wave functions are

$$\psi_{2,1,0} = C_{2,1,0}\frac{Zr}{a_0}e^{-Zr/2a_0}\cos\theta \qquad \text{37-37}$$

$$\psi_{2,1\pm1} = C_{2,1,1}\frac{Zr}{a_0}e^{-Zr/2a_0}\sin\theta\, e^{\pm i\phi} \qquad \text{37-38}$$

where $C_{2,0,0}$, $C_{2,1,0}$, and $C_{2,1,1}$ are normalization constants. The probability densities are given by

$$\psi^2_{2,0,0} = C^2_{2,0,0}\left(2 - \frac{Zr}{a_0}\right)^2 e^{-Zr/a_0} \qquad \text{37-39}$$

$$\psi^2_{2,1,0} = C^2_{2,1,0}\left(\frac{Zr}{a_0}\right)^2 e^{-Zr/a_0}\cos^2\theta \qquad \text{37-40}$$

$$|\psi_{2,1\pm1}|^2 = C^2_{2,1,1}\left(\frac{Zr}{a_0}\right)^2 e^{-Zr/a_0}\sin^2\theta \qquad \text{37-41}$$

The wave functions and probability densities for $\ell \neq 0$ are not spherically symmetric, but instead depend on the angle θ. The probability densities do not depend on ϕ. Figure 37-11 shows the probability density $|\psi|^2$ for $n = 2$, $\ell = 0$, $m = 0$ (Figure 37-11*a*); for $n = 2$, $\ell = 1$, $m = 0$ (Figure 37-11*b*); and for $n = 2$, $\ell = 1$, $m = \pm1$ (Figure 37-11*c*). An important feature of these plots is that the electron cloud is spherically symmetric for $\ell = 0$ and is not spherically symmetric for $\ell \neq 0$. These angular distributions of the electron charge density depend only on the values of ℓ and m and not on the radial part of the wave function. Similar charge distributions for the valence electrons of more complicated atoms play an important role in the chemistry of molecular bonding.

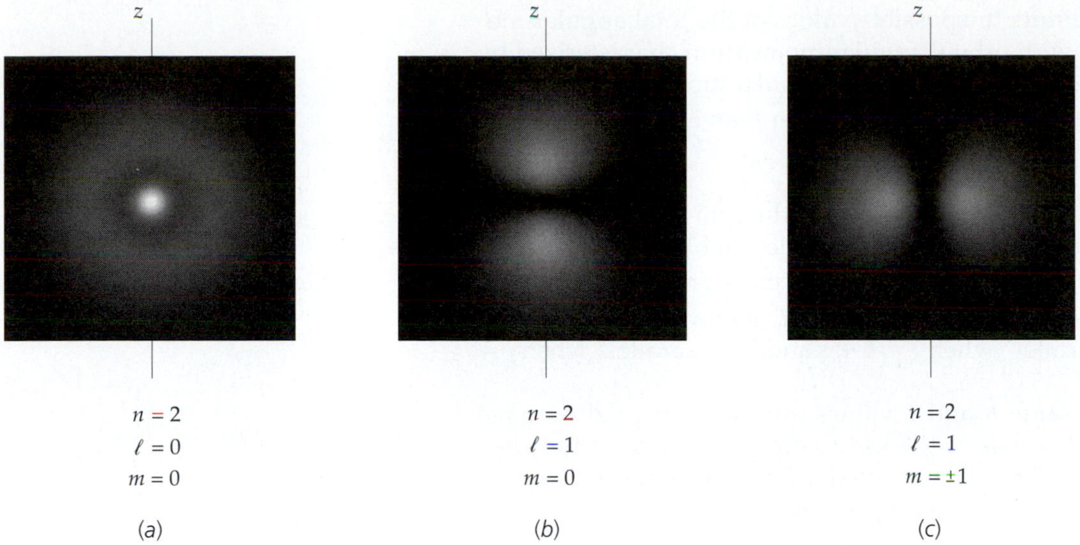

Figure 37-11 Computer-generated picture of the probability densities $|\psi|^2$ for the electron in the $n = 2$ states of hydrogen. (*a*) For $\ell = 0$, $|\psi|^2$ is spherically symmetric. (*b*) For $\ell = 1$ and $m = 0$, $|\psi|^2$ is proportional to $\cos^2 \theta$. (*c*) For $\ell = 1$ and $m = +1$ or -1, $|\psi|^2$ is proportional to $\sin^2\theta$.

Figure 37-12 shows the probability of finding the electron at a distance r as a function of r for $n = 2$ when $\ell = 1$ and when $\ell = 0$. We can see from the figure that the probability distribution depends on ℓ as well as on n.

For $n = 1$, we found that the most likely distance between the electron and the nucleus is a_0, the first Bohr radius, whereas for $n = 2$, $\ell = 1$, it is $4a_0$. These are the orbital radii for the first and second Bohr orbits (Equation 37-11). For $n = 3$ (and $\ell = 2$),* the most likely distance between the electron and nucleus is $9a_0$, the radius of the third Bohr orbit.

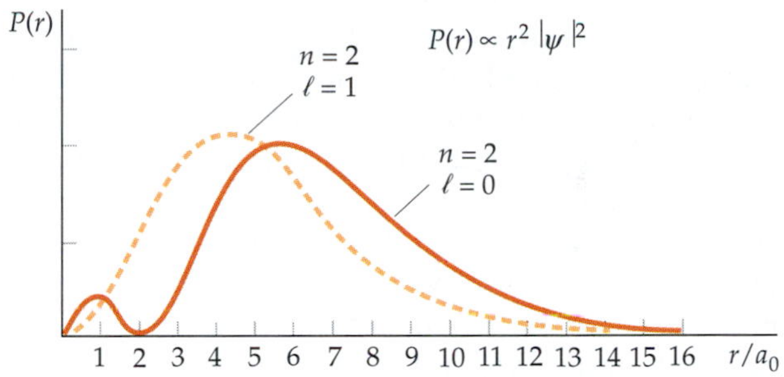

Figure 37-12 Radial probability density $P(r)$ versus r/a_0 for the $n = 2$ states of hydrogen. For $\ell = 1$, $P(r)$ is maximum at the Bohr value $r = 2^2a_0$. For $\ell = 0$, there is a maximum near this value and a smaller submaximum near the origin.

* The correspondence with the Bohr model is closest for the maximum value of ℓ, which is $n - 1$.

37-5 The Spin–Orbit Effect and Fine Structure

In general, an electron in an atom has both orbital angular momentum characterized by the quantum number ℓ and spin angular momentum characterized by the quantum number s. Analogous classical systems that have two kinds of angular momentum are the earth, which is spinning about its axis of rotation in addition to revolving about the sun, and a precessing gyroscope that has angular momentum of precession in addition to its spin. The total angular momentum $\vec{J}$ is the sum of the orbital angular momentum $\vec{L}$ and the spin angular momentum $\vec{S}$

$$\vec{J} = \vec{L} + \vec{S} \tag{37-42}$$

Classically $\vec{J}$ is an important quantity because the resultant torque on a system equals the rate of change of the total angular momentum, and in the case of central forces, the total angular momentum is conserved. For a classical system, the magnitude of the total angular momentum J can have any value between $L + S$ and $L - S$. In quantum mechanics, angular momentum is more complicated. Both L and S are quantized, and their directions are restricted.

Quantum mechanics also limits the possible values of the total angular momentum J. For an electron with orbital angular momentum characterized by the quantum number ℓ and spin $s = \frac{1}{2}$, the total angular momentum J has the magnitude $\sqrt{j(j+1)}\,\hbar$, where the quantum number j can be either

$$j = \ell + \tfrac{1}{2} \qquad \text{or} \qquad j = \ell - \tfrac{1}{2}, \qquad \ell \neq 0 \qquad \text{37-43}$$

(For $\ell = 0$, the total angular momentum is simply the spin and $j = \frac{1}{2}$.) Figure 37-13 is a vector model illustrating the two possible combinations $j = \frac{3}{2}$ and $j = \frac{1}{2}$ for the case of $\ell = 1$. The lengths of the vectors are proportional to $\sqrt{\ell(\ell+1)}\,\hbar$, $\sqrt{s(s+1)}\,\hbar$, and $\sqrt{j(j+1)}\,\hbar$. The spin and orbital angular momentum are said to be "parallel" when $j = \ell + s$ and "antiparallel" when $j = \ell - s$.

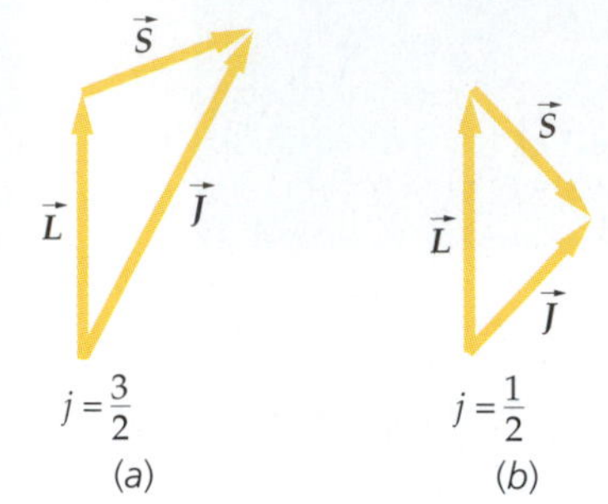

Figure 37-13 Vector diagrams illustrating the addition of orbital and spin angular momentum for the case $\ell = 1$ and $s = \frac{1}{2}$. There are two possible values of the quantum number for the total angular momentum: $j = \ell + s = \frac{3}{2}$ and $j = \ell - s = \frac{1}{2}$.

Atomic states with the same n and ℓ values but different j values have slightly different energies because of the interaction of the spin of the electron with its orbital motion. This effect is called the **spin–orbit effect.** The resulting splitting of spectral lines is called **fine-structure splitting.**

In spectroscopic notation, the total angular-momentum quantum number of an atomic state is written as a subscript after the code letter describing the orbital angular momentum. For example, the ground state of hydrogen is written $1S_{1/2}$, where the 1 indicates the value of n. The $n = 2$ states can have either $\ell = 0$ or $\ell = 1$, and the $\ell = 1$ state can have either $j = \frac{3}{2}$ or $j = \frac{1}{2}$. These states are thus denoted by $2S_{1/2}$, $2P_{3/2}$, and $2P_{1/2}$. Because of the spin–orbit effect, the $2P_{3/2}$ and $2P_{1/2}$ states have slightly different energies resulting in the fine-structure splitting of the transitions $2P_{3/2} \to 2S_{1/2}$ and $2P_{1/2} \to 2S_{1/2}$.

We can understand the spin–orbit effect qualitatively from a simple Bohr-model picture as shown in Figure 37-14. In this picture, the electron moves in a circular orbit around a fixed proton. In Figure 37-14a, the orbital angular momentum $\vec{L}$ is up. In the reference frame of the electron (Figure 37-14b), the proton is moving in a circle around it, thus constituting a circular loop of current that produces a magnetic field $\vec{B}$ at the position of the electron. The direction of $\vec{B}$ is up, parallel to $\vec{L}$. The energy of the electron depends on its spin because of the magnetic moment $\vec{\mu}_s$ associated with its spin. The energy is lowest when $\vec{\mu}_s$ is parallel to $\vec{B}$ and highest when it is antiparallel. This energy is given by (Equation 28-16)

$$U = -\vec{u}_s \cdot \vec{B} \qquad \text{37-44}$$

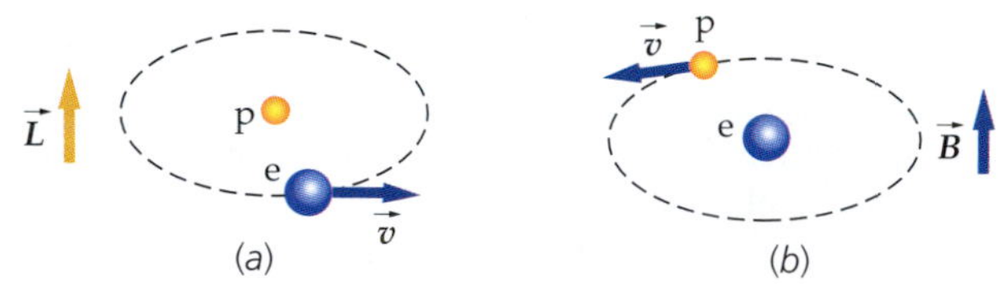

Figure 37-14 (a) An electron moving about a proton in a circular orbit in the horizontal plane with angular momentum $\vec{L}$ up. (b) The magnetic field $\vec{B}$ seen by the electron due to the apparent (relative) motion of the proton is also up. When the electron spin is parallel to $\vec{L}$, the magnetic moment is antiparallel to $\vec{L}$ and $\vec{B}$, so the spin–orbit energy is at its greatest.

Since $\vec{\mu}_s$ is directed opposite to its spin (because the electron has a negative charge), the energy is lowest when the spin is antiparallel to $\vec{B}$ and thus to $\vec{L}$. The energy of the $2P_{1/2}$ state in hydrogen, in which $\vec{L}$ and $\vec{S}$ are antiparallel (Figure 37-15),

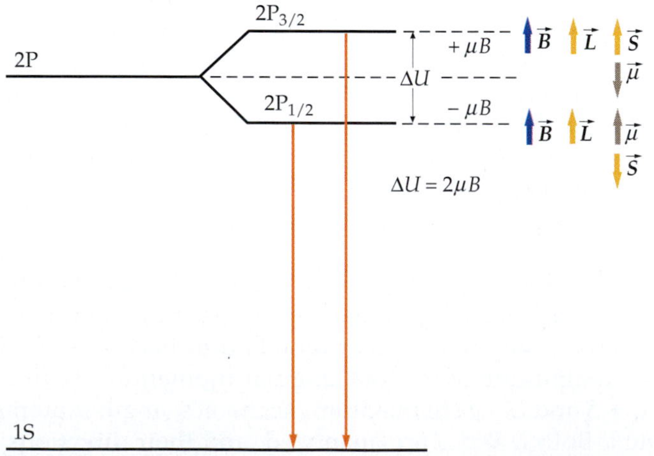

Figure 37-15 Fine-structure energy-level diagram. On the left, the levels in the absence of a magnetic field are shown. The effect of an applied field is shown on the right. Because of the spin–orbit interaction, the magnetic field splits the 2P level into two energy levels, with the $j = \frac{3}{2}$ level having slightly greater energy than the $j = \frac{1}{2}$ level. The spectral line due to the transition $2P \to 1S$ is therefore split into two lines of slightly different wavelengths.

is therefore slightly lower than that of the $2P_{3/2}$ state, in which $\vec{L}$ and $\vec{S}$ are parallel.

Example 37-5

As a consequence of fine-structure splitting, the energies of the $2P_{3/2}$ and $2P_{1/2}$ levels in hydrogen differ by 4.5×10^{-5} eV. If the 2p electron sees an internal magnetic field B, the spin–orbit energy splitting will be of the order of $\Delta E = 2\mu_B B$, where μ_B is the Bohr magneton.* From this, estimate the magnetic field that the 2p electron in hydrogen experiences.

* The Bohr magneton unit is discussed in Section 29-5.

1. Write the energy splitting in terms of the magnetic moment:

$$\Delta E = 2\mu_B B = 4.5 \times 10^{-5}\ \text{eV}$$

2. Solve for the magnetic field B:

$$B = \frac{4.5 \times 10^{-5}\ \text{eV}}{2\mu_B} = \frac{4.5 \times 10^{-5}\ \text{eV}}{2(5.79 \times 10^{-5}\ \text{eV/T})} = 0.389\ \text{T}$$

37-6 The Periodic Table

For atoms with more than one electron, the Schrödinger equation cannot be solved exactly. However, powerful approximation methods allow us to determine the energy levels of the atoms and wave functions of the electrons to a high degree of accuracy. As a first approximation, the Z electrons in an atom are assumed to be noninteracting. The Schrödinger equation can then be solved, and the resulting wave functions used to calculate the interaction of the electrons, which in turn can be used to better approximate the wave functions.† Because the spin of an electron can have two possible components along an axis, there is an additional quantum number m_s, which can have the possible values $+\frac{1}{2}$ or $-\frac{1}{2}$. The state of each electron is thus described by the four quantum numbers n, ℓ, m, and m_s. The energy of the electron is determined mainly by the principal quantum number n (which is related to the radial dependence of the wave function) and by the orbital angular-momentum quantum number ℓ. Generally, the lower the values of n and ℓ, the lower the energy. The dependence of the energy on ℓ is due to the interaction of the electrons in the atom with each other. In hydrogen, of course, there is only one electron, and the energy is independent of ℓ. The specification of n and ℓ for each electron in an atom is called the **electron configuration.** Customarily, ℓ is specified according to the same code used to label the states of the hydrogen atom rather than by its numerical value. The code is

	s	p	d	f	g	h
ℓ value	0	1	2	3	4	5

The n values are sometimes referred to as shells, which are identified by another letter code: $n = 1$ denotes the K shell; $n = 2$, the L shell; and so on.

The electron configuration of atoms is governed by the Pauli exclusion principle, which states that no two electrons in an atom can be in the same quantum state; that is, no two electrons can have the same set of values for

† This approximation method, called perturbation theory, is similar to that used to find the orbits of planets around the sun. The orbits are first found neglecting any interaction between them, then these orbits are used to calculate the perturbation in the orbit due to the forces exerted by one planet on another.

the quantum numbers n, ℓ, m, and m_s. Using the exclusion principle and the restrictions on the quantum numbers discussed in the previous sections (n is an integer, ℓ is an integer that ranges from 0 to $n-1$, m can have $2\ell+1$ values from $-\ell$ to ℓ in integral steps, and m_s can be either $+\frac{1}{2}$ or $-\frac{1}{2}$), we can understand much of the structure of the periodic table.

We have already discussed the lightest element, hydrogen, which has just one electron. In the ground (lowest energy) state, the electron has $n = 1$ and $\ell = 0$, with $m = 0$ and $m_s = +\frac{1}{2}$ or $-\frac{1}{2}$. We call this a 1s electron. The 1 signifies that $n = 1$, and the s signifies that $\ell = 0$.

As electrons are added to make the heavier atoms, the electrons go into those states that will give the lowest total energy consistent with the Pauli exclusion principle.

Helium ($Z = 2$)

The next element after hydrogen is helium ($Z = 2$), which has two electrons. In the ground state, both electrons are in the K shell with $n = 1$, $\ell = 0$, and $m = 0$; one electron has $m_s = +\frac{1}{2}$ and the other has $m_s = -\frac{1}{2}$. This configuration is lower in energy than any other two-electron configuration. The resultant spin of the two electrons is zero. Since the orbital angular momentum is also zero, the total angular momentum is zero. The electron configuration for helium is written $1s^2$. The 1 signifies that $n = 1$, the s signifies that $\ell = 0$, and the superscript 2 signifies that there are two electrons in this state. Since ℓ can be only 0 for $n = 1$, these two electrons fill the K ($n = 1$) shell. The energy required to remove the most loosely bound electron from an atom in the ground state is called the **ionization energy**. This energy is the binding energy of the last electron placed in the atom. For helium, the ionization energy is 24.6 eV, which is relatively large. Helium is therefore basically inert.

Example 37-6

(*a*) Use the measured ionization energy to calculate the energy of interaction of the two electrons in the ground state of the helium atom. (*b*) Use your result to estimate the average separation of the two electrons.

Picture the Problem The energy of one electron in the ground state of helium is E_1 (which is negative) given by Equation 37-27 with $n = 1$ and $Z = 2$. If the electrons did not interact, the energy of the second electron would also be E_1, the same as that of the first electron, and the ground–state energy would be $E_{non} = 2E_1$. This is represented by the lowest level in Figure 37-16. Because of the interaction energy, the ground–state energy is greater than this as represented by the higher level labeled E_g in the figure. When we add $E_{ion} = 24.6$ eV to ionize He, we obtain ionized helium, written He^+, which has just one electron and therefore energy E_1.

Figure 37-16

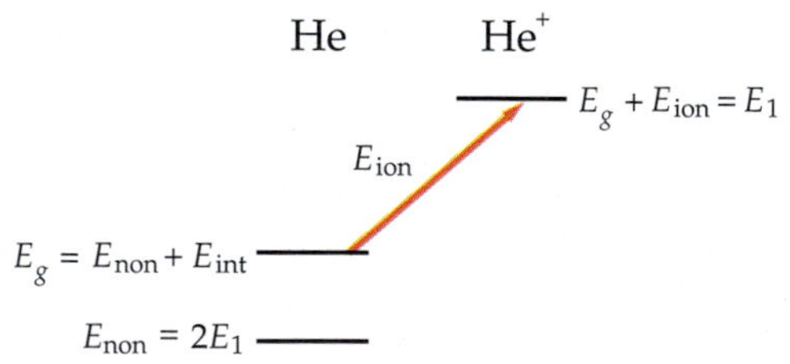

(*a*) 1. The energy of interaction plus the energy of two noninteracting electrons equals the ground–state energy of helium:

$$E_{int} + E_{non} = E_g$$

2. Solve for E_{int} and substitute $E_{non} = 2E_1$:

$$E_{int} = E_g - E_{non} = E_g - 2E_1$$

3. Use Equation 37-27 to calculate the energy E_1 of one electron in the ground state:

$$E_1 = -Z^2\frac{E_0}{n^2} = -(2)^2\frac{13.6 \text{ eV}}{1^2} = -54.4 \text{ eV}$$

4. Substitute this value for E_1:

$$E_{int} = E_g - 2E_1 = E_g - (2)(-54.4 \text{ eV})$$
$$= E_g + 108.8 \text{ eV}$$

5. The ground-state energy of He, E_g, plus the ionization energy equals the energy of He^+, which is E_1:

$$E_g + E_{ion} = E_1 = -54.4 \text{ eV}$$

6. Substitute $E_{ion} = 24.6$ eV to calculate E_g:

$$E_g = -54.4 \text{ eV} - E_{ion} = -54.4 \text{ eV} - 24.6 \text{ eV}$$
$$= -79 \text{ eV}$$

7. Substitute this result for E_g to obtain E_{int}:

$$E_{int} = E_g + 108.8 \text{ eV} = -79 \text{ eV} + 108.8 \text{ eV}$$
$$= 29.8 \text{ eV}$$

(*b*)1. The energy of interaction of two electrons a distance r apart is the potential energy:

$$U = +\frac{ke^2}{r}$$

2. Set U equal to 29.8 eV, and solve for r. It is convenient to express r in terms of a_0, the radius of the first Bohr orbit in hydrogen:

$$r = \frac{ke^2}{U} = \frac{ke^2/a_0}{U}a_0 = \frac{13.6 \text{ eV}}{29.8 \text{ eV}}a_0 = 0.456a_0$$

Check the Result This separation is approximately equal to the radius of the first Bohr orbit for an electron in helium, which is $r_1 = a_0/Z = 0.50a_0$.

Lithium ($Z = 3$)

The next element, lithium, has three electrons. Since the K shell is completely filled with two electrons, the third electron must go into a higher energy shell. The next lowest energy shell after $n = 1$ is the $n = 2$ or L shell. The outer electron is much farther from the nucleus than are the two inner, $n = 1$ electrons. It is most likely to be found at the radius of the second Bohr orbit, which is four times the radius of the first Bohr orbit.

The nuclear charge is partially screened from the outer electron by the two inner electrons. Recall that the electric field outside a spherically symmetric charge density is the same as if all the charge were at the center of the sphere. If the outer electron were completely outside of the charge cloud of the two inner electrons, the electric field it would see would be that of a single charge $+e$ at the center due to the nuclear charge of $+3e$ and the charge $-2e$ of the inner electron cloud. However, the outer electron does not have a well-defined orbit; instead, it is itself a charge cloud that penetrates the charge cloud of the inner electrons to some extent. Because of this penetration, the effective nuclear charge $Z'e$ is somewhat greater than $+1e$. The energy of the outer electron at a distance r from a point charge $+Z'e$ is given by Equation 37-6 with the nuclear charge $+Ze$ replaced by $+Z'e$:

$$E = -\frac{1}{2}\frac{kZ'e^2}{r} \qquad \text{37-45}$$

The greater the penetration of the inner electron cloud, the greater the effective nuclear charge $Z'e$ and the lower the energy. Because the penetration is greater for lower ℓ values (see Figure 37-12), the energy of the outer electron in lithium is lower for the s state ($\ell = 0$) than for the p state ($\ell = 1$). The electron configuration of lithium in the ground state is therefore $1s^2 2s$. The ionization energy of lithium is only 5.39 eV. Because its outer electron is so loosely bound to the atom, lithium is very active chemically. It behaves like a "one-electron atom," similar to hydrogen.

Example 37-7

If the outer electron in lithium moved in the $n = 2$ Bohr orbit, the nuclear charge would be shielded by the two inner electrons and the effective nuclear charge would be $Z'e = 1e$. Then the energy of the electron would be $-13.6\text{ eV}/2^2 = -3.4$ eV. However, the ionization energy of lithium is 5.39 eV, not 3.4 eV. Use this fact to calculate the effective nuclear charge Z' seen by the outer electron in lithium.

Picture the Problem Because the outer electron moves in the $n = 2$ shell, we will take $r = 4a_0$ for its average distance from the nucleus. We can then calculate Z' from Equation 37-45. Since r is given in terms of a_0, it will be convenient to use the fact that $E_0 = ke^2/2a_0 = 13.6$ eV (Equation 37-16).

1. Equation 37-45 relates the energy of the outer electron to its average distance r and the effective nuclear charge Z':

$$E = \frac{1}{2}\frac{kZ'e^2}{r}$$

2. Substitute the given values $r = 4a_0$ and $E = -5.39$ eV:

$$-5.39\text{ eV} = -\frac{1}{2}\frac{kZ'e^2}{4a_0} = -\frac{Z'}{4}\left(\frac{ke^2}{2a_0}\right)$$

3. Use $ke^2/2a_0 = 13.6$ eV and solve for Z':

$$-5.39\text{ eV} = -\frac{Z'}{4}\left(\frac{ke^2}{2a_0}\right) = -\frac{Z'}{4}(13.6\text{ eV})$$

$$Z' = 4\frac{5.39\text{ eV}}{13.6\text{ eV}} = 1.59$$

Remark This calculation is interesting but not very rigorous. We essentially used the circular orbit from the semiclassical Bohr model and the measured ionization energy to calculate the effective inner charge seen by the outer electron. We know, of course, that this outer electron does not move in a circular orbit of constant radius, but is better represented by a stationary charged cloud of charge density $|\psi|^2$ that penetrates the charged clouds of the inner electrons.

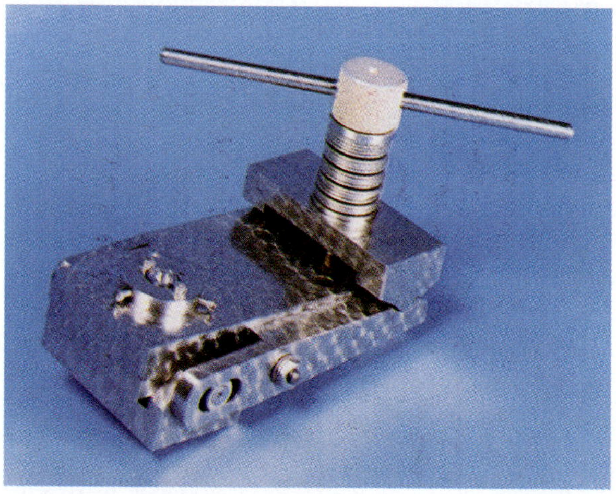

(a)

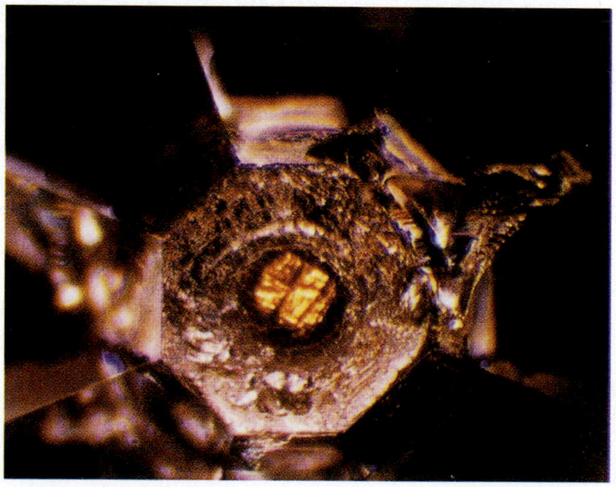

(b)

(*a*) A diamond anvil cell, in which the facets of two diamonds (about 1 mm^2 each) are used to compress a sample substance, subjecting it to very high pressure. (*b*) Samarium monosulfide (SmS) is normally a black, dull-looking semiconductor. When it is subjected to pressure above 7000 atm, an electron from the 4f state is dislocated into the 5d state. The resulting compound glitters like gold and behaves like a metal.

Beryllium ($Z = 4$)

The fourth electron has the least energy in the 2s state. There can be two electrons with $n = 2$, $\ell = 0$, and $m = 0$ because of the two possible values for the spin quantum number m_s. The configuration of beryllium is thus $1s^2 2s^2$.

Boron to Neon ($Z = 5$ to $Z = 10$)

Since the 2s subshell is filled, the fifth electron must go into the next available (lowest energy) subshell, which is the 2p subshell, with $n = 2$ and $\ell = 1$. Since there are three possible values of m ($+1$, 0, and -1) and two values of m_s for each value of m, there can be six electrons in this subshell. The electron configuration for boron is $1s^2 2s^2 2p$. The electron configurations for the elements carbon ($Z = 6$) to neon ($Z = 10$) differ from that for boron only in the number of electrons in the 2p subshell. The ionization energy increases with Z for these elements, reaching the value of 21.6 eV for the last element in the group, neon. Neon has the maximum number of electrons allowed in the $n = 2$ shell. Its electron configuration is $1s^2 2s^2 2p^6$. Because of its very high ionization energy, neon, like helium, is basically chemically inert. The element just before neon, fluorine, has a "hole" in the 2p subshell; that is, it has room for one more electron. It readily combines with elements such as lithium that have one outer electron. Lithium, for example, will donate its single outer electron to the fluorine atom to make an F^- ion and an Li^+ ion. These ions then bond together to form a molecule of lithium fluoride.

Sodium to Argon ($Z = 11$ to $Z = 18$)

The eleventh electron must go into the $n = 3$ shell. Since this electron is very far from the nucleus and from the inner electrons, it is weakly bound in the sodium ($Z = 11$) atom. The ionization energy of sodium is only 5.14 eV. Sodium therefore combines readily with atoms such as fluorine. With $n = 3$, the value of ℓ can be 0, 1, or 2. Because of the lowering of the energy due to penetration of the electron shield formed by the other ten electrons (similar to that discussed for lithium) the 3s state is lower than the 3p or 3d states. This energy difference between subshells of the same n value becomes greater as the number of electrons increases. The electron configuration of sodium is $1s^2 2s^2 2p^6 3s^1$. As we move to elements with higher values of Z, the 3s subshell and then the 3p subshell fill. These two subshells can accommodate $2 + 6 = 8$ electrons. The configuration of argon ($Z = 18$) is $1s^2 2s^2 2p^6 3s^2 3p^6$. One might expect the nineteenth electron to go into the third subshell (the d subshell with $\ell = 2$), but the penetration effect is now so strong that the energy of the next electron is lower in the 4s subshell than in the 3d subshell. There is thus another large energy difference between the eighteenth and nineteenth electrons, and so argon, with its full 3p subshell, is basically stable and inert.

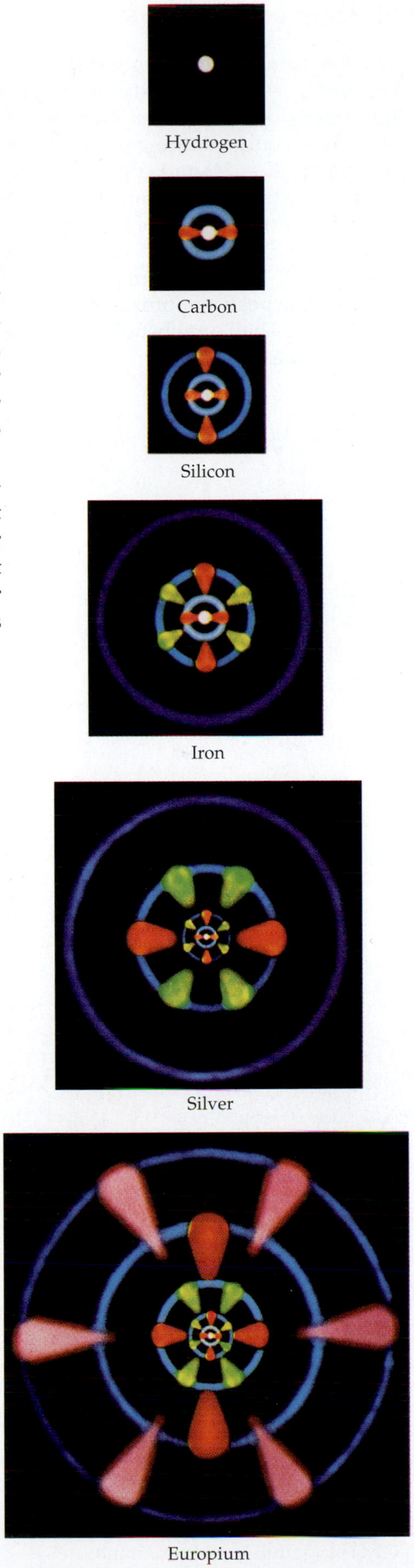

A schematic depiction of the electron configurations in atoms. The spherically symmetric s states can contain 2 electrons and are colored white and blue. The dumbbell-shaped p states can contain up to 6 electrons and are colored orange. The d states can contain up to 10 electrons and are colored yellow-green. The f states can contain up to 14 electrons and are colored purple.

Elements with $Z > 18$

The nineteenth electron in potassium ($Z = 19$) and the twentieth electron in calcium ($Z = 20$) go into the 4s rather than the 3d subshell. The electron configurations of the next ten elements, scandium ($Z = 21$) through zinc ($Z = 30$), differ only in the number of electrons in the 3d shell, except for chromium ($Z = 24$) and copper ($Z = 29$), each of which has only one 4s electron. These ten elements are called **transition elements.** Since their chemical properties are mainly due to their 4s electrons, they are quite similar chemically.

Figure 37-17 shows a plot of the ionization energy versus Z for $Z = 1$ to $Z = 60$. The peaks in ionization energy at $Z = 2$, 10, 18, 36, and 54 mark the closing of a shell or subshell. Table 37-1 gives the electron configurations of all the elements.

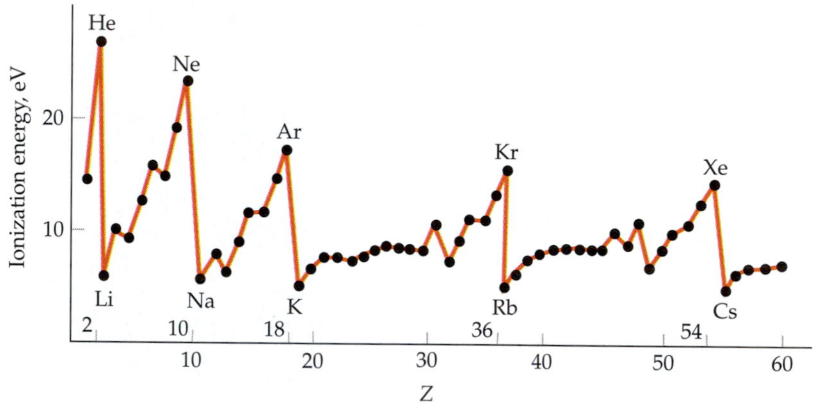

Figure 37-17 Ionization energy versus Z for $Z = 1$ to $Z = 60$. This energy is the binding energy of the last electron in the atom. The binding energy increases with Z until a shell is closed at $Z = 2$, 10, 18, 36, and 54. Elements with a closed shell plus one outer electron, such as sodium ($Z = 11$), have very low binding energies because the outer electron is very far from the nucleus and is shielded by the inner core electrons.

Table 37-1

Electron Configurations of the Atoms in Their Ground States

For some of the rare-earth elements ($Z = 57$ to 71) and the heavy elements ($Z > 89$) the configurations are not firmly established.

		Shell (n):	K (1)	L (2)		M (3)			N (4)				O (5)				P (6)			Q (7)
			s	s	p	s	p	d	s	p	d	f	s	p	d	f	s	p	d	s
Z		Element / Subshell (ℓ):	(0)	(0)	(1)	(0)	(1)	(2)	(0)	(1)	(2)	(3)	(0)	(1)	(2)	(3)	(0)	(1)	(2)	(1)
1	H	hydrogen	1																	
2	He	helium	2																	
3	Li	lithium	2	1																
4	Be	beryllium	2	2																
5	B	boron	2	2	1															
6	C	carbon	2	2	2															
7	N	nitrogen	2	2	3															
8	O	oxygen	2	2	4															
9	F	fluorine	2	2	5															
10	Ne	neon	2	2	6															
11	Na	sodium	2	2	6	1														
12	Mg	magnesium	2	2	6	2														
13	Al	aluminum	2	2	6	2	1													
14	Si	silicon	2	2	6	2	2													
15	P	phosphorus	2	2	6	2	3													
16	S	sulfur	2	2	6	2	4													
17	Cl	chlorine	2	2	6	2	5													
18	Ar	argon	2	2	6	2	6													

Table 37-1 *(Continued)*

Electron Configurations of the Atoms in Their Ground States
For some of the rare-earth elements (Z = 57 to 71) and the heavy elements (Z > 89) the configurations are not firmly established.

		Shell (n):	K (1)	L (2)		M (3)			N (4)				O (5)				P (6)			Q (7)
			s	s	p	s	p	d	s	p	d	f	s	p	d	f	s	p	d	s
Z		Element Subshell (ℓ):	(0)	(0)	(1)	(0)	(1)	(2)	(0)	(1)	(2)	(3)	(0)	(1)	(2)	(3)	(0)	(1)	(2)	(1)
19	K	potassium	2	2	6	2	6	.	1											
20	Ca	calcium	2	2	6	2	6	.	2											
21	Sc	scandium	2	2	6	2	6	1	2											
22	Ti	titanium	2	2	6	2	6	2	2											
23	V	vanadium	2	2	6	2	6	3	2											
24	Cr	chromium	2	2	6	2	6	5	1											
25	Mn	manganese	2	2	6	2	6	5	2											
26	Fe	iron	2	2	6	2	6	6	2											
27	Co	cobalt	2	2	6	2	6	7	2											
28	Ni	nickel	2	2	6	2	6	8	2											
29	Cu	copper	2	2	6	2	6	10	1											
30	Zn	zinc	2	2	6	2	6	10	2											
31	Ga	gallium	2	2	6	2	6	10	2	1										
32	Ge	germanium	2	2	6	2	6	10	2	2										
33	As	arsenic	2	2	6	2	6	10	2	3										
34	Se	selenium	2	2	6	2	6	10	2	4										
35	Br	bromine	2	2	6	2	6	10	2	5										
36	Kr	krypton	2	2	6	2	6	10	2	6										
37	Rb	rubidium	2	2	6	2	6	10	2	6	.	.	1							
38	Sr	strontium	2	2	6	2	6	10	2	6	.	.	2							
39	Y	yttrium	2	2	6	2	6	10	2	6	1	.	2							
40	Zr	zirconium	2	2	6	2	6	10	2	6	2	.	2							
41	Nb	niobium	2	2	6	2	6	10	2	6	4	.	1							
42	Mo	molybdenum	2	2	6	2	6	10	2	6	5	.	1							
43	Tc	technetium	2	2	6	2	6	10	2	6	6	.	1							
44	Ru	ruthenium	2	2	6	2	6	10	2	6	7	.	1							
45	Rh	rhodium	2	2	6	2	6	10	2	6	8	.	1							
46	Pd	palladium	2	2	6	2	6	10	2	6	10	.	.							
47	Ag	silver	2	2	6	2	6	10	2	6	10	.	1							
48	Cd	cadmium	2	2	6	2	6	10	2	6	10	.	2							
49	In	indium	2	2	6	2	6	10	2	6	10	.	2	1						
50	Sn	tin	2	2	6	2	6	10	2	6	10	.	2	2						
51	Sb	antimony	2	2	6	2	6	10	2	6	10	.	2	3						
52	Te	tellurium	2	2	6	2	6	10	2	6	10	.	2	4						
53	I	iodine	2	2	6	2	6	10	2	6	10	.	2	5						
54	Xe	xenon	2	2	6	2	6	10	2	6	10	.	2	6						
55	Cs	cesium	2	2	6	2	6	10	2	6	10	.	2	6	.	.	1			
56	Ba	barium	2	2	6	2	6	10	2	6	10	.	2	6	.	.	2			
57	La	lanthanum	2	2	6	2	6	10	2	6	10	.	2	6	1	.	2			
58	Ce	cerium	2	2	6	2	6	10	2	6	10	1	2	6	1	.	2			
59	Pr	praseodymium	2	2	6	2	6	10	2	6	10	3	2	6	.	.	2			
60	Nd	neodymium	2	2	6	2	6	10	2	6	10	4	2	6	.	.	2			
61	Pm	promethium	2	2	6	2	6	10	2	6	10	5	2	6	.	.	2			
62	Sm	samarium	2	2	6	2	6	10	2	6	10	6	2	6	.	.	2			
63	Eu	europium	2	2	6	2	6	10	2	6	10	7	2	6	.	.	2			
64	Gd	gadolinium	2	2	6	2	6	10	2	6	10	7	2	6	1	.	2			
65	Tb	terbium	2	2	6	2	6	10	2	6	10	9	2	6	.	.	2			
66	Dy	dysprosium	2	2	6	2	6	10	2	6	10	10	2	6	.	.	2			
67	Ho	holmium	2	2	6	2	6	10	2	6	10	11	2	6	.	.	2			
68	Er	erbium	2	2	6	2	6	10	2	6	10	12	2	6	.	.	2			
69	Tm	thulium	2	2	6	2	6	10	2	6	10	13	2	6	.	.	2			
70	Yb	ytterbium	2	2	6	2	6	10	2	6	10	14	2	6	.	.	2			
71	Lu	lutetium	2	2	6	2	6	10	2	6	10	14	2	6	1	.	2			
72	Hf	hafnium	2	2	6	2	6	10	2	6	10	14	2	6	2	.	2			
73	Ta	tantalum	2	2	6	2	6	10	2	6	10	14	2	6	3	.	2			
74	W	tungsten (wolfram)	2	2	6	2	6	10	2	6	10	14	2	6	4	.	2			

Continued on page 1192

Table 37-1 *(Continued)*

Electron Configurations of the Atoms in Their Ground States
For some of the rare-earth elements (Z = 57 to 71) and the heavy elements (Z > 89) the configurations are not firmly established.

		Shell (n):	K (1)	L (2)		M (3)			N (4)				O (5)				P (6)			Q (7)
			s	s	p	s	p	d	s	p	d	f	s	p	d	f	s	p	d	s
Z		Element / Subshell (ℓ):	(0)	(0)	(1)	(0)	(1)	(2)	(0)	(1)	(2)	(3)	(0)	(1)	(2)	(3)	(0)	(1)	(2)	(1)
75	Re	rhenium	2	2	6	2	6	10	2	6	10	14	2	6	5	.	2			
76	Os	osmium	2	2	6	2	6	10	2	6	10	14	2	6	6	.	2			
77	Ir	iridium	2	2	6	2	6	10	2	6	10	14	2	6	7	.	2			
78	Pt	platinum	2	2	6	2	6	10	2	6	10	14	2	6	9	.	1			
79	Au	gold	2	2	6	2	6	10	2	6	10	14	2	6	10	.	1			
80	Hg	mercury	2	2	6	2	6	10	2	6	10	14	2	6	10	.	2			
81	Tl	thallium	2	2	6	2	6	10	2	6	10	14	2	6	10	.	2	1		
82	Pb	lead	2	2	6	2	6	10	2	6	10	14	2	6	10	.	2	2		
83	Bi	bismuth	2	2	6	2	6	10	2	6	10	14	2	6	10	.	2	3		
84	Po	polonium	2	2	6	2	6	10	2	6	10	14	2	6	10	.	2	4		
85	At	astatine	2	2	6	2	6	10	2	6	10	14	2	6	10	.	2	5		
86	Rn	radon	2	2	6	2	6	10	2	6	10	14	2	6	10	.	2	6		
87	Fr	francium	2	2	6	2	6	10	2	6	10	14	2	6	10	.	2	6	.	1
88	Ra	radium	2	2	6	2	6	10	2	6	10	14	2	6	10	.	2	6	.	2
89	Ac	actinium	2	2	6	2	6	10	2	6	10	14	2	6	10	.	2	6	1	2
90	Th	thorium	2	2	6	2	6	10	2	6	10	14	2	6	10	.	2	6	2	2
91	Pa	protactinium	2	2	6	2	6	10	2	6	10	14	2	6	10	1	2	6	2	2
92	U	uranium	2	2	6	2	6	10	2	6	10	14	2	6	10	3	2	6	1	2
93	Np	neptunium	2	2	6	2	6	10	2	6	10	14	2	6	10	4	2	6	1	2
94	Pu	plutonium	2	2	6	2	6	10	2	6	10	14	2	6	10	6	2	6	.	2
95	Am	americium	2	2	6	2	6	10	2	6	10	14	2	6	10	7	2	6	.	2
96	Cm	curium	2	2	6	2	6	10	2	6	10	14	2	6	10	7	2	6	1	2
97	Bk	berkelium	2	2	6	2	6	10	2	6	10	14	2	6	10	8	2	6	1	2
98	Cf	californium	2	2	6	2	6	10	2	6	10	14	2	6	10	10	2	6	.	2
99	Es	einsteinium	2	2	6	2	6	10	2	6	10	14	2	6	10	11	2	6	.	2
100	Fm	fermium	2	2	6	2	6	10	2	6	10	14	2	6	10	12	2	6	.	2
101	Md	mendelevium	2	2	6	2	6	10	2	6	10	14	2	6	10	13	2	6	.	2
102	No	nobelium	2	2	6	2	6	10	2	6	10	14	2	6	10	14	2	6	.	2
103	Lw	lawrencium	2	2	6	2	6	10	2	6	10	14	2	6	10	14	2	6	1	2

37-7 Optical and X-Ray Spectra

When an atom is in an excited state (that is, when it is in an energy state above the ground state), it makes transitions to lower energy states, and in doing so emits electromagnetic radiation. The wavelength of the electromagnetic radiation emitted is related to the initial and final states by the Bohr formula (Equation 37-17), $\lambda = hc/(E_i - E_f)$, where E_i and E_f are the initial and final energies and h is Planck's constant. The atom can be excited to a higher energy state by bombarding it with a beam of electrons as in a spectral tube with a high voltage across it. Since the excited energy states of an atom form a discrete (rather than continuous) set, only certain wavelengths are emitted. These wavelengths of the emitted radiation constitute the emission spectrum of the atom.

Optical Spectra

To understand atomic spectra we thus need to understand the excited states of the atom. The situation for an atom with many electrons is, in general, much more complicated than that of hydrogen with just one electron. An excited state of the atom may involve a change in the state of any one of the electrons, or even two or more electrons. Fortunately, in most cases, an excited state of an atom involves the excitation of just one of the electrons in the atom. The energies of excitation of the outer, valence electrons of an atom are of the order of a few electron volts. Transitions involving these electrons result in photons in or near the visible or **optical spectrum**. (Recall that the energies of visible photons range from about 1.5 to 3 eV.) The excitation energies can often be calculated from a simple model in which the atom is pictured as a single electron plus a stable core consisting of the nucleus plus the other, inner electrons. This model works particularly well for the alkali metals: Li, Na, K, Rb, and Cs. These elements are in the first column of the periodic table. The optical spectra of these elements are similar to that of hydrogen.

Figure 37-18 shows an energy-level diagram for the optical transitions in sodium, whose electrons form a neon core plus one outer electron. Since the spin angular momentum of the core adds up to zero, the spin of each state of sodium is $\frac{1}{2}$. Because of the spin–orbit effect, the states with $j = \ell - \frac{1}{2}$ have a slightly lower energy than those with $j = \ell + \frac{1}{2}$. Each state (except for the S states) is therefore split into two states, called a doublet. The doublet splitting is very small and not evident on the energy scale of this diagram. The states are labeled by the usual spectroscopic notation with the superscript 2 before the letter indicating that the state is a doublet. Thus $^2P_{3/2}$, read as

A neon sign outside a Chinatown restaurant in Paris. Neon atoms in the tube are excited by an electron current passing through the tube. The excited neon atoms emit light in the visible range as they decay toward their ground states. The colors of neon signs result from the characteristic red-orange spectrum of neon plus the color of the glass tube itself.

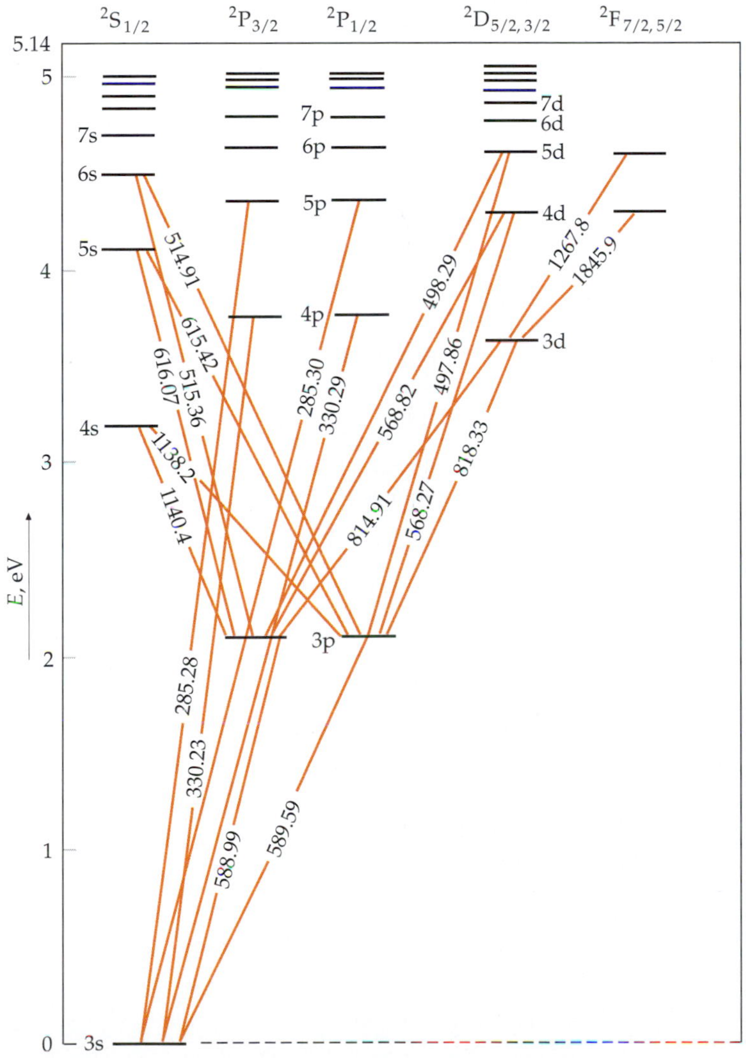

Figure 37-18 Energy-level diagram for sodium. The diagonal lines show observed optical transitions, with wavelengths given in nanometers. The energy of the ground state has been chosen as the zero point for the scale on the left.

"doublet P three halves," denotes a state in which $\ell = 1$ and $j = \frac{3}{2}$. (The S states are customarily labeled as if they were doublets even though they are not.) In the first excited state, the outer electron is excited from the 3s level to the 3p level, which is about 2.1 eV above the ground state. The energy difference between the $P_{3/2}$ and $P_{1/2}$ states due to the spin–orbit effect is about 0.002 eV. Transitions from these states to the ground state give the familiar sodium yellow doublet:

$$3p(^2P_{1/2}) \rightarrow 3s(^2S_{1/2}), \qquad \lambda = 589.6 \text{ nm}$$

$$3p(^2P_{3/2}) \rightarrow 3s(^2S_{1/2}), \qquad \lambda = 589.0 \text{ nm}$$

The energy levels and spectra of other alkali metal atoms are similar to those for sodium. The optical spectrum for atoms such as helium, beryllium, and magnesium that have two outer electrons is considerably more complex because of the interaction of the two outer electrons.

X-Ray Spectra

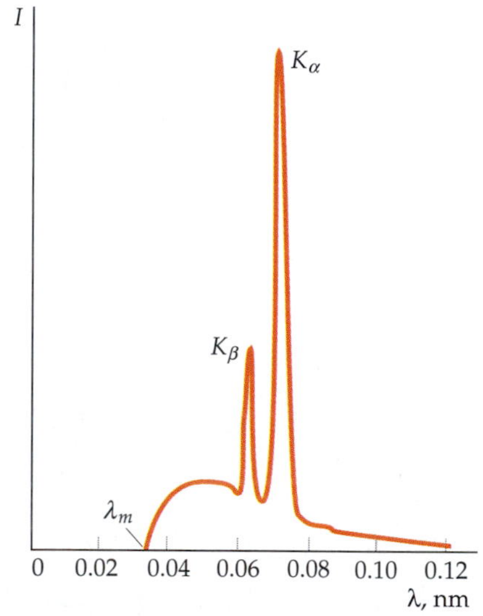

Figure 37-19 X-ray spectrum of molybdenum. The sharp peaks labeled K_α and K_β are characteristic of the element. The cutoff wavelength λ_m is independent of the target element and is related to the voltage V of the X-ray tube by $\lambda_m = hc/eV$.

X rays are usually produced by bombarding a target element with a high-energy beam of electrons in an X-ray tube. The result (Figure 37-19) consists of a continuous spectrum that depends only on the energy of the bombarding electrons, and a line spectrum that is characteristic of the target element. The characteristic spectrum results from excitation of the inner core electrons in the target element.

The energy needed to excite an inner core electron—for example, an electron in the $n = 1$ state (K shell)—is much greater than that required to excite an outer, valence electron. An inner electron cannot be excited to any of the filled states (for example, the $n = 2$ states in sodium) because of the exclusion principle. The energy required to excite an inner core electron to an unoccupied state is typically of the order of several keV. If an electron is knocked out of the $n = 1$ state (K shell), there is a vacancy left in this shell. This vacancy can be filled if an electron in the L shell (or in a higher shell) makes a transition into the K shell. The photons emitted by electrons making such transitions also have energies of the order of keV and produce the sharp peaks in the X-ray spectrum, as shown in Figure 37-18. The K_α line arises from transitions from the $n = 2$ (L) shell to the $n = 1$ (K) shell. The K_β line arises from transitions from the $n = 3$ shell to the $n = 1$ shell. These and other lines arising from transitions ending at the $n = 1$ shell make up the K series of the characteristic X-ray spectrum of the target element. Similarly, a second series, the L series, is produced by transitions from higher energy states to a vacated place in the $n = 2$ (L) shell.

We can use the Bohr theory to calculate approximately the frequencies of the characteristic X-ray spectra. According to the Bohr model, the energy of a single electron in a state n is given by

$$E_n = -Z^2 \frac{13.6 \text{ eV}}{n^2}$$

Since for any atom other than hydrogen, there are two electrons in the innermost shell, the K shell, the effective charge seen by one of the electrons is less than Ze because of the shielding due to the other electron. Assuming that the effective charge is $(Z - 1)e$, the energy of an electron in the K shell is given by this equation with $n = 1$ and Z replaced by $Z - 1$:

$$E_1 = -(Z - 1)^2 (13.6 \text{ eV})$$

The energy of an electron in the state n (assuming the same effective charge) is given by

$$E_n = -(Z-1)^2 \frac{13.6\text{ eV}}{n^2}$$

When an electron from state n drops into the vacated state in the $n = 1$ shell, a photon of energy $E_n - E_1$ is emitted. The wavelength of this photon is

$$\lambda = \frac{hc}{E_n - E_1} = \frac{hc}{(Z-1)^2(13.6\text{ eV})(1 - 1/n^2)} \qquad \text{37-46}$$

In 1913 the English physicist H. Moseley measured the wavelengths of the characteristic X-ray spectra for about 40 elements. From his data he was able to accurately determine the atomic number Z for each element.

Example 37-8

The wavelength of the K_α X-ray line for a certain element is $\lambda = 0.0721$ nm. What is the element?

Picture the Problem The K_α line corresponds to a transition from $n = 2$ to $n = 1$. The wavelength is related to the atomic number Z by Equation 37-46 with $n = 2$.

1. Solve Equation 37-46 for $(Z-1)^2$:

$$\lambda = \frac{hc}{(Z-1)^2(13.6\text{ eV})(1 - 1/n^2)}$$

$$(Z-1)^2 = \frac{hc}{\lambda(13.6\text{ eV})(1 - 1/n^2)}$$

2. Substitute the given data and solve for Z:

$$(Z-1)^2 = \frac{1240\text{ eV}\cdot\text{nm}}{(0.0721\text{ nm})(13.6\text{ eV})(1 - 1/4)} = 1686$$

$$Z - 1 = \sqrt{1686} = 41.06$$

$$Z = 42$$

Remark Since Z is an integer, we round to the nearest integer. The element is molybdenum.

Summary

1. The Bohr model is important historically because it was the first successful model to explain the discrete optical spectrum of atoms in terms of the quantization of energy.
2. The quantum theory of atoms results from the application of the Schrödinger equation to a bound system consisting of nucleus of charge $+Ze$ and Z electrons of charge $-e$.
3. For the simplest atom, hydrogen, consisting of one proton and one electron, the Schrödinger equation can be solved exactly to obtain the wave functions ψ, which depend on the quantum numbers n, ℓ, m, and m_s.

4. The electron configuration of atoms is governed by the Pauli exclusion principle, which states that no two electrons in an atom can have the same set of values for the quantum numbers n, ℓ, m, and m_s. Using the exclusion principle and the restrictions on the quantum numbers, we can understand much of the structure of the periodic table.

Topic	Remarks and Relevant Equations	
1. Bohr Model		
Postulates for the hydrogen atom		
Nonradiating orbits	The electron moves in a circular nonradiating orbit around the proton.	
Frequency of radiation related to energy states	$f = \frac{E_i - E_f}{h}$	**37-7**
Quantization of angular momentum	$mvr = \frac{nh}{2\pi} = n\hbar$	**37-9**
Radius of allowed orbits	$r = n^2 \frac{\hbar^2}{mkZe^2} = n^2 \frac{a_0}{Z}$	**37-11**
Bohr radius	$a_0 = \frac{\hbar^2}{mke^2} \approx 0.0529 \text{ nm}$	**37-12**
Energy levels in the H atom	$E_n = -\frac{mk^2e^4}{2\hbar^2}\frac{Z^2}{n^2} = -Z^2 \frac{E_0}{n^2}, \quad n = 1, 2, \ldots$	**37-15**
	$E_0 = \frac{mk^2e^4}{2\hbar^2} = \frac{1}{2}\frac{ke^2}{a_0} \approx 13.6 \text{ eV}$	**37-16**
Wavelengths emitted by the H atom	$\lambda = \frac{c}{f} = \frac{hc}{E_i - E_f} = \frac{1240 \text{ eV}\cdot\text{nm}}{E_i - E_f}$	**37-17, 37-18**
2. Quantum Theory of Atoms	The electron is described by a wave function ψ that obeys the Schrödinger equation. Energy quantization arises from standing wave conditions. ψ is described by the quantum numbers n, ℓ, m, and the spin quantum number $m_s = \pm\frac{1}{2}$.	
Schrödinger equation in spherical coordinates	$-\frac{\hbar^2}{2mr^2}\frac{\partial}{\partial r}\left(r^2 \frac{\partial \psi}{\partial r}\right) - \frac{\hbar^2}{2mr^2}\left[\frac{1}{\sin\theta}\frac{\partial}{\partial\theta}\left(\sin\theta \frac{\partial\psi}{\partial\theta}\right) + \frac{1}{\sin^2\theta}\frac{\partial^2\psi}{\partial\phi^2}\right] + U(r)\psi = E\psi$	**37-21**
The solutions can be written as products of functions of r, θ, and ϕ separately.	$\psi(r, \theta, \phi) = R(r)f(\theta)g(\phi)$	**37-22**
Quantum numbers		
Principal quantum number	$n = 1, 2, 3, \ldots$	
Orbital quantum number	$\ell = 0, 1, 2, 3, \ldots, n - 1$	
Magnetic quantum number	$m = -\ell, (-\ell + 1), \ldots, 0, 1, 2, \ldots, \ell$	**37-23**
Orbital angular momentum	$L = \sqrt{\ell(\ell + 1)}\,\hbar$	**37-24**

z component of angular momentumn	$L_z = m\hbar$	37-25

3. Quantum Theory of the Hydrogen Atom

Energy levels (same as for Bohr model)	$E_n = -\frac{mk^2e^4}{2\hbar^2}\frac{Z^2}{n^2} = -Z^2\frac{E_0}{n^2}, \quad n = 1, 2, 3, \ldots$	37-27
	where $E_0 = mk^2e^4/2\hbar^2 \approx 13.6\text{ eV}$	37-28
Wavelengths emitted by H atom (same as for Bohr model)	$\lambda = \frac{c}{f} = \frac{hc}{E_i - E_f} = \frac{1240\text{ eV}\cdot\text{nm}}{E_i - E_f}$	37-17, 37-18
Wave functions		
Ground state	$\psi_{1,0,0} = C_{1,0,0}e^{-Zr/a_0} = \frac{1}{\sqrt{\pi}}\left(\frac{Z}{a_0}\right)^{3/2}e^{-Zr/a_0}$	37-31, 37-33
First excited state	$\psi_{2,0,0} = C_{2,0,0}\left(2 - \frac{Zr}{a_0}\right)e^{-Zr/2a_0}$	37-36
	$\psi_{2,1,0} = C_{2,1,0}\frac{Zr}{a_0}e^{-Zr/2a_0}\cos\theta$	37-37
	$\psi_{2,1\pm1} = C_{2,1,1}\frac{Zr}{a_0}e^{-Zr/2a_0}\sin\theta\, e^{\pm i\phi}$	37-38
Probability densities	For $\ell = 0$, $\lvert\psi\rvert^2$ is spherically symmetric. For $\ell \neq 0$, $\lvert\psi\rvert^2$ depends on the angle θ.	
Radial probability density	$P(r) = 4\pi r^2\lvert\psi\rvert^2$	37-34
	The radial probability density is maximum at the distances corresponding roughly to the Bohr orbits.	

4. Spin–Orbit Effect and Fine Structure

The total angular momentum of an electron in an atom is a combination of the orbital angular momentum and spin angular momentum. It is characterized by the quantum number j, which can be either $\ell - \frac{1}{2}$ or $\ell + \frac{1}{2}$. Because of the interaction of the orbital and spin magnetic moments, the state $j = \ell - \frac{1}{2}$ has lower energy than the state $j = \ell + \frac{1}{2}$. This small splitting of the energy states gives rise to a small splitting of the spectral lines called fine structure.

5. Periodic Table

Beginning with hydrogen, each larger neutral atom adds one electron. The electrons go into those states that will give the lowest energy consistent with the Pauli exclusion principle.

The state of an atom is described by its electron configuration, which gives the values of n and ℓ for each electron. The ℓ values are specified by a code:

	s	p	d	f	g	h
ℓ values	0	1	2	3	4	5

Pauli exclusion principle	No two electrons in an atom can have the same set of values for the quantum numbers n, ℓ, m, and m_s.

6. Atomic Spectra

Atomic spectra include optical spectra and X-ray spectra. Optical spectra result from transitions between energy levels of a single outer electron moving in the field of the nucleus and core electrons of the atom. Characteristic X-ray spectra result from the excitation of an inner core electron and the subsequent filling of the vacancy by other electrons in the atom.

Selection rules	Transitions between energy states with the emission of a photon are governed by the following selection rules: $\Delta m = 0 \quad \text{or} \quad \pm 1$ **37-29** $\Delta \ell = \pm 1$

Problem-Solving Guide

Begin by drawing a neat diagram that includes the important features of the problem.

Summary of Worked Examples

Type of Calculation	Procedure and Relevant Examples
1. Calculate Wavelengths of Emitted Radiation	
Find the wavelength of radiation emitted by an atom.	Use $\lambda = hc/(E_{\text{i}} - E_{\text{f}})$, where $hc = 1240$ eV·nm and E_{i} and E_{f} are the energies of the initial and final states. **Example 37-2**
2. Angular Momentum	
Find the possible values of L_z given ℓ.	The possible values are $m\hbar$, where m ranges in integral steps from $-\ell$ to $+\ell$. **Example 37-3**
Find the angle between $\vec{L}$ and the z axis.	Use $\cos\theta = L_z/L$, where $L_z = m\hbar$ and $L = \sqrt{\ell(\ell+1)}\,\hbar$. **Example 37-3**
3. Hydrogen Atom Wave Functions	
Determine the probability of finding the electron in some range Δr in the hydrogen atom.	For small Δr, the probability is given by $P(r)\,\Delta r = 4\pi r^2\|\psi\|^2\,\Delta r$, where ψ is the appropriate wave function. **Example 37-4**
4. Spin–Orbit Effect	
Estimate the magnetic field seen by an electron given the fine-structure splitting of the energy levels.	Use $\Delta E = 2\mu_{\text{B}}B$. **Example 37-5**
5. Periodic Table	
Estimate the energy of interaction of two electrons in an atom.	Compare the ionization energy with the total energy of the electrons, neglecting their interaction. **Example 37-6**
Find the effective nuclear charge seen by an outer electron.	Choose the orbit radius from the Bohr model and calculate the energy from $E = -kZ'e^2/2r$ using the value of E determined from the ionization energy. **Example 37-7**
6. X-Ray Spectra	
Find the element given the wavelength of a line in the K series.	Use $\lambda = \dfrac{hc}{E_n - E_1} = \dfrac{hc}{(Z-1)^2(13.6\text{ eV})(1 - 1/n^2)}$ **Example 37-8**

Problems

Conceptual Problems

Problems from Optional and Exploring sections

In a few problems, you are given more data than you actually need; in a few other problems, you are required to supply data from your general knowledge, outside sources, or informed estimates.

• Single-concept, single-step, relatively easy

•• Intermediate-level, may require synthesis of concepts

••• Challenging, for advanced students

The Bohr Model

1 • As n increases, does the spacing of adjacent energy levels increase or decrease?

2 • The energy of the ground state of doubly ionized lithium ($Z = 3$) is _______, where $E_0 = 13.6$ eV.

(*a*) $-9E_0$
(*b*) $-3E_0$
(*c*) $-E_0/3$
(*d*) $-E_0/9$

3 • Bohr's quantum condition on electron orbits requires

(*a*) that the angular momentum of the electron about the hydrogen nucleus equal $n\hbar$.
(*b*) that no more than one electron occupy a given stationary state.
(*c*) that the electrons spiral into the nucleus while radiating electromagnetic waves.
(*d*) that the energies of an electron in a hydrogen atom be equal to nE_0, where E_0 is a constant energy and n is an integer.
(*e*) none of the above.

4 •• If an electron moves to a larger orbit, does its total energy increase or decrease? Does its kinetic energy increase or decrease?

5 •• The kinetic energy of the electron in the ground state of hydrogen is 13.6 eV $= E_0$. The kinetic energy of the electron in the state $n = 2$ is _______.

(*a*) $4E_0$
(*b*) $2E_0$
(*c*) $E_0/2$
(*d*) $E_0/4$

6 • The radius of the $n = 1$ orbit in the hydrogen atom is $a_0 = 0.053$ nm. What is the radius of the $n = 5$ orbit?

(*a*) $5a_0$
(*b*) $25a_0$
(*c*) a_0
(*d*) $\frac{1}{5}a_0$
(*e*) $\frac{1}{25}a_0$

7 • Use the known values of the constants in Equation 37-11 to show that a_0 is approximately 0.0529 nm.

8 • The longest wavelength of the Lyman series was calculated in Example 37-2. Find the wavelengths for the transitions (*a*) $n_1 = 3$ to $n_2 = 1$ and (*b*) $n_1 = 4$ to $n_2 = 1$.

9 • Find the photon energy for the three longest wavelengths in the Balmer series and calculate the wavelengths.

10 • (*a*) Find the photon energy and wavelength for the series limit (shortest wavelength) in the Paschen series ($n_2 = 3$). (*b*) Calculate the wavelengths for the three longest wavelengths in this series and indicate their positions on a horizontal linear scale.

11 • Repeat Problem 10 for the Brackett series ($n_2 = 4$).

12 • A hydrogen atom is in its tenth excited state according to the Bohr model ($n = 11$). (*a*) What is the radius of the Bohr orbit? (*b*) What is the angular momentum of the electron? (*c*) What is the electron's kinetic energy? (*d*) What is the electron's potential energy? (*e*) What is the electron's total energy?

13 •• The binding energy of an electron is the minimum energy required to remove the electron from its ground state to a large distance from the nucleus. (*a*) What is the binding energy for the hydrogen atom? (*b*) What is the binding energy for He^+? (*c*) What is the binding energy for Li^{2+}? [Singly ionized helium (He^+) and doubly ionized lithium (Li^{2+}) are "hydrogen-like" in that the system consists of a positively charged nucleus and a single bound electron.]

14 •• The electron of a hydrogen atom is in the $n = 2$ state. The electron makes a transition to the ground state. (*a*) What is the energy of the photon according to the Bohr model? (*b*) The linear momentum of the emitted photon is related to its energy by $p = E/c$. If we assume conservation of linear momentum, what is the recoil velocity of the atom? (*c*) Find the recoil kinetic energy of the atom in electron volts. By what percentage must the energy of the photon calculated in part (*a*) be corrected to account for this recoil energy?

15 •• Show that the speed of an electron in the nth Bohr orbit of hydrogen is given by $v_n = e^2/2\epsilon_0 hn$.

16 •• In this problem you will estimate the radius and the energy of the lowest stationary state of the hydrogen atom using the uncertainty principle. The total energy of the electron of momentum p and mass m a distance r from the proton in the hydrogen atom is given by $E = p^2/2m - ke^2/r$, where k is the Coulomb constant. Assume that the minimum value of p^2 is $p^2 \approx (\Delta p)^2 = \hbar^2/r^2$, where Δp is the uncertainty in p and we have taken $\Delta r \sim r$ for the order of magnitude of the uncertainty in position; the energy is then $E = \hbar^2/2mr^2 - ke^2/r$. Find the radius r_m for which this energy is a minimum, and calculate the minimum value of E in electron volts.

17 •• In a reference frame with the origin at the center of mass of an electron and the nucleus of an atom, the electron and nucleus have equal and opposite momenta of magnitude p. (*a*) Show that the total kinetic energy of the electron and nucleus can be written $K = p^2/2m_r$, where $m_r = m_eM/(M + m_e)$ is called the reduced mass, m_e is the mass of the electron, and M is the mass of the nucleus. It can be shown that the motion of the nucleus can be accounted for by replacing the mass of the electron by the reduced mass. In general, the reduced mass for a two-body problem with masses m_1 and m_2 is given by

$$m_r = \frac{m_1 m_2}{m_1 + m_2} \qquad 37\text{-}47$$

(*b*) Use Equation 37-14 with m replaced by m_r to calculate the Rydberg constant for hydrogen ($M = m_p$) and for a very massive nucleus ($M = \infty$). (*c*) Find the percentage correction for the ground-state energy of the hydrogen atom due to the motion of the proton.

Quantum Numbers in Spherical Coordinates

18 • For the principal quantum number $n = 4$, how many different values can the orbital quantum number ℓ have?

(*a*) 4
(*b*) 3
(*c*) 7
(*d*) 16
(*e*) 6

19 • For the principal quantum number $n = 4$, how many different values can the magnetic quantum number m have?

(*a*) 4
(*b*) 3
(*c*) 7
(*d*) 16
(*e*) 6

20 • For $\ell = 1$, find (*a*) the magnitude of the angular momentum L and (*b*) the possible values of m. (*c*) Draw to scale a vector diagram showing the possible orientations of $\vec{L}$ with the z axis.

21 • Work Problem 20 for $\ell = 3$.

22 • A compact disk has a moment of inertia of about 2.3×10^{-5} kg·m^2. (*a*) Find its angular momentum L when it is rotating at 500 rev/min. (*b*) Find the approximate value of the quantum number ℓ for this angular momentum.

23 • If $n = 3$, (*a*) what are the possible values of ℓ? (*b*) For each value of ℓ in (*a*), list the possible values of m. (*c*) Using the fact that there are two quantum states for each value of ℓ and m because of electron spin, find the total number of electron states with $n = 3$.

24 • Find the total number of electron states with (*a*) $n = 2$ and (*b*) $n = 4$. (See Problem 23.)

25 • Find the minimum value of the angle θ between $\vec{L}$ and the z axis for (*a*) $\ell = 1$, (*b*) $\ell = 4$, and (*c*) $\ell = 50$.

26 • What are the possible values of n and m if (*a*) $\ell = 3$, (*b*) $\ell = 4$, and (*c*) $\ell = 0$?

27 • What are the possible values of n and ℓ if (*a*) $m = 0$, (*b*) $m = -1$, and (*c*) $m = 2$?

Quantum Theory of the Hydrogen Atom

28 • For the ground state of the hydrogen atom, find the values of (*a*) ψ, (*b*) ψ^2, and (*c*) the radial probability density $P(r)$ at $r = a_0$. Give your answers in terms of a_0.

29 • (*a*) If spin is not included, how many different wave functions are there corresponding to the first excited energy level $n = 2$ for hydrogen? (*b*) List these functions by giving the quantum numbers for each state.

30 • For the ground state of the hydrogen atom, calculate the probability of finding the electron in the range $\Delta r = 0.03a_0$ at (*a*) $r = a_0$ and (*b*) $r = 2a_0$.

31 • The value of the constant $C_{2,0,0}$ in Equation 37-36 is

$$C_{2,0,0} = \frac{1}{4\sqrt{2\pi}}\left(\frac{Z}{a_0}\right)^{3/2}$$

Find the values of (*a*) ψ, (*b*) ψ^2, and (*c*) the radial probability density $P(r)$ at $r = a_0$ for the state $n = 2$, $\ell = 0$, $m = 0$ in hydrogen. Give your answers in terms of a_0.

32 •• Show that the radial probability density for the $n = 2$, $\ell = 1$, $m = 0$ state of a one-electron atom can be written as $P(r) = A\cos^2\theta r^4 e^{-Zr/a_0}$, where A is a constant.

33 •• Calculate the probability of finding the electron in the range $\Delta r = 0.02a_0$ at (*a*) $r = a_0$ and (*b*) $r = 2a_0$ for the state $n = 2$, $\ell = 0$, $m = 0$ in hydrogen. (See Problem 31 for the value of $C_{2,0,0}$.)

34 •• The radial probability distribution function for a one-electron atom in its ground state can be written $P(r) = Cr^2e^{-2Zr/a_0}$, where C is a constant. Show that $P(r)$ has its maximum value at $r = a_0/Z$.

35 •• Find the expectation value of r, $\langle r\rangle = \int_0^\infty rP(r)\,dr$ for hydrogen in its ground state.

36 ••• Show that the number of states in the hydrogen atom for a given n is $2n^2$.

37 ••• Calculate the probability that the electron in the ground state of a hydrogen atom is in the region $0 < r < a_0$.

The Spin–Orbit Effect

38 • The potential energy of a magnetic moment in an external magnetic field is given by $U = -\vec{\mu}\cdot\vec{B}$. (*a*) Calculate the difference in energy between the two possible orientations of an electron in a magnetic field $\vec{B} = 0.600\text{ T}\,\hat{k}$. (*b*) If these electrons are bombarded with photons of energy equal to this energy difference, "spin flip" transitions can be induced. Find the wavelength of the photons needed for such transitions. This phenomenon is called *electron spin resonance*.

39 • The total angular momentum of a hydrogen atom in a certain excited state has the quantum number $j = \frac{1}{2}$. What can you say about the orbital angular-momentum quantum number ℓ?

40 • The total angular momentum of a hydrogen atom in a certain excited state has the quantum number $j = 1\frac{1}{2}$. What can you say about the orbital angular-momentum quantum number ℓ?

41 • A hydrogen atom is in the state $n = 3$, $\ell = 2$. What are the possible values of j?

The Periodic Table

42 • The p state of an electronic configuration corresponds to

(*a*) $n = 2$.
(*b*) $\ell = 2$.
(*c*) $\ell = 1$.
(*d*) $n = 0$.
(*e*) $\ell = 0$.

43 •• Why is the energy of the 3s state considerably lower than that of the 3p state for sodium, whereas in hydrogen these states have essentially the same energy?

44 •• Discuss the evidence from the periodic table of the need for a fourth quantum number. How would the properties of He differ if there were only three quantum numbers, n, ℓ, and m?

45 •• The properties of iron (Z = 26) and cobalt (Z = 27), which have adjacent atomic numbers, are similar, whereas the properties of neon (Z = 10) and sodium (Z = 11), which also have adjacent atomic numbers, are very different. Explain why.

46 •• Separate the following six elements—potassium, calcium, titanium, chromium, manganese, and copper—into two groups of three each such that those in a group have similar properties.

47 • What element has the electron configuration (*a*) $1s^2 2s^2 2p^6 3s^2 3p^2$ and (*b*) $1s^2 2s^2 2p^6 3s^2 3p^6 4s^2$?

48 • The total number of quantum states of hydrogen with quantum number $n = 4$ is ______.

(*a*) 4
(*b*) 16
(*c*) 32
(*d*) 36

49 • How many of oxygen's eight electrons are found in the p state?

(*a*) 0
(*b*) 2
(*c*) 4
(*d*) 6
(*e*) 8

50 • Write the electron configuration of (*a*) carbon and (*b*) oxygen.

51 • Write the electron configuration of (*a*) aluminum and (*b*) chromium.

52 • Give the possible values of the z component of the orbital angular momentum of (*a*) a d electron and (*b*) an f electron.

53 •• If the outer electron in sodium moves in the $n = 3$ Bohr orbit, the effective nuclear charge would be $Z'e = 1e$, and the energy of the electron would be $-13.6\text{ eV}/3^2 = -1.51\text{ eV}$. However, the ionization energy of sodium is 5.14 eV, not 1.51 eV. Use this fact and Equation 37-45 to calculate the effective nuclear charge Z' seen by the outer electron in sodium. Assume that $r = 9a_0$ for the outer electron.

Optical and X-Ray Spectra

54 • The optical spectra of atoms with two electrons in the same outer shell are similar, but they are quite different from the spectra of atoms with just one outer electron because of the interaction of the two electrons. Separate the following elements into two groups such that those in each group have similar spectra: lithium, beryllium, sodium, magnesium, potassium, calcium, chromium, nickel, cesium, barium.

55 • Write down the possible electron configurations for the first excited state of (*a*) hydrogen, (*b*) sodium, and (*c*) helium.

56 • Indicate which of the following elements should have optical spectra similar to hydrogen and which should be similar to helium: Li, Ca, Ti, Rb, Hg, Ag, Cd, Ba, Fr, Ra.

57 • (*a*) Calculate the next two longest wavelengths in the K series (after the K_α line) of molybdenum. (*b*) What is the wavelength of the shortest wavelength in this series?

58 • The wavelength of the K_α line for a certain element is 0.3368 nm. What is the element?

59 • The wavelength of the K_α line for a certain element is 0.0794 nm. What is the element?

60 • Calculate the wavelength of the K_α line of rhodium.

61 • Calculate the wavelength of the K_α line in (*a*) magnesium (Z = 12) and (*b*) copper (Z = 29).

General Problems

62 • For the principal quantum number $n = 3$, what are the possible values of the quantum numbers ℓ and m?

63 • An electron in the L shell means that

(*a*) $\ell = 0$.
(*b*) $\ell = 1$.
(*b*) $n = 1$.
(*c*) $n = 2$.
(*d*) $m = 2$.
(*e*) none of the above are true.

64 •• The Bohr theory and the Schrödinger theory of the hydrogen atom give the same results for the energy levels. Discuss the advantages and disadvantages of each model.

65 •• In Figure 37-17, there are small dips in the ionization-energy curve at Z = 31 (gallium) and Z = 49 (indium) that are not labeled. Explain these dips using the electron configurations of these atoms given in Table 37-1.

66 • What is the energy of the shortest wavelength photon emitted by the hydrogen atom?

67 • The wavelength of a spectral line of hydrogen is 97.254 nm. Identify the transition that results in this line.

68 • The wavelength of a spectral line of hydrogen is 1093.8 nm. Identify the transition that results in this line.

69 • Spectral lines of the following wavelengths are emitted by singly ionized helium: 164 nm, 230.6 nm, and 541 nm. Identify the transitions that result in these spectral lines.

70 •• We are often interested in finding the quantity ke^2/r in electron volts when r is given in nanometers. Show that $ke^2 = 1.44$ eV·nm.

71 •• The wavelengths of the photons emitted by potassium corresponding to transitions from the $4P_{3/2}$ and $4P_{1/2}$ states to the ground state are 766.41 and 769.90 nm. (*a*) Calculate the energies of these photons in electron volts. (*b*) The difference in the energies of these photons equals the difference in energy ΔE between the $4P_{3/2}$ and $4P_{1/2}$ states in potassium. Calculate ΔE. (*c*) Estimate the magnetic field that the 4p electron in potassium experiences.

72 •• To observe the characteristic K lines of the X-ray spectrum, one of the $n = 1$ electrons must be ejected from the atom. This is generally accomplished by bombarding the target material with electrons of sufficient energy to eject this tightly bound electron. What is the minimum energy required to observe the K lines of (*a*) tungsten, (*b*) molybdenum, and (*c*) copper?

73 •• The combination of physical constants $\alpha = e^2k/\hbar c$, where k is the Coulomb constant, is known as the *fine-structure constant*. It appears in numerous relations in atomic physics. (*a*) Show that α is dimensionless. (*b*) Show that in the Bohr model of hydrogen $v_n = c\alpha/n$, where v_n is the speed of the electron in the stationary state of quantum number n.

74 •• The *positron* is a particle identical to the electron except that it carries a positive charge of e. *Positronium* is the bound state of an electron and positron. (*a*) Calculate the energies of the five lowest energy states of positronium using the reduced mass as given by Equation 37-47 in Problem 17. (*b*) Do transitions between any of the levels found in (*a*) fall in the visible range of wavelengths? If so, which transitions are these?

75 •• The deuteron, the nucleus of deuterium ("heavy hydrogen"), was first recognized from the spectrum of hydrogen. The deuteron has a mass twice that of the proton. (*a*) Calculate the Rydberg constant for hydrogen and for deuterium using the reduced mass as given by Equation 37-47 in Problem 17. (*b*) Using the result obtained in (*a*), determine the wavelength difference between the longest wavelength Balmer lines of hydrogen and deuterium.

76 •• The *muonium* atom is a hydrogen atom with the electron replaced by a μ^- particle. The μ^- is identical to an electron but has a mass 207 times as great as the electron. (*a*) Calculate the energies of the five lowest energy levels of muonium using the reduced mass as given by Equation 37-47 in Problem 17. (*b*) Do transitions between any of the levels found in (*a*) fall in the visible range of wavelengths, i.e., between $\lambda = 700$ nm and 400 nm? If so, which transitions are these?

77 •• The triton, a nucleus consisting of a proton and two neutrons, is unstable with a fairly long half-life of about 12 years. *Tritium* is the bound state of an electron and a triton. (*a*) Calculate the Rydberg constant of tritium using the reduced mass as given by Equation 37-47 in Problem 17. (*b*) Using the result obtained in (*a*) and in part (*a*) of Problem 75 determine the wavelength difference between the longest wavelength Balmer lines of tritium and deuterium and between tritium and hydrogen.

78 ••• Suppose that the interaction between an electron and proton were of the form $F = -Kr$, where K is a constant, rather than $1/r^2$. If the stationary state orbits are again limited by the angular momentum condition $L = n\hbar$, what are then the radii of these orbits? Show that for this case the total energies of the stationary states are given by $E = n\hbar\omega$, where ω is the angular frequency of the electron about the proton.

79 ••• The frequency of revolution of an electron in a circular orbit of radius r is $f_{\text{rev}} = v/2\pi r$, where v is the speed. (*a*) Show that in the nth stationary state

$$f_{\text{rev}} = \frac{k^2Z^2e^4m}{2\pi\hbar^3}\frac{1}{n^3}$$

(*b*) Show that when $n_1 = n$, $n_2 = n - 1$, and n is much greater than 1,

$$\frac{1}{n_2^2} - \frac{1}{n_1^2} \approx \frac{2}{n^3}$$

(*c*) Use your result in part (*b*) and Equation 37-13 to show that in this case the frequency of radiation emitted equals the frequency of motion. This result is an example of Bohr's correspondence principle: When n is large, so that the energy difference between adjacent states is a small fraction of the total energy, classical and quantum physics must give the same results.

CHAPTER 38

Molecules and Solids

Molten tin solidifies in a pattern of tree-shaped crystals called dendrites as it cools under controlled circumstances.

Most atoms bond together to form molecules or solids. Molecules may exist as separate entities as in gaseous O_2 or N_2 or they, too, may bond together to form liquids or solids. A molecule is the smallest constituent of a substance that retains its chemical properties.

In this chapter we use our understanding of quantum mechanics to discuss molecular bonding, the energy levels and spectra of diatomic molecules, the structure of solids, and solid-state semiconducting devices. Much of our discussion will be qualitative because, as in atomic physics, the quantum-mechanical calculations are very difficult. Before studying semiconductor devices you should review the material in Chapter 27 on the microscopic theory of electrical conduction.

38-1 Molecular Bonding

There are two extreme views that we can take of a molecule. Consider, for example, H_2. We can think of it either as two H atoms joined together, or as a quantum-mechanical system of two protons and two electrons. The latter picture is more fruitful in this case because neither of the electrons in the H_2 molecule can be identified as belonging to either proton. Instead, the wave function for each electron is spread out in space throughout the whole molecule. For more complicated molecules, however, an intermediate picture is useful. For example, the nitrogen molecule N_2 consists of 14 protons and 14 electrons, but only two of the electrons take part in the bonding. We therefore can consider this molecule as two N^+ ions and two electrons that belong to the molecule as a whole. The molecular wave functions for these bonding electrons are called **molecular orbitals.** In many cases these molecular wave functions can be constructed from combinations of the atomic wave functions with which we are familiar.

The two principal types of bonds responsible for the formation of molecules are the ionic bond and the covalent bond. Other types of bonds that are important in the bonding of liquids and solids are van der Waals bonds, metallic bonds, and hydrogen bonds. In many cases, bonding is a mixture of these mechanisms.

The Ionic Bond

The simplest type of bond is the **ionic bond,** which is found in salts such as sodium chloride (NaCl). The sodium atom has one 3s electron outside a stable core. The energy needed to remove this electron, the ionization energy, is just 5.14 eV (see Figure 37-15). The removal of this electron leaves a positive ion with a spherically symmetric, closed-shell electron core. Chlorine, on the other hand, is one electron short of having a closed shell. The energy released by an atom's acquisition of one electron is called its **electron affinity,** which in the case of chlorine is 3.62 eV. The acquisition of one electron by chlorine results in a negative ion with a spherically symmetric, closed-shell electron core. Thus, the formation of an Na^+ ion and a Cl^- ion by the donation of one electron of sodium to chlorine requires only $5.14\text{ eV} - 3.62\text{ eV} = 1.52\text{ eV}$ at infinite separation. The electrostatic potential energy of the two ions when they are a distance r apart is $-ke^2/r$. When the separation of the ions is less than about 0.95 nm, the negative potential energy of attraction is of greater magnitude than the 1.52 eV of energy needed to create the ions. Thus, at separation distances less than 0.95 nm it is energetically favorable (that is, the total energy of the system is reduced) for the sodium atom to donate an electron to the chlorine atom to form NaCl.

Since the electrostatic attraction increases as the ions get closer together, it might seem that equilibrium could not exist. However, when the separation of the ions is very small, there is a strong repulsion that is quantum mechanical in nature and is related to the exclusion principle. This **exclusion-principle repulsion** is responsible for the repulsion of the atoms in all molecules (except H_2)* for all bonding mechanisms. We can understand it qualitatively as follows. When the ions are very far apart, the wave function for a core electron in one of the ions does not overlap that of any electron in the other ion. We can distinguish the electrons by the ion to which they belong. This means that electrons in the two ions can have the same quantum numbers because they occupy different regions of space. However, as the distance between the ions decreases, the wave functions of the core electrons begin to overlap; that is, the electrons in the two ions begin to occupy the same region of space. Because of the exclusion principle, some of these electrons must go into higher energy quantum states.† But energy is required to shift the elec-

---

* In H_2, the repulsion is simply that of the two positively charged protons.

† Recall from our discussion in Chapter 36 that the exclusion principle is related to the fact that the wave function for two identical electrons is antisymmetric on the exchange of the electrons and that an antisymmetric wave function for two electrons with the same quantum numbers is zero if the space coordinates of the electrons are the same.

trons into higher energy quantum states. This increase in energy when the ions are pushed closer together is equivalent to a repulsion of the ions. It is not a sudden process. The energy states of the electrons change gradually as the ions are brought together. A sketch of the potential energy of the Na^+ and Cl^- ions versus separation is shown in Figure 38-1. The energy is lowest at an equilibrium separation of about 0.236 nm. At smaller separations, the energy rises steeply as a result of the exclusion principle. The energy required to separate the ions and form neutral sodium and chlorine atoms is called the **dissociation energy,** which is about 4.27 eV for NaCl.

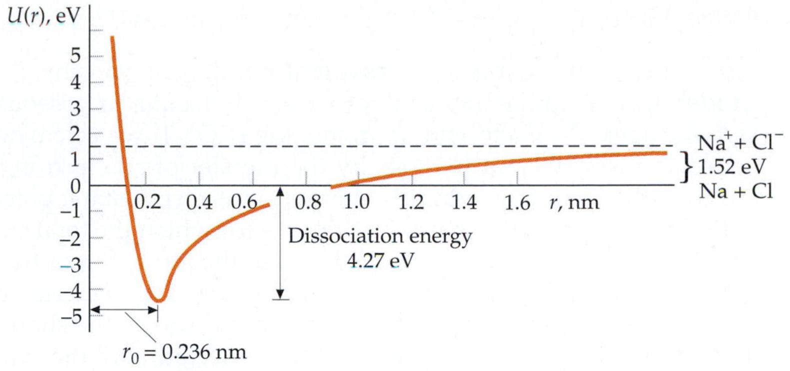

Figure 38-1 Potential energy for Na^+ and Cl^- ions as a function of separation distance r. The energy at infinite separation was chosen to be 1.52 eV, corresponding to the energy needed to form the ions from neutral atoms. The minimum energy is at the equilibrium separation $r_0 = 0.236$ nm for the ions in the molecule.

The equilibrium separation distance of 0.236 nm is for gaseous diatomic NaCl, which can be obtained by evaporating solid NaCl. Normally, NaCl exists in a cubic crystal structure, with the Na^+ and Cl^- ions at the alternate corners of a cube. The separation of the ions in a crystal is somewhat larger, about 0.28 nm. Because of the presence of neighboring ions of opposite charge, the electrostatic energy per ion pair is lower when the ions are in a crystal.

Example 38-1

The electron affinity of fluorine is 3.40 eV, and the equilibrium separation of sodium fluoride (NaF) is 0.193 nm. (*a*) How much energy is needed to form Na^+ and F^- ions from neutral sodium and fluorine atoms? (*b*) What is the electrostatic potential energy of the Na^+ and F^- ions at their equilibrium separation? (*c*) The dissociation energy of NaF is 5.38 eV. What is the energy due to repulsion of the ions at the equilibrium separation?

Picture the Problem (*a*) The energy ΔE needed to form Na^+ and F^- ions from the neutral sodium and fluorine atoms is the difference between the ionization energy of sodium (5.14 eV) and the electron affinity of fluorine. (*b*) The electrostatic potential energy with $U = 0$ at infinity is $U_e = \frac{-ke^2}{r}$. (*c*) If we choose the potential energy at infinity to be ΔE, the total potential energy is $U_{tot} = U_e + \Delta E + U_{rep}$, where U_{rep} is the energy of repulsion, which is found by setting the dissociation energy equal to $-U_{tot}$.

(*a*) Calculate the energy needed to form Na^+ and F^- ions from the neutral sodium and fluorine atoms:

$$\Delta E = 5.14\text{ eV} - 3.40\text{ eV} = 1.74\text{ eV}$$

(*b*)1. Calculate the electrostatic potential energy at the equilibrium separation of $r = 0.193$ nm:

$$U_e = -\frac{ke^2}{r} = -\frac{(8.99\times10^9\text{ N}\cdot\text{m}^2/\text{C}^2)(1.60\times10^{-19}\text{ C})^2}{1.93\times10^{-10}\text{ m}}$$

$$= -1.19\times10^{-18}\text{ J}$$

2. Convert from joules to electron volts:

$$U_e = -1.19\times10^{-18}\text{ J}\,\frac{1\text{ eV}}{1.60\times10^{-19}\text{ J}} = -7.45\text{ eV}$$

(*c*)1. The dissociation energy equals the negative of the total potential energy:

$$E_d = -U_{tot} = -(U_e + \Delta E + U_{rep})$$

$$= -(-7.45\text{ eV} + 1.74\text{ eV} + U_{rep}) = 5.38\text{ eV}$$

2. Solve for U_{rep}:

$$U_{rep} = 7.45\text{ eV} - 1.74\text{ eV} - 5.38\text{ eV} = 0.33\text{ eV}$$

The Covalent Bond

A completely different mechanism, the **covalent bond,** is responsible for the bonding of identical or similar atoms to form such molecules as gaseous hydrogen (H_2), nitrogen (N_2), and carbon monoxide (CO). If we calculate the energy needed to form H^+ and H^- ions by the transfer of an electron from one atom to the other and then add this energy to the electrostatic potential energy, we find that there is no separation distance for which the total energy is negative. The bond thus cannot be ionic. Instead, the attraction of two hydrogen atoms is an entirely quantum-mechanical effect. The decrease in energy when two hydrogen atoms approach each other is due to the sharing of the two electrons by both atoms. It is intimately connected with the symmetry properties of the wave functions of electrons.

We can gain some insight into covalent bonding by considering a simple, one-dimensional quantum-mechanics problem of two finite square wells. We first consider a single electron that is equally likely to be in either well. Since the wells are identical, the probability distribution, which is proportional to $|\psi^2|$, must be symmetric about the midpoint between the wells. Then ψ must be either symmetric or antisymmetric with respect to the two wells. The two possibilities for the ground state are shown in Figure 38-2*a* for the case in which the wells are far apart and in Figure 38-2*b* for the case in which the wells are close together. An important feature of Figure 38-2*b* is that in the region between the wells the symmetric wave function is large and the antisymmetric wave function is small.

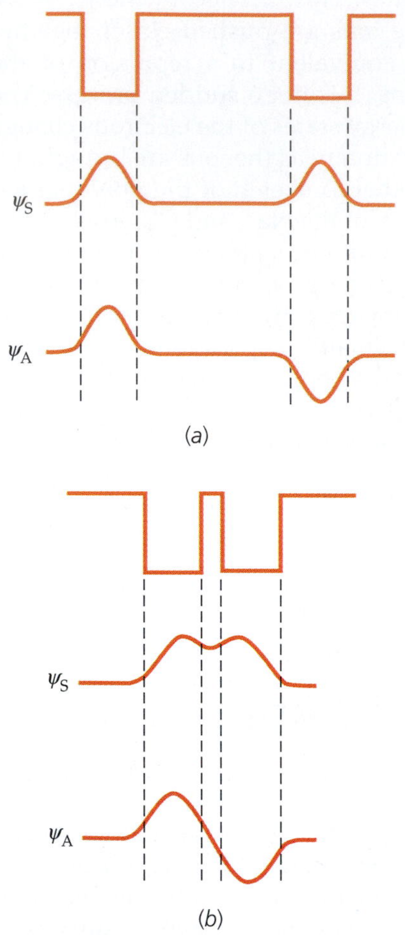

Figure 38-2 (*a*) Two square wells far apart. The electron wave function can be either symmetric (ψ_S) or antisymmetric (ψ_A) in space. The probability distributions and energies are the same for the two wave functions when the wells are far apart. (*b*) Two square wells that are close together. Between the wells the symmetric space wave function is larger than the antisymmetric space wave function.

Now consider adding a second electron to the two wells. We saw in Chapter 36 that the wave functions for particles that obey the exclusion principle are antisymmetric on exchange of the particles. Thus the total wave function for the two electrons must be antisymmetric on exchange of the electrons. Note that exchanging the electrons in the wells here is the same as exchanging the wells.The total wave function for two electrons can be written as a product of a space part and a spin part. So an antisymmetric wave function can be the product of a symmetric space part and an antisymmetric spin part or of a symmetric spin part and an antisymmetric space part.

To understand the symmetry of the total wave function, we must therefore understand the symmetry of the spin part of the wave function. The spin of a single electron can have two possible values for its quantum number m_s: $m_s = +\frac{1}{2}$, which we call spin up, or $m_s = -\frac{1}{2}$, which we call spin down. We will use arrows to designate the spin wave function for a single electron: $\uparrow_1$ or $\uparrow_2$ for electron 1 or electron 2 with spin up, and $\downarrow_1$ or $\downarrow_2$ for electron 1 or electron 2 with spin down. The total spin quantum number for two electrons can be $S = 1$, with $m_S = +1, 0,$ or -1, or $S = 0$, with $m_S = 0$. We use ϕ_{S,m_S} to denote the spin wave function for two electrons. The spin state $\phi_{1,+1}$, corresponding to $S = 1$ and $m_S = +1$, can be written

$$\phi_{1,+1} = \uparrow_1\uparrow_2, \qquad S = 1, m_S = +1 \tag{38-1}$$

Similarly, the spin state for $S = 1, m_S = -1$ is

$$\phi_{1,-1} = \downarrow_1\downarrow_2, \qquad S = 1, m_S = -1 \tag{38-2}$$

Note that both of these states are symmetric upon exchange of the electrons. The spin state corresponding to $S = 1$ and $m_S = 0$ is not quite so obvious. It turns out to be proportional to

$$\phi_{1,0} = \uparrow_1\downarrow_2 + \uparrow_2\downarrow_1, \qquad S = 1, m_S = 0 \tag{38-3}$$

This spin state is also symmetric upon exchange of the electrons. The spin state for two electrons with antiparallel spins ($S = 0$) is

$$\phi_{0,0} = \uparrow_1\downarrow_2 - \uparrow_2\downarrow_1, \qquad S = 0, m_S = 0 \tag{38-4}$$

This spin state is antisymmetric upon exchange of electrons.

We thus have the important result that the *spin* part of the wave function is symmetric for parallel spins ($S = 1$) and antisymmetric for antiparallel spins ($S = 0$). Because the total wave function is the product of the space function and spin function, we have the following important result:

> For the total wave function of two electrons to be antisymmetric, the space part of the wave function must be antisymmetric for parallel spins ($S = 1$) and symmetric for antiparallel spins ($S = 0$).

We can now consider the problem of two hydrogen atoms. Figure 38-3*a* shows a spatially symmetric wave function ψ_S and a spatially antisymmetric wave function ψ_A for two hydrogen atoms that are far apart, and Figure 38-3*b* shows the same two wave functions for two hydrogen atoms that are close together. The squares of these two wave functions are shown in Figure 38-3*c*. Note that the probability distribution $|\psi|^2$ in the region between the protons is large for the symmetric wave function and small for the antisymmetric wave function. Thus, when the space part of the wave function is symmetric ($S = 0$), the electrons are often found in the region between the protons. The negatively charged electron cloud representing these electrons is concentrated in the space between the protons, as shown in the upper part of Figure 38-3*c*, and the protons are bound together by this negatively charged cloud. Conversely, when the space part of the wave function is antisymmetric ($S = 1$), the electrons spend little time between the protons, and the atoms do not bind together to form a molecule. In this case, the electron cloud is not concentrated in the space between the protons, as shown in the lower part of Figure 38-3*c*.

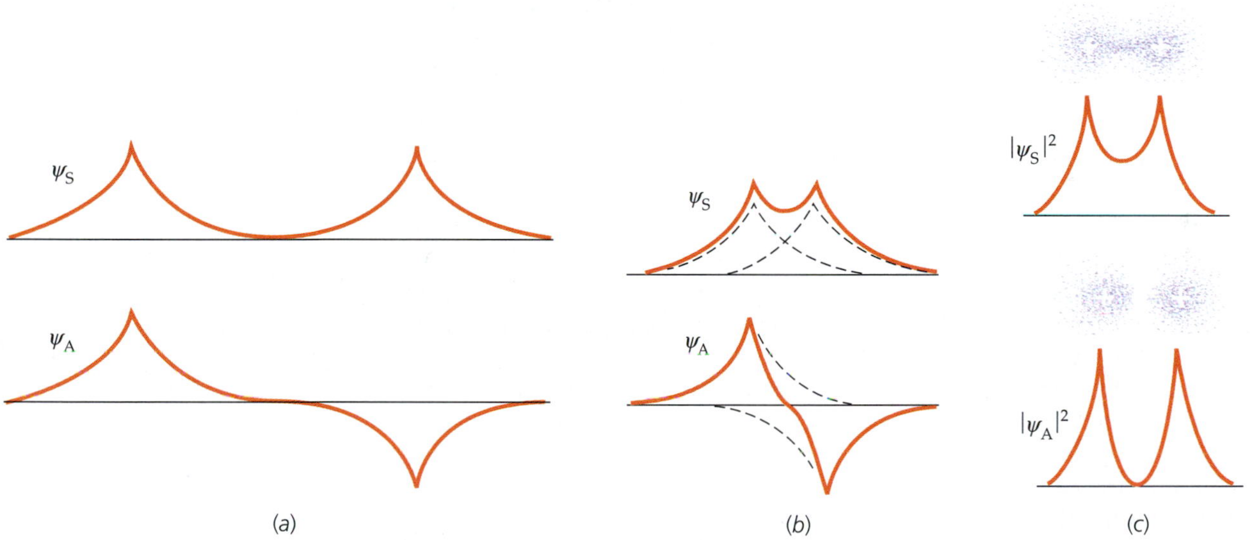

Figure 38-3 One-dimensional symmetric and antisymmetric wave functions for two hydrogen atoms (*a*) far apart and (*b*) close together. (*c*) Electron probability distributions ($|\psi|^2$) for the wave functions in (*b*). For the symmetric wave function, the electron charge density is large between the protons. This negative charge density holds the protons together in the hydrogen molecule H_2. For the antisymmetric wave function, the electron charge density is not large between the protons.

The total electrostatic potential energy for the H_2 molecule consists of the positive energy of repulsion of the two electrons and the negative potential energy of attraction of each electron for each proton. Figure 38-4 shows the electrostatic potential energy for two hydrogen atoms versus separation for the case in which the space part of the electron wave function is symmetric (U_S) and for the case in which it is antisymmetric (U_A). We can see that the potential energy for the symmetric state is the lower of the two and that the shape of this potential-energy curve is similar to that for ionic bonding. The equilibrium separation for H_2 is r_0 = 0.074 nm, and the binding energy is 4.52 eV. For the antisymmetric state, the potential energy is never negative and there is no bonding.

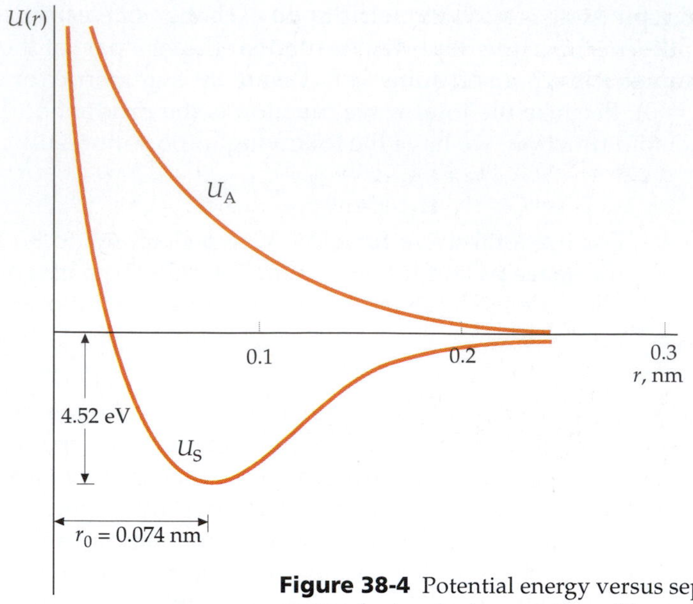

Figure 38-4 Potential energy versus separation for two hydrogen atoms. The curve labeled U_S is for a wave function with a symmetric space part, and the curve labeled U_A is for a wave function with an antisymmetric space part.

We can now see why three hydrogen atoms do not bond to form H_3. If a third hydrogen atom is brought near an H_2 molecule, the third electron cannot be in a 1s state and have its spin antiparallel to the spin of both of the other electrons. If this electron is in an antisymmetric space state with respect to exchange with one of the electrons, the repulsion of this atom is greater than the attraction of the other. As the three atoms are pushed together, the third electron is, in effect, forced into a higher quantum-energy state by the exclusion principle. The bond between two hydrogen atoms is called a **saturated bond** because there is no room for another electron. The two shared electrons essentially fill the 1s states of both atoms.

We can also see why two helium atoms do not normally bond together to form the He_2 molecule. There are no valence electrons that can be shared. The electrons in the closed shells are forced into higher energy states when the two atoms are brought together. At low temperatures or high pressures, helium atoms do bond together due to van der Waals forces, which we will discuss next. This bonding is so weak that at atmospheric pressure helium boils at 4 K, and it does not form a solid at any temperature unless the pressure is greater than about 20 atm.

When two identical atoms bond, as in O_2 or N_2, the bonding is purely covalent. However, the bonding of two dissimilar atoms is often a mixture of covalent and ionic bonding. Even in NaCl, the electron donated by sodium to chlorine has some probability of being at the sodium atom because its wave function does not suddenly fall to zero. Thus, this electron is partially shared in a covalent bond, although this bonding is only a small part of the total bond, which is mainly ionic.

A measure of the degree to which a bond is ionic or covalent can be obtained from the electric dipole moment of the molecule. For example, if the bonding in NaCl were purely ionic, the center of positive charge would be at the Na^+ ion and the center of negative charge would be at the Cl^- ion. The electric dipole moment would have the magnitude

$$p_{\text{ionic}} = er_0 \tag{38-5}$$

where r_0 is the equilibrium separation of the ions. Thus, the dipole moment of NaCl would be (from Figure 38-1)

$$p_{\text{ionic}} = er_0$$
$$= (1.60 \times 10^{-19}\ \text{C})(2.36 \times 10^{-10}\ \text{m}) = 3.78 \times 10^{-29}\ \text{C}\cdot\text{m}$$

The actual measured electric dipole moment of NaCl is

$$p_{\text{measured}} = 3.00 \times 10^{-29}\ \text{C}\cdot\text{m}$$

We can define the ratio of p_{measured} to p_{ionic} as the fractional amount of ionic bonding. For NaCl, this ratio is $3.00/3.78 = 0.79$. Thus, the bonding in NaCl is about 79% ionic.

Exercise The equilibrium separation of HCl is 0.128 nm and its measured electric dipole moment is 3.60×10^{-30} C·m. What is the percentage of ionic bonding in HCl? (*Answer* 18%)

Other Bonding Types

The van der Waals Bond Any two separated molecules will be attracted to one another by electrostatic forces called van der Waals forces. So will any two atoms that do not form ionic or covalent bonds. The **van der Waals bonds** due to these forces are much weaker than the bonds already discussed. At high enough temperatures, these forces are not strong enough to overcome the ordinary thermal agitation of the atoms or molecules, but at sufficiently low temperatures, thermal agitation becomes negligible, and the van der Waals forces will cause virtually all substances to condense into a liquid and then a solid form.* The van der Waals forces arise from the interaction of the instantaneous electric dipole moments of the molecules.

Figure 38-5 shows how two polar molecules—molecules with *permanent* electric dipole moments, such as H_2O—can bond. The electric field due to the dipole moment of one molecule orients the other molecule such that the two dipole moments attract. Nonpolar molecules also attract other nonpolar molecules via the van der Waals forces. Although nonpolar molecules have zero electric dipole moments on the average, they have instantaneous dipole moments that are generally not zero because of fluctuations in the positions of the charges. When two nonpolar molecules are near each other, the fluctuations in the instantaneous dipole moments tend to become correlated so as to produce attraction. This is illustrated in Figure 38-6.

Figure 38-5 Bonding of H_2O molecules because of the attraction of the electric dipoles. The dipole moment of each molecule is indicated by $\vec{p}$. The field of one dipole orients the other dipole so the moments tend to be parallel. When the dipole moments are approximately parallel, the center of negative charge of one molecule is close to the center of positive charge of the other molecule, and the molecules attract.

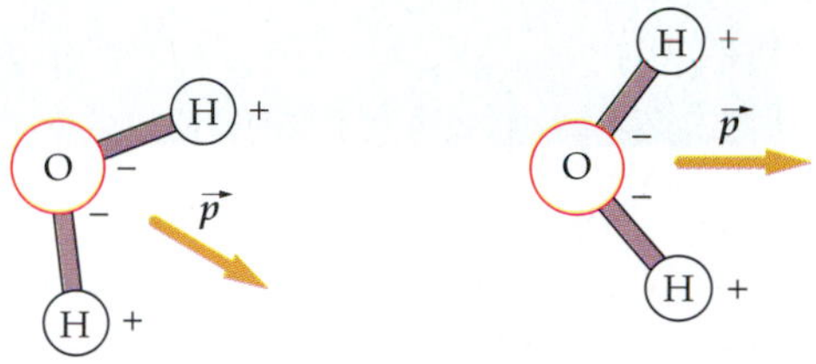

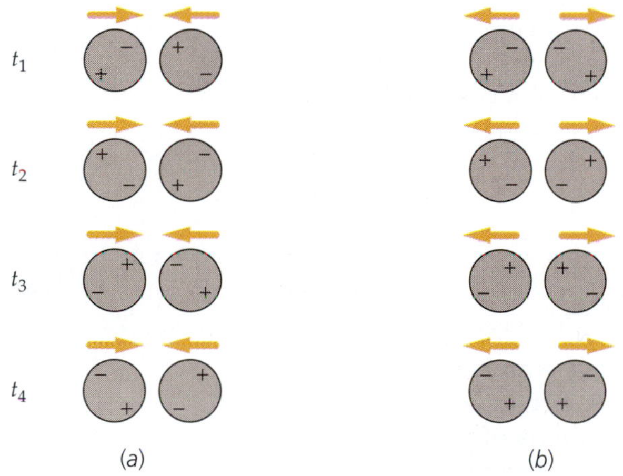

Figure 38-6 van der Waals attraction of molecules with zero average dipole moments. (*a*) Possible orientations of instantaneous dipole moments at different times leading to attraction. (*b*) Possible orientations leading to repulsion. The electric field of the instantaneous dipole moment of one molecule tends to polarize the other molecule; thus the orientations leading to attraction (*a*) are much more likely than those leading to repulsion (*b*).

The Hydrogen Bond Another bonding mechanism of great importance is the hydrogen bond, which is formed by the sharing of a proton (the nucleus of the hydrogen atom) between two atoms, frequently two oxygen atoms. This sharing of a proton is similar to the sharing of electrons responsible for the covalent bond already discussed. It is facilitated by the small mass of the proton and by the absence of inner-core electrons in hydrogen. The hydrogen bond often holds groups of molecules together and is responsible for the

* Helium is the only element that does not solidify at any temperature at atmospheric pressure.

cross-linking that allows giant biological molecules and polymers to hold their fixed shapes. The well-known helical structure of DNA is due to hydrogen-bond linkages across turns of the helix (Figure 38-7).

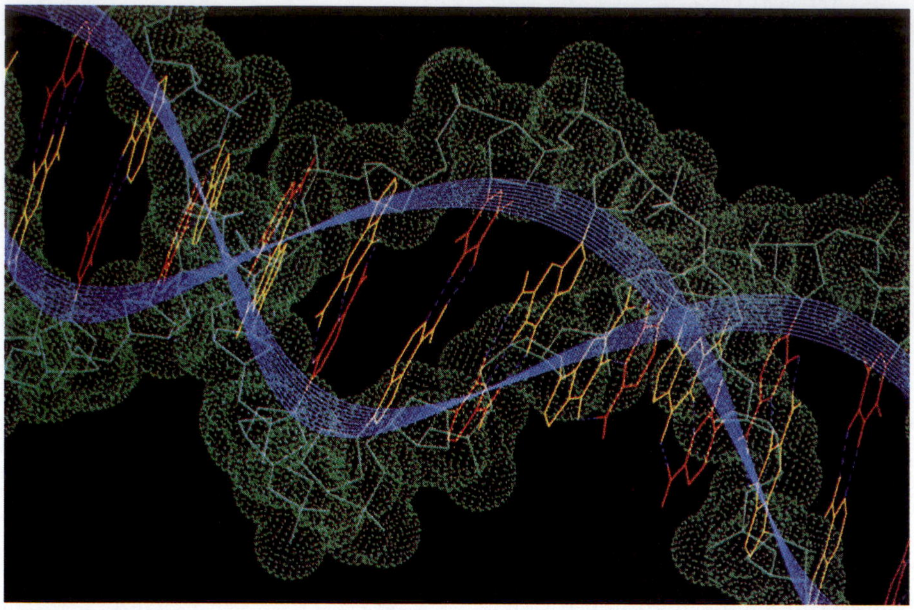

Figure 38-7 The DNA molecule.

The Metallic Bond In a metal, two atoms do not bond together by exchanging or sharing an electron to form a molecule. Instead, each valence electron is shared by many atoms. The bonding is thus distributed throughout the entire metal. A metal can be thought of as a lattice of positive ions held together by a "gas" of essentially free electrons that roam throughout the solid. In the quantum-mechanical picture, these free electrons form a cloud of negative charge density between the positively charged lattice ions that holds the ions together. In this respect, the metallic bond is somewhat similar to the covalent bond. However, with the metallic bond, there are far more than just two atoms involved, and the negative charge is distributed uniformly throughout the volume of the metal. The number of free electrons varies from metal to metal but is of the order of one per atom.

optional

38-2 Polyatomic Molecules

Molecules with more than two atoms range from such relatively simple molecules as water, which has a molecular mass number of 18, to such giants as proteins and DNA, which can have molecular masses of hundreds of thousands up to many millions. As with diatomic molecules, the structure of polyatomic molecules can be understood by applying basic quantum mechanics to the bonding of individual atoms. The bonding mechanisms for most polyatomic molecules are the covalent bond and the hydrogen bond. We will discuss only some of the simplest polyatomic molecules—H_2O, NH_3, and CH_4—to illustrate both the simplicity and complexity of the application of quantum mechanics to molecular bonding.

The basic requirement for the sharing of electrons in a covalent bond is that the wave functions of the valence electrons in the individual atoms must overlap as much as possible. As our first example, we will consider the water molecule. The ground-state configuration of the oxygen atom is $1s^22s^22p^4$. The 1s and 2s electrons are in closed-shell states and do not contribute to the bonding. The 2p shell has room for six electrons, two in each of the three space states corresponding to $\ell = 1$. In an isolated atom, we describe these space states by the hydrogen-like wave functions corresponding to $\ell = 1$ and $m = +1, 0,$ and -1. Since the energy is the same for these three space states, we could equally well use any linear combination of these wave functions. When an atom participates in molecular bonding, certain combinations of these atomic wave functions are important. These combinations are called the p_x, p_y, and p_z **atomic orbitals.** The angular dependence of these orbitals is

$$p_x \propto \sin\theta\cos\phi \qquad \text{38-6}$$

$$p_y \propto \sin\theta\sin\phi \qquad \text{38-7}$$

$$p_z \propto \cos\theta \qquad \text{38-8}$$

The electron charge distribution is maximum along the x, y, or z axis, respectively, for these orbitals as shown in Figure 38-8.

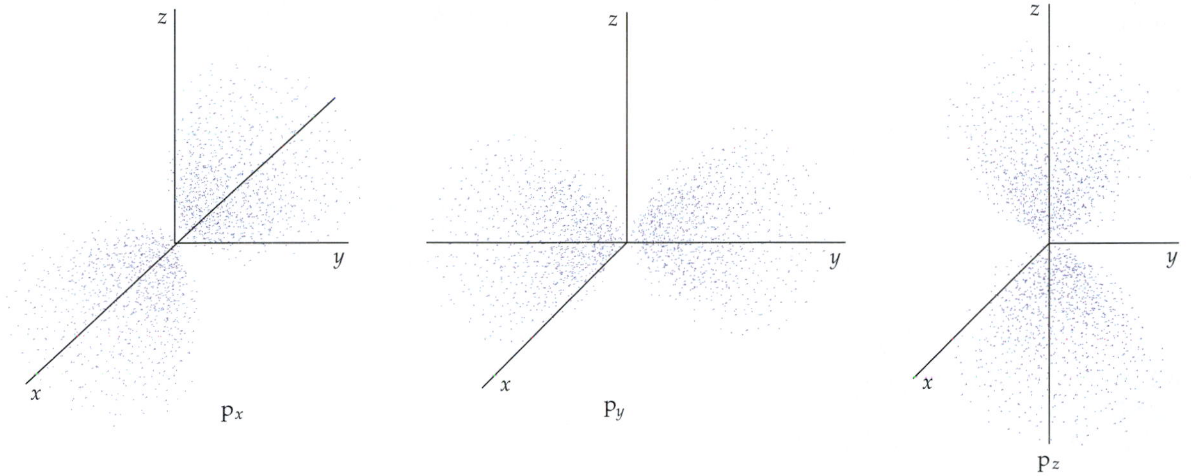

Figure 38-8 Computer-generated dot plot illustrating the spatial dependence of the electron charge distribution in the p_x, p_y, and p_z atomic orbitals.

For the oxygen in an H_2O molecule, maximum overlap of the electron wave functions occurs when two of the four 2p electrons are paired with their spins antiparallel in one of the atomic orbitals (for this example, assume the p_z orbital), one of the other electrons is in a second orbital (the p_x orbital), and the other electron is in the third orbital (the p_y orbital). Each of the unpaired electrons (in the p_x and p_y orbitals, in this illustration) forms a bond with the electron of a hydrogen atom as shown in Figure 38-9. Because of the repulsion of the two hydrogen atoms, the angle between the O–H bonds is actually greater than 90°. The effect of this repulsion can be calculated, and the result is in agreement with the measured angle of 104.5°.

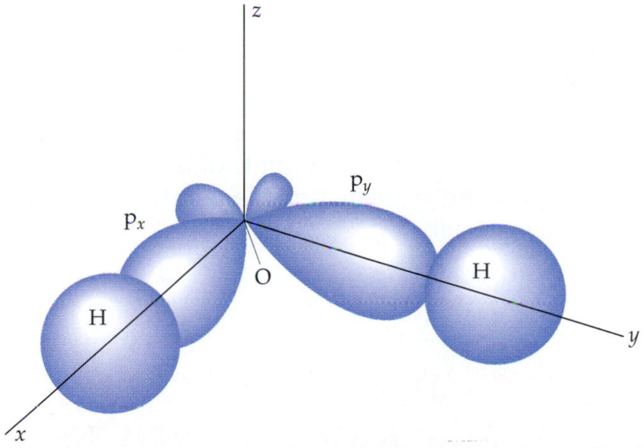

Figure 38-9 Electron charge distribution in the H_2O molecule.

Similar reasoning leads to an understanding of the bonding in NH_3. In the ground state, nitrogen has three electrons in the 2p state. When these three electrons are in the p_x, p_y, and p_z atomic orbitals, they bond to the electrons of hydrogen atoms. Again, because of the repulsion of the hydrogen atoms, the angles between the bonds are somewhat larger than 90°.

The bonding of carbon atoms is somewhat more complicated. Carbon forms a wide variety of different types of molecular bonds, leading to a great diversity in the kinds of organic molecules. The ground-state configuration of carbon is $1s^22s^22p^2$. From our previous discussion, we might expect carbon to be divalent—that is, bonding only through its two 2p electrons—with the two bonds forming at approximately 90°. However, one of the most important features of the chemistry of carbon is that tetravalent carbon compounds, such as CH_4, are overwhelmingly favored.

The observed valence of 4 for carbon comes about in an interesting way. One of the first excited states of carbon occurs when a 2s electron is excited to a 2p state, giving a configuration of $1s^22s^12p^3$. In this excited state, we can have four unpaired electrons, one each in the 2s, $2p_x$, $2p_y$, and $2p_z$ atomic orbitals. We might expect there to be three similar bonds corresponding to the three p orbitals and one different bond corresponding to the s orbital. However, when carbon forms tetravalent bonds, these four atomic orbitals become mixed and form four new *equivalent* molecular orbitals called **hybrid orbitals.** This mixing of atomic orbitals, called **hybridization,** is among the most important features involved in the physics of complex molecular bonds. Figure 38-10 shows the tetrahedral structure of the methane molecule (CH_4), and Figure 38-11 shows the structure of the ethane molecule (CH_3–CH_3), which is similar to two joined methane molecules in which one of the C–H bonds is replaced with a C–C bond.

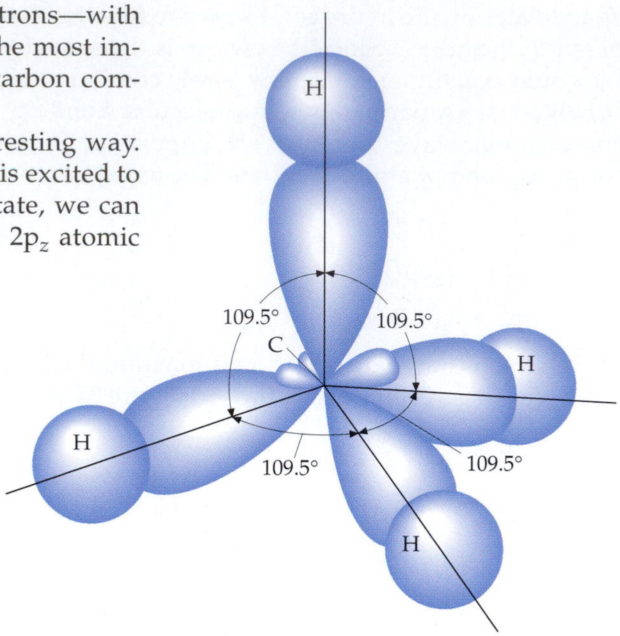

Figure 38-10 Electron charge distribution in the CH_4 (methane) molecule.

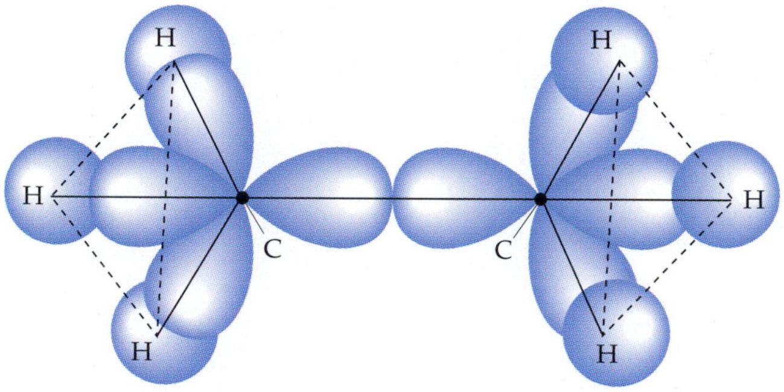

Figure 38-11 Electron charge distribution in the CH_3–CH_3 (ethane) molecule.

Carbon orbitals can also hybridize, with the s, p_x, and p_y orbitals combining to form three hybrid orbitals in the *xy* plane with 120° bonds and the p_z orbital remaining unmixed. An example of this configuration is graphite, in which the bonds in the *xy* plane provide the strongly layered structure characteristic of the material.

38-3 Energy Levels and Spectra of Diatomic Molecules

As is the case with an atom, a molecule often emits electromagnetic radiation when it makes a transition from an excited energy state to a state of lower energy. Conversely, a molecule can absorb radiation and make a transition from a lower energy state to a higher energy state. The study of molecular emission and absorption spectra thus provides us with information about the energy states of molecules. For simplicity, we will consider only diatomic molecules here.

The energy of a molecule can be conveniently separated into three parts: electronic, due to the excitation of the electrons of the molecule; vibrational, due to the oscillations of the atoms of the molecule; and rotational, due to the rotation of the molecule about its center of mass. The magnitudes of these energies are sufficiently different that they can be treated separately. The energies due to the electronic excitations of a molecule are of the order of magnitude of 1 eV, the same as for the excitation of an atom. The energies of vibration and rotation are much smaller than this.

Rotational Energy Levels

Figure 38-12 shows a simple schematic model of a diatomic molecule consisting of a mass m_1 and a mass m_2 separated by a distance r and rotating about its center of mass. Classically, the kinetic energy of rotation (see Section 9-5) is

$$E = \frac{1}{2}I\omega^2 \tag{38-9}$$

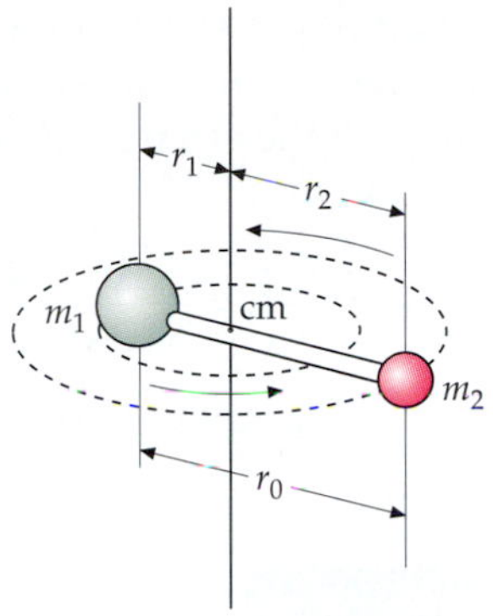

Figure 38-12 Diatomic molecule rotating about an axis through its center of mass.

where I is the moment of inertia and ω is the angular frequency of rotation. If we write this in terms of the angular momentum $L = I\omega$, we have

$$E = \frac{(I\omega)^2}{2I} = \frac{L^2}{2I} \tag{38-10}$$

The solution of the Schrödinger equation for rotation leads to quantization of the angular momentum with values given by

$$L^2 = \ell(\ell + 1)\hbar^2, \qquad \ell = 0, 1, 2, \ldots \tag{38-11}$$

where ℓ is the **rotational quantum number.** This is the same quantum condition on angular momentum that holds for the orbital angular momentum of an electron in an atom. Note, however, that L in Equation 38-10 refers to the angular momentum of the entire molecule rotating about its center of mass. The energy levels of a rotating molecule are therefore given by

$$E = \frac{\ell(\ell + 1)\hbar^2}{2I} = \ell(\ell + 1)E_{0r} \qquad \ell = 0, 1, 2, \ldots \tag{38-12}$$

Rotational energy levels

where E_{0r} is the characteristic rotational energy of a particular molecule, which is inversely proportional to its moment of inertia:

$$E_{0r} = \frac{\hbar^2}{2I} \tag{38-13}$$

Characteristic rotational energy

A measurement of the rotational energy of a molecule from its rotational spectrum can be used to determine the moment of inertia of the molecule, which can then be used to find the separation of the atoms in the molecule. The moment of inertia about an axis through the center of mass of a diatomic molecule (see Figure 38-12) is

$$I = m_1r_1^2 + m_2r_2^2$$

Using $m_1r_1 = m_2r_2$, which relates the distances r_1 and r_2 from the atoms to the center of mass, and $r_0 = r_1 + r_2$ for the separation of the atoms, we can write the moment of inertia as (see Problem 26)

$$I = \mu r_0^2 \tag{38-14}$$

where μ, called the **reduced mass,** is

$$\mu = \frac{m_1 m_2}{m_1 + m_2} \qquad 38\text{-}15$$

Definition—Reduced mass

If the masses are equal ($m_1 = m_2 = m$), as in H_2 and O_2, the reduced mass $\mu = \frac{1}{2}m$ and

$$I = \tfrac{1}{2}mr_0^2 \qquad 38\text{-}16$$

A unit of mass convenient for discussing atomic and molecular masses is the **unified mass unit,** u, which is defined as one-twelfth the mass of the neutral carbon-12 (^{12}C) atom. The mass of one ^{12}C atom is thus 12 u. The mass of an atom in unified mass units is therefore numerically equal to the molar mass of the atom in grams. The unified mass unit is related to the gram and kilogram by

$$1\text{ u} = \frac{1\text{ g}}{N_A} = \frac{10^{-3}\text{ kg}}{6.0221 \times 10^{23}} = 1.6606 \times 10^{-27}\text{ kg} \qquad 38\text{-}17$$

where N_A is Avogadro's number.

Example 38-2

Find the reduced mass of the HCl molecule.

Picture the Problem We find the masses of the hydrogen and chlorine atoms in the periodic table* in Appendix C and use the definition in Equation 38-15.

1. The reduced mass μ is related to the individual masses m_H and m_{Cl} $\qquad \mu = \frac{m_H m_{Cl}}{m_H + m_{Cl}}$

2. Find the masses in the periodic table: $\qquad m_H = 1.01\text{ u}, \quad m_{Cl} = 35.5\text{ u}$

3. Substitute to calculate the reduced mass: $\qquad \mu = \frac{m_H m_{Cl}}{m_H + m_{Cl}} = \frac{(1.01\text{ u})(35.5\text{ u})}{1.01\text{ u} + 35.5\text{ u}} = 0.982\text{ u}$

Remark Note that the reduced mass is less than the mass of either atom in the molecule and that it is approximately equal to the mass of the hydrogen atom. When one atom of a diatomic molecule is much more massive than the other, the center of mass of the molecule is approximately at the center of the more massive atom, and the reduced mass is approximately equal to the mass of the lighter atom.

*The masses in these tables are weighted according to the natural isotopic distribution. Thus, the mass of carbon is given as 12.011 rather than 12.000 because natural carbon consists of about 98.9% ^{12}C and 1.1% ^{13}C. Similarly, natural chlorine consists of about 76% ^{35}Cl and 24% ^{37}Cl.

Example 38-3

Estimate the characteristic rotational energy of an O_2 molecule, assuming that the separation of the atoms is 0.1 nm.

1. The characteristic rotational energy is inversely proportional to the moment of inertia: $\qquad E_{0r} = \frac{\hbar^2}{2I}$

2. Calculate the moment of inertia: $I = \mu r_0^2 = \frac{1}{2}mr_0^2$

3. Substitute this expression for I into the expression for E_{0r}: $E_{0r} = \frac{\hbar^2}{mr_0^2}$

4. Use $m = 16$ u for the mass of oxygen and the given values of the constants to calculate E_{0r}: $E_{0r} = \frac{\hbar^2}{mr_0^2} = \frac{(1.05 \times 10^{-34}\,\mathrm{J\cdot s})^2}{(16\,\mathrm{u})(10^{-10}\,\mathrm{m})^2} \times \frac{1\,\mathrm{u}}{1.66 \times 10^{-27}\,\mathrm{kg}}$
$= 4.15 \times 10^{-23}\,\mathrm{J} = 2.59 \times 10^{-4}\,\mathrm{eV}$

We can see from Example 38-3 that the rotational energy levels are several orders of magnitude smaller than energy levels due to electron excitation, which have energies of the order of 1 eV or higher. Transitions within a given set of rotational energy levels yield photons in the microwave region of the electromagnetic spectrum. The rotational energies are also small compared with the typical thermal energy kT at normal temperatures. For $T = 300$ K, for example, kT is about 2.6×10^{-2} eV, which is about 100 times the characteristic rotational energy as calculated in Example 38-3, and about 1% of the typical electronic energy. Thus, at ordinary temperatures, a molecule can be easily excited to the lower rotational energy levels by collisions with other molecules. But such collisions cannot excite the molecule to its electronic energy levels above the ground state.

Vibrational Energy Levels

The quantization of energy in a simple harmonic oscillator was one of the first problems solved by Schrödinger in his paper proposing his wave equation. Solving the Schrödinger equation for a simple harmonic oscillator gives

$$E_\nu = (\nu + \tfrac{1}{2})hf, \qquad \nu = 0, 1, 2, \ldots \tag{38-18}$$

Vibrational energy levels

where f is the frequency of the oscillator and ν is the **vibrational quantum number.*** An interesting feature of this result is that the energy levels are equally spaced with intervals equal to hf. The frequency of vibration of a diatomic molecule can be related to the force exerted by one atom on the other. Consider two objects of mass m_1 and m_2 connected by a spring of force constant K. The frequency of oscillation of this system can be shown to be (see Problem 30)

$$f = \frac{1}{2\pi}\sqrt{\frac{K}{\mu}} \tag{38-19}$$

where μ is the reduced mass given by Equation 38-15. The effective force constant of a diatomic molecule can thus be determined from a measurement of the frequency of oscillation of the molecule.

A selection rule on transitions between vibrational states (of the same electronic state) requires that ν can change only by ± 1, so the energy of a photon emitted by such a transition is hf and the frequency is f, the same as the frequency of vibration. There is a similar selection rule that ℓ must change by ± 1 for transitions between rotational states.

A typical measured frequency of a transition between vibrational states is 5×10^{13} Hz, which gives for the order of magnitude of vibrational energies

$$E \sim hf = (4.14 \times 10^{-15}\,\mathrm{eV\cdot s})(5 \times 10^{13}\,\mathrm{s^{-1}}) = 0.2\,\mathrm{eV}$$

*We use ν (the Greek letter nu) here rather than n so as not to confuse the vibrational quantum number with the principal quantum number n for electronic energy levels.

This typical vibrational energy is about 1000 times greater than the typical rotational energy E_{0r} of the O_2 molecule we found in Example 38-3 and about 8 times greater than the typical thermal energy $kT = 0.026$ eV at $T = 300$ K. Thus the vibrational levels cannot be excited by molecular collisions at ordinary temperatures.

Example 38-4

The frequency of vibration of the CO molecule is 6.42×10^{13} Hz. What is the effective force constant for this molecule?

Picture the Problem We use Equation 38-19 to relate K to the frequency and reduced mass, and calculate μ from its definition.

1. The effective force constant is related to the frequency and reduced mass by Equation 38-19:

$$f = \frac{1}{2\pi}\sqrt{\frac{K}{\mu}}$$

$$K = (2\pi f)^2\,\mu$$

2. Calculate the reduced mass using 12 u for the mass of the carbon atom and 16 u for the mass of the oxygen atom:

$$\mu = \frac{m_1 m_2}{m_1 + m_2} = \frac{(12\text{ u})(16\text{ u})}{12\text{ u} + 16\text{ u}} = 6.86\text{ u}$$

3. Substitute this value of μ into the equation for K in step 1 and convert to SI units:

$$K = (2\pi f)^2\,\mu = 4\pi^2\,(6.42 \times 10^{13}\text{ Hz})^2(6.86\text{ u})\,\frac{1.66 \times 10^{-27}\text{ kg}}{1\text{ u}}$$

$$= 1.85 \times 10^3\text{ N/m}$$

Emission Spectra

Figure 38-13 shows schematically some electronic, vibrational, and rotational energy levels of a diatomic molecule. The vibrational levels are labeled with the quantum number ν and the rotational levels are labeled with ℓ. The lower vibrational levels are evenly spaced, with $\Delta E = hf$. For higher vibrational levels, the approximation that the vibration is simple harmonic is not valid and the levels are not quite evenly spaced. Note that the potential-energy curves representing the force between the two atoms in the molecule do not have exactly the same shape for the electronic ground and excited states. This implies that the fundamental frequency of vibration f is different for different electronic states. For transitions between vibrational states of different electronic states, the selection rule $\Delta\nu = \pm 1$ does not hold. Such transitions result in the emission of photons of wavelengths in or near the visible spectrum, so the emission spectrum of a molecule for electronic transitions is also sometimes called the optical spectrum.

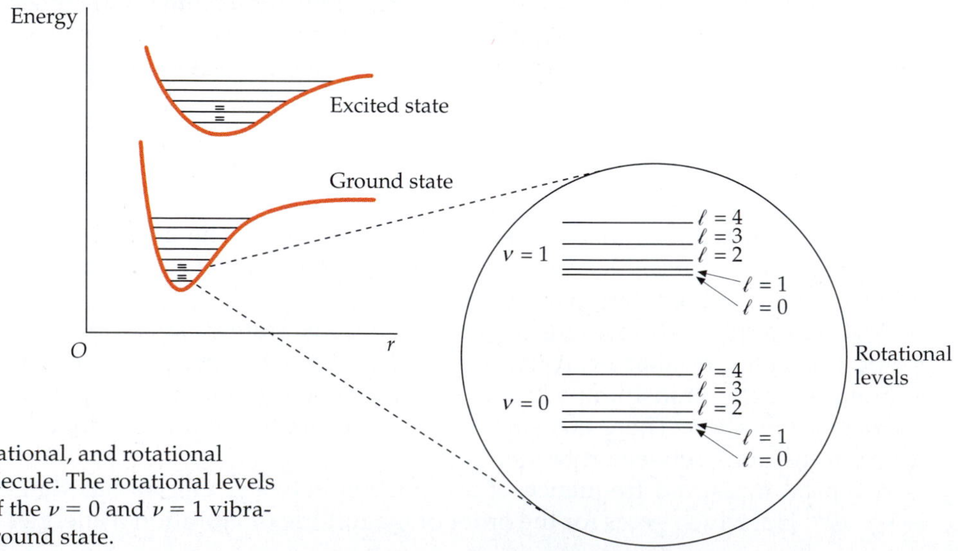

Figure 38-13 Electronic, vibrational, and rotational energy levels of a diatomic molecule. The rotational levels are shown in an enlargement of the $\nu = 0$ and $\nu = 1$ vibrational levels of the electronic ground state.

The spacing of the rotational levels increases with increasing values of ℓ. Since the energies of rotation are so much smaller than those of vibrational or electronic excitation of a molecule, molecular rotation shows up in optical spectra as a fine splitting of the spectral lines. When the fine structure is not resolved, the spectrum appears as bands as shown in Figure 38-14*a*. Close inspection of these bands reveals that they have a fine structure due to the rotational energy levels, as shown in the enlargement in Figure 38-14*b*.

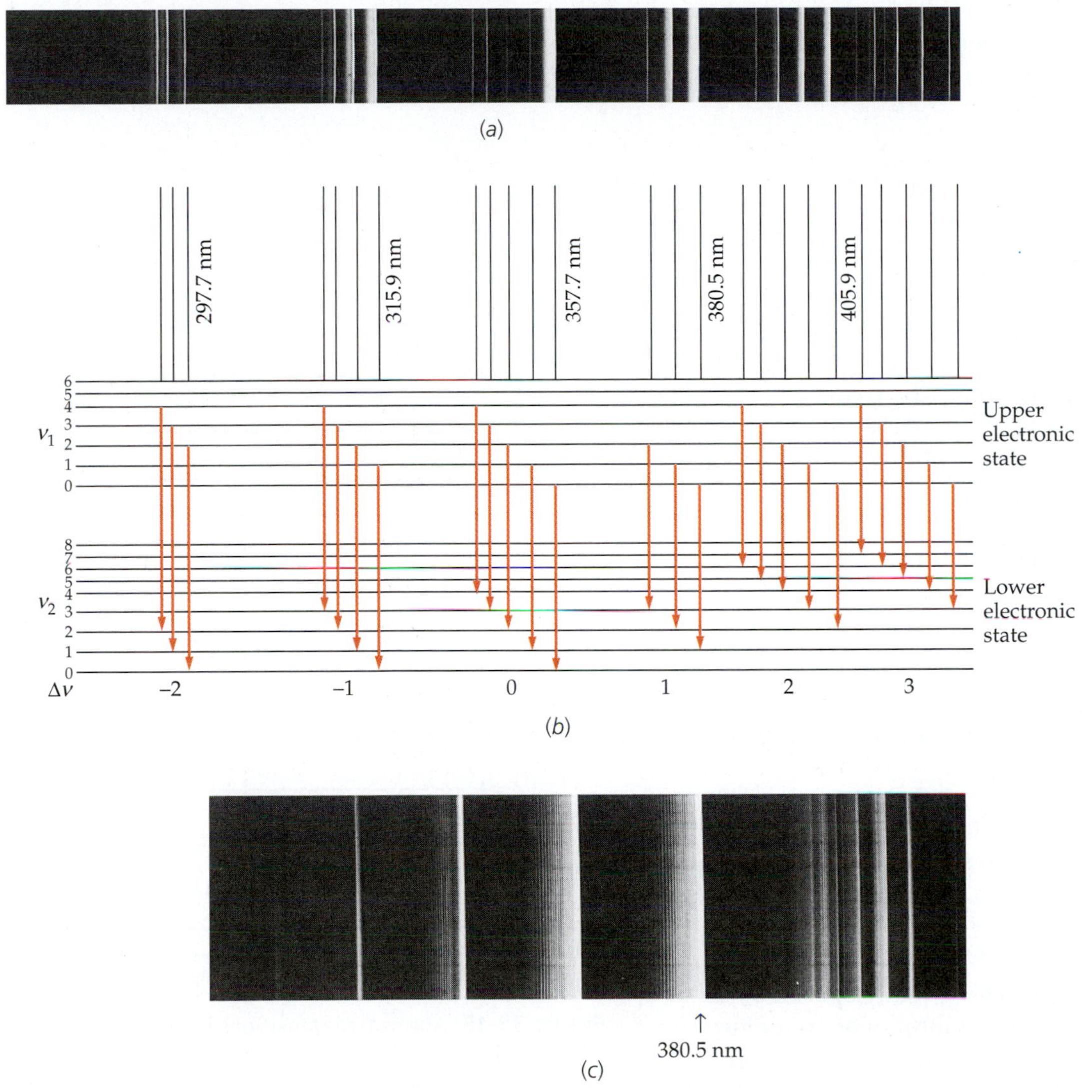

Figure 38-14 (*a*) Part of the emission spectrum of N_2. The spectral lines are due to transitions between the vibrational levels of two electronic states, as indicated in the energy level diagram (*b*). (*c*) An enlargement of part of (*a*) shows that the apparent lines in (*a*) are in fact bands with structure caused by rotational levels.

Absorption Spectra

Much molecular spectroscopy is done using infrared absorption techniques in which only the vibrational and rotational energy levels of the ground-state electronic level are excited. For ordinary temperatures, the vibrational energies are sufficiently large in comparison with the thermal energy kT that most of the molecules are in the lowest vibrational state $\nu = 0$, for which the energy is $E_0 = \frac{1}{2}hf$. The transition from $\nu = 0$ to $\nu = 1$ is the predominant transition in absorption. The rotational energies, however, are sufficiently less than kT that the molecules are distributed among several rotational energy states. If the molecule is originally in a vibrational state characterized by

$\nu = 0$ and a rotational state characterized by the quantum number ℓ, its initial energy is

$$E_\ell = \tfrac{1}{2}hf + \ell(\ell + 1)E_{0r} \qquad \text{38-20}$$

where E_{0r} is given by Equation 38-13. From this state, two transitions are permitted by the selection rules. For a transition to the next higher vibrational state $\nu = 1$ and a rotational state characterized by $\ell + 1$, the final energy is

$$E_{\ell+1} = \tfrac{3}{2}hf + (\ell + 1)(\ell + 2)E_{0r} \qquad \text{38-21}$$

For a transition to the next higher vibrational state and to a rotational state characterized by $\ell - 1$, the final energy is

$$E_{\ell-1} = \tfrac{3}{2}hf + (\ell - 1)\ell E_{0r} \qquad \text{38-22}$$

The energy differences are

$$\Delta E_{\ell\to\ell+1} = E_{\ell+1} - E_\ell = hf + 2(\ell + 1)E_{0r} \qquad \text{38-23}$$

where $\ell = 0, 1, 2, \ldots,$ and

$$\Delta E_{\ell\to\ell-1} = E_{\ell-1} - E_\ell = hf - 2\ell E_{0r} \qquad \text{38-24}$$

where $\ell = 1, 2, 3, \ldots$. (In Equation 38-24, ℓ begins at $\ell = 1$ because from $\ell = 0$ only the transition $\ell \to \ell + 1$ is possible.) Figure 38-15 illustrates these transitions. The frequencies of these transitions are given by

$$f_{\ell\to\ell+1} = \frac{\Delta E_{\ell\to\ell+1}}{h}$$

$$= f + \frac{2(\ell + 1)E_{0r}}{h}, \qquad \ell = 0, 1, 2, \ldots \qquad \text{38-25}$$

and

$$f_{\ell\to\ell-1} = \frac{\Delta E_{\ell\to\ell-1}}{h}$$

$$= f - \frac{2\ell E_{0r}}{h}, \qquad \ell = 1, 2, 3, \ldots \qquad \text{38-26}$$

The frequencies for the transitions $\ell \to \ell + 1$ are thus $f + 2(E_{0r}/h)$, $f + 4(E_{0r}/h)$, $f + 6(E_{0r}/h)$, and so forth; those corresponding to the transition $\ell \to \ell - 1$ are $f - 2(E_{0r}/h)$, $f - 4(E_{0r}/h)$, $f - 6(E_{0r}/h)$, and so forth. We thus expect the absorption spectrum to contain frequencies equally spaced by $2E_{0r}/h$ except for a gap of $4E_{0r}/h$ at the vibrational frequency f as shown in Figure 38-16. A measurement of the position of the gap gives f, and a measurement of the spacing of the absorption peaks gives E_{0r}, which is inversely proportional to the moment of inertia of the molecule.

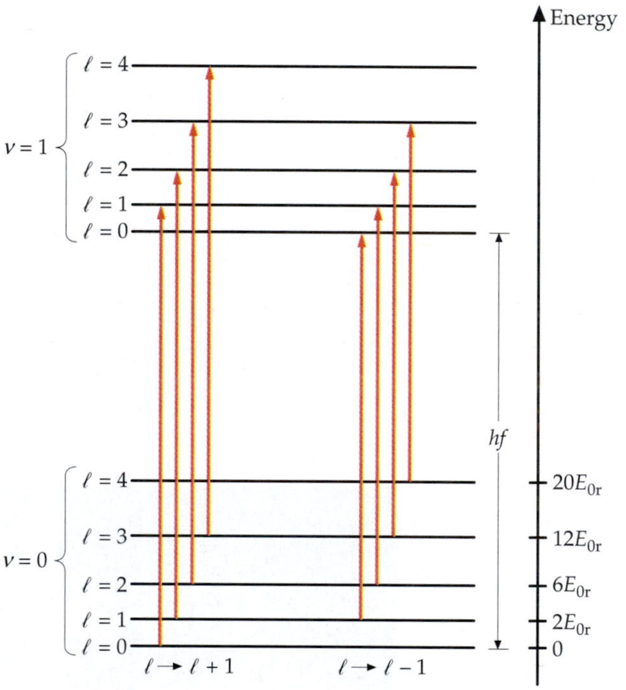

Figure 38-15 Absorptive transitions between the lowest vibrational states $\nu = 0$ and $\nu = 1$ in a diatomic molecule. These transitions obey the selection rule $\Delta\ell = \pm 1$ and fall into two bands. The energies of the $\ell \to \ell + 1$ band are $hf + 2E_{0r}$, $hf + 4E_{0r}$, $hf + 6E_{0r}$, and so forth, whereas the energies of the $\ell \to \ell - 1$ band are $hf - 2E_{0r}$, $hf - 4E_{0r}$, $hf - 6E_{0r}$, and so forth.

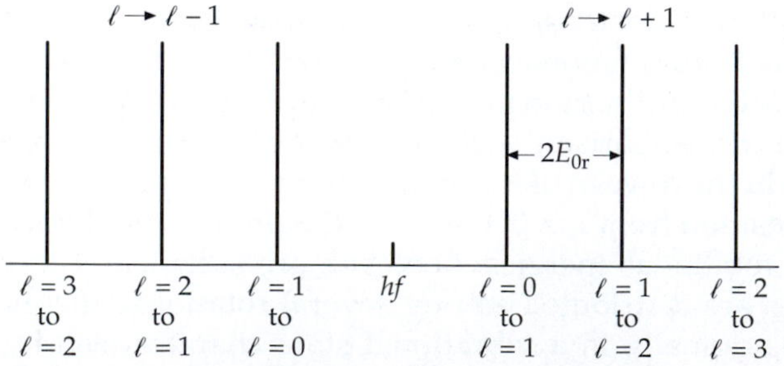

Figure 38-16 Expected absorption spectrum of a diatomic molecule. The right branch corresponds to transitions $\ell \to \ell + 1$ and the left branch to $\ell \to \ell - 1$. The lines are equally spaced by $2E_{0r}$. The energy midway between the branches is hf where f is the frequency of vibration of the molecule.

Figure 38-17 shows the absorption spectrum of HCl. The double-peak structure results from the fact that chlorine occurs naturally in two isotopes, ^{35}Cl and ^{37}Cl, which have different moments of inertia. If all the rotational levels were equally populated initially, we would expect the intensities of each absorption line to be equal. However, the population of a rotational level ℓ is proportional to the degeneracy of the level, that is, to the number of states with the same value of ℓ, which is $2\ell + 1$, and to the Boltzmann factor $e^{-E/kT}$, where E is the energy of the state. For low values of ℓ, the population increases slightly because of the degeneracy factor, whereas for higher values of ℓ, the population decreases because of the Boltzmann factor. The intensities of the absorption lines therefore increase with ℓ for low values of ℓ and then decrease with ℓ for high values of ℓ, as can be seen from the figure.

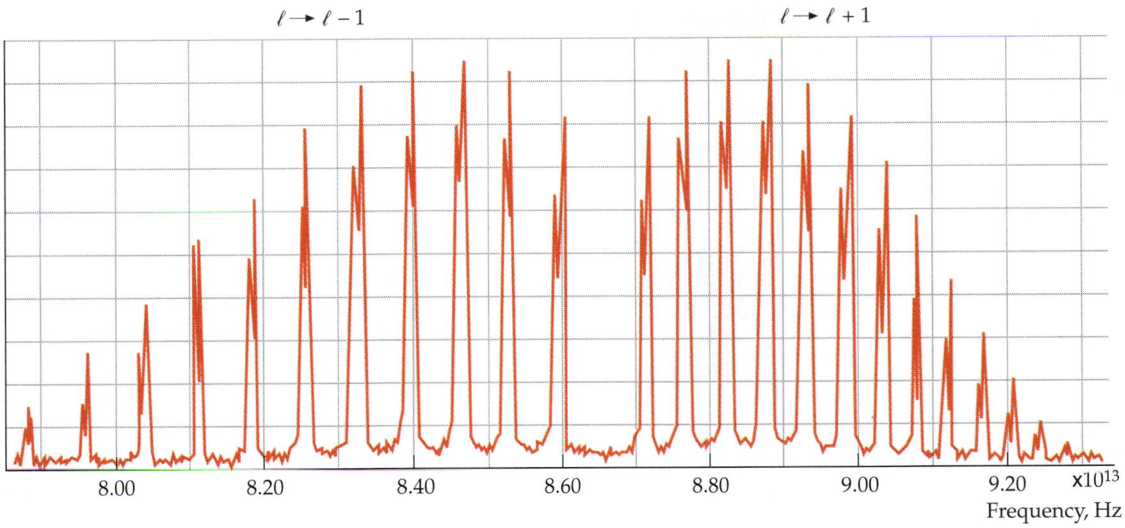

Figure 38-17 Absorption spectrum of the diatomic molecule HCl. The double-peak structure results from the two isotopes of chlorine, ^{35}Cl (abundance 75.5%) and ^{37}Cl (abundance 24.5%). The intensities of the peaks vary because the population of the initial state depends on ℓ.

38-4 The Structure of Solids

The three phases of matter we observe—gas, liquid, and solid—result from the relative strengths of the attractive forces between molecules and the thermal energy of the molecules. Molecules in the gas phase have a high thermal kinetic energy and have little influence on one another except during their frequent but brief collisions. At sufficiently low temperatures (depending on the type of molecule), van der Waals forces will cause practically every substance to condense into a liquid and then a solid. In liquids the molecules are close enough—and their kinetic energy is low enough—that they can develop a temporary **short-range order**. As thermal kinetic energy is further reduced, the molecules form solids, characterized by a lasting order.

If a liquid is cooled slowly so that the kinetic energy of its molecules is reduced slowly, the molecules (or atoms or ions) may arrange themselves in a regular crystalline array, producing the maximum number of bonds and leading to a minimum potential energy. However, if the liquid is cooled rapidly so that its internal energy is removed before the molecules have a chance to arrange themselves, the solid formed is often not crystalline but instead resembles a snapshot of the liquid. Such a solid is called an **amorphous solid.** It displays short-range order but not the **long-range order** (over many molecular diameters) that is characteristic of a crystal. Glass is a typical amorphous solid. A characteristic result of the long-range ordering of a crystal is that it has a well-defined melting point, whereas an amorphous solid merely softens as its temperature is increased. Many materials may solidify into either an amorphous or a crystalline state depending on how they are prepared; others exist only in one form or the other.

Most common solids are polycrystalline; that is, they consist of many single crystals that meet at grain boundaries. The size of a single crystal is typically a fraction of a millimeter. However, large single crystals do occur naturally and can be produced artificially. The most important property of a single crystal is the symmetry and regularity of its structure. It can be thought of as having a single unit structure that is repeated throughout the crystal. This smallest unit of a crystal is called the **unit cell;** its structure depends on the type of bonding—ionic, covalent, metallic, hydrogen, van der Waals—between the atoms, ions, or molecules. If more than one kind of atom is present, the structure will also depend on the atoms' relative sizes.

Figure 38-18 shows the structure of the ionic crystal sodium chloride (NaCl). The Na^+ and Cl^- ions are spherically symmetric, and the Cl^- ion is approximately twice as large as the Na^+ ion. The minimum potential energy for this crystal occurs when an ion of either kind has 6 nearest neighbors of the other kind. This structure is called *face-centered-cubic* (fcc). Note that the Na^+ and Cl^- ions in solid NaCl are *not* paired into NaCl molecules.

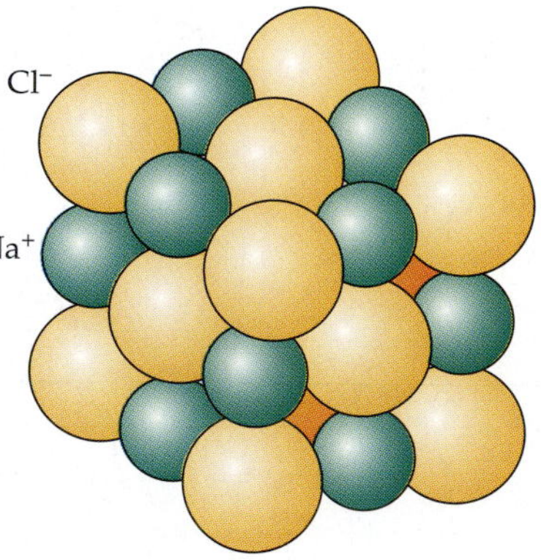

Figure 38-18 Face-centered-cubic structure of the NaCl crystal.

The net attractive part of the potential energy of an ion in a crystal can be written

$$U_{att} = -\alpha \frac{ke^2}{r} \tag{38-27}$$

where r is the separation distance between neighboring ions (0.281 nm for the Na^+ and Cl^- ions in crystalline NaCl), and α, called the **Madelung constant,** depends on the geometry of the crystal. If only the 6 nearest neighbors of each ion were important, α would be 6. However, in addition to the 6 neighbors of the opposite charge at a distance r, there are 12 ions of the same charge at a distance $\sqrt{2}r$, 8 ions of opposite charge at a distance $\sqrt{3}r$, and so on. The Madelung constant is thus an infinite sum:

$$\alpha = 6 - \frac{12}{\sqrt{2}} + \frac{8}{\sqrt{3}} - \cdots \tag{38-28}$$

The result for face-centered-cubic structures is $\alpha = 1.7476$.*

When Na^+ and Cl^- ions are very close together, they repel each other because of the overlap of their electrons and the exclusion-principle repulsion

* A large number of terms are needed to calculate the Madelung constant accurately because the sum converges very slowly.

B

A

D

C

Crystal Structure

A The hexagonal symmetry of a snowflake arises from a hexagonal symmetry in its lattice of hydrogen and oxygen atoms.

B NaCl (salt) crystals, magnified about 30 times. The crystals are built up from a cubic lattice of sodium and chloride ions. In the absence of impurities, an exact cubic crystal is formed. This (false-color) scanning electron micrograph shows that in practice the basic cube is often disrupted by dislocations, giving rise to crystals with a wide variety of shapes. The underlying cubic symmetry, though, remains evident.

C A crystal of quartz (SiO_2, silicon dioxide), the most abundant and widespread mineral on earth. If molten quartz is allowed to solidify without crystallizing, it will form glass.

D A soldering iron tip, ground down to reveal the copper core within its iron sheath. Visible in the iron is its underlying microcrystalline structure.

discussed in Section 38-1. A simple empirical expression for the potential energy associated with this repulsion that works fairly well is

$$U_{\text{rep}} = \frac{A}{r^n}$$

where A and n are constants. The total potential energy of an ion is then

$$U = -\alpha\frac{ke^2}{r} + \frac{A}{r^n} \qquad \text{38-29}$$

The equilibrium separation $r = r_0$ is that at which the force $F = -dU/dr$ is zero. Differentiating and setting $dU/dr = 0$ at $r = r_0$, we obtain

$$A = \frac{\alpha ke^2 r_0^{n-1}}{n} \qquad \text{38-30}$$

The total potential energy can thus be written

$$U = -\alpha\frac{ke^2}{r_0}\left[\frac{r_0}{r} - \frac{1}{n}\left(\frac{r_0}{r}\right)^n\right] \qquad \text{38-31}$$

At $r = r_0$, we have

$$U(r_0) = -\alpha\frac{ke^2}{r_0}\left(1 - \frac{1}{n}\right) \qquad \text{38-32}$$

If we know the equilibrium separation r_0, the value of n can be found approximately from the *dissociation energy* of the crystal, which is the energy needed to break up the crystal into atoms.

Example 38-5

Calculate the equilibrium spacing r_0 for NaCl from the measured density of NaCl, which is $\rho = 2.16\ \text{g/cm}^3$.

Picture the Problem We consider each ion to occupy a cubic volume of side r_0. The mass of 1 mol of NaCl is 58.4 g, which is the sum of the atomic masses of sodium and chlorine. The ions occupy a volume of $2N_A r_0^3$, where $N_A = 6.02 \times 10^{23}$ is Avogadro's number.

1. Relate r_0 to the density ρ: $\qquad \rho = \frac{m}{V} = \frac{m}{2N_A r_0^3}$

2. Solve for r_0^3 and substitute the known values: $\qquad r_0^3 = \frac{m}{2N_A\rho} = \frac{58.4\ \text{g}}{2(6.02 \times 10^{23})(2.16\ \text{g/cm}^3)} = 2.25 \times 10^{-23}\ \text{cm}^3$

3. Solve for r_0: $\qquad r_0 = 2.82 \times 10^{-8}\ \text{cm} = 0.282\ \text{nm}$

The measured dissociation energy of NaCl is 770 kJ/mol. Using 1 eV = 1.602×10^{-19} J, and the fact that 1 mol of NaCl contains N_A pairs of ions, we can express the dissociation energy in electron volts per ion pair. The conversion between electron volts per ion pair and kilojoules per mole is

$$1\frac{\text{eV}}{\text{ion pair}} \times \frac{6.022 \times 10^{23}\ \text{ion pairs}}{\text{mol}} \times \frac{1.602 \times 10^{-19}\ \text{J}}{1\ \text{eV}}$$

The result is

$$1\frac{\text{eV}}{\text{ion pair}} = 96.47\frac{\text{kJ}}{\text{mol}} \tag{38-33}$$

Thus 770 kJ/mol = 7.98 eV per ion pair. Substituting -7.98 eV for $U(r_0)$, 0.282 nm for r_0, and 1.75 for α in Equation 38-32, we can solve for n. The result is $n = 9.35 \approx 9$.

Most ionic crystals, such as LiF, KF, KCl, KI, and AgCl, have a face-centered-cubic structure. Some elemental solids that have this structure are silver, aluminum, gold, calcium, copper, nickel, and lead.

Figure 38-19 shows the structure of CsCl, which is called the *body-centered-cubic* (bcc) structure. In this structure, each ion has 8 nearest neighbor ions of the opposite charge. The Madelung constant for these crystals is 1.7627. Elemental solids with this structure include barium, cesium, iron, potassium, lithium, molybdenum, and sodium.

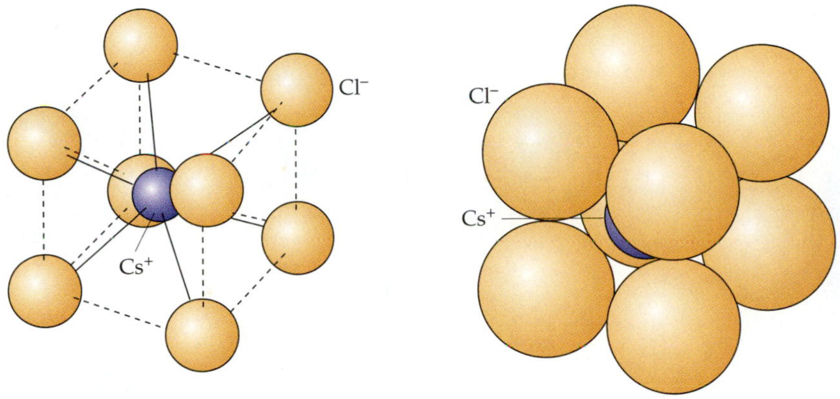

Figure 38-19 Body-centered-cubic structure of the CsCl crystal.

Figure 38-20 shows another important crystal structure: the *hexagonal close-packed* (hcp) structure. It is obtained by stacking identical spheres, such as bowling balls. In the first layer, each ball touches 6 others; thus, the name *hexagonal.* In the next layer, each ball fits into a triangular depression of the first layer. In the third layer, each ball fits into a triangular depression of the second layer, so it lies directly over a ball in the first layer. Elemental solids with hcp structure include beryllium, cadmium, cerium, magnesium, osmium, and zinc.

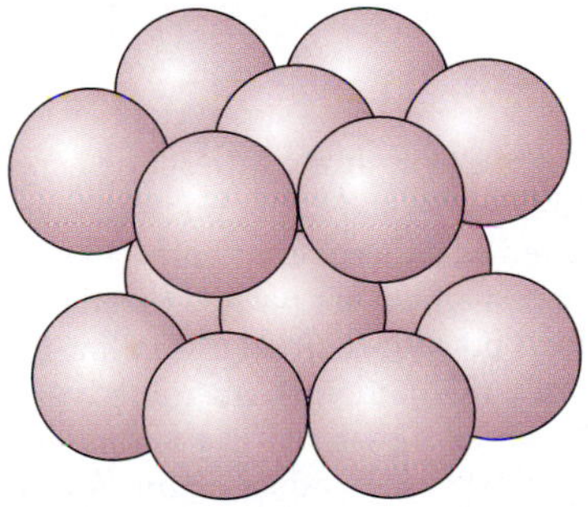

Figure 38-20 Hexagonal close-packed crystal structure.

In some solids with covalent bonding, the crystal structure is determined by the directional nature of the bonds. Figure 38-21 illustrates the diamond structure of carbon, in which each atom is bonded to four others as a result of

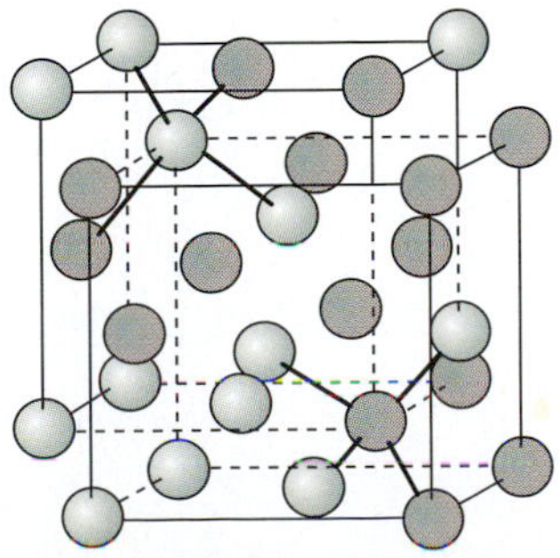

Figure 38-21 Diamond crystal structure. This structure can be considered to be a combination of two interpenetrating face-centered-cubic structures.

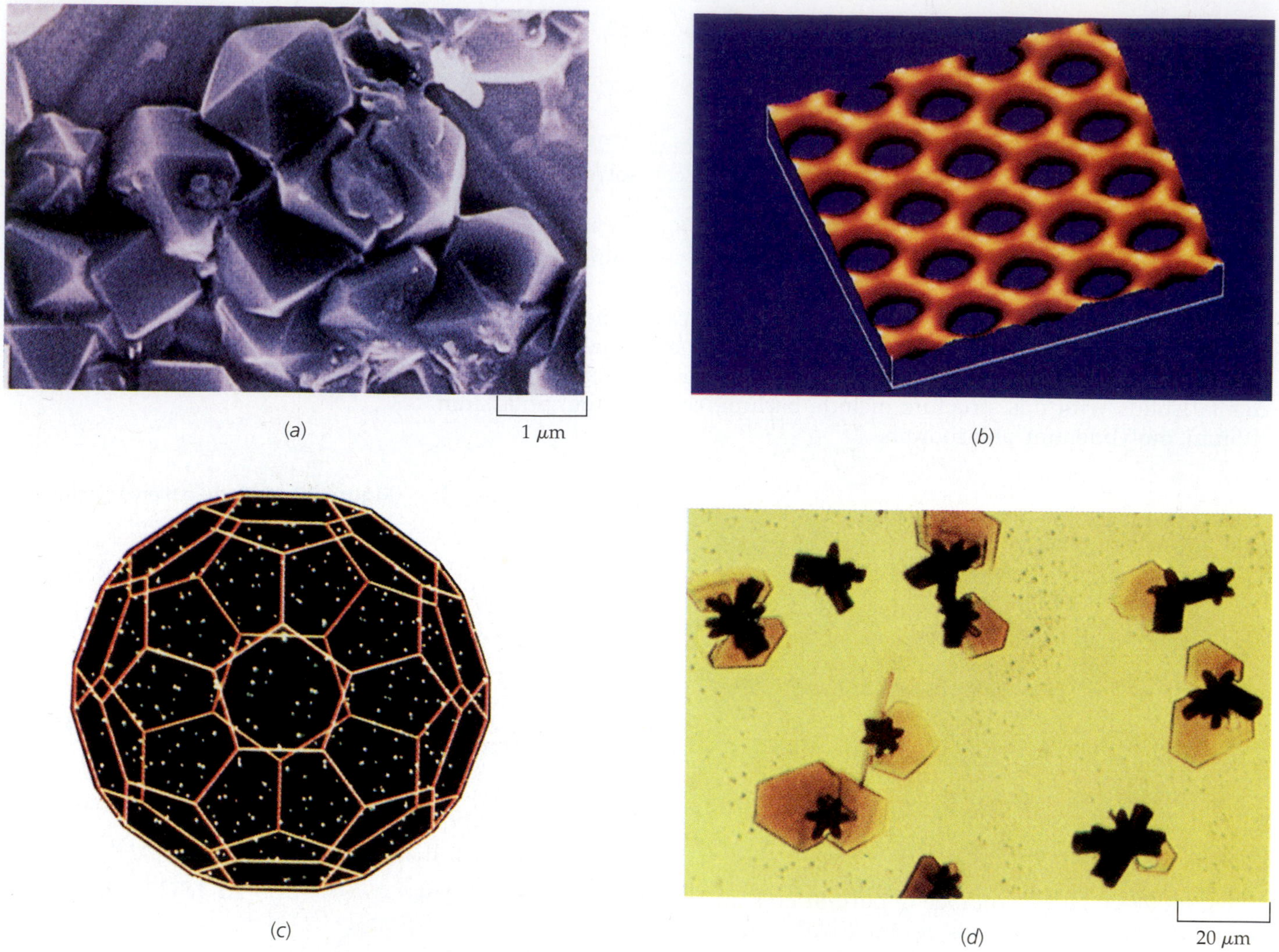

Carbon exists in three well-defined crystalline forms: diamond, graphite, and fullerenes (short for "buckminsterfullerenes"), the third of which was predicted and discovered only a few years ago. The forms differ in how the carbon atoms are packed together in a lattice. A fourth form of carbon, in which no well-defined crystalline form exists, is common charcoal. (*a*) Synthetic diamonds, magnified about 75,000 times. In diamond, each carbon atom is centered in a tetrahedron of four other carbon atoms. The strength of these bonds accounts for the hardness of a diamond. (*b*) An atomic-force micrograph of graphite. In graphite, carbon atoms are arranged in sheets, each sheet made up of atoms in hexagonal rings. The sheets slide easily across one another, a property that allows graphite to function as a lubricant. (*c*) A single sheet of carbon rings can be closed on itself if certain rings are allowed to be pentagonal, instead of hexagonal. A computer-generated image of the smallest such structure, C_{60}, is shown here. Each of the 60 vertices corresponds to a carbon atom; 20 of the faces are hexagons and 12 are pentagons. The same geometric pattern is encountered in a soccer ball. (*d*) Fullerene crystals, in which C_{60} molecules are close-packed. The smaller crystals tend to form thin brownish platelets; larger crystals are usually rod-like in shape. Fullerenes exist in which more than 60 carbon atoms appear. In the crystals shown here, about one-sixth of the molecules are C_{70}.

hybridization, discussed in Section 38-2. This is also the structure of germanium and silicon.

38-5 Semiconductors

In Section 27-4, it was shown that certain materials, such as silicon and germanium, behave as *intrinsic semiconductors* because of the small energy gap between their filled valence band and empty conduction band. The semiconducting property of such materials makes them useful as a basis for electronic circuit components whose resistivity can be controlled by application of an external voltage or current. Most such *solid-state devices,* however, such as the semiconductor diode and the transistor, make use of **impurity semiconductors,** which are created through the controlled addition of certain impurities to intrinsic semiconductors. This process is called **doping.** Figure 38-22*a* is a schematic illustration of silicon doped with a small amount of arsenic such that arsenic atoms replace a few of the silicon atoms in the crystal lattice. Arsenic has five valence electrons rather than the four of silicon. Four of these electrons take part in bonds with the four neighboring silicon atoms, and the fifth electron is very loosely bound to the atom. This extra electron occupies an energy level that is just slightly below the conduction band in the solid, and it is easily excited into the conduction band, where it can contribute to electrical conduction.

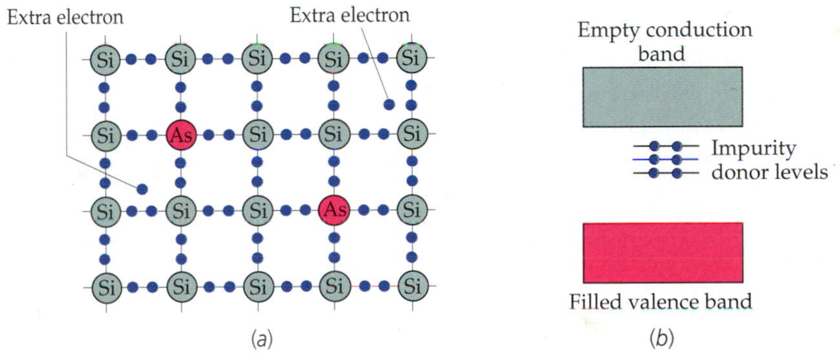

Figure 38-22 (*a*) A two-dimensional schematic illustration of silicon doped with arsenic. Because arsenic has five valence electrons, there is an extra, weakly bound electron that is easily excited to the conduction band, where it can contribute to electrical conduction. (*b*) Band structure of an *n*-type semiconductor such as silicon doped with arsenic. The impurity atoms provide filled energy levels that are just below the conduction band. These levels donate electrons to the conduction band.

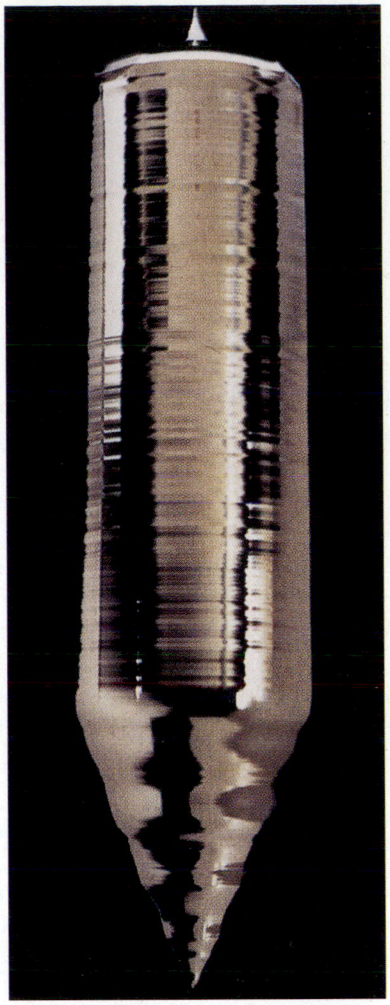

Synthetic crystal silicon is produced beginning with a raw material containing silicon (for instance, common beach sand), separating out the silicon, and melting it. From a seed crystal, the molten silicon grows into a cylindrical crystal, such as the one shown here. The crystals (typically about 1.3 m long) are formed under highly controlled conditions to ensure that they are flawless, and sliced into thousands of thin wafers, onto which the layers of an integrated circuit are etched.

The effect on the band structure of a silicon crystal achieved by doping it with arsenic is shown in Figure 38-22*b*. The levels shown just below the conduction band are due to the extra electrons of the arsenic atoms. These levels are called **donor levels** because they donate electrons to the conduction band without leaving holes in the valence band. Such a semiconductor is called an ***n*-type semiconductor** because the major charge carriers are negative electrons. The conductivity of a doped semiconductor can be controlled by controlling the amount of impurity added. The addition of just one part per million can increase the conductivity by several orders of magnitude.

Another type of impurity semiconductor can be made by replacing a silicon atom with a gallium atom, which has 3 valence electrons (Figure 38-23*a*).

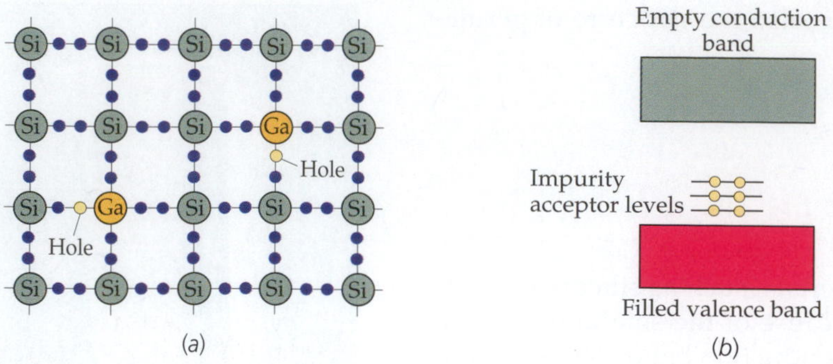

Figure 38-23 (*a*) A two-dimensional schematic illustration of silicon doped with gallium. Because gallium has only three valence electrons, there is a hole in one of its bonds. As electrons move into the hole, the hole moves about, contributing to the conduction of electrical current. (*b*) Band structure of a *p*-type semiconductor such as silicon doped with gallium. The impurity atoms provide empty energy levels just above the filled valence band that accept electrons from the valence band.

The gallium atom accepts electrons from the valence band to complete its four covalent bonds, thus creating a hole in the valence band. The effect on the band structure of silicon achieved by doping it with gallium is shown in Figure 38-23*b*. The empty levels shown just above the valence band are due to the holes from the ionized gallium atoms. These levels are called **acceptor levels** because they accept electrons from the filled valence band when these electrons are thermally excited to a higher energy state. This creates holes in the valence band that are free to propagate in the direction of an electric field. Such a semiconductor is called a ***p*-type semiconductor** because the charge carriers are positive holes. The fact that conduction is due to the motion of holes can be verified by the Hall effect.*

*The Hall effect is discussed in Chapter 28.

Example 38-6 *try it yourself*

The number of free electrons in pure silicon is about 10^{10} electrons/cm^3 at ordinary temperatures. If one silicon atom out of every million atoms is replaced by an arsenic atom, how many free electrons per cubic centimeter are there? (The density of silicon is 2.33 g/cm^3 and its molar mass is 28.1 g/mol.)

Picture the Problem The number of silicon atoms per cubic centimeter, n_{Si} can be found from $n_{\mathrm{s}} = N_{\mathrm{A}}\rho/M$. Then, since each arsenic atom contributes one free electron, the number of electrons contributed by the arsenic atoms is $10^{-6}\, n_{\mathrm{Si}}$.

Cover the column to the right and try these on your own before looking at the answers.

Steps	***Answers***
1. Calculate the number of silicon atoms per cubic centimeter.	$n_{\mathrm{Si}} = \frac{\rho N_{\mathrm{A}}}{M} = \frac{(2.33\ \mathrm{g/cm^3})(6.02 \times 10^{23}\ \mathrm{atoms/mol})}{28.1\ \mathrm{g/mol}}$ $= 4.99 \times 10^{22}\ \mathrm{atoms/cm^3}$
2. Multiply by 10^{-6} to obtain the number of arsenic atoms per cubic centimeter, which equals the added number of free electrons per cubic centimeter.	$n_{\mathrm{e}} = 10^{-6}\, n_{\mathrm{s}} = 4.99 \times 10^{16}\ \mathrm{electrons/cm^3}$

Remark Because silicon has so few free electrons per atom, the number of conduction electrons is increased by a factor of about 5 million by doping silicon with just one arsenic atom per million silicon atoms.

Exercise How many free electrons are there per silicon atom in pure silicon? (*Answer* 2×10^{-13})

38-6 Semiconductor Junctions and Devices

Semiconductor devices such as diodes and transistors make use of *n*-type and *p*-type semiconductors joined together as shown in Figure 38-24. In practice, the two types of semiconductors are often incorporated into a single silicon crystal doped with donor impurities on one side and acceptor impurities on the other. The region in which the semiconductor changes from *p*-type to *n*-type is called a **junction.**

When an *n*-type and a *p*-type semiconductor are placed in contact, the initially unequal concentrations of electrons and holes result in the diffusion of electrons across the junction from the *n* side to the *p* side and holes from the *p* side to the *n* side until equilibrium is established. The result of this diffusion is a net transport of positive charge from the *p* side to the *n* side. Unlike the case when two different metals are in contact, the electrons cannot travel very far from the junction region because the semiconductor is not a particularly good conductor. The diffusion of electrons and holes therefore creates a double layer of charge at the junction similar to that on a parallel-plate capacitor. There is thus a potential difference *V* across the junction, which tends to inhibit further diffusion. In equilibrium, the *n* side with its net positive charge will be at a higher potential than the *p* side with its net negative charge. In the junction region, between the charge layers, there will be very few charge carriers of either type, so the junction region has a high resistance. Figure 38-25 shows the energy level diagram for a *pn* junction. The junction region is also called the **depletion region** because it has been depleted of charge carriers.

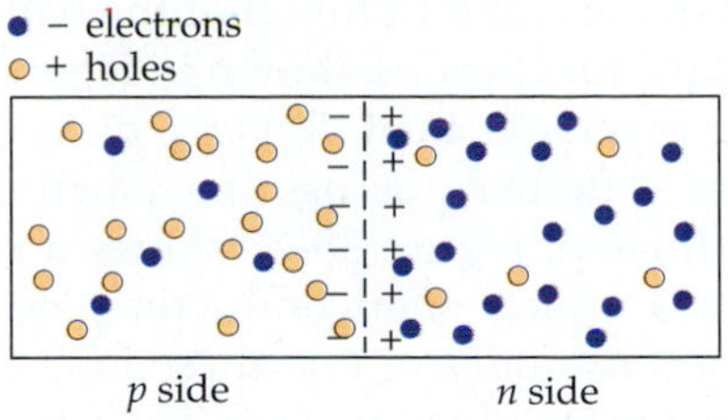

Figure 38-24 A *pn* junction. Because of the difference in their concentrations, holes diffuse from the *p* side to the *n* side and electrons diffuse from the *n* side to the *p* side. As a result, there is a double layer of charge at the junction, with the *p* side being negative and the *n* side being positive.

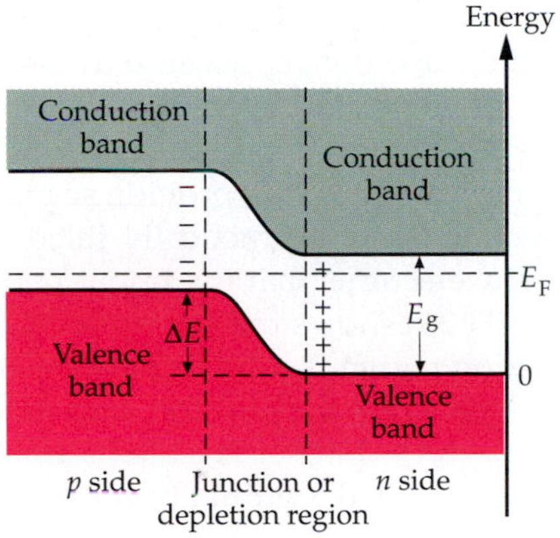

Figure 38-25 Electron energy levels for a *pn* junction.

Diodes

In Figure 38-26 an external potential difference has been applied across a *pn* junction by connecting a battery and resistor to the semiconductor. When the positive terminal of the battery is connected to the *p* side of the junction, as shown in Figure 38-26*a*, the junction is said to be **forward biased.** Forward biasing lowers the potential across the junction. The diffusion of electrons and holes is thereby increased as they attempt to reestablish equilibrium, resulting in a current in the circuit.

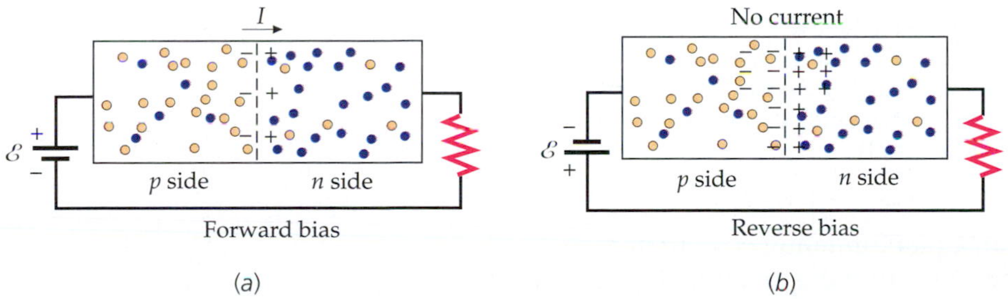

Figure 38-26 A *pn*-junction diode. (*a*) Forward-biased *pn* junction. The applied potential difference enhances the diffusion of holes from the *p* side to the *n* side and of electrons from the *n* side to the *p* side, resulting in a current *I*. (*b*) Reverse-biased *pn* junction. The applied potential difference inhibits the further diffusion of holes and electrons, so there is no current.

optional

If the positive terminal of the battery is connected to the *n* side of the junction, as shown in Figure 38-26*b*, the junction is said to be **reverse biased.** Reverse biasing tends to increase the potential difference across the junction, thereby further inhibiting diffusion. Figure 38-27 shows a plot of current versus voltage for a typical semiconductor junction. Essentially, the junction conducts only in one direction. A single-junction semiconductor device is called a **diode.*** Diodes have many uses. One is to convert alternating current into direct current, a process called rectification.

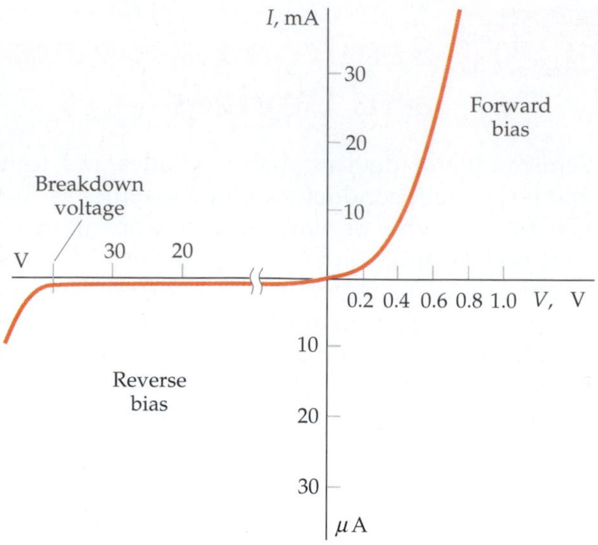

Figure 38-27 Current versus applied voltage across a *pn* junction. Note the different scales on both axes for the forward and reverse bias conditions.

Note that the current in Figure 38-27 suddenly increases in magnitude at extreme values of reverse bias. In such large electric fields, electrons are stripped from their atomic bonds and accelerated across the junction. These electrons, in turn, cause others to break loose. This effect is called **avalanche breakdown.** Although such a breakdown can be disastrous in a circuit where it is not intended, the fact that it occurs at a sharply defined voltage makes it of use in a special voltage reference standard known as a **Zener diode.**

An interesting effect that we can discuss only qualitatively occurs if both the *n* side and *p* side of a *pn*-junction diode are so heavily doped that the donors on the *n* side provide so many electrons that the lower part of the conduction band is practically filled and the acceptors on the *p* side accept so many electrons that the upper part of the valence band is nearly empty. Figure 38-28*a* shows the energy-level diagram for this situation. Because the depletion region is now so narrow, electrons can easily penetrate the potential barrier across the junction, and tunnel to the other side. The flow of electrons through the barrier is called a **tunneling current,** and such a heavily doped diode is called a **tunnel diode.**

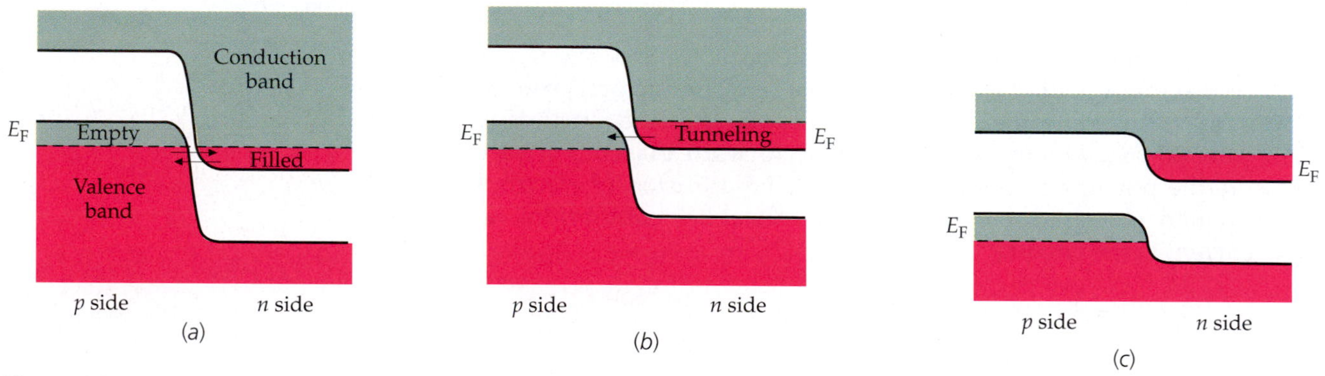

Figure 38-28 Electron energy levels for a heavily doped *pn*-junction tunnel diode. (*a*) With no bias voltage, some electrons tunnel in each direction. (*b*) With a small bias voltage, the tunneling current is enhanced in one direction, making a sizable contribution to the net current. (*c*) With further increases in the bias voltage, the tunneling current decreases dramatically.

At equilibrium with no bias, there is an equal tunneling current in each direction. When a small bias voltage is applied across the junction, the energy-level diagram is as shown in Figure 38-28*b* and the tunneling of electrons from the *n* side to the *p* side is increased whereas that in the opposite direction is decreased. This tunneling current in addition to the usual current due to diffusion results in a considerable net current. When the bias voltage is increased slightly, the energy-level diagram is as shown in Figure 38-28*c* and the tunneling current is decreased. Although the diffusion current is in-

*The name diode originates from a vacuum tube device consisting of just two electrodes that also conducts electric current in one direction only.

creased, the net current is decreased. At large bias voltages, the tunneling current is completely negligible, and the total current increases with increasing bias voltage due to diffusion as in an ordinary *pn*-junction diode. Figure 38-29 shows the current-versus-voltage curve for a tunnel diode. Such diodes are used in electric circuits because of their very fast response time. When operated near the peak in the current-versus-voltage curve, a small change in bias voltage results in a large change in the current.

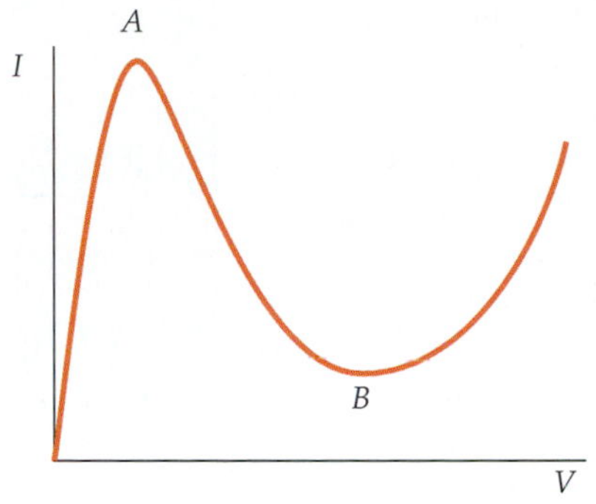

Figure 38-29 Current versus applied voltage for a tunnel diode. Up to point *A*, an increase in the bias voltage enhances tunneling. Between points *A* and *B*, an increase in the bias voltage inhibits tunneling. After point *B*, the tunneling is negligible, and the diode behaves like an ordinary *pn*-junction diode.

Another use for the *pn*-junction semiconductor is the **solar cell,** which is illustrated schematically in Figure 38-30. When a photon of energy greater than the gap energy (1.1 eV in silicon) strikes the *p*-type region, it can excite an electron from the valence band into the conduction band, leaving a hole in the valence band. This region is already rich in holes. Some of the electrons created by the photons will recombine with holes, but some will migrate to the junction. From there they are accelerated into the *n*-type region by the electric field between the double layer of charge. This creates an excess negative charge in the *n*-type region and excess positive charge in the *p*-type region. The result is a potential difference between the two regions, which in practice is about 0.6 V. If a load resistance is connected across the two regions, a charge flows through the resistance. Some of the incident light energy is thus converted into electrical energy. The current in the resistor is proportional to the number of incident photons, which is in turn proportional to the intensity of the incident light.

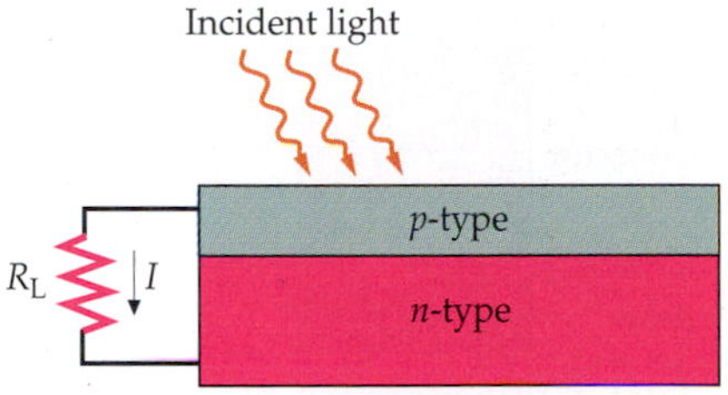

Figure 38-30 A *pn*-junction semiconductor as a solar cell. When light strikes the *p*-type region, electron–hole pairs are created, resulting in a current through the load resistance R_L.

There are many other applications of semiconductors with *pn* junctions. Particle detectors called **surface-barrier detectors** consist of a *pn*-junction semiconductor with a large reverse bias so that there is ordinarily no current. When a high-energy particle, such as an electron, passes through the semiconductor, it creates many electron–hole pairs as it loses energy. The resulting current pulse signals the passage of the particle. **Light-emitting diodes (LEDs)** are *pn*-junction semiconductors with a large forward bias that produces a large excess concentration of electrons on the *p* side and holes on the *n* side of the junction. Under these conditions, the diode emits light as the electrons and holes recombine. This is essentially the reverse of the process that occurs in a solar cell, in which electron–hole pairs are created by the absorption of light. LEDs are commonly used in displays for digital watches and calculators.

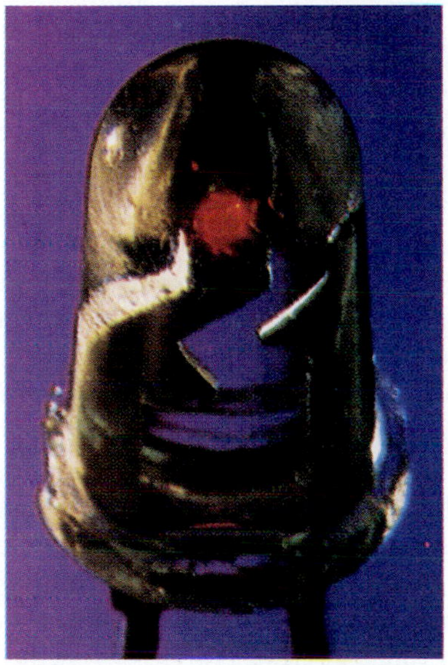

A light-emitting diode (LED).

Transistors

The transistor, a semiconducting device that is used to produce a desired output signal in response to an input signal, was invented in 1948 by William Shockley, John Bardeen, and Walter H. Brattain and has revolutionized the electronics industry and our everyday world. A simple *bipolar junction transistor** consists of three distinct semiconductor regions called the **emitter,** the **base,** and the **collector.** The base is a very thin region of one type of semiconductor sandwiched between two regions of the opposite type. The emitter semiconductor is much more heavily doped than either the base or the col-

* Besides the bipolar junction transistor, there are other categories of transistors, notably, the field-effect transistor.

lector. In an *npn* transistor, the emitter and collector are *n*-type semiconductors and the base is a *p*-type semiconductor; in a *pnp* transistor, the base is an *n*-type semiconductor and the emitter and collector are *p*-type semiconductors.

Figures 38-31 and 38-32 show, respectively, a *pnp* transistor and an *npn* transistor with the symbols used to represent each transistor in circuit diagrams. We see that either transistor consists of two *pn* junctions. We will discuss the operation of a *pnp* transistor. The operation of an *npn* transistor is similar.

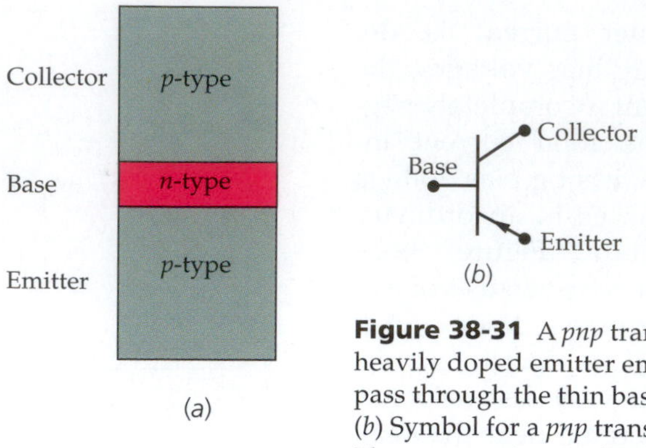

Figure 38-31 A *pnp* transistor. (*a*) The heavily doped emitter emits holes that pass through the thin base to the collector. (*b*) Symbol for a *pnp* transistor in a circuit. The arrow points in the direction of the conventional current, which is the same as that of the emitted holes.

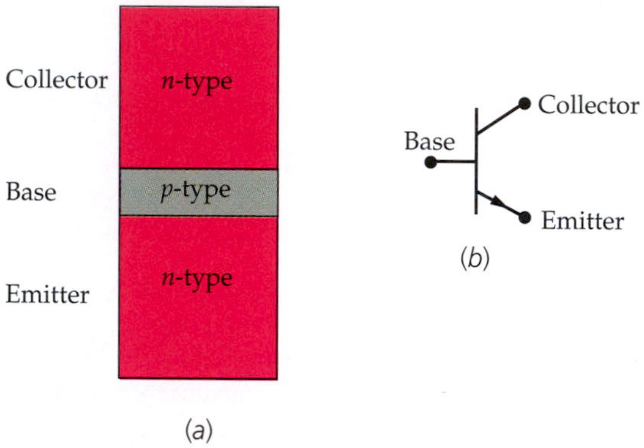

Figure 38-32 An *npn* transistor. (*a*) The heavily doped emitter emits electrons that pass through the thin base to the collector. (*b*) Symbol for an *npn* transistor. The arrow points in the direction of the conventional current, which is opposite the direction of the emitted electrons.

In normal operation, the emitter–base junction is forward biased, and the base–collector junction is reverse biased, as shown in Figure 38-33. The heavily doped *p*-type emitter emits holes that flow across the emitter–base junction into the base. Because the base is very thin, most of these holes flow across the base into the collector. This flow constitutes a current I_c from the emitter to the collector. However, some of the holes recombine in the base producing a positive charge that inhibits the further flow of current. To prevent this, some of the holes that do not reach the collector are drawn off the base as a base current I_b in a circuit connected to the base. In Figure 38-33, therefore, I_c is almost but not quite equal to I_e, and I_b is much smaller than either I_c or I_e. It is customary to express I_c as

$$I_c = \beta I_b \tag{38-34}$$

where β is called the **current gain** of the transistor. Transistors can be designed to have values of β as low as 10 or as high as several hundred.

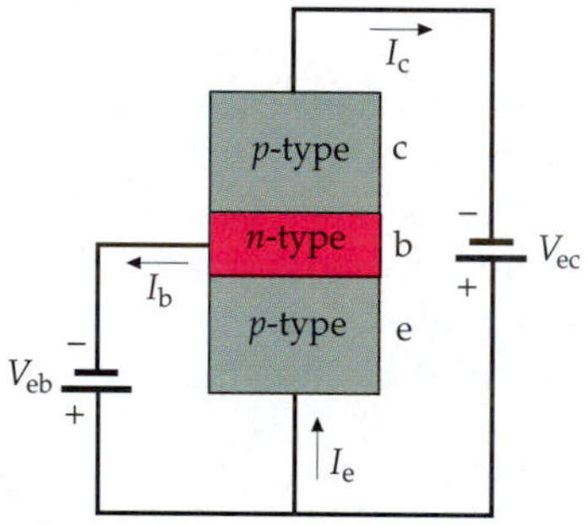

Figure 38-33 A *pnp* transistor biased for normal operation. Holes from the emitter can easily diffuse across the base, which is only tens of nanometers thick. Most of the holes flow to the collector, producing the current I_c.

Figure 38-34 shows a simple *pnp* transistor used as an amplifier. A small time-varying input voltage v_s is connected in series with a bias voltage V_{eb}. The base current is then the sum of a steady current I_b produced by the bias voltage V_{eb} and a varying current i_b due to the signal voltage v_s. Because v_s may at any instant be either positive or negative, the bias voltage V_{eb} must be large enough to ensure that there is always a forward bias on the emitter–base junction. The collector current will consist of two parts: a direct current $I_c = \beta I_b$ and an alternating current $i_c = \beta i_b$. We thus have a current amplifier in which the time-varying output current i_c is β times the input current i_b. In such an amplifier, the steady currents I_c and I_b, although essential to the operation of the transistor, are usually not of interest. The input signal voltage v_s is related to the base current by Ohm's law:

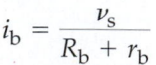

$$i_b = \frac{v_s}{R_b + r_b} \tag{38-35}$$

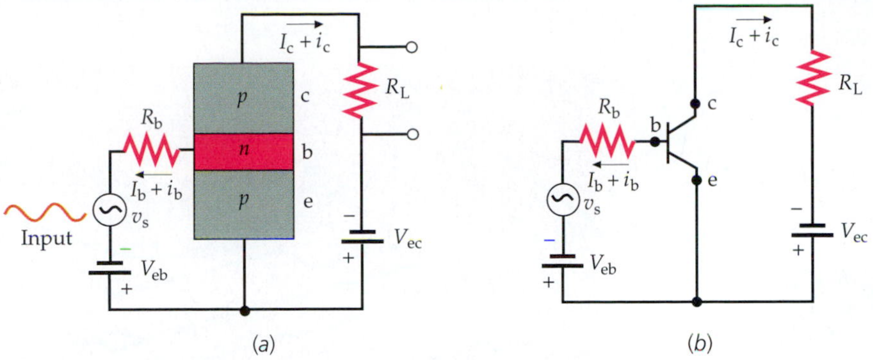

Figure 38-34 (*a*) A *pnp* transistor used as an amplifier. A small change i_b in the base current results in a large change i_c in the collector current. Thus, a small signal in the base circuit results in a large signal across the load resistor R_L in the collector circuit. (*b*) The same circuit as (*a*) with the conventional symbol for the transistor.

where r_b is the internal resistance of the transistor between the base and emitter. Similarly, the collector current i_c produces a voltage v_L across the output or load resistance R_L given by

$$v_L = i_c R_L \qquad \text{38-36}$$

Using Equations 38-34 and 38-35, we have

$$i_c = \beta i_b = \beta \frac{v_s}{R_b + r_b}$$

The output voltage is thus related to the input voltage by

$$v_L = \beta \frac{R_L}{R_b + r_b} v_s \qquad \text{38-37}$$

The ratio of the output voltage to the input voltage is the **voltage gain** of the amplifier:

$$\text{Voltage gain} = \frac{v_L}{v_s} = \beta \frac{R_L}{R_b + r_b} \qquad \text{38-38}$$

A typical amplifier, such as that in a tape player, has several transistors similar to the one in Figure 38-34 connected in series so that the output of one transistor serves as the input for the next. Thus, the very small voltage produced by the passage of the magnetized tape past the pickup heads controls the large amounts of power required to drive the loudspeakers. The power delivered to the speakers is supplied by the dc sources connected to each transistor.

The technology of semiconductors extends well beyond individual transistors and diodes. Many of the electronic devices we now take for granted, such as laptop computers and the processors that govern the operation of vehicles and appliances, rely on large-scale integration of many transistors and other circuit components on a single "chip." Large-scale integration combined with advanced concepts in semiconductor theory has created remarkable new instruments for scientific research; see the Exploring sections that follow.

Exploring

Integrated Circuits

Integrated circuits (ICs; often called chips) combine "active" electronic devices (transistors and diodes) with "passive" ones (capacitors and resistors) on a single semiconductor crystal. Chips containing the equivalent of hundreds of thousands of transistors may be as small as a few millimeters square and can be connected to dozens of leads (Figures 1 and 2). Although resistors, capacitors, and conductors (Figure 3) can be incorporated into the chip, no means have been found to directly fabricate inductors (the remaining passive circuit component) on ICs; they are simulated with other circuitry or appended to a chip as discrete components.

Figure 2 Scanning electron micrograph showing two conductor leads precision bonded to the edge of a chip (magnification: ×163).

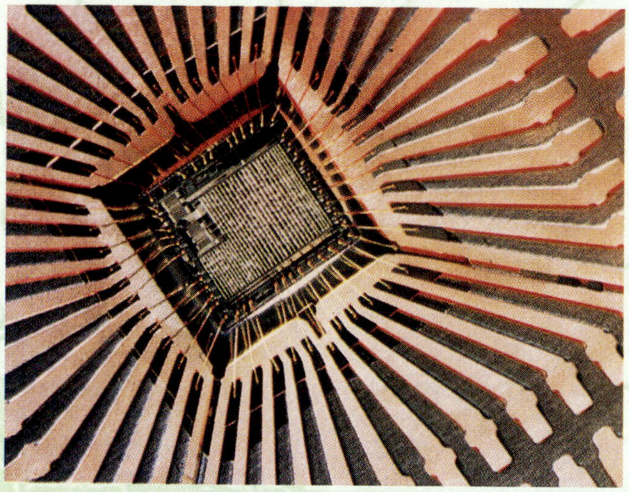

Figure 1 A chip used to format digitized voice and data signals so they can share a single transmission line. This chip is connected to 44 conductor leads and has an actual size of 6.4 mm squared.

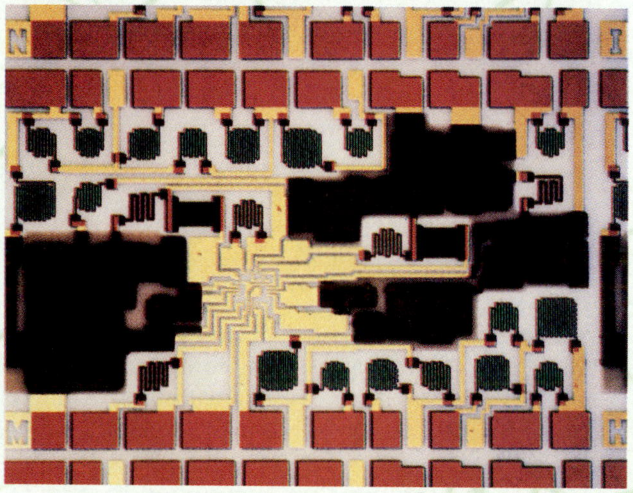

Figure 3 Capacitors (orange blocks), resistors (brown blocks and meandering black lines), and conductors (gold lines) on a ceramic base, formed here by metal films only a few tenths of a micrometer thick.

A particular solid-state technology that especially lends itself to very large scale integration is metal-oxide-semiconductor (MOS) technology. MOS transistors are manufactured by heating an original silicon wafer to about 1000°C, causing a layer of silicon dioxide (SiO_2) to form on its surface (Figures 4 and 5). This is coated with a photoresist and exposed to light through a mask. Unexposed (masked) windows of photoresist are etched away with a developer, exposing the silicon dioxide, which is etched away with acid. The exposed (unmasked) areas are resistant to the developer and are not affected. The wafer is again heated and this time doped, via a diffusion process, with a *p*-type impurity, forming *pn* junctions

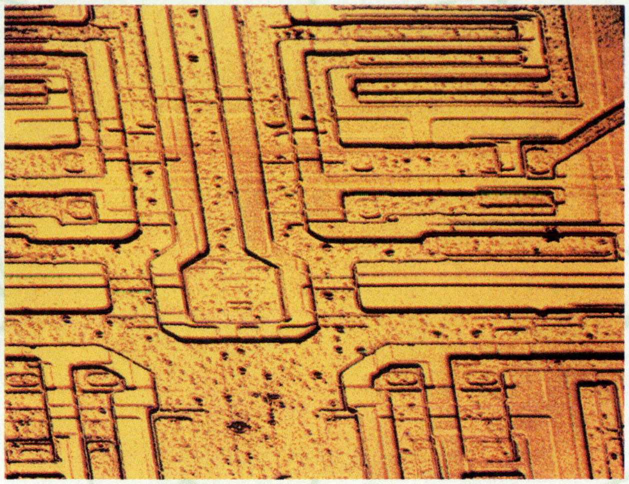

Figure 4 Scanning electron micrograph of metal-oxide-semiconductor (MOS) transistors in patterned layers (magnification ×106).

Integrated circuits making use of the magnetic properties of materials can serve as nonvolatile digital memory. Magnetic bubble memory chips are the integrated-circuit analog to magnetic recording tape and disks (Figure 6). In a thin-film garnet memory crystal, magnetic "bubbles" are created when the garnet is placed between two permanent magnets. They represent regions whose magnetic polarity points in a direction opposite to that of the surrounding crystal. An additional external magnetic field manipulates the position of the bubbles. (Garnet is easy to magnetize, up or down, along a particular axis, but hard to magnetize perpendicular to that axis. This property is necessary for the formation and movement of bubbles.) Storage sites for bubbles are established using a layer of ferromagnetic material deposited on the surface of the crystal; the presence or absence of a bubble at a site can be used to represent a bit of data.

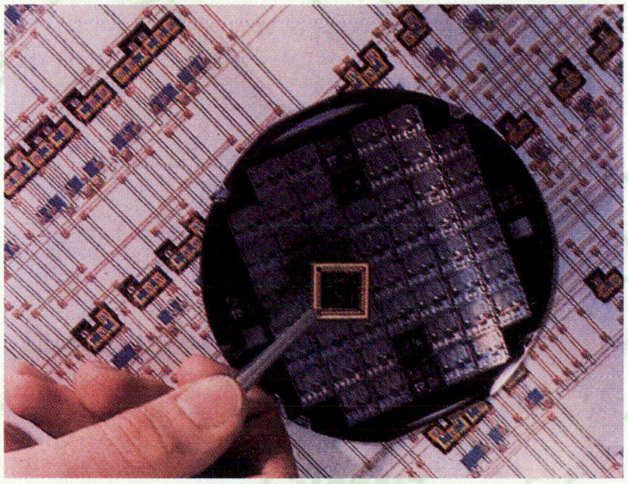

Figure 5 The chip in the tweezers holds 150,000 transistors. Beneath it is a 4-inch wide silicon wafer, awaiting dicing, on which a group of chips have been fabricated simultaneously. In the background is a detail of the "stare plot," the layout of the chip's circuits.

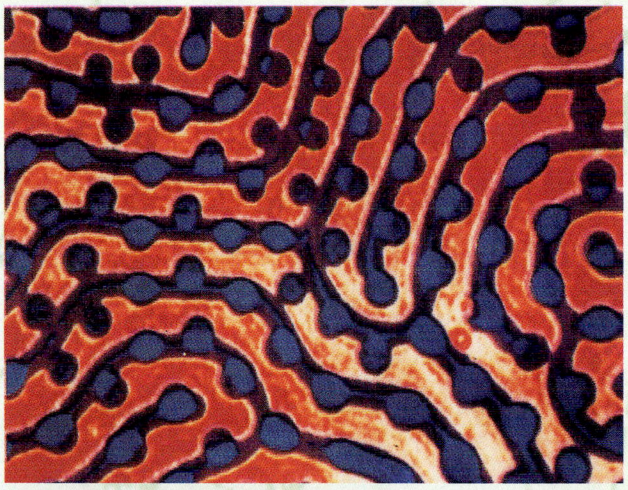

Figure 6 Magnetized domains ("bubbles," actually cylinders seen in cross section), blue in this video micrograph, flow along channels in a thin-film garnet memory crystal. The bubbles are created when the garnet is placed between two permanent magnets. Storage sites for bubbles are established using a layer of ferromagnetic material deposited on the surface of the crystal; the presence or absence of a bubble at a site can be used to represent a bit of data.

in the *n*-type silicon. The chip is covered with a contact metal (typically aluminum), which bonds to the SiO_2 that has re-formed in windows while the chip was heated and doped. The contact metal itself is patterned in a final photo-etching process. Entire microchips are fabricated by an elaboration, using many masks, of this process.

Exploring

Charge-Coupled Devices

Charge-coupled devices (CCDs) are light-sensitive semiconductors at the forefront of imaging technology. They are efficient and fast, and their output is easily stored electronically for processing by computer. Typically 40 to 80% of the photons incident on a CCD surface are converted into a stored electrical signal, allowing for short exposure times and a very low detection threshold. This compares with the 2 or 3% of incoming photons that react with a film's light-sensitive atoms to produce exposed film grains. Also, unlike a photographic film, the response of a CCD is directly proportional to the amount of incoming light, making possible a much more precise measurement of data. Arrays of CCDs used as receptors can greatly expand the capabilities of optical telescopes.

Figure 2 Platinum silicide CCD chip that responds to infrared wavelengths. It contains pixels in a 320 by 244 array.

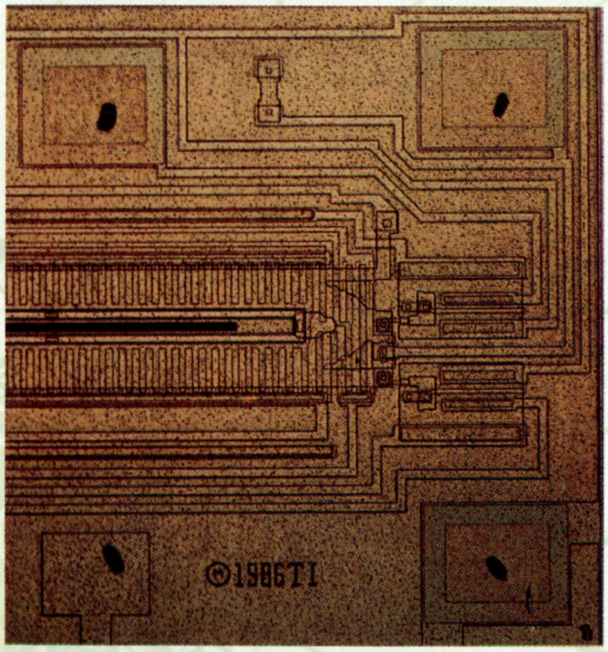

Figure 1 Close-up of part of a CCD. The horizontal bar emerging from the left is the photosensitive area (called a "pixel"). The vertical segments above and below it (called a "transfer register") contain the succession of electrodes that transfer accumulated charge packets along a line of potential wells, from left to right, eventually depositing them in an amplifier located in the central right portion of the chip.

A CCD is a three-layer semiconductor: The top layer is a series of metallic electrodes (see Figure 1), the bottom layer is a silicon crystal, and the middle layer is an insulator separating the two. Light striking silicon in the semiconductor frees electrons, which accumulate in potential wells at the surface of the silicon (Figure 2). Each well in the two-dimensional array on the silicon surface stores an amount of charge that is proportional to the number of photons that strike the surface in the region of the well. The charge is dumped electronically into a computer that records the location and amount of charge in each well. A conventional TV monitor can then be used to reconstruct the original image.

Figure 3 shows a CCD image of two galaxies presented with minimal computer processing. In Figures 4 through 6, the same data that yielded the image in Figure 3 have been processed to reveal different degrees and kinds of detail.

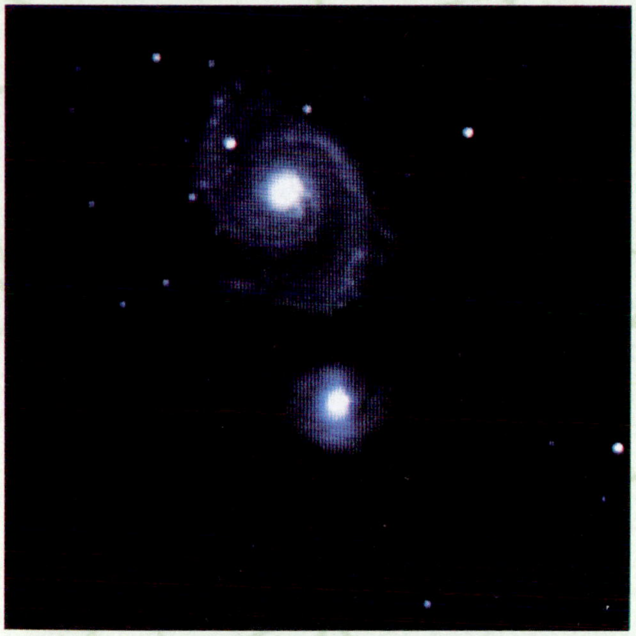

Figure 3 Unprocessed CCD image of spiral galaxy Messier 51 and companion galaxy.

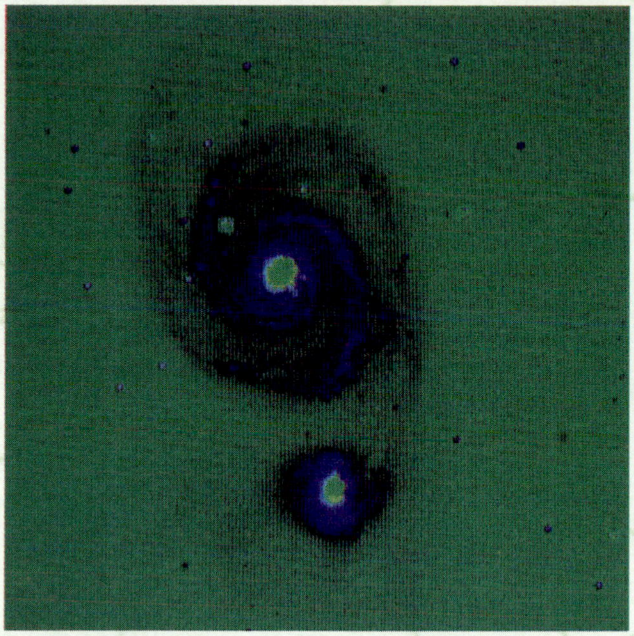

Figure 5 An image that again, like Figure 4, has been generated from the data in Figure 3 and enhanced and colorized by computer.

Figure 4 An image generated from the data contained in Figure 3 in which false colors have been assigned, corresponding to different intensity ranges.

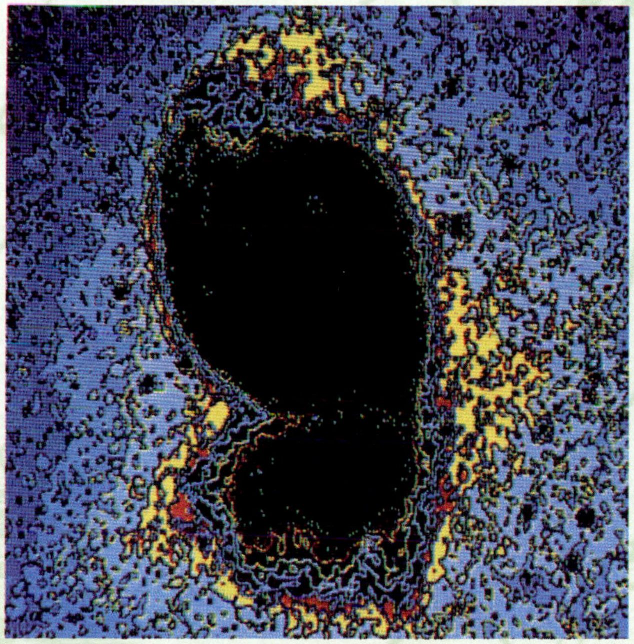

Figure 6 This time the image has been processed for maximum contrast and contoured to show detail in the outer rims of the galaxies.

Summary

1. Atoms are usually found in nature bonded to form molecules or in the lattices of crystalline solids.
2. Ionic bonds and covalent bonds are the principal mechanisms responsible for forming molecules. van der Waals bonds and metallic bonds are important in the formation of liquids and solids. Hydrogen bonds enable large biological molecules to maintain their shape.
3. Like atoms, molecules emit electromagnetic radiation when making a transition from a higher to a lower energy state. The internal energy of a molecule can be separated into three parts: electronic, vibrational, and rotational energy.
4. The molecules in liquids are characterized by a temporary short-range order. The molecules or ions in solids have a more lasting order. Amorphous solids maintain a short-range order similar to the short-range order of a liquid. Crystalline solids display a long-range order determined by their minimum potential energy state.

Topic	Remarks and Relevant Equations	
1. Bonding Mechanisms		
Ionic	Ionic bonds result when an electron is transferred from one atom to another, resulting in a positive ion and a negative ion that bond together.	
Covalent	The covalent bond is a quantum-mechanical effect that arises from the sharing of one or more electrons by atoms.	
van der Waals	The van der Waals bonds are weak bonds that result from the interaction of the instantaneous electric dipole moments of molecules.	
Hydrogen	The hydrogen bond results from the sharing of a hydrogen atom by other atoms.	
Metallic	In the metallic bond, the positive lattice ions of the metal are held together by a cloud of negative charge comprised of free electrons.	
Mixed	A diatomic molecule formed from two identical atoms, such as O_2, must bond by covalent bonding. The bonding of two nonidentical atoms is often a mixture of covalent and ionic bonding. The percentage of ionic bonding can be found from the ratio of the measured electric dipole moment to the ionic electric dipole moment defined by	
	$p_{\text{ionic}} = er_0$	38-5
	where r_0 is the equilibrium separation of the ions.	
2. Polyatomic Molecules (optional)	The shapes of such polyatomic molecules as H_2O and NH_3 can be understood from the spatial distribution of the atomic-orbital or molecular-orbital wave functions. The tetravalent nature of the carbon atom is a result of the hybridization of the 2s and 2p atomic orbitals.	
3. Diatomic Molecules		
Moment of inertia	$I = \mu r_0^2$	38-14
	where r_0 is the equilibrium separation, and μ is the reduced mass:	
Reduced mass	$\mu = \dfrac{m_1 m_2}{m_1 + m_2}$	38-15

Rotational energy levels	$E = \frac{\ell(\ell + 1)\hbar^2}{2I} = \ell(\ell + 1)E_{0r} \qquad \ell = 0, 1, 2, \ldots$	38-12
	where $E_{0r} = \hbar^2/2I$.	38-13
Vibrational energy levels	$E_\nu = (\nu + \frac{1}{2})hf \qquad \nu = 0, 1, 2, 3, \ldots$	38-18
Effective force constant K	$f = \frac{1}{2\pi}\sqrt{\frac{K}{\mu}}$	38-19

4. Molecular Spectra

The optical spectra of molecules have a band structure due to transitions between rotational levels. Information about the structure and bonding of a molecule can be found from its rotational and vibrational absorption spectrum involving transitions from one vibrational–rotational level to another. These transitions obey the selection rules

$$\Delta\nu = \pm 1, \qquad \Delta\ell = \pm 1$$

5. Crystal Structure of Solids

Solids are often found in crystalline form in which a small structure called the unit cell is repeated over and over. A crystal may have a face-centered-cubic, body-centered-cubic, hexagonal-close-packed, or other structure depending on the type of bonding between the atoms, ions, or molecules in the crystal and on the relative sizes of the atoms if there are more than one kind as in NaCl.

Potential energy

$$U = -\alpha\frac{ke^2}{r} + \frac{A}{r^n} \qquad \textbf{38-29}$$

where r is the separation distance between neighboring ions, α is the Madelung constant, which depends on the geometry of the crystal and is of the order of 1.8, and n is about 9.

6. Impurity Semiconductors

The conductivity of a semiconductor can be greatly increased by doping. In an n-type semiconductor, the doping adds electrons just below the conduction band. In a p-type semiconductor, holes are added just above the valence band.

7. Semiconductor Devices (optional)

Junction	Semiconductor devices such as diodes and transistors make use of n-type and p-type semiconductors joined together. The two types of semiconductors are often a single silicon crystal doped with donor impurities on one side and acceptor impurities on the other. The region in which the semiconductor changes from a p-type to an n-type is called a junction. Junctions are used in diodes, solar cells, surface barrier detectors, LEDs, and transistors
Diode	A diode is a single-junction device that carries current in one direction only.
Zener diode	A Zener diode is a diode with a very high reverse bias. It breaks down suddenly at a distinct voltage and is therefore used as a voltage reference standard.
Tunnel diode	A tunnel diode is a diode that is heavily doped so that electrons tunnel through the depletion barrier. At normal operation, a small change in bias voltage results in a large change in current.
Transistors	A transistor consists of a very thin semiconductor of one type sandwiched between two semiconductors of the opposite type. Transistors are used in amplifiers because a small variation in the base current results in a large variation in the collector current.

Problem-Solving Guide

Summary of Worked Examples

Type of Calculation	Procedure and Relevant Examples	
1. Molecular Bonding		
Calculate the energy due to repulsion of two ions in a ionic molecule.	The dissociation energy equals the total potential energy, which is the sum of the electrostatic energy of attraction plus the energy of repulsion plus the energy required to form the ions.	**Example 38-1**
2. Diatomic Molecules		
Find the reduced mass.	Use $\mu = m_1 m_2/(m_1 + m_2)$.	**Example 38-2**
Find the characteristic rotational energy.	Use $E_{0r} = \hbar^2/2I$ with $I = \frac{1}{2} m_0^2$.	**Example 38-3**
Find the effective force constant given the vibration frequency.	Use $f = \frac{1}{2\pi}\sqrt{\frac{K}{\mu}}$.	**Example 38-4**
3. Solids		
Estimate the equilibrium spacing of ions in an ionic crystal.	Assume each ion occupies a cubic volume of side r_0. Then calculate r_0 from the density and molecular mass.	**Example 38-5**
Find the number of free electrons per cubic centimeter in a doped semiconductor.	Calculate the number of doped atoms per cubic centimeter.	**Example 38-6**

Problems

Conceptual Problems

Problems from Optional and Exploring sections

- • Single-concept, single-step, relatively easy
- •• Intermediate-level, may require synthesis of concepts
- ••• Challenging, for advanced students

In a few problems, you are given more data than you actually need; in a few other problems, you are required to supply data from your general knowledge, outside sources, or informed estimates.

Molecular Bonding

1 • Would you expect the NaCl molecule to be polar or nonpolar?

2 • Would you expect the N_2 molecule to be polar or nonpolar?

3 • Does neon occur naturally as Ne or Ne_2? Why?

4 • What type of bonding mechanism would you expect for (*a*) HF, (*b*) KBr, (*c*) N_2?

5 • What kind of bonding mechanism would you expect for (*a*) the N_2 molecule, (*b*) the KF molecule, (*c*) Ag atoms in a solid?

6 • Calculate the separation of Na^+ and Cl^- ions for which the potential energy is -1.52 eV.

7 • The dissociation energy of Cl_2 is 2.48 eV. Consider the formation of an NaCl molecule according to the reaction

$$Na + \tfrac{1}{2}Cl_2 \rightarrow NaCl$$

Does this reaction absorb energy or release energy? How much energy per molecule is absorbed or released?

8 • The dissociation energy is sometimes expressed in kilocalories per mole (kcal/mol). (*a*) Find the relation between the units eV/molecule and kcal/mol. (*b*) Find the dissociation energy of molecular NaCl in kcal/mol.

9 • The equilibrium separation of the HF molecule is 0.0917 nm and its measured electric dipole moment is 6.40×10^{-30} C·m. What percentage of the bonding is ionic?

10 • Do Problem 9 for CsCl, for which the equilibrium separation is 0.291 nm and the measured electric dipole moment is 3.48×10^{-29} C·m.

11 •• The dissociation energy of LiCl is 4.86 eV and the equilibrium separation is 0.202 nm. The electron affinity of chlorine is 3.62 eV, and the ionization energy of lithium is 5.39 eV. Determine the core-repulsion energy of LiCl.

12 •• The equilibrium separation of the K^+ and Cl^- ions in KCl is about 0.267 nm. (*a*) Calculate the potential energy of attraction of the ions assuming them to be point charges at this separation. (*b*) The ionization energy of potassium is 4.34 eV and the electron affinity of Cl is 3.62 eV. Find the dissociation energy neglecting any energy of repulsion. (See Figure 38-1.) The measured dissociation energy is 4.49 eV. What is the energy due to repulsion of the ions at the equilibrium separation?

13 •• Indicate the mean value of r for two vibration levels in the potential-energy curve for a diatomic molecule and show that because of the asymmetry in the curve, r_{av} increases with increasing vibration energy, and therefore solids expand when heated.

14 •• Calculate the potential energy of attraction between the Na^+ and Cl^- ions at the equilibrium separation $r_0 = 0.236$ nm and compare this result with the dissociation energy given in Figure 38-1. What is the energy due to repulsion of the ions at the equilibrium separation?

15 •• The equilibrium separation of the K^+ and F^- ions in KF is about 0.217 nm. (*a*) Calculate the potential energy of attraction of the ions, assuming them to be point charges at this separation. (*b*) The ionization energy of potassium is 4.34 eV and the electron affinity of F is 3.40 eV. Find the dissociation energy neglecting any energy of repulsion. (*c*) The measured dissociation energy is 5.07 eV. Calculate the energy due to repulsion of the ions at the equilibrium separation.

16 ••• Assume that the core repulsion can be represented by a potential energy of the form $U_{rep} = C/r^n$ so the total potential energy is

$$U = -\frac{ke^2}{r} + \frac{C}{r^n} + \Delta E$$

Use the fact that $dU/dr = 0$ at $r = r_0$, and the results for U_{rep} at $r = r_0$ from Problem 11, to calculate C and n.

17 ••• (*a*) Find U_{rep} at $r = r_0$ for NaCl. (*b*) Assume $U_{rep} = C/r^n$ and find C and n for NaCl. (See Problem 16.)

Polyatomic Molecules (optional)

18 • Find other elements with the same subshell electron configuration in the two outermost orbitals as carbon. Would you expect the same type of hybridization for these elements as for carbon?

Energy Levels and Spectra of Diatomic Molecules

19 • How does the effective force constant calculated for the CO molecule in Example 38-4 compare with the force constant of an ordinary spring?

20 • Explain why the moment of inertia of a diatomic molecule increases slightly with increasing angular momentum.

21 • The characteristic rotational energy E_{0r} for the rotation of the N_2 molecule is 2.48×10^{-4} eV. From this find the separation distance of the N atoms in N_2.

22 • The separation of the O atoms in O_2 is actually slightly greater than the 0.1 nm used in Example 38-3, and the characteristic energy of rotation E_{0r} is 1.78×10^{-4} eV rather than the result obtained in that example. Use this value to calculate the separation distance of the O atoms in O_2.

23 •• Show that the reduced mass is smaller than either mass in a diatomic molecule, and calculate it for (*a*) H_2, (*b*) N_2, (*c*) CO, and (*d*) HCl. Express your answers in unified mass units.

24 • The equilibrium separation between the nuclei of the LiH molecule is 0.16 nm. Determine the energy separation between the $\ell = 3$ and $\ell = 2$ rotational levels of this diatomic molecule.

25 •• Repeat Problem 24 for LiD, where D is the symbol for deuterium. Note that replacing the proton by the deuteron does not change the equilibrium separation between the nuclei of the molecule.

26 •• Derive Equations 38-14 and 38-15 for the moment of inertia in terms of the reduced mass of a diatomic molecule.

27 •• Use the separation of the K^+ and Cl^- ions given in Problem 12 and the reduced mass of KCl to calculate the characteristic rotational energy E_{0r}.

28 •• The central frequency for the absorption band of HCl shown in Figure 38-17 is at $f = 8.66 \times 10^{13}$ Hz, and the absorption peaks are separated by about $\Delta f = 6 \times 10^{11}$ Hz. Use this information, to find (*a*) the lowest (zero-point) vibrational energy for HCl, (*b*) the moment of inertia of HCl, and (*c*) the equilibrium separation of the atoms.

29 •• Calculate the effective force constant for HCl from its reduced mass and the fundamental vibrational frequency obtained from Figure 38-17.

30 •• Two objects of mass m_1 and m_2 are attached to a spring of force constant K and equilibrium length r_0. (*a*) Show that when m_1 is moved a distance Δr_1 from the center of mass, the force exerted by the spring is

$$F = -K\left(\frac{m_1 + m_2}{m_2}\right)\Delta r_1$$

(*b*) Show that the frequency of oscillation is $f = (1/2\pi)\sqrt{K/\mu}$, where μ is the reduced mass.

31 ••• Calculate the reduced mass for the $H^{35}Cl$ and $H^{37}Cl$ molecules and the fractional difference $\Delta\mu/\mu$. Show that the mixture of isotopes in HCl leads to a fractional difference in the frequency of a transition from one rotational state to another given by $\Delta f/f = -\Delta\mu/\mu$. Compute $\Delta f/f$ and compare your result with Figure 38-17.

32 ••• In calculating the rotational energy levels of a diatomic molecule, we did not consider rotation of the molecule about the line joining the atoms. (*a*) Estimate the moment of inertia of the H_2 molecule about this line. (*b*) Use your results of (*a*) to estimate the typical rotational energy E_{0r} for rotation about the line joining the atoms. (*c*) Compare your answer in (*b*) with the typical thermal energy kT at T = 300 K.

The Structure of Solids

33 • Suppose that hard spheres of radius R are located at the corners of a unit cell with a simple cubic structure. (*a*) If the hard spheres touch so as to take up the minimum volume possible, what is the size of the unit cell? (*b*) What fraction of the volume of the cubic structure is occupied by the hard spheres?

34 • Calculate the distance r_0 between the K^+ and the Cl^- ions in KCl, assuming that each ion occupies a cubic volume of side r_0. The molar mass of KCl is 74.55 g/mol and its density is 1.984 g/cm^3.

35 • The distance between the Li^+ and Cl^- ions in LiCl is 0.257 nm. Use this and the molecular mass of LiCl (42.4 g/mol) to compute the density of LiCl.

36 • Find the value of n in Equation 38-32 that gives the measured dissociation energy of 741 kJ/mol for LiCl, which has the same structure as NaCl and for which $r_0 = 0.257$ nm.

37 •• Suppose identical bowling balls of radius R are packed into a hexagonal close-packed structure. What fraction of the available volume of the unit cell is filled by the bowling balls?

Semiconductors

38 • Which of the following elements are most likely to act as acceptor impurities in germanium?

(*a*) Bromine (*b*) Gallium
(*c*) Silicon (*d*) Phosphorus
(*e*) Magnesium

39 • Which of the following elements are most likely to serve as donor impurities in germanium?

(*a*) Bromine (*b*) Gallium
(*c*) Silicon (*d*) Phosphorus
(*e*) Magnesium

40 • What type of semiconductor is obtained if silicon is doped with (*a*) aluminum and (*b*) phosphorus? (See Table 37-1 for the electron configurations of these elements.)

41 • What type of semiconductor is obtained if silicon is doped with (*a*) indium and (*b*) antimony? (See Table 37-1 for the electron configurations of these elements.)

42 • The donor energy levels in an n-type semiconductor are 0.01 eV below the conduction band. Find the temperature for which $kT = 0.01$ eV.

43 •• The relative binding of the extra electron in the arsenic atom that replaces an atom in silicon or germanium can be understood from a calculation of the first Bohr orbit of this electron in these materials. Four of arsenic's outer electrons form covalent bonds, so the fifth electron sees a singly charged center of attraction. This model is a modified hydrogen atom. In the Bohr model of the hydrogen atom, the electron moves in free space at a radius a_0 given by

$$a_0 = \frac{\epsilon_0 h^2}{\pi m_e e^2}$$

When an electron moves in a crystal, we can approximate the effect of the other atoms by replacing ϵ_0 with $\kappa\epsilon_0$ and m_e with an effective mass for the electron. For silicon, κ is 12 and the effective mass is about $0.2m_e$. For germanium, κ is 16 and the effective mass is about $0.1m_e$. Estimate the Bohr radii for the outer electron as it orbits the impurity arsenic atom in silicon and germanium.

44 •• The ground-state energy of the hydrogen atom is given by

$$E_1 = -\frac{mk^2e^4}{2\hbar^2} = -\frac{e^4 m_e}{8\epsilon_0^2 h^2}$$

Modify this equation in the spirit of Problem 43 by replacing ϵ_0 by $\kappa\epsilon_0$ and m_e by an effective mass for the electron to estimate the binding energy of the extra electron of an impurity arsenic atom in (*a*) silicon and (*b*) germanium.

45 •• A doped n-type silicon sample with 10^{16} electrons per cubic centimeter in the conduction band has a resistivity of 5×10^{-3} $\Omega \cdot$m at 300 K. Find the mean free path of the electrons. Use the effective mass of $0.2m_e$ for the mass of the electrons. (See Problem 43.) Compare this mean free path with that of conduction electrons in copper at 300 K.

46 •• The measured Hall coefficient of a doped silicon sample is 0.04 V·m/A·T at room temperature. If all the doping impurities have contributed to the total charge carriers of the sample, find (*a*) the type of impurity (donor or acceptor) used to dope the sample and (*b*) the concentration of these impurities.

Semiconductor Junctions and Devices (optional)

47 • When a *pnp* junction transistor is used as an amplifier, a small signal in the ________ current results in a large signal in the ________ current.

(*a*) collector; base (*b*) base; collector
(*c*) emitter; base (*d*) emitter; collector
(*e*) collector; base

48 • When light strikes the p-type semiconductor in a pn junction solar cell,

(*a*) only free electrons are created.
(*b*) only positive holes are created.
(*c*) both electrons and holes are created.
(*d*) positive protons are created.
(*e*) none of these is correct.

49 •• Simple theory for the current versus the bias voltage across a *pn* junction yields the equation

$$I = I_0(e^{eV_b/kT} - 1)$$

Sketch I versus V_b for both positive and negative values of V_b using this equation.

50 •• For a temperature of 300 K, use the equation in Problem 49 to find the bias voltage V_b for which the exponential term has the value (*a*) 10 and (*b*) 0.1.

51 •• In Figure 38-34 for the *pnp*-transistor amplifier, suppose $R_b = 2\ k\Omega$ and $R_L = 10\ k\Omega$. Suppose further that a 10-μA ac base current generates a 0.5-mA ac collector current. What is the voltage gain of the amplifier?

52 •• Germanium can be used to measure the energy of incident particles. Consider a 660-keV gamma ray emitted from ^{137}Cs. (*a*) Given that the band gap in germanium is 0.72 eV, how many electron–hole pairs can be generated as this gamma ray travels through germanium? (*b*) The number of pairs N in part (*a*) will have statistical fluctuations given by $\pm\sqrt{N}$. What then is the energy resolution of this detector in this photon energy region?

53 •• Make a sketch showing the valence and conduction band edges and Fermi energy of a *pn*-junction diode when biased (*a*) in the forward direction and (*b*) in the reverse direction.

54 •• A "good" silicon diode has the current–voltage characteristic given in Problem 49. Let $kT = 0.025$ eV (room temperature) and the saturation current $I_0 = 1$ nA. (*a*) Show that for small reverse-bias voltages, the resistance is 25 MΩ. *Hint:* Do a Taylor expansion of the exponential function, or use your calculator and enter small values for V_b. (*b*) Find the dc resistance for a reverse bias of 0.5 V. (*c*) Find the dc resistance for a 0.5-V forward bias. What is the current in this case? (*d*) Calculate the ac resistance dV/dI for a 0.5-V forward bias.

55 •• A slab of silicon of thickness $t = 1.0$ mm and width $w = 1.0$ cm is placed in a magnetic field $B = 0.4$ T. The slab is in the xy plane, and the magnetic field points in the positive z direction. When a current of 0.2 A flows through the sample in the positive x direction, a voltage difference of 5 mV develops across the width of the sample with the electric field in the sample pointing in the positive y direction. Determine the semiconductor type (n or p) and the concentration of charge carriers.

General Problems

56 • Why would you expect the separation distance between the two protons to be greater in the H_2^+ ion than in the H_2 molecule?

57 • What kind of bonding mechanism would you expect for (*a*) the HCl molecule (*b*) the O_2 molecule, and (*c*) Cu atoms in a solid?

58 • Why does an atom usually absorb radiation only from the ground state, whereas a diatomic molecule can absorb radiation from many different rotational states?

59 • The equilibrium separation of CsF is 0.2345 nm. If its bonding is 70% ionic, what is its electric dipole moment.

60 • Show that when one atom in a diatomic molecule is much more massive than the other the reduced mass is approximately equal to the mass of the lighter atom.

61 •• The equilibrium separation between the nuclei of the CO molecule is 0.113 nm. Determine the energy difference between the $\ell = 2$ and $\ell = 1$ rotational energy levels of this molecule.

62 •• When a thin slab of semiconducting material is illuminated with monochromatic light most of the light is transmitted through the slab if the wavelength is greater than 1.85 μm. For wavelengths less than 1.85 μm, most of the incident light is absorbed. Determine the energy gap of this semiconductor.

63 •• Show that when an intrinsic semiconductor carries a current in a transverse magnetic field no Hall voltage is developed across the sample.

64 •• The semiconducting compound CdSe is widely used for light emitting diodes (LEDs). The energy gap in CdSe is 1.8 eV. What is the frequency of the light emitted by a CdSe LED?

65 •• The resistivity of a sample of pure silicon diminishes drastically when it is irradiated with infrared light of wavelength less than 1.13 μm. What is the energy gap of silicon?

66 •• The effective force constant for the HF molecule is 970 N/m. Find the frequency of vibration for this molecule.

67 •• The frequency of vibration of the NO molecule is 5.63×10^{13} Hz. Find the effective force constant for NO.

68 •• The force constant of the hydrogen bond in the H_2 molecule is 580 N/m. Obtain the energies of the four lowest vibrational levels of H_2, HD, and D_2 molecules, and the wavelengths of photons resulting from transitions between adjacent vibrational levels of these molecules.

69 •• The potential energy between two atoms in a molecule can often be described rather well by the Lenard-Jones potential, which can be written

$$U = U_0\left[\left(\frac{a}{r}\right)^{12} - 2\left(\frac{a}{r}\right)^{6}\right]$$

where U_0 and a are constants. Find the interatomic separation r_0 in terms of a for which the potential energy is a minimum. Find the corresponding value of U_{min}. Use Figure 38-4 to obtain numerical values of r_0 and U_0 for the H_2 molecule and express your answers in nanometers and electron volts.

70 •• In this problem you are to find how the van der Waals force between a polar and a nonpolar molecule depends on the distance between the molecules. Let the dipole moment of the polar molecule be in the x direction and the nonpolar molecule be a distance x away. (*a*) How does the electric field due to an electric dipole depend on distance x? (*b*) Use the fact that the potential energy of an electric dipole of moment $\vec{p}$ in an electric field $\vec{E}$ is $U = -\vec{p}\cdot\vec{E}$ and that the induced dipole moment of the nonpolar molecule is proportional to E to find how the potential energy of interaction of the two molecules depends on separation distance. (*c*) Using $F_x = -dU/dx$, find the x dependence of the force between the two molecules.

71 •• Find the dependence of the force on separation distance between two polar molecules. (See Problem 70.)

72 •• Use the infrared absorption spectrum of HCl in Figure 38-17 to obtain (*a*) the characteristic rotational energy E_{0r} (in eV) and (*b*) the vibrational frequency f and the vibrational energy hf (in eV).

73 •• For a molecule such as CO, which has a permanent electric dipole moment, radiative transitions obeying the selection rule $\Delta\ell = \pm 1$ between two rotational energy levels of the same vibrational level are allowed. (That is, the selection rule $\Delta\nu = \pm 1$ does not hold.) (*a*) Find the moment of inertia of CO and calculate the characteristic rotational energy E_{0r} (in eV). (*b*) Make an energy level diagram for the rotational levels for $\ell = 0$ to $\ell = 5$ for some vibrational level. Label the energies in electron volts starting with $E = 0$ for $\ell = 0$. (*c*) Indicate on your diagram transitions that obey $\Delta\ell = -1$ and calculate the energy of the photon emitted. (*d*) Find the wavelength of the photons emitted for each transition in (*c*). In what region of the electromagnetic spectrum are these photons?

74 ••• Use the results of Problem 16 to calculate the vibrational frequency of the LiCl molecule. To do this, expand the potential about $r = r_0$, where r_0 is the equilibrium separation, in a Taylor series. Retain only the term proportional to $(r - r_0)^2$. Recall that the potential energy of a simple harmonic oscillator is given by $V = \frac{1}{2}m\omega^2x^2$. What is the wavelength resulting from transitions between adjacent harmonic oscillator levels of this molecule?

75 ••• Repeat Problem 74 for the NaCl molecule.

76 ••• A one-dimensional model of an ionic crystal consists of a line of alternating positive and negative ions with distance r_0 between each ion. (*a*) Show that the potential energy of attraction of one ion in the line is

$$V = -\frac{2ke^2}{r_0}\left(1 - \frac{1}{2} + \frac{1}{3} - \frac{1}{4} + \frac{1}{5} - \cdots\right)$$

(*b*) Using the result that

$$\ln(1 + x) = x - \frac{x^2}{2} + \frac{x^3}{3} - \frac{x^4}{4} + \cdots$$

show that the Madelung constant for this one-dimensional model is $\alpha = 2 \ln 2 = 1.386$.

CHAPTER 39

Relativity

Albert Einstein in 1916.

The theory of relativity consists of two rather different theories, the special theory and the general theory. The special theory, developed by Einstein and others in 1905, concerns the comparison of measurements made in different inertial reference frames moving with constant velocity relative to one another. Its consequences, which can be derived with a minimum of mathematics, are applicable in a wide variety of situations encountered in physics and engineering. On the other hand, the general theory, also developed by Einstein and others around 1916, is concerned with accelerated reference frames and gravity. A thorough understanding of the general theory requires sophisticated mathematics, and the applications of this theory are chiefly in the area of gravitation. It is of great importance in cosmology, but it is rarely encountered in other areas of physics or in engineering. In this chapter we concentrate on the special theory (often referred to as *special relativity*). General relativity will be discussed briefly in an Exploring section near the end of the chapter.

39-1 Newtonian Relativity

Newton's first law does not distinguish between a particle at rest and one moving with constant velocity. If there is no net external force acting, the particle will remain in its initial state—either at rest or moving with its initial velocity. A particle at rest relative to you is moving with constant velocity relative to an observer who is moving with constant velocity relative to you. How might we distinguish whether you and the particle are at rest and the second observer is moving with constant velocity, or the second observer is at rest and you and the particle are moving?

Let us consider some simple experiments. Suppose we have a railway boxcar moving along a straight, flat track with a constant velocity V. We note that a ball at rest in the boxcar remains at rest. If we drop the ball, it falls straight down relative to the boxcar with an acceleration g due to gravity. Of course, when viewed from the track the ball moves along a parabolic path because it has an initial velocity V to the right. No mechanics experiment that we can do—measuring the period of a pendulum, observing the collisions between two objects, or whatever—will tell us whether the boxcar is moving and the track is at rest or the track is moving and the boxcar is at rest. If we have a coordinate system attached to the track and another attached to the boxcar, Newton's laws hold in either system.

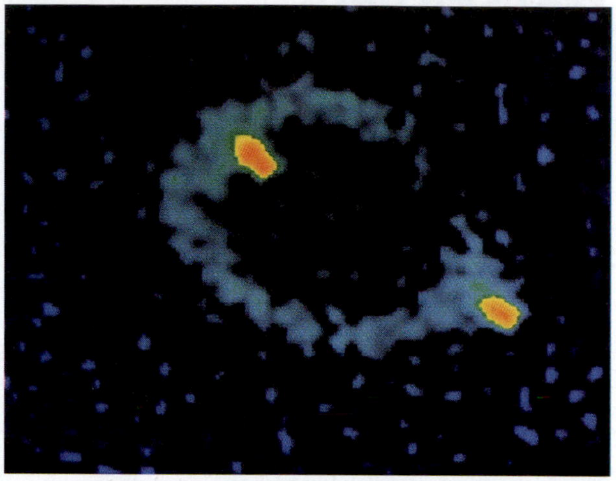

This ringlike structure of the radio source MG1131 + 0456 is thought to be due to "gravitational lensing," first proposed by Einstein in 1936, in which a source is imaged into a ring by a large, massive object in the foreground.

A set of coordinate systems at rest relative to each other is called a *reference frame.** A reference frame in which Newton's laws hold is called an *inertial reference frame.* All reference frames moving at constant velocity relative to an inertial reference frame are also inertial reference frames. If we have two inertial reference frames moving with constant velocity relative to each other, there are no mechanics experiments that can tell us which is at rest and which is moving or if they are both moving. This result is known as the principle of **Newtonian relativity:**

> Absolute motion cannot be detected.

Principle of Newtonian relativity

This principle was well known by Galileo, Newton, and others in the seventeenth century. By the late nineteenth century, however, this view had changed. It was then generally thought that Newtonian relativity was not valid and that absolute motion could be detected in principle by a measurement of the speed of light.

Ether and the Speed of Light

We saw in Chapter 15 that the velocity of a wave depends on the properties of the medium in which the wave travels and not on the velocity of the source of the waves. For example, the velocity of sound relative to still air depends on the temperature of the air. Light and other electromagnetic waves (radio, X rays, etc.) travel through a vacuum with a speed $c \approx 3 \times 10^8$ m/s that is predicted by Maxwell's equations for electricity and magnetism. But what is this speed relative to? What is the equivalent of still air for a vacuum? A proposed medium for the propagation of light was called the *ether;* it was

* Inertial reference frames were discussed in Section 4-1.

thought to pervade all space. The velocity of light relative to the ether was assumed to be c as predicted by Maxwell's equations. The velocity of any object relative to the ether was considered its absolute velocity.

Albert Michelson, first in 1881 and then again with Edward Morley in 1887, set out to measure the velocity of the earth relative to the ether by an ingenious experiment in which the velocity of light relative to the earth was compared for two light beams, one in the direction of the earth's motion relative to the sun and the other perpendicular to the direction of the earth's motion. Despite painstakingly careful measurements, they could detect no difference. The experiment has since been repeated under various conditions by a number of people, and no difference has ever been found. The absolute motion of the earth relative to the ether cannot be detected.

39-2 Einstein's Postulates

In 1905, at the age of 26, Albert Einstein published a paper on the electrodynamics of moving bodies.* In this paper, he postulated that absolute motion cannot be detected by any experiment. That is, there is no ether. The earth can be considered to be at rest and the velocity of light will be the same in any direction.† His theory of special relativity can be derived from two postulates. Simply stated, these postulates are:

> Postulate 1. Absolute uniform motion cannot be detected.
>
> Postulate 2. The speed of light is independent of the motion of the source.

Einstein's postulates

Postulate 1 is merely an extension of the Newtonian principle of relativity to include all types of physical measurements (not just those that are mechanical). Postulate 2 describes a common property of all waves. For example, the speed of sound waves does not depend on the motion of the sound source. The sound waves from a car horn travel through the air with the same velocity independent of whether the car is moving or not. The speed of the waves depends only on the properties of the air, such as its temperature.

Although each postulate seems quite reasonable, many of the implications of the two together are quite surprising and contradict what is often called common sense. For example, one important implication of these postulates is that every observer measures the same value for the speed of light independent of the relative motion of the source and the observer. Consider a light source S and two observers, R_1 at rest relative to S and R_2 moving toward S with speed v, as shown in Figure 39-1*a*. The speed of light measured by R_1 is $c = 3 \times 10^8$ m/s. What is the speed measured by R_2? The answer is *not* $c + v$. By postulate 1, Figure 39-1*a* is equivalent to Figure 39-1*b*, in which R_2 is at rest and the source S and R_1 are moving with speed v. That is, since absolute motion cannot be detected, it is not possible to say

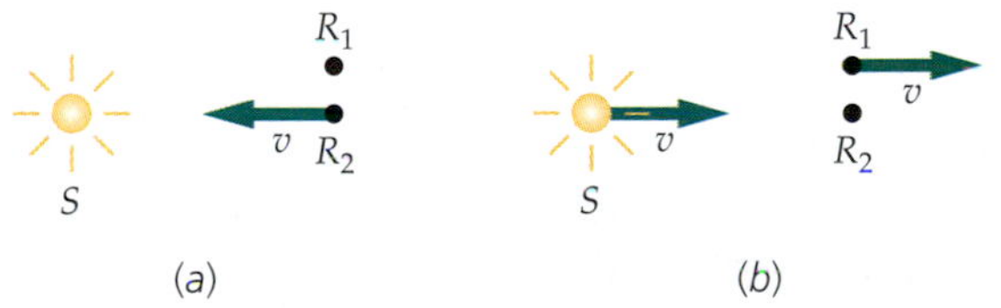

Figure 39-1 (*a*) A stationary light source S and a stationary observer R_1, with a second observer R_2 moving toward the source with speed v. (*b*) In the reference frame in which the observer R_2 is at rest, the light source S and observer R_1 move to the right with speed v. If absolute motion cannot be detected, the two views are equivalent. Since the speed of light does not depend on the motion of the source, observer R_2 measures the same value for that speed as observer R_1.

* *Annalen der Physik*, vol. 17, 1905, p. 841. For a translation from the original German, see W. Perrett and G.B. Jeffery (trans.), *The Principle of Relativity: A Collection of Original Memoirs on the Special and General Theory of Relativity* by H. A. Lorentz, A. Einstein, H. Minkowski, and W. Weyl, Dover, New York, 1923.

† Einstein did not set out to explain the results of the Michelson–Morley experiment. His theory arose from his considerations of the theory of electricity and magnetism and the unusual property of electromagnetic waves that they propagate in a vacuum. In his first paper, which contains the complete theory of special relativity, he made only a passing reference to the Michelson–Morley experiment, and in later years he could not recall whether he was aware of the details of this experiment before he published his theory.

which is really moving and which is at rest. By postulate 2, the speed of light from a moving source is independent of the motion of the source. Thus, looking at Figure 39-1*b*, we see that R_2 measures the speed of light to be c, just as R_1 does. This result is often considered as an alternative to Einstein's second postulate:

> Postulate 2 (Alternate). Every observer measures the same value c for the speed of light.

This result contradicts our intuitive ideas about relative velocities. If a car moves at 50 km/h away from an observer and another car moves at 80 km/h in the same direction, the velocity of the second car relative to the first car is 30 km/h. This result is easily measured and conforms to our intuition. However, according to Einstein's postulates, if a light beam is moving in the direction of the cars, observers in both cars will measure the same speed for the light beam. Our intuitive ideas about the combination of velocities are approximations that hold only when the speeds are very small compared with the speed of light. Even in an airplane moving with the speed of sound, it is not possible to measure the speed of light accurately enough to distinguish the difference between the results c and $c + v$, where v is the speed of the plane. To perceive such a distinction, we would have to either move with a very great velocity (much greater than that of sound) or make extremely accurate measurements.

39-3 The Lorentz Transformation

Einstein's postulates have important consequences for measuring time intervals and space intervals as well as relative velocities. Throughout this chapter we will be comparing measurements of the positions and times of events (such as lightning flashes) made by observers who are moving relative to each other. We will use a rectangular coordinate system xyz with origin O, called the S reference frame, and another system $x'y'z'$ with origin O', called the S' frame, that is moving with a constant velocity $\vec{V}$ relative to the S frame. Relative to the S' frame, the S frame is moving with a constant velocity $-\vec{V}$. For simplicity, we will consider the S' frame to be moving along the x axis in the positive x direction relative to S. In each frame, we will assume that there are as many observers as are needed who are equipped with measuring devices, such as clocks and metersticks, that are identical when compared at rest (see Figure 39-2).

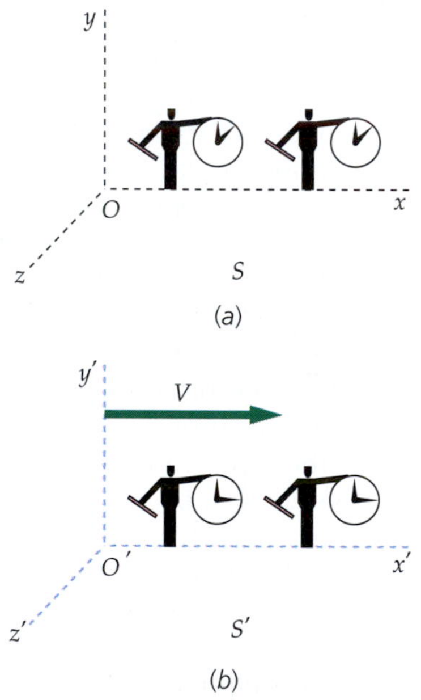

Figure 39-2 Coordinate reference frames S and S' moving with relative speed V. In each frame, there are observers with metersticks and clocks that are identical when compared at rest.

We will use Einstein's postulates to find the general relation between the coordinates x, y, and z and the time t of an event as seen in reference frame S and the coordinates x', y', and z' and the time t' of the same event as seen in reference frame S', which is moving with uniform velocity relative to S. We assume that the origins are coincident at time $t = t' = 0$. The classical relation, called the **Galilean transformation,** is

$$x = x' + Vt', \quad y = y', \quad z = z', \quad t = t' \qquad \text{39-1}a$$

Galilean transformation

The inverse transformation is

$$x' = x - Vt, \quad y' = y, \quad z' = z, \quad t' = t \qquad \text{39-1}b$$

Galilean transformation

These equations are consistent with experimental observations as long as V is much less than c. They lead to the familiar classical addition law for velocities. If a particle has velocity $u_x = dx/dt$ in frame S, its velocity in frame S' is

$$u'_x = \frac{dx'}{dt'} = \frac{dx'}{dt} = \frac{dx}{dt} - V = u_x - V \qquad \text{39-2}$$

If we differentiate this equation again, we find that the acceleration of the particle is the same in both frames:

$$a_x = \frac{du_x}{dt} = \frac{du'_x}{dt'} = a'_x$$

It should be clear that the Galilean transformation is not consistent with Einstein's postulates of special relativity. If light moves along the x axis with speed $u'_x = c$ in S', these equations imply that the speed in S' is $u_x = c + V$ rather than $u_x = c$, which is consistent with Einstein's postulates and with experiment. The classical transformation equations must therefore be modified to make them consistent with Einstein's postulates. We will give a brief outline of one method of obtaining the relativistic transformation.

We assume that the relativistic transformation equation for x is the same as the classical equation (Equation 39-1*a*) except for a constant multiplier on the right side. That is, we assume the equation is of the form

$$x = \gamma(x' + Vt') \qquad \text{39-3}$$

where γ is a constant that can depend on V and c but not on the coordinates. The inverse transformation must look the same except for the sign of the velocity:

$$x' = \gamma(x - Vt) \qquad \text{39-4}$$

Let us consider a light pulse that starts at the origin of S at $t = 0$. Since we have assumed that the origins are coincident at $t = t' = 0$, the pulse also starts at the origin of S' at $t' = 0$. Einstein's postulates require that the equation for the x component of the wave front of the light pulse is $x = ct$ in frame S and $x' = ct'$ in frame S'. Substituting ct for x and ct' for x' in Equations 39-3 and 39-4, we obtain

$$ct = \gamma(ct' + Vt') = \gamma(c + V)t' \qquad \text{39-5}$$

and

$$ct' = \gamma(ct - Vt) = \gamma(c - V)t \qquad \text{39-6}$$

We can eliminate either t' or t from these two equations and determine γ. We get

$$\gamma^2 = \left(1 - \frac{V^2}{c^2}\right)^{-1}$$

$$\gamma = \frac{1}{\sqrt{1 - V^2/c^2}} \qquad \text{39-7}$$

Note that γ is always greater than 1, and that when V is much less than c, $\gamma \approx 1$. The relativistic transformation for x and x' is therefore given by Equations 39-3 and 39-4 with γ given by Equation 39-7. We can obtain equations for t and t' by combining Equation 39-3 with the inverse transformation given by Equation 39-4. Substituting $x = \gamma(x' + Vt')$ for x in Equation 39-4, we obtain

$$x' = \gamma[\gamma(x' + Vt') - Vt] \qquad \text{39-8}$$

which can be solved for t in terms of x' and t'. The complete relativistic transformation is

$$x = \gamma(x' + Vt'), \qquad y = y', \quad z = z' \tag{39-9}$$

$$t = \gamma\left(t' + \frac{Vx'}{c^2}\right) \tag{39-10}$$

Lorentz transformation

The inverse transformation is

$$x' = \gamma(x - Vt), \qquad y' = y, \quad z' = z \tag{39-11}$$

$$t' = \gamma\left(t - \frac{Vx}{c^2}\right) \tag{39-12}$$

The transformation described by Equations 39-9 through 39-12 is called the **Lorentz transformation.** It relates the space and time coordinates x, y, z, and t of an event in frame S to the coordinates x', y', z', and t' of the same event as seen in frame S', which is moving along the x axis with speed V relative to frame S.

We will now look at some applications of the Lorentz transformation.

Time Dilation

Consider two events that occur at a single point x_0' at times t_1' and t_2' in frame S′. We can find the times t_1 and t_2 for these events in S from Equation 39-10. We have

$$t_1 = \gamma\left(t_1' + \frac{Vx_0'}{c^2}\right)$$

and

$$t_2 = \gamma\left(t_2' + \frac{Vx_0'}{c^2}\right)$$

so

$$t_2 - t_1 = \gamma(t_2' - t_1')$$

The time between events that happen at the *same place* in a reference frame is called **proper time** t_{p}. In this case, the time interval $t_2' - t_1'$ measured in frame S' is proper time. The time interval Δt measured in any other reference frame is always longer than the proper time. This expansion is called **time dilation:**

$$\Delta t = \gamma\,\Delta t_{\mathrm{p}} \tag{39-13}$$

Time dilation

Example 39-1

Two events occur at the same point x_0' at times t_1' and t_2' in frame S', which is traveling at speed V relative to frame S. What is the spatial separation of these events in frame S?

Picture the Problem The spatial separation in S is $x_2 - x_1$, where x_2 and x_1 are the coordinates of the events in S, which are found using Equation 39-9.

1. The position x_1 in S at time t_1' is given by Equation 39-9: $x_1 = \gamma(x_0' + Vt_1')$

2. Similarly, at time t'_2, the position is x_2, given by: $x_2 = \gamma(x'_0 + Vt'_2)$

3. Subtract to find the spatial separation: $x_2 - x_1 = \gamma V(t'_2 - t'_1) = V(t_2 - t_1)$

Remarks The spatial separation of these events in S is the distance a single point, such as x'_0 in S', moves in S during the time interval between the events.

We can understand time dilation directly from Einstein's postulates without using the Lorentz transformation. Figure 39-3*a* shows an observer A' a distance D from a mirror. The observer and the mirror are in a spaceship that is at rest in frame S'. The observer explodes a flash gun and measures the time interval $\Delta t'$ between the original flash and his seeing the return flash from the mirror. Since light travels with speed c, this time is

$$\Delta t' = \frac{2D}{c}$$

We now consider these same two events, the original flash of light and the receiving of the return flash, as observed in reference frame S, in which observer A' and the mirror are moving to the right with speed V as shown in Figure 39-3*b*. The events happen at two different places x_1 and x_2 in frame S.

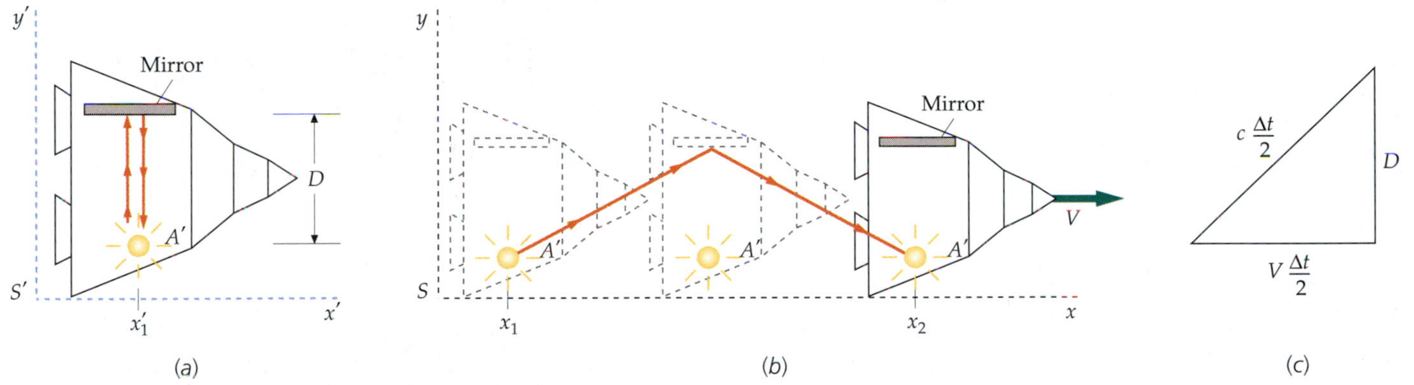

Figure 39-3 (*a*) Observer A' and the mirror are in a spaceship at rest in frame S'. The time it takes for the light pulse to reach the mirror and return is measured by A' to be $2D/c$. (*b*) In frame S, the spaceship is moving to the right with speed V. If the speed of light is the same in both frames, the time it takes for the light to reach the mirror and return is longer than $2D/c$ in S because the distance traveled is greater than $2D$. (*c*) A right triangle for computing the time Δt in frame S.

During the time interval Δt (as measured in S) between the original flash and the return flash, observer A' and his spaceship have moved a horizontal distance $V\,\Delta t$. In Figure 39-3*b*, we can see that the path traveled by the light is longer in S than in S'. However, by Einstein's postulates, light travels with the same speed c in frame S as it does in frame S'. Since it travels farther in S at the same speed, it takes longer in S to reach the mirror and return. The time interval in S is thus longer than it is in S'. From the triangle in Figure 39-3*c*, we have

$$\left(\frac{c\,\Delta t}{2}\right)^2 = D^2 + \left(\frac{V\,\Delta t}{2}\right)^2$$

or

$$\Delta t = \frac{2D}{\sqrt{c^2 - V^2}} = \frac{2D}{c}\frac{1}{\sqrt{1 - V^2/c^2}}$$

Using $\Delta t' = 2D/c$, we obtain

$$\Delta t = \frac{\Delta t'}{\sqrt{1 - V^2/c^2}} = \gamma\,\Delta t'$$

Example 39-2 *try it yourself*

Astronauts in a spaceship traveling at $V = 0.6c$ relative to the earth sign off from space control, saying that they are going to nap for 1 h and then call back. How long does their nap last as measured on earth?

Picture the Problem Since the astronauts go to sleep and wake up at the same place in their reference frame, the time interval for their nap of 1 h as measured by them is proper time. In the earth's reference frame, they move a considerable distance between these two events. The time interval measured in the earth's frame (using two clocks located at those events) is longer by the factor γ.

Cover the column to the right and try these on your own before looking at the answers.

Steps	*Answers*
1. Relate the time interval measured on earth Δt to the proper time Δt_{p}.	$\Delta t = \gamma\,\Delta t_{\mathrm{p}}$
2. Calculate γ for $V = 0.6c$.	$\gamma = 1.25$
3. Substitute to calculate the time of the nap in the earth's frame.	$\Delta t = \gamma\,\Delta t_{\mathrm{p}} = 1.25\text{ h}$

Exercise If the spaceship is moving at $V = 0.8c$, how long would a 1-h nap last as measured on earth? (*Answer* 1.67 h)

Length Contraction

A phenomenon closely related to time dilation is **length contraction.** The length of an object measured in the reference frame in which the object is at rest is called its **proper length** L_{p}. In a reference frame in which the object is moving, the measured length is shorter than its proper length. Consider a rod at rest in frame S' with one end at x_2' and the other end at x_1'. The length of the rod in this frame is its proper length $L_{\mathrm{p}} = x_2' - x_1'$. Some care must be taken to find the length of the rod in frame S. In this frame, the rod is moving to the right with speed V, the speed of frame S'. The length of the rod in frame S is *defined* as $L = x_2 - x_1$, where x_2 is the position of one end at some time t_2, and x_1 is the position of the other end *at the same time* $t_1 = t_2$ as measured in frame S. Equation 39-11 is convenient to use to calculate $x_2 - x_1$ at some time t because it relates x and x' to t, whereas Equation 39-9 is not convenient because it relates x and x' to t':

$$x_2' = \gamma(x_2 - Vt_2)$$

and

$$x_1' = \gamma(x_1 - Vt_1)$$

Since $t_2 = t_1$, we obtain

$$x_2' - x_1' = \gamma(x_2 - x_1)$$

$$x_2 - x_1 = \frac{1}{\gamma}(x_2' - x_1') = \sqrt{1 - V^2/c^2}\,(x_2' - x_1')$$

or

$$L = \frac{1}{\gamma}L_p = \sqrt{1 - V^2/c^2}\,L_p \qquad \text{39-14}$$

Length contraction

Thus, the length of a rod is smaller when it is measured in a frame in which it is moving. Before Einstein's paper was published, Lorentz and FitzGerald tried to explain the null result of the Michelson–Morley experiment by assuming that distances in the direction of motion contracted by the amount given in Equation 39-14. This contraction is now known as the **Lorentz–FitzGerald contraction.**

Example 39-3

A stick that has a proper length of 1 m moves in a direction along its length with speed V relative to you. The length of the stick as measured by you is 0.914 m. What is the speed V?

Picture the Problem Since both L and L_p are given, we can find V directly from Equation 39-14.

1. Equation 39-14 relates the lengths L and L_p and the speed V: $\quad L = \sqrt{1 - V^2/c^2}\,L_p$

2. Solve for V: $\quad 1 - V^2/c^2 = (L/L_p)^2 = (0.914\text{ m}/1\text{ m})^2 = 0.835$

$\quad V = c\sqrt{1 - 0.835} = 0.406c$

An interesting example of time dilation or length contraction is afforded by the appearance of muons as secondary radiation from cosmic rays. Muons decay according to the statistical law of radioactivity:

$$N(t) = N_0\,e^{-t/\tau} \qquad \text{39-15}$$

where N_0 is the original number of muons at time $t = 0$, $N(t)$ is the number remaining at time t, and τ is the mean lifetime, which is about 2 μs for muons at rest. Since muons are created (from the decay of pions) high in the atmosphere, usually several thousand meters above sea level, few muons should reach sea level. A typical muon moving with speed $0.9978c$ would travel only about 600 m in 2 μs. However, the lifetime of the muon measured in the earth's reference frame is increased by the factor $1/\sqrt{1 - V^2/c^2}$, which is 15 for this particular speed. The mean lifetime measured in the earth's reference frame is therefore 30 μs, and a muon with speed $0.9978c$ travels about 9000 m in this time. From the muon's point of view, it lives only 2 μs, but the atmosphere is rushing past it with a speed of $0.9978c$. The distance of 9000 m in the earth's frame is thus contracted to only 600 m in the muon's frame, as indicated in Figure 39-4.

It is easy to distinguish experimentally between the classical and relativistic predictions of the observation of muons at sea level. Suppose that we observe 10^8 muons at an altitude of 9000 m in some time interval with a muon detector. How many would we expect to observe at sea level in the same time interval? According to the nonrelativistic prediction, the time it takes for these muons to travel 9000 m is $(9000\ \text{m})/0.998c \approx 30\ \mu\text{s}$, which is 15 lifetimes. Substituting $N_0 = 10^8$ and $t = 15\tau$ into Equation 39-15, we obtain

$$N = 10^8 e^{-15} = 30.6$$

We would thus expect all but about 31 of the original 100 million muons to decay before reaching sea level.

According to the relativistic prediction, the earth must travel only the contracted distance of 600 m in the rest frame of the muon. This takes only $2\ \mu\text{s} = 1\tau$. Therefore, the number of muons expected at sea level is

$$N = 10^8 e^{-1} = 3.68 \times 10^7$$

Thus, relativity predicts that we would observe 36.8 million muons in the same time interval. Experiments of this type have confirmed the relativistic predictions.

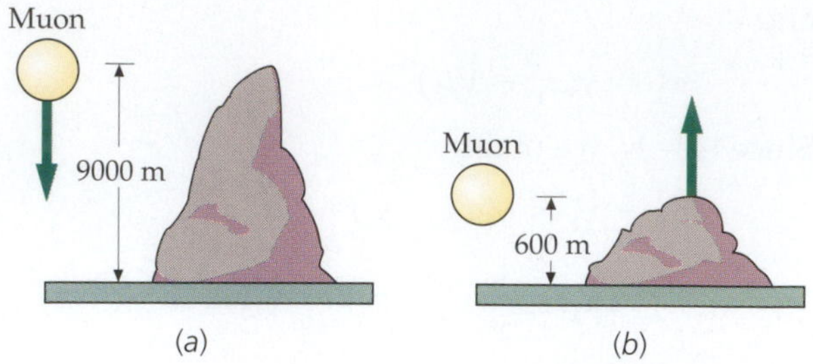

Figure 39-4 Although muons are created high above the earth and their mean lifetime is only about 2 μs when at rest, many appear at the earth's surface. (*a*) In the earth's reference frame, a typical muon moving at 0.998*c* has a mean lifetime of 30 μs and travels 9000 m in this time. (*b*) In the reference frame of the muon, the distance traveled by the earth is only 600 m in the muon's lifetime of 2 μs.

The Relativistic Doppler Effect

For light or other electromagnetic waves in a vacuum a distinction between motion of source and receiver cannot be made. Therefore, the expressions we derived in Chapter 15 for the Doppler effect cannot be correct for light. The reason is that in that derivation, we assumed the time intervals in the reference frames of the source and receiver to be the same.

Consider a source moving toward a receiver with velocity V, relative to the receiver. If the source emits N electromagnetic waves in a time Δt_R (measured in the frame of the receiver), the first wave will travel a distance $c\,\Delta t_R$ and the source will travel a distance $V\,\Delta t_R$ measured in the frame of the receiver. The wavelength will be

$$\lambda' = \frac{c\,\Delta t_R - V\,\Delta t_R}{N}$$

The frequency f' observed by the receiver will therefore be

$$f' = \frac{c}{\lambda'} = \frac{c}{c - V}\frac{N}{\Delta t_R} = \frac{1}{1 - V/c}\frac{N}{\Delta t_R}$$

If the frequency of the source is f_0, it will emit $N = f_0\,\Delta t_S$ waves in the time Δt_S measured by the source. Then

$$f' = \frac{1}{1 - V/c}\frac{N}{\Delta t_R} = \frac{1}{1 - V/c}\frac{f_0\,\Delta t_S}{\Delta t_R} = \frac{f_0}{1 - V/c}\frac{\Delta t_S}{\Delta t_R}$$

Here Δt_S is the proper time interval (the first wave and the Nth wave are emitted at the same place in the source's reference frame). Times Δt_S and Δt_R are related by Equation 39-13 for time dilation:

$$\Delta t_R = \gamma\,\Delta t_S = \frac{\Delta t_S}{\sqrt{1 - V^2/c^2}}$$

Thus, when the source and receiver are moving toward one another we obtain

$$f' = \frac{f_0}{1 - V/c}\frac{1}{\gamma} = \frac{\sqrt{1 - V^2/c^2}}{1 - V/c} f_0 = \sqrt{\frac{1 + V/c}{1 - V/c}} f_0 \qquad \text{approaching} \qquad \text{39-16}a$$

This differs from our classical equation only in the time-dilation factor.

When the source and receiver are moving away from one another, the same analysis shows that the observed frequency is given by

$$f' = \frac{\sqrt{1 - V^2/c^2}}{1 + V/c} f_0 = \sqrt{\frac{1 - V/c}{1 + V/c}} f_0 \qquad \text{receding} \qquad \text{39-16}b$$

It is left as a problem (Problem 36) for you to show that the same results are obtained if the calculations are done in the reference frame of the source.

An application of the relativistic Doppler effect is the **redshift** observed in the light from distant galaxies. Because the galaxies are moving away from us, the light they emit is shifted toward the longer, red wavelengths. The speed of the galaxies relative to us can be determined by measuring this shift.

Example 39-4 *try it yourself*

The longest wavelength of light emitted by hydrogen in the Balmer series is λ_0 = 656 nm. In light from a distant galaxy, this wavelength is measured to be λ' = 1458 nm. Find the speed at which the distant galaxy is receding from the earth.

Cover the column to the right and try these on your own before looking at the answers.

Steps	*Answers*
1. Use Equation 39-16*b* to relate the speed V to the received frequency f' and the emitted frequency f_0.	$f' = \sqrt{\frac{1 - V/c}{1 + V/c}} f_0$
2. Substitute $f' = c/\lambda'$ and $f_0 = c/\lambda_0$ and solve for V/c.	$\frac{1 - V/c}{1 + V/c} = \left(\frac{f'}{f_0}\right)^2 = \left(\frac{\lambda_0}{\lambda'}\right)^2 = 0.202$ $V/c = 0.664$

39-4 Clock Synchronization and Simultaneity

We saw in Section 39-3 that proper time is the time interval between two events that occur at the same point in some reference frame. It can therefore be measured on a single clock. However, in another reference frame moving relative to the first, the same two events occur at different places, so two clocks are needed to record the times. The time of each event is measured on a different clock, and the interval is found by subtraction. This procedure requires that the clocks be **synchronized.** We will show in this section that

> Two clocks that are synchronized in one reference frame are not synchronized in any other frame moving relative to the first frame.

A corollary to this result is:

> Two events that are simultaneous in one reference frame are not simultaneous in another frame moving relative to the first.*

Comprehension of these facts usually resolves all relativity paradoxes. Unfortunately, the intuitive (and incorrect) belief that simultaneity is an absolute relation is difficult to overcome.

Suppose we have two clocks at rest at points A and B a distance L apart in frame S. How can we synchronize these two clocks? If an observer at A looks at the clock at B and sets her clock to read the same time, the clocks will not be synchronized because of the time L/c it takes light to travel from one clock to another. To synchronize the clocks, the observer at A must set her clock ahead by the time L/c. Then she will see that the clock at B reads a time that is L/c behind the time on her clock, but she will calculate that the clocks are synchronized when she allows for the time L/c for the light to reach her. Any other observers except those equidistant from the clocks will see the clocks reading different times, but they will also calculate that the clocks are synchronized when they correct for the time it takes the light to reach them. An equivalent method for synchronizing two clocks would be for an observer C at a point midway between the clocks to send a light signal and for observers at A and B to set their clocks to some prearranged time when they receive the signal.

We now examine the question of **simultaneity.** Suppose A and B agree to explode flashguns at t_0 (having previously synchronized their clocks). Observer C will see the light from the two flashes at the same time, and because he is equidistant from A and B, he will conclude that the flashes were simultaneous. Other observers in frame S will see the light from A or B first, depending on their location, but after correcting for the time the light takes to reach them, they also will conclude that the flashes were simultaneous. We can thus define simultaneity as follows:

> Two events in a reference frame are simultaneous if light signals from the events reach an observer halfway between the events at the same time.

Definition—Simultaneity

To show that two events that are simultaneous in frame S are not simultaneous in another frame S' moving relative to S, we will use an example introduced by Einstein. A train is moving with speed V past a station platform. We will consider the train to be at rest in S' and the platform to be at rest in S. We have observers A', B', and C' at the front, back, and middle of the train. We now suppose that the train and platform are struck by lightning at the front and back of the train and that the lightning bolts are simultaneous in the frame of the platform S (Figure 39-5). That is, an observer C on the platform halfway between the positions A and B, where the lightning strikes, sees the two flashes at the same time. It is convenient to suppose that the lightning scorches the

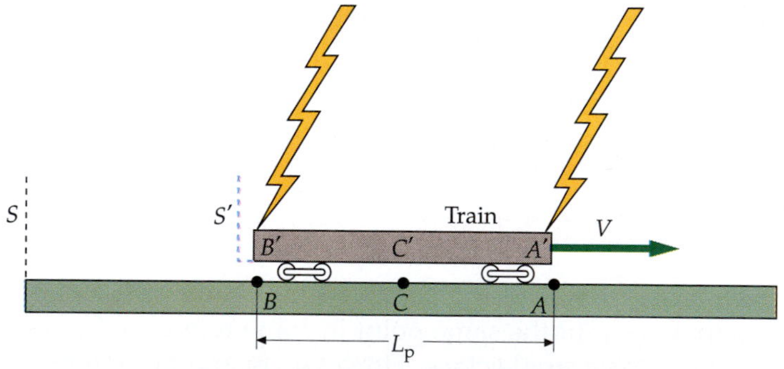

Figure 39-5 Simultaneous lightning bolts strike the ends of a train traveling with speed V in frame S attached to the platform. The light from these simultaneous events reaches observer C midway between the events at the same time. The distance between the bolts is $L_{p,\text{platform}}$.

* This is true unless the events and clocks are in the same plane and are perpendicular to the relative motion.

train and platform so that the events can be easily located. Since C' is in the middle of the train, halfway between the places on the train that are scorched, the events are simultaneous in S' only if C' sees the flashes at the same time. However, the flash from the front of the train is seen by C' before the flash from the back of the train. We can understand this by considering the motion of C' as seen in frame S (Figure 39-6). By the time the light from the front flash reaches C', C' has moved some distance toward the front flash and some distance away from the back flash. Thus, the light from the back flash has not yet reached C' as indicated in the figure. Observer C' must therefore conclude that the events are not simultaneous and that the front of the train was struck before the back. Furthermore, all observers in S' on the train will agree with C' when they have corrected for the time it takes the light to reach them.

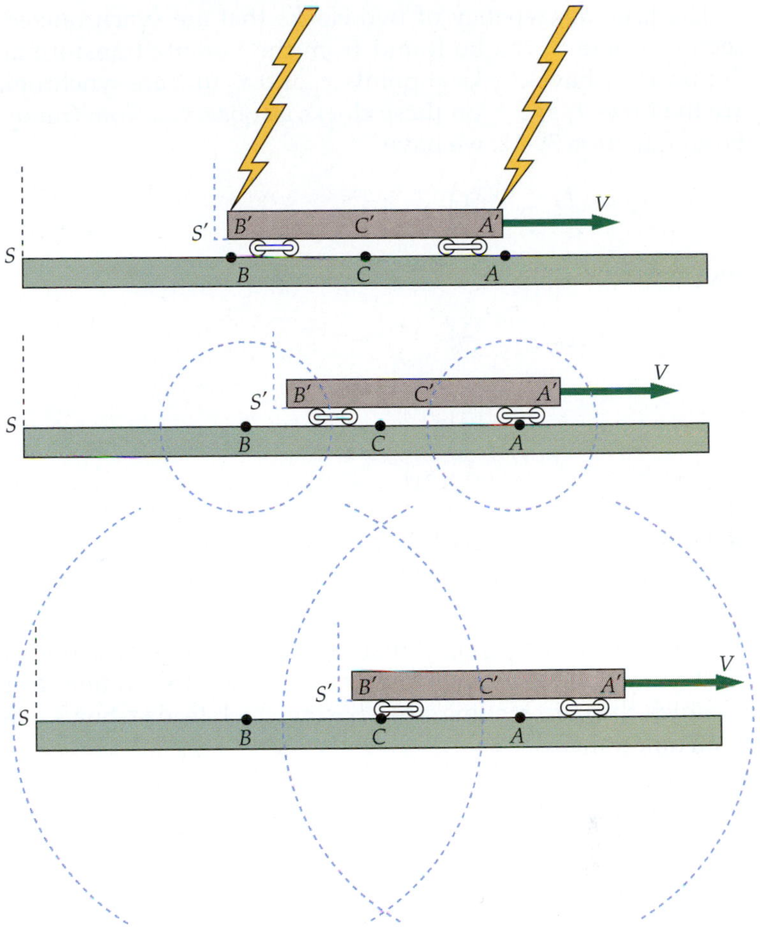

Figure 39-6 The light from the lightning bolt at the front of the train reaches observer C' at the middle of the train before that from the bolt at the back of the train. Since C' is midway between the events (which occur at the front and rear of the train), these events are not simultaneous for him.

Figure 39-7 shows the events of the lightning bolts as seen in the reference frame of the train (S'). In this frame, the platform is moving so the distance between the burns on the platform is contracted. The platform is shorter than it is in S, and, since the train is at rest, the train is longer than its contracted length in S. When the lightning bolt strikes the front of the train at A', the front of the train is at point A, and the back of the train has not yet reached point B. Later, when the lightning bolt strikes the back of the train at B', the back has reached point B on the platform.

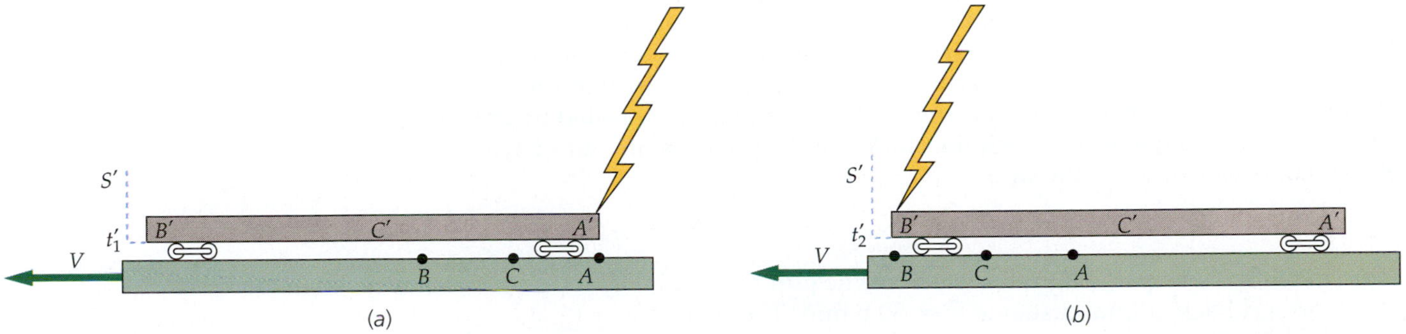

Figure 39-7 The lightning bolts of Figure 39-5 as seen in frame S' of the train. In this frame, the distance between A and B on the platform is less than $L_{p,platform}$, and the proper length of the train $L_{p,train}$ is longer than $L_{p,platform}$. The first lightning bolt strikes the front of the train when A' and A are coincident. The second bolt strikes the rear of the train when B' and B are coincident.

The time discrepancy of two clocks that are synchronized in frame S as seen in frame S' can be found from the Lorentz transformation equations. Suppose we have clocks at points x_1 and x_2 that are synchronized in S. What are the times t_1 and t_2 on these clocks as observed from frame S' at a time t'_0? From Equation 39-12, we have

$$t'_0 = \gamma\left(t_1 - \frac{Vx_1}{c^2}\right)$$

and

$$t'_0 = \gamma\left(t_2 - \frac{Vx_2}{c^2}\right)$$

Then

$$t_2 - t_1 = \frac{V}{c^2}(x_2 - x_1)$$

Note that the chasing clock (at x_2) leads the other (at x_1) by an amount that is proportional to their proper separation $L_p = x_2 - x_1$.

If two clocks are synchronized in the frame in which they are at rest, they will be out of synchronization in another frame. In the frame in which they are moving, the chasing clock leads (shows a later time) by an amount

$$\Delta t_S = L_p \frac{V}{c^2} \qquad \text{39-17}$$

where L_p is the proper distance between the clocks.

A numerical example should help clarify time dilation, clock synchronization, and the internal consistency of these results.

Example 39-5

An observer in a spaceship has a flash gun and a mirror as shown in Figure 39-3. The distance from the gun to the mirror is 15 light-minutes (written 15c·min) and the spaceship in frame S' travels with speed $V = 0.8c$ relative to a very long space platform in frame S that has two synchronized clocks, one at the position x_1 of the spaceship when the observer explodes the flash gun and the other at the position x_2 of the spaceship when the light returns to the gun from the mirror. Find the time intervals between the events (exploding the flash gun and receiving the return flash from the mirror) (*a*) in the frame of the ship and (*b*) in the frame of the platform. (*c*) Find the distance traveled by the ship, and (*d*) the amount by which the clocks on the platform are out of synchronization as viewed by the ship.

(*a*)1. In the spaceship, the light travels from the gun to the mirror and back, a total distance $D = 30\ c{\cdot}\text{min}$. The time required is D/c:

$$\Delta t' = \frac{D}{c} = \frac{30\ c{\cdot}\text{min}}{c} = 30\ \text{min}$$

2. Since these events happen at the same place in the spaceship, the time interval is proper time:

$$\Delta t_p = 30\ \text{min}$$

(*b*)1. In frame *S*, the time between the events is longer by the factor γ: $\Delta t = \gamma\,\Delta t' = \gamma(30\text{ min})$

2. Calculate γ: $\gamma = \dfrac{1}{\sqrt{1 - V^2/c^2}} = \dfrac{1}{\sqrt{1-(0.8)^2}} = \dfrac{1}{\sqrt{0.36}} = \dfrac{5}{3}$

3. Use this value of γ to calculate the time between the events as observed in frame *S*: $\Delta t = \gamma\,\Delta t_{\rm p} = \dfrac{5}{3}(30\text{ min}) = 50\text{ min})$

(*c*)1. The distance traveled by the ship in *S* is $V\,\Delta t$: $D = V\,\Delta t = (0.8c)(50\text{ min}) = 40\;c\cdot\text{min}$

2. This distance is the proper distance between the clocks on the platform: $L_{\rm p} = D = 40\;c\cdot\text{min}$

(*d*) The amount that the clocks on the platform are out of synchronization is related to the proper distance between the clocks $L_{\rm p}$: $\Delta t_{\rm s} = L_{\rm p}\dfrac{V}{c^2} = \dfrac{(40\;c\cdot\text{min})(0.8c)}{c^2} = 32\text{ min}$

Remarks Observers on the platform would say that the spaceship's clock is running slow because it records a time of only 30 min between the events, whereas the time measured on the platform is 50 min.

Figure 39-8 shows the situation viewed from the spaceship in S'. The platform is traveling past the ship with speed $0.8c$. There is a clock at point x_1, which coincides with the ship when the flash gun is exploded, and another at point x_2, which coincides with the ship when the return flash is received from the mirror. We assume that the clock at x_1 reads 12:00 noon at the time of the light flash. The clocks at x_1 and x_2 are synchronized in *S* but not in S'. In S', the clock at x_2, which is chasing the one at x_1, leads by 32 min; it would thus read 12:32 to an observer in S'. When the spaceship coincides with x_2, the clock there reads 12:50. The time between the events is therefore 50 min in *S*. Note that according to observers in S', this clock ticks off 50 min − 32 min = 18 min for a trip that takes 30 min in S'. Thus, observers in S' see this clock run slow by the factor $30/18 = 5/3$.

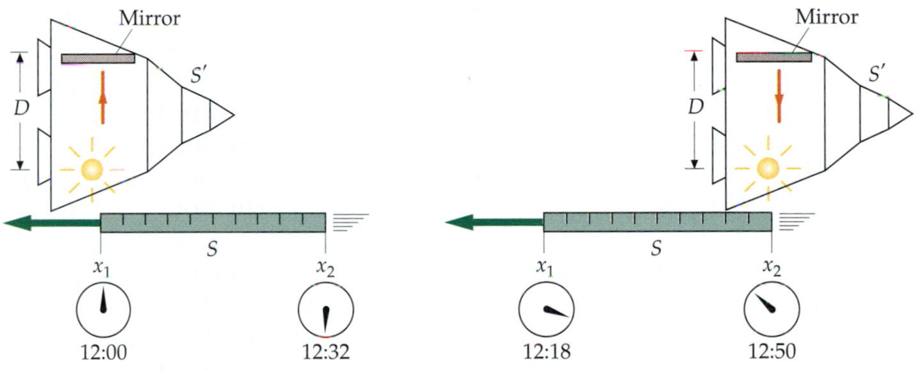

Figure 39-8 Clocks on a platform as observed from the spaceship's frame of reference S'. During the time $\Delta t' = 30$ min it takes for the platform to pass the spaceship, the clocks on the platform run slow and tick off $(30\text{ min})/\gamma = 18$ min. But the clocks are unsynchronized, with the chasing clock leading by $L_{\rm p}V/c^2$, which for this case is 32 min. The time it takes for the spaceship to pass as measured on the platform is therefore 32 min + 18 min = 50 min.

Every observer in one frame sees the clocks in the other frame run slow. According to observers in *S*, who measure 50 min for the time interval, the time interval in S' (30 min) is too small, so they see the single clock in S' run too slow by the factor 5/3. According to the observers in S', the observers in *S* measure a time that is too *long* despite the fact that their clocks run too slow because the clocks in *S* are out of synchronization. The clocks tick off only 18 min, but the second one leads the first by 32 min, so the time interval is 50 min.

Exploring

The Twin Paradox

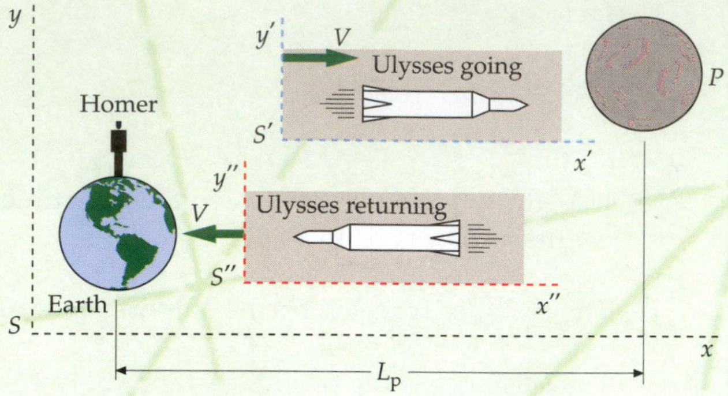

Figure 1 The twin paradox. The earth and a distant planet are fixed in frame S. Ulysses coasts in frame S' to the planet and then coasts back in frame S''. His twin Homer stays on earth. When Ulysses returns, he is younger than his twin. The roles played by the twins are not symmetric. Homer remains in one inertial reference frame, but Ulysses must accelerate if he is to return home.

Homer and Ulysses are identical twins. Ulysses travels at high speed to a planet beyond the solar system and returns while Homer remains at home. When they are together again, which twin is older, or are they the same age? The correct answer is that Homer, the twin who stays at home, is older. This problem, with variations, has been the subject of spirited debate for decades, though there are very few who disagree with the answer. The problem is a paradox because of the seemingly symmetric roles played by the twins with the asymmetric result in their aging. The paradox is resolved when the asymmetry of the twins' roles is noted. The relativistic result conflicts with common sense based on our strong but incorrect belief in absolute simultaneity. We will consider a particular case with some numerical magnitudes that, though impractical, make the calculations easy.

Let planet P and Homer on earth be at rest in reference frame S a distance L_p apart, as illustrated in Figure 1. We neglect the motion of the earth. Reference frames S' and S'' are moving with speed V toward and away from the planet, respectively. Ulysses quickly accelerates to speed V, then coasts in S' until he reaches the planet, where he stops and is momentarily at rest in S. To return he quickly accelerates to speed V toward earth and then coasts in S'' until he reaches earth, where he stops. We can assume that the acceleration times are negligible compared with the coasting times. We use the following values for illustration: $L_p = 8$ light-years and $V = 0.8c$. Then $\sqrt{1 - V^2/c^2} = 3/5$ and $\gamma = 5/3$.

It is easy to analyze the problem from Homer's point of view on earth. According to Homer's clock, Ulysses coasts in S' for a time $L_p/V = 10$ y and in S'' for an equal time. Thus, Homer is 20 y older when Ulysses returns. The time interval in S' between Ulysses' leaving earth and his arriving at the planet is shorter because it is proper time. The time it takes to reach the planet by Ulysses' clock is

$$\Delta t' = \frac{\Delta t}{\gamma} = \frac{10\text{ y}}{5/3} = 6\text{ y}$$

Since the same time is required for the return trip, Ulysses will have recorded 12 y for the round trip and will be 8 y younger than Homer upon his return.

From Ulysses' point of view, the distance from the earth to the planet is contracted and is only

$$L' = \frac{L_p}{\gamma} = \frac{8\text{ light-years}}{5/3} = 4.8\text{ light-years}$$

At $V = 0.8c$, it takes only 6 y each way.

The real difficulty in this problem is for Ulysses to understand why his twin aged 20 y during his absence. If we consider Ulysses as being at rest and Homer as moving away, Homer's clock should run slow and measure only $3/5(6) = 3.6$ y. Then why shouldn't Homer age only 7.2 y during the round trip? This, of course, is the paradox. The difficulty with the analysis from the point of view of Ulysses is that he does not remain in an inertial frame. What happens while Ulysses is stopping and starting? To investigate this problem in detail, we would need to treat accelerated reference frames, a subject dealt with in the study of general relativity and beyond the scope of this book. However, we can get some insight into the problem by having the twins send regular signals to each other so that they can record the other's age continuously. If they arrange to send a signal once a year, each can determine the age of the other merely by counting the signals received. The arrival frequency of the signals will not be 1 per year because of the Doppler shift. The frequency observed will be given by Equations 39-16*a* and *b*. Using $V/c = 0.8$ and $V^2/c^2 = 0.64$, we have for the case in which the twins are receding from each other

$$f' = \frac{\sqrt{1 - V^2/c^2}}{1 + V/c} f_0 = \frac{\sqrt{1 - 0.64}}{1 + 0.8} f_0 = \frac{1}{3} f_0$$

When they are approaching, Equation 39-16*a* gives $f' = 3f_0$.

Consider the situation first from the point of view of Ulysses. During the 6 y it takes him to reach the planet (remember that the distance is contracted in his frame), he receives signals at the rate of $\frac{1}{3}$ signal per year, and so he receives 2 signals. As soon as he turns around and starts back to earth, he begins to receive 3 signals per year. In the 6 y it takes him to return he receives 18 signals, giving a total of 20 for the trip. He accordingly expects his twin to have aged 20 years.

We now consider the situation from Homer's point of view. He receives signals at the rate of $\frac{1}{3}$ signal per year not only for the 10 y it takes Ulysses to reach the planet but also for the time it takes for the last signal sent by Ulysses before he turns around to get back to earth. (He cannot know that Ulysses has turned around until the signals begin reaching him with increased frequency.) Since the planet is 8 light-years away, there is an additional 8 y of receiving signals at the rate of $\frac{1}{3}$ signal per year. During the first 18 y, Homer receives 6 signals. In the final 2 y before Ulysses arrives, Homer receives 6 signals, or 3 per year. (The first signal sent after Ulysses turns around takes 8 y to reach earth, whereas Ulysses, traveling at 0.8*c*, takes 10 y to return and therefore arrives just 2 y after Homer begins to receive signals at the faster rate.) Thus, Homer expects Ulysses to have aged 12 y. In this analysis, the asymmetry of the twins' roles is apparent. When they are together again, both twins agree that the one who has been accelerated will be younger than the one who stayed home.

The predictions of the special theory of relativity concerning the twin paradox have been tested using small particles that can be accelerated to such large speeds that γ is appreciably greater than 1. Unstable particles can be accelerated and trapped in circular orbits in a magnetic field, for example, and their lifetimes can then be compared with those of identical particles at rest. In all such experiments, the accelerated particles live longer on the average than those at rest, as predicted. These predictions have also been confirmed by the results of an experiment in which high-precision atomic clocks were flown around the world in commercial airplanes, but the analysis of this experiment is complicated due to the necessity of including gravitational effects treated in the general theory of relativity.

39-5 The Velocity Transformation

We can find how velocities transform from one reference frame to another by differentiating the Lorentz transformation equations. Suppose a particle has velocity $u'_x = dx'/dt'$ in frame S', which is moving to the right with speed V relative to frame S. Its velocity in frame S is

$$u_x = \frac{dx}{dt}$$

From the Lorentz transformation equations (Equations 39-9 and 39-10), we have

$$dx = \gamma(dx' + V\,dt')$$

and

$$dt = \gamma\left(dt' + \frac{V\,dx'}{c^2}\right)$$

The velocity in S is thus

$$u_x = \frac{dx}{dt} = \frac{\gamma(dx' + V\,dt')}{\gamma\left(dt' + \dfrac{V\,dx'}{c^2}\right)} = \frac{\dfrac{dx'}{dt'} + V}{1 + \dfrac{V}{c^2}\dfrac{dx'}{dt'}} = \frac{u'_x + V}{1 + \dfrac{Vu'_x}{c^2}}$$

If a particle has components of velocity along the y or z axes, we can use the same relation between dt and dt', with $dy = dy'$ and $dz = dz'$, to obtain

$$u_y = \frac{dy}{dt} = \frac{dy'}{\gamma\left(dt' + \dfrac{V\,dx'}{c^2}\right)} = \frac{\dfrac{dy'}{dt'}}{\gamma\left(1 + \dfrac{V}{c^2}\dfrac{dx'}{dt'}\right)} = \frac{u'_y}{\gamma\left(1 + \dfrac{Vu'_x}{c^2}\right)}$$

and

$$u_z = \frac{u'_z}{\gamma\left(1 + \dfrac{Vu'_x}{c^2}\right)}$$

The complete relativistic velocity transformation is

$$u_x = \frac{u'_x + V}{1 + Vu'_x/c^2} \qquad \text{39-18}a$$

$$u_y = \frac{u'_y}{\gamma(1 + Vu'_x/c^2)} \qquad \text{39-18}b$$

$$u_z = \frac{u'_z}{\gamma(1 + Vu'_x/c^2)} \qquad \text{39-18}c$$

Relativistic velocity transformation

The inverse velocity transformation equations are

$$u'_x = \frac{u_x - V}{1 - Vu_x/c^2} \qquad \text{39-19}a$$

$$u'_y = \frac{u_y}{\gamma(1 - Vu_x/c^2)} \qquad \text{39-19}b$$

$$u'_z = \frac{u_z}{\gamma(1 - Vu_x/c^2)} \qquad \text{39-19}c$$

Relativistic velocity transformation

These equations differ from the classical and intuitive result $u_x = u'_x + V$, $u_y = u'_y$, and $u_z = u'_z$ because the denominators in the equations are not equal to 1. When V and u'_x are small compared with the speed of light c, $\gamma \approx 1$ and $Vu'_x/c^2 \ll 1$. Then the relativistic and classical expressions are the same.

Example 39-6

A supersonic plane moves away from you along the *x* axis with speed 1000 m/s (about 3 times the speed of sound) relative to you. Another plane moves along the *x* axis away from you and away from the second plane at speed 500 m/s relative to the first plane. How fast is the second plane moving relative to you?

Picture the Problem These speeds are so small compared with c that we expect the classical equations for combining velocities to be accurate. We show this by calculating the correction term in the denominator of Equation 39-18*a*. Let frame S be your rest frame and frame S' be moving with velocity $V = 1000$ m/s. The first plane is then at rest in frame S' and the second has velocity $u'_x = 500$ m/s in S'.

1. The classical formula for combining velocities, gives for the velocity of the second plane relative to you: $u_x = u'_x + V = 500\text{ m/s} + 1000\text{ m/s} = 1500\text{ m/s}$

2. Calculate the correction term in the denominator of Equation 39-18*a*: $\dfrac{Vu'_x}{c^2} = \dfrac{(1000)(500)}{(3\times 10^8)^2} \approx 5.6\times 10^{-12}$

Remark This correction term is so small that the classical and relativistic results are essentially the same.

Example 39-7

Work Example 39-6 if the first plane moves with speed $V = 0.8c$ relative to you and the second plane moves with the same speed $0.8c$ relative to the first plane.

Picture the Problem These speeds are not small compared with c so we use the relativistic expression (Equation 39-18*a*). We again assume that you are at rest in frame S and the first plane is at rest in frame S' that is moving at $V = 0.8c$ relative to you. The velocity of the second plane in S' is $u'_x = 0.8c$.

1. Use Equation 39-18*a* to calculate the speed of the second plane relative to you: $u_x = \dfrac{u'_x + V}{1 + Vu'_x/c^2} = \dfrac{0.8c + 0.8c}{1 + (0.8c)(0.8c)/c^2} = \dfrac{1.6c}{1.64} = 0.98c$

The result in Example 39-7 is quite different from the classically expected result of $0.8c + 0.8c = 1.6c$. In fact, it can be shown from Equation 39-18 that if the speed of an object is less than c in one frame, it is less than c in all other frames moving relative to that frame with a speed less than c. (See Problem 20.) We will see in Section 39-7 that it takes an infinite amount of energy to accelerate a particle to the speed of light. The speed of light c is thus an upper, unattainable limit for the speed of a particle having mass.*

* There are massless particles, such as photons, that always move at the speed of light.

Example 39-8

Light moves along the x axis in frame S' with speed $u'_x = c$. What is its speed in frame S?

1. The speed in S' is given by Equation 39-18*a*:

$$u_x = \frac{u'_x + V}{1 + Vu'_x/c^2} = \frac{c + V}{1 + Vc/c^2} = \frac{c(1 + V/c)}{1 + V/c} = c$$

Remark The speed in both frames is c as required by Einstein's postulates.

Example 39-9

Two spaceships, each 100 m long when measured at rest, travel toward each other with speeds of 0.85c relative to the earth. (*a*) How long is each ship as measured by someone on earth? (*b*) How fast is each ship traveling as measured by an observer on the other? (*c*) How long is one ship when measured by an observer on the other? (*d*) At time $t = 0$ on earth, the fronts of the ships are together as they just begin to pass each other. At what time on earth are their ends together?

Picture the Problem (*a*) The length of each ship as measured on earth is the contracted length $\sqrt{1 - V^2/c^2}\,L_p$ (Equation 39-14). To solve part (*b*) let the earth be in frame S, and the ship on the left be in frame S' moving with velocity $V = 0.85c$ relative to the earth. Then the ship on the right moves with velocity $u_x = -0.85c$ as shown in Figure 39-9. (*c*) The length of one ship as seen by the other is $\sqrt{1 - V_2^2/c^2}\,L_p$, where V_2 is the velocity of one ship relative to the other.

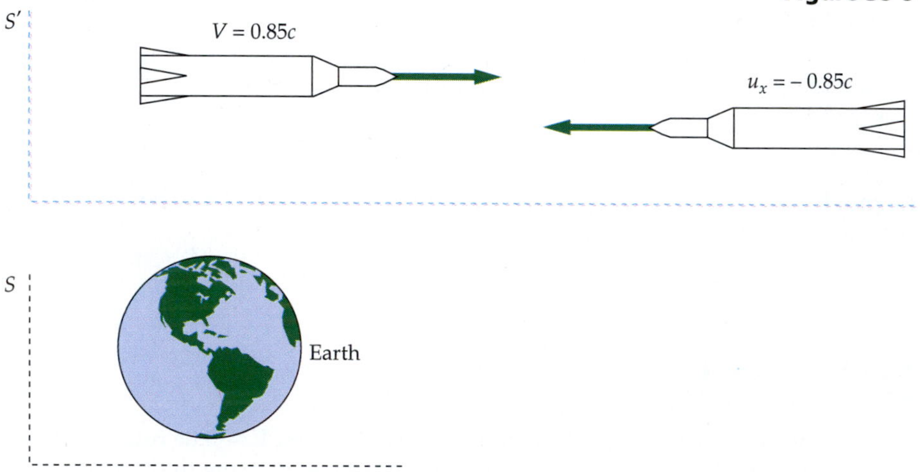

Figure 39-9

(*a*) The length of each ship in the earth's frame is the proper length divided by γ:

$$L = \sqrt{1 - V^2/c^2}\,L_p = \sqrt{1 - (0.85c)^2/c^2}\,(100\text{ m}) = 52.7\text{ m}$$

(*b*) Use the velocity transformation formula (Equation 39-19*a*) to find the velocity u'_x of the ship on the right as seen in frame S':

$$u'_x = \frac{u_x - V}{1 - Vu_x/c^2} = \frac{-0.85c - 0.85c}{1 - (0.85c)(-0.85c)/c^2} = \frac{-1.70c}{1 + 0.7225} = -0.987c$$

(*c*) In the frame of the left ship, the right ship is moving with velocity $V_2 = -0.987c$. Use this to calculate the contracted length of the ship on the right:

$$L = \sqrt{1 - V_2^2/c^2}\,L_p = \sqrt{1 - (0.987c)^2/c^2}\,(100\text{ m}) = 16.1\text{ m}$$

(*d*) If the heads of the ships are together at $t = 0$ on earth, their ends will be together after the time it takes either ship to move the length of the ship as seen on earth:

$$t = \frac{L}{V} = \frac{52.7\text{ m}}{0.85c} = \frac{52.7\text{ m}}{(0.85)(3 \times 10^8\text{ m/s})} = 2.07 \times 10^{-7}\text{ s}$$

39-6 Relativistic Momentum

We have seen in previous sections that Einstein's postulates require important modifications in our ideas of simultaneity and in our measurements of time and length. Perhaps more importantly, they also require modifications in our concepts of mass, momentum, and energy. In classical mechanics, the momentum of a particle is defined as the product of its mass and its velocity, $\vec{p} = m\vec{u}$, where $\vec{u}$ is the velocity. In an isolated system of particles, with no net force acting on the system, the total momentum of the system remains constant.

We can see from a simple thought experiment that the quantity $\vec{p} = m\vec{u}$ is not conserved in an isolated system. We consider two observers: observer *A* in reference frame *S* and observer *B* in frame *S'*, which is moving to the right in the *x* direction with speed *V* with respect to frame *S*. Each has a ball of mass *m*. The two balls are identical when compared at rest. One observer throws his ball up with a speed u_0 relative to him and the other throws his ball down with a speed u_0 relative to him, so that each ball travels a distance *L*, makes an elastic collision with the other ball, and returns. Figure 39-10 shows how the collision looks in each reference frame. Classically, each ball has vertical momentum of magnitude mu_0. Since the vertical components of the momenta are equal and opposite, the total vertical component of momentum is zero before the collision. The collision merely reverses the momentum of each ball, so the total vertical momentum is zero after the collision.

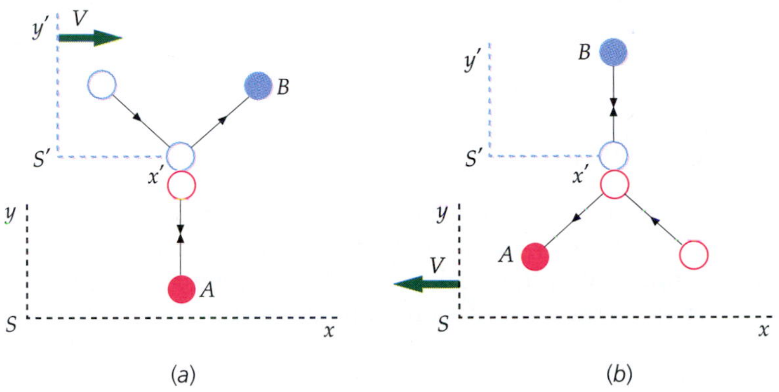

Figure 39-10 (*a*) Elastic collision of two identical balls as seen in frame *S*. The vertical component of the velocity of ball *B* is u_0/γ in *S* if it is u_0 in *S'*. (*b*) The same collision as seen in *S'*. In this frame, ball *A* has vertical component of velocity equal to u_0/γ.

Relativistically, however, the vertical components of the velocities of the two balls as seen by either observer are not equal and opposite. Thus, when they are reversed by the collision, classical momentum is not conserved. Consider the collision as seen by *A* in frame *S*. The velocity of his ball is $u_{Ay} = +u_0$. Since the velocity of *B*'s ball in frame *S'* is $u'_{Bx} = 0$, $u'_{By} = -u_0$, the *y* component of the velocity of *B*'s ball in frame *S* is $u_{By} = -u_0/\gamma$ (Equation 39-18*b*). Thus, if the classical expression $\vec{p} = m\vec{u}$ is taken as the definition of momentum, the vertical components of momentum of the two balls are not equal and opposite as seen by observer *A*. Since the balls are reversed by the collision, classical momentum is not conserved. Of course, the same result is observed by *B*. In the classical limit, when *u* is much less than *c*, γ is approximately 1, and the momentum of the system is conserved as seen by either observer.

The reason that the total momentum of a system is important in classical mechanics is that it is conserved when there are no external forces acting on the system, as is the case in collisions. But we have just seen that $\Sigma m\vec{u}$ is conserved only in the approximation that $u \ll c$. We will define the relativistic momentum $\vec{p}$ of a particle to have the following properties:

1. In collisions, $\vec{p}$ is conserved.
2. As u/c approaches zero, $\vec{p}$ approaches $m\vec{u}$.

We will show below that the quantity

$$\vec{p} = \frac{m\vec{u}}{\sqrt{1 - u^2/c^2}} \qquad 39\text{-}20$$

is conserved in the elastic collision shown in Figure 39-10. Since this quantity also approaches $m\vec{u}$ as u/c approaches zero, we take this equation for the definition of the **relativistic momentum** of a particle.

One interpretation of Equation 39-20 is that the mass of an object increases with speed. The quantity $m/\sqrt{1-u^2/c^2}$ is called the **relativistic mass** of a particle. The mass of a particle when it is at rest in some reference frame is called its **rest mass** m_0. The mass thus increases from m_0 at rest to $m_r = m_0/\sqrt{1-u^2/c^2}$ when it is moving at speed u. To avoid confusion, we will label the rest mass m_0 and use $m_0/\sqrt{1-u^2/c^2}$ for the relativistic mass in this chapter. The rest mass of a particle is the same in all reference frames. Using this notation, the relativistic momentum of a particle is written

$$\vec{p} = \frac{m_0\vec{u}}{\sqrt{1-u^2/c^2}} \qquad \text{39-21}$$

Relativistic momentum

Illustration of Conservation of Relativistic Momentum

optional

We will compute the y component of the relativistic momentum of each particle in the reference frame S for the collision of Figure 39-10 and show that the y component of the total relativistic momentum is zero. The speed of ball A in S is u_0, so the y component of its relativistic momentum is

$$p_{Ay} = \frac{mu_0}{\sqrt{1-u_0^2/c^2}}$$

The speed of ball B in S is more complicated. Its x component is V and its y component is $-u_0/\gamma$. Thus,

$$u_B^2 = u_{Bx}^2 + u_{By}^2 = V^2 + (-u_0\sqrt{1-V^2/c^2})^2 = V^2 + u_0^2 - \frac{u_0^2V^2}{c^2}$$

Using this result to compute $\sqrt{1-u_B^2/c^2}$, we obtain

$$1 - \frac{u_B^2}{c^2} = 1 - \frac{V^2}{c^2} - \frac{u_0^2}{c^2} + \frac{u_0^2V^2}{c^4} = \left(1 - \frac{V^2}{c^2}\right)\left(1 - \frac{u_0^2}{c^2}\right)$$

and

$$\sqrt{1-u_B^2/c^2} = \sqrt{1-V^2/c^2}\,\sqrt{1-u_0^2/c^2} = (1/\gamma)\,\sqrt{1-u_0^2/c^2}$$

The y component of the relativistic momentum of ball B as seen in S is therefore

$$p_{By} = \frac{mu_{By}}{\sqrt{1-u_B^2/c^2}} = \frac{-mu_0/\gamma}{(1/\gamma)\sqrt{1-u_0^2/c^2}} = \frac{-mu_0}{\sqrt{1-u_0^2/c^2}}$$

Since $p_{By} = -p_{Ay}$, the y component of the total momentum of the two balls is zero. If the speed of each ball is reversed by the collision, the total momentum will remain zero and momentum will be conserved.

39-7 Relativistic Energy

In classical mechanics, the work done by an unbalanced force acting on a particle equals the change in the kinetic energy of the particle. In relativistic mechanics, we equate the unbalanced force to the rate of change of the relativistic momentum. The work done by such a force can then be calculated and set equal to the change in kinetic energy.

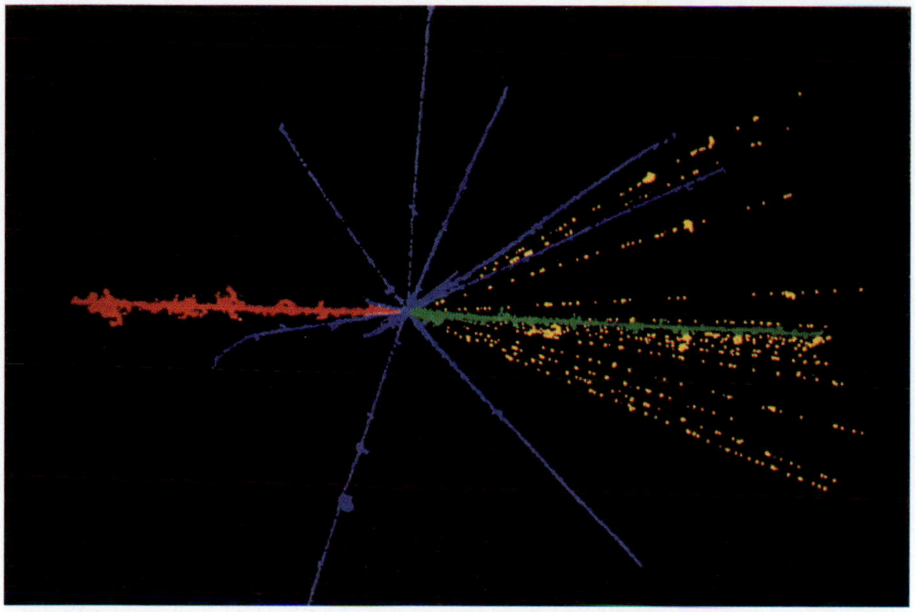

The creation of elementary particles demonstrates the conversion of kinetic energy to rest energy. In this 1950 photograph of a cosmic ray shower, a high-energy sulfur nucleus (red) collides with a nucleus in a photographic emulsion and produces a spray of particles, including a fluorine nucleus (green), other nuclear fragments (blue), and about 16 pions (yellow).

As in classical mechanics, we will define kinetic energy as the work done by an unbalanced force in accelerating a particle from rest to some velocity. Considering one dimension only, we have

$$K = \int_{u=0}^{u} \sum F\, ds = \int_0^u \frac{dp}{dt}\, ds = \int_0^u u\, dp$$
$$= \int_0^u u\, d\left(\frac{m_0 u}{\sqrt{1 - u^2/c^2}}\right) \qquad \text{39-22}$$

where we have used $u = ds/dt$. It is left as a problem (Problem 51) for you to show that

$$d\left(\frac{m_0 u}{\sqrt{1 - u^2/c^2}}\right) = m_0\left(1 - \frac{u^2}{c^2}\right)^{-3/2} du$$

If we substitute this expression into the integrand in Equation 39-22, we obtain

$$K = \int_0^u u\, d\left(\frac{m_0 u}{\sqrt{1 - u^2/c^2}}\right) = \int_0^u m_0\left(1 - \frac{u^2}{c^2}\right)^{-3/2} u\, du$$
$$= m_0 c^2\left(\frac{1}{\sqrt{1 - u^2/c^2}} - 1\right)$$

or

$$K = \frac{m_0 c^2}{\sqrt{1 - u^2/c^2}} - m_0 c^2 \qquad \text{39-23}$$

Relativistic kinetic energy

The expression for kinetic energy consists of two terms. The first term depends on the speed of the particle. The second, m_0c^2, is independent of the speed. The quantity m_0c^2 is called the **rest energy** E_0 of the particle. The rest energy is the product of the rest mass and c^2:

$$E_0 = m_0 c^2 \qquad \text{39-24}$$

Rest energy

The total **relativistic energy** E is then defined to be the sum of the kinetic energy and the rest energy:

$$E = K + m_0c^2 = \frac{m_0c^2}{\sqrt{1 - u^2/c^2}} \qquad \text{39-25}$$

Relativistic energy

Thus, the work done by an unbalanced force increases the energy from the rest energy m_0c^2 to the final energy $m_0c^2/\sqrt{1 - u^2/c^2} = m_rc^2$ where $m_r = m_0/\sqrt{1 - u^2/c^2}$ is the relativistic mass. We can obtain a useful expression for the velocity of a particle by multiplying Equation 39-21 for the relativistic momentum by c^2 and comparing the result with Equation 39-25 for the relativistic energy. We have

$$pc^2 = \frac{m_0c^2u}{\sqrt{1 - u^2/c^2}} = Eu$$

or

$$\frac{u}{c} = \frac{pc}{E} \qquad \text{39-26}$$

Energies in atomic and nuclear physics are usually expressed in units of electron volts (eV) or mega-electron-volts (MeV):

$$1\ \text{eV} = 1.6 \times 10^{-19}\ \text{J}$$

A convenient unit for the masses of atomic particles is eV/c^2 or MeV/c^2, which is just the rest energy of the particle divided by c^2. The rest energies of some elementary particles and light nuclei are given in Table 39-1.

Table 39-1

Rest Energies of Some Elementary Particles and Light Nuclei

Particle	Symbol	Rest energy, MeV
Photon	γ	0
Electron (positron)	e or e^- (e^+)	0.5110
Muon	$\mu^\pm$	105.7
Pion	π^0	135
	$\pi^\pm$	139.6
Proton	p	938.280
Neutron	n	939.573
Deuteron	^{2}H or d	1875.628
Triton	^{3}H or t	2808.944
Helium-3	^{3}He	2808.41 MeV
Alpha particle	^{4}He or α	3727.409

Example 39-10

An electron (rest energy 0.511 MeV) moves with speed $u = 0.8c$. Find (*a*) its total energy, (*b*) its kinetic energy, and (*c*) the magnitude of its momentum.

(*a*) The total energy is given by Equation 39-25:

$$E = \frac{m_0c^2}{\sqrt{1 - u^2/c^2}} = \frac{m_0c^2}{\sqrt{1 - 0.64}} = \frac{m_0c^2}{0.6}$$

$$= \frac{0.511\ \text{MeV}}{0.6} = 0.852\ \text{MeV}$$

(*b*)1. The kinetic energy is the total energy minus the rest energy:

$$K = E - m_0c^2 = 0.852\ \text{MeV} - 0.511\ \text{MeV} = 0.341\ \text{MeV}$$

2. The magnitude of the momentum is found from Equation 39-21:

$$p = \frac{m_0u}{\sqrt{1 - u^2/c^2}} = \frac{m_0(0.8c)}{0.6}$$

3. We can simplify by multiplying both numerator and denominator by c:

$$p = \frac{(0.8)m_0c^2}{(0.6)c} = \frac{(1.33)(0.511\ \text{MeV})}{c} = 0.680\ \text{MeV}/c$$

Remark The unit MeV/c is a convenient unit for momentum.

The expression for kinetic energy given by Equation 39-23 doesn't look much like the classical expression $\frac{1}{2}m_0u^2$. However, when u is much less than c, we can approximate $1/\sqrt{1 - u^2/c^2}$ using the binomial expansion

$$(1 + x)^n = 1 + nx + n(n-1)\frac{x^2}{2} + \cdots \approx 1 + nx \qquad \text{39-27}$$

Then

$$\frac{1}{\sqrt{1-u^2/c^2}} = \left(1 - \frac{u^2}{c^2}\right)^{-1/2} \approx 1 + \frac{1}{2}\frac{u^2}{c^2}$$

From this result, when u is much less than c, the expression for relativistic kinetic energy becomes

$$K = m_0c^2\left(\frac{1}{\sqrt{1 - u^2/c^2}} - 1\right) \approx m_0c^2\left(1 + \frac{1}{2}\frac{u^2}{c^2} - 1\right) = \frac{1}{2}m_0u^2$$

Thus, at low speeds, the relativistic expression is the same as the classical expression.

We note from Equation 39-25 that as the speed u approaches the speed of light c, the energy of the particle becomes very large because $1/\sqrt{1 - u^2/c^2}$ becomes very large. At $u = c$, the energy becomes infinite. For u greater than c, $\sqrt{1 - u^2/c^2}$ is the square root of a negative number and is therefore imaginary. A simple interpretation of the result that it takes an infinite amount of energy to accelerate a particle to the speed of light is that no particle that is ever at rest in any inertial reference frame can travel as fast or faster than the speed of light c. As we noted in Example 39-7, if the speed of a particle is less than c in one reference frame, it is less than c in all other reference frames moving relative to that frame at speeds less than c.

In practical applications, the momentum or energy of a particle is often known rather than the speed. Equation 39-21 for the relativistic momentum and Equation 39-25 for the relativistic energy can be combined to eliminate the speed u. (See Problem 52.) The result is

$$E^2 = p^2c^2 + (m_0c^2)^2 \qquad \text{39-28}$$

Relation for total energy, momentum, and rest energy

This useful equation can be conveniently remembered from the right triangle shown in Figure 39-11. If the energy of a particle is much greater than its rest energy m_0c^2, the second term on the right side of Equation 39-28 can be neglected, giving the useful approximation

$$E \approx pc \qquad \text{for } E \gg m_0c^2 \qquad \text{39-29}$$

Equation 39-29 is an exact relation between energy and momentum for particles with no rest mass, such as photons.

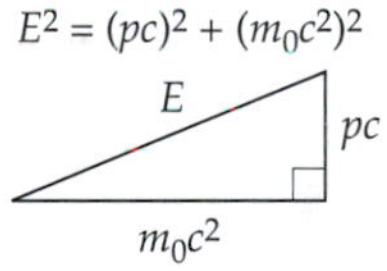

Figure 39-11 Right triangle for remembering Equation 39-28.

Exercise A proton (rest mass 938 MeV/c^2) has a total energy of 1400 MeV. Find (*a*) $1/\sqrt{1 - u^2/c^2}$, (*b*) the momentum of the proton, and (*c*) the speed u of the proton. (*Answers* (*a*) 1.49, (*b*) $p = 1.04 \times 10^3$ MeV/c, (*c*) $u = 0.74c$)

Rest Mass and Energy

Einstein considered Equation 39-24 relating the energy of a particle to its mass to be the most significant result of the theory of relativity. Energy and inertia, which were formerly two distinct concepts, are related through this famous equation. As discussed in Chapter 7, the conversion of rest energy to kinetic energy with a corresponding loss in rest mass is a common occurrence in radioactive decay and nuclear reactions, including nuclear fission and nuclear fusion. We illustrated this in Chapter 7 with the deuteron, whose rest mass is 2.22 MeV/c^2 less than the rest mass of its parts—a proton and neutron. When a neutron and proton combine to form a deuteron, 2.22 MeV of energy is released. The breaking up of a deuteron into a neutron and proton requires 2.22 MeV of energy input. The proton and neutron are thus bound together in a deuteron by a binding energy of 2.22 MeV. Any stable composite particle, such as a deuteron or a helium nucleus (2 neutrons plus 2 protons), that is made up of other particles has a rest mass and rest energy that are less than the sum of the rest masses and rest energies of its parts. The difference in rest energy is the binding energy of the composite particle. The binding energies of atoms and molecules are of the order of a few electron volts, which leads to a negligible difference in mass between the composite particle and its parts. The binding energies of nuclei are of the order of several MeV, which leads to a noticeable difference in mass. Some very heavy nuclei, such as radium, are radioactive and decay into a lighter nucleus plus an alpha particle. In this case, the original nucleus has a rest energy greater than that of the decay particles. The excess energy appears as the kinetic energy of the decay products.

To further illustrate the interrelation of rest mass and energy, we consider a perfectly inelastic collision of two particles. Classically, kinetic energy is lost in such a collision. Relativistically, this loss in kinetic energy shows up as an increase in rest energy of the system, that is, the total energy of the system is conserved. Consider a particle of rest mass $m_{0,1}$ moving with initial speed u_1 that collides with a particle of rest mass $m_{0,2}$ moving with initial speed u_2. The particles collide and stick together, forming a particle of rest mass M_0 that moves with speed u_f, as shown in Figure 39-12. The initial total energy of particle 1 is

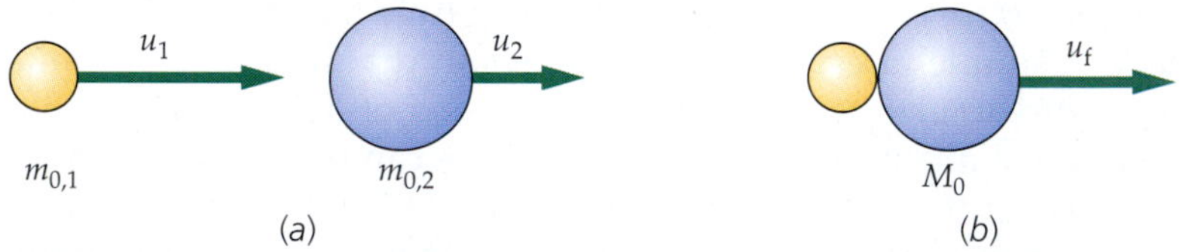

Figure 39-12 A perfectly inelastic collision between two particles. One particle of rest mass $m_{1,0}$ collides with another particle of rest mass $m_{2,0}$. After the collision, the particles stick together, forming a composite particle of rest mass M_0 that moves with speed u_f such that relativistic momentum is conserved. Kinetic energy is lost in this process. If we assume that the total energy is conserved, the loss in kinetic energy must equal c^2 times the increase in the rest mass of the system.

$$E_1 = K_1 + m_{0,1}c^2$$

where K_1 is its initial kinetic energy. Similarly the initial total energy of particle 2 is

$$E_2 = K_2 + m_{0,2}c^2$$

The total initial energy of the system is

$$E_i = E_1 + E_2 = K_1 + m_{0,1}c^2 + K_2 + m_{0,2}c^2 = K_i + (m_{0,1} + m_{0,2})c^2$$

where $K_i = K_1 + K_2$ is the initial kinetic energy of the system. The final total energy of the system is

$$E_f = K_f + M_0c^2$$

If we set the final total energy equal to the initial total energy, we obtain

$$K_f + M_0c^2 = K_i + (m_{0,1} + m_{0,2})c^2$$

The change in kinetic energy is thus

$$K_f - K_i = [M_0 - (m_{0,1} + m_{0,2})]c^2 = (\Delta m_0)c^2 \qquad 39\text{-}30$$

where $\Delta m_0 = M_0 - (m_{0,1} + m_{0,2})$ is the increase in rest mass of the system.

Example 39-11

A particle of rest mass 2 MeV/c^2 and kinetic energy 3 MeV collides with a stationary particle of rest mass 4 MeV/c^2. After the collision, the two particles stick together. Find *(a)* the initial momentum of the system, *(b)* the final velocity of the two-particle system, and *(c)* the rest mass of the two-particle system.

Picture the Problem *(a)* The initial momentum of the system is that of the incoming particle, which can be found from the total energy of the particle. *(b)* The final velocity of the system can be found from its total energy and momentum using $u/c = pc/E$ (Equation 39-26). The energy is found from conservation of energy, and the momentum from conservation of momentum. *(c)* Since the final energy and momentum are known, the final rest mass can be found from $E^2 = p^2c^2 + (M_0c^2)^2$.

(*a*)1. The initial momentum of the incoming particle is related to its energy and rest mass:

$$E^2 = p^2c^2 + (m_0c^2)^2$$
$$pc = \sqrt{E_1^2 - (m_0c^2)^2}$$

2. The total energy of the moving particle is the sum of its kinetic energy and rest energy:

$$E_1 = 3\text{ MeV} + 2\text{ MeV} = 5\text{ MeV}$$

3. Use this total energy to calculate the momentum:

$$pc = \sqrt{E_1^2 - (m_0c^2)^2} = \sqrt{(5\text{ MeV})^2 - (2\text{ MeV})^2} = \sqrt{21}\text{ MeV}$$
$$p = 4.58\text{ MeV}/c$$

(*b*)1. We can find the final velocity of the two-particle system from its total energy E and its momentum p using Equation 39-26:

$$\frac{u}{c} = \frac{pc}{E}$$

2. By the conservation of total energy, the final energy of the system equals the initial total energy of the two particles:

$$E_f = E_i = E_1 + E_2 = 5\text{ MeV} + 4\text{ MeV} = 9\text{ MeV}$$

3. By the conservation of momentum, the final momentum of the two-particle system equals the initial momentum:

$$p = 4.58\text{ MeV}/c$$

4. Calculate the velocity of the two-particle system from its total energy and momentum using $u/c = pc/E$:

$$\frac{u}{c} = \frac{pc}{E} = \frac{4.58\text{ MeV}}{9\text{ MeV}} = 0.509$$
$$u = 0.509c$$

(*c*) We can find the rest mass of the final two-particle system from Equation 39-28 using $pc = 4.58$ MeV and $E = 9$ MeV:

$$E^2 = (pc)^2 + (M_0c^2)^2$$
$$(9\text{ MeV})^2 = (4.58\text{ MeV})^2 + (M_0c^2)^2$$
$$M_0 = 7.75\text{ MeV}/c^2$$

Remarks Note that the rest mass of the system increased from 6 MeV/c^2 to 7.75 MeV/c^2. This increase times c^2 equals the loss in kinetic energy of the system, as you will show in the exercise below.

Exercise *(a)* Find the final kinetic energy of the two-particle system in Example 39-11. *(b)* Find the loss in kinetic energy, K_{loss}, in the collision. *(c)* Show that $K_{loss} = \Delta Mc^2$, where ΔM is the increase in rest mass of the system found in part *(c)* of that example. (*Answers* *(a)* $K_f = E - M_0c^2 = 9\text{ MeV} - 7.75\text{ MeV} = 1.25\text{ MeV}$, *(b)* $K_{loss} = K_i - K_f = 3\text{ MeV} - 1.25\text{ MeV} = 1.75\text{ MeV}$, *(c)* $\Delta Mc^2 = M_0c^2 - M_i\,c^2 = 7.75\text{ MeV} - (2\text{ MeV} + 4\text{ MeV}) = 1.75\text{ MeV} = K_{loss}$)

Example 39-12

A rocket of mass $m_r = 10^6$ kg is coasting through space when it suddenly becomes necessary to accelerate. The rocket ejects 10^3 kg of burned fuel in a very short time at a speed of $c/2$ relative to the rocket. (*a*) Neglecting any change in the rest mass of the system, calculate the speed of the rocket u_r in the frame in which it was initially at rest. (*b*) Use your results from (*a*) to estimate the change in the rest mass of the system. (*c*) Calculate the speed of the rocket using classical, newtonian mechanics.

Picture the Problem The speed of the rocket is calculated from its relativistic momentum. In the frame in which the rocket is initially at rest the total momentum of rocket plus fuel is zero. After the burn, the magnitude of the momentum of the rocket equals that of the ejected fuel. Let m_r be the mass of the rocket *after* ejecting the burned fuel and m_f be the mass of the fuel. Since $m_r \approx 1000m_f$, the loss in fuel has a negligible effect on the mass of the rocket. In part (*b*) we can neglect the change in rest mass of the ejected fuel compared to the change in rest mass of the remaining rocket.

(*a*)1. The speed of the rocket is related to its momentum:

$$p_r = \frac{m_r u_r}{\sqrt{1 - u_r^2/c^2}}$$

2. Set the magnitude of the momentum of the rocket equal to the magnitude of the momentum of the ejected fuel:

$$\frac{m_r u_r}{\sqrt{1 - u_r^2/c^2}} = \frac{m_f u_f}{\sqrt{1 - u_f^2/c^2}}$$

$$= \frac{m_f(0.5c)}{\sqrt{1 - (0.5c)^2/c^2}} = \frac{m_f(0.5c)}{\sqrt{1 - 0.25}} = \frac{m_f c}{\sqrt{3}}$$

3. Solve for u_r:

$$\sqrt{3}m_r u_r = \sqrt{1 - u_r^2/c^2}\, m_f c = m_f\sqrt{c^2 - u_r^2}$$

$$c^2 - u_r^2 = 3(m_r/m_f)^2 u_r^2 = 3 \times 10^6 u_r^2$$

$$u_r = (1/\sqrt{3}) \times 10^{-3}c = 1.73 \times 10^5 \text{ m/s}$$

(*b*)1. The rest mass is related to the total energy. Write the initial energy E_i in terms of m_r and m_f:

$$E_i = m_r c^2 + m_f c^2$$

2. Write the final energy in terms of the final rest mass of the rocket minus fuel m_r' and the final rest mass of the fuel m_f':

$$E_f = \frac{m_r' c^2}{\sqrt{1 - u_r^2/c^2}} + \frac{m_f' c^2}{\sqrt{1 - (0.5c)^2/c^2}}$$

3. Simplify using $u_r \ll c$ and $m_f' \approx m_f$:

$$E_f \approx m_r' c^2 + \frac{2}{\sqrt{3}} m_f c^2$$

4. Apply conservation of energy and solve for m_r':

$$E_f = E_i$$

$$m_r' c^2 + \frac{2}{\sqrt{3}} m_f c^2 = m_r c^2 + m_f c^2$$

$$m_r' = m_r - \left(\frac{2}{\sqrt{3}} - 1\right) m_f = m_r - 0.155 m_f$$

$$\Delta m_r = -0.155 m_f = -155 \text{ kg}$$

(*c*)1. Set the magnitude of the classical momentum of the rocket equal to the magnitude of the classical momentum of the ejected fuel and solve for u_r:

$$m_r u_r = m_f u_f$$

$$u_r = \frac{m_f}{m_r} u_f = \frac{10^3 \text{ kg}}{10^6 \text{ kg}}(0.5c) = 1.5 \times 10^5 \text{ m/s}$$

Remarks The decrease in rest mass is extremely small as it usually is in macroscopic problems. Note that the classical calculation of the speed gives an error of about 13% in this case.

Exploring

General Relativity

The generalization of the theory of relativity to noninertial reference frames by Einstein in 1916 is known as the general theory of relativity. It is much more difficult mathematically than the special theory of relativity, and there are fewer situations in which it can be tested. Nevertheless, its importance calls for a brief qualitative discussion.

The basis of the general theory of relativity is the **principle of equivalence:**

> A homogeneous gravitational field is completely equivalent to a uniformly accelerated reference frame.

Principle of equivalence

This principle arises in Newtonian mechanics because of the apparent identity of gravitational mass and inertial mass. In a uniform gravitational field, all objects fall with the same acceleration $\vec{g}$ independent of their mass because the gravitational force is proportional to the (gravitational) mass, whereas the acceleration varies inversely with the (inertial) mass. Consider a compartment in space undergoing a uniform acceleration $\vec{a}$, as shown in Figure 1*a.* No mechanics experiment can be performed *inside* the compartment that will distinguish whether the compartment is actually accelerating in space or is at rest (or is moving with uniform velocity) in the presence of a uniform gravitational field $\vec{g} = -\vec{a}$, as shown in Figure 1*b*. If objects are dropped in the compartment, they will fall to the "floor" with an acceleration $\vec{g} = -\vec{a}$. If people stand on a spring scale, it will read their "weight" of magnitude ma.

Einstein assumed that the principle of equivalence applies to all physics and not just to mechanics. In effect, he assumed that there is no experiment of any kind that can distinguish uniformly accelerated motion from the presence of a gravitational field.

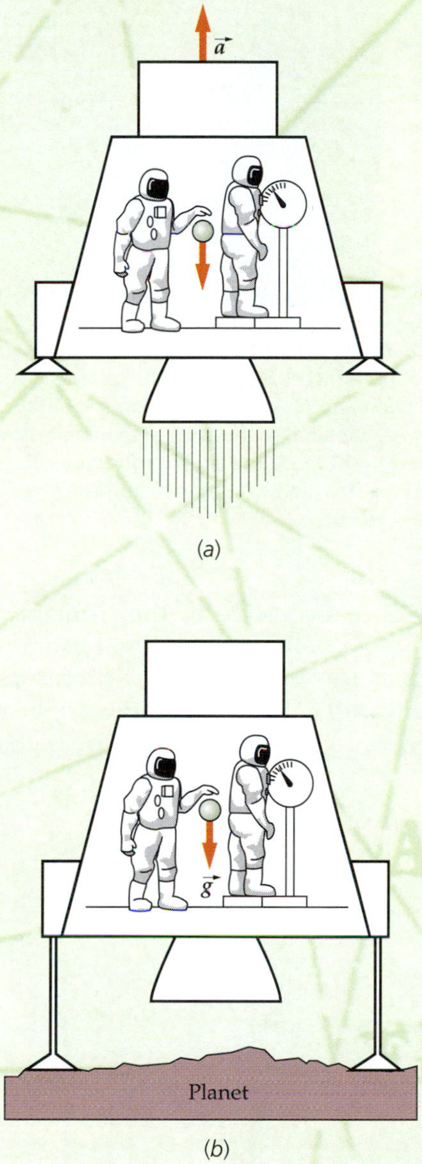

Figure 1 The results of experiments in a uniformly accelerated reference frame (*a*) cannot be distinguished from those in a uniform gravitational field (*b*) if the acceleration $\vec{a}$ and the gravitational field $\vec{g}$ have the same magnitude.

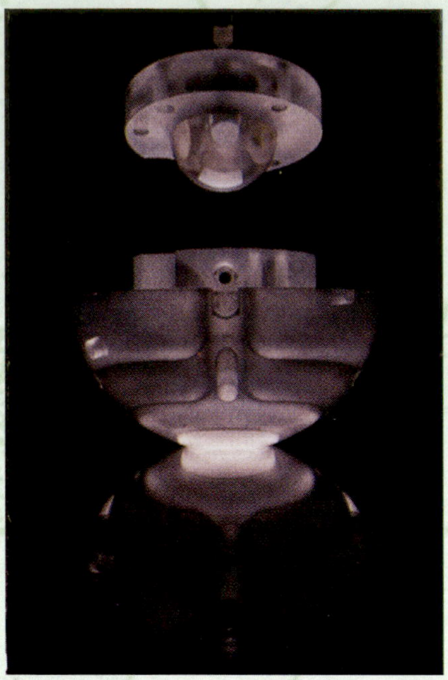

The quartz sphere in the top part of the container is probably the world's most perfectly round object. It is designed to spin as a gyroscope in a satellite orbiting the earth. General relativity predicts that the rotation of the earth will cause the axis of rotation of the gyroscope to precess in a circle at a rate of about 1 revolution in 100,000 years.

One consequence of the principle of equivalence—the deflection of a light beam in a gravitational field—was one of the first to be tested experimentally. Figure 2 shows a beam of light entering a compartment that is accelerating. Successive positions of the compartment at equal time intervals are shown in Figure 2*a*. Because the compartment is accelerating, the distance it moves in each time interval increases with time. The path of the beam of light as observed from inside the compartment is therefore a parabola, as shown in Figure 2*b*. But according to the principle of equivalence, there is no way to distinguish between an accelerating compartment and one moving with uniform velocity in a uniform gravitational field. We conclude, therefore, that a beam of light will accelerate in a gravitational field, just like objects that have mass. For example, near the surface of the earth, light will fall with an acceleration of 9.81 m/s^2. This is difficult to observe because of the enormous speed of light. For example, in a distance of 3000 km, which takes light about 0.01 s to traverse, a beam of light should fall about 0.5 mm. Einstein pointed out that the deflection of a light beam in a gravitational field might be observed when light from a distant star passes close to the sun, as illustrated in Figure 3. Because of the brightness of the sun, such a star cannot ordinarily be seen. Such a deflection was first observed in 1919 during an eclipse of the sun. This well-publicized observation brought instant worldwide fame to Einstein.

A second prediction from Einstein's theory of general relativity, which we will not discuss in detail, is the excess precession of the perihelion of the orbit of Mercury of about 0.01° per century. This effect had been known and unexplained for some time, so, in a sense, explaining it constituted an immediate success of the theory.

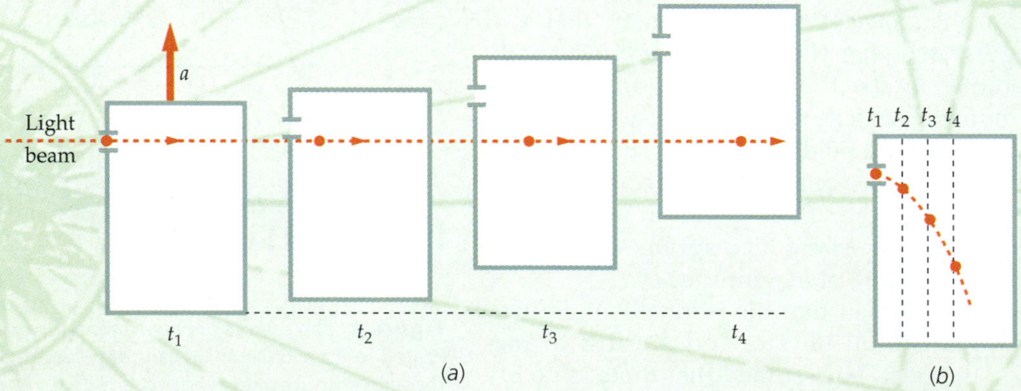

Figure 2 (*a*) A light beam moving in a straight line through a compartment that is undergoing uniform acceleration. The position of the beam is shown at equally spaced times t_1, t_2, t_3, and t_4. (*b*) In the reference frame of the compartment, the light travels in a parabolic path as a ball would if it were projected horizontally. The vertical displacements are greatly exaggerated in both (*a*) and (*b*) for emphasis.

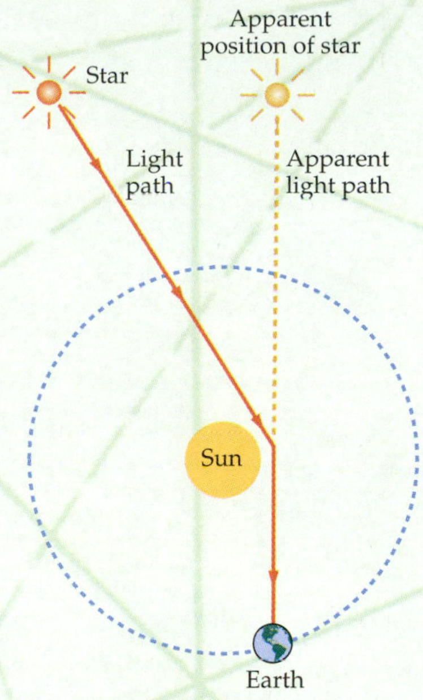

Figure 3 The deflection (greatly exaggerated) of a beam of light due to the gravitational attraction of the sun.

This extremely accurate hydrogen maser clock was launched in a satellite in 1976, and its time was compared to that of an identical clock on earth. In accordance with the prediction of general relativity, the clock on earth, where the gravitational potential was lower, "lost" about 4.3×10^{-10} s each second compared with the clock orbiting the earth at an altitude of about 10,000 km.

A third prediction of general relativity concerns the change in time intervals and frequencies of light in a gravitational field. In Chapter 11, we found that the gravitational potential energy between two masses M and m a distance r apart is

$$U = -\frac{GMm}{r}$$

where G is the universal gravitational constant, and the point of zero potential energy has been chosen to be when the separation of the masses is infinite. The potential energy per unit mass near a mass M is called the *gravitational potential* ϕ:

$$\phi = -\frac{GM}{r} \tag{1}$$

According to the general theory of relativity, clocks run more slowly in regions of low gravitational potential. (Since the gravitational potential is negative, as can be seen from Equation 1, low gravitational potential occurs near the mass, where the *magnitude* of the potential is large.) If Δt_1 is a time interval between two events measured by a clock where the gravitational potential is ϕ_1 and Δt_2 is the interval between the same events as measured by a clock where the gravitational potential is ϕ_2, general relativity predicts that the fractional difference between these times will be approximately*

$$\frac{\Delta t_2 - \Delta t_1}{\Delta t} = \frac{1}{c^2}(\phi_2 - \phi_1) \tag{2}$$

A clock in a region of low gravitational potential will therefore run slower than one in a region of high potential. Since a vibrating atom can be considered to be a clock, the frequency of vibration of an atom in a region of low potential, such as near the sun, will be lower than that of the same atom on earth. This shift toward a lower frequency and therefore a longer wavelength is called the **gravitational redshift.**

As our final example of the predictions of general relativity, we mention **black holes,** which were first predicted by Oppenheimer and Snyder in 1939. According to the general theory of relativity, if the density of an object such as a star is great enough, its gravitational attraction will be so great that once inside a critical radius, nothing

*Since this shift is usually very small, it does not matter by which interval we divide on the left side of the equation.

can escape, not even light or other electromagnetic radiation. (The effect of a black hole on objects outside the critical radius is the same as that of any other mass.) A remarkable property of such an object is that nothing that happens inside it can be communicated to the outside. As sometimes occurs in physics, a simple but incorrect calculation gives the correct results for the relation between the mass and the critical radius of a black hole. In Newtonian mechanics, the speed needed for a particle to escape from the surface of a planet or star of mass M and radius R is given by Equation 11-19:

$$v_e = \sqrt{\frac{2GM}{R}}$$

If we set the escape speed equal to the speed of light and solve for the radius, we obtain the critical radius R_S, called the **Schwarzschild radius:**

$$R_S = \frac{2GM}{c^2} \qquad 3$$

For an object with a mass equal to that of our sun to be a black hole, its radius would have to be about 3 km. Since no radiation is emitted from a black hole and its radius is expected to be small, the detection of a black hole is not easy. The best chance of detection would occur if a black hole were a companion to a normal star in a binary star system. Then both stars would revolve around their center of mass, and the existence of the black hole could be inferred from the motion of the normal star. At present there are several excellent candidates—one in the constellation Cygnus, one in the Small Magellanic Cloud, and perhaps one in our own galaxy—but the evidence is not conclusive.

Summary

Topic	Remarks and Relevant Equations	
1. Einstein Postulates	The special theory of relativity is based on two postulates of Albert Einstein. All of the results of special relativity can be derived from these postulates. Postulate 1. Absolute, uniform motion cannot be detected. Postulate 2. The speed of light is independent of the motion of the source. An important implication of these postulates is Postulate 2 (Alternate). Every observer measures the same value for the speed of light independent of the relative motion of the source and observer.	
2. Lorentz Transformation	$x = \gamma(x' + Vt'), \quad y = y', \quad z = z'$	**39-9**
	$t = \gamma\left(t' + \frac{Vx'}{c^2}\right)$	**39-10**
Inverse transformation	$x' = \gamma(x - Vt), \quad y' = y, \quad z' = z$	**39-11**
	$t' = \gamma\left(t - \frac{Vx}{c^2}\right)$	**39-12**
3. Proper Time and Time Dilation	The time interval measured between two events that occur at the same point in space in some reference frame is called the proper time. In another reference frame in which the events occur at different places, the time interval between the events is longer by the factor γ: $\Delta t = \gamma \Delta t_p$	**39-13**
4. Length Contraction	The length of an object measured in a frame in which it is at rest is called its proper length L_p. When measured in another reference frame, the length of the object is $L = \frac{L_p}{\gamma}$	**39-14**
5. The Relativistic Doppler Effect	$f' = \frac{\sqrt{1 - V^2/c^2}}{1 - V/c} f_0$ approaching	**39-16*a***
	$f' = \frac{\sqrt{1 - V^2/c^2}}{1 + V/c} f_0$ receding	**39-16*b***
6. Clock Synchronization	Two events that are simultaneous in one reference frame are not simultaneous in another frame that is moving relative to the first. If two clocks are synchronized in the frame in which they are at rest, they will be out of synchronization in another frame. In the frame in which they are moving, the chasing clock leads by an amount $\Delta t_S = L_p(V/c^2)$	**39-17**
	where L_p is the proper distance between the clocks.	
7. Velocity Transformation	$u_x = \frac{u'_x + V}{1 + Vu'_x/c^2}$	**39-18*a***
	$u_y = \frac{u'_y}{\gamma(1 + Vu'_x/c^2)}$	**39-18*b***
	$u_z = \frac{u'_z}{\gamma(1 + Vu'_x/c^2)}$	**39-18*c***

	Inverse velocity transformation	$u'_x = \dfrac{u_x - V}{1 - Vu_x/c^2}$	**39-19*a***
		$u'_y = \dfrac{u_y}{\gamma(1 - Vu_x/c^2)}$	**39-19*b***
		$u'_z = \dfrac{u_z}{\gamma(1 - Vu_x/c^2)}$	**39-19*c***
8.	**Relativistic Momentum**	$\vec{p} = \dfrac{m_0\vec{u}}{\sqrt{1 - u^2/c^2}}$ where m_0 is the rest mass of the particle.	**39-21**
9.	**Relativistic Energy**		
	Kinetic energy	$K = \dfrac{m_0c^2}{\sqrt{1 - u^2/c^2}} - m_0c^2 = \dfrac{m_0c^2}{\sqrt{1 - u^2/c^2}} - E_0$	**39-23**
	Rest energy	$E_0 = m_0c^2$	**39-24**
	Total energy	$E = K + E_0 = \dfrac{m_0c^2}{\sqrt{1 - u^2/c^2}}$	**39-25**
10.	**Useful Formulas for Speed, Energy, and Momentum**	$\dfrac{u}{c} = \dfrac{pc}{E}$	**39-26**
		$E^2 = p^2c^2 + (m_0c^2)^2$	**39-28**
		$E \approx pc \qquad \text{for } E \gg m_0c^2$	**39-29**
11.	**Binding Energy**	The total rest mass of bound systems of particles, such as nuclei or atoms, is less than the sum of the rest masses of the particles making up the system. The difference in mass times c^2 equals the binding energy of the system. The binding energy is the energy that must be added to break up the system into its parts. The binding energies of electrons in atoms are of the order of eV or keV, leading to a negligible difference in rest mass. The binding energies of nuclei are of the order of several MeV, and the difference in rest mass is noticeable.	

Problem-Solving Guide

Summary of Worked Examples

	Type of Calculation	Procedure and Relevant Examples
1.	**Lorentz Transformation**	
	Find the spatial separation or time interval between two events in another reference frame.	Use the Lorentz transformation. If the events happen at the same place in one frame, the time interval in that frame is proper time and in another frame $\Delta t = \gamma\,\Delta t_{\text{p}}$. **Examples 39-1, 39-2, 39-3, 39-5**
2.	**Doppler Effect**	
	Find the speed of a moving source from its Doppler shift.	Use $f' = \dfrac{\sqrt{1 - V^2/c^2}}{1 \pm V/c} f_0$ and solve for V/c. **Example 39-4**

3. Clock Synchronization		
Find the amount by which two clocks are unsynchronized.	The chasing clock leads the other by $\Delta t_s = \gamma\, VL_p/c^2$, where L_p is the proper separation of the clocks.	**Example 39-5**
4. Velocity Transformation		
Find the velocity of an object relative to you given its velocity in another frame.	Use the velocity transformation equations.	**Examples 39-6, 39-7, 39-8, 39-9**
5. Momentum and Energy		
Find the total energy, kinetic energy, and momentum of a particle given its rest energy and speed.	Use $E = m_0c^2/\sqrt{1 - u^2/c^2}$, $K = E - E_0$, $p = m_0u/\sqrt{1 - u^2/c^2}$.	**Examples 39-10, 39-11**
Find the velocity and rest energy of a system after an inelastic collision.	Use $E^2 = p^2c^2 + (m_0c^2)^2$ to find the initial momentum of the particles. Then use conservation of momentum and energy. The velocity can be found from $u/c = pc/E$.	**Example 39-11**
Find the speed of a rocket given the mass and speed of its exhaust.	Use conservation of momentum.	**Example 39-12**

Problems

Conceptual Problems

Problems from Optional and Exploring sections

In a few problems, you are given more data than you actually need; in a few other problems, you are required to supply data from your general knowledge, outside sources, or informed estimates.

- • Single-concept, single-step, relatively easy
- •• Intermediate-level, may require synthesis of concepts
- ••• Challenging, for advanced students

Time Dilation and Length Contraction

1 • You are standing on a corner and a friend is driving past in an automobile. Both of you note the times when the car passes two different intersections and determine from your watch readings the time that elapses between the two events. Which of you has determined the proper time interval?

2 • The proper mean lifetime of pions is 2.6×10^{-8} s. If a beam of pions has a speed of $0.85c$, (*a*) what would their mean lifetime be as measured in the laboratory? (*b*) How far would they travel, on average, before they decay? (*c*) What would your answer be to part (*b*) if you neglect time dilation?

3 • (*a*) In the reference frame of the pion in Problem 2, how far does the laboratory travel in a typical lifetime of 2.6×10^{-8} s? (*b*) What is this distance in the laboratory's frame?

4 • The proper mean lifetime of a muon is 2 μs. Muons in a beam are traveling at $0.999c$. (*a*) What is their mean lifetime as measured in the laboratory? (*b*) How far do they travel, on average, before they decay?

5 • (*a*) In the reference frame of the muon in Problem 4, how far does the laboratory travel in a typical lifetime of 2 μs? (*b*) What is this distance in the laboratory's frame?

6 • Jay has been posted to a remote region of space to monitor traffic. Toward the end of a quiet shift, a spacecraft goes by, and he measures its length using a laser device, which reports a length of 85 m. He flips open his handy reference catalogue and identifies the craft as a CCCNX-22, which has a proper length of 100 m. When he phones in his report, what speed should Jay give for this spacecraft?

7 • A spaceship travels to a star 95 light-years away at a speed of 2.2×10^8 m/s. How long does it take to get there (*a*) as measured on earth and (*b*) as measured by a passenger on the spaceship?

8 • The mean lifetime of a pion traveling at high speed is measured to be 7.5×10^{-8} s. Its lifetime when measured at rest is 2.6×10^{-8} s. How fast is the pion traveling?

9 • A meterstick moves with speed $V = 0.8c$ relative to you in the direction parallel to the stick. (*a*) Find the length of the stick as measured by you. (*b*) How long does it take for the stick to pass you?

10 • The half-life of charged pions, π^+ and π^-, is 1.8×10^{-8} s; i.e., in the rest frame of the pions if there are N pions at time $t = 0$, there will be only $N/2$ pions at time $t = 1.8 \times 10^{-8}$ s. Pions are produced in an accelerator and emerge with a speed of $0.998c$. How far do these particles travel in the laboratory before half of them have decayed?

11 •• A friend of yours who is the same age as you travels to the star Alpha Centauri, which is 4 light-years away and returns immediately. He claims that the entire trip took just 6 y. How fast did he travel?

12 •• Two spaceships pass each other traveling in opposite directions. A passenger in ship A, which she knows to be 100 m long, notes that ship B is moving with a speed of $0.92c$ relative to A and that the length of B is 36 m. What are the lengths of the two spaceships as measured by a passenger in ship B?

13 •• In the Stanford linear collider, small bundles of electrons and positrons are fired at each other. In the laboratory's frame of reference, each bundle is about 1 cm long and 10 μm in diameter. In the collision region, each particle has an energy of 50 GeV, and the electrons and positrons are moving in opposite directions. (*a*) How long and how wide is each bundle in its own reference frame? (*b*) What must be the minimum proper length of the accelerator for a bundle to have both its ends simultaneously in the accelerator in its own reference frame? (The actual length of the accelerator is less than 1000 m.) (*c*) What is the length of a positron bundle in the reference frame of the electron bundle?

The Lorentz Transformation

14 • Use the binomial expansion

$$(1 + x)^n = 1 + nx + n(n-1)\frac{x^2}{2} + \cdots \approx 1 + nx$$

to derive the following results for the case when V is much less than c, and use the results when applicable in the following problems:

(*a*) $\gamma \approx 1 + \frac{1}{2}\frac{V^2}{c^2}$

(*b*) $\frac{1}{\gamma} \approx 1 - \frac{1}{2}\frac{V^2}{c^2}$

(*c*) $\gamma - 1 \approx 1 - \frac{1}{\gamma} \approx \frac{1}{2}\frac{V^2}{c^2}$

15 •• Show that when $V \ll c$ the transformation equations for x, t, and u reduce to the Galilean equations.

16 •• Supersonic jets achieve maximum speeds of about $(3 \times 10^{-6})c$. (*a*) By what percentage would you see a jet traveling at this speed contracted in length? (*b*) During a time of 1 y $= 3.15 \times 10^7$ s on your clock, how much time would elapse on the pilot's clock? How many minutes are lost by the pilot's clock in 1 y of your time?

17 •• How great must the relative speed of two observers be for the time-interval measurements to differ by 1%? (See Problem 14.)

18 •• A spaceship of proper length $L' = 400$ m moves past a transmitting station at a speed of $0.76c$. At the instant that the nose of the ship passes the transmitter, clocks at the transmitter and in the nose of the ship are synchronized to $t = t' = 0$. The instant that the tail of the ship passes the transmitter a signal is sent and subsequently detected by the receiver in the nose of the ship. (*a*) When, according to the clock in the ship, is the signal sent? (*b*) When, according to the clock at the transmitter, is the signal received by the spaceship? (*c*) When, according to the clock in the ship, is the signal received? (*d*) Where, according to an observer at the transmitter, is the nose of the spaceship when the signal is received?

19 •• A beam of unstable particles emerges from the exit slit of an accelerator with a speed of $0.89c$. Particle detectors 3.0 and 6.0 m from the exit slit measure beam intensities of 2×10^8 particles/cm^2·s and 5×10^7 particles/cm^2·s, respectively. (*a*) Find the proper half-life of the particles. (*b*) Determine the beam intensity at the exit slit of the accelerator. (*c*) The accelerator is adjusted so that the particles emerge from the exit slit with a speed of $0.96c$. The beam intensity at the farther detector is again 5×10^7 particles/cm^2·s. Find the beam intensity at the exit slit of the accelerator.

20 •• Show that if u'_x and V in Equation 39-18*a* are both less than c, then u_x is less than c. (*Hint:* Let $u'_x = (1 - \epsilon_1)c$ and $V = (1 - \epsilon_2)c$, where ϵ_1 and ϵ_2 are small positive numbers that are less than 1.)

21 ••• Two events in S are separated by a distance $D = x_2 - x_1$ and a time $T = t_2 - t_1$. (*a*) Use the Lorentz transformation to show that in frame S', which is moving with speed V relative to S, the time separation is $t'_2 - t'_1 = \gamma(T - VD/c^2)$. (*b*) Show that the events can be simultaneous in frame S' only if D is greater than cT. (*c*) If one of the events is the *cause* of the other, the separation D must be less than cT, since D/c is the smallest time that a signal can take to travel from x_1 to x_2 in frame S. Show that if D is less than cT, t'_2 is greater than t'_1 in all reference frames. This shows that if the cause precedes the effect in one frame, it must precede it in all reference frames. (*d*) Suppose that a signal could be sent with speed $c' > c$ so that in frame S the cause precedes the effect by the time $T = D/c'$. Show that there is then a reference frame moving with speed V less than c in which the effect precedes the cause.

Clock Synchronization and Simultaneity

22 • If event A occurs before event B in some frame, might it be possible for there to be a reference frame in which event B occurs before event A?

23 • Two events are simultaneous in a frame in which they also occur at the same point in space. Are they simultaneous in other reference frames?

24 •• Two observers are in relative motion. In what circumstances can they agree on the simultaneity of two different events?

Problems 25 through 29 refer to the following situation: An observer in S' lays out a distance $L' = 100$ light-minutes between points A' and B' and places a flashbulb at the midpoint C'. She arranges for the bulb to flash and for clocks at A' and B' to be started at zero when the light from the flash reaches them (see Figure 39-13). Frame S' is moving to the right with speed $0.6c$ relative *to an observer C in S who is at the midpoint between A′ and B′ when the bulb flashes. At the instant he sees the flash, observer C sets his clock to zero.*

Figure 39-13
Problems 25 through 29

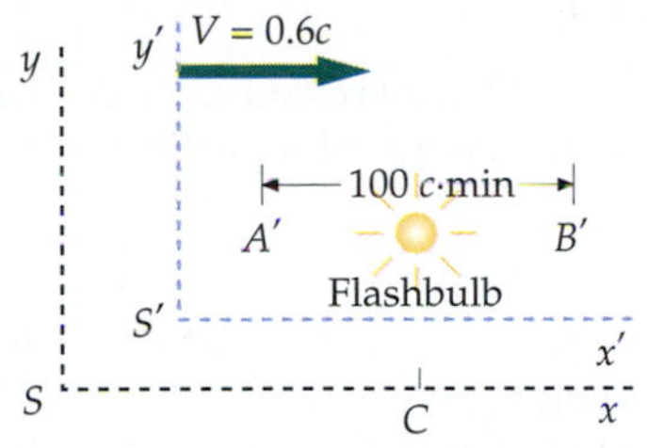

25 •• What is the separation distance between clocks A' and B' according to the observer in S?

26 •• As the light pulse from the flashbulb travels toward A' with speed c, A' travels toward C with speed $0.6c$. Show that the clock in S reads 25 min when the flash reaches A'. (*Hint*: In time t, the light travels a distance ct and A' travels $0.6ct$. The sum of these distances must equal the distance between A' and the flashbulb as seen in S.)

27 •• Show that the clock in S reads 100 min when the light flash reaches B', which is traveling away from C with speed $0.6c$. (See the hint for Problem 26.)

28 •• The time interval between the reception of the flashes at A' and B' in Problems 26 and 27 is 75 min according to the observer in S. How much time does he expect to have elapsed on the clock at A' during this 75-min interval?

29 •• The time interval calculated in Problem 28 is the amount that the clock at A' leads that at B' according to the observer in S. Compare this result with L_pV/c^2.

30 •• In frame S, event B occurs 2 μs after event A, which occurs at $\Delta x = 1.5$ km from event A. How fast must an observer be moving along the $+x$ axis so that events A and B occur simultaneously? Is it possible for event B to precede event A for some observer?

31 •• Observers in reference frame S see an explosion located at $x_1 = 480$ m. A second explosion occurs 5 μs later at $x_2 = 1200$ m. In reference frame S', which is moving along the $+x$ axis at speed V, the explosions occur at the same point in space. What is the separation in time between the two explosions as measured in S'?

The Doppler Effect

32 • How fast must you be moving toward a red light ($\lambda = 650$ nm) for it to appear green ($\lambda = 525$ nm)?

33 • A distant galaxy is moving away from us at a speed of 1.85×10^7 m/s. Calculate the fractional redshift $(\lambda' - \lambda_0)/\lambda_0$ in the light from this galaxy.

34 • Sodium light of wavelength 589 nm is emitted by a source that is moving toward the earth with speed V. The wavelength measured in the frame of the earth is 620 nm. Find V.

35 • A student on earth hears a tune on her radio that seems to be coming from a record that is being played too fast. She has a 33-rev/min record of that tune and determines that the tune sounds the same as when her record is played at 78 rev/min, that is, the frequencies are all too high by a factor of 78/33. If the tune is being played correctly, but is being broadcast by a spaceship that is approaching the earth at speed V, determine V.

36 •• Derive Equation 39-16*a* for the frequency received by an observer moving with speed V toward a stationary source of electromagnetic waves.

Exploring . . . The Twin Paradox

37 • Herb and Randy are twin jazz musicians who perform as a trombone–saxophone duo. At the age of twenty, however, Randy got an irresistible offer to join a road trip to perform on a star 15 light-years away. To celebrate his bounteous luck, he bought a new vehicle for the trip—a deluxe space-coupe which could do $0.999c$. Each of the twins promises to practice diligently, so they can reunite afterward. Randy's gig goes so fabulously well, however, that he stays for a full 10 years before returning to Herb. After their reunion, (*a*) how many years of practice will Randy have? (*b*) how many years of practice will Herb have?

38 •• A clock is placed in a satellite that orbits the earth with a period of 90 min. By what time interval will this clock differ from an identical clock on earth after 1 y? (Assume that special relativity applies and neglect general relativity.)

39 •• A and B are twins. A travels at $0.6c$ to Alpha Centauri (which is 4 $c\cdot$y from earth as measured in the reference frame of the earth) and returns immediately. Each twin sends the other a light signal every 0.01 y as measured in her own reference frame. (*a*) At what rate does B receive signals as A is moving away from her? (*b*) How many signals does B receive at this rate? (*c*) How many total signals are received by B before A has returned? (*d*) At what rate does A receive signals as B is receding from her? (*e*) How many signals does A receive at this rate? (*f*) How many total signals are received by A? (*g*) Which twin is younger at the end of the trip, and by how many years?

The Velocity Transformation

40 • A light beam moves along the y' axis with speed c in frame S', which is moving to the right with speed V relative to frame S. (*a*) Find the x and y components of the velocity of the light beam in frame S. (*b*) Show that the magnitude of the velocity of the light beam in S is c.

41 • A spaceship is moving east at speed $0.90c$ relative to the earth. A second spaceship is moving west at speed $0.90c$ relative to the earth. What is the speed of one spaceship relative to the other?

42 •• Two spaceships are approaching each other. (*a*) If the speed of each is $0.6c$ relative to the earth, what is the speed of one relative to the other? (*b*) If the speed of each relative to the earth is 30,000 m/s (about 100 times the speed of sound), what is the speed of one relative to the other?

43 •• A particle moves with speed $0.8c$ along the x'' axis of frame S'', which moves with speed $0.8c$ along the x' axis relative to frame S'. Frame S' moves with speed $0.8c$ along the x axis relative to frame S. (*a*) Find the speed of the particle relative to frame S'. (*b*) Find the speed of the particle relative to frame S.

Energy and Momentum

44 • The approximate total energy of a particle of mass m moving at speed $u \ll c$ is

(*a*) mc^2.
(*b*) $\frac{1}{2}mu^2$.
(*c*) cmu.
(*d*) $\frac{1}{2}mc^2$.
(*e*) $\frac{1}{2}cmu$.

45 • Find the ratio of the total energy to the rest energy of a particle of rest mass m_0 moving with speed (*a*) $0.1c$, (*b*) $0.5c$, (*c*) $0.8c$, and (*d*) $0.99c$.

46 • A proton (rest energy 938 MeV) has a total energy of 1400 MeV. (*a*) What is its speed? (*b*) What is its momentum?

47 • How much energy would be required to accelerate a particle of mass m_0 from rest to (*a*) $0.5c$, (*b*) $0.9c$, and (*c*) $0.99c$? Express your answers as multiples of the rest energy.

48 • If the kinetic energy of a particle equals its rest energy, what error is made by using $p = m_0u$ for its momentum?

49 • What is the energy of a proton whose momentum is $3m_0c$?

50 •• A particle with momentum of 6 MeV/c has total energy of 8 MeV. (*a*) Determine the rest mass of the particle. (*b*) What is the energy of the particle in a reference frame in which its momentum is 4 MeV/c? (*c*) What are the relative velocities of the two reference frames?

51 •• Show that

$$d\left(\frac{m_0u}{\sqrt{1 - u^2/c^2}}\right) = m_0\left(1 - \frac{u^2}{c^2}\right)^{-3/2} du$$

52 •• Use Equations 39-21 and 39-25 to derive the equation $E^2 = p^2c^2 + (m_0c^2)^2$.

53 •• Use the binomial expansion (Equation 39-27) and Equation 39-28 to show that when $pc << m_0c^2$, the total energy is given approximately by

$$E \approx m_0c^2 + \frac{p^2}{2m_0}$$

54 •• (*a*) Show that the speed u of a particle of mass m_0 and total energy E is given by

$$\frac{u}{c} = \left[1 - \frac{(m_0c^2)^2}{E^2}\right]^{1/2}$$

and that when E is much greater than m_0c^2, this can be approximated by

$$\frac{u}{c} \approx 1 - \frac{(m_0c^2)^2}{2E^2}$$

Find the speed of an electron with kinetic energy of (*b*) 0.51 MeV and (*c*) 10 MeV.

55 •• The rest energy of a proton is about 938 MeV. If its kinetic energy is also 938 MeV, find (*a*) its momentum and (*b*) its speed.

56 •• What percentage error is made in using $\frac{1}{2}m_0u^2$ for the kinetic energy of a particle if its speed is (*a*) $0.1c$ and (*b*) $0.9c$?

57 •• The K^0 particle has a rest mass of 497.7 MeV/c^2. It decays into a π^- and π^+, each with rest mass 139.6 MeV/c^2. Following the decay of a K^0, one of the pions is at rest in the laboratory. Determine the kinetic energy of the other pion and of the K^0 prior to the decay.

58 •• The sun radiates energy at the rate of about 4×10^{26} W. Assume that this energy is produced by a reaction whose net result is the fusion of 4 H nuclei to form 1 He nucleus, with the release of 25 MeV for each He nucleus formed. Calculate the sun's loss of rest mass per day.

59 •• Two protons approach each other head on at $0.5c$ relative to reference frame S'. (*a*) Calculate the total kinetic energy of the two protons as seen in frame S'. (*b*) Calculate the total kinetic energy of the protons as seen in reference frame S, which is moving with speed $0.5c$ relative to S' such that one of the protons is at rest.

60 •• An antiproton $\bar{p}$ has the same rest energy as a proton. It is created in the reaction $p + p \rightarrow p + p + p + \bar{p}$. In an experiment, protons at rest in the laboratory are bombarded with protons of kinetic energy K_L, which must be great enough so that kinetic energy equal to $2m_0c^2$ can be converted into the rest energy of the two particles. In the frame of the laboratory, the total kinetic energy cannot be converted into rest energy because of conservation of momentum. However, in the zero-momentum reference frame in which the two initial protons are moving toward each other with equal speed u, the total kinetic energy can be converted into rest energy. (*a*) Find the speed of each proton u such that the total kinetic energy in the zero-momentum frame is $2m_0c^2$. (*b*) Transform to the laboratory's frame in which one proton is at rest, and find the speed u' of the other proton. (*c*) Show that the kinetic energy of the moving proton in the laboratory's frame is $K_L = 6m_0c^2$.

61 ••• A particle of rest mass 1 MeV/c^2 and kinetic energy 2 MeV collides with a stationary particle of rest mass 2 MeV/c^2. After the collision, the particles stick together. Find (*a*) the speed of the first particle before the collision, (*b*) the total energy of the first particle before the collision, (*c*) the initial total momentum of the system, (*d*) the total kinetic energy after the collision, and (*e*) the rest mass of the system after the collision.

Exploring . . . General Relativity

62 • A set of twins work in an office building. One works on the top floor and the other works in the basement. Considering general relativity, which one will age more quickly?

(*a*) They will age at the same rate.

(*b*) The twin who works on the top floor will age more quickly.
(*c*) The twin who works in the basement will age more quickly.
(*d*) It depends on the speed of the office building.
(*e*) None of these is correct.

63 ••• A horizontal turntable rotates with angular speed ω. There is a clock at the center of the turntable and one at a distance r from the center. In an inertial reference frame, the clock at distance r is moving with speed $u = r\omega$. (*a*) Show that from time dilation according to special relativity, time intervals Δt_0 for the clock at rest and Δt_r for the moving clock are related by

$$\frac{\Delta t_r - \Delta t_0}{\Delta t_0} \approx -\frac{r^2\omega^2}{2c^2} \qquad \text{if } r\omega \ll c$$

(*b*) In a reference frame rotating with the table, both clocks are at rest. Show that the clock at distance r experiences a pseudoforce $F_r = mr\omega^2$ in this accelerated frame and that this is equivalent to a difference in gravitational potential between r and the origin of $\phi_r - \phi_0 = \frac{1}{2}r^2\omega^2$. Use this potential difference in Equation 2 to show that in this frame the difference in time intervals is the same as in the inertial frame.

General Problems

64 • True or false:
(*a*) The speed of light is the same in all reference frames.
(*b*) Proper time is the shortest time interval between two events.
(*c*) Absolute motion can be determined by means of length contraction.
(*d*) The light-year is a unit of distance.
(*e*) Simultaneous events must occur at the same place.
(*f*) If two events are not simultaneous in one frame, they cannot be simultaneous in any other frame.
(*g*) If two particles are tightly bound together by strong attractive forces, the rest mass of the system is less than the sum of the masses of the individual particles when separated.

65 • An observer sees a system consisting of a mass oscillating on the end of a spring moving past at a speed u and notes that the period of the system is T. Another observer, who is moving with the mass–spring system, also measures its period. The second observer will find a period that is

(*a*) equal to T.
(*b*) less than T.
(*c*) greater than T.
(*d*) either (*a*) or (*b*) depending on whether the system was approaching or receding from the first observer.
(*e*) There is not sufficient information to answer the question.

66 • The Lorentz transformation for y and z is the same as the classical result: $y = y'$ and $z = z'$. Yet the relativistic velocity transformation does not give the classical result $u_y = u'_y$ and $u_z = u'_z$. Explain.

67 • A spaceship departs from earth for the star Alpha Centauri, which is 4 light-years away. The spaceship travels at $0.75c$. How long does it take to get there (*a*) as measured on earth and (*b*) as measured by a passenger on the spaceship?

68 • The total energy of a particle is twice its rest energy. (*a*) Find u/c for the particle. (*b*) Show that its momentum is given by $p = \sqrt{3}\, m_0c$.

69 • How fast must a muon travel so that its mean lifetime is 46 μs if its mean lifetime at rest is 2 μs?

70 • A distant galaxy is moving away from the earth with a speed that results in each wavelength received on earth being shifted such that $\lambda' = 2\lambda_0$. Find the speed of the galaxy relative to the earth.

71 • How fast must a meterstick travel relative to you in the direction parallel to the stick so that its length as measured by you is 50 cm?

72 • Show that if V is much less than c, the Doppler shift is given approximately by $\Delta f/f \approx \pm V/c$.

73 •• If a plane flies at a speed of 2000 km/h, for how long must it fly before its clock loses 1 s because of time dilation?

74 • The radius of the orbit of a charged particle in a magnetic field is related to the momentum of the particle by

$$p = BqR \qquad \text{39-41}$$

This equation holds classically for $p = mu$ and relativistically for $p = m_0u/\sqrt{1 - u^2/c^2}$. An electron with kinetic energy of 1.50 MeV moves in a circular orbit perpendicular to a uniform magnetic field $B = 5 \times 10^{-3}$ T. (*a*) Find the radius of the orbit. (*b*) What result would you obtain if you used the classical relations $p = mu$ and $K = p^2/2m$?

75 •• Oblivious to economics and politics, Professor Spenditt proposes building a circular accelerator around the earth's circumference using bending magnets that provide a magnetic field of magnitude 1.5 T. (*a*) What would be the kinetic energy of protons orbiting in this field in a circle of radius R_E? (See Problem 74.) (*b*) What would be the period of rotation of these protons?

76 •• Frames S and S' are moving relative to each other along the x and x' axis. Observers in the two frames set their clocks to $t = 0$ when the origins coincide. In frame S, event 1 occurs at $x_1 = 1.0\ c\cdot$y and $t_1 = 1$ y and event 2 occurs at $x_2 = 2.0\ c\cdot$y and $t_2 = 0.5$ y. These events occur simultaneously in frame S'. (*a*) Find the magnitude and direction of the velocity of S' relative to S. (*b*) At what time do both these events occur as measured in S'?

77 •• An interstellar spaceship travels from the earth to a distant star system 12 light-years away (as measured in the earth's frame). The trip takes 15 y as measured on the ship. (*a*) What is the speed of the ship relative to the earth? (*b*) When the ship arrives, it sends a signal to the earth. How long after the ship leaves the earth will it be before the earth receives the signal?

78 •• The neutral pion π^0 has a rest mass of 135 MeV/c^2. This particle can be created in a proton–proton collision:

$$\text{p} + \text{p} \rightarrow \text{p} + \text{p} + \pi^0$$

Determine the threshold kinetic energy for the creation of a π^0 in a collision of a moving and stationary proton. (See Problem 60.)

79 •• A rocket with a proper length of 1000 m moves in the $+x$ direction at $0.6c$ with respect to an observer on the ground. An astronaut stands at the rear of the rocket and fires a bullet toward the front of the rocket at $0.8c$ relative to the rocket. How long does it take the bullet to reach the front of the rocket (*a*) as measured in the frame of the rocket, (*b*) as measured in the frame of the ground, and (*c*) as measured in the frame of the bullet?

80 ••• In a simple thought experiment, Einstein showed that there is mass associated with electromagnetic radiation. Consider a box of length L and mass M resting on a frictionless surface. At the left wall of the box is a light source that emits radiation of energy E, which is absorbed at the right wall of the box. According to classical electromagnetic theory, this radiation carries momentum of magnitude $p = E/c$ (Equation 32-13). (*a*) Find the recoil velocity of the box such that momentum is conserved when the light is emitted. (Since p is small and M is large, you may use classical mechanics.) (*b*) When the light is absorbed at the right wall of the box, the box stops, so the total momentum remains zero. If we neglect the very small velocity of the box, the time it takes for the radiation to travel across the box is $\Delta t = L/c$. Find the distance moved by the box in this time. (*c*) Show that if the center of mass of the system is to remain at the same place, the radiation must carry mass $m = E/c^2$.

81 ••• A rocket with a proper length of 700 m is moving to the right at a speed of $0.9c$. It has two clocks, one in the nose and one in the tail, that have been synchronized in the frame of the rocket. A clock on the ground and the nose clock on the rocket both read $t = 0$ as they pass. (*a*) At $t = 0$, what does the tail clock on the rocket read as seen by an observer on the ground? When the tail clock on the rocket passes the ground clock, (*b*) what does the tail clock read as seen by an observer on the ground, (*c*) what does the nose clock read as seen by an observer on the ground, and (*d*) what does the nose clock read as seen by an observer on the rocket? (*e*) At $t = 1$ h, as measured on the rocket, a light signal is sent from the nose of the rocket to an observer standing by the ground clock. What does the ground clock read when the observer receives this signal? (*f*) When the observer on the ground receives the signal, he sends a return signal to the nose of the rocket. When is this signal received at the nose of the rocket as seen on the rocket?

82 ••• An observer in frame S standing at the origin observes two flashes of colored light separated spatially by $\Delta x = 2400$ m. A blue flash occurs first, followed by a red flash 5 μs later. An observer in S' moving along the x axis at speed V relative to S also observes the flashes 5 μs apart and with a separation of 2400 m, but the red flash is observed first. Find the magnitude and direction of V.

83 ••• Reference frame S' is moving along the x' axis at $0.6c$ relative to frame S. A particle that is originally at $x' = 10$ m at $t_1' = 0$ is suddenly accelerated and then moves at a constant speed of $c/3$ in the $-x'$ direction until time $t_2' = 60$ m/c, when it is suddenly brought to rest. As observed in frame S, find (*a*) the speed of the particle, (*b*) the distance and direction the particle traveled from t_1' to t_2', and (*c*) the time the particle traveled.

84 ••• In reference frame S the acceleration of a particle is $\vec{a} = a_x\hat{i} + a_y\hat{j} + a_z\hat{k}$. Derive expressions for the acceleration components a_x', a_y', and a_z' of the particle in reference frame S' that is moving relative to S in the x direction with velocity V.

85 ••• When a projectile particle with kinetic energy greater than the threshold kinetic energy K_{th} strikes a stationary target particle, one or more particles may be created in the inelastic collision. Show that the threshold kinetic energy of the projectile is given by

$$K_{th} = \frac{(\Sigma\, m_{in} + \Sigma\, m_{fin})(\Sigma\, m_{fin} - \Sigma\, m_{in})c^2}{2m_{target}}$$

Here Σm_{in} is the sum of the rest masses of the projectile and target particles, $\Sigma\, m_{fin}$ is the sum of the rest masses of the final particles, and m_{target} is the rest mass of the target particle. Use this expression to determine the threshold kinetic energy of protons incident on a stationary proton target for the production of a proton–antiproton pair; compare your result with that of Problem 60.

86 ••• A particle of rest mass M_0 decays into two identical particles of rest mass m_0, where $m_0 = 0.3M_0$. Prior to the decay, the particle of rest mass M_0 has an energy of $4M_0c^2$ in the laboratory. The velocities of the decay products are along the direction of motion of M_0. Find the velocities of the decay products in the laboratory.

87 ••• A stick of proper length L_p makes an angle θ with the x axis in frame S. Show that the angle θ' made with the x' axis in frame S', which is moving along the $+x$ axis with speed V, is given by $\tan\theta' = \gamma\tan\theta$ and that the length of the stick in S' is

$$L' = L_p\left(\frac{1}{\gamma^2}\cos^2\theta + \sin^2\theta\right)^{1/2}$$

88 ••• Show that if a particle moves at an angle θ with the x axis with speed u in frame S, it moves at an angle θ' with the x' axis in S' given by

$$\tan\theta' = \frac{\sin\theta}{\gamma(\cos\theta - V/u)}$$

89 ••• For the special case of a particle moving with speed u along the y axis in frame S, show that its momentum and energy in frame S' are related to its momentum and energy in S by the transformation equations

$$P_x' = \gamma\left(p_x - \frac{VE}{c^2}\right), \quad p_y' = p_y, \quad p_z' = p_z$$

$$\frac{E'}{c} = \gamma\left(\frac{E}{c} - \frac{Vp_x}{c^2}\right)$$

Compare these equations with the Lorentz transformation for x', y', z', and t'. These equations show that the quantities p_x, p_y, p_z, and E/c transform in the same way as do x, y, z, and ct.

90 ••• The equation for the spherical wavefront of a light pulse that begins at the origin at time $t = 0$ is $x^2 + y^2 + z^2 -$

$(ct)^2 = 0$. Using the Lorentz transformation, show that such a light pulse also has a spherical wavefront in frame S' by showing that $x'^2 + y'^2 + z'^2 - (ct')^2 = 0$ in S'.

91 ••• In Problem 90, you showed that the quantity $x^2 + y^2 + z^2 - (ct)^2$ has the same value (0) in both S and S'. Such a quantity is called an *invariant*. From the results of Problem 89, the quantity $p_x^2 + p_y^2 + p_z^2 - (E/c)^2)$ must also be an invariant. Show that this quantity has the value $-m_0c^2$ in both the S and S' reference frames.

92 ••• Two identical particles of rest mass m_0 are each moving toward the other with speed u in frame S. The particles collide inelastically with a spring that locks shut (Figure 39-14) and come to rest in S, and their initial kinetic energy is transformed into potential energy. In this problem you are going to show that the conservation of momentum in reference frame S', in which one of the particles is initially at rest, requires that the total rest mass of the system after the collision be $2m_0/\sqrt{1 - u^2/c^2}$. (*a*) Show that the speed of the particle not at rest in frame S' is $u' = 2u/(1 + u^2/c^2)$ and use this result to show that

$$\sqrt{1 - \frac{u'^2}{c^2}} = \frac{1 - u^2/c^2}{1 + u^2/c^2}$$

(*b*) Show that the initial momentum in frame S' is $p' = 2m_0u/(1 - u^2/c^2)$. (*c*) After the collision, the composite particle moves with speed u in S' (since it is at rest in S). Write the total momentum after the collision in terms of the final rest mass M_0, and show that the conservation of momentum implies that $M_0 = 2m_0/\sqrt{1 - u^2/c^2}$. (*d*) Show that the total energy is conserved in each reference frame.

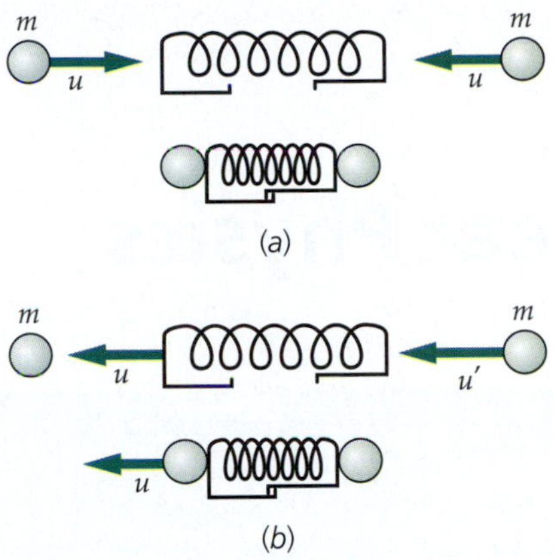

Figure 39-14 An inelastic collision between two identical objects (*a*) in the zero-momentum reference frame S and (*b*) in frame S', which is moving to the right with speed $V = u$ relative to frame S such that one of the particles is initially at rest. The spring, which is assumed to be massless, is merely a device for visualizing the storage of potential energy.

CHAPTER 40

Nuclear Physics

A nuclear power plant in Germany. The fission reactor core is housed in a hemispherical containment structure (center). Two large cooling towers are to its left.

To the chemist, the atomic nucleus is essentially a point charge that contains most of the mass of the atom. It plays a negligible role in the structure of atoms and molecules. When examined closely, the nucleus is found to contain protons and neutrons, whose interactions have played important roles in our everyday life as well as in the history and structure of the universe. The fission of very heavy nuclei such as uranium is a major source of power today, while the fusion of very light nuclei is the energy source that powers the stars, including our sun, and may hold the key to our energy needs of the future.

40-1 Properties of Nuclei

The nucleus of an atom contains just two kinds of particles: protons and neutrons,* which have approximately the same mass (the neutron is about 0.2% more massive). The proton has a charge of $+e$ and the neutron is uncharged. The number of protons, Z, is the atomic number of the atom, which also equals the number of electrons in the atom. The number of neutrons, N, is ap-

* The normal hydrogen nucleus contains a single proton.

proximately equal to Z for light nuclei, and for heavier nuclei is increasingly greater than Z. The total number of nucleons* $A = N + Z$ is called the mass number of the nucleus. A particular nuclear species is called a **nuclide.** Two or more nuclides with the same atomic number Z but different N and A numbers are called **isotopes.** A particular nuclide is designated by its atomic symbol (H for hydrogen, He for helium, etc.) with the mass number A as a presuperscript. The lightest element, hydrogen, has three isotopes: ordinary hydrogen, ^{1}H, whose nucleus is just a single proton; deuterium, ^{2}H, whose nucleus contains one proton and one neutron; and tritium, ^{3}H, whose nucleus contains one proton and two neutrons. Although the mass of the deuterium atom is about twice that of the hydrogen atom and that of the tritium atom about three times that of hydrogen, these three atoms have nearly identical chemical properties because they each have one electron. On the average, there are about three stable isotopes for each atom, although some atoms have only one while others have five or six. The most common isotope of the second lightest atom, helium, is ^{4}He. The ^{4}He nucleus is also known as an α particle. Another isotope of helium is ^{3}He.

Inside the nucleus, the nucleons exert a strong attractive force on their nearby neighbors. This force, called the **strong nuclear force** or the **hadronic force,** is much stronger than the electrostatic force of repulsion between the protons and is very much stronger than the gravitational forces between the nucleons. (Gravity is so weak that it can always be neglected in nuclear physics.) The strong nuclear force is roughly the same between two neutrons, two protons, or a neutron and a proton. Two protons, of course, also exert a repulsive electrostatic force on each other due to their charges, which tends to weaken the attraction between them somewhat. The strong nuclear force decreases rapidly with distance, and it is negligible when two nucleons are more than a few femtometers apart.

Size and Shape

The size and shape of the nucleus can be determined by bombarding it with high-energy particles and observing the scattering. The results depend somewhat on the kind of experiment. For example, a scattering experiment using electrons measures the charge distribution of the nucleus, whereas one using neutrons determines the region of influence of the strong nuclear force. Despite these differences, a wide variety of experiments suggest that most nuclei are approximately spherical, with radii given approximately by

$$R = R_0 A^{1/3} \qquad \text{40-1}$$

Nuclear radius

where R_0 is about 1.5 fm. The fact that the radius of a spherical nucleus is proportional to $A^{1/3}$ implies that the volume of the nucleus is proportional to A. Since the mass of the nucleus is also approximately proportional to A, the densities of all nuclei are approximately the same. This is analogous to a drop of liquid, which also has constant density independent of its size. The **liquid-drop model** of the nucleus has proved quite successful in explaining nuclear behavior, especially the fission of heavy nuclei.

N and *Z* Numbers

For light nuclei, the greatest stability is achieved when the numbers of protons and neutrons are approximately equal, $N \approx Z$. For heavier nuclei, insta-

* The word nucleon refers to either a neutron or a proton.

bility caused by the electrostatic repulsion between the protons is minimized when there are more neutrons than protons. We can see this by looking at the N and Z numbers for the most abundant isotopes of some representative elements: for $^{16}_{8}O$, $N = 8$ and $Z = 8$; for $^{40}_{20}Ca$, $N = 20$ and $Z = 20$; for $^{56}_{26}Fe$, $N = 30$ and $Z = 26$; for $^{207}_{82}Pb$, $N = 125$ and $Z = 82$; and for $^{238}_{92}U$, $N = 146$ and $Z = 92$. (The atomic number Z has been included here as a presubscript of the atomic symbol for emphasis. It is not actually needed because the atomic number is implied by the atomic symbol.)

Figure 40-1 shows a plot of N versus Z for the known stable nuclei. The curve follows the straight line $N = Z$ for small values of N and Z. We can understand this tendency for N and Z to be equal by considering the total energy of A particles in a one-dimensional box. Figure 40-2 shows the energy levels for eight neutrons and for four neutrons and four protons. Because of the exclusion principle, only two identical particles (with opposite spins) can be in the same space state. Since protons and neutrons are not identical, we can put two each in a state as in Figure 40-2*b*. Thus, the total energy for four protons and four neutrons is less than that for eight neutrons (or eight protons) as in Figure 40-2*a*. When the Coulomb energy of repulsion, which is proportional to Z^2, is included, this result changes somewhat. For large values of A and Z, the total energy may be increased less by adding two neutrons than by adding one neutron and one proton because of the electrostatic repulsion involved in the latter case. This explains why $N > Z$ for the heavier nuclei.

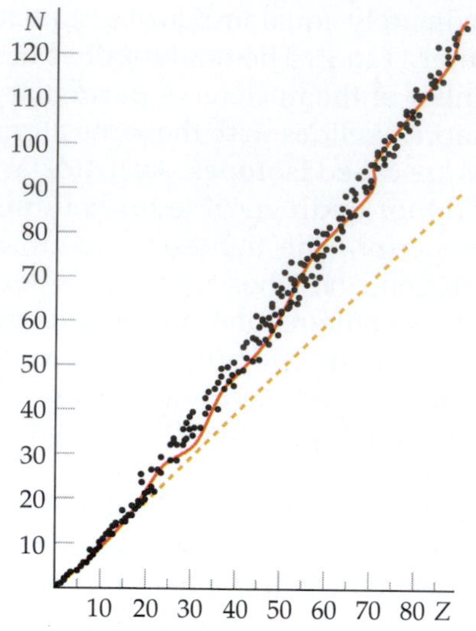

Figure 40-1 Plot of number of neutrons N versus number of protons Z for the stable nuclides. The dashed line is $N = Z$.

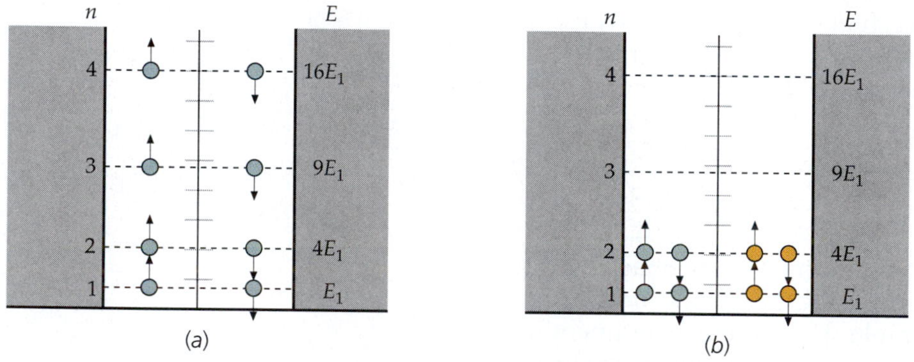

Figure 40-2 (*a*) Eight neutrons in a one-dimensional box. In accordance with the exclusion principle, only two neutrons (with opposite spins) can be in a given energy level. (*b*) Four neutrons and four protons in a one-dimensional box. Because protons and neutrons are not identical particles, two of each can be in each energy level. The total energy is much less for this case than for that in (*a*).

Mass and Binding Energy

The mass of a nucleus is less than the mass of its parts by $\Delta E/c^2$, where ΔE is the binding energy and c is the speed of light. When two or more nucleons fuse together to form a nucleus, the total mass decreases and energy is given off. Conversely, to break up a nucleus into its parts, energy must be put into the system to produce the increase in rest mass.

Atomic and nuclear masses are often given in unified mass units (u), defined as one-twelfth the mass of the neutral ^{12}C atom. The rest energy of one unified mass unit is

$$(1\ \text{u})c^2 = 931.5\ \text{MeV} \tag{40-2}$$

Consider ^4He, for example, which consists of two protons and two neutrons. The mass of an atom can be accurately measured in a mass spectrometer. The mass of the ^4He atom is 4.002 603 u. This includes the masses of the two electrons in the atom. The mass of the ^{1}H atom is 1.007 825 u, and that of the neutron is 1.008 665 u. The sum of the masses of two ^{1}H atoms plus two neutrons is 2(1.007 825 u) + 2(1.008 665 u) = 4.032 98 u, which is greater than the mass of the ^{4}He atom by 0.030 377 u.* We can find the binding energy of the ^{4}He nucleus from this mass difference of 0.030 377 u by using the mass conversion factor $(1\text{ u})c^2 = 931.5$ MeV from Equation 40-2. Then

$$(0.030\,377\text{ u})c^2 = (0.030\,377\text{ u})c^2 \times \frac{931.5\text{ MeV}/c^2}{1\text{ u}}$$

$$= 28.30\text{ MeV}$$

The total binding energy of ^{4}He is thus 28.3 MeV. In general, the binding energy of a nucleus of an atom of atomic mass M_A containing Z protons and N neutrons is found by calculating the difference between the mass of the parts and the mass of the nucleus and then multiplying by c^2:

$$E_b = (ZM_H + Nm_n - M_A)c^2 \qquad \textbf{40-3}$$

Total nuclear binding energy

where M_H is the mass of the ^{1}H atom and m_n that of the neutron. (Note that the mass of the Z electrons in the term ZM_H is canceled by the mass of the Z electrons in the term M_A.) The atomic masses of the neutron and of some selected isotopes are listed in Table 40-1.

* Note that by using the masses of two ^{1}H atoms rather than two protons, the masses of the electrons in the atom are accounted for. We do this because it is atomic masses, not nuclear masses, that are measured directly and listed in mass tables.

Table 40-1

Atomic Masses of the Neutron and Selected Isotopes

Element	Symbol	Z	Atomic mass, u
Neutron	n	0	1.008 665
Hydrogen	^{1}H	1	1.007 825
Deuterium	^{2}H or D	1	2.014 102
Tritium	^{3}H or T	1	3.016 050
Helium	^{3}He	2	3.016 030
	^{4}He	2	4.002 603
Lithium	^{6}Li	3	6.015 125
	^{7}Li	3	7.016 004
Boron	^{10}B	5	10.012 939
Carbon	^{12}C	6	12.000 000
	^{13}C	6	13.003 354
	^{14}C	6	14.003 242
Nitrogen	^{13}N	7	13.005 738
	^{14}N	7	14.003 074
Oxygen	^{16}O	8	15.994 915
Sodium	^{23}Na	11	22.989 771
Potassium	^{39}K	19	38.963 710
Iron	^{56}Fe	26	55.939 395
Copper	^{63}Cu	29	62.929 592
Silver	^{107}Ag	47	106.905 094
Gold	^{197}Au	79	196.966 541
Lead	^{208}Pb	82	207.976 650
Polonium	^{212}Po	84	211.989 629
Radon	^{222}Rn	86	222.017 531
Radium	^{226}Ra	88	226.025 360
Uranium	^{238}U	92	238.048 608
Plutonium	^{242}Pu	94	242.058 725

Example 40-1

Find the binding energy of the last neutron in ^{4}He.

Picture the Problem The binding energy is c^2 times the difference in mass of ^{3}He plus a neutron and ^{4}He. We find these masses from Table 40-1 and convert to energy using Equation 40-3.

1. Add the mass of the neutron to that of ^{3}He:

$$m_{^3\text{He}} + m_n = 3.016\,030\text{ u} + 1.008\,665\text{ u} = 4.024\,695\text{ u}$$

2. Subtract the mass of ^{4}He from the result:

$$(m_{^3\text{He}} + m_n) - m_{^4\text{He}} = 4.024\,695\text{ u} - 4.002\,603\text{ u}$$
$$= 0.022\,092\text{ u}$$

3. Multiply this mass difference by c^2 and convert to MeV:

$$E_b = (\Delta m)c^2 = (0.022\,092\text{ u})c^2 \times \frac{931.5\text{ MeV}/c^2}{1\text{ u}}$$
$$= 20.58\text{ MeV}$$

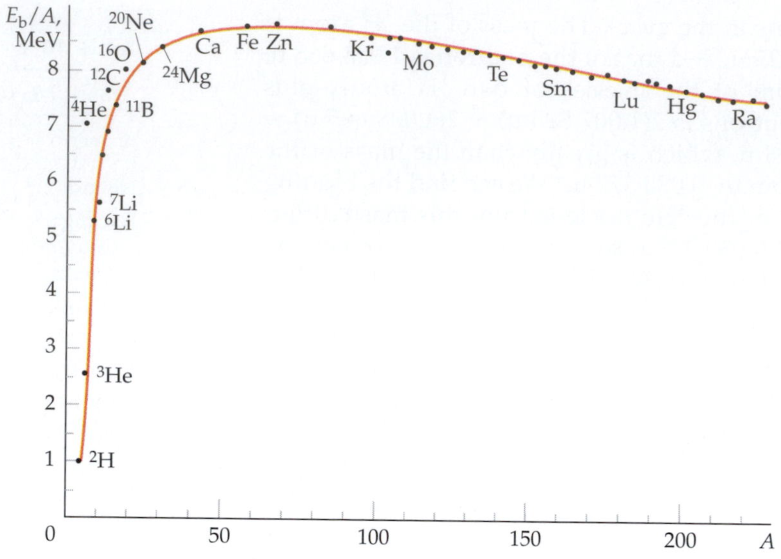

Figure 40-3 The binding energy per nucleon versus the mass number A. For nuclei with values of A greater than about 50, the curve is approximately flat, indicating that the total binding energy is approximately proportional to A.

Figure 40-3 shows the binding energy per nucleon E_b/A versus A. The mean value is about 8.3 MeV. The flatness of this curve for $A > 50$ shows that E_b is approximately proportional to A. This indicates that there is saturation of nuclear forces in the nucleus as would be the case if each nucleon were attracted only to its nearest neighbors. Such a situation also leads to a constant nuclear density consistent with the measurements of the radius. If, for example, there were no saturation and each nucleon bonded to each other nucleon, there would be $A - 1$ bonds for each nucleon and a total of $A(A - 1)$ bonds altogether. The total binding energy, which is a measure of the energy needed to break all these bonds, would then be proportional to $A(A - 1)$, and E_b/A would not be approximately constant. The steep rise in the curve for low A is due to the increase in the number of nearest neighbors and therefore to the increased number of bonds per nucleon. The gradual decrease at high A is due to the Coulomb repulsion of the protons, which increases as Z^2 and decreases the binding energy. Eventually, for very large A this Coulomb repulsion becomes so great that a nucleus with A greater than about 300 is unstable and undergoes spontaneous fission.

40-2 Radioactivity

Many nuclei are radioactive; that is, they decay into other nuclei by the emission of particles, such as photons, electrons, neutrons, or α particles. The terms α decay, β decay, and γ decay were used before it was known that α particles are ^{4}He nuclei, β particles are either electrons (β^-) or positrons* (β^+), and γ rays are photons. The rate of decay is not constant over time, but decreases exponentially. *This exponential time dependence is characteristic of all radioactivity and indicates that radioactive decay is a statistical process.* Because each nucleus is well shielded from others by the atomic electrons, pressure and temperature changes have little or no effect on the rate of radioactive decay or other nuclear properties.

Let N be the number of radioactive nuclei at some time t. If the decay of an individual nucleus is a random event, we expect the number of nuclei that decay in some time interval dt to be proportional to N and to dt. Because of these decays, the number N will decrease. The change in N is given by

$$dN = -\lambda N\, dt \qquad \text{40-4}$$

where λ is a constant of proportionality called the **decay constant.** The rate of change of N, dN/dt, is proportional to N. This is characteristic of exponential decay. To solve Equation 40-4 for N, we first divide each side by N, thus separating the variables N and t:

$$\frac{dN}{N} = -\lambda\, dt$$

* The positron is identical to an electron except it has a charge of $+e$.

Integrating, we obtain

$$\ln N = -\lambda t + C \qquad \text{40-5}$$

where C is some constant of integration. Taking the exponential of each side we obtain

$$N = e^{-\lambda t + C} = e^C e^{-\lambda t}$$

or

$$N = N_0 e^{-\lambda t} \qquad \text{40-6}$$

where $N_0 = e^C$ is the number of nuclei at $t = 0$. The number of radioactive decays per second is called the decay rate R:

$$R = -\frac{dN}{dt} = \lambda N = \lambda N_0 e^{-\lambda t} = R_0 e^{-\lambda t} \qquad \text{40-7}$$

Decay rate

where

$$R_0 = \lambda N_0 \qquad \text{40-8}$$

is the rate of decay at time $t = 0$. The decay rate R is the quantity that is determined experimentally.

The average or **mean lifetime** τ is the reciprocal of the decay constant:

$$\tau = \frac{1}{\lambda} \qquad \text{40-9}$$

(See Problem 33.) The mean lifetime is analogous to the time constant in the exponential decrease in the charge on a capacitor in an RC circuit that we discussed in Section 26-6. After a time equal to the mean lifetime, the number of radioactive nuclei and the decay rate have each decreased to 37% of their original values. The **half-life** $t_{1/2}$ is defined as the time it takes for the number of nuclei and the decay rate to decrease by half. Setting $t = t_{1/2}$ and $N = N_0/2$ in Equation 40-6 gives

$$\frac{N_0}{2} = N_0 e^{-\lambda t_{1/2}} \qquad \text{40-10}$$

or

$$e^{+\lambda t_{1/2}} = 2$$

Solving for $t_{1/2}$ gives

$$t_{1/2} = \frac{\ln 2}{\lambda} = \frac{0.693}{\lambda} = 0.693\tau \qquad \text{40-11}$$

Figure 40-4 shows a plot of N versus t. If we multiply the numbers on the N axis by λ, this graph becomes a plot of R versus t. After each time interval of one half-life, the number of nuclei left and the decay rate have decreased to half of their previous values. For example, if the decay rate is R_0 initially, it will be $\frac{1}{2}R_0$ after one half-life, $(\frac{1}{2})(\frac{1}{2})R_0$ after two half-lives, and so forth. After n half-lives, the decay rate will be

$$R = (\tfrac{1}{2})^n R_0 \qquad \text{40-12}$$

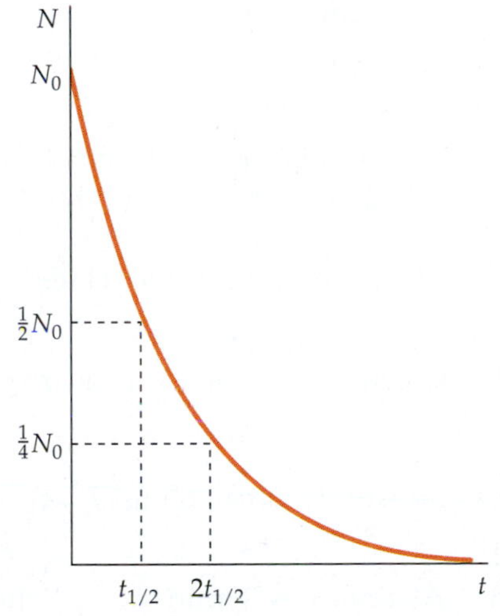

Figure 40-4 Exponential radioactive decay. After each half-life $t_{1/2}$, the number of nuclei remaining has decreased by one-half. The decay rate $R = \lambda N$ has the same time dependence.

The half-lives of radioactive nuclei vary from very small times (less than 1 μs) to very large times (up to 10^{16} y).

Example 40-2

A radioactive source has a half-life of 1 min. At time $t = 0$, it is placed near a detector, and the counting rate (the number of decay particles detected per unit time) is observed to be 2000 counts/s. Find the counting rate at times $t = 1$ min, 2 min, 3 min, and 10 min.

Picture the Problem The counting rate decreases by a factor of 2 each minute.

1. Since the half-life is 1 min, the counting rate will be half as great at $t = 1$ min as at $t = 0$:

$$r_1 = \tfrac{1}{2} r_0 = \tfrac{1}{2}(2000 \text{ counts/s}) = 1000 \text{ counts/s at 1 min}$$

2. At $t = 2$ min, the rate is half that at 1 min. It decreases by one-half each minute:

$$r_2 = \tfrac{1}{2} r_1 = \tfrac{1}{2}(1000 \text{ counts/s}) = 500 \text{ counts/s at 2 min}$$

$$r_3 = \tfrac{1}{2} r_2 = \tfrac{1}{2}(500 \text{ counts/s}) = 250 \text{ counts/s at 3 min}$$

3. At $t = 10$ min, the rate will be $(\frac{1}{2})^{10}$ times the initial rate:

$$r_{10} = (\tfrac{1}{2})^{10} r_0 = (\tfrac{1}{2})^{10} (2000 \text{ counts/s})$$

$$= 1.95 \text{ counts/s} \approx 2 \text{ counts/s at 10 min}$$

Example 40-3

If the detection efficiency in Example 40-2 is 20%, (*a*) how many radioactive nuclei are there at time $t = 0$? (*b*) At time $t = 1$ min? (*c*) How many nuclei decay in the first minute?

Picture the Problem The detection efficiency depends on the probability that a radioactive decay particle will enter the detector and the probability that upon entering the detector it will produce a count. If the efficiency is 20%, the decay rate must be 5 times the counting rate.

(*a*) 1. The number of radioactive nuclei is related to the decay rate R, and the decay constant λ:

$$R = \lambda N$$

2. The decay constant is related to the half-life:

$$\lambda = \frac{0.693}{t_{1/2}} = \frac{0.693}{1 \text{ min}}$$

3. Calculate the decay rate from the counting rate:

$$R_0 = 5 \times 2000 \text{ counts/s} = 10^4 \text{ s}^{-1}$$

4. Substitute to calculate N_0 at $t = 0$:

$$N_0 = \frac{R_0}{\lambda} = \frac{10{,}000 \text{ s}^{-1}}{0.693 \text{ min}^{-1}} \times \frac{60 \text{ s}}{1 \text{ min}} = 8.66 \times 10^5$$

(*b*) At time $t = 1 \text{ min} = t_{1/2}$, there are half as many radioactive nuclei as at $t = 0$:

$$N_1 = \tfrac{1}{2}(8.66 \times 10^5) = 4.33 \times 10^5$$

(*c*) The number of nuclei that decay in the first minute is $N_0 - N_1$:

$$\Delta N = N_0 - N_1 = 8.66 \times 10^5 - 4.33 \times 10^5 = 4.33 \times 10^5$$

The SI unit of radioactive decay is the **becquerel** (Bq), which is defined as one decay per second:

$$1\ \text{Bq} = 1\ \text{decay/s} \qquad \text{40-13}$$

A historical unit that applies to all types of radioactivity is the **curie** (Ci), which is defined as

$$1\ \text{Ci} = 3.7 \times 10^{10}\ \text{decays/s} = 3.7 \times 10^{10}\ \text{Bq} \qquad \text{40-14}$$

The curie is the rate at which radiation is emitted by 1 g of radium. Since this is a very large unit, the millicurie (mCi) or microcurie (μCi) are often used.

Beta Decay

Beta decay occurs in nuclei that have too many or too few neutrons for stability. In β decay, A remains the same while Z either increases by 1 (β^- decay) or decreases by 1 (β^+ decay).

The simplest example of β decay is the decay of the free neutron into a proton plus an electron. (The half-life of a free neutron is about 10.8 min.) The energy of decay is 0.782 MeV, which is the difference between the rest energy of the neutron and that of the proton plus electron. More generally, in β^- decay, a nucleus of mass number A and atomic number Z decays into a nucleus, referred to as the **daughter nucleus,** of mass number A and atomic number $Z' = Z + 1$ with the emission of an electron. If the decay energy were shared by only the daughter nucleus and the emitted electron, the energy of the electron would be uniquely determined by the conservation of energy and momentum. Experimentally, however, the energies of the electrons emitted in the β^- decay of a nucleus are observed to vary from zero to the maximum energy available. A typical energy spectrum for these electrons is shown in Figure 40-5.

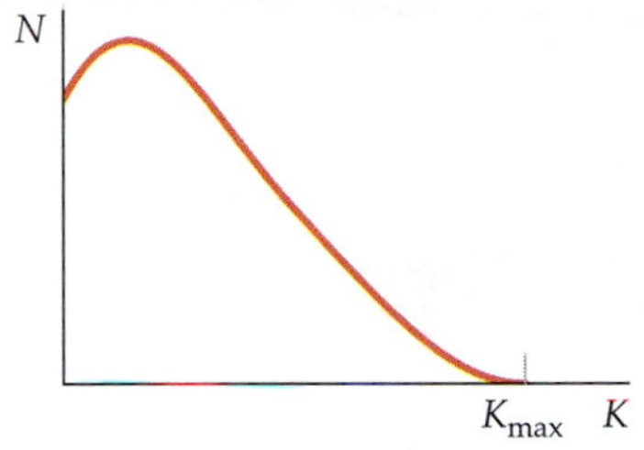

Figure 40-5 Number of electrons emitted in β^- decay versus kinetic energy. The fact that all the electrons do not have the same energy K_{max} suggests that another particle that shares the energy available for decay is emitted.

To explain the apparent nonconservation of energy in β decay, W. Pauli in 1930 suggested that a third particle, which he called the **neutrino,** is also emitted. Because the measured maximum energy of the emitted electrons is equal to the total available for the decay, the rest energy and therefore the mass of the neutrino was assumed to be zero. (It is now believed that the mass of the neutrino is very small but not zero.) In 1948, measurements of the momenta of the emitted electron and the recoiling nucleus showed that the neutrino was also needed for the conservation of linear momentum in β decay. The neutrino was first observed experimentally in 1957. It is now known that there are at least three kinds of neutrinos, one (ν_e) associated with electrons, one (ν_μ) associated with muons, and one (ν_τ) not yet observed experimentally, associated with the newly discovered tau particle, τ. Moreover, each neutrino has an antiparticle, written $\bar{\nu}_e$, $\bar{\nu}_\mu$, and $\bar{\nu}_\tau$. It is the electron antineutrino that is emitted in the decay of a neutron, which is written*

$$\text{n} \rightarrow \text{p} + \beta^- + \bar{\nu}_e \qquad \text{40-15}$$

In β^+ decay, a proton changes into a neutron with the emission of a positron (and a neutrino). A free proton cannot decay by positron emission because of conservation of energy (the rest mass of the neutron plus the positron is greater than that of the proton), but because of binding-energy effects, a proton inside a nucleus can decay. A typical β^+ decay is

$$^{13}_{7}\text{N} \rightarrow {}^{13}_{6}\text{C} + \beta^+ + \nu_e \qquad \text{40-16}$$

The electrons or positrons emitted in β decay do not exist inside the nucleus. They are created in the process of decay, just as photons are created when an atom makes a transition from a higher to a lower energy state.

* This reaction is also written $\text{n} \rightarrow \text{p} + \text{e}^- + \bar{\nu}_e$.

An important example of β decay is that of ^{14}C, which is used in **radioactive carbon dating:**

$$^{14}C \rightarrow {}^{14}N + \beta^- + \bar{\nu}_e \qquad \text{40-17}$$

The half-life for this decay is 5730 y. The radioactive isotope ^{14}C is produced in the upper atmosphere in nuclear reactions caused by cosmic rays. The chemical behavior of carbon atoms with ^{14}C nuclei is the same as those with ordinary ^{12}C nuclei. For example, atoms with these nuclei combine with oxygen to form CO_2 molecules. Since living organisms continually exchange CO_2 with the atmosphere, the ratio of ^{14}C to ^{12}C in a living organism is the same as the equilibrium ratio in the atmosphere, which is about 1.3×10^{-12}. After an organism dies, it no longer absorbs ^{14}C from the atmosphere, so the ratio of ^{14}C to ^{12}C continually decreases due to the radioactive decay of ^{14}C. The number of ^{14}C decays per minute per gram of carbon in a living organism can be calculated from the known half-life of ^{14}C and the number of ^{14}C nuclei in a gram of carbon. The result is that there are about 15.0 decays per minute per gram of carbon in a living organism. Using this result and the measured number of decays per minute per gram of carbon in a nonliving sample of bone, wood, or other object containing carbon, we can determine the age of the sample. For example, if the measured rate were 7.5 decays per minute per gram, the sample would be one half-life = 5730 years old.

Example 40-4

A bone containing 200 g of carbon has a β-decay rate of 400 decays/min. How old is the bone?

Picture the Problem We first obtain a rough estimate of the age of the bone. If the bone were from a living organism, we would expect the decay rate to be (15 decays/min·g)(200 g) = 3000 decays/min. Since 400/3000 is roughly 1/8 (actually 1/7.5), the sample must be about three half-lives old, which is about 3(5730) y = 17,190 y. To find the age of the bone more accurately, we note that after n half-lives, the decay rate will have decreased by a factor of $(\frac{1}{2})^n$.

1. Write the decay rate after n half-lives in terms of the initial decay rate:

$$R_n = (\tfrac{1}{2})^n R_0$$

2. Calculate the initial decay rate for 200 g:

$$R_0 = (15\text{ decays/min}\cdot\text{g})(200\text{ g}) = 3000\text{ decays/min}$$

3. Substitute the measured decay rate and simplify:

$$R_n = (\tfrac{1}{2})^n\, 3000\text{ decays/min} = 400\text{ decays/min}$$

$$(\tfrac{1}{2})^n = \frac{400}{3000}$$

$$2^n = \frac{3000}{400} = 7.5$$

4. We solve for n by taking the logarithm of each side:

$$n \ln 2 = \ln 7.5$$

$$n = \frac{\ln 7.5}{\ln 2} = 2.91$$

5. The age of the bone is $nt_{1/2}$:

$$t = nt_{1/2} = 2.91(5730\text{ y}) = 1.67 \times 10^4\text{ y}$$

Gamma Decay

In γ decay a nucleus in an excited state decays to a lower-energy state by the emission of a photon. This is the nuclear counterpart of spontaneous emission of photons by atoms and molecules. Unlike β or α decay, the radioactive nucleus remains the same nucleus after γ decay. Since the spacing of the nuclear energy levels is of the order of 1 MeV (as compared with spacing of the order of 1 eV in atoms), the wavelengths of the emitted photons are of the order of 1 pm (1 pm = 10^{-12} m):

$$\lambda = \frac{hc}{E} \approx \frac{1240 \text{ eV}\cdot\text{nm}}{1 \text{ MeV}} = 0.00124 \text{ nm} = 1.24 \text{ pm}$$

The mean lifetime for γ decay is often very short. Usually it is observed only because it follows either α or β decay. For example, if a radioactive parent nucleus decays by β decay to an excited state of the daughter nucleus, the daughter nucleus then decays to its ground state by γ emission. Direct measurements of mean lifetimes as short as about 10^{-11} s are possible. Measurements of mean lifetimes shorter than 10^{-11} s are difficult, but they can sometimes be made by indirect methods.

A few γ emitters have very long lifetimes, of the order of hours. Nuclear energy states that have such long lifetimes are called **metastable states.**

Alpha Decay

All very heavy nuclei ($Z > 83$) are theoretically unstable to α decay because the mass of the original radioactive nucleus is greater than the sum of the masses of the decay products—an α particle and the daughter nucleus. Consider the decay of ^{232}Th ($Z = 90$) into ^{228}Ra ($Z = 88$) plus an α particle. This is written as

$$^{232}\text{Th} \rightarrow {}^{228}\text{Ra} + \alpha = {}^{228}\text{Ra} + {}^{4}\text{He} \qquad \text{40-18}$$

The mass of the ^{232}Th atom is 232.038 124 u. The mass of the daughter atom ^{228}Ra is 228.031 139 u. Adding 4.002 603 u to this for the mass of ^{4}He, we get 232.033 742 u for the total mass of the decay products. This is less than the mass of ^{232}Th by 0.004 382 u, which multiplied by 931.5 MeV/c^2 gives 4.08 MeV/c^2 for the excess rest mass of ^{232}Th over that of the decay products. The isotope ^{232}Th is therefore theoretically unstable to α decay. This decay does in fact occur in nature with the emission of an α particle of kinetic energy 4.08 MeV. (The kinetic energy of the α particle is actually somewhat less than 4.08 MeV because some of the decay energy is shared by the recoiling ^{228}Ra nucleus.)

In general, when a nucleus emits an α particle, both N and Z decrease by 2 and A decreases by 4. The daughter of a radioactive nucleus is often itself radioactive and decays by either α or β decay or both. If the original nucleus has a mass number A that is 4 times an integer, the daughter nucleus and all those in the chain will also have mass numbers equal to 4 times an integer. Similarly, if the mass number of the original nucleus is $4n + 1$, where n is an integer, all the nuclei in the decay chain will have mass numbers given by $4n + 1$, with n decreasing by one at each decay. We can see, therefore, that there are four possible α-decay chains, depending on whether A equals $4n$, $4n + 1$, $4n + 2$, or $4n + 3$, where n is an integer. All but one of these decay chains are found on Earth. The $4n + 1$ series is not found because its longest lived member (other than the stable end product ^{209}Bi) is ^{237}Np, which has a half-life of only 2×10^6 y. Because this is much less than the age of the earth, this series has disappeared.

Figure 40-6 shows the thorium series, for which $A = 4n$. It begins with an α decay from ^{232}Th to ^{228}Ra. The daughter nuclide of an α decay is on the left or neutron-rich side of the stability curve (the dashed line in the figure), so it often decays by β^- decay. In the thorium series, ^{228}Ra decays by β^- decay to ^{228}Ac, which in turn decays by β^- decay to ^{228}Th. There are then four α decays to ^{212}Pb, which decays by β^- decay to ^{212}Bi. The series branches at ^{212}Bi, which decays either by α decay to ^{208}Tl or by β^- decay to ^{212}Po. The branches meet at the stable lead isotope ^{208}Pb.

The energies of α particles from natural radioactive sources range from about 4 to 7 MeV, and the half-lives of the sources range from about 10^{-5} s to 10^{10} y. In general, the smaller the energy of the emitted α particle, the longer the half-life. As we discussed in Section 36-4, the enormous variation in half-lives was explained by George Gamow in 1928. He considered α decay to be a process in which an α particle is first formed inside a nucleus and then tunnels through the Coulomb barrier (Figure 40-7). A slight increase in the energy of the α particle reduces the relative height $U - E$ of the barrier and also the thickness. Because the probability of penetration is so sensitive to the relative height and thickness of the barrier, a small increase in E leads to a large increase in the probability of barrier penetration and therefore to a shorter lifetime. Gamow was able to derive an expression for the half-life as a function of E that is in excellent agreement with experimental results.

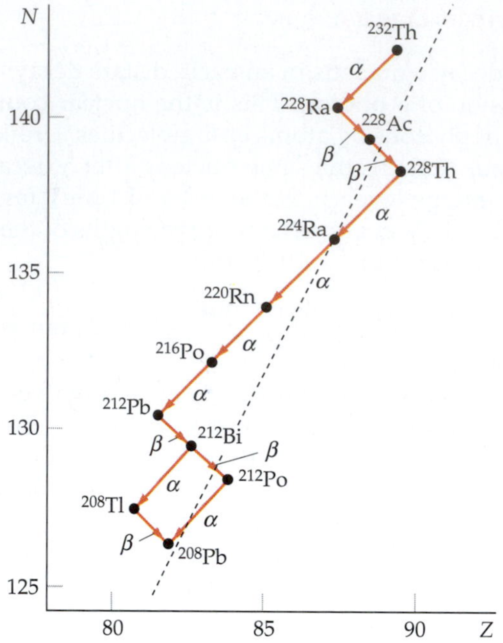

Figure 40-6 The thorium (4n) α decay series. The dashed line is the curve of stability.

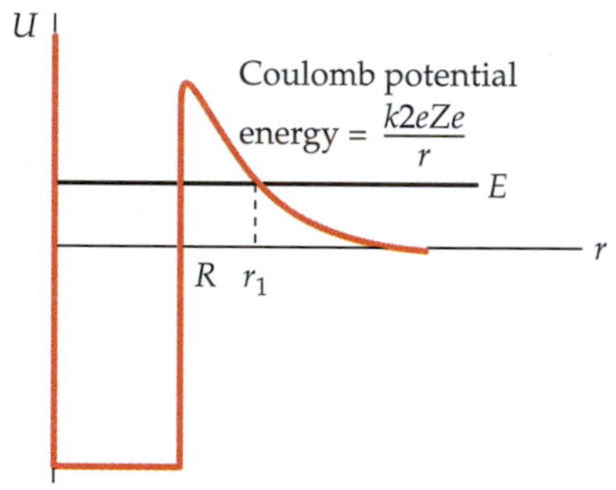

Figure 40-7 A model of the potential energy for an α particle and a nucleus. The strong attractive nuclear force that exists for values of r less than the nuclear radius R is indicated by the potential well. Outside the nucleus, the nuclear force is negligible, and the potential energy is given by Coulomb's law $U = +k2eZe/r$, where Ze is the nuclear charge and $2e$ is the charge of the α particle. The energy E is the kinetic energy of the α particle when it is far away from the nucleus. A small increase in E reduces the relative height of the barrier and also its thickness, leading to a much greater chance of penetration. An increase in the energy of the emitted α particles by a factor of 2 results in a reduction of the half-life by a factor of more than 10^{20}.

40-3 Nuclear Reactions

Information about nuclei is typically obtained by bombarding them with various particles and observing the results. Although the first experiments of this type were limited by the need to use naturally occurring radiation, they produced many important discoveries. In 1932 J. D. Cockcroft and E. T. S. Walton succeeded in producing the reaction

$$p + {}^7Li \rightarrow {}^8Be \rightarrow {}^4He + {}^4He$$

using artificially accelerated protons. At about the same time, the Van de Graaff electrostatic generator was built (by R. Van de Graaff in 1931), as was the first cyclotron (by E. O. Lawrence and M. S. Livingston in 1932). Since then, enormous advances in the technology for accelerating and detecting particles have been made, and many nuclear reactions have been studied.

When a particle is incident on a nucleus, several different things can happen. The incident particle may be scattered elastically or inelastically, or the incident particle may be absorbed by the nucleus, and another particle or particles may be emitted. In inelastic scattering the nucleus is left in an excited state and subsequently decays by emitting photons (or other particles).

The amount of energy released or absorbed in a reaction (in the center of mass reference frame) is called the **Q value** of the reaction. The Q value equals c^2 times this mass difference. When energy is released by a nuclear reaction, the reaction is said to be an **exothermic reaction.** In an exothermic reaction, the total mass of the incoming particles is greater than that of the outgoing particles, and the Q value is positive. If the total mass of the incoming particles is less than that of the outgoing particles, energy is required for the reaction to take place, and the reaction is said to be an **endothermic reaction.** The Q value of an endothermic reaction is negative. In general, if Δm is the *increase* in mass, the Q value is

$$Q = -(\Delta m)c^2 \qquad \text{40-19}$$

Q value

An endothermic reaction cannot take place below a certain threshold energy. In the laboratory reference frame in which stationary particles are bombarded by incoming particles, the threshold energy is somewhat greater than $|Q|$ because the outgoing particles must have some kinetic energy to conserve momentum.

A measure of the effective size of a nucleus for a particular nuclear reaction is the **cross section σ.** If I is the number of incident particles per unit time per unit area (the incident intensity) and R is the number of reactions per unit time per nucleus, the cross section is

$$\sigma = \frac{R}{I} \qquad \text{40-20}$$

The cross section σ has the dimensions of area. Since nuclear cross sections are of the order of the square of the nuclear radius, a convenient unit for them is the **barn,** which is defined as

$$1 \text{ barn} = 10^{-28} \text{ m}^2 \qquad \text{40-21}$$

The cross section for a particular reaction is a function of energy. For an endothermic reaction, it is zero for energies below the threshold energy.

Example 40-5

Find the Q value of the reaction $p + {}^7\text{Li} \rightarrow {}^4\text{He} + {}^4\text{He}$ and state whether the reaction is exothermic or endothermic.

Picture the Problem We find the masses of the atoms from Table 40-1 and calculate the difference in the total mass of the outgoing particles and incomming particles. The Q value is $-(\Delta m)c^2$. If we use the mass of hydrogen rather than the mass of the proton, there will be 4 electrons on each side of the reaction so the electron masses will cancel.

1. Find the mass of each atom from Table 40-1:

	${}^1\text{H}$	1.007 825 u
	${}^7\text{Li}$	7.016 004 u
	${}^4\text{He}$	4.002 603 u

2. Calculate the initial mass m_i of the incoming particles: $m_i = 1.007\,825\text{ u} + 7.016\,004\text{ u} = 8.023\,829\text{ u}$

3. Calculate the final mass m_f: $m_f = 2(4.002\,603\text{ u}) = 8.005\,206\text{ u}$

4. Calculate the increase in mass:
$$\Delta m = m_f - m_i = 8.005\,206\text{ u} - 8.023\,829\text{ u}$$
$$= -0.018\,623\text{ u}$$

5. Calculate the Q value:
$$Q = -(\Delta m)c^2 = (+0.018\,623\text{ u})c^2(931.5\text{ MeV/u }c^2)$$
$$= 17.35\text{ MeV}$$
Q is positive so the reaction is exothermic.

Remarks Since the initial mass is greater than the final mass, mass is converted into energy and the reaction is exothermic, yeilding 17.35 MeV.

Reactions With Neutrons

Nuclear reactions involving neutrons are important for understanding nuclear reactors. The most likely reaction between a nucleus and a neutron having an energy of more than about 1 MeV is scattering. However, even if the scattering is elastic, the neutron loses some energy to the nucleus because the nucleus recoils. If a neutron is scattered many times in a material, its energy decreases until it is of the order of the energy of thermal motion kT, where k is Boltzmann's constant and T is the absolute temperature. (At ordinary room temperatures, kT is about 0.025 eV.) The neutron is then equally likely to gain or lose energy from a nucleus when it is elastically scattered. A neutron with energy of the order of kT is called a **thermal neutron.**

At low energies, a neutron is likely to be captured, with the emission of a γ ray from the excited nucleus. Figure 40-8 shows the neutron-capture cross section for silver as a function of the energy of the neutron. The large peak in this curve is called a **resonance.** Except for the resonance, the cross section varies fairly smoothly with energy, decreasing with increasing energy roughly as $1/v$, where v is the speed of the neutron. We can understand this energy dependence as follows: Consider a neutron moving with speed v near a nucleus of diameter $2R$. The time it takes the neutron to pass the nucleus is $2R/v$. Thus, the neutron-capture cross section is proportional to the time spent by the neutron in the vicinity of the nucleus. The dashed line in Figure 40-8 indicates this $1/v$ dependence. At the maximum of the resonance, the value of the cross section is very large ($\sigma > 5000$ barns) compared with a value of only about 10 barns just past the resonance. Many elements show similar resonances in their neutron-capture cross sections. For example, the maximum cross section for ^{113}Cd is about 57,000 barns. This material is thus very useful for shielding against low-energy neutrons.

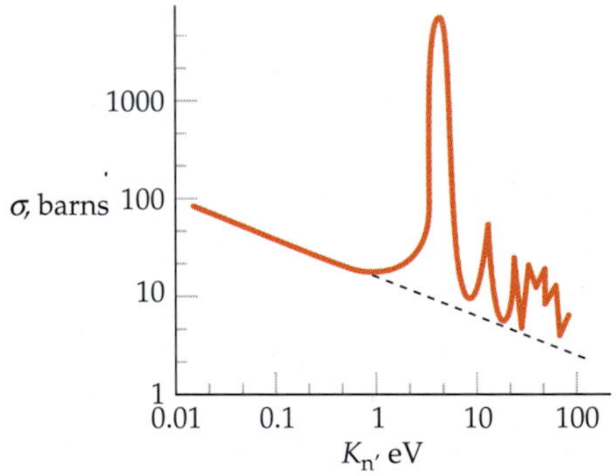

Figure 40-8 Neutron-capture cross section for silver versus energy of the neutron. The straight line indicates the $1/v$ dependence of the cross section, which is proportional to the time spent by the neutron near the silver nucleus. Superimposed on this dependence are a large resonance and several smaller resonances.

An important nuclear reaction that involves neutrons is fission, which is discussed in the next section.

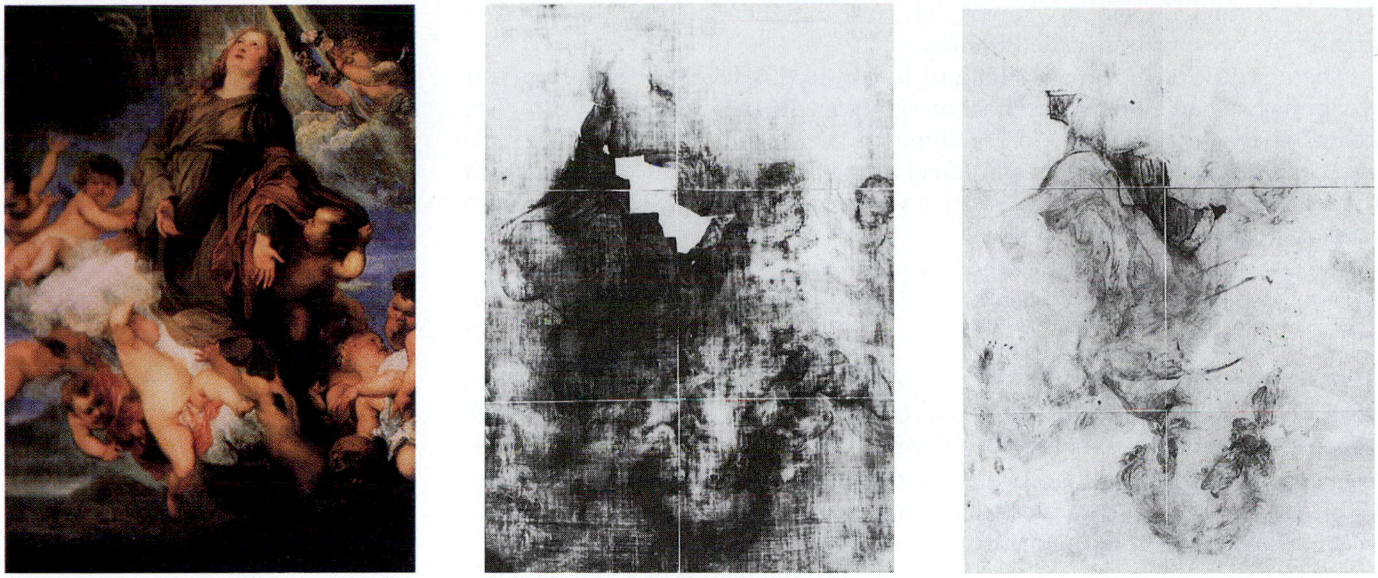

(a) (b) (c)

Hidden layers in paintings are analyzed by bombarding the painting with neutrons and observing the radiative emissions from nuclei that have captured a neutron. Different elements used in the painting have different half-lives. (*a*) Van Dyck's painting "Saint Rosalie Interceding for the Plague-Stricken of Palermo." The black-and-white images in (*b*) and (*c*) were formed using a special film sensitive to electrons emitted by the radioactively decaying elements. Image (*b*), taken a few hours after the neutron irradiation, reveals the presence of manganese, found in umber, a dark earth-pigment used for the painting's base layer. (Blank areas show where modern repairs, free of manganese, have been made.) The image in (*c*) was taken four days later, after the umber emissions had died away and when phosphorus, found in charcoal and boneblack, was the main radiating element. Upside down is revealed a sketch of Van Dyck himself. The self-portrait, executed in charcoal, had been overpainted by the artist.

40-4 Fission and Fusion

Figure 40-9 shows a plot of the nuclear mass difference per nucleon $(M - Zm_p - Nm_n)/A$ in units of MeV/c^2 versus A. This is just the negative of the binding-energy curve shown in Figure 40-3. From Figure 40-9, we can see that the rest mass per nucleon for both very heavy ($A \approx 200$) and very light ($A \lesssim 20$) nuclides is more than that for nuclides of intermediate mass. Thus, energy is released when a very heavy nucleus, such as ^{235}U, breaks up into two lighter nuclei—a process called **fission**—or when two

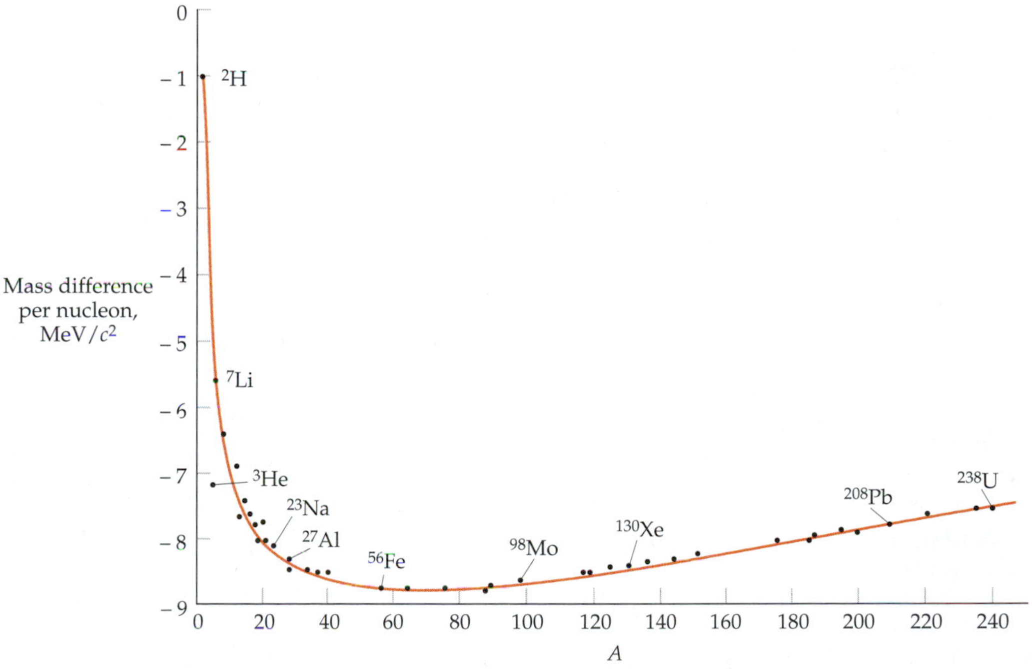

Figure 40-9 Plot of mass difference per nucleon $(M - Zm_p - Nm_n)/A$ in units of MeV/c^2 versus A. The rest mass per nucleon is less for nuclei of intermediate mass than for either very light or very heavy nuclei.

very light nuclei, such as ^{2}H and ^{3}H, fuse together to form a nucleus of greater mass—a process called **fusion.**

The application of both fission and fusion to the development of nuclear weapons has had a profound effect on our lives during the past 50 years. The peaceful application of these reactions to the development of energy resources may have an even greater effect in the future. We will look at some of the features of fission and fusion that are important for their application in reactors to generate power.

Fission

Very heavy nuclei ($Z > 92$) are subject to spontaneous fission. They break apart into two nuclei even if left to themselves with no outside disturbance. We can understand this by considering the analogy of a charged liquid drop. If the drop is not too large, surface tension can overcome the repulsive forces of the charges and hold the drop together. There is, however, a certain maximum size beyond which the drop will be unstable and will spontaneously break apart. Spontaneous fission puts an upper limit on the size of a nucleus and therefore on the number of elements that are possible.

Some heavy nuclei—uranium and plutonium, in particular—can be induced to fission by the capture of a neutron. In the fission of ^{235}U, for example, the uranium nucleus is excited by the capture of a neutron, causing it to split into two nuclei and emit several neutrons. The Coulomb force of repulsion drives the fission fragments apart, with the energy eventually showing up as thermal energy. Consider, for example, the fission of a nucleus of mass number $A = 200$ into two nuclei of mass number $A = 100$. Since the rest energy for $A = 200$ is about 1 MeV per nucleon greater than that for $A = 100$, about 200 MeV per nucleus is released in such a fission. This is a large amount of energy. By contrast, in the chemical reaction of combustion, only about 4 eV of energy is released per molecule of oxygen consumed.

Example 40-6

Calculate the total energy in kilowatt-hours released in the fission of 1 g of ^{235}U, assuming that 200 MeV is released per fission.

Picture the Problem We need to find the number of uranium nuclei in one gram of ^{235}U, which we do using the fact that there are Avogadro's number ($N_A = 6.02 \times 10^{23}$) of nuclei in 235 grams.

1. The total energy is the number of nuclei times the energy per nucleus:

$$E = NE_{\text{nucleus}} = N(200\ \text{MeV/nucleus})$$

2. Calculate N:

$$N = \frac{6.02 \times 10^{23}\ \text{nuclei/mol}}{235\ \text{g/mol}} \times 1\ \text{g} = 2.56 \times 10^{21}\ \text{nuclei}$$

3. Calculate the energy per gram in eV and convert to kW·h:

$$E = \frac{200 \times 10^{6}\ \text{eV}}{\text{nucleus}} \times 2.56 \times 10^{21}\ \text{nuclei}$$

$$\times \frac{1.6 \times 10^{-19}\ \text{J}}{1\ \text{eV}} \times \frac{1\ \text{h}}{3600\ \text{s}} \times \frac{1\ \text{kW}}{1000\ \text{J/s}}$$

$$= 2.28 \times 10^{4}\ \text{kW·h}$$

The fission of uranium was discovered in 1939 by Hahn and Strassmann, who found, by careful chemical analysis, that medium-mass elements (such as barium and lanthanum) were produced in the bombardment of uranium with neutrons. The discovery that several neutrons are emitted in the fission process led to speculation concerning the possibility of using these neutrons to cause further fissions, thereby producing a chain reaction. When ^{235}U captures a neutron, the resulting ^{236}U nucleus emits γ rays as it deexcites to the ground state about 15% of the time and undergoes fission about 85% of the time. The fission process is somewhat analogous to the oscillation of a liquid drop, as shown in Figure 40-10. If the oscillations are violent enough, the drop splits in two. Using the liquid-drop model, Bohr and Wheeler calculated the critical energy E_c needed by the ^{236}U nucleus to undergo fission. (^{236}U is the nucleus formed momentarily by the capture of a neutron by ^{235}U.) For this nucleus, the critical energy is 5.3 MeV, which is less than the 6.4 MeV of excitation energy produced when ^{235}U captures a neutron. The capture of a neutron by ^{235}U therefore produces an excited state of the ^{236}U nucleus that has more than enough energy to break apart. On the other hand, the critical energy for fission of the ^{239}U nucleus is 5.9 MeV. The capture of a neutron by a ^{238}U nucleus produces an excitation energy of only 5.2 MeV. Therefore, when a neutron is captured by ^{238}U to form ^{239}U, the excitation energy is not great enough for fission to occur. In this case, the excited ^{239}U nucleus deexcites by γ emission and then decays to $^{239}N_p$ by β decay, and then again to ^{239}Pu by β decay.

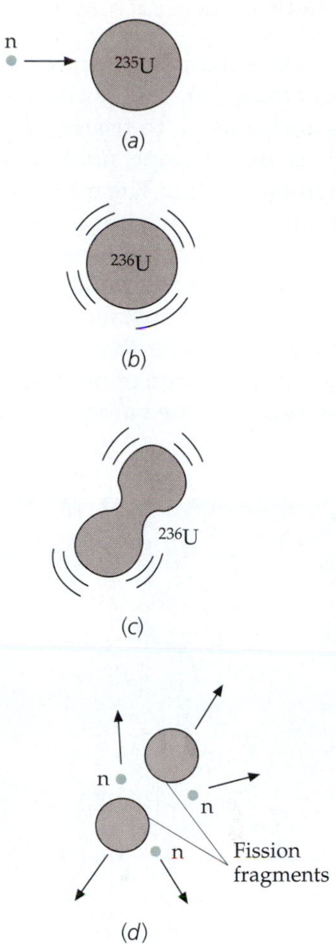

Figure 40-10 Schematic illustration of nuclear fission. (*a*) The absorption of a neutron by ^{235}U leads to (*b*) ^{236}U in an excited state. In (*c*), the oscillation of ^{236}U has become unstable. (*d*) The nucleus splits apart into two nuclei of medium mass and emits several neutrons that can produce fission in other nuclei.

A fissioning nucleus can break into two medium-mass fragments in many different ways, as shown in Figure 40-11. Depending on the particular reaction, 1, 2, or 3 neutrons may be emitted. The average number of neutrons emitted in the fission of ^{235}U is about 2.5. A typical fission reaction is

$$n + {}^{235}U \rightarrow {}^{141}Ba + {}^{92}Kr + 3n$$

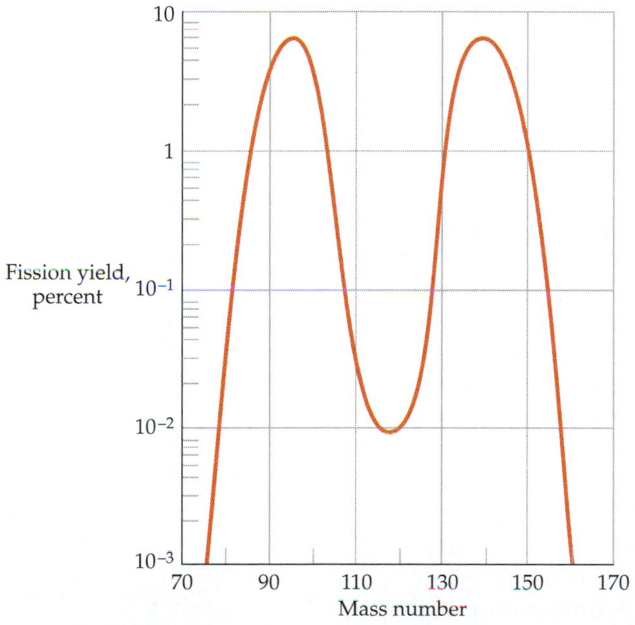

Figure 40-11 Distribution of the possible fission fragments of ^{235}U. The splitting of ^{235}U into two fragments of unequal mass is more likely than its splitting into fragments of equal mass.

Nuclear Fission Reactors

To sustain a chain reaction in a fission reactor, one of the neutrons (on the average) emitted in the fission of ^{235}U must be captured by another ^{235}U nucleus and cause it to fission. The **reproduction constant** k of a reactor is defined as the average number of neutrons from each fission that cause a subsequent fission. The maximum possible value of k is 2.5, but it is normally less than this for two important reasons: (1) Some of the neutrons may escape from the region containing fissionable nuclei, and (2) some of the neutrons may be captured by nonfissioning nuclei in the reactor. If k is exactly 1, the reaction will be self-sustaining. If it is less than 1, the reaction will die out. If k is significantly greater than 1, the reaction rate will increase rapidly and "run away." In the design of nuclear bombs, such a runaway reaction is desired. In power reactors, the value of k must be kept very nearly equal to 1.

The inside of a nuclear power plant in Kent, England. A technician is standing on the reactor charge transfer plate, into which uranium fuel rods fit.

Since the neutrons emitted in fission have energies of the order of 1 MeV, whereas the chance for neutron capture leading to fission in ^{235}U is largest at small energies, the chain reaction can be sustained only if the neutrons are slowed down before they escape from the reactor. At high energies (1 to 2 MeV), neutrons lose energy rapidly by inelastic scattering from ^{238}U, the principal constituent of natural uranium. (Natural uranium contains 99.3% ^{238}U and only 0.7% fissionable ^{235}U.) Once the neutron energy is below the excitation energies of the nuclei in the reactor (about 1 MeV), the main process of energy loss is by elastic scattering, in which a fast neutron collides with a nucleus at rest and transfers some of its kinetic energy to that nucleus. Such energy transfers are efficient only if the masses of the two bodies are comparable. A neutron will not transfer much energy in an elastic collision with a heavy uranium nucleus. Such a collision is like one between a marble and a billiard ball. The marble will be deflected by the much more massive billiard ball, and very little of its kinetic energy will be transferred to the billiard ball. A **moderator** consisting of material such as water or carbon that contains light nuclei is therefore placed around the fissionable material in the core of the reactor to slow down the neutrons. The neutrons are slowed down by elastic collisions with the nuclei of the moderator until they are in thermal equilibrium with the moderator. Because of the relatively large neutron-capture cross section of the hydrogen nucleus in water, reactors using ordinary water as a moderator cannot easily achieve $k \approx 1$ unless they use enriched uranium, in which the ^{235}U content has been increased from 0.7% to between 1 and 4%. Natural uranium can be used if heavy water (D_2O) is used instead of ordinary (light) water (H_2O) as the moderator. Although heavy water is expensive, most Canadian reactors use it for a moderator to avoid the cost of constructing uranium-enrichment facilities.

Figure 40-12 shows some of the features of a pressurized-water reactor commonly used in the United States to generate electricity. Fission in the core heats the water to a high temperature in the primary loop, which is closed. This water, which also serves as the moderator, is under high pressure to prevent it from boiling. The hot water is pumped to a heat exchanger, where it heats the water in the secondary loop and converts it to steam, which is then used to drive the turbines that produce electrical power. Note that the water in the secondary loop is isolated from that in the primary loop to prevent its contamination by the radioactive nuclei in the reactor core.

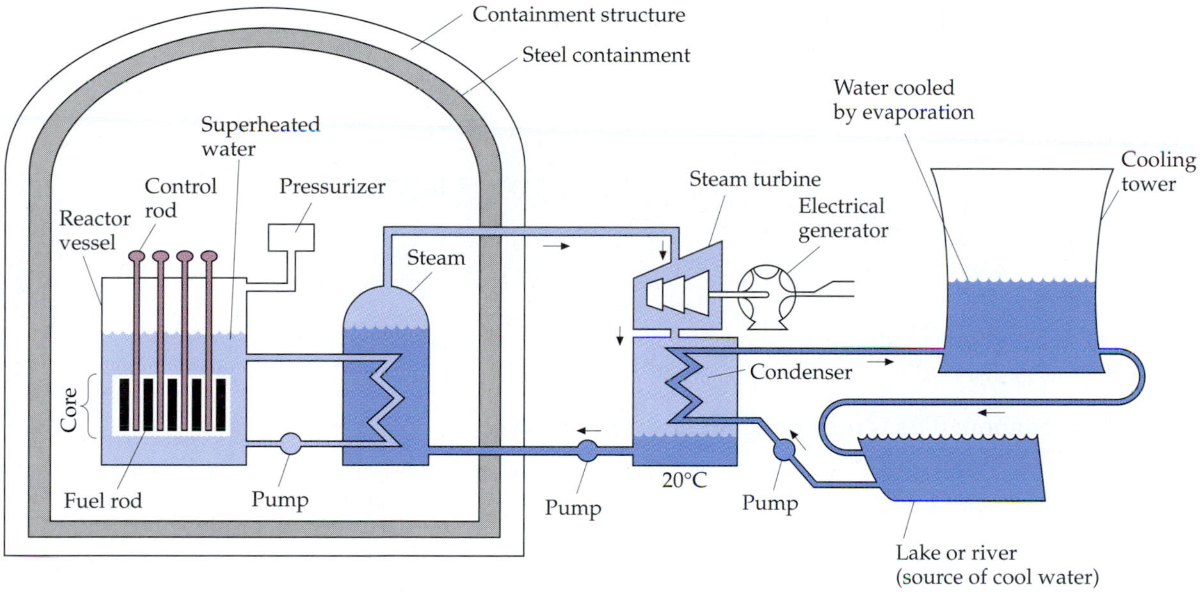

Figure 40-12 Simplified drawing of a pressurized-water reactor. The water in contact with the reactor core serves as both the moderator and the heat-transfer material. It is isolated from the water used to produce the steam that drives the turbines. Many features, such as the backup cooling mechanisms, are not shown here.

The ability to control the reproduction factor k precisely is important if a power reactor is to be operated safely. Both natural negative-feedback mechanisms and mechanical methods of control are used. If k is greater than 1 and the reaction rate increases, the temperature of the reactor increases. If water is used as a moderator, its density decreases with increasing temperature and it becomes a less effective moderator. A second important control method is the use of control rods made of a material, such as cadmium, that has a very large neutron-capture cross section. When the reactor is started up, the control rods are inserted so that k is less than 1. As the rods are gradually withdrawn from the reactor, fewer neutrons are captured by them and k increases to 1. If k becomes greater than 1, the rods are inserted again.

Mechanical control of the reaction rate of a nuclear reactor using control rods is possible only because some of the neutrons emitted in the fission process are **delayed neutrons.** The time needed for a neutron to slow down from 1 or 2 MeV to the thermal-energy level and then be captured is only of the order of a millisecond. If all the neutrons emitted in fission were prompt neutrons, that is, emitted immediately in the fission process, mechanical control would not be possible because the reactor would run away before the rods could be inserted. However, about 0.65% of the neutrons emitted are delayed by an average time of about 14 s. These neutrons are emitted not in the fission process itself but in the decay of the fission fragments. The effect of the delayed neutrons can be seen in the following examples.

Example 40-7

If the average time between fission generations (the time it takes for a neutron emitted in one fission to cause another) is $t_1 = 1$ ms $= 0.001$ s and the reproduction constant is 1.001, how long will it take for the reaction rate to double?

Picture the Problem The time to double is the number of generations N needed to double times the generation time. If $k = 1.001$, the reaction rate after N generations is 1.001^N. We find the number of generations by setting 1.001^N equal to 2 and solving for N.

1. Set 1.001^N equal to 2 and solve for N:

$$(1.001)^N = 2$$

$$N \ln 1.001 = \ln 2$$

$$N = \frac{\ln 2}{\ln 1.001} = 693$$

2. Multiply the number of generations by the generation time:

$$t = Nt_1 = 693(0.001\text{ s}) = 0.693\text{ s}$$

Remark The doubling time of about 0.7 s is not enough time for insertion of control rods.

Example 40-8 *try it yourself*

Assuming that 0.65% of the neutrons emitted are delayed by 14 s, find the average generation time and the doubling time if $k = 1.001$.

Picture the Problem The doubling time is Nt_{av}, where t_{av} is the average time between generations. Since 99.35% of the generation times are 0.001 s and 0.65% are 14 s, the average generation time is 0.9935(0.001 s) + 0.0065(14 s).

Cover the column to the right and try these on your own before looking at the answers.

Steps	*Answers*
1. Compute the average generation time.	1. $t_{av} = 0.092$ s
2. Use your result to find the time for 693 generations.	2. $t = 63.8$ s

Remarks Even though the number of delayed neutrons is less than 1%, they have a large effect on the doubling time. Here they increase the generation time by a factor of 92, resulting in a doubling time of about 64 s, which is plenty of time for mechanical insertion of control rods.

Because of the limited supply of natural uranium, the small fraction of ^{235}U in natural uranium, and the limited capacity of enrichment facilities, reactors based on the fission of ^{235}U cannot meet our energy needs for very long. A promising alternative is the **breeder reactor.** When the relatively plentiful but nonfissionable ^{238}U nucleus captures a neutron, it decays by β decay (with a half-life of 20 min) to ^{239}Np, which in turn decays by β decay (with a half-life of 2.35 days) to the fissionable nuclide ^{239}Pu. Since ^{239}Pu fissions with fast neutrons, no moderator is needed. A reactor initially fueled with a mixture of ^{238}U and ^{239}Pu will breed as much fuel as it uses or more if one or more of the neutrons emitted in the fission of ^{239}Pu is captured by ^{238}U. Practical studies indicate that a typical breeder reactor can be expected to double its fuel supply in 7 to 10 years.

There are two major safety problems inherent with breeder reactors. The fraction of delayed neutrons is only 0.3% for the fission of ^{239}Pu, so the time between generations is much less than that for ordinary reactors. Mechanical control is therefore much more difficult. Also, since the operating temperature of a breeder reactor is relatively high and a moderator is not desired, a heat-transfer material such as liquid sodium metal is used rather than water (which is the moderator as well as the heat-transfer material in an ordinary reactor). If the temperature of the reactor increases, the resulting decrease in the density of the heat-transfer material leads to positive feedback, since it will absorb fewer neutrons than before. Because of these safety considerations, breeder reactors are not yet in commercial use in the United States. There are, however, several in operation in France, Great Britain, and the Soviet Union.

Fusion

In fusion, two light nuclei such as deuterium (^{2}H) and tritium (^{3}H) fuse together to form a heavier nucleus. A typical fusion reaction is

$$^2\text{H} + {}^3\text{H} \rightarrow {}^4\text{He} + \text{n} + 17.6\text{ MeV}$$

The energy released in fusion depends on the particular reaction. For the ^{2}H + ^{3}H reaction, it is 17.6 MeV. Although this is less than the energy released in a fission reaction, it is a greater amount of energy per unit mass. The energy released in this fusion reaction is (17.6 MeV)/(5 nucleons) = 3.52 MeV per nucleon. This is about 3.5 times as great as the 1 MeV per nucleon released in fission.

The production of power from the fusion of light nuclei holds great promise because of the relative abundance of the fuel and the absence of some of the dangers inherent in fission reactors. Unfortunately, the technology necessary to make fusion a practical source of energy has not yet been developed. We will consider the ^{2}H + ^{3}H reaction; other reactions present similar problems.

Because of the Coulomb repulsion between the ^{2}H and ^{3}H nuclei, very large kinetic energies, of the order of 1 MeV, are needed to get the nuclei close enough together for the attractive nuclear forces to become effective and cause fusion. Such energies can be obtained in an accelerator, but since the scattering of one nucleus by the other is much more probable than fusion, the bombardment of one nucleus by another in an accelerator requires the input of more energy than is recovered. To obtain energy from fusion, the particles must be heated to a temperature great enough for the fusion reaction to occur as the result of random thermal collisions. Because a significant number of particles have kinetic energies greater than the mean kinetic energy, $\frac{3}{2}kT$, and because some particles can tunnel through the Coulomb barrier, a temperature T corresponding to $kT \approx 10$ keV is adequate to ensure that a reasonable number of fusion reactions will occur if the density of particles is sufficiently high. The temperature corresponding to $kT = 10$ keV is of the order of 10^8 K. Such temperatures occur in the interiors of stars, where such reactions are common. At these temperatures, a gas consists of positive ions and negative electrons and is called a **plasma.** One of the problems arising in attempts to produce controlled fusion reactions is that of confining the plasma long enough for the reactions to take place. In the interior of the sun the plasma is confined by the enormous gravitational field of the sun. In a laboratory on earth, confinement is a difficult problem.

The energy required to heat a plasma is proportional to the density of its ions, n, whereas the collision rate is proportional to n^2, the square of the density. If τ is the confinement time, the output energy is proportional to $n^2\tau$. If the output energy is to exceed the input energy, we must have

$$C_1 n^2 \tau > C_2 n$$

where C_1 and C_2 are constants. In 1957, the British physicist J. D. Lawson evaluated these constants from estimates of the efficiencies of various hypothetical fusion reactors and derived the following relation between density and confinement time, known as **Lawson's criterion:**

$$n\tau > 10^{20}\ \text{s}\cdot\text{particles/m}^3 \qquad \text{40-22}$$

Lawson's criterion

If Lawson's criterion is met and the thermal energy of the ions is great enough ($kT \sim 10$ keV), the energy released by a fusion reactor will just equal the energy input; that is, the reactor will just break even. For the reactor to be practical, much more energy must be released.

Two schemes for achieving Lawson's criterion are currently under investigation. In one scheme, **magnetic confinement,** a magnetic field is used to confine the plasma (see Section 28-2). In the most common arrangement, first developed in the USSR and called the Tokamak, the plasma is confined in a large toroid. The magnetic field is a combination of the doughnut-shaped magnetic field due to the windings of the toroid and the self-field due to the current of the circulating plasma. The break-even point has been achieved recently using magnetic confinement, but we are still a long way from building a practical fusion reactor.

In a second scheme, called **inertial confinement,** a pellet of solid deuterium and tritium is bombarded from all sides by intense pulsed laser beams of energies of the order of 10^4 J lasting about 10^{-8} s. (Intense beams of ions are also used.) Computer simulation studies indicate that the pellet should be compressed to about 10^4 times its normal density and heated to a temperature greater than 10^8 K. This should produce about 10^6 J of fusion energy in 10^{-11} s, which is so brief that confinement is achieved by inertia alone.

Because the break-even point is just barely being achieved in magnetic-confinement fusion, and because the building of a fusion reactor involves many practical problems that have not yet been solved, the availability of fusion to meet our energy needs is not expected for at least several decades. However, fusion holds great promise as an energy source for the future.

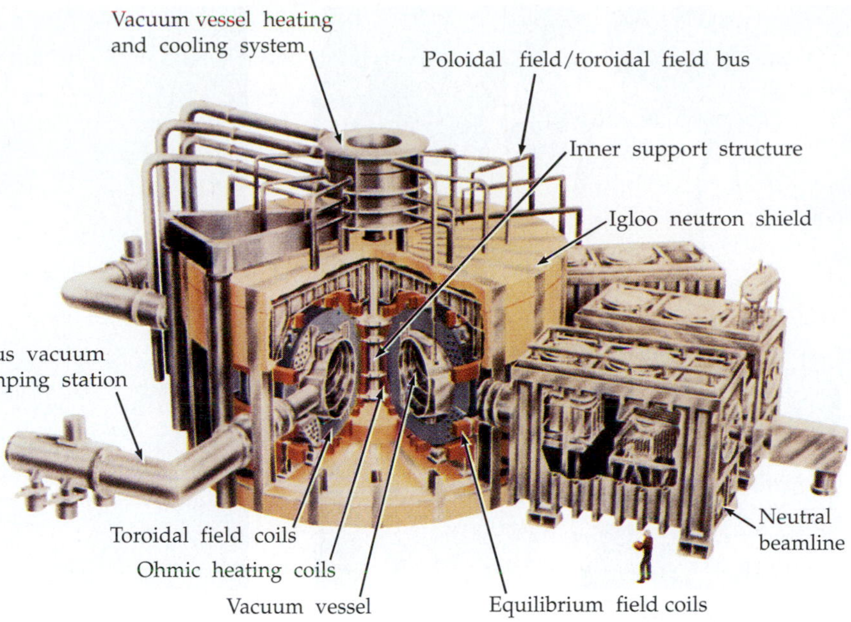

(*a*)

(*b*)

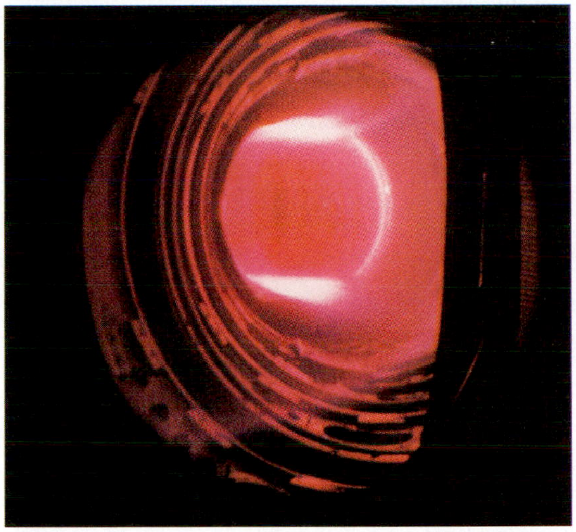

(*c*)

(*a*) Schematic of the Tokamak Fusion Test Reactor (TFTR). The toroidal coils, surrounding the doughnut-shaped vacuum vessel, are designed to conduct current for 3-s pulses, separated by waiting times of 5 min. Pulses peak at 73,000 A, producing a magnetic field of 5.2 T. This field is the principal means of confining the deuterium–tritium plasma that circulates within the vacuum vessel. Current for the pulses is delivered by converting the rotational energy of two 600-ton flywheels. Sets of poloidal coils, perpendicular to the toroidal coils, carry an oscillating current that generates a current through the confined plasma itself, heating it ohmically. Additional poloidal fields help stabilize the confined plasma. Between four and six neutral-beam injection systems (only one of which is shown in the schematic) are used to inject high-energy deuterium atoms into the deuterium–tritium plasma, heating beyond what could be obtained ohmically—ultimately to the point of fusion. (*b*) The TFTR itself. The diameter of the vacuum vessel is 7.7 m (see also photo on page 900). (*c*) An 800-kA plasma, lasting 1.6 s, as it discharges within the vacuum vessel.

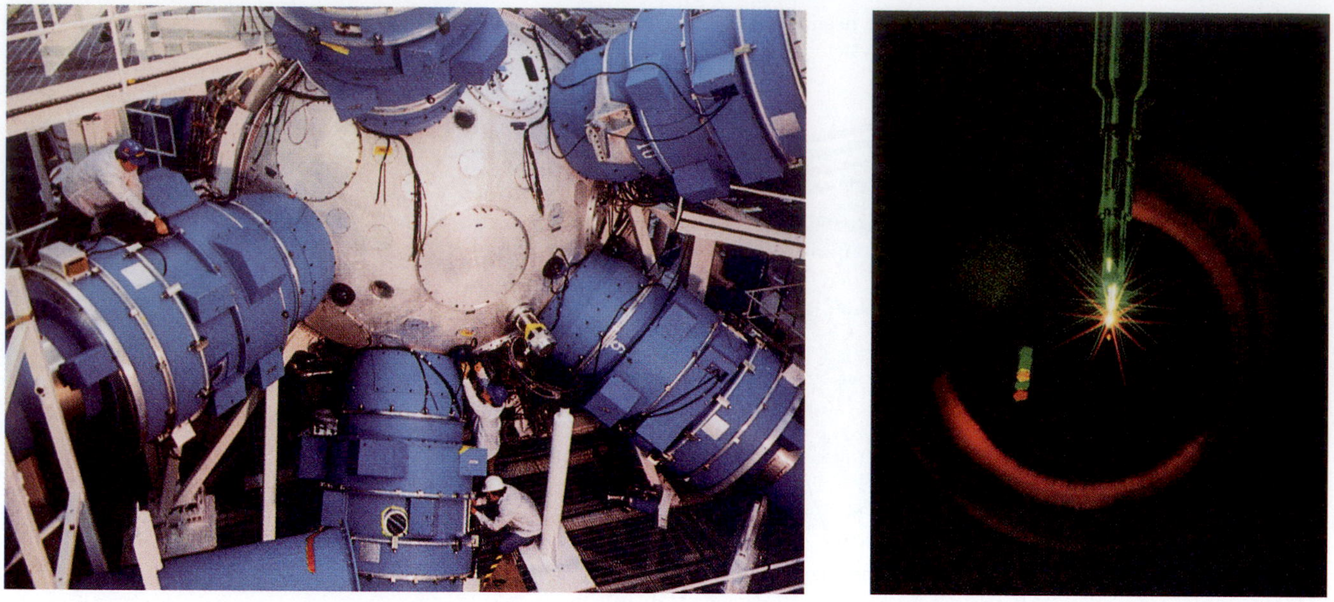

(a) (b)

(*a*) The Nova target chamber, an aluminum sphere approximately 5 m in diameter, inside which 10 beams from the world's most powerful laser converge onto a hydrogen-containing pellet 0.5 mm in diameter. The resulting fusion reaction is visible as a tiny star (*b*), lasting 10^{-10} s, releasing 10^{13} neutrons.

Summary

Topic	Relevant Equations and Remarks	
1. Nuclear Properties	Nuclei have N neutrons, Z protons, and a mass number $A = N + Z$. For light nuclei, N and Z are approximately equal, whereas for heavy nuclei, N is greater than Z.	
Isotopes	Isotopes consist of two or more nuclei having the same value of Z but different values of N and A.	
Size and shape	Most nuclei are approximately spherical in shape and have a volume that is proportional to A. Because the mass is proportional to A, nuclear density is independent of A.	
Radius	$R = R_0 A^{1/3} \approx (1.5\text{ fm})A^{1/3}$	**40-1**
Mass and binding energy	The mass of a stable nucleus is less than the sum of the masses of its nucleons. The mass difference times c^2 equals the binding energy of the nucleus. The binding energy is approximately proportional to the mass number A.	
2. Radioactivity	Unstable nuclei are radioactive and decay by emitting α particles (^{4}He nuclei), β particles (electrons or positrons), or γ rays (photons). All radioactivity is statistical in nature and follows an exponential decay law:	
Decay law	$N = N_0 e^{-\lambda t}$	**40-6**
Decay rate	$R = \lambda N = R_0 e^{-\lambda t}$	**40-7**
Mean life	$\tau = \frac{1}{\lambda}$	**40-9**

Half-life	$$t_{1/2} = \frac{0.693}{\lambda}$$ Half-lives of α decay range from a fraction of a second to millions of years. For β decay they range up to hours or days and for γ decay half-lives are usually less than a microsecond.	40-11
Curie	The number of decays per second of 1 g of radium is the curie, which equals 3.7×10^{10} decays/s $= 3.7 \times 10^{10}$ Bq.	
3. Nuclear Reactions		
Q value	The Q value equals c^2 times the difference in the total mass of the incoming particles and the total mass of the outgoing particles. If the net mass increase is Δm, the Q value is $$Q = -(\Delta m)c^2$$	40-19
Exothermic reaction	The mass decreases, Q is positive and measures the energy released.	
Endothermic reaction	The mass increases, Q is negative. Then $\lvert Q \rvert$ is the threshold energy for the reaction in the center of mass reference frame.	
3. Fission	Fission occurs when some heavy elements, such as ^{235}U or ^{239}Pu, capture a neutron and split apart into two medium-mass nuclei. The two nuclei then fly apart because of electrostatic repulsion, releasing a large amount of energy. A chain reaction is possible because several neutrons are emitted by a nucleus when it undergoes fission. A chain reaction can be sustained in a reactor if, on the average, one of the emitted neutrons is slowed down by scattering in the reactor and is then captured by another fissionable nucleus. Very heavy nuclei ($Z > 92$) are subject to spontaneous fission.	
4. Fusion	A large amount of energy is released when two light nuclei, such as ^2H and ^3H, fuse together. Fusion takes place spontaneously inside the sun and other stars, where the temperature is great enough (about 10^8 K) for thermal motion to bring the charged hydrogen ions close enough together to fuse. Although controlled fusion holds great promise as a future energy source, practical difficulties have thus far prevented its development.	
Lawson criterion	The minimum product of particle density n and confinement time τ to get more energy out of a fusion reactor than is put in is $n\tau > 10^{20}$ s·particles/m^3.	

Problem-Solving Guide

Summary of Worked Examples

Type of Calculation	Procedure and Relevant Examples
1. Nuclear Properties	
Find the binding energy of a nucleon in a nucleus.	Compute the difference in mass between the final nucleus plus nucleon and the original nucleus and multiply Δm by c^2. **Example 40-1**

2. Radioactivity		
Find the counting rate of a radioactive source at various times.	The rate decreases by a factor of 2 for each half-life.	**Examples 40-2, 40-3**
Find the number of nuclei that decay during a given time interval.	Find the number of nuclei at the beginning and end of the interval and subtract.	**Example 40-3**
Find the age of an organic sample.	Compare the actual decay rate of carbon in the sample with the original rate of 15 decays per gram per minute. You can compute the number of half-lives from $R = (0.5)^n R_0$ and then find t from $t_{1/2} = 5730$ y.	**Example 40-4**
3. Find the Q value of a reaction.	Calculate the difference in the total mass of the incoming particles and outgoing particles and multiply by c^2.	**Example 40-5**
4. Nuclear Fission		
Calculate the total energy released in fission given the mass decrease.	Use 200 MeV per nuclei and find the number of nuclei in the sample.	**Example 40-6**
Find the time for the reaction rate in a reactor to double given the reproduction constant.	Multiply the generation time by the number of generations. If there are delayed neutrons, you need to compute the average generation time.	**Examples 40-7, 40-8**

Problems

In a few problems, you are given more data than you actually need; in a few other problems, you are required to supply data from your general knowledge, outside sources, or informed estimates.

Conceptual Problems

Problems from Optional and Exploring sections

• Single-concept, single-step, relatively easy

•• Intermediate-level, may require synthesis of concepts

••• Challenging, for advanced students

Properties of Nuclei

1 • Give the symbols for two other isotopes of (*a*) ^{14}N, (*b*) ^{56}Fe, and (*c*) ^{118}Sn.

2 • Calculate the binding energy and the binding energy per nucleon from the masses given in Table 40-1 for (*a*) ^{12}C, (*b*) ^{56}Fe, and (*c*) ^{238}U.

3 • Repeat Problem 2 for (*a*) ^{6}Li, (*b*) ^{39}K, and (*c*) ^{208}Pb.

4 • Use Equation 40-1 to compute the radii of the following nuclei: (*a*) ^{16}O, (*b*) ^{56}Fe, and (*c*) ^{197}Au.

5 • (*a*) Given that the mass of a nucleus of mass number A is approximately $m = CA$, where C is a constant, find an expression for the nuclear density in terms of C and the constant R_0 in Equation 40-1. (*b*) Compute the value of this nuclear density in grams per cubic centimeter using the fact that C has the approximate value of 1 g per Avogadro's number of nucleons.

6 • Derive Equation 40-2; that is, show that the rest energy of one unified mass unit is 931.5 MeV.

7 • Use Equation 40-1 for the radius of a spherical nucleus and the approximation that the mass of a nucleus of mass number A is A u to calculate the density of nuclear matter in grams per cubic centimeter.

8 •• The electrostatic potential energy of two charges q_1 and q_2 separated by a distance r is $U = kq_1q_2/r$, where k is the Coulomb constant. (*a*) Use Equation 40-1 to calculate the radii of ^{2}H and ^{3}H. (*b*) Find the electrostatic potential energy when these two nuclei are just touching, that is, when their centers are separated by the sum of their radii.

9 •• (*a*) Calculate the radii of $^{141}_{56}$Ba and $^{92}_{36}$Kr from Equation 40-1. (*b*) Assume that after the fission of ^{235}U into ^{141}Ba and ^{92}Kr, the two nuclei are momentarily separated by a distance r equal to the sum of the radii found in (*a*), and calculate the electrostatic potential energy for these two nuclei at this separation. (See Problem 8.) Compare your result with the measured fission energy of 175 MeV.

Radioactivity

10 • Why is the decay series $A = 4n + 1$ not found in nature?

11 • A decay by α emission is often followed by β decay. When this occurs, it is by β^- and not β^+ decay. Why?

12 • The half-life of ^{14}C is much less than the age of the universe, yet ^{14}C is found in nature. Why?

13 • What effect would a long-term variation in cosmic-ray activity have on the accuracy of ^{14}C dating?

14 • Homer enters the visitors' chambers, and his geiger-beeper goes off. He shuts off the beep, removes the device from his shoulder patch and holds it near the only new object in the room: an orb which is to be presented as a gift from the visiting Cartesians. Pushing a button marked "monitor," Homer reads that the orb is a radioactive source with a counting rate of 4000 counts/s. After 10 min, the counting rate has dropped to 1000 counts/s. The source's half-life appears on the geiger-beeper display. (*a*) What is the half-life? (*b*) What will the counting rate be 20 min after the monitoring device was switched on?

15 • A certain source gives 2000 counts/s at time $t = 0$. Its half-life is 2 min. (*a*) What is the counting rate after 4 min? (*b*) After 6 min? (*c*) After 8 min?

16 • The counting rate from a radioactive source is 8000 counts/s at time $t = 0$, and 10 min later the rate is 1000 counts/s. (*a*) What is the half-life? (*b*) What is the decay constant? (*c*) What is the counting rate after 20 min?

17 • The half-life of radium is 1620 y. Calculate the number of disintegrations per second of 1 g of radium, and show that the disintegration rate is approximately 1 Ci.

18 • A radioactive silver foil ($t_{1/2} = 2.4$ min) is placed near a Geiger counter and 1000 counts/s are observed at time $t = 0$. (*a*) What is the counting rate at $t = 2.4$ min and at $t = 4.8$ min? (*b*) If the counting efficiency is 20%, how many radioactive nuclei are there at time $t = 0$? At time $t = 2.4$ min? (*c*) At what time will the counting rate be about 30 counts/s?

19 • Use Table 40-1 to calculate the energy in MeV for the α decay of (*a*) ^{226}Ra and (*b*) ^{242}Pu.

20 • Suppose that two billion years ago 10% of the mass of the earth was ^{14}C. Approximately what percentage of the mass of the earth today would be ^{14}C, neglecting formation of ^{14}C in the atmosphere?

21 • At the scene of the crime, in the museum's west wing, Angela found some wood chips, so she slipped them into her purse for future analysis. They were allegedly from an old wooden mask, which the guard said he threw at the would-be thief. Later, in the lab, she determined the age of the chips, using a sample that contained 10 g of carbon and showed a ^{14}C decay rate of 100 counts/min. How old are they?

22 • The thief in Problem 21 had been after a valuable carving made from a 10,000-year-old bone. The guard said that he chased the thief away, but Angela suspects that the guard is an accomplice, and that the bone in the display case is, in fact, a fake. If a sample of the bone containing 15 g of carbon were to be analyzed, what should the decay rate of ^{14}C be if it is a 10,000-year-old bone?

23 • Through a friend in security at the museum, Angela got a sample having 175 g of carbon. The decay rate of ^{14}C was 8.1 Bq. (*a*) How old is it? (*b*) Is it from the carving described in Problem 22?

24 • A sample of a radioactive isotope is found to have an activity of 115.0 Bq immediately after it is pulled from the reactor that formed it. Its activity 2 h 15 min later is measured to be 85.2 Bq. (*a*) Calculate the decay constant and the half-life of the sample. (*b*) How many radioactive nuclei were there in the sample initially?

25 •• Derive the result that the activity of 1 g of natural carbon due to the β decay of ^{14}C is 15 decays/min = 0.25 Bq.

26 •• Measurements of the activity of a radioactive sample have yielded the following results. Plot the activity as a function of time, using semilogarithmic paper, and determine the decay constant and half-life of the radioisotope.

Time, min	Activity	Time, min	Activity
0	4287	20	880
5	2800	30	412
10	1960	40	188
15	1326	60	42

27 •• (*a*) Show that if the decay rate is R_0 at time $t = 0$ and R_1 at some later time t_1, the decay constant is given by $\lambda = t_1^{-1} \ln(R_0/R_1)$ and the half-life is given by $t_{1/2} = 0.693t_1/\ln(R_0/R_1)$. (*b*) Use these results to find the decay constant and the half-life if the decay rate is 1200 Bq at $t = 0$ and 800 Bq at $t_1 = 60$ s.

28 •• A wooden casket is thought to be 18,000 years old. How much carbon would have to be recovered from this object to yield a ^{14}C counting rate of no less than 5 counts/min?

29 •• A 1.00-mg sample of substance of atomic mass 59.934 u emits β particles with an activity of 1.131 Ci. Find the decay constant for this substance in s^{-1} and its half-life in years.

30 •• The counting rate from a radioactive source is measured every minute. The resulting counts per second are 1000, 820, 673, 552, 453, 371, 305, 250. Plot the counting rate versus time on semilog graph paper, and use your graph to find the half-life of the source.

31 •• A sample of radioactive material is initially found to have an activity of 115.0 decays/min. After 4 d 5 h, its activity is measured to be 73.5 decays/min. (*a*) Calculate the half-life of the material. (*b*) How long (from the initial time) will it take for the sample to reach an activity level of 10.0 decays/min?

32 •• The rubidium isotope ^{87}Rb is a β emitter with a half-life of 4.9×10^{10} y that decays into ^{87}Sr. It is used to determine the age of rocks and fossils. Rocks containing the fossils of early animals contain a ratio of ^{87}Sr to ^{87}Rb of 0.0100. Assuming that there was no ^{87}Sr present when the rocks were formed, calculate the age of these fossils.

33 ••• If there are N_0 radioactive nuclei at time = 0, the number that decay in some time interval dt at time t is $-dN = \lambda N_0 e^{-\lambda t}\, dt$. If we multiply this number by the lifetime t of these nuclei, sum over all the possible lifetimes from $t = 0$ to $t = \infty$, and divide by the total number of nuclei, we get the mean lifetime τ:

$$\tau = \frac{1}{N_0}\int_0^\infty t|dN| = \int_0^\infty t\lambda e^{-\lambda t}\, dt$$

Show that $\tau = 1/\lambda$.

Nuclear Reactions

34 • Using Table 40-1, find the Q values for the following reactions: (*a*) ${}^{1}H + {}^{3}H \rightarrow {}^{3}He + n + Q$ and (*b*) ${}^{2}H + {}^{2}H \rightarrow {}^{3}He + n + Q$.

35 • Using Table 40-1, find the Q values for the following reactions: (*a*) ${}^{2}H + {}^{2}H \rightarrow {}^{3}H + {}^{1}H + Q$, (*b*) ${}^{2}H + {}^{3}He \rightarrow {}^{4}He + {}^{1}H + Q$, and (*c*) ${}^{6}Li + n \rightarrow {}^{3}H + {}^{4}He + Q$.

36 •• (*a*) Use the atomic masses $m = 14.00324$ u for ${}^{14}_{6}C$ and $m = 14.00307$ u for ${}^{14}_{7}N$ to calculate the Q value (in MeV) for the β decay

$$^{14}_{6}C \rightarrow {}^{14}_{7}N + \beta^- + \bar{\nu}_e$$

(*b*) Explain why you do not need to add the mass of the β^- to that of atomic ${}^{14}_{7}N$ for this calculation.

37 •• (*a*) Use the atomic masses $m = 13.00574$ u for ${}^{13}_{7}N$ and $m = 13.003354$ u for ${}^{13}_{6}C$ to calculate the Q value (in MeV) for the β decay

$$^{13}_{7}N \rightarrow {}^{13}_{6}C + \beta^+ + \nu_e$$

(*b*) Explain why you need to add two electron masses to the mass of ${}^{13}_{6}C$ in the calculation of the Q value for this reaction.

Fission and Fusion

38 • Why isn't there an element with $Z = 130$?

39 • Why is a moderator needed in an ordinary nuclear fission reactor?

40 • Explain why water is more effective than lead in slowing down fast neutrons.

41 • What happens to the neutrons produced in fission that do not produce another fission?

42 • What is the advantage of a breeder reactor over an ordinary one? What are the disadvantages?

43 • Assuming an average energy of 200 MeV per fission, calculate the number of fissions per second needed for a 500-MW reactor.

44 • If the reproduction factor in a reactor is $k = 1.1$, find the number of generations needed for the power level to (*a*) double, (*b*) increase by a factor of 10, and (*c*) increase by a factor of 100. Find the time needed in each case if (*d*) there are no delayed neutrons, so the time between generations is 1 ms, and (*e*) there are delayed neutrons that make the average time between generations 100 ms.

45 • Compute the temperature T for which $kT = 10$ keV, where k is Boltzmann's constant.

46 •• In 1989, researchers claimed to have achieved fusion in an electrochemical cell at room temperature. They claimed a power output of 4 W from deuterium fusion reactions in the palladium electrode of their apparatus. If the two most likely reactions are

$$^{2}H + {}^{2}H \rightarrow {}^{3}He + n + 3.27 \text{ MeV}$$

and

$$^{2}H + {}^{2}H \rightarrow {}^{3}H + {}^{1}H + 4.03 \text{ MeV}$$

with 50% of the reactions going by each branch, how many neutrons per second would we expect to be emitted in the generation of 4 W of power?

47 •• A fusion reactor using only deuterium for fuel would have the two reactions in Problem 46 taking place in it. The ${}^{3}H$ produced in the second reaction reacts immediately with another ${}^{2}H$ to produce

$$^{3}H + {}^{2}H \rightarrow {}^{4}He + n + 17.6 \text{ MeV}$$

The ratio of ${}^{2}H$ to ${}^{1}H$ atoms in naturally occurring hydrogen is 1.5×10^{-4}. How much energy would be produced from 4 L of water if all of the ${}^{2}H$ nuclei undergo fusion?

48 ••• The fusion reaction between ${}^{2}H$ and ${}^{3}H$ is

$$^{3}H + {}^{2}H \rightarrow {}^{4}He + n + 17.6 \text{ MeV}$$

Using the conservation of momentum and the given Q value, find the final energies of both the ${}^{4}He$ nucleus and the neutron, assuming the initial kinetic energy of the system is 1.00 MeV and the initial momentum of the system is zero.

49 ••• Energy is generated in the sun and other stars by fusion. One of the fusion cycles, the proton–proton cycle, consists of the following reactions:

$$^{1}H + {}^{1}H \rightarrow {}^{2}H + \beta^+ + \nu_e$$

$$^{1}H + {}^{2}H \rightarrow {}^{3}He + \gamma$$

followed by

$$^{1}H + {}^{3}He \rightarrow {}^{4}He + \beta^+ + \nu_e$$

(*a*) Show that the net effect of these reactions is

$$4\,{}^{1}H \rightarrow {}^{4}He + 2\beta^+ + 2\nu_e + \gamma$$

(*b*) Show that rest energy of 24.7 MeV is released in this cycle (not counting the energy of 1.02 MeV released when each positron meets an electron and the two annihilate). (*c*) The sun radiates energy at the rate of about 4×10^{26} W. Assuming that this is due to the conversion of four protons into helium plus γ rays and neutrinos, which releases 26.7 MeV, what is the rate of proton consumption in the sun? How long will the sun last if it continues to radiate at its present level? (Assume that protons constitute about half of the total mass [2×10^{30} kg] of the sun.)

General Problems

50 • True or false:

(*a*) The atomic nucleus contains protons, neutrons, and electrons.
(*b*) The mass of ${}^{2}H$ is less than the mass of a proton plus a neutron.
(*c*) After two half-lives, all the radioactive nuclei in a given sample have decayed.
(*d*) In a breeder reactor, fuel can be produced as fast as it is consumed.

51 • Why do extreme changes in the temperature or pressure of a radioactive sample have little or no effect on the radioactivity?

52 • The stable isotope of sodium is ${}^{23}Na$. What kind of radioactivity would you expect of (*a*) ${}^{22}Na$ and (*b*) ${}^{24}Na$?

53 • Why does fusion occur spontaneously in the sun but not on earth?

54 • (*a*) Show that $ke^2 = 1.44$ MeV·fm, where k is the Coulomb constant and e is the electron charge. (*b*) Show that $hc = 1240$ MeV·fm.

55 • The counting rate from a radioactive source is 6400 counts/s. The half-life of the source is 10 s. Make a plot of the counting rate as a function of time for times up to 1 min. What is the decay constant for this source?

56 • Find the energy needed to remove a neutron from (*a*) ^{4}He and (*b*) ^{7}Li.

57 • The isotope ^{14}C decays according to $^{14}\text{C} \rightarrow {}^{14}\text{N} + e^- + \bar{\nu}_e$. The atomic mass of ^{14}N is 14.003074 u. Determine the maximum kinetic energy of the electron. (Neglect recoil of the nitrogen atom.)

58 • A neutron star is an object of nuclear density. If our sun were to collapse to a neutron star, what would be the radius of that object?

59 • Nucleus A has a half-life that is twice that of nucleus B. At $t = 0$ the number of B nuclei in a sample is twice that of A nuclei. If the half-life of A is 1 h, will there ever be an instant when the number of A and B nuclei are equal? If so, when will this moment occur?

60 • Calculate the nuclear radii of ^{19}F, ^{145}La, and ^{246}Cm.

61 • The relative abundance of ^{40}K (molecular mass 40.0 g/mol) is 1.2×10^{-4}. The isotope ^{40}K is radioactive with a half-life of 1.3×10^9 y. Potassium is an essential element of every living cell. In the human body the mass of potassium constitutes approximately 0.36% of the total mass. Determine the activity of this radioactive source in a student whose mass is 60 kg.

62 •• A 0.05394-kg sample of ^{144}Nd (atomic mass 143.91 u) emits an average of 2.36 α particles each second. Find the decay constant in s^{-1} and the half-life in years.

63 •• The isotope ^{24}Na is a β emitter with a half-life of 15 h. A saline solution containing this radioactive isotope with an activity of 600 kBq is injected into the bloodstream of a patient. Ten hours later, the activity of 1 mL of blood from this individual yields a counting rate of 60 Bq. Determine the volume of blood in this patient.

64 •• (*a*) Determine the closest distance of approach of an 8-MeV α particle in a head-on collision with a nucleus of ^{197}Au and a nucleus of ^{10}B, neglecting the recoil of the struck nuclei. (*b*) Repeat the calculation taking into account the recoil of the struck nuclei.

65 •• Twelve nucleons are in a one-dimensional infinite square well of length $L = 3$ fm. (*a*) Using the approximation that the mass of a nucleon is 1 u, find the lowest energy of a nucleon in the well. Express your answer in MeV. What is the ground-state energy of the system of 12 nucleons in the well if (*b*) all the nucleons are neutrons so that there can be only 2 in each state and (*c*) 6 of the nucleons are neutrons and 6 are protons so that there can be 4 nucleons in each state? (Neglect the energy of Coulomb repulsion of the protons.)

66 •• The helium nucleus or α particle is a very tightly bound system. Nuclei with $N = Z = 2n$, where n is an integer, such as ^{12}C, ^{16}O, ^{20}Ne, and ^{24}Mg, may be thought of as agglomerates of α particles. (*a*) Use this model to estimate the binding energy of a pair of α particles from the atomic masses of ^{4}He and ^{16}O. Assume that the four α particles in ^{16}O form a regular tetrahedron with one α particle at each vertex. (*b*) From the result obtained in part (*a*) determine, on the basis of this model, the binding energy of ^{12}C and compare your result with that obtained from the atomic mass of ^{12}C.

67 •• Radioactive nuclei with a decay constant of λ are produced in an accelerator at a constant rate R_p. The number of radioactive nuclei N then obeys the equation $dN/dt = R_p - \lambda N$. (*a*) If N is zero at $t = 0$, sketch N versus t for this situation. (*b*) The isotope ^{62}Cu is produced at a rate of 100 per second by placing ordinary copper (^{63}Cu) in a beam of high-energy photons. The reaction is

$$\lambda + {}^{63}\text{Cu} \rightarrow {}^{62}\text{Cu} + \text{n}$$

^{62}Cu decays by β decay with a half-life of 10 min. After a time long enough so that $dN/dt \approx 0$, how many ^{62}Cu nuclei are there?

68 •• The total energy consumed in the United States in 1 y is about 7.0×10^{19} J. How many kilograms of ^{235}U would be needed to provide this amount of energy if we assume that 200 MeV of energy is released by each fissioning uranium nucleus, that all of the uranium atoms undergo fission, and that all of the energy-conversion mechanisms used are 100% efficient?

69 •• (*a*) Find the wavelength of a particle in the ground state of a one-dimensional infinite square well of length $L = 2$ fm. (*b*) Find the momentum in units of MeV/c for a particle with this wavelength. (*c*) Show that the total energy of an electron with this wavelength is approximately $E \approx pc$. (*d*) What is the kinetic energy of an electron in the ground state of this well? This calculation shows that if an electron were confined in a region of space as small as a nucleus, it would have a very large kinetic energy.

70 •• (*a*) How many α decays and how many β decays must a ^{222}Rn nucleus undergo before it becomes a ^{210}Pb nucleus? (*b*) Calculate the total energy released in the decay of one ^{222}Rn nucleus to ^{210}Pb. (The mass of ^{210}Pb is 209.984187 u.)

71 ••• Assume that a neutron decays into a proton plus an electron without the emission of a neutrino. The energy shared by the proton and electron is then 0.782 MeV. In the rest frame of the neutron, the total momentum is zero, so the momentum of the proton must be equal and opposite that of the electron. This determines the relative energies of the two particles, but because the electron is relativistic, the exact calculation of these relative energies is somewhat difficult. (*a*) Assume that the kinetic energy of the electron is 0.782 MeV and calculate the momentum p of the electron in units of MeV/c. (*Hint:* Use Equation 39-28.) (*b*) From your result for (*a*), calculate the kinetic energy $p^2/2m_p$ of the proton. (*c*) Since the total energy of the electron plus proton is 0.782 MeV, the calculation in (*b*) gives a correction to the

assumption that the energy of the electron is 0.782 MeV. What percentage of 0.782 MeV is this correction?

72 ••• Consider a neutron of mass m moving with speed v_L and making an elastic head-on collision with a nucleus of mass M that is at rest in the laboratory frame of reference. (*a*) Show that the speed of the center of mass in the lab frame is $V = mv_L/(m + M)$. (*b*) What is the speed of the nucleus in the center-of-mass frame before the collision? After the collision? (*c*) What is the speed of the nucleus in the lab frame after the collision? (*d*) Show that the energy of the nucleus after the collision in the lab frame is

$$\frac{1}{2}M(2V)^2 = \frac{4mM}{(m + M)^2}\left(\frac{1}{2}mv_L^2\right)$$

(*e*) Show that the fraction of the energy lost by the neutron in this elastic collision is

$$\frac{-\Delta E}{E} = \frac{4mM}{(m + M)^2} = \frac{4(m/M)}{(1 + m/M)^2} \qquad \text{40-23}$$

73 ••• (*a*) Use the result of part (*e*) of Problem 72 (Equation 40-23) to show that after N head-on collisions of a neutron with carbon nuclei at rest, the energy of the neutron is approximately $(0.714)^N E_0$, where E_0 is its original energy. (*b*) How many head-on collisions are required to reduce the energy of the neutron from 2 MeV to 0.02 eV, assuming stationary carbon nuclei?

74 ••• On the average, a neutron loses 63% of its energy in a collision with a hydrogen atom and 11% of its energy in a collision with a carbon atom. Calculate the number of collisions needed to reduce the energy of a neutron from 2 MeV to 0.02 eV if the neutron collides with (*a*) hydrogen atoms and (*b*) carbon atoms. (See Problem 73.)

75 ••• Frequently, the "daughter" of a radioactive "parent" is itself radioactive. Suppose the parent, designated by A, has a decay constant λ_A, while the daughter, designated B, has a decay constant λ_B. The number of nuclei of B are then given by the solution to the differential equation

$$dN_B/dt = \lambda_A N_A - \lambda_B N_B$$

(*a*) Justify this differential equation. (*b*) Show that the solution for this equation is

$$N_B(t) = \frac{N_{A0}\lambda_A}{\lambda_B - \lambda_A}(e^{-\lambda_A t} - e^{-\lambda_B t})$$

where N_{A0} is the number of A nuclei present at $t = 0$ when there are no B nuclei. (*c*) Show that $N_B(t) > 0$ whether $\lambda_A > \lambda_B$ or $\lambda_B > \lambda_A$. (*d*) Make a plot of $N_A(t)$ and $N_B(t)$ as a function of time when $\tau_B = 3\tau_A$.

76 ••• Suppose isotope A decays to isotope B with a decay constant λ_A, and isotope B in turn decays with a decay constant λ_B. Suppose a sample contains, at $t = 0$, only isotope A. Derive an expression for the time at which the number of isotope B nuclei will be a maximum. (See Problem 75.)

77 ••• An example of the situation discussed in Problem 75 is the radioactive isotope ^{229}Th, an α emitter with a half-life of 7300 y. Its daughter, ^{225}Ra, is a β emitter with a half-life of 14.8 d. In this, as in many instances, the half-life of the parent is much longer than that of the daughter. Using the expression given in Problem 75 (*b*), show that, starting with a sample of pure ^{229}Th containing N_{A0} nuclei, the number, N_B, of ^{225}Ra nuclei will, after several years, be a constant, given by

$$N_B = \frac{\lambda_A}{\lambda_B}N_A$$

The number of daughter nuclei are said to be in "secular equilibrium."

CHAPTER 41

Elementary Particles and the Beginning of the Universe

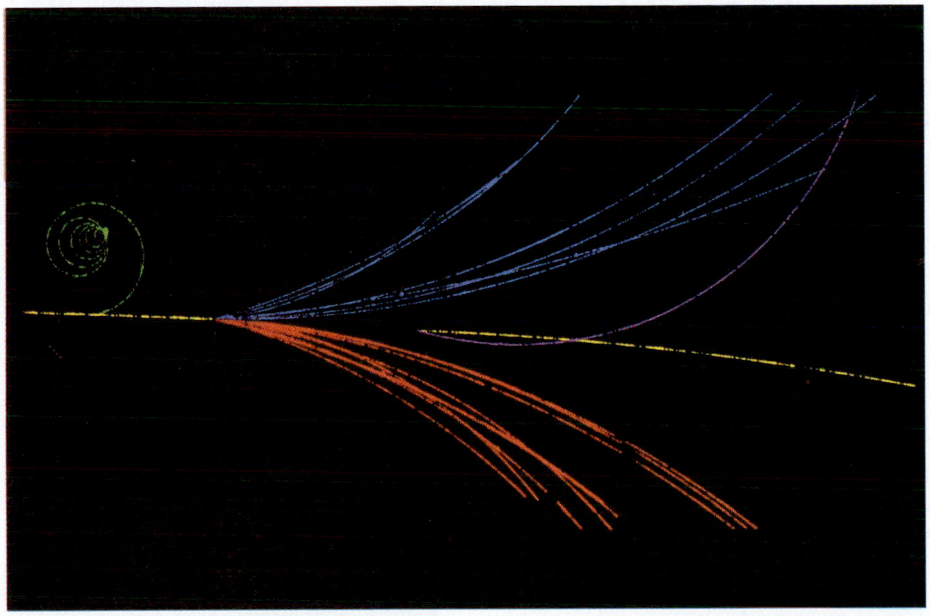

Tracks in a bubble chamber produced by an incoming high-energy proton (yellow) incident from the left, colliding with a proton at rest. The small green spiral is an electron knocked out of an atom. It curves to the left because of an external magnetic field in the chamber. The collision produces seven negative particles (π^-)(blue); a neutral particle Λ^0 that leaves no track; and nine positive particles (red) including seven π^+, a K^+, and a proton. The Λ^0 travels in the original direction of the incoming proton before decaying into a proton (yellow) and a π^- (purple).

In Dalton's atomic theory of matter (1808), the atom was considered to be the smallest indivisible constituent of matter, that is, an elementary particle. Then, with the discovery of the electron by Thomson (1897), the Bohr theory of the nuclear atom (1913), and the discovery of the neutron (1932), it became clear that atoms and even nuclei have considerable structure. For a time, it was thought that there were just four "elementary" particles: proton, neutron, electron, and photon. However, the positron or antielectron was discovered in 1932, and shortly thereafter the muon, the pion, and many other particles were predicted and discovered.

Since the 1950s, enormous sums of money have been spent constructing particle accelerators of greater and greater energies in hopes of finding particles predicted by various theories. At present, we know of several hundred particles that at one time or another have been considered to be elementary, and research teams at the giant accelerator laboratories around the world are searching for and finding new particles. Some of these have such short lifetimes (of the order of 10^{-23} s) that they can be detected only indirectly. Many are observed only in nuclear reactions with high-energy accelerators. In addition to the usual particle properties of mass, charge, and spin, new properties have been found and given whimsical names such as strangeness, charm, color, topness, and bottomness.

In this chapter, we will first look at the various ways of classifying the multitude of particles that have been found. We will then describe the current theory of elementary particles, called the *standard model,* in which all matter in nature—from the exotic particles produced in the giant accelerator laboratories to ordinary grains of sand—is considered to be constructed from just two families of elementary particles, leptons and quarks. In the final section we will use our knowledge of elementary particles to discuss the Big Bang theory of the origin of the universe.

41-1 Hadrons and Leptons

All the different forces observed in nature, from ordinary friction to the tremendous forces involved in supernova explosions, can be understood in terms of the four basic interactions: (1) the strong nuclear interaction (also called the hadronic interaction), (2) the electromagnetic interaction, (3) the weak (nuclear) interaction, and (4) the gravitational interaction. The four basic interactions provide a convenient structure for the classification of particles. Some particles participate in all four interactions, whereas others participate in only some of them. For example, all particles participate in gravity, the weakest of the interactions. All particles that carry electric charge participate in the electromagnetic interaction.

Particles that interact via the strong interaction are called **hadrons.** There are two kinds of hadrons: **baryons,** which have spin $\frac{1}{2}$ (or $\frac{3}{2}$, $\frac{5}{2}$, and so on), and **mesons,** which have zero or integral spin. Baryons, which include nucleons, are the most massive of the elementary particles. Mesons have intermediate masses between the mass of the electron and the mass of the proton. Particles that decay via the strong interaction have very short lifetimes of the order of 10^{-23} s, which is about the time it takes light to travel a distance equal to the diameter of a nucleus. On the other hand, particles that decay via the weak interaction have much longer lifetimes of the order of 10^{-10} s. Table 41-1 lists some of the properties of those hadrons that are stable against decay via the strong interaction.

Hadrons are rather complicated entities with complex structures. If we use the term "elementary particle" to mean a point particle without structure that is not constructed from some more elementary entities, hadrons do not fit the bill. It is now believed that all hadrons are composed of more fundamental entities called *quarks,* which are truly elementary particles.

Particles that participate in the weak interaction but not in the strong interaction are called **leptons.** These include electrons, muons, and neutrinos, which are all less massive than the lightest hadron. The word *lepton,* meaning "light particle," was chosen to reflect the relatively small mass of these particles. However, the most recently discovered lepton, the *tau,* found by Perl in 1975, has a mass of 1784 MeV/c^2, nearly twice that of the proton (938 MeV/c^2), so we now have a "heavy lepton." As far as we know, leptons are point particles with no structure and can be considered to be truly elementary in the sense that they are not composed of other particles.

There are six leptons, each of which has an antiparticle. They are the electron, the muon, and the tau, and a distinct neutrino associated with each of these three particles. (The neutrino associated with the tau has not yet been observed experimentally.) The masses of these particles are quite different. The mass of the electron is 0.511 MeV/c^2, the mass of the muon is 106 MeV/c^2, and that of the tau is 1784 MeV/c^2. The neutrinos were originally thought to be massless, but there is now strong evidence that their mass,

The Super-Kamiokande detector, built in Japan in 1996 as a joint Japanese–American experiment, is essentially a water tank the size of a large cathedral installed in a deep zinc mine one mile inside a mountain. When neutrinos pass through the tank, one of them occasionally collides with an atom, sending blue light through the water to an array of detectors. This is a picture of the detector wall and top with about 9000 photomultiplier tubes which help detect the neutrinos. Experimental results reported in June 1998 indicate that the mass of the neutrino cannot be zero.

Table 41-1

Hadrons That Are Stable Against Decay via the Strong Nuclear Interaction

Name	Symbol	Mass MeV/c^2	Spin, $\hbar$	Charge, e	Antiparticle	Mean Lifetime, s	Typical Decay Products[a]
Baryons							
Nucleon	p (proton)	938.3	$\frac{1}{2}$	+1	$\overline{p}$	Infinite	
	n (neutron)	939.6	$\frac{1}{2}$	0	$\overline{n}$	930	$p + e^- + \overline{\nu}_e$
Lambda	Λ^0	1116	$\frac{1}{2}$	0	$\overline{\Lambda^0}$	2.5×10^{-10}	$p + \pi^-$
Sigma[b]	Σ^+	1189	$\frac{1}{2}$	+1	$\overline{\Sigma^-}$	0.8×10^{-10}	$n + \pi^+$
	Σ^0	1193	$\frac{1}{2}$	0	$\overline{\Sigma^0}$	10^{-20}	$\Lambda^0 + \gamma$
	Σ^-	1197	$\frac{1}{2}$	−1	$\overline{\Sigma^+}$	1.7×10^{-10}	$n + \pi^-$
Xi	Ξ^0	1315	$\frac{1}{2}$	0	$\overline{\Xi^0}$	3.0×10^{-10}	$\Lambda^0 + \pi^0$
	Ξ^-	1321	$\frac{1}{2}$	−1	$\overline{\Xi^+}$	1.7×10^{-10}	$\Lambda^0 + \pi^-$
Omega	Ω^-	1672	$\frac{3}{2}$	−1	Ω^+	1.3×10^{-10}	$\Xi^0 + \pi^-$
Mesons							
Pion	π^+	139.6	0	+1	π^-	2.6×10^{-8}	$\mu^+ + \nu_\mu$
	π^0	135	0	0	π^0	0.8×10^{-16}	$\gamma + \gamma$
	π^-	139.6	0	−1	π^+	2.6×10^{-8}	$\mu^- + \overline{\nu}_\mu$
Kaon[c]	K^+	493.7	0	+1	K^-	1.24×10^{-8}	$\pi^+ + \pi^0$
	K^0	497.7	0	0	$\overline{K^0}$	0.88×10^{-10} and 5.2×10^{-8}	$\pi^+ + \pi^-$ / $\pi^+ + e^- + \overline{\nu}_e$
Eta	η^0	549	0	0		2×10^{-19}	$\gamma + \gamma$

[a]Other decay modes also occur for most particles.
[b]The Σ^0 is included here for completeness even though it does decay via the strong interaction.
[c]The K^0 has two distinct lifetimes, sometimes referred to as K^0_{short} and K^0_{long}. All other particles have a unique lifetime.

though very small, is not zero. There is no present measurement of the neutrino mass, but it is expected that it is of the order of a few eV/c^2. Experiments designed to detect neutrinos emitted from the sun have found a much smaller number than expected, which could be explained if the mass of the neutrino were not zero. In addition, a mass even as small as 40 eV/c^2 for the neutrino would have great cosmological significance. The answer to the question of whether the universe will continue to expand indefinitely or will reach a maximum size and begin to contract depends on the total mass in the universe. Thus, the answer could depend on whether the rest mass of the neutrino is merely small rather than zero since the cosmic density of each species of neutrino is about 100 cm^3. The observation of electron neutrinos from the supernova 1987A puts an upper limit on the mass of these neutrinos. Since the velocity of a particle with mass depends on its energy, the arrival time of a burst of neutrinos with mass from a supernova would be spread out in time. The fact that the electron neutrinos from the 1987 supernova all arrived at the earth within 13 s of one another results in an upper limit of about 16 eV/c^2 for their mass. Note that an upper limit does not imply that the mass is not zero. Recent measurements of the relative number of muon neutrinos and electron neutrinos entering a huge underground detector in Japan called Super-Kamiokande suggest that at least one type of neutrino can oscillate between types (for example, between a mu neutrino and a tau neutrino). Such oscillation is only possible if the neutrino has mass.

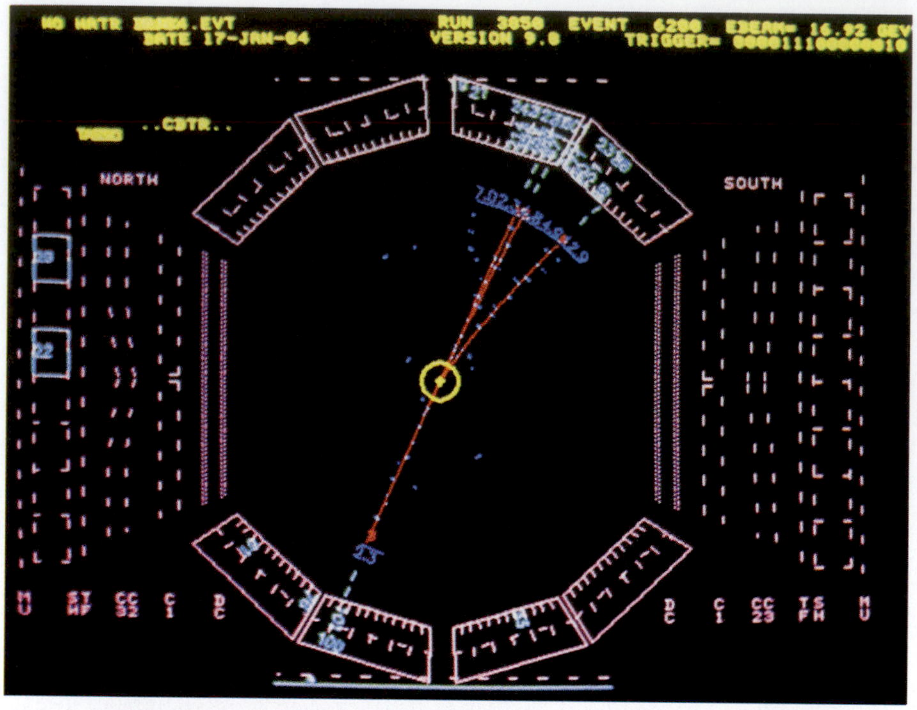

(a)

(b)

(a) A computer display of the production and decay of a τ^+ and τ^- pair. An electron and positron annihilate at the center marked by the yellow cross, producing a τ^+ and τ^- pair, which travel in opposite directions, but quickly decay while still inside the beam pipe (yellow circle). The τ^+ decays into two invisible neutrinos and a μ^+, which travels toward the bottom left. Its track in the drift chamber is calculated by a computer and indicated in red. It penetrates the lead–argon counters outlined in purple and is detected at the blue dot near the bottom blue line that marks the end of a muon detector. The τ^- decays into three charged pions (red tracks moving upward) plus invisible neutrinos. (b) The Mark I detector, built by a team from the Stanford Linear Accelerator Center (SLAC) and the Lawrence Berkeley Laboratory, became famous for many discoveries, including the ψ/J meson and the τ lepton. Tracks of particles are recorded by wire spark chambers wrapped in concentric cylinders around the beam pipe extending out to the ring where physicist Carl Friedberg has his right foot. Beyond this are two rings of protruding tubes, housing photomultipliers that view various scintillation counters. The rectangular magnets at the left guide the counterrotating beams that collide in the center of the detector.

41-2 Spin and Antiparticles

One important characteristic of a particle is its intrinsic spin angular momentum. We have already discussed the fact that the electron has a quantum number m_s that corresponds to the z component of its intrinsic spin characterized by the quantum number $s = \frac{1}{2}$. Protons, neutrons, neutrinos, and the various other particles that also have an intrinsic spin characterized by the quantum number $s = \frac{1}{2}$ are called **spin-$\frac{1}{2}$ particles.** Particles that have spin $\frac{1}{2}$

(or $\frac{3}{2}, \frac{5}{2}, \ldots$) are called fermions and obey the Pauli exclusion principle. Particles such as pions and other mesons have zero spin or integral spin ($s = 0, 1, 2, \ldots$). These particles are called bosons and do not obey the Pauli exclusion principle. Any number of these particles can be in the same quantum state.

Spin-$\frac{1}{2}$ particles are described by the Dirac equation, an extension of the Schrödinger equation that includes special relativity. One feature of Dirac's theory, proposed in 1927, is the prediction of the existence of antiparticles. In special relativity, the energy of a particle is related to the mass and momentum of the particle by $E = \pm\sqrt{p^2c^2 + m^2c^4}$ (Equation 39-28). We usually choose the positive solution and dismiss the negative-energy solution with a physical argument. However, the Dirac equation requires the existence of wave functions that correspond to the negative-energy states. Dirac got around this difficulty by postulating that all the negative-energy states were filled and would therefore not be observable. Only holes in the "infinite sea" of negative-energy states would be observed. For example, a hole in the negative sea of electron energy states would appear as a particle identical to the electron except with positive charge. When such a particle came in the vicinity of an electron the two particles would annihilate, releasing energy of $2m_ec^2$. This interpretation received little attention until the positron with just these properties was discovered in 1932 by Carl Anderson.

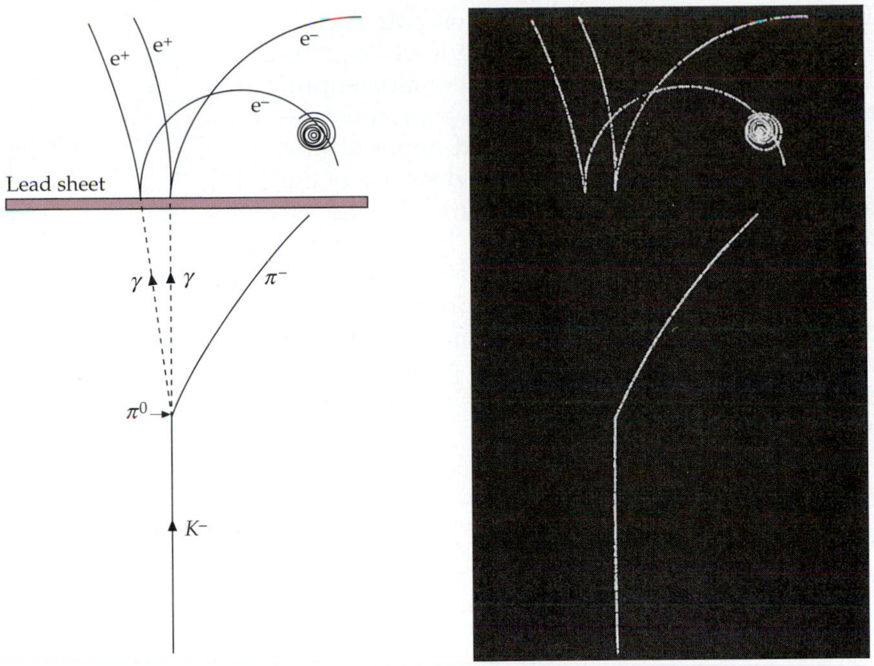

A negative kaon (K^-) enters a bubble chamber from the bottom and decays into a π^-, which moves off to the right, and a π^0, which immediately decays into two photons whose paths are indicated by the dashed lines in the drawing. Each photon interacts in the lead sheet, producing an electron–positron pair. The spiral at the right is another electron that has been knocked out of an atom in the chamber. (Other extraneous tracks have been removed from the photograph.)

Antiparticles are never created alone but always in particle–antiparticle pairs. In the creation of an electron–positron pair by a photon, the energy of the photon must be greater than the rest energy of the electron plus that of the positron, which is $2m_ec^2 \approx 1.02$ MeV, where m_e is the mass of the electron. Although the positron is stable, it has only a short-term existence in our universe because of the large supply of electrons in matter. The fate of a positron is annihilation according to the reaction

$$e^+ + e^- \rightarrow \gamma + \gamma \qquad 41\text{-}1$$

The probability of this reaction is large only if the positron is at rest or nearly at rest. Two photons moving in opposite directions are needed to conserve linear momentum.

The fact that we call electrons *particles* and positrons *antiparticles* does not imply that positrons are less fundamental than electrons. It merely reflects the nature of our part of the universe. If our matter were made up of negative protons and positive electrons, then positive protons and negative electrons would suffer quick annihilation and would be called antiparticles.

The antiproton (p^-) was discovered in 1955 by E. Segré and O. Chamberlain using a beam of protons in the Bevatron at Berkeley to produce the reaction*

$$p^+ + p^+ \rightarrow p^+ + p^+ + p^+ + p^- \qquad 41\text{-}2$$

*The antiproton is sometimes denoted by $\bar{p}$ rather than p^-. For neutral particles, such as the neutron, the bar must be used to denote the antiparticle. Thus the antineutron is denoted by $\bar{n}$. The normal electron and proton are often denoted by e and p without the minus or plus superscripts.

The creation of a proton–antiproton pair (Figure 41-1) requires kinetic energy of at least $2m_pc^2$ = 1877 MeV = 1.877 GeV in the zero-momentum reference frame in which the two protons approach each other with equal and opposite momenta. In the laboratory frame in which one of the protons is initially at rest, the kinetic energy of the incoming proton must be at least $6m_pc^2$ = 5.63 GeV (see Problem 60 of Chapter 39). This energy was not available in laboratories before the development of high-energy accelerators in the 1950s. Antiprotons annihilate with protons to produce two gamma rays in a reaction similar to that in Equation 41-1.

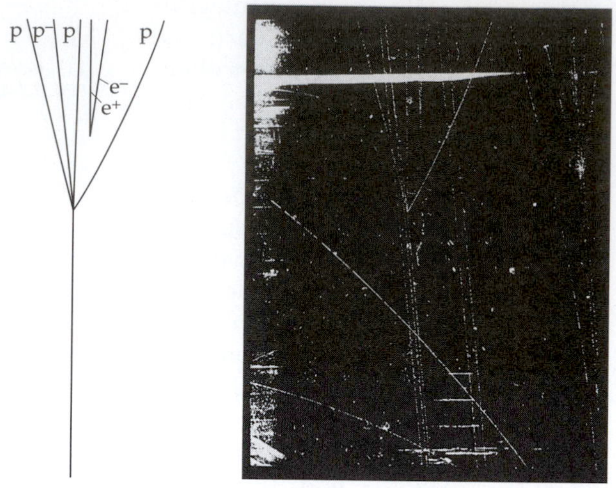

Figure 41-1 Bubble-chamber tracks showing the creation of a proton–antiproton pair in the collision of an incident 25-GeV proton with a stationary proton in liquid hydrogen.

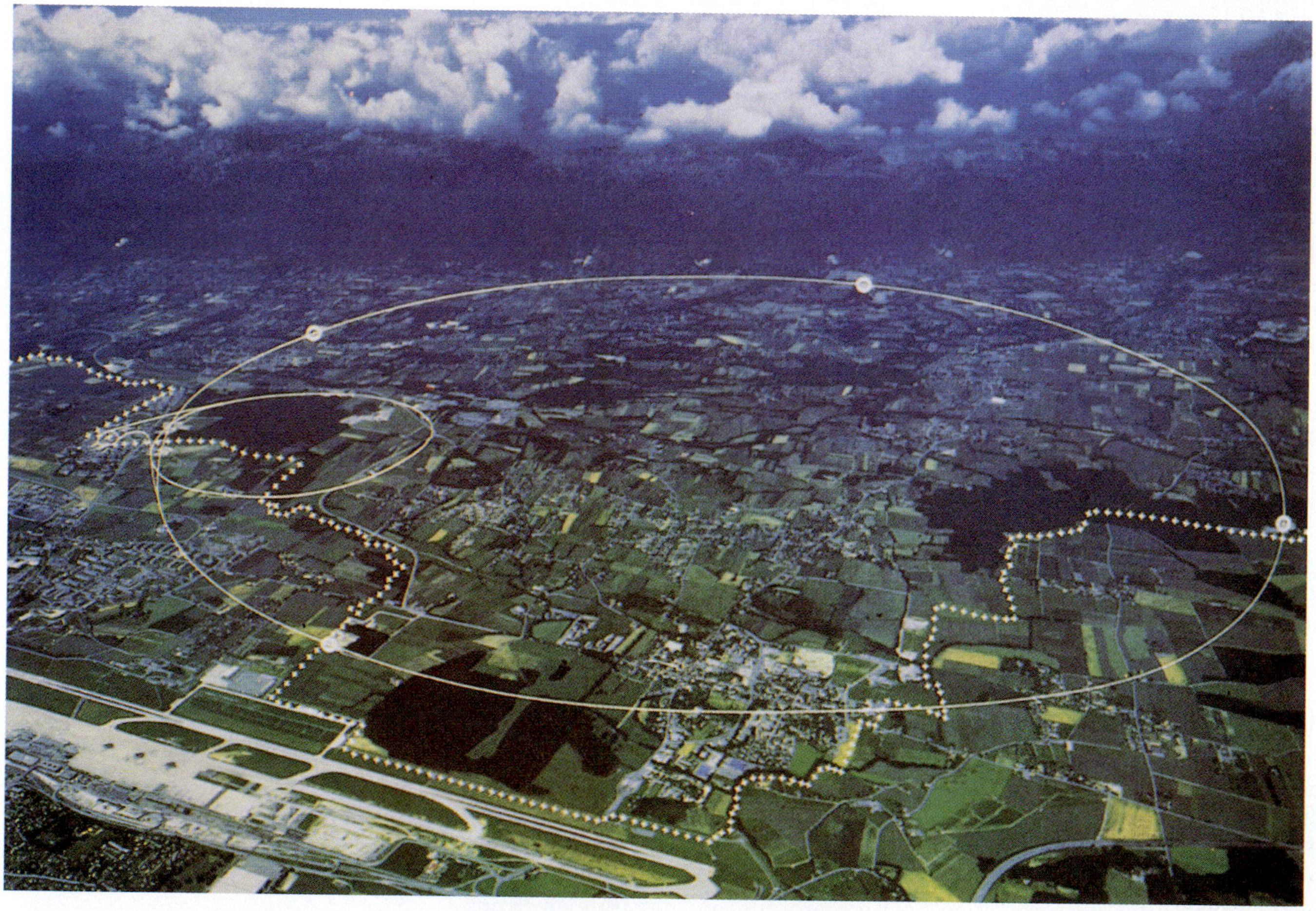

Air view of the European Laboratory for Particle Physics (CERN) just outside of Geneva, Switzerland. The large circle shows the Large Electron–Positron collider (LEP) tunnel, which is 27 km in circumference. The irregular dashed line is the border between France and Switzerland.

The tunnel of the proton–antiproton collider at CERN. The same bending magnets and focusing magnets can be used for protons or antiprotons moving in opposite directions. The rectangular box in the foreground is a focusing magnet; the next four boxes are bending magnets.

Example 41-1

A proton and an antiproton at rest annihilate according to the reaction $p^+ + p^- \rightarrow \gamma + \gamma$. Find the energies and wavelengths of the photons.

Picture the Problem Because the proton and the antiproton are at rest, conservation of momentum requires that the two photons created in their annihilation have equal and opposite momenta and therefore equal energies. Conservation of energy implies that the total energy of the photons equals the rest energy of the proton plus that of the antiproton (approximately 938 MeV each).

1. Set the total energy of the two photons, $2E_\gamma$, equal to the rest energy of the proton plus antiproton and solve for E_γ:

$$2E_\gamma = 2m_pc^2$$
$$E_\gamma = m_pc^2 = 938\text{ MeV}$$

2. Set the energy of the photon equal to $hf = hc/\lambda$ and solve for the wavelength λ:

$$E_\gamma = hf = \frac{hc}{\lambda}$$
$$\lambda = \frac{hc}{E_\gamma} = \frac{1240\text{ eV}\cdot\text{nm}}{938\text{ MeV}} = 1.32 \times 10^{-6}\text{ nm}$$
$$= 1.32\text{ fm}$$

41-3 The Conservation Laws

One of the maxims of nature is "anything that can happen does." If a conceivable decay or reaction does not occur, there must be a reason. The reason is usually expressed in terms of a conservation law. The conservation of energy rules out the decay of any particle for which the total rest mass of the decay products would be greater than the initial rest mass of the particle before decay. The conservation of linear momentum requires that when an electron and positron at rest annihilate, two photons must be emitted. Angular momentum must also be conserved in a reaction or decay. A fourth conservation law that restricts the possible particle decays and reactions is that of electric charge. The net electric charge before a decay or reaction must equal the net charge after the decay or reaction.

There are two additional conservation laws that are important in the reactions and decays of elementary particles: the conservation of baryon number and the conservation of lepton number. Consider the possible decay

$$p \to \pi^0 + e^+$$

This decay would conserve charge, energy, angular momentum, and linear momentum, but it does not occur. It does not conserve either lepton number or baryon number. The conservation of lepton number and baryon number implies that whenever a lepton or baryon is created, an antiparticle of the same type is also created. We assign the **lepton number** $L = +1$ to all leptons, $L = -1$ to all antileptons, and $L = 0$ to all other particles. Similarly, the **baryon number** $B = +1$ is assigned to all baryons, $B = -1$ to all antibaryons, and $B = 0$ to all other particles. The baryon and lepton numbers cannot change in a reaction or decay. The conservation of baryon number along with the conservation of energy implies that the least massive baryon, the proton, must be stable.

The conservation of lepton number implies that the neutrino emitted in the β decay of the free neutron is an antineutrino:

$$n \to p^+ + e^- + \bar{\nu}_e \qquad \text{41-3}$$

The fact that neutrinos and antineutrinos are different is illustrated by an experiment in which ^{37}Cl is bombarded with an intense antineutrino beam from the decay of reactor neutrons. If neutrinos and antineutrinos were the same, we would expect the following reaction:

$$^{37}Cl + \bar{\nu}_e \to {}^{37}Ar + e^- \qquad \text{41-4}$$

This reaction is not observed. However, if *protons* are bombarded with antineutrinos, the reaction

$$p + \bar{\nu}_e \to n + e^+ \qquad \text{41-5}$$

is observed. Note that the lepton number is -1 on the left side of reaction 41-4 and $+1$ on the right side. But the lepton number is -1 on both sides of reaction 41-5.

Not only are neutrinos and antineutrinos distinct particles, but the neutrinos associated with electrons are distinct from the neutrinos associated with muons. Electron-like leptons (e and ν_e), muon-like leptons (μ and ν_μ), and tau-like leptons (τ and ν_τ) are each separately conserved, so we assign separate lepton numbers L_e, L_μ, and L_τ to the particles. For e and ν_e, $L_e = +1$; for their antiparticles, $L_e = -1$; and for all other particles, $L_e = 0$. The lepton numbers L_μ and L_τ are similarly assigned.

Example 41-2

What conservation laws (if any) are violated by the following decays?
(*a*) $n \to p + \pi^-$ (*b*) $\Lambda^0 \to p^- + \pi^+$ (*c*) $\mu^- \to e^- + \gamma$

(*a*) There are no leptons in this decay, so there is no problem with the conservation of lepton number. The net charge is zero before and after the decay, so charge is conserved. Also, the baryon number is $+1$ before and after the decay. However, the rest energy of the proton (938.3 MeV) plus that of the pion (139.6 MeV) is greater than the rest energy of the neutron (939.6 MeV). Thus, this decay violates the conservation of energy.

(*b*) Again, there are no leptons involved, and the net charge is zero before and after the decay. Also, the rest energy of the Λ^0 (1116 MeV) is greater than the rest energy of the antiproton (938.3 MeV) plus that of the pion (139.6 MeV), so energy is conserved, with the loss in rest energy equaling the gain in kinetic energy of the decay products. However, this decay does not conserve baryon number, which is $+1$ for the Λ^0 and -1 for the antiproton.

(*c*) This reaction does not conserve muon lepton number or electron lepton number. The muon does decay via

$$\mu^- \rightarrow e^- + \bar{\nu}_e + \nu_\mu$$

which does conserve both muon and electron lepton numbers.

There are some conservation laws that are not universal but apply only to certain kinds of interactions. In particular, there are quantities that are conserved in decays and reactions that occur via the strong interaction but not in decays or reactions that occur via the weak interaction. One of these quantities that is particularly important is **strangeness,** introduced by M. Gell-Mann and K. Nishijima in 1952 to explain the strange behavior of some of the heavy baryons and mesons. Consider the reaction

$$p + \pi^- \rightarrow \Lambda^0 + K^0 \qquad \text{41-6}$$

The proton and pion interact via the strong interaction. Both the Λ^0 and K^0 decay into hadrons

$$\Lambda^0 \rightarrow p + \pi^- \qquad \text{41-7}$$

and

$$K^0 \rightarrow \pi^+ + \pi^- \qquad \text{41-8}$$

However, the decay times for both the Λ^0 and K^0 are of the order of 10^{-10} s, which is characteristic of the weak interaction, rather than 10^{-23} s, which would be expected for the strong interaction. Other particles showing similar behavior were called **strange particles.** These particles are always produced in pairs and never singly, even when all other conservation laws are met. This behavior is described by assigning a new property called strangeness to these particles. In reactions and decays that occur via the strong interaction, strangeness is conserved. In those that occur via the weak interaction the strangeness can change by ± 1. The strangeness of the ordinary hadrons—the nucleons and pions—was arbitrarily taken to be zero. The strangeness of the K^0 was arbitrarily chosen to be $+1$. The strangeness of the Λ^0 particle must then be -1 so that strangeness is

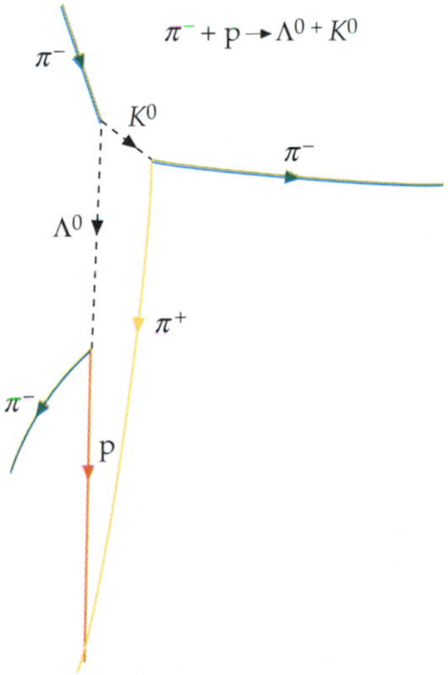

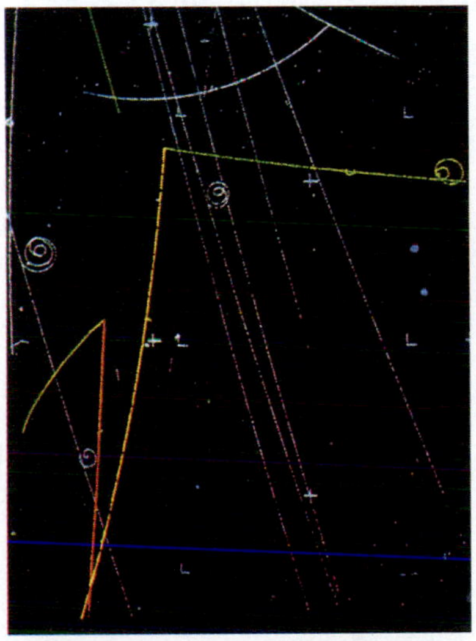

An early photograph of bubble-chamber tracks at the Lawrence Berkeley Laboratory, showing the production and decay of two strange particles, the K^0 and the Λ^0. These neutral particles are identified by the tracks of their decay particles. The lambda particle was named because of the similarity of the tracks of its decay particles to the Greek letter Λ. (The blue tracks are particles not involved in the reaction of Equation 41-6.)

conserved in reaction 41-6. The strangeness of other particles could then be assigned by looking at their various reactions and decays. In those that occur via the weak interaction, the strangeness can change by ±1.

Figure 41-2 shows the masses of the baryons and mesons that are stable against decay via the strong interaction versus strangeness. We can see from this figure that these particles cluster in multiplets of one, two, or three particles of approximately equal mass, and that the strangeness of a multiplet of particles is related to the "center of charge" of the multiplet.

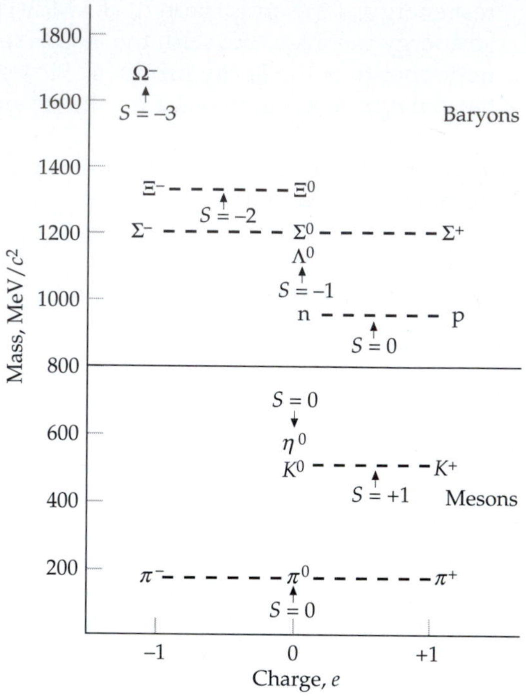

Figure 41-2 The strangeness of hadrons shown on a plot of rest mass versus charge. The strangeness of a baryon-charge multiplet is related to the number of places the center of charge of the multiplet is displaced from that of the nucleon doublet. For each displacement of $\frac{1}{2}e$, the strangeness changes by ±1. For mesons, the strangeness is related to the number of places the center of charge is displaced from that of the pion triplet. Because of the unfortunate original assignment of +1 for the strangeness of kaons, all of the baryons that are stable against decay via the strong interaction have negative or zero strangeness.

Example 41-3

State whether the following decays can occur via the strong interaction, via the weak interaction, or not at all:
(*a*) $\Sigma^+ \rightarrow p + \pi^0$ (*b*) $\Sigma^0 \rightarrow \Lambda^0 + \gamma$ (*c*) $\Xi^0 \rightarrow n + \pi^0$

Picture the Problem We first note that the mass of each decaying particle is greater than that of the decay products, so there is no problem with energy conservation in any of the decays. In addition, there are no leptons involved in any of the decays, and charge and baryon number are both conserved in all the decays. The decay will occur via the strong interaction if strangeness is conserved. If $\Delta S = \pm 1$, it will occur via the weak interaction. If S changes by more than 1, the decay will not occur.

(*a*) From Figure 41-2, we can see that the strangeness of the Σ^+ is -1, whereas the strangeness of both the proton and the pion is zero. This decay is possible via the weak interaction but not the strong interaction. It is, in fact, one of the decay modes of the Σ^+ particle with a lifetime of the order of 10^{-10} s.

(*b*) Since the strangeness of both the Σ^0 and Λ^0 is -1, this decay can proceed via the strong interaction. It is, in fact, the dominant mode of decay of the Σ^0 particle with a lifetime of about 10^{-20} s.

(*c*) The strangeness of the Ξ^0 is -2, whereas the strangeness of both the neutron and pion is zero. Since strangeness cannot change by 2 in a decay or reaction, this decay cannot occur.

41-4 Quarks

Leptons appear to be truly elementary particles in that they do not break down into smaller entities and they seem to have no measurable size or structure. Hadrons, on the other hand, are complex particles with size and structure, and they decay into other hadrons. Furthermore, at the present time, there are only six known leptons, whereas there are many more hadrons. Except for the Σ^0 particle, Table 41-1 includes only hadrons that are stable against decay via the strong interaction. Hundreds of other hadrons have been discovered, and their properties, such as charge, spin, mass, strangeness, and decay schemes, have been measured.

The most important advance in our understanding of elementary particles was the quark model proposed by M. Gell-Mann and G. Zweig in 1963 in which all hadrons consist of combinations of two or three truly elementary particles called **quarks.*** In the original model, quarks came in three types, called **flavors,** labeled u, d, and s (for *up*, *down*, and *strange*). An unusual property of quarks is that they carry fractional electron charges. The charge of the u quark is $+\frac{2}{3}e$ and that of the d and s quarks is $\frac{1}{3}e$. Each quark has spin $\frac{1}{2}$ and a baryon number of $\frac{1}{3}$. The strangeness of the u and d quark is 0 and that of the s quark is -1. Each quark has an antiquark with the opposite electric charge, baryon number, and strangeness. Baryons consist of three quarks (or three antiquarks for antiparticles), whereas mesons consist of a quark and an antiquark, giving them a baryon number $B = 0$, as required. The proton consists of the combination uud and the neutron, udd. Baryons with a strangeness $S = -1$ contain one s quark. All the particles listed in Table 41-1 can be constructed from these three quarks and three antiquarks.† The great strength of the quark model is that all the allowed combinations of three quarks or quark–antiquark pairs result in known hadrons. Strong evidence for the existence of quarks inside a nucleon is provided by high-energy scattering experiments called *deep inelastic scattering.* In these experiments, a nucleon is bombarded with electrons, muons, or neutrinos of energies from 15 to 200 GeV. Analyses of particles scattered at large angles indicate that inside the nucleon are 3 spin-$\frac{1}{2}$ particles of sizes much smaller than that of the nucleon. These experiments are analogous to Rutherford's scattering of α particles by atoms in which the presence of a tiny nucleus in the atom was inferred from the large-angle scattering of the α particles.

*The name *quark* was chosen by Gell-Mann from a quotation from *Finnegans Wake* by James Joyce.

†The correct quark combinations of hadrons are not always obvious, because of the symmetry requirements on the total wave function. For example, the π^0 meson is represented by a linear combination of $u\bar{u}$ and $d\bar{d}$.

Example 41-4

What are the properties of the particles made up of the following quarks: (*a*) $u\bar{d}$, (*b*) $\bar{u}d$, (*c*) dds, and (*d*) uss?

Picture the Problem Baryons are made up of 3 quarks, whereas mesons consist of a quark and an antiquark. We add the electric charges of the quarks to find the total charge of the hadron. We also find the strangeness of the hadron by adding the strangeness of the quarks.

(*a*) Since $u\bar{d}$ is a quark–antiquark combination, it has baryon number 0 and is therefore a meson. There is no strange quark here, so the strangeness of the meson is zero. The charge of the up quark is $+\frac{2}{3}e$ and that of the antidown quark is $+\frac{1}{3}e$, so the charge of the meson is $+1e$. This is the quark combination of the π^+ meson.

(*b*) The particle $\bar{u}d$ is also a meson with zero strangeness. Its electric charge is $-\frac{2}{3}e + (-\frac{1}{3}e) = -1e$. This is the quark combination of the π^- meson.

(*c*) The particle *dds* is a baryon with strangeness −1 since it contains one strange quark. Its electric charge is $-\frac{1}{3}e - \frac{1}{3}e - \frac{1}{3}e = -1e$. This is the quark combination for the Σ^- particle.

(*d*) The particle *uss* is a baryon with strangeness −2. Its electric charge is $+\frac{2}{3}e - \frac{1}{3}e - \frac{1}{3}e = 0$. This is the quark combination for the Ξ^0 particle.

In 1967, a fourth quark was proposed to explain some discrepancies between experimental determinations of certain decay rates and calculations based on the quark model. The fourth quark is labeled *c* for a new property called **charm.** Like strangeness, charm is conserved in strong interactions but changes by ±1 in weak interactions. In 1975, a new heavy meson called the **ψ/J particle** (or simply the **ψ particle**) was discovered that has the properties expected of a $c\bar{c}$ combination. Since then other mesons with combinations such as $c\bar{d}$ and $\bar{c}d$, as well as baryons containing the charmed quark, have been discovered. Two more quarks labeled *t* and *b* (for *t*op and *b*ottom) were proposed in the 1970s. In 1977, a massive new meson called the **Υ meson** or **bottomonium,** which is considered to have the quark combination $b\bar{b}$, was discovered. The top quark was observed in 1995. The properties of the six quarks are listed in Table 41-2.

Table 41-2

Properties of Quarks and Antiquarks

Flavor	Spin	Charge	Baryon Number	Strangeness	Charm	Topness	Bottomness
Quarks							
u (up)	$\frac{1}{2}\hbar$	$+\frac{2}{3}e$	$+\frac{1}{3}$	0	0	0	0
d (down)	$\frac{1}{2}\hbar$	$-\frac{1}{3}e$	$+\frac{1}{3}$	0	0	0	0
s (strange)	$\frac{1}{2}\hbar$	$-\frac{1}{3}e$	$+\frac{1}{3}$	−1	0	0	0
c (charmed)	$\frac{1}{2}\hbar$	$+\frac{2}{3}e$	$+\frac{1}{3}$	0	+1	0	0
t (top)	$\frac{1}{2}\hbar$	$+\frac{2}{3}e$	$+\frac{1}{3}$	0	0	+1	0
b (bottom)	$\frac{1}{2}\hbar$	$-\frac{1}{3}e$	$+\frac{1}{3}$	0	0	0	+1
Antiquarks							
$\bar{u}$	$\frac{1}{2}\hbar$	$-\frac{2}{3}e$	$-\frac{1}{3}$	0	0	0	0
$\bar{d}$	$\frac{1}{2}\hbar$	$+\frac{1}{3}e$	$-\frac{1}{3}$	0	0	0	0
$\bar{s}$	$\frac{1}{2}\hbar$	$+\frac{1}{3}e$	$-\frac{1}{3}$	+1	0	0	0
$\bar{c}$	$\frac{1}{2}\hbar$	$-\frac{2}{3}e$	$-\frac{1}{3}$	0	−1	0	0
$\bar{t}$	$\frac{1}{2}\hbar$	$-\frac{2}{3}e$	$-\frac{1}{3}$	0	0	−1	0
$\bar{b}$	$\frac{1}{2}\hbar$	$+\frac{1}{3}e$	$-\frac{1}{3}$	0	0	0	−1

The six quarks and six leptons (and their antiparticles) are thought to be the fundamental, elementary particles of which all matter is composed. Table 41-3 lists the masses of the fundamental particles. In this table, the masses given for neutrinos are upper limits. The masses given for quarks are educated guesses. There is experimental evidence for the existence of each of these particles.

Table 41-3

Masses of Fundamental Particles

Particle	Mass
Quarks	
u (up)	336 MeV/c^2
d (down)	338 MeV/c^2
s (strange)	540 MeV/c^2
c (charmed)	1,500 MeV/c^2
t (top)	174,000 MeV/c^2
b (bottom)	500 MeV/c^2
Leptons	
e^- (electron)	0.511 MeV/c^2
ν_e (electron neutrino)	<7 eV/c^2
μ^- (muon)	105.659 MeV/c^2
ν_μ (muon neutrino)	<0.27 MeV/c^2
τ^- (tau)	1,784 MeV/c^2
ν_τ (tau neutrino)	<31 MeV/c^2

Quark Confinement

Despite considerable experimental effort, no isolated quark has ever been observed. It is now believed that it is impossible to obtain an isolated quark. Although the force between quarks is not known, it is believed that the potential energy of two quarks increases with increasing separation distance so that an infinite amount of energy would be needed to separate the quarks completely. This would be true, for example, if the force of attraction between two quarks remains constant or increases with separation distance, rather than decreasing with increasing separation distance as is the case for other fundamental forces, such as the electric force between two charges, the gravitational force between two masses, and the strong nuclear force between two hadrons.

When a large amount of energy is added to a quark system such as a nucleon, a quark–antiquark pair is created and the original quarks remain confined within the original system. Because quarks cannot be isolated, but are always bound in a baryon or meson, the mass of a quark cannot be accurately known, which is why the masses listed in Table 41-3 are merely educated guesses.

41-5 Field Particles

In addition to the six fundamental leptons and six fundamental quarks, there are other particles, called *field particles* or *field quanta,* that are associated with the forces exerted by one elementary particle on another. In **quantum electrodynamics** the electromagnetic field of a single charged particle is described by **virtual photons** that are continuously being emitted and reabsorbed by the particle. If we put energy into the system by accelerating the charge, some of these virtual photons can be "shaken off" and become real, observable photons. The photon is said to mediate the electromagnetic interaction. Each of the four basic interactions can be described in this way.

The field quantum associated with the gravitational interaction, called the **graviton,** has not yet been observed. The gravitational "charge" analogous to electric charge is mass.

The weak interaction is thought to be mediated by three field quanta called **vector bosons:** W^+, W^-, and Z^0. These particles were predicted by S. Glashow, A. Salam, and S. Weinberg in a theory called the *electroweak theory,* which we discuss in the next section. The *W* and *Z* particles were first observed in 1983 by a group of over a hundred scientists led by C. Rubbia using the high-energy accelerator at CERN in Geneva, Switzerland. The masses of the $W^{\pm}$ particles (about 80 GeV/c^2) and the *Z* particle (about 91 GeV/c^2) measured in this experiment were in excellent agreement with those predicted by the electroweak theory. (The W^- particle is the antiparticle of the W^+ particle, so they must have identical masses.)

The field quanta associated with the strong force between quarks are called **gluons.** Isolated gluons have not been observed experimentally. The "charge" responsible for the strong interactions comes in three varieties, labeled *red, green,* and *blue* (analogous with the three primary colors), and the strong charge is called the **color charge.** The field theory for strong interactions, analogous to quantum electrodynamics for electromagnetic interactions, is called **quantum chromodynamics (QCD).**

Table 41-4 lists the bosons responsible for mediating the basic interactions.

Table 41-4

Bosons That Mediate the Basic Interactions

Interaction	Boson	Spin	Mass	Electric Charge
Strong	*g* (gluon)[a]	1	0	0
Weak	$W^{\pm}$	1	80.22 GeV/c^2	$\pm 1e$
	Z^0	1	91.19 GeV/c^2	0
Electromagnetic	γ (photon)	1	0	0
Gravitational	Graviton[a]	2	0	0

[a] Not yet observed.

41-6 The Electroweak Theory

In the **electroweak theory,** the electromagnetic and weak interactions are considered to be two different manifestations of a more fundamental electroweak interaction. At very high energies ($\gg$100 GeV), the electroweak interaction would be mediated by four bosons. From symmetry considerations, these would be a triplet consisting of W^+, W^0, and W^-, all of equal mass, and a singlet B^0 of some other mass. Neither the W^0 nor the B^0 would be observed directly, but one linear combination of the W^0 and the B^0 would be the Z^0 and another would be the photon. At ordinary energies, the symmetry is broken. This leads to the separation of the electromagnetic interaction mediated by the massless photon and the weak interaction mediated by the W^+, W^-, and Z^0 particles. The fact that the photon is massless and that the *W* and *Z* particles have masses of the order of 100 GeV/c^2 shows that the symmetry assumed in the electroweak theory does not exist at lower energies.

The symmetry-breaking mechanism is called a **Higgs field,** which requires a new boson, the **Higgs boson,** whose rest energy is expected to be of the order of 1 TeV (1 TeV = 10^{12} eV). The Higgs boson has not yet been observed. Calculations show that Higgs bosons (if they exist) should be produced in head-on collisions between protons of energies of the order of 20 TeV. Such energies are not presently available.

41-7 The Standard Model

The combination of the quark model, electroweak theory, and quantum chromodynamics is called the **standard model.** In this model, the fundamental particles are the leptons and quarks, each of which comes in six flavors as shown in Table 41-3, and the force carriers are the photon, the $W^{\pm}$ and Z particles, and the gluons (of which there are eight types). The leptons and quarks are all spin-$\frac{1}{2}$ fermions, which obey the Pauli exclusion principle, and the force carriers are integral-spin bosons, which do not obey the Pauli exclusion principle. Every force in nature is due to one of the four basic interactions: strong, electromagnetic, weak, and gravitational. A particle experiences one of the basic interactions if it carries a charge associated with that interaction. Electric charge is the familiar charge that we have studied previously. Weak charge, also called flavor charge, is carried by leptons and quarks. The charge associated with the strong interaction is called color charge and is carried by quarks and gluons but not by leptons. The charge associated with the gravitational force is mass. It is important to note that the photon, which mediates the electromagnetic interaction, does not carry electric charge. Similarly, the $W^{\pm}$ and Z particles, which mediate the weak interaction, do not carry weak charge. However, the gluons, which mediate the strong interaction, do carry color charge. This fact is related to the confinement of quarks as discussed in Section 41-4.

All matter is made up of leptons or quarks. There are no known composite particles consisting of leptons bound together by the weak force. Leptons exist only as isolated particles. Hadrons (baryons and mesons) are composite particles consisting of quarks bound together by the color charge. A result of QCD theory is that only color-neutral combinations of quarks are allowed. Three quarks of different colors can combine to form color-neutral baryons, such as the neutron and proton. Mesons contain a quark and an antiquark and are also color-neutral. Excited states of hadrons are considered to be different particles. For example, the Δ^+ particle is an excited state of the proton. Both are made up of the *uud* quarks, but the proton is in the ground state with spin $\frac{1}{2}$ and a rest energy of 938 MeV, whereas the Δ^+ particle is in the first excited state with spin $\frac{3}{2}$ and a rest energy of 1232 MeV. The two u quarks can be in the same spin state in the Δ^+ without violating the exclusion principle, because they have different color. All baryons eventually decay to the lightest baryon, the proton. The proton cannot decay because of conservation of energy and conservation of baryon number.

The strong interaction has two parts, the fundamental or color interaction and what is called the *residual strong interaction.* The fundamental interaction is responsible for the force exerted by one quark on another and is mediated by gluons. The residual strong interaction is responsible for the force between color-neutral nucleons, such as the neutron and proton. This force is due to the residual strong interactions between the color-charged quarks that make up the nucleons and can be viewed as being mediated by the exchange of mesons. The residual strong interaction between color-neutral nucleons

Table 41-5

Properties of the Basic Interactions

	Gravitational	Weak	Electromagnetic	Strong	
				Fundamental	Residual
Acts on	Mass	Flavor	Electric charge	Color charge	
Particles experiencing	All	Quarks, leptons	Electrically charged	Quarks, gluons	Hadrons
Particles mediating	Graviton	$W^{\pm}$, Z	γ	Gluons	Mesons
Strength for two quarks at 10^{-18} m[a]	10^{-41}	0.8	1	25	(not applicable)
Strength for two protons in nucleus[a]	10^{-36}	10^{-7}	1	(not applicable)	20

[a] Strengths are relative to electromagnetic strength.

can be thought of as analogous to the residual electromagnetic interaction between neutral atoms that bind them together to form molecules. Table 41-5 lists some of the properties of the basic interactions.

For each particle there is an antiparticle. A particle and its antiparticle have identical mass and spin but opposite electric charge. For leptons, the lepton numbers L_e, L_μ, and L_τ of the antiparticles are the negatives of the corresponding numbers for the particles. For example, the lepton number for the electron is $L_e = +1$ and that for the positron is $L_e = -1$. For hadrons, the baryon number, strangeness, charm, topness, and bottomness are the sums of those quantities for the quarks that make up the hadron. The number of each antiparticle is the negative of the number for the corresponding particle. For example, the lambda particle Λ^0, which is made up of the *uds* quarks, has $B = 1$ and $S = -1$, whereas its antiparticle $\overline{\Lambda}^0$, which is made up of the $\overline{u}\overline{d}\overline{s}$ quarks, has $B = -1$ and $S = +1$. A particle such as the photon γ or the Z^0 particle that has zero electric charge, $B = 0$, $L = 0$, $S = 0$, and zero charm, topness, and bottomness, is its own antiparticle. Note that the K^0 meson ($d\overline{s}$) has a zero value for all of these quantities except strangeness, which is +1. Its antiparticle, the $\overline{K}^0$ meson ($\overline{d}s$), has strangeness −1, which makes it distinct from the K^0. The π^+ ($u\overline{d}$) and π^- ($\overline{u}d$) are somewhat special in that they have electric charge but zero values for L, B, and S. They are antiparticles of each other, but since there is no conservation law for mesons, it is impossible to say which is the particle and which is the antiparticle. Similarly, the W^+ and W^- are antiparticles of each other.

Grand Unification Theories

With the success of the electroweak theory, attempts have been made to combine the strong, electromagnetic, and weak interactions in various **grand unification theories** known as **GUTs**. In one of these theories, leptons and quarks are considered to be two aspects of a single class of particles. Under certain conditions, a quark could change into a lepton and vice versa, even though this would appear to violate the conservation of lepton number and baryon number. One of the exciting predictions of this theory is that the proton is not stable but merely has a very long lifetime of the order of 10^{31} y. Such a long lifetime makes proton decay difficult to observe. However, projects are ongoing in which detectors monitor very large numbers of protons in search of an event indicating the decay of a proton.

41-8 Evolution of the Universe

In the presently accepted model, the universe began with a singular cataclysmic event called the **Big Bang** and is expanding. The first evidence that the universe is expanding was the astronomer E. P. Hubble's discovery of the relation between the redshifts in the spectra of galaxies and their distances from us. This relation is illustrated in Figure 41-3 for a group of spiral galaxies used by astronomers for calibrating distances. Provided that the redshift is due to the Doppler effect, the recession velocity v of a galaxy is related to its distance r from us by **Hubble's law,**

$$v = Hr \tag{41-9}$$

where H is the **Hubble constant.** In principle, the value of H is easy to obtain since it relies on the direct calculation of v from redshift measurements. However, astronomical distances are very difficult to obtain and they have been computed for only a fraction of the 10^{10} or so galaxies in the observable universe. Thus, the value of H changes as distance calibration data are refined. The currently accepted value of the Hubble constant is

$$H = \frac{23\ \text{km/s}}{10^6\, c\cdot\text{y}} \tag{41-10}$$

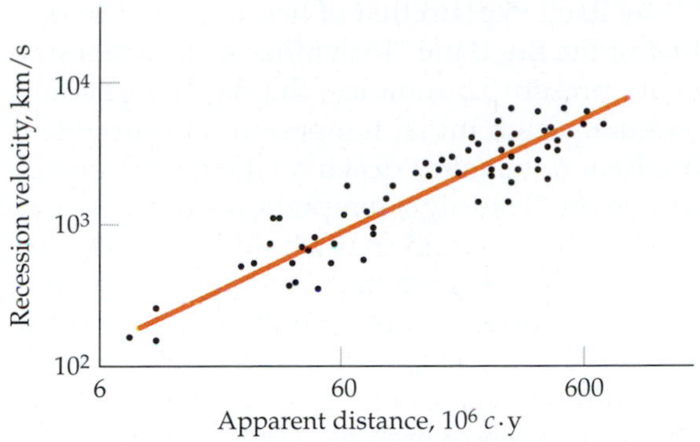

Figure 41-3 A plot of the recession velocities of individual galaxies versus apparent distance.

Hubble's law tells us that the galaxies are all rushing away from us, with those the farthest away moving the fastest. However, there is no reason why our location should be special. An observer in any galaxy would make the same observations and compute the same Hubble constant. Thus, Hubble's law suggests that all of the galaxies are receding from each other at an average speed of 23 km/s per $10^6\ c\cdot$y of separation. In other words, the universe is expanding. Notice that the basic dimension of H is reciprocal time. The quantity $1/H$ is called the **Hubble age** and equals about 1.3×10^{10} y. This would correspond to the age of the universe if the gravitational pull on the receding galaxies were ignored.

Example 41-5

Redshift measurements of a galaxy in the constellation Virgo yield a recession velocity of 1200 km/s. How far is it to that galaxy?

Picture the Problem We calculate the distance from Hubble's law.

1. Use Hubble's law to find r: $\quad r = \frac{v}{H} = (1200\ \text{km/s})\frac{10^6\ c\cdot\text{y}}{23\ \text{km/s}} = 52 \times 10^6\ c\cdot\text{y}$

Exercise Show that $1/H = 1.3 \times 10^{10}$ y.

The 2.7-K Background Radiation

In investigating ways of accounting for the cosmic abundance of elements heavier than hydrogen, cosmologists recognized that nucleosynthesis in stars could explain the abundance of elements heavier than helium but could not by itself explain that of helium. Helium must therefore have been formed during the Big Bang. To synthesize an amount of helium sufficient to account for its present abundance, the Big Bang would have to have occurred at an extremely high initial temperature to provide the necessary reaction rate before fusion was shut down by the decreasing density of the very rapid initial expansion. The high temperature implies a corresponding thermal (blackbody) radiation field that would cool as the expansion progressed. Theoretical analysis predicted that from the estimated time of the Big Bang to the present, the remnants of the radiation field should have cooled to a temperature of about 3 K, corresponding to a blackbody spectrum with peak wavelength λ_{max} in the microwave region. In 1965, the predicted cosmic background radiation was discovered by Arno Penzias and Robert Wilson at the Bell Labs. Since this landmark discovery, careful analysis has established that the temperature of the background field is 2.7 ± 0.1 K and has shown that it has an isotropic distribution in space.

The Big Bang

The singular event that initiated the expansion of the universe is thought to have been a huge explosion. Initially, the four forces of nature (strong, electromagnetic, weak, and gravity) were unified into a single force. Physicists have been successful in developing theoretical descriptions that unify the first three, but a theory of quantum gravity, needed for the extreme densities of the single-force period, does not yet exist. Consequently, until the cooling universe "froze" or "condensed out" the gravitational force at about 10^{-43} s after the Big Bang, when the temperature was still 10^{32} K, we have no means of describing what was occurring. At this point, the average energy of the particles created would have been about 10^{19} GeV. As the universe continued to cool below 10^{32} K, the three forces other than gravity remained unified and are described by the grand unification theories. Quarks and leptons were indistinguishable and particle quantum numbers were not conserved. It was during this period that a slight excess of quarks over antiquarks occurred, roughly 1 in 10^9, that ultimately resulted in the predominance of matter over antimatter that we now observe in the universe.

At 10^{-35} s, the universe had expanded sufficiently to cool to about 10^{27} K, at which point another phase transition occurred as the strong force condensed out of the GUTs group, leaving only the electromagnetic and weak forces still unified as the **electroweak force.** During this period, the previously free quarks in the dense mixture of roughly equal numbers of quarks, leptons, their antiparticles, and photons began to combine into hadrons and their antiparticles, including the nucleons. By the time the universe had cooled to about 10^{13} K, at about $t = 10^{-6}$ s, the hadrons had mostly disappeared. This is because 10^{13} K corresponds to $kT \sim 1$ GeV, which is the minimum energy needed to create nucleons and antinucleons from the photons present via the reactions

$$\gamma \rightarrow \mathrm{p}^+ + \mathrm{p}^- \qquad \text{41-11}a$$

and

$$\gamma \rightarrow \mathrm{n}^+ + \bar{\mathrm{n}} \qquad \text{41-11}b$$

The particle–antiparticle pairs annihilated and there was no new production to replace them. Only the slight earlier excess of quarks over antiquarks led to a slight excess of protons and neutrons over their antiparticles. The annihilations resulted in photons and leptons, and after about $t = 10^{-4}$ s, those particles in roughly equal numbers dominated the universe. This was the **lepton era.** At about $t = 10$ s, the temperature had fallen to 10^{10} K ($kT \sim 1$ MeV).

Further expansion and cooling dropped the average photon energy below that needed to form an electron–positron pair. Annihilation then removed all of the positrons as it had the antiprotons and antineutrons earlier, leaving only the small excess of electrons arising from charge conservation, and the **radiation era** began. The particles present were primarily photons and neutrinos.

Within a few more minutes, the temperature dropped sufficiently to enable fusing protons and neutrons to form nuclei that were not immediately photodisintegrated. The nuclei of deuterium, helium, and lithium were produced in this **nucleosynthesis period,** but the rapid expansion soon dropped the temperature too low for the fusion to continue and the formation of heavier elements had to await the birth of stars.

A long time later, when the temperature had dropped to about 3000 K as the universe grew to about 1/1000 of its present size, kT dropped below typical atomic ionization energies and atoms were formed. By then the expansion had redshifted the radiation field so that the total radiation energy was about equal to the energy represented by the remaining mass. As expansion and cooling continued, the energy of the steadily redshifting radiation steadily declined until, at $t = 10^{10}$ y (now), matter came to dominate the universe, with its energy density exceeding that of the 2.7-K radiation remaining from the Big Bang by a factor of about 1000.

Summary

Topic	Remarks and Relevant Equations
1. Basic Interactions	There are four basic interactions: strong, electromagnetic, weak, and gravitational.
Gravitational	All particles with mass experience the force due to the gravitational interaction.
Electromagnetic	All particles with electric charge experience the force due to the electromagnetic interaction.
Weak	The "charge" associated with the weak interaction is called flavor. Quarks and leptons have flavor and experience the weak interaction. Decay times via weak interaction are typically 10^{-10} s.
Strong	The "charge" associated with the strong interaction is called color. Quarks and gluons have color and experience the strong interaction. Hadrons (baryons and mesons) experience a residual strong interaction resulting from the fundamental strong interaction between the quarks that make up the hadrons. Decay times via strong interaction are typically 10^{-23} s.
2. Fundamental Particles	There are two families of fundamental particles, leptons and quarks, each containing six members. It is thought that these particles have no size and no internal structure.
Leptons	Leptons are spin-$\frac{1}{2}$ fermions: the electron e and its neutrino ν_e, the muon μ and its neutrino ν_μ, and the tau τ and its neutrino ν_τ. The electron, muon, and tau have mass, electric charge, and flavor but not color, so they participate in the gravitational, electromagnetic, and weak interactions but not the strong interaction. The neutrinos have flavor but no electric charge and no color. They appear to have a very small mass.

	Quarks	There are six quarks, called up u, down d, strange s, charmed c, top t, and bottom b. Each is a spin-$\frac{1}{2}$ fermion. The quarks participate in all of the basic interactions. Because they are always confined in mesons or baryons, their masses can only be estimated.
3.	**Hadrons**	Hadrons are composite particles that are made up of quarks. There are two types, baryons and mesons. Baryons, which include the neutron and proton, are fermions of half-integral spin consisting of three quarks. Mesons, which include pions and kaons, have zero or integral spin. Hadrons interact with each other via the residual strong interaction.
4.	**Field Particles**	In addition to the six fundamental leptons and six fundamental quarks, there are field particles that are associated with the basic interactions.

Interaction	*Field Particle*
Gravity	Graviton
Electromagnetic	Photon
Weak	W^+, W^-, Z^0
Strong	Gluons

5.	**Conservation Laws**	Some quantities, such as energy, momentum, electric charge, angular momentum, baryon number, and each of the three lepton numbers, are strictly conserved in all reactions and decays. Others, such as strangeness and charm, are conserved in reactions and decays that proceed via the strong interaction but not in those that proceed via the weak interaction.
6.	**Particles and Antiparticles**	Particles and their antiparticles have identical masses but opposite values for their other properties, such as charge, lepton number, baryon number, and strangeness. Particle–antiparticle pairs can be produced in various nuclear reactions if the energy available is greater than $2mc^2$, where m is the mass of the particle.
7.	**Hubble's Law**	Hubble's law relates the recession velocity of a galaxy, determined from the redshift of its spectrum, to the distance of the galaxy from us:

$$v = Hr \qquad \textbf{41-9}$$

where the Hubble constant $H = 23$ km/s per million light-years. From Hubble's law, we conclude that the universe is expanding and that the expansion began approximately $1/H$ years ago.

8.	**The Big Bang**	According to the model currently used to describe the evolution of the universe, the universe began with a Big Bang approximately 10^{10} years ago. The Big Bang model is supported by substantial experimental observations, including the isotropic, 2.7-K, background blackbody radiation spectrum.

Problem-Solving Guide

Summary of Worked Examples

Type of Calculation	Procedure and Relevant Examples
1. Spin and Antiparticles	
Find the energy of the photons emitted when a particle and an antiparticle annihilate.	When the particle and antiparticle annihilate from rest, two photons of equal energy and opposite momentum are emitted. The energy of each photon is mc^2, where m is the mass of the particle or antiparticle. **Example 41-1**

2. Conservation Laws

Check a possible decay for violation of conservation laws.	Check the spin, electric charge, baryon number, and lepton number of each side of the equation. For energy conservation, check that the initial rest energy is greater than the final rest energy.	Example 41-2
Determine if a decay occurs via the weak or strong interaction or not at all.	First check the rest masses for conservation of energy. Then compute the change in strangeness. If $\Delta S = 0$, the decay will be via the strong interaction. If $\Delta S = \pm 1$, the decay will be via the weak interaction. If $\lvert\Delta S\rvert \geq 2$ the decay will not occur.	Example 41-3

3. Quarks

Determine the properties of a particle given the quark combination.	Baryons contain 3 quarks; mesons contain a quark and an antiquark. Add the charges of the quarks to determine the charge of the particle and add the strangenesses of the quarks to determine the strangeness of the particle.	Example 41-4

4. Hubble's Law

Given the recession velocity, find the distance to a galaxy.	Compute r from Hubble's law $v = Hr$.	Example 41-5

Problems

In a few problems, you are given more data than you actually need; in a few other problems, you are required to supply data from your general knowledge, outside sources, or informed estimates.

Conceptual Problems

Problems from Optional and Exploring sections

• Single-concept, single-step, relatively easy

•• Intermediate-level, may require synthesis of concepts

••• Challenging, for advanced students

Hadrons and Leptons

1 • How are baryons and mesons similar? How are they different?

2 • The muon and the pion have nearly the same mass. How do these particles differ?

3 • How can you tell whether a decay proceeds via the strong interaction or the weak interaction?

4 • True or false:

(*a*) All baryons are hadrons.
(*b*) All hadrons are baryons.

Spin and Antiparticles

5 • True or false:

Mesons are spin-$\frac{1}{2}$ particles.

6 • Two pions at rest annihilate according to the reaction $\pi^+ + \pi^- \rightarrow \gamma + \gamma$. (*a*) Why must the energies of the two γ rays be equal? (*b*) Find the energy of each γ ray. (*c*) Find the wavelength of each γ ray.

7 • Find the minimum energy of the photon needed for the following pair-production reactions: (*a*) $\gamma \rightarrow \pi^+ + \pi^-$, (*b*) $\gamma \rightarrow p + p^-$, and (*c*) $\gamma \rightarrow \mu^- + \mu^+$.

The Conservation Laws

8 • State which of the decays or reactions that follow violate one or more of the conservation laws, and give the law or laws violated in each case: (*a*) $p^+ \rightarrow n + e^+ + \bar{\nu}_e$, (*b*) $n \rightarrow p^+ + \pi^-$, (*c*) $e^+ + e^- \rightarrow \gamma$, (*d*) $p + p^- \rightarrow \gamma + \gamma$, and (*e*) $\bar{\nu}_e + p \rightarrow n + e^+$.

9 • Determine the change in strangeness in each reaction that follows, and state whether the reaction can proceed via the strong interaction, the weak interaction, or not at all: (*a*) $\Omega^- \rightarrow \Xi^0 + \pi^-$, (*b*) $\Xi^0 \rightarrow p + \pi^- + \pi^0$, and (*c*) $\Lambda^0 \rightarrow p^+ + \pi^-$.

10 • Determine the change in strangeness for each decay, and state whether the decay can proceed via the strong interaction, the weak interaction, or not at all: (*a*) $\Omega^- \rightarrow \Lambda^0 + K^-$ and (*b*) $\Xi^0 \rightarrow p + \pi^-$.

11 • Determine the change in strangeness for each decay, and state whether the decay can proceed via the strong interaction, the weak interaction, or not at all: (*a*) $\Omega^- \to \Lambda^0 + \bar{\nu}_e + e^-$ and (*b*) $\Sigma^+ \to p + \pi^0$.

12 • (*a*) Which of the following decays of the τ particle is possible?

$$\tau \to \mu^- + \bar{\nu}_\mu + \nu_\tau$$

$$\tau \to \mu^- + \nu_\mu + \bar{\nu}_\tau$$

(*b*) Explain why the other is not possible. (*c*) Calculate the kinetic energy of the decay products for the decay that is possible.

13 •• Consider the following decay chain:

$$\Omega^- \to \Xi^0 + \pi^-$$
$$\Xi^0 \to \Sigma^+ + e^- + \bar{\nu}_e$$
$$\pi^- \to \mu^- + \bar{\nu}_\mu$$
$$\Sigma^+ \to n + \pi^+$$
$$\pi^+ \to \mu^+ + \nu_\mu$$
$$\mu^+ \to e^+ + \bar{\nu}_\mu + \nu_e$$
$$\mu^- \to e^- + \bar{\nu}_e + \nu_\mu$$

(*a*) Are all the final products shown stable? If not, finish the decay chain. (*b*) Write the overall decay reaction for Ω^- to the final products. (*c*) Check the overall decay reaction for the conservation of electric charge, baryon number, lepton number, and strangeness.

14 •• Test the following decays for violation of the conservation of energy, electric charge, baryon number, and lepton number: (*a*) $n \to \pi^+ + \pi^- + \mu^+ + \mu^-$; (*b*) $\pi^0 \to e^+ + e^- + \gamma$. Assume that linear and angular momentum are conserved. State which conservation laws (if any) are violated in each decay.

The Quark Model

15 • How can you tell whether a particle is a meson or a baryon by looking at its quark content?

16 • Are there any quark–antiquark combinations that result in a nonintegral electric charge?

17 • Find the baryon number, charge, and strangeness for the following quark combinations and identify the hadron: (*a*) *uud*, (*b*) *udd*, (*c*) *uus*, (*d*) *dds*, (*e*) *uss*, and (*f*) *dss*.

18 • Repeat Problem 17 for the following quark combinations: (*a*) $u\bar{d}$, (*b*) $\bar{u}d$, (*c*) $u\bar{s}$, and (*d*) $\bar{u}s$.

19 • The Δ^{++} particle is a baryon that decays via the strong interaction. Its strangeness, charm, topness, and bottomness are all zero. What combination of quarks gives a particle with these properties?

20 • Find a possible combination of quarks that gives the correct values for electric charge, baryon number, and strangeness for (*a*) K^+ and (*b*) K^0.

21 • The D^+ meson has no strangeness, but it has charm of +1. (*a*) What is a possible quark combination that will give the correct properties for this particle? (*b*) Repeat (*a*) for the D^- meson, which is the antiparticle of the D^+.

22 • Find a possible combination of quarks that gives the correct values for electric charge, baryon number, and strangeness for (*a*) K^- (the K^- is the antiparticle of the K^+) and (*b*) $\bar{K}^0$.

23 •• Find a possible quark combination for the following particles: (*a*) Λ^0, (*b*) $\bar{p}^-$, and (*c*) Σ^-.

24 •• Find a possible quark combination for the following particles: (*a*) $\bar{n}$, (*b*) Ξ^0, and (*c*) Σ^+.

25 •• Find a possible quark combination for the following particles: (*a*) Ω^- and (*b*) Ξ^-.

26 •• State the properties of the particles made up of the following quarks: (*a*) *ddd*, (*b*) $u\bar{c}$, (*c*) $u\bar{b}$, and (*d*) $\bar{s}\bar{s}\bar{s}$.

General Problems

27 • True or false:

(*a*) Leptons consist of three quarks.
(*b*) The times for decays via the weak interaction are typically longer than those for decays via the strong interaction.
(*c*) Electrons interact with protons via the strong interaction.
(*d*) Strangeness is not conserved in weak interactions.
(*e*) Neutrons have no charm.

28 • (*a*) What conditions are necessary for a particle and its antiparticle to be the same? Find the antiparticle for (*b*) π^0 and (*c*) Ξ^0.

29 •• Consider the following decay chain:

$$\Xi^0 \to \Lambda^0 + \pi^0$$
$$\Lambda^0 \to p + \pi^-$$
$$\pi^0 \to \gamma + \gamma$$
$$\pi^- \to \mu^- + \bar{\nu}_\mu$$
$$\mu^- \to e^- + \bar{\nu}_e + \nu_\mu$$

(*a*) Are all the final products shown stable? If not, finish the decay chain. (*b*) Write the overall decay reaction for Ξ^0 to the final products. (*c*) Check the overall decay reaction for the conservation of electric charge, baryon number, lepton number, and strangeness. (*d*) In the first step of the chain, could the Λ^0 have been a Σ^0?

30 •• Test the following decays for violation of the conservation of energy, electric charge, baryon number, and lepton number: (*a*) $\Lambda^0 \to p + \pi^-$, (*b*) $\Sigma^- \to n + p^-$, (*c*) $\mu^- \to e^- + \bar{\nu}_e + \nu_\mu$. Assume that linear and angular momentum are conserved. State which conservation laws (if any) are violated in each decay.

31 ••• (*a*) Calculate the total kinetic energy of the decay products for the decay $\Lambda^0 \to p + \pi^-$. Assume the Λ^0 is initially at rest. (*b*) Find the ratio of the kinetic energy of the pion to the kinetic energy of the proton. (*c*) Find the kinetic energies of the proton and the pion for this decay.

32 ••• A Σ^0 particle at rest decays into a Λ^0 plus a photon. (*a*) What is the total energy of the decay products? (*b*) Assuming that the kinetic energy of the Λ^0 is negligible compared with the energy of the photon, calculate the approximate momentum of the photon. (*c*) Use your result for (*b*) to calculate the kinetic energy of the Λ^0. (*d*) Use your result for (*c*) to obtain a better estimate of the momentum and the energy of the photon.

33 ••• In this problem, you will calculate the difference in the time of arrival of two neutrinos of different energy from a supernova that is 170,000 light-years away. Let the energies of the neutrinos be $E_1 = 20$ MeV and $E_2 = 5$ MeV, and assume that the rest mass of a neutrino is 20 eV/c^2. Because their total energy is so much greater than their rest energy, the neutrinos have speeds that are very nearly equal to c and energies that are approximately $E \approx pc$. (*a*) If t_1 and t_2 are the times it takes for neutrinos of speeds u_1 and u_2 to travel a distance x, show that

$$\Delta t = t_2 - t_1 = x\frac{u_1 - u_2}{u_1 u_2} \approx \frac{x\,\Delta u}{c^2}$$

(*b*) The speed of a neutrino of rest mass m_0 and total energy E can be found from Equation 39-25. Show that when $E >> m_0c^2$, the speed u is given approximately by

$$\frac{u}{c} \approx 1 - \frac{1}{2}\left(\frac{m_0c^2}{E}\right)^2$$

(*c*) Use the results for (*b*) to calculate $u_1 - u_2$ for the energies and rest mass given, and calculate Δt from the result for (*a*) for $x = 170{,}000c\cdot$y. (*d*) Repeat the calculation in (*c*) using $m_0c^2 = 40$ eV for the rest energy of a neutrino.

APPENDIX A

SI Units and Conversion Factors

Basic Units	
Length	The *meter* (m) is the distance traveled by light in a vacuum in 1/299,792,458 s.
Time	The *second* (s) is the duration of 9,192,631,770 periods of the radiation corresponding to the transition between the two hyperfine levels of the ground state of the ^{133}Cs atom.
Mass	The *kilogram* (kg) is the mass of the international standard body preserved at Sèvres, France.
Current	The *ampere* (A) is that current in two very long parallel wires 1 m apart that gives rise to a magnetic force per unit length of 2×10^{-7} N/m.
Temperature	The *kelvin* (K) is 1/273.16 of the thermodynamic temperature of the triple point of water.
Luminous intensity	The *candela* (cd) is the luminous intensity, in the perpendicular direction, of a surface of area 1/600,000 m^2 of a blackbody at the temperature of freezing platinum at a pressure of 1 atm.

Derived Units		
Force	newton (N)	$1\ N = 1\ kg{\cdot}m/s^2$
Work, energy	joule (J)	$1\ J = 1\ N{\cdot}m$
Power	watt (W)	$1\ W = 1\ J/s$
Frequency	hertz (Hz)	$1\ Hz = s^{-1}$
Charge	coulomb (C)	$1\ C = 1\ A{\cdot}s$
Potential	volt (V)	$1\ V = 1\ J/C$
Resistance	ohm (Ω)	$1\ \Omega = 1\ V/A$
Capacitance	farad (F)	$1\ F = 1\ C/V$
Magnetic field	tesla (T)	$1\ T = 1\ N/A{\cdot}m$
Magnetic flux	weber (Wb)	$1\ Wb = 1\ T{\cdot}m^2$
Inductance	henry (H)	$1\ H = 1\ J/A^2$

Conversion Factors

Conversion factors are written as equations for simplicity; relations marked with an asterisk are exact.

Length

1 km = 0.6215 mi
1 mi = 1.609 km
1 m = 1.0936 yd = 3.281 ft = 39.37 in
*1 in = 2.54 cm
*1 ft = 12 in = 30.48 cm
*1 yd = 3 ft = 91.44 cm
1 lightyear = 1 $c \cdot y$ = 9.461×10^{15} m
*1 Å = 0.1 nm

Area

*1 $m^2 = 10^4\ cm^2$
1 $km^2 = 0.3861\ mi^2$ = 247.1 acres
*1 $in^2 = 6.4516\ cm^2$
1 $ft^2 = 9.29 \times 10^{-2}\ m^2$
1 $m^2 = 10.76\ ft^2$
*1 acre = 43,560 ft^2
1 mi^2 = 640 acres = 2.590 km^2

Volume

*1 $m^3 = 10^6\ cm^3$
*1 L = 1000 cm^3 = $10^{-3}\ m^3$
1 gal = 3.786 L
1 gal = 4 qt = 8 pt = 128 oz = 231 in^3
1 $in^3 = 16.39\ cm^3$
1 $ft^3 = 1728\ in^3$ = 28.32 L = $2.832 \times 10^4\ cm^3$

Time

*1 h = 60 min = 3.6 ks
*1 d = 24 h = 1440 min = 86.4 ks
1 y = 365.24 d = 31.56 Ms

Speed

1 km/h = 0.2778 m/s = 0.6215 mi/h
1 mi/h = 0.4470 m/s = 1.609 km/h
1 mi/h = 1.467 ft/s

Angle and Angular Speed

*π rad = 180°
1 rad = 57.30°
1° = 1.745×10^{-2} rad
1 rev/min = 0.1047 rad/s
1 rad/s = 9.549 rev/min

Mass

*1 kg = 1000 g
*1 tonne = 1000 kg = 1 Mg
1 u = 1.6606×10^{-27} kg
1 kg = 6.022×10^{23} u
1 slug = 14.59 kg
1 kg = 6.852×10^{-2} slug
1 u = 931.50 MeV/c^2

Density

*1 g/cm^3 = 1000 kg/m^3 = 1 kg/L
$(1\ g/cm^3)g$ = 62.4 lb/ft^3

Force

1 N = 0.2248 lb = 10^5 dyn
1 lb = 4.4482 N
$(1\ kg)g$ = 2.2046 lb

Pressure

*1 Pa = 1 N/m^2
*1 atm = 101.325 kPa = 1.01325 bars
1 atm = 14.7 lb/in^2 = 760 mmHg
= 29.9 inHg = 33.8 ftH_2O
1 lb/in^2 = 6.895 kPa
1 torr = 1 mmHg = 133.32 Pa
1 bar = 100 kPa

Energy

*1 kW·h = 3.6 MJ
*1 cal = 4.1840 J
1 ft·lb = 1.356 J = 1.286×10^{-3} Btu
*1 L·atm = 101.325 J
1 L·atm = 24.217 cal
1 Btu = 778 ft·lb = 252 cal = 1054.35 J
1 eV = 1.602×10^{-19} J
1 $u \cdot c^2$ = 931.50 MeV
*1 erg = 10^{-7} J

Power

1 horsepower = 550 ft·lb/s = 745.7 W
1 Btu/min = 17.58 W
1 W = 1.341×10^{-3} horsepower
= 0.7376 ft·lb/s

Magnetic Field

*1 G = 10^{-4} T
*1 T = 10^4 G

Thermal Conductivity

1 W/m·K = 6.938 Btu·in/h·ft^2·F°
1 Btu·in/h·ft^2·F° = 0.1441 W/m·K

APPENDIX B

Numerical Data

Terrestrial Data	
Acceleration of gravity g	$9.80665\ \mathrm{m/s^2}$
Standard value	$32.1740\ \mathrm{ft/s^2}$
At sea level, at equator†	$9.7804\ \mathrm{m/s^2}$
At sea level, at poles†	$9.8322\ \mathrm{m/s^2}$
Mass of earth M_E	5.98×10^{24} kg
Radius of earth R_E, mean	6.37×10^{6} m; 3960 mi
Escape speed $\sqrt{2R_E g}$	1.12×10^{4} m/s; 6.95 mi/s
Solar constant‡	$1.35\ \mathrm{kW/m^2}$
Standard temperature and pressure (STP):	
Temperature	273.15 K
Pressure	101.325 kPa; 1.00 atm
Molar mass of air	28.97 g/mol
Density of air (STP), ρ_{air}	$1.293\ \mathrm{kg/m^3}$
Speed of sound (STP)	331 m/s
Heat of fusion of H_2O (0°C, 1 atm)	333.5 kJ/kg
Heat of vaporization of H_2O (100°C, 1 atm)	2.257 MJ/kg

†Measured relative to the earth's surface.
‡Average power incident normally on 1 m^2 outside the earth's atmosphere at the mean distance from the earth to the sun.

Astronomical Data	
Earth	
Distance to moon†	3.844×10^{8} m; 2.389×10^{5} mi
Distance to sun, mean†	1.496×10^{11} m; 9.30×10^{7} mi; 1.00 AU
Orbital speed, mean	2.98×10^{4} m/s
Moon	
Mass	7.35×10^{22} kg
Radius	1.738×10^{6} m
Period	27.32 d
Acceleration of gravity at surface	$1.62\ \mathrm{m/s^2}$
Sun	
Mass	1.99×10^{30} kg
Radius	6.96×10^{8} m

† Center to center.

Physical Constants

Gravitational constant	G	$6.672\,6 \times 10^{-11}$ N·m²/kg²
Speed of light	c	$2.997\,924\,58 \times 10^{8}$ m/s
Fundamental charge	e	$1.602\,177\,33 \times 10^{-19}$ C
Avogadro's number	N_A	$6.022\,136\,7 \times 10^{23}$ particles/mol
Gas constant	R	8.314 51 J/mol·K
		1.987 22 cal/mol·K
		$8.205\,78 \times 10^{-2}$ L·atm/mol·K
Boltzmann's constant	$k = R/N_A$	$1.380\,658 \times 10^{-23}$ J/K
		$8.617\,385 \times 10^{-5}$ eV/K
Unified mass unit	$u = (1/N_A)$ g	$1.660\,540 \times 10^{-24}$ g
Coulomb constant	$k = 1/4\pi\epsilon_0$	$8.987\,551\,788 \times 10^{9}$ N·m²/C²
Permittivity of free space	ϵ_0	$8.854\,187\,817 \times 10^{-12}$ C²/N·m²
Permeability of free space	μ_0	$4\pi \times 10^{-7}$ N/A²
		$1.256\,637 \times 10^{-6}$ N/A²
Planck's constant	h	$6.626\,075\,5 \times 10^{-34}$ J·s
		$4.135\,669\,2 \times 10^{-15}$ eV·s
	$\hbar = h/2\pi$	$1.054\,572\,66 \times 10^{-34}$ J·s
		$6.582\,122\,0 \times 10^{-16}$ eV·s
Mass of electron	m_e	$9.109\,389\,7 \times 10^{-31}$ kg
		510.999 1 keV/c^2
Mass of proton	m_p	$1.672\,623\,1 \times 10^{-27}$ kg
		938.272 3 MeV/c^2
Mass of neutron	m_n	$1.674\,929 \times 10^{-27}$ kg
		939.565 6 MeV/c^2
Bohr magneton	$m_B = e\hbar/2m_e$	$9.274\,015\,4 \times 10^{-24}$ J/T
		$5.788\,382\,63 \times 10^{-5}$ eV/T
Nuclear magneton	$m_n = e\hbar/2m_p$	$5.050\,786\,6 \times 10^{-27}$ J/T
		$3.152\,451\,66 \times 10^{-8}$ eV/T
Magnetic flux quantum	$\phi_0 = h/2e$	$2.067\,834\,6 \times 10^{-15}$ T·m²
Quantized Hall resistance	$R_K = h/e^2$	$2.581\,280\,7 \times 10^{4}$ Ω
Rydberg constant	R_H	$1.097\,373\,153\,4 \times 10^{7}$ m^{-1}
Josephson frequency–voltage quotient	$2e/h$	$4.835\,979 \times 10^{14}$ Hz/V
Compton wavelength	$\lambda_C = h/m_e c$	$2.426\,310\,58 \times 10^{-12}$ m

For additional data, see the last four pages in the book and the following tables in the text.

APPENDIX C

Periodic Table of Elements

1	2	3	4	5	6	7	8	9	10	11	12	13	14	15	16	17	18
1 **H** 1.00797																	2 **He** 4.003
3 **Li** 6.941	4 **Be** 9.012											5 **B** 10.81	6 **C** 12.011	7 **N** 14.007	8 **O** 15.9994	9 **F** 19.00	10 **Ne** 20.179
11 **Na** 22.990	12 **Mg** 24.31											13 **Al** 26.98	14 **Si** 28.09	15 **P** 30.974	16 **S** 32.064	17 **Cl** 35.453	18 **Ar** 39.948
19 **K** 39.102	20 **Ca** 40.08	21 **Sc** 44.96	22 **Ti** 47.88	23 **V** 50.94	24 **Cr** 52.00	25 **Mn** 54.94	26 **Fe** 55.85	27 **Co** 58.93	28 **Ni** 58.69	29 **Cu** 63.55	30 **Zn** 65.38	31 **Ga** 69.72	32 **Ge** 72.59	33 **As** 74.92	34 **Se** 78.96	35 **Br** 79.90	36 **Kr** 83.80
37 **Rb** 85.47	38 **Sr** 87.62	39 **Y** 88.906	40 **Zr** 91.22	41 **Nb** 92.91	42 **Mo** 95.94	43 **Tc** (98)	44 **Ru** 101.1	45 **Rh** 102.905	46 **Pd** 106.4	47 **Ag** 107.870	48 **Cd** 112.41	49 **In** 114.82	50 **Sn** 118.69	51 **Sb** 121.75	52 **Te** 127.60	53 **I** 126.90	54 **Xe** 131.29
55 **Cs** 132.905	56 **Ba** 137.33	57–71 **Rare Earths**	72 **Hf** 178.49	73 **Ta** 180.95	74 **W** 183.85	75 **Re** 186.2	76 **Os** 190.2	77 **Ir** 192.2	78 **Pt** 195.09	79 **Au** 196.97	80 **Hg** 200.59	81 **Tl** 204.37	82 **Pb** 207.19	83 **Bi** 208.98	84 **Po** (210)	85 **At** (210)	86 **Rn** (222)
87 **Fr** (223)	88 **Ra** (226)	89–103 **Actinides**	104 **Rf** (261)	105 **Ha** (260)	106 (263)	107 (262)	108 (265)	109 (266)									

Rare Earths (Lanthanides)	57 **La** 138.91	58 **Ce** 140.12	59 **Pr** 140.91	60 **Nd** 144.24	61 **Pm** (147)	62 **Sm** 150.36	63 **Eu** 152.0	64 **Gd** 157.25	65 **Tb** 158.92	66 **Dy** 162.50	67 **Ho** 164.93	68 **Er** 167.26	69 **Tm** 168.93	70 **Yb** 173.04	71 **Lu** 174.97
Actinides	89 **Ac** 227.03	90 **Th** 232.04	91 **Pa** 231.04	92 **U** 238.03	93 **Np** 237.05	94 **Pu** (244)	95 **Am** (243)	96 **Cm** (247)	97 **Bk** (247)	98 **Cf** (251)	99 **Es** (252)	100 **Fm** (257)	101 **Md** (258)	102 **No** (259)	103 **Lr** (260)

The 1–18 group designation has been recommended by the International Union of Pure and Applied Chemistry (IUPAC).

Atomic Numbers and Atomic Masses

Name	Symbol	Atomic Number	Mass	Name	Symbol	Atomic Number	Mass
Actinium	Ac	89	227.03	Mercury	Hg	80	200.59
Aluminum	Al	13	26.98	Molybdenum	Mo	42	95.94
Americium	Am	95	(243)	Neodymium	Nd	60	144.24
Antimony	Sb	51	121.75	Neon	Ne	10	20.179
Argon	Ar	18	39.948	Neptunium	Np	93	237.05
Arsenic	As	33	74.92	Nickel	Ni	28	58.69
Astatine	At	85	(210)	Niobium	Nb	41	92.91
Barium	Ba	56	137.3	Nitrogen	N	7	14.007
Berkelium	Bk	97	(247)	Nobelium	No	102	(259)
Beryllium	Be	4	9.012	Osmium	Os	76	190.2
Bismuth	Bi	83	208.98	Oxygen	O	8	15.9994
Boron	B	5	10.81	Palladium	Pd	46	106.4
Bromine	Br	35	79.90	Phosphorus	P	15	30.974
Cadmium	Cd	48	112.41	Platinum	Pt	78	195.09
Calcium	Ca	20	40.08	Plutonium	Pu	94	(244)
Californium	Cf	98	(251)	Polonium	Po	84	(210)
Carbon	C	6	12.011	Potassium	K	19	39.102
Cerium	Ce	58	140.12	Praseodymium	Pr	59	140.91
Cesium	Cs	55	132.905	Promethium	Pm	61	(147)
Chlorine	Cl	17	35.453	Protactinium	Pa	91	231.04
Chromium	Cr	24	52.00	Radium	Ra	88	(226)
Cobalt	Co	27	58.93	Radon	Rn	86	(222)
Copper	Cu	29	63.55	Rhenium	Re	75	186.2
Curium	Cm	96	(247)	Rhodium	Rh	45	102.905
Dysprosium	Dy	66	162.50	Rubidium	Rb	37	85.47
Einsteinium	Es	99	(252)	Ruthenium	Ru	44	101.1
Erbium	Er	68	167.26	Rutherfordium	Rf	104	(261)
Europium	Eu	63	152.0	Samarium	Sm	62	150.36
Fermium	Fm	100	(257)	Scandium	Sc	21	44.96
Fluorine	F	9	19.00	Selenium	Se	34	78.96
Francium	Fr	87	(223)	Silicon	Si	14	28.09
Gadolinium	Gd	64	157.25	Silver	Ag	47	107.870
Gallium	Ga	31	69.72	Sodium	Na	11	22.990
Germanium	Ge	32	72.59	Strontium	Sr	38	87.62
Gold	Au	79	196.97	Sulfur	S	16	32.064
Hafnium	Hf	72	178.49	Tantalum	Ta	73	180.95
Hahnium	Ha	105	(260)	Technetium	Tc	43	(98)
Helium	He	2	4.003	Tellurium	Te	52	127.60
Holmium	Ho	67	164.93	Terbium	Tb	65	158.92
Hydrogen	H	1	1.00797	Thallium	Tl	81	204.37
Indium	In	49	114.82	Thorium	Th	90	232.04
Iodine	I	53	126.90	Thulium	Tm	69	168.93
Iridium	Ir	77	192.2	Tin	Sn	50	118.69
Iron	Fe	26	55.85	Titanium	Ti	22	47.88
Krypton	Kr	36	83.80	Tungsten	W	74	183.85
Lanthanum	La	57	138.91	Uranium	U	92	238.03
Lawrencium	Lr	103	(260)	Vanadium	V	23	50.94
Lead	Pb	82	207.19	Xenon	Xe	54	131.29
Lithium	Li	3	6.941	Ytterbium	Yb	70	173.04
Lutetium	Lu	71	174.97	Yttrium	Y	39	88.906
Magnesium	Mg	12	24.31	Zinc	Zn	30	65.38
Manganese	Mn	25	54.94	Zirconium	Zr	40	91.22
Mendelevium	Md	101	(258)				

APPENDIX D

Review of Mathematics

In this appendix, we will review some of the basic results of algebra, geometry, trigonometry, and calculus. In many cases, we will merely state results without proof. Table D-1 lists some mathematical symbols.

Equations

The following operations can be performed on mathematical equations to facilitate their solution:

1. The same quantity can be added to or subtracted from each side of the equation.
2. Each side of the equation can be multiplied or divided by the same quantity.
3. Each side of the equation can be raised to the same power.

It is important to understand that the preceding rules apply to each *side* of the equation and not to each *term* in the equation.

Table D-1

Mathematical Symbols

$=$	is equal to
$\neq$	is not equal to
$\approx$	is approximately equal to
$\sim$	is of the order of
$\propto$	is proportional to
$>$	is greater than
$\geq$	is greater than or equal to
$\gg$	is much greater than
$<$	is less than
$\leq$	is less than or equal to
$\ll$	is much less than
Δx	change in x
$\lvert x \rvert$	absolute value of x
$n!$	$n(n-1)(n-2)\cdots 1$
Σ	sum
lim	limit
$\Delta t \to 0$	Δt approaches zero
$\frac{dx}{dt}$	derivative of x with respect to t
$\frac{\partial x}{\partial t}$	partial derivative of x with respect to t
$\int$	integral

Example D-1

Solve the following equation for x: $(x-3)^2 + 7 = 23$.

1. Subtract 7 from each side: $(x-3)^2 = 16$
2. Take the square root of each side: $x - 3 = \pm 4$
3. Add 3 to each side: $x = 4 + 3 = 7$ or $x = -4 + 3 = -1$

Remark Note that in step 2 we do not need to write $\pm(x-3) = \pm 4$ because all possibilities are included in $x - 3 = \pm 4$.

Check the Result We check our result by substituting each value into the original equation: $(7-3)^2 + 7 = 16 + 7 = 23$ and $(-1-3)^2 + 7 = 16 + 7 = 23$.

Example D-2

Solve the following equation for x:

$$\frac{1}{x} + \frac{1}{4} = \frac{1}{3}$$

1. Subtract $\frac{1}{4}$ from each side: $\frac{1}{x} = \frac{1}{3} - \frac{1}{4} = \frac{4}{12} - \frac{3}{12} = \frac{1}{12}$
2. Multiply each side by $12x$: $x = 12$

Remark This type of equation occurs both in geometric optics and in analyses of electric circuits. Although it is easy to solve, errors are often made. A typical mistake is to take the reciprocal of each *term*, obtaining $x + 4 = 3$. Taking the reciprocal of each term is not allowed; taking the reciprocal of each *side* of an equation is allowed. Note that multiplying each side by $12x$ in step 2 is equivalent to taking the reciprocal of each side of the equation.

Direct and Inverse Proportion

The relationships of direct proportion and inverse proportion are so important in physics that they deserve special consideration. Often much algebraic manipulation can be avoided through a simple knowledge of these relationships. Suppose, for example, that you work for 5 days at a certain pay rate and earn \$400. How much would you earn at the same pay rate if you worked 8 days? In this problem, the money earned is *directly proportional* to the time worked. We can write an equation relating the money earned M to the time worked t using a constant of proportionality R:

$$M = Rt$$

The constant of proportionality in this case is the pay rate. We can express R in dollars per day. Since \$400 was earned in 5 d, the value of R is $\$400/(5\text{ d}) = \$80/\text{d}$. In 8 d, the amount earned is therefore

$$M = (\$80/\text{d})(8\text{ d}) = \$640$$

However, we do not have to find the pay rate explicitly to work the problem. Since the amount earned in 8 d is $\frac{8}{5}$ times that earned in 5 d, this amount is

$$M = \tfrac{8}{5}(\$400) = \$640$$

We can use a similar example to illustrate inverse proportion. If you get a 25% raise, how long would you need to work to earn $400? Here we consider R to be a variable and we wish to solve for t:

$$t = \frac{M}{R}$$

In this equation, the time t is *inversely proportional* to the pay rate R. Thus, if the new rate is $\frac{5}{4}$ times the old rate, the new time will be $\frac{4}{5}$ times the old time or 4 d.

There are some situations in which one quantity varies as the square or some other power of another quantity where the ideas of proportionality are also very useful. Suppose, for example, that a 10-in diameter pizza costs $8.50. How much would you expect a 12-in diameter pizza to cost? We expect the cost of a pizza to be approximately proportional to the amount of its contents, which is proportional to the area of the pizza. Since the area is in turn proportional to the square of the diameter, the cost should be proportional to the square of the diameter. If we increase the diameter by a factor of 12/10, the area increases by a factor of $(12/10)^2 = 1.44$, so we should expect the cost to be $(1.44)(\$8.50) = \$12.24$.

Example D-3

The intensity of light from a point source varies inversely with the square of the distance from the source. If the intensity is 3.20 W/m² at 5 m from a source, what is it at 6 m from the source?

1. Write an equation expressing the fact that the intensity varies inversely with the square of the distance:

$$I = \frac{C}{r^2}$$

where C is some constant.

2. Let I_1 be the intensity at $r_1 = 5$ m and I_2 be the intensity at $r_2 = 6$ m, and express the ratio I_2/I_1 in terms of r_1 and r_2:

$$\frac{I_2}{I_1} = \frac{C/r_2^2}{C/r_1^2} = \frac{r_1^2}{r_2^2} = \left(\frac{r_1}{r_2}\right)^2 = \left(\frac{5}{6}\right)^2 = 0.694$$

3. Solve for I_2:

$$I_2 = 0.694I_1 = (0.694)(3.20\ \text{W/m}^2) = 2.22\ \text{W/m}^2$$

Linear Equations

An equation in which the variables occur only to the first power is said to be linear. A linear equation relating y and x can always be put into the standard form

$$y = mx + b \tag{D-1}$$

where m and b are constants that may be either positive or negative. Figure D-1 shows a graph of the values of x and y that satisfy Equation D-1. The constant b, called the **intercept,** is the value of y at $x = 0$. The constant m is the **slope** of the line, which equals the ratio of the change in y to the correspond-

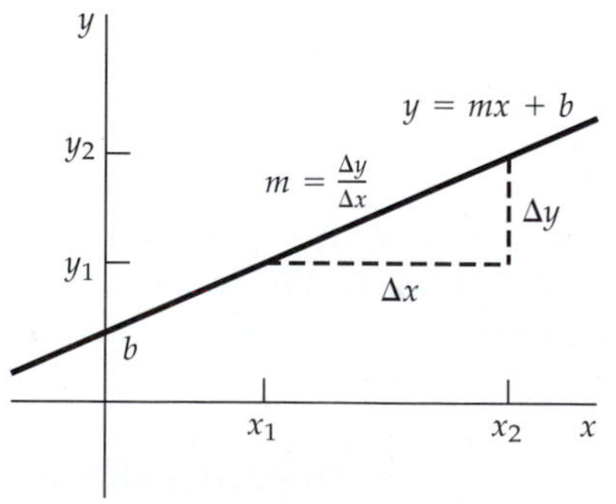

Figure D-1 Graph of the linear equation $y = mx + b$, where b is the intercept and $m = \Delta y/\Delta x$ is the slope.

ing change in x. In the figure, we have indicated two points on the line, x_1, y_1 and x_2, y_2, and the changes $\Delta x = x_2 - x_1$ and $\Delta y = y_2 - y_1$. The slope m is then

$$m = \frac{y_2 - y_1}{x_2 - x_1} = \frac{\Delta y}{\Delta x}$$

If x and y are both unknown, there is no unique solution for their values. Any pair of values x_1, y_1 on the line in Figure D-1 will satisfy the equation. If we have two equations, each with the same two unknowns x and y, the equations can be solved simultaneously for the unknowns.

Example D-4

Find the values of x and y that satisfy

$$3x - 2y = 8 \qquad \text{D-2}$$

and

$$y - x = 2 \qquad \text{D-3}$$

Picture the Problem Figure D-2 shows a graph of each of these equations. At the point where the lines intersect, the values of x and y satisfy both equations. We can solve two simultaneous equations by first solving either equation for one variable in terms of the other variable and then substituting the result into the other equation. An alternative method is to multiply one equation by a constant such that one of the unknown terms is eliminated when the equations are added or subtracted.

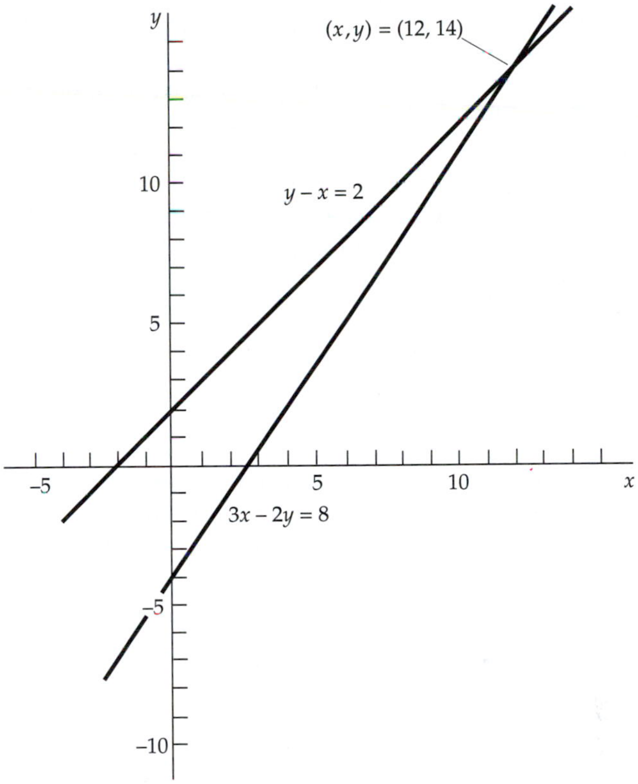

Figure D-2 Graph of Equations D-2 and D-3. At the point where the lines intersect, the values of x and y satisfy both equations.

1. Solve Equation D-3 for y: $\quad y = x + 2$

2. Substitute this value for y into Equation D-2: $\quad 3x - 2(x + 2) = 8$

3. Simplify and solve for x: $\quad 3x - 2x - 4 = 8$

   $x = 12$

4. To solve these equations using the alternative method, we first multiply Equation D-3 by 2: $\quad 2y - 2x = 4$

5. Add this equation to Equation D-2:

$$\begin{array}{r} 2y - 2x = 4 \\ 3x - 2y = 8 \\ \hline 3x - 2x = 12 \end{array}$$

   $x = 12$

Factoring

Equations can often be simplified by factoring. Three important examples are

1. Common factor: $2ax + 3ay = a(2x + 3y)$
2. Perfect square: $x^2 \pm 2xy + y^2 = (x \pm y)^2$
3. Difference of squares: $x^2 - y^2 = (x + y)(x - y)$

The Quadratic Formula

An equation that contains a variable to the second power is called a *quadratic equation.* The standard form for a quadratic equation is

$$ax^2 + bx + c = 0 \qquad \text{D-4}$$

where a, b, and c are constants. The general solution of this equation is

$$x = -\frac{b}{2a} \pm \frac{1}{2a}\sqrt{b^2 - 4ac} \qquad \text{D-5}$$

When b^2 is greater than $4ac$, there are two solutions corresponding to the + and − signs. Figure D-3 shows a graph of y versus x where $y = ax^2 + bx + c$. The curve, called a **parabola,** crosses the x axis twice. The values of x for which $y = 0$ are the solutions to Equation D-4. When $b^2 < 4ac$, the graph of y versus x does not intersect the x axis, as is shown in Figure D-4, and there are no real solutions to Equation D-4. When $b^2 = 4ac$, the graph of y versus x is tangent to the x axis at the point $x = -b/2a$.

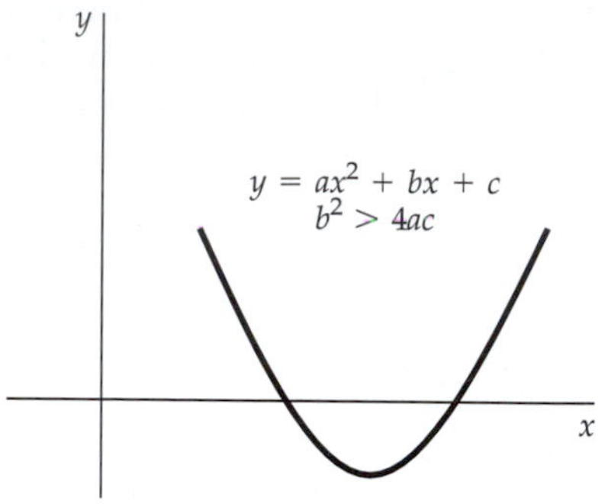

Figure D-3 Graph of y versus x when $y = ax^2 + bx + c$ for the case $b^2 > 4ac$. The two values of x for which $y = 0$ satisfy the quadratic equation (Equation D-4).

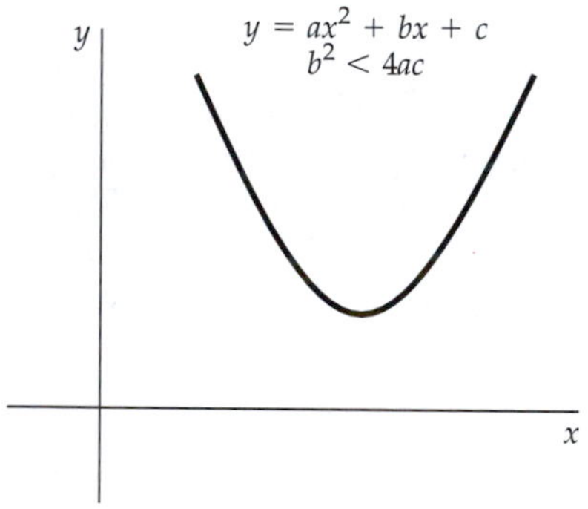

Figure D-4 Graph of y versus x when $y = ax^2 + bx + c$ for the case $b^2 < 4ac$. In this case, there are no (real) values of x for which $y = 0$.

Exponents

The notation x^n stands for the quantity obtained by multiplying x times itself n times. For example, $x^2 = x \cdot x$ and $x^3 = x \cdot x \cdot x$. The quantity n is called the **power,** or the **exponent,** of x. When two powers of x are multiplied, the exponents are added:

$$(x^m)(x^n) = x^{m+n} \qquad \text{D-6}$$

This can be readily seen from an example:

$$x^2x^3 = (x \cdot x)(x \cdot x \cdot x) = x^5$$

Any number raised to the 0 power is defined to be 1:

$$x^0 = 1 \qquad \text{D-7}$$

Then

$$x^n x^{-n} = x^0 = 1$$

$$x^{-n} = \frac{1}{x^n} \qquad \text{D-8}$$

When two powers are divided, the exponents are subtracted:

$$\frac{x^n}{x^m} = x^n x^{-m} = n^{n-m} \qquad \text{D-9}$$

Using these rules, we have

$$x^{1/2} \cdot x^{1/2} = x$$

so

$$x^{1/2} = \sqrt{x}$$

When a power is raised to another power, the exponents are multiplied:

$$(x^n)m = x^{nm} \qquad \text{D-10}$$

Logarithms

When y is related to x by $y = a^x$, the number x is said to be the logarithm of y to the base a and is written

$$x = \log_a y$$

If $y_1 = a^n$ and $y_2 = a^m$, then

$$y_1 y_2 = a^n a^m = a^{n+m}$$

and

$$\log_a y_1 y_2 = n + m = \log_a y_2 + \log_a y_1 \qquad \text{D-11}$$

It then follows that

$$\log_a y^n = n \log_a y \qquad \text{D-12}$$

Since $a^1 = a$ and $a^0 = 1$,

$$\log_a a = 1 \qquad \text{D-13}$$

and

$$\log_a 1 = 0 \qquad \text{D-14}$$

There are two bases in common use: base 10, called **common logarithms,** and base e ($e = 2.728...$), called **natural logarithms.** When no base is specified, the base is understood to be 10. Thus, $\log 100 = \log_{10} 100 = 2$ since $100 = 10^2$.

The symbol ln is used for natural logarithms. Thus,

$$y = \ln x \qquad \text{D-15}$$

implies

$$x = e^y \qquad \text{D-16}$$

Logarithms can be changed from one base to another. Suppose that

$$z = \log x \qquad \text{D-17}$$

Then

$$10^z = x \qquad \text{D-18}$$

Taking the natural logarithm of both sides of Equation D-18, we obtain

$$z \ln 10 = \ln x$$

or

$$\ln x = (\ln 10)\log x \qquad \text{D-19}$$

The Exponential Function

When the rate of change of a quantity is proportional to the quantity itself, the quantity increases or decreases exponentially. An example of *exponential decrease* is nuclear decay. If N is the number of radioactive nuclei at some time, then the change dN in some very small time interval dt will be proportional to N and to dt:

$$dN = -\lambda N\, dt$$

where the constant of proportionality λ is the decay rate. The function N satisfying this equation is

$$N = N_0 e^{-\lambda t} \qquad \text{D-20}$$

where N_0 is the number at time $t = 0$. Figure D-5 shows N versus t. A characteristic of exponential decay is that N decreases by a constant factor in a given time interval. The time interval for N to decrease to half its original value is its half-life $t_{1/2}$, which is related to the decay rate by

$$t_{1/2} = \frac{\ln 2}{\lambda} = \frac{0.693}{\lambda} \qquad \text{D-21}$$

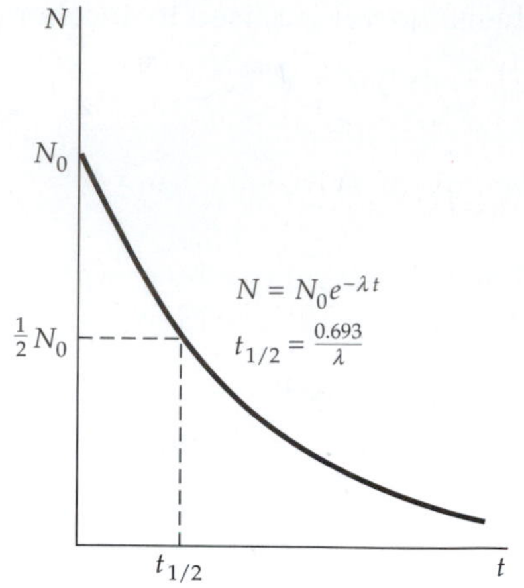

Figure D-5 Graph of N versus t when N decreases exponentially. The time $t_{1/2}$ is the time it takes for N to decrease by one-half.

An example of *exponential increase* is population growth. If the number of organisms is N, the change in N after a small time interval dt is given by

$$dN = +\lambda N\, dt$$

where λ is a constant that characterizes the rate of increase. The function N satisfying this equation is

$$N = N_0 e^{\lambda t} \qquad \text{D-22}$$

A graph of this function is shown in Figure D-6. An exponential increase is characterized by a doubling time T_2, which is related to λ by

$$T_2 = \frac{\ln 2}{\lambda} = \frac{0.693}{\lambda} \qquad \text{D-23}$$

If the rate of increase λ is expressed as a percentage, $r = \lambda/100\%$, the doubling time is

$$T_2 = \frac{69.3}{r} \qquad \text{D-24}$$

For example, if the population increases by 2% per year, the population will double every $69.3/2 \approx 35$ years. Table D-2 lists some useful relations for exponential and logarithmic functions.

Table D-2

Exponential and Logarithmic Functions

$e = 2.71828; \quad e^0 = 1$

If $y = e^x$, then $x = \ln y$.

$e^{\ln x} = x$

$e^x e^y = e^{(x+y)}$

$(e^x)^y = e^{xy} = (e^y)^x$

$\ln e = 1; \quad \ln 1 = 0$

$\ln xy = \ln x + \ln y$

$\ln \frac{x}{y} = \ln x - \ln y$

$\ln e^x = x; \quad \ln a^x = x \ln a$

$\ln x = (\ln 10) \log x$

$\quad = 2.3026 \log x$

$\log x = \log e \ln x = 0.43429 \ln x$

$e^x = 1 + x + \frac{x^2}{2!} + \frac{x^3}{3!} + \cdots$

$\ln(1 + x) = x - \frac{x^2}{2} + \frac{x^3}{3} - \frac{x^4}{4} + \cdots$

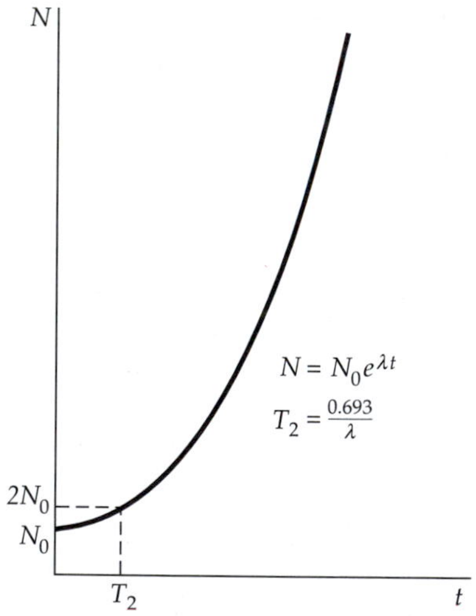

Figure D-6 Graph of N versus t when N increases exponentially. The time T_2 is the time it takes for N to double.

Geometry

The ratio of the circumference of a circle to its diameter is a natural number π, which has the approximate value

$$\pi = 3.141592$$

The circumference C of a circle is thus related to its diameter d and its radius r by

$$C = \pi d = 2\pi r \qquad \text{circumference of circle} \qquad \text{D-25}$$

The area of a circle is

$$A = \pi r^2 \qquad \text{area of circle} \qquad \text{D-26}$$

The area of a parallelogram is the base b times the height h (Figure D-7) and that of a triangle is one-half the base times the height (Figure D-8). A sphere of radius r (Figure D-9) has a surface area given by

$$A = 4\pi r^2 \qquad \text{spherical surface area} \qquad \text{D-27}$$

and a volume given by

$$V = \tfrac{4}{3}\pi r^3 \qquad \text{spherical volume} \qquad \text{D-28}$$

A cylinder of radius r and length L (Figure D-10) has surface area (not including the end faces) of

$$A = 2\pi rL \qquad \text{cylindrical surface} \qquad \text{D-29}$$

and volume of

$$V = \pi r^2 L \qquad \text{cylindrical volume} \qquad \text{D-30}$$

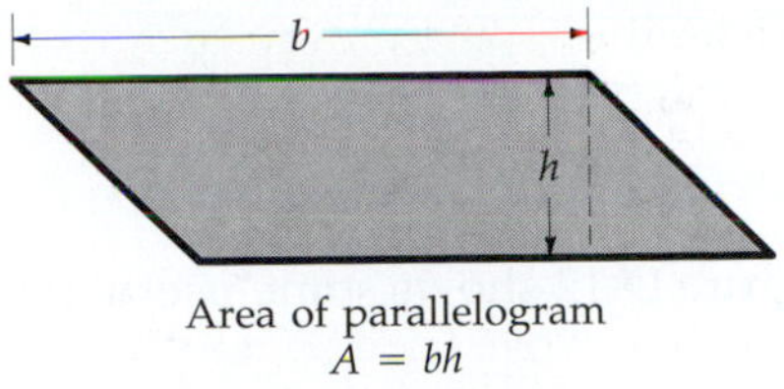

Figure D-7 Area of a parallelogram.

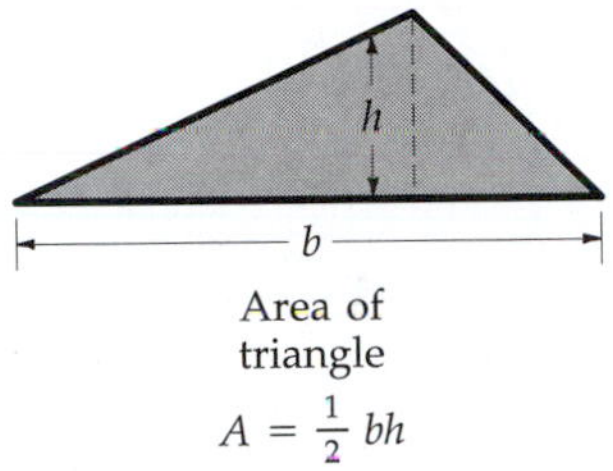

Figure D-8 Area of a triangle.

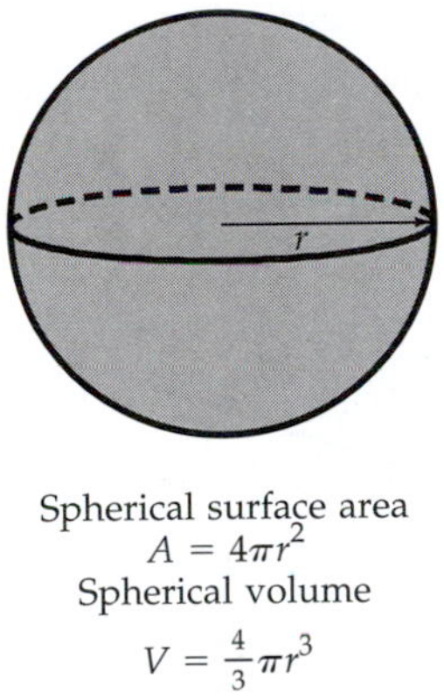

Figure D-9 Surface area and volume of a sphere.

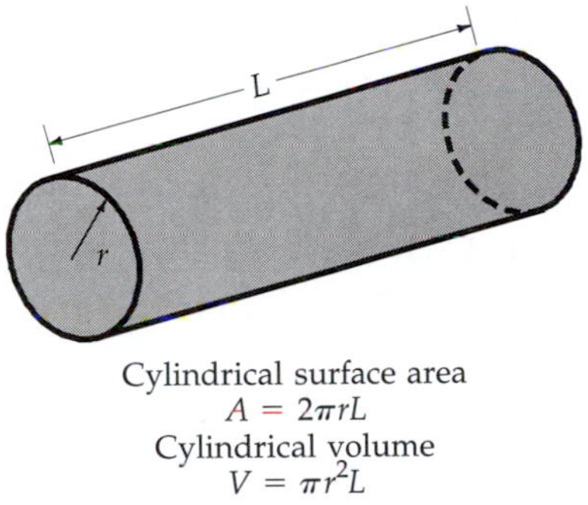

Figure D-10 Surface area (not including the end faces) and volume of a cylinder.

Trigonometry

The angle between two intersecting straight lines is measured as follows. A circle is drawn with its center at the intersection of the lines, and the circular arc is divided into 360 parts called **degrees.** The number of degrees in the arc between the lines is the measure of angle between the lines. For very small angles, the degree is divided into minutes (′) and seconds (″) with $1' = 1°/60$ and $1'' = 1'/60 = 1°/3600$. For scientific work, a more useful measure of an angle is the radian (rad), which is defined as the length of the circular arc between the lines divided by the radius of the circle (Figure D-11). If s is the arc length and r is the radius of the circle, the angle θ measured in radians is

$$\theta = \frac{s}{r} \qquad \text{D-31}$$

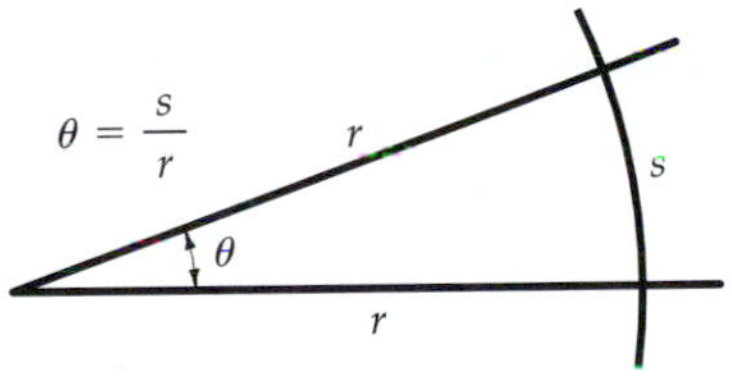

Figure D-11 The angle θ in radians is defined to be the ratio s/r, where s is the arc length intercepted on a circle of radius r.

Since the angle measured in radians is the ratio of two lengths, it is dimensionless. The relation between radians and degrees is

$$360° = 2\pi \text{ rad}$$

or

$$1\ \text{rad} = \frac{360°}{2\pi} = 57.3° \qquad \text{D-32}$$

Figure D-12 shows some useful relations for angles.

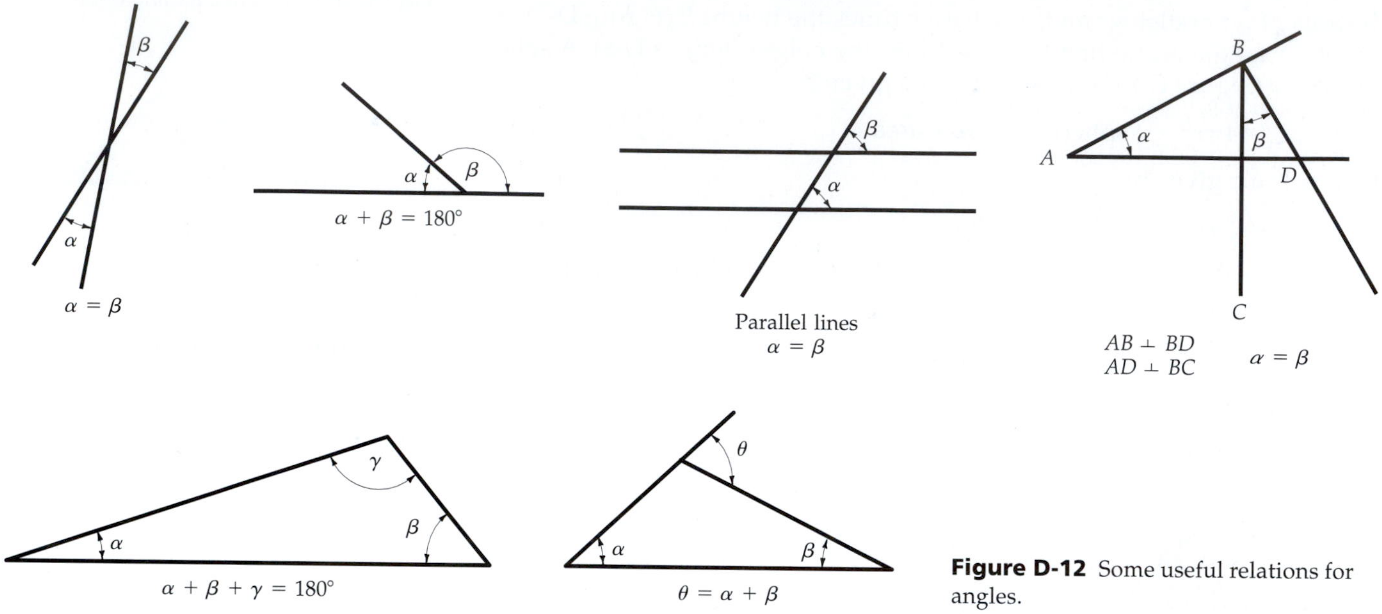

Figure D-12 Some useful relations for angles.

Figure D-13 shows a right triangle formed by drawing the line BC perpendicular to AC. The lengths of the sides are labeled a, b, and c. The trigonometric functions $\sin\theta$, $\cos\theta$, and $\tan\theta$ for an acute angle θ are defined as

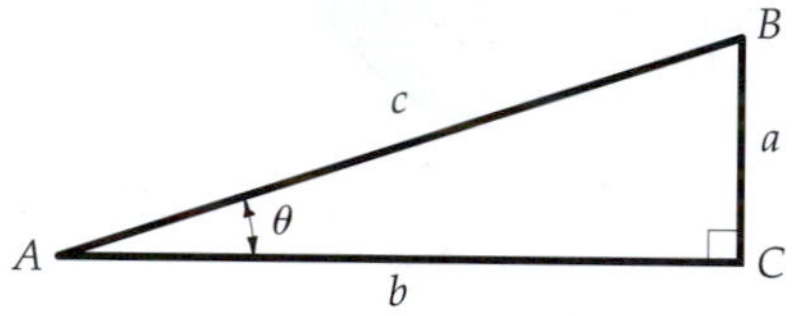

Figure D-13 A right triangle with sides of length a and b and a hypotenuse of length c.

$$\sin\theta = \frac{a}{c} = \frac{\text{opposite side}}{\text{hypotenuse}} \qquad \text{D-33}$$

$$\cos\theta = \frac{b}{c} = \frac{\text{adjacent side}}{\text{hypotenuse}} \qquad \text{D-34}$$

$$\tan\theta = \frac{a}{b} = \frac{\text{opposite side}}{\text{adjacent side}} = \frac{\sin\theta}{\cos\theta} \qquad \text{D-35}$$

Three other trigonometric functions are the reciprocals of these functions and are defined as

$$\sec\theta = \frac{c}{b} = \frac{1}{\cos\theta} \qquad \text{D-36}$$

$$\csc\theta = \frac{c}{a} = \frac{1}{\sin\theta} \qquad \text{D-37}$$

$$\cot\theta = \frac{b}{a} = \frac{1}{\tan\theta} = \frac{\cos\theta}{\sin\theta} \qquad \text{D-38}$$

The angle θ whose sine is x is called the arcsine of x, and is written $\sin^{-1} x$. That is, if

$$\sin\theta = x$$

then

$$\theta = \arcsin x = \sin^{-1} x \qquad \text{D-39}$$

The arcsine is the inverse of the sine. The inverse of the cosine and tangent are defined similarly. The angle whose cosine is y is the arccosine of y. That is, if

$$\cos \theta = y$$

then

$$\theta = \arccos y = \cos^{-1} y \qquad \text{D-40}$$

The angle whose tangent is z is the arctangent of z. That is, if

$$\tan \theta = z$$

$$\theta = \arctan z = \tan^{-1} z \qquad \text{D-41}$$

The Pythagorean theorem

$$a^2 + b^2 = c^2 \qquad \text{D-42}$$

gives some useful identities. If we divide each term in this equation by c^2, we obtain

$$\frac{a^2}{c^2} + \frac{b^2}{c^2} = 1$$

or, from the definitions of $\sin \theta$ and $\cos \theta$,

$$\sin^2 \theta + \cos^2 \theta = 1 \qquad \text{D-43}$$

Similarly, we can divide each term in Equation D-42 by a^2 or b^2 and obtain

$$1 + \cot^2 \theta = \csc^2 \theta \qquad \text{D-44}$$

and

$$1 + \tan^2 \theta = \sec^2 \theta \qquad \text{D-45}$$

These and other useful trigonometric formulas are listed in Table D-3.

Table D-3

Trigonometric Formulas

$\sin^2 \theta + \cos^2 \theta = 1; \quad \sec^2 \theta - \tan^2 \theta = 1; \quad \csc^2 \theta - \cot^2 \theta = 1$

$\sin 2\theta = 2 \sin \theta \cos \theta$

$\cos 2\theta = \cos^2 \theta - \sin^2 \theta = 2 \cos^2 \theta - 1 = 1 - 2 \sin^2 \theta$

$$\tan 2\theta = \frac{2 \tan \theta}{1 - \tan^2 \theta}$$

$$\sin \frac{1}{2}\theta = \sqrt{\frac{1 - \cos \theta}{2}}; \quad \cos \frac{1}{2}\theta = \sqrt{\frac{1 + \cos \theta}{2}}; \quad \tan \frac{1}{2}\theta = \sqrt{\frac{1 - \cos \theta}{1 + \cos \theta}}$$

$\sin(A \pm B) = \sin A \cos B \pm \cos A \sin B$

$\cos(A \pm B) = \cos A \cos B \mp \sin A \sin B$

$$\tan(A \pm B) = \frac{\tan A \pm \tan B}{1 \mp \tan A \tan B}$$

$\sin A \pm \sin B = 2 \sin[\frac{1}{2}(A \pm B)] \cos[\frac{1}{2}(A \mp B)]$

$\cos A + \cos B = 2 \cos[\frac{1}{2}(A + B)] \cos[\frac{1}{2}(A - B)]$

$\cos A - \cos B = 2 \sin[\frac{1}{2}(A + B)] \sin[\frac{1}{2}(B - A)]$

$$\tan A \pm \tan B = \frac{\sin(A \pm B)}{\cos A \cos B}$$

Example D-5

Use the isosceles right triangle shown in Figure D-14 to find the sine, cosine, and tangent of 45°.

Picture the Problem It is clear from the figure that the two acute angles of this triangle are equal. Since the sum of the three angles in a triangle must equal 180°, and the right angle is 90°, each acute angle must be 45°. If we multiply each side of any triangle by a common factor, we obtain a similar triangle with the same angles as the first. We can therefore choose any convenient length for one side. We choose the equal sides to have a length of 1 unit.

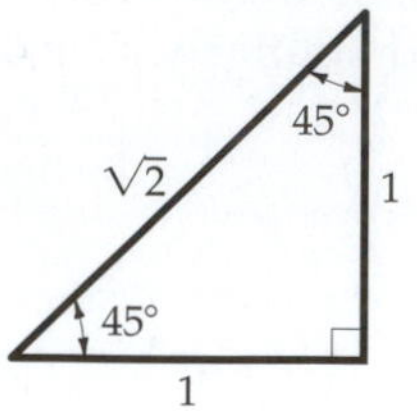

Figure D-14 An isosceles right triangle.

1. Find the length of the hypotenuse from the Pythagorean theorem: $c = \sqrt{a^2 + b^2} = \sqrt{1^2 + 1^2} = \sqrt{2}$ units

2. Calculate sin 45° from its definition: $\sin 45° = \frac{a}{c} = \frac{1}{\sqrt{2}} = 0.707$

3. Calculate cos 45° from its definition: $\cos 45° = \frac{b}{c} = \frac{1}{\sqrt{2}} = 0.707$

4. Calculate tan 45° from its definition: $\tan 45° = \frac{a}{b} = \frac{1}{1} = 1$

Example D-6

The sine of 30° is exactly 0.5. Find the ratios of the sides of a 30–60° right triangle.

Picture the Problem This common triangle is shown in Figure D-15. We choose a length of 1 unit for the side opposite the 30° angle.

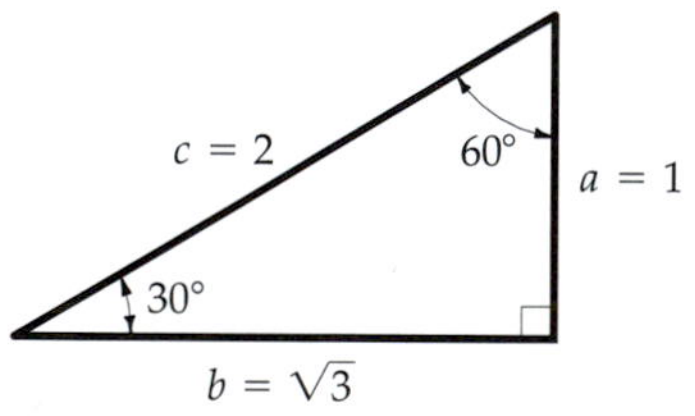

Figure D-15 A 30–60° right triangle.

1. Calculate the hypotenuse from the definition of the sine and the choice of 1 unit for the opposite side: $\sin 30° = \frac{a}{c} = \frac{1}{c} = 0.5$

   $c = \frac{1}{0.5} = 2$

2. Use the Pythagorean theorem to find the length b of the side opposite the 60° angle: $b = \sqrt{c^2 - a^2} = \sqrt{2^2 - 1^2} = \sqrt{3}$

3. Use these results to calculate cos 30°, tan 30°, sin 60°, cos 60°, and tan 60°:

$$\cos 30° = \frac{b}{c} = \frac{\sqrt{3}}{2} = 0.866$$

$$\tan 30° = \frac{a}{b} = \frac{1}{\sqrt{3}} = 0.577$$

$$\sin 60° = \frac{b}{c} = \cos 30° = 0.866$$

$$\cos 60° = \frac{a}{c} = \sin 30° = 0.500$$

$$\tan 60° = \frac{b}{a} = \frac{\sqrt{3}}{1} = 1.732$$

For small angles, the length a is nearly equal to the arc length s, as can be seen in Figure D-16. The angle $\theta = s/c$ is therefore nearly equal to $\sin\theta = a/c$:

$$\sin\theta \approx \theta \qquad \text{for small values of } \theta \qquad \text{D-46}$$

Similarly, the lengths c and b are nearly equal, so $\tan\theta = a/b$ is nearly equal to both θ and $\sin\theta$ for small values of θ:

$$\tan\theta \approx \theta \sin\theta \approx \theta \qquad \text{for small values of } \theta \qquad \text{D-47}$$

Equations D-46 and D-47 hold only if θ is measured in radians. Since $\cos\theta = b/c$ and these lengths are nearly equal for small values of θ, we have

$$\cos\theta \approx 1 \qquad \text{for small values of } \theta \qquad \text{D-48}$$

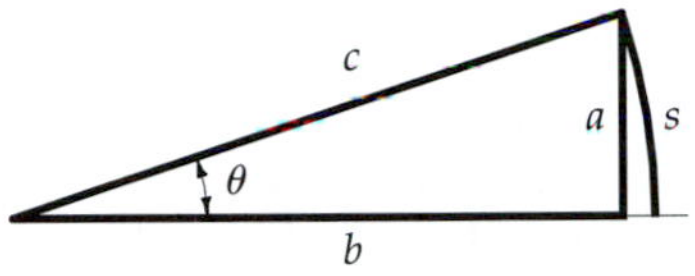

Figure D-16 For small angles, $\sin\theta = a/c$, $\tan\theta = a/b$, and the angle $\theta = s/c$ are all approximately equal.

Example D-7

By how much do sin θ, tan θ, and θ differ when θ = 15°?

1. Convert 15° to radians:

$$\theta = 15° \frac{2\pi \text{ rad}}{360°} = 0.262 \text{ rad}$$

2. Find sin 15° and tan 15° using a calculator:

$$\sin 15° = 0.259$$

$$\tan 15° = 0.268$$

3. Compute the percentage difference between θ and $\sin\theta$:

$$\frac{|\sin\theta - \theta|}{\theta} = \frac{|0.259 - 0.262|}{0.262} = \frac{0.003}{0.262} = 0.011 \approx 1\%$$

4. Compute the percentage difference between θ and $\tan\theta$:

$$\frac{|\tan\theta - \theta|}{\theta} = \frac{|0.268 - 0.262|}{0.262} = \frac{0.006}{0.262} = 0.023 \approx 2\%$$

Remark For smaller angles, the approximation $\theta \approx \sin\theta \approx \tan\theta$ is even more accurate.

Example D-7 shows that if accuracy of a few percent is needed, small angle approximations can be used only for angles of about 15° or less. Figure D-17 shows graphs of θ, $\sin\theta$, and $\tan\theta$ versus θ, for small values of θ.

Figure D-18 shows an obtuse angle with its vertex at the origin and one side along the x axis. The trigonometric functions for a general angle such as this are defined by

$$\sin\theta = \frac{y}{c} \tag{D-49}$$

$$\cos\theta = \frac{x}{c} \tag{D-50}$$

$$\tan\theta = \frac{y}{x} \tag{D-51}$$

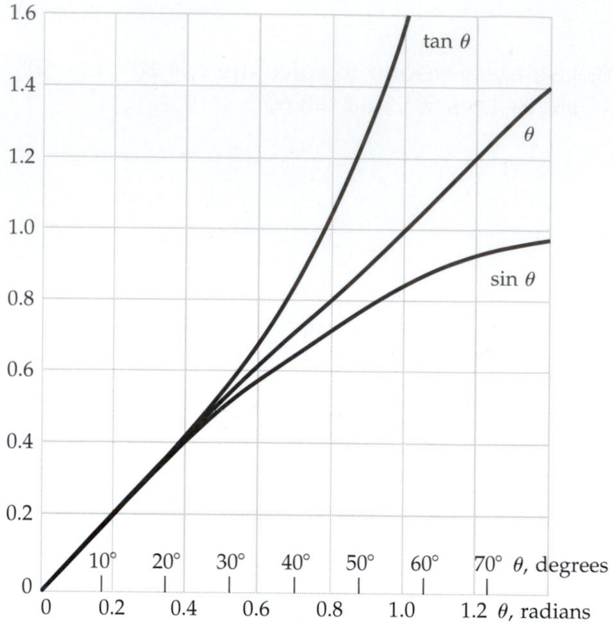

Figure D-17 Graphs of $\tan\theta$, θ, and $\sin\theta$ versus θ for small values of θ.

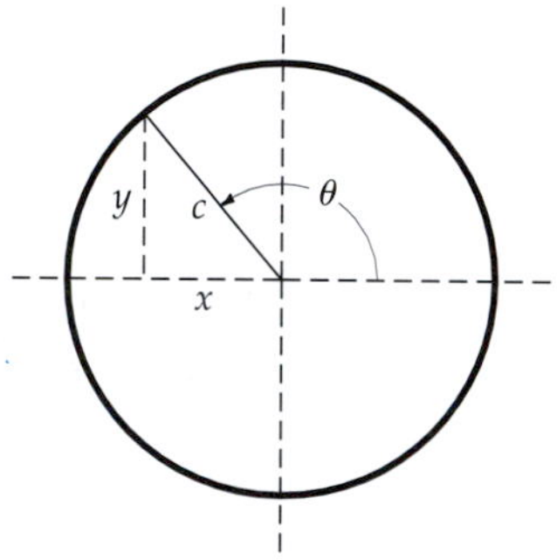

Figure D-18 Diagram for defining the trigonometric functions for an obtuse angle.

Figure D-19 shows plots of these functions versus θ. All trigonometric functions have a period of 2π. That is, when an angle changes by 2π rad, the function returns to its original value. Thus, $\sin(\theta + 2\pi) = \sin\theta$ and so forth. Some other useful relations are

$$\sin(\pi - \theta) = \sin\theta \tag{D-52}$$

$$\cos(\pi - \theta) = -\cos\theta \tag{D-53}$$

$$\sin(\pi/2 - \theta) = \cos\theta \tag{D-54}$$

$$\cos(\pi/2 - \theta) = \sin\theta \tag{D-55}$$

The trigonometric functions can be expressed as power series in θ. The series for $\sin\theta$ and $\cos\theta$ are

$$\sin\theta = \theta - \frac{\theta^3}{3!} + \frac{\theta^5}{5!} - \frac{\theta^7}{7!} + \cdots \tag{D-56}$$

$$\cos\theta = 1 - \frac{\theta^2}{2!} + \frac{\theta^4}{4!} - \frac{\theta^6}{6!} + \cdots \tag{D-57}$$

When θ is small, good approximations are obtained using only the first few terms in the series.

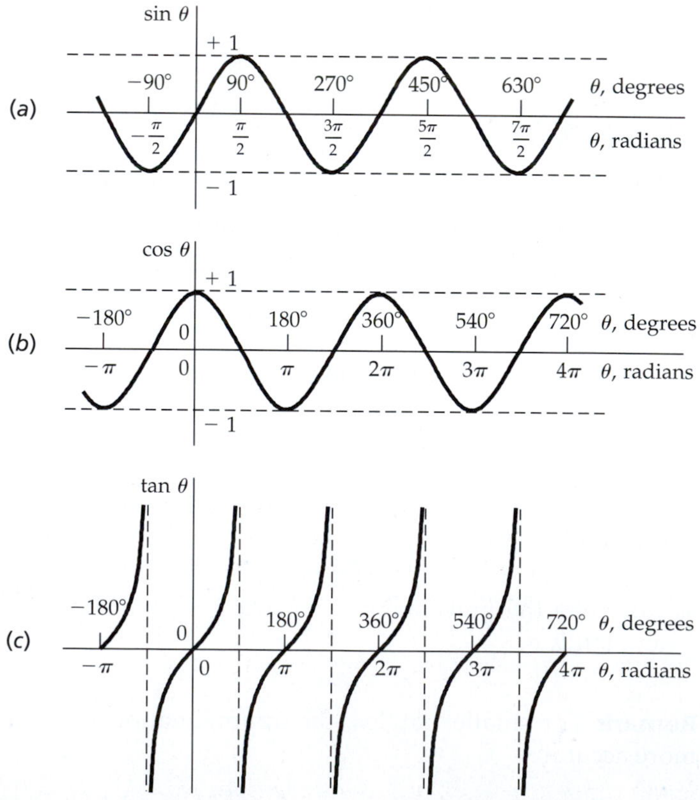

Figure D-19 The trigonometric functions $\sin\theta$, $\cos\theta$, and $\tan\theta$ versus θ.

The Binomial Expansion

The binomial theorem is very useful for making approximations. One form of this theorem is

$$(1 + x)^n = 1 + nx + \frac{n(n-1)}{2!}x^2 + \frac{n(n-1)(n-2)}{3!}x^3 + \frac{n(n-1)(n-2)(n-3)}{4!}x^4 + \cdots \qquad \text{D-58}$$

If n is a positive integer, there are $n + 1$ terms in this series. If n is a real number other than a positive integer, there are an infinite number of terms. The series is valid for any value of n if x^2 is less than 1. It is also valid for $x^2 = 1$ if n is positive. The series is particularly useful if $|x|$ is much less than 1. Then each term is much smaller than the previous term and we can drop all but the first two or three terms in the equation. If $|x|$ is much less than 1, we have

$$(1 + x)^n \approx 1 + nx, \qquad |x| \ll 1 \qquad \text{D-59}$$

Example D-8

Use Equation D-59 to find an approximate value for the square root of 101.

1. Write $(101)^{1/2}$ so it is in the form $(1 + x)^n$ with x much less than 1: $(101)^{1/2} = (100 + 1)^{1/2} = (100)^{1/2}(1 + 0.01)^{1/2} = 10(1 + 0.01)^{1/2}$
2. Use Equation D-59 with $n = \frac{1}{2}$ and $x = 0.1$ to expand $(1 + 0.01)^{1/2}$: $(1 + 0.01)^{1/2} \approx 1 + \frac{1}{2}(0.01) = 1.005$
3. Substitute this result into the equation in step 1: $(101)^{1/2} = 10(1 + 0.01)^{1/2} \approx 10.05$

Remarks We can assess the accuracy of this result by computing the first term in Equation D-58 that was neglected. This term is

$$\frac{n(n-1)}{2}x^2 = \frac{\frac{1}{2}(-\frac{1}{2})}{2}(0.01)^2 = -\frac{0.0001}{8} \approx -0.00001 = -0.001\%$$

We therefore expect our answer to be correct to within about 0.001%. The value of $(101)^{1/2}$ to eight significant figures is 10.049875, which differs from 10.05 by 0.000124 or about 0.001% of 10.05.

Complex Numbers

A general complex number z can be written

$$z = a + bi \qquad \text{D-60}$$

where a and b are real numbers and $i = \sqrt{-1}$. The quantity a is called the real part and the quantity ib is called the imaginary part of z. We can represent a complex number in a plane as shown in Figure D-20, where the x axis is the real axis and the y axis is the imaginary axis. We can use the relations $a = r\cos\theta$ and $b = r\sin\theta$ from Figure D-20 to write the complex number z in polar coordinates:

$$z = r\cos\theta + (r\sin\theta)i \qquad \text{D-61}$$

where $r = \sqrt{a^2 + b^2}$ is called the magnitude of z.

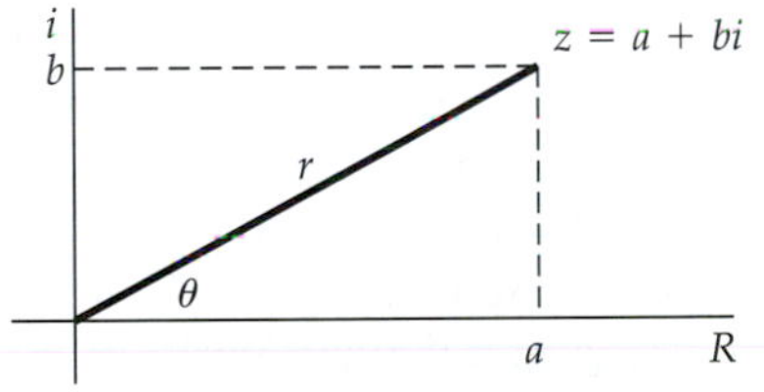

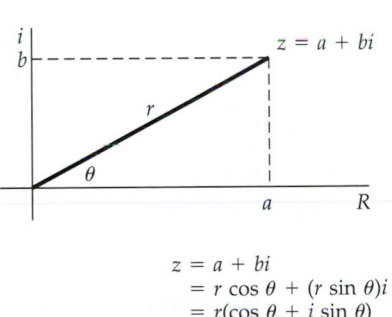

$$z = a + bi = r\cos\theta + (r\sin\theta)i = r(\cos\theta + i\sin\theta)$$

Figure D-20 Representation of a complex number in a plane. The real part of the complex number is plotted along the horizontal axis, and the imaginary part is plotted along the vertical axis.

When complex numbers are added or subtracted, the real and imaginary parts are added or subtracted separately:

$$z_1 + z_2 = (a_1 + b_1 i) + (a_2 + b_2 i) = (a_1 + a_2) + (b_1 + b_2)i \qquad \text{D-62}$$

However, when two complex numbers are multiplied, each part of one number is multiplied by each part of the other number:

$$z_1 z_2 = (a_1 + b_1 i)(a_2 + b_2 i) = a_1 a_2 + b_1 b_2 i^2 + (a_1 b_2 + a_2 b_1)i \qquad \text{D-63}$$
$$= a_1 a_2 - b_1 b_2 + (a_1 b_2 + a_2 b_1)i$$

where we have used $i^2 = -1$.

The complex conjugate of a complex number, z^*, is that number obtained by replacing i with $-i$:

$$z^* = (a + bi)^* = a - bi \qquad \text{D-64}$$

The product of a complex number and its complex conjugate equals the square of the magnitude of the number:

$$zz^* = (a + bi)(a - bi) = a^2 + b^2 \qquad \text{D-65}$$

A particularly useful function of a complex number is the exponential $e^{i\theta}$. Using the expansion for e^x given in Table D-2, we have

$$e^{i\theta} = 1 + i\theta + \frac{(i\theta)^2}{2!} + \frac{(i\theta)^3}{3!} + \frac{(i\theta)^4}{4!} + \cdots$$

Using $i^2 = -1$, $i^3 = -i$, $i^4 = +1$, and so forth and separating the real parts from the imaginary parts, this expansion can be written

$$e^{i\theta} = 1 - \frac{\theta^2}{2!} + \frac{\theta^4}{4!} + \cdots + i\left(\theta - \frac{\theta^3}{3!} + \cdots\right)$$

Comparing this result with Equations D-56 and D-57, we can see that

$$e^{i\theta} = \cos\theta + i\sin\theta \qquad \text{D-66}$$

Using this result, we can express a general complex number as an exponential:

$$z = a + bi = r\cos\theta + (r\sin\theta)i = re^{i\theta} \qquad \text{D-67}$$

where $r = \sqrt{a^2 + b^2}$.

Differential Calculus

When we say that x is a function of t, we mean that for each value of t there is a corresponding value of x. An example is $x = At^2$, where A is a constant. To indicate that x is a function of t, we sometimes write $x(t)$ for x. Figure D-21 is a graph of x versus t for a typical function $x(t)$. At a particular value $t = t_1$, x has the value of x_1 as indicated. At another value t_2, x has the value x_2. The change in t, $t_2 - t_1$, is written $\Delta t = t_2 - t_1$ and the corresponding change in x is written $\Delta x = x_2 - x_1$. The ratio

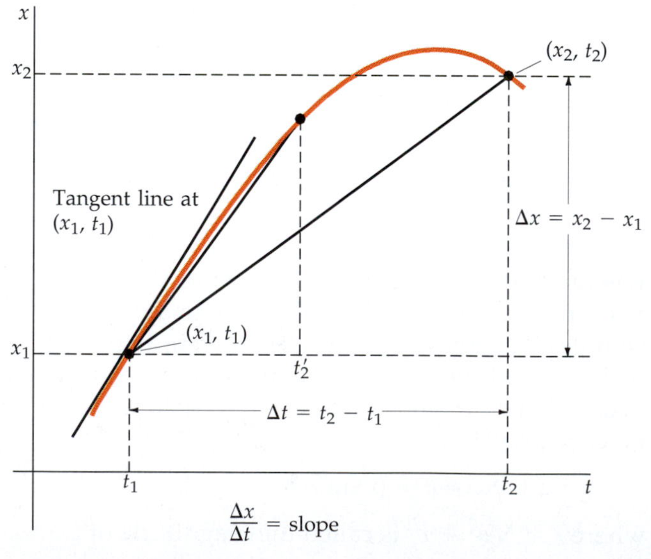

Figure D-21 Graph of a typical function $x(t)$. The points (x_1, t_1) and (x_2, t_2) are connected by a straight line. The slope of this line is $\Delta x/\Delta t$. As the time interval beginning at t_1 is decreased, the slope for that interval approaches the slope of the line tangent to the curve at time t_1, which is the derivative of x with respect to t.

$\Delta x/\Delta t$ is the slope of the straight line connecting (x_1, t_1) and (x_2, t_2). If we make Δt smaller and smaller, the line connecting (x_1, t_1) and (x_2, t_2) approaches the line that is tangent to the curve at the point (x_1, t_1). The slope of this tangent line is called the derivative of x with respect to t and is written dx/dt:

$$\frac{dx}{dt} = \lim_{\Delta t \to 0} \frac{\Delta x}{\Delta t} \qquad \text{D-68}$$

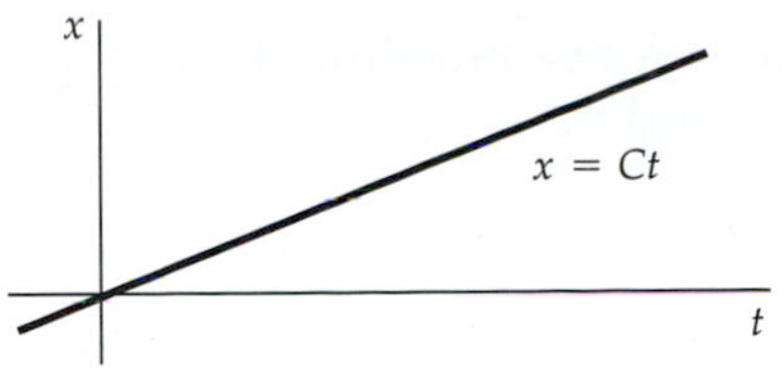

Figure D-22 Graph of the linear function $x = Ct$. This function has a constant slope C.

The derivative of a function of t is another function of t. If x is a constant, the graph of x versus t is a horizontal line with zero slope. The derivative of a constant is thus zero. In Figure D-22, x is proportional to t:

$$x = Ct$$

This function has a constant slope equal to C. Thus the derivative of Ct is C. Table D-4 lists some properties of derivatives and the derivatives of some particular functions that occur often in physics. It is followed by comments aimed at making these properties and rules clearer. More detailed discussion can be found in most calculus books.

Table D-4

Properties of Derivatives and Derivatives of Particular Functions

Linearity

1. The derivative of a constant times a function equals the constant times the derivative of the function:

$$\frac{d}{dt}[Cf(t)] = C\frac{df(t)}{dt}$$

2. The derivative of a sum of functions equals the sum of the derivatives of the functions:

$$\frac{d}{dt}[f(t) + g(t)] = \frac{df(t)}{dt} + \frac{dg(t)}{dt}$$

Chain rule

3. If f is a function of x and x is in turn a function of t, the derivative of f with respect to t equals the product of the derivative of f with respect to x and the derivative of x with respect to t:

$$\frac{d}{dt}f(x) = \frac{df}{dx}\frac{dx}{dt}$$

Derivative of a product

4. The derivative of a product of functions $f(t)g(t)$ equals the first function times the derivative of the second plus the second function times the derivative of the first:

$$\frac{d}{dt}[f(t)g(t)] = f(t)\frac{dg(t)}{dt} + \frac{df(t)}{dt}g(t)$$

Reciprocal derivative

5. The derivative of t with respect to x is the reciprocal of the derivative of x with respect to t, assuming that neither derivative is zero:

$$\frac{dx}{dt} = \left(\frac{dt}{dx}\right)^{-1} \quad \text{if} \quad \frac{dt}{dx} \neq 0$$

Derivatives of particular functions

6. $\dfrac{dC}{dt} = 0$ where C is a constant

7. $\dfrac{d(t^n)}{dt} = nt^{n-1}$

8. $\dfrac{d}{dt}\sin\omega t = \omega\cos\omega t$

9. $\dfrac{d}{dt}\cos\omega t = -\omega\sin\omega t$

10. $\dfrac{d}{dt}\tan\omega t = \omega\sec^2\omega t$

11. $\dfrac{d}{dt}e^{bt} = be^{bt}$

12. $\dfrac{d}{dt}\ln bt = \dfrac{1}{t}$

Example D-9

Find the derivative of $x = at^2 + bt + c$, where a, b, and c are constants.

Picture the Problem From rule 2, we can differentiate each term separately and add the results.

1. Use rules 1 and 7 to find the derivative of the first term: $\frac{d(at^2)}{dt} = 2at^1 = 2at$

2. Compute the derivatives of the second and third terms: $\frac{d(bt)}{dt} = b, \quad \frac{d(c)}{dt} = 0$

3. Add these results: $\frac{dx}{dt} = 2at + b$

Comments on Rules 1 Through 5

Rules 1 and 2 follow from the fact that the limiting process is linear. We can understand rule 3, the chain rule, by multiplying $\Delta f/\Delta t$ by $\Delta x/\Delta x$ and noting that, since x is a function of t, both Δx and Δf approach zero as Δt approaches zero. Since the limit of a product of two functions equals the product of their limits, we have

$$\lim_{\Delta t \to 0} \frac{\Delta f}{\Delta t} = \lim_{\Delta t \to 0} \frac{\Delta f}{\Delta x}\frac{\Delta x}{\Delta t} = \left(\lim_{\Delta x \to 0} \frac{\Delta f}{\Delta x}\right)\left(\lim_{\Delta t \to 0} \frac{\Delta x}{\Delta t}\right) = \frac{df}{dx}\frac{dx}{dt}$$

Rule 4 is not immediately apparent. The derivative of a product of functions is the limit of the ratio

$$\frac{f(t + \Delta t)g(t + \Delta t) - f(t)g(t)}{\Delta t}$$

If we add and subtract the quantity $f(t + \Delta t)g(t)$ in the numerator, we can write this ratio as

$$\frac{f(t + \Delta t)g(t + \Delta t) - f(t + \Delta t)g(t) + f(t + \Delta t)g(t) - f(t)g(t)}{\Delta t}$$

$$= f(t + \Delta t)\left[\frac{g(t + \Delta t) - g(t)}{\Delta t}\right] + g(t)\left[\frac{f(t + \Delta t) - f(t)}{\Delta t}\right]$$

As Δt approaches zero, the terms in brackets become $dg(t)/dt$ and $df(t)/dt$, respectively, and the limit of the expression is

$$f(t)\frac{dg(t)}{dt} + g(t)\frac{df(t)}{dt}$$

Rule 5 follows directly from the definition:

$$\frac{dx}{dt} = \lim_{\Delta t \to 0} \frac{\Delta x}{\Delta t} = \lim_{\Delta t \to 0}\left(\frac{\Delta t}{\Delta x}\right)^{-1} = \lim_{\Delta x \to 0}\left(\frac{\Delta t}{\Delta x}\right)^{-1} = \left(\lim_{\Delta t \to 0} \frac{\Delta t}{\Delta x}\right)^{-1}$$

Comments on Rule 7

We can obtain this important result using the binomial expansion. We have

$$f(t) = t^n$$

$$f(t + \Delta t) = (t + \Delta t)^n = t^n\left(1 + \frac{\Delta t}{t}\right)^n$$

$$= t^n\left[1 + n\frac{\Delta t}{t} + \frac{n(n-1)}{2!}\left(\frac{\Delta t}{t}\right)^2 + \frac{n(n-1)(n-2)}{3!}\left(\frac{\Delta t}{t}\right)^3 + \cdots\right]$$

Then

$$f(t - \Delta t) - f(t) = t^n\left[n\frac{\Delta t}{t} + \frac{n(n-1)}{2!}\left(\frac{\Delta t}{t}\right)^2 + \cdots\right]$$

and

$$\frac{f(t - \Delta t) - f(t)}{\Delta t} = nt^{n-1} + \frac{n(n-1)}{2!}t^{n-2}\,\Delta t + \cdots$$

The next term omitted from the last sum is proportional to $(\Delta t)^2$, the following to $(\Delta t)^3$, and so on. Each term except the first approaches zero as Δt approaches zero. Thus

$$\frac{df}{dt} = \lim_{\Delta x \to 0}\frac{f(t + \Delta t) + f(t)}{\Delta t} = nt^{n-1}$$

Comments on Rules 8 to 10

We first write $\sin \omega t = \sin \theta$ with $\theta = \omega t$ and use the chain rule,

$$\frac{d \sin \theta}{dt} = \frac{d \sin \theta}{d\theta}\frac{d\theta}{dt} = \omega\frac{d \sin \theta}{d\theta}$$

We then use the trigonometric formula for the sine of the sum of two angles θ and $\Delta\theta$:

$$\sin(\theta + \Delta\theta) = \sin \Delta\theta \cos \theta + \cos \Delta\theta \sin \theta$$

Since $\Delta\theta$ is to approach zero, we can use the small-angle approximations

$$\sin \Delta\theta \approx \Delta\theta \qquad \text{and} \qquad \cos \Delta\theta \approx 1$$

Then

$$\sin(\theta + \Delta\theta) \approx \Delta\theta \cos \theta + \sin \theta$$

and

$$\frac{\sin(\theta + \Delta\theta) - \sin \theta}{\Delta\theta} \approx \cos \theta$$

Similar reasoning can be applied to the cosine function to obtain rule 9.

Rule 10 is obtained by writing $\tan \theta = \sin \theta/\cos \theta$ and applying Rule 4 along with Rules 8 and 9.

$$\frac{d}{dt}(\tan \theta) = \frac{d}{dt}(\sin \theta)(\cos \theta)^{-1} = \sin \theta\frac{d}{dt}(\cos \theta)^{-1} + \frac{d(\sin \theta)}{dt}(\cos \theta)^{-1}$$

$$= \sin \theta(-1)(\cos \theta)^{-2}(-\sin \theta) + (\cos \theta)(\cos \theta)^{-1}$$

$$= \frac{\sin^2 \theta}{\cos^2 \theta} + 1 = \tan^2 \theta + 1 = \sec^2 \theta$$

Comments on Rule 11

Again we use the chain rule

$$\frac{de^{\theta}}{dt} = b\frac{de^{\theta}}{d\theta} \qquad \text{with} \qquad \theta = bt$$

and the series expansion for the exponential function:

$$e^{\theta+\Delta\theta} = e^{\theta}e^{\Delta\theta} = e^{\theta}\left[1 + \Delta\theta + \frac{(\Delta\theta)^2}{2!} + \frac{(\Delta\theta)^3}{3!} + \cdots\right]$$

Then

$$\frac{e^{\theta+\Delta\theta} - e^{\theta}}{\Delta\theta} = e^{\theta} + e^{\theta}\frac{\Delta\theta}{2!} + e^{\theta}\frac{(\Delta\theta)^2}{3!} + \cdots$$

As $\Delta\theta$ approaches zero, the right side of the equation above approaches e^{θ}.

Comments on Rule 12

Let

$$y = \ln bt$$

Then

$$e^{y} = bt \qquad \text{and} \qquad \frac{dt}{dy} = \frac{1}{b}e^{y} = t$$

Then using rule 5, we obtain

$$\frac{dy}{dt} = \left(\frac{dt}{dy}\right)^{-1} = \frac{1}{t}$$

Integral Calculus

Integration is related to the problem of finding the area under a curve. It is also the inverse of differentiation. Figure D-23 shows a function $f(t)$. The area of the shaded element is approximately $f_i\,\Delta t_i$, where f_i is evaluated anywhere in the interval Δt_i. This approximation improves if Δt_i is very small. The total area from t_1 to t_2 is found by summing all the area elements from t_1 to t_2 and taking the limit as each Δt_i approaches zero. This limit is called the integral of f over t and is written

$$\int_{t_1}^{t_2} f\,dt = \text{Area} = \lim_{\Delta t_i \to 0} \sum_i f_i\,\Delta t_i$$

If we integrate some function $f(t)$ from t_1 to some general value of t, we obtain another function of t. Let us call this function y:

$$y = \int_{t_1}^{t} f\,dt$$

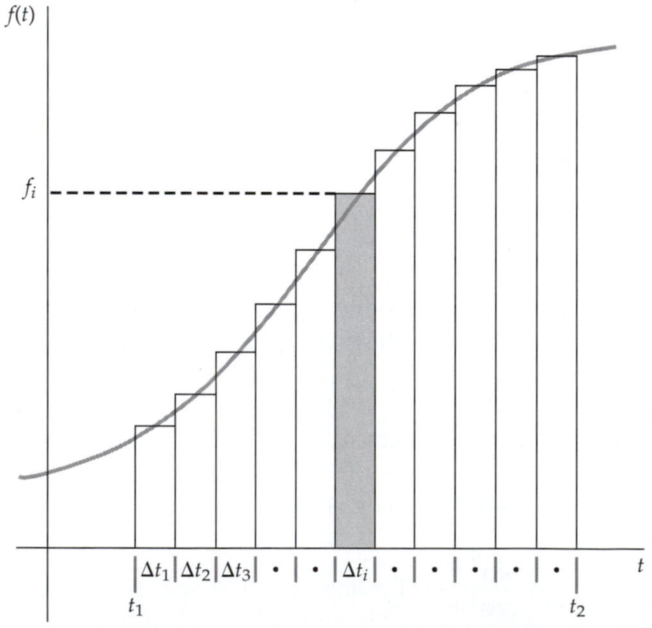

Figure D-23 A general function $f(t)$. The area of the shaded element is approximately $f_i\,\Delta t_i$, where f_i is evaluated anywhere in the interval.

The function y is the area under the f-versus-t curve from t_1 to a general value t. For a small interval Δt, the change in the area Δy is approximately $f\,\Delta t$.

$$\Delta y \approx f\,\Delta t$$

$$f \approx \frac{\Delta y}{\Delta t}$$

If we take the limit as Δt approaches 0, we can see that f is the derivative of y:

$$f = \frac{dy}{dt}$$

The relation between y and f is often written

$$y = \int f\,dt$$

where $\int f\,dt$ is called an **indefinite integral**. To evaluate an indefinite integral, we find the function y whose derivative is f. The definite integral of f from t_1 to t_2 is $y(t_1) - y(t_2)$, where $df/dt = y$:

$$\int_{t_1}^{t_2} f\,dt = y(t_2) - y(t_1)$$

Example D-10

Find the indefinite integral of $f(t) = t$.

1. The function whose derivative is t is $\frac{1}{2}t^2$ plus any constant:

$$\int t\,dt = \tfrac{1}{2}t^2 + C$$

where C is any constant

Table D-5 lists some important integration formulas. More extensive lists of differentiation and integration formulas can be found in handbooks such as Herbert Dwight's *Tables of Integrals and Other Mathematical Data*, fourth edition, Macmillan Publishing Company, Inc., New York, 1961.

Table D-5

Integration Formulas†

1. $\int A\,dt = At$	5. $\int e^{bt}\,dt = \frac{1}{b}e^{bt}$
2. $\int At\,dt = \frac{1}{2}At^2$	6. $\int \cos \omega t\,dt = \frac{1}{\omega}\sin \omega t$
3. $\int At^n\,dt = A\frac{t^{n+1}}{n+1}\quad n \neq -1$	7. $\int \sin \omega t\,dt = -\frac{1}{\omega}\cos \omega t$
4. $\int At^{-1}\,dt = A\ln t$	

†In these formulas, A, b, and ω are constants. An arbitrary constant C can be added to the right side of each equation.

ILLUSTRATION CREDITS

Part Openers

Part I p. 18 James Sugar/Black Star
Part II p. 402 Michael Freeman
Part III p. 540 Sandia National Laboratory
Part IV p. 656 © 1988 Richard Megna/Fundamental Photographs
Part V p. 1026 Courtesy AT&T Archives
Part VI p. 1142 Courtesy AT&T Archives

Chapter 1

Opener p. 1 The Granger Collection; **p. 4 (a)** The Granger Collection; **(b)** Frequency Electronics, Inc.; **p. 5 (a)** McDonald Observatory; **(b)** Bruce Coleman; **p. 7** Owen Franken/Stock, Boston; **p. 12 (a)** IBM Almaden Research Center; **(b)** from *The Body Victorius,* The Delacourt Press/Lennart Nilsson; **(c)** Kent and Donnan Dannen/Photo Researchers; **(d)** N.A.S.A. (68-HC-74); **(e)** Smithsonian Institution.

Chapter 2

Opener p. 19 Barry O'Rourke/The Stock Market; **p. 26** George Hall/Woodfin Camp and Assoc.; **p. 27 (top)** Estate of Harold E. Edgerton/Palm Press, Inc.; **(bottom)** Gunter Ziesler/Peter Arnold Inc.; **p. 29** Sidney Harris; **p. 32** Courtesy General Motors Corporation; **p. 34 (a)** Stanford Linear Accelerator, U.S. Department of Energy; **(b)** John Seeman/Stanford Linear Accelerator Center.

Chapter 3

Opener p. 54 Lewis Portnoy/Spectra-Action, Inc.

Chapter 4

Opener p. 83 The Granger Collection; **p. 84** Arthur Tilley/FPG International; **p. 86** N.A.S.A./Science Source/Photo Researchers; **p. 88** Berenice Abbott/Commerce Graphics, Ltd., Inc.; **p. 90** Culver Pictures; **p. 91 (a)** Cotton Coulson/Woodfin Camp and Assoc.; **(b)** Gary Ladd; **(c)** Los Alamos Scientific Laboratory; **(d)** Science Photo Library/Photo Researchers; **p. 93 (b)** Fundamental Photographs; **Figure 4-6** David J. Phillip/AP Wide World; **p. 110** Bruce Hayes/Photo Researchers.

Chapter 5

Opener p. 114 Courtesy Center for Engineering Design, University of Utah, Salt Lake City; **p. 115 (left)** F.P. Bowden and D. Tabor, *Friction and Lubrication of Solids,* Oxford University Press; **(right)** Uzi Landman, David W. Luedtke/Georgia Institute of Technology; **p. 118** Jean-Claude LeJeune/Stock, Boston; **p. 129** N.A.S.A. (85-HC-345); **p. 130** Malraux Photography/The Image Bank; **p. 131** Sandia National Laboratories; **p. 136** Richard Megna/Fundamental Photographs; **p. 143** David de Lossy/The Image Bank.

Chapter 6

Opener p. 148 Estate of Harold E. Edgerton/Palm Press Inc.; **p. 152** Courtesy of Dr. Roger Craig; **p. 165** David J. Phillip/AP Wide World.

Chapter 7

Opener p. 178 M.C. Escher Heirs/Cordon Art, De Baarn, Holland; **p. 180** Berenice Abbott/Photo Researchers; **p. 188 (a)** Visual Horizons/FPG International; **(b)** New York State Commerce Dept.; **(c)** L. Grant/FPG International; **p. 193** Stan Sholik/FPG International; **p. 195 (top left)** Courtesy of Omega Engineering, Inc.; **(top right)** Lennart Nilsson; **(bottom left)** Bruce Iverson; **(bottom right)** Lennart Nilsson; **p. 196** Leicester University/Science Photo Library/Photo Researchers.

Chapter 8

Opener p. 212 Daniel Wray/The Image Works; **p. 218** Estate of Harold E. Edgerton/Palm Press Inc.; **p. 223** N.A.S.A.; **p. 227** Robert R. Edwards/ BOB-E Photography; **p. 228** Romilly Lockyer/The Image Bank; **p. 230** Courtesy of Mercedes-Benz of North America, Montvale, NJ; **p. 233** Estate of Harold E. Edgerton/Palm Press Inc.; **p. 236 (left and right)** Estate of Harold E. Edgerton/Palm Press Inc.; **p. 238 (left)** Berenice Abbot/Photo Researchers; **(right)** Brookhaven National Laboratory; **p. 239 (top left)** Courtesy Central Scientific Company; **(top right)** David Parker/Science Photo Library/Photo Researchers; **(bottom left)** Alar and Juri Toomre, from *Cambridge Atlas of Astronomy,* Cambridge University Press, 1985; **(bottom right)** Kitt Peak National Observatory, from *Cambridge Atlas of Astronomy,* Cambridge University Press, 1985; **p. 243** N.A.S.A./Shostal Superstock.

Chapter 9

Opener p. 257 David Malin/Anglo-Australian Telescope Board; **p. 258** University of Utah/Center for Engineering Design; **p. 262** Culver Pictures; **p. 267** Michael Furman/The Stock Market; **p. 271** David Malin/Anglo-Australian Telescope Board; **p. 274** PAR/NYC, Inc. Archives; **p. 285** © Treë.

Chapter 10

Opener p. 295 National Oceanographic and Atmospheric Administration; **p. 301** Dick Luria/Science Source/Photo Researchers; **p. 303** Richard Minnix; **p. 304 (Figure 10-19)** Estate of Harold E. Edgerton/Palm Press Inc.; **(Figure 10-20)** Focus on Sports; **p. 305** Dick Luria/FPG International; **p. 307 (left and right)** N.A.S.A./Goddard Space Flight Center; **p. 318** Chris Sorenson/The Stock Market.

Chapter 11

Opener p. 321 (top) Collection of Historical Scientific Instruments, Harvard University; **(bottom)** Erich Lessing/Magnum; **p. 325**

Courtesy Central Scientific Company; **p. 331** N.A.S.A. (69-HC-905); **p. 341** N.A.S.A.

Chapter 12

Opener p. 351 Thomas Zimmerman/FPG International.

Chapter 13

Opener p. 374 Michael Dunn/The Stock Market; **p. 381** Chuck O'Rear/Woodfin Camp and Assoc.; **p. 382 (a)** PAR/NYC, Inc. Archives; **(b)** David Burnett/Woodfin Camp and Assoc.; **p. 384 (a)** Estate of Harold E. Edgerton/Palm Press, Inc.; **(b)** Richard Megna/Fundamental Photographs; **p. 388 (a) and (b)** Office National d'Etudes et de Recherches Areospatiales; **p. 392 (top)** Picker International; **(bottom)** Dr. Owen M. Griffin, Naval Research Laboratory.

Chapter 14

Opener p. 403 Citibank; **p. 405** N.A.S.A. (73-HC-787); **p. 409** Institute for Marine Dynamics; **p. 417** Berenice Abbott/Commerce Graphics Ltd, Inc.; **p. 422** Estate of Harold E. Edgerton/Palm Press Inc.; **p. 423** Monroe Auto Equipment; **p. 426** Royal Swedish Academy of Music.

Chapter 15

Opener p. 442 (top) Four by Five Inc.; **(bottom)** Berenice Abbott/Photo Researchers; **p. 443** Berenice Abbott/Photo Researchers; **p. 454 (Figure 15-11)** *PSSC Physics,* 2nd ed., 1965. D.C. Heath & Co., and Education Development Center Inc., Newton, MA; **(Figure 15-12)** From Winston E. Kock, *Lasers and Holography,* Dover Publications, New York, 1981; **p. 459 (a) and (b)** *PSSC Physics,* 2nd ed., 1965. D.C. Heath & Co., and Education Development Center Inc., Newton, MA; **p. 461 (Figure 15-22)** *PSSC Physics,* 2nd ed., 1965. D.C. Heath & Co., and Education Development Center Inc., Newton, MA; **p. 462 (left)** Naval Research Laboratory; **(right)** Courtesy San Francisco Symphony; **(Figures 15-23 and 15-25)** *PSSC Physics,* 2nd ed., 1965. D.C. Heath & Co., and Education Development Center Inc., Newton, MA; **p. 463** Courtesy of the author; **p. 464** *PSSC Physics,* 2nd ed., 1965. D.C. Heath & Co., and Education Development Center Inc., Newton, MA; **p. 467 (left)** Sandia National Laboratory; **(right)** Robert de Gast/Photo Researchers; **(bottom)** Estate of Harold E. Edgerton/Palm Press Inc.; **p. 468 (Figure 15-28)** *PSSC Physics,* 2nd ed., 1965. D.C. Heath & Co., and Education Development Center Inc., Newton, MA; **(right)** Department of Energy.

Chapter 16

Opener p. 480 Dr. John S. Shelton; **p. 486** Berenice Abbott(8J 1328)/Photo Researchers; **p. 489** David Yost/Steinway & Sons; **p. 490 (bottom)** University of Washington; **p. 492** *PSSC Physics,* 2nd ed., 1965. D.C. Heath & Co., and Education Development Center Inc., Newton, MA; **p. 496 (top)** David Hathaway/N.A.S.A.; **(bottom)** Professor Thomas D. Rossing, Northern Illinois University, DeKalb.

Chapter 17

Opener p. 509 (a–d) Courtesy of Akira Tonomura, Advanced Research Laboratory, Hitachi, Ltd.; **p. 519 (a, b)** *PSSC Physics,* 2nd ed., 1965. D.C. Heath & Co., and Education Development Center Inc., Newton, MA; **(c)** C.G. Shull; **(d)** Claus Jönsson; **(bottom)** Jack Griffith/University of North Carolina.

Chapter 18

Opener p. 541 Lockheed Corporation; **p. 544** Courtesy Honeywell, Inc.; **p. 545** Dr. William Magnum/National Bureau of Standards; **p. 559 (top)** N.A.S.A.; **(bottom)** Jet Propulsion Laboratory/N.A.S.A.

Chapter 19

Opener p. 566 Phoenix Pipe and Tube/Lana Berkovich; **p. 572 (b)** Science Museum, London; **p. 588** Will and Deni McIntyre/Photo Researchers.

Chapter 20

Opener p. 600 Sandia National Laboratory; **p. 603** J.M. Mejuto/FPG International; **p. 610** Michael Collier/Stock, Boston; **p. 611 (b)** Jean-Pierre Horlin/The ImageBank; **(c)** Sandia National Laboratories; **(d)** Peter Miller/The Image Bank; **(e)** Sandia National Laboratories.

Chapter 21

Opener p. 633 Michael Melford/The Image Bank.

Chapter 22

Opener p. 657 Lawrence Berkeley Laboratory/Science Photo Library/Photo Researchers; **p. 658** *PSSC* Physics, 2nd ed., 1965. D.C. Heath & Co. and Education Development Center, Inc., Newton, MA; **p. 659** Bruce Terris/IBM Almaden Research Center; **p. 662 (left)** © Grant Heilman; **(right)** Ann Ronan Picture Library; **p. 663** Burndy Library, Norwalk, Connecticut; **p. 672 (Figure 22-17)** Harold M. Waage; **(Figure 22-18)** Harold M. Waage; **p. 673 (Figure 22-19)** Harold M. Waage; **p. 675** Courtesy of Hulon Forrester/Video Display Corporation, Tucker, Georgia.

Chapter 23

Opener p. 688 © Dagmar Hailer-Hamann/Peter Arnold, Inc.; **p. 691** Ben Damsky Electric Power Research Institute; **p. 702 (Figure 23-21)** Runk/Schoenberger from Grant Heilman; **p. 708** Harold M. Waage.

Chapter 24

Opener p. 720 © 1990 Richard Megna/Fundamental Photographs; **p. 739 (Figure 24-18) (b)** © Karen R. Preuss; **(c)** © Hank Morgan/Rainbow; **p. 741 (Figure 24-20)** Harold M. Waage.

Chapter 25

Opener p. 752 © Picturepoint, Ltd., London; **p. 756 (Figure 25-2)** Harold M. Waage; **p. 758 (left)** © Mark Antman/The Image Works; **(right)** © Bruce Iverson; **p. 759 (left)** © Bruce Iverson; **(right)** © Paul Brierly; (bottom) Courtesy Tusonix, Tucson, Arizona; **p. 764** © Lawrence Livermore National Laboratory; **p. 767** © Manfred Kage/Peter Arnold, Inc.

Chapter 26

Opener p. 786 © Rabbit System, Santa Monica, California. Photo caption by kind cooperation of Larry Albright; **p. 793** © Chris Rogers/The Stock Market; **p. 795 (Figure 26-9)** © Paul Silverman/Fundamental Photographs; **p. 798** Ann Ronan Picture Library.

Chapter 28

Opener p. 855 © Stadler GmbH, Courtesy Transrapid International, Munich; **(Figure 28-1)** Fred Weiss; **p. 859 (Figure 28-8)** © 1990 Richard Megna/Fundamental Photographics; **p. 861 (top)** Larry Langrill; **(bottom)** © Lawrence Berkeley Laboratory/Science Photo Library; **p. 862 (Figure 28-13)** Carl E. Nielsen.

Chapter 29

Opener p. 883 Bob Williamson, Oakland University, Rochester, Michigan; **p. 885 (left)** © 1990 Richard Megna/Fundamental Photographs; **(right)** © 1990 Richard Megna/Fundamental Photographs; **p. 887 (Figure 29-6)** © 1990 Richard Megna/ Fundamental Photographs; **p. 890 (Figure 29-10)** © 1990 Richard Megna/Fundamental Photographs; **p. 891** © Bruce Iverson; **p. 893 (Figure 29-14)** © 1990 Richard Megna/Fundamental Photographs; **p. 894** Courtesy F.W. Bell Co.; **p. 896 (Figure 29-19)** Clarence Bennett, Oakland University, Rochester, Michigan; **p. 900 (left)** Courtesy Princeton University Plasma Physics Laboratory; **(right)** Courtesy Princeton University Plasma Physics Laboratory; **p. 909 (top)** J. F. Allen, St. Andrews University, Scotland; **(bottom)** © Paul Silverman/Fundamental Photographs; **p. 910 (Figure 29-34)** Robert J. Celotta, National Institute of Standards and Technology; **p. 911 (left)** Akira Tonomura, Hitachi Advanced Research Laboratory, Hatomaya, Japan; **(right)** © Bruce Iverson; **p. 912 (left)** Courtesy Seagate Technology; **(right)** Tom Chang/IBM Storage Systems Division, San Jose, CA; **p. 914** © Bill Pierce/Time Magazines, Inc.

Chapter 30

Opener p. 927 © 1990 Richard Megna/Fundamental Photographs; **p. 941 (Figure 30-19)** © Michael Holford, Collection of the Science Museum, London; **p. 948 (Figure 30-26)** A. Leitner/Renselaer Polytechnic Institute.

Chapter 31

Opener p. 959 © 1990 C. J. Allen/New England Hydro Transmission Corporation; **p. 961 (left)** Courtesy U.S. Dept. of the Interior, Dept. of Reclamation; **(right)** © Lee Langum/Photo Researchers; **p. 979** © George H. Clark Radioana Collection-Archive Center, National Museum of American History; **p. 982 (top)** © Yoav/Phototake; **(right)** © Daniel S. Brody/Stock Boston.

Chapter 32

Opener p. 999 N.A.S.A.; **p. 1004** Wabash Instrument Corporation; **p. 1018** Courtesy Cavendish Laboratory, University of Cambridge; **p. 1025 (Figure 32-13)** Courtesy Central Scientific Company.

Chapter 33

Opener p. 1027 © Robert Greenler; **p. 1028 (Figure 33-1)** Corbis-Bettmann; **p. 1029 (Figure 33-2)** Corbis-Bettmann; **(bottom)** Adapted from Eastman Kodak and Wabash Instrument Corporation; **p. 1032 (left)** © 1991 Paul Silverman/Fundamental Photographs; **(right)** © 1991 Paul Silverman/Fundamental Photographs; **p. 1037 (a)** © Chuck O'Rear/West Light; **(b)** Courtesy of Ahmed H. Zewail, California Institute of Technology; **(c)** © Chuck O'Rear/West Light; **(d)** © Michael W. Berns/Scientific American; **(e)** © David Scharf, 1988. All rights reserved; **p. 1042 (Figure 33-18) (a)** © 1987 Ken Kay/ Fundamental Photographs; **(b)** Courtesy Battelle-Northwest Laboratories; **(bottom)** © 1990 Richard Megna/ Fundamental Photographs; **p. 1043 (Figure 33-19)** Macduff Everton/Corbis; **(Figure 33-20)** © 1987 Pete Saloutos/The Stock Market; **p. 1044 (Figure 33-22)** © 1987 Ken Kay/Fundamental Photographs; **p. 1046 (Figure 33-24)** © Ted Horowitz/The Stock Market; **(bottom left)** © Dan Boyd/Courtesy Naval Research Laboratory; **(bottom right)** Courtesy AT&T Archives; **p. 1047 (Figure 33-25)** © Robert Greenler; **p. 1048 (Figure 33-27)** David Parker/Science Photo Library/Photo Researchers; **p. 1049 (top)** © Robert Greenler; **(bottom)** Giovanni DeAmici, NSF, Lawrence Berkeley Laboratory; **p. 1051** © Robert Greenler; **p. 1053 (Figure 33-34) (a)** Larry Langrill; **(b)** Larry Langrill; **p. 1054 (left)** © 1970 Fundamental Photographs; **(right)** © 1990 PAR/NYC, Inc./Photo by Elizabeth Algieri; **p. 1056 (Figure 33-40)** © 1987 Paul Silverman Photographs; **p. 1057 (top left)** Glen A. Izett, U.S. Geological Survey, Denver Colorado; **(top right)** Glen A. Izett, U.S. Geological Survey, Denver Colorado; **(center left)** Dr. Anthony J. Gow/Cold Regions Research and Engineering Laboratory, Hanover, New Hampshire; **(center right)** Dr. Anthony J. Gow/Cold Regions Research and Engineering Laboratory, Hanover, New Hampshire; **(bottom)** © Sepp Seitz/Woodfin Camp & Associates.

Chapter 34

Opener p. 1070 © Dagmar Heiler-Hamann/Peter Arnold, Inc.; **p. 1071 (Figure 34-2)** Demetrios Zangos; **p. 1077 (left)** © 1990 Richard Megna/Fundamental Photographs; **(right)** © 1990 Richard Megna/Fundamental Photographs; **p. 1080** © 1990 Richard Megna/Fundamental Photographs; **p. 1082 (Figure 34-26) (a)** Nils Abramson; **(b)** © 1974 Fundamental Photographs; **p. 1083 (Figure 34-27)** © Fundamental Photographs; **p. 1085** © Bohdan Hrynewych/Stock Boston; **p. 1094 (top left)** © Lennart Nilsson; **(top right)** © Lennart Nilsson; **(bottom left)** © Lennart Nilsson; **(bottom right)** Courtesy IMEC and University of Pennsylvania Department of Electrical Engineering; **p. 1098 (top left)** Scala/Art Resource; **(top center)** © Royal Astronomical Society Library; **(top right)** © Lick Observatory, University of California Regents; **(bottom left)** California Institute of Technology; **(bottom right)** © 1980 Gary Ladd; **p. 1099 (top left)** © California Association for Research in Astronomy; **(top right)** © California Association for Research in Astronomy; **(center)** © California Association for Research in Astronomy; **(bottom)** NASA/Corbis.

Chapter 35

Opener p. 1110 © Ken Kay/Fundamental Photographs; **p. 1112** © 1990 Richard Megna/Fundamental Photographs; **p. 1113 (Figure 35-3)** Courtesy Bausch & Lomb; **p. 1114 (Figure 35-5) (a)** Courtesy T. A. Wiggins; **(b)** Courtesy T. A. Wiggins; **p. 1115 (Figure 35-6)** *PSSC Physics*, 2nd ed., 1965. D.C. Heath & Co. and Education Development Center, Newton, MA; **p. 1116 (Figure 35-8)** Courtesy Michel Cagnet; **p. 1118 (Figure 35-10)** Courtesy Michel Cagnet; **p. 1120 (Figure 35-13)** Courtesy Michel Cagnet; **p. 1123 (Figure 35-20) (a)** Courtesy Michel Cagnet; **(b)** Courtesy Michel Cagnet; **(c)** Courtesy Michel Cagnet; **p. 1127 (Figure 35-28) (a)** M. Cagnet, M. Françon, J. C. Thrierr, *Atlas of Optical Phenomena*; **(b)** M. Cagnet, M. Françon, J. C. Thrierr, *Atlas of Optical Phenomena*; **(Figure 35-29)** Courtesy Battelle-Northwest Laboratories; **(Figure 35-30)** Courtesy Michel Cagnet; **p. 1128 (Figure 35-31)** Courtesy Michel Cagnet; **(Figure 35-33) (a)** Courtesy Michel Cagnet; **(b)** Courtesy Michel Cagnet; **p. 1130** Kevin R. Morris/Corbis; **p. 1131 (Figure 35-35) (a)** Clarence Bennett/Oakland University, Rochester, Michigan; **(b)** NRAO/AUI/Science Photo Library/Photo Researchers; **p. 1133 (left)** © Ronald R. Erickson, 1981. Hologram by Nicklaus Phillips, 1978, for Digital Equipment Corporation; **(right)** © Ronald R. Erickson, 1981. Hologram by Nicklaus Phillips, 1978, for Digital Equipment Corporation.

Chapter 36

Opener p.1143 Lawrence Livermore/Photo Researchers; **p.1155 (right)** Education Development Center; **(left)** Education

Development Center; **p.1158 (left)** R. Trump/IBM Research; **(right)** Courtesy of E. D. Williams; **(bottom)** Courtesy E. D. Williams; **p.1159 (top left)** D. M. Eigler, and E. K. Schweizer, IBM Almaden Research Center; **(top right)** D. M. Eigler, and E. K. Schweizer, IBM Almaden Research Center; **(bottom left)** D. M. Eigler, and E. K. Schweizer, IBM Almaden Research Center; **(bottom right)** D. M. Eigler, and E. K. Schweizer, IBM Almaden Research Center.

Chapter 37

Opener p.1169 Dr. Bruce Schardt/Courtesy Digital Instruments, Inc.; **p.1170 (Figure 37–01) (a)** Adapted from Eastman Kodak and Wabash Instrument Corporation; **(b)** Adapted from Eastman Kodak and Wabash Instrument Corporation; **p.1188 (left)** A. Jayaraman/AT&T Bell Labs; **(right)** A. Jayaraman/AT&T Bell Labs; **p.1189 (all images)** David Parker/Photo Researchers; **p.1193** © Robert Landau/Westlight.

Chapter 38

Opener p.1203 Courtesy AT&T Archives; **p.1210 (Figure 38–07)** Will and Deni McIntyre/Photo Researchers; **p.1217 (Figure 38–14) (a)** Courtesy Dr. J. A. Marquisee; **(b)** Courtesy Dr. J. A. Marquisee; **p.1221 (top left)** Richard Walters 2/89, p.52/*Discover*; **(top right)** Dr. Jeremy Burgess/Science Photo Library/Photo Researchers; **(bottom left)** Thomas R. Taylor /Photo Researchers; **(bottom right)** Courtesy AT&T Archives; **p.1224 (top left)** Chris Kovach 3/91, p.69/*Discover*; **(top right)** Srinivas Manne, University of California-Santa Barbara; **(bottom left)** Dr. F. A. Quiocho and J. C. Spurlino/Howard Hughes Medical Institute, Baylor College of Medicine; **(bottom right)** W. Krätschmer/Max-Planck Institute for Nuclear Physics; **p.1225** Museum of Modern Art; **p.1229** C. Falco/Photo Researchers; **p.1232 (top)** Dr. Jeremy Burgess/Science Photo Library/Photo Researchers; **(left)** Courtesy AT&T Archives; **(right)** AT&T Archives; **p.1233 (top)** Dr. Jeremy Burgess/Science Photo Library /Photo Researchers; **(left)** AT&T Archives; **(right)** AT&T Archives; **p.1234 (bottom)** Courtesy of Texas Instruments; **(top)** Courtesy of David Sarnoff Research Center; **p.1235 (top left)** Images by Patterson Electronics and computer processing by John Sanford; **(top right)** Images by Patterson Electronics and computer processing by John Sanford; **(bottom left)** Images by Patterson Electronics and computer processing by John Sanford; **(bottom right)** Images by Patterson Electronics and computer processing by John Sanford.

Chapter 39

Opener p.1243 Albert Einstein Archives/AIP Niels Bohr Library; **p.1244** Courtesy NRAO/AUI; **p.1265** C. Powell, P. Fowler & D. Perkins/Science Photo Library/Photo Researchers; **p.1272** © Michael Freeman; **p.1273** N.A.S.A.

Chapter 40

Opener p.1284 Hans Wolf /The Image Bank; **p.1297 (left)** © 1991 by the Metropolitan Museum of Art; **(center)** Courtesy of Paintings Conservation Dept., Metropolitan Museum of Art; **(right)** Paintings Conservation Dept., Metropolitan Museum of Art; **p.1300** Jerry Mason/Photo Researchers; **p.1305 (left)** Courtesy of Princeton Plasma Physics Laboratory; **(right)** Courtesy of Princeton Plasma Physics Laboratory; **p.1306 (left)** Courtesy of Lawrence Livermore National Laboratory/U. S. Department of Energy; **(right)** Courtesy of Lawrence Livermore National Laboratory/U. S. Department of Energy.

Chapter 41

Opener p.1313 Lawrence Livermore Lab/Science Photo Library/Photo Researchers; **p.1315** ICRR (Institute for Cosmic Ray Research), The University of Tokyo; **p.1316 (top)** Science Photo Library /Photo Researchers; **(bottom)** Lawrence Berkeley Laboratory/Science Photo Library/Photo Researchers; **p.1317** Lawrence Berkeley Lab/Science Photo Library/Photo Researchers; **p.1318 (Figure 41–01)** Richard Ehrlich; **(bottom)** CERN; **p.1319** CERN; **p.1321** Lawrence Berkeley Lab/Science Photo Library/Photo Researchers.

ANSWERS

ANSWERS

Problem answers are calculated using $g = 9.81\ \text{m/s}^2$ unless otherwise specified in the Problem. Differences in the last figure can easily result from differences in rounding the input data and are not important.

To help you master the techniques in Examples and to solve the intermediate-level problems at the end of each chapter, the problem maps preceding the answers for the chapters indicate which Examples and odd-numbered intermediate-level Problems deal with similar material.

Chapter 1

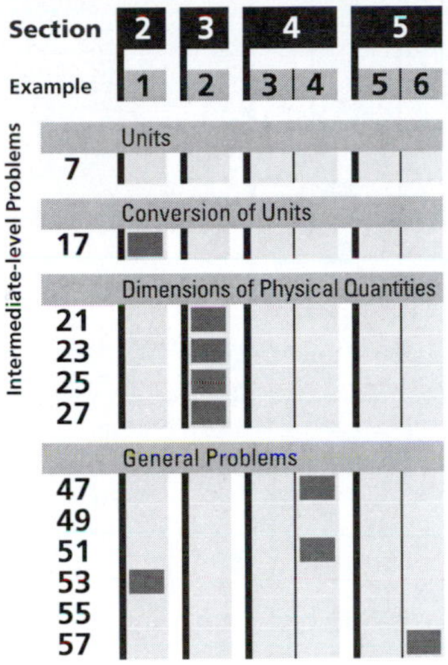

1. (*c*)

3. (*a*) 1 MW; (*b*) 2 mg; (*c*) 3 μm; (*d*) 30 ks

5. (*a*) 1 picoboo; (*b*) 1 gigalow; (*c*) 1 microphone; (*d*) 1 attoboy; (*e*) 1 megaphone; (*f*) 1 nanogoat; (*g*) 1 terabull

7. (*a*) C_1: ft, C_2: ft/s; (*b*) C_1: ft/s^2; (*c*) C_1: m/s^2; (*d*) C_1: ft, C_2: s^{-1}; (*e*) C_1: ft/s, C_2: s^{-1}

9. 2450 km/h, 1520 mi/h

11. (*a*) 62.1 mi/h; (*b*) 23.6 in; (*c*) 91.4 m

13. 1.61 km/mi

15. (*a*) 3.784 L; (*b*) 0.1589 m^3

17. (*a*) 0.505 ft^3; (*b*) 0.0143 m^3; (*c*) 14.3 L

19. (*a*) C_1: L, C_2: L/T; (*b*) C_1: L/T^2; (*c*) C_1: L/T^2; (*d*) C_1: L, C_2: $1/T$; (*e*) C_1: L/T, C_2: $1/T$

21. $L^3/(MT^2)$, m^3/kg·s^2

23. $M(L/T^2)(L/T) = ML^2/T^3$

25. $(ML/T^2)(L/T)$

27. $T = Cr^{3/2}/\sqrt{M_S G}$

29. (*d*)

31. (*c*)

33. (*a*) 30,000; (*b*) 0.0062; (*c*) 0.000004; (*d*) 217,000

35. (*a*) 1.14×10^5; (*b*) 2.24×10^{-8}; (*c*) 8.27×10^3; (*d*) 6.27×10^2

37. 4×10^6 membranes

39. (*a*) 1.69×10^3; (*b*) 4.8; (*c*) 5.6; (*d*) 10

41. One advantage of using your arm is that a measure of standard length is always available. The disadvantage is that arm lengths are not uniform. For example, if you wish to purchase a board of "two arm lengths," it may be shorter or longer than you expected, depending on the owner's arm length.

43. (*a*) True; (*b*) False; (*c*) True

45. 31.7 yr

47. 2.0×10^{23}

49. 3.51 Mm

51. 3.86×10^2 mi^2

53. (*a*) 1.8×10^2; (*b*) 3.4; (*c*) 2.9×10^8; (*d*) 0.45

55. (*a*) 6.6×10^3 electrons/m^3; (*b*) 3.6 protons/m^3

57. (*a*) 1×10^{11} cans/yr; (*b*) 2×10^9 kg/yr; (*c*) \$2 billion

59. (*a*) In a plot of T versus m^n, $n = 1/2$ provides the best fit to a straight line.

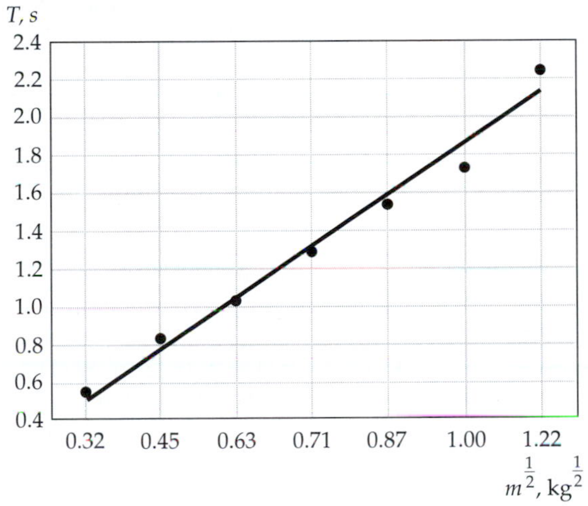

The slope of the line $C \approx 1.77$ s/kg$^{1/2}$. (*b*) The points with the greatest deviation are $T = 1.75$ s and $T = 2.22$ s.

61. (*a*) $C\sqrt{L/g}$; (*b*) Check by using pendulums of lengths 1 m and 0.5 m; the periods should be about 2 s and 1.4 s.

63. (*a*) 1.6×10^{10} bits; (*b*) 350 books

Chapter 2

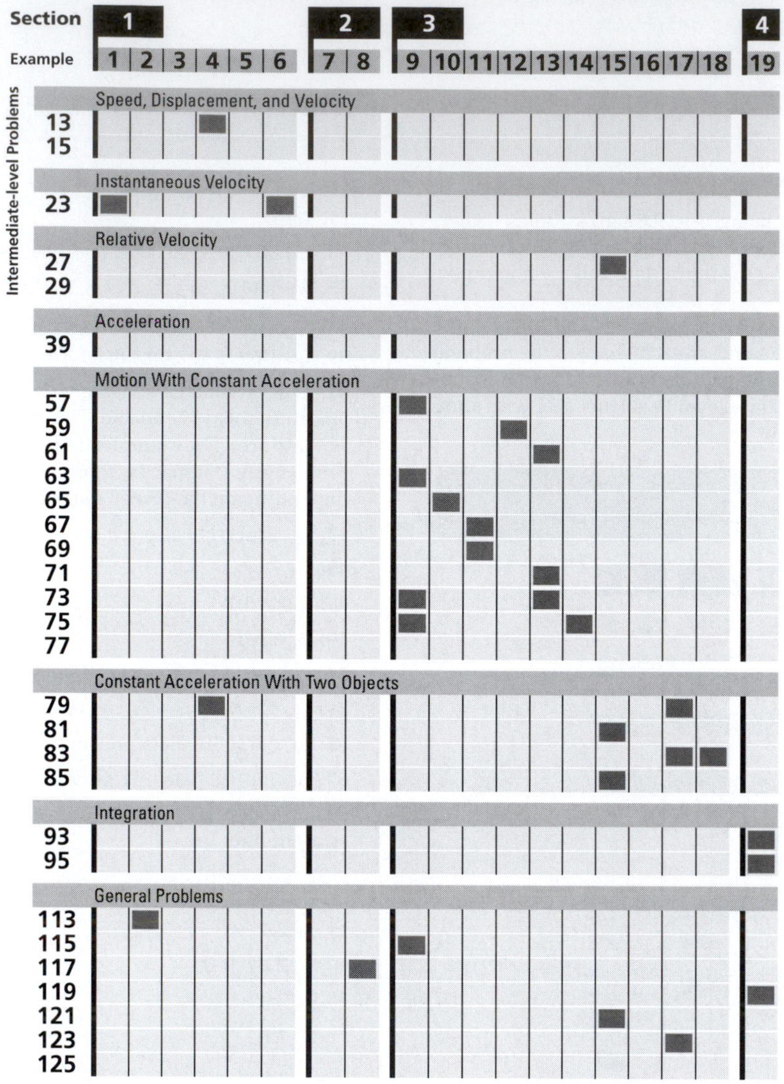

1. Since the cars go round a closed circuit and return nearly to the starting point, the displacement is nearly zero, and the average velocity is zero.

3. Yes. In a round trip, A to B and back to A, the average velocity is zero; the average velocity between A and B is not zero.

5. (*a*) 4 ns; (*b*) 4 ks

7. (*a*) 260 km; (*b*) 65 km/h

9. (*a*) 73.3 mi/h; (*b*) 44.3

11. (*a*) 434 yr; (*b*) 4.34×10^6 yr; he need not pay for pizza.

13. (*a*) 13 m; (*b*) 2.2 s

15. (*a*) 7.9×10^4 m/s; (*b*) 3.16×10^7 m/s; (*c*) 2×10^{10} yr

17. No, because for constant velocity, the instantaneous velocity and average velocities are equal.

19. (*b*)

21. (*a*) 1 m/s; (*b*) 2 m/s

23. (*a*) 2 m, 2 m/s; (*b*) $\Delta x = [(2t - 5)\,\Delta t + \Delta t^2]$ m; (*c*) $(2t - 5)$ m/s

25. (*a*)

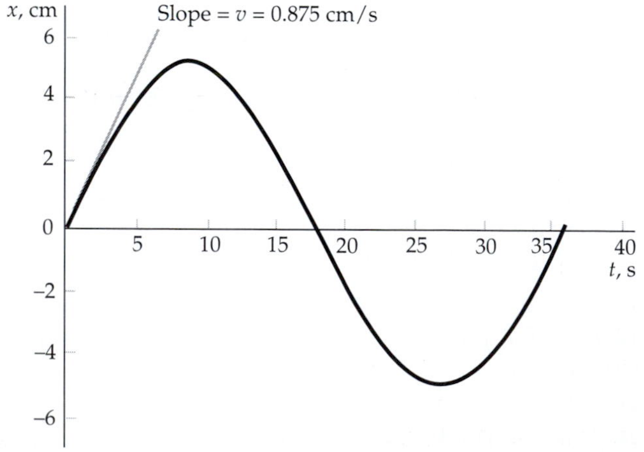

(*b*) 0.875 cm/s; (*c*) 0.723, 0.835, 0.857, 0.871, 0.874, 0.875; (*d*) 0.875 cm/s

27. 120 km

29. 6 h

31. (*a*) After getting started, walk in the opposite (negative) direction. Gradually slow the speed of walking, until the other end of the room is reached.

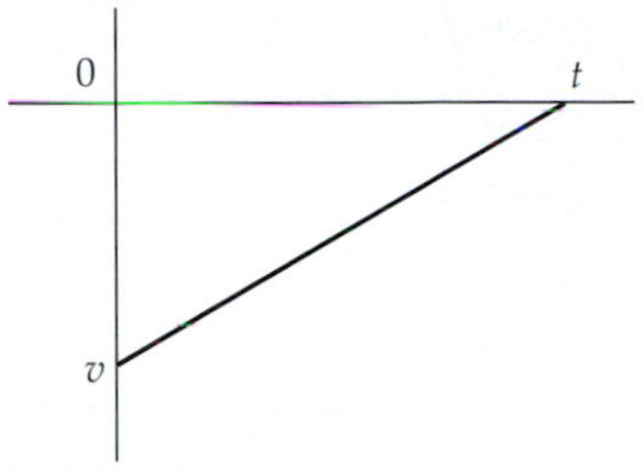

33. Yes. An object tossed up has a constant downward acceleration; at the top its instantaneous velocity is zero.

35. (*a*) $a = 0$; (*b*) $a > 0$; (*c*) $a < 0$; (*d*) $a = 0$

37. (*a*) 2.42 m/s^2, (*b*) 89.2 km/h

39. (*a*) 8 m/s^2, 8 m/s^2; (*b*) 8 m/s^2

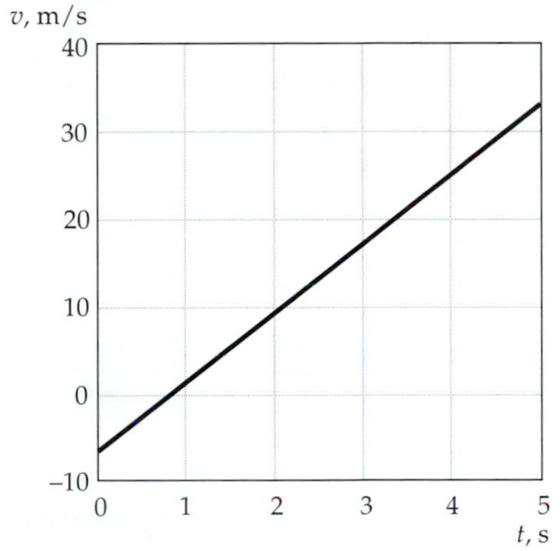

41. The initial downward velocities of the two rocks are not the same.

43. (*b*)

45. (*a*)

47. (*d*)

49. (*c*)

51. (*a*) 80 m/s; (*b*) 400 m; (*c*) 40 m/s

53. 15.6 m/s^2

55. 4.59×10^3 m

57. (*a*) 4.08 s; (*b*) 20.4 m; (*c*) 0.99 s or 3.09 s are both acceptable.

59. (*a*) -5.1×10^5 m/s^2; (*b*) 0.686 ms

61. 1467 m

63. (*a*)

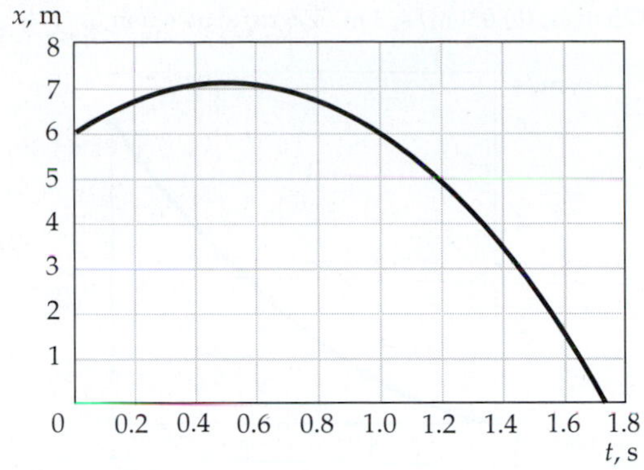

(*b*) 7.27 m; (*c*) 1.73 s; (*d*) 11.9 m/s

65. 1.62 m

67. 43.6 m

69. ±68 m/s; the stone may be thrown either up or down.

71. (*a*) 666 m; (*b*) 13.6 m/s

73. (*a*) 19.0 km; (*b*) 2 min 18 s; (*c*) 610 m/s

75. (*a*) 2.24 s; (*b*) 6.17 m

77. 145.7 m

79. 15.6 m

81. 100.4 km/h

83. Deceleration ≥ 0.754 m/s^2, v_{rel} = 3.77 m/s, distance = 518 m

85. 4.8 m/s

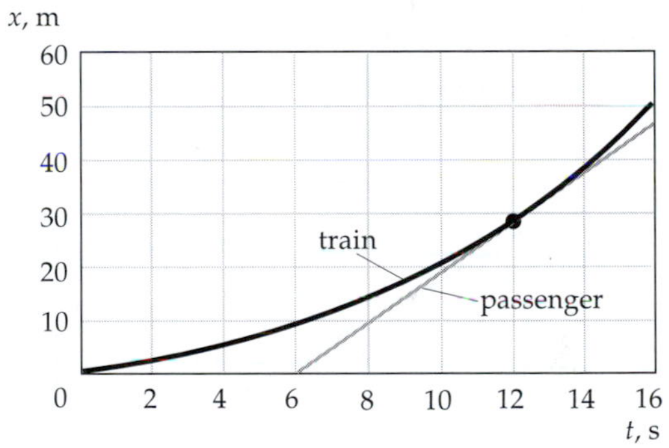

87. $2h/3$

89. 32.4 km

91. (*a*)

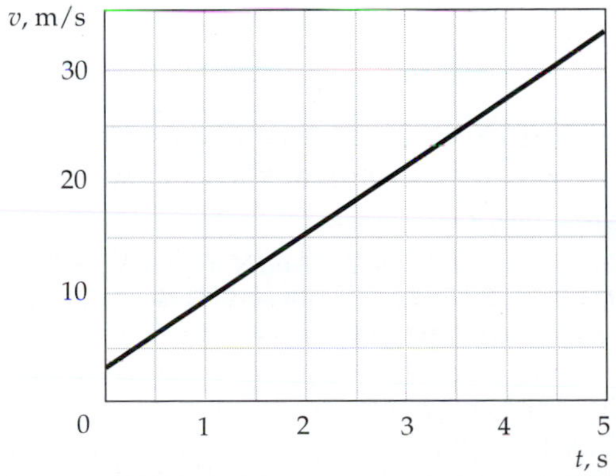

(*b*) 90 m

93. $x(t) = \int (7t^2 - 5)\,dt = (7/3)\,t^3 - 5t + C$

95. (*a*) 0.25 m/s; (*b*) 0.9 m/s, 3 m/s, 6 m/s; (*c*) 6.5 m

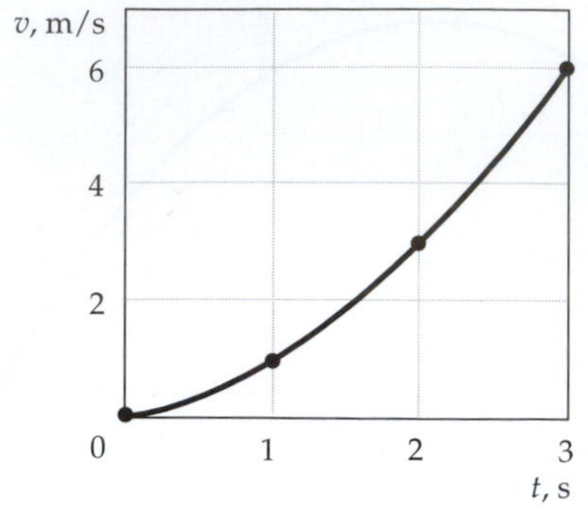

97.

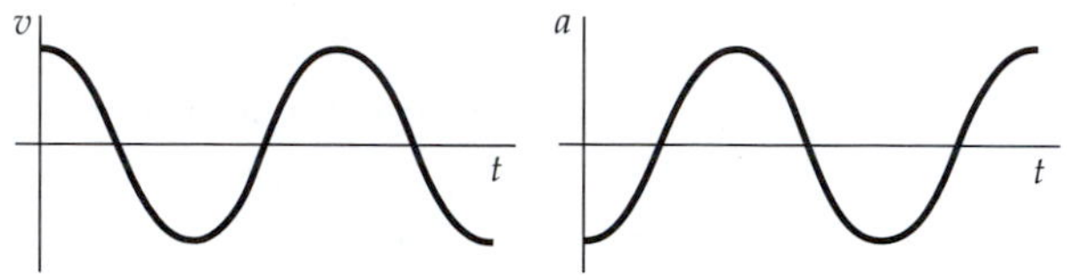

99. (*c*)

101. (*a*)

103. (*d*)

105. (*d*)

107. (*a*) t_0, t_1; (*b*) t_3, t_1, t_6, t_7; (*c*) t_2, t_5; (*d*) t_4; (*e*) t_2, t_6; (*f*) t_0, t_1, t_3, t_5, t_7

109. (*a*) *a, f, i*; (*b*) *c, d*; (*c*) *a, d, e, f, h, i*; (*d*) *b, c, g*. The graphs *d* and *h* and the graphs *f* and *i* are mutually consistent.

111. (*a*) 36 m; (*b*) −36 m; (*c*) −9 m/s

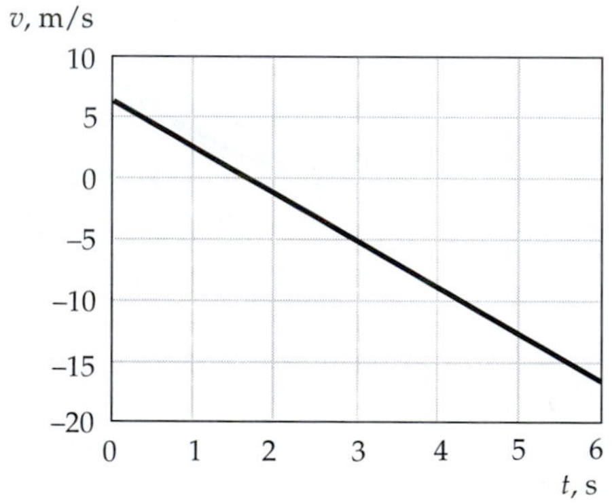

113. 134 m/s

115. 24 m, 1.4 s

117. $v(t) = (50 - 10t)$ m/s, $a = -10$ m/s^2, $x(t) = 50t - 5t^2$

119. (*a*) $x(t) = Ct^3/6 + At + B$; (*b*) 37.5 m/s, 62.5 m

121. (*a*) 6.67 s; (*b*) 6.67 s

123. (*a*) $\Delta x_s = 100.3$ m; $\Delta x_p = 232.3$ m; Note that $\Delta x_s + 100\text{ m} < x_p$, so the cars collide. (*b*) 5.52 s. (*c*) With added reaction time, the collision will occur sooner and be more severe.

125. The running time of the ball player is 2.42 s and the flight time of the ball is 2.19 s. A good umpire will call him out!

127. (*a*) Initially, $v = 0$ and increases as gt, but as v becomes finite, the acceleration diminishes to $g - bv$. Ultimately, the acceleration approaches zero and v remains constant, $v_{term} = g/b$.

(*b*)

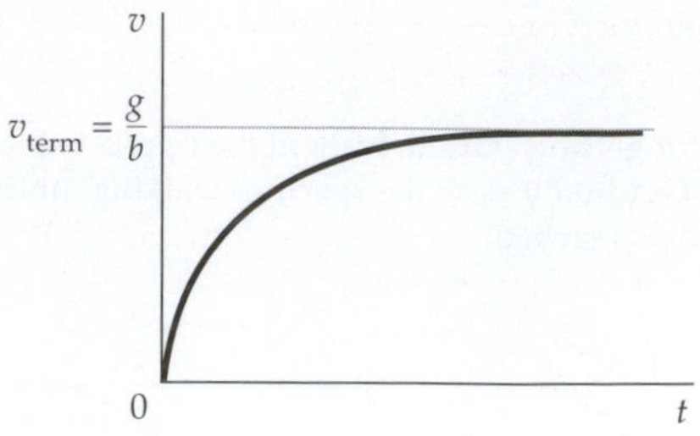

129. $x(t) = x_0\,e^{t - t_0}$.

131. (*a*)

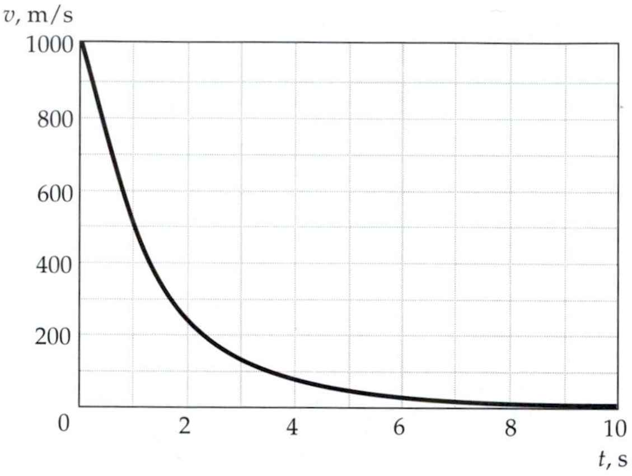

(*b*) 144.1 m/s

Chapter 3

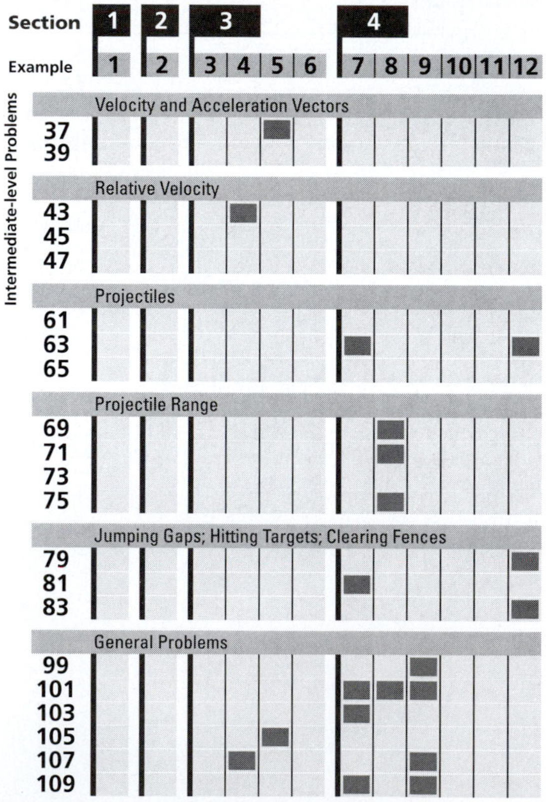

1. The magnitude of the displacement can be less than but never more than the distance traveled.

3. (*e*)

5. $D = (-5\hat{i} + 5\hat{j})$ m

7.

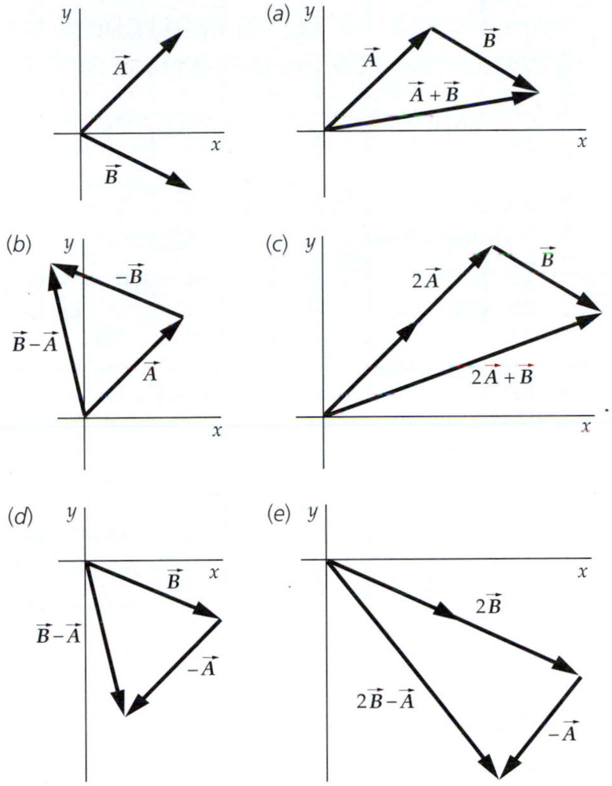

9. No, the magnitudes can be equal if the vector is along the component direction.

11. No, if $\vec{B} = -\vec{A}$, then $\vec{C} = 0$.

13. (*b*)

15. (*a*) $A_x = 8.66$ m, $A_y = 5$ m; (*b*) $A_x = 3.54$ m, $A_y = 3.54$ m; (*c*) $A_x = 3.5$ km, $A_y = 6.06$ km; (*d*) $A_x = 0$ km, $A_y = 5$ km; (*e*) $A_x = -13$ km/s, $A_y = 7.5$ km/s; (*f*) $A_x = -5$ m/s, $A_y = -8.66$ m/s; (*g*) $A_x = 0$, $A_y = -8\ \text{m/s}^2$

17. (*a*) $A = 5.83$, $\theta = 31°$; (*b*) $B = 12.2$, $\theta = -35°$; (*c*) $C = 5.39$, $\theta = 42°$, $\phi = 236°$

19. (*a*) $5\ \text{m/s}\ \hat{i} + 8.66\ \text{m/s}\ \hat{j}$; (*b*) $-3.54\ \text{m}\ \hat{i} - 3.54\ \text{m}\ \hat{j}$; (*c*) $14\ \text{m}\ \hat{i} - 6\ \text{m}\ \hat{j}$

21. $\vec{B} = -1.5\vec{A}$

23. No

25. A particle moving at constant speed in a circular path is accelerating (the direction of the velocity vector is changing). If a particle is moving at constant velocity, it is not accelerating.

27. (*a*) The velocity vector is always tangent to the path.

(*b*)

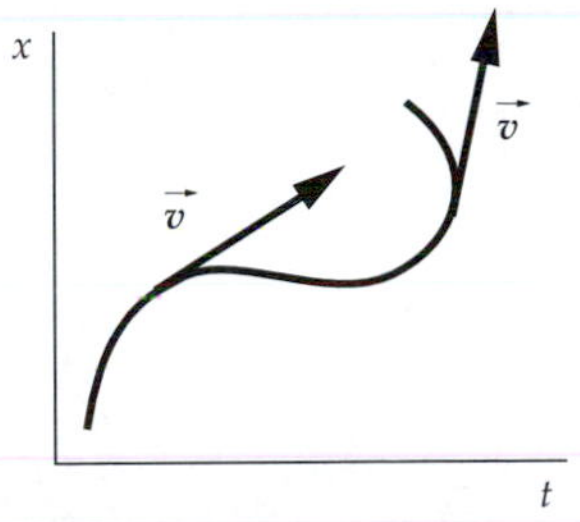

29.

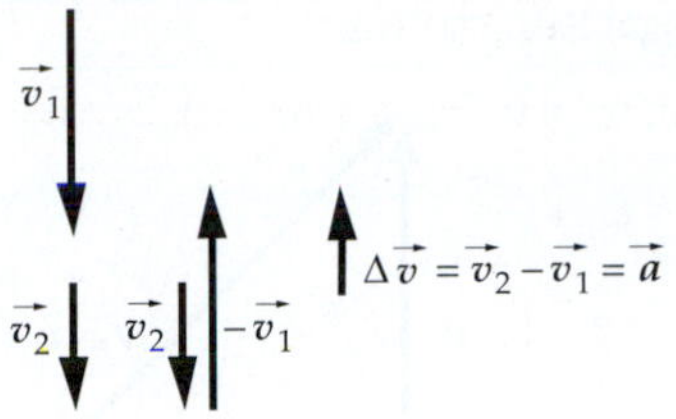

31. $14.1\ \text{km/h}\ \hat{i} - 4.1\ \text{km/h}\ \hat{j}$

33. (*b*)

35. (*a*) -10 m/s; (*b*) 90°; (*c*) $40\ \text{m/s}\ \hat{i} + 30\ \text{m/s}\ \hat{j}$; (*d*) $10\ \text{m/s}^2$ at 37° with x axis

37. (*a*) $10\ \text{m/s}\ \hat{i} - 3\ \text{m/s}\ \hat{j}$; (*b*) $44\ \text{m}\ \hat{i} - 9\ \text{m}\ \hat{j}$; $r = 44.9$ m, $\theta = -11.6°$

39. (*a*) $\vec{v} = [6t\,\hat{i} + 4t\,\hat{j}]$ m/s, $\vec{r} = [(10 + 3t^2)\,\hat{i} + 2t^2\,\hat{j}]$ m

(*b*)

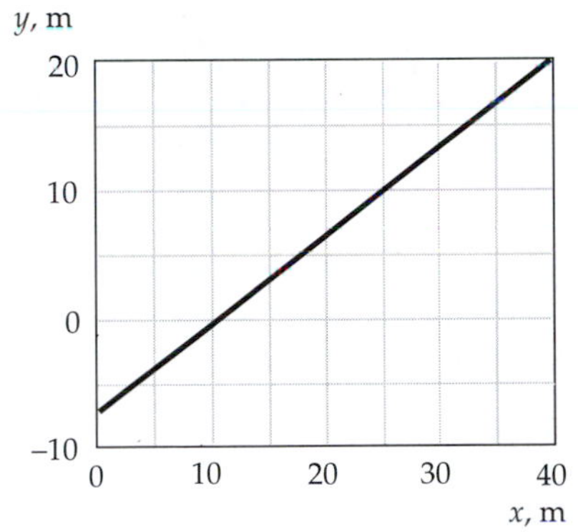

41. (*d*)

43. (*a*) 0.8 m/s; (*b*) 1.79 m/s; (*c*) 30° upstream

45. 5.18 km/h

47. (*a*) $-120\ \text{m}\ \hat{i} + 4\ \text{m}\ \hat{j}$; (*b*) $-20\ \text{m/s}\ \hat{i} - 12\ \text{m/s}\ \hat{j}$; (*c*) $-2\ \text{m/s}^2\ \hat{j}$

49. 261.7 km/h, 6.58° west of north

51. True

53. (*e*)

55. (*a*) *A* and *E*; (*b*) *C*; (*c*) *A* and *E* have the same speed, but not the same velocities.

57. 1.1 m

59. 15 m/s, $9.81\ \text{m/s}^2$ downward

61. 33.8 m/s

63. 20.3 m/s, 36.2°

65. 69.3°

67. (*a*) 49.5 s; (*b*) 12.4 km; (*c*) 12.0 km

69. (*a*) 8.14 m/s; (*b*) 23.2 m/s

71. −63.4°

73. $d[(v_0^2 \sin 2\theta)/g]/d\theta = (v_0^2/g)\, d(\sin 2\theta)/d\theta = (2v_0^2/g) \cos 2\theta$; set equal to 0; $2\theta = 90°$, $\theta = 45°$

75. $2v^2 \tan\phi/(g \cos\phi)$

77. 18.2 m from wall

79. −1.93°

81. 3.19 m/s

83. (*a*) 44.3 m/s, 9.33° above horizontal; (*b*) 0.733 s; (*c*) 3.23 m

85. 0.785 m

87. (*a*) 11.0 m; (*b*) $v_{0x} = 10$ m/s; $v_{0y} = 14.7$ m/s; (*c*) 17.8 m/s; (*d*) 55.8°

89. (*a*) False; (*b*) False; (*c*) True

91.

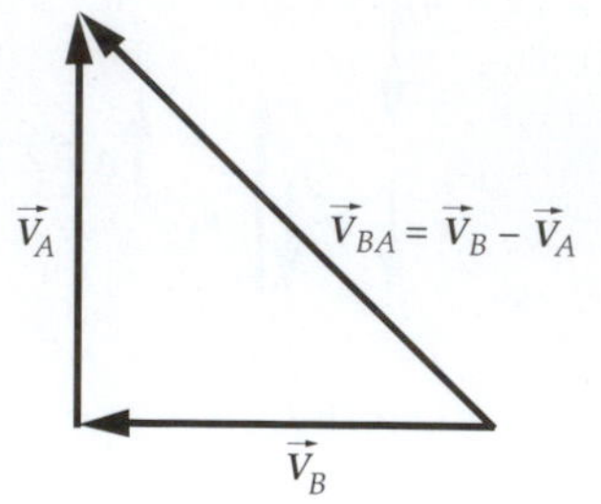

93. (*a*) north, northeast, east, southeast, south; (*b*) AB north, BC southeast, CD $\vec{a} = 0$, DE southwest, EF north; (*c*) equal

95. $4.9\ \text{m/s}^2\ \hat{i} + 8.5\ \text{m/s}^2\ \hat{j}$

97. (*a*)

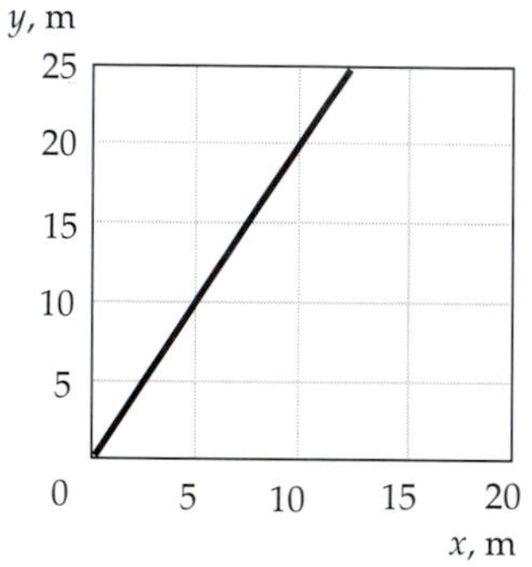

(*b*) $(5\hat{i} + 10\hat{j})$ m/s, 11.2 m/s

99. 4.29 m

101. Assuming an initial height of 1.5 m and an initial speed of 25 min/s: (*a*) 14 m; (*b*) 64 m; (*c*) 41 m; (*d*) 75 m

103. 14.8 m/s

105. (*a*) $(\hat{i} + \hat{j})$ m/s; (*b*) $(2\hat{i} - 3.5\hat{j})$ m/s^2;
(*c*) $[\hat{i} + \hat{j} + (2\hat{i} - 3.5\hat{j})t]$ m/s;
(*d*) $[4\hat{i} + 3\hat{j} + (\hat{i} + \hat{j})t + \frac{1}{2}(2\hat{i} - 3.5\hat{j})t^2]$ m

107. (*a*) $(25\hat{i} + 7.07\hat{j} + 7.07\hat{k})$ m/s; (*b*) $(39.9\hat{i} + 11.3\hat{j})$ m

109. Yes, if it is a high fly.

111. (*a*) 0.98 km/h; (*b*) 39.6 west of north; (*c*) 3 h 19 min

113. $v_y^2 = v_{0y}^2 + 2gh$, regardless of direction (up or down). $v^2 = v_x^2 + v_y^2$, and since $v_x = v_{0x}$, $v^2 = v_{0x}^2 + v_{0y}^2 + 2gH = v_0^2 + 2gH$, for any angle. $v = (v_0^2 + 2gh)^{1/2}$

Chapter 4

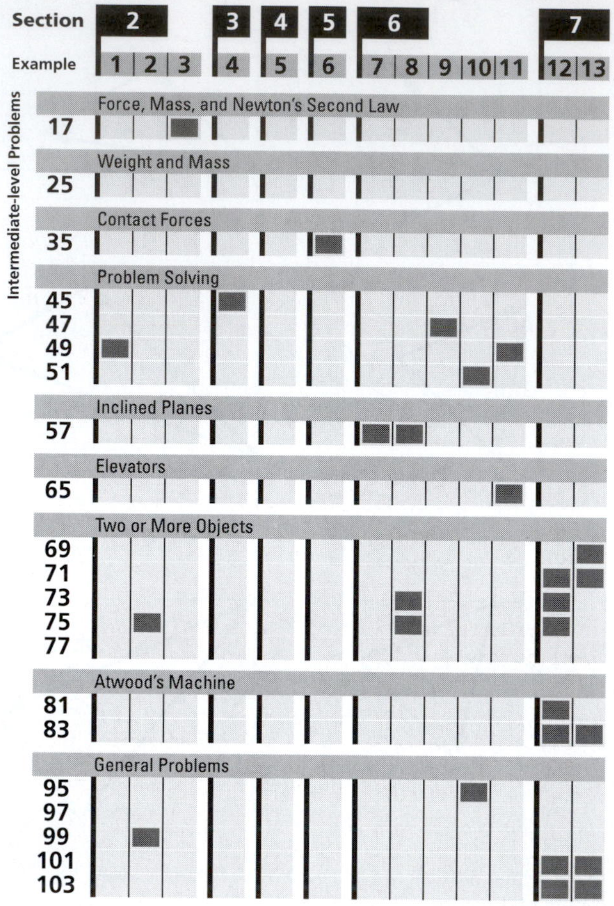

1. If Newton's first law is obeyed, the reference frame is an inertial reference frame.

3. No

5. No

7. (*d*)

9. (*b*)

11. $F_2 = 3F_1$

13. 4.46 m/s^2

15. (*a*) 250 N; (*b*) 400 N

17. (*a*) $(1.5\hat{i} - 3.5\hat{j})$ m/s^2; (*b*) $(4.5\hat{i} - 10.5\hat{j})$ m/s;
(*c*) $(6.75\hat{i} - 15.75\hat{j})$ m

19. To accelerate, she must exert a force proportional to her mass.

21. (*c*)

23. 74.8 kg

25. (*a*) 395 N; (*b*) No

27. (*b*)

29. (*c*)

31.

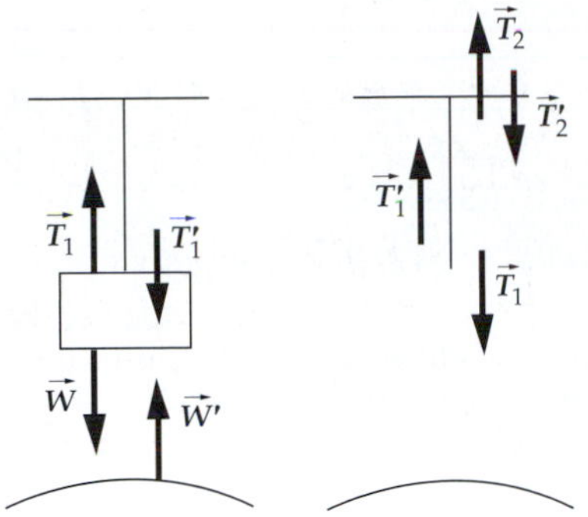

33. (*a*) 60 N; (*b*) 57.7 N

35. (*a*)

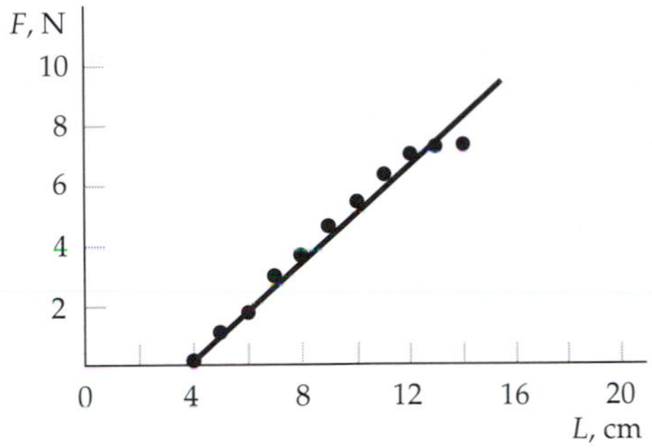

(*b*) 7.15 N; (*c*) 5.3 cm

37. No, the tension in the line must have a vertical component to support the weight of the towel.

39. (*b*)

41. (*c*)

43. (*a*) -8.81 m/s^2; (*b*) -7.81 m/s^2; (*c*) 10.19 m/s^2

45. (*a*) $v = -10.8\text{ m/s}$; (*b*) $v = -7.7\text{ m/s}$

47. (*a*) $T = w/2\,(\sin\theta)$; $\theta = 90°$, T least; $\theta \rightarrow 0°$, T greatest; (*b*) 19.6 N

49. (*a*) 11,810 N; (*b*) 9810 N; (*c*) 7810 N

51. (*a*) $T_1 = 60$ N, $T_2 = 52$ N, $M = 5.3$ kg; (*b*) $T_1 = 46.2$ N, $T_2 = 46.2$ N, $M = 4.71$ kg; (*c*) $T_1 = T_3 = 34$ N, $T_2 = 58.9$ N, $M = 3.46$ kg

53.

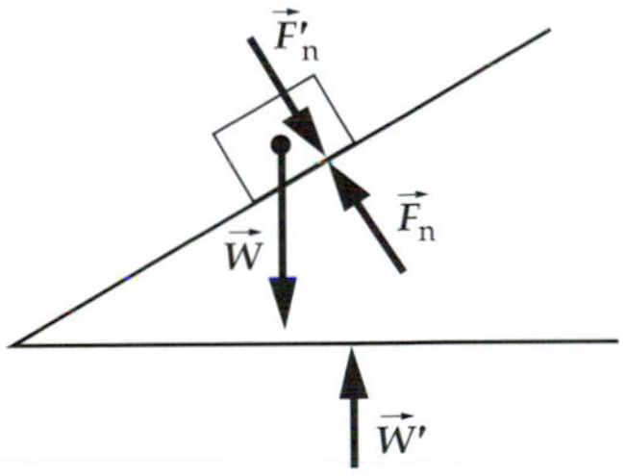

55. (*a*) and (*b*) = 98.1 N; (*c*) 49 N; (*d*) 49 N

57. (*a*) 149 N; (*b*) 3.41 m/s^2

59. (*a*)

61. high

63. 5.19 m/s^2

65. (*a*) 19.6 N; (*b*) 19.6 N; (*c*) 39.6 N; for $0 \le t \le 2$ s, $F = 19.6$ N; for $2\text{ s} \le t \le 4$ s, $F = 9.62$ N

67. (*d*)

69. (*a*) $a = F/(m_1 + m_2)$; $F_c = F\,[m_2/(m_1 + m_2)]$; (*b*) $a = 0.4\text{ m/s}^2$; $F_c = 2.4$ N

71. $F_A = 160$ N, $F_B = 161.2$ N, $F_C = 321.6$ N

73. (*a*) 492 N, 3.71 m/s; (*b*) 20.7 m/s

75. (*a*) $m_1 = m_2(g - a)/(a + g\sin\theta) = 48$ kg; (*b*) 424 N

77. (*a*) 10 m/s^2; (*b*) $60 + 8x$ N

79. $a_{10} = 1.12\text{ m/s}^2$, $a_{20} = -1.12\text{ m/s}^2$, $T = 44.8$ N

81. $\Sigma F_x = m_1 a = T - m_1 g\sin\theta$, $\Sigma F_y = m_2 a = m_2 g - T$. Add the two equations and solve for a, using $\theta = 90°$. Use the expression for a to solve for T.

83. $2\,m_1 m g/(m_1 + m_2)$

85. (*a*) $(2L/t^2)\,[(m_1 + m_2)/(m_1 - m_2)]$; (*b*) Differentiate with respect to t; $dg/dt = -2g/t$ or $dg/g = -2dt/t$; $m_2 = 0.926$ kg or 1.08 kg

87. (*a*)

89. 3 kg

91. 14.3 m/s^2

93. 20.8 lb

95. (*a*) The object will swing backward; (*b*) $T_x = ma$, $T_y = mg$, $\tan\theta = T_x/T_y = ma/mg$, $a = g\tan\theta$; (*c*) 1.6 m/s^2, 9.3°

97. (*a*) $a = F/(m_1 + m_2)$; (*b*) $F_{net} = Fm_2/(m_1 + m_2)$; (*c*) $Fm_1/(m_1 + m_2)$

99. 4000 N

101. (*a*) 55 g; (*b*) 2.45 m/s^2

103. (*a*) $T = (F_2 + 2F_1)/3$; (*b*) $3T_0/4C$

105. $a_1 = \dfrac{m_2 - m_1}{m_1 + m_2}g + \dfrac{2m_2}{m_1 + m_2}a$

$$a_2 = \frac{m_1 - m_2}{m_1 + m_2}g + \frac{2m_1}{m_1 + m_2}a$$

$$T = \frac{2m_1 m_2}{m_1 + m_2}(g + a)$$

Chapter 5

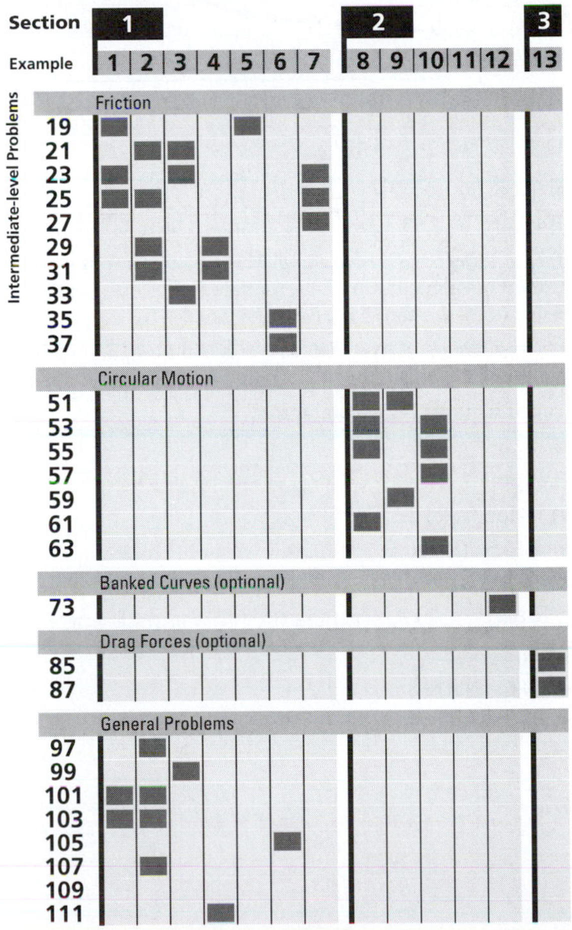

1. The force of friction between the objects and the floor of the truck

3. (*a*) False; (*b*) True; (*c*) True; (*d*) True

5. (*d*)

7. (*e*)

9. (*b*)

11. 0.417

13. (*a*) $\mu_s > \mu_k$; therefore f is greater if wheels do not spin; (*b*) $\mu_s = 0.425$

15. (*a*) 49.05 N; (*b*) 123 N

17. (*a*) $\theta > 0$ is preferable; it reduces F_n and therefore f_s; (*b*) 294 N

19. (*b*)

21. (*a*) 0.614 kg, 3.39 kg; (*b*) 9.81 N

23. 83.9 m

25. (*a*) 49.1 m; (*b*) 110 m

27. (*a*) 2.75 m/s^2; (*b*) 10.1 s

29. (*a*) -0.809 m/s^2; (*b*) $T = 0.176$ N

31. (*a*) -0.944 m/s^2; (*b*) -0.425 N

33. (*a*) One expects that F will increase with increasing magnitude of the angle since the normal component increases and the tangential component decreases.
(*b*) A force applied at a 0° angle will be most efficient.

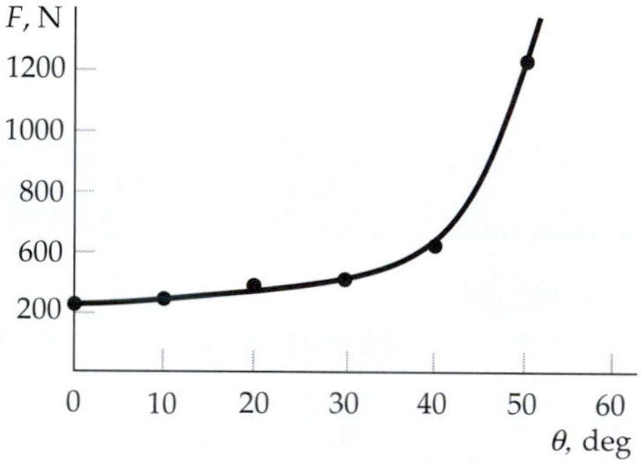

35. (*a*) 0.238; (*b*) 1.4 m/s^2

37. (*a*) 17.7 N; (*b*) 2.95 m/s^2, 5.9 N; (*c*) $a_1 = 1.96$ m/s^2, $a_2 = 7.87$ m/s^2

39. (*a*) $F = \mu_s mg/(\cos\theta + \mu_s \sin\theta)$; (*b*) F is a minimum when the denominator is a maximum. Differentiate $(\cos\theta + \mu_s\sin\theta)$ and set to 0. Solve for θ: $\theta = \tan^{-1}\mu_s$. For $\mu_s = 0.6$, $\theta = 31°$

41. $a_{min} = g(\sin\theta - \mu_s\cos\theta)/(\cos\theta + \mu_s\sin\theta)$, $a_{max} = g(\sin\theta + \mu_s\cos\theta)/(\cos\theta - \mu_s\sin\theta)$

43. True, it requires centripetal force.

45. (*e*)

47. 33.4 rpm

49. (*a*) 1.41 m/s; (*b*) 8.5 N

51. (*a*) 4660 N; (*b*) 110 m

53. $R = m_1v^2/m_2g$

55. $T_1 = [m_1L_1 + m_2(L_1 + L_2)](2\pi/T)^2$; $T_2 = m_2(L_1 + L_2)(2\pi/T)^2$

57. 410 N, 53.3°

59. 640 N to 1000 N

61. (*a*) 2.61 N; (*b*) 1.21 m/s

63. 51.6°

65. (*a*) $a_c = \dfrac{v_0^2}{r}\left[\dfrac{1}{1 + (\mu_k v_0/r)t}\right]^2$; (*b*) $-\mu_k a_c$; (*c*) $a = a_c\,(1 + \mu_k^2)^{1/2}$

67. (*d*)

69. Bonita

71. (*a*) 7.25 m/s; (*b*) 0.536

73. (*a*) 8245 N; (*b*) 1565 N; (*c*) 0.19

75. 176 m

77. The constant b should increase with density as more air molecules collide with the object as it falls.

79. (*d*)

81. 3.27×10^{-9} kg/s

83. (*a*) 589 N; (*b*) 0.942 kg/m

85. 88.2 km/h

87. (*a*) 21.8 kg/m; (*b*) 78.48 kN. This initial acceleration would cause internal damage.

89. (*a*) 2.42 cm/s; (*b*) 1.15 h

91. (*d*)

93. (*b*)

95. (*a*) 10.7 m/s; (*b*) 8.0 N

97. (*a*) 0.289; (*b*) 600 N

99. 1486 N

101. (*d*)

103. $\mu_s = 0.577$, $\mu_k = 0.342$

105. $(m_1 + m_2)g\mu_s\cos\theta$

107. 0.433

109. 23.6 rpm

111. 1.22 kg, 0.672

Chapter 6

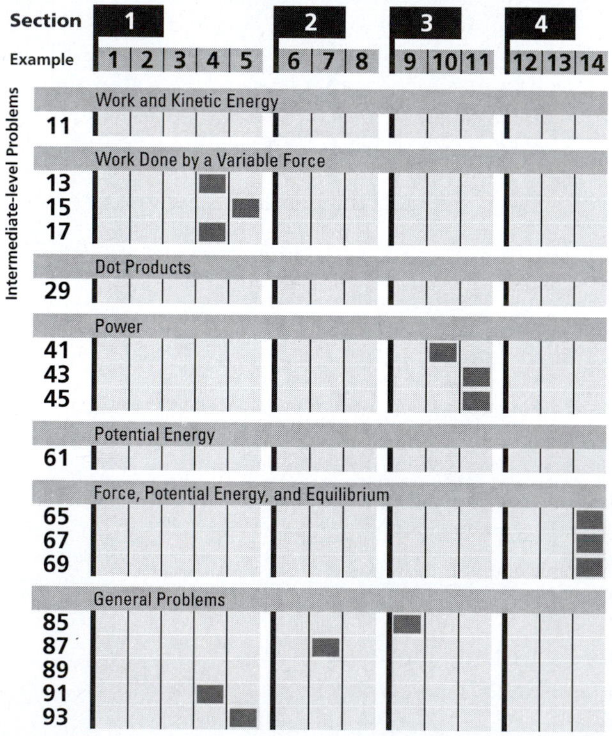

1. (*a*) False; (*b*) True; (*c*) True

3. Yes, if your center of mass is higher when standing than when lying in bed. Otherwise, no.

5. No, $dW = \vec{F}\cdot d\vec{r}$ and here $\vec{F}$ is perpendicular to $d\vec{r}$

7. (*a*) 10.8 kJ; (*b*) 2.7 kJ; (*c*) 43.2 kJ

9. (*a*) 240 J; (*b*) −177 J; (*c*) 63 J

11. 54.4 kg

13. (*a*) 6 J; (*b*) 12 J; (*c*) 3.46 m/s

15. 19C J

17. (*a*) 2.75 J; (*b*) 11.4 J; (*c*) 2.76 m/s; (*d*) 3.5 J; (*e*) 2.84 m/s

19. No. Since the force is perpendicular to the motion, it will cause the particle to depart from straight-line motion.

21. (*a*) $F_{3m} = 417$ N, $F_{4m} = 313$ N, $F_{5m} = 250$ N; (*b*) 1.25 kJ for each L; (*c*) Choosing a longer length means one can exert a smaller force.

23. 180°

25. (*a*) −24; (*b*) −10; (*c*) 0

27. (*a*) 1 J; (*b*) 0.213 N

29. (*a*) $(d/dt)(\vec{r}\cdot\vec{r}) = \vec{r}\cdot(d\vec{r}/dt) + (d\vec{r}/dt)\cdot\vec{r} = 2\vec{v}\cdot\vec{r} = 0$. Therefore $\vec{v} \perp \vec{r}$. (*b*) $(d/dt)(\vec{v}\cdot\vec{v}) = 2\vec{a}\cdot\vec{v} = 0$. Therefore $\vec{a} \perp \vec{v}$. The above impiies that the component of $\vec{a}$ in the plane formed by $\vec{r}$ and $\vec{v}$ is colinear with $\vec{r}$. (*c*) $(d/dt)(\vec{v}\cdot\vec{r}) = \vec{v}\cdot(d\vec{r}/dt) + \vec{r}\cdot(d\vec{v}/dt) = v^2 + \vec{r}\cdot\vec{a} = 0$. Therefore, $a_r = -v^2/r$.

31. (*d*)

33. (*a*)

35. F_B

37. (*a*) 2 m/s; (*b*) 24 J

39. (*a*) 24 W; (*b*) −50 W; (*c*) 24 W

41. 26.7 kW

43. From Example 6-11, $x = (8P/9m)^{1/2}t^{3/2}$. From $v = (2P/m)^{1/2}t^{1/2}$, solve for $t^{3/2}$ and substitute this into the expression for x. Simplifying, one obtains $x = (m/3P)v^3$.

45. 4.9 m/s, 19.6 m

47. (*c*)

49. (*a*)

51. (*c*)

53. No. The steeper trail requires fewer steps, but more effort per step.

55. 4.71 kJ

57. 639 MW

59. (*a*) $U(x) = -6(x - x_0)$; (*b*) $U(x) = 24 - 6x$; (*c*) $U(x) = 50 - 6x$

61. $m_1 = 5.68$ kg, $m_2 = 4.32$ kg

63. (*a*) *A*, *E*: $F_x < 0$; *B*, *D*, *E*: $F_x = 0$; *C*: $F_x > 0$; (*b*) *A*; (*c*) *B*: unstable equilibrium; *D*: stable equilibrium; and *F*: neutral equilibrium

65. (*a*) $F_x = C/x^2$; (*b*) directed away from the origin; (*c*) $U(x)$ decreases as x increases; (*d*) If C is a negative constant, F_x is directed toward the origin and $U(x)$ increases as x increases.

67. $U(x) = a/x + U_0$

69. (*a*) $U(y) = 2Mg\,[(d^2 + y^2)^{1/2} - d] - mgy$; (*b*) $y = d\,(m/2M)/\sqrt{1 - (m/2M)^2}$; (*c*) $y_0 = d(m/2M)/\sqrt{1 - (m/2M)^2}$

71. Stable equilibrium at $x = 0$, unstable equilibrium at $x = \pm 2$. $U(x)$ is local minimum at $x = 0$, local maximum at $x = \pm 2$.

73. (*a*) $U(x)$ decreases; (*b*) $U(x) = 4/x^2$ J;

(*c*)

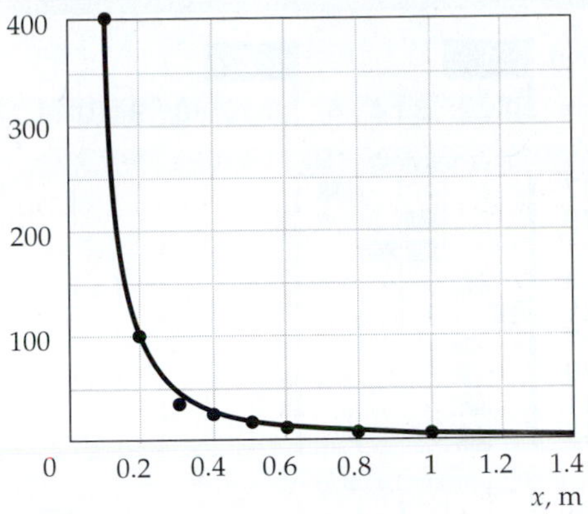

75. (*d*)

77. (*a*) $2h$; (*b*) wh; (*c*) $F \times 2h = wh$

79. (*a*) 706 MJ; (*b*) 11.8 MW

81. 0.5 m

83. 39.3 m

85. (*a*) 12 MW; (*b*) 1261 citizens

87. (*a*) 0; (*b*) 78 J

89. (*a*) 9 J; (*b*) −22 W

91. (*a*)

x, m	−4	−3	−2	−1	0	1	2	3	4
W, J	−11	−10	−7	−3	0	1	0	−2	−3

(*b*)

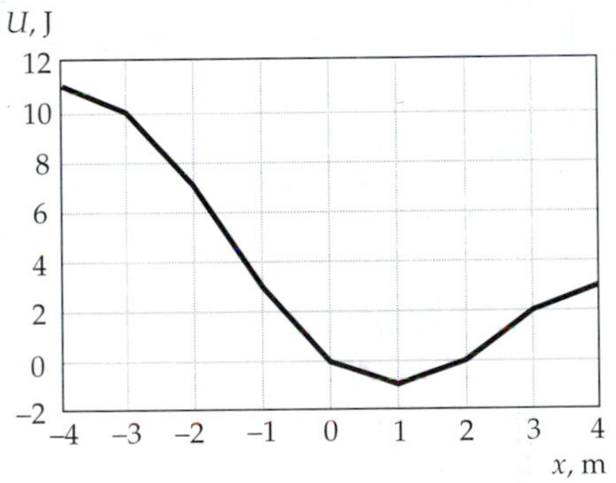

93. (*a*) μyg; (*b*) $\mu g(l^2/2)$

95. (*a*) Answer given in the problem. (*b*) $W = 10\pi F_0$ for clockwise rotation, $W = -10\pi F_0$ for counterclockwise rotation. The force is not conservative.

(*c*) $P = T\sqrt{2[(T/M) - g\sin\theta]x}$

Chapter 7

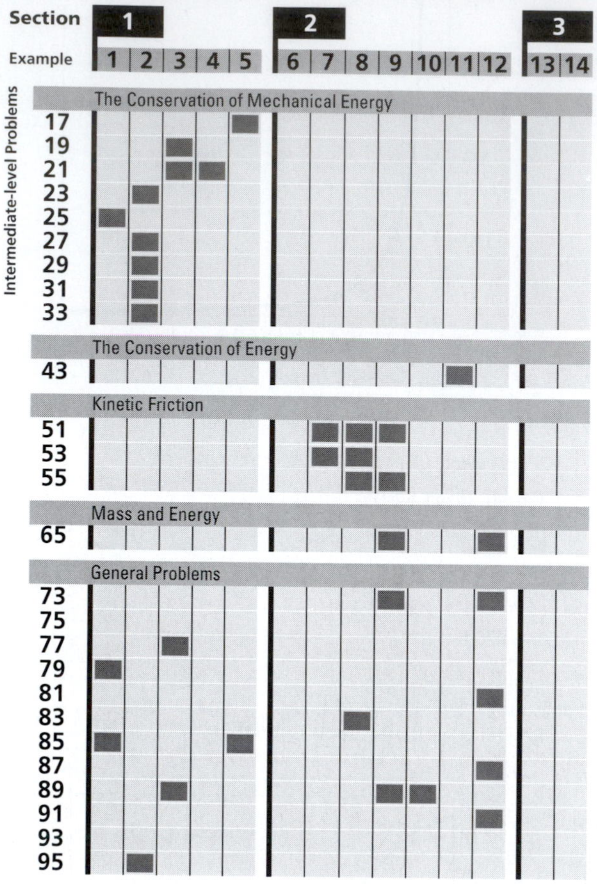

1. Using conservation of mechanical energy is generally simpler, involving only scalars; however, some details can't be obtained (e.g., trajectories).

3. (*c*)

5. (*c*)

7. (*a*) 1.215 J; (*b*) 3.49 m/s; (*c*) 2.34 m

9. (*a*) 0.858 m; (*b*) The spring will accelerate the mass and it will then retrace the path, rising to a height of 5 m.

11. (*a*)

13. 26.7 m/s

15. 25.6°

17. 1.40 m/s

19. $[mg(\sin\theta + \mu_s \cos\theta)]^2/2k$

21. 8.0 m/s

23. $6mg$

25. 27.2 m/s

27. (*a*) 60°; (*b*) 51.3°

29. (*a*) 3.52 m/s; (*b*) 7.89 J; (*c*) 25.3 N; (*d*) 49°

31. (*a*) $(5/2)\,mgL$; (*b*) $6\,mg$

33. Assuming your weight is 650 N: (*a*) 20.2°; (*b*) 6.4 m/s

35. $v = L\theta\sqrt{g/L + (k/m)\theta^2}$

37. The kinetic energy of the man increases at the expense of metabolic (chemical) energy.

39. (*b*)

41. (*a*) 0; (*b*) 560 J; (*c*) The increase in kinetic energy is at the expense of a loss in metabolic (chemical) energy.

43. (*a*) 94.2 kJ; (*b*) The energy comes from metabolic energy. (*c*) 471 kJ

45. Metabolic (chemical) energy converted to thermal energy is released through friction.

47. (*a*) 104 J; (*b*) 70.2 J; (*c*) 33.8 J; (*d*) 2.91 m/s

49. (*a*) 7.67 m/s; (*b*) 58.9 J; (*c*) 1/3

51. (*a*) $13.73y$ J; (*b*) $-13.73y$ J; (*c*) 1.98 m/s

53. (*a*) $(3/8)\,mv_0^2$; (*b*) $(3v_0^2)/(16\pi gr)$; (*c*) 1/3 revolution

55. (*a*) 2.15 m; (*b*) 0.63 m/s; (*c*) 133 J

57. (*a*) $d = (mg/k)(\sin\theta + \mu_s\cos\theta)$; (*b*) $\mu_k = \tan\theta - \frac{1}{2}(1 + \mu_s\cot\theta)$

59. (*a*) 9×10^{13} J; (*b*) \$$2.5 \times 10^6$; (*c*) 9×10^{11} s = 28,400 yr

61. 3.56×10^{14} reactions

63. 0.782 MeV

65. 1.05 kg; (*b*) 3.06×10^9 kg

67. (*b*)

69. (*a*) 8.64 MJ; (*b*) 2065 kcal

71. 57.6 MJ

73. 47 kW

75. 45.1 kW

77. $\Delta E_{\text{therm}} = 0.602$ J; $v = 17.3$ m/s

79. (*a*) 1600 J; (*b*) 619 J; (*c*) 16 m/s

81. 18 kW

83. (*a*) $y = y_0 = 0$; (*b*) $F = mg - ky$; (*c*) $y = y_{\max} = 2mg/k$; (*d*) $y_{\text{eq}} = mg/k$; (*e*) $W_f = \frac{1}{2}(m^2g^2/k)$

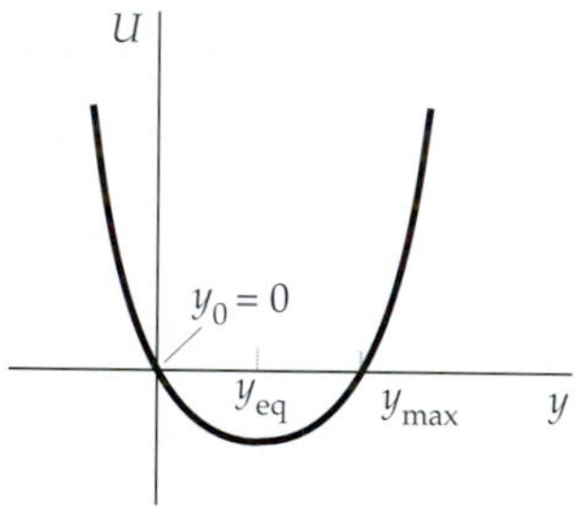

85. (*a*) 17.34 m; (*b*) 4905 N; (*c*) 4.905 m/s^2; (*d*) 13.41 kN; (*e*) 5461 N; (*f*) 1440 N

87. (*a*) $F_{20} = 491$ N; $F_{30} = 981$ N; (*b*) $P_{20} = 9.82$ kW; $P_{30} = 29.4$ kW; (*c*) 8.85°; (*d*) 6.36 km/L

89. (*a*) 0.989 m; (*b*) 0.783 m; (*c*) 1.54 m

91. (*a*) 115.7 kJ; (*b*) 90 kJ; (*c*) 1.58

93. 10.96 m/s, 9.12 s. The pole vaulter uses additional metabolic energy to raise himself on the pole.

95. $\theta_2 = \cos^{-1}\left[1 - \dfrac{L}{L - x}(1 - \cos\theta_1)\right]$

97. (*a*) $F_{\tan} = -mg\sin\theta$; therefore $a = dv/dt = -g\sin\theta$;
(*b*) For circular motion, $v = r\omega = L\,d\theta/dt$; $d\theta/dt = v/L$;
(*c*) $dv/dt = (dv/d\theta)(d\theta/dt) = (v/L)(dv/d\theta)$;
(*d*) $dv/d\theta = (L/v)(dv/dt) = -(L/v)\,g\sin\theta$; $v\,dv = -gL\sin\theta\,d\theta$;

(*e*) $\int_0^v v\,dv = \int_{\theta_0}^0 -gL\sin\theta\,d\theta$; $\frac{1}{2}v^2 = gL(1 - \cos\theta_0)$;

Note that $L(1 - \cos\theta_0) = h$; consequently, $v = \sqrt{2gh}$.

Chapter 8

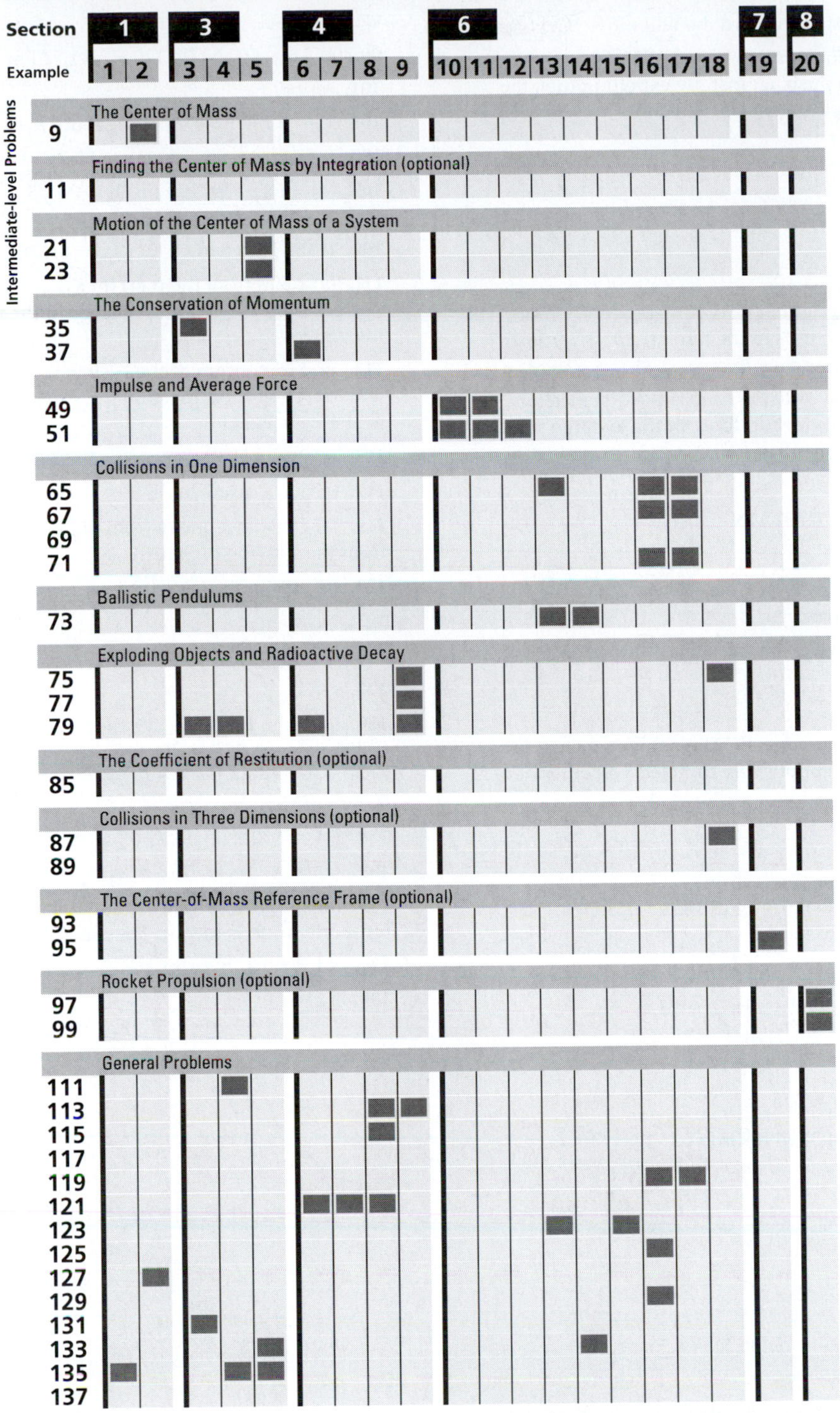

1. hollow sphere

3. 15.6 m

5. 3.36 m

7. (2 m, 1.4 m)

9. (1.5 m, 1.36 m)

11. $(9/16)L$

13. $(0, 0, R/2)$

15. The external force is the force of the static friction between the tires and the road.

17. (*b*)

19. 4 m/s east

21. (*a*) Yes. Initially the scale reads $(M + m)g$. After m drops, the scale reads Mg. (*b*) $mg/(M + m)$ directed downward; (*c*) $F_{net} = (M + m)g - (M + m)\,a_{cm} = Mg$.

23. (*a*) $[(m_1 - m_2)^2 g]/[(m_1 + m_2)(m_1 + m_2 + m_c)]$; (*b*) $4gm_1m_2/(m_1 + m_2) + m_c g$; (*c*) Since $T = 2gm_1m_2/(m_1 + m_2)$, then $F = 2T + m_c g$

25. (*a*) True; (*b*) True; (*c*) True

27. If the man throws something forward, he will move backward.

29. Because of the conservation of momentum, the rocket can move without having something to push against. By expelling fuel, the rocket increases in velocity in the opposite direction.

31. 1.83 m/s to the left

33. (*c*)

35. 40 m

37. $\sqrt{gh/3}$

39. rolling ball

41. (*a*) 60 J; (*b*) 3.75 m/s to the right; (*c*) v of 3-kg block = 1.25 m/s to the right, v of 5-kg block = 0.75 m/s to the left; (*d*) 3.75 J; (*e*) 56.25 J

43. Assume that the ball travels 80 mi/h = 35 m/s and that the ball stops in a distance of 1 cm. The collision time is the distance for which the ball and bat are in contact divided by the average speed. Thus the collision time ≈ 0.02 m/18 m/s = 1 ms.

45. (*a*) 10.75 N·s; (*b*) 1344 N

47. $I = 1.81 \times 10^6$ MN·s; $F_{av} = 0.602$ MN

49. 1.15 kN

51. (*a*) If $d = 0.7$ m, then $F_{av} = 84$ N. (*b*) Yes, the weight of the ball is much less than the average force.

53. (*a*) 4.95×10^{-5}N; (*b*) The weight of the water droplet is about six times the average force due to 10 drops.

55. (*a*) False; (*b*) True; (*c*) True

57. (*a*) The energy losses are the same. (*b*) The percentage loss is greater for the two objects with oppositely directed velocities of $v/2$.

59. 0.56 m/s

61. 3.13 m/s

63. $3v$

65. (*a*) 1.5 m/s to the right; (*b*) v of 0.3-kg muffin = 2.5 m/s, v of 0.5-kg muffin = 4.5 m/s

67. v of 2-kg mass = 4.8 m/s, v of 3-kg mass = 0.8 m/s

69. (*a*) 5 m/s; (*b*) 25 cm; (*c*) $v_1 = 0$ m/s, $v_2 = 7$ m/s

71. (*a*) $0.2v_0$; (*b*) 0.4 v_0

73. $[2(m_1 + m_2)/m_1]\sqrt{gL}$

75. (*a*) 9 m/s $\hat{i}$ − 2 m/s $\hat{j}$; (*b*) 6 m/s $\hat{i}$

77. $v_p = 1.74 \times 10^7$ m/s, $v_\alpha = 4.34 \times 10^6$ m/s

79. (*a*) $v_x = 312$ m/s, $v_y = 66.6$ m/s; (*b*) 5610 m; (*c*) 35.8 kJ

81. 0.913

83. 0.894

85. (*a*) 1.7 m/s; (*b*) 0.83

87. (*a*) $\frac{1}{2}v_0\hat{i} + v_0\hat{j}$; (*b*) $mv_0^2/16$

89. (*a*) $v_2 = \sqrt{2}v$, $\theta_2 = 45°$; (*b*) $K_i = \frac{1}{2}m(3v_c)^2 = 4.5\,mv_0^2$, $K_f = \frac{1}{2}m(\sqrt{5}v_0)^2 + \frac{1}{2}(2m)(\sqrt{2}v_0)^2 = 4.5mv_0^2$

91. (*a*) Let m_1 and m_2 be the incoming and struck particles; assume $\vec{v}_0$ is in the x direction. From $\vec{p}_i = \vec{p}_f$ one obtains $m_1v_0 = m_1v\cos\phi + m_2v_2\cos\theta$ (1) and $m_1v\sin\phi = m_2v_2\sin\theta$ (2). Rearrange the first equation to $m_1(v_0 - v\cos\phi) = m_2v_2\cos\phi$ (1a). Divide (2) by (1a) to obtain $\tan\theta = (v\sin\phi)/(v_0 - v\cos\phi)$. (*b*) Did not use energy conservation; therefore, the result is valid for elastic and inelastic collisions.

93. $K_i = p_1^2/2m_1 + p_1^2/2m_2$; $K_f = p_1'^2/2m_1 + p_1'^2/2m_2$. In an elastic collision, $K_f = K_i$, thus $p_1'^2 = p_1^2$ and $p_1' = \pm p_1$. If $p_1' = +p_1$, the particles do not collide.

95. (*a*) 3.75 m/s; (*b*) $u_{3kg} = 1.25$ m/s, $u_{5kg} = -0.75$ m/s; (*c*) $u'_{3kg} = -1.25$ m/s, $u'_{5kg} = 0.75$ m/s; (*d*) $v'_{3kg} = 2.5$ m/s, $v'_{5kg} = 4.5$ m/s; (*e*) $K_i = K_f = 60$ J

97. 15 km/s

99. (*a*) 3.6×10^5 N; (*b*) 120 s; (*c*) 1.72 km/s

101. (*e*)

103. $v_f = 0.192$ m/s, $K_i = 31.25$ mJ, $K_f = 12.0$ mJ

105. 0.46 m/s

107. (*a*) -1.1×10^5 kg·km/h $\hat{i}$ + 1.05×10^5 kg·km/h $\hat{j}$; (*b*) $v_f = 43.4$ km/h, $\theta = 43°$ north of west

109. (*a*) 81 J; (*b*) 0 m/s; (*c*) $K_{cm} = 0$; (*d*) $K_{rel} = 81$ J

111. (*a*) 2.5 m from the pier; (*b*) $K_{total} = 330$ J, $K_{on\ land} = 270$ J; (*c*) 60 J derives from the chemical energy of the woman. (*d*) Shot will land in the water.

113. (*a*) Momentum of the system is not conserved because the rails will exert a vertical reaction force. (*b*) 4.33 m/s in direction opposite to shell; (*c*) 346 kJ

115. (*a*) 2.3 s; (*b*) The pumpers hit the ground at 24 km/h.

117. $0.75v\hat{i} - v\hat{j}$

119. (*a*) 29.6 km/s away from planet; (*b*) 8.1: This energy comes from the slowing of Saturn.

121. (*a*) −0.6 m/s; (*b*) 960 N

123. No

125. 8.85 kg

127. $r/14\hat{k}$

129. (*a*) From Problem 128, the fraction of energy retained after collision is $-\Delta K/K = 4(m/M)/(1 + m/M)^2 = 0.284$. The fraction of energy retained = 1 − 0.284 = 0.716. After N collisions, $K_f = K_i \times 0.716^N$. (*b*) 55

131. (*a*) $v^2t^2/2L$; (*b*) v^2/L; (*c*) $Mv^3t/L^2 + Mvtg/L$

133. (*a*) $(m_p + m_b)g$; (*b*) $F = g[m_p + m_b\sqrt{(2kh)/g(m_c + m_b)}]$

135. (*a*) 4670 km; (*b*) gravitational force of sun and other planets; (*c*) acceleration is toward the sun; (*d*) 9330 km

137. -1.5×10^{-22} kg·m/s $\hat{j}$

139. $1.2\sqrt{L}$

Chapter 9

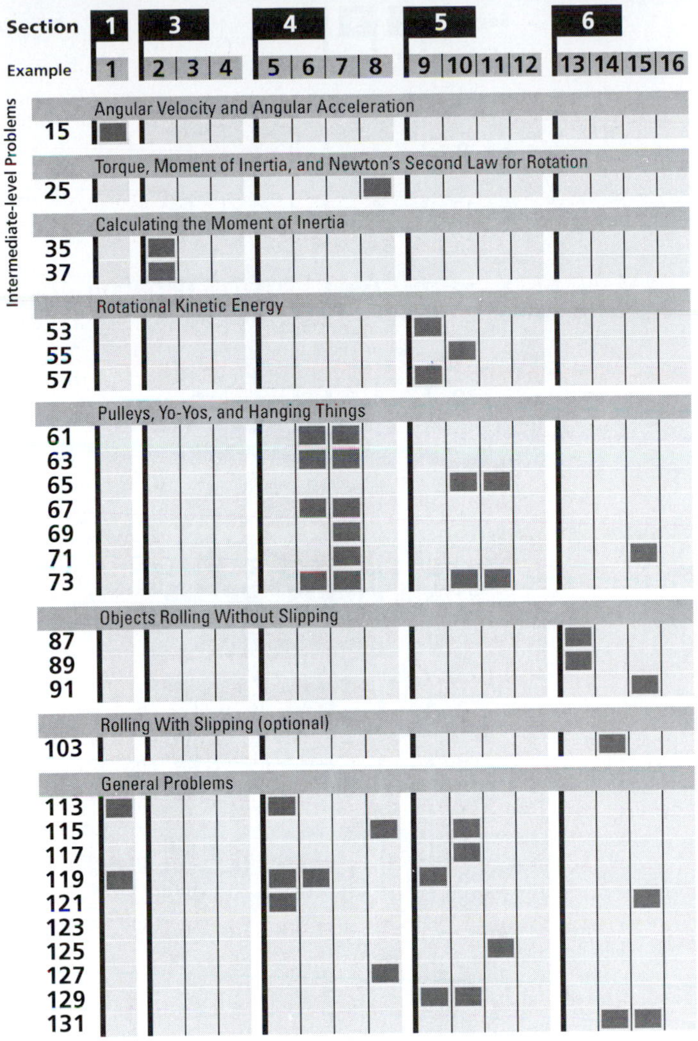

1. The point on the rim moves the greater distance. Both turn through the same angle. The point on the rim has the greater speed. Both have the same angular velocity. Both have zero tangential acceleration. Both have zero angular acceleration. The point on the rim has the greater centripetal acceleration.

3. (*c*)

5. (*a*) 15.6 rad/s; (*b*) 46.8 rad; (*c*) 7.45 rev; (*d*) 4.68 m/s, 73 m/s^2

7. (*a*) 40 rad/s; (*b*) 0.96 m/s^2; 192 m/s^2

9. (*a*) 0.233 rad/s; (*b*) 2.8 m/s; 0.65 m/s^2

11. 7.27×10^{-5} rad/s

13. (*a*)

15. 1.04 rad/s, 9.93 rev/min

17. (*d*)

19. No, it may cause a rotating object to come to rest.

21. (*b*)

23. (*a*) 1.87 N·m; (*b*) 124 rad/s^2; (*c*) 620 rad/s

25. (*a*) $g \sin\theta$; (*b*) $mgL \sin\theta$; (*c*) $g \sin\theta$

27. (*a*) $2(M/R^2)\,\mu_k g r^2\,dr$; (*b*) $(2/3)\,MR\mu_k g$; (*c*) $3Rw/4\mu_k g$

29. 4.66×10^{-5} kg·m^2

31. 28 kg·m^2

33. $(7/5)MR^2$

35. (*a*) $m_1x^2 + m_2(L-x)^2$; (*b*) $2(m_1x + m_2x - m_2L)$, $dI/dx = 0$ when $x = m_2L/(m_1+m_2)$

37. (*a*) $0.04\ \text{kg·m}^2/0.04145\ \text{kg·m}^2 = 0.965$; (*b*) The moment of inertia would increase because I_{cm} of the hollow sphere $> I_{cm}$ of the solid sphere.

39. Let the element of mass be $dm = \rho\,dV = 2\pi\rho h r\,dr$, where h is the height of the cylinder. The mass m of the hollow cylinder is $m = \pi\rho h(R_2^2 - R_1^2)$, so $\rho = m/[\pi h(R_2^2 - R_1^2)]$. The element $dI = r^2\,dm = 2\pi\rho h r^3\,dr$. Integrate dI from R_1 to R_2 and obtain $I = \frac{1}{2}\pi\rho h(R_2^4 - R_1^4) = \frac{1}{2}\pi\rho h(R_2^2 + R_1^2)(R_2^2 - R_1^2) = \frac{1}{2}m(R_2^2 + R_1^2)$.

41. (*a*) $0.509\ M/R^3$; (*b*) $0.329\ MR^2$

43. $(\frac{1}{2})MR^2$

45. $(\frac{1}{2})MR^2$

47. (*b*)

49. (*a*) 52 kg·m^2; (*b*) 20.9 rev/min

51. 155 kW

53. $K_{rot} = 2.6 \times 10^{29}$ J, $K_{orb} = 2.7 \times 10^{33}$ J; $K_{orb} \simeq 10^4\,K_{rot}$

55. (*a*) $\sqrt{(8mg/R)/(2m+M)}$; (*b*) $mg\,[1 + 8m/(2m+M)]$

57. 1.95 m

59. $\sqrt{(1 + M/2m)/(1 + M/3m)}$

61. 3.11 m/s^2; $T_1 = 12.4$ N, $T_2 = 13.4$ N

63. 1.56 m/s^2; $T_1 = 16.0$ N, $T_2 = 16.5$ N

65. 8.2 m/s

67. (*a*) $g/[1 + (2M/5m)]$; (*b*) $2mMg/(5m + 2M)$

69. (*a*) 72 kg; (*b*) 1.37 rad/s^2, $T_1 = 294$ N, $T_2 = 745$ N

71. (*a*) mg; (*b*) $2g/R$; (*c*) $2g$

73. (*a*) $(g \sin\theta)/(1 + m_1/2m_2)$; (*b*) $(\frac{1}{2}m_1g \sin\theta)/(1 + m_1/2m_2)$; (*c*) m_2gh; (*d*) m_2gh; (*e*) $\sqrt{(2gh)/(1 + m_1/2m_2)}$; (*f*) For $\theta = 0$: $a = T = 0$; for $\theta = 90°$: $a = g/(1 + m_1/2m_2)$, $T = \frac{1}{2}m_1 a$, $v = \sqrt{(2gh)/(1 + m_1/2m_2)}$; for $m_1 = 0$: $a = g \sin\theta$, $T = 0$, $v = \sqrt{2gh}$

75. True

77. (*b*)

79. (*c*)

81. Assume the ball is rolling to the right. If f acts to the right along the direction of motion, the center of mass will then accelerate to the right and its linear speed will increase. But the torque produced by f about the center of mass will be counterclockwise tending to decrease the angular speed. But $v = R\omega$, so ω cannot decrease while v is increasing. If f acts to the left so that v decreases, it produces a clockwise torque that tends to increase ω. The only possibility consistent with $v = R\omega$ is $f = 0$.

83. 1125 J

85. 45.9 m

87. 3/2

89. $1.09L$

91. 0.325°

93. 233 J

95. (*a*) From the figure it is evident that $x = r_0 \cos\theta$ and $y = r_0 \sin\theta$ relative to the center of the wheel. Therefore, if the coordinates of the center are X and R, those of point P are as stated.
(*b*) $v_{Px} = d(X + r_0 \cos\theta)/dt = dX/dt - r_0 \sin\theta\, d\theta/dt$. Note that $dX/dt = V$ and $d\theta/dt = -\omega = -V/R$; therefore, $v_{Px} = V + (r_0V \sin\theta)/R$. $v_{Py} = d(R + r_0 \sin\theta)/dt = r_0 \cos\theta\, d\theta/dt$ $(dR/dt = 0)$. Again, $d\theta/dt = -\omega$, so $v_{Py} = -(r_0V \cos\theta)/R$.

(c) $\vec{v}\cdot\vec{r} = v_{Px}r_x + v_{Py}r_y = (V + r_0V\sin\theta/R)(r_0\cos\theta) - (r_0V\cos\theta/R)(R + r_0\sin\theta) = 0$
(d) $v^2 = v_x^2 + v_y^2 = V^2[1 + (2r_0/R)\sin\theta + r_0^2/R^2]$; $r^2 = r_x^2 + r_y^2 = R^2[1 + (2r_0/R)\sin\theta + r_0^2/R^2]$; so $v/r = V/R = \omega$.

97. $2F/R(M + 3m)$, counterclockwise; (b) $F/(M + 3m)$; (c) $-2F/(M + 3m)$.

99. 54°

101. (a)

103. The translational impulse $P_t = P_0\,\Delta t = mv_0$. The rotational impulse about the center of mass is $P_\tau = P_t(h - r) = I\omega_0$. With $I = (2/5)mr^2$ one then obtains $\omega_0 = 5v_0(h - r)/2r^2$.

105. (a) 200 m/s; (b) 8000 rad/s; (c) 257 m/s; (d) 11.6 s

107. (a) $5v_0/3R$; (b) $(5/21)v_0$; (c) $1.056\,mv_0^2$; (d) $-1.016\,mv_0^2$

109. (a) $(2/3)\,v_0$; (b) $5v_0^2/18\mu_k g$; (c) 1/3

111. 2.7×10^{-6} rad/s

113. (a) 0.0873 rad/s²; (b) 572 N·m; (c) 6550 kg·m²

115. (a) 7.36 m/s²; (b) 14.7 m/s²; (c) 2.43 m/s

117. $(6gL/7)^{1/2}$

119. (a) 780 kJ; (b) 211 N·m, 151 N; (c) 1375 rev

121. (a) 15 m; (b) 15.4 rad/s

123. (a), (b) $I_z = \int r^2\,dm = \int(x^2 + y^2)\,dm = \int x^2\,dm + \int y^2\,dm = I_y + I_x$; (c) Let the z axis be the axis of rotation of the disk. By symmetry, $I_x = I_y$. So $I_x = \frac{1}{2}I_z = (1/4)MR^2$. See Table 9-1.

125. (a) The spool will move down the plane at a constant acceleration, spinning in a counterclockwise direction as the string unwinds. $v = \sqrt{(2MgD\sin\theta)/(M + I/r^2)}$; (b) The direction of the friction force is up along the plane. $f_s = (Mg\sin\theta)/(1 + R/r)$

127. $0.75mg$, $0.5mg$, $0.25mg$, 0, 0, 0

129. (a) 230.4 N/m; (b) 122.7 J

131. (a) The only force is F; therefore, $a_{cm} = F/M$. The torque about the center of mass is $\tau = FR$ and $I = MR^2/2$. Thus $\alpha = \tau/I = 2F/MR$. If the cylinder rolls without slipping, $a_{cm} = \alpha R$. Here, $\alpha = 2a_{cm}/R$. (b) $f = F/3$, same direction as F; $a_{cm} = 4F/3M$

133. (a) $T(2r/R - 1)/3$; (b) $(2T/3m)(1 + r/R)$; (c) $r > R/2$ (d) Same direction as T.

Chapter 10

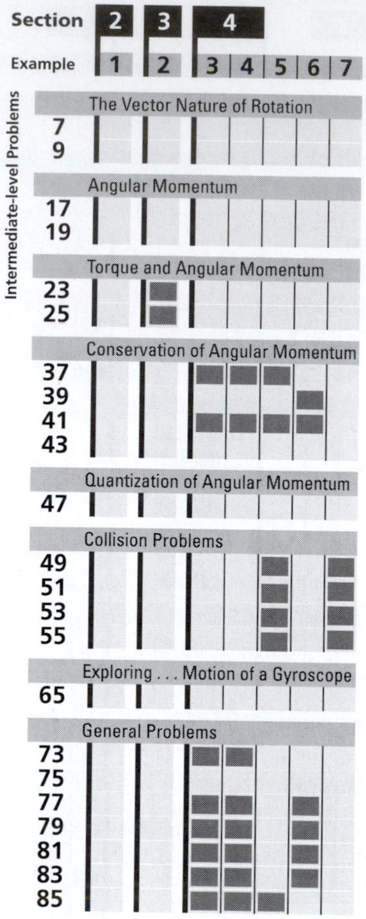

1. (a) True; (b) True; (c) True

3. $\vec{F} = -F\hat{i}$, $\vec{r} = R\hat{j}$, $\tau = +FR\hat{k}$

5. (a) $24\hat{k}$; (b) $-24\hat{j}$; (c) $13\hat{k}$

7. (a) Let $\vec{r}$ be in the x-y plane. If $\vec{\omega}$ points in the positive z direction, i.e., $\vec{\omega} = \omega\hat{k}$, the particle's velocity is in the $\hat{j}$ direction when $\vec{r} = r\hat{i}$ (see Figure) and has the magnitude $r\omega$. Thus, $\vec{v} = \vec{\omega} \times \vec{r} = r\omega\hat{j}$.

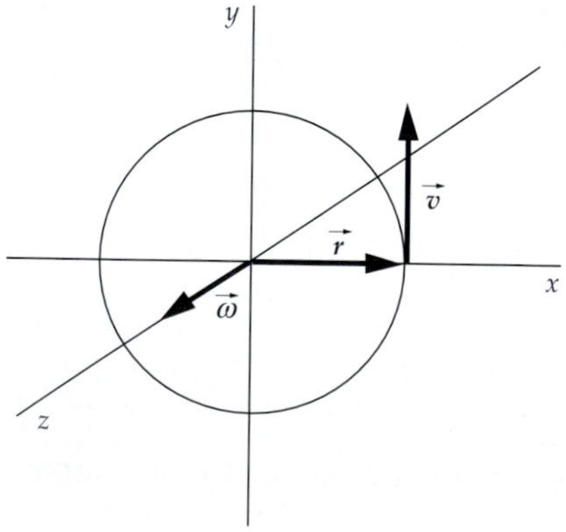

(b) $\vec{a} = d\vec{v}/t = (d\vec{\omega}/dt) \times \vec{r} + \vec{\omega} \times (d\vec{r}/dt) = (d\vec{\omega}/dt) \times \vec{r} + \vec{\omega} \times \vec{v} = \vec{a}_t + \vec{\omega} \times (\vec{\omega} \times \vec{r}) = \vec{a}_t + \vec{a}_c$, where $\vec{a}_t$ and $\vec{a}_c$ are the tangential and centripetal accelerations, respectively.

9. $4\hat{j} + 3\hat{k}$

11. (*b*)

13. constant

15. (*a*) 28 kg·m²/s; (*b*) 32 kg·m²; (*c*) 0.875 rad/s²

17. The area at $t = t_1$ is $A_1 = \frac{1}{2}br_1 \cos\theta_1 = \frac{1}{2}bx_1$, where θ_1 is the angle between $\vec{r}_1$ and $\vec{v}$ and x_1 is the component of $\vec{r}_1$ in the direction of $\vec{v}$. At $t = t_1 + dt$, $A = A_1 + dA = \frac{1}{2}b(x + dx) = \frac{1}{2}b(x + v\,dt)$. Thus, $dA/dt = \frac{1}{2}bv$ = constant. Note that $r \sin\theta = b$; consequently, $\frac{1}{2}bv = \frac{1}{2}L/m$.

19. $\vec{\tau} = \vec{r}_1 \times \vec{F}_1 + \vec{r}_2 \times \vec{F}_2 = (\vec{r}_1 - \vec{r}_2) \times \vec{F}_1$ since $\vec{F}_2 = -\vec{F}_1$. But $\vec{r}_1 - \vec{r}_2$ points along $-\vec{F}_1$ so $(\vec{r}_1 - \vec{r}_2) \times \vec{F}_1 = 0$.

21. (*a*) 4 N·m; (*b*) $0.192t$ rad/s

23. (*a*) $rg\,(m_2 \sin\theta - m_1)$ into the page; (*b*) $vr(I/r^2 + m_1 + m_2)$; (*c*) $(m_2 \sin\theta - m_1)g/(I^2/r + m_1 + m_2)$

25. (*a*) 0.785 N·m; (*b*) 2.75 kg·m²/s; (*c*) 0.131 kg·m²/s; (*d*) 6.64, 13.3

27. False

29. (*e*)

31. No

33. (*b*)

35. (*c*)

37. (*a*) $\omega_i/(1 + mR^2/I_0)$; (*b*) ω_f

39. (*a*) $r_0 m v_0$; (*b*) $\frac{1}{2}mv_0^2$; (*c*) mv_0^2/r_0, $(3/2)\,mv_0^2$

41. 0.11 ms or 1.31×10^{-9} days

43. (*a*) 4 m/s; (*b*) 3.2 N/m

45. 54.7°

47. (*a*) 3.46×10^{-47} kg·m²; (*b*) 2.0 meV, 6.0 meV, 12 meV

49. 14 J·s

51. $L[(M - m)/12m]^{1/2}$

53. 7.75 m/s

55. 0.88

57. 7.36 rad/s; 11.1 J

59. (*a*) False; (*b*) True; (*c*) False

61. (*d*)

63. First, imagine that $\vec{L}$ points vertically upward and the car is moving directly away from you, rounding the top of a hill. The original $\vec{L}$ is now altered by the addition of a ΔL pointing to the left, hence $\vec{L}$ is rotated counterclockwise and the car rolls to its left. If the original $\vec{L}$ had pointed downward, the same ΔL would have been added, but now it causes a clockwise rotation of $\vec{L}$, and hence the car rolls to the right. Similarly, imagine $\vec{L}$ points forward, the car is moving away from you, and is turning to the left. The torque exerted by the road on the car in this case points upward. This means an upward component of angular momentum is added to $\vec{L}$, thus tending to lift the front of the car. If $\vec{L}$ had pointed backward, the same left turn would add the same angular momentum increment, this time raising the rear of the car.

65. (*a*) 3.27 rad/s; (*b*) 0.163 m/s; (*c*) 0.535 m/s²; (*d*) 24.5 N, 1.34 N

67. To prevent body of helicopter from rotating. If rear rotor fails, body of helicopter will tend to rotate on its main axis.

69. (*e*) The ball rotates counterclockwise. The torque about the center of the pole is clockwise and of magnitude RT, where R is the pole's radius and T is the tension. So L must decrease.

71. (*a*) $-47.7\hat{k}$; (*b*) $15.9\hat{k}$ N·m

73. 4.2 rev/s with $m_{arm} = 4$ kg

75. (*a*) 1. Use Newton's laws to determine v; $T\cos\theta = mg$; $T\sin\theta = mv^2/(r\sin\theta)$; $v = (rg\sin\theta\tan\theta)^{1/2} = 2.06$ m/s;
$\vec{v} = d\vec{r}/dt = -2.06 \sin\omega t\,\hat{i} + 2.06\cos\omega t\,\hat{j}$
2. Find $\vec{L} = m\vec{r} \times \vec{v} = 6.18[\sin 30°(\cos\omega t\,\hat{i} + \sin\omega t\,\hat{j}) - \cos 30°\,\hat{k}] \times (-\sin\omega t\,\hat{i} + \cos\omega t\,\hat{j}) = [(3.09\hat{k} + 5.35(\sin\omega t\,\hat{j} + \cos\omega t\,\hat{i})]$ J·s
(*b*) $dL/dt = 14.7$ N·m, $\tau = mgr\sin 30° = 14.7$ N·m

77. $\omega_f = [(M + 5m\ell^2/L^2)/(M + 5m)]\omega$; $K_i = (ML^2 + 5m\ell^2)\omega^2/20$; $K_f = [(ML^2 + 5m\ell^2)^2/(ML^2 + 5mL^2)]\omega^2/20$

79. 30 rad/s; 10.5 rad/s; 176 J, 61.7 J

81. $\omega_i = 30$ rad/s, $\omega_f = 14.13$ rad/s; 93.3 J

83. 6.32 rad/s; 1.54 J

85. 0.55 s

87. $\omega = 12.5$ rad/s in both instances

89. (*a*) 21.4 rad/s; (*b*) $0.378\,e^{1.14t}$ J·s

Chapter 11

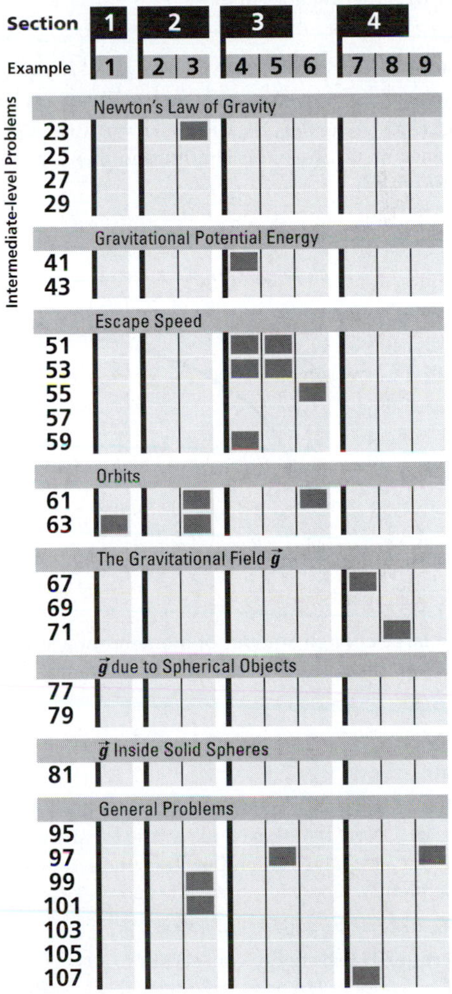

1. (*a*) False; (*b*) True

3. 4.39×10^{11}m

5. 3.91×10^{13}m

7. 11.7 y

9. The mass of the building is negligible compared to the mass of the earth.

11. (*d*)

13. (*a*) 1.88×10^9 m; (*b*) 1.9×10^{27} kg

15. 6.02×10^{24} kg

17. $2.45\ \mathrm{m/s^2}$

19. $39.2\ \mathrm{m/s^2}$

21. 2.63×10^6 m/s

23. (*a*) 37.6 N; (*b*) 2.66 km/s; (*c*) 1.33×10^5 s

25. $F \propto 1/R^3$

27. (*a*) 0.0027 N; (*b*) 0.48 N; (*c*) −0.061%

29. 0.605

31. Because gravity is weak compared to other forces of nature, either an extremely massive object or an extremely accurate instrument is needed to see its effects.

33. (*a*) 2.45×10^{-9} N; (*b*) $\tau = 4.41 \times 10^{-10}$ N·m

35. (*a*) His effectiveness would depend on his mass rather than his weight. (*b*) The power requirement would not be determined by the car's weight, but by its mass. (*c*) No significant effect.

37. (*a*) 5.77 kg; (*b*) gravitational mass

39. GMm_0/R

41. 6.95 km/s

43. (*a*) Outside: $\vec{F} = -(GMm_0/r^2)\hat{r}$. Inside: $\vec{F} = 0$. (*b*) $U(r) = -GMm_0/r$; $U(R) = -GMm_0/R$. (*c*) Since $F = 0$ for $r < R$, $dU/dr = 0$ and $U =$ constant. (*d*) Since U is continuous, then for $r < R$, $U(r) = U(R) = -GMm_0/R$.

(*e*)

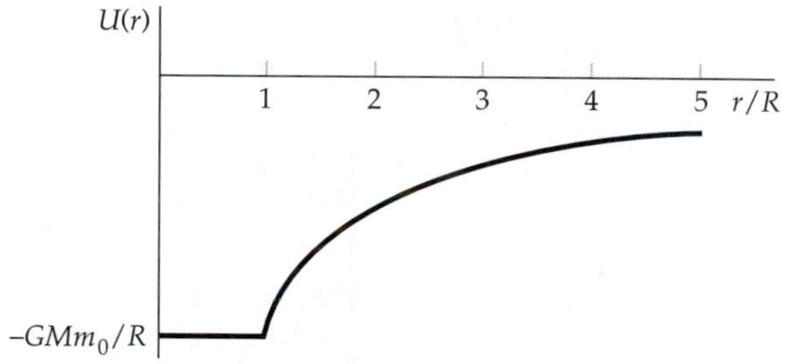

45. (*a*) $M/2\pi R$; (*b*) $-\dfrac{GM}{R}\ln\left(\dfrac{R + \sqrt{x^2 + R^2}}{x}\right)$
(*c*) $GM/(x\sqrt{x^2 + R^2})$

47. Yes, if a large enough amount of momentum was imparted to the earth, say through a collision with a huge comet.

49. 35.5 km/s

51. 19.4 km/s

53. 61.0 km/s

55. From Example 11-6, the kinetic energy of a mass m in a circular orbit is $K = \frac{1}{2}|U|$. Note that this result is true for any circular orbit of a mass m about a massive center. But $|U|$ is the escape energy of $\frac{1}{2}mv_e^2$. Consequently, $K = \frac{1}{2}mv_c^2 = \frac{1}{2}(\frac{1}{2}mv_e^2)$ and $v_c = v_e/\sqrt{2}$.

57. $v_{eS} = 42.2$ km/s, $v_{e,\,solar} = 43.7$ km/s

59. To determine H use energy conservation: $\frac{1}{2}mv^2 = -\Delta U = GMm[1/R_E - 1/(R_E + H)]$ or $v^2 = 2gR_E^2[1/R_E - 1/(R_E + H)]$. So $H' = R_EH/(R_E + H)$ and, solving for H, $H = R_EH'/(R_E - H')$.

61. (*a*) 7.3 h; (*b*) 1.04×10^9 J; (*c*) 8.73×10^{10} J·s

63. $v = \sqrt{GM_S/r - GM_Er/(D - r)^2}$

65. $1 \times 10^{-8}\ \mathrm{N}\ \hat{j}$

67. (*a*) $9.66 \times 10^{-8}\ \mathrm{N}\ \hat{j}$; (*b*) $4.83 \times 10^{-8}\ \mathrm{N/kg}\ \hat{j}$

69. (*a*) g of opposite elements of mass $R\lambda\, d\theta$ cancel. By symmetry, $g = 0$ at center. (*b*) 1. $m_1/m_2 = r_1/r_2$; 2. $g_1 > g_2$; 3. toward m_1; (*c*) toward m_1; (*d*) $g = 0$; (*e*) 1. $m_1/m_2 = r_1^2/r_2^2$; 2. $g_1 = g_2$; 3. $g = 0$; $g = 0$

71. (*a*) $CL^2/2$; (*b*) $-GC\int_0^L \dfrac{x\,dx}{(x_0 - x)^2} = \dfrac{2GM}{L^2}\left[\ln\left(\dfrac{x_0}{x_0 - L}\right) - \left(\dfrac{L}{x_0 - L}\right)\right]\hat{i}$

73. g is proportional to the mass within the sphere and inversely proportional to the radius, i.e., proportional to $r^3/r^2 = r$.

75. Zero. The gravitational field inside the 2-m shell is zero; therefore, it exerts no force on the 1-m shell, and, by Newton's third law, that shell exerts no force on the larger shell.

77. $g_1\,(R_1^2/R_2^2)$

79. (*a*) $(Gm/a^2)(M_1/4.84 + M_2/9)$; (*b*) $GmM_1/1.21\,a^2$; (*c*) 0

81. $G\left(\dfrac{4\pi\rho_0 R_3}{3}\right)\left[\dfrac{1}{x^2} - \dfrac{1}{8(x - \frac{1}{2}R)^2}\right]$

83. $\sqrt{4\pi\rho_0 G/3}$

85. 1.04×10^{-5} m/s

87. (*a*) $-(GMm/d^2)[1 - (d^3/4)/(d^2 + R^2/4)^{3/2}]\hat{i}$; (*b*) $-0.821(GMm/R^2)\hat{i}$

89. (*d*)

91. 1.9×10^{27} kg

93. (*a*) 1.16×10^6 s; (*b*) 8.79×10^{25} kg

95. (*a*) $GM_Em\left(\dfrac{1}{r_1} - \dfrac{1}{r_2}\right)$; (*b*) In the above expression, replace GM_Em by mgR_E^2, r_1 by R_E, and r_2 by $R_E + h$ to obtain the result given. (*c*) $[(1/R_E) - 1/(R_E + h)] = h/[R_E(R_E + h)]$; if $h \ll R_E$, the denominator $\approx R_E^2$ and $W = mgh$.

97. (*a*) $-0.56/r^2$; (*b*) $-(5.6 \times 10^{-7})r$

99. 8.94×10^7 m

101. 1700 m

103. Take the coordinate origin at the center of mass. Then $r_1m_1 = r_2m_2$ and $r = r_1 + r_2$. The force holding m_2 in orbit is $Gm_1m_2/(r_1 + r_2)^2 = m_2r_2\omega^2$. $\omega^2 = Gm_1/r_2(r_1 + r_2)^2$. Now $r_2 = rm_1/(m_1 + m_2)$, so $\omega^2 = 4\pi^2/T^2 = G(m_1 + m_2)/r^3$ and $T^2 = 4\pi^2r^3/G(m_1 + m_2)$.

105. (*a*) $gmR_E/2$; (*b*) $\sqrt{gR_E}$; (*c*) $\sqrt{3gR_E}$

107. (*a*)

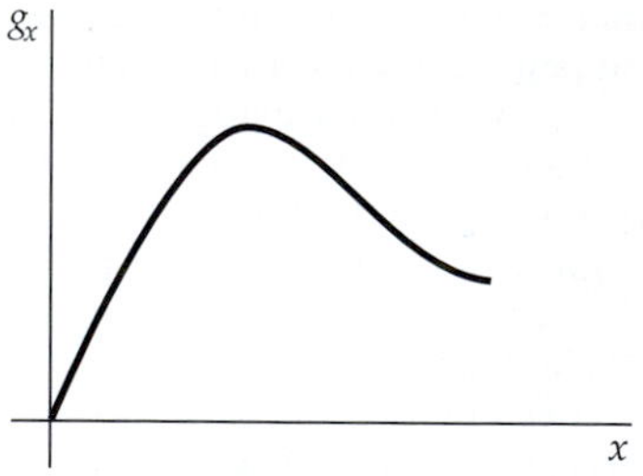

(*b*) $x = R/\sqrt{2}$

109. $Gm^2/[a(a + L)]$

111. (*a*) Since force on a mass m is $F = GMm/r^2$, then $F_s/F_m = M_sr_m^2/M_mr_s^2 = 179$. (*b*) $dF/dr = -2Gm_1m_2/r^3 = -2F/r$; $dF/F = -2dr/r$. (*c*) Since $\Delta F = -2(F/r)\,\Delta r$, then $\Delta F_s/\Delta F_m = (F_s/F_m)(r_m/r_s) = (M_sr_m^3)/(M_mr_s^3) = 0.46$.

Chapter 12

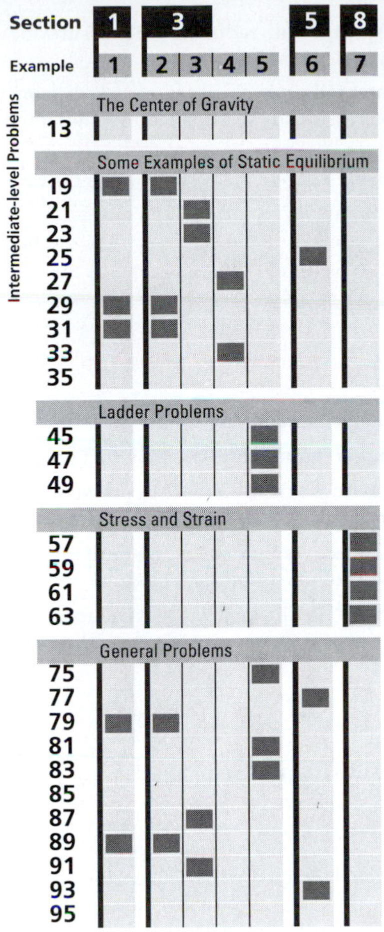

1. (*a*) False; (*b*) True; (*c*) True; (*d*) False

3. 318 N

5. 101 N

7. False. Location depends on mass distribution.

9. center of gravity

11. 3.4 m

13. $(0.94a, a)$

15. Left support, $F = 245$ N; right support, $F = 736$ N

17. $T_B = 692$ N, $F_{horizontal} = 0$, $F_{vertical} = 2540$ N, No, so no block is needed

19. 4.17 m

21. (*a*) 182 N; (*b*) 456 N; (*c*) $\vec{F} = 157$ N, $\vec{F}_{hinge} = 508$ N

23. (*a*) $\vec{T}_1, \vec{T}_2, \vec{F}_H$ (the force exerted on the strut by the hinge); (*b*) $\Sigma\tau$ about hinge = 0, so $T_{2y} = T_1 = 80$ N; (*c*) 139 N

25. 27°

27. $\dfrac{Mg\sqrt{2Rh - h^2}}{h - R}\hat{i} + Mg\hat{j}$

29. Force on middle support = 2920 N, compression. Force on end support = 1940 N, tension.

31. If her boss is 1 m from the left end, Julie must stand 2.33 m or less from the right end.

33. (*a*) $Mg - F\sqrt{(2R - h)/h}$; (*b*) $-F$; (*c*) $F\sqrt{(2R - h)/h}$

35. (*a*) 49 N; (*b*) 73.6 N·m; (*c*) 736 N, acting upward; (*d*) 687 N, acting downward

37. 636 N, 21.5°

39. $F_{2m} = 49.1$ N, $F_{6m} = 147$ N; both forces act in a direction 60° above the x axis.

41. (*a*) Cube is stationary, thus $f_s = -F$. $\tau = Fa$. (*b*) $x = a/3$ from center of cube. (*c*) $Mg/2$

43. No. The wall exerts a horizontal force on the ladder, so there must be friction between the floor and the ladder to balance that horizontal force on the ladder.

45. $\mu_s L \sin\theta \tan\theta$

47. $2h/L \sin\theta \tan\theta$

49. 59°

51. (*a*) $0.361mg$; (*b*) $0.313mg$; (*c*) $0.820mg$

53. 0.98 mm

55. 0.83 mm

57. 3.00181 m

59. (*a*) 34.7 cm; (*b*) 0.0776 J

61. 0.515 mm

63. 0.686

65. $A = A_0 - T/Y$

67. (*c*)

69. The body's center of gravity must be above the feet.

71. 99 cm

73. (*a*) 392 N; (*b*) 3

75. 0.577

77. 0.5

79. 304 kg

81. $(\cot\theta - 1)/2$

83. (*a*) 147 N; (*b*) 72.4% of total length

85. 0.207

87. (*a*) $T = 10{,}260$ N, $F_x = 5130$ N, $F_y = -4520$ N; (*b*) $T = 5924$ N, $F_x = 5130$ N, $F_y = 1403$ N

89. $2g$

91. (*a*) Since the center of gravity of the picture is in front of the wall, the torque due to mg about the nail must be balanced by an opposing torque due to the force of the wall on the picture, acting horizontally. So that $\Sigma F_x = 0$, the tension in the wire must have a horizontal component, and the picture must therefore tilt forward. (*b*) 3.43 N

93. (*a*) 83 km/h; (*b*) 105 km/h

95. 1.5 m

97. (*a*) 7.9 N; (*b*) 29.4 N; (*c*) 7.9 N

99. (*a*) Forces that act on the beam: weight = 49.1 N; force of the cylinder on the beam, $F_c = 28.3$ N; normal force of the ground, $F_{nb} = 24.5$ N; and friction force, $f_{sb} = 14.2$ N. Forces that act on the cylinder: weight = 78.5 N; force of the beam on the cylinder, $F_b = 28.3$ N; normal force of the ground, $F_{nc} = 103$ N; and friction force, $f_{sc} = 14.2$ N. (*b*) μ_s(beam–floor) = 0.58; μ_s(cylinder–floor) = 0.14.

101. $d/a = 1$

Chapter 13

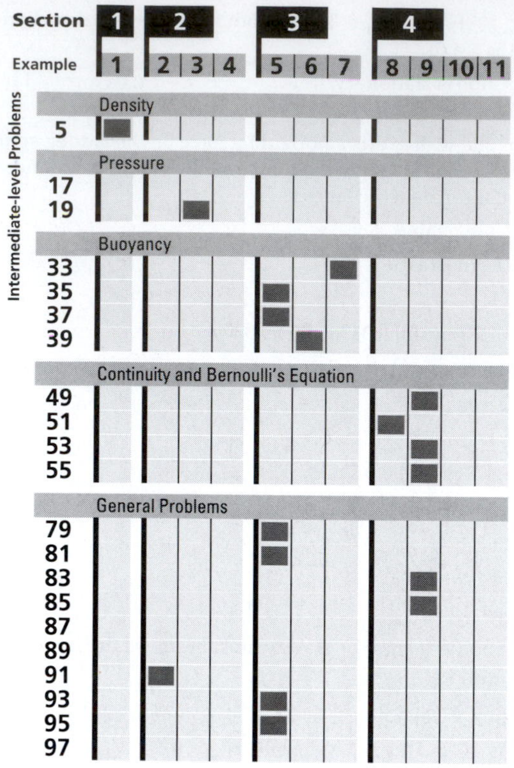

1. 0.673 kg

3. 103 kg

5. 13,621 kg/m^3

7. 29.8 inHg

9. (*a*) 1.5 atm; (*b*) 0.5 atm

11. 230 N

13. 200 atm

15. (*a*) 14,800 N; (*b*) 340 g

17. 45 cm

19. $W = Fh$; $F_1 = F_2\,(A_1/A_2)$; $h_1A_1 = h_2A_2$; $h_1 = h_2\,(A_2/A_1)$; $W_1 = F_1h_1 = F_2\,(A_1/A_2)h_2\,(A_2/A_1) = F_2h_2 = W_2$

21. (*a*) $V = 5.89 \times 10^{-3}$ m^3; $w = 57.8$ N; (*b*) $F = 173$ N; the water exerts a force that has an upward component on the slanting walls of the cone, so the walls exert a downward force on the water. This downward force plus the weight of the water must be balanced by the upward force exerted by the base on the water, which equals the downward force of the water on the base.

23. (*c*)

25. Nothing. The fish is in neutral buoyancy, so the upward acceleration of the fish is balanced by the downward acceleration of the displaced water.

27. (*b*) Volume of copper > volume of lead.

29. 4.36 N

31. (*a*) 11.1×10^3 kg/m^3; (*b*) lead

33. $\rho = 800$ kg/m^3; sp. gr. = 1.11

35. 250 kg/m^3

37. 183 m^3

39. 3.89 kg in freshwater

41. 2.06×10^7 kg

43. It blows over the ball, reducing the pressure above the ball below atmospheric pressure.

45. As the water falls, its velocity increases, causing the internal pressure to decrease; the pressure of air pushes it into a smaller stream.

47. (*a*) 12 m/s; (*b*) 132.5 kPa; (*c*) The volume flows are equal.

49. (*a*) 4.58 L/min; (*b*) 763 cm^2

51. $2\sqrt{h(H-h)}$

53. 1.85 m/s

55. 241 kPa

57. 1.47 kPa

59. 3.98 mPa·s

61. The water level remains constant.

63. The density of humid air is less than that of dry air.

65. The force acting is the difference in pressure between the wide and narrow parts times the area of the narrow part.

67. (*a*) Water spills over because the amount displaced is now greater than the volume of the lead block.

69. $F = 6.48 \times 10^4$ N. The table doesn't collapse because the atmosphere also exerts an upward force on the bottom surface of the table.

71. 1061 kg/m^3

73. 5000 kg/m^3

75. 72% of the block will be submerged.

77. 0.06 m^3 if mass = 60 kg

79. 40.2 g

81. The additional weight on the beaker side equals the weight of the displaced water, i.e., 64 g. That is the mass that must be placed on the other cup to maintain balance.

83. $h_A = 12.6$ m; $h_B = 5.3$ m

85. 6.5 mm

87. Since the object floats, the volume of displaced liquid is $m/\rho_0 = 4A\,\Delta h$; $\Delta h = m/4\rho_0 A$

89. (*a*) $dF = \rho g y L\,dy$; (*b*) 9.20×10^7 N; (*c*) Atmospheric pressure is exerted on each side of the dam and can therefore be neglected.

91. 1.37

93. (*a*) 70 m^3; (*b*) 6.61 m/s^2

95. 150 N (mass ≃ 15 kg)

97. (*a*) 2.4%; (*b*) 59 m^3

99. (*a*) 5.3 g; (*b*) 8 cm; (*c*) 1.17

Chapter 14

Intermediate-level Problems	Section 1				Section 2	Section 3						Section 4	Section 5
Example	1	2	3	4	5	6	7	8	9	10	11	12	13
Simple Harmonic Motion													
11		■											
13		■	■										
15	■			■									
Springs													
35			■	■	■								
37					■								
39					■	■							
41						■	■						
43							■						
Energy of an Object on a Vertical Spring													
45								■					
Simple Pendulums													
55													
Physical Pendulums (optional)													
61										■	■		
63										■			
65										■	■		
Clocks													
71									■				
Damped Oscillations													
77													
79												■	
81												■	
Driven Oscillations and Resonance													
89													■
91													
General Problems													
105													
107							■						
109													
111													
113			■	■	■		■						
115					■		■						
117							■						
119							■						

1. $4A$, 0

3. 0, $A(2\pi f)^2$

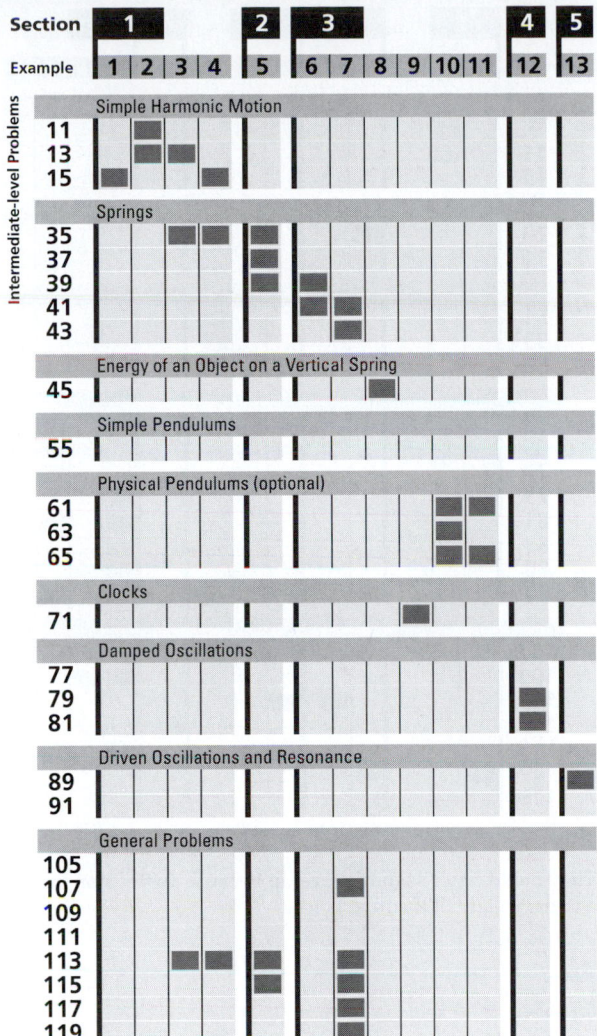

5. (*a*) False; (*b*) True; (*c*) True

7. (*a*) 1.3 m/s; (*b*) 25 m/s

9. (*a*) $x(t) = (0.25\text{ m})\cos(4\pi/3)t$;
(*b*) $v(t) = dx/dt = -(\pi/3\text{ m/s})\sin(4\pi/3)t$;
(*c*) $a(t) = dv/dt = -(4\pi^2/9\text{ m/s}^2)\cos(4\pi/3)t$

11. (*a*) $x(t) = (0.277\text{ m})\cos(4\pi t/3 - 0.445)$; (*b*) $v(t) = dx/dt = -(1.16\text{ m/s})\sin(4\pi t/3 - 0.445)$; (*c*) $a(t) = dv/dt = -(4.86\text{ m/s}^2)\cos(4\pi t/3 - 0.445)$

13. (*a*) $x(t) = 10\cos(\pi t/4)$

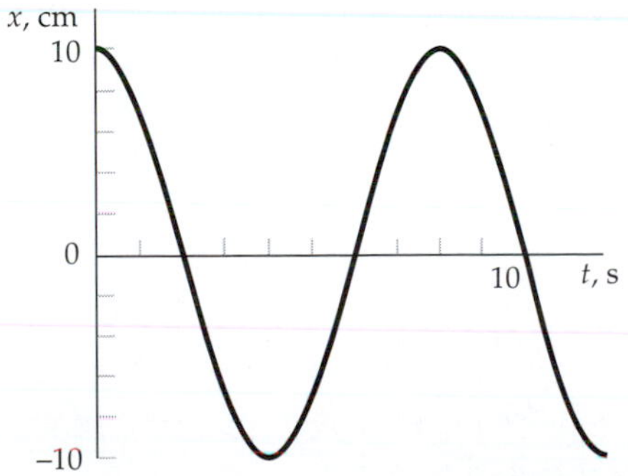

(*b*) 2.9 cm, 7.1 cm, 7.1 cm, 2.9 cm

15. (*a*) $v_{max} = 7.85$ m/s, $a_{max} = 24.7$ m/s^2; (*b*) $|v| = 6.28$ m/s, $a = -14.8$ m/s^2

17. (*a*) $f = 0.32$ Hz, $\omega = 2$ rad/s; (*b*) 3.1 s; (*c*) $x(t) = (40\text{ cm})\cos(2t)$

19. 9

21. 23 J

23. (*a*) 0.37 J; (*b*) 3.8 cm

25. 1.4×10^3 N/m

27. (*a*) True; (*b*) True

29. (*a*) 6.89 Hz; (*b*) 0.145 s; (*c*) 0.1 m; (*d*) 4.33 m/s; (*e*) 187 m/s^2; (*f*) $t = T/4 = 36.3$ ms; $a = 0$

31. (*a*) 680 N/m; (*b*) 0.42 s; (*c*) 1.5 m/s; (*d*) 23 m/s^2

33. (*a*) 3.1 kN/m; (*b*) 4.2 Hz; (*c*) 0.24 s

35. (*a*) 0.44 m/s; (*b*) 0.38 m/s, 1.2 m/s^2; (*c*) 0.095 s

37. 0.26 s

39. 10 kJ

41. 25 cm

43. (*a*) yes; (*b*) $A_{max} = 0.25$ m

45. (*a*) 0.13 J; (*b*) −0.32 J; (*c*) 0.45 J; (*d*) 0.13 J

47. False, the amplitude must be small.

49. The clock would run too slowly.

51. 12 s

53. 11.7 s

55. $T = 2\pi\sqrt{L/g(1 - \sin\theta)}$

57. 1.1 s

59. 0.5 kg·m^2

61. $0.0918L$

63. 2.4 s

65. (*a*) 0.25 m; (*b*) 0.57 m, 2.1 s

67. (*a*) $T = 2\pi\sqrt{I/mgL}$. Since $I = \frac{2}{5}mr^2 + mL^2$, $T = 2\pi\sqrt{L/g + 2r^2/5gL} = 2\pi\sqrt{L/g}\,(\sqrt{1 + 2r^2/5L^2}) = T_0\sqrt{1 + 2r^2/5L^2}$; (*b*) $\sqrt{1 + x} \approx 1 + x/2$ for small x. When $r << L$, $T = T_0\sqrt{1 + 2r^2 5L^2} \approx T_0(1 + r^2/5L^2)$; (*c*) 0.008%, 22 cm

69. (*a*)

71. 6.43°

73. True

75. 3.1%

77. $A(t + T)/A(t) = e^{-T/2\tau}$

79. (*a*) 31%; (*b*) 0.031%

81. (*a*) 3.6 cm, 2.2 cm; (*b*) 38 J, 14 J

83. (*a*) 2.6 s; (*b*) 3.6 s; (*c*) 6.0 J

85. Pendulum of a clock, violin string when bowed

87. (*a*) $f_0 = 1.01$ Hz; (*b*) $f_0 = 2.01$ Hz; (*c*) $f_0 = 0.352$ Hz

89. (*a*) 0.498 m; (*b*) 14.1 rad/s; (*c*) 0.354 m; (*d*) 1 rad/s

91. 180 J; (*b*) 57 J; (*c*) 19 W

93. (*a*) 0; (*b*) 4 m/s

95. (*a*) 0.14 m, 0.44 s; (*b*) 0.23 m, 0.36 s; (*c*) For inelastic collisions, $x(t) = (0.14\text{ m})\sin[(14\text{ rad/s})t]$; for elastic collisions, $x(t) = (0.231\text{ m})\sin[(17\text{ rad/s})t]$

97. 1. B; 2. D; 3. A, C

99. (*d*)

101. (*c*)

103. (*a*) $v(t) = -(1.2\text{ m/s})\sin(3t + \pi/4)$; (*b*) −0.85 m/s; (*c*) 1.2 m/s; (*d*) 0.26 s or 1.3 s

105. (*a*) Since there is no friction, the only forces acting on the particle are mg and the normal force acting radially inward; the normal force is identical to the tension in the string that keeps the particle moving in a circular path if attached to a string of length r. (*b*) The particles meet at the bottom of the bowl because in simple harmonic motion, the period is independent of amplitude.

107. (*a*) 2.3 kg; (*b*) 0.65

109. $T = (2\pi/a)\sqrt{m/\rho g}$

111. The error is greater when the clock is elevated.

113. (*a*) $\mu_s = kA/(m_1 + m_2)g$; (*b*) A and E are unchanged, $\omega_f = \sqrt{m_1/(m_1 + m_2)}\ \omega_i$, $T_f = \sqrt{(m_1 + m_2)m_1}\ T_i$

115. (*a*) 2.3 Hz, 43 J; (*b*) maximum compression: 3.2 Hz, 0.60 m, 22 J; maximum extension: 3.2 Hz, 0.60 m, 43 J.

117. (*a*) If $\omega^2 A > g$, then the piston's acceleration is greater than the maximum acceleration of the block. (*b*) 2.4×10^{-2} s

119. (*a*) 2.5; (*b*) 6.4 cm

121. (*a*)

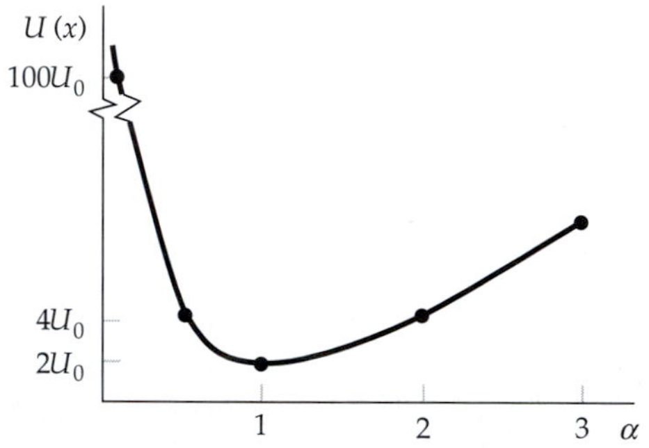

(*b*) $x_0 = a$; (*c*) $U(x_0 + \epsilon) = U_0[(1 + \epsilon/a)^2 + (1 + \epsilon/a)^{-2}]$; (*d*) $U(x_0 + \epsilon) \approx \text{constant} + 4U_0\epsilon^2/a^2$; (*e*) $f = (1/\pi a)\sqrt{2U_0/m}$

123. $T = 8.59\sqrt{R/g}$

125. (*a*) $F = -(Gm\,M_E/R_E^3)r$; $F_x = (\sin\theta)F = (x/r)F = -(Gm\,M_E/R_E^3)x$; (*b*) $T = 2\pi\sqrt{m/k}$; $k = GmM_E/R_E^3 = gm/R_E$. Substituting k, $T = 2\pi\sqrt{R_E/g}$; $T = 84.4$ min.

127. Answer given in the problem.

129. Answer given in the problem.

Chapter 15

Section	1	1	1	2	2	3	3	3	4	5	5	5	5
Example	**1**	**2**	**3**	**4**	**5**	**6**	**7**	**8**	**9**	**10**	**11**	**12**	**13**
Intermediate-level Problems													
Speed of Waves													
9		■											
11		■											
13													
Harmonic Waves on a String													
29				■									
31													
Intensity Level (optional)													
51								■					
53								■					
55								■					
The Doppler Effect													
73												■	
75												■	
77												■	
79										■			
81										■		■	
83										■			
85										■			
87												■	
General Problems													
107	■												
109										■	■		
111					■	■							
113									■				
115										■		■	
117	■			■									
119													

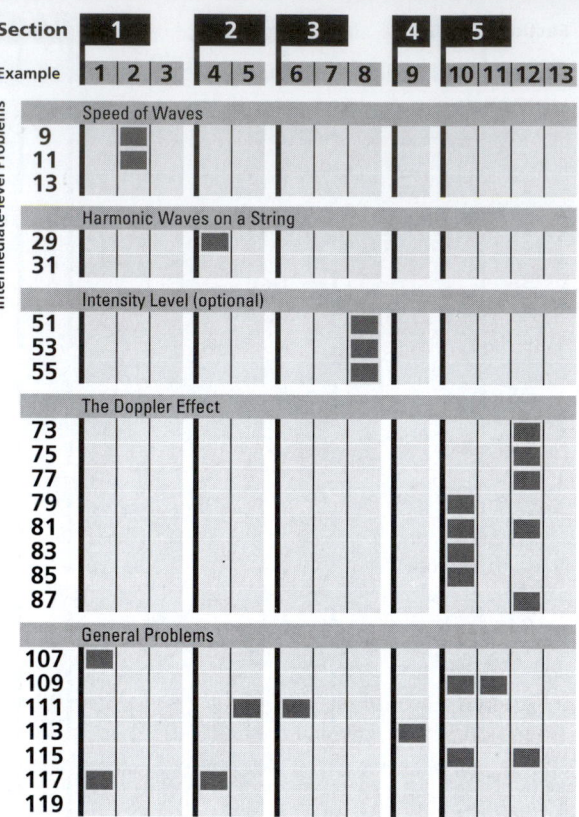

1. They move faster as they move up because the tension increases due to the weight of the rope below.

3. 1.32 km/s

5. 19.6 g

7. (*a*) 265 m/s; (*b*) 15 g

9. (*a*) 0.340 km/s; (*b*) 98% accurate; (*c*) No, because the speed of light is much greater than the speed of sound.

11. (*a*) 78 m; (*b*) 0.23 s, 70 m; (*c*) 70 m

13. (*a*) $v = \sqrt{\gamma RT/M}$; $dv/dt = \frac{1}{2}\sqrt{\gamma R/MT} = \sqrt{\gamma RT/M}/2T = v/2T$. Thus $dv/v = dT/2T$; (*b*) 4.95%; (*c*) $v_{\text{approximate}} = 347.4$ m/s, $v_{\text{exact}} = 347$ m/s

15. The lightning struck 680 m from the ballpark, 58.4° W or E of North.

17. (*a*) $d^2y/dx^2 = 6k(x + vt)$, $d^2y/dt^2 = 6kv^2(x + vt)$; thus $d^2y/dx^2 = (1/v^2)(d^2y/dt^2)$; (*b*) $d^2y/dx^2 = A(ik)^2e^{ik(x-vt)} = -Ak^2e^{ik(x-vt)}$, $d^2y/dt^2 = A(ikv)^2e^{ik(x-vt)} = -Ak^2v^2e^{ik(x-vt)}$; thus $d^2y/dx^2 = (1/v^2)(d^2y/dt^2)$; (*c*) $d^2y/dx^2 = -k^2/[k(x - vt)]^2 = -1/(x - vt)^2$ $d^2y/dt^2 = -(-kv)^2/[k(x - vt)]^2 = -v^2/(x - vt)^2$; thus $d^2y/dx^2 = (1/v^2)(d^2y/dt^2)$

19. For $y(x,t) = A\sin(kx - \omega t)$: $d^2y/dx^2 + i\,dy/dt = -Ak^2\sin(kx - wt) - Ai\omega\cos(kx - \omega t) \neq 0$. For $y(x,t) = Ae^{i(kx \pm \omega t)}$: $d^2y/dx^2 + i\,dy/dt = -k^2y + \omega y = 0$ if $k^2 = \alpha\omega$.

21. True

23. 20 cm

25. (*a*) 7.5×10^{14} to 4.3×10^{14} Hz; (*b*) 10 GHz

27. 171 Hz

29. (*a*) $y(x,t) = (0.025\text{ m}) \sin(42x - 500t)$; (*b*) 13 m/s; (*c*) 6.3 km/s

31. (*a*) $P_0 = \frac{1}{2}\mu\omega^2 A_0^2 v$; (*b*) $P(x) = \frac{1}{2}\mu\omega^2 A_0^2 v e^{-2bx}$

33. (*a*) 0.75 Pa; (*b*) 4 m; (*c*) 85 Hz; (*d*) 340 m/s

35. (*a*) 36.7×10^{-5} m; (*b*) 8.27×10^{-2} Pa

37. (*a*) 0; (*b*) 3.67×10^{-7} m

39. (*a*) 138 Pa; (*b*) 21.7 W/m^2; (*c*) 0.217 W

41. (*a*) 50 W; (*b*) 2 m; (*c*) 4.4×10^{-3} W/m^2

43. False

45. (*a*) 10^{-11} W/m^2; (*b*) 2×10^{-12} W/m^2; (*c*) for (*a*) 9.4×10^{-5} Pa, for (*b*) 4.2×10^{-5} Pa

47. (*a*)

49. 99%

51. (*a*) 100 m; (*b*) 0.13 W

53. (*a*) 100 dB; (*b*) 50 W; (*c*) 20 m; (*d*) 96 dB

55. (*a*) 81 dB; (*b*) Eliminating the two least intense sources will not help much in reducing the intensity level of the noise. Most of the intensity comes from the 80-dB source.

57. 88 dB

59. 57 dB

61. (*a*)

63. (*a*) 15 m, 20 pies/min; (*b*) 13.5 m, 22.2 pies/min; (*c*) 15 m, 22 pies/min

65. (*a*) 1.3 m; (*b*) 262 Hz

67. (*a*) 2.1 m; (*b*) 162 Hz

69. (*a*) 80 m/s toward listener; (*b*) 420 m/s; (*c*) 1.7 m; (*d*) 247 Hz

71. (*a*) 24°; (*b*) 50 km high and 11.5 km over

73. 22 m/s = 79.2 km/h

75. $\Delta f = 9.2 \times 10^4$ Hz, $f = 30.00092$ GHz

77. 4.45 kHz

79. (*a*) 8.1×10^{-2} J; (*b*) 29 cm, 0.14 s

81. Both students must walk 0.75 m/s away from each other.

83. $f' = 1000 + 9.4 \sin 4t$ Hz

85. (*a*) 823.5 Hz; (*b*) 848.5 Hz

87. 183 m, 713 Hz.

89. 6.73×10^{10} m, 4.3×10^{34} kg

91. No; the frequencies are the same, but since the propagation speeds differ, so do the wavelengths.

93. (*a*) True; (*b*) False; (*c*) False

95. Segments 2 cm $< x \leq 3$ cm are moving up. Segments 1 cm $\leq x <$ 2 cm are moving down. $x = 2$ cm is instantaneously at rest.

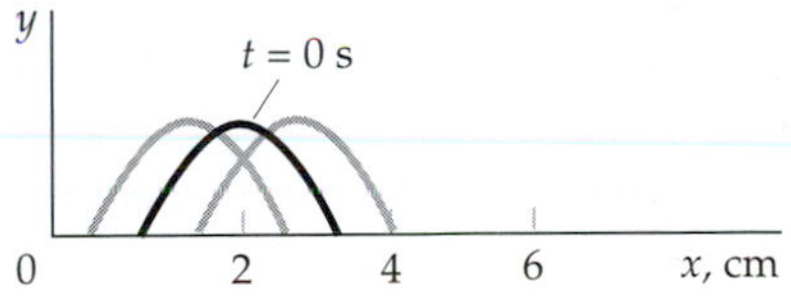

97. Cars behind slow down and then speed up in turn, as each reacts to the reduced distance between them and the car in front of them. This gives rise to a longitudinal wave pulse propagating backwards along the line of cars. No energy is transported. The speed of propagation is proportional to the length of a car and inversely proportional to the average driver's reaction time.

99. 21 cm

101. (*a*) 34 m/s to the left; (*b*) 20 m/s to the right; (*c*) 10 m/s to the right

103. (*a*) 110 Hz; (*b*) 1.1×10^4 Hz; (*c*) 570 Hz, 5.7×10^4 Hz

105. (*a*) 0.59 Hz; (*b*) 1.59 Hz

107. 8.0 m from the left end

109. 65 km/h

111. 1.07 kHz

113. (*a*) $A_r = A/2$, $A_t = 3A/2$; (*b*) $P_r = P_{in}/4$, $P_t = 3P_{in}/4$; (*c*) $A_1 = A + A_r = A_t$ so the amplitudes of displacement are the same.

115. (*a*) 4.5 m/s; (*b*) 887 Hz; (*c*) 842 Hz

117. (*a*) $T = 2.5 \times 10^{-3}$ s, $f = 400$ Hz; (*b*) 316 m/s; (*c*) $\lambda = 79$ cm, $k = 7.9\text{ m}^{-1}$ (*d*) $y(x,t) = [5 \times 10^{-4} \sin(7.9x - 2.5 \times 10^3 t)]$ m; (*e*) $v_{max} = 1.26$ m/s, $a_{max} = 3.16$ km/s^2; (*f*) 2.5 W

119. (*a*) $F = \mu v_0^2$; (*b*) $v_0 = \sqrt{F/\mu}$; (*c*) With respect to a fixed point on the chain, the pulse travels through 360°.

121. (*a*) The speed is given by $v = \sqrt{F/\mu}$. At a distance y from the bottom, $F = \mu g y$. Thus, $v = \sqrt{gy}$. (*b*) 2.2 s

123. (*a*) For $\Delta y/\Delta x \ll 1$, $\Delta\ell = \Delta x[1 + \frac{1}{2}(\Delta y/\Delta x)^2]$; so, $\Delta\ell - \Delta x = \frac{1}{2}(\Delta y/\Delta x)^2\,\Delta x$ and $\Delta U = \frac{1}{2}F(\Delta y/\Delta x)^2\,\Delta x$.
(*b*) $(dy/dx)^2 = A^2k^2\cos^2(kx - \omega t)$. So $\Delta U = \frac{1}{2}FA^2k^2\,\Delta x\cos^2(kx - \omega t)$.
(*c*) Replace F by $\mu v^2 = \mu\omega^2/k^2$. This gives Eq. 15-16*b*.

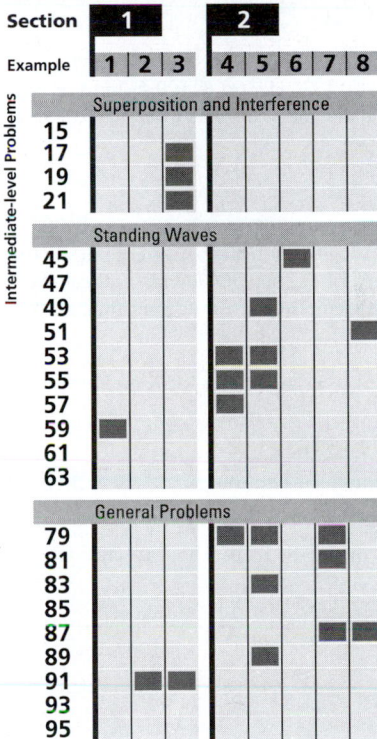

Chapter 16

Section	1			2				
Example	1	2	3	4	5	6	7	8
Intermediate-level Problems								
Superposition and Interference								
15								
17			■					
19			■					
21			■					
Standing Waves								
45						■		
47								
49					■			
51								■
53				■	■			
55				■	■			
57				■				
59	■							
61								
63								
General Problems								
79				■	■		■	
81							■	
83					■			
85								
87							■	■
89					■			
91		■	■					
93								
95								

1. (*a*) False; (*b*) False; (*c*) True

3.

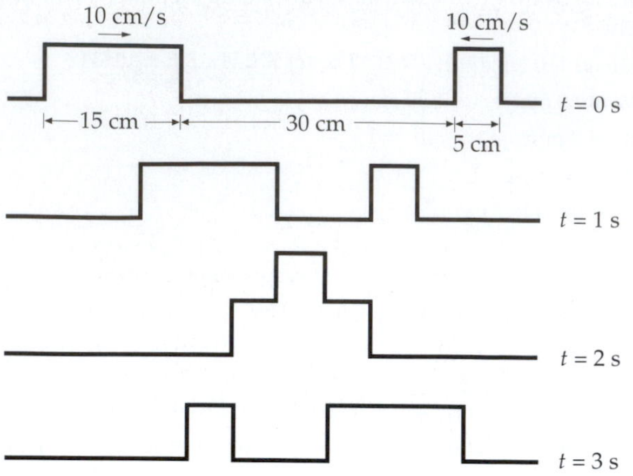

5. (*a*) 3.86 cm; (*b*) 3.46 cm

7. 0.071 m

9. (*a*) $\pi/2$ rad; (*b*) $\sqrt{2}A$

11. (*a*) 85 Hz, 255 Hz; (*b*) Some sound will reflect off walls, thus giving a variety of path differences.

13. (*a*) $4I_0$; (*b*) $2I_0$; (*c*) 0

15. For destructive interference, $\delta = \pi = 2\pi\Delta x/\lambda$. $\Delta x = d \sin\theta \leq d$, where d is the source separation. So if $\Delta x < \lambda/2$, $\delta < \pi$ and there is no complete destructive interference in any direction.

17. 0.8π rad

19. 2361 Hz, 7083 Hz

21. 1.81 m; $\theta = 51°$

23. (*a*) 0.279 m; (*b*) 1220 Hz; (*c*) $\theta_3 = 0.432$ rad, $\theta_4 = 0.592$ rad, $\theta_5 = 0.772$ rad, $\theta_6 = 0.992$ rad, $\theta_7 = 1.354$ rad; (*d*) 0.07 rad

25. 1.98 rad

27. True

29. 4 Hz

31. (*a*) 496 Hz, 504 Hz; (*b*) If the beat frequency is diminished, the second fork has a frequency of 496 Hz, whereas if the beat frequency is increased, the frequency of the second fork is 504 Hz.

33. (*b*)

35. (*c*)

37. 180 m/s

39. (*a*) 17 Hz; (*b*) 8.5 Hz

41. (*a*) 1.25 Hz; (*b*) No second harmonic is excited. (*c*) 3.75 Hz

43. (*a*) 4.25 m; (*b*) 8.5 m

45. (*a*) $\lambda = 31$ cm, $f = 48$ Hz; (*b*) 15 m/s; (*c*) 63 cm

47. (*a*)

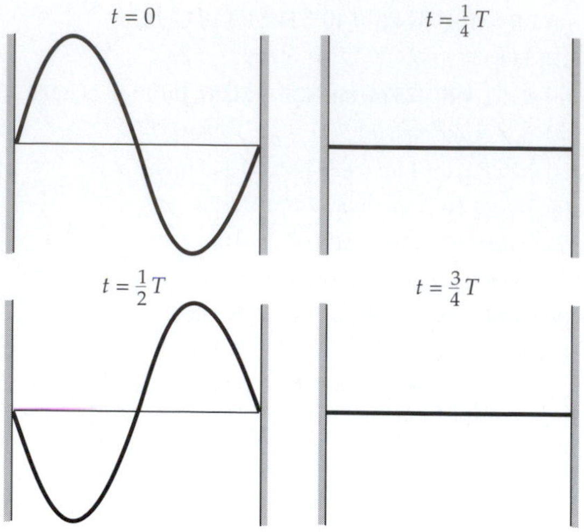

(*b*) 0.0126 s; (*c*) All the energy is kinetic.

49. (*a*) 75 Hz/125 Hz = 3/5, 125 Hz/175 Hz = 5/7; (*b*) The even harmonics are missing. (*c*) 25 Hz; (*d*) 3rd, 5th, 7th harmonics; (*e*) 4 m

51. (*a*) 345 m/s; (*b*) 0.45 cm

53. (*a*) 0.8 m; (*b*) 480 N; (*c*) 9.2 cm from the end

55. (*a*) 75 Hz; (*b*) 5th, 6th; (*c*) 2 m

57. 16.2 Hz

59. (*a*) The two sounds produce a beat because the third harmonic of 440 Hz equals the second harmonic of 660 Hz, and the original frequency of the E string is slightly greater than 660 Hz. If $f_E = (660 + \Delta f)$ Hz, a beat of $2\Delta f$ will be heard. (*b*) 661.5 Hz; (*c*) 79.6 N

61. 12.5 cm

63. (*a*) $y = 0.03 \sin(3\pi/4)x \cos 200\pi t$;
(*b*) $dK = \frac{1}{2}\mu\,[6\pi \sin(3\pi/4)x \sin 200\pi t]^2\,dx$, $t = 2.5 \times 10^{-3}$ s, straight;
(*c*) $89m$ J

65. (*a*) $N \simeq f_0\,\Delta t$; (*b*) $\lambda \simeq \Delta x/N$; (*c*) $k \simeq 2\pi N/\Delta x$; (*d*) N is uncertain because the waveform dies out gradually, rather than stopping abruptly at some time; hence, where the pulse starts and stops is not well defined. (*e*) $\Delta k \approx 2\pi\,\Delta N/\Delta x$. Since $\Delta N = 1$, $\Delta k \approx 2\pi/\Delta x$.

67. No

69. When the edges of the glass vibrate, sound waves are produced in the air in the glass. The resonance frequency of the air columns depends on the length of the air column, which depends on how much water is in the glass.

71. (*b*)

73. The pitch is determined in large part by the resonant cavity of the mouth. Since $v_{He} > v_{air}$, the resonance frequency is higher if helium is the gas in the cavity.

75. (*a*) $3400n$ Hz, $n = 1, 3, 5\ldots$; (*b*) Frequencies near 3400 Hz will be the most readily perceived.

77. (*a*) 2270 Hz; (*b*) 8th harmonic ($f_8 = 18.1$ kHz) or 9th harmonic ($f_9 = 20.4$ kHz)

79. (*a*) $f_n = n(0.66)$ Hz, $n = 1, 2, 3\ldots$; (*b*) $f_n = n(0.33)$ Hz, $n = 1, 3, 5\ldots$

81. (*a*) 4 m; (*b*) $\pi/2\ \text{m}^{-1}$; (*c*) 800π rad/s;
(*d*) $y = 0.03[\sin(\pi/2)x][\cos 800\pi t]$

83. (*a*) $n = 5$; (*b*) 5.5 Hz, 11.0 Hz, 16.4 Hz

85. (*a*) 720 N; (*b*) 2880 N, 6480 N, 11,520 N

87. 338 m/s. This method is not very accurate because the antinode generally does not occur exactly at the end of an open pipe.

89. 9.1 g

91. (*a*) $I_1 = 1.99 \times 10^{-5}$ W/m², $I_2 = 8.84 \times 10^{-6}$ W/m²; (*b*) 5.53×10^{-5} W/m²; (*c*) 2.21×10^{-6} W/m²; (*d*) 2.87×10^{-5} W/m²

93. (*a*) Answer given in the problem. (*b*) 203.4 Hz

95. (*a*) $K = \frac{1}{4}\mu L\omega_n^2 A_n^2 \sin^2 \omega_n t$; (*b*) $K_{max} = \frac{1}{4}\mu L\omega_n^2 A_n^2$; (*c*) $y = 0$; (*d*) Since $\omega_n = n\omega_1$, $K_{max} \propto n^2 A_n^2$.

97. $y_1 = A_0 \cos(kx - \omega t)$, $y_2 = A_0 \cos(kx - \omega t + k\,\Delta x + \delta_0)$; (*b*) $y = 2A_0 \cos \frac{1}{2}(k\,\Delta x + \delta_0) \cos\frac{1}{2}[kx - \omega t + \frac{1}{2}(k\,\Delta x + \delta_0)]$; $A = 2A_0 \cos \frac{1}{2}(\delta + \delta_0)$, where $\delta = 2\pi\,\Delta x/\lambda$.

(*c*)

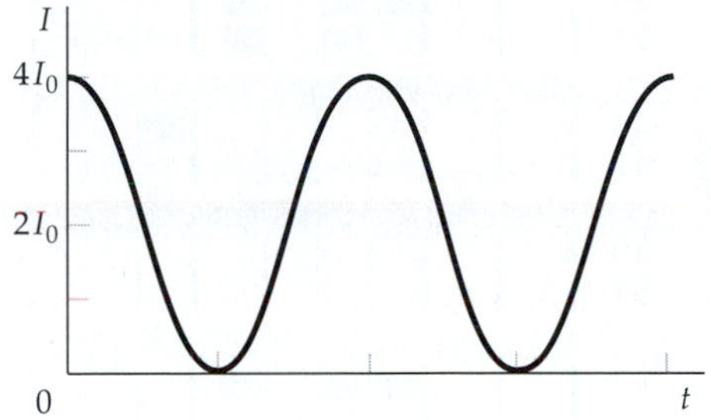

The time-average intensity = $2I_0$.

(*d*)

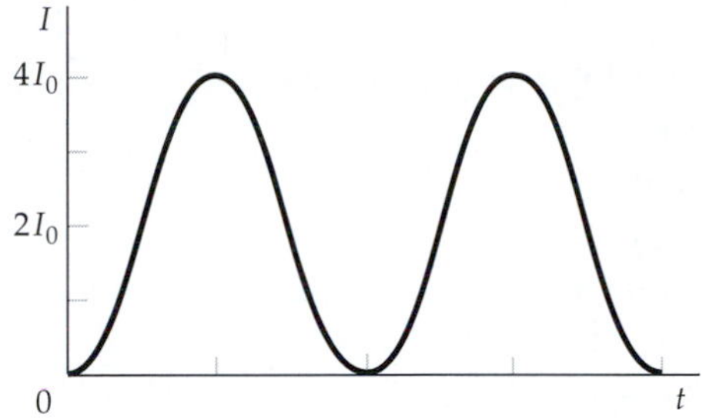

99. (*a*) 0.02 J; (*b*) 0; (*c*) dK_{max} at midpoint: $x = 1$ m; (*d*) dU_{max} at midpoint: $x = 1$ m.

Chapter 17

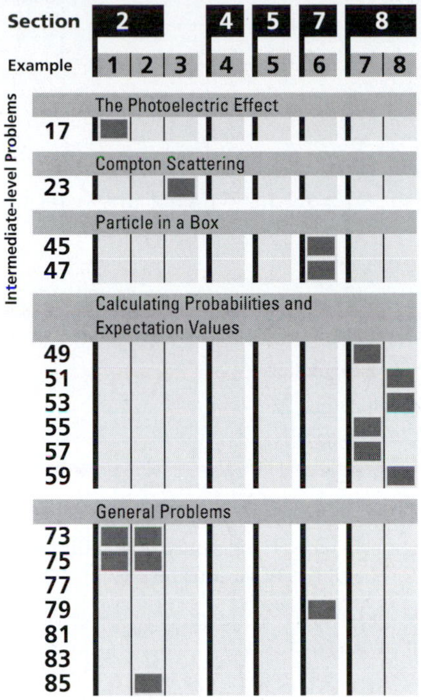

1. (*c*)

3. (*a*) 6.63×10^{-26} J, 4.14×10^{-7} eV; (*b*) 5.96×10^{-28} J, 3.72×10^{-9} eV

5. (*a*) 2.42×10^{14} Hz; (*b*) 2.42×10^{17} Hz; (*c*) 2.42×10^{20} Hz

7. (*a*) 12 keV; (*b*) 1.2 GeV

9. (*a*) True; (*b*) False; (*c*) True; (*d*) True

11. (*a*)

13. (*a*) 1.11×10^{15} Hz, 270 nm; (*b*) 1.62 eV; (*c*) 0.38 eV

15. (*a*) 4.73 eV; (*b*) 2.36 eV

17. 1.52 eV

19. 30°

21. (*a*) 17.4 keV; (*b*) 0.0760 nm; (*c*) 16.3 keV

23. (*a*) 2.43 pm; (*b*) 59.6 keV

25. (*a*) True; (*b*) True; (*c*) True; (*d*) False

27. (*c*)

29. 2.9 nm

31. (*a*) 0.061 eV; (*b*) 15 keV

33. 2.02×10^{-14} m

35. (*a*) 8.2×10^{-4} eV; (*b*) 8.2×10^{8} eV

37. 0.167 nm

39. 4.6 pm

41. Aperture $\approx 1.7 \times 10^{-33}$ m. The size of an atomic nucleus is on the order of 10^{-15}, which is much larger than the size of the aperture.

43. 0.087 nm, atoms in crystals

45. (*a*) $E_1 = 205$ MeV, $E_2 = 821$ MeV, $E_3 = 1846$ MeV

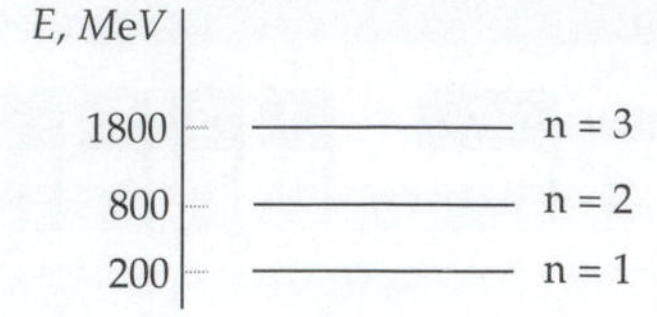

(*b*) 2.01 fm; (*c*) 1.21 fm; (*d*) 0.758 fm

47. (*a*) 5.6×10^{-55} J, 2.2×10^{-54} J, 5.0×10^{-54} J; (*b*) 5×10^{-16} J, 3×10^{19}

49. (*a*) 0; (*b*) 0.003; (*c*) 0

51. $L/2$, $L^2/3$ for box between 0 and L

53. (*a*) $L/2$; (*b*) $0.328L^2$

55. (*a*) $A = \sqrt{1/a}$; (*b*) 0.865

57. (*a*) 0.5; (*b*) 0.402; (*c*) 0.75

59. $\langle x \rangle = 0$, $\langle x^2 \rangle = 0.0327L^2$

61. Yes

63. (*a*) 0; (*b*) 1.5

65. Quantum mechanics predicts only probabilities. Change "results" to "probabilities of various results."

67. 3×10^{19}

69. 1.0×10^{22} photons

71. 4.2×10^{-20} nm

73. 6800 km

75. 1.7×10^{-14} W/m^2

77. 1.9 eV

79. 5

81. 1.04 eV, 554 nm

83. (*a*) $\dfrac{E_{n+1} - E_n}{E_n} = \dfrac{(n+1)^2 - n^2}{n^2} = \dfrac{2}{n} + \dfrac{1}{n^2} \approx \dfrac{2}{n}$ for large n.
(*b*) 0.2%; (*c*) For large quantum numbers n, the fractional difference between adjacent energy levels here is proportional to $1/n$, so as n approaches infinity, the quantized energy levels approach a continuum of levels as in classical physics.

85. (*a*) 6.25×10^{-4} eV/s; (*b*) 3200 s

Chapter 18

Section	2	3	4				5			
Example	1	2	3	4	5	6	7	8	9	10
Intermediate-level Problems										
The Ideal-Gas Law										
25			■							
27			■							
29			■	■		■				
31				■		■				
Kinetic Theory of Gases										
45							■			
47										
The Distribution of Molecular Speeds (optional)										
49										
51										■
General Problems										
67			■	■		■				
69										
71				■		■				
73				■		■				
75										

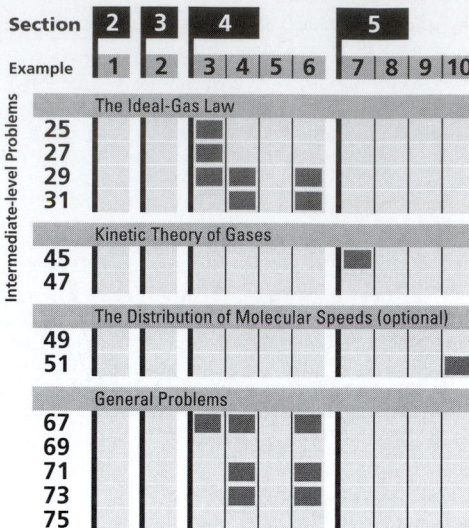

1. (*a*) False; (*b*) False; (*c*) True; (*d*) False

3. 1 C° > 1 F°

5. 10°F to 19°F

7. 57°C, −62°C

9. (*a*) 8.4 cm; (*b*) 107°C

11. −320.4°F

13. (*a*) 54.9 torr; (*b*) 3700 K

15. −40°

17. −183°C, −297°F

19. (*a*) $R_0 = 3.89$ mΩ, $B = 3.94 \times 10^3$ K; (*b*) 1.32 kΩ; (*c*) −389 Ω/K at ice point, −4.37 Ω/K at steam point; (*d*) more sensitive at ice point

21. Pressure increases

23. 1.15

25. 3.2×10^8 molecules

27. (*a*) n = 3.66×10^3 mol; (*b*) 60 mol

29. 0.86 L; (*b*) 60 L

31. (*a*) 230 kPa; (*b*) 200 kPa

33. (*a*) 776 mol; (*b*) 4.86 km; (*c*) yes; (*d*) 6.0 km

35. 4

37. helium molecules

39. (*a*) 276 m/s; (*b*) 872 m/s

41. 5×10^5 m/s, 1.29 keV

43. $\lambda = 1/\sqrt{2}n_v\pi d^2$; since $n_v = N/V = P/KT$, $\lambda = KT/P\sqrt{2}\pi d^2$

45. (*a*) 1.24 km/s; (*b*) 0.31 km/s; (*c*) 0.26 km/s; (*d*) All three gases are likely to be present.

47. $K/mgh = 8.0 \times 10^4$

49. Answer given in the problem.

51. $v_{av} = \sqrt{8kT/\pi m}$

53. False

55. The Fahrenheit and Celsius scales emphasize biologically significant temperatures; 0° is "cold" and 100° is "hot" in everyday experience.

57. (*d*)

59. (*d*)

61. $v_2/v_1 = \sqrt{M_1/M_2}$, $K_{2,av}/K_{1,av} = 1$

63. The rate of molecule–wall collisions increases with increased density.

65. 8.79 K

67. (*a*) 122 K; (*b*) 244 K; (*c*) 1.43 atm

69. 110 mol of H_2, 55 mol of O_2

71. $m_N = 7m_H$

73. $(18/11)P_0$

75. 400.5 K

Chapter 19

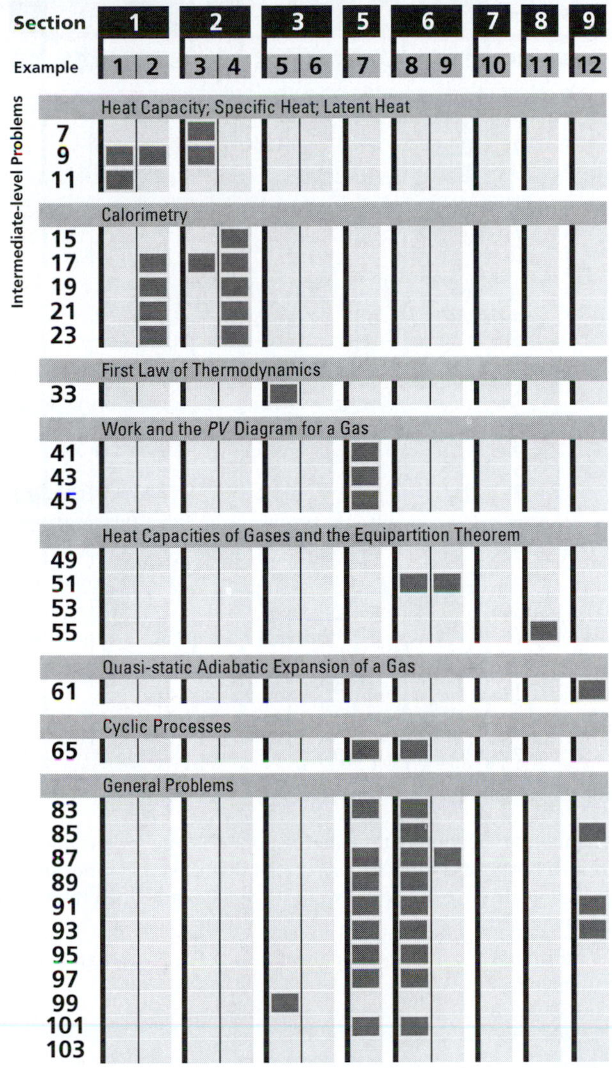

1. $\Delta T_A/\Delta T_B = 1/4$

3. (*c*)

5. 5×10^5 kJ

7. 311 kJ

9. 100 g

11. 6.25 kg

13. 0.092 cal/g·K

15. (*a*) 0°C; (*b*) 125 g

17. (*a*) 4.9°C; (*b*) no

19. (*a*) 3.01°C; (*b*) 200 g; (*c*) no

21. 18 ice cubes

23. (*a*) 28.5°C; (*b*) 15.5°C

25. Yes

27. 2.2 kJ

29. 176°C

31. 54 J

33. (*a*) 34 km; (*b*) yes; (*c*) The snowflakes reach terminal velocity and hit without melting.

35. For an ideal gas, $\Delta U = 0$ implies $\Delta T = 0$ because U is a function of T alone. A real gas is likely to have small, long-range, attractive interactions, which may introduce a dependence on volume (so that molecular kinetic energies decrease slightly as volume increases).

37. Since $Q = 0$ and $W = 0$, $\Delta U = 0$. Thus the temperature must increase to offset the decrease of potential energy from the increase in average distance between the ions.

39. (*a*) 608 J; (*b*) 1060 J

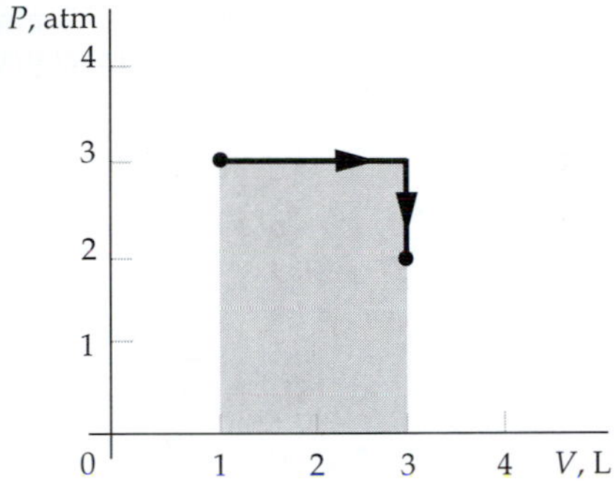

41. (*a*) 334 J; (*b*) 790 J

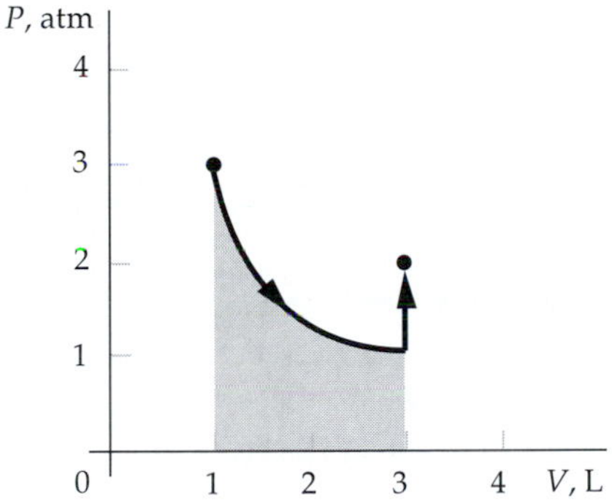

43. 100 L·atm

45. (*a*) −1.57 kJ; (*b*) −1.57 kJ

47. (*a*) 4.00 mol; (*b*) 14.9 kJ; (*c*) 83.0 J/K

49. (*a*) 0.713 J/g·K; (*b*) 252 J/L

51. $5P_0V$

53. (*a*) 3.50 mol; (*b*) $C_v = 43.7$ J/K, $C_p = 72.7$ J/K; (*c*) $C_v = 72.7$ J/K, $C_p = 102$ J/K

55. 1740 K

57. (*a*) $V_i = 2.2$ L, $V_f = 5.9$ L; (*b*) 143 K; (*c*) 1.62 kJ

59. 739 K, 10.1 atm

61. (*a*) 300 K, 7.79×10^{-3} m^3, $W = 1.14$ kJ, $Q = 1.14$ J; (*b*) 231 K, 6.00×10^{-3} m^3, $W = 287$ J, $Q = 0$

63. 575 J

65. (*a*)

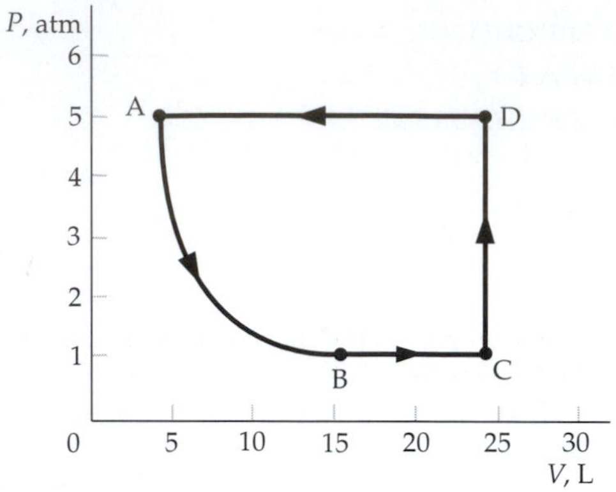

(*b*) $W \approx -64$ L·atm; (*c*) -65.1 L·atm; (*d*) $W = -65.1$ L·atm

67. D to A: $W = 0$, $Q = 89$ L·atm; A to B: $W = 130$ L·atm, $Q = 130$ L·atm; B to C: $W = 0$, $Q = -89$ L·atm; C to D: $W = -65$ L·atm, $Q = -65$ L·atm; Total $W = 65$ L·atm, total $Q = 65$ L·atm

69. D to A: $W = 0$, $Q = 148$ L·atm; A to B: $W = 105$ L·atm, $Q = 0$; B to C: $W = 0$, $Q = -95$ L·atm; C to D: $W = -52$ L·atm, $Q = 0$; Total $W = 53$ L·atm, Total $Q = 53$ L·atm

71. (*a*)

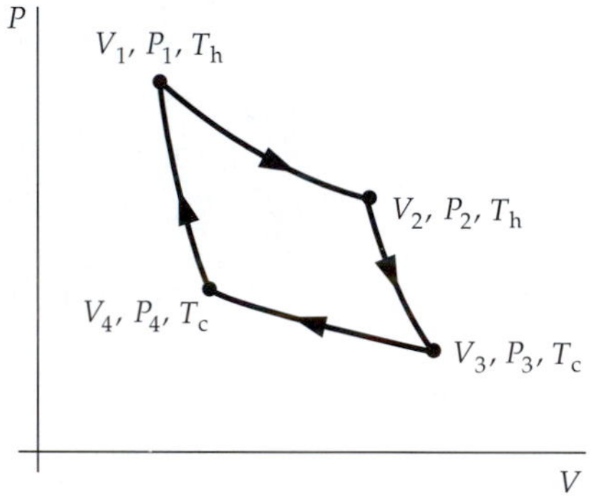

(*b*)–(*f*) Answer given in the problem.

73. (*a*) False; (*b*) False; (*c*) False; (*d*) True; (*e*) True; (*f*) False; (*g*) True

75. (*d*)

77. The temperature decreases.

79. 43 kcal

81. (*a*) $T_A = 65.2$ K, $T_C = 81.3$ K; (*b*) 16.0 L·atm; (*c*) 21.9 L·atm

83. (*a*) $T_A = 65.2$ K, $T_C = 81.3$ K; (*b*) 26.2 L·atm; (*c*) 32.1 L·atm

85. $W = 21.0$ L·atm, $Q = 26.9$ L·atm

87. $W = -170$ cal, $\Delta U = 0$, $T_f = T_i = 53$ K

89. $W = -170$ cal, $\Delta U = 0$, $T_f = T_i = 53$ K

91. (*a*) 3.91 kJ; (*b*) 5.49 kJ

93. (*a*) 2.74 kJ; (*b*) 4.91 kJ

95. (*a*) For an ideal gas, ΔU depends only on the states of the gas (n, T) and not on the path taken to go between the states. (*b*) Answer given in the problem.

97. (*a*) 8.6 K; (*b*) 143 J; (*c*) 1.03

99. 0.20 g

101. $4RT$

103. 4650 J/mol

105. For adiabatic expansion, $PV^\gamma =$ constant. Taking the derivative of both sides, $V^\gamma dP + \gamma P V^{\gamma-1} dV = 0$. Thus $dP/dV = -\gamma P V^{\gamma-1}/V^\gamma = -\gamma P/V$.

For isothermal expansion, $PV = nRT =$ constant. Taking the derivative, $V\,dP + P\,dV = 0$. Thus $dP/dV = -P/V$. $(-P/V)_{\text{isothermal}} = \gamma(-P/V)_{\text{adiabatic}}$.

107. (*a*) $T = 7418$ K, $V = 1832$ L; (*b*) $P = 0.0403$ atm

109. (*a*) $P = 218.2$ kPa; (*b*) $T = 3426$ K, $P = 2136$ kPa; (*c*) P = 170.2 kPa. Since N_2 is a diatomic gas, the contribution to specific heat from vibrations is R only, not $3R$ as for the triatomic gases.

111. (*a*) Answer given in the problem. (*b*) 4650 J

Chapter 20

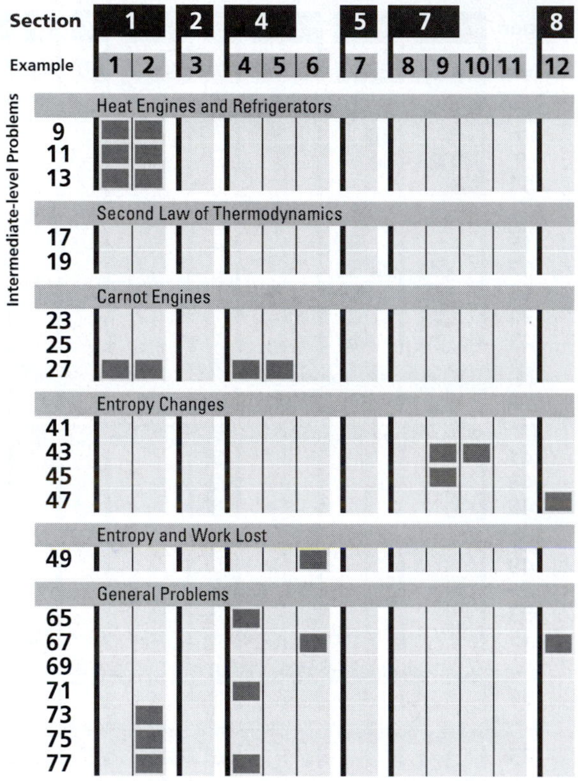

1. In an internal combustion engine, fuel is burned internally to produce hot gases which press a piston. In a steam engine, an external heat source converts water to hot steam, which then presses against a piston.

3. A refrigerator exhausts more heat into the kitchen than it absorbs from its interior.

5. (*a*) 500 J; (*b*) 400 J

7. (*a*) 40%; (*b*) 80 W

9.

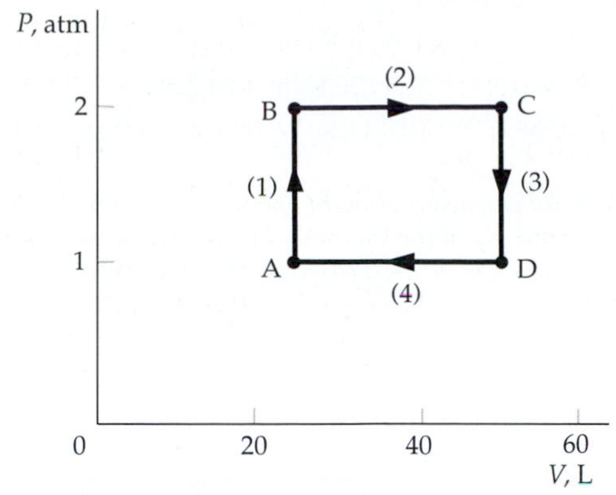

(*a*)

Step	W, L·atm	Q, L·atm	ΔU, L·atm
1	0	36.9	36.9
2	49.2	123.6	74.4
3	0	−74.4	−74.4
4	−24.6	−61.5	−36.9

(*b*) 15.3%

11.

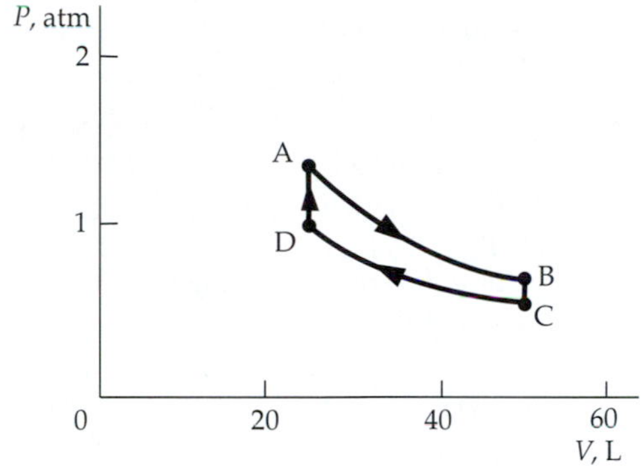

$\varepsilon = 13\%$

13. (*a*) 600 K, 1800 K, 600 K; (*b*) 15.4%

15. $\varepsilon = \varepsilon_C/\{1 + C_v\varepsilon_C/[R\ln(V_a/V_b)]\}$, where ε_C = Carnot efficiency = $(T_h - T_c)/T_h$.

17. This engine takes 200 J from a hot reservoir, does 60 J of work, and rejects the remaining 140 J to a cold reservoir. If the refrigerator statement of the second law of thermodynamics is false, then it would be possible to build a perfect refrigerator that would transfer the 140 J of heat from the cold to the hot reservoir with no other effects. The net result is that 60 J of heat is taken from the hot reservoir, 60 J of work is done, and no heat is rejected to the cold reservoir, a violation of the heat-engine statement of the second law.

19. If the proposed cycle is performed in the clockwise sense, then positive net work is done during the cycle. On the other hand, heat is exchanged only during the isothermal expansion. In this process, heat is drawn in from a high-temperature source. Note, however, that no heat is ever rejected to a low-temperature reservoir in the cycle, so the cycle would violate the heat-engine statement of the second law, namely, that it is impossible to have a cycle that uses heat to do work without exhausting waste heat in the process.

21. (*a*) 13.7; (*b*) 8.77

23. Answer given in the problem.

25. (*a*) 33.3%; (*b*) Answer given in the problem.

27. (*a*) 373 K; (*b*) 3.12 kJ, 0, −2.91 kJ; (*c*) 7%; (*d*) 35%

29. (*a*) 5.26; (*b*) 3.2 kW; (*c*) 4.8 kW

31. (*a*) 303 kJ/min; (*b*) 212 kJ

33. (*c*)

35. (*a*) 11.5 J/K; (*b*) 0

37. 6.05 kJ/K

39. 1.22 kJ/K

41. (*a*) 0; (*b*) 267 K

43. (*a*) 244 kJ/K; (*b*) −244 kJ/K; (*c*) 0

45. (*a*) −24 J/K; (*b*) 138 J/K; (*c*) 114 J/K

47. 1.98 kJ/K

49. (*a*) 0.42 J/K; (*b*) 125 J

51. (*c*)

53. (*d*)

55. The diagram is an Otto cycle, which represents an internal combustion engine.

57.

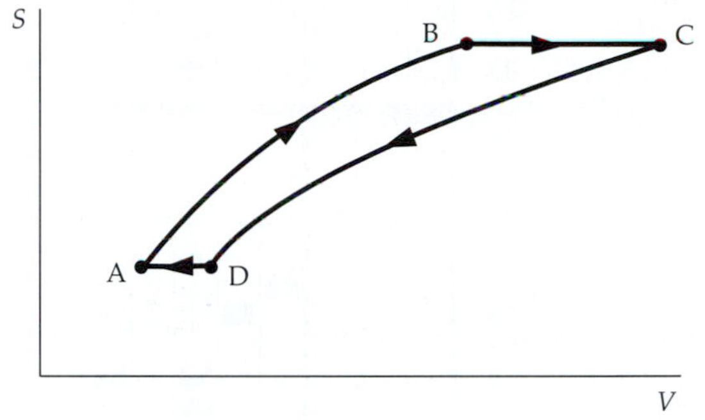

59.

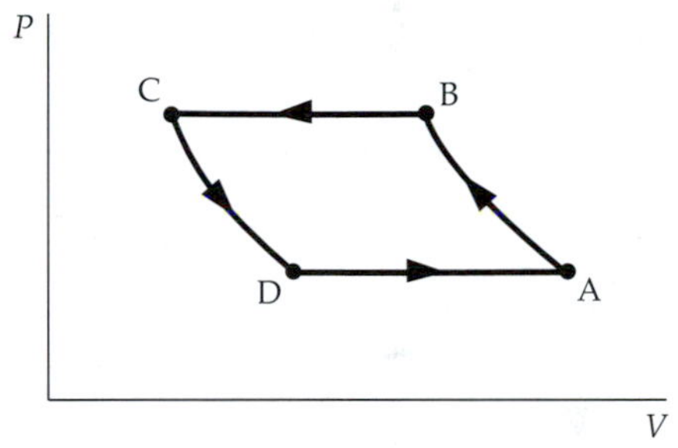

AB, CD are adiabatic; BC, DA are isobaric.

61. The 5-K decrease in the temperature of the cold reservoir

63. (*a*) 51%; (*b*) 102 kJ; (*c*) 98 kJ

65. 60%

67. (*a*) (1); (*b*) 7.66 J/K; 0.83 J/K

69. 313 K

71. 10 W

73. (*a*) 2.5 atm; (*b*) 462 K; (*c*) 68.7 L·atm, 26%

75. (*a*) 2.5 atm; (*b*) 415 K; (*c*) 65.3 L·atm, 35%

77. $\varepsilon = \varepsilon_C/\{1 + C_v\varepsilon_C/[R\ln(V_a/V_b)]\}$, which is less than the Carnot efficiency ε_C.

79. (*a*) Perfect refrigerator = $\Delta S_u = Q(T_h - T_c)/(T_hT_c) < 0$
(*b*) Perfect heat engine: $\Delta S_u = -Q/T_h < 0$; (*c*) $\Delta S \geq 0$ implies both statements, but the converse has not been proven here.

81. $\varepsilon = \varepsilon_1 + \varepsilon_2(1 - \varepsilon_1)$. Substitute $W/Q_h = (1 - T_m/T_h) + (1 - T_c/T_m)(T_m/T_h) = 1 - T_c/T_h$.

83. 2.30×10^{-22} J/K

Chapter 21

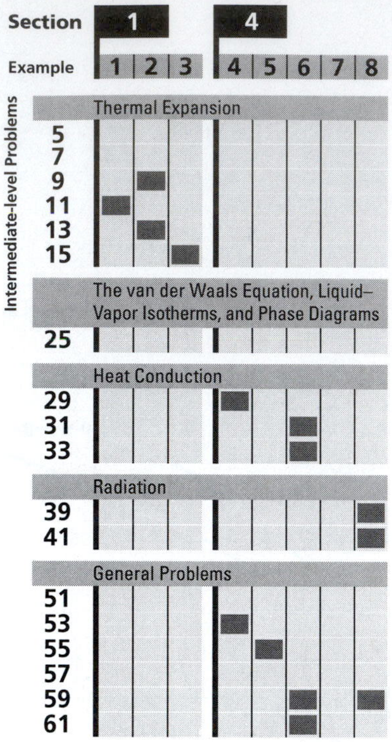

1. The glass of the thermometer expands first, causing an increased volume of space and a corresponding dip in the level of mercury.

3. 30.026 cm

5. (*a*) $(1/A)(dA/dT)$; (*b*) 2α in each case

7. 217°C

9. $15 \times 10^{-6}\ \mathrm{K}^{-1}$

11. (*a*) 1.99935 m; (*b*) 2.00065 m

13. 0.78 L

15. 6.54 cm^3

17. 395°C

19. (*c*)

21. (*a*) For boiling, the temperature and pressure values must lie on the *OF* curve. As the altitude increases, the pressure decreases. The *OF* curve shows that at lower pressures, lower temperatures are required for boiling to occur. The *OH* curve shows the temperature and pressure values required for melting ice to water. As the pressure decreases (with increased altitude), higher temperatures are required for melting. (*b*) Because at higher altitudes water boils at lower temperatures, foods boiled in water would take a longer time to heat thoroughly. Also, more heat would have to be applied to heat foods to a certain temperature.

23. (*a*) 30.6 L; (*b*) 374 K

25. $3.94 \times 10^{-23}\ \mathrm{cm}^3$, 2.1×10^{-8} cm

27. (*a*) 16 K/W; (*b*) 6.3 W; (*c*) 50 K/m; (*d*) 87.5°C

29. (*a*) Cu: 0.083 K/W, Al: 0.14 K/W; (*b*) 0.22 K/W; (*c*) 360 W; (*d*) 70°C

31. (*a*) Because of conservation of energy, the thermal current into an element must equal the thermal current out of the element. (*b*) $I = 4\pi r^2 k\, dT/dr$; (*c*) $dT = (I/4\pi k)(dr/r^2)$; integrating from r_1 to r_2 gives $I = 4\pi k r_1 r_2 (T_2 - T_1)/(r_2 - r_1)$; (*d*) Answer given in the problem.

33. 665 m

35. 4.4×10^{-4}

37. 9470 nm

39. 0.022 K/s

41. 97 g/h

43. (*a*) True; (*b*) True; (*c*) False; (*d*) False; (*e*) True

45. (*d*) assuming that the atmosphere is part of the earth.

47. radiation

49. 2.1 km

51. Rearranging $\beta = (1/V)(\Delta V/\Delta T)$ gives $\Delta V/V = \beta\,\Delta T$; $\rho = m/V$. Since m = constant, $\Delta\rho/\rho = -\Delta V/V = -\beta\,\Delta T$; hence $\Delta\rho = -\rho\beta\,\Delta T$

53. 360 s

55. 1.2 cm

57. (*a*) $\beta = (1/V)(dV/dT)$, where $V = nRT/P$ and $dV/dT = nR/P$. Thus $\beta = (P/nRT)/(nR/P) = 1/T$; (*b*) $\beta_t = 3.663 \times 10^{-3}\ \mathrm{K}^{-1}$

59. 1.26×10^{13} W; $P_{lost}/P_{absorbed} = 7.3 \times 10^{-5}$ (Note that the effective area for absorption of energy from the sun is πR^2 whereas the loss is through area $4\pi R^2$.)

61. 142 W

63. $\omega_2 = \omega_1/(1 + \alpha\,\Delta T)^2$, $K_2 = MR_1^2\,\omega_1^2/4\,(1 + \alpha\,\Delta T)^2$, $L_2 = MR_1^2\,\omega_1/2$

65. Answer given in the problem.

67. 3.1×10^{-6} W/m·K

Chapter 22

Section	1	3				4			5	6		7
Example	1	2	3	4	5	6	7	8	9	10	11	12
Intermediate-level Problems												
Coulomb's Law												
13					■							
15					■							
17					■							
19					■							
The Electric Field												
29							■					
31							■					
33							■					
Motion of Point Charges in Electric Fields												
45										■		
47												
49										■	■	
Electric Dipoles												
53												
55												
General Problems												
67						■						
69	■	■										
71				■								
73				■								
75												
77		■										
79				■		■						
81												
83												■
85												

1. Yes

3. 5×10^{12} electrons

5. 4.82×10^{7} C

7. (*c*)

9. (*a*)

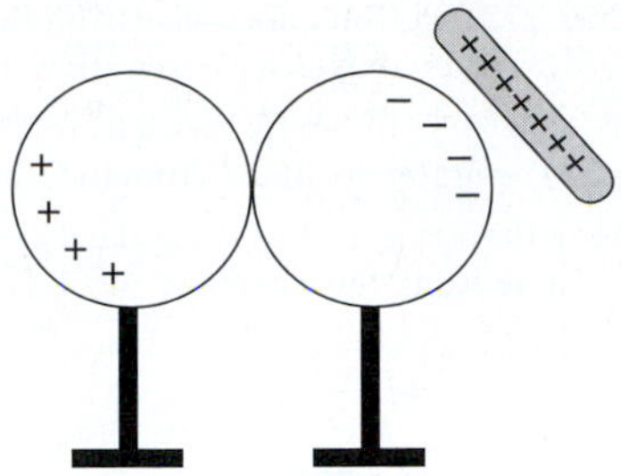

(*b*)

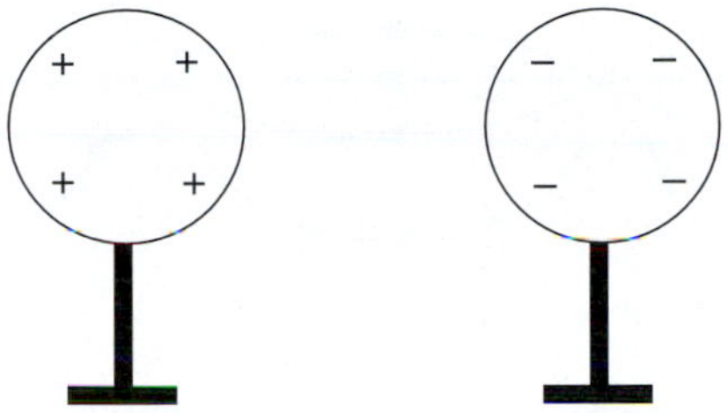

11. (*a*) $0.024\ \mathrm{N}\ \hat{i}$; (*b*) $-0.024\ \mathrm{N}\ \hat{i}$; (*c*) The direction of both forces would be reversed.

13. $1.27 \times 10^{-3}\ \mathrm{N}\ \hat{i} - 3.24 \times 10^{-4}\ \mathrm{N}\ \hat{j}$

15. $-8.66\ \mathrm{N}\ \hat{j}$

17. $0.899\ \mathrm{N}\ \hat{i} + 1.80\ \mathrm{N}\ \hat{j}$, $-1.29\ \mathrm{N}\ \hat{i} - 1.16\ \mathrm{N}\ \hat{j}$, $0.391\ \mathrm{N}\ \hat{i} - 0.643\ \mathrm{N}\ \hat{j}$

19. $(\sqrt{2} + 1)\dfrac{kqQ}{R^2}\hat{i}$

21. (*d*)

23. (*d*)

25. (*a*) $-9.35 \times 10^{3}\ \mathrm{N/C}\ \hat{i}$; (*b*) $7.99 \times 10^{3}\ \mathrm{N/C}\ \hat{i}$; (*c*) $-7.99 \times 10^{3}\ \mathrm{N/C}\ \hat{i}$; (*d*) $9.35 \times 10^{3}\ \mathrm{N/C}\ \hat{i}$; (*e*) 4 m

(*f*)

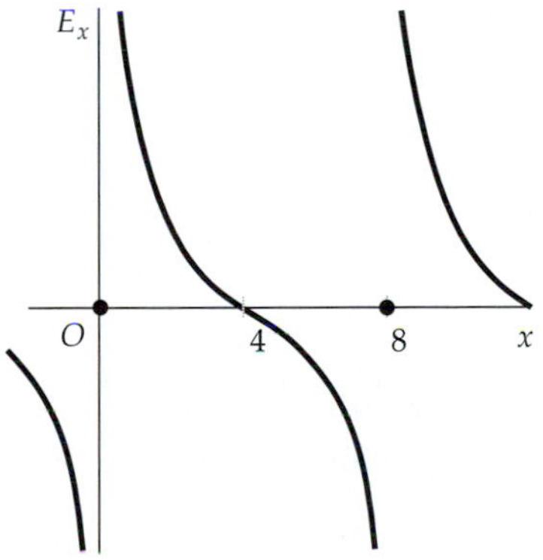

27. 8.18×10^{5} N/C, upward

29. (*a*) $3.45 \times 10^{4}\ \mathrm{N/C}\ \hat{i}$; (*b*) $6.90 \times 10^{-5}\ \mathrm{N/C}\ \hat{i}$

31. (*a*) 1.30×10^4 N/C, 231° counterclockwise from the x axis; (*b*) 2.07×10^{-15} N, 51.3° counterclockwise from the x axis

33. (*a*) 1.91×10^3 N/C, 234° counterclockwise from the x axis; (*b*) 3.06×10^{-16} N, 234° counterclockwise from the x axis

35. (*a*) Unstable along the x axis, stable along the y axis; (*b*) Stable along the x axis, unstable along the y axis; (*c*) $-q/4$; (*d*) The system will be unstable.

37. (*d*)

39.

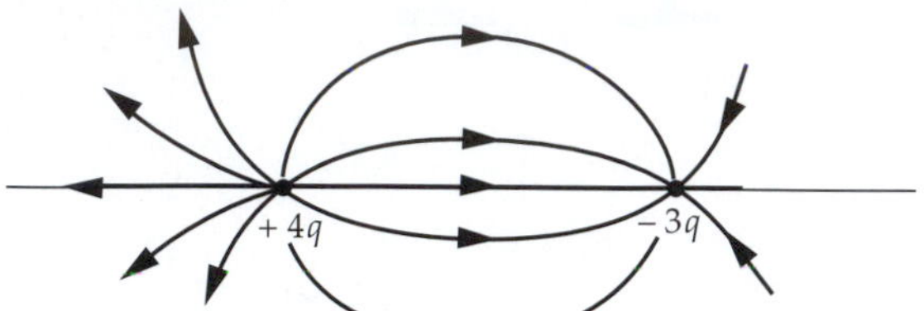

41.

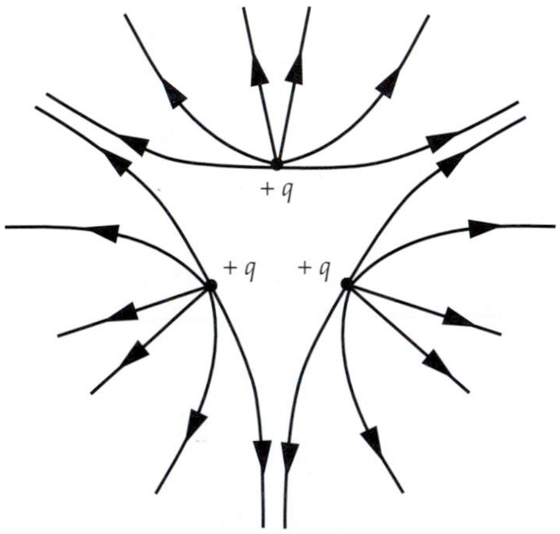

43. (*a*) 9.58×10^7 C/kg, 9.58×10^9 m/s^2; (*b*) 3.13×10^{-4} s

45. 3.75×10^7 m/s

47. 8×10^{-4} C

49. Bottom plate, 4.07 cm from the starting point

51. (*a*) 8×10^{-18} C·m
(*b*)

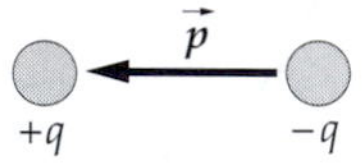

53. Answer given in the problem.

55. (*a*) The force on the positive charge is $qC(x_1 + a)\hat{i}$ and the force on the negative charge is $-qC(x_1 - a)\hat{i}$. Therefore, the net force on the dipole is $2aqC\hat{i} = pC\hat{i}$. (*b*) Answer given in the problem.

57. (*a*) $-3\,kqa^2x^{-4}\hat{i}$; (*b*) $6\,kqa^2y^{-4}\hat{j}$

59. (*a*)

61. The dipole will oscillate about its equilibrium orientation $\theta = 0$. In the nonuniform field, the dipole will accelerate in the x direction as it oscillates about $\theta = 0$.

63. The nonconducting ball is polarized with the side closest to the positive charge becoming negative. This produces a net attractive force. With a negative charge, the ball is polarized with the side closest to the charge becoming positive.

65. Yes. A positively charged ball will induce a dipole on the metal ball, and if the two are in close proximity, the net force can be attractive.

67. 1.14×10^8 N/C, 1.74×10^6 N/C, 0.46 cm

69. 3.3×10^{-7}%, 32.4 N

71. $3kq^2/d^2$, $9kq^2/d^2$

73. 3.0 μC

75. qE/g

77. (*a*) 28 μC, 172 μC; (*b*) 250 N

79. (*a*) −97 μC; (*b*) Between 0.16 m and 0.17 m

81. (*a*) 10.2°; (*b*) 9.9°

83. Answer given in the problem.

85. $\sqrt{ke^2/2mr}$

87. (*a*), (*b*) Answer given in the problem.

89. (*a*) 8.5°; (*b*) 7.8° and 8.6°

91. 1.1×10^4 N/C, upward

93. (*a*) $kQ[(L/2 + x)^{-2} - (L/2 - x)^{-2}]\hat{i}$; (*b*) Answer given in the problem; (*c*) Answer given in the problem; (*d*) $2\pi\sqrt{mL^3/32kqQ}$

Chapter 23

Section	1			2	3				5
Example	1	2	3	4	5	6	7	8	9
Intermediate-level Problems									
Calculate $\vec{E}$ from Coulomb's Law									
11									
13	■								
Gauss's Law									
27									
Spherical Charge Distributions									
33						■			■
35							■		
37							■		
Cylindrical Charge Distributions									
43						■			
45								■	
47								■	
49							■	■	
Charge and Field at Conductor Surfaces									
57									
59									■
General Problems									
71						■			■
73									■
75					■				
77					■				
79									
81									
83					■	■			
85									
87									
89			■						
91									

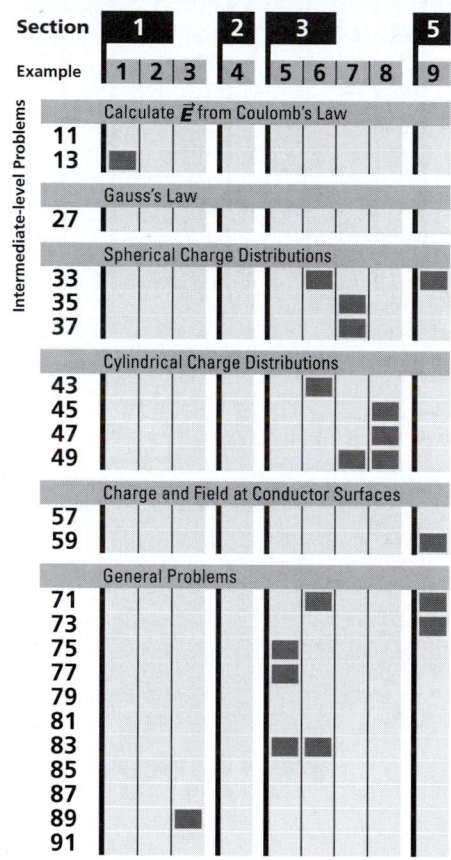

1. (*a*) 17.5 nC; (*b*) 26.2 N/C; (*c*) 4.37 N/C; (*d*) 2.57×10^{-3} N/C; (*e*) 2.52×10^{-3} N/C

3. (*a*) 4.69×10^5 N/C; (*b*) 1.13×10^6 N/C; (*c*) 1.54×10^3 N/C; (*d*) 1.55×10^3 N/C

5. (*a*) 2.00×10^5 N/C; (*b*) 2.54 N/C

7. $a/\sqrt{3}$

9. (*a*) $0.804(2\pi k\sigma)$; (*b*) $0.553(2\pi k\sigma)$; (*c*) $0.427(2\pi k\sigma)$; (*d*) $0.293(2\pi k\sigma)$; (*e*) $0.106(2\pi k\sigma)$

(*f*)

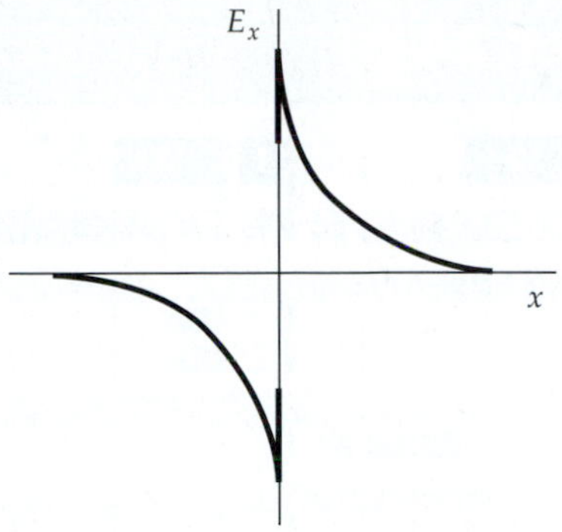

11.

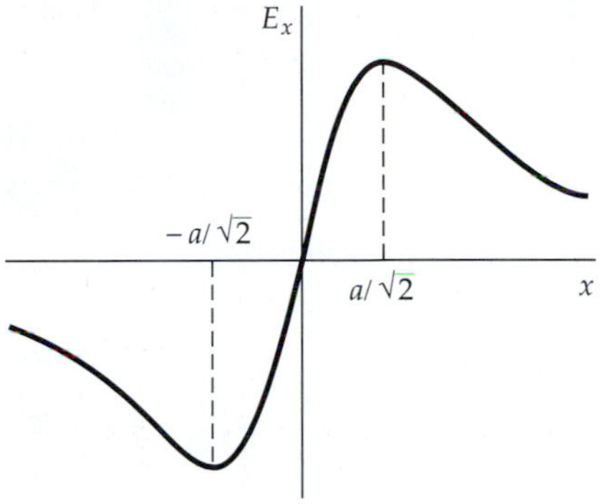

13. (*a*), (*b*) Answer given in the problem.

15. $(\sigma/2\epsilon_0)(1 - 1/\sqrt{2})$

17. (*a*) False; (*b*) False

19. E in Gauss's law is the total electric field due to all charges.

21. (*a*) Three lines enter the sphere.

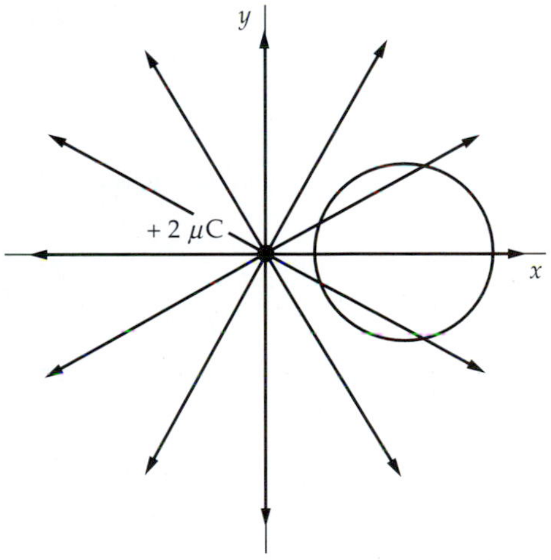

(*b*) 0; (*c*) 0

23. (*a*) N; (*b*) $N/6$; (*c*) q/ϵ_0; (*d*) $q/6\epsilon_0$; (*e*) (*b*) and (*d*)

25. (*a*) $3.14\ m^2$; (*b*) 7.19×10^4 N/C; (*c*) $2.26 \times 10^5\ N{\cdot}m^2/C$; (*d*) No; (*e*) $2.26 \times 10^5\ N{\cdot}m^2/C$

27. $1.13 \times 10^5\ N{\cdot}m^2/C$

29. The field at a distance r is proportional to Q'/r^2, where Q' is the charge within a sphere of radius r. The charge Q' is proportional to the volume of the sphere and therefore is proportional to r^3. Thus the field is proportional to $r^3/r^2 = r$.

31. (*a*) 0.407 nC; (*b*) 0; (*c*) 0; (*d*) 984 N/C; (*e*) 366 N/C

33. (*a*) $-5Q$, $-2Q$; (*b*) $-5Q$, $-5Q$, 0, goes to 0; (*c*) $-2Q$, $-2Q$, 0

35. (*a*) πAR^4; (*b*) $Ar^2/4\epsilon_0$, $AR^4/4\epsilon_0 r^2$

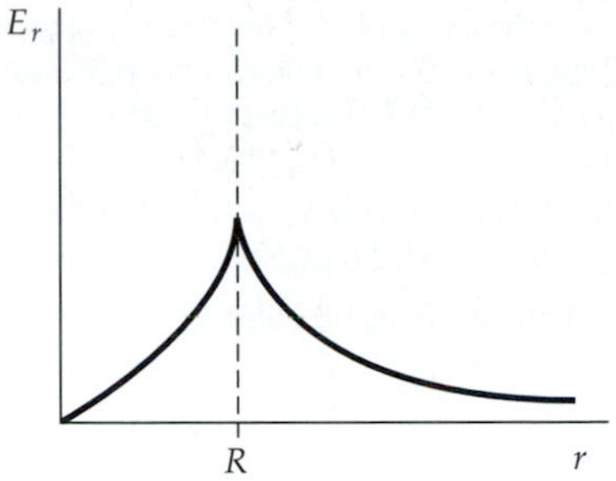

37. (*a*) $4\pi CR$; (*b*) $C/\epsilon_0 r$, $CR/\epsilon_0 r^2$

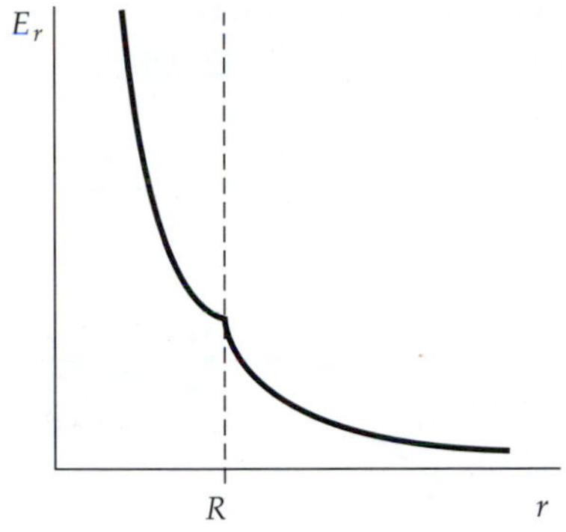

39. (*a*) $(4\pi\rho/3)(b^3 - a^3)$; (*b*) 0, $(\rho/3\epsilon_0 r^2)(r^3 - a^3)$, $(\rho/3\epsilon_0 r)(r^3 - a^3)$

41. Answer given in the problem.

43. (*a*) 6.79×10^{-7} C; (*b*) 0; (*c*) 0; (*d*) 1.00×10^3 N/C; (*e*) 6.11×10^2 N/C

45. (*a*) 6.79×10^{-7} C; (*b*) 3.39×10^2 N/C; (*c*) 1.00×10^3 N/C; (*d*) 1.00×10^3 N/C; (*e*) 6.11×10^2 N/C; The fields for the shell charge and solid cylinder of charge with the same total charge are identical outside the charges. Inside the shell, the field is zero.

47. (*a*) zero for $0 \le r \le 1.5$ cm and 4.5 cm $\le r < 6.5$ cm, $(108/r)$ N/C for 1.5 cm $< r <$ 4.5 cm and 6.5 cm $\le r$; (*b*) $2.10 \times 10^{-8}\ C/m^2$, $1.47 \times 10^{-8}\ C/m^2$

49. Answer given in the problem.

51. (*a*) 1.885×10^{-8} C/m; (*b*) 2.26×10^4 N/C for $0 \le r \le 1.5$ cm, $(339.3/r)$ N/C for 1.5 cm $< r <$ 4.5 cm, zero for 4.5 cm $< r <$ 6.5 cm, $(339.3/r)$ N/C for 6.5 cm $< r$

53. 9.41×10^3 N/C

55. (*a*) kq/r^2, 0, kq/r^2

(*b*)

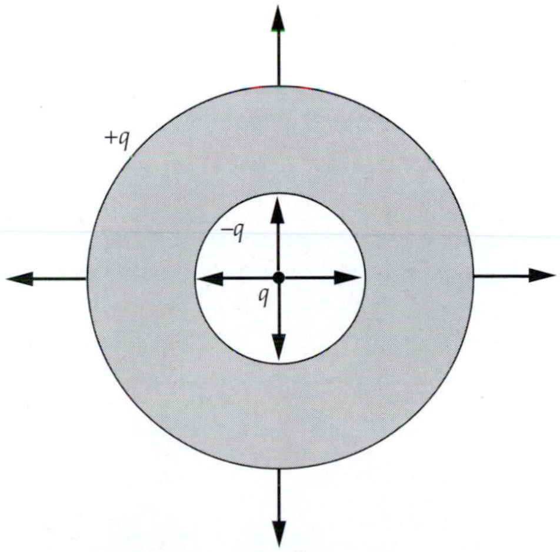

(c) $-q/4\pi a^2, q/4\pi b^2$

57. (a) $-0.553\ \mu C/m^2, 0.246\ \mu C/m^2, 2.5\ \mu C$ for outer surface and $-2.5\ \mu C$ for inner surface; (b) kq/r^2 for $0 \le r \le 60$ cm, zero for 60 cm $< r <$ 90 cm, kq/r^2 for $r \ge 90$ cm; (c) $-0.553\ \mu C/m^2$, $0.589\ \mu C/m^2$, $6\ \mu C$ for outer surface and $-2.5\ \mu C$ for inner surface; $kq/r^2, 0, (6\ \mu C)k/r^2$

59. (a) $1.60 \times 10^{-6}\ C/m^2, 1.81 \times 10^5$ N/C; (b) 6.78×10^4 N/C, 2.94×10^5 N/C, $0.6\ \mu C/m^2, 2.6\ \mu C/m^2$

61. (a) False; (b) True; (c) True; (d) False

63. (a)

65. (b)

67. (c)

69. Gauss's law cannot be used to find the field of a finite line charge because there is not enough symmetry to find a Gaussian surface from which to calculate the flux.

71. (a) toward; (b) $+Q_0$; (c) $-Q_0$; (d) $+Q_0$; (e) 0

(f)

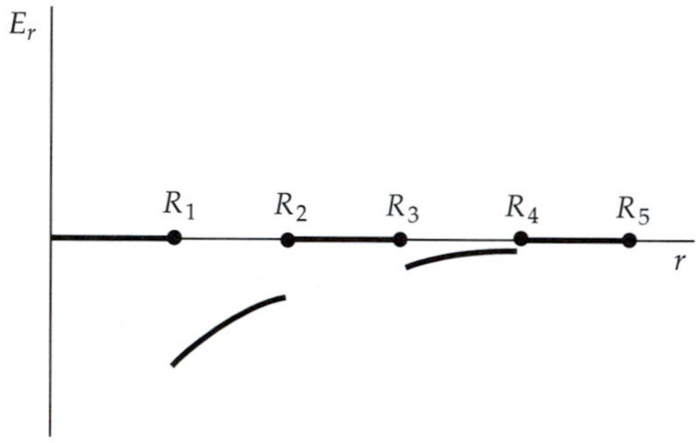

73. 1.15×10^5 N/C

75. (a) -1.41×10^5 N/C $\hat{i}$; (b) -5.37×10^5 N/C $\hat{i}$; (c) 1.41×10^5 N/C $\hat{i}$

77. (a) 1.27×10^3 N/C $\hat{i} + 1.47 \times 10^3$ N/C $\hat{j}$; (b) -1.27×10^3 N/C $\hat{i} + 5.87 \times 10^3$ N/C $\hat{j}$

79. (a) $3\sqrt{3}\,gR^2/k|Q|$; (b) $1.896R$

81. $(2kQ\lambda/L)\ln[(d + L)/d]$

83. (a) $(1.13\hat{i} + 1.69\hat{j}) \times 10^5$ N/C; (b) $(2.43\hat{i} + 2.17\hat{j}) \times 10^5$ N/C

85. $2\pi(mR^2/2k|q\lambda|)^{1/2}$

87. 7.42 rad/s

89. (a) 1.19×10^5 N/C; (b) 1.03×10^5 N/C; (c) 1.12×10^3 N/C

91. 3.27×10^{-8} C/m

93. (a) $k\lambda^2\ln[(d + L)^2/d(2L + d)]$
(b) Answer given in the problem.

95. (a) $12Q/L^3$; (b) $dF = [2ka\lambda y^2/(y + b)]\,dy$; (c) $F = 2ka\lambda(L/2 + d)\{(L/2 + d)\ln[(d + L)/d] - L\}$

Chapter 24

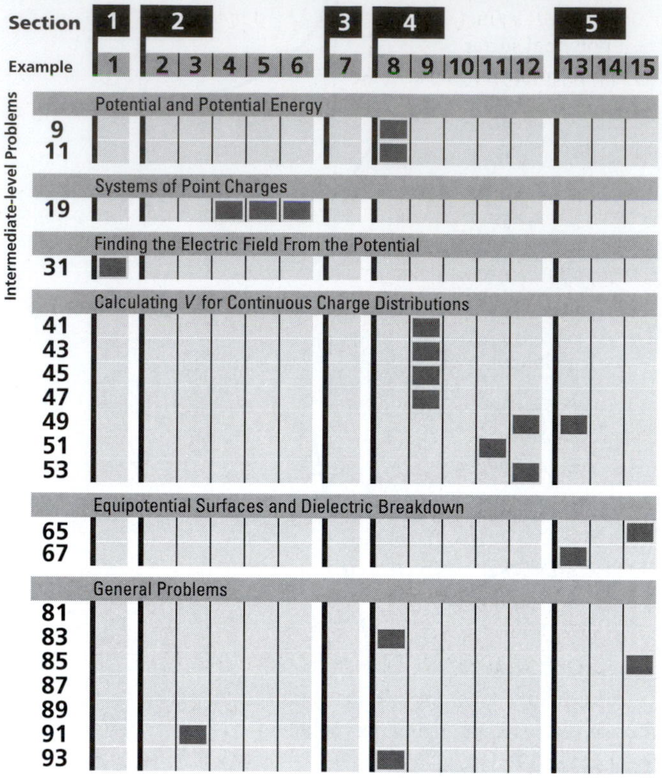

1. (a) -8000 V; (b) -2.4×10^{-2} J; (c) 2.4×10^{-2} J; (d) $(-2\ kV/m)x$; (e) 4000 V $-$ (2 kV/m)x; (f) 2000 V $-$ (2 kV/m)x

3. (a) 5 kV/m, positive; (b) 500 eV, 8×10^{-17} J; (c) -500 eV, 500 eV

5. smaller

7. (a) 4.5×10^3 V; (b) 1.35×10^{-2} J; (c) 1.35×10^{-2} J

9. Answer given in the problem.

11. (a) 30,000 eV; (b) 4.8×10^{-15} J; (c) 1.03×10^8 m/s

13. (a) 25.4 kV; (b) 12.7 kV; (c) 0

15. (a) 12,000 V; (b) 0.0599 J; (c) 0, 0

17. $-1/2$

19. (a)

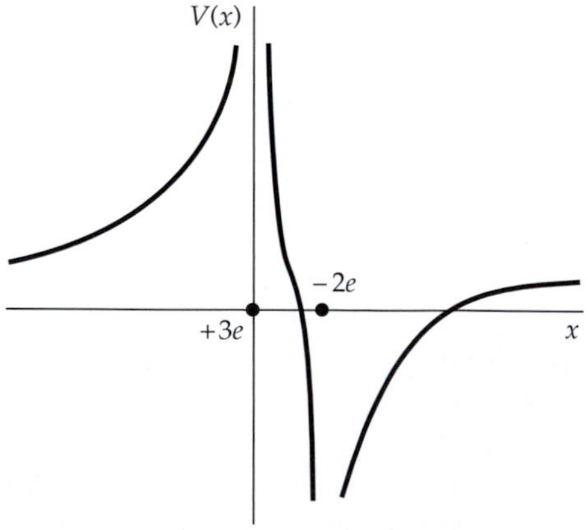

(b) $3a/5$; (c) $2ke^2/a$

21. No

23. (*a*) positive; (b) 25,000 V/m

25. (*a*) 8.99×10^3 V, 8.96×10^3 V; (*b*) decreases, 2.97×10^3 V/m; (*c*) 3.00×10^3 V/m; (*d*) 8.99×10^3 V. The displacement Δy is on an equipotential surface.

27. (*a*) positive; (*b*) 25,000 V/m

29. $-2C_2x$, $-\hat{i}$ for $x > 0$ and $\hat{i}$ for $x < 0$

31. -7500 V

33. $(-8\hat{i} - 2\hat{j} - \hat{k})$ V/m

35. No, both V and E_x would be the same if Q were not uniformly distributed.

37. (*a*) 6.24×10^3 V/m, 0; (*b*) 749 V, 749 V; (*c*) 749 V, 0

39. (*a*) 6.02×10^3 V; (*b*) -1.27×10^4 V; (*c*) -4.23×10^4 V

41. (*a*) $\frac{kQ}{L}\ln\left(\frac{\sqrt{x^2 + L^2/4} + L/2}{\sqrt{x^2 + L^2/4} - L/2}\right)$
(*b*) Answer given in the problem.

43. (*a*) $\pi\sigma_0 R^2/2$; (*b*) $\sigma_0 k$

45. (*a*) $2\pi\sigma_0 k[2\sqrt{x^2 + R^2/2} - x - \sqrt{x^2 + R^2}]$; (*b*) $\pi\sigma_0 kR^4/8x^3$

47. (*a*) $(kQ/L)\ln[(x + L/2)/(x - L/2)]$; (*b*) Answer given in the problem.

49. $-2kq/L\ln(b/a)$

51. (*a*) zero for $0 < x < a$, $-(\sigma/\epsilon_0)(x - a)$ for $x > a$, $\sigma x/\epsilon_0$ for $x < 0$; (*b*) $-\sigma x/\epsilon_0$ for $0 < x < a$, $-\sigma a/\epsilon_0$ for $x > a$, zero for $x < 0$

53. (*a*) Qr^3/R^3, kQr^2/R^3; (*b*) $3kQr'\,dr'/R^3$; (*c*) $(3kQ/2R^3)(R^2 - r^2)$; (*d*) $(kQ)/2R)(3 - r^2/R^2)$

55. (*c*)

57.

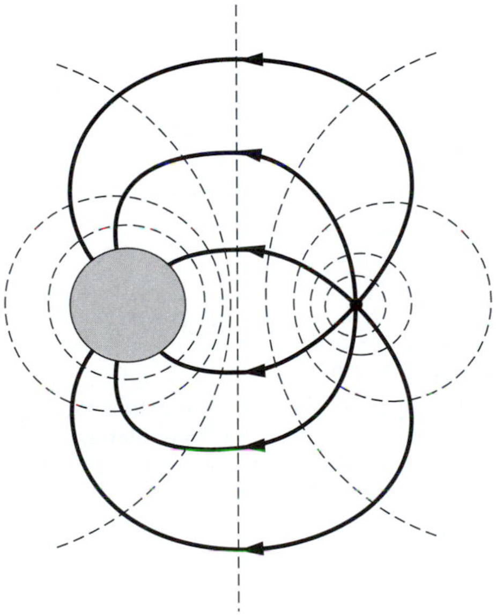

59.

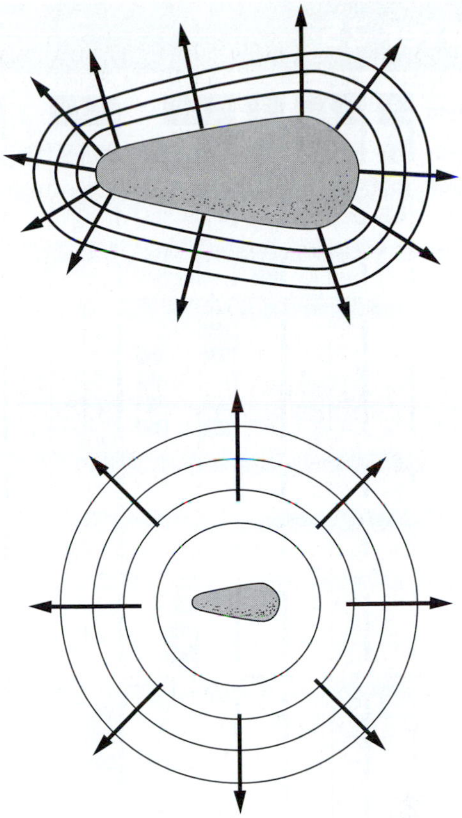

61. 5.06×10^{-4} m

63. (*a*) ± 8.54 N/C; (*b*) $\pm 4.80 \times 10^5$ V

65. The smaller sphere has the larger field by a factor of 2.

67. $kq(1/a - 1/b)$

69. (*b*)

71. (*c*)

73. (*a*) No; (*b*) Yes, yes

75. (*a*) kg/Cs^2; (*b*) $q_0ax^2/2$; (*c*) $-ax^2/2$

77. 3.33×10^{-3} m

79. 20 μC

81. 250 W

83. 1.45×10^{-7} J, 9.03×10^{11} eV

85. $2R_1/3$

87. 50 cm

89. 6000 V

91. 4.46×10^{-14} m

93. 55.1 J

95. (*a*) $kq_1/x^2 + kq_2/(x - a)^2$ for $x > a$, $kq_1/x^2 - kq_2/(a - x)^2$ for $0 < x < a$, $-kq_1/x^2 - kq_2/(x - a)^2$ for $x < 0$; (*b*) $kq_1/|y| + kq_2/(y^2 + a^2)^{1/2}$; (*c*) $\pm kq_1/y^2 + kq_2y/(y^2 + a^2)^{3/2}$ (positive for $y > 0$ and negative for $y < 0$)

97. (*a*) $V(a) = kQ(1/b - 1/c)$, $V(b) = kQ(1/b - 1/c)$, $V(c) = 0$; (*b*) $V(a) = 0$, $V(b) = kQ(b - a)(c - b)/b^2(c - a)$, $V(c) = 0$, $Q_a = -Q\frac{a}{b}\left(\frac{c - b}{c - a}\right)$, $Q_b = Q$, $Q_c = -Q\frac{c}{b}\left(\frac{b - a}{c - a}\right)$

Chapter 25

Section	1	2	2	3	4	4	4	5	5	5	6
Example	1	2	3	4	5	6	7	8	9	10	11
Electrostatic Potential Energy											
5	X										
Storage of Electrical Energy											
23				X							
Combinations of Capacitors											
31					X						
33					X		X				
35							X				
37					X		X				
39					X		X				
Parallel-Plate Capacitors											
43											
Spherical Capacitors											
49											
Disconnected and Reconnected Capacitors											
53						X					
55						X					
57						X					
Dielectrics											
63								X			
65								X			
67								X			
69											
71									X	X	
General Problems											
79											
81								X			
83				X							
85											
87					X	X					
89											
91								X			
93											
95								X			
97								X	X		
99									X		

Intermediate-level Problems

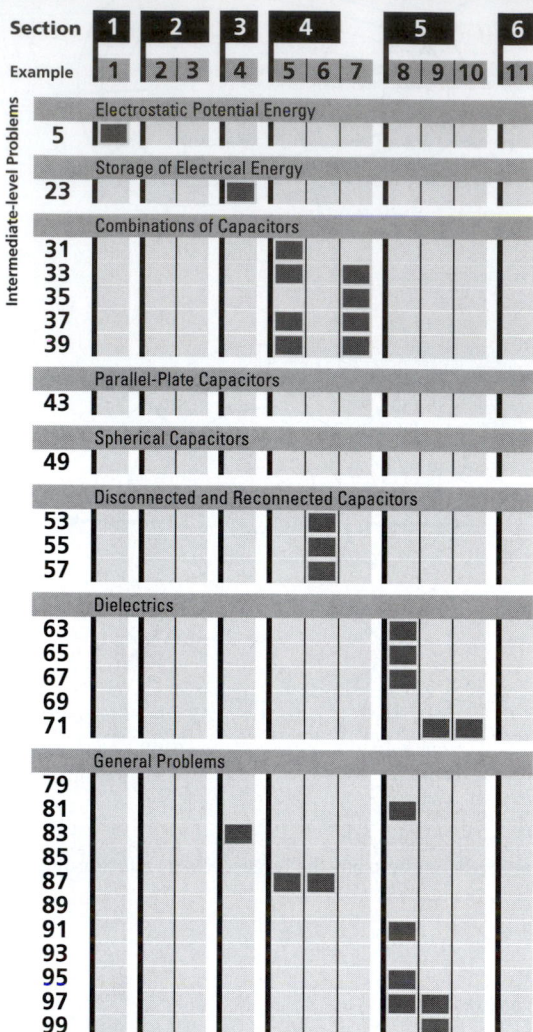

1. (*a*) 3.00×10^{-2} J; (*b*) -5.99×10^{-3} J; (*c*) -1.80×10^{-2} J

3. 2.23×10^{-5} J

5. $(6\sqrt{2}\,k/ma)^{1/2}\,q$

7. (*c*)

9. (*a*) 2.22×10^{-8} C; (*b*) 1.11×10^{-11} F; (*c*) no change

11. (*a*) 1.69×10^{7} m^2; (*b*) 4117 m

13. 2

15. (*d*)

17. (*a*) 1.5×10^{-2} J; (*b*) 4.5×10^{-2} J

19. (*a*) 0.625 J; (*b*) 1.875 J

21. (*a*) 10^5 V/m; (*b*) 4.43×10^{-2} J/m^3; (*c*) 8.85×10^{-5} J; (*d*) 1.77×10^{-8} F; (*e*) same

23. (*a*) 1.11×10^{-8} C; (*b*) 5.53×10^{-7} J

25. (*a*) True; (*b*) True

27. (*a*) 100; (*b*) 10 V; (*c*) 10^{-5} C

29. $C_2 + C_1C_3/(C_1 + C_3)$

31. (*a*) 4.00×10^{-5} C; (*b*) $V_{10} = 4$ V, $V_{20} = 2$ V

33. (*a*) 15.2 μF; (*b*) $Q_{12} = 2400\ \mu$C, $Q_4 = Q_{15} = 632\ \mu$C; (*c*) 0.303 J

35. (*a*) 0.242 μF; (*b*) $Q_{0.30} = 2.42\ \mu$C, $Q_{1.0} = 1.94\ \mu$C, $Q_{0.25} = 0.484\ \mu$C; (*c*) 1.21×10^{-5} J

37. (*a*) 5.0 μF; (*b*) 133 V

39. 3 μF, 5 μF, 6 μF, 7 μF (parallel); (4/7) μF, (4/5) μF, (4/3) μF, (2/3) μF (series); (6/7) μF, (10/7) μF, (12/7) μF (1 in series, 2 in parallel); (14/5) μF, (7/3) μF, (14/3) μF (2 in series, 1 in parallel)

41. (*a*) 40 V; (*b*) 4.24 m

43. (*a*) 0.333 mm; (*b*) 3.77 m^2

45. (*a*) 1.55×10^{-12} F; (*b*) 1.55×10^{-8} C/m

47. 1.79×10^{-10} F/m

49. (*a*) kQ/r^2 between spheres, zero elsewhere, $u_e = \epsilon_0 k^2Q^2/2r^4$; (*b*) $(kQ^2/2r^2)\,dr$; (*c*) $\frac{1}{2}kQ^2\,(R_2 - R_1)/R_1R_2$

51. $5R/3$

53. (*a*) 2.0 kV; (*b*) 0

55. (*a*) 2.4 μF; (*b*) 3.60×10^{-4} J

57. (*a*) 6 V; (*b*) 1.15×10^{-3} J, 2.88×10^{-4} J

59. (*a*) 200 V, 200 V, 200 V; (*b*) $Q_1 = -533\ \mu$C, $Q_2 = -133\ \mu$C, $Q_3 = 267\ \mu$C; (*c*) $V_1 = -267$ V, $V_2 = -133$ V, $V_3 = 400$ V

61. 2.71 nF

63. (*a*) $kQ/\kappa r^2$ between spheres, zero elsewhere, $u_e = \epsilon_0 k^2Q^2/2\kappa^2 r^4$; (*b*) $(kQ^2/2\kappa r^2)dr$; (*c*) $kQ^2(R_2 - R_1)/2\kappa R_1 R_2$

65. $4\kappa_1\kappa_2 C_0/(3\kappa_1 + \kappa_2)$

67. (*a*) 1.67×10^{-8} F; (*b*) 1.17×10^{-9} C; (*c*) 7×10^6 V/m

69. (*a*) 2.08; (*b*) 45.2 cm^2; (*c*) 5.2 nC

71. (*a*) 2.5×10^4 V/m, 6.64×10^{-7} J; (*b*) 6.25×10^3 V/m; (*c*) 25 V; (*d*) 1.66×10^{-7} J

73. (*a*) False; (*b*) False; (*c*) False

75. (*a*) 14 μF; (*b*) (8/7) μF

77. 1.0 mm

79. $C_1C_4 = C_2C_3$

81. $2/(\kappa + 2)$

83. (*a*) $\epsilon_0 A/2d$; (*b*) 2 V; (*c*) $\epsilon_0 AV^2/d$; (*d*) $\epsilon_0 AV^2/2d$

85. For a balloon with a radius of 3 m, $C = 0.3$ nF.

87. (*a*) 40 V; (*b*) 1.49×10^{-5} m^2; (*c*) 6

89. 6.67 cm

91. (*a*) $Q/\kappa_1 A\epsilon_0$, $Q/\kappa_2 A\epsilon_0$; (*b*) $(Q/2C_0)\,(1/\kappa_1 + 1/\kappa_2)$; (*c*), (*d*) Answer given in the problem.

93. (*a*) Since the capacitor plates are conductors, the potentials are the same across the entire upper and lower plates. So the system is equivalent to two capacitors in parallel, each of area $A/2$. (*b*) Answer given in the problem.

95. (*a*) $(\epsilon_0 b/d)[(\kappa - 1)\,x + a]$; (*b*) Answer given in the problem.

97. 2.55×10^{-6} J

99. (*a*) $(\epsilon_0 bV^2/2d)[(\kappa - 1)x + a]$; (*b*) $(\epsilon_0 bV^2/2d)(\kappa - 1)$

101. (*a*) 2.51×10^3 m^3; (*b*) 5.02×10^{-2} m^3

103. (*a*) $2kQ/\kappa L\,\ln(b/a)$; (*b*) $Q/2\pi aL$, $-Q/2\pi bL$; (*c*) $-Q(\kappa - 1)/2\pi aL\kappa$, $Q(\kappa - 1)/2\pi bL\kappa$; (*d*) $kQ^2/\kappa L\,\ln(b/a)$; (*e*) $(kQ^2/L\kappa)\,(\kappa - 1)\,\ln(b/a)$

105. 3.89×10^{-11} F

107. 0.1 μF, 32 μC

109. (*a*) 0.001 J; (*b*) 47.6 μC, 152 μC; (*c*) 4.76×10^{-4} J

111. (*a*) $C(V) = C_0[1 + (\kappa\epsilon_0\,V^2/2Yd^2)]$ if $\Delta x \ll d$; $C(V) = 0.133(1 + 6.64 \times 10^{-11}\,V^2)\mu$ F/m^2; (*b*) 7.97 kV; (*c*) 99.8%, 0.2%

113. (*a*) 5×10^{-4} J; (*b*) 1.59×10^{-4} N·m

Chapter 26

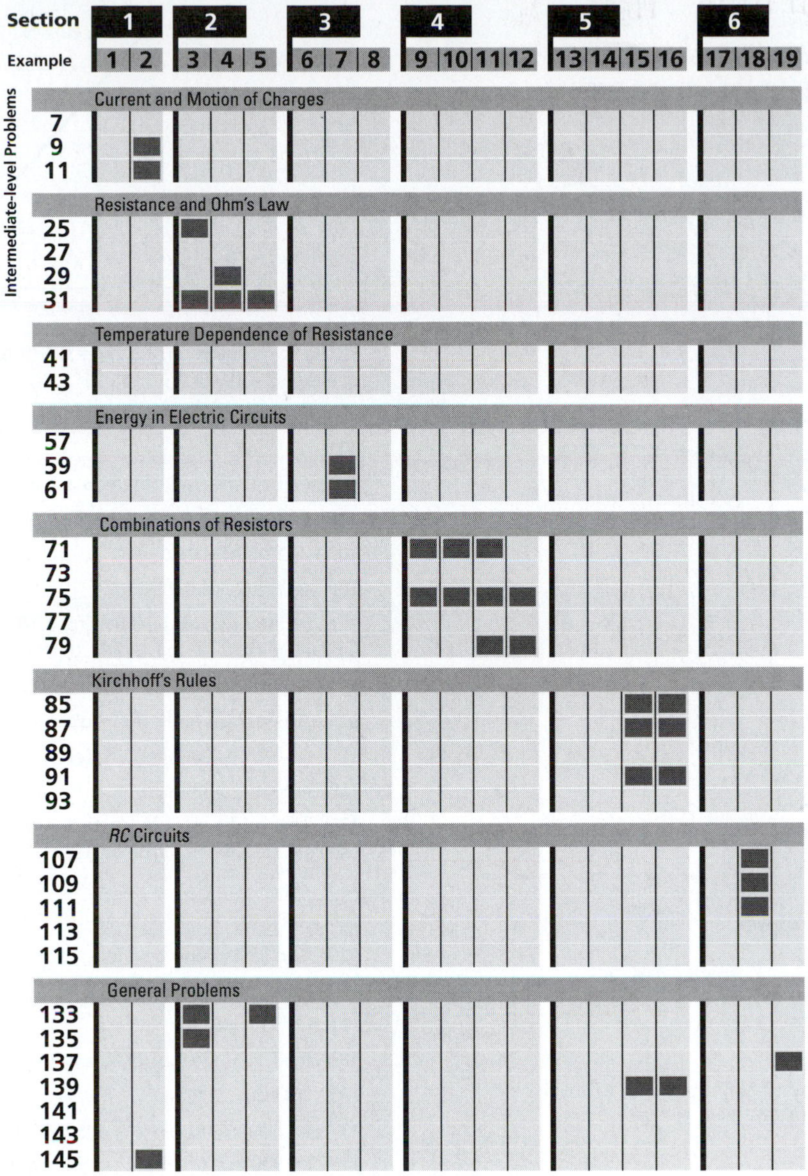
Section
1
2
3
4
5
6
Example
1 2
3 4 5
6 7 8
9 10 11 12
13 14 15 16
17 18 19
Intermediate-level Problems
Current and Motion of Charges
7
9
11
Resistance and Ohm's Law
25
27
29
31
Temperature Dependence of Resistance
41
43
Energy in Electric Circuits
57
59
61
Combinations of Resistors
71
73
75
77
79
Kirchhoff's Rules
85
87
89
91
93
RC Circuits
107
109
111
113
115
General Problems
133
135
137
139
141
143
145

1. Electric fields inside a conductor cause free electrons in the conductor to move; the conductor is no longer in static equilibrium.

3. 2.81×10^{-4} m/s

5. (*a*) 5.93×10^{7} m/s; (*b*) 3.73×10^{-5} A

7. $\lambda \omega a$

9. (*a*) 3.21×10^{13} protons/m^3; (*b*) 3.75×10^{17} protons; (*c*) (10^{-3} C/s)t

11. (*a*) 1.04×10^{8} protons/m; (*b*) 1.04×10^{14} protons/m^3

13. (*e*)

15. battery (chemical), generator (mechanical)

17. (*b*)

19. (*a*) 33.3 Ω; (*b*) (3/4) A

21. 8.98 mm

23. 1.95 V

25. 5.88×10^{17} m

27. (*a*) 5.88; (*b*) iron

29. (*a*) 79.6 Ω; (*b*) 318 Ω

31. (*a*) 0.323 Ω; (*b*) 9.69×10^{-2} V/m; (*c*) 2.38×10^{5} s

33. (*a*) 1.28 Ω; (*b*) 0.029 Ω

35. $\rho L / \pi ab$

37. (*a*) $(\rho/2\pi L)\ln(b/a)$; (*b*) 2.05 A

39. 45.6°C

41. (*a*) 15 A; (*b*) 11.1 Ω; (*c*) 1.29 kW

43. 1750 K

45. (*d*)

47. decrease

49. (*a*) 2.88 kW; (*b*) 1.44 kW

51. (*a*) 57.6 Ω, 4.17 A; (*b*) 250 W

53. 0.03 Ω

55. $77.76

57. (*a*) 8 min 13 s; (*b*) 52 min 13 s

59. (*a*) 6.91×10^6 J; (*b*) 12 h 48 min

61. (*a*) 26.7 kW; (*b*) 5.76×10^6 C; (*c*) 6.91×10^7 J; (*d*) 57.6 km; (*e*) 3 cents/km

63. (*b*)

65. (*e*)

67. (*a*) 4.5 Ω; (*b*) $I_3 = 2.67$ A, $I_6 = 0.67$ A, $I_2 = 2.0$ A

69. (*a*) 16 V; (*b*) 2 A

71. (*a*) I_2 (series) = 2.14 A, $I_4 = 0.43$ A, I_2 (parallel) = 0.86 A; (*b*) 12.8 W

73. (*a*) $R_{eq} = xR_1/(1 + x)$
(*b*)

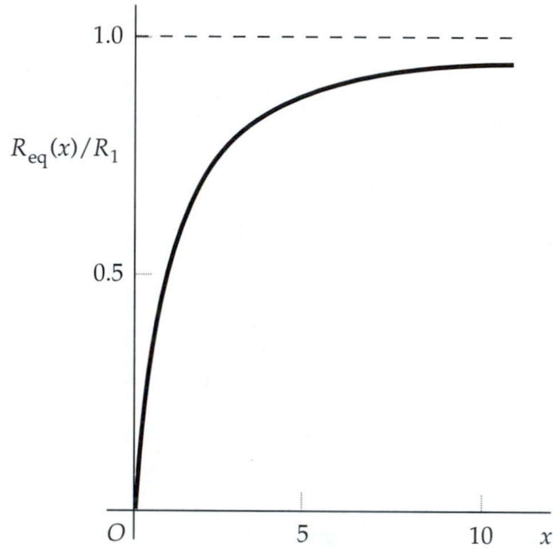

75. (*a*) 4.10 Ω; (*b*) $I_6 = 1.43$ A, I_4 (upper) = 0.86 A, $I_{2,4} = 0.57$ A, I_4 (lower) = 1.50 A, $I_8 = 0.75$ A

77. 11.3 Ω

79. Answer given in the problem.

81. (*b*)

83. (*a*) 1 A; (*b*) 12 W delivered by 12-V battery, 6 W absorbed by 6-V battery; (*c*) 2 W in the 2 Ω, 4 W in the 4 Ω

85. 600 Ω

87. (*a*) $I_4 = \frac{2}{3}$ A, $I_3 = \frac{8}{9}$ A, $I_6 = \frac{14}{9}$ A; (*b*) $\frac{28}{3}$ V; (*c*) 8 W (left battery), $\frac{32}{3}$ W (right battery)

89. Greater when $R < r$ (parallel) and $R > r$ (series)

91. 2.4 V

93. (*a*) For $R > 0.4$ Ω connect in series and for $R < 0.4$ Ω connect in parallel; (*b*) 10.7 A; (*c*) 6.67 A; (*d*) 5.45 A; (*e*) 4.44 A

95. (*a*) $I_1 = -57.0\text{ A} + (10\text{ A/h})t$, $I_2 = 63.0\text{ A} - (10\text{ A/h})t$
(*b*)

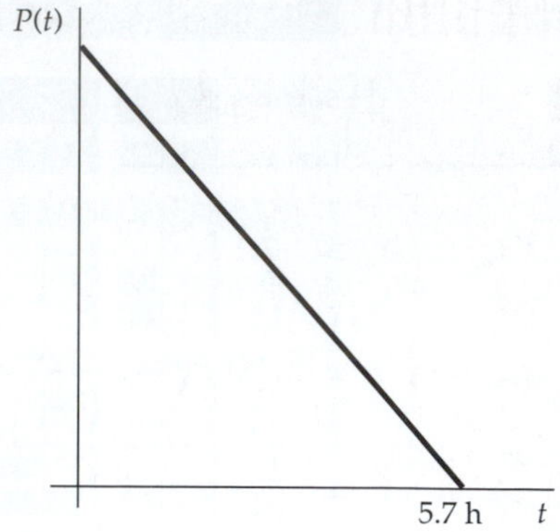

97. I_3 (parallel) = 2.13 A, $I_4 = 0.678$ A, $I_5 = 0.735$ A, $I_2 = 2.807$ A, I_3 (series) = 0.057 A

99. (*b*)

101. (*e*)

103. (*a*) not affected; (*b*) decreases; (*c*) doubles

105. (*a*) 6×10^{-4} C; (*b*) 0.2 A; (*c*) 3×10^{-3} s; (*d*) 8.12×10^{-5} C

107. 2.18×10^6 Ω

109. (*a*) 8 μC; (*b*) 0.0737 s

111. (*a*) 5.69 μC; (*b*) 1.10 μA; (*c*) 1.10 μA; (*d*) 6.62×10^{-6} W; (*e*) 2.44×10^{-6} W; (*f*) 4.19×10^{-6} W

113. (*a*) 25 A; (*b*) 0.4 Ω, 10 Ω, 6.67 Ω

115. Answer given in the problem.

117. (*a*) 0.25 A; (*b*) 0.0625 A; (*c*) $I_{600} = 0.0625(1 - e^{-t/0.00075})$

119. (*a*) 0; (*b*) never

121. To avoid a sizable fraction of the energy being dumped into the internal resistance of the battery

123. Yes

125. (*c*)

127. The 25-W bulb because it has the higher operating resistance

129. (*a*) 0.707 A; (*b*) 7.07 V

131. (*a*) 30 A; (*b*) 4 V

133. (*a*) 3.11 V; (*b*) 0.0779 V/m; (*c*) 18.7 W

135. 2.03 m

137. (*a*) 3.42 A; (*b*) 0.962 A; (*c*) 260 μC (10 μF), 130 μC (5 μF)

139. (*a*) 9800 Ω; (*b*) For $x = 98.2$ cm, $R_x = 10{,}911$ Ω. For $x = 97.8$ cm, $R_x = 8891$ Ω. The errors are 11.3% and 9.28% respectively. (*c*) Increase to 9800 Ω

141. 0.2 Ω

143. $359

145. (*a*) 0.05 A; (*b*) 5000 W

147. (*a*) 0.01 s; (*b*) 0.2 V + 79,980 (V/s)t; (*c*) 10^9 Ω; (*d*) 6.09×10^{-11} s; (*e*) 1.28 W, 1.76×10^4 W

149. $I_{10} = 0.740$ A, $I_{40} = 0.472$ A, $I_{30} = 0.383$ A, $I_{80} = 0.357$ A, $I_{20} = 0.855$ A

151. (*a*) $QC_1/(C_1 + C_2)$, $QC_2/(C_1 + C_2)$; (*b*) $\frac{1}{2}Q^2/C_1$, $\frac{1}{2}Q^2/(C_1 + C_2)$; (*c*) Heat generated in resistor R

153. (*a*) 0.12 A; (*b*) 0.04 A; (*c*) 8 V; (*d*) 6 V; (*e*) $0.04e^{-133.3t}$ A

155. (*a*) 10^{12} electrons; (*b*) 0.16 mA; (*c*) 6.4×10^4 W; (*d*) 6.4×10^8 W; (*e*) 10^{-4}

Chapter 27

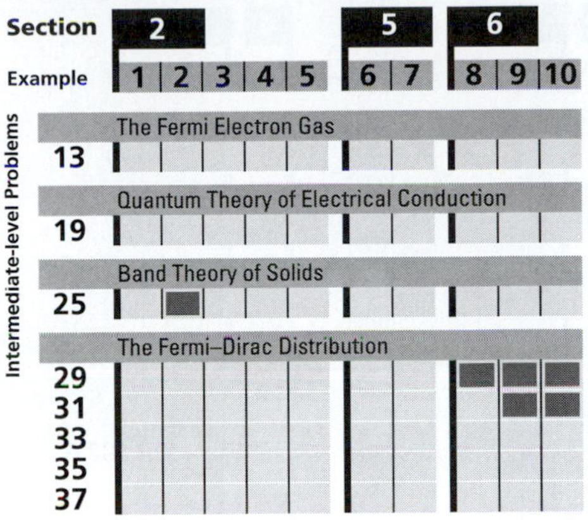

1. as heat energy

3. (*a*) $1.23 \times 10^{-7}\ \Omega{\cdot}\text{m}$; (*b*) $7.10 \times 10^{-8}\ \Omega{\cdot}\text{m}$

5. 3 electrons/Al atoms

7. (*a*) 1.36×10^5 K; (*b*) 2.46×10^4 K; (*c*) 1.18×10^5 K

9. (*a*) 11.7 eV; (*b*) 2.12 eV; (*c*) 10.2 eV

11. (*a*) 1.30×10^5 K; (*b*) 11.2 eV

13. (*a*), (*b*) Answer given in the problem.
(*c*) $6.36 \times 10^{10}\ \text{N/m}^2$, less than half the measured value. In this calculation, the electron gas is treated as an ideal gas and thus the effect of electron-electron interactions are not taken into account.

15. (*a*) Au, Ag; (*b*) 0.1 eV

17. The impurity effect in brass is higher.

19. (*a*) 66.1 nm; (*b*) $1.77 \times 10^{-20}\ \text{m}^2$

21. (*b*)

23. 1.68×10^{-6} m

25. (*a*) 0.37 eV; (*b*) 4300 K

27. (*a*) 2.17×10^{-3} eV; (*b*) 0.455 mm

29. Answer given in the problem.

31. Approximately 1.0

33. Answer given in the problem.

35. 60%

37. Answer given in the problem.

39. Answer given in the problem.

41. (*a*) 4.57×10^{45} neutrons/m^3; (*b*) 5.48×10^8 eV

43. The resistivity of copper increases and the resistivity of silicon decreases.

45. (*a*) 8.62×10^{22} electrons/cm^3; (*b*) 13.1×10^{22} electrons/cm^3

47. (*a*) 3.12×10^{15} pairs/s; (*b*) 3.12×10^{15} pairs/s; (*c*) 8×10^{-4} W

Chapter 28

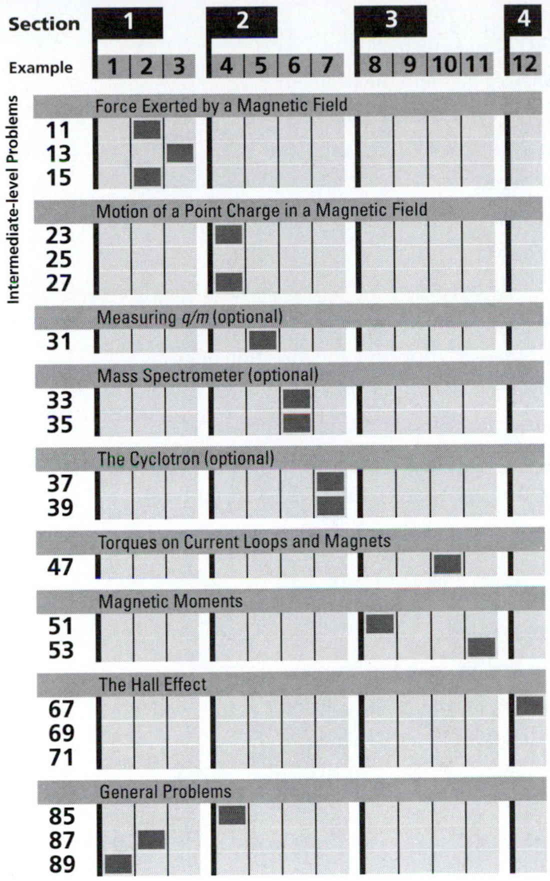

1. (*b*)

3. $-1.25 \times 10^{-12}\ \text{N}\ \hat{j}$

5. (*a*) $-6.39 \times 10^{-13}\ \text{N}\ \hat{j}$; (*b*) $8.76 \times 10^{-13}\ \text{N}\ \hat{i}$; (*c*) 0;
(*d*) $(7.10\hat{i} - 9.47\hat{j}) \times 10^{-13}$ N

7. 0.962 N

9. $(-1.92\hat{i} - 1.28\hat{j} - 5.76\hat{k}) \times 10^{-13}$ N

11. 1.48 A

13. Answer given in the problem.

15. $10\ \text{T}\ \hat{i} + 10\ \text{T}\ \hat{j} - 15\ \text{T}\ \hat{k}$

17. False

19. (*a*) 8.74×10^{-8} s; (*b*) 4.67×10^7 m/s; (*c*) 1.82×10^{-12} J

21. (*a*) 142 m; (*b*) 2.85 m

23. (*a*) $v_p/v_\alpha = 2$; (*b*) $K_p/K_\alpha = 1$; (*c*) $L_p/L_\alpha = 1/2$

25. Answer given in the problem.

27. (*a*) 12.6×10^6 m/s, 24°; (*b*) 6.28×10^6 m/s, 24°

29. (*a*) 1.64×10^6 m/s; (*b*) 14.1 keV; (*c*) 7.68 eV

31. (*a*) 7.35 mm; (*b*) 6.64×10^{-5} T

33. (*a*) 63.5 cm; (*b*) 2.59 cm

35. 1.57×10^{-5} s, 1.63×10^{-5} s

37. (*a*) 2.13×10^7 Hz; (*b*) 46.0 MeV; (*c*) *f* and *K* are reduced by a factor of 2.

39. Answer given in the problem.

41. If the plane of current loop is parallel to the field.

43. 2.83×10^{-5} N·m

45. (*a*) 0; (*b*) 2.08×10^{-3} N·m

47. (*a*) 37° below the *x*-axis; (*b*) $0.799\hat{i} - 0.602\hat{j}$; (*c*) $[0.335\hat{i} - 0.253\hat{j}]$ A·m²; (*d*) 0.503 N·m $\hat{k}$; (*e*) 0.380 J

49. Answer given in the problem.

51. 0.377 A·m² into the page

53. Answers given in the problem.

55. 0.02 A·m²

57. Answers given in the problem.

59. $\frac{4}{3}\pi\sigma\omega R^4$

61. $2^{-1/4}R$

63. $(\pi\omega L\rho_0/4)(2R_s^4 - R_i^4 - R_o^4)$

65. $(\pi/4)\omega\rho_0 R^5$, opposite the direction of $\vec{\omega}$.

67. (*a*) 3.69×10^{-5} m/s; (*b*) 1.48 μV

69. 1.02×10^{-3} V

71. 3.5 electrons/Al atom

73. (*a*) True; (*b*) True; (*c*) True; (*d*) False; (*e*) True

75. (*c*)

77. (*e*)

79. (*c*)

81. 8.78 N/m $\hat{k}$

83. Answer given in the problem.

85. Answers given in the problem.

87. (*a*) $(mg/I\ell)\tan\theta$; (*b*) $g\sin\theta$

89. (*a*) 1.6×10^{-18} N $\hat{j}$; (*b*) 10 V/m $\hat{j}$; (*c*) 20 V

91. 5.10 m

93. $2\pi(M/\pi IB)^{1/2}$

Chapter 29

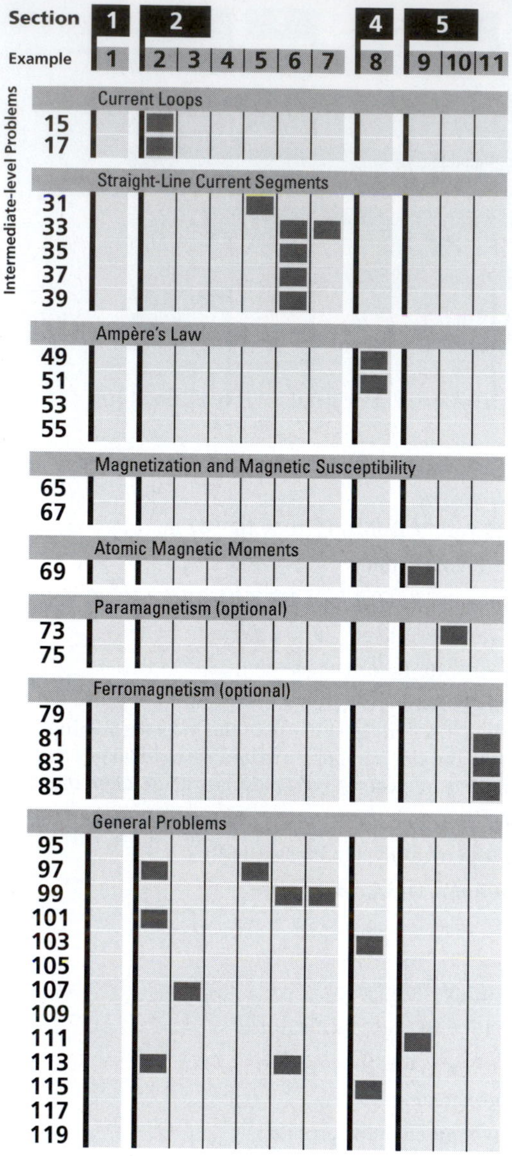

1. (*a*) opposite; (*b*) same

3. (*a*) 1.27×10^{-11} T $\hat{k}$; (*b*) 0; (*c*) 3.22×10^{-12} T $\hat{k}$

5. 12.5 T

7. (*a*), (*b*)

9. -9.6×10^{-12} T $\hat{i}$

11. No

13. 11.1 A

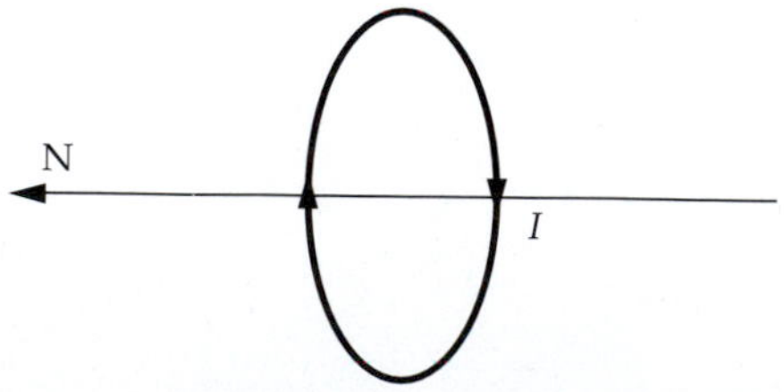

15. 9.47 A

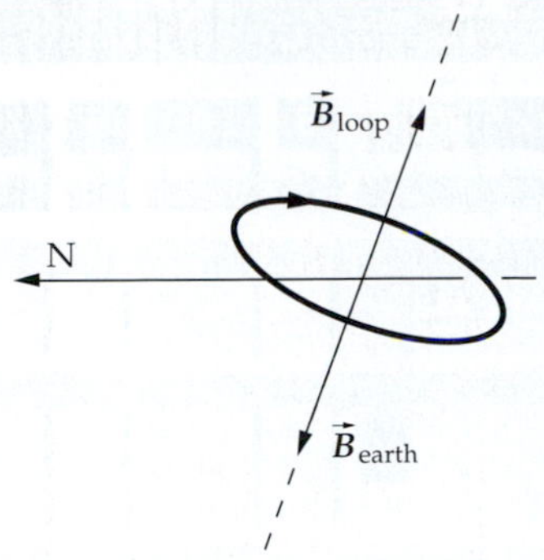

17. (*a*) 0.0540 T, 0.0539 T, 0.0526 T, 0.0486 T

(*b*)

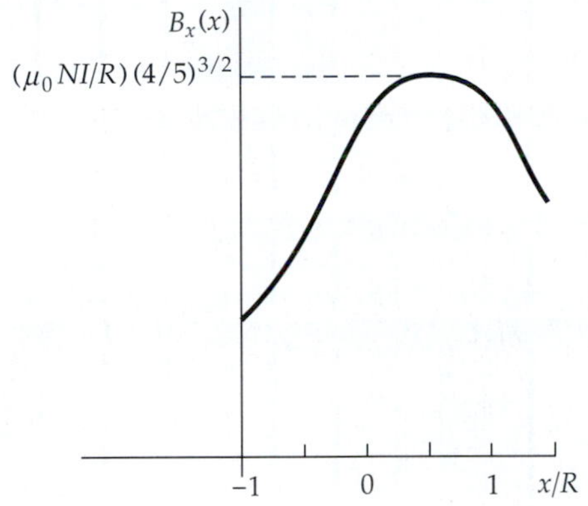

19. (*c*)

21. (*c*)

23. (*a*) 2×10^{-5} T; (*b*) 4×10^{-6} T; (*c*) 10^{-6} T

25.

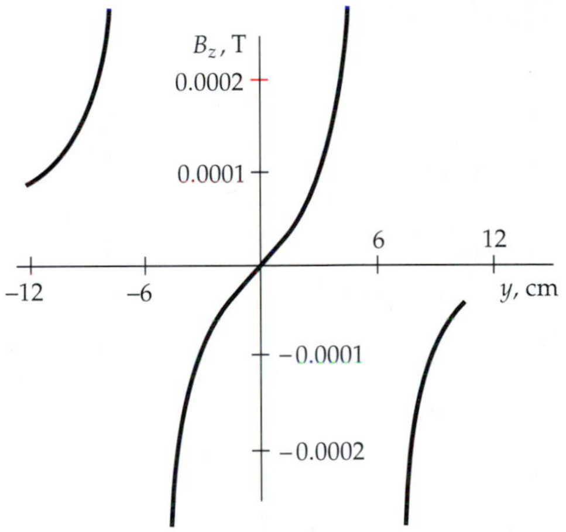

27.

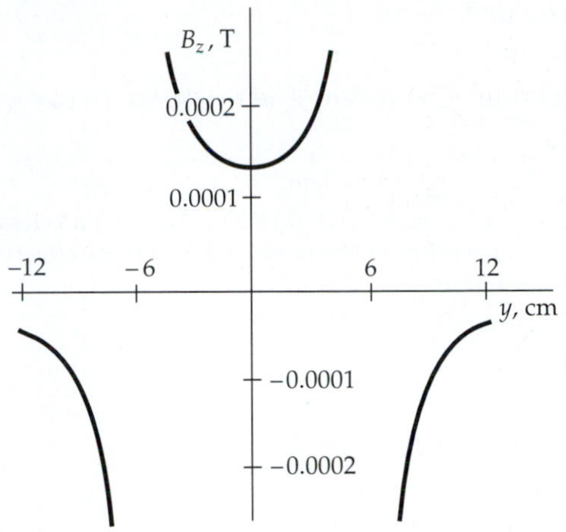

29. 6.67×10^{-4} N/m

31. $B_{1\text{ cm}} = 56.6\ \mu$T, into the page; $B_{2\text{ cm}} = 113\ \mu$T, into the page; $B_{total} = 226\ \mu$T

33. (*a*) 7.79×10^{-4} N/m, upward; (*b*) 5.20×10^{-5} T, to the right

35. Along line $y = x$ except at $x = y = 0$ the field is not defined.

37. (*a*) $(3\mu_0 I/4\pi L)(\hat{i} - \hat{j})$; (*b*) $(\mu_0 I/4\pi L)(\hat{i} - \hat{j})$; (*c*) $(\mu_0 I/4\pi L)(-\hat{i} - 3\hat{j})$

39. $[(1 + \sqrt{2})\mu_0 I/2\pi R]\hat{i}$

41. 6.98×10^{-4} T

43. Answers given in the problem.

45. (*e*)

47. (*a*) C_1: 1.01×10^{-5} T·m, C_2: 0, C_3: -1.01×10^{-5} T·m; (*b*) none

49. (*a*) 8×10^{-4} T; (*b*) 4×10^{-3} T; (*c*) 2.86×10^{-3} T

(*d*)

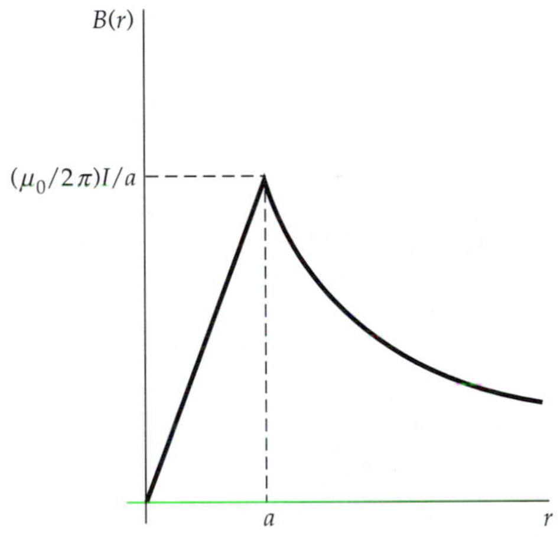

51. (*a*) 2.26×10^{-5} T·m; (*b*) 1.24×10^{-5} T·m; (*c*) 0

53. Answer given in the problem.

55. (*a*) $-\hat{i}$; (*b*) $-\hat{i}$; (*c*) $\hat{i}$; (*d*) Answer given in the problem.

57. (*a*)

59. Diamagnetic: hydrogen, carbon dioxide, nitrogen. Paramagnetic: oxygen.

61. 1.01×10^{-2} T, 0.547 A/m, 1.01×10^{-2} T

63. -4.00×10^{-5}

65. 2.38×10^{-3}

67. (*a*) 8.12×10^3 A/m; (*b*) 1.62×10^{21} electrons; (*c*) 24.4 A

69. $0.587\mu_B$

71. Answer given in the problem.

73. (*a*) 5.58×10^5 A/m, 0.701 T; (*b*) 5.23×10^{-4}; (*c*) μ is less than μ_B. Diamagnetic effects have been neglected, and these effects tend to reduce susceptibility.

75. (*a*) 95.5 A/m; (*b*) 0.0301 T; (*c*) 0.4%

77. 0.864 T, 6.87×10^5 A/m

79. 11.0 A

81. (*a*) 1.26×10^{-3} T; (*b*) 1.26×10^6 A/m; (*c*) 1250

83. (*a*) 0.096 T; (*b*) 0.96 N/m

85. 0.755 T

87. (*a*) False; (*b*) True; (*c*) False; (*d*) False; (*e*) True

89. No

91. If the current in the tube is uniform and parallel to the axis, the magnetic field is zero inside. In a solenoid the current encircles the axis.

93. 2.36×10^{-5} T

95. Answer given in the problem.

97. (*a*) $\pi\mu_0 I/\ell$; (*b*) $(8\sqrt{2}/\pi)\mu_0 I/\ell$; (*c*) $(27/2\pi)\mu_0 I/\ell$; The greatest field is produced by the triangle.

99. (*a*) The force on each of the horizontal segments is 0.251×10^{-4} N, down on the upper segment and up on the lower; the force on the left vertical segment is 1×10^{-4} N to the right, and the force on the right vertical segment is 0.286×10^{-4} N to the left.
(*b*) 0.714×10^{-4} N, to the right

101. 7.07×10^{-6} T, into the page

103. (*a*) 1.64 mT; (*b*) 1.05 mT; (*c*) 17.2 A

105. 0.246 N·m/rad, 0.523 s

107. 0.512 s

109. (*a*) 0.0524 A·m²; (*b*) 7.70×10^5 A/m; (*c*) 2.31×10^4 A

111. (*a*) 70.5 A·m²; (*b*) 17.6 N·m

113. 3.18 cm

115. (*a*) 10^{-5} T; (*b*) 10^{-5} T; (*c*) 5×10^{-6} T

117. 2.32×10^{-22} J $= 1.45 \times 10^{-3}$ eV

119. 6.26 T

121. (*a*) $\vec{B} = -\dfrac{\mu_0 I}{2\pi(R^2 - a^2)}\left[\dfrac{R}{2} - \dfrac{a^2}{2R - b}\right]\hat{j}$;

(*b*) $\vec{B} = \dfrac{\mu_0 I}{2\pi(R^2 - a^2)}\left(\dfrac{R}{2} - \dfrac{2a^2R}{4R^2 + b^2}\right)\hat{i} - \dfrac{\mu_0 I}{2\pi(R^2 - a^2)}\left(\dfrac{a^2b}{4R^2 + b^2}\right)\hat{j}$

123. $\vec{B} = 2\dfrac{\mu_0}{4\pi x^3}\left(\dfrac{I\ell^2}{[1 + (\ell^2/4x^2)][1 + (\ell^2/2x^2)]^{1/2}}\right)\hat{i}$.

In the limit $x \gg 1$, this reduces to

$$\vec{B} = 2\frac{\mu_0 I\ell^2}{4\pi x^3}\hat{i} = \frac{\mu_0 2\vec{\mu}}{4\pi x^3} \text{ where } \vec{\mu} = I\ell^2\hat{i}.$$

125. $r < R$: $B = \dfrac{\mu_0 I_0}{2\pi r}(1 - e^{-r/a})$; $r > R$: $B = \dfrac{\mu_0 I_0}{2\pi r}(1 - e^{-R/a})$

Chapter 30

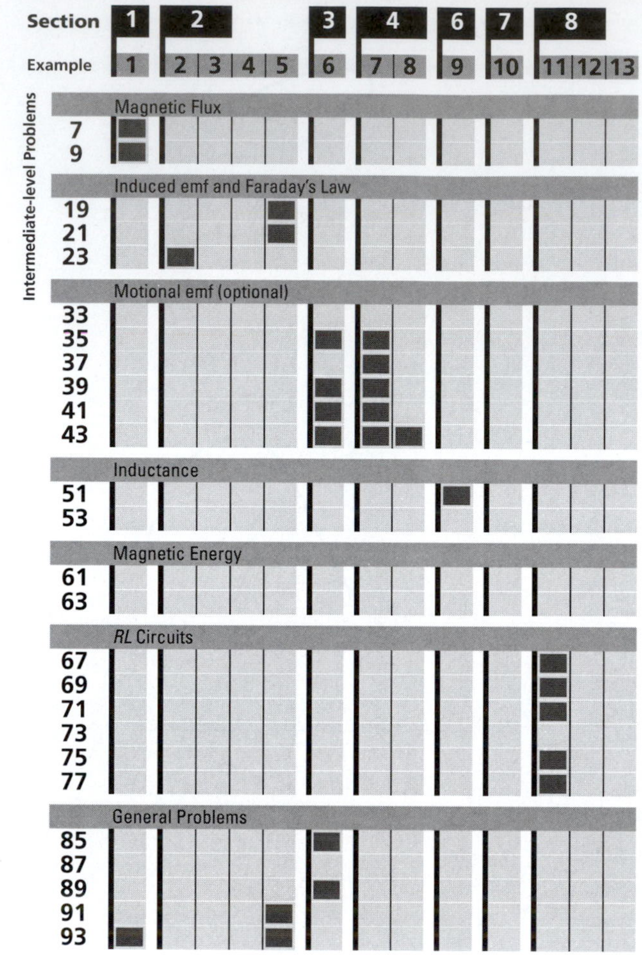

1. (*a*) 5×10^{-4} Wb; (*b*) 4.33×10^{-4} Wb; (*c*) 2.5×10^{-4} Wb; (*d*) 0

3. (*a*) 4.2×10^{-2} Wb; (*b*) 2.1×10^{-2} Wb

5. $\pi R^2 B$

7. 6.74×10^{-3} Wb

9. (*a*) $\mu_0 nNI\pi R_1^2$; (*b*) $\mu_0 nNI\pi R_3^2$

11. $\mu_0 I/4\pi$

13. (*d*)

15. (*a*) $(-0.2t + 0.4)$ V; (*b*) $t = 0$ s: 0 Wb and 0.4 V, $t = 2$ s: -0.4 Wb and 0 V, $t = 4$ s: 0 Wb and -0.4 V, $t = 6$ s: 1.2 Wb and -0.8 V

17. 1.88×10^{-2} V

19. (*a*) 1.26×10^{-3} C; (*b*) 12.6 mA; (*c*) 0.628 V

21. 0.798 G

23. (*a*) $\left(\dfrac{\mu_0}{4\pi}\right)4a \ln 3$; (*b*) 6.59 $\mu\Omega$, counterclockwise

25. (*a*) counterclockwise; (*b*) clockwise

27. (*a*)

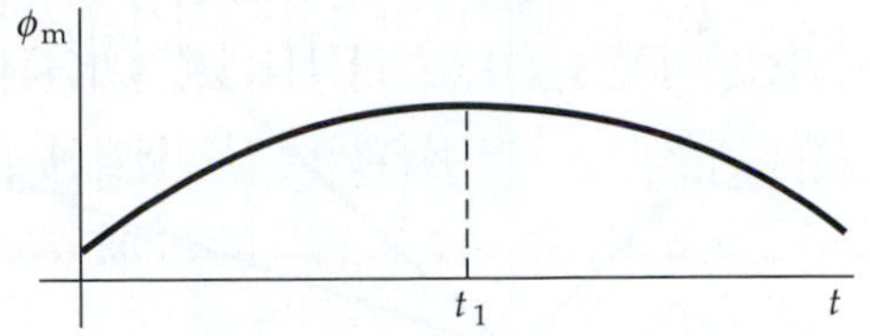

(*b*)

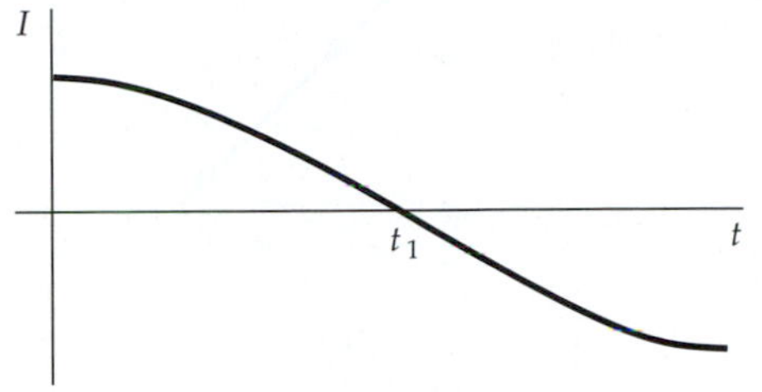

29. (*a*) 6.4×10^{-20} N; (*b*) 0.4 V/m; (*c*) 0.12 V

31. (*a*) 1.6 V; (*b*) 0.8 A; (*c*) 0.128 N; (*d*) 1.28 W; (*e*) 1.28 W

33. (*a*)

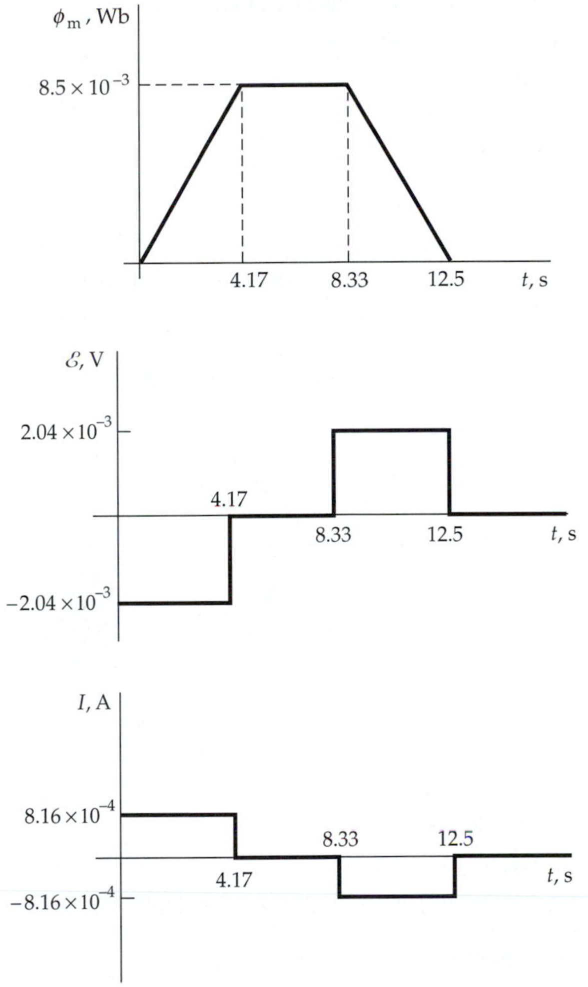

35. (*a*) $[(\mathcal{E} - B\ell v)\ell B]/R = ma$; (*b*) $|v| = \mathcal{E}/B\ell$; (*c*) 0

37. $v_0 mR/B^2\ell^2$

39. Answer given in the problem.

41. (*a*) 1.38×10^{-4} V/m; (*b*) 5.51×10^{-6} V

43. $m^2g^2R \sin^2\theta/B^2\ell^2 \cos^2\theta$

45. Answers given in the problem.

47 Answers given in the problem.

49. (*a*) 24 Wb; (*b*) -1600 V

51. (*a*) 6.03×10^{-3} T; (*b*) 7.58×10^{-4} Wb; (*c*) 253 μH; (*d*) $|\mathcal{E}| = 37.9$ mV

53. 162 Ω, 0

55. 12 mH

57. (*c*)

59. (*a*) 4.43×10^{-4} J; (*b*) 9.95×10^4 J; (*c*) 9.95×10^4 J

61. (*a*) 0.0536 J; (*b*) 447 J/m^3; (*c*) 0.0335 T; (*d*) 447 J/m^3

63. (*a*) 7958; (*b*) 2.55 T; (*c*) 2.58×10^6 J/m^3, 5093 J

65. (*a*) 13.5 mA; (*b*) 7.44×10^{-44} A

67. (*a*) 47.7 W, 48.0 W; (*b*) 47.4 W, 48.0 W; (*c*) 0.321 W, 0

69. (*a*) 2.30; (*b*) 4.61; (*c*) 6.91

71. (*a*) 2.41 s; (*b*) 20.1 s

73. Answer given in the problem.

75. 20 Ω, 4 H

77. 2.31×10^{-4} s

79. (*a*)

81. In the metal tube, the moving magnet creates eddy currents, which produce a retarding force.

83. (*a*) 7.07×10^{-3} V; (*b*) 6.64×10^{-3} V

85. (*a*) Answer given in the problem; (*b*) 275 rad/s

87. Answer given in the problem.

89. $(0.35 \sin 2t)$ A

91. Answer given in the problem.

93. (*a*) $-\frac{1}{2}\mu_0 nrI_0\omega \cos \omega t$; (*b*) $-\frac{1}{2}\mu_0 n(R^2/r)I_0\omega \cos \omega t$. In each case, the electric field is tangent to the circle of radius r. The minus sign indicates the direction of $\vec{E}$ relative to the current.

95. (*a*) $\mu_0 I/2\pi r$, no current is enclosed for $r < r_1$ and $r > r_2$; (*b*)–(*d*) Answer given in the problem

97. (*a*) $0.600v$ A, counterclockwise; (*b*) $-0.180v$ N, up ($+y$ is down); (*c*) $(4.90 - 0.180v)$ N; (*d*) $a_y = \frac{-1}{2.78}(v - 27.2)$ m/s^2; (*e*) $v = 27.2(1 - e^{-t/2.78})$ m/s; (*f*) $y = 27.2[t - 2.78(1 - e^{-t/2.78})]$ m; (*g*) 0.553 s; (*h*) 4.91 m/s; (*i*) 5.24 m/s

99. Answer given in the problem.

Chapter 31

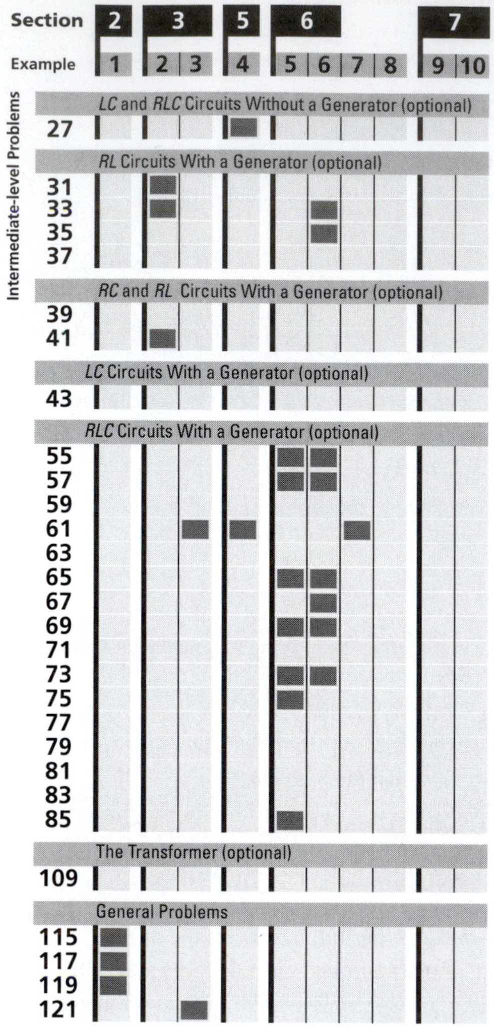
Section
2
3
5
6
7
Example
1
2 3
4
5 6 7 8
9 10
Intermediate-level Problems
LC and RLC Circuits Without a Generator (optional)
27
RL Circuits With a Generator (optional)
31
33
35
37
RC and RL Circuits With a Generator (optional)
39
41
LC Circuits With a Generator (optional)
43
RLC Circuits With a Generator (optional)
55
57
59
61
63
65
67
69
71
73
75
77
79
81
83
85
The Transformer (optional)
109
General Problems
115
117
119
121

1. (*a*) 39.8 Hz; (*b*) 15.1 V

3. (*a*) 13.6 V; (*b*) 486 Hz

5. (*b*)

7. (*a*) 0.833 A; (*b*) 1.18 A; (*c*) 200 W

9. (*a*) 21.2 A; (*b*) 1.8 kW

11. (*a*)

13. Yes, yes

15. (*a*) 0.377 Ω; (*b*) 3.77 Ω; (*c*) 37.7 Ω

17. 1.59 kHz

19. (*a*) 0.025 A; (*b*) 0.0178 A

21.

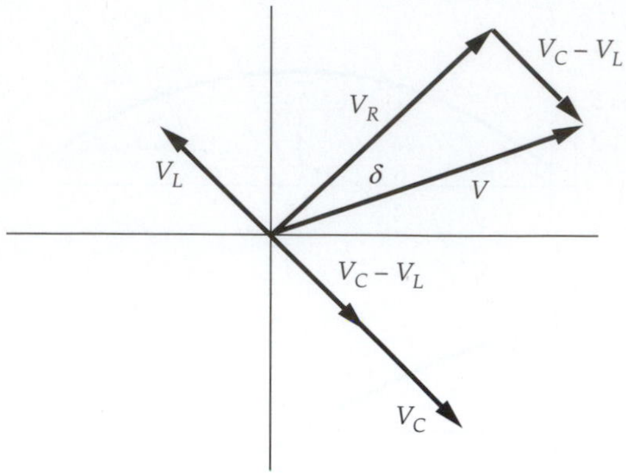

The net emf V lags the current I, which is in the same direction as V_R, by the angle δ. In the diagram, it is clear that $\tan\delta = (V_C - V_L)/V$.

23. (*a*)

25. Answer given in the problem.

27. Answer given in the problem. (*b*) third circuit

29. (*a*) 0.553; (*b*) 0.663 A; (*c*) 44 W

31. 29.2 mH

33. (*a*) 0.333; (*b*) 26.7 Ω; (*c*) 0.2 H; (*d*) lags, 70.5°

35. 0.397

37. (*a*) 8.81 A, 7.87 A, 3.95 A; (*b*) 4.64 A, 2.07 A, 4.15 A; (*c*) 0.502, 0.800

39. 60 V

41. (*a*) 6 Ω; (*b*) 35.5 mH

43. (*a*) $I_C = \omega CV$ leads voltage by 90°, $I_L = V/\omega L$ lags voltage by 90°; (*b*) 100 rad/s; (*c*) 0.25 A, 0.25 A;
(*d*)

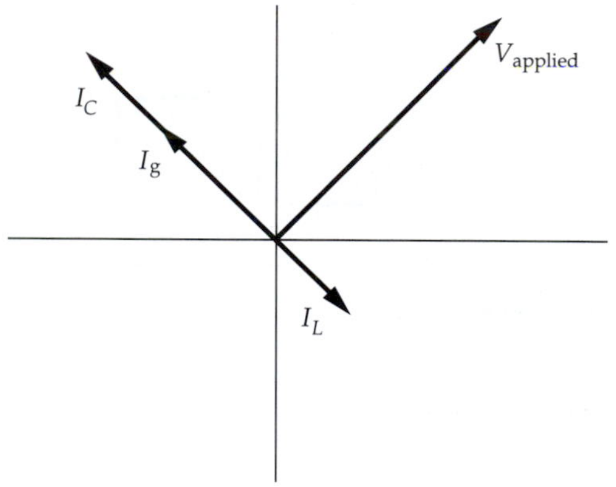

45. 1.24×10^9 N/m^2

47. (*a*) True; (*b*) True; (*c*) True

49. Yes, the freqency needs to be very precisely tuned to get a good signal.

51. 9.89×10^{-9} F to 1.01×10^{-7} F

53. (*a*) 0.539; (*b*) 95.3 mA; (*c*) 0.727 W

55. (*a*) 7.07×10^3 rad/s; (*b*) 14.1 A; (*c*) 62.5 Ω, 80 Ω; (*d*) 18.2 Ω, 3.86 A; (*e*) 74.1°

57. (*a*) 0.155, −81.1°; (*b*) 0.839, −33.0°; (*c*) 0.237, 76.3°

59. 2002

61. (*a*) 18.8 μF, 0.531 A; (*b*) 25 V

63. (*a*) Answer given in the problem.
(*b*)

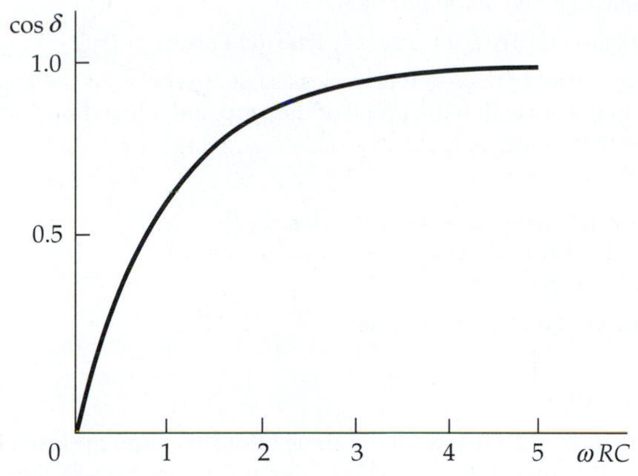

65. (*a*) 450 Hz; (*b*) 0.141; (*c*) 3.25 kHz, 62.4 Hz

67. (*a*) 933 W; (*b*) 7.71 Ω; (*c*) 99.9 μF; (*d*) 40.9 μF, 20.5 mH

69. (*a*) 120 Hz; (*b*) 356 Hz, 2.96; (*c*) 40.7 Hz, 0.338

71. (*a*) 15 W; (*b*) 15 Ω; (*c*) 0.235 H

73. (*a*) 0.8 mH, 12.5 μF; (*b*) 1.6; (*c*) 2 A

75. (*a*) 12 Ω; (*b*) 7.2 Ω of *R*, 9.6 Ω of *X*; (*c*) capacitive

77. Answers given in the problem.

79. (*a*) 40 Ω; (*b*) 41.1 Ω; (*c*) closed: $0.75\mathcal{E}$, open: $0.73\mathcal{E}$; (*d*) 23.7 Ω, 22.7 Ω, closed: $0.701\mathcal{E}$, open: $0.028\mathcal{E}$ (*e*) open S

81. Assuming the resonance to be sharp, $Q = 1/\omega RC = \omega L/R$

83. 5.66 kHz, 5.66, 1.14 kΩ

85. (*a*) 13.3 MHz; (*b*) 0.5 A, $V_L = 1.5$ kV, $V_C = 1.5$ kV; (*c*) 0.354 A, $V_L = 1.07$ kV, $V_C = 1.05$ kV

87. (*a*), (*b*) Answer given in the problem.
(*c*)

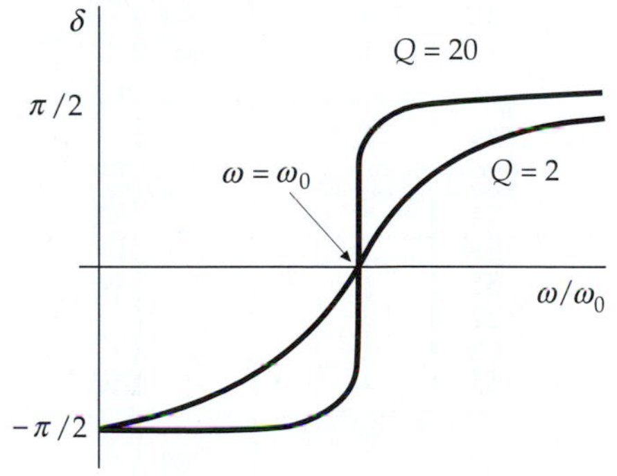

89. Answers given in the problem.

91. Answer given in the problem.

93. (*a*) 3.55×10^{-5} H; (*b*) 1.98×10^{-11} F; (*c*) 3.67×10^{-4}

95. (*a*) 19.8 MHz; (*b*) 1.10 mm

97. 7.96 kHz

99. (*a*) 1975 rad/s; (*b*) 16.9 Ω, 23.7 Ω;
(*c*)

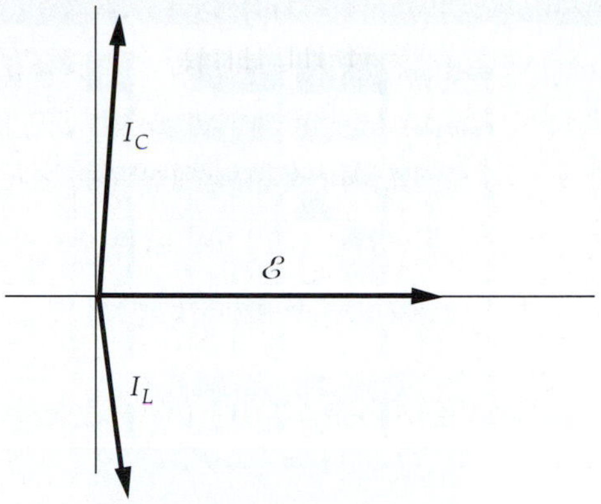

(*d*) 45.3 Ω

101. True

103. (*a*) 5; (*b*) 50 A

105. (*a*) 1.5 A; (*b*) 19

107. 3,333 turns

109. Answer given in the problem.

111. (*a*) 20.8 A; (*b*) 29.5 A; (*c*) 41.7 A, 58.9 A

113.

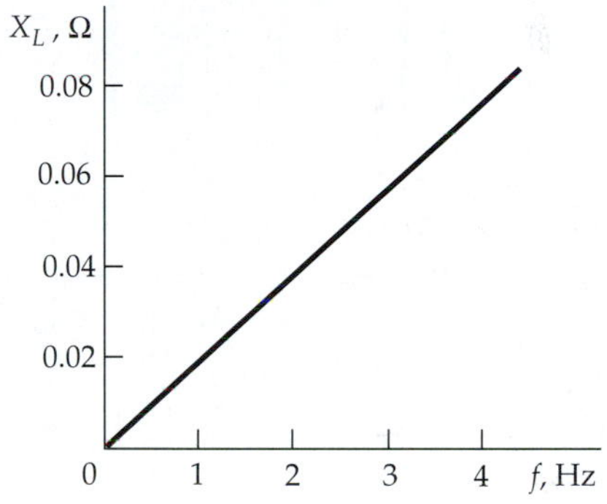

115. (*a*) 6.08 A; (*b*) 444 W; (*c*) 73 V

117. (*a*) 4.74 A; (*b*) 150 W

119. (*a*) 2.00 A, 2.31 A; (*b*) 2.00 A, 2.83 A

121. 45.2 mA, −45.2 mA, 0, 32.0 mA

Chapter 32

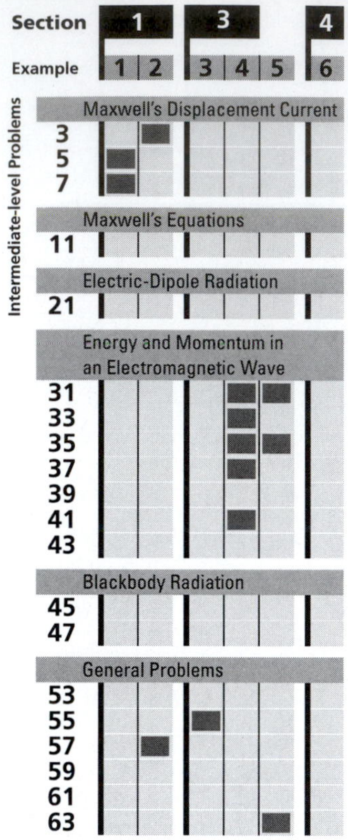

1. (*a*) 3.4×10^{14} V/m·s; (*b*) 5 A

3. Answer given in the problem.

5. (*a*) 10 A; (*b*) 2.26×10^{12} V/m·s; (*c*) 7.89×10^{-7} T·m

7. (*a*) $(A/100\rho d)t$; (*b*) $\kappa\epsilon_0 A/100d$; (*c*) $\kappa\epsilon_0\rho$

9. Answers given in the problem.

11. Answer given in the problem.

13. greater

15. (*a*) 300 m; (*b*) 3 m

17. 3×10^{18} Hz

19. Horizontal and normal to the direction from the antenna to the receiver.

21. (*a*) 30°; (*b*) 7.07 m

23. 4.14×10^{-6} W/m^2, 5.21×10^{17} photons/s·cm^2

25. 1.51×10^{-7} W/m^2

27. (*a*) 283 V/m; (*b*) 943 nT; (*c*) 212 W/m^2; (*d*) 707 nPa

29. Answer given in the problem.

31. (*a*) 4×10^{-8} N; (*b*) 8×10^{-8} N

33. (*a*) 12 V/m, 4×10^{-8} T; (*b*) 1.2 V/m, 4×10^{-9} T; (*c*) 0.12 V/m, 4×10^{-10} T

35. (*a*) 3×10^{-12} N; (*b*) 6×10^{-12} N

37. 7.52×10^4 V/m, 2.51×10^{-4} T

39. (*a*) positive *x* direction; (*b*) 0.628 m, 4.77×10^8 Hz; (*c*) $\vec{E} = (194\ \text{V/m})\cos[10x - (3 \times 10^9)t]\,\hat{j}$, $\vec{B} = (0.647 \times 10^{-6}\ \text{T}) \times \cos[10x - (3 \times 10^9)t]\hat{k}$

41. 3.42×10^6 W/m^2

43. 7.59×10^{-4} kg/m^2, the ratio does not change

45. (*a*) 1425 W/m^2; (*b*) 907 W/m^2; (*c*) 585 V/m; (*d*) 1.95×10^{-6} T

47. (*a*) 278 K; (*b*) 245 K

49. Answer given in the problem.

51. (*a*) False; (*b*) True; (*c*) True; (*d*) True; (*e*) False; (*f*) True

53. The output of the receiver varies as cos θ, where θ is the angle between the normal to the plane of the loop and a line directed toward the source.

55. 7.25×10^{-3} V

57. (*a*) $V_0[(1/R)\sin\omega t + (\epsilon_0\pi a^2\omega/d)\cos\omega t]$; (*b*) $(\mu_0/2\pi)[(V_0/aR)\sin\omega t + (\epsilon_0\omega\pi V_0 a/d)\cos\omega t]$; (*c*) $\tan\delta = \epsilon_0\pi a^2\omega R/d$

59. Answer given in the problem.

61. 111 m^2

63. 247 mW

65. (*a*) $-\mu_0 nar/2$, tangent to the circle of radius *r* and opposite the sense of the current; (*b*) $\mu_0 n^2a^2Rt/2$, points inward toward the axis of the solenoid; (*c*) $\mu_0\pi R^2n^2a^2Lt$

67. Answers given in the problem.

69. Answers given in the problem.

Chapter 33

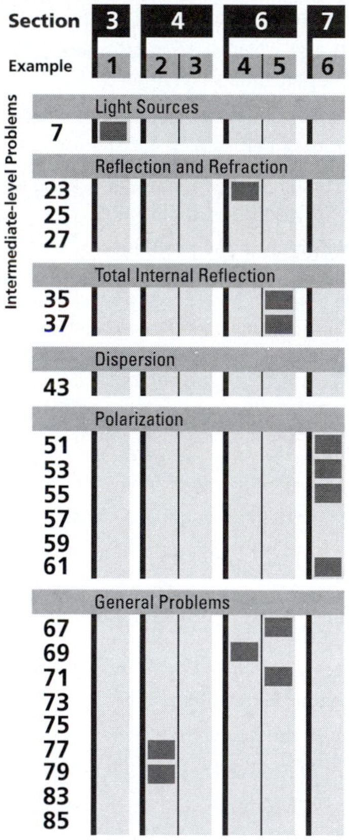

1. Helium atoms are needed to obtain the inverted energy population in neon atoms to produce a laser beam.

3. (*a*) 0.015 J; (*b*) 5.24×10^{16} photons

5. (*a*) 435 nm; (*b*) 1210 nm

7. (*a*) 387.5 nm; 1138 nm and 588 nm; (*b*) 285 nm; 1078 nm, 1140 nm, 554 nm, 387.5 nm, 588 nm

9. 20 μs

11. 2.1×10^6 light years

13. ± 0.3 m

15. You will see less light from your headlights reflected off the road, but you will see more reflected light from an oncoming car.

17. Because of refraction due to the density variation of air, a ray from the bottom of the sun curves in the direction of the curvature of the earth so the apparent position of the bottom of the sun is raised and can be seen even after it is below the horizon. The curvature of the rays from the top of the sun is less so its position is raised less than that of the bottom, which results in the observed flattening.

19. (*a*) 14.9°; (*b*) 22.1°; (*c*) 32.1°; (*d*) 40.6°

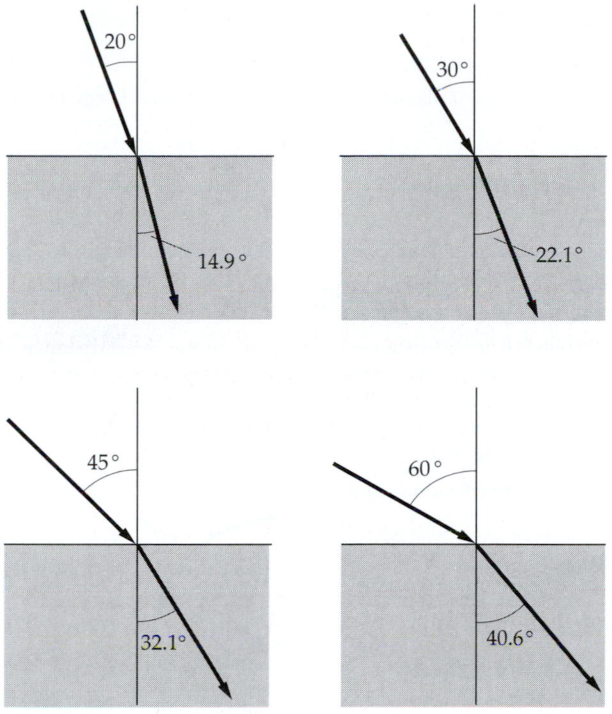

21. 2.26×10^8 m/s, 2×10^8 m/s

23. (*a*) 50.2°; (*b*) 38.8°; (*c*) 26.3°

25. $\approx 92\%$

27.

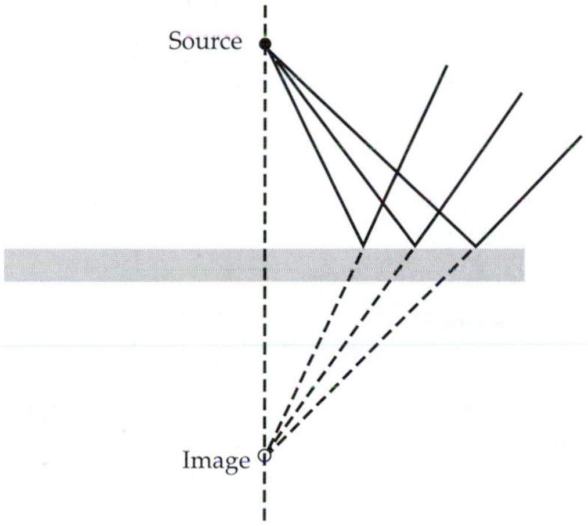

29. (*a*) $\sin^{-1}(n^2 - n\sqrt{n^2 - 1})^{1/2}$; (*b*) 48.5°, 2.81 cm

31. (*a*) Yes; (*b*) If the ball travels with constant speed, the path of least distance will be the path of least time, which is the path that would be taken by a light beam according to Fermat's principle.

33. 48.6°

35. 102 m^2

37. 1.30

39. (*a*) 1.41; (*b*) between 1.63 and 1.88

41. 2.18 cm

43. 1.1°

45. (*b*)

47. (*a*) 53.1°; (*b*) 56.3°

49. (*a*) $I_0/8$; (*b*) $3I_0/32$

51. $I = (1/8)I_0 \sin^2 2\theta$

53. (*a*) $I_0 [\cos(\pi/2N)]^{2N}$; (*b*) $I_0/4$; (*c*) $0.976I_0$; (*d*) Perpendicular to the original polarization.

55. $0.211I_0$, which is greater than $I = 0.125I_0$ for a single sheet in the middle at 45°.

57. Answer given in the problem.

59. $\vec{E} = E_0 \cos(kz + \omega t)\hat{i} + E_0 \sin(kz + \omega t)\hat{j}$, $\vec{B} = (E_0/c)\sin(kz + \omega t)\hat{i} - (E_0/c)\cos(kz + \omega t)\hat{j}$

61. Horizontally polarized, $I_0 [\cos(\pi/2N)]^{2N}$

63. (*c*)

65. (*a*) 526 nm; (*b*) the same color

67. 35.3°

69. 1.45 m

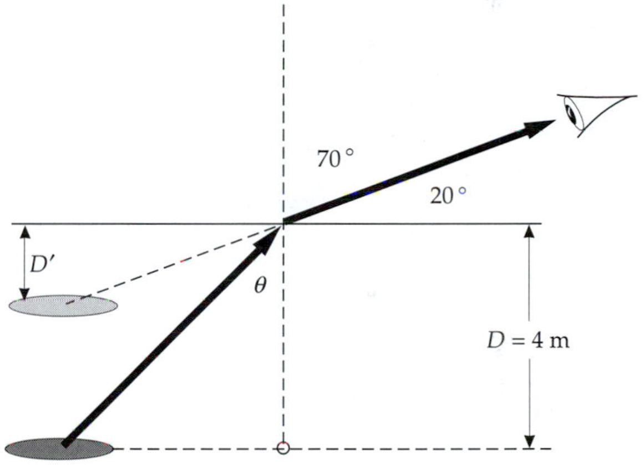

71. Answer given in the problem.

73. Answer given in the problem.

75. Answer given in the problem.

77. (*a*) Answer given in the problem. (*b*) 3.47°

79. (*a*) 1.60; (*b*) 38.7°

81. 1.68°

83. (*a*) $[4n/(n + 1)^2]^{2N}$; (*b*) 0.783; (*c*) 28

85. (*a*) 24.0°; (*b*) The cloud appears to move in a circle of angular radius 24.0°. (*c*) The angular radius θ is related to the earth's speed v by $\tan\theta = v/c$. (*d*) 2.99×10^8 m/s

87. (*a*) Answer given in the problem. (*b*) 5.12 cm

89. Answer given in the problem.

Chapter 34

Section	1		2									4			
Example	1	2	3	4	5	6	7	8	9	10	11	12	13	14	15
Intermediate-level Problems															
Plane Mirrors															
7															
Spherical Mirrors															
19		■													
21	■														
23	■														
25															
Images Formed by Refraction															
31				■											
33				■											
35				■											
Thin Lenses															
51									■	■					
53						■	■								
55									■	■					
The Eye (optional)															
71													■		
73															
The Simple Magnifier (optional)															
81														■	
83															
Microscopes (optional)															
85															■
87															■
Telescopes (optional)															
93															
95															
General Problems															
107	■	■													
109								■							
111									■	■					
113				■			■								
115								■							

1. Yes

3.

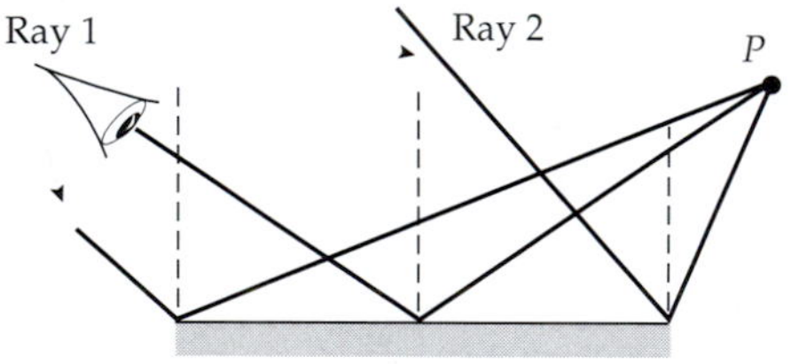

5.

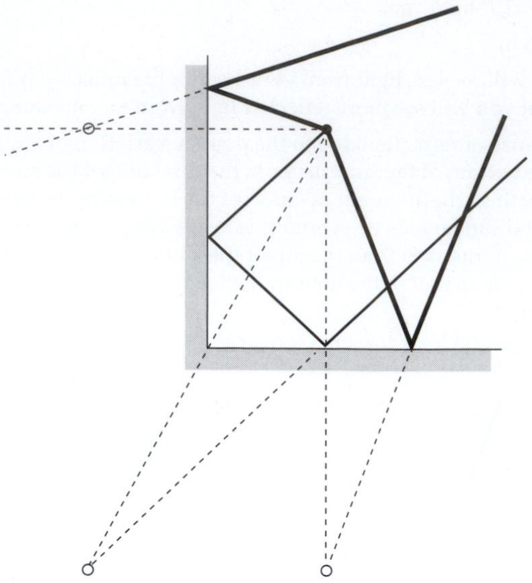

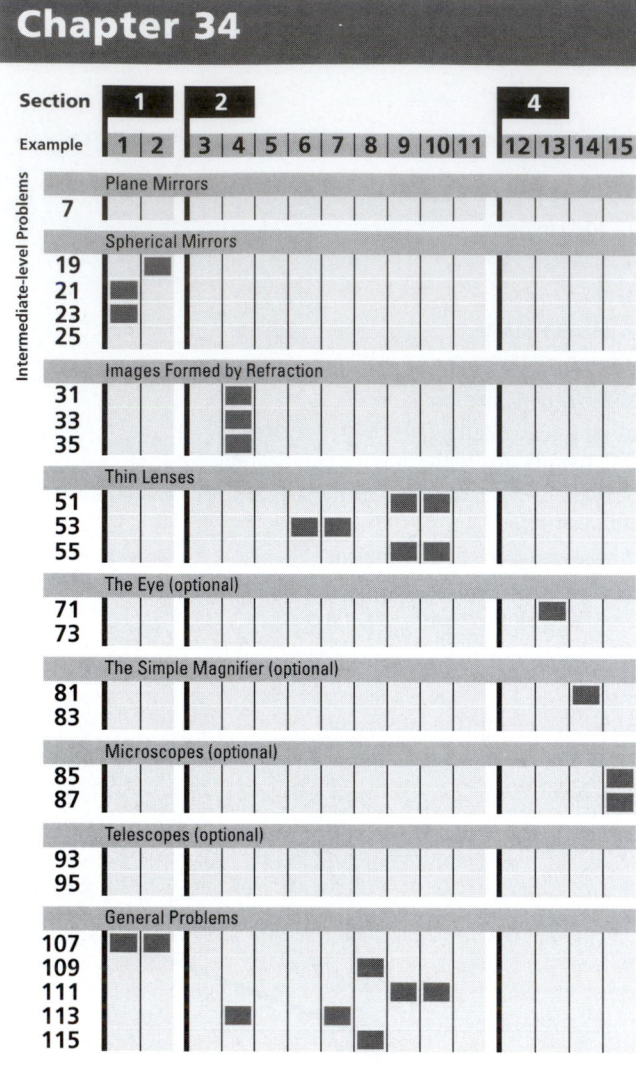

7. (*a*) 10 cm, 50 cm, 70 cm, 110 cm; (*b*) 20 cm, 40 cm, 80 cm, 100 cm

9. The image will be virtual and erect if $s < F$; the image will be smaller if $s > 2F$ and larger if $s < 2F$

11. While the image distance is smaller than the object distance, the size of the image is also smaller than the object. The second effect outweighs the first so that the angular size presented to the driver is smaller for the image than for the actual object. Thus the image appears more distant.

13. (*a*) real, inverted, reduced

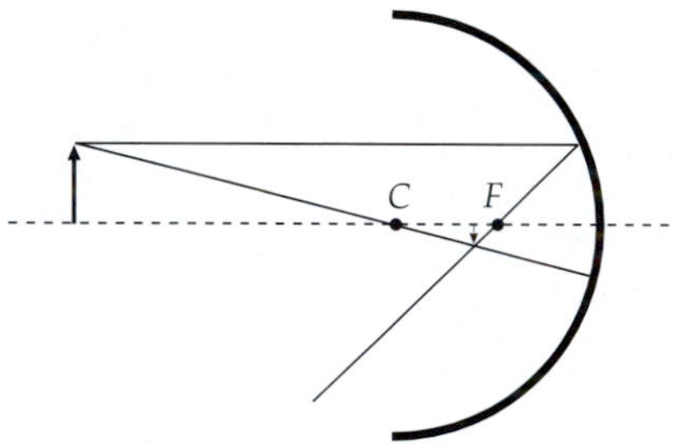

(*b*) real, inverted, same size

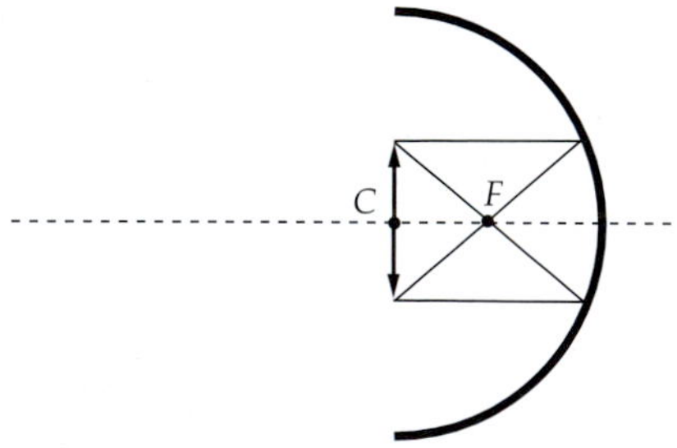

(*c*) no image is formed

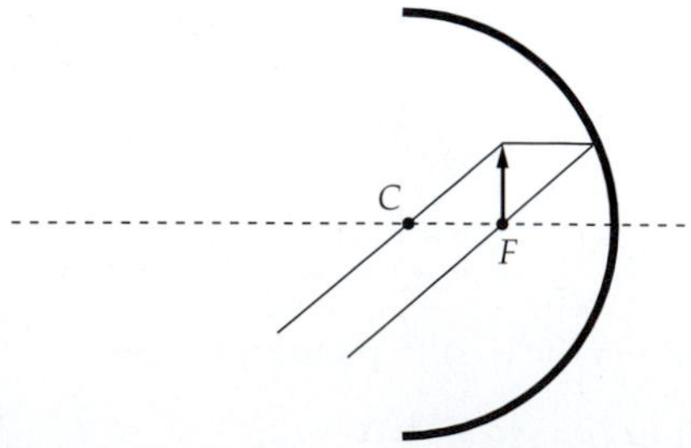

(*d*) virtual, erect, enlarged

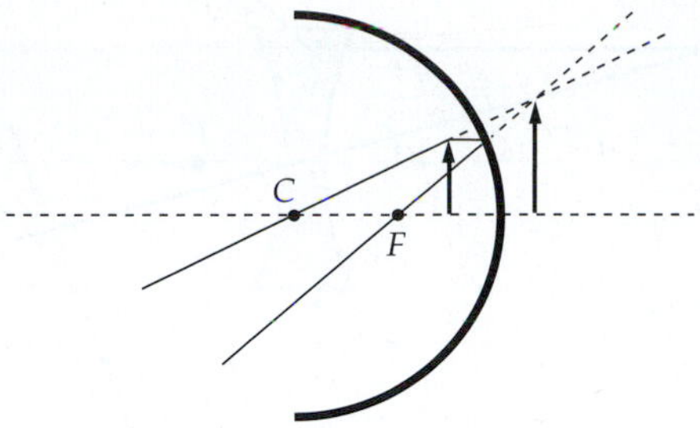

15. (*a*) virtual, erect, reduced

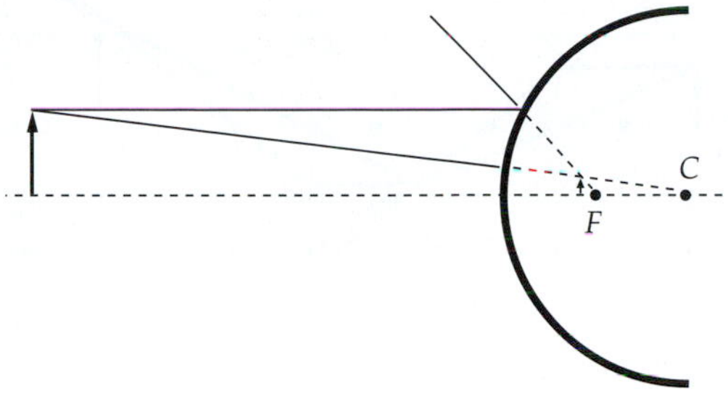

(*b*) virtual, erect, reduced

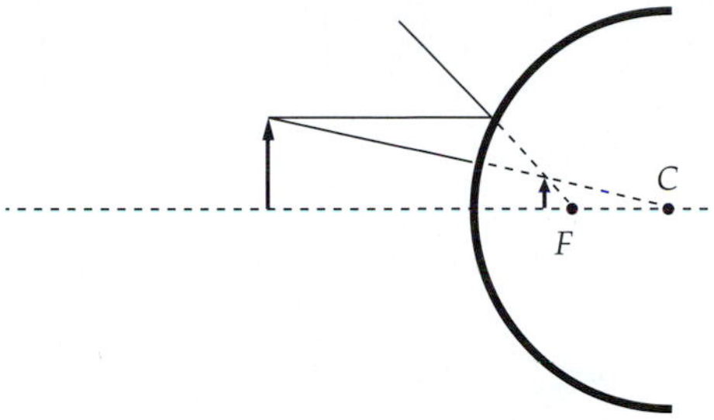

(*c*) virtual, erect, reduced

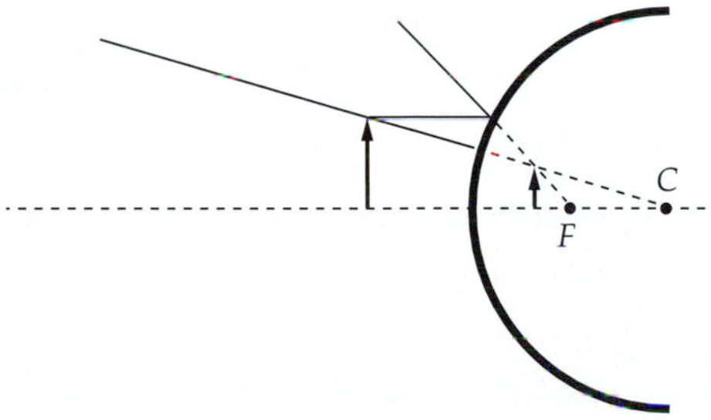

(*d*) virtual, erect, reduced

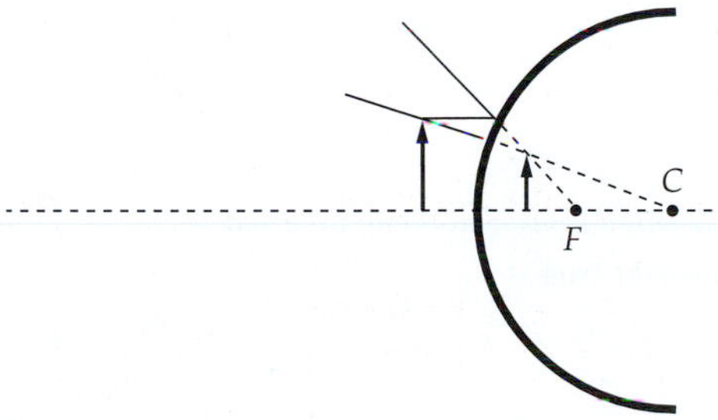

17. Answer given in the problem.

19. (*a*) 0.566 m; (*b*) behind; (*c*) 0.113 m

21. 1.5 cm

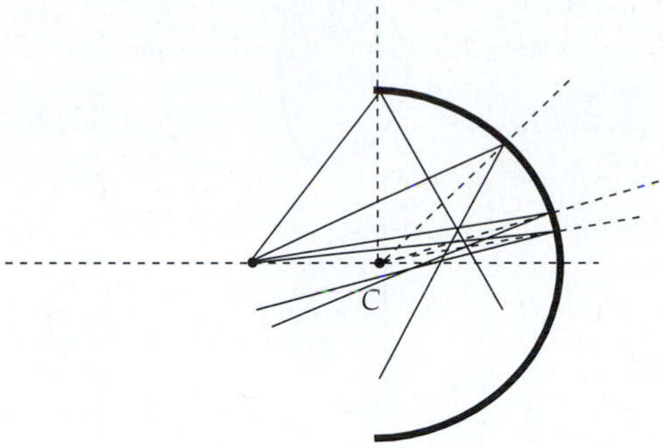

23. (*a*) −66.7 cm; (*b*) virtual

25. (*a*) −1.33 m; (*b*) convex

27. farther

29. (*a*) −8.58 cm; (*b*) −35.9 cm

31. 10 cm

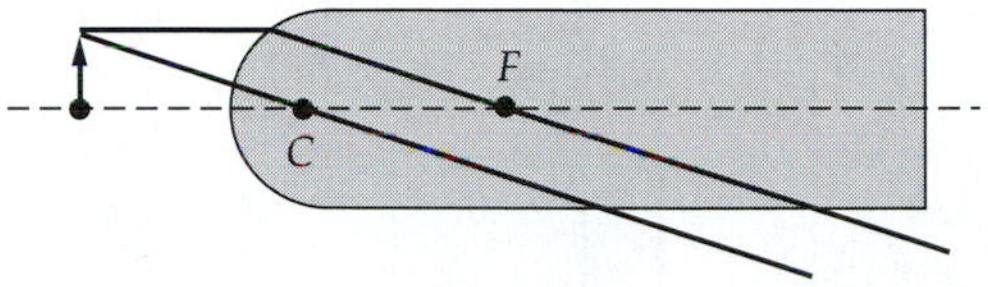

33. (*a*) −46.2 cm, virtual

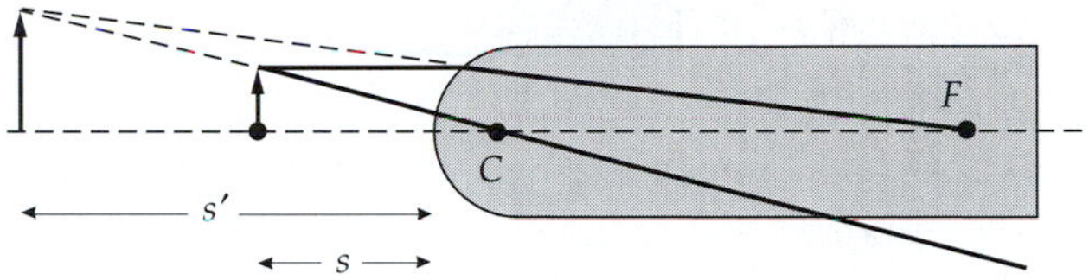

(*b*) −6.47 cm, virtual

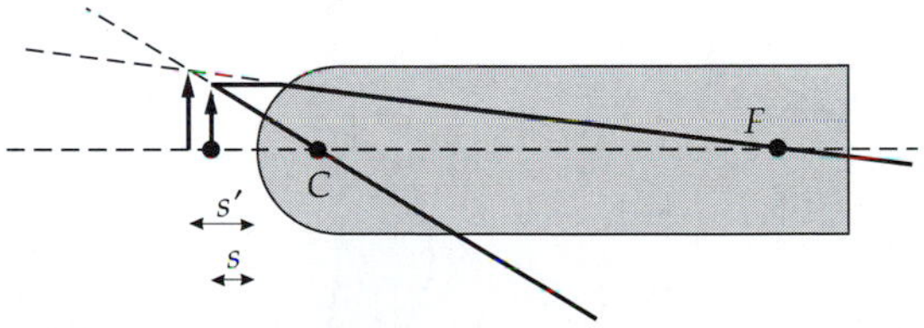

(*c*) 44.1 cm, real

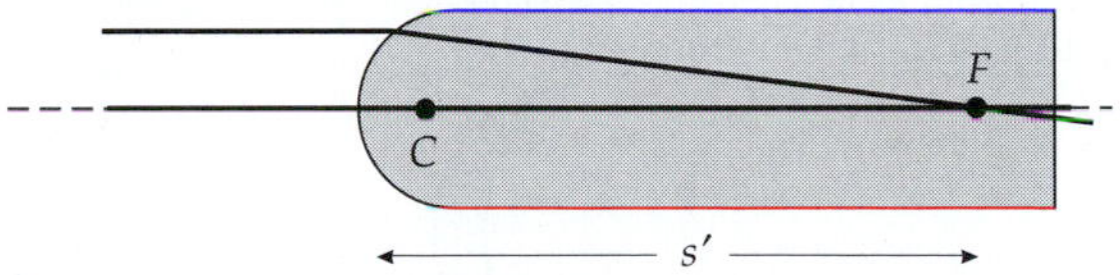

35. (*a*) 64 cm; (*b*) −80 cm; (*c*) virtual

37. Assuming that *n* is greater than 1, a thin lens will have a positive focal length if it is double convex, plano convex, or if the radius of curvature of the convex side is smaller in magnitude than the radius of curvature of the concave side. The focal length will be negative if the lens is double concave, plano concave, or if the radius of curvature of the concave side is smaller in magnitude than the radius of curvature of the convex side.

39. (*d*)

41. (*a*) 13.5 cm

(*b*) 20 cm

(*c*) −10 cm

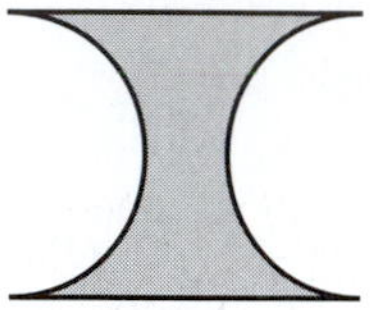

(*d*) −40 cm

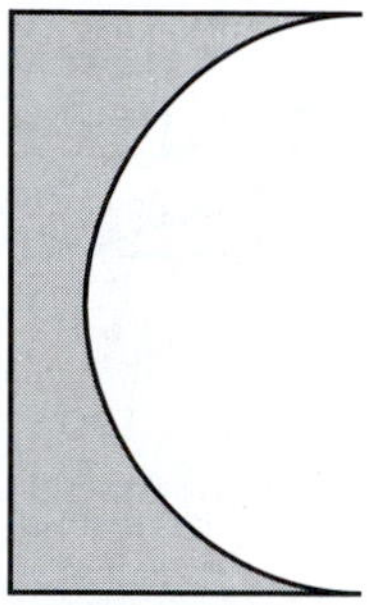

43. 108 cm

45. (*a*) −33.3 cm

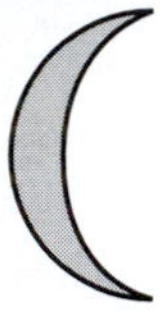

(*b*) 33.3 cm

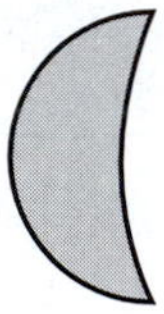

(*c*) −33.3 cm

47. $s' = 6.67$ cm, $h = 1$ cm

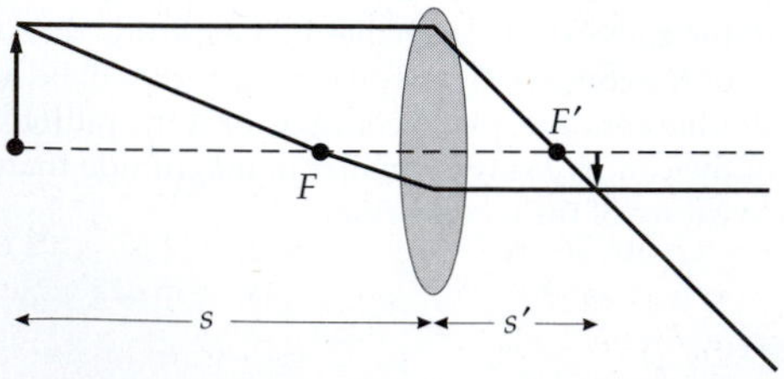

49. $s' = -3.33$ cm, $h = 0.333$ cm

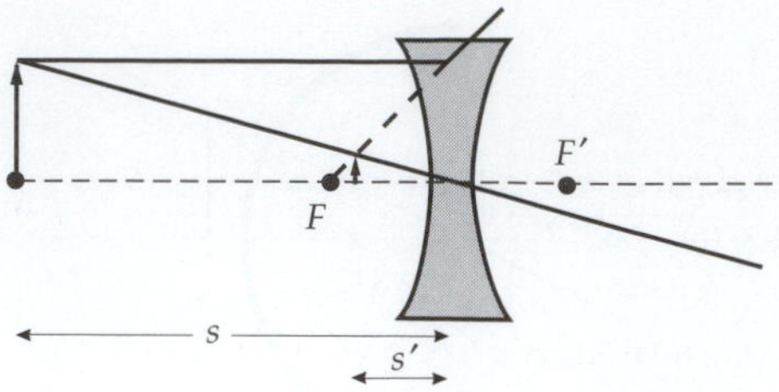

51. (*a*) 30 cm to the far side of the second lens. (*b*) real, erect; (*c*) 2

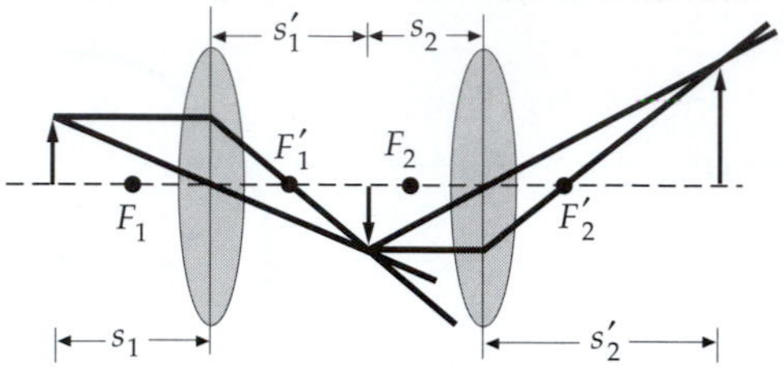

53. (*a*) 35.0 cm, concave

(*b*)

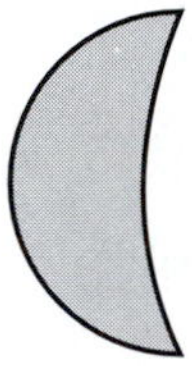

55. 15 cm to right of second lens, real, inverted, magnification of −1

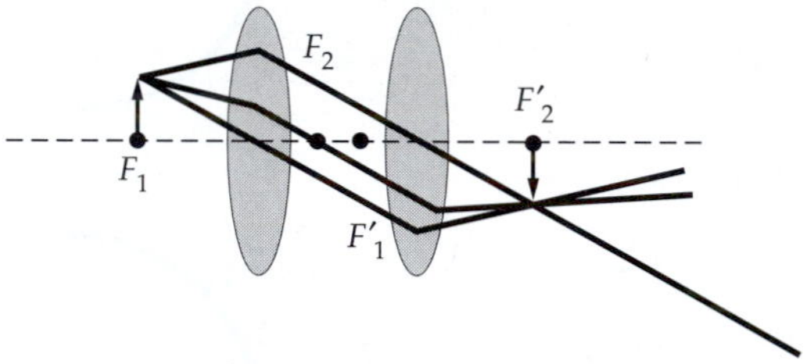

57. Answer given in the problem.

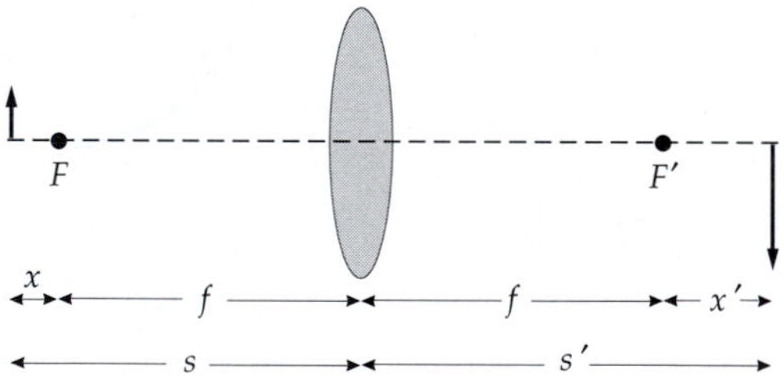

59. (*a*) 41.2 cm; (*b*) −1.53; (*c*) real, inverted

61. (*a*) False; (*b*) True

63. (*a*)

65. (*b*)

67. 0.172 cm

69. 44.4 cm

71. (*a*) 103 cm; (*b*) 0.972 diopters

73. 0.714, smaller

75. (*a*) 46.0 cm; (*b*) 186 cm; (*c*) 3.84 diopters

77. 6

79. 5

81. (*a*) 3; (*b*) 4

83. Answer given in the problem.

85. (*a*) -11.3; (*b*) -56.3

87. (*a*) 30 cm; (*b*) -6; (*c*) -30; (*d*) 5.83 cm

89. (*a*) 1.67 cm; (*b*) 0.508 cm; (*c*) 0.496 cm

91. 17.6 cm

93. 4 cm (eyepiece), 28 cm

95. (*a*) 103.4 cm; (*b*) -6.25 cm; (*c*) 97.2 cm; (*d*) 20.7 cm, 25.6

97. 0.00667

99. (*a*)

101. Answer given in the problem.

103. 0.158 mm

105. (*a*) 5 cm, -10 cm

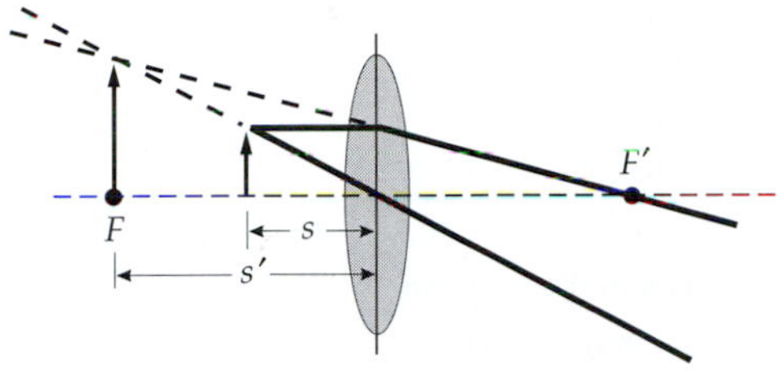

(*b*) 15 cm, 30 cm

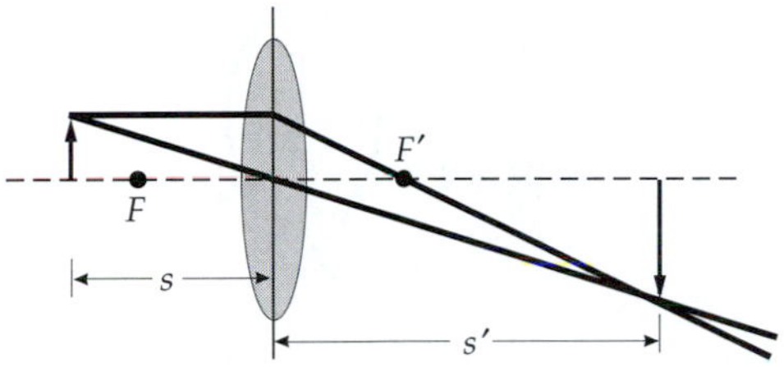

107. concave, 90 cm

109. 3.7 m

111. (*a*) 9.52 cm; (*b*) -1.19

(*c*)

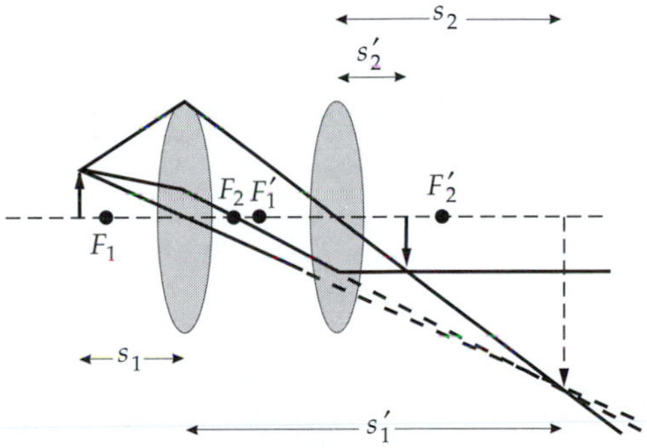

113. (*a*) 19.3 cm; (*b*) 77.3 cm

115. 9.7 cm/s

117. (*a*) 18 cm to left of lens; (*b*) real, erect

(*c*)

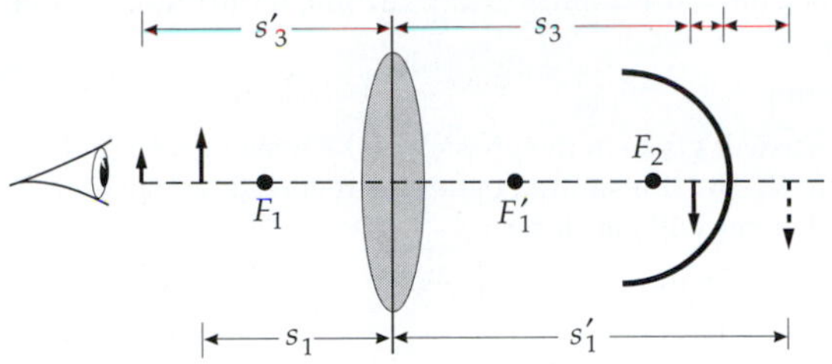

119. 1.6

121. 1.36

123. (*a*) Answer given in the problem. (*b*) 17.4 cm

125. Answer given in the problem.

Chapter 35

Intermediate-level Problems	Section 1 (Example 1)	Section 2 (Example 2)	Section 3 (Example 3)	Section 4 (Example 4)	Section 4 (Example 5)	Section 5 (Example 6)	Section 7 (Example 7)	Section 8 (Example 8)
Phase Difference and Coherence								
5	■							
Interference in Thin Films								
11		■						
13		■						
15		■						
Newton's Rings								
17								
19								
Two-Slit Interference Pattern								
25								
27				■				
Diffraction Pattern of a Single Slit								
33				■				
Interference–Diffraction Pattern of Two Slits								
35					■			
37					■			
Using Phasors to Add Harmonic Waves (optional)								
41						■		
43						■		
Diffraction and Resolution								
53						■	■	
55						■	■	
Diffraction Gratings (optional)								
63								
65								■
67								
69								
General Problems								
81		■						
83		■						
85			■					

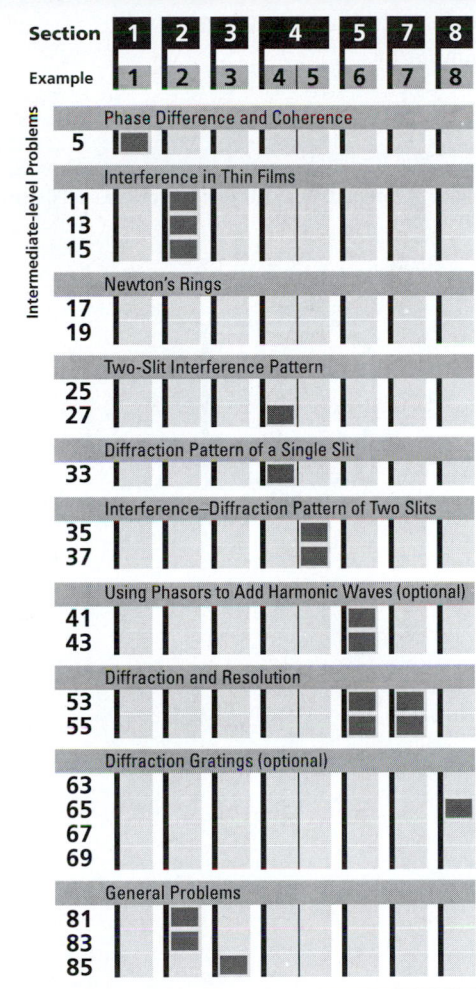

1. It goes to where constructive interference occurs.

3. (*a*) 300 nm; (*b*) 135°

5. 164°

7. The fringes are too close together to be resolved by the eye.

9. (*a*) The top of the film approaches zero thickness, so the phase difference approaches 180°. (*b*) violet; (*c*) The top of the film is white; the color of the first band is red.

11. $5.46\ \mu m < d < 5.75\ \mu m$

13. (*a*) 602 nm; (*b*) 401 nm, 516 nm, 722 nm; (*c*) 401 nm, 516 nm, 722 nm

15. 476 nm

17. (*a*) Answer given in the problem. (*b*) Answer given in the problem. (*c*) The transmitted pattern is reversed. (*d*) 68; (*e*) 1.14 cm; (*f*) The fringes will get closer.

19. 534 μm, 926 μm

21. 8.33 fringes/cm

23. (*a*) 9.29 μm; (*b*) 29

25. Answer given in the problem.

27. (*a*) 0.30°; (*b*) 8

29. The equation $d \sin\theta = m\lambda$ describes the angles at which the *interference* between *two slits separated by a distance d* produces maximum intensity. The equation $a \sin\theta = m\lambda$ describes the angles at which the *diffraction of light from a single slit of width a* results in zero intensity.

31. (*a*) 2 cm; (*b*) 20 cm; (*c*) 231 cm

33. 32.2 km

35. (*a*) 20 μm; (*b*) 9

37. 8

39. $3.61 \sin(\omega t - 56.3°)$

41. $0.0162 I_0$

43. (*a*) Answer given in the problem. (*b*) 4 sources: 6.00 mm, 2 sources: 12.0 mm

45. (*a*)

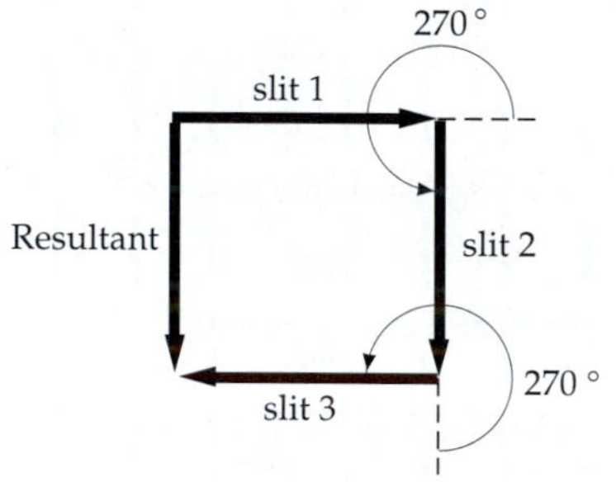

(*b*) $5.56 \times 10^{-3}\ W/m^2$

47. (*a*) $3\pi, 5\pi, 7\pi$; (*b*) $2.86\pi, 4.92\pi, 6.94\pi$

49. 8.54 cm

51. 8.35 km

53. (*a*) 55.6 km; (*b*) 55.6 m

55. 5×10^6 km

57. (*a*)

59. 485 nm, 658 nm

61. 500 nm

63. 2, The second-order long wavelength overlaps the third-order short wavelength.

65. (*a*) 0.353 m, 0.707 m; (*b*) 88.4 μm; (*c*) 8000

67. 3.09×10^5, 5.14×10^4 lines/cm

69. (*a*) 750 lines/cm; (*b*) 4.21 cm, 12.6 cm

71. (*a*) Answer given in the problem. (*b*) 491; (*c*) 0.988 mm

73. Answer given in the problem.

75. 695 nm

77. 1.68 cm

79. 1.30×10^{-4} rad

81. Answer given in the problem.

83. (*a*) 97.8 nm; (*b*) No; (*c*) 0.273, 0.124

85. 12.3 m

87. Answers given in the problem.

89. Answer given in the problem.

91. (*a*) $I_{max} \cos^2[(\pi/2)\sin\theta]$

(*b*)

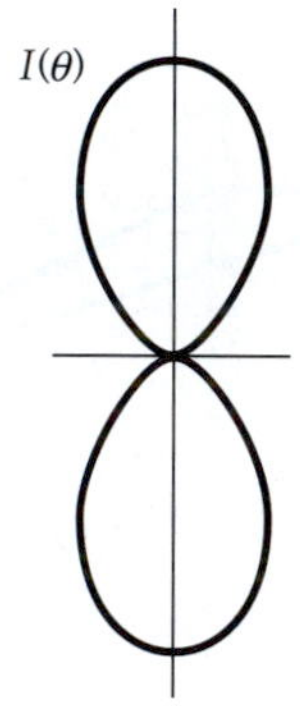

Chapter 36

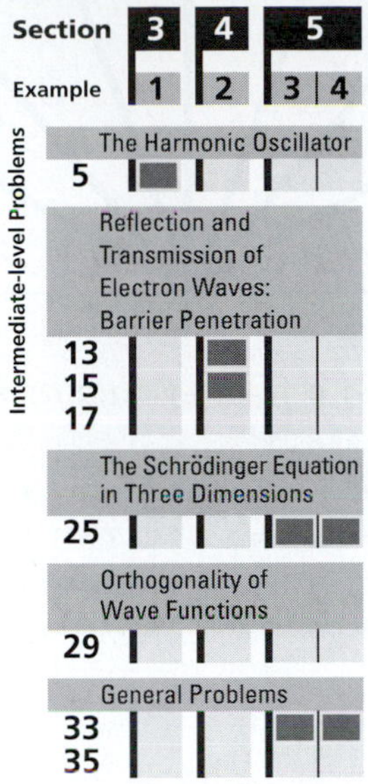

1. True

3. (*a*)

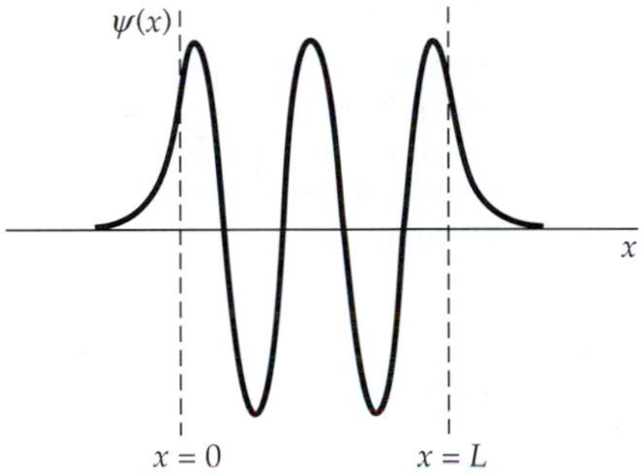

(*b*)

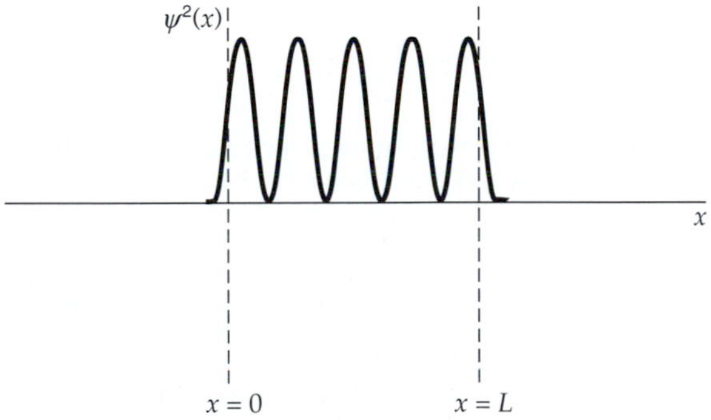

5. Answer given in the problem.

7. $[(2/\sqrt{\pi})(m\omega_0/\hbar)^{3/2}]^{1/2}$

9. Answer given in the problem.

11. $m\omega_0\hbar/2$

13. (*a*) $k_1/\sqrt{2}$, $(2mU_0)^{1/2}/\hbar$; (*b*) 0.0294; (*c*) 0.971; (*d*) 9.71×10^5 particles. Classically, 100% of the particles are transmitted.

15. (*a*) $0.1k_1$, $(0.02mU_0)^{1/2}/\hbar$; (*b*) 0.671; (*c*) 0.329; (*d*) 3.29×10^5 particles. Classically, 100% of the particles are transmitted.

17. 0.342

19. (1, 1, 1), (1, 1, 2), (1, 2, 1), (1, 1, 3), (1, 2, 2), (1, 2, 3), (1, 1, 4), (1, 3, 1), (1, 3, 2), (1, 2, 4)

21. (*a*) (1, 1, 1), (1, 1, 2), (1, 1, 3), (1, 2, 1), (1, 1, 4), (1, 2, 2), (1, 2, 3), (1, 1, 5), (1, 2, 4), (1, 3, 1), (1, 1, 6), (1, 3, 2); (*b*) (1, 1, 4) and (1, 2, 2); (1, 1, 6) and (1, 3, 2)

23. (*a*) $\psi(x, y, z) = (8/L^3)^{1/2}\cos(\pi x/L)\sin(\pi y/L)\sin(\pi z/L)$; (*b*) The energy levels are the same.

25. $50h^2/8mL^2$

27. $10E_1$

29. Answer given in the problem.

31. Answer given in the problem.

33. $20\,\hbar^2\pi^2/mL^2$, $21\hbar^2\pi^2/mL^2$, $21\hbar^2\pi^2/mL^2$

35. (*a*)

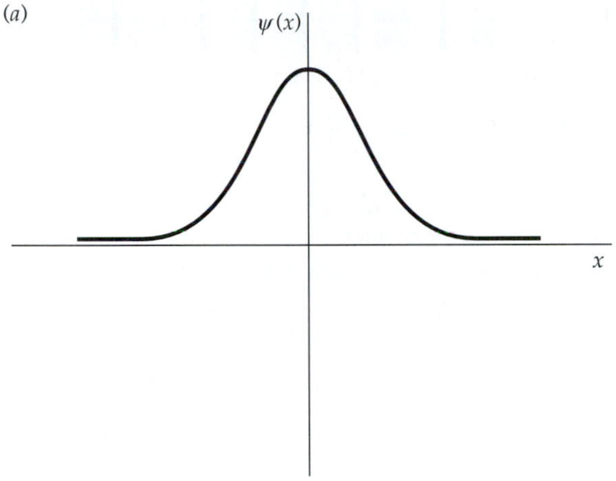

(*b*)

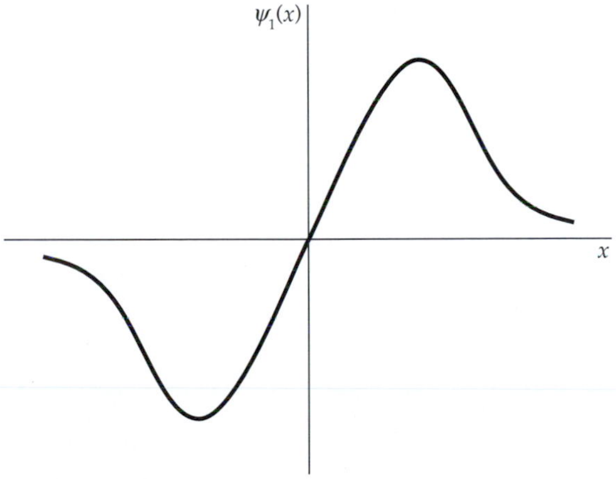

37. Answer given in the problem.

39. $\sqrt{2}\,[m\omega_0/(\pi\hbar)]^{1/4}$

41. Answer given in the problem.

Chapter 37

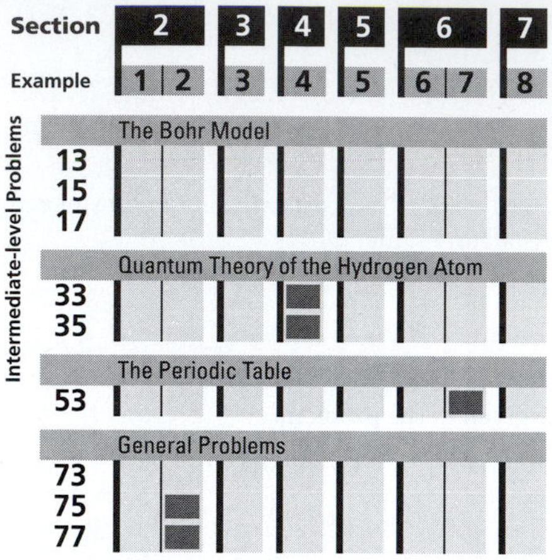

Section
2
3
4
5
6
7
Example
1 2
3
4
5
6 7
8
Intermediate-level Problems
The Bohr Model
13
15
17
Quantum Theory of the Hydrogen Atom
33
35
The Periodic Table
53
General Problems
73
75
77

1. decrease

3. (*a*)

5. (*d*)

7. Answer given in the problem.

9. 1.89 eV, 656 nm; 2.55 eV, 486 nm; 2.86 eV, 434 nm

11. (*a*) 0.85 eV, 1459 nm; (*b*) 4052 nm, 2627 nm, 2168 nm

13. (*a*) 13.6 eV; (*b*) 54.4 eV; (*c*) 122.4 eV

15. Answer given in the problem.

17. (*a*) Answer given in the problem; (*b*) $R_H = 1.096\,776 \times 10^7\ \text{m}^{-1}$, $R_\infty = 1.097\,373 \times 10^7\ \text{m}^{-1}$; (*c*) 0.0544%

19. (*c*)

21. (*a*) $(2\sqrt{3})\hbar$; (*b*) $-3, -2, -1, 0, 1, 2, 3$;

(*c*)

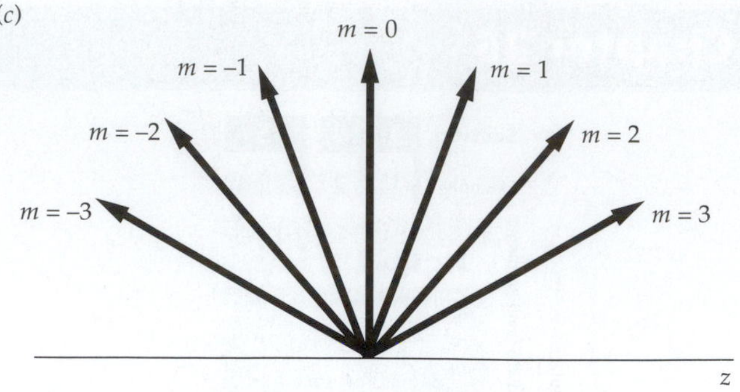

23. (*a*) 0, 1, 2; (*b*) 0; −1, 0, 1; −2, −1, 0, 1, 2; (*c*) 18

25. (*a*) 45°; (*b*) 26.6°; (*c*) 8.05°

27. (*a*) $n \geq 1, \ell \geq 0$; (*b*) $n \geq 2, \ell \geq 1$; (*c*) $n \geq 3, \ell \geq 2$

29. (*a*) 4; (*b*) $n = 2, \ell = 0, m = 0$; $n = 2, \ell = 1, m = 1$; $n = 2, \ell = 1, m = 0$; $n = 2, \ell = 1, m = -1$

31. (*a*) $0.0605a_0^{-3/2}$; (*b*) $0.00366a_0^{-3}$; (*c*) $0.046a_0^{-1}$

33. (*a*) 0.00092; (*b*) 0

35. $(3/2)a_0$

37. 0.323

39. $\ell = 0$ or $\ell = 1$

41. 5/2 or 3/2

43. In sodium, the screening of the nuclear charge by the inner electrons is less for the 3s state than for the 3p state; there is no screening in hydrogen.

45. Iron and cobalt each have two valence electrons in the 4s shell so they have similar properties. They differ only in that cobalt has one more electron in the 3d shell. Neon has a closed shell structure so it is inert. Sodium has one more electron, which is in the outer 3s shell, so its properties are very different.

47. (*a*) silicon; (*b*) calcium

49. (*c*)

51. (*a*) $1s^2\,2s^2\,2p^6\,3s^2\,3p^1$; (*b*) $1s^2\,2s^2\,2p^6\,3s^2\,3p^6\,3d^5\,4s^1$

53. 3.4

55. (*a*) $2s^1$ or $2p^1$; (*b*) $1s^2\,2s^2\,2p^6\,3p^1$; (*c*) $1s^1 2s^1$

57. (*a*) 0.0611 nm, 0.0580 nm; (*b*) 0.0543 nm

59. zirconium

61. (*a*) 1.01 nm; (*b*) 0.155 nm

63. (*c*)

65. These atoms have a single outer p-shell electron, which is shielded from the nucleus more than an s-shell electron.

67. $n = 4$ to $n = 1$.

69. $n = 3$ to $n = 2$, $n = 9$ to $n = 3$, $n = 7$ to $n = 4$

71. (*a*) 1.61774 eV, 1.61041 eV; (*b*) 7.33×10^{-3} eV; (*c*) 63.3 T

73. Answer given in the problem.

75. (*a*) $R_H = 1.096\,776 \times 10^{-7}\ \text{m}^{-1}$, $R_D = 1.097\,074 \times 10^{-7}\ \text{m}^{-1}$; (*b*) 0.178 nm

77. (*a*) $R_T = 1.097\,174 \times 10^{-7}\ \text{m}^{-1}$; (*b*) 0.0598 nm, 0.238 nm

79. Answer given in the problem.

Chapter 38

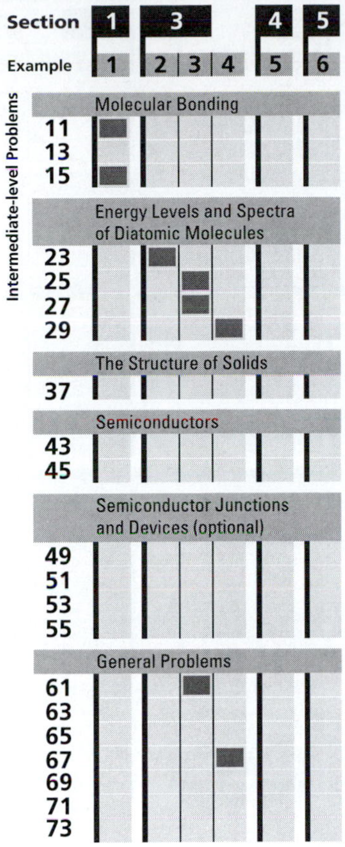

1. polar
3. Ne
5. (*a*) covalent; (*b*) ionic; (*c*) metallic
7. releases energy, 3.02 eV
9. 43.6%
11. 0.499 eV
13.

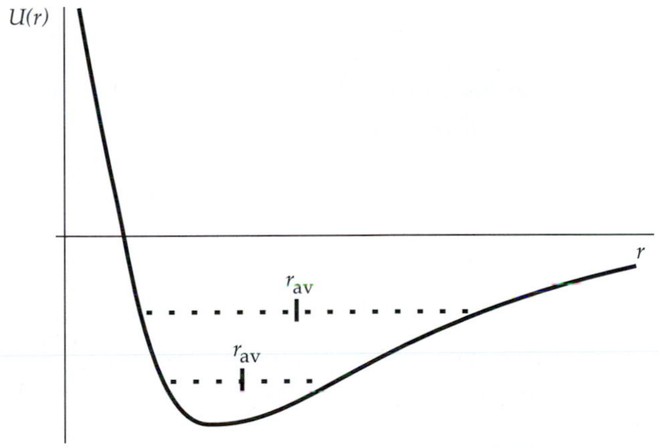

From the graph we see that r_{av} increases with increasing vibration energy. As a result, such a molecule expands when heated.

15. (*a*) −6.64 eV; (*b*) 5.70 eV; (*c*) 0.64 eV
17. (*a*) 3.1 eV; (*b*) $C = 1.37 \times 10^{-13}$ eV·nm^n, $n = 19.7$
19. The force constant is similar to that of a stiff ordinary spring.
21. 0.110 nm
23. (*a*) 0.504 u; (*b*) 7.00 u; (*c*) 6.86 u; (*d*) 0.980 u
25. 0.00314 eV
27. 1.58×10^{-5} eV
29. 477 N/m
31. (*a*) ^{35}Cl: 0.9722 u, ^{37}Cl: 0.9737 u, $\Delta\mu/\mu = 0.00153$ u; (*b*) $\Delta f/f = -0.00153$
33. (*a*) cube with sides of length $2R$; (*b*) 0.524
35. 2.07 g/cm^3
37. 0.740
39. (*d*)
41. (*a*) *p*-type; (*b*) *n*-type
43. 3.17 nm, 8.46 nm
45. 342×10^{-10} m
47. (*b*)
49.

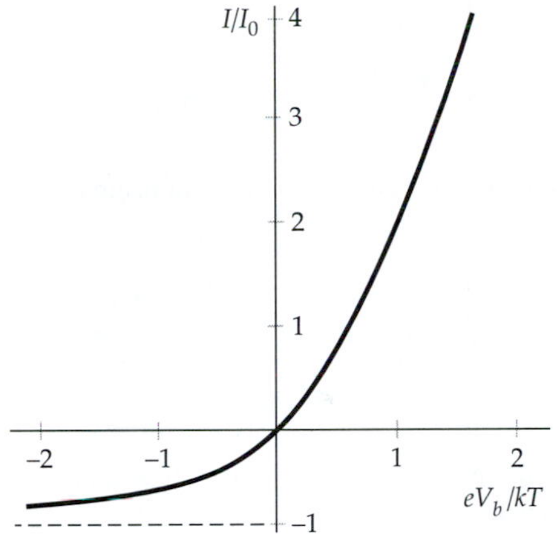

51. 250
53.

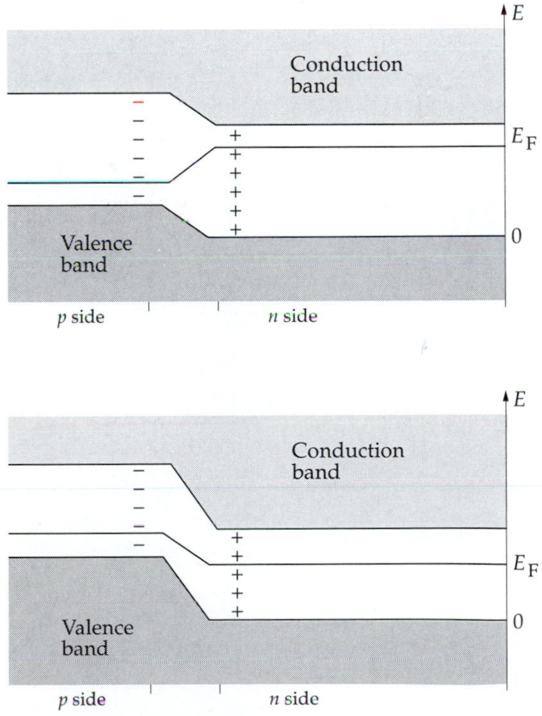

55. p-type, 10^{23} holes/m^3

57. (*a*) ionic; (*b*) covalent; (*c*) metallic

59. 2.63×10^{-29} C·m

61. 9.55×10^{-4} eV

63. Answer given in the problem.

65. 1.1 eV

67. 1551 N/m

69. $r_0 = a$, $U_{min} = -U_0$, $r_0 = 0.074$ nm, $U_0 = 4.52$ eV

71. $F_x \propto 1/x^4$

73. (*a*) $I = 1.45 \times 10^{-46}$ kg·m^2, $E_{0r} = 2.39 \times 10^{-4}$ eV

(*b*)

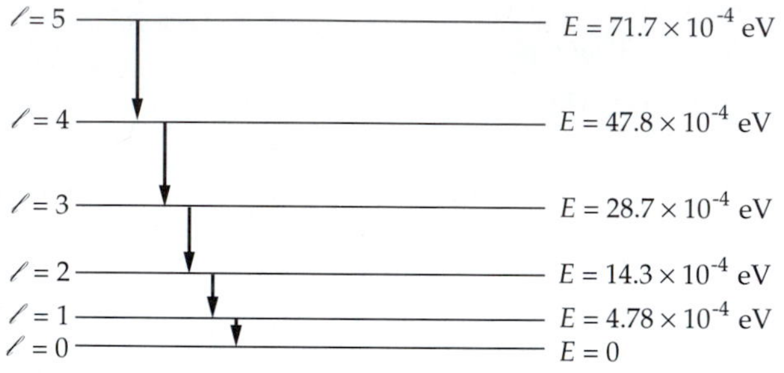

(*c*), (*d*) Transitions that obey $\Delta\ell = -1$ are indicated by the arrows in the diagram.

$\Delta\ell$	ΔE	λ
$\ell = 1 \to \ell = 0$	4.78×10^{-4} eV	2.60 mm
$\ell = 2 \to \ell = 1$	9.55×10^{-4} eV	1.30 mm
$\ell = 3 \to \ell = 2$	14.4×10^{-4} eV	0.862 mm
$\ell = 4 \to \ell = 3$	19.1×10^{-4} eV	0.650 mm
$\ell = 5 \to \ell = 4$	23.9×10^{-4} eV	0.519 mm

These photons fall in the microwave and short radio wave portion of the electromagnetic spectrum.

75. $f = 1.9 \times 10^{13}$ Hz, $\lambda = 1.6 \times 10^{-5}$ m

Chapter 39

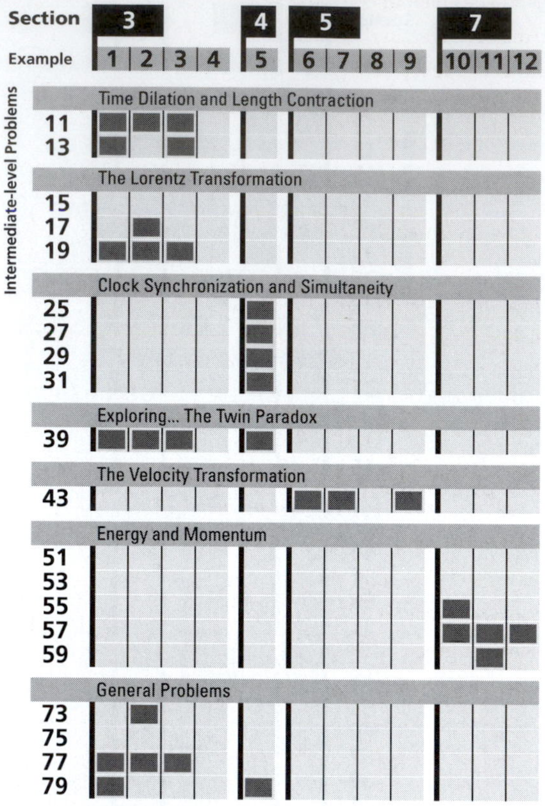

1. The friend in the car.

3. (*a*) 6.63 m; (*b*) 12.6 m

5. (*a*) 599 m; (*b*) 13.4 km

7. (*a*) 130 y; (*b*) 88.1 y

9. (*a*) 0.6 m; (*b*) 2.5 ns

11. $0.8c$

13. (*a*) 978 m; (*b*) 9.57×10^7 m; (*c*) 1.02×10^{-7} m

15. Answer given in the problem.

17. $0.141c$

19. (*a*) 2.56×10^{-9} s; (*b*) 8×10^8 particles/cm^2·s; (*c*) 2.4×10^8 particles/cm^2·s

21. Answers given in the problem.

23. Yes

25. 80 c·min

27. Answer given in the problem.

29. $L_pV/c^2 = 60$ min, which is the same as the time interval found in Problem 28.

31. 4.39 μs

33. 0.0637

35. $0.696c$

37. (*a*) 11.3 y; (*b*) 40 y

39. (*a*) 50 signals/y; (*b*) 533.3 signals; (*c*) 1067 signals; (*d*) 50 signals/y; (*e*) 267 signals; (*f*) 1333 signals; (*g*) A by 2.67 y

41. $0.994c$

43. (*a*) $0.976c$; (*b*) $0.997c$

45. (*a*) 1.005; (*b*) 1.15; (*c*) 1.67; (*d*) 7.09

47. (*a*) $0.155E_0$; (*b*) $1.29E_0$; (*c*) $6.09E_0$

49. 2.97 GeV

51. Answer given in the problem.

53. Answer given in the problem.

55. (*a*) 1625 MeV/*c*; (*b*) 0.866*c*

57. 608 MeV, 389 MeV

59. (*a*) 290 MeV; (*b*) 625 MeV

61. (*a*) 0.943*c*; (*b*) 3 MeV; (*c*) 2.83 MeV/*c*; (*d*) 0.877 MeV; (*e*) 4.12 MeV/c^2

63. Answers given in the problem.

65. (*b*)

67. (*a*) 5.33 y; (*b*) 3.53 y

69. 0.999*c*

71. 0.866*c*

73. 1.85×10^4 y

75. (*a*) 2.87×10^9 MeV; (*b*) 0.133 s

77. (*a*) 0.625*c*; (*b*) 31.2 y

79. (*a*) 4.17 μs; (*b*) 7.71 μs; (*c*) 2.5 μs

81. (*a*) 2.1 μs; (*b*) 2.59 μs; (*c*) 0.493 μs; (*d*) 2.59 μs; (*e*) 4.36 h; (*f*) 19 h

83. (*a*) $c/3$; (*b*) 20 m; (*c*) 0.20 μs

85. Answer given in the problem.

87. Answer given in the problem.

89. Answer given in the problem.

91. Answer given in the problem.

Chapter 40

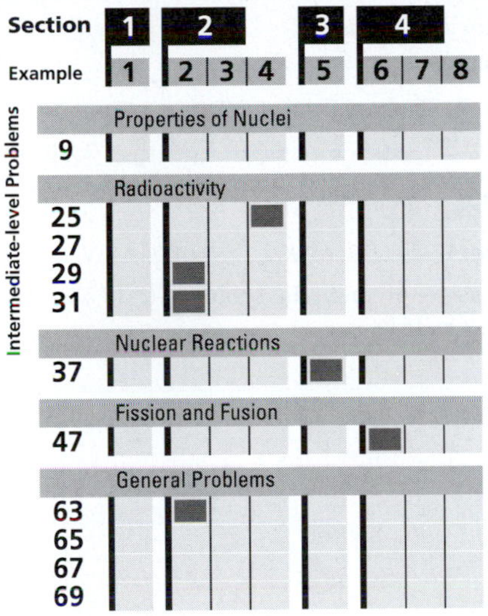

1. (*a*) ^{15}N, ^{13}N; (*b*) ^{57}Fe, ^{58}Fe; (*c*) ^{117}Sn, ^{119}Sn

3. (*a*) 31.99 MeV, 5.332 MeV/nucleon; (*b*) 333.7 MeV, 8.557 MeV/nucleon; (*c*) 1636 MeV, 7.868 MeV/nucleon

5. (*a*) $3C/4\pi R_0^3$; (*b*) 1.17×10^{14} g/cm^3

7. 1.17×10^{14} g/cm^3

9. (*a*) 7.81 fm, 6.77 fm; (*b*) 199 MeV

11. α decay is often followed by β^- decay (not β^+ decay) because the daughter nucleus is neutron rich.

13. It would cause inaccuracies since we assume the initial ratio of ^{14}C/^{12}C was the same as today's ratio.

15. (*a*) 500 counts/s; (*b*) 250 counts/s; (*c*) 125 counts/s

17. 3.61×10^{10} disintegrations/s

19. (*a*) 4.868 MeV; (*b*) 6.999 MeV

21. 3350 y

23. (*a*) 13,950 y; (*b*) Yes

25. Answer given in the problem.

27. (*a*) Answer given in the problem. (*b*) 6.76×10^{-3} s^{-1}, 103 s

29. (*a*) 4.16×10^{-9}/s^{-1}; (*b*) 5.28 y

31. (*a*) 156 h; (*b*) 551 h

33. Answer given in the problem.

35. (*a*) 4.032 MeV; (*b*) 18.35 MeV; (*c*) 4.785 MeV

37. (*a*) 1.20 MeV; (*b*) The mass of $^{13}_{6}$C includes one less electron than that of $^{13}_{7}$N so one electron mass plus the mass of the emitted electron must be added.

39. The neutrons emitted in fission are fast (energy of the order of 1 MeV), whereas the probability of the capture varies inversely with neutron speed, so a moderator is needed to slow down the emitted neutrons to thermal energies (of the order of 0.02 eV).

41. They escape from the reactor or are captured by nuclei other than ^{235}U.

43. 1.56×10^{19} fissions/s

45. 1.16×10^8 K

47. 3.20×10^{10} J

49. (*a*), (*b*) Answer given in the problem. (*c*) 3.74×10^{38} protons/s, 5.07×10^{10} y

51. Pressure changes affect the electron clouds surrounding the nucleus and therefore affect the spacing of the nuclei slightly, and temperature changes affect the vibration energy of nuclei, but neither affects the internal structure of the nucleus responsible for radioactivity.

53. The temperature of the sun is so great that the thermal energy of positively charged nuclei such as ^{1}H and ^{2}He can overcome the Coulomb repulsion and bring the nuclei close enough to fuse. On earth, the temperature is much too small for thermal energy to overcome the Coulomb energy of repulsion.

55. $\lambda = 0.0693$ s^{-1}

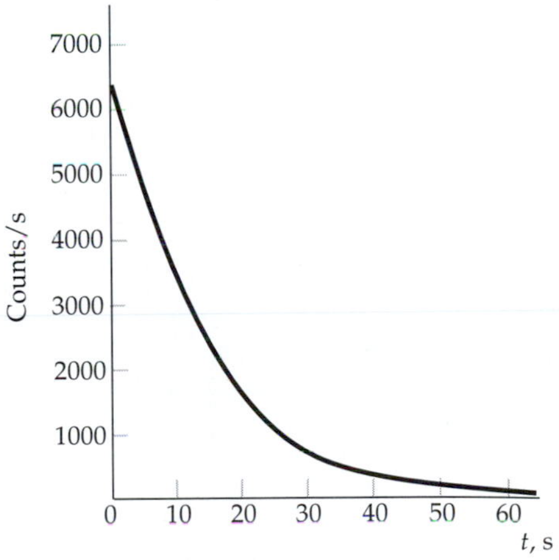

57. 0.156 MeV

59. Yes, 1h

61. 6.60×10^3 decays/s

63. 6.3 L

65. (*a*) 22.96 MeV; (*b*) 4179 MeV; (*c*) 1286 MeV

67. (*a*)

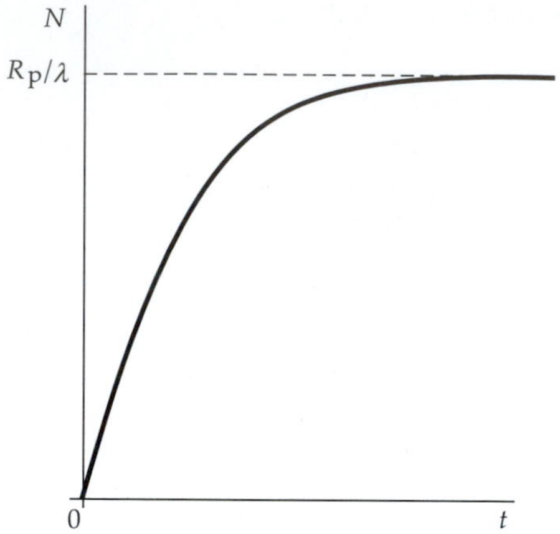

(*b*) 8.66×10^4 nuclei

69. (*a*) 4 fm; (*b*) 310 MeV/*c*; (*c*) Answer given in the problem. (*d*) 310 MeV

71. (*a*) 1.19 MeV/*c*; (*b*) 0.752 keV; (*c*) 0.0962%

73. Answer given in the problem. (*b*) 55 collisions

75. (*a*)–(*c*) Answer given in the problem.
(*d*)

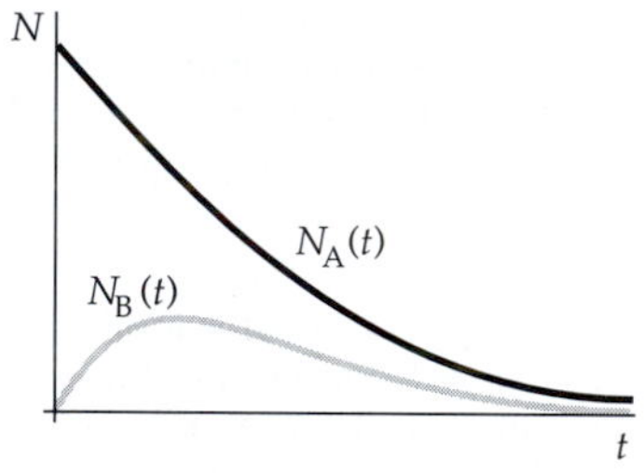

77. Answer given in the problem.

Chapter 41

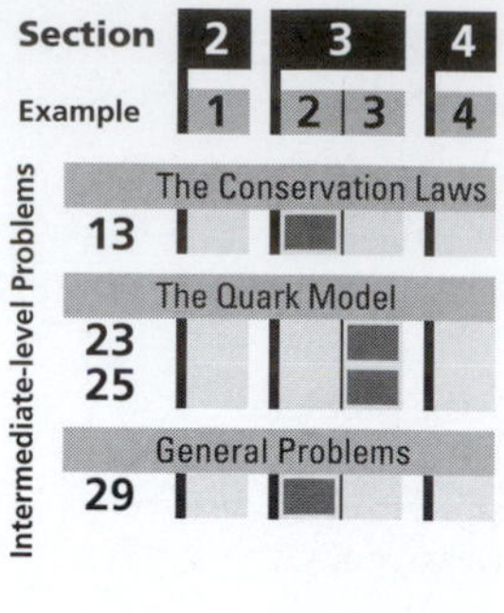

1. Both are hadrons, that is, they interact via strong nuclear interactions and are composed of quarks. Baryons have $\frac{1}{2}$-integral spins, whereas mesons have zero or integral spin. Mesons are less massive.

3. from the lifetime

5. False

7. (*a*) 279.2 MeV; (*b*) 1876.6 MeV; (*c*) 211.4 MeV

9. (*a*) $\Delta S = +1$, weak interaction; (*b*) $\Delta S = +2$, not allowed; (*c*) $\Delta S = +1$, weak interaction

11. (*a*) $\Delta S = +2$, not allowed; (*b*) $\Delta S = +1$, weak interaction

13. (*a*) No, $n \rightarrow p + e^- + \bar{\nu}_e$; (*b*) $\Omega^- \rightarrow p + 3e^- + 3\bar{\nu}_e + 2\bar{\nu}_\mu + 2\nu_\mu + e^+ + \nu_e$ (c) Only strangeness is not conserved.

15. Two quarks make a meson and three quarks make a baryon.

17. (*a*) +1, +1, 0, p^+; (*b*) +1, 0, 0, n; (*c*) +1, +1, −1, Σ^+; (*d*) +1, −1, −1, Σ^-; (*e*) +1, 0, −2, Ξ^0; (*f*) +1, −1, −2, Ξ^-

19. *uuu*

21. (*a*) $c\bar{d}$; (*b*) $\bar{c}d$

23. (*a*) *uds*; (*b*) $\bar{u}\,\bar{u}\,\bar{d}$; (*c*) *dds*

25. (*a*) *sss*; (*b*) *dss*

27. (*a*) False; (*b*) True; (*c*) False; (*d*) True; (*e*) True

29. (*a*) Yes; (*b*) $\Xi^0 \rightarrow p + e^- + \bar{\nu}_e + \nu_\mu + \bar{\nu}_\mu + 2\gamma$; (*c*) Only strangeness is not conserved. (*d*) No, the rest mass of $(\Lambda^0 + \pi^0)$ is greater than that of Σ^0.

31. (*a*) 38.1 MeV; (*b*) 6.72; (*c*) $K_{proton} = 4.93$ MeV, $K_{pion} = 33.2$ MeV

33. (*a*), (*b*) Answer given in the problem. (*c*) $(7.5 \times 10^{-12})c$, 40.3 s; (*d*) $(30 \times 10^{-12})c$, 161 s

INDEX

Numbers in **bold** indicate additional display material, such as diagrams; *n* indicates a footnote; AP indicates material in the Appendixes.

Pages 1–655 are found in Volume 1; pages 656–1141 are found in Volume 2; pages 1142–1335 are found in Volume 3.

Prefixes for Powers of 10

Multiple	Prefix	Abbreviation
10^{18}	exa	E
10^{15}	peta	P
10^{12}	tera	T
10^{9}	giga	G
10^{6}	mega	M
10^{3}	kilo	k
10^{2}	hecto	h
10^{1}	deka	da
10^{-1}	deci	d
10^{-2}	centi	c
10^{-3}	milli	m
10^{-6}	micro	μ
10^{-9}	nano	n
10^{-12}	pico	p
10^{-15}	femto	f
10^{-18}	atto	a

Some Physical Data

Acceleration of gravity at earth's surface	g	$9.81\ \text{m/s}^2 = 32.2\ \text{ft/s}^2$
Radius of earth	R_E	6370 km = 3960 mi
Mass of earth	M_E	5.98×10^{24} kg
Mass of sun		1.99×10^{30} kg
Mass of moon		7.36×10^{22} kg
Escape speed at earth's surface		11.2 km/s = 6.95 mi/s
Standard temperature and pressure (STP)		0°C = 273.15 K 1 atm = 101.3 kPa
Earth–moon distance		3.84×10^{8} m = 2.39×10^{5} mi
Earth–sun distance (mean)		1.50×10^{11} m = 9.30×10^{7} mi
Speed of sound in dry air (at STP)		331 m/s
Density of air		$1.29\ \text{kg/m}^3$
Density of water		$1000\ \text{kg/m}^3$
Heat of fusion of water	L_f	333.5 kJ/kg
Heat of vaporization of water	L_v	2.257 MJ/kg

The Greek Alphabet

Alpha	A	α	Iota	I	ι	Rho	P	ρ
Beta	B	β	Kappa	K	κ	Sigma	Σ	σ
Gamma	Γ	γ	Lambda	Λ	λ	Tau	T	τ
Delta	Δ	δ	Mu	M	μ	Upsilon	Υ	υ
Epsilon	E	ϵ	Nu	N	ν	Phi	Φ	ϕ
Zeta	Z	ζ	Xi	Ξ	ξ	Chi	X	χ
Eta	H	η	Omicron	O	o	Psi	Ψ	ψ
Theta	Θ	θ	Pi	Π	π	Omega	Ω	ω

Abbreviations for Units

A	ampere
Å	angstrom (10^{-10} m)
atm	atmosphere
Btu	British thermal unit
Bq	becquerel
C	coulomb
°C	degree Celsius
cal	calorie
Ci	curie
cm	centimeter
dyn	dyne
eV	electron volt
°F	degree Fahrenheit
fm	femtometer, fermi (10^{-15} m)
ft	foot
Gm	gigameter (10^9 m)
G	gauss
Gy	gray
g	gram
H	henry
h	hour
Hz	hertz
in	inch
J	joule
K	kelvin
kg	kilogram
km	kilometer
keV	kilo-electron volt
lb	pound
L	liter
m	meter
MeV	mega-electron volt
Mm	megameter (10^6 m)
mi	mile
min	minute
mm	millimeter
ms	millisecond
N	newton
nm	nanometer (10^{-9} m)
pt	pint
qt	quart
rev	revolution
R	roentgen
Sv	seivert
s	second
T	tesla
u	unified mass unit
V	volt
W	watt
Wb	weber
y	year
yd	yard
μm	micrometer (10^{-6} m)
μs	microsecond
μC	microcoulomb
Ω	ohm

Some Conversion Factors

1 m = 39.37 in = 3.281 ft = 1.094 yd

1 m = 10^{15} fm = 10^{10} Å = 10^9 nm

1 km = 0.6215 mi

1 mi = 5280 ft = 1.609 km

1 lightyear = 1 $c\cdot$y = 9.461×10^{15} m

1 in = 2.540 cm

1 L = 10^3 cm^3 = 10^{-3} m^3 = 1.057 qt

1 h = 3.6 ks

1 y = 365.24 d = 3.156×10^7 s

1 km/h = 0.278 m/s = 0.6215 mi/h

1 ft/s = 0.3048 m/s = 0.6818 mi/h

1 rev = 2π rad = 360°

1 rad = 57.30°

1 rev/min = 0.1047 rad/s

1 slug = 14.59 kg

1 tonne = 10^3 kg = 1 Mg

1 atm = 101.3 kPa = 1.013 bar = 76.00 cmHg = 14.70 lb/in^2

1 N = 10^5 dyn = 0.2248 lb

1 lb = 4.448 N

1 Pa·s = 10 poise

1 J = 10^7 erg = 0.7373 ft·lb = 9.869×10^{-3} L·atm

1 kW·h = 3.6 MJ

1 cal = 4.184 J = 4.129×10^{-2} L·atm

1 L·atm = 101.3 J = 24.22 cal

1 eV = 1.602×10^{-19} J

1 Btu = 778 ft·lb = 252 cal = 1054 J

1 horsepower = 550 ft·lb/s = 746 W

1 W/m·K = 6.938 Btu·in/h·ft^2·°F

1 T = 10^4 G

1 kg weighs about 2.205 lb

Some Physical Constants

Avogadro's number	N_A	$6.022\ 136\ 7 \times 10^{23}$ particles/mol
Boltzmann's constant	k	$1.380\ 658 \times 10^{-23}$ J/K
Bohr magneton	$m_B = e\hbar/2m_e$	$9.274\ 015\ 4 \times 10^{-24}$ J/T
Coulomb constant	$k = 1/4\pi\epsilon_0$	$8.987\ 551\ 788 \times 10^{9}$ N·m²/C²
Compton wavelength	$\lambda_C = h/2e$	$2.426\ 310\ 58 \times 10^{-12}$ m
Fundamental charge	e	$1.602\ 177\ 33 \times 10^{-19}$ C
Gas constant	$R = N_A k$	8.314 51 J/mol·K = 1.987 22 cal/mol·K $= 8.205\ 78 \times 10^{-2}$ L·atm/mol·K
Gravitational constant	G	$6.672\ 6 \times 10^{-11}$ N·m²/kg²
Mass, of electron	m_e	$9.109\ 389\ 7 \times 10^{-31}$ kg = 510.999 1 keV/c^2
of proton	m_p	$1.672\ 623\ 1 \times 10^{-27}$ kg = 938.272 3 MeV/c^2
of neutron	m_n	$1.674\ 929 \times 10^{-27}$ kg = 939.565 6 MeV/c^2
Permeability of free space	μ_0	$4\pi \times 10^{-7}$ N/A²
Planck's constant	h	$6.626\ 075\ 5 \times 10^{-34}$ J·s $= 4.135\ 669\ 2 \times 10^{-15}$ eV·s
	$\hbar$	$1.054\ 572\ 66 \times 10^{-34}$ J·s $6.582\ 122\ 0 \times 10^{-16}$ eV·s
Speed of light	c	$2.997\ 924\ 58 \times 10^{8}$ m/s
Unified mass unit	u	$1.660\ 540 \times 10^{-27}$ kg = 931.494 32 MeV/c^2

Mathematical Symbols

$=$	is equal to	Δx	change in x
$\neq$	is not equal to	$\|x\|$	absolute value of x
$\approx$	is approximately equal to	$n!$	$n(n-1)(n-2)\cdots 1$
$\sim$	is of the order of	Σ	sum
$\propto$	is proportional to	lim	limit
$>$	is greater than	$\Delta t \to 0$	Δt approaches zero
$\geq$	is greater than or equal to	$\frac{dx}{dt}$	derivative of x with respect to t
$\gg$	is much greater than		
$<$	is less than	$\frac{\partial x}{\partial t}$	partial derivative of x with respect to t
$\leq$	is less than or equal to		
$\ll$	is much less than	$\int$	integral